ORGANIC COMPOUNDS

Family						
Ether	Amine	Aldehyde	Ketone	Carboxylic Acid	Ester	Amide
CH_3OCH_3	CH_3NH_2	$\overset{O}{\overset{\|}{CH_3CH}}$	$\overset{O}{\overset{\|}{CH_3C\,CH_3}}$	$\overset{O}{\overset{\|}{CH_3COH}}$	$\overset{O}{\overset{\|}{CH_3COCH_3}}$	$\overset{O}{\overset{\|}{CH_3CNH_2}}$
Methoxy-methane	Methan-amine	Ethanal	Propanone	Ethanoic Acid	Methyl ethanoate	Ethanamide
Dimethyl ether	Methyl-amine	Acetal-dehyde	Acetone	Acetic acid	Methyl acetate	Acetamide
ROR	RNH_2 R_2NH R_3N	$\overset{O}{\overset{\|}{RCH}}$	$\overset{O}{\overset{\|}{RCR}}$	$\overset{O}{\overset{\|}{RCOH}}$	$\overset{O}{\overset{\|}{RCOR}}$	$\overset{O}{\overset{\|}{RCNH_2}}$ $\overset{O}{\overset{\|}{RCNHR}}$ $\overset{O}{\overset{\|}{RCNR_2}}$
$-\overset{\|}{\underset{\|}{C}}-O-\overset{\|}{\underset{\|}{C}}-$	$-\overset{\|}{\underset{\|}{C}}-N-$	$\overset{O}{\overset{\|}{-\overset{\|}{\underset{\|}{C}}-H}}$	$\overset{O}{\overset{\|}{-\overset{\|}{\underset{\|}{C}}-\overset{\|}{C}-\overset{\|}{\underset{\|}{C}}-}}$	$\overset{O}{\overset{\|}{-\overset{\|}{\underset{\|}{C}}-OH}}$	$\overset{O}{\overset{\|}{-\overset{\|}{\underset{\|}{C}}-O-\overset{\|}{\underset{\|}{C}}-}}$	$\overset{O}{\overset{\|}{-\overset{\|}{\underset{\|}{C}}-N-}}$

ORGANIC CHEMISTRY

revised printing

T. W. Graham Solomons
University of South Florida

JOHN WILEY & SONS

New York Chichester Brisbane Toronto

This book was set in Times Roman by York Graphic
Services Inc. It was printed and bound by Halliday
Lithograph Corp. The designer was Jerry Wilke. The
drawings were designed and executed by John Balbalis with
the assistance of the Wiley Illustration Department.
Susan Giniger was the copyeditor.

Permission for the publication herein of Sadtler Standard
Spectra® has been granted, and all rights are reserved by
Sadtler Research, Laboratories, Inc.

Library of Congress Cataloging in Publication Data

Solomons, T. W. Graham.
  Organic chemistry, revised printing.

  Includes index, bibliographies
  1. Chemistry, Organic. I. Title.
QD251.2.S66   547   78-4880
ISBN 0-471-03561-O

Printed in the United States of America

10 9 8 7 6 5 4

For Judith

PREFACE TO
REVISED PRINTING

With this revised printing of *Organic Chemistry* I have tried to improve the book in two important ways: by providing many additional problems, and by adding an introduction to mass spectroscopy. As with earlier reprintings, I have also tried to eliminate errors.

The new problems have been added to the end-of-chapter sets. These problem sets are of graded difficulty and the new problems range from problems of moderate difficulty to those that offer students a considerable challenge. Problems that are especially challenging have been designated as such by an asterisk. Thus, if instructors wish to do so they can make these problems optional.

The new problems provide students with many opportunities to apply principles they have mastered to new situations. Naturally, these problems cover a wide range of subjects. Wherever possible, however, I have tried to develop problems from material that will stimulate students' interest—especially from molecules of biological importance or from processes used in the chemical industry. A revised printing of the solutions manual that gives answers to all of the problems in the revised printing of the text is also available.

The introduction to mass spectroscopy has been added as an appendix. This section can be brought into the course at a variety of places but it will probably fit best if used to supplement Chapter 14 or some chapter thereafter. The treatment of mass spectroscopy begins with very simple spectra that illustrate basic principles. From there students move gradually to more and more complex spectra. They begin to see how things learned earlier about carbocations and free radicals can help them use mass spectroscopy in determining the structures of organic molecules.

Again I thank all the people who helped me prepare the first edition of this text and whose names are given in the acknowledgments section. Not listed there are three people who have been a great help in preparing the revised printing. Two colleagues at the University of South Florida, George R. Jurch and Rebecca O'Malley, reviewed the section on mass spectroscopy, and Professor O. C. Dermer, Oklahoma State University, proofread the entire revised printing. I am especially grateful to them.

T. W. Graham Solomons

PREFACE

A new textbook almost always has its origin in the writer's perception of the needs of his own students, and this book is no exception. When I began I felt that there were specific features that a new organic text should have. These features and how I have attempted to provide them are described in topical form below.

A clear and unified development of important concepts. Students in a first-year organic chemistry course usually have had little exposure to most of the ideas that they will encounter. Furthermore, they often find themselves in large classes where, even under the best of circumstances, their opportunities for individualized instruction are limited. Recognizing this, I have tried to pay special attention to the language of the science by defining or illustrating each new term when it is first presented and by using these terms consistently thereafter. I have tried to avoid placing the student in a situation where his understanding a concept or his solving a problem requires a knowledge of facts that he has not yet encountered. Also I have tried to develop an internal order of the basic concepts so that, as the student progresses from chapter to chapter, he will be able to apply the knowledge he has gained to new situations.

An early presentation of important functional groups. The traditional organization of organic texts—based on the structures and functional groups of organic molecules—is clearly the one that students find most accessible. Therefore, I have used this organization here. A disadvantage of it, however, in other texts is that it delays treatment of many of the important functional groups until late in the course. Chapter 2 should provide a solution to this problem. It gives the student an introduction to the most important functional groups and to the major types of reactions. Recognizing that too much diversity at this stage should be avoided, my treatment in Chapter 2 is nonmechanistic, and it is limited to simple examples that use only C_1, C_2, and C_3 compounds. Since the presentation at this level is intended primarily as an overview, almost all of the material is repeated in later chapters when mechanisms are discussed.

A broad and logical presentation of subsequent topics. Because the student has had an introduction to the important functional groups in Chapter 2, subsequent chapters can be broader and topics can be dealt with in a more logical way. Consider an example. A modern treatment of alkenes (Chapters 5 and 6) requires a discussion of hydroboration-oxidation, oxymercuration-demercuration, epoxidation, and ozonolysis, for instance. These topics require some knowledge of the chemistry of alcohols, aldehydes, ketones, ethers, and acids. Since the student will have encountered examples of these functional groups in Chapter 2, a reference to them in Chapters 5 and 6 will not be confusing. The early presentation of important functional groups also allows the introduction of many bio-organic applications at appropriate points.

An emphasis on bio-organic chemistry. Those of us who teach organic chemistry have known for a number of years that most of the students in our one-year course have a special interest in biochemical and biological applications of the subject. That this is true of premedical and predental students and of biology and biochemistry majors is obvious. We know now, however, that an emphasis on health-related research has made this true of our chemistry majors as well, and that concern with environmental problems has even made it true of our engineering students.

One of the things I have tried to do, here, therefore, is to address myself to the needs and interests of these students; I have attempted to do this in a way that will enhance their comprehension of basic organic chemistry. Thus, the many diversions into biochemical topics have been chosen not only to stimulate the students' interest but also to help to teach them the fundamentals of organic chemistry and to reinforce their understanding of it.

Examples of biochemical applications are given early in the book and increase in number as the students master more and more organic chemistry. These examples take a variety of forms. Where appropriate, parallels have been drawn between the reactions that occur in the glassware of the organic laboratory and those that occur in living cells. Three examples that illustrate this approach are in the two chapters on the chemistry of alkenes. Here I relate laboratory hydrogenation-dehydrogenation reactions to those catalyzed by the enzyme *succinate dehydrogenase;* hydration-dehydration reactions to those catalyzed by the enzyme *aconitase;* and carbocation alkylations to the remarkable series of reactions catalyzed by *squalene oxide cyclase.* Biological applications also form the basis for many problems, and they are the subject of many special topics.

Special topics and an organization that allows for a flexible approach. The topics that are fundamental to any organic course are systematically developed in twenty-one basic chapters. The remaining five chapters are designated as special topics. These special topics are interspersed throughout the text where they can be used to augment the immediately preceding chapters. The special topics stand by themselves and can be omitted at the instructor's choice. An understanding of the material in the basic chapters does not require a knowledge of any of the special topics. Among the subjects treated as special topics are divalent carbon compounds, addition polymers, the photochemistry of vision, electrocyclic and cycloaddition reactions, alkaloids, condensation polymers, and lipids.

A modern and easily understood treatment of molecular orbital theory. Chemists now recognize that the plus and minus phase signs associated with orbitals can be of great importance in explaining many organic reactions. Consequently, I have introduced these signs into my treatment of molecular orbital theory from the outset; and, by doing so, I believe that I have made many aspects of simple molecular orbital theory much more understandable. In Chapter 1, for example, orbital signs are used to account for the bonding and antibonding orbitals of hydrogen. In Chapter 3, they are used to make the idea of orbital hybridization clearer, and in Chapter 5 they are applied to the bonding and antibonding orbitals of alkenes. These early and simple introductions will prepare the student for a better understanding of conjugated systems and visible-ultraviolet spectroscopy in Chapter 10 and of aromaticity in Chapter 12. They also make possible a presentation of the theory of electrocyclic and cycloaddition reactions as a special topic in Chapter 11.

Extensive use of resonance theory. Although molecular orbital theory is presented in a new and clear way, resonance theory is not neglected. Resonance theory is introduced in Chapter 1 as an explanation of bonding in simple molecules, and it is amplified, developed, applied, and reapplied as the student progresses through subsequent chapters.

Problems and examples. Almost 850 problems provide more than 2600 separate exercises. Problems are interspersed throughout the text material; others are collected in sets at the ends of chapters. The intratext problems are specifically designed to test a student's understanding of the material up to that point. The end-of-chapter problems are graded, beginning with simpler exercises and progressing to more difficult ones. The examples used to illustrate reactions are actual examples taken from the literature of organic chemistry and usually have percentage yields and conditions included.

Chapter summaries. Summaries of the reactions of important functional groups are included at the end of appropriate chapters.

T. W. Graham Solomons

ACKNOWLEDGMENTS

I am grateful to many people for their help, direct or indirect, in preparing the manuscript for this book. Let me begin by acknowledging my debt to those who have brought organic chemistry to its present stage of development, to those who have done the research, and to those who have written the articles, monographs, and texts that were my sources of information. Some of the former are mentioned in the book; many, but not all of the latter, are listed in the bibliography at the end.

I am also grateful to colleagues who were kind enough to read my manuscript at different stages of its development and to offer comments and suggestions that were invaluable in my revisions. At the University of South Florida there were: Sotirios A. Barber, Jack E. Fernandez, Douglas J. Raber, Stewart W. Schneller, and George R. Wenzinger, all were extremely helpful. Other critical reviewers of the manuscript included: Paul A. Barks (North Hennepin State Junior College), Edward M. Burgess (Georgia Institute of Technology), Philip L. Hall (Virginia Polytechnic Institute and State University), Harold Hart (Michigan State University), John R. Holum (Augsburg College), Stanley N. Johnson (Orange Coast College), Philip W. LeQuesne (Northeastern University), Jerry March (Adelphi University), William A. Pryor (Louisiana State University), and Thomas R. Riggs (University of Michigan).

I also thank Fred W. Clough for checking all of the problems in the book and Linda Hardee for help in proofreading and indexing.

Part of the manuscript was written while I was a visiting member of the faculty at the University of Sussex, England. I am indebted to many people there for making my stay such a pleasant one, especially to Dean Aubrey D. Jenkins and to James R. Hanson.

I also thank the people at Wiley who transformed my manuscript into this book, especially; Gary Carlson and Robert Rogers, the present and former chemistry editors, Marion Palen, the production editor, Susan Giniger, the staff editor, John Balbalis, who executed the illustrations, and special thanks to Jerry Wilke for his fine interior design.

My greatest debt is to Judith Taylor Solomons, who took my illegible hand-written copy and somehow miraculously turned it into a manuscript, who edited and proofread, and who did a thousand other tasks that gave me the time to write, erase, and write again.

T. W. Graham Solomons

CONTENTS

1
A STUDY OF THE COMPOUNDS OF CARBON

2 FUNCTIONAL GROUPS AND FAMILIES OF ORGANIC COMPOUNDS: THE MAJOR REACTION TYPES

3 ALKANES AND CYCLOALKANES: THEIR STRUCTURES, PROPERTIES, AND SYNTHESES

4 CHEMICAL REACTIVITY: REACTIONS OF ALKANES AND CYCLOALKANES

5 ALKENES: STRUCTURE AND SYNTHESIS

6 REACTIONS OF ALKENES: ADDITION REACTIONS OF THE CARBON-CARBON DOUBLE BOND

7 STEREOCHEMISTRY

8 SPECIAL TOPICS 1

9 ALKYNES

10 CONJUGATED UNSATURATED SYSTEMS VISIBLE-ULTRAVIOLET SPECTROSCOPY

11 SPECIAL TOPICS II

12 AROMATIC COMPOUNDS I: THE PHENOMENON OF AROMATICITY

13 AROMATIC COMPOUNDS II: REACTIONS OF AROMATIC COMPOUNDS WITH ELECTROPHILES

14 PHYSICAL METHODS OF STRUCTURE DETERMINATION
NUCLEAR MAGNETIC RESONANCE SPECTROSCOPY INFRARED SPECTROSCOPY

15 ORGANIC HALIDES AND ORGANOMETALLIC COMPOUNDS

16 ALCOHOLS, PHENOLS, AND ETHERS

19 CARBOXYLIC ACIDS AND THEIR DERIVATIVES: NUCLEOPHILIC SUBSTITUTION AT ACYL CARBON

20 SYNTHESIS AND REACTIONS OF β-DICARBONYL COMPOUNDS

21 AMINES

22 SPECIAL TOPICS III

23 SPECIAL TOPICS IV
LIPIDS

24 CARBOHYDRATES

25 AMINO ACIDS AND PROTEINS

26 SPECIAL TOPICS V NUCLEIC ACIDS: PROTEIN SYNTHESIS

1 A STUDY OF THE COMPOUNDS OF CARBON

1.1 INTRODUCTION

Organic chemistry is the chemistry of the compounds of carbon. The compounds of carbon are the "stuff" of which all living things on this planet are made. Carbon compounds include DNA, the giant molecules that, for a given species, contain all its genetic information—the molecules that determine whether we are men or women, humans or frogs, whether we have blue eyes or brown, whether we have black hair or blonde, and whether in our old age we have gray hair or none. Carbon compounds make up the proteins of our muscle and skin. They make up the enzymes that catalyze the reactions that occur in our bodies. Together with oxygen in the air we breathe, carbon compounds in our diets furnish the energy that sustains life.

Most of the carbon atoms from which these molecules are constructed have been on this planet since it was formed. In the billions of years that have passed these same carbon atoms have, at one time or another, been part of billions of different molecules and of billions of different organisms. We now know that living organisms are not static in relation to their environment, but are constantly taking in molecules (and thus atoms) from that environment. At the same time, organisms are constantly expelling other atoms (in the form of molecules) back into their surroundings. When we reflect on this it is not too disturbing to realize that some of the carbon atoms that are part of molecules of our body now were, a few hours ago, a part of the egg we ate for breakfast. Before that, these same atoms were a part of a chicken, and before that they were a part of the grain the chicken ate. What is more surprising, and also more pleasing, is to realize that thousands of the carbon atoms that are a part of our bodies now were once a part of the bodies of Socrates, Plato, or Aristotle.

There is considerable evidence indicating that several billion years ago most of the carbon atoms on this planet were present in the form of the gas, methane. This simple organic molecule, CH_4, along with water, ammonia, and hydrogen, were the main constituents of the primordial atmosphere. It is believed that as lightning and highly energetic radiation passed through this atmosphere many of these simple molecules were fragmented into highly reactive pieces. These pieces then recombined to form more complex arrangements. Compounds called amino acids, formaldehyde, purines, and pyrimidines were formed in this way. These and other compounds, produced in the same way, were carried by rain into the sea. The sea became richer and richer in organic molecules until it became a vast storehouse containing all of the compounds necessary for the emergence of life. Amino acids reacted with each other to form proteins, formaldehyde became sugars, and some of these sugars, together with purines and pyrimidines, became simple molecules of DNA. At some point, and in a manner still not understood, these larger molecules collected together to form the first primitive living cells. From these first cells, through the long process of natural selection, evolved man and all the other living things present on this earth today.

1.2 THE DEVELOPMENT OF ORGANIC CHEMISTRY AS A SCIENCE

Man has used organic compounds and their reactions for thousands of years. His first deliberate experience with an organic reaction probably dates from his discovery of fire. The ancient Egyptians used organic compounds (indigo and alizarin) to dye cloth. The famous "royal purple" used by the Phoenicians was also an organic substance, obtained from mollusks. The fermentation of grapes to produce ethyl alcohol and the acidic qualities of "soured wine" are both described in the Bible and were probably known earlier.

The science of organic chemistry is less than 200 years old. Most historians of science date its origin to the early part of the nineteenth century. For a period of time prior to 1850 chemists believed that there was something distinctive about organic compounds that would prevent their preparation outside of living organisms. All of the organic compounds that were known before 1828 had been obtained from living organisms and, for this reason, it was believed that the intervention of a "vital force" was necessary for their creation. During the 1780s the old distinction was first made between *organic compounds* (obtained from living organisms) and *inorganic compounds* (derived from nonliving sources).

Between 1828 and 1850 a number of compounds that were clearly "organic" were synthesized from sources that were clearly "inorganic." The first of these syntheses was accomplished by Friedrich Wöhler in 1828. Wöhler found that the organic compound urea (a constituent of urine) could be made by heating the inorganic compound ammonium cyanate. Although "vitalism" died slowly and did

$$NH_4{}^+NCO^- \xrightarrow{\text{heat}} H_2N-\overset{\displaystyle O}{\underset{\displaystyle \|}{C}}-NH_2$$

Ammonium cyanate   Urea

not disappear completely from scientific circles until 1850, its passing made possible the flowering of the science of organic chemistry that has occurred since 1850.

Even while vitalism persisted* extremely important advances were made in the development of qualitative and quantitative methods for analyzing organic substances. In 1784 Antoine Lavoisier first showed that organic compounds were composed primarily of carbon, hydrogen, and oxygen. Between 1811 and 1831, *quantitative* methods for determining the composition of organic compounds were developed by Justus Liebig, J. J. Berzelius, and J. B. A. Dumas. A great confusion was dispelled in 1860 when Stanislao Cannizzaro showed that the earlier hypothesis

---

* It is still held today by some groups. While there are sound arguments made against foods contaminated with pesticides, it is impossible to argue that "natural" vitamin C, for example, is healthier than the "synthetic" vitamin, since they are identical.

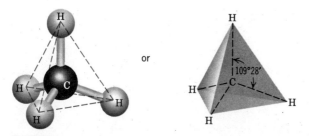

or

FIG. 1.1
The tetrahedral structure of methane.

of Amedeo Avogadro (1811) could be used to distinguish between *empirical* and *molecular formulas*. As a result, many molecules that had appeared earlier to have the same formula were seen to be composed of different number of atoms. For example, ethylene, cyclopentane, and cyclohexane all have the same empirical formula: CH_2. However they have molecular formulas of C_2H_4, C_5H_{10}, and C_6H_{12}, respectively.

Between 1858 and 1861, August Kekulé, Archibald Scott Couper and Alexander M. Butlerov, working independently, laid the basis for one of the most fundamental ideas in organic chemistry: *the structural theory*. Kekulé, Couper, and Butlerov proposed that carbon is *tetravalent* (i.e., forms four bonds) in its compounds and, most importantly, that a carbon atom can use one or more of its valences to form bonds to other carbon atoms. In his original publication Couper represented these bonds by lines much in the same way that most of the formulas in this book are drawn. In his textbook (published in 1861), Kekulé gave the science of organic chemistry its modern definition: *a study of the compounds of carbon*.

In 1874, the structural formulas originated by Kekulé, Couper, and Butlerov were expanded into three dimensions by the independent work of J. H. van't Hoff and J. A. Le Bel. Van't Hoff and Le Bel demonstrated that the four bonds of the carbon atom in methane, for example, are arranged in such a way that they would point toward the corners of a regular tetrahedron if the carbon atom were placed at its center (Fig. 1.1). The necessity for determining the arrangement of the atoms in space, taken together with an understanding of the order in which they are connected, is central to an understanding of organic chemistry.

1.3 BONDING IN ORGANIC MOLECULES

The first explanations of the nature of chemical bonds were advanced by W. Kössel and G. N. Lewis in 1916. They proposed two major types of chemical bonds:

   1. The *ionic* or *electrovalent* bond (formed by the transfer of one or more electrons from one atom to another).

   2. The *covalent* bond (a bond that results when atoms share electrons).

The concepts and explanations that arise from the original propositions of Lewis and Kössel are satisfactory for explanations of many of the problems we deal with in organic chemistry today. For this reason we will review these two types of bonds in more modern terms.

TABLE 1.1 Pauling Electronegativities of Some of the Elements

			H			
			2.2			
Li	Be	B	C	N	O	F
1.0	1.5	2.0	2.5	3.0	3.5	4.0
Na	Mg	Al	Si	P	S	Cl
0.9	1.2	1.5	1.8	2.1	2.5	3.0
K						Br
0.8						2.8

Ionic or Electrovalent Bonds

An electrovalent or ionic bond is formed when two atoms of widely differing electronegativities (Table 1.1) unite. We can see an example of electrovalent bond formation in the reaction of lithium atoms with fluorine atoms. Lithium, a typical metal, has a very low electronegativity; fluorine, a nonmetal, is the most electronegative element of all. The loss of an electron (a negatively charged species) by the lithium atom results in the formation of a lithium cation, Li^+; the gain of an electron by the fluorine atom results in the formation of a fluoride anion, F^-. Why

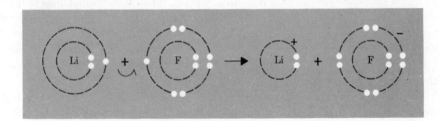

do these ions form? In terms of the Lewis-Kössel theory both ions achieve the electronic structure of a noble gas. The lithium ion is now like the noble gas helium, and the fluoride ion is like the noble gas neon (Fig. 1.2). Moreover, crystalline lithium fluoride (Fig. 1.3) forms from the individual lithium and fluoride ions; in this process negative fluoride ions become surrounded by positive lithium ions, and positive lithium ions by negative fluoride ions. In this crystalline state, the ions have substantially lower energies than the atoms from which they were formed. Lithium and fluorine are thus "stabilized" when they react to form crystalline lithium fluoride.

FIG. 1.2
The electronic structure of helium and neon.

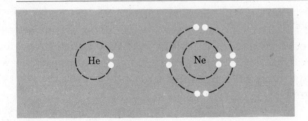

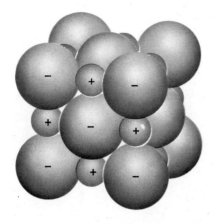

FIG. 1.3

Crystal structure of lithium fluoride, Li⁺ F⁻. The colored spheres represent lithium ions, Li⁺. The grey spheres represent fluoride ions, F⁻.

It is questionable whether an ionic bond should be called a bond at all, since it is essentially omnidirectional. It is impossible to say that any two ions (i.e., Li^+ and F^-) are bonded to each other in the sense that they are attached to each other. Crystals of ionic compounds are simply orderly arrangements of the oppositely charged ions that mutually surround each other (Fig. 1.3).

Ionic substances, because of their strong internal electrostatic forces, tend to be very high melting solids, often melting above 1000°C. In polar solvents, such as water, the solvated ions usually conduct an electric current.

Problem 1.1

Write equations and account for the formation of an ionic bond when
(a) Sodium reacts with chlorine atoms
(b) Magnesium reacts with fluorine atoms
(c) Potassium reacts with bromine atoms

Covalent Bonds

When two or more atoms of the same or similar electronegativities react, a complete transfer of electrons does not occur. In these instances the atoms achieve the noble gas structure by sharing electrons, and *covalent* bonds are formed. *Covalent* molecules can be represented by electron-dot formulas but, more conveniently, by dash formulas where each dash represents a pair of electrons bonding two atoms together. Some examples are shown below.

H_2 H:H   or   H—H Cl_2 :C̈l:C̈l: or   :C̈l—C̈l:

HCl H:C̈l: or   H—C̈l: CH_4 H
                                                    H:C̈:H   or   H—C—H
                                                    H   H

In certain cases, multiple covalent bonds are formed, for example,

N_2 :N::N: or   :N≡N:

and ions themselves may contain covalent bonds.

$$
\overset{+}{N}H_4 \qquad H\!:\!\overset{\displaystyle H}{\underset{\displaystyle \ddot{H}}{\overset{..}{\underset{..}{N}}}}\!:\!H \quad \text{or} \quad H\!-\!\!\!\underset{\displaystyle H}{\overset{\displaystyle H}{\overset{|}{\underset{|}{N^{\pm}}}}}\!\!\!-\!H
$$

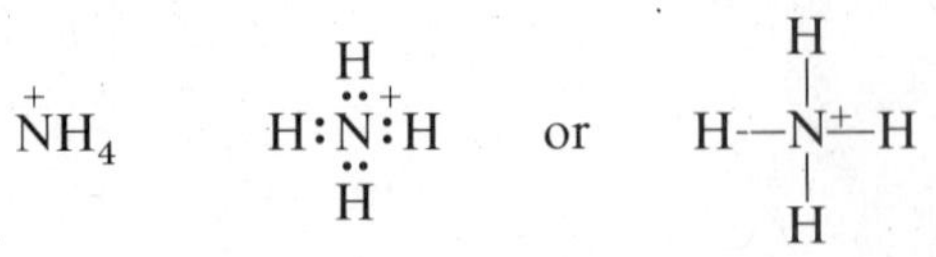

Problem 1.2

Write electron-dot and line formulas for each of the following molecules or ions. In each case show how the atoms achieve the noble gas structure.

(a) HBr
(b) Br_2
(c) CO_2
(d) CH_4
(e) HNO_3 ($HONO_2$)
(f) HNO_2 (HONO)
(g) H_2O_2
(h) SiH_4
(i) NH_3

(j) PCl_3
(k) NF_3
(l) CH_3Cl
(m) H_2O
(n) OH^-
(o) NO_3^-
(p) NO_2^-
(q) NH_4Cl ($NH_4^+Cl^-$)
(r) $MgSO_4$

1.4 COVALENT IONS AND FORMAL CHARGE

Most of the covalent substances that we have considered so far have consisted of neutral molecules. Some of these compounds, however, have not been neutral but have been positive or negative ions that contain covalent bonds. In problem 1.2, for example, you were asked to write electron-dot formulas for the ammonium ion, NH_4^+, the hydroxide ion, OH^-, the nitrite ion, NO_2^-, and the nitrate ion, NO_3^-.* It will be helpful at this point, as a review, to show how the charges on covalent ions can be calculated. Let us begin with a simple example: the ammonium ion.

There are several ways to calculate the charge on the ammonium ion. The most fundamental is to calculate the arithmetic sum of all of the nuclear protons (positive charges) and of all of the extranuclear electrons (negative charges). This approach is shown below:

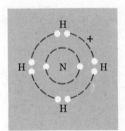

One nitrogen atom = 7 protons
Four hydrogen atoms = 4 protons
Total number = 11 protons

Number of valence shell electrons = 8
Number of inner shell electrons = 2
Total number = 10 electrons

Charge on the ion = $(+11) + (-10) = +1$

---

* These ions do not exist alone, of course, but are always associated with oppositely charged ions that balance the charge; the ammonium ion as ammonium *chloride*, $NH_4^+Cl^-$, for example; the hydroxide ion as *sodium* hydroxide, Na^+OH^-, and so on.

A faster way to demonstrate that the ammonium ion has one positive charge is to use the equation showing its formation in a chemical reaction:

$$H{-}\overset{\cdot\cdot}{N}{-}H \quad + \quad H^+ \quad \longrightarrow \quad H{-}\overset{\overset{\textstyle H}{|}}{\underset{\underset{\textstyle H}{|}}{N^{\pm}}}{-}H$$

Ammonia   +   A proton $\longrightarrow$ Ammonium ion

(an electrically neutral molecule)   +   (one positive charge)   =   (one positive charge)

0   +   1   =   +1

This approach is based on the principle that in a correctly balanced chemical equation the electrical charges must also balance.

Formal charge

Still a third way of calculating the charge is based on the concept of *formal charge*, a procedure that is really nothing more than a method for electron book-keeping. In this approach, the formal charge on each atom is calculated first, then the arithmetic sum of all of the formal charges is taken, giving the charge on the

$\overset{\textstyle H}{\underset{\textstyle H}{H\!:\!\overset{\cdot\cdot}{\underset{\cdot\cdot}{N}}\!:\!H}}^{+}$	Formal charge on each hydrogen $= 0 \times 4 = 0$ Formal charge on nitrogen $\quad = +1 \times 1 = +1$ Total charge on the ion $\quad\quad = +1$

ion as a whole. The formal charge of each atom is calculated by taking the group number of that atom (from the Periodic Table) and subtracting the number of electrons associated with it: one for each shared pair and two for each unshared pair. For the ammonium ion this is done as follows.

Formal charge = group number − [½ (number of shared electrons)
+ (number of unshared electrons)]

Formal charge on hydrogen $= +1 - [\frac{1}{2}\,(2) + (0)]$

Formal charge on hydrogen $= +1 - 1 = 0$

Formal charge on nitrogen $= +5 - [\frac{1}{2}\,(8) + (0)]$

Formal charge on nitrogen $= +5 - 4 = +1$

$+7 - 2 = +5$

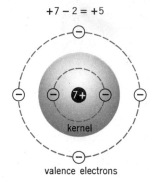

valence electrons

A moment's reflection will reveal why this formula works: the group number of an element is nothing more than the *kernel charge* on that element; that is, it is the sum of the nuclear protons and the inner shell electrons. For nitrogen the kernel charge is $+5$.

Taking the sum [½ (number of shared electrons) + (number of unshared electrons)] is nothing more than a way of apportioning the valence electrons.

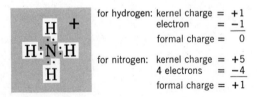

for hydrogen: kernel charge = +1
                 electron   = −1
                 formal charge =   0

for nitrogen: kernel charge = +5
               4 electrons   = −4
               formal charge = +1

Let us consider several other examples:

The nitrate ion, $NO_3{}^-$:

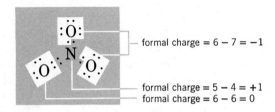

formal charge = 6 − 7 = −1

formal charge = 5 − 4 = +1
formal charge = 6 − 6 = 0

Charge on ion = 2(−1) + 1 + 0 = −1

or

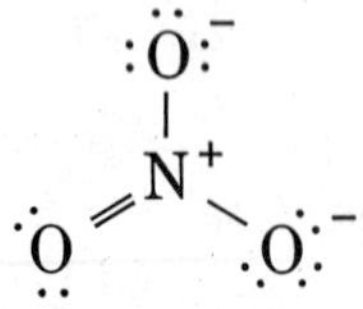

The sulfate ion, $SO_4^=$, can be written with double bonds to two oxygens and with single bonds to the other two, that is,

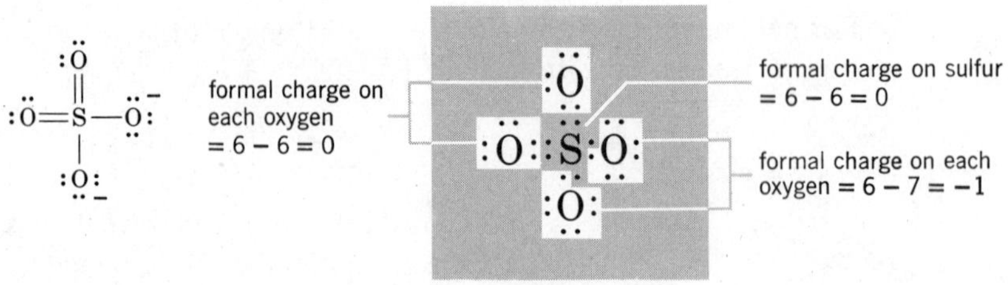

formal charge on each oxygen = 6 − 6 = 0

formal charge on sulfur = 6 − 6 = 0

formal charge on each oxygen = 6 − 7 = −1

Charge on ion = 0 + 2(0) + 2(−1) = −2

Such a structure is possible even though it places 12 electrons around sulfur, because sulfur is in the third row of the Periodic Table and can accommodate more than 8 electrons. This is how we write sulfate compounds in this text.

In many references the sulfate ion will be written with single bonds to all four oxygens, that is,

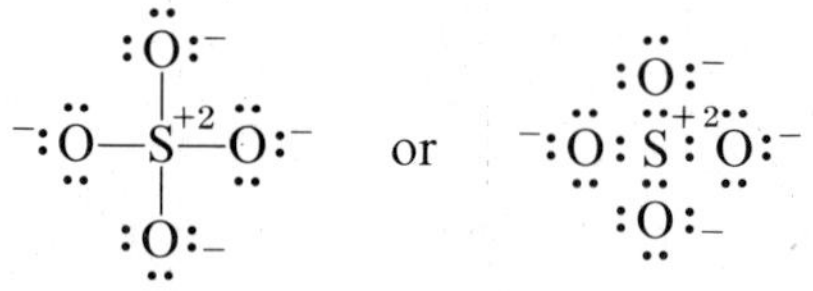

$$
\begin{array}{cc}
\ddot{:}\ddot{O}\!:^- & \ddot{:}\ddot{O}\!:^- \\
\ddot{:}\ddot{O}\!-\!\overset{+2}{S}\!-\!\ddot{O}\!:^- & -:\ddot{O}\!:\overset{+2}{S}\!:\ddot{O}\!:^- \\
\ddot{:}\ddot{O}\!:_- & :\ddot{O}\!:_-
\end{array}
$$

Charge on ion = $4(-1) + (+2) = -2$

Either way we write the ion, however, we find that it has a charge of -2.

Electrically Neutral Molecules

The sum of the formal charges of each atom of a (neutral) molecule must, of course, equal zero. Consider the following examples.

Ammonia:

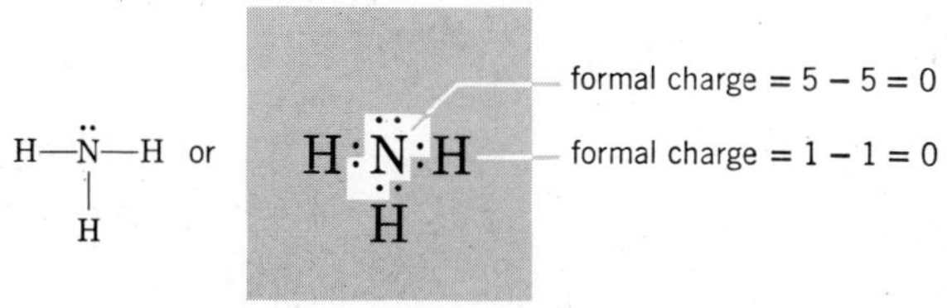

formal charge = 5 − 5 = 0
formal charge = 1 − 1 = 0

Charge on molecule = $0 + 3(0) = 0$

Water:

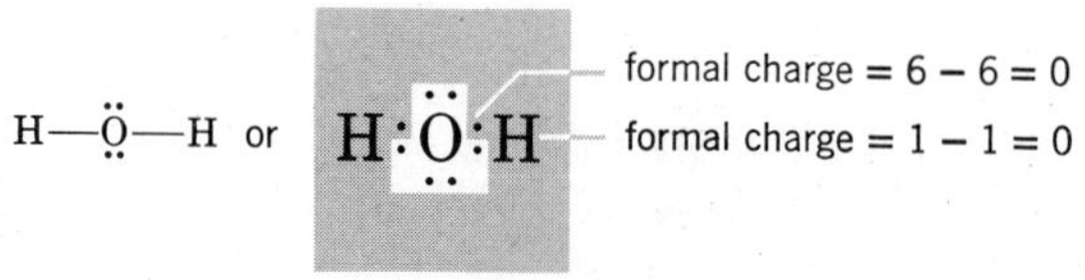

formal charge = 6 − 6 = 0
formal charge = 1 − 1 = 0

Charge on molecule = $0 + 2(0) = 0$

Sulfuric acid:

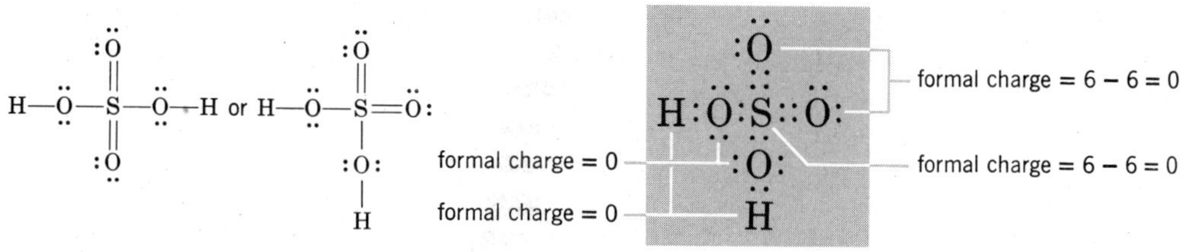

formal charge = 0
formal charge = 0
formal charge = 6 − 6 = 0
formal charge = 6 − 6 = 0

Charge on molecule = $2(0) + 2(0) + 2(0) + 0 = 0$

Problem 1.3

Calculate both the formal charge on each atom, and the total charge on the molecule or ion, for each of the following species on page 10.

(a) BH_4

(b) H_2SO_3

(c) BF_4

(d) H_3O

(e) $:NF_3$

(f) $:CH_3$ (a carbanion)

(g) CH_3 (a carbocation)

(h) $\cdot CH_3$ (a free radical)

(i) $:CH_2$ (a carbene)

Summary

With this background it should now be clear that each time an oxygen atom of the type, $-\ddot{O}:$ appears in a molecule or ion it will have a formal charge of -1, and that each time an oxygen atom of the type $=\ddot{O}:$ or $-\ddot{O}-$ appears it will have a formal charge of zero. Similarly; $-\overset{|}{N}-$ will be $+1$, and $-\overset{|}{\dot{N}}-$ will be zero.

It is much easier to memorize these common structures rather than to calculate the formal charge each time they are encountered. These common structures are summarized below.

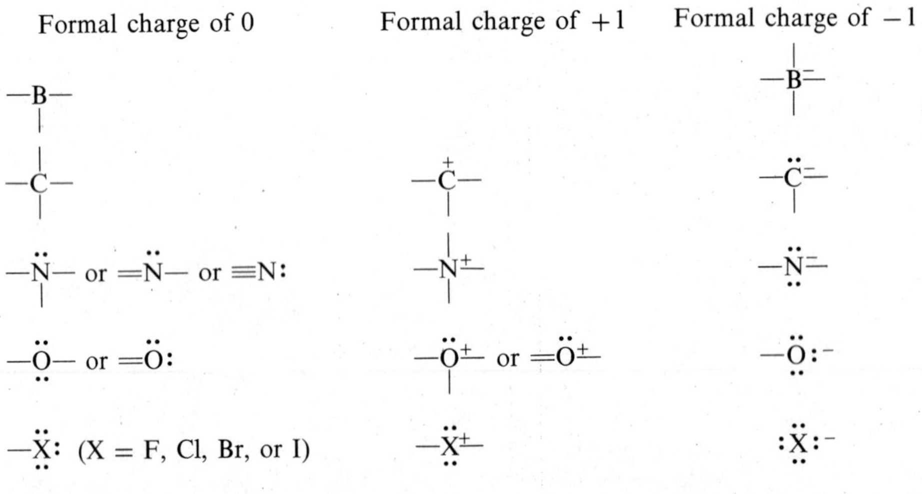

Formal charge of 0	Formal charge of +1	Formal charge of -1
$-B-$		$-\overset{|}{B}=$
$-\overset{|}{\underset{|}{C}}-$	$-\overset{+}{C}-$	$-\overset{\cdot\cdot}{C}=$
$-\ddot{N}-$ or $=\ddot{N}-$ or $\equiv N:$	$-\overset{|}{N}{}^{\pm}$	$-\ddot{N}=$
$-\ddot{O}-$ or $=\ddot{O}:$	$-\overset{|}{\ddot{O}}{}^{\pm}$ or $=\ddot{O}{}^{\pm}$	$-\ddot{O}:^-$
$-\ddot{X}:$ (X = F, Cl, Br, or I)	$-\overset{|}{\ddot{X}}{}^{\pm}$	$:\ddot{X}:^-$

Problem 1.4

Using the chart given above, determine the formal charge on each colored atom of the following molecules and ions. (Remember the formal charge of $-\ddot{O}-$ = $=\ddot{O}:$, that of $-\ddot{N}-$ = $\equiv N:$, and so on.)

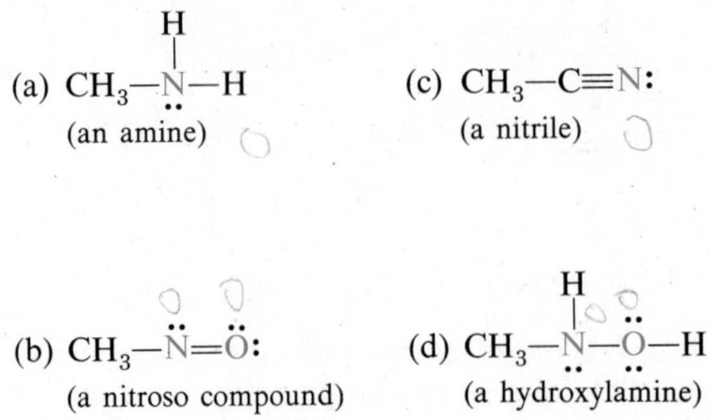

(a) $CH_3-\overset{H}{\underset{\cdot\cdot}{N}}-H$

(an amine)

(c) $CH_3-C\equiv N:$

(a nitrile)

(b) $CH_3-\ddot{N}=\ddot{O}:$

(a nitroso compound)

(d) $CH_3-\overset{H}{\underset{\cdot\cdot}{N}}-\ddot{O}-H$

(a hydroxylamine)

(e) CH_3—$\overset{\overset{\displaystyle CH_3}{|}}{\underset{\underset{\displaystyle :\overset{..}{\underset{..}{O}}: \ -}{|}}{N}}$—$CH_3$

(an amine oxide)

(f) CH_3—$\overset{+}{N}\overset{\displaystyle :\overset{..}{O}:}{\underset{\displaystyle \overset{..}{\underset{..}{O}}: \ -}{\diagup\diagdown}}$

(a nitro compound)

1.5 POLAR COVALENT BONDS

When two atoms of different electronegativities form a covalent bond the electrons are not shared equally between them. The atom with greater electronegativity draws the electron pair closer to it, and a *polar covalent bond* results. (One definition of *electronegativity* is *the ability of an element to attract electrons that it is sharing in a covalent bond.*) An example of such a polar covalent bond is the one in hydrogen chloride. The chlorine atom, with its greater electronegativity, pulls the bonding electrons closer to it. This makes the hydrogen atom somewhat electron

$$\overset{\delta+}{H} \ \overset{\delta-}{:\overset{..}{\underset{..}{Cl}}:}$$

deficient and gives it a *partial* positive charge ($\delta+$). The chlorine atom becomes somewhat electron rich and bears a *partial* negative charge ($\delta-$). Because the hydrogen chloride molecule has a partially positive end and a partially negative end, it is a dipole.

a dipole

Because the positive and negative charges are *separated*, the HCl molecule has a *dipole moment*. The dipole moment (μ) is expressed in Debye units (D), and it is the product of the magnitude of the charge in electrostatic units and the distance that separates them in Angstrom units. (One Angstrom unit equals 10^{-8} cm.)

Dipole moment = charge (in esu) $\times$ distance (in Angstrom units)

$$\mu = e \times d$$

The dipole moment of HCl is 1.08 D.

The direction of polarity of a polar bond is sometimes symbolized by $\longmapsto$. In HCl this is expressed in the following way:

(positive end) $\underset{\text{H—Cl}}{\longmapsto}$ (negative end)

Dipole moments, as we will see, are very useful quantities in accounting for physical properties of compounds.

Problem 1.5

Predict the direction of the dipole (if any) in the following molecules.

(a) HBr   (c) H_2
(b) ICl   (d) Cl_2

1.6 STRUCTURES OF MOLECULES: VALENCE SHELL ELECTRON PAIR REPULSION

In later chapters we discuss the structures of molecules on the basis of theories that arise from quantum mechanics. It is possible, however, to make good guesses about the structures of many molecules simply from the fact that in most compounds, electrons exist as pairs and *electron pairs repel one another*.* This is true whether electrons are in bonds or in lone (nonbonding) pairs. Consequently, within the confines of the molecule, *electron pairs of the valence shell tend to stay as far apart as possible*. Let us apply this principle to a few simple examples.

Methane

As we pointed out earlier, the structure of a molecule of methane is tetrahedral. If we assume that the bonding electron pairs of methane molecules are located somewhere between the central carbon atom and each of the hydrogen atoms, we see that the tetrahedral arrangement allows the electron pairs to be as far apart as possible (Fig. 1.4).

Any other arrangement of the atoms, for example, a square planar structure (Fig. 1.5), places the electron pairs closer together.

The bond angles for any atom that has a regular tetrahedral structure are 109.5°. One way of representing a tetrahedral atom is shown in Fig. 1.6. This representation is derived from the familiar ball-and-stick model. The lines that intersect the circle are directed out of the plane of the paper, and the lines that do not intersect the circle are directed behind the plane of the paper.

Water

The H—O—H bond angle of water molecules is 105°, an angle that is quite close to the 109.5° bond angles of molecules that have a tetrahedral structure.

$$\overset{\displaystyle \cdot\cdot}{\underset{H \qquad H}{O}}$$

105°

* An explanation of electron pairing will be given in Section 1.12.

FIG. 1.4

Tetrahedral structure of methane showing the maximum separation of the bonding electron pairs.

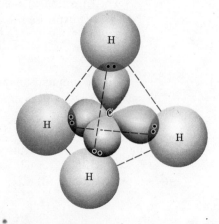

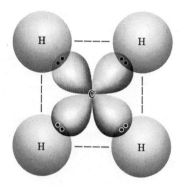

FIG. 1.5

Hypothetical square planar structure for methane. When comparing this structure to that in Fig. 1.4, remember all of the atoms in the square structure are in the same plane (of the paper), while in the tetrahedral structure the atoms are in three dimensions.

We can write a tetrahedral structure for a water molecule if we place the two nonbonding electron pairs at corners. Such a structure is shown below.

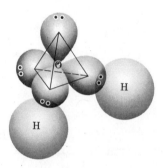

A regular tetrahedral structure for water molecules would, of course, result in a H—O—H bond angle of 109.5°, not the 105° bond angle that is actually found. We can account for this deviation, however, because of a *difference in the repulsive force between the nonbonding electron pairs and that between the bonding electron pairs.* Bonding pairs, because they are under the influence of two nuclei, are smaller than nonbonding electron pairs. Since electron pairs repel each other, and since these "repulsions" are caused by interactions of electron pairs in space, it is clear that

FIG. 1.6

(a) *Line-and-circle representation of a tetrahedral atom.*

(b) *Ball-and-stick model.*

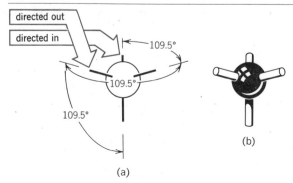

(a)

(b)

the repulsive force between the electrons of the smaller bonding pairs should be less than that between the larger nonbonding pairs. The H—O—H bond angle of water is, as a result, compressed to 105°.

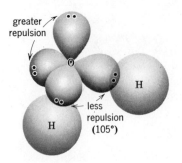

Ammonia

The bond angles in ammonia are 107°, a value even closer to the tetrahedral

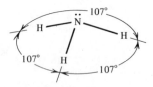

angle (109.5°) than that of water. We write a tetrahedral structure for ammonia by placing the nonbonding pair at one corner. We can account for the deviation of

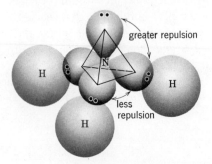

the H—N—H bond angles from those of a regular tetrahedron by assuming that greater repulsion occurs in the interaction of the electrons of the nonbonding pair with those of the bonding pairs than in the interaction of electrons of the bonding pairs with each other.

Boron Trifluoride

Boron, a Group III element, has only three outer shell electrons. In the compound boron trifluoride (BF_3) these three electrons are shared with three fluorine atoms. As a result, the boron atom in BF_3 has only six electrons (three bonding pairs) around it. The boron trifluoride molecule is known to have a shape such that the three fluorines lie at the corners of an equilateral triangle. The bond angles are 120°.

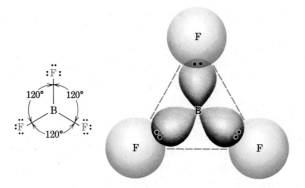

This trigonal (triangular) planar structure allows for the maximum separation (120°) of the bonding pairs of electrons.

Problem 1.6

Predict the general shapes of the following molecules and ions.

(a) SiH_4 (d) BF_4^-

(b) $CH_3{:}^-$ (e) CCl_4

(c) CH_3^+ (f) $BeCl_2$

1.7 POLAR AND NONPOLAR MOLECULES

In our discussion of dipole moments in Section 1.5, we restricted our attention to simple diatomic molecules. Any *diatomic* molecule in which the two atoms are *different* (and thus have different electronegativities), will, of necessity, have a dipole moment. If we examine Table 1.2, however, we find that there are a number of molecules (e.g., CCl_4, CO_2) that consist of more than two atoms and that have *polar* bonds, *but they have no dipole moment.* Now that we have an understanding of the shapes of molecules we can understand how this can occur.

TABLE 1.2 Dipole Moments of Some Simple Molecules

FORMULA	μ IN D	FORMULA	μ IN D
H_2	0	CH_4	0
Cl_2	0	CH_3Cl	1.87
HF	1.91	CH_2Cl_2	1.55
HCl	1.08	$CHCl_3$	1.02
HBr	0.80	CCl_4	0
HI	0.42	NH_3	1.47
BF_3	0	NF_3	0.23
CO_2	0	H_2O	1.85

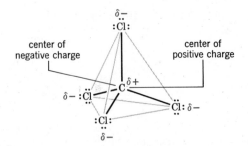

FIG. 1.7

Charge distribution in carbon tetrachloride.

Let us consider carbon tetrachloride (CCl_4) as an example. Because the electronegativity of chlorine is greater than that of carbon, each of the carbon-chlorine bonds in CCl_4 is polar. Each chlorine has a partial negative charge, and the carbon is considerably positive. Because carbon tetrachloride is tetrahedral (Fig. 1.7), however, *the center of positive charge and the center of negative charge coincide, and the molecule has no net dipole moment.*

This can be illustrated in a slightly different way: if we use arrows ($\leftrightarrow$) to represent the direction of polarity of each bond we get the arrangement of bond moments shown below.

If we then consider the bond moments to be vectors of equal magnitude we can, with the application of geometry, see that a tetrahedral orientation of the equal bond moments causes them to exactly cancel and thus result in *no net dipole moment.*

The chloromethane molecule (CH_3Cl) has a net dipole moment of 1.87 D. Since carbon and hydrogen have electronegativities (Table 1.1) that are nearly the same, the contribution of three C—H bonds to the net dipole is negligible. The electronegativity difference between carbon and chlorine is large, however, and it is this highly polar bond that accounts for most of the dipole moment of CH_3Cl.

Problem 1.7

Carbon dioxide (CO_2) is a linear molecule. Show how this fact accounts for the fact that CO_2 has no dipole moment.

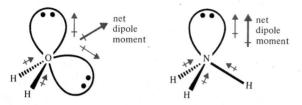

FIG. 1.8
*Bond moments and the resulting dipole moment of
water and ammonia.*

Unshared electron pairs make large contributions to the dipole moments of water and ammonia. The unshared pairs have no atom attached to them to partially neutralize their negative charge; and as a result, these unshared electron pairs contribute a large moment away from the central atom (Fig. 1.8). (The O—H and N—H moments are also appreciable.)

Problem 1.8

Nitrogen trifluoride ($:NF_3$) has a shape very much like that of ammonia. It has, however, a very low dipole moment ($\mu = 0.24$ D). How can you explain this?

Problem 1.9

BF_3 has no dipole moment. How can this be explained?

1.8 THE CARBON-CARBON COVALENT BOND

Carbon's ability to form strong covalent bonds to other carbon atoms is the single property of the carbon atom that—more than any other—accounts for the very existence of a field of study called organic chemistry. It is this property, too, that accounts in part for carbon being the element around which most of the molecules of living organisms are constructed. Carbon's ability to form strong bonds to other carbon atoms and to form strong bonds to hydrogen, oxygen, sulfur, and nitrogen atoms as well, provides the necessary versatility of structure that makes possible the vast number of different molecules required for complex living organisms.

There are molecules that contain literally thousands of carbon atoms joined in a single chain. The familiar plastic, *polyethylene*, is an example of a substance whose molecules contain chains of this great length. A portion of a polyethylene chain is shown in Fig. 1.9. Each carbon atom in the polyethylene chain is covalently bonded to a carbon atom on either side and to two hydrogen atoms.

Most of the molecules that we will study in this book have chains of atoms much shorter than those of polyethylene. Compounds containing chains of great length are usually difficult to isolate in a chemically pure state. Polyethylene plastic is a mixture of molecules containing chains ranging in length from a few hundred atoms up to several thousand. While compounds whose molecules have very long chains often have certain practical uses, (and in living cells very important ones), they are not so chemically complex as they might appear. Pentane (Fig. 1.10), a compound with much simpler molecules, shows almost all of the chemical reactions

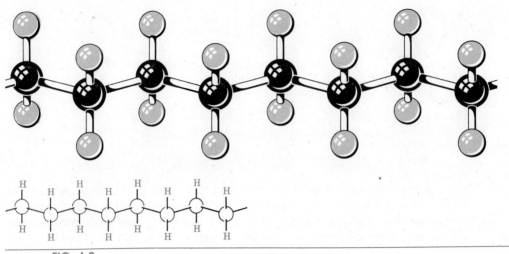

FIG. 1.9

A portion of the carbon chain of polyethylene.

that polyethylene does, and pentane is much easier to deal with experimentally and conceptually. After we understand the structure and reactions of smaller molecules, such as those of pentane, we can then move on to the larger ones.

Carbon is not unique in its ability to form covalent bonds to itself. Other atoms (e.g., oxygen, nitrogen, and silicon) have the same property. Carbon *is unique*, however, in its ability to form *very stable* bonds to other carbon atoms while at the same time being able to form strong bonds to atoms of other elements. The average amount of energy required to break a carbon-carbon bond is $\sim$ 83 kcal/mole.*

*A kilocalorie of energy is the amount of energy (in the form of heat) required to raise by 1°C the temperature of one kilogram (1000 grams) of water at 15°C.

FIG. 1.10

A pentane molecule.

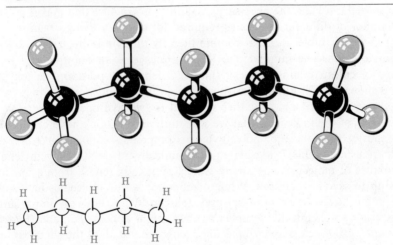

Silicon, the element most like carbon in its atomic structure, forms silicon-silicon covalent bonds but, because of the larger size of the silicon atom, the silicon-silicon bond is considerably longer and weaker. Only 53 kcal/mole are required to break a silicon-silicon bond. There also exist compounds that contain several nitrogen-nitrogen covalent bonds. Those containing as few as five linearly bonded nitrogen atoms are highly unstable, however, and this instability seems to be caused by the

$$-\overset{\displaystyle ..}{\underset{\displaystyle |}{N}}-\overset{\displaystyle ..}{\underset{\displaystyle |}{N}}-$$

Typical bond energy = 50 kcal/mole

repulsive forces between the nonbonding electron pairs of nitrogen. When only two oxygen atoms are linked the bond is quite weak. Once again, the explanation for the instability of the oxygen-oxygen bond lies in the large repulsive forces between the *two* nonbonding pairs on each atom.

$$-\overset{\displaystyle ..}{\underset{\displaystyle ..}{O}}-\overset{\displaystyle ..}{\underset{\displaystyle ..}{O}}-$$

Typical bond energy = 34 kcal/mole

1.9 EMPIRICAL AND MOLECULAR FORMULAS

In Section 1.2, we discussed briefly the pioneering work of Berzelius, Dumas, Liebig, and Cannizzaro in devising methods for determining the formulas of organic compounds. Although the experimental procedures for these analyses have been refined, the basic processes for determining the elemental composition of an organic compound today are not substantially different from those used in the nineteenth century. A carefully weighed quantity of the compound to be analyzed is oxidized completely to carbon dioxide and water. The weights of carbon dioxide and water that are formed are then carefully measured and from these weights the percentages of carbon and hydrogen in the compound can be calculated. The percentage of nitrogen is usually determined by measuring the volume of nitrogen (N_2) produced in a similar analysis.

Special techniques for determining the percentage composition of other elements typically found in organic compounds have also been developed, but the determination of the percentage of oxygen is usually difficult. However, if the percentage composition of all the other elements present in a compound is known, then the percentage of oxygen can be determined by difference. The following examples will illustrate how these calculations can be carried out.

Example A

A new organic compound is found to have the following elemental analysis.

Carbon	67.95%
Hydrogen	5.69
Nitrogen	26.20
Total:	99.84%

Since the total of these percentages is very close to 100% (within experimental error) we can assume that no other element is present. For the purpose of our calculation it is convenient to assume that we have a 100-g sample. If we did, it would contain the following:

67.95 g of carbon
5.69 g of hydrogen
26.20 g of nitrogen

We can divide each of these quantities by the atomic weight of each element and obtain the number of moles of each element, respectively.

$$\text{C:} \quad \frac{67.95g}{12.01 \text{ g/mole}} = 5.66 \text{ moles}$$

$$\text{H:} \quad \frac{5.69 \text{ g}}{1.008 \text{ g/mole}} = 5.64 \text{ moles}$$

$$\text{N:} \quad \frac{26.20 \text{ g}}{14.01 \text{ g/mole}} = 1.87 \text{ moles}$$

Since it is not possible for a molecule to contain fractions of atoms, we then attempt to convert these fractional numbers of moles to a series of whole numbers. Dividing each by the smallest number gives:

$$\text{C:} \quad \frac{5.66}{1.87} = 3.03 \text{ which is} \sim 3$$

$$\text{H:} \quad \frac{5.64}{1.87} = 3.02 \text{ which is} \sim 3$$

$$\text{N:} \quad \frac{1.87}{1.87} = 1.00$$

These values are very close to C_3H_3N (once again, within experimental error), and thus C_3H_3N is the *simplest formula* for the compound. The simplest formula for a compound is called the *empirical formula*. By simplest formula we mean the formula in which the subscripts are the smallest integers that give the ratio of atoms in the compound. The molecular formula of this particular compound could be C_3H_3N or some whole-number multiple of C_3H_3N; that is, $C_6H_6N_2$, $C_9H_9N_3$, $C_{12}H_{12}N_4$, and so on. If, in a separate determination, we find that the molecular weight of the compound is 108 ± 3, we can be certain that the *molecular formula* of the compound is $C_6H_6N_2$.

Formula	Molecular Weight
C_3H_3N	53.06
$C_6H_6N_2$	106.13 (which is within the range 108 ± 3)
$C_9H_9N_3$	159.19
$C_{12}H_{12}N_4$	212.26

The most accurate method of determining formula weights is by a technique called mass spectral analysis, but a variety of other methods based on freezing point depression, boiling point elevation, osmotic pressure and vapor density, are used as well.

Example B

Histidine, an amino acid isolated from protein, has the following elemental analysis:

Carbon	46.38%
Hydrogen	5.90
Nitrogen	27.01
Total:	79.29
Difference	20.71 (assumed to be oxygen)
	100.00%

Since no elements, other than oxygen, are found to be present in histidine the difference is assumed to be oxygen. Again, we assume a 100-g sample and divide the weight of each element by its gram-atomic weight. This gives us the ratio of moles (A).

$$\text{(A) moles} \qquad \text{(B)} \qquad \text{(C)}$$

$$\frac{46.38}{12.01} = 3.86 \qquad \frac{3.86}{1.29} = 2.99 \times 2 = 5.98 = 6$$

$$\frac{5.90}{1.008} = 5.85 \qquad \frac{5.85}{1.29} = 4.53 \times 2 = 9.06 = 9$$

$$\frac{27.11}{14.01} = 1.94 \qquad \frac{1.94}{1.29} = 1.50 \times 2 = 3.00 = 3$$

$$\frac{20.71}{16.00} = 1.29 \qquad \frac{1.29}{1.29} = 1.00 \times 2 = 2.00 = 2$$

Dividing each of the moles (A) by the smallest, does not give a set of numbers (B) that is close to a set of whole numbers. Multiplying each of the numbers in column (B) by 2 does, however, in column (C). The empirical formula of histidine is, therefore, $C_6H_9N_3O_2$.

In a separate determination the molecular weight of histidine was found to be 158 ± 5. The empirical formula weight of $C_6H_9N_3O_2$ (155.15) is within this range; thus the molecular formula for histidine is the same as the empirical formula.

Problem 1.10

The widely used antibiotic, Penicillin G, gave the following elemental analysis: C, 57.45%; H, 5.40%; N, 8.45%; S, 9.61%. The molecular weight of Penicillin G is 330 ± 10. Assume that no other elements except oxygen, are present and calculate the empirical and molecular formulas for Penicillin G.

1.10 ISOMERS AND THE REPRESENTATION OF STRUCTURAL FORMULAS

More than two million organic compounds have now been isolated in a pure state and have been characterized on the basis of their physical and chemical properties. Additional compounds are added to this list by the tens of thousands each year. A look into *Chemical Abstracts* or Beilstein's *Handbuch der Organischen Chemie*,

TABLE 1.3 Properties of Ethyl Alcohol and Dimethyl Ether

	ETHYL ALCOHOL	DIMETHYL ETHER
Boiling point	78.5°	−24.9°
Melting point	−117.3°	−138°
Reaction with sodium	Displaces H_2	No reaction

where known organic compounds are catalogued, shows that there are dozens and sometimes hundreds of *different compounds that have the same molecular formula.* Such compounds are called *isomers.* Different compounds with the same molecular formula are said to be *isomeric,* and this phenomenon is called *isomerism.* Before going further, let us consider a simple example of two isomeric compounds.

There are two substances that have the molecular formula C_2H_6O. One of these isomers is called dimethyl ether, the other ethyl alcohol. These two compounds are strikingly different in both their physical and chemical properties (Table 1.3).

One glance at the structural formulas for these two compounds reveals their difference (Fig. 1.11). The atoms of ethyl alcohol are connected in a way that is different from those of dimethyl ether. In ethyl alcohol there is a C—C—O linkage; in dimethyl ether the linkage is C—O—C. Ethyl alcohol has a hydrogen attached to oxygen; in dimethyl ether all of the hydrogens are attached to carbon. It is the hydrogen atom covalently bonded to oxygen in ethanol that is displaced when ethanol reacts with sodium:

$$\text{H}-\overset{\overset{\displaystyle H}{|}}{\underset{\underset{\displaystyle H}{|}}{C}}-\overset{\overset{\displaystyle H}{|}}{\underset{\underset{\displaystyle H}{|}}{C}}-\ddot{\underset{\displaystyle\cdot\cdot}{O}}-\text{H} + \text{Na} \longrightarrow \text{H}-\overset{\overset{\displaystyle H}{|}}{\underset{\underset{\displaystyle H}{|}}{C}}-\overset{\overset{\displaystyle H}{|}}{\underset{\underset{\displaystyle H}{|}}{C}}-\overset{\cdot\cdot}{\underset{\displaystyle\cdot\cdot}{\overset{-}{O}}}\text{:}\text{Na}^+ + \tfrac{1}{2}\text{H}_2$$

Hydrogen atoms that are covalently bonded to carbon are normally unreactive toward sodium. As a result, none of the hydrogens in dimethyl ether is displaced when one treats dimethyl ether with sodium.

Ethanol and dimethyl ether are examples of *structural isomers.* Structural isomers are different compounds that have the same molecular formula, but *differ in the order in which their atoms are bonded together.* Structural isomers always have different physical properties (e.g., melting point, boiling point, density). The differences, however, may not always be as large as those between ethanol and dimethyl ether.

Problem 1.11

Write formulas for all of the structural isomers of C_3H_8O.

Organic chemists use a variety of ways to write structural formulas. The representation that is chosen in a particular instance is the one that best illustrates the property being considered or the one that represents the structure most conveniently. The *dot structure* for propanol (Fig. 1.12) shows us clearly the

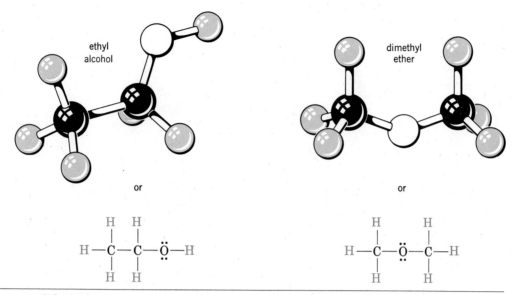

FIG. 1.11
*Ball-and-stick models and molecular formulas
for ethyl alcohol and dimethyl ether.*

number of valence electrons and the way that they are shared. The *dash formula* is easier to write and shows us, as does the dot structure, the order in which the atoms are attached. The condensed formula is still easier to write, and when we become more familiar with it, it will impart all the information that is contained in either the dot or dash structure. In condensed formulas, the atoms that are attached to a particular carbon atom are written immediately after that atom. For example:

FIG. 1.12
Structural formulas for propanol.

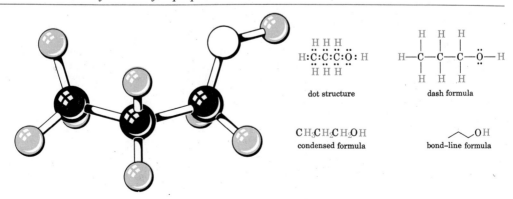

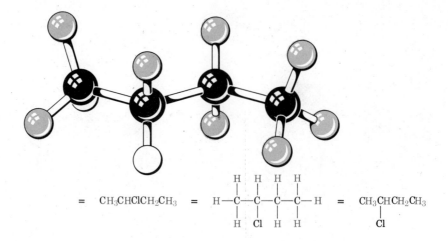

$$= \quad CH_3CHClCH_2CH_3 \quad = \quad H-\overset{\overset{\displaystyle H}{|}}{\underset{\underset{\displaystyle H}{|}}{C}}-\overset{\overset{\displaystyle H}{|}}{\underset{\underset{\displaystyle Cl}{|}}{C}}-\overset{\overset{\displaystyle H}{|}}{\underset{\underset{\displaystyle H}{|}}{C}}-\overset{\overset{\displaystyle H}{|}}{\underset{\underset{\displaystyle H}{|}}{C}}-H \quad = \quad CH_3\underset{\underset{\displaystyle Cl}{|}}{CH}CH_2CH_3$$

and

$$= \quad CH_3CH(CH_3)CH_2CH_3 \quad = \quad H-\overset{\overset{\displaystyle H}{|}}{\underset{\underset{\displaystyle H}{|}}{C}}-\overset{\overset{\displaystyle H}{|}}{\underset{\underset{\displaystyle |}{C}}{C}}-\overset{\overset{\displaystyle H}{|}}{\underset{\underset{\displaystyle H}{|}}{C}}-\overset{\overset{\displaystyle H}{|}}{\underset{\underset{\displaystyle H}{|}}{C}}-H \quad = \quad CH_3\underset{\underset{\displaystyle CH_3}{|}}{CH}CH_2CH_3$$

$$H-\overset{\overset{\displaystyle H}{|}}{\underset{\underset{\displaystyle H}{|}}{C}}-H$$

The bond-line representation is the easiest of all to write because it shows only the carbon skeleton. The number of hydrogen atoms necessary to fulfill the carbon atoms' valences are assumed to be present, but we do not write them in. Other atoms (e.g., O, Cl, N) *are* written in:

$$CH_3CHClCH_2CH_3 = \quad = CH_3\underset{\underset{\displaystyle Cl}{|}}{CH}CH_2CH_3$$

$$CH_3—CH(CH_3)CH_2CH_3 = \quad = CH_3\underset{\underset{\displaystyle CH_3}{|}}{CH}CH_2CH_3$$

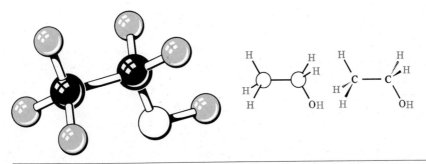

FIG. 1.13

The circle-and-line structure and the dash-line-wedge structure for ethanol.

None of the structures that we have described so far tell us how the atoms are arranged in space. There are, however, two frequently used formulations that do impart the three-dimensional structure of a molecule. They are the Alexander (or circle-and-line structure) and the dash-line-wedge structure (Fig. 1.13). In the dash-line-wedge formula, atoms that project out of the plane of the paper are connected by a wedge (◄), those that lie behind the plane are connected with a dash (----), and those atoms in the plane of the paper are connected by a line.

1.11 QUANTUM MECHANICS

In 1926 a new theory of atomic and molecular structure was advanced independently and almost simultaneously by three men: Erwin Schrödinger, Werner Heisenberg, and Paul Dirac. This theory, called *wave mechanics* by Schrödinger or *quantum mechanics* by Heisenberg, has become the basis from which we derive our modern understanding of bonding in molecules. The language of quantum mechanics is largely a mathematical language and, as such, it has a formidable appearance to those unfamiliar with differential equations, matrix algebra, and wave theory. Yet, in a sense, wave mechanics is as central to modern chemistry as the theory of natural selection is to modern biology. For this reason we have attempted to give an essentially nonmathematical account of the development of quantum mechanics as a special topic in Chapter 8. Now, however, we will discuss only those essential ideas of quantum mechanics that are necessary to develop a modern understanding of bonding in covalent molecules.

The formulation of quantum mechanics that Schrödinger advanced is the form that is most often used by chemists. In Schrödinger's publication the motion of electrons was described in terms that took into account the wave nature of the electron.* Schrödinger developed mathematical expressions for the energy of the

---

*This idea—that electrons have the properties of a wave as well as those of a particle—was proposed by Louis de Broglie in 1923. According to de Broglie the wavelength of an electron, λ, is given by the following equation:

$$\lambda = \frac{h}{mv_e}$$

where h is Planck's constant (6.625×10^{-27} erg-sec), m is the mass of the electron, and v_e is the velocity of the electron.

electron called *wave equations*. These wave equations were then manipulated to yield a series of solutions called *wave functions*.

Wave functions are most often denoted by the Greek letter psi (ψ) and each wave function (ψ function) corresponds to a different energy level for the electron.

1.12 ATOMIC ORBITALS

For a short time after Schrödinger's proposal in 1926, a precise physical interpretation for the wave function eluded early practitioners of quantum mechanics. It remained for Max Born, a few months later, to point out that the square of ψ *could* be given a precise physical meaning. According to Born, ψ^2 for a particular location (x,y,z), expresses the *probability* of finding an electron at that particular location in space. If ψ^2 is large in a unit volume of space, the probability of finding an electron in that volume is great; conversely if ψ^2 for some other unit volume of space is small, the probability of finding an electron there is low. Plots of ψ^2 in three dimensions generate the shapes of the familiar s, p, and d atomic orbitals.

The f orbitals are practically never used in organic chemistry, so we will not concern ourselves with them in this book. The d orbitals will be discussed briefly later when we discuss compounds in which d orbital interactions are important. The s and p orbitals are, by far, the most important in the formation of organic molecules and, at this point, we will limit our discussion to them.

The shapes of s and p orbitals are shown in Fig. 1.14.* Both the $1s$ and $2s$

---

* There is a finite, but very small, probability of finding an electron at greater distances from the nucleus. The volumes that we typically use to illustrate an orbital are those volumes that would contain the electron 90 to 95% of the time.

FIG. 1.14
The shapes of some s *and* p *orbitals.*

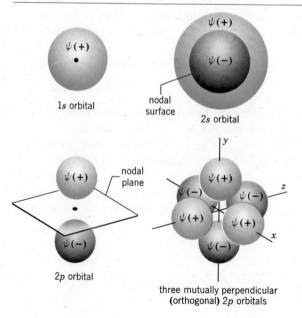

1s orbital

nodal surface

2s orbital

nodal plane

2p orbital

three mutually perpendicular (orthogonal) 2p orbitals

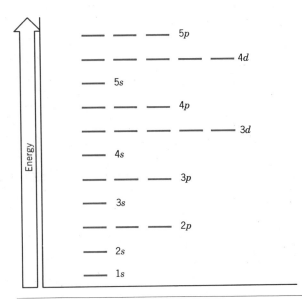

FIG. 1.15
The relative energy levels of some atomic orbitals.

orbitals (as are all higher *s* orbitals) are spherically symmetrical. The sign of the wave function, ψ_{1s}, is positive (+) over the entire 1*s* orbital. The 2*s* orbital contains a nodal surface, that is, an area where $\psi = 0$. In the inner portion of the 2*s* orbital, ψ_{2s} is negative.

The 2*p* orbitals have the shape of two almost-touching spheres. The sign of the wave function, ψ_{2p}, is positive in one lobe (or sphere) and negative in the other.* A nodal plane separates the two lobes of a *p* orbital, and the three *p* orbitals are arranged in space so that they are mutually perpendicular.

The relative energies of the orbitals are shown in Fig. 1.15. Electrons in 1*s* orbitals have the lowest energy because they are closest to the positive nucleus. Electrons in 2*s* orbitals are next lowest in energy. Electrons of 2*p* orbitals have equal but still higher energy. (Orbitals of equal energy are said to be *degenerate orbitals*.)

We can use this chart to arrive at the electron configuration of the lowest energy state (ground state) of any atom we choose. We need only follow a few simple rules.

1. *The aufbau principle:* orbitals are filled so that those of lowest energy are filled first. (*Aufbau* is German for "building up.")

---

* You should not associate the sign of the wave function with anything having to do with charge. The (+) and (−) signs, associated with ψ are simply the arithmetic signs of the mathematical expression for the wave function in that region of space (cf. Section 8.1). The (+) and (−) signs do not imply a greater or lesser probability of finding an electron because the probability of finding the electron is ψ^2, and ψ^2 is positive in either lobe. The significance of the (+) and (−) signs of the wave function in atomic orbitals will become clear later, when we see how atomic orbitals combine to form molecular orbitals.

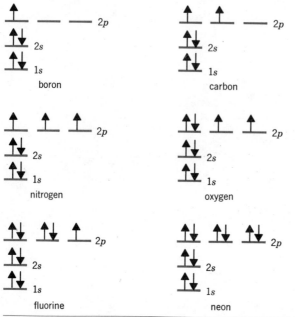

FIG. 1.16

The electron configuration of some second-row elements.

2. *The Pauli exclusion principle:* a maximum of two electrons may be placed in each orbital but only when the spins of the electrons are opposed.*

3. *Hund's rule:* when we come to orbitals of equal energy (degenerate orbitals) we add one electron to each orbital until all the degenerate orbitals contain one electron. We then begin adding electrons to the degenerate orbitals so that the spins are opposed.

If we apply these principles to some of the second row elements of the Periodic Table we get the results shown in Fig. 1.16.

Before we leave the subject of atomic orbitals we should point out that the electronic configurations of the atoms given above are obtained from solutions of the Schrödinger equation calculated for the hydrogen atom. The hydrogen atom, consisting of only two particles—a proton and an electron—is the only atom for which an exact solution to the Schrödinger equation has been obtained. When more complex systems are considered, approximations have to be made in order to simplify the mathematics. Even when these approximations are made, however, quantum mechanics gives surprisingly accurate correlations with results obtained experimentally for many complex molecules. In less complex molecules (consisting of only three or four interacting particles), the correlation obtained between theory and experiment is remarkable.

---

*Electrons spin about an axis. Since the electron is a charged particle, its spin produces a magnetic field, and this gives rise to the spin quantum number. For reasons that we cannot go into here, the electron is permitted only two equal and opposite spin orientations corresponding to two equal and opposite spin states, $+\frac{1}{2}$ and $-\frac{1}{2}$. We designate these two spin states by an arrow up ↑ or down ↓.

1.13 MOLECULAR ORBITALS

For the organic chemist the greatest utility of atomic orbitals is in using them to understand how atoms combine to form molecules. We will have much more to say about this subject in subsequent chapters for, as we have already said, covalent bonds are central to the study of organic chemistry. First, however, we concern ourselves with a very simple case: the covalent bond that is formed when two hydrogen atoms combine to form a hydrogen molecule. We will see that the description of the formation of the H—H bond is the same, or at least very similar, to the description of bonds in more complex molecules.

Let us begin by examining what happens to the total energy of two hydrogen atoms with electrons of opposite spins when they are brought closer and closer together. This can best be shown with the curve shown in Fig. 1.17.

When the atoms of hydrogen are only a few Angstrom units apart (I) their total energy is simply that of two isolated hydrogen atoms. As the hydrogen atoms move closer together (II), each nucleus increasingly attracts the other's electron. This attraction more than compensates for the repulsive force between the two electrons (or the two nuclei), and the result of this attraction *is to lower the energy of the total system.* When the two nuclei are 0.74 Å (III) apart, the most stable (lowest energy) state is obtained. This distance, 0.74 Å, corresponds to the *bond distance* for the hydrogen molecule. If the nuclei are moved closer together (IV) the repulsion of the two positively charged nuclei predominates, and the energy of the system rises.

Figure 1.17 shows us why hydrogen atoms do not remain in the atomic state very long when there are other hydrogen atoms nearby. Hydrogen atoms will inevitably collide with each other, and as they do so they will assume a state of lower energy by forming a hydrogen molecule.

Figure 1.17 also shows why we have to supply energy to fragment a hydrogen molecule into its two constituent atoms. We must do so in order to break the H—H covalent bond. The amount of energy required to fragment one mole (6.02×10^{23}) hydrogen molecules is equal to 104.2 kcal at 25°C, and is called the *bond dissociation energy* for the hydrogen-hydrogen bond.

FIG. 1.17

The potential energy of the hydrogen molecule as a function of internuclear distance.

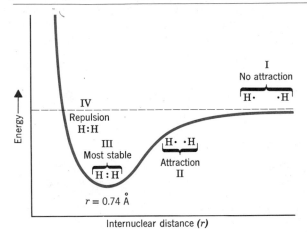

When hydrogen atoms combine to form hydrogen molecules 104.2 kcal/mole of energy are evolved.

$$H:H \rightarrow 2H\cdot \qquad \Delta H = +104.2 \text{ kcal/mole}$$

$$2H\cdot \rightarrow H:H \qquad \Delta H = -104.2 \text{ kcal/mole}$$

A quantum mechanical explanation for what happens when two hydrogen atoms combine to form a hydrogen molecule is given in the following way: as the two hydrogen atoms approach each other their $1s$ atomic orbitals begin to overlap. This overlap increases as they move closer together. As a result of orbital overlap, the *atomic orbitals* (ψ_{1s}) combine and become molecular orbitals (ψ molec). One molecular orbital that is formed (Fig. 1.18) encompasses both nuclei, and in it the two electrons are free to move about both nuclei. They are not restricted to the vicinity of only one nucleus as they were in the separate atomic orbitals.

What we have just described can also be described mathematically by the method of *linear combination of atomic orbitals* (called the LCAO method). In this treatment, wave functions of the atomic orbitals are combined in a linear fashion (either added or subtracted) in order to obtain new wave functions for molecular orbitals. *The number of molecular orbitals that are obtained equals the number of atomic orbitals that are combined.* Thus, for the hydrogen molecule, we obtain two molecular orbitals. The first molecular orbital is obtained by adding the wave functions for the two atomic orbitals:

$$\psi_{\text{molec}} = N_1 \left(\psi_{1s}^{H} + \psi_{1s}^{H} \right) \text{ (bonding molecular orbital)}$$

(Where N_1 is a constant called the normalization constant) This molecular orbital is called the *bonding* molecular orbital, and in actual calculations it can be shown that the energy of two electrons in this molecular orbital is lower than that of two electrons in separate $1s$ atomic orbitals.

The second molecular orbital is obtained by subtraction of the atomic orbitals:

$$\psi_{\text{molec}}^* = N_2 \left(\psi_{1s}^{H} - \psi_{1s}^{H} \right) \text{ (antibonding molecular orbital)}$$

Calculations show that the energy of two electrons in this orbital will be greater than their total energy in two separate $1s$ atomic orbitals. Consequently, this molecular orbital is called an *antibonding* molecular orbital.

An energy diagram for the molecular orbitals of hydrogen is shown in Fig. 1.19. Notice that electrons are placed in molecular orbitals in the same way that they were in atomic orbitals. Two electrons (with their spins opposed) occupy the bonding molecular orbital, where their total energy is less than in the separate atomic

FIG. 1.18

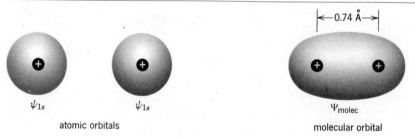

ψ_{1s} ψ_{1s}

atomic orbitals

Ψ_{molec}

molecular orbital

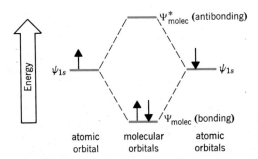

FIG. 1.19
Energy diagram for the hydrogen molecule. Combination of two atomic orbitals. ψ_{1s} gives two molecular orbitals. Ψ_{molec} and Ψ^*_{molec}. The energy of Ψ_{molec} is lower than that of the separate atomic orbitals, and in the lowest electronic energy state of molecular hydrogen it contains both electrons.

orbitals. This is the *lowest electronic energy state* or *ground state* of the hydrogen molecule.

A way of visualizing the formation of the bonding and antibonding molecular orbitals that takes into account the mathematical signs of the wave functions of the orbitals is shown in Fig. 1.20. Addition of two 1s orbitals yields a molecular orbital in which the wave function is positive over the entire region. This is the *bonding molecular orbital*. Subtraction of the two wave functions yields the antibonding orbital. The antibonding orbital has a nodal plane halfway between the two nuclei. When electrons occupy the antibonding molecular orbital they repel each other strongly and, as a result, the antibonding orbital is of high energy.

Now that we have seen several different ways of explaining how the reaction $H\cdot + H\cdot \longrightarrow H_2$ occurs; we can also explain why the reaction $He: + He: \longrightarrow He_2$ does not occur. We begin by assuming that the molecular orbitals for He_2 are similar to those for H_2. With the electrons included, the energy diagram for the formation of He_2 is given in Fig. 1.21. In the case of He_2 the bonding orbital is fully occupied just as it is in the case of H_2; but in He_2 the antibonding orbital is also fully occupied. The two electrons in the antibonding orbital strongly repel each other. Detailed calculations for He_2 show that the stability gained from the electrons in the bonding orbital is insufficient to counterbalance the instability that arises from the electrons in the antibonding orbital. As a result, He_2 is not predicted to form from separate

FIG. 1.20

Two ways that 1s atomic orbitals can combine to form molecular orbitals. Addition of the atomic orbitals gives the bonding molecular orbital, Ψ_{molec}. Subtraction of the atomic orbitals gives the antibonding orbital, Ψ^*_{molec}.

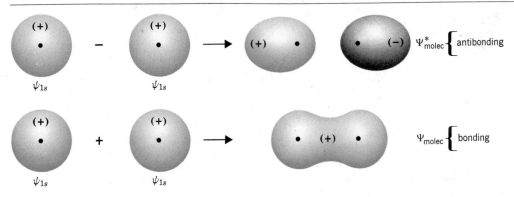

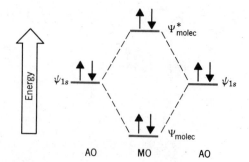

FIG. 1.21
Energy diagram for the formation of a hypothetical He$_2$ molecule.

helium atoms. The theory, thus, accommodates what we know from experiments to be true. Helium exists as a monoatomic species, He, not as He$_2$.

At this point it may seem that these quantum mechanically based ideas are unnecessarily complicated and that the much simpler Lewis theory would do just as well. For certain features of simple molecules this is true. As we go on to more complicated molecules, however, we will find that a great many features of their structures and reactions require more detailed explanations than the Lewis theory permits. For many of these we will find the molecular orbital treatment to be invaluable.

1.14 VALENCE BOND THEORY AND RESONANCE

During the 1920s two mathematical methods for describing covalent bonds were developed: molecular orbital theory, which we have just discussed, and *valence bond theory*, which we will now describe in a limited way. Both theories are approximate ways of solving the Schrödinger equations for molecules; and both theories, when appropriately applied, give descriptions of complex molecules that correspond very closely to the experimentally determined properties of these molecules. The molecular orbital method is usually more amenable to detailed calculations, particularly those carried out with computers, and is more widely used today. Valence bond theory, however, corresponds more closely to the Lewis structures for molecules and ions that we have written earlier in this chapter. Valence bond theory has also given rise to a very useful qualitative approach to molecules and ions called *resonance theory*. Since we will make extensive use of resonance theory at many points in this text, it is appropriate to introduce some of the basic concepts of resonance theory at this point.

Valence bond theory differs from molecular orbital theory in this way: in molecular orbital theory, atomic orbitals are used mathematically to calculate molecular orbitals. In valence bond theory, *we use Lewis-type structures as a starting point for our calculations instead*. The hydrogen molecule, for example, is treated in valence bond theory as though it were a combination of the following structures:

$$H : H \quad \text{and} \quad H^{+ \ -}:H \quad \text{and} \quad H:^{- \ +}H$$

In its mathematical application, wave functions for each of these structures are combined and, in this way, wave functions for the molecular orbitals of hydrogen can be obtained. These calculations are beyond the scope of this text, but the concept of *resonance* that is employed in valence bond theory is very important.

In resonance theory the three structures that we have just written are said to be three *resonance* (or canonical) *contributors* to the actual structure of the hydrogen molecule. We indicate the fact that these structures *are* resonance structures by connecting them with *double-headed arrows:*

$$H : H \longleftrightarrow {}^{+}H \; {}^{-}{:}H \longleftrightarrow H{:}^{-} \; H^{+}$$

Resonance structures are defined as structures that differ from each other only in the positions of the electrons. In the structures above only the electrons have been moved, the nuclei of the hydrogen atoms (the protons) remain in the same relative position with respect to each other.

In the first structure, H : H, the two electrons are an equal distance from either nucleus. We can understand how this structure would help hold the molecule together, for in it, each nucleus attracts the centrally located electrons, and this attraction is greater than the repulsive force between the two nuclei. In the other two structures, $H^{+} \; {}^{-}{:}H$ and $H{:}^{-} \; {}^{+}H$, the electrons are closer to one nucleus than the other, and the nuclei, as a result, bear formal positive and negative charges. The hydrogen ions in these structures attract each other because of their opposite charges, and these structures also contribute to bonding in the molecule.

Resonance structures are not actual discrete structures for the molecule; *they exist only in theory.* No single contributor adequately represents the molecule. In resonance theory, we view the hydrogen molecule, which is of course a real entity, as being a *hybrid* of these three *hypothetical* resonance structures.

Two analogies that have been used may help make this clear. One might, for example, describe a *real* person as being like a combination of Don Quixote, Sir Galahad, and Robin Hood (three fictional characters). Or one might describe a rhinoceros (a real animal) as resembling a combination (hybrid) of a unicorn and a dragon (both mythical animals). Resonance structures, therefore, are like dragons, unicorns, and Sir Galahads. While they have no real existence of their own, they are useful in helping describe real molecules.

Resonance structures do not necessarily make equal contributions to the hybrid. In the example of the hydrogen molecule, the structure H : H makes a greater contribution than either of the other structures. $H^{+} \; {}^{-}{:}H$ and $H{:}^{-} \; {}^{+}H$. (In terms of one of our analogies, this is like saying that a real person is more like Don Quixote, but with a little of Sir Galahad and Robin Hood thrown in.)

We can only estimate the contribution that a particular resonance structure will make on the basis of its "reasonableness." We illustrate what we mean by this with the three structures for the hydrogen molecule. The structure, H : H is the most reasonable because in it the electrons are equidistant from two atoms of necessarily equal electronegativity. The structures, $H^{+} \; {}^{-}{:}H$ and $H{:}^{-} \; {}^{+}H$, are less reasonable because they presuppose that one hydrogen nucleus will exert a greater pull on the electron pair than the other nucleus. Therefore, the most reasonable structure, H : H, makes the greatest contribution to the hybrid.

Resonance theory is particularly helpful in explaining molecules or ions for which more than one *equivalent* Lewis structure can be written. Consider the carbonate ion, for example. Three different but *equivalent* structures can be written.

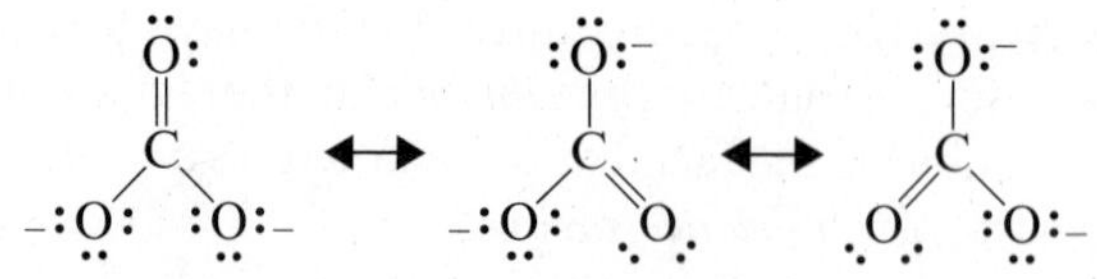

These structures differ from each other in only one way: in the position of the electrons. They are, therefore, *resonance structures*, and the carbonate ion itself is not described by any one of these structures but, instead, is a *hybrid* of all of them. The bonds of the carbonate ion are not discrete single and double bonds but are one and one-third bonds and are all equivalent. This prediction of resonance theory

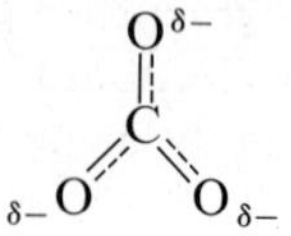

is confirmed by physical evidence. All of the bonds of the carbonate ion are of equal length, and the length is in between that of a carbon-oxygen single bond and that of a carbon-oxygen double bond. The oxygens of the carbonate ion all bear an equal partial negative charge, that is, two-thirds of a negative charge.

Problem 1.12

The ozone molecule, O_3, is a bent molecule with a 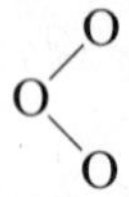 bond angle of 117°. The oxygen-oxygen bonds are of equal length, however. Can you account for this on the basis of resonance theory?

Additional Problems

1.13
Rewrite each of the condensed structural formulas given below, as *dash formulas* and as *bond-line formulas*.

(a) $CH_3CCl_2CH_2CH_3$

(b) $CH_3CH(CH_2Cl)CH_2CH_3$

(c) $CH_3C(CH_3)_2CH_2CH_3$

(d) $CH_3CHClCHClCH_3$

(e) $CH_3CH(OH)CH_2CH_3$

(f) $CH_3CH_2CH_2CH_2OH$

(g) $CH_3CCH_2CH(CH_3)_2$ (with O double-bonded)

(h) $CH_3CH_2CH(OH)CH(CH_3)_2$

1.14
Are any of the compounds listed in problem 1.13 structural isomers? If so, which ones?

1.15
Calculate the percentage composition of each of the following compounds

(a) $C_6H_{12}O_6$

(b) $CH_3CH_2NO_2$

(c) $CH_3CH_2CBr_3$

1.16
Show an electron-dot formula, including any formal charge, for each of the following molecules.

(a) CH_3NCS

(b) CH_3CNO

(c) CH_3ONO_2

(d) CH_3NCO

(e) CH_2CO

(f) CH_2N_2

1.17
Write dash formulas for each of the following bond-line formulas.

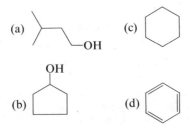

1.18
(a) Write out the electron configuration for each of the following atoms.
(b) Make a sketch of the atom showing the orbital arrangement, shape, and the disposition of the electrons in s and p orbitals.
(a) Be   (c) C   (e) O
(b) B   (d) N

1.19
What are the empirical formulas of each of the following compounds?
(a) Hydrazine, N_2H_4 (d) Nicotine, $C_{10}H_{14}N_2$
(b) Benzene, C_6H_6 (e) Cyclodecane, $C_{10}H_{20}$
(c) Dioxane, $C_4H_8O_2$ (f) Acetylene, C_2H_2

1.20
The empirical formulas and molecular weights of several compounds are given below. In each case calculate the molecular formula for the compound.

Empirical Formula	Molecular Weight
(a) CH_2O	179 ± 5
(b) CHN	80 ± 5
(c) CCl_2	410 ± 10

1.21
A gaseous compound gave the following analysis: C, 40.04%; H, 6.69%. At standard temperature and pressure, 1.00 g of the gas occupied a volume of 746 ml. What is the molecular formula of the compound?

1.22
Ethene, a gaseous hydrocarbon, has a density of 1.251 g/liter at standard temperature and pressure. When subjected to complete combustion, a 1.000-liter sample of ethene gave 3.926 g of carbon dioxide and 1.608 g of water. What is the molecular formula for ethene?

1.23
Nicotinamide, a vitamin that prevents the occurrence of pellagra, gave the following analysis: C, 59.10%; H, 4.92%; N, 22.91%. The molecular weight of nicotinamide was shown in a separate determination to be 120±5. What is the molecular formula for nicotinamide?

1.24
The antibiotic, chloramphenicol gave the following analysis: C, 40.88%; H, 3.74%; Cl, 21.95%; N, 8.67%. The molecular weight was found to be 300±30. What is the molecular formula for chloramphenicol?

1.25
Tetramethyllead, $Pb(CH_3)_4$, a compound used as an "antiknock" component of gasoline is

a liquid that boils at 106°. Lead fluoride, by contrast, is a high-melting solid, mp 824°.
(a) What kinds of bonds do you think are present in the two compounds?
(b) How can you explain the differences?

1.26

Resonance structures that can be written for hydrogen chloride are:

$$\text{H} : \overset{\cdot\cdot}{\underset{\cdot\cdot}{\text{Cl}}}: \longleftrightarrow \text{H}:^{-}\ \overset{\cdot\cdot}{\underset{\cdot\cdot}{\text{Cl}}}:^{+} \longleftrightarrow \text{H}^{+}\ :\overset{\cdot\cdot}{\underset{\cdot\cdot}{\text{Cl}}}:^{-}$$

(a) Of the two charged structures, which is the more reasonable and why?
(b) Which charged structure would make a greater contribution to the hybrid?
(c) Can you account for the direction of the dipole moment of HCl in terms of resonance theory?

1.27

When we wrote the structure of the nitrate ion, NO_3^-, on p. 8, we wrote it with single bonds to two oxygens and with a double bond to the third, for example,

$$\begin{array}{c} :\overset{\cdot\cdot}{\text{O}}:^{-} \\ \| \\ \overset{+}{\text{N}} \\ \diagdown \\ :\overset{\cdot\cdot}{\text{O}}\cdot\quad \cdot\overset{\cdot\cdot}{\underset{\cdot\cdot}{\text{O}}}:^{-} \end{array}$$

There is, however, considerable physical evidence indicating that all three nitrogen-oxygen bonds are equivalent and that they have a bond distance in between that expected for a nitrogen-oxygen single bond and that expected for a nitrogen-oxygen double bond. Explain this in terms of resonance theory.

1.28

Write eleven resonance structures for the sulfate ion and show the formal charge on each atom in each structure. Do these structures account for the fact that the sulfur-oxygen bonds are of equal length?

1.29

Multiplying dipole moments expressed in Debye units (Table 1.2) by 10^{-18} converts them to cgs units. The dipole moment of HCl, for example, is 1.08×10^{-18} esu-cm. We can assume that the distance d in the equation $\mu = e \times d$ is equal to the bond length; for HCl this is 1.27×10^{-8} cm. (a) What is the magnitude of the charge (in esu) on each atom in HCl? (b) Given that the electron charge is 4.8×10^{-10} esu, what fraction of an electron does the chlorine atom of HCl have in excess?

1.30

Chloromethane has a larger dipole moment ($\mu = 1.87$ D) than fluoromethane ($\mu = 1.81$ D) even though fluorine is more electronegative than chlorine. Explain.

1.31

(a) Write a structural formula for formaldehyde, CH_2O. (b) Formaldehyde has an unusually large dipole moment ($\mu = 2.27$ D), larger even than that of CH_3F ($\mu = 1.81$ D). Explain.

1.32

An organometallic compound called *ferrocene* contains 30.02% iron. What is the minimum molecular weight of ferrocene?

1.33

Cyanic acid, H—O—C≡N, and isocyanic acid, H—N=C=O, differ in the positions of their electrons but their structures do not represent resonance structures. (a) Explain. (b) Loss of a proton from cyanic acid yields the same anion as that obtained by loss of a proton from isocyanic acid. Explain.

1.34

The valence shell electron pair repulsion method (Sect. 1.6) can also be used to predict the shapes of molecules containing multiple bonds if we assume that the electrons of a multiple bond are located in the region of space between the two atoms joined by a multiple bond. That is, *we assume that the electrons of a multiple bond behave as though they were a single electron pair*. Use this method to predict the shapes and bond angles of (a) CO_2, (b) SO_2, and (c) SO_3.

1.35

Experiments show that the sulfur-oxygen bonds of SO_2 (cf. problem 1.34) are the same length. (a) How can you account for this? (b) What would you expect for the bonds of SO_3?

2

FUNCTIONAL GROUPS AND FAMILIES OF ORGANIC COMPOUNDS: THE MAJOR REACTION TYPES

2.1 INTRODUCTION

One great advantage of the structural theory is that it enables us to organize the vast number of organic compounds into structure-based families. The compounds that fall into a particular family are characterized by the presence, as a part of their overall structure, of certain arrangements of atoms called *functional groups*. When organized in this way, the manifold features and reactions of organic molecules become readily comprehensible. The necessity of memorizing vast numbers of unrelated facts disappears, for we find that the reactions of the simplest member of a family of organic compounds are often the same for hundreds, and sometimes thousands, of compounds in the same family. Methane (CH_4) for example, a very simple molecule, belongs to the family of molecules called *alkanes*. Molecules of alkanes contain only carbon-hydrogen bonds and carbon-carbon single bonds. Ethane (C_2H_6), propane (C_3H_8), butane (C_4H_{10}) pentane (C_5H_{12}), and heptacontane ($C_{70}H_{142}$) are also alkanes. Ethane, propane, butane, pentane, and heptacontane show the same kinds of reactions that methane does. We find, then, that an understanding of the chemistry of methane helps us to understand much of the chemistry of all the alkanes.

Let us consider another example. Acetaldehyde ($CH_3\overset{\text{O}}{\overset{\|}{C}}H$) is a simple member of a broad class of organic compounds known as *aldehydes;* aldehydes contain the functional group, $-\overset{\text{O}}{\overset{\|}{C}}-H$. Later, we will find that an understanding of the chemistry of acetaldehyde will give us a general understanding of the chemistry of most aldehydes.

In this chapter we will consider, in limited detail, representative members from some of the most important families encountered in a study of organic chemistry. We will concern ourselves only with the fundamental structural features of molecules containing carbon-carbon single bonds, carbon-carbon double bonds, and carbon-carbon triple bonds. We will also examine the structures of molecules in which carbon is singly and doubly bonded to oxygen, and singly bonded to nitrogen. As we do this, we will use these simple molecules to illustrate some of the major reaction types that are associated with organic compounds.

This chapter, then, is an overview of much of the material in the remainder of the book. In this initial presentation, our descriptions of structures and reactions are largely nontheoretical and mastery of them will require some memorization. Later, we will examine all of them again in a more theoretical context.

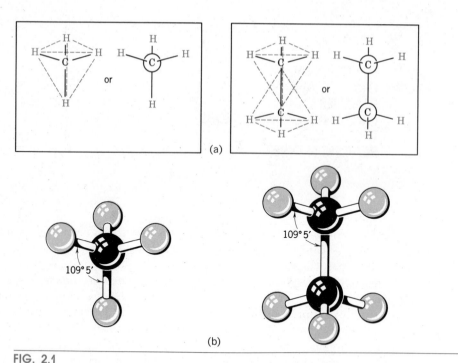

FIG. 2.1

(a) *Two ways of representing the structures of methane and ethane that show the tetrahedral arrangements of the atoms around carbon.* (b) *Ball-and-stick models of methane and ethane.*

2.2 METHANE AND ETHANE

Methane, CH_4, and ethane, C_2H_6, are two members of a broad family of organic compounds called *hydrocarbons*. Hydrocarbons, as the name implies, are compounds whose molecules contain only carbon and hydrogen atoms. Methane and ethane also belong to a subgroup of hydrocarbons known as *alkanes* because methane and ethane do not contain multiple bonds between carbon atoms. Hydrocarbons that contain a carbon-carbon double bond are called *alkenes,* and those that contain a carbon-carbon triple bond are called *alkynes.*

We saw earlier that the bonds of the carbon atom in methane are directed in space toward the corners of a tetrahedron. The bond angles of the carbon atoms of ethane, and of all alkanes, are also tetrahedral. In the case of ethane (Fig. 2.1), each carbon is at one corner of the other carbon's tetrahedron; hydrogen atoms are situated at the other three corners.

We have also seen (Section 1.6) how the tetrahedral structure of methane can be explained on the basis of a maximum possible separation of the pairs of bonding electrons. We can explain the tetrahedral arrangement of the four atoms surrounding each carbon atom in ethane in the same way.

We know from a variety of evidence that groups bonded by carbon-carbon single bonds are capable of essentially free rotation at temperatures near or above room temperature. The different arrangements of the atoms in space that result from rotations of groups about single bonds are called *conformations,* and an analysis

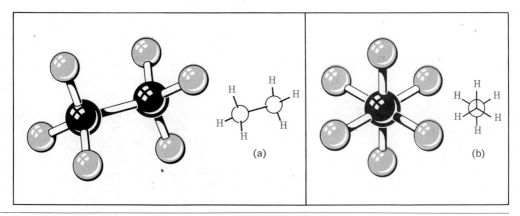

FIG. 2.2

(a) *The staggered conformation of ethane.* (b) *The Newman projection formula for the staggered conformation. This formulation is obtained by viewing the molecule end-on along the carbon-carbon bond axis. Bonds of the front carbon are depicted as* $\curlyvee$, *those of the back carbon as* $\curlywedge$.

of the energy changes that a molecule undergoes as groups rotate about single bonds is called a *conformational analysis.*

Let us consider the ethane molecule as an example: there are obviously an infinite number of different conformations that could result from rotations about the carbon-carbon bond. These different conformations, however, are not all of equal stability. The conformation (Fig. 2.2) in which the hydrogen atoms attached to each carbon are perfectly staggered when viewed from one end of the molecule along the carbon-carbon bond axis is the most stable conformation. This is easily explained in terms of repulsive interactions between bonding pairs of electrons: the staggered conformation allows the maximum possible separation of the electron pairs of the six carbon-hydrogen bonds.

The least stable conformation of ethane is the *eclipsed* conformation (Fig. 2.3). When viewed from one end along the carbon-carbon bond axis, the hydrogen

FIG. 2.3

(a) *The eclipsed conformation of ethane.* (b) *The Newman projection formula for the eclipsed conformation.*

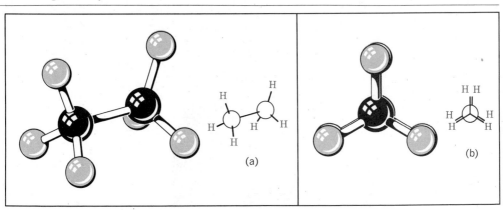

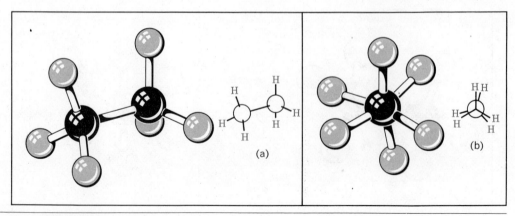

FIG. 2.4

(a) *A skew conformation of ethane.* (b) *The Newman projection formula for a skew conformation.*

atoms attached to each carbon in the eclipsed conformation are in direct opposition to each other. This conformation permits *minimum* separation of the electrons of the six carbon-hydrogen bonds; it is, therefore, of highest energy and has the lowest stability.

Conformations between the eclipsed conformation and the staggered conformation are called skew conformations (Fig. 2.4). All of the skew conformations have stabilities in between that of the staggered conformation and that of the eclipsed conformation, that is, the relative stabilities are: staggered > skew > eclipsed.

We represent this situation graphically by plotting the potential energy of an ethane molecule as a function of rotation about the carbon-carbon bond. The energy changes that occur are illustrated in Fig. 2.5.

FIG. 2.5

*Potential energy changes that accompany rotation
of the carbon-carbon bond of ethane.*

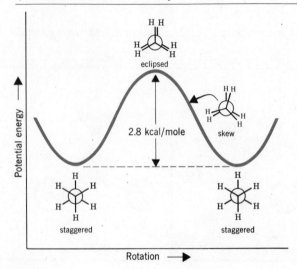

The difference in potential energy between the staggered and eclipsed conformations is 2.8 kcal/mole of ethane. This barrier to rotation is called the *torsional energy* and, because of it, the eclipsed conformation is said to have *torsional strain*. At room temperature, however, molecules have an average energy of 15–20 kcal/mole because of their thermal motion. Thus, at temperatures even considerably below room temperature ethane molecules have more than enough energy to surmount this small barrier and, for all practical purposes, the groups rotate freely about the single bond.

What does all this mean about ethane? We can answer this question in two different ways. If we consider a single molecule of ethane, we can say, for example, that it will spend most of its time in the lowest energy, staggered conformation or in a skew conformation very close to being staggered. It will spend very little time in an eclipsed conformation. If we speak in terms of a large number of ethane molecules (a more realistic situation), we can say that at any given moment most of the molecules will be in staggered or nearly-staggered conformations.

2.3 COMBUSTION: THE REACTIONS OF METHANE AND ETHANE WITH OXYGEN

By far the most widely used reactions of alkanes are their reactions with atmospheric oxygen to form carbon dioxide and water. Methane, for example, is the major constituent of natural gas. It reacts with oxygen according to the following equation.

$$CH_4 + 2O_2 \longrightarrow CO_2 + 2H_2O + \text{heat}$$

Combustion of methane is useful, of course, because of the heat that is produced. Methane, because it is a gas, is easily transported through vast networks of pipelines that cross the entire continent from its sources in oil fields to the places where it is burned.

When methane is subjected to complete combustion in a calorimeter, a device used to determine accurately the heat produced by a reaction, the heat of reaction, ΔH, can be measured. The values of ΔH for the combustion of three common hydrocarbons are given in Table 2.1.

On a molar basis, propane ($CH_3CH_2CH_3$) liberates much more heat than methane, since a mole of propane weighs almost three times as much as a mole of methane. On a gram basis, however, all three hydrocarbons liberate approximately the same quantity of heat, ~ 12 kcal/g.

TABLE 2.1 The Values of ΔH for the Combustion of Three Common Hydrocarbons

REACTION	ΔH
(measured at 25° and 1 atm)	kcal/mole
$CH_{4(g)} + 2O_{2(g)} \longrightarrow CO_{2(g)} + 2H_2O_{(g)}$	-192
$CH_3—CH_{3(g)} + \frac{7}{2}O_{2(g)} \longrightarrow 2CO_{2(g)} + 3H_2O_{(g)}$	-341
$CH_3CH_2CH_{3(g)} + 5O_{2(g)} \longrightarrow 3CO_{2(g)} + 4H_2O_{(g)}$	-531

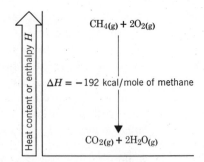

FIG. 2.6

Enthalpy (heat content) change for the complete combustion of methane.

By convention, the sign of ΔH for *exothermic* reactions (those evolving heat) is negative. *Endothermic* reactions (those which absorb heat) have a positive ΔH. The heat of reaction, ΔH, measures the change in heat content (or enthalpy) of the atoms of the reactants as they are converted to products. For an exothermic reaction the atoms have a smaller heat content as products than they do as reactants. For endothermic reactions, the reverse is true. Figure 2.6 shows that a mole of methane and two moles of oxygen have a much higher heat content than do the relatively stable products, carbon dioxide (one mole) and water (two moles). As a consequence, the heat content of the atoms decreases by a very large amount as reactants are converted to products.*

2.4 THE REACTIONS OF METHANE AND ETHANE WITH HALOGENS: SUBSTITUTION REACTIONS

Methane, ethane, and other alkanes react with the first three members of the halogen family: fluorine, chlorine, and bromine. Alkanes do not react appreciably with iodine. With methane the reactions that take place produce a mixture of halomethanes and a hydrogen halide.

$$\begin{array}{ccc} & H & \\ & | & \\ H - & C - H & + \quad X_2 \quad \longrightarrow \\ & | & \\ & H & \end{array}$$

Methane   Halogen

$$\begin{array}{ccccc} H & X & X & X & \\ | & | & | & | & \\ H-C-X & + \quad H-C-X & + \quad H-C-X & + \quad X-C-X & + \quad H-X \\ | & | & | & | & \\ H & H & X & X & \end{array}$$

Halomethane   Dihalomethane   Trihalomethane   Tetrahalomethane   Hydrogen halide

X = F, Cl, or Br

* A discussion of ΔH and its relation to other thermodynamic quantities is given as a special topic in Section 8.2.

The reaction of an alkane with fluorine is so vigorous that special precautions have to be taken to prevent the reaction mixture from exploding. The reaction of alkanes with chlorine or bromine is more easily controlled and is found to require an energy input, either heat or light. Chlorine, however, is considerably more reactive toward alkanes than is bromine.

In all of these reactions a halogen atom replaces one or more of the hydrogen atoms of the alkane. Reactions of this type, *where one group replaces another,* are called *substitution reactions.*

$$-\overset{|}{\underset{|}{C}}-H + Cl_2 \longrightarrow -\overset{|}{\underset{|}{C}}-Cl + H-Cl \qquad \textbf{A substitution reaction}$$

Substitution reactions are not limited to the replacement of hydrogen. When chloromethane, for example, is treated with sodium hydroxide, a substitution reaction takes place in which an —OH group replaces the —Cl.

$$H-\overset{\overset{\displaystyle H}{|}}{\underset{\underset{\displaystyle H}{|}}{C}}-Cl \ + NaOH \xrightarrow[\text{heat}]{\text{solvent}} H-\overset{\overset{\displaystyle H}{|}}{\underset{\underset{\displaystyle H}{|}}{C}}-OH + NaCl \qquad \textbf{A substitution reaction}$$

Chloromethane Methyl alcohol

When bromomethane reacts with potassium iodide, —I replaces —Br; this, too, is a substitution reaction.

$$H-\overset{\overset{\displaystyle H}{|}}{\underset{\underset{\displaystyle H}{|}}{C}}-Br \ + KI \xrightarrow{\text{solvent}} H-\overset{\overset{\displaystyle H}{|}}{\underset{\underset{\displaystyle H}{|}}{C}}-I + KBr \qquad \textbf{A substitution reaction}$$

Bromomethane

Substitution reactions are among the most common and most useful reactions that we will encounter in our study of organic chemistry. We will see many other examples in succeeding chapters.

Halogenation Reactions

The substitution reactions that take place when alkanes and halogens react are often called alkane *halogenations.* One characteristic of alkane halogenations is that multiple substitution reactions almost always occur. As we saw at the beginning of this section, the halogenation of methane produces a mixture of monohalomethane, dihalomethane, trihalomethane, and tetrahalomethane.

We can understand how this happens if we recognize that all hydrogens attached to carbon are capable of reacting with fluorine, chlorine, or bromine.

Let us consider the reaction that takes place between chlorine and methane as an example. If we bring together a mixture of methane and chlorine (both substances are gases at room temperature) and then either heat the mixture or irradiate it with light, a reaction begins to occur vigorously. At the outset, the only compounds that are present in the mixture are chlorine and methane, and the only reaction that can take place is one that produces chloromethane and hydrogen chloride.

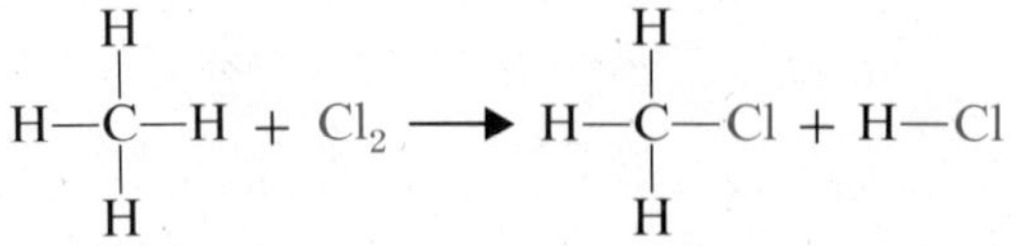

$$\underset{\underset{\displaystyle H}{|}}{\overset{\overset{\displaystyle H}{|}}{H-C-H}} + Cl_2 \longrightarrow \underset{\underset{\displaystyle H}{|}}{\overset{\overset{\displaystyle H}{|}}{H-C-Cl}} + H-Cl$$

As the reaction progresses, however, the concentration of chloromethane in the mixture increases, and a second substitution reaction begins to occur. Chloromethane reacts with chlorine to produce dichloromethane.

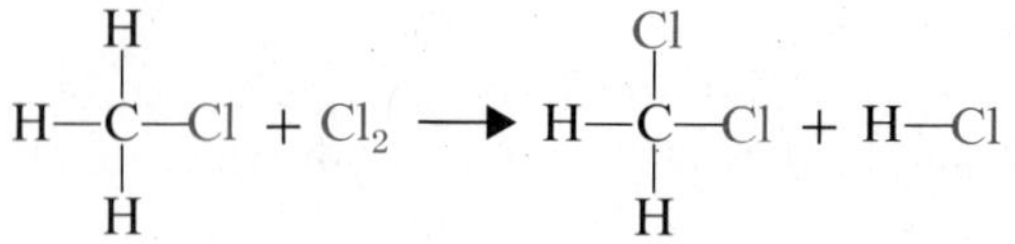

$$\underset{\underset{\displaystyle H}{|}}{\overset{\overset{\displaystyle H}{|}}{H-C-Cl}} + Cl_2 \longrightarrow \underset{\underset{\displaystyle H}{|}}{\overset{\overset{\displaystyle Cl}{|}}{H-C-Cl}} + H-Cl$$

Dichloromethane can then produce trichloromethane,

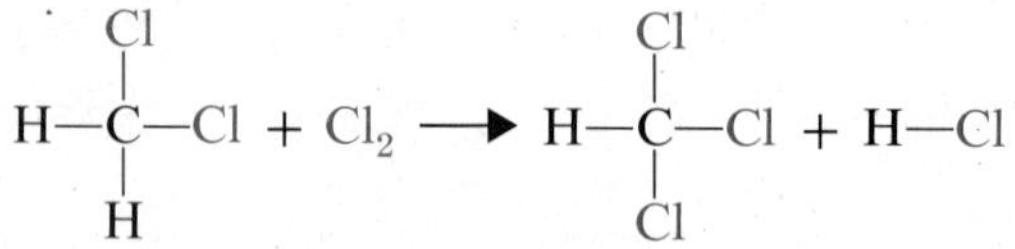

$$\underset{\underset{\displaystyle H}{|}}{\overset{\overset{\displaystyle Cl}{|}}{H-C-Cl}} + Cl_2 \longrightarrow \underset{\underset{\displaystyle Cl}{|}}{\overset{\overset{\displaystyle Cl}{|}}{H-C-Cl}} + H-Cl$$

and trichloromethane, as it accumulates in the mixture, can react with chlorine to produce tetrachloromethane.

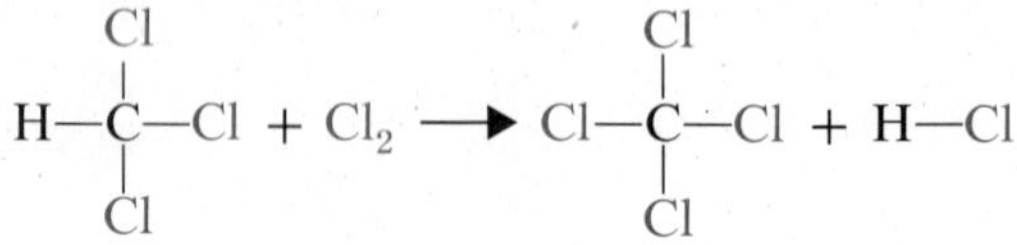

$$\underset{\underset{\displaystyle Cl}{|}}{\overset{\overset{\displaystyle Cl}{|}}{H-C-Cl}} + Cl_2 \longrightarrow \underset{\underset{\displaystyle Cl}{|}}{\overset{\overset{\displaystyle Cl}{|}}{Cl-C-Cl}} + H-Cl$$

Each time a substitution of —Cl for —H takes place a molecule of H—Cl is produced.

When ethane and chlorine react similar substitution reactions occur. Ultimately all six hydrogen atoms of ethane may be replaced. We notice below that

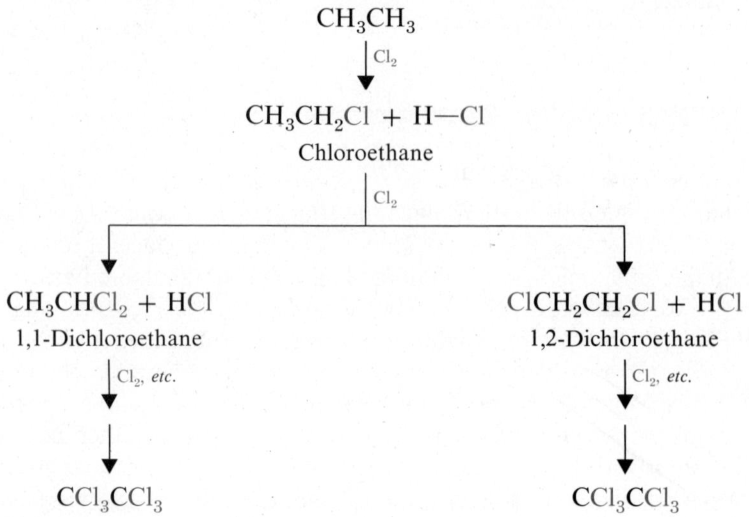

$$CH_3CH_3$$
$$\downarrow Cl_2$$
$$CH_3CH_2Cl + H-Cl$$
Chloroethane
$$\downarrow Cl_2$$

$$CH_3CHCl_2 + HCl \qquad\qquad ClCH_2CH_2Cl + HCl$$
1,1-Dichloroethane $\qquad\qquad$ 1,2-Dichloroethane
$$\downarrow Cl_2, etc. \qquad\qquad\qquad\qquad \downarrow Cl_2, etc.$$

$$CCl_3CCl_3 \qquad\qquad\qquad\qquad CCl_3CCl_3$$

the second substitution reaction of ethane results in the formation of two different molecules: 1,1-dichloroethane and 1,2-dichloroethane.* These two molecules have the same molecular formula, $C_2H_4Cl_2$, but they have *different structures*. They are *structural isomers*.

Problem 2.1

(a) How many trichloroethanes would be produced when 1,1-dichloroethane reacts with chlorine? (b) How many trichloroethanes would be produced when 1,2-dichloroethane reacts with chlorine? (c) Write structural formulas for these trichloroethanes. (d) Are they isomers? (e) How many tetrachloroethanes might be produced in the next chlorination step? (f) Write structural formulas for these tetrachloroethanes. (g) How many pentachloroethanes are theoretically possible?

An observation emerges from the preceding discussion and problem: what appears, at first, to be a straightforward reaction may often be complicated by additional reactions. The additional substitution reactions that occur when methane and ethane are halogenated would be disadvantageous if our goal was the preparation of chloromethane or chloroethane only. Conversely, these additional reactions would be essential if our aim was the preparation of tetrachloromethane or hexachloroethane.

We often find that we can achieve some measure of control over the course of reactions of this type by varying the relative proportions of halogen and alkane in our reaction mixture. Chloromethane or chloroethane can be obtained as major products if a large molar excess of methane or ethane is used in the individual reaction mixture. The excess methane or ethane can be recovered and reused after the reaction is over. On the other hand, the presence of an excess of chlorine in the appropriate reaction mixture will assure the formation of larger amounts of more highly substituted alkanes. An illustration of these effects can be seen in the two examples given below.

$$CH_4 \ + \ Cl_2 \ \xrightarrow{440°} \ CH_3Cl \ + \ CH_2Cl_2 \ + \ CHCl_3 \ + \ CCl_4$$

| (2 moles) | (1 mole) | (0.62 mole fraction) | (0.30 mole fraction) | (0.07 mole fraction) | (0.01 mole fraction) |

$$CH_4 \ + \ Cl_2 \ \xrightarrow{440°} \ CH_3Cl \ + \ CH_2Cl_2 \ + \ CHCl_3 \ + \ CCl_4$$

| (1 mole) | (1.10 mole) | (0.37 mole fraction) | (0.41 mole fraction) | (0.19 mole fraction) | (0.03 mole fraction) |

We can easily account for the origin of these results. When methane is present in the mixture *in excess* the opportunities for it to react with chlorine will be much greater than for similar reactions of chloromethane, dichloromethane, and trichloromethane. As a result, chloromethane will be the major product. On the other hand, when chlorine is present *in excess,* most of the methane will be quickly converted to chloromethane. At this point there will still be excess chlorine in the mixture that can produce subsequent chlorinations. The result: a greater yield of dichloromethane, trichloromethane, and tetrachloromethane.

---

*These names are obtained by assigning numbers to the carbons of ethane.

2.5 ALKENES: COMPOUNDS CONTAINING THE CARBON-CARBON DOUBLE BOND; ETHENE AND PROPENE

The carbon atoms of all of the molecules that we have considered so far have used their four valence electrons to form four single covalent bonds to four other atoms. We find, however, that there are many important organic compounds in which carbon atoms share more than two electrons with another atom; in these compounds the bonds that are formed are multiple covalent bonds. When two carbon atoms share four electrons, for example, the result is a carbon-carbon double bond. Hydrocarbons whose molecules contain a carbon-carbon double bond are called *alkenes*.

$$\ddot{C}::\ddot{C} \quad \text{or} \quad \diagdown C{=}C \diagup$$

Ethene, C_2H_4, and propene, C_3H_6, are both alkenes. (Ethene is also called ethylene, and propene is sometimes called propylene.)

$$\underset{\text{Ethene}}{\overset{\displaystyle H\diagdown \quad \diagup H}{\underset{\displaystyle H\diagup \quad \diagdown H}{C{=}C}}} \qquad \underset{\text{Propene}}{\overset{\displaystyle H\diagdown \quad \diagup H}{\underset{\displaystyle CH_3\diagup \quad \diagdown H}{C{=}C}}}$$

In ethene the only carbon-carbon bond is a double bond; in propene, there is one carbon-carbon single bond and one carbon-carbon double bond.

The spatial arrangement of the atoms of alkenes is different from that of alkanes. The six atoms of ethene are coplanar, and the arrangement of atoms around each carbon atom is triangular (Fig. 2.7).

Problem 2.2

(a) Explain the trigonal-planar structure of ethene on the basis of a maximum separation of the bonding pairs of electrons. (Assume that the two pairs of the double bond are located in a region of space between the two carbon atoms.) (b) Explain why the bond angles of ethene deviate somewhat from the ideal 120° bond angles

that would result from a regular triangular structure. The $C\diagup^{\displaystyle H}_{\diagdown H}$ bond angle of ethene

FIG. 2.7
The trigonal-planar-structure of ethene.

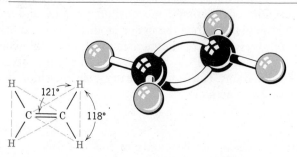

is 118° and the $\text{C} \underset{\text{C}}{\overset{\text{H}}{\diagup}}$ bond angle is 121°.

Cis-Trans Isomerism

Another property of groups joined by a carbon-carbon double bond is that they do not rotate with respect to each other at temperatures near and considerably above room temperature. This restricted rotation of groups attached by double-bonded carbon atoms is in marked contrast to the behavior of groups attached by single-bonded carbon atoms; the latter rotate quite freely at these temperatures. The restricted rotation of groups joined by a carbon-carbon double bond results in a new type of isomerism that we illustrate with the two dichloroethenes written below.

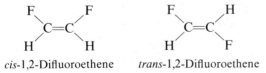

cis-1,2-Dichloroethene   *trans*-1,2-Dichloroethene

These two compounds are isomers because they both have the same molecular formula, $C_2H_2Cl_2$, and because they are not superimposable. They are not structural isomers, however, because the order of attachment of atoms is the same in each. *Cis*-1,2-dichloroethene and *trans*-1,2-dichloroethene differ from each other only in the arrangement of their atoms in space. Isomers of this general type are classified as *stereoisomers*. *Cis*- and *trans*-1,2-dichloroethene are commonly called *cis-trans* isomers (*cis*, Latin: on this side; *trans*, Latin: across).*

The structural requirements for *cis-trans* isomerism will become clear if we consider a few additional examples. 1,1-Dichloroethene and 1,1,2-trichloroethene do not show this type of isomerism.

1,1-Dichloroethene
(no *cis-trans* isomerism)   1,1,2-Trichloroethene
(no *cis-trans* isomerism)

1,2-Difluoroethene and 1,2-difluoro-1,2-dichloroethene do exist as *cis-trans* isomers.

cis-1,2-Difluoroethene   *trans*-1,2-Difluoroethene

* *Cis-trans* isomers of this type belong to one class of stereoisomers known as *diastereomers*. We will not refer to them in this way until we have had the opportunity to study stereoisomerism in detail in Chapter 7. An older term, used to describe *cis-trans* isomerism, is "geometric isomerism."

TABLE 2.2 Physical Properties of Cis-Trans Isomers

COMPOUND	MELTING POINT	BOILING POINT	DIPOLE MOMENT
cis-1,2-Dichloroethene	$-80°$	$60°$	1.85 D
trans-1,2-Dichloroethene	$-50°$	$48°$	0
cis-1,2-Dibromoethene	$-53°$	$110°$	1.35 D
trans-1,2-Dibromoethene	$-6°$	$108°$	0

$$
\begin{array}{cc}
\underset{Cl}{\overset{F}{\diagdown}}C=C\underset{Cl}{\overset{F}{\diagup}} & \underset{Cl}{\overset{F}{\diagdown}}C=C\underset{F}{\overset{Cl}{\diagup}}
\end{array}
$$

cis-1,2-Difluoro-1,2-dichloroethene   *trans*-1,2-Difluoro-1,2-dichloroethene

Clearly, then, *cis-trans isomerism of this type is not possible if one carbon atom of the double bond bears two identical groups.*

   Cis-trans isomers have different physical properties. They have different melting points and boiling points, and often *cis-trans* isomers differ markedly in the magnitude of their dipole moments. Table 2.2 summarizes some of the physical properties of two pairs of *cis-trans* isomers.

Problem 2.3

(a) How do you explain the fact that *trans*-1,2-dichloroethene and *trans*-1,2-dibromoethene have no dipole moments ($\mu = 0$); whereas the corresponding *cis* isomers have rather large dipole moments (for *cis*-1,2-dichloroethene, $\mu = 1.85$, and for *cis*-1,2-dibromoethene, $\mu = 1.35$)? (b) Account for the fact that *cis*-1,2-dichloroethene has a larger dipole moment than *cis*-1,2-dibromoethene? (c) Would you expect to be able to separate a mixture of *cis*- and *trans*-1,2-dichloroethene isomers by fractional distillation? (d) By fractional crystallization?

FIG. 2.8
The structure of propene. Carbon atoms 1 and 2 have the same geometry as those of ethene; carbon atom 3 is tetrahedral like that of methane.

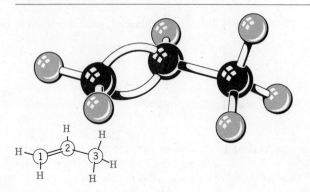

Propene

Two of the carbon atoms of propene (Fig. 2.8) have a trigonal planar structure, the third carbon atom is tetrahedral.

Problem 2.4

Write structural formulas for: (a) all of the compounds that could be obtained by replacing one hydrogen of propene with chlorine; (b) all of the compounds that could be obtained by replacing two hydrogens of propene with chlorine; (c) three hydrogens; (d) four hydrogens; (e) five hydrogens. (f) In each instance [(a)–(e)] designate the isomeric relationships (structural or *cis-trans*) that exist.

2.6 ADDITION REACTIONS

The most characteristic reaction of compounds containing multiple bonds is an *addition reaction*. When a reagent X-Y *adds* to a carbon-carbon double bond, X becomes attached to one atom of the double bond and Y to the other. The double bond changes to a single bond.

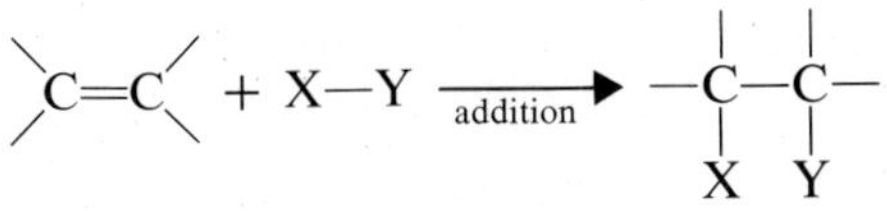

$$\underset{/}{\overset{\backslash}{}}C{=}C\underset{\backslash}{\overset{/}{}} + X{-}Y \xrightarrow{\text{addition}} -\underset{|X}{\overset{|}{C}}-\underset{|Y}{\overset{|}{C}}-$$

A specific example is the addition of hydrogen bromide to ethene.

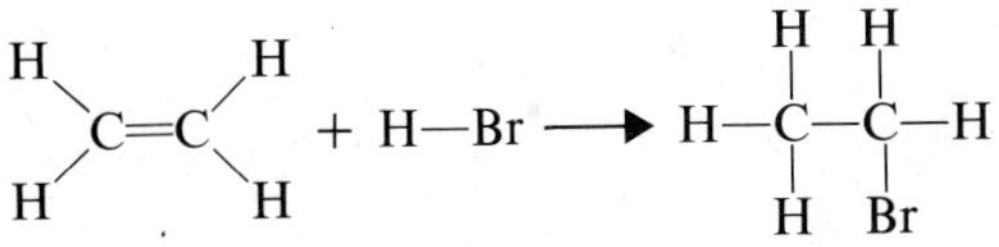

$$\underset{H}{\overset{H}{}}{>}C{=}C{<}\underset{H}{\overset{H}{}} + H{-}Br \longrightarrow H-\underset{|H}{\overset{|H}{C}}-\underset{|Br}{\overset{|H}{C}}-H$$

Bromoethane

The reagent that adds may also be a symmetrical reagent, as in the addition of bromine to ethene.

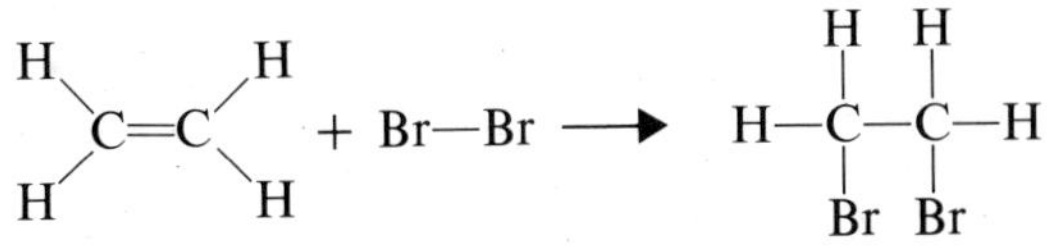

$$\underset{H}{\overset{H}{}}{>}C{=}C{<}\underset{H}{\overset{H}{}} + Br{-}Br \longrightarrow H-\underset{|Br}{\overset{|H}{C}}-\underset{|Br}{\overset{|H}{C}}-H$$

1,2-Dibromoethane

Further examples of addition reactions of both symmetrical and unsymmetrical reagents to ethene are shown below.

$$CH_2{=}CH_2 + Cl_2 \longrightarrow CH_2ClCH_2Cl \qquad \text{1,2-Dichloroethane}$$

$$CH_2{=}CH_2 + HCl \longrightarrow CH_3CH_2Cl \qquad \text{Chloroethane}$$

$$CH_2{=}CH_2 + HI \longrightarrow CH_3CH_2I \qquad \text{Iodoethane}$$

$$CH_2{=}CH_2 + HOH \xrightarrow{H^+} CH_3CH_2OH \qquad \text{Ethanol*}$$

Propene shows the same addition reactions. If the adding reagent is symmetrical only one product can result:

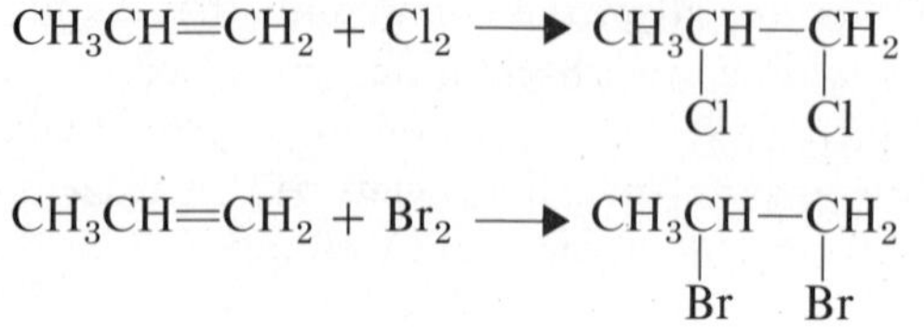

$$CH_3CH{=}CH_2 + Cl_2 \longrightarrow CH_3\underset{\underset{Cl}{|}}{CH}{-}\underset{\underset{Cl}{|}}{CH_2} \qquad \text{1,2-Dichloropropane}$$

$$CH_3CH{=}CH_2 + Br_2 \longrightarrow CH_3\underset{\underset{Br}{|}}{CH}{-}\underset{\underset{Br}{|}}{CH_2} \qquad \text{1,2-Dibromopropane}$$

On the other hand, if the adding reagent is unsymmetrical, two products are theoretically possible. Only one, however, is usually obtained in actual practice.

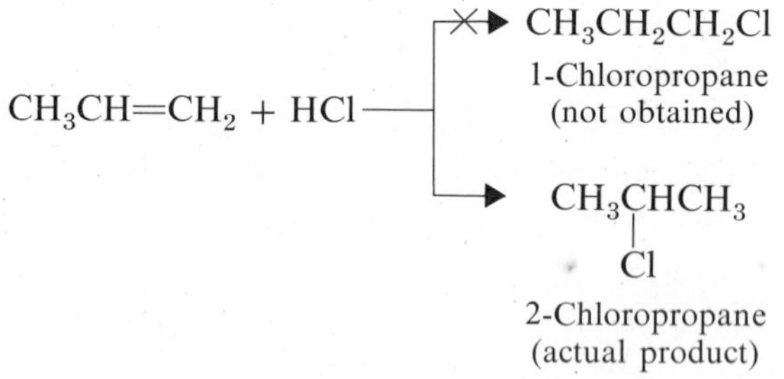

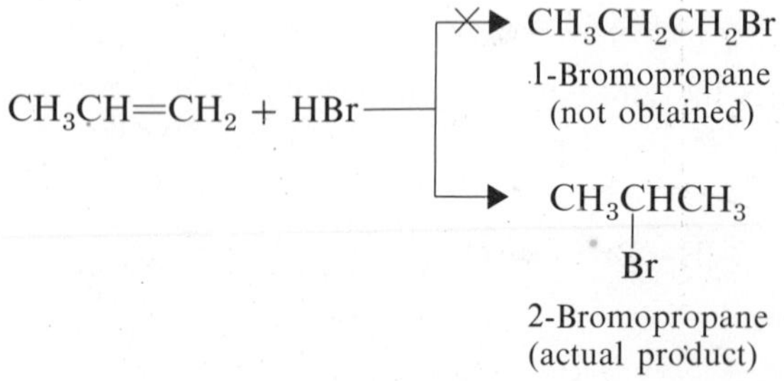

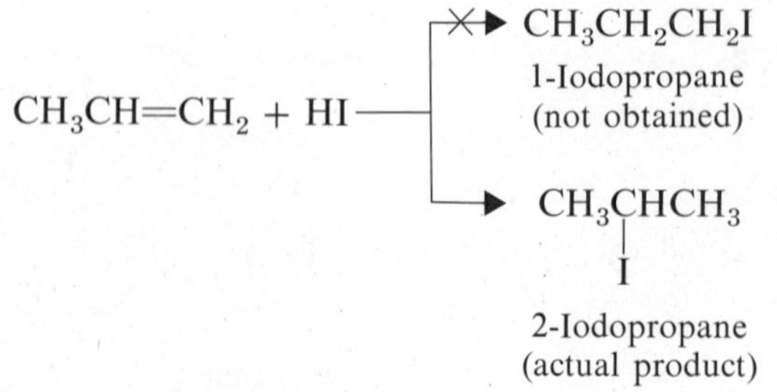

$$CH_3CH{=}CH_2 + HCl$$

⤫► $CH_3CH_2CH_2Cl$
1-Chloropropane
(not obtained)

► $CH_3\underset{\underset{Cl}{|}}{CH}CH_3$
2-Chloropropane
(actual product)

$$CH_3CH{=}CH_2 + HBr$$

⤫► $CH_3CH_2CH_2Br$
1-Bromopropane
(not obtained)

► $CH_3\underset{\underset{Br}{|}}{CH}CH_3$
2-Bromopropane
(actual product)

$$CH_3CH{=}CH_2 + HI$$

⤫► $CH_3CH_2CH_2I$
1-Iodopropane
(not obtained)

► $CH_3\underset{\underset{I}{|}}{CH}CH_3$
2-Iodopropane
(actual product)

---

*The H^+ over the arrow in this reaction indicates that the reaction is *acid catalyzed*. That is, acid is required to make the reaction proceed at a reasonable rate, but it is not consumed or transformed in the conversion of reactants to products. When we examine the mechanisms of this reaction and others that require catalysts in later chapters, the actual role of the catalyst will become clear.

There is a recognizable pattern in these examples. In each case the halogen of the halogen acid becomes attached to the central carbon of the product. On the basis of observations similar to these, but with many more examples, the Russian chemist Vladimir Markovnikov formulated in 1870 what is now known as Markovnikov's rule. Markovnikov stated that when a halogen acid adds to a carbon-carbon double bond "the halogen adds itself to the less hydrogenated carbon atom." Using propene as an illustration, Markovnikov's rule takes the form shown below. We will

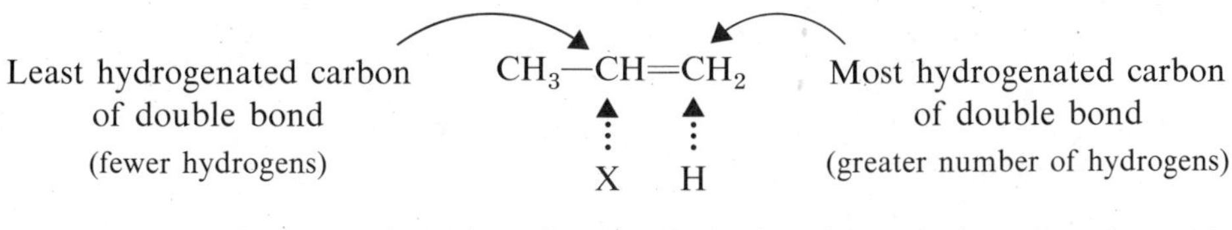

Least hydrogenated carbon $CH_3—CH=CH_2$ Most hydrogenated carbon
     of double bond   of double bond
    (fewer hydrogens)   X   H   (greater number of hydrogens)

give a modern version of Markovnikov's rule and a theoretical explanation of it in Chapter 6.

Problem 2.5

(a) On the basis of Markovnikov's rule, predict the product that would be obtained from the addition reaction of hydrogen chloride and $CH_3—\overset{\underset{\textstyle CH_3}{|}}{C}=CH_2$; (b) from the addition reaction of hydrogen iodide and $CH_2=CHCH_2CH_3$.

In the presence of a variety of finely divided metals, molecular hydrogen also adds to carbon-carbon double bonds. The catalysts most commonly used are finely

$$CH_2=CH_2 + H_2 \xrightarrow[\text{or Pt}]{\text{Ni, Pd}} CH_3—CH_3$$

$$CH_3CH=CH_2 + H_2 \xrightarrow[\text{or Pt}]{\text{Ni, Pd}} CH_3—CH_2—CH_3$$

divided nickel, palladium, or platinum.

The product that results from the addition of hydrogen to an alkene is an alkane. Alkanes have only single bonds and contain the maximum number of hydrogen atoms that a hydrocarbon can possess. For this reason, alkanes are said to be "saturated" compounds. Alkenes, because they contain a double bond and possess fewer than the maximum number of hydrogens, are capable of adding hydrogen and are said to be "unsaturated." The process of adding hydrogen to an alkene is sometimes described as being one of *saturation*. Most often, however, the term used to describe this process, that is, the addition of hydrogen, is *hydrogenation*.

2.7 ALKYNES: COMPOUNDS CONTAINING THE CARBON-CARBON TRIPLE BOND; ETHYNE AND PROPYNE

Hydrocarbons in which two carbon atoms share six electrons and are thus bonded by a triple bond, are called alkynes. The two simplest alkynes are ethyne and propyne.

$$H:C::C:H \qquad CH_3:C::C:H$$
$$H-C\equiv C-H \qquad CH_3-C\equiv C-H$$

    Ethyne   Propyne
    (C_2H_2) (C_3H_4)

Ethyne, a compound that is also called acetylene, consists of linear molecules. The H—C≡C bond angles of ethyne molecules are 180°.

$$H-C\equiv C-H$$
      180° 180°

Problem 2.6

(a) Assuming that three bonding pairs of electrons are located generally between the two carbon atoms of ethyne, are bond angles of 180° consistent with the principle of maximum possible separation of the bonding pairs of electrons? (b) Which atoms in propyne would lie in a straight line? (c) Write out a complete three-dimensional structure for propyne.

The carbon-carbon triple bond is shorter than the carbon-carbon double bond, and the carbon-carbon double bond is shorter than the carbon-carbon single bond. The carbon-hydrogen bonds of ethyne are shorter, too, than the corresponding bonds of ethene, and carbon-hydrogen bonds of ethene are shorter than those of ethane. The differences in bond lengths and bond angles of ethyne, ethene, and ethane are summarized below (Fig. 2.9).

Because they have a multiple covalent bond, alkynes also undergo addition reactions. Because an alkyne has a triple bond, there is the possibility that two moles of the adding reagent will react. The addition of two moles bromine to ethyne is an example. When the proper experimental conditions are chosen the reaction can

$$H-C\equiv C-H + 2Br_2 \longrightarrow H-\overset{\overset{\displaystyle Br}{|}}{C}-\overset{\overset{\displaystyle Br}{|}}{C}-H$$
$$\underset{Br \quad Br}{}$$

FIG. 2.9
Bond angles and bond distances of ethyne, ethene, and ethane.

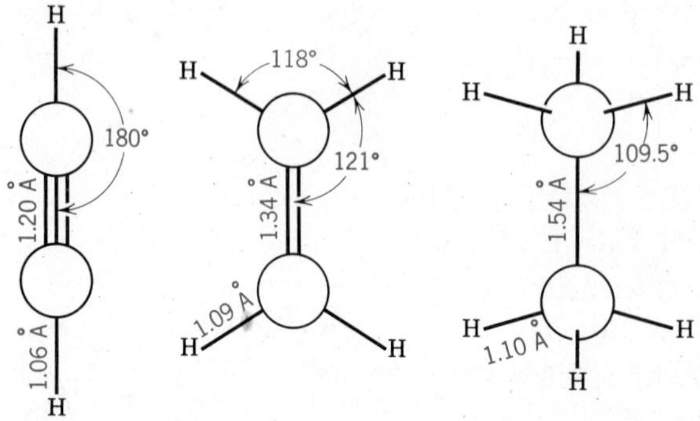

usually be controlled, and either the single addition product or the double addition product can be obtained.

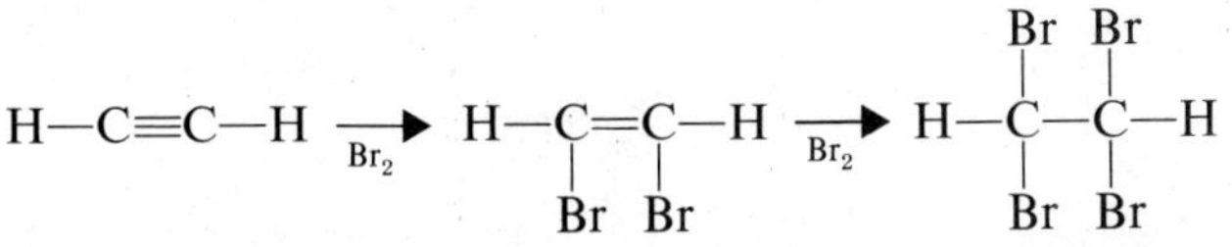

$$H-C\equiv C-H \xrightarrow{Br_2} H-\underset{\underset{Br}{|}}{C}=\underset{\underset{Br}{|}}{C}-H \xrightarrow{Br_2} H-\underset{\underset{Br}{|}}{\overset{\overset{Br}{|}}{C}}-\underset{\underset{Br}{|}}{\overset{\overset{Br}{|}}{C}}-H$$

When we consider the addition of hydrogen chloride to ethyne there is no difficulty predicting the structure of the first addition product. However the second addition could in theory, at least, yield two products; 1,2-dichloroethane or 1,1-dichloroethane. The product that is actually obtained is 1,1-dichloroethane.

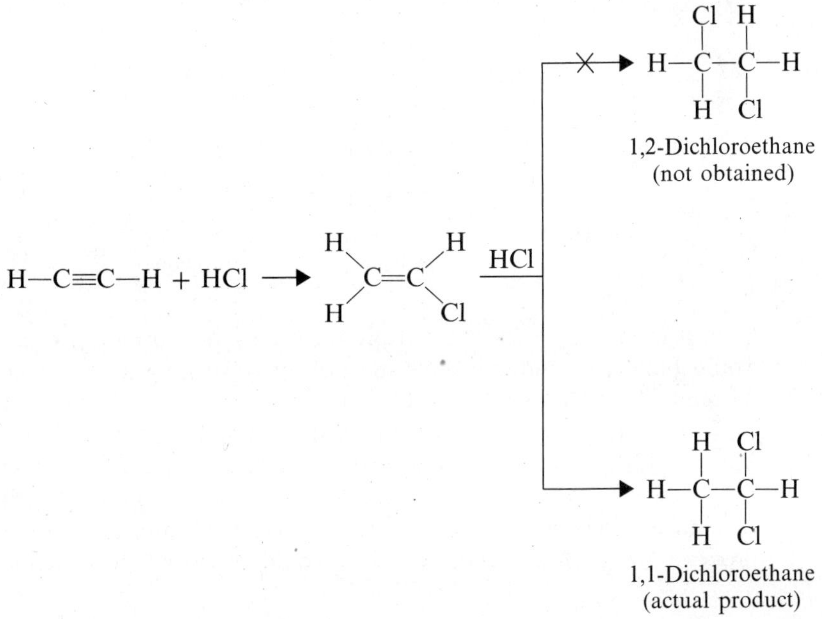

1,2-Dichloroethane
(not obtained)

1,1-Dichloroethane
(actual product)

Problem 2.7

(a) In the reaction just given, state whether or not the product of the second addition of hydrogen chloride is the one that you would expect on the basis of Markovnikov's rule? (b) What products would you expect from the single and double addition of hydrogen chloride to propyne?

2.8 SYNTHESIS

We have now seen a sufficient number of examples of organic reactions to consider the kind of problems faced by an organic chemist when he wants to prepare a specific organic compound. Let us assume that we have available to us as starting materials the seven hydrocarbons that we have studied so far: methane, ethane, propane, ethene, propene, ethyne, and propyne. Let us assume further that our goal is the development of a practical synthesis of 1,2-dichloropropane, $CH_2ClCHClCH_3$.

We might think first of a direct substitution reaction of propane with chlorine. With a little reflection, however, we conclude that this would be a poor method; for as we see below, the direct chlorination of propane would be quite complicated:

$$CH_3CH_2CH_3$$

$$\xrightarrow[\text{}]{Cl_2 \mid \text{heat or light}}$$

$$CH_3\underset{\underset{Cl}{|}}{C}HCH_3 \qquad\qquad CH_3CH_2CH_2Cl$$

2-Chloropropane   1-Chloropropane

$$\xrightarrow[\text{heat or light} \mid Cl_2]{} \qquad\qquad \xrightarrow[Cl_2 \mid \text{heat or light}]{}$$

$$CH_3-\underset{\underset{Cl}{|}}{\overset{\overset{Cl}{|}}{C}}-CH_3 \qquad CH_3\underset{\underset{Cl}{|}}{C}H\underset{\underset{Cl}{|}}{C}H_2 \qquad CH_3CH_2\underset{\underset{Cl}{|}}{C}HCl \qquad CH_2ClCH_2CH_2Cl$$

2,2-Dichloro-   1,2-Dichloro-   1,1-Dichloro-   1,3-Dichloro-
propane   propane   propane   propane

Even if more highly chlorinated products were not formed (as they most certainly would be), we would still have the task of separating a very complex mixture at the end of the reaction. In addition, a large portion of our starting compound, propane, would be consumed by the formation of the other dichloropropane isomers. This would lead to a low yield of the required compound.

Let us now give our attention to a synthesis beginning with propene. Here we find a much better prospect. The addition of chlorine to the double bond of propene occurs in a specific way. We could carry out the reaction in the dark at

$$CH_3CH\!=\!CH_2 + Cl_2 \xrightarrow[\substack{\text{Room} \\ \text{temperature} \\ \text{(no light)}}]{CCl_4} CH_3\underset{\underset{Cl}{|}}{C}H-\underset{\underset{Cl}{|}}{C}H_2$$

room temperature by simply dissolving the propene in an unreactive solvent such as carbon tetrachloride. We could then bubble gaseous chlorine into the solution until all of the propene had been consumed. Under these conditions (the absence of light and at room temperature), substitution reactions would not occur to any appreciable extent and our product, 1,2-dichloropropane, would be free from undesired side products.*

---

* Under other conditions (i.e., at elevated temperatures) in the gas phase, and in the presence of light, one gets *substitution* at the methyl group of propene to the virtual exclusion of addition to the double bond. Thus, specifying of the conditions of a chemical reaction is often just as important as specifying the reactants.

Problem 2.8

Assume that you have available, as organic starting reagents; methane, ethane, propane, ethene, propene, ethyne, and propyne. Outline a practical synthesis of the following compounds. (You may use any needed inorganic reagents.)

(a) 1,1-Dichloroethane; CH_3CHCl_2

(b) 1,2-Dichloroethane; CH_2ClCH_2Cl

(c) 1,2-Dibromoethane; CH_2BrCH_2Br

(d) 1,1,2,2-Tetrabromoethane; $CHBr_2CHBr_2$

(e) 1-Bromo-1-chloroethane; $CH_3CHClBr$

(f) 2-Bromo-2-chloropropane; $CH_3CClBrCH_3$

(g) 2,2-Dibromopropane; $CH_3CBr_2CH_3$

(h) Bromoethane (two ways)

2.9 ELIMINATION REACTIONS

One of the most convenient methods for the syntheses of alkenes is the general type of reaction called an *elimination reaction*. We will study the elimination reaction in detail later, but we can illustrate three examples here.

When any of the bromoalkanes written below are heated with a solution of potassium hydroxide dissolved in ethanol, C_2H_5OH, the elements of hydrogen bromide are *eliminated* from the bromoalkane and an alkene is produced. These

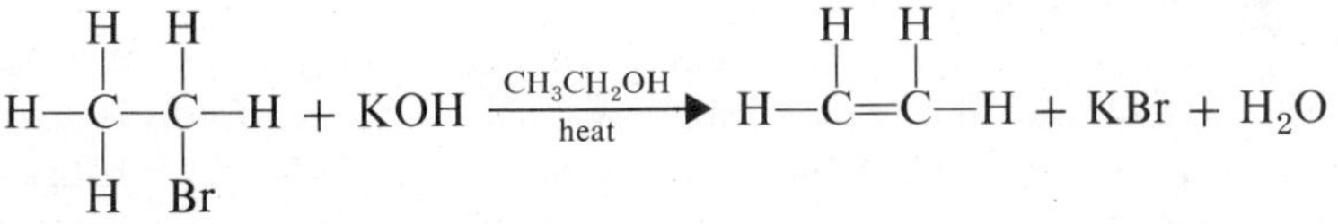

Bromoethane → Ethene

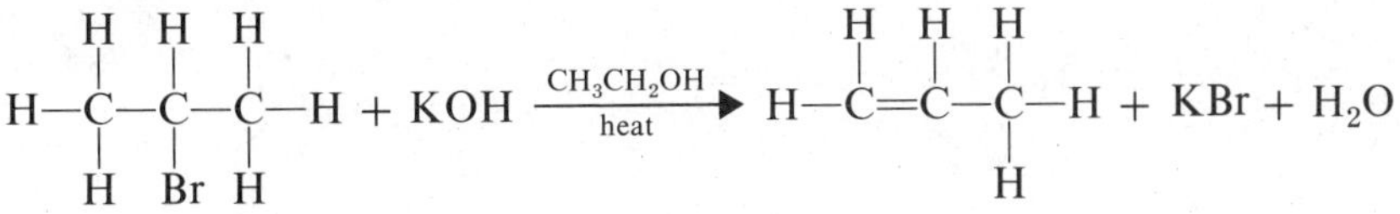

2-Bromopropane → Propene

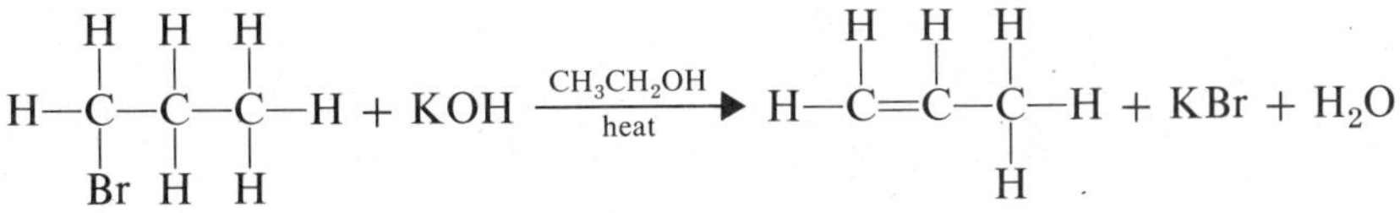

1-Bromopropane → Propene

reactions are not limited to the elimination of hydrogen bromide. Chloroalkanes also undergo the elimination of hydrogen chloride, iodoalkanes undergo the elimination of hydrogen iodide and, in all cases, alkenes are produced. When the elements of a hydrogen halide are eliminated from a haloalkane in this way the reaction is often called *dehydrohalogenation*.

2.10 ETHYL ALCOHOL, ACETALDEHYDE, AND ACETIC ACID: ORGANIC OXIDATION AND REDUCTION REACTIONS

Ethyl alcohol (more systematically called ethanol) has the structural formula CH_3CH_2OH and is one member of a broad family of organic compounds known as alcohols. The hydroxyl group (—OH) is the characteristic functional group of this family. Other examples of alcohols are methyl alcohol, CH_3OH (a compound that is also called methanol); n-propyl alcohol, $CH_3CH_2CH_2OH$ (also called 1-propanol); and isopropyl alcohol, $CH_3CHOHCH_3$ (also called 2-propanol).

There are two structure-based ways of viewing alcohols: (1) as hydroxy-derivatives of alkanes, and (2) as alkyl derivatives of water. Ethyl alcohol, for example, can be seen as an ethane molecule in which one hydrogen has been replaced by a hydroxyl group, or as a water molecule in which one hydrogen has been replaced by an ethyl group. That these ways of regarding ethyl alcohol are valid is shown

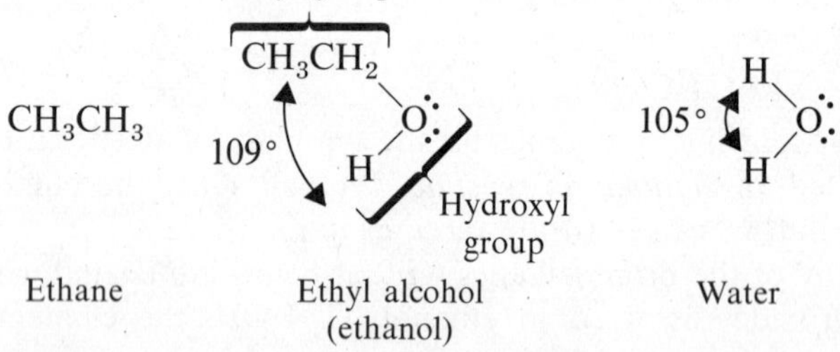

Ethane   Ethyl alcohol (ethanol)   Water

by the fact that the C—O—H bond angle of ethyl alcohol is very close to the H—O—H bond angle of water.

Alcohols, generally, are classified into three groups: as being primary (1°), secondary (2°), or tertiary (3°) alcohols. *This classification is based on the condition of the carbon to which the hydroxyl group is directly attached.* If the carbon to which the hydroxyl group is attached is itself attached to only one other carbon, the carbon is said to be a primary carbon and the alcohol a primary alcohol.

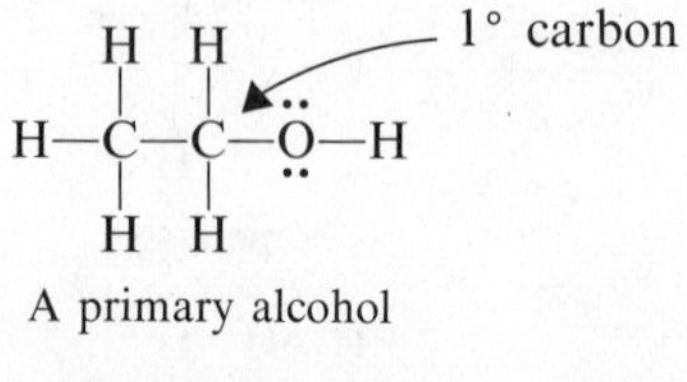

A primary alcohol

If the carbon that bears the hydroxyl group also has two other carbons attached to it, this carbon is called a secondary carbon, and the alcohol is a secondary alcohol.

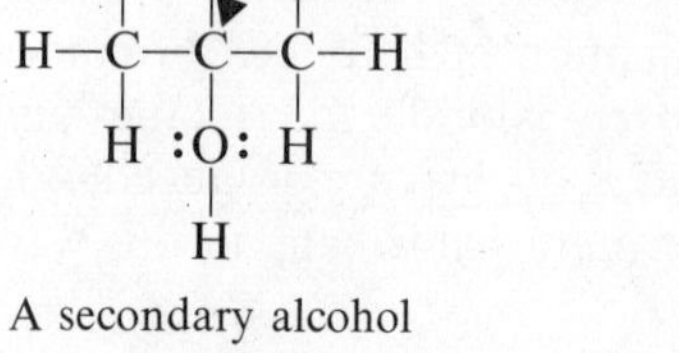

A secondary alcohol

Finally, if the carbon with the hydroxyl group is attached to three other carbons, the alcohol is called a tertiary alcohol.

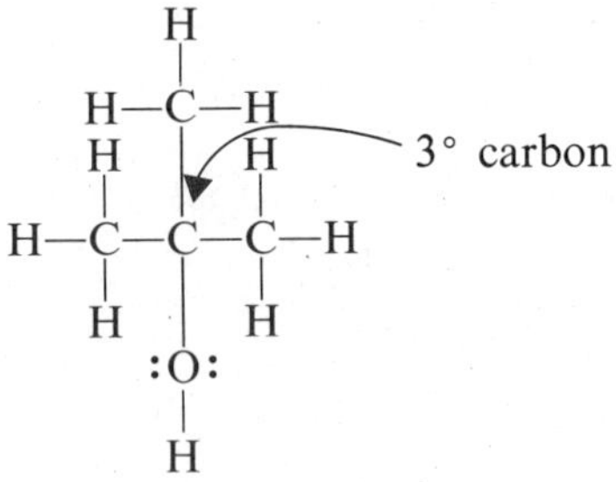

A tertiary alcohol

A characteristic reaction that primary and secondary alcohols undergo is oxidation to compounds known as *aldehydes* and *ketones,* respectively. For example, the primary ethyl alcohol, can be oxidized to the aldehyde known as acetaldehyde by the action of chromic acid.

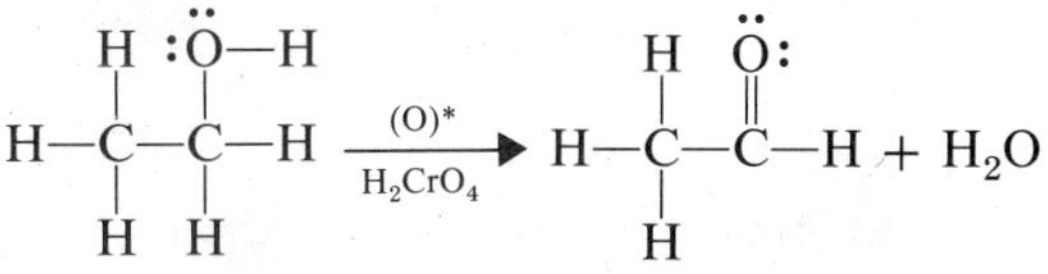

Ethyl alcohol   Acetaldehyde
(a 1° alcohol)   (an aldehyde)

The secondary alcohol, isopropyl alcohol, can be oxidized in the same way to the ketone known as acetone.

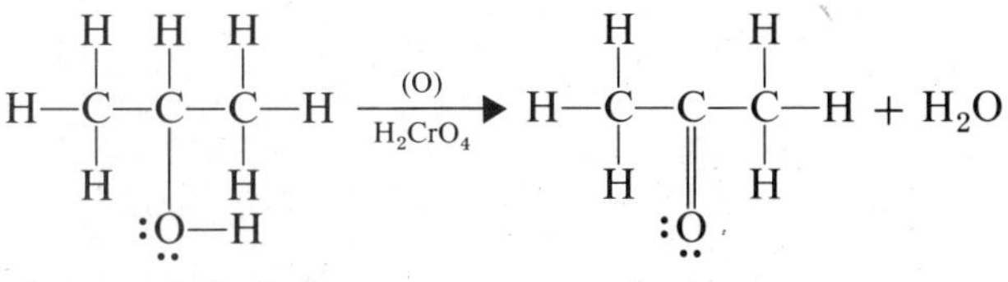

Isopropyl alcohol   Acetone
(a 2° alcohol)   (a ketone)

Aldehydes and ketones both contain the carbonyl functional group:

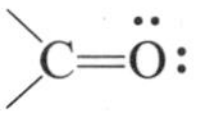

The carbonyl group

The carbonyl group in aldehydes is bonded to at least one *hydrogen atom,* and in ketones it is bonded to *two carbon atoms.*

---

*(O) means "oxidation" (in general) by some oxidizing agent that may or may not be specified in a given case. Here, chromic acid is specified as the oxidizing agent, but there are others that would serve as well.

Aldehydes   Ketones

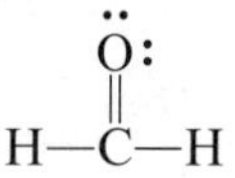

$$\overset{\overset{\displaystyle ..}{O:}}{\underset{}{\|}}$$

H—C—H CH_3—C—CH_3

Formaldehyde   Acetone

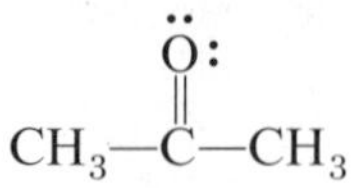

$$\overset{..}{O:}$$

CH_3—C—H CH_3CH_2—C—CH_3

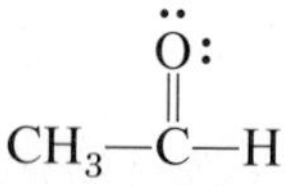

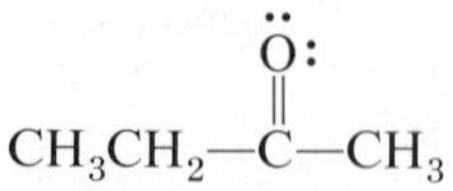

Acetaldehyde   Methyl ethyl ketone

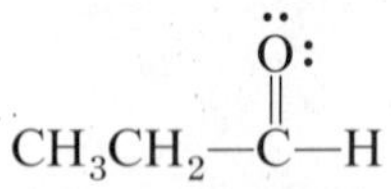

$$\overset{..}{O:}$$

CH_3CH_2—C—H

Propionaldehyde

We can simplify much of our future discussion if, at this point, we introduce a symbol that is widely used in designating general structures of organic molecules: the symbol R. *R is used as a general symbol to represent any alkyl group.* R, for example, might be a methyl group, an ethyl group, a propyl group, or an isopropyl group.

CH_3—	Methyl	All of	
CH_3CH_2—	Ethyl	these	by R
$CH_3CH_2CH_2$—	Propyl	can be	
CH_3CHCH_3	Isopropyl	designated	

Using R in this way, we see that we can designate the general formula for an aldehyde as

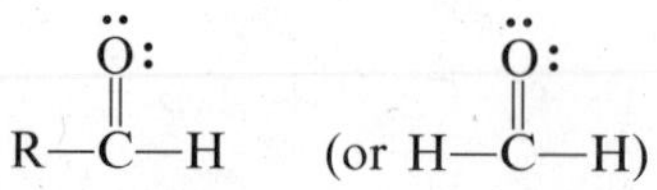

$$\overset{..}{O:} \qquad \overset{..}{O:}$$

R—C—H   (or H—C—H)

and the general formula for a ketone as

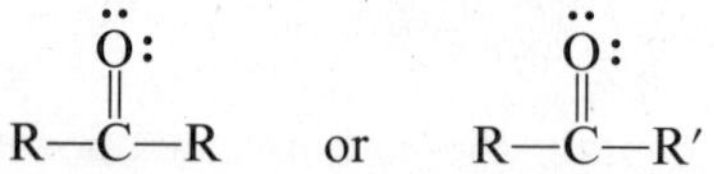

$$\overset{..}{O:} \qquad \overset{..}{O:}$$

R—C—R   or   R—C—R′

where R′ is an alkyl group different from R.

In the same way an alkane can be designated as R—H and an alcohol as R—OH.

Further oxidation of ketones is difficult, for further oxidation requires oxidative cleavage of a carbon-carbon bond, a bond that is not easily oxidized. Aldehydes, on the other hand, have a carbon-hydrogen bond and C—H bonds are easily oxidized. Aldehydes can be oxidized to compounds known as carboxylic acids.

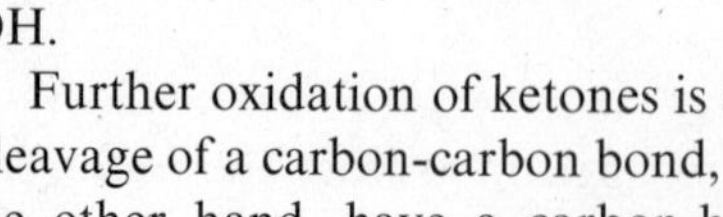

$$\overset{..}{O:} \qquad\qquad \overset{..}{O:}$$

R—C—H $\xrightarrow{(O)}$ R—C—ÖH

Aldehyde   A carboxylic acid

For example, the mild oxidizing agent, silver oxide, oxidizes acetaldehyde to the carboxylic acid, acetic acid.

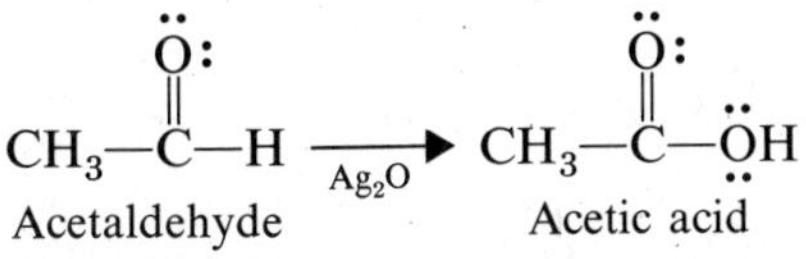

$$\underset{\text{Acetaldehyde}}{CH_3-\overset{\overset{\displaystyle \ddot{O}:}{\|}}{C}-H} \xrightarrow{\;Ag_2O\;} \underset{\text{Acetic acid}}{CH_3-\overset{\overset{\displaystyle \ddot{O}:}{\|}}{C}-\ddot{O}H}$$

The reverse of an oxidation reaction is a reduction reaction, and all of the oxidation reactions that we have considered in this section have counterparts in reduction. Using indirect methods carboxylic acids can be reduced to aldehydes.

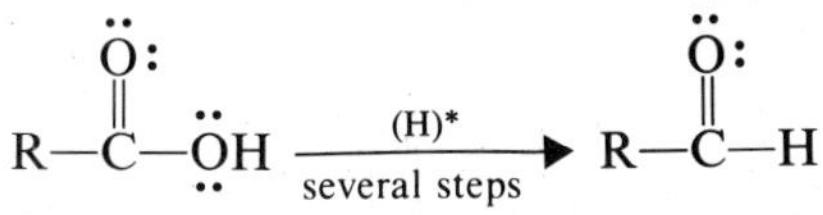

$$R-\overset{\overset{\displaystyle \ddot{O}:}{\|}}{C}-\ddot{O}H \xrightarrow[\text{several steps}]{(H)^*} R-\overset{\overset{\displaystyle \ddot{O}:}{\|}}{C}-H$$

The very powerful reducing agent, lithium aluminum hydride (LiAlH$_4$), reduces carboxylic acids to primary alcohols. Aldehydes are also reduced to primary

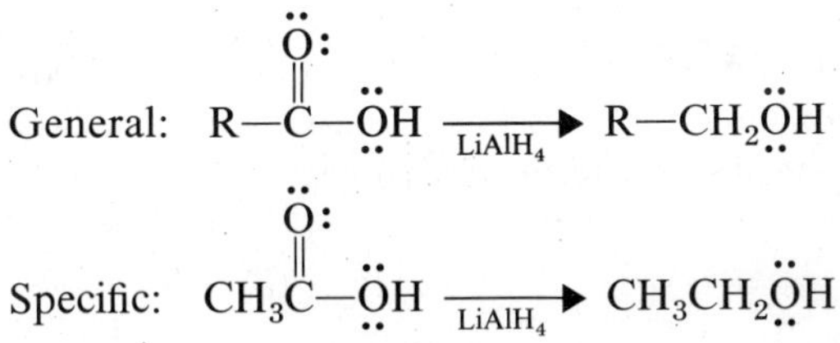

General: $\quad R-\overset{\overset{\displaystyle \ddot{O}:}{\|}}{C}-\ddot{O}H \xrightarrow{\;LiAlH_4\;} R-CH_2\ddot{O}H$

Specific: $\quad CH_3\overset{\overset{\displaystyle \ddot{O}:}{\|}}{C}-\ddot{O}H \xrightarrow{\;LiAlH_4\;} CH_3CH_2\ddot{O}H$

alcohols by lithium aluminum hydride, but the milder reducing agent, sodium borohydride (NaBH$_4$) is used more often.

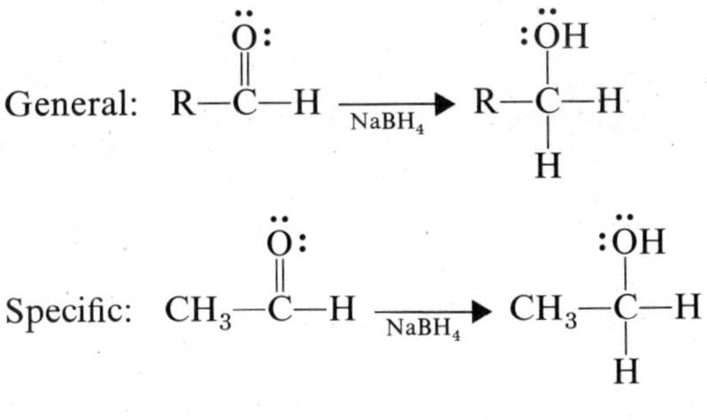

General: $\quad R-\overset{\overset{\displaystyle \ddot{O}:}{\|}}{C}-H \xrightarrow{\;NaBH_4\;} R-\overset{\overset{\displaystyle :\ddot{O}H}{|}}{\underset{\underset{\displaystyle H}{|}}{C}}-H$

Specific: $\quad CH_3-\overset{\overset{\displaystyle \ddot{O}:}{\|}}{C}-H \xrightarrow{\;NaBH_4\;} CH_3-\overset{\overset{\displaystyle :\ddot{O}H}{|}}{\underset{\underset{\displaystyle H}{|}}{C}}-H$

Ketones can be reduced to secondary alcohols in the same way:

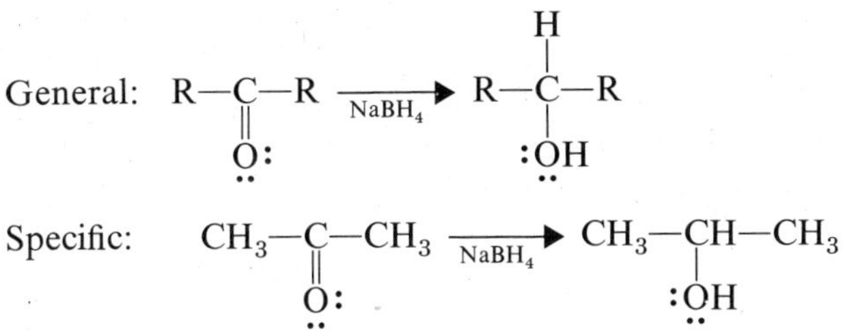

General: $\quad R-\overset{\overset{\displaystyle H}{|}}{\underset{\underset{\displaystyle \ddot{O}:}{\|}}{C}}-R \xrightarrow{\;NaBH_4\;} R-\overset{\overset{\displaystyle H}{|}}{\underset{\underset{\displaystyle :\ddot{O}H}{|}}{C}}-R$

Specific: $\quad CH_3-\underset{\underset{\displaystyle \ddot{O}:}{\|}}{C}-CH_3 \xrightarrow{\;NaBH_4\;} CH_3-\underset{\underset{\displaystyle :\ddot{O}H}{|}}{CH}-CH_3$

*(H) is a general way of designating that a reduction takes place.

Finally, alcohols can also be reduced, through indirect methods, to alkanes.

General: $R-\ddot{O}H \xrightarrow{(H)} R-H$

Specific: $CH_3CH_2-\ddot{O}H \xrightarrow{(H)} CH_3CH_2-H$

We can summarize all of the oxidation-reduction reactions that we have described for C_2 compounds in the following way. When seen in this way, a progres-

$$CH_3-CH_3 \underset{(H)}{\overset{(O)}{\rightleftharpoons}} CH_3-CH_2-\ddot{O}-H \underset{(H)}{\overset{(O)}{\rightleftharpoons}}$$

Lowest oxidation state

$$CH_3-\overset{\overset{\displaystyle \ddot{O}:}{\|}}{C}-H \underset{(H)}{\overset{(O)}{\rightleftharpoons}} CH_3-\overset{\overset{\displaystyle \ddot{O}:}{\|}}{C}-\ddot{O}H$$

Highest oxidation state

sion of oxidation states of the C_2 molecules becomes clear. The lowest oxidation state is the alkane; and the highest oxidation state is the carboxylic acid. Since ketones occupy the same general position as aldehydes, we can express the progression of oxidation states in general terms as follows:

$$R-CH_2-H \underset{(H)}{\overset{(O)}{\rightleftharpoons}} R-CH_2-\ddot{O}H \underset{(H)}{\overset{(O)}{\rightleftharpoons}} R-\overset{\overset{\displaystyle \ddot{O}:}{\|}}{C}-H \underset{(H)}{\overset{(O)}{\rightleftharpoons}} R-\overset{\overset{\displaystyle \ddot{O}:}{\|}}{C}-\ddot{O}H$$

Alkane   Alcohol   Aldehyde   Carboxylic acid

or

$$\left(R-\overset{\overset{\displaystyle \ddot{O}:}{\|}}{C}-R \right)$$

Ketone

2.11 CONDENSATION REACTIONS: ESTERS AND ETHERS

Another characteristic reaction of aldehydes, ketones, alcohols, and carboxylic acids is a reaction known as the *condensation reaction*. A condensation reaction is one in which two (or more) components become covalently bonded through the loss of a molecule of water or some other simple molecule. The condensation reaction that occurs when an alcohol and a carboxylic acid are heated in the presence of a strong acid catalyst is an example. The alcohol and the acid *condense* to form a new type of organic compound called an *ester*. (Esters occur widely in nature.)

General: $R-\ddot{O}-H + H\ddot{O}-\overset{\overset{\displaystyle \ddot{O}:}{\|}}{C}-R' \xrightarrow{H^+} R-\ddot{O}-\overset{\overset{\displaystyle \ddot{O}:}{\|}}{C}-R' + H-\ddot{O}H$

      Alcohol   Carboxylic   Ester   Water
                     acid

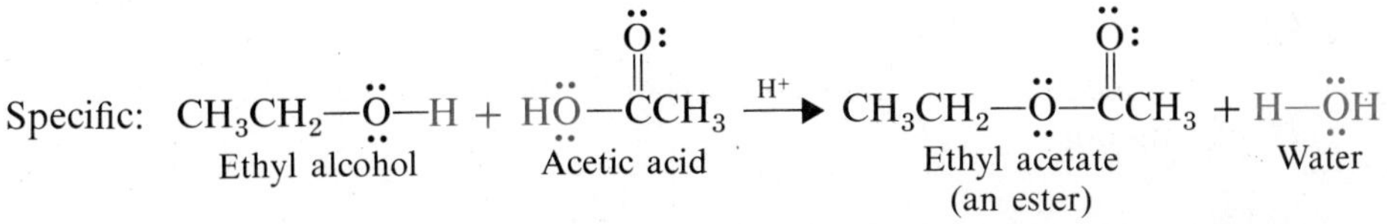

Specific: $CH_3CH_2-\overset{..}{\underset{..}{O}}-H + H\overset{..}{\underset{..}{O}}-\overset{\overset{\overset{..}{O}:}{\|}}{C}CH_3 \xrightarrow{H^+} CH_3CH_2-\overset{..}{\underset{..}{O}}-\overset{\overset{\overset{..}{O}:}{\|}}{C}CH_3 + H-\overset{..}{O}H$

Ethyl alcohol   Acetic acid   Ethyl acetate   Water
(an ester)

When esters are heated with aqueous sodium hydroxide they undergo a reaction known as *hydrolysis*. Hydrolysis is a process in which the cleavage (*lysis*) of a molecule is accompanied by the addition of the pieces of a water molecule, H— and —OH. In the hydrolysis of the ester shown below the H— and —OH groups appear in the alcohol and carboxylic acid.

The initial products of the basic (NaOH) hydrolysis of an ester are an acid salt and an alcohol. Acidification of the acid salt produces a carboxylic acid, and the overall result of both reactions is to convert the ester back into the carboxylic acid and alcohol from which it was originally synthesized.

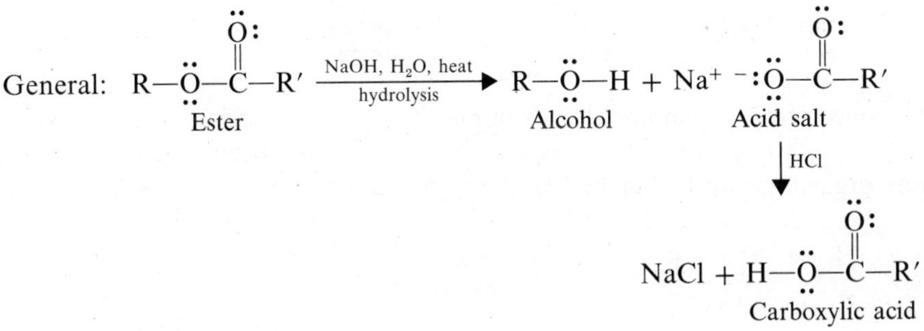

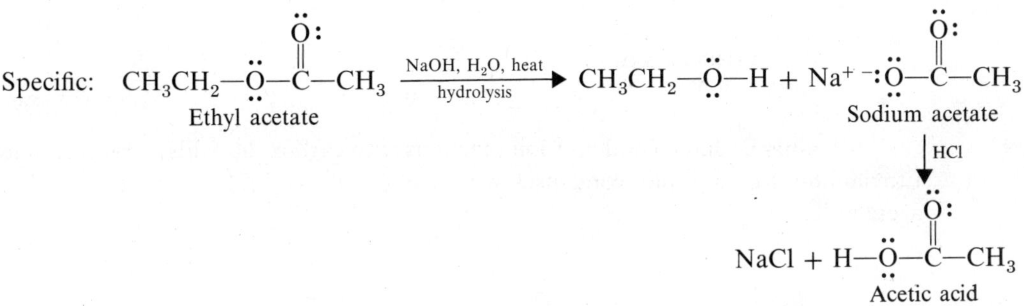

General: $R-\overset{..}{\underset{..}{O}}-\overset{\overset{\overset{..}{O}:}{\|}}{C}-R' \xrightarrow[\text{hydrolysis}]{\text{NaOH, H}_2\text{O, heat}} R-\overset{..}{\underset{..}{O}}-H + Na^+ \ ^-:\overset{..}{\underset{..}{O}}-\overset{\overset{\overset{..}{O}:}{\|}}{C}-R'$

Ester   Alcohol   Acid salt

$\downarrow$ HCl

$NaCl + H-\overset{..}{\underset{..}{O}}-\overset{\overset{\overset{..}{O}:}{\|}}{C}-R'$

Carboxylic acid

Specific: $CH_3CH_2-\overset{..}{\underset{..}{O}}-\overset{\overset{\overset{..}{O}:}{\|}}{C}-CH_3 \xrightarrow[\text{hydrolysis}]{\text{NaOH, H}_2\text{O, heat}} CH_3CH_2-\overset{..}{\underset{..}{O}}-H + Na^+ \ ^-:\overset{..}{\underset{..}{O}}-\overset{\overset{\overset{..}{O}:}{\|}}{C}-CH_3$

Ethyl acetate   Sodium acetate

$\downarrow$ HCl

$NaCl + H-\overset{..}{\underset{..}{O}}-\overset{\overset{\overset{..}{O}:}{\|}}{C}-CH_3$

Acetic acid

Alcohols are found to undergo *self-condensation reactions* to form compounds known as *ethers*. Ethers are derivatives of water in which both hydrogens have been

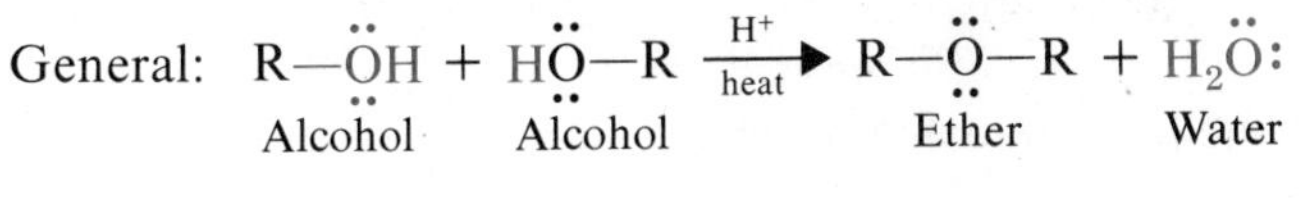

General: $R-\overset{..}{O}H + H\overset{..}{O}-R \xrightarrow[\text{heat}]{H^+} R-\overset{..}{\underset{..}{O}}-R + H_2\overset{..}{\underset{..}{O}}:$

Alcohol   Alcohol   Ether   Water

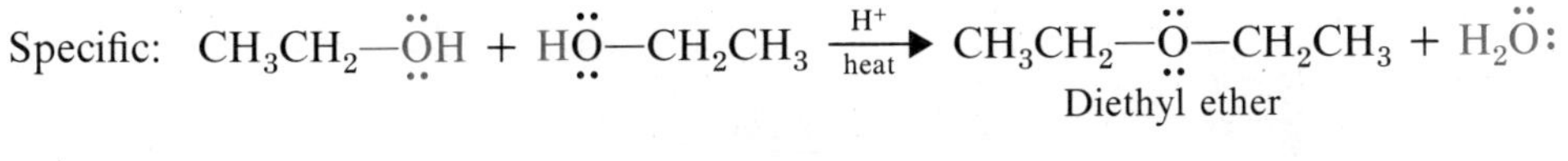

Specific: $CH_3CH_2-\overset{..}{\underset{..}{O}}H + H\overset{..}{\underset{..}{O}}-CH_2CH_3 \xrightarrow[\text{heat}]{H^+} CH_3CH_2-\overset{..}{\underset{..}{O}}-CH_2CH_3 + H_2\overset{..}{\underset{..}{O}}:$

Diethyl ether

replaced by alkyl groups. The reaction given above is catalyzed by strong acids and usually takes place at temperatures greater than 150°.

2.12 AMINES AND AMIDES

Just as alcohols and ethers are alkyl derivatives of water, amines are alkyl derivatives of ammonia.

$$H\overset{\cdot\cdot}{-N}-H \qquad R\overset{\cdot\cdot}{-N}-H$$
$$\underset{H}{\mid} \qquad\qquad \underset{H}{\mid}$$

Ammonia   An amine

Amines are classified as being either primary amines, secondary amines, or tertiary amines. This classification is based on *the number of organic groups that are attached to the nitrogen atom*. Notice that this is quite different from the way alcohols

$$R\overset{\cdot\cdot}{-N}-H \qquad R\overset{\cdot\cdot}{-N}-H \qquad R\overset{\cdot\cdot}{-N}-R$$
$$\underset{H}{\mid} \qquad\qquad \underset{R}{\mid} \qquad\qquad \underset{R}{\mid}$$

A primary (1°)   A secondary (2°)   A tertiary (3°)
amine   amine   amine

were classified. Isopropylamine, for example, is a primary amine even though its —NH$_2$ group is attached to a secondary carbon atom. It is a primary amine because only one organic group is attached to the nitrogen atom.

$$\begin{array}{ccc} & H & H & H \\ & \mid & \mid & \mid \\ H- & C- & C- & C-H \\ & \mid & \mid & \mid \\ & H & :NH_2 & H \end{array}$$

Isopropylamine
(a 1° amine)

Amines undergo condensation reactions with carboxylic acids to form *amides*. Ethylamine, for example, condenses with acetic acid to form the amide, *N*-ethylacetamide.

$$CH_3CH_2\overset{\cdot\cdot}{-N}-H + H\overset{\cdot\cdot}{O}-\overset{\overset{\textstyle \overset{\cdot\cdot}{O}:}{\|}}{C}-CH_3 \xrightarrow{heat} CH_3CH_2\overset{\cdot\cdot}{-N}-\overset{\overset{\textstyle \overset{\cdot\cdot}{O}:}{\|}}{C}-CH_3 + H_2\overset{\cdot\cdot}{O}:$$

Ethylamine   Acetic acid   *N*-ethylacetamide
(an amine)   (an acid)   (an amide)

This reaction is a general one for ammonia, primary amines, and secondary amines.

$$H\overset{\cdot\cdot}{-N}-H + H\overset{\cdot\cdot}{O}-\overset{\overset{\textstyle \overset{\cdot\cdot}{O}:}{\|}}{C}-R \xrightarrow{heat} H\overset{\cdot\cdot}{-N}-\overset{\overset{\textstyle \overset{\cdot\cdot}{O}:}{\|}}{C}-R + H_2\overset{\cdot\cdot}{O}:$$

Ammonia   An acid   An amide

$$R'-\overset{\cdot\cdot}{\underset{H}{N}}-H \;+\; H\overset{\cdot\cdot}{\underset{\cdot\cdot}{O}}-\overset{\overset{\displaystyle \overset{\cdot\cdot}{\underset{\cdot\cdot}{O}}}{\|}}{C}-R \;\xrightarrow{\;heat\;}\; R'-\overset{\overset{\displaystyle \overset{\cdot\cdot}{O}:}{}}{\underset{H}{N}}-\overset{\overset{\displaystyle \overset{\cdot\cdot}{\underset{}{O}}:}{\|}}{C}-R \;+\; H_2\overset{\cdot\cdot}{\underset{\cdot\cdot}{O}}:$$

| 1° amine | An acid | An *N*-substituted amide |

$$R'-\overset{\cdot\cdot}{\underset{R}{N}}-H \;+\; H\overset{\cdot\cdot}{\underset{\cdot\cdot}{O}}-\overset{\overset{\displaystyle \overset{\cdot\cdot}{\underset{\cdot\cdot}{O}}}{\|}}{C}-R \;\xrightarrow{\;heat\;}\; R'-\overset{}{\underset{R}{N}}-\overset{\overset{\displaystyle \overset{\cdot\cdot}{\underset{}{O}}:}{\|}}{C}-R \;+\; H_2\overset{\cdot\cdot}{\underset{\cdot\cdot}{O}}:$$

| 2° amine | An acid | An *N,N*-disubstituted amide |

Amides are like esters in that they, too, can be hydrolyzed. Refluxing an amide with aqueous sodium hydroxide produces an amine and the salt of the acid. After the amine is separated the acid salt can be acidified to produce the acid itself.

$$R-\overset{\cdot\cdot}{N}H\overset{\overset{\displaystyle \overset{\cdot\cdot}{O}:}{\|}}{C}-R' + NaOH \xrightarrow[\;heat\;]{H_2\overset{\cdot\cdot}{O}:} R-\overset{\cdot\cdot}{N}H_2 + R'-\overset{\overset{\displaystyle \overset{\cdot\cdot}{O}:}{\|}}{C}-\overset{\cdot\cdot}{\underset{\cdot\cdot}{O}}:^-Na^+ \xrightarrow{\;H^+\;} R'-\overset{\overset{\displaystyle \overset{\cdot\cdot}{O}:}{\|}}{C}-\overset{\cdot\cdot}{\underset{\cdot\cdot}{O}}-H$$

| | | Amine | Acid salt | Acid |

The amide group is one of the most important functional groups in living organisms, for it is the amide group that bonds individual amino acids together to form proteins.

$$H-\overset{}{\underset{H}{N}}:\overset{\overset{\displaystyle \overset{\displaystyle R}{|}}{\overset{\displaystyle CH}{}}}{}\;\overset{}{C}-\overset{\cdot\cdot}{O}H$$

An amino acid

A portion of a protein molecule

Hydrolysis of proteins produces the individual amino acids.

2.13 ACID-BASE REACTIONS: BRØNSTED-LOWRY ACIDS AND BASES

Involved at some point in the vast majority of reactions that occur with organic compounds are acid-base reactions. For this reason, we need to review some of the essential principles of acid-base chemistry.

Strong Acids and Bases

According to the Brønsted-Lowry theory an acid is a substance that can donate a proton, and a base is a substance that can accept a proton. Let us consider, as an example of this concept, the reaction that occurs when gaseous hydrogen chloride dissolves in water.

$$H-\overset{\cdot\cdot}{\underset{\cdot\cdot}{Cl}}:_{(g)} + H-\overset{|}{\underset{|}{\overset{\cdot\cdot}{O}}}: \longrightarrow H-\overset{\overset{H}{|}}{\underset{H}{\overset{\cdot\cdot}{O}}}:^{+} + :\overset{\cdot\cdot}{\underset{\cdot\cdot}{Cl}}:^{-}_{(aq)}$$

| Acid | Base | Conjugate acid | Conjugate base |

Hydrogen chloride, a very strong acid, transfers its proton to water. Water acts as a base and accepts the proton. The products that result from this reaction are the hydronium ion (H_3O^+, the conjugate acid of water) and the chloride ion (Cl^-, the conjugate base of hydrogen chloride). The reaction, for all practical purposes, goes to completion.

Other strong acids that show essentially complete proton transfer when dissolved in water are hydrogen bromide, hydrogen iodide, nitric acid, perchloric acid, and sulfuric acid.*

$$HBr + H_2O \longrightarrow H_3O^+ + Br^-$$
$$HI + H_2O \longrightarrow H_3O^+ + I^-$$
$$HNO_3 + H_2O \longrightarrow H_3O^+ + NO_3^-$$
$$HClO_4 + H_2O \longrightarrow H_3O^+ + ClO_4^-$$
$$H_2SO_4 + H_2O \longrightarrow H_3O^+ + HSO_4^-$$
$$HSO_4^- + H_2O \longrightarrow H_3O^+ + SO_4^=$$

Because sulfuric acid has two protons that it can transfer to a base, it is called a dibasic acid. The proton transfer is stepwise; the first proton transfer occurs completely and the second nearly so.

Hydronium ions, H_3O^+, and hydroxide ions, OH^-, are the strongest acids and bases, respectively, that are capable of existence in aqueous solutions. When sodium hydroxide (a crystalline compound consisting of sodium ions and hydroxide ions) dissolves in water, the result is a solution consisting of solvated sodium ions and hydroxide ions.

$$NaOH_{solid} + H_2O \longrightarrow Na^+_{aq} + OH^-_{aq}$$

When an aqueous solution of sodium hydroxide is mixed with an aqueous solution of hydrogen chloride (hydrochloric acid) the reaction that occurs is between hydroxide ions and hydronium ions. The sodium ions and chloride ions are called spectator ions because they play no part in the acid-base reaction.

---

* The extent to which an acid transfers the protons to a base like water is a measure of its "strength" as an acid. Acid strength is thus a measure of the percent of ionization and not of concentration.

Net reaction: $:\ddot{O}H^- + H{-}\overset{\displaystyle H}{\underset{|}{\ddot{O}^+}}{-}H \longrightarrow 2H:\ddot{O}H$

Spectator ions: $Na^+ +$ Cl^- $Na^+ + Cl^-$

What we have just related about hydrochloric acid and aqueous sodium hydroxide is true when aqueous solutions of all strong acids and strong bases are mixed. The net ionic reaction that occurs is simply

$$H_3O:^+ + :\ddot{O}H^- \longrightarrow 2H_2\ddot{O}:$$

Weak Acids and Bases

In contrast to the strong acids, such as HCl and H_2SO_4, acetic acid is a much weaker acid. When acetic acid dissolves in water, the reaction below does not proceed to completion.

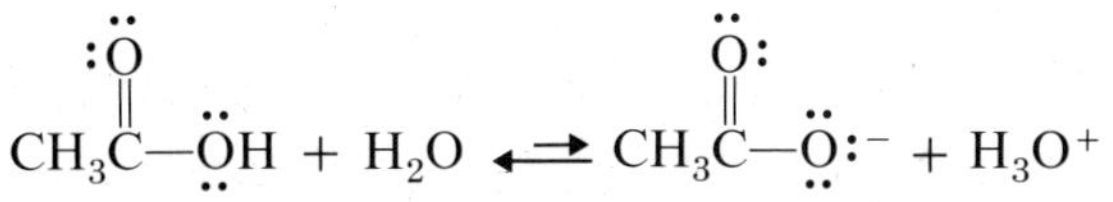

$$CH_3\overset{\displaystyle :\ddot{O}}{\underset{\|}{C}}{-}\ddot{O}H + H_2O \rightleftharpoons CH_3\overset{\displaystyle \ddot{O}:}{\underset{\|}{C}}{-}\ddot{O}:^- + H_3O^+$$

Experiments show that in a 0.1-molar solution of acetic acid at 25°C only 1% of the acetic acid molecules transfer their protons to water.

Because the reaction that occurs in an aqueous solution of acetic acid is an equilibrium, we can describe it with an expression for the equilibrium constant.

$$K_{eq} = \frac{[H_3O^+][CH_3COO^-]}{[CH_3COOH][H_2O]}$$

For dilute aqueous solutions the concentration of water is essentially constant ($\sim$55 moles/liter), so we can rewrite the expression for the equilibrium constant in terms of a new constant, K_a, called *the acidity constant.*

$$K_a = K_{eq}[H_2O] = \frac{[H_3O^+][CH_3COO^-]}{[CH_3COOH]}$$

At 25°C, the acidity constant for acetic acid is 1.8×10^{-5}.

We can write similar expressions for any acid dissolved in water. Using a generalized hypothetical acid, HA, the reaction in water is

$$HA + H_2\ddot{O}: \rightleftharpoons H_3O:^+ + A^-$$

and the expression for the acidity constant is

$$K_a = \frac{[H_3O^+][A^-]}{[HA]}$$

In this standard form the molar concentrations of the products of the proton transfer reaction are written in the numerator of the expression and the molar concentration of the undissociated acid is written in the denominator. For this reason, a large value of K_a means the acid is a strong acid, and a small value of K_a means the acid is a weak acid. If the acidity constant is greater than 10 the acid will be, for all practical purposes, completely dissociated in water.

Problem 2.9

Trifluoroacetic acid (CF_3COOH) has a $K_a = 1$ at 25°. (a) What are the molar concentrations of hydronium ion and trifluoroacetate ion (CF_3COO^-) in a 0.1 M aqueous solution of trifluoroacetic acid? (b) What percentage of the trifluoroacetic acid is ionized?

Although acetic acid and other carboxylic acids containing fewer than five carbon atoms are water soluble, many other carboxylic acids of higher molecular weight are not appreciably soluble in water. Because of their acidity, however, *water-insoluble acids dissolve in aqueous sodium hydroxide;* they do so by reacting to form *water-soluble* sodium salts.

$$C_7H_{15}COOH + NaOH_{(aq)} \longrightarrow C_7H_{15}COO^-Na^+ + H_2O$$
Water insoluble $\qquad\qquad\qquad$ Water soluble

Amines, are like ammonia in that they are weak bases. The reaction of an aqueous solution of ethylamine with a strong aqueous acid, for example, is very similar to that of aqueous ammonia.

$$:NH_3 \ + H_3O:^+ \rightleftarrows NH_4^+ \ + H_2\ddot{O}:$$
Ammonia $\qquad\qquad$ Ammonium
$\qquad\qquad\qquad$ ion

$$CH_3CH_2\ddot{N}H_2 + H_3O:^+ \rightleftarrows CH_3CH_2\overset{+}{N}H_3 \ + H_2\ddot{O}:$$
Ethylamine $\qquad\qquad$ Ethylammonium
$\qquad\qquad\qquad\quad$ ion

While ethylamine and most amines of low molecular weight are very soluble in water, higher molecular-weight amines, such as hexylamine ($C_6H_{13}NH_2$), have limited water solubility. However, such water-insoluble amines dissolve readily in hydrochloric acid because the acid-base reaction produces a soluble salt.

$$C_6H_{13}\ddot{N}H_2 + HCl_{(aq)} \longrightarrow C_6H_{13}\overset{+}{N}H_3Cl^-$$
Hexylamine $\qquad\qquad\qquad$ Hexylammonium chloride
Slight water $\qquad\qquad\qquad\quad$ *Water-soluble*
solubility $\qquad\qquad\qquad\qquad$ *salt*

Water, itself, is a very weak acid and undergoes self-ionization even in the absence of added acids or bases.

$$H_2\ddot{O}: + H_2\ddot{O}: \rightleftarrows H_3O:^+ + :\ddot{O}H^-$$

The self-ionization of pure water produces concentrations of hydronium and hydroxide ions equal to 10^{-7} moles/liter at 25°C. Since the concentration of water in pure water is 55.5 moles/liter we can calculate the K_a for water.

$$K_a = \frac{[H_3O^+][OH^-]}{[H_2O]} \qquad K_a = \frac{(10^{-7})(10^{-7})}{55.5} = 1.8 \times 10^{-16}*$$

In our discussion so far we have dealt only with the strengths of acids. Arising as a natural corollary of all that we have related about acids is a principle that allows us to estimate the strengths of bases as well. Simply stated, *the conjugate base of a strong acid will be a weak base, and the conjugate base of a weak acid will be a strong base.* Moreover, *the stronger the acid, the weaker will be its conjugate base* and vice versa.

As examples of this principle consider the following;

1. The chloride ion (Cl^-), bromide ion (Br^-), iodide ion (I^-), nitrate ion (NO_3^-), and perchlorate ion (ClO_4^-) are all conjugate bases of very strong acids (K_a's > 10) and, thus, all are very weak bases. In aqueous solution these anions have virtually no affinity for protons and exist as simple solvated ions.

2. The hydroxide ion (OH^-) is the conjugate base of water; water is a weak acid ($K_a \simeq 10^{-16}$). The hydroxide ion is therefore a strong base. It is, as we have said, the strongest base that can exist in aqueous solutions.

3. The acetate ion (CH_3COO^-) is the conjugate base of the moderately weak acid, acetic acid ($K_a \simeq 10^{-5}$). As a result, the acetate ion is a moderately strong base.

Acids and Bases in Nonaqueous Solutions

Many of the organic acid-base reactions that we describe in this text do not occur in aqueous solution. Many of the acids that we do encounter are much weaker acids than water. The conjugate bases of these very weak acids are exceedingly powerful bases. Reactions of organic compounds, for example, are often carried out in liquid ammonia (the liquefied gas, not the aqueous solution commonly used in your general chemistry laboratory). The acidity constant, K_a, of ammonia is 10^{-36}, thus liquid ammonia is a weaker acid than water by a factor of 10^{20}. The most powerful base that can be employed in liquid ammonia is the amide ion, $:\ddot{N}H_2^-$, the conjugate base of ammonia. The amide ion, because it is the conjugate base of an extremely weak acid, is an extremely powerful base. In liquid ammonia, and in the presence of amide ion, even the hydrocarbon ethyne becomes an acid.

$$H-C\equiv C-H + :\ddot{N}H_2^- \xrightarrow[NH_3]{liq.} H-C\equiv C:^- + :NH_3$$

| Stronger acid | Stronger base | Weaker base | Weaker acid |

Cyclohexane (C_6H_{12}) is an even weaker acid ($K_a = 10^{-45}$) than ammonia. The cyclohexyl anion ($C_6H_{11}:^-$) is, thus, an even stronger base than the amide ion. If ammonia were added to a solution containing cyclohexyl anions in cyclohexane the following reaction would take place.

$$C_6H_{11}:^- + :NH_3 \longrightarrow C_6H_{12} + :\ddot{N}H_2^-$$

| Stronger base | Stronger acid | Weaker acid | Weaker base |

---

*The ion-product constant for water, that is, $[H_3O^+] \times [OH^-]$, is 10^{-14}.

TABLE 2.3 Scale of Acidities and Basicities

	ACID	APPROXIMATE K_a	CONJUGATE BASE	
INCREASING STRENGTH OF ACID	C_6H_{12}	10^{-45}	$C_6H_{11}^-$	INCREASING STRENGTH OF CONJUGATE BASE
	CH_3CH_3	10^{-42}	$CH_3CH_2^-$	
	$CH_2{=}CH_2$	10^{-36}	$CH_2{=}CH^-$	
	NH_3	10^{-36}	NH_2^-	
	$HC{\equiv}CH$	10^{-26}	$HC{\equiv}C^-$	
	CH_3CH_2OH	10^{-18}	$CH_3CH_2O^-$	
	H_2O	10^{-16}	HO^-	
	CH_3COOH	10^{-5}	CH_3COO^-	
	CF_3COOH	1	CF_3COO^-	
	HNO_3	20	NO_3^-	
	H_3O^+	50	H_2O	
	HCl	10^7	Cl^-	
	H_2SO_4	10^9	HSO_4^-	
	HI	10^{10}	I'	
	$HClO_4$	10^{10}	ClO_4^-	
	$SbF_5{\cdot}FSO_3H$	$>10^{12}$	$SbF_5{\cdot}FSO_3^-$	

Later, we will encounter nonaqueous acid-base reactions that involve extremely powerful acids reacting with extremely weak bases. Ethene, for example, is a much weaker base than ammonia, but in concentrated sulfuric acid ethene accepts a proton.

$$H_2SO_4 + CH_2{=}CH_2 \longrightarrow CH_3CH_2^+ + HSO_4^-$$

In the presence of the substance called "superacid," $SbF_5 \cdot FSO_3H$, *even methane becomes a base*. It accepts a proton to form a highly unstable and short-lived intermediate, CH_5^+.

$$SbF_5 \cdot FSO_3H + CH_4 \longrightarrow CH_5^+ + SbF_5 \cdot FSO_3^-$$

The acidity constants of some of the molecules we have considered so far are listed in Table 2.3. All of the proton acids that we consider in this book have strengths in between that of cyclohexane and that of the "superacid" $SbF_5 \cdot FSO_3H$. On the other hand, all of the bases that we will consider will be weaker in strength than the cyclohexyl anion. As you examine Table 2.3, however, take care not to lose sight of the vast spectrum of acidities and basicities that it represents.

2.14 LEWIS ACID-BASE THEORY

Acid-base theory was broadened immensely by G. N. Lewis in 1923. Striking at what he called "the cult of the proton," Lewis proposed that acids be defined as *electron-pair acceptors* and bases as *electron-pair donors*. In the Lewis theory, not only is the proton an acid, but many other species are as well. Aluminum chloride and boron trifluoride, for example, react with amines in much the same way a proton does.

$$
H^+ + \overset{\overset{\displaystyle H}{|}}{\underset{\underset{\displaystyle H}{|}}{:N}}\!-R \longrightarrow H-\overset{\overset{\displaystyle H}{|}}{\underset{\underset{\displaystyle H}{|}}{N^{\pm}}}\!-R
$$

$$
\overset{\overset{\displaystyle :\ddot{C}l:}{|}}{\underset{\underset{\displaystyle :\ddot{C}l:}{|}}{:\ddot{C}l}}\!-Al + \overset{\overset{\displaystyle H}{|}}{\underset{\underset{\displaystyle H}{|}}{:N}}\!-R \longrightarrow \overset{\overset{\displaystyle :\ddot{C}l:}{|}}{\underset{\underset{\displaystyle :\ddot{C}l:}{|}}{:\ddot{C}l}}\!-Al\!-\!\overset{\overset{\displaystyle H}{|}}{\underset{\underset{\displaystyle H}{|}}{N^{\pm}}}\!-R
$$

$$
\overset{\overset{\displaystyle :\ddot{F}:}{|}}{\underset{\underset{\displaystyle :\ddot{F}:}{|}}{:\ddot{F}}}\!-B + \overset{\overset{\displaystyle H}{|}}{\underset{\underset{\displaystyle H}{|}}{:N}}\!-R \longrightarrow \overset{\overset{\displaystyle :\ddot{F}:}{|}}{\underset{\underset{\displaystyle :\ddot{F}:}{|}}{:\ddot{F}}}\!-B\!-\!\overset{\overset{\displaystyle H}{|}}{\underset{\underset{\displaystyle H}{|}}{N^{\pm}}}\!-R
$$

In the examples above, aluminum chloride and boron trifluoride accept the electron pair of the amine just as the proton does. They do this because the central aluminum and boron atoms have only a sextet of electrons and are thus electron deficient. When they accept an electron pair, aluminum chloride and boron trifluoride are, in the Lewis definition, *acting as acids*.

Bases are much the same in both the Lewis theory and the Bronsted-Lowry theory.

The Lewis theory, by virtue of its broader definition of acids, allows acid-base theory to include all of the Lowry-Brønsted reactions and, as we will see, a great many others.

2.15 PHYSICAL PROPERTIES AND MOLECULAR STRUCTURE

So far, we have said little about one of the most obvious characteristics of organic compounds, that is, *their physical state*. Whether a particular substance is a solid, or a liquid, or a gas, would certainly be one of the first observations that we would note in any experimental work. The temperatures at which transitions occur between physical states, that is, melting points and boiling points, are also among the more easily measured physical properties. Melting points and boiling points are also useful in identifying and isolating organic compounds.

Suppose, for example, we have just carried out the synthesis of an organic compound that is known to be a liquid at room temperature and one-atmosphere pressure. If we know the boiling point of our desired product, and the boiling points of other by-products and solvents that may be present in the reaction mixture, we can decide whether or not simple distillation will be a feasible method for isolating our product.

In another instance our product might be a solid. In this case, in order to isolate the substance by crystallization, we need to know its melting point and its solubility in different solvents.

TABLE 2.4

COMPOUND	STRUCTURE	mp °C	bp °C at 1 atm
Methane	CH_4	−183	−162
Ethane	CH_3CH_3	−172	−88.2
Ethene	$CH_2{=}CH_2$	−169	−102
Ethyne	$HC{\equiv}CH$	−82	−75
Chloromethane	CH_3Cl	−24	5
Chloroethane	CH_3CH_2Cl	12.5	38
Ethyl alcohol	CH_3CH_2OH	−115	78.3
Acetaldehyde	CH_3CHO	−121	20
Acetic acid	CH_3COOH	16.6	118
Sodium acetate	CH_3COONa	324	Dec
Ethylamine	$CH_3CH_2NH_2$	−80	17
Ethyl ether	$(CH_3CH_2)_2O$	−116	34.6
Ethyl acetate	$CH_3COOCH_2CH_3$	−84	77

The physical constants of known organic substances are easily found in handbooks and journals.* Table 2.4 lists the melting and boiling points of some of the compounds that we have discussed in this chapter.

Often in the course of research work, however, the product of a synthesis is a new compound—one that has never been described before. In these instances, our success in isolating the new compound depends on our making reasonably accurate estimates of its melting point, boiling point, and solubilities. Our estimations of these macroscopic physical properties are based on the structure of the substance at the molecular or ionic level and on the forces that act between molecules and ions. What are these forces? How do they affect the melting point, boiling point, and solubilities of a compound? These are questions that we will attempt to answer.

The *melting point* of a substance is the temperature at which an equilibrium exists between the well-ordered crystalline state and the more random liquid state.

* Two useful handbooks are: *Handbook of Chemistry,* N. A. Lange, McGraw-Hill, New York and *Handbook of Chemistry and Physics,* Chemical Rubber Co., Cleveland.

FIG. 2.10

The melting of sodium acetate.

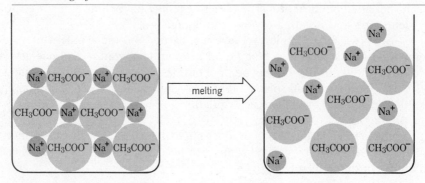

If the substance is an ionic compound, such as sodium acetate (Table 2.4), the forces that hold the ions together in the crystalline state are the strong electrostatic lattice forces that act between the positive and negative ions in the orderly crystalline structure. In Fig. 2.10 each sodium ion is surrounded by negatively charged acetate ions, and each acetate ion is surrounded by positive sodium ions. A large amount of thermal energy is required to break up the orderly structure of the crystal into the disorderly open structure of a liquid. As a result, the temperature at which sodium acetate melts is quite high, 324°. The *boiling points* of ionic compounds are higher still, so high that most ionic organic compounds decompose before they boil. Sodium acetate shows this behavior.

If we consider a nonpolar molecule like methane, however, we find that the melting point and boiling point are *much* lower: −183° and −162°, respectively. Rather than ask, "Why does methane melt and boil at low temperatures?" a more appropriate question might be: "Why does methane, a nonionic, nonpolar substance, become a liquid or a solid at all?" The answer to this question can be given in terms of attractive intermolecular forces called *van der Waals forces*.

An accurate account of the nature of van der Waals forces requires the use of quantum mechanics. We can, however, visualize the origin of these forces in the following way. The average distribution of charge in a nonpolar molecule (like methane) over a period of time is uniform. At any given instant, however, *because electrons move,* the electrons and thus the charge may not be uniformly distributed. Electrons may, in one instant, be slightly accumulated on one side of the molecule and, as a consequence, *a small temporary dipole will occur* (Fig. 2.11). This temporary dipole in one molecule can induce opposite (attractive) dipoles in surrounding molecules. These temporary dipoles change constantly, but the net result of their existence is to produce small attractive forces between nonpolar molecules, and thus make possible the existence of their liquid and solid states.

The *boiling point* of a liquid is the temperature at which the vapor pressure of the liquid equals the pressure of the atmosphere above it. For this reason, the boiling points of liquids are *pressure dependent,* and boiling points are always reported as occurring at a particular pressure, at 1 atm and at 10 torrs,* for example. A substance that boils at 150° at one atmosphere pressure will boil at a substantially lower temperature if the pressure is reduced to, for example, 0.01 torrs (a pressure easily obtained with a vacuum pump). The *standard boiling point* of a liquid is its boiling point at 1 atm.

* One torr is the pressure exerted by a column of mercury 1 mm high. One atmosphere (1 atm) is equal to 760 torrs.

FIG. 2.11
Temporary dipoles and induced dipoles in nonpolar molecules resulting from a non-uniform distribution of electrons at a given instant.

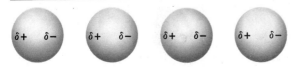

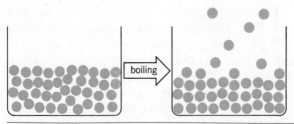

FIG. 2.12
The boiling of a liquid.

In passing from a liquid to a gaseous state the individual molecules (or ions) of the substance must separate considerably (Fig. 2.12). Because of this, we can understand why ionic organic compounds often decompose before they boil. The thermal energy required to completely separate (volatilize) the ions is so great that chemical reactions (decompositions) occur first.

Nonpolar compounds, where the intermolecular forces are very weak, usually boil at low temperatures even at 1 atm pressure. This is not always true, however, because of other factors that we have not yet mentioned: the effects of molecular weight and molecular size. Heavier molecules require greater thermal energy in order to acquire velocities sufficiently great to escape the liquid surface. In addition, because their surface areas are usually much greater, intermolecular van der Waals attractions are also much larger. These factors explain why nonpolar ethane (bp, $-88.2°$) boils higher than methane (bp, $-162°$) at a pressure of 1 atm. It also explains why, at 1 atm, the even heavier and larger nonpolar molecule *n*-decane ($C_{10}H_{22}$) boils at $+174°$.

Most organic molecules are neither ionic nor nonpolar but have *permanent dipoles* resulting from a nonuniform distribution of the bonding electrons. Chloromethane and chloroethane are examples of molecules with permanent dipoles. In these compounds, the attractive forces between molecules are much easier to visualize. In the liquid or solid state the molecules orient themselves so that the positive end of one molecule is directed toward, and thus attracted by, the negative end of another (Fig. 2.13).

Very strong dipole-dipole attractions occur between hydrogen atoms bonded to strongly electronegative elements and nonbonding electron pairs on other strongly electronegative elements (Fig. 2.14). This type of intermolecular force is called a *hydrogen bond*. The hydrogen bond (bond dissociation energy about 5 kcal/mole) is weaker than an ordinary covalent bond, but is much stronger than the dipole-dipole interactions that occur in chloromethane.

FIG. 2.13

Dipole-dipole interactions between chloromethane molecules.

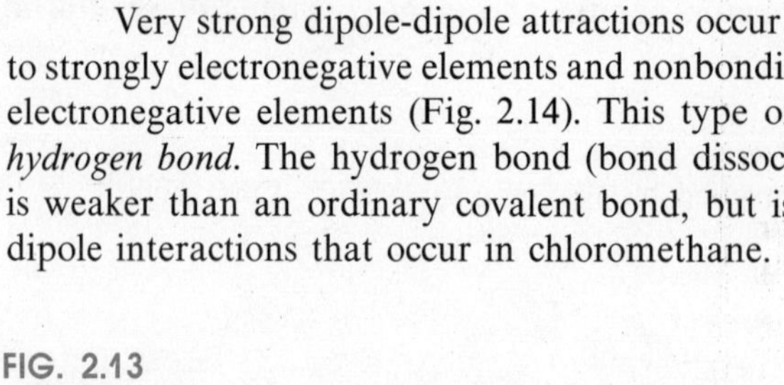

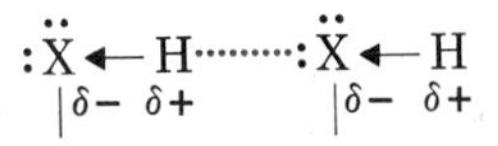

FIG. 2.14

$$:\overset{\displaystyle..}{\text{X}} \leftarrow \text{H} \cdots\cdots :\overset{\displaystyle..}{\text{X}} \leftarrow \text{H}$$
$$|_{\delta-\ \ \delta+} \qquad |_{\delta-\ \ \delta+}$$

The hydrogen bond. X is a strongly electronegative element, usually oxygen or nitrogen.

Hydrogen bonding accounts for the fact that ethyl alcohol has a much higher boiling point ($+78.3°$) than dimethyl ether ($-25°$) even though the two compounds have the same molecular weight. Ethyl alcohol, with one hydrogen atom covalently bonded to oxygen, can form strong hydrogen bonds. Dimethyl ether, because it lacks

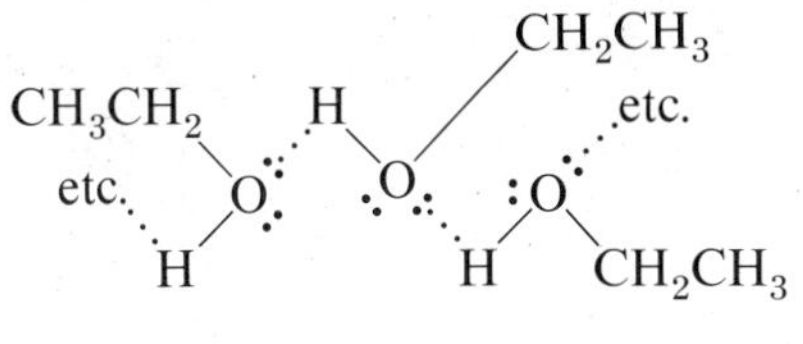

a hydrogen attached to a strongly electronegative element, cannot form strong hydrogen bonds. In dimethyl ether the intermolecular forces are weaker dipole-dipole interactions.

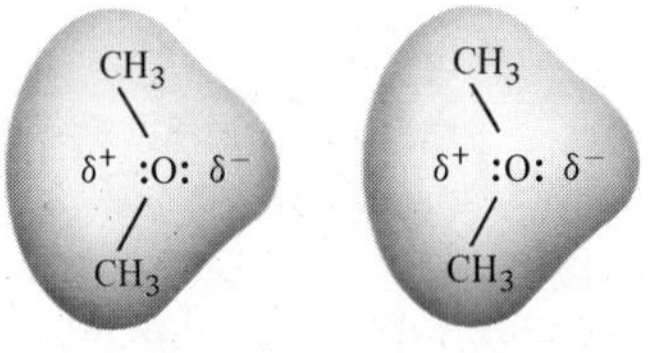

Problem 2.10

Explain why trimethylamine $(CH_3)_3N$ has a considerably lower boiling point ($3°$) than *n*-propylamine, $CH_3CH_2CH_2NH_2$, ($49°$) even though these two compounds have the same molecular weight.

Intermolecular forces are of primary importance in explaining the *solubilities* of substances. Dissolution of a solid in a liquid is, in many respects, like the melting of a solid. The orderly crystal structure of the solid is destroyed, and the result is the formation of the more disorderly arrangement of the molecules (or ions) in solution. In the process of dissolving, too, the molecules or ions must be separated from each other, and energy must be supplied to do both of these things. The energy required to overcome lattice energies and intermolecular or interionic attractions comes from the formation of new attractive forces between solute and solvent.

Consider the dissolution of an ionic substance as an example. Here both the lattice energy and interionic attractions are large. We find that water and only a few other very polar solvents are capable of dissolving ionic compounds. These solvents dissolve ionic compounds by *hydrating* or *solvating* the ions (Fig. 2.15).

Water molecules, by virtue of their great polarity, as well as their very small compact shapes, can very effectively surround the individual ions as they are freed from the crystal surface. Positive ions are surrounded by water molecules with the negative end of the water dipole pointed toward the positive ion; negative ions are solvated in exactly the opposite way. Because water is highly polar, and because water is capable of forming strong hydrogen bonds, the *dipole-ion* attractive forces

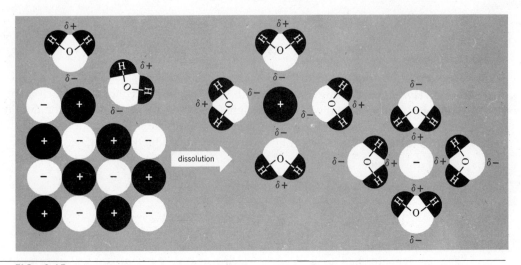

FIG. 2.15
The dissolution of an ionic solid in water showing the hydration of positive and negative ions by the very polar water molecules.

are also large. The energy supplied by their formation is great enough to overcome both the lattice energy and interionic attractions of the crystal.

A rule of thumb for predicting solubilities is that "like dissolves like." Polar and ionic compounds tend to dissolve in polar solvents. Polar liquids are generally miscible with each other in all proportions. Nonpolar solids are usually soluble in nonpolar solvents. On the other hand, nonpolar solids are insoluble in polar solvents. Nonpolar liquids are usually mutually miscible, but nonpolar liquids and polar liquids "like oil and water" do not mix.

We can understand why this is true if we understand that, when substances of similar polarities are mixed, the "new" intermolecular forces that form in the solution are very much like those that existed in the separate substances. The miscibility of nonpolar carbon tetrachloride with a nonpolar alkane would be an example. Very polar water molecules are probably capable of inducing polarities in alkane molecules that are sufficiently large to form attractive forces between them. Water and alkanes are not soluble in each other, however, because dissolution of the alkane in water requires the separation of strongly attractive water molecules from each other.

Ethanol and water, by contrast, are miscible in all proportions. In this example, both molecules are polar and the new attractive forces are as strong as those they replace and, in this instance, both compounds are capable of forming strong hydrogen bonds.

Additional Problems

2.11
Classify each of the following compounds as to whether it is an alkane, alkene, alkyne, alcohol, or aldehyde, and so forth.

(a) $CH_3C{\equiv}CCH_3$ (b) $CH_3{-}\underset{\underset{CH_3}{|}}{C}HCH_2\overset{\overset{O}{\|}}{C}OH$

(c)

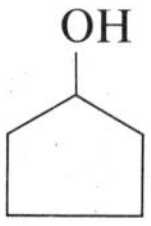

OH

(e) $CH_3CHCH_2CH_2CH_2CH_2CH_2CH_2CH_2CH_2CH_2CH_2CH_2CH_2CH_2CH_3$
$\quad\ |$
$\quad CH_3$

(a sex attractant of the female tiger moth)

(d) 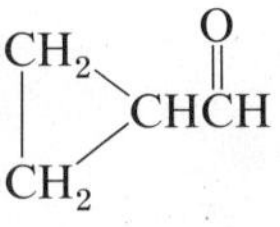 CH_2
$\qquad\diagdown$
$\qquad\quad CHCH$
$\qquad\diagup\ \quad\ \parallel$
$CH_2\qquad\quad O$

(f)

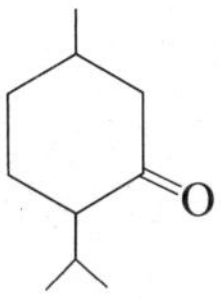

(obtained from peppermint oil)

2.12
Identify all of the functional groups in each of the following compounds.

(a) 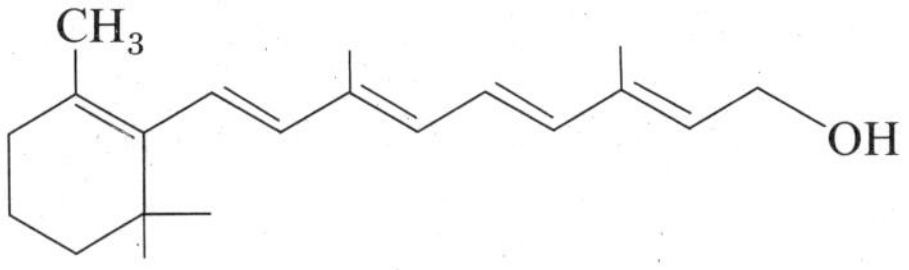 CH_3

OH

(vitamin A$_1$)

(b)

OH

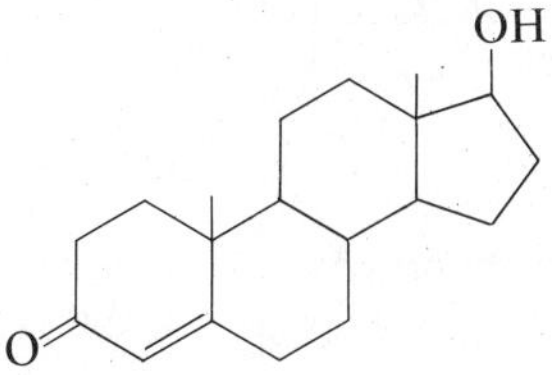

(testosterone, a male sex hormone)

O

(c) 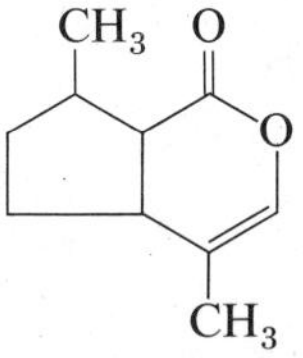 CH_3 $\quad$ O

(nepetalactone, one constituent of catnip)

CH_3

(d) $[-NH-(CH_2)_6-NH\overset{\overset{\displaystyle O}{\parallel}}{C}-(CH_2)_4-\overset{\overset{\displaystyle O}{\parallel}}{C}-]_n$ $\quad$ (nylon)

(e)

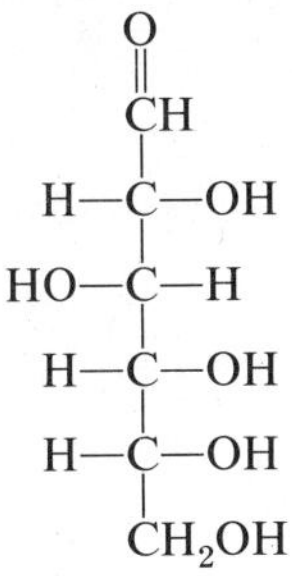

$\qquad\ \overset{\overset{\displaystyle O}{\parallel}}{CH}$
$\quad H-C-OH$
$HO-C-H \qquad$ (glucose, a sugar)
$\quad H-C-OH$
$\quad H-C-OH$
$\qquad CH_2OH$

(f) $CH_2{=}CH-O-CH{=}CH_2$ $\quad$ (an anesthetic)

(g) 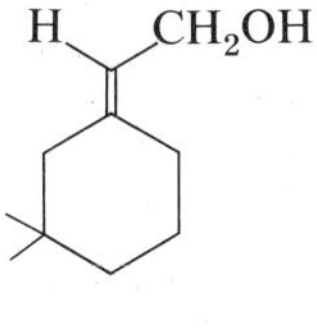 H $\quad CH_2OH$

(male boll-weevil sex attractant)

2.13

Classify each of the following reactions as to whether it is a substitution reaction, addition reaction, elimination reaction, condensation reaction, or oxidation reaction, and so forth.

(a)

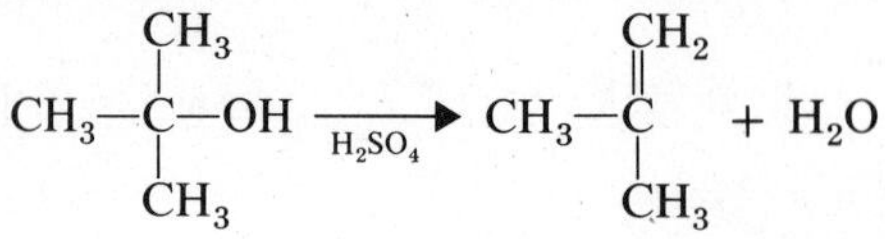

$$CH_3-\underset{\underset{CH_3}{|}}{\overset{\overset{CH_3}{|}}{C}}-OH \xrightarrow{H_2SO_4} CH_3-\underset{\underset{CH_3}{|}}{\overset{\overset{CH_2}{||}}{C}} + H_2O$$

(b) $CH_3CH_2CH_2Br + :CN^- \longrightarrow CH_3CH_2CH_2-CN + Br^-$

(c) $CH_2{=}CHCH_3 + Br_2 \xrightarrow{heat} CH_2{=}CHCH_2Br + HBr$

(d)

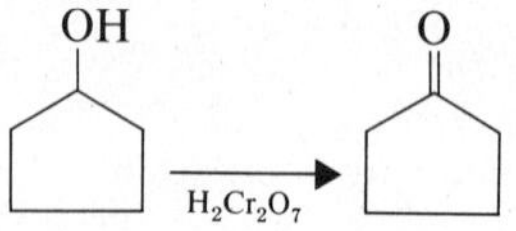

(e)

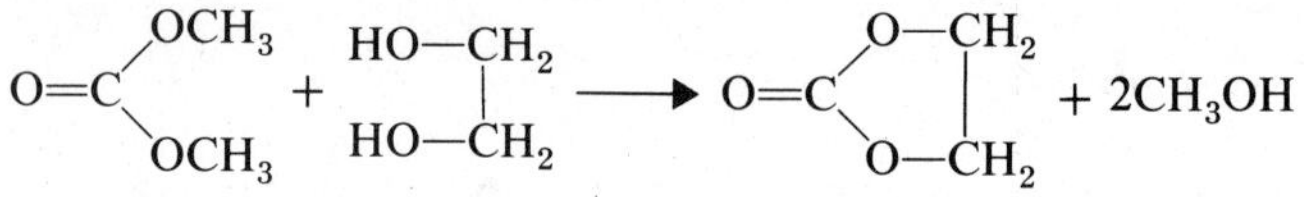

$$O{=}C\underset{OCH_3}{\overset{OCH_3}{\big\langle}} + \begin{matrix} HO-CH_2 \\ | \\ HO-CH_2 \end{matrix} \longrightarrow O{=}C\underset{O-CH_2}{\overset{O-CH_2}{\big\langle}}\Big| + 2CH_3OH$$

(f)

$$CH_3CH_2\overset{\overset{O}{||}}{C}H + CH_3NHNH_2 \longrightarrow CH_3CH_2\underset{\underset{H}{|}}{C}{=}NNHCH_3 + H_2O$$

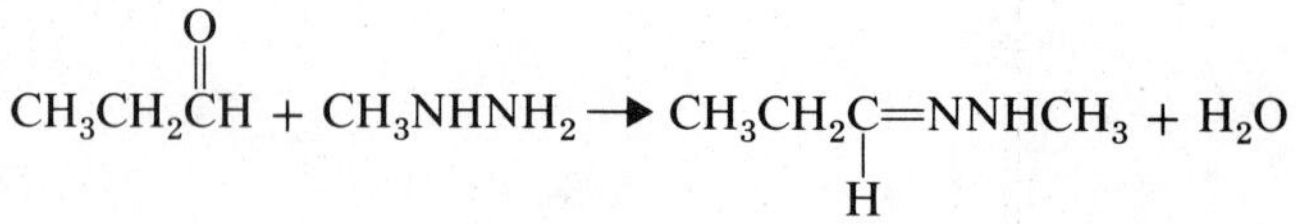

2.14

On p. 54 we saw all of the monochloro- and dichloropropanes that can be formed when propane is chlorinated. (a) How many different trichloropropanes are possible? (b) How many tetrachloropropanes?

2.15

(a) What product would you expect to be formed when isobutyl bromide is heated with

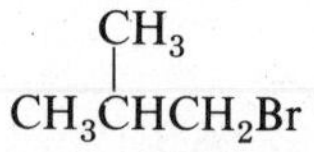

$$CH_3\overset{\overset{CH_3}{|}}{C}HCH_2Br$$
isobutyl bromide

potassium hydroxide in ethanol? (b) Can you suggest a method for the conversion of isobutyl bromide into *tert*-butyl bromide?

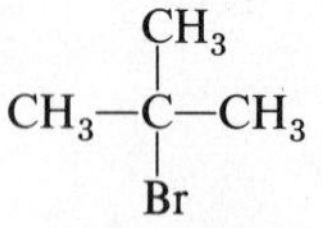

$$CH_3-\underset{\underset{Br}{|}}{\overset{\overset{CH_3}{|}}{C}}-CH_3$$
tert-butyl bromide

2.16

Classify the following alcohols as to whether they are primary alcohols, secondary alcohols, or tertiary alcohols.

(a) $CH_3CH_2CH_2CH_2OH$

(b) $CH_3CH_2CH_2CH(OH)CH_3$

(c) $CH_3CH_2C(OH)(CH_3)CH_3$

(d) $CH_3CH(CH_3)CH(OH)CH_3$

(e)

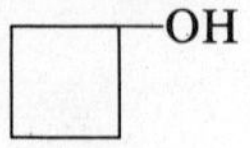

(f)

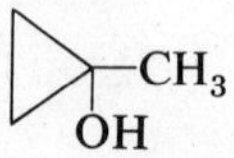

2.17
Classify the following amines as to whether they are primary amines, secondary amines, or tertiary amines.

(a) $CH_3CH_2NHCH_2CH_3$

(b) $CH_3CH(NH_2)CH_2CH_3$

(c) $CH_3NCH_2CH_2CH_3$
     |
    CH_3

(d) 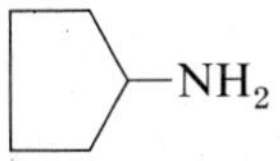 ⬠—NH_2

(e) 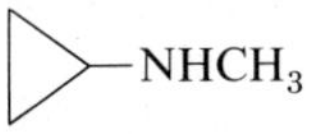 ▷—$NHCH_3$

2.18
Write net ionic equations for the following acid-base reactions.

(a) $HCl_{(aq)} + Na_2CO_{3(aq)} \longrightarrow H_2O + CO_2 + NaCl_{(aq)}$

(b) $HBr_{(aq)} + CH_3\overset{\overset{O}{\|}}{C}ONa_{(aq)} \longrightarrow CH_3\overset{\overset{O}{\|}}{C}OH + NaBr_{(aq)}$

(c) $Na_2CO_{3(aq)} + H_2O \longrightarrow NaHCO_3 + NaOH$

(d) $NaH + H_2O \longrightarrow H_2 + NaOH_{(aq)}$

(e) $CH_3Li + H_2O \longrightarrow CH_4 + LiOH$

(f) $CH_3Li + HC\equiv CH \longrightarrow LiC\equiv CH + CH_4$

(g) $HCl_{(aq)} + NH_{3(aq)} \longrightarrow NH_4Cl_{(aq)}$

(h) $NH_4Cl + NaNH_2 \xrightarrow{NH_3} 2NH_3 + NaCl$

(i) $CH_3CH_2ONa + H_2O \longrightarrow CH_3CH_2OH + NaOH$

2.19
Explain why almost all oxygen-containing organic compounds dissolve in concentrated sulfuric acid.

2.20
Which compound in each of the following pairs would have the higher boiling point?
(a) Ethyl alcohol, CH_3CH_2OH, or methyl ether, CH_3OCH_3
(b) Ethylene glycol, CH_2OHCH_2OH, or ethyl alcohol, CH_3CH_2OH
(c) Pentane, C_5H_{12}, or heptane, C_7H_{16}

(d) Acetone, $CH_3\overset{\overset{O}{\|}}{C}CH_3$ or 1-propanol, $CH_3CH_2CH_2OH$

(e) *cis*-1,2-Dichloroethene, $\underset{H}{\overset{Cl}{}}C=C\underset{H}{\overset{Cl}{}}$ or *trans*-1,2-Dichloroethene, $\underset{H}{\overset{Cl}{}}C=C\underset{Cl}{\overset{H}{}}$

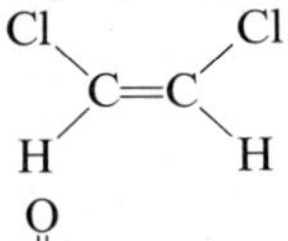

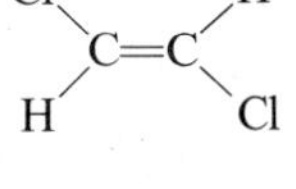

(f) Propionic acid, $CH_3CH_2\overset{\overset{O}{\|}}{C}OH$, or methyl acetate, $CH_3\overset{\overset{O}{\|}}{C}-OCH_3$

2.21
How would you carry out the following reactions?

(a) $CH_3CH_2CH_2\overset{\overset{O}{\|}}{C}OH \xrightarrow{?} CH_3CH_2CH_2CH_2OH$

(b) $CH_3CH_2CH_2CH_2OH \xrightarrow{?} CH_3CH_2CH_2\overset{\overset{O}{\|}}{C}-H$

(c) $CH_3\overset{\overset{}{\underset{\underset{O}{\|}}{}}}{C}CH_2CH_3 \xrightarrow{?} CH_3\overset{}{\underset{\underset{OH}{|}}{C}}HCH_2CH_3$

(d) $\underset{\underset{OH}{|}}{\overset{\overset{H}{|}}{CH_3CCH_2CH_3}} \xrightarrow{?} CH_3\overset{\overset{O}{\|}}{C}CH_2CH_3$

(e) $CH_3CH_2CH_2\overset{\overset{O}{\|}}{C}OH \xrightarrow{?} CH_3CH_2CH_2\overset{\overset{O}{\|}}{C}OCH_3$

(f) $CH_3OH \xrightarrow{?} CH_3OCH_3$

2.22

Write structural formulas for each of the following:

(a) An ether with the formula C_3H_8O
(b) A primary alcohol with the formula C_3H_8O
(c) A secondary alcohol with the formula C_3H_8O
(d) Two esters with the formula $C_4H_8O_2$
(e) A primary alkyl halide (R—X) with the formula C_4H_9X (alkyl halides are classified in the same way as alcohols)
(f) A secondary alkyl halide with the formula C_4H_9X
—(g) A tertiary alkyl halide with the formula C_4H_9X
(h) An aldehyde with the formula C_4H_8O
(i) A ketone with the formula C_4H_8O
(j) A primary amine with the formula $C_4H_{11}N$
(k) A secondary amine with the formula $C_4H_{11}N$
(l) A tertiary amine with the formula $C_4H_{11}N$
(m) An amide with the formula C_4H_9NO
(n) An N-substituted amide with the formula C_4H_9NO

2.23

Write Newman projection formulas for the staggered and eclipsed forms of propane.

2.24

$(CF_3)_3N$ is a weaker base than $(CH_3)_3N$. Explain.

2.25

Compounds of the general type shown below are called lactones. What functional group do they contain?

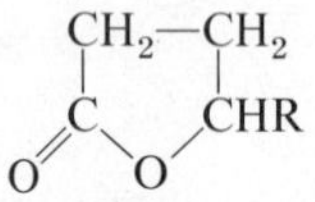

$$\underset{\underset{O}{\nearrow}\underset{O}{\nwarrow}}{\overset{\overset{CH_2-CH_2}{|\qquad\ |}}{C\qquad\ CHR}}$$

2.26

(a) What products will be obtained initially from a sodium hydroxide hydrolysis of each of the following compounds? (b) What products will be formed if the reaction mixture is acidified with hydrochoric acid?

(a) $CH_3CH_2\overset{\overset{O}{\|}}{C}{-}OCH_2CH_3$ (d) $CH_3\overset{\overset{O}{\|}}{C}{-}\overset{\overset{\cdot\cdot}{}}{\underset{\underset{CH_3}{|}}{N}}{-}CH_3$

(b) $CH_3\overset{\overset{O}{\|}}{C}{-}O{-}\langle\text{cyclopentyl}\rangle$ (e) $CH_3CH_2\overset{\overset{O}{\|}}{C}{-}NHCH_2CH_3$

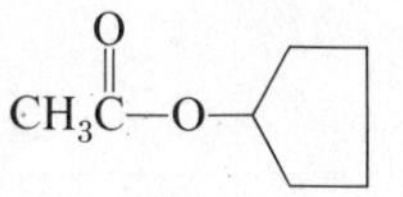

(c) $CH_3CH_2CH_2\overset{\overset{O}{\|}}{C}{-}NH_2$

2.27
Hydrogen fluoride has a dipole moment of 1.9 D; its boiling point is 19°. Ethyl fluoride, CH_3CH_2F, has an almost identical dipole moment and has a larger molecular weight, yet its boiling point is −32°. Explain.

* **2.28**
What factors might account for the fact that tetrafluoromethane, CF_4, has a much lower boiling point (bp = −129°) than hexane, C_6H_{14}, (bp = 68°) even though both compounds are nonpolar and both have approximately the same molecular weight?

* **2.29**
The boiling points of nonpolar molecules increase with increasing molecular size and increasing numbers of electrons. The boiling points of the halogens and of the noble gases, for example, show the following orders: $I_2 > Br_2 > Cl_2 > F_2$, and $Xe > Kr > Ar > Ne > He$. Explain.

* **2.30**
One method for assigning an oxidation state to a carbon atom of an organic compound is to base that assignment on the groups attached to the carbon; a bond to hydrogen makes it −1, a bond to oxygen, nitrogen, or halogen makes it +1, and a bond to another carbon 0. Thus the carbon of methane is assigned an oxidation state of −4, and that of carbon dioxide +4. (a) Use this method to assign oxidation states to the carbons of methyl alcohol (CH_3OH),

formic acid (HCOH), and formaldehyde (HCH). (b) Arrange the compounds methane, carbon dioxide, methyl alcohol, formic acid, and formaldehyde in order of increasing oxidation state. (c) What change in oxidation state accompanies the reaction, methyl alcohol ⟶ formaldehyde? (d) Is this an oxidation or a reduction? (e) When H_2CrO_4 acts as an oxidizing agent the chromium of H_2CrO_4 becomes Cr^{+3}. What change in oxidation state does chromium undergo? (f) How many moles of H_2CrO_4 will be required to oxidize one mole of methyl alcohol to formaldehyde?

* **2.31**
(a) Use the method described in the preceding problem to assign oxidation states to each carbon of ethyl alcohol and to each carbon of acetaldehyde. (b) What do these numbers reveal about the site of oxidation when ethyl alcohol is oxidized to acetaldehyde? (c) Repeat this procedure for the oxidation of acetaldehyde to acetic acid. (d) When silver oxide oxidizes acetaldehyde to acetic acid, the silver oxide is reduced to metallic silver. How many moles of Ag_2O will be required to oxidize one mole of acetaldehyde?

* **2.32**
(a) Although we have described the hydrogenation of an alkene (p. 51) as an addition reaction, organic chemists often refer to it as a "reduction." Refer to the method described in problem 2.30 and explain. (b) Make similar comments about the reversible reaction:

$$CH_3-\overset{\overset{\textstyle O}{\|}}{C}-H + H_2 \overset{Ni}{\rightleftharpoons} CH_3CH_2OH$$

* An asterisk beside a problem indicates that it is somewhat more challenging. Your instructor may tell you that these problems are optional.

3

ALKANES AND CYCLOALKANES: THEIR STRUCTURES, PROPERTIES, AND SYNTHESES

3.1 INTRODUCTION

We noted earlier that the family of organic compounds called hydrocarbons can be divided into several groups based on the type of bond that exists between the individual carbon atoms. Those hydrocarbons in which all of the carbon-carbon bonds are single bonds are called *alkanes*; those hydrocarbons that contain a carbon-carbon double bond are called *alkenes*; and those with a carbon-carbon triple bond are called *alkynes*. In Chapter 12, we will study a special group of cyclic hydrocarbons known as *aromatic hydrocarbons*.

Cycloalkanes are alkanes in which the carbon atoms are arranged in a ring. Alkanes have the general formula C_nH_{2n+2}; cycloalkanes have two fewer hydrogen atoms and thus have the general formula C_nH_{2n}. Several examples of alkanes and cycloalkanes are given below.

Alkanes	Cycloalkanes

$$CH_3CH_2CH_3$$
Propane
(C_3H_8)

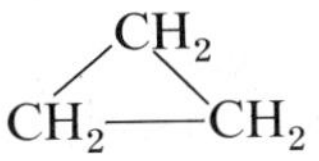

$$\begin{array}{c} CH_2 \\ \diagup \quad \diagdown \\ CH_2 \!\!-\!\! CH_2 \end{array}$$
Cyclopropane
(C_3H_6)

$$CH_3CH_2CH_2CH_3$$
Butane
(C_4H_{10})

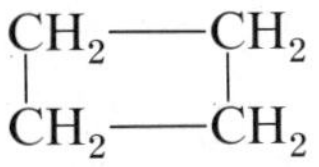

$$\begin{array}{c} CH_2 \!\!-\!\! CH_2 \\ | \qquad | \\ CH_2 \!\!-\!\! CH_2 \end{array}$$
Cyclobutane
(C_4H_8)

$$CH_3CH_2CH_2CH_2CH_3$$
Pentane
(C_5H_{12})

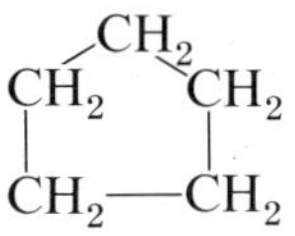

$$\begin{array}{c} CH_2 \\ \diagup \quad \diagdown \\ CH_2 \qquad CH_2 \\ | \qquad | \\ CH_2 \!\!-\!\! CH_2 \end{array}$$
Cyclopentane
(C_5H_{10})

$$CH_3CH_2CH_2CH_2CH_2CH_3$$
Hexane
(C_6H_{14})

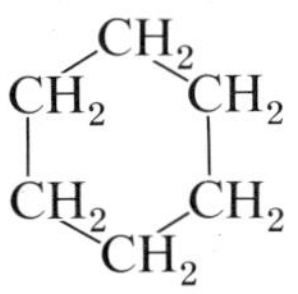

$$\begin{array}{c} CH_2 \\ \diagup \quad \diagdown \\ CH_2 \qquad CH_2 \\ | \qquad | \\ CH_2 \qquad CH_2 \\ \diagdown \quad \diagup \\ CH_2 \end{array}$$
Cyclohexane
(C_6H_{12})

Alkanes and cycloalkanes are so similar that many of their basic properties can be considered side by side. There are differences between these two groups of compounds, however, and certain structural features arise from the rings of cyclo- alkanes that are more conveniently dealt with separately. We will point out the chemical and physical similarities of alkanes and cycloalkanes as we go along.

3.2 ORBITAL HYBRIDIZATION AND THE STRUCTURE OF METHANE

The s and p orbitals used in the quantum mechanical description of the carbon atom, given in Section 1.12, were based on calculations for hydrogen atoms. These simple s and p orbitals do not, when taken alone, provide a satisfactory explanation for the *tetravalent-tetrahedral* carbon of methane. However, a satisfactory description of methane's structure that is based on quantum mechanics *can* be obtained through an approach called *orbital hybridization*. Orbital hybridization, in its simplest terms, is nothing more than a mathematical approach that involves the combining of individual s and p orbitals to obtain new orbitals. The new orbitals have, *in varying proportions*, the properties of the original orbitals taken separately. These new orbitals are called *hybrid orbitals*.

In order to see why the idea of orbital hybridization is a necessary addition to our store of quantum mechanical ideas, let us begin with two "thought" experi- ments. In the first of these we will attempt to arrive at a satisfactory description of the structure of methane using four hydrogen atoms *and a carbon atom in its lowest energy state (or ground state)*. In the second thought experiment we will use, instead, a carbon atom *in an excited state* (i.e., a state in which one electron is promoted to a higher orbital). In both instances we will fail to account satisfactorily for methane's structure and thus demonstrate the need for some new idea—for orbital hybridization.

We begin with the thought experiment using carbon in its ground state. According to quantum mechanical calculations for hydrogen the electronic structure of the ground state of carbon should be that given below.

$$C \quad \underset{1s}{\uparrow\downarrow} \quad \underset{2s}{\uparrow\downarrow} \quad \underset{2p}{\uparrow} \quad \underset{2p}{\uparrow} \quad \underset{2p}{\underline{}}$$

ground-state carbon

The valence electrons of carbon (those used in bonding) are those of the *outer shell*, that is, the $2s$ and $2p$ electrons. If we recall the shapes of these orbitals, we see that

FIG. 3.1
The electronic configuration of a ground state carbon atom. The p *orbitals are designated* $2p_x$, $2p_y$, *and* $2p_z$ *to indicate their respective orientations along the* x, y, *and* z *axes. The assignment of the unpaired electrons to the* $2p_y$ *and* $2p_x$ *orbitals is arbitrary. They could also have been placed in the* $2p_x$ *and* $2p_z$ *or* $2p_y$ *and* $2p_z$ *orbitals. (To have placed them both in the same orbital would not have been correct, however, for this would have violated Hund's Rule.) (Section 1.12)*

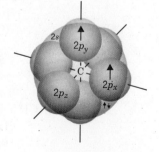

a picture of the valence electrons of a carbon atom in its ground state might resemble that given in Fig. 3.1.

The formation of the covalent bonds of methane *from individual atoms* requires that the carbon atom overlap its orbitals containing *single electrons* with 1s orbitals of hydrogen atoms (which also contain a single electron). If a ground-state carbon atom were to combine with hydrogen atoms in this way, the result would be that depicted below. *Only two carbon-hydrogen bonds would be formed, and these would be at right angles to each other.*

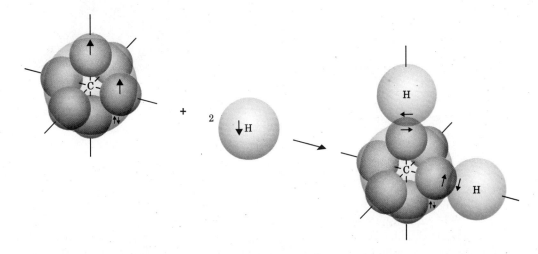

The molecule that we have described here, CH_2 (called carbene), has been detected, but only as a highly reactive substance capable of momentary existence. Clearly, then, a carbon atom in this ground state is *neither tetravalent nor tetrahedral*, and thus, it does not offer a proper model for the carbon of methane.

What, then, of carbon in an excited state? Let us imagine one electron being promoted from the 2s orbital to the 2p_z orbital. Such a carbon atom could be tetravalent, because all four orbitals would each contain a single electron.

$$C^* \quad \underset{1s}{\uparrow\downarrow} \quad \underset{2s}{\uparrow} \quad \underset{2p_x}{\uparrow} \quad \underset{2p_y}{\uparrow} \quad \underset{2p_z}{\uparrow}$$

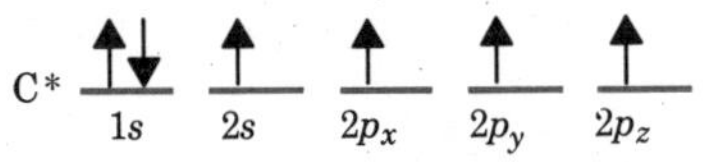

excited-state carbon

An excited-state carbon atom might combine with four hydrogen atoms as shown in Fig. 3.2.

The promotion of an electron from the 2s orbital to the 2p_z orbital requires energy. The amount of energy required has been determined and is equal to 96 kcal/mole. This energy expenditure can be rationalized by arguing that the energy released when two additional covalent bonds are formed would more than compensate for that required to excite the electron. No doubt this is true, but this solves only one problem. The problems that cannot be solved by using an excited-state carbon as a basis for a model of methane are the problems of the carbon-hydrogen bond angles and the apparent equivalence of all four carbon-hydrogen bonds. Three of the hydrogens—those overlapping p orbitals—would, in this model, be at angles

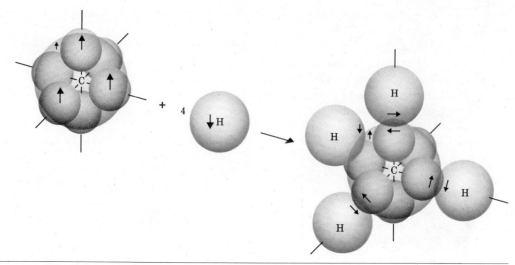

FIG. 3.2
The hypothetical formation of CH$_4$ from excited-state carbon.

of 90° with respect to each other; the fourth hydrogen, the one overlapping its 1s orbital with the 2s orbital of carbon, would be at some other angle, probably as far from the other bonds as the confines of the molecule would allow. Basing our model of methane on this excited state of carbon gives us a carbon that is tetravalent *but one that is not tetrahedral*, and it predicts a structure for methane in which one carbon-hydrogen bond differs from the other three.

We see then, that still another model is needed to account for the tetrahedral bond angles of methane and its four equivalent hydrogens. This model *can be provided by the theory of orbital hybridization.*

Hybrid orbitals that account for methane's structure can be obtained by combining the wave functions of the 2s orbital of carbon with those of the three 2p orbitals. The mathematical process of hybridization can be approximated by the illustration that is shown in Fig. 3.3.

In this process, four orbitals are mixed—or hybridized—and four new hybrid orbitals are obtained. The hybrid orbitals are called sp^3 orbitals to indicate that they have one part the character of an s orbital and three parts the character of a p orbital. The mathematical treatment of orbital hybridization also shows that *the four sp^3 orbitals should be oriented at angles of 109.5° with respect to each other*. This is, of course, precisely the spatial orientation of the four hydrogen atoms of methane.

If, in our imagination, we visualize the formation of methane from an sp^3-hybridized carbon atom and four hydrogen atoms, the process might be like that shown in Fig. 3.4. We see that an sp^3 hybridized carbon gives a *tetrahedral structure for methane, and one with four equivalent C—H bonds*.

In addition to accounting properly for the shape of methane, the theory of orbital hybridization also explains the very strong bonds that are formed between carbon and hydrogen. To see how this is so, consider the shape of the individual

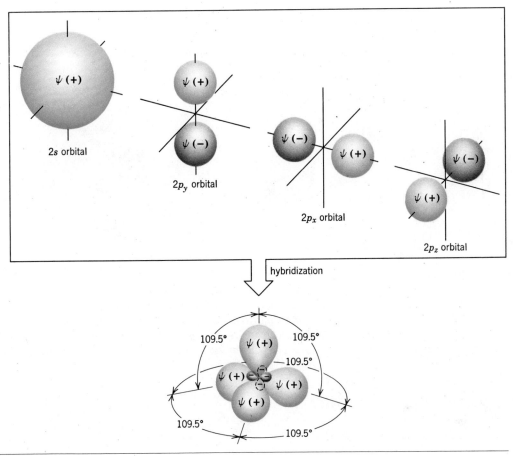

FIG. 3.3
Hybridization of the atomic orbitals of carbon to produce sp^3-*hybrid orbitals.*

sp^3 orbital shown below:

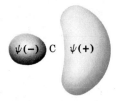

Because the sp^3 orbital has the character of a p orbital, the positive lobe of the sp^3 orbital is large and is extended quite far into space.* We see why this is true by examining the following illustration.

---

* Remember, the + and − signs associated with orbitals are the arithmetic signs of the wave functions that describe them. They are not electrical charges.

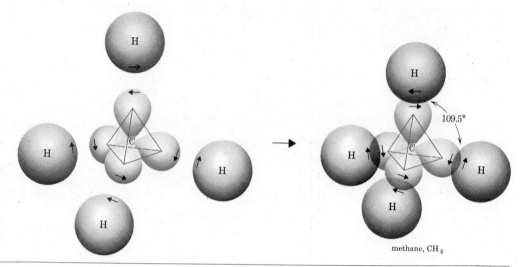

methane, CH$_4$

FIG. 3.4

The formation of methane from an sp^3*-hybridized carbon atom. [In orbital hybridization we combine orbitals* not *electrons. The electrons can then be replaced in the hybrid orbitals as necessary for bond formation, but always in accordance with the Pauli principle of no more than two electrons (with opposite spin) in each orbital. In this illustration we have placed one electron in each of the hybrid carbon orbitals.]*

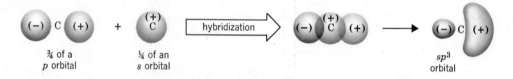

We see that on one side of the carbon atom the signs of the wave functions of both the *s* and *p* orbitals are positive. On this side the wave functions *add* to give a large *positive* lobe to the *sp^3* orbital. On the other side the wave functions are of opposite signs and they *subtract*. This results in a small *negative* lobe.

It is the positive lobe of the *sp^3* orbital that overlaps with the positive 1*s* orbital of hydrogen to form a carbon-hydrogen bond.

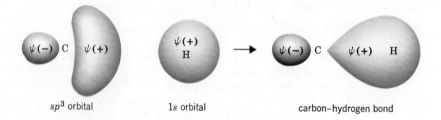

Because the positive lobe of the *sp^3* orbital is large and is extended into space, the overlap between it and the 1*s* orbital of hydrogen is also large, and the resulting carbon-hydrogen bond is quite strong.

The bond formed from the overlap of an *sp^3* orbital and a 1*s* orbital is called a *sigma bond*. The term *sigma bond* is a general term applied to those bonds in which orbital overlap gives a bond that is *circularly symmetrical in cross section when viewed along the bond axis.*

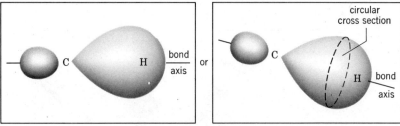

a sigma (σ) bond

3.3 SHAPES OF OTHER ALKANES

The sp^3-hybridized carbon atom not only provides a satisfactory model for methane, it provides a satisfactory model for the carbon atoms of other alkanes as well. Let us consider ethane as an example. We know from experiments that the arrangement of atoms around each carbon of ethane is tetrahedral. We can now see how this arrangement can be explained in terms of sp^3-hybridized carbon atoms. Let us imagine two sp^3-hybridized carbon atoms and six hydrogen atoms overlapping their orbitals in the manner shown below.

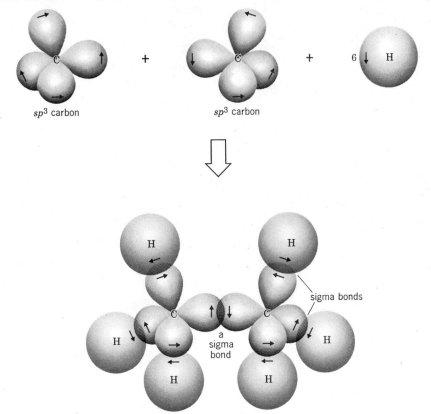

We see that the carbon-carbon bond of ethane is a *sigma bond*, formed as the result of two overlapping sp^3 orbitals. We also see that remaining sp^3 orbitals on each carbon atom can form sigma bonds to hydrogen, and that the spatial arrangement of the four atoms that surround each carbon of ethane is tetrahedral.[*]

General tetrahedral orientation of the four atoms around carbon, and thus sp^3 hybridization, is the rule for all saturated carbons. Using line-and-circle formulas, we can represent the shapes of several alkanes in the following way.

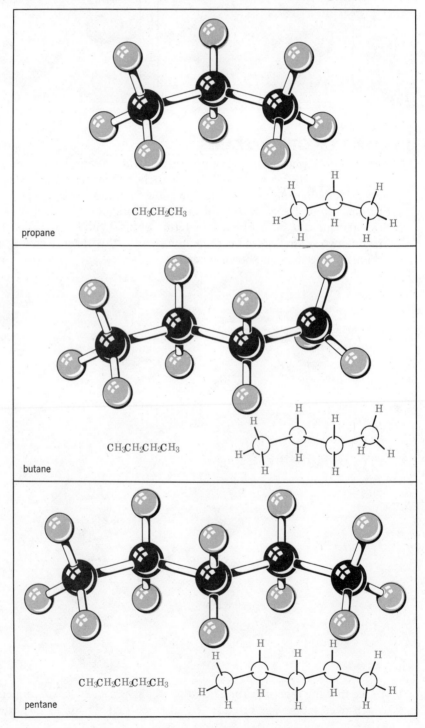

$CH_3CH_2CH_3$

propane

$CH_3CH_2CH_2CH_3$

butane

$CH_3CH_2CH_2CH_2CH_3$

pentane

Butane and pentane are examples of alkanes that are sometimes called "straight-chain" alkanes. One glance at their three-dimensional structures shows that because of the tetrahedral carbon atoms the chain is zigzagged and not at all straight. Indeed, the formulas that we have depicted above are the straightest possible arrangements of the chains, for rotations about the carbon-carbon single bonds create arrangements of the chains that are even less straight. It is more accurate to refer to the chains of butane and pentane as being *unbranched*. This means that each carbon of the chain is bonded to no more than two other carbon atoms. Unbranched alkanes are also often called "normal" alkanes.

Isobutane and isopentane are examples of branched-chain alkanes.

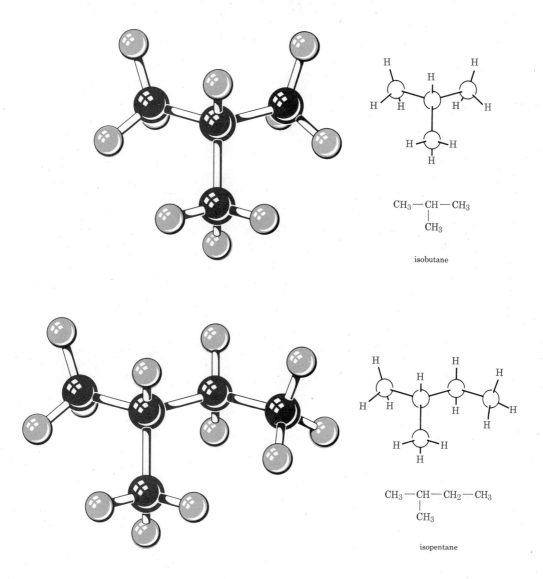

$$CH_3 - CH - CH_3$$
$$|$$
$$CH_3$$

isobutane

$$CH_3 - CH - CH_2 - CH_3$$
$$|$$
$$CH_3$$

isopentane

In each of the compounds above one carbon is attached to three other carbon atoms.

In neopentane (below) the central carbon is bonded to four carbon atoms:

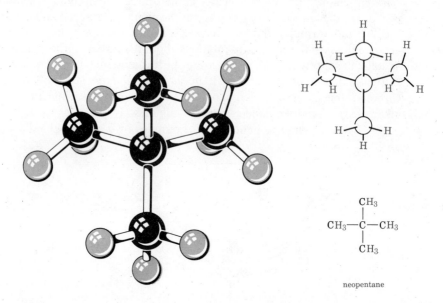

neopentane

Butane and isobutane have the same molecular formula: C_4H_{10}. The two compounds have different structures and are, therefore, *structural isomers*. Pentane, isopentane, and neopentane are also structural isomers. They, too, have the same molecular formula (C_5H_{12}) but have different structures.

Problem 3.1

Write line-and-circle and condensed structural formulas for all of the structural isomers of C_6H_{14}. Compare your answers with the condensed structural formulas given in Table 3.1.

Structural isomers, as stated earlier, have different physical properties. The differences may not always be large, but structural isomers will always be found to have different melting points, boiling points, densities, indices of refraction, and so forth. Table 3.1 gives some of the physical properties of the C_4H_{10}, C_5H_{12}, and C_6H_{14} isomers.

3.4 NOMENCLATURE OF ALKANES

Thus far, we have used the names of a number of organic compounds without specifying any systematic origin for these names. You have certainly noticed patterns and structural relationships in the names of some of the compounds we have studied. For example, you have probably noticed the relation between ethane, ethanol, and ethylamine, between hexane and cyclohexane, and between 1,1-dichloroethane and 1,2-dichloroethane. You may also have noticed the iso- structure common to both isobutane and isopentane.

TABLE 3.1 Physical Constants of the Butane, Pentane, and Hexane Isomers

MOLECULAR FORMULA	STRUCTURAL FORMULA	mp °C	bp °C	DENSITY	INDEX OF REFRACTION n_D 20°C
C_4H_{10}	$CH_3CH_2CH_2CH_3$	-138.3	-0.5	0.6012_4^0	1.3543
C_4H_{10}	CH_3CHCH_3 $\quad\ \ \|$ $\quad\ \ CH_3$	-159	-12	0.603_4^0	—
C_5H_{12}	$CH_3CH_2CH_2CH_2CH_3$	-129.72	36	0.6262_4^{20}	1.3579
C_5H_{12}	$CH_3CHCH_2CH_3$ $\quad\ \ \|$ $\quad\ \ CH_3$	-160	27.9	0.6197_4^{20}	1.3537
C_5H_{12}	$\qquad CH_3$ $\qquad\ \|$ CH_3-C-CH_3 $\qquad\ \|$ $\qquad CH_3$	-20	9.45	0.61350_4^{20}	1.3476
C_6H_{14}	$CH_3CH_2CH_2CH_2CH_2CH_3$	-95	68	0.65937_4^{20}	1.3748
C_6H_{14}	$CH_3CHCH_2CH_2CH_3$ $\quad\ \ \|$ $\quad\ \ CH_3$	-153.67	60.3	0.6532_4^{20}	1.3714
C_6H_{14}	$CH_3CH_2CHCH_2CH_3$ $\qquad\ \ \ \|$ $\qquad\ \ \ CH_3$	-118	63.265	0.6643_4^{20}	1.3765
C_6H_{14}	$CH_3CH-CHCH_3$ $\qquad\ \|\qquad\ \|$ $\qquad CH_3\ \ CH_3$	-128.8	58	0.6616_4^{20}	1.3750
C_6H_{14}	$\qquad CH_3$ $\qquad\ \|$ $CH_3-C-CH_2CH_3$ $\qquad\ \|$ $\qquad CH_3$	-98	49.7	0.6492_4^{20}	1.3688

$$CH_3CH-$$
$$|$$
$$CH_3$$

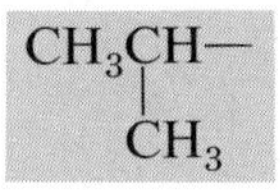

The iso- structure

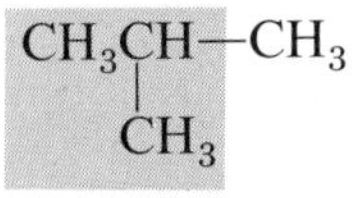

CH_3CH-CH_3
$\quad\ \ |$
$\quad\ \ CH_3$

Isobutane

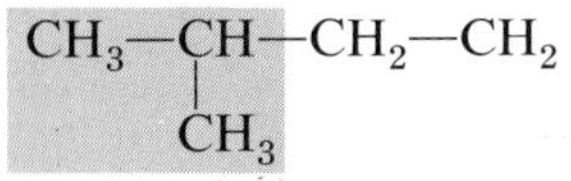

$CH_3-CH-CH_2-CH_2$
$\qquad\ \ |$
$\qquad\ \ CH_3$

Isopentane

$CH_3-CH-CH_2-CH_2-CH_3$
$\qquad\ \ |$
$\qquad\ \ CH_3$

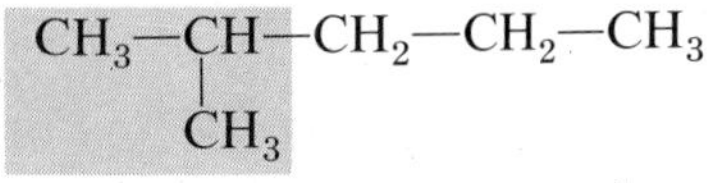

Isohexane

Isohexane, and all isoalkanes, have a one-carbon branch on the second carbon of the chain.

You may even have noticed the relationship between *ace*tic acid, and *ace*tal-dehyde, when we discussed them in Chapter 2.

$$\underset{\text{Acetic acid}}{CH_3\overset{\displaystyle O}{\overset{\|}{C}}-OH} \qquad \underset{\text{Acetaldehyde}}{CH_3\overset{\displaystyle O}{\overset{\|}{C}}-H}$$

The development of a formal system for naming organic compounds did not come about until near the end of the nineteenth century. Prior to that time many organic compounds had already been discovered. The names given these compounds sometimes reflected a source of the compound. Acetic acid, for example, can be obtained from vinegar; it got its name from the Latin word for vinegar, *acetum*. Formic acid can be obtained from ants; it got its name from the Latin word for ants, *formicae*. Ethanol (or ethyl alcohol) was at one time called grain alcohol because it was obtained by the fermentation of grains. Methanol was called wood alcohol because it was obtained by the destructive distillation of wood. The names of other compounds were based on their method of preparation. Ethylene dichloride (CH_2ClCH_2Cl), for example, got its name because it was made by allowing ethylene to react with chlorine.

These older names for organic compounds are now called "common" or "trivial" names. Trivial though they may be, many of these names are still widely used by chemists, biochemists, and the companies that sell chemicals. For these reasons it is still necessary to learn the common names for some of the common compounds. We will point out these common names as we go along, and we will use them occasionally. Most of the time, however, the names that we will use will be those called IUPAC names.

The formal system of nomenclature that is used today is one proposed by the International Union of Pure and Applied Chemistry (IUPAC). This system was first developed in 1892 and has been revised at regular intervals to keep it up to date. Underlying the IUPAC system of nomenclature for organic compounds is a fundamental principle: *each different compound should have a different name*. Thus, through a systematic set of rules, the IUPAC system provides different names for the more than two million known organic compounds, and names can be devised

TABLE 3.2 The Normal Alkanes

NAME	NO. OF CARBONS	STRUCTURE	NAME	NO. OF CARBONS	STRUCTURE
Methane	1	CH_4	Heptadecane	17	$CH_3(CH_2)_{15}CH_3$
Ethane	2	CH_3CH_3	Octadecane	18	$CH_3(CH_2)_{16}CH_3$
Propane	3	$CH_3CH_2CH_3$	Nonadecane	19	$CH_3(CH_2)_{17}CH_3$
Butane	4	$CH_3(CH_2)_2CH_3$	Eicosane	20	$CH_3(CH_2)_{18}CH_3$
Pentane	5	$CH_3(CH_2)_3CH_3$	Heneicosane	21	$CH_3(CH_2)_{19}CH_3$
Hexane	6	$CH_3(CH_2)_4CH_3$	Docosane	22	$CH_3(CH_2)_{20}CH_3$
Heptane	7	$CH_3(CH_2)_5CH_3$	Tricosane	23	$CH_3(CH_2)_{21}CH_3$
Octane	8	$CH_3(CH_2)_6CH_3$	Triacontane	30	$CH_3(CH_2)_{28}CH_3$
Nonane	9	$CH_3(CH_2)_7CH_3$	Hentriacontane	31	$CH_3(CH_2)_{29}CH_3$
Decane	10	$CH_3(CH_2)_8CH_3$	Tetracontane	40	$CH_3(CH_2)_{38}CH_3$
Undecane	11	$CH_3(CH_2)_9CH_3$	Pentacontane	50	$CH_3(CH_2)_{48}CH_3$
Dodecane	12	$CH_3(CH_2)_{10}CH_3$	Hexacontane	60	$CH_3(CH_2)_{58}CH_3$
Tridecane	13	$CH_3(CH_2)_{11}CH_3$	Heptacontane	70	$CH_3(CH_2)_{68}CH_3$
Tetradecane	14	$CH_3(CH_2)_{12}CH_3$	Octacontane	80	$CH_3(CH_2)_{78}CH_3$
Pentadecane	15	$CH_3(CH_2)_{13}CH_3$	Nonacontane	90	$CH_3(CH_2)_{88}CH_3$
Hexadecane	16	$CH_3(CH_2)_{14}CH_3$	Hectane	100	$CH_3(CH_2)_{98}CH_3$

for any one of millions of other compounds yet to be synthesized. In addition, the IUPAC system is simple enough to allow any chemist familiar with the rules (or with the rules at hand) to write the name for any compound he might encounter. In the same way, one is also able to derive the structure of a given compound from its IUPAC name.

The IUPAC system for naming alkanes is quite simple, and the principles involved are used in naming compounds in other families as well. For these reasons we begin our study of the IUPAC system with the rules for naming alkanes.

The names for several of the unbranched alkanes are listed in Table 3.2. The ending for all of the names of alkanes is *ane*. The prefixes of the names of most of the alkanes (above C_4) are of Greek and Latin origin. Learning the prefixes is like learning to count in organic chemistry. Thus, one, two, three, four, five, becomes meth-, eth-, prop-, but-, pent-. If you take the time now to memorize only the first 10 prefixes, you will, through application of the rules that follow, have learned the names for nearly 100,000 alkanes. In addition, you will have formed the basis for the names of perhaps 100,000 alkenes, 100,000 alkynes, 100,000 alcohols, and 100,000 aldehydes. (The specific rules for alkenes, alkynes, alcohols and aldehydes will be given in subsequent chapters.) One has few opportunities to gain such a large body of knowledge from such a small investment of time.

Nomenclature of Alkyl Groups

The names of groups derived from alkanes by removal of one hydrogen end in *yl* and these groups are called *alkyl groups*. When the alkane is *unbranched*, and the hydrogen that is removed is a *terminal* hydrogen the names are straightforward:

Alkane		Alkyl Group
CH_4 Methane	becomes	CH_3- Methyl
CH_3CH_3 *Ethane*	becomes	CH_3CH_2- Ethyl
$CH_3CH_2CH_3$ Propane	becomes	$CH_3CH_2CH_2-$ Propyl
$CH_3CH_2CH_2CH_3$ Butane	becomes	$CH_3CH_2CH_2CH_2-$ Butyl

For alkanes with more than two carbons, however, more than one derived group is possible. Two groups can be derived from propane; the propyl group (above) is derived by removal of a terminal hydrogen, and the isopropyl group (below) is derived by removal of an inner hydrogen.

$$\overset{\displaystyle CH_3}{\underset{\displaystyle }{|}}$$
$$CH_3CH- \qquad or \qquad CH_3-\overset{\displaystyle }{\underset{\displaystyle |}{CH}}-CH_3$$
Isopropyl group

There are four alkyl groups that contain four carbon atoms:

$$CH_3CH_2CH_2CH_2- \qquad \overset{\displaystyle CH_3}{\underset{\displaystyle }{|}}$$
$$\qquad\qquad\qquad CH_3CHCH_2-$$

Butyl (or *n*-butyl)   Isobutyl

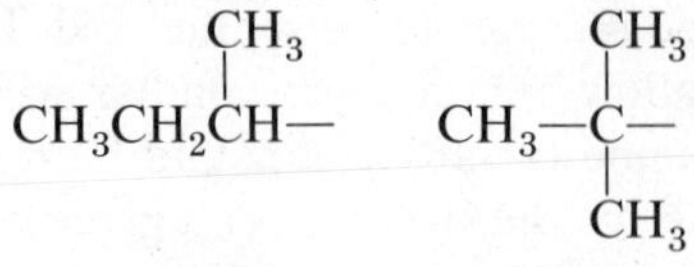

$$CH_3CH_2\overset{\overset{\displaystyle CH_3}{|}}{CH}- \qquad CH_3-\overset{\overset{\displaystyle CH_3}{|}}{\underset{\underset{\displaystyle CH_3}{|}}{C}}-$$

sec-Butyl   tert-Butyl
                   (or t-butyl)

All of these groups should be memorized so well that you are able to recognize them when they are written backward or upside down. As an aid in learning these groups it may be helpful to point out the following characteristics.

1. The base name of any group relates to the total number of carbon atoms in the group. (The propyl and isopropyl groups have *three* carbons and all of the "butyl" groups have *four* carbons.)

2. The isopropyl and isobutyl groups both have an iso structure.*

$$CH_3-\overset{\overset{\displaystyle CH_3}{|}}{CH}- \qquad CH_3\overset{\overset{\displaystyle CH_3}{|}}{CH}-CH_2-$$

Iso structures

3. The incomplete valence of the *sec*-butyl group is directed away from a *secondary* carbon.

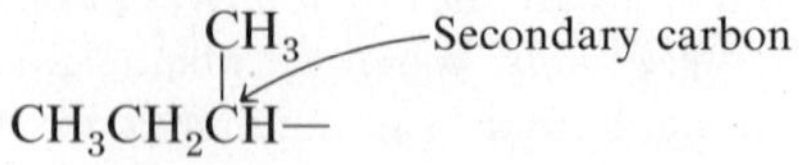

$$CH_3CH_2\overset{\overset{\displaystyle CH_3}{|}}{CH}-\quad\text{—Secondary carbon}$$

sec-butyl

4. The incomplete valence of the *tert*-butyl group is directed away from a *tertiary* carbon.

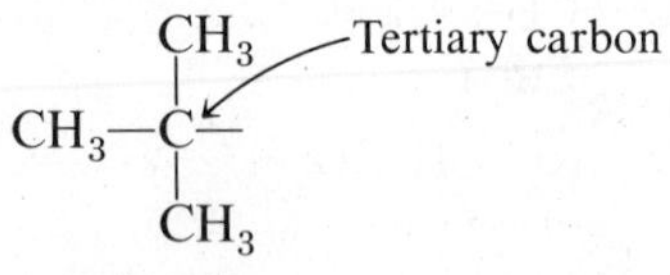

$$CH_3-\overset{\overset{\displaystyle CH_3}{|}}{\underset{\underset{\displaystyle CH_3}{|}}{C}}-\quad\text{—Tertiary carbon}$$

tert-butyl

Nomenclature of Branched-Chain Alkanes

Branched-chain alkanes are named according to the following rules:

1. *Locate the longest continuous chain of carbon atoms; this chain determines the base name for the alkane.*

We designate the compound listed below, for example, as a *hexane* because the longest continuous chain contains six carbon atoms.

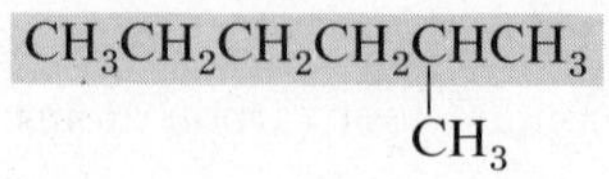

$$CH_3CH_2CH_2CH_2\overset{}{\underset{\underset{\displaystyle CH_3}{|}}{CH}}CH_3$$

*So too, do the isopentyl and isohexyl groups, $CH_3\overset{\overset{\displaystyle CH_3}{|}}{CH}CH_2CH_2-$ and

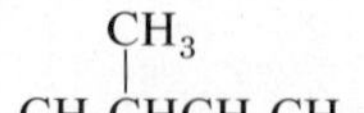

$CH_3\overset{\overset{\displaystyle CH_3}{|}}{CH}CH_2CH_2CH_2-$, respectively. These names are also provided in the IUPAC system.

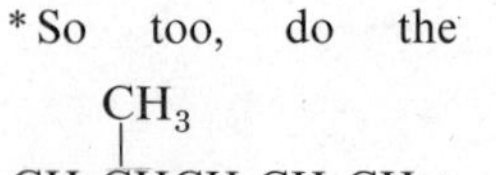

The longest continuous chain may not always be obvious from the way the formula is written. Notice, for example, that the alkane written below,

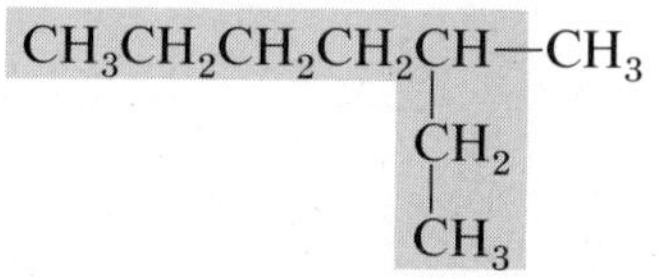

$$CH_3CH_2CH_2CH_2CH-CH_3$$
$$|$$
$$CH_2$$
$$|$$
$$CH_3$$

is designated as a *heptane* because the longest chain contains seven carbon atoms.

2. *Number the longest chain beginning with the end of the chain nearest the branching.*

Applying this rule, we number the two alkanes that we illustrated above in the following way.

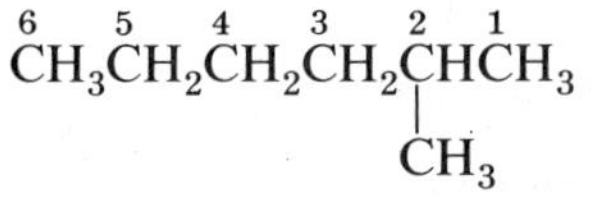

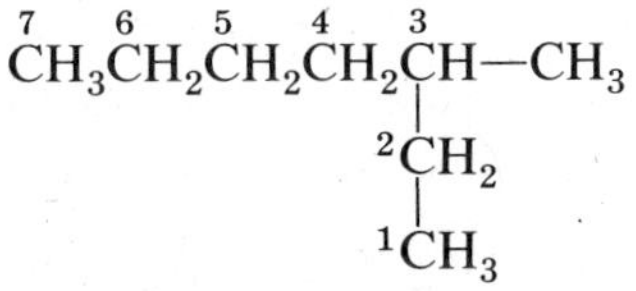

$$\overset{6}{C}H_3\overset{5}{C}H_2\overset{4}{C}H_2\overset{3}{C}H_2\overset{2}{C}HC\overset{1}{H}_3$$
$$|$$
$$CH_3$$

$$\overset{7}{C}H_3\overset{6}{C}H_2\overset{5}{C}H_2\overset{4}{C}H_2\overset{3}{C}H-CH_3$$
$$|$$
$$\overset{2}{C}H_2$$
$$|$$
$$\overset{1}{C}H_3$$

3. *Use the numbers obtained by application of rule 2 to designate the location of the substituent group.* The base name is placed last; and the substituent group, preceded by the number designating its location on the chain, is placed first. Numbers are separated from words by a dash. Our two examples are 2-methylhexane and 3-methylheptane, respectively.

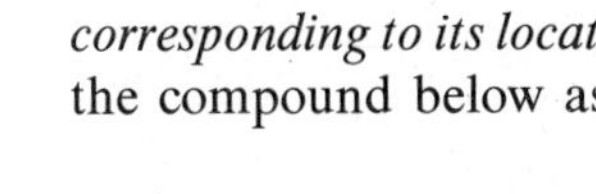

$$\overset{6}{C}H_3\overset{5}{C}H_2\overset{4}{C}H_2\overset{3}{C}H_2\overset{2}{C}HC\overset{1}{H}_3$$
$$|$$
$$CH_3$$

$$\overset{7}{C}H_3\overset{6}{C}H_2\overset{5}{C}H_2\overset{4}{C}H_2\overset{3}{C}HCH_3$$
$$|$$
$$\overset{2}{C}H_2$$
$$|$$
$$\overset{1}{C}H_3$$

2-Methylhexane 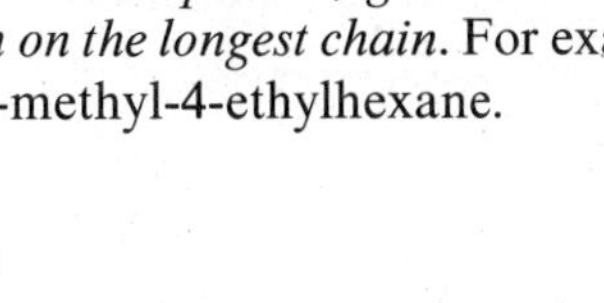   3-Methylheptane

4. *When two or more substituents are present, give each substituent a number corresponding to its location on the longest chain.* For example, we designate the compound below as 2-methyl-4-ethylhexane.

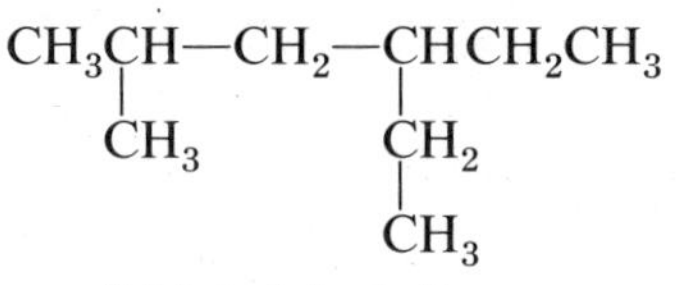

$$CH_3CH-CH_2-CHCH_2CH_3$$
$$| \qquad\qquad |$$
$$CH_3 \qquad\quad CH_2$$
$$|$$
$$CH_3$$

2-Methyl-4-ethylhexane
or
4-Ethyl-2-methylhexane

The groups can be listed in order of complexity (i.e., methyl before ethyl) or alphabetically (ethyl before methyl). The latter system is the one used in the important publication *Chemical Abstracts*.

5. *When two or more substituents are present on the same carbon, use that number twice.*

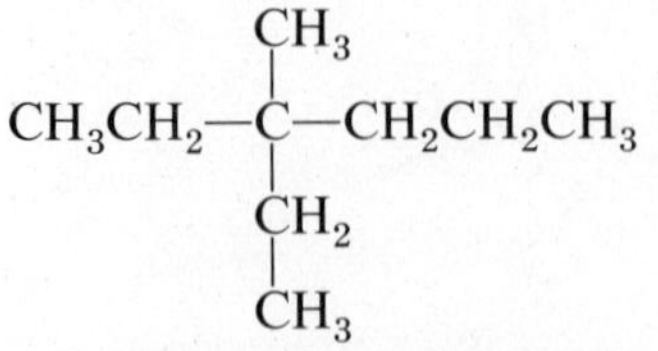

$$CH_3CH_2-\underset{\underset{CH_3}{\overset{|}{\underset{|}{CH_2}}}}{\overset{\overset{CH_3}{|}}{C}}-CH_2CH_2CH_3$$

3-Ethyl-3-methylhexane

6. *When two or more substituents are identical indicate this by the use of the prefixes di-, tri-, tetra-, and so on. Commas are used to separate numbers from each other.*

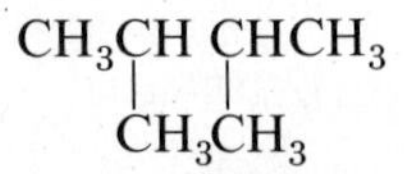

$$CH_3CH\;CHCH_3 \qquad CH_3CHCHCHCH_3$$
$$\underset{CH_3CH_3}{|\;\;|} \qquad\qquad \overset{\overset{CH_3}{|}}{}\underset{CH_3\;\;CH_3}{|\;\;\;\;|}$$

2,3-Dimethylbutane 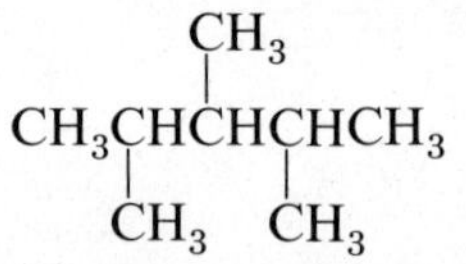   2,3,4-Trimethylpentane

Application of these six rules allows us to name most of the alkanes that we will encounter. Two other rules, however, may be required occasionally.

7. *When two chains of equal length compete for selection as the base chain, choose the chain with the greater number of substituents.*

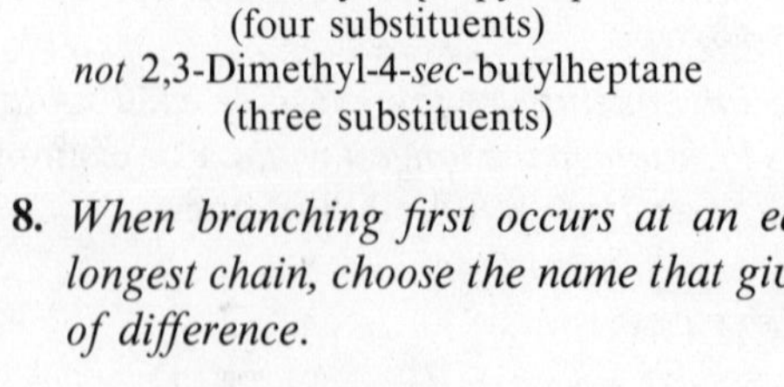

$$\overset{7}{CH_3}\overset{6}{CH_2}-\overset{5}{CH}-\overset{4}{CH}-\overset{3}{CH}-\overset{2}{CH}-\overset{1}{CH_3}$$

with CH₃ CH₂ CH₃ CH₃, CH₂, CH₃ substituents

2,3,5-Trimethyl-4-propylheptane
(four substituents)
not 2,3-Dimethyl-4-*sec*-butylheptane
(three substituents)

8. *When branching first occurs at an equal distance from either end of the longest chain, choose the name that gives the lowest number at the first point of difference.*

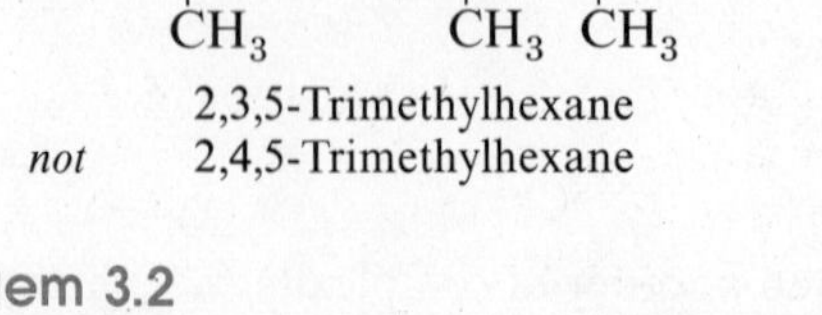

$$\overset{6}{CH_3}-\overset{5}{CH}-\overset{4}{CH_2}-\overset{3}{CH}-\overset{2}{CH}-\overset{1}{CH_3}$$

2,3,5-Trimethylhexane
not   2,4,5-Trimethylhexane

Problem 3.2

(a) Give correct IUPAC names for all of the C_6H_{14} isomers in Table 3.1. (b) Write structural formulas for the nine isomers of C_7H_{16} and give IUPAC names for each.

(Hint: You may find it helpful to name each compound as you write its structure. This will help you to decide whether or not two structures are really different. If their IUPAC names are different then so are the structures.)

Alkanes bearing halogen substituents are named, in the IUPAC system, as *haloalkanes;* for example,

$$CH_3CH_2Cl \qquad CH_3\underset{\underset{Br}{|}}{C}HCH_3$$

Chloroethane   2-Bromopropane

Common names for many simple haloalkanes are still widely used, however. In this common nomenclature system haloalkanes are named as if they were alkyl halides.

$$CH_3CH_2Cl \qquad CH_3\underset{\underset{Br}{|}}{C}HCH_3$$

Ethyl chloride   Isopropyl bromide

3.5 PHYSICAL PROPERTIES OF ALKANES

If we examine the unbranched alkanes in Table 3.2 we notice that each alkane differs from the preceding one by one —CH_2— group. Butane, for example, is $CH_3(CH_2)_2CH_3$ and pentane is $CH_3(CH_2)_3CH_3$. A series of compounds like this, where each member differs from the next member by a constant unit, is called a *homologous* series. Members of a homologous series are called *homologs*.

At room temperature (25°C) and one atmosphere pressure the first four members of the homologous series of normal alkanes (Table 3.3) are gases; the C_5 to C_{17} *n*-alkanes (pentane to heptadecane) are liquids; and the *n*-alkanes with 18 and more carbon atoms are solids.

Boiling Points. The boiling points of the *n*-alkanes show a regular increase with increasing molecular weight (Fig. 3.5). Branching of the alkane chain, however, dramatically lowers the boiling point. Examples of the effect of chain branching that contrast the boiling points of the pentane and hexane isomers are shown in Table 3.1.

Melting Points. The unbranched alkanes do not show the same smooth increase in melting points with increasing molecular weight (black line Fig. 3.6) that they show in their boiling points. There is an alternation as one progresses from an unbranched alkane with an even number of carbon atoms to the next one with an odd number of carbon atoms. For example, propane (mp −188°) melts lower than ethane (mp −172°) and even lower than methane (mp −183°). Butane (mp −135°) melts 53° higher than propane and only 5° lower than pentane (mp −130°). If, however, the even- and odd-numbered alkanes are plotted on *separate* curves (white and colored lines in Fig. 3.6), there *is* a smooth increase in melting point with increasing molecular weight.

X-ray-diffraction studies have revealed the source of this apparent anomaly. Alkane chains with an even number of carbon atoms pack more closely in the

TABLE 3.3 Physical Constants of *n*-Alkanes

NUMBER OF CARBON ATOMS	NAME	bp °C (1 atm.)	mp °C	DENSITY d_4^{20}	REFRACTIVE INDEX n_D^{20}
1	Methane	−161.5	−183		
2	Ethane	−88.6	−172		
3	Propane	−42.1	−188		
4	Butane	−0.5	−135		
5	Pentane	36.1	−130	0.626	1.3575
6	Hexane	68.7	−95	0.659	1.3749
7	Heptane	98.4	−91	0.684	1.3876
8	Octane	125.7	−57	0.703	1.3974
9	Nonane	150.8	−54	0.718	0.4054
10	Decane	174.1	−30	0.730	1.4119
11	Undecane	195.9	−26	0.740	1.4176
12	Dodecane	216.3	−10	0.749	1.4216
13	Tridecane	243	−5.5	0.756	1.4233
14	Tetradecane	253.5	6	0.763	1.4290
15	Pentadecane	270.5	10	0.769	1.4315
16	Hexadecane	287	18	0.773	1.4345
17	Heptadecane	303	22	0.778	1.4369
18	Octadecane	305–307	28	0.777	1.4349
19	Nonadecane	330	32	0.777	1.4409
20	Eicosane	343	36.8	0.789	1.4425

FIG. 3.5
Boiling points of n-*alkanes.*

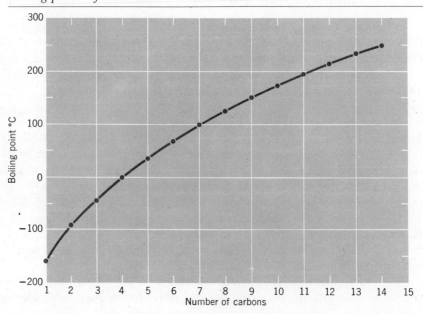

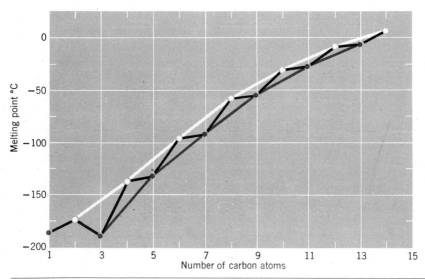

FIG. 3.6

Melting points of unbranched alkanes.

crystalline state. As a result, attractive forces between individual chains are greater and melting points are higher.

The effect of chain branching on the melting points of alkanes is more difficult to predict. Generally, however, branching that produces highly symmetrical structures results in abnormally high melting points. The compound 2,2,3,3-tetramethylbutane, for example, melts at 100.7°. Its boiling point is only six degrees higher, 106.3°.

$$CH_3-\underset{\underset{CH_3}{|}}{\overset{\overset{CH_3}{|}}{C}}-\underset{\underset{CH_3}{|}}{\overset{\overset{CH_3}{|}}{C}}-CH_3$$

2,2,3,3-Tetramethylbutane

Density. As a class, the alkanes are the least dense of all groups of organic molecules. All alkanes have densities considerably less than 1.00 g/cc (the density of water at 4°C). As a result, oil (a mixture of hydrocarbons rich in alkanes) floats on water.

Solubility. Alkanes are almost totally insoluble in water because of their very low polarity and their inability to form hydrogen bonds. Liquid alkanes are miscible with each other, and alkanes generally dissolve in solvents of low polarity. Good solvents for alkanes are benzene, carbon tetrachloride, chloroform, and other alkanes.

3.6 CONFORMATIONAL ANALYSIS OF BUTANE

A study of the energy changes that occur in a molecule when groups are rotated about single bonds is called a *conformational analysis*. We saw the results of such a study for ethane in Chapter 2 (p. 39). At that time we found that in ethane there is a slight barrier (2.8 kcal/mole) to free rotation about the carbon-carbon single bond. This barrier causes the energy of the ethane molecule to rise to a maximum when rotation brings the hydrogen atoms into an eclipsed arrangement. This barrier to free rotation, which can be accounted for by quantum mechanics but for which no simple physical picture can be presently given, was called the *torsional energy* of the molecule.

If we consider rotation about one C-2—C-3 bond of butane, torsional energy plays a part, too. There are, however, additional factors. To see what these are, we should look at the important conformations of *n*-butane shown below.

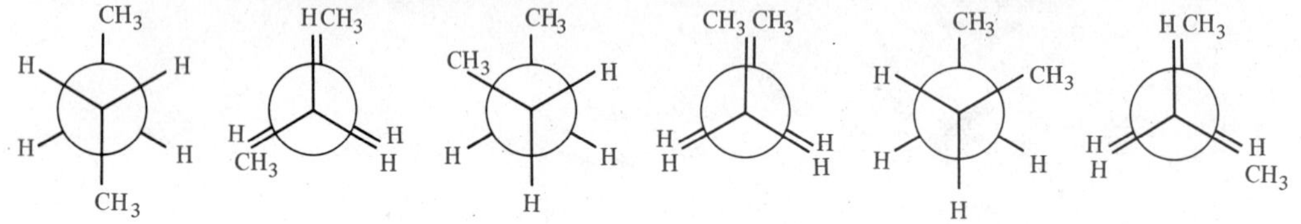

I	II	III	IV	V	VI
anti conformation	an eclipsed conformation	a *gauche-* conformation	an eclipsed conformation	a *gauche* conformation	an eclipsed conformation

The *anti* conformation (I) and the *gauche* conformations (III and V) do not have torsional strain. Of these, the *anti* conformation is the most stable. The methyl groups in the *gauche* conformations are close enough to each other that the van der Waals forces between them are *repulsive*: the electron clouds of the two groups are so close that they repel each other. This repulsion causes the *gauche* conformations to have approximately 0.8 kcal/mole more energy than the *anti* conformation.

The eclipsed conformations (II, IV, and VI) represent energy maxima in the potential energy diagram (Fig. 3.7). Eclipsed conformations II and VI not only have torsional strain, they have van der Waals repulsions arising from the eclipsed methyls and hydrogens. Eclipsed conformation IV has the greatest energy of all because, in addition to torsional strain, there is the large van der Waals repulsive force between the eclipsed methyl groups.

We saw earlier (p. 71) that van der Waals forces can be *attractive*. Here, however, we find that they can also be *repulsive*. Whether or not van der Waals interactions lead to attraction or repulsion depends on the distance that separates the two groups. As two nonpolar groups are brought closer and closer together, the first effect is one in which a momentarily unsymmetrical distribution of electrons in one group induces an opposite polarity in the other. The opposite charges induced in those portions of the two groups that are in closest proximity leads to attraction between them. This attraction increases to a maximum as the internuclear distance of the two groups decreases. The internuclear distance at which the attractive force is at a maximum is equal to the sum of what are called the *van der Waals radii* of the two groups. The van der Waals radius of a group is, in effect, a measure of its size. If the two groups are brought still closer—closer than the sum of their van der Waals radii—the interaction between them becomes repulsive: their electron clouds begin to penetrate each other, and strong electron-electron interactions begin to occur.

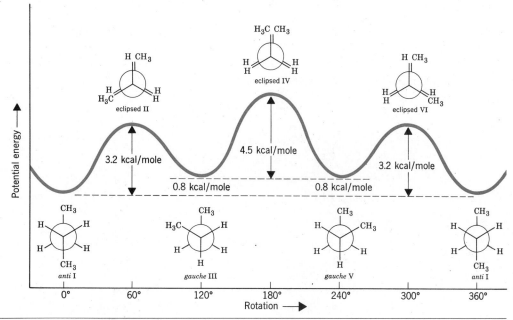

FIG. 3.7

Energy changes that arise from rotation of the C-2—C-3 bond of butane.

3.7 THE STRUCTURES OF CYCLOALKANES: ANGLE STRAIN

Depicted below are the structures of cyclopropane, cyclobutane, cyclopentane, and cyclohexane. If we assume that all of these molecules are planar—an assumption that is not justified in every case, as we will see later—another type of strain, called *angle strain,* can be considered.

	cyclopropane	cyclobutane
C–C–C bond angles (if planar)	60°	90°
Angle strain (deviation from 109.5° if planar)	49.5°	19.5°

	cyclopentane	cyclohexane
C–C–C bond angles (if planar)	108°	120°
Angle strain (deviation from 109.5° if planar)	1.5°	10.5°

In each case, except cyclopentane, a planar-ring structure causes the carbon-carbon bond angles to deviate considerably from that of a normal tetrahedral carbon atom (109.5°). Planar rings larger than cyclohexane would also have angle strain. For example, a planar molecule of cycloheptane would be strained by 19.1°.

$$CH_2$$

$$CH_2 \qquad CH_2$$

$$CH_2 \qquad\qquad CH_2$$

$$CH_2 \xleftarrow{128.6°} \quad CH_2$$

$$CH_2 \longrightarrow CH_2$$

Angle strain 19.1°
(if planar)

Considerations such as these led Adolf von Baeyer (in 1885) to conclude that, of all the cycloalkanes, cyclopentane should be the most stable. The evidence available to von Baeyer justified his conclusions: cyclopropane and cyclobutane were known to undergo reactions where the ring broke open, (Section 4.12), and rings larger than cyclohexane had proved to be very difficult to synthesize.

Subsequently, experimental evidence has shown that except for cyclopropane, von Baeyer's assumption of ring planarity was *incorrect,* and that the large predicted deviations from 109.5° bond angles are strikingly incorrect for rings of six members and more. Part of this experimental evidence came from a consideration of the heats of combustion of cycloalkanes.

Heats of Combustion of Cycloalkanes

The heat of combustion of a compound (Section 2.3) is the quantity of heat evolved when one mole of a substance is completely oxidized to carbon dioxide and water. The cycloalkanes constitute a homologous series; each member of the series differs from the one immediately preceding it, by the constant amount of one —CH_2— group. Thus, the general equation for combustion of a cycloalkane can be formulated as follows:

$$(CH_2)_n + \tfrac{3}{2}n\, O_2 \longrightarrow nCO_2 + nH_2O + heat$$

From the total amount of heat evolved in the combustion of a cycloalkane, we can calculate the amount of heat evolved *per CH_2 group*. On this basis, the stabilities of the cycloalkanes become directly comparable. The results of such an investigation are given in Table 3.4.

Several observations emerge from a consideration of these results.

1. Cyclohexane, *not cyclopentane* as von Baeyer had predicted, is found to be the most stable cycloalkane. It has the lowest heat of combustion per CH_2 group. Moreover, the heat of combustion of cyclohexane (157.4 kcal/mole/CH_2) does not differ from that of *n*-alkanes where there is assumed to be no angle strain at all.

2. Cyclopropane evolves the greatest amount of heat per CH_2 group. This shows that cyclopropane is the least stable (has the greatest potential energy) of all the cycloalkanes and this is in accord with von Baeyer's prediction. It is reasonable, then, to assume that cyclopropane has the greatest angle strain.

TABLE 3.4 Heats of Combustion of Cycloalkanes

CYCLOALKANE $(CH_2)_n$	n	HEAT OF COMBUSTION kcal/mole	HEAT OF COMBUSTION/CH_2 GROUP kcal/mole
Cyclopropane	3	499.83	166.6
Cyclobutane	4	655.86	164.0
Cyclopentane	5	793.52	158.7
Cyclohexane	6	944.48	157.4
Cycloheptane	7	1108.2	158.3
Cyclooctane	8	1269.2	158.6
Cyclononane	9	1429.5	158.8
Cyclodecane	10	1586.0	158.6
Cyclopentadecane	15	2362.5	157.5
n-alkane			157.4

3. Cyclobutane, too, is found to deviate considerably (164.0 kcal/mole versus 157.4 kcal/mole) from the heat of combustion of an unstrained n-alkane. An assumption of considerable angle strain in cyclobutane is thus also justified.

4. Rings of seven-members and more are not found to be exceptionally unstable. This suggests that little, if any, angle strain is present in these large rings. The lack of angle strain in rings of seven-members or more is contrary to von Baeyer's prediction based on ring planarity.

We can explain all of these results if we examine carefully the actual conformations of the cycloalkanes.

3.8 CONFORMATIONAL ANALYSIS OF CYCLOHEXANE

There is considerable evidence that the most stable form of the cyclohexane ring is the "chair" form illustrated in Fig. 3.8.* In this nonplanar structure the carbon-carbon bond angles are all 109.5° and are thus free of angle strain. The chair conformation is free of torsional strain as well. When viewed along the carbon-carbon bonds of any side (viewing the structure from an end, Fig. 3.9) the atoms are seen to be perfectly staggered. (Figures 3.8, 3.9 are on the following page.) Moreover, the hydrogen atoms on carbons at opposite corners of the cyclohexane ring are maximally separated. The separation between two such hydrogens is illustrated in Fig. 3.10 on page 103.

Another possible conformation of cyclohexane that is also free of angle strain is the "boat" form (Fig. 3.11).

The boat form, however, is not free of torsional strain. When a model of the boat form is sighted along side carbon atoms (Fig. 3.12) the hydrogen atoms

---

*An understanding of this and subsequent discussions of conformational analysis can be aided immeasurably through the use of appropriate molecular models.

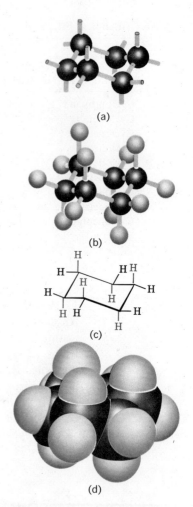

FIG. 3.8

Representations of the chair form of cyclohexane:
(a) *carbon skeleton only;* (b) *carbon and hydrogen*
atoms; (c) *line drawing;* (d) *space filling model of*
cyclohexane.

of the side carbons are found to be eclipsed. Additionally, two of the hydrogen atoms on C-1 and C-4 are close enough to each other to cause van der Waals repulsion (Fig. 3.13). This latter effect has been called the "flagpole" interaction of the boat form. The flagpole interaction is an example of a *transannular interaction*—one that occurs across a ring. Torsional strain and flagpole interactions cause the boat form to have considerably higher energy than the chair form. (Figs. 3.12, 3.13 are on p. 104.)

Although it is more stable, the chair conformation is much more rigid than the boat form. The boat conformation appears to be quite flexible. By flexing to a new form—the twist conformation (Fig. 3.14, p. 105)—the boat form can relieve some of its torsional strain and, at the same time, reduce the flagpole interactions.

FIG. 3.9

A Newman projection of cyclohexane. (Comparison with
an actual molecular model will make this formulation
clearer, and will show that similar staggered arrangements
are seen when other carbon-carbon bonds are chosen for
sighting.)

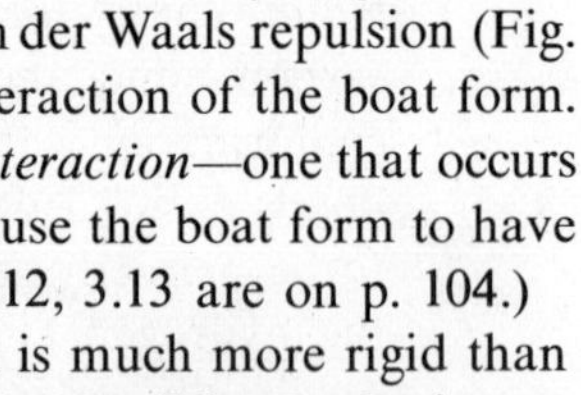

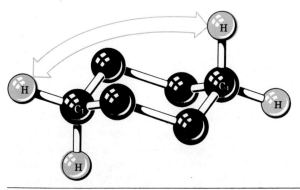

FIG. 3.10
*Illustration of large separation between hydrogens
at opposite corners of the ring (designated as C-1
and C-4) in the chair form of cyclohexane.*

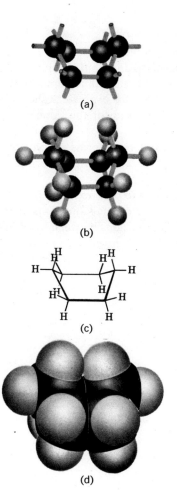

(a)

(b)

(c)

(d)

FIG. 3.11
*The boat-conformation cyclohexane: (a) carbon
skeleton only; (b) carbon and hydrogen atoms; (c)
line drawing; (d) space-filling model.*

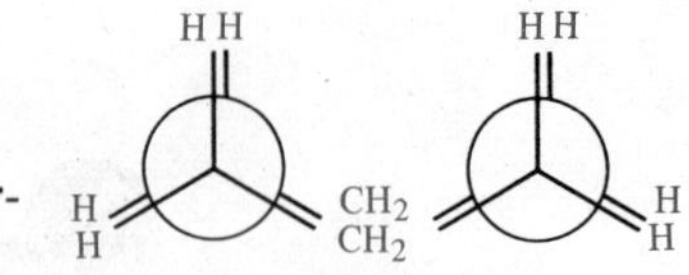

FIG. 3.12
Illustration of eclipsed hydrogens of the boat confor-
mation of cyclohexane.

This flexing causes the twist conformation to have lower energy than the boat
form. *The stability gained by flexing is insufficient, however, to cause the twist form*
of cyclohexane to be more stable than the chair form. The chair form is estimated
to be lower in energy than the twist form by approximately 5 kcal/mole.

The energy barriers between the chair, boat, and twist forms of cyclohexane
are low enough to make separation of the different conformations impossible at room
temperature. At room temperature, the thermal energies of the molecules are great
enough to cause approximately one million interconversions to occur each second
and, *because of its greater stability, more than 99% of the molecules are estimated*
to be in the chair form at any given moment.

3.9 CONFORMATIONS OF OTHER CYCLOALKANES

Cyclopropane is, of necessity, planar. (Three points must lie in a single plane.)
Cyclobutane, however, is known not to be exactly planar, but is slightly folded (Fig.
3.15a). This folding increases the angle strain of cyclobutane slightly (the carbon-
carbon bonds are 88° rather than 90°), but allows for a greater relief of torsional
strain. Cyclopentane is similarly bent (Fig. 3.15b) and for the same reasons. (Figure 3.15a,b is on page 105.)

There are, of course, different folded and bent forms of cyclobutane and
cyclopentane. At room temperature, as with cyclohexane, thermal energy causes these
forms to be interconverted rapidly.

Cycloheptane, cyclooctane, and cyclononane also exist in nonplanar conformations. The small instabilities of these higher cycloalkanes appear to be due,
primarily, to torsional strain and van der Waals repulsions between hydrogens across
rings. The nonplanar conformations of these rings, however, are essentially free

FIG. 3.13
Flagpole interaction of the C-1 and C-4 hydrogens of
the boat conformation of cyclohexane.

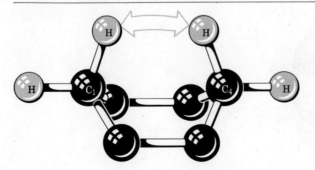

(a)

(b)

FIG. 3.14
(a) *Carbon skeleton and* (b) *line drawing of the twist conformation of cyclohexane.*

of angle strain. Although they are not known with certainty, the most stable conformations of cycloheptane and cyclooctane appear to be those shown below.

cycloheptane cyclooctane

X-ray crystallographic studies of cyclodecane reveal that the most stable conformation has carbon-carbon bond angles of 117°. This indicates some angle strain. The wide bond angles apparently allow the molecule to expand and thereby minimize unfavorable transannular repulsions between hydrogens.

FIG. 3.15
The most stable conformations of cyclobutane (a) *and cyclopentane* (b).

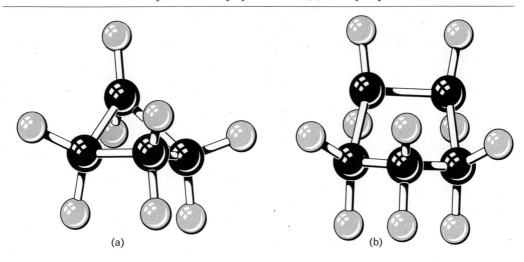

(a)

(b)

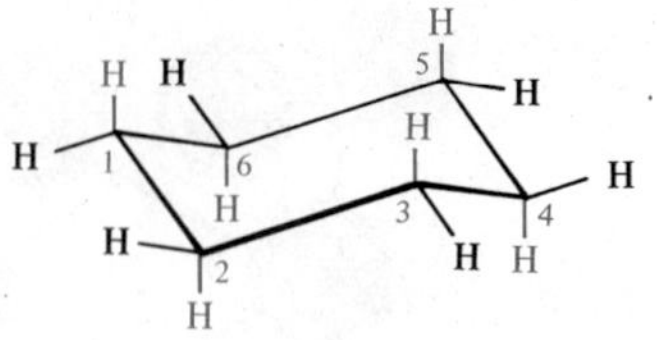

FIG. 3.16
The chair conformation of cyclohexane. The axial hydrogens are shown in color.

3.10 SUBSTITUTED CYCLOHEXANE COMPOUNDS: AXIAL AND EQUATORIAL HYDROGENS

The six-membered ring is the most common ring found among nature's organic molecules. For this reason, we will give it special attention. We have already seen that the chair conformation of cyclohexane is the most stable one and that it is the predominant conformation of the molecules in a sample of cyclohexane. With this fact in mind, we are in a position to undertake a limited analysis of the conformations of substituted cyclohexanes.

If we look carefully at the chair conformation of cyclohexane (Fig. 3.16) we can see that there are only two different kinds of hydrogens. One hydrogen attached to each of the six carbons lies in a plane generally defined by the ring of carbon atoms. These hydrogens, by analogy with the equator of the earth, are called *equatorial* hydrogens. Six other hydrogens, one on each carbon, are oriented in a direction that is generally perpendicular to the average plane of the ring. These hydrogens, again by analogy with the earth, are called *axial* hydrogens. There are three axial hydrogens on each face of the cyclohexane ring and their orientation (up or down) alternates from one carbon to the next.

The question one might next ask is the following: what is the most stable conformation of a cyclohexane derivative *in which one hydrogen has been replaced by a substituent*? That is, what is the most stable conformation of a *monosubstituted* cyclohexane? We can answer this question by considering methylcyclohexane as an example.

There are two possible chair conformations of methylcyclohexane (Fig. 3.17), and these two forms are interconvertible (via a twist conformation) through partial rotations about the single bonds of the ring. In one conformation (Fig. 3.17a) the methyl group occupies an *axial* position, and in the other (Fig. 3.17b) the methyl group occupies an *equatorial* position. Studies indicate that the conformation with the methyl group equatorial is more stable than the conformation with the methyl group axial by about 1.6 kcal/mole. Thus, in the equilibrium mixture, the conformation

FIG. 3.17
The conformations of methylcyclohexane with the methyl group axial (a) and equatorial (b).

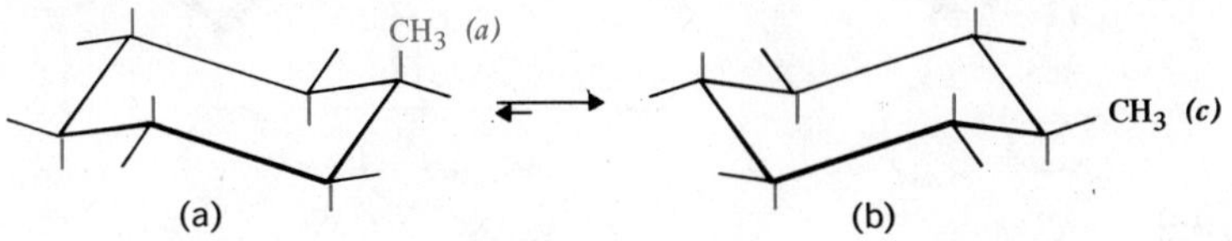

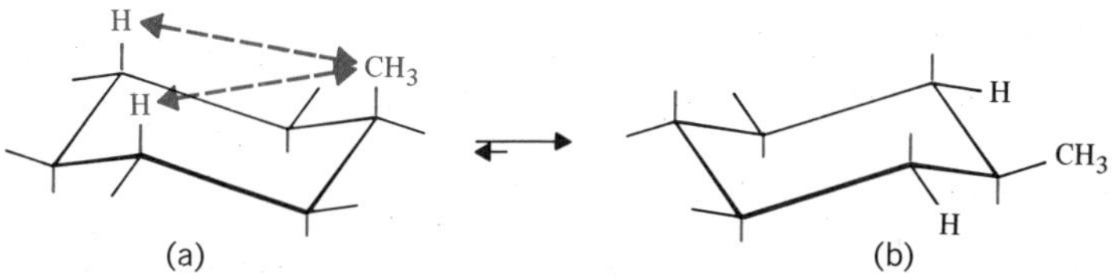

FIG. 3.18

(a) *1,3-Diaxial interactions between the two axial hydrogens and the axial methyl group in the axial conformation of methylcyclohexane.* (b) *Less crowding occurs in the equatorial conformation.*

with the methyl group in the equatorial position is the predominant one; calculations show that it constitutes about 93% of the equilibrium mixture.

The greater stability of methylcyclohexane with an equatorial methyl group can be understood through an inspection of the two forms as they are shown in Fig. 3.18.

Studies done with scale models of the two conformations show that when the methyl group is axial, it is so close to the two axial hydrogens on the same side of the molecule that the van der Waals forces between them are repulsive. Similar studies also reveal that *any group has considerably more room when it occupies an equatorial position.*

In cyclohexane derivatives with larger substituents, this effect is even more pronounced. The conformation of *tert*-butylcyclohexane with the *tert*-butyl group equatorial is estimated to be more than 5 kcal/mole more stable than the axial form. This large energy difference between the two conformations means that, at room temperature, virtually 100% of the molecules of *tert*-butylcyclohexane have the *tert*-butyl group in the equatorial position.

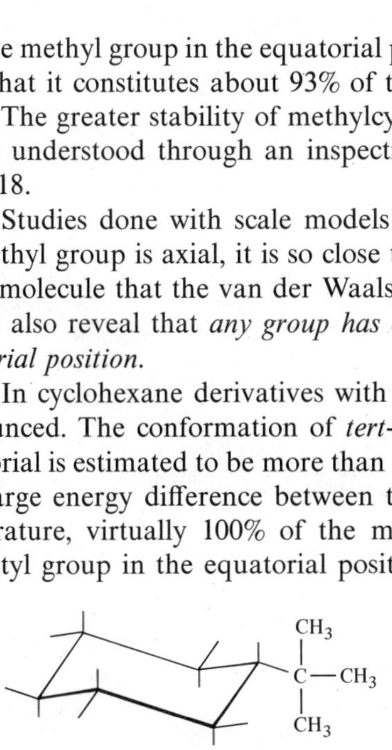

Equatorial *tert*–butylcyclohexane

3.11 DISUBSTITUTED CYCLOALKANES: CIS-TRANS ISOMERISM

The presence of two substituents on a ring of a cycloalkane allows for the possibility of *cis-trans* isomerism. We can see this most easily if we begin by examining cyclopentane derivatives because the cyclopentane ring is essentially planar. (At any given moment the ring of cyclopentane is, of course, slightly bent, but we know that the various bent conformations are rapidly interconverted. Over a period of time, the average conformation of the cyclopentane ring is planar.) Since the planar representation is much more convenient for an initial presentation of *cis-trans* isomerism in cycloalkanes we will use it here.

Let us consider 1,2-dimethylcyclopentane as an example. We can write the structures shown in Fig. 3.19. In the first structure the methyl groups are on the same

FIG. 3.19

Cis- *and* trans-*1,2-dimethylcyclopen-*
tanes.

cis–1, 2–dimethylcyclopentane
bp 130°

trans–1, 2–dimethylcyclopentane
bp 123.7°

side of the ring, that is, they are *cis*. In the second structure the methyl groups are on opposite sides of the ring; they are *trans*.

The *cis*- and *trans*- 1,2-dimethylcyclopentanes are stereoisomers: they differ from each other only in the arrangement of the atoms in space. Because of the ring structure, the two forms cannot be interconverted without breaking carbon-carbon bonds. This requires a great deal of energy, and does not occur at temperatures even considerably above room temperature. As a result, the *cis* and *trans* forms can be separated, placed in separate bottles, and kept indefinitely.

> Actually, three different structures can be written for 1,2-dimethylcyclo-pentanes. One structure can be written for *cis*-1,2-dimethylcyclopentane and two structures can be written for *trans*-1,2-dimethylcyclopentanes:

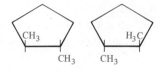

cis–1, 2–dimethylcyclopentane

trans–1, 2–dimethylcyclopentane

> Compounds representing all three structures are known and all three compounds are stereoisomers. The stereoisomeric relation between *cis*-1,2-dimethylcyclopentane and either of the *trans*-1,2-dimethylcyclopen-tanes is called a *diastereomeric* one. The stereoisomeric relation between the *trans*-1,2-dimethylcyclopentanes is called an *enantiomeric* one. We will discuss the meaning of these terms and their application to cyclic molecules in detail in Chapter 7.

1,3-dimethylcyclopentanes show *cis-trans* isomerism as well:

cis–1, 3–dimethylcyclopentane

trans–1, 3–dimethylcyclopentane

The physical properties of *cis-trans* isomers are different; they have different melting points, boiling points, etc. Table 3.5 lists some of the physical constants of the dimethylcyclopentanes.

Problem 3.3

Write structures for the *cis* and *trans* isomers of: (a) 1,2-dimethylcyclopropane; (b) 1,2-dibromocyclobutane.

TABLE 3.5 Physical Constants of *cis*- and *trans*-**Cyclopentane Derivatives**

SUBSTITUENTS	ISOMER	mp °C	bp °C*
1,2-Dimethyl	*cis*	−50.1	130.04[760]
1,2-Dimethyl	*trans*	−89.4	123.7[760]
1,3-Dimethyl	*cis*	−85	124.9[760]
1,3-Dimethyl	*trans*	−79.4	123.5[760]
1,2-Dichloro	*cis*	−6	93.5[22]
1,2-Dichloro	*trans*	−7	74[16]

*The pressures (in units of torr) at which the boiling points were measured are given as superscripts.

The cyclohexane ring is, of course, not planar. A "time average" of the various interconverting chair conformations would, however, be planar and, as with cyclopentane, this planar representation is convenient for introducing the topic of *cis-trans* isomerism of cyclohexane derivatives. The planar representations of the 1,2-, 1,3-, and 1,4-dimethylcyclohexane isomers are shown below.

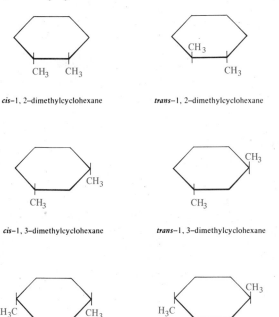

cis–1, 2–dimethylcyclohexane   *trans*–1, 2–dimethylcyclohexane

cis–1, 3–dimethylcyclohexane   *trans*–1, 3–dimethylcyclohexane

cis–1, 4–dimethylcyclohexane   *trans*–1, 4–dimethylcyclohexane

If we consider the *actual* conformations of these isomers, the structures are somewhat more complex. Beginning with *trans*-1,4-dimethylcyclohexane, since it is easiest to visualize, we find there are two possible chair conformations (Fig. 3.20). In one conformation both methyl groups are axial; in the other both are equatorial. The diequatorial conformation is, as we would expect it to be, the more stable conformation, and it represents the structure of at least 99% of the molecules at equilibrium.

That the diaxial form of *trans*-1,4-dimethylcyclohexane is a *trans* isomer is easy to see; the two methyl groups are clearly on opposite sides of the ring. The

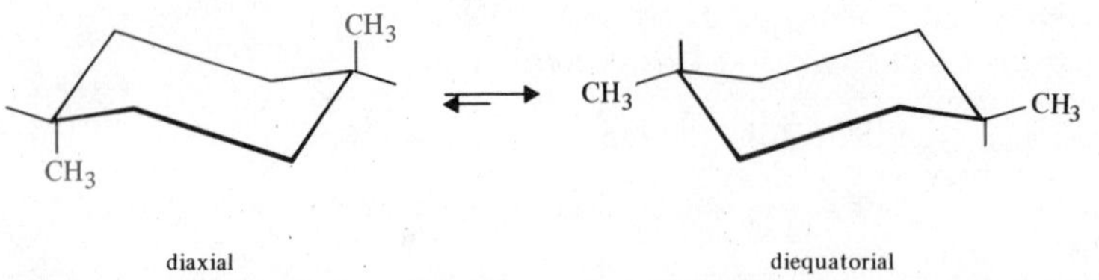

diaxial diequatorial

FIG. 3.20
The two chair conformations of trans-*1,4-dimethylcyclohexane.*
(Note: all other C—H bonds have been omitted for clarity.)

trans relationship of the methyl groups in the diequatorial isomer is not as obvious, however. The *trans* relationship of the methyls becomes more apparent if we imagine ourselves "flattening" the molecule by turning one end up and the other down.

A second way to recognize a *trans*-disubstituted cyclohexane is to notice that one group is attached by the *upper* bond (of the two to its carbon) and one by the *lower* bond.

Cis-1,4-dimethylcyclohexane also exists in two conformations. The *cis* relationship of the methyl groups, however, precludes the possibility of a structure with both groups in an equatorial position. There are, thus, two equivalent conformations (Fig. 3.21) of *cis*-1,4-dimethylcyclohexane.

Problem 3.4

(a) Write structural formulas for the two chair conformations of *cis*-1-*tert*-butyl-4-methylcyclohexane. (b) Are these two conformations equivalent? (c) If not, which would be more stable? (d) Which would be the preferred conformation at equilibrium?

Trans-1,3-dimethylcyclohexane is like the *cis*-1,4-compound in that no conformation is possible with both methyl groups in the favored equatorial position. The two conformations below are of equal energy and are equally populated at equilibrium.

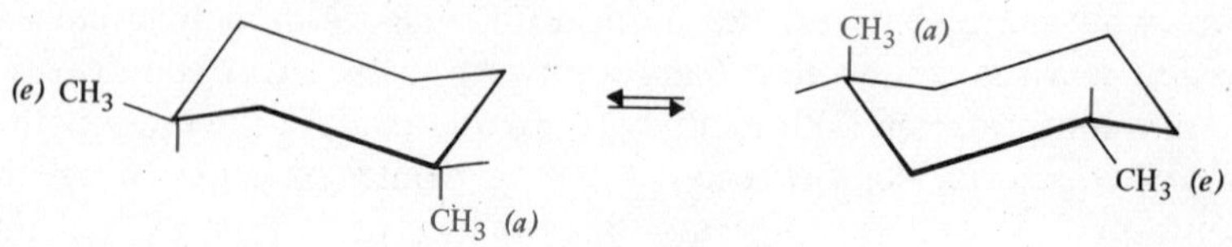

trans-1, 3-dimethylcyclohexane

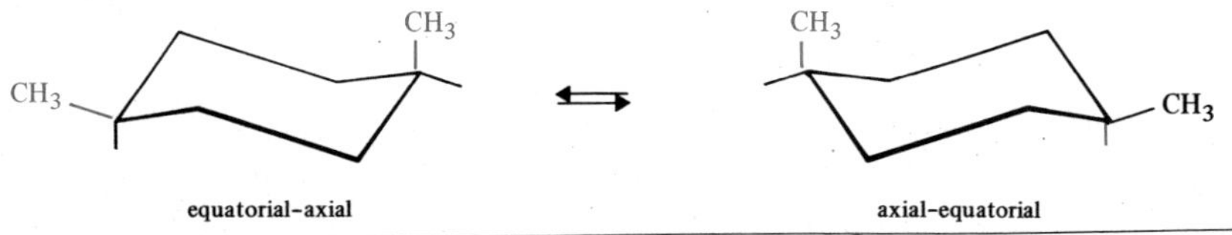

equatorial–axial   axial–equatorial

FIG. 3.21
Equivalent conformations of cis-*1,4-dimethylcyclohexane.*

If, however, we consider some other *trans*-1,3-disubstituted cyclohexane in which one group is larger than the other, the lower energy conformation is the one that has the larger group in the equatorial position. For example, the more stable conformation of *trans*-1-*tert*-butyl-3-methylcyclohexane is the one shown below. The large *tert*-butyl group occupies the equatorial position.

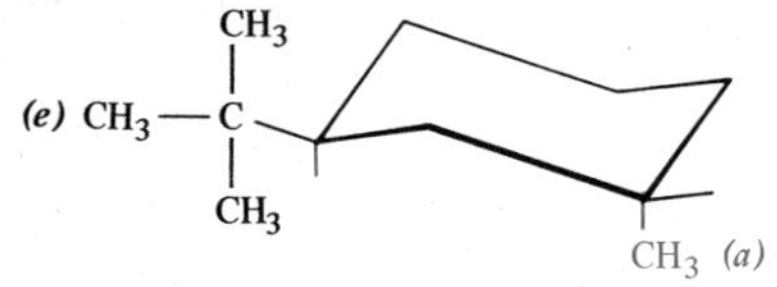

Problem 3.5

(a) Write chair conformations for *cis*- and *trans*-1,2-dimethylcyclohexane. (b) For which isomer (*cis* or *trans*) are the two conformations equivalent? (c) For the isomer where the two conformations are not equivalent, which conformation is more stable? (d) Which conformation would be more highly populated at equilibrium? (Check your answer with Table 3.6.)

The different conformations of the dimethylcyclohexanes are summarized in Table 3.6. The more stable conformation, where one exists, is set in heavy type.

TABLE 3.6 Conformations of Dimethylcyclohexanes

COMPOUND	*cis* ISOMER			*trans* ISOMER		
1,2-Dimethyl-	*a,e*	or	*e,a*	**e,e**	or	*a,a*
1,3-Dimethyl-	**e,e**	or	*a,a*	*a,e*	or	*e,a*
1,4-Dimethyl-	*a,e*	or	*e,a*	**e,e**	or	*a,a*

3.12 BICYCLIC AND POLYCYCLIC ALKANES

The molecules of many of the compounds that we will encounter will contain more than one ring. A molecule with two rings that share two or more carbon atoms is called a *bicyclic system*. The rings of such a system are said to be "fused," and the atoms that are common to both rings are called "bridgehead atoms." One of the most important bicyclic systems is the one that has been given the common name decalin.

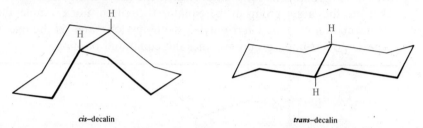

Decalin
(carbons 1 and 6 are bridgehead carbons)

Decalin shows *cis-trans* isomerism:

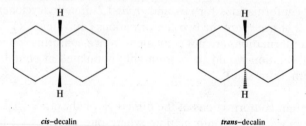

cis–decalin   *trans*–decalin

In *cis*-decalin the two hydrogens attached to the bridgehead atoms lie on the same side of the ring; in *trans*-decalin they are on opposite sides. We often indicate this by writing their structures in the following way.

cis–decalin   *trans*–decalin

Simple rotations of groups about carbon-carbon bonds do not interconvert *cis*- and *trans*-decalins. In this respect they resemble the isomeric *cis*- and *trans*-disubstituted cyclohexanes. (We can, in fact, regard them as being *cis*- or *trans*-1,2-disubstituted cyclohexanes in which the 1,2-substituents are the two ends of a four carbon bridge, that is, —$CH_2CH_2CH_2CH_2$—.)

The *cis*- and *trans*-decalins can be separated. *Cis*-decalin boils at 195° (at 760 torrs) and *trans*-decalin boils at 185.5° (at 760 torrs).

Norbornane (below) is another bicyclic system. In norbornane the cyclohexane ring is held in a boat form by a methylene bridge (—CH_2—) between the 1- and 4-carbons.

norbornane

Adamantane (below) is a tricyclic system that contains a three-dimensional array of cyclohexane rings, all of which are in the chair form. Extending the structure of adamantane in three dimensions gives the structure of diamond. The great hardness of diamond results from the fact that the entire diamond crystal is actually one very large molecule—a molecule that is held together by millions of strong covalent bonds.

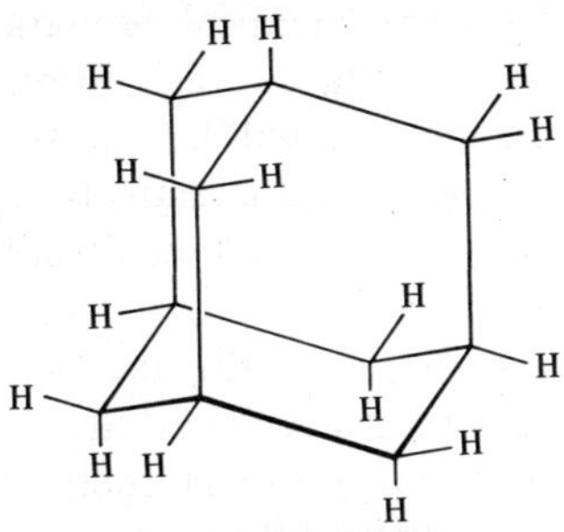

adamantane 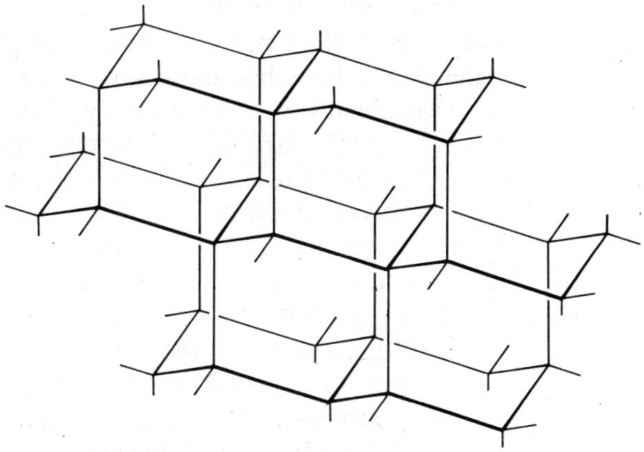   a portion of the diamond structure

The tetracyclic ring system of perhydrocyclopentanophenanthrene (below) is incorporated into a broad group of biologically potent compounds known as steroids. Many steroids function as hormones: they regulate numerous biochemical processes including sexual development, fertility, and growth. Cholesterol, a steroid that is found in nearly all body tissues, apparently serves as a precursor for many steroidal hormones. Accumulation of cholesterol in the arteries of the heart, however, has been strongly implicated in the development of heart disease.

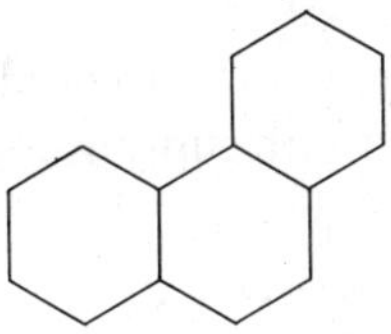

Perhydrophenanthrene 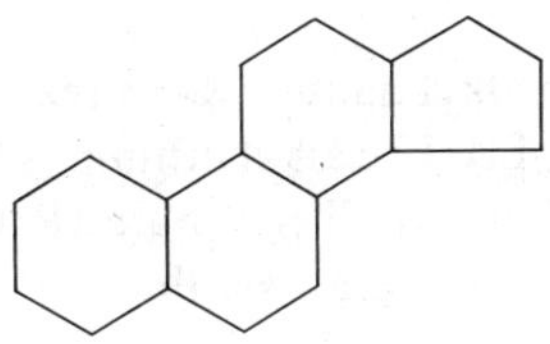   Perhydrocyclopentanophenanthrene

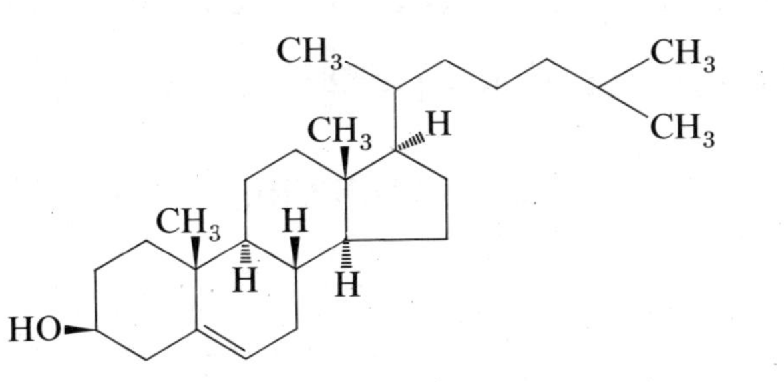

Cholesterol

3.13 SYNTHESIS OF ALKANES AND CYCLOALKANES

The primary source of alkanes is petroleum. The primary use of alkanes is as a source of energy through combustion. Most of the fuels that we commonly use—natural gas, gasoline, kerosene, and diesel fuel, for example—are composed of hydrocarbons containing mixtures of alkanes of similar but varying chain lengths. Mixtures of alkanes are perfectly suitable as fuels, and separation of petroleum into individual pure alkanes is unnecessary. This is fortunate because such a separation is extremely difficult. The boiling points of many components of petroleum are so close that complete separations become almost impossible.

Occasionally, in our laboratory work, however, we may need to have a pure sample of a particular alkane. For these purposes, the chemical preparation—or synthesis—of that particular alkane is often the most reliable way of obtaining it. The preparative method that we choose should be one that will lead to the desired product alone or, at least, to products that can be easily and effectively separated.

Several such methods are available, and four are outlined here. In subsequent chapters we will encounter others.

1. *The Corey-House alkane synthesis.* A highly versatile method for the synthesis of alkanes from alkyl halides has been developed independently by Professors E. J. Corey (Harvard University) and Herbert O. House (Georgia Institute of Technology). In this synthesis lithium dialkylcuprates, R_2CuLi, are treated with alkyl halides, $R'-X$; the product of the reaction is an alkane, $R-R'$.

$$R_2CuLi + R'-X \longrightarrow R-R' + RCu + LiX$$

Lithium dialkylcuprates are prepared in the following way.

$$R{-}X \xrightarrow[\text{Ether}]{2Li} \underset{\substack{\text{Alkyl}\\ \text{lithium}}}{2RLi} \xrightarrow{\text{CuI}} \underset{\substack{\text{Lithium}\\ \text{dialkylcuprate}}}{R_2CuLi}$$

An alkyl halide (R-X) reacts with lithium metal to produce an alkyllithium (R-Li). The alkyl lithium is then treated with cuprous iodide to produce the lithium dialkylcuprate (R_2CuLi).

The overall synthesis,

$$R_2CuLi + R'-X \longrightarrow R-R' + RCu + LiX$$

gives excellent yields when the alkyl halide (R'-X) is a primary halide and when the alkyl groups of the dialkylcuprate are either primary or secondary. Several specific examples are outlined below.

$$(CH_3)_2CuLi + CH_3CH_2CH_2CH_2CH_2{-}I \xrightarrow[\text{3.5 hr, 25°}]{\text{diethyl ether}} CH_3{-}CH_2CH_2CH_2CH_2CH_3$$
$$(98\%)$$

$$(CH_3CH_2CH_2CH_2)_2CuLi + CH_3CH_2CH_2CH_2CH_2CH_2CH_2{-}Cl$$

$$\xrightarrow[\text{5 days, 0°}]{\text{diethyl ether}}$$

$$CH_3CH_2CH_2CH_2{-}CH_2CH_2CH_2CH_2CH_2CH_2CH_3$$
$$(75\%)$$

$$
\underset{(CH_3CH_2\overset{\overset{\displaystyle CH_3}{|}}{CH})_2CuLi} + CH_3CH_2CH_2CH_2CH_2\!\!-\!\!Cl \xrightarrow[\text{1 hr, 25}^{\circ}]{}
$$

$$
CH_3CH_2\overset{\overset{\displaystyle CH_3}{|}}{CH}\!\!-\!\!CH_2CH_2CH_2CH_2CH_3
$$
$$(64\%)$$

Problem 3.6

Outline a Corey-House synthesis of the following compounds from the alkyl halides given.

(a) $CH_3CH_2CH_2CH_2CH_3$ from $CH_3\!\!-\!\!I$ and $CH_3CH_2CH_2CH_2\!\!-\!\!Cl$

(b) $CH_3CH_2\overset{\overset{\displaystyle CH_3}{|}}{C}HCH_3$ from $CH_3CH_2\!\!-\!\!Br$ and $CH_3\overset{\overset{\displaystyle}{|}}{\underset{\underset{\displaystyle Br}{|}}{C}}HCH_3$

(c) ⬡—CH_3 from $CH_3\!\!-\!\!I$ and   ⬡—Br

(d) ⬠—$CH_2CH_2CH_3$ from $CH_3CH_2CH_2\!\!-\!\!Br$ and   ⬠—Br

---

2. *The Wurtz reaction.* In the Wurtz reaction an alkyl halide is allowed to react with sodium metal; this produces an alkane with twice the number of carbon atoms as the starting alkyl halide.

$$
2\,CH_3CH_2CH_2\!\!-\!\!Br + 2\,Na \longrightarrow CH_3CH_2CH_2\!\!-\!\!CH_2CH_2CH_3 + 2\,NaBr
$$
1-Bromopropane   Hexane

The Wurtz reaction is of limited usefulness, however, because numerous "side products" are often obtained. This is particularly true when secondary and tertiary alkyl halides are employed as starting materials.

"Mixed" Wurtz reactions, even with primary halides, are not generally practical because several alkanes are formed:

$$
CH_3\!\!-\!\!CH_3 \qquad \text{Ethane}
$$
$$
CH_3Cl + CH_3CH_2Cl \xrightarrow{\text{Na}} CH_3\!\!-\!\!CH_2CH_3 \qquad \text{Propane}
$$
$$
CH_3CH_2\!\!-\!\!CH_2CH_3 \qquad \text{Butane}
$$

3. *Hydrogenation of alkenes.* Alkenes add hydrogen in the presence of metal catalysts to produce alkanes. Examples are:

$$
CH_3CH\!\!=\!\!CHCH_3 + H_2 \xrightarrow[\substack{C_2H_5OH \\ (25^{\circ},\ 50\ \text{atm})}]{\text{Ni}} CH_3\overset{\overset{\displaystyle}{|}}{\underset{\underset{\displaystyle H}{|}}{C}}H\!\!-\!\!\overset{\overset{\displaystyle}{|}}{\underset{\underset{\displaystyle H}{|}}{C}}HCH_3
$$

2-butene   Butane

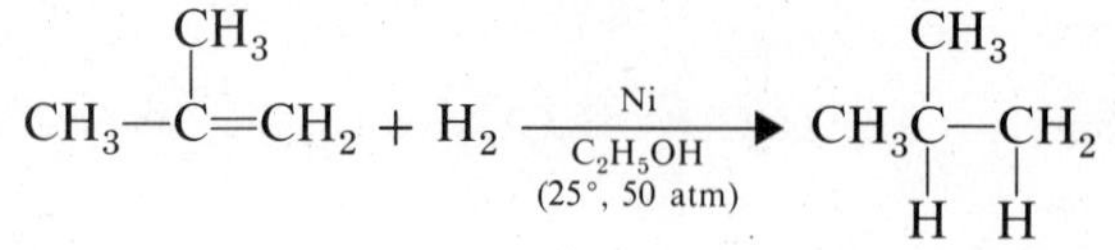

$$CH_3-\underset{\displaystyle CH_3}{C}=CH_2 + H_2 \xrightarrow[\substack{C_2H_5OH \\ (25°,\ 50\ atm)}]{Ni} CH_3\underset{\displaystyle \underset{H}{|}}{\overset{\displaystyle \overset{CH_3}{|}}{C}}-\underset{\displaystyle \underset{H}{|}}{CH_2}$$

2-Methylpropene · Isobutane

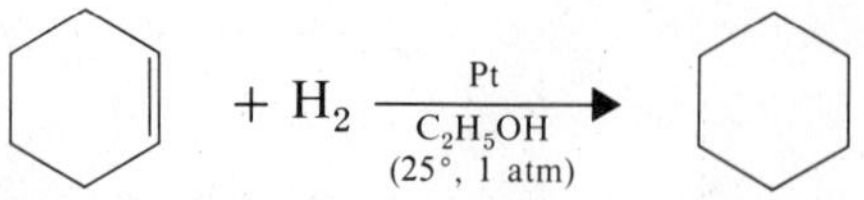

$$\text{Cyclohexene} + H_2 \xrightarrow[\substack{C_2H_5OH \\ (25°,\ 1\ atm)}]{Pt} \text{Cyclohexane}$$

Cyclohexene · Cyclohexane

These reactions are very easy to carry out in the laboratory. We will see several applications of hydrogenation in Chapter 6.

4. *Reduction of alkyl halides.* Most alkyl halides react with zinc and aqueous acid to produce an alkane. In this reaction the more electronegative halogen is replaced by the less electronegative hydrogen. For this reason the reaction is one of reduction.

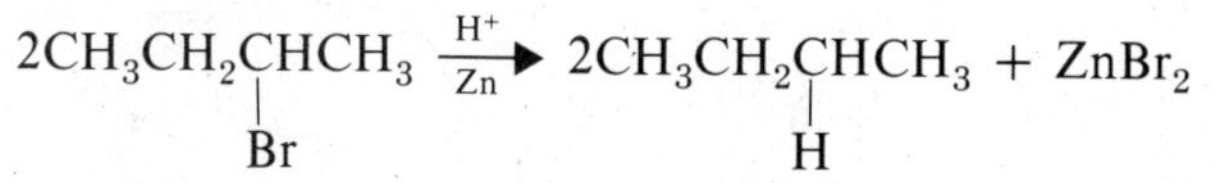

$$2CH_3CH_2\underset{\displaystyle \underset{Br}{|}}{CH}CH_3 \xrightarrow[Zn]{H^+} 2CH_3CH_2\underset{\displaystyle \underset{H}{|}}{CH}CH_3 + ZnBr_2$$

2-Bromobutane · · · · · · · · · · · · · · Butane

$$2CH_3\underset{\displaystyle \overset{CH_3}{|}}{CH}CH_2CH_2-Br \xrightarrow[Zn]{H^+} 2CH_3\underset{\displaystyle \overset{CH_3}{|}}{CH}CH_2CH_2-H + ZnBr_2$$

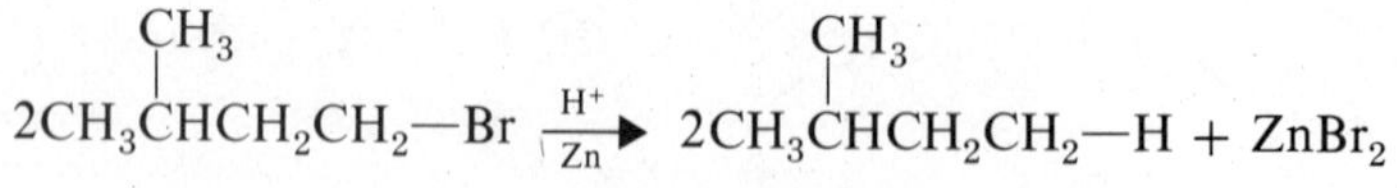

Cyclopropane, but not higher cycloalkanes, can be prepared by a similar reaction—one that results in the formation of a new carbon-carbon bond.

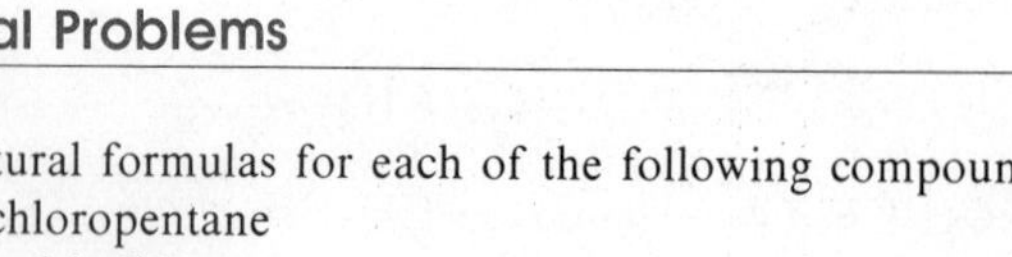

$$\begin{matrix} I-CH_2 \\ \\ I-CH_2 \end{matrix}\!\!\!\! CH_2 \xrightarrow[\text{alcohol}]{Zn} \begin{matrix} CH_2 \\ \\ CH_2 \end{matrix}\!\!\!\! CH_2 + ZnI_2$$

1,3-Diiodopropane · · · · · · · · · · · · · · (70%)

Additional Problems

3.7

Write structural formulas for each of the following compounds:
(a) 2,3-Dichloropentane
(b) *tert*-Butyl iodide
(c) 3-Ethylpentane
(d) 2,3,4-Trimethyldecane
(e) 4-Isopropylnonane
(f) 1,1-Dimethylcyclopropane
(g) *cis*-1,2-Dimethylcyclobutane
(h) *trans*-1,3-Dimethylcyclobutane
(i) Isopropylcyclohexane (more stable conformation)
(j) *trans*-1-Isopropyl-3-methylcyclohexane (more stable conformation)
(k) Isohexyl chloride

(l) 2,2,4,4-Tetramethyloctane
(m) Neopentyl chloride
(n) Isopentane

3.8
Name each of the following compounds by the IUPAC system.
(a) $CH_3CH(C_2H_5)CH(CH_3)CH_2CH_3$
(b) $CH_3CH(CH_3)CH_2CH_3$

(c)
$$CH_3\overset{|}{C}H—CH_2—CH\overset{CH_3}{\underset{CH_3}{|}}$$

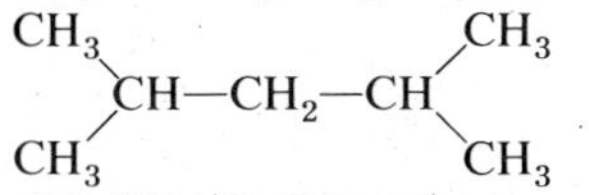

(d) $CH_3CH_2CH(C_2H_5)CH_3$

(e) $CH_3CH_2—\hexagon$

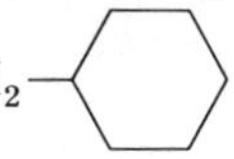

(f) [bicyclic structure: cyclobutane joined to cyclohexane]

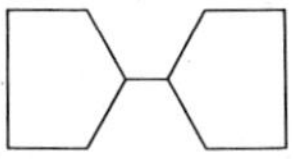

(g)
$$CH_3CH_2CH_2CH_2\overset{CH_3}{\underset{|}{C}}HCH_2CH_2CH_2\overset{|}{C}HCH_3$$
with branch $CH_2—CH—CH_3,CH_3$

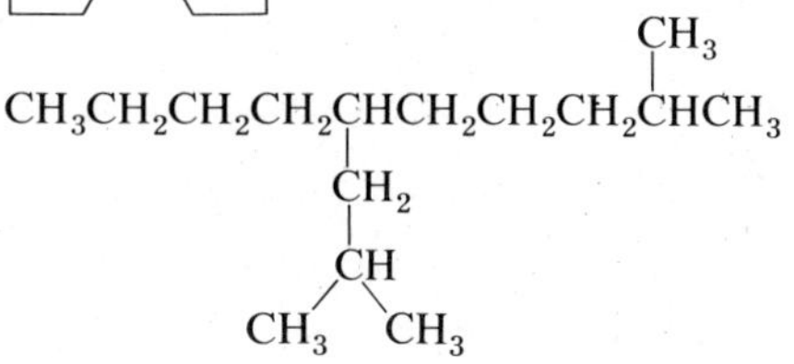

3.9
Give both common and IUPAC names for the isomers of
(a) C_3H_7Cl
(b) C_4H_9Br.

3.10
Write the structure and give the IUPAC name of an alkane or cycloalkane with the formula:
(a) C_5H_{12} that has only primary hydrogens
(b) C_5H_{12} that has only one tertiary hydrogen
(c) C_5H_{12} that has only primary and secondary hydrogens
(d) C_5H_{10} that has only secondary hydrogens
(e) C_6H_{14} that has only primary and tertiary hydrogens

3.11
An alternative system of nomenclature that is occasionally used for naming alkanes with a high degree of symmetry is based on naming them as alkyl-substituted methanes. In this system, 3-ethylpentane would be called triethylmethane:

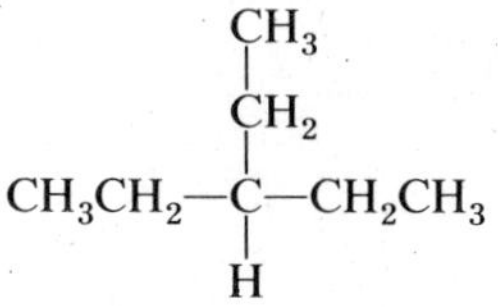

$$CH_3CH_2—\overset{\overset{\displaystyle CH_3}{|}\overset{\displaystyle CH_2}{|}}{\underset{|}{C}}—CH_2CH_3$$
$$H$$

3-Ethylpentane
(or triethylmethane)

The following names are based on this system. Give their structures and their IUPAC names.
(a) Triisopropylmethane   (d) Tri-*sec*-butylmethane
(b) Tetraethylmethane   (e) Di-*tert*-butylmethane
(c) Tetraisobutylmethane   (f) Tetra-*n*-butylmethane

3.12
What alkane of molecular weight 72 would yield:
(a) Only one monochloro derivative?
(b) Three different monochloro derivatives?
(c) Four different monochloro derivatives?
(d) Only two dichloro derivatives?
(The same alkane may serve as the answer to more than one part.)

3.13

The carbon-carbon bond angles of isobutane are $\sim 111.5°$. These angles are larger than those expected from purely tetrahedral carbon (i.e., 109.28°). Explain.

3.14

Sketch approximate potential-energy diagrams for rotations about
(a) One carbon-carbon bond of propane
(b) The C-2—C-3 bond of 2,3-dimethylbutane
(c) The C-2—C-3 bond of 2,2,3,3-tetramethylbutane

3.15

Without referring to tables, decide which member of each of the following pairs has the higher boiling point.
(a) Hexane or isohexane
(b) Hexane or pentane
(c) Pentane or neopentane
(d) Ethane or chloroethane

3.16

Cis-1,2-dimethylcyclopropane has a larger heat of combustion than trans-1,2-dimethylcyclopropane. (a) Which compound is more stable? (b) Give a reason that would explain your answer to part (a).

3.17

Write structural formulas for (a) the two chair conformations of cis-3-isopropyl-1-methylcyclohexane, (b) the two chair conformations of trans-3-isopropyl-1-methylcyclohexane. (c) Designate which conformation in part (a) and (b) is more stable.

3.18

Which member of each of the following pairs of compounds is more stable?
(a) cis- or trans-1,2-Dimethylcyclohexane
(b) cis- or trans-1,3-Dimethylcyclohexane
(c) cis- or trans-1,4-Dimethylcyclohexane

3.19

Professor Norman L. Allinger of the University of Georgia has obtained evidence indicating that while cis-1,3-di-tert-butylcyclohexane exists predominantly in a chair conformation, trans-1,3-di-tert-butylcyclohexane adopts a twist conformation. Explain.

3.20

The important sugar glucose exists in the cyclic form shown below:

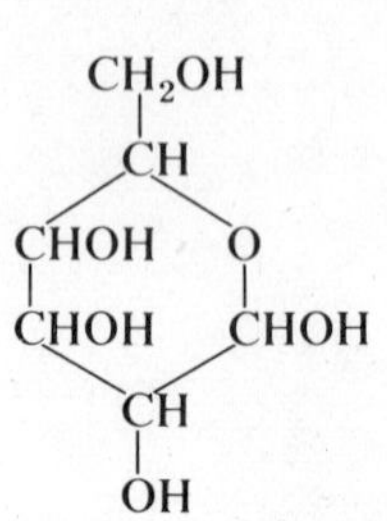

$$CH_2OH$$
$$|$$
$$CH$$

CHOH   O

CHOH   CHOH

CH
|
OH

The six-membered ring of glucose has the chair conformation. All of the hydroxyl groups and the —CH$_2$OH group are equatorial. Write a structure for glucose.

3.21
Outline methods showing how *n*-hexane could be prepared from
(a) A propyl chloride (two ways)   (b) A hexylbromide
(c) An alkene

3.22
When 1,2-dimethylcyclohexene (below) is allowed to react with hydrogen in the presence of a platinum catalyst, the product that is obtained has a melting point of −50° and a boiling point of 130° (at 760 torr). (a) What is the structure of the product of this reaction? (b) Consult an appropriate table and tell which stereoisomer it is. (c) What does this experiment suggest about the mode of addition of hydrogen to the double bond?

1,2-Dimethylcyclohexene

3.23
When cyclohexene is dissolved in an appropriate solvent and allowed to react with chlorine, the product of the reaction, C$_6$H$_{10}$Cl$_2$, has a melting point of −7° and a boiling point (at 16 torr) of 74°. (a) Which stereoisomer is this? (b) What does this experiment suggest about the mode of addition of chlorine to the double bond?

3.24
Compounds with rings containing atoms other than carbon are called *heterocyclic compounds*. Molecules of the heterocyclic compound given below have been shown to exist predominantly in a chair conformation with the hydroxyl group axial. (a) Write this conformation. (b) What factor will account for the molecule preferring to have the hydroxyl group axial rather than equatorial?

* 3.25
(Use models in solving this problem) In contrast to most cyclohexane compounds *trans*-decalin is *conformationally rigid*, that is, its rings do not exhibit the usual chair ⇄ chair "flipping." Thus, it is possible for a substituent of *trans*-decalin to be unequivocally defined as being axial or equatorial and we will see how this is useful when we study steroids in Chapter 23. (a) Consider what would happen to the carbons of one ring were the other to be "flipped" into another chair conformation and explain why *trans*-decalin is conformationally rigid. (b) Would you expect *cis*-decalin to show the same rigidity?

* 3.26
The following data have been advanced to support the assertion that *trans*-1,2-dibromo-cyclohexane exists to a considerable extent in a diaxial conformation. (1) *Cis*-1,2-dibromo-

cyclohexane has a dipole moment equal to 3.09 D, a value that is quite close to that of *cis*-3-bromo-*trans*-4-bromo-1-*tert*-butylcyclohexane, which is assumed to exist primarily in the conformation shown below.

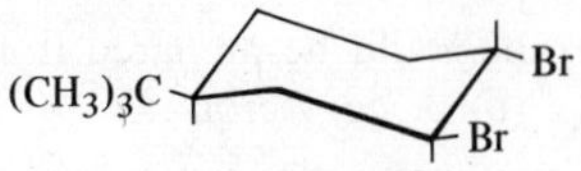

$$\mu = 3.28 \text{ D}$$

cis—3—bromo—*trans*—4—bromo—1—*tert*—butylcyclohexane

(2) *Trans*-1,2-dibromocyclohexane has a much lower dipole moment, $\mu = 2.11$ D. (a) Assume that C—H and C—C bonds make negligible contributions to the dipole moments of the compounds given, and explain how the assertion is justified. (b) We saw earlier that *trans*-1,2-dimethylcyclohexane exists almost exclusively in a diequatorial conformation. What factor might account for the different behavior of *trans*-1,2-dibromocyclohexane?

4 CHEMICAL REACTIVITY: REACTIONS OF ALKANES AND CYCLOALKANES

4.1 INTRODUCTION: HOMOLYSIS AND HETEROLYSIS OF COVALENT BONDS

Reactions of organic compounds almost inevitably involve the making and breaking of covalent bonds. If we consider a hypothetical molecule A:B, there are three possible ways in which the covalent bond may break:

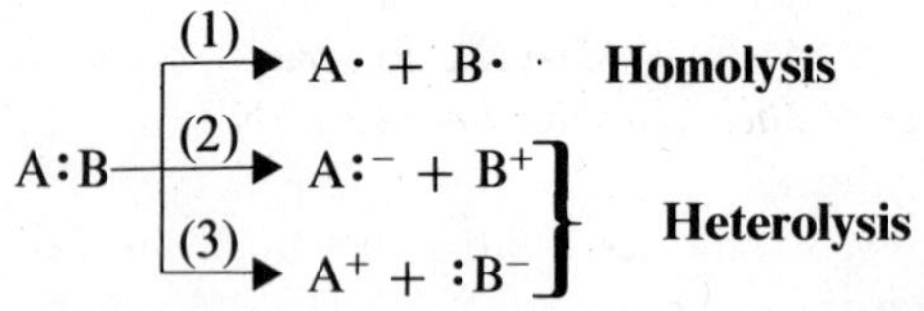

$$A:B \begin{cases} \xrightarrow{(1)} A\cdot + B\cdot & \textbf{Homolysis} \\ \xrightarrow{(2)} A:^- + B^+ \\ \xrightarrow{(3)} A^+ + :B^- \end{cases} \Bigg\} \quad \textbf{Heterolysis}$$

In (1) above the bond breaks so that A and B both retain one of the electrons of the bond and cleavage leads to the neutral fragments $A\cdot$ and $B\cdot$. This type of bond breaking is called *homolysis* (Gr: *homo-*, the same, + *lysis*, loosening or cleavage): the bond is said to have broken *homolytically*. The neutral fragments $A\cdot$ and $B\cdot$ are called *radicals*. Radicals always contain an odd number of electrons.

In (2) and (3) bond cleavage leads to charged fragments or *ions* ($A:^-$ and B^+ or A^+ and $^-:B$). This kind of bond cleavage is called *heterolysis* (Gr: *hetero,* different, + *lysis*); the bond is said to have broken *heterolytically*.

4.2 REACTIVE INTERMEDIATES IN ORGANIC CHEMISTRY

Many organic reactions take place through the formation of a *highly reactive intermediate*—one that results either from homolysis or heterolysis of a bond to carbon. Homolysis of a bond to carbon leads to an intermediate known as a *free radical* (or simply a *radical*).

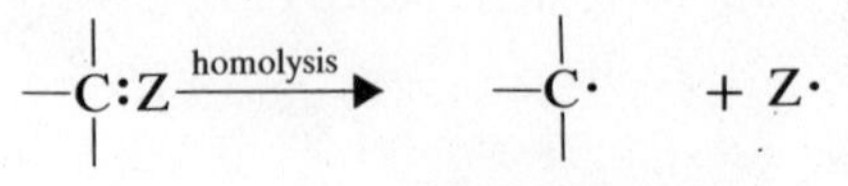

$$-\overset{|}{\underset{|}{C}}{:}Z \xrightarrow{\text{homolysis}} -\overset{|}{\underset{|}{C}}\cdot \quad + Z\cdot$$

A free radical
(or radical)

Heterolysis of a bond to carbon can lead either to a carbon cation or carbon anion.

$$\overset{|}{\underset{|}{-C}}:Z \xrightarrow{\text{heterolysis}} \begin{cases} \overset{|}{-C^+} \quad + \; :Z^- \\[2em] \overset{|}{\underset{|}{-C}}:^- \; + \; Z^+ \end{cases}$$

Carbocation
(or *carbonium ion*)

Carbanion

Carbon cations are called either *carbocations* or *carbonium ions*. The term *carbocation* has a clear and distinct meaning; the older term *carbonium ion* has taken on a different meaning in some of the chemical literature.* Because of this, we always refer to trivalent positively charged species such as $-\overset{|}{C}{}^+$ as carbocations.

Carbon anions are called *carbanions*.

Free radicals and carbocations are electron-deficient species. A free radical has seven electrons in its valence shell; a carbocation has only six. As a consequence, both species are *electrophiles: in their reactions they seek the extra electron or electrons that will give them a stable octet.*

Carbanions are usually strong *bases* and strong *nucleophiles*. *They seek either a proton* or *a positively charged center to neutralize their negative charge.*

Free radicals, carbocations, and carbanions are usually highly reactive species; in most instances they exist only as transient intermediates in an organic reaction. Under certain conditions however, these species may exist long enough for chemists to study them using special techniques. We will have more to say about how this is done in later chapters. A few free radicals, carbocations, and carbanions are stable enough to be isolated. This only happens, however, when there are special groups attached to the central carbon that allow the charge or the odd electron to be stabilized.

4.3 BOND DISSOCIATION ENERGIES

We saw in Section 1.11 that energy is released when atoms combine to form molecules. This energy release results from the formation of a covalent bond. The molecules of the product have lower enthalpy (heat content) than the separate atoms. When hydrogen atoms combine to form hydrogen molecules, for example, the reaction is *exothermic;* it evolves 104 kcal of heat for every mole of hydrogen that is produced. Similarly, when chlorine atoms combine to form chlorine molecules the reaction evolves 58 kcal/mole of chlorine produced.

$$\begin{array}{ll} H\cdot \; + \; H\cdot \; \longrightarrow \; H{-}H & \Delta H = -104 \text{ kcal/mole} \\ Cl\cdot \; + \; Cl\cdot \; \longrightarrow \; Cl{-}Cl & \Delta H = -58 \text{ kcal/mole} \end{array} \bigg\} \begin{array}{l} \textit{Bond formation} \\ \textit{is an exothermic process} \end{array}$$

---

* For example, the highly reactive species, CH_5^+, that we mentioned earlier, is called a carbonium ion by some chemists. CH_5^+ is clearly a different kind of chemical entity than $-\overset{|}{C}{}^+$, so the term carbonium ion has become ambiguous.

When covalent bonds are broken, energy must be supplied. Reactions in which only bond breaking occurs are always endothermic. The energy required to break the covalent bonds of hydrogen or chlorine homolytically is exactly equal to that evolved when the separate atoms combine to form molecules. In the bond cleavage reaction, however, ΔH is positive.

$$H\text{—}H \longrightarrow H\cdot + H\cdot \quad \Delta H = +104 \text{ kcal/mole}$$

$$Cl\text{—}Cl \longrightarrow Cl\cdot + Cl\cdot \quad \Delta H = +58 \text{ kcal/mole}$$

The energies required to break covalent bonds homolytically have been determined experimentally for many types of covalent bonds. These energies are called *bond dissociation energies,* and they are usually abbreviated by the letter D. The bond dissociation energies of hydrogen and chlorine, for example, might be written in the following way.

$$\begin{array}{cc} H\text{—}H & Cl\text{—}Cl \\ (D = 104 \text{ kcal/mole}) & (D = 58 \text{ kcal/mole}) \end{array}$$

The bond dissociation energies of a variety of covalent bonds are listed in Table 4.1.

Bond Dissociation Energies and Heats of Reaction

Bond dissociation energies have, as we will see, a variety of uses. They can be used, for example to calculate the enthalpy change (ΔH) for a reaction. To make such a calculation (below) we must only remember that for bond breaking ΔH is $+$ and for bond formation ΔH is $-$. Let us consider, for example, the reaction of hydrogen and chlorine to produce two moles of hydrogen chloride. From Table 4.1 we get the following values of D.

$$\begin{array}{ccc} H\text{—}H \ + \ Cl\text{—}Cl & \longrightarrow & 2H\text{—}Cl \\ (D = 104) \quad (D = 58) & & (D = 103) \times 2 \end{array}$$

$$\begin{array}{cc} +162 \text{ kcal/mole is required} & -206 \text{ kcal/mole is evolved} \\ \text{for bond cleavage} & \text{in bond formation} \end{array}$$

Overall, the reaction is exothermic: $\Delta H = (-206 \text{ kcal/mole} + 162 \text{ kcal/mole}) = -44 \text{ kcal/mole}$.

For the purpose of our calculation, we have assumed a particular pathway, that is,

$$H\text{—}H \longrightarrow 2H\cdot$$
$$\text{and} \qquad Cl\text{—}Cl \longrightarrow 2Cl\cdot$$
$$\text{then} \qquad 2H\cdot + 2Cl\cdot \longrightarrow 2H\text{—}Cl$$

This is not the way the reaction actually occurs. Nonetheless, the heat of reaction, ΔH, is a thermodynamic quantity that is dependent *only* on the initial and final states of the reacting molecules. ΔH is independent of the path followed and, for this reason, our calculation is valid.

TABLE 4.1 Single-Bond Dissociation Energies in kcal/mole

A:B $\longrightarrow$ A· + B·			
	D		D
H—H	104	$(CH_3)_2CH—H$	94.5
D—D	106	$(CH_3)_2CH—F$	105
F—F	38	$(CH_3)_2CH—Cl$	81
Cl—Cl	58	$(CH_3)_2CH—Br$	68
Br—Br	46	$(CH_3)_2CH—I$	53
I—I	36	$(CH_3)_2CH—OH$	92
H—F	136	$(CH_3)_2CH—OCH_3$	80.5
H—Cl	103		
H—Br	87.5	$(CH_3)_3C—H$	91
H—I	71	$(CH_3)_3C—Cl$	78.5
$CH_3—H$	104	$(CH_3)_3C—Br$	63
$CH_3—F$	108	$(CH_3)_3C—I$	49.5
$CH_3—Cl$	83.5	$(CH_3)_3C—OH$	90.5
$CH_3—Br$	70	$(CH_3)_3C—OCH_3$	78
$CH_3—I$	56	$C_6H_5CH_2—H$	85
$CH_3—OH$	91.5	$CH_2=CHCH_2—H$	85
$CH_3—OCH_3$	80	$CH_2=CH—H$	103
$CH_3CH_2—H$	98	$C_6H_5—H$	103
$CH_3CH_2—F$	106	$HC\equiv C—H$	125
$CH_3CH_2—Cl$	81.5	$CH_3—CH_3$	88
$CH_3CH_2—Br$	69	$CH_3CH_2—CH_3$	85
$CH_3CH_2—I$	53.5	$CH_3CH_2CH_2—CH_3$	85
$CH_3CH_2—OH$	91.5	$CH_3CH_2—CH_2CH_3$	82
$CH_3CH_2—OCH_3$	80	$(CH_3)_2CH—CH_3$	84
		$(CH_3)_3C—CH_3$	80
$CH_3CH_2CH_2—H$	98	HO—H	119
$CH_3CH_2CH_2—F$	106	HOO—H	90
$CH_3CH_2CH_2—Cl$	81.5	HO—OH	51
$CH_3CH_2CH_2—Br$	69	$CH_3CH_2O—OCH_3$	44
$CH_3CH_2CH_2—I$	53.5		
$CH_3CH_2CH_2—OH$	91.5		
$CH_3CH_2CH_2—OCH_3$	80		

From S. W. Benson, "Bond Energies," *J. Chem. Ed.*, *42*, 502 (1965).

Problem 4.1

Calculate the heat of reaction, ΔH, for the following reactions.

(a) $CH_4 + F_2 \longrightarrow CH_3F + HF$

(b) $CH_4 + Cl_2 \longrightarrow CH_3Cl + HCl$

(c) $CH_4 + Br_2 \longrightarrow CH_3Br + HBr$

(d) $CH_4 + I_2 \longrightarrow CH_3I + HI$

Bond Dissociation Energies and the Relative Stabilities of Free Radicals

Bond dissociation energies also provide us with a convenient way to estimate the relative stabilities of free radicals. If we examine the data given in Table 4.1

for methane, ethane, propane, and isobutane, we find that homolysis of the C—H bonds of these alkanes requires different amounts of energy. The amount of energy required is related to the type of C—H bond being broken; that is, it is related to whether the carbon bearing the hydrogen is 3°, 2°, or 1° or methyl carbon. Free radicals are classified as being 3°, 2°, or 1° on the basis of the carbon that has the odd electron.

$$
(CH_3)_3C—H \longrightarrow \begin{array}{c} CH_3 \\ | \\ CH_3—C\cdot \\ | \\ CH_3 \end{array} + H\cdot \qquad \Delta H = +91 \text{ kcal/mole}
$$

Isobutane   *tert*-Butyl radical
       (a 3° radical)

$$
(CH_3)_2CH—H \longrightarrow \begin{array}{c} CH_3 \\ | \\ CH_3—C\cdot \\ | \\ H \end{array} + H\cdot \qquad \Delta H = +94.5 \text{ kcal/mole}
$$

Propane   Isopropyl radical
      (a 2° radical)

$$
CH_3CH_2—H \longrightarrow CH_3CH_2\cdot + H\cdot \qquad \Delta H = +98 \text{ kcal/mole}
$$

Ethane   Ethyl radical
     (a 1° radical)

$$
CH_3—H \longrightarrow CH_3\cdot + H\cdot \qquad \Delta H = +104 \text{ kcal/mole}
$$

All of these reactions resemble one another in two respects: they all begin with an alkane, and they all produce an alkyl radical and a hydrogen atom. They differ, however, in the amount of energy required and in the type of free radical being produced. These differences must be related.

Less energy is required, for example, to produce a *tert*-butyl radical from isobutane than is required to produce the isopropyl radical from propane. This must mean that relative to the alkane from which it is formed, the *tert*-butyl radical contains less energy and, thus, it has a *lower potential energy*. Since the relative stability of a chemical species is inversely related to its relative potential energy, the *tert*-butyl radical must be relatively *more stable*. A comparison of the energy changes involved in these two reactions is given in Fig. 4.1.

We can make a similar comparison of the isopropyl radical and the ethyl radical (a 1° radical). We can also compare the relative potential energies of the ethyl radical and the methyl radical. When we do this we find that, overall, their relative stabilities are

$$
\begin{array}{c} CH_3 \\ | \\ CH_3—C\cdot \\ | \\ CH_3 \end{array} > \begin{array}{c} CH_3 \\ | \\ CH_3—C\cdot \\ | \\ H \end{array} > \begin{array}{c} H \\ | \\ CH_3—C\cdot \\ | \\ H \end{array} > \begin{array}{c} H \\ | \\ H—C\cdot \\ | \\ H \end{array}
$$

Bond dissociation energies of other alkanes also show that this pattern is a general one and the relative stabilities of alkyl radicals is

$$3° > 2° > 1° > \text{methyl}$$

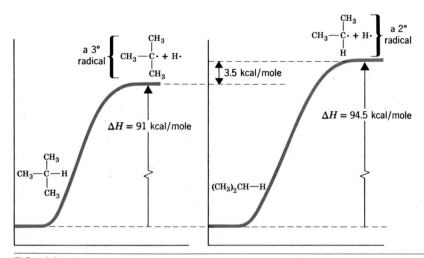

FIG. 4.1

A comparison of the potential energies of the tert-*butyl radical* (+ H·)
and the isopropyl radical (+ H·) *relative to the alkanes from which they
were formed. The relative energy of the tertiary radical is lower than the
secondary radical and, therefore, its relative stability is greater.*

Problem 4.2

Sketch diagrams similar to those in Fig. 4.1 showing the relative stabilities of (a)
$(CH_3)_2CH\cdot$ and $CH_3CH_2\cdot$; (b) $CH_3CH_2\cdot$ and $CH_3\cdot$; (c) $CH_3CH_2\cdot$ and
$CH_3CH_2CH_2\cdot$. (d) Account for the similarity of the two energy diagrams in part
(c).

Problem 4.3

One can also estimate the relative stabilities of free radicals by comparing the bond
dissociation energies of the C—X bonds of haloalkanes. Show how this can be done
with CH_3—Cl, CH_3CH_2—Cl, $(CH_3)_2CH$—Cl, and $(CH_3)_3C$—Cl.

4.4 CHEMICAL REACTIONS OF ALKANES

Alkanes, as a class, are characterized by a general inertness to many chemical
reagents. Carbon-carbon and carbon-hydrogen bonds are quite strong; they do not
break unless alkanes are heated to very high temperatures. Because carbon and
hydrogen have nearly the same electronegativity, the carbon hydrogen bonds of
alkanes are only slightly polarized. As a consequence, they are generally unaffected
by most bases. There are no unshared electrons in alkanes to offer sites for attack
by acids. This low reactivity of alkanes toward many reagents accounts for the fact
that alkanes were originally called *paraffins* (Latin: *parum affinis,* low affinity).

The term paraffin, however, is probably not an appropriate one. We all know
that alkanes react vigorously with oxygen when an appropriate mixture is ignited.
This combustion occurs in the cylinders of automobiles, and in oil furnaces, for
example. When heated, alkanes also react with chlorine, and they react explosively
with fluorine.

Most of the reactions of alkanes are characterized by an attack on the alkane by a reagent with unpaired electrons. Oxygen, the reagent responsible for combustion of alkanes, has unpaired electrons:

$$\cdot \ddot{O} : \ddot{O} \cdot$$

The reactions of methane with halogens, reactions that we will soon consider, involve the attack of a halogen atom on the methane molecule.

$$X \cdot + \underset{\overset{\displaystyle |}{H}}{\overset{\overset{\displaystyle H}{|}}{H : C}} - H \longrightarrow X : H + \underset{\overset{\displaystyle |}{H}}{\overset{\overset{\displaystyle H}{|}}{\cdot C}} - H$$

Exceptions to what we have just related are the recently discovered reactions of alkanes with very powerful acids, called superacids. Superacids react with alkanes by donating protons to their carbon-carbon and carbon-hydrogen sigma bonds. We will describe these reactions later in this chapter.

4.5 THE HALOGENATION OF METHANE: EXPERIMENTAL FACTS AND THE REACTION MECHANISM

One of the most useful concepts in the study of the reactions of organic compounds is the concept of a *reaction mechanism*. A reaction mechanism is a step-by-step description of the events that take place at the molecular level as reacting molecules are converted to products.

A reaction mechanism gives us a theoretical framework on which to hang our experimental facts. It helps us to account for the fate of the bonding electrons, to see which bonds are broken and which are formed and, in so doing, it helps us to remember the reaction. Later, as we discuss mechanisms for other reactions we will find that many reactions, which at first seem unrelated, actually have similar mechanisms. In this way mechanisms help us to correlate reactions in an understandable way. Finally, a knowledge of mechanisms is of great utility in making predictions about new reactions and syntheses. They assist us in eliminating from consideration reactions whose occurrence would be unlikely and help us in choosing experimental conditions that will give us the greatest probability of success.

At the same time it should be kept in mind that a reaction mechanism is only a hypothesis. It is a theoretical construction that we use to account for experimental observations. As such, it is difficult to say that a reaction mechanism is ever *proved*. Early in our study of a particular reaction, the mechanism that we propose may be as tenuous as a sketch of steps that seem reasonable. What we consider to be reasonable will be based on our knowledge of similar reactions and on our examination of the structures of the reacting substances and products. At this point several reasonable mechanisms can often be written. Later, as more experimental facts about the reaction become known, we may modify one mechanistic pathway, and we may discard others altogether.

The reactions of methane with the halogens occur in the gas phase. As such they are particularly suited as vehicles for introducing the reasoning that goes into fashioning a reaction mechanism out of experimental observations. Solvent effects, which complicate reactions that occur in solution, do not play a role in reactions that take place in the gaseous state. Moreover, many of the ideas that can be developed in a discussion of the halogenation of methane, are common to most of the reactions that we will discuss later. For this reason, we shall describe the mechanism for the halogenation of methane in some detail.

We have already seen that methane and chlorine react to produce, successively, mono-, di-, tri-, and tetrachloromethanes. We have also seen that each of

$$CH_4 \xrightarrow{Cl_2} CH_3Cl \xrightarrow{Cl_2} CH_2Cl_2 \xrightarrow{Cl_2} CHCl_3 \xrightarrow{Cl_2} CCl_4$$

$$+ \qquad + \qquad + \qquad +$$

$$HCl \qquad HCl \qquad HCl \qquad HCl$$

these reactions falls into the general class of reactions called substitution reactions. Other experimental observations that can be made about this reaction are as follows:

1. *The reaction is promoted by heat or light.* For all practical purposes, methane and chlorine do not react in the dark at room temperature. Methane and chlorine do react, however, at room temperature if the reaction mixture is irradiated by ultraviolet light, and methane and chlorine do react in the dark, if the mixture is heated to temperatures greater than 250°C.

2. The light-promoted reaction has a very high quantum yield; that is, one photon of light is found to initiate thousands of chlorination reactions.

A mechanism that is consistent with these facts is a stepwise one. The first step involves the fragmentation of a chlorine molecule, by heat or light, into two chlorine atoms.

Step (1): $:\overset{..}{\underset{..}{Cl}}:\overset{..}{\underset{..}{Cl}}: \xrightarrow[\substack{or \\ light}]{heat} 2:\overset{..}{\underset{..}{Cl}}\cdot$

Chlorine is known, from other evidence, to undergo such reactions. It can be shown, moreover, that the frequency of light that promotes the chlorination of methane is a frequency that is absorbed by chlorine molecules and not by methane molecules.

The second and third steps of the mechanism are as follows:*

Step (2): $:\overset{..}{\underset{..}{Cl}}\cdot \; + \; H:\overset{H}{\underset{H}{\overset{|}{C}}}-H \longrightarrow H:\overset{..}{\underset{..}{Cl}}: \; + \; \cdot\overset{H}{\underset{H}{\overset{|}{C}}}-H$

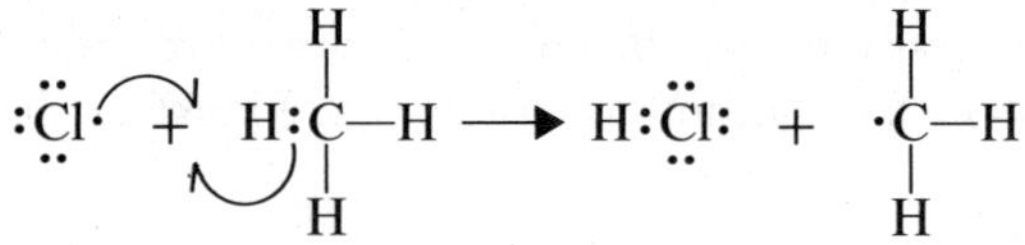

Step (3): $H-\overset{H}{\underset{H}{\overset{|}{C}}}\cdot \; + \; :\overset{..}{\underset{..}{Cl}}:\overset{..}{\underset{..}{Cl}}: \longrightarrow H-\overset{H}{\underset{H}{\overset{|}{C}}}:\overset{..}{\underset{..}{Cl}}: \; + \; \cdot\overset{..}{\underset{..}{Cl}}:$

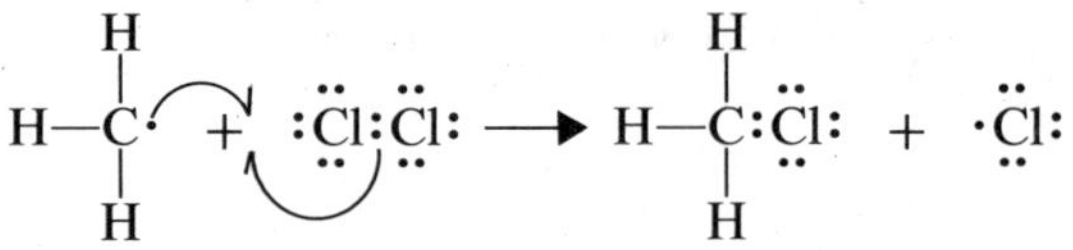

*These conventions are used in illustrating reaction mechanisms in this text.
 1. Arrows ⌒ or ⌒ always show the direction of movement of electrons.
 2. Single-barbed arrows ⌒ show the attack (or movement) of an unpaired electron.
 3. Double-barbed arrows, ⌒ show the attack (or movement) of an electron pair.

Step 2 involves the abstraction of a hydrogen atom from the methane molecule by a chlorine atom. This step results in the formation of a molecule of hydrogen chloride and a methyl radical.

In step 3 the highly reactive methyl radical reacts with a chlorine molecule by abstracting a chlorine atom. This step results in the formation of a molecule of chloromethane (one of the ultimate products of the reaction) and a *chlorine atom*. This latter product is particularly significant, for the chlorine atom formed in step 3 can attack another methane molecule and cause a repetition of step 2.

$$:\ddot{Cl}\cdot \ + \ H:\overset{\overset{\displaystyle H}{|}}{\underset{\underset{\displaystyle H}{|}}{C}}-H \longrightarrow H:\ddot{Cl}: \ + \ \cdot\overset{\overset{\displaystyle H}{|}}{\underset{\underset{\displaystyle H}{|}}{C}}-H \qquad \text{Step 2 is repeated}$$

Then,

$$H-\overset{\overset{\displaystyle H}{|}}{\underset{\underset{\displaystyle H}{|}}{C}}\cdot \ + \ :\ddot{Cl}:\ddot{Cl}: \longrightarrow H-\overset{\overset{\displaystyle H}{|}}{\underset{\underset{\displaystyle H}{|}}{C}}:\ddot{Cl}: \ + \ \cdot\ddot{Cl}: \qquad \text{Step 3 is repeated}$$

Step 2 is repeated again, then step 3, and so forth. With each repetition of step 3 a molecule of chloromethane is produced. This type of sequential, stepwise reaction, in which each step generates the reactive substance that causes the next step to occur, is called a *chain reaction*.

It is the chain nature of the reaction that accounts for the observation that one photon of light promotes the formation of thousands of molecules of products. In step 1 two chlorine atoms are produced by a single photon of light. Each of the two chlorine atoms thus generated can initiate chains; and each chain can ultimately produce thousands of molecules of chloromethane before it is terminated.

What causes the chains to terminate? Why does one photon of light not promote the chlorination of all of the methane molecules present? We know that this does not happen because we find that at low temperatures continuous irradiation is required or the reaction slows and stops. The answer to these questions can be found in existence of *chain-terminating steps;* steps that occur infrequently, but occur often enough to use up one or both of the reactive intermediates, and thus explain the requirement for continuous irradiation. Plausible chain-terminating steps are

$$H-\overset{\overset{\displaystyle H}{|}}{\underset{\underset{\displaystyle H}{|}}{C}}\cdot \ + \ \cdot\ddot{Cl}: \longrightarrow H-\overset{\overset{\displaystyle H}{|}}{\underset{\underset{\displaystyle H}{|}}{C}}:\ddot{Cl}:$$

$$H-\overset{\overset{\displaystyle H}{|}}{\underset{\underset{\displaystyle H}{|}}{C}}\cdot \ + \ \cdot\overset{\overset{\displaystyle H}{|}}{\underset{\underset{\displaystyle H}{|}}{C}}-H \longrightarrow H-\overset{\overset{\displaystyle H}{|}}{\underset{\underset{\displaystyle H}{|}}{C}}:\overset{\overset{\displaystyle H}{|}}{\underset{\underset{\displaystyle H}{|}}{C}}-H$$

and

$$:\ddot{C}l\cdot + \cdot\ddot{C}l: \longrightarrow :\ddot{C}l:\ddot{C}l:*$$

How do we account for the formation of such compounds as dichloromethane, trichloromethane and tetrachloromethane during the course of the chlorination of methane? As the reaction proceeds molecules of chloromethane accumulate in the reaction mixture. As they do, chlorine atoms begin to react with chloromethane molecules as well. This reaction, also of the chain type, ultimately produces dichloromethane.

Step 4
$$:\ddot{C}l\cdot + \begin{matrix} :\ddot{C}l: \\ | \\ H:C-H \\ | \\ H \end{matrix} \longrightarrow H:\ddot{C}l: + \begin{matrix} :\ddot{C}l: \\ | \\ \cdot C-H \\ | \\ H \end{matrix}$$

Chloromethane

Step 5
$$\begin{matrix} :\ddot{C}l: \\ | \\ H-C\cdot + :\ddot{C}l:\ddot{C}l: \\ | \\ H \end{matrix} \longrightarrow \begin{matrix} :\ddot{C}l: \\ | \\ H-C:\ddot{C}l: \\ | \\ H \end{matrix} + \cdot\ddot{C}l:$$

Dichloromethane

As dichloromethane accumulates in the reaction mixture it undergoes similar chain reactions with chlorine atoms and produces trichloromethane. And then, trichloromethane reacts with chlorine atoms to produce tetrachloromethane.

4.6 THE CHLORINATION OF METHANE: ENERGY CHANGES

We saw earlier that we can calculate the overall heat of reaction from bond dissociation energies. We can also calculate the heat of reaction for each individual step of a mechanism.

Chain-initiating step	$Cl-Cl \longrightarrow 2Cl\cdot$ $(D = 58)$	$\Delta H = +58$ kcal/mole
Chain-propagating steps	$Cl\cdot + CH_3-H \longrightarrow CH_3\cdot + H-Cl$ $(D = 104) \qquad\qquad (D = 103)$	$\Delta H = +1$ kcal/mole
	$CH_3\cdot + Cl-Cl \longrightarrow CH_3-Cl + Cl\cdot$ $(D = 58) \qquad (D = 83.5)$	$\Delta H = -25.5$ kcal/mole

* This last step probably occurs least frequently. The two chlorine atoms are highly energetic; as a result, the simple diatomic chlorine molecule that is formed has to dissipate its excess energy rapidly by colliding with some other molecule or the walls of the container. Otherwise it simply flies apart again. By contrast, chloromethane and ethane, formed in the other two chain-terminating steps, can dissipate their excess energy through vibrations of their C—H bonds.

$$\text{Chain-terminating steps} \begin{cases} CH_3\cdot + Cl\cdot \longrightarrow CH_3\text{—}Cl & \Delta H = -83.5 \text{ kcal/mole} \\ \qquad\qquad (D = 83.5) \\ CH_3\cdot + \cdot CH_3 \longrightarrow CH_3\text{—}CH_3 & \Delta H = -88 \text{ kcal/mole} \\ \qquad\qquad (D = 88) \\ Cl\cdot + Cl\cdot \longrightarrow Cl\text{—}Cl & \Delta H = -58 \text{ kcal/mole} \\ \qquad\qquad (D = 58) \end{cases}$$

In the chain-initiating step (p. 129) only one bond is broken—the bond between two chlorine atoms—and no bonds are formed. The heat of reaction for this step is simply the bond dissociation energy for a chlorine molecule and it is highly endothermic.

In the chain-terminating steps bonds are formed, but no bonds are broken. As a result, all of the chain-terminating steps are highly exothermic.

Each of the chain-propagating steps, on the other hand, requires the breaking of one bond and the formation of another. The value of ΔH for each of these steps is the difference between the bond dissociation energy of the bond that is broken and the bond dissociation energy for the bond that is formed. The first chain-propagating step is slightly endothermic ($\Delta H = +1$ kcal/mole), but the second is exothermic by a large amount ($\Delta H = -25.5$ kcal/mole).

Problem 4.4

Assuming the same mechanism occurs, calculate ΔH for the chain-initiating, chain-propagating, and chain-terminating steps involved in the bromination of methane.

The addition of the chain-propagating steps yields the overall equation for the chlorination of methane:

$$\begin{aligned} Cl\cdot + CH_3\text{—}H &\longrightarrow CH_3\cdot + H\text{—}Cl & \Delta H &= +1 \text{ kcal/mole} \\ CH_3\cdot + Cl\text{—}Cl &\longrightarrow CH_3\text{—}Cl + Cl\cdot & \Delta H &= -25.5 \text{ kcal/mole} \\ \hline CH_3\text{—}H + Cl\text{—}Cl &\longrightarrow CH_3\text{—}Cl + H\text{—}Cl & \Delta H &= -24.5 \text{ kcal/mole} \end{aligned}$$

and the addition of the values of ΔH for the individual chain-propagating steps yields the overall value of ΔH for the reaction.

Problem 4.5

Why would it be incorrect to include the chain-initiating and chain-terminating steps in the calculation of the overall value of ΔH given above?

We will return to the individual steps of the reaction of methane with halogens soon. Before doing so, however, we will need to introduce some of the basic ideas of the collision theory of reaction rates.

4.7 REACTION RATES: COLLISION THEORY

Reaction rates are important in guiding our predictions about the eventual outcome of chemical reactions. Many reactions that have favorable energy changes occur so slowly as to be imperceptible. In other instances several reaction pathways may compete with each other, and the actual distribution of products of such a reaction may not be governed by a position of equilibrium—or by the magnitude of ΔH—but *by the rate of the reaction that occurs most rapidly.*

In order to understand the factors that affect the rate of a chemical reaction, let us consider the hypothetical example:

$$A + B \longrightarrow C$$

Let us assume that the reaction occurs in the gas phase where interactions of the reacting substances with solvent molecules are not involved, and let us further assume that A and B react to produce C by colliding with each other.

The rate of the reaction $A + B \longrightarrow C$ can be determined experimentally by measuring the rates at which A and/or B disappear from the reaction mixture, or by measuring the rate at which C is formed in the reaction mixture. We can make either of these measurements in the laboratory by simply withdrawing small samples of the reaction mixture at measured intervals of time and by analyzing the samples for the concentrations of A, B, or C.

After determining the reaction rate in this way, we can analyze our data to find certain relationships. In this instance, we would find that the overall rate of the reaction is proportional to the concentrations of A *and* B present in the reaction at any given moment, that is,

$$\text{Rate } \alpha \ [A]$$
$$\text{Rate } \alpha \ [B]$$

therefore, $\quad\quad\quad\quad\quad$ Rate α [A][B]

This proportionality can be expressed as an equation by the introduction of a proportionality constant, k, called the rate constant:

$$\text{Rate} = k \, [A][B]$$

We can discuss the reaction rate in other terms: since the reaction of $A + B \longrightarrow C$ requires collisions between molecules of A and B, the collision theory of reaction rates tells us that the rate of the reaction will be determined by three factors:

1. *The collision frequency,* that is, the number of collisions between A and B that occur in each unit volume of the reaction mixture in each second of time.

2. *An energy factor,* that is, the fraction of collisions that occur with energies that are sufficient to bring about a reaction between the colliding molecules.

3. *A probability (or orientation) factor,* that is, the fraction of collisions in which the orientation of the colliding molecules, with respect to each other, allows a reaction to take place.

The overall rate of the reaction is the product of these three quantities:

$$\text{Rate of reaction} = \frac{\text{collision}}{\text{frequency}} \times \frac{\text{an energy}}{\text{factor}} \times \frac{\text{a probability}}{\text{factor}}$$

We now examine carefully each of these terms.

The Collision Frequency

Many experiments show that the rates of chemical reactions are directly proportional to *the collision frequency*—the greater the collision frequency, the faster the reaction. Two major factors that determine the magnitude of the collision frequency are *concentration and temperature.*

The more concentrated the reacting molecules are (or for a gas phase reaction, the greater is the pressure), the greater will be the number of collisions occurring in each unit volume of the reaction mixture each second. A simple analogy will help make this clear. Consider, for example, a room in which blindfolded people walk about randomly. It is easy to see that if the room is crowded (a high concentration of people), people will bump into each other more often (have a higher collision frequency) than they will if only a few are present.

The higher the temperature, the faster the molecules move and, as a result, the greater are the number of collisions in a unit volume per unit of time. In our analogy, this would correspond to having our blindfolded people run rather than walk. Clearly in this situation, more collisions will occur in each unit of time, at any given population (concentration) of the room.

The Orientation or Probability Factor

A great deal of experimental evidence indicates that collisions between molecules must occur in a particular way in order to be effective. While the probability factor for a particular reaction is difficult to estimate, it is easy to see why it should exist. Consider, for example, the reaction of a chlorine atom with a molecule of hydrogen bromide to produce a bromine atom and a molecule of hydrogen chloride: Cl· + H—Br ⟶ H—Cl + Br·

FIG. 4.2
Some of the possible orientations of collisions between chlorine atoms and molecules of hydrogen bromide.

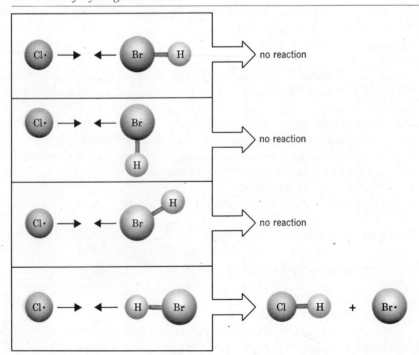

There are, naturally, an infinite number of ways that collisions between chlorine atoms and molecules of hydrogen bromide could occur. Some of these possibilities are shown in Fig. 4.2. Since the reaction that we are considering is one in which the chlorine atom abstracts a hydrogen atom from hydrogen bromide, it is reasonable to assume that in order to be effective, the chlorine atom must collide with the hydrogen end of the hydrogen bromide molecule. Collisions in which the chlorine atom collides with the bromine end of the hydrogen bromide molecule are not likely to be effective.

Another example is the reaction of a chlorine atom and a molecule of methane. This reaction is one of the chain–propagating steps in the chlorination of methane.

$$\text{Cl}\cdot \; + \; \text{H}-\overset{\displaystyle H}{\underset{\displaystyle H}{\text{C}}}-\text{H} \; \longrightarrow \; \text{Cl}-\text{H} \; + \; \cdot\overset{\displaystyle H}{\underset{\displaystyle H}{\text{C}}}-\text{H}$$

Here the probability that collisions will occur with the proper orientation is greater (Fig. 4.3). There are four hydrogens and collisions between a chlorine atom and any one of them will be effective. Even here, however, not all collisions will produce reactions. Chlorine atoms might collide with that part of the methane molecule that is between hydrogens, and these collisions would not be effective.

The Energy Factor and the Energy of Activation

We have already seen that one effect of increasing the temperature of a reaction is to increase the number of collisions that occur between molecules. In-

FIG. 4.3

Two possible orientations for a collision between a chlorine atom and a methane molecule.

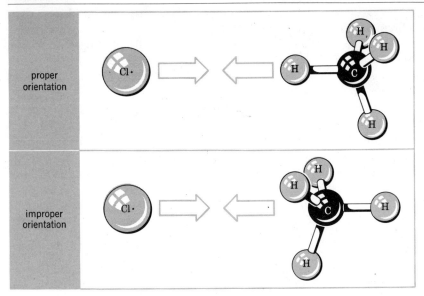

creasing the temperature of a reaction, however, produces another effect that is even more important. We find, for example, that for most reactions, a rather small increase in temperature results in a very large increase in reaction rate. The rate of some reactions is doubled and sometimes trebled by a temperature increase of only 10%. This effect is far too large to be explained solely on the basis of an increase in the number of collisions that occur in each unit volume each second. The explanation lies, instead, *in the large increase in the fraction of collisions that are sufficiently energetic to produce reactions*. To see how this happens we need to introduce a new concept at this point: the *energy of activation* of a reaction.

Not all collisions between molecules produce a chemical reaction even when the colliding molecules have the proper orientation. According to the collision theory, a certain minimum amount of energy, called the *energy of activation,* must be brought to a properly aligned collision for it to be an effective collision. The source of the activation energy (abbreviated E_{act}) is the kinetic energy* of the colliding molecules. If the molecules bring insufficient energy to a collision they apparently bounce apart without reacting.

How do we explain this requirement for a minimum amount of energy for a collision to be effective? Let us answer by considering a specific example: the reaction of a fluorine atom with a methane molecule.

$$\text{F} \cdot + \text{CH}_3\text{—H} \longrightarrow \text{H—F} + \cdot\text{CH}_3 \qquad \Delta H = -32 \text{ kcal/mole}$$
$$\qquad (104) \qquad\qquad (136) \qquad\qquad\qquad E_{act} = +1.2 \text{ kcal/mole}$$

This reaction, the first chain-propagating step in the reaction of fluorine with methane, is a highly exothermic one. The energy released in forming the H—F bond is considerably greater than the energy required to break the C—H bond of methane. If bond formation and bond rupture could occur simultaneously as the fluorine atom and methane molecule collide, then the energy released in bond formation ought to be more than adequate to bring about bond rupture. We find experimentally, however, that this is not the case. On a molar basis, an additional 1.2 kcal of energy (the energy of activation) must be brought to the collision for a reaction to occur. *Bond rupture and bond formation do not occur simultaneously.* Bond formation appears to lag behind bond rupture, and it is this lag that is the origin of the requirement for an energy of activation.

We can illustrate this situation graphically by plotting the potential energy of the reacting particles as they are transformed from reactants to products during the course of a collision against what is called the *reaction coordinate*. Such an illustration is given in Fig. 4.4.

The reaction coordinate is a measure of the *progress of the reaction* during the collision. For most reactions a point along the reaction coordinate represents a collection of distances of all of the particles of the reactants at a given moment during the reaction. We may think of it as representing the changes in geometry that the colliding particles undergo as atoms are forced apart in bond breaking and as atoms are drawn together in bond formation.

In this illustration (Fig. 4.4), we can see that what we call the energy of activation of the reaction is, effectively, an energy barrier between reactants and

---

* The kinetic energy of a molecule is proportional to its mass and to the square of its velocity ($\text{KE} = \frac{1}{2} mv^2$); and the temperature of a gas is a measure of the average kinetic energy of the molecules present in the gas.

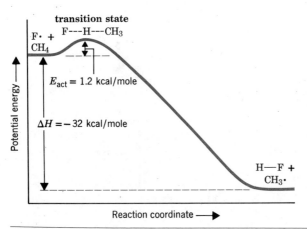

FIG. 4.4

The energy changes that occur in the reaction of a fluorine atom and a methane molecule.

products. It is, in a sense, an energy hill that the reacting species must traverse in order to become products. The height of this barrier (in kcal/mole) above the level of reactants is the energy of activation.

The top of the energy hill, called the *transition state,* corresponds to a particular arrangement of the atoms of the reacting species as they are converted from reactants to products. It represents a configuration in which bonds are partially broken and partially formed. In the present example, the bond between the hydrogen atom and the carbon of the methyl group is partially broken; the bond between the fluorine and the hydrogen atom is partially formed. Although we know that the transition state has only fleeting existence, we conceive of it, at times, as though it were an actual molecule. And we will soon see that the idea of a transition state is one of the most useful concepts that arises from theories of reaction rates.

The difference in potential energy between the reactants and the transition state is the energy of activation, E_{act}. The difference in potential energy between the reactants and products is the heat of reaction, ΔH. For the reaction, $F\cdot + CH_4 \longrightarrow H{-}F + CH_3\cdot$, the energy level of the products is 32 kcal/mole lower than that of the reactants. In terms of our analogy, we can say that the reactants on one energy plateau must traverse an energy hill (the energy of activation) in order to arrive at the lower energy plateau of products.

When a three-dimensional plot of potential energy versus the reaction coordinate is made, the transition state is found to resemble a mountain pass or *col* (Fig. 4.5) rather than the top of an energy hill as we have shown in our two-dimensional plot above. That is, the reactants and products appear to be separated by an energy barrier resembling a mountain range. While there are an infinite number of possible routes from reactants to products, the transition state lies at the top of the route that requires the lowest (energy) climb.

How does one know what the energy of activation for a reaction will be? Could we, for example, have predicted that the energy of activation for the reaction, $F\cdot + CH_4 \longrightarrow HF + CH_3\cdot$, would be precisely 1.2 kcal/mole? The answer is *no*.

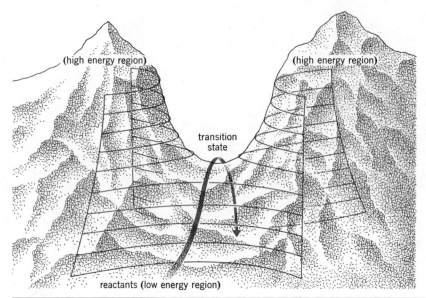

FIG. 4.5

Mountain pass or col *analogy for the transition state. (Adapted with permission from J. E. Leffler and E. Grunwald,* Rates and Equilibria of Organic Reactions, *Wiley, New York, 1963, page 6.)*

The energy of activation must be determined experimentally, and even then, by indirect methods. Certain principles can be established, however, that enable one to arrive at estimates of the energies of activation of different types of reactions:

1. Any reaction in which *bonds are broken* will have an energy of activation.
2. The energy of activation of a reaction can be estimated from a consideration of the transition state and the stabilities of reactants and products. *Factors that tend to stabilize the transition state relative to the reactants lower the energy of activation.* Conversely, *factors that increase the energy of the transition state, relative to reactants, increase the energy of activation.* We will see how these estimates can be made later.
3. Activation energies of *endothermic reactions that involve bond formation and bond rupture must be larger than the heat of reaction,* ΔH. Two examples illustrate this principle: the first chain-propagating step in the chlorination of methane and the corresponding step in the bromination of methane:

$$Cl\cdot + CH_3-H \longrightarrow H-Cl + CH_3\cdot \qquad \Delta H = +1\,\text{kcal/mole}$$
$$\text{(104)} \qquad\qquad \text{(103)} \qquad\qquad E_{act} = +3.8\,\text{kcal/mole}$$

$$Br\cdot + CH_3-H \longrightarrow H-Br + CH_3\cdot \qquad \Delta H = +16.5\,\text{kcal/mole}$$
$$\text{(104)} \qquad\qquad \text{(87.5)} \qquad\qquad E_{act} = +18.6\,\text{kcal/mole}$$

In both of these reactions the energy released in bond formation is less than that required for bond rupture; both reactions are, therefore, endothermic. The first reaction, that of a chlorine atom with methane, is only slightly endothermic, $\Delta H = +1$ kcal/mole. The energy of activation is somewhat higher, $+3.8$ kcal/mole. The second reaction, that of a bromine atom with methane, is much more endothermic, $\Delta H = +16.5$ kcal/mole.

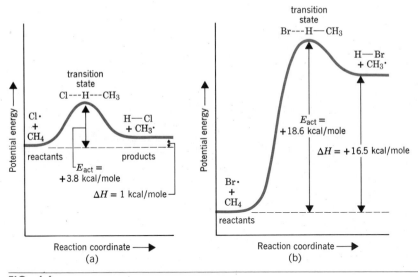

FIG. 4.6

Potential energy diagrams for (a) *the reaction of a chlorine atom with methane and* (b) *the reaction of a bromine atom with methane.*

The energy of activation for this latter reaction is quite large, +18.6 kcal/mole. We can easily see why the energy of activation for each reaction is greater than the heat of reaction by looking at the potential energy diagrams in Fig. 4.6. In each case the path from reactants to products is from a lower energy plateau to a higher one. In each case the intervening energy hill (or *col*) is higher still, and since the energy of activation is the vertical (energy) distance between the plateau of reactants and the top of this hill the energy of activation exceeds the heat of reaction.

4. *The energy of activation of a reaction where bonds are broken but no bonds are formed is equal to ΔH.* An example of this type of reaction is the chain-initiating step in the chlorination of methane; the dissociation of chlorine molecules into chlorine atoms.

$$Cl—Cl \longrightarrow 2Cl\cdot \qquad \Delta H = +58 \text{ kcal/mole}$$
$$\qquad (58) \qquad\qquad E_{act} = +58 \text{ kcal/mole}$$

The potential energy diagram for this reaction is shown in Fig. 4.7.

5. The energy of activation for a reaction in which *bonds are formed but no bonds are broken is usually zero.* In reactions of this type the problem of nonsimultaneous bond formation and bond rupture does not exist; only one process occurs; that of bond formation. All of the chain-terminating steps in the chlorination of methane fall into this category. An example is the combination of two methyl radicals to form a molecule of ethane.

$$2CH_3\cdot \longrightarrow CH_3—CH_3 \qquad \Delta H = -88 \text{ kcal/mole}$$
$$\qquad (88) \qquad\qquad E_{act} = 0$$

Figure 4.8 illustrates the potential energy changes that occur in this reaction.

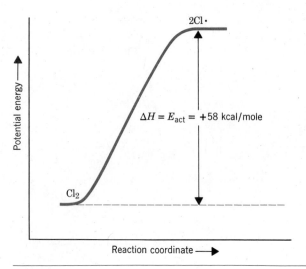

FIG. 4.7
The potential energy diagram for the dissociation
of a chlorine molecule into chlorine atoms.

Problem 4.6

Sketch potential energy diagrams for the following reactions. Label the heat of reaction (ΔH) and the energy of activation (E_{act}) in each case.

(a) $CH_3 \cdot + HCl \longrightarrow CH_3-H + Cl \cdot$
(b) $CH_3 \cdot + HBr \longrightarrow CH_3-H + Br \cdot$
(c) $CH_3-CH_3 \longrightarrow 2CH_3 \cdot$
(d) $Br-Br \longrightarrow 2Br \cdot$
(e) $2Cl \cdot \longrightarrow Cl-Cl$

FIG. 4.8
The potential energy diagram for the combination of
two methyl radicals to form a molecule of ethane.

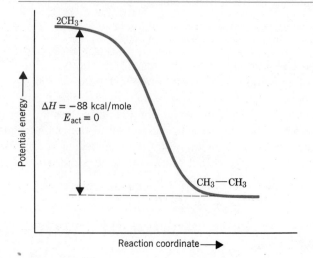

Reaction Rates: an Explanation of the Effect of Temperature Changes

With an understanding of the origin and nature of the energy of activation behind us, we now return to our original problem: that of providing an explanation for the remarkable increase in the rate of most reactions that accompanies a modest increase in temperature.

In a given sample of a gas at a given temperature the molecules move with widely varying velocities. If we could actually see the molecules we would observe that, at a given instant, some molecules would be moving quite slowly and others very rapidly. Most of the molecules would have velocities at or near the average velocity of all of the molecules present. Since the kinetic energy of the individual molecules is proportional to the square of their velocities ($KE = \frac{1}{2} mv^2$), the distribution of the kinetic energies of the molecules should parallel the distribution of their velocities. This distribution is given in the curve shown in Fig. 4.9a. Note that while a few of the molecules have very low kinetic energies and a few have very high kinetic energies, most of the molecules have kinetic energies near the average.

The energies brought to collisions between molecules in the gas phase show the same general distribution (Fig. 4.9b). Occasionally molecules having very low kinetic energies collide and these collisions have, as a result, very low energy. Occasionally, too, molecules possessing very high kinetic energies collide and these give very high energy collisions; most of the collisions, however, are between molecules with kinetic energies at or near the average and have collision energies at or near the average.

If we designate a particular collision energy (E_{act}) as that required, as *a minimum* to bring about a reaction between the colliding molecules, the number of collisions having sufficient energy to cause a reaction is proportional to the area under that portion of the curve that represents collision energies greater than or equal to E_{act}. If, at the same time, we also examine the distribution of collision energies at two different temperatures, T_1 and T_2, where T_2 is a higher temperature than T_1, we get the result shown in Fig. 4.10.

Because of the shapes of the curves, and because the curve for the higher temperature is generally shifted to higher kinetic energies, the number of collisions

FIG. 4.9

The distributions of (a) *kinetic energies among gaseous molecules, and* (b) *kinetic energies of collisions among collisions between gaseous molecules.*

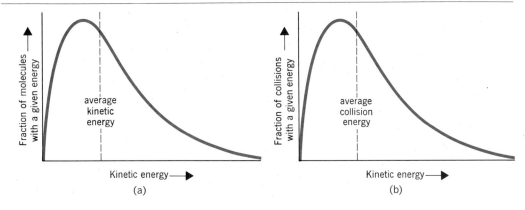

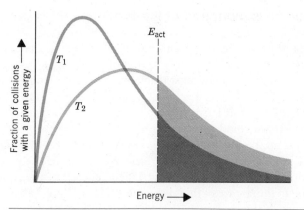

FIG. 4.10

The distribution of collision energies at two different temperatures, T_1 *and* T_2 *(*$T_2 > T_1$*). The number of collisions with energies greater than the activation energy is indicated by the appropriately shaded area under each curve.*

with energies great enough to cause a reaction, that is, with energies greater than E_{act}, *is very much larger at the higher temperature.* The area under curve T_1 with energy greater than E_{act} is quite small, but that under curve T_2 with energy greater than E_{act} is quite large. It is this difference that accounts, primarily, for the very dramatic increase in reaction rate that can occur with a very modest increase in temperature.

What we have just illustrated graphically can also be explained mathematically. The relation between the rate constant for a reaction (p. 131), the energy of activation and the absolute temperature is given by the Arrhenius equation:

$$k = A\, e^{-E_{act}/RT}$$

where
- k = the rate constant
- A = the Arrhenius factor
- e = 2.718 (the base of natural logarithms)
- R = 1.986 cal/°K-mole (the gas constant)

and
- T = the absolute temperature

The Arrhenius factor, A, is a measure of both the collision frequency and the probability factor; it measures the probability that a reaction will take place independently of energy requirements. The term,

$$e^{-E_{act}/RT}$$

is the *energy factor,* and it equals the fraction of collisions with energies greater than the energy of activation.

It is the exponential nature of the energy factor that accounts for the remarkable increase in the rate of a reaction that usually accompanies an increase in temperature. This point can be illustrated in the following way. If we assume that a particular reaction has an energy of activation of 10 kcal/mole, we can show through a simple calculation that the small change from room temperature, $300°K$, to the temperature of a steambath, $373°K$ should cause *a 20-fold increase* in the number of collisions with the energy greater than the activation energy.

Reaction Rates: The effect of the Energy of Activation

The energy factor also accounts for another important observation: *the relation between the reaction rate and the size of the energy of activation for different reactions of the same type occurring at the same temperature.* The nature of this relationship can be understood by considering, as examples, three different reactions—all having very different energies of activation—but all occurring at the same temperature. The reactions that we have chosen are the first chain propagating steps that occur when methane is fluorinated, chlorinated, and brominated.

$$F\cdot + CH_3\text{—}H \longrightarrow H\text{—}F + CH_3\cdot \qquad E_{act} = 1.2 \text{ kcal/mole}$$
$$Cl\cdot + CH_3\text{—}H \longrightarrow H\text{—}Cl + CH_3\cdot \qquad E_{act} = 3.8 \text{ kcal/mole}$$
$$Br\cdot + CH_3\text{—}H \longrightarrow H\text{—}Br + CH_3\cdot \qquad E_{act} = 18.6 \text{ kcal/mole}$$

Substituting these values for E_{act} into the expression $e^{-E_{act}/RT}$ gives the following results. (The temperature that we have chosen for our calculation is $300°C$ or $573°K$.)

REACTION	E_{act}	FRACTION OF COLLISIONS WITH ENERGY GREATER THAN E_{act}
$F\cdot + CH_4 \longrightarrow HF + CH_3\cdot$	1.2 kcal/mole	0.34
$Cl\cdot + CH_4 \longrightarrow HCl + CH_3\cdot$	3.8 kcal/mole	0.035
$Br\cdot + CH_4 \longrightarrow H\text{—}Br + CH_3\cdot$	18.6 kcal/mole	0.00000008

The meaning of these results can be more fully appreciated if we express them in the following way: out of every 100 million collisions, 34 million will be sufficiently energetic in the fluorination reaction, 3.5 million will be sufficiently energetic in the chlorination reaction, but only 8 collisions will be sufficiently energetic in the bromination reaction!

The same kind of explanation can be given graphically. To see how this is so, consider Fig. 4.11.

We see from the shape of the curve that the area under the curve representing the fraction of collisions with sufficient energy to allow a reaction is quite small when the energy of activation is large. On the other hand, when the energy of activation is small, the area under the curve is quite large. The first situation might represent the reaction, $Br\cdot + CH_4 \longrightarrow HBr + CH_3\cdot$ —a reaction that at room temperature occurs very slowly. The latter might represent the reaction, $F\cdot + CH_4 \longrightarrow HF + CH_3\cdot$ —a reaction, that under the same conditions, occurs very rapidly.

It thus becomes clear that in planning experiments, or in explaining their results, we need to give careful consideration to the relative sizes of the energies of activation.

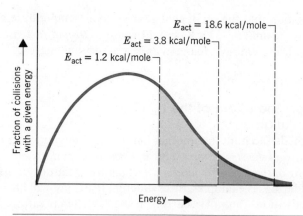

FIG. 4.11
The fraction of collisions with energies greater than the energy of activation at three different activation energies.

4.8 THERMODYNAMICS AND KINETICS OF THE REACTIONS OF METHANE WITH HALOGENS

The *relative reactivity* of one substance toward another is measured by the *rate* at which the two substances react. A reagent that reacts very rapidly with a particular substance is said to be highly reactive toward that substance. One that reacts slowly or not at all under the same experimental conditions (e.g., concentration, pressure, and temperature) is said to have a low relative reactivity or to be unreactive. The reactions of the halogens (fluorine, chlorine, bromine, and iodine) with methane, show a wide spread of relative reactivities. Fluorine is most reactive; so reactive, in fact, that without special precautions mixtures of fluorine and methane often explode. Chlorine is next most reactive. However, the chlorination of methane is easily controlled by the judicious control of heat and light. Bromine is much less reactive toward methane than chlorine, and iodine is so unreactive that for all practical purposes we can say that no reaction takes place at all.

If we assume that the mechanisms for fluorination, bromination, and iodination are the same as for the chlorination of methane, we can explain the wide variation in reactivity of the halogens by a careful examination of the thermodynamic quantities ΔH and E_{act} for each step.

Fluorination	ΔH (kcal/mole)	E_{act} (kcal/mole)
Chain initiating:		
$F_2 \longrightarrow 2F\cdot$	$+ 38$	$+38$
Chain propagating		
$F\cdot + CH_4 \longrightarrow HF + CH_3\cdot$	$- 32$	$+ 1.2$
$CH_3\cdot + F_2 \longrightarrow CH_3F + F\cdot$	$- 70$	small
Overall $\Delta H =$	$\overline{-102}$	

The chain-initiating step in fluorination is highly endothermic and thus has a high energy of activation.

If we did not know otherwise, we might carelessly conclude from the energy of activation of the chain-initiating step alone that fluorine would be quite unreactive toward methane. (If we then proceeded to try the reaction, as a result of this careless assessment, the results would be literally disastrous.) We know, however, that the chain-initiating step occurs only infrequently relative to the chain-propagating steps. One initiating step is able to produce thousands of fluorination reactions. As a result, the high activation energy for this step is not an impediment to the reaction.

Chain-propagating steps, by contrast, cannot afford to have high energies of activation. If they do, the highly reactive intermediates are consumed by chain-terminating steps before the chains progress very far. Both of the chain-propagating steps in fluorination have very small energies of activation. This allows a relatively large fraction of energetically favorable collisions even at room temperature. Moreover, the overall heat of reaction, ΔH, is very large. This means that as the reaction occurs, a large quantity of heat is evolved. This heat may accumulate in the reaction medium causing the temperature to rise and with it a rapid occurrence of additional chain-initiating steps and thus additional chains. These two factors, the low energy of activation for the chain-propagating steps and the large overall heat of reaction, account for the high reactivity of fluorine towards methane.*

Chlorination	ΔH (kcal/mole)	E_{act} (kcal/mole)
Chain initiating:		
$Cl_2 \longrightarrow 2Cl\cdot$	+58	+58
Chain propagating:		
$Cl\cdot + CH_4 \longrightarrow HCl + CH_3\cdot$	+ 1	+ 3.8
$CH_3\cdot + Cl_2 \longrightarrow CH_3Cl + Cl\cdot$	−25.5	small
Overall $\Delta H =$	−24.5	

The higher energy of activation of the first chain-propagating step in chlorination of methane (+3.8 kcal/mole), versus the lower energy of activation (+1.2 kcal/mole) in fluorination, partly explains the lower reactivity of chlorine. The greater energy required to break the chlorine-chlorine bond in the initiating step (+58 kcal/mole for Cl_2 versus +38 kcal/mole for F_2) has some effect, too. However, the much greater overall heat of reaction in fluorination probably plays the greatest role in accounting for the much greater reactivity of fluorine.

Bromination	ΔH (kcal/mole)	E_{act} (kcal/mole)
Chain initiating:		
$Br_2 \longrightarrow 2Br\cdot$	+46	46
Chain propagating:		
$Br\cdot + CH_4 \longrightarrow HBr + CH_3\cdot$	+ 16.5	+18.6
$CH_3\cdot + Br_2 \longrightarrow CH_3Br + Br\cdot$	−24	small
Overall $\Delta H =$	− 7.5	

In contrast to chlorination, the first chain-propagating step in bromination has a very high energy of activation ($E_{act} = 18.6$ kcal/mole). This means that only a very tiny fraction of all of the collisions between bromine atoms and methane

---

*Fluorination reactions can be controlled. This is usually accomplished by diluting both the hydrocarbon and the fluorine with an inert gas like helium before bringing them together. The reaction is also carried out in a reactor packed with copper shot. The copper, by absorbing the heat produced, moderates the reaction.

molecules will be energetically effective even at a temperature of 300°C. Bromine, as a result, is much less reactive toward methane than chlorine even though the net reaction is slightly exothermic.

Iodination ΔH (kcal/mole) E_{act} (kcal/mole)
Chain initiating:

$$I_2 \longrightarrow 2I\cdot \qquad\qquad +36 \qquad\qquad +36$$

Chain propagating:

$$I\cdot + CH_4 \longrightarrow HI + CH_3\cdot \qquad +31 \qquad\qquad +33.5$$
$$CH_3\cdot + I_2 \longrightarrow CH_3I + I\cdot \qquad -20 \qquad\qquad \text{small}$$
$$\text{Overall } \Delta H = \overline{+11}$$

The thermodynamic quantities for iodination of methane make it clear that the chain initiating step is not responsible for the observed order of reactivities: $F_2 > Cl_2 > Br_2 > I_2$. The iodine-iodine bond is even weaker than the fluorine-fluorine bond. On this basis alone, one would predict that iodine would be the most reactive of the halogens. This clearly is not the case. Once again, it is the first chain-propagating step that correlates with the experimentally determined order of reactivities. The energy of activation of this step in the iodine reaction (33.5 kcal/mole) is so large that only two collisions out of every 10^{12} have sufficient energy to produce reactions at 300°C. As a result, iodination is not a feasible reaction experimentally.

Before we leave this topic one further point needs to be made. We have given explanations of the relative reactivities of the halogens toward methane that have been based on energy considerations alone. This has been possible *only because the reactions are quite similar and thus have similar probability factors*. Had the reactions been of different types, this kind of analysis would not have been proper and might have given incorrect explanations.

Mathematically, we say it this way: the A terms in the Arrhenius equations for these similar reactions have all been of the same order of magnitude because the A term is a function of collision frequency and the probability factor. Comparisons made on the basis of the energy term, $e^{-E_{act}/RT}$ alone, have been valid because the similar A terms have canceled out. We make similar comparisons at many other points in this book, but we are justified in doing this *only* when the reactions are quite similar.

4.9 HALOGENATION OF HIGHER ALKANES

The reactions of higher alkanes with halogens proceed by the same mechanism as the chlorination of methane. Using R—H as the designation for an alkane, we can state the steps for the general reaction as follows:

Chain initiating: (1) $X_2 \xrightarrow{\text{heat or light}} 2X\cdot$

Chain propagating: (2) $X\cdot + R-H \longrightarrow R\cdot + H-X$
                     (3) $R\cdot + X_2 \longrightarrow R-X + X\cdot$

Chlorination of most alkanes containing more than three carbons gives a mixture of isomeric monochloroproducts as well as more highly halogenated compounds. An example is the light-promoted chlorination of isobutane.

$$CH_3\underset{\underset{\displaystyle CH_3}{|}}{C}HCH_3 \xrightarrow[\text{light}]{Cl_2} CH_3\underset{\underset{\displaystyle CH_3}{|}}{C}HCH_2Cl + CH_3\underset{\underset{\displaystyle Cl}{|}}{\overset{\overset{\displaystyle CH_3}{|}}{C}}CH_3 + \text{Polychlorinated products} + HCl$$

| Isobutane | Isobutyl chloride (48%) | *tert*-Butyl chloride (29%) | (23%) |

The ratio of products that we obtain from chlorination reactions of higher alkanes are not identical with what we would expect if all the hydrogens of the alkane were equally reactive. We find that there is a correlation between reactivity of different hydrogens and the type of hydrogen (1°, 2°, or 3°) being replaced. The tertiary hydrogens of an alkane are most reactive, secondary hydrogens are next most reactive, and primary hydrogens are the least reactive.

Problem 4.7

If we examine the products of the reaction of isobutane with chlorine given above, we find that isobutyl chloride represents 62.5% of the monochlorinated product, while *tert*-butyl chloride represents 37.5%. Explain how this demonstrates that the tertiary hydrogen is more reactive. (Hint: Consider what percentages of the butyl chlorides would be obtained if the nine primary hydrogens and the single tertiary hydrogen were all equally reactive.)

We can account for the relative reactivities of the primary, secondary, and tertiary hydrogens in a chlorination reaction on the basis of bond dissociation energies we saw earlier. Of the three types, breaking a tertiary C—H bond requires the least energy, and breaking a primary C—H bond requires the most. Since the step in which these C—H bonds are broken (i.e., the hydrogen abstraction step) determines the net orientation of the chlorination reaction, we would expect tertiary hydrogens to be most reactive, secondary hydrogens to be next most reactive, and primary hydrogens to be the least reactive.

The differences in the rates with which primary, secondary, and tertiary hydrogens are replaced by chlorine are not large, however. Chlorine, as a result, does not discriminate between the different types of hydrogens in a way that makes chlorination of higher alkanes a generally useful laboratory procedure. We will find later that there are much better laboratory methods for introducing chlorine into a molecule. (Alkane chlorinations do find use in some industrial processes, especially in those instances where mixtures of alkyl chlorides can be used.)

Problem 4.8

Chlorination reactions of certain higher alkanes can be used for laboratory preparations. Examples are the preparation of neopentyl chloride from neopentane and cyclopentyl chloride from cyclopentane. What structural feature of these molecules makes this possible?

Problem 4.9

The hydrogen abstraction steps for most alkane chlorinations are exothermic. Show that this is true by calculating ΔH for the reaction by which Cl· abstracts

(a) a primary hydrogen of ethane.

(b) a secondary hydrogen of propane.

(c) a primary hydrogen of propane.

Reactivity and Selectivity

Bromine is less reactive toward alkanes than chlorine, but bromine is more *selective*. Bromine shows a much greater ability to discriminate among the different types of hydrogens. The reaction of isobutane and bromine, for example, gives almost exclusive replacement of the tertiary hydrogens.

$$
\underset{\underset{H}{|}}{\overset{\overset{CH_3}{|}}{CH_3-C-CH_3}} \xrightarrow[\text{light, 127°}]{Br_2} \underset{\underset{\underset{>99\%}{Br}}{|}}{\overset{\overset{CH_3}{|}}{CH_3-C-CH_3}} + \underset{\underset{\underset{\text{trace}}{H}}{|}}{\overset{\overset{CH_3}{|}}{CH_3-C-CH_2Br}}
$$

The greater selectivity of bromine can be explained in terms of transition state theory and bromine's greater selectivity is directly related to its lower reactivity. According to a postulate made by Professor G. S. Hammond (of the California Institute of Technology) *the transition states of endothermic reactions resemble products more than those of exothermic reactions.*

This principle can be better understood through consideration of the potential energy versus reaction coordinate diagrams given in Fig. 4.12.

Highly exothermic and highly endothermic reactions are shown in Fig. 4.12 because they illustrate Hammond's postulate most dramatically.* In the highly

---

* Hammond's postulate is quite general, however, and applies to reactions where values of ΔH are not so exaggerated.

FIG. 4.12

Energy diagrams for a highly exothermic and a highly endothermic reaction. (From William A. Pryor, Introduction to Free Radical Chemistry *© 1966. Reprinted by permission of Prentice-Hall, Inc., Englewood Cliffs, N.J., p. 53.)*

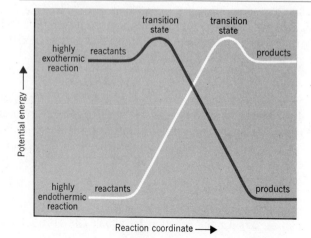

exothermic reaction the energy levels of the reactants and the transition state are close to each other. The transition state also lies close to the reactants along the *reaction coordinate*. This means that in the highly exothermic reaction relatively little bond-breaking has occurred when the transition state is reached. In the exothermic reactions of isobutane with chlorine, for example, relatively little carbon-hydrogen bond breaking has occurred in the transition states. The transition states for the two chlorination reactions might resemble those shown below.

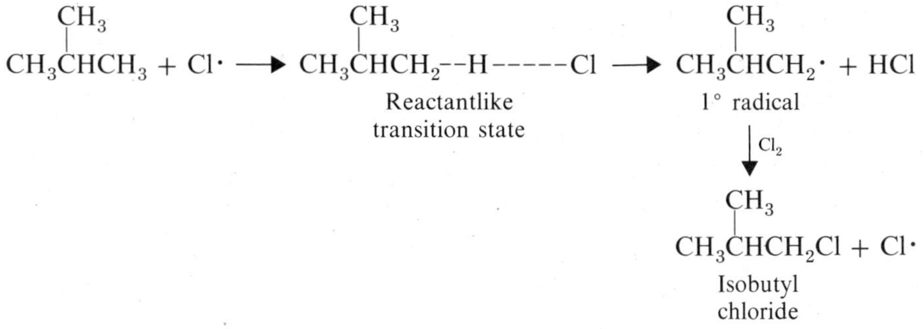

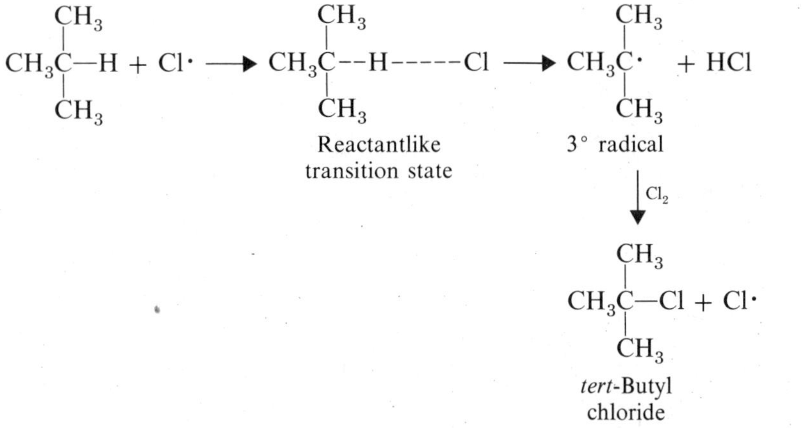

$$CH_3\overset{\overset{\displaystyle CH_3}{|}}{C}HCH_3 + Cl\cdot \longrightarrow CH_3\overset{\overset{\displaystyle CH_3}{|}}{C}HCH_2\text{-}\text{-}H\text{-}\text{-}\text{-}\text{-}\text{-}Cl \longrightarrow CH_3\overset{\overset{\displaystyle CH_3}{|}}{C}HCH_2\cdot + HCl$$

Reactantlike transition state   1° radical

$$\Big\downarrow Cl_2$$

$$CH_3\overset{\overset{\displaystyle CH_3}{|}}{C}HCH_2Cl + Cl\cdot$$

Isobutyl chloride

$$CH_3\overset{\overset{\displaystyle CH_3}{|}}{\underset{\underset{\displaystyle CH_3}{|}}{C}}\text{-}H + Cl\cdot \longrightarrow CH_3\overset{\overset{\displaystyle CH_3}{|}}{\underset{\underset{\displaystyle CH_3}{|}}{C}}\text{-}\text{-}H\text{-}\text{-}\text{-}\text{-}\text{-}Cl \longrightarrow CH_3\overset{\overset{\displaystyle CH_3}{|}}{\underset{\underset{\displaystyle CH_3}{|}}{C}}\cdot + HCl$$

Reactantlike transition state   3° radical

$$\Big\downarrow Cl_2$$

$$CH_3\overset{\overset{\displaystyle CH_3}{|}}{\underset{\underset{\displaystyle CH_3}{|}}{C}}\text{-}Cl + Cl\cdot$$

tert-Butyl chloride

Since the transition states in both cases are reactantlike in both structure and energy, they show little resemblance to the products of the hydrogen abstraction step, a 1° radical and a 3° radical. And, since the reactants in both cases are the same, the exact type of C—H bond being broken (primary or tertiary) has a relatively small influence on the relative rates of the reactions. The two reactions proceed with similar (but not identical) rates because their respective activation energies are quite similar (Fig. 4.13).

The transition states of highly endothermic reactions lie close to products on the potential energy coordinate and *along the reaction coordinate*. In highly endothermic reactions considerable bond breaking has occurred when the transition state is reached. The two reactions of isobutane with bromine are both highly endothermic. In these reactions considerable carbon-hydrogen bond breaking has occurred when the transition state is reached. These transitions states might be depicted in the way shown on the top half of page 148.

$$\begin{array}{c} CH_3 \\ | \\ CH_3CHCH_3 + Br\cdot \end{array} \longrightarrow \begin{array}{c} CH_3 \\ | \\ CH_3CHCH_2\text{-----}H\text{--}Br \end{array} \longrightarrow \begin{array}{c} CH_3 \\ | \\ CH_3CHCH_2\cdot + HBr \end{array}$$

Productlike   1° radical
transition state

$$\downarrow Br_2$$

$$\begin{array}{c} CH_3 \\ | \\ CH_3CHCH_2Br + Br\cdot \end{array}$$

Isobutyl
bromide

$$\begin{array}{c} CH_3 \\ | \\ CH_3C\text{--}H + Br\cdot \\ | \\ CH_3 \end{array} \longrightarrow \begin{array}{c} CH_3 \\ | \\ CH_3C\text{-----}H\text{--}Br \\ | \\ CH_3 \end{array} \longrightarrow \begin{array}{c} CH_3 \\ | \\ CH_3C\cdot \; + HBr \\ | \\ CH_3 \end{array}$$

Productlike   3° radical
transition state

$$\downarrow Br_2$$

$$\begin{array}{c} CH_3 \\ | \\ CH_3C\text{--}Br + Br\cdot \\ | \\ CH_3 \end{array}$$

tert-Butyl
bromide

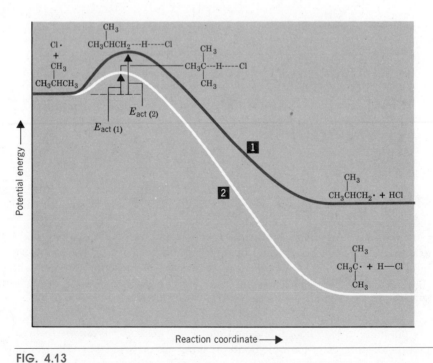

FIG. 4.13

Potential energy diagrams for the two hydrogen abstraction steps in the reaction of isobutane with chlorine. Both reactions are exothermic and both transition states resemble the reactants. The activation energies are similar; but, because 3° C—H bonds are broken more easily than 1° C—H bonds, reaction (2) has a lower activation energy and proceeds at a somewhat faster rate.

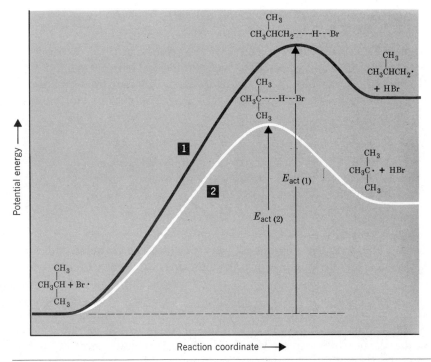

FIG. 4.14

Potential energy diagrams for the two hydrogen abstraction steps in the reaction of isobutane with bromine. Both reactions are highly endothermic and, in both, the transition states resemble the products. Since the products—a 3° radical and a 1° radical—have quite different potential energies (stabilities) the transition states for the two reactions are also quite different. The transition state for reaction (1) resembles a 1° radical. It occurs at a much higher potential energy than the transition state for reaction (2) because the state for reaction (2) resembles a much more stable 3° radical. The activation energy for reaction (2) is much lower than that reaction for (1). Reaction (2), consequently, proceeds at a much faster rate. The ultimate product that arises from reaction (2) is tert-*butyl bromide, and this is the predominant product of the reaction.*

In bromination, since the transition states in both cases are productlike in both structure and energy, and since the products of the hydrogen abstraction step are, in fact, quite different (a 1° radical versus a 3° radical), the type of C—H bond being broken will have a marked influence on the relative rates of the reactions. In fact, they proceed with very different rates. Bromine, as a result, discriminates more effectively between the primary and tertiary hydrogens. A comparison of potential energy diagrams for the abstraction of the primary and tertiary hydrogens by bromine is given in Fig. 4.14.

Problem 4.10

Fluorine is far less selective than bromine and is even less selective than chlorine. The products that one obtains from alkane fluorinations are, in fact, almost those that one would expect if the different types of hydrogen were equally reactive. Explain.

4.10 REACTIONS OF ALKANES WITH SUPERACIDS

Alkanes, as we noted earlier, are normally unreactive towards acids. Alkanes can be boiled for days with concentrated hydrochloric acid, for example, and no detectable reaction will occur. However, Professor George Olah (of Case Western Reserve University) has discovered that when alkanes are treated with very powerful acids, called superacids, sigma bonds break *heterolytically,* and the reactions give carbocations.*

$$\underset{\substack{\text{Heterolytic} \\ \text{cleavage of} \\ \text{sigma bond}}}{-\overset{|}{\underset{|}{C}}\!:\!H} \;+\; \underset{\substack{\text{From a} \\ \text{superacid}}}{H^+} \;\longrightarrow\; \underset{\substack{\text{Carbo-} \\ \text{cation}}}{-\overset{|}{\underset{|}{C}}{}^+} \;+\; H\!:\!H$$

Superacids are mixtures of very powerful Lewis acids and very powerful proton donors. One of the strongest superacids is a mixture of antimony pentafluoride, SbF_5, (the Lewis acid) and fluorosulfonic acid, FSO_3H (the proton donor).

The reaction between an alkane and a superacid is a *protolysis* (cleavage by a proton). The process begins with the donation of a proton to a sigma bond of the alkane and the formation of an intermediate with a pentacoordinate carbon. In the reaction of isobutane with SbF_5—FSO_3H (below), three of the bonds to the central carbon of the intermediate are ordinary carbon-carbon sigma bonds; the two hydrogens, however, are attached through a three-center bond, that is, *a bond in which three nuclei are held together by only two electrons.*

$$\underset{\underset{CH_3}{|}}{\overset{\overset{CH_3}{|}}{CH_3\!-\!C\!-\!H}} + FSO_3H\cdot SbF_5 \longrightarrow \left[\underset{\underset{CH_3}{|}}{\overset{\overset{CH_3}{|}}{CH_3\!-\!C}}\!\!\overset{H}{\underset{H}{\diagup\!\diagdown}} \right]^+ \longrightarrow \underset{\underset{CH_3}{|}}{\overset{\overset{CH_3}{|}}{CH_3\!-\!C^+}} + H_2$$

Pentacoordinate carbon   *tert*-Butyl
compound containing a   cation
three-center bond

The intermediate then loses a molecule of hydrogen through cleavage of three-center bond, this gives a *tert*-butyl cation.

Three-center bonds are not uncommon. The inorganic compound diborane, B_2H_6, contains two three-center bonds. When we examine the reactions of carbo-

$$\underset{\text{Diborane}}{\overset{\displaystyle H\diagdown\;\;\;H\diagdown\;\;\;\diagup H}{\underset{\displaystyle H\diagup\;\;\;{}^{\backprime}H\diagup\;\;\;{}^{\backprime}H}{B\quad\;\;\;B}}}$$

cations in subsequent chapters we will see other examples of intermediates containing three-center bonds.

Carbocations, like free radicals, are highly reactive species. In most reactions

---

* Professor Olah and his co-workers have isolated and studied the properties of a number of carbocations. In doing so, they have contributed greatly to our understanding of them.

in which they are encountered, they have only transient existence. In superacids, however, carbocations have lifetimes long enough to allow them to be analyzed spectroscopically (Chapter 14).

4.11 THE STRUCTURE OF CARBOCATIONS AND FREE RADICALS: sp^2 HYBRIDIZATION

Carbocations

There is considerable experimental evidence indicating that the structure of carbocations is *trigonal planar*. Just as the tetrahedral structure of methane can be

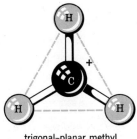

trigonal–planar methyl
carbocation

accounted for on the basis of orbital hybridization so, too, can the trigonal-planar structure of carbocations. In the case of carbocations, however, the central carbon is sp^2 hybridized.

Mathematically, the model for sp^2 hybridized carbon is obtained by mixing or hybridizing, one s orbital with two p orbitals. In the illustration below, we hybridize the s orbital with the p_x and p_z orbitals.

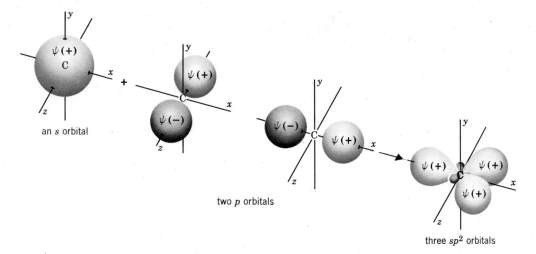

an s orbital

two p orbitals

three sp^2 orbitals

The three sp^2 orbitals that result, are directed toward the corners of a regular triangle (with angles of 120° between them). The carbon p orbital (the p_y orbital) that is not hybridized with the *s* orbital is perpendicular to the plane of the triangle formed by the hybrid *sp²* orbitals.

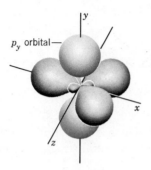

The central carbon in a carbocation is electron deficient; it has only six electrons in its outside energy level. In our model (Fig. 4.15) these six electrons are

FIG. 4.15

(a) *Orbital structure of the methyl cation. The bonds to hydrogen are sigma bonds (σ) formed by overlap of the carbon's* sp² *orbitals with the 1s orbital of hydrogen. The* p *orbital is vacant.* (b) *A line-dash-wedge representation of the* tert-*butyl cation. The bonds between carbon atoms are formed by overlap of* sp³ *orbitals of the methyl groups.*

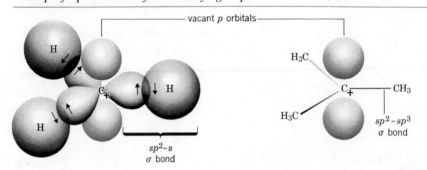

FIG. 4.16

sp²-*Hybridized methyl radical showing the odd electron in one lobe of the vacant* p *orbital. It may also occupy the other lobe.*

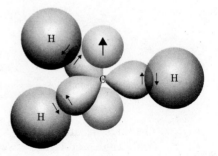

used to form sigma-type covalent bonds to hydrogens (or to alkyl groups). The p orbital contains no electrons.

Free Radicals

Experimental evidence indicates that the structure of most alkyl radicals is also trigonal planar at the carbon having the unpaired electron. This structure, too, can be accommodated by an sp^2 hybridized central carbon. In an alkyl radical, however, the p orbital is not vacant, but contains the unpaired electron (Fig. 4.16).

Some free radicals, such as the trifluoromethyl radical, appear to be sp^3 hybridized (Fig. 4.17). Experiments also indicate sp^3 hybridized radicals undergo rapid inversion about the central carbon.

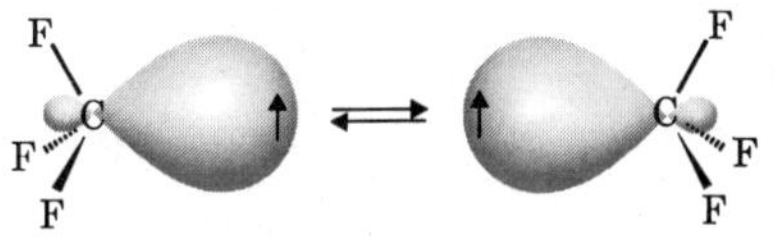

FIG. 4.17
Inverting trifluoromethyl radical.

4.12 REACTIONS OF CYCLOALKANES: RING-OPENING REACTIONS OF CYCLOPROPANE

The reactions of cycloalkanes with rings of five carbon atoms or larger are similar to those of alkanes. Cyclopentane and cyclohexane, for example, undergo halogen-

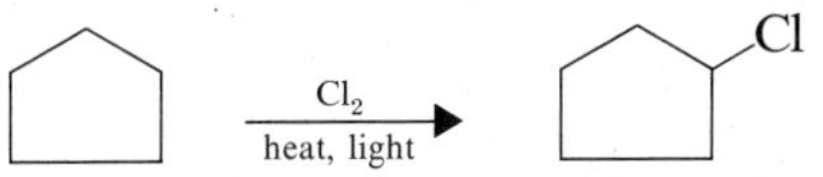

Cyclopentane → Chlorocyclopentane

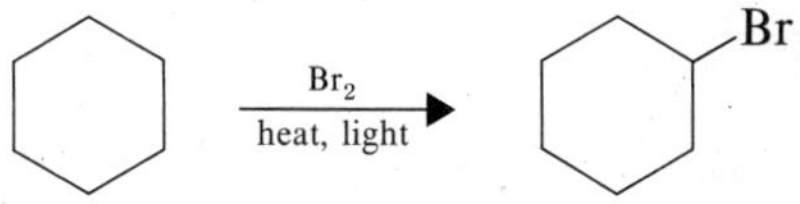

Cyclohexane → Bromocyclohexane

ation by substitution. Cyclopropane, however, undergoes an unusual ring opening reaction when treated with bromine (and a Lewis acid) in carbon tetrachloride.

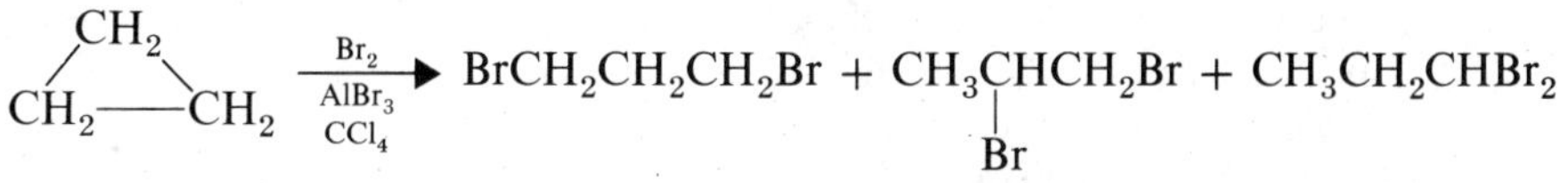

$$\text{BrCH}_2\text{CH}_2\text{CH}_2\text{Br} + \text{CH}_3\text{CHCH}_2\text{Br} + \text{CH}_3\text{CH}_2\text{CHBr}_2$$

Ring opening reactions also occur when cyclopropane is treated with hydroiodic acid or aqueous sulfuric acid.

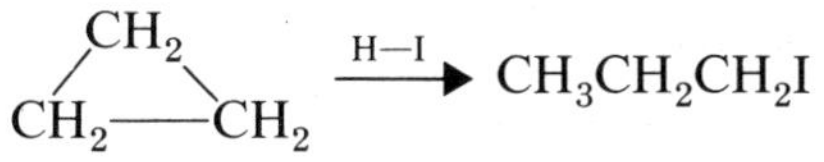

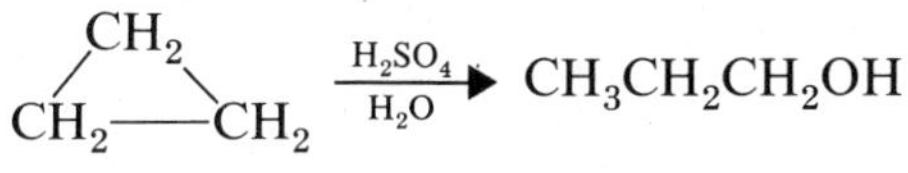

$$\xrightarrow{\text{H—I}} \text{CH}_3\text{CH}_2\text{CH}_2\text{I}$$

$$\xrightarrow[\text{H}_2\text{O}]{\text{H}_2\text{SO}_4} \text{CH}_3\text{CH}_2\text{CH}_2\text{OH}$$

Although we will not consider their mechanisms in detail, the reactions of cyclopropane with HI and H_2SO_4 both involve heterolytic cleavage of a carbon-carbon sigma bond of the cyclopropane ring through protolysis. The ring opening reaction of cyclopropane with bromine (given above) apparently proceeds through a similar mechanism; in it, the agent responsible for bond cleavage is either the Lewis acid or a positive bromine ion.

An explanation for the ring-opening reactions of cyclopropane can be found in the great strain of the three-membered ring. The carbon-carbon bonds of cyclopropane are, as a result of this strain, considerably weakened. An indication of the strain present can be obtained by considering the value of ΔH for opening the cyclopropane ring by homolytic cleavage of one of the carbon-carbon bonds:

$$\underset{\overset{|}{CH_2}\text{---}\overset{|}{CH_2}}{\overset{CH_2}{\diagup \diagdown}} \longrightarrow \cdot CH_2CH_2CH_2\cdot \qquad \Delta H = 55 \text{ kcal/mole}$$

By contrast, the bond dissociation energy for the carbon-carbon bonds of most alkanes is ~82 kcal/mole.

The cyclobutane ring is strained, as well, but not nearly so much as the cyclopropane ring. Cyclobutane does not undergo a ring-opening reaction when it is treated with bromine in carbon tetrachloride.

Additional Problems

4.11

Define and give an example of each of the following.
(a) Homolysis of a covalent bond
(b) Heterolysis of a covalent bond
(c) Bond dissociation energy
(d) A free radical
(e) A carbocation
(f) A carbanion

4.12

Arrange the following free radicals in order of decreasing stability.

$$\underset{CH_3\overset{\overset{\displaystyle CH_3}{|}}{C}HCH_2CH_2\cdot}{} \qquad \underset{CH_3\overset{\overset{\displaystyle CH_3}{|}}{C}HCHCH_3}{} \qquad \underset{CH_3\overset{\overset{\displaystyle CH_3}{|}}{C}CH_2CH_3}{} \qquad \underset{\cdot CH_2\overset{\overset{\displaystyle CH_3}{|}}{C}HCH_2CH_3}{}$$

4.13

(a) Bromination of 2-methylbutane gives a predominance of one product. What is it? (b) Chlorination of 2-methylbutane gives (in addition to polychloro compounds) a mixture of monochloro products in the following proportions: 1-chloro-2-methylbutane (34%), 2-chloro-2-methylbutane (22%), 3-chloro-2-methylbutane (28%), and 1-chloro-3-methylbutane (16%). Show how each of these products is formed. (c) Account for the different results obtained from the bromination and chlorination reactions.

4.14

Give an example and sketch a potential energy diagram for:
(a) A reaction in which a covalent bond is broken but no bond is formed
(b) A reaction in which a covalent bond is formed but no bond is broken
(c) A highly exothermic reaction
(d) A highly endothermic reaction

4.15

The reaction $CH_3CH_3 + Cl_2 \xrightarrow[\text{or}\atop\text{light}]{\text{heat}} CH_3CH_2Cl + HCl$ proceeds by a mechanism similar to that for the chlorination of methane. (a) Write equations for the chain-initiating, chain-propagating, and chain-terminating steps. (b) Calculate values of ΔH for each of the chain-propagating steps. (c) Ethane undergoes free radical chlorination more readily than methane. This can be demonstrated by allowing a mixture of equimolar amounts of ethane and methane to react with chlorine; the reaction produces more chloroethane than chloromethane. What factor accounts for this?

4.16

A possible reaction of ethane with chlorine is

$$Cl_2 + CH_3CH_3 \longrightarrow 2CH_3—Cl$$

This reaction could conceivably occur by the following chain reaction:

Initiating   (1) $Cl—Cl \longrightarrow 2Cl\cdot$

Propagating $\begin{cases} (2)\ Cl\cdot + CH_3—CH_3 \longrightarrow CH_3Cl + CH_3\cdot \\ (3)\ CH_3\cdot + Cl—Cl \longrightarrow CH_3Cl + Cl\cdot \end{cases}$
then (2), (3), (2), (3), etc.

(a) Do you think this overall reaction would compete effectively with the one you gave as answer to problem 4.15? (Base your answer on the ΔH and E_{act} for each step.)
(b) Speculate about the possibility of a similar reaction of ethane with fluorine, that is,

Initiating   (1) $F—F \longrightarrow 2F\cdot$

Propagating $\begin{cases} (2)\ F\cdot + CH_3—CH_3 \longrightarrow CH_3—F + CH_3\cdot \\ (3)\ CH_3\cdot + F—F \longrightarrow CH_3—F + F\cdot \end{cases}$
then (2), (3), (2), (3), etc.

4.17

Free radical fluorination of methane occurs in the absence of light. A mechanism that has been proposed for the dark reaction is

$$CH_4 + F_2 \xrightarrow{\text{slow}} CH_3\cdot + HF + F\cdot$$

$$CH_3\cdot + F\cdot \xrightarrow{\text{fast}} CH_3F$$

(a) Basing your answer on bond dissociation energies, assess the likelihood of the reaction occurring by this mechanism.
(b) What is the likelihood of a similar mechanism occurring when a mixture of methane and chlorine is heated in the dark?

4.18

We saw on p. 150 that treating isobutane with $FSO_3H—SbF_5$ gives hydrogen gas and a solution containing a *tert*-butyl cation. A side reaction also takes place in the mixture; this reaction produces methane and an isopropyl cation, $CH_3\overset{+}{C}HCH_3$. Propose a mechanism that accounts for the side reaction.

4.19

The carbon-carbon bonds of cyclopropane are sometimes described as being "bent" because overlap of the carbon sp^3 orbitals occurs in the manner shown on p. 156. While bent bonds probably relieve some angle strain they also make the carbon-carbon sigma bonds more susceptible to attack by an electrophile such as a proton. (a) Show how protolysis of cyclopropane by HI (p. 153) could lead to an *n*-propyl cation, $CH_3CH_2CH_2^+$ and (b) show how a propyl cation could stabilize itself by reacting with an iodide ion to form *n*-propyl iodide.

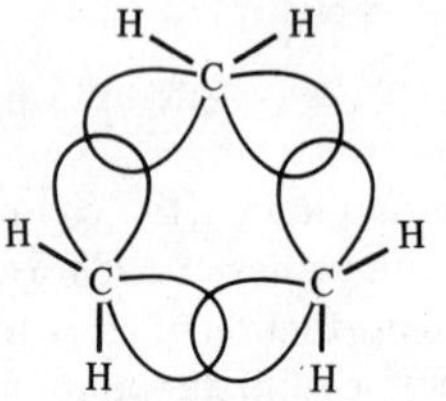

bent bonds of cyclopropane

4.20

When alkanes are heated to very high temperatures (500°) their carbon-carbon and carbon-hydrogen bonds break homolytically. This process is called thermal "cracking." [Cracking is very important in the petroleum industry because it can convert long-chain alkanes with low-octane ratings (p. 221) to shorter-chain alkanes with higher-octane ratings.] Use bond dissociation energies in Table 4.1 to account for the following: (a) Thermal cracking of a C—H bond of methane requires a higher temperature ($\sim$1200°) than does a similar breaking of a C—H bond of ethane (500–600°). (b) When ethane undergoes homolysis at high temperatures, the C—C bond breaks more readily than the C—H bonds. (c) When *n*-butane "cracks," the reaction, $CH_3CH_2CH_2CH_3 \longrightarrow 2CH_3CH_2\cdot$, occurs more readily than the reaction $CH_3CH_2CH_2CH_3 \longrightarrow CH_3CH_2CH_2\cdot + CH_3\cdot$.

4.21

When propane is heated to a very high temperature it undergoes thermal cracking (cf. problem 4.20) through homolysis of C—C and C—H bonds. The major products of the reaction are methane and ethene. A chain mechanism has been proposed for this reaction. (a) Which of the following reactions is most likely to be the major chain initiating step? Explain your answer by estimating activation energies for each reaction.

$$CH_3CH_2CH_3 \longrightarrow CH_3CH_2\cdot + CH_3\cdot$$
$$CH_3CH_2CH_3 \longrightarrow CH_3CH_2CH_2\cdot + H\cdot$$
$$CH_3CH_2CH_3 \longrightarrow CH_3\overset{\cdot}{C}HCH_3 + H\cdot$$

Possible chain-propagating steps are:

(1) $CH_3\cdot + CH_3CH_2CH_3 \longrightarrow CH_4 + \cdot CH_2CH_2CH_3$

(2) $\cdot CH_2{-}CH_2{:}CH_3 \longrightarrow CH_2{=}CH_2 + \cdot CH_3$

(b) Both reactions have reasonably low activation energies (low enough to occur at very high temperatures). Show that this is likely for step 1 by calculating ΔH for step 1.

(c) An alternative to step 1 is:

$$CH_3\cdot + CH_3CH_2CH_3 \longrightarrow CH_4 + CH_3\overset{\cdot}{C}HCH_3$$

Comment on the likelihood of this reaction occurring in terms of energy and probability factors.

4.22

Carbocations are powerful Lewis acids. (a) What factor accounts for this? (b) Write the Lewis acid-base reactions that take place when an ethyl cation, $CH_3CH_2^+$, reacts with a chloride ion. (c) With a hydrogen sulfate ion. (d) With a molecule of water. (e) Carbocations often undergo reactions in which they are transformed into alkenes. What is involved here? (f) Illustrate your answer to (e) by showing the reaction in which an ethyl cation is converted to ethene.

* 4.23

Compounds with an oxygen-oxygen single bond—called peroxides—are often used to initiate free radical chain reactions such as alkane halogenations. (a) Examine the bond energies in Table 4.1 and give reasons that will explain why peroxides are especially effective as free-radical initiators. (b) Illustrate your answer by outlining how di-*tert*-butyl peroxide, $(CH_3)_3CO{-}OC(CH_3)_3$, might initiate an alkane halogenation.

* **4.24**

The reactions below show comparisons between two sets of similar reactions. In each set we compare reactions in which a hydrogen atom is abstracted from methane and from ethane. In the first set **(A)** the abstracting agent is a methyl radical; in the second set **(B)** it is a bromine atom. (a) Sketch potential energy diagrams for each set of reactions that take Hammond's postulate into account. Take care to locate each transition state properly not only along the energy axis but along the reaction coordinate as well. For convenience in making comparisons, you should align the curves so that the potential energies of the reactants are the same. (b) For which reaction will bond-breaking have occurred to the *least* extent when the transition state is reached? (c) To the *greatest* extent? (d) To what approximate extent will bond breaking have occurred in reaction **A**(1)? (e) For which set of reactions will the transition states most resemble products? (f) Notice that the difference in ΔH for the two sets of reactions is the same (6 kcal/mole). Why is this so? (g) The difference in E_{act} for the first set of reactions is relatively small (2.8 kcal/mole). For the second set of reactions, however, the difference in E_{act} is large (5.0 kcal/mole); it is nearly as large as the difference in ΔH. Explain.

		ΔH (KCAL/MOLE)	E_{act} (KCAL/MOLE)
(A)	(1) $CH_3 \cdot + H{-}CH_3 \longrightarrow CH_3{-}H + \cdot CH_3$	0	14.5
	(2) $CH_3 \cdot + H{-}CH_2CH_3 \longrightarrow CH_3{-}H + \cdot CH_2CH_3$	−6.0	11.7
	difference	6.0	2.8
(B)	(1) $Br \cdot + H{-}CH_3 \longrightarrow Br{-}H + \cdot CH_3$	16.5	18.6
	(2) $Br \cdot + H{-}CH_2CH_3 \longrightarrow Br{-}H + \cdot CH_2CH_3$	10.5	13.6
	difference	6.0	5.0

5

ALKENES: STRUCTURE AND SYNTHESIS

5.1 INTRODUCTION

Alkenes are hydrocarbons whose molecules contain the carbon-carbon double bond; they are also called *unsaturated* compounds.

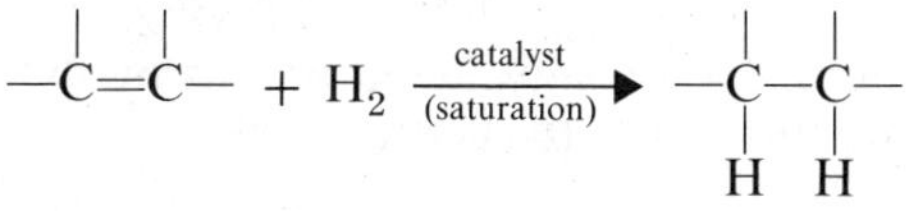

$$-\overset{|}{\underset{|}{C}}=\overset{|}{\underset{|}{C}}- + H_2 \xrightarrow[\text{(saturation)}]{\text{catalyst}} -\overset{|}{\underset{|}{C}}-\overset{|}{\underset{|}{C}}- \\ \qquad\qquad\qquad\qquad\qquad H\ \ H$$

$\qquad$ An alkene $\qquad\qquad\qquad\qquad\qquad$ An alkane

An old name for this family of compounds that is still often used is the name *olefins*. Ethylene, the simplest olefin (alkene), was called olefiant gas (Latin: *oleum*, oil + *facere*, to make) because gaseous ethylene (C_2H_4) reacts with chlorine to form $C_2H_4Cl_2$, a liquid (oil).

$\qquad$ The double bond is a reactive site or a *functional group*; it is the double bond that primarily determines the properties of alkenes.

$\qquad$ Alkenes whose molecules contain only one double bond have the general formula C_nH_{2n}. They are, therefore, isomeric with cycloalkanes. Propene and cyclopropane, for example, have the same molecular formula, C_3H_6, and are structural isomers.

$$CH_3CH=CH_2 \qquad\qquad \overset{\displaystyle CH_2}{\underset{\displaystyle CH_2——CH_2}{\diagup\ \diagdown}}$$

$\qquad$ Propene $\qquad\qquad\qquad$ Cyclopropane
$\qquad$ C_3H_6 $\qquad\qquad\qquad\qquad$ C_3H_6

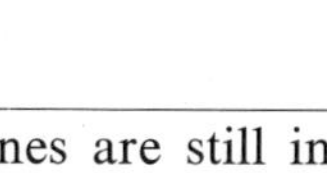

5.2 NOMENCLATURE

Many older names for alkenes are still in common use. Propene is often called propylene, and 2-methylpropene frequently bears the name isobutylene.

$$CH_2=CH_2 \qquad\qquad CH_3CH=CH_2 \qquad\qquad \overset{\displaystyle CH_3}{\overset{|}{CH_3-C=CH_2}}$$

$\qquad$ *IUPAC:* Ethene $\qquad$ *IUPAC:* Propene $\qquad$ *IUPAC:* 2-Methylpropene
$\qquad$ or ethylene* $\qquad$ *Common:* Propylene $\qquad$ *Common:* Isobutylene

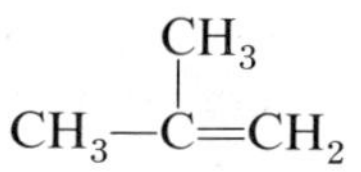

*The IUPAC system also retains the name ethylene.

The IUPAC rules for naming alkenes are similar in many respects to those for naming alkanes:

1. We determine the base name by selecting the longest chain *that contains the double bond* and by changing the ending of the name of the alkane of identical length from *ane* to *ene*. Thus, if this longest chain contains five carbon atoms, the base name for the alkene is *pentene*; if it contains six carbons, the base name is hexene, and so on.

2. We number the chain so as to include both carbons of the double bond, and *we begin numbering at the end of the chain nearest the double bond.* We designate the location of the double bond by using the number of the first atom of the double bond as a prefix:

$$\overset{1}{C}H_2=\overset{2}{C}H\overset{3}{C}H_2\overset{4}{C}H_3 \qquad CH_3CH=CHCH_2CH_2CH_3$$

1-Butene   2-Hexene

3. We indicate the locations of the substituent groups by the number of the carbon atom to which they are attached.

$$\underset{\substack{1\;\;\;\;2\;\;\;3\;\;\;4}}{CH_3-\overset{\overset{\displaystyle CH_3}{|}}{C}=CHCH_3} \qquad \underset{\substack{1\;\;2\;\;\;\;\;3\;\;\;4\;\;\;5\;\;\;6}}{CH_3\overset{\overset{\displaystyle CH_3}{|}}{C}=CHCH_2\overset{\overset{\displaystyle CH_3}{|}}{C}HCH_3}$$

2-Methyl-2-butene   2,5-Dimethyl-2-hexene

$$\underset{\substack{1\;\;\;\;2\;\;\;\;\;3\;\;\;4\;\;|5\;\;6}}{CH_3CH=CHCH_2\overset{\overset{\displaystyle CH_3}{|}}{\underset{\underset{\displaystyle CH_3}{|}}{C}}-CH_3}$$

5,5-Dimethyl-2-hexene

4. We number substituted cycloalkenes in the way that gives the carbon atoms of the double bond the 1- and 2- positions and that also gives the substituent groups the lowest numbers at the first point of difference. With substituted cycloalkenes it is not necessary to specify the position of the double bond. The two examples listed below illustrate the application of these rules.

1-Methylcyclopentene   3,5-Dimethylcyclohexene
(not 2-methylcyclopentene)   (not 4,6-dimethylcyclohexene)

5. Two frequently encountered alkenyl groups are often named as though they were substituents. These groups are the *vinyl group* and the *allyl group*.

$$\overset{H}{\underset{H}{>}}C=C\overset{H}{\underset{\;}{<}} \quad \text{or} \quad CH_2=CH- \qquad \overset{H}{\underset{H}{>}}C=C\overset{H}{\underset{CH_2^-}{<}} \quad \text{or} \quad CH_2=CHCH_2-$$

The vinyl group   The allyl group

The following examples illustrate how these names are employed.

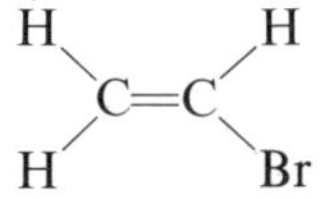

$$\begin{array}{cc} H & H \\ \diagdown \quad \diagup \\ C{=}C \\ \diagup \quad \diagdown \\ H & Br \end{array} \qquad \begin{array}{cc} H & H \\ \diagdown \quad \diagup \\ C{=}C \\ \diagup \quad \diagdown \\ H & CH_2Cl \end{array}$$

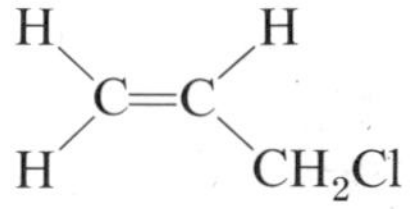

Bromoethene   3-Chloropropene
or   or
Vinyl bromide   Allyl chloride

5.3 ORBITAL HYBRIDIZATION AND THE STRUCTURE OF ALKENES

We have already seen that the quantum-mechanically based concept of orbital hybridization provides satisfactory models for the structures of alkanes, free radicals, and carbocations. We will now find that we can also account for the structure of the carbon-carbon double bond in terms of orbital hybridization and here, as with free radicals and carbocations, the basis for our model is *sp²-hybridized carbon*.

Consider, for example, the simplest alkene, ethene. Ethene is known to be a coplanar molecule; that is, all of the nuclei of ethene lie in the same plane as shown in Fig. 5.1. The H—C—C bond angles of ethene are known to be ~121°, and the H—C—H bond angles are ~118°.

In our model for ethene (Fig. 5.2) we see that two *sp²*-hybridized carbon atoms form a sigma(σ)bond between them by overlapping one *sp²*-orbital from each atom. The remaining *sp²*-orbitals of each carbon form σ bonds to four hydrogens through overlap with the 1s orbitals of the hydrogen atoms. These five bonds account for 10 of the 12 bonding electrons of ethene, and they are called the *σ-bond framework*. The bond angles that we would predict on the basis of *sp²*-hybridized carbon atoms (120° all around) are quite close to the bond angles that are actually found (Fig. 5.2) for ethene.

The two bonding electrons that we have not yet accounted for in our model are located in the *p* orbitals of each carbon. We can better visualize how these *p* orbitals interact with each other if we imagine the σ-bond framework shown in Fig. 5.2. as being rotated perpendicular to the plane of this page. (It is also helpful, in this regard, to replace the σ bonds by lines.) This new perspective is shown in Fig. 5.3. We see that the parallel *p* orbitals *overlap above and below the plane of the*

FIG. 5.1
The structure and bond angles of ethene.

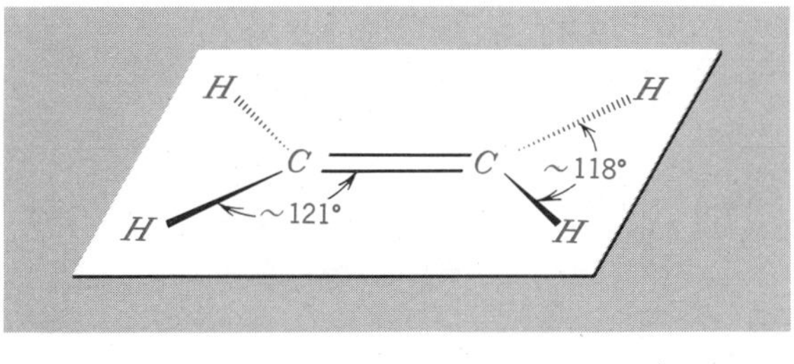

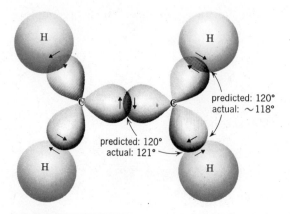

FIG. 5.2
The sigma-bond framework for ethene based on
sp²-*hybridized carbon atoms.*

σ framework. This sideways overlap of the *p* orbitals results in a new type of covalent bond, known as a *pi(π)bond.**

According to molecular orbital theory both bonding and antibonding π-molecular orbitals are formed when *p* orbitals interact in this way to form a π bond. The bonding orbital (Fig. 5.4) results when *p* orbital lobes of like sign overlap; the antibonding orbital is formed when *p* orbital lobes of opposite signs overlap.

The bonding orbital is the lower energy orbital and contains both electrons (with opposite spins) in the ground state of the molecule. The region of greatest probability of finding the electrons in the bonding orbital is a region generally situated between the two carbon atoms above and below the plane of the σ-bond framework. The antibonding orbital is of high energy, and it is not occupied by electrons when the molecule is in the ground state. It can become occupied, however, if the molecule absorbs light of the right frequency, and an electron is promoted from the lower energy level to the higher one. The antibonding orbital has a nodal plane between the two carbon atoms.

To summarize: In our model based on orbital hybridization, the carbon-carbon double bond is seen as consisting of two different kinds of bonds, *a σ bond and a π bond.* The σ bond is formed by two overlapping *sp²* orbitals and is symmet-

---

* Note the difference in shape of the bonding molecular orbital of a π bond as contrasted to that of a σ bond. A σ bond has cylindrical symmetry about a line connecting the two bonded nuclei. A π bond has a shape similar to that of a *p* orbital. In each case it is the shape of the bonding molecular orbital that is related to the name.

FIG. 5.3
The overlapping p *orbitals of ethene.*

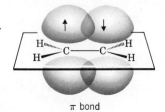

π bond

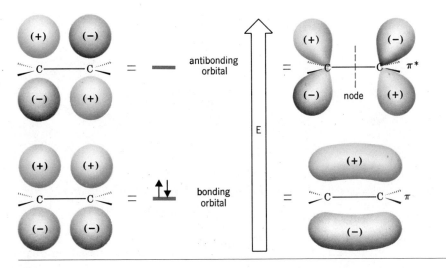

FIG. 5.4

The interaction of p *orbitals to form the bonding and antibonding molecular orbitals of a π bond.*

rical about an axis linking the two carbon atoms. The π bond is formed by a sideways overlap of two *p* orbitals; it has a shape similar to that of a *p* orbital. In the ground state the electrons of the π bond are located between the two carbon atoms but generally above and below the plane of the σ-bond framework.

Our model also accounts for the fact that the carbon-carbon double bond is shorter than the carbon-carbon single bond (Fig. 5.5). We explain the shorter carbon-carbon double bond primarily in terms of the closer proximity of the carbon atoms required for effective overlap of the *p* orbitals.

We can also explain the fact that the carbon-hydrogen bonds of ethene are slightly shorter than those of ethane. The sp^2 orbitals of ethene contain less *p* character than the sp^3 orbitals of ethane. As a consequence, sp^2 orbitals are shorter than sp^3 orbitals, and carbon-hydrogen bonds formed by sp^2-orbitals are shorter as well (Fig. 5.6).

The σ-π model for the carbon-carbon double bond also accounts for a property of the double bond that we encountered in Chapter 2: *the large barrier to free rotation possessed by groups joined by a double bond*. Maximum overlap between the *p* orbitals of a π bond occurs when *p* orbitals are exactly parallel. Rotating one carbon of the

FIG. 5.5

The bond distances of ethene (a), *and ethane* (b).

1.34 Å

|←1.54 Å→|

1.10 Å

H

H

H

H⋯C — C⋯H —1.09 Å
H   H

H⋯C   C⋯H
H   H

(a)

(b)

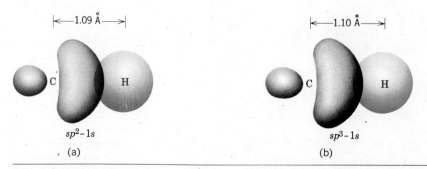

$|\leftarrow\!\!-1.09\ \text{Å}\!\!-\!\!\rightarrow|$ $|\leftarrow\!\!\!-1.10\ \text{Å}\!\!-\!\!\rightarrow|$

C   H   C   H

sp^2-1s sp^3-1s

(a)   (b)

FIG. 5.6
Carbon-hydrogen bonds formed by sp^2 *(a), and* sp^3 *(b) carbon atoms.*

double bond 90° (Fig. 5.7) breaks the π bond, for then the p orbitals are perpendicular and there is no overlap between them. Estimates based on quantum mechanical calculations indicate that the strength of the π bond is 60 kcal/mole. This, then, is the barrier to rotation of the double bond as well, and it contrasts markedly with the rotational barrier of groups joined by carbon-carbon single bonds (3-6 kcal/ mole). As a result, groups joined by single bonds rotate freely at room temperature while those joined by double bonds do not. This also explains why the separation of *anti-* and *gauche* isomers of butane (p. 98) is *impossible* at room temperature while the separation of *cis-* and *trans*-2-butene is not only possible but why, once separated, *cis* and *trans* isomers can be stored indefinitely at the normal laboratory temperature. *Cis-trans* isomers *can* be interconverted, but only if they are heated (to temperatures usually greater than 300°), for under these conditions the molecules acquire sufficient thermal energy to surmount the rotational barrier. *Cis-trans-* isomers can also be interconverted by exposure to light of a wavelength absorbed by the double bond or by chemical means.

Problem 5.1

Can you explain how the light-catalyzed interconversion of *cis-trans* isomers might occur?

The σ bond of ethene has been estimated to have a bond strength of 85 kcal/ mole. Taken together, the σ bond and π bond give a total strength for the double bond of ~145 kcal/mole.

FIG. 5.7
Rotation of a carbon atom of a double bond through an angle of 90° results in the breaking of the π *bond.*

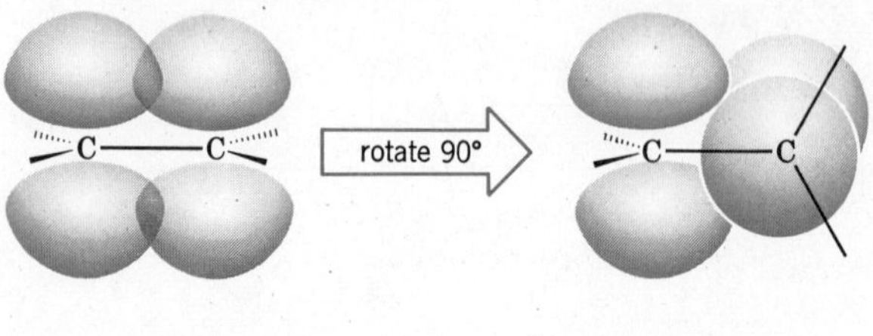

C——C   rotate 90°   C——C

σ-bond strength $\simeq$ 85 kcal/mole

π-bond strength $\simeq$ 60 kcal/mole

Double-bond strength $\simeq$ 145 kcal/mole

This is the bond strength for a homolytic scission of the ethene molecule, that is,

$$CH_2\text{:}\,|\,\text{:}CH_2 \longrightarrow 2CH_2\text{:} \quad \Delta H = 145 \text{ kcal/mole}$$

In the next chapter, when we discuss the reactions of alkenes, we will find that the relative weakness of the π bond ($\sim$60 kcal/mole) *and its exposed electrons* are important in explaining the reactivity of alkenes.

5.4 HEATS OF HYDROGENATION: THE STABILITY OF ALKENES

The reaction between alkenes and hydrogen (hydrogenation) can often be used to determine their relative stabilities. Hydrogenation of an alkene is an exothermic reaction, and most alkenes show heats of hydrogenation of approximately 30 kcal/mole. Individual alkenes, however, have heats of hydrogenation that differ from this average value by more than 2 kcal/mole.

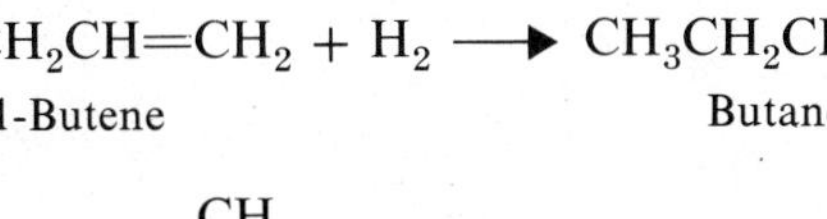

$$\mathrm{C{=}C} + H{-}H \longrightarrow {-}\underset{H}{\overset{|}{C}}{-}\underset{H}{\overset{|}{C}}{-} \quad \Delta H \simeq -30 \text{ kcal/mole}$$

These differences allow us to measure the relative stabilities of alkenes *when hydrogenation converts them to the same product.*

Consider, as examples, the three butene isomers shown below:

$$CH_3CH_2CH{=}CH_2 + H_2 \longrightarrow CH_3CH_2CH_2CH_3 \quad \Delta H = -30.3 \text{ kcal/mole}$$

1-Butene   Butane

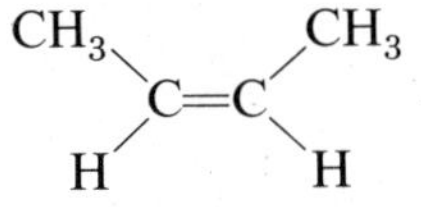

$$\underset{H}{\overset{CH_3}{\diagup}}C{=}C\underset{H}{\overset{CH_3}{\diagdown}} + H_2 \longrightarrow CH_3CH_2CH_2CH_3 \quad \Delta H = -28.6 \text{ kcal/mole}$$

cis-2-Butene   Butane

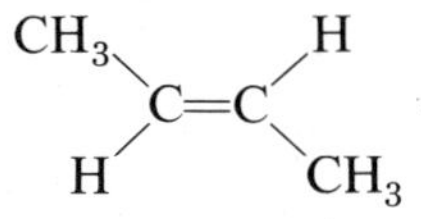

$$\underset{H}{\overset{CH_3}{\diagup}}C{=}C\underset{CH_3}{\overset{H}{\diagdown}} + H_2 \longrightarrow CH_3CH_2CH_2CH_3 \quad \Delta H = -27.6 \text{ kcal/mole}$$

trans-2-Butene   Butane

In each case the product of the reaction (butane) is the same. In each case, too, one of the reactants (hydrogen) is the same. Different amounts of *heat* are evolved in each reaction, however, and these differences must be related to different relative stabilities (different heat contents) of the individual butenes. 1-Butene evolves the greatest amount of heat when hydrogenated; and *trans*-2-butene evolves the least. Therefore 1-butene must have the greatest potential energy and be the least stable

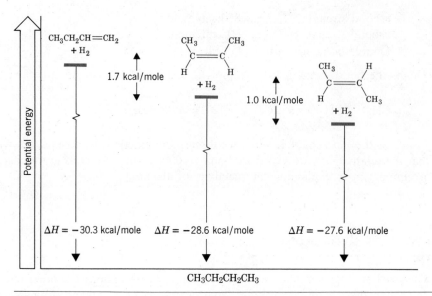

FIG. 5.8
A potential energy diagram for the three butene isomers. The order of stability is trans-*2-butene* > cis-*2-butene* > *1-butene.*

isomer. *Trans*-2-butene must have the lowest potential energy and be the most stable isomer. The potential energy (and stability) of *cis*-2-butene falls in between. The order of stabilities of the butenes is easier to see if we examine the potential energy diagram in Fig. 5.8.

The greater stability of the *trans*-2-butene when compared to *cis*-2-butene illustrates a general pattern found in *cis-trans* alkene pairs. The 2-pentenes, for example, show the same stability relationship: *trans* isomer > *cis* isomer.

$$\begin{array}{c}CH_3CH_2 \quad\quad CH_3 \\ \diagdown\;\; C=C\;\;\diagup \\ H \quad\quad\quad H \end{array} + H_2 \longrightarrow CH_3CH_2CH_2CH_2CH_3 \quad\quad \Delta H = -28.6 \text{ kcal/mole}$$

cis-2-Pentene   Pentane

$$\begin{array}{c}CH_3CH_2 \quad\quad H \\ \diagdown\;\; C=C\;\;\diagup \\ H \quad\quad\quad CH_3 \end{array} + H_2 \longrightarrow CH_3CH_2CH_2CH_2CH_3 \quad\quad \Delta H = -27.6 \text{ kcal/mole}$$

trans-2-Pentene   Pentane

The greater potential energy of *cis* isomers can be attributed to strain caused by the crowding of two alkyl groups on the same side of the double bond (Fig. 5.9)—a strain that is relieved when the molecule becomes saturated and therefore more flexible.

Studies of numerous alkenes also reveal a pattern of stabilities that is related to the number of alkyl groups attached to the carbon atoms of the double bond. *The greater the number of attached alkyl groups, (i.e., the more highly substituted is*

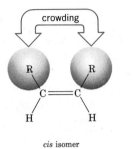

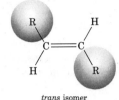

FIG. 5.9

Cis-*and* trans-*Alkene isomers. The less stable* cis *isomer has greater strain.*

the double bond) *the greater is the alkene's stability.* We can represent this order of stabilities in general terms, as follows:*

Relative stabilities of alkenes

$$\underset{\text{Tetrasubstituted}}{\overset{R}{\underset{R}{>}}C=C\overset{R}{\underset{R}{<}}} > \underset{\text{Trisubstituted}}{\overset{R}{\underset{R}{>}}C=C\overset{R}{\underset{H}{<}}} >$$

$$\underset{\text{Disubstituted}}{\overset{R}{\underset{H}{>}}C=C\overset{H}{\underset{R}{<}}} > \overset{R}{\underset{H}{>}}C=C\overset{R}{\underset{H}{<}} \simeq \overset{R}{\underset{R}{>}}C=C\overset{H}{\underset{H}{<}} > \underset{\text{Monosubstituted}}{\overset{R}{\underset{H}{>}}C=C\overset{H}{\underset{H}{<}}} > \underset{\text{Unsubstituted}}{\overset{H}{\underset{H}{>}}C=C\overset{H}{\underset{H}{<}}}$$

A portion of this overall pattern can be seen by inspecting the heats of hydrogenation of the following pentenes.

$$\overset{CH_3}{\underset{CH_3}{>}}C=C\overset{H}{\underset{CH_3}{<}} \qquad \overset{CH_3CH_2}{\underset{CH_3}{>}}C=C\overset{H}{\underset{H}{<}} \qquad \overset{CH_3\overset{|}{C}H}{\underset{H}{>}}C=C\overset{H}{\underset{H}{<}}$$

$$\Delta H = -26.9 \text{ kcal/mole} \qquad \Delta H = -28.5 \qquad \Delta H = -30.3 \text{ kcal/mole}$$

$$\underset{\text{Trisubstituted}}{\overset{R}{\underset{R}{>}}C=C\overset{H}{\underset{R}{<}}} > \underset{\text{Disubstituted}}{\overset{R}{\underset{R}{>}}C=C\overset{H}{\underset{H}{<}}} > \underset{\text{Monosubstituted}}{\overset{R}{\underset{H}{>}}C=C\overset{H}{\underset{H}{<}}}$$

Direct comparisons can be made for these pentenes because hydrogenation of each yields the same alkane, isopentane.

$$\underset{\overset{|}{CH_3}}{CH_3CH-CH_2CH_3}$$

Isopentane

* This order of stabilities may seem contradictory when compared with explanation given for the relative stabilities of *cis* and *trans* isomers. Although a detailed explanation of the trend given here is beyond our scope, the relative stabilities of substituted alkenes can be rationalized. Part of the explanation can be given in terms of the electron releasing effect of alkyl groups; an effect that satisfies the electron-withdrawing properties of the sp^2-hybridized carbons of the double bond.

Problem 5.2

Another C_4H_8 isomer, 2-methylpropene, cannot be compared directly with the butene isomers that we have already discussed (*cis*-2-butene, *trans*-2-butene, and 1-butene) because 2-methylpropene yields a different product on hydrogenation. (a) What is this product? (b) What products would all four C_4H_8 isomers yield on complete combustion? (c) Could heats of combustion be used to determine the relative stabilities of these four C_4H_8 isomers? (d) What are two other C_4H_8 isomers? (e) Could their relative stabilities be determined in the same way?

5.5 CYCLOALKENES

Rings containing six carbon atoms or fewer exist only in the *cis* form (Fig. 5.10). The introduction of a *trans* double bond into rings this small would, if it were possible, introduce greater strain than the bonds of the ring atoms could accommodate. If it existed, *trans*-cyclohexene might resemble the structure shown in Fig. 5.11.

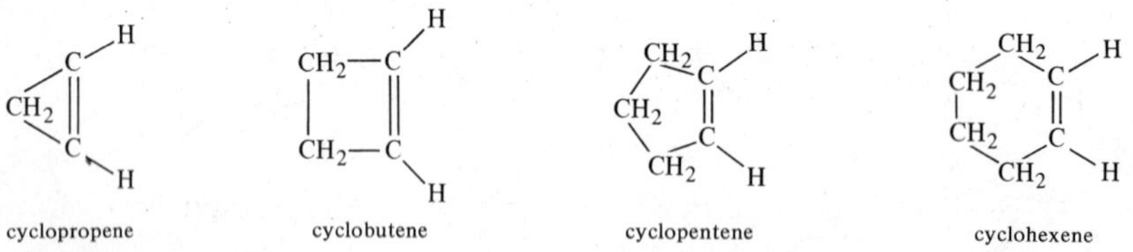

cyclopropene · cyclobutene · cyclopentene · cyclohexene

FIG. 5.10
Cis-*cycloalkenes.*

Trans-cycloheptene has been observed with instruments called spectrometers, but it is a substance with a very short lifetime and has not been isolated.

Trans-cyclooctene (Fig. 5.12) has been isolated, however. Here the ring is large enough to accommodate the geometry required by a *trans* double bond.

FIG. 5.11
Hypothetical trans-*cyclohexene. This molecule is apparently too highly strained to exist.*

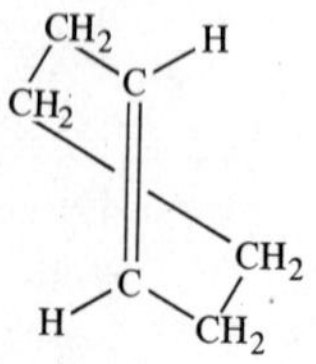

cis–cyclooctene · *trans*–cyclooctene

FIG. 5.12
Cis- *and* trans- *forms of cyclooctene.*

5.6 SYNTHESIS OF ALKENES: ELIMINATION REACTIONS

The most widely used methods for synthesizing alkenes are based on elimination reactions of the general type shown below.

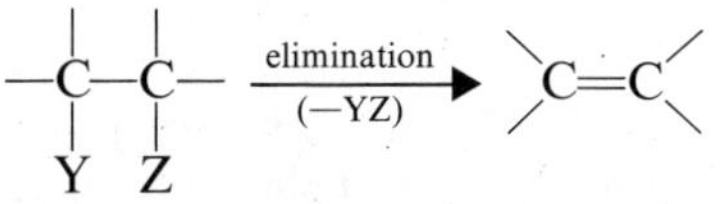

$$-\underset{Y}{\underset{|}{C}}-\underset{Z}{\underset{|}{C}}- \xrightarrow[\text{(—YZ)}]{\text{elimination}} \quad \overset{}{C}=\overset{}{C}$$

Alkene

In these reactions the pieces of some molecule, YZ, are eliminated from adjacent carbons of the reactant. Y and Z may be the same as, for example, in *dehydrogenation* and *dehalogenation*. Or Y and Z may be different as in *dehydration* or *dehydrohalogenation*.

Dehydrogenation

General: $\text{R}-\underset{H}{\underset{|}{\text{CH}}}-\underset{H}{\underset{|}{\text{CH}}}-\text{R} \xrightarrow[\text{(—H}_2)]{\overset{\text{catalyst}}{\text{heat}}} \text{RCH}=\text{CHR}$

Specific example: $\text{CH}_3\text{CH}_2\text{CH}_2\text{CH}_3 \xrightarrow[\text{(—H}_2)]{\overset{\text{Cr}_2\text{O}_3 \cdot \text{Al}_2\text{O}_3}{500°}}$

$$\underset{H}{\overset{CH_3}{\diagdown}}C=C\underset{CH_3}{\overset{H}{\diagup}} \quad + \quad \text{other butenes}$$

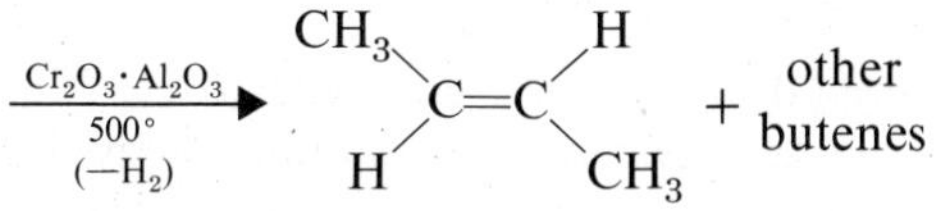

Dehalogenation

General: $\text{R}-\underset{X}{\underset{|}{\text{CH}}}-\underset{X}{\underset{|}{\text{CH}}}-\text{R} \xrightarrow[\text{(—ZnX}_2)]{\overset{\text{Zn}}{\text{acetone}}} \text{R}-\text{CH}=\text{CH}-\text{R} \quad (X = \text{halogen})$

Specific example:

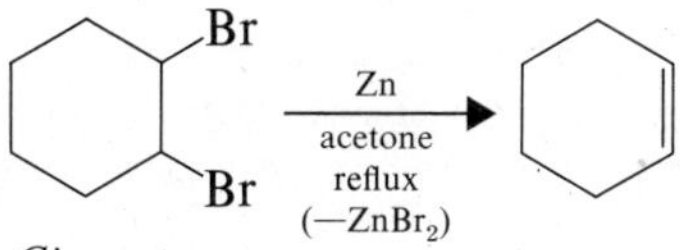

Cis or *trans*

Dehydration

General: $\text{R}-\underset{H}{\underset{|}{\text{CH}}}-\underset{OH}{\underset{|}{\text{CH}}}-\text{R} \xrightarrow[\text{(—H}_2\text{O})]{\overset{\text{acid}}{\text{heat}}} \text{R}-\text{CH}=\text{CH}-\text{R}$

Specific example: $\text{CH}_3\text{CH}_2\text{OH} \xrightarrow[\text{(—H}_2\text{O})]{\overset{\text{conc. H}_2\text{SO}_4}{180°}} \text{CH}_2=\text{CH}_2$

Dehydrohalogenation

General: $\text{R}-\underset{H}{\underset{|}{\text{CH}}}-\underset{X}{\underset{|}{\text{CH}}}-\text{R} \xrightarrow[\text{(—HX)}]{\overset{\text{base}}{\text{heat}}} \text{R}-\text{CH}=\text{CH}-\text{R} \quad (X = \text{halogen})$

Specific example: $\text{CH}_3\underset{Br}{\underset{|}{\text{CHCH}}}_3 \xrightarrow[\substack{55° \\ \text{(—HBr)}}]{\overset{\text{C}_2\text{H}_5\text{O}^- \text{Na}^+}{\text{C}_2\text{H}_5\text{OH}}} \underset{(79\%)}{\text{CH}_2=\text{CHCH}_3}$

In this chapter we will examine in detail the four methods for alkene synthesis given above. In Chapters 17 and 18 we will discuss other methods for alkene synthesis that involve elimination reactions, and in Chapter 9 we will see how alkenes can be prepared through *an addition reaction*—through the addition of one mole of hydrogen to an alkyne.

5.7 HYDROGENATION AND DEHYDROGENATION: THE FUNCTION OF THE CATALYST

We have just seen (Section 5.4) that alkenes can be converted to alkanes by hydrogenation. The reverse of a hydrogenation reaction is called *dehydrogenation*. Hydrogenation of an alkene is an exothermic reaction ($\Delta H \cong -30$ kcal/mole). Dehydrogenation of the same alkene is *endothermic* and by the same amount.

$$\text{R—CH=CH—R} + \text{H}_2 \underset{\text{dehydrogenation}}{\overset{\text{hydrogenation}}{\rightleftharpoons}} \text{R—CH}_2\text{—CH}_2\text{—R} + \text{heat}$$

Both hydrogenation and dehydrogenation reactions usually have high energies of activation. The reaction of an alkene with molecular hydrogen does not take place at room temperature in the absence of a catalyst, but often *does* take place at room temperature when a metal catalyst is added. The catalyst provides a reaction pathway with a *lower energy of activation* (Fig. 5.13).

The most commonly used hydrogenation catalysts are finely divided platinum, nickel, palladium, rhodium and ruthenium. (These same catalysts are often used in carrying out dehydrogenation reactions.) In a hydrogenation reaction the metal

FIG. 5.13

Potential energy diagram for the hydrogenation of an alkene in the presence of and in the absence of a catalyst. The energy of activation for the uncatalyzed reaction [$E_{act(1)}$] is very much larger than the largest energy of activation for the catalyzed reaction [$E_{act(2)}$].

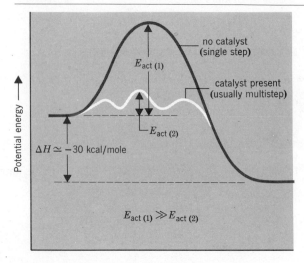

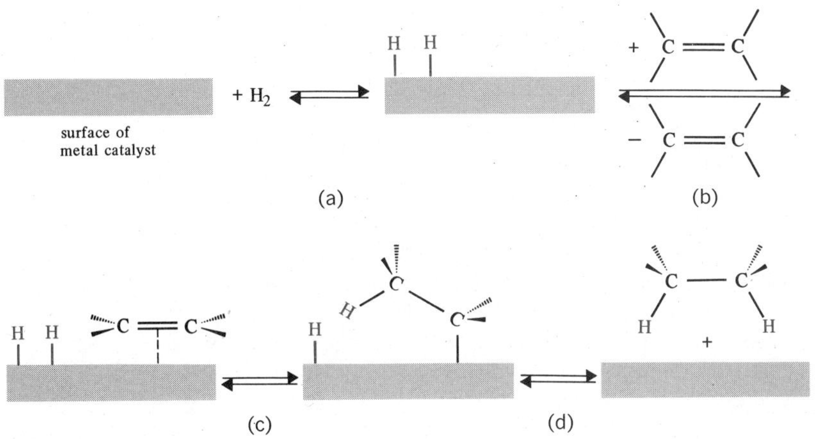

FIG. 5.14

The mechanism for the hydrogenation of an alkene (forward reaction) and the dehydrogenation of an alkane (reverse reaction) as catalyzed by finely divided platinum metal.

catalyst apparently serves to adsorb hydrogen molecules on its surface. This adsorption of hydrogen is essentially a chemical reaction; unpaired electrons on the surface of the metal *pair* with the electrons of hydrogen (Fig. 5.14*a*) and bind the hydrogen to the surface. The collision of an alkene with the surface-bearing adsorbed hydrogen causes adsorption of the alkene as well (Fig. 5.14*b*). A stepwise transfer of hydrogen atoms takes place, and this produces an alkane before the organic molecule leaves the catalyst surface (Fig. 5.14*c* and *d*). As a consequence, the addition of hydrogen to an alkene is usually a *syn addition; that is, both hydrogens usually add from the same side of the molecule.*

The mechanism for a dehydrogenation reaction is simply the reverse of that of a hydrogenation reaction. The overall direction of this reversible reaction can often be determined by the experimental conditions we employ. We usually carry out hydrogenation with a large excess of hydrogen and at lower temperatures. We bring about dehydrogenation by using higher temperatures because the reaction is endothermic. In dehydrogenations we usually sweep the reaction mixture with an inert gas (nitrogen) to remove the hydrogen as it is formed. This minimizes the concentration of hydrogen in the reaction flask and on the catalyst surface.

Reversible hydrogenation-dehydrogenation reactions occur in living cells as well as in the apparatus of an organic laboratory. An example of a biochemical hydrogenation-dehydrogenation equilibrium is the one catalyzed by the enzyme, *succinate dehydrogenase* (Fig. 5.15).

Enzymes are very large complex protein molecules that catalyze biochemical reactions. The enzyme *succinate dehydrogenase* has a molecular weight of 175,000. *Succinate dehydrogenase* also contains four atoms of iron, and it appears likely that there is an $Fe^{++} \rightleftharpoons Fe^{+++}$ oxidation-state change as the enzyme is oxidized and reduced. The cell's reliance on enzymatically bound metals, shows a remarkable resemblance to the chemist's reliance on catalytic metals in laboratory hydrogenations and dehydrogenations.

In theory, dehydrogenation of succinic acid could produce either the *trans* isomer, fumaric acid, or the *cis* isomer, maleic acid.

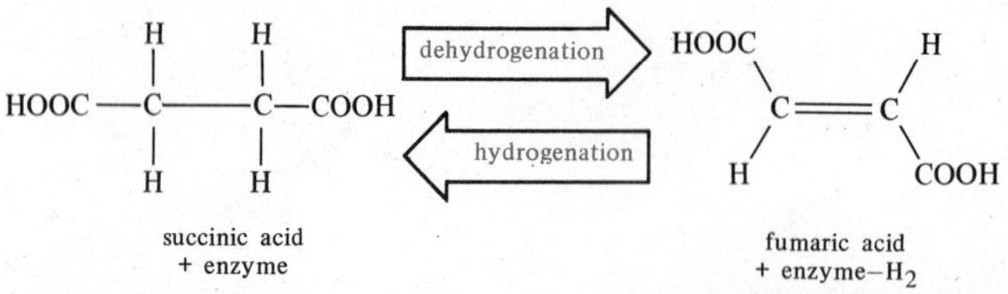

$$HOOC-\overset{\overset{\displaystyle H}{|}}{\underset{\underset{\displaystyle H}{|}}{C}}-\overset{\overset{\displaystyle H}{|}}{\underset{\underset{\displaystyle H}{|}}{C}}-COOH \underset{\xleftarrow{\text{hydrogenation}}}{\xrightarrow{\text{dehydrogenation}}} \overset{HOOC}{\underset{H}{}}C=C\overset{H}{\underset{COOH}{}}$$

succinic acid   fumaric acid
+ enzyme   + enzyme—H_2

FIG. 5.15

The succinic dehydrogenase *equilibrium. The enzyme serves as a hydrogen acceptor in the dehydrogenation reaction and, as a result, a reduced form of the enzyme (enzyme—H_2) is produced. In the hydrogenation reaction hydrogens are transferred to fumaric acid producing succinic acid and the oxidized enzyme.*

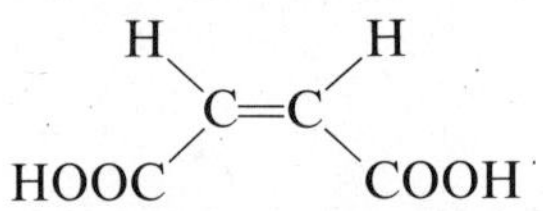

$$\overset{H}{\underset{HOOC}{}}C=C\overset{H}{\underset{COOH}{}}$$

Maleic acid

The enzymatically catalyzed dehydrogenation reaction produces the single isomer, fumaric acid. Moreover, the enzymatically catalyzed hydrogenation reaction utilizes only fumaric acid to produce succinic acid. This type of specificity is a general rule in biochemical reactions.

5.8 SYNTHESIS OF ALKENES BY DEHYDRATION OF ALCOHOLS

Most alcohols dehydrate and form alkenes when they are heated in the presence of a strong acid. The most commonly used acids are proton donors such as sulfuric acid and phosphoric acid, or Lewis acids such as alumina (Al_2O_3).

Dehydration reactions of alcohols show several important characteristics:

1. The experimental conditions—temperature and acid concentration—that are required to bring about dehydration are closely related to the structure of individual alcohols. Alcohols in which the hydroxyl group is attached to a primary carbon (primary alcohols) are the most difficult to dehydrate. Dehydration of ethyl alcohol, for example, requires concentrated sulfuric acid and a temperature of 180°.

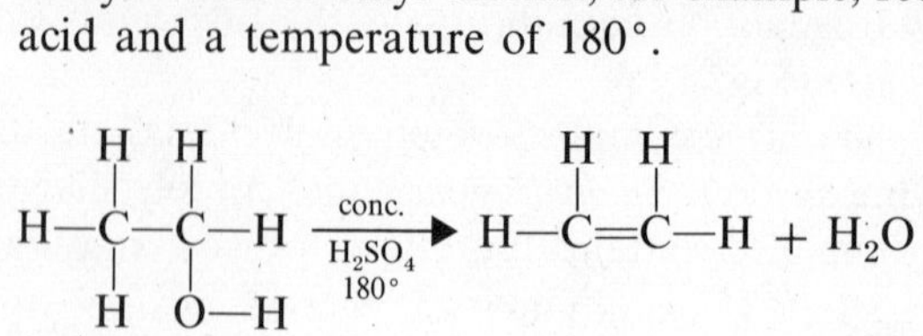

$$H-\overset{\overset{\displaystyle H}{|}}{\underset{\underset{\displaystyle H}{|}}{C}}-\overset{\overset{\displaystyle H}{|}}{\underset{\underset{\displaystyle O-H}{|}}{C}}-H \xrightarrow[\substack{H_2SO_4 \\ 180°}]{\text{conc.}} H-\overset{\overset{\displaystyle H}{|}}{C}=\overset{\overset{\displaystyle H}{|}}{C}-H + H_2O$$

Ethyl alcohol
(a 1° alcohol)

Dehydration of *n*-propyl alcohol, also a primary alcohol, requires concentrated acid at a similarly high temperature.

$$CH_3CH_2CH_2OH \xrightarrow[170°]{\text{conc.} \atop H_2SO_4} CH_3CH{=}CH_2 + H_2O$$

n-Propyl alcohol   Propene

Secondary alcohols usually dehydrate under milder conditions. Cyclohexanol, for example, dehydrates in 85% phosphoric acid at 165 to 170°

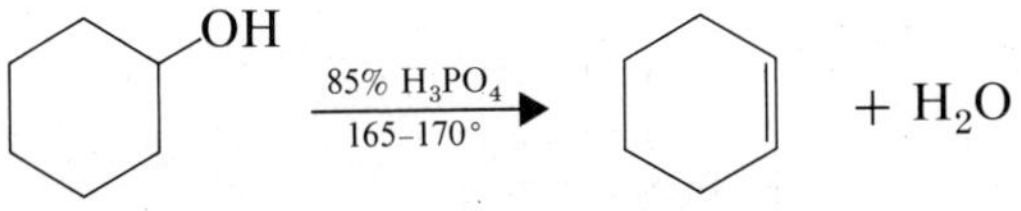

Cyclohexanol   Cyclohexene
                   (80%)

Tertiary alcohols are usually so easily dehydrated that extremely mild conditions can be used. *Tert*-butyl alcohol, for example, dehydrates in 20% aqueous sulfuric acid at a temperature of 85°.

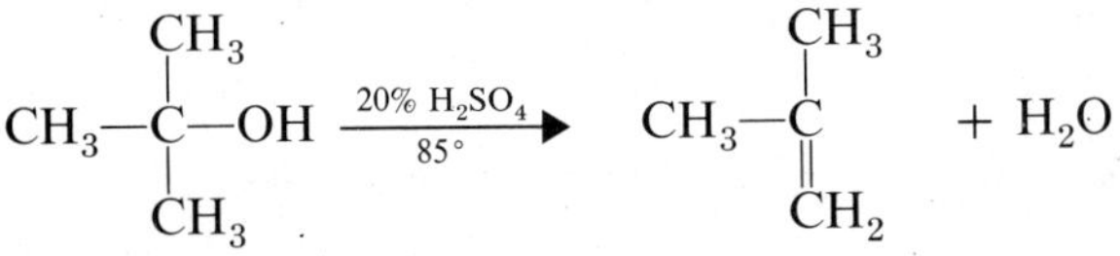

tert-Butyl   2-Methylpropene
alcohol   (84%)

Thus, overall, the relative ease with which alcohols undergo dehydration shows the following order.

Ease of dehydration:

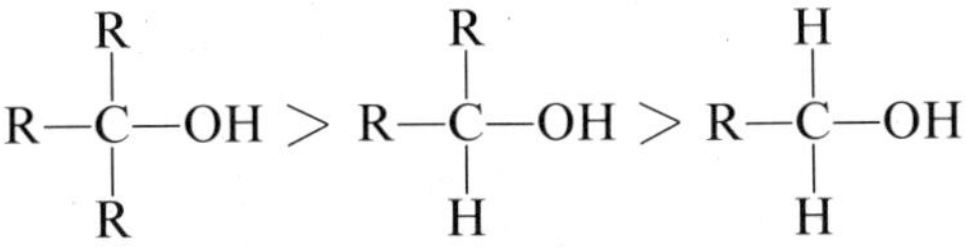

       3° Alcohol   2° Alcohol   1° Alcohol

2. Certain alcohols dehydrate to give more than one product: when 1-butanol dehydrates, for example, the reaction gives three products: *trans*-2-butene, *cis*-2-butene, and 1-butene.

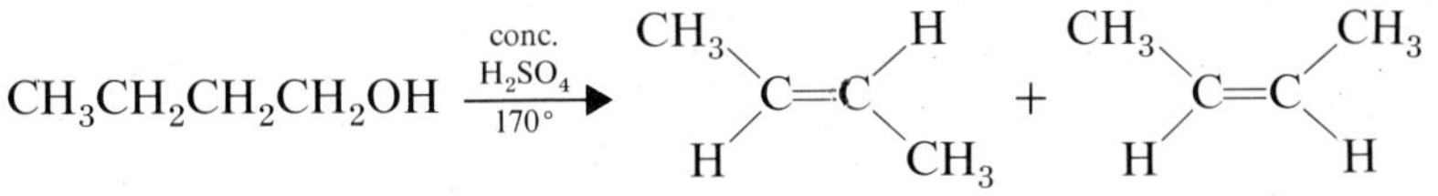

$$CH_3CH_2CH_2CH_2OH \xrightarrow[170°]{\text{conc.} \atop H_2SO_4}$$

trans-2-Butene   *cis*-2-Butene
(major product)   (minor product)

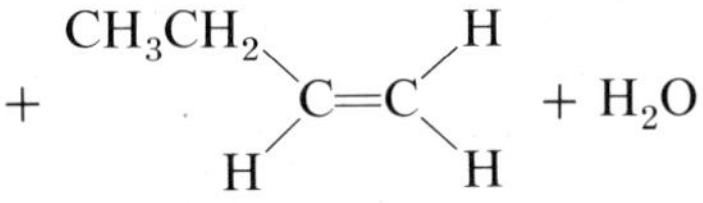

               1-Butene
            (minor product)

The dehydration of 2-methylcyclohexanol gives both 1-methylcyclohexene and 3-methylcyclohexene.

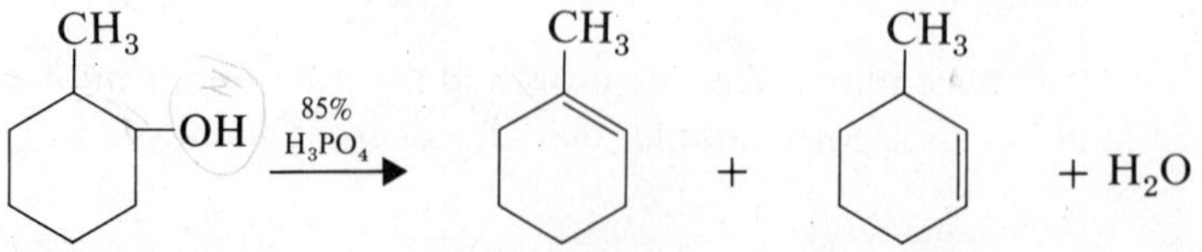

| 2-Methyl-cyclohexanol | 1-Methyl-cyclohexene (major product) | 3-Methyl-cyclohexene (minor product) |

3. Some primary and secondary alcohols also undergo *rearrangement of their carbon skeleton* when they are subjected to dehydration. Such a rearrangement occurs in the dehydration of 3,3-dimethyl-2-butanol.

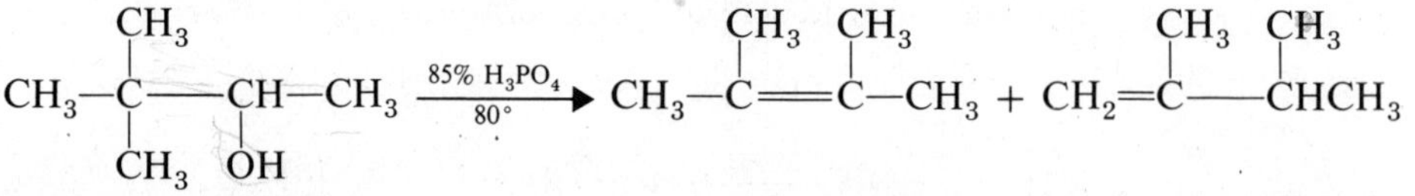

3,3-Dimethyl-2-butanol   2,3-Dimethyl-2-butene (80%)   2,3-Dimethyl-1-butene (20%)

Notice that the carbon skeleton of the reactant is

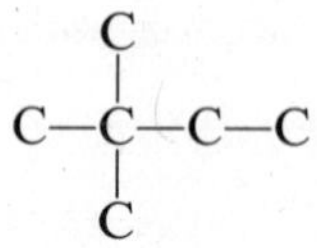

while that of the products is

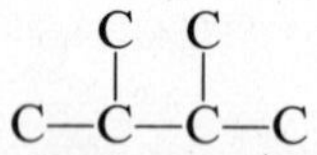

Explanations for all of these observations can be given on the basis of a mechanism originally postulated by Professor F. Whitmore of Pennsylvania State University. The mechanism is a stepwise one, and it is useful to consider each step in some detail. We use the dehydration of *tert*-butyl alcohol as an example.

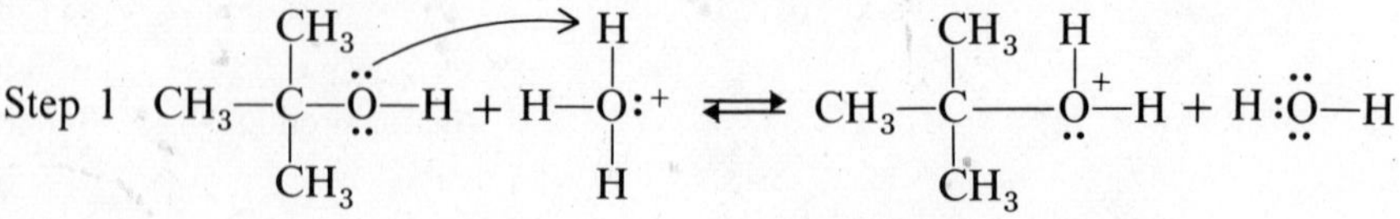

Step 1

Protonated alcohol
or oxonium ion

The first step, an acid-base reaction, involves the transfer of a proton from the acid to one of the unshared electron pairs of the alcohol. In dilute sulfuric acid the acid is a hydronium ion; in concentrated sulfuric acid the acid is sulfuric acid itself. The protonated alcohol is an *oxonium* ion. The structure of the protonated alcohol is similar to that of a hydronium ion (H_3O^+), but it has an alkyl group in the place of one hydrogen.

$$R-\overset{\overset{\displaystyle H}{|}}{\underset{\cdot\cdot}{O}}{}^{\pm}\!-H \qquad H-\overset{\overset{\displaystyle H}{|}}{\underset{\cdot\cdot}{O}}{}^{\pm}\!-H$$

Protonated alcohol   Hydronium ion

The presence of the positive charge on the oxygen of the protonated alcohol weakens all bonds from oxygen including the carbon-oxygen bond and in step 2 the carbon-oxygen bond breaks.

Step 2
$$CH_3-\overset{\overset{\displaystyle CH_3}{|}}{\underset{\underset{\displaystyle CH_3}{|}}{C}}\frown\overset{\overset{\displaystyle H}{|}}{\underset{\cdot\cdot}{O}}{}^{\pm}\!-H \rightleftharpoons CH_3-\overset{\overset{\displaystyle CH_3}{|}}{\underset{\underset{\displaystyle CH_3}{|}}{C}}{}^{+} \;+\; :\overset{\overset{\displaystyle H}{|}}{\underset{\cdot\cdot}{O}}-H$$

A carbocation

The carbon-oxygen bond breaks *heterolytically*. The bonding electrons depart with the water molecule and leave behind a carbocation.* The carbocation is, of course, highly reactive because it bears a positive charge and because the central carbon atom has only six electrons around it.

Finally, in step 3, the carbocation stabilizes itself by transferring a proton to a molecule of water. The result is the formation of a hydronium ion and an alkene.

Step 3
$$CH_3-\overset{\overset{\displaystyle H-\overset{\overset{\displaystyle H}{|}}{\underset{\displaystyle}{C}}-H}{}}{\underset{\underset{\displaystyle CH_3}{|}}{C}}{}^{+} \;+\; :\overset{}{\underset{\cdot\cdot}{O}}-H \rightleftharpoons CH_3-\overset{\overset{\displaystyle CH_2}{\|}}{\underset{\underset{\displaystyle CH_3}{|}}{C}} \;+\; H-\overset{\overset{\displaystyle H}{|}}{\underset{\cdot\cdot}{O}}{}^{\pm}\!-H$$

2-Methylpropene

Step 3 is also an acid-base reaction, and in it any one of the nine protons of the three methyl groups can be transferred to a molecule of water. The electron pair that bonded the proton to carbon in the carbocation becomes the second bond of the double bond of the alkene.

By itself, the Whitmore mechanism does not explain the observed order of reactivity of alcohols: tertiary > secondary > primary. Taken alone, it does not explain the formation of more than one product in the dehydration of certain

---

* Notice that this type of reaction is different from the radical reactions that we saw in Chapter 4. In this reaction a *pair* of electrons leaves with the oxygen atom.

$$CH_3-\overset{\overset{\displaystyle CH_3}{|}}{\underset{\underset{\displaystyle CH_3}{|}}{C}}\frown\overset{\overset{\displaystyle H}{|}}{\underset{\cdot\cdot}{O}}{}^{\pm}\!-H \xrightarrow[\text{cleavage}]{\text{heterolytic}} CH_3-\overset{\overset{\displaystyle CH_3}{|}}{\underset{\underset{\displaystyle CH_3}{|}}{C}}{}^{+} \;+\; :\overset{\overset{\displaystyle H}{|}}{\underset{\cdot\cdot}{O}}-H$$

In our illustrations a double-barbed arrow, $\frown$, indicates the movement of a pair of electrons.

alcohols nor the occurrence of a rearranged carbon skeleton in the dehydration of others. But, when coupled with what is known about *the stability of carbocations*, the Whitmore mechanism *does* eventually account for all of these observations.

5.9 THE STABILITY OF CARBOCATIONS

A large body of experimental evidence indicates that the relative stabilities of carbocations parallel those of free radicals. Tertiary carbocations are the most stable, and the methyl cation is the least stable. The overall order of stability is

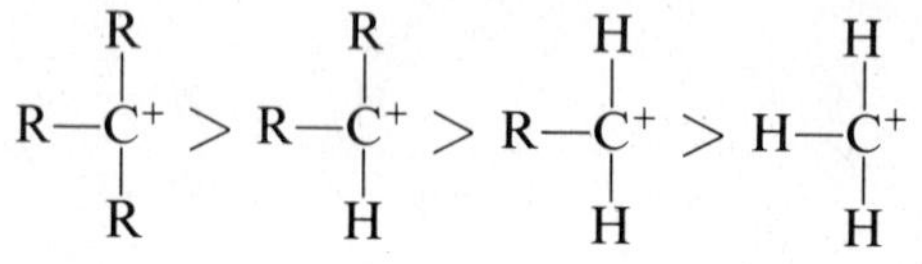

$$
\underset{\substack{\text{R} \\ | \\ \text{R}-\overset{\displaystyle |}{\underset{\displaystyle |}{\text{C}}}{}^{+} \\ \text{R}}}{} >
\underset{\substack{\text{R} \\ | \\ \text{R}-\overset{\displaystyle |}{\underset{\displaystyle |}{\text{C}}}{}^{+} \\ \text{H}}}{} >
\underset{\substack{\text{H} \\ | \\ \text{R}-\overset{\displaystyle |}{\underset{\displaystyle |}{\text{C}}}{}^{+} \\ \text{H}}}{} >
\underset{\substack{\text{H} \\ | \\ \text{H}-\overset{\displaystyle |}{\underset{\displaystyle |}{\text{C}}}{}^{+} \\ \text{H}}}{}
$$

Most　　　　　　　　**Least**
stable　　　　　　　 **stable**

This order of stability of carbocations can be explained on the basis of a law of physics that states that *a charged system is stabilized when the charge is dispersed or delocalized*. Alkyl groups, in contrast to hydrogen atoms, are *electron releasing*. This means that alkyl groups have a tendency to shift electron density (negative charge) toward a positive charge. Through electron release, *alkyl groups* attached to the positive carbon of a carbocation *delocalize* the positive charge. In doing so, the attached alkyl groups assume part of the positive charge themselves and thus *stabilize* the carbocation. We can see how this occurs by inspecting Fig. 5.16.

In the *tert*-butyl cation three electron-releasing methyl groups surround the central carbon atom and assist in delocalizing the positive charge. In the isopropyl cation there are only two attached methyl groups that can serve to delocalize the charge. In the ethyl cation there is only one attached methyl group, and in the methyl cation there is none at all. As a result, *the delocalization of charge and the order of stability of the carbocations parallel the number of attached methyl groups.*

FIG. 5.16

The stabilization of carbocations through delocalization of charge. Arrows (→) indicate the direction of electron flow from the electron-releasing methyl groups. (Hydrogens, by contrast, are less electron releasing.) Partial charges are indicated by delta (δ).

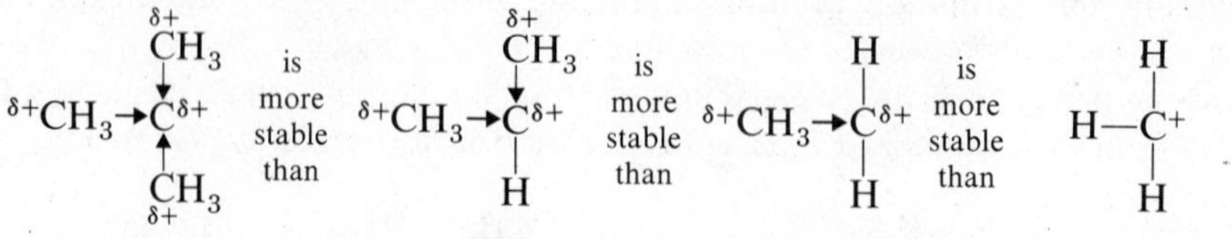

tert-Butyl cation	Isopropyl cation	Ethyl cation	Methyl cation
(3°)	(2°)	(1°)	
Most stable			**Least stable**

5.10 CARBOCATION STABILITY AND THE TRANSITION STATE

An important principle that governs all *multistep reactions* can be stated as follows: *the overall rate of a multistep reaction cannot exceed the rate of its slowest step.* Thus the slowest step of a multistep reaction is often called the *rate-limiting* or *rate-determining step.* In the dehydration of alcohols, (i.e., following the equations below in the forward direction) the slowest step is step 2: the formation of the carbocation from the protonated alcohol. The first step is a simple acid-base reaction. Proton-transfer reactions of this type occur very rapidly. The third step, the loss of a proton by the carbocation, is also very rapid, because the highly reactive carbocation stabilizes itself as quickly as possible by losing a proton and becoming an alkene.

Mechanism for the Acid-Catalyzed Dehydration of an Alcohol

$$\text{Step 1} \quad -\overset{|}{\underset{H}{C}}-\overset{|}{C}-\overset{..}{O}H + H_3O:^+ \rightleftharpoons -\overset{|}{\underset{H}{C}}-\overset{|}{C}-\overset{\overset{H}{|}}{\underset{..}{O}}{}^{+}\!\!-H + H_2\overset{..}{O}: \quad \text{fast}$$

$$\text{Step 2} \quad -\overset{|}{\underset{H}{C}}-\overset{|}{C}-\overset{\overset{H}{|}}{\underset{..+}{O}}-H \rightleftharpoons -\overset{|}{\underset{H}{C}}-\overset{|}{\underset{+}{C}}- + H_2\overset{..}{O}: \quad \begin{array}{c}\text{slow}\\ \text{(rate-limiting)}\end{array}$$

$$\text{Step 3} \quad -\overset{|}{\underset{|\ H}{C}}-\overset{|}{\underset{+}{C}}- + H_2\overset{..}{O}: \rightleftharpoons -\overset{|}{C}=\overset{|}{C}- + H_3O:^+ \quad \text{fast}$$

Step 2 is, then, the rate-limiting step, and step 2 is the step that determines the reactivity of alcohols toward dehydration. With this in mind, we can now understand why tertiary alcohols are the most easily dehydrated: the formation of a tertiary carbocation is easiest because the energy of activation (Fig. 5.17) for a reaction leading to a tertiary carbocation is lowest.

The reactions by which carbocations are formed are all highly *endothermic.* According to Hammond's postulate (Section 4.9), there should be a strong resemblance between the transition state and the product in each case. Of the three, *the transition state that leads to the tertiary carbocation is lowest in potential energy because it resembles the most stable product.* By contrast, the transition state that leads to the primary carbocation occurs at highest potential energy because it resembles the least stable product. In each instance, moreover, the transition state is stabilized by the same factor that stabilizes the carbocation itself: *delocalization of charge.* We can understand this point if we examine the process by which the transition state is formed.

The oxygen atom of the protonated alcohol bears a full positive charge. As the transition state develops this oxygen begins to separate from the carbon to which it is attached. The carbon, because it is losing the electrons that bonded it to the oxygen, begins to develop a partial positive charge. This developing positive charge

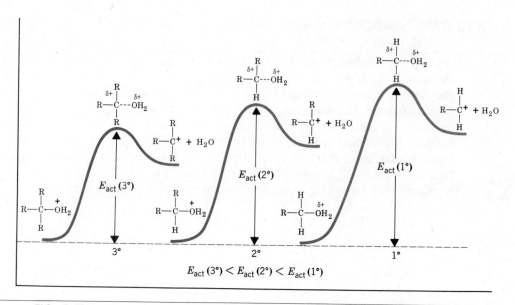

FIG. 5.17

Potential energy diagrams for the formation of carbocations from protonated tertiary, secondary, and primary alcohols. The relative energies of activation are tertiary < secondary < primary.

is most effectively delocalized in the transition state leading to a tertiary carbocation because of the presence of three electron releasing alkyl groups. The positive charge

is less effectively delocalized in the transition state leading to a secondary carbocation (*two* electron releasing groups) and is least effectively delocalized in the transition state leading to a primary carbocation (*one* electron-releasing group).

5.11 CARBOCATION STABILITY AND THE OCCURRENCE OF MOLECULAR REARRANGEMENTS

With an understanding of carbocation stability and its effect on transition states behind us, we now proceed to explain the rearrangements of carbon skeletons that occur in some alcohol dehydrations. For example, let us consider again the rearrangement that occurs when 3,3-dimethyl-2-butanol is dehydrated.

$$CH_3-\underset{\underset{CH_3}{|}}{\overset{\overset{CH_3}{|}}{C}}-\underset{\underset{OH}{|}}{CH}-CH_3 \xrightarrow[\text{heat}]{85\% \; H_3PO_4} CH_3-\underset{\overset{|}{CH_3}}{\overset{\overset{CH_3}{|}}{C}}=\overset{\overset{CH_3}{|}}{C}-CH_3 + CH_2=\underset{\overset{|}{CH_3}}{\overset{\overset{CH_3}{|}}{C}}-CHCH_3$$

3,3-Dimethyl-2-butanol   (major product)   (minor product)

2,3-Dimethyl-2-butene   2,3-Dimethyl-1-butene

The first step of this dehydration is the formation of the protonated alcohol in the usual way:

(1) $CH_3-\underset{\underset{CH_3}{|}}{\overset{\overset{CH_3}{|}}{C}}-CH-CH_3 + H-\overset{\overset{H}{|}}{\underset{\underset{H}{|}}{O}}:^+ \rightleftharpoons CH_3-\underset{\underset{CH_3}{|}}{\overset{\overset{CH_3}{|}}{C}}-CHCH_3 + H_2\overset{..}{O}:$

Protonated alcohol

In the second step the protonated alcohol loses water and a secondary carbocation forms:

(2) $CH_3-\underset{\underset{CH_3}{|}}{\overset{\overset{CH_3}{|}}{C}}-CH-CH_3 \rightleftharpoons CH_3-\underset{\overset{|}{CH_3}}{\overset{\overset{CH_3}{|}}{C}}-\overset{+}{C}HCH_3 + H_2\overset{..}{O}:$

A 2° carbocation

It is at this point that the rearrangement occurs: *a less stable, secondary carbocation rearranges to a more stable, tertiary carbocation.*

(3) $CH_3-\underset{\overset{|}{CH_3}}{\overset{\overset{CH_3}{|}}{C}}-\overset{+}{C}HCH_3 \longrightarrow CH_3-\underset{\overset{|}{CH_3}}{\overset{\overset{CH_3}{|}}{C}}\cdots\overset{+}{\cdots}CHCH_3 \longrightarrow CH_3-\underset{\overset{|}{CH_3}}{\overset{+}{C}}-\overset{\overset{CH_3}{|}}{CH}-CH_3$

2° Carbocation   Transition state   3° Carbocation
(less stable)   (more stable)

The rearrangement occurs through the migration of a methyl group from the carbon adjacent to the carbon that bears the positive charge. The methyl group migrates *with its pair of electrons,* that is, as a methanide, $CH_3:^-$, ion. After the migration is complete the carbon that the methanide ion left has become a carbocation, and the positive charge on the carbon to which it migrated has been neutralized.

The transition state for the methyl migration is one with a *three-center bond.* In the transition state the shifting methyl is partly bonded to both carbon atoms by the pair of electrons with which it migrates.

The final step of the reaction is the loss of proton by the new carbocation and the formation of an alkene. This step, however, can occur in two ways.

$$(a) \quad H-CH_2-\overset{+}{\underset{\underset{CH_3}{|}}{C}}-\overset{\underset{H\,(b)}{|}}{\underset{\underset{CH_3}{|}}{C}}-CH_3 \quad
\begin{cases}
(a) & CH_2=\underset{\underset{CH_3}{|}}{C}-\underset{\underset{CH_3}{|}}{C}HCH_3 \quad \text{Less stable alkene} \\
& \text{(minor product)} \\
(b) & CH_3-\underset{\underset{CH_3}{|}}{C}=\underset{\underset{CH_3}{|}}{C}-CH_3 \quad \text{More stable alkene} \\
& \text{(major product)}
\end{cases}$$

The choice here is dictated by the type of the alkene that is being formed. Path (b) leads to the highly stable tetrasubstituted alkene and this is the path followed by most of the carbocations. Path (a), on the other hand, leads to a less stable, disubstituted alkene and produces the minor product of the reaction. *The formation of the more stable alkene is the general rule in the acid-catalyzed dehydration reactions of alcohols.*

When we examine the rearrangement that occurs in the dehydration of n-butyl alcohol we find that the mechanism is similar to that for 3,3-dimethyl-2-butanol. In this instance, however, the migrating group is a hydride ion, $H:^-$—*a group that also migrates with a pair of electrons.*

(1) $CH_3CH_2CH_2CH_2OH + H^+ \rightleftharpoons CH_3CH_2CH_2CH_2\overset{+}{O}H_2$

(2) $CH_3CH_2CH_2CH_2\overset{+}{O}H_2 \rightleftharpoons CH_3CH_2CH_2CH_2{}^+ + H_2O$

$\quad\quad\quad\quad\quad\quad$ A 1° carbocation

(3) $CH_3CH_2\underset{\underset{H}{|}}{C}H-\overset{+}{C}H_2 \longrightarrow CH_3CH_2CH\underset{\underset{\overset{+}{H}}{\diagdown\;\diagup}}{\quad\quad}CH_2 \longrightarrow CH_3CH_2\overset{+}{C}H-\underset{\underset{H}{|}}{C}H_2$

$\quad$ 1° Carbocation $\quad\quad\quad\quad$ Transition state $\quad\quad\quad\quad$ 2° Carbocation

$$(4) \quad CH_3\underset{\underset{\underset{(a)\; H}{|}}{|}}{C}H-\overset{+}{\underset{\underset{\overset{..}{O}}{\diagup\diagdown}}{C}}H-\underset{\underset{\overset{|}{H\,(b)}}{|}}{C}H_2 \quad
\begin{cases}
(a) & CH_3CH=CHCH_3 + CH_3CH=CHCH_3 + H_3O^+ \\
& \quad\quad \textit{trans} \quad\quad\quad\quad\quad\quad\quad \textit{cis} \\
& \text{(major product)} \quad\quad\quad\quad \text{(minor product)} \\
& \textit{Most stable} \quad\quad\quad\quad\quad \textit{Less stable} \\
& \textit{alkene} \quad\quad\quad\quad\quad\quad\quad \textit{alkene} \\
(b) & CH_3CH_2CH=CH_2 + H_3O^+ \\
& \text{(minor product)} \\
& \textit{Least stable alkene}
\end{cases}$$

In the final step, three different alkenes are possible products: *trans*-2-butene, *cis*-2-butene, and 1-butene. Of these, *trans*-2-butene is the most stable and it is, as a result, the major product of the reaction.

Studies of thousands of reactions involving carbocations show that rearrangements like those just described are general phenomena. They occur almost inevitably when the migration of an alkyl group or hydride ion can lead to a more stable carbocation.

Problem 5.3

On page 172 we illustrated the dehydration of 2-methylcyclohexanol. The products of this reaction were 1-methylcyclohexene (major product) and 3-methylcyclohexene (minor product).

(a) Is a rearrangement involved here?

(b) If not, how can you account for the formation of two products?

(c) How do you account for the fact that 1-methylcyclohexene is the major product?

5.12 THE FORMATION OF ALKENES BY DEHYDROHALOGENATION OF ALKYL HALIDES

The elimination of the elements of a hydrogen halide (H-X) from an alkyl halide is a commonly used method for the preparation of an alkene. The overall elimination reaction is of the following type:

$$-\overset{\displaystyle H}{\underset{\displaystyle :X:}{\overset{|}{\underset{|}{C^\beta}}}}-\overset{|}{\underset{|}{C^\alpha}}- \xrightarrow[\text{base}]{B:^-} \quad -\overset{|}{C}=\overset{|}{C}- \ + \ H:B \ + \ :\overset{..}{\underset{..}{X}}:^-$$

The hydrogen that is eliminated in dehydrohalogenation is one from a carbon adjacent to the carbon bearing the halogen. This carbon is called a *β carbon* and, as a result, these reactions are often called *β eliminations*.

Several specific examples of dehydrohalogenation reactions are given below.

Specific examples:

$$\underset{\underset{\displaystyle Br}{|}}{CH_3CHCH_3} \xrightarrow[C_2H_5OH,\ 55°]{C_2H_5ONa} CH_2{=}CH{-}CH_3 + NaBr + C_2H_5OH$$

(79%)

$$\underset{\underset{\displaystyle Br}{|}}{CH_3CHCH_2CH_3} \xrightarrow[C_2H_5OH,\ 80°]{KOH} CH_3CH{=}CHCH_3 + CH_3CH_2CH{=}CH_2 + KBr + H_2O$$

cis and *trans*   (20%)
(80%)

$$\underset{\underset{\displaystyle CH_3}{|}}{\overset{\overset{\displaystyle CH_3}{|}}{CH_3{-}C{-}Br}} \xrightarrow[C_2H_5OH,\ 25°]{C_2H_5ONa} \underset{}{\overset{\overset{\displaystyle CH_3}{|}}{CH_3{-}C{=}CH_2}} + NaBr + C_2H_5OH$$

(91%)

A variety of strong bases have been used for dehydrohalogenations: potassium hydroxide dissolved in ethyl alcohol is a reagent often encountered, but the sodium salts of alcohols are also commonly used and, in some instances, they offer distinct advantages.

The sodium salt of an alcohol (a sodium alkoxide) can be prepared by treating an alcohol with sodium metal:

$$R-\ddot{O}H + Na \longrightarrow R-\ddot{O}:^-Na^+ + \tfrac{1}{2}H_2$$

Alcohol   Sodium
alkoxide

This reaction involves the displacement of hydrogen from the alcohol and is, thus, an *oxidation-reduction reaction*. Sodium, an alkali metal, is a very powerful reducing agent and always displaces hydrogen atoms that are bonded to oxygen atoms. The vigorous (at times explosive) reaction of sodium with water is of the same type.

$$H-\ddot{O}H + Na \longrightarrow H\ddot{O}:^-Na^+ + \tfrac{1}{2}H_2$$

Sodium
hydroxide

Sodium (and potassium) alkoxides are usually prepared by using an excess of the alcohol, and the excess alcohol is used as a solvent for the reaction. Sodium ethoxide is frequently employed in this way.

$$CH_3CH_2OH + Na \longrightarrow CH_3CH_2O^-Na^+/CH_3CH_2OH + \tfrac{1}{2}H_2$$

Ethyl alcohol   Sodium ethoxide in ethyl alcohol
(excess)

Potassium *tert*-butoxide is also a highly effective dehydrohalogenating reagent.

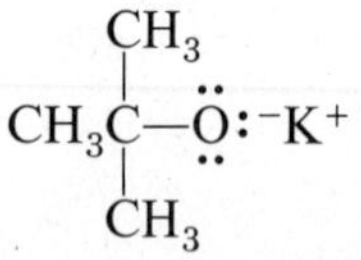

$$CH_3\overset{\overset{\displaystyle CH_3}{|}}{\underset{\underset{\displaystyle CH_3}{|}}{C}}-\ddot{O}:^-K^+$$

Potassium *tert*-butoxide

When dehydrohalogenations carried out with hydroxide ion or ethoxide ion can lead to more than one alkene, the major product of the reaction is usually *the most highly substituted* (the most stable) *alkene*. An example that illustrates this principle is the following:

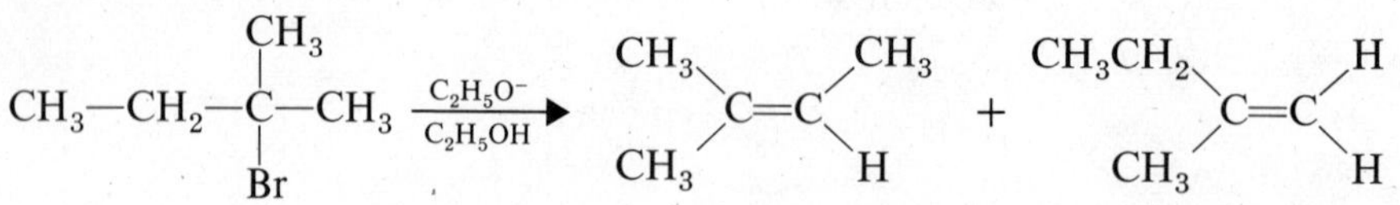

(62%)   (38%)
A trisubstituted alkene   A disubstituted alkene

This principle does not apply when the base used is *tert*-butoxide ion. When potassium *tert*-butoxide is employed, the *less* substituted alkene is usually the predominant product (see pp.664 for an explanation).

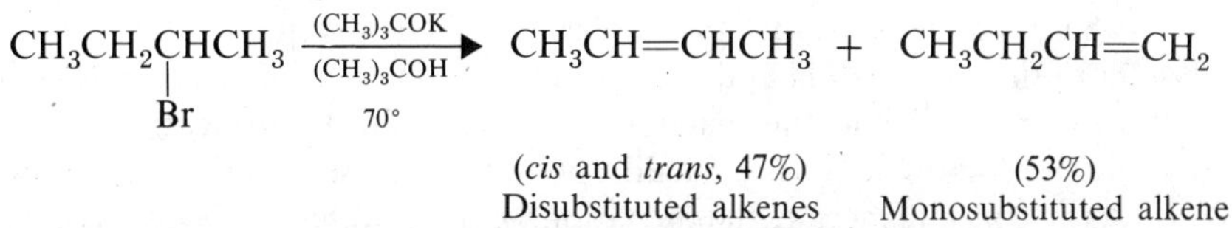

$$CH_3CH_2CHCH_3 \xrightarrow[\text{(CH}_3)_3\text{COH}]{\text{(CH}_3)_3\text{COK}} CH_3CH{=}CHCH_3 + CH_3CH_2CH{=}CH_2$$

at position: Br, 70°

(*cis* and *trans*, 47%) (53%)
Disubstituted alkenes   Monosubstituted alkene

A reaction that often competes with a dehydrohalogenation reaction is a *substitution reaction*. The reaction of 2-bromopropane with sodium ethoxide in ethyl alcohol, for example, not only yields the elimination product, propene, but also produces ethyl isopropyl ether, a *substitution product*.

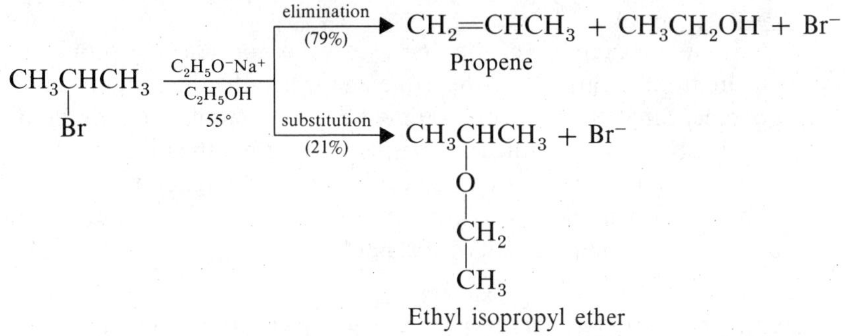

$$CH_3CHCH_3 \xrightarrow[\text{C}_2\text{H}_5\text{OH}]{\text{C}_2\text{H}_5\text{O}^-\text{Na}^+}$$

Br, 55°

elimination (79%) → $CH_2{=}CHCH_3 + CH_3CH_2OH + Br^-$
Propene

substitution (21%) → $CH_3CHCH_3 + Br^-$

$$\begin{array}{c} O \\ | \\ CH_2 \\ | \\ CH_3 \end{array}$$

Ethyl isopropyl ether

We will see other examples of substitution reactions of this type in Section 9.10 and the entire topic of substitution and elimination reactions will be considered in Chapter 17.

5.13 THE FORMATION OF ALKENES BY DEHALOGENATION OF VICINAL DIHALIDES

Vicinal (or *vic*) dihalides are dihalo compounds in which the halogens are situated on adjacent carbon atoms. The name *geminal* dihalide is used for those dihalides where both halogens are attached to the same carbon atom.

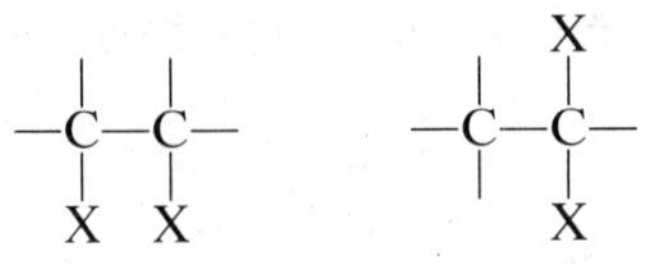

$$\begin{array}{cc} -\overset{|}{C}-\overset{|}{C}- & -\overset{|}{C}-\overset{X}{\underset{|}{C}}- \\ \overset{|}{X}\ \overset{|}{X} & \overset{|}{X} \end{array}$$

A *vic*-dihalide   A *gem*-dihalide

Vic-dihalides undergo the loss of halogen (dehalogenation) when they are treated with zinc. The organic products of these reactions are alkenes. A specific example is the dehalogenation of 1,2-dibromocyclohexane.

$$\text{(1,2-dibromocyclohexane)} + Zn \xrightarrow[\text{CH}_3\overset{O}{\overset{\|}{C}}\text{CH}_3]{} \text{(cyclohexene)} + ZnBr_2$$

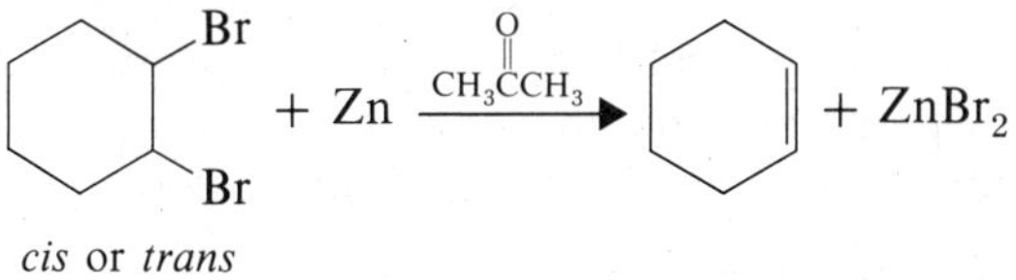

cis or *trans*

These reactions are usually carried out in a solvent such as acetone that dissolves both the *vic*-dihalide and the alkene product. The zinc (as a powder) is usually held in suspension and the reaction takes place on its surfaces.

Vic-dihalides are usually prepared by the addition of halogen to an alkene. As a consequence, dehalogenation of a *vic*-dihalide is of little use as a general preparative reaction. This method is useful, however, in the purification of alkenes, and as a "protecting group" for the double bond.

$$R-CH=CH-R + X_2 \longrightarrow R-\underset{X}{CH}-\underset{X}{CH}-R$$

As an example of the first use, let us consider the problem of separating an alkene from an alkane of the same chain length: the separation of 1-pentene from pentane. Since the two compounds have very similar boiling points, separation by simple distillation is difficult. However, if we treat the mixture of the two compounds with chlorine in carbon tetrachloride at room temperature and in the absence of light, a reaction takes place that produces 1,2-dichloropentane from the 1-pentene but no new product results from pentane.

$$CH_3CH_2CH_2CH_2CH_3 \xrightarrow[CCl_4]{Cl_2} CH_3CH_2CH_2CH_2CH_3$$

Pentane, bp 36°   no reaction   bp 36°

$$CH_3CH_2CH_2CH=CH_2 \xrightarrow[CCl_4]{Cl_2} CH_3CH_2CH_2\underset{Cl}{CH}\underset{Cl}{CH_2}$$

addition

1-Pentene, bp 31°   bp 148°

Difficult to separate by   **Easily separated by**
distillation   **distillation**

These two compounds, pentane and 1,2-dichloropentane, can be separated easily by distillation, and the pure 1,2-dichloropentane can be converted to pure 1-pentene by dechlorination.

$$CH_3CH_2CH_2\underset{Cl}{CH}-\underset{Cl}{CH_2} + Zn \xrightarrow{acetone} CH_3CH_2CH_2CH=CH_2 + ZnCl_2$$

Halogenation-dehalogenation can also be used to protect a double bond while some other reaction is carried out on another part of the molecule. We find an example of this in the synthesis of the *steroid* shown at the top of page 183. The overall reaction that is desired in this synthesis is the oxidation of a tertiary hydrogen to a tertiary alcohol by chromium trioxide. Double bonds, unfortunately, are not stable toward this oxidation. Direct oxidation (dotted arrow) would cause oxidation of both the double bond and the tertiary hydrogen. However, conversion of the double bond to a *vic* dihalide "protects" the double bond by transforming it temporarily into a saturated grouping, and the desired oxidation can be carried out. Afterward, the double bond can be regenerated by allowing the *vic* dihalide to react with zinc.

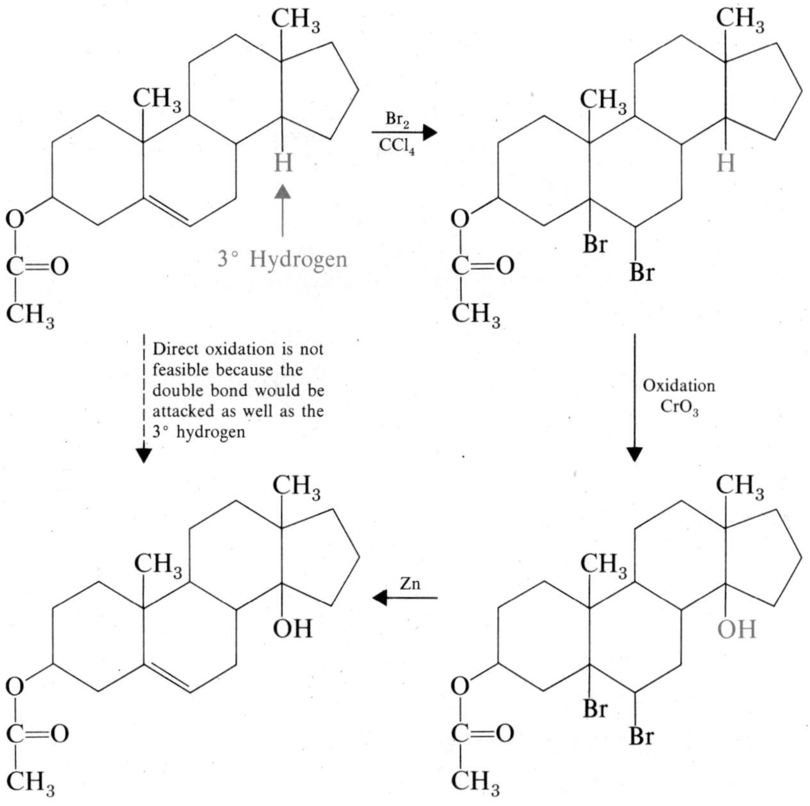

3° Hydrogen

Direct oxidation is not feasible because the double bond would be attacked as well as the 3° hydrogen

Oxidation
CrO$_3$

Zn

5.14 SUMMARY OF METHODS FOR THE PREPARATION OF ALKENES

In this chapter we described four general methods for the preparation of alkenes. All are based on elimination reactions.

1. Dehydrogenation of alkanes (Section 5.7)

General:

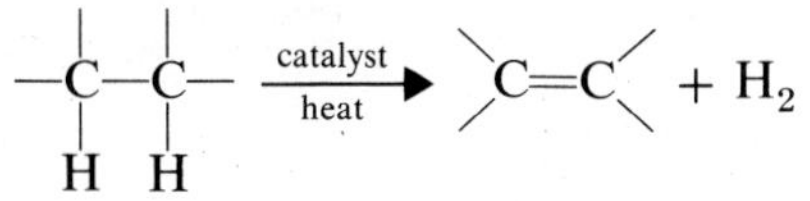

$$-\underset{\underset{H}{|}}{C}-\underset{\underset{H}{|}}{C}- \xrightarrow[\text{heat}]{\text{catalyst}} \quad \underset{}{C}{=}\underset{}{C} + H_2$$

Specific
example: $CH_3CH_2CH_2CH_3 \xrightarrow[500°]{Cr_2O_3 \cdot Al_2O_3} CH_3CH{=}CHCH_3$
(*cis* and *trans*)

+ other butenes + H$_2$

2. Dehydration of alcohols (Section 5.8)

General:

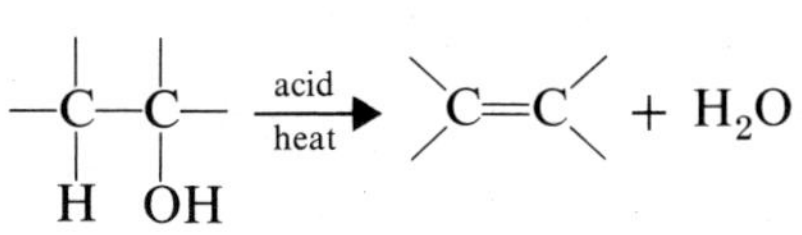

$$-\underset{\underset{H}{|}}{C}-\underset{\underset{OH}{|}}{C}- \xrightarrow[\text{heat}]{\text{acid}} \quad \underset{}{C}{=}\underset{}{C} + H_2O$$

Specific examples: $CH_3CH_2OH \xrightarrow[180°]{\text{conc. } H_2SO_4} CH_2{=}CH_2 + H_2O$

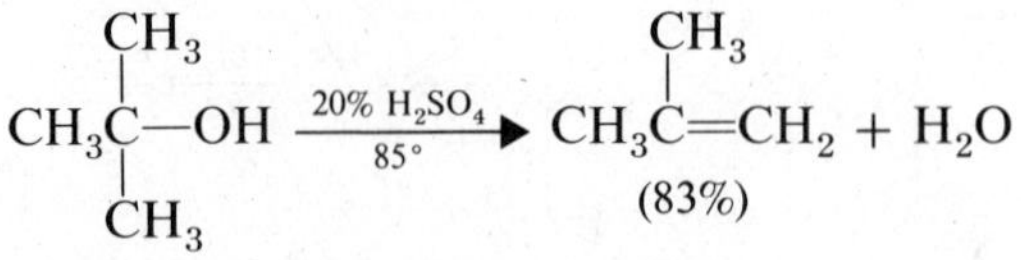

$$CH_3\underset{\underset{CH_3}{|}}{\overset{\overset{CH_3}{|}}{C}}{-}OH \xrightarrow[85°]{20\% \text{ } H_2SO_4} CH_3\overset{\overset{CH_3}{|}}{C}{=}CH_2 + H_2O$$
$$(83\%)$$

3. **Dehydrohalogenation of alkyl halides (Section 5.12)**

General: $-\underset{\underset{H}{|}}{C}{-}\underset{\underset{X}{|}}{C}{-} \xrightarrow[\substack{\text{heat} \\ (-HX)}]{\text{base}} \overset{}{C}{=}\overset{}{C}$

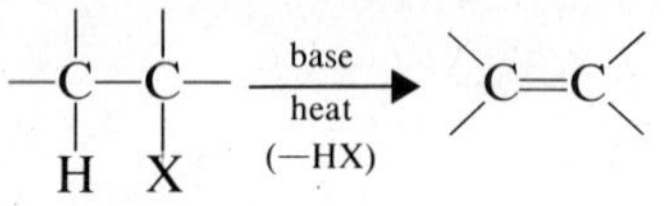

Specific example:

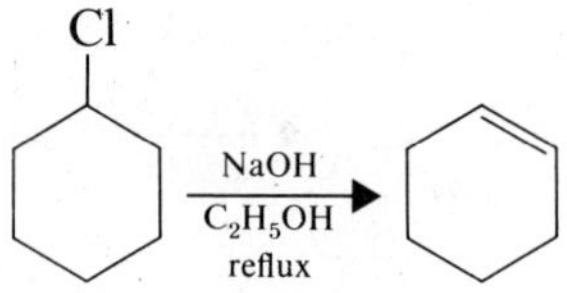

4. **Dehalogenation of *vic*-dihalides (Section 5.13)**

General: $-\underset{\underset{X}{|}}{C}{-}\underset{\underset{X}{|}}{C}{-} \xrightarrow[\text{acetone}]{Zn} \overset{}{C}{=}\overset{}{C} + ZnX_2$

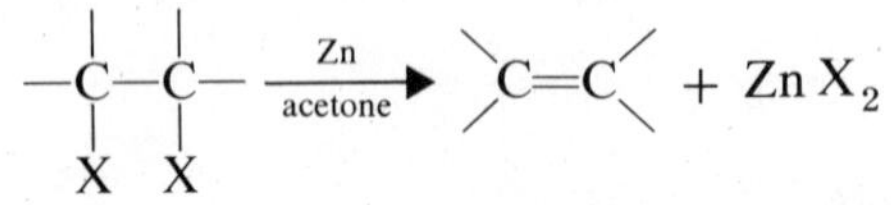

Specific example:

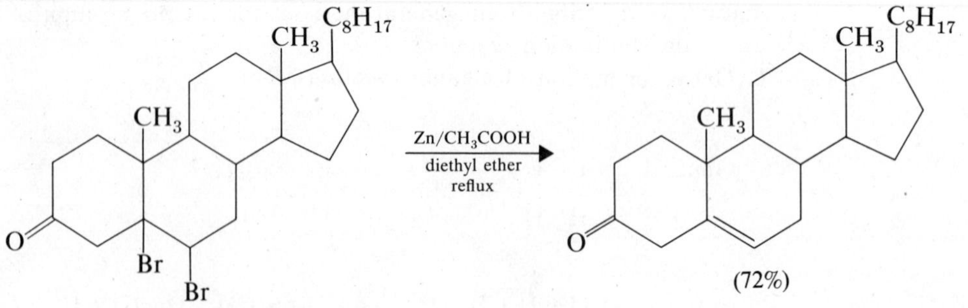

In subsequent chapters we will see a number of related methods for alkene synthesis.

5. **Hofmann elimination (Section 17.9)**

General: $-\underset{\underset{HO^-}{}}{\overset{}{C}}\underset{\underset{H}{|}}{}-\underset{\underset{\overset{+}{N}(CH_3)_3}{|}}{C}- \xrightarrow[\text{heat}]{} \overset{}{C}{=}\overset{}{C} + H_2O + (CH_3)_3N{:}$

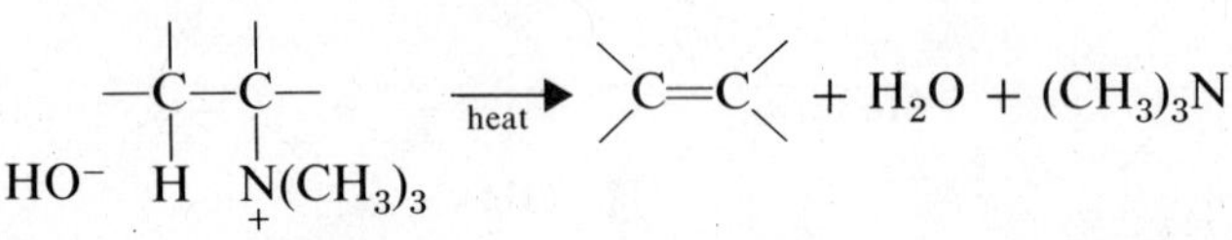

Specific
example:

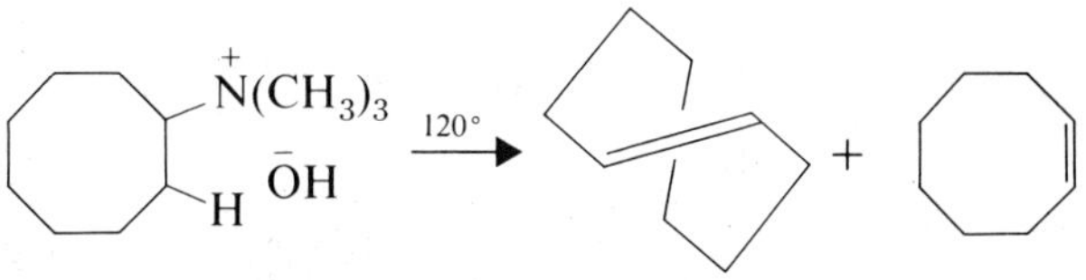

trans-Cyclo-
octene (60%)   *cis*-Cyclo-
octene (40%)

6. Cope elimination (Section 17.10)

General:

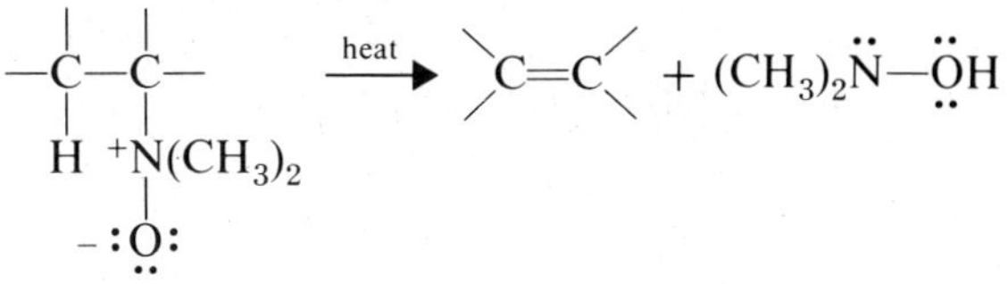

Specific
example:

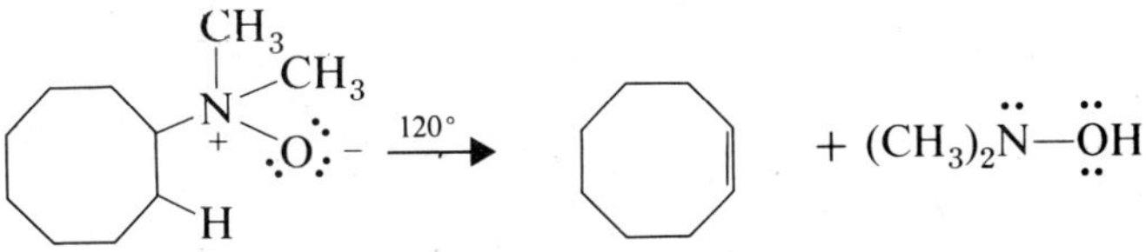

cis-Cyclo-octene
(81%)

7. Acetate pyrolysis (Section 17.10)

General:

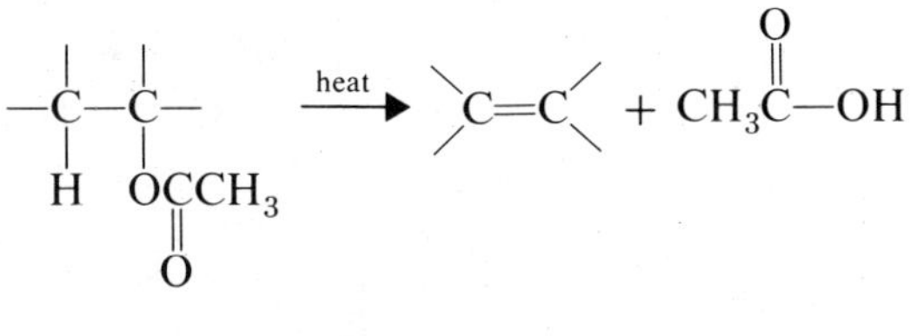

Specific
example: $CH_3CH_2CHCH_3$

$+ CH_3CH_2CH{=}CH_2 + CH_3\overset{\text{O}}{\underset{\|}{C}}{-}OH$

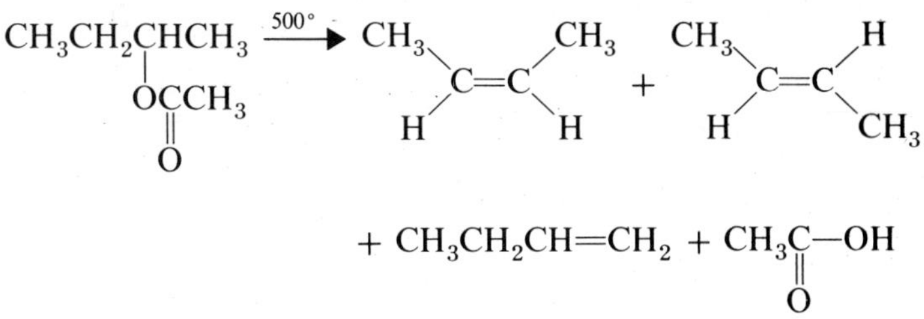

8. Tosylate elimination (Section 17.9)

General:

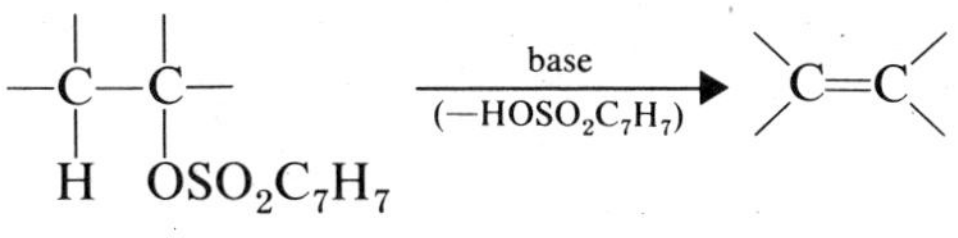

Specific
example: $CH_3CH_2\underset{\underset{OSO_2C_7H_7}{|}}{C}HCH_3$ $+ (CH_3)_3COK$ $\xrightarrow[55°]{\text{dimethylsulfoxide}}$ $CH_3CH_2CH{=}CH_2$

(61%)

$+$ $\underset{H}{\overset{CH_3}{>}}C{=}C\underset{CH_3}{\overset{H}{<}}$ $+$ $\underset{H}{\overset{CH_3}{>}}C{=}C\underset{H}{\overset{CH_3}{<}}$ $+ C_7H_7SO_3K + (CH_3)_3COH$

(28%)   (11%)

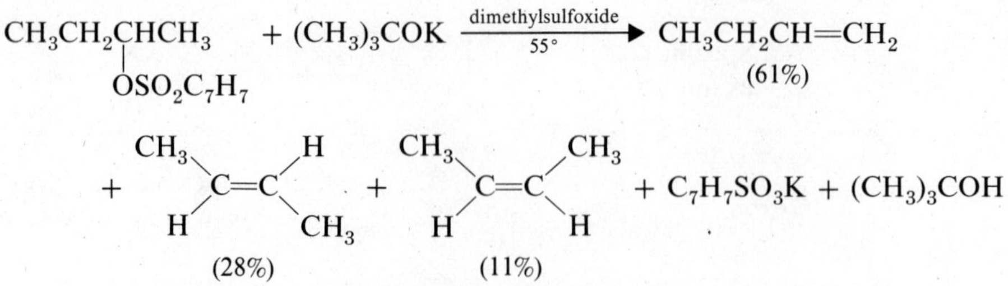

9. Wittig reaction (Section 18.11)

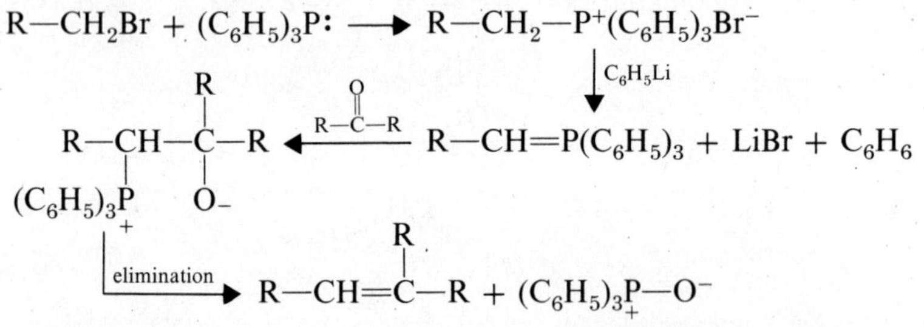

$$R{-}CH_2Br + (C_6H_5)_3P: \longrightarrow R{-}CH_2{-}P^+(C_6H_5)_3Br^-$$

$\downarrow C_6H_5Li$

$\underset{\underset{+}{(C_6H_5)_3P}}{\overset{R}{\underset{|}{R{-}CH{-}}}}\overset{O}{\underset{O^-}{\underset{|}{C}}}{-}R$ $\xleftarrow{R{-}\overset{O}{\overset{||}{C}}{-}R}$ $R{-}CH{=}P(C_6H_5)_3 + LiBr + C_6H_6$

$\xrightarrow{\text{elimination}}$ $R{-}CH{=}\overset{R}{\underset{|}{C}}{-}R + (C_6H_5)_3\underset{+}{P}{-}O^-$

All of the foregoing methods for the preparation of alkenes involve elimination. One other general method for the preparation of alkenes is based on the *addition of hydrogen* to an alkyne.

10. Reduction of alkynes (Section 9.6)

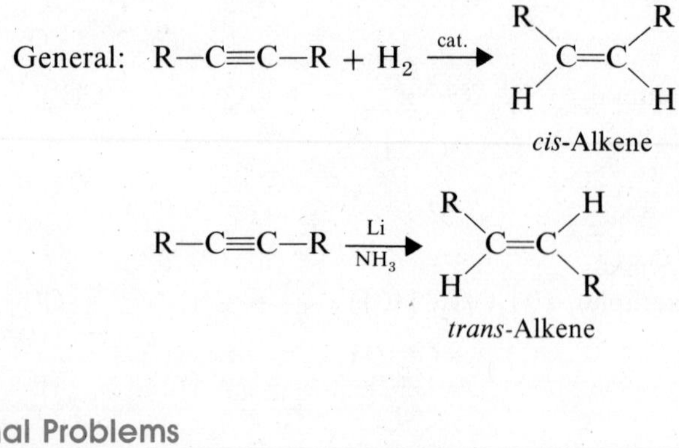

General: $R{-}C{\equiv}C{-}R + H_2 \xrightarrow{\text{cat.}}$ $\underset{H}{\overset{R}{>}}C{=}C\underset{H}{\overset{R}{<}}$

cis-Alkene

$R{-}C{\equiv}C{-}R \xrightarrow[NH_3]{Li}$ $\underset{H}{\overset{R}{>}}C{=}C\underset{R}{\overset{H}{<}}$

trans-Alkene

Additional Problems

5.4

Give IUPAC names for the following alkenes:

(a) $\underset{CH_3}{\overset{CH_3}{>}}C{=}C\underset{CH_3}{\overset{H}{<}}$

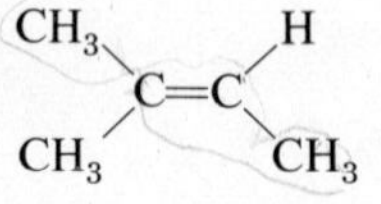

(c) $\underset{CH_3}{\overset{CH_3}{>}}C{=}C\underset{Br}{\overset{H}{<}}$

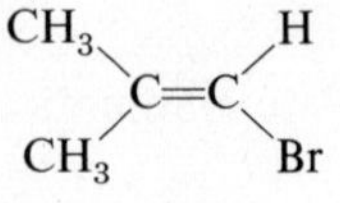

(b) $CH_3CH_2CH_2\underset{H}{\overset{}{>}}C{=}C\underset{H}{\overset{}{<}}CH_2CH_2CH_3$

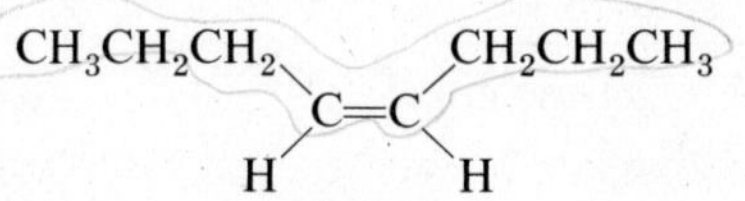

(d) [cyclohexene with CH₃ substituent]

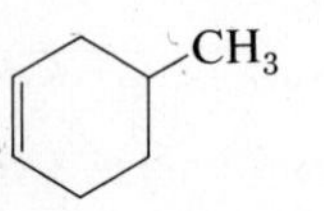

5.5
Write structural formulas for each of the following:
(a) 1-Methylcyclobutene
(b) 3-Methylcyclopentene
(c) 2,3-Dimethyl-2-pentene
(d) *trans*-2-Hexene
(e) *cis*-3-Heptene
(f) 3,3,3-Trichloropropene
(g) Isobutylene
(h) Propylene
(i) 4-Cyclopentyl-1-pentene
(j) Cyclopropylethene

5.6
Write structural formulas and give IUPAC names for all alkene isomers of
(a) C_5H_{10} and (b) C_6H_{12}. (c) What other isomers are possible for C_5H_{10} and C_6H_{12}? Write their structures.

5.7
Give structural formulas for the alcohols that, on dehydration, would yield each of the following alkenes as major products.
(a) 2-Methylpropene   (c) Cyclopentene
(b) 2,3-Dimethyl-2-butene   (d) Cyclohexene

5.8
Cyclohexane and 1-hexene have the same molecular formula. Suggest a simple chemical reaction that will distinguish one from the other.

5.9
For which of the following compounds is *cis-trans* isomerism possible? In each case, where *cis-trans* isomerism is possible, write structural formulas for the isomeric compounds.

(a) 1-Butene
(b) 2-Methylpropene
(c) 2-Heptene
(d) 2-Methyl-2-heptene
(e) CH_2

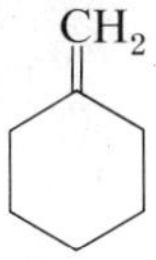

(f) $CHCH_3$

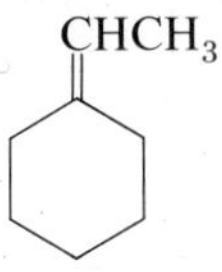

(g) $CHCH_3$ CH_3

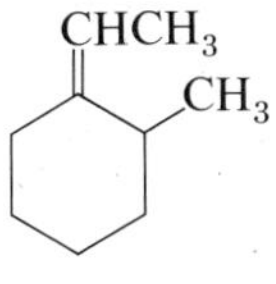

(h) 1-Chloro-1-butene
(i) 1,1-Dichloro-1-butene
(j) CH_3

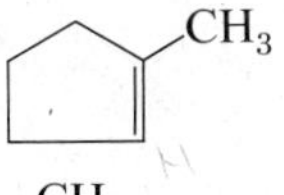

(k) CH_2 —CH_3

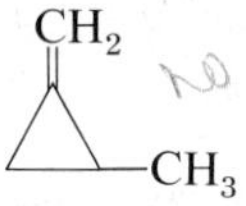

(l) $CHCH_3$ —CH_3

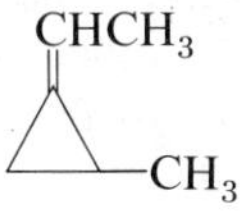

5.10
(a) Arrange the following alkenes in order of their relative stabilities:

   trans-3-hexene; 1-hexene; 2-methyl-2-pentene; *cis*-2-hexene; 2,3-dimethyl-2-butene

(b) For which of the alkenes listed in part (a) could comparative heats of hydrogenation be used to measure their relative stabilities?

5.11

Which compound would you expect to have the larger heat of hydrogenation (in kcal/mole): *cis*-cyclooctene or *trans*-cyclooctene? Explain.

5.12

When *cis*-2-butene is heated to a temperature greater than 300°, a mixture of two 2-butene isomers results. (a) What chemical change takes place? (b) Which butene isomer would you expect to predominate in the mixture when equilibrium is established between them?

5.13

The sp^2-hybridized carbons of alkenes are often said to be "more electronegative" than the sp^3-hybridized carbons of alkanes. (a) Explain this statement. (b) Can you account for the fact that ethane ($K_a = 10^{-42}$) is a weaker acid than ethene ($K_a = 10^{-36}$)?

5.14

Write structural formulas for the alkenes that could be formed when each of the following alkyl halides is subjected to dehydrohalogenation by the action of ethoxide ion in ethanol. When more than one product results, designate the product that is most likely to be the major product.
(a) $CH_3CHBrCH(CH_3)CH_3$
(b) $CH_3CH_2CH(CH_3)CH_2Br$
(c) $CH_3CH_2CHBrCH_2CH_3$
(d) $CH_3CHBrCH_2CH_2CH_3$

(e) 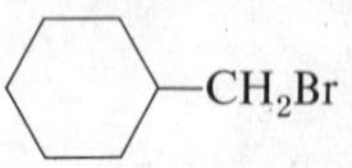 ⬡—CH_2Br (f) 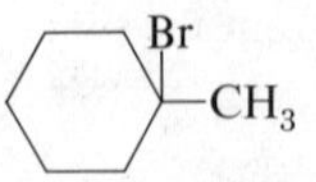 ⬡ with Br and CH_3

5.15

Arrange the following alcohols in order of their reactivity toward acid-catalyzed dehydration and explain your reasoning.

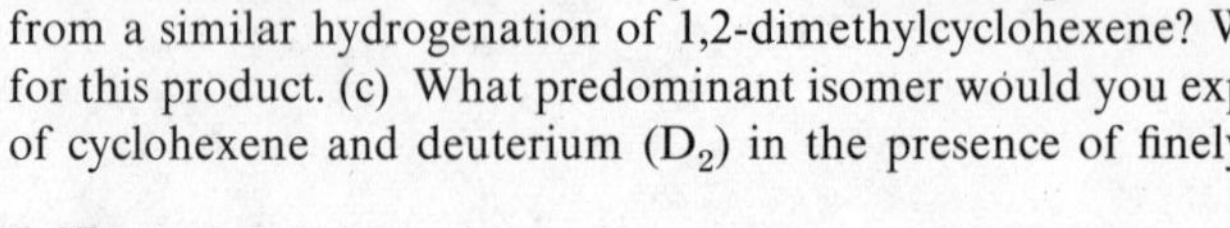

$$CH_3CHCH_2CH_2OH \qquad CH_3\overset{OH}{\underset{CH_3}{C}}CH_2CH_3 \qquad CH_3\overset{OH}{\underset{CH_3}{CH}}CHCH_3$$
$$\underset{CH_3}{|}$$

5.16

(a) When 1,2-dimethylcyclopentene reacts with hydrogen in the presence of finely divided platinum, a single 1,2-dimethylcyclopentane isomer is formed. What is it? (Assume a mechanism similar to that in Fig. 5.14) (b) What predominant product would you expect from a similar hydrogenation of 1,2-dimethylcyclohexene? Write a conformational formula for this product. (c) What predominant isomer would you expect to obtain from the reaction of cyclohexene and deuterium (D_2) in the presence of finely divided platinum?

5.17

Write step-by-step mechanisms that account for each product of the following reactions and explain the relative proportions of the isomers obtained in each instance.

(a) 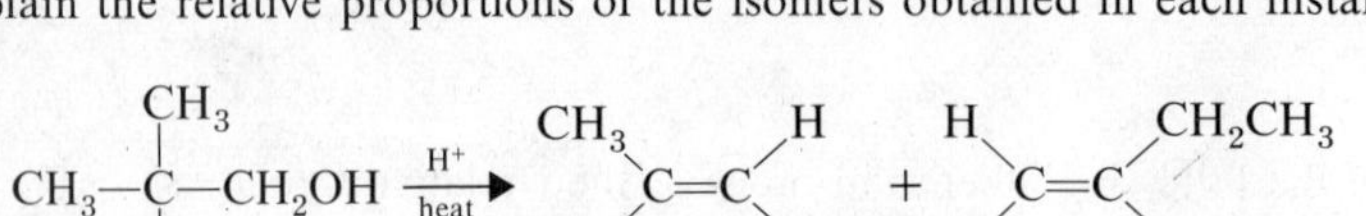 $CH_3-\overset{CH_3}{\underset{CH_3}{\overset{|}{C}}}-CH_2OH \xrightarrow[\text{heat}]{H^+}$ $\overset{CH_3}{\underset{CH_3}{C}}=\overset{H}{\underset{CH_3}{C}}$ $+$ $\overset{H}{\underset{H}{C}}=\overset{CH_2CH_3}{\underset{CH_3}{C}}$

(major product)   (minor product)

(b)

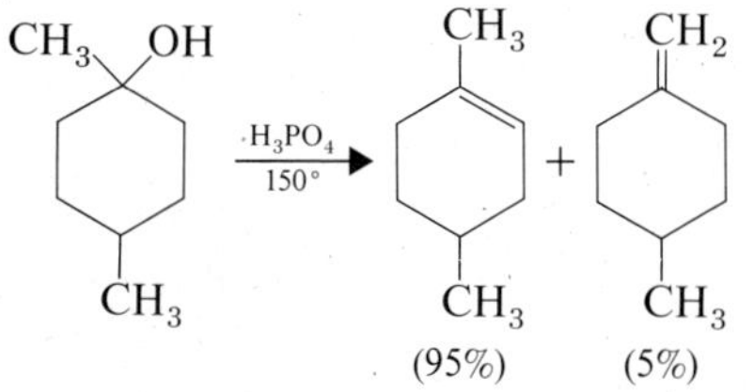

(95%)   (5%)

(c)

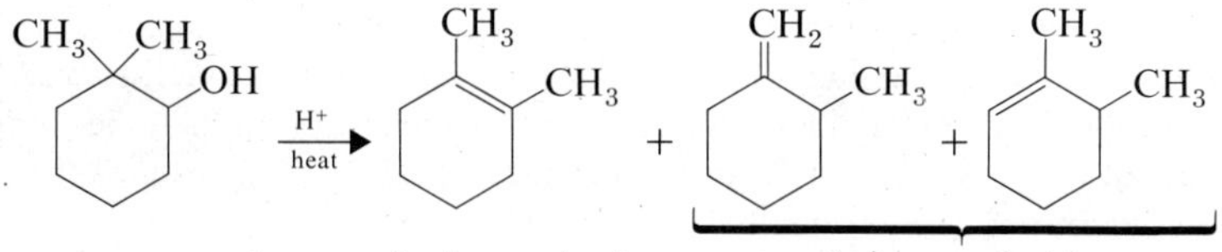

(major product)   (minor products)

5.18

Cholesterol (below) is an important steroid found in nearly all body tissues; it is also the major component of gallstones. Impure cholesterol can be obtained from gallstones through a simple extraction with an organic solvent. The cholesterol thus obtained can be purified through (a) treatment with Br_2 in $CHCl_3$, (b) careful crystallization of the product, and (c) treatment of the latter with zinc in ethyl alcohol. What reactions are involved in the purification procedure?

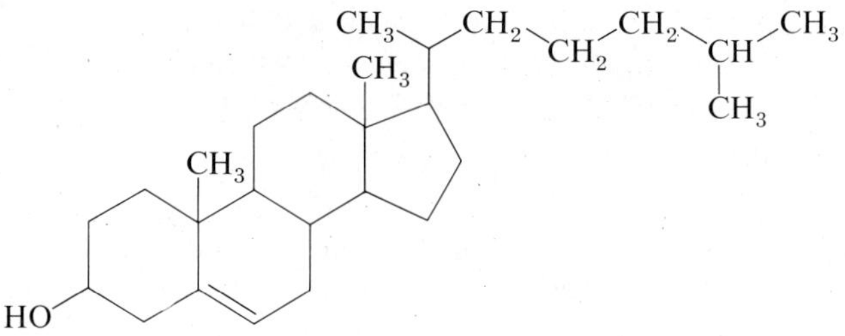

Cholesterol

5.19

A spectacular example of the kind of rearrangement that can occur in an alcohol dehydration is found in the example given below. When 3β-friedelanol is treated with acid, a carbocation is formed, and then a total of *seven* methyl and hydrogen migrations occur before the loss of a proton produces 13(18)-oleanene. Each migration is a 1,2-shift, that is, a methide ion $(CH_3:^-)$ or hydride ion $(H:^-)$ migrates from one atom to an adjacent atom. Show how each of these migrations occur.

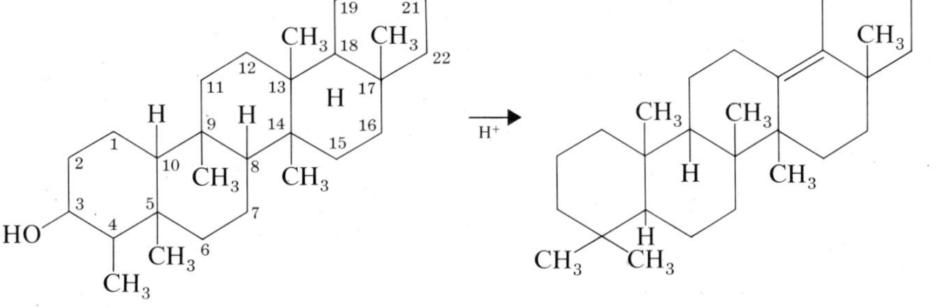

3β-Friedelanol   13(18)-Oleanene

5.20
Reconsider the interconversion of *cis*-2-butene and *trans*-2-butene given in problem 5.12. (a) What is the value of ΔH for the reaction, *cis*-2-butene $\longrightarrow$ *trans*-2-butene? (b) What minimum value of E_{act} would you expect for this reaction? (c) Sketch a potential energy diagram for the reaction and label ΔH and E_{act}.

5.21
Considerable information about the structure of a hydrocarbon can be obtained from its molecular formula and a number called *the index of hydrogen deficiency*. The index of hydrogen deficiency is defined as the *number of pairs* of hydrogen atoms that must be subtracted from the formula of a saturated alkane to give the formula of the compound under consideration. (a) What is the index of hydrogen deficiency of 1-hexene? (b) Of 2-hexene? (c) Of cyclohexane? (d) Of methylcyclopentane? (e) Of 1-methylcyclopentene? (f) Of 1,3-hexadiene, that is, $CH_2{=}CHCH{=}CHCH_2CH_3$? (g) Of 1,3,5-hexatriene, that is, $CH_2{=}CHCH{=}CHCH{=}CH_2$? (h) Of 13(18)-oleanene (cf. problem 5.19)? (i) What does the index of hydrogen deficiency reveal about the number of double bonds and rings in a molecule of the compound? (j) About the sizes of the rings? (k) About the locations of the double bonds? (l) In general terms, what structural possibilities exist for a compound with the molecular formula $C_{10}H_{16}$?

5.22
(Work this problem after problem 5.21.) Zingiberene, a fragrant compound isolated from ginger, has the molecular formula $C_{15}H_{24}$. (a) What is the index of hydrogen deficiency of zingiberene? (b) When zingiberene is subjected to catalytic hydrogenation using an excess of hydrogen, one mole of zingiberene absorbs three moles of hydrogen and produces a compound with the formula $C_{15}H_{30}$. How many double bonds does a molecule of zingiberene have? (c) How many rings?

5.23
(Work this problem after problem 5.21.) Caryophyllene, a compound found in oil of cloves, has the molecular formula $C_{15}H_{24}$. Reaction of caryophyllene with an excess of hydrogen in the presence of a platinum catalyst produces a compound with the formula $C_{15}H_{28}$. How many (a) double bonds, and (b) rings does a molecule of caryophyllene have?

5.24
(Work this problem after 5.21.) Squalene, an important intermediate in the biosynthesis of steroids, has the molecular formula $C_{30}H_{50}$. (a) What is the index of hydrogen deficiency of squalene? (b) Squalene undergoes catalytic hydrogenation to yield a compound with the molecular formula $C_{30}H_{62}$. How many double bonds does a molecule of squalene have? (c) How many rings?

6

REACTIONS OF ALKENES: ADDITION REACTIONS OF THE CARBON-CARBON DOUBLE BOND

6.1 INTRODUCTION

The most commonly encountered reaction of compounds containing a carbon-carbon double bond is one of addition. As we described in Chapter 2, addition may involve a symmetrical reagent:

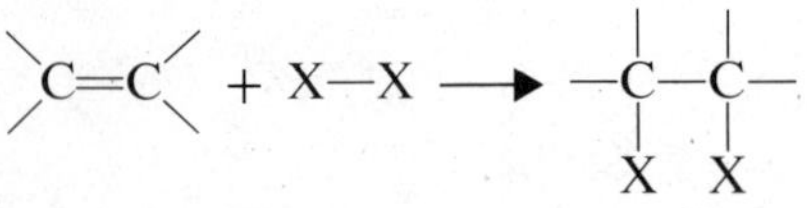

$$\text{C}=\text{C} + \text{X}-\text{X} \longrightarrow -\overset{|}{\underset{|}{\text{C}}}-\overset{|}{\underset{|}{\text{C}}}-$$
$$\phantom{\text{C}=\text{C} + \text{X}-\text{X} \longrightarrow }\;\; \text{X} \;\; \text{X}$$

or an unsymmetrical one:

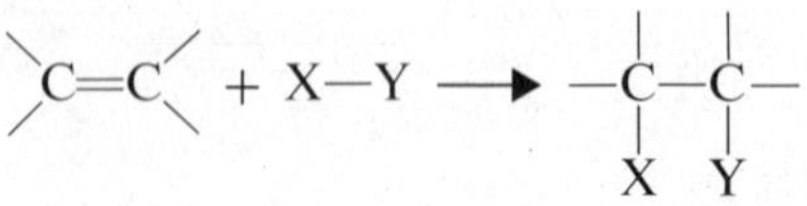

$$\text{C}=\text{C} + \text{X}-\text{Y} \longrightarrow -\overset{|}{\underset{|}{\text{C}}}-\overset{|}{\underset{|}{\text{C}}}-$$
$$\phantom{\text{C}=\text{C} + \text{X}-\text{Y} \longrightarrow }\;\; \text{X} \;\; \text{Y}$$

Examples of additions of symmetrical reagents are the additions of hydrogen or bromine:

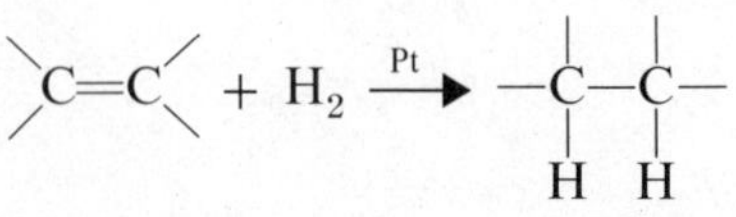

$$\text{C}=\text{C} + \text{H}_2 \xrightarrow{\text{Pt}} -\overset{|}{\underset{|}{\text{C}}}-\overset{|}{\underset{|}{\text{C}}}-$$
$$\phantom{\text{C}=\text{C} + \text{H}_2 \xrightarrow{\text{Pt}} }\;\; \text{H} \;\; \text{H}$$

Alkene   Alkane

$$\text{C}=\text{C} + \text{Br}_2 \xrightarrow{\text{CCl}_4} -\overset{|}{\underset{|}{\text{C}}}-\overset{|}{\underset{|}{\text{C}}}-$$
$$\phantom{\text{C}=\text{C} + \text{Br}_2 \xrightarrow{\text{CCl}_4} }\;\; \text{Br} \;\; \text{Br}$$

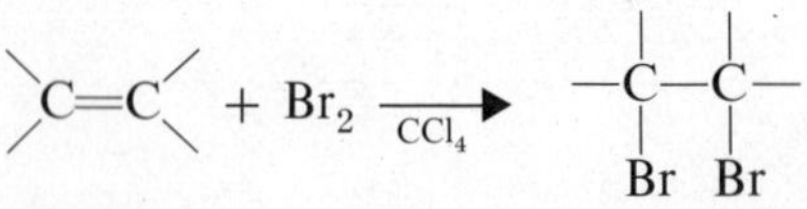

Alkene   *vic*-Dibromide

Examples of additions of unsymmetrical reagents are the additions of hydrogen halides or the acid-catalyzed addition of water:

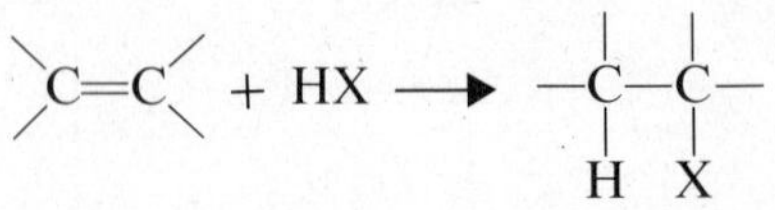

$$\text{C}=\text{C} + \text{HX} \longrightarrow -\overset{|}{\underset{|}{\text{C}}}-\overset{|}{\underset{|}{\text{C}}}-$$
$$\phantom{\text{C}=\text{C} + \text{HX} \longrightarrow }\;\; \text{H} \;\; \text{X}$$

Alkene   Alkyl halide

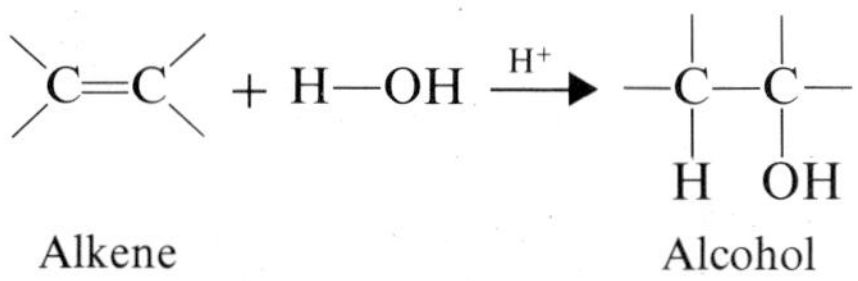

$$\text{C=C} + \text{H—OH} \xrightarrow{\text{H}^+} -\underset{\underset{\text{H}}{|}}{\text{C}}-\underset{\underset{\text{OH}}{|}}{\text{C}}-$$

Alkene   Alcohol

The large numbers of reactions of alkenes that fall into the category of addition reactions can be accounted for on the basis of two characteristics of the carbon-carbon double bond:

1. An addition reaction results in the conversion of one π bond and one σ bond into two σ bonds. The result of this change is usually energetically

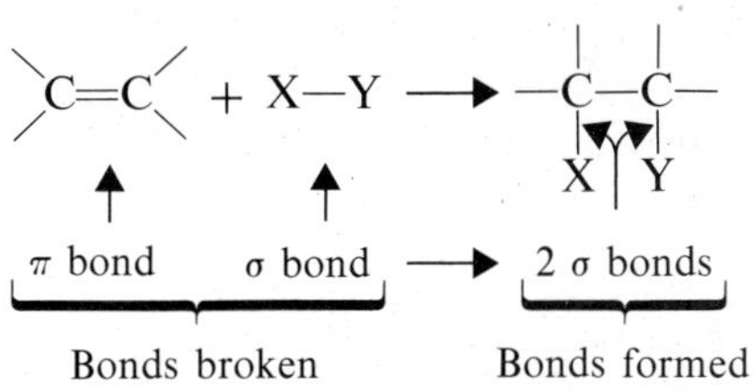

$$\text{C=C} + \text{X—Y} \longrightarrow -\underset{\underset{\text{X}}{|}}{\text{C}}-\underset{\underset{\text{Y}}{|}}{\text{C}}-$$

$\quad\quad\pi$ bond$\quad\quad\sigma$ bond $\longrightarrow$ 2 σ bonds

Bonds broken$\quad\quad\quad$ Bonds formed

favorable. The heat evolved in bond formation exceeds that necessary for bond cleavage, and addition reactions are usually exothermic.

2. The electrons of the π bond are exposed because the π orbital has considerable p character; moreover, the overlap between the orbitals of a π bond is less than that of a σ bond. Thus, a π bond is particularly susceptible to electron-seeking reagents. Such electron-seeking reagents are said to be *electrophilic* (electron loving) and are called *electrophiles*. Electrophiles include positive reagents like the proton, H^+, or neutral reagents such as bromine (because it can be polarized), and the Lewis acids BF_3 and $AlCl_3$. Metal ions that contain vacant orbitals—the silver ion, Ag^+, the mercuric ion, Hg^{++}, and the platinum ion, Pt^{++}, for example—also act as electrophiles.

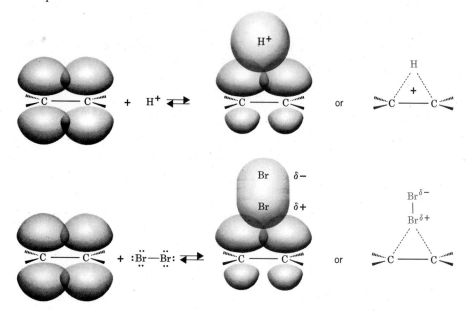

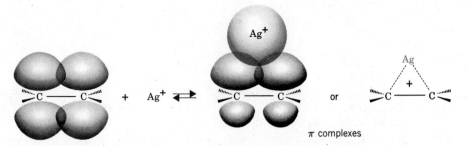

π complexes

The reaction of an electrophile with an alkene nearly always begins with interaction between the exposed π bond of the alkene and the electrophile, to form a π complex.

These π complexes often undergo further change. A protonated alkene may become a carbocation, for example, and then react with a *nucleophile* (a reagent

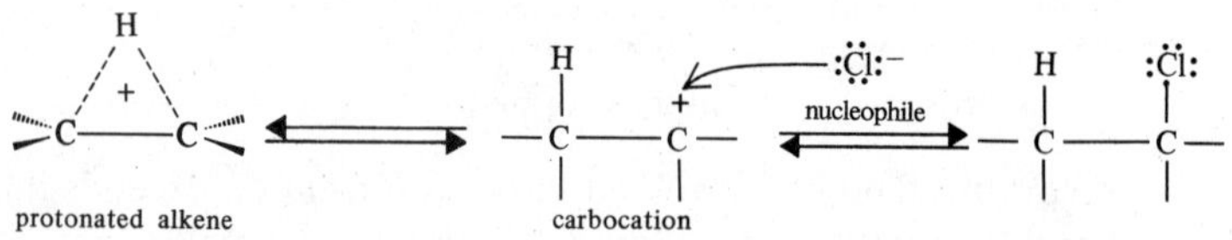

that seeks a positive center) to form an addition product.

Electrophiles are really Lewis acids: they are molecules or ions that are capable of accepting an electron pair. Nucleophiles are molecules or ions that are capable of furnishing an electron pair (i.e., Lewis bases). Any reaction of an electrophile also involves a nucleophile. In the protonation of an alkene the electrophile is the proton donated by an acid; the nucleophile is the alkene.

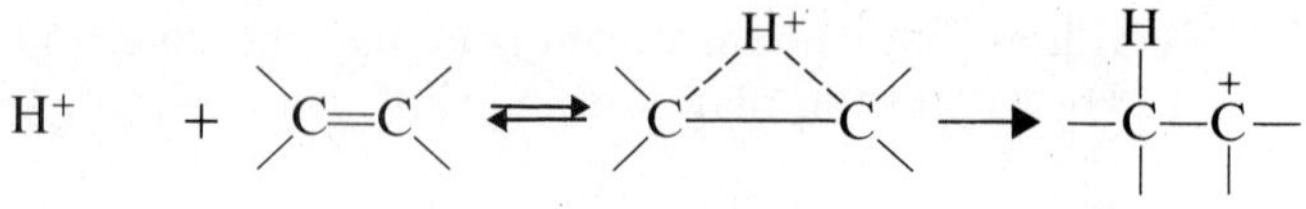

Electrophile   nucleophile

In the next step, the reaction of the carbocation with a chloride ion, the carbocation is the electrophile and the chloride ion is the nucleophile.

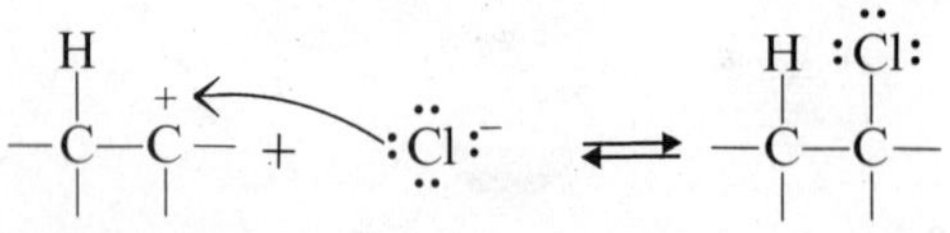

Electrophile   Nucleophile

Either reaction could be described in two ways: the first step could be described as an electrophilic attack by a proton on the alkene, or as a nucleophilic attack by the alkene on the proton. Chemists generally consider the organic molecule as the substance being attacked, and would usually describe this reaction as an *electrophilic attack by a proton on the alkene.*

"Since electrophile-nucleophile reactions resemble Lewis acid-base reactions, why do chemists use two different sets of words for them?" To answer this question, one must first recall that the terms acidity and basicity are ways of talking about relative equilibrium constants. A strong acid, in this sense, is one that produces a high concentration of the conjugate acid (and conjugate base) at equilibrium. Electrophilicity and nucleophilicity are ways of talking about relative rates of reactions. These terms refer to the rates at which the electrophiles and nucleophiles react. Ethoxide ion, $CH_3CH_2\ddot{O}{:}^-$, for example, is a stronger base than the ethyl mercaptide ion, $CH_3CH_2\ddot{S}{:}^-$. (The K_a for C_2H_5OH is $\sim 10^{-18}$; that for C_2H_5SH is $\sim 10^{-12}$). The ethyl mercaptide ion, however, is in many cases the stronger nucleophile; it reacts faster with electrophiles—particularly with electrophilic carbon atoms.

6.2 THE ADDITION OF HYDROGEN HALIDES TO ALKENES: AN EXPLANATION OF MARKOVNIKOV'S RULE

Hydrogen chloride, hydrogen bromide, and hydrogen iodide add readily to the double bond of alkenes.

$$CH_3CH{=}CHCH_3 + H{-}Cl \longrightarrow CH_3CH_2\underset{\underset{Cl}{|}}{C}HCH_3$$

2-Butene   2-Chlorobutane
(*cis* or *trans*)

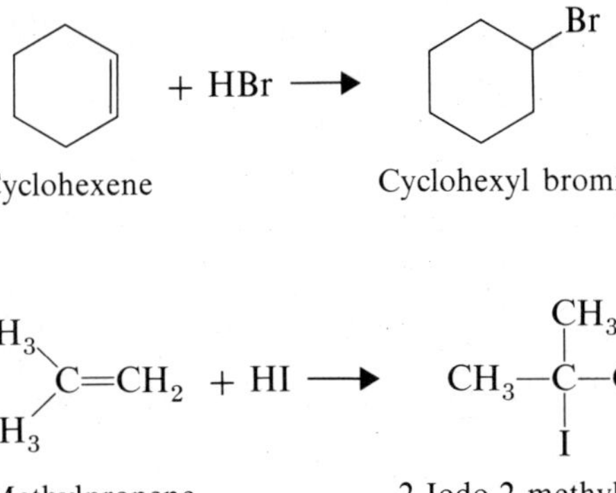

Cyclohexene   + HBr   Cyclohexyl bromide

$$\underset{CH_3}{\overset{CH_3}{{>}}}C{=}CH_2 + HI \longrightarrow CH_3{-}\underset{\underset{I}{|}}{\overset{\overset{CH_3}{|}}{C}}{-}CH_3$$

2-Methylpropene   2-Iodo-2-methylpropane
(isobutylene)   (*tert*-butyl iodide)

In carrying out these reactions, the hydrogen halide may be dissolved in acetic acid and mixed with the alkene, or gaseous hydrogen halide may be bubbled directly into the alkene with the alkene, itself, being used as the solvent.

As we noted earlier (Chapter 2), the addition of H-X to an unsymmetrical alkene can conceivably occur in two ways. With propene, for example, addition could lead to a 2-halopropane, by Markovnikov addition or to a 1-halopropane, by an anti-Markovnikov addition.

$$CH_3CH{=}CH_2 + HX \longrightarrow CH_3\underset{\underset{X}{|}}{C}HCH_3 \qquad \text{Markovnikov addition}$$

Propene

2-Halopropane

$$CH_3CH=CH_2 + HX \xrightarrow{\quad\times\quad} CH_3CH_2CH_2X$$

Propene   1-Halopropane

Anti-Markovnikov addition

In most instances, the only product that is obtained from the reaction is the Markovnikov addition product.* The reason for Markovnikov addition becomes clear when we examine the reaction mechanism below.

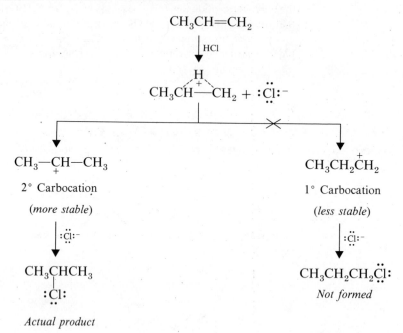

We see that Markovnikov addition takes place because it leads to the more stable carbocation in the first step. Of the two possibilities below, the first reaction

(1) $CH_3CH=CH_2 + H^+ \xrightarrow{\text{slow}} CH_3\overset{+}{C}HCH_3$ 2° Carbocation

(2) $CH_3CH=CH_2 + H^+ \xrightarrow[\text{slow}]{\text{very}} CH_3CH_2CH_2^+ +$ 1° Carbocation

has the lower energy of activation because it leads to the more stable product, a 2° carbocation. The second reaction, because of its higher energy of activation—the transition state resembles a less stable 1° carbocation—is very much slower and does not compete with the first reaction.

The reaction of H-I with 2-methylpropene produces only 2-iodo-2-methylpropane, and for the same reason. Here, in the first step (i.e., the attachment of the proton) the choice is even more pronounced—between a 3° carbocation and a 1° carbocation.

Thus, 1-iodo-2-methylpropane is *not* obtained as a product of the reaction because its formation would require the formation of a primary carbocation. Such a reaction would have a much higher energy of activation than that leading to a tertiary carbocation.

_____

* The addition of hydrogen bromide to an alkene occurs in an anti-Markovnikov fashion when the reaction is promoted by peroxides. We will discuss this topic in Section 6.9.

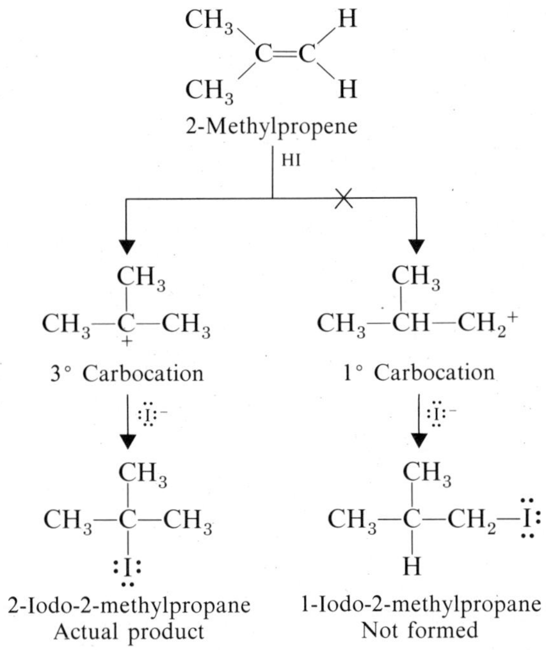

2-Methylpropene

HI

| 3° Carbocation | 1° Carbocation |

$$CH_3-\underset{+}{\overset{\overset{\displaystyle CH_3}{|}}{C}}-CH_3 \qquad CH_3-\overset{\overset{\displaystyle CH_3}{|}}{CH}-CH_2^+$$

2-Iodo-2-methylpropane
Actual product

1-Iodo-2-methylpropane
Not formed

A Modern Statement of Markovnikov's Rule

With this understanding of the mechanism for the ionic addition of hydrogen halides to alkenes behind us, we are now in a position to give the following modern statement of Markovnikov's rule: *in the ionic addition of an unsymmetrical reagent to a double bond, the positive portion of the adding reagent attaches itself to a carbon of the double bond so as to yield the more stable carbocation.* Since this is the step that occurs first (prior to the addition of the negative portion of the adding reagent), it is the step that determines the overall orientation of the reaction.

Notice that this formulation of Markovnikov's rule allows us to predict the outcome of the addition of a reagent such as I-Cl. Because of the greater electronegativity of chlorine, the positive portion of this molecule is iodine. The addition of I-Cl to 2-methylpropene takes place in the following way and produces 1-iodo-2-chloro-2-methylpropane.

2-Methylpropene

1-Iodo-2-chloro-2-methylpropane

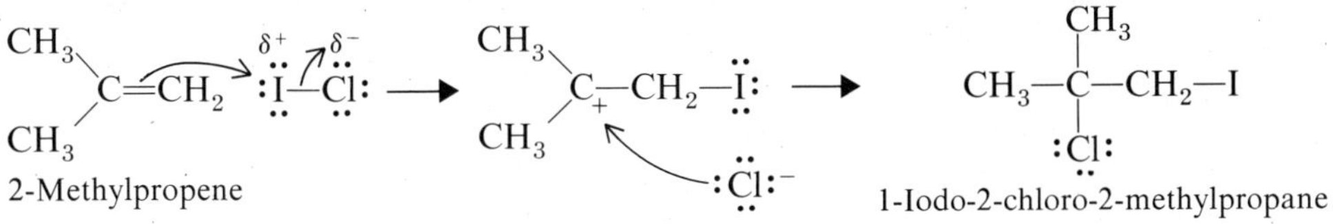

Problem 6.1

Give the structure and name of the product that would be obtained from Markovnikov addition of I-Cl to propene.

Problem 6.2

Outline mechanisms for the ionic additions of (a) hydrogen iodide to 1-butene, (b) of I-Br to 2-methyl-2-butene, (c) of hydrogen chloride to 1-methylcyclohexene.

Problem 6.3

The addition of hydrogen chloride to 3,3-dimethyl-1-butene (below) yields two products: 3-chloro-2,2-dimethylbutane and 2-chloro-2,3-dimethylbutane. Write mechanisms that account for the formation of each product.

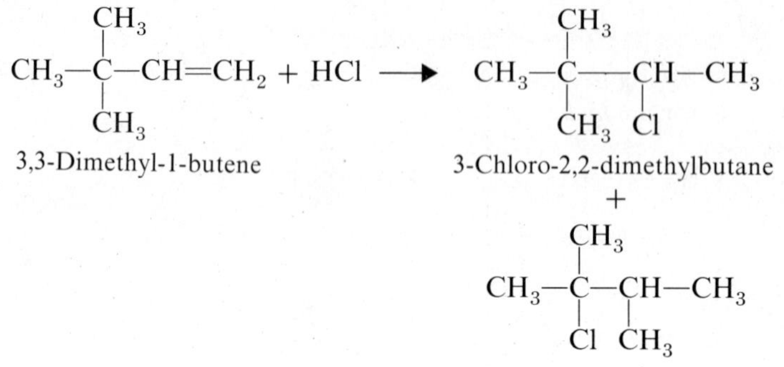

$$CH_3-\underset{\underset{CH_3}{|}}{\overset{\overset{CH_3}{|}}{C}}-CH=CH_2 + HCl \longrightarrow CH_3-\underset{\underset{CH_3}{|}}{\overset{\overset{CH_3}{|}}{C}}-\underset{\underset{Cl}{|}}{CH}-CH_3$$

3,3-Dimethyl-1-butene   3-Chloro-2,2-dimethylbutane

+

$$CH_3-\underset{\underset{Cl}{|}}{\overset{\overset{CH_3}{|}}{C}}-\underset{\underset{CH_3}{|}}{CH}-CH_3$$

2-Chloro-2,3-dimethylbutane

Regiospecific Reactions

We can now introduce a new term that chemists use to describe reactions like the Markovnikov additions of hydrogen halides to alkenes: the term is *regiospecific*. *Regio* comes from the Latin word meaning to rule or to govern. A reaction is said to be regiospecific if *from a standpoint of orientation* it tends to give predominantly *one* of two or more possible isomeric products. The ionic additions of hydrogen halides to alkenes are regiospecific.

6.3 THE ADDITION OF WATER TO ALKENES: ACID-CATALYZED HYDRATION

The acid-catalyzed addition of water to the double bond of an alkene (hydration of an alkene) is a convenient method for the preparation of secondary and tertiary alcohols. The acids most commonly used to catalyze the hydration of alkenes are sulfuric acid and phosphoric acid. These reactions, too, are usually regiospecific and the addition of water to the double bond follows Markovnikov's rule. Primary alcohols are not generally formed.

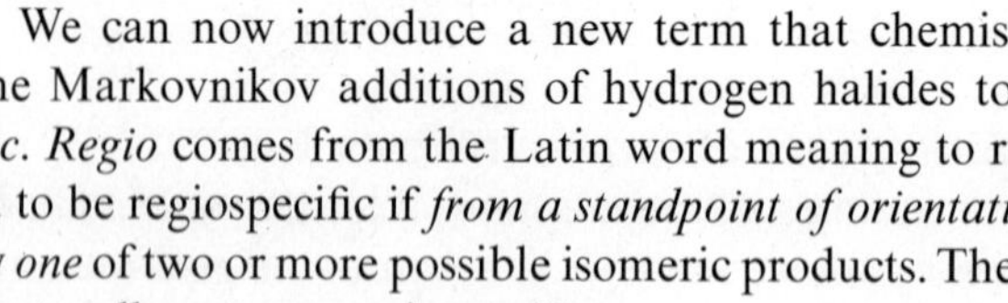

$$CH_3-\overset{\overset{CH_3}{|}}{C}=CH_2 + HOH \xrightarrow[25°]{H^+} CH_3-\underset{\underset{OH}{|}}{\overset{\overset{CH_3}{|}}{C}}-CH_3$$

2-Methylpropene   *tert*-Butyl alcohol

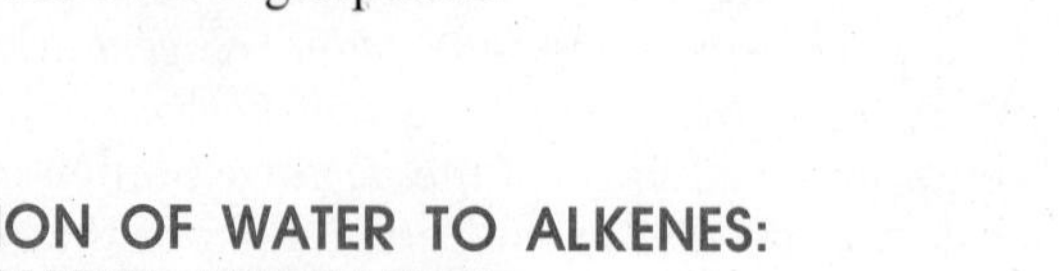

$$+ HOH \xrightarrow[25°]{H^+}$$

Methylene cyclopentane   1-Methylcyclopentanol

A primary alcohol (ethanol) is formed, of course, in the acid-catalyzed hydration of ethene.

$$\underset{\text{Ethene}}{CH_2\!=\!CH_2} + HOH \xrightarrow[240°]{H^+} \underset{\text{Ethanol}}{CH_3CH_2OH}$$

The mechanism for the hydration of an alkene is simply the reverse of the mechanism for the dehydration of an alcohol. We can illustrate this by giving the mechanism for the *hydration* of isobutylene and by comparing it with the mechanism for the *dehydration* of *tert*-butyl alcohol given in p. 172–173.

Step 1

$$CH_3\!-\!\overset{\overset{\displaystyle CH_2}{\|}}{\underset{\underset{\displaystyle CH_3}{|}}{C}} + H\!-\!\overset{H}{\underset{\overset{+}{..}}{O}}\!-\!H \rightleftarrows CH_3\!-\!\overset{\overset{\displaystyle CH_3}{|}}{\underset{\underset{\displaystyle CH_3}{|}}{C^+}} + :\overset{H}{\underset{..}{O}}\!-\!H$$

Step 2

$$CH_3\!-\!\overset{\overset{\displaystyle CH_3}{|}}{\underset{\underset{\displaystyle CH_3}{|}}{C^+}} + :\overset{H}{\underset{..}{O}}\!-\!H \rightleftarrows CH_3\!-\!\overset{\overset{\displaystyle CH_3}{|}}{\underset{\underset{\displaystyle CH_3}{|}}{C}}\!-\!\overset{H}{\underset{\overset{+}{..}}{O}}\!-\!H$$

Step 3

$$CH_3\!-\!\overset{\overset{\displaystyle CH_3}{|}}{\underset{\underset{\displaystyle CH_3}{|}}{C}}\!-\!\overset{H}{\underset{..}{O}}\!\overset{+}{-}H + :\overset{H}{\underset{..}{O}}\!-\!H \rightleftarrows CH_3\!-\!\overset{\overset{\displaystyle CH_3}{|}}{\underset{\underset{\displaystyle CH_3}{|}}{C}}\!-\!\underset{..}{\overset{..}{O}}\!-\!H + H\!-\!\overset{H}{\underset{\overset{+}{..}}{O}}\!-\!H$$

The relation between the mechanisms for hydration of an alkene and dehydration of an alcohol is an illustration of the *principle of microscopic reversibility*. According to this principle, a reaction and its reverse proceed through the same path but in opposite directions.

The rate-limiting step in the *hydration* mechanism is step 1: the formation of the carbocation. It is this step, too, that accounts for the Markovnikov addition of water to the double bond.

Problem 6.4

(a) Show all steps in the acid-catalyzed hydration of propene. (b) Account for the fact that the product of the reaction is isopropyl alcohol (in accordance with Markovnikov's rule) and not *n*-propyl alcohol, that is,

$$CH_3\!-\!CH\!=\!CH_2 \longrightarrow \underset{\substack{| \\ OH}}{CH_3CHCH_3} \quad \text{but not} \quad CH_3CH_2CH_2OH$$

$$\qquad\qquad\qquad\qquad\underset{\text{alcohol}}{\text{Isopropyl}} \qquad\qquad\qquad \underset{\text{alcohol}}{n\text{-Propyl}}$$

A spectacular biochemical example of both hydration of a double bond and dehydration of an alcohol occurs in the reactions catalyzed by the enzyme *aconitase*. Aconitase is an important enzyme in the biochemical pathway called the tricarboxylic acid or Krebs cycle—a biological pathway by which much of the energy content of carbohydrates is "extracted" and made available for various energy-requiring

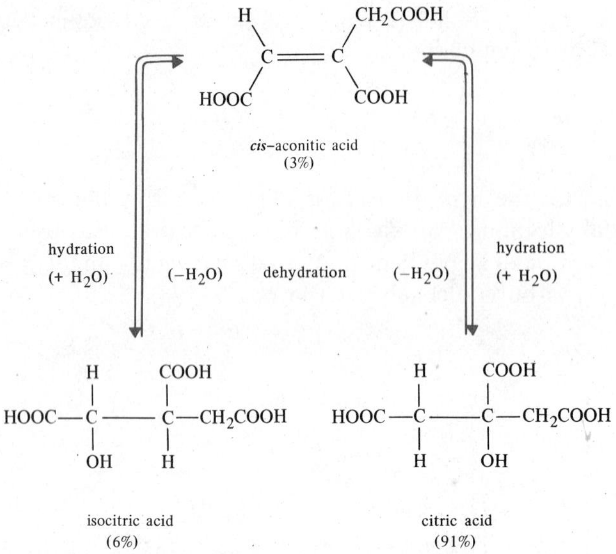

FIG. 6.1
The aconitase *equilibrium.*

processes in living organisms. The reactions that are catalyzed by aconitase are shown above in Fig. 6.1.

There are, in the aconitase equilibrium, four reactions that are catalyzed by a single enzyme. Two involve dehydration, and two involve hydration. The equilibrium mixture at 25° and pH 7.4 is indicated by the percentages given in parentheses in Fig. 6.1.

Addition of Sulfuric Acid to Alkenes

When alkenes are treated with cold concentrated sulfuric acid they react by addition to form alkyl hydrogen sulfates. In the first step of this reaction the alkene accepts a proton from sulfuric acid to form a carbocation; in the second step the carbocation reacts with a hydrogen sulfate ion to form an alkyl hydrogen sulfate:

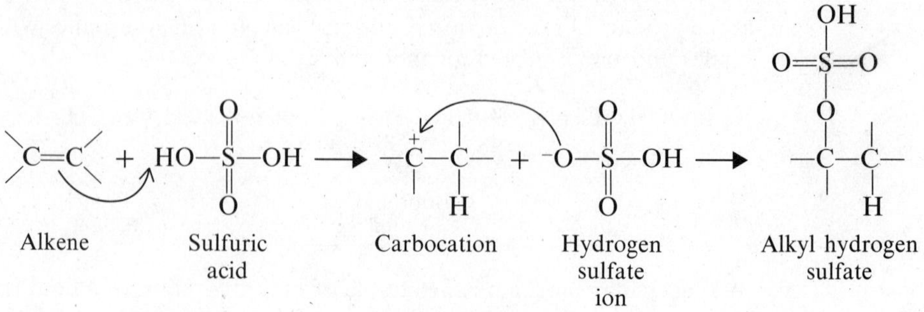

The addition of sulfuric acid is also regiospecific, and it follows Markovnikov's rule. Propene, for example, reacts to yield isopropyl hydrogen sulfate rather than *n*-propyl hydrogen sulfate.

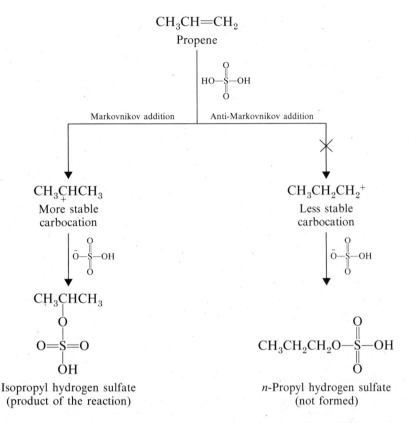

Alkyl hydrogen sulfates can be hydrolyzed to alcohols. Thus, the overall result of the addition of sulfuric acid to an alkene followed by hydrolysis is the same as that obtained by acid-catalyzed hydration: Markovnikov addition of H— and —OH.

$$CH_3CH{=}CH_2 \xrightarrow[H_2SO_4]{cold} CH_3CHCH_3 \xrightarrow{H_2O, \text{ heat}} CH_3CHCH_3 + H_2SO_4$$
$$\qquad\qquad\qquad\qquad\quad \underset{OSO_3H}{|} \qquad\qquad\qquad \underset{OH}{|}$$

6.4 ALCOHOLS FROM ALKENES THROUGH OXYMERCURATION-DEMERCURATION (SOLVOMERCURATION-DEMERCURATION)

A highly useful procedure for synthesizing alcohols from alkenes, one that complements acid-catalyzed hydration, is a two-step process called oxymercuration-demercuration.

Alkenes react with mercuric acetate in a mixture of tetrahydrofuran* and

---

* Tetrahydrofuran, or THF, is a cyclic ether with the structure

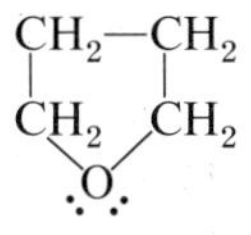

$$CH_2{-}CH_2$$
$$CH_2\quad CH_2$$
$$\ddot{\cdot}O\ddot{\cdot}$$

water to produce (hydroxyalkyl) mercury compounds. These (hydroxyalkyl) mercury compounds can be reduced to alcohols with sodium borohydride:

$$-\overset{|}{C}=\overset{|}{C}- + H_2O + Hg(O_2CCH_3)_2 \xrightarrow[\text{oxymercuration}]{\text{THF}} -\overset{|}{\underset{HO}{C}}-\overset{|}{\underset{HgO_2CCH_3}{C}}-$$

$$+ CH_3\overset{O}{\overset{||}{C}}OH$$

$$-\overset{|}{\underset{HO}{C}}-\overset{|}{\underset{HgO_2CCH_3}{C}}- + OH^- + NaBH_4 \xrightarrow{\text{demercuration}} -\overset{|}{\underset{HO}{C}}-\overset{|}{\underset{H}{C}}- + Hg + CH_3\overset{O}{\overset{||}{C}}O^-$$

In the first step, *oxymercuration*, water and mercuric acetate add to the double bond; in the second step, *demercuration*, sodium borohydride reduces the mercuriacetate group and replaces it with hydrogen.

Both steps can be carried out in the same reaction vessel, and both reactions take place very rapidly at room temperature or below. The first step—oxymercuration—usually goes to completion within a period of 20 seconds to 10 minutes; the second step—demercuration—normally requires less than an hour. The overall reaction gives alcohols in very high yields, usually greater than 90%.

The oxymercuration-demercuration synthesis is also highly regiospecific. The net orientation of the addition of the elements of water is in accordance with Markovnikov's rule.

$$n\text{-}C_3H_7CH=CH_2 \xrightarrow[\text{THF-H}_2\text{O}]{\text{Hg(OAc)}_2^*} n\text{-}C_3H_7\overset{|}{\underset{OH}{C}}H-\overset{|}{\underset{HgOAc}{C}}H_2 \xrightarrow[\text{OH}^-]{\text{NaBH}_4} n\text{-}C_3H_7\overset{|}{\underset{OH}{C}}HCH_3 + Hg$$
(15 sec)   (1 hr)

1-Pentene   2-Pentanol
                                                              (93%)

$$CH_3CH_2\overset{CH_3}{\overset{|}{C}}=CH_2 \xrightarrow[\text{THF-H}_2\text{O}]{\text{Hg(OAc)}_2} CH_3CH_2\overset{CH_3}{\underset{OH}{\overset{|}{C}}}-\overset{|}{\underset{HgOAc}{C}}H_2 \xrightarrow[\text{OH}^-]{\text{NaBH}_4} CH_3CH_2\overset{CH_3}{\underset{OH}{\overset{|}{C}}}CH_3 + Hg$$
(10 sec)   (5 min)

2-Methyl-1-   2-Methyl-2-
butene   butanol
                                                              (90%)

1-Methylcyclo-   (20 sec.)   (6 min.)   1-Methylcyclo-
pentene   pentanol

$$* Hg(OAc)_2 = Hg(O\overset{O}{\overset{||}{C}}CH_3)_2$$

Rearrangements of the carbon skeleton seldom occur in oxymercuration-demercuration syntheses. A striking example that illustrates this characteristic of an oxymercuration-demercuration synthesis is the oxymercuration-demercuration of 3,3-dimethyl-1-butene shown below.

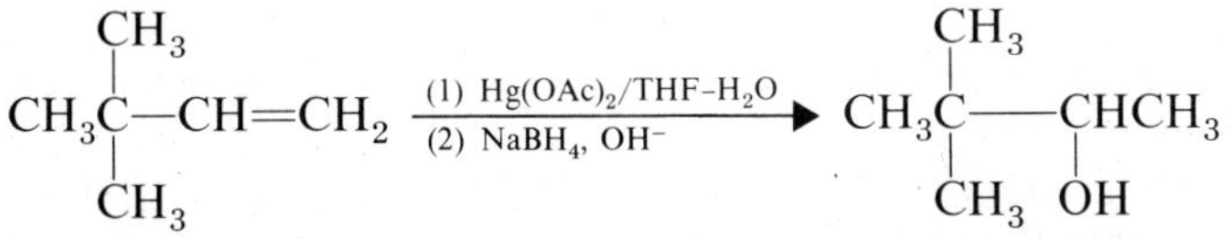

$$
\begin{array}{c}
CH_3 \\
| \\
CH_3\overset{|}{\underset{|}{C}}-CH{=}CH_2 \\
| \\
CH_3
\end{array}
\quad\xrightarrow[\text{(2) NaBH}_4,\ \text{OH}^-]{\text{(1) Hg(OAc)}_2/\text{THF–H}_2\text{O}}\quad
\begin{array}{c}
CH_3 \\
| \\
CH_3\overset{|}{C}{-}{-}\overset{}{C}HCH_3 \\
|\quad | \\
CH_3\ \ OH
\end{array}
$$

3,3-Dimethyl-1-butene → 3,3-Dimethyl-2-butanol (94%)

Analysis of the reaction mixture by gas-liquid chromatography failed to reveal the presence of any 2,3-dimethyl-2-butanol. The acid-catalyzed hydration of 3,3-dimethyl-1-butene, by contrast, gives 2,3-dimethyl-2-butanol as the major product.

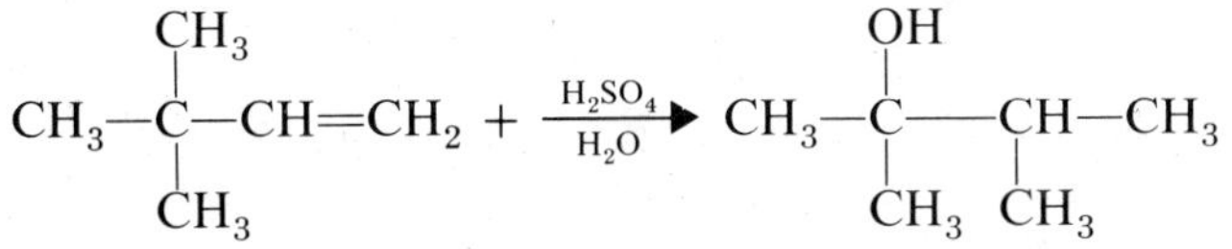

$$
\begin{array}{c}
CH_3 \\
| \\
CH_3-\overset{|}{\underset{|}{C}}-CH{=}CH_2 \\
| \\
CH_3
\end{array}
+\xrightarrow[\text{H}_2\text{O}]{\text{H}_2\text{SO}_4}
\begin{array}{c}
OH \\
| \\
CH_3-\overset{|}{C}{-}{-}CH{-}CH_3 \\
|\quad\ | \\
CH_3\ CH_3
\end{array}
$$

3,3-Dimethyl-1-butene → 2,3-Dimethyl-2-butanol (major product)

A mechanism that accounts for the orientation of addition in the oxymercuration stage, and one that also explains the general lack of accompanying rearrangements, has been proposed by Professor H. C. Brown of Purdue University. According to this mechanism, the first step of the oxymercuration reaction is an electrophilic attack by the mercury species, $\overset{+}{H}gOAc$, at the less substituted carbon of the double bond (i.e., at that carbon that bears the greater number of hydrogens). We can illustrate this step using 1-pentene in the following example.

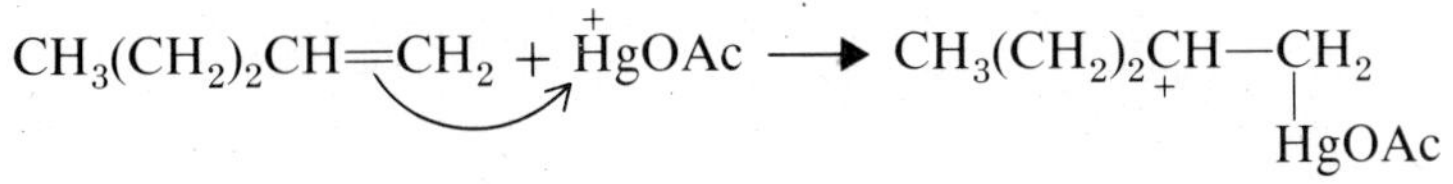

$$
CH_3(CH_2)_2CH{=}CH_2 + \overset{+}{H}gOAc \longrightarrow CH_3(CH_2)_2\overset{+}{C}H{-}CH_2 \\
\qquad\qquad\qquad\qquad\qquad\qquad\qquad\qquad\qquad\quad |\\
\qquad\qquad\qquad\qquad\qquad\qquad\qquad\qquad\qquad HgOAc
$$

1-Pentene   Mercury-substituted carbocation

The mercury-substituted carbocation produced in this way then reacts very rapidly with water to produce a (hydroxyalkyl) mercury compound.

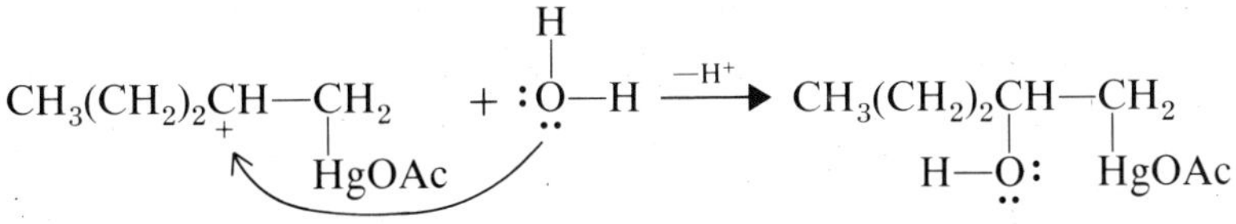

$$
CH_3(CH_2)_2\overset{+}{C}H{-}CH_2 \ + \ :\overset{H}{\underset{..}{O}}{-}H \xrightarrow{-H^+} CH_3(CH_2)_2\overset{}{C}H{-}CH_2 \\
\qquad\qquad\quad |\qquad\qquad\qquad\qquad\qquad\qquad\qquad\qquad |\qquad\quad |\\
\qquad\qquad HgOAc\qquad\qquad\qquad\qquad\qquad\qquad H{-}\underset{..}{O}:\ \ HgOAc
$$

Mercury-substituted carbocation   Hydroxyalkylmercury compound

Calculations indicate that mercury-substituted carbocations like those formed in this reaction retain much of the positive charge on the mercury moiety:

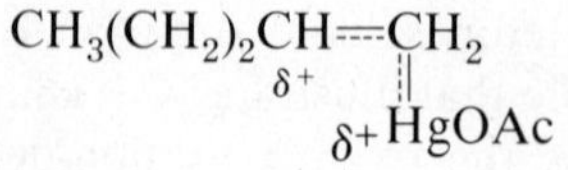

$$CH_3(CH_2)_2\overset{\delta^+}{CH}\!=\!\!\overset{|}{\underset{\delta^+ HgOAc}{CH_2}}$$

As a result, only a small portion of the positive charge resides on carbon. The charge on carbon is large enough to account for the observed Markovnikov addition, but it is too small to allow the usual rapid carbon-skeleton rearrangements that take place with full carbocations.

The mechanism for the demercuration stage is not well understood. Free radicals are thought to be involved and rearrangements occasionally occur.

Solvomercuration-Demercuration Reactions

When mercuration-demercuration reactions are carried out in solvents other than H_2O-THF, they can be used to synthesize ethers. H. C. Brown, who has developed these general procedures, has proposed that these reactions (including oxymercuration) be called *solvomercuration-demercuration* reactions.

In oxymercuration, water acts as a nucleophile and attacks the mercury-substituted carbocation; after demercuration the product is an *alcohol*. When the reaction is carried out in an alcohol solvent, however, the alcohol acts as a nucleophile and attacks the mercury-substituted carbocation; after demercuration the product is an *ether*:

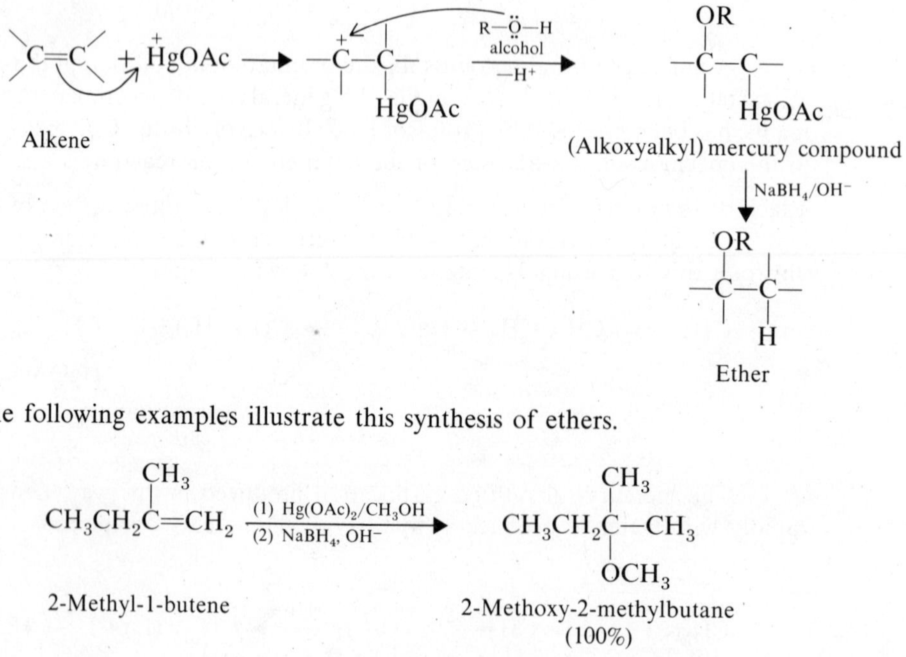

The following examples illustrate this synthesis of ethers.

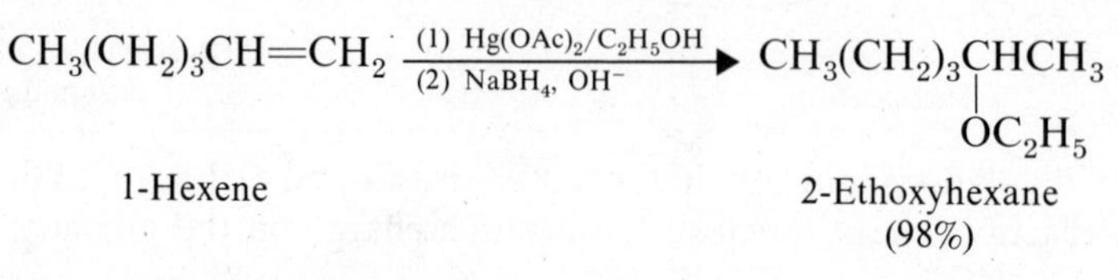

2-Methyl-1-butene → 2-Methoxy-2-methylbutane (100%)

1-Hexene → 2-Ethoxyhexane (98%)

When tertiary alcohols are used in this synthesis, best results are obtained by replacing mercuric acetate by mercuric trifluoroacetate, $Hg(OOCCF_3)_2$.

$$CH_3(CH_2)_3CH=CH_2 \xrightarrow[\text{(2) NaBH}_4,\ \text{OH}^-]{\text{(1) Hg(OOCCF}_3)_2/(CH_3)_3COH}} CH_3(CH_2)_3\underset{\underset{\underset{C(CH_3)_3}{|}}{\overset{|}{O}}}{C}HCH_3$$

1-Hexene	2-*tert*-Butoxyhexane
	(100%)

Problem 6.5

Outline all steps in the mechanism for the solvomercuration-demercuration synthesis of 2-ethoxyhexane from 1-hexene.

Problem 6.6

Show how solvomercuration (including oxymercuration)-demercuration syntheses could be used to prepare the following alcohols and ethers from the alkenes given.

(a) $CH_3-\underset{\underset{CH_3}{|}}{C}=CH-CH_3 \longrightarrow CH_3-\underset{\underset{CH_3}{|}}{\overset{\overset{OH}{|}}{C}}-CH_2-CH_3$

(b)

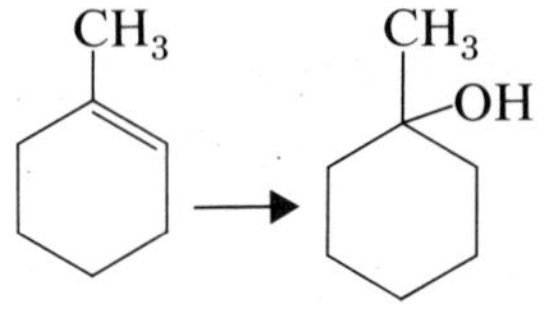

(c)

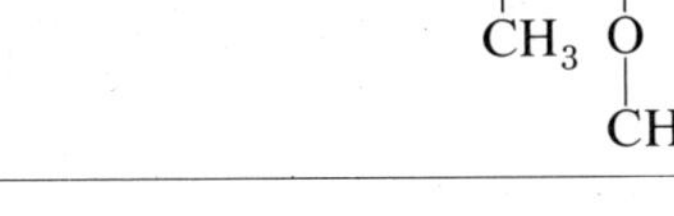

(d) $CH_3-\underset{\underset{CH_3}{|}}{\overset{\overset{CH_3}{|}}{C}}-CH=CH_2 \longrightarrow CH_3-\underset{\underset{CH_3}{|}}{\overset{\overset{CH_3}{|}}{C}}-\underset{\underset{O}{|}}{C}H-CH_3$
$\underset{CH_3}{|}$

HYDROBORATION-OXIDATION

Addition of the elements of water to a double bond can also be achieved in the laboratory through the use of the boron hydride $(BH_3)_2$, called diborane. The addition of water is an indirect one and two reactions are involved. The first reaction is the addition of a boron hydride to the double bond (hydroboration); the second reaction is the oxidation of the organoboron derivative to an alcohol and boric acid. We can illustrate these steps with the hydroboration-oxidation of propene.*

$$3CH_3CH{=}CH_2 + \tfrac{1}{2}(BH_3)_2 \longrightarrow (CH_3CH_2CH_2)_3B$$

$$\downarrow {\scriptstyle H_2O_2,\ OH^-}$$

$$3CH_3CH_2CH_2OH + H_3BO_3$$

n-Propyl alcohol   Boric acid

Both diborane and alkylboranes ignite spontaneously when exposed to air. To avoid this difficulty, diborane is usually generated in the presence of the alkene (*in situ*) by the addition of boron trifluoride in ether to a mixture of sodium borohydride and the alkene. Boron trifluoride and sodium borohydride react to produce diborane according to the following equation:

$$3Na^+BH_4^- + 4BF_3 \longrightarrow 2(BH_3)_2 + 3Na^+BF_4^-$$

Boron trifluoride and sodium borohydride are much more easily handled. Boron trifluoride is usually obtained in ethyl ether solution, where it forms the stable complex $(C_2H_5)_2O{:}BF_3$. Hydroboration reactions are usually carried out in ethers, either in diethyl ether $(C_2H_5)_2O$ itself or in some higher-molecular-weight ether such as "diglyme" $(CH_3OCH_2CH_2)_2O$, *di*ethylene *glyc*ol *di*methyl ether.

The alkylboranes produced by hydroboration are not usually isolated. They are oxidized to alcohols in the same reaction vessel by the addition of hydrogen peroxide in an aqueous base.

$$R_3B \xrightarrow[\substack{NaOH\ 25°\\ \text{Oxidation}}]{H_2O_2} 3R{-}OH + H_3BO_3$$

Hydroboration reactions are regiospecific and the net result of hydroboration-oxidation is an apparent *anti-Markovnikov addition of water*. As a consequence, *hydroboration-oxidation gives us a method for the preparation of alcohols that cannot normally be obtained through the acid-catalyzed hydration of alkenes or by oxymercuration-demercuration.* The following examples illustrate this.

$$CH_3CH_2CH_2CH_2CH{=}CH_2 \xrightarrow[\substack{(2)\ H_2O_2,\ OH^-}]{(1)\ (BH_3)_2} CH_3CH_2CH_2CH_2CH_2CH_2OH$$

1-Hexene   1-Hexanol (90%)

$$\underset{\displaystyle \text{3,3-Dimethyl-1-butene}}{\overset{\displaystyle CH_3}{\underset{\displaystyle CH_3}{CH_3{-}\overset{|}{\underset{|}{C}}{-}CH{=}CH_2}}} \xrightarrow[\substack{(2)\ H_2O_2,\ OH^-}]{(1)\ (BH_3)_2} \underset{\displaystyle \text{3,3-Dimethyl-1-butanol (67\%)}}{\overset{\displaystyle CH_3}{\underset{\displaystyle CH_3}{CH_3{-}\overset{|}{\underset{|}{C}}{-}CH_2CH_2OH}}}$$

$$\underset{\displaystyle \text{2-Methyl-2-butene}}{\overset{\displaystyle CH_3}{CH_3{-}\overset{|}{C}{=}CHCH_3}} \xrightarrow[\substack{(2)\ H_2O_2,\ OH^-}]{(1)\ (BH_3)_2} \underset{\displaystyle \text{3-Methyl-2-butanol (59\%)}}{\overset{\displaystyle CH_3}{CH_3{-}\overset{|}{CH}{-}\underset{\displaystyle OH}{\overset{|}{CH}CH_3}}}$$

---

* Hydroboration reactions were discovered by H. C. Brown (p. 201). They have proved to be of great synthetic versatility. We will see other examples in subsequent chapters.

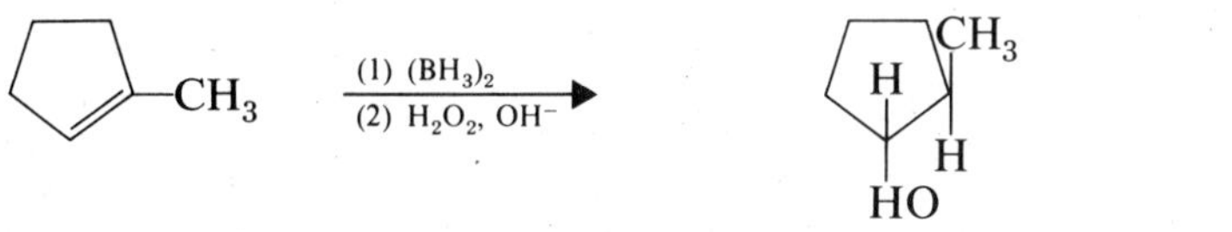

$$\text{1-Methylcyclopentene} \xrightarrow[\text{(2) H}_2\text{O}_2,\ \text{OH}^-]{\text{(1) (BH}_3)_2} \text{trans-2-Methylcyclopentanol (86\%)}$$

1-Methylcyclopentene   *trans*-2-Methylcyclopentanol (86%)

Mechanism of Hydroboration

Hydroboration, the first step of the hydroboration-oxidation synthesis, occurs in a stepwise manner. Diborane $(BH_3)_2$, acts as if it were in a monomeric form, as if it were borane, BH_3. Borane adds successively to the double bonds of three molecules of the alkene until all of its hydrogens are replaced by alkyl groups.

$$CH_3CH=CH_2 \longrightarrow CH_3CH_2CH_2-BH_2 \xrightarrow{CH_3CH=CH_2} (CH_3CH_2CH_2)_2BH$$
$$+ \qquad\qquad\qquad\qquad\qquad\qquad \downarrow CH_3CH=CH_2$$
$$H-BH_2 \qquad\qquad\qquad\qquad\qquad\qquad (CH_3CH_2CH_2)_3B$$

Tripropylboron

In each addition step the boron atom becomes attached to the *less substituted carbon of the double bond* and a hydrogen is transferred to the other carbon of the double bond.

The addition of borane is consistent with what we have learned about the stabilities of carbocations. To see how this can be said let us consider mechanism for the addition of borane.

Experimental evidence indicates that the addition of borane to the double bond occurs *in a single step* through a four-center transition state. The orientation predicted by this single-step mechanism is in accord with the observed attachment of boron to the less substituted carbon of the double bond. An addition occurring through a four-center transition state is shown in Fig. 6.2 using propene as an example.

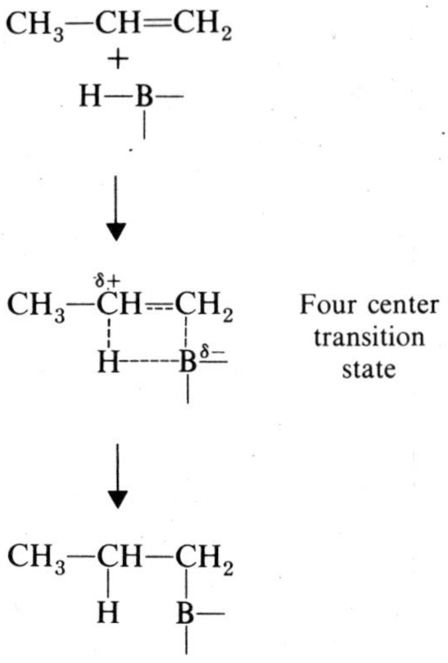

$$CH_3-CH=CH_2$$
$$+$$
$$H-B-$$
$$\downarrow$$
$$CH_3-\overset{\delta+}{C}H\cdots CH_2 \quad \text{Four center}$$
$$H\cdots\cdots\overset{\delta-}{B} \quad\quad \text{transition}$$
$$\text{state}$$
$$\downarrow$$
$$CH_3-CH-CH_2$$
$$H \quad B-$$

FIG. 6.2
The one-step mechanism for the addition of a boron hydride to propene. In the transition state the formation of the carbon- boron bond has taken place to a greater extent than the formation of the carbon-hydrogen bond. Consequently, a partial positive charge develops on the secondary carbon of propene.

FIG. 6.3

A hypothetical transition state for the reverse addition of a boron hydride to propene. This transition state is of higher energy than that shown in Fig. 6.2, because here, the partially positive carbon resembles a primary carbocation.

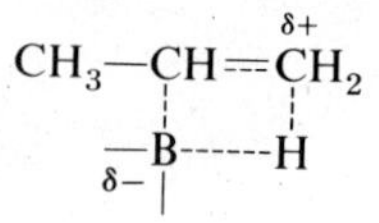

$$CH_3-CH\!\!=\!\!\overset{\delta+}{CH_2}$$
$$\underset{\delta-}{-}\!\!-B\text{-----}H$$

The driving force for hydroboration is the formation of the carbon-boron covalent bond. The carbon-boron bond forms most rapidly and formation of the carbon-hydrogen bond lags behind. Since the carbon-boron bond is formed at the expense of the π electrons of the double bond, a positive charge begins to develop on the central carbon of propene and the transition state resembles a *partial secondary carbocation*.

Were the addition to occur in the reverse way the transition state would resemble a less stable primary carbocation (Fig. 6.3).

Syn and Anti Additions

An addition reaction that places the parts of the adding reagent on the same side (or face) of the reactant is called a *syn* addition.

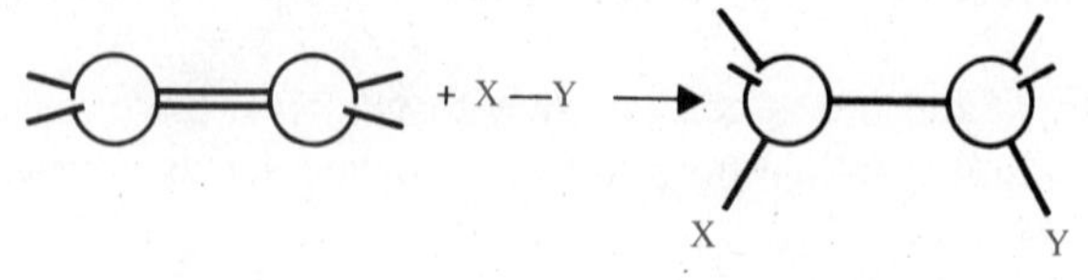

a syn addition

The opposite of a syn addition is an *anti* addition. An antiaddition places the parts of the adding reagent on opposite faces of the reactant.

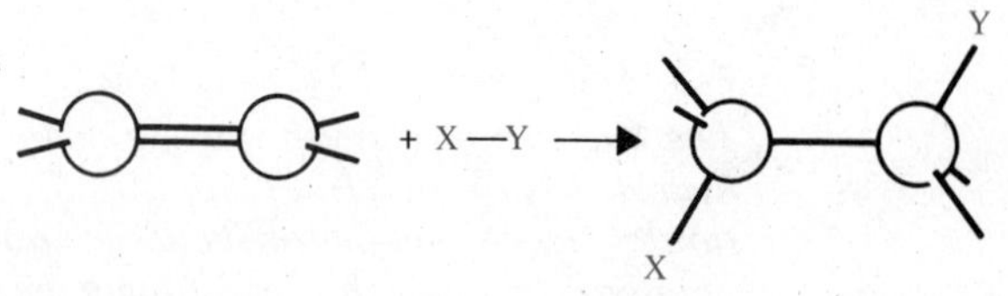

an anti addition

Syn addition of Boron Hydrides

The four-center transition state for the addition of boron hydrides to a double bond requires the addition of both boron and hydrogen to the same face of the molecule. The addition of boron hydrides is then, of necessity, a syn addition. With 1-methylcyclopentene, for example, both boron and hydrogen add to the same face of the ring (Equation 1, Fig. 6.4).

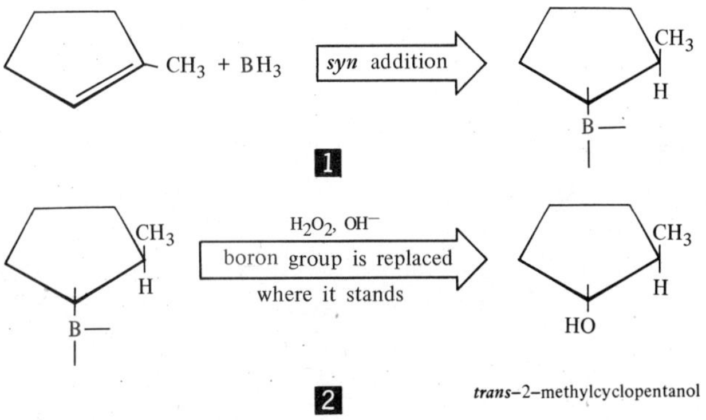

FIG. 6.4
The hydroboration-oxidation of 1-methylcyclopentene. The first reaction is a syn *addition of boron hydride. (In this illustration we have shown the boron and hydrogen both entering from the bottom side of 1-methylcyclopentene. We could also have shown the reaction with both groups entering from the top.) In the second reaction the boron atom is replaced by a hydroxyl group on the same side of the molecule. The product is a* trans *compound (*trans-*2-methylcyclopentanol) and the overall result is the* syn *addition of —H and —OH.*

Oxidation of Alkylboranes

The mechanism for the oxidation reaction in the hydroboration-oxidation synthesis of alcohols is beyond the scope of our present discussion. One important point can be made about the oxidation reaction, however: that *the hydroxyl group replaces the boron atom where it stands in the organoboron compound.* The net result of the two reactions (hydroboration and oxidation) is the syn addition of —H and —OH, that is, the pieces of a water molecule. We can see this if we examine the hydroboration-oxidation of 1-methylcyclopentene (Fig. 6.4).

Problem 6.7

Show how you might employ hydroboration-oxidation reactions to carry out the following syntheses.

(a) 1-Butene $\longrightarrow$ $CH_3CH_2CH_2CH_2OH$

(b) 2-Methyl-2-butene $\longrightarrow$ $CH_3\overset{\displaystyle CH_3}{\underset{\displaystyle OH}{CHCHCH_3}}$

(c) 1-Methylcyclohexene $\longrightarrow$

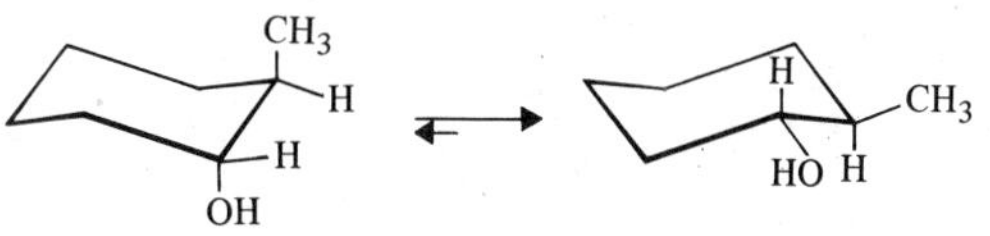

Problem 6.8

Alkylboranes react with acetic acid in the following way:

$$R-B- \xrightarrow[\text{heat}]{CH_3COOH} R-H + CH_3\overset{\text{O}}{\underset{\|}{C}}-O-B-$$

Alkylborane   Alkane

In this reaction hydrogen replaces boron where it stands in the alkylborane. With this fact in mind, and assuming you have deuterioacetic acid (CH_3COOD) available, can you suggest syntheses of the following deuterium-labeled compounds?

(a) $CH_3\overset{CH_3}{\underset{|}{C}HCH_2D}$

(b)

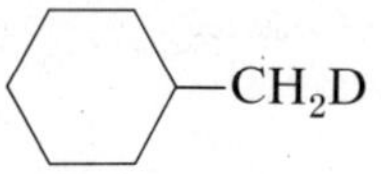

—CH_2D

(c)

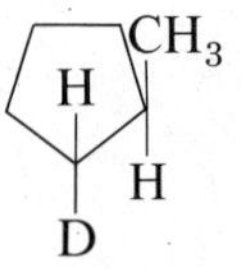

(d) Assume you also have available $(BD_3)_2$. Can you suggest a synthesis of the following?

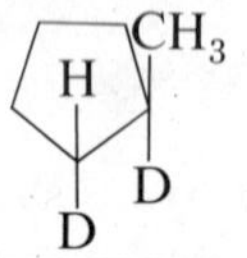

6.6 ADDITION OF HALOGENS TO ALKENES

In the absence of light, *alkanes* do not react appreciably with chlorine or bromine at room temperature. If we add an alkane to a solution of bromine in carbon tetrachloride, the red-brown color of the bromine will persist in the solution as long as we keep the mixture away from sunlight and as long as the solution is not heated.

$$R-H \quad + \quad Br_2 \quad \xrightarrow[\text{in the dark, } CCl_4]{\text{room temperature}} \text{ no appreciable reaction}$$

Alkane   Bromine
(colorless)   (red-brown)

On the other hand, if we expose the reactants to sunlight, the bromine color will fade rapidly. If we now place a small piece of moist blue litmus in the region above the liquid, the litmus will turn red because of the hydrogen bromide that evolves as the alkane and bromine react. (Hydrogen bromide is not very soluble in carbon tetrachloride.)

$$R-H \quad + \quad Br_2 \quad \xrightarrow[\text{sunlight, } CCl_4]{\text{room temperature}} \quad R-Br \quad + \quad HBr$$

Alkane   Bromine   Alkyl halide   (detected by
(colorless)   (red-brown)   (colorless)   moist blue litmus)

The behavior of *alkenes* toward bromine in carbon tetrachloride contrasts markedly with that of alkanes. Alkenes react rapidly with bromine at room temperature and in the *absence of light*. If we add bromine to an alkene the red-brown color of the bromine disappears almost instantly as long as the alkene is present in excess. If we test the atmosphere above the solution with moist blue litmus, we will find that no hydrogen bromide is present. The reaction is one of addition.

$$\text{C=C} + Br_2 \xrightarrow[\text{in the dark, } CCl_4]{\text{room temperature}} -\underset{Br}{\underset{|}{C}}-\underset{Br}{\underset{|}{C}}-$$

An alkene   *vic*-Dibromide
(colorless)   (colorless)

Because of these differences, bromine in carbon tetrachloride is a useful reagent for distinguishing between alkanes and alkenes.

The addition reaction between alkenes and chlorine or bromine is a general one. The products are vicinal dihalides.

$$CH_3CH=CHCH_3 + Cl_2 \xrightarrow[-9°]{O_2^*} CH_3\underset{Cl}{\underset{|}{C}}H\underset{Cl}{\underset{|}{C}}HCH_3 \quad (100\%)$$

$$CH_3CH_2CH=CH_2 + Cl_2 \xrightarrow[-9°]{O_2} CH_3CH_2\underset{Cl}{\underset{|}{C}}H-\underset{Cl}{\underset{|}{C}}H_2 \quad (97\%)$$

trans-1,2-Dibromocyclohexane

Mechanism of Halogen Addition

One mechanism that has been proposed for halogen addition is an ionic mechanism.† In the first step the exposed electrons of the π bond of the alkene attack the halogen to form a π complex. In a bromine addition, for example, π complex formation probably takes place in the following way:

Step 1

As the π electrons of the alkene approach the bromine molecule the electrons of the bromine-bromine bond drift in the direction of the bromine atom more distant from the approaching alkene. The bromine molecule becomes *polarized* as a result. The more distant bromine develops a partial negative charge; the nearer bromine becomes partially positive:

*Oxygen acts as a free-radical inhibitor.

† There is evidence that in the absence of oxygen some reactions between alkenes and chlorine proceed through a free-radical mechanism. We will not discuss this mechanism here, however.

Polarization weakens the bromine-bromine bond, and in the next step *it breaks heterolytically*. A bromide ion departs, and a *bromonium ion* is formed. The

Step 2

π complex   Bromonium ion   Bromide ion

bromonium ion has a bromine atom that is bonded to two carbon atoms by *two pairs of electrons*. This is possible because of the unshared electron pairs on bromine.

Bromonium ion

In the third step, one of the bromide ions produced in step 2 attacks one of the carbon atoms of the bromonium ion. This nucleophilic attack results in the formation of a *vic*-dibromide.

Step 3

vic-Dibromide

Bromohydrin Formation

If the bromination of an alkene is carried out in aqueous solution (rather than carbon tetrachloride) the major product of the overall addition reaction is not a *vic*-dibromide, but rather a *bromoalcohol*.

$$\text{RCH=CHR} + \text{Br}_2 \xrightarrow{\text{H}_2\text{O}} \underset{\substack{\text{Bromohydrin} \\ \text{(major product)}}}{\overset{\displaystyle \text{RCH—CHR}}{\underset{\displaystyle \text{Br} \quad \text{OH}}{\big| \quad \big|}}} + \text{HBr} + \underset{\substack{\textit{vic}\text{-Dibromide} \\ \text{(minor product)}}}{\overset{\displaystyle \text{RCH—CHR}}{\underset{\displaystyle \text{Br} \quad \text{Br}}{\big| \quad \big|}}}$$

Bromoalcohols of this type are commonly called *bromohydrins,* and the formation of bromohydrins can be easily explained in terms of the mechanism we have just proposed. The first two steps of the mechanism for bromohydrin formation are the same as for bromine addition. However, in the third step water—acting as a nucleophile—attacks the bromonium ion. The three-membered ring of the bromonium ion opens, and a protonated alcohol is produced. Loss of a proton from the protonated alcohol leads to the formation of the bromohydrin.

Water, because of its unshared electron pairs, acts as a nucleophile in this and in many other reactions. In this instance water molecules far outnumber bromide ions because water is the solvent for the reaction. This accounts for the bromohydrin being the major product.

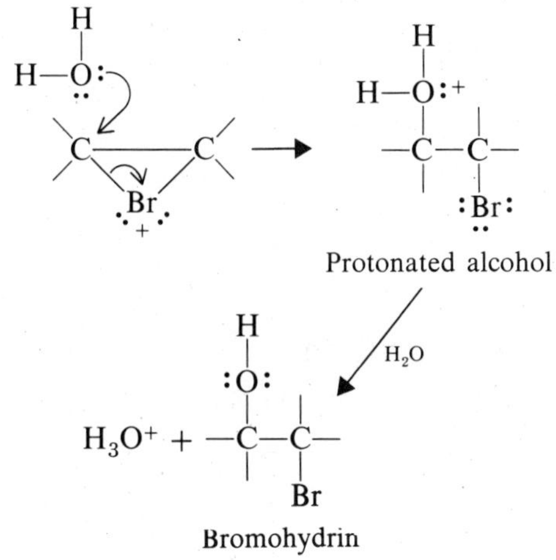
H
H—O:
C
C
Br
+
H
H—O:+
—C—C—
:Br:
Protonated alcohol
H
:O:
H_2O
H_3O^+ +
—C—C—
Br
Bromohydrin

Anti Addition of Bromine

The addition of bromine to cyclic alkenes provides additional evidence for bromonium ion intermediates in bromine additions. When cyclopentene reacts with bromine in carbon tetrachloride, *anti addition* occurs, and the product of the reaction is *trans*-1,2-dibromocyclopentane. The anti addition of bromine to cyclopentene can be accounted for by the mechanism given in Fig. 6.5—one in which a bromide ion attacks a carbon of the ring from the side opposite that of the bromonium ion.

FIG. 6.5
Anti *addition of bromine to cyclopentene. (The formation of the* π *complex is omitted for convenience.) In this reaction the bromide ion attacks a carbon atom of the bromonium ion from a side opposite the one to which the bromine is already attached because this provides the most accessible route.*

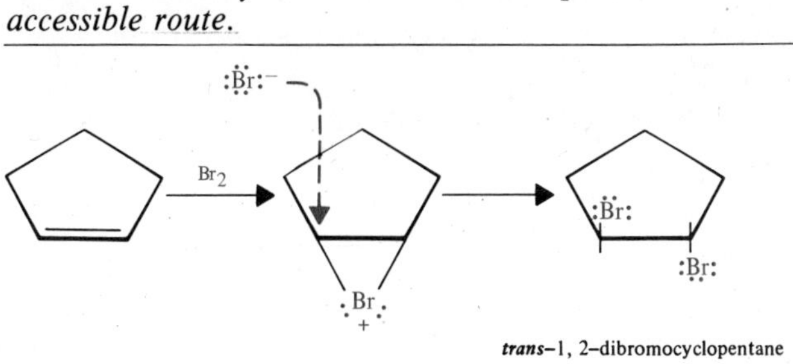
:Br:⁻
Br_2
Br
+
:Br:
:Br:
trans–1, 2–dibromocyclopentane

Problem 6.9

Outline a mechanism that accounts for the formation of *trans*-2-chlorocyclopentanol from cyclopentene and chlorine in aqueous solution.

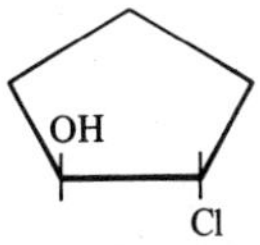
OH
Cl

trans–2–chlorocyclopentanol

All halogen additions to double bonds are not antiadditions, nor do all additions proceed through a halonium ion intermediate. We will see, in Chapter 13, that syn additions occur frequently when phenyl groups (C_6H_5 groups) are attached to one or both carbons of the double bond. In these instances it appears that carbocations and carbocationlike intermediates are formed instead of halonium ions, because the carbocations are quite stable.

6.7 EPOXIDES: EPOXIDATION OF ALKENES

Epoxides are cyclic ethers with three-membered rings:

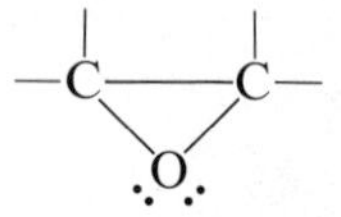

An epoxide

In IUPAC nomenclature epoxides are called *oxiranes*. More commonly, however, they are called *alkene oxides*.

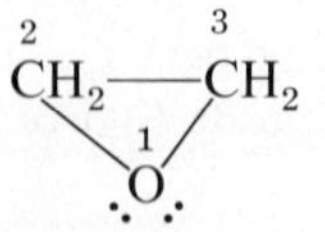

IUPAC name: oxirane
Common name: ethylene oxide

The common names for epoxides originate from the most widely used method for synthesizing them: *epoxidation*, the reaction of an alkene and an organic peracid.

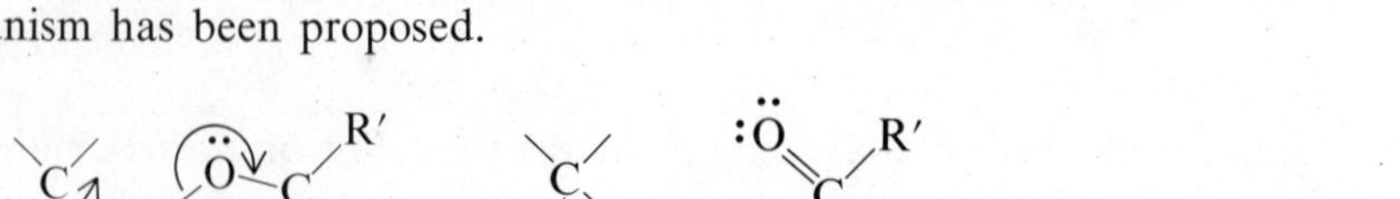

$$RCH=CHR + R'\overset{O}{\overset{\|}{C}}-O-OH \xrightarrow{epoxidation} RCH-CHR + R'\overset{O}{\overset{\|}{C}}-OH$$

An alkene   A peracid   An alkene oxide
                                               (or oxirane)

In this reaction the peracid transfers an oxygen atom to the alkene. The following mechanism has been proposed.

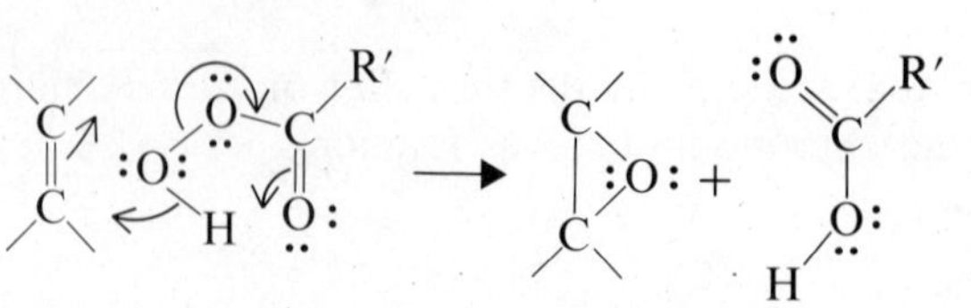

The addition of oxygen to the double bond in an epoxidation reaction is, of necessity, a *syn* addition. With cyclopentene, for example, epoxidation gives the following result.

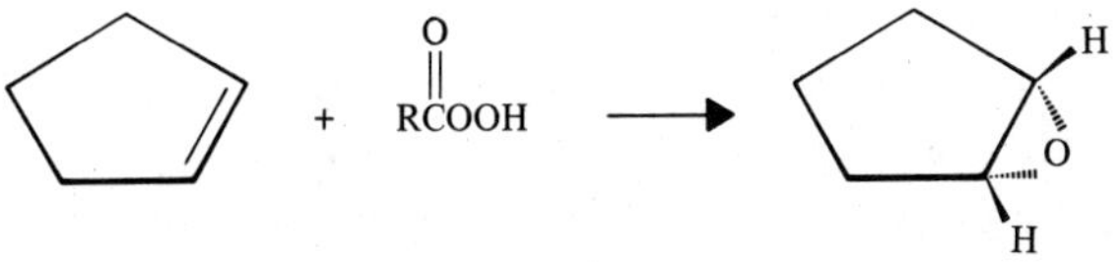

cyclopentene   cyclopentene oxide

The peracids most commonly used are performic acid (HĊOOH), peracetic acid (CH$_3$ĊOOH), and perbenzoic acid (C$_6$H$_5$ĊOOH). Cyclohexene, for example, reacts with perbenzoic acid to give cyclohexene oxide in a quantitative yield.

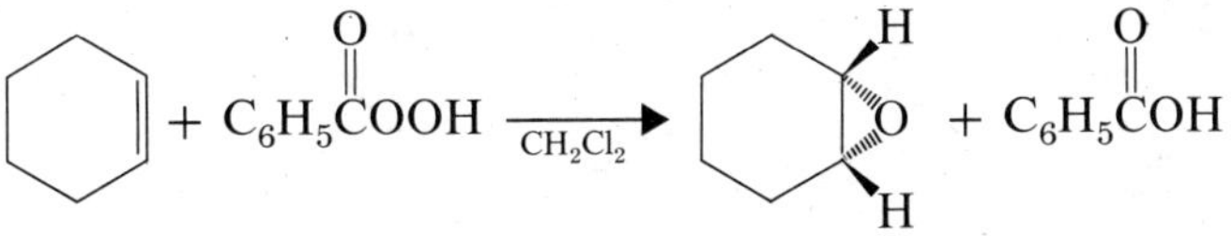

Perbenzoic   Cyclohexene
acid   oxide
                 (100%)

Acid-Catalyzed Hydrolysis of Epoxides

Although most ethers react with few reagents, the strained three-membered ring of epoxides makes them highly susceptible to ring-opening reactions. Ring opening takes place through cleavage of one of the carbon-oxygen bonds. It can be initiated by either electrophiles or nucleophiles, or catalyzed by either acids or bases. The acid-catalyzed hydrolysis of an epoxide, for example, is a useful procedure for preparing vicinal-dihydroxy compounds called glycols.

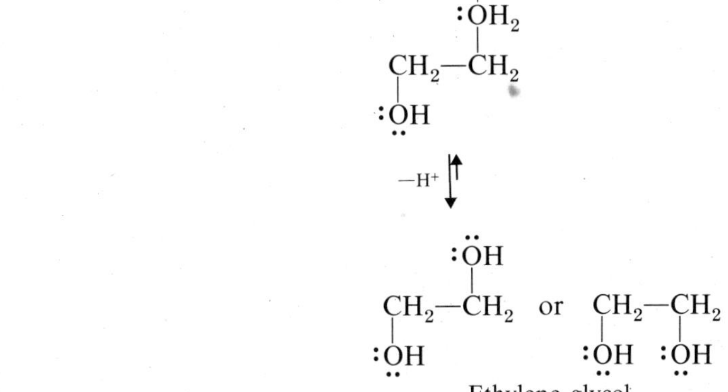

Ethylene glycol

Antihydroxylation of Alkenes

When cyclopentene oxide is subjected to acid-catalyzed hydrolysis, water attacks from the side opposite the epoxide ring. The product of the reaction is a *trans*-glycol (*trans*-1,2-cyclopentanediol). Glycols are "dialcohols" or diols.

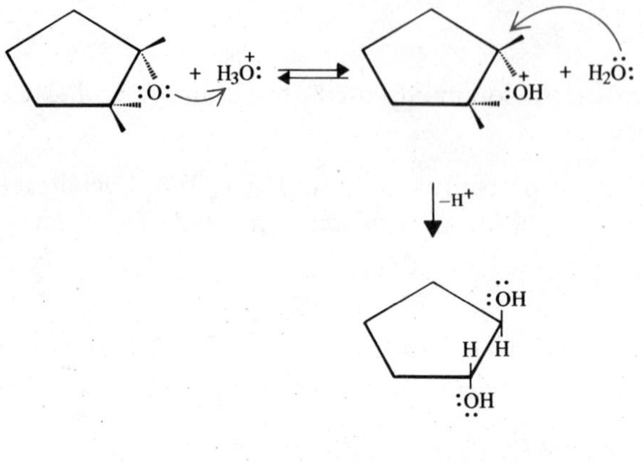

trans—1, 2—cyclopentanediol

Epoxidation followed by acid-catalyzed hydrolysis gives us, therefore, a method for *hydroxylating* a double bond (i.e., a method for adding a hydroxyl group to each carbon). This hydroxylation technique, moreover, amounts to a net anti-hydroxylation, and its mechanism parallels closely the mechanism for bromination of an alkene given earlier (p. 211).

6.8 OXIDATION OF ALKENES

Alkenes undergo a number of other reactions in which the carbon-carbon double bond is oxidized. Potassium permanganate or osmium tetroxide, for example, can also be used to oxidize alkenes to glycols.

$$CH_2{=}CH_2 + KMnO_4 \xrightarrow[OH^-]{cold} \underset{\underset{\displaystyle OH \quad\;\; OH}{|\qquad\;\; |}}{CH_2{-}CH_2}$$

Ethylene · · · · · · · · · · · · · · · · Ethylene glycol

$$CH_3CH{=}CH_2 \xrightarrow[\text{(2) Na}_2\text{SO}_3]{\text{(1) OsO}_4} \underset{\underset{\displaystyle OH \quad\;\; OH}{|\qquad\;\; |}}{CH_3CH{-}CH_2}$$

Propylene · · · · · · · · · · · · · · · · Propylene glycol

Syn Hydroxylation of Alkenes

The mechanisms for glycol formation by permanganate ion and osmium tetroxide oxidations involve the formation of cyclic intermediates.

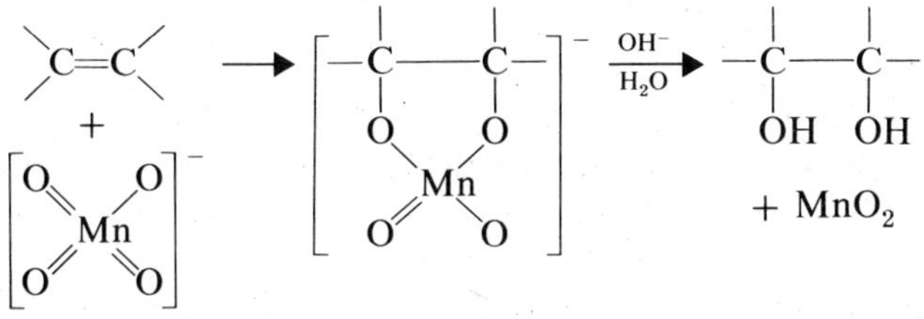

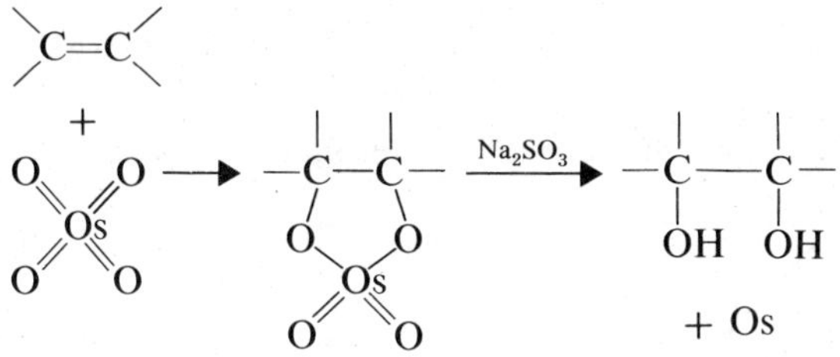

An osmate
ester

The course of these reactions is *syn hydroxylation*. This can be seen, readily, when cyclopentene reacts with cold dilute potassium permanganate (in base) or with osmium tetroxide (followed by treatment with Na_2SO_3). The product in either case is *cis*-1,2-cyclopentanediol.

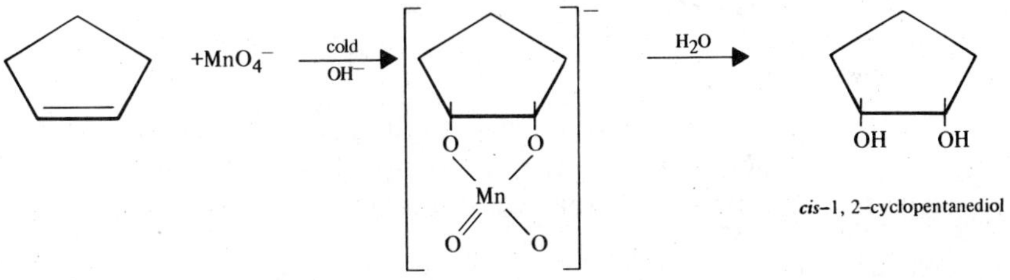

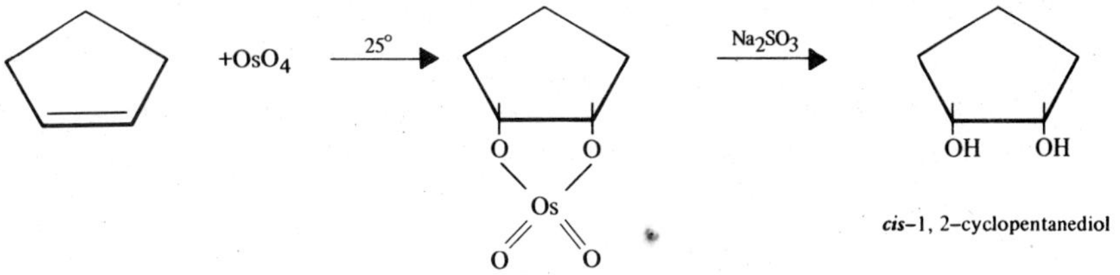

Of the two reagents used for syn hydroxylation, osmium tetroxide gives the highest yields. Unfortunately, however, osmium tetroxide is highly toxic and very expensive. Potassium permanganate is a very powerful oxidizing agent and is easily capable of causing further oxidation of the glycol. Limiting the reaction to hydroxylation alone is often difficult, but is usually attempted by using cold, dilute, and basic solutions of potassium permanganate.

Oxidative Cleavage of Alkenes

Alkenes are oxidatively cleaved to carboxylic acid salts by hot permanganate solutions. We can illustrate this reaction with the oxidative cleavage of either *cis*- or *trans*-2-butene to two moles of acetate ion.

$$CH_3CH{=}CHCH_3 \xrightarrow[\text{heat}]{MnO_4^-} 2CH_3C{\overset{O}{\underset{O^-}{\diagdown}}} \xrightarrow{H^+} 2CH_3C{\overset{O}{\underset{OH}{\diagdown}}}$$

| (*cis* or *trans*) | Acetate ion | Acetic acid |

Acidification of the reaction mixture, after the oxidation is complete, produces two moles of acetic acid.

The terminal CH_2 group of a 1-alkene is completely oxidized to carbon dioxide and water by hot permanganate.

$$CH_3CH_2CH{=}CH_2 \xrightarrow[\text{(2) } H^+]{\substack{\text{(1) } KMnO_4, H_2O, \\ \text{heat}}} CH_3CH_2\overset{O}{\overset{\|}{C}}OH + CO_2 + H_2O$$

Propanoic acid

The oxidative cleavage of alkenes has frequently been used to locate the position of the double bond in an alkene chain or ring. We can see how this might be done with the following example:

Example

An unknown alkene with the formula C_8H_{16} was found, on oxidation with hot permanganate, to yield a five-carbon carboxylic acid (pentanoic acid) and a three-carbon carboxylic acid (propanoic acid).

$$C_8H_{16} \xrightarrow[\text{(2) } H^+]{\substack{\text{(1) } KMnO_4, H_2O, \\ OH^-, \Delta}} CH_3CH_2CH_2CH_2\overset{O}{\overset{\|}{C}}{-}OH + CH_3CH_2\overset{O}{\overset{\|}{C}}{-}OH$$

Pentanoic acid   Propanoic acid

We can see, then, from what we know about oxidative cleavage by permanganate, that the original alkene must have been either *cis*- or *trans*-3-octene

$$CH_3CH_2CH{=}CHCH_2CH_2CH_2CH_3$$

3-Octene
(*cis* or *trans*)

Ozonization (Ozonolysis) of Alkenes

A more widely used method for locating the double bond of an alkene involves the use of ozone, O_3. Ozone reacts with alkenes to form unstable compounds called molozonides.

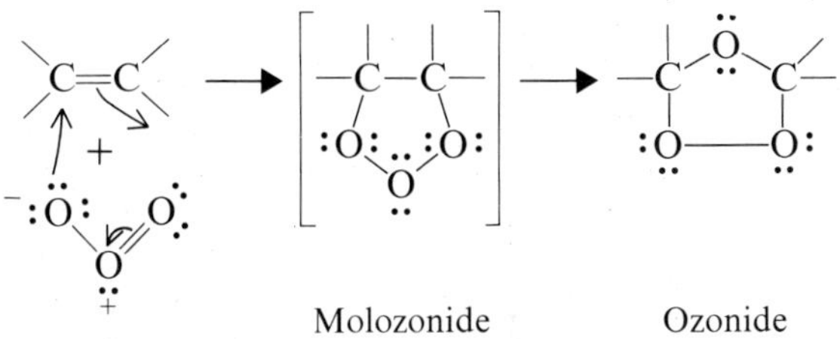

Molozonide Ozonide

The molozonide that is formed initially rearranges spontaneously to produce a compound known as an *ozonide*. This rearrangement is thought to occur through dissociation of the molozonide into reactive fragments that recombine to yield the ozonide.

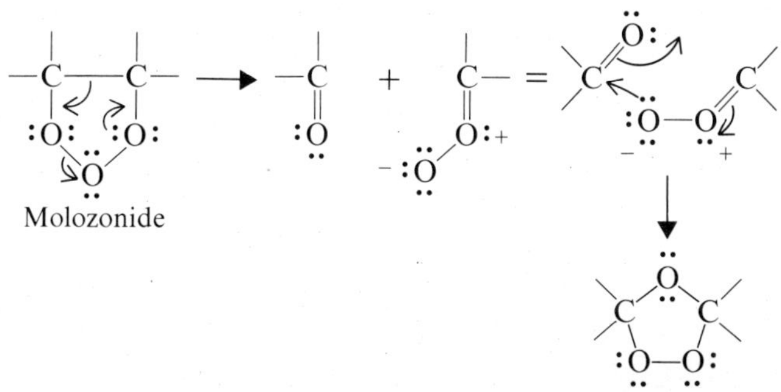

Molozonide

Ozonide

Ozonides, themselves, are relatively unstable compounds and may often explode violently. Because of this property they are not usually isolated, but are reduced directly by treatment with zinc and water. The reduction produces carbonyl compounds (either aldehydes or ketones) that can be safely isolated and identified.

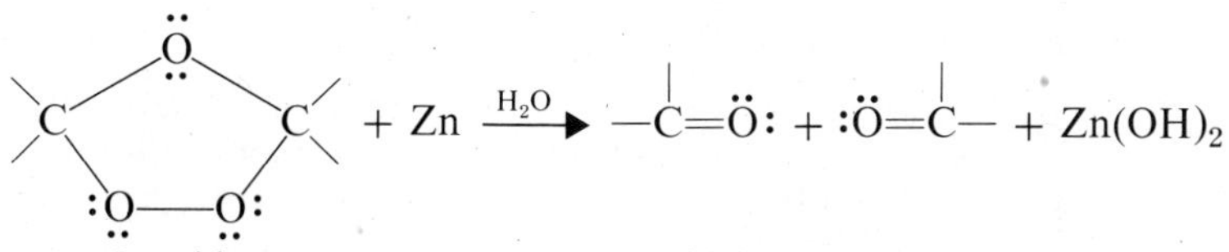

Ozonide

Aldehydes and/or ketones

Once the identities of the aldehydes and/or ketones are established the location of the double bond in the original alkene can be ascertained. The examples listed below will illustrate the kinds of products that result from ozonization and subsequent treatment with zinc and water.

$$CH_3\overset{\overset{\displaystyle CH_3}{|}}{C}{=}CHCH_3 \xrightarrow[(2)\ Zn,\ H_2O]{(1)\ O_3} CH_3\overset{\overset{\displaystyle CH_3}{|}}{C}{=}O + CH_3\overset{\overset{\displaystyle O}{||}}{C}H$$

2-Methyl-2-butene (*cis* or *trans*) — Acetone — Acetaldehyde

$$CH_3\overset{\overset{\displaystyle CH_3}{|}}{C}H{-}CH{=}CH_2 \xrightarrow[(2)\ Zn,\ H_2O]{(1)\ O_3} CH_3\overset{\overset{\displaystyle CH_3}{|}}{C}H{-}\overset{\overset{\displaystyle O}{||}}{C}H + H\overset{\overset{\displaystyle O}{||}}{C}H$$

3-Methyl-1-butene — Isobutyraldehyde — formaldehyde

Problem 6.10

Write the general structures of the alkenes that would produce the following products when treated with ozone and then with zinc and water.

(a) $CH_3CH_2\overset{\overset{\displaystyle O}{\|}}{C}H$ + $CH_3\overset{\overset{\displaystyle O}{\|}}{C}H$

(b) $CH_3-\underset{\underset{\displaystyle CH_3}{|}}{C}=O$ only (Two moles are produced from one mole of alkene)

(c) $CH_3CH_2\underset{\underset{\displaystyle CH_3}{|}}{C}H-\overset{\overset{\displaystyle O}{\|}}{C}H$ + $H\overset{\overset{\displaystyle O}{\|}}{C}H$

(d) $H-\overset{\overset{\displaystyle O}{\|}}{C}CH_2CH_2CH_2CH_2\overset{\overset{\displaystyle O}{\|}}{C}-H$ only

6.9 FREE-RADICAL ADDITION TO ALKENES: THE ANTI-MARKOVNIKOV ADDITION OF HYDROGEN BROMIDE

Before 1933, the orientation of hydrogen bromide additions to alkenes was the subject of much confusion. At times addition of hydrogen bromide occurred in accordance with Markovnikov's rule; at other times it occurred in just the opposite manner. Many instances were reported where, under what seemed to be the same experimental conditions, Markovnikov additions were obtained in one laboratory and anti-Markovnikov additions in another. At times even the same chemist would obtain different results using the same conditions but on different occasions.

The mystery was solved in 1933 by the research of M. S. Kharasch and F. R. Mayo. The culprit turned out to be organic peroxides present in the alkenes—peroxides that were formed by the action of atmospheric oxygen on the alkenes. Kharasch and Mayo found that when alkenes that contained peroxides reacted with hydrogen bromide, anti-Markovnikov addition of hydrogen bromide occurred.

$$R-\ddot{\underset{\cdot\cdot}{O}}-\ddot{\underset{\cdot\cdot}{O}}-R$$

An organic peroxide

Under these conditions, for example, propene yields 1-bromopropane. In the absence of peroxides, or in the presence of compounds that would "trap" free radicals, normal Markovnikov addition occurs.

$$CH_3CH=CH_2 + HBr \xrightarrow{\text{ROOR}} CH_3CH_2CH_2Br \quad \text{Anti-Markovnikov addition}$$

$$CH_3CH=CH_2 + HBr \xrightarrow[\text{peroxides}]{\text{no}} CH_3\underset{\underset{\displaystyle Br}{|}}{C}HCH_3 \quad \text{Markovnikov addition}$$

2-Bromopropane

We have already seen the carbocation mechanism (Section 6.2) that accounts for Markovnikov addition of hydrogen bromide, so we need not concern ourselves with it here. The mechanism of anti-Markovnikov addition is of concern, however.

According to Kharasch and Mayo, the mechanism for anti-Markovnikov addition of hydrogen bromide is a free-radical chain reaction initiated by peroxides.

Chain-initiating steps

(1) $R—\overset{..}{\underset{..}{O}}:\overset{..}{\underset{..}{O}}—R \xrightarrow{\text{heat}} 2R—\overset{..}{\underset{..}{O}}·$

(2) $R—\overset{..}{\underset{..}{O}}· + H:\overset{..}{\underset{..}{Br}}: \longrightarrow R—\overset{..}{\underset{..}{O}}:H + :\overset{..}{\underset{..}{Br}}·$

Chain-propagating steps

(3) $:\overset{..}{\underset{..}{Br}}· + CH_2—CHCH_3 \longrightarrow :\overset{..}{\underset{..}{Br}}:CH_2CHCH_3$

2° Free radical

(4) $:\overset{..}{\underset{..}{Br}}—CH_2CHCH_3 + H:\overset{..}{\underset{..}{Br}}: \longrightarrow BrCH_2CHCH_3 + ·\overset{..}{\underset{..}{Br}}:$
$\underset{H}{}$

1-Bromopropane

then (3), (4), (3), (4), and so on.

Step 1 is the simple homolytic cleavage of the peroxide molecule to produce two peroxy free radicals. The oxygen-oxygen bond of peroxides is weak and such reactions are known to occur readily.

$$R—O:O—R \rightarrow 2R—O· \qquad \Delta H \cong +35 \text{ kcal/mole}$$

A question arises about step 2 of the mechanism, however. Why does not the following reaction occur instead?

$$R—\overset{..}{\underset{..}{O}}· + H:\overset{..}{\underset{..}{Br}}: \longrightarrow R—\overset{..}{\underset{..}{O}}:\overset{..}{\underset{..}{Br}}: + H·$$

The answer can be found in the thermodynamic quantities associated with the two possibilities. Step 2 of the mechanism, abstraction of a hydrogen atom by the peroxyradical, is exothermic, and has a low energy of activation. The alternative

$$R—\overset{..}{\underset{..}{O}}· + H:\overset{..}{\underset{..}{Br}}: \longrightarrow R—\overset{..}{\underset{..}{O}}:H + :\overset{..}{\underset{..}{Br}}· \qquad \begin{array}{l}\Delta H \cong -13 \text{ kcal/mole} \\ E_{act} \text{ is low}\end{array}$$

reaction, the abstraction of a bromine atom by the peroxyradical, is highly endothermic and, consequently, has a high energy of activation.

$$R—\overset{..}{\underset{..}{O}}· + H:\overset{..}{\underset{..}{Br}}: \longrightarrow R—\overset{..}{\underset{..}{O}}—\overset{..}{\underset{..}{Br}}: + H· \qquad \begin{array}{l}\Delta H \cong +39 \text{ kcal/mole} \\ E_{act} > +39 \text{ kcal/mole}\end{array}$$

Step 3 of the mechanism determines the final orientation of bromine in the product. It occurs, as it does, because a *more stable secondary radical* is produced. Had the bromine attacked propene at the central carbon atom, a less stable, primary radical would have been the result,

$$Br\cdot + CH_2{=}CHCH_3 \;\xcancel{\longrightarrow}\; \cdot CH_2\underset{\underset{Br}{|}}{C}HCH_3$$

1° Free radical

.and this reaction would have had a higher energy of activation.

Step 4 of the mechanism is simply the abstraction of a hydrogen atom from hydrogen bromide by the radical produced in step 3. This hydrogen abstraction produces a bromine atom that can bring about step 3 again, then step 4 occurs again—a chain reaction.

Problem 6.11

You may have noticed that step 1 of the mechanism is almost as endothermic as the reaction that we disallowed for step 2, that is,

$$R{-}\ddot{\underset{..}{O}}{:}\ddot{\underset{..}{O}}{-}R \longrightarrow 2R\ddot{\underset{..}{O}}\cdot \qquad\qquad \Delta H \cong +35 \text{ kcal/mole}$$

$$R{-}\ddot{\underset{..}{O}}\cdot + HBr \;\xcancel{\longrightarrow}\; R{-}\ddot{\underset{..}{O}}{-}Br + H\cdot \qquad \Delta H \cong +39 \text{ kcal/mole}$$

Yet, to invoke the first reaction as a chain-initiating step, and to preclude the second from any important role, is perfectly reasonable. How can you explain this?

Many molecules, other than hydrogen bromide, add to alkenes under the influence of a peroxide catalyst. The following reactions are examples.

$$CH_3CH_2CH_2CH{=}CH_2 + HCCl_3 \xrightarrow{\text{peroxides}} CH_3CH_2CH_2CH_2CH_2CCl_3$$

1,1,1-Trichlorohexane

$$\underset{\underset{CH_3C=CH_2}{|}}{CH_3} + CH_3CH_2SH \xrightarrow{\text{peroxides}} CH_3{-}\underset{\underset{CH}{|}}{\overset{\overset{CH_3}{|}}{}}{-}CH_2{-}S{-}CH_2CH_3$$

$$\underset{\underset{CH_3CH_2C=CH_2}{|}}{CH_3} + CCl_4 \xrightarrow{\text{peroxides}} CH_3CH_2\underset{\underset{Cl}{|}}{\overset{\overset{CH_3}{|}}{C}}{-}CH_2{-}CCl_3$$

1,1,1,3-Tetrachloro-3-methylpentane

Problem 6.12

Write free-radical, chain-reaction mechanisms that account for the products formed in each of the reactions listed above.

Free radicals can also cause alkenes to add to each other to form large molecules called addition polymers. These reactions are described as a special topic in Section 8.3.

6.10 DIMERIZATION OF ALKENES: ALKYLATION OF ALKENES BY CARBOCATIONS

Heating isobutylene (2-methylpropene) in 60% sulfuric acid at 70° causes the formation of two main products—two isomeric compounds called "diisobutylenes." The diisobutylenes are examples of dimers (di- = two + Gr., *meros* = part); isobutylene is said to have been dimerized by the reaction.

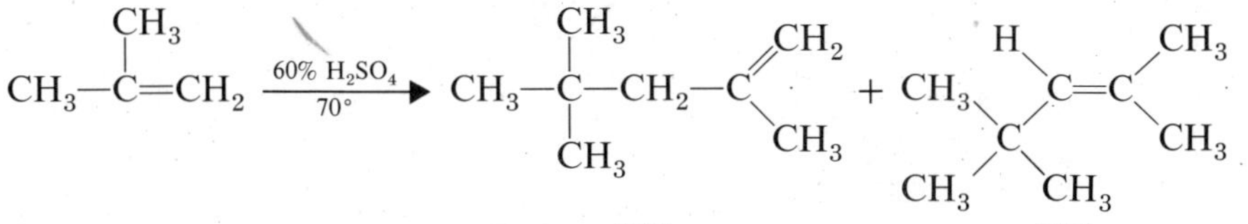

80%   20%

Diisobutylenes

In this reaction the less highly substituted alkene is the major product because it is apparently the more stable. The more highly substituted alkene has considerable internal repulsion arising from the large *tert*-butyl group being *cis* to the methyl group. This is an exception to the general rule concerning alkene stability (p. 164).

The dimerization of isobutylene is important in the petroleum industry because hydrogenation of the mixture of dimers produces the single product called "isooctane."*

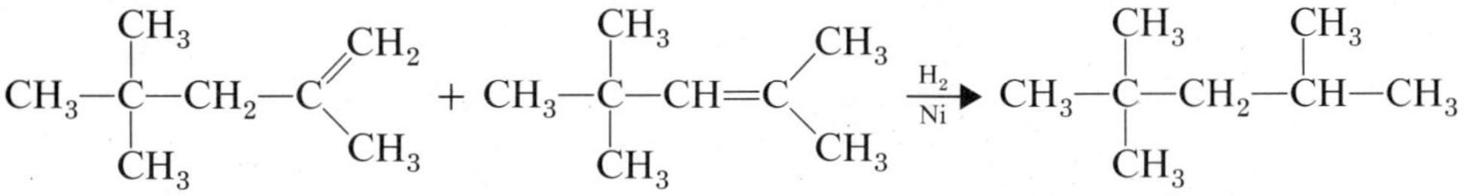

Isooctane
(2,2,4-trimethylpentane)

Isooctane burns very smoothly in internal combustion engines (without "knocking") and is used as one of the standards by which the octane rating of gasolines is established. According to this scale isooctane has an octane rating of 100. Heptane, a compound that produces much knocking when it is burned in an internal combustion engine, is given an octane rating of zero. Mixtures of isooctane and heptane are used as standards for octane ratings in between zero and 100. A gasoline, for example, that has the same characteristics in an engine as a mixture of 90% isooctane-10% heptane would be rated as 90-octane gasoline.

The mechanism by which isobutylene dimerizes (below), is straightforward. The second step shows us another property of carbocations: *their ability to react with alkenes.*

---

* Although calling this compound isooctane is incorrect, the name has been used for many years in the petroleum industry. The correct IUPAC name is 2,2,4-trimethylpentane.

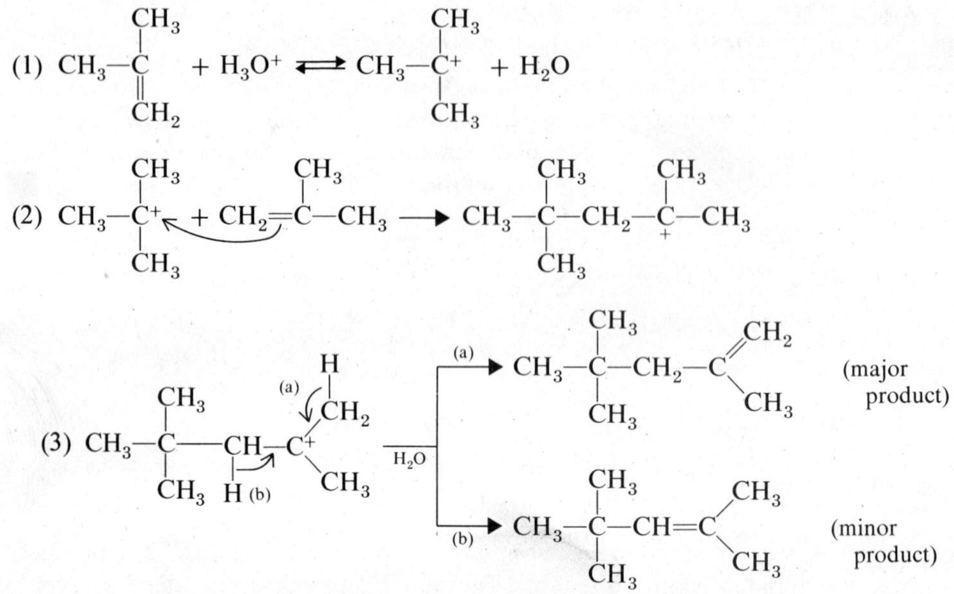

(1) $CH_3-\overset{\overset{\displaystyle CH_3}{|}}{\underset{\underset{\displaystyle CH_2}{\|}}{C}} + H_3O^+ \rightleftharpoons CH_3-\overset{\overset{\displaystyle CH_3}{|}}{\underset{\underset{\displaystyle CH_3}{|}}{C^+}} + H_2O$

(2) $CH_3-\overset{\overset{\displaystyle CH_3}{|}}{\underset{\underset{\displaystyle CH_3}{|}}{C^+}} + CH_2{=}\overset{\overset{\displaystyle CH_3}{|}}{C}-CH_3 \longrightarrow CH_3-\overset{\overset{\displaystyle CH_3}{|}}{\underset{\underset{\displaystyle CH_3}{|}}{C}}-CH_2-\overset{\overset{\displaystyle CH_3}{|}}{\underset{+}{C}}-CH_3$

(3)

(major product)

(minor product)

In step 1 the acid donates a proton to isobutylene; this leads to the formation of a *tert*-butyl cation. We have seen similar reactions several times before. In step 2 the carbocation—acting as an electrophile—attacks another molecule of isobutylene. Isobutylene acts as a nucleophile; its π electrons form a σ bond to the *tert*-butyl cation. The larger carbocation that is formed as a result may lose a proton in two different ways (step 3) to form the mixture of diisobutylenes.

This type of reaction, *the alkylation of an alkene by a carbocation*, has a number of counterparts in biochemistry. An extraordinary example is the conversion of squalene-2,3-epoxide into lanosterol (below and on next page).

The first step, ring opening of the three-membered epoxide ring, produces a tertiary carbocation. The second step—a series of alkylation reactions occurring in concert—is initiated by the original carbocation. This series of electron shifts brings about the *closure of four rings*. Reactions involving hydride and methanide shifts occur next and these ultimately produce lanosterol. All of these steps are catalyzed by a single enzyme—*squalene oxide cyclase*.

Squalene-2,3-epoxide

$H^+ \downarrow$ Squalene oxide cyclase

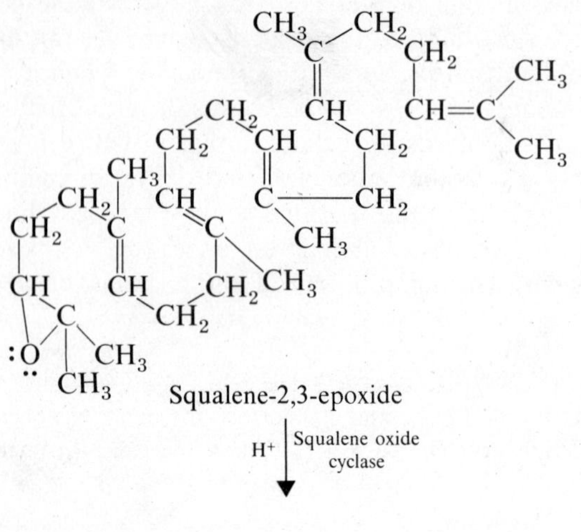

(continued on next page)

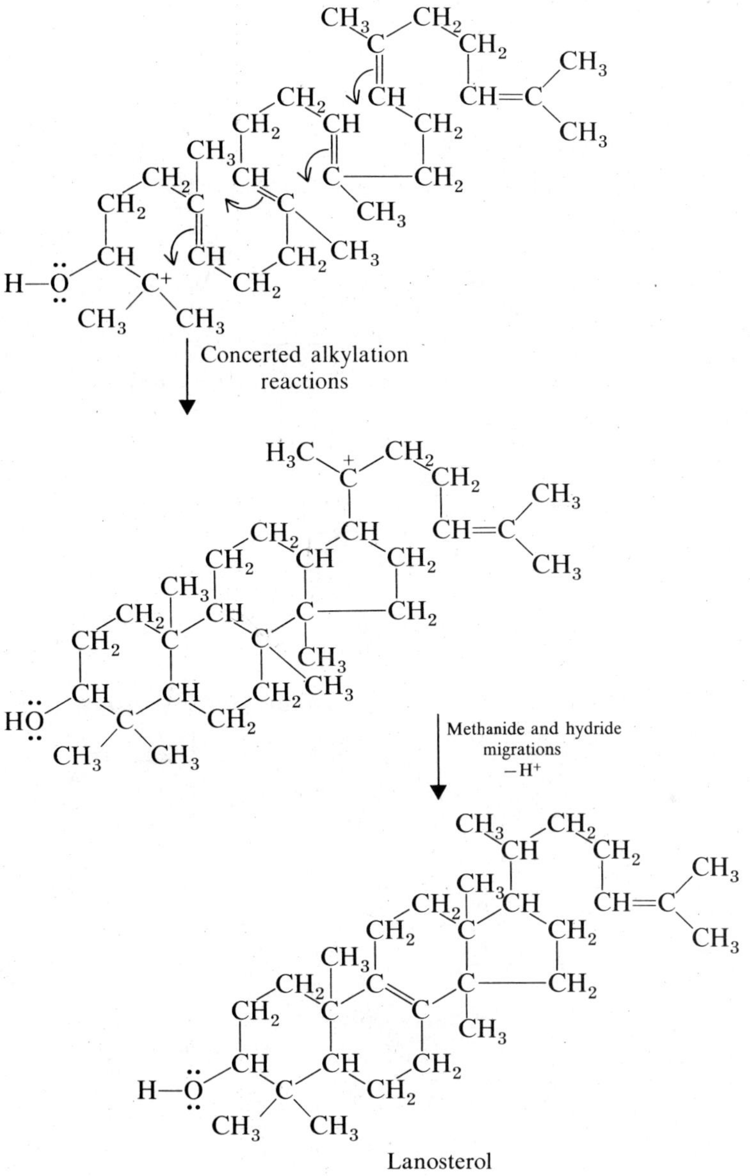

Concerted alkylation reactions

Methanide and hydride migrations
$-H^+$

Lanosterol

6.11 SUMMARY OF ADDITION REACTIONS OF ALKENES

1. Addition of hydrogen (Section 5.7)

General: $RCH{=}CHR + H_2 \xrightarrow{\text{cat.}} RCH_2{-}CH_2R$

Specific
examples: $CH_3CH_2CH{=}CH_2 + H_2 \xrightarrow[25°]{Pt} CH_3CH_2CH_2CH_3$ (100%)

   $CH_3CH{=}CHCH_3 + H_2 \xrightarrow[25°]{Pt} CH_3CH_2CH_2CH_3$ (100%)
   cis or *trans*

2. Addition of hydrogen halides (Sections 6.2 and 6.9)

General: $R-CH=CH_2 + HX \longrightarrow R-\underset{\underset{X}{|}}{CH}-CH_3$ Markovnikov addition

$R-CH=CH_2 + HBr \xrightarrow{ROOR} R-CH_2-CH_2Br$ Anti-Markovnikov addition

Specific examples: $CH_3-\underset{\underset{CH_3}{|}}{C}=CH_2 + HCl \xrightarrow{-80°} CH_3-\underset{\underset{CH_3}{|}}{\overset{\overset{Cl}{|}}{C}}-CH_3$ (94%)

$CH_3CH_2CH_2CH_2CH=CH_2 + HBr \xrightarrow{peroxides} CH_3CH_2CH_2CH_2CH_2CH_2Br$ (83%)

3. Addition of water (Section 6.3)

General: $R-CH=CH_2 + HOH \xrightarrow{H^+} R-\underset{\underset{OH}{|}}{CH}-CH_3$

Specific examples: $CH_2=CH_2 \xrightarrow[\text{(2) }H_2O]{\text{(1) 98\% }H_2SO_4} CH_3CH_2OH$

$\underset{CH_3}{\overset{CH_3}{\diagdown\diagup}}C=CH_2 \xrightarrow[25°]{10\% H_2SO_4} CH_3-\underset{\underset{OH}{|}}{\overset{\overset{CH_3}{|}}{C}}-CH_3$

4. Solvomercuration-demercuration (Section 6.4)

General: $RCH=CH_2 \xrightarrow{Hg(OAc)_2 \atop THF-H_2O} R\underset{\underset{OH}{|}}{CH}CH_2Hg(OAc) \xrightarrow[OH^-]{NaBH_4} R\underset{\underset{OH}{|}}{CH}CH_3 + Hg$

Alcohol

$RCH=CH_2 \xrightarrow{Hg(OAc)_2 \atop R'-OH} R\underset{\underset{OR'}{|}}{CH}CH_2Hg(OAc) \xrightarrow[OH^-]{NaBH_4} R\underset{\underset{OR'}{|}}{CH}CH_3 + Hg$

Ether

Specific examples:

$CH_3(CH_2)_3CH=CH_2 \xrightarrow{Hg(OAc)_2 \atop THF-H_2O} CH_3(CH_2)_3\underset{\underset{OH}{|}}{CH}CH_2Hg(OAc) \xrightarrow[OH^-]{NaBH_4}$

$CH_3(CH_2)_3\underset{\underset{OH}{|}}{CH}CH_3 + Hg$

(95%)

$(CH_3)_3C-CH=CH_2 \xrightarrow{Hg(OAc)_2 \atop CH_3OH} (CH_3)_3C-\underset{\underset{OCH_3}{|}}{CH}CH_2Hg(OAc) \xrightarrow[OH^-]{NaBH_4} (CH_3)_3C-\underset{\underset{OCH_3}{|}}{CH}CH_3$

(83%)

5. Addition of boron hydrides (Section 6.5)

General: $3R-CH=CH_2 + \frac{1}{2}(BH_3)_2 \longrightarrow (RCH_2CH_2)_3B$

Specific example:

$CH_3CH_2CH_2CH=CH_2 \xrightarrow[\text{(an ether solvent)}]{(BH_3)_2} (CH_3CH_2CH_2CH_2CH_2)_3B$
(94%)

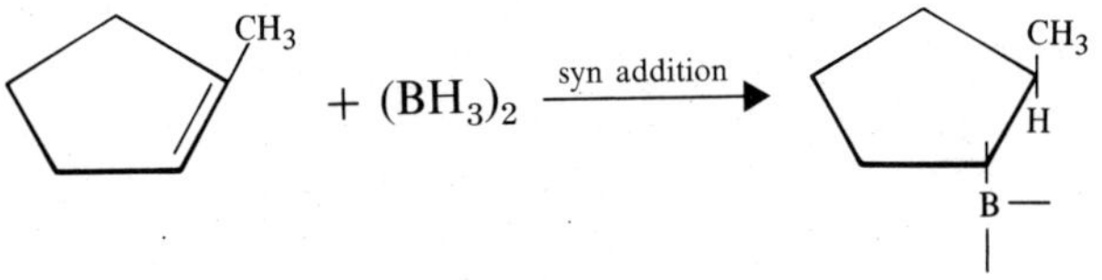

Reactions of alkyl boranes:

$CH_3(CH_2)_7CH=CH_2 \xrightarrow[\substack{\text{diglyme} \\ 25°}]{(BH_3)_2} [CH_3(CH_2)_7CH_2CH_2]_3B \xrightarrow[H_2O_2]{NaOH} CH_3(CH_2)_7CH_2CH_2OH$
(93%)
Alcohol

$CH_3(CH_2)_3CH=CH_2 \xrightarrow[\substack{\text{diglyme} \\ 25°}]{(BH_3)_2} [CH_3(CH_2)_3CH_2CH_2]_3B \xrightarrow{CH_3\overset{O}{\overset{\|}{C}}OH} CH_3(CH_2)_3CH_2CH_3$
(91%)
Alkane

6. Addition of halogens (Section 6.6)

General: $R-CH=CH-R + X_2 \xrightarrow{CCl_4} \underset{\underset{X\ \ \ X}{|\quad\ |}}{R-CH-CH-R}$

Specific examples:

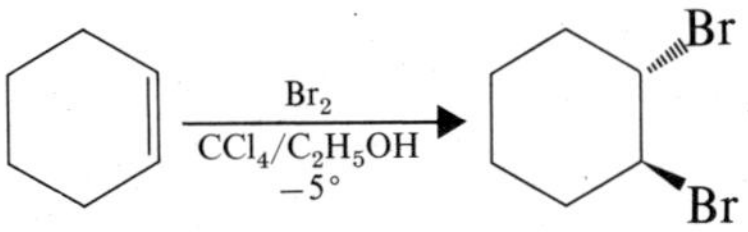

(95%)

$CH_3CH=CHCH_3 \xrightarrow[\substack{CH_3COOH \\ 25°}]{Cl_2} \underset{\underset{Cl\ \ Cl}{|\ \ |}}{CH_3CHCHCH_3}$ (100%)

$(CH_3)_2C=CH_2 + Br_2 + H_2O \longrightarrow \underset{\underset{OH\ Br}{|\quad\ \ |}}{(CH_3)_2C-CH_2}$
(73%)
A bromohydrin

7. Anti hydroxylation: epoxidation (Section 6.7)

General: $\overset{}{\underset{}{C}}=\overset{}{\underset{}{C} + R\overset{O}{\overset{\|}{C}}OH \longrightarrow} \underset{O}{-\overset{|}{C}\diagup\diagdown\overset{|}{C}-} \xrightarrow[H_2O]{H^+} \underset{\underset{OH}{|}}{-\overset{\overset{OH}{|}}{C}-\overset{|}{C}-}$

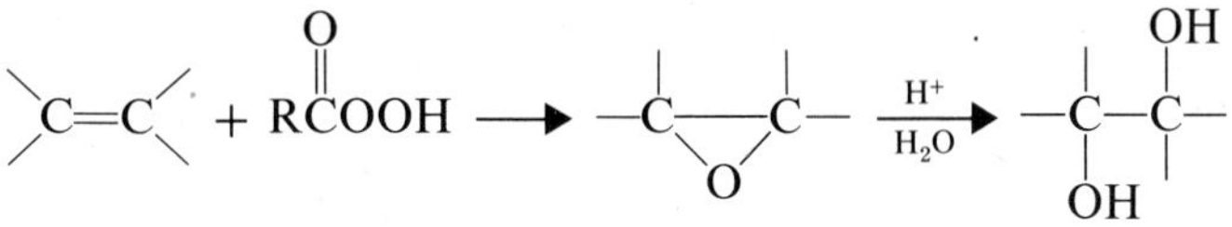

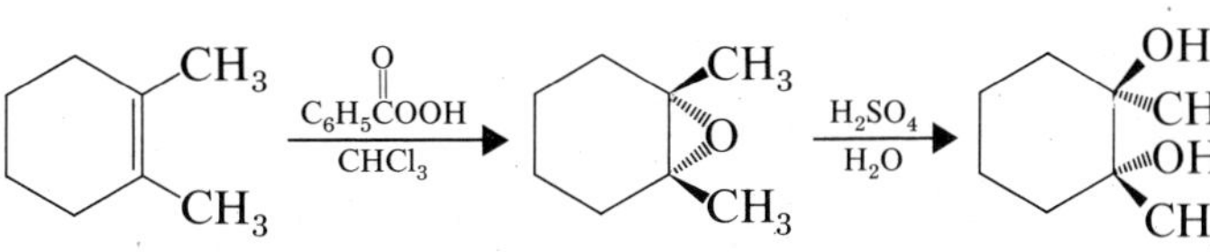

(60%)

8. Syn hydroxylation (Section 6.8)

General:

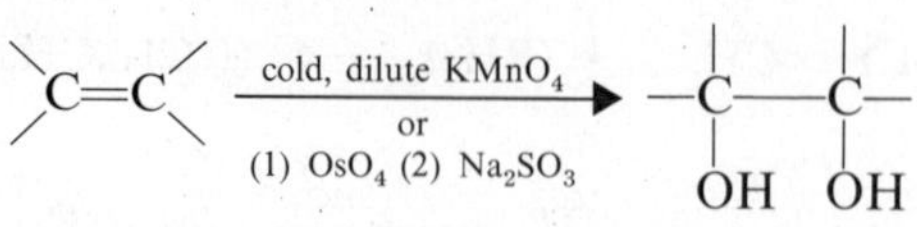

$$\text{C}=\text{C} \xrightarrow[\text{(1) OsO}_4 \text{ (2) Na}_2\text{SO}_3]{\overset{\text{cold, dilute KMnO}_4}{\text{or}}} \underset{\underset{\text{OH}}{|}}{-\text{C}}-\underset{\underset{\text{OH}}{|}}{\text{C}}-$$

Specific example:

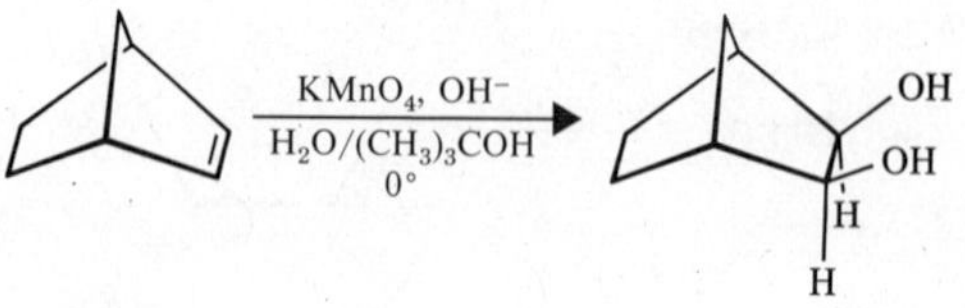

(40%)

9. Ozonolysis

General:

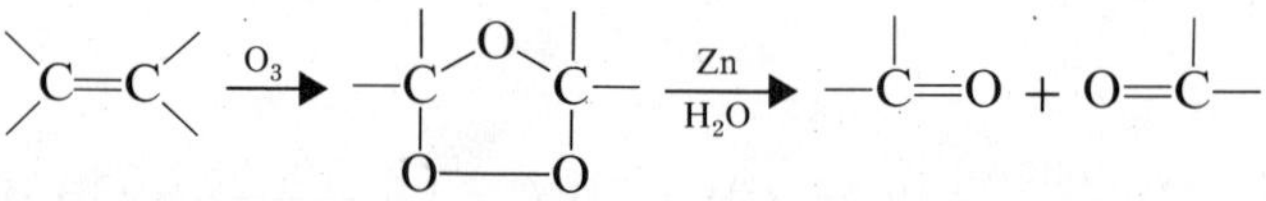

$$\text{C}=\text{C} \xrightarrow{\text{O}_3} \underset{\underset{\text{O}-\text{O}}{|}}{-\text{C}}\underset{}{\overset{\text{O}}{<}}\underset{}{\text{C}}- \xrightarrow[\text{H}_2\text{O}]{\text{Zn}} -\text{C}=\text{O} + \text{O}=\text{C}-$$

Specific example:

$$\underset{\text{CH}_3\text{CH}_2}{\overset{\text{CH}_3\text{CH}_2}{>}}\text{C}=\text{C}\underset{\text{H}}{\overset{\text{CH}_3}{<}} \xrightarrow[\text{(2) Zn, H}_2\text{O}]{\text{(1) O}_3} \underset{\text{CH}_3\text{CH}_2}{\overset{\text{CH}_3\text{CH}_2}{>}}\text{C}=\text{O} + \text{O}=\text{C}\underset{\text{H}}{\overset{\text{CH}_3}{<}}$$

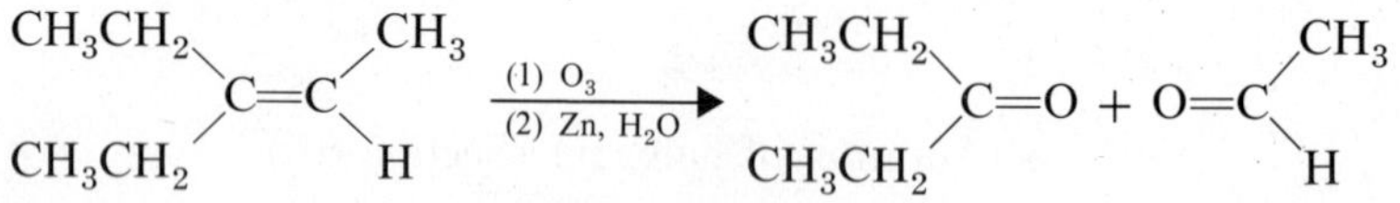

10. Addition of other alkenes (Section 6.10)

General:

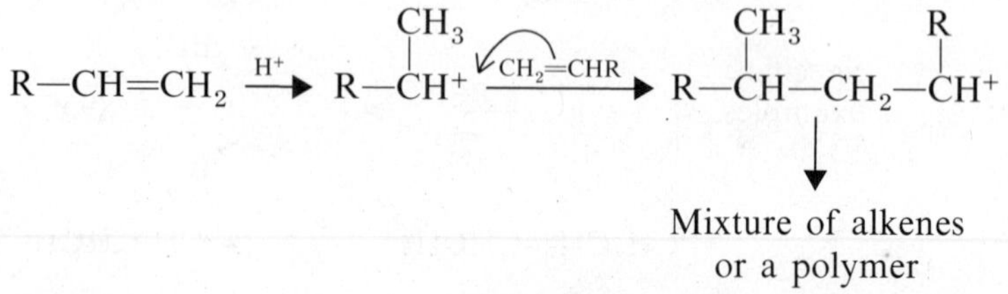

$$\text{R}-\text{CH}=\text{CH}_2 \xrightarrow{\text{H}^+} \text{R}-\overset{\overset{\text{CH}_3}{|}}{\text{CH}^+} \xoverset{\text{CH}_2=\text{CHR}}{\longrightarrow} \text{R}-\overset{\overset{\text{CH}_3}{|}}{\text{CH}}-\text{CH}_2-\overset{\overset{\text{R}}{|}}{\text{CH}^+}$$

Mixture of alkenes
or a polymer

Specific example:

$$\underset{}{\overset{\overset{\text{CH}_3}{|}}{\text{CH}_3\text{C}}}=\text{CH}_2 \xrightarrow[80°]{60\% \text{ H}_2\text{SO}_4} \text{CH}_3-\overset{\overset{\text{CH}_3}{|}}{\underset{\underset{\text{CH}_3}{|}}{\text{C}}}-\text{CH}_2-\text{C}\overset{\text{CH}_2}{\underset{\text{CH}_3}{<}} + \underset{(\text{CH}_3)_3\text{C}}{\overset{\text{H}}{>}}\text{C}=\text{C}\underset{\text{CH}_3}{\overset{\text{CH}_3}{<}}$$

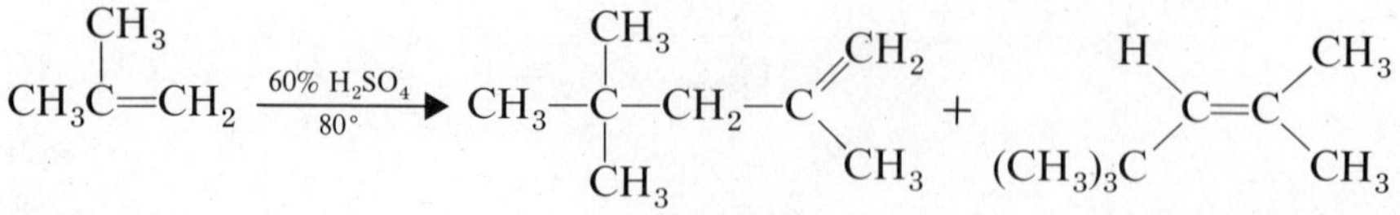

(80%) (20%)

11. Addition of carbenes and carbenoids (Section 8.5)

General: $\text{R}-\text{CH}=\text{CH}-\text{R} + :\text{CH}_2 \longrightarrow \text{R}-\text{CH}\underset{\underset{\text{CH}_2}{\diagdown \diagup}}{-}\text{CH}-\text{R}$

$$\underset{\text{CH}_2\text{N}_2}{\overset{}{\underset{\text{light}}{\upharpoonleft}} \xrightarrow{} \text{N}_2}$$

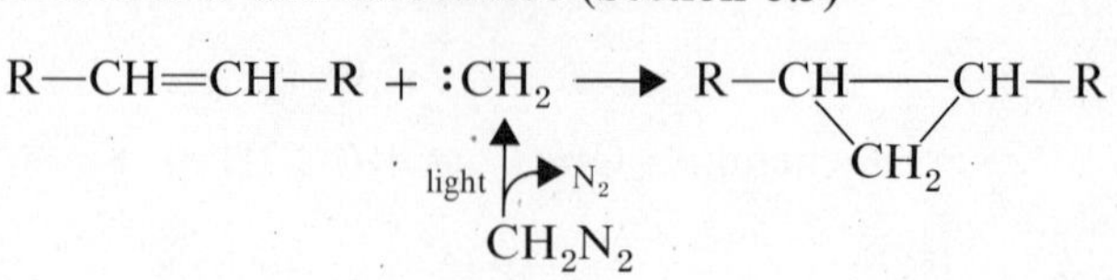

Additional Problems

6.13

Write structural formulas for the products that would be formed in the following reactions:

(a) 1-Butene + Br$_2$ $\xrightarrow{\text{CCl}_4}$

(b) Cyclohexene + dilute KMnO$_4$ $\xrightarrow[\text{cold}]{\text{OH}^-}$

(c) Cyclohexene + OsO$_4$ $\xrightarrow{\text{(2) Na}_2\text{SO}_3}$

(d) Cyclohexene + C$_6$H$_5\overset{\overset{\displaystyle O}{\|}}{\text{C}}$OOH $\longrightarrow$ $\xrightarrow[\text{H}_2\text{O}]{\text{H}^+}$

(e) Ethene + conc. H$_2$SO$_4$ $\xrightarrow{\text{cold}}$.

(f) Product of (e) + H$_2$O $\longrightarrow$
(g) 2-Methylpropene + (BH$_3$)$_3$ $\longrightarrow$
(h) Product of (g) + NaOH/H$_2$O$_2$ $\longrightarrow$

(i) 2-Butanol + hot conc. H$_2$SO$_4$ $\longrightarrow$

(j) Bromocyclopentane + KOH $\xrightarrow[\text{heat}]{\text{alcohol}}$

(k) 1,2-Dibromopentane + Zn $\longrightarrow$

(l) 2,3-Dimethyl-1-butene + HBr $\xrightarrow[\text{heat}]{\text{(free-radical inhibitor)}}$

(m) 1-Hexene + HCl $\xrightarrow{\text{heat}}$

(n) 1-Hexene + HBr $\xrightarrow{\text{peroxides}}$

(o) *cis*-3-Hexene + O$_3$ $\longrightarrow$ then

(p) Product of (o) + Zn $\xrightarrow{\text{H}_2\text{O}}$

(q) *cis*-3-Hexene + KMnO$_4$ $\xrightarrow[\text{heat}]{\text{OH}^-}$

(r) 1-Methylcyclopentene + O$_3$ $\longrightarrow$ then

(s) Product of (r) + Zn $\xrightarrow{\text{H}_2\text{O}}$

(t) 1-Methylcyclopentene + KMnO$_4$ $\xrightarrow{\text{heat}}$

(u) 1-Methylcyclopentene + H$_2$O $\xrightarrow{\text{H}^+}$

(v) 1-Methylcyclopentene + H$_2$ $\xrightarrow{\text{Ni}}$

(w) 2-Methyl-2-pentene + HI $\xrightarrow{\text{heat}}$

(x) 1,2-Dimethylcyclopentene + HCl $\longrightarrow$

6.14
(a) Write all steps of a mechanism for the acid-catalyzed dimerization of propene.
(b) What would be the major product?

6.15
When either *cis*- or *trans*-2-butene is treated with hydrogen chloride in ethyl alcohol, one of the products of the reaction is

$$CH_3CH_2 \diagdown$$
$$ CH-OCH_2CH_3$$
$$CH_3 \diagup$$

Write a mechanism that accounts for the formation of this product.

6.16

Write three-dimensional formulas for the products of the following reactions. In each case, designate the location of deuterium atoms.

(a) 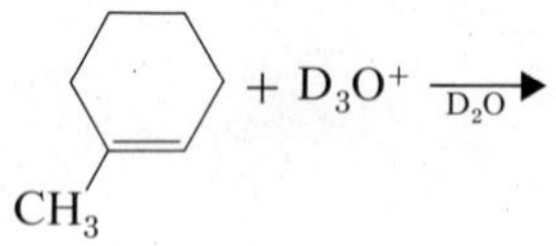 [cyclohexene ring with CH_3 substituent] $+ D_3O^+ \xrightarrow[D_2O]{}$

(b) 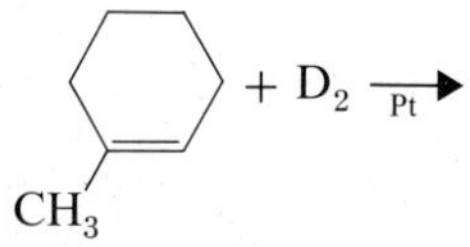 [cyclohexene ring with CH_3 substituent] $+ D_2 \xrightarrow[Pt]{}$

(c) 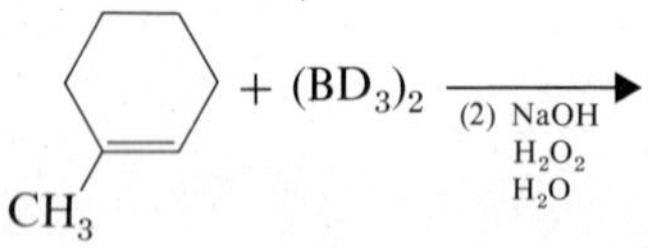 [cyclohexene ring with CH_3 substituent] $+ (BD_3)_2 \xrightarrow[\substack{(2)\ NaOH \\ H_2O_2 \\ H_2O}]{}$

6.17

Write the products and show how many moles of each would be formed when squalene is subjected to ozonolysis and the ozonide is subsequently treated with zinc and water?

$$CH_3\overset{\overset{\displaystyle CH_3}{|}}{C}=CHCH_2CH_2\overset{\overset{\displaystyle CH_3}{|}}{C}=CHCH_2CH_2\overset{\overset{\displaystyle CH_3}{|}}{C}=CHCH_2CH_2CH=\overset{\overset{\displaystyle CH_3}{|}}{C}CH_2CH_2CH=\overset{\overset{\displaystyle CH_3}{|}}{C}CH_2CH_2CH=\overset{\overset{\displaystyle CH_3}{|}}{C}CH_3$$

Squalene

6.18

Arrange the following alkenes in order of their reactivity toward acid-catalyzed hydration and explain your reasoning.

$$CH_2=CH_2 \qquad CH_3CH=CH_2 \qquad CH_3\overset{\overset{\displaystyle CH_3}{|}}{C}=CH_2$$

6.19

(a) When treated with strong acid at 25°, either cis-2-butene or trans-2-butene is converted to a mixture of trans-2-butene, cis-2-butene, and 1-butene. Trans-2-butene predominates in the mixture. (The mixture contains 74% trans-2-butene, 23% cis-2-butene, and 3% 1-butene.) Write mechanisms for the reactions that occur, and account for the relative amounts of the alkene isomers that are formed.

(b) When treated with strong acid, 1-butene is converted to the same mixture of alkenes referred to in part (a). How can you explain this?

(c) Can you also explain why 2-methylpropene is *not* formed in either of the reactions referred to in parts (a) or (b) even though 2-methylpropene is more stable than 1-butene?

6.20

Write a mechanism that explains the course of the following reaction.

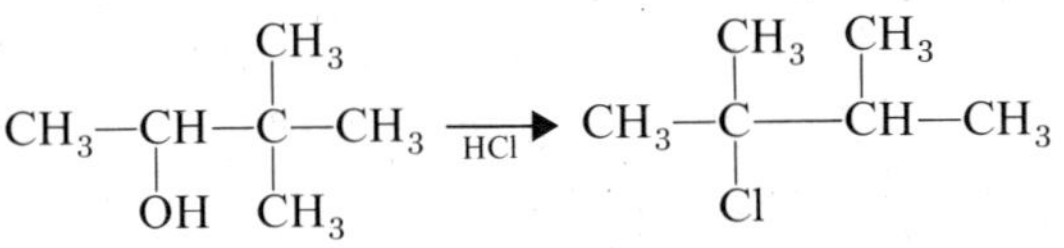

$$CH_3-CH-\underset{\underset{CH_3}{|}}{\overset{\overset{CH_3}{|}}{C}}-CH_3 \xrightarrow{HCl} CH_3-\underset{\underset{Cl}{|}}{\overset{\overset{CH_3}{|}}{C}}-\overset{\overset{CH_3}{|}}{CH}-CH_3$$
$$\overset{|}{OH}\ \overset{|}{CH_3}$$

6.21

A cycloalkene reacts with hydrogen and a catalyst to yield methylcyclohexane. On vigorous oxidation with potassium permanganate the cycloalkene yields only

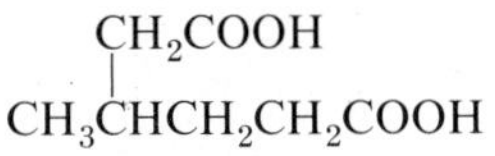

$$CH_2COOH$$
$$CH_3\overset{|}{CHCH_2CH_2COOH}$$

What is the structure of the cycloalkene?

6.22

Outline all steps in a laboratory synthesis of each of the following compounds. You should begin with the organic compound indicated and you may use any needed solvents or inorganic compounds. These syntheses may require more than one step and should be designed to give reasonably good yields of reasonably pure products.
(a) Propene from propane
(b) 2-Bromopropane from propane
(c) 1-Bromopropane from propane
(d) 2-Methylpropene from 2-methylpropane
(e) *tert*-Butyl alcohol from 2-methylpropane
(f) 1,2-Dichlorobutane from 1-chlorobutane
(g) Ethylene bromohydrin from ethyl bromide
(h) 2,5-Dimethylhexane from 2-methylpropene

(i) 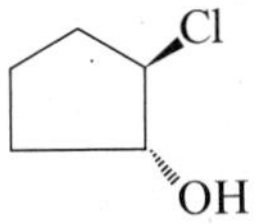 from cyclopentane

trans

(j) 2-Bromobutane from 1-bromobutane
(k) 3,4-Dimethylhexane from 1-chlorobutane

(l) 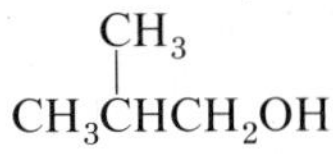 $CH_3\overset{\overset{\overset{CH_3}{|}}{}}{CHCH_2OH}$ from $CH_3\underset{\underset{CH_3}{|}}{\overset{\overset{CH_3}{|}}{C}}-OH$

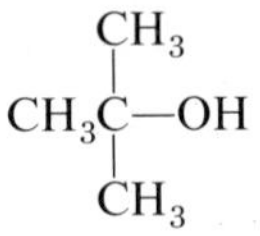

6.23

(a) How many grams of bromine will react with 7.0 g of 1-pentene?
(b) What is the molecular weight of the alkene 2.24 g of which reacts with 3.20 g of bromine?

6.24

Write a mechanism that accounts for the following cyclization reaction:

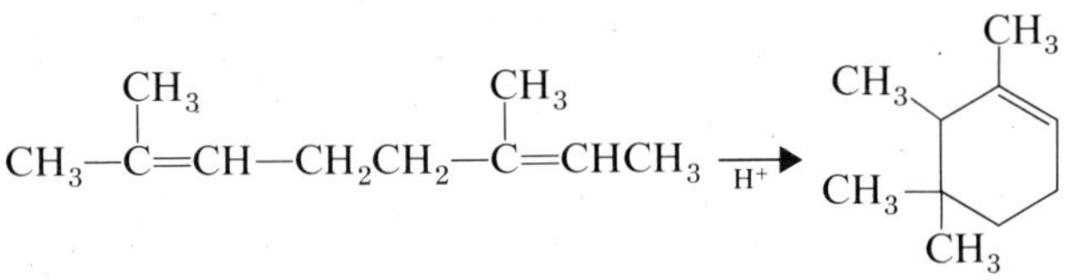

$$CH_3-\underset{\underset{CH_3}{|}}{\overset{\overset{CH_3}{|}}{C}}=CH-CH_2CH_2-\underset{\underset{CH_3}{|}}{\overset{\overset{CH_3}{|}}{C}}=CHCH_3 \xrightarrow{H^+}$$

6.25

When cyclopentene is allowed to react with bromine in an aqueous solution of sodium chloride, the products of the reaction are *trans*-1,2-dibromocyclopentane, the *trans*-bromohydrin of cyclopentene, *and trans-1-bromo-2-chlorocyclopentane*. Write a mechanism that explains the formation of this last product.

6.26

The relative rate of addition of hydrogen chloride to the following alkenes increases in the order given below.

Increasing rate of addition of H—Cl

Ethene
Propene
2-Butene (*cis* or *trans*)
2-Methylpropene

Explain.

6.27

Pheromones are substances secreted by animals (especially insects) that produce a specific behavioral reaction in other members of the same species. Pheromones are effective at very low concentrations and include sex attractants, warning substances, and "aggregation" compounds. After many years of research, the sex attractant of the gypsy moth has been identified and synthesized in the laboratory. This sex pheromone is unusual in that it appears to be equally attractive to male and female gypsy moths. (It promises to be useful in their control even though this may seem somewhat unfair.) The final step in the synthesis of the pheromone involves treatment of *cis*-2-methyl-7-octadecene with a peracid. What is the structure of the gypsy moth sex pheromone?

6.28

The green peach aphid is repelled by its own defensive pheromone. (It is also repelled by other squashed aphids.) This alarm pheromone has been isolated and has been shown to have the molecular formula $C_{15}H_{24}$. On catalytic hydrogenation it absorbs four moles of hydrogen and yields 2,6,10-trimethyldodecane, that is,

$$
\overset{\underset{\textstyle |}{CH_3}}{} \qquad \overset{\underset{\textstyle |}{CH_3}}{} \qquad \overset{\underset{\textstyle |}{CH_3}}{}
$$
$$
CH_3CHCH_2CH_2CH_2CHCH_2CH_2CH_2CHCH_2CH_3
$$

When subjected to ozonolysis followed by treatment with zinc and water one mole of the alarm pheromone produces:

Two moles of formaldehyde, $H-\overset{O}{\overset{\|}{C}}-H$;

One mole of acetone, $CH_3\overset{O}{\overset{\|}{C}}CH_3$

One mole of $CH_3\overset{O}{\overset{\|}{C}}CH_2CH_2\overset{O}{\overset{\|}{CH}}$

One mole of $H\overset{O}{\overset{\|}{C}}CH_2CH_2\overset{O}{\overset{\|}{C}}-\overset{O}{\overset{\|}{CH}}$

Neglecting *cis-trans* isomerism, propose a structure for the green peach aphid alarm pheromone.

6.29

Write a mechanism that accounts for the following reaction.

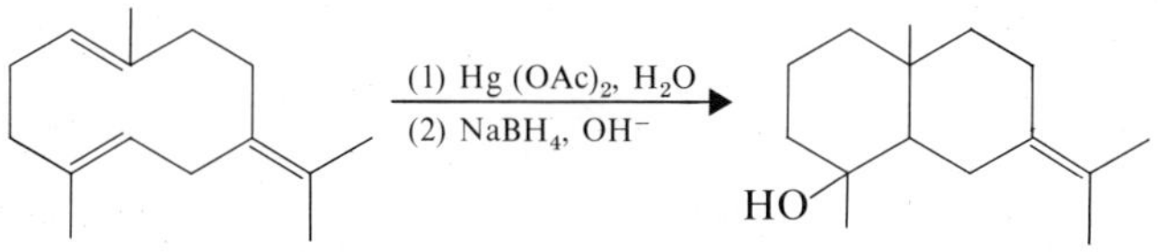

$$\xrightarrow[\text{(2) NaBH}_4,\ \text{OH}^-]{\text{(1) Hg (OAc)}_2,\ \text{H}_2\text{O}}$$

6.30

When an alkene, $RCH{=}CH_2$, and hydrogen sulfide are irradiated with light (of wavelength that can be absorbed by H_2S), a chain reaction takes place producing a thiol, RCH_2CH_2SH. (a) Outline a possible mechanism for this reaction. (b) A side-product of the reaction is a thioether, $(RCH_2CH_2)_2S$. Suggest how it is formed.

6.31

The structures of the diisobutylene dimers (Sect. 6.10) were determined by F. C. Whitmore and his students on the basis of the products formed when each dimer was subjected to ozonolysis. Show how this might have been done.

6.32

In an industrial process, propene is heated with phosphoric acid at $205°$ under a pressure of 1000 atm. The major products of the reaction are two isomers with the molecular formula $C_{12}H_{24}$. Propose structures for the isomers and write a mechanism that explains their formation. (Note: at one time these isomers were used extensively in the synthesis of a non-biodegradable detergent. This synthesis is given in problem 13.33.)

* 6.33

When hydrogen chloride adds to the double bond of 3,3,3-trifluoropropene, $CF_3CH{=}CH_2$, the product is $CF_3CH_2CH_2Cl$ rather than $CF_3CHClCH_3$. Although the orientation of addition in this reaction is not the one that would have been predicted on the basis of the original version of Markovnikov's rule (p. 51), it is consistent with the modern version. Explain.

7 STEREOCHEMISTRY

7.1 INTRODUCTION

In 1877, Hermann Kolbe, one of the most eminent organic chemists of the time, wrote the following:

"Not long ago, I expressed the view that the lack of general education and of thorough training in chemistry was one of the causes of the deterioration of chemical research in Germany. . . . Will anyone to whom my worries seem exaggerated please read, if he can, a recent memoir by a Herr van't Hoff on 'The Arrangements of Atoms in Space' a document crammed to the hilt with the outpourings of a childish fantasy . . . This Dr. J. H. van't Hoff, employed by the Veterinary College at Utrecht, has, so it seems, no taste for accurate chemical research. He finds it more convenient to mount his Pegasus (evidently taken from the stables of the Veterinary College) and to announce how, on his bold flight to Mount Parnassus, he saw the atoms arranged in space."

Kolbe, nearing the end of his career, was reacting to a publication of a 22-year-old Dutch scientist. This publication had appeared two years earlier in September 1874, and in it, van't Hoff had argued that the spatial arrangement of groups around the carbon atom was tetrahedral. A young French scientist, J. A. Le Bel, had independently advanced the same idea in a publication in November 1874. Within 10 years after Kolbe's comments, however, abundant evidence accumulated to substantiate the "childish fantasy" of van't Hoff, and in 1901 he was named the first recipient of the Nobel Prize for chemistry.

Together, the publications of van't Hoff and Le Bel marked the beginning of a field of study that is concerned with the structures of molecules in three-dimensions: *stereochemistry*.

Until now in our study of organic chemistry, we have been concerned primarily with the order in which the atoms of molecules are attached to each other. We will find, however, that an understanding of the properties of many organic compounds requires that we also concern ourselves with the arrangement of their atoms in space.

7.2 ISOMERISM: STRUCTURAL ISOMERS AND STEREOISOMERS

Isomers are different compounds that have the same molecular formula. In our study of carbon compounds, thus far, most of our attention has been directed toward those isomers that we have called structural isomers.

Structural isomers are isomers that differ because their atoms are joined in a different order.* Several examples of structural isomers are shown below.

Molecular Formula	Structural Isomers

C_4H_{10}

$$CH_3CH_2CH_2CH_3 \text{ and } \overset{\overset{\displaystyle CH_3}{|}}{CH_3CHCH_3}$$

Butane   Isobutane

C_3H_7Cl

$$CH_3CH_2CH_2Cl \text{ and } \underset{\underset{\displaystyle Cl}{|}}{CH_3CHCH_3}$$

1-Chloropropane   2-Chloropropane

C_2H_6O

$$CH_3CH_2OH \text{ and } CH_3OCH_3$$

Ethanol   Dimethyl ether

Structural isomers are sometimes classified into subcategories. Butane and isobutane, for example, are sometimes called *chain isomers,* 1-chloropropane and 2-chloropropane are called *position isomers,* and ethanol and dimethyl ether are sometimes called *functional group isomers.* The origin of these subclassifications is self-evident: Butane and isobutane differ in the structure of their carbon chains, 1-chloropropane and 2-chloropropane differ in the position of attachment of the chlorine atom, and ethanol and dimethyl ether differ in their functional groups.

Stereoisomers are not structural isomers—they have their constituent atoms attached in the same order—*stereoisomers differ only in arrangement of their atoms in space.* The *cis* and *trans* isomers of alkenes are stereoisomers; we can see that this is true if we examine the *cis*- and *trans*-2-butenes shown below.

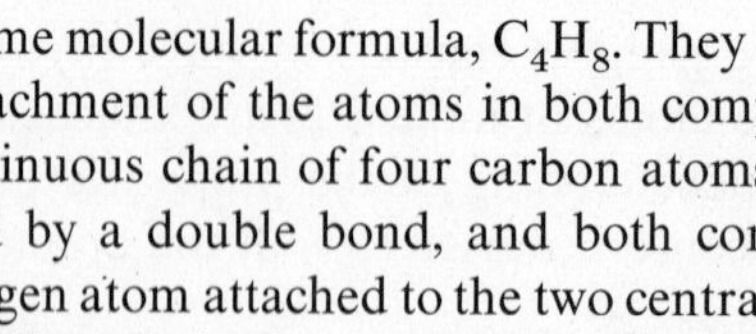

cis-2-Butene   *trans*-2-Butene

Cis-2-butene and *trans*-2-butene are isomers because both compounds have the same molecular formula, C_4H_8. They are *not* structural isomers, because the order of attachment of the atoms in both compounds is the same. Both compounds have a continuous chain of four carbon atoms, both compounds have two central atoms joined by a double bond, and both compounds have one methyl group and one hydrogen atom attached to the two central atoms. *Cis*- and *trans*-2-butene are isomers that differ only in the arrangement of their atoms in space. In *cis*-2-butene the methyl groups are on the same side of the molecule, and in *trans*-2-butene the methyl groups are on opposite sides. Thus, *cis*- and *trans*-2-butene are stereoisomers.

---

* Another term that is used for what we have called structural isomers is the term *constitutional isomers*. Both terms are defined in the same way. The term constitutional isomers may eventually become widely adopted, but at present most texts and monographs use the term structural isomers. For this reason we have retained the latter terminology in this book. It would be wise, however, to learn both terms.

Stereoisomers can be subdivided into two general categories: *enantiomers* and *diastereomers*. Enantiomers are stereoisomers that *are* mirror reflections of each other. Diastereomers are stereoisomers that *are not* mirror reflections of each other.

Cis- and *trans*-2-butene *are not* mirror reflections of each other. If one holds a model of *cis*-2-butene up to a mirror, the model that one sees in the mirror is not *trans*-2-butene. But, *cis-* and *trans*-2-butene *are* stereoisomers and, since they are not related to each other as an object and its mirror reflection, they are diastereomers.

7.3 ENANTIOMERS AND CHIRAL MOLECULES

Enantiomers occur only with those compounds whose molecules are *chiral*. A chiral molecule can be defined as *one that is not superposable on its mirror reflection.*

The word chiral comes from the Greek word *cheir,* meaning "hand." Chiral objects (including molecules) are said to possess "handedness." The term chiral is used to describe molecules because enantiomers are related to each other in the same way that a left hand is related to a right hand. When you view your left hand in a mirror, the mirror reflection of your left hand is a right hand (Fig. 7.1). Your left and right hands, moreover, are not *superposable* (Fig. 7.2). (This fact becomes obvious when one attempts to put a "left-handed" glove on a right hand or vice versa.)

Many familiar objects are chiral and the chirality of some of these objects is clear because we normally speak of them as having "handedness." We speak, for example, of nuts and bolts as having right- or left-handed threads or of a propeller as having a right- or left-handed pitch. The chirality of many other objects is not obvious in this sense, but becomes obvious when we apply the test of nonsuperposability of the object and its mirror reflection.

Objects (and molecules) that *are* superposable on their mirror images are *achiral*. Socks, for example, are achiral whereas gloves are chiral.

FIG. 7.1

The mirror reflection of a left hand is a right hand.

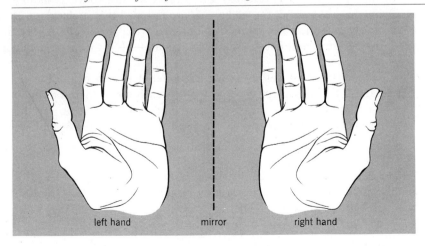

left hand   mirror   right hand

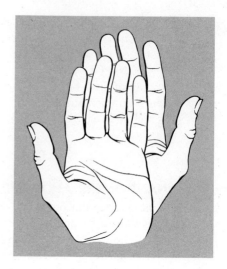

FIG. 7.2
Left and right hands are not superposable.

Problem 7.1

Classify the following objects as to whether they are chiral or achiral.

    (a) Screw   (e) Foot
    (b) Plain spoon   (f) Ear
    (c) Fork   (g) Shoe
    (d) Cup   (h) Spiral staircase

The chirality of molecules can be demonstrated with relatively simple compounds. Consider, for example, 2-butanol.

$$CH_3CHCH_2CH_3$$
$$|$$
$$OH$$

2-Butanol

Until now, we have presented the formula written above as though it represented only one compound, for we have not mentioned that molecules of 2-butanol are chiral. Because they are, there are actually two different 2-butanols and these two 2-butanols are enantiomers. We can understand this if we examine the drawings and models in Fig. 7.3.

If model I is held before a mirror, model II is seen in the mirror and vice versa. Models I and II are not superposable on each other, therefore they represent different, but isomeric, molecules. Because models I and II are nonsuperposable mirror reflections of each other, the molecules that they represent are enantiomers.

Problem 7.2

(a) If models are available, construct the 2-butanols represented in Fig. 7.3 and demonstrate for yourself that they are not mutually superposable. (b) Make similar models of 2-propanol, $CH_3CHOHCH_3$. Are they superposable? (c) Is 2-propanol chiral? (d) Would you expect to find enantiomeric forms of 2-propanol?

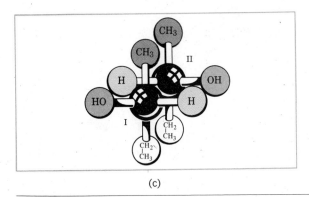

FIG. 7.3
(a) *Three-dimensional drawings of the 2-butanol enantiomers I and II.* (b) *Models of the 2-butanol enantiomers.* (c) *An unsuccessful attempt to superpose models of I and II.*

The possibility of the existence of a pair of enantiomers will be found for all molecules that contain *a single* chiral carbon. *A chiral carbon is a carbon atom that has four different groups attached to it.** In 2-butanol (Fig. 7.4) the chiral carbon is carbon-2. The four different groups that are attached to carbon-2 of 2-butanol are a hydroxyl group, a hydrogen, a methyl group, and an ethyl group.

---

* Any tetrahedral atom with four different groups attached to it is called a *chiral center.*

FIG. 7.4

The chiral carbon of 2-butanol. (By convention chiral carbons are often designated with an asterisk.)

(hydrogen)

$$\text{(methyl)} \quad \overset{1}{CH_3}-\overset{\overset{\displaystyle H}{|}}{\underset{\underset{\displaystyle OH}{|}}{\overset{2}{C}{}^{*}}}-\overset{3}{CH_2}\overset{4}{CH_3} \quad \text{(ethyl)}$$

(hydroxyl)

FIG. 7.5

A demonstration of chirality of a generalized molecule containing one chiral carbon. (a) The four different groups around the carbon atom in III and IV are arbitrary. (b) III is rotated and placed in front of a mirror. III and IV are found to be related as an object and its mirror reflection. (c) III and IV are not superposable; therefore, the molecules that they represent are chiral and are enantiomers.

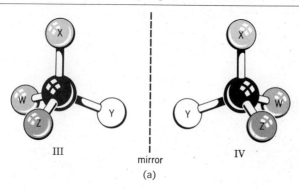

III mirror IV

(a)

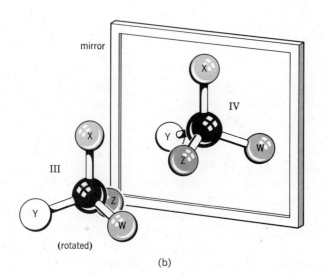

(b)

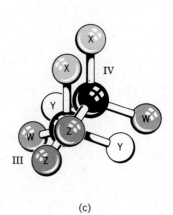

(c)

Figure 7.5 demonstrates the validity of the generalization that enantiomeric compounds are possible whenever a molecule contains a single chiral carbon.*

Problem 7.3

(a) If models are not available, demonstrate the validity of what we have represented in Fig. 7.5 by constructing models made from gumdrops and toothpicks. Select one gumdrop of a particular color for the central atoms of III and IV and insert toothpicks so that they point generally toward the corners of a tetrahedron. Then arrange four different-colored gumdrops at each corner. Demonstrate for yourself that III and IV are related as an object and its mirror reflection and that they are not superposable (i.e., III and IV are chiral molecules and are enantiomers). (b) Replace one gumdrop on each model so that each model has two gumdrops of the same color arranged around the central atom. Are the molecules that these models represent chiral? (c) Are they enantiomers?

If two or more of the groups that are attached to a tetrahedral atom *are the same,* the molecule is superposable on its mirror image and is achiral. An example of a molecule of this type is 2-propanol, since two identical methyl groups are attached to the central atom. If we write three-dimensional formulas for 2-propanol we find (Fig. 7.6) that one structure can be superposed on its mirror reflection.

Thus, we would not predict the existence of enantiomeric forms of 2-propanol. Our prediction correlates with what is known from experience to be true: only one form of 2-propanol has ever been found.

Problem 7.4

Some of the molecules listed below have chiral carbons; some do not. Write three-dimensional formulas for the enantiomers of those molecules that do have chiral carbons.

    (a) 1-Chloropropane
    (b) Chlorobromoiodomethane
    (c) 2-Methyl-1-chloropropane
    (d) 2-Methyl-2-chloropropane
    (e) 2-Bromobutane
    (f) 1-Chloropentane
    (g) 2-Chloropentane
    (h) 3-Chloropentane

It was reasoning of the type that we have just presented that led van't Hoff to the conclusion that the spatial orientation of groups around carbon atoms is tetrahedral when carbon is bonded to four other atoms. The following information was available to van't Hoff.

1. Only one compound with the general formula CH_3X is ever found.

---

*This generalization is not necessarily true of molecules that contain more than one chiral carbon. It is also not necessary for a molecule to have a chiral carbon in order to exist in enantiomeric forms. We shall see examples of both instances later.

(a)   (b)

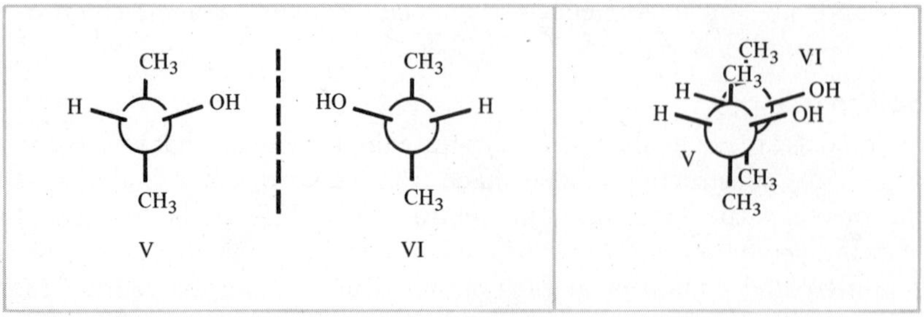

FIG. 7.6
(a) *2-Propanol* (*V*) *and its mirror reflection* (*VI*). (b) *When one is rotated,
the two structures* are *superposable and thus do not represent enantiomers.
They represent two molecules of the same compound.*

2. Only one compound with the formula CH_2X_2 or CH_2XY is ever found.
3. Two enantiomeric compounds with the formula CHXYZ are found.

Problem 7.5

(a) Prove to yourself the correctness of van't Hoff's reasoning by writing tetrahedral
representations for carbon compounds of the three types given above. (b) How many
isomers would be possible in each instance if the carbon atom were at the center
of a square? (c) At the center of a rectangle? (d) At one corner of a regular pyramid?

7.4 SYMMETRY ELEMENTS: PLANES OF SYMMETRY

The ultimate way to test for molecular chirality is to construct a model of the
molecule and a model of its mirror image and then attempt to superpose the two
models. If the two models are superposable, the molecule that they represent is
achiral. If the models are not superposable, then the molecules that they represent
are chiral.* We can apply this test with actual models, as we have just described,
or we can apply it by drawing three-dimensional structures and attempting to
superpose them in our minds.

There are other aids, however, that will assist us in recognizing chiral molecules. We have mentioned one already: the presence of a *single* chiral center. The
other aids that can be used are based on the presence in the molecule of certain
symmetry elements. A molecule *will not be chiral,* for example, if it possesses (1)
a plane of symmetry, (2) a center of symmetry, or (3) any *n*-fold (*n* = even number)
alternating axis of symmetry. The latter two symmetry elements are beyond the scope
of our discussion at this point, but an ability to recognize planes of symmetry will
serve us very well. An ability to recognize planes of symmetry will, in fact, enable
us to make a decision about the existence or nonexistence of chirality in most of
the molecules that we will encounter.

A plane of symmetry is defined as an imaginary plane that bisects a molecule

---

*Two older terms that you may encounter in other books that have the same meaning as
  chiral and achiral are the terms *dissymmetric* and *nondissymmetric*. A chiral object is
  dissymmetric, and an achiral object is *nondissymmetric*.

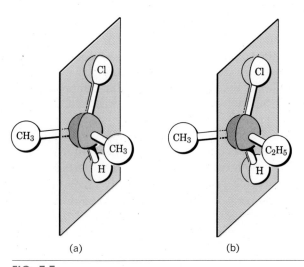

(a)   (b)

FIG. 7.7
(a) *2-Chloropropane has a plane of symmetry and is achiral.* (b) *2-Chlorobutane does not possess a plane of symmetry and is chiral.*

in such a way that the two halves of the molecule are mirror reflections of each other. For example, 2-chloropropane has a plane of symmetry (Fig. 7.7*a*), while 2-chlorobutane does not (Fig. 7.7*b*).

Problem 7.6

Which of the objects listed in problem 7.1 possess a plane of symmetry?

Problem 7.7

Write three-dimensional formulas and designate a plane of symmetry for all of the achiral molecules in problem 7.4. (In order to be able to designate a plane of symmetry you may have to write the molecule in an appropriate conformation. This is permissible in all of these molecules because they have only single bonds and groups joined by single bonds are capable of essentially free rotation at room temperature.)

7.5 NOMENCLATURE OF ENANTIOMERS: THE *R-S* SYSTEM

The two enantiomers of 2-butanol are shown below.

$$HO \overset{\displaystyle CH_3}{\underset{\displaystyle CH_2}{\overset{|}{\underset{|}{C}}}} H \qquad H \overset{\displaystyle CH_3}{\underset{\displaystyle CH_2}{\overset{|}{\underset{|}{C}}}} OH$$

$$CH_3 \qquad\qquad CH_3$$

I   II

If we name these two enantiomers using the IUPAC system of nomenclature both enantiomers will have the same name—2-butanol (Sec. 16.1). This is an undesirable situation because, as we pointed out earlier, *each compound should have its own name*. Moreover, the name that is given a compound should allow a chemist who is familiar with the rules of nomenclature to write the structure of the compound from its name alone. Given the name 2-butanol a chemist could write either structure I or structure II.

Three chemists, Professors R. S. Cahn (England), C. K. Ingold (England), and V. Prelog (Switzerland) have devised a system of nomenclature that, when added to the IUPAC system, solves both of these problems. This system, called the *R-S* system, or the Cahn-Ingold-Prelog system, is now widely used.

According to the Cahn-Ingold-Prelog system one 2-butanol enantiomer should be designated *R*-2-butanol and the other enantiomer should be designated *S*-2-butanol. (*R* and *S* are from the Latin words *rectus* and *sinister* meaning right and left, respectively.)

R and *S* designations are made on the following basis.

1. Each group attached to the chiral carbon is assigned a priority number from 1 to 4. Priority is first assigned on the basis of the atomic number of the atom that is directly attached to the chiral carbon. The group with the lowest atomic number is given the lowest priority number, 1; the group with next highest atomic number is given the next higher number, 2, and so on.

We can illustrate the application of this rule with the 2-butanol enantiomer, I.

$$CH_3 \ (2 \ or \ 3)$$
$$(4) \ HO \!\!-\!\!\! C \!\!-\!\! H \ (1)$$
$$CH_2 \ (2 \ or \ 3)$$
$$CH_3$$

I

Oxygen has the highest atomic number and is assigned the highest priority number, 4. Hydrogen has lowest atomic number and is assigned the lowest priority number, 1. A priority assignment cannot be made for the methyl group and the ethyl group because in both groups the atom that is directly attached to the chiral carbon is a carbon atom.

2. When a priority assignment cannot be made on the basis of the atomic number of the atoms that are directly attached to the chiral carbon then the next sets of atoms in the groups are examined and this process is continued until a decision can be made.

When we examine the methyl group of enantiomer I, we find that the next set of atoms consists of three hydrogens (H, H, H). In the ethyl group of I the next set of atoms consists of one carbon and two hydrogens (C, H, H). Carbon has a higher atomic number than hydrogen so we assign the ethyl group the higher priority number, 3, and the methyl group the lower priority number, 2.

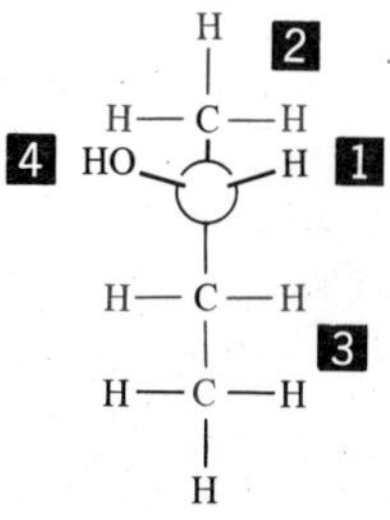

I

3. We now rotate the formula (or model) so that the group with lowest priority number is directed away from us.

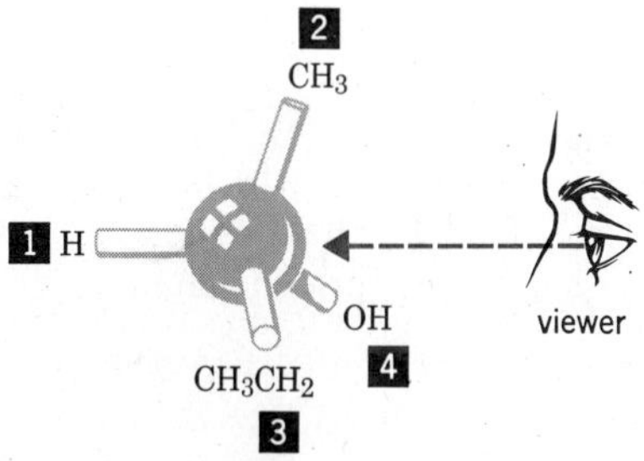

Then we trace a circle starting with the group numbered 4, to 3, to 2. If, as we do this, the direction of our finger (or pencil) is *clockwise* the enantiomer is designated *R*. If the direction is *counterclockwise,* the enantiomer is designated *S*.
On this basis the 2-butanol enantiomer I is *R*-2-butanol.

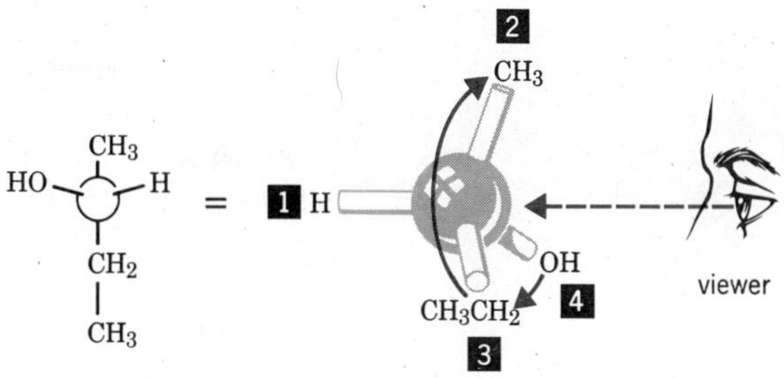

arrows are clockwise ·

Problem 7.8

Apply the procedure, just given, to the 2-butanol enantiomer II and show that it is *S*-2-butanol.

Problem 7.9

Give *R* and *S* designations for each pair of enantiomers given as answers to problem 7.4.

The first three rules of the Cahn-Ingold-Prelog system allow us to make an *R* or *S* designation for compounds containing single bonds only. For compounds containing multiple bonds one other rule is necessary.

    4. Groups containing double or triple bonds are assigned priorities as if both atoms were duplicated or triplicated, that is,

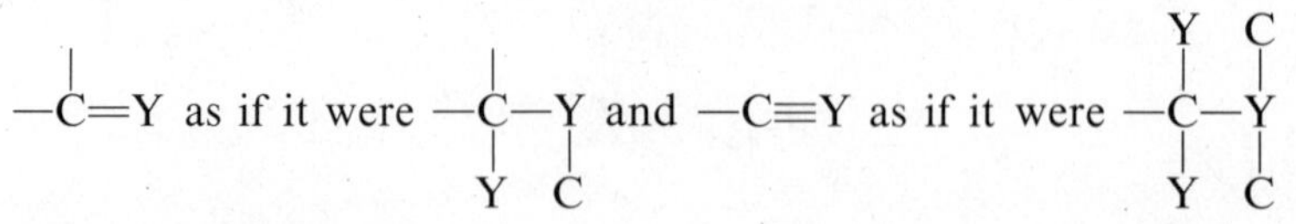

$$-\overset{|}{C}=Y \quad \text{as if it were} \quad -\overset{|}{\underset{Y}{C}}-Y \quad \text{and} \quad -C\equiv Y \quad \text{as if it were} \quad -\overset{Y}{\underset{Y}{C}}-\overset{C}{\underset{C}{Y}}$$

Thus, the vinyl group, $-CH=CH_2$, is of higher priority than the isopropyl group, $-CH(CH_3)_2$.

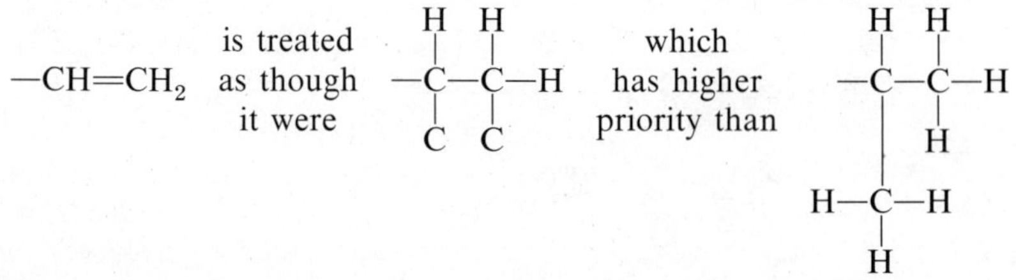

$$-CH=CH_2 \quad \begin{matrix}\text{is treated}\\\text{as though}\\\text{it were}\end{matrix} \quad -\overset{H}{\underset{C}{C}}-\overset{H}{\underset{C}{C}}-H \quad \begin{matrix}\text{which}\\\text{has higher}\\\text{priority than}\end{matrix} \quad -\overset{H}{\underset{H-\overset{|}{\underset{H}{C}}-H}{C}}-\overset{H}{\underset{H}{C}}-H$$

The ethynyl group, $-C\equiv CH$, is of higher priority than the tert-butyl group, $-C(CH_3)_3$.

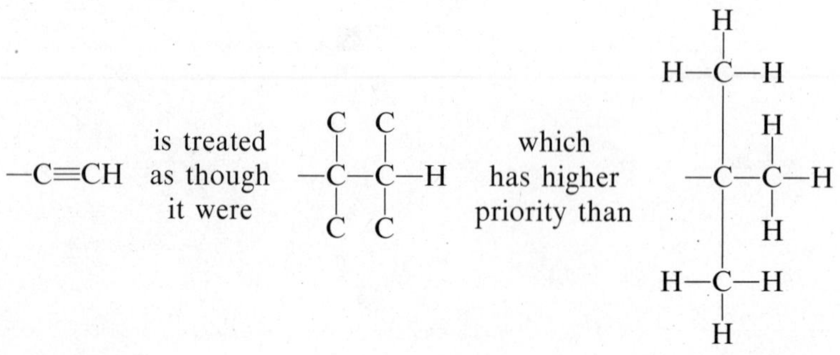

$$-C\equiv CH \quad \begin{matrix}\text{is treated}\\\text{as though}\\\text{it were}\end{matrix} \quad -\overset{C}{\underset{C}{C}}-\overset{C}{\underset{C}{C}}-H \quad \begin{matrix}\text{which}\\\text{has higher}\\\text{priority than}\end{matrix} \quad -\overset{H-\overset{H}{\underset{H}{C}}-H}{\underset{H-\overset{|}{\underset{H}{C}}-H}{C}}-\overset{H}{\underset{H}{C}}-H$$

Problem 7.10

Assign *R* and *S* designations to each of the following compounds.

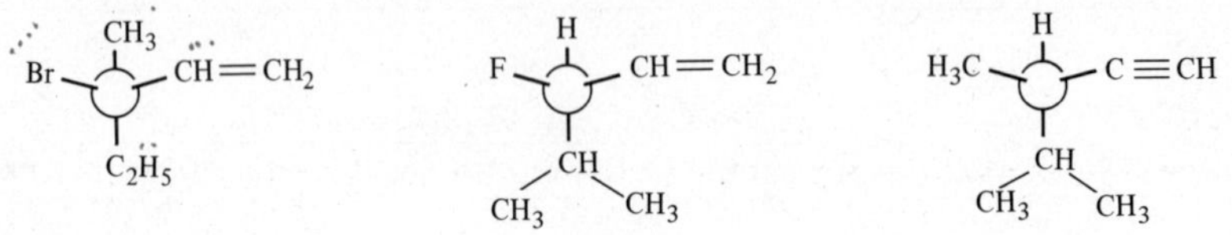

      (a)   (b)   (c)

TABLE 7.1 Physical Properties of R- and S-2-Butanol

PHYSICAL PROPERTY	R-2-BUTANOL	S-2-BUTANOL
Boiling point (1 atm)	99.5°	99.5°
Density (20/4)	0.808	0.808
Index of refraction (20°)	1.397	1.397

7.6 PROPERTIES OF ENANTIOMERS: OPTICAL ACTIVITY

Enantiomers are not superposable one on the other, and on this basis alone, we have concluded that enantiomers are different compounds. How are they different? Do enantiomers resemble structural isomers and diastereomers in having different melting and boiling points? The answer to this question is *no*. Enantiomers have *identical* melting and boiling points. Do enantiomers have different indexes of refraction, different solubilities in common solvents, different infrared spectra, and different rates of reaction with ordinary reagents? The answer to each of these questions is also no.*

We can see examples if we examine Table 7.1 where some of the physical properties of the 2-butanol enantiomers are listed.

Enantiomers differ only when they interact with other chiral substances or phenomena. One easily observable way in which enantiomers differ is in *their behavior toward plane-polarized light*. Plane-polarized light has chiral properties. When a beam of plane-polarized light passes through an enantiomer, the plane of polarization *rotates*. Moreover, separate enantiomers rotate the plane of plane-polarized light equal amounts *but in opposite directions*. Because of their effect on plane-polarized light, separate enantiomers are said to be *optically active compounds*.

In order to understand this behavior of enantiomers we need to understand the nature of plane-polarized light. We also need to understand how an instrument called a *polarimeter* operates.

Plane-Polarized Light

Light is an electromagnetic phenomenon. A beam of light consists of two mutually perpendicular oscillating fields: an oscillating electric field and an oscillating magnetic field. The planes in which the electrical and magnetic oscillations occur are also perpendicular to the direction of propagation of the beam of light (Fig. 7.8).

If we were to view a beam of ordinary light from one end, and if we could actually see the planes in which the electrical oscillations were occurring, we would find that oscillations of the electric field were occurring in all possible planes perpendicular to the direction of propagation (Fig. 7.9). (The same would be true of the magnetic field.)

---

*There are exceptions to what we have just stated. Enantiomers show different solubilities in chiral solvents; that is, in solvents that consist of a single enantiomer or an excess of a single enantiomer. Enantiomers also show different rates of reaction toward other chiral compounds; that is, toward reagents that consist of a single enantiomer or an excess of a single enantiomer.

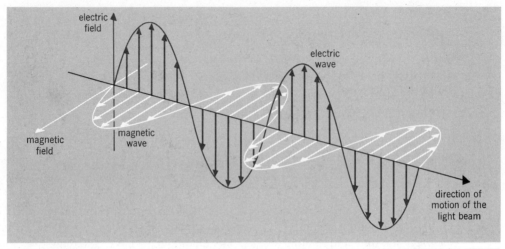

FIG. 7.8
The oscillating electric and magnetic fields of a beam of ordinary light.

When ordinary light is passed through a polarizer, the polarizer interacts with the electrical field so that the electrical field of the light that emerges from the polarizer is oscillating only in one plane.* Such light is called plane-polarized light (Fig. 7.10).

The Polarimeter

The device that is used for measuring the effect of plane-polarized light on optically active compounds is a polarimeter. A sketch of a polarimeter is shown in Fig. 7.11. The principal working parts of a polarimeter are (1) a light source (usually a sodium lamp), (2) a polarizer, (3) a tube for holding the optically active substance (or solution) in the light beam, (4) an analyzer, and (5) a scale for measuring the number of degrees that the plane of polarized light has been rotated.

The analyzer of a polarimeter (Fig. 7.11) is nothing more than another polarizer. If the tube of the polarimeter is empty, or if an optically *inactive* substance is present, the axes of the plane-polarized light and the analyzer will be exactly parallel when the instrument reads 0°, and the observer will detect the maximum amount of light passing through. If, by contrast, the tube contains an optically active substance, a solution of one enantiomer, for example, the plane of polarization of

*The lenses of Polaroid sunglasses have this effect.

FIG. 7.9
Oscillation of the electrical field of ordinary light occurs in all possible planes perpendicular to the direction of propagation.

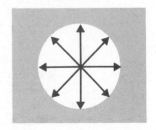

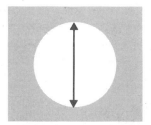

FIG. 7.10
The plane of oscillation of the electrical field of plane-polarized light. In this example the plane of polarization is vertical.

FIG. 7.11
The principal working parts of a polarimeter and the measurement of optical rotation. (From John R. Holum, Organic Chemistry: A Brief Course, *Wiley, New York, 1975, p. 316.)*

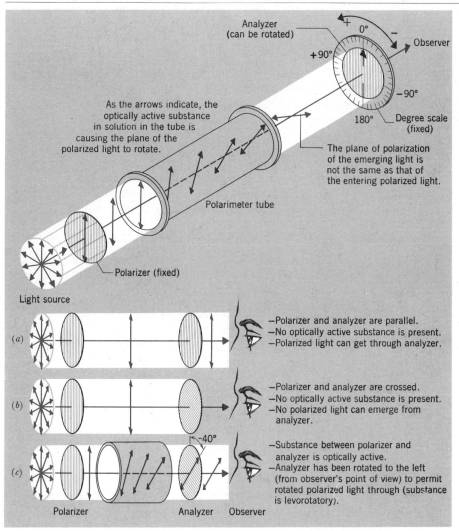

the light will be rotated as it passes through the tube. In order to detect the maximum brightness of light the observer will have to rotate the axis of the analyzer in either a clockwise or counterclockwise direction. If the analyzer is rotated in a clockwise direction, the rotation, α, (measured in degrees) is said to be positive ($+$). If the rotation is counterclockwise, the rotation is said to be negative ($-$). A substance that rotates plane-polarized light in the clockwise direction is also said to be *dextrorotatory,* and one that rotates plane-polarized light in a counterclockwise direction is said to be *levorotatory.*

Specific Rotation

The number of degrees that the plane of polarization is rotated as it passes through a solution of an enantiomer depends on the number of chiral molecules that it encounters. This, of course, depends on the length of the tube and the concentration of the enantiomer. In order to place measured rotations on a standard basis, chemists calculate a quantity called the specific rotation, $[\alpha]$. This is done as follows:

$$[\alpha] = \frac{\alpha}{c \cdot l}$$

where $[\alpha] =$ the specific rotation
  $\alpha\ =$ the observed rotation
  $c\ =$ the concentration of the solution in grams per cubic
    centimeter
  $l\ =$ the length of the tube in decimeters (1 dm $=$ 10 cm)

The specific rotation also depends on the temperature and the wavelength of light that is employed. Specific rotations are reported so as to incorporate these quantities as well. A specific rotation might be given as follows:

$$[\alpha]_D^{25} = +\,3.12°$$

This means that, using the D line* of a sodium lamp, at a temperature of 25°C, a sample containing 1.00 g/cc of the optically active substance, in a 1-dm tube, produces a rotation of 3.12° in a clockwise direction.

The specific rotation of *R*-2-butanol and *S*-2-butanol are given below.

R–2–butanol
$[\alpha]_D^{25} = -13.52°$

S–2–butanol
$[\alpha]_D^{25} = +13.52°$

The direction of rotation of plane-polarized light is often incorporated into the names of optically active compounds. The two sets of enantiomers at the top of the next page show how this is done.

* Wavelength $=$ 5896 Å.

R-(+)-2-methyl-1-butanol
$[\alpha]_D^{25} = +5.756°$

S-(−)-2-methyl-1-butanol
$[\alpha]_D^{25} = -5.756°$

R-(−)-1-chloro-2-methylbutane
$[\alpha]_D^{25} = -1.64°$

S-(+)-1-chloro-2-methylbutane
$[\alpha]_D^{25} = +1.64°$

These compounds (above) also illustrate an important principle: *there is no obvious correlation between the configurations of enantiomers and the direction in which they rotate plane-polarized light.*

R-(+)-2-methyl-1-butanol and R-(−)-1-chloro-2-methylbutane have the same *configuration,* that is, they have the same general arrangement of their atoms in space.

R-(+)-2-methyl-1-butanol   R-(−)-1-chloro-2-methylbutane

They have, however, an opposite effect on plane-polarized light.*

These same compounds also illustrate a second important principle: *there is no necessary correlation between the R and S designation and the direction of rotation of plane-polarized light.* R-2-methyl-1-butanol is dextrorotatory, (+), and R-1-chloro-2-methylbutane is levorotatory, (−).

7.7 THE ORIGIN OF OPTICAL ACTIVITY

It is not possible to give a complete, condensed account of the origin of the optical activity observed for separate enantiomers. An insight into the source of this phenomenon can be obtained, however, by comparing what occurs when a beam of plane-polarized light passes through a solution of *achiral* molecules, with what occurs when a beam of plane-polarized light passes through a solution of *chiral* molecules.

Almost all *individual* molecules, whether chiral or achiral, are theoretically capable of producing a slight rotation of the plane of plane-polarized light. The direction and magnitude of the rotation produced by an individual molecule depends, in part, on its orientation at the precise moment that it encounters the beam. In a solution there are, of course, billions of molecules in the path of the light beam

---

*A method based on the measurement of optical rotation measured at many different wavelengths, called optical rotary dispersion, has been used to correlate configurations of chiral molecules. A discussion of the technique of optical rotatory dispersion, however, is beyond the scope of this text.

and at any given moment these molecules will be present in all possible orientations. If the beam of plane-polarized light passes through a solution of the achiral compound 2-propanol, for example, it should encounter at least two molecules in the exact orientations shown below in Fig. 7.12. The effect of the first encounter might be to produce a very slight rotation of the plane of polarization to the right. Before the beam emerges from the solution, however, it should encounter at least one molecule of 2-propanol that is in exactly the mirror reflection orientation of the first. The effect of this second encounter will be to produce an equal and opposite rotation of the plane: a rotation that exactly cancels the first rotation. The beam, therefore, emerges with no net rotation.

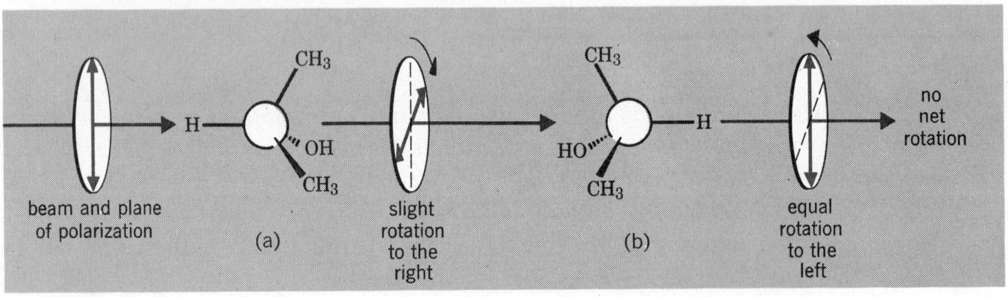

FIG. 7.12

A beam of plane-polarized light encountering a molecule of 2-propanol (an achiral molecule) in orientation (a) and then a second molecule in the mirror-reflection orientation (b). The beam emerges from these two encounters with no net rotation of its plane of polarization.

What we have just described for the two encounters shown in Fig. 7.12 can be said of all possible encounters of the beam with molecules of 2-propanol. *For each encounter with a particular orientation there will be an encounter with a molecule that is in a mirror-reflection orientation.* The result of all of these encounters will be such that all of the rotations produced by individual molecules will be canceled and 2-propanol will be found to be *optically inactive.*

We see from this illustration that optical inactivity is not a property of individual molecules but, instead, optical inactivity is a property of the random distribution of individual molecules serving in mirror reflection orientations of each other.

What, then, is the situation when a beam of plane-polarized light passes through a solution of one enantiomer of a chiral compound? We can answer this question by considering what might occur when plane-polarized light passes through a solution of pure *R*-2-butanol. Figure 7.13 illustrates one possible encounter of a beam of plane-polarized light with a molecule of *R*-2-butanol.

When a beam of plane-polarized light passes through a solution of *R*-2-butanol, *no molecule is present that can serve in an exact mirror-reflection orientation of any given orientation of an R-2-butanol molecule.* The only molecules that could do this would be molecules of *S*-2-butanol, and *S*-2-butanol is not present. Exact cancellation of the rotations produced by all of the encounters of the beam with random orientations of *R*-2-butanol does not happen and, as a result, a net rotation of the plane of polarization is observed. *R*-2-butanol is found to be *optically active.*

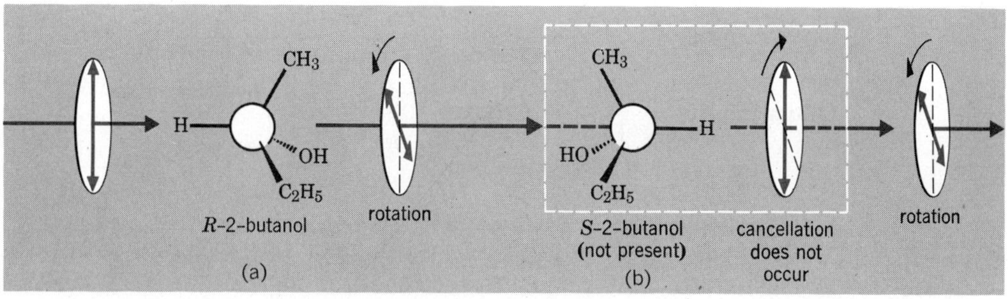

FIG. 7.13

(a) *A beam of plane-polarized light encounters a molecule of* R-*2-butanol (a chiral molecule) in a particular orientation. This encounter produces a slight rotation of the plane of polarization.* (b) *Exact cancellation of this rotation requires that a second molecule be present in an exact mirror-reflection orientation. This cancellation does not occur because the only molecule that can serve in an exact mirror reflection orientation of the encounter is a molecule of* S-*2-butanol and* S-*2-butanol is not present. There is, as a result, a net rotation of the plane of polarization.*

Racemic Modifications

The net rotation of the plane of polarization that we observe for a solution consisting of molecules of *R*-2-butanol alone would not be observed if we passed the beam through a solution that contained equimolar amounts of *R*-2-butanol and *S*-2-butanol. In the latter instance, molecules of *S*-2-butanol would be present in a quantity equal to those of *R*-2-butanol and for every possible orientation of one enantiomer, a molecule of the other enantiomer would be in a mirror-reflection orientation. Exact cancellations of all rotations would occur, and the solution of the equimolar mixture of enantiomers would be *optically inactive*.

Equimolar mixtures of two enantiomers are called *racemic modifications*. Racemic modifications show no rotation of plane-polarized light; as such, they are often designated as being ($\pm$). A racemic mixture of *R*-($-$)-2-butanol and *S*-($+$)-2-butanol might be indicated as

$$(\pm)\text{-2-butanol}$$

or as $\qquad\qquad (\pm) \; CH_3CH_2CHOHCH_3$

Optical Purity

A sample of an optically active substance that consists of a single enantiomer is said to be *optically pure*. An optically pure sample of *S*-($+$)-2-butanol shows a specific rotation of $+13.52°$ ($[\alpha]_D^{25} = +13.52°$). On the other hand, a sample of *S*-($+$)-2-butanol that contains less than an equimolar amount of *R*-($-$)-2-butanol will show a specific rotation that is less than $+13.52°$ but greater than $0°$. Such a sample is said to have an *optical purity* less than 100%.

Let us suppose, for example, that the sample showed a specific rotation of $+6.76°$. We would then say that the optical purity of the *S*-($+$)-2-butanol is 50%.*

---

*The term optical purity is applied to a single enantiomer or to mixtures of enantiomers only. It should not be applied to mixtures in which some other compound is present.

$$\text{Optical purity} = \frac{\text{observed specific rotation}}{\text{specific rotation of the pure enantiomer}} \times 100$$

$$\text{Optical purity} = \frac{+6.76°}{13.52°} \times 100 = 50\%$$

Problem 7.11

What relative molar proportions of S-(+)-2-butanol and R-(−)-2-butanol would give a specific rotation, $[\alpha]_D^{25}$, equal to +6.76°?

7.8 THE SYNTHESIS OF ENANTIOMERS

There are many times in the course of working in the organic laboratory when a reaction carried out with reactants whose molecules are achiral results in the formation of products whose molecules are chiral. The usual outcome of such a reaction is the formation of a optically inactive racemic modification. The reason: the chiral molecules of the product are obtained as a 50:50 mixture of enantiomers.

An example is the synthesis of 2-butanol by the nickel-catalyzed hydrogenation of 2-butanone.

$$\text{CH}_3\text{CH}_2\overset{\text{O}}{\underset{\|}{\text{C}}}\text{CH}_3 + \quad \text{H—H} \quad \xrightarrow{\text{Ni}} \quad (\pm)\ \text{CH}_3\text{CH}_2\overset{*}{\underset{\text{OH}}{\text{C}}}\text{HCH}_3$$

| 2-Butanone (achiral molecules) | Hydrogen (achiral molecules) | (±) 2-Butanol (chiral molecules but 50:50 mixture R and S) |

Molecules of neither reactant—2-butanone nor hydrogen—are chiral. The molecules of the product—2-butanol—are chiral. The product, however, is obtained as a racemic modification because the two enantiomers, R-(−)-2-butanol and S-(+)-2-butanol, are obtained in equal amounts.*

Figure 7.14 shows why a racemic modification of 2-butanol is obtained. Hydrogen, absorbed on the surface of the nickel catalyst, adds with equal facility at either face of 2-butanone. Reaction at one face produces one enantiomer; reaction at the other face produces the other enantiomer, and the two reactions occur at the same rate.

Another example, of the same type, is the catalytic hydrogenation of pyruvic acid to produce a racemic modification of lactic acid.

$$\text{CH}_3\overset{\text{O}}{\underset{\|}{\text{C}}}\text{COOH} + \quad \text{H}_2 \quad \xrightarrow{\text{Pt}} \quad (\pm)\ \text{CH}_3\overset{\text{OH}}{\underset{|}{\text{C}}}\text{HCOOH}$$

| Pyruvic acid (achiral molecules) | Hydrogen (achiral molecules) | (±) Lactic acid (chiral molecules but 50:50 mixture of R and S forms) |

---

* This is not the result if reactions like this are carried out in the presence of a chiral influence such as an optically active solvent or, as we will see later, an enzyme. The nickel catalyst used in this reaction does not exert a chiral influence.

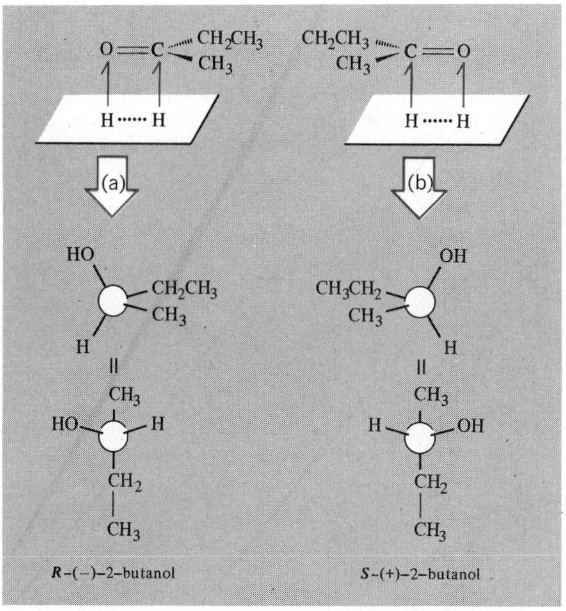

FIG. 7.14

The reaction of 2-butanone with hydrogen in the presence of a nickel catalyst. The reaction rate by path (a) *is equal to that by path* (b). *R-(−)-2-butanol and S-(+)-2-butanol are produced in equal amounts:* as a racemic modification.

Problem 7.12

What products would you expect from each of the following reactions? What relative amounts of each product would be formed?

(a) $CH_3CH_2CCH_3 + NaBH_4 \longrightarrow$
     $\quad\quad\quad\quad \overset{||}{O}$

(b) $CH_3CH_2CCH_3 + LiAlH_4 \longrightarrow$
     $\quad\quad\quad\quad \overset{||}{O}$

(c) 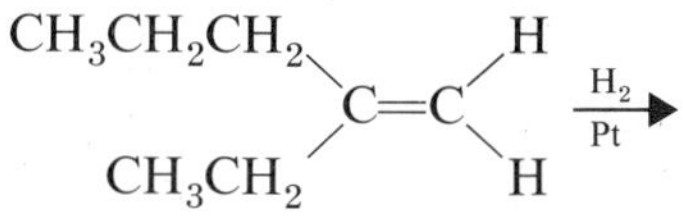 $\begin{array}{c} CH_3CH_2CH_2 \\ \\ CH_3CH_2 \end{array} \!\!\! C{=}C \!\!\! \begin{array}{c} H \\ \\ H \end{array} \xrightarrow[Pt]{H_2}$

(d) $CH_3CH_2CH{=}CH_2 \xrightarrow[H_2O]{H^+}$

(e) $CH_3CH{=}CHCH_3 \xrightarrow[\text{(2) OH}^-,\, H_2O_2,\, \text{heat}]{\text{(1) } (BH_3)_2}$

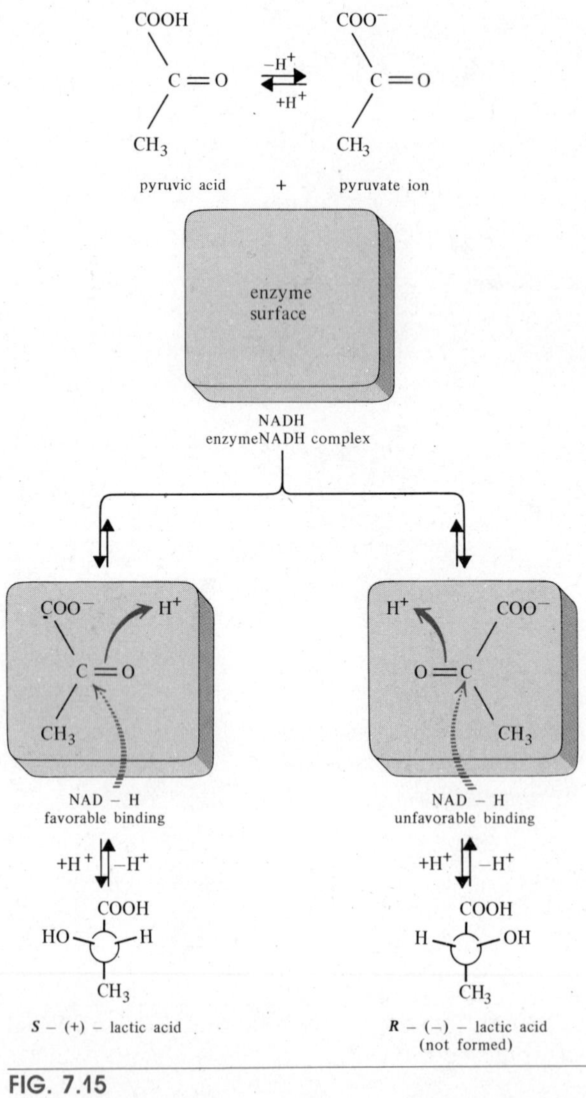

FIG. 7.15

A mechanism that accounts for stereoselectivity in the enzyme-catalyzed reduction of pyruvic acid to S-(+)-lactic acid. Pyruvate ion can bind itself to the surface of the enzyme-NADH complex in only one favorable way. As a result, NADH can transfer hydrogen (H:⁻) to only one surface of the pyruvate ion; this leads to the formation of S-(+)-lactic acid. R-(−)-lactic acid is not produced because its formation would require an unfavorable binding between the NADH-enzyme complex.

The stereochemical course of reactions that occur in living cells are often dramatically different from those that are carried out in the glassware of the laboratory. The laboratory hydrogenation of pyruvic acid in the presence of a platinum catalyst as we have just seen produces a racemic modification of lactic acid. In muscle cells, however, the conversion of pyruvic acid to lactic acid is catalyzed by an enzyme called *lactic acid dehydrogenase*.* The enzymatically catalyzed reaction results in the formation of a single enantiomer, S-($+$)-lactic acid:

$$\underset{\substack{\text{pyruvic acid}}}{CH_3\overset{\displaystyle\underset{\displaystyle O}{\|}}{C}COOH} \xrightarrow[\substack{\text{dehydrogenase}\\ \text{NADH}\dagger}]{\text{lactic acid}} \underset{\substack{S\text{-}(+)\text{-lactic acid}\\ [\alpha] = +3.82}}{\overset{\displaystyle COOH}{\underset{\displaystyle CH_3}{HO\text{---}\overset{|}{\underset{|}{C}}\text{---}H}}}$$

The enzymatically catalyzed reaction is said to be *stereoselective; Only one stereoisomer* (S-($+$)-lactic acid) *results from a reactant* (pyruvic acid) *that is not capable of existing in stereoisomeric forms.* The stereoselectivity of the enzyme catalyzed reaction is a result of a special relation between the reacting molecule and the enzyme surface (Fig. 7.15).

7.9 COMPOUNDS WITH MORE THAN ONE CHIRAL CENTER

Thus far all of the chiral molecules that we have considered have contained only one chiral carbon. Many organic molecules, however, contain more than one chiral center. In Chapter 24, for example, we will consider the chemistry of carbohydrates and, in doing so, we will pay considerable attention to the important sugar, glucose. A molecule of glucose in its simplest form contains *four* chiral carbons.

$$\underset{\text{Glucose}}{\overset{\displaystyle O}{\underset{\substack{|\\ *CHOH\\ |\\ *CHOH\\ |\\ *CHOH\\ |\\ *CHOH\\ |\\ CH_2OH}}{\overset{\|}{C}H}}}$$

We can begin our study of molecules containing more than one chiral center with simpler examples, however. Let us start by examining a compound whose molecules contain two chiral carbons, 2,3-dibromopentane.

$$\underset{\substack{\text{2,3-Dibromopentane}}}{CH_3\text{---}\overset{*}{\underset{\underset{\displaystyle Br}{|}}{C}H}\text{---}\overset{*}{\underset{\underset{\displaystyle Br}{|}}{C}H}\text{---}CH_2CH_3}$$

*The name of the enzyme is derived from the fact that this enzyme also catalyzes the reverse reaction, that is, the dehydrogenation of S-($+$)-lactic acid to produce pyruvic acid.

†The reducing agent in this reaction is a compound, NADH, called the reduced form of nicotinamide adenine dinucleotide, NAD^+ (Sec. 12.10).

There is a useful rule that helps us to know how many stereoisomers to expect from compounds like this one: *the total number of stereoisomers will not exceed 2^n where n is equal to the number of chiral centers.* For 2,3-dibromopentane we should not expect more than four stereoisomers ($2^2 = 4$).

Our next task is to write three-dimensional formulas for the stereoisomers of 2,3-dibromopentane. We begin by writing a three-dimensional formula for one stereoisomer and then by writing the formula for *its* mirror reflection. This is shown below.

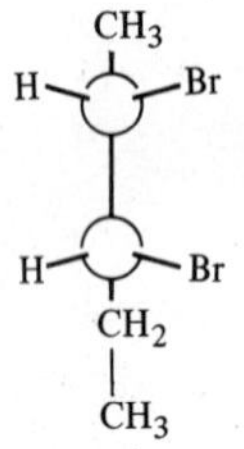

It is helpful to follow certain conventions when we write these three-dimensional formulas. For example, we usually write our structures in eclipsed conformations. When we do this we do not mean to imply that eclipsed conformations are the most stable ones—they most certainly are not. We write eclipsed conformations because, as we shall see on page 256, they make it easy for us to recognize planes of symmetry when they are present. We also write the longest carbon chain in a generally vertical orientation on the page; this makes the structures that we write directly comparable. As we do these things, however, *we must remember that molecules can rotate in their entirety* and that *at normal temperatures groups attached by single bonds rotate freely.* If rotations of the structure itself, or rotations of groups joined by single bonds make one structure superposable with another, then *the structures do not represent different compounds*; rather, they represent different orientations or different conformations of two molecules of the same compound.

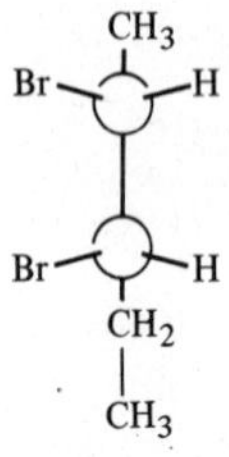

Since structures **1** and **2** are not superposable, they represent different compounds. Since structures **1** and **2** differ *only* in the arrangement of their atoms in space, they represent stereoisomers. Structures **1** and **2** are also mirror reflections of each other, thus **1** and **2** represent enantiomers.

We find, moreover, that we can also write a structure **3** (below) that is different from either **1** or **2**. We can also write a structure **4** that is a nonsuperposable mirror reflection of structure **3**.

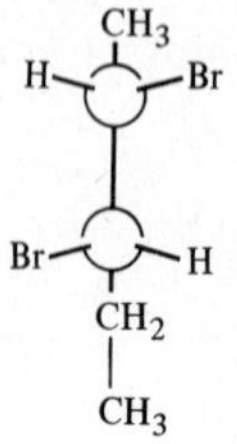

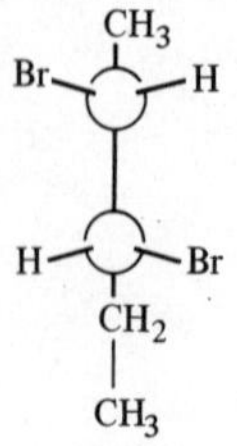

Structures **3** and **4** represent another pair of enantiomers. Structures **1–4** are all different, so there are a total of four stereoisomers of 2,3-dibromopentane. At this point you should convince yourself that there are no other stereoisomers of

2,3-dibromopentane by writing other structural formulas. You will find that rotation of the single bonds (or of the entire structure) of any other arrangement of the atoms will cause the structure to become superposable with one of the structures that we have written here.

The compounds represented by structures **1–4** are all optically active compounds. Any one of them, if placed separately in a polarimeter, would show optical activity.

The compounds represented by formulas **1** and **2** are enantiomers. The compounds represented by formulas **3** and **4** are also enantiomers. But, what is the isomeric relation between the compounds represented by formulas **1** and **3**?

We can answer this question by observing that **1** and **3** are stereoisomers and that they are *not* mirror reflections of each other. They are, therefore, *diastereomers*.

Problem 7.13

(a) What is the stereoisomeric relation between compounds **2** and **3**? (b) Between **1** and **4**? (c) Between **2** and **4**? (d) Make a table showing all of the stereoisomeric relations between compounds **1** to **4**. (e) Would compounds **1** and **2** have the same boiling point? (f) Would compounds **1** and **3**?

Let us consider another example of a compound whose molecules contain two chiral carbons, 2,3-dibromobutane.

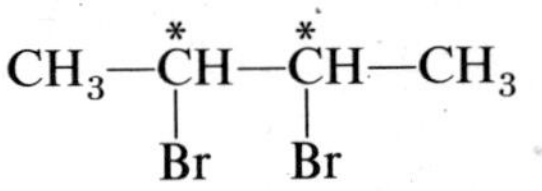

$$CH_3 - \overset{*}{C}H - \overset{*}{C}H - CH_3$$
$$\quad\quad | \quad\quad |$$
$$\quad\quad Br \quad Br$$

In this example we will find that there are not four stereoisomers, as is the case with 2,3-dibromopentane, but only *three*.

We begin in the same way as we did before, by writing the structural formula of one stereoisomer and then its mirror image.

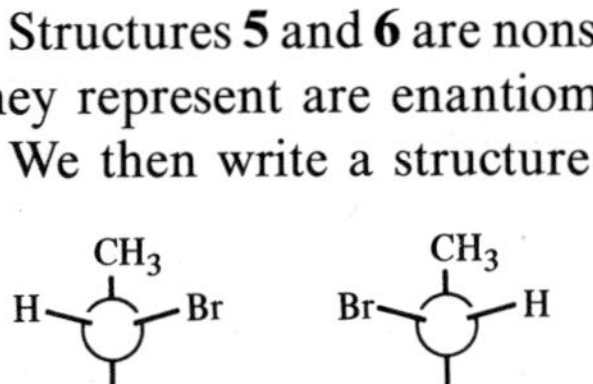

5   **6**

Structures **5** and **6** are nonsuperposable mirror reflections, and the compounds that they represent are enantiomers.

We then write a structure **7** that is different from either structure **5** or **6**.

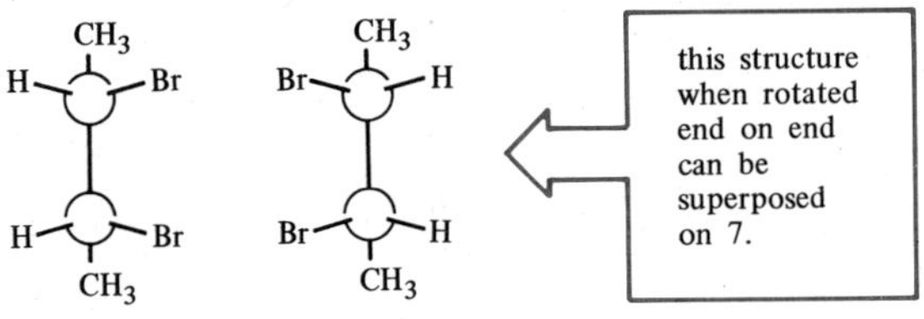

this structure when rotated end on end can be superposed on 7.

7   **8**

a *meso* compound
(optically inactive)

We also write a structure **8** that is the mirror reflection of **7**. We find, however, in this instance the two structures *are superposable*. Structures **7** and **8** are simply two different orientations of a molecule of the same compound.

The molecule represented by structure **7** is not chiral even though it contains chiral carbons. *Achiral molecules that contain chiral centers* are called *meso compounds*. *Meso* compounds are optically inactive.

The ultimate test for molecular chirality is to construct a model (or write the structure) of the molecule and then test whether or not the model (or structure) is superposable on its mirror reflection. If it is, the molecule is achiral: if it *is not* the molecule is chiral.

We have already carried out this test with structure **7** and found that it is achiral. We can also demonstrate that **7** is achiral in another way. Figure 7.16 shows that structure **7** *has a plane of symmetry*.

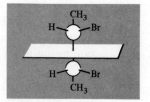

FIG. 7.16
The plane of symmetry of meso-*2,3-dibromobutane. This plane divides the molecule into two halves that are mirror reflections of each other.*

Problem 7.14

Which of the following would be optically active?
  (a) **5** alone
  (b) **6** alone
  (c) **7** alone
  (d) An equimolar mixture of **5** and **6**.

Problem 7.15

Write three-dimensional formulas for all of the stereoisomers of each of the following compounds.
  (a) $CH_3CH(OH)CH(OH)CH_3$
  (b) $CH_3CHBrCHClCH_3$
  (c) $CH_3CHBrCHBrCH_2Br$
  (d) $CH_2BrCHBrCHBrCH_2Br$
  (e) $CH_3CHClCHClCHClCH_3$
  (f) In answers to parts (a) to (e) label pairs of enantiomers
      and *meso* compounds.

Stereoisomerism of Cyclic Compounds

Cyclic compounds also exist in stereoisomeric forms. For example, 1,2-cyclopentanediol has two chiral carbons and exists in three stereoisomeric forms, **9, 10,** and **11.**

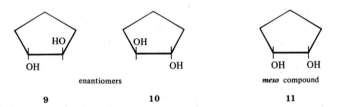

enantiomers   *meso* compound

9   10   11

The *trans* compound exists as a pair of enantiomers **9** and **10**. *Cis*-1,2-cyclo-pentanediol is a *meso* compound. It has a plane of symmetry that is perpendicular to the plane of the ring.

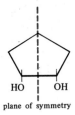

HO   OH

plane of symmetry

Problem 7.16

(a) Is the *trans*-1,2-cyclopentanediol, **9**, superposable on its mirror image (i.e., compound **10**)? (b) Is the *cis*-1,2-cyclopentanediol, **11**, superposable on its mirror image? (c) Is the *cis*-1,2-cyclopentanediol a chiral molecule? (d) Would *cis*-1,2-cyclopentanediol show optical activity? (e) What is the stereoisomeric relation between **9** and **11**? (f) Between **10** and **11**?

In Chapter 6 we saw that *cis*-1,2-cyclopentanediol could be synthesized by the syn hydroxylation of cyclopentene.

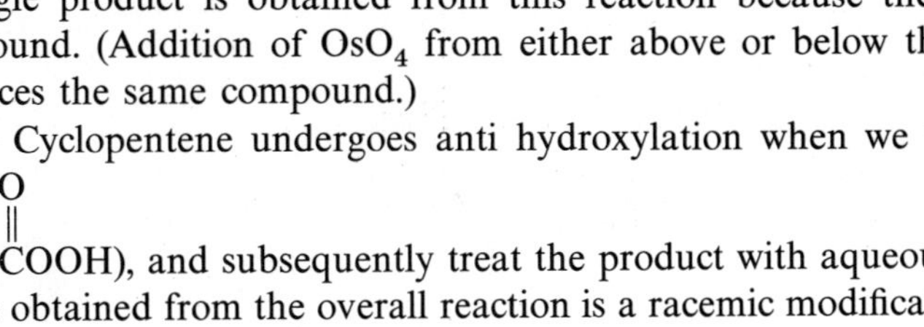

$$\text{cyclopentene} \xrightarrow[25°]{OsO_4} \xrightarrow{Na_2SO_3} \text{cis-1,2-cyclopentanediol}$$

OH   OH

cis–1, 2–cyclopentanediol
11

A single product is obtained from this reaction because the product is a *meso* compound. (Addition of OsO_4 from either above or below the plane of the ring produces the same compound.)

Cyclopentene undergoes anti hydroxylation when we use perbenzoic acid

O
‖
(C_6H_5COOH), and subsequently treat the product with aqueous acid. The product that is obtained from the overall reaction is a racemic modification of the *trans*-1,2-cyclopentanediols.

$$\text{Cyclopentene} \xrightarrow[\text{(2) H}^+, \text{ H}_2\text{O}]{\text{(1) } C_6H_5\overset{O}{\overset{\|}{C}}OOH, \text{ CHCl}_3, \text{ 25°}} (\pm)\text{-}trans\text{-1,2-cyclopentanediol}$$

We can see why a racemic modification is obtained if we examine the two steps in detail.

Cyclopentene reacts with perbenzoic acid and forms an epoxide.

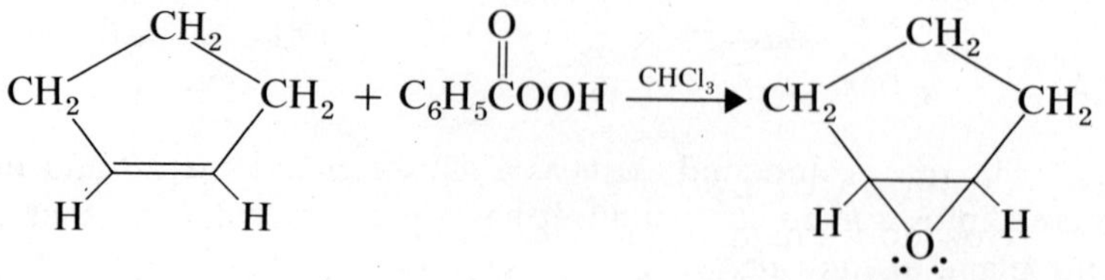

Cyclopentene
oxide
(an epoxide)

The epoxide ring is highly strained. When cyclopentene oxide is treated with aqueous acid the epoxide ring opens. The new C—O bond is always formed from the side opposite that of the original epoxide ring.

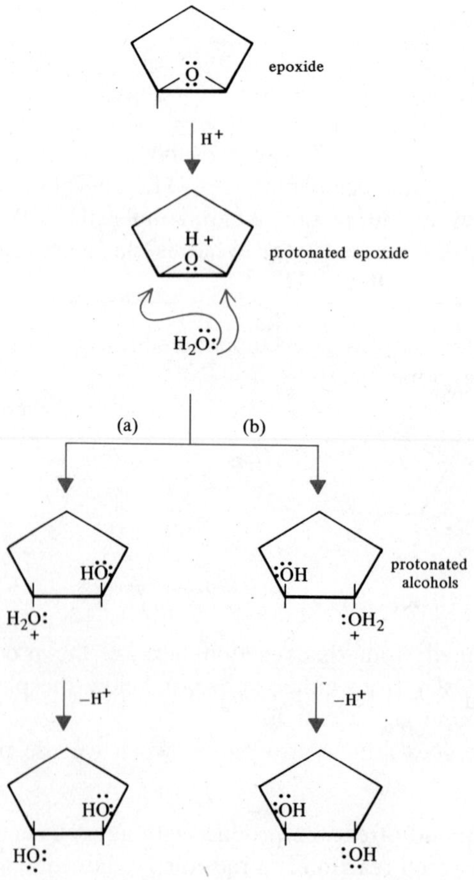

Since the rates of ring opening by paths (a) and (b) are equal, the enantiomers **9** and **10** are obtained in equal amounts.

Problem 7.17

On p. 211 we gave a mechanism for the addition of bromine to cyclopentene. (a) What products would you expect from this reaction? (b) In what form would they be obtained? (c) If a syn addition of bromine to cyclopentene were to occur, what product would be obtained?

Naming Compounds with More than One Chiral Center

How do we name compounds that contain more than one chiral carbon so as to specify the configuration at each chiral center? The answer is straightforward: we analyze each center separately using the rules specified in Section 7.5 and decide whether it is R or S. Then, using numbers, we tell which designation refers to which carbon.

Compound **1**, (p. 254) for example, is named in the following way.

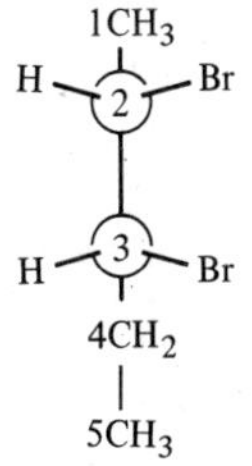

$$1CH_3$$

$$H \text{—} \overset{}{\underset{2}{\bigcirc}} \text{—} Br$$

$$H \text{—} \overset{}{\underset{3}{\bigcirc}} \text{—} Br$$

$$4CH_2$$

$$|$$

$$5CH_3$$

1

When this formula is rotated so that the lowest priority group attached to carbon-2 is directed away from the viewer it resembles that shown below.

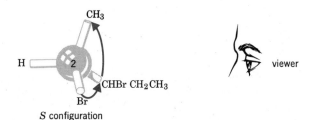

S configuration

The order of progression from the group of highest priority, to that of next highest priority, (from -Br, to -CHBrCH$_2$CH$_3$, to CH$_3$) is counterclockwise. So carbon -2 has the S configuration.

When we repeat this procedure with carbon-3, we find that carbon-3 has the R configuration. Compound **1** is then (2S, 3R)-2,3-dibromopentane.

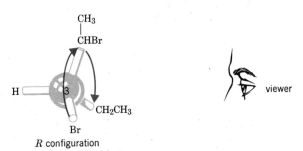

R configuration

Problem 7.18

Give names that include R and S designations for compounds **2** to **7** (p. 254–255).

7.10 STEREOSPECIFIC REACTIONS

Reactions in which stereoisomeric reactants give diastereomerically different products are called *stereospecific reactions*.

The distinction between the terms stereoselective and stereospecific is a subtle one. A stereoselective reaction is one that yields a predominance of one diastereomer *regardless* of the stereochemistry of the reactant. On p. 253 we described the enzymatic reduction of pyruvic acid as being stereoselective. This is true because molecules of pyruvic acid are achiral and do not exist in different stereoisomeric forms. When they associate with the chiral molecules of the enzyme they produce a predominance of one diastereomeric form of the pyruvic acid-enzyme complex. This complex then goes on to yield a single enantiomeric form of lactic acid, that is, S-(+)-lactic acid.

On the other hand, when we say that a reaction is *stereospecific* we mean that *a single stereoisomeric form of the starting material* reacts in such a way that it gives a *specific diastereomeric form of the product*. It does this because the reaction mechanism requires that the configurations of the atoms involved change in a characteristic way.

We should notice, too, the terms stereoselective and stereospecific are not mutually exclusive. All reactions that are stereospecific are, of necessity, stereoselective. The reverse, however, is not true; the enzymatic reduction of pyruvic acid is stereoselective, but it is not stereospecific. It is not stereospecific because the reactant cannot exist in stereoisomeric forms. Other reactions may be stereoselective but not be stereospecific, because reactants of different stereochemistry yield one predominant stereoisomeric product.

The reaction of alkenes and peracids takes place in a stereospecific way: *cis*-2-butene, for example, yields only *cis*-2,3-dimethyloxirane, and *trans*-2-butene yields only the *trans*-2,3-dimethyloxiranes.

(*a*)

cis-2-Butene

cis-2,3-Dimethyloxirane
(a *meso* compound)

(*b*)

trans-2-Butene

Enantiomeric *trans*-2,3-dimethyloxiranes

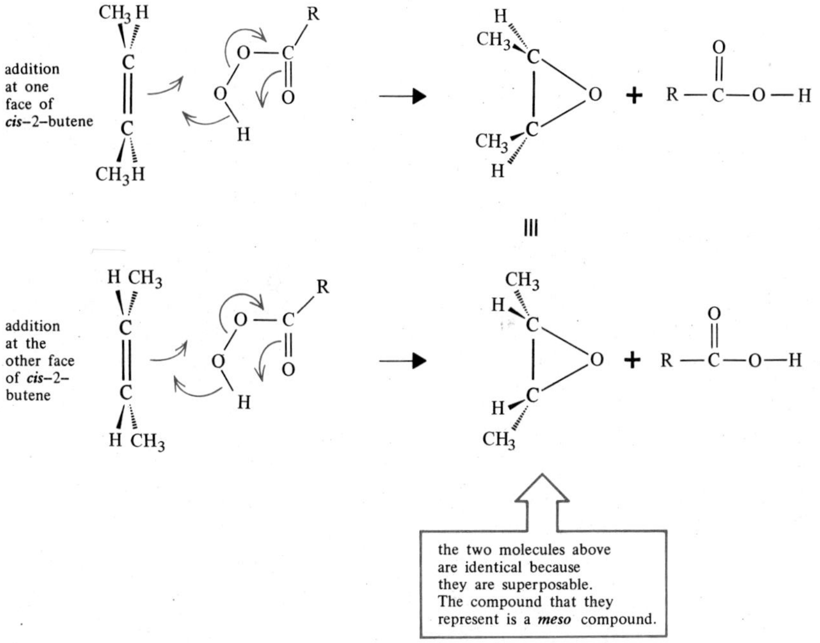

FIG. 7.17

Epoxidation of cis-*2-butene in a stereospecific manner yields* cis-*2,3-dimethyloxirane (a* meso *compound).*

FIG. 7.18

Epoxidation of trans-*2-butene yields enantiomeric* trans-*2,3-dimethyloxiranes.*

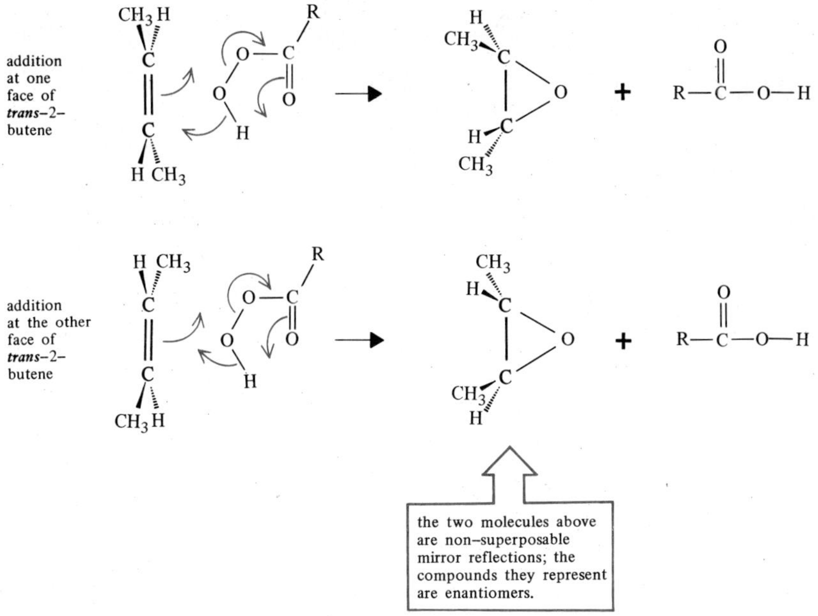

The reactants *cis*-2-butene and *trans*-2-butene are stereoisomers; they are *diastereomers*. The product of reaction (*a*), *cis*-2,3-dimethyloxirane is a *meso* compound, and it is a diastereomer of either of the products of reaction (*b*), that is, the enantiomeric *trans*-2,3-dimethyloxiranes. Thus, by definition, both reactions are stereospecific.

We can better understand the results of these two reactions if we examine their mechanisms. *Cis*-2-butene (Fig. 7.17) reacts with the peracid through a syn addition. A syn addition of oxygen to either face of *cis*-2-butene produces the same compound: the *meso* compound, *cis*-2,3-dimethyloxirane.

On the other hand, when *trans*-2-butene undergoes a syn addition of oxygen at one face it produces one enantiomeric form of *trans*-2,3-dimethyloxirane; when it undergoes syn addition at the other face it produces the other enantiomeric form of *trans*-2,3-dimethyloxirane. These reactions are shown in Fig. 7.18.

The reactions that lead to the enantiomeric *trans*-2,3-dimethyloxiranes, moreover, have identical energies of activation. As a result, they occur at identical rates and produce the enantiomers in equal amounts: as a racemic modification.

When either *cis*-2,3-dimethyloxirane or the racemic modification of *trans*-2,3-dimethyloxirane is subjected to acid-catalyzed hydrolysis, stereospecific reactions also take place.

The *meso* compound, *cis*-2,3-dimethyloxirane (Fig. 7.19), hydrolyzes to yield (2*R*, 3*R*)-2,3-butanediol and (2*S*, 3*S*)-2,3-butanediol. These products are enantiomers. Since the attack by water at either carbon [path (*a*) or path (*b*)] occurs at the same rate, the product is obtained as a racemic modification.

When either of the *trans*-2,3-dimethyloxirane enantiomers undergoes acid catalyzed hydrolysis, the only product that is obtained is the *meso* compound, (2*R*, 3*S*)-2,3-butanediol. The hydrolysis of one enantiomer is shown in Fig. 7.20. (You might construct a similar diagram showing the hydrolysis of the other enantiomer to convince yourself that it, too, yields the same product.)

Since both steps in this method for the conversion of an alkene to a diol (glycol) are stereospecific (i.e., both the epoxidation step and the acid-catalyzed hydrolysis), the net result is a stereospecific anti hydroxylation of the double bond.

(*cis*–2–butene)   enantiomeric 2, 3–butanediols

(*trans*–2–butene)   (*meso*–2, 3–butanediol)

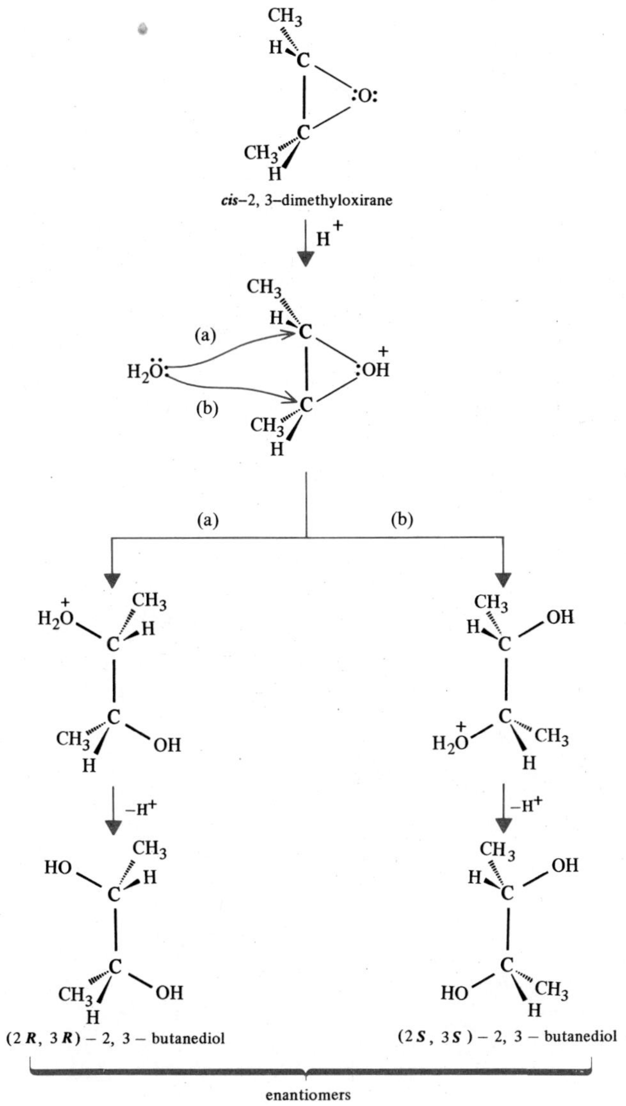

FIG. 7.19
Acid-catalyzed hydrolysis of cis-*2,3-dimethyloxirane yields (2R, 3R)-2,3-butanediol by path* (a) *and (2S, 3S)-2,3-butanediol by path* (b).

Problem 7.19

What product(s) would you expect to obtain from the addition of bromine to (a) *cis*-2-butene and (b) *trans*-2-butene? Show all steps in each addition reaction in ways that clearly illustrate their stereochemistry.

Problem 7.20

What products would you expect to obtain from the syn hydroxylation of (a) *cis*-2-butene and (b) *trans*-2-butene using either $KMnO_4$ or OsO_4 and Na_2SO_3?

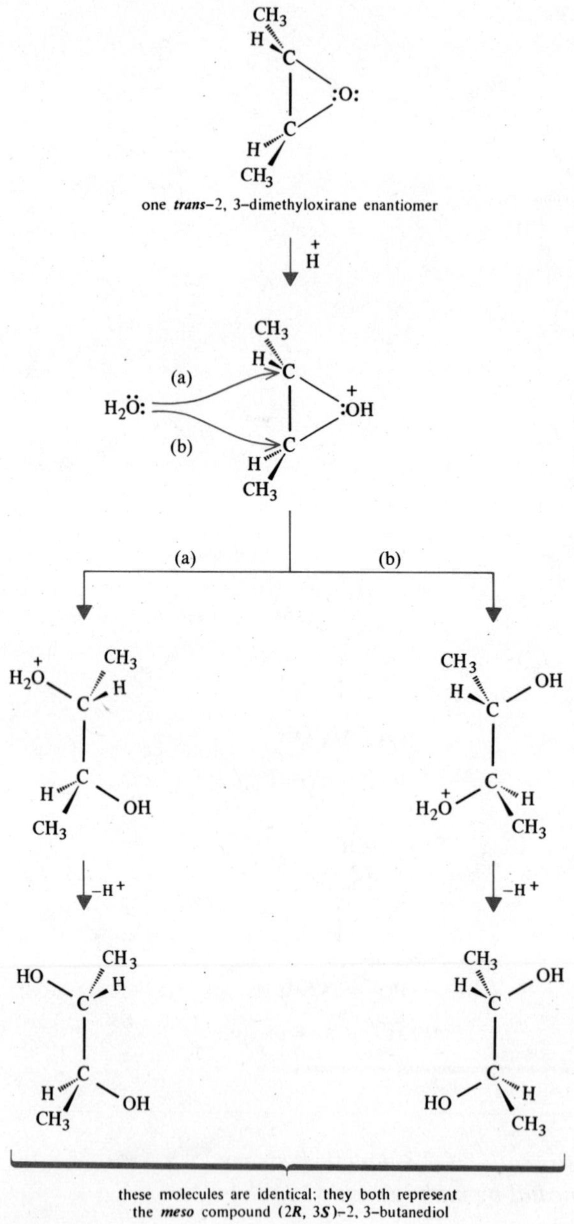

one *trans*-2, 3-dimethyloxirane enantiomer

these molecules are identical; they both represent
the *meso* compound (2**R**, 3**S**)-2, 3-butanediol

FIG. 7.20

The acid-catalyzed hydrolysis of one trans-2,3-
dimethyloxirane enantiomer produces the meso *compound,* (2R, 3S)-2,3-butanediol *by path* (a) *or* (b).
*Hydrolysis of the other enantiomer (or the racemic
modification) would yield the same product.*

Other stereospecific and stereoselective reactions of alkenes—those that lead to addition polymers and those that take place with divalent carbon compounds—are described as special topics in Sections 8.4 and 8.5.

7.11 REACTIONS OF CHIRAL MOLECULES

The reactions that chiral molecules undergo can be divided into two general categories: (1) those reactions in which no bonds to the chiral carbon are broken, and (2) those reactions in which a bond to the chiral carbon *is* broken. We discuss reactions of the first type in detail in this chapter because they are simpler and because they can be used *to relate configurations* of chiral molecules. Reactions of the second type are more complicated. Except for pointing out the possible results of these reactions, we will postpone a detailed discussion of them until later.

Reactions in Which No Bonds to the Chiral Carbon Are Broken

Reactions in which no bonds to the chiral carbon are broken must proceed with *retention of configuration.* We can illustrate what we mean by this by examining the reaction of *R*-(−)-2-butanol with acetic anhydride shown in Fig. 7.21.

The reaction of anhydrides with alcohols (Sec. 19.7) is an easy way to synthesize esters. The reaction of acetic anhydride with *R*-(−)-2-butanol produces the ester, *R*-(−)-2-butyl acetate. Since the reaction does not involve cleavage of any of the bonds to the chiral carbon we can be certain that the general spatial arrangement of the groups in the product will be the same as that of the reactant. The configuration of the reactant is said to be *retained* in the product; the reaction is said to proceed with *retention of configuration.*

In the example that we just cited, the optically active reactant and product

FIG. 7.21

The reaction of R-(−)-2-butanol with acetic anhydride. The bonds that are broken are shown with dashed lines. A bond between one of the carbonyl groups and the oxygen of acetic anhydride is broken. The only bond that is broken in R-(−)-2-butanol is the one between the hydrogen and oxygen of the hydroxyl group. No bonds to the chiral carbon (shown in color) are broken.*

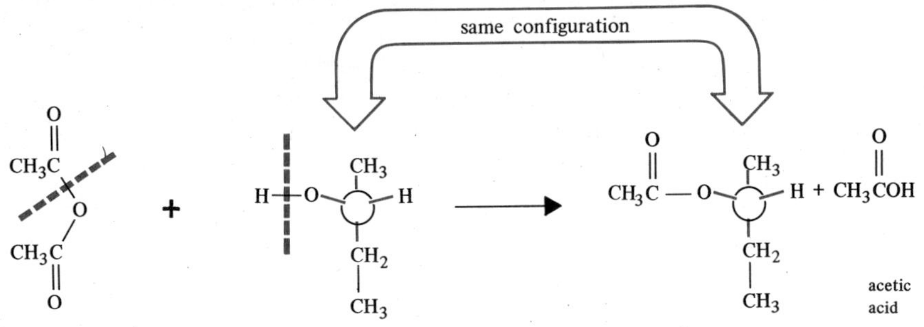

acetic anhydride   *R*-(−)-2-butanol   *R*-(−)-2-butyl acetate

$[\alpha]_D^{25} = -13.52$ $[\alpha]_D^{25} = -25.43$

both rotate plane-polarized light in the same direction. This is not always the case as the following example illustrates.

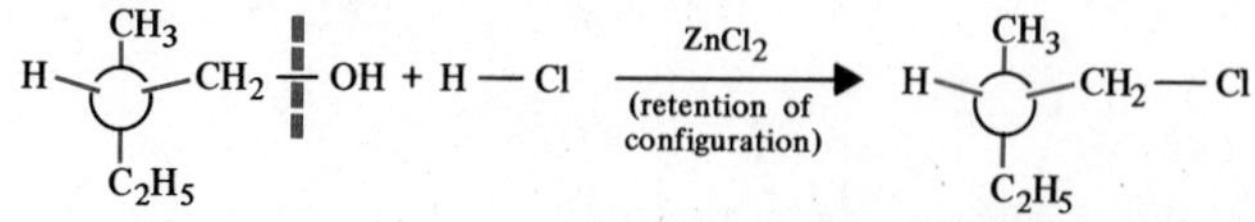

S–(–)–2–methyl–1–butanol
$[\alpha] = -5.756$

S–(+)–1–chloro–2–methylbutane
$[\alpha] = +1.64$

And in some instances the R or S designation may change even though the reaction proceeds with retention of configuration.

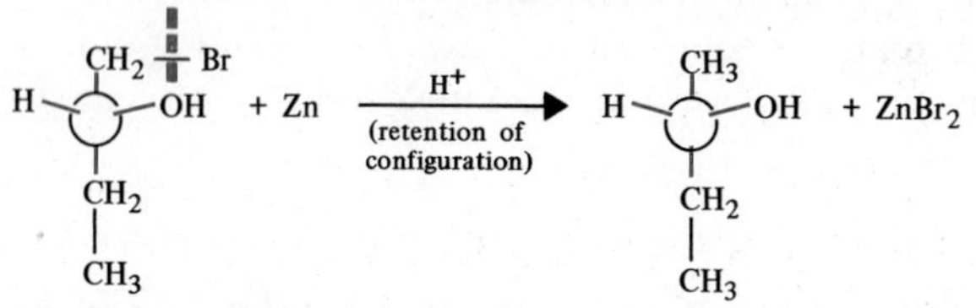

R–1–bromo–2–butanol

S–2–butanol

In this example the R-S designation changes because the —CH$_2$Br group of the reactant (—CH$_2$Br has a higher priority number than —CH$_2$CH$_3$) changes to a —CH$_3$ group in the product (—CH$_3$ has a lower priority number than —CH$_2$CH$_3$).

Relative and Absolute Configurations

Reactions in which no bonds to the chiral carbon are broken are useful in relating configurations of chiral molecules. That is, they allow us to demonstrate that certain compounds have the same *relative configuration*. In each of the three examples that we have just cited, the products of the reactions have the same *relative configurations* as the reactants. (They also, we now know, have the same *absolute configuration*.)

Before 1949, only relative configurations of chiral molecules were known. No one, prior to that time, had been able to demonstrate with certainty what the actual spatial arrangement of groups was in any chiral molecule. To say this another way, no one had been able to determine the *absolute configuration* of an optically active compound.

Configurations of chiral molecules were related to each other *through reactions of known stereochemistry*. Attempts were also made to relate all configurations back to a single compound that had been chosen as a standard. This standard compound was glyceraldehyde.

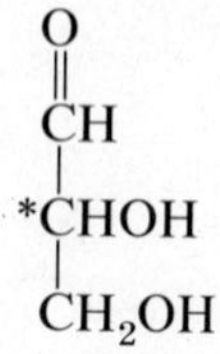

$$
\begin{array}{c}
O \\
\parallel \\
CH \\
\mid \\
*CHOH \\
\mid \\
CH_2OH
\end{array}
$$
Glyceraldehyde

Glyceraldehyde molecules have one chiral carbon; therefore, glyceraldehyde exists as a pair of enantiomers.

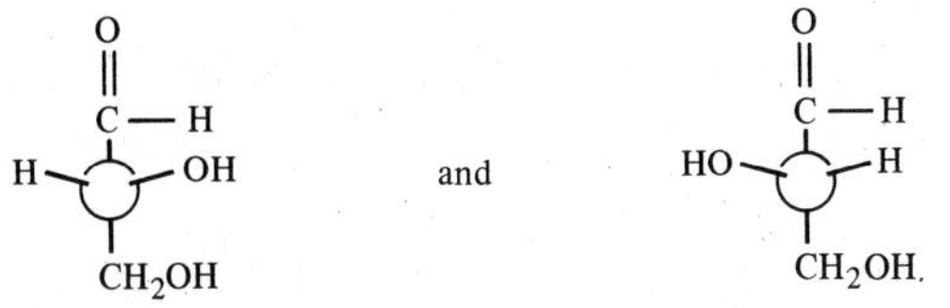

R–glyceraldehyde*   and   S–glyceraldehyde

One glyceraldehyde enantiomer is dextrorotatory (+) and the other, of course, is levorotatory (−). Before 1949, no one could be sure, however, which configuration belonged to which enantiomer. Chemists decided arbitrarily to assign the R configuration to the (+)-enantiomer. Then configurations of other molecules were related to one glyceraldehyde enantiomer or the other through reactions of known stereochemistry.

For example, the configuration of (−)-lactic acid can be related to (+)-glyceraldehyde through the following sequence of reactions.

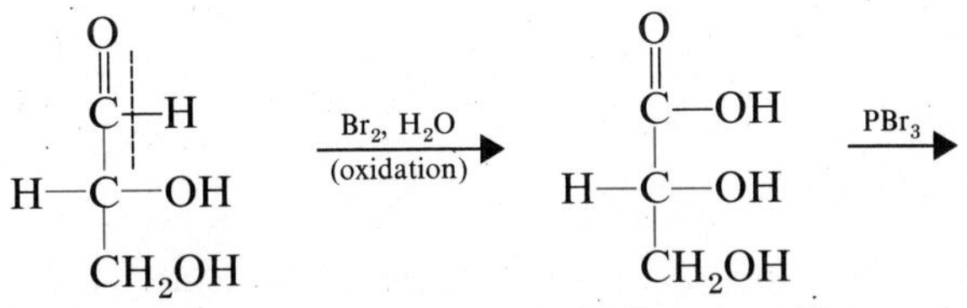

(+)-Glyceraldehyde   (−)-Glyceric acid

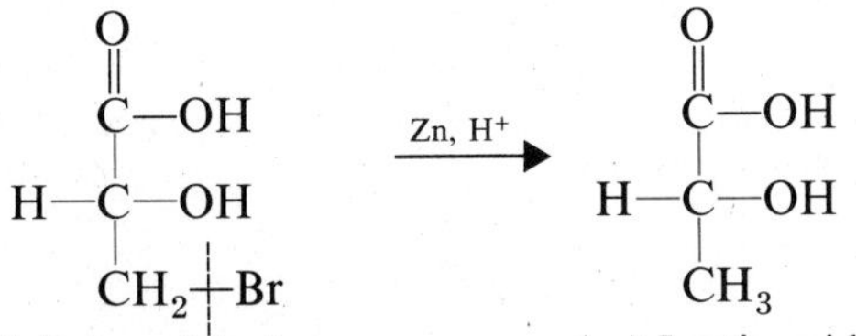

(−)-3-Bromo-2-hydroxy-   (−)-Lactic acid
propanoic acid

The stereochemistry of all of these reactions is known. Because bonds to the chiral carbon are not broken in any of them, they all proceed with retention of configuration. If the assumption is made that the configuration of (+)-glyceraldehyde is

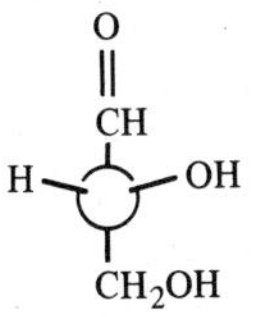

---

*In the older system for designating configurations R-glyceraldehyde was called D-glyceraldehyde and S-glyceraldehyde was called L-glyceraldehyde.

then the configuration of $(-)$-lactic acid is

$$
\begin{array}{c}
\text{O} \\
\| \\
\text{C} - \text{OH} \\
\text{H} - \overset{}{\underset{}{\text{C}}} - \text{OH} \\
\text{CH}_3
\end{array}
$$

Problem 7.21

(a) Write three-dimensional structures for the relative configurations of $(-)$ glyceric acid and $(-)$-3-bromo-2-hydroxypropanoic acid. (b) What is the R-S designation of $(-)$ glyceric acid? (c) Of $(-)$-3-bromo-2-hydroxypropanoic acid? (d) Of $(-)$-lactic acid?

The configuration of $(-)$-glyceraldehyde was also related through reactions of known stereochemistry to $(+)$-tartaric acid.

$$
\begin{array}{c}
\text{COOH} \\
\text{H} - \text{C} - \text{OH} \\
\\
\text{HO} - \text{C} - \text{H} \\
\text{COOH}
\end{array}
$$

(+)–tartaric acid

In 1949, J. M. Bijvoet, the director of the van't Hoff Laboratory of the University of Utrecht in Holland, using a special technique of X-ray diffraction, was able to show conclusively that $(+)$-tartaric acid had the absolute configuration given above. This meant that the original guess about the configurations of $(+)$- and $(-)$-glyceraldehyde was also correct. It also meant that the configurations of all of the compounds that had been related to one glyceraldehyde enantiomer or the other were now absolute configurations.*

Problem 7.22

Lactic acid with a specific rotation of $+3.82°$ can be converted to the ester, methyl lactate, $\text{CH}_3\text{CHOHCO}_2\text{CH}_3$. The methyl lactate that is produced has a specific rotation of $-8.25°$. What is the absolute configuration of $(-)$ methyl lactate?

Reactions in which Bonds to the Chiral Carbons are Broken

In subsequent chapters we will encounter many reactions that involve cleavage of a bond to a chiral carbon. When we do so we will find that such reactions can give one of the following results:

*It is interesting to notice that Bijvoet was director of a laboratory named for van't Hoff, one of the discoverers of the tetrahedral arrangement of carbon, and that Bijvoet used tartaric acid as the basis for his determination. Tartaric acid, as we will see, also provided Louis Pasteur the vehicle for the discovery of enantiomerism 101 years earlier.

1. Retention of configuration.
2. Inversion of configuration.
3. Racemization (or partial racemization).

At this point we will examine briefly three examples that illustrate each of these situations.

When an aqueous solution of R-2-bromopropanoic acid is treated with silver

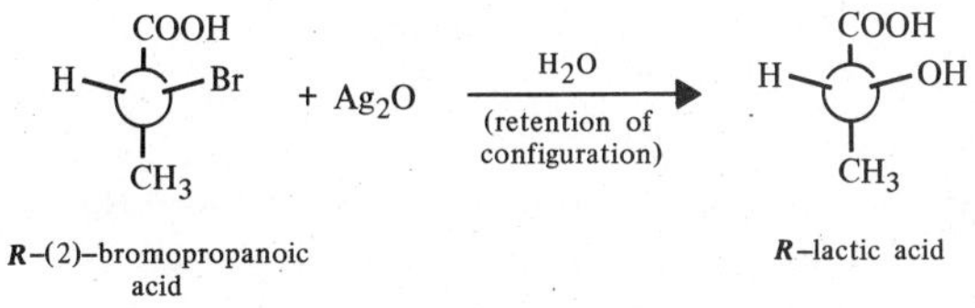

R–(2)–bromopropanoic acid   R–lactic acid

oxide the product that is obtained is *R*-lactic acid. Both the product and the reactant have the same configuration so this reaction proceeds with *retention of configuration*.

When *R*-2-bromopropanoic acid is heated with concentrated sodium hydroxide the product that is obtained is *S*-lactic acid. In this example the reactant and

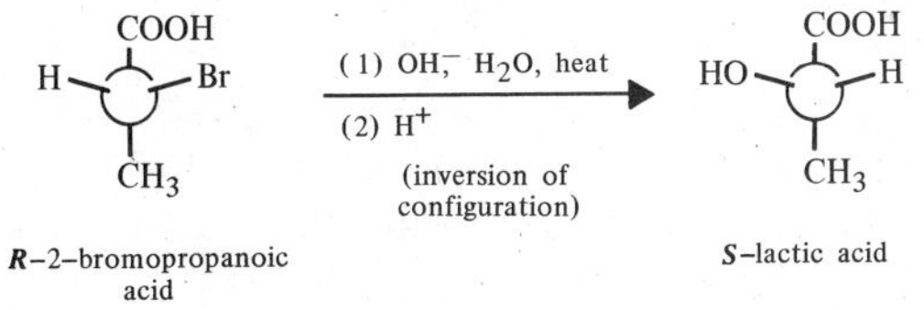

R–2–bromopropanoic acid   S–lactic acid

the product have opposite configurations. This reaction is said to have occurred with *inversion of configuration*.

An example of a reaction that proceeds with *racemization* can be seen in the following:

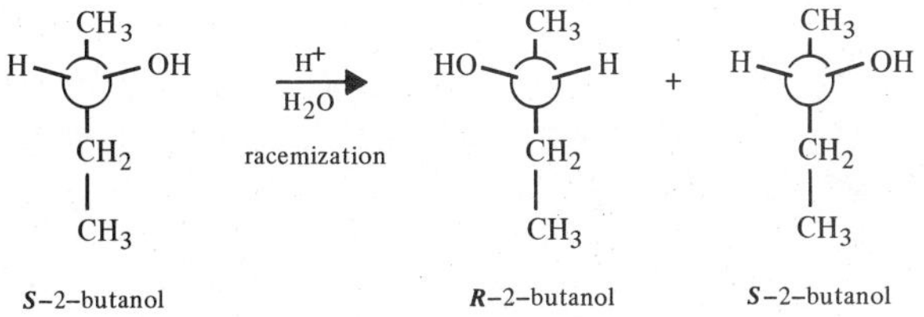

S–2–butanol   R–2–butanol   S–2–butanol

When *S*-2-butanol is heated in an aqueous solution of a strong acid the optical activity of the solution gradually disappears. This occurs because some of the *S*-2-butanol molecules are converted to *R*-2-butanol. If heating is continued long enough the solution will contain the racemic modification (±)-2-butanol, and the reaction is said to have occurred with complete *racemization*.

When we study these and other examples of reactions like these in subsequent chapters, we will present their mechanisms.

7.12 SEPARATION OF ENANTIOMERS: RESOLUTION

So far we have left unanswered an important question about optically active compounds and racemic modifications: How are enantiomers separated? Enantiomers have identical solubilities in ordinary solvents and they have identical boiling points.

Consequently, the conventional methods for separating organic compounds (crystallization and distillation) fail when applied to a racemic modification.

It was in fact, Louis Pasteur's separation of a racemic modification in 1848 that led to the discovery of the phenomenon called enantiomerism and to the later founding of the field of stereochemistry.

Although Pasteur's name is best known for his discoveries in bacteriology and their applications in medicine, Pasteur was trained as a chemist. He received his doctorate in chemistry from the Royal College of Besançon. Pasteur was also interested in wine making, and it was an outgrowth of this interest that led to his discovery of enantiomerism.

Tartaric acid is one of the by-products of wine manufacture. Pasteur had obtained a sample of racemic tartaric acid* from the owner of a chemical plant. In the course of his investigation Pasteur began examining the crystal structure of the sodium ammonium salt of racemic tartaric acid. He noticed that two types of crystals were present. One was identical with crystals of the sodium ammonium salt of (+)-tartaric acid.† The other type of crystals were *non*superposable mirror reflections of the first kind. Pasteur separated the two kinds of crystals under a microscope, dissolved them in water, and placed the solutions in a polarimeter. The solution of crystals of the first type was dextrorotatory, and the crystals themselves proved to be identical with the sodium ammonium salt of (+)-tartaric acid that was already known. The solution of crystals of the second type were levorotatory: they rotated plane-polarized light in the opposite direction and by an equal amount. The crystals of the second type were the sodium ammonium salt of (−)-tartaric acid.

Pasteur's discovery of enantiomerism and his demonstration that the optical activity of the two forms of tartaric acid was a property of the molecules themselves led, in 1874, to the proposal of the tetrahedral structure of carbon by van't Hoff and Le Bel.

Problem 7.23

On p. 268 we gave the absolute configuration of (+)-tartaric acid. (a) What is the absolute configuration of (−)-tartaric acid? (b) What other isomer of tartaric acid is possible? (c) Would it be optically active?

Unfortunately, few organic compounds show the same behavior when crystallized that the (+)- and (−)-tartaric acid salts do. Few organic compounds crystallize into separate crystals (containing separate enantiomers) that are visibly chiral like the crystals of the sodium ammonium salt of tartaric acid. Pasteur's method, therefore, is not one that is generally applicable.

The most useful procedure for separating enantiomers is based on allowing a racemic modification to react with a single enantiomer of some other compound. By such a reaction *a racemic modification is converted into a mixture of diastereomers;* and *diastereomers, because they have different melting and boiling points, can be separated by conventional means.* The process of separating the enantiomers of a racemic modification is called *resolution.*

We can illustrate this procedure in general terms by showing how a racemic

---

*At the time racemic tartaric acid was simply called racemic acid (L. *racemis,* a bunch of grapes).

†(+)-Tartaric acid had been discovered earlier and had been shown to be dextrorotatory.

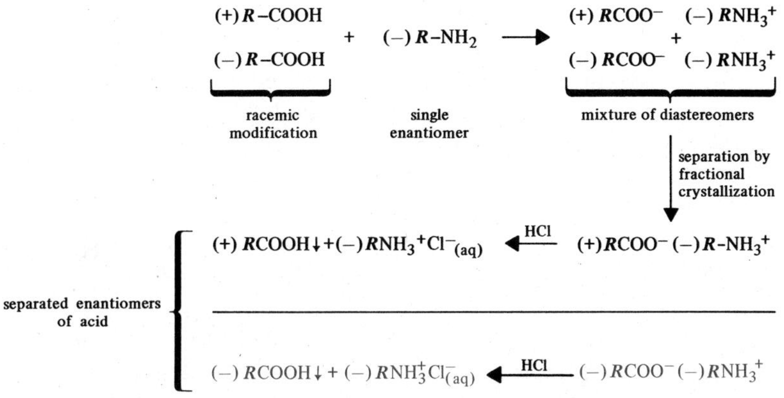

$$(+)\,R\text{–COOH}$$
$$(-)\,R\text{–COOH} \quad + \quad (-)\,R\text{–NH}_2 \quad \longrightarrow \quad \begin{array}{l}(+)\,RCOO^- \quad (-)\,RNH_3^+ \\ \qquad\qquad + \\ (-)\,RCOO^- \quad (-)\,RNH_3^+\end{array}$$

racemic   single   mixture of diastereomers
modification   enantiomer

separation by
fractional
crystallization

separated enantiomers of acid

$$(+)\,RCOOH\downarrow + (-)\,RNH_3{}^+Cl^-{}_{(aq)} \quad \xleftarrow{\text{HCl}} \quad (+)\,RCOO^-\,(-)\,R\text{-}NH_3{}^+$$

$$(-)\,RCOOH\downarrow + (-)\,RNH_3^+Cl^-_{(aq)} \quad \xleftarrow{\text{HCl}} \quad (-)\,RCOO^-\,(-)\,RNH_3{}^+$$

FIG. 7.22

The resolution of the racemic modification of an organic acid.

modification of an organic acid might be resolved (separated) using the single enantiomer of an amine (Fig. 7.22) as a resolving agent.

In this procedure the single enantiomer of an amine [$(-)RNH_2$] is added to the racemic modification of the acid. The salts that form—$(+)RCOO^-(-)RNH_3^+$ and $(-)RCOO^-(-)RNH_3^+$—are not enantiomers: they are diastereomers. (The chiral carbon of the acid portion of the salts are enantiomerically related to each other, but the chiral carbons of the amine portions are not.) The diastereomers have different solubilities and can be separated by careful crystallization. The separated salts are then acidified with hydrochloric acid and the enantiomeric acids precipitate from the separate solutions. The amines remain in solution as hydrochloride salts (p. 66).

The single enantiomers that are employed in resolutions of the type that we have just illustrated are readily available from natural sources. Since most of the organic molecules that occur in living organisms are synthesized by enzymatically catalyzed reactions, most of them occur as a single enantiomer. Naturally occurring optically active amines such as $(-)$-quinine, $(-)$-strychnine, and $(-)$-brucine are often employed as resolving agents for racemic acids.* Acids such as $(+)$- or $(-)$ tartaric acid are often used for resolving racemic bases.

Racemic compounds that are neither acids nor bases are often resolved by first attaching an acidic "handle." A racemic alcohol, for example, can react with a cyclic anhydride to produce a product that is both an ester and *an acid*.

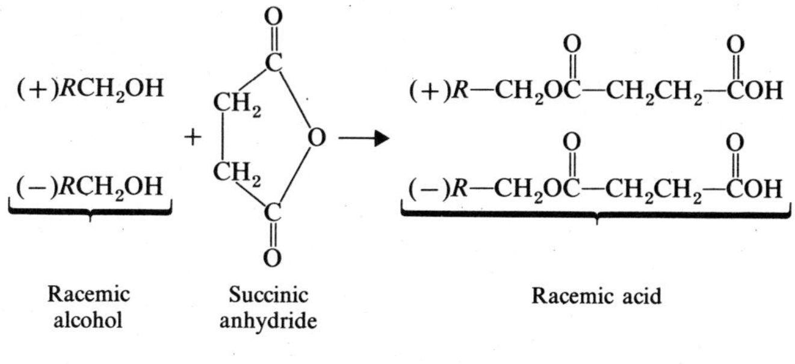

$$\begin{array}{l}(+)RCH_2OH \\ (-)RCH_2OH\end{array} \;+\; \underset{\substack{\text{Succinic}\\\text{anhydride}}}{\begin{array}{c}O\\ \parallel\\ C\\ CH_2\\ | \;\;\; O\\ CH_2\\ C\\ \parallel\\ O\end{array}} \longrightarrow \begin{array}{l}(+)R\text{—}CH_2O\overset{O}{\overset{\parallel}{C}}\text{—}CH_2CH_2\text{—}\overset{O}{\overset{\parallel}{C}}OH \\ (-)R\text{—}CH_2O\overset{O}{\overset{\parallel}{C}}\text{—}CH_2CH_2\text{—}\overset{O}{\overset{\parallel}{C}}OH\end{array}$$

Racemic   Succinic   Racemic acid
alcohol   anhydride

*These compounds are called alkaloids; alkaloids are discussed as a special topic in Section 22.2.

The racemic acid that is produced can then be converted to a mixture of diastereomeric salts through its reaction with an optically active amine such as $(-)$ quinine. The diastereomeric salts are separated and converted back to the enantiomeric acids in the way outlined in Fig. 7.22. The acidic "handle" is then removed by hydrolysis of the ester group (p. 61) and the separate alcohol enantiomers are obtained.

$$(+)R—CH_2OCCH_2CH_2COH \xrightarrow[\substack{H_2O \\ heat}]{OH^-} (+)R—CH_2OH + \begin{matrix} COO^- \\ | \\ CH_2 \\ | \\ CH_2 \\ | \\ COO^- \end{matrix}$$

$$(-)R—CH_2OCCH_2CH_2COH \xrightarrow[\substack{H_2O \\ heat}]{OH^-} (-)RCH_2OH + \begin{matrix} COO^- \\ | \\ CH_2 \\ | \\ CH_2 \\ | \\ COO^- \end{matrix}$$

7.13 COMPOUNDS WITH CHIRAL CENTERS OTHER THAN CARBON

Any tetrahedral atom with four different groups attached to it is a chiral center. Listed below are examples of compounds whose molecules contain chiral centers other than carbon.

$$R_4—\underset{\underset{R_3}{|}}{\overset{\overset{R_1}{|}}{Si}}—R_2 \qquad R_4—\underset{\underset{R_3}{|}}{\overset{\overset{R_1}{|}}{Ge}}—R_2 \qquad R_4—\underset{\underset{R_3}{|}}{\overset{\overset{R_1}{|}}{N}}{}^+—R_2 \quad X^-$$

Silicon and germanium are in the same group of the Periodic Table as carbon. They form tetrahedral compounds as carbon does. When four different groups are situated around the central atom in silicon and germanium compounds, the molecules are chiral and the enantiomers can be separated.

Nitrogen forms four covalent bonds in ammonium compounds. Numerous examples of chiral ammonium compounds have been resolved; listed below are two.

$$C_6H_5—\underset{\underset{CH_2}{|}}{\overset{\overset{CH_3}{|}}{N}}{}^+—CH_2C_6H_5 \quad Cl^- \qquad C_6H_5CH_2—\underset{\underset{CH_2}{|}}{\overset{\overset{CH_3}{|}}{N}}{}^+—C_6H_5$$
$$\quad\quad CH_3 \qquad\qquad\qquad Cl^- \quad CH_3$$

$$C_6H_5—\underset{\underset{CH_2}{|}}{\overset{\overset{CH_3}{|}}{N}}{}^+—O^- \qquad {}^-O—\underset{\underset{CH_2}{|}}{\overset{\overset{CH_3}{|}}{N}}{}^+—C_6H_5$$
$$\quad\quad CH_3 \qquad\qquad\qquad CH_3$$

The nitrogen atom of a tertiary amine is also tetrahedral if one considers the unshared pair as occupying one corner of the tetrahedron.

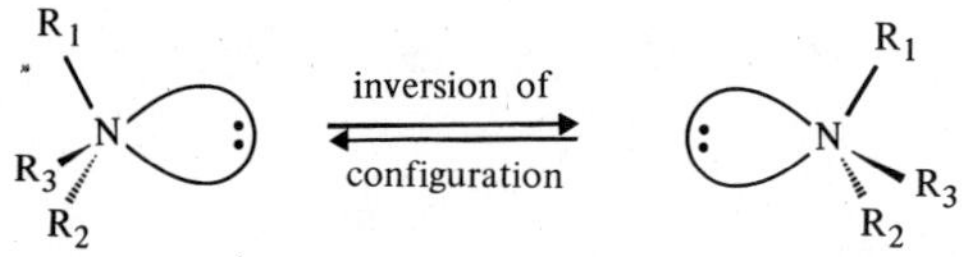

The nitrogen atoms of most tertiary amines invert their configuration so rapidly at room temperature that they cannot be resolved into separate enantiomers even when three different groups are attached at the other corners.*

Sulfur compounds with three different attached groups have been resolved. The sulfonium compounds shown below are examples.

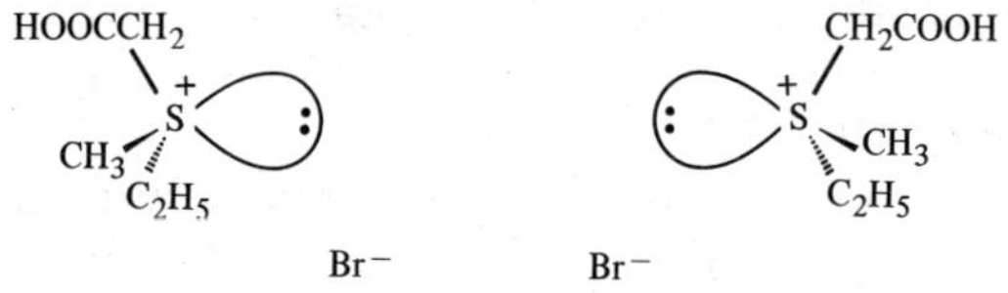

Molecules containing sulfur atoms of this type retain their configuration at room temperature and do not undergo the characteristic rapid inversion associated with tertiary amines.

7.14 CHIRAL MOLECULES THAT DO NOT POSSESS A CHIRAL CENTER

A molecule is chiral if it is not superposable on its mirror reflection. The presence of a chiral center is only one focus that will confer chirality on a molecule. Most of the molecules that we will encounter do have chiral centers (usually chiral carbons). Many chiral molecules are known, however, that do not. Two examples of chiral molecules that do not have chiral centers are 1,3-dichloroallene and *trans*-cyclooctene.

Allenes are compounds that contain the double bond sequence shown below.

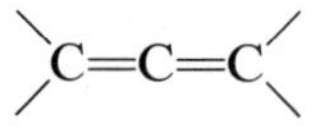

(The double bonds of an allene are said to be *cumulated*.) The π bonds of allenes are perpendicular to each other.

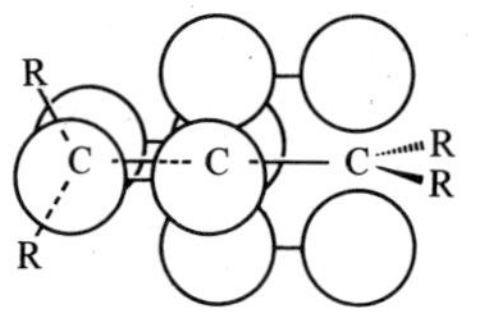

---

*Tertiary amines with chiral centers have been resolved when the tertiary nitrogen is a part of a ring that prevents inversion of its configuration.

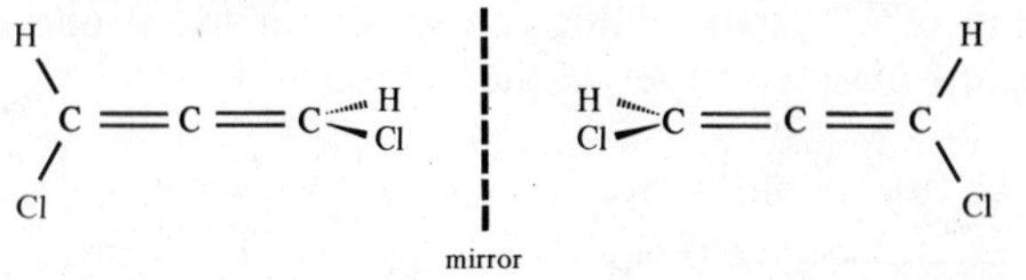

FIG. 7.23
Enantiomeric forms of 1,3-dichloroallene. These two molecules are nonsuperposable mirror reflections of each other and are, therefore, chiral. They do not possess a chiral center, however.

This geometry of the π bonds causes the groups attached to the end carbons to lie in perpendicular planes and, because of this, allenes with different substituents on the end carbons are chiral (Fig. 7.23).

Trans-cyclooctene (Fig. 7.24) also exists in enantiomeric forms because of the geometry of the *trans*-cycloalkene ring.

FIG. 7.24
The trans-*cyclooctene enantiomers.*

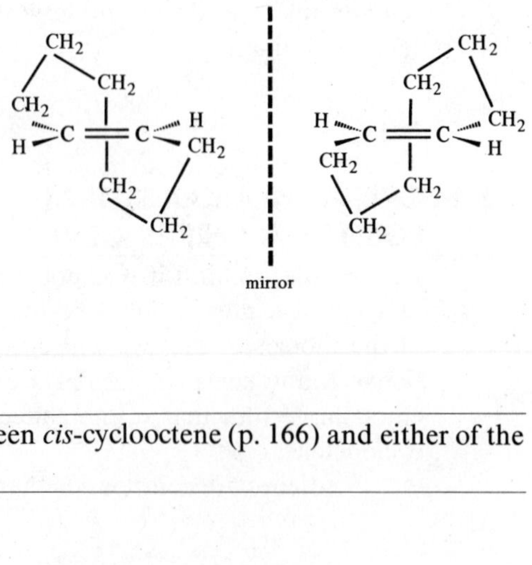

Problem 7.24

What is the stereoisomeric relation between *cis*-cyclooctene (p. 166) and either of the *trans*-cyclooctene enantiomers?

7.15 THE *E-Z* SYSTEM FOR DESIGNATING ALKENE DIASTEREOMERS

The terms *cis* and *trans*, when used to designate the stereochemistry of alkene diastereomers, are unambiguous only when applied to disubstituted alkenes. If the alkene is trisubstituted or tetrasubstituted the terms *cis* and *trans* are either ambiguous or do not apply at all. Consider the alkenes shown below as examples.

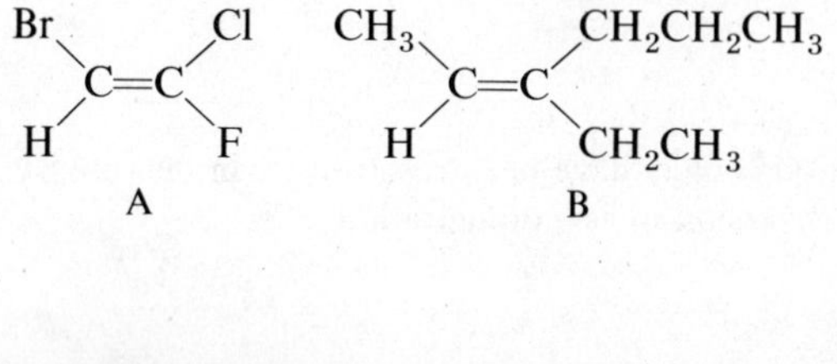

It is impossible to decide whether A is *cis* or *trans* since no two groups are the same. The same is true of compound B.*

A system of designations for diastereomeric alkenes that is based on the priorities of groups in the Cahn-Ingold-Prelog convention has been proposed. This system, called the *E-Z* system, applies to alkene diastereomers of all types. In the *E-Z* system, we examine the two groups attached to each carbon of the double bond and arrange them in order of priority.

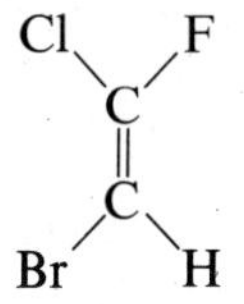

Cl > F
Br > H

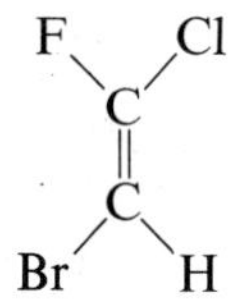

We then take the group of higher priority on one carbon and compare it with the group of higher priority on the other carbon. If the two groups of higher priority are on the same side of the double bond the alkene is designated *Z* (from the German word, *Zusammen*, meaning together). If the two groups of higher priority are on opposite sides of the double bond the alkene is designated *E* (from the German word, *entgegen*, meaning opposite).†

Most compounds that we would normally designate *cis* are designated *Z*. Similarly most compounds that we would designate *trans* are designated *E*.

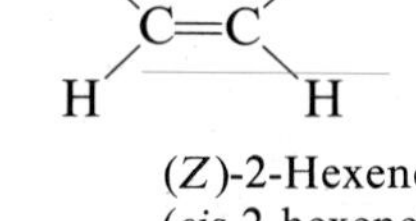

(*Z*)-2-Butene
(*cis*-2-butene)

CH₃ > H
CH₃ > H

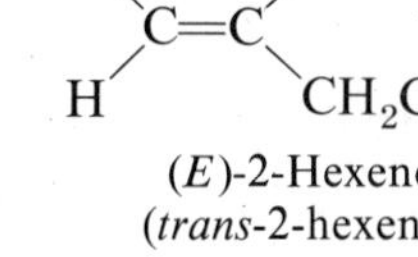

(*E*)-2-Butene
(*trans*-2-butene)

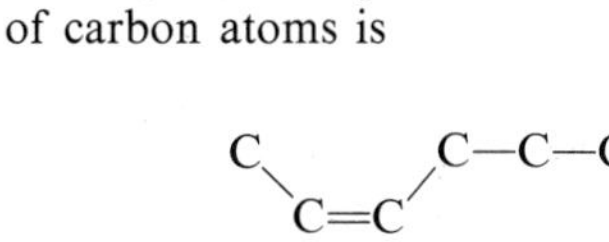

(*Z*)-2-Hexene
(*cis*-2-hexene)

CH₃ > H
C₃H₇ > H

(*E*)-2-Hexene
(*trans*-2-hexene)

* In an older system compound B is said to be *cis* because the configuration of the longest chain of carbon atoms is

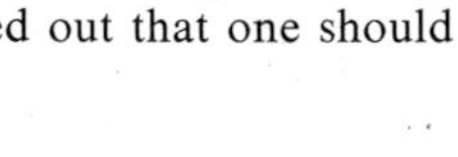

rather than

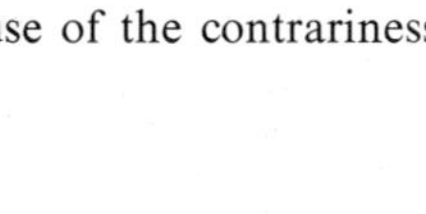

Even this older system is ambiguous, however, when it is applied to certain compounds.
† If the German derivations of these letters are difficult for you to remember, it may be helpful to notice that in a sense, their meanings *are the opposite* of the appearance of the letters *E* and *Z*. Intuitively the letter *E* appears "cisoid," and the letter *Z* "transoid." It has been pointed out that one should make *some* use of the contrariness of these letters.

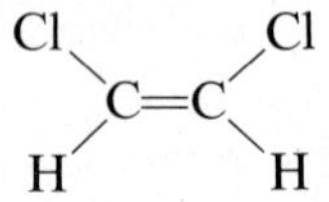

$$Cl > H$$
$$Cl > H$$

(Z)-1,2-Dichloroethene
(cis-1,2-dichloroethene)

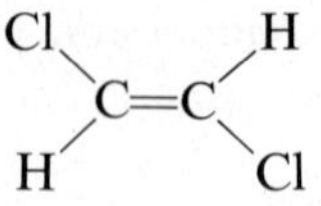

(E)-1,2-Dichloroethene
(trans-1,2-dichloroethene)

There are exceptions, however.

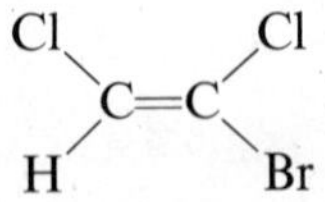

$$Cl > H$$
$$Br > Cl$$

(E)-1-Bromo-1,2-dichloroethene
(cis-1-bromo-1,2-dichloroethene)

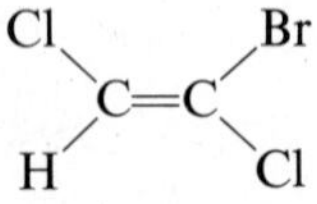

(Z)-1-Bromo-1,2-dichloroethene
(trans-1-bromo-1,2-dichloroethene)

Problem 7.25

Give the *E-Z* designation and name of each of the following.

(a)

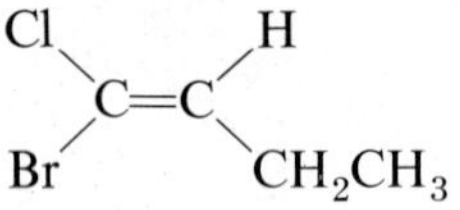

(b)

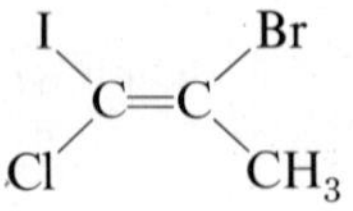

(c)

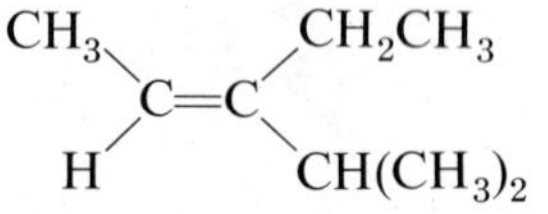

(d)

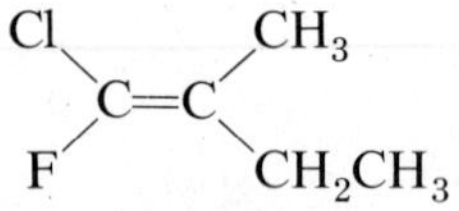

Additional Problems

7.26
Give definitions of each of the following terms and examples that illustrate their meaning.

(a) Isomers
(b) Structural isomers
(c) Stereoisomers
(d) Diastereomers
(e) Enantiomers
(f) *Meso* compound
(g) Racemic modification

(h) Plane of symmetry
(i) Chiral center
(j) Chiral molecule
(k) Achiral molecule
(l) Optical activity
(m) Dextrorotatory
(n) Retention of configuration

7.27
Listed on pp. 277–278 are pairs of structures. Identify the relation between them by describing them as representing enantiomers, diastereomers, structural isomers, or two molecules of the same compound.

(a) 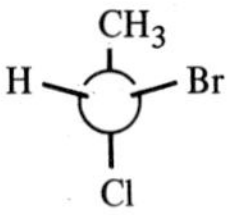 and 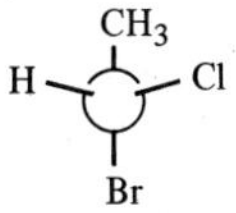

(b) 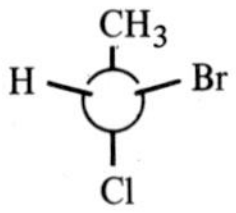 and 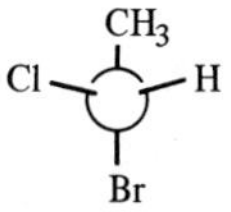

(c) 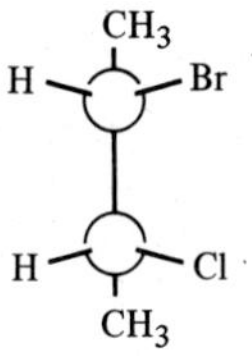 and 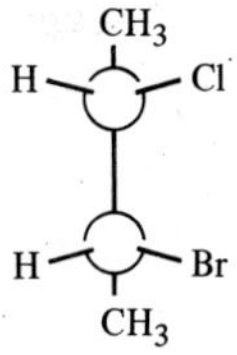

(d) 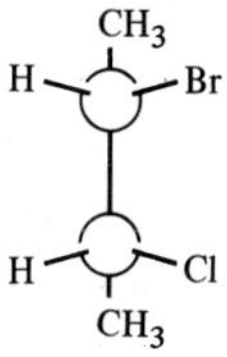 and 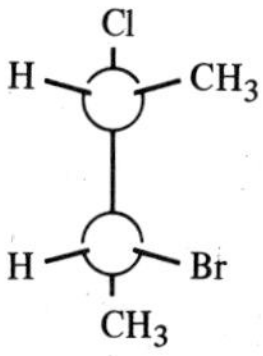

(e) 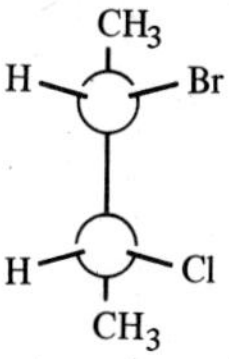 and 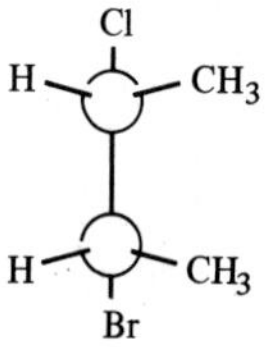

(f) 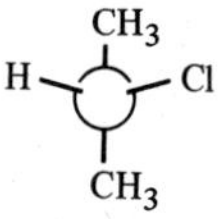 and 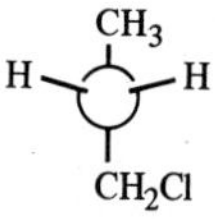

(g) 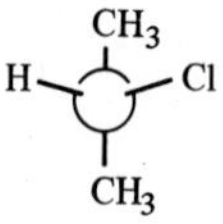 and 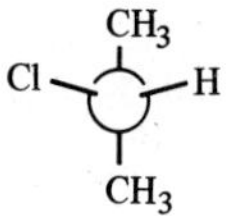

(h) 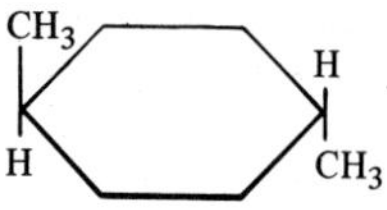 and 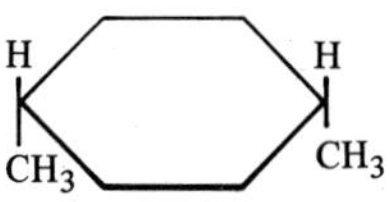

(i) 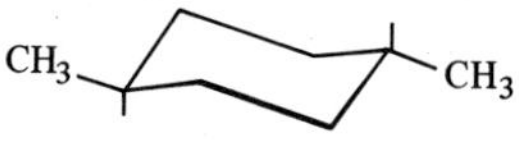 and 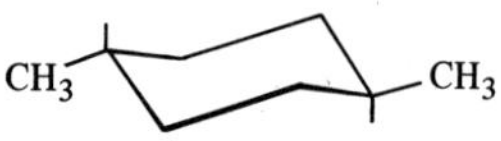

(j) 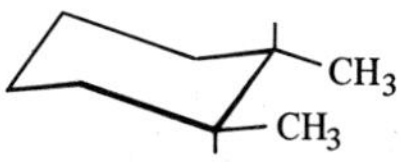 and 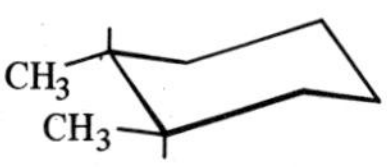

(k) 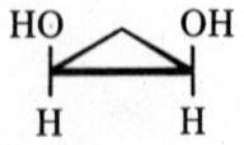   and

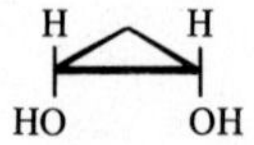

(l) 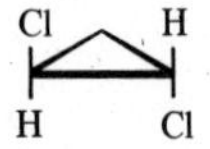   and

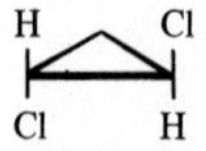

(m) 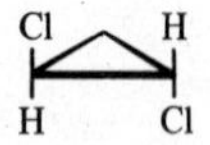   and

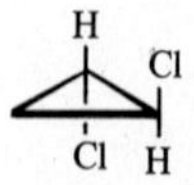

(n) 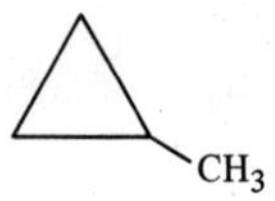   and

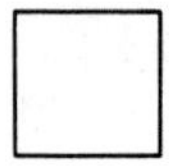

(o) 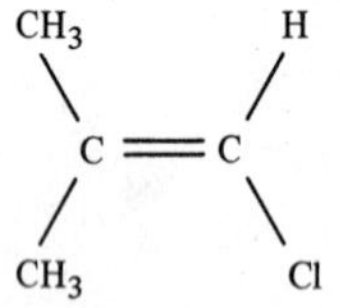   and

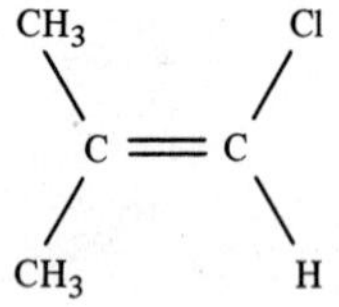

(p) 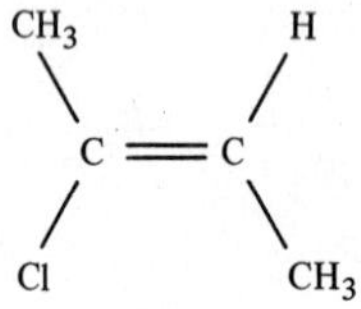   and

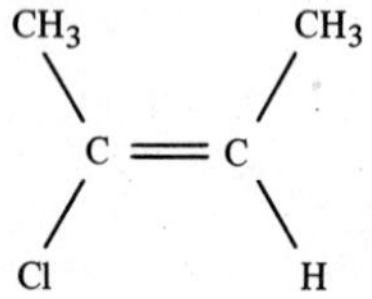

(q) 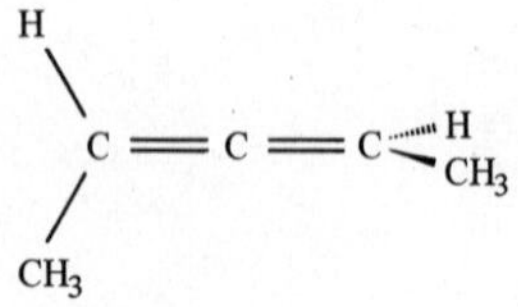   and

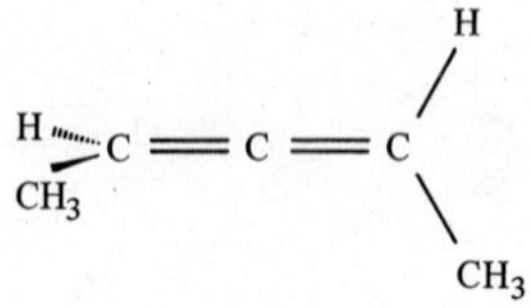

7.28

There are four dimethylcyclopropane isomers. (a) Write three-dimensional formulas for them. (b) Which dimethylcyclopropane isomers would, if taken separately, show optical activity? (c) Which dimethylcyclopropane isomer is a *meso* compound? (d) If a mixture consisting of one mole of each of the four dimethylcyclopropane isomers were subjected to fractional distillation, how many fractions would be obtained? (e) How many of these fractions would show optical activity?

7.29

Write stereochemical formulas for all of the products that you would expect from each of the following reactions.

(a) 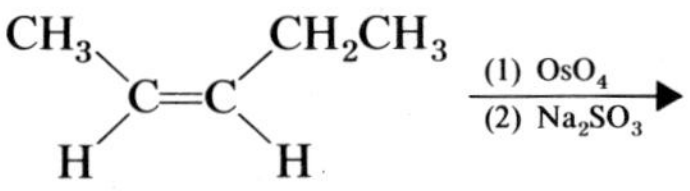
$$CH_3, \quad CH_2CH_3 \atop H \atop C{=}C \atop H \xrightarrow[(2)\ Na_2SO_3]{(1)\ OsO_4}$$

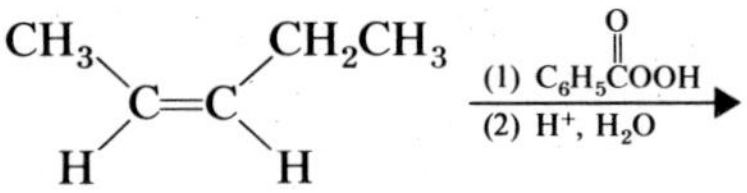

(b)
$$CH_3, \quad CH_2CH_3 \atop H \atop C{=}C \atop H \xrightarrow[(2)\ H^+,\ H_2O]{(1)\ C_6H_5\overset{O}{\underset{\|}{C}}OOH}$$

(c)
$$CH_3, \quad H \atop H \atop C{=}C \atop CH_2CH_3 \xrightarrow[(2)\ Na_2SO_3]{(1)\ OsO_4}$$

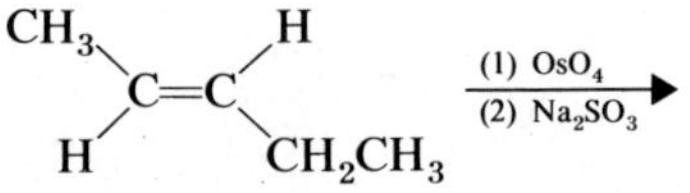

(d) 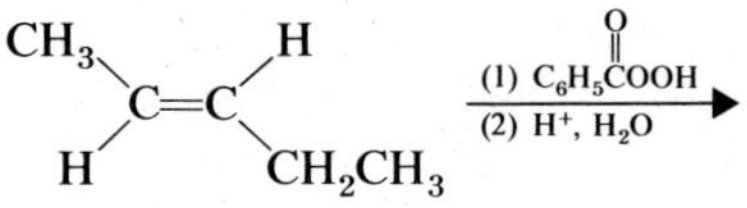
$$CH_3, \quad H \atop H \atop C{=}C \atop CH_2CH_3 \xrightarrow[(2)\ H^+,\ H_2O]{(1)\ C_6H_5\overset{O}{\underset{\|}{C}}OOH}$$

(e) 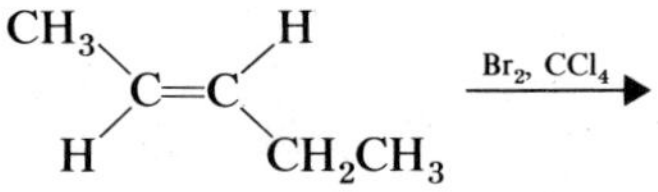
$$CH_3, \quad H \atop H \atop C{=}C \atop CH_2CH_3 \xrightarrow{Br_2,\ CCl_4}$$

(f) 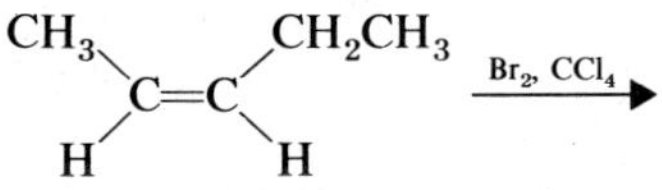
$$CH_3, \quad CH_2CH_3 \atop H \atop C{=}C \atop H \xrightarrow{Br_2,\ CCl_4}$$

7.30

Give *R-S* designations for each different compound given as an answer to problem 7.29.

7.31

Which of the following reactions *must* occur with retention of configuration?

(a) 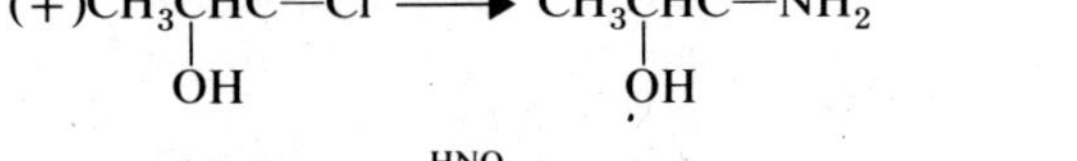 $(+)CH_3CH_2\underset{OH}{CH}CH_3 + PCl_3 \longrightarrow CH_3CH_2\underset{Cl}{CH}CH_3$

(b) 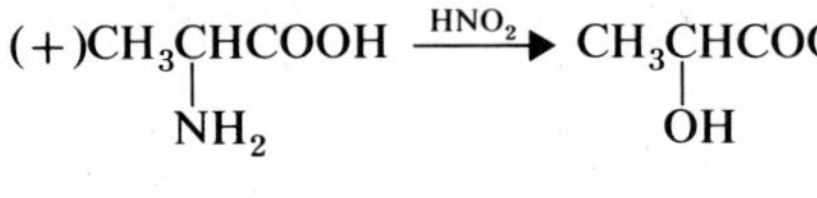 $(-)CH_3\underset{OH}{CH}CHO \xrightarrow[H_2O]{Br_2} CH_3\underset{OH}{CH}COOH$

(c) 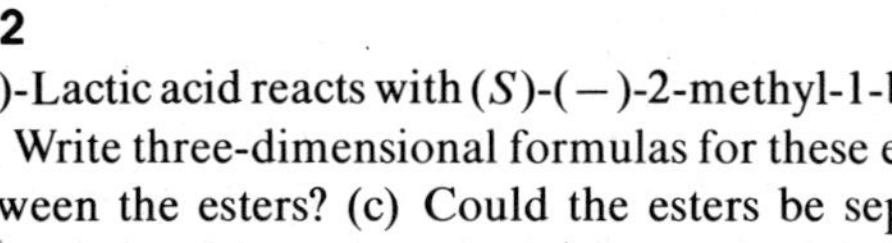 $(+)CH_3\underset{OH}{CH}\overset{O}{\underset{\|}{C}}{-}Cl \xrightarrow{NH_3} CH_3\underset{OH}{CH}\overset{O}{\underset{\|}{C}}{-}NH_2$

(d) 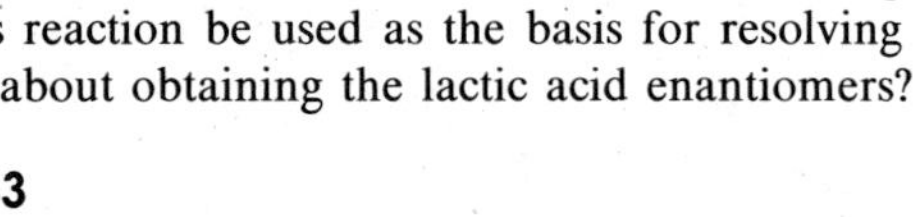 $(+)CH_3\underset{NH_2}{CH}COOH \xrightarrow{HNO_2} CH_3\underset{OH}{CH}COOH$

7.32

$(\pm)$-Lactic acid reacts with (S)-$(-)$-2-methyl-1-butanol (p. 247) to produce a mixture of esters. (a) Write three-dimensional formulas for these esters. (b) What is the stereochemical relation between the esters? (c) Could the esters be separated by fractional distillation? (d) Could this reaction be used as the basis for resolving $(\pm)$ lactic acid? (e) If so, how would you go about obtaining the lactic acid enantiomers?

7.33

(a) What product would be obtained from the following Wurtz reaction (cf. p. 115)?
(b) Would the product be optically active?

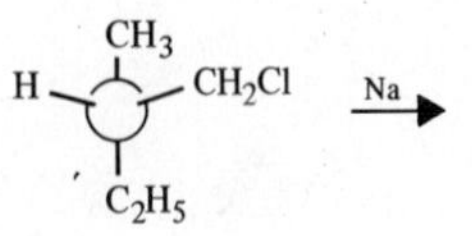

$$\text{H} - \overset{\overset{\displaystyle CH_3}{|}}{\underset{\underset{\displaystyle C_2H_5}{|}}{\bigcirc}} - CH_2Cl \quad \xrightarrow{\text{Na}}$$

7.34

(a) What stereoisomer would you expect to obtain from the following reaction?

1,2-Dimethylcyclopentene $\xrightarrow[\text{Pt}]{\text{H}_2}$

(b) Would the product be optically active?

7.35

When 1-methylcyclopentene is subjected to hydroboration-oxidation (p. 207) the product is obtained as a racemic modification. (a) Write three-dimensional structures for the enantiomers that are formed. (b) Write a mechanism that shows how they are formed.

7.36

Ricinoleic acid, a compound that can be isolated from castor oil, has the structure $CH_3(CH_2)_5CHOHCH_2CH=CH(CH_2)_7COOH$. (a) How many stereoisomers of this structure are possible? (b) Write these structures.

7.37

There are two dicarboxylic acids with the general formula $HOOCCH=CHCOOH$. One dicarboxylic acid is called maleic acid; the other is called fumaric acid. In 1880, Kekulé found that on treatment with cold dilute $KMnO_4$, maleic acid yields *meso*-tartaric acid and that fumaric yields ($\pm$)-tartaric acid. Show how this information allows one to arrive at stereochemical formulas for maleic acid and fumaric acid.

7.38

Use your answers to the preceding problem to predict the stereochemical outcome of the addition of bromine to maleic acid and to fumaric acid. (a) Which dicarboxylic acid would add bromine to yield a *meso*-compound? (b) Which would yield a racemic modification?

7.39

An optically active compound **A** (assume that it is dextrorotatory) has the molecular formula $C_7H_{11}Br$. **A** reacts with hydrogen bromide, in the absence of peroxides, to yield isomeric products, **B** and **C**, with molecular formula $C_7H_{12}Br_2$. **B** is optically active; **C** is not. Treating **B** with one mole of potassium *tert*-butoxide yields (+)**A**. Treating **C** with one mole of potassium *tert*-butoxide yields ($\pm$)**A**. Treating **A** with potassium *tert*-butoxide yields **D** (C_7H_{10}). Subjecting one mole of **D** to ozonolysis followed by treatment with zinc and water yields two moles of formaldehyde and one mole of 1,3-cyclopentanedione.

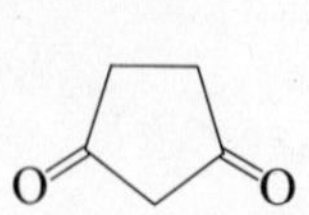

$$O \diagdown\!\!\bigtriangleup\!\!\diagup O$$

1,3-Cyclopentanedione

Propose stereochemical formulas for **A, B, C,** and **D** and outline the reactions involved in these transformations.

7.40

Propose structures for compounds **E–H.**
(a) Compound **E** has the molecular formula C_5H_8 and is optically active. On catalytic hydrogenation **E** yields **F**. **F** has the molecular formula C_5H_{10}, is optically inactive, and cannot be resolved.

(b) Compound **G** has the molecular formula C_6H_{10} and is optically active. **G** contains no triple bonds. On catalytic hydrogenation **G** yields **H**. **H** has the molecular formula C_6H_{14}, is optically inactive, and cannot be resolved.

7.41

Compounds **I** and **J** both have the molecular formula C_7H_{14}. **I** and **J** are both optically active and both rotate plane-polarized light in the same direction. On catalytic hydrogenation **I** and **J** yield the same compound **K**, C_7H_{16}. **K** is optically active. Propose possible structures for **I**, **J**, and **K**.

7.42

Compounds **L** and **M** have the molecular formula C_7H_{14}. **L** and **M** are optically inactive, are nonresolvable, and are diastereomers of each other. Catalytic hydrogenation of either **L** or **M** yields **N**. **N** is optically inactive but can be resolved. Propose possible structures for **L**, **M**, and **N**.

7.43

One stereoisomeric form of 1,3-di-*sec*-butylcyclohexane is optically inactive. Write a conformational structure for this isomer.

* 7.44

When the 3-bromo-2-butanol with the stereochemical structure **A** is treated with concentrated HBr it yields *meso*-2,3-dibromobutane; a similar reaction of the 3-bromo-2-butanol **B** yields ($\pm$)-2,3-dibromobutane. This classic experiment performed in 1939 by S. Winstein and H. J. Lucas was the starting point for a series of investigations of what are called *neighboring group effects* (see also Sect. 17.8). Propose mechanisms that will account for the stereochemistry of these reactions.

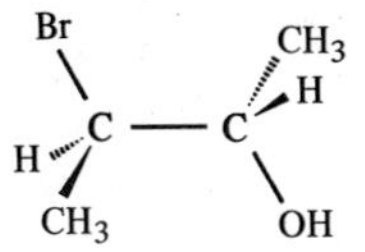

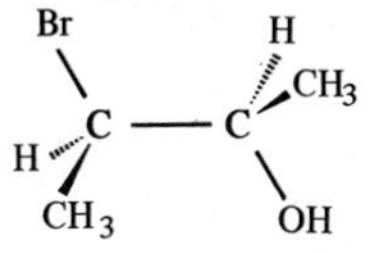

8 SPECIAL TOPICS I

THE DEVELOPMENT OF QUANTUM MECHANICS
ELEMENTARY THERMODYNAMICS: ΔH, ΔS, AND ΔG
POLYMERIZATION OF ALKENES: ADDITION POLYMERS
STEREOCHEMISTRY OF ADDITION POLYMERS
DIVALENT CARBON COMPOUNDS: CARBENES

8.1 THE DEVELOPMENT OF QUANTUM MECHANICS

At the end of the last century, and early in this one, revolutionary ideas about the nature of matter and energy were advanced by Max Planck, J. J. Thomson, Ernest Rutherford, Albert Einstein, and Niels Bohr. These ideas were essential to the later development of quantum mechanics, therefore, we consider them first.

Planck's Contribution

Planck's contribution came as a consequence of his studies of the energy radiated from very hot but nonreflective objects called "black bodies." (Black only because they were not reflecting energy; actually they were red hot.) Planck found that the radiation emitted from these objects, particularly its intensity at different wavelengths (or colors), could not be explained in terms of the theories that existed at that time. In order to explain his results, Planck made an imaginative proposal. Planck postulated that matter is not capable of emitting or absorbing energy (radiation) in continuous and infinitely divisible amounts but that *matter emits or absorbs energy in small bundles* (or packets) *called quanta.*

According to Planck's theory, the energy emitted by the very hot objects is emitted by the very rapidly oscillating atoms on their surfaces. Moreover, the energy of these atoms is proportional to the frequency (number of oscillations per second) with which they oscillate, so that those atoms that oscillate very rapidly (with high frequency) have high energies, while those that oscillate very slowly (low frequency) have low energies. This relationship is expressed mathematically in Planck's famous equation:

$$E = h\nu$$

where E is the energy of the oscillation, ν is the frequency, and h is a proportionality constant known as Planck's constant. Most importantly, according to Planck, the energy of the oscillating atoms is *quantized*. As a result, atoms can only gain or lose energy in units (or packets) of $h\nu$. Planck's constant—called one quantum of action—has the value of 6.625×10^{-27} erg-sec and is apparently one of the most fundamental constants in the universe. It appears in all equations where the granular or quantized nature of matter and energy are expressed.

Thomson and Rutherford

Just before Planck's proposals were published in 1900, J. J. Thomson had discovered the electron as a fundamental constituent of matter. Later, in 1904, Thomson postulated a "pudding" model of the atom. According to this model, negative electrons were embedded like "plums" in a sphere of positive charge. Soon afterwards, however, Rutherford, as a result of his experiments with the scattering of alpha particles, postulated a model of the atom where the positive charge and most of the mass of the atom is intensely concentrated in a tiny nucleus. The electrons, Rutherford proposed, move in orbits about the nucleus much as the planets do around the sun.

Bohr's Model of the Atom

In 1912, Niels Bohr, extended Rutherford's ideas and simultaneously introduced Planck's quantum conditions with a new model of the atom. According to Bohr, the small, massive, and positively charged nucleus is surrounded by electrons moving in circular orbits. The hydrogen atom, for example, would have its electron moving in the orbit closest to the nucleus in its lowest energy state. Moreover, according to the Bohr model, the electron can be excited to higher orbits by absorbing

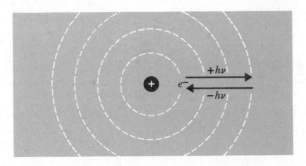

one or more quanta of energy. By losing energy, again in units of $h\nu$, the electron can fall back to lower orbits. The position of the electron is restricted to the discrete orbits and, according to Bohr, can not assume an energy state in between. This *quantized* state of the atom, introduced by Bohr quite arbitrarily was, at the time, a brilliant step forward. Bohr's mathematical treatment of the hydrogen atom, using his model as a basis, explained for the first time the spectrum of atomic hydrogen.

As time passed, however, and further experiments were done, it became clear that the Bohr model was inadequate. The Bohr treatment was successful in explaining only the spectrum of one-electron systems, and although many attempts were made to refine the Bohr model (including A. Sommerfeld's introduction of elliptical orbits), by 1920 it became clear that a new model was necessary.

Before we discuss the next steps in the development of quantum mechanics we must return to a slightly earlier time and relate an important new theory about the nature of light that had been advanced by Einstein.

Einstein and Light Photons

For almost 100 years, prior to 1905, scientists had conceived of light as a *wave phenomenon*. All of the properties of light known in the nineteenth century could be explained by assuming that light and other forms of electromagnetic

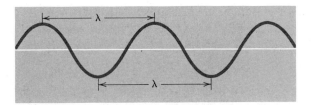

FIG. 8.1
A simple wave.

radiation were waves occurring simultaneously in perpendicular electrical and magnetic fields. Properties of light, such as reflection and diffraction, were easily explained on that basis. Diffraction, the bending of light waves as they pass sharp edges (accounting for the fact that shadows having fuzzy edges rather than sharp ones), can be explained *only on the basis of the wave nature of light*. A new phenomenon, known as the photoelectric effect, had been discovered, however, and this new property of light, associated with light's ability to cause electrons to be emitted from the surfaces of metals, could *not* be accounted for on the basis of light's wave nature. An explanation for the photoelectric effect required that light have the nature of a particle as well. In 1905, Albert Einstein proposed that light, too, was *quantized* and that light had *both the properties of a wave and a particle*. According to this new conception we observe light's wave character when we observe diffraction; we observe its particlelike or granular nature when we observe the photoelectric effect. Einstein called these wave particles of light *photons*.

We can aid our subsequent discussion if we introduce, at this point, some of the mathematical relationships used to describe waves.

A simple wave (Fig 8.1) can be described in terms of its wavelength, λ, and its frequency, ν. The distance between two consecutive crests (or troughs) is the *wavelength;* the number of full cycles (wavelengths) that the wave completes each second, as it is propagated, is the *frequency*. Frequency has the units of cycles/sec or hertz (Hz). Since the wave (consider a wave moving across a pond or in a long, plucked string) moves one wavelength in each cycle, the product of the wavelength, λ, and frequency, ν, gives the wave velocity, v. Thus:

$$v = \lambda \cdot \nu$$

$$v\left(\text{wave velocity in } \frac{\text{cm}}{\text{sec}}\right) = \lambda\left(\text{wavelength in } \frac{\text{cm}}{\text{cycle}}\right) \cdot \nu\left(\text{frequency in } \frac{\text{cycle}}{\text{sec}}\right)$$

According to Einstein's theory of relativity, the velocity of a light wave is constant, and is equal to 3×10^{10} cm/sec. The velocity of light is usually abbreviated c, thus for a light wave:

$$c = \lambda \nu$$

Moreover, according to Einstein, light has associated with it the property of momentum because of its particlelike nature. This momentum can be expressed by the following formula:

$$\text{Momentum} = \frac{h\nu}{c} = \frac{h}{\lambda}$$

where once again, h is Planck's constant.

The energy of a photon of light, in Einstein's theory, is the product of Planck's constant and its frequency:

$$E = h\nu = \frac{hc}{\lambda}$$

Thus a photon of light of high frequency and short wavelength (blue light would be an example) will have greater energy than light of lower frequency and longer wavelength (i.e., red light).

De Broglie's Contribution

We now return to 1923, and with these simple mathematical relationships about light in mind, we can fully appreciate the bold proposal about the nature of the electron that was made by a young French graduate student, Louis de Broglie.

De Broglie was familiar with the insights gained by Planck and Einstein; he apparently asked himself the following question. If light—which until 1905 had been conceived of as purely a wave—has associated with it the properties of a particle, could the electron—explained in 1923 as being purely a particle—have associated with it *the properties of a wave?* De Broglie then proposed that the answer to this question was *yes,* and that the relationship between the wave and particle properties of the electron is expressed in the following way:

$$\text{Momentum } (mv_e) = \frac{h\nu}{v} = \frac{h}{\lambda}$$

where m = the mass of the electron
v_e = the velocity of the electron
h = Planck's constant
ν = the frequency of the wave associated with the electron
λ = the wavelength of the wave associated with the electron
and v = the velocity of the associated wave

De Broglie's hypothesis was seen, at the time, as being so wild that he almost missed getting his degree. A few years later, however, the wave nature of the electron was confirmed when Davisson and Germer observed the diffraction of a beam of electrons. De Broglie's flight of the imagination set the stage for the last step in the development of quantum mechanics and with it our modern understanding of atoms and molecules.

Schrödinger and the Quantum Mechanical Description of the Atom

This last act was performed independently and almost simultaneously by three men: Erwin Schrödinger, Werner Heisenberg, and Paul Dirac. Because Schrödinger's treatment of the atom is the most easily visualized, we will deal with it alone.

Schrödinger saw in De Broglie's hypothesis of the wave nature of the electron a way to introduce a quantized model for the atom quite naturally. (In Bohr's model, you will recall, the quantized state of the electron had been introduced arbitrarily.) Schrödinger knew that waves of a particular type, called "standing waves," were quantized, that is, in the mathematical equations that describe standing waves, a series of integers (whole numbers) inevitably appear.

To see what we mean by this, let us consider some very simple standing waves; those that move in only one dimension. Such standing waves would occur in the plucked string of a violin, or any string where the ends are fixed and cannot move. The kind of waves that occur in a string of length, l, are shown below:

fundamental wave

first overtone

second overtone

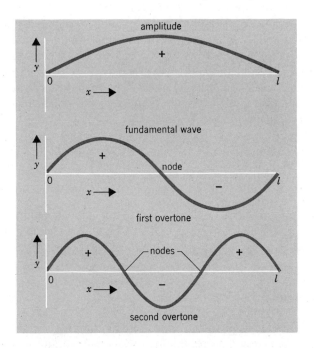

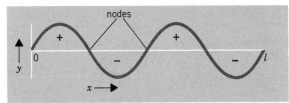

third overtone

The height of the string along the y axis, as it moves periodically up and down, is called the *amplitude* of the wave. In the first, second, third, and higher overtones, there are parts of the string that do not move at all. These are called *nodes*. The sign of the amplitude, y, is positive (+) above the midpoint of the wave and negative (−) below it.

The mathematical equation that describes all of these waves is

$$y_n = a \sin n\pi(x/l)$$

where a is a constant and $n = 1$ for the fundamental, $n = 2$ for the first overtone, $n = 3$ for the second overtone, $n = 4$ for the third overtone, and so on. The fact that n is always a whole number and the fact that only certain vibrations occur allow us to say that the standing waves are quantized.

Standing waves occurring in two dimensions (as in the head of a vibrating drum) are described by somewhat more complicated equations, but the basic properties of such waves are the same as those in a plucked string.

In order to describe the motion of the electron in an atom in terms of the standing waves associated with it, Schrödinger had to find wave equations describing standing waves occurring in three dimensions. Fortunately for him, these equations had already been written.

One hundred years earlier, William Hamilton, a mathematical genius, had considered the problem of the standing waves that might occur on a planet uniformly flooded with water. While, at the time, these efforts by Hamilton must have appeared somewhat useless in any practical sense, it turned out that the equations that Hamilton developed for his flooded planet were precisely those that Schrödinger needed for the atom. Indeed, three-dimensional plots of the solutions obtained by Hamilton for his flooded planet resemble very closely the *s, p,* and *d* orbitals for the electron that we use so commonly today.

Schrödinger used a wave function, ψ, to describe the electron wave. Psi, ψ, is a function of the amplitude of the electron wave much as *y* is an amplitude function in the equation for a simple wave. It was Schrödinger's genius that led him to substitute into Hamilton's equations for the flooded planet, the de Broglie wavelength ($\lambda = h/mv_e$), for the electron.

From that point on, straightforward mathematical manipulations of the equations produced solutions for the hydrogen atom where the quantum conditions of discrete orbitals and energies for the electron appeared quite naturally.

In its most concise form, the famous Schrödinger equation can be written as follows:

$$H\psi = E\psi$$

H, called the Hamiltonian operator, tells us to do certain mathematical operations on ψ. Solutions to the equation are obtained in terms of ψ and *E*, for a particular *set of integers* called *quantum numbers. E* represents the energy of the electron. If ψ is the wave function for the 1*s* orbital, *E* is the energy of the electron in the 1*s* orbital. If ψ is the wave function for the 2*s* orbital, then the value of *E* that is obtained is the energy of the electron in the 2*s* orbital.

Wave functions—*or ψ functions*—can be plotted in a number of ways. One way of plotting the wave functions of the 2*s* and 2*p* orbitals of a hydrogen atom is shown in Fig. 8.2. The contours given are for constant values of ψ at different distances from the nucleus.

The contours for the 2*s* orbital are a series of concentric circles centered on the nucleus. (Three-dimensional plots of the 2*s* orbital show that it is spherical.) The 2*s* orbital has a *nodal surface;* that is, it has a contour for which $\psi = 0$. Inside the nodal surface, ψ is negative; outside the nodal surface, ψ is positive.

The 2*p* orbital has two lobes that resemble two slightly squashed circles. (In three dimensions these lobes resemble slightly squashed spheres.) The lobes lie on opposite sides of the nucleus. In one lobe ψ is negative; in the other ψ is positive. A nodal plane separates the two lobes.

A 1*s* orbital (not shown) is shaped like a 2*s* orbital. In the 1*s* orbital, however, there is *no* nodal surface. The ψ function is positive throughout.

The positive and negative signs associated with ψ have nothing to do with electric charge. They are simply arithmetic signs of the wave function and in this sense they resemble the $(+)$ and $(-)$ signs associated with the amplitude of a simple wave (p. 285). It is the square of ψ, that is, ψ^2, that expresses the probability of finding an electron at a particular region of space (cf. Section 1.12). The quantity, ψ^2, of course,

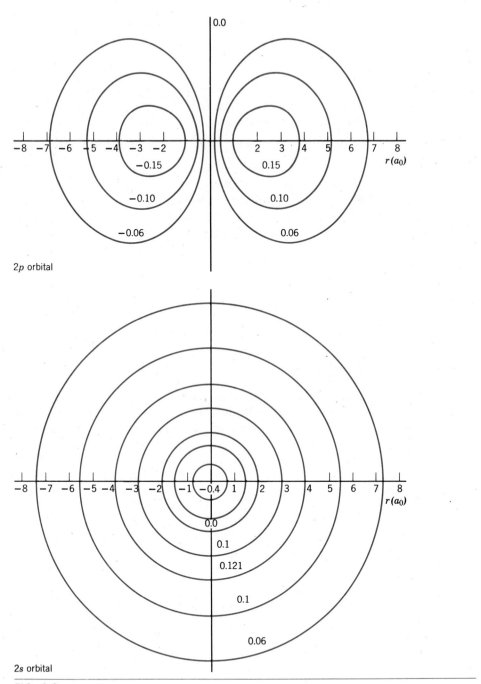

FIG. 8.2
Shapes of a 2s *orbital and a* 2p *orbital. The numbers on the contours give the value of* ψ; r *is the distance from the nucleus in units of* a_0 (a_0 *is 0.529 Å). (From K. B. Wiberg,* Physical Organic Chemistry, *Wiley, New York, 1964, pp. 29 and 30. Used with permission.)*

is always positive. Thus in the $2p$ orbital, for example, there is an equal probability of finding the electron in either lobe.

8.2 ELEMENTARY THERMODYNAMICS: ΔH, ΔS, AND ΔG

The reaction of methane with oxygen (p. 42) is a highly exothermic reaction. The enthalpy change (ΔH) for the reaction is -192 kcal/mole of methane.

$$CH_{4(g)} + 2O_{2(g)} \longrightarrow CO_{2(g)} + 2H_2O_{(g)} \qquad \Delta H = -192 \text{ kcal/mole}$$

Many studies of a large number of chemical reactions have shown that reactions that have a negative enthalpy change as large as this always go to completion at equilibrium. A *small* negative value for ΔH, however, is not a guarantee that a reaction will produce substantial amounts of products at equilibrium. To account for this we need to consider two other thermodynamic quantities; ΔG, the *free-energy change,* and ΔS, *the entropy change.* To see how these thermodynamic quantities determine the position of equilibrium of a chemical reaction let us consider the general reaction written below:

$$A + B \rightleftharpoons C$$

An expression can be written for the equilibrium constant, K_{eq},

$$K_{eq} = \frac{[C]}{[A][B]}$$

where $[A]$ and $[B]$ are the equilibrium concentrations of the reactants A and B in moles/liter, and $[C]$ is the equilibrium concentration of the product C in moles/liter.

By convention, the equilibrium concentration of the products of a reaction are written in the numerator of this expression and those of the reactants are written in the denominator. This means that a large value of K_{eq} is associated with a reaction that goes essentially to completion, and a small value of K_{eq} is associated with a reaction that produces only small amounts of products at equilibrium.

Problem 8.1

Consider the simple reaction:

$$X \longrightarrow Y$$

Assume an initial concentration of X equal to 1.0 mole/liter, and calculate the equilibrium concentration of Y that is formed if the equilibrium constant is (a) 10, (b) 1, (c) 10^{-3}. In each case, what percentage of the reactant is converted to product?

The equilibrium constant for a reaction is directly related to the thermodynamic quantity called the free-energy change for the reaction. The free-energy change is symbolized as ΔG and the relationship between ΔG and the equilibrium constant, K_{eq}, is:

$$\log K_{eq} = \frac{\Delta G}{-2.3RT}$$

where R is the gas constant (1.986 cal/degree-mole) and T is the absolute temperature.

It is easy to show with this relationship that a large *negative* free-energy change results in a large value for the equilibrium constant; and thus, a large negative free-energy change favors the formation of products at equilibrium.

The free-energy change (ΔG) and the enthalpy change (ΔH) are related to one another through another equation that involves a third thermodynamic quantity, the entropy change (ΔS) and the absolute temperature, T. This important equation is:

$$\Delta G = \Delta H - T\Delta S$$

The entropy of any system is a measure of the relative order or randomness of that system. The more random (less ordered) a system is, the greater is its entropy. Thus, a *positive* entropy change ($+\Delta S$) is always associated with a change from a more ordered arrangement to a more random one. A negative entropy change ($-\Delta S$) always accompanies the reverse process.

For a chemical system the relative order (or randomness) of the molecules can be related to the number of *degrees of freedom* available to the molecules and their constituent atoms. Degrees of freedom are associated with ways in which *movement* or *changes in relative position* can occur. Molecules have three sorts of degrees of freedom; translational degrees of freedom associated with movements of the whole molecule through space, rotational degrees of freedom associated with the tumbling motions of the molecule, and vibrational degrees of freedom associated with the stretching and bending motion of atoms about the bonds that connect them.

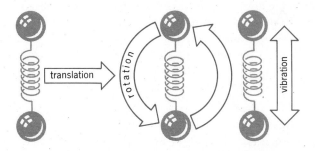

If the atoms of the products of a reaction have more degrees of freedom available than they did as reactants, the entropy change for the reaction will be positive. If, on the other hand, the atoms of the products are more constrained (have fewer degrees of freedom) than they were as reactants, a negative ΔS will result.

With these ideas and relationships in mind, we are now in a position to understand how these three thermodynamic quantities, ΔH, ΔG, and ΔS, account for the course of a chemical reaction. For reactions occurring near room tempera-ture*, if ΔH has a large negative value ($\Delta H \simeq -15$ kcal/mole), a small negative entropy change (associated with the atoms being more constrained in the products) will not generally result in a positive free-energy change ($+\Delta G$). For these reactions ΔG will be negative and the formation of products at equilibrium will be favored:

---

*At very high temperatures the entropy term ($T\Delta S$) in the equation $\Delta G = \Delta H - T\Delta S$, becomes much more important because T becomes very large. Most of the reactions that we will consider, however, are carried out between 250°K and 450°K. Under these conditions ΔH is the important term if it is larger than -15 kcal/mole.

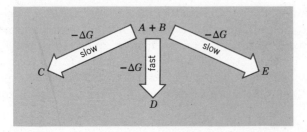

FIG. 8.3
The hypothetical reaction of A *and* B *to form differ-ent products,* C, D, *and* E. *All of the reactions have favorable free-energy changes.* D *is the product that is actually obtained, however, because it is produced by the reaction that proceeds most rapidly.* E *and* C *are not isolated in appreciable amounts because the reactions by which they are formed are slow.*

$\Delta G = \qquad \Delta H \qquad - \qquad T \qquad \times \qquad \Delta S$
$\Delta G = $ (large negative quantity) $-$ (positive quantity) $\times$ (small negative quantity)
$\Delta G = $ (large negative quantity) $-$ (smaller negative quantity)
$\Delta G = $ negative quantity

We can now see how, for some reactions that are endothermic (have a $+\Delta H$), the formation of products at equilibrium will still be favored. For these reactions there must be a larger accompanying positive entropy change. This latter situation would result when the atoms of the products are much less constrained than those of the reactants. In such cases the large positive entropy term $(T) \times (+\Delta S)$ more than compensates for the unfavorable $+\Delta H$.

$\Delta G = \qquad \Delta H \qquad - \qquad T \qquad \times \qquad \Delta S$
$\Delta G = $ (small positive quantity) $-$ (positive quantity) $\times$ (large positive quantity)
$\Delta G = $ (small positive quantity) $-$ (larger positive quantity)
$\Delta G = $ negative quantity

A negative free-energy change for a reaction only ensures that substantial amounts of products will be formed *when the reaction reaches equilibrium*. The overall free-energy change for a reaction tells us nothing, however, about how long it will take that particular reaction to reach equilibrium. The reaction of hydrogen and oxygen to produce water, for example,

$$H_2 + O_2 \longrightarrow 2H_2O$$

has associated with it a very large negative free-energy change. In the absence of a catalyst (or applied heat), however, mixtures of hydrogen and oxygen react so slowly that the formation of water is essentially imperceptible.

For many other reactions (Fig. 8.3) there may be several pathways to different products that have favorable free-energy changes. In such cases, the products that we actually obtain in greatest amounts are the ones that are formed most rapidly. We can see from this that a knowledge of the factors that determine reaction rates is of considerable importance as well (cf. Section 4.7).

Problem 8.2

The reaction of ethyne with hydrogen to produce ethene,

$$H-C\equiv C-H_{(g)} + H_{2(g)} \longrightarrow CH_2=CH_{2(g)}$$

has the following values of ΔH and ΔS at 27°C (300°K): $\Delta H = -41.7$ kcal/mole, $\Delta S = -26.6$ cal deg K^{-1} mole^{-1}. (a) Calculate the value of ΔG for this reaction. (Remember: 1000 cal = 1.0 kcal.) (b) Would you expect the position of equilibrium to favor the formation of ethene? (c) Does the entropy change for this reaction, taken alone, favor the formation of ethene? (d) How can you account for the negative entropy change?

Problem 8.3

The reaction of ethene with hydrogen to produce ethane,

$$CH_2=CH_{2(g)} + H_{2(g)} \longrightarrow CH_3-CH_{3(g)}$$

has the following thermodynamic parameters at 27°C.

$$\Delta H = -32.7 \text{ kcal/mole}$$
$$\Delta G = -24.14 \text{ kcal/mole}$$
$$\Delta S = +2.9 \text{ cal deg K}^{-1} \text{mole}^{-1}$$

(a) Will the formation of ethane be favored at equilibrium?
(b) Does the entropy term, taken alone, favor the formation of products?
(c) How can you account for the fact that the entropy change associated with this reaction is positive while the entropy change for the reaction in Problem 8.2 was negative?

8.3 POLYMERIZATION OF ALKENES: ADDITION POLYMERS

The names *Orlon, Plexiglas, Lucite, Polyethylene,* and *Teflon* are now familiar names to most of us. These "plastics" or polymers are used in the construction of many objects that surround us—from the clothing we wear, to portions of the houses we live in. Yet, all of these compounds were unknown 50 years ago. The development of the processes by which synthetic polymers are made, more than any other single factor, has been responsible for the remarkable growth of chemical industry in this century.

At the same time, some scientists are now expressing concern about the reliance we have placed on these "man-made" materials. Because they are the products of laboratory and industrial processes rather than processes that occur in nature, nature often has no way of disposing of many of them. Although progress has been made in the development of "biodegradable plastics" in recent years, many materials are still used that are not biodegradable. Although all of these objects are combustible, incineration is not always a feasible method of disposal because of attendant air pollution.

All polymers are not synthetic. Many naturally occurring compounds are polymers as well. Silk and wool are polymers that we call proteins. The starches of our diet are polymers and so is the cellulose of cotton and wood.

Polymers are compounds that consist of very large molecules made up of many repeating subunits. The molecular subunits that are used to synthesize polymers are called *monomers,* and the reactions by which monomers are joined together are called polymerization reactions.

Propylene, for example, can be polymerized to form *polypropylene.* The reaction by which this polymerization occurs is an addition reaction and, as a consequence, polymers such as polypropylene are called *addition polymers.*

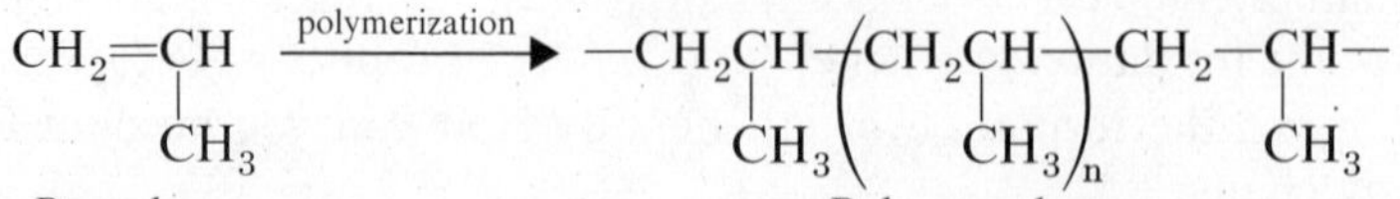

$$CH_2{=}CH \xrightarrow{\text{polymerization}} -CH_2CH{-}\!\!\left(\!CH_2CH{-}\!\right)\!\!-CH_2{-}CH{-}$$

$$\underset{\text{Propylene}}{\overset{|}{\underset{CH_3}{}}} \qquad \underset{}{\overset{|}{CH_3}}\left(\overset{|}{\underset{CH_3}{}}\right)_{n} \quad \overset{|}{CH_3}$$

Propylene   Polypropylene

Alkenes, as we might expect, are convenient starting materials for the preparation of addition polymers. The addition reactions occur through free-radical, cationic, or anionic mechanisms depending on how they are initiated. The following examples illustrate these mechanisms.

Free-Radical Polymerization

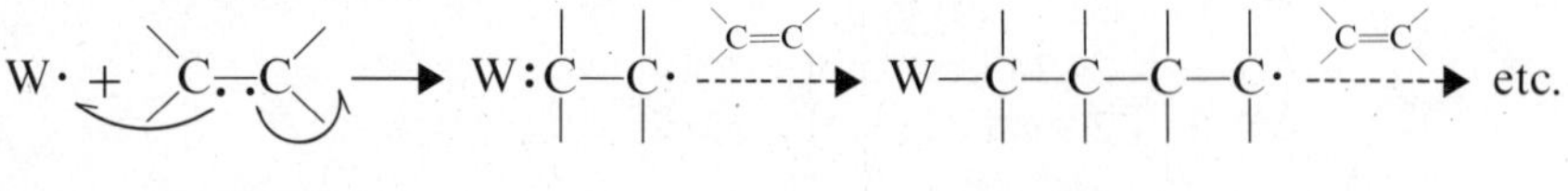

Cationic Polymerization

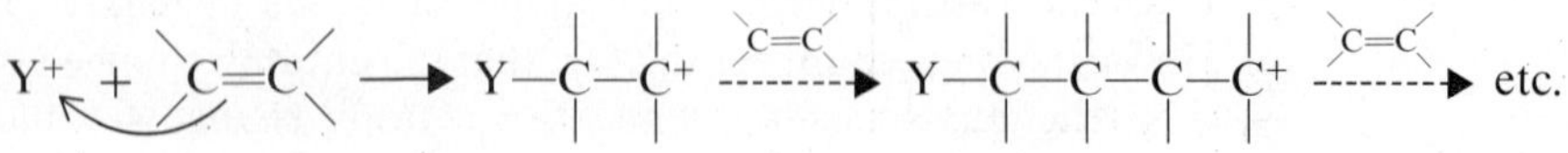

Anionic Polymerization

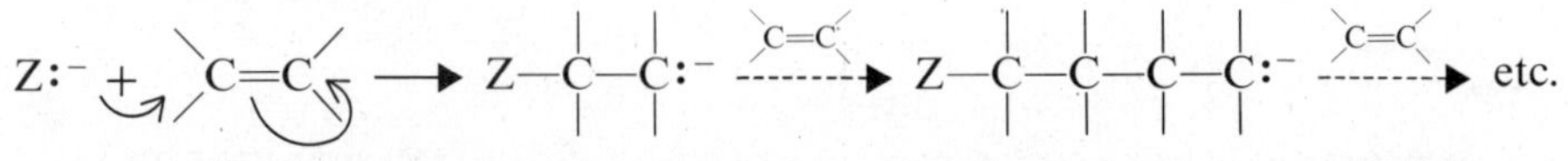

All of these reactions are chain reactions. Ethylene, for example, polymerizes by a free-radical mechanism when it is heated at a pressure of 1000 atm with a small amount of an organic peroxide. The peroxide dissociates to produce free radicals which in turn initiate chains.*

Chain Initiation

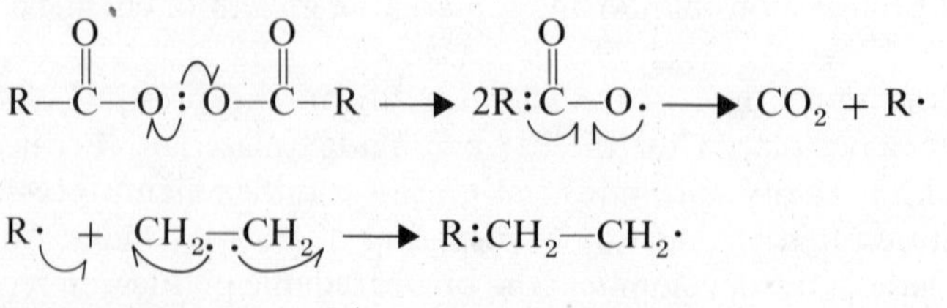

$$R{-}\overset{\overset{O}{\|}}{C}{-}O \!:\! O{-}\overset{\overset{O}{\|}}{C}{-}R \longrightarrow 2R\!:\!\overset{\overset{O}{\|}}{C}{-}O\cdot \longrightarrow CO_2 + R\cdot$$

$$R\cdot + CH_2{-}CH_2 \longrightarrow R\!:\!CH_2{-}CH_2\cdot$$

Chain Propagation

$$R{-}CH_2CH_2\cdot + nCH_2{=}CH_2 \longrightarrow R{-}(CH_2CH_2)_n{-}CH_2CH_2\cdot$$

Chains propagate by adding successive ethylene units, until their growth is stopped by combination or disproportionation.

---

*Oxygen is also often used as an initiator because it, too, produces free radicals.

Chain Termination

$$2R—(CH_2CH_2)_nCH_2CH_2· \begin{cases} \xrightarrow{\text{combination}} \left[R—(CH_2CH_2)_nCH_2CH_2\right]_2 \\ \\ \xrightarrow{\text{disproportionation}} R—(CH_2CH_2)_nCH=CH_2 + \\ \qquad\qquad\qquad\quad R—(CH_2CH_2)_nCH_2CH_3 \end{cases}$$

The polyethylene produced in this way is not generally useful unless it has a molecular weight of nearly 1,000,000. Very high molecular weight polyethylene can be obtained by using a low concentration of the initiator. This initiates the growth of only a few chains and ensures that each chain will have a large excess of the monomer available. More initiator may be added as chains terminate during the polymerization and, in this way, new chains are begun.

Polyethylene has been produced commercially since 1943. It is used in manufacturing flexible bottles, films, sheets, and insulation for electric wires. Polyethylene, produced by free radical polymerization, has a softening point of about 110°.

Free radical polymerization of chloroethene (vinyl chloride) produces a polymer called polyvinyl chloride.

$$\underset{\text{Vinyl chloride}}{\underset{|}{CH_2{=}CH}\atop Cl} \longrightarrow \underset{\text{Polyvinyl chloride}}{\left(\underset{Cl}{\underset{|}{CH_2{-}CH}}\right)_n}$$

This reaction produces a polymer that has a molecular weight of 1,500,000 and that is a hard, brittle, and rigid material. In this form it is often used to make pipes, rods and phonograph records. Polyvinyl chloride can be softened by mixing it with esters (called plasticizers). The softer material is used for making "vinyl leather," plastic raincoats, shower curtains, and garden hoses.

A mixture of vinyl chloride and vinylidene chloride polymerizes to form what is known as a *copolymer*. The familiar *Saran Wrap* used in food packaging is made by polymerizing a mixture in which the vinylidene chloride predominates.

$$m\underset{\underset{Cl}{|}}{\overset{\overset{Cl}{|}}{CH_2{=}C}} + n\underset{\underset{Cl}{|}}{CH_2{=}CH} \xrightarrow{R·} \left[\left(\underset{\underset{Cl}{|}}{\overset{\overset{Cl}{|}}{CH_2{-}C}}\right)_m\left(\underset{\underset{Cl}{|}}{CH_2CH}\right)_n\right]$$

(excess)

Vinylidene chloride   Vinyl chloride   Saran Wrap

The subunits do not necessarily alternate regularly along the polymer chain.

Acrylonitrile ($CH_2{=}CHCN$) polymerizes to form polyacrylonitrile or *Orlon*. The initiator for the polymerization is a mixture of ferrous sulfate and hydrogen peroxide; these two compounds react to produce hydroxyl radicals, ·OH, which act as chain initiators.

$$CH_2\!\!=\!\!\underset{\underset{CN}{|}}{CH} \xrightarrow[\text{H—O—O—H}]{FeSO_4} \left(\!\!\begin{array}{c} CH_2\!\!-\!\!CH \\ | \\ CN \end{array}\!\!\right)$$

Acrylonitrile   Polyacrylonitrile
Orlon

Polyacrylonitrile decomposes before it melts, thus melt spinning cannot be used for the production of fibers. Polyacrylonitrile, however, is soluble in dimethylformamide, and these solutions can be used to spin fibers. Fibers produced in this way are used in making carpets and clothing.

   Teflon is made by polymerizing tetrafluoroethylene in aqueous suspension.

$$n\,CF_2\!\!=\!\!CF_2 \xrightarrow[\substack{H_2O_2 \\ H_2O}]{Fe^{++}} \left(\!CF_2\!\!-\!\!CF_2\!\right)_n$$

The reaction is highly exothermic and water helps dissipate the heat that is produced. Teflon has a melting point (327°) that is unusually high for an addition polymer. It is also highly resistant to chemical attack and has a low coefficient of friction. Because of these properties Teflon is used in greaseless bearings, in liners for pots and pans, and in many special situations that require a substance that is highly resistant to corrosive chemicals.

   Vinyl alcohol is an unstable compound that rearranges spontaneously to acetaldehyde. Consequently, the water-soluble polymer, polyvinyl alcohol can not be

$$CH_2\!\!=\!\!\underset{\underset{OH}{|}}{CH} \;\rightleftharpoons\; CH_3\!\!-\!\!\underset{\underset{O}{\|}}{CH}$$

Vinyl alcohol   Acetaldehyde

made directly. It can be made, however, by an indirect method. This method first involves the polymerization of vinyl acetate to produce polyvinyl acetate and then hydrolysis of the polyvinyl acetate to polyvinyl alcohol. Hydrolysis is rarely carried

$$n\,CH_2\!\!=\!\!\underset{\substack{| \\ O \\ | \\ C=O \\ | \\ CH_3}}{CH} \longrightarrow \left(\!\!\begin{array}{c} CH_2CH \\ | \\ O \\ | \\ C=O \\ | \\ CH_3 \end{array}\!\!\right)_n \xrightarrow[H_2O]{OH^-} \left(\!\!\begin{array}{c} CH_2CH \\ | \\ OH \end{array}\!\!\right)_n$$

Vinyl acetate   Polyvinyl   Polyvinyl
                acetate   alcohol

to completion, however, because the presence of a few ester groups helps confer water solubility on the product. The ester groups apparently help keep the polymer chains apart and this permits hydration of the hydroxyl groups. Polyvinyl alcohol, in which 10% of the ester groups remain, dissolves readily in water. Polyvinyl alcohol is used to manufacture water-soluble films and adhesives. Polyvinyl acetate is used as an emulsion in water-base paints.

   A polymer with excellent optical properties can be made by the free-radical

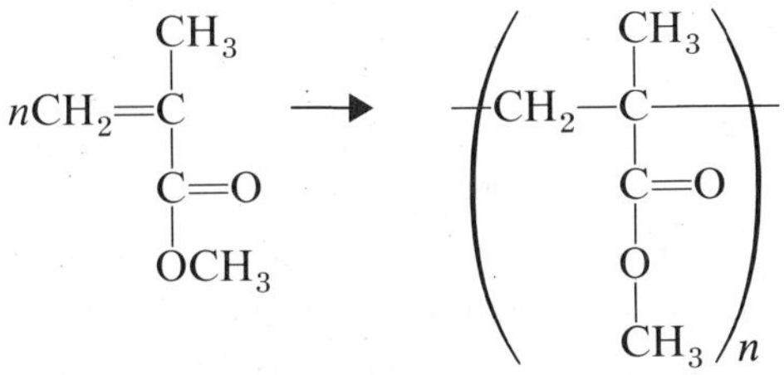

$$n CH_2 = \overset{\overset{\displaystyle CH_3}{|}}{\underset{\underset{\displaystyle OCH_3}{|}}{\underset{\displaystyle C=O}{\overset{|}{C}}}} \longrightarrow \left(CH_2 - \overset{\overset{\displaystyle CH_3}{|}}{\underset{\underset{\displaystyle CH_3}{|}}{\underset{\displaystyle O}{\underset{\displaystyle C=O}{\overset{|}{C}}}}} \right)_n$$

Poly(methyl methacrylate)

polymerization of methyl methacrylate. Poly(methyl methacrylate) is marketed under the names *Lucite, Plexiglas,* and *Perspex.*

Problem 8.4

Can you suggest an explanation that accounts for the fact that the radical polymerization of propylene occurs in a head-to-tail fashion

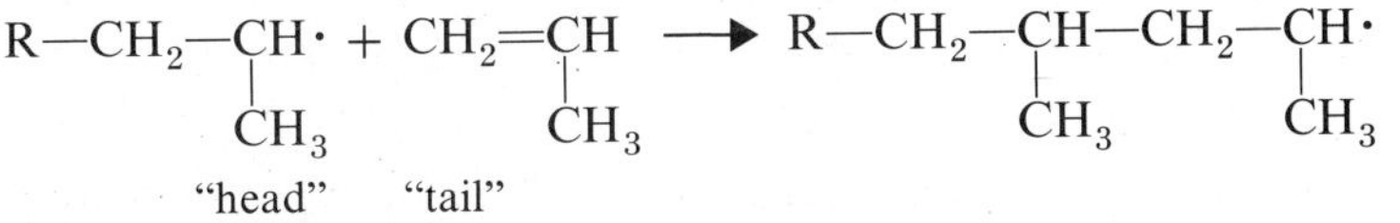

$$R-CH_2-\underset{\underset{\displaystyle CH_3}{|}}{CH}\cdot + CH_2 = \underset{\underset{\displaystyle CH_3}{|}}{CH} \longrightarrow R-CH_2-\underset{\underset{\displaystyle CH_3}{|}}{CH}-CH_2-\underset{\underset{\displaystyle CH_3}{|}}{CH}\cdot$$

"head"   "tail"

rather than the head-to-head manner, shown below?

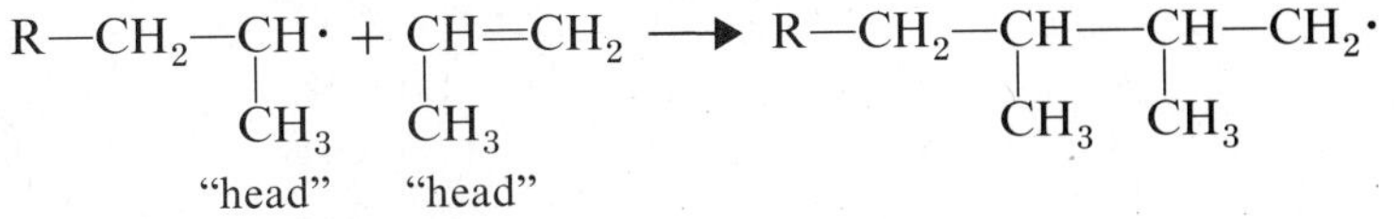

$$R-CH_2-\underset{\underset{\displaystyle CH_3}{|}}{CH}\cdot + \underset{\underset{\displaystyle CH_3}{|}}{CH} = CH_2 \longrightarrow R-CH_2-\underset{\underset{\displaystyle CH_3}{|}}{CH}-\underset{\underset{\displaystyle CH_3}{|}}{CH}-CH_2\cdot$$

"head"   "head"

Problem 8.5

Outline general methods for the synthesis of each of the following polymers by free radical polymerization. Assume that the appropriate monomers are available.
  (a) Polyvinylfluoride (Tedlar), $(CH_2CHF)_n$
  (b) Poly(chlorotrifluoroethylene) (Kel—F), $(CF_2-CFCl)_n$
  (c) *Viton,* a copolymer of hexafluoropropene, $CF_2 = CFCF_3$ and vinylidene fluoride, $CH_2 = CF_2$

Alkenes also polymerize when they are treated with strong acids. The growing chains in acid-catalyzed polymerizations are *cations* rather than free radicals. The cationic polymerization of isobutene is illustrated below.

(1) $H_2O + BF_3 \rightleftharpoons H^+ + BF_3(OH)^-$

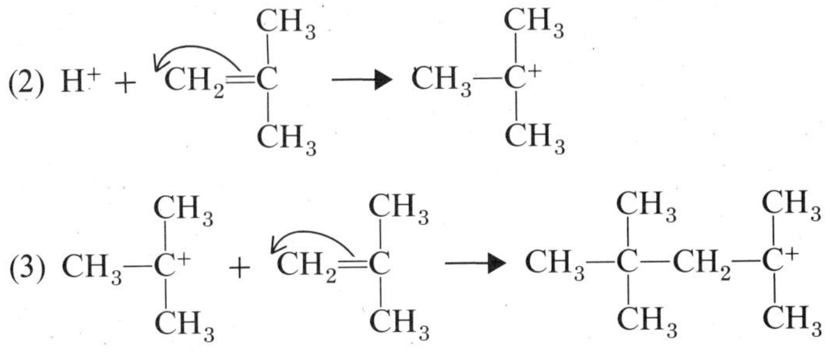

$$(2)\ H^+ + CH_2 = \underset{\underset{\displaystyle CH_3}{|}}{\overset{\overset{\displaystyle CH_3}{|}}{C}} \longrightarrow CH_3 - \underset{\underset{\displaystyle CH_3}{|}}{\overset{\overset{\displaystyle CH_3}{|}}{C}}{}^+$$

$$(3)\ CH_3 - \underset{\underset{\displaystyle CH_3}{|}}{\overset{\overset{\displaystyle CH_3}{|}}{C}}{}^+ + CH_2 = \underset{\underset{\displaystyle CH_3}{|}}{\overset{\overset{\displaystyle CH_3}{|}}{C}} \longrightarrow CH_3 - \underset{\underset{\displaystyle CH_3}{|}}{\overset{\overset{\displaystyle CH_3}{|}}{C}} - CH_2 - \underset{\underset{\displaystyle CH_3}{|}}{\overset{\overset{\displaystyle CH_3}{|}}{C}}{}^+$$

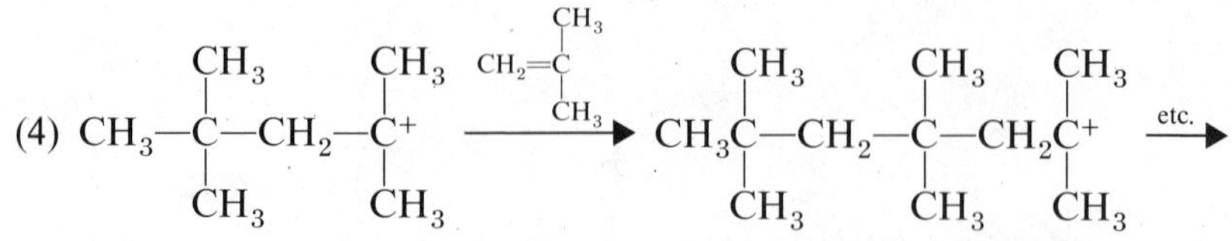

(4) $CH_3-\overset{\overset{\displaystyle CH_3}{|}}{\underset{\underset{\displaystyle CH_3}{|}}{C}}-CH_2-\overset{\overset{\displaystyle CH_3}{|}}{\underset{\underset{\displaystyle CH_3}{|}}{C^+}}$ $\xrightarrow{\quad CH_2=\overset{\overset{\displaystyle CH_3}{|}}{\underset{\underset{\displaystyle CH_3}{|}}{C}} \quad}$ $CH_3\overset{\overset{\displaystyle CH_3}{|}}{\underset{\underset{\displaystyle CH_3}{|}}{C}}-CH_2-\overset{\overset{\displaystyle CH_3}{|}}{\underset{\underset{\displaystyle CH_3}{|}}{C}}-CH_2\overset{\overset{\displaystyle CH_3}{|}}{C^+}$ $\xrightarrow{etc.}$

The catalysts used for cationic polymerizations are usually Lewis acids that contain a small amount of water. The polymerization of isobutene illustrates how the catalyst (BF_3 and H_2O) function to produce growing cationic chains.

Problem 8.6

How can you account for the fact that isobutene polymerizes in the way we indicated above, rather than in the manner shown below?

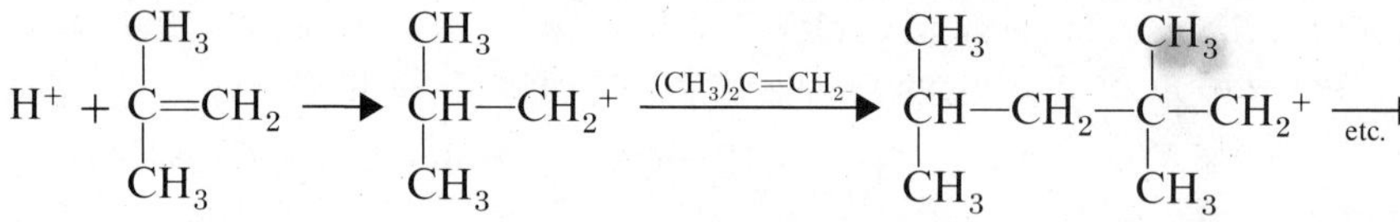

$H^+ + \overset{\overset{\displaystyle CH_3}{|}}{\underset{\underset{\displaystyle CH_3}{|}}{C}}=CH_2 \longrightarrow \overset{\overset{\displaystyle CH_3}{|}}{\underset{\underset{\displaystyle CH_3}{|}}{CH}}-CH_2{}^+ \xrightarrow{(CH_3)_2C=CH_2} \overset{\overset{\displaystyle CH_3}{|}}{\underset{\underset{\displaystyle CH_3}{|}}{CH}}-CH_2-\overset{\overset{\displaystyle CH_3}{|}}{\underset{\underset{\displaystyle CH_3}{|}}{C}}-CH_2{}^+ \xrightarrow{etc.}$

Alkenes containing electron-withdrawing groups polymerize in the presence of strong bases. Acrylonitrile, for example, polymerizes when it is treated with sodium amide ($Na^+N\bar{H}_2$) in liquid ammonia. The growing chains in this polymerization are anions.

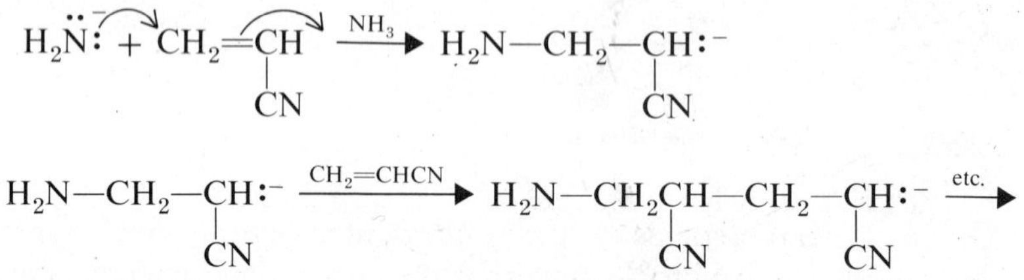

$H_2\ddot{N}{:}^- + CH_2{=}\overset{\displaystyle CH}{\underset{\underset{\displaystyle CN}{|}}{}} \xrightarrow{NH_3} H_2N-CH_2-\overset{\displaystyle CH}{\underset{\underset{\displaystyle CN}{|}}{}}{:}^-$

$H_2N-CH_2-\overset{\displaystyle CH}{\underset{\underset{\displaystyle CN}{|}}{}}{:}^- \xrightarrow{CH_2=CHCN} H_2N-CH_2\overset{\displaystyle CH}{\underset{\underset{\displaystyle CN}{|}}{}}-CH_2-\overset{\displaystyle CH}{\underset{\underset{\displaystyle CN}{|}}{}}{:}^- \xrightarrow{etc.}$

Anionic polymerization of acrylonitrile is less important in commercial production than the free radical process we illustrated earlier.

Problem 8.7

If alkene monomers used in anionic polymerization are extremely pure, the chains continue growing until all of the monomer is consumed. Even then, however, most of the chains ends are still anions. These chains are said to be "living" chains. The "living" chains can be terminated—"killed"—by the addition of water. (a) How does water terminate the chain? (b) Speculate about what happens when one adds first ethylene oxide and then water to a "living" anionic polymer.

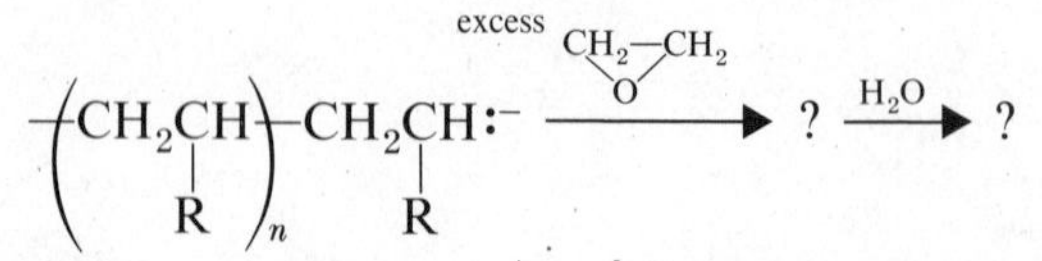

$\left(\overset{\displaystyle CH_2CH}{\underset{\underset{\displaystyle R}{|}}{}}\right)_n \overset{\displaystyle CH_2CH}{\underset{\underset{\displaystyle R}{|}}{}}{:}^- \xrightarrow[\quad O \quad]{\overset{excess}{\quad CH_2-CH_2 \quad}} ? \xrightarrow{H_2O} ?$

Hint: Ethylene oxide, $\underset{\underset{\displaystyle O}{\diagdown\diagup}}{CH_2-CH_2}$, can also be polymerized by anions. The reaction involves ring opening of the highly strained three-membered ring.

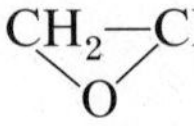

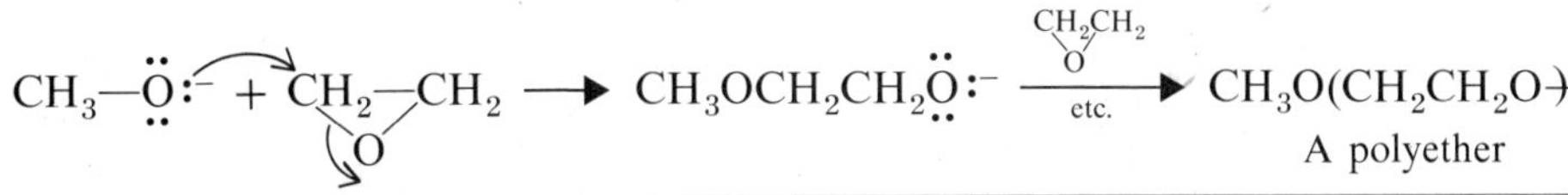

$$CH_3-\overset{..}{\underset{..}{O}}:^- + \overset{CH_2-CH_2}{\underset{O}{\diagdown}} \longrightarrow CH_3OCH_2CH_2\overset{..}{\underset{..}{O}}:^- \xrightarrow[\text{etc.}]{\overset{CH_2CH_2}{\overset{\diagup\diagdown}{O}}} CH_3O(CH_2CH_2O)_n$$

A polyether

8.4 STEREOCHEMISTRY OF ADDITION POLYMERS

Head-to-tail polymerization of propylene produces a polymer in which every other atom is chiral. Many of the physical properties of the polypropylene produced in this way depend on the stereochemistry of these chiral centers.

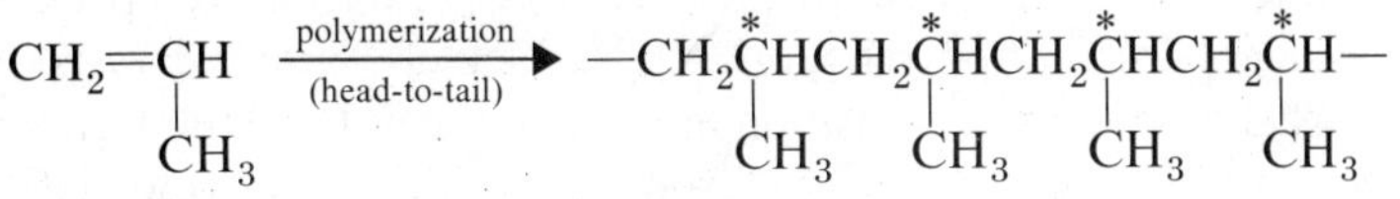

$$CH_2=\underset{\underset{CH_3}{|}}{CH} \xrightarrow[\text{(head-to-tail)}]{\text{polymerization}} -CH_2\overset{*}{\underset{\underset{CH_3}{|}}{CH}}CH_2\overset{*}{\underset{\underset{CH_3}{|}}{CH}}CH_2\overset{*}{\underset{\underset{CH_3}{|}}{CH}}CH_2\overset{*}{\underset{\underset{CH_3}{|}}{CH}}-$$

There are three general arrangements of the methyl groups and hydrogens along the chain. These arrangements are described as being *atactic, syndyotactic,* and *isotactic.*

If the stereochemistry at the chiral centers is random (Fig. 8.4) the polymer is said to be atactic (a, without + Gr: *tactikos,* order).

We see that in atactic polypropylene the methyl groups are randomly disposed on either side of the stretched carbon chain. If we were to arbitrarily designate one end of the chain as having higher priority than the other we could give *R-S* designations (Section 7.5) to the chiral carbons. In atactic polypropylene the *R-S* designations would also be random, that is, *R, R, S, R, S, S,* and so on.

Polypropylene produced by free-radical polymerization at high pressures is atactic. Because the polymer is atactic, it is noncrystalline, has a low softening point, and has weak mechanical properties.

A second possible arrangement of the groups along the carbon chain is that of *syndyotactic* polypropylene. In syndyotactic polypropylene the methyl groups alternate regularly from one side of the stretched chain to the other (Fig. 8.5). If we were to arbitrarily designate one end of the chain of syndyotactic polypropylene as having higher priority, the configuration of the chiral carbons would alternate, *R, S, R, S, R, S, R, S,* and so on.

The third possible arrangement of chiral centers is the *isotactic* arrangement shown in Fig. 8.6. In the isotactic arrangement all of the methyl groups are on the same side of the stretched chain. The configurations of the chiral centers are either all *R* or all *S* depending on which end of the chain is assigned higher priority.

The names isotactic and syndyotactic come from the Greek terms *taktikos* (order) plus *iso* (same) and *syndyo* (two together).

FIG. 8.4

Atactic polypropylene. (In this illustration a "stretched" carbon chain is used for clarity.)

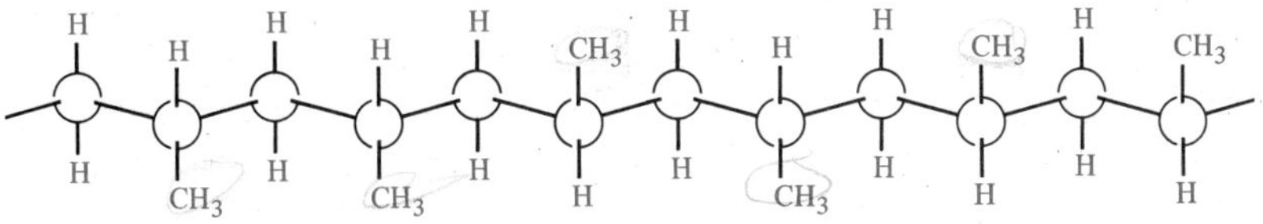

FIG. 8.5

Syndyotactic polypropylene.

Before 1953, isotactic and syndyotactic addition polymers were unknown. In that year, however, a German chemist, Karl Ziegler, and an Italian chemist, Giulio Natta, announced independently the discovery of catalysts that permit stereochemical control of polymerization reactions.* The Ziegler-Natta catalysts, as they are now called, are prepared from transition metal halides and a reducing agent. The catalysts most commonly used are prepared from titanium tetrachloride (TiCl$_4$) and a trialkyl aluminum (R$_3$Al).

Ziegler-Natta catalysts are generally employed as suspended solids and polymerization probably occurs at metal atoms on the surfaces of the particles. The mechanism for the polymerization is an ionic mechanism but its details are not fully understood. There is evidence that polymerization occurs through an insertion of the alkene monomer between the metal and the growing polymer chain.

Both syndyotactic and isotactic polypropylene have been made using Ziegler-Natta catalysts. The polymerizations occur at much lower pressures and the polymers that are produced are much higher melting than atactic polypropylene. Isotactic polypropylene, for example, melts at 175°. Isotactic and syndyotactic polymers are also much more crystalline than atactic polymers. The regular arrangement of groups along the chains apparently allows them to fit together better in a crystal structure.

Atactic, syndyotactic, and isotactic forms of poly(methyl methacrylate) (p. 295) are also known. The atactic form is a noncrystalline glass. The crystalline syndyotactic and isotactic forms melt at 160° and 200°, respectively.

Problem 8.8

Write structural formulas for portions of the chain of:
   (a) Atactic poly(methyl methacrylate)
   (b) Syndyotactic poly(methyl methacrylate)
   (c) Isotactic poly(methyl methacrylate)

* Ziegler and Natta were awarded the Nobel Prize for their discoveries in 1963.

FIG. 8.6

Isotactic polypropylene.

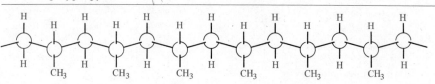

8.5 DIVALENT CARBON COMPOUNDS: CARBENES

In recent years considerable research has been devoted to investigating the structures and reactions of a group of compounds in which carbon forms only *two bonds*. These neutral divalent carbon compounds are called *carbenes*.

Most carbenes are highly unstable compounds that are capable of only fleeting existence. Soon after carbenes are formed in a reaction mixture, they usually react with another molecule. The reactions of carbenes are especially interesting because, in many instances, the reactions show a remarkable degree of stereospecificity. The reactions of carbenes are also of great synthetic use in the preparation of compounds that have three-membered rings.

Structure of Methylene

The simplest carbene is the compound called methylene, CH_2. Methylene can be prepared by the decomposition of diazomethane,* CH_2N_2. This decomposition can be accomplished by heating diazomethane (thermolysis) or by irradiating it with light of a wavelength that it can absorb (photolysis).

$$:\overset{-}{C}H_2-\overset{+}{N}\equiv N: \xrightarrow[\text{or light}]{\text{heat}} \quad :CH_2 \quad + \quad :N\equiv N:$$

Diazomethane   Methylene   Nitrogen

The structure of methylene is more complicated than its simple formula, CH_2, suggests. There are actually three different forms of methylene. These forms, **1, 2,** and **3,** differ in the arrangement of their nonbonding electrons.

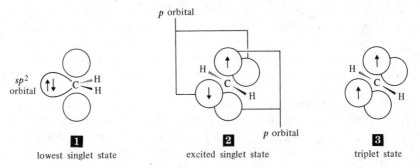

1	**2**	**3**
lowest singlet state	excited singlet state	triplet state

The two states, **1** and **2,** that have spin-paired electrons are called *singlet states*. State **3,** in which the electron spins are *not* paired, is called a *triplet state*.† In the lowest singlet state **1,** the *spin-paired* nonbonding electrons occupy an sp^2 orbital of an

---

* Diazomethane is resonance hybrid of the three structures, I, II, and III shown below.

$$:\overset{-}{C}H_2-\overset{+}{N}\equiv N: \longleftrightarrow CH_2=\overset{+}{N}=\overset{-}{N}: \longleftrightarrow \overset{..}{\overset{-}{C}}H_2-\overset{..}{N}=\overset{+}{N}:$$
$$\text{I} \qquad\qquad\qquad \text{II} \qquad\qquad\qquad \text{III}$$

We have chosen resonance structure I to illustrate the decomposition of diazomethane because with I it is readily apparent that heterolytic cleavage of the carbon-nitrogen bond results in the formation of methylene and molecular nitrogen.

$$:\overset{-}{C}H_2 \overset{|}{\underset{|}{\,}} :\overset{+}{N}\equiv N: \longrightarrow :CH_2 + :N\equiv N:$$

† The names singlet and triplet are derived from spectroscopic designations for species of this general type.

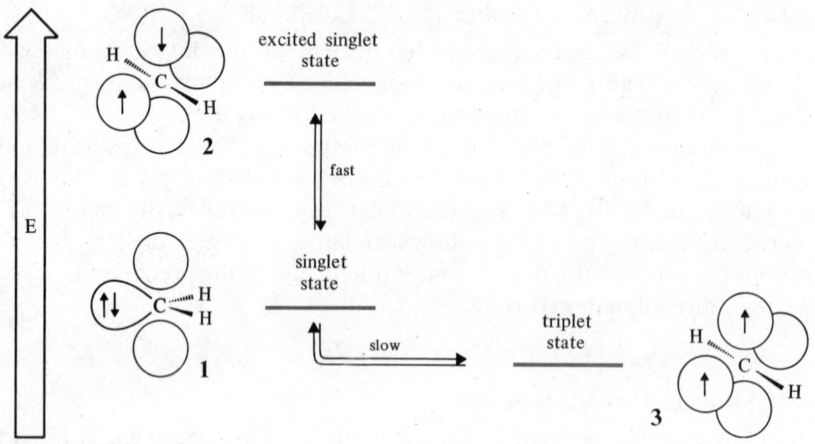

FIG. 8.7
The relative energies of the three states of methylene.

essentially sp^2-hybridized carbon. (The C $\overset{H}{\underset{H}{<}}$ bond angle is close to 120°.) The p orbital of **1** is vacant.

The carbon atoms of the excited singlet state **2** and the triplet state **3** are sp hybridized and both are linear molecules. The electrons of the excited singlet state occupy separate p orbitals *but their spins are paired.* The electrons of the triplet state also occupy separate p orbitals but their spins are *unpaired.* The triplet state of methylene resembles a diradical and, as we will see, it reacts like one.

The relative energies of the three states of methylene are shown in Fig. 8.7. There is evidence that photolysis of diazomethane produces singlet methylene initially, even though triplet methylene has slightly lower energy. Transitions between singlet methylene and triplet methylene, however, do not occur rapidly. The transition from singlet methylene to triplet methylene requires that the singlet methylene lose energy. It can only do this by colliding with some other molecule or the walls of the container, to which it transfers the excess energy.

Reactions of Methylene

Singlet and triplet methylene both react with alkenes by addition to the double bond. These addition reactions produce cyclopropanes. The addition reactions of singlet and triplet methylene, however, give different stereochemical results.

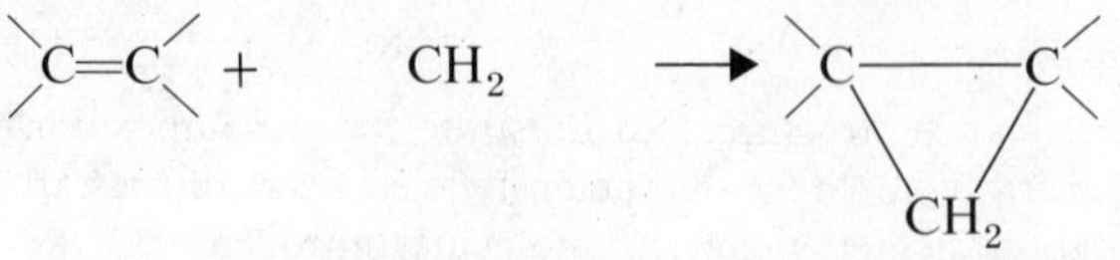

$$\overset{}{\underset{}{>}}C=C\overset{}{\underset{}{<}} \; + \quad CH_2 \quad \longrightarrow \quad \overset{}{\underset{}{>}}C\text{---}C\overset{}{\underset{}{<}}$$
$$\underset{CH_2}{\diagup\diagdown}$$

Alkene	Methylene	Cyclopropane
	(singlet or triplet)	

Singlet methylene reacts *stereospecifically.* Singlet methylene reacts with *cis*-2-butene, for example, to give *cis*-1,2-dimethylcyclopropane.

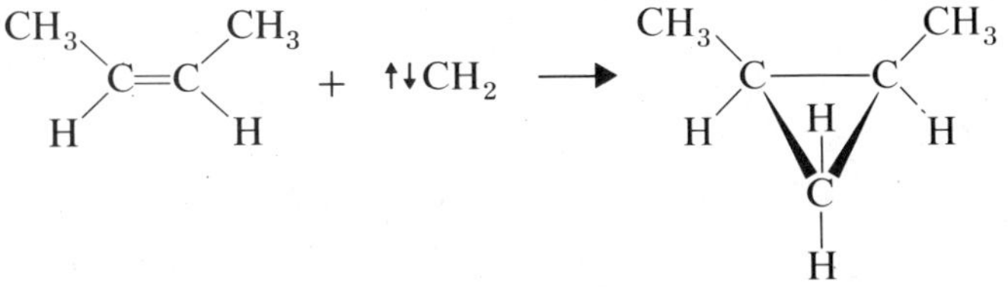

cis-2-Butene   Singlet   cis-1,2-Dimethyl-
                      methylene   cyclopropane
                                       (a *meso* compound)

Singlet methylene reacts with *trans*-2-butene to give a racemic modification of the *trans*-1,2-dimethylcyclopropane enantiomers.*

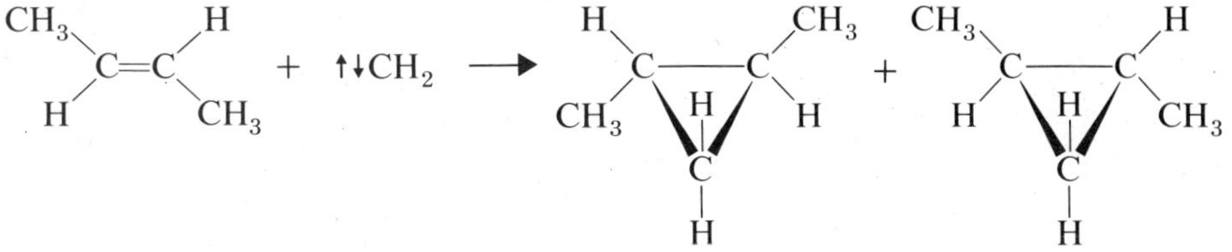

trans-2-Butene   Singlet   Racemic-*trans*-1,2-dimethylcyclopropane
                      methylene

When triplet methylene reacts with either *cis*- or *trans*-2-butene the reactions are *nonstereospecific*. The reaction of either *cis*-2-butene, or *trans*-2-butene with triplet methylene gives a mixture of the *cis*- and *trans*-1,2-dimethylcyclopropanes.

$$CH_3CH=CHCH_3 + \overset{\uparrow}{\uparrow}CH_2 \longrightarrow$$
cis- or *trans*-   Triplet
2-Butene   methylene

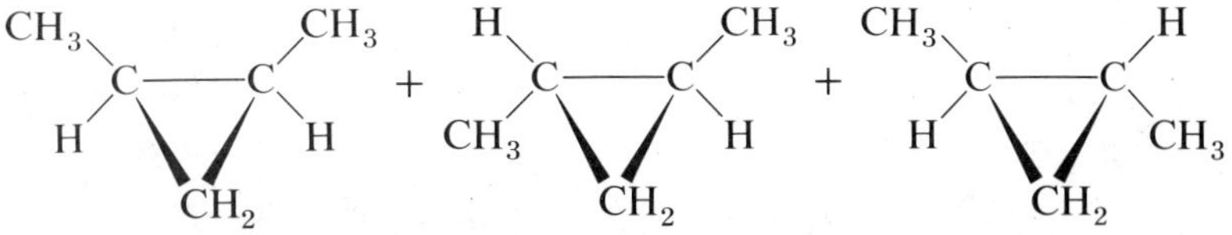

Mixture of *cis*- or *trans*-1,2-dimethylcyclopropanes

As we might expect, different stereochemical results are obtained from the reactions of singlet and triplet methylene because the mechanisms of the two reactions are different. Singlet methylene adds to the double bond in a single step; bonds form between the carbon of singlet methylene and the carbons of the alkene at the

cis-Alkene   *cis*-Dialkylcyclopropane

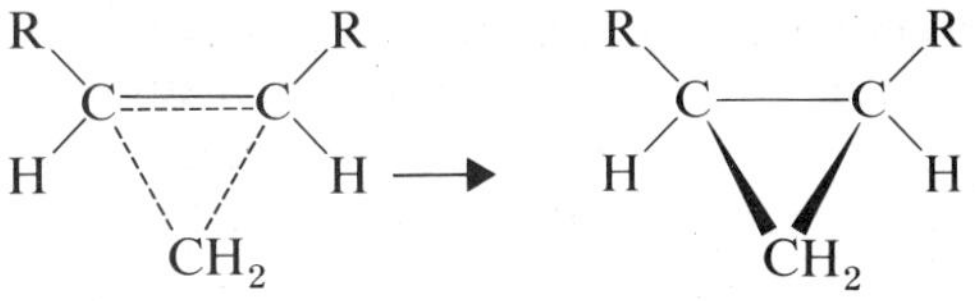

Singlet methylene

*These reactions are stereospecific (rather than stereoselective) because in them, diastereomerically different reactants (*cis*- and *trans*-2-butene) give diastereomerically different products (*cis*-1,2-dimethylcyclopropane and the *trans*-1,2-dimethylcyclopropanes, respectively).

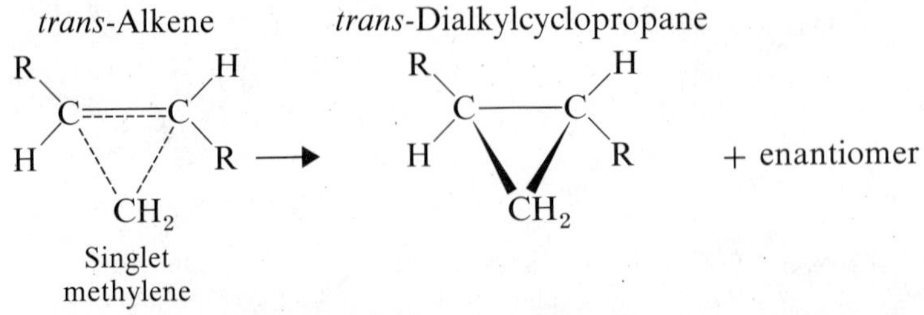

trans-Alkene   *trans*-Dialkylcyclopropane

Singlet
methylene

+ enantiomer

same time. As a result, the stereochemistry of the reactant is preserved in the product: *cis*-alkenes yield *cis*-dialkylcyclopropanes; *trans*-alkenes yield *trans*-dialkylcyclopropanes.

Problem 8.9

How can you account for the fact that the addition of singlet methylene to *trans*-2-butene produces equal amounts of the two enantiomeric *trans*-1,2-dimethylcyclopropanes.

Triplet methylene, because its electrons are not paired, reacts in a two-step mechanism. Triplet methylene (a diradical itself) reacts with the alkene to form an

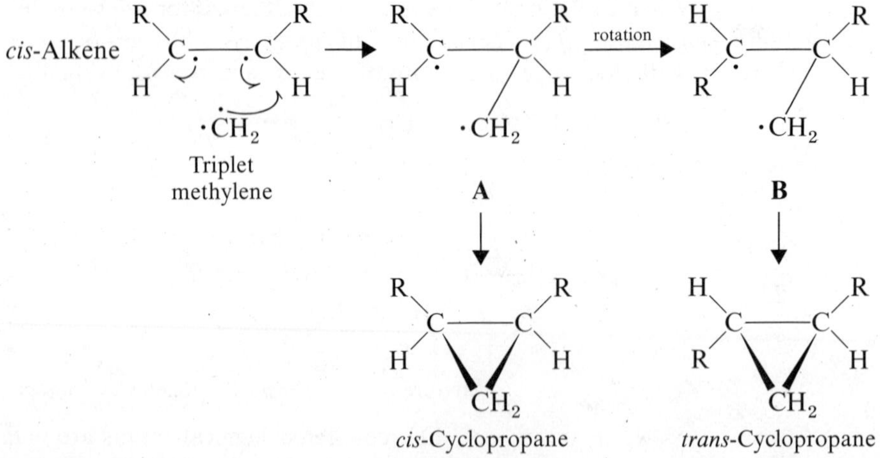

intermediate diradical in conformation **A.** The intermediate diradical has a lifetime long enough to allow rotation of groups joined by single bonds. Rotations lead to a mixture of conformations **A** and **B,** and ring closures of **A** and **B** give diastereomerically different cyclopropanes.

When we plan experiments we must take into account the stereochemical differences between the reactions of singlet and triplet methylene. But how do we know which form of methylene will be the predominant one in a reaction? The answer is that the predominant form will depend on the reaction conditions that we use.

If we carry out the reaction in the liquid phase the reaction of singlet methylene will predominate because it forms first in the decomposition of diazomethane and because *it reacts with the alkene before it can undergo transition to triplet methylene.* In the liquid state the singlet methylene is in close proximity to many alkene molecules. The reaction of singlet methylene and an alkene, therefore, occurs rapidly. The transition of singlet to triplet methylene, by contrast, is slow and does not occur to an appreciable extent. We find that reactions of methylene in the liquid state are largely stereospecific.

When we carry out methylene additions in the gas phase, particularly when we dilute the mixture by the addition of an inert gas, the molecules will be widely separated. In gas-phase reactions singlet methylene will have an opportunity to undergo transition to the lower energy triplet state through collisions with molecules of the inert gas. Thus, gas-phase reactions give primarily those products expected from triplet methylene; the reactions are largely nonstereospecific.

Problem 8.10

What products would you expect from the following reactions?

(a) *cis*-2-Pentene + CH_2N_2 $\xrightarrow[\text{(liquid state)}]{\text{light}}$

(b) *trans*-2-Pentene + CH_2N_2 $\xrightarrow[\text{(liquid state)}]{\text{light}}$

(c) *cis*-2-Pentene + CH_2N_2 $\xrightarrow[\substack{\text{He} \\ \text{(gas phase)}}]{\text{light}}$

(d) *trans*-2-Pentene + CH_2N_2 $\xrightarrow[\substack{\text{He} \\ \text{(gas phase)}}]{\text{light}}$

Other Reactions of Methylene

Singlet methylene not only reacts with double bonds to form three-membered rings, it also reacts with carbon-hydrogen bonds by *insertion*. This reaction is partic-

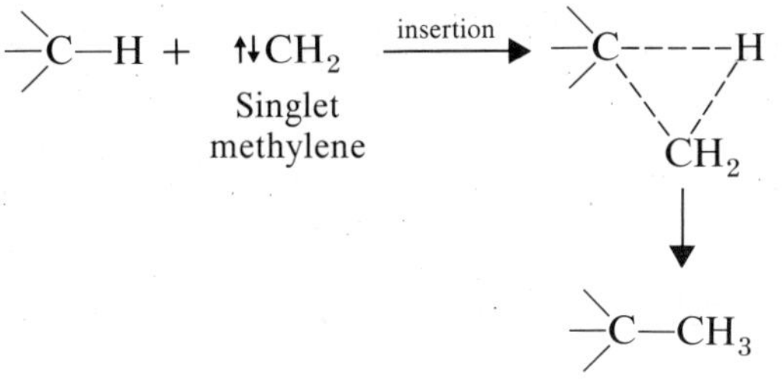

$$\text{>\!\!C-H} + \text{↑↓CH}_2 \xrightarrow{\text{insertion}} \text{>\!\!C} \overset{\cdots\cdots\text{H}}{\underset{\underset{\text{CH}_2}{\Big\downarrow}}{\diagdown}}$$

$$\underset{\text{Singlet}}{\underset{\text{methylene}}{}}$$

$$\text{>\!\!C-CH}_3$$

ularly important if the reactant does not contain a double bond. Butane, for example, reacts with singlet methylene to give a mixture of pentane and isopentane. Insertion

$$\text{CH}_3\text{CH}_2\text{CH}_2\text{CH}_3 \xrightarrow[\text{light}]{\text{CH}_2\text{N}_2} \text{CH}_3\text{CH}_2\text{CH}_2\text{CH}_2\text{CH}_3 + \text{CH}_3\text{CH}_2\underset{\underset{\text{CH}_3}{|}}{\text{CHCH}_3}$$

reactions of methylene are highly exothermic. As a consequence, singlet methylene is relatively unselective. It does not discriminate between primary, secondary, and tertiary hydrogens to an appreciable extent.

When triplet methylene reacts with alkanes, it reacts primarily by abstracting hydrogens.

$$CH_3CH_2CH_3 + \ \ ^\uparrow\ddot{C}H_2 \longrightarrow CH_3CH_2CH_2\cdot + CH_3\dot{C}HCH_3$$

Triplet methylene

$+CH_3\cdot$

Problem 8.11

What eventual products would you expect from the reaction of propane with triplet methylene given above?

Reactions of Other Carbenes

Dihalocarbenes are also frequently employed in the synthesis of cyclopropane derivatives from alkenes. The singlet state of dihalocarbenes appears to be the more stable and most reactions of dihalocarbenes are stereospecific.

Dichlorocarbene can be synthesized by the α *elimination* of the elements of hydrogen chloride from chloroform. This reaction resembles the β elimination

$$R-\ddot{O}:^- K^+ + H:CCl_3 \rightleftharpoons R-\ddot{O}:H + \ ^-:CCl_3 + K^+$$

$$^-:CCl_3 \xrightarrow{\text{slow}} :CCl_2 + :\ddot{C}l:^-$$

Dichlorocarbene

reactions by which alkenes are synthesized (p. 179). Compounds *with a β hydrogen* react by β elimination preferentially. Compounds with no β hydrogen (such as chloroform) react by α elimination.

A variety of cyclopropane derivatives have been prepared by generating dichlorocarbene in the presence of alkenes. Cyclohexene, for example, reacts with

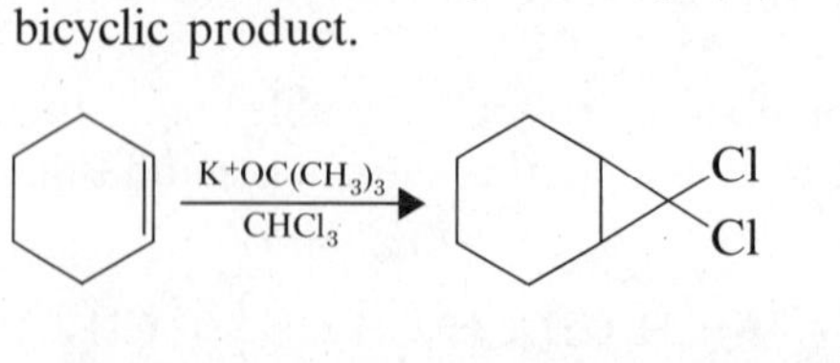

$$\overset{\diagdown}{\underset{\diagup}{C}}=\overset{\diagdown}{\underset{\diagup}{C}} + :CCl_2 \longrightarrow \overset{\diagdown}{\underset{\diagup}{C}}\!\!-\!\!\overset{\diagdown}{\underset{\diagup}{C}}$$

dichlorocarbene generated by treating chloroform with potassium *tert*-butoxide to give a bicyclic product.

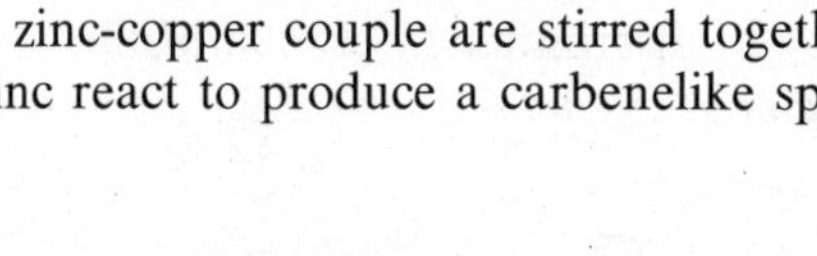

(59%)

Carbenoids: The Simmons-Smith Cyclopropane Synthesis

An extremely useful cyclopropane synthesis has been developed by H. E. Simmons and R. D. Smith of the du Pont Company. In this synthesis diiodomethane and a zinc-copper couple are stirred together with an alkene. The diiodomethane and zinc react to produce a carbenelike species called a *carbenoid*.

$$CH_2I_2 + Zn(Cu) \longrightarrow ICH_2ZnI$$
$$\text{A carbenoid}$$

The carbenoid then brings about the stereospecific addition of a CH_2 group directly to the double bond.

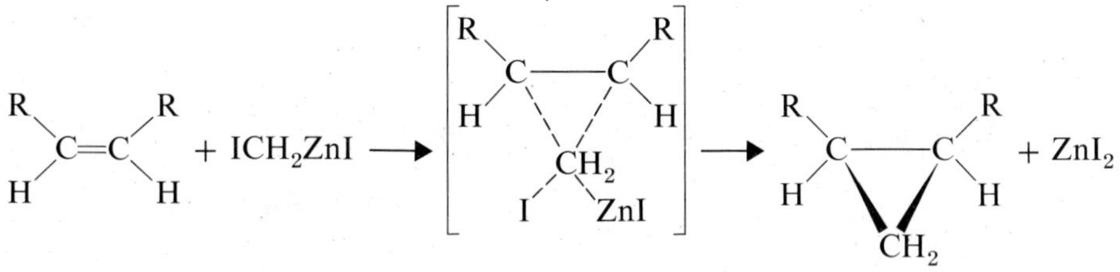

This synthesis has been used widely. One example is the synthesis of methyl dihydrosterculate from methyl oleate.

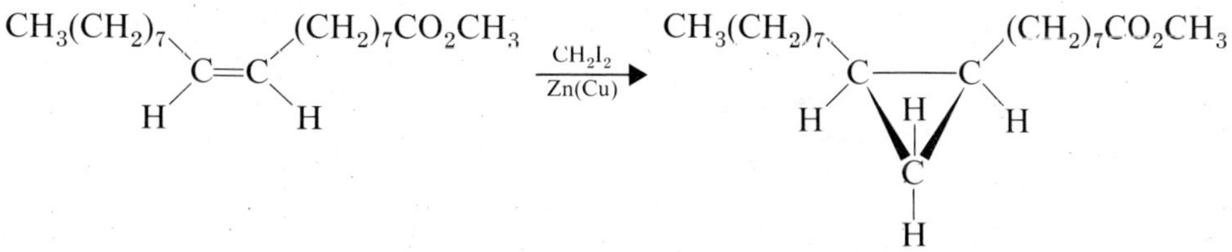

Methyl oleate · · · · · · · · · · · · · · · Methyl dihydrosterculate

Methyl dihydrosterculate is of interest because it is related to sterculic acid, a compound that has been isolated from the kernel oil of the tropical tree *Sterculia Foetida*. Sterculic acid was the first naturally occurring compound found to have the highly strained cyclopropene ring.

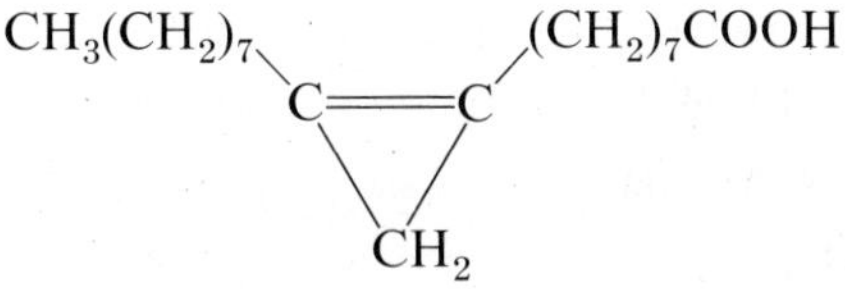

Sterculic acid

Sterculic acid, itself, has been synthesized using the Simmons-Smith method.

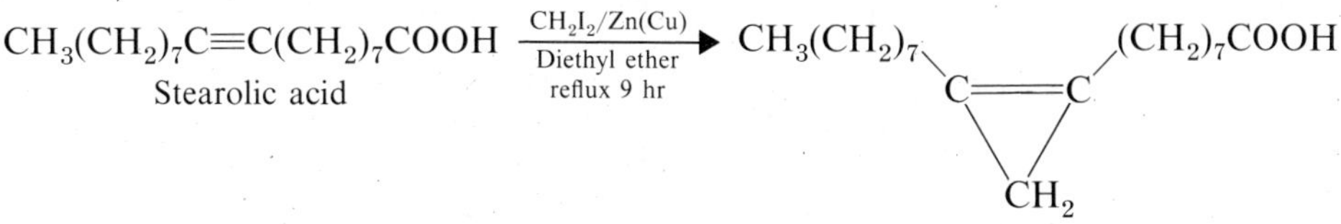

$$CH_3(CH_2)_7C \equiv C(CH_2)_7COOH \xrightarrow[\substack{\text{Diethyl ether} \\ \text{reflux 9 hr}}]{CH_2I_2/Zn(Cu)}$$

Stearolic acid

(4%)

This reaction illustrates the addition of a carbenoid to a carbon-carbon triple bond. (In Sect. 9.11 we will see how stearolic acid is prepared.)

The zinc-copper couple used in the Simmons-Smith synthesis can also be prepared *in situ* (in the reaction mixture) as the following example illustrates.

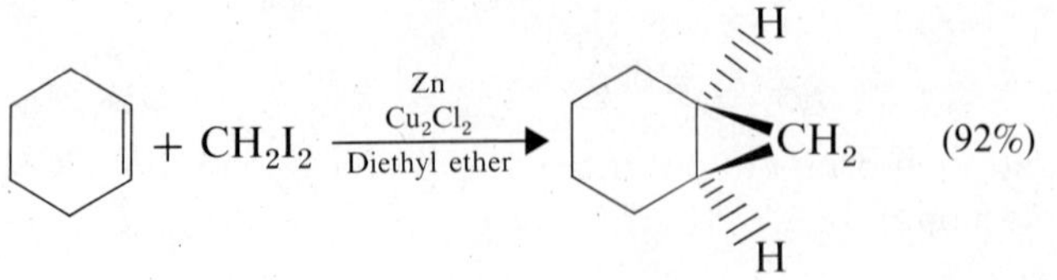

$$\text{(cyclohexene)} + CH_2I_2 \xrightarrow[\text{Diethyl ether}]{\substack{\text{Zn} \\ \text{Cu}_2\text{Cl}_2}} \text{(bicyclic product)} \quad (92\%)$$

Problem 8.12

How might the following compounds be synthesized?

(a)

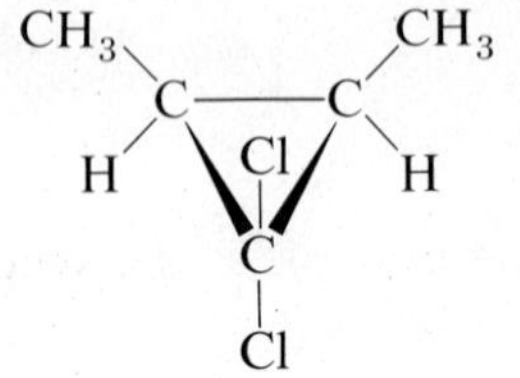

(b)

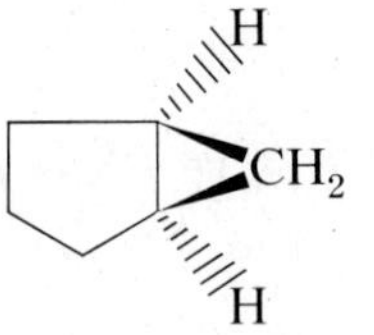

(c)

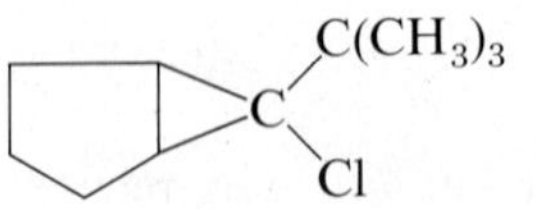

*Problem 8.13

[Problem for Section 8.3] Studies have shown that when 3-methyl-1-butene, $CH_2=CHCH(CH_3)_2$, undergoes cationic polymerization the product is not the expected 1,2-linked polymer that follows,

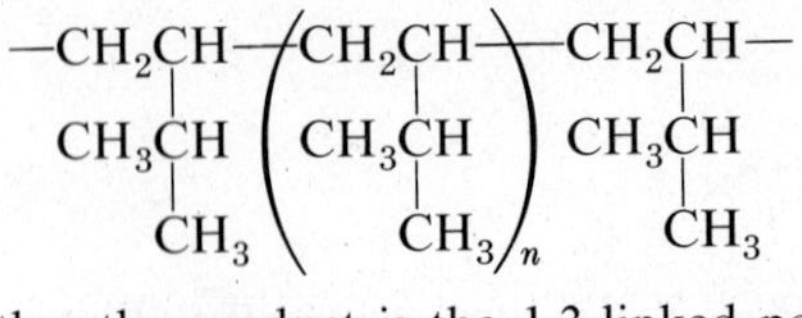

$$-CH_2CH-\left(CH_2CH\right)-CH_2CH- \\ \quad CH_3CH \quad \left(CH_3CH\right)_n \quad CH_3CH \\ \qquad CH_3 \qquad CH_3 \qquad CH_3$$

but rather the product is the 1,3-linked polymer shown below:

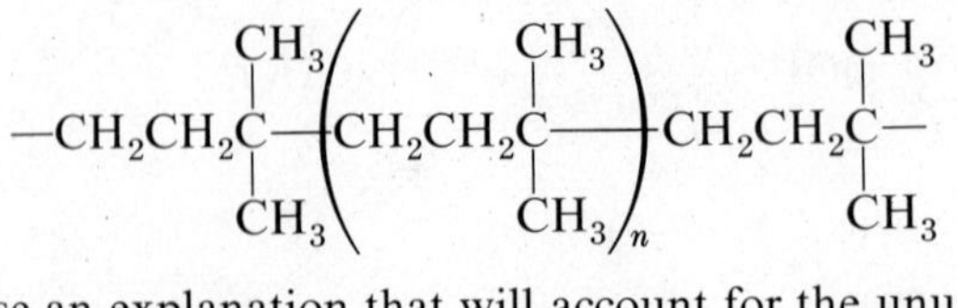

$$\quad CH_3 \qquad CH_3 \qquad CH_3 \\ -CH_2CH_2C-\left(CH_2CH_2C\right)-CH_2CH_2C- \\ \quad CH_3 \qquad \left(CH_3\right)_n \qquad CH_3$$

Propose an explanation that will account for the unusual course of this polymerization.

9

ALKYNES

9.1 INTRODUCTION

Hydrocarbons whose molecules contain the carbon-carbon triple bond are called alkynes. Alkynes have the general formula C_nH_{2n-2} and therefore contain a smaller proportion of hydrogen than alkenes or alkanes.

Alkane	Alkene	Alkyne
C_nH_{2n+2}	C_nH_{2n}	C_nH_{2n-2}

$$CH_3-CH_3 \qquad \overset{H}{\underset{H}{\diagdown}}C=C\overset{H}{\underset{H}{\diagup}} \qquad H-C\equiv C-H$$

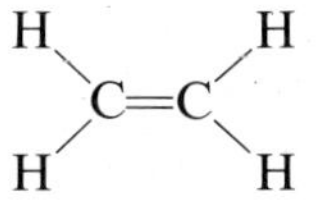

Ethane	Ethene	Ethyne or acetylene
C_2H_6	C_2H_4	C_2H_2

9.2 NOMENCLATURE OF ALKYNES

IUPAC Nomenclature

Alkynes are named in much the same way as alkenes. One names unbranched alkynes, for example, by replacing the -*ane* of the name of the corresponding alkane with the ending -*yne*. The chain is numbered in order to give the carbon atoms of the triple bond the lowest possible numbers. The lower number of the two carbons of the triple bond is used to designate the location of the triple bond. The IUPAC names of several unbranched alkynes are shown below.

$HC\equiv CH$	Ethyne or acetylene*
$CH_3C\equiv CH$	Propyne
$CH_3CH_2C\equiv CH$	1-Butyne
$CH_3C\equiv CCH_3$	2-Butyne
$CH_3CH_2CH_2C\equiv CH$	1-Pentyne
$CH_3CH_2C\equiv CCH_3$	2-Pentyne

The location of substituent groups of branched alkynes and substituted alkynes are also indicated with numbers.

*The name acetylene is retained by the IUPAC system for the compound $HC\equiv CH$. The name acetylene is used much more frequently than the name ethyne.

$$Cl—\overset{3}{C}H_2\overset{2}{C}{\equiv}\overset{1}{C}H$$

3-Chloropropyne

$$\overset{4}{C}H_3\overset{3}{C}{\equiv}\overset{2}{C}\overset{1}{C}H_2Cl$$

1-Chloro-2-butyne

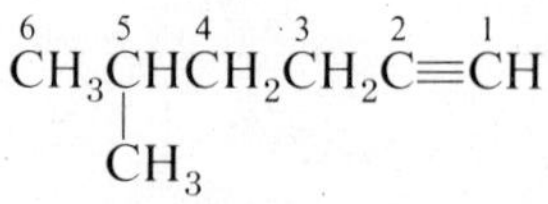

$$\overset{6}{C}H_3\overset{5}{C}H\overset{4}{C}H_2\overset{3}{C}H_2\overset{2}{C}{\equiv}\overset{1}{C}H$$
$$\underset{CH_3}{|}$$

5-Methyl-1-hexyne

$$\overset{CH_3}{\underset{|}{}}$$
$$CH_3\overset{|}{C}CH_2C{\equiv}CH$$
$$\underset{CH_3}{|}$$

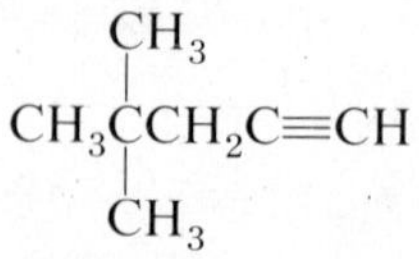

4,4-Dimethyl-1-pentyne

Problem 9.1

Give the IUPAC names of all of the alkyne isomers of (a) C_4H_6, (b) C_5H_8, (c) C_6H_{10}.

Common Names

In an older system of nomenclature, alkynes are named as though they were substituted acetylenes. Propyne, for example, a monosubstituted acetylene, is sometimes called methylacetylene. The disubstituted acetylene, 2-butyne, is called dimethylacetylene. The following examples illustrate this older system of nomenclature.

$$CH_3C{\equiv}CH$$

Propyne
(methylacetylene)

$$CH_3CH_2C{\equiv}CH$$

1-Butyne
(ethylacetylene)

$$CH_3C{\equiv}CCH_3$$

2-Butyne
(dimethylacetylene)

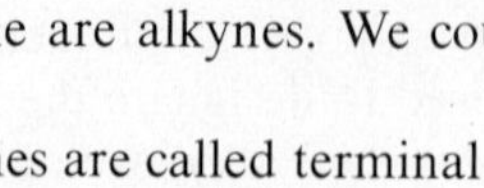

$$\overset{CH_3}{\underset{|}{}}$$
$$CH_3C{\equiv}C\overset{|}{C}HCH_3$$

4-Methyl-2-pentyne
(methylisopropylacetylene)

The name acetylene is not only used for the compound $HC{\equiv}CH$, it is also used as a general name for all of the alkynes. We could say, for example, that propyne, 1-butyne, and 2-butyne are alkynes. We could also say that they are all acetylenes.

Monosubstituted acetylenes are called terminal acetylenes, and the hydrogen attached to the carbon of the triple bond is called the acetylenic hydrogen.

$$R—C{\equiv}C—H$$ ⟵ acetylenic hydrogen

A terminal
acetylene

The anion obtained when the acetylenic hydrogen is removed is known as *an alkynide ion* or an acetylide ion.

$$R—C{\equiv}C{:}^-$$

An alkynide ion
(an acetylide ion)

9.3 *sp*-HYBRIDIZED CARBON: THE STRUCTURE OF ACETYLENE

Acetylene is a linear molecule. The carbon-carbon triple bond (1.20 Å) is considerably shorter than the carbon-carbon double bond of ethene (1.34 Å). The carbon-hydrogen bonds of acetylene (1.08 Å) are also shorter than those of ethene (1.09 Å).

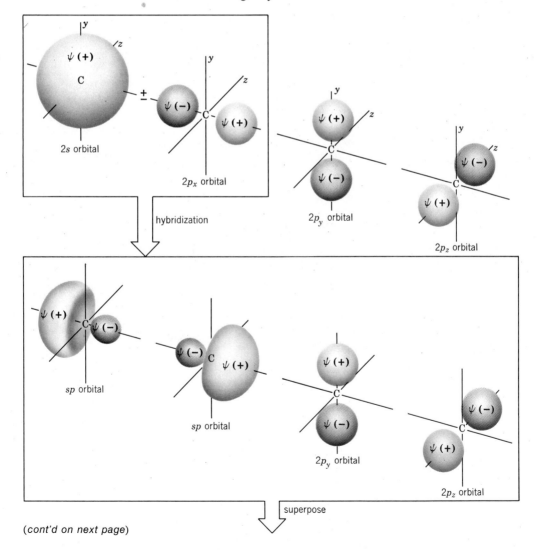

We can account for the structure of acetylene on the basis of orbital hybridization in much the same way that we did for ethane and ethene. In our model for ethane (p. 85) the carbon orbitals are *sp*³ hybridized, and in our model for ethene (p. 160) they are *sp*² hybridized. In our model of acetylene the carbon atoms are *sp hybridized*.

The mathematical process for obtaining the *sp*-hybrid orbitals of acetylene can be visualized in the following way.

(cont'd on next page)

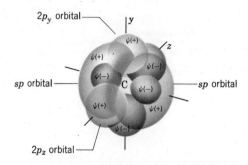

The 2s orbital and one 2p orbital of carbon are hybridized to form two sp orbitals. (In our illustration we have used the $2p_x$ orbital for this purpose but we could have used either of the other two.) The remaining two 2p orbitals (in this illustration the $2p_z$ and $2p_y$ orbitals) are not hybridized. When the centers of the orbitals are superposed, the sp hybrid orbitals are found to have their large positive lobes oriented at an angle of 180° with respect to each other. The 2p orbitals that were not hybridized are perpendicular to the axis that passes through the center of the two sp orbitals.

If we include the valence electrons we can see how two sp hybridized carbons and two hydrogen atoms might be used to construct a model for acetylene.

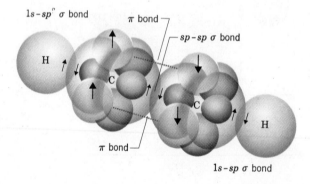

We see that the triple bond consists of one σ bond and two π bonds. If we replace the σ bonds of the illustration above with lines, it is easier to see how the p orbitals overlap.

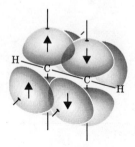

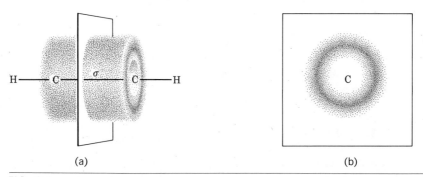

(a)   (b)

FIG. 9.1

(a) *The shape of the π-electron cloud in acetylene.* (b) *A cross section. The electrons of the carbon-carbon σ bond are located in the "hollow" part of the cylinder where the π-electron density is low. The σ and π bonds, however, share some space in common.*

The *p* orbitals overlap both above and below the carbon atoms and in front of and behind them as well. Because the *p* orbitals overlap in this way, the π-electron density of the triple bond is actually cylindrical in cross section (Fig. 9.1).

In addition to accounting for the linear structure of acetylene, the model based on *sp*-hybridized carbons also helps us to understand why the carbon-carbon and carbon-hydrogen bonds of acetylene are shorter than those of ethane. Orbitals that are *sp*-hybridized contain more *s* character than orbitals that are *sp²* hybridized and, consequently, *sp* orbitals themselves are shorter than *sp²* orbitals. This effect, along with the greater strength of the acetylene triple bond (198 kcal/mole versus 145 kcal/mole for ethene), accounts for the fact that carbon-carbon triple bond is shorter. The greater *s* character of *sp* orbitals accounts for the shorter carbon-hydrogen bonds of acetylene as well.

sp–s C—H bond   *sp²–s* C—H bond

The C—H bonds of acetylene are *more difficult* to break *homolytically* than are the C—H bonds of ethene or ethane

$$\equiv\text{C:H} \xrightarrow[\text{cleavage}]{\text{homolytic}} \equiv\text{C}\cdot \ + \text{H}\cdot$$

but they are *easier* to break in the following *heterolytic* way.

$$\equiv\text{C:H} \xrightarrow[\text{cleavage}]{\text{heterolytic}} \equiv\text{C:}^- \ + \text{H}^+$$

As a result of this ease of heterolytic cleavage, the hydrogens of acetylene are considerably more acidic than those of ethene or ethane.

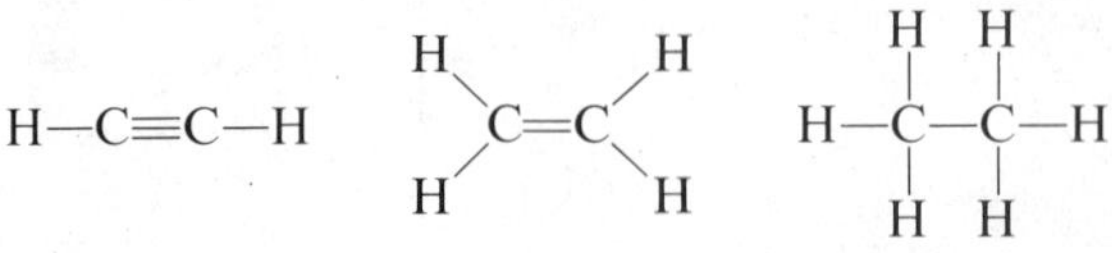

$$H-C\equiv C-H \qquad \underset{H}{\overset{H}{\diagdown}}C=C\underset{H}{\overset{H}{\diagup}} \qquad H-\underset{\underset{H}{|}}{\overset{\overset{H}{|}}{C}}-\underset{\underset{H}{|}}{\overset{\overset{H}{|}}{C}}-H$$

$$K_a \simeq 10^{-26} \qquad\qquad K_a \simeq 10^{-36} \qquad\qquad K_a \simeq 10^{-42}$$

We can account for this order of acidities on the basis of the hybridization state of carbon in each compound. An sp-hybridized carbon of acetylene acts as though it were more electronegative than an sp^2-hybridized carbon of ethene; an sp^2-hybridized carbon of ethene acts as though it were more electronegative than an sp^3-hybridized carbon of ethane. In each instance the carbon with the greater amount of s character in its hybrid orbitals exerts a greater pull on the electrons of the C—H bond; this permits an easier removal of the hydrogen *as a proton*.

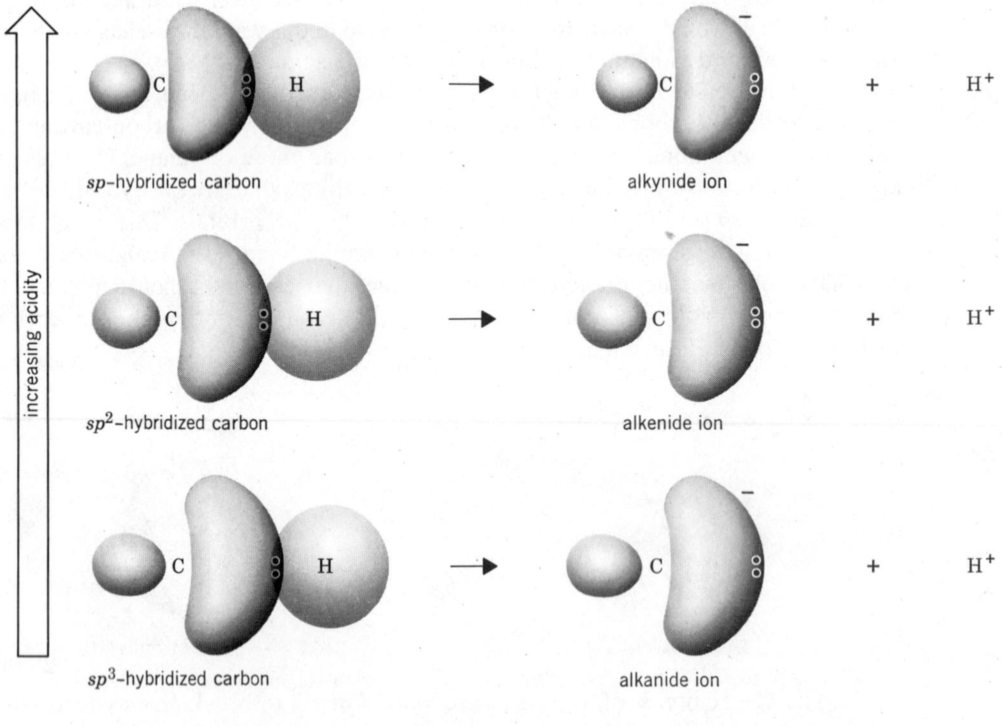

sp-hybridized carbon   alkynide ion

sp^2-hybridized carbon   alkenide ion

sp^3-hybridized carbon   alkanide ion

If we include other hydrogen compounds of the first row of the periodic table we can write the following orders of relative acidities and basicities.

Relative acidity: $H-\overset{..}{\underset{..}{O}}H > H-C\equiv CR > H-\overset{..}{N}H_2 > H-CH=CH_2 > H-CH_2CH_3$

Relative basicity: $^-\!:\overset{..}{\underset{..}{O}}H < {}^-\!:C\equiv CR < {}^-\!:\overset{..}{N}H_2 < {}^-\!:CH=CH_2 < {}^-\!:CH_2CH_3$

Problem 9.2

Predict the products of the following acid-base reactions. If the equilibrium would not result in the formation of appreciable amounts of products you should so indicate. In each case label the stronger acid, the stronger base, the weaker acid, and the weaker base.

(a) $H—C\equiv C—H + NaNH_2$
(b) $CH_2=CH_2 + NaNH_2$
(c) $CH_3CH_3 + NaNH_2$
(d) $H—C\equiv C:^-Na^+ + CH_3CH_2OH$
(e) $H—C\equiv C:^-Na^+ + H_2O$

9.4 ACETYLENE

Acetylene, an important industrial chemical, is used as a starting material for the preparation of many useful compounds. Acetylene can be prepared from relatively inexpensive starting materials. In one process coke and lime are heated in an electric furnace to produce calcium carbide. The calcium carbide, thus produced, is allowed to react with water and in this way acetylene is generated.

$$3C + CaO \xrightarrow{2500°} CaC_2 + CO$$

$\quad$ Coke $\quad$ Lime $\qquad$ Calcium
$\qquad\qquad\qquad\qquad$ carbide

$$CaC_2 + 2H_2O \xrightarrow[\text{temperature}]{\text{room}} H—C\equiv C—H + Ca(OH)_2$$

In another process methane (from natural gas) is heated briefly to a very high temperature; acetylene is one of the products.

$$2CH_4 \xrightarrow[0.1 - 0.01 \text{ sec}]{1500°} H—C\equiv C—H + 3H_2$$

Acetylene is a colorless gas at room temperature. It is thermodynamically unstable; and when shocked, acetylene explodes to form elemental carbon and hydrogen.

$$C_2H_2 \longrightarrow 2C + H_2 \qquad \Delta H = -56 \text{ kcal/mole}$$

The explosive decomposition of acetylene is more likely if the acetylene is at a high pressure but even liquid acetylene (bp $-84°$) has to be handled with extreme care. Special techniques have been developed, however, that allow for the relatively safe manipulation of acetylene even in large quantities and at pressures up to 30 atm. When large diameter pipes are used for transporting acetylene, these pipes are filled with smaller pipes to minimize the free volume. Acetylene is also often diluted with an unreactive gas such as water vapor or nitrogen.

$\qquad$ Acetylene is used extensively in welding because it burns with an extremely hot flame (up to $2800°$). When we examine the heats of combustion of ethane, ethene, and acetylene below, we might expect that any one of these gases could be used for welding. We might even expect that ethane and ethene would produce hotter flames than acetylene.

$$CH_3CH_3 + 3\tfrac{1}{2}O_2 \longrightarrow 2CO_2 + 3H_2O \qquad \Delta H = -373 \text{ kcal/mole}$$

$$CH_2{=}CH_2 + 3O_2 \longrightarrow 2CO_2 + 2H_2O \qquad \Delta H = -337 \text{ kcal/mole}$$

$$CH{\equiv}CH + 2\tfrac{1}{2}O_2 \longrightarrow 2CO_2 + H_2O \qquad \Delta H = -317 \text{ kcal/mole}$$

Ethane and ethene produce more heat on a molar basis but they also require more oxygen, and heat is consumed in bringing the oxygen up to the flame temperature. More water is produced in the combustion of ethane and ethene as well, so the number of product molecules, among which the heat must be distributed, is also greater. Because of these factors, of the three hydrocarbons, acetylene produces the highest flame temperature.

Acetylene for use in welding is dissolved in acetone and placed under a pressure of 300 atm in cylinders filled with a porous but inert substance.

9.5 REACTIONS OF ALKYNES

Many of the reactions of alkynes are characterized by addition reactions of the triple bond. Depending on the conditions used, the additions can occur once or twice.

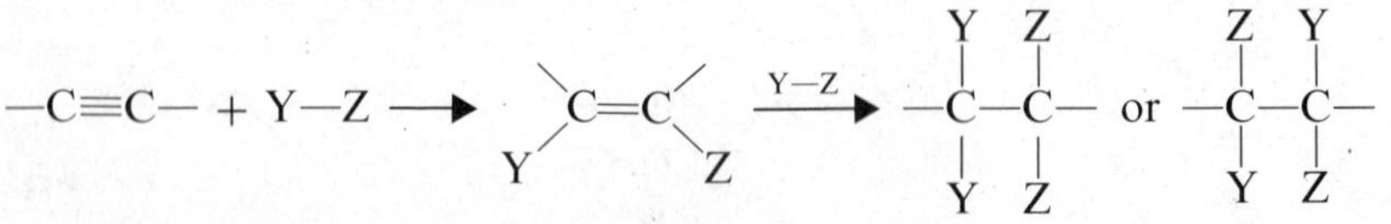

We will see several examples of reactions of this general type in Sections 9.6–9.9.

Other reactions of alkynes result from the acidity of acetylenic hydrogen. In the presence of strong bases the following reaction takes place.

$$-C{\equiv}C-H + \ :B^- \ \longrightarrow \ -C{\equiv}C{:}^- + H{:}B$$

$$\text{strong base} \qquad\qquad \text{alkynide ion}$$

When we study these reactions in Section 9.10 we will find that alkynide ions are important reagents for the synthesis of other alkynes.

9.6 ADDITION OF HYDROGEN

Depending on the reaction conditions and catalyst employed, one or two moles of hydrogen can be added to a carbon-carbon triple bond. When a platinum catalyst is used the alkyne generally reacts with two moles of hydrogen and produces an alkane.

$$CH_3C{\equiv}CCH_3 \xrightarrow[\text{H}_2]{\text{Pt}} \left[CH_3CH{=}CHCH_3\right] \xrightarrow[\text{H}_2]{\text{Pt}} CH_3CH_2CH_2CH_3$$

However, reduction of an alkyne to the alkene stage can be accomplished through the use of special catalysts or reagents. Moreover, these special methods allow the preparation of either *cis*- or *trans*-alkenes from disubstituted acetylenes.

A catalyst that permits hydrogenation of an alkyne to an alkene is the nickel-boride catalyst called P-2 catalyst. This catalyst can be prepared by the reduction of nickel acetate with sodium borohydride.

$$\underset{\text{(P-2)}}{Ni(O\overset{\overset{\displaystyle O}{\|}}{C}CH_3)_2 \xrightarrow[C_2H_5OH]{NaBH_4} Ni_2B}$$

When alkynes with an internal triple bond are hydrogenated in the presence of P-2 catalyst *syn addition of hydrogen occurs* and the alkene that is formed has the *cis* configuration. The hydrogenation of 3-hexyne (below) illustrates this method.

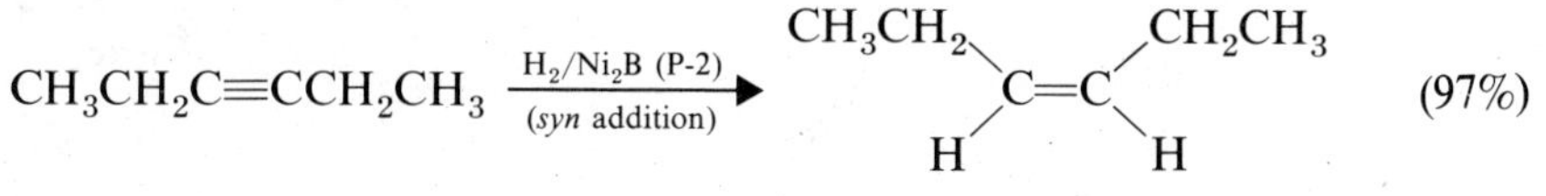

$$CH_3CH_2C\equiv CCH_2CH_3 \xrightarrow[(syn\ addition)]{H_2/Ni_2B\ (P-2)}$$

(97%)

3-Hexyne   cis-3-Hexene

Other specially conditioned catalysts can be used to prepare *cis*-alkenes from disubstituted acetylenes. Metallic palladium deposited on calcium carbonate can be used in this way after it has been conditioned with lead acetate and quinoline

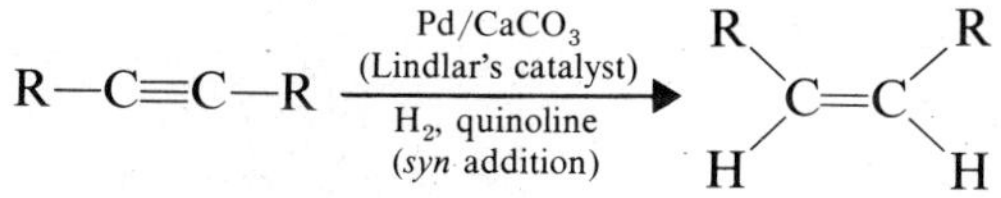

$$R-C\equiv C-R \xrightarrow[\substack{H_2,\ quinoline \\ (syn\ addition)}]{\substack{Pd/CaCO_3 \\ (Lindlar's\ catalyst)}}$$

(Sec. 21.1). This special catalyst is known as Lindlar's catalyst. Palladium deposited on barium sulfate gives similar results.

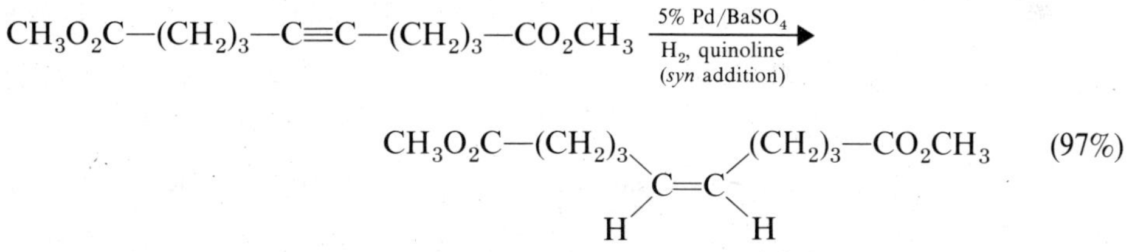

$$CH_3O_2C-(CH_2)_3-C\equiv C-(CH_2)_3-CO_2CH_3 \xrightarrow[\substack{H_2,\ quinoline \\ (syn\ addition)}]{5\%\ Pd/BaSO_4}$$

(97%)

Diborane can also be used to prepare *cis*-alkenes from appropriate alkynes. Diborane adds once to the triple bond of 3-hexyne to form the vinylic borane shown below.

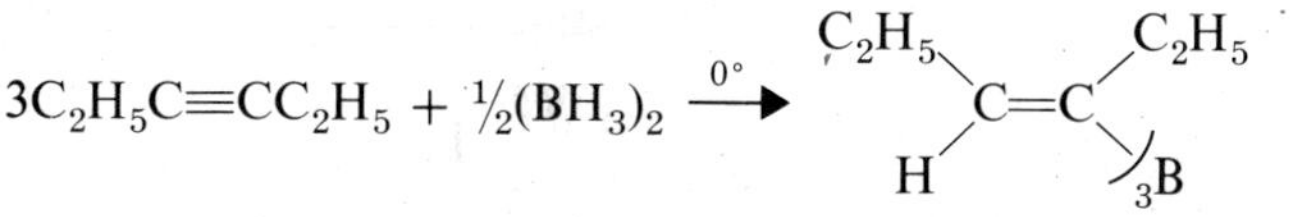

$$3C_2H_5C\equiv CC_2H_5 + \tfrac{1}{2}(BH_3)_2 \xrightarrow{0°}$$

A vinylic borane

The vinylic borane can be converted to *cis*-2-hexene by treating the solution with acetic acid (cf. problem 6.8, p. 208).

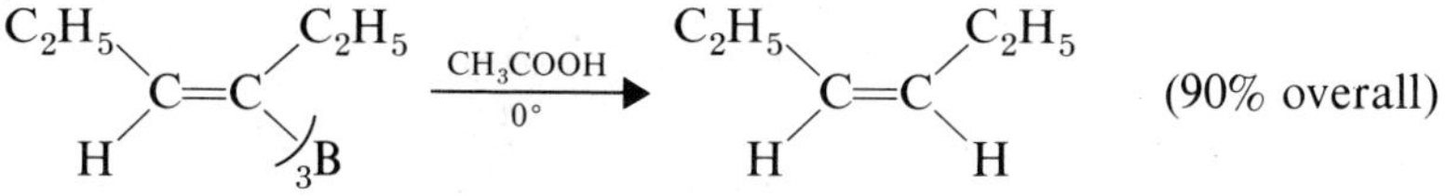

$$\xrightarrow[0°]{CH_3COOH}$$

(90% overall)

cis-3-Hexene

Unlike the methods based on the use of nickel boride and the specially conditioned palladium catalysts, the addition of diborane cannot be used to reduce terminal alkynes to alkenes because diborane adds twice to the triple bond of terminal alkynes (p. 319).

Trans-alkenes are obtained when alkynes are reduced with lithium metal in ammonia or ethylamine at low temperatures.

$$CH_3(CH_2)_2—C≡C—(CH_2)_2CH_3 \xrightarrow[\substack{C_2H_5NH_2 \\ -78°}]{Li} \begin{array}{c} CH_3(CH_2)_2 \\ \diagdown \\ \diagup \\ H \end{array} C=C \begin{array}{c} H \\ \diagup \\ \diagdown \\ (CH_2)_2CH_3 \end{array}$$ (52%)

4-Octyne   *trans*-4-Octene

9.7 ADDITION OF HALOGENS

Alkynes show the same kind of reactions toward chlorine and bromine that alkenes do: *they react by addition*. However, with alkynes the addition may occur twice. The

$$—C≡C— \xrightarrow[CCl_4]{Br_2} \begin{array}{c} \diagdown \\ Br \end{array} C=C \begin{array}{c} Br \\ \diagup \end{array} \xrightarrow[CCl_4]{Br_2} \begin{array}{c} Br \ Br \\ | \ \ | \\ —C—C— \\ | \ \ | \\ Br \ Br \end{array}$$

Dibromoalkene   Tetrabromoalkane

$$—C≡C— \xrightarrow[CCl_4]{Cl_2} \begin{array}{c} \diagdown \\ Cl \end{array} C=C \begin{array}{c} Cl \\ \diagup \end{array} \xrightarrow[CCl_4]{Cl_2} \begin{array}{c} Cl \ Cl \\ | \ \ | \\ —C—C— \\ | \ \ | \\ Cl \ Cl \end{array}$$

Dichloroalkene   Tetrachloroalkane

dihaloalkenes that result from the first addition are usually less reactive toward further addition than the alkyne itself. Thus, it is usually possible to prepare a dihaloalkene by simply adding one mole of the halogen.

$$CH_3CH_2CH_2CH_2C≡CCH_2OH \xrightarrow[\substack{CCl_4 \\ 0°}]{Br_2 \ (one \ mole)} CH_3CH_2CH_2CH_2CBr=CBrCH_2OH$$
(80%)

Most additions of chlorine and bromine to alkynes are anti additions and yield *trans*-dihaloalkenes. Addition of bromine to acetylene dicarboxylic acid, for example, gives the *trans* isomer in 70% yield.

$$HOOC—C≡C—COOH \xrightarrow{Br_2} \begin{array}{c} HOOC \\ \diagdown \\ \diagup \\ Br \end{array} C=C \begin{array}{c} Br \\ \diagup \\ \diagdown \\ COOH \end{array}$$

Acetylene dicarboxylic   (70%)
acid

9.8 ADDITION OF HYDROGEN HALIDES

Alkynes react with hydrogen chloride and hydrogen bromide to form haloalkenes or geminal dihalides depending, once again, on whether one or two moles of the hydrogen halide are used. Both additions follow Markovnikov's rule: the hydrogen

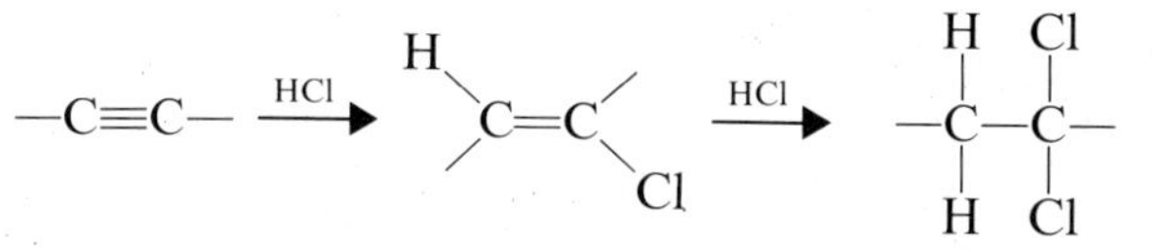

Chloroalkene   *gem*-Dichloride

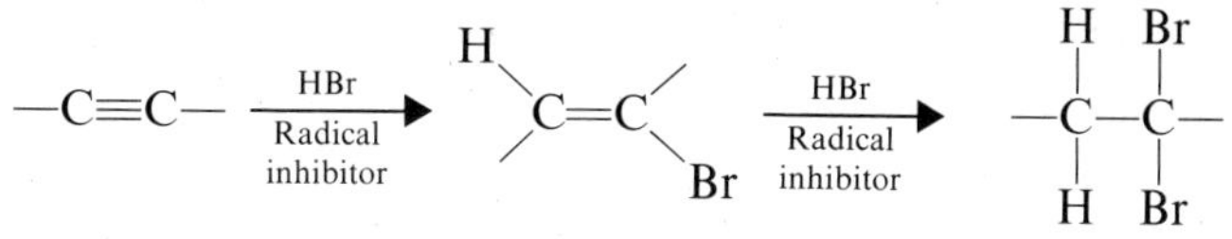

Bromoalkene   *gem*-Dibromide

of the hydrogen halide becomes attached to the carbon that has the greater number of hydrogens. Propyne, for example, reacts with one mole of hydrogen chloride to yield 2-chloropropene and with two moles to yield 2,2-dichloropropane.

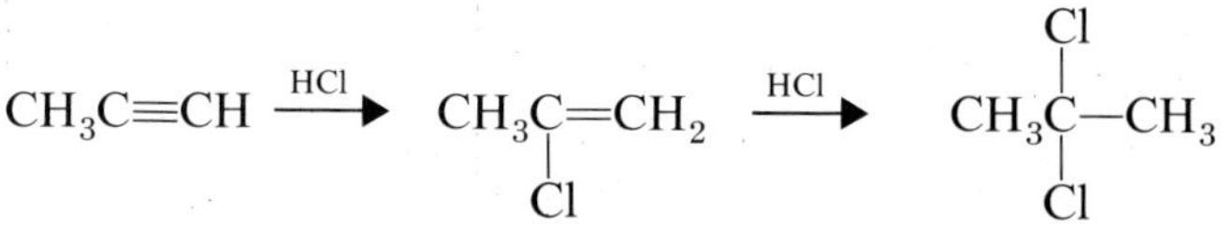

2-Chloropropene   2,2-Dichloropropane

The initial addition of a hydrogen halide to an alkyne usually occurs in an anti manner. This is especially likely if an ionic halide corresponding to the halogen of the hydrogen halide is present in the reaction mixture.

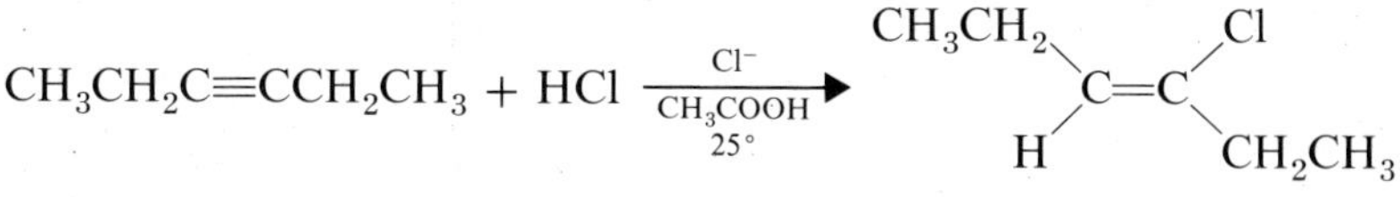

(97%)

Anti-Markovnikov addition of hydrogen bromide to alkynes occurs when peroxides are present in the reaction mixture.

$$CH_3CH_2CH_2CH_2C{\equiv}CH \xrightarrow[\text{peroxides}]{HBr} CH_3CH_2CH_2CH_2CH{=}CHBr$$
(74%)

9.9 ADDITION OF WATER

Alkynes add water readily when the reaction is catalyzed by strong acids and mercuric (Hg^{++}) ions. Aqueous solutions of sulfuric acid and mercuric sulfate are often used for this purpose. The vinyl alcohol that is initially produced is usually unstable, and it rearranges rapidly to an aldehyde or ketone. The rearrangement

$$-C\equiv C- + H-OH \xrightarrow[\text{H}_2\text{SO}_4]{\text{HgSO}_4} \left[\begin{array}{c} -CH=C- \\ | \\ OH \end{array} \right] \longrightarrow \begin{array}{c} H \\ | \\ -C-C- \\ | \parallel \\ H\ O \end{array}$$

A vinyl   Aldehyde
alcohol   or
      ketone

involves a shift of a proton from the hydroxyl group to the adjacent carbon. This kind of rearrangement, known as a *tautomerization*, is acid catalyzed and occurs in the following way.

$$\begin{array}{c} -C=C- \\ \\ H-O: \quad H^+ \end{array} \rightleftharpoons \begin{array}{c} -C-C- + H^+ \\ \parallel \\ :O\ H \end{array}$$

Enol form   Keto form

Vinyl alcohols are often called *enols* (-*ene*, the ending for alkenes plus -*ol*, the ending for alcohols). The product of the rearrangement is often a ketone and these rearrangements are known as *keto-enol tautomerizations*. We describe this phenomenon in greater detail in Chapter 18.

Mercuric ion probably forms a complex with the π electrons of the triple bond and thus activates the triple bond toward the addition of water.

$$\begin{array}{c} -C\equiv C- \\ \downarrow \\ Hg^{+2} \end{array}$$

When acetylene itself undergoes addition of water the product is an aldehyde.

$$H-C\equiv C-H + H_2O \xrightarrow[\text{H}_2\text{SO}_4]{\text{HgSO}_4} \left[\begin{array}{c} H \qquad H \\ \diagdown \quad \diagup \\ C=C \\ \diagup \qquad \diagdown \\ H \qquad OH \end{array} \right] \longrightarrow \begin{array}{c} H \\ | \\ H-C-C-H \\ | \parallel \\ H\ O \end{array}$$

Acetaldehyde

This method has been important in the commercial production of acetaldehyde.

The addition of water to alkynes also follows Markovnikov's rule. When higher alkynes are hydrated, ketones, rather than aldehydes, are the products.

$$R-C\equiv C-H \xrightarrow[\text{H}_2\text{O, H}^+]{\text{Hg}^{++}} \left[\begin{array}{c} RC=CH_2 \\ | \\ OH \end{array} \right] \longrightarrow \begin{array}{c} R-C-CH_3 \\ \parallel \\ O \end{array}$$

A ketone

Several examples of this ketone synthesis are listed below.

$$CH_3C\equiv CH \xrightarrow[\text{H}_2\text{O, H}^+]{\text{Hg}^{++}} \left[\begin{array}{c} CH_3-C=CH_2 \\ | \\ OH \end{array} \right] \longrightarrow \begin{array}{c} CH_3-C-CH_3 \\ \parallel \\ O \end{array}$$

Acetone

$$CH_3CH_2CH_2CH_2C{\equiv}CH \xrightarrow[\substack{H_2SO_4 \\ H_2O}]{HgSO_4} CH_3CH_2CH_2CH_2\underset{\underset{O}{\|}}{C}CH_3$$

(80%)

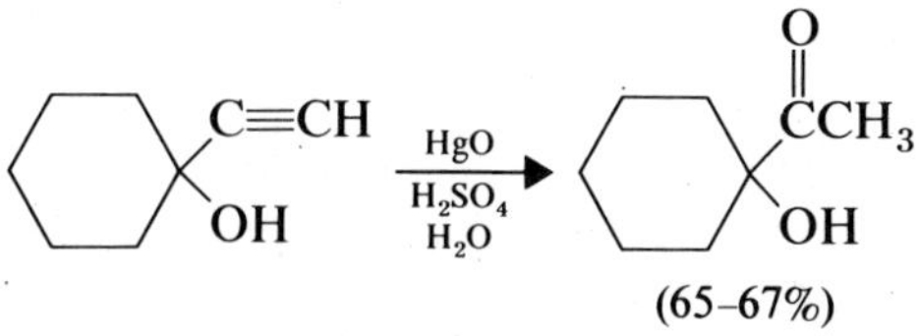

(65–67%)

Boron hydrides can also be used to synthesize ketones from alkynes. The overall method is very similar to the hydroboration-oxidation of alkenes that we saw in Chapter 6. Diborane adds only once to alkynes that have an internal triple bond (p. 315).

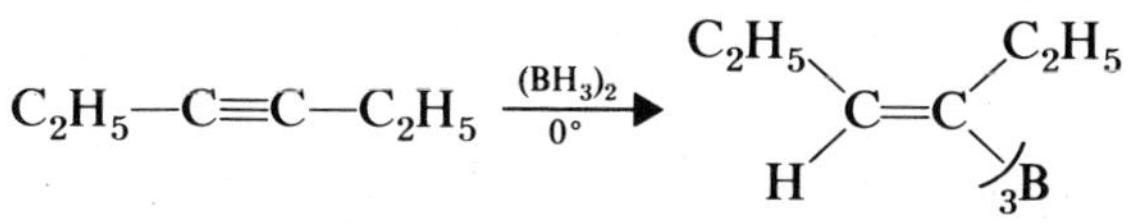

A vinylic borane

The vinylic borane formed in the reaction above can be treated with hydrogen peroxide and aqueous sodium hydroxide to produce an enol. The enol rearranges spontaneously to a ketone.

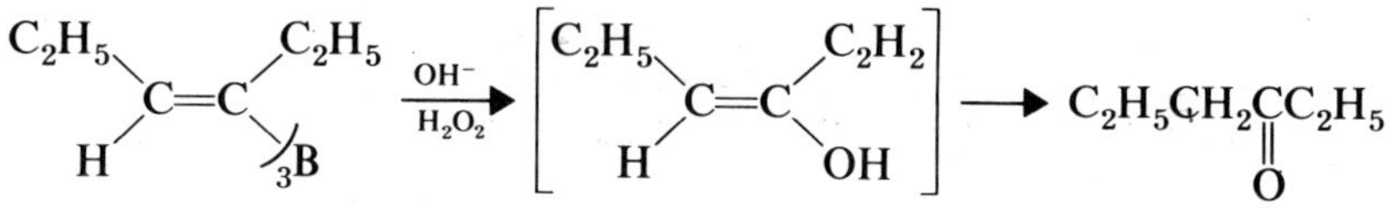

(62% overall)

Terminal alkynes generally react with two moles of diborane to produce *gem*-dibora derivatives:

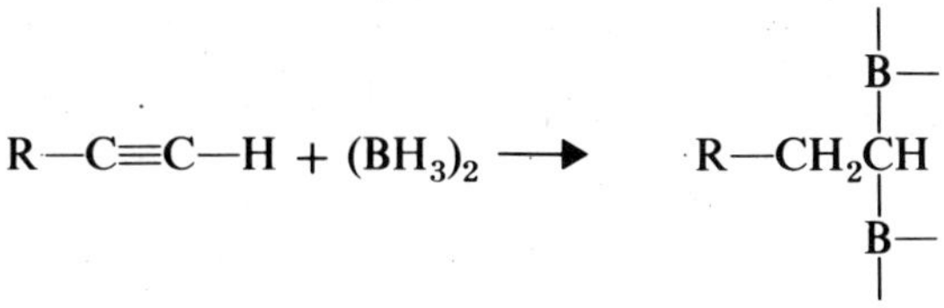

gem-Dibora compound

However, terminal alkynes can be converted to vinylic boranes through the use of

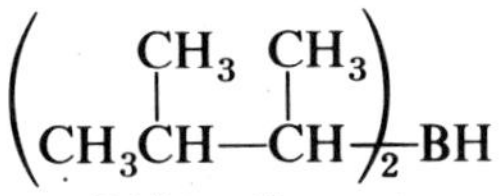

Disiamylborane
(Sia$_2$BH)

disiamylborane. The bulky alkyl groups of disiamylborane hinder* its reaction with terminal acetylenes and the addition occurs only once.

$$n-C_4H_9C\equiv CH + Sia_2BH \longrightarrow n-C_4H_9CH=CH-BSia_2 \quad (100\%)$$

Oxidation of the product with alkaline hydrogen peroxide produces an aldehyde.

$$n-C_4H_9CH=CH-Sia_2B \xrightarrow[OH^-/H_2O]{H_2O_2} n-C_4H_9CH_2\overset{\overset{O}{\|}}{C}H$$

Hexanal

Problem 9.3

What other organic product would you expect to obtain from the oxidation of $n-C_4H_9CH=CH-BSia_2$ with hydrogen peroxide in base?

This procedure provides an anti-Markovnikov-addition of H— and —OH to an alkyne, and thus it is a convenient synthesis of aldehydes.

Problem 9.4

Show how each of the following transformations could be carried out. [Hint: In parts (c) and (d) assume that you have CH_3COOD available.]

(a) $CH_3CH_2C\equiv CH \longrightarrow CH_3CH_2\overset{\overset{O}{\|}}{C}CH_3$

(b)

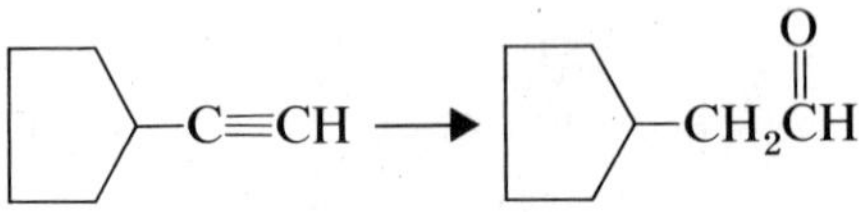

(c) 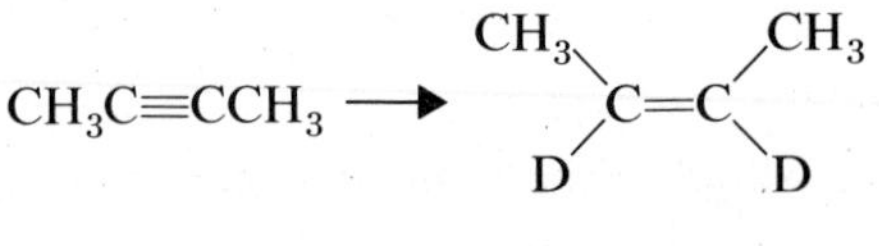 $CH_3C\equiv CCH_3 \longrightarrow$

(d) 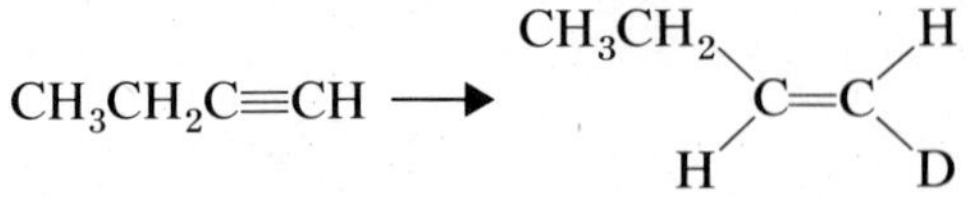 $CH_3CH_2C\equiv CH \longrightarrow$

9.10 REPLACEMENT OF THE ACETYLENIC HYDROGEN

Alkali Metal Acetylides

Alkali metals such as sodium and potassium displace the hydrogen of acetylene.

---

* This is an example of what is often called *steric hindrance*. Steric hindrance arises from the effect of the spatial bulk of groups on the transition state of a reaction. We will see another example of steric hindrance in Section 9.10.

$$H-C\equiv C-H + Na \xrightarrow[\text{NH}_3]{\text{liq.}} H-C\equiv C:^-Na^+ + \tfrac{1}{2}H_2$$

Sodium
acetylide

This reaction is often carried out by bubbling acetylene gas into liquid ammonia that contains dissolved sodium.* The reaction that occurs when the acetylene is added is an oxidation-reduction reaction similar to the reaction of alcohols with sodium (p. 180).

$$R-\ddot{O}H + Na \longrightarrow R-\ddot{O}:^-Na^+ + \tfrac{1}{2}H_2$$

Sodium
alkoxide

Sodium acetylide and other sodium alkynides can also be prepared by treating terminal alkynes with sodium amide in liquid ammonia.

$$R-C\equiv C-H + Na^+ {}^-:\ddot{N}H_2 \longrightarrow R-C\equiv C:^-Na^+ + \ddot{N}H_3$$

$$CH_3C\equiv C-H + Na^+ {}^-:\ddot{N}H_2 \longrightarrow CH_3C\equiv C:^-Na^+ + \ddot{N}H_3$$

$$CH_3CH_2C\equiv CH + Na^+ {}^-:\ddot{N}H_2 \longrightarrow CH_3CH_2C\equiv C:^-Na^+ + \ddot{N}H_3$$

These are acid-base reactions. The amide ion, by virtue of its being the anion of the very weak acid, ammonia ($K_a \simeq 10^{-36}$), is able to remove the acetylenic protons of terminal alkynes ($K_a \simeq 10^{-26}$). These reactions, for all practical purposes, go to completion.

Sodium alkynides are useful intermediates for the synthesis of other alkynes. Such a synthesis can be accomplished by treating the sodium alkynide with a primary alkyl halide.

$$(H)R-C\equiv C:^-Na^+ + \quad R'-Br \quad \longrightarrow (H)R-C\equiv C-R' + NaBr$$

Sodium   Primary   Mono- or di-
alkynide   alkyl halide   substituted
acetylene

The following examples illustrate this synthesis of alkynes.

$$HC\equiv C:^-Na^+ + CH_3-Br \xrightarrow[\text{5 hr}]{\text{liq. NH}_3} H-C\equiv C-CH_3$$

Propyne
(84%)

$$H-C\equiv C:^-Na^+ + CH_3CH_2CH_2Br \xrightarrow[\text{5 hr}]{\text{liq. NH}_3} HC\equiv CCH_2CH_2CH_3$$

1-Pentyne
(85%)

$$H-C\equiv C:^-Na^+ + CH_3\overset{\overset{\displaystyle CH_3}{|}}{C}HCH_2CH_2Br \xrightarrow[\text{6 hr}]{\text{liq. NH}_3} H-C\equiv CCH_2CH_2\overset{\overset{\displaystyle CH_3}{|}}{C}HCH_3$$

5-Methyl-1-hexyne
(68%)

* Sodium dissolves in liquid ammonia to produce an intensely blue-colored solution that is thought to contain sodium ions and "solvated electrons."

$$CH_3CH_2C{\equiv}C{:}^-Na^+ + CH_3CH_2Br \xrightarrow[\text{6 hr}]{\text{liq. NH}_3} CH_3CH_2C{\equiv}CCH_2CH_3$$

3-Hexyne
(75%)

In all of these examples the sodium alkynides act as nucleophiles and carry out a displacement of the halogen from the primary alkyl halide. The result is a *nucleophilic substitution reaction.* The unshared electron pair of the alkynide ion attacks the

$$(H)R{-}C{\equiv}C{:}^- \rightsquigarrow \overset{R'}{\underset{\underset{H}{\overset{|}{H}}}{\overset{\delta+}{C}}}{-}\overset{\cdot\cdot}{\underset{\cdot\cdot}{Br}}{:} \xrightarrow[\text{substitution}]{\text{nucleophilic}} (H)R{-}C{\equiv}C{-}CH_2R' + {:}\overset{\cdot\cdot}{\underset{\cdot\cdot}{Br}}{:}^-$$

Na⁺   Na⁺

Sodium   1° Alkyl
alkynide   halide

carbon that bears the halogen and forms a bond to it. The halogen departs as a halide ion.

When secondary or tertiary halides are treated with sodium alkynides the alkynide ion acts as a base rather than as a nucleophile, and the major result is an *elimination reaction.* The products of the elimination reaction are an alkene and the alkyne from which the sodium alkynide was originally formed.

$$R{-}C{\equiv}C{:}^- \quad R'{-}CH{-}CH{-}\overset{\cdot\cdot}{\underset{\cdot\cdot}{Br}}{:} \xrightarrow[\text{reaction}]{\text{Elimination}} R{-}C{\equiv}C{-}H + R'CH{=}CHR'' + {:}\overset{\cdot\cdot}{\underset{\cdot\cdot}{Br}}{:}^-$$

Na⁺   Na⁺

2° Alkyl
halide

Elimination reactions nearly always compete with nucleophilic substitution reactions because all nucleophiles are also potential bases. Primary alkyl halides and methyl halides are more susceptible to nucleophilic substitution because attack at the carbon bearing the halogen is relatively unhindered.

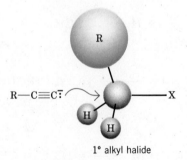

1° alkyl halide

Alkyl groups are large in comparison to hydrogens, and in secondary and tertiary alkyl halides the additional alkyl groups hinder attack at the carbon that bears the halogen. With secondary and tertiary halides the alkynide ion attacks a

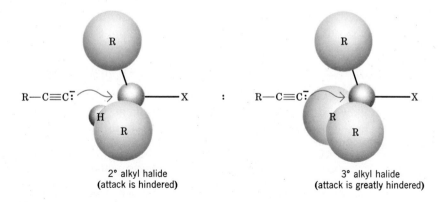

2° alkyl halide
(attack is hindered)

3° alkyl halide
(attack is greatly hindered)

β hydrogen instead, and this brings about an elimination reaction. We have more to say about elimination and substitution reactions in Chapter 17.

Problem 9.5

In addition to sodium amide and liquid ammonia, assume that you have the following four compounds available and want to carry out a synthesis of 2,2-dimethyl-3-

$$CH_3CH_2C{\equiv}CH \qquad CH_3-\underset{\underset{CH_3}{|}}{\overset{\overset{CH_3}{|}}{C}}-C{\equiv}CH \qquad CH_3CH_2Br \qquad CH_3-\underset{\underset{CH_3}{|}}{\overset{\overset{CH_3}{|}}{C}}-Br$$

hexyne. Which synthetic route would you choose?

Other Metal Acetylides

Acetylene and terminal alkynes also form metal derivatives with silver and cuprous ions.

$$RC{\equiv}CH + Cu(NH_3)_2^+ + OH^- \xrightarrow{H_2O} R-C{\equiv}CCu + H_2O + 2NH_3$$

$$RC{\equiv}CH + Ag(NH_3)_2^+ + OH^- \xrightarrow{H_2O} RC{\equiv}CAg + H_2O + 2NH_3$$

Silver and copper alkynides differ from sodium alkynides in several ways. The metal-carbon bond in silver and copper alkynides is largely covalent; as a result, silver and copper alkynides are poor bases and poor nucleophiles. Silver and copper alkynides can be prepared in water, whereas sodium alkynides react vigorously with water.

$$R—C{\equiv}CNa + H_2O \longrightarrow R—C{\equiv}CH + NaOH$$

Silver and copper alkynides are also quite insoluble in water and precipitate when they are prepared. This affords a convenient test for terminal alkynes as well as a method for separating terminal alkynes from alkynes that have an internal triple bond.

$$R—C{\equiv}C—R + Ag(NH_3)_2^+OH^- \longrightarrow \text{no reaction}$$

$$R—C{\equiv}CH + Ag(NH_3)_2^+OH^- \longrightarrow RC{\equiv}CAg \downarrow$$

Once a separation has been carried out the terminal alkyne can be regenerated by treating it with a solution of sodium cyanide.

$$R—C{\equiv}CAg + 2CN^- + H_2O \longrightarrow R—C{\equiv}CH + Ag(CN)_2^- + OH^-$$

Silver and copper alkynides must be handled cautiously; when dry they are likely to explode.

Coupling Reactions

Terminal alkynes undergo an interesting coupling (or dimerization) reaction when they are treated with copper salts and amines. One method for dimerizing terminal alkynes uses cuprous chloride, ammonia, and oxygen. The general reaction can be illustrated as follows.

$$R—C{\equiv}C—H + CuCl \xrightarrow[O_2, \ CH_3OH]{NH_3} R—C{\equiv}C—C{\equiv}C—R$$
$$\text{A diyne}$$

Propyne has been dimerized in this way to 2,4-hexadiyne.

$$CH_3C{\equiv}CH + CuCl \xrightarrow[O_2, \ CH_3OH]{NH_3} CH_3—C{\equiv}C—C{\equiv}C—CH_3$$
$$(74\%)$$

The mechanism of the reaction is not understood completely, but it is thought that cuprous ions form a π complex with the alkyne and that, at some point in the coupling reaction, oxygen causes a $Cu^+ \longrightarrow Cu^{++}$ oxidation.

Professor F. Sondheimer (of University College, London) has made extensive use of a modification of this procedure—one that utilizes cupric ion in pyridine (p. 436) in the absence of oxygen. This method is particularly useful because diacetylenes are dimerized, trimerized, and so on, to cyclic compounds. This has proved to be

$$
\begin{array}{ccc}
& C{\equiv}C—C{\equiv}C & \\
(CH_2)_5 & & (CH_2)_5 \qquad (10\%) \\
& C{\equiv}C—C{\equiv}C & \\
\end{array}
$$
$$\text{Cyclic dimer}$$
$$+$$

$$HC{\equiv}C(CH_2)_5C{\equiv}CH \xrightarrow[\text{pyridine}]{\text{cupric acetate}}$$

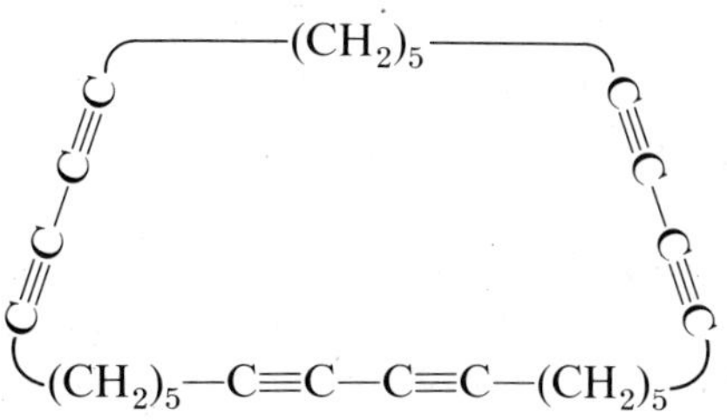

(13%)

Cyclic trimer
+
cyclic tetramer   (11%)
+
cyclic pentamer   (4%)

one of the most valuable methods for the synthesis of compounds whose molecules have large carbon rings.

9.11 SYNTHESIS OF ALKYNES

From Other Alkynes

Alkynes can be prepared from other terminal alkynes through the nucleophilic substitution reaction that we saw in Section 9.10.

$$R—C≡CH \xrightarrow[\text{liq. } NH_3]{NaNH_2} R—C≡C:^- Na^+ \xrightarrow[\substack{\text{Primary} \\ \text{halide}}]{R'—X} R—C≡C—R'$$

By Elimination Reactions

Alkynes can also be synthesized from alkenes. In this method an alkene is first treated with bromine to form a *vic*-dibromo compound, and the *vic*-dibromide

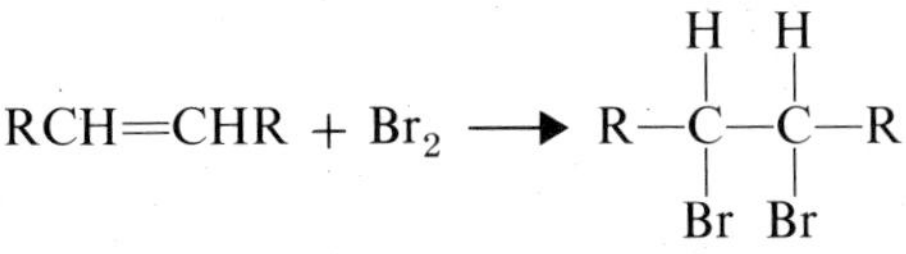

$$RCH{=}CHR + Br_2 \longrightarrow R{-}\underset{\underset{Br}{|}}{\overset{\overset{H}{|}}{C}}{-}\underset{\underset{Br}{|}}{\overset{\overset{H}{|}}{C}}{-}R$$

vic-Dibromide

is dehydrohalogenated through its reaction with a strong base. The dehydrohalogenation occurs in two steps. The first step yields a bromoalkene.

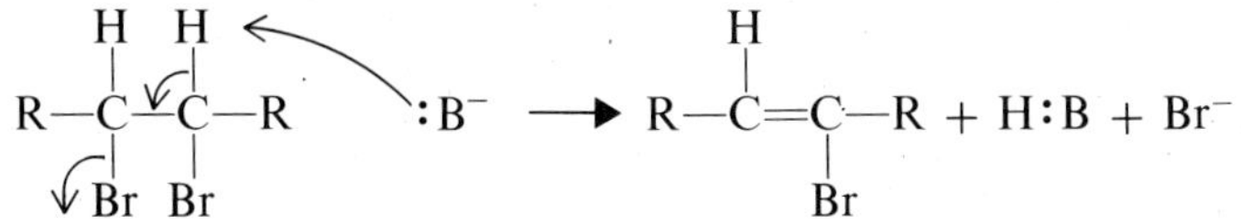

vic-Dibromide   Strong   Bromoalkene
                          base

The second dehydrohalogenation yields an alkyne.

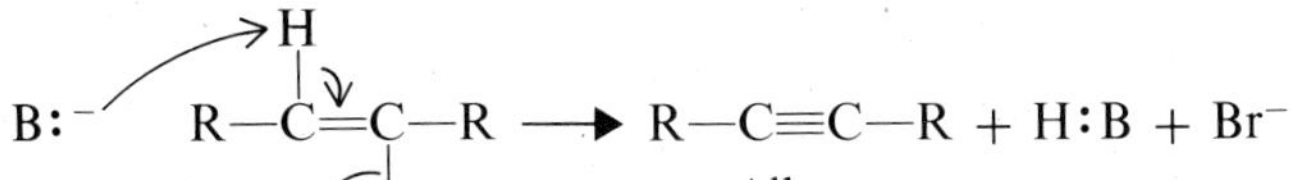

Depending on the reaction conditions employed, these two dehydrohalogenations may be carried out in separate reactions or they may be carried out consecutively in a single reaction mixture. Potassium hydroxide in ethanol usually brings about a single dehydrohalogenation unless the reaction is carried out at a very high temperature. The stronger base, sodium amide, is capable of effecting both dehydrohalogenations in a single reaction mixture. (Two moles of sodium amide per mole of the dihalide must be used.) Dehydrohalogenations with sodium amide are usually carried out in liquid ammonia or in an inert solvent such as mineral oil.

The following examples illustrate both methods.

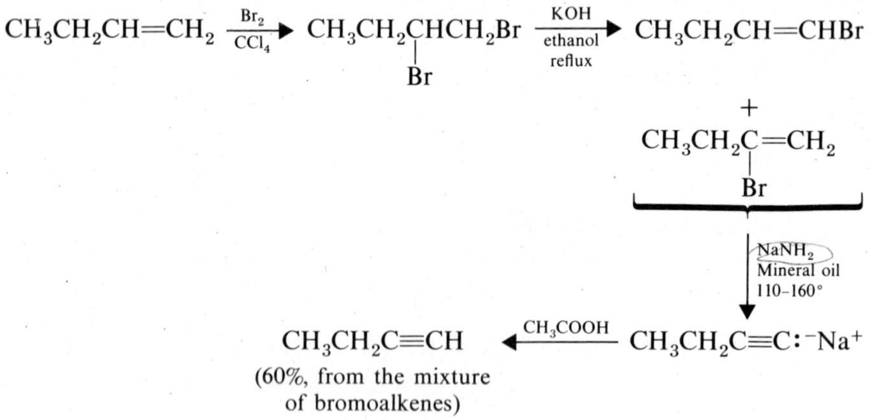

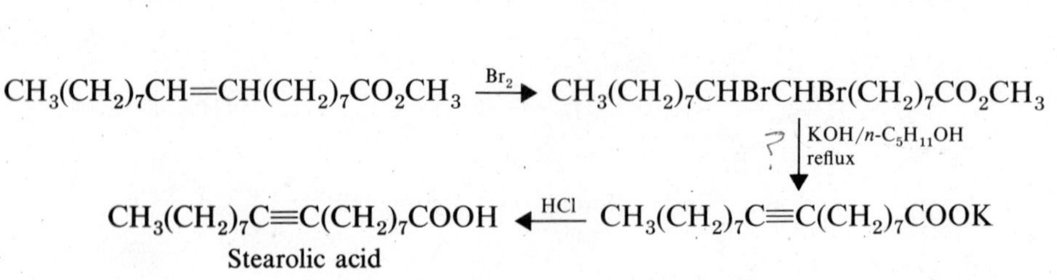

$$CH_3CH_2CH{=}CH_2 \xrightarrow[CCl_4]{Br_2} CH_3CH_2\underset{\underset{Br}{|}}{C}HCH_2Br \xrightarrow[\substack{ethanol \\ reflux}]{KOH} CH_3CH_2CH{=}CHBr$$

$$+$$

$$CH_3CH_2\underset{\underset{Br}{|}}{C}{=}CH_2$$

$$\xrightarrow[\substack{NaNH_2 \\ Mineral\ oil \\ 110-160°}]{}$$

$$CH_3CH_2C{\equiv}CH \xleftarrow{CH_3COOH} CH_3CH_2C{\equiv}C{:}^-Na^+$$

(60%, from the mixture
of bromoalkenes)

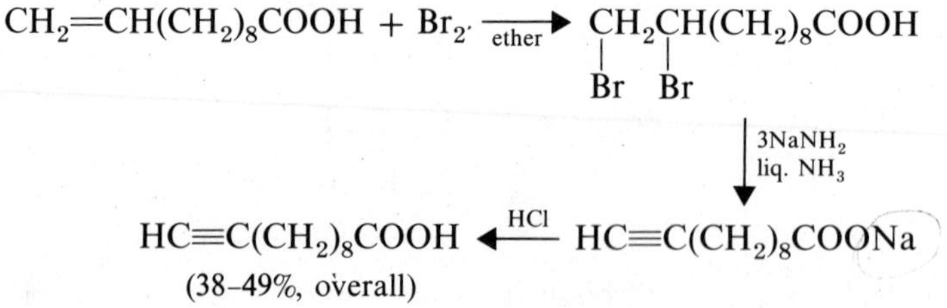

$$CH_2{=}CH(CH_2)_8COOH + Br_2 \cdot \xrightarrow{ether} CH_2\underset{\underset{Br}{|}}{C}H(CH_2)_8COOH$$
with Br on first carbon as well: $\underset{Br}{|}$

$$\xrightarrow[\substack{3NaNH_2 \\ liq.\ NH_3}]{}$$

$$HC{\equiv}C(CH_2)_8COOH \xleftarrow{HCl} HC{\equiv}C(CH_2)_8COONa$$

(38–49%, overall)

$$CH_3(CH_2)_7CH{=}CH(CH_2)_7CO_2CH_3 \xrightarrow{Br_2} CH_3(CH_2)_7CHBrCHBr(CH_2)_7CO_2CH_3$$

$$\xrightarrow[\substack{KOH/n\text{-}C_5H_{11}OH \\ reflux}]{}$$

$$CH_3(CH_2)_7C{\equiv}C(CH_2)_7COOH \xleftarrow{HCl} CH_3(CH_2)_7C{\equiv}C(CH_2)_7COOK$$
Stearolic acid

Stearolic acid (above) has been used to synthesize sterculic acid, a naturally occurring compound that contains a cyclopropene ring (p. 305).

Ketones can be converted to *gem*-dichlorides through their reaction with phosphorus pentachloride and *gem*-dichlorides can also be used to synthesize alkynes.

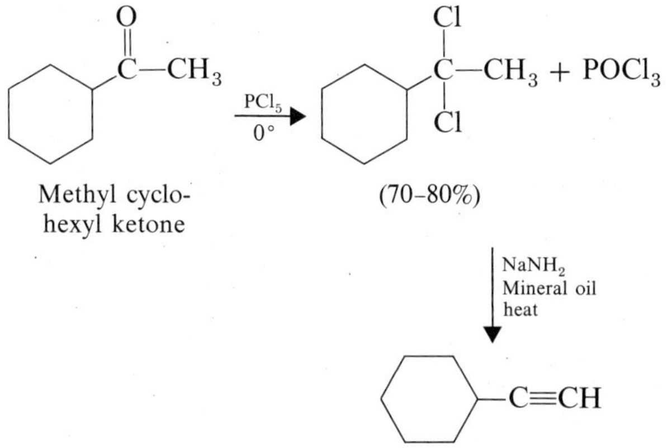

Methyl cyclo-
hexyl ketone

(70–80%)

Cyclohexylacetylene
(46%)

Problem 9.6

(a) Suggest a method for converting cyclohexylacetylene back into methyl cyclo-hexyl ketone. (b) Suggest a method for carrying out the following transformation:

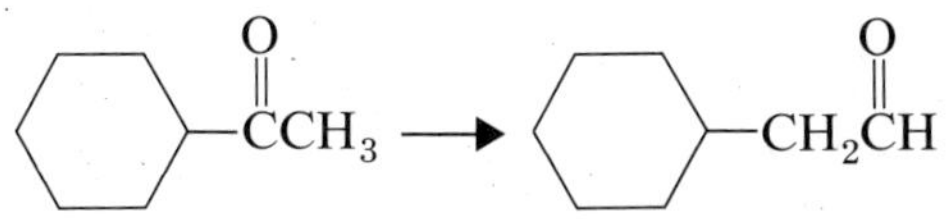

9.12 PHYSICAL PROPERTIES OF ALKYNES

As we might expect, the physical properties of alkynes are very similar to those of alkenes and alkanes. Alkynes are only very slightly soluble in water although they are somewhat more soluble than alkenes and alkanes. Alkynes, like alkenes and alkanes, are soluble in solvents of low polarity such as carbon tetrachloride, ether, and alkanes. Alkynes, like alkanes and alkenes are less dense than water.

The first three alkynes (Table 9.1) are gases at room temperature.

9.13 OXIDATION OF ALKYNES

When alkynes are treated with basic potassium permanganate they undergo oxidative cleavage in much the same way as alkenes do.

$$R—C{\equiv}C—R' \xrightarrow[\text{OH}^-,\ 25°]{\text{KMnO}_4} RCOO^- + R'COO^- \xrightarrow{\text{H}^+} RCOOH + R'COOH$$

The products of the reaction (after acidification) are carboxylic acids. Carboxylic acids are easy to identify and their identities allow one to locate the position of the triple bond of an alkyne. For example, 3-hexyne yields *only* a three-carbon acid (propanoic acid) when it is subjected to oxidative cleavage.

TABLE 9.1 Physical Constants of Alkynes

NAME	FORMULA	mp °C	bp °C	DENSITY d_4^{20}
Acetylene	$HC{\equiv}CH$	-81.8	-83.6	—
Propyne	$CH_3C{\equiv}CH$	-101.51	-23.2	—
1-Butyne	$CH_3CH_2C{\equiv}CH$	-122.5	8.1	—
2-Butyne	$CH_3C{\equiv}CCH_3$	-32.3	27	0.691
1-Pentyne	$CH_3(CH_2)_2C{\equiv}CH$	-90	39.3	0.695
2-Pentyne	$CH_3CH_2C{\equiv}CCH_3$	-101	55.5	0.714
1-Hexyne	$CH_3(CH_2)_3C{\equiv}CH$	-132	71	0.715
2-Hexyne	$CH_3(CH_2)_2C{\equiv}CCH_3$	-88	84	0.730
3-Hexyne	$CH_3CH_2C{\equiv}CCH_2CH_3$	-51	81.8	0.724
1-Heptyne	$CH_3(CH_2)_4C{\equiv}CH$	-81	100	0.734
2-Heptyne	$CH_3(CH_2)_3C{\equiv}CCH_3$	—	112	0.748
3-Heptyne	$CH_3(CH_2)_2C{\equiv}CCH_2CH_3$	—	105	0.752
1-Octyne	$CH_3(CH_2)_5C{\equiv}CH$	-80	125.2	0.746
2-Octyne	$CH_3(CH_2)_4C{\equiv}CCH_3$	-60.2	137.2	0.759
3-Octyne	$CH_3(CH_2)_3C{\equiv}CCH_2CH_3$	-105	133	0.752
4-Octyne	$CH_3(CH_2)_2C{\equiv}C(CH_2)_2CH_3$	-102	131	0.751
1-Nonyne	$CH_3(CH_2)_6C{\equiv}CH$	-65	160.7	0.760
2-Nonyne	$CH_3(CH_2)_5C{\equiv}CCH_3$	—	155^{747}	0.769
3-Nonyne	$CH_3(CH_2)_4C{\equiv}CCH_2CH_3$	—	154^{745}	0.762
4-Nonyne	$CH_3(CH_2)_3C{\equiv}C(CH_2)_2CH_3$	—	152^{752}	0.757
1-Decyne	$CH_3(CH_2)_7C{\equiv}CH$	-44	174	0.765
3-Decyne	$CH_3(CH_2)_5C{\equiv}CCH_2CH_3$	—	175	0.765

$$CH_3CH_2C{\equiv}CCH_2CH_3 \xrightarrow[\text{(2) H}^+]{\overset{\text{(1) KMnO}_4}{\text{OH}^-,\ 25°}} 2CH_3CH_2COOH$$

3-Hexyne   Propanoic acid

The alkynes, 1-hexyne and 2-hexyne both yield different products.

$$CH_3CH_2CH_2CH_2C{\equiv}CH \xrightarrow[\text{(2) H}^+]{\overset{\text{(1) KMnO}_4}{\text{OH}^-,\ 25°}} CH_3CH_2CH_2CH_2COOH + CO_2$$

1-Hexyne   Pentanoic acid

$$CH_3CH_2CH_2C{\equiv}CCH_3 \xrightarrow[\text{(2) H}^+]{\overset{\text{(1) KMnO}_4}{\text{OH}^-,\ 25°}} CH_3CH_2CH_2COOH + CH_3COOH$$

2-Hexyne   Butanoic acid   Acetic acid

The ${\equiv}CH$ group of a terminal alkyne (e.g., 1-hexyne) is oxidized to carbon dioxide.

Problem 9.7

Give the name and structure of each of the following alkynes.

(a) $C_7H_{12} \xrightarrow[\text{(2) H}^+]{\overset{\text{(1) KMnO}_4}{\text{OH}^-,\ 25°}} CH_3\underset{\underset{\displaystyle CH_3}{|}}{CH}COOH + CH_3CH_2COOH$

(b) C_8H_{12} $\xrightarrow[\text{(2) } H^+]{\substack{\text{(1) } KMnO_4 \\ OH^-, 25°}}$ $HOOC-(CH_2)_6-COOH$ (only)

(c) C_7H_{12} $\xrightarrow{\substack{2H_2 \\ Pt}}$ $CH_3CH_2CH_2CH_2CH_2CH_2CH_3$

$\xrightarrow[OH^-]{Ag(NH_3)_2{}^+}$ $C_7H_{11}Ag\downarrow$

Ozone can also be used to locate the position of the triple bond of an alkyne. Stearolic acid, for example, reacts with ozone to yield an ozonide that decomposes to a nine-carbon carboxylic acid and a nine-carbon dicarboxylic acid.

$$CH_3(CH_2)_7C\equiv C(CH_2)_7COOH \xrightarrow[\text{(2) } H_2O]{\text{(1) } O_3,\ CCl_4} CH_3(CH_2)_7COOH + HOOC(CH_2)_7COOH$$
$$\text{Stearolic acid}$$

9.14 CHEMICAL ANALYSIS OF ALKANES, ALKENES, ALKYNES, ALKYL HALIDES, AND ALCOHOLS

Chemical Tests

Very often in the course of laboratory work we need to decide what functional groups are present in a compound that we have isolated. We may have isolated a compound from a synthesis, for example, and the presence of a particular functional group may tell us whether our synthesis has succeeded or failed. Or, we may have isolated a compound from some natural material; before we go on to more elaborate procedures for structure determination it is often desirable to know just what kind of compound we are dealing with.

Spectroscopic methods are available that will do all of these things for us and we will study these procedures in Chapters 10 and 14. Spectrometers are expensive instruments, however, and spectroscopic procedures are sometimes time consuming. It is helpful, therefore, to have simpler and more rapid means to identify the presence or absence of a particular functional group.

Very often this can be done by a simple chemical test. Such a test will often consist of a single reagent that, when mixed with the compound in question, will indicate the presence of a particular functional group. Not all reactions of a functional group will serve as chemical tests, however. To be useful the reaction ought to proceed with a clear signal; a color change, the evolution of a gas, or the appearance of a precipitate.

Let us consider an example. Suppose we have just finished carrying out what we hope was an elimination reaction of an alkyl halide to produce an alkene. We have isolated a colorless liquid by distillation of the reaction mixture. Both alkenes and alkyl halides are colorless when pure. The question that needs an answer at this point is, "Is this colorless liquid our product (an alkene), or is it our starting material (an alkyl halide)?". So we begin to search our minds for a way to decide. We might first recall that alkenes undergo addition of hydrogen halides and that alkyl halides, because they are already saturated, do not. But clearly this reaction will not help us very much. For even if our compound is an alkene, bubbling a colorless gas (hydrogen chloride) into a colorless liquid (an alkene) will simply produce another colorless liquid (an alkyl halide). When our test is finished we will not be more enlightened than we were before we carried it out.

At this point we might remember that alkenes also undergo rapid addition of bromine, and that if the alkene is present in excess, the red-brown color of the bromine will disappear within seconds.

$$R—CH{=}CHR \; + \quad Br_2 \quad \xrightarrow{CCl_4} \; R—CHBrCHBr—R$$

Colorless   Red-brown   Colorless

Alkyl halides, on the other hand, do not react with bromine in carbon tetrachloride at room temperature and in the absence of steady irradiation by a strong light source.

$$R—Br \; + \quad Br_2 \quad \xrightarrow{CCl_4} \; \text{no reaction}$$

Colorless   Red-brown   (Red-brown color persists as long as exposure to a strong light source is avoided.)

So by simply adding a drop of bromine in carbon tetrachloride to a small amount of our compound, we will be able to make a decision. If the red-brown color of bromine disappears within a few seconds we can be reasonably certain that our elimination reaction has succeeded. If the color persists, then the elimination reaction did not occur. Now we may decide that we need to use a stronger base or a higher temperature or a different solvent to bring about the reaction.

The test that we have just described will serve us very well if the elimination reaction that we were attempting was one in which an alkyl halide was being converted to an alkene. But if the elimination reaction were one in which we were attempting to convert a *vic*-dihalide to an alkyne (below), testing with bromine in carbon tetrachloride would not be definitive.

$$R—CHBrCHBr—R \xrightarrow{\text{base}} R—CH{=}CHBr—R \xrightarrow{\text{base}} R—C{\equiv}C—R$$

vic-Dibromide   Vinylic bromide   Alkyne

or

$$R—CHBrCH_2Br \xrightarrow{\text{base}} R—CHBr{=}CH_2 \xrightarrow{\text{base}} R—C{\equiv}CH$$

vic-Dibromide   Vinylic bromide   Terminal alkyne

In either of the reactions above, the material that we isolate might be the starting *vic*-dihalide, it might be the intermediate vinylic bromide, or it might be an alkyne. Decolorization of bromine in carbon tetrachloride by the liquid that we isolate would only tell us that our product is not our starting material but it would not tell us whether it was the vinylic halide or the alkyne. Both compounds react with bromine, and both would decolorize bromine in carbon tetrachloride.

$$-\overset{|}{C}{=}\overset{|}{C}- \; + \; Br_2 \quad \xrightarrow{CCl_4} \; -\overset{|}{\underset{|}{C}}-\overset{\overset{Br}{|}}{\underset{|}{C}}-$$

        Br   Br   Br

Vinylic bromide

$$-C{\equiv}C- \;+\; 2Br_2 \xrightarrow{\text{CCl}_4} \; \underset{\underset{Br \;\; Br}{|\;\;\;\;|}}{\overset{\overset{Br \;\; Br}{|\;\;\;\;|}}{-C-C-}}$$

Alkyne

An additional test is required. If the reaction that we were carrying out was one that would yield a terminal alkyne, then we could use ammoniacal silver nitrate.

$$RC{\equiv}CH + Ag(NH_3)_2{}^+OH^- \xrightarrow[\text{H}_2\text{O}]{} RC{\equiv}CAg\downarrow$$

$$RCBr{=}CHBr + Ag(NH_3)_2{}^+OH^- \xrightarrow[\text{H}_2\text{O}]{} \text{no reaction (and no precipitate)}$$

The formation of a precipitate would tell us that our liquid was a terminal acetylene. If no precipitate appeared and if the liquid had also given a positive test with bromine in carbon tetrachloride we could be reasonably sure that we had isolated the vinylic bromide.

If, however, the reaction we were carrying out was one that would yield a disubstituted acetylene, testing with ammoniacal silver nitrate would not allow us to make a decision.

$$R-C{\equiv}C-R + Ag(NH_3)_2{}^+OH^- \longrightarrow \text{no precipitate}$$

$$R-CH{=}CBrR + Ag(NH_3)_2{}^+OH^- \longrightarrow \text{no precipitate}$$

Neither compound would give a precipitate.

In this instance we would need still another method. If our product gave a positive test with bromine in carbon tetrachloride, then we would be reasonably safe in assuming that it was either the vinylic bromide or the disubstituted acetylene. We could then distinguish between these two possibilities by adding a drop of the compound to a small amount of molten sodium. This reaction would reduce the bromine of the vinylic bromide to sodium bromide.

$$R-CH{=}CBrR + Na \xrightarrow[\text{fusion}]{} Na^+Br^- + \text{other products}$$

Acidification of the resulting mixture with nitric acid, followed by treatment with aqueous silver nitrate, would give a precipitate of silver bromide.

$$Na^+Br^- + Ag^+NO_3{}^- \xrightarrow{\text{H}_2\text{O}} AgBr\downarrow + Na^+NO_3{}^-$$

The disubstituted acetylene, since it does not contain a halogen, would not give a positive test.

Chemical Tests

A number of reagents that are used as tests for some of the functional groups that we have studied so far are summarized below. We are restricting our attention at this point to alkanes, alkenes, alkynes, alkyl halides, and alcohols.

Concentrated Sulfuric Acid

Alkenes, alkynes, and alcohols are protonated and thus dissolve when they are added to cold concentrated sulfuric acid.

$$\text{C=C} + H_2SO_4 \longrightarrow -\overset{|}{\underset{H}{C}}-\overset{|}{\underset{+}{C}}- + HSO_4^-$$

Soluble

$$-C\equiv C- + H_2SO_4 \longrightarrow -C\equiv C-\ \text{or}\ -C=\overset{|}{\underset{+}{C}} + HSO_4^-$$

Soluble

$$-\overset{|}{\underset{|}{C}}-\overset{..}{\underset{..}{O}}-H + H_2SO_4 \longrightarrow -\overset{|}{\underset{|}{C}}-\overset{H}{\underset{+}{O}}-H\ \text{or}\ -\overset{|}{C}{}^+ + HSO_4^-$$

Soluble

Alkanes and alkyl halides are insoluble in cold concentrated sulfuric acid.

Bromine in Carbon Tetrachloride

Alkenes and alkynes both add bromine at room temperature and in the absence of light. Alkanes, alkyl halides, and alcohols do not react with bromine unless the reaction mixture is heated or exposed to strong irradiation. Thus, rapid decolorization of bromine in carbon tetrachloride at room temperature and in the absence of strong irradiation by light indicates the presence of a carbon-carbon double bond or a carbon-carbon triple bond.

Cold Dilute Potassium Permanganate

Alkenes and alkynes are oxidized by cold dilute solutions of potassium permanganate.

$$\text{C=C} + MnO_4^- \xrightarrow[H_2O]{25°} -\overset{|}{\underset{OH}{C}}-\overset{|}{\underset{OH}{C}}- + MnO_2 + \text{other oxidation products}$$

Purple   Brown

$$-C\equiv C-\ MnO_4^- \xrightarrow[H_2O]{25°} -\overset{|}{\underset{O}{C}}-\overset{|}{\underset{O}{C}}- + MnO_2 + \text{other oxidation products}$$

Purple   Brown

If the alkene or alkyne is present in excess, the deep-purple color of the permanganate solution disappears and is replaced by the brown color of precipitated manganese dioxide.

Alkanes, alkyl halides, and pure alcohols do not react with cold dilute potassium permanganate. When these compounds are tested the purple color is not discharged and a precipitate of manganese dioxide does not appear. (Impure alcohols often contain aldehydes and aldehydes give a positive test with cold dilute potassium permanganate.)

Cold dilute potassium permanganate is often called Baeyer's reagent (after Adolf von Baeyer, p. 100).

Alcoholic Silver Nitrate

Silver nitrate in alcohol reacts with alkyl and allyl halides to form a precipitate of silver halide.

$$R\!-\!X \; + \; AgNO_3 \xrightarrow[\text{alcohol}]{} AgX\!\downarrow \; + \; R^+ \longrightarrow \text{ other products}$$
Alkyl
halide

$$R\!-\!CH\!=\!CHCH_2X \; + \; AgNO_3 \xrightarrow[\text{alcohol}]{} AgX\!\downarrow \; + \; RCH\!=\!CH\overset{+}{C}H_2 \longrightarrow \text{ other products}$$
Allyl halide

Vinyl halides and phenyl halides (Chapter 12) do not give a silver halide precipitate when treated with silver nitrate in alcohol.

$$R\!-\!CH\!=\!CHX \; + \; AgNO_3 \xrightarrow[\text{alcohol}]{} \text{ no reaction}$$
A vinyl halide

$$C_6H_5X \quad + \; AgNO_3 \xrightarrow[\text{alcohol}]{} \text{ no reaction}$$
Phenyl halide

Ammoniacal Silver Nitrate

Silver nitrate in ammonia reacts with terminal alkynes to form a precipitate of the silver alkynide.

$$R\!-\!C\!\equiv\!CH \; + \; Ag(NH_3)_2^+OH^- \longrightarrow R\!-\!C\!\equiv\!CAg\!\downarrow \; + \; HOH \; + \; 2NH_3$$

Nonterminal alkynes do not give a precipitate. *Silver alkynides can be distinguished from silver halides on the basis of their solubility in nitric acid; silver alkynides dissolve whereas silver halides do not.*

Sodium Fusion

The halogens of all organic halides including vinyl halides and phenyl halides are reduced to halide ions when these compounds are fused with sodium metal. Afterwards, the solutions can be acidified with nitric acid and treated with aqueous silver nitrate. A precipitate of silver halide indicates the presence of halogen in the original compound.

$$\begin{array}{c} R\!-\!X \\ \text{or} \\ R\!-\!CH\!=\!CHX \; + \; Na \xrightarrow[\text{fuse}]{} Na^+X^- \; + \; \text{other products} \end{array}$$
Vinyl halide
or
$$C_6H_5Br$$
Phenyl halide

(1) HNO_3
(2) $AgNO_3$

$$AgX\!\downarrow$$
Silver halide

Chromic Oxide

Primary and secondary alcohols are oxidized by a solution of chromic oxide, CrO_3, in aqueous sulfuric acid.

$$R—CH_2OH$$

or $+ HCrO_4^-$ ⟶ blue-green opaque mixture

$$R—\underset{\underset{R}{|}}{C}HOH$$

Clear
orange
solution

The clear orange solution containing $HCrO_4^-$ ions becomes opaque and a blue-green color appears within seconds. Tertiary alcohols do not give this test because they are more difficult to oxidize.

Hydrogenation

This test is more time consuming than the others that we have mentioned. It also requires more elaborate equipment. None of the tests that we have described thus far will distinguish between an alkene and a nonterminal alkyne. If the approximate formula weight of the compound is known, however, quantitative hydrogenation will distinguish between an alkene and a nonterminal alkyne because the alkyne will ultimately absorb twice as much hydrogen.

$$R—CH{=}CH—R + H_2 \xrightarrow{Pt} R—CH_2CH_2—R$$

$$R—C{\equiv}C—R + 2H_2 \xrightarrow{Pt} R—CH_2CH_2—R$$

This test is usually carried out by hydrogenating a known weight of the substance in an apparatus that permits an accurate measurement of the initial and final volumes of hydrogen at the same pressure. From the volume of hydrogen consumed, and the number of moles of the unknown compound used, a simple calculation will tell whether it is an alkene or an alkyne.

Quantitative hydrogenation, of course, will not distinguish between an alkyne and an alkadiene; both compounds absorb two moles of hydrogen.

$$R—CH{=}CH—CH{=}CH—R + 2H_2 \xrightarrow{Pt} RCH_2CH_2CH_2CH_2R$$

$$R—C{\equiv}C—R + 2H_2 \xrightarrow{Pt} RCH_2CH_2R$$

Table 9.2 summarizes, in tabular form, all of the simple chemical tests that we have presented. Hydrogenation is not included.

9.15 SUMMARY OF THE CHEMISTRY OF ALKYNES

Synthesis of Alkynes

1. From other alkynes (Sections 9.10 and 9.11)
   General:

$$R—C{\equiv}C—H \xrightarrow[\text{liq. } NH_3]{NaNH_2} R—C{\equiv}C{:}^- Na^+ \xrightarrow[\substack{\text{primary} \\ \text{halide}}]{R'—X} R—C{\equiv}C—R' + NaX$$

TABLE 9.2 Simple Chemical Tests

	CONC. H_2SO_4	Br_2/CCl_4	DILUTE $KMnO_4$	$AgNO_3/$ C_2H_5OH	$Ag(NH_3)_2^+OH^-$	(1) SODIUM (2) DIL. HNO_3 (3) $AgNO_3$	CrO_3 in H_2SO_4
Alkane	Insol.	—	—	—	—	—	—
$\diagdown C{=}C \diagup$	Sol.	*Rapid Br_2 addition:* red-brown ↓ colorless	*Oxidation:* purple soln. ↓ brown ppt.	—	—		
$RC{\equiv}CH$	Sol.	*Rapid Br_2 addition:* red-brown ↓ colorless	*Oxidation:* purple soln. ↓ brown ppt.	*	Precipitation of $R{-}C{\equiv}CAg$ Ppt. dissolves in dil. HNO_3	—	—
$RC{\equiv}CR$	Sol.	*Rapid Br_2 addition:* red-brown ↓ colorless	*Oxidation:* purple soln. ↓ brown ppt.	—	—	—	—
$R{-}X$	Insol.	—	—	Precipitation of AgX	—	Precipitation of AgX	—
$-\underset{\vert}{\overset{\vert}{C}}{=}\underset{\vert}{\overset{\vert}{C}}-\underset{\vert}{\overset{\vert}{C}}-X$	Sol.	*Rapid Br_2 addition:* red-brown ↓ colorless	*Oxidation:* purple soln. ↓ brown ppt.	Precipitation of AgX	—	Precipitation of AgX	—
$-\underset{\vert}{\overset{\vert}{C}}{=}\underset{\vert}{C}-X$	Sol.	*Rapid Br_2 addition:* red-brown ↓ colorless	*Oxidation:* purple soln. ↓ brown ppt.	—	—	Precipitation of AgX	—
$-\underset{H}{\overset{H}{C}}-OH$	Sol.	—	—	—	—	—	*Oxidation:* orange soln. ↓ blue-green opaque
$C-\underset{H}{\overset{C}{C}}-OH$	Sol.	—	—	—	—	—	*Oxidation:* orange soln. ↓ blue-green opaque
$C-\underset{C}{\overset{C}{C}}-OH$	Sol.	—	—	—	—	—	—

*Some terminal alkynes may give a precipitate ($R{-}C{\equiv}CAg$) when treated with $AgNO_3$ in alcohol. Silver alkynides, however, can be distinguished from silver halides on the basis of their solubility in dilute HNO_3, see p. 333.

Specific example:

$$CH_3CH_2C{\equiv}CH \xrightarrow[\text{liq. } NH_3]{NaNH_2} CH_3CH_2C{\equiv}C{:}^- \ Na^+ \xrightarrow{CH_3CH_2Br} CH_3CH_2C{\equiv}CCH_2CH_3$$
$$(75\%)$$

2. Through elimination reactions (Section 9.11)

General:

$$R-\underset{\underset{X}{\vert}}{\overset{\overset{H}{\vert}}{C}}-\underset{\underset{X}{\vert}}{\overset{\overset{H}{\vert}}{C}}-R \xrightarrow[(-2HX)]{\text{base}} R-C{\equiv}C-R$$

vic-Dihalide

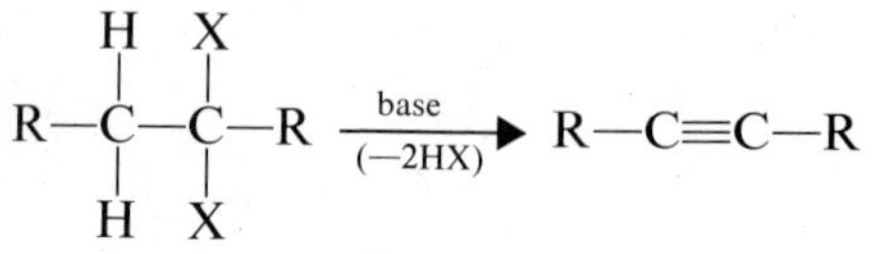

$$R-\underset{\underset{H}{|}}{\overset{\overset{H}{|}}{C}}-\underset{\underset{X}{|}}{\overset{\overset{X}{|}}{C}}-R \xrightarrow[(-2HX)]{base} R-C\equiv C-R$$

gem-Dihalide

Specific examples:

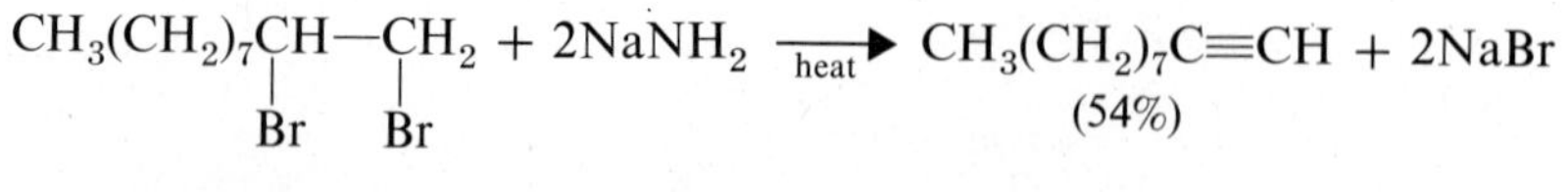

$$CH_3(CH_2)_7\underset{\underset{Br}{|}}{CH}-\underset{\underset{Br}{|}}{CH_2} + 2NaNH_2 \xrightarrow{heat} CH_3(CH_2)_7C\equiv CH + 2NaBr$$
$$(54\%)$$

$$CH_3CH_2CHBr_2 + 2KOH \xrightarrow[heat]{C_2H_5OH} CH_3C\equiv CH + 2KBr$$

Reactions of Alkynes

1. Addition of hydrogen (Section 9.6)
   General:

$$R-C\equiv C-H \xrightarrow{\underset{Cat.}{H_2}} R-CH=CH_2 \xrightarrow{\underset{Cat.}{H_2}} R-CH_2CH_3$$

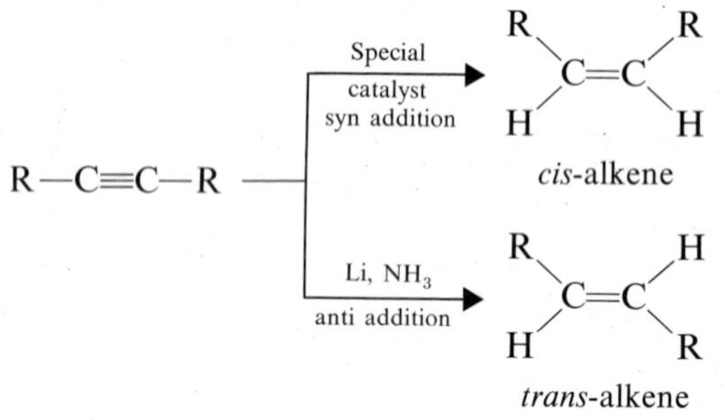

$$R-C\equiv C-R \begin{cases} \xrightarrow[\substack{\text{catalyst} \\ \text{syn addition}}]{\text{Special}} & \underset{\text{cis-alkene}}{\overset{R \diagdown \quad \diagup R}{\underset{H \diagup \quad \diagdown H}{C=C}}} \\[3em] \xrightarrow[\text{anti addition}]{\text{Li, NH}_3} & \underset{\text{trans-alkene}}{\overset{R \diagdown \quad \diagup H}{\underset{H \diagup \quad \diagdown R}{C=C}}} \end{cases}$$

Specific examples:

$$CH_3-\underset{\underset{CH_3}{|}}{\overset{\overset{CH_3}{|}}{C}}-CH_2C\equiv CH \xrightarrow{\underset{Ni}{H_2}} \left[CH_3-\underset{\underset{CH_3}{|}}{\overset{\overset{CH_3}{|}}{C}}-CH_2-CH=CH_2 \right] \xrightarrow{\underset{Ni}{H_2}} CH_3-\underset{\underset{CH_3}{|}}{\overset{\overset{CH_3}{|}}{C}}-CH_2CH_2CH_3$$

$$CH_3(CH_2)_2C\equiv C(CH_2)_2CH_3 \xrightarrow[\text{liq. NH}_3]{Li} \underset{(52\%)}{\overset{CH_3CH_2CH_2 \diagdown \qquad \diagup H}{\underset{H \diagup \qquad \diagdown CH_2CH_2CH_3}{C=C}}}$$

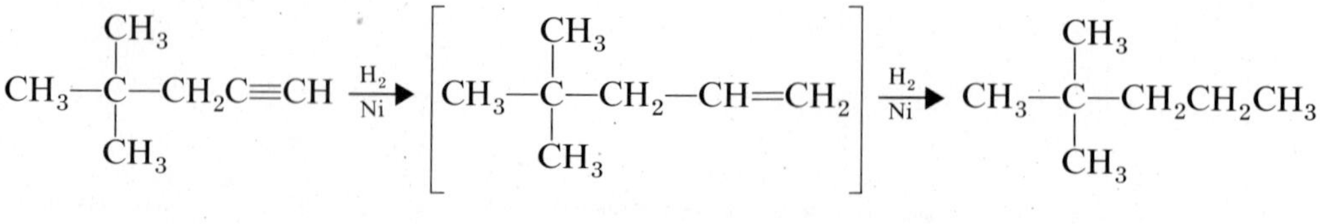

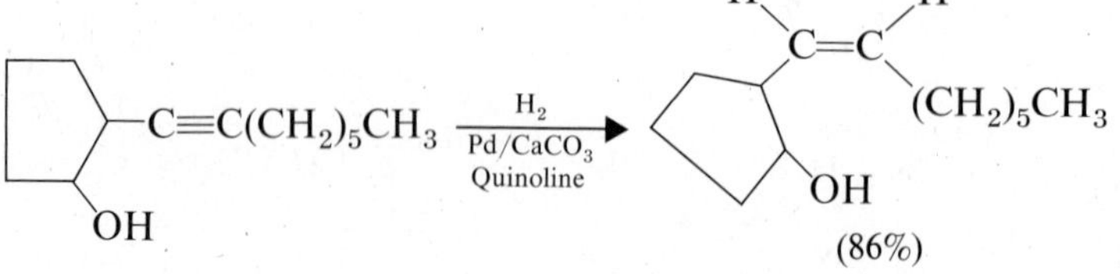

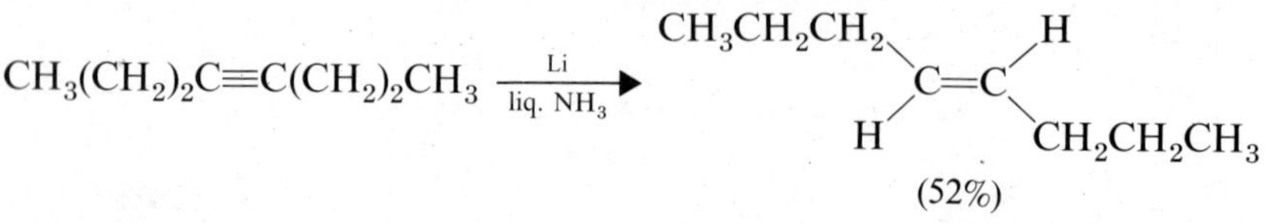

2. Addition of halogens (Section 9.7)
   General:

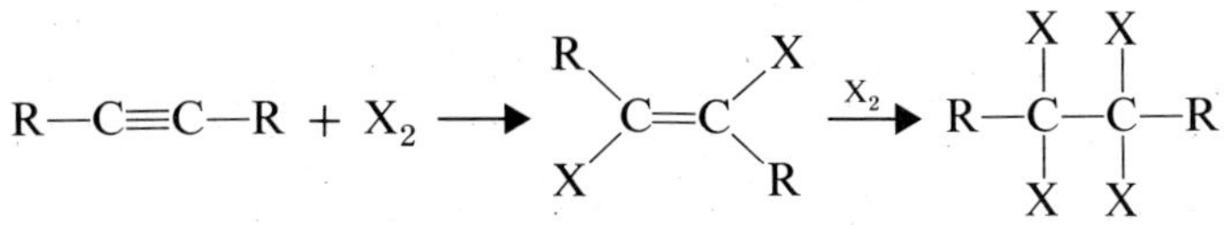

$$R-C\equiv C-R + X_2 \longrightarrow \underset{X}{\overset{R}{\diagup}} C=C \overset{X}{\underset{R}{\diagdown}} \xrightarrow{X_2} R-\underset{\underset{X}{|}}{\overset{\overset{X}{|}}{C}}-\underset{\underset{X}{|}}{\overset{\overset{X}{|}}{C}}-R$$

   Specific example:

$$C_6H_5-C\equiv C-CH_3 + Br_2 \xrightarrow[\substack{CH_3COOH \\ 25°}]{LiBr} \underset{Br}{\overset{C_6H_5}{\diagup}} C=C \overset{Br}{\underset{CH_3}{\diagdown}}$$

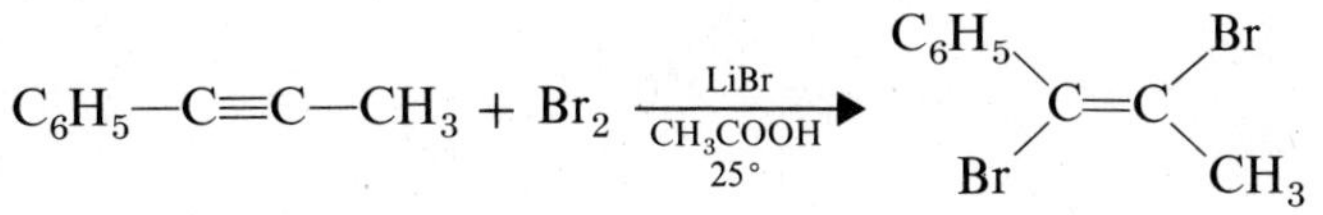

   (98%)

3. Addition of hydrogen halides (Section 9.8)
   General:

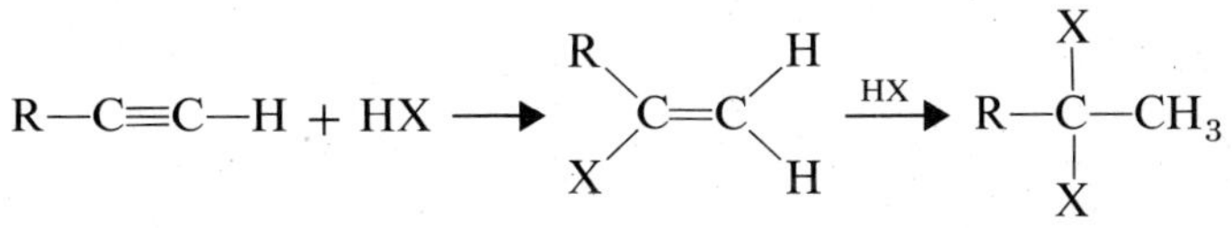

$$R-C\equiv C-H + HX \longrightarrow \underset{X}{\overset{R}{\diagup}} C=C \overset{H}{\underset{H}{\diagdown}} \xrightarrow{HX} R-\underset{\underset{X}{|}}{\overset{\overset{X}{|}}{C}}-CH_3$$

   Specific example:

$$CH_3-C\equiv CH + HCl \longrightarrow \underset{Cl}{\overset{CH_3}{\diagup}} C=C \overset{H}{\underset{H}{\diagdown}} \xrightarrow{HCl} CH_3-\underset{\underset{Cl}{|}}{\overset{\overset{Cl}{|}}{C}}-CH_3$$

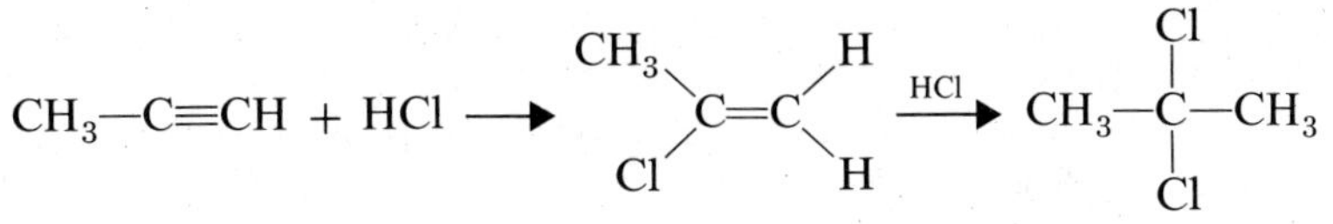

4. Addition of water (Section 9.9)
   General:

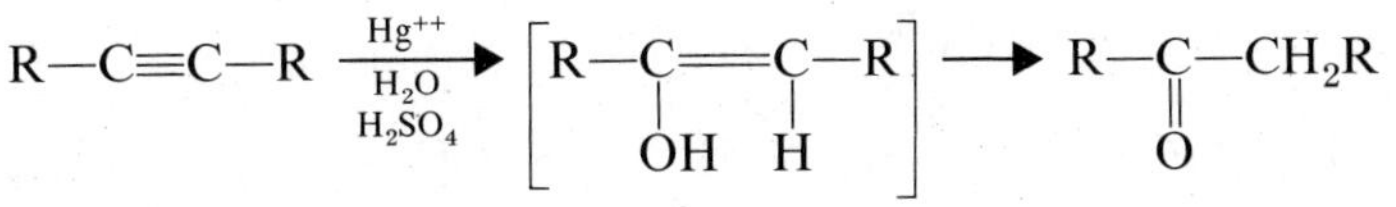

$$R-C\equiv C-R \xrightarrow[\substack{H_2O \\ H_2SO_4}]{Hg^{++}} \left[R-\underset{\underset{OH}{|}}{C}=\underset{\underset{H}{|}}{C}-R \right] \longrightarrow R-\underset{\underset{O}{\|}}{C}-CH_2R$$

   Specific examples:

$$H-C\equiv C-H + H_2O \xrightarrow[\substack{18\% \ H_2SO_4 \\ 90°}]{Hg^{++}} CH_3-\underset{\underset{O}{\|}}{C}-H$$

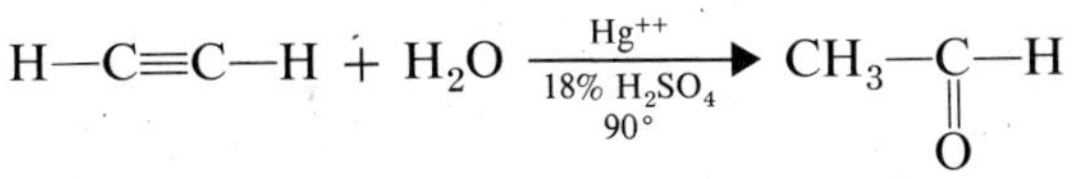

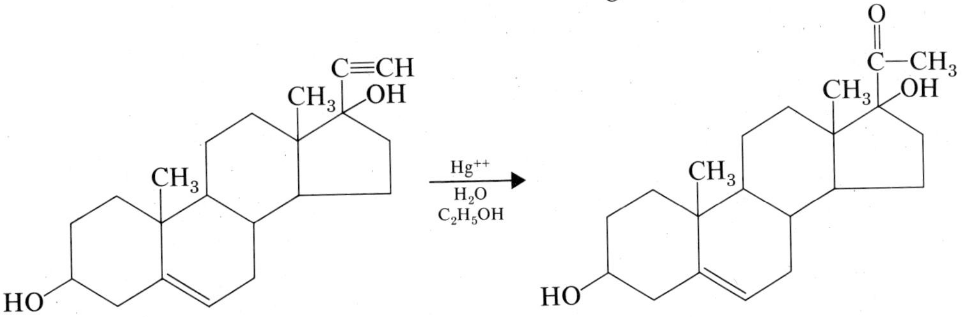

(95%)

5. Addition of boron hydrides (Sections 9.6 and 9.9)
   General:

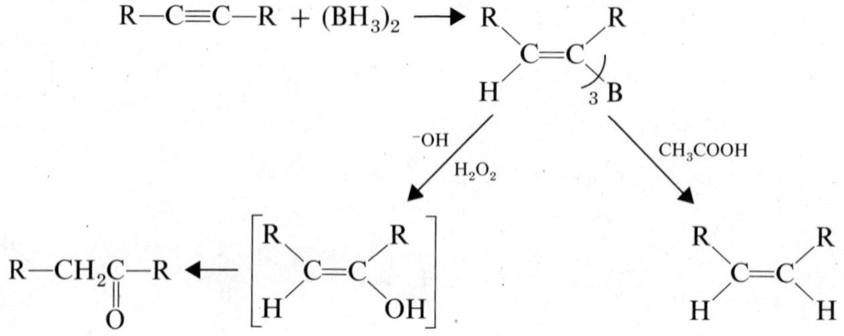

$$R-C \equiv C-R + (BH_3)_2 \longrightarrow \underset{H}{\overset{R}{\diagdown}} C = C \underset{3 \ B}{\overset{R}{\diagup}}$$

$$\xrightarrow[\substack{H_2O}]{^-OH}$$

$$\xrightarrow{CH_3COOH}$$

$$R-CH_2\underset{\underset{O}{\|}}{C}-R \longleftarrow \left[\underset{H}{\overset{R}{\diagdown}} C = C \underset{OH}{\overset{R}{\diagup}} \right] \qquad \underset{H}{\overset{R}{\diagdown}} C = C \underset{H}{\overset{R}{\diagup}}$$

Specific example:

$$C_3H_7-C \equiv C-C_3H_7 + (BH_3)_2 \xrightarrow{0°} \underset{H}{\overset{C_3H_7}{\diagdown}} C = C \underset{B^-}{\overset{C_3H_7}{\diagup}}$$

$$\xrightarrow{OH^-, \ H_2O} \qquad 0° \Big| CH_3COOH$$

$$\underset{\underset{O}{\|}}{C_3H_7C}CH_2C_3H_7 \qquad \underset{H}{\overset{C_3H_7}{\diagdown}} C = C \underset{H}{\overset{C_3H_7}{\diagup}}$$

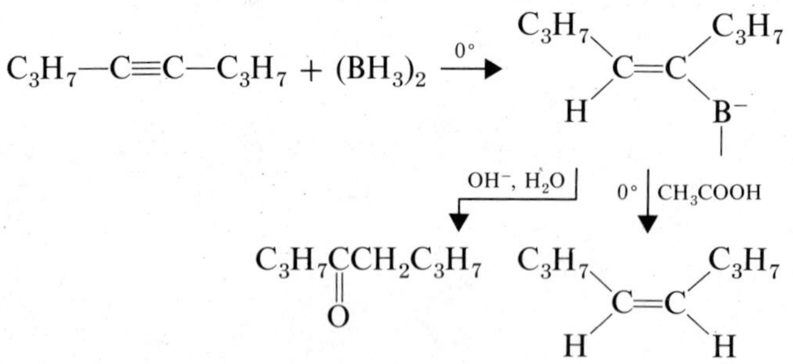

6. Replacement of the acetylenic hydrogen (Section 9.10)
   Specific examples:
   Replacement with Metals

$$H-C \equiv C-H + Na \xrightarrow[NH_3]{liq.} H-C \equiv C:^-Na^+ + \tfrac{1}{2}H_2$$

$$H-C \equiv C-H + NaNH_2 \xrightarrow[NH_3]{liq.} H-C \equiv C:^-Na^+ + NH_3$$

$$CH_3(CH_2)_6C \equiv CH + Ag(NH_3)_2{}^+OH^- \xrightarrow[H_2O]{} CH_3(CH_2)_6C \equiv C:^-Ag^+ + 2NH_3 + H_2O$$

   Oxidative Coupling

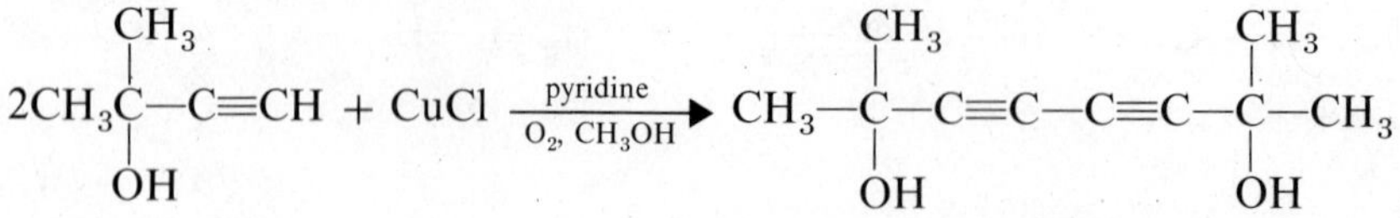

$$2CH_3\underset{OH}{\overset{CH_3}{\underset{|}{\overset{|}{C}}}}-C \equiv CH + CuCl \xrightarrow[O_2, \ CH_3OH]{pyridine} CH_3-\underset{OH}{\overset{CH_3}{\underset{|}{\overset{|}{C}}}}-C \equiv C-C \equiv C-\underset{OH}{\overset{CH_3}{\underset{|}{\overset{|}{C}}}}-CH_3$$

$$(80\%)$$

7. Oxidative cleavage of alkynes (Section 9.13)
   General:

$$R-C \equiv C-H \xrightarrow[\substack{(2) \ H^+}]{(1) \ KMnO_4, \ OH^-} R-COOH + CO_2$$

$$R-C \equiv C-R' \xrightarrow[\substack{(2) \ H^+}]{(1) \ KMnO_4, \ OH^-} R-COOH + R'-COOH$$

$$R-C \equiv C-R' \xrightarrow[\substack{(2) \ H_2O}]{(1) \ O_3, \ CCl_4} R-COOH + R'-COOH$$

Additional Problems

9.8
Give IUPAC names for each of the following alkynes.

(a) $CH_3CHC \equiv CH$
     $|$
     CH_3

(b) $(CH_3)_3CC \equiv CCH_2CH_3$

(c) $CH_3CH_2C \equiv C(CH_2)_4CH_3$

(d) $CH_3CH_2C \equiv CCH_2CH_3$

(e) di-*tert*-Butylacetylene

(f) diisopropylacetylene

(g) Methylpropylacetylene

(h) Methyl *sec*-butylacetylene

(i) Diisobutylacetylene

(j) *n*-Hexylacetylene

9.9
Which of the compounds in problem 9.8 would give a positive test with $Ag(NH_3)_2^+OH^-$?

9.10
(a) Which of the compounds in problem 9.8 would yield *n*-hexane when reduced with two moles of hydrogen and a platinum catalyst?

(b) Which would yield *n*-heptane?

9.11
Starting only with coke, lime, water, methane, and any other needed inorganic reagents show how you might synthesize each of the following compounds. (You need not show the synthesis of the same compound twice.)

(a) Acetylene

(b) Ethene

(c) Propyne

(d) Propene

(e) Acetone

(f) 2-Butyne

(g) 1-Butyne

(h) $CH_3CH_2CCH_3$
         $\|$
         O

(i) Bromoethane

(j) Methylethylacetylene

(k) Ethanol

(l) 1,2-Dibromoethane

(m) 2,2-Dichloropropane

(n) *cis*-2-Butene

(o) *trans*-2-Butene

(p) 1-Butene

(q) 2-Bromobutane

(r) 1-Bromobutane

(s) *n*-Butyl alcohol

(t) *sec*-Butyl alcohol

(u) *meso*-2,3-Dibromobutane

(v) Racemic-2,3-dibromobutane

(w) (Z)-2-chloro-2-butene

(x) 1-Bromopropane

(y) 3-Hexyne

(z) 1-Bromo-1-butene

9.12
Give the structure of the products that you would expect from the reaction of 1-pentyne with:

(a) One mole Br_2

(b) One mole HCl

(c) Two moles HCl

(d) One mole HBr and peroxides

(e) H_2O, H^+, Hg^{++}

(f) H_2, Pd-$CaCO_3$, quinoline

(g) $(Sia)_2BH$ then CH_3COOH

(h) $(Sia)_2BH$ then OH^-, H_2O_2

(i) $NaNH_2$ in liq. NH_3

(j) $NaNH_2$ in liq. NH_3, then CH_3I

(k) CuCl, NH_3, O_2, CH_3OH

(l) $Ag(NH_3)_2OH$

(m) $Cu(NH_3)_2OH$

(n) Hot $KMnO_4$

9.13

Give the structure of the products you would expect from the reaction (if any) of 3-hexyne with:

(a) One mole HCl
(b) Two moles HCl
(c) One mole Br_2
(d) Two moles Br_2
(e) $Ni_2B(P-2)$, H_2
(f) $Pd/BaSO_4$, quinoline, H_2
(g) Li/NH_3
(h) H_2O, H^+, Hg^{++}
(i) $Ag(NH_3)_2OH$
(j) $Cu(NH_3)_2OH$
(k) $(BH_3)_2$ then CH_3COOH
(l) $(BH_3)_2$ then OH^-, H_2O_2
(m) CuCl, NH_3, O_2, CH_3OH
(n) Two moles H_2, Pt
(o) Hot $KMnO_4$
(p) O_3, H_2O
(q) $NaNH_2$, liq. NH_3

9.14

Show how each of the following compounds might be transformed into 1-pentyne.
(a) 1-Pentene
(b) 1-Chloropentane
(c) 1-Chloro-1-pentene
(d) 1,1-Dichloropentane
(e) 1-Bromopropane

9.15

Describe with equations a simple test that would distinguish between each of the following pairs of compounds. (In each case tell what you would see.)
(a) Propane and propyne
(b) Propene and propyne
(c) 1-Bromopropene and 2-bromopropane
(d) 2-Bromo-2-butene and 1-butyne
(e) 2-Bromo-2-butene and 2-butyne
(f) 2-Butyne and n-butyl alcohol
(g) 2-Butyne and 2-bromobutane
(h) $CH_3C{\equiv}CCH_2OH$ and $CH_3CH_2CH_2CH_2OH$
(i) $CH_3CH{=}CHCH_2OH$ and $CH_3CH_2CH_2CH_2OH$

9.16

Three compounds A, B, and C all have the formula C_5H_8. All three compounds rapidly decolorize bromine in carbon tetrachloride, all three give a positive test with Baeyer's reagent, and all three are soluble in cold concentrated sulfuric acid. Compound A gives a precipitate when treated with ammoniacal silver nitrate but compounds B and C do not. Compounds A and B both yield n-pentane (C_5H_{12}) when they are treated with excess hydrogen in the presence of a platinum catalyst. Under these same conditions, compound C absorbs only one mole of hydrogen and gives a product with the formula C_5H_{10}.
(a) Suggest a possible structure for A, B, and C.
(b) Are other structures possible for B and C?
(c) On oxidation with hot potassium permanganate B gave (after acidification) acetic acid and CH_3CH_2COOH. What is the structure of B?

(d) On oxidation with hot potassium permanganate, C gave (after acidification) $HOOCCH_2CH_2CH_2COOH$. What is the structure of C?

9.17

Starting with 3-methyl-1-butyne and any inorganic reagents show how the following compounds could be synthesized.

(a) 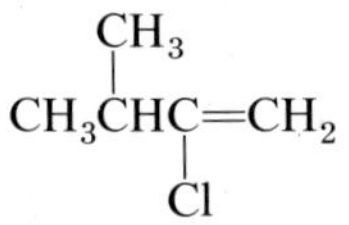 $CH_3\overset{\overset{\displaystyle CH_3}{|}}{C}H\underset{\underset{\displaystyle Cl}{|}}{C}=CH_2$

(d) $CH_3\overset{\overset{\displaystyle CH_3}{|}}{C}HCCl_2CH_2Cl$

(b) $CH_3\overset{\overset{\displaystyle CH_3}{|}}{C}HCH_2CH_2Br$

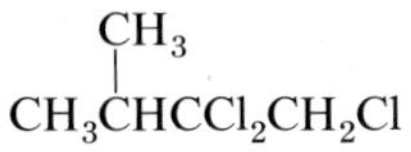

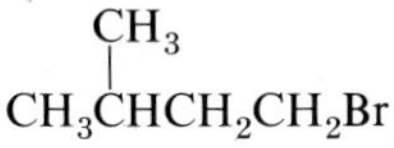

(e) $CH_3\overset{\overset{\displaystyle CH_3}{|}}{C}HCClBrCH_3$

(c) $CH_3\overset{\overset{\displaystyle CH_3}{|}}{C}H\underset{\underset{\displaystyle Cl}{|}}{C}HCH_3$

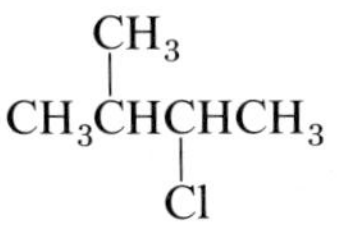

(f) 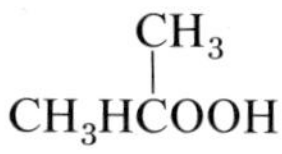 $CH_3\overset{\overset{\displaystyle CH_3}{|}}{H}COOH$

9.18

In addition to 1-pentyne and ordinary inorganic reagents, assume that you also have available the following deuterium compounds: D_2, DCl, CH_3COOD, and $(Sia)_2BD$. Show how you might prepare the following deuterium-labeled compounds.

(a) 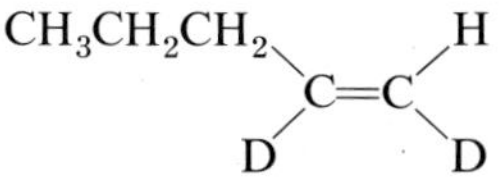 $CH_3CH_2CH_2$ and H on one carbon, $C=C$, D and D on other

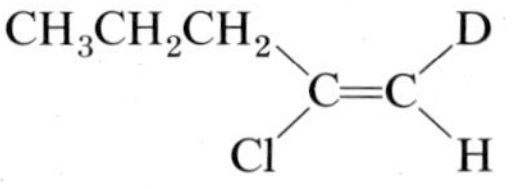

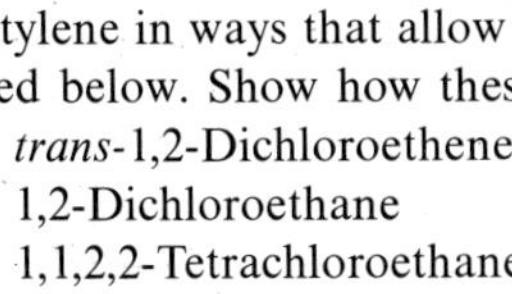

(b) $CH_3CH_2CH_2$ and H on one carbon, $C=C$, H and D on other

(c) $CH_3CH_2CH_2$ and D on one carbon, $C=C$, Cl and H on other

(d) $CH_3CH_2CH_2\underset{\underset{\displaystyle D}{|}}{C}H\overset{\overset{\displaystyle O}{\|}}{C}H$

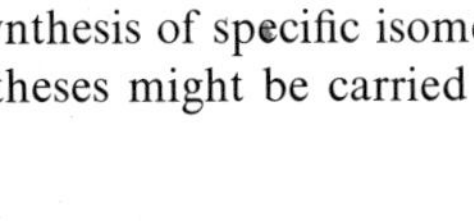

9.19

A number of industrially important chloroethenes and chloroethanes can be prepared from acetylene in ways that allow the synthesis of specific isomers. Examples are the compounds listed below. Show how these syntheses might be carried out.
(a) *trans*-1,2-Dichloroethene
(b) 1,2-Dichloroethane
(c) 1,1,2,2-Tetrachloroethane
(d) 1,1,2-Trichloroethene
(e) 1,1,2-Trichloroethane
(f) 1,1,1,2,2-Pentachloroethane
(g) 1,1,2,2-Tetrachloroethene

9.20

Disiamylborane (p. 319) is prepared by allowing diborane to react with an alkene. (a) Give the structure of this alkene and show how it reacts with diborane. (b) What factor favors the formation of disiamylborane, Sia_2BH, in this reaction by preventing the formation of Sia_3B?

9.21

An optically active compound **D** has the molecular formula C_6H_{10}. The compound gives a precipitate when treated with a solution containing $Ag(NH_3)_2OH$. On catalytic hydrogenation **D** yields **E**, C_6H_{14}. **E** is optically inactive and cannot be resolved. Propose structures for **D** and **E**.

* 9.22

Alkenes are more reactive than alkynes toward addition of electrophilic reagents (i.e., Br_2, Cl_2, HCl, etc). Yet when alkynes are treated with these same electrophilic reagents it is easy to stop the addition at the "alkene stage." This appears to be a paradox and yet it is not. Explain.

* 9.23

A number of compounds containing carbon-carbon triple bonds have been found in nature. One such compound obtained from the seeds of an African tree is called *erythrogenic acid*. *Erythro* comes from a Greek word meaning red, and when exposed to light erythrogenic acid forms a bright red polymer. Erythrogenic acid is a monocarboxylic acid (i.e., each molecule has one —COOH group) and has the molecular formula $C_{18}H_{26}O_2$. When treated with excess hydrogen in the presence of a platinum catalyst, erythrogenic acid absorbs five moles of hydrogen and yields a compound with the structural formula $CH_3(CH_2)_{16}COOH$. (a) What important clues does this experiment give us about the structure of erythrogenic acid? (b) At this point many possible structures can be written for erythrogenic acid. Write at least four. (c) Erythrogenic acid reacts with ozone and when the ozonide is subsequently treated with

$$\overset{O}{\overset{\|}{}}$$

water, four products can be isolated: $H—C—H$, $HOOC(CH_2)_4COOH$, $HOOC—COOH$, and $HOOC(CH_2)_7COOH$. This experiment narrows the structural possibilities for erythrogenic acid down to two formulas. What are they? (d) When an alcoholic solution of $CH_2{=}CH(CH_2)_4C{\equiv}CH$ and $HC{\equiv}C(CH_2)_7COOH$ is treated with cuprous chloride, ammonia, and oxygen, three products can be isolated from the reaction mixture; one is non-acidic, one is a monocarboxylic acid, and one is a dicarboxylic acid. This monocarboxylic acid is erythrogenic acid. What is the structure of erythrogenic acid?

10 CONJUGATED UNSATURATED SYSTEMS. VISIBLE-ULTRAVIOLET SPECTROSCOPY

10.1 ALLYLIC SUBSTITUTION AND THE ALLYL RADICAL

When propene reacts with bromine or chlorine at low temperatures the reaction that takes place is the usual addition of halogen to the double bond.

$$CH_2{=}CH{-}CH_3 + X_2 \xrightarrow[\text{CCl}_4 \text{ (addition reaction)}]{\text{low temperature}} \underset{\substack{| \\ X}}{CH_2}{-}\underset{\substack{| \\ X}}{CH}{-}CH_3$$

However, when propene reacts with chlorine or bromine at very high temperatures or under conditions in which the concentration of the halogen is very small, the reaction that occurs is a *substitution reaction.* These two examples illustrate how we can often change the course of an organic reaction by simply changing the conditions. (They also illustrate the need for specifying the conditions of a reaction carefully when we report experimental results.)

$$\underset{\text{Propene}}{CH_2{=}CH{-}CH_3} + X_2 \xrightarrow[\substack{\text{or} \\ \text{low conc. of } X_2 \\ \text{(substitution reaction)}}]{\text{high temperature}} CH_2{=}CH{-}CH_2X + HX$$

In the substitution reaction a halogen replaces one of the hydrogens of the methyl group of propene. These hydrogens are called the *allylic hydrogens* and the substitution reaction is known as an *allylic substitution.**

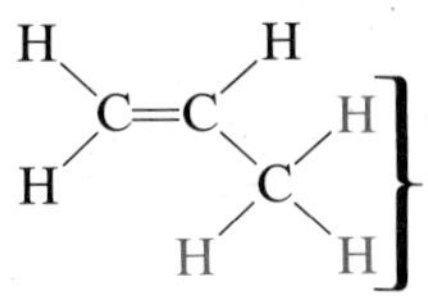

Allylic hydrogens

* These are general terms as well. The hydrogens of any saturated carbon adjacent to a double bond, that is,

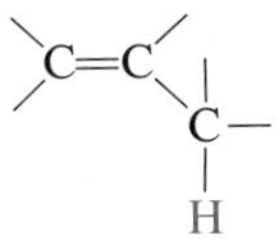

are called *allylic* hydrogens and any reaction in which an allylic hydrogen is replaced is called an *allylic substitution.*

Propene undergoes allylic chlorination when propene and chlorine react in the gas phase at 500°.

$$CH_2{=}CH{-}CH_3 + Cl_2 \xrightarrow[\text{Gas phase}]{500°} CH_2{=}CH{-}CH_2Cl + HCl$$
Allyl chloride

Propene undergoes allylic bromination when it is treated with *N*-bromosuccinimide in carbon tetrachloride at room temperature.

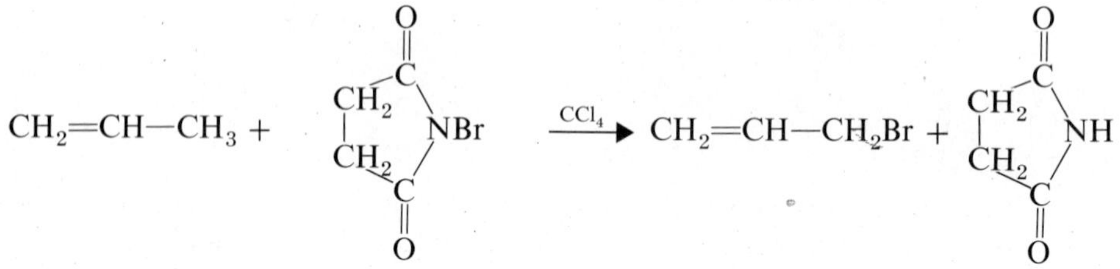

N-bromosuccinimide   Allyl bromide   Succinimide
(NBS)

N-bromosuccinimide provides a constant but low concentration of bromine in the reaction mixture. It does this by reacting with the hydrogen bromide formed in the substitution reaction. Each molecule of HBr that is formed is replaced by one molecule of Br_2.

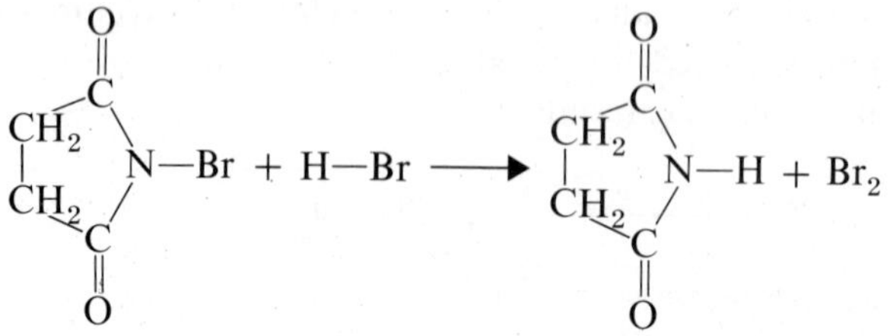

The mechanism for these substitution reactions is much the same as the chain mechanism for alkane halogenations that we saw in Chapter 4. In the chain-initiating step, the halogen (chlorine or bromine) dissociates into halogen atoms.

Chain-initiating step
$$:\ddot{X}:\ddot{X}: \longrightarrow 2:\ddot{X}\cdot$$

In the first chain-propagating step the halogen atom abstracts one of the allylic hydrogens.

First chain-propagating step

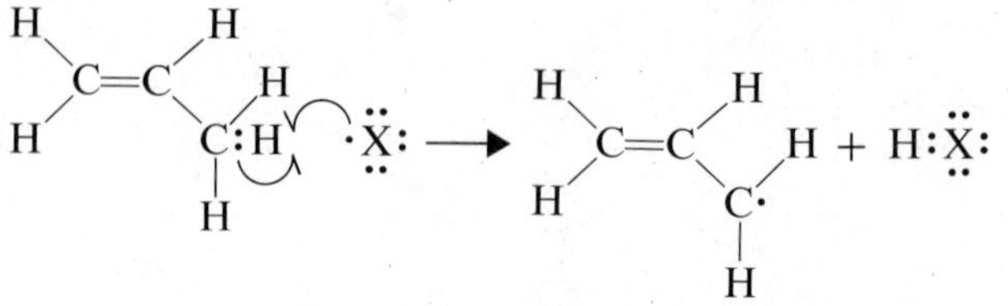

Allyl radical

The radical that is produced in this step is called an *allyl radical*.

In the second chain-propagating step the allyl radical reacts with a molecule of the halogen.

Second chain-propagating step

$$\underset{H}{\overset{H}{\diagdown}}C=C\underset{CH_2}{\overset{H}{\diagup}} \quad :\overset{..}{\underset{..}{X}}:\overset{..}{\underset{..}{X}}: \longrightarrow \underset{H}{\overset{H}{\diagdown}}C=C\underset{CH_2:X}{\overset{H}{\diagup}} + :\overset{..}{\underset{..}{X}}\cdot$$

Allyl halide

This step results in the formation of a molecule of allyl halide and a halogen atom. The halogen atom then brings about the repetition of the first chain-propagating step. The chain reaction continues until the usual chain-terminating steps consume the reactive species.

The allylic substitution reactions of propene will seem somewhat less surprising if we examine the bond dissociation energy of an allylic carbon-hydrogen bond and compare it with the bond dissociation energies of other carbon hydrogen bonds (cf. Table 4.1.)

$$CH_2=CHCH_2-H \longrightarrow CH_2=CHCH_2\cdot + H\cdot \qquad \Delta H = 85 \text{ kcal/mole}$$
$$\underset{\text{Propene}}{} \qquad \underset{\text{Allyl radical}}{}$$

$$(CH_3)_3C-H \longrightarrow (CH_3)_3C\cdot + H\cdot \qquad \Delta H = 91 \text{ kcal/mole}$$
$$\underset{\text{Isobutane}}{} \qquad \underset{3° \text{ Radical}}{}$$

$$(CH_3)_2CH-H \longrightarrow (CH_3)_2CH\cdot + H\cdot \qquad \Delta H = 94.5 \text{ kcal/mole}$$
$$\underset{\text{Propane}}{} \qquad \underset{2° \text{ Radical}}{}$$

$$CH_3CH_2CH_2-H \longrightarrow CH_3CH_2CH_2\cdot + H\cdot \qquad \Delta H = 98 \text{ kcal/mole}$$
$$\underset{\text{Propane}}{} \qquad \underset{1° \text{ Radical}}{}$$

We see that an allylic carbon-hydrogen bond of propene is broken with greater ease than even the tertiary carbon-hydrogen bond of isobutane and with far greater ease

FIG. 10.1

The relative stability of the allyl radical compared to 1°, 2°, and 3° radicals. (The stabilities of the radicals are relative to the hydrocarbon from which each was formed and the overall order of stability is: allyl > 3° > 2° > 1°.

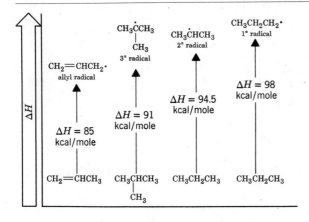

than either a secondary carbon-hydrogen bond or a primary carbon-hydrogen bond of propane.

The ease with which an allylic carbon-hydrogen bond is broken means that relative to primary, secondary, and tertiary free radicals the allyl radical is the *most stable* (Fig. 10.1).

10.2 THE STABILITY OF THE ALLYL RADICAL

The unusual stability of the allyl radical can be accounted for in two ways: in terms of molecular-orbital theory and in terms of resonance theory. The molecular-orbital explanation is easier to visualize so let us begin with it.

Molecular-Orbital Description of the Allyl Radical

As an allylic hydrogen is abstracted from propene, the sp^3-hybridized carbon of the methyl group changes its hybridization state to sp^2.

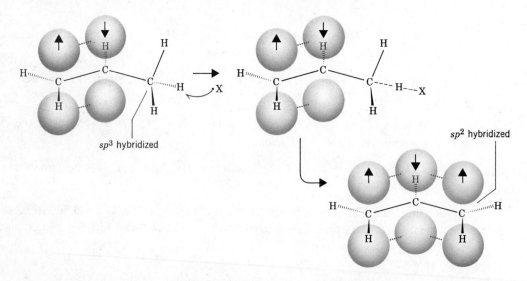

The p orbital of this new sp^2-hybridized carbon overlaps with the p orbital of the central carbon. Thus, in the allyl radical three p orbitals overlap to form π-molecular orbitals that encompass all three carbons. The new p-orbital of the allyl radical is said to be *conjugated* with those of the double bond and the allyl radical is said to be a *conjugated unsaturated system*.

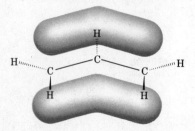

The unpaired electron of the allyl radical and the two electrons of the π bond are *delocalized* over all three carbons. This delocalization of the unpaired electron

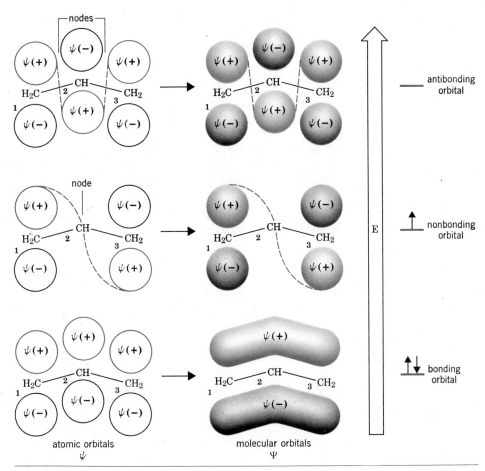

FIG. 10.2
The combination of atomic orbitals to form molecular orbitals in the allyl radical. The bonding molecular orbital is formed by the combination of three p orbitals with lobes of the same sign overlapping above and below the plane of the atoms. The non-bonding molecular orbital is formed by a combination of p orbitals from only two carbon atoms: carbons 1 and 3. This nonbonding molecular orbital has a node at carbon 2. The antibonding molecular orbital has two nodes: between carbons 1 and 2, and between carbons 2 and 3.

accounts for the greater stability of the allyl radical when compared to primary, secondary, and tertiary radicals. Although some delocalization occurs in primary, secondary, and tertiary radicals, delocalization is not as effective because it occurs through σ bonds.

The diagram in Figure 10.2 illustrates how the three *p* orbitals of the allyl radical combine to form three π-molecular orbitals.

The bonding molecular orbital of the allyl radical encompasses all three carbons and is occupied by two spin-paired electrons. The nonbonding molecular orbital is occupied by one unpaired electron. Because the nonbonding molecular orbital has a node located at the central carbon, the unpaired electron spends its time in the vicinity of carbons **1** and **3** only. The antibonding molecular orbital of the allyl radical has two nodes, and it is empty in the ground state.

We can illustrate the picture of the allyl radical given by molecular orbital theory in simpler terms with the structure shown below.

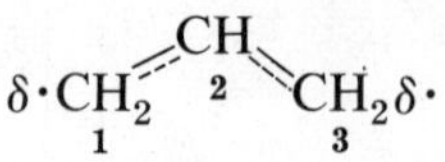

$$\delta \cdot \underset{1}{CH_2} \overset{CH}{\underset{2}{=\!\!\!=}} \underset{3}{CH_2} \delta \cdot$$

We indicate with dotted lines that both carbon-carbon bonds are partial double bonds. This accommodates one of the things that molecular-orbital theory tells us: *that there is a π bond encompassing all three atoms.* We also place the symbol $\delta \cdot$ beside carbons **1** and **3**. This denotes a second thing molecular-orbital theory tells us: *that the unpaired electron spends its time in the vicinity of carbons **1** and **3**.* Finally, implicit in the molecular orbital picture of the allyl radical is this: the two ends of the allyl radical are *equivalent.* This aspect of the molecular-orbital description is also implicit in the formula given above.

Resonance Description of the Allyl Radical

Earlier in this section we wrote the structure of the allyl radical as though it were **A.**

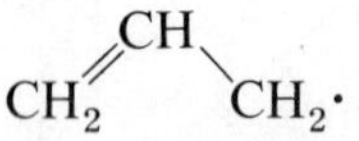

$$\underset{\textbf{A}}{CH_2 \overset{CH}{=\!\!\!=} CH_2 \cdot}$$

However, we might just as well have written the equivalent structure **B.**

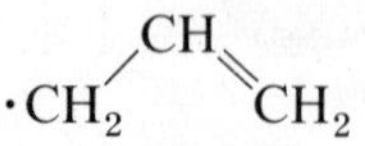

$$\underset{\textbf{B}}{\cdot CH_2 \overset{CH}{=\!\!\!=} CH_2}$$

In writing structure **B** we do not mean to imply that we have simply taken structure **A** and turned it over. What we have done is moved the electrons in the following way.

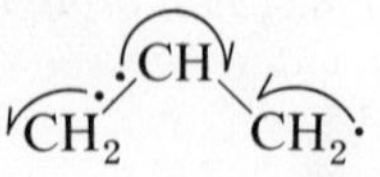

$$\overset{\curvearrowleft CH \curvearrowright}{CH_2 \qquad CH_2 \cdot}$$

We have not moved the atoms themselves.

Resonance theory (Section 1.13) tells us that whenever we can write two structures for a chemical entity *that differ only in the positions of the electrons,* the chemical entity is not represented by either structure alone but is a *hybrid* of both. We can represent the hybrid in two ways: we can write both structures **A** and **B,** and connect them with a double-headed arrow to indicate that they are resonance structures.

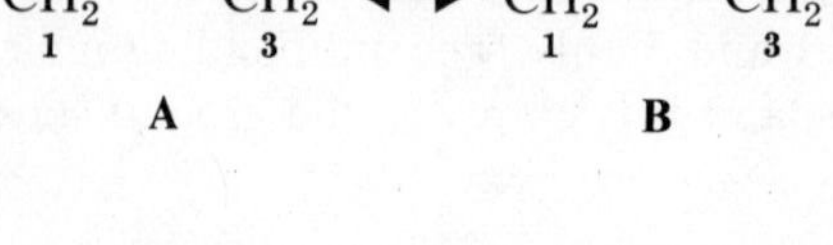

$$\underset{\textbf{A}}{\underset{1}{CH_2} \overset{CH}{\underset{2}{=\!\!\!=}} \underset{3}{CH_2} \cdot} \quad \longleftrightarrow \quad \underset{\textbf{B}}{\cdot \underset{1}{CH_2} \overset{CH}{\underset{2}{=\!\!\!=}} \underset{3}{CH_2}}$$

Or, we can write a single structure, **C**, that summarizes the features of both resonance structures.

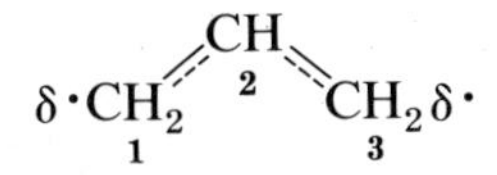

$$\delta \cdot \underset{1}{CH_2} \overset{CH}{\underset{2}{\diagup \ \diagdown}} \underset{3}{CH_2} \delta \cdot$$

C

Structure **C** describes the carbon-carbon bonds of the allyl radical as partial double bonds. This summarizes one feature of both **A** and **B**. In **A** the bond between carbon **1** and carbon **2** is a double bond and the bond between carbon **2** and carbon **3** is a single bond. In structure **B** just the reverse is true.

The resonance structures **A** and **B** also tell us that the unpaired electron is associated only with carbons **1** and **3**. We indicate this in structure **C** by placing a $\delta \cdot$ beside carbons **1** and **3**.*

Resonance theory also tells us that whenever *equivalent* resonance structures can be written for a chemical species, *the chemical species is much more stable than either resonance structure (when taken alone) would indicate.* If we were to examine either **A** or **B** alone we might decide that they resemble primary radicals. Thus, we might estimate the stability of the allyl radical as approximately that of a primary radical. In doing so, we would greatly underestimate the stability of the allyl radical. Resonance theory tells us, however, that since **A** and **B** are *equivalent resonance structures,* the allyl radical should be much more stable than either, that is, much more stable than a primary radical. This correlates with what experiments have shown to be true: the allyl radical is even more stable than a tertiary radical.

Problem 10.1

(a) What product(s) would you expect to obtain if propene labeled with ^{14}C at carbon **1** were subjected to allylic chlorination or bromination? (b) Explain your answer.

$$^{14}CH_2{=}CHCH_3 + X_2 \xrightarrow[\substack{\text{or} \\ \text{low conc. of } X_2}]{\text{high temperature}} \ ?$$

(c) If more than one product would be obtained what relative proportions would you expect?

10.3 THE ALLYL CATION

Although we cannot go into the experimental evidence here, the allyl cation, $CH_2{=}CHCH_2^+$, is an unusually stable carbocation. It is even more stable than a

---

* A resonance structure such as the one shown below would indicate that an unpaired electron is associated with carbon **2**. This structure is not a proper resonance structure because resonance theory dictates that *all resonance structures must have the same number of unpaired electrons.*

$$\cdot CH_2{-}\overset{\displaystyle \cdot}{CH}{-}CH_2\cdot$$
(an incorrect resonance structure)

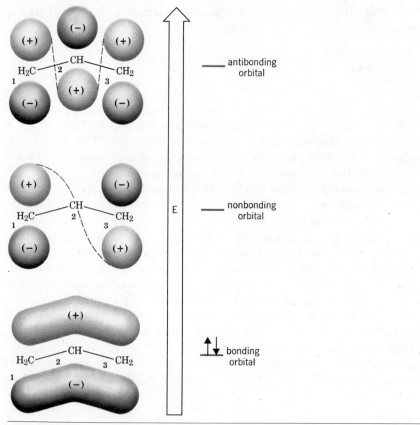

FIG. 10.3

The molecular orbitals of the allyl cation. The allyl cation, like the allyl radical (Fig. 10.2), is a conjugated unsaturated system.

secondary carbocation and is almost as stable as a tertiary carbocation. In general terms, the relative order of stabilities of carbocations is that given below.

Relative order of carbocation stability

$$
\begin{matrix}
& C & & & & C & & H \\
& | & & | & | & & | & & | \\
C-C^+ & > & -C{=}C-\overset{+}{C}H_2 & > & C-C^+ & > & C-C^+ \\
& | & & & & | & & | \\
& C & & & & H & & H \\
\\
& 3° & & \text{Allyl} & & 2° & & 1°
\end{matrix}
$$

As we might expect, the unusual stability of the allyl cation can also be accounted for in terms of molecular orbital or resonance theory.

The molecular orbital description of the allyl cation is shown in Fig. 10.3.

The bonding molecular orbital of the allyl cation, like that of the allyl radical (Fig. 10.2), contains two spin-paired electrons. The nonbonding molecular orbital of the allyl cation, however, is empty. Since an allyl cation is what we would get if we removed an electron from an allyl radical, we can say, that, in effect, we remove the electron from the nonbonding molecular orbital.

$$CH_2{=}CHCH_2 \cdot \xrightarrow{-e^-} CH_2{=}CHCH_2^+$$

Removal of an electron from a nonbonding orbital is known to require less energy than removal of an electron from a bonding orbital. In addition, the positive charge that forms on the allyl cation is *effectively delocalized* between carbons **1** and **3**. Thus, in molecular orbital theory these two factors, the ease of removal of a nonbonding electron, and the delocalization of charge account for the unusual stability of the allyl cation.

Resonance theory depicts the allyl cation as a hybrid of structures **D** and **E** below.

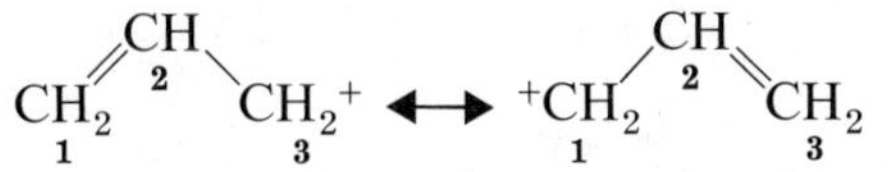

Because **D** and **E** are equivalent resonance structures, resonance theory predicts that the allyl cation should be unusually stable. Since the positive charge is located on carbon **3** in **D** and on carbon **1** in **E**, resonance theory also tells us that the positive charge should be delocalized over both carbons. The hybrid structure **F** (below) includes charge and bond features of both **D** and **E**.

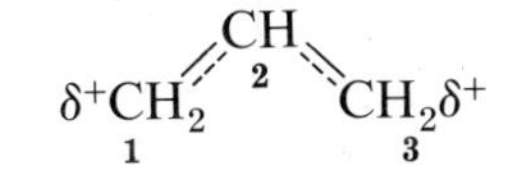

Problem 10.2

(a) Write structures corresponding to D, E, and F for the carbocation shown below.

$$CH_3\text{—}\overset{+}{CH}\text{—}CH\text{=}CH_2 \longleftrightarrow$$ *CH₃CH=CH—CH₂   CH₃—CH=CH=CH₂*

(b) This carbocation appears to be even more stable than a tertiary carbocation; how can you explain this?

(c) What product(s) would you expect to be formed if this carbocation reacted with a chloride ion?

CH₃CHCl—CH=CH₂   CH₃CH=CH—CH₂Cl

10.4 ALKADIENES AND POLYUNSATURATED HYDROCARBONS

Many hydrocarbons are known whose molecules contain more than one double or triple bond. A hydrocarbon whose molecules contain two double bonds is called an *alkadiene;* one whose molecules contain three double bonds is called a *alkatriene,* and so on. Colloquially, these compounds are often referred to simply as "dienes" or "trienes." A hydrocarbon with two triple bonds is called an *alkadiyne,* and a hydrocarbon with a double and a triple bond is called a *alkenyne.*

The following examples of polyunsaturated hydrocarbons illustrate how specific compounds are named.

$$CH_2\text{=}C\text{=}CH_2$$
1,2-Propadiene
(Allene)

$$\overset{1}{CH_2}\text{=}\overset{2}{CH}\text{—}\overset{3}{CH}\text{=}\overset{4}{CH_2}$$
1,3-Butadiene

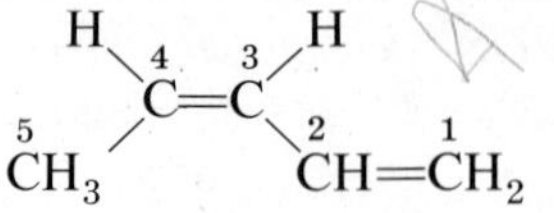

cis-1,3-Pentadiene

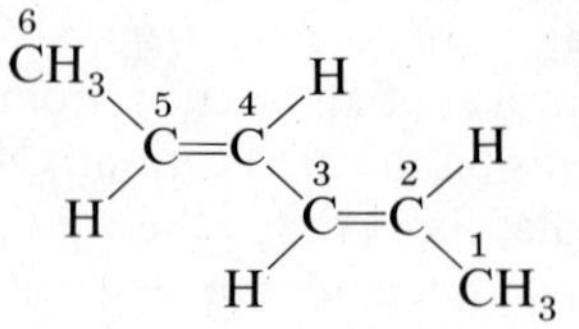

trans, trans-2,4-Hexadiene

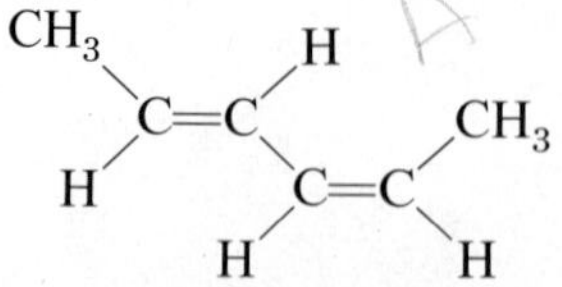

cis, trans-2,4-Hexadiene

$HC\equiv C-CH_2CH=CH_2$

1-Penten-4-yne*

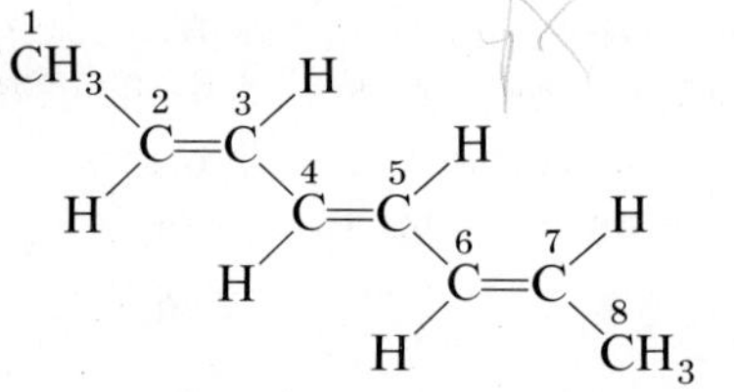

trans, trans, trans-2,4,6-Octatriene

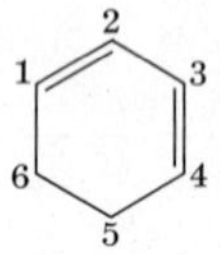

1,3-Cyclohexadiene 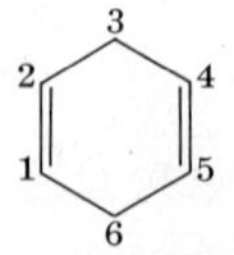   1,4-Cyclohexadiene

The multiple bonds of polyunsaturated compounds are classified as being *cumulated, conjugated,* or *isolated.* The double bonds of allene are said to be cumulated, because one carbon (the central carbon) participates in two double bonds.

$CH_2=C=CH_2$

Allene

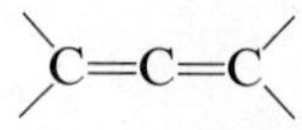

A cumulated
diene

An example of a conjugated diene is 1,3-butadiene. In conjugated polyenes the double and single bonds *alternate* along the chain.

$CH_2=CH-CH=CH_2$

1,3-Butadiene

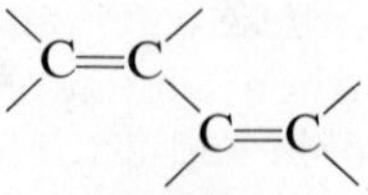

A conjugated diene

Trans, trans, trans-2,4,6-Octatriene is an example of a conjugated alkatriene.

If one or more saturated carbons intervene between the double bonds of an

---

* Where there is a choice, the double bond is given the lower number.

alkadiene, the double bonds are said to be *isolated*. An example of an isolated diene is 1,4-pentadiene.

$$CH_2=CH-CH_2-CH=CH_2$$

1,4-Pentadiene

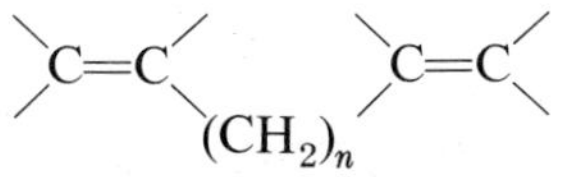

An isolated diene
$(n \neq 0)$

Problem 10.3

(a) Which other compounds on p. 352 are conjugated dienes? (b) Which other compound is an isolated diene? (c) Which compound is an isolated enyne?

In Chapter 7 we saw that appropriately substituted allenes give rise to chiral molecules even though the molecules do not have chiral centers.

Allenes are often obtained by the isomerization of alkynes.

$$-\overset{|}{C}-C\equiv C- \longrightarrow \quad C=C-C$$

An alkyne   An allene

Allenes are of some commercial importance and cumulated double bonds are occasionally found in naturally occurring compounds.

The double bonds of isolated dienes behave just as their name suggests—as isolated "enes." They undergo all of the reactions of alkenes; and, except for the fact that they are capable of reacting twice, their behavior is not unusual. Conjugated dienes are far more interesting because we find that their double bonds interact with each other. This interaction leads to unexpected properties and reactions. We will, therefore, consider the chemistry of conjugated dienes in detail.

10.5 1,3-BUTADIENE: ELECTRON DELOCALIZATION

Bond Lengths of 1,3-Butadiene

The carbon-carbon bond lengths of 1,3-butadiene have been determined and are shown below.

$$\overset{1}{CH_2}=\overset{2}{CH}-\overset{3}{CH}=\overset{4}{CH_2}$$
1.34 Å   1.47 Å   1.34 Å

The C-1—C-2 bond and the C-3—C-4 bond are, (within experimental error) the same length as the carbon-carbon double bond of ethene. The central bond of 1,3-butadiene (1.47 Å), however, is considerably shorter than the single bond of ethane (1.54 Å).

This should not be surprising. All of the carbons of 1,3-butadiene are sp^2 hybridized and, as a result, the central bond of butadiene results from overlapping sp^2 orbitals. The carbon-carbon bond of ethane, by contrast, results from overlapping

TABLE 10.1 Carbon-Carbon Single Bond Lengths and Hybridization State

COMPOUND	HYBRIDIZATION STATE	BOND LENGTH Å
$H_3C{-}CH_3$	$sp^3\text{-}sp^3$	1.54
$CH_2{=}CH{-}CH_3$	$sp^2\text{-}sp^3$	1.50
$CH_2{=}CH{-}CH{=}CH_2$	$sp^2\text{-}sp^2$	1.47
$HC{\equiv}C{-}CH_3$	$sp\text{-}sp^3$	1.46
$HC{\equiv}C{-}C{=}CH_2$	$sp\text{-}sp^2$	1.43
$HC{\equiv}C{-}C{\equiv}CH$	$sp\text{-}sp$	1.37

sp^3 orbitals. And, as we know, sp^3 *orbitals are longer*. There is, in fact, a steady decrease in bond length of carbon-carbon single bonds as the hybridization state of the bonded atoms changes from sp^3 to sp (Table 10.1).

Conformations of 1,3-Butadiene

There are two possible planar conformations of 1,3-butadiene: the *s-cis* and the *s-trans* conformation.

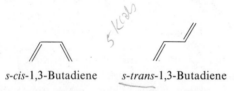

    s-cis-1,3-Butadiene   *s-trans*-1,3-Butadiene

These are not true *cis* and *trans* forms since the *s-cis* and *s-trans* conformations of 1,3-butadiene can be interconverted through rotation about the central bond. The barrier to rotation is approximately 5 kcal/mole. The *s-trans* conformation is the predominant one at room temperature.

Molecular Orbitals of 1,3-Butadiene

The central carbons of 1,3-butadiene (Fig. 10.4) are close enough so that there is overlap between the *p* orbitals of carbon-2 and carbon-3. The overlap between the orbitals of C-2 and C-3 is not as great as that between the orbitals of C-1 and C-2 (or those of C-3 and C-4). The C-2, C-3 orbital overlap, however, gives the central bond partial double bond character and allows the four π electrons of 1,3-butadiene to be delocalized over all four atoms.

Figure 10.5 shows how the four *p* orbitals of 1,3-butadiene combine to form a set of four molecular orbitals.

Two of the molecular orbitals of 1,3-butadiene are bonding molecular orbitals and in the ground state these orbitals hold the four π electrons with two spin-paired electrons in each. The other two molecular orbitals are antibonding molecular orbitals; in the ground state these orbitals are empty. An electron can be excited from the highest occupied molecular orbital to the lowest empty molecular orbital when 1,3-butadiene absorbs light with a wavelength of 217 nanometers (2170 Å). (We will study the absorption of light by unsaturated molecules in Section 10.9).

The delocalized bonding that we have just described for 1,3-butadiene is characteristic of all conjugated polyenes.

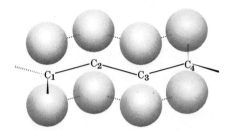

FIG. 10.4
The p orbitals of 1,3-butadiene.

FIG. 10.5
The molecular orbitals of 1,3-butadiene.

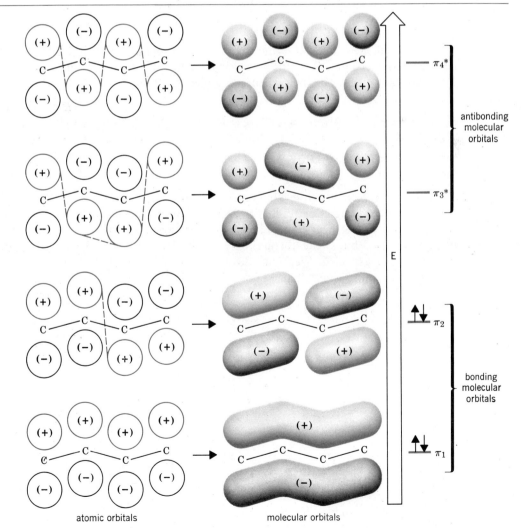

atomic orbitals   molecular orbitals

TABLE 10.2 Heats of Hydrogenation of Alkenes and Alkadienes

COMPOUND	MOLES H_2	ΔH, KCAL/MOLE
1-Butene	1	-30.3
1-Pentene	1	-30.1
trans-2-Pentene	1	-27.6
1,3-Butadiene	2	-57.1
trans-1,3-Pentadiene	2	-54.1
1,4-Pentadiene	2	-60.8
1,5-Hexadiene	2	-60.5

10.6 THE STABILITY OF CONJUGATED DIENES

The double bonds of conjugated alkadienes are thermodynamically more stable than those of alkenes or those of isolated alkadienes. Two examples of this extra stability of conjugated dienes can be seen if we analyze the heats of hydrogenation given in Table 10.2.

In itself, 1,3-butadiene cannot be compared directly to an isolated diene of the same chain length. We can, however, compare the heat of hydrogenation of 1,3-butadiene to that obtained when two moles of 1-butene are hydrogenated.

$$\Delta H, \text{kcal/mole}$$

$$2CH_2{=}CHCH_2CH_3 + 2H_2 \longrightarrow 2CH_3CH_2CH_2CH_3 \ (2 \times -30.3) = -60.6$$
$$\text{1-butene}$$

$$CH_2{=}CHCH{=}CH_2 + 2H_2 \longrightarrow CH_3CH_2CH_2CH_3 \qquad \underline{-57.1}$$
$$\text{Difference} \qquad 3.5 \text{ kcal/mole}$$

Since 1-butene has the same kind of monosubstituted double bond as those of 1,3-butadiene, we might expect that hydrogenation of 1,3-butadiene would liberate the same amount of heat (-60.6 kcal/mole) as two moles of 1-butene. We find, however, that 1,3-butadiene liberates only 57.1 kcal/mole. This is 3.5 kcal/mole *less* than what we expected. We conclude, therefore, that the double bonds of 1,3-butadiene are more stable than the double bond of 1-butene (Fig. 10.6).

With the pentadienes we can make a direct comparison between conjugated and isolated dienes of the same chain length:

$$CH_2{=}CH{-}CH_2{-}CH{=}CH_2 \qquad \Delta H = -60.8 \text{ kcal/mole}$$
$$\text{1,4-Pentadiene}$$

$$CH_2{=}CH \diagdown$$
$$\qquad\quad C{=}C \diagup^{H}_{}$$
$$H \diagup \qquad \diagdown CH_3 \qquad \qquad \Delta H = -54.1 \text{ kcal/mole}$$
$$\text{trans-1,3-Pentadiene} \qquad \overline{\text{Difference} = \quad 6.7 \text{ kcal/mole}}$$

It is not really appropriate, however, to make a direct comparison of these two compounds in assessing the degree of stabilization that conjugation affords the double bonds of *trans*-1,3-pentadiene. Both double bonds of 1,4-pentadiene are monosubstituted double bonds, while one of the double bonds of *trans*-1,3-

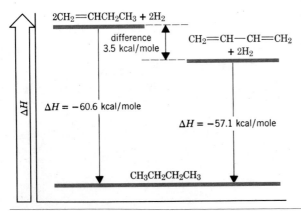

FIG. 10.6
*Heats of hydrogenation of two moles of 1-butene and
one mole of butadiene.*

pentadiene is a *trans*-disubstituted double bond. Disubstituted double bonds are
known to be more stable (p. 165).

A better assessment of the stabilization that conjugation provides the double
bonds of *trans*-1,3-pentadiene can be made by comparing the heat of hydrogenation
of *trans*-1,3-pentadiene to the sum of the heats of hydrogenation of 1-pentene and
trans-2-pentene. This way we will be comparing double bonds of comparable types.

$$CH_2{=}CHCH_2CH_2CH_3 \qquad\qquad \Delta H = -30.1 \text{ kcal/mole}$$
1-Pentene

$$\Delta H = -27.6 \text{ kcal/mole}$$
$$\overline{\text{Sum} = -57.7 \text{ kcal/mole}}$$
trans-2-Pentene

$$\Delta H = -54.1 \text{ kcal/mole}$$
$$\overline{\text{Difference} = \quad 3.6 \text{ kcal/mole}}$$
trans-1,3-Pentadiene

We see from these calculations that conjugation affords the double bonds of
trans-1,3-pentadiene an extra stability of 3.6 kcal/mole, a value that is very close
to the one we calculated for 1,3-butadiene (3.5 kcal/mole).

When calculations like these are carried out for other conjugated dienes,
similar results are obtained; conjugated dienes are found to be more stable than
isolated dienes. The question, then, is this: What is the source of the extra stability
associated with conjugated dienes? This is a question that is still being debated. Some
chemists argue that the extra stability of conjugated dienes arises solely from the
stronger sp^2-sp^2 bond that they contain. Others argue that the extra stability arises
from the additional delocalization of the π-electrons that occurs in conjugated dienes.
It is likely that both arguments are correct and that the additional stability is derived
from both sources.

10.7 ELECTROPHILIC ATTACK AND CONJUGATED DIENES: 1,4 ADDITION

Not only are conjugated dienes somewhat more stable than we might expect them to be, they also display unusual behavior when they react with electrophilic reagents. For example, 1,3-butadiene reacts with one mole of hydrogen chloride to produce two products: 3-chloro-1-butene and 1-chloro-2-butene.

$$CH_2=CH-CH=CH_2 \xrightarrow[25°]{HCl} CH_2=CH-\underset{\underset{Cl}{|}}{CH}-CH_3 + ClCH_2-CH=CH-CH_3$$

1,3-Butadiene   3-Chloro-1-butene   1-Chloro-2-butene
                                   (78%)   (22%)

If only the first product (3-chloro-1-butene) were formed, we would not be particularly surprised. We would conclude that hydrogen chloride had added to one double bond of 1,3-butadiene in the usual way.

$$\overset{1}{C}H_2=\overset{2}{C}H-\overset{3}{C}H=\overset{4}{C}H_2 \atop + \atop H-Cl \xrightarrow{\text{1,2-addition}} \underset{\underset{H}{|}}{CH_2}-\underset{\underset{Cl}{|}}{CH}-CH=CH_2$$

                                                 3-Chloro-1-butene

It is the second product (1-chloro-2-butene) that is unusual. In it the double bond is between the central atoms and the elements of hydrogen chloride have added to carbons-1 and -4.

$$\overset{1}{C}H_2=\overset{2}{C}H-\overset{3}{C}H=\overset{4}{C}H_2 \atop + \atop H-Cl \xrightarrow{\text{1,4-addition}} \underset{\underset{H}{|}}{CH_2}-CH=CH-\underset{\underset{Cl}{|}}{CH_2}$$

                                                 1-Chloro-2-butene

This unusual behavior of 1,3-butadiene can be attributed directly to the stability and the delocalized nature of an allylic cation. In order to see how this can be done, let us consider a mechanism for the addition of hydrogen chloride.

Step 1 $H^+ + CH_2=CH-CH=CH_2 \longrightarrow CH_3-\underset{+}{CH}-CH=CH_2 \longleftrightarrow CH_3-CH=CH-\underset{+}{CH_2}$

                                                  An allylic cation
                                                  equivalent to

$$CH_3-\underset{\delta^+}{CH}\cdots CH \cdots \underset{\delta^+}{CH_2}$$

Step 2 $CH_3\underset{\delta^+}{CH}\cdots CH \cdots \underset{\delta^+}{CH_2} + :\ddot{C}l:^-$ ⟶

                              ⟶ $CH_3\underset{\underset{Cl}{|}}{CH}-CH=CH_2$ 1,2 Addition

                              ⟶ $CH_3CH=CHCH_2Cl$ 1,4 Addition

In step 1 a proton adds to one of the terminal carbons of 1,3-butadiene to form a highly stable allylic cation. Addition to one of the inner carbons would have produced a much less stable primary cation.

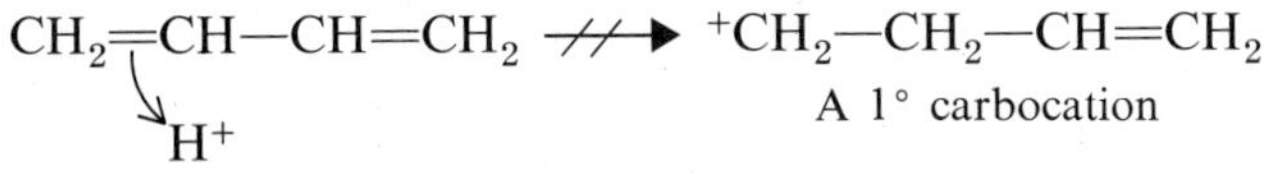

$$CH_2=CH-CH=CH_2 \xrightarrow{H^+} {}^+CH_2-CH_2-CH=CH_2$$

A 1° carbocation

In step 2 a chloride ion forms a bond to one of the carbons of the allylic cation that bears a partial positive charge. Reaction at one carbon results in the 1,2-addition product, reaction at the other gives the 1,4-addition product.

Problem 10.4

(a) What products would you expect to obtain if hydrogen chloride were allowed to react with a 2,4-hexadiene, $CH_3CH=CHCH=CHCH_3$? (b) With a 1,3-pentadiene, $CH_2=CHCH=CHCH_3$? (Neglect *cis-trans* isomerism.)

Butadiene shows 1,4-addition reactions with electrophilic reagents other than hydrogen chloride. Two examples are shown below, the addition of hydrogen bromide and the addition of bromine.

$$CH_2=CHCH=CH_2 + HBr \xrightarrow{40°} CH_3CHBrCH=CH_2 + CH_3CH=CHCH_2Br$$

$$(20\%) \qquad\qquad (80\%)$$

$$CH_2=CHCH=CH_2 + Br_2 \xrightarrow[\text{hexane}]{-15°} CH_2BrCHBrCH=CH_2 + CH_2BrCH=CHCH_2Br$$

$$(54\%) \qquad\qquad\qquad (46\%)$$

Reactions of this type are quite general with other conjugated dienes. Conjugated trienes often show 1,6 addition. An example is the 1,6 addition of bromine to 1,3,5-cyclooctatriene:

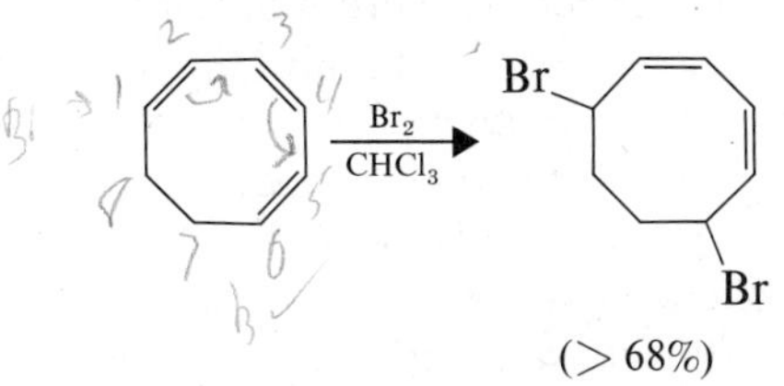

($> 68\%$)

Rate Versus Equilibrium Control of a Chemical Reaction

The addition of hydrogen bromide to 1,3-butadiene is interesting in another respect: the relative amounts of 1,2- and 1,4-addition product that we obtain are dependent on the temperature at which we carry out the reaction.

When 1,3-butadiene and hydrogen bromide are allowed to react at a low temperature ($-80°$), the major reaction pathway is 1,2 addition; we obtain about 80% of the 1,2 product and only about 20% of the 1,4 product. At a higher temperature ($40°$) the situation is reversed. The major reaction pathway is 1,4 addition; we obtain about 80% of the 1,4 product and only about 20% of the 1,2 product.

When the reaction mixture that is formed at the lower temperature is brought to the higher temperature, moreover, the relative amounts of the two products change. Ultimately, this reaction mixture contains the same proportion of products given by the reaction carried out at the higher temperature.

It can also be shown that at the higher temperature and in the presence of

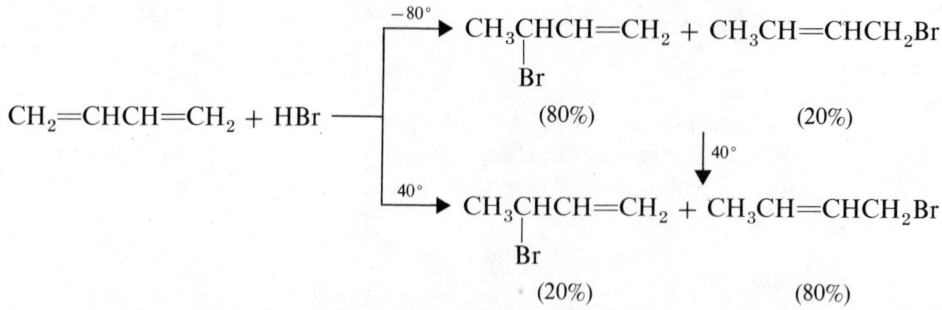

$$\text{CH}_2\text{=CHCH=CH}_2 + \text{HBr} \xrightarrow{\quad} \begin{array}{l} \xrightarrow{-80°} \text{CH}_3\underset{\underset{\text{Br}}{|}}{\text{CHCH=CH}_2} + \text{CH}_3\text{CH=CHCH}_2\text{Br} \\ \qquad\qquad (80\%) \qquad\qquad\qquad (20\%) \\[2em] \xrightarrow{40°} \text{CH}_3\underset{\underset{\text{Br}}{|}}{\text{CHCH=CH}_2} + \text{CH}_3\text{CH=CHCH}_2\text{Br} \\ \qquad\qquad (20\%) \qquad\qquad\qquad (80\%) \end{array}$$

(middle arrow: 40°)

hydrogen bromide, the 1,2-addition product rearranges to the 1,4-product and that an equilibrium exists between them.

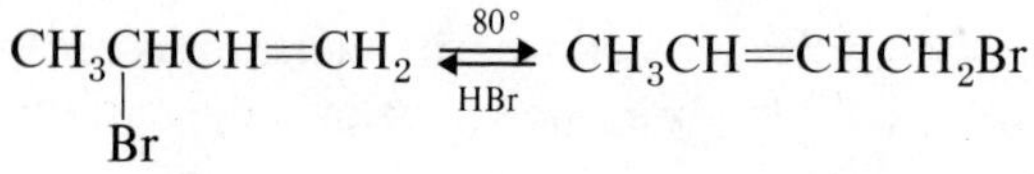

$$\text{CH}_3\underset{\underset{\text{Br}}{|}}{\text{CHCH=CH}_2} \underset{\text{HBr}}{\overset{80°}{\rightleftharpoons}} \text{CH}_3\text{CH=CHCH}_2\text{Br}$$

| 1,2-Addition product | 1,4-Addition product |

Since this equilibrium favors the 1,4-addition product, *it must be more stable.*

The reactions of hydrogen bromide and 1,3-butadiene serve as a striking illustration of the way that the outcome of a chemical reaction can be determined, in one instance, by relative rates of competing pathways and, in another, by the relative stabilities of the final products themselves. At the lower temperature, the relative amounts of the products of the addition are determined by the relative rates at which the two additions occur: 1,2 addition occurs faster so the 1,2-addition product is the major product. At the higher temperature, the relative amounts of the products are determined by the position of an equilibrium: the 1,4-addition product is the more stable so it is the major product.

This behavior of 1,3-butadiene and hydrogen bromide can be more fully understood if we examine the diagram shown in Fig. 10.7.

The step that determines the overall outcome of the reaction is the step in which the hybrid allylic cation combines with a bromide ion, that is,

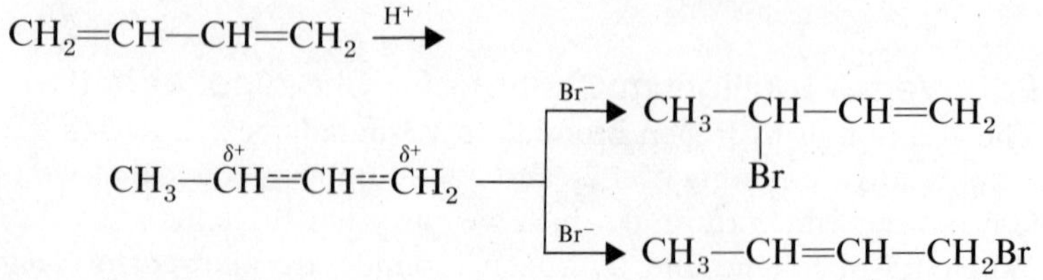

$$\text{CH}_2\text{=CH—CH=CH}_2 \xrightarrow{\text{H}^+}$$

$$\text{CH}_3\overset{\delta+}{\text{—CH}}\text{=\!=\!=CH}\overset{\delta+}{\text{=\!=\!=CH}_2} \xrightarrow{\quad} \begin{array}{l} \xrightarrow{\text{Br}^-} \text{CH}_3\text{—CH—CH=CH}_2 \qquad \text{1,2 product} \\ \qquad\qquad\quad \underset{}{|} \\ \qquad\qquad\ \ \text{Br} \\[1em] \xrightarrow{\text{Br}^-} \text{CH}_3\text{—CH=CH—CH}_2\text{Br} \qquad \text{1,4 product} \end{array}$$

This step determines the orientation of the reaction

We see in Fig. 10.7 that, for this step, the energy of activation leading to the 1,2-addition product is less than the energy of activation leading to the 1,4-addition product, even though the 1,4 product is more stable. At low temperatures, a larger fraction of collisions between the intermediate ions will have enough energy to cross the lower barrier (leading to the 1,2 product), and only a very small fraction of collisions will have enough energy to cross the higher barrier (leading to the 1,4 product). In either case (and this is the *key point*) whichever barrier is crossed, product formation is *irreversible* because there is not enough energy available to

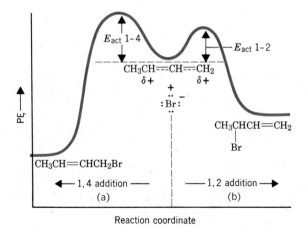

FIG. 10.7

Potential energy versus *reaction coordinate diagram for the reactions of an allyl cation with a bromide ion. One reaction pathway (a) leads to the 1,4 addition product and the other (b) leads to the 1,2 addition product.*

lift either product out of its deep potential energy well. Since 1,2 addition occurs faster, the 1,2 product predominates and the reaction is said to be under *rate control.*

At higher temperatures, the intermediate ions have sufficient energy to cross both barriers with relative ease. More importantly, however, *both reactions are reversible.* Sufficient energy is also available to take the products back over their energy barriers to the intermediate level of allylic cations and bromide ions. The 1,2 product is still formed faster, but being less stable than the 1,4 product, it also reverts to the allylic cation faster. Under these conditions, that is, at higher temperatures, the relative proportions of the products *do not reflect* the relative heights of the energy barriers leading from allylic cation to products. Instead, *they reflect the relative stabilities of the products themselves.* Since the 1,4 product is more stable, it is formed at the expense of the 1,2 product because the overall change from 1,2 product to 1,4 product is energetically favorable. Such a reaction is said to be under *equilibrium control.*

Before we leave this subject one final point should be made. This example clearly demonstrates that relative reaction rate predictions made on the basis of product stabilities alone can conflict with the results we actually obtain. This is not always the case, however. For many reactions in which a common intermediate leads to two or more products, the most stable product is formed fastest.

Problem 10.5

(a) Can you suggest a possible explanation for the fact that the 1,2-addition reaction of 1,3-butadiene and hydrogen bromide occurs faster? (Hint: Consider the relative contributions that the two forms $CH_3\overset{+}{C}HCH=CH_2$ and $CH_3CH=CH\overset{+}{C}H_2$ make to the resonance hybrid of the allylic cation.) (b) How can you account for the fact that the 1,4-addition product is more stable?

10.8 THE DIELS-ALDER REACTION: A 1,4-CYCLOADDITION REACTION OF DIENES

In 1928 two German chemists, Otto Diels and Kurt Alder, discovered a 1,4-cyclo-addition reaction of dienes that has since come to bear their names. The reaction proved to be one of such great versatility and synthetic utility that Diels and Alder were awarded the Nobel Prize in 1950.

An example of the Diels-Alder reaction is the reaction that takes place when 1,3-butadiene and maleic anhydride are heated together at 100°. The product is obtained in quantitative yield.

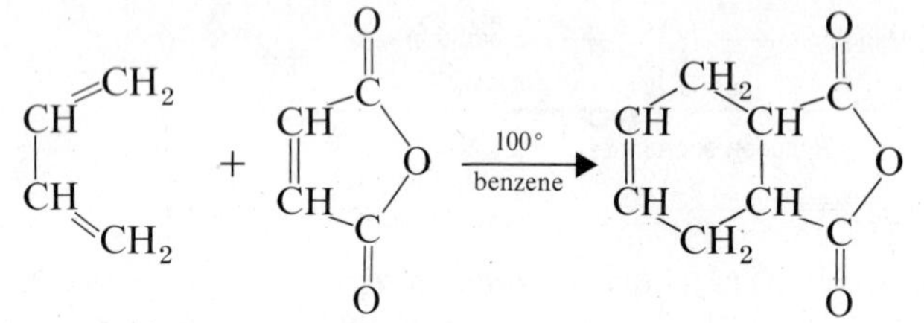

| 1,3-Butadiene (diene) | Maleic anhydride (dienophile) | (100%) (adduct) |

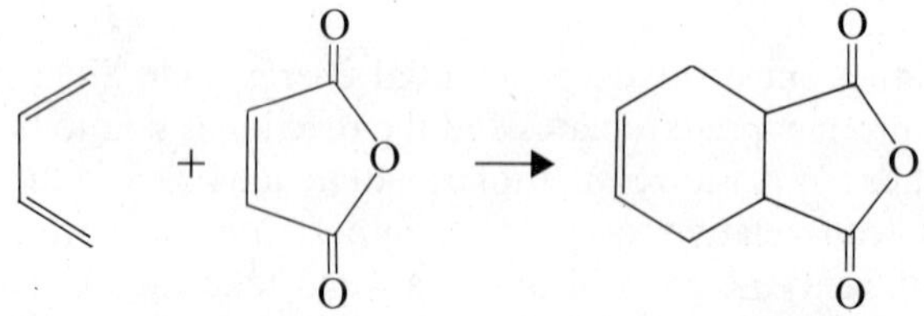

In general terms, the reaction is one between a conjugated *diene* (a 4π-electron system) and a double-bond containing compound (a 2π-electron system) called a *dienophile* (diene + Gr: *philein,* to love). The product of a Diels-Alder reaction is often called the *adduct.* In the Diels-Alder reaction, two new σ bonds are formed at the expense of two π bonds of the diene and dienophile. Since σ bonds are usually stronger than π bonds, formation of the adduct is usually favored energetically, *but most Diels-Alder reactions are reversible.*

The simplest example of a Diels-Alder reaction is the one that takes place between 1,3-butadiene and ethylene.

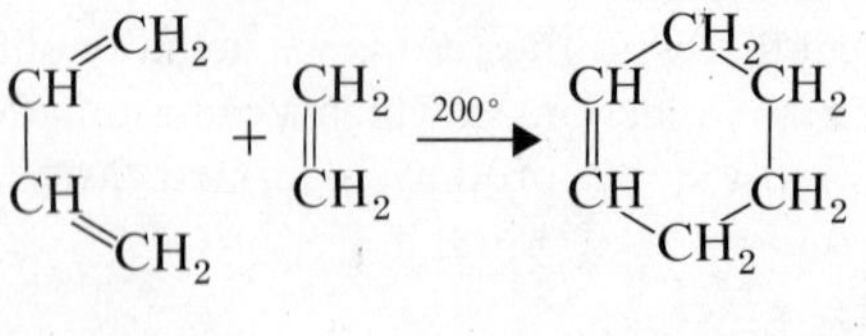

This is also one of the poorest examples because the product, cyclohexene, is obtained in only 20% yield.

Alder originally stated that the Diels-Alder reaction is favored by the presence of electron-withdrawing groups in the dienophile and by electron-releasing groups in the diene. Maleic anhydride, a very potent dienophile, has two carbonyl groups on carbons adjacent to the double bond. Carbonyl groups are electron withdrawing because of the electronegativity of their oxygen atoms and because resonance structures such as those shown below contribute to the hybrid.

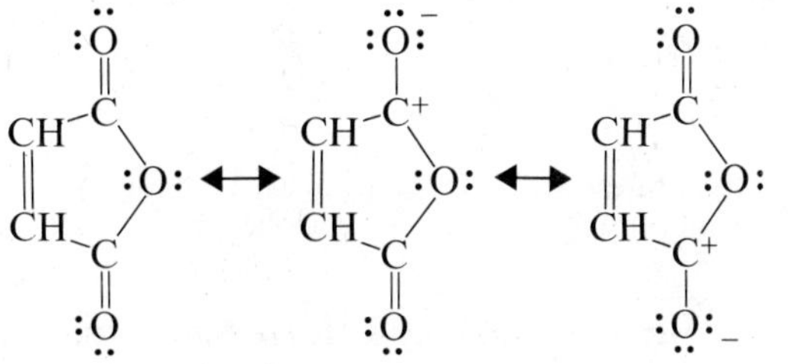

The two examples that we have given (1,3-butadiene + maleic anhydride and 1,3-butadiene + ethylene) illustrate the effect that the presence of electron-withdrawing groups in the dienophile has on the outcome of the reaction.

The effect of electron-releasing groups in the diene can also be demonstrated; 2,3-dimethyl-1,3-butadiene, for example, is nearly five times as reactive in Diels-Alder reactions as is 1,3-butadiene. When 2,3-dimethyl-1,3-butadiene reacts with acrolein at only 30° the adduct is obtained in quantitative yield.

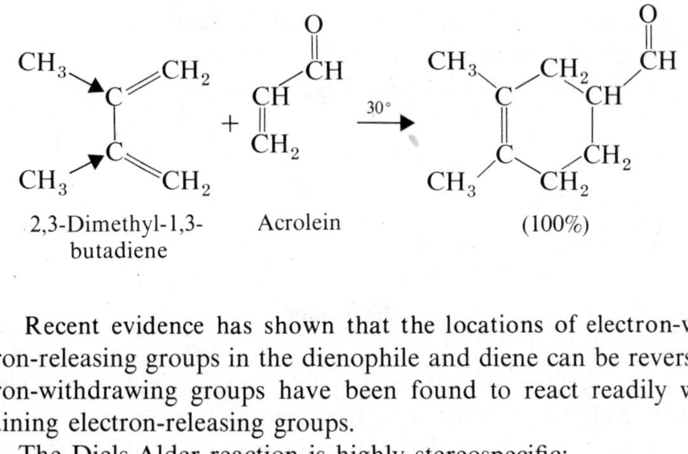

2,3-Dimethyl-1,3-butadiene   Acrolein   (100%)

Recent evidence has shown that the locations of electron-withdrawing and electron-releasing groups in the dienophile and diene can be reversed. Dienes with electron-withdrawing groups have been found to react readily with dienophiles containing electron-releasing groups.

The Diels-Alder reaction is highly stereospecific:

1. The reaction is a syn addition and the configuration of the dienophile is *retained* in the products. Two examples that illustrate this aspect of the reaction are shown below.

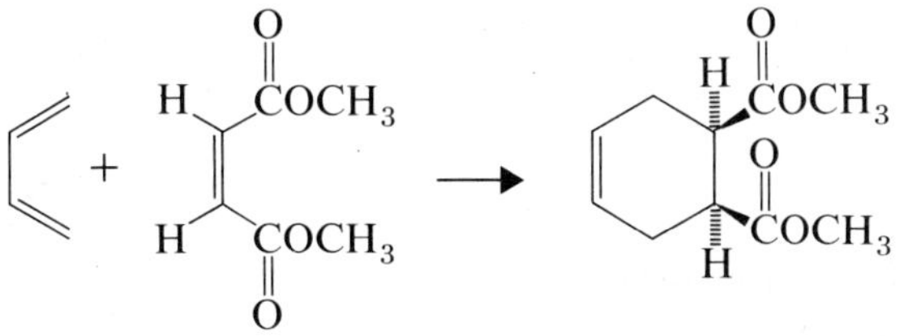

Dimethyl maleate   cis-4,5-Dicarbomethoxy-
(a cis-dienophile)   cyclohexene

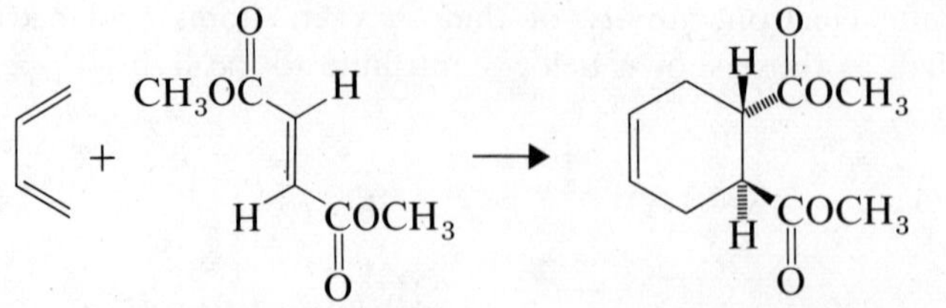

Dimethyl fumarate
(a *trans*-dienophile)

trans-4,5-Dicarbomethoxy-
cyclohexene

In the first example, a dienophile with *cis* ester groups reacts with 1,3-butadiene to give an adduct with *cis* ester groups. In the second example just the reverse is true. A *trans* dienophile gives a *trans* adduct.

2. The diene component of the Diels-Alder reaction must, of necessity, react in the *s-cis* conformation rather than the *s-trans*.

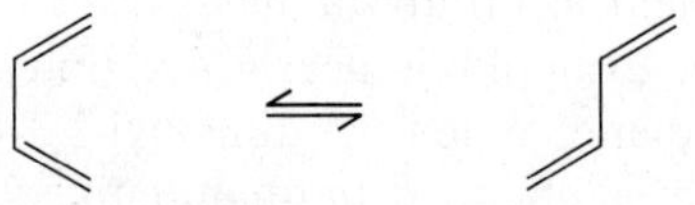

s-cis Conformation   *s-trans* Conformation

Reaction in the *s-trans* conformation would, if it occurred, produce a six-membered ring with a highly strained *trans* double bond. This course of the Diels-Alder reaction has never been observed.

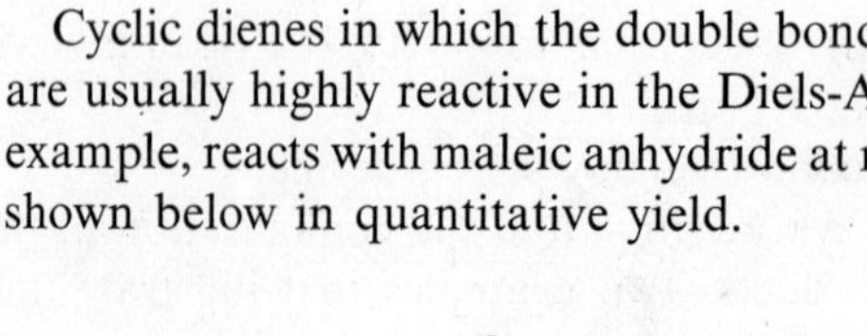

Highly strained

Cyclic dienes in which the double bonds are held in the *s-cis* configuration are usually highly reactive in the Diels-Alder reaction. Cyclopentadiene, for example, reacts with maleic anhydride at room temperature to give the adduct shown below in quantitative yield.

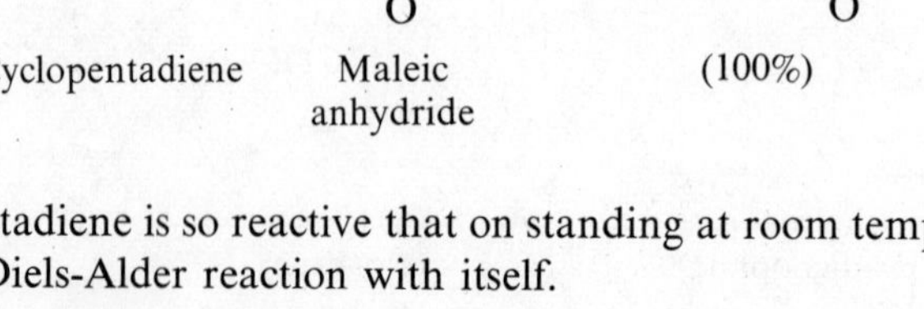

Cyclopentadiene   Maleic
anhydride

(100%)

Cylopentadiene is so reactive that on standing at room temperature it slowly undergoes a Diels-Alder reaction with itself.

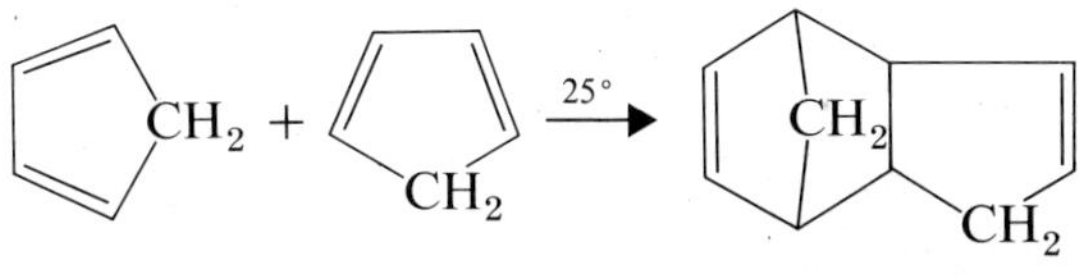

"Dicyclopentadiene"

The reaction is reversible, however. When "dicyclopentadiene" is distilled, it dissociates into two moles of cyclopentadiene.

The reactions of cyclopentadiene illustrate a third stereochemical characteristic of the Diels-Alder reaction.

> **3.** The Diels-Alder reaction occurs primarily in an *endo* rather than an *exo* fashion.

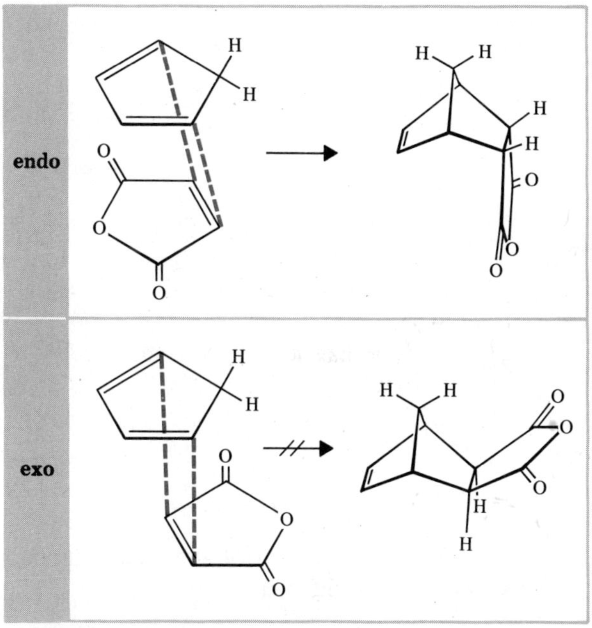

The favored *endo* stereochemistry seems to arise from favorable interactions between the π electrons of the developing double bond in the diene and the π electrons of unsaturated groups of the dienophile. In the example above, the π electrons of the —C—O—C— linkage of the anhydride interact with the π electrons of the devel-
$$\underset{O}{\|}\qquad\underset{O}{\|}$$
oping double bond in cyclopentadiene.

Problem 10.6

The dimerization of cyclopentadiene also occurs in an *endo* way. (a) Show how this happens. (b) Which π electrons interact? (c) What is the three-dimensional structure of the product?

Problem 10.7

What products would you expect from the following reactions?

(a)

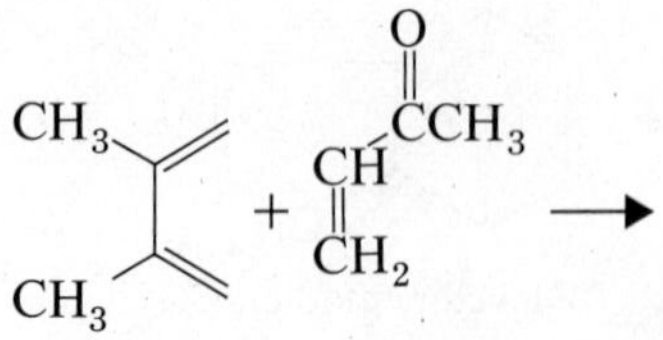

(b)

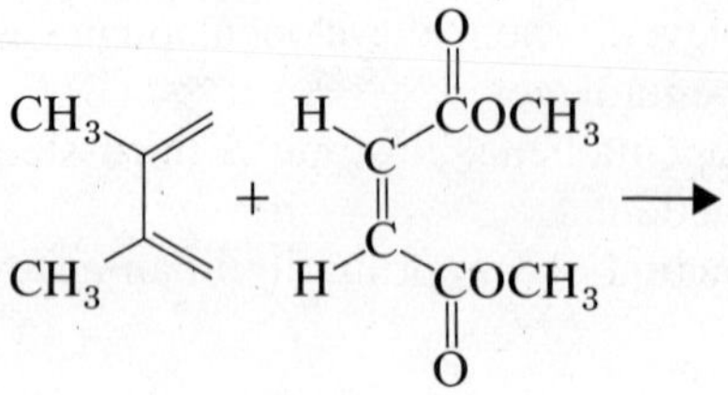

(c)

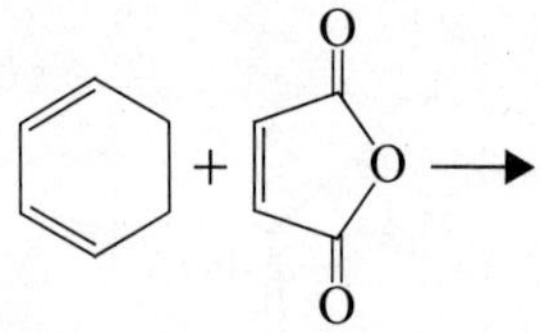

Problem 10.8

What diene and dienophile would you employ to synthesize the following compound?

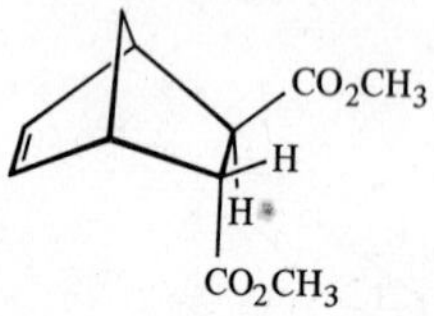

Problem 10.9

Diels-Alder reactions also take place with triple-bonded (acetylenic) dienophiles. What diene and what dienophile would you use to prepare:

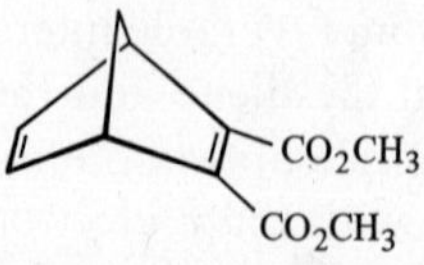

10.9 VISIBLE AND ULTRAVIOLET SPECTROSCOPY

The Electromagnetic Spectrum

The names of most forms of electromagnetic energy have become familiar terms. The *X rays* used in medicine, the *light* that we see, the *ultraviolet* rays that produce sunburns, and the *radio* and *radar* waves used in communication are all different forms of the same phenomenon: electromagnetic energy.

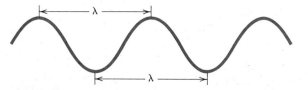

FIG. 10.8

A simple wave and the wavelength, λ.

According to quantum mechanics, electromagnetic energy has a dual and seemingly contradictory nature. Electromagnetic energy has both the properties of a wave and a particle (p. 283). Electromagnetic energy can be described as though it were a wave occurring simultaneously in electrical and magnetic fields. It can also be described as though it consisted of particles called quanta or photons.

A wave is usually described in terms of its wavelength (λ) or its frequency (*v*). A simple wave is shown in Fig. 10.8. The distance between consecutive crests (or troughs) is the wavelength. The number of full cycles of the wave that pass a given point each second, as the wave moves through space, is called the frequency.

An analogy may make the term frequency more understandable. Imagine that you are anchored offshore in a small boat. Since your boat is anchored, your position on the surface of the water is fixed. As waves pass by, the boat rises to one crest and falls into another trough. Now suppose that you decide to count the number of times the boat rises from one crest to the next each minute. In doing this you will be measuring the frequency of the waves in cycles per minute.

The frequencies of most electromagnetic waves are much higher than those of a wave on an ocean or lake. As a result, the frequencies of electromagnetic waves are usually reported in cycles per second.* The wavelengths of electromagnetic radiation are expressed in either meters (m), millimeters (1 mm = 10^{-3}m), microns (1 μ = 10^{-6}m), or nanometers (1 nm = 10^{-9}m).

The energy of a quantum of electromagnetic energy is directly related to its frequency.

$$E = hv$$

where *h* = Planck's constant, 6.625 × 10^{-27} erg-sec,

and

v = the frequency in cycles/sec.

This means that the higher the frequency of radiation the greater is its energy. X rays, for example, are much more energetic than rays of visible light. The frequencies of X rays are of the order of 10^{19} cycles/sec, while those of visible light are of the order of 10^{15} cycles/sec.

The energy of electromagnetic radiation is inversely proportional to its wavelength,

$$E = \frac{hc}{\lambda} \qquad (c = \text{the velocity of light})$$

---

* The term Hertz (after the German physicist H. R. Hertz) is now often used in place of *cycles per second.* Frequency of electromagnetic radiation is also sometimes expressed in *wave numbers,* that is, the number of waves per centimeter.

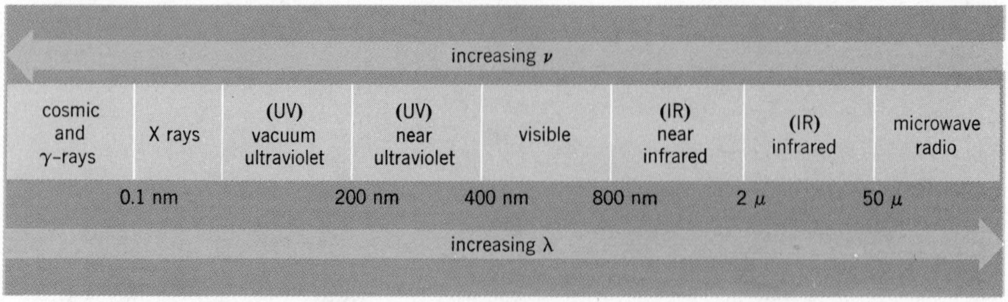

increasing ν							
cosmic and γ-rays	X rays	(UV) vacuum ultraviolet	(UV) near ultraviolet	visible	(IR) near infrared	(IR) infrared	microwave radio
0.1 nm		200 nm		400 nm	800 nm	2 μ	50 μ
increasing λ							

FIG. 10.9
The electromagnetic spectrum.

Thus, electromagnetic radiation of long wavelength has low energy, while that of short wavelength has high energy. X rays have wavelengths of the order of 0.1 nm, while visible light has wavelengths between 400 nm and 750 nm.*

It may be helpful to point out, too, that for visible light, wavelengths (and, thus, frequencies) are related to what we perceive as colors. The light that we call red light has a wavelength of approximately 750 nm. The light we call violet light has a wavelength of approximately 400 nm. All of the other colors of the visible spectrum (the rainbow) lie in between these wavelengths.

The different regions of the electromagnetic spectrum are shown in Fig. 10.9. Nearly every portion of the electromagnetic spectrum from the region of X rays to those of microwave and radio wave has been used in elucidating structures of atoms and molecules. In Chapter 14 we discuss the use that can be made of the infrared and radio regions when we take up infrared spectroscopy and nuclear magnetic resonance spectroscopy. At this point we will direct our attention to electromagnetic radiation in the near ultraviolet and visible regions and see how it interacts with conjugated polyenes.

Visible and Ultraviolet Spectra

When electromagnetic radiation in the ultraviolet and visible regions passes through a compound containing multiple bonds, a portion of the radiation is usually absorbed by the compound. Just how much of the radiation is absorbed depends on the wavelength of the radiation and the structure of the compound. The absorption of radiation is caused by the subtraction of energy from the radiation beam when electrons in orbitals of lower energy are excited into orbitals of higher energy.

Instruments, called visible-ultraviolet spectrometers, are used to measure the amount of light absorbed at each wavelength of the visible and ultraviolet region. In these instruments a beam of light is split; half the beam (the sample beam) is directed through a transparent cell containing a sample of the compound being analyzed, and half (the reference beam) is directed through an identical cell that

---

* A convenient formula for calculating the energy of electromagnetic radiation in kcal/mole is the following.

$$E \text{ (in kcal/mole)} = \frac{28,600}{\text{wavelength in nanometers}}$$

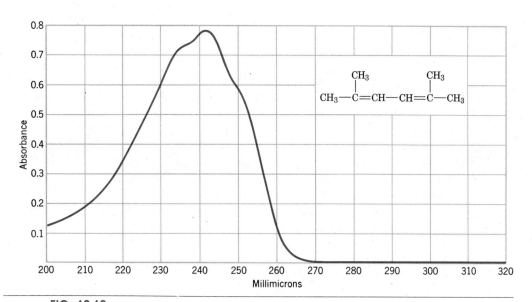

FIG. 10.10

The ultraviolet absorption spectrum of 2,5-dimethyl-2,4-hexadiene. (Spectrum courtesy of Sadtler Research Laboratories, Philadelphia, Pa.)

does not contain the compound. The instrument is designed so that it can make a comparison of the intensities of the two beams at each wavelength of the region. If the compound absorbs light at a particular wavelength, the intensity of the sample beam (I_S) will be less than that of the reference beam (I_R). The instrument indicates this by producing a graph—a plot of the wavelength of the entire region *versus* the absorbance (A) of light at each wavelength. [The absorbance at a particular wavelength is defined by the equation: $A_\lambda = \log (I_R/I_S)$.] Such a graph is called an *absorption spectrum*.

A typical ultraviolet absorption spectrum, that of 2,5-dimethyl-2,4-hexadiene, is shown in Fig. 10.10.

The spectrum of 2,5-dimethyl-2,4-hexadiene shows a broad absorption band in the region between 210 and 260 nm. The absorption is at a maximum at 241.5 nm. It is this wavelength that is usually reported in the chemical literature.

In addition to reporting the wavelength of maximum absorption (λ_{max}), chemists often report another quantity called the molar absorptivity, ϵ.*

The molar absorptivity is simply the proportionality constant that relates the observed absorbance at a particular wavelength, λ, to the molar concentration, C, of the sample and the path length, l, (in centimeters) of the light beam through the sample cell.

$$A = \epsilon \times C \times l \quad \text{or} \quad \epsilon = \frac{A}{C \times l}$$

For 2,5-dimethyl-2,4-hexadiene dissolved in methanol the molar absorptivity at the wavelength of maximum absorbance (241.5 nm) is 13,100 M^{-1} cm^{-1}. In the chemical literature this would be reported as

2,5-dimethyl-2,4-hexadiene, $\lambda_{max}^{methanol}$ 241.5 nm ($\epsilon = 13,100$)

---

*The molar absorptivity, ϵ, is often referred to as the molar extinction coefficient.

As we noted earlier, when compounds absorb light in the ultraviolet and visible regions, electrons are excited from lower electronic energy levels to higher ones. For this reason, visible and ultraviolet spectra are often called *electronic spectra.* The absorption spectrum of 2,5-dimethyl-2,4-hexadiene is a typical electronic spectrum because the absorption band (or peak) is very broad. Most absorption bands in the visible and ultraviolet region are broad because each electronic energy level has, associated with it, vibrational and rotational levels. Thus, electron transitions may occur between any of several vibrational and rotational states of one electronic level to any of several vibrational and rotational states of a higher level.

Compounds whose molecules contain only one carbon-carbon double bond, and those whose molecules contain isolated carbon-carbon double bonds usually absorb light of a wavelength shorter than 200 nm. Ethene, for example, gives an absorption maximum at 171 nm; 1,4-pentadiene gives an absorption maximum at 178 nm. These absorptions occur at such short wavelengths that they are out of the range of operation of most visible-ultraviolet spectrometers. Special vacuum techniques must be employed in measuring them.

Compounds whose molecules contain *conjugated* multiple bonds absorb light at wavelengths longer than 200 nm. For example, 1,3-butadiene, absorbs at 217 nm. This longer wavelength absorption by conjugated dienes is a direct consequence of the conjugated double bonds.

We can understand how conjugation of multiple bonds brings about longer wavelength absorption of light if we examine Fig. 10.11.

When a molecule absorbs light at its longest wavelength, an electron is excited from its highest occupied molecular orbital (HOMO) to the lowest unoccupied molecular orbital (LUMO). For alkenes and alkadienes the highest occupied molecular orbital is a bonding π orbital and the lowest unoccupied molecular orbital is an antibonding π^* orbital. The wavelength of the absorption maximum is determined by the difference in energy between these two levels. The energy gap between the highest occupied molecular orbital and lowest unoccupied molecular orbital of ethene is greater than that between the corresponding orbitals of 1,3-butadiene. Thus, the $\pi \longrightarrow \pi^*$ electron excitation of ethene requires absorption of light of greater energy (shorter wavelength) than the corresponding $\pi_2 \longrightarrow \pi_3^*$ excitation in 1,3-butadiene. The energy difference between the highest occupied molecular orbitals and the lowest

FIG. 10.11

The relative energies of the π-molecular orbitals of ethene and 1,3-butadiene.

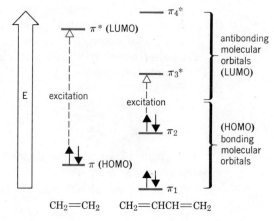

unoccupied molecular orbitals of the two compounds is reflected in their absorption spectra. Ethene has its longest wavelength maxima at 171 nm; 1,3-butadiene has a λ_{max} at 217 nm.

The narrowing of the gap between the highest occupied molecular orbital and the lowest unoccupied molecular orbital in 1,3-butadiene results from the conjugation of the double bonds. Molecular orbital calculations indicate that a much larger gap should occur in isolated alkadienes. This is borne out experimentally. Isolated alkadienes give absorption spectra similar to those of alkenes. They absorb at shorter wavelengths, usually below 200 nm. As we mentioned, 1,4-pentadiene, has its longest wavelength absorption at 178 nm.

Conjugated alkatrienes absorb at longer wavelengths than conjugated alkadienes, and this too can be accounted for in molecular orbital calculations. The energy gap between the highest occupied molecular orbital and the lowest unoccupied molecular orbital of an alkatriene is even smaller than that of an alkadiene. In fact, there is a general rule that states that *the greater the number of conjugated multiple bonds a compound contains, the longer will be the wavelength at which the compound absorbs light.*

Polyenes with eight or more conjugated double bonds absorb light in the visible region of the spectrum. For example, β-carotene, a precursor of Vitamin A, and a compound that imparts its orange color to carrots, has 11 conjugated double bonds; β-carotene has an absorption maximum at 497 nm.

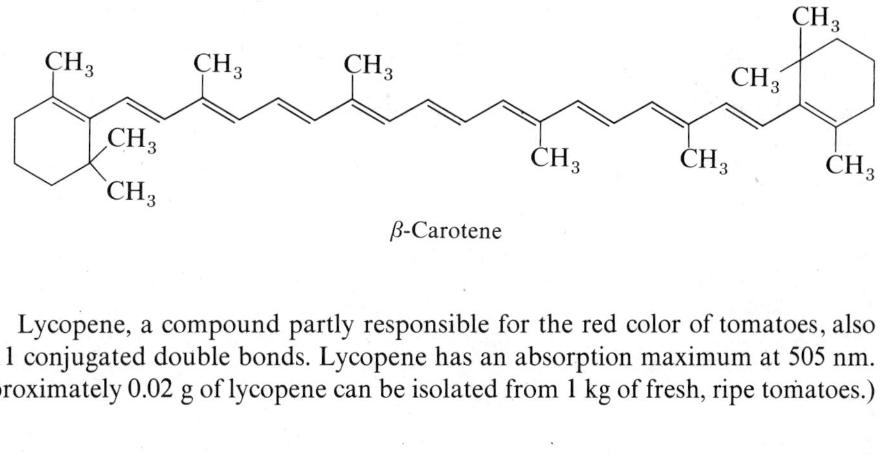

β-Carotene

Lycopene, a compound partly responsible for the red color of tomatoes, also has 11 conjugated double bonds. Lycopene has an absorption maximum at 505 nm. (Approximately 0.02 g of lycopene can be isolated from 1 kg of fresh, ripe tomatoes.)

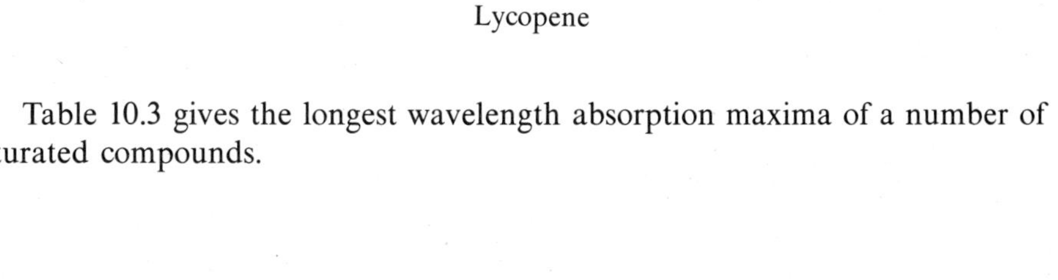

Lycopene

Table 10.3 gives the longest wavelength absorption maxima of a number of unsaturated compounds.

TABLE 10.3 Long Wavelength Absorption Maxima of Unsaturated Hydrocarbons

COMPOUND	STRUCTURE	λ_{max} (nm)	ϵ_{max}
Ethene	$CH_2\!\!=\!\!CH_2$	171	15,530
trans-2-Hexene	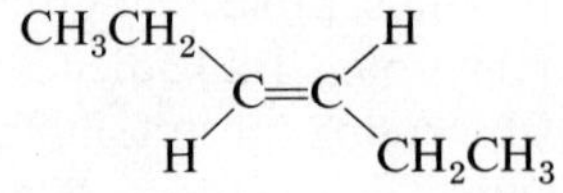 CH_3CH_2 H $C\!\!=\!\!C$ H CH_2CH_3	184	10,000
Cyclohexene	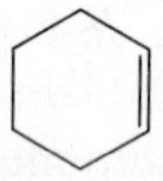	182	7,600
1-Octene	$CH_3(CH_2)_5CH\!\!=\!\!CH_2$	177	12,600
1-Octyne	$CH_3(CH_2)_5C\!\!\equiv\!\!CH$	185	2,000
1,3-Butadiene	$CH_2\!\!=\!\!CHCH\!\!=\!\!CH_2$	217	21,000
cis-1,3-Pentadiene	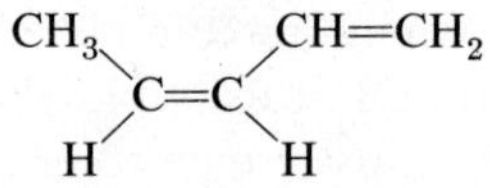 CH_3 $CH\!\!=\!\!CH_2$ $C\!\!=\!\!C$ H H	223	22,600
trans-1,3-Pentadiene	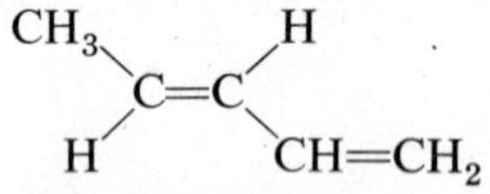 CH_3 H $C\!\!=\!\!C$ H $CH\!\!=\!\!CH_2$	223.5	23,000
1-Butene-3-yne	$CH_2\!\!=\!\!CHC\!\!\equiv\!\!CH$	228	7,800
1,4-Pentadiene	$CH_2\!\!=\!\!CHCH_2CH\!\!=\!\!CH_2$	178	
1,3-Cyclopentadiene		239	3,400
1,3-Cyclohexadiene		256	8,000
trans-1,3,5-Hexatriene	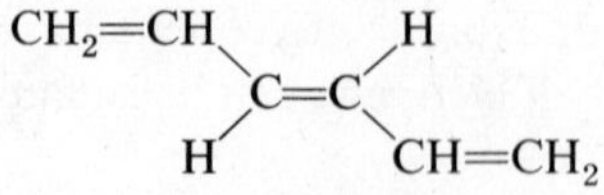 $CH_2\!\!=\!\!CH$ H $C\!\!=\!\!C$ H $CH\!\!=\!\!CH_2$	274	50,000

Additional Problems

10.10

Outline a synthesis of 1,3-butadiene starting from

(a) 1,4-dibromobutane   (e) $CH_2\!\!=\!\!CHCHClCH_3$
(b) $HOCH_2(CH_2)_2CH_2OH$ (f) $CH_2\!\!=\!\!CHCHOHCH_3$
(c) $CH_2\!\!=\!\!CHCH_2CH_2OH$ (g) $HC\!\!\equiv\!\!CCH\!\!=\!\!CH_2$
(d) $CH_2\!\!=\!\!CHCH_2CH_2Cl$

10.11

What product would you expect from the following reaction?

$$(CH_3)_2\underset{\underset{Cl}{|}}{C}-\underset{\underset{Cl}{|}}{C}(CH_3)_2 + 2KOH \xrightarrow[\text{heat}]{\text{ethanol}}$$

10.12

What products would you expect from the reaction of 1,3-butadiene and each of the following reagents. (If no reaction would occur you should indicate that as well.)

(a) One mole Cl_2
(b) Two moles Cl_2
(c) Two moles Br_2
(d) Two moles H_2, Ni

(e) $Ag(NH_3)_2^+OH^-$
(f) One mole Cl_2 in H_2O
(g) Hot $KMnO_4$
(h) H^+, H_2O

10.13

Show how you might carry out each of the following transformations. (In some transformations several steps may be necessary.)

(a) 1-Butene $\longrightarrow$ 1,3-butadiene
(b) 1-Pentene $\longrightarrow$ 1,3-pentadiene
(c) $CH_3CH_2CH_2CH_2OH \longrightarrow CH_2BrCH=CHCH_2Br$
(d) $CH_3CH=CHCH_3 \longrightarrow CH_3CH=CHCH_2Br$

(e)

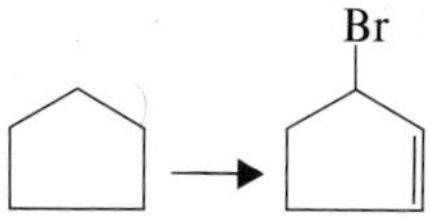

(f)

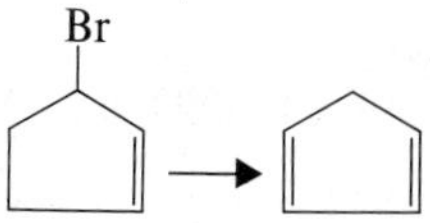

10.14

Conjugated dienes react with free radicals by both 1,2 and 1,4 addition. Account for this fact by using the peroxide catalyzed addition of one mole of HBr to 1,3-butadiene as an illustration.

10.15

Outline a simple chemical test that would distinguish between each of the following pairs of compounds.

(a) 1,3-Butadiene and 1-butyne
(b) 1,3-Butadiene and *n*-butane
(c) 1,3-Butadiene and $CH_2=CHCH_2CH_2OH$
(d) 1,3-Butadiene and $CH_2=CHCH_2CH_2Br$
(e) $CH_2BrCH=CHCH_2Br$ and $CH_3CBr=CBrCH_3$

10.16

(a) The protons of carbon-3 of 1,4-pentadiene are unusually susceptible to abstraction by free radicals. How can you account for this? (b) Can you also provide an explanation for the fact that the protons of carbon-3 of 1,4-pentadiene are more acidic than the methyl hydrogens of propene?

10.17

When 2-methyl-1,3-butadiene (isoprene) undergoes a 1,4-addition of hydrogen chloride, the major product that is formed is 1-chloro-3-methyl-2-butene. Little or no 1-chloro-2-methyl-2-butene is formed. How can you explain this?

10.18

Which diene and dienophile would you employ in a synthesis of each of the following?

(a)

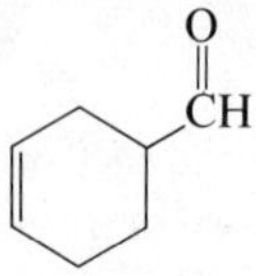

(d)

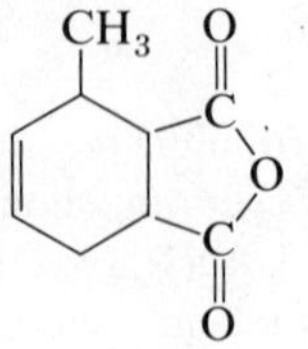

(b)

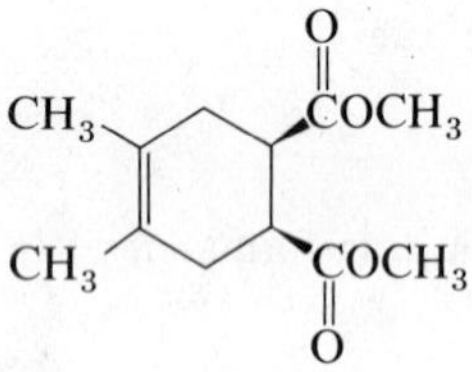

(e)

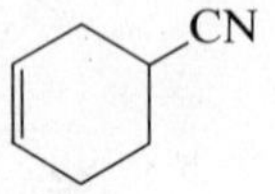

(c)

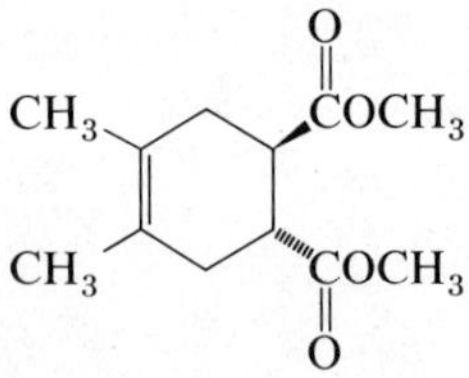

10.19

Account for the fact that neither of the following compounds undergoes a Diels-Alder reaction with maleic anhydride.

$HC\equiv C-C\equiv CH$ or.

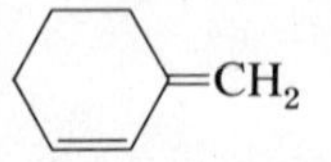

10.20

Acetylenic compounds may be used as dienophiles in the Diels-Alder reaction (cf. problem 10.9). Write structures for the adducts that you expect from the reaction of 1,3-butadiene with:

(a) $CH_3OCC\equiv CCOCH_3$ (dimethyl acetylenedicarboxylate)
(b) $CF_3C\equiv CCF_3$ (hexafluoro-2-butyne)

10.21

Two compounds, **A** and **B**, have the same molecular formula C_6H_8. Both **A** and **B** decolorize bromine in carbon tetrachloride and both give positive tests with cold dilute potassium permanganate. Both **A** and **B** react with two moles of hydrogen in the presence of platinum to yield cyclohexane. **A** shows an absorption maximum at 256 nm, while **B** shows no absorption maximum beyond 200 nm. What are the structures of **A** and **B**?

10.22

Cyclopentadiene undergoes a dimerization reaction at room temperature that results in the formation of "dicyclopentadiene" (p. 365). Can you suggest a simple method for following the rate of this reaction?

10.23

Three compounds, **D**, **E**, and **F**, have the same molecular formula C_5H_6. In the presence of a platinum catalyst, all three compounds absorb three moles of hydrogen and yield

terminal

n-pentane. Compounds **E** and **F** give a precipitate when treated with ammoniacal silver nitrate; compound **D** gives no reaction. Compounds **D** and **E** show an absorption maximum near 230 nm. Compound **F** shows no absorption maximum beyond 200 nm. Propose structures for **D, E,** and **F**.

10.24

The rotational barrier that separates *s-cis*-1,3-butadiene and *s-trans*-1,3-butadiene is considerably higher than the rotational barrier of an ordinary carbon-carbon single bond. Explain.

10.25

When furan and maleimide (below) undergo a Diels-Alder reaction at 25°, the major product is the *endo* adduct **G**. When the reaction is carried out at 90°, however, the major product is the *exo* isomer **H**. The *endo* adduct isomerizes to the *exo* adduct when it is heated to 90°. Propose an explanation that will account for these results.

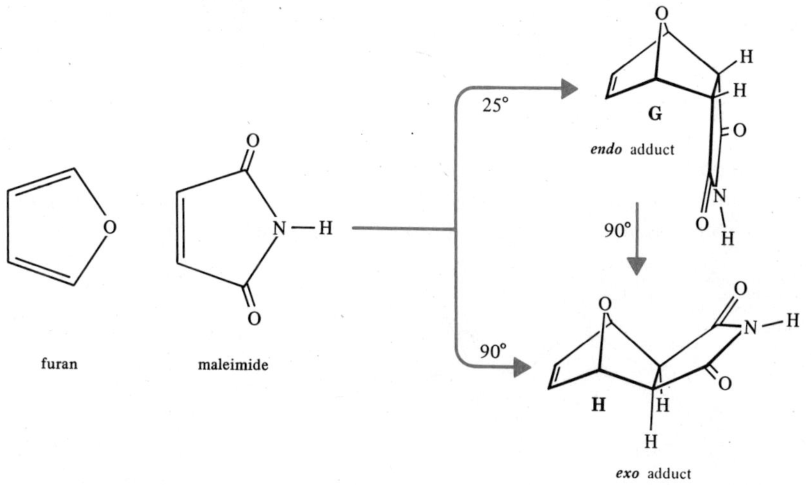

furan   maleimide

25°

G

endo adduct

90°

90°

H H H

exo adduct

10.26

Two controversial "hard" insecticides are Aldrin and Dieldrin (below). (The Environmental Protection Agency has recommended discontinuance of use of these insecticides because of possible harmful side effects and because they are not biodegradable.) The commercial synthesis of Aldrin begins with hexachlorocyclopentadiene and norbornadiene. Dieldrin is synthesized from Aldrin. Show how these syntheses might be carried out.

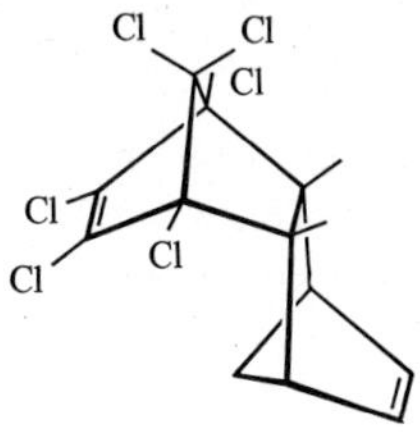

Per acid

aldrin

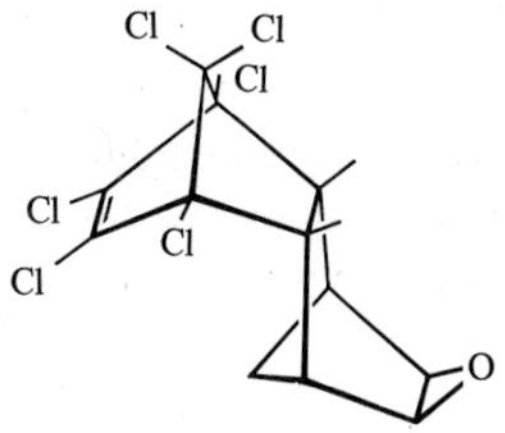

dieldrin

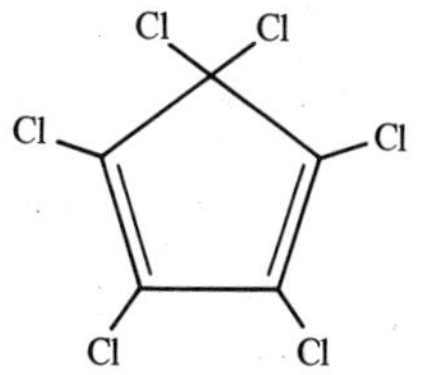

hexachlorocyclopentadiene

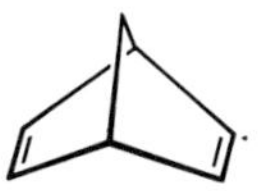

norbornadiene

10.27

Norbornadiene for the Aldrin synthesis (problem 10.26) can be prepared from cyclopenta-diene and acetylene. Show the reaction involved.

10.28

Two other hard insecticides are chlordan and heptachlor. Their commercial syntheses begin with cyclopentadiene and hexachlorocyclopentadiene. Show how these syntheses might be carried out.

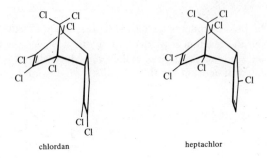

chlordan   heptachlor

10.29

Endrin, an isomer of Aldrin, is obtained when cyclopentadiene reacts with the hexachloro-norbornadiene shown below. Propose a structure for Endrin

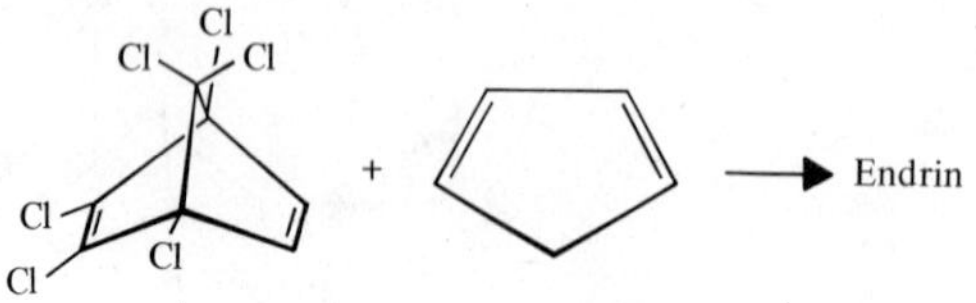

10.30

When $CH_3CH=CHCH_2OH$ is treated with concentrated HCl, two products are produced, $CH_3CH=CHCH_2Cl$ and $CH_3CHClCH=CH_2$. Outline a mechanism that will explain this.

10.31

When a solution of 1,3-butadiene in CH_3OH is treated with chlorine, the products are $ClCH_2CH=CHCH_2OCH_3$ (30%) and $ClCH_2\underset{\overset{|}{OCH_3}}{CH}CH=CH_2$ (70%). Write a mechanism that accounts for their formation.

10.32

Dehydrohalogenation of *vic*-dihalides (with the elimination of two moles of HX) normally leads to an alkyne rather than to a conjugated diene. However, when 1,2-dibromocyclohex-ane is dehydrohalogenated, 1,3-cyclohexadiene is produced in good yield. What factor accounts for this?

10.33

When 1-pentene reacts with *N*-bromosuccinimide, two products with the formula $C_5H_{11}Br$ are obtained. What are these products and how are they formed?

* **10.34**

Mixing furan (problem 10.25) with maleic anhydride in ether yields a crystalline solid with a melting point of 125°. When melting of this compound takes place, however, one can notice that the melt evolves a gas. If the melt is allowed to resolidify, one finds that it no longer melts at 125° but instead it melts at 56°. Consult an appropriate handbook and provide an explanation for what is taking place.

* **10.35**

Propose explanations for the following facts: (a) The reactivity of *trans*-1,3-pentadiene in Diels–Alder reactions is much like that of 1,3-butadiene. On the other hand, *cis*-1,3-pentadiene reacts much more slowly. (b) 2-*tert*-Butyl-1,3-butadiene undergoes a Diels–Alder reaction much more rapidly than 1,3-butadiene, while *cis*-5,5-dimethyl-1,3-hexadiene (*cis*-1-*tert*-butyl-1,3-butadiene) does not react at all.

* **10.36**

Treating hexachloropropene with aluminum chloride leads to the formation of a stable organic salt. Propose a reasonable structure for this salt and account for its formation.

11 SPECIAL TOPICS II

PHOTOCHEMISTRY OF VISION
ELECTROCYCLIC REACTIONS
CYCLOADDITION REACTIONS
NATURALLY OCCURRING ALKENES

11.1 THE PHOTOCHEMISTRY OF VISION

The chemical changes that occur when light impinges on the retina of the eye involve several of the phenomena that we have studied in earlier chapters. Central to an understanding of the visual process are two phenomena in particular: the absorption of light by conjugated polyenes and the interconversion of *cis-trans* isomers.

The retina of the human eye contains two types of receptor cells. Because of their shapes, these cells have been named *rods* and *cones*. Rods are located primarily at the periphery of the retina and are responsible for vision in dim light. Rods, however, are color-blind and "see" only in shades of gray. Cones are found mainly in the center of the retina and are responsible for vision in bright light. Cones also possess the pigments that are responsible for color vision.

Some animals do not possess both rods and cones. The retinas of pigeons contain only cones, and so, while pigeons have color vision; they see only in the bright light of day. The retinas of owls, on the other hand, have only rods; owls see very well in dim light, but are color blind.

The chemical processes that occur in rods are much better understood than those in cones. For this reason we will concern ourselves here with rod vision alone.

When light strikes rod cells, it is absorbed by a compound called rhodopsin. This absorption of a photon of light by a molecule of rhodopsin initiates a series of chemical events that ultimately results in the transmission of a nerve impulse to the brain.

Our understanding of the chemical nature of rhodopsin and the conformational changes that occur when rhodopsin absorbs light have resulted largely from the research of Professor George Wald and his co-workers at Harvard University. Wald's research began in 1933 when he was a graduate student in Berlin; work with rhodopsin, however, began much earlier.

Rhodopsin was discovered in 1877 by the German physiologist Franz Boll. Boll noticed that the initial red-purple color of a pigment in the retina of frogs was "bleached" by the action of light. The bleaching process led first to a yellow retina and then to a colorless one. A year later, another German scientist, Willy Kuhne isolated the red-purple pigment and named it, because of its color, *sehpurpur* or "visual purple." The name visual purple is still commonly used for rhodopsin.

In 1952, Wald and one of his students, Ruth Hubbard, showed that the chromophore (light absorbing group) of rhodopsin is the polyunsaturated aldehyde, 11-*cis*-retinal.

Rhodopsin is formed in the retina by a condensation reaction between 11-*cis*-retinal and a protein called opsin. The condensation reaction is one between the aldehyde group of 11-*cis*-retinal and an amine group on the surface of the protein.

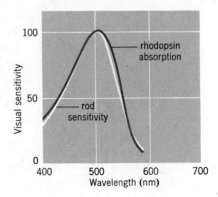

Other secondary interactions involving —SH groups of the protein probably also hold the *cis*-retinal in place. The site on the surface of the protein is one on which *cis*-retinal fits precisely.

The conjugated polyunsaturated chain of 11-*cis*-retinal gives rhodopsin the ability to absorb light over a broad region of the visible spectrum. Figure 11.1 shows

FIG. 11.1
A comparison of the visible absorption spectrum of rhodopsin and the sensitivity curve for rod vision. [*Adapted from S. Hecht, S. Shlaer, and M. H. Pirenne,* J. Gen. Chem. Physiol., 25, *819 (1942).*]

the absorption curve of rhodopsin in the visible region and compares it with the sensitivity curve for human rod vision. The fact that these two curves coincide provides strong evidence that rhodopsin is the light-sensitive material in rod vision.

When rhodopsin absorbs a photon of light the visual process begins. Two very important phenomena accompany the absorption of the photon: a nerve impulse is generated, and 11-*cis*-retinal of rhodopsin is ultimately isomerized to the all-*trans* form of metarhodopsin II. The all-*trans* configuration of retinal does not fit the site on the surface of the protein and, because it does not, hydrolysis of the —CH=N— group occurs. Hydrolysis produces opsin and all-*trans*-retinal. These steps are illustrated on page 380.

Rhodopsin has an absorption maximum at 498 nm. This gives rhodopsin its red-purple color. Together, all-*trans*-retinal and opsin have an absorbance maximum at 387 nm and thus, are yellow. The light-initiated transformation of rhodopsin to all-*trans*-retinal and opsin corresponds to the initial bleaching that Boll observed in the retinas of frogs. Further bleaching to a colorless form occurs when all-*trans*-retinal is reduced enzymatically to all-*trans*-vitamin A. This reduction converts the aldehyde group of retinal to the primary alcohol group of vitamin A.

all-*trans*–retinal

(H) | enzyme

all-*trans*–vitamin A

Regeneration of Rhodopsin

If the retina of a live animal is subjected to constant irradiation, a steady-state concentration of rhodopsin in the retina develops: rhodopsin is created at the same rate that it is destroyed. This is important because rhodopsin is essential to vision.

If an animal is placed in the dark for about 25 minutes a process of "dark adaptation" takes place and retinal rhodopsin reaches its maximum value.

Rhodopsin regeneration occurs in two important ways: one (in light) occurs in the retina itself; the other (in the dark) involves enzymes of the liver. Let us begin by describing how rhodopsin regeneration occurs in light.

Photoregeneration of Rhodopsin

Several different chemical species that occur between rhodopsin and metarhodopsin II have been identified on the basis of their absorption spectra. These intermediates are shown in Table 11.1.

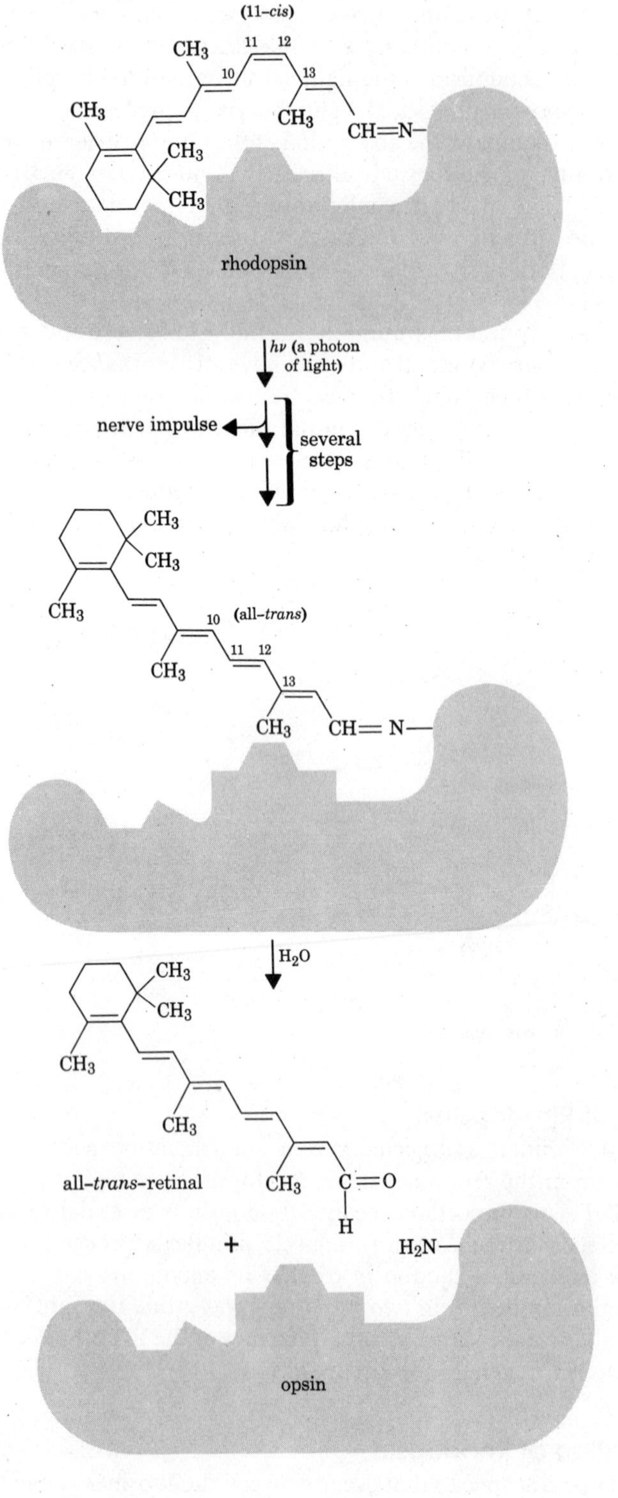

(11-*cis*)

rhodopsin

hν (a photon of light)

nerve impulse ← several steps

(all-*trans*)

H₂O

all-*trans*-retinal

+

opsin

TABLE 11.1

COMPOUND	λ_{MAX}
Rhodopsin	498 nm
↓ *hv*	
Prelumirhodopsin	534 nm
↓	
Lumirhodopsin	500 nm
↓	
Metarhodopsin I	478 nm
↓	
Metarhodopsin II	380 nm

Isomerization of the 11-*cis* double bond of rhodopsin occurs in the first step, rhodopsin ⟶ prelumirhodopsin. The changes from prelumirhodopsin to meta-rhodopsin II probably involve conformational changes of the protein, particularly at the site of its attachment to the all-*trans*-retinal. Prelumirhodopsin and lumirhodopsin have lifetimes so short that they cannot be detected in solution at room temperature.

Metarhodopsin I absorbs light in the same general region as rhodopsin itself, and there is evidence that when the retina is exposed to strong irradiation (i.e., when a large number of photons fall on the rod cells) metarhodopsin I is converted back to rhodopsin. This regeneration of rhodopsin from metarhodopsin I is a photochemical process, too. Metarhodopsin I absorbs a photon of light and its all-*trans* double bond structure is reconverted to the 11-*cis* form of rhodopsin. We can summarize the reactions that occur in the retina in the following way (Figure 11.2).

FIG. 11.2
A summary of the reactions occurring in the retina. Wavy arrows, ⟿, are used to represent reactions involving light; and straight arrows, ⟶, are used to represent reactions that occur in the dark.

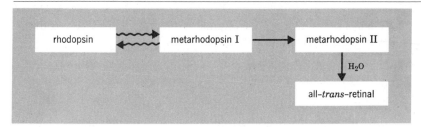

The bleaching sequence begins when rhodopsin absorbs a photon of light and is converted to metarhodopsin I. If, however, metarhodopsin I absorbs a second photon of equal energy it can be converted back to rhodopsin before it is converted to metarhodopsin II. This is not likely to occur in dim light when the number of photons striking the retina is low. However, in bright light photoreversal becomes important because the number of photons striking the retina is large, and the probability that the same molecule will absorb a second photon before it is further transformed is also large. Under these conditions the amount of rhodopsin being bleached reaches a limiting value.

The all-*trans*-retinal that is formed when metarhodopsin II is hydrolyzed can be isomerized to 11-*cis*-retinal by an enzyme in the retina. This enzymatic reaction also requires light but it requires light of shorter wavelength than that for the photoregeneration of rhodopsin from metarhodopsin I. Rhodopsin is resynthesized in the retina when the 11-*cis*-retinal recombines with opsin.

Regeneration of Rhodopsin in the Dark

The all-*trans* retinal that is not is isomerized to 11-*cis*-retinal in the retina is reduced to all-*trans*-vitamin A by an enzyme, retinal reductase. The all-*trans*-vitamin A is then transported to the liver and the next step in the dark synthesis of rhodopsin occurs there.

Enzymes in the liver convert all-*trans*-vitamin A into 11-*cis*-vitamin A.

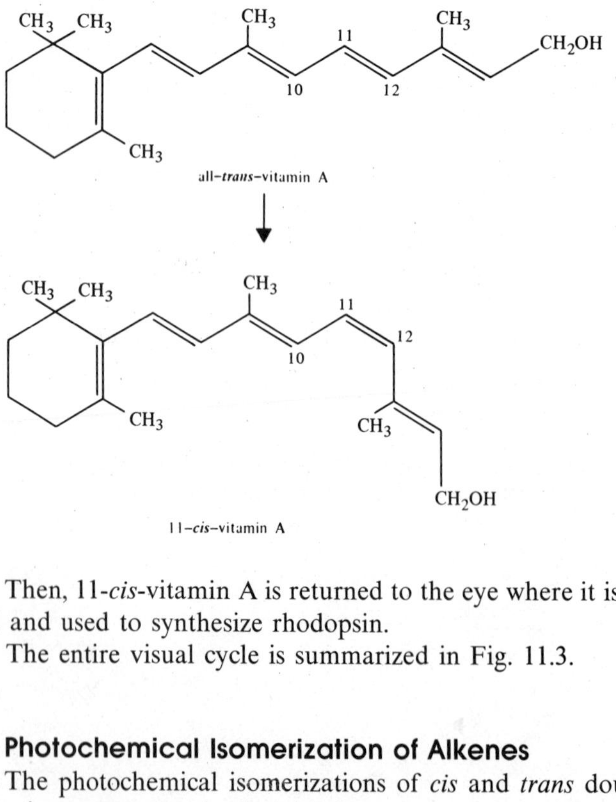

all-*trans*-vitamin A

11-*cis*-vitamin A

Then, 11-*cis*-vitamin A is returned to the eye where it is reoxidized to 11-*cis*-retinal and used to synthesize rhodopsin.

The entire visual cycle is summarized in Fig. 11.3.

Photochemical Isomerization of Alkenes

The photochemical isomerizations of *cis* and *trans* double bonds that play such an important part in the visual process require extra comment. The barrier to rotation of groups joined by a carbon-carbon double bond is quite large. The energy of activation for the interconversion of 11-*cis* retinal and all-*trans*-retinal, for example, has been estimated to be approximately 25 kcal/mole. Yet, when rhodopsin absorbs light of the proper wavelength, the reaction occurs very rapidly.

We can understand how light absorption brings about the rapid interconversion of *cis-trans* isomers if we examine the molecular orbitals given in Fig. 11.4.

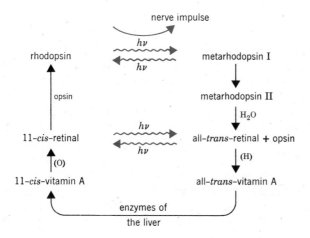

FIG. 11.3
The visual cycle. (All of the photoreactions require light of a wavelength that can be absorbed by reacting molecules.)

In the ground state of an alkene, both electrons are in the bonding molecular orbital, π. In this bonding orbital the electrons are located in a region of space above and below but generally between the two carbon atoms. In the excited state, however, one electron is promoted to the antibonding orbital, π^*. The antibonding orbital has four lobes that are directed generally away from the area between the two carbon

FIG. 11.4
The ground state and an excited state of an alkene.

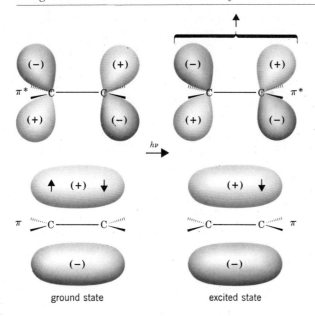

ground state   excited state

atoms. Thus in the excited state the carbon-carbon bond is much more like that of a single bond and has, as a result, a very low barrier to rotation.

This does not mean, however, that in photoisomerization we have avoided the energy barrier. The energy for the isomerization is provided by the photon, and it is supplied just where it is needed: at the carbon-carbon double bond.

The Sensitivity of the Eye

The human eye is truly an amazing instrument. Although each rod contains at least 10 billion molecules of rhodopsin, it has been shown that the absorption of as few as *five* photons of light by five molecules of rhodopsin, is detectable if the rhodopsin molecules are in different rods. Each rod apparently possesses an extraordinary mechanism for amplification of the nerve impulse generated by absorption of the photon.

11.2 ELECTROCYCLIC REACTIONS

There are a number of reactions like the one shown below in which a conjugated polyene is transformed into a cyclic compound.

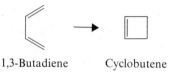

1,3-Butadiene Cyclobutene

There are also many reactions in which the ring of a cyclic compound opens and a conjugated polyene is formed.

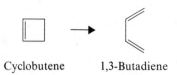

Cyclobutene 1,3-Butadiene

Reactions of either type are called *electrocyclic reactions*.

In electrocyclic reactions σ and π bonds are interconverted. In our first example, one π bond of 1,3-butadiene becomes a σ bond in cyclobutene. In our second example, the reverse is true: a σ bond of cyclobutene becomes a π bond in 1,3-butadiene.

Electrocyclic reactions have several characteristic features:

1. They require only heat or light for initiation.

2. Their mechanisms do not involve free-radical or ionic intermediates.

3. Bonds are made and broken in *a single concerted and cyclic transition state*.

4. The reactions are *highly stereospecific*.

The examples that follow demonstrate this last characteristic of electrocyclic reactions.

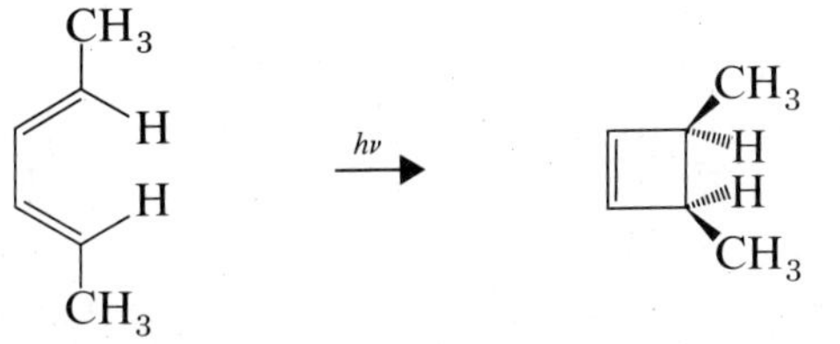

trans, trans-2,4-Hexadiene   cis-3,4-Dimethylcyclobutene

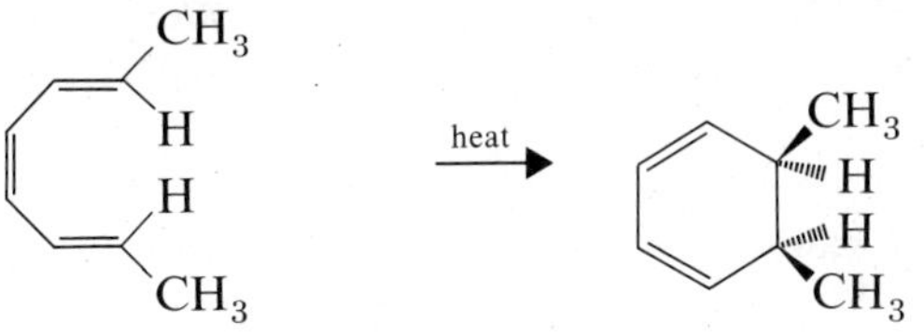

trans, cis, trans-2,4,6-Octatriene   cis-5,6-Dimethyl-1,3-
                                              cyclohexadiene

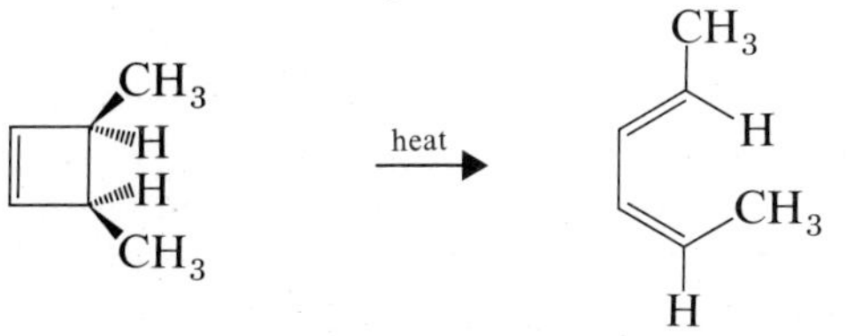

cis-3,4-Dimethylcyclobutene   cis, trans-2,4-Hexadiene

In each of these three examples a single stereoisomeric form of the reactant yields a single stereoisomeric form of the product. The concerted photochemical cyclization of *trans-trans*-2,4-hexadiene, for example, yields only *cis*-3,4-dimethylcyclobutene; it does not yield *trans*-3,4-dimethylcyclobutene.

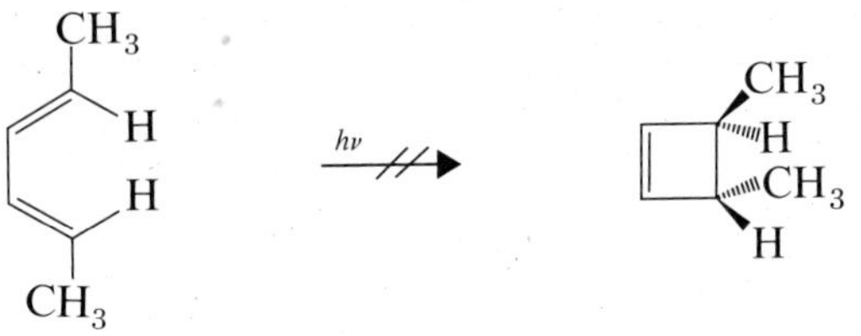

trans, trans-2,4-Hexadiene   trans-3,4-Dimethylcyclobutene

The other two concerted reactions are characterized by the same stereospecificity.

The electrocyclic reactions that we will study here and the concerted cyclo-addition reactions that we will study in Section 11.3 were poorly understood by chemists before 1960. Since then, several scientists, most notably K. Fukui in Japan, H. C. Longuet-Higgins in England, and R. B. Woodward and R. Hoffmann at Harvard University, have provided us with a basis for understanding how these reactions occur and why they take place with such remarkable stereospecificity.

All of these men worked from molecular orbital theory. In 1965, Woodward and Hoffmann formulated their theoretical insights into a set of rules that not only allowed chemists to understand reactions that were already known but that correctly predicted the outcome of many reactions that had not been attempted.

The Woodward–Hoffmann rules are formulated for concerted reactions only. Concerted reactions are reactions in which bonds are broken and formed simultaneously and thus no intermediates occur. The Woodward–Hoffman rules are based on this hypothesis: *in concerted reactions molecular orbitals of the reactant are continuously converted into molecular orbitals of the product.* This conversion of molecular orbitals is not a random one, however. Molecular orbitals have symmetry characteristics and, because they do, restrictions are placed on just which molecular orbitals of the reactant may be transformed into particular molecular orbitals of the product.

According to Woodward and Hoffmann certain reaction paths are said to be *symmetry allowed* while others are said to be *symmetry forbidden.* To say that a particular path is symmetry forbidden does not necessarily mean, however, that the reaction will not occur. It simply means that if the reaction were to occur through a symmetry-forbidden path, the concerted reaction would have a much higher energy of activation. The reaction may occur, but it will probably do so in a different way: through another path that is symmetry allowed or through a nonconcerted path.

A complete analysis of electrocyclic reactions using the Woodward–Hoffman rules requires a correlation of symmetry characteristics of *all* of the molecular orbitals of the reactants and products. Such analyses are beyond the scope of our discussion here. We will find, however, that a simplified approach can be undertaken; one that will be easy to visualize and, at the same time, will be accurate in most instances. In this simplified approach to electrocyclic reactions we focus our attention only on the *highest occupied molecular orbital* (HOMO) *of the conjugated polyene.* We are justified in doing this because this is the orbital that interacts first in the reaction; in a sense, it contains the "valence" electrons.

Electrocyclic Reactions of 4n π-Electron Systems

Let us begin such an approach with an analysis of the thermal interconversion of *cis*-3,4-dimethylcyclobutene and *cis,trans*-2,4-hexadiene shown below.

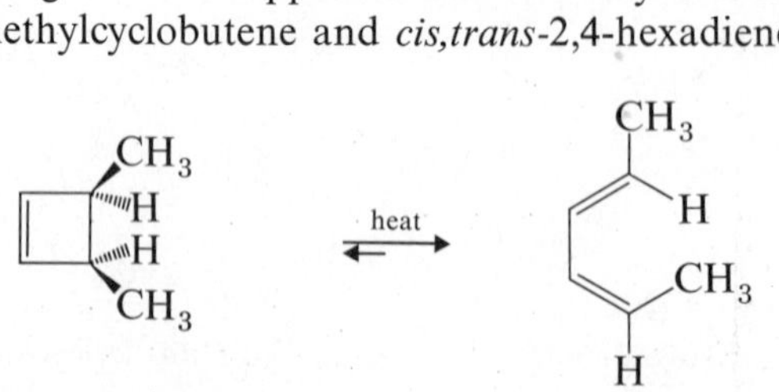

cis-3,4-Dimethylcyclobutene   cis, trans-2,4-Hexadiene

Electrocyclic reactions are reversible; thus, according to the principle of microscopic reversibility, the path for the forward reaction is the same as that for the reverse reaction. In this example it is easier to see what happens to the orbitals, if we follow the *cyclization* reaction, *cis,trans*-2,4-hexadiene ⟶ *cis*-3,4-dimethylcyclobutene.

In the cyclization reaction one π bond of the hexadiene is transformed into a σ bond of the cyclobutene. But which π bond? And, how does the conversion occur?

Let us begin to answer these questions by examining the π molecular orbitals of a hexadiene and, in particular, let us look at *the highest occupied molecular orbital of the ground state* (Fig. 11.5).

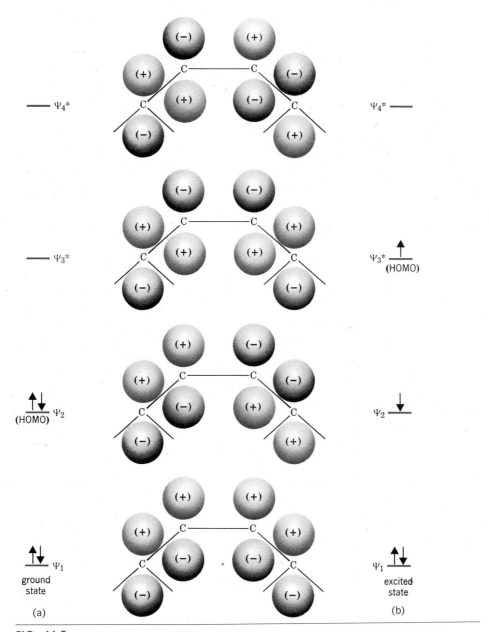

FIG. 11.5
*The π-molecular orbitals of a 2,4-hexadiene. (a) The electron distribution
of the ground state. (b) The electronic distribution of the first excited state.
(The first excited state is formed when the molecule absorbs a photon of
light of the proper wavelength.) Notice that the orbitals of a 2,4-hexadiene
are like those of 1,3-butadiene shown in Figure 10.5.)*

The cyclization reaction that we are concerned with now, *cis,trans*-2,4-
cyclohexadiene ⇌ *cis*-3,4-dimethylcyclobutene, is one that requires heat alone.
We can conclude, therefore, that excited states of the hexadiene are not involved,
for these would require the intervention of light. If we focus our attention on Ψ_2—the
highest occupied molecular orbital of the ground state—we can see how the *p* orbitals

at carbon-2 and carbon-5 can be transformed into a σ bond in the cyclobutene.

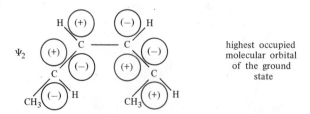

highest occupied
molecular orbital
of the ground
state

A bonding σ- molecular orbital between C-2 and C-5 is formed when the *p* orbitals *rotate in the same direction* (both clockwise, as shown below, or both counter-clockwise, which leads to an equivalent result). The term *conrotatory* is used to describe this type of motion of the two *p* orbitals relative to each other.

Conrotatory motion allows *p*-orbital lobes of the *same sign* to overlap. It also

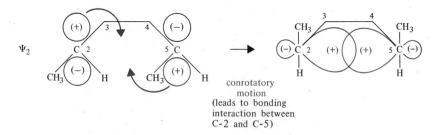

conrotatory
motion
(leads to bonding
interaction between
C-2 and C-5)

places the two methyl groups on the same side of the molecule in the product, that is, in the *cis* configuration.*

The pathway with conrotatory motion of the methyl groups is consistent with what we know from experiments to be true: the *thermal reaction* results in the interconversion of *cis*-3,4-dimethylcyclobutene and *cis,trans*-2,4-hexadiene.

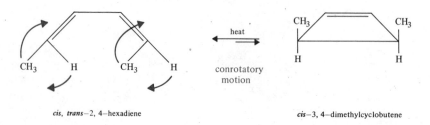

*cis, **trans**-2, 4-hexadiene*   *cis*-3, 4-dimethylcyclobutene

*Notice that if conrotatory motion occurs in the opposite (counterclockwise) direction, lobes of the same sign still overlap, and the methyl group are still *cis*.

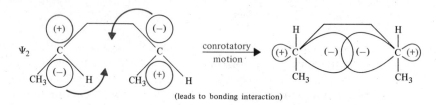

(leads to bonding interaction)

We can now examine another 2,4-hexadiene $\rightleftharpoons$ dimethylcyclobutene inter-conversion: one that takes place under the influence of light. This reaction is shown below.

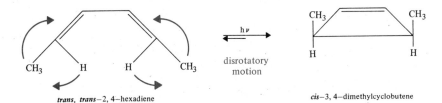

trans, trans—2, 4—hexadiene

cis—3, 4—dimethylcyclobutene

In the photochemical reaction *cis*-3,4-dimethylcyclobutene and *trans, trans*-2,4-hexadiene are interconverted. The photochemical interconversion occurs with the methyl groups rotating in *opposite directions,* that is, with the methyl groups under-going *disrotatory motion.*

The photochemical reaction can also be understood by considering orbitals of the 2,4-hexadiene. In this reaction, however—since the absorption of light is involved—we want to look at the first *excited state* of the hexadiene. We want to examine Ψ_3^*, because in the first excited state, Ψ_3^* *is the highest occupied molecular orbital.*

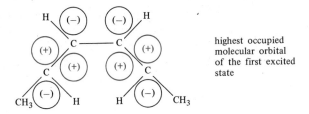

highest occupied
molecular orbital
of the first excited
state

We find that disrotatory motion of the orbitals at carbons-2 and -5 of Ψ_3^* allows lobes of the same sign to overlap and form a bonding sigma molecular orbital between them.

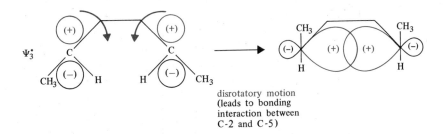

disrotatory motion
(leads to bonding
interaction between
C-2 and C-5)

Disrotatory motion of the orbitals, of course, also requires disrotatory motion of the methyl groups and, once again, this is consistent with what we find experimentally: the *photochemical reaction* results in the interconversion of *cis*-3,4-dimethylcyclo-butene and *trans, trans*-2,4-hexadiene.

Since both of the interconversions that we have presented, so far, involve *cis*-3,4-dimethylcyclobutene, we can summarize them in the following way.

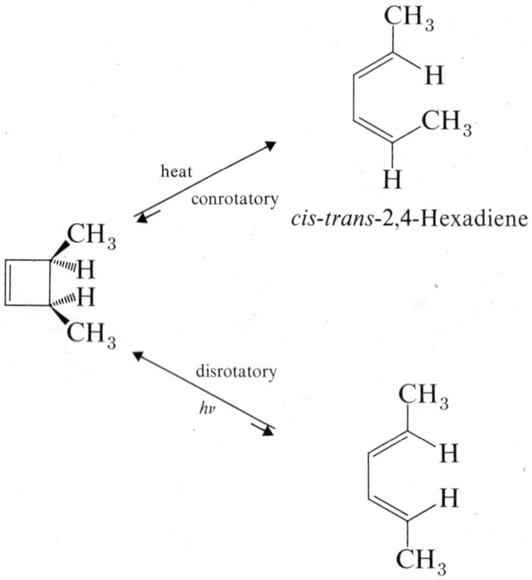

cis-trans-2,4-Hexadiene

trans-trans-2,4-Hexadiene

We see that these two interconversions occur with precisely opposite stereochemistry. We also see that the stereochemistry of the interconversions depends on whether or not the reaction is brought about by the application of heat or light.

The first Woodward–Hoffmann rule can be stated as follows:

   1. A thermal electrocyclic reaction involving $4n$ π electrons (where $n = 1, 2, 3, \ldots$) proceeds with conrotatory motion; the photochemical reaction proceeds with disrotatory motion.

Both of the interconversions that we have studied involve systems of 4π electrons and both follow this rule. Many other $4n$ π electron systems have been studied and many of them since Woodward and Hoffmann stated this rule: all concerted electrocyclic reactions of $4n$ π electron systems have been found to follow it.

   Before we leave the subject of $4n$ π electron systems let us illustrate an application of the rule with one other example.

   When *trans*-3,4-dimethylcyclobutene is heated, ring opening occurs and *trans,trans*-2,4 hexadiene is formed.

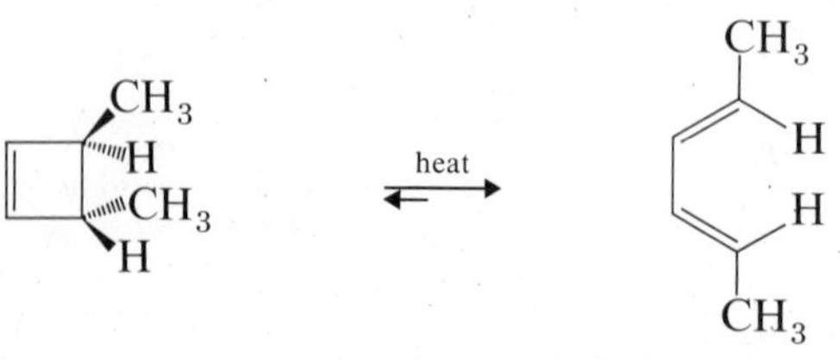

trans-3,4-Dimethylcyclobutene   *trans, trans*-2,4-Hexadiene

According to the Woodward–Hoffmann rule this thermal reaction of a 4π electron system should occur with *conrotatory motion; and this is precisely what happens.* *Trans*-3,4-dimethylcyclobutene is transformed into *trans,trans*-2,4-hexadiene.

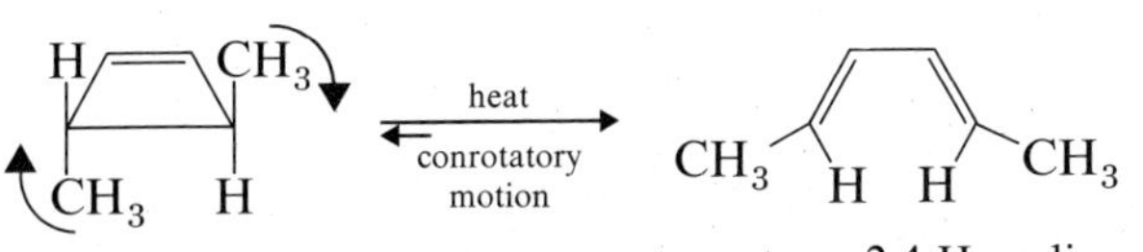

trans-trans-2,4-Hexadiene

Problem 11.1

In the example given above, another conrotatory path is available. This path would produce *cis,cis*-2,4-hexadiene. Can you suggest a reason that will account for the fact that this path is not followed to any appreciable extent?

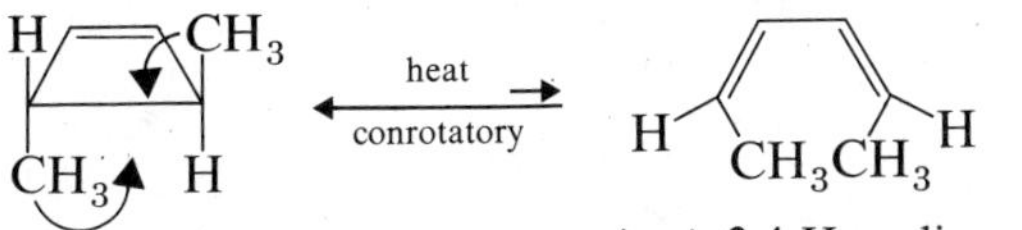

cis, cis-2,4-Hexadiene

Problem 11.2

What product would you expect from a concerted photochemical cyclization of *cis,- trans*-2,4-hexadiene?

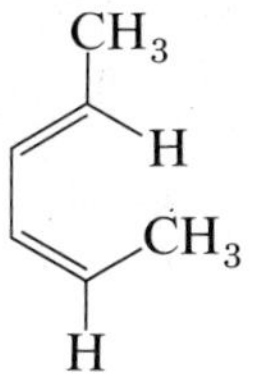

cis, trans-2,4-Hexadiene

Problem 11.3

(a) Show the orbitals involved in the following thermal electrocyclic reaction.

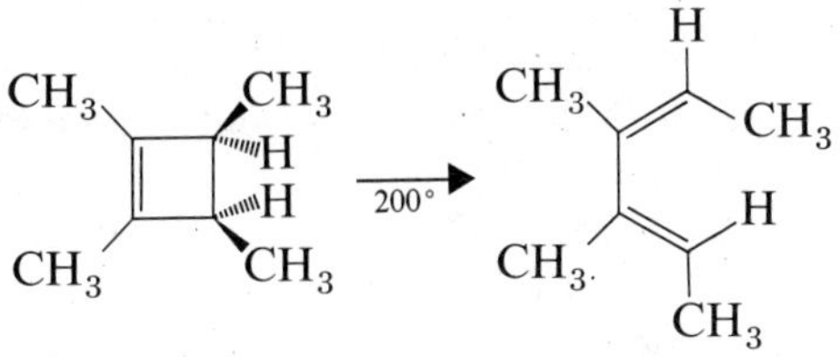

(b) Do the groups rotate in a conrotatory or disrotatory manner?

Problem 11.4

Can you suggest a method for carrying out a stereospecific conversion of *trans,- trans*-2,4-hexadiene into *cis,trans*-2,4-hexadiene?

Problem 11.5

The 2,4,6,8-tetradecaenes on the next page undergo ring closure to dimethylcyclo-octatrienes when heated. What product would you expect from each reaction?

(a) 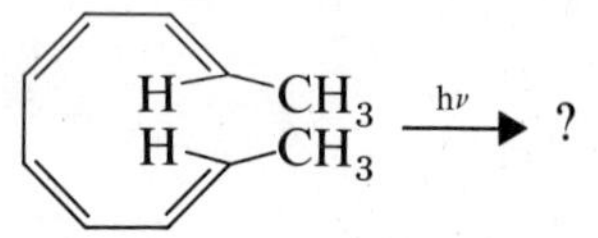 $\xrightarrow{h\nu}$?

(b) 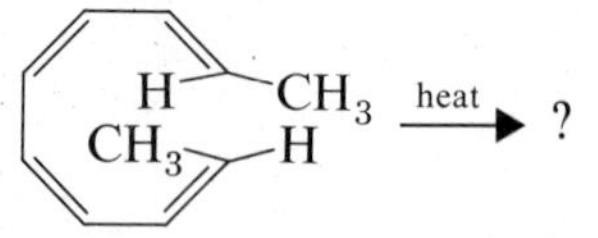 $\xrightarrow{\text{heat}}$?

Problem 11.6

(a) For each of the following reactions, state whether conrotatory or disrotatory motion of the groups is involved and (b) state whether you would expect the reaction to occur under the influence of heat or light.

(a)

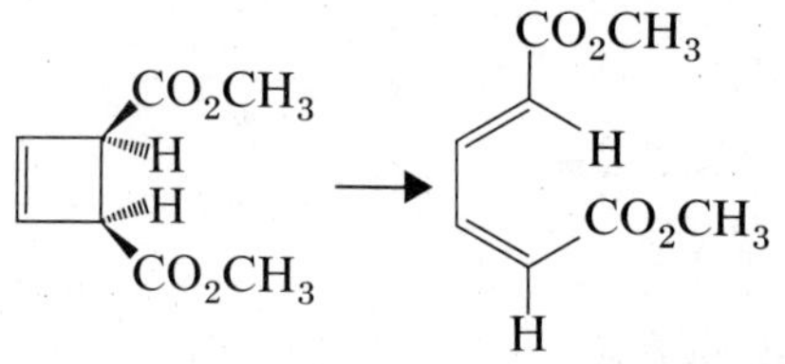

(b)

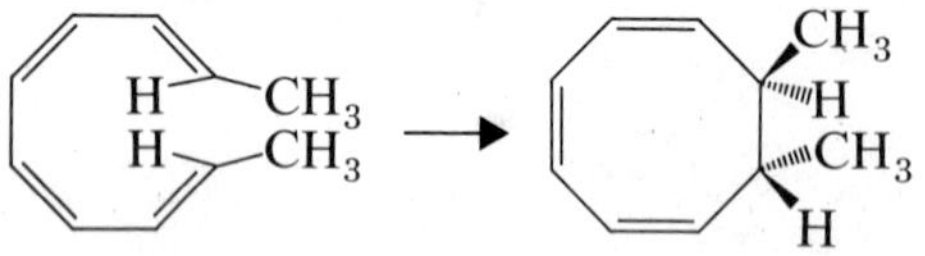

(c)

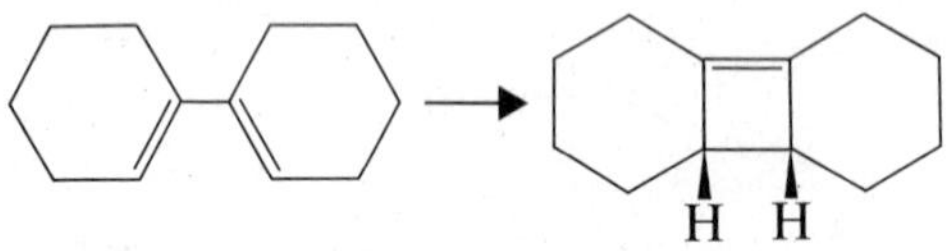

Electrocyclic Reactions of 4n + 2 π-Electron Systems

The second Woodward–Hoffmann rule for electrocyclic reactions is stated as follows:

2. **A thermal electrocyclic reaction involving $(4n + 2)$ π electrons (where $n = 0, 1, 2, \ldots$) proceeds with disrotatory motion; the photochemical reaction proceeds with conrotatory motion.**

According to this rule the direction of rotation of the thermal and photochemical reactions of $(4n + 2)$ π electron systems is the opposite of that for corresponding $4n$ systems. Thus, we can summarize both systems in the following way (Table 11.2).

TABLE 11.2 Woodward-Hoffmann Rules for Electrocyclic Reactions

NUMBER OF ELECTRONS	MOTION	RULE
$4n$	Conrotatory	Thermally allowed, photochemically forbidden
$4n$	Disrotatory	Photochemically allowed, thermally forbidden
$4n + 2$	Disrotatory	Thermally allowed, photochemically forbidden
$4n + 2$	Conrotatory	Photochemically allowed, thermally forbidden

The interconversions of *trans*-5,6-dimethyl-1,3-cyclohexadiene and the two different 2,4,6-octatrienes shown below illustrate thermal and photochemical interconversions of six π-electron systems ($4n + 2$ where $n = 1$).

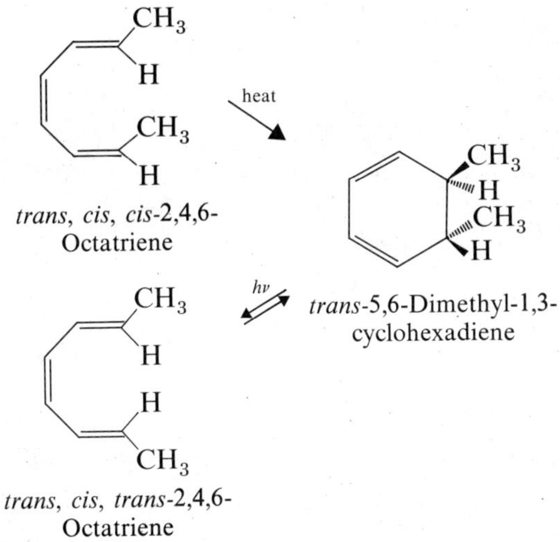

trans, cis, cis-2,4,6-Octatriene

trans-5,6-Dimethyl-1,3-cyclohexadiene

trans, cis, trans-2,4,6-Octatriene

In the thermal reaction (below) the methyl groups rotate in a disrotatory fashion.

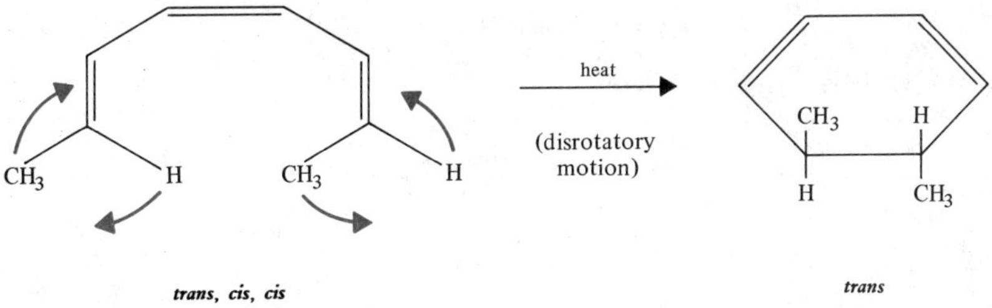

heat

(disrotatory motion)

trans, cis, cis

trans

In the photochemical reaction the groups rotate in a conrotatory way.

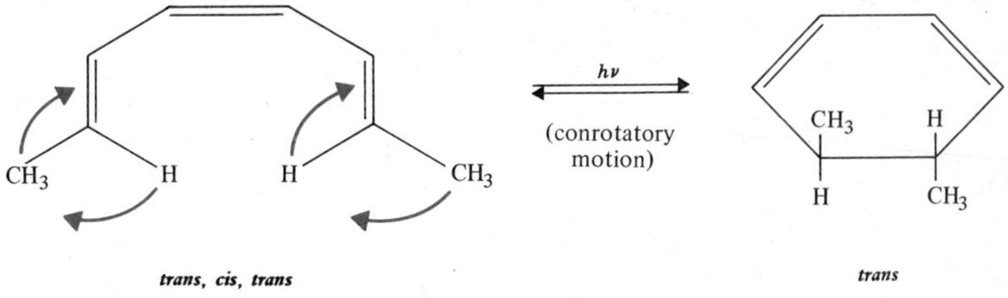

$h\nu$

(conrotatory motion)

trans, cis, trans

trans

We can understand how these reactions occur if we examine the π-molecular orbitals shown in Fig. 11.6. Once again we want to pay attention to the highest occupied molecular orbitals. For the thermal reaction of a 2,4,6-octatriene, the highest occupied orbital will be Ψ_3 because the molecule reacts in its ground state.

We see on p. 395 that disrotatory motion of orbitals at carbons-2 and -7 of Ψ_3 allows the formation of a bonding sigma molecular orbital between them. Disrotatory motion of the orbitals, of course, also requires disrotatory motion of the groups attached to carbons-2 and -7. And, disrotatory motion of the groups is what

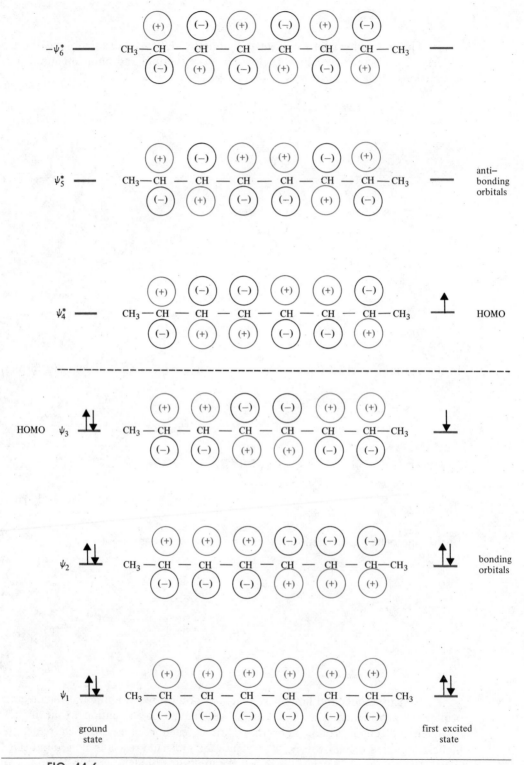

FIG. 11.6

The π-molecular orbitals of 2,4,6-octatriene. The first excited state is formed when the molecule absorbs light of the proper wavelength. (These molecular orbitals are obtained from calculations that are beyond the scope of our discussions.)

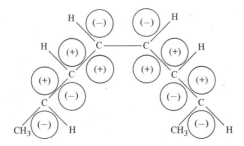

Ψ_3 of **trans, cis, cis,**$-2, 6-$octatriene

we observe in the thermal reaction: *trans,cis,cis*-2,4,6-octatriene $\longrightarrow$ *trans*-5,6-dimethyl-1,3-cyclohexadiene.

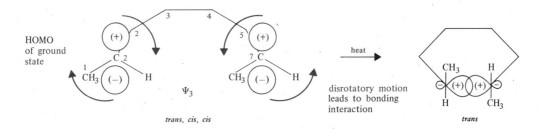

When we consider the photochemical reaction, *trans,cis,trans*-2,4-6-octatriene $\rightleftharpoons$ *trans*-5,6-dimethyl-1,3-cyclohexadiene, we want to focus our attention on Ψ_4^*. In the photochemical reaction, light causes the promotion of an electron from Ψ_3 to Ψ_4^*, and thus, Ψ_4^* becomes the highest occupied molecular orbital. We also want to look at the symmetry of the orbitals at C–2 and C–7 of Ψ_4^*, for these are the orbitals that form a σ bond. In the interconversion shown below, conrotatory motion

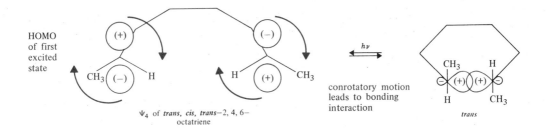

of the orbitals allows lobes of the same sign to overlap. Thus, we can understand why conrotatory motion of the groups is what we observe in the photochemical reaction.

Problem 11.7

Give the stereochemistry of the product that you would expect from each of the following electrocyclic reactions.

(a) 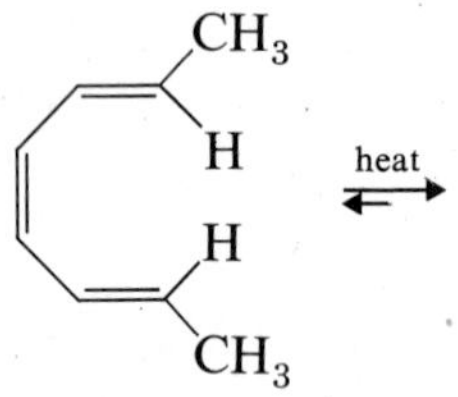 heat

(b) 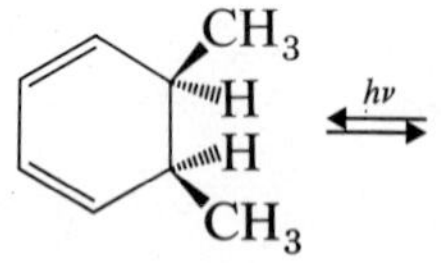 hv

Problem 11.8

Can you suggest a stereospecific method for converting *trans*-5,6-dimethyl-1,3-cyclohexadiene into *cis*-5,6-dimethyl-1,3-cyclohexadiene?

Problem 11.9

When compound **A** is heated, compound **B** can be isolated from the reaction mixture. A sequence of two electrocyclic reactions occurs; the first involves a 4π-electron

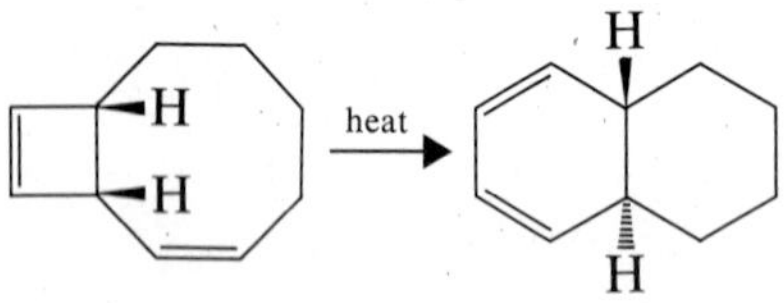

system, the second involves a 6π-electron system. Outline both electrocyclic reactions and give the structure of the intermediate that intervenes.

Problem 11.10

Cyclopropyl cations undergo rapid disrotatory ring-opening reactions when heated. A general example is shown below.

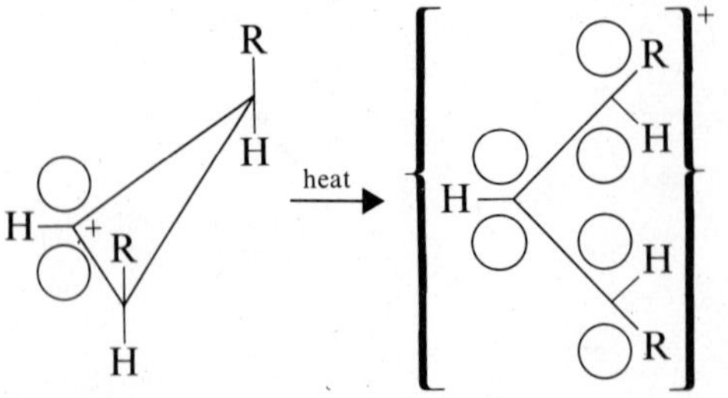

(a) What kind of π system is involved in this reaction, $4n$ or $4n + 2$?
(b) What kind of cation is formed?
(c) What kind of rotation would you expect the cyclopropyl *anion* to show in a thermal reaction, conrotatory or disrotatory?

Problem 11.11

Would you expect the following reactions to occur under the influence of heat or light?

(a)

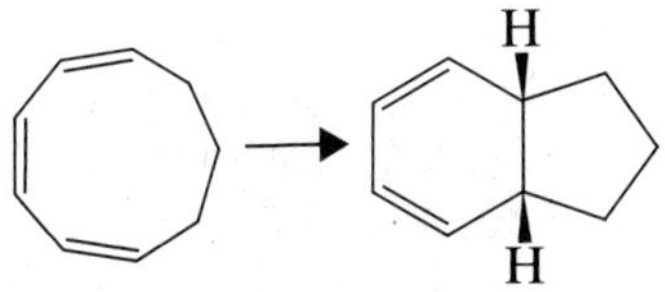

(b)

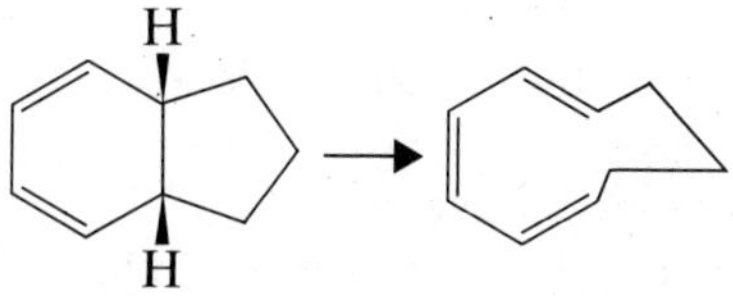

(c)

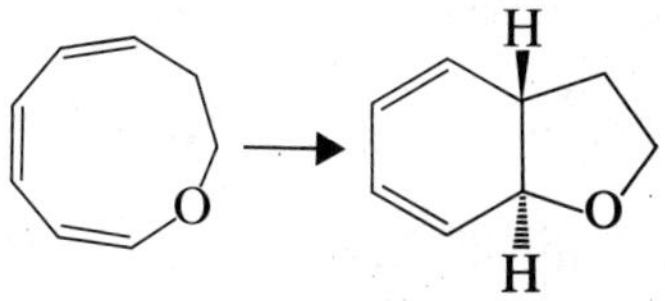

(d)

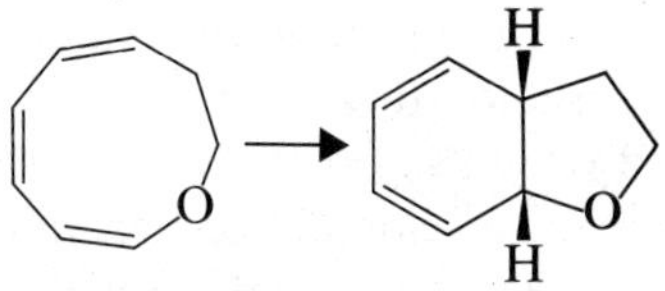

Problem 11.12

What are the structures of compounds **1** and **2** below?

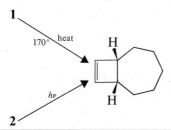

11.3 CYCLOADDITION REACTIONS

There are a number of reactions of alkenes and polyenes in which two molecules react to form a cyclic product. These reactions, called *cycloaddition* reactions, are shown below.

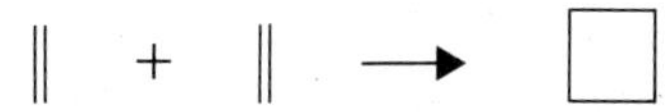

|| + || ⟶ ▢   A [2 + 2] cycloaddition

Alkene   Alkene   Cyclobutane

Diene + || ⟶ ⬡   A [4 + 2] cycloaddition

Diene   Alkene   Cyclohexene
(dienophile)   (adduct)

Chemists classify cycloaddition reactions on the basis of the number of π electrons involved in each component. The reaction of two alkenes to form a cyclobutane is a [2 + 2] cycloaddition; the reaction of a diene and an alkene to form a cyclohexene is called a [4 + 2] cycloaddition. We are already familiar with the [4 + 2] cycloaddition, for it is the Diels-Alder reaction that we studied in Section 10.8.

Cycloaddition reactions resemble electrocyclic reactions in the following important ways:

1. Sigma and pi bonds are interconverted.
2. Cycloaddition reactions require only heat or light for initiation.
3. Free radicals and ionic intermediates are not involved in the mechanisms for cycloadditions.
4. Bonds are made and broken in a single, concerted, and cyclic transition state.
5. Cycloaddition reactions are highly stereospecific.

As we might expect, concerted cycloaddition reactions resemble electrocyclic reactions in still another important way: the symmetry elements of the interacting molecular orbitals allow us to account for their stereochemistry. The symmetry elements of the interacting molecular orbitals also allow us to account for two other observations that have been made about cycloaddition reactions:

1. *Photochemical [2 + 2] cycloaddition reactions occur readily while thermal [2 + 2] cycloadditions only take place under extreme conditions.* When thermal [2 + 2] cycloadditions do take place they occur through free radical (or ionic) mechanisms not through a concerted process.
2. *Thermal [4 + 2] cycloaddition reactions occur readily and photochemical [4 + 2] cycloadditions are difficult.*

[2 + 2] Cycloadditions

Let us begin with an analysis of the [2 + 2] cycloaddition of two ethylene molecules to form a molecule of cyclobutane.

$$2 \ \begin{matrix} CH_2 \\ \| \\ CH_2 \end{matrix} \longrightarrow \begin{matrix} CH_2-CH_2 \\ | \qquad | \\ CH_2-CH_2 \end{matrix}$$

In this reaction we see that two π bonds are converted into two σ bonds. But how does this conversion take place? In a concerted reaction we would expect that molecular orbitals of one reactant would overlap with molecular orbitals of the other. But which molecular orbitals overlap?

We can answer this last question if we consider two factors: the Pauli exclusion principle and effect of orbital symmetry.

According to the Pauli exclusion principle, a molecular orbital may not contain more than *two* (spin paired) electrons. This means that if a molecular orbital of one reactant already contains two electrons it must overlap with an *unoccupied* molecular orbital of the other. (Otherwise, the Pauli principle would be violated, for the new σ molecular orbital would contain more than two electrons.)

We can see how orbital interactions come into play if we examine the possibility of a *concerted thermal* conversion of two ethene molecules into cyclobutane.

Thermal reactions involve molecules reacting in their ground states. The orbital diagram for ethene in its ground state is shown below.

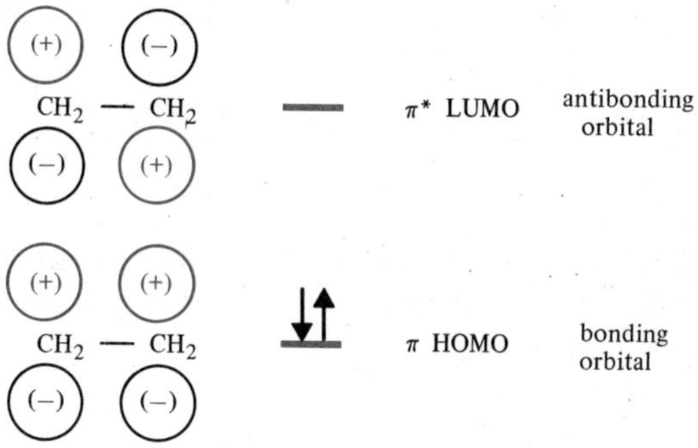

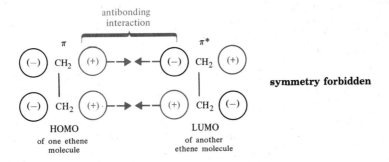

the ground state of ethene

The highest occupied molecular orbital (HOMO) of ethene in its ground state is the π orbital. Since this orbital contains two electrons, it must interact with an *unoccupied* molecular orbital of another ethene molecule. The lowest unoccupied molecular orbital (LUMO) of the ground state of ethene is, of course, π^*.

We see from the diagram above, however, that overlapping the π orbital of one ethene molecule with the π^* orbital of another does not lead to bonding between both sets of carbons. Moreover, complete correlation diagrams of all the molecular orbitals show that this reaction is *symmetry forbidden*. What does this mean? It means that a thermal (or ground state) cycloaddition of ethene would be unlikely to occur in a concerted process. This is exactly what we find experimentally; thermal cyclo-additions of ethene, when they occur, take place through nonconcerted, free-radical mechanisms.

What, then, can we decide about the other possibility—a photochemical [2 + 2] cycloaddition? If an ethene molecule absorbs a photon of light of the proper wavelength, an electron is promoted from π to π^*. In this excited state the highest occupied molecular orbital of an ethene molecule is π^*. The diagram on p. 400 shows how the highest occupied molecular orbital of an excited state ethene molecule interacts with the lowest unoccupied molecular orbital of a ground-state ethene molecule.

Here we find that bonding interactions occur between both CH_2 groups, that is, lobes of the same sign overlap between both sets of carbons. Complete

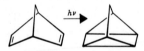

bonding
interaction
(−)
CH2
(+)
(+)
CH2
(−)
(+)
CH2
(−)
(−)
CH2
(+)
bonding
interaction
HOMO
of an excited
state ethene
molecule
LUMO
of a ground
state ethene
molecule

symmetry allowed

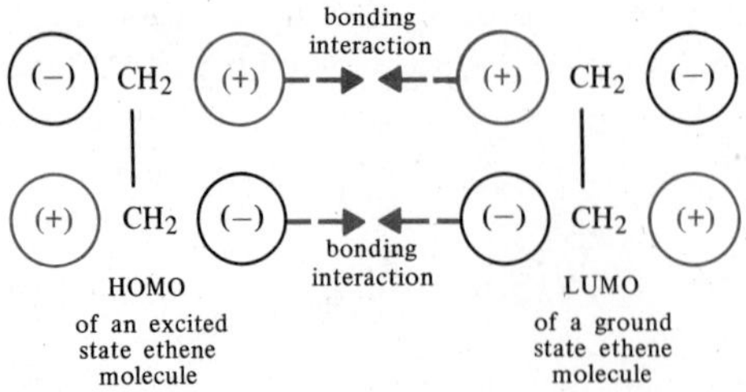

correlation diagrams also show that the photochemical reaction is *symmetry allowed* and should occur readily through a concerted process. This, moreover, is what we observe experimentally: ethene reacts readily in a *photochemical* cycloaddition.

The analysis that we have given for the [2 + 2] ethene cycloaddition can be made for any alkene [2 + 2] cycloaddition because the symmetry elements of the π and π^* orbitals of all alkenes are the same.

Problem 11.13

What products would you expect from the following concerted cycloaddition reactions? (Give stereochemical formulas.)

(a) *cis*-2-butene $\xrightarrow{h\nu}$

(b) *trans*-2-butene $\xrightarrow{h\nu}$

Problem 11.14

Show what happens in the following reaction.

hν

Problem 11.15

When allene is heated to 150° it undergoes a "head-to-head" cyclization to give 1,2-dimethylenecyclobutane.

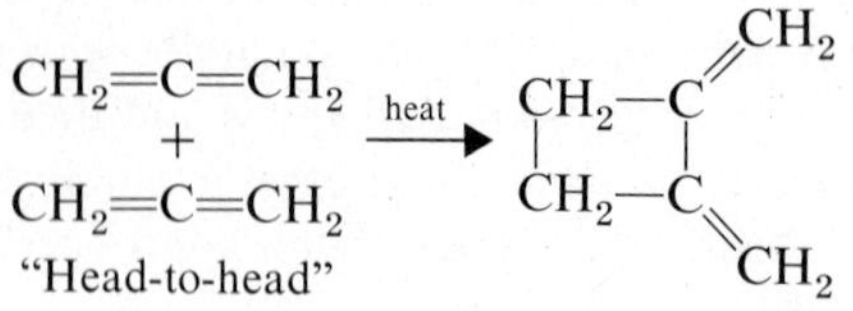

CH2=C=CH2
+
CH2=C=CH2
"Head-to-head"
heat
CH2
CH2—C
CH2—C
CH2

(a) Is it likely that this reaction follows a concerted path?

(b) Can you propose a free-radical mechanism for the reaction that will account for the fact that most of the product results from a "head-to-head" addition rather than a "head-to-tail" addition?

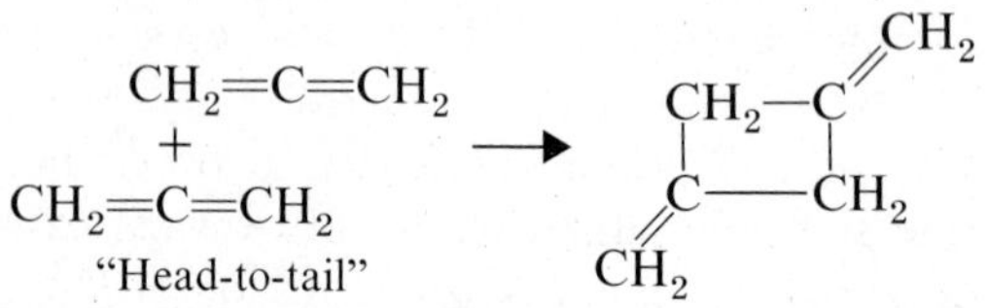

CH2=C=CH2
+
CH2=C=CH2
"Head-to-tail"
CH2
CH2—C
C—CH2
CH2

minor product

[4 + 2] Cycloadditions

Concerted [4 + 2] cycloadditions—Diels-Alder reactions— are *thermal reactions*. Considerations of orbital interactions allow us to account for this fact as well. To see how, let us consider the diagrams shown in Figure 11.7.

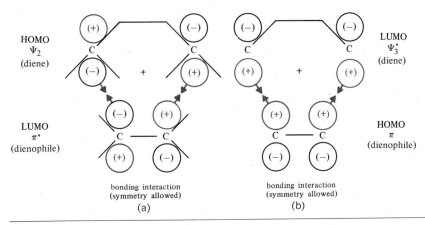

FIG. 11.7

Two symmetry-allowed thermal [4 + 2] cycloadditions. (a) Bonding interaction between the highest occupied molecular orbital of a diene and the lowest unoccupied orbital of a dienophile. (b) Bonding interaction between the lowest unoccupied molecular orbital of the diene and the highest occupied molecular orbital of the dienophile.

Both modes of orbital overlap shown in Fig. 11.7 lead to bonding interactions and both involve *ground states* of the reactants. The ground state of a diene has two electrons in Ψ_2 (its highest occupied molecular orbital). The overlap shown in part (*a*) allows these two electrons to flow into the lowest unoccupied molecular orbital of the dienophile, π^*. The overlap shown in part (*b*) allows two electrons to flow from the highest occupied molecular orbital of the dienophile, π, into the lowest unoccupied molecular orbital of the diene, Ψ_3^*. Complete correlation diagrams also show that the $4n + 2$ thermal reaction is symmetry allowed.

In Section 10.8 we saw that the Diels-Alder reaction proceeds with retention

retention of
configuration of
the dienophile

of configuration of the dienophile. Because the Diels-Alder reaction is usually concerted, it also proceeds with retention of configuration of the diene.

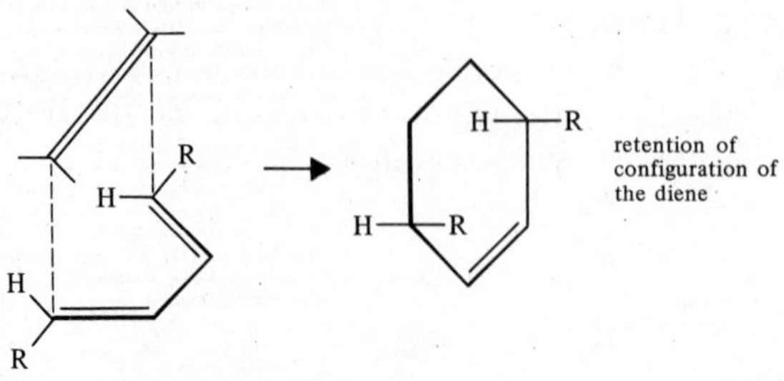
R
H
H
R
H
R
H
R
retention of
configuration of
the diene

Problem 11.16

What products would you expect from the following reaction?

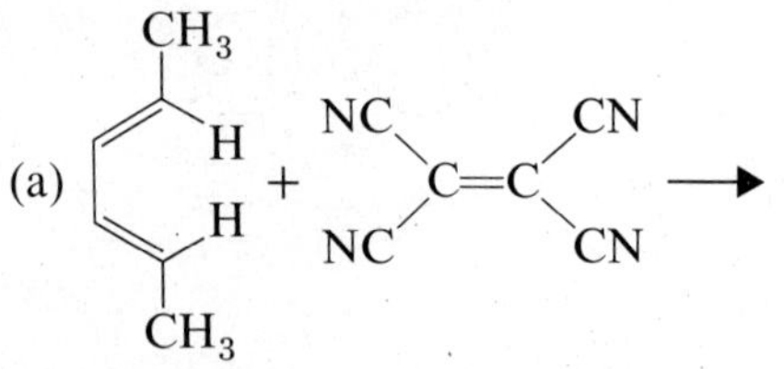
(a)
CH3
H
H
CH3
NC
CN
NC
CN
C=C

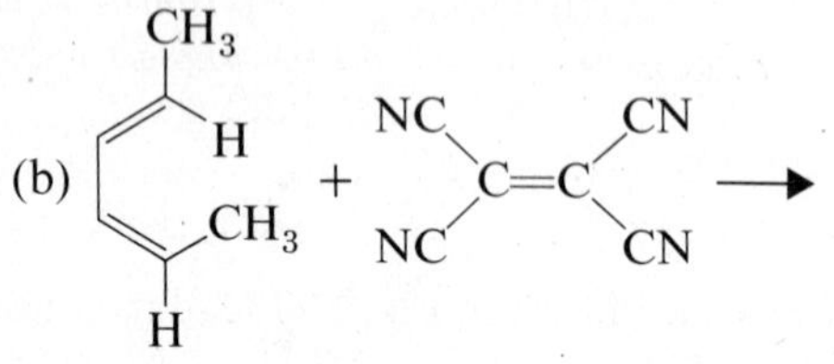
(b)
CH3
H
CH3
H
NC
CN
NC
CN
C=C

Problem 11.17

What are compounds **3**, **4**, and **5** in the following reaction sequence?

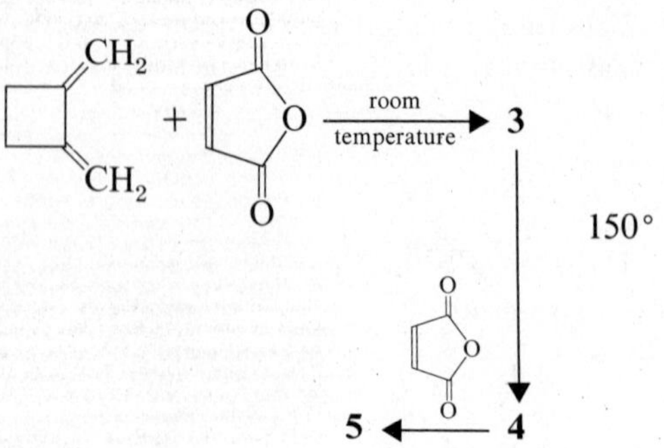
CH2
CH2
O
O
O
room
temperature
3
150°
5
4

Problem 11.18

What is compound **7** in the following reaction sequence?

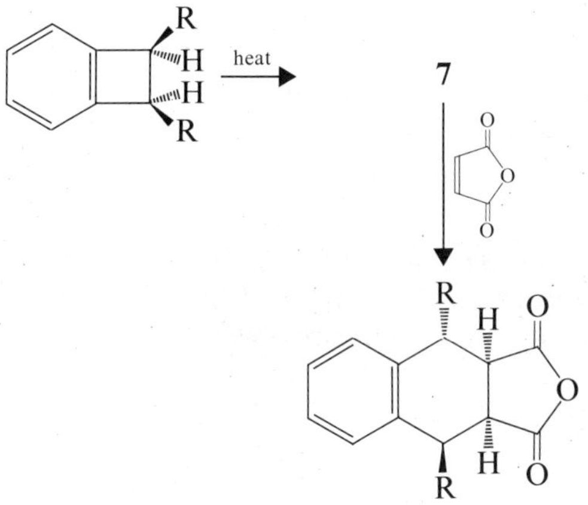

11.4 NATURALLY OCCURRING ALKENES

Man has isolated organic compounds from plants since antiquity. By gently heating and later by steam distilling certain plant materials, one can obtain mixtures of odoriferous compounds known as "essential oils." These compounds have had a variety of uses, particularly in early medicine and in the making of perfumes.

As the science of organic chemistry developed, chemists separated the various components of these mixtures, determined their molecular formulas and then later their structural formulas. Even today these compounds offer challenging problems for chemists who are interested in structure determination and synthesis. Research in this area has also given us important information about the ways the plants, themselves, synthesize these compounds.

Compounds known generally as *terpenes* and *terpenoids* are the most important constituents of essential oils. Most terpenes have carbon skeletons of ten, fifteen, twenty, or thirty atoms and are classified in the following way.

NUMBER OF CARBONS	CLASS
10	Monoterpenes
15	Sesquiterpenes
20	Diterpenes
30	Triterpenes

One can view terpenes as being built up from two or more five-carbon units known as *isoprene units*. Isoprene is 2-methyl-1,3-butadiene. Isoprene and the isoprene unit can be represented in various ways.

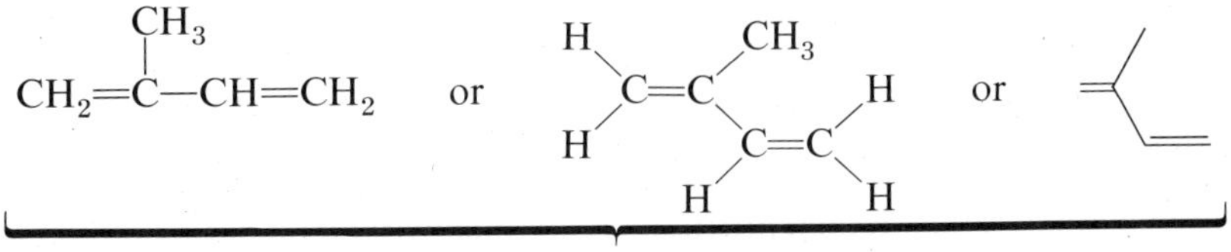

Isoprene

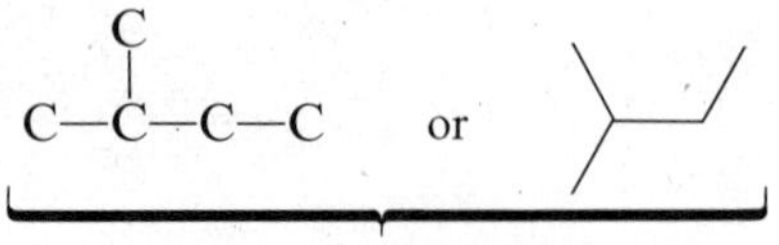

An isoprene unit

We now know that plants do not synthesize terpenes from isoprene. However, recognition of the isoprene unit as a component of the structure of terpenes has been a great aid in elucidating their structures. We can see how, if we examine the structures shown below.

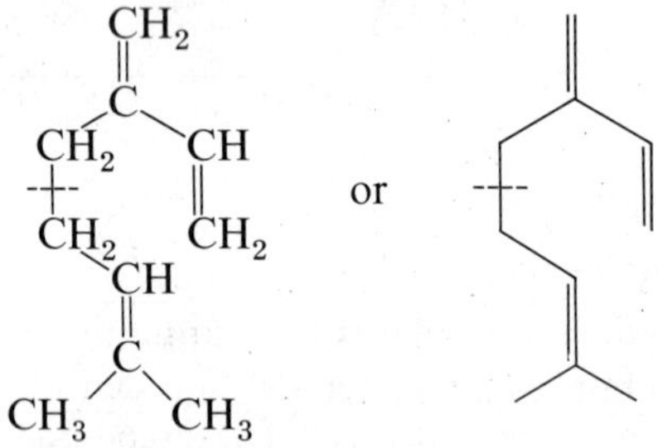

Myrcene
(isolated from bay oil)

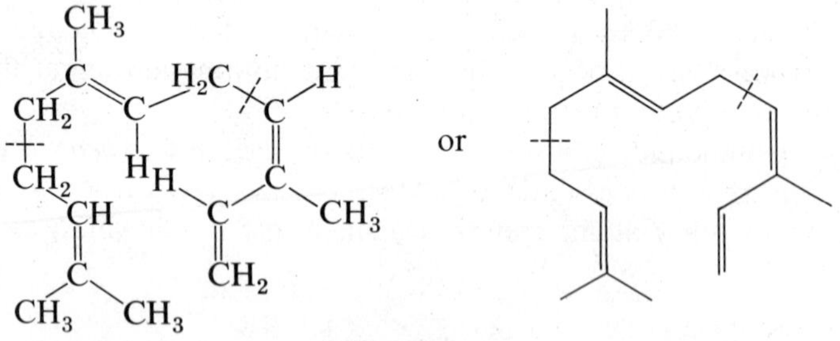

α-Farnesene
(from oil of citronella)

Using dashed lines to separate isoprene units we can see that the monoterpene, myrcene, has two isoprene units; and that the sesquiterpene, α-farnesene, has three. In both compounds the isoprene units are linked head to tail.

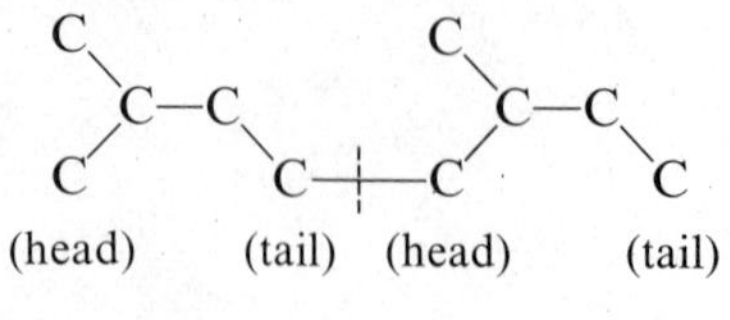

(head)   (tail)   (head)   (tail)

Many terpenes also have isoprene units linked in rings, and others (terpenoids) contain oxygen.

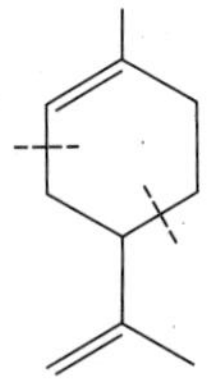

Limonene
(from oil of lemon or orange)

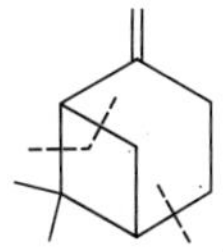

β-Pinene
(from oil of turpentine)

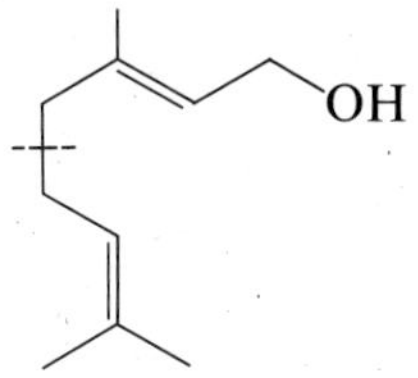

Geraniol
(from roses and other flowers)

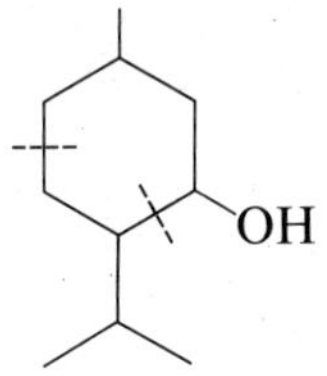

Menthol
(from peppermint)

Problem 11.19

(a) Show the isoprene units in each of the following terpenes. (b) Classify each as a monoterpene, sesquiterpene, diterpene, and so on.

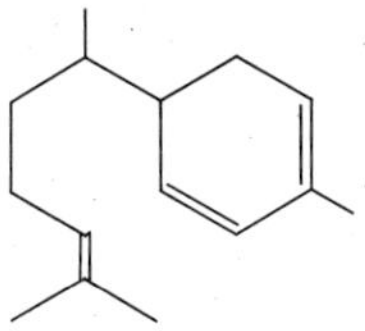

Zingiberene
(from oil of ginger)

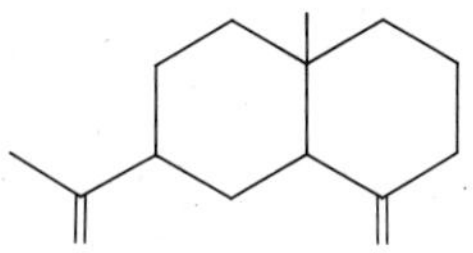

β-Selinene
(from oil of celery)

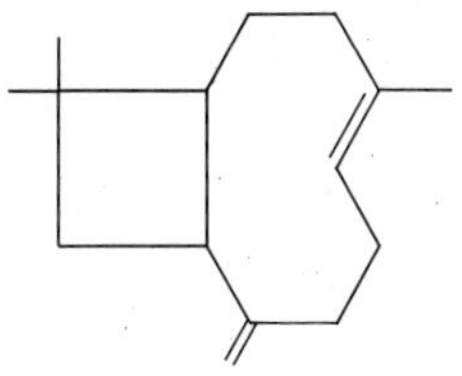

Caryophyllene
(from oil of cloves)

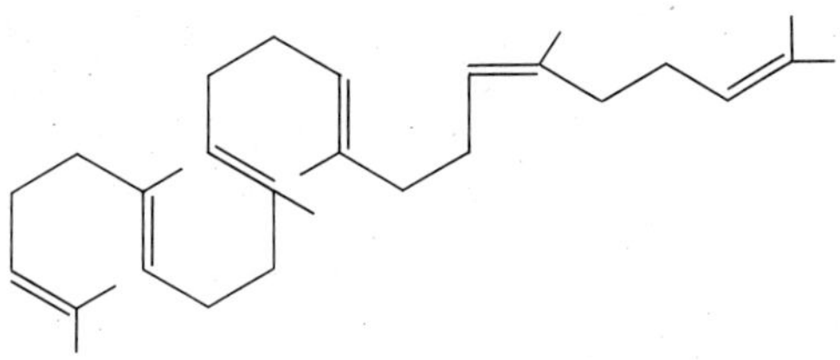

Squalene
(from shark liver oil)

Problem 11.20

What products would you expect to obtain if each of the following terpenes were subjected to ozonization and subsequent treatment with zinc and water?

    (a) Myrcene   (d) Geraniol
    (b) Limonene   (e) Squalene
    (c) α-Farnesene

Problem 11.21

Give structural formulas for the products that you would expect from the following reactions.

(a) β-Pinene + hot $KMnO_4 \longrightarrow$

(b) Zingiberene + $H_2 \xrightarrow{\text{Pt}}$

(c) Caryophyllene + HCl $\longrightarrow$

(d) β-Selinene + $2(BH_3)_2 \xrightarrow[\text{(2) } H_2O_2, \text{ OH}^-]{}$

Problem 11.22

What simple chemical test could you use to distinguish between geraniol and menthol?

The carotenes are tetraterpenes. They can be thought of as two diterpenes linked in tail-to-tail fashion.

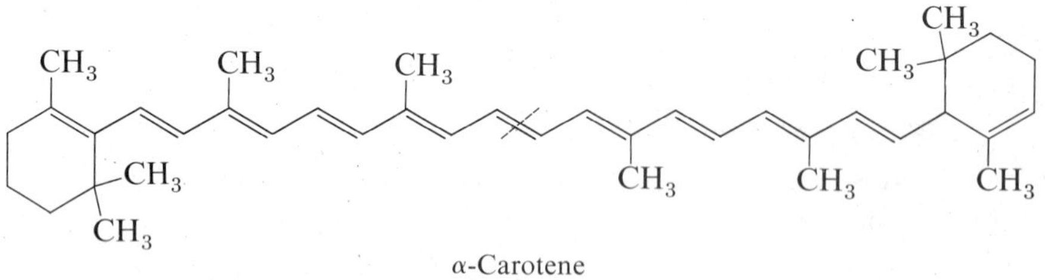

α-Carotene

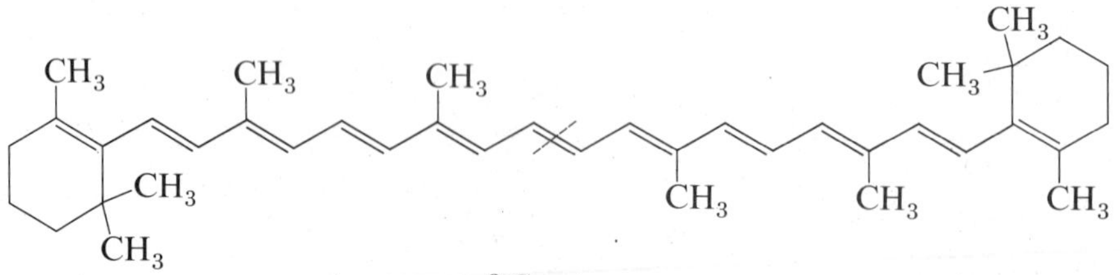

β-Carotene

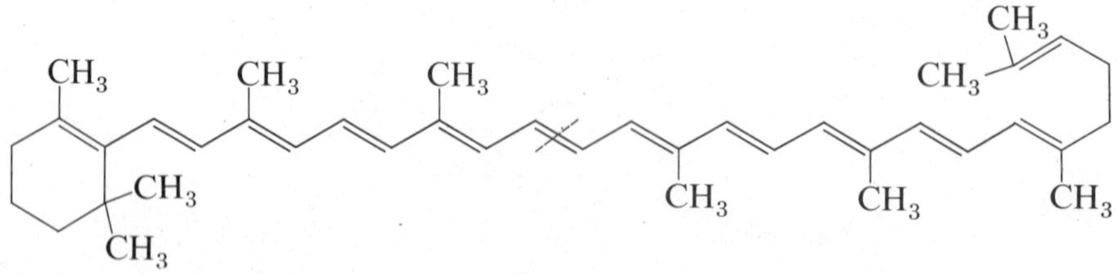

γ-Carotene

The carotenes are present in almost all green plants. All three carotenes serve as precursors for vitamin A, for they all can be converted to vitamin A by enzymes in the liver.

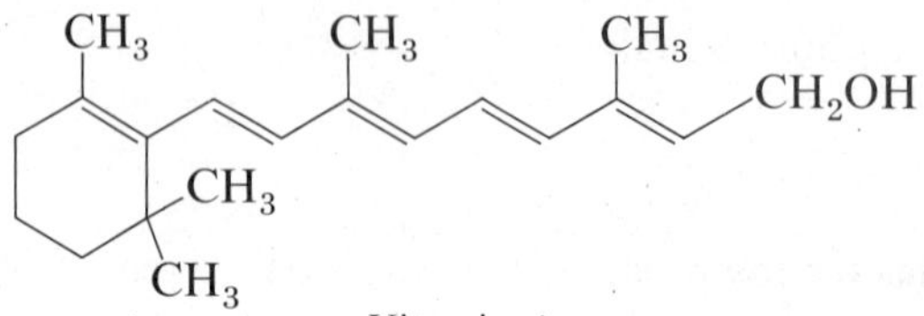

Vitamin A

In this conversion, β-carotene yields two molecules of vitamin A; α- and γ-carotene give only one. Because β-carotene became available before vitamin A; β-carotene has been adopted as the standard for vitamin A content in foods. One international unit (I.U.) of vitamin A is equivalent to the Vitamin A activity of 0.6 μg of crystalline β-carotene. Vitamin A is important not only in vision but in many other ways as well. For example, young animals whose diets are deficient in Vitamin A fail to grow.

Natural Rubber

Natural rubber can be viewed as a 1,4-addition polymer of isoprene. In fact, heat degrades natural rubber to isoprene when the heating is carried out in the absence of air. The isoprene units of natural rubber are all linked in a head-to-tail fashion and all of the double bonds are *cis*.

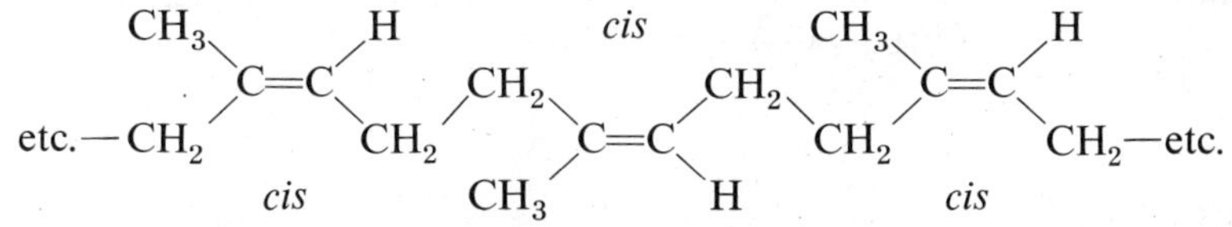

Natural rubber
(*cis*-1,4-Polyisoprene)

Using Ziegler-Natta catalysts (p. 298), it is now possible to polymerize isoprene and obtain a synthetic product that is identical with the rubber obtained from natural sources.

Pure natural rubber is soft and tacky. For it to be useful, natural rubber has to be "vulcanized." In the vulcanization process, natural rubber is heated with sulfur. A reaction takes place that produces cross-links between the *cis*-polyisoprene chains and makes the rubber much harder. Sulfur reacts both at the double bonds and at allylic hydrogens.

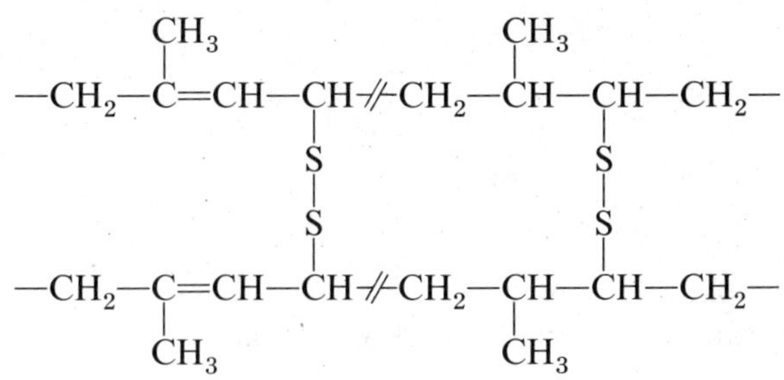

Vulcanized rubber

Biosynthesis of Isoprenoids

One of the most interesting areas of research today is the discovery of the pathways—the specific reaction sequences—by which plants and animals synthesize the compounds that they need. We will, of course, have much more to say about this area in later chapters. We can show now, in outline form, how some of those compounds composed of isoprene units are constructed in living organisms.

The basic building block for the synthesis of terpenes and terpenoids is

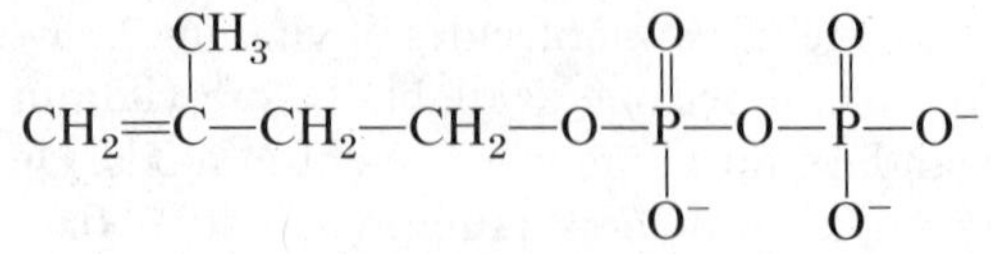

$$CH_2\!=\!\overset{\overset{\textstyle CH_3}{|}}{C}\!-\!CH_2\!-\!CH_2\!-\!O\!-\!\overset{\overset{\textstyle O}{\|}}{\underset{\underset{\textstyle O^-}{|}}{P}}\!-\!O\!-\!\overset{\overset{\textstyle O}{\|}}{\underset{\underset{\textstyle O^-}{|}}{P}}\!-\!O^-$$

3-Isopentenylpyrophosphate

3-isopentenylpyrophosphate. The five carbons of this compound are the source of the "isoprene units" of all "isoprenoids." The pyrophosphate group is a group that nature relies upon for a vast number of chemical processes. In the reactions that we will show below the pyrophosphate group functions as a natural "leaving group" in enzymatic reactions that resemble the alkylation reactions of alkenes which we studied in Section 6.10.

3-Isopentenylpyrophosphate (3-IPP) is isomerized by an enzyme to 2-iso-pentenylpyrophosphate (2-IPP). The isomerization is an equilibrium, and because of this, the cell has available both 3-IPP and 2-IPP.

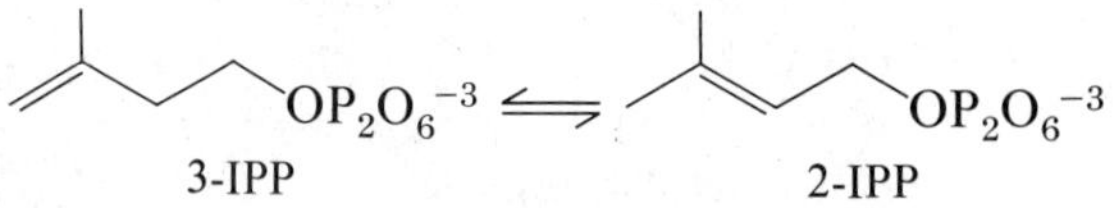

These two 5-carbon compounds condense with each other in another enzymatic reaction to yield the 10-carbon compound, geranyl pyrophosphate.

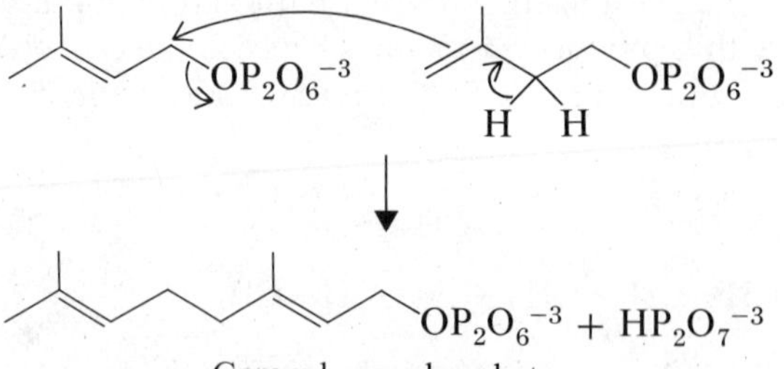

Geranyl pyrophosphate

Geranyl pyrophosphate is the precursor of the monoterpenes; hydrolysis of geranyl pyrophosphate, for example, yields geraniol.

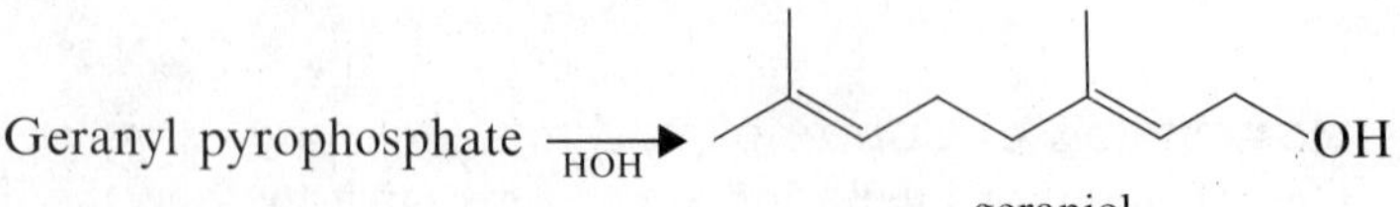

geraniol

Geranyl pyrophosphate can also condense with 3-isopentenylpyrophosphate to form the 15 carbon precursor for sesquiterpenes, farnesyl pyrophosphate.

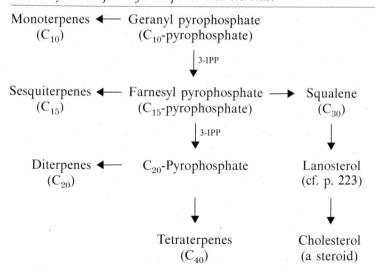

Geranyl pyrophosphate

$-HP_2O_7^{-3}$

Farnesyl pyrophosphate

Farnesol

Other sesquiterpenes

Farnesol has been isolated from ambrette seed oil. It has the odor of lily of the valley. Farnesol also functions as a hormone in insects and causes the change from caterpillar to cocoon to moth.

Similar condensation reactions yield the precursors for all of the other terpenes (Fig. 11.8). In addition, a tail-to-tail reductive coupling of two molecules of farnesyl pyrophosphate produces squalene, the precursor for another important group of isoprenoids known as *steroids* (cf. Chapter 23).

FIG. 11.8

The biosynthetic paths for terpenes and steroids.

Monoterpenes ◄─── Geranyl pyrophosphate
(C_{10})   (C_{10}-pyrophosphate)

│ 3-IPP

Sesquiterpenes ◄─── Farnesyl pyrophosphate ───► Squalene
(C_{15})   (C_{15}-pyrophosphate)   (C_{30})

│ 3-IPP   │

Diterpenes ◄─── C_{20}-Pyrophosphate   Lanosterol
(C_{20})   (cf. p. 223)

│   │

Tetraterpenes   Cholesterol
(C_{40})   (a steroid)

Problem 11.23

(a) Show how reductive tail-to-tail coupling of two molecules of farnesyl pyro-phosphate would yield squalene.

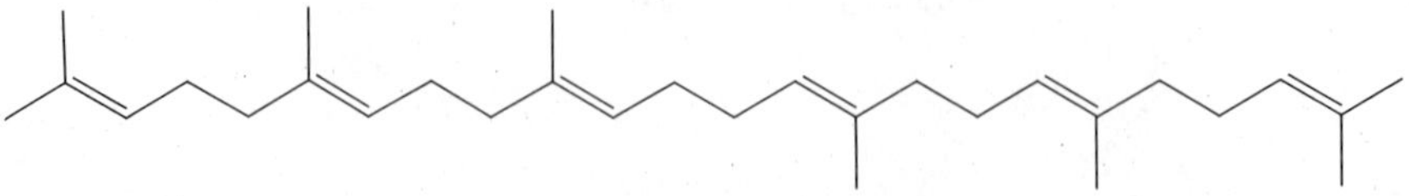

Squalene

(b) Show how an oxidative tail-to-tail coupling of two of the C_{20}-pyrophosphates would yield a precursor for the carotenes.

Problem 11.24

When farnesol is treated with sulfuric acid it is converted to bisabolene. Outline a possible mechanism for this reaction.

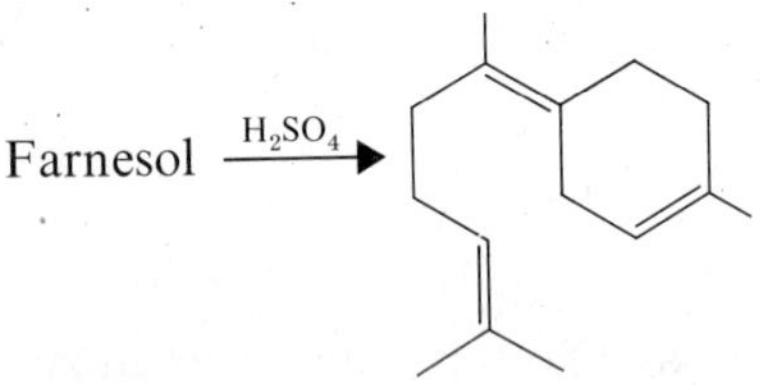

Farnesol $\xrightarrow{\text{H}_2\text{SO}_4}$

Bisabolene

Problem 11.25

When limonene (p. 405) is heated strongly, it yields two moles of isoprene. What kind of reaction is involved here?

*Problem 11.26

α-Phellandrene and β-phellandrene are isomeric compounds that are minor constit-uents of spearmint oil; they have the molecular formula $C_{10}H_{16}$. Each compound has an ultraviolet absorption maximum in the 230–270 nm range. On catalytic hydro-genation each compound yields 1-isopropyl-4-methylcyclohexane. On vigorous oxidation with potassium permanganate, α-phellandrene yields $CH_3\overset{\overset{\displaystyle O}{\|}}{C}COOH$ and $CH_3CHCH(COOH)CH_2COOH$. A similar oxidation of β-phellandrene yields
$\quad\quad\;\; |$
$\quad\;\; CH_3$
$CH_3CHCH(COOH)CH_2CH_2\overset{\overset{\displaystyle O}{\|}}{C}COOH$ as the only isolable product. Propose structures
$\quad\quad\;\; |$
$\quad\;\; CH_3$
for α- and β-phellandrene.

*Problem 11.27

The double bonds of vitamin A (below) all have a trans-configuration. An isomer of vitamin A, called neovitamin A, that can be obtained from sharkliver oil has a 13-*cis*-double bond. Both compounds react with maleic anhydride at the 11, 13-diene unit; however, neovitamin A reacts much more slowly. (The difference in their reactivities is so great that the Diels–Alder reaction can be used as an assay for neovitamin A in fish liver oils once the total concentration of vitamin A isomers is known.) Propose an explanation for the difference in reactivity of vitamin A and neovitamin A.

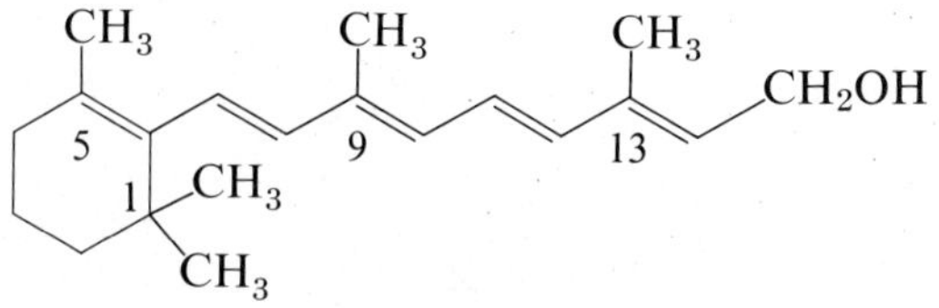

Vitamin A

* **Problem 11.28**

[Problem for Section 11.2] (a) What is the product of the following reaction?

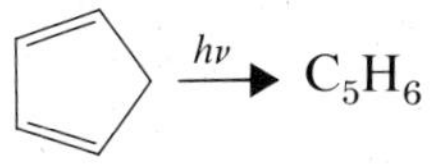

(b) Although the product of this reaction is *highly strained,* it is remarkably stable to heat. (At room temperature it has a half-life of about 2 hours and it can be heated to 50° for short periods without extensive degradation.) What factor might account for this unusual thermal stability?

* Problem 11.29

[Problem for Section 11.3] Propose structures for compounds **A**, **B**, and **C**.

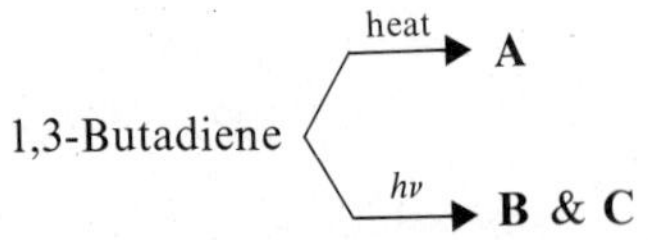

1,3-Butadiene
$\xrightarrow{\text{heat}}$ **A**

$\xrightarrow{hv}$ **B & C**

* Problem 11.30

[Problem for Section 11.3] What are the intermediates **A** and **B** in the synthesis of basketene given below?

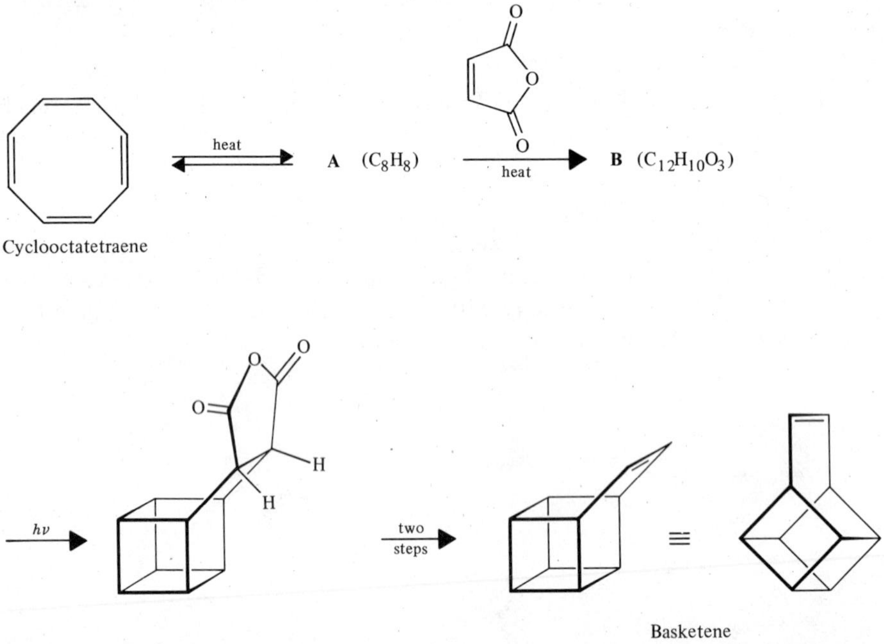

Cyclooctatetraene $\xrightarrow{\text{heat}}$ **A** (C_8H_8) $\xrightarrow[\text{heat}]{}$ **B** $(C_{12}H_{10}O_3)$

$\xrightarrow{hv}$ $\xrightarrow[\text{steps}]{\text{two}}$ $\equiv$

Basketene

* Problem 11.31

[Problem for Section 11.3] Account for the products of the following reaction with a step-by-step mechanism.

$$CH_2=C=CH_2 + CH_2=CH-C\equiv N \xrightarrow{\text{heat}}$$

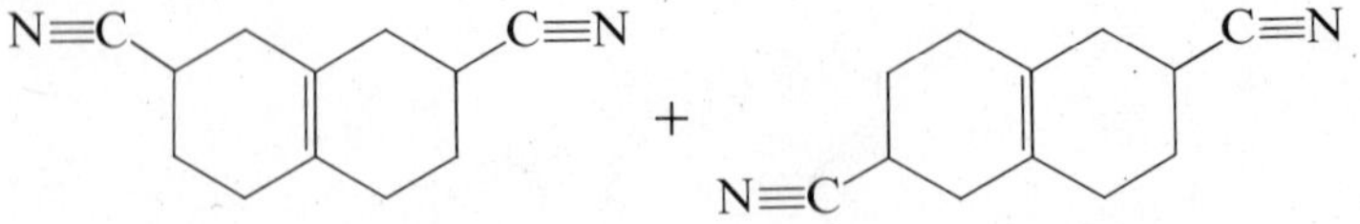

+

12 AROMATIC COMPOUNDS I: THE PHENOMENON OF AROMATICITY

12.1 INTRODUCTION

In 1825, Michael Faraday, a scientist most often remembered for his work with electricity, isolated a new substance from a gas used at that time for illumination. This compound, now called *benzene,* was an example of a new class of organic substances called *aromatic compounds.*

In 1834, benzene was shown to have the empirical formula CH. Later its molecular formula was shown to be C_6H_6. This in itself was a surprising discovery. Benzene has *only* as many hydrogen atoms as it has carbon atoms! Most compounds that were then known had a far greater proportion of hydrogen atoms, usually twice as many. Benzene, with a formula of C_6H_6, or C_nH_{2n-6}, must consist of *highly unsaturated* molecules. However, as we will see, benzene and other aromatic compounds are characterized by a tendency to undergo the substitution reactions characteristic of saturated compounds, rather than the addition reactions characteristic of unsaturated compounds.

During the latter part of the nineteenth century the Kekulé-Couper-Butlerov theory of valence was systematically applied to all known organic compounds. One result of this effort was the placing of organic compounds in either of two broad categories; compounds were classified as being either *aliphatic* or *aromatic.* To be classified as aliphatic meant then that the chemical behavior of a compound was "fat-like". (Now it means that the compound reacts like an alkane, an alkene, an alkyne, or one of their cyclic counterparts.) To be classified as aromatic meant then that the compound had a low hydrogen to carbon ratio and that it was "fragrant". Most of the early aromatic compounds were obtained from balsams, resins, or essential oils. Included among these were benzaldehyde (from oil of bitter almonds), benzoic acid and benzyl alcohol (from gum benzoin), and toluene (from tolu balsam).

Kekulé was the first to recognize that these early aromatic compounds all contain a six-carbon unit and that they retain this six-carbon unit through most chemical transformations and degradations. Benzene was eventually recognized as being the parent compound of this new series.

Since this new group of compounds proved to be distinctive in ways that are far more important than their odors, the term "aromatic" began to take on a purely chemical connotation. We will see, in this chapter, that the meaning of the term aromatic has evolved as chemists have learned more about the reactions and properties of aromatic compounds.

12.2 REACTIONS OF BENZENE

The benzene molecule is highly unsaturated, and because of this we might expect that it would react accordingly. We might expect, for example, that benzene would decolorize bromine in carbon tetrachloride by *adding* bromine, that it would decolorize aqueous potassium permanganate by being *oxidized*, that it would *add* hydrogen easily in the presence of a catalyst, and that it would *add* water in the presence of acids.

Benzene does none of these. When benzene is treated with bromine in carbon tetrachloride in the dark or with aqueous potassium permanganate or with dilute acids, no reaction occurs (cf. Table 12.1). Benzene does add hydrogen in the presence of finely divided nickel, but only at high temperatures and under high pressures.

TABLE 12.1 Reactions of Benzene

REAGENT	EXPECTED OF AN UNSATURATED COMPOUND	OBSERVED
Br_2/CCl_4	Addition of Br_2	No reaction
$KMnO_4/H_2O$	Oxidation	No reaction
H_2/Ni	Rapid addition of hydrogen	Slow addition at high temperature and pressure
H_2O/acid catalyst	Addition of water	No reaction

Benzene reacts with bromine but only in the presence of an iron salt catalyst such as ferric bromide. Most surprisingly, however, it reacts not by addition but by *substitution*.

Substitution: $C_6H_6 + Br_2 \xrightarrow{FeBr_3} C_6H_5Br + HBr$ Observed

Addition: $C_6H_6 + Br_2 \xrightarrow{\times} C_6H_6Br_2 + C_6H_6Br_4 + C_6H_6Br_6$ Not observed

Benzene undergoes similar substitution reactions with chlorine in the presence of ferric ion, with nitric acid in the presence of sulfuric acid, with SO_3 in the presence of sulfuric acid, and with alkyl halides in the presence of aluminum chloride.

$$C_6H_6 + Cl_2 \xrightarrow{FeCl_3} \underset{\text{Chlorobenzene}}{C_6H_5Cl} + HCl \qquad \text{Chlorination}$$

$$C_6H_6 + HONO_2 \xrightarrow{H_2SO_4} \underset{\text{Nitrobenzene}}{C_6H_5NO_2} + H_2O \qquad \text{Nitration}$$

$$C_6H_6 + SO_3 \xrightarrow{H_2SO_4} \underset{\text{Benzenesulfonic acid}}{C_6H_5SO_3H} \qquad \text{Sulfonation}$$

$$C_6H_6 + CH_3Cl \xrightarrow{AlCl_3} \underset{\text{Toluene}}{C_6H_5CH_3} + HCl \qquad \text{Alkylation}$$

When benzene reacts with bromine *only one monobromobenzene* is formed. That is, only one compound with the formula C_6H_5Br is found among the products.

Similarly, when benzene is chlorinated *only one monochlorobenzene* results, and when benzene is nitrated *only one mononitrobenzene* is formed. The same behavior is true of the sulfonation and alkylation reactions of benzene.

There are obviously two possible explanations that can be given for these observations. The first is that only one of the six hydrogens in benzene is reactive toward these reagents. The second is that all six hydrogens in benzene are equivalent so that replacing any one of them with a substituent results in the same product. As we will see, it is the second explanation that is correct.

Problem 12.1

Listed below are several compounds that have the formula C_6H_6. (a) For which of these compounds, if any, would a substitution of bromine for hydrogen yield only one *mono*bromo product? (b) Which of these compounds would you expect to react with bromine by substitution alone?

$$H—C\equiv C—CH_2CH_2—C\equiv C—H$$
$$(a)$$

$$CH_2=CH—CH=CH—C\equiv CH$$
$$(b)$$

$$CH_3CH_2—C\equiv C—C\equiv C—H$$
$$(c)$$

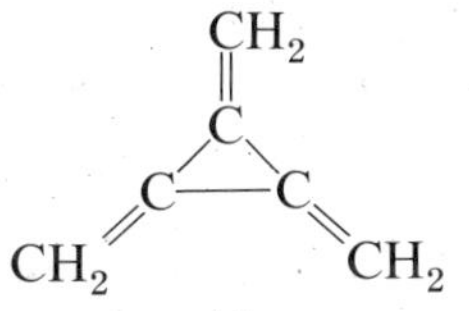

$$(d)$$

$$CH_2=CH—C\equiv C—CH=CH_2$$
$$(e)$$

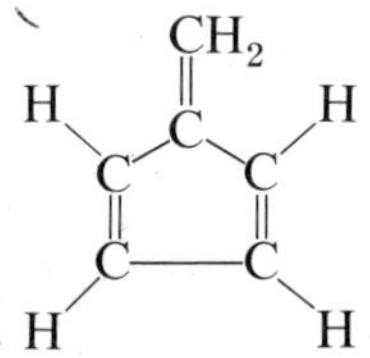

$$(f)$$

12.3 THE KEKULE STRUCTURE FOR BENZENE

In 1865, August Kekulé proposed a structure for benzene. Kekulé suggested that the carbon atoms of benzene are in a ring, that they are bonded to each other by alternating single and double bonds, and that one hydrogen atom is attached to each carbon. This structure satisfied the requirements that carbon atoms form four bonds, and that all of the hydrogen atoms of benzene are equivalent.

or

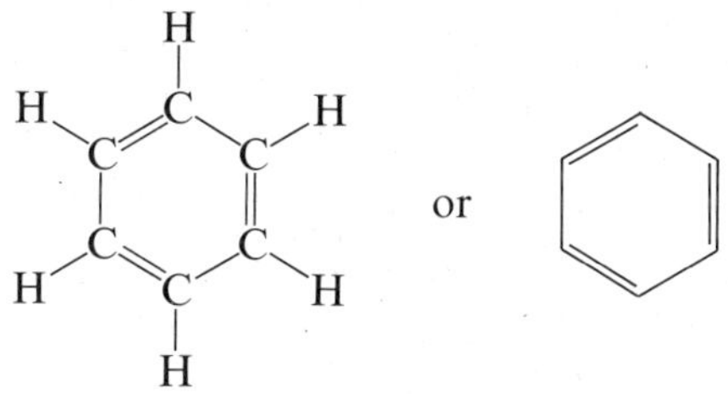

The Kekulé formula for benzene

A problem soon arose with the Kekulé structure, however. The Kekulé structure predicts that there should be two different 1,2-dibromobenzenes. In one of these hypothetical compounds (page 414), the carbons that bear the bromines are separated by a single bond, and in the other they are separated by a double bond. *Only one 1,2-dibromobenzene, however, has ever been found.*

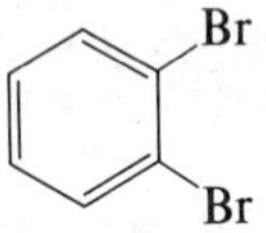

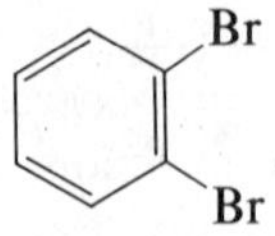

In order to accommodate this objection, Kekulé proposed that the two forms of benzene (and of benzene derivatives) are in a state of equilibrium, and that this equilibrium is so rapidly established that it prevents isolation of the separate molecules.

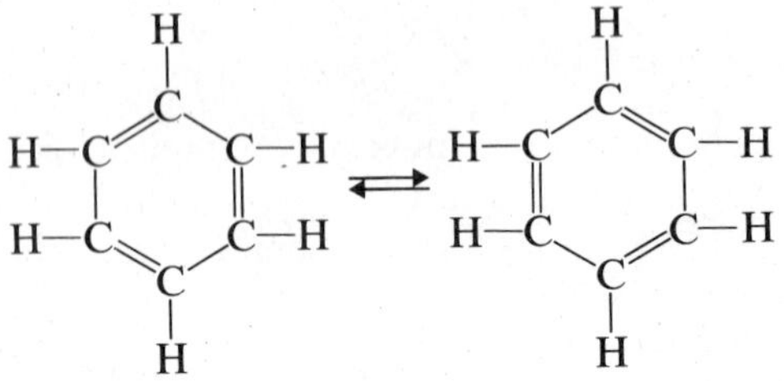

Thus, the two 1,2-dibromobenzenes would also be in rapidly equilibrated, and this would explain why chemists had not been able to isolate the two forms.

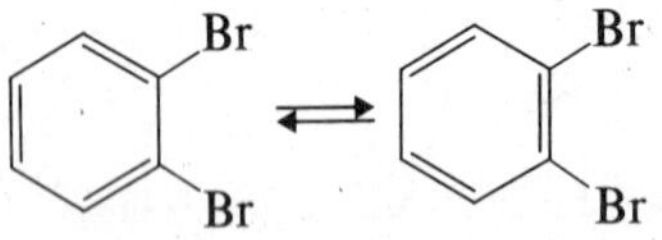

We now know that this proposal was wrong and that *no such equilibrium exists.* Nonetheless, the Kekulé formulation of benzene's structure was an important step forward and, for very practical reasons, it is still used today. We conceptualize its meaning differently, however.

With the Kekulé formula we are able to reformulate the substitution reactions of benzene in a clearer way.

Bromination: $+ Br_2 \xrightarrow{FeBr_3}$ $+ HBr$

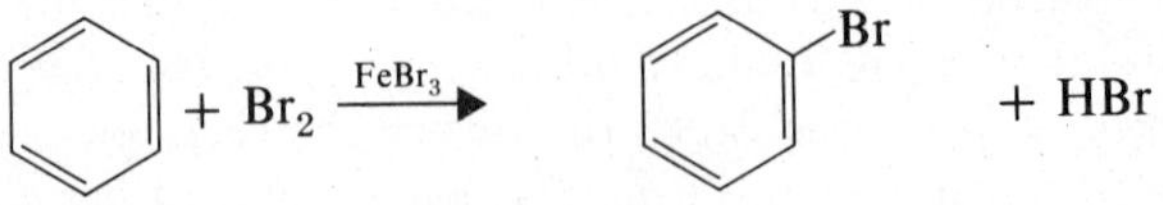

Nitration: $+ HONO_2 \xrightarrow{H_2SO_4}$ $+ H_2O$

Sulfonation: $+ SO_3 \xrightarrow{H_2SO_4}$

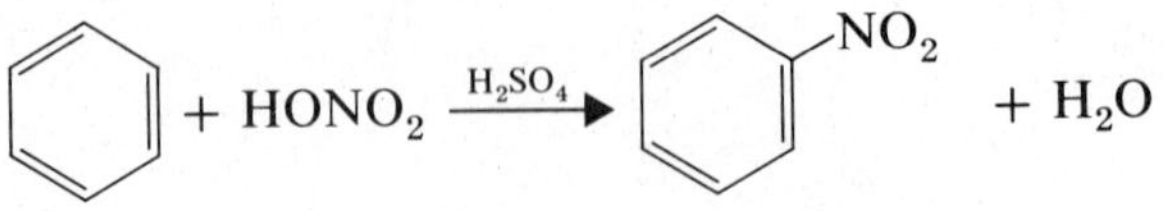

Alkylation: $+ CH_3Cl \xrightarrow{AlCl_3}$ $+ HCl$

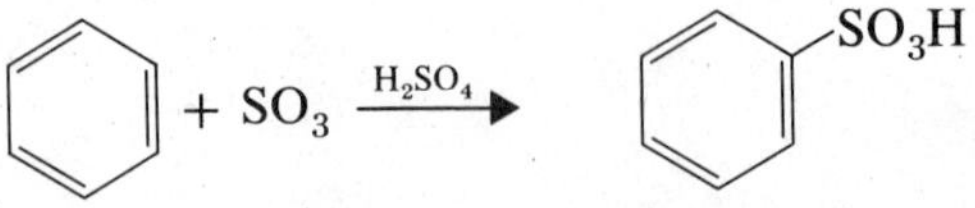

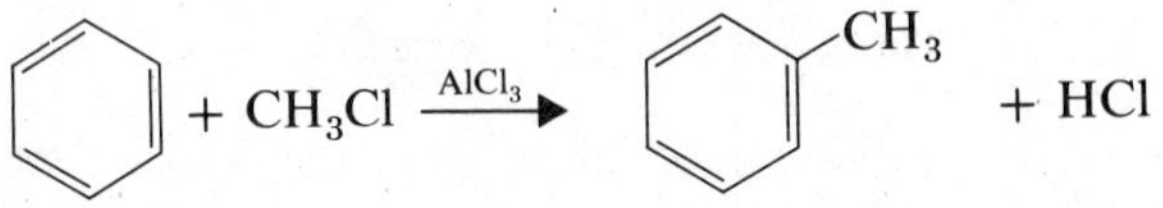

In each of these substitution reactions the benzene ring with its alternating single and double bonds is retained in the product. This would not have been true if benzene had reacted by addition. Had an addition reaction occurred, the ring of alternating single and double bonds would no longer exist in the products. Addition of bromine, for example, would have resulted in the following reaction.

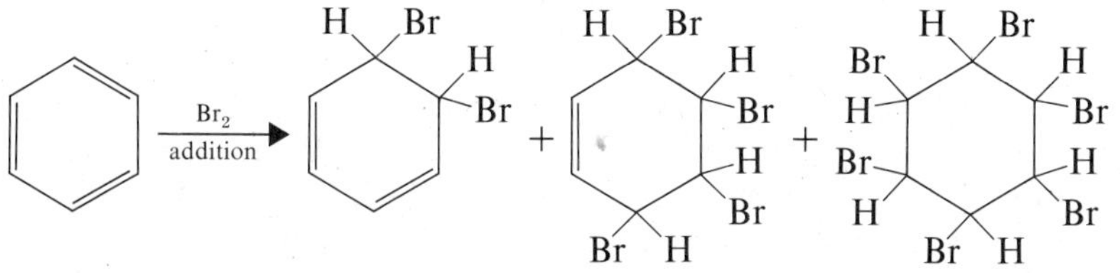

The tendency of benzene to react by substitution rather than addition gave rise to another concept of aromaticity. Since benzene reacts by substitution, and thereby retains its ring of alternating single and double bonds, aromatic compounds were seen as those molecules that retain their cyclic, unsaturated structure in most of their reactions. For a compound to be called aromatic meant, experimentally, that it gave substitution reactions rather than addition reactions even though it was highly unsaturated.

Before 1900, chemists assumed that the ring of alternating single and double bonds was the structural feature that gave rise to the aromatic properties. Since benzene and benzene derivatives (that is, compounds with six-membered rings) were the only aromatic compounds known, chemists naturally sought other examples. The compound cyclooctatetraene seemed to be a likely candidate.

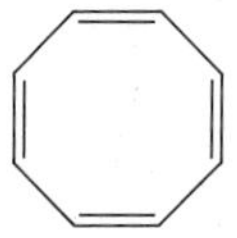

Cyclooctatetraene

In 1911, through a brilliant synthesis that we will discuss later, Richard Willstätter succeeded in synthesizing cyclooctatetraene. Willstätter found, however, that cyclooctatetraene is not at all like benzene. Cyclooctatetraene reacts with bromine by addition, it adds hydrogen readily, it decolorizes solutions of potassium permanganate, and thus it is clearly *not aromatic*. While these findings must have been a keen disappointment to Willstätter, they were very significant for what they did not prove. Chemists, as a result, had to look deeper to discover the origin of benzene's aromaticity.

12.4 THE STABILITY OF BENZENE

Benzene not only undergoes substitution reactions when, on the basis of its Kekulé structure, we would expect that it should undergo addition, benzene is also *more stable* than the Kekulé structure suggests that it should be. To see how this is so, consider the following thermochemical results.

Cyclohexene, a six-membered ring containing one double bond, can be hydrogenated easily to cyclohexane. When the ΔH for this reaction is measured it is found to be -28.6 kcal/mole, very much like that of any alkene.

$$\text{Cyclohexene} + H_2 \xrightarrow{\text{Pt}} \text{Cyclohexane} \qquad \Delta H = -28.6 \text{ kcal/mole}$$

We would expect that hydrogenation of 1,3-cyclohexadiene would liberate roughly twice as much heat and thus have a ΔH equal to about -57.2 kcal/mole. When this experiment is done, the result is a $\Delta H = -55.4$ kcal/mole. This result is quite close to what we calculated, and the difference can be explained by taking

$$\text{1,3-Cyclohexadiene} + 2H_2 \xrightarrow{\text{Pt}} \text{Cyclohexane}$$

Calculated
$\Delta H = (2 \times -28.6) = -57.2$ kcal/mole

Observed
$\Delta H = -55.4$

into account the fact that compounds containing conjugated double bonds are usually somewhat more stable than those that contain isolated double bonds.

If we extend this kind of thinking, and if benzene is simply 1,3,5-cyclohexatriene, we would predict that benzene would liberate approximately 85.8 kcal/mole (3×-28.6) when it is hydrogenated. When the experiment is actually done the result is surprisingly different. The reaction is exothermic, but only by 49.8 kcal/mole.

$$\text{Benzene} + 3H_2 \xrightarrow{\text{Pt}} \text{Cyclohexane}$$

Calculated
$\Delta H = (3 \times -28.6) = -85.8$ kcal/mole

Observed	ΔH	=	-49.8
Difference		=	36 kcal/mole

When these results are represented in Fig. 12.1 it becomes clear that benzene is much more stable than we calculated it to be. Indeed, it is more stable than the hypothetical 1,3,5-cyclohexatriene by 36 kcal/mole. This difference between the amount of heat actually released and that calculated on the basis of the Kekulé structure is now called the *resonance energy* or *delocalization energy* of the compound.

12.5 MODERN THEORIES OF THE STRUCTURE OF BENZENE

It was not until the development of quantum mechanics in the 1920s that the unusual behavior and stability of benzene began to be understood. Quantum mechanics, as we have seen, produced two approximate ways of viewing the bonding in molecules: resonance theory and molecular orbital theory. We now look at both of these as they apply to benzene.

The Resonance Explanation of the Structure of Benzene

It is a basic postulate of resonance theory (Sections 1.13 and 10.2) that whenever two or more structures can be written for a molecule *that differ only in the positions of the electrons,* none of the structures will be in complete accord with

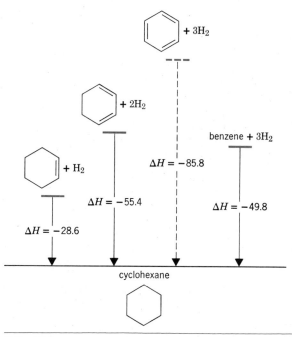

$\Delta H = -28.6$

$\Delta H = -55.4$

$\Delta H = -85.8$

$\Delta H = -49.8$

benzene $+ 3H_2$

$+ H_2$

$+ 2H_2$

$+ 3H_2$

cyclohexane

FIG. 12.1

Relative stabilities of cyclohexene, 1,3-cyclohexa-diene, 1,3,5-cyclohexatriene (hypothetically), and benzene.

the compound's chemical and physical properties. If we recognize this, we can now understand the true nature of the two Kekulé structures (I and II) for benzene. The two Kekulé structures differ only in the positions of the electrons. Structures I and II, then, do not represent two separate molecules in equilibrium as Kekulé had proposed; instead, they are the closest we can get to a structure for benzene within the limitations of its molecular formula, the classical rules of valence, and the fact that the six hydrogens are chemically equivalent. That they are false structures is not the fault of nature; it is the fault of earlier theories of molecular structure. Resonance theory, fortunately, does not stop with telling us when to expect this kind of trouble; it also instructs us in a way out. Resonance theory tells us to use structures I and II; to treat them simply as "contributors"—calling them resonance structures—contributors to a picture of the real molecule of benzene. As such they should be connected with a double-headed arrow and not with two separate ones (because we must reserve the symbol of two separate arrows for chemical equilibria). Resonance contributors, we emphasize again, are not in equilibrium. They are not structures of real molecules. They are the closest we can get if we are bound by simple rules of valence, and they are very useful in helping us visualize the actual molecule as a hybrid.

(not ⇌)

I   II

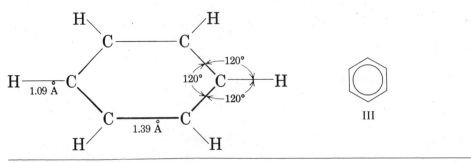

FIG. 12.2
Bond lengths and angles in benzene.

Look at the structures carefully. All of the single bonds in structure I are double bonds in structure II. If we blend I and II, that is, fashion a hybrid of them, then the carbon-carbon bonds in benzene are neither single bonds nor double bonds but have a bond order between that of a single bond and that of a double bond. This is exactly what we find experimentally. Spectroscopic measurements show that benzene is a planar molecule and that all of the carbon-carbon bonds are of equal length. Moreover, the carbon-carbon bond lengths in benzene are 1.39 Å, a value in between that for a carbon-carbon single bond (1.54 Å) and that for a carbon-carbon double bond (1.33 Å) (Fig. 12.2).

This situation is represented by inscribing a circle in the hexagon, and it is this new formula, III, that is most often used for benzene today. There are times, however, when an accounting of the electrons must be made, and for these purposes we may use one or the other of the Kekulé structures. We do this simply because the electron count in Kekulé structures is obvious, while the number of electrons represented by a circle or portion of a circle is more difficult to estimate.

Problem 12.2

If benzene were 1,3,5-cyclohexatriene what carbon-carbon bond lengths would you expect the molecule to have?

FIG. 12.3

A six-membered ring of sp^2 carbon atoms. Sigma bonds are shown as lines.

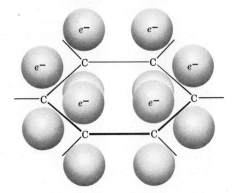

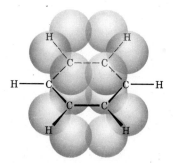

FIG. 12.4
Overlapping p orbitals in benzene.

than any of the resonance structures individually. It is in this way that resonance theory accounts for the much greater stability of benzene when compared to the hypothetical 1,3,5-cyclohexatriene. And it is for this reason that the difference between them is sometimes called the *resonance energy*.

The Molecular Orbital Explanation of the Structure of Benzene

The fact that the bond angles of the carbon atoms in the benzene ring are 120° strongly suggests that the carbon atoms are sp^2 *hybridized*. If we accept this and construct a planar six-membered ring from sp^2 carbon atoms as shown in Fig. 12.3, another picture of benzene begins to emerge.

Since the carbon-carbon bond distances are all 1.39 Å, the p orbitals are close enough to overlap effectively. The p orbitals also overlap equally all around the ring (Fig. 12.4).

According to molecular-orbital theory, the six overlapping p orbitals combine to form a set of six molecular orbitals. Molecular-orbital theory also allows us to calculate the relative energies of the π-molecular orbitals. These calculations are beyond the scope of our discussion but the energy levels are shown in Fig. 12.5.

FIG. 12.5
The energy levels of the π-molecular orbitals of benzene. Energy levels below the dotted line are bonding while those above the dotted line are antibonding.

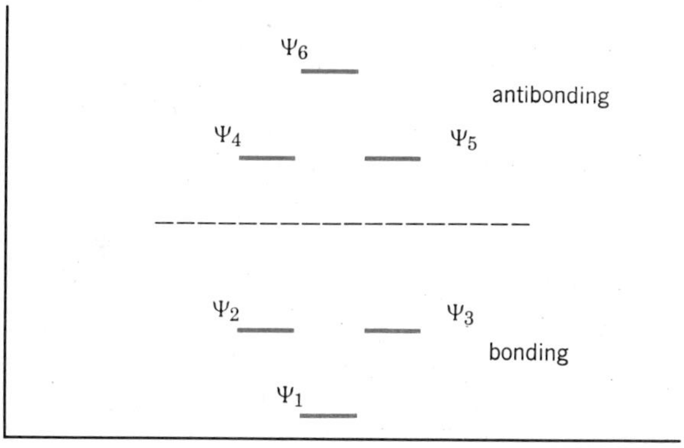

Ψ_6

antibonding

Ψ_4 Ψ_5

Ψ_2 Ψ_3

bonding

Ψ_1

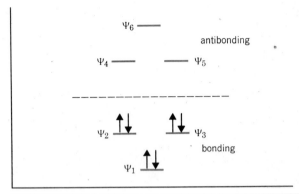

FIG. 12.6
The ground state of benzene.

Molecular orbitals, as we have seen, can accommodate two electrons as long as their spins are opposed. Thus, the electronic structure of the ground state of benzene is obtained by adding the six electrons to the molecular orbitals so that those of lowest energy are filled first. This is shown in Fig. 12.6. Notice that in benzene, all of the bonding orbitals are filled, all of the electrons have their spins paired, and there are no electrons in anti-bonding orbitals. Benzene is, thus, said to have a *closed bonding shell* of electrons. It is this closed bonding shell that accounts, in part, for the stability of benzene.

The shapes of the molecular orbitals of benzene are given in Fig. 12.7.

Further insight into the unusual stability of benzene can be obtained by inspecting the lowest energy molecular orbital, ψ_1 (Fig. 12.7). In quantum mechanics electrons are described in terms of their *wave nature* and the energy of an electron is inversely proportional to its wavelength (Section 8.1). That is *the longer the wavelength* of the electron; *the lower is its energy.* Molecular orbital ψ_1 encompasses the entire ring of carbon atoms and has no nodes between carbon atoms; the electrons that occupy ψ_1 can move about this entire distance. Thus, these electrons can have a wavelength much longer than electrons in the molecular orbital of an isolated double bond (where the motion of the electrons is restricted to the vicinity of only two atoms). As a result, the electrons in the lowest molecular orbital of benzene have unusually low energies, and it is this factor, as well, that helps us account for the exceptional stability of benzene.

12.6 HÜCKEL'S RULE

In 1931, the German physicist E. Hückel carried out a series of mathematical calculations based on the kind of theory that we have just described. Hückel concerned himself with the general situation of *coplanar monocyclic* rings in which each atom of the ring has an available *p* orbital as in benzene. His calculations indicate that coplanar rings containing $(4n + 2)$ π electrons, where $n = 0, 1, 2, 3 \ldots$ etc., have closed shells of electrons like benzene, and should have substantial resonance or delocalization energies. In other words, *coplanar monocyclic rings with 2, 6, 10, 14, 18, and 22, delocalized π electrons should be aromatic.*

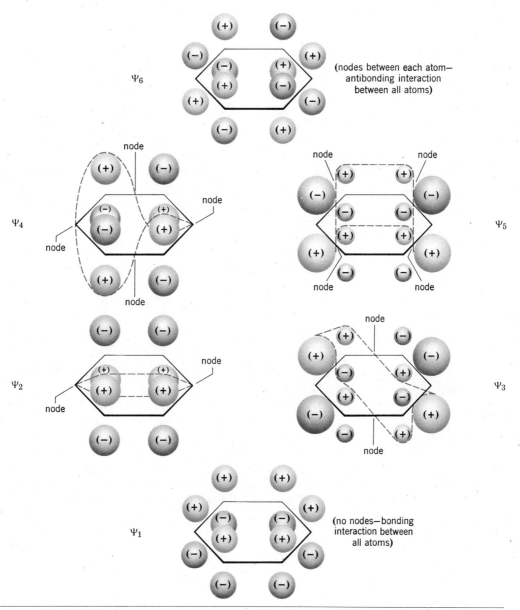

FIG. 12.7
Shapes of the molecular orbitals in benzene.

The fact that cyclooctatetraene lacks aromatic properties correlates with Hückel's rule. Cyclooctatetraene has a total of 8 π electrons; on the basis of Hückel's calculations, monocyclic rings with 8 π electrons are not predicted to have a closed shell of π electrons or substantial resonance energies and they are not expected to be aromatic. Cyclooctatetraene, it is now known, is not a coplanar molecule, but is shaped like a tub (Fig. 12.8).

The bonds of cyclooctatetraene are alternately long and short; X-ray crystallographic studies indicate that they are 1.54 Å and 1.34 Å, respectively. These values

FIG. 12.8

The actual (tub) shape of cyclooctatetraene.

are those of single bonds and double bonds, and are not at all like the hybrid bonds of benzene. Cyclooctatetraene is then, clearly, a simple cyclic polyene.

The Annulenes

The name annulene has been proposed as a general name for monocyclic compounds that can be represented by structures having alternating single and double bonds. The ring size of an annulene is indicated by a number in brackets. Thus, benzene is [6] annulene and cyclooctatetraene is [8] annulene.* Hückel's rule

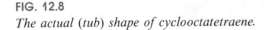

Benzene   Cyclooctatetraene
[6] Annulene   [8] Annulene

predicts that annulenes will be aromatic, provided they contain $(4n + 2)$ π electrons and have a planar carbon skeleton.

Before 1960, the only annulenes that were available to test Hückel's predictions were benzene and cyclooctatetraene. During the 1960s, and largely as a result of research by Professor Franz Sondheimer (now at University College, London), annulenes up to [24] annulene were synthesized and the predictions of Hückel's rule have been verified in every instance.†

Sondheimer's syntheses of large-ring annulenes involve reactions that we have seen before. Consider, as an example, his synthesis of [18] annulene.

$$3HC \equiv CCH_2CH_2C \equiv CH \xrightarrow[\text{pyridine}]{Cu^{++}}$$

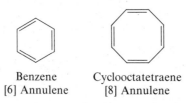

1

KOC(CH₃)₃ →

---

*These names are seldom used for benzene and cyclooctatetraene. They are often used, however, for conjugated rings of 10 or more carbon atoms.

† A [30] annulene has also been synthesized. However, recent calculations by Professor M. J. S. Dewar of the University of Texas indicate that Hückel's rule has an upper limit lying between [22] annulene and [26] annulene. Thus, even though this [30] annulene obeys Hückel's rule ($4n + 2$, where $n = 7$), it is above this limit. The [30] annulene is not aromatic.

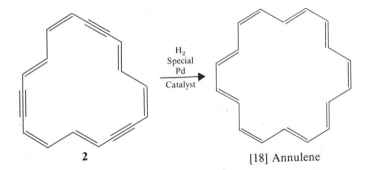

2   [18] Annulene

In the first step, oxidative coupling of 1,5-hexadiyne with cupric ion in pyridine (p. 324) gave the 18-membered cyclic "trimer" **1** in ~6% yield. When **1** was treated with potassium *tert*-butoxide, three of the triple bonds migrated to produce the six double bonds of the tridehydro [18] annulene* **2**. In the last step, hydrogenation of the triple bonds of **2** using a specially prepared palladium catalyst (p. 315) gave [18] annulene.

Examples of [14]-, [16]-, [20]-, [22]-, and [24]-annulenes have been synthesized by similar methods. Of these, *as Hückel's rule predicts,* the [14]-, [18]-, and [22] annulenes ($4n + 2$ when $n = 3, 4, 5$ respectively) have been found to be aromatic. The [16] annulene and the [24] annulene are not aromatic. They are $4n$ compounds, not $4n + 2$ compounds.

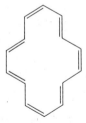

[14] Annulene
(aromatic)

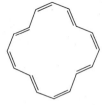

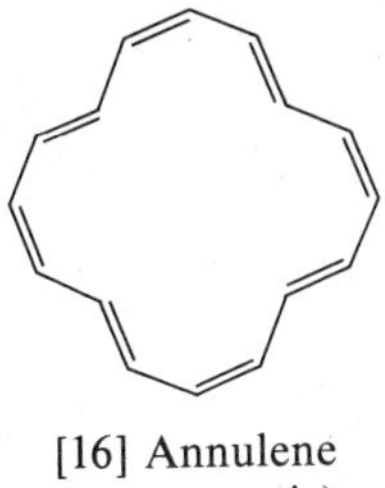

[16] Annulene
(nonaromatic)

Examples of [10] and [12] annulenes have been synthesized by other methods and none are aromatic. We would not expect [12] annulenes to be aromatic since they have 12 π electrons and, thus, do not obey Hückel's rule. The [10] annulenes (next page) would be expected to be aromatic on the basis of electron count, but their rings are not planar. The [10] annulene, **4**, has two *trans* double bonds.

* The prefix *tridehydro* indicates that the annulene has three triple bonds.

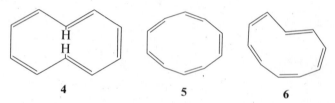

4   5   6

[10] Annulenes
(none are aromatic because none are planar)

The carbons of its ring are prevented from becoming coplanar because the two hydrogens in the center of the ring interfere with each other. The [10] annulene with all *cis* double bonds, **5**, and the [10] annulene with one *trans* double bond, **6**, are not planar because of ring strain.

After many unsuccessful attempts [4] annulene (or cyclobutadiene) has also been synthesized. As we would expect, it is a highly reactive compound and *it is not aromatic.*

Cyclobutadiene
or [4] annulene
(not aromatic)

Aromatic Ions

In addition to the neutral molecules that we have discussed so far, there are a number of monocyclic species that bear either a positive or negative charge. Some of these ions show unexpected stabilities that suggest that they, too, are aromatic. Hückel's rule is helpful in accounting for the properties of these ions as well. We will consider two examples: the cyclopentadienide anion and the cyclo-heptatrienyl cation.

Cyclopentadiene is not aromatic; however, it is unusually acidic for a hydro-carbon. (The K_a for cyclopentadiene is 10^{-15} and, by contrast, the K_a for cyclo-heptatriene is 10^{-36}.) Because of its acidity, cyclopentadiene can be converted to its anion by treatment with moderately strong bases. The cyclopentadienide anion, moreover, is unusually stable and nuclear magnetic resonance (NMR) spectroscopy (Chapter 14) shows that all five hydrogens in the cyclopentadienide ion are equivalent.

Cyclopentadiene   Cyclopentadienyl anion

When we look at the orbital structure of cyclopentadiene (Fig. 12.9*a*) it becomes clear why cyclopentadiene, itself, is not aromatic. Not only does it not have the proper number of π electrons, but the π electrons cannot be delocalized about the entire ring because of the intervening saturated —CH_2— group.

On the other hand, if the —CH_2— group becomes sp^2 hybridized as it loses a proton (Fig. 12.9*b*), the two electrons that are left behind can occupy the new

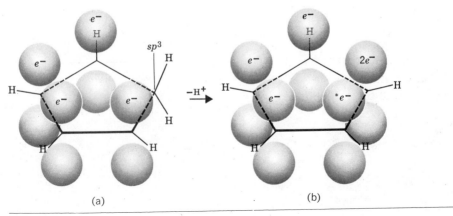

FIG. 12.9
(a) p *Orbitals of cyclopentadiene and* (b) *the* p *orbitals of the cyclopentadienide ion.*

p orbital that is produced. Moreover, this new *p* orbital can overlap with the *p* orbitals on either side of it and give rise to a ring with *six* delocalized π electrons (Fig. 12.10). Since the electrons are delocalized, all of the hydrogens are equivalent and this is what NMR spectroscopy tells us. Six is, of course, a Hückel number ($4n + 2$, where $n = 1$), and the cyclopentadienide ion is, in fact, an *aromatic anion*. The unusual acidity of cyclopentadiene is a result of the unusual stability of its anion.

Problem 12.3

(a) Write all of the resonance structures for the cyclopentadienide anion. (b) Can the equivalence of the five hydrogen atoms be explained by resonance?

Cycloheptatriene has six π electrons. However, the six π electrons of cycloheptatriene cannot be fully delocalized because of the presence of the —CH$_2$— group—a group that does not have an available *p* orbital (Fig. 12.11).

When cycloheptatriene is treated with a reagent that can abstract a hydride ion, it is converted to the cycloheptatrienyl (or tropylium) cation. The loss of a

FIG. 12.10
Overlapping p *orbitals in the cyclopentadienide ion.*

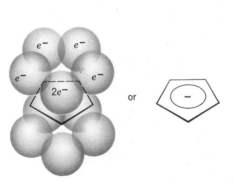

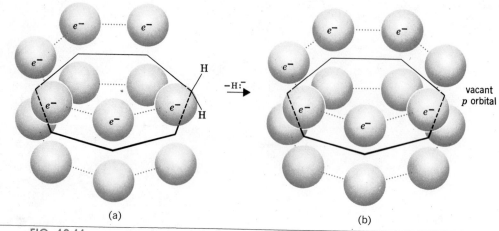

FIG. 12.11

p *orbitals of cycloheptatriene* (a) *and the cycloheptatrienyl* (*tropylium*) *cation* (b).

Cycloheptatriene

hydride ion from cycloheptatriene occurs with unexpected ease, and the cycloheptatrienyl cation is found to be unusually stable. The NMR spectrum of the cycloheptatrienyl cation indicates that all seven hydrogens are equivalent. If we look closely at Figs. 12.11 and 12.12 we will see how we can account for these observations.

As a hydride ion is removed from the —CH_2— group of cycloheptatriene, a vacant p orbital is created, and the carbon atom becomes sp^2 hybridized. The cation that results has seven overlapping p orbitals containing *six* delocalized π electrons. The cycloheptatrienyl cation is, therefore, an aromatic cation and all of its hydrogens should be equivalent; again this is exactly what we find experimentally.

FIG. 12.12

Overlapping p *orbitals in the cyclo-heptatrienyl cation.*

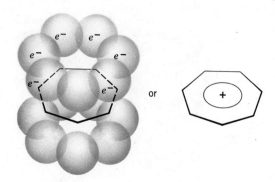

Problem 12.4

The conversion of cycloheptatriene to the cycloheptatrienyl cation can be accomplished by treating cycloheptatriene with triphenylcarbonium perchlorate $[(C_6H_5)_3C^+ClO_4^-]$. (a) Write the structure of triphenylcarbonium perchlorate and show how it abstracts a hydride ion from cycloheptatriene. (b) What other product is formed in this reaction? (c) What anion is associated with the cycloheptatrienyl cation that is produced?

Problem 12.5

Tropylium bromide, C_7H_7Br, is insoluble in nonpolar solvents but dissolves readily in water. When an aqueous solution of tropylium bromide is treated with silver nitrate a precipitate of AgBr forms immediately. The melting point of tropylium bromide is above 200°, quite high for an organic molecule. How can you account for these facts?

Aromatic, Antiaromatic, and Nonaromatic Compounds

When we say that cyclopentadienide anion is aromatic we do not mean that we can brominate, nitrate, or sulfonate it the way we can benzene. If we were, for example, to place the cyclopentadienide anion in a mixture of nitric and sulfuric acids in an attempt to nitrate it, the anion would immediately accept a proton and become cyclopentadiene. From that point on any reactions that took place would not be reactions of the cyclopentadienide anion but would be reactions of cyclopentadiene.

Many of the higher annulenes that we have classified as being aromatic are too highly reactive to nitrate, sulfonate, and so on.

What do we mean, then, when we say that these compounds are aromatic? We mean that their π electrons are *delocalized* over the entire ring and that they are *stabilized* by the π electron delocalization.

One of the best ways to determine whether or not the π electrons of a cyclic system are delocalized is through the use of nuclear magnetic resonance spectroscopy. NMR spectroscopy gives us a direct physical measurement of whether or not the π electrons are delocalized. We will have more to say about how this is done in Chapter 14.

But, what do we mean by saying that a compound is stabilized by π electron delocalization? We have an idea of what this means from our comparison of heats of hydrogenation of benzene with the hypothetical 1,3,5-cyclohexatriene. We saw that benzene—in which the π electrons are delocalized—is much more stable than 1,3,5-cyclohexatriene (a model in which the π electrons are not delocalized). We call the energy difference between them the resonance energy or delocalization energy.

In order to make similar comparisons for other aromatic compounds we need to choose proper models. But what should these models be?

One proposal that has been made is that we should compare the π electron energy of the cyclic system with that of the corresponding open-chain compound. This approach is particularly useful because it furnishes us with models not only for annulenes but for aromatic cations and anions as well.

When we use this approach we take as our model a linear chain of sp^2-hybridized carbon atoms that carries the same number of π electrons as our cyclic compound. Then we imagine ourselves removing two hydrogens from the end of this chain and joining the ends to form a ring. If the ring has *lower* π-electron energy than the open chain then the ring is *aromatic*. If the ring and chain have *the same* π-electron energy then the ring is *nonaromatic*. If the ring has *greater* π-electron energy than the open chain then the ring is *antiaromatic*.

The actual calculations and experiments used in determining π-electron energies are beyond our scope, but we can examine four examples that illustrate how this approach has been used.

Cyclobutadiene. For cyclobutadiene we consider the change in π-electron energy for the following transformation.

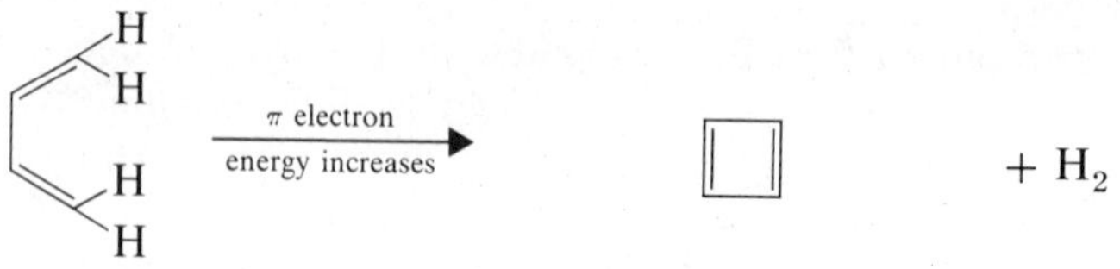

1,3-Butadiene
4 π electrons

Cyclobutadiene
(4 π electrons—**antiaromatic**)

Calculations indicate and experiments appear to confirm that the π-electron energy of cyclobutadiene is higher than that of its open-chain counterpart. Thus cyclobutadiene is classified as being antiaromatic.

Benzene. Here our comparison is based on the following transformation.

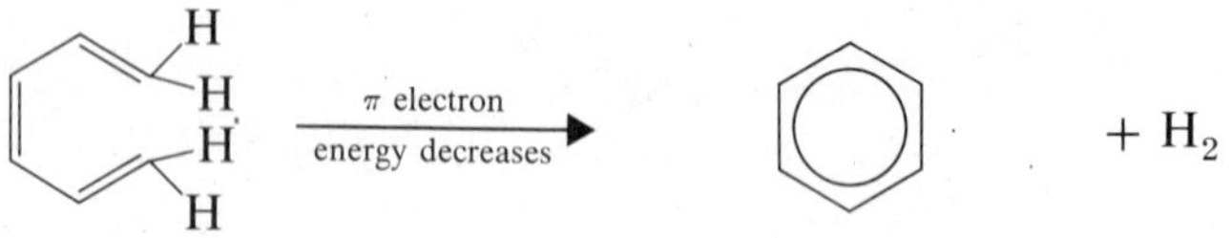

1,3,5-Hexatriene
6 π electrons

Benzene
(6 π electrons, **aromatic**)

Calculations indicate and experiments confirm that benzene has a much lower π-electron energy than 1,3,5-hexatriene. Benzene is classified as being aromatic on the basis of this comparison as well.

Cyclopentadienide Anion. Here we use a linear anion:

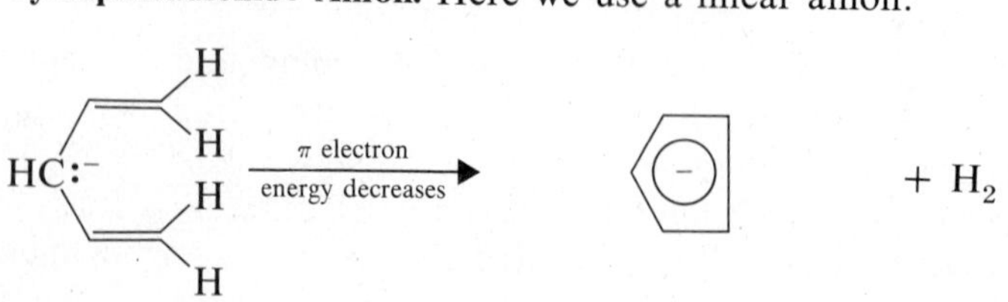

6 π electrons

Cyclopentadienide anion
(6 π electrons, **aromatic**)

Both calculations and experiments confirm that the cyclic anion has a lower π-electron energy than its open-chain counterpart. Therefore the cyclopentadienyl anion is classified as being aromatic.

[30] Annulene. Calculations indicate that when the number of π electrons is greater than 22–26, the open-chain conjugated polyene and its cyclic counterpart have the same energy. Thus, we would expect [30] annulene to be *nonaromatic*. Experiments indicate that this is true.

Problem 12.6

(a) What open-chain compound would you use as a comparison for the cyclohepta-trienyl cation? (b) Both theory and experiments indicate that the cycloheptatrienyl cation has a lower π-electron energy than its open-chain counterpart. What conclusion does this justify?

Problem 12.7

(a) The cyclopentadienyl cation is apparently *antiaromatic*. Show what this means through a comparison of π-electron energies of the cyclic and open-chain compounds. (b) Make a similar comparison for the cyclopropenyl cation (below)—a

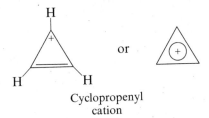

Cyclopropenyl
cation

system that is known from experiments to be *aromatic*. (c) What does Hückel's rule predict for the cyclopropenyl cation? (d) Calculations and experiments indicate that the cyclopropenyl *anion* is *antiaromatic*. What does this mean?

OTHER AROMATIC COMPOUNDS

Benzenoid Aromatic Compounds

In addition to the molecules that we have seen so far, there are many other examples of aromatic compounds. Representatives of one broad class of aromatic molecules, called *polycyclic benzenoid aromatic hydrocarbons,* are illustrated on the next page (Fig. 12.13).

All of these molecules consist simply of two or more benzene rings that are *fused* together. A close look at one of them, naphthalene, will illustrate what we mean by this.

According to resonance theory, naphthalene can be considered to be a hybrid of three Kekulé structures. One of these Kekulé structures is shown in Fig. 12.14. There are two carbon atoms in naphthalene (C-9 and C-10) that are common to both rings. These two atoms are said to be at the points of *ring fusion*. These two atoms direct all of their bonds toward other carbon atoms and do not bear hydrogen atoms.

Problem 12.8

(a) Write the three resonance structures for naphthalene. (b) The C-1—C-2 bond of naphthalene is shorter than the C-2—C-3 bond. Do the resonance structures you have written account for this?

Molecular orbital calculations for naphthalene begin with the model shown in Fig. 12.15. The p orbitals overlap around the periphery of both rings *and* across the points of ring *fusion*.

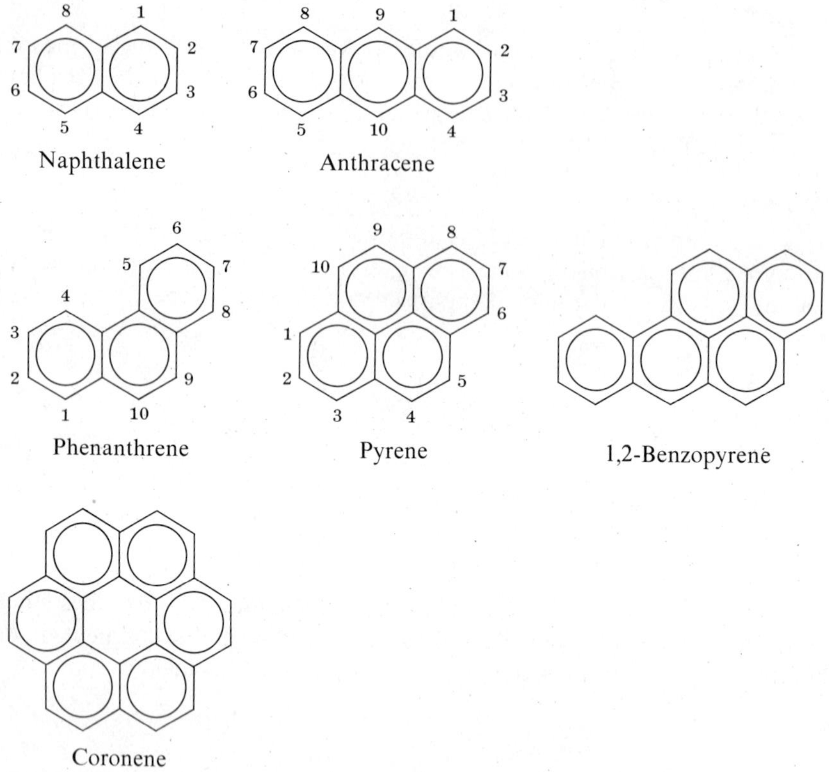

FIG. 12.13
Benzenoid aromatic hydrocarbons.

When molecular-orbital calculations are carried out for naphthalene using the model shown in Fig. 12.15, the results of the calculations correlate well with our experimental knowledge of naphthalene. The calculations indicate that delocalization of the 10 π electrons over the two rings produces a structure with considerably lower energy than that calculated for any individual Kekulé structure. Naphthalene, consequently, has a substantial resonance energy. Based on what we know about benzene, moreover, naphthalene's tendency to react by substitution rather than addition, and to show other properties associated with aromatic compounds, is understandable.

FIG. 12.14
One Kekulé structure for naphthalene.

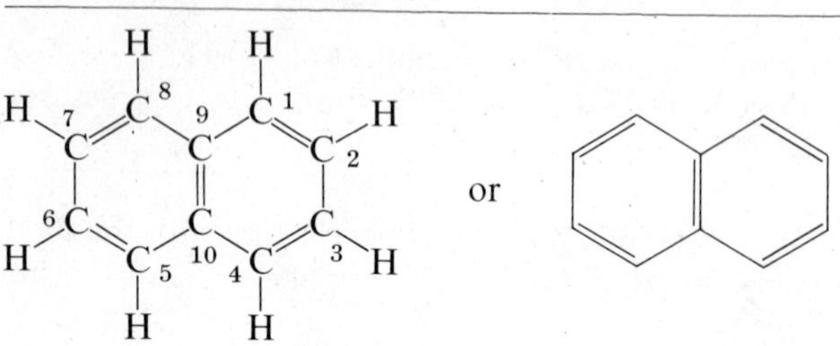

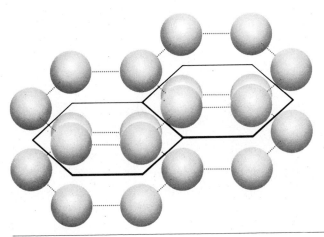

FIG. 12.15
The p *orbitals of naphthalene.*

Anthracene and phenanthrene are isomers. In anthracene the three rings are fused in a linear way, and in phenanthrene they are fused so as to produce an angular molecule. Both of these molecules also show large resonance energies and chemical properties typical of aromatic compounds.

Pyrene and coronene are also aromatic. Pyrene itself has been known for a long time; a pyrene derivative, however, has been the object of recent research that shows another interesting application of Hückel's rule.

In order to understand this particular piece of research we need to pay special attention to the Kekulé structure for pyrene (Fig. 12.16). The total number of π electrons in pyrene is 16 (8 double bonds = 16 π electrons). Sixteen is a non-Hückel number, but Hückel's rule is intended to be applied only to monocyclic compounds and pyrene is clearly tetracyclic. If we disregard the internal double bond of pyrene, however, and look only at the periphery, we see that the periphery resembles a monocyclic ring with 14 π electrons. The periphery is, in fact, very much like that of the [14] annulene shown below. Fourteen *is* a Hückel number ($4n + 2$, where $n = 3$) and one might then predict that the periphery of pyrene would be aromatic by itself, in the absence of the internal double bond.

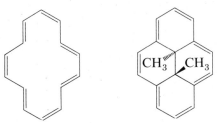

[14] Annulene   *trans*-15,16-Dimethyldihydropyrene

This prediction was confirmed when Professor V. Boekelheide (of the University of Oregon) synthesized the very unusual *trans*-15,16-dimethyldihydropyrene and showed that it is aromatic.

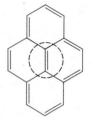

FIG. 12.16

One Kekulé structure for pyrene. The internal double bond is enclosed in a dotted circle for emphasis.

Nonbenzenoid Aromatic Compounds

Naphthalene, phenanthrene, and anthracene are examples of *benzenoid* aromatic compounds. On the other hand, the cyclopentadienyl anion, the cyclo-heptatrienyl cation, *trans*-15,16-dimethyldihydropyrene, and the aromatic annulenes (except for [6]-annulene) are classified as *nonbenzenoid* aromatic compounds.

Another example of a *nonbenzenoid* aromatic hydrocarbon is the compound azulene.

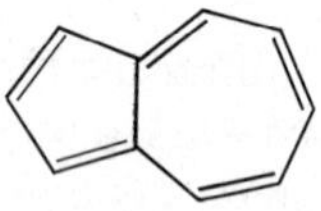

Azulene

This deep-blue hydrocarbon (its name is derived from the word azure) is an isomer of naphthalene. It has the same number of π electrons as naphthalene and, for this reason, azulene is also said to be *isoelectronic* with naphthalene. In addition to its deep-blue color (naphthalene by contrast is colorless), azulene differs from naphthalene in another respect that seems, at first, to be peculiar. Azulene is found to have a substantial dipole moment. The dipole moment of azulene is 1.0 D, whereas the dipole moment of naphthalene is zero.

That azulene has a dipole moment at all indicates that charge separation exists in the molecule. If we recognize this and begin writing resonance structures for azulene that involve charge separation, we find that we can write a number of structures like the one shown in Fig. 12.17.

When we inspect this resonance structure we see that the five-membered ring is very much like the *aromatic* cyclopentadienide ion and that the seven-membered ring resembles the *aromatic* cycloheptatrienyl cation. If resonance structures of this type contribute to the overall hybrid for azulene then we not only understand why azulene has a dipole moment, but we also have some insight into why it is aromatic. Such speculation is strengthened by the results of studies done with substituted azulenes that show quite conclusively that the five-membered ring is negatively charged and the seven-membered ring is positive.

FIG. 12.17

One resonance structure for azulene that has separated charges.

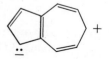

Problem 12.9

Diphenylcyclopropenone (I) has a much larger dipole moment than benzophenone (II). Can you think of an explanation that would account for this?

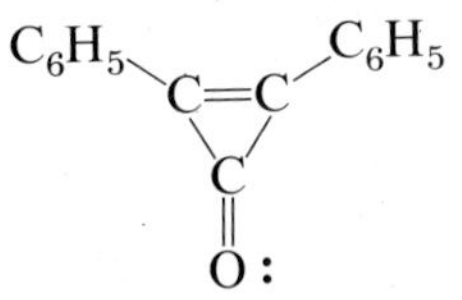

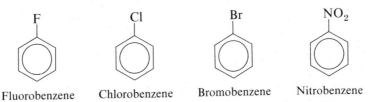

12.8 NOMENCLATURE OF BENZENE DERIVATIVES

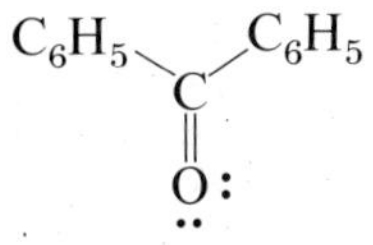

Two systems are used in naming monosubstituted benzenes. In compounds of one type, *benzene* is the base name and the substituent is simply indicated by a prefix. We have, for example,

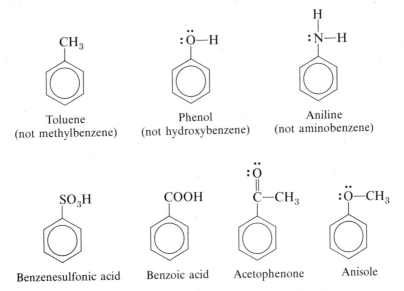

Fluorobenzene Chlorobenzene Bromobenzene Nitrobenzene

For compounds of the other type, the substituent and the benzene ring taken together form a new base name. Methylbenzene is always called *toluene*, hydroxybenzene is always called *phenol*, and aminobenzene is always called *aniline*. These and other examples are indicated below.

Toluene
(not methylbenzene)

Phenol
(not hydroxybenzene)

Aniline
(not aminobenzene)

Benzenesulfonic acid Benzoic acid Acetophenone Anisole

When two substituents are present, their relative positions are indicated by the use of *ortho*, *meta*, and *para* (abbreviated *o-*, *m-*, and *p-*). For the dibromobenzenes we have

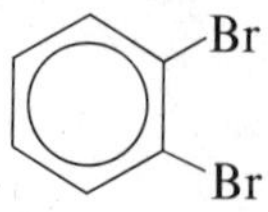

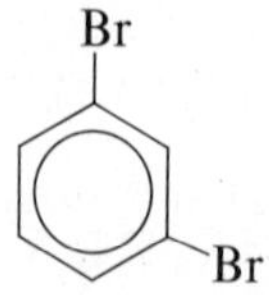

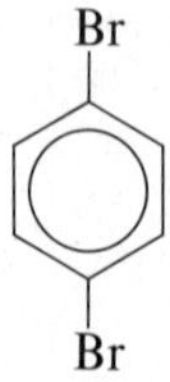

| o-Dibromobenzene | m-Dibromobenzene | p-Dibromobenzene |
| ortho | meta | para |

For the chlorotoluenes we have

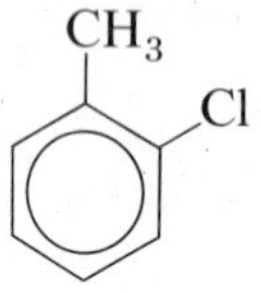

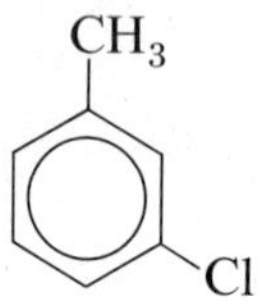

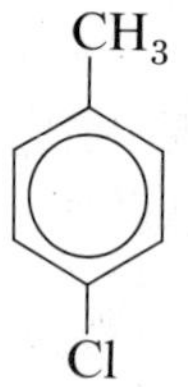

o-Chlorotoluene   m-Chlorotoluene   p-Chlorotoluene

and for the nitrobenzoic acids:

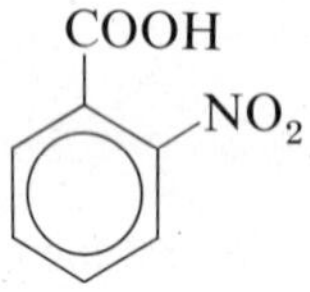

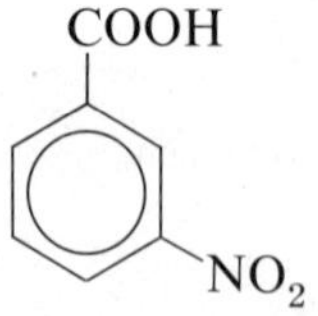

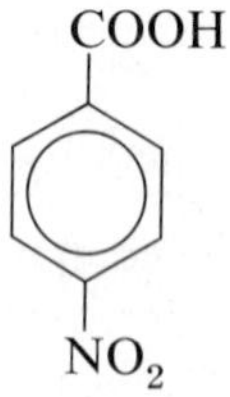

o-Nitrobenzoic acid   m-Nitrobenzoic acid   p-Nitrobenzoic acid

The dimethylbenzenes are called *xylenes*.

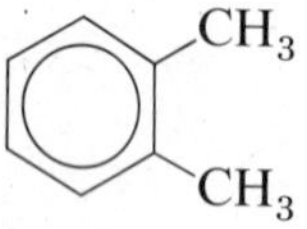

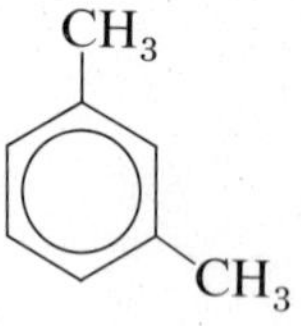

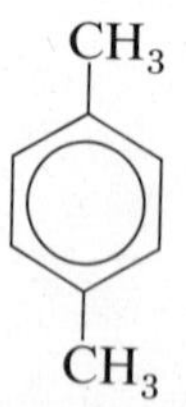

o-Xylene   m-Xylene   p-Xylene

If more than two groups are present on the benzene ring their positions must be indicated by the use of *numbers*. As examples, consider the two compounds below.

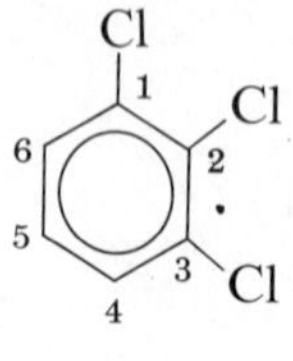

1,2,3-Trichlorobenzene 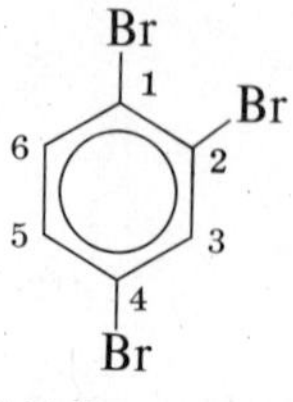   1,2,4-Tribromobenzene
(not 1,3,4-tribromobenzene)

We notice, too, that the benzene ring is numbered so as to give *the lowest possible numbers to the substituents*.

When more than two substituents are present, and the substituents are different, they are usually listed in alphabetical order:

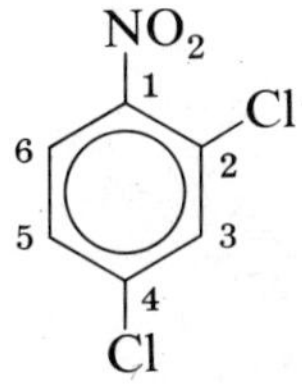

2,4-Dichloro-1-nitrobenzene 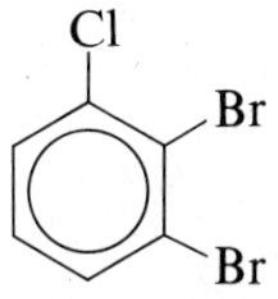   2,3-Dibromo-1-chlorobenzene

When a substituent is one that when taken together with the benzene ring gives a new base name, that substituent is assumed to be in position −1 and the new base name is used:

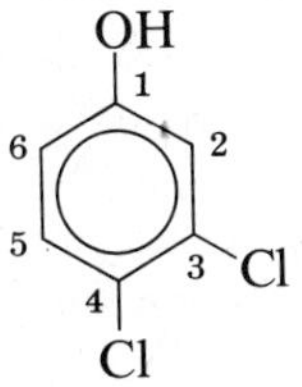

3,4-Dichlorophenol 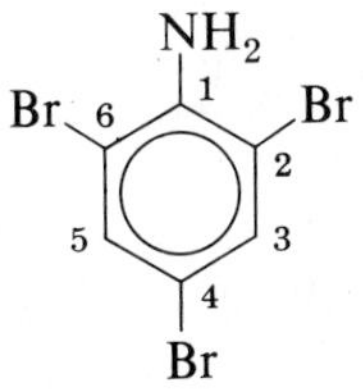   2,4,6-Tribromoaniline

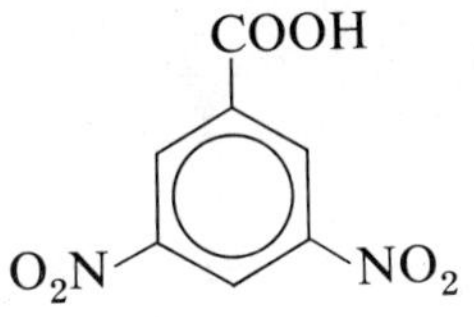

3,5-Dinitrobenzoic acid 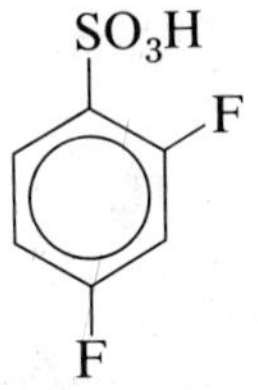   2,4-Difluorobenzenesulfonic acid

When the benzene ring itself is a substituent it is called a *phenyl group*. For example,

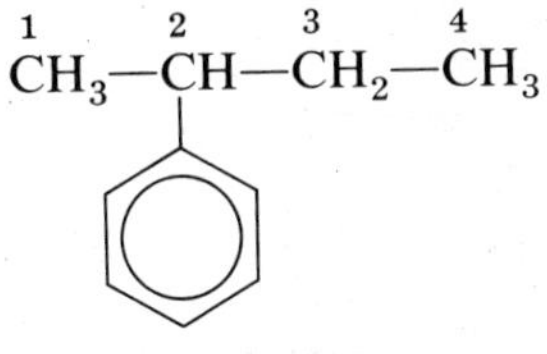

$$\overset{1}{C}H_3-\overset{2}{C}H-\overset{3}{C}H_2-\overset{4}{C}H_3$$

2-Phenylbutane
(not 2-benzylbutane)

The name benzyl is reserved for the group derived from toluene in which one of the hydrogens of the methyl group is replaced.

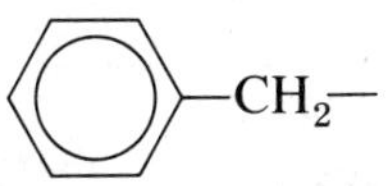

—CH₂—   —CH₂Cl

The benzyl group 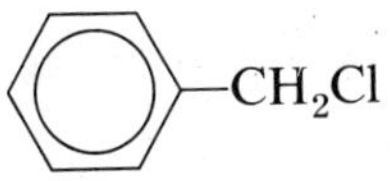   Benzyl chloride
(or α-chlorotoluene)

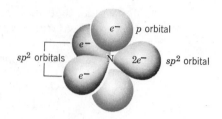

FIG. 12.18

The arrangement of electrons in an sp² *nitrogen atom of the pyridine type.*

12.9 HETEROCYCLIC AROMATIC COMPOUNDS

All of the aromatic molecules that we have discussed so far have had rings composed solely of carbon atoms. There are, however, many aromatic compounds in which an element other than carbon is present in the ring. These compounds are called *heterocyclic* aromatic compounds. Heterocyclic molecules are quite commonly encountered in nature. For this reason, and because the structures of some of these molecules are closely related to the compounds that we have discussed earlier, we will describe a few of the most important heterocyclic aromatic compounds at this point.

Heterocyclic compounds containing nitrogen, oxygen, and sulfur are by far the most common. Four important examples are listed below in their Kekulé forms.

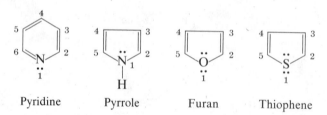

Pyridine   Pyrrole   Furan   Thiophene

If we examine these structures we will see that pyridine is structurally related to benzene, and that pyrrole, furan, and thiophene are related to the cyclopentadienide ion.

The nitrogen atoms in both pyridine and pyrrole are sp² hybridized. In pyridine the electrons of the sp² nitrogen are arranged in the way shown in Fig. 12.18.

FIG. 12.19

Orbital structure of pyridine.

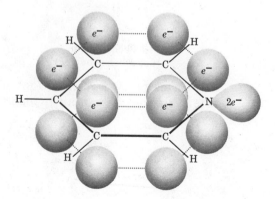

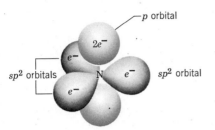

FIG. 12.20
The arrangement of electrons in an sp² *nitrogen atom of the pyrrole type.*

In the pyridine ring the two nitrogen sp^2 orbitals that contain only one electron overlap with sp^2 orbitals of the 2- and 6-carbon atoms to form σ bonds. The remaining sp^2 orbital (with two electrons) does not form a bond and is in the same plane as the ring but directed away from it (Fig. 12.19). The p orbital of nitrogen (containing one electron) is perpendicular to the plane of the ring and overlaps with the p orbitals of the 2- and 6-carbon atoms to form a π system of six π electrons. The π system of pyridine is isoelectronic with the π system of benzene. Pyridine, consequently, shows aromatic properties.

In pyrrole the nitrogen atom is sp^2 hybridized but the electrons are arranged differently. The sp^2 orbitals each contain one electron and *the p orbital contains two electrons* (Fig. 12.20). The reason for this different arrangement of the electrons becomes clear when we look at the orbital structure of the entire pyrrole ring (Fig. 12.21). Since only four π electrons are contributed by the carbon atoms in pyrrole, the nitrogen atom must contribute two electrons to give an aromatic sextet. Two of the sp^2 orbitals of the pyrrole nitrogen form bonds to the 2- and 5-carbon atoms; the remaining sp^2 bond overlaps with the $1s$ orbital of a hydrogen atom.

Furan and thiophene are structurally quite similar to pyrrole. The oxygen atom in furan and the sulfur atom in thiophene are sp^2 hybridized. In both compounds the p orbital of the heteroatom donates two electrons to the π system. The oxygen and sulfur atoms of furan and thiophene carry an unshared pair of electrons in an sp^2 orbital (Fig. 12.22).

FIG. 12.21
The orbital structure of pyrrole. (Compare with the orbital structure of the cyclopentadienide anion in Fig. 12.9.)

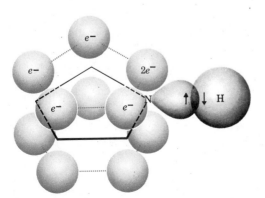

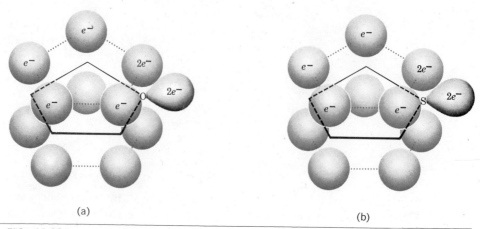

(a)   (b)

FIG. 12.22
The orbital structures of furan (a) and thiophene (b).

12.10 AROMATIC COMPOUNDS IN BIOCHEMISTRY

Compounds with aromatic rings occupy numerous and important positions in reactions that occur in living systems. It would be impossible to describe them all in this chapter. We will, however, point out a few examples now and we will see others later.

Two amino acids necessary for protein synthesis contain the benzene ring:

Phenylalanine   Tyrosine

A third aromatic amino acid, tryptophan, contains a benzene ring fused to a pyrrole ring. (This aromatic ring system is called an indole system, cf. Section 22.1.)

Tryptophan

It appears that man, through the course of evolution, has lost the biochemical ability to synthesize the benzene ring. As a result, phenylalanine and tryptophan derivatives are essential in the human diet. Because tyrosine can be synthesized from phenylalanine by an enzyme known as *phenylalanine hydroxylase,* it is not essential in the diet as long as phenylalanine is present.

Heterocyclic aromatic compounds are also present in many biochemical systems. Derivatives of purine and pyrimidine are essential parts of DNA and RNA.

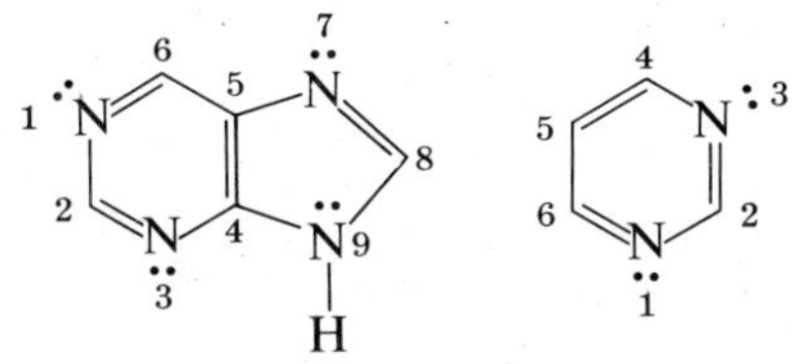

Purine   Pyrimidine

DNA is the molecule responsible for the storage of genetic information and RNA is prominently involved in the synthesis of enzymes (Chapter 26).

Problem 12.10

Classify each nitrogen atom in the purine molecule as to whether it is of the pyridine type or of the pyrrole type.

Both a pyridine derivative (nicotinamide) and a purine derivative (adenine) are present in one of the most important coenzymes in biological oxidations. This molecule, *nicotinamide-adenine-dinucleotide* (NAD$^+$), is shown in Fig. 12.23.

NAD$^+$, together with another compound in the liver (an apoenzyme), is capable of oxidizing alcohols to aldehydes. (We saw an example of this in the oxidation of vitamin A to retinal in Section 11.1). While the overall change is quite complex, a look at one aspect of it will illustrate a *biological use* of the extra stability (resonance or delocalization energy) associated with an aromatic ring.

A simplified version of the oxidation of an alcohol to an aldehyde is illustrated below:

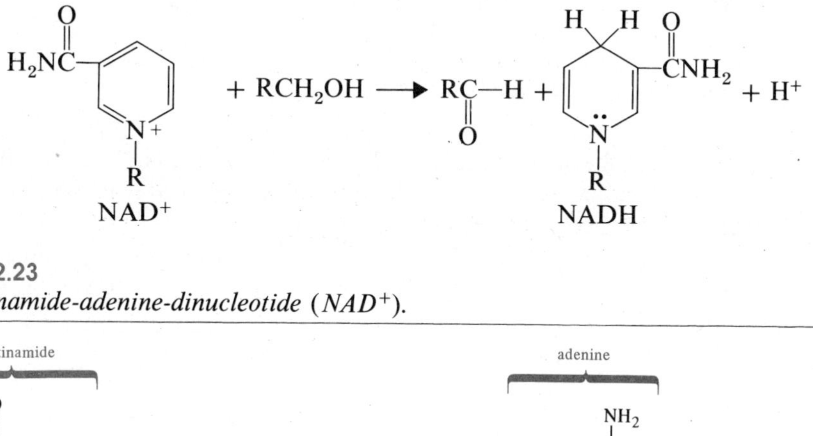

FIG. 12.23
Nicotinamide-adenine-dinucleotide (NAD$^+$).

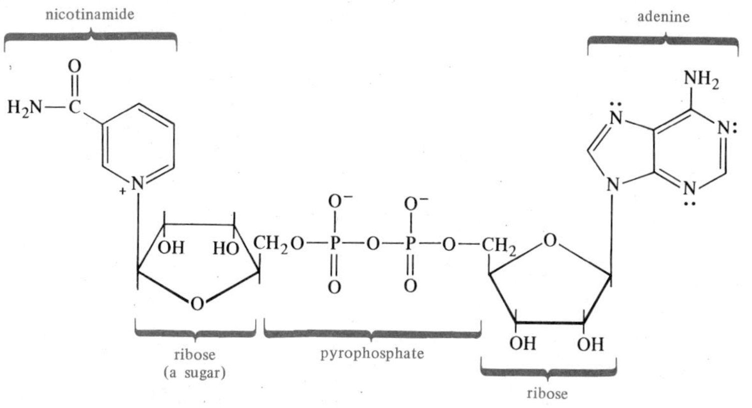

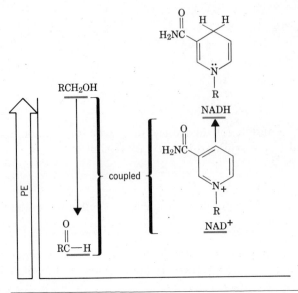

FIG. 12.24
Potential energy diagram for the biologically coupled oxidation of an alcohol and reduction of nicotinamideadenine dinucleotide.

The *aromatic* pyridine ring (actually *pyridinium* because it is positively charged) in NAD^+ is converted to a *nonaromatic* ring in NADH. The extra stability of the pyridine ring is lost in this process; and, as a result, the potential (chemical) energy of NADH is greater than that of NAD^+. The conversion of the alcohol to the aldehyde, however, occurs with a decrease in potential energy. Because these reactions are coupled in biological systems (Fig. 12.24), a portion of the potential energy contained in the alcohol becomes chemically contained in NADH. This stored energy in NADH can be used to bring about other biochemical reactions that require energy and that are necessary to life.

While many aromatic compounds are essential to life, others are hazardous. Many benzene compounds are quite toxic and many polycyclic benzenoid compounds are *carcinogenic* (i.e., cancer causing). Two examples of carcinogenic hydrocarbons are 1,2-benzopyrene and 10-methyl[1,2]benzanthracene:

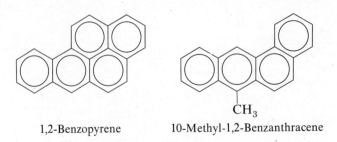

1,2-Benzopyrene   10-Methyl-1,2-Benzanthracene

The hydrocarbon 1,2-benzopyrene has been found in cigarette smoke and in the exhaust from automobiles. It is also formed in the incomplete combustion of any fossil fuel. It is found on charcoal-broiled steaks and exudes from asphalt streets on a hot summer day. 1,2-Benzopyrene is so carcinogenic that one can induce skin cancers in mice with almost total certainty by simply shaving an area of the body of the mouse and applying a coating of 1,2-benzopyrene.

12.11 VISIBLE AND ULTRAVIOLET ABSORPTION SPECTRA OF AROMATIC COMPOUNDS

Benzene and alkyl benzenes have two absorption bands in the ultraviolet region. One band lies near 200 nm and is, by far, the more intense. The second band, called the *benzenoid band,* is found at longer wavelengths, near 260 nm. The benzenoid band is quite weak.

Both of these bands are shifted to longer wavelengths when groups are attached to the ring in a way that causes the conjugated system to be extended. We can see this effect if we compare the absorption maxima of benzene and styrene.

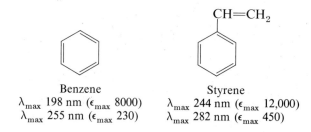

Benzene
λ_{max} 198 nm (ϵ_{max} 8000)
λ_{max} 255 nm (ϵ_{max} 230)

Styrene
λ_{max} 244 nm (ϵ_{max} 12,000)
λ_{max} 282 nm (ϵ_{max} 450)

The benzenoid band of polycyclic compounds is shifted to longer wavelengths in a dramatic way when benzene rings are fused in a linear fashion.

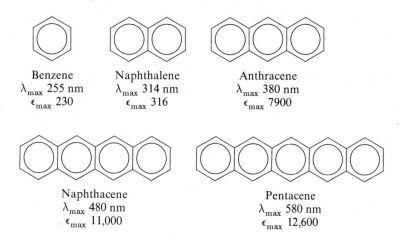

Benzene
λ_{max} 255 nm
ϵ_{max} 230

Naphthalene
λ_{max} 314 nm
ϵ_{max} 316

Anthracene
λ_{max} 380 nm
ϵ_{max} 7900

Naphthacene
λ_{max} 480 nm
ϵ_{max} 11,000

Pentacene
λ_{max} 580 nm
ϵ_{max} 12,600

Benzene, naphthalene, and anthracene are colorless because their absorption maxima lie in the ultraviolet region. Naphthacene and pentacene, however, are colored because their absorption maxima lie in the visible region. Naphthacene is orange-yellow and pentacene is purple.

Additional Problems

12.11

Draw structural formulas for the following:
(a) *p*-Nitrobenzenesulfonic acid
(b) *o*-Chlorotoluene
(c) *m*-Dichlorobenzene
(d) *p*-Dinitrobenzene
(e) *p*-Bromoanisole
(f) *m*-Nitrobenzoic acid
(g) *p*-Iodophenol
(h) *o*-Chloroacetophenone
(i) 2-Bromonaphthalene
(j) 9-Chloroanthracene
(k) 3-Nitrophenanthrene
(l) 4-Nitropyridine
(m) 2-Methylpyrrole
(n) 2,4-Dichloronitrobenzene
(o) *p*-Nitrobenzyl bromide
(p) *o*-Chloroaniline
(q) 2,5-Dibromo-3-nitrobenzoic acid
(r) 1,3,5-Trimethylbenzene (mesitylene)
(s) *p*-Hydroxybenzoic acid
(t) Vinylbenzene (styrene)
(u) 1,2-Benzopyrene
(v) 2-Phenylcyclohexanol
(w) 2,4,6-Trinitrotoluene (TNT)
(x) A [12] annulene
(y) A [14] annulene
(z) An [18] annulene

12.12

Write structural formulas and give names for all of the following:
(a) Trichlorobenzenes
(b) Dibromonitrobenzenes
(c) Dichlorotoluenes
(d) Monochloronaphthalenes
(e) Nitropyridines
(f) Methylfurans
(g) Chlorodinitrobenzenes
(h) Chloroxylenes
(i) Cresols (hydroxytoluenes)

12.13

(a) Write all of the resonance structures for phenanthrene. (There are five.)
(b) On the basis of these can you speculate about the length of the 9,10 bond?
(c) About its double-bond character?
(d) Phenanthrene, in contrast to most aromatic molecules, tends to *add* one mole of bromine to form a molecule with the formula $C_{14}H_{10}Br_2$. How can you account for this?

12.14

The following problem illustrates the method used by W. Koerner (in 1875) for determining the structures of *ortho, meta,* and *para-* disubstituted benzene derivatives.

There are three dibromobenzenes, melting at $+87°$, $+6°$, and $-7°$. Without con-

sulting any tables, consider the following results and decide which compound is *o*-dibromobenzene, which is *m*-dibromobenzene, and which is *p*-dibromobenzene.
(a) The dibromobenzene melting at $+87°$ yields (on nitration) a single mononitrodibromobenzene.
(b) The dibromobenzene melting at $+6°$ yields two different mononitrodibromobenzenes.
(c) The dibromobenzene melting at $-7°$ yields three different mononitrodibromobenzenes.

12.15

How might you use the method described in problem 12.14 to determine the structure of the xylenes?

12.16

How many trimethylbenzenes are there? Which one of these would yield only one mononitro product? Which one could yield two mononitro products? Which one could yield three?

12.17

The methoxycyclopropene (III) has been found to react with fluoroboric acid to yield methanol and a compound with the formula $C_6H_9{}^+BF_4{}^-$ (IV).

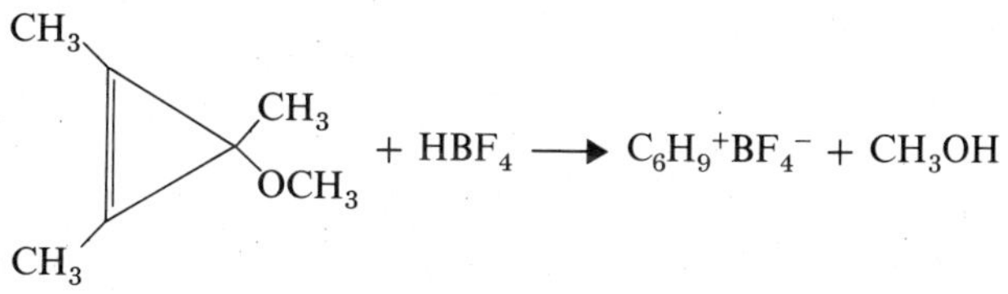

$$+ HBF_4 \longrightarrow C_6H_9{}^+BF_4{}^- + CH_3OH$$

III   IV

(a) What is the structure of $C_6H_9{}^+BF_4{}^-$?
(b) How can you account for its formation?

12.18

Diphenylcyclopropenone (cf. Prob. 12.9) reacts with hydrogen bromide to form a stable crystalline hydrobromide. What is its structure?

12.19

Cyclooctatetraene has been found to react with 2 moles of potassium to yield an unusually stable compound with the formula $2K^+C_8H_8{}^=$ (V). The NMR spectrum of V indicates that all of its hydrogens are equivalent.
(a) What is the structure of V?
(b) How can you account for the formation of V?

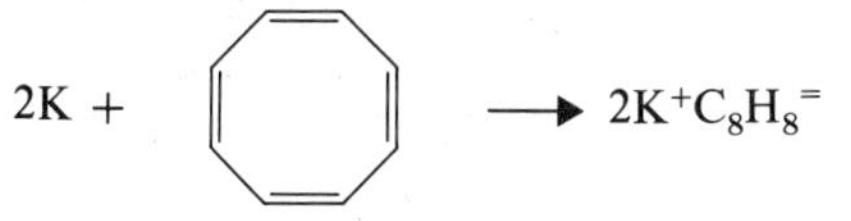

$$2K + \quad \text{(cyclooctatetraene)} \quad \longrightarrow 2K^+C_8H_8{}^=$$

Cyclooctatetraene   V

12.20

The bicyclic molecule, VI, reacts with 2 moles of lithium to yield lithium chloride and a compound with the formula $Li^+C_9H_9{}^-$ (VII).

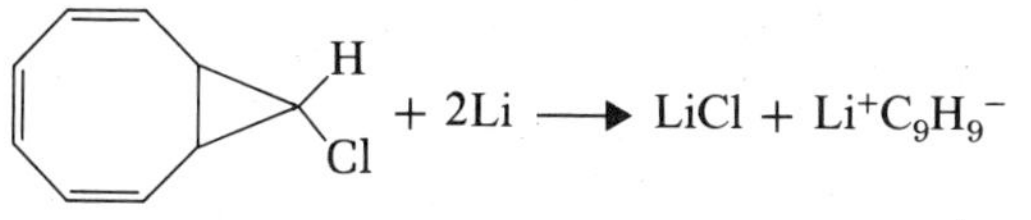

$$+ 2Li \longrightarrow LiCl + Li^+C_9H_9{}^-$$

VI   VII

(a) What is VII?

(b) Account for its formation.

12.21

Using the data in Section 12.4 and Fig. 12.1 calculate ΔH for the following reactions.

(a)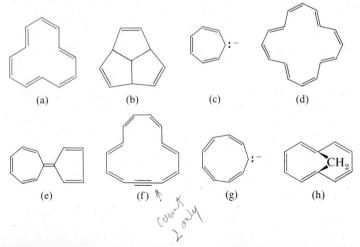

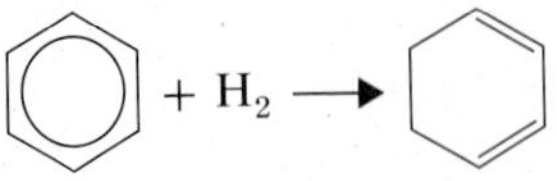

(b) 2

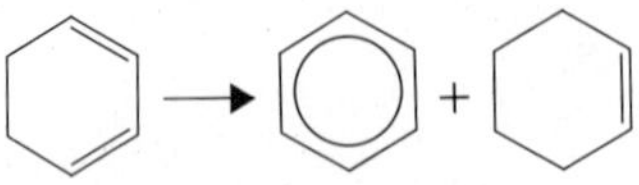

(c) On the basis of what you have found in (a) and (b), what is the likelihood of reaction (a) occurring successfully?

12.22

Under certain conditions, benzoic acid derivatives can be caused to lose a molecule of CO_2 by a process called decarboxylation. For benzoic acid itself this reaction occurs as follows:

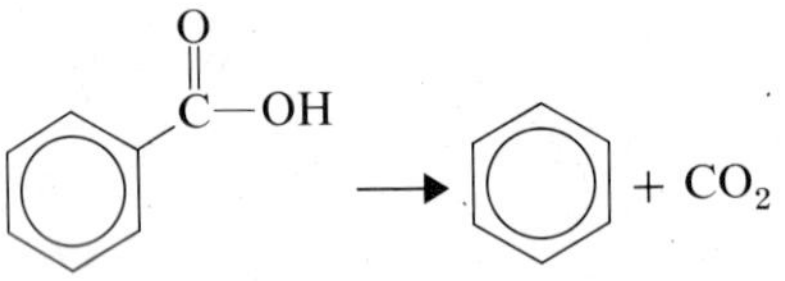

In 1874, a year before Koerner published his method (problem 12.14), Griess arrived at the structures for the diaminobenzenes from the following data:

1. Six diaminobenzoic acids (i.e., molecules with the formula $C_6H_3(NH_2)_2COOH$) were known.

2. Three of these acids, on decarboxylation, produced a single diaminobenzene, mp 63°.

3. Two diaminobenzoic acids produced the same diaminobenzene, mp 104°.

4. One diaminobenzoic acid gave a diaminobenzene, mp 142°.

    (a) What are the structures of the three diaminobenzenes melting at 63°, 104°, and 142°, respectively?

    (b) What are the structures of the diaminobenzoic acids that yield each of the diaminobenzenes?

12.23

Consider whether each of the following molecules or ions would or would not be aromatic. Explain your answer in each instance.

(a)	(b)	(c)	(d)

(e)	(f)	(g)	(h)

* **12.24**

Cycloheptatrienone, **I**, is a very stable molecule. Cyclopentadienone, **II**, by contrast is quite unstable and rapidly undergoes a Diels–Alder reaction with itself. (a) Propose an explanation for the different stabilities of these two compounds. (b) Write the structure of the Diels–Alder adduct of cyclopentadienone.

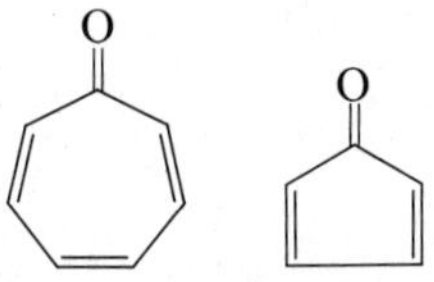

I   **II**

* **12.25**

The final product **C** of the following synthesis is an aromatic compound. What is its structure and what are **A** and **B?**

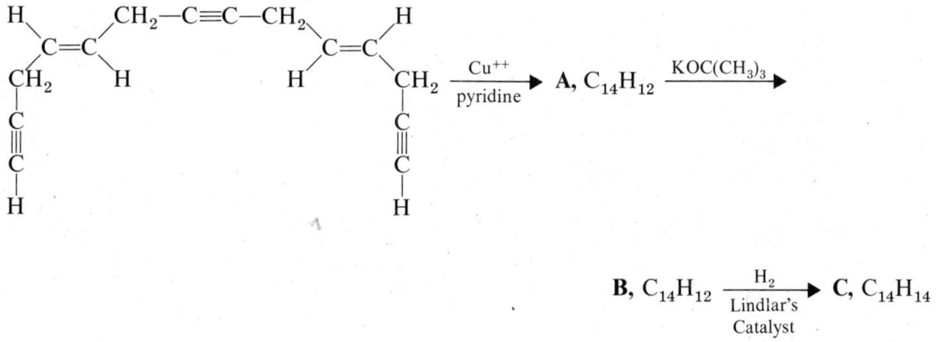

$$\xrightarrow[\text{pyridine}]{\text{Cu}^{++}} \textbf{A, C}_{14}\text{H}_{12} \xrightarrow{\text{KOC(CH}_3)_3}$$

$$\textbf{B, C}_{14}\text{H}_{12} \xrightarrow[\substack{\text{Lindlar's}\\ \text{Catalyst}}]{\text{H}_2} \textbf{C, C}_{14}\text{H}_{14}$$

13

AROMATIC COMPOUNDS II: REACTIONS OF AROMATIC COMPOUNDS WITH ELECTROPHILES

13.1 ELECTROPHILIC AROMATIC SUBSTITUTION REACTIONS

The most characteristic reactions of benzenoid aromatic compounds are the *substitution* reactions that occur when they react with *electrophilic* reagents.

$$ArH + E^+ \longrightarrow Ar\text{-}E + H^+$$

Electrophilic aromatic substitution reactions allow us to introduce a wide variety of groups into aromatic rings and because of this, they give us synthetic access to a vast number of aromatic compounds that would not be available otherwise. Listed below are examples that illustrate some of these reactions.

Electrophilic Aromatic Substitution Reactions

1. Halogenation (Section 13.3)

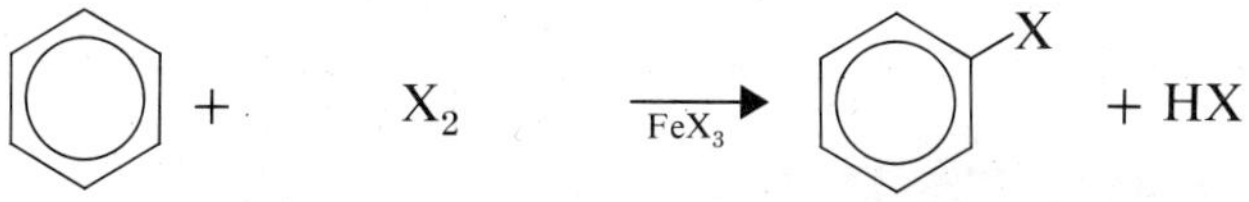

$(X_2 = Cl_2 \text{ or } Br_2)$ Halobenzene
$(X = Cl, 90\%)$
$(X = Br, 75\%)$

2. Nitration (Section 13.4)

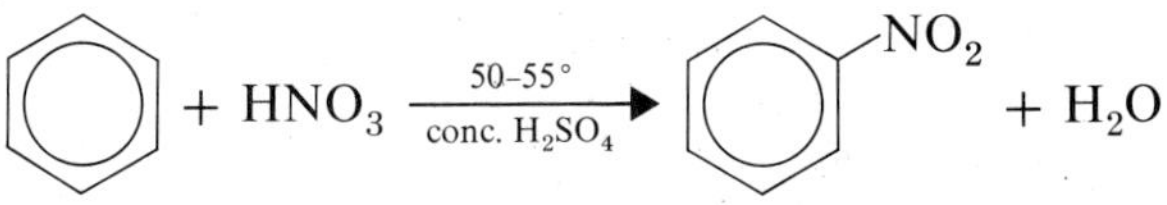

Nitrobenzene
(85%)

3. Sulfonation (Section 13.5)

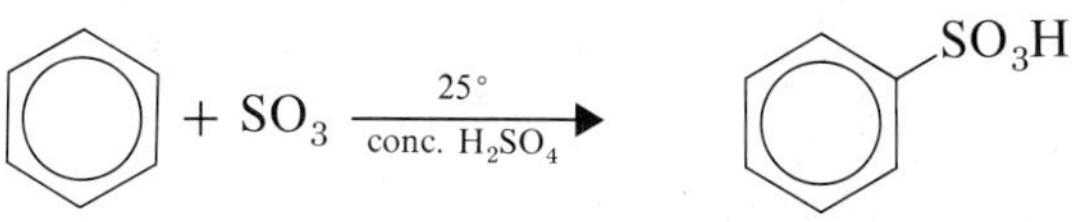

Benzenesulfonic acid (56%)

4. Friedel-Crafts alkylation (Section 13.6)

$$\text{benzene} + CH_3CH_2X \xrightarrow{AlCl_3} \text{Ethylbenzene (83\%)} + HX$$

5. Friedel-Crafts acylation (Section 13.7)

$$\text{benzene} + CH_3\overset{O}{\overset{\|}{C}}Cl \xrightarrow[AlCl_3]{80°} \text{Acetophenone (97\%)} + HCl$$

All of these reactions involve an attack on the benzene ring by an electron-deficient species—an electrophile.

We will examine the specific mechanisms for each of these electrophilic aromatic substitution reactions later (Section 13.3 through 13.7). When we do this we will find that in some reactions the electrophile is a positive ion, that in others it is a species that bears only a partial positive charge, and that in still others it is a neutral Lewis acid. In nitration reactions, for example, the electrophile is a nitronium ion, NO_2^+; in halogenation reactions the electrophiles are halogen-Lewis acid complexes in which a halogen is partially positive; and in sulfonation reactions the electrophile is usually the Lewis acid, SO_3.

13.2 THE MECHANISM FOR ELECTROPHILIC AROMATIC SUBSTITUTION: σ COMPLEXES

Benzene is susceptible to electrophilic attack primarily because of its exposed π electrons. In this respect benzene resembles an alkene, for in the reaction of an alkene with an electrophile the site of attack is the exposed π bond. Like an alkene, too, benzene forms π complexes with electrophiles, and there is evidence that in some

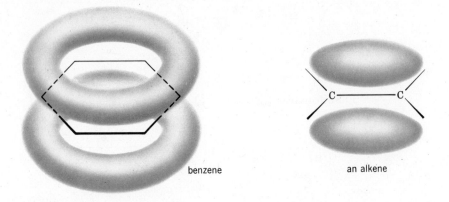

benzene   an alkene

reactions benzene forms intermediates resembling the alkene bromonium ions we saw in Section 6.6.

But, as we saw in Chapter 12, benzene differs from an alkene in a very significant way. Benzene's closed shell of six π electrons gives it a special stability. So, while benzene is susceptible to electrophilic attack, benzene undergoes *substitution reactions* rather than *addition reactions*. Substitution reactions allow the closed shell of π electrons of benzene to be regenerated after attack by the electrophile has occurred. We can see how this happens if we examine a general mechanism for electrophilic aromatic substitution.

There is a considerable body of experimental evidence indicating that electrophiles attack the π system of benzene to form a delocalized carbocation known as a σ complex.*

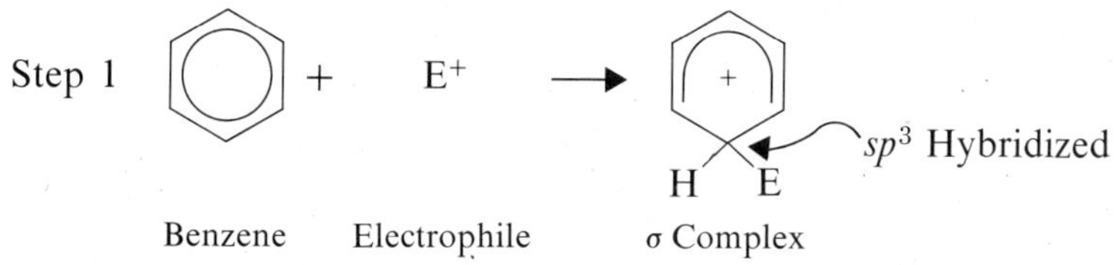

Step 1

Benzene   Electrophile   σ Complex

In step 1 the electrophile uses two electrons of the π system to form a σ bond to one carbon of the benzene ring. This interrupts the cyclic system of π electrons, for in the formation of the σ complex one carbon becomes sp^3 hybridized and thus, no longer has an available p orbital. The four remaining π electrons of the σ complex are delocalized over the five remaining sp^2 carbons.

In step 2 the σ complex loses a proton from the carbon that bears the electrophile. The two electrons that bonded this proton to carbon become a part of the π system. The carbon that bears the electrophile becomes sp^2 hybridized and a benzene derivative with six fully delocalized π electrons is formed.

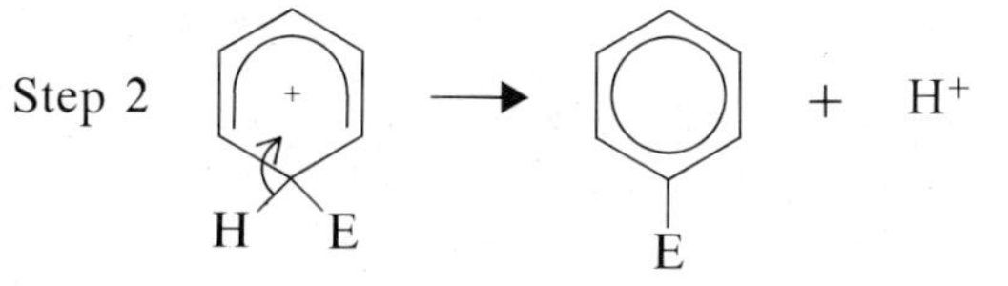

Step 2

We can represent both steps of this mechanism with one of the Kekulé formulas for benzene. When we do this it becomes much easier to account for the π electrons. The first step can be shown in the following way:

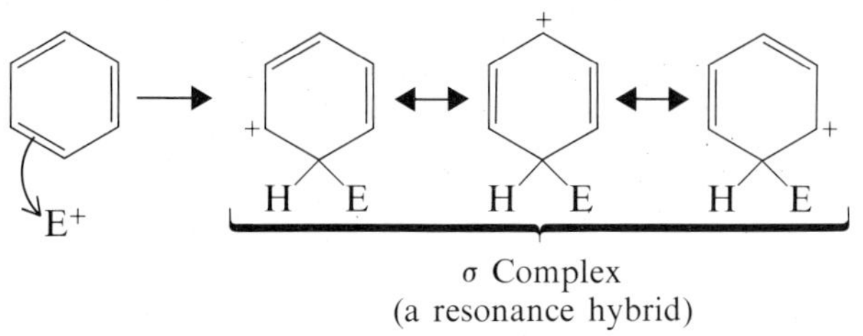

σ Complex
(a resonance hybrid)

---

* Pi complexes may be formed first in electrophilic aromatic substitution reactions but these subsequently become σ complexes. The formation of the σ complex is usually the rate-limiting step.

By using a Kekulé structure we can also see that the σ complex is a hybrid of three resonance structures each of which has a positive charge *on an ortho or para carbon.*

We can represent the second step, the loss of a proton, with any one of the resonance structures for the σ complex.

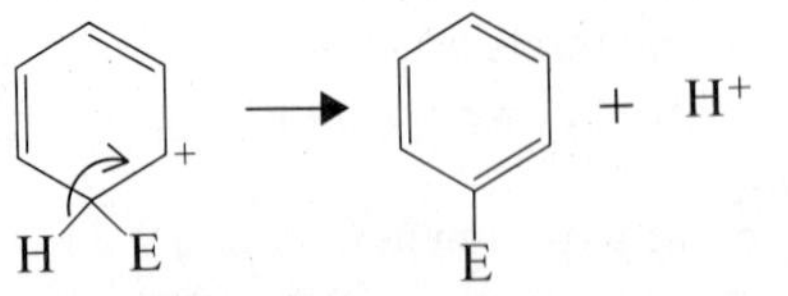

When we do this we can see clearly that two electrons from the carbon-hydrogen bond serve to regenerate the ring of alternating single and double bonds.

Problem 13.1

Show the second step of this mechanism using each of the other two resonance structures for the σ complex.

Part of the evidence for the existence of σ complexes as intermediates in electrophilic substitution reactions comes from experiments in which they have actually been isolated. In one experiment a σ complex was isolated from the reaction of mesitylene and fluoroboric acid, HBF_4.

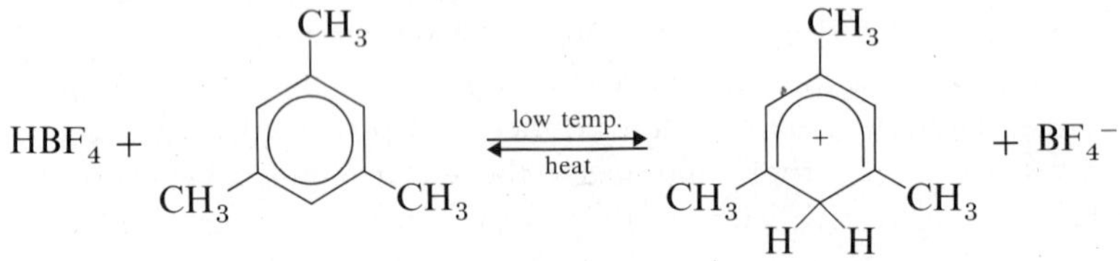

The formation of this σ complex takes place at low temperatures and the electrophile is a proton. The σ complex can be isolated as a bright yellow ionic solid, but heating the σ complex causes it to lose fluoroboric acid and regenerate mesitylene.

Problem 13.2

(a) What σ complex would you expect to obtain from a similar reaction using DBF_4 instead of HBF_4? (b) What products would you expect to obtain when this σ complex is heated?

The σ complex is a true *intermediate* in electrophilic substitution reactions. It is not a transition state. This means that in a potential-energy diagram (Fig. 13.1) the σ complex lies in an energy valley between two transition states.

The energy of activation for the reaction leading from benzene and the electrophile, E^+, to the σ complex ($E_{act(1)}$) has been shown to be much greater than the energy of activation ($E_{act(2)}$) leading from the σ complex to the substituted benzene. This is consistent with what we would expect. The reaction leading from benzene and an electrophile to the σ complex is highly endothermic, for in it the

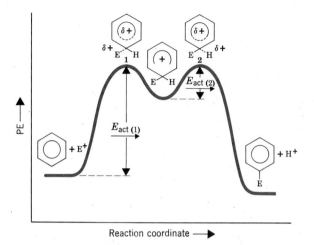

FIG. 13.1
*The potential energy diagram for an electrophilic aromatic substitution reaction. The σ complex is a true intermediate lying between transition states **1** and **2**. In transition state **1** the bond between the electrophile and one carbon of the benzene ring is only partially formed. In transition state **2** the bond between the same benzene carbon and its hydrogen is partially broken.*

benzene ring loses its resonance (or delocalization) energy. The reaction leading from the σ complex to the substituted benzene, by contrast, is highly exothermic because in it the benzene ring regains its resonance energy.

This means that of the two steps shown below, step 1—the formation of the σ complex—is the rate-limiting step in electrophilic aromatic substitution.

Step 1   ⬡ + E⁺ ⟶ ⬡(+) H E   **Slow, rate limiting**

Step 2   ⬡(+) H E ⟶ ⬡ E + H⁺   **Fast**

Step 2, the loss of a proton, occurs rapidly relative to step 1 and has no effect on the overall rate of reaction.

There is also evidence that the height of the energy barriers on either side of the σ complex varies from one reaction to another and depends partly on the nature of the electrophile.

In nitration reactions the electrophile is a nitronium ion, NO_2^+. It appears that in nitration reactions (Fig. 13.2) the energy of activation, $E_{act(1)}$, leading back to

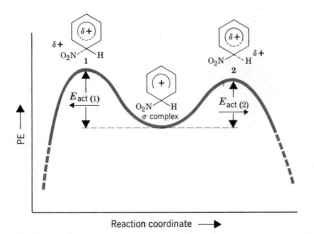

FIG. 13.2

The σ complex for a nitration reaction and the energies of activation leading from it through transition states **1** *and* **2.** $\underset{\longleftarrow}{E_{act(1)}}$ *is greater than* $\underset{\longrightarrow}{E_{act(2)}}$.

reactants through transition state **1** is considerably *greater* than the energy of activation, $\underset{\longrightarrow}{E_{act(2)}}$, leading to products through transition state **2.**

This difference in the two energies of activation is consistent with the fact that nitration reactions are essentially irreversible. Once the σ complex is reached in a nitration reaction, the reaction proceeds over the lower energy barrier ($\underset{\longrightarrow}{E_{act(2)}}$) to the product, nitrobenzene.

Sulfonation reactions, on the other hand, *are* reversible and it appears that in sulfonation reactions the energy barriers on either side of the σ complex are of roughly the same height (Fig. 13.3).

FIG. 13.3

The σ complex for a sulfonation reaction and the energies of activation leading through transition states **1** *and* **2.** $\underset{\longleftarrow}{E_{act(1)}}$ *is approximately equal to* $\underset{\longrightarrow}{E_{act(2)}}$.

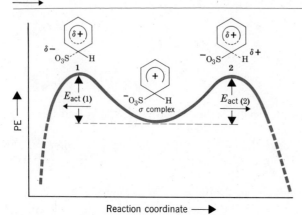

13.3 HALOGENATION OF BENZENE

Benzene does not react with bromine or chlorine unless a Lewis acid is present in the reaction mixture. As a consequence, benzene does not decolorize a solution of bromine in carbon tetrachloride. When Lewis acids are present, however, benzene reacts readily with bromine or chlorine and the reactions give bromobenzene and chlorobenzene in good yields.

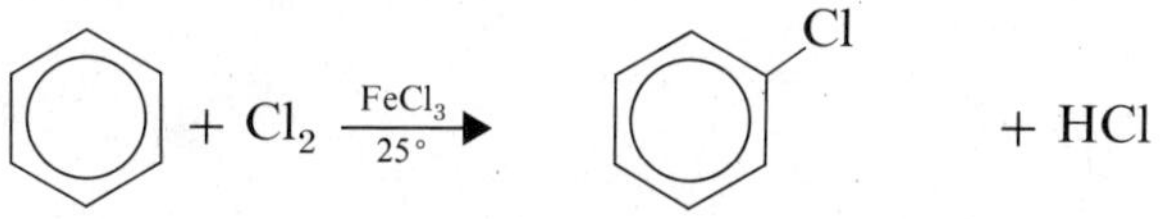

Chlorobenzene (90%)

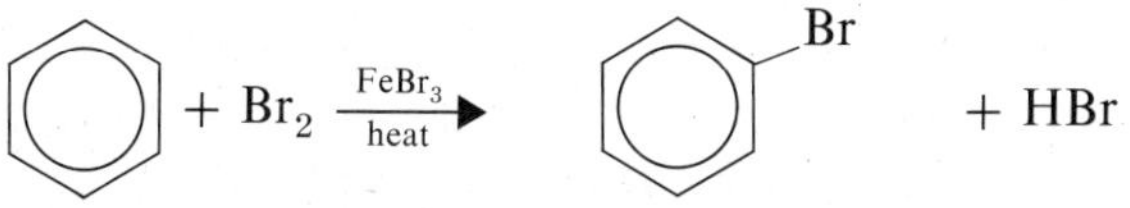

Bromobenzene (75%)

The Lewis acids most commonly used to effect chlorination and bromination reactions are $FeCl_3$, $FeBr_3$, and $AlCl_3$.

The mechanism for aromatic bromination is shown below.

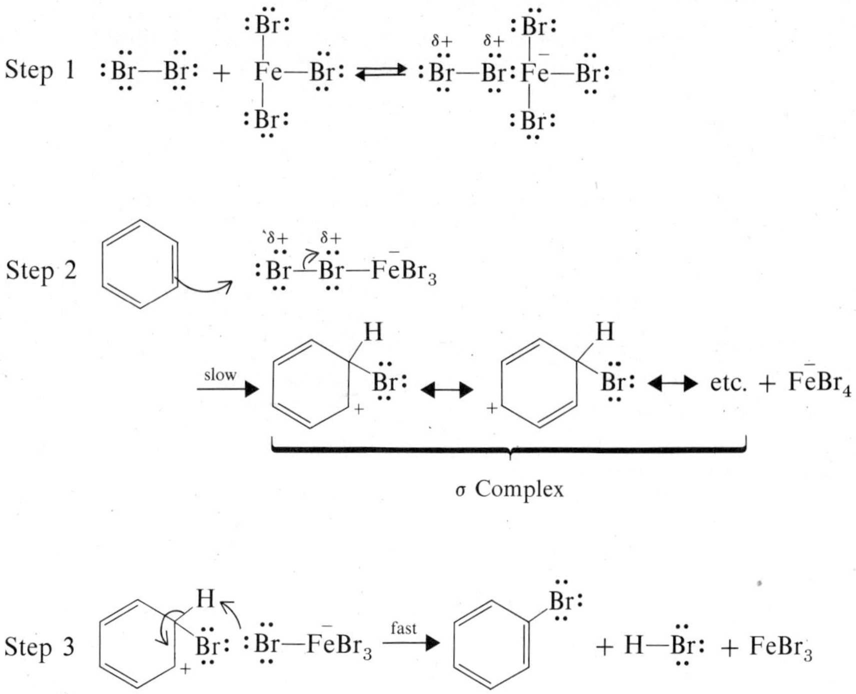

Step 1

Step 2

σ Complex

Step 3

The function of the Lewis acid can be seen by examining steps 1 and 2. In step 1 the Lewis acid forms a complex with bromine; in step 2 this complex acts as an electrophile and transfers a positive bromine ion, Br^+, to the benzene ring. The result of both steps is the formation of a σ complex and an $FeBr_4^-$ ion.

In step 3 the σ complex transfers a proton to $FeBr_4^-$. This results in the formation of bromobenzene and hydrogen bromide—the products of the reaction. At the same time this step regenerates the catalyst—$FeBr_3$.

One might ask, why do aromatic brominations require the presence of a Lewis acid, whereas alkene brominations do not? The answer can be given in terms of the relative strength with which the π electrons are held. The π electrons of alkenes are rather loosely held; so loosely, in fact, they can polarize the bromine-bromine bond and cause it, ultimately, to break.

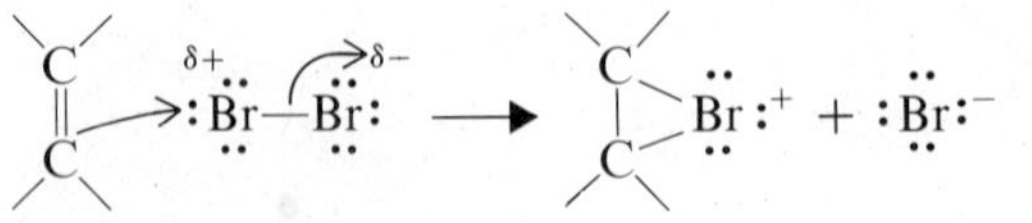

An alkene

The π electrons of benzene, by contrast, are held more tightly. The π electrons of benzene are not able to polarize the bromine-bromine bond to an extent that is sufficient to cause it to break. In the bromination of benzene, the Lewis acid assists in this process by weakening the bromine-bromine bond through complex formation.

We will see aromatic compounds later (e.g., aniline and phenol) in which the π electrons are more loosely held. These compounds react with bromine directly, without the intervention of a Lewis acid.

The chlorination of benzene in the presence of ferric chloride proceeds through a mechanism analogous to the one we have shown here for bromination. The ferric chloride serves the same purpose in aromatic chlorinations as ferric bromide does in aromatic brominations. It assists in the transfer of a positive halogen ion.

> It should be mentioned here that in *alkene* chlorinations Lewis acids are often added to the reaction mixture. The chlorine-chlorine bond is stronger than the bromine-bromine bond, and in alkene chlorinations Lewis acids ensure the transfer of $:\ddot{C}l^+$ ions.

Problem 13.3

Very often in aromatic chlorinations and brominations iron metal is added to the reaction mixture instead of ferric bromide or ferric chloride. Can you suggest what happens in these reactions?

Fluorine reacts so rapidly with benzene that aromatic fluorination requires special conditions and special types of apparatus. Even then, it is difficult to limit the reaction to monofluorination. Fluorobenzene can be made, however, through an indirect method that we will see in Chapter 21.

Iodine, on the other hand, is so unreactive that special techniques have to be used to effect direct iodination. One way this can be done chemically is by adding silver perchlorate to the reaction mixture.

(80%)

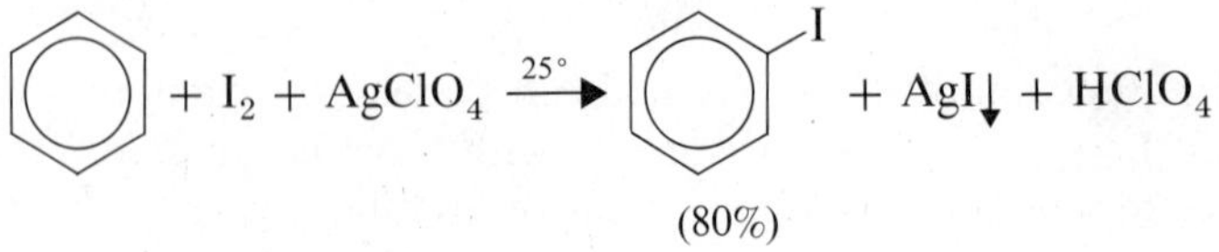

The mechanism of this reaction is not understood.

13.4 NITRATION OF BENZENE

Benzene reacts slowly with hot concentrated nitric acid to yield nitrobenzene. The reaction is much faster if it is carried out by heating benzene with a mixture of concentrated nitric acid and concentrated sulfuric acid.

$$\text{C}_6\text{H}_6 + \text{HNO}_3 + \text{H}_2\text{SO}_4 \xrightarrow{50-55°} \text{C}_6\text{H}_5\text{NO}_2 + \text{H}_3\text{O}^+ + \text{HSO}_4^-$$

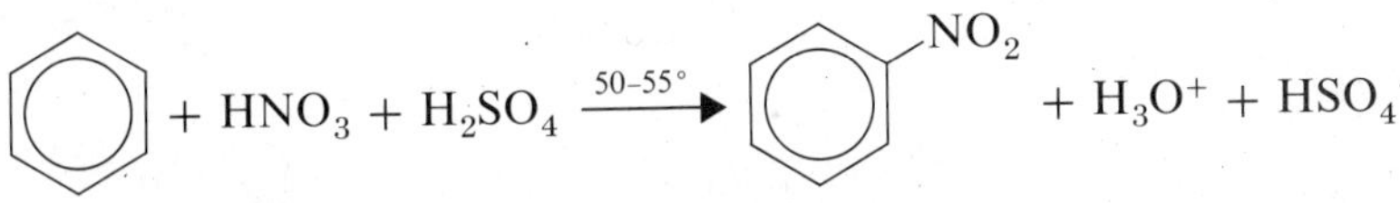

(85%)

Concentrated sulfuric acid increases the rate of the reaction by increasing the concentration of the electrophile—the nitronium ion.

Step 1 $\quad$ H—Ö—NO$_2$ + HOSO$_3$H $\rightleftharpoons$ H—Ö$^+$—NO$_2$ + HSO$_4^-$
$\qquad\qquad\qquad\qquad$ (H$_2$SO$_4$) $\qquad\qquad\qquad$ H

Step 2 $\quad$ H—Ö$^+$—NO$_2$ + H$_2$SO$_4$ $\rightleftharpoons$ $\overset{+}{\text{N}}$O$_2$ + H$_3$O$^+$ + HSO$_4^-$
$\qquad\qquad$ H $\qquad\qquad\qquad\qquad\qquad$ Nitronium
$\qquad\qquad\qquad\qquad\qquad\qquad\qquad\qquad\qquad$ ion

In the first step (above) nitric acid acts as a base and accepts a proton from the stronger acid, sulfuric acid. In the second step protonated nitric acid dissociates and produces a nitronium ion.

The nitronium ion reacts with benzene by attacking the π cloud and forming a σ complex.

Step 3

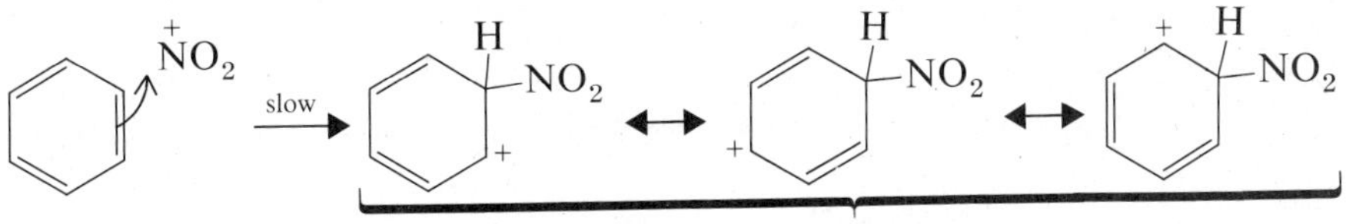

σ complex

The σ complex then loses a proton and becomes nitrobenzene.

Step 4

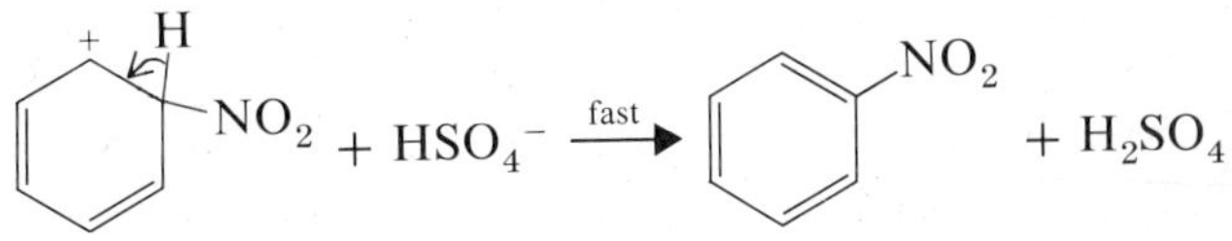

The nitronium ion has been observed spectroscopically in mixtures of nitric and sulfuric acids. Further evidence for the nitronium ion as the electrophile in aromatic nitration reactions comes from research in which stable nitronium salts such as NO$_2^+$ClO$_4^-$ and NO$_2^+$PF$_6^-$ have been prepared and used to carry out nitration reactions. We will see examples of the use of NO$_2^+$PF$_6^-$ in Section 13.8.

Problem 13.4

Write equations that show how nitronium ions might be formed in nitration reactions in which concentrated nitric acid is used by itself.

13.5 SULFONATION OF BENZENE

Benzene reacts with fuming sulfuric acid at room temperature to produce benzenesulfonic acid. Fuming sulfuric acid is sulfuric acid that contains added sulfur trioxide, SO_3. Sulfonation also takes place in concentrated sulfuric acid alone, but more slowly.

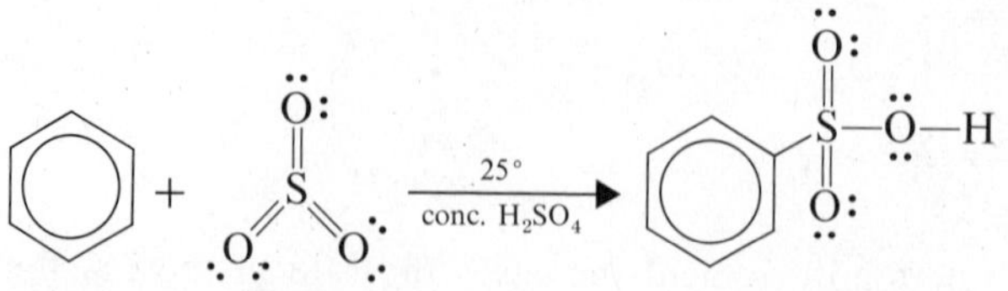

Sulfur trioxide

Benzenesulfonic acid (56%)

In either reaction the electrophile appears to be sulfur trioxide. In concentrated sulfuric acid, sulfur trioxide is produced by the following equilibrium.

Step 1 $\quad 2H_2SO_4 \rightleftharpoons SO_3 + H_3O^+ + HSO_4^-$

When sulfur trioxide reacts with benzene the following steps occur.

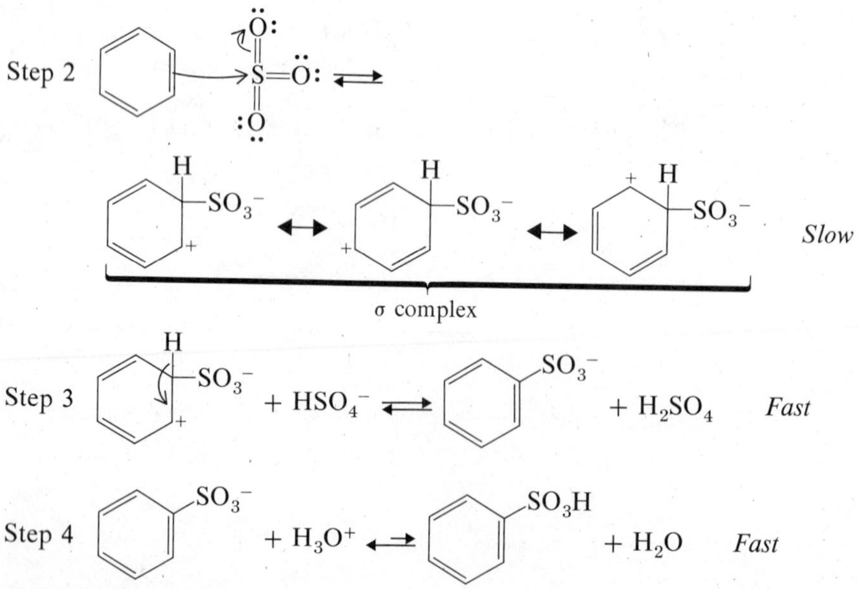

Step 2

σ complex

Slow

Step 3 $\quad + HSO_4^- \rightleftharpoons \quad + H_2SO_4 \quad$ *Fast*

Step 4 $\quad + H_3O^+ \rightleftharpoons \quad + H_2O \quad$ *Fast*

All of the steps in sulfonation reactions, including the step in which sulfur trioxide is formed from sulfuric acid, are equilibria. This means that the overall reaction is an equilibrium as well. In concentrated sulfuric acid, the overall equilibrium is the sum of steps 1 to 4.

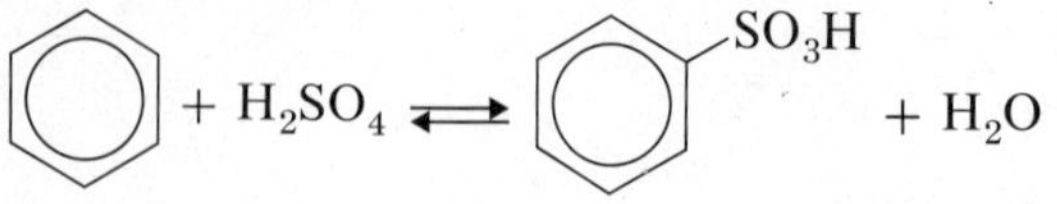

$+ H_2SO_4 \rightleftharpoons \quad + H_2O$

In fuming sulfuric acid, step 1 is unimportant because the dissolved sulfur trioxide reacts directly.

The fact that all of the steps in sulfonation reactions are equilibria means that the position of equilibrium can be influenced by the reaction conditions that we employ. If we want to sulfonate benzene we use concentrated sulfuric acid—or better yet—fuming sulfuric acid. Under these conditions the position of equilibrium lies appreciably to the right and we obtain benzenesulfonic acid in good yield.

On the other hand, there are times when we may want to remove a sulfonic acid group from a benzene ring. To do this we employ dilute sulfuric acid and usually pass steam through the reaction mixture. Under these conditions—with a high concentration of water—the equilibrium lies appreciably to the left and desulfonation occurs. The equilibrium is shifted even further to the left with volatile aromatic compounds because the aromatic compound distills with the steam.

We will see later that sulfonation and desulfonation reactions are often used in synthetic work. We may, for example, introduce a sulfonic acid group into a benzene ring to influence the course of some further reaction. Later, we may remove the sulfonic acid group by desulfonation.

Problem 13.5

In most desulfonation reactions the electrophile is a proton. Other electrophiles may be used, however. (a) Show all steps of the desulfonation reaction that would occur when benzenesulfonic acid is desulfonated with deuterium sulfate, D_2SO_4, dissolved in D_2O. (b) When benzenesulfonic acid reacts with bromine in the presence of ferric bromide, bromobenzene is obtained from the reaction mixture. What is the electrophile in this reaction? Show all steps in the mechanism for this desulfonation.

13.6 FRIEDEL-CRAFTS ALKYLATION

In 1877 a French chemist, Charles Friedel, and his American collaborator, James M. Crafts, discovered new methods for the preparation of alkylbenzenes, (ArR) and acylbenzenes, (ArCOR). These reactions are now called the Friedel-Crafts alkylation and acylation reactions. We will study the Friedel-Crafts alkylation reaction here and take up the Friedel-Crafts acylation reaction in Section 13.7.

Friedel-Crafts alkylations result from the reaction of a carbocation—or carbocationlike species—with an aromatic ring. An example is the reaction of benzene with ethyl bromide in the presence of aluminum chloride.

$$\text{(large excess)} + CH_3CH_2Br \xrightarrow[\text{AlCl}_3]{80°} \text{Ethylbenzene} + HBr$$

Ethylbenzene
(83%)

This reaction appears to occur through an electrophilic attack of a complex formed from ethyl bromide and aluminum chloride.

$$\text{Step 1} \quad CH_3CH_2-\overset{..}{\underset{..}{Br}}: + AlCl_3 \rightleftharpoons CH_3CH_2-\overset{\delta+}{\underset{..}{Br}}-\overset{\delta+}{\overline{Al}}Cl_3$$

Complex formation makes the —CH_2— group quite positive and allows the complex to transfer an ethyl group to the benzene ring.

Step 2

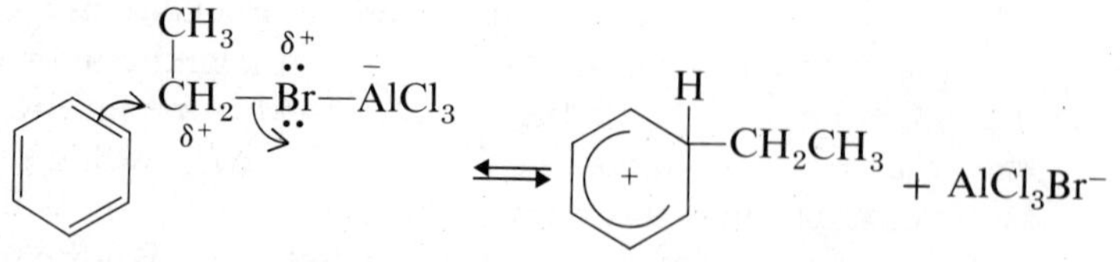

Then the σ complex that is produced loses a proton and yields ethylbenzene.

Step 3

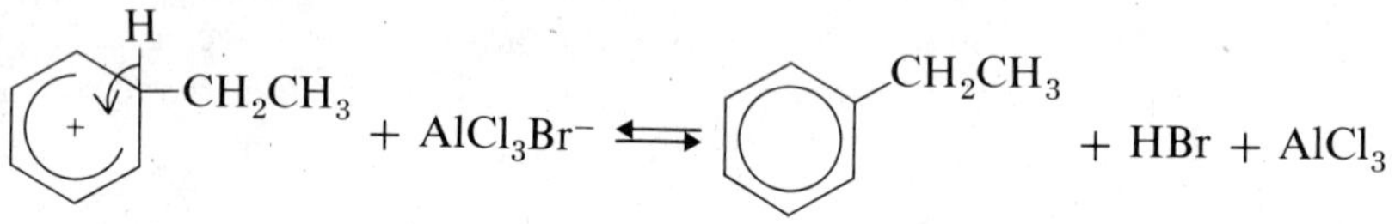

When similar alkylations are carried out with secondary and tertiary halides, the reactions appear to occur through the formation of carbocations.

Step 1

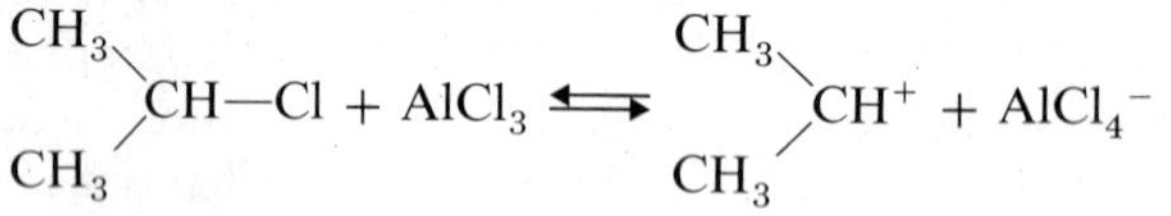

Isopropyl
chloride

Step 2

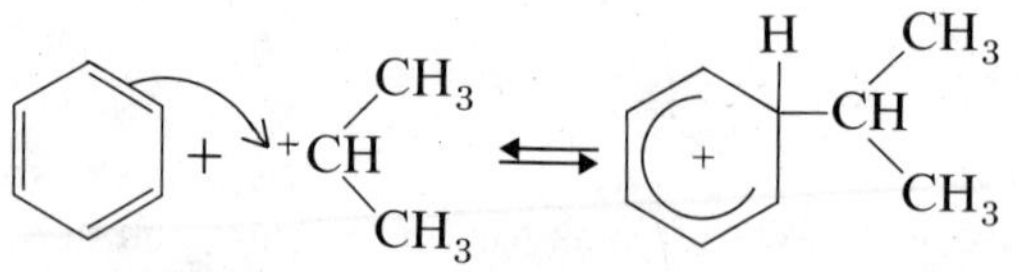

Step 3

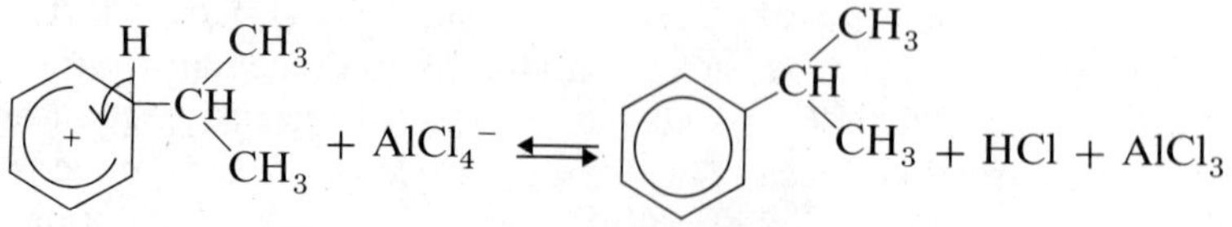

Isopropylbenzene

Friedel-Crafts alkylations are not restricted to the use of alkyl halides and aluminum chloride. Many other pairs of reagents that form carbocations (or carbocation-like species) may be used as well. These possibilities include the use of a mixture of an alkene and an acid.

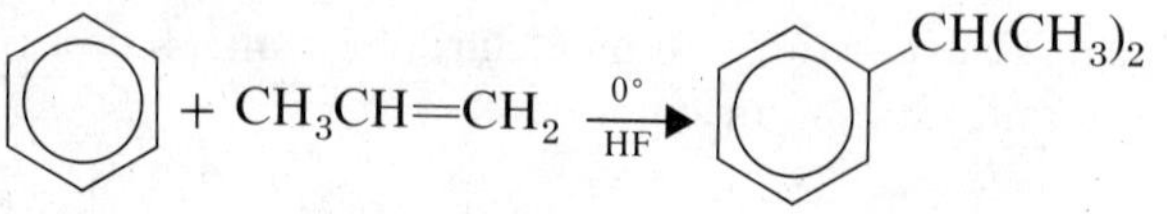

Propene   Isopropylbenzene
                     (84%)

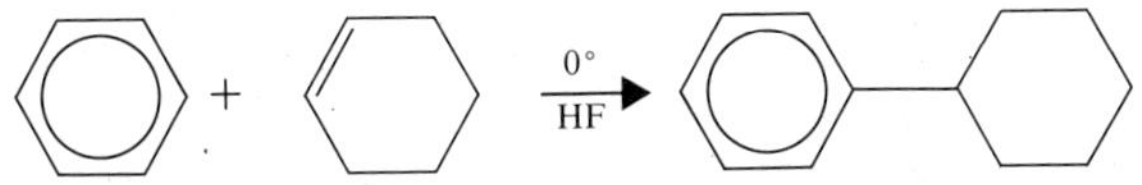

Cyclohexene   Cyclohexylbenzene
(62%)

A mixture of an alcohol and an acid may also be used

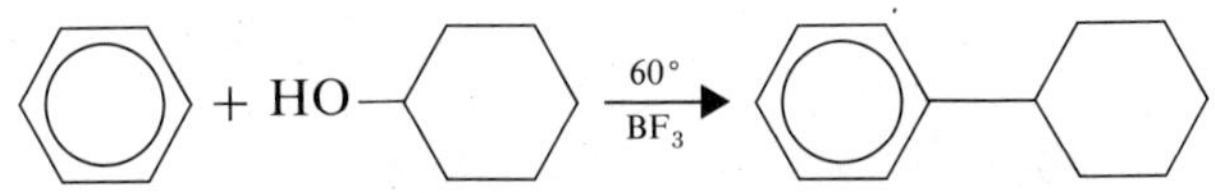

Cyclohexanol   Cyclohexylbenzene
(56%)

Problem 13.6

Assume that carbocations are involved and propose step-by-step mechanisms for both of the syntheses of cyclohexylbenzene given above.

There are several restrictions that limit the usefulness of the Friedel-Crafts alkylation reaction:

1. Aryl and vinyl halides cannot be used as the halide component because they do not form carbocations readily.

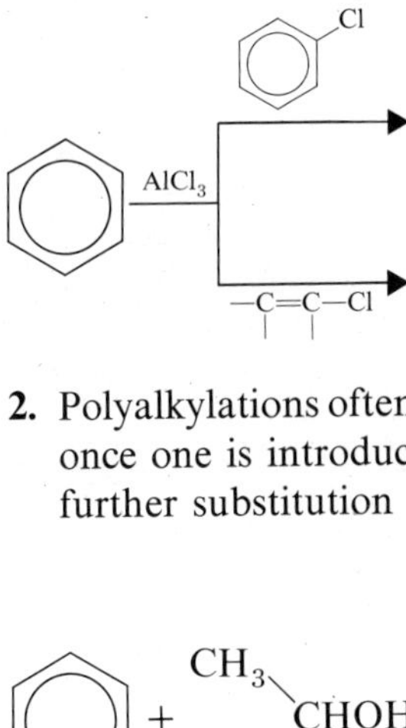

2. Polyalkylations often occur. Alkyl groups are electron-releasing groups and once one is introduced into the benzene ring it activates the ring toward further substitution (cf. Section 13.8).

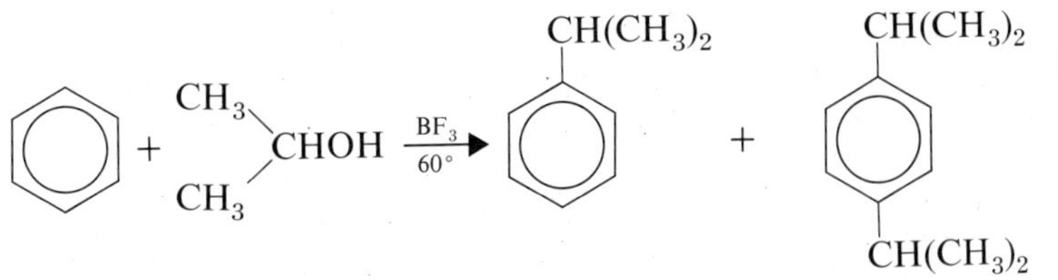

Isopropyl-   p-Diisopropylbenzene
benzene   (14%)
(24%)

3. When the carbocation formed from an alkyl halide, alkene, or alcohol can rearrange to a more stable carbocation, it does so and the major product obtained from the reaction is usually the one from the more stable carbocation.

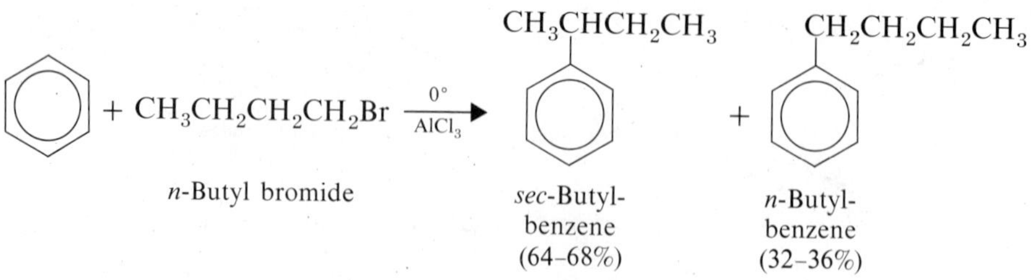

$$CH_3CH_2CH_2CH_2Br \xrightarrow[AlCl_3]{0°}$$

n-Butyl bromide

sec-Butyl-
benzene
(64–68%)

n-Butyl-
benzene
(32–36%)

4. Alkyl group rearrangements often take place in Friedel-Crafts reactions.

$$\xrightarrow[\cdot \, HCl]{AlCl_3}$$

+ AlCl_3 + HCl

1,2,4-Trimethylbenzene

Mesitylene

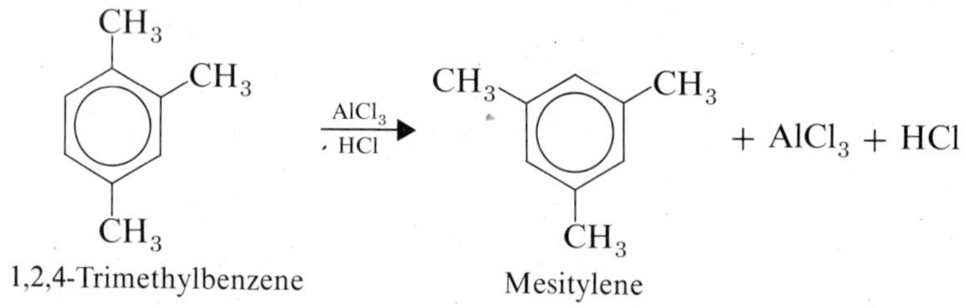

5. Friedel-Crafts reactions do not occur when powerful electron-withdrawing groups (Section 13.8) are present on the aromatic ring or when the ring bears an —NH$_2$, —NHR, or —NR$_2$ group.

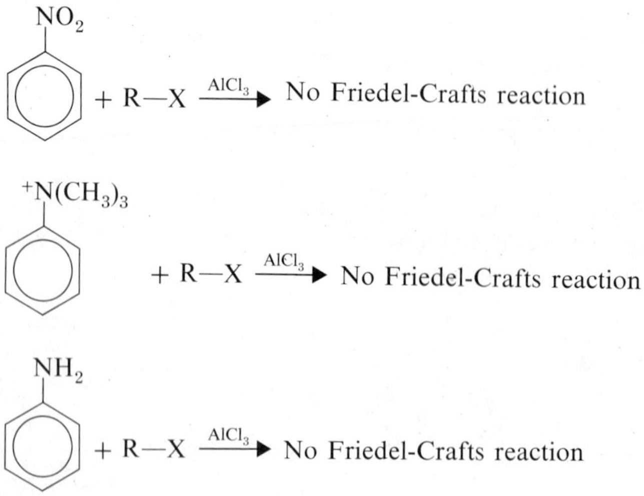

+ R—X $\xrightarrow{AlCl_3}$ No Friedel-Crafts reaction

+ R—X $\xrightarrow{AlCl_3}$ No Friedel-Crafts reaction

+ R—X $\xrightarrow{AlCl_3}$ No Friedel-Crafts reaction

Any substituent more electron withdrawing (or deactivating) than a halogen (cf. Table 13.3) usually makes an aromatic ring too electron deficient to undergo electrophilic substitution by a carbocation or a carbocation-like electrophile. The amino groups, —NH$_2$, —NHR, and —NR$_2$, become powerful electron-withdrawing groups when compounds containing them are placed in Friedel-Crafts reaction mixtures. They do so by reacting with the Lewis acid as follows.

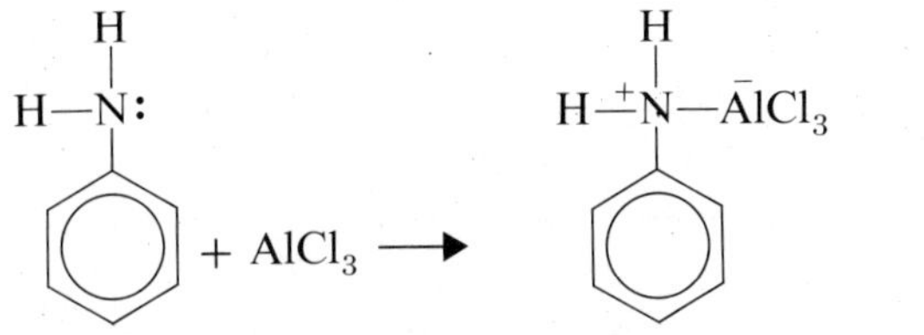

Does not undergo
a Friedel-Crafts reaction

Problem 13.7

One of our illustrations of the limitations of the Friedel-Crafts reaction shows the reaction of benzene and *n*-butyl bromide in the presence of aluminum chloride producing both *n*-butylbenzene and *sec*-butylbenzene. (a) Write mechanisms that account for the formation of both products. (b) When benzene reacts with *n*-propyl alcohol in the presence of boron trifluoride, both *n*-propylbenzene and isopropyl benzene are obtained as products. How can you account for this?

13.7 FRIEDEL-CRAFTS ACYLATION

The $R\overset{\overset{O}{\|}}{C}-$ group is called an *acyl group,* and a reaction whereby an acyl group is introduced into a compound is called an *acylation* reaction. Two common acyl groups are the acetyl group and the propionyl group. The benzoyl group is an aroyl group, but it is often called an acyl group as well.

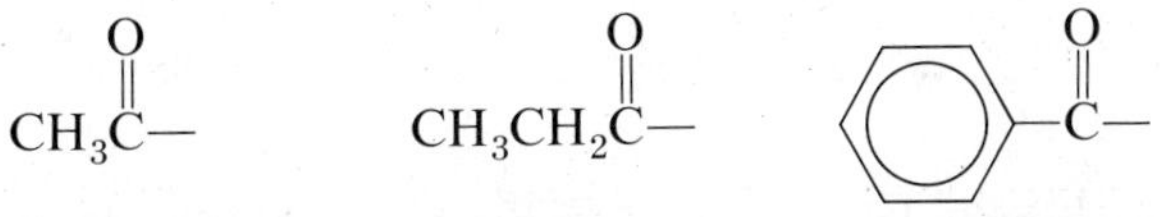

Acetyl
group
(ethanoyl group)

Propionyl
group
(propanoyl group)

Benzoyl
group

The Friedel-Crafts acylation reaction is an effective means of introducing an acyl or an aroyl group into an aromatic ring. The reaction is often carried out by treating the aromatic compound with an acyl or aroyl halide. Unless the aromatic compound is one that is highly reactive, the reaction requires the addition of at least one equivalent of a Lewis acid (such as $AlCl_3$) as well. The product of the reaction is an aryl ketone.

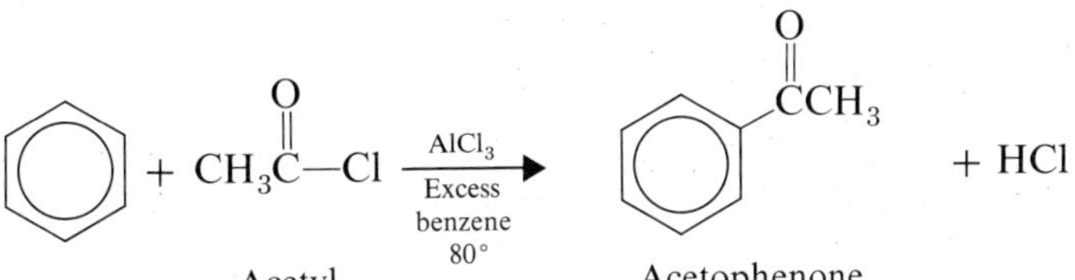

Acetyl
chloride

Acetophenone
(methyl phenyl ketone)
(97%)

Acyl and aroyl chlorides are also called *acid chlorides*. They are prepared easily by treating carboxylic acids with thionyl chloride ($SOCl_2$) or phosphorus pentachloride (PCl_5).

$$CH_3\overset{O}{\overset{\|}{C}}OH + SOCl_2 \xrightarrow{80°} CH_3\overset{O}{\overset{\|}{C}}Cl + SO_2 + HCl$$

Acetic acid   Thionyl chloride   Acetyl chloride (80–90%)

$$\text{Benzoic acid} \quad + \quad PCl_5 \quad \longrightarrow \quad \text{Benzoyl chloride} + POCl_3 + HCl$$

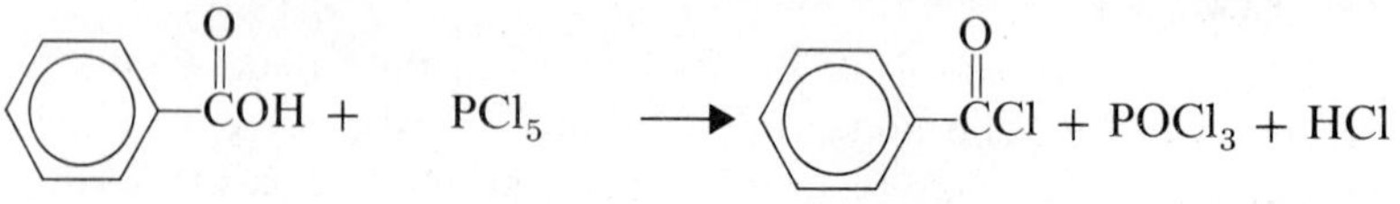

Benzoic acid   Phosphorus pentachloride   Benzoyl chloride (90%)

Friedel-Crafts acylations can also be carried out using acid anhydrides. The following reaction is an example.

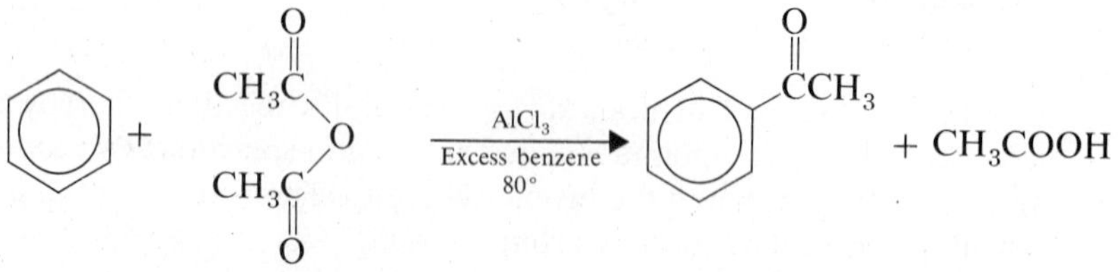

Acetic anhydride $\xrightarrow[\text{Excess benzene} \atop 80°]{AlCl_3}$ Acetophenone (82–85%) $+ CH_3COOH$

In most Friedel-Crafts reactions the electrophile appears to be an *acylium ion*. An acylium ion can be formed from an acyl halide in the following way.

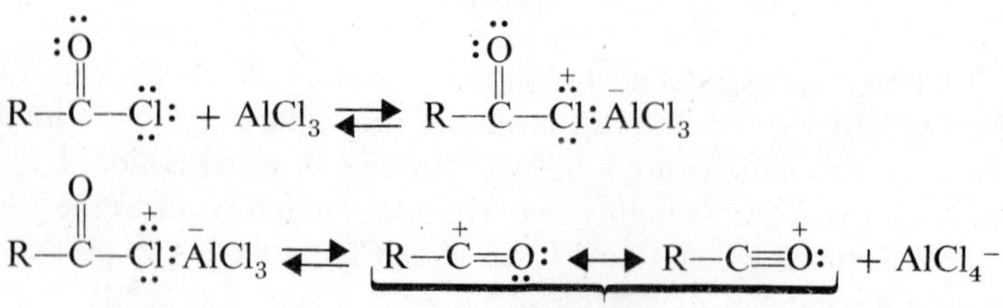

$$R-\overset{:O:}{\overset{\|}{C}}-\overset{..}{\underset{..}{Cl}}: + AlCl_3 \rightleftarrows R-\overset{:O:}{\overset{\|}{C}}-\overset{+}{\underset{..}{Cl}}:\bar{A}lCl_3$$

$$R-\overset{O}{\overset{\|}{C}}-\overset{+}{\underset{..}{Cl}}:\bar{A}lCl_3 \rightleftarrows \underline{R-\overset{+}{C}=\overset{..}{O}: \longleftrightarrow R-C\equiv\overset{+}{O}:} + AlCl_4^-$$

An acylium ion
(a resonance hybrid)

Problem 13.8

Show how an acylium ion could be formed from an acid anhydride.

The remaining steps in the Friedel-Crafts acylation of benzene are shown on the top of the next page.

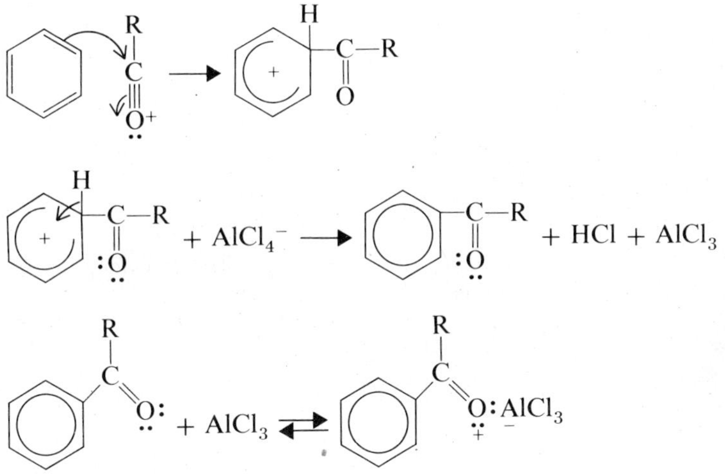

In the last step (above) aluminum chloride forms a complex with the ketone. After the reaction is over, treating the complex with water liberates the ketone.

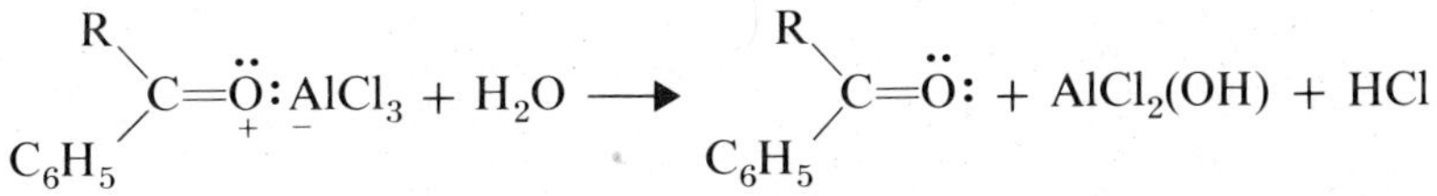

Polyacylations are not a problem in Friedel-Crafts acylation reactions. The acyl group is an electron-withdrawing group by itself; and once introduced, it deactivates the benzene ring toward further electrophilic attack (cf. Section 13.8). As we have just seen, the acyl group also forms a complex with aluminum chloride. Complex formation makes the acyl group an even more powerful electron-withdrawing group and this makes the ring even less reactive.

Synthetic Applications of the Friedel-Crafts Acylation Reaction

Rearrangements of the carbon chain do not occur in Friedel-Crafts acylations. The acylium ion, because it is stabilized by resonance, is more stable than most other carbocations. Thus, there is no driving force for a rearrangement. Because rearrangements do not occur, Friedel-Crafts acylation reactions give us much better synthetic routes to *n*-alkylbenzenes than do Friedel-Crafts alkylation reactions.

As an example, let us consider the problem of synthesizing *n*-propylbenzene. If we attempt this synthesis through a Friedel-Crafts alkylation reaction, a rearrangement occurs and the major product is isopropylbenzene.

Isopropylbenzene (major product)   *n*-Propylbenzene (minor product)

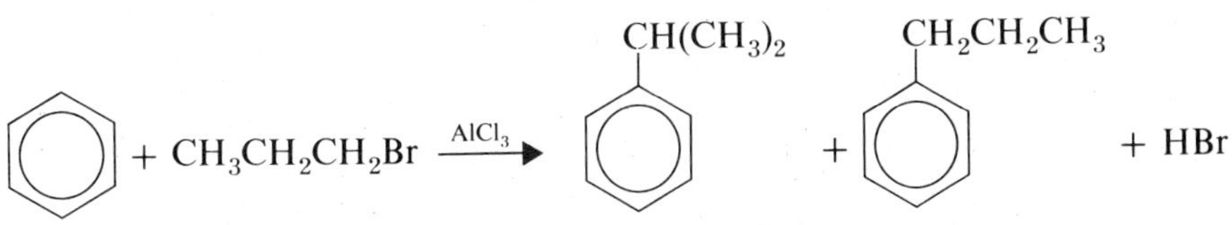

By contrast, the Friedel-Crafts acylation of benzene with propionyl chloride produces a ketone with an unrearranged carbon chain in excellent yield.

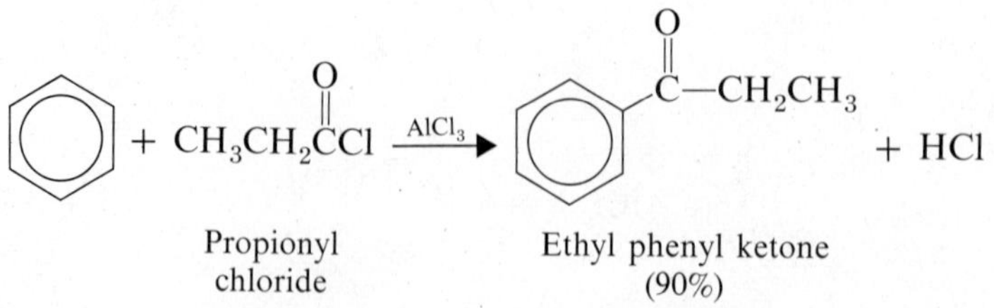

Propionyl chloride

Ethyl phenyl ketone (90%)

This ketone can then be reduced to *n*-propylbenzene by several methods. One method—called the Clemmensen reduction—consists of refluxing the ketone with amalgamated zinc and hydrochloric acid.

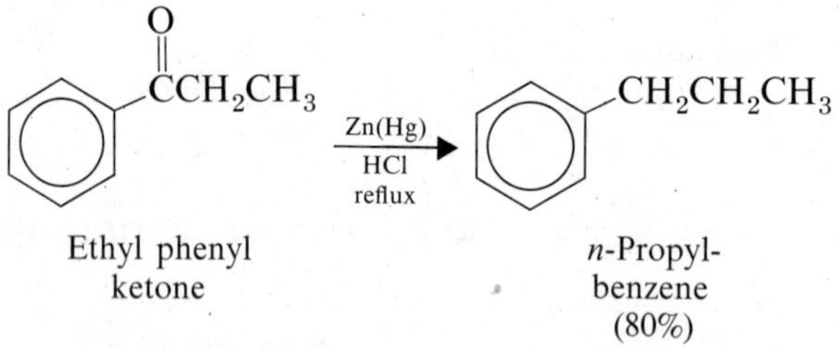

Ethyl phenyl ketone

n-Propyl-benzene (80%)

Friedel-Crafts acylations can be used in a variety of other ways. Diaryl ketones are formed when the acid chloride of an aromatic acid is used as one component. The reaction of benzene with benzoyl chloride is an example.

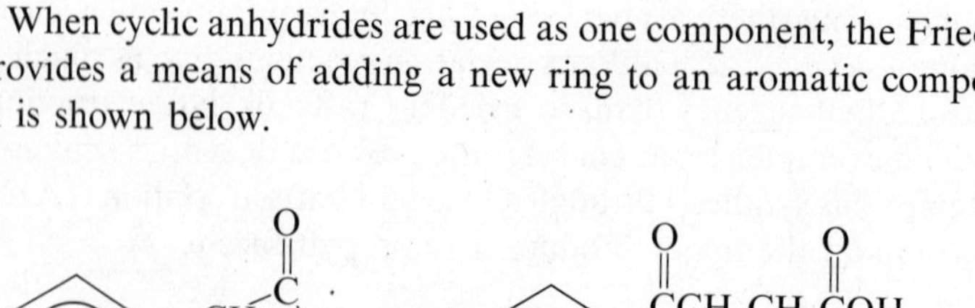

Benzoyl chloride

Diphenyl ketone (benzophenone) (82%)

When cyclic anhydrides are used as one component, the Friedel-Crafts acylation provides a means of adding a new ring to an aromatic compound. One illustration is shown below.

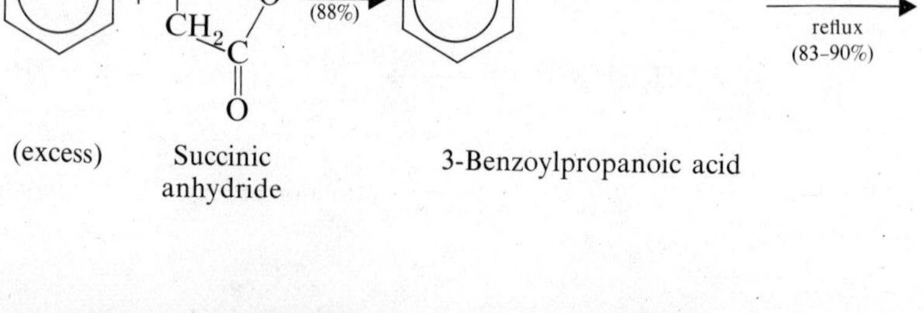

(excess)   Succinic anhydride

3-Benzoylpropanoic acid

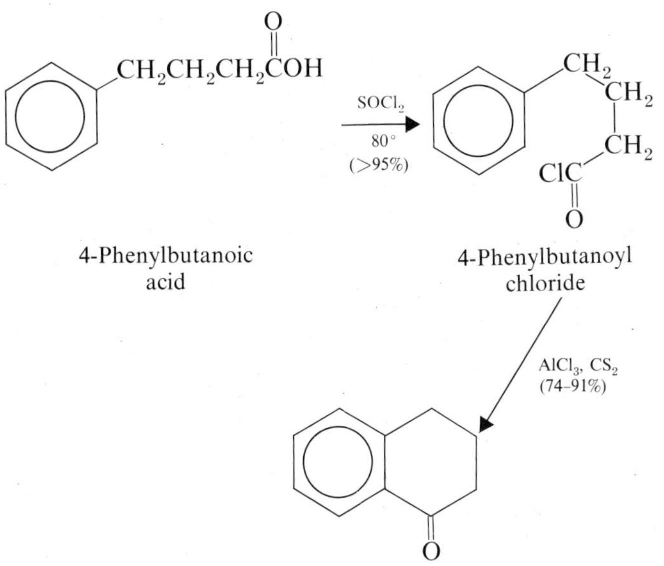

4-Phenylbutanoic
acid

4-Phenylbutanoyl
chloride

α-Tetralone

Problem 13.9

Starting with benzene and the appropriate acid chloride or anhydride, outline a synthesis of each of the following.

    (a) *n*-Hexylbenzene

    (b) Isobutylbenzene

    (c) Anthrone,

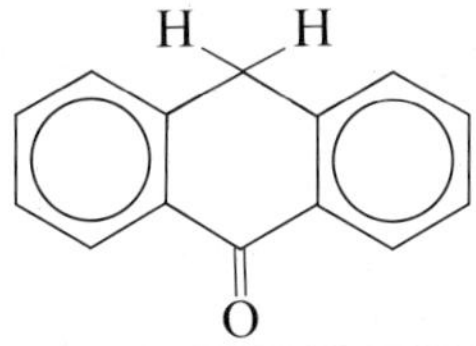

13.8 EFFECT OF SUBSTITUENTS: REACTIVITY AND ORIENTATION

Monosubstituted Benzenes

We find that when substituted benzenes undergo electrophilic attack, groups already present on the ring affect both the rate of the reaction and the site of attack. We can say, therefore, that substituent groups affect both the *reactivity* and *orientation* in electrophilic aromatic substitution.

We can divide substituent groups into two categories depending on how they influence the reactivity of the ring. Those groups that cause the ring to be more reactive than benzene itself we call *activating groups*. Those groups that cause the ring to be less reactive than benzene we call *deactivating groups*.

We also find that we can divide substituent groups into two categories depending on how they influence the orientation of attack by the incoming electrophile.

Groups in one category tend to bring about electrophilic substitution primarily at the *ortho* and *para* positions. We call these groups *ortho-para directors* because they tend to *direct* the incoming group into the ortho and para positions. Groups in the second category tend to direct the incoming electrophile toward substitution at the *meta* position. We call these groups *meta directors.*

Several examples will illustrate more clearly what we mean by these terms.

The methyl is an *activating* group and an ortho-para director. Toluene reacts considerably faster than benzene does in all electrophilic substitution reactions.

More reactive toward   Less reactive toward
electrophilic substitution   electrophilic substitution

We observe the greater reactivity of toluene in several different ways. We find, for example, that toluene generally requires milder conditions—lower temperatures and lower concentrations of the electrophile—in electrophilic substitution reactions than benzene does. We also find that under the same conditions, toluene reacts faster than benzene. In nitration, for example, toluene reacts 25 times as fast as benzene.

We find, moreover, that when toluene undergoes electrophilic substitution, most of the substitution takes place at the ortho and para positions. When we nitrate toluene using nitric and sulfuric acid we get mononitrotoluenes in the following relative proportions.

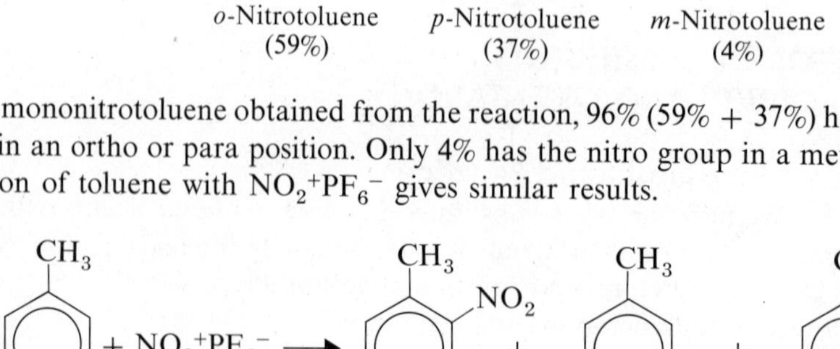

o-Nitrotoluene   *p*-Nitrotoluene   *m*-Nitrotoluene
(59%)   (37%)   (4%)

Of the mononitrotoluene obtained from the reaction, 96% (59% + 37%) has the nitro group in an ortho or para position. Only 4% has the nitro group in a meta position. Nitration of toluene with $NO_2^+PF_6^-$ gives similar results.

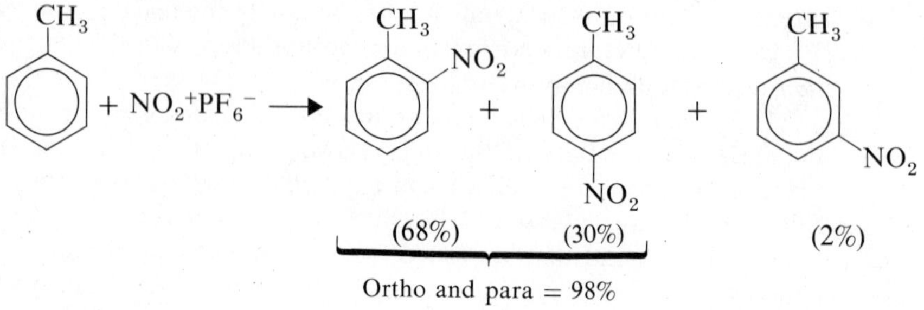

(68%)   (30%)   (2%)

Ortho and para = 98%

Problem 13.10
What percentage of each nitrotoluene would you expect if substitution were to take place on a purely *statistical* basis?

Predominant substitution at the ortho and para positions of toluene is not restricted to nitration reactions as the following examples show.

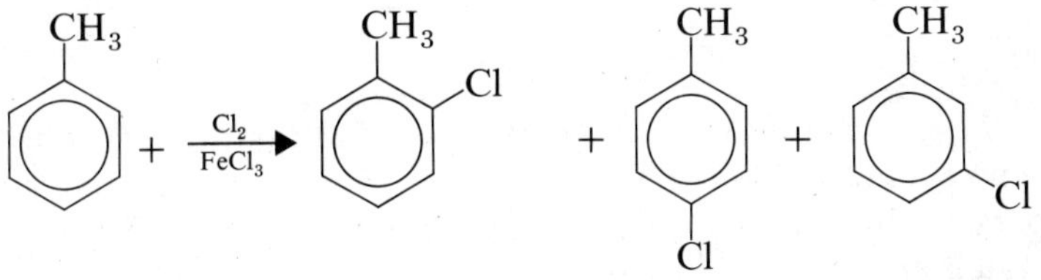

(75%)
o-Chloro-
toluene

(23%)
p-Chloro-
toluene

(2%)
m-Chloro-
toluene

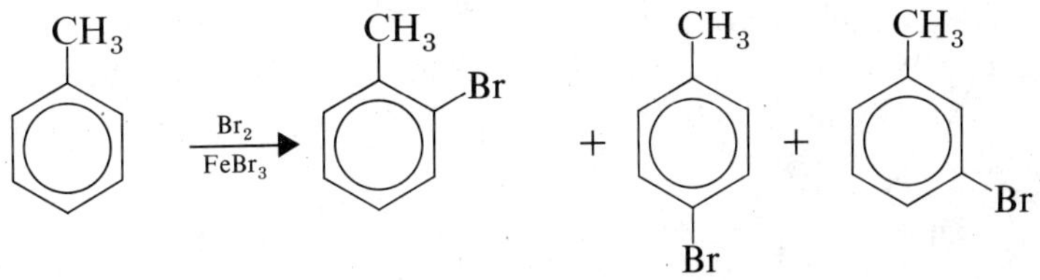

(40%)
o-Bromo-
toluene

(60%)
p-Bromo-
toluene

(trace)
m-Bromo-
toluene

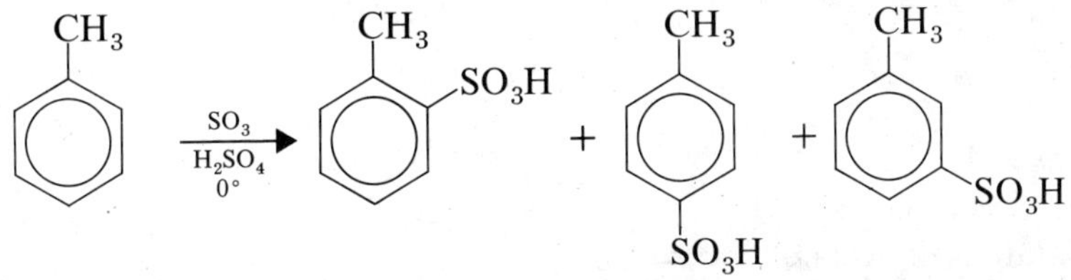

(43%)
o-Toluene-
sulfonic
acid

(53%)
p-Toluene-
sulfonic
acid

(4%)
m-Toluene-
sulfonic
acid

Problem 13.11
When toluene is sulfonated at 100° the following proportions of monosubstituted products are obtained: *o*-toluenesulfonic acid (13%), *m*-toluenesulfonic acid (8%), *p*-toluenesulfonic acid (79%). (a) Keeping in mind that sulfonation reactions are reversible, how can you account for the different relative percentages formed at the lower and higher temperature? (b) Which toluenesulfonic acid appears to be more stable, *o*-toluenesulfonic acid or *p*-toluenesulfonic acid?

All alkyl groups are activating groups, and they are all also ortho-para directors. Nitration of *tert*-butylbenzene, for example, takes place at a rate that is roughly 16 times as fast as benzene, and gives the following percentages of mononitro products.

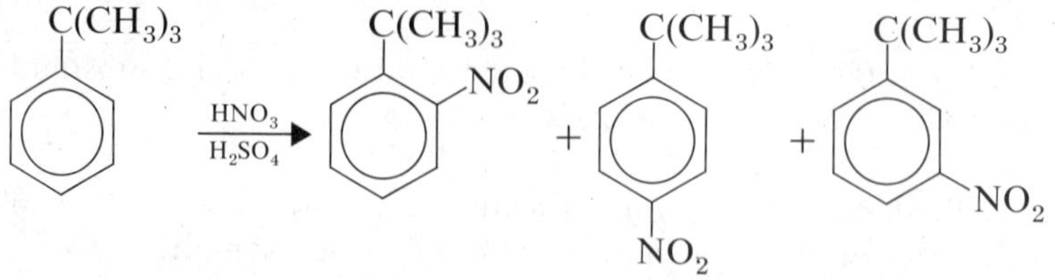
C(CH3)3
HNO3
H2SO4
C(CH3)3
NO2
+
C(CH3)3
NO2
+
C(CH3)3
NO2

(12%) Ortho | (80%) Para | (8%) Meta

If we compare the nitration reaction of *tert*-butylbenzene with the nitration reaction of toluene (p. 464), we can see an effect on the relative amounts of ortho and para product produced by the steric bulk of the *tert-butyl* group. As we would expect, this effect is most pronounced in ortho substitution. *Tert*-butylbenzene gives much less ortho-nitro product (12% versus 59%) and gives much more para-nitro product (80% versus 37%). Steric effects are, as we might expect, always greater at positions ortho to the group on the ring.

The methoxyl group, $CH_3O—$, and the acetamido group $CH_3CONH—$, are strong activating groups and both are ortho-para directors. Bromination of anisole proceeds readily in the absence of a catalyst.

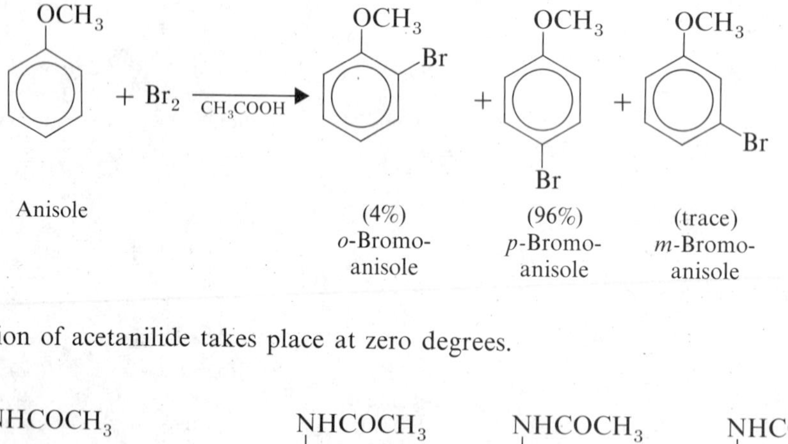
OCH3
+ Br2
CH3COOH
OCH3
Br
+
OCH3
Br
+
OCH3
Br

Anisole

(4%) *o*-Bromoanisole | (96%) *p*-Bromoanisole | (trace) *m*-Bromoanisole

Nitration of acetanilide takes place at zero degrees.

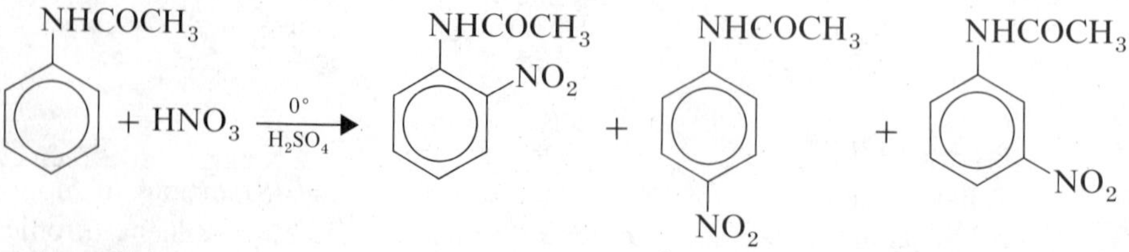
NHCOCH3
+ HNO3
0°
H2SO4
NHCOCH3
NO2
+
NHCOCH3
NO2
+
NHCOCH3
NO2

Acetanilide

(19%) *o*-Nitroacetanilide | (79%) *p*-Nitroacetanilide | (2%) *m*-Nitroacetanilide

The hydroxyl group and the amino group are very powerful activating groups and are also powerful ortho-para directors. Phenol and aniline react with bromine in water (no catalyst is required) to produce products in which both ortho positions and the para position are substituted. These tribromo products are obtained in nearly quantitative yield.

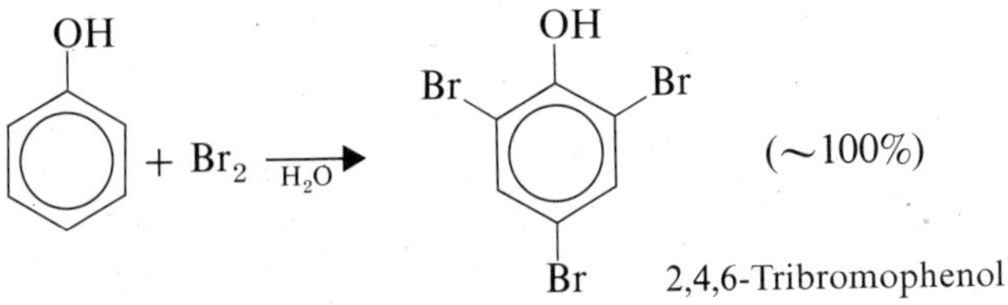

$(\sim 100\%)$

2,4,6-Tribromophenol

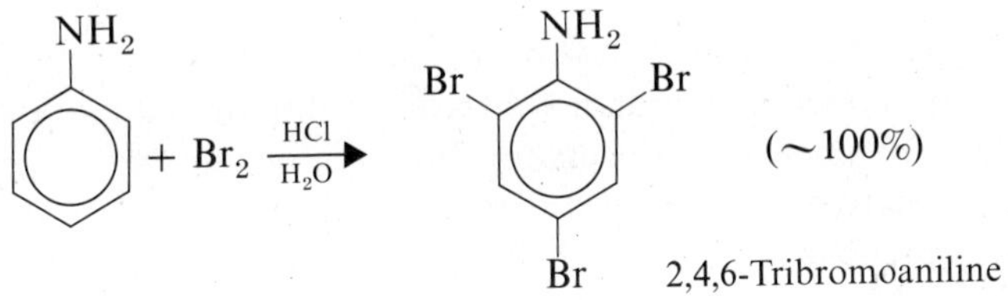

$(\sim 100\%)$

2,4,6-Tribromoaniline

The nitro group is a very strong *deactivating* group. Nitrobenzene undergoes nitration at a rate that is 10^4 times slower than benzene. The nitro group is a meta director. When nitrobenzene is nitrated with nitric and sulfuric acid, 93% of the substitution occurs at the meta position.

(6%)　　(1%)　　(93%)

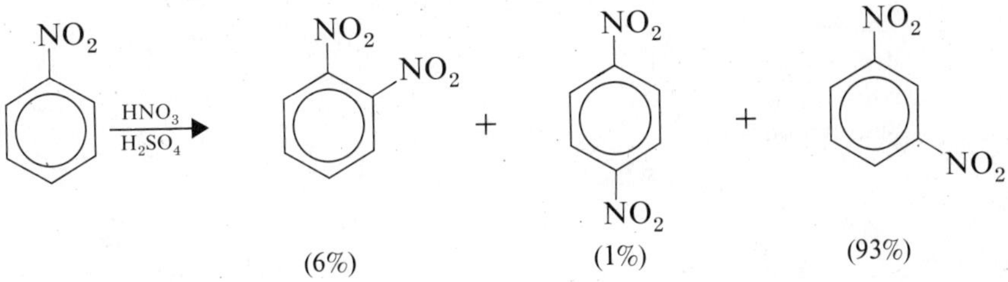

When $NO_2^+PF_6^-$ is used as the nitrating agent, 88.5% of the nitration occurs at the meta position

(10%)　　(1.5%)　　(88.5%)

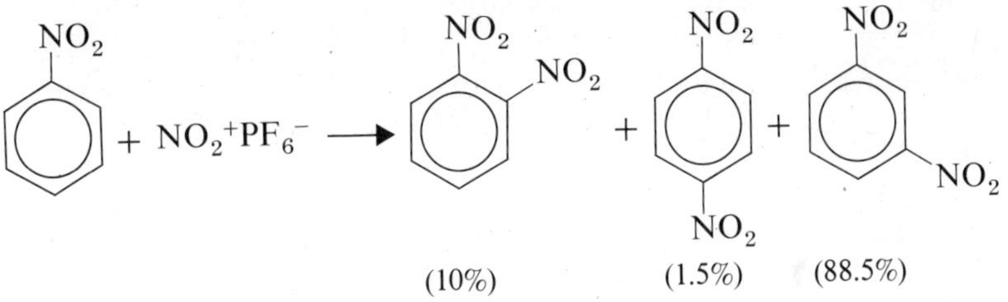

The carboxyl group (—COOH) and the trifluoromethyl group (—CF$_3$) are also strong deactivating groups; they are also meta directors.

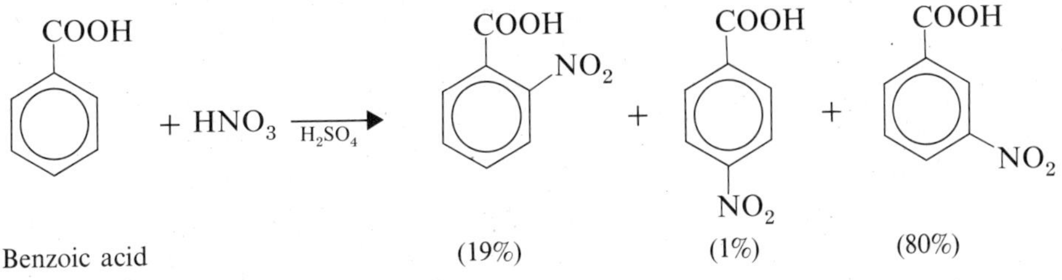

Benzoic acid　　　　(19%)　　(1%)　　(80%)

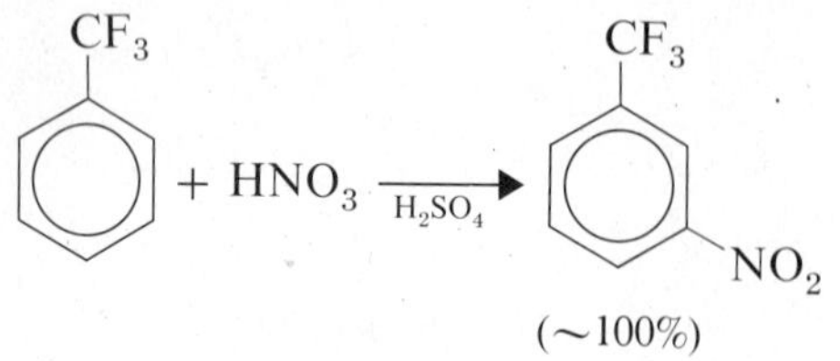

$(\sim 100\%)$

The chloro and bromo groups are moderate deactivating groups. Chlorobenzene and bromobenzene undergo nitration at rates that are, respectively, 33 and 30 times slower than benzene. The chloro and bromo groups are ortho-para directors, however. The relative percentages of monosubstituted products that are obtained when chlorobenzene is chlorinated, brominated, nitrated, and sulfonated are shown in Table 13.1.

TABLE 13.1 Electrophilic Substitution of Chlorobenzene

REACTION	ORTHO PRODUCT (percent)	PARA PRODUCT (percent)	TOTAL ORTHO AND PARA (percent)	META PRODUCT (percent)
Chlorination	39	55	94	6
Bromination	11	87	98	2
Nitration	30	70	100	—
Sulfonation	—	100	100	—

The corresponding results from electrophilic substitution reaction of bromobenzene are given in Table 13.2.

TABLE 13.2 Electrophilic Substitution of Bromobenzene

REACTION	ORTHO PRODUCT (percent)	PARA PRODUCT (percent)	TOTAL ORTHO AND PARA (percent)	META PRODUCT (percent)
Chlorination	45	53	98	2
Bromination	13	85	98	2
Nitration	38	62	100	—
Sulfonation	—	100	100	—

Classification of Substituents

Studies like the ones that we have presented in this section have been done for a number of other substituted benzenes. The effects of these substituents on reactivity and orientation are included in Table 13.3.

Problem 13.12

What would be the major monochloro product (or products) formed when each of the following compounds reacts with chlorine in the presence of ferric chloride?

(a) Ethylbenzene, $C_6H_5CH_2CH_3$
(b) Trifluoromethylbenzene, $C_6H_5CF_3$
(c) Phenyltrimethylammonium chloride, $C_6H_5\overset{+}{N}(CH_3)_3$ Cl^-
(d) Methyl benzoate, $C_6H_5COOCH_3$

TABLE 13.3 Effect of Substituents on Electrophilic Aromatic Substitution

ORTHO-PARA DIRECTORS	META DIRECTORS
Strongly Activating	*Strongly Deactivating*
$-\ddot{N}H_2$, $-\ddot{N}HR$, $-\ddot{N}R_2$	$-NO_2$
$-\ddot{O}H$	$-NR_3^+$
	$-CF_3$, $-CCl_3$
Moderately Activating	*Moderately Deactivating*
$-\ddot{N}HCOCH_3$, $-\ddot{N}HCOR$	$-C\equiv N$
$-\ddot{O}CH_3$, $-\ddot{O}R$	$-SO_3H$
Weakly Activating	$-COOH$, $-COOR$
$-CH_3$, $-C_2H_5$, $-R$	$-CHO$, $-COR$
$-C_6H_5$	
Weakly Deactivating	
$-\ddot{\underset{..}{F}}:$, $-\underset{..}{\ddot{C}l}:$, $-\underset{..}{\ddot{B}r}:$, $-\ddot{\underset{..}{I}}:$	

Disubstituted Benzenes

The problem of orientation is somewhat more complicated when two substituents are present on a benzene ring. We find, however, that in many instances we can make very good estimates of the outcome of the reaction by relatively simple analyses.

If two groups are located so that their directive effects reinforce each other then predicting the outcome is easy. Consider the examples shown below. In each case the entering substituent is directed by both groups into the position indicated by the arrows.

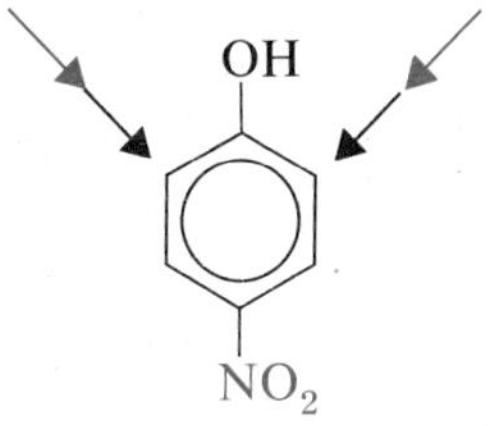

1 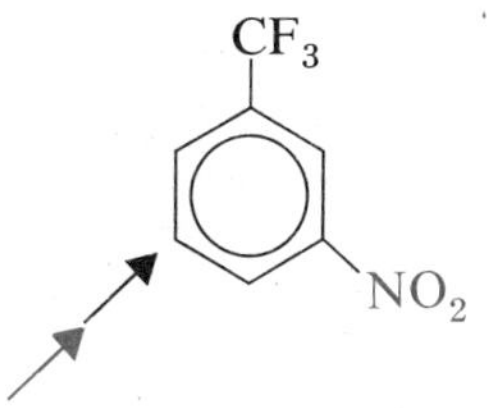   2 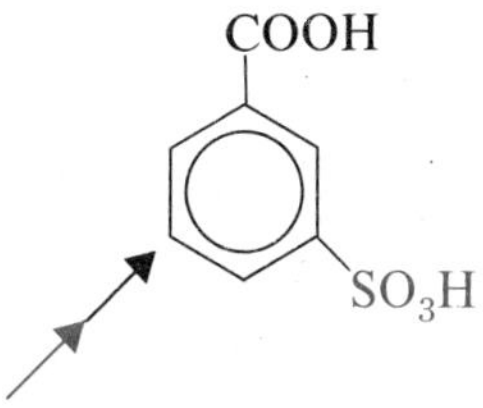   3

When the directive effect of two groups oppose each other, *the most powerful activating group generally determines the outcome of the reaction.* Let us consider, as an example, the orientation of electrophilic substitution of *p*-methylacetanilide.

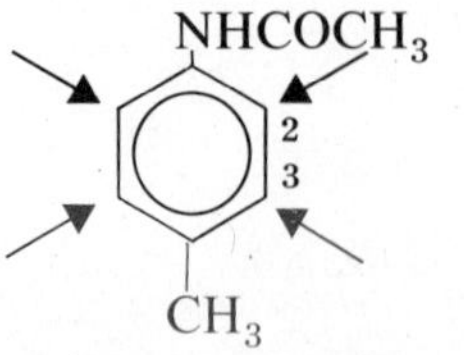

p-Methylacetanilide

The acetamido group is a much stronger activating group than the methyl group. The following examples show that the acetamido group determines the outcome of the reaction. Substitution occurs primarily at position 2.

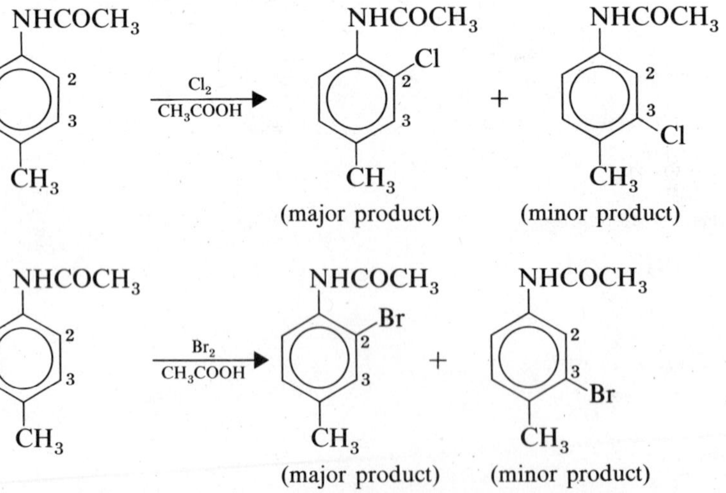

(major product)   (minor product)

(major product)   (minor product)

Other examples that illustrate the greater directive influence of the more powerful activating group are shown below.

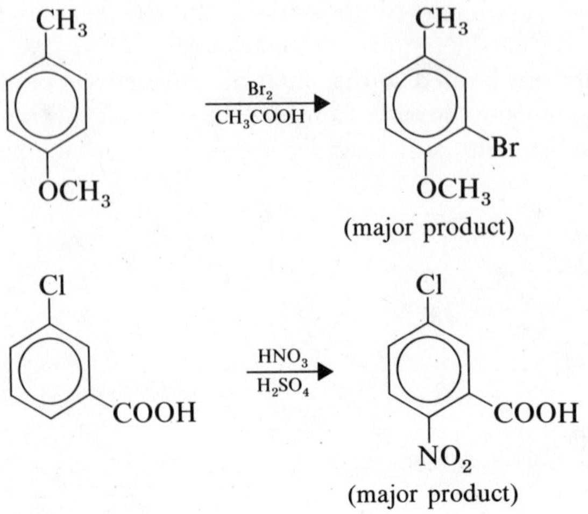

(major product)

(major product)

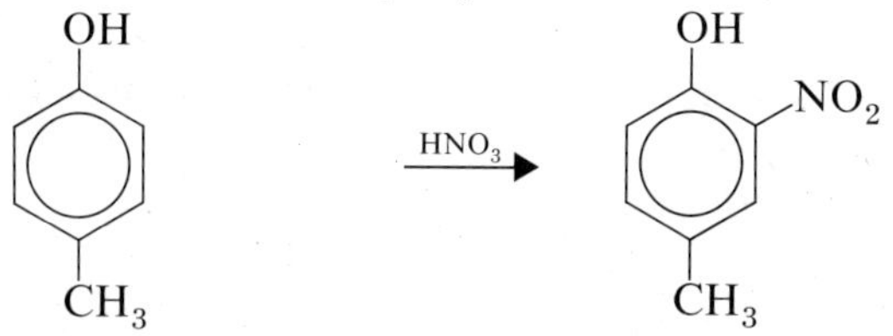

(major product)

When two opposing groups have approximately the same activating effect the results are not nearly so clear-cut. The following reaction is a typical example.

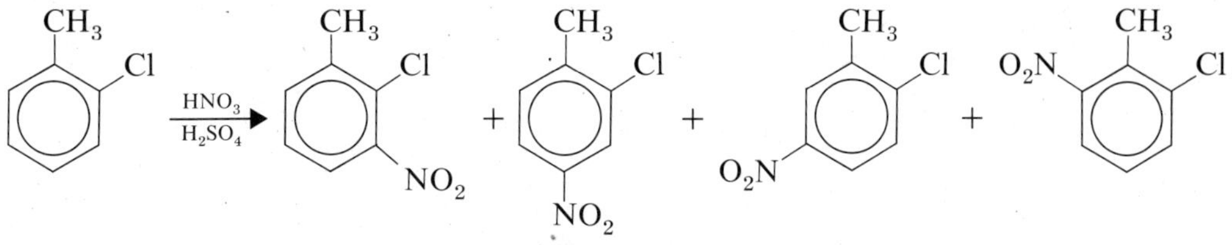

(19%)   (17%)   (43%)   (21%)

One exception to this pattern occurs in compounds that have an acetamido group and a methoxyl group. Even though both groups are moderately activating, the acetamido group apparently has a much greater directive influence.

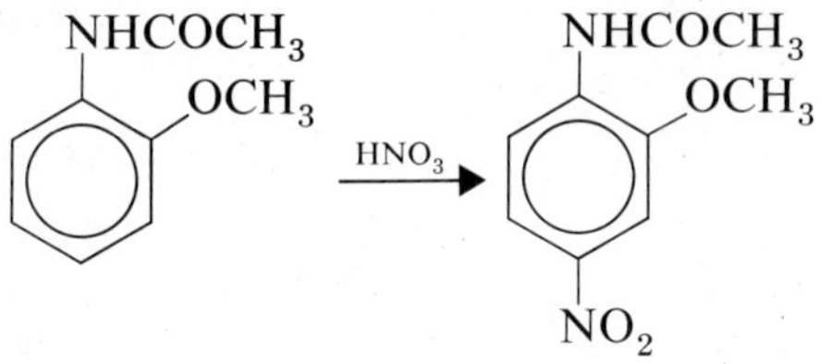

(major product)

Steric effects are also important in aromatic substitutions. We saw one example in the nitration of *tert*-butylbenzene, where we found that substitution ortho to the bulky *tert*-butyl group is inhibited. Another example of a steric effect is in the inhibition of substitution between meta substituents. *Substitution does not occur to an appreciable extent between meta substituents if another position is open.* A good example of this effect can be seen in the nitration of *m*-bromochlorobenzene.

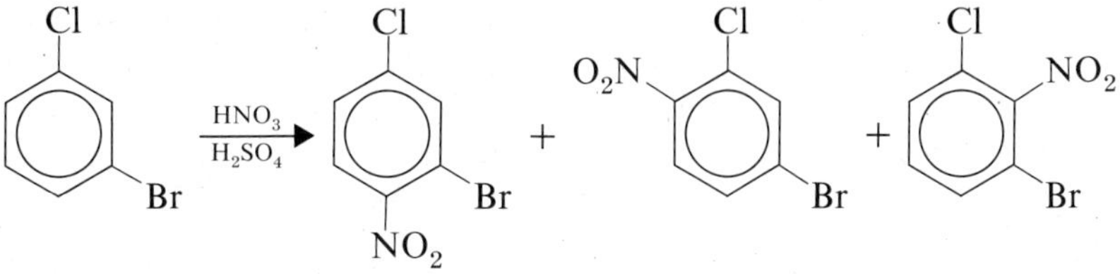

(62%)   (37%)   (1%)

Only 1% of the mononitro product obtained has the nitro group between the bromine and chlorine.

Problem 13.13

Predict the major product (or products) that would be obtained when each of the following compounds is nitrated.

(a) 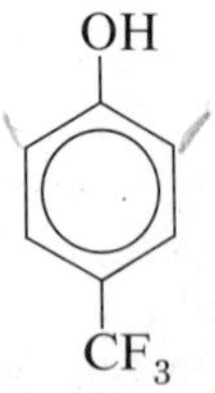

(b) 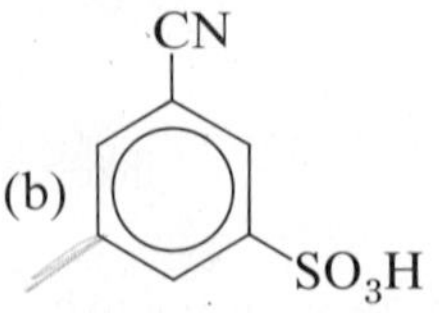

(c) 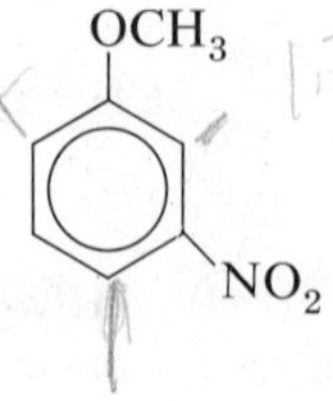

(d) 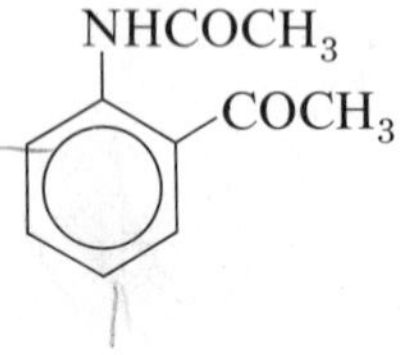

(e) 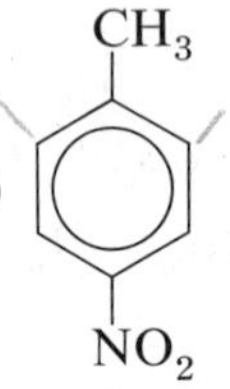

(f) 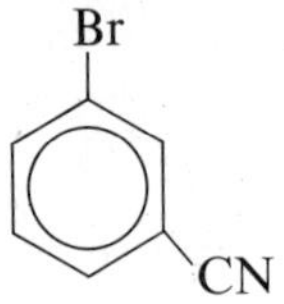

(g) 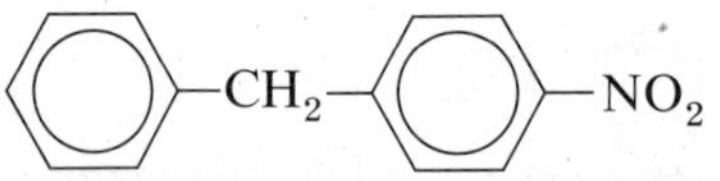

13.9 THEORY OF ELECTROPHILIC AROMATIC SUBSTITUTION

Theory of Reactivity

We have now seen that certain groups *activate* the benzene ring toward electrophilic substitution, while other groups *deactivate* the ring. When we say that a group activates the ring, what we mean, of course, is that the group increases the relative rate of the reaction. We mean that an aromatic compound with an activating group reacts faster in electrophilic substitutions than benzene does. When we say that a group deactivates the ring, we mean that an aromatic compound with a deactivating group reacts slower than benzene does.

We have also seen that we can account for relative reaction rates by examining the transition states for the rate-limiting steps of the reactions involved. We know that any factor that increases the energy of the transition state relative to that of the reactants decreases the relative rate of the reaction. It does this because it increases the energy of activation of the reaction. In the same way, any factor that decreases the energy of the transition state relative to that of the reactants, lowers the energy of activation, and increases the relative rate of the reaction.

The rate-limiting step in most electrophilic substitution reactions of substituted benzenes is the step that results in the formation of the σ complex. We can write the formula for a substituted benzene in a generalized way if we use the letter **S** to represent any ring substituent including hydrogen. (If **S** is hydrogen the compound is benzene itself.)

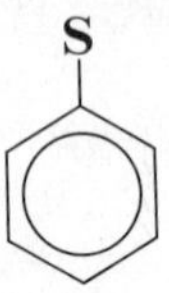

We can write the structure for the σ complex in the way shown below. By this formula we mean that **S** can be in any position—ortho, meta, or para—relative to the electrophile, E.

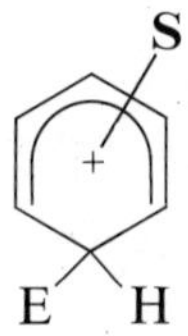

Using these conventions, then, we are able to write the rate-determining step for electrophilic aromatic substitution in the following general way.

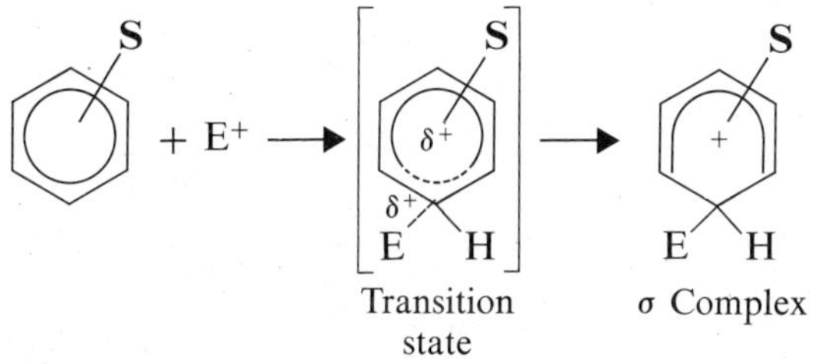

When we examine this step for a large number of reactions we find that the relative rates of the reactions depend on whether **S** *withdraws* or *releases* electrons. If **S** is an electron-releasing group (relative to hydrogen) the reaction occurs faster than the corresponding reaction of benzene. If **S** is an electron-withdrawing group, the reaction is slower than that of benzene.

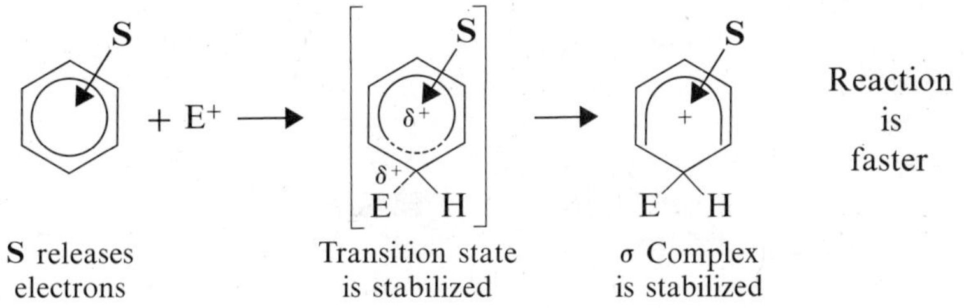

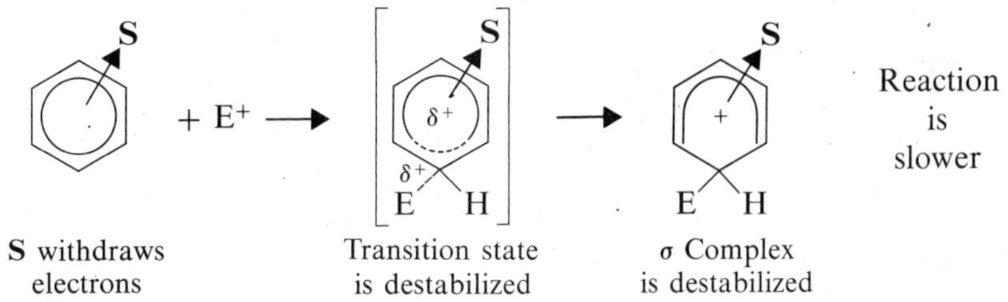

It appears then, that the substituent, **S**, must affect the stability of the transition state relative to that of the reactants. Electron releasing groups apparently make the transition state more stable, while electron-withdrawing groups make it less stable. That this is so, is entirely reasonable, for the transition states resemble the σ complex, and the σ complex is a delocalized *carbocation*.

Since the σ complex is positively charged, we would expect an electron-releasing group to stabilize it *and the transition state leading to the σ complex*, for

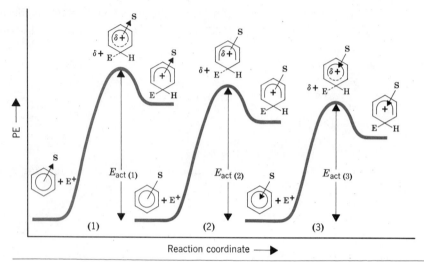

FIG. 13.4

*Energy profiles for the formation of the σ complex in three electrophilic aromatic substitution reactions. In (1), **S** is an electron-withdrawing group. In (2) **S** = H. In (3) **S** is an electron-releasing group.* $E_{act(1)} > E_{act(2)} > E_{act(3)}$.

the transition state is a developing delocalized carbocation. We can make the same kind of arguments about the effect of electron-withdrawing groups. An electron-withdrawing group should make the σ complex *less* stable and in a corresponding way it should make the transition state leading to the σ complex less stable.

Figure 13.4 shows how the electron-withdrawing abilities of substituents affect the relative energies of activation of electrophilic aromatic substitution reactions.

Inductive and Resonance Effects: Theory of Orientation

We can account for the electron-withdrawing and electron-releasing properties of groups on the basis of two factors: *an inductive effect and a resonance effect.* We will also see that these two factors determine orientation in aromatic substitution reactions.

An inductive effect is an effect that results from electronegativity differences between two groups. Inductive effects are important in accounting for a great many phenomena in organic chemistry. In Chapter 19 we will see how inductive influences affect the strengths of acids. Another example of an inductive effect—one that we will soon be able to relate to orientation in electrophilic aromatic substitution reactions—occurs in the addition of hydrogen chloride to trifluoromethylethylene (3,3,3-trifluoropropene). Hydrogen chloride adds to 3,3,3-trifluoropropene more slowly than to propene itself and the addition takes place in the manner shown below.

$$CF_3CH=CH_2 \xrightarrow{HCl} CF_3CH_2CH_2Cl$$

At first inspection this addition appears to occur in an anti-Markovnikov fashion. If we examine the two possible carbocations that could be formed by addition of a proton to trifluoromethylethylene we will see that the addition is consistent with a modern statement of Markovnikov's rule. *The reaction proceeds through the formation of the more stable carbocation.*

$$CF_3 \leftarrow CH=CH_2 \xrightarrow{H^+} \begin{cases} \text{(1)} \quad \cancel{\times} \rightarrow CF_3 \leftarrow \underset{+}{CH}-CH_3 \quad \text{(not formed)} \\ \\ \text{(2)} \quad \rightarrow CF_3 \leftarrow CH_2-\underset{+}{CH_2} \quad \text{(more stable)} \end{cases}$$

The trifluoromethyl group, because of the three highly electronegative fluorines, is strongly electron withdrawing. The carbocation that would be formed by path 1 would be highly unstable because even though it is a secondary carbocation, the positive charge is on the carbon adjacent to the highly electronegative trifluoromethyl group. Consequently, this carbocation does not form, and the reaction follows path 2. The carbocation formed by path 2 is more stable even though it is a primary carbocation, because the positive charge is separated from the trifluoromethyl group by an intervening —CH$_2$— group, and inductive effects are not transmitted very effectively through σ bonds.

Meta-Directing Groups

The trifluoromethyl group is a strong deactivating group and a powerful meta director in electrophilic aromatic substitution reactions. We can account for both of these characteristics of the trifluoromethyl group on the basis of its inductive effect.

The trifluoromethyl group affects reactivity by causing the transition state leading to the σ complex to be highly unstable. It does this by withdrawing electrons from the developing carbocation.

Trifluoromethylbenzene   Transition state σ Complex

We can understand how the trifluoromethyl group affects orientation in electrophilic aromatic substitution if we examine the resonance structures for the σ complex that would be formed when an electrophile attacks the ortho, meta, and para positions of trifluoromethylbenzene.

Ortho attack:

Highly unstable

Meta attack:

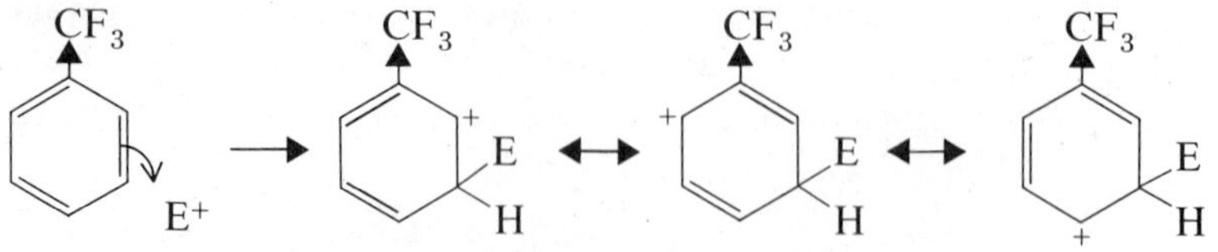

Para attack:

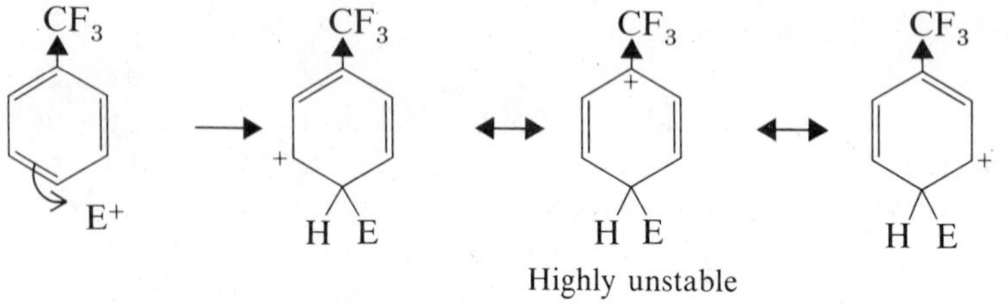

Highly unstable

We see in the resonance structures for the σ complex arising from ortho and para attack, that *one contributing structure is highly unstable because the positive charge is located on the ring carbon that bears the electron-withdrawing group.* We see *no* such highly unstable resonance structure in the σ complex arising from *meta* attack. This means that the σ complex formed by meta attack should be the most stable of the three. By the usual reasoning we would also expect the transition state leading to the meta σ complex to be the most stable and, therefore, that meta attack would be favored. This is exactly what we find experimentally. The trifluoromethyl group is a powerful meta director.

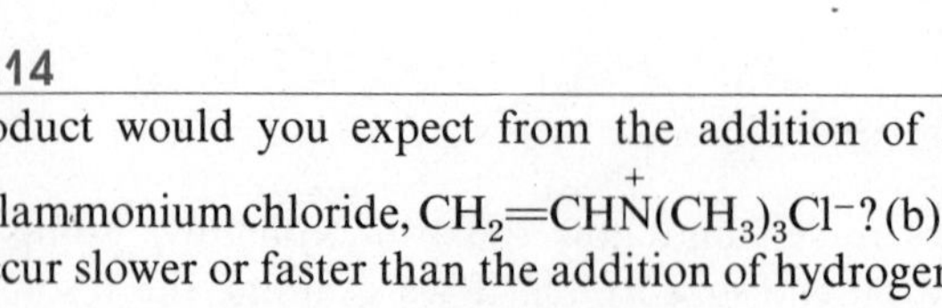

Trifluoromethylbenzene
(benzotrifluoride)

(100%)

Problem 13.14

(a) What product would you expect from the addition of hydrogen chloride to vinyltrimethylammonium chloride, $CH_2{=}CH\overset{+}{N}(CH_3)_3Cl^-$? (b) Would you expect this addition to occur slower or faster than the addition of hydrogen chloride to propene?

(c) How do you account for the fact that the trimethylammonium group, $(CH_3)_3\overset{+}{N}-$, is a strong deactivating group in electrophilic aromatic substitution? (d) Write resonance structures that account for the fact that the trimethylammonium group is also a meta-directing group.

Problem 13.15

(a) When the following trimethylammonium compounds were subjected to nitration they gave the following percentages of meta product: **A,** 100%; **B,** 88%; **C,** 19%; **D,** 5%.

How can you account for these results?

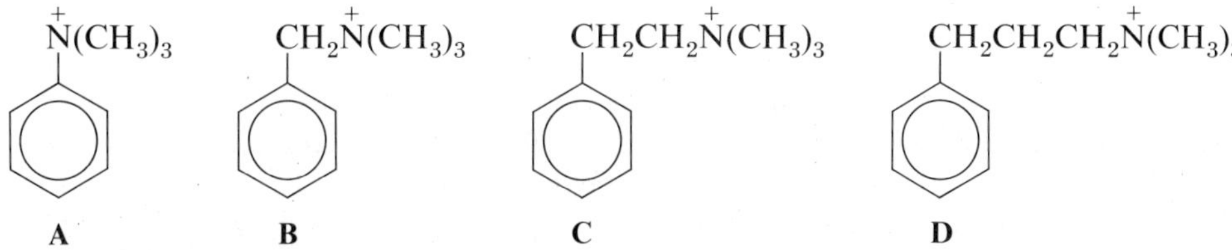

$\overset{+}{N}(CH_3)_3$ $CH_2\overset{+}{N}(CH_3)_3$ $CH_2CH_2\overset{+}{N}(CH_3)_3$ $CH_2CH_2CH_2\overset{+}{N}(CH_3)_3$

A   B   C   D

(b) How do you account for the percentages of meta product obtained from nitration of the following compounds?

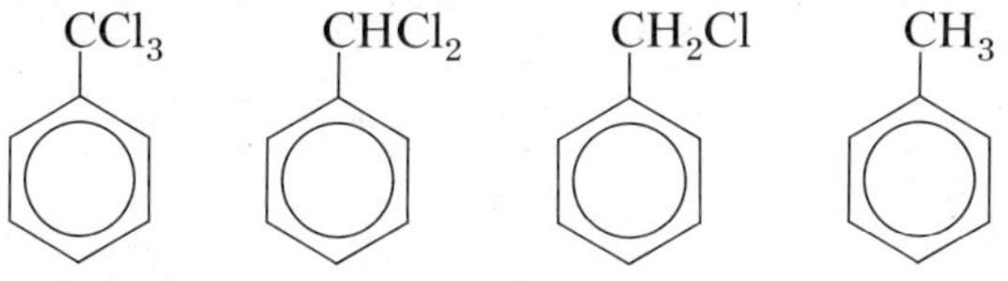

CCl_3 $CHCl_2$ CH_2Cl CH_3

| Benzotri-chloride (64%) | Benzal-dichloride (34%) | Benzyl chloride (14%) | Toluene (4%) | % meta product |

The nitro group is another powerful electron-withdrawing group. The nitro group withdraws electrons through an inductive effect *and a resonance effect*. These two effects combine to cause a large deactivation of the ring.

The inductive effect of the nitro group arises from three electronegative atoms—a nitrogen and two oxygens. The electronegativities of these atoms combine to make the nitro group as a whole strongly electronegative. The inductive effect of the nitro group increases the energy of the transition state leading to σ complex by withdrawing electrons from it. This accounts, in part, for the deactivating effect of the nitro group.

In order to see how the nitro group asserts its resonance effect and how this affects the energy of activation in electrophilic aromatic substitution, we need to examine some of the resonance structures for nitrobenzene, itself, and then some of the resonance structures for the σ complex.

Resonance structures for nitrobenzene

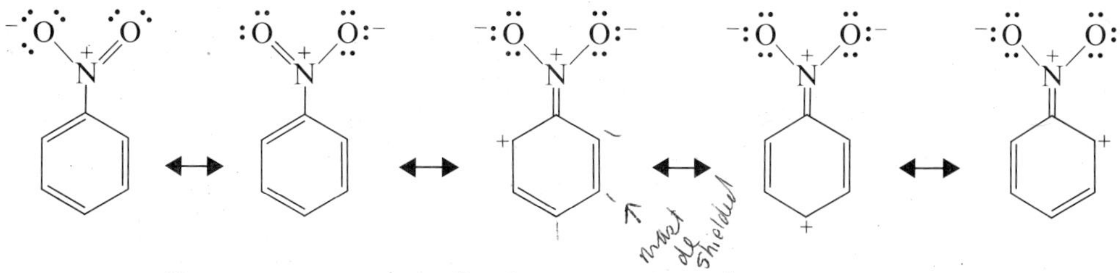

Resonance structures for the meta σ complex

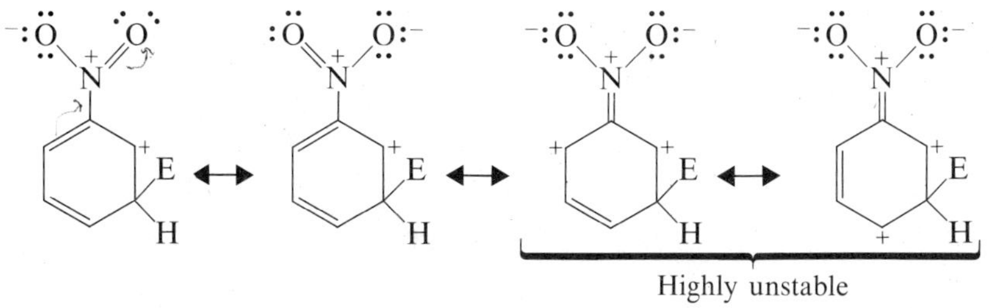

Highly unstable

When we examine the last three resonance structures for nitrobenzene, itself, we see that the nitro group withdraws electrons from the benzene ring. In each of these three structures an ortho or para carbon of the ring bears a positive charge. We know that these structures make a substantial contribution to the overall hybrid of nitrobenzene because nitrobenzene has an unusually high dipole moment (3.95 D). Another important effect of these structures is one that bears directly on nitrobenzene's reactivity in electrophilic substitution: the last three structures stabilize nitrobenzene. That is, they reduce its potential energy relative to that of benzene.

When we examine the last two structures for the σ complex we see that they correspond to the last three structures for nitrobenzene in the sense that in them electrons are withdrawn from the ring. These structures make only a *small contribution* to the hybrid for the σ complex, however, because *they involve electron withdrawal from a ring that is already positively charged*. As a consequence, resonance stabilization *is also of little importance in the transition state leading to the σ complex*.

Thus, the resonance effect and the inductive effect of the nitro group combine to make the energy of activation for an electrophilic attack on nitrobenzene much greater than that for benzene (Fig. 13.5). The resonance effect lowers the potential energy of the reactants; the inductive effect raises the potential energy of the transition state. Thus, we can understand why the electrophilic substitution of nitrobenzene occurs at a rate that is approximately 10,000 times slower than that of benzene.

We can account for the meta-directing property of the nitro group in much the same way as we did for trifluoromethylbenzene. Since the nitro group is an electron withdrawing group we can write resonance structures (below) for the σ complexes formed by ortho, meta, and para attack. When we do, we find that the

Ortho attack:

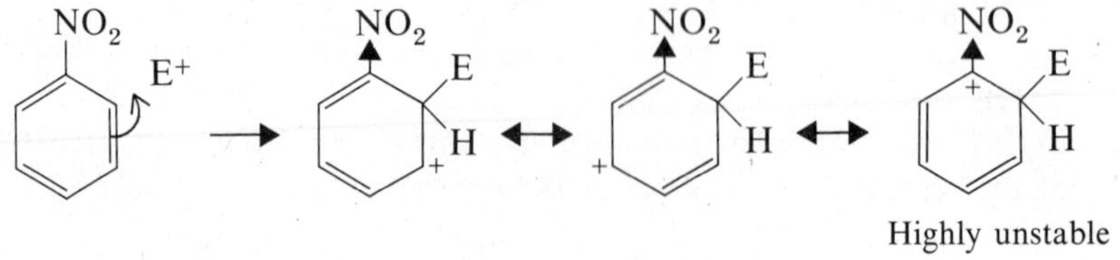

Highly unstable

Meta attack:

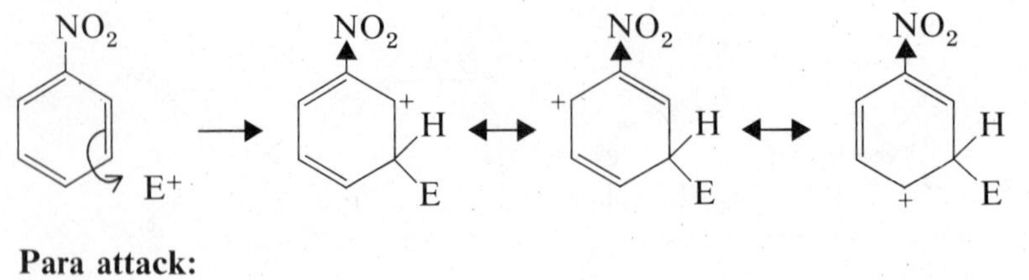

Para attack:

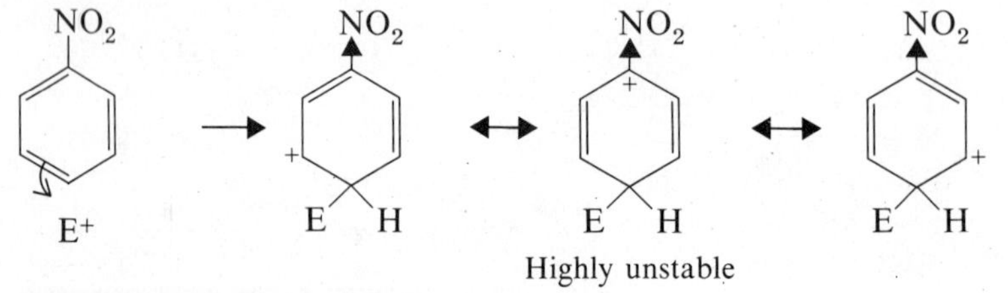

Highly unstable

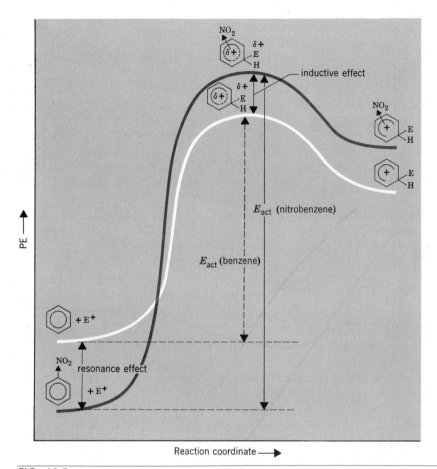

FIG. 13.5

Resonance and inductive effects combine to make the energy of activation for an electrophilic attack on nitrobenzene substantially greater than that for an electrophilic attack on benzene.

meta σ complex is most stable of the three because no highly unstable structure contributes to its hybrid. Thus, the transition state leading to the meta σ complex occurs at a lower potential energy than those for ortho and para attack, and meta substitution is favored. Meta substitution is favored, however, only in the sense that it is the least unfavorable of three unfavorable pathways.

The resonance explanation that we have given for nitrobenzene and trifluoromethylbenzene can be given for the directive influence of all meta-directing groups. All meta-directing groups are electron withdrawing; therefore, they all destabilize the transition states leading to ortho and para substitution to a greater extent than they destabilize the transition state leading to meta substitution.

Resonance effects similar to those we saw in nitrobenzene also play a part in determining the reactivity of other aromatic compounds. Greater resonance stabilization of the reactant relative to that of the transition state leading to the σ complex will be found for all those compounds containing meta directing groups of the following type.

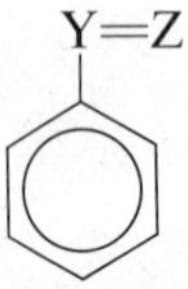

Y=Z

Z is an electronegative
element

All of these groups make stabilizing resonance contributions to the hybrid of the reactant like those that follow:

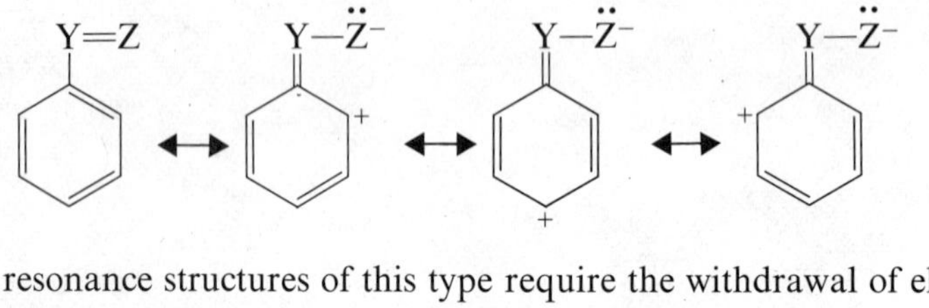

Since resonance structures of this type require the withdrawal of electrons from the ring, *similar stabilizing structures do not make significant contributions to the developing delocalized carbocation of the transition state*. This results in a large energy of activation for the electrophilic substitution reaction.

Problem 13.16

All of the following groups are of the type shown above: $-\overset{\overset{\displaystyle \ddot{O}:}{\|}}{C}-OH$, $-\overset{\overset{\displaystyle \ddot{O}:}{\|}}{C}-R$,

$-\overset{\overset{\displaystyle \ddot{O}:}{\|}}{C}-H$, $-C\equiv N:$. Write resonance structures that show how benzene compounds containing each of these groups would be stabilized by resonance.

Ortho-Para Directing Groups

Except for the alkyl and phenyl substituents, all of the ortho-para directing groups in Table 13.3 are of the general type shown below.

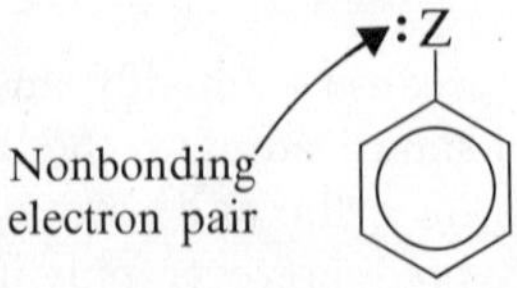

Nonbonding
electron pair :Z

All of these ortho-para directors have at least one pair of nonbonding electrons on the atom adjacent to the benzene ring. The following compounds illustrate this structural characteristic.

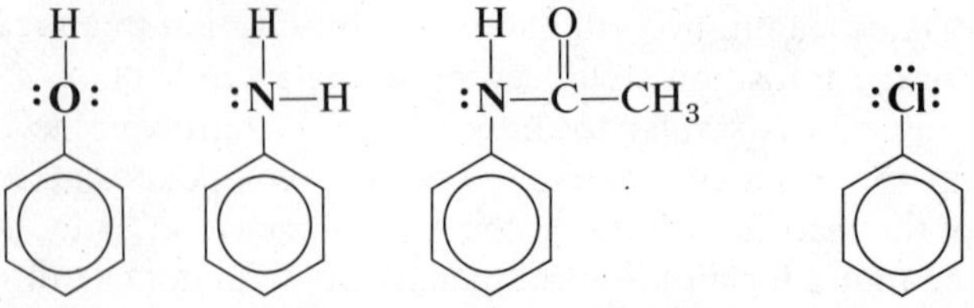

| Phenol | Aniline | Acetanilide | Chlorobenzene |

This structural feature—an unshared electron pair on the atom adjacent to the ring—allows us to understand how these groups determine the orientation and influence reactivity in electrophilic substitution reactions.

The directive *effect* of these groups is predominantly a resonance effect. This resonance effect, moreover, operates primarily in the σ complex and, consequently, in the transition state leading to it.

Except for the halogens, which represent a special case, the primary *reactivity* effect of these groups is also a resonance effect. And, again, this effect operates primarily in the transition state leading to the σ complex.

In order to understand these resonance effects let us begin by recalling the effect of the amino group on electrophilic aromatic substitution reactions. The amino group is not only a powerful activating group, it is also a powerful ortho-para director. We saw earlier (p. 467) that aniline reacts with bromine at room temperature and in the absence of a catalyst to yield a product in which both ortho positions and the para position are substituted.

The inductive effect of the amino group makes it slightly electron withdrawing. Nitrogen, as we know, is more electronegative than carbon. The difference between the electronegativities of nitrogen and carbon in aniline is not large, however, because the carbon of the benzene ring is sp^2 hybridized and thus somewhat electronegative itself.

The resonance effect of the amino group is far more important in electrophilic aromatic substitution, and this resonance effect makes the amino group electron releasing. We can understand this effect if we write the resonance structures for the aniline σ complexes that would arise from ortho, meta and para attack.

Ortho attack:

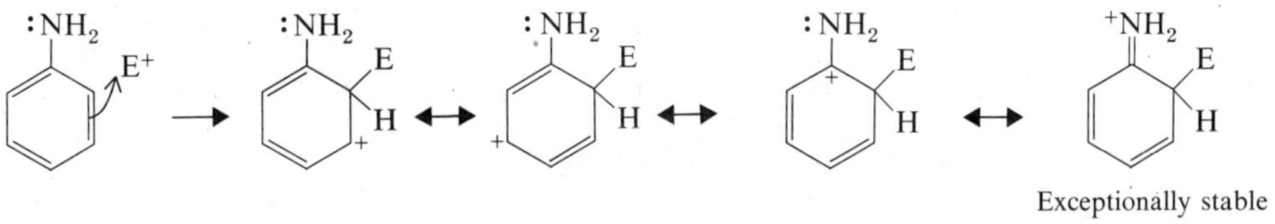

Exceptionally stable

Meta attack:

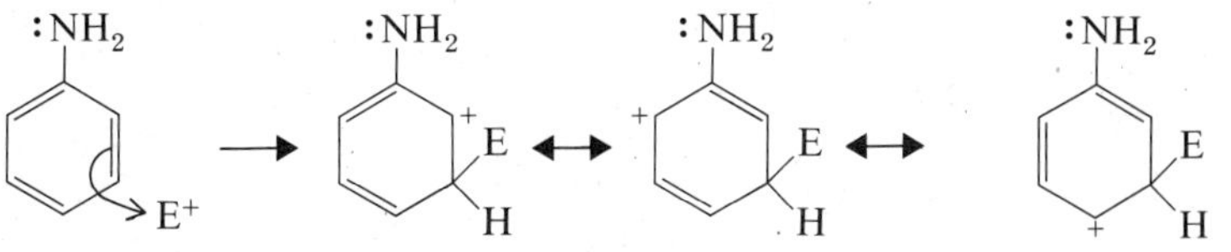

Para attack:

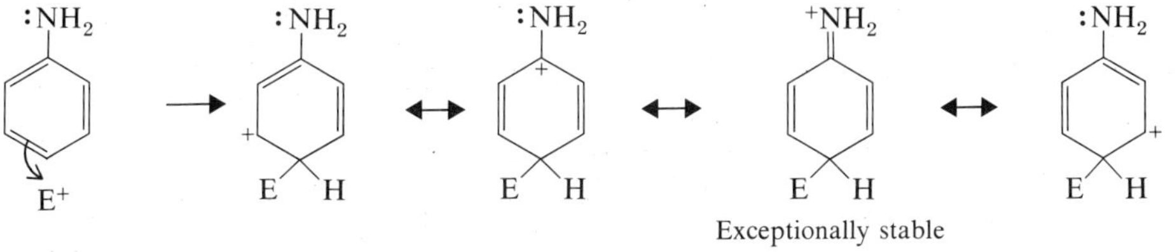

Exceptionally stable

We see that four resonance structures can be written for the σ complexes resulting from ortho and para attack, whereas only three can be written for the σ complex that results from meta attack. This, in itself, suggests that the ortho and para σ complexes should be more stable. Of greater importance, however, are the exceptionally stable structures that contribute to the hybrid for the ortho and para σ complexes. In exceptionally stable structures, nonbonding pairs of electrons from nitrogen form an extra bond to the carbon of the ring. This extra bond—and the fact that every atom in each of these structures has a complete valence shell of electrons—makes these structures unusually stable. Because these structures are exceptionally stable, they make a large—*and stabilizing*—contribution to the hybrid. This means, of course, that the ortho and para σ complexes themselves are considerably more stable than the σ complex that results from meta attack. The transition states leading to the ortho and para σ complexes occur at unusually low potential energies. As a result, electrophiles react at the ortho and para positions very rapidly.

Problem 13.17

(a) Write resonance structures for the σ complexes that would result from electrophilic attack on the ortho, meta, and para positions of phenol. (b) Can you account for the fact that phenol is highly susceptible to electrophilic attack? (c) Can you account for the fact that the hydroxyl group is an ortho and para director? (d) Would you expect the phenoxide ion, C_6H_5—O^-, to be more or less reactive than phenol in electrophilic substitution? (e) Explain.

Problem 13.18

(a) Ignore resonance structures involving electrons of the ring and write *one* other resonance structure for acetanilide. (Your structure will contain + and − charges.)

Acetanilide

(b) Acetanilide is less reactive toward electrophilic substitution than aniline. How can you explain this on the basis of the resonance structure you have just written? (c) Acetanilide, however, is much more reactive than benzene and the acetamido group is an ortho-para director. Can you account for these facts in terms of resonance structures that involve the ring? (d) What relative reactivity would you expect phenyl acetate to show, that is, would it be *more* or *less* reactive than phenol? Explain.

Phenyl acetate

(e) What kind of directional influence would you expect the acetoxy group,

$CH_3\overset{\displaystyle \overset{..}{\underset{..}{O}}}{\underset{\displaystyle \|}{C}}-\overset{..}{\underset{..}{O}}-$, to show? (f) Would you expect phenyl acetate to be *more* or *less* reactive in electrophilic substitution than benzene? Explain.

The directive and reactivity effects of halo substituents are, in a sense, contradictory. The halo groups are the only ortho-para directors that are also deactivating groups. All other deactivating groups (Table 13.3) are meta directors.

We find that we can account for this peculiar behavior of halo substituents if we assume that their inductive effect influences reactivity and their resonance effect governs orientation.

Let us apply this kind of analysis specifically to chlorobenzene. The chloro group is highly electronegative. Thus, we would expect a chloro group to withdraw electrons from the benzene ring and thereby deactivate it.

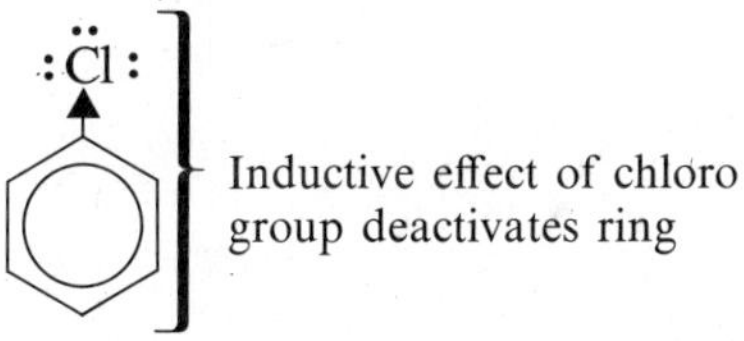

Inductive effect of chloro group deactivates ring

On the other hand, when electrophilic attack does take place, the chloro group stabilizes the σ complexes resulting from ortho and para attack. The chloro group does this in the same way as amino groups and hydroxyl groups do—*by donating*

Ortho attack:

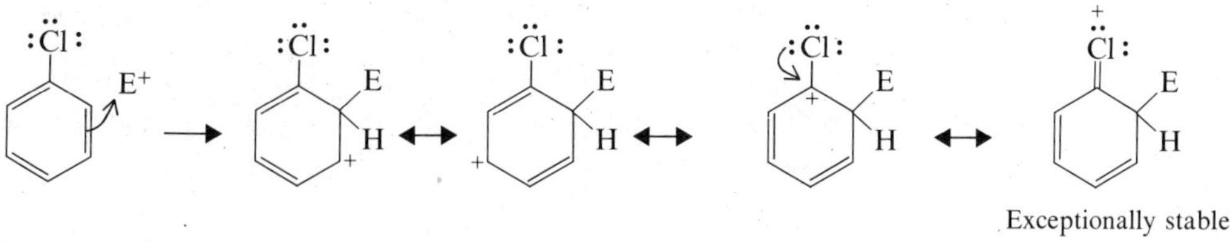

Exceptionally stable

Meta attack:

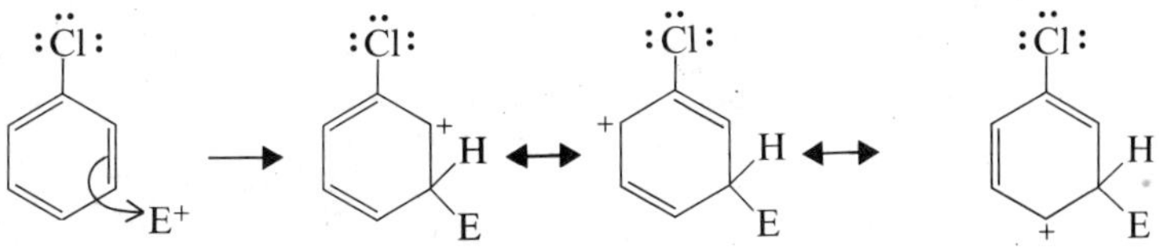

Para attack:

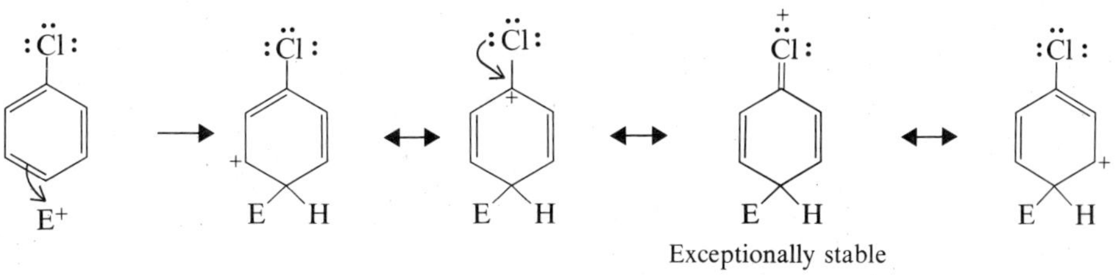

Exceptionally stable

a pair of nonbonding electrons. These nonbonding electrons give rise to exceptionally stable resonance structures in the hybrids for the ortho and para σ complexes.

What we have said about chlorobenzene is, of course, true of bromobenzene.

We can summarize the inductive and resonance effects of halo substituents in the following way. Through their inductive effect halo groups make the ring more positive than that of benzene. This causes the energy of activation for any electrophilic aromatic substitution reaction to be greater than that for benzene, and, therefore, halo groups are deactivating. Through their resonance effect, however, halo substituents cause the energies of activation leading to ortho and para substitution to be lower than the energy of activation leading to meta substitution. This makes halo substituents ortho-para directors.

Problem 13.19

The trifluoromethyl group and the chloro group both withdraw electrons inductively, and both trifluoromethylethene and chloroethene add hydrogen chloride more slowly than ethene does. The mode of addition of hydrogen chloride to chloroethene, however, is opposite that of 3,3,3-trifluoropropene. How can you account for this?

$$CF_3CH=CH_2 \xrightarrow{\text{HCl}} CF_3\underset{\underset{\text{H}}{|}}{C}H-\underset{\underset{\text{Cl}}{|}}{C}H_2$$

$$ClCH=CH_2 \xrightarrow{\text{HCl}} Cl-\underset{\underset{\text{Cl}}{|}}{C}H-\underset{\underset{\text{H}}{|}}{C}H_2$$

You may have noticed an apparent contradiction between the rationale offered for the unusual effects of the halogens and that offered earlier for amino or hydroxyl groups. That is, oxygen is *more* electronegative than chlorine or bromine (and especially iodine). Yet, the hydroxyl group is an activating group while these halogens are deactivating groups. An explanation for this can be obtained if we consider the relative stabilizing contributions made to the transition state leading to the σ complex by resonance structures involving a group $-\ddot{Z}$ ($-\ddot{Z} =$ $-\ddot{N}H_2$, $-\ddot{O}-H$, $-\ddot{F}:$, $-\ddot{C}l:$, $-\ddot{B}r:$, $-\ddot{I}:$) that is directly attached to the benzene ring in which $\ddot{Z}$ donates an electron pair. If $-\ddot{Z}$ is $-\ddot{O}H$ or $-\ddot{N}H_2$, these resonance structures arise from the overlap of a $2p$ orbital of carbon with that of oxygen or nitrogen. Such overlap is favorable because the atoms are almost the same size. With chlorine, however, donation of an electron pair to the benzene ring requires overlap of a carbon $2p$ orbital with a chlorine $3p$ orbital. Such overlap is less effective; the chlorine atom is much larger and its $3p$ orbital is much further from its nucleus. With bromine and iodine overlap is even less effective. Justification for this explanation can be found in the observation that fluorobenzene ($Z = -\ddot{F}:$) is the most reactive halobenzene in spite of the high electronegativity of fluorine and that $-\ddot{F}:$ is the most powerful ortho-para director of the halogens. With fluorine, donation of an electron pair arises from overlap of a $2p$ orbital of fluorine with a $2p$ orbital of carbon (as with $-\ddot{N}H_2$ and $-\ddot{O}H$). This overlap is effective because the atoms $=\overset{|}{C}-$ and $-F$ are of the same relative size.

Reactivity and Orientation of Alkylbenzenes

Alkyl groups can be much better electron-releasing groups than hydrogen. Because of this they can activate a benzene ring toward electrophilic substitution by stabilizing the transition state leading to the σ complex:

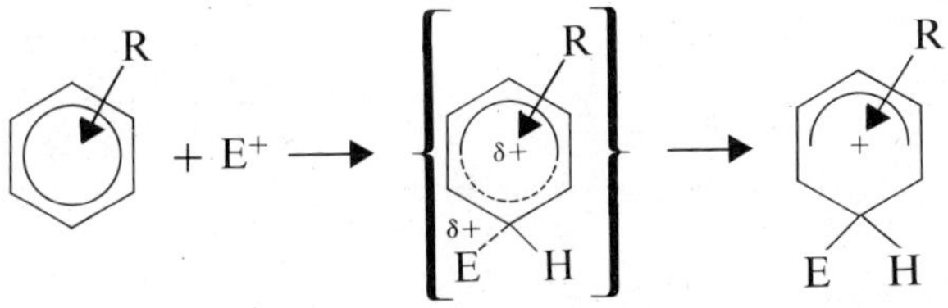

Transition state   σ Complex
is stabilized   is stabilized

For an alkylbenzene the energy of activation of the step leading to the σ complex (above) is lower than that for benzene, and alkylbenzenes react faster.

Alkyl groups are ortho-para directors. We can also account for this property of alkyl groups on the basis of their ability to release electrons—an effect that is particularly important when the alkyl group is attached directly to a carbon that bears a positive charge. (Recall the ability of alkyl groups to stabilize carbocations that we discussed in Section 5.9.)

If, for example, we write resonance structures for the σ complexes formed when toluene undergoes electrophilic substitution we get the following result:

Ortho attack:

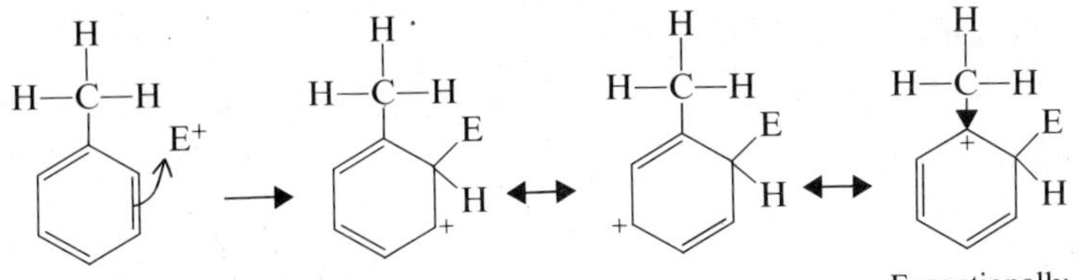

Exceptionally
stable

Meta attack:

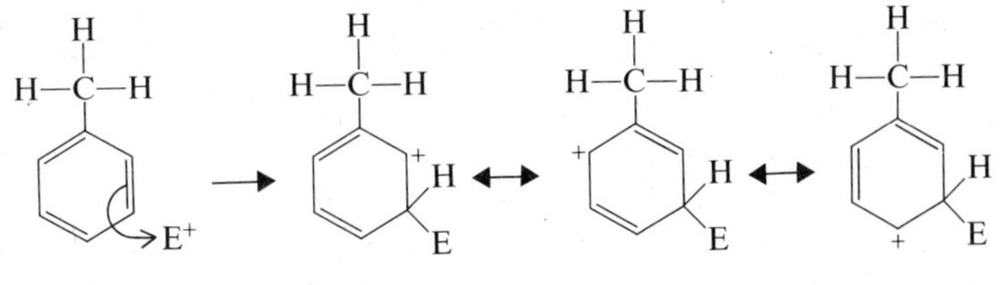

Para attack:

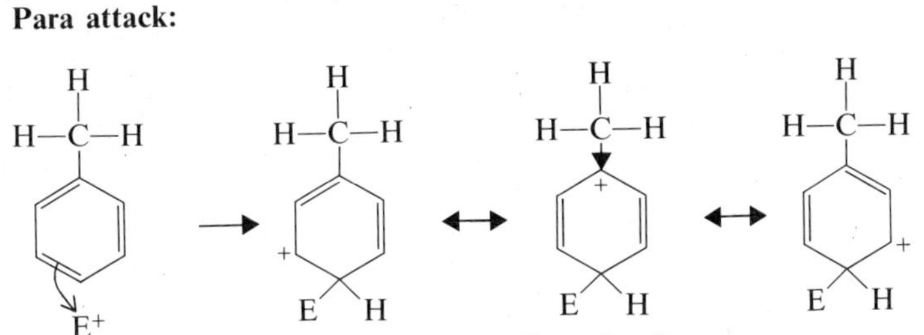

Exceptionally
stable

In ortho attack and para attack we find that we can write resonance structures in which the methyl group is directly attached to a positively charged carbon of the ring. These structures are exceptionally *stable* because in them the stabilizing influence of the methyl group (by electron release) is most effective. These structures, therefore, make a large (stabilizing) contribution to the overall hybrid for ortho and para σ complexes. No such exceptionally stable structure contributes to the hybrid for the meta σ complex and as a result, it is less stable than the ortho or para σ complex. Since the ortho and para σ complexes are more stable, the transition states leading to them occur at lower energy and ortho and para substitution takes place most rapidly.

Problem 13.20

(a) Write resonance structures for the σ complexes formed when ethyl benzene undergoes electrophilic attack. (b) Do these structures account for the fact that the ethyl group is an ortho-para director? (c) How can you account for the fact that the ethyl group is an activating group?

Problem 13.21

Resonance structures can also be used to account for the fact that phenyl group is an ortho-para director and that it is an activating group. Show how this is possible.

13.10 ARENES: REACTIONS OF THE SIDE CHAIN OF ALKYLBENZENES

Hydrocarbons that consist of both aliphatic and aromatic groups are known as *arenes*. Toluene, ethylbenzene, isopropylbenzene, and styrene are all arenes.

| Toluene | Ethylbenzene | Isopropyl benzene | Styrene (phenylethene or vinylbenzene) |

We can think of arenes in two ways. We can see these compounds as benzene compounds with alkyl or alkenyl substituents or we can see them as alkanes and alkenes with phenyl substituents. In Section 13.9 we examined the reactions that take place in the ring of alkylbenzenes. When we did this we found it was convenient to treat them as alkyl substituted benzenes. In doing so we found that the alkyl group had a distinct effect on both the reactivity and orientation of the substitution reactions that take place in the ring.

We will now examine some of the reactions that take place in the side chain of alkylbenzenes. Here it will be convenient to treat these compounds as phenyl substituted alkanes. As we do this, we should not be surprised to find that phenyl substituents affect the reactivity and orientation of reactions that occur in the side chain.

Halogenation of the Side Chain

We have seen that bromine and chlorine replace hydrogens of the ring of toluene when the reaction takes place in the presence of a Lewis acid. In ring halogenations the electrophiles are *positive* chlorine or bromine ions or they are Lewis-acid complexes that have positive halogens. These positive electrophiles attack the π electrons of the benzene ring and aromatic substitution takes place.

Chlorine and bromine can also be made to replace hydrogens of the methyl group of toluene. Side-chain halogenation takes place when the reaction is carried out *in the absence of Lewis acids* and under conditions that favor the formation of free radicals. When toluene reacts with *N*-bromosuccinimide in the presence of peroxides, for example, the major product is benzyl bromide. (*N*-bromosuccinimide furnishes a low concentration of Br_2 (cf. p. 344.)

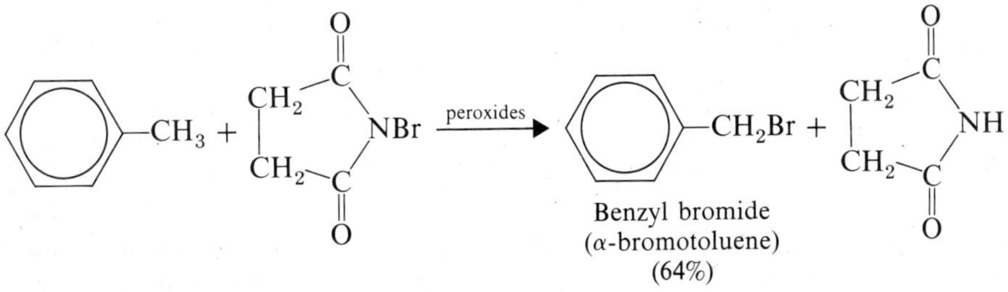

Benzyl bromide
(α-bromotoluene)
(64%)

Side-chain chlorination of toluene also takes place in the gas phase at 400 to 600° or in the presence of ultraviolet light. When an excess of chlorine is used multiple chlorinations of the side chain occur.

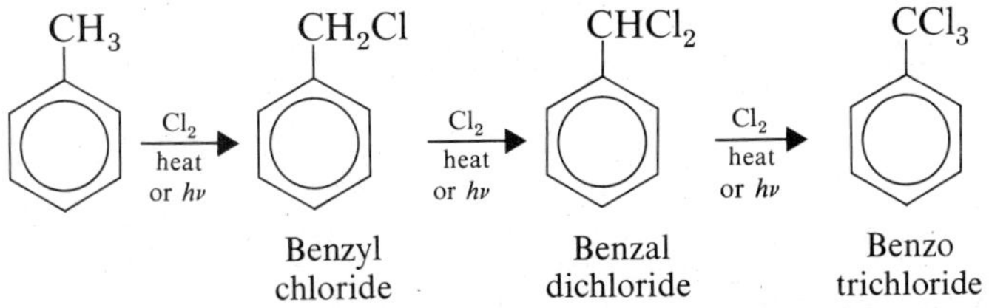

CH₃	CH₂Cl	CHCl₂	CCl₃
	Benzyl chloride	Benzal dichloride	Benzo trichloride

These halogenations take place through free-radical mechanisms like those we saw for alkanes in Section 4.9. The halogens dissociate to produce halogen atoms and then the halogen atoms initiate chains by abstracting hydrogens of the methyl group.

Chain initiating step

$$(1) \quad X_2 \xrightarrow[\text{or } h\nu]{\substack{\text{peroxides,} \\ \text{heat}}} 2X \cdot$$

Chain propagating steps

$$(2) \quad C_6H_5CH_3 + X\cdot \longrightarrow \underset{\substack{\text{Benzyl} \\ \text{radical}}}{C_6H_5CH_2\cdot} + HX$$

$$(3) \quad \underset{\substack{\text{Benzyl} \\ \text{radical}}}{C_6H_5CH_2\cdot} + X_2 \longrightarrow \underset{\substack{\text{Benzyl} \\ \text{halide}}}{C_6H_5CH_2X} + X\cdot$$

Abstraction of a hydrogen from the methyl group of toluene produces *a benzyl radical*. The benzyl radical then reacts with a halogen molecule to produce a benzyl halide and a halogen atom. The halogen atom then brings about a repetition of step 2, then step 3 occurs again, and so on.

The name benzyl radical is used not only as a specific name for the radical produced in the reaction above, it is also used as a general name (benzylic radical) for all radicals that have an unpaired electron on the side chain carbon that is directly attached to the benzene ring. The hydrogens of the carbon directly attached to the benzene ring are called benzylic hydrogens.

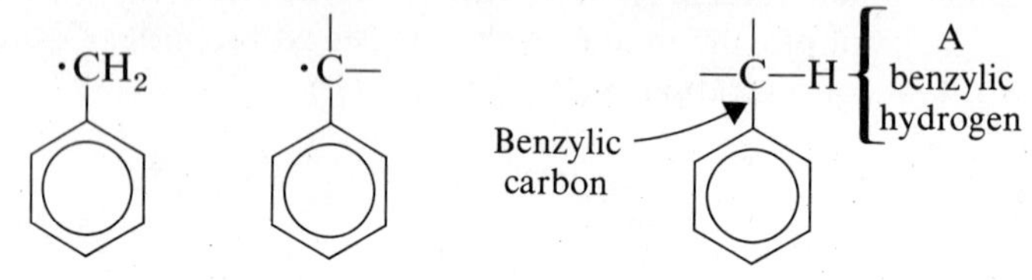

The benzyl radical   A benzylic radical

Benzylic halogenations are similar to allylic halogenations in that they involve the formation of *unusually stable free radicals*. Benzyl and allyl radicals are even more stable than tertiary radicals. We can see this if we examine the bond dissociation energies given below.

$$C_6H_5CH_2{-}H \longrightarrow C_6H_5CH_2\cdot + H\cdot \qquad \Delta H = 85 \text{ kcal/mole}$$

Benzyl
radical

$$CH_2{=}CHCH_2{-}H \longrightarrow CH_2{=}CHCH_2\cdot + H\cdot \quad \Delta H = 85 \text{ kcal/mole}$$

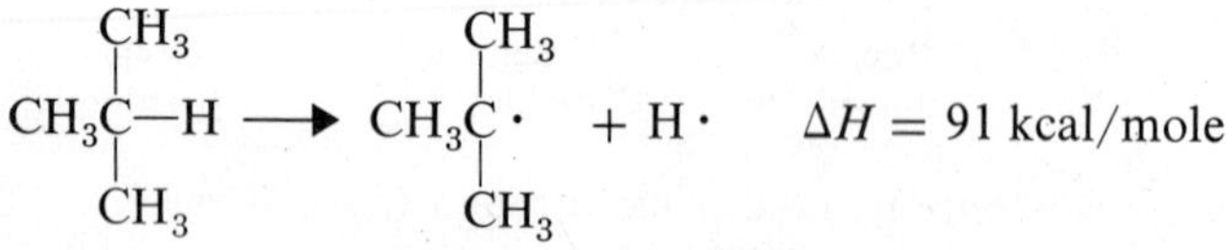

Allyl
radical

3° Radical

We can account for the unusual stability of the benzyl radical in much the same way that we did for the allyl radical: through resonance theory and molecular-orbital theory.

We can understand how resonance theory accounts for this stability of the benzyl radical by examining the following structures.

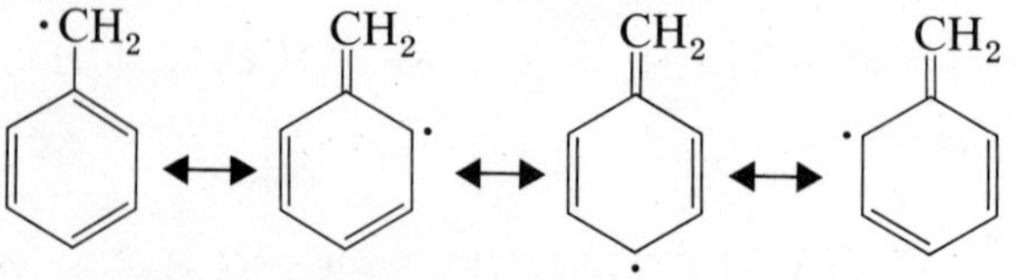

We see that four "reasonable" resonance structures contribute to the hybrid for the benzyl radical. Since all of these structures are "reasonable" structures, they all make stabilizing contributions to the hybrid. They also tell us that the unpaired electron is *delocalized* over the benzylic carbon and the ortho and para carbons of the ring.

In molecular-orbital theory the benzyl radical is treated by combining the wave functions for the *p* orbitals of the ring with the *p* orbital of the benzylic carbon.

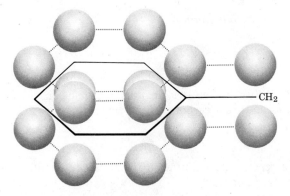

When this is done, calculations give energy levels for each of the π-molecular orbitals. These are shown in Fig. 13.6.

The molecular orbital that contains the unpaired electron is π_4. Orbital π_4 is nonbonding and the calculations show that it involves only the *p* orbitals of the benzylic carbon and the ortho and para carbons of the ring. It has nodes located at the other benzene carbons (Fig. 13.7).

FIG. 13.6

The energy levels of the π-molecular orbitals of the benzyl radical.

π_7 ———	
π_6 ———	antibonding molecular orbitals
π_5 ———	
π_4 ———	nonbonding molecular orbitals
π_3 ———	
π_2 ———	bonding molecular orbitals
π_1 ———	

E

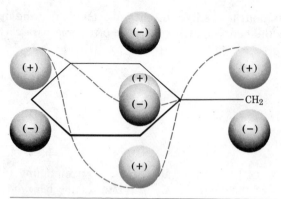

FIG. 13.7

The p *orbitals of the nonbonding orbital* (π_4) *of the benzyl radical.*

Thus, molecular orbital theory tells us the same thing that resonance theory does: the unpaired electron of the benzyl radical is delocalized over the benzylic carbon and the ortho and para carbons of the ring.

Problem 13.22

When ethylbenzene reacts with chlorine in the presence of ultraviolet irradiation the following results are obtained.

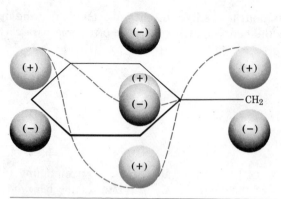

| Ethylbenzene | 1-Chloro-1-phenyl-ethane (91%) | 2-Chloro-1-phenylethane (9%) |

(a) How do you account for the fact that 1-chloro-1-phenylethane is the major product? (b) What kind of free radical is involved in the formation of 2-chloro-1-phenylethane? (c) What would you expect the major product to be if *n*-propylbenzene were subjected to a similar chlorination?

Problem 13.23

(a) Write resonance structures for the *benzyl cation* $C_6H_5CH_2^+$. (b) The benzyl cation is exceptionally stable; it is even more stable than a tertiary cation. Can you account for this? (c) Would you expect the positive charge of the benzyl cation to be delocalized? (d) If so, over which carbons? (e) Which molecular orbitals of the benzyl cation would be occupied? (f) What does molecular orbital theory tell you about the location of the positive charge?

Oxidation of the Side Chain

Strong oxidizing agents oxidize toluene to benzoic acid. The oxidation can be carried out by heating toluene with dilute nitric acid in a sealed tube, by the action of hot chromic acid, or by the action of hot alkaline potassium permanganate. The last method gives benzoic acid in almost quantitative yield.

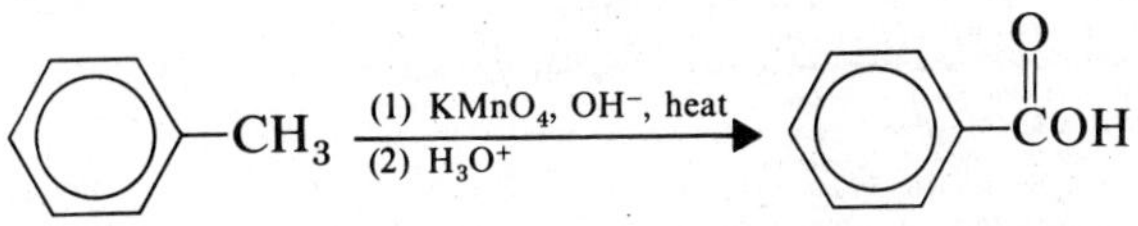

Benzoic acid
(~100%)

Other alkylbenzenes can be oxidized in the same way. Hot alkaline potassium permanganate oxidizes *o*-chlorotoluene to *o*-chlorobenzoic acid.

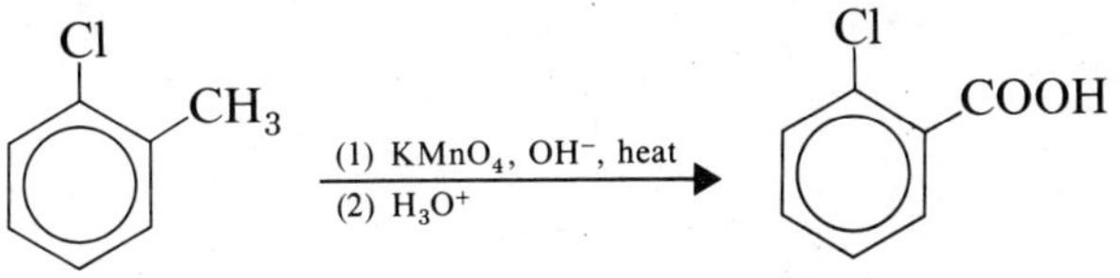

o-Chlorotoluene

o-Chlorobenzoic acid
(77%)

Hot nitric acid oxidizes one methyl group of *o*-xylene to give *o*-toluic acid.

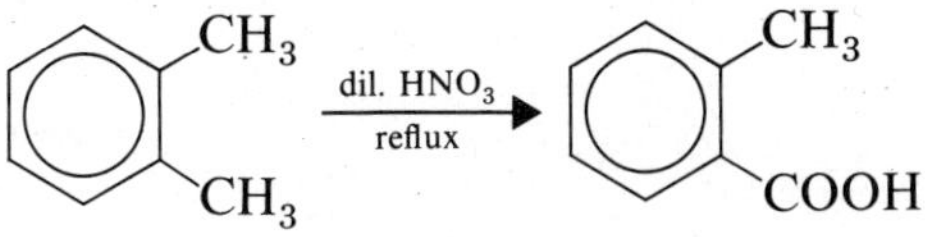

o-Toluic acid
(55%)

Hot nitric acid also selectively oxidizes the isopropyl group of *p*-isopropyltoluene to give *p*-toluic acid.

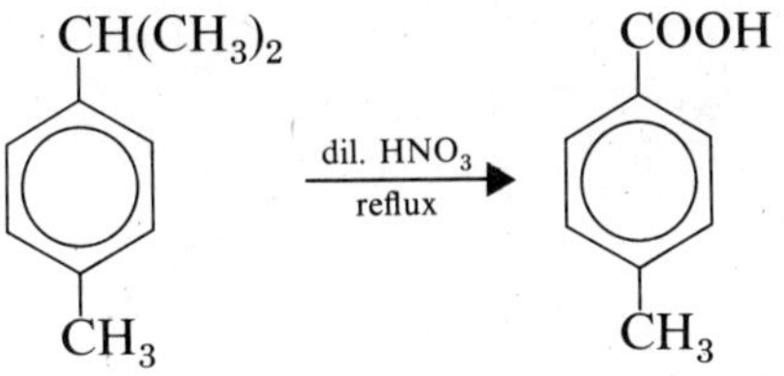

p-Toluic acid
(51%)

This last reaction illustrates another characteristic of side-chain oxidation: oxidation takes place at the benzylic carbon and, consequently, alkyl groups longer than methyl are degraded to benzoic acids.

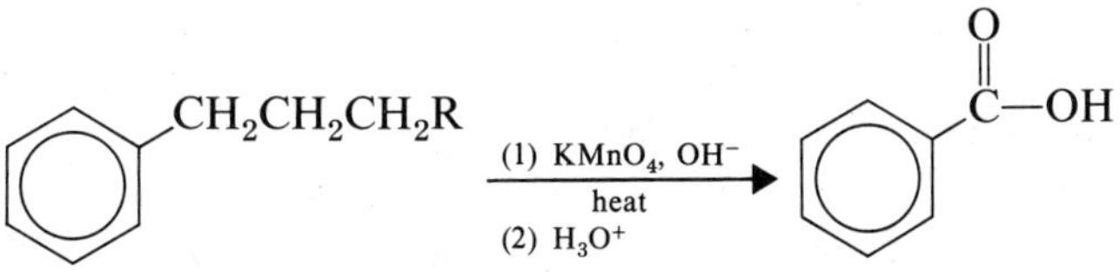

Alkylbenzene

Benzoic acid

Side-chain oxidations are similar to benzylic halogenations, for in the first step the oxidizing agent abstracts a benzylic hydrogen. Once oxidation is begun at the benzylic carbon, it continues at that site. Ultimately, the oxidizing agent oxidizes the benzylic carbon to a carboxyl group and, in the process, it scissors the remaining carbons of the side chain.

Benzene compounds with acyl side chains can also be oxidized to benzoic acids with strong oxidizing agents. Solutions of bromine in sodium hydroxide oxidize acetyl groups to carboxylic acids but do not affect alkyl groups that are present on the ring (cf. p. 721).

$$CH_3- \bigcirc -\overset{\overset{\displaystyle O}{\|}}{C}CH_3 \xrightarrow[\text{(2) } H_3O^+]{\text{(1) } Br_2, \text{ } OH^-} CH_3- \bigcirc -\overset{\overset{\displaystyle O}{\|}}{C}OH$$

p-Toluic acid

13.11 ALKENYLBENZENES: ADDITIONS TO THE DOUBLE BOND

The phenyl group of *conjugated* alkenylbenzenes affects the reactivity, orientation, and *stereochemistry* of reactions that take place at the double bond.

For example, bromine adds to styrene three times as fast as it does to ethene.

$$\bigcirc -CH{=}CH_2 \xrightarrow{Br_2} \bigcirc -\underset{\underset{\displaystyle Br}{|}}{C}H-\underset{\underset{\displaystyle Br}{|}}{C}H_2$$

Styrene

In the presence of peroxides, hydrogen bromide adds to the double bond of 1-phenylpropene to give 1-phenyl-2-bromopropane as the major product.

$$\bigcirc -CH{=}CHCH_3 \xrightarrow[\text{peroxides}]{HBr} \bigcirc -CH_2\underset{\underset{\displaystyle Br}{|}}{C}HCH_3$$

1-Phenylpropene   1-Phenyl-2-bromopropane

In the absence of peroxides, bromine adds in just the opposite way.

$$\bigcirc -CH{=}CHCH_3 \xrightarrow[\text{(no peroxides)}]{HBr} \bigcirc -\underset{\underset{\displaystyle Br}{|}}{C}HCH_2CH_3$$

1-Phenylpropene   1-Phenyl-1-bromopropane

Reactivity and orientation in all of these reactions are consequences of the unusual stabilities of benzylic cations (cf. problem 13.23) and benzylic radicals.

The addition of bromine to styrene appears to occur through a bromonium ion, *but one in which the benzylic carbon has considerable cationic character.*

$$C_6H_5-CH\!=\!CH_2 \xrightarrow{\ Br_2\ } C_6H_5-\overset{\delta+}{C}H\!-\!CH_2 + :\ddot{B}r:^- $$

$$\downarrow$$

$$C_6H_5-\underset{\underset{\textstyle Br}{|}}{C}H\!-\!CH_2Br$$

Because the bromonium ion resembles a very stable benzyl cation, bromination of styrene proceeds faster than the corresponding reaction of ethene.

The addition of hydrogen bromide to 1-phenylpropene proceeds through a benzylic radical in the presence of peroxides, and through a benzylic cation in their absence (cf. (1) and (2) below).

1. Hydrogen bromide addition in the presence of peroxides.

Chain-initiating steps
$$\begin{cases} R-O-O-R \longrightarrow 2R-O\cdot \\ RO\cdot + H-Br \longrightarrow R-O-H + Br\cdot \end{cases}$$

Chain-propagating steps
$$\begin{cases} Br\cdot + C_6H_5CH\!=\!CHCH_3 \longrightarrow C_6H_5\overset{\cdot}{C}H\!-\!\underset{\underset{\textstyle Br}{|}}{C}HCH_3 \\ \qquad\qquad\qquad\qquad\qquad\qquad\text{A benzyl radical} \\ \\ C_6H_5\overset{\cdot}{C}H\underset{\underset{\textstyle Br}{|}}{C}HCH_3 + H-Br \longrightarrow C_6H_5CH_2\underset{\underset{\textstyle Br}{|}}{C}HCH_3 + Br\cdot \\ \qquad\qquad\qquad\qquad\qquad\qquad\qquad\text{1-Phenyl-2-bromopropane} \end{cases}$$

The mechanism for the addition of hydrogen bromide to 1-phenylpropene in the presence of peroxides is a chain mechanism analogous to the one we discussed when we described anti-Markovnikov addition in Section 6.9. The step that determines the orientation of the reaction is the first chain-propagating step. Bromine attacks the second carbon of the chain because by doing so the reaction produces a more stable benzylic radical. Had the bromine atom attacked the double bond in the opposite way a less stable secondary radical would have been formed.

$$C_6H_5CH\!=\!CHCH_3 + Br\cdot \;/\!\!/\!\!\longrightarrow C_6H_5\underset{\underset{\textstyle Br}{|}}{C}H\!-\!\overset{\cdot}{C}HCH_3$$

A secondary radical

2. Hydrogen bromide addition in the absence of peroxides.

$$C_6H_5CH\!=\!CHCH_3 + HBr \longrightarrow C_6H_5\overset{+}{C}HCH_2CH_3 + Br^-$$

A benzylic cation

$$\downarrow$$

$$C_6H_5\underset{\underset{\textstyle Br}{|}}{C}HCH_2CH_3$$

In the absence of peroxides hydrogen bromide adds through an ionic mechanism. The step that determines the orientation in the ionic mechanism is the first step. The proton attacks the double bond as it does because a more stable benzylic cation is formed. Had the proton attacked the double bond in the opposite way a less stable secondary cation would have been formed.

$$C_6H_5CH=CHCH_3 + HBr \xrightarrow{\quad\quad} C_6H_5\underset{\underset{H}{|}}{C}H-\overset{+}{C}HCH_3$$

A secondary cation

Syn Additions

Studies of the peroxide-free addition using DBr rather than HBr show that the addition is primarily (88%) *a syn addition* (Fig. 13.8). This means that in the addition, the benzyl cation and the bromide ion are not completely independent

FIG. 13.8
Syn *addition of DBr to cis-1-phenylpropene. The
product of* syn *addition is a racemic modification
of (1S, 2S) and (1R, 2R)-1-phenyl-1-bromo-2-
deuteriopropane.*

cis–1–phenylpropene

DBr, CH$_2$Cl$_2$, 0°

addition from below

addition from above

ion pairs

a racemic modification

(1S, 2S)-1-phenyl-1-bromo–
2–deuteriopropane

(1R, 2R)-1-phenyl-1-bromo
2–deuteriopropane

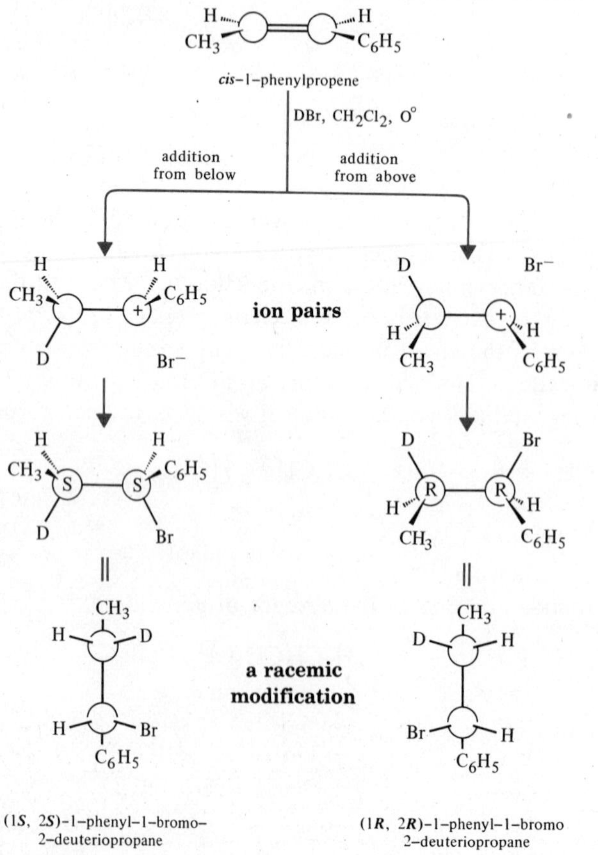

of each other. They remain in close proximity as an *ion pair* and the bromide ion adds from the same side of the double bond.

Problem 13.24

(a) What products would be obtained from the anti addition of DBr to *cis*-1-phenylpropene? (b) Would knowing the stereochemistry of the products of the addition of DBr to *cis*-1-phenylpropene allow you to decide whether or not the addition occurred in a syn or anti fashion?

Problem 13.25

(a) What products would be obtained from the syn addition of DBr to *trans*-1-phenylpropene? (b) From the anti addition?

Syn additions also predominate when chlorine and fluorine add to 1-phenylpropene (Table 13.4). When bromine adds, however, the addition is predominantly *anti*.

These experimental results and others like them allow us to place the mechanisms for the electrophilic additions to alkenes that we have studied in a unified scheme. We can demonstrate how this can be done with the addition of reagents X-Y to *cis*-alkenes of the general type, RCH=CHR'.

The overall stereochemical course of electrophilic additions to alkenes appears to depend on the relative stabilities of the onium ions and carbocations that can be formed. These stabilities, then, depend on the nature of the adding reagent and the structure of the alkene undergoing addition.

A number of studies indicate that the relative stabilities of *onium ions* depend on the nature of X.

Relative stability: $X = Br > X = Cl > X = F > X = H$ or D

When R and R' are aliphatic groups the onium ion appears to be more stable than the carbocation even when X = H or D. Thus, in most additions to aliphatic alkenes (Section 6.6) the initially formed π complex **1** collapses to the onium ion **2** and this ultimately gives anti addition.

TABLE 13.4 Addition Reactions to *cis* and *trans* 1-Phenylpropenes

ADDING REAGENT	CONDITIONS	ALKENE	PERCENT SYN ADDITION	PERCENT ANTI ADDITION
DBr	CH_2Cl_2, 0°	1-Phenylpropene	88	12
F_2	CCl_3F, −126°	*cis*-1-Phenylpropene	78	22
F_2	CCl_3F, −126°	*trans*-1-Phenylpropene	73	27
Cl_2	CH_2Cl_2, 0°	*cis*-1-Phenylpropene	75	25
Cl_2	CH_2Cl_2, 0°	*trans*-1-Phenylpropene	67	33
Br_2	CCl_4, 2–5°	*cis*-1-Phenylpropene	17	83
Br_2	CCl_4, 2–5°	*trans*-1-Phenylpropene	12	88

(From an article by W. R. Dolbier, Jr., *J. Chem. Education*, *46*, 342 (1969). The experimental results were obtained by M. J. S. Dewar, R. C. Fahey, C. Schubert, H. J. Schneider, and R. F. Merrit.)

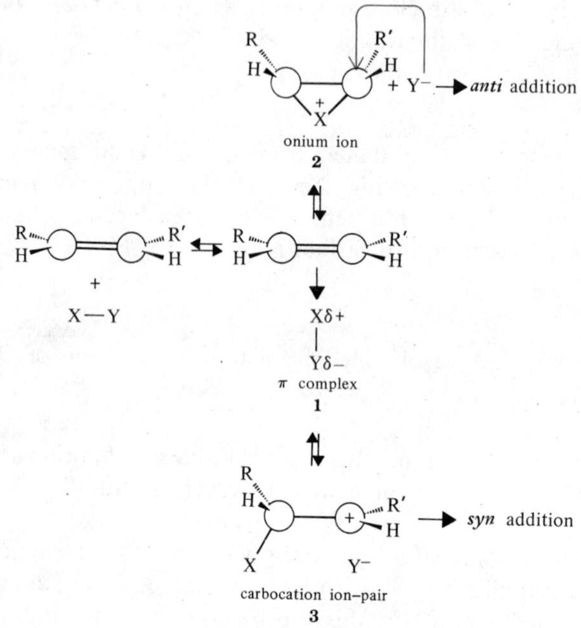

Anti addition is also the result when R or R′ = C_6H_5 and when X = Br. Here the onium ion **2** is still more stable than the carbocation **3**.

However, when R or R′ = C_6H_5 and X = Cl, F, H, or D, the carbocation **3** is formed because it is more stable than the onium ion. Since **3** is an ion pair in which the charged groups do not separate, it collapses to give syn addition.

13.12 SYNTHETIC APPLICATIONS

The substitution reactions of aromatic rings and the reactions of the side chains of alkyl and alkenylbenzenes, when taken together offer us a powerful set of reactions for organic synthesis. By using these reactions skillfully, we will be able to synthesize a vast number of benzene derivatives.

Part of the skill in planning a synthesis is in deciding the order in which reactions should be carried out. Let us suppose, for example, that we want to synthesize *o*-bromonitrobenzene. We can see very quickly that we should introduce the bromine into the ring first because it is an ortho-para director.

o-Bromonitro-benzene

p-Bromonitro-benzene

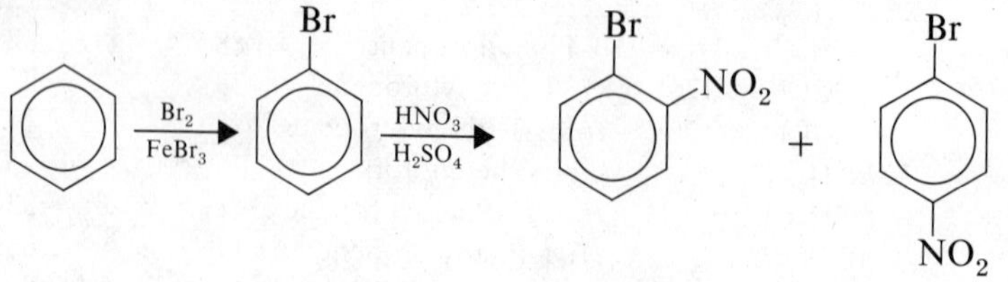

The ortho and para compounds that we get as products can be separated by fractional distillation. However, had we introduced the nitro group first, we would have obtained *m*-nitrobromobenzene as the major product.

Other examples in which choosing the proper order for the reactions are important are the syntheses of the ortho, meta, and para nitrobenzoic acids. We can synthesize the ortho and para nitrobenzoic acids from toluene by nitrating it, separating the ortho and para nitrotoluenes, and then oxidizing the methyl groups to carboxyl groups.

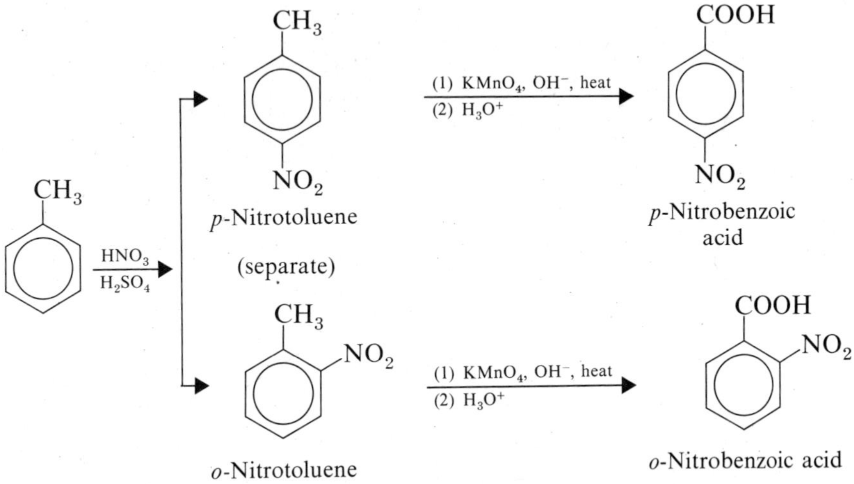

We can synthesize *m*-nitrobenzoic acid by reversing the order of the reactions.

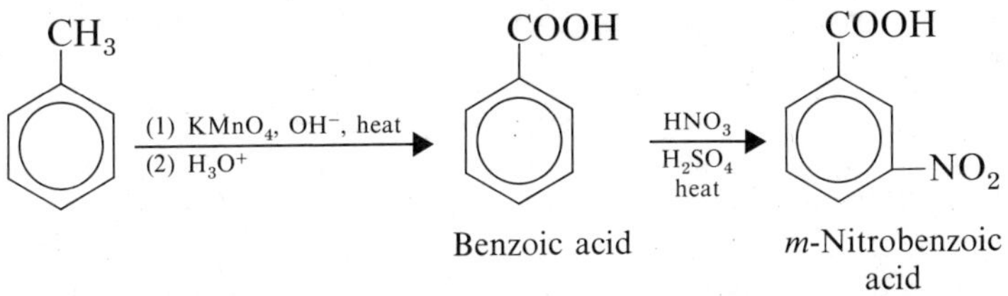

Problem 13.26

Suppose you needed to synthesize 1-(*p*-chlorophenyl)-propene from propylbenzene.

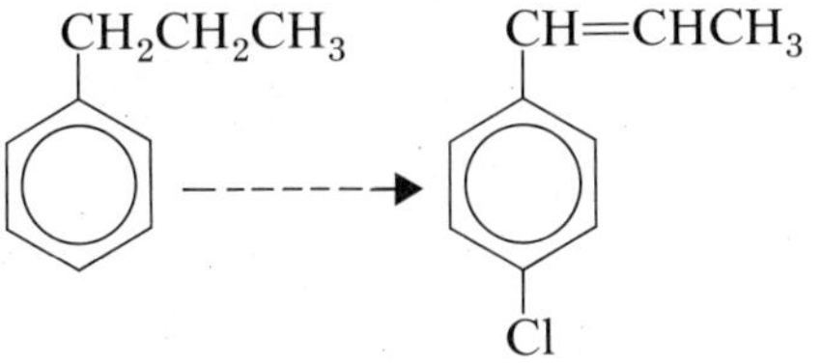

You could introduce the double bond into the side chain through a benzylic halogenation and subsequent dehydrohalogenation. You could introduce the chlorine into the benzene ring through a Lewis-acid catalyzed chlorination. Which reaction would you carry out first? Why?

Very powerful activating groups like the amino group and the hydroxyl group cause the benzene ring to be so reactive that undesirable reactions take place.

Nitration of aniline, for example, results in considerable destruction of the benzene ring because it is oxidized by the nitric acid. Direct nitration of aniline, consequently, is not a satisfactory method for the preparation of *o*- and *p*-nitroaniline.

Acetylation of the amino group (below) converts it to a group that is only moderately activating and one that does not make the ring highly susceptible to oxidation. With acetanilide direct nitration becomes possible.

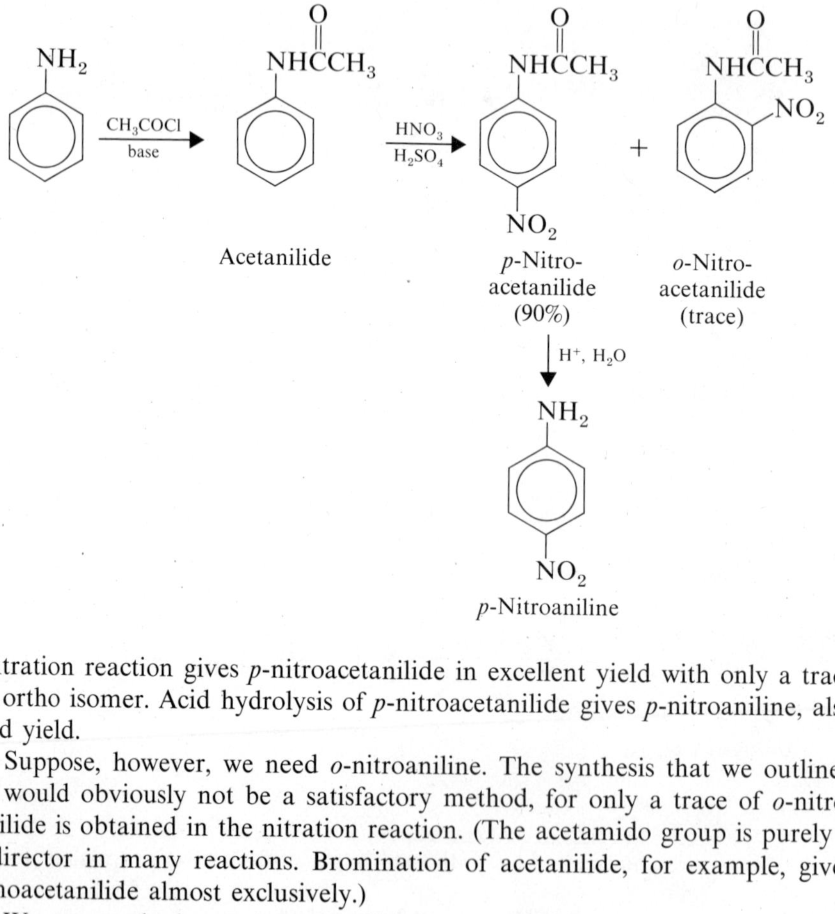

Acetanilide   *p*-Nitro-acetanilide (90%)   *o*-Nitro-acetanilide (trace)

p-Nitroaniline

The nitration reaction gives *p*-nitroacetanilide in excellent yield with only a trace of the ortho isomer. Acid hydrolysis of *p*-nitroacetanilide gives *p*-nitroaniline, also in good yield.

Suppose, however, we need *o*-nitroaniline. The synthesis that we outlined above would obviously not be a satisfactory method, for only a trace of *o*-nitro-acetanilide is obtained in the nitration reaction. (The acetamido group is purely a para director in many reactions. Bromination of acetanilide, for example, gives *p*-bromoacetanilide almost exclusively.)

We can synthesize *o*-nitroacetanilide, however, through the reaction shown below.

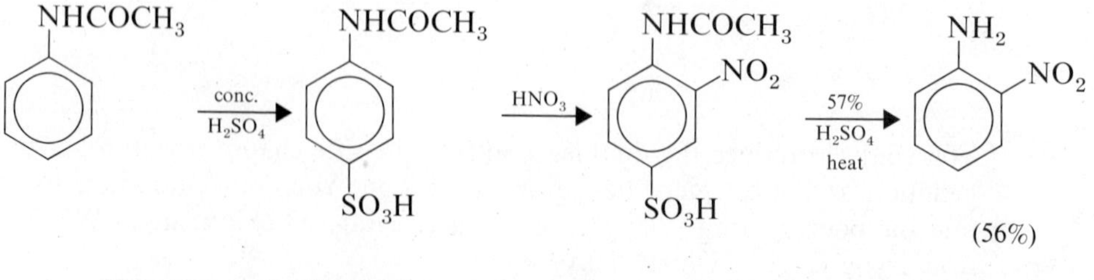

(56%)

Here we see an illustration of the utility of introducing a sulfonic acid group as a "blocking group" at one stage in the synthetic sequence. We can remove the sulfonic acid group by desulfonation at a later stage. In this example, the desulfonation reaction also conveniently removes the acetyl group that we employed to "protect" the benzene ring from oxidation.

Additional Problems

13.27

Outline ring bromination, nitration, and sulfonation reactions of the following compounds. In each case give the structure of the major reaction product or products. Also indicate whether the reaction would occur faster or slower than the corresponding reaction of benzene.

(a) Anisole, $C_6H_5OCH_3$
(b) Benzal difluoride, $C_6H_5CHF_2$
(c) Ethylbenzene
(d) Nitrobenzene
(e) Chlorobenzene
(f) Benzenesulfonic acid
(g) Ethyl benzoate, $C_6H_5\overset{O}{\overset{\|}{C}}OC_2H_5$
(h) Phenoxybenzene, $C_6H_5OC_6H_5$
(i) Biphenyl, $C_6H_5-C_6H_5$
(j) *tert*-Butylbenzene
(k) Fluorobenzene
(l) Propanoylbenzene, $C_6H_5\overset{O}{\overset{\|}{C}}C_2H_5$
(m) Benzonitrile (cyanobenzene), C_6H_5CN
(n) Phenylacetate, $C_6H_5O\overset{O}{\overset{\|}{C}}CH_3$
(o) Benzamide, $C_6H_5\overset{O}{\overset{\|}{C}}NH_2$
(p) Iodobenzene

13.28

Predict the major products of the following reactions.
(a) Sulfonation of *p*-methylacetophenone
(b) Nitration of *m*-dichlorobenzene
(c) Nitration of 1,3-dimethoxybenzene
(d) Monobromination of *p*-$CH_3CONHC_6H_4NH_2$
(e) Nitration of *p*-$HSO_3C_6H_4OH$
(f) Nitration of ⬡—CH_2—⬡—COOH
(g) Chlorination of $C_6H_5CCl_3$

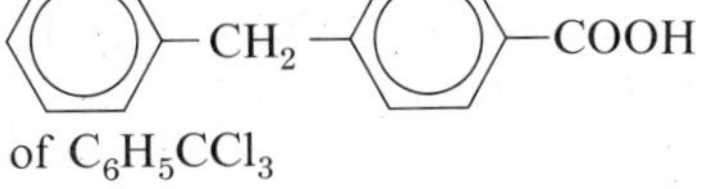

13.29

Starting with benzene, toluene, or aniline, show how you might synthesize each of the following compounds.
(a) *m*-Chlorobenzoic acid
(b) *p*-Bromoaniline
(c) *o*-Bromoaniline
(d) 2-Bromo-4-nitroaniline
(e) 4-Bromo-2-nitroaniline
(f) *o*-Acetyltoluene
(g) *o*-Toluic acid, *o*-$CH_3C_6H_4COOH$

(h) *p*-Iodobenzenesulfonic acid
(i) 2-Bromo-4-nitrotoluene
(j) *p*-Bromobenzoic acid
(k) *n*-Butylbenzene
(l) 1-(*p*-Bromophenyl)-1-butene
(m) *m*-Nitrobenzotrichloride
(n) 2,4,6-Trinitrotoluene (TNT)

13.30

(a) Which ring of benzanilide would you expect to undergo electrophilic substitution more readily? (b) Write resonance structures that explain your choice.

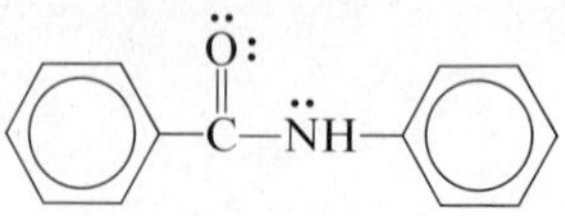

Benzanilide

13.31

What products would you expect from the nitration of phenyl benzoate?

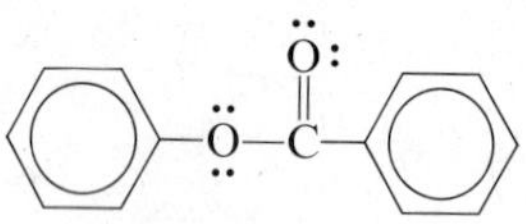

Phenyl benzoate

13.32

Hexadeuteriobenzene, C_6D_6, undergoes ring nitration at a rate that is indistinguishable from the rate at which benzene itself undergoes ring nitration. Yet, experiments with other compounds show that carbon-deuterium bonds are broken at a rate that is 5 to 10 times slower than carbon-hydrogen bonds. (We find that many reactions of deuterio-compounds show an "isotope effect"; that is, those reactions in which C-D bonds are broken occur more slowly than those in which C-H bonds are broken.) (a) How can you explain the absence of an isotope effect in the nitration of C_6D_6? (b) When C_6D_6 is sulfonated, an isotope effect *is* observed. Can you account for this as well?

13.33

We have seen that benzene undergoes ring substitution when it reacts with chlorine in the presence of a Lewis acid. However, benzene can be made to undergo *addition* of chlorine by irradiating a mixture of benzene and chlorine with ultraviolet light. The addition reaction produces a mixture of 1,2,3,4,5,6-hexachlorocyclohexanes. One of these hexachlorocyclohexanes is *Lindane*, a very effective (but potentially hazardous) insecticide. The chloro groups of Lindane at carbons 1, 2, and 3 are equatorial, those at 4, 5, and 6 are axial. (a) Write the structure of Lindane. (b) Would you expect Lindane to exist in enantiomeric forms? (c) If not, why not? (d) One 1,2,3,4,5,6-hexachlorocyclohexane isomer does exist in enantiomeric forms. Write its structure.

13.34

Naphthalene undergoes electrophilic attack at the 1 position much more rapidly than it does at the 2 position.

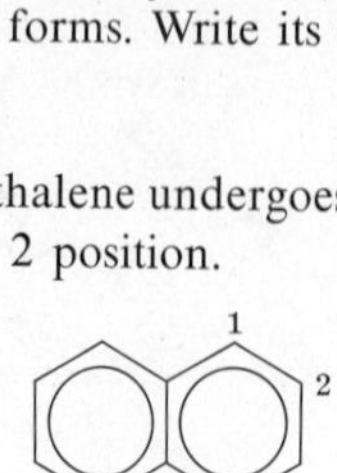

Naphthalene

The greater reactivity at the 1 position can be accounted for by writing resonance structures for the ring that undergoes electrophilic attack. Show how this is possible.

13.35

Naphthalene can be synthesized from benzene through the sequence of reactions shown below. Write the structures of each intermediate.

$$\text{Benzene + succinic anhydride} \xrightarrow{\text{AlCl}_3} \underset{(C_{10}H_{10}O_3)}{A} \xrightarrow[\text{HCl}]{\text{Zn(Hg)}} \underset{(C_{10}H_{12}O_2)}{B} \xrightarrow{\text{SOCl}_2}$$

$$\underset{(C_{10}H_{11}ClO)}{C} \xrightarrow{\text{AlCl}_3} \underset{(C_{10}H_{10}O)}{D} \xrightarrow{\text{H}_2;\ \text{Pt}} \underset{(C_{10}H_{12}O)}{E} \xrightarrow[\text{heat}]{\text{H}^+} \underset{(C_{10}H_{10})}{F} \xrightarrow[\text{heat}]{\text{Pt}}$$

$$\text{naphthalene} + H_2$$

13.36

Anthracene and many other polycyclic aromatic compounds have been synthesized by a cyclization reaction known as the *Bradsher reaction* or *aromatic cyclodehydration*. This method, discovered by Professor C. K. Bradsher of Duke University, can be illustrated by the conversion of an *o*-benzylphenyl ketone to a substituted anthracene.

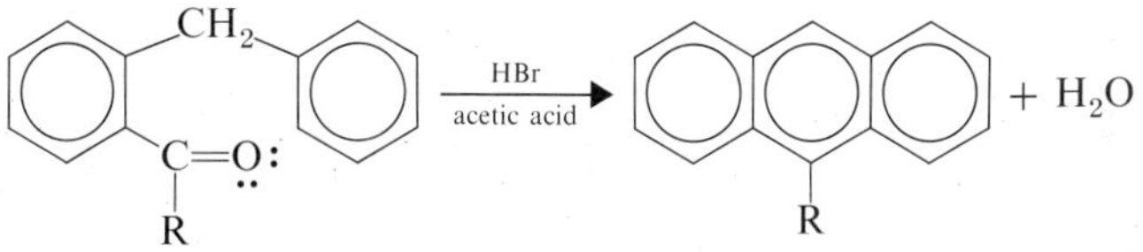

An *o*-benzylphenyl ketone → Substituted anthracene

A σ complex is an intermediate in this reaction and the last step involves the dehydration of an alcohol. Propose a mechanism for the Bradsher reaction.

13.37

Propose structures for compounds **G–I.**

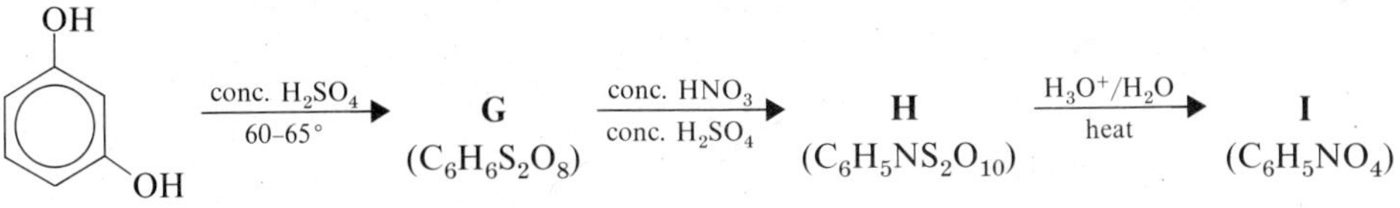

13.38

2,6-Dichlorophenol has been isolated from the females of two species of ticks (*A. Americanum* and *A. maculatum*) where it apparently serves as a sex attractant. Each female tick yields about 5 nanograms of 2,6-dichlorophenol. Assume that you need larger quantities than this, and outline a synthesis of 2,6-dichlorophenol from phenol. (Hint: when phenol is sulfonated at 100°, the product is chiefly *p*-hydroxybenzenesulfonic acid.)

* **13.39**

Treating propene with acetyl chloride in the presence of aluminum chloride gives two isomeric compounds **A** and **B** with the molecular formula C_5H_8O. Carrying the reaction out at low temperatures permits the isolation of a compound **C** with the molecular formula C_5H_9ClO. Propose structures for **A, B,** and **C** and write mechanisms that explain their formation.

* **13.40**

The addition of a hydrogen halide (hydrogen bromide or hydrogen chloride) to 1-phenyl-1, 3-butadiene produces (only) 1-phenyl-3-halo-1-butene. (a) Write a mechanism that accounts for the formation of this product. (b) Is this 1,4-addition or 1,2-addition to the butadiene system? (c) Is the product of the reaction consistent with the formation of the most stable intermediate carbocation? (d) Does the reaction appear to be under kinetic control or equilibrium control? Explain.

* 13.41

Treating *cis*-stilbene (*cis*-$C_6H_5CH{=}CHC_6H_5$) with chlorine in 1,2-dichloroethane yields an optically inactive addition product. All attempts to resolve the addition product into separate enantiomers fail. What does this suggest about (a) the stereochemistry of the addition product and (b) the mode of addition? (c) What result would you expect from a similar reaction starting with *trans*-stilbene?

* 13.42

Write mechanisms that account for the products of the following reactions:

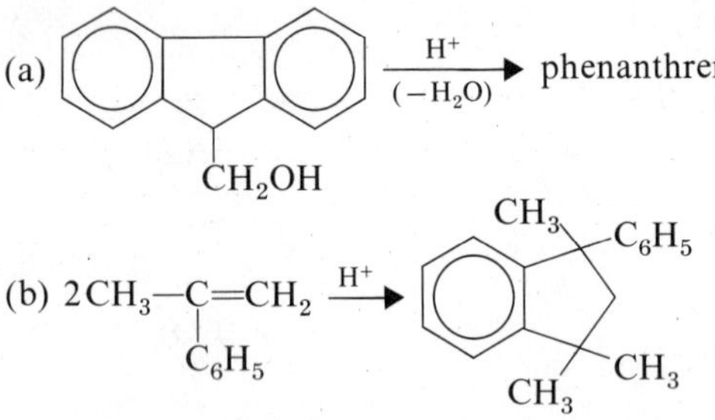

(a)

$$\xrightarrow[(-H_2O)]{H^+}$$ phenanthrene

(b) $2CH_3{-}\underset{\underset{C_6H_5}{|}}{C}{=}CH_2 \xrightarrow{H^+}$

* 13.43

At one time extensive use was made of detergents manufactured from the propene tetramers whose synthesis we saw in problem 6.32. In the industrial process, heating benzene and a mixture of the propene tetramers with aluminum chloride at 35–45° gave a mixture of isomers with the molecular formula $C_{18}H_{30}$. This mixture was then heated with sulfuric acid and the products of this reaction (isomers with the molecular formula $C_{18}H_{30}SO_3$) were treated with aqueous sodium hydroxide to give the detergent (also a mixture of isomers). Propose structures for the detergent isomers and write the reactions involved in their formation. (Note: use of these detergents was discontinued a number of years ago because they proved not to be biodegradable, cf. Sect. 23.2.)

* 13.44

The compound phenylbenzene, $C_6H_5{-}C_6H_5$, is called *biphenyl* and the rings are numbered in the manner shown below.

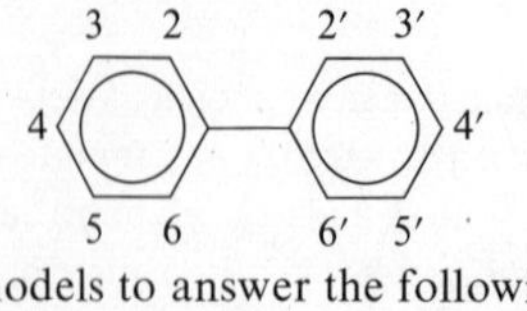

Use models to answer the following questions about substituted biphenyls. (a) When certain large groups occupy three or four of the *ortho* positions (i.e., 2,6,2′ and 6′) the substituted biphenyl may exist in enantiomeric forms. An example of a biphenyl that exists in enantiomeric forms is the compound in which the following substituents are present: 2-NO_2, 6-COOH, 2′-NO_2, 6′-COOH. What factors account for this? (b) Would you expect a biphenyl with 2-Br, 6-COOH, 2′-COOH, 6′-H to exist in enantiomeric forms? (c) The biphenyl with 2-NO_2, 6-NO_2, 2′-COOH, 6′-Br cannot be resolved into enantiomeric forms. Explain.

14

PHYSICAL METHODS OF STRUCTURE DETERMINATION
NUCLEAR MAGNETIC RESONANCE SPECTROSCOPY
INFRARED SPECTROSCOPY

14.1 THE NUCLEAR MAGNETIC RESONANCE SPECTRUM

The hydrogen nucleus, or proton, has magnetic properties. When one places a compound containing hydrogen in a very strong magnetic field and simultaneously irradiates it with electromagnetic energy, the hydrogen nuclei of the compound may absorb energy through a process known as *magnetic resonance*.* This absorption of energy, like all processes that occur on the atomic and molecular scale, is *quantized*. Absorption of energy does not occur until the strength of the magnetic field and the frequency of electromagnetic radiation are at specific values.

Instruments, known as nuclear magnetic resonance (NMR) spectrometers (Fig. 14.1), allow chemists to measure the absorption of energy by hydrogen nuclei. These instruments use very powerful magnets and irradiate the sample with electromagnetic radiation in the radio frequency region.

Nuclear magnetic resonance spectrometers are usually designed so that they irradiate the compound with electromagnetic energy of a constant frequency while the magnetic field strength is varied. When the magnetic field reaches the correct strength, the nuclei absorb energy and resonance occurs. This generates a tiny electrical current in an antenna coil surrounding the sample. The instrument then

*Magnetic resonance is an entirely different phenomenon from the resonance theory that we have discussed in earlier chapters.

FIG. 14.1

Essential parts of a nuclear magnetic resonance spectrometer.

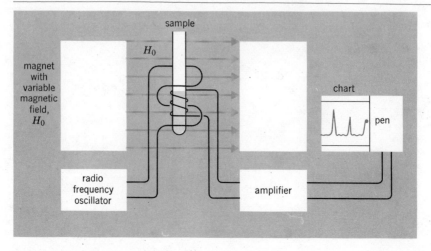

amplifies this current and displays it as a signal (a peak or series of peaks) on a strip of calibrated chart paper.

If hydrogen nuclei were stripped of their electrons and isolated from other nuclei, all hydrogen nuclei (protons) would absorb at the same magnetic field strength for a given frequency of electromagnetic radiation. If this were the case, nuclear magnetic resonance spectrometers would only be very expensive instruments for hydrogen analysis.

Fortunately, the nuclei of hydrogen atoms of compounds of interest to the organic chemist are not stripped of their electrons, and they are not isolated from each other. Some hydrogen nuclei are in regions of greater electron density than others. As a result, the protons of these compounds absorb at *slightly different* magnetic field strengths. The actual field strength at which absorption occurs is highly dependent on the magnetic environment of each proton. This magnetic environment depends on two factors: magnetic fields generated by circulating electrons and magnetic fields that result from other nearby protons.

We will discuss the theory of nuclear magnetic resonance in more detail later. Before we do, however, it will be helpful to examine the proton magnetic resonance spectra of some simple compounds. ·

Figure 14.2 shows the proton magnetic resonance spectra of *p*-xylene and *tert*-butyl acetate.

Magnetic field strength is measured along the bottom of the spectra on a delta (δ) scale in units of parts per million (ppm) and along the top in cycles per second. We will have more to say about these units later; for the moment, we need only to point out that the magnetic field strength increases from left to right. A signal that occurs at $\delta = 7$ ppm occurs at a lower magnetic field strength than one that occurs at $\delta = 2$ ppm. Signals on the left of the spectrum are also said to occur *downfield* and those on the right are said to be *upfield*.

Both spectra in Fig. 14.2 show a small signal at $\delta = 0$ ppm. This arises from a compound that has been added to the sample to allow calibration of the instrument.

The first thing on which we want to focus our attention is the relation between the number of signals in each spectrum and the number of different types of hydrogens in each compound.

Both *p*-xylene and *tert*-butyl acetate have only *two* different types of hydrogens, and each compound gives only *two* signals in its nuclear magnetic resonance spectrum.

The two different types of hydrogens of *p*-xylene are the hydrogens of the methyl groups and the hydrogens of the benzene ring. The six methyl hydrogens of *p*-xylene are all chemically *equivalent* and they are in a different chemical environment from the four hydrogens of the ring. The six methyl hydrogens give rise to the signal that occurs at $\delta = 2.30$ ppm. The four hydrogens of the benzene ring are also chemically equivalent; they give rise to the signal at $\delta = 7.05$ ppm.

The two different kinds of hydrogens of *tert*-butyl acetate are the three equivalent hydrogens of the methyl group (*b*) and the nine equivalent hydrogens of the *tert*-butyl group (*a*). The *tert*-butyl hydrogens absorb at $\delta = 1.45$ ppm and the methyl hydrogens absorb at $\delta = 1.97$ ppm.

The next aspect of these two spectra that we want to examine are the relative magnitudes of the peaks (or signals), for these are often helpful in assigning peaks to particular groups of hydrogens. What is important here is not necessarily the height of each peak but *the area underneath it*. This area, when accurately measured

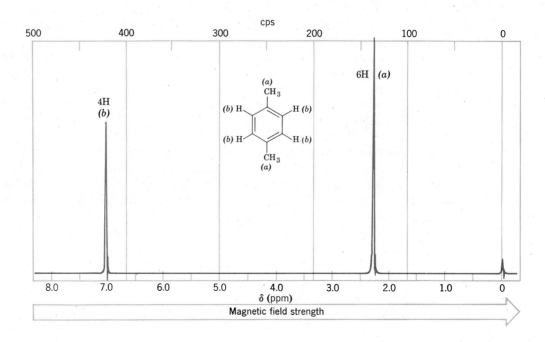

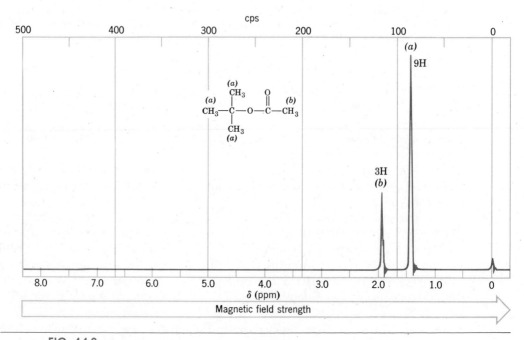

FIG. 14.2
The proton magnetic resonance spectra of p-*xylene and* tert-*butyl acetate. (Spectrum courtesy of Varian Associates, Palo Alto, Calif.)*

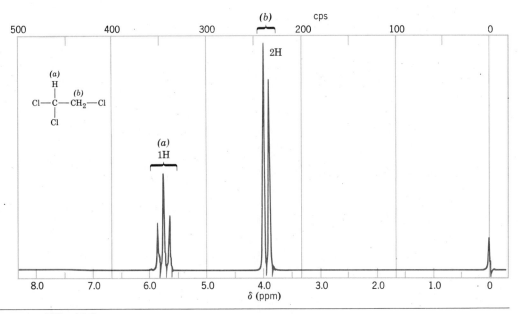

FIG. 14.3

The proton magnetic resonance (pmr) spectrum of 1,1,2-trichloroethane. (Spectrum courtesy of Varian Associates, Palo Alto, Calif.)

(the spectrometers do this automatically), is proportional to the number of hydrogens giving rise to the signal. We can see however, without measuring, that the area under the signal for the methyl groups of *p*-xylene (6H) is larger than that for the benzene hydrogens (4H). When these areas are measured accurately they are found to be in ratio of 3:2 or 6:4. When measurements are made of the areas under the signals from *tert*-butyl acetate, they are found to be in a ratio of 3:1 or 9:3 for the *tert*-butyl and methyl hydrogens, respectively.

A third feature of proton magnetic resonance (pmr) spectra that provides us with information about the structure of a compound can be illustrated if we examine the spectrum for 1,1,2-trichloroethane (Fig. 14.3).

In Fig. 14.3 we have an example of signal splitting. Signal splitting is a phenomenon that arises from magnetic influences of hydrogens on adjacent atoms. The signal (*b*) from the two equivalent hydrogens of the -CH_2Cl group is split into two peaks (a doublet) by the magnetic influence of the hydrogen of the $CHCl_2$-group. Conversely, the signal (*a*) from the hydrogen of the $CHCl_2$- group is split into three peaks (a triplet) by the magnetic influences of the two equivalent hydrogens of the -CH_2Cl group.

At this point signal splitting may seem like an unnecessary complication. As we gain experience in interpreting pmr spectra we will find that because signal splitting occurs in a predictable way, it often provides us with important information about the structure of the compound.

Now that we have had an introduction to the important features of pmr spectra we are in a position to consider them in greater detail.

14.2 NUCLEAR SPIN: THE ORIGIN OF THE SIGNAL

We are already familiar with the concept of electron spin and with the fact that the spins of electrons confer on them the spin quantum states of $+\frac{1}{2}$ or $-\frac{1}{2}$. Electron spin is the basis for the Pauli exclusion principle (p. 28); it allows us to understand how two electrons with paired spins may occupy the same atomic or molecular orbital.

The nuclei of certain isotopes also spin and because of this spin, nuclei possess spin quantum numbers, I. The nucleus of ordinary hydrogen, ^1H (i.e., a proton), is like the electron; its spin quantum number I is $\frac{1}{2}$ and it can assume either of two spin states: $+\frac{1}{2}$ or $-\frac{1}{2}$. These correspond to the magnetic moments allowed for $I = \frac{1}{2}$, $m = +\frac{1}{2}$ or $-\frac{1}{2}$. Other nuclei with spin quantum numbers $I = \frac{1}{2}$ are ^{13}C, ^{19}F, and ^{31}P. Some nuclei, such as ^{12}C, ^{16}O, and ^{32}S have no spin ($I = 0$); other nuclei have spin quantum numbers greater than $\frac{1}{2}$. In our treatment here, we will limit ourselves to discussions of the magnetic spectra that arise from protons alone. These spectra are called proton magnetic resonance (pmr) spectra to distinguish them from the spectra associated with other nuclei.

Since the proton is electrically charged, the spinning proton generates a tiny magnetic moment—one that coincides with the axis of spin (Fig. 14.4). This tiny magnetic moment confers on the spinning proton the properties of a tiny bar magnet.

When a compound containing hydrogen (and thus protons) is placed in an external magnetic field the protons may assume one of two possible orientations with respect to the external magnetic field. The magnetic moment of the proton may be aligned "with" the external field or "against" it (Fig. 14.5).

As we might expect, the two alignments of the proton are not of equal energy. When the proton is aligned with the magnetic field its energy is lower than when it is aligned against the magnetic field (Fig. 14.6).

Energy is required to "flip" the proton from its lower energy state (with the

FIG. 14.4

The magnetic field associated with a spinning proton (a). *The spinning proton resembles a tiny bar magnet* (b).

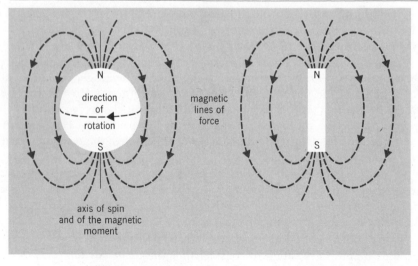

(a)   (b)

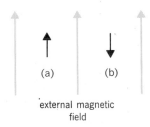

(a)   (b)

external magnetic
field

FIG. 14.5
The two orientations of the magnetic moment of a proton in an external magnetic field. Proton (a) is aligned with the magnetic field; proton (b) is aligned against it.

field) to its higher-energy state (against the field). In a nuclear magnetic resonance spectrometer this energy is supplied by electromagnetic radiation in the radio frequency region. The energy required is proportional to the strength of the magnetic field. One can show by relatively simple calculations, that in a magnetic field of approximately 14,092 gauss, electromagnetic radiation of 60×10^6 cycles per second (60 MHz) supplies the correct amount of energy.*

As we mentioned earlier, pmr spectrometers are designed so that the radio frequency is kept constant (at 60 MHz, for example) and the magnetic field is varied. When the magnetic field is at precisely the right strength the protons flip from one state to the other and in doing so they absorb radio frequency energy. This flipping of the protons generates a small electric current in a coil of wire surrounding the sample. After being amplified, the current is displayed as a signal in the spectrum.

14.3 THE POSITION AND NUMBER OF SIGNALS

Shielding and Deshielding Effects

All protons do not absorb energy at the same magnetic field strength. The three spectra that we examined earlier demonstrate this for us. The aromatic protons of *p*-xylene absorb at lower field strength ($\delta = 7.05$ ppm); the various alkyl protons of *p*-xylene, *tert*-butyl acetate, and 1,1,2-trichloroethane all absorb at higher magnetic field strengths.

---

* The relationship between the frequency of the radiation, v, and the strength of the magnetic field, H_0, is,

$$v = \frac{\gamma H_0}{2\pi}$$

where γ is the *gyromagnetic ratio*. For a proton, $\gamma = 26,750$.

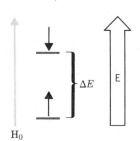

H_0

FIG. 14.6
The two energy states of a proton in an external magnetic field. The difference between these two energy states is ΔE.

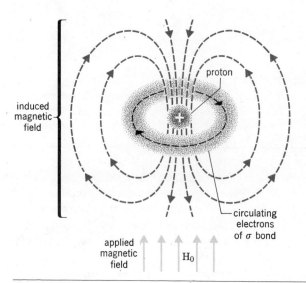

FIG. 14.7

*The circulations of the electrons of a C—H bond
under the influence of an applied magnetic field. The
electron circulations generate a small magnetic field*
(an induced field) *that* shields *the proton from the
applied field.*

The general position of a signal in a nuclear magnetic resonance spectrum
—that is, the strength of the magnetic field required to bring about absorption of
energy—can be related to electron densities and electron circulations in the compounds. Under the influence of an applied magnetic field the electrons circulate in
certain preferred paths. Because they do, and because electrons are charged particles,
the circulating electrons generate tiny magnetic fields.

We can see how this happens if we consider the circulations of the electrons
around the proton of σ bond of a C-H group. In doing so, we will oversimplify the
situation by assuming that σ electrons move in generally circular paths. The magnetic field generated by these circulating σ electrons is shown in Fig. 14.7.

The small magnetic field generated by the circulating electrons is called *an
induced field. At the proton, the induced magnetic field opposes the applied magnetic
field.* This means that the actual magnetic field sensed by the proton is slightly less
than the applied field. This proton is said to be *shielded.*

A shielded proton will not, of course, absorb at the same applied field strength
as a proton that is stripped of its electrons. A shielded proton will absorb *at higher
applied field strengths;* the applied field must be made larger by the spectrometer
in order to compensate for the small induced field.

The extent to which a proton is shielded by the circulations of σ electrons
depends on the relative electron density around the proton. This electron density
depends largely on the presence or absence of electronegative groups. Electronegative
groups withdraw electrons from the C-H bond, particularly if they are attached to
the same carbon. We can see an example of this effect in the spectrum of
1,1,2-trichloroethane (Fig. 14.3). The proton of carbon -1, absorbs at a lower magnetic
field strength ($\delta = 5.77$ ppm) than the protons of carbon -2 ($\delta = 3.95$ ppm). Carbon

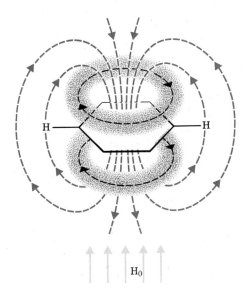

FIG. 14.8
The induced magnetic field of the π electrons of benzene deshields the benzene protons.

-1 bears two highly electronegative chloro groups whereas carbon -2 bears only one. Carbon -2, consequently, is more effectively shielded because the σ-electron density around it is greater.

The circulations of delocalized π electrons generate magnetic fields that can either *shield or deshield* nearby protons. Whether shielding or deshielding occurs, depends on the location of the proton. The aromatic protons of benzene derivatives (Fig. 14.8) are *deshielded* because their locations are such that the induced magnetic field reinforces the applied magnetic field.

Because of this deshielding effect the absorption of energy by benzene protons occurs at relatively low magnetic field strength. The protons of benzene, itself, absorb at $\delta = 7.27$ ppm. The aromatic protons of *p*-xylene (Fig. 14.2) absorb at $\delta = 7.05$ ppm.

The deshielding of external aromatic protons that results from the circulating π electrons is one of the best pieces of physical evidence that we have for π electron delocalization in aromatic rings. In fact, low field strength proton absorption is often used as a criterion for aromaticity in newly synthesized conjugated cyclic compounds.

Not all aromatic protons absorb at low magnetic field strengths, however. Large ring aromatic compounds have been synthesized that have hydrogens *in the*

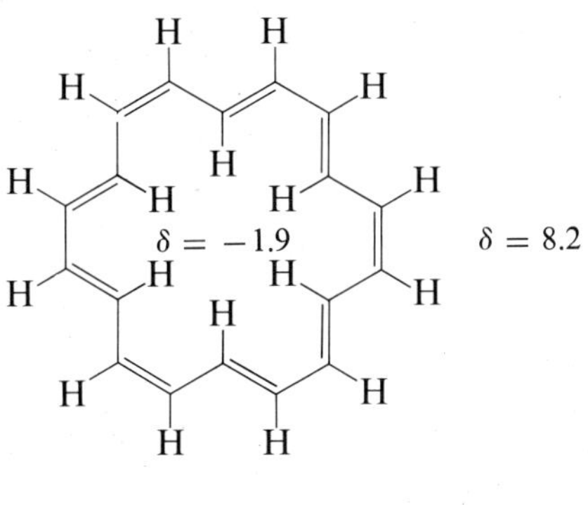

FIG. 14.9
18-Annulene. The internal protons are highly shielded and absorb at $\delta = -1.9$. The external protons are highly deshielded and absorb at $\delta = 8.2$.

center of the ring (in the π electron cavity). The protons of these internal hydrogens absorb at unusually high magnetic field strengths because they are highly shielded by the opposite induced field in the center of the ring. These internal protons often absorb at field strengths greater than that used for the reference point, $\delta = 0$. The internal protons of 18-annulene (Fig. 14.9) absorb at $\delta = -1.9$ ppm.

Problem 14.1

The methyl protons of 15,16-dimethylpyrene (p. 431) absorb at very high magnetic field strengths, $\delta = -4.2$. Can you account for this?

Pi-electron circulations also *shield* the protons of acetylene causing them to absorb at higher magnetic field strengths than we might otherwise expect. If we were to consider *only* the relative electronegativities of carbon in its three hybridization states we might expect the following order of protons attached to each type of carbon:

(low field strength) $sp < sp^2 < sp^3$ (high field strength)

In fact, acetylenic protons absorb between $\delta = 2.0$ and $\delta = 3.0$ and the order is:

(low field strength) $sp^2 < sp < sp^3$ (high field strength)

This upfield shift of the absorption of acetylenic protons is a result of shielding produced by the circulating π electrons of the triple bond. The origin of this shielding is illustrated in Fig. 14.10.

The Chemical Shift

We see now that shielding and deshielding effects cause the absorptions of protons to be shifted from the position that a bare proton would absorb (i.e., a proton stripped of its electrons). Since these shifts result from the circulations of electrons in *chemical* bonds, they are called *chemical shifts*.

Chemical shifts are measured with reference to the absorption of protons of standard compounds. This is done because it is impractical to measure the actual value of the magnetic field at which absorptions occur. The reference compound

FIG. 14.10

The shielding of acetylenic protons by π-electron circulations. Shielding causes acetylenic protons to absorb further upfield than vinyl protons.

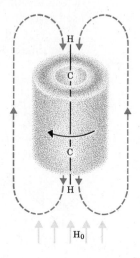

most often used is tetramethylsilane (TMS). A small amount of tetramethylsilane is usually added to the sample whose spectrum is being measured, and the signal from the 12 equivalent protons of tetramethylsilane is used to establish the zero point on the delta scale.

$$CH_3-\underset{\underset{CH_3}{|}}{\overset{\overset{CH_3}{|}}{Si}}-CH_3$$

Tetramethylsilane

Tetramethylsilane was chosen as a reference compound for several reasons: it has 12 hydrogens and, therefore, a very small amount of tetramethylsilane gives a measurable signal. Because the hydrogens are all equivalent they give a *single signal*. Since silicon is less electronegative than carbon the protons of tetramethylsilane are in regions of high electron density. They are, as a result, highly shielded and the signal from tetramethylsilane occurs in a region of the spectrum where few other hydrogens absorb. Thus, their signal seldom interferes with the signals from other hydrogens. Tetramethylsilane, like an alkane, is also unreactive. Finally, it is volatile; its boiling point is 27°C. After the spectrum has been determined the tetramethylsilane can be removed easily by evaporation.

Chemical shifts are measured in Hertz (cycles per second), as if the frequency of the electromagnetic radiation were being varied. In actuality it is the magnetic field that is changed. But, since the values of frequency and the strength of the magnetic field are mathematically related, this amounts to the same thing.

When expressed in Hertz, the magnitudes of chemical shifts are proportional to the strength of the applied magnetic field. Since pmr spectrometers with different magnetic field strengths are commonly used, it is desirable to express chemical shifts in a form that is independent of the strength of the applied field. This can be done easily by expressing the chemical shift *as a fraction of the total field*, with both numerator and denominator of the fraction expressed in frequency units (Hertz). Since chemical shifts are always very small (typically less than 500 Hz) compared with the total field strength (commonly 30, 60, or 100 *million* Hz), it is convenient to express these fractions in units of *parts per million* (ppm). This is the origin of the delta scale for the expression of chemical shifts relative to TMS.

$$\delta = \frac{\text{(observed shift from TMS in Hertz)}}{\text{(operating frequency of the pmr instrument in megahertz)}}$$

Table 14.1 gives the values of proton chemical shifts for some common hydrogen-containing groups.

Equivalent and Nonequivalent Protons

Protons that are in the same electrical environment have the same chemical shift and, therefore, give only one pmr signal. How do we know when protons are in the same electrical environment? For most purposes, protons that are in the same electrical environment are also equivalent in chemical reactions. That is, chemically equivalent protons are *chemical shift equivalent* in pmr spectra.

We saw, for example, that the six methyl protons of *p*-xylene give a single

TABLE 14.1 Typical Proton Chemical Shifts

TYPE OF PROTON	CHEMICAL SHIFT, PPM
Cyclopropane	0.2–0.9
Primary alkyl RCH_3	0.9
Secondary alkyl R_2CH_2	1.3
Tertiary alkyl R_3CH	1.5
Vinylic $\diagdown\!C\!=\!C\diagup$ H	4.6–5.9
Acetylenic $-C\!\equiv\!C-H$	2–3
Aromatic Ar—H	6–8.5
Benzylic Ar—CH_3	2–3
Allylic $C\!=\!C\!-\!CH_3$	1.7–1.8
Aldehydes $-\overset{\overset{\displaystyle O}{\|}}{C}-H$	9–10
Ketones $-\overset{\overset{\displaystyle O}{\|}}{C}-\overset{\|}{C}-H$	2–2.7
Esters $-\overset{\overset{\displaystyle O}{\|}}{C}-O-C-H$	2–2.2
Ethers $R-O-\overset{\|}{C}-H$	3.3–4
Alcohols $HO-\overset{\|}{C}-H$	3.4–4
Alkyl fluorides $F-\overset{\|}{C}-H$	4–4.5
Alkyl chlorides $Cl-\overset{\|}{C}-H$	3–4
Alkyl bromides $Br-\overset{\|}{C}-H$	2.7–4
Alkyl iodides $I-\overset{\|}{C}-H$	2–4
Hydroxylic* R—OH	1–6
Phenolic* Ar—OH	4–12
Carboxylic* R—COOH	10–12
Amino* R—NH_2	1–5

*The chemical shift of these groups vary in different solvents and with temperature and concentration.

pmr signal. We probably recognize, intuitively, that these six hydrogens are chemically equivalent. We can demonstrate their equivalence, however, by replacing each hydrogen in turn with some other group. If in making these substitutions we get the same compound from each replacement then the protons are chemically equivalent and are chemical shift equivalent. The replacements can be replacements that occur in an actual chemical reaction or they can be purely imaginary. For the methyl hydrogens of *p*-xylene we can think of an actual chemical reaction that demonstrates their equivalence, *benzylic bromination*.

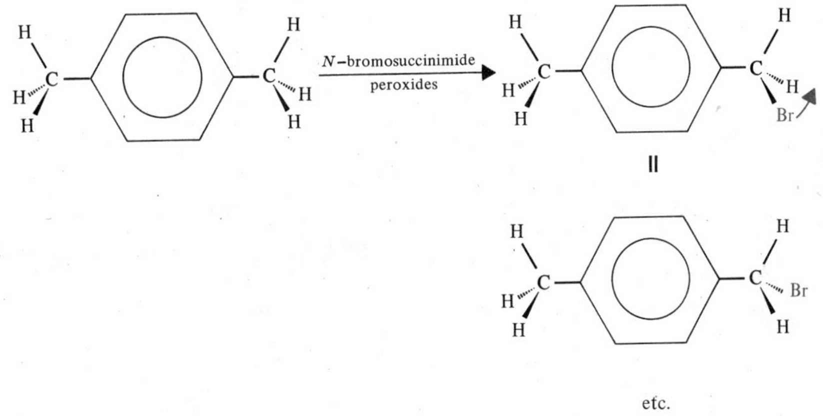

II

etc.

Benzylic bromination produces the same monobromo product regardless of which of the six hydrogens is replaced.

We can also think of a chemical reaction that demonstrates the equivalence of the four aromatic hydrogens of *p*-xylene, *ring bromination*. Once again, we get the same compound regardless of which of the four hydrogens is replaced.

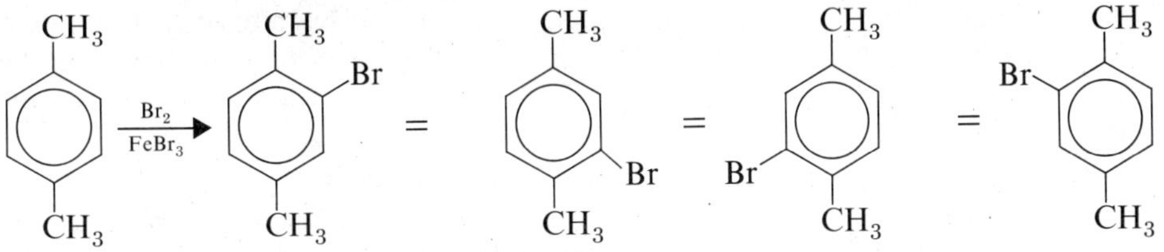

With *tert*-butyl acetate we can demonstrate that the three hydrogens of the methyl group are equivalent by replacing each with an imaginary group **W.**

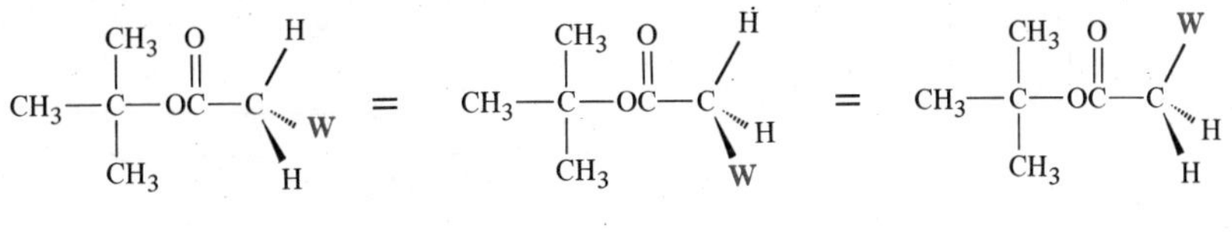

We can demonstrate the equivalence of the nine *tert*-butyl hydrogens by the same process.

Problem 14.2

How many different sets of equivalent protons do each of the following compounds have? How many signals would each compound give in its pmr spectrum?

(a) CH_3CH_3　　　　(b) $CH_3CH_2CH_3$

(c) CH_3OCH_3

(e) 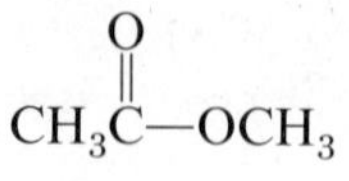 $CH_3\overset{\overset{\textstyle O}{\|}}{C}{-}OCH_3$

(d) 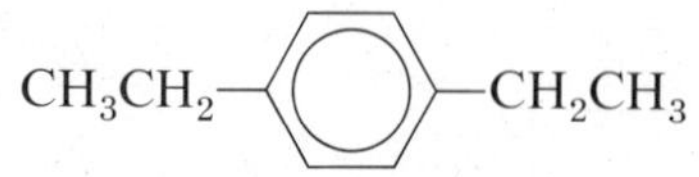 CH_3CH_2—⬡—CH_2CH_3

(f) 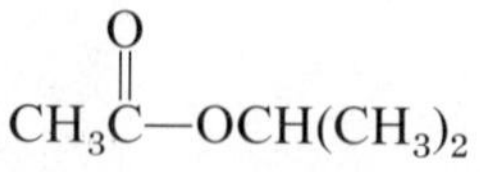 $CH_3\overset{\overset{\textstyle O}{\|}}{C}{-}OCH(CH_3)_2$

Enantiotopic and Diastereotopic Hydrogens

If replacement of each of two hydrogens by the same group yields compounds that are enantiomers, the two hydrogens are said to be *enantiotopic. The protons of enantiotopic hydrogens have the same chemical shift and give only one pmr signal.**

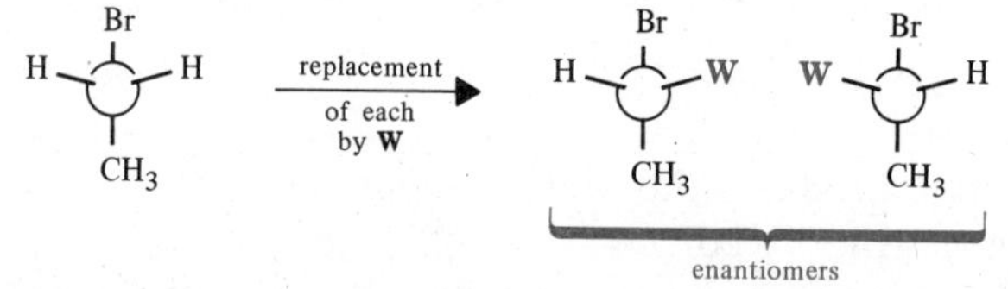

enantiomers

The two hydrogens of the -CH_2Br group of ethyl bromide are enantiotopic. Ethyl bromide, then, gives two signals in its pmr spectrum. The three equivalent protons of the CH_3- group give one signal; the two enantiotopic protons of the -CH_2Br group give the other signal. [The pmr spectrum of ethyl bromide as we will see, actually consists of seven peaks. This is a result of signal splitting. One signal (from the CH_3- group) is split into three peaks; the other signal (from the -CH_2Br group) is split into four peaks.]

If replacement of each of two hydrogens by a group, **W**, gives compounds that are diastereomers, the two hydrogens are said to be *diastereotopic. Except for accidental coincidence, diastereotopic protons do not have the same chemical shift and give rise to different pmr signals.*

The two protons of the $=CH_2$ group of chloroethene are diastereotopic.

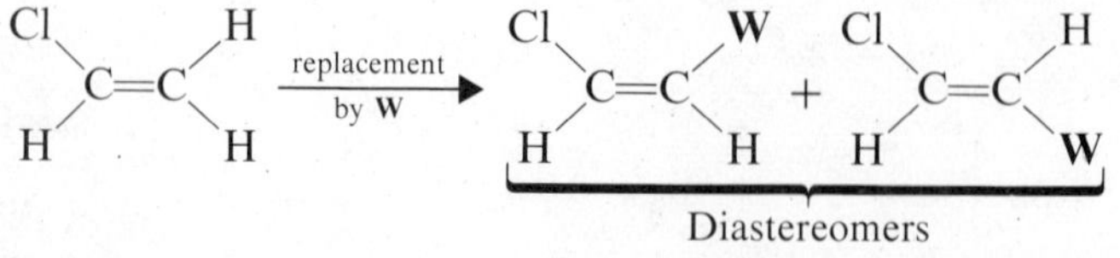

Diastereomers

Chloroethene, then, should give signals from three nonequivalent protons; one for the proton of the $ClCH=$ group, and one for each of the diastereotopic protons of the $=CH_2$ group.

The two methylene (—CH_2—) protons of *sec*-butyl alcohol (p. 515) are also diastereotopic. We can illustrate this with one enantiomer of *sec*-butyl alcohol in the following way:

---

*Enantiotopic hydrogens may not have the same chemical shift if the compound is dissolved in an optically active solvent. However, most pmr spectra are determined using optically inactive solvents and in this situation enantiotopic protons have the same chemical shift.

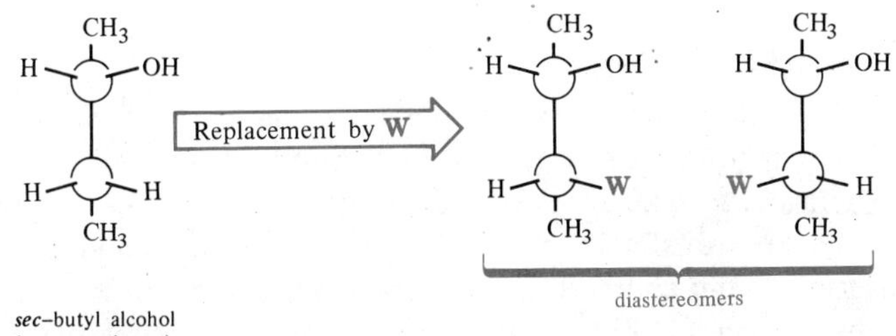

diastereomers

sec-butyl alcohol
(one enantiomer)

Problem 14.3

(a) Show that replacing each of the two methylene protons of the other *sec*-butyl alcohol enantiomer by **W** also leads to a pair of diastereomers. (b) How many chemically different kinds of protons are there in *sec*-butyl alcohol? (c) How many pmr signals would you expect to find in the spectrum of *sec*-butyl alcohol?

Problem 14.4

How many pmr signals would you expect from each of the following compounds?

(a) $CH_3CH_2CH_2CH_3$ (f) 1,1-Dimethylcyclopropane
(b) CH_3CH_2OH (g) *trans*-1,2-Dimethylcyclopropane
(c) $CH_3CH=CH_2$ (h) *cis*-1,2-Dimethylcyclopropane
(d) *trans*-2-Butene   (i) 1-Pentene
(e) 1,2-Dibromopropane

14.4 SIGNAL SPLITTING: SPIN-SPIN COUPLING

Signal splitting reflects the environment of a proton with respect to magnetic fields of protons on nearby atoms. We have seen an example of signal splitting in the spectrum of 1,1,2-trichloroethane (Fig. 14.3). The signal from the two magnetically equivalent protons of the —CH_2Cl group of 1,1,2-trichloroethane is split into two peaks by the single proton of the $CHCl_2$— group. The signal from the proton of the $CHCl_2$— group is split into three peaks by the two protons of the —CH_2Cl group.

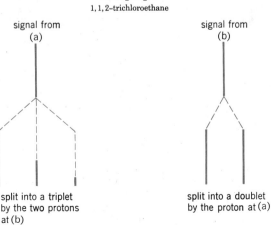

(a)   (b)
$CHCl_2CH_2Cl$
1, 1, 2–trichloroethane

signal from (a)   signal from (b)

split into a triplet by the two protons at (b)   split into a doublet by the proton at (a)

Signal splitting arises from a phenomenon known as *spin-spin coupling*. This spin-spin coupling occurs because protons tend to align themselves in particular ways in external magnetic fields. Since protons have magnetic moments themselves, the magnetic moment of one proton can affect the magnetic environment of a proton on an adjacent atom. (The spins of the protons are said to be coupled.) Spin-spin coupling effects are transferred primarily through the bonding electrons and are not usually observed if the coupled protons are separated by more than three σ bonds. Thus, we observe signal splitting primarily from the protons of *adjacent σ-bonded* atoms. We would not, for example, expect to observe splitting of the signals from the protons (a) and (b) of *tert*-butyl acetate (Fig. 14.2).

$$\overset{(a)}{CH_3} \quad \overset{O}{\parallel}$$
$$(a)\ CH_3-\overset{|}{\underset{|}{C}}-O-\overset{\parallel}{C}-\overset{(b)}{CH_3}$$
$$\underset{(a)}{CH_3}$$

tert-Butyl acetate
(no signal splitting)

Spin-spin splitting is not observed for protons that are *chemically equivalent or enantiotopic*. That is, spin-spin splittings do not occur between protons that have *exactly the same chemical shift*. Thus, we would not expect, and do not find, signal splitting in the signal from the six equivalent hydrogens of ethane.

CH_3CH_3 (no signal splitting)

FIG. 14.11

The pmr spectrum of methoxyacetonitrile. No signal splitting occurs with the enantiotopic protons (b). (Spectrum courtesy of Varian Associates, Palo Alto, Calif.)

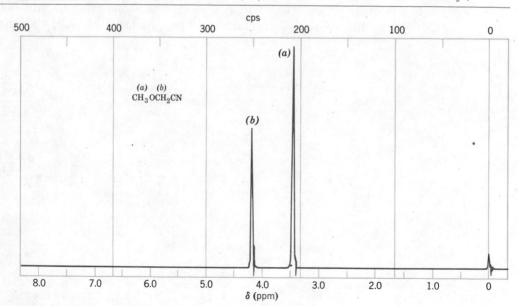

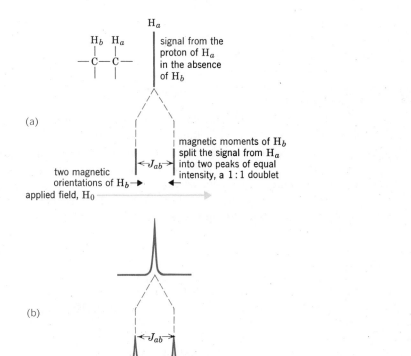

FIG. 14.12
*Signal splitting arising from spin-spin coupling
with one nonequivalent proton of a neighboring
hydrogen. A theoretical spectrum is shown in* (a)
and the actual appearance of the spectrum in (b).
*The distance between the centers of the peaks of
the doublet is called the* coupling constant, J_{ab}.
J_{ab} *is measured in Hertz (cycles per second). The
magnitudes of coupling constants are not dependent upon the magnitude of the applied field
and their values, in Hertz will be the same regardless of the operating frequency of the spectrometer used.*

Nor do we find signal splitting occurring from enantiotopic protons of methoxy
acetonitrile (Fig. 14.11).

We do, however, observe signal splitting from *protons that are not chemical
shift equivalent.*

We have seen that protons can be aligned in only two ways in an external
magnetic field: with the field or against it. Therefore, the magnetic moment of a
proton on an adjacent atom may affect the magnetic field of the proton whose signal
we are observing in only two ways: the proton of the adjacent atom may *add* its
small magnetic field to that of the applied field, or it may *subtract* its small magnetic
field from that of the applied field.

Figure 14.12 shows how two possible orientations of the proton of a neighboring hydrogen, H_b, splits the signal of the proton H_a. (H_b and H_a are not equivalent.)

We see, in Fig. 14.12*a,* that in the first instance, the magnetic moment of the proton of H_b is aligned with the applied magnetic field (H_0). Thus, the small magnetic moment of the proton of H_b *adds* to that of the large applied field. The proton of H_a feels the *sum* of these magnetic fields and absorbs energy at a slightly *lower* applied field strength than it would if H_b were not present.

In the second instance, the proton of H_b is aligned against the applied field. Its small magnetic moment *subtracts* from that of the large applied field. The proton H_a feels the *difference* of these fields and, therefore, does not absorb energy until the applied field strength is slightly greater than that required if H_b were not present.

When we determine pmr spectra we are, of course, observing effects produced by billions of molecules. Since the difference in energy between the two possible orientations of the proton of H_b is small, the two orientations will be present in roughly (but not exactly) equal amounts. The signal that we observe from H_a is, therefore, split into two peaks of roughly equal intensity, *a 1:1 doublet.*

Problem 14.5

Sketch the pmr spectrum of $CHBr_2CHCl_2$. Which signal would you expect to occur at lowest magnetic field strength; that of the proton of the $CHBr_2$— group or of the —$CHCl_2$ group? Why?

Two equivalent protons on an adjacent carbon (or carbons) split the signal from an absorbing proton into a 1:2:1 triplet. The following diagram illustrates how this occurs.

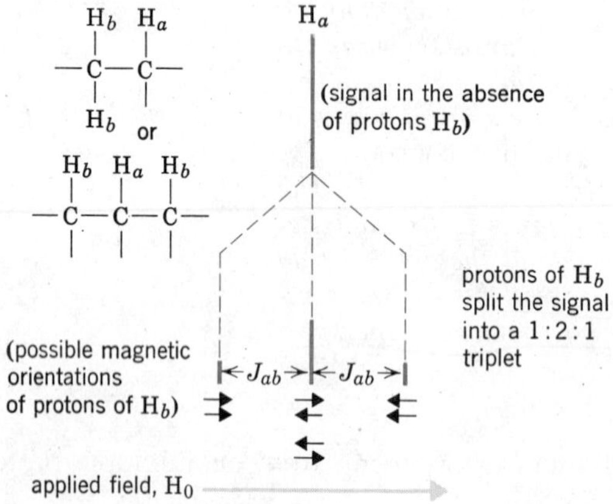

In compounds of either type, both protons may be aligned with the applied field and thus add their small magnetic fields to it. This causes a peak to appear at lower applied field strength than would occur in the absence of the two hydrogens H_b. Conversely, both protons may be aligned against the applied field and thus subtract their small magnetic fields from the applied field. This orientation of the protons of H_b causes a peak to appear at higher applied field strengths than would occur in their absence. Finally there are two ways in which the two protons may be aligned in which one opposes the applied field and one reinforces it. These arrangements do not displace the signal. Since the probability of this last arrangement is twice that of either of the other two, the center peak of the triplet is twice as intense.

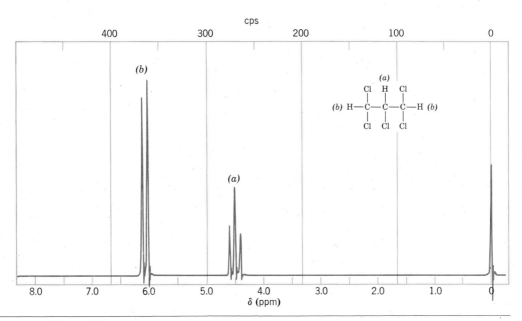

FIG. 14.13
The pmr spectrum of 1,1,2,3,3-pentachloropropane. (Spectrum courtesy of Varian Associates, Palo Alto, Calif.)

The proton of the —$CHCl_2$ group of 1,1,2-trichloroethane is an example of a proton of the type having two equivalent protons on an adjacent carbon. The signal from the —$CHCl_2$ group (Fig. 14.3) appears as a 1:2:1 triplet and, as we would expect, the protons of the —CH_2Cl group of 1,1,2-trichloroethane are split into a 1:1 doublet by the proton of the —$CHCl_2$ group.

The spectrum of 1,1,2,3,3-pentachloropropane (Fig. 14.13) is similar to that of 1,1,2-trichloroethane in that it also consists of a 1:2:1 triplet and a 1:1 doublet. The hydrogens H_b of 1,1,2,3,3-pentachloropropane are equivalent even though they are on separate carbons.

Problem 14.6

The relative positions of the doublet and triplet of 1,1,2-trichloroethane (Fig. 14.3) and 1,1,2,3,3-pentachloropropane (Fig. 14.13) are reversed. Explain this.

Three equivalent protons (H_b) on a neighboring carbon split the signal from the H_a into a 1:3:3:1 quartet. This is shown on p. 520.

The signal from two equivalent protons of the —CH_2Br group of ethyl bromide (Fig. 14.14) appears as a 1:3:3:1 quartet because of this type of signal splitting. The three equivalent protons of the CH_3— group are split into a 1:2:1 triplet by the two protons of the —CH_2Br group.

The kind of analysis that we have just given can be extended to compounds with even larger numbers of equivalent protons on adjacent atoms. These analyses show that *a set of n equivalent protons on adjacent atoms will split a pmr signal into n + 1 peaks.* (We may not always see all of these peaks in actual spectra, however, because some of them may be very small.)

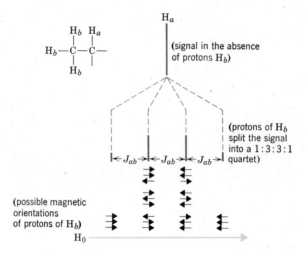

Problem 14.7

What kind of pmr spectrum would you expect the following compound to give?

$(Cl_2CH)_3CH$

Sketch the spectrum showing the splitting patterns and relative position of each signal.

FIG. 14.14

The pmr spectrum of ethyl bromide.

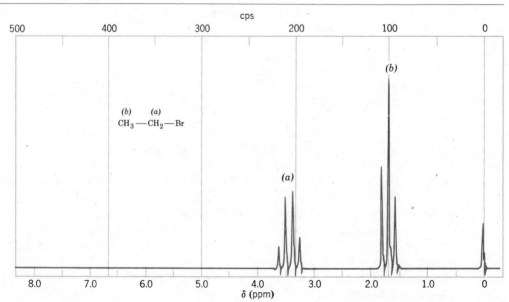

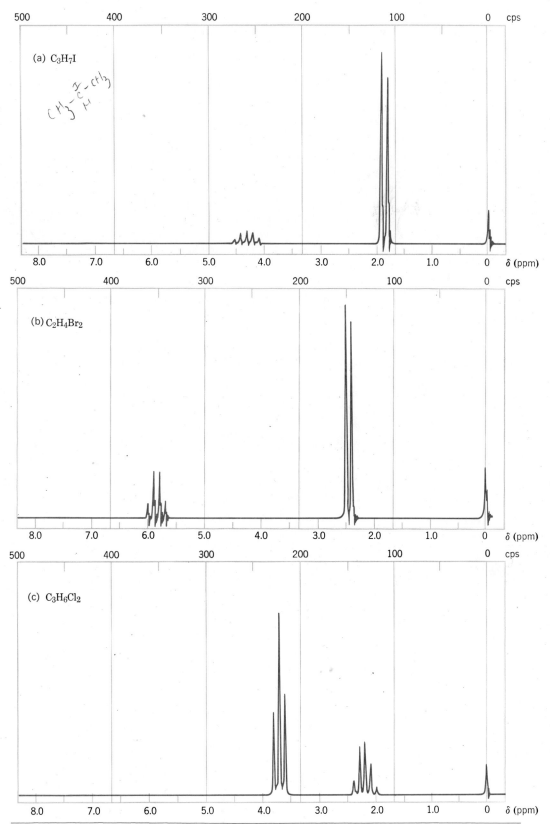

FIG. 14.15 *Pmr spectra for problem 14.8. (Spectrum courtesy of Varian Associates, Palo Alto, Calif.)*

Problem 14.8

Propose structures for each of the compounds shown in Fig. 14.15, and account for the splitting pattern of each signal.

The splitting patterns shown in Fig. 14.14 are fairly easy to recognize because in each compound there are only two sets of nonequivalent hydrogens. Two features are present in pmr spectra, however, that will often help us recognize splitting patterns in more complicated spectra.

1. The separation of the peaks in cycles per second gives us the value of the coupling constants. Thus, if we look for doublets, triplets, quartets, and so on, that have the same coupling constants the chances are good that these multiplets are related to each other because they arise from reciprocal spin-spin couplings.

The two sets of protons of an ethyl group, for example, appear as a triplet and a quartet as long as the ethyl group is attached to a carbon that does not bear

FIG. 14.16

(*Top*) *A theoretical splitting pattern for an ethyl group.* (*Bottom*) *The spectrum of ethyl iodide.* (*Spectrum courtesy of Varian Associates, Palo Alto, Calif.*)

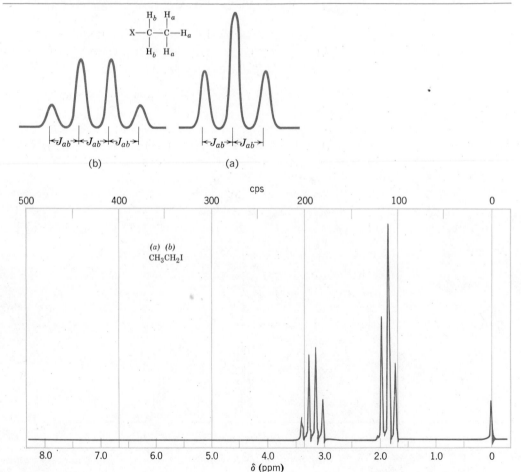

any hydrogens. The spacings of the peaks of the triplet and the quartet will be the same because the coupling constants, J_{ab}, are the same (Fig. 14.16).

2. Actual multiplets are not exactly symmetrical. We have drawn them symmetrically in our theoretical spectra. However, if we look closely at the actual quartet and triplet of ethyl iodide in Fig. 14.16 we see that the peaks are really like this:

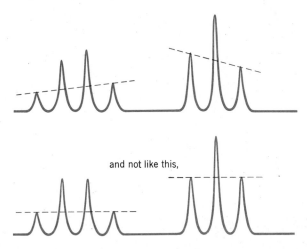

and not like this,

In the actual spectrum, a line connecting the top of the outermost peak of each multiplet *slopes upward in the direction of the set of peaks with which it is coupled.*

The phenomenon occurs because the spin states of the protons are not populated in an exactly statistical way when actual spectra are measured.

Proton magnetic spectra have other features, however, that are not at all helpful when we try to determine the structures of a compound.

1. Signals may overlap. This happens when the chemical shifts of the signals are very nearly the same. In the spectrum of ethyl chloroacetate (Fig. 14.17) we see that the singlet of the CH_2Cl- group falls directly on top of one of the outermost peaks of the ethyl quartet.

We see a more dramatic example of signal overlapping in the spectrum of *n*-octane (Fig. 14.18). The methyl protons absorb at $\delta = 0.88$, but all of the three different sets of methylene ($-CH_2-$) protons have approximately the same chemical shift, $\delta = 1.27 \pm 0.03$.

2. Spin-spin couplings between the protons of nonadjacent atoms may occur. This happens frequently when π-bonded atoms intervene between the atoms bearing the coupled protons. We can see an example of "long-range" spin-spin coupling in the expanded signals of the spectrum of methyl

formate (Fig. 14.19). Even though the proton of the formyl group, $-\overset{\displaystyle O}{\overset{\|}{C}}-H$, is separated by two atoms from the proton of the methyl group, the signal

from the proton of the $-\overset{\displaystyle O}{\overset{\|}{C}}H$ is split into a quartet and the signal from the protons of the methyl is split into a doublet. The coupling constants are small, however, and in the "unexpanded" spectrum both peaks simply look like "ragged" singlets. (We also see in the expanded signal that the

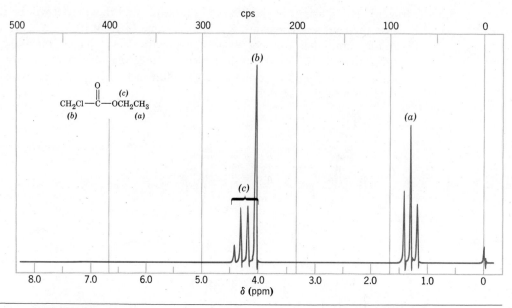

FIG. 14.17

The pmr spectrum of ethyl chloroacetate. The singlet from the protons of (b) falls on one of the outermost peaks of the quartet from (c). (Spectrum courtesy of Varian Associates, Palo Alto, Calif.)

FIG. 14.18

The pmr spectrum of n-octane. (Spectrum courtesy of Varian Associates, Palo Alto, Calif.)

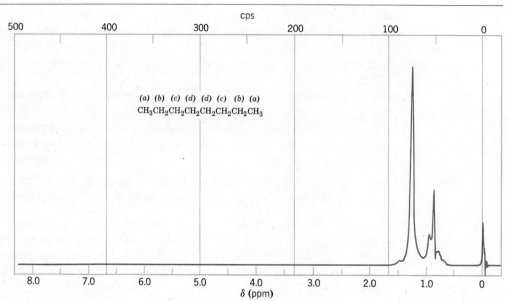

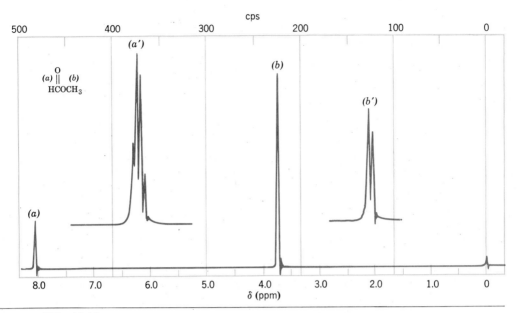

FIG. 14.19
The pmr spectrum of methyl formate. The signals (a') and (b') are expanded versions of (a) and (b). The expanded signals are offset on the δ scale for clarity and their relative areas are not related to the number of protons giving rise to each signal. (Spectrum courtesy of Varian Associates, Palo Alto, Calif.)

outermost peaks of the formyl quartet do not slope upwards in the direction of the signal from the methyl protons.This happens because the chemical shifts of the groups are very different.)

3. The splitting patterns of aromatic groups are difficult to analyze. A monosubstituted benzene ring (a phenyl group) has three different kinds of protons.

The chemical shifts of these protons may be so similar that the phenyl group will give a signal that resembles a singlet. This happens, for example, in the spectrum of toluene (Fig. 14.20). Or the chemical shifts may be different and because of long-range couplings, the phenyl group appears as a very complicated multiplet. The signal from the aromatic protons of anisole (Fig. 14.21) gives us an example of this.

Disubstituted benzenes show a range of complicated splitting patterns. In many instances these patterns can be analyzed using techniques that are beyond the scope of our discussion here. We will see later in this chapter that infrared spectroscopy gives us a relatively easy method for deciding whether the substituents of disubstituted benzenes are ortho, meta or para to each other.

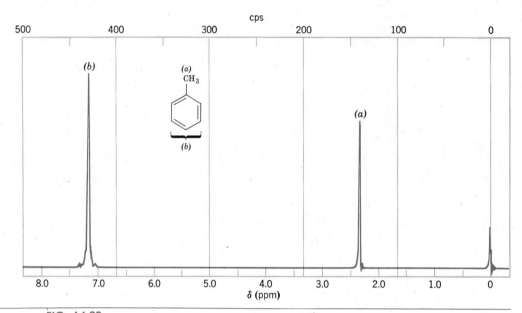

FIG. 14.20

The pmr spectrum of toluene. (Spectrum courtesy of Varian Associates, Palo Alto, Calif.)

FIG. 14.21

The pmr spectrum of anisole (methoxybenzene). (Spectrum courtesy of Varian Associates, Palo Alto, Calif.)

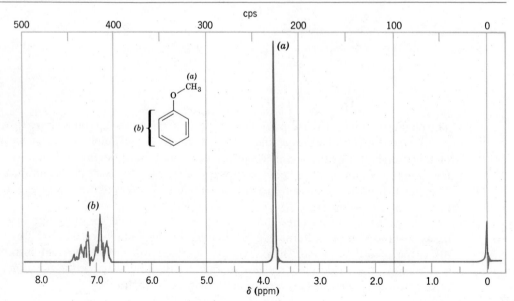

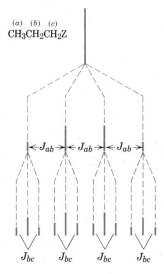

(a) (b) (c)
$CH_3CH_2CH_2Z$

FIG. 14.22
The splitting pattern that would occur for the (b) *protons of* $CH_3CH_2CH_2Z$ *if* J_{ab} *is much larger than* J_{bc}. *Here* $J_{ab} = 4J_{bc}$.

In all of the pmr spectra that we have considered so far, we have restricted our attention to signal splittings arising from interactions of only two sets of equivalent protons on adjacent atoms. What kind of patterns should we expect from compounds in which more than two sets of equivalent protons are interacting? We cannot answer this question completely but we can give an example that illustrates the kind of analysis that is involved. Let us consider a 1-substituted propane.

$$(a) \quad (b) \quad (c)$$
$$CH_3—CH_2—CH_2—Z$$

Here, there are three sets of equivalent protons. We have no problem in deciding what kind of signal splitting to expect from the protons of the $CH_3—$ group or the $—CH_2Z$ group. The methyl group is spin-spin coupled only to the two protons of the central $—CH_2—$ group. Therefore, the methyl group should appear as a triplet. The protons of the $—CH_2Z$ group are similarly coupled only to the two protons of the central $—CH_2—$ group. Thus, the protons of the $—CH_2Z$ group should also appear as a triplet.

But what about the protons of the central $—CH_2—$ group (b)? They are spin-spin coupled with the three protons at (a) and with two protons at (c). The protons at (a) and (c), moreover, are not equivalent. If the coupling constants J_{ab} and J_{bc} have quite different values, then, the protons at (b) could give as many as 12 peaks. That is, the signal from the protons (b) could be split into a quartet by the three protons (a) and each line of the quartet could be split into a triplet by the two protons (c) (Fig. 14.22).

It is unlikely, however, that we would observe as many as 12 peaks in an actual spectrum because the coupling constants are such that peaks usually fall on top of peaks. The pmr spectrum of 1-nitropropane (Fig. 14.23) is typical of 1-substituted propyl compounds. We see that the (b) protons are split into six major peaks, each of which shows a slight sign of further splitting.

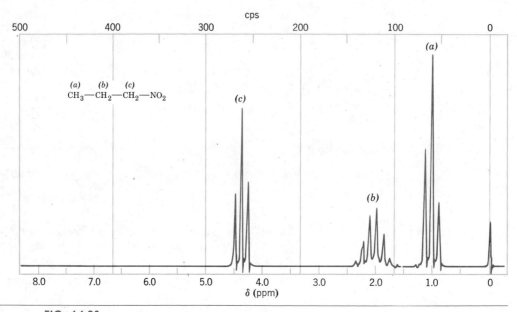

FIG. 14.23

The pmr spectrum of 1-nitropropane. (Spectrum courtesy of Varian Associates, Palo Alto, Calif.)

Problem 14.9

Carry out an analysis like that shown in Fig. 14.22 and show how many peaks the signal from (b) would be split into if $J_{ab} = 2J_{bc}$ and if $J_{ab} = J_{bc}$. (Hint: in both cases peaks will fall on top of peaks so that the total numbers of peaks in the signal is less than 12.)

Problem 14.10

Assign structures to each of the compounds in Fig. 14.24.

Problem 14.11

The pmr spectrum of 1,2-dibromo-1-phenylethane is given in Fig. 14.25. Make assignments for each signal and account for the signal splittings that occur. [Hint: Hydrogens (a) and (b) are diastereotopic but there is no discernible signal splitting due to spin-spin coupling between (a) and (b). Spin-spin coupling does occur, however, between (a) and (c) and between (b) and (c). J_{ac} is greater than J_{bc}.]

14.5 PROTON MAGNETIC RESONANCE SPECTRA OF COMPOUNDS CONTAINING FLUORINE AND DEUTERIUM

The fluorine (^{19}F) nucleus has spin quantum numbers of $+\frac{1}{2}$ and $-\frac{1}{2}$. In this respect ^{19}F nuclei resemble protons and fluorine magnetic spectra can be measured. The signals from ^{19}F absorptions occur at considerably different magnetic field strengths than those of protons, so we do not see peaks due to ^{19}F absorption in proton magnetic resonance spectra. We do, however, see splitting of proton signals due to

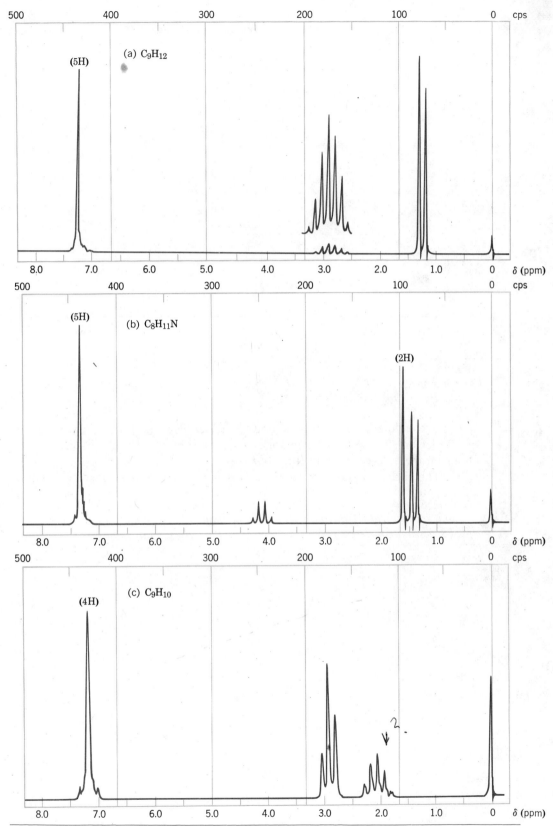

FIG. 14.24 *Pmr spectra for problem 14.10. (Spectra courtesy of Varian Associates, Palo Alto, Calif.)*

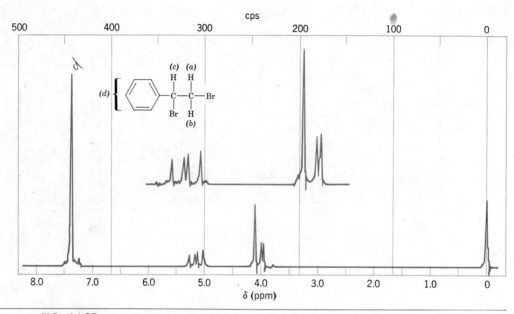

FIG. 14.25

The pmr spectrum of 1,2-dibromo-1-phenylethane. (Spectrum courtesy of Varian Associates, Palo Alto, Calif.)

spin-spin couplings between protons and ^{19}F nuclei. The signal from the two protons of 1,2-dichloro-1,1-difluoroethane, for example, is split into a triplet by the fluorine atoms on the adjacent carbon (Fig. 14.26).

The nucleus of a deuterium atom (a deuteron) has a much smaller magnetic

FIG. 14.26

The pmr spectrum of 1,2-dichloro-1,1-difluoroethane. The J_{HF} coupling constant is ~ 12 cps. (Spectrum courtesy of Varian Associates, Palo Alto, Calif.)

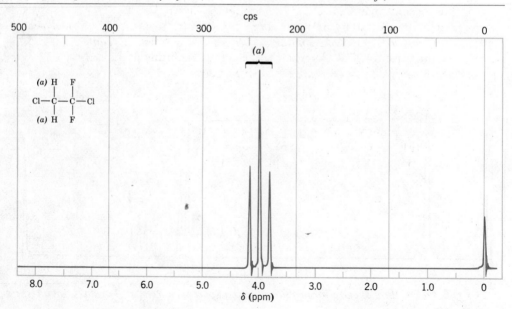

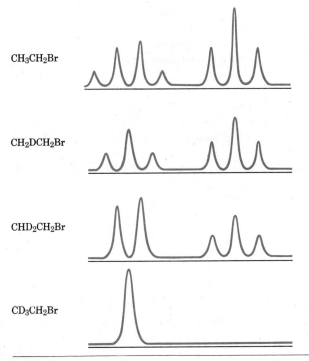

FIG. 14.27
The splitting patterns of ethyl bromide with deuterium labeling in the methyl group.

moment than a proton. Deuterons, therefore, absorb at much higher magnetic fields than protons, and signals from deuteron absorption do not occur in proton magnetic resonance spectra.

Spin-spin couplings between deuterons and protons are small and the presence of deuterium on an adjacent atom does not usually cause splitting of the proton signal. (It may broaden the signal from the proton slightly, however.) *Thus, when deuterium atoms replace protons the splitting of proton signals is usually reduced.*

Figure 14.27 shows what happens to the splitting patterns of an ethyl group as the methyl hydrogens are successively replaced by deuterium.

Problem 14.12

Sketch the pmr spectra that you would expect from each of the following compounds.

 (a) $CH_3CF_2CH_3$ (c) $CH_3CD_2CH_3$
 (b) CH_3CHDCH_3 (d) CH_3CHDBr

14.6 PROTON MAGNETIC RESONANCE SPECTRA AND RATE PROCESSES

Professor J. D. Roberts (of the California Institute of Technology), a pioneer in the application of NMR spectroscopy to problems of organic chemistry, has compared the nuclear magnetic resonance spectrometer to a camera with a relatively slow shutter speed. Just as a camera with a slow shutter speed blurs photographs of objects that are moving rapidly, the nuclear magnetic resonance spectrometer blurs its picture of molecular processes that are occurring rapidly.

What are some of the rapid processes that occur in organic molecules?

At temperatures near room temperature, groups connected by carbon-carbon single bonds rotate very rapidly. Because of this, when we determine pmr spectra of compounds with single bonds, the spectra that we obtain often show us the individual hydrogens in an average environment. That is, in an environment that is an average of all the environments that the protons have as a result of the group rotations.

To see an example of this effect, let us consider the pmr spectrum of ethyl iodide again. The most stable conformation of ethyl iodide is the one in which the groups are perfectly staggered. In this staggered conformation the hydrogen *anti*

to the iodine is in a different environment from that of the other two hydrogens of the methyl group. If the NMR spectrometer were to detect this particular conformation of ethyl iodide it would show the proton of the antihydrogen of the methyl group at *a different chemical shift*. We know, however, that in the spectrum of ethyl iodide (Fig. 14.16), the three protons of the methyl group give a single signal (a signal that is split into a triplet by spin-spin coupling with the two protons of the adjacent carbon).

The methyl protons of ethyl iodide give a single signal because at room temperature the groups connected by the carbon-carbon single bond rotate approximately one million times each second. The "shutter speed" of the NMR spectrometer is too slow to "photograph" this rapid rotation; instead, it photographs the methyl hydrogens in their average environments, and in this sense, it gives us a blurred picture of the methyl group.

Rotations about single bonds slow down as the temperature of the compound is lowered. Sometimes, this slowing of rotations allows us to "see" the different conformations of a molecule when we determine the spectrum at a sufficiently low temperature.

An example of this, and one that also shows the usefulness of deuterium labeling, can be seen in the low temperature pmr spectra of cyclohexane and of undecadeuteriocyclohexane.

undecadeuteriocyclohexane

At room temperature, ordinary cyclohexane gives a single signal because interconversions between the various chair forms occur very rapidly. At low temperatures, however, ordinary cyclohexane gives a very complex pmr spectrum. At low temperatures interconversions are slow; the axial and equatorial protons have different chemical shifts; and complex spin-spin couplings occur.

At $-100°$, however, undecadeuteriocyclohexane gives only two unsplit signals of equal intensity. These signals correspond to the axial and equatorial hydrogens of the two chair conformations shown below.

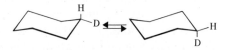

Interconversions between these conformations occur at this low temperature, but they happen slowly enough for the NMR spectrometer to detect the individual conformations.

Problem 14.13

What kind of pmr spectrum would you expect to obtain from undecadeuteriocyclohexane at room temperature?

Another example of a rapidly occurring process can be seen in pmr spectra of ethanol. The pmr spectrum of ordinary ethanol shows the hydroxyl proton as a singlet and the protons of the $-CH_2-$ group as a quartet. (Fig. 14.28). In ordinary ethanol we observe *no signal splitting arising from coupling between the hydroxyl proton and the protons of the $-CH_2-$ group even though they are on adjacent atoms.*

If we examine a pmr spectrum of *very pure* ethanol, however, we find that the signal from the hydroxyl proton is split into a triplet, and that the signal from the protons of $-CH_2-$ group is split into a multiplet of eight peaks. Clearly, in

FIG. 14.28
The pmr spectrum of ordinary ethanol. (Spectrum courtesy of Varian Associates, Palo Alto, Calif.)

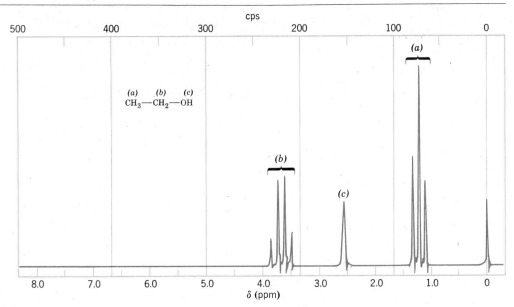

very pure ethanol the spin of the proton of the hydroxyl group couples with the spins of the protons of the —CH_2— groups.

Whether or not coupling occurs between the hydroxyl protons and the methylene protons depends on the length of time the proton spends on a particular ethanol molecule. Protons attached to very electronegative atoms such as oxygen can undergo rapid *chemical exchange*. That is they can be transferred rapidly from one molecule to another. The rate of chemical exchange in very pure ethanol is slow and, as a consequence, we see the signal splitting of and by the hydroxyl proton in the spectrum. In ordinary ethanol acidic and basic impurities catalyze the rate of chemical exchange; the exchange occurs so rapidly that the hydroxyl proton gives an unsplit signal and the methylene protons are split only by coupling with the protons of the methyl group. We say, then, that rapid exchange causes *spin decoupling*.

Spin decoupling is often found in the pmr spectra of alcohols, amines, and carboxylic acids.

14.7 PMR SPECTRA OF CARBOCATIONS

Professor Olah (p. 150) has developed methods for preparing carbocations under conditions where they are stable enough to be studied by NMR spectroscopy. Olah has found, for example, that in liquid sulfur dioxide, alkyl fluorides react with antimony pentafluoride to yield solutions of carbocations.

$$R—F + SbF_5 \xrightarrow[\text{liq. } SO_2]{} R^+SbF_6^-$$

Antimony pentafluoride is a powerful Lewis acid.

When the pmr spectrum of *tert*-butyl fluoride is measured in liquid sulfur dioxide the nine protons appear as a doublet centered at δ 1.3. (Why do the protons of *tert*-butyl fluoride appear as a doublet?) When antimony pentafluoride is added to the solution, the doublet at δ 1.35 is replaced by a singlet at δ 4.35. Both the change in the splitting pattern of the methyl protons and the downfield shift are consistent with the formation of a *tert*-butyl cation.

$$(CH_3)_3C—F + SbF_5 \xrightarrow[\text{SO}_2]{\text{liq.}} (CH_3)_3C^+ \quad SbF_6^-$$

$$\begin{matrix} \delta\,1.35 & & \delta\,4.35 \\ \text{(doublet)} & & \text{(singlet)} \end{matrix}$$

When a solution of isopropyl fluoride in liquid sulfur dioxide is treated with antimony pentafluoride an even more remarkable downfield shift occurs.

$$(CH_3)_2—CHF + SbF_5 \longrightarrow (CH_3)_2—CH^+ \quad SbF_6^-$$

$$\begin{matrix} \delta\,1.23 & \delta\,4.64 & & \delta\,5.03 & \delta\,13.50 \\ \text{(quartet)} & \text{(multiplet)} & & \text{(doublet)} & \text{(septet)} \end{matrix}$$

The pmr spectrum of the σ complex that results when mesitylene is protonated (p. 448) has also been observed. In this experiment a solution of mesitylene in liquid sulfur dioxide was treated with hydrogen fluoride and antimony pentafluoride.

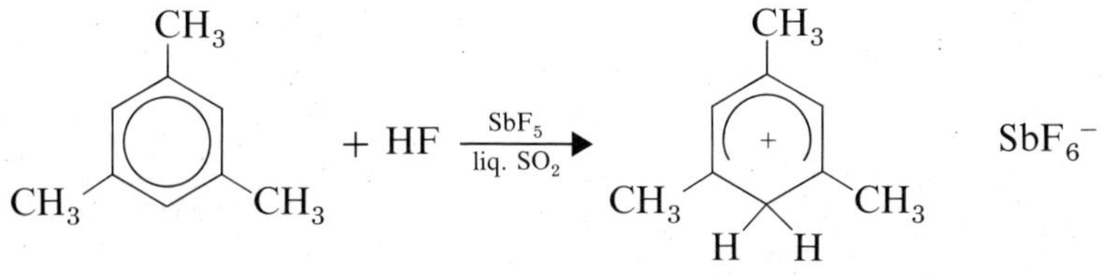

(a) Singlet, δ 2.35 (9H)   (a) Singlet, δ 2.8 (6H)
(b) Singlet, δ 6.70 (3H)   (b) Singlet, δ 2.9 (3H)
                                    (c) Singlet, δ 4.6 (2H)
                                    (d) Singlet, δ 7.7 (2H)

Problem 14.14

Make assignments for each of the pmr signals of mesitylene and the σ complex given above.

Problem 14.15

When 1,1-diphenyl-2-methyl-2-propanol in liquid sulfur dioxide was treated with the "superacid" FSO_3H-SbF_5 the solution showed the pmr absorptions given below.

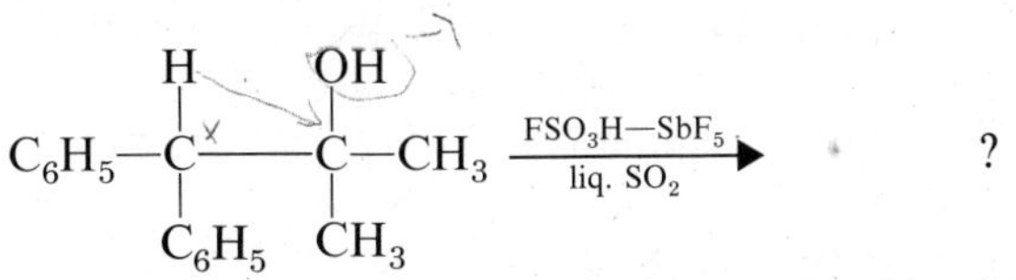

1,1-Diphenyl-2-methyl-
2-propanol

Doublet, δ 1.48 (6H)
Multiplet, δ 4.45 (1H)
Multiplet, δ 8.0 (10H)

What carbocation is formed in this reaction?

14.8 INFRARED SPECTROSCOPY

We saw in Section 10.9 that many organic compounds absorb light in the visible and ultraviolet regions of the electromagnetic spectrum. We also saw that when compounds absorb light of the visible and ultraviolet regions, electrons are excited from lower energy molecular orbitals to higher ones.

Organic compounds also absorb electromagnetic energy in the infrared region. Infrared radiation does not have sufficient energy to cause the excitation of electrons but it does cause atoms and groups of organic compounds to vibrate about the covalent bonds that connect them. The vibrations are *quantized* and as they occur, the compounds absorb infrared energy in particular regions of the spectrum.*

In their vibrational absorptions, covalent bonds behave as though they were elastic—as though there were tiny springs connecting the atoms. When covalently bonded atoms vibrate they do so only at certain frequencies, as though the bonds

---

* Radiation in the far infrared region causes rotations of the molecules and these absorptions give rise to what are known as rotational spectra. We will limit our discussion, however, to the region of the infrared spectrum in which the primary molecular absorptions are a result of molecular vibrations.

were "tuned." Because of this, covalently bonded atoms have only particular vibrational energy levels and excitation of a molecule from one vibrational energy level to another occurs only when the compound absorbs infrared radiation of a particular wavelength or frequency.

The infrared spectra of even relatively simple compounds contain many absorption peaks. It can be shown that a nonlinear molecule of n atoms has $3n - 6$ possible *fundamental* vibrational modes that can be responsible for the absorption of infrared radiation. This means that, theoretically, methane has 9 possible fundamental absorption peaks and benzene has 30.

All molecular vibrations do not result in the absorption of energy, however. In order for a vibration to occur with the absorption of infrared energy, *the dipole moment of the molecule must change as the vibration occurs.* Thus when the four hydrogens of methane vibrate symmetrically, methane does not absorb infrared energy. Symmetrical vibrations of the carbon-carbon double and triple bonds of ethene and ethyne do not result in the absorption of infrared radiation, either.

Vibrational absorption may occur outside the region measured by a particular infrared spectrophotometer and vibrational absorptions may occur so closely together that peaks fall on top of peaks. These factors, together with the absence of absorptions because of vibrations that have no dipole moment change, cause infrared spectra to contain fewer peaks than the formula $3n - 6$ would predict.

However, other factors bring about even more absorption peaks. Overtones (harmonics) of fundamental absorption bands may be seen in infrared spectra even though these overtones occur with greatly reduced intensity. Combination bands and difference bands also appear in infrared spectra.

Because infrared spectra contain so many peaks, the possibility that two compounds will have the same infrared spectrum is exceedingly small. In this sense an infrared spectrum has been compared to a "fingerprint" of a molecule. Thus, with organic compounds, if two pure samples give different infrared spectra, one

FIG. 14.29

The infrared spectrum of n-*octane. (Notice that, in infrared spectra, the peaks are "upside down." This is simply a result of the way infrared spectrophotometers operate.) (Spectrum by author.)*

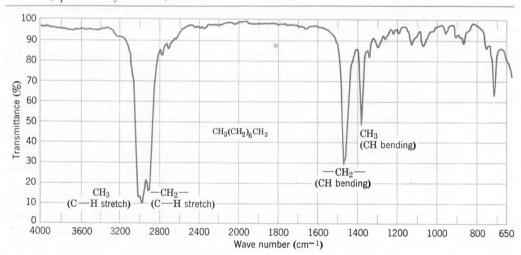

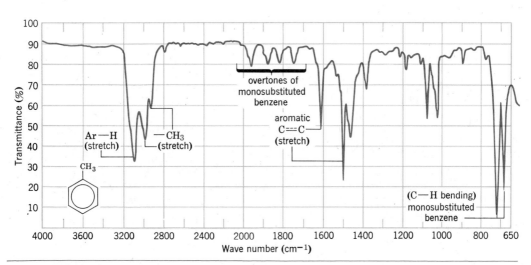

FIG. 14.30
The infrared spectrum of toluene. (Spectrum by author.)

can be certain that they are different compounds. If they give the same infrared spectrum then they are the same compound.

In the hands of one skilled in their interpretation, infrared spectra contain a wealth of information about the structures of compounds. We show some of the information that can be gathered from the spectra of *n*-octane and toluene in Figs. 14.29 and 14.30. We have neither the time nor the space here to develop the skill that would lead to complete interpretations of infrared spectra, but we can learn how to recognize the presence of absorption peaks in the infrared spectrum that result from vibrations of characteristic functional groups in the compound. By doing only this, however, we will be able to use the information we gather from infrared spectra in a powerful way, particularly when we couple it with the information we gather from proton magnetic resonance spectra.

Infrared absorption peaks are usually measured in either *wave numbers* (cm^{-1}) or in microns (μ). Wave numbers are frequency units that correspond to the number of cycles of the wave in each centimeter. Microns are measurements of wave length (1 $\mu = 10^{-4}$ cm). Table 14.2 gives the positions of infrared peaks of some important functional groups.

Let us now see how we can apply the data given in Table 14.2 to the interpretation of infrared spectra.

Hydrocarbons

All hydrocarbons give absorption peaks in the 2800–3300 cm^{-1} region that are associated with carbon-hydrogen stretching vibrations. We can use these peaks in interpreting infrared spectra because the exact location of the peak depends on the hybridization state of the carbon that bears the hydrogen.

The carbon-hydrogen stretching peaks of hydrogens attached to *sp*-hybridized carbons occur at highest frequencies, ~3300 cm^{-1}. Thus, $\equiv$C—H groups of terminal alkynes give peaks in this region. We can see the absorption of the acetylenic hydrogen of 1-hexyne at 3320 cm^{-1} in Fig. 14.31.

TABLE 14.2 Characteristic Infrared Absorptions of Functional Groups

GROUP		FREQUENCY RANGE cm^{-1}	INTENSITY*
A. Alkyl			
C—H (Stretching)		2853–2962	(m–s)
Isopropyl, —CH(CH$_3$)$_2$		1380–1385	(s)
	and	1365–1370	(s)
tert-Butyl, —C(CH$_3$)$_3$		1385–1395	(m)
	and	~1365	(s)
B. Alkenyl			
C—H (stretching)		3010–3095	(m)
C=C (stretching)		1620–1680	(v)
R—CH=CH$_2$		985–1000	(s)
	and	905–920	(s)
R$_2$C=CH$_2$	(out-of-plane C—H bendings)	880–900	(s)
cis-RCH=CHR		675–730	(s)
trans-RCH=CHR		960–975	(s)
C. Alkynyl			
≡C—H (stretching)		~3300	(s)
C≡C (stretching)		2100–2260	(v)
D. Aromatic			
Ar—H (stretching)		~3030	(v)
Aromatic substitution type (C—H out-of-plane bendings)			
Monosubstituted		690–710	(v, s)
	and	730–770	(v, s)
o-Disubstituted		735–770	(s)
m-Disubstituted		680–725	(s)
	and	750–810	(s)
p-Disubstituted		790–840	(s)
E. Alcohols, Phenols, Carboxylic Acids			
OH (alcohols, phenols)		3200–3600	(broad, s)
OH (carboxylic acids)		2500–3000	(broad, s)
F. Aldehydes, Ketones, Esters, and Carboxylic Acids			
C=O stretch		1690–1750	(s)
G. Amines			
N—H		3300–3500	(m)
H. Nitriles			
C≡N		2220–2260	(m)

*Abbreviations: s = strong, m = medium, w = weak, v = variable, ~ = approximately.

The carbon-hydrogen stretching peaks of hydrogens attached to sp^2-hybridized carbons occur in the 3000–3100 cm^{-1} region. Thus, vinyl groups and the C—H groups of aromatic rings give absorption peaks in this region. We can see the vinyl C—H absorption peak at 3080 cm^{-1} in the spectrum of 1-hexene (Fig. 14.32) and we can see the C—H absorption of the aromatic hydrogens at 3090 cm^{-1} in the spectrum of toluene (Fig. 14.30).

The carbon hydrogen stretching bands of hydrogens attached to sp^3-hybridized carbons occur in the 2800–3000 cm^{-1} region. We can see methyl and methylene

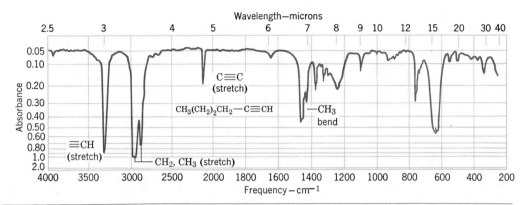

FIG. 14.31

The infrared spectrum of 1-hexyne. (Spectrum courtesy of Sadtler Research Laboratories, Inc., Philadelphia, Pa.)

absorption peaks in the spectra of *n*-octane (Fig. 14.29), toluene (Fig. 14.30), 1-hexyne (Fig. 14.31), and 1-hexene (Fig. 14.32).

Hydrocarbons also give absorption peaks in their infrared spectra that result from carbon-carbon bond stretchings. Carbon-carbon single bonds normally give rise to very weak peaks that are usually of little use in assigning structures. More useful peaks arise from multiple carbon-carbon bonds, however. Carbon-carbon double bonds give absorption peaks in the 1620–1680 cm^{-1} region and carbon-carbon triple bonds give absorption peaks between 2100 cm^{-1} and 2260 cm^{-1}. These absorptions are not usually strong ones and they will not be present at all if the double or triple bond is symmetrically substituted. (No dipole moment change will be associated with the vibration.) The stretchings of the carbon-carbon bonds of benzene rings usually give a set of characteristic sharp peaks in the 1450–1600 cm^{-1} region.

Absorptions arising from carbon-hydrogen bending vibrations of alkenes occur in the 600–1000 cm^{-1} region. The exact location of these peaks can often be used to determine the *nature of the double bond and its stereochemistry.*

Monosubstituted alkenes give two strong peaks in the 905–920 cm^{-1} and the 985–1000 cm^{-1} regions. Disubstituted alkenes of the type $R_2C{=}CH_2$ give a strong

FIG. 14.32

The infrared spectrum of 1-hexene. (Spectrum courtesy of Sadtler Research Laboratories, Inc., Philadelphia, Pa.)

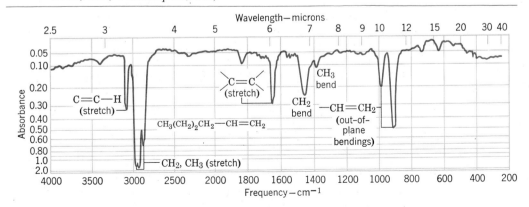

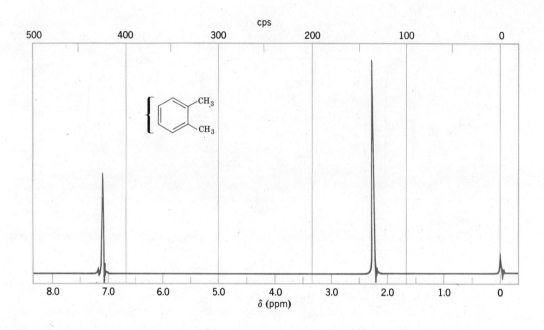

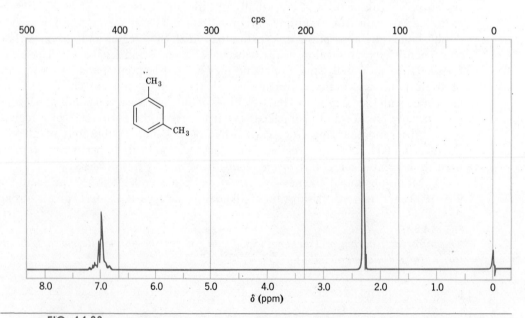

FIG. 14.33

The pmr spectra of ortho- *and* meta-*xylene. (Spectrum courtesy of Varian Associates, Palo Alto, Calif.)*

peak in the 880–900 cm^{-1} range. *Cis*-alkenes give an absorption peak in the 675–730 cm^{-1} region and *trans*-alkenes give a peak between 960 cm^{-1} and 975 cm^{-1}. These ranges for the carbon-hydrogen bending vibrations can be used with fair reliability for alkenes that do not have an electron releasing or withdrawing substituent (other than an alkyl group) on one of the carbons of the double bond. When

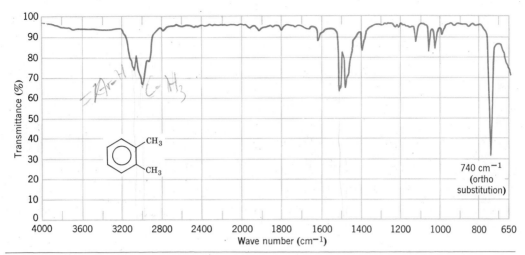

FIG. 14.34
The infrared spectrum of o-*xylene.*

electron-releasing or withdrawing substituents are present on a double-bond carbon the bending absorption peaks may be shifted out of the regions we have given.

Substituted Benzenes

If we compare the pmr spectra of the three xylenes (Fig. 14.2 and Fig. 14.33) we find that they are very similar. Using pmr spectra alone, we would have little difficulty in deciding whether an unknown compound was a xylene or not. We would, however, have considerable difficulty in deciding just which xylene it was, *o*-xylene, *m*-xylene, or *p*-xylene.

FIG. 14.35
The infrared spectrum of m-*xylene.*

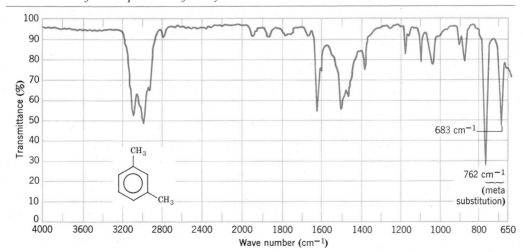

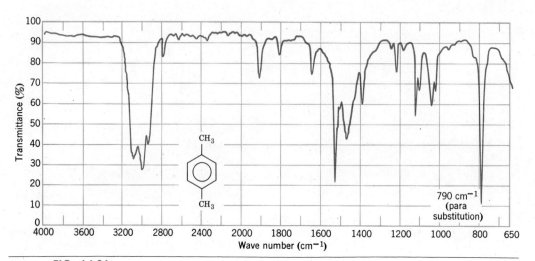

FIG. 14.36
The infrared spectrum of p-*xylene.* (*Spectra by author.*)

Infrared spectroscopy provides us with a way around this difficulty. Ortho-, meta-, and para-substituted benzenes give absorption peaks in the 680–840 cm^{-1} region that characterize their substitution patterns. Ortho-substituted benzenes show an absorption peak due to bending motions of the aromatic hydrogens between 735 and 770 cm^{-1}. Meta-substituted benzenes show two peaks; one between 680 and 725 cm^{-1} and one between 750 and 810 cm^{-1}. Para-substituted benzenes give a single absorption band between 790 and 840 cm^{-1}.

These characteristic peaks are designated in the infrared spectra of ortho-, meta-, and para-xylene (Figs. 14.34–36).

Monosubstituted benzenes give two peaks, between 690–710 cm^{-1} and 730–770 cm^{-1}.

Problem 14.16

Using Table 14.2, make assignments to as many other peaks as you can in the spectra of *o*-, *m*-, and *p*-xylene.

Problem 14.17

Four benzenoid compounds, all with the formula C_7H_7Br, gave the following infrared peaks in the 680–840 cm^{-1} region.

        A, 740 cm^{-1} **C,** 680 cm^{-1} and 760 cm^{-1}
        B, 795 cm^{-1} **D,** 693 cm^{-1} and 765 cm^{-1}

Propose structures for **A**, **B**, **C**, and **D**.

Other Functional Groups

We will study the infrared spectra of other functional groups in more detail in subsequent chapters. Table 14.2, however, lists some characteristic absorption peaks that may be helpful in interpreting some of the spectra at the end of this chapter.

Oxygen-hydrogen stretching vibrations give absorption peaks in the 3200–3600 cm^{-1} range for alcohols and phenols and in the 2500–3000 cm^{-1} region for carboxylic acids. Nitrogen-hydrogen stretching vibrations of amines give absorption in the 3300–3500 cm^{-1} region. All of the bands may be very broad because of hydrogen bonding.

Carbonyl groups of aldehydes, ketones, esters, and carboxylic acids give strong and very characteristic absorption peaks in the 1690–1750 cm^{-1} region.

Additional Problems

14.18

Listed below are pmr absorption peaks for several compounds. Propose structures that are consistent with each set of data. (In some cases characteristic infrared absorptions are given as well.)

(a) $C_4H_{10}O$ pmr spectrum
        singlet, $\delta1.28$ (9H)
        singlet, $\delta1.35$ (1H)

(b) C_3H_7Br pmr spectrum
        doublet, $\delta1.71$ (6H)
        septet, $\delta4.32$ (1H)

(c) C_4H_8O pmr spectrum   IR spectrum
        triplet, $\delta1.05$ (3H)   1720 cm^{-1}
        singlet, $\delta2.13$ (3H)
        quartet, $\delta2.47$ (2H)

(d) C_7H_8O pmr spectrum   IR spectrum
        singlet, $\delta2.43$ (1H)   broad peak in
        singlet, $\delta4.58$ (2H)   3200–3600 cm^{-1}
        multiplet, $\delta7.28$ (5H)   region

(e) C_4H_9Cl pmr spectrum
        doublet, $\delta1.04$ (6H)
        multiplet, $\delta1.95$ (1H)
        doublet, $\delta3.35$ (2H)

(f) $C_{15}H_{14}O$ pmr spectrum   IR spectrum
        singlet, $\delta2.20$ (3H)   strong peak
        singlet, $\delta5.08$ (1H)   near 1720 cm^{-1}
        multiplet, $\delta7.25$ (10H)

(g) $C_4H_7BrO_2$ pmr spectrum   IR spectrum
        triplet, $\delta1.08$ (3H)   broad peak in
        multiplet, $\delta2.07$ (2H)   2500–3000 cm^{-1}
        triplet, $\delta4.23$ (1H)   region
        singlet, $\delta10.97$ (1H)

(h) C_8H_{10} pmr spectrum
        triplet, $\delta1.25$ (3H)
        quartet, $\delta2.68$ (2H)
        multiplet, $\delta7.23$ (5H)

(i) $C_4H_8O_3$ pmr spectrum   IR spectrum
        triplet, $\delta1.27$ (3H)   broad peak in
        quartet, $\delta3.66$ (2H)   2500–3000 cm^{-1}
        singlet, $\delta4.13$ (2H)   region
        singlet, $\delta10.95$ (1H)

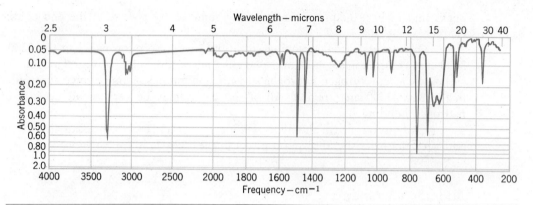

FIG. 14.37

The infrared spectrum of compound E. (Spectrum courtesy of Sadtler Research Laboratories, Philadelphia, Pa.)

(j) $C_3H_7NO_2$ pmr spectrum
          doublet, $\delta 1.55$ (6H)
          septet, — $\delta 4.67$ (1H)

(k) $C_4H_{10}O_2$ pmr spectrum
          singlet, $\delta 3.25$ (6H)
          singlet, $\delta 3.45$ (4H)

(l) $C_5H_{10}O$ pmr spectrum   IR spectrum
          doublet, $\delta 1.10$ (6H)   strong peak
          singlet, $\delta 2.10$ (3H)   near 1720 cm^{-1}
          septet, $\delta 2.50$ (1H)

(m) C_8H_9Br pmr spectrum
          doublet, $\delta 2.0$ (3H)
          quartet, $\delta 5.15$ (1H)
          multiplet, $\delta 7.35$ (5H)

14.19

The infrared spectrum of compound **E**, C_8H_6, is shown in Fig. 14.37. **E** decolorizes bromine in carbon tetrachloride and gives a precipitate when treated with ammoniacal silver nitrate. What is the structure of **E**?

14.20

The pmr spectrum of cyclooctatetraene consists of a single line located at $\delta 5.78$. What does this suggest about electron delocalization in cyclooctatetraene?

14.21

The [14] annulene shown below obeys Huckel's rule. Its pmr spectrum shows signals at $\delta 7.78$

[14] Annulene   Dehydro [14] annulene

(10H) and δ−0.61 (4H). Dehydro [14] annulene gives pmr signals at δ8.0 (10H) and δ0.0 (2H). How can you account for the relative intensities of the signals given by the two compounds?

14.22

Give a structure for compound **F** that is consistent with the pmr and IR spectra below.

FIG. 14.38

The pmr and IR spectra of compound F, problem 14.22. (pmr spectrum adapted from Varian Associates, Palo Alto, Calif. IR spectrum adapted from Sadtler Research Laboratories, Philadelphia, Pa.)

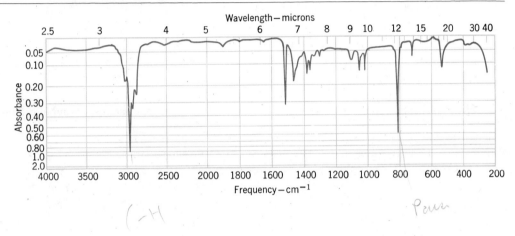

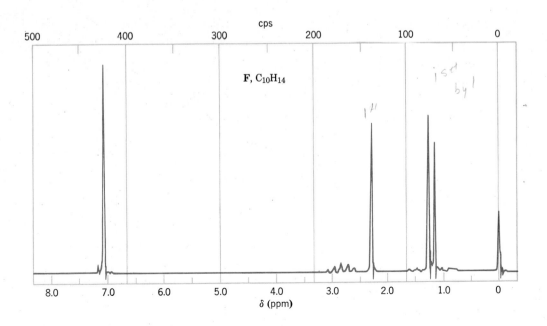

F, $C_{10}H_{14}$

14.23

(a) When either *n*-butyl fluoride or *sec*-butyl fluoride is treated with excess SbF_5, their solutions give identical pmr spectra. These spectra, moreover, are the same as that obtained when *tert*-butyl fluoride is treated with SbF_5. Explain these results. (b) Treating the seven isomeric fluoropentanes with excess SbF_5 furnishes solutions that give identical pmr spectra. What species is formed in these reactions? Sketch the spectrum that you would expect to obtain.

14.24

(a) How many peaks would you expect to find in the pmr spectrum of caffeine?

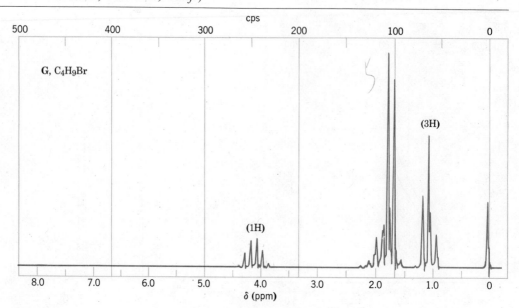

Caffeine

(b) What characteristic peaks would you expect to find in the infrared spectrum of caffeine?

14.25

Propose structures for the compounds **G** and **H** whose pmr spectra are shown in Fig. 14.39 and Fig. 14.40.

FIG. 14.39

The pmr spectrum of compound **G** (*problem 14.25*). (*Spectrum courtesy of Varian Associates, Palo Alto, Calif.*)

14.26

Propose a structure for compound **I** whose pmr and infrared spectra are given in Fig. 14.41 and Fig. 14.42.

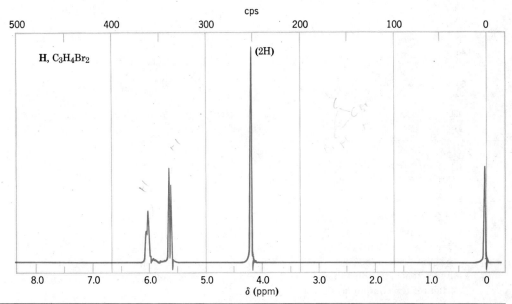

FIG. 14.40

*The pmr spectrum of compound **H** (problem 14.25). (Spectrum courtesy of Varian Associates, Palo Alto, Calif.)*

FIG. 14.41

*The pmr spectrum of compound **I** (problem 14.26). (Spectrum courtesy of Varian Associates, Palo Alto, Calif.)*

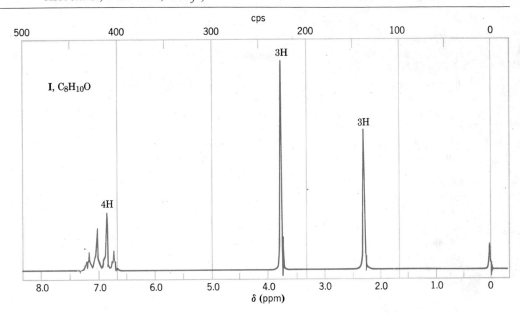

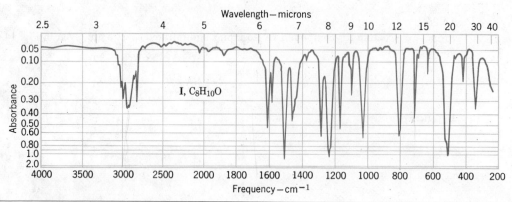

FIG. 14.42

The infrared spectrum of compound **I** (*problem 14.26*). (*Spectrum courtesy of Sadtler Research Laboratories, Philadelphia, Pa.*)

14.27

A two-carbon compound, **J**, contains only carbon, hydrogen, and chlorine. Its infrared spectrum is relatively simple and shows the following absorbance peaks: 3125 cm^{-1}(m), 1625 cm^{-1}(m), 1280 cm^{-1}(m), 820 cm^{-1}(s), 695 cm^{-1}(s). The pmr spectrum of **J** consists of a singlet at $\delta6.3$. Using Table 14.2 make as many infrared assignments as you can and propose a structure for compound **J**.

14.28

When dissolved in $CDCl_3$, a compound, **K**, with the molecular formula $C_4H_8O_2$ gives a pmr spectrum that consists of a doublet at $\delta1.35$, a singlet at $\delta2.15$, a broad singlet at $\delta3.75$ (1H) and a quartet at $\delta4.25$(1H). When dissolved in D_2O, the compound gives a similar pmr spectrum with the exception that the signal at $\delta3.75$ has disappeared. The infrared spectrum of the compound shows a strong absorption peak near 1720 cm^{-1}. (a) Propose a structure for compound **K** and (b) explain why the pmr signal at $\delta3.75$ disappears when D_2O is used as the solvent.

14.29

A compound, **L**, with the molecular formula C_9H_{10} decolorizes bromine in carbon tetrachloride and gives an infrared absorption spectrum that includes the following absorption peaks: 3035 cm^{-1}(m), 3020 cm^{-1}(m), 2925 cm^{-1}(m), 2853 cm^{-1}(w), 1640 cm^{-1}(m), 990 cm^{-1}(s), 915 cm^{-1}(s), 740 cm^{-1}(s), 695 cm^{-1}(s). The pmr spectrum of **L** consists of:

doublet $\delta3.1$(2H)   multiplet $\delta5.1$ multiplet $\delta7.1$(5H)
multiplet $\delta4.8$ multiplet $\delta5.8$

The ultraviolet spectrum shows a maximum at 255 nm. Propose a structure for compound **L** and make assignments for each of the infrared peaks. *not conjugated w/* $=$

14.30

Assume that in a certain pmr spectrum, you find two peaks of roughly equal intensity. You are not certain whether these two peaks are *singlets* arising from uncoupled protons at different chemical shifts, or whether they are two peaks of a *doublet* that arises from protons coupling with a single adjacent proton. What simple experiment would you perform to distinguish between these two possibilities?

14.31

Compound **M** has the molecular formula C_9H_{12}. The pmr spectrum of **M** is given in Fig. 14.43 and the infrared spectrum in Fig. 14.44. Propose a structure for **M**.

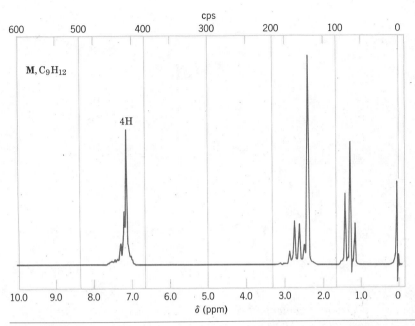

FIG. 14.43
*The pmr spectrum of compound **M**, problem 14.31. (Spectrum courtesy of*
Aldrich Chemical Co., Milwaukee, Wisconsin.)

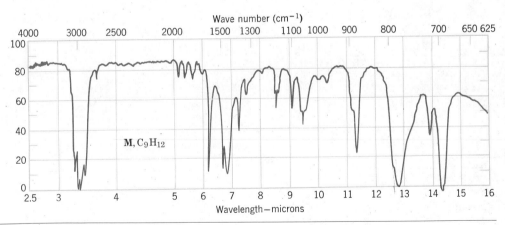

FIG. 14.44
*The infrared spectrum of compound **M**, problem 14.31. (Spectrum courtesy of Aldrich*
Chemical Co., Milwaukee, Wis.)

14.32

A compound, **N**, with the molecular formula $C_9H_{10}O$ gives a positive test with cold dilute
aqueous potassium permanganate. The pmr spectrum of **N** is shown in Fig. 14.45 and the
infrared spectrum of **N** is shown in Fig. 14.46. Propose a structure for **N**.

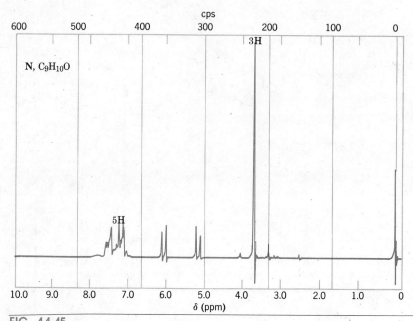

FIG. 14.45

*The pmr spectrum of compound **N**, problem 14.32. (Spectrum courtesy of Aldrich Chemical Co., Milwaukee, Wis.)*

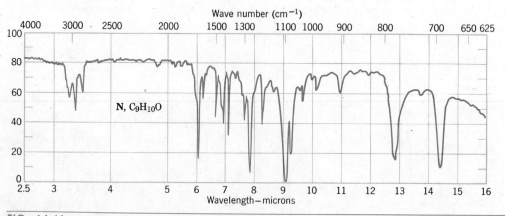

FIG. 14.46

*The infrared spectrum of compound **N**, problem 14.32. (Spectrum courtesy of Aldrich Chemical Co., Milwaukee, Wis.)*

* 14.33

When 2,3-dibromo-2,3-dimethylbutane is treated with SbF_5 in liquid SO_2 at $-60°$, the pmr spectrum does not show the two signals that would be expected of a carbocation like $CH_3\overset{+}{C}Br\overset{|}{C}CH_3$. Instead only one signal (at $\delta 2.9$) is observed. What carbocation is formed in $\underset{CH_3CH_3}{}$

this reaction and of what special significance is this experiment?

* **14.34**

1,2,3,4-Tetramethyl-3,4-dichlorocyclobutene, I, gives a pmr signal at $\delta1.15$ corresponding to the protons labeled **A** and a signal at $\delta1.26$ corresponding to the protons labeled **B**.

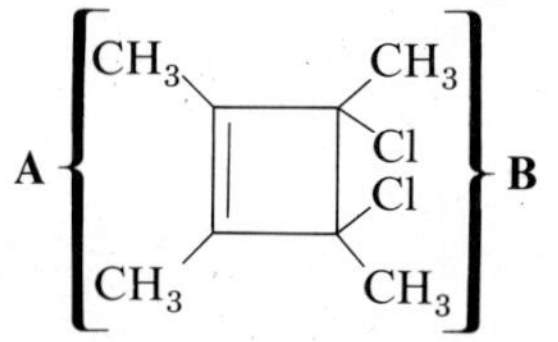

I

When I is added to SbF_5—SO_2 at $-78°$ a pale-yellow solution is formed whose pmr spectrum shows the following singlets: $\delta2.05$ (3H), $\delta2.20$ (3H), $\delta2.65$ (6H). After several minutes, these peaks begin to be replaced by a sharp singlet at $\delta3.68$. Recall that SbF_5 is a powerful Lewis acid and explain what is taking place.

* **14.35** .

Given the following information predict the appearance of the pmr spectrum given by the vinyl hydrogens of p-chlorostyrene.

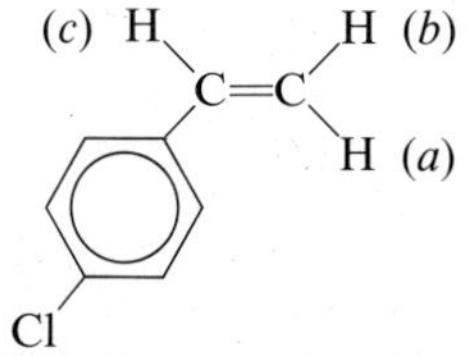

Deshielding by the induced magnetic field of the ring is greatest at proton c ($\delta6.7$) and is least at proton b ($\delta5.3$). The chemical shift of b is $\sim \delta5.7$. The coupling constants have the following approximate magnitudes: $J_{ac} \simeq 18$ Hz, $J_{bc} \simeq 11$ Hz, and $J_{ab} \simeq 2$ Hz. (These coupling constants are typical of those given by vinylic systems: coupling constants for *trans* hydrogens are larger than those for *cis* hydrogens, and coupling constants for geminal hydrogens are very small.)

15

ORGANIC HALIDES AND ORGANOMETALLIC COMPOUNDS

15.1 INTRODUCTION

We have already encountered organic halogen compounds at many points in this text and we will see them again at many others. The reason for this is simple. Halogen substituents, particularly chlorine and bromine, can be introduced into organic compounds by a variety of methods and they can usually be easily replaced by other groups. Organic halogen compounds, therefore, are used frequently in synthetic procedures.

Halogen substituents on saturated carbon atoms can be replaced in substitution reactions. We saw an example of this in the synthesis of alkynes from sodium acetylides and primary alkyl halides in Section 9.10.

$$R-C{\equiv}C{:}^- \, Na^+ + R'-CH_2-X \longrightarrow R-C{\equiv}C-CH_2-R' + NaX$$

Halogen substituents also give us the possibility of carrying out elimination reactions and thus provide a method for introducing carbon-carbon multiple bonds. We saw examples of this synthetic application in the preparations of alkenes and alkynes (Sections 5.12 and 9.11).

$$\underset{\underset{X}{|}}{R-CH_2-CH-R} + NaOR'' \longrightarrow R-CH{=}CH-R + R''OH + NaX$$

$$\underset{\underset{X}{|} \quad \underset{X}{|}}{R-CH-CH-R'} + 2NaNH_2 \longrightarrow R-C{\equiv}C-R' + 2NaX$$

Halogen substituents also allow us to prepare compounds in which organic groups are bonded to metals. *Organometallic compounds*, as we will see, are of considerable synthetic importance because the metal portion of many organometallic compounds can be replaced by a variety of other groups. Organometallic compounds also give us valuable synthetic methods for creating new carbon-carbon single bonds.

Organometallic compounds are almost always prepared from organic halides. Therefore, we will study organometallic compounds in detail in this chapter. As we do this, we will focus our attention particularly on the organic compounds of magnesium, lithium, and thallium.

Substitution and elimination reactions are reactions that organic halides share with a number of other compounds that contain good "leaving groups." For this reason, we will postpone a detailed study of substitution and elimination reactions until we reach Chapter 17.

15.2 PREPARATION OF ALIPHATIC HALIDES

Alkyl, allyl, and vinyl halides can be prepared by a variety of methods, many of which we have already seen.

Alkyl Halides From Alkanes (discussed in Section 4.9)

We saw in Chapter 4 that alkanes and cycloalkanes undergo substitution reactions when they react with halogens. While alkane halogenations find considerable use in industrial preparations they are less important in laboratory work because halogens react with most alkanes in relatively nonspecific ways.

Alkyl and Allyl Halides From Alkenes

Alkyl halides can be prepared from alkenes by the addition reactions of hydrogen halides (Section 6.2).

$$R—CH{=}CH—R + HX \longrightarrow R—CH_2\underset{\underset{X}{|}}{C}H—R$$

Cyclohexyl fluoride, for example, can be prepared through the addition reaction of hydrogen fluoride and cyclohexene.

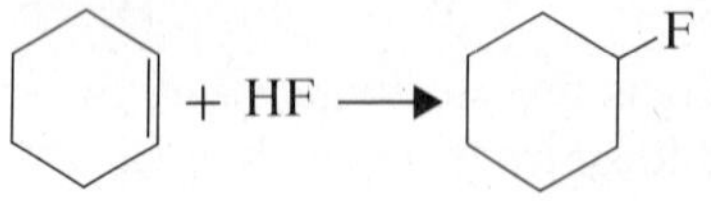

(60%)

Cyclohexene also reacts with a mixture of potassium iodide and phosphoric acid to produce cyclohexyl iodide.

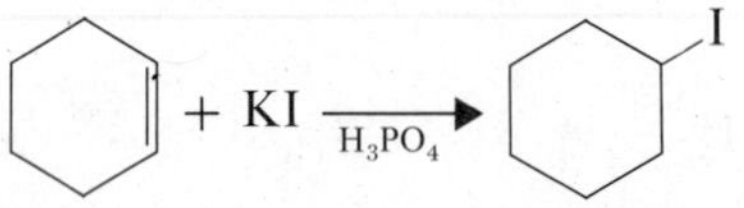

(88–90%)

Hydrogen bromide adds to alkenes in a Markovniknov fashion in the absence of peroxides and in an anti-Markovniknov way in their presence (Section 6.9).

$$CH_3CH_2CH_2CH{=}CH_2 + HBr \xrightarrow[CH_3COOH]{} CH_3CH_2CH_2\underset{\underset{Br}{|}}{C}HCH_3$$

(84%)

$$CH_3CH_2CH_2CH{=}CH_2 + HBr \xrightarrow[peroxides]{} CH_3CH_2CH_2CH_2CH_2Br$$

(81%)

Allyl halides can be prepared by allowing alkenes to react with chlorine or bromine at high temperatures or under conditions in which the concentration of halogen is low (Section 10.1).

$$CH_2=CHCH_3 + Cl_2 \xrightarrow{500°} CH_2=CHCH_2Cl + HCl$$

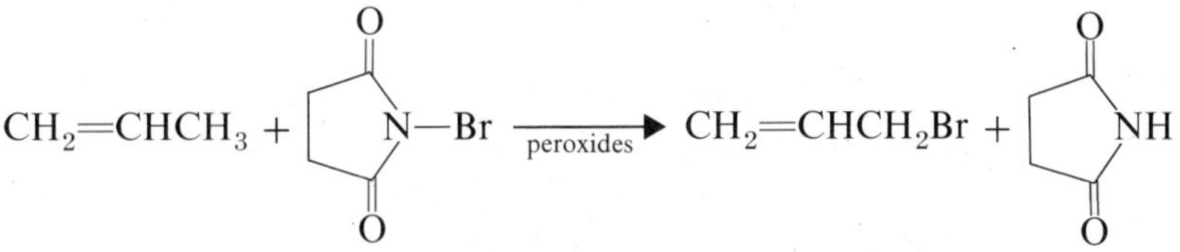

$$CH_2=CHCH_3 + \underset{O}{\overset{O}{\underset{\|}{\overset{\|}{N}}}}-Br \xrightarrow{\text{peroxides}} CH_2=CHCH_2Br + \underset{O}{\overset{O}{\underset{\|}{\overset{\|}{N}}}}NH$$

Benzyl halides can be prepared from alkyl benzenes in this same way (Section 13.10).

Vinyl Halides From Alkynes

Vinyl halides can be synthesized by the addition of one mole of a hydrogen halide to an alkyne (Section 9.8).

$$HC\equiv CH + HX \longrightarrow CH_2=CHX$$
$$RC\equiv CH + HX \longrightarrow RCX=CH_2$$

Addition of two moles of the hydrogen halide gives a *gem*-dihalide.

Alkyl, Allyl, and Benzyl Halides From Alcohols

Alcohols react with a variety of reagents to produce alkyl halides. Two commonly used reagents are phosphorus tribromide (PBr_3) and thionyl chloride ($SOCl_2$). One mole of phosphorus tribromide generally converts three moles of an alcohol to the alkyl bromide.

$$3ROH + PBr_3 \longrightarrow 3RBr + P(OH)_3$$

The other product is phosphorous acid, $P(OH)_3$.

One mole of thionyl chloride ($SOCl_2$) converts one mole of an alcohol to an alkyl chloride. The other products are SO_2 and HCl.

$$ROH + SOCl_2 \longrightarrow RCl + SO_2 + HCl$$

Phosphorus tribromide and thionyl chloride can also be used to synthesize bromides and chlorides from benzyl and allyl alcohols.

Listed below are several specific examples that illustrate the use of these two reagents.

$$3(CH_3)_2CHCH_2OH + PBr_3 \xrightarrow[\text{4 hrs.}]{-10° \text{ to } 0°} 3(CH_3)_2CHCH_2Br + P(OH)_3$$
$$(55-60\%)$$

$$3C_2H_5OCH_2CH_2OH + PBr_3 \xrightarrow{\text{reflux}} 3C_2H_5OCH_2CH_2Br + P(OH)_3$$
$$(66\%)$$

$$CH_3SCH_2CH_2OH + SOCl_2 \xrightarrow[\text{CHCl}_3]{\text{6 hrs.}} CH_3SCH_2CH_2Cl + SO_2 + HCl$$
$$(75-85\%)$$

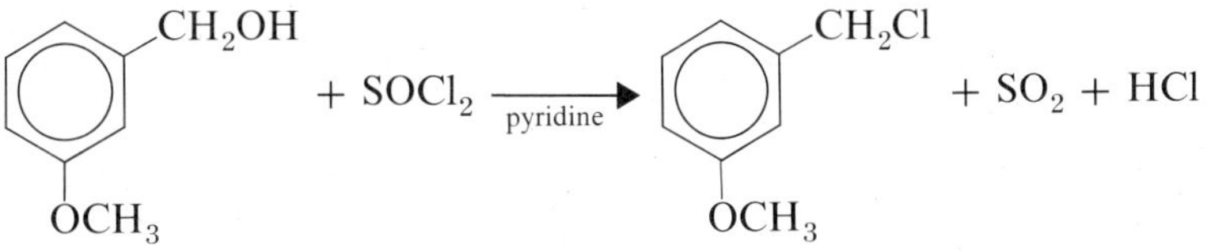

$$+ SOCl_2 \xrightarrow{\text{pyridine}} + SO_2 + HCl$$
$$(91\%)$$

Hydrogen halides also convert alcohols to alkyl halides:

$$ROH + HX \longrightarrow RX + H_2O$$

Several specific examples are shown below.

$$(C_2H_5)_3COH + HCl \longrightarrow (C_2H_5)_3CCl + H_2O$$
$$(94\%)$$
$$C_{12}H_{25}OH + HBr \longrightarrow C_{12}H_{25}Br + H_2O$$
$$(91\%)$$
$$HOCH_2CH_2CH_2CH_2CH_2CH_2OH \xrightarrow[H_3PO_4]{KI} ICH_2CH_2CH_2CH_2CH_2CH_2I$$
$$(85\%)$$

We will discuss the mechanisms of these reactions in Section 16.6.

Triphenylphosphine and bromine react to form a reagent that can also be used to synthesize bromides from alcohols. The synthesis of cinnamyl bromide (below) from cinnamyl alcohol illustrates the use of this reagent.

$$(C_6H_5)_3P\colon\; + Br_2 \xrightarrow{0-5°} (C_6H_5)_3PBr_2$$

Triphenylphos-
phine

Dibromotriphenylphosphine

CH=CHCH_2OH

+ (C_6H_5)_3PBr_2 $\xrightarrow[CH_3CN]{room\ temp.}$ CH=CHCH_2Br

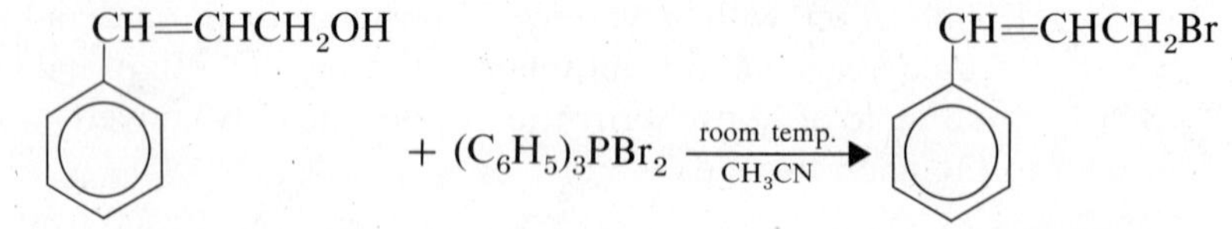

Cinnamyl alcohol

Cinnamyl bromide
(75%)

The reagent $(C_6H_5O)_3PX_2$ also converts alcohols to alkyl halides in excellent yields.

$$R—OH + (C_6H_5O)_3PX_2 \xrightarrow{1\ hr.} RX + (C_6H_5O)_2\overset{O}{\overset{\|}{P}}X + C_6H_5OH$$
$$(65-95\%)$$

15.3 ARYL HALIDES FROM ARENES: THALLATION REACTIONS

Aryl halides can be prepared through the direct introduction of a halogen in an electrophilic substitution reaction.

$$Ar—H + X_2 \xrightarrow{FeX_3} Ar—X + HX$$

We saw many examples of this method in Chapter 13. (We will also see how aryl halides can be prepared from aryl amines in Chapter 21.)

Recent experimental work done by Professors Edward C. Taylor (of Princeton University) and Alexander McKillop (of the University of East Anglia) has shown that compounds of the metal *thallium* can be of great use in organic synthesis. We will see other examples of organothallium chemistry later in this chapter, but we begin our study of organothallium chemistry here by examining the reaction of thallium trifluoroacetate with benzene compounds. As we do this, we will see that the use of thallium trifluoroacetate gives a method for introducing halogens (and other groups) into a benzene ring in highly specific ways.

Arenes react with thallium trifluoroacetate* by electrophilic aromatic substitution. The reaction is usually carried out in trifluoroacetic acid and takes place in the manner shown below.

$$ArH + Tl(OOCCF_3)_3 \xrightarrow{CF_3COOH} ArTl(OOCCF_3)_2 + CF_3COOH$$

Thallium trifluoroacetate   Arylthallium bis-trifluoroacetate

This reaction usually gives the arylthallium bis-trifluoroacetate in very high yields.

Because the carbon-thallium bond of arylthallium bis-trifluoroacetates is very weak (25–30 kcal/mole), it is easily cleaved homolytically and heterolytically. Cleavage of the carbon-thallium bond allows us to replace the thallium bis-trifluoroacetate group by a number of other substituents.

Replacement by Iodine

Arylthallium bis-trifluoroacetates react with potassium iodide to give aryl iodides.

$$ArTl(OOCCF_3)_2 + KI \longrightarrow ArI$$

Aryl iodide

Replacement by a Cyano Group

The thallium bis-trifluoroacetate group can be replaced by a cyano group in two ways: through the reaction of arylthallium bis-trifluoroacetates with cuprous cyanide in dimethylformamide (DMF) or through their reaction with aqueous potassium cyanide under the influence of ultraviolet light.

$$ArTl(OOCCF_3)_2 \quad \begin{array}{l} \xrightarrow{CuCN} ArCN \\[2mm] \xrightarrow[h\nu]{KCN} ArCN \end{array}$$

Aryl cyanide

Replacement by a Hydroxyl Group

The thallium bis-trifluoroacetate group can be replaced by a hydroxyl group through the treatment of arylthallium bis-trifluoroacetates with lead tetraacetate and triphenylphosphine followed by treatment with sodium hydroxide. This gives us a method for the synthesis of phenols.

---

* Thallium compounds are extremely toxic and must be handled with great care.

$$ArTl(OOCCF_3)_2 \xrightarrow[\text{(2) OH}^-]{\text{(1) } Pb(OAc)_4/(C_6H_5)_3P} ArOH$$

A phenol

The most remarkable aspect of aromatic thallation reactions is that they are highly *regiospecific*.

Para Substitution

Thallation takes place almost exclusively at the para position of benzene compounds with a single alkyl group, halo group, or methoxyl group. The regiospecificity of these thallation reactions is attributed to the steric bulk of the electrophile. Thallium trifluoroacetate, because of its size, reacts faster at the uncrowded para position.

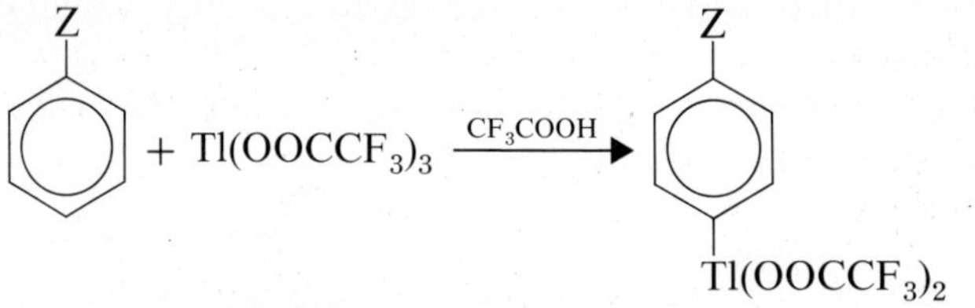

$$\text{Z-C}_6\text{H}_5 + Tl(OOCCF_3)_3 \xrightarrow{CF_3COOH} \text{para-Z-C}_6\text{H}_4\text{-}Tl(OOCCF_3)_2 \qquad \text{Para thallation}$$

Z— = R—, X—, or CH_3O—

The examples listed below show how thallation reactions have been used to synthesize para-disubstituted benzene compounds in a regiospecific way.

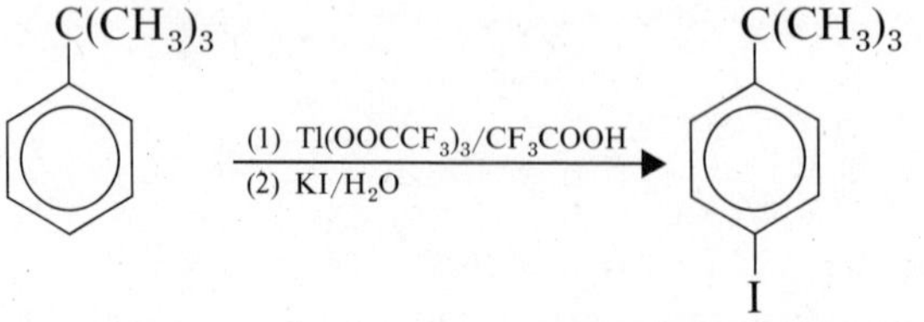

$$\text{C(CH}_3)_3\text{-C}_6\text{H}_5 \xrightarrow[\text{(2) KI/H}_2\text{O}]{\text{(1) } Tl(OOCCF_3)_3/CF_3COOH} \text{para-C(CH}_3)_3\text{-C}_6\text{H}_4\text{-I} \qquad (93\%)$$

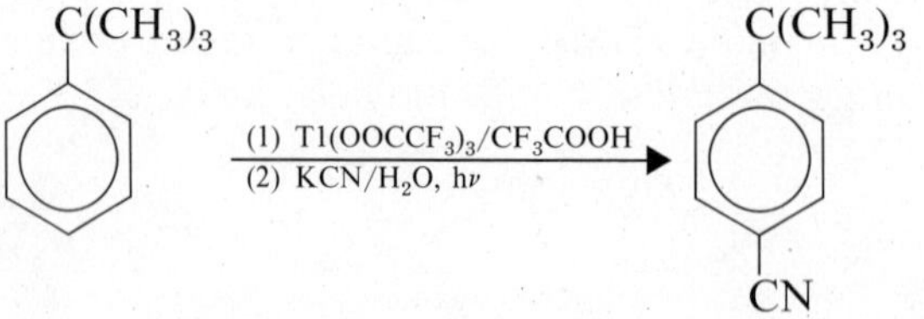

$$\text{C(CH}_3)_3\text{-C}_6\text{H}_5 \xrightarrow[\text{(2) KCN/H}_2\text{O, } h\nu]{\text{(1) } Tl(OOCCF_3)_3/CF_3COOH} \text{para-C(CH}_3)_3\text{-C}_6\text{H}_4\text{-CN} \qquad (80\%)$$

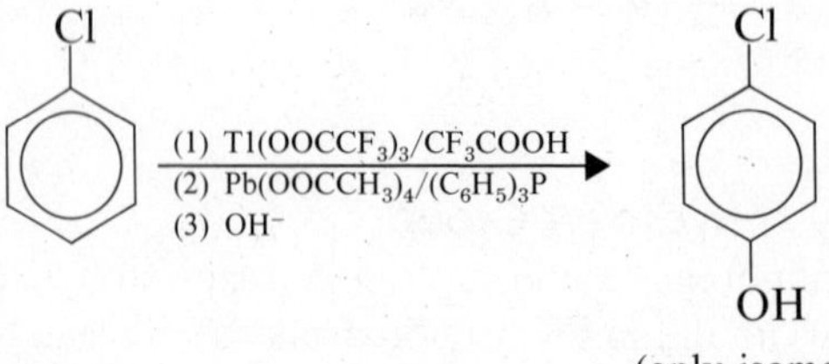

$$\text{Cl-C}_6\text{H}_5 \xrightarrow[\substack{\text{(2) } Pb(OOCCH_3)_4/(C_6H_5)_3P \\ \text{(3) OH}^-}]{\text{(1) } Tl(OOCCF_3)_3/CF_3COOH} \text{para-Cl-C}_6\text{H}_4\text{-OH} \qquad (56\%)$$

(only isomer)

Para-bromobenzene compounds can be prepared by using thallium trifluoroacetate as the Lewis acid in bromination reactions. Although ring thallation does not take place in these reactions, para orientation is again a result of the steric bulk of the electrophile. The electrophile is a bromine-thallium trifluoroacetate complex, and it reacts faster at the more-open para position.

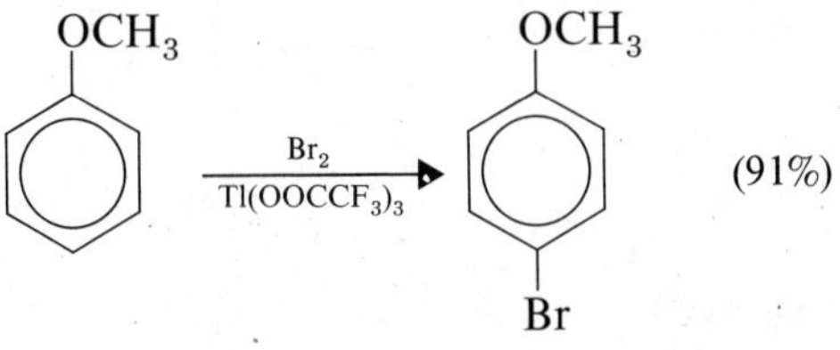

(91%)

Para-acetylanisole can be prepared in a similar way by using thallium trifluoroacetate and acetyl chloride.

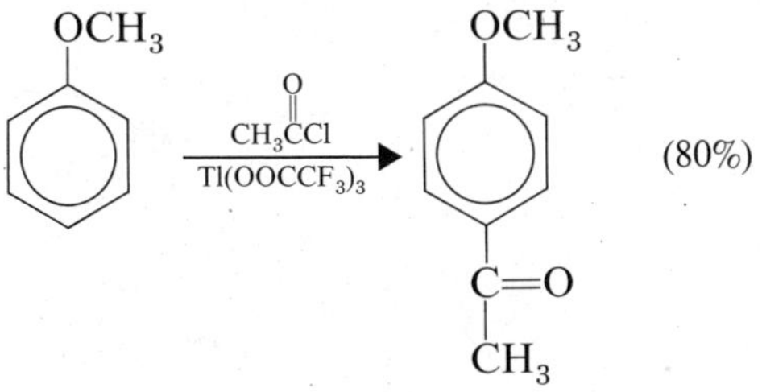

(80%)

Ortho Substitution

Ortho thallation takes place when benzoic acid and methyl benzoate are treated with thallium trifluoroacetate. Ortho thallation also occurs with benzyl alcohol, benzyl methyl ether, 2-phenylethanol, 2-phenylethyl methyl ether, 2-phenylacetic acid, and methyl 2-phenylacetate.

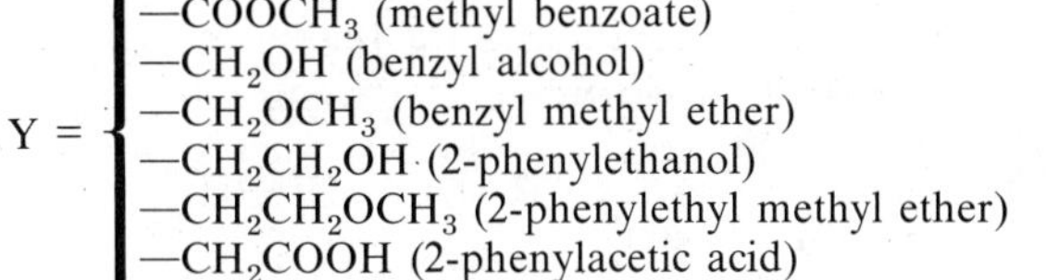

$$Y = \begin{cases} -COOH \text{ (benzoic acid)} \\ -COOCH_3 \text{ (methyl benzoate)} \\ -CH_2OH \text{ (benzyl alcohol)} \\ -CH_2OCH_3 \text{ (benzyl methyl ether)} \\ -CH_2CH_2OH \text{ (2-phenylethanol)} \\ -CH_2CH_2OCH_3 \text{ (2-phenylethyl methyl ether)} \\ -CH_2COOH \text{ (2-phenylacetic acid)} \\ -CH_2COOCH_3 \text{ (methyl 2-phenylacetate)} \end{cases}$$

Ortho substitution is the preferred course in all of these reactions because each compound contains an oxygen that can form a complex with thallium trifluoroacetate and deliver it to the ortho position.

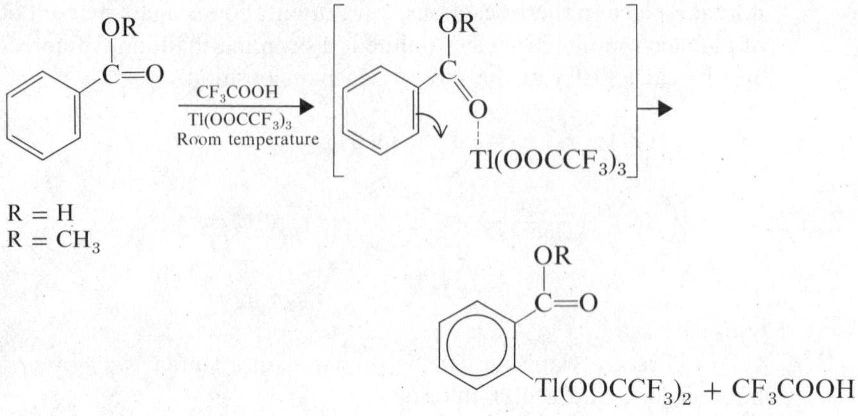

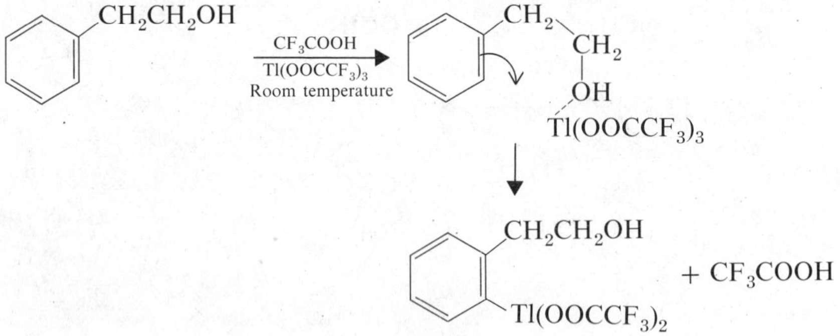

We see that in each case the intermediate complex resembles a stable five- or six-membered ring.

Problem 15.1

Write the structure of complexes that would account for ortho thallation of the following compounds:

    (a) Benzyl alcohol
    (b) Benzyl methyl ether
    (c) 2-Phenylethyl methyl ether
    (d) Phenylacetic acid
    (e) Methyl phenylacetate

Problem 15.2

The ester, 2-phenylethyl acetate undergoes para thallation when it is treated with thallium trifluoroacetate in trifluoroacetic acid at room temperature.

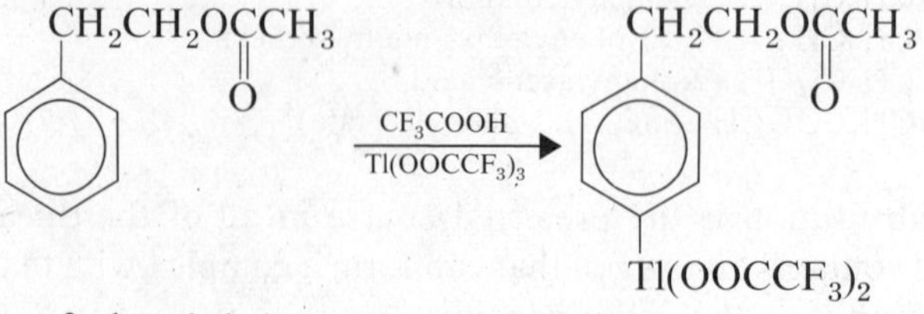

Can you suggest a reason that will account for the orientation of this reaction? (Hint: Complex-formation probably occurs with carbonyl oxygen of the ester group.)

The two examples listed below illustrate how thallation reactions can be used for the regiospecific synthesis of ortho-disubstituted benzenes.

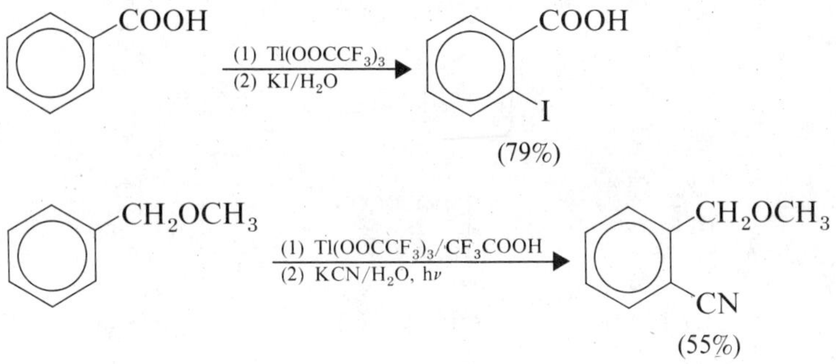

(79%)

(55%)

Meta Substitution

Thallation reactions are reversible. In the absence of complex formation, thallation occurs at the para position because reaction at the para position takes place faster. Meta-thallated benzenes appear to be more stable, however. So at higher temperatures meta thallation is the preferred course of the reaction. An example that illustrates the effect of a temperature change on the course of a thallation reaction is shown below.

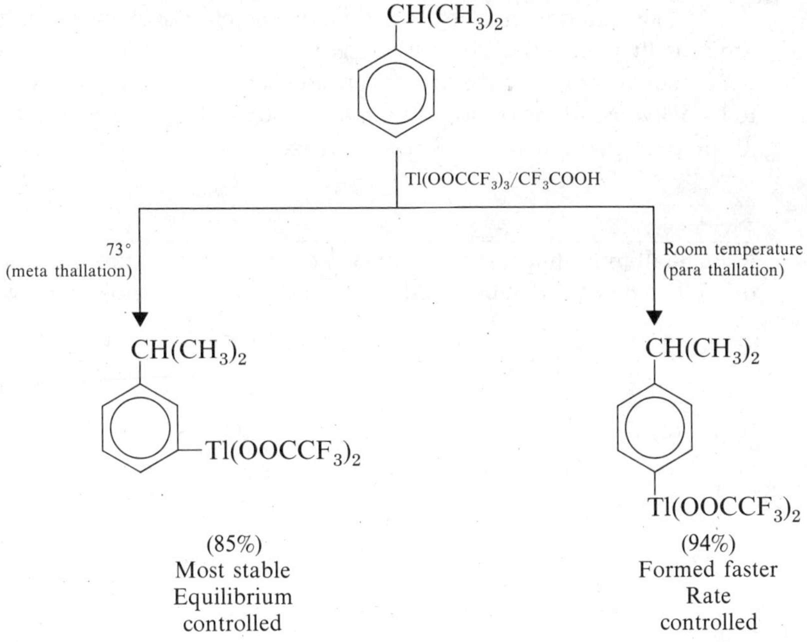

Tl(OOCCF$_3$)$_3$/CF$_3$COOH

73°
(meta thallation)

Room temperature
(para thallation)

(85%)
Most stable
Equilibrium
controlled

(94%)
Formed faster
Rate
controlled

This example also furnishes us with another illustration of rate versus equilibrium control of an organic reaction (cf. Section 10.7).

Ortho, Meta, and Para Substitution of the Same Compound

The versatility and control offered by thallation reactions can be illustrated by one other example—one that is truly spectacular. Thallation reactions allow 2-phenylethanol to be *selectively converted* to ortho-, meta-, and para-iodo isomers.

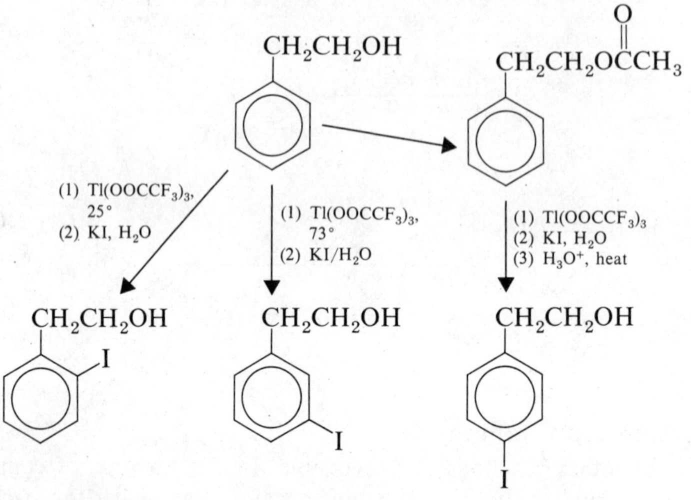

A (83%)   **B (54%)**   **C (84%)**

The ortho isomer results from reaction **A** because the oxygen of the alcohol substituent delivers the thallium to that position.

The meta isomer results from reaction **B** because the reaction is carried out at a higher temperature and the meta-thallated compound is *more stable.*

The para isomer is obtained from the reaction sequence **C** because the alcohol group is first converted to an acetate group. Complex formation of thallium with the carbonyl oxygen of the acetate group places the thallium at too great a distance to be delivered to the ortho position (cf. problem 15.2). Instead, thallium substitutes at the para position where it *reacts faster.*

Problem 15.3

Using thallium trifluoroacetate and any other necessary starting materials, show how the following compounds could be synthesized in a regiospecific way.

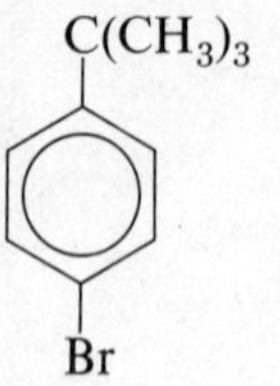

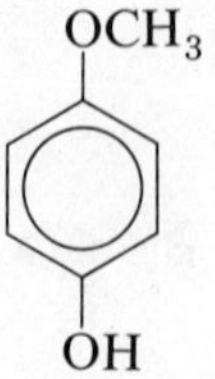

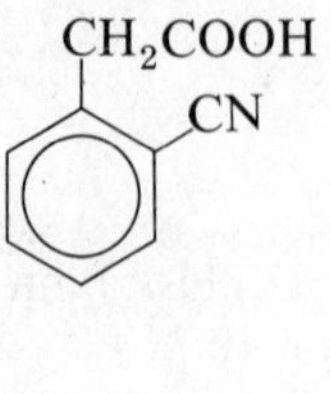

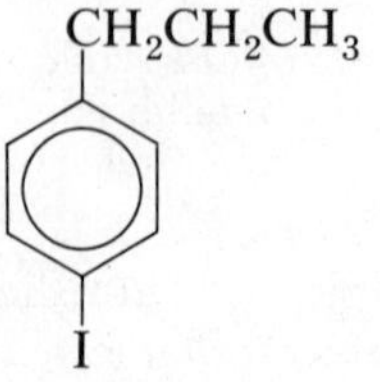

(e) $CH_2CH_2CH_3$ ... I

(g) CH_2CH_3 ... CN

(f) CH_3 ... CH_3 ... I

15.4 PHYSICAL PROPERTIES OF ORGANIC HALIDES

Alkyl and aryl halides have very low water solubilities, but as we might expect, they are miscible with each other and with other relatively nonpolar solvents. Methylene chloride (CH_2Cl_2), chloroform ($CHCl_3$), and carbon tetrachloride are often used as solvents for nonpolar and moderately polar organic compounds. Many chlorohydrocarbons have a cumulative toxicity, however, and should, therefore, be used only in fume hoods.

Methyl iodide (bp 42°) is the only monohalomethane that is a liquid at room temperature and atmospheric pressure. Ethyl bromide (bp 38°) and ethyl iodide (bp 72°) are both liquids but ethyl chloride (bp 13.1°) is a gas. The propyl chlorides,

TABLE 15.1 Organic Halides

NAME OF GROUP	FLUORIDE bp °C	FLUORIDE DENSITY	CHLORIDE bp °C	CHLORIDE DENSITY	BROMIDE bp °C	BROMIDE DENSITY	IODIDE bp °C	IODIDE DENSITY
Methyl	−78.4	0.84^{-60}	−23.8	0.92^{20}	3.6	1.73^{0}_{0}	42.5	2.28^{20}_{4}
Ethyl	−37.7		13.1	0.91^{15}_{4}	38.4	1.46^{20}_{4}	72	1.95^{20}_{20}
n-Propyl	−2.5	0.78^{-3}	46.6	0.89^{20}_{4}	70.8	1.35^{20}_{4}	102	1.74^{20}_{4}
Isopropyl			34	0.86^{20}_{4}	59.4	1.31^{20}_{4}	89.4	1.70^{20}_{4}
n-Butyl	32	0.78^{20}_{4}	78.4	0.89^{20}_{4}	101	1.27^{20}_{4}	130	1.61^{20}_{4}
sec-Butyl			68	0.87^{20}_{4}	91.2	1.26^{20}_{4}	120	1.60^{20}_{4}
Isobutyl			69	0.87^{20}_{4}	91	1.26^{20}_{4}	119	1.60^{20}_{4}
tert-Butyl			51	0.84^{20}_{4}	73.3	1.22^{20}_{4}	100 dec.	1.57^{0}_{0}
n-Pentyl	62	0.79^{20}_{4}	108.2	0.88^{20}_{4}	129.6	1.22^{20}_{4}	155^{740}	1.52^{20}_{4}
Neopentyl			84.4	0.87^{20}_{4}	93		127 dec.	1.53^{13}
Vinyl	−51		13.9		16	1.52^{14}_{4}	56	2.04^{20}
Allyl	−3		45	0.94^{20}_{4}	70	1.40^{20}_{4}	102–3	1.84^{22}_{4}
Phenyl (halo-benzene)	85	1.02^{20}_{4}	132	1.10^{20}_{4}	155	1.52^{20}_{4}	189	1.82^{20}_{4}
Benzyl	140	1.02^{25}_{4}	179	1.10^{25}_{4}	201	1.44^{22}_{0}	93^{10}	1.73^{25}

bromides, and iodides are all liquids. In general alkyl chlorides, bromides, and iodides tend to have boiling points near those of alkanes of similar molecular weights.

Polyfluoroalkanes, however, tend to have unusually low boiling points. Hexafluoroethane boils at $-79°$, even though its molecular weight (138) is near that of decane (bp $174°$).

Table 15.1 lists some of the physical properties of organic halides.

15.5 OTHER ORGANOMETALLIC COMPOUNDS

Compounds that contain carbon-metal bonds are called *organometallic compounds*. The nature of carbon-metal bonds vary widely, from bonds that are essentially ionic to those that are primarily covalent. While the structure of the organic portion of the organometallic compound has some effect on the nature of the carbon-metal bond, the identity of the metal itself is of far greater importance. Carbon-sodium and carbon-potassium bonds are largely ionic in character; carbon-lead, carbon-tin, carbon-thallium (III), and carbon-mercury bonds are essentially covalent. Carbon-lithium and carbon-magnesium bonds fall in between these extremes.

$$-\overset{|}{\underset{|}{C}}\!:^- \overset{+}{M} \qquad\qquad -\overset{|}{C}\!:\overset{|\delta^-\ \ \delta^+|}{M} \qquad\qquad -\overset{|}{\underset{|}{C}}\!-M$$

primarily ionic   (M = Mg or Li)   primarily covalent
(M = Na⁺ or K⁺)   (M = Pb, Sn, Hg, Tl)

As we might expect, the reactivity of organometallic compounds increases with the percent ionic character of the carbon metal bond. Alkylsodium and alkylpotassium compounds are highly reactive and are among the most powerful of bases. They react explosively with water and burst into flame when exposed to air. Organomercury and lead compounds are much less reactive; they are volatile and are stable in air. They are generally soluble in nonpolar solvents. Tetraethyllead, for example, is used as an "antiknock" compound in gasoline.

Organometallic compounds of lithium and magnesium are of great importance in organic synthesis. They are relatively stable in ether solutions but their carbon-metal bonds have considerable ionic character. Because of this, the carbon atom of organolithium and organomagnesium compounds that is bonded to the metal is a strong base and a powerful nucleophile. We will see reactions that illustrate both of these properties in Section 15.7.

15.6 PREPARATION OF ORGANOLITHIUM AND ORGANOMAGNESIUM COMPOUNDS

Organolithium Compounds

Organolithium and organomagnesium compounds are usually prepared through the reaction of an organic halide with lithium or magnesium metal. These reactions are usually carried out in ether solvents, and since organolithium and organomagnesium compounds are strong bases, care must be taken to exclude moisture. (Why?) The ethers most commonly used as solvents are diethyl ether and tetrahydrofuran. (Tetrahydrofuran is a cyclic ether.)

$$CH_3CH_2\overset{..}{\underset{..}{O}}CH_2CH_3$$

$$\begin{array}{c} CH_2{-}CH_2 \\ | \qquad | \\ CH_2 \quad CH_2 \\ \diagdown \; \underset{..}{\overset{..}{O}} \; \diagup \end{array}$$

Diethyl ether (Et$_2$O)   Tetrahydrofuran (THF)

For example, *n*-butylbromide reacts with lithium metal in diethyl ether to give a solution of *n*-butyllithium.

$$CH_3CH_2CH_2CH_2Br + 2Li \xrightarrow[\text{Et}_2\text{O}]{-10°} CH_3CH_2CH_2CH_2Li + LiBr$$

n-Butyl bromide   (80–90%)
                              n-Butyllithium

Other organolithium compounds, such as methyllithium, ethyllithium, and phenyllithium, can be prepared in the same general way.

$$\begin{array}{ccc} R{-}X & + 2Li \xrightarrow{\text{ether}} & RLi & + LiX \\ (\text{or } Ar{-}X) & & (\text{or } ArLi) \end{array}$$

The order of reactivity of halides is RI > RBr > RCl. (Alkyl and aryl fluorides are seldom used in the preparation of organolithium compounds.)

Vinyllithium can be prepared by treating vinyl chloride with a lithium-sodium mixture.

$$CH_2{=}CHCl + 2Li \xrightarrow{\text{Et}_2\text{O}} CH_2{=}CHLi + LiCl$$
$$\qquad\qquad (2\% \text{ Na}) \qquad\qquad (60\%)$$

Most organolithium compounds slowly attack ethers by bringing about an elimination reaction.

$$\overset{\delta-}{R}{:}\overset{\delta+}{Li} + \overset{}{H}{-}CH_2{-}CH_2{-}OCH_2CH_3 \longrightarrow RH + CH_2{=}CH_2 + \overset{+}{Li}\,\overset{-}{O}CH_2CH_3$$

For this reason, ether solutions of organolithium reagents are not usually stored but are used directly in some other reaction. Organolithium compounds are much more stable in hydrocarbon solvents. Several alkyl and aryllithium reagents are commercially available in paraffin wax or mineral oil.

Grignard Reagents

Organomagnesium halides were discovered by the French chemist Victor Grignard in 1900. Grignard received the Nobel Prize for his discovery in 1912 and organomagnesium halides are now called Grignard reagents in his honor. Grignard reagents probably have a greater use in organic synthesis than any other single analogous organic reagent.

Grignard reagents are usually prepared through the reaction of an organic halide and magnesium metal (turnings) in an ether solvent.

$$\left.\begin{array}{l} \text{R—X} + \text{Mg} \xrightarrow{\text{ether}} \text{RMgX} \\ \text{Ar—X} + \text{Mg} \xrightarrow{\text{ether}} \text{ArMgX} \end{array}\right\} \begin{array}{l}\text{Grignard} \\ \text{reagents}\end{array}$$

The order of reactivity of halides with magnesium is also RI > RBr > RCl. Very few organomagnesium fluorides have been successfully prepared. Aryl Grignard reagents are more easily prepared from aryl bromides and aryl iodides. Aryl chlorides react sluggishly.

Grignard reagents are seldom isolated but are used for further reactions in ether solution. The ether solutions can be analyzed for the content of the Grignard reagent, however, and the yields of Grignard reagent are almost always very high (85–95%). Several examples are shown below.

$$\text{CH}_3\text{I} + \text{Mg} \xrightarrow[35°]{\text{Et}_2\text{O}} \quad \text{CH}_3\text{MgI} \qquad (95\%)$$
Methylmagnesium
iodide

$$\text{C}_2\text{H}_5\text{Br} + \text{Mg} \xrightarrow[35°]{\text{Et}_2\text{O}} \quad \text{C}_2\text{H}_5\text{MgBr} \qquad (93\%)$$
Ethylmagnesium
bromide

$$\text{C}_6\text{H}_5\text{Br} + \text{Mg} \xrightarrow[35°]{\text{Et}_2\text{O}} \quad \text{C}_6\text{H}_5\text{MgBr} \qquad (95\%)$$
Phenylmagnesium
bromide

$$\text{CH}_2\text{=CHBr} + \text{Mg} \xrightarrow[66°]{\text{THF}} \text{CH}_2\text{=CHMgBr} \qquad (90\%)$$
Vinylmagnesium
bromide

The actual structure of Grignard reagents are more complex than the general formula RMgX indicates. Experiments done with radioactive magnesium have established that, for most Grignard reagents, there is an equilibrium between an alkylmagnesium halide and a dialkylmagnesium.

$$2\text{RMgX} \rightleftharpoons \text{R}_2\text{Mg} + \text{MgX}_2$$
Alkylmagnes-   Dialkylmagnesium
ium halide

For convenience in this text, however, we will write the formula for the Grignard reagent as though it were simply RMgX. As we do this we should always remember that the formula RMgX is an oversimplification.

Grignard reagents also form complexes with their ether solvents like the one shown below.

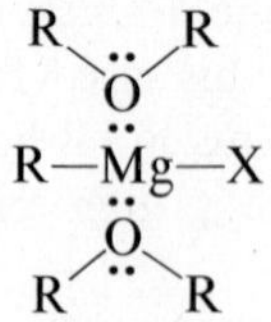

$$\begin{array}{c} \text{R} \quad \text{R} \\ \ddot{\text{O}} \\ \text{R—Mg—X} \\ \ddot{\text{O}} \\ \text{R} \quad \text{R} \end{array}$$

Complex formation with ether solvents is an important factor in the formation and stability of Grignard reagents. Organomagnesium compounds can be prepared in nonethereal solvents but the preparations are more difficult.

The mechanism by which Grignard reagents are formed is still not fully understood. The most likely path, however, appears to be the two-step radical mechanism shown below.

$$R-X + :Mg \longrightarrow R\cdot + \cdot MgX$$
$$R\cdot + \cdot MgX \longrightarrow RMgX$$

15.7 REACTIONS OF ORGANOLITHIUM AND ORGANOMAGNESIUM COMPOUNDS

Reactions With Less Electropositive Metal Halides

Grignard reagents and alkyllithium reagents react with a number of halides of less electropositive metals to produce new organometallic compounds. Since Grignard reagents and alkyllithiums are easily prepared, these reactions furnish us with useful syntheses of alkyl derivatives of mercury, zinc, cadmium, copper, silicon, and phosphorus, for example. In general terms the reactions using Grignard reagents take the form shown below.

$$n RMgX + MX_n \longrightarrow R_nM + n MgX_2$$
(M is less electropositive than Mg)

Several specific examples are shown below.

$$2CH_3MgCl + HgCl_2 \longrightarrow (CH_3)_2Hg + 2MgCl_2$$
Dimethylmercury

$$2CH_3CH_2MgBr + ZnCl_2 \longrightarrow (CH_3CH_2)_2Zn + 2MgClBr$$
(100%)
Diethylzinc

$$2C_6H_5MgBr + CdCl_2 \longrightarrow (C_6H_5)_2Cd + 2MgClBr$$
(>83%)
Diphenylcadmium

$$3CH_3CH_2CH_2CH_2MgBr + PCl_3 \longrightarrow (CH_3CH_2CH_2CH_2)_3P + 3MgClBr$$
(57%)
Tributylphosphine

$$4CH_3CH_2MgCl + SiCl_4 \longrightarrow (CH_3CH_2)_4Si + 4MgCl_2$$
(80%)
Tetraethylsilane

We will see in Chapter 18 that organo cadmium compounds are useful reagents for preparing ketones.

Alkyllithiums are used in the preparation of lithium dialkylcuprates for the Corey-House synthesis of alkanes (cf. Section 3.13) and alkenes.

$$2CH_3Li + CuI \xrightarrow[Et_2O]{0°} (CH_3)_2CuLi$$

Lithium dimethylcuprate

Coupling reactions of lithium dimethylcuprate and an alkyl and an allylic halide are shown below,

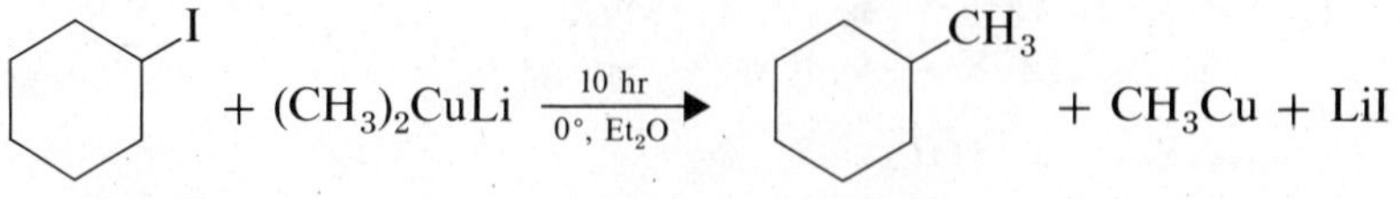

(75%)
Methylcyclohexane

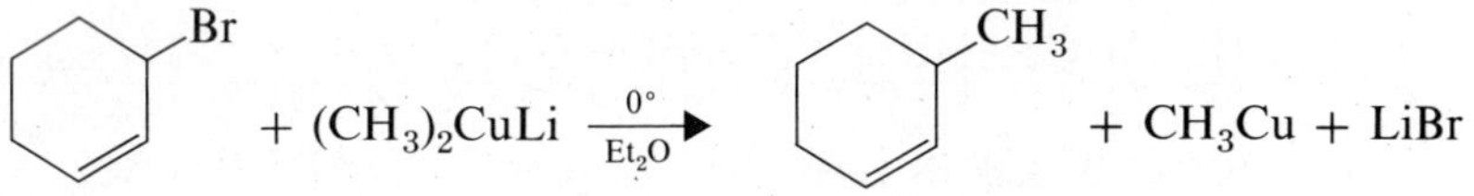

(75%)
3-Methylcyclohexene

Vinylic halides also couple with lithium dialkylcuprates. They do so in a stereo-specific way as the following example illustrates.

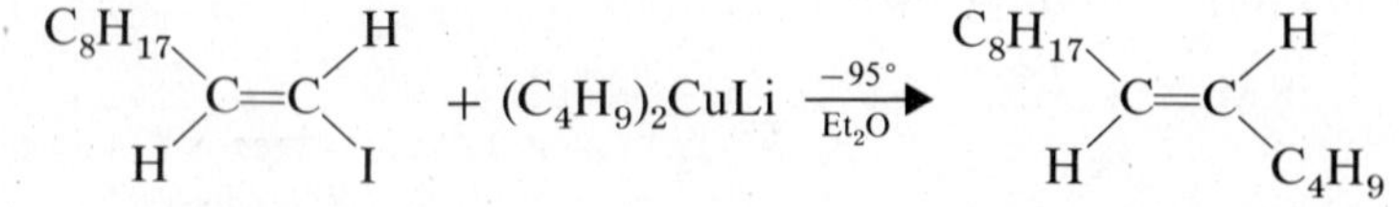

trans-1-Iodo-1-decene

(74%)
trans-5-Tetradecene

Aryl Grignard reagents undergo coupling reactions in very high yield when they are treated with thallium(I) bromide. Two examples are shown below.

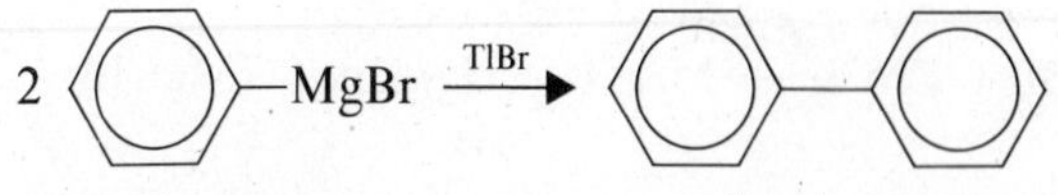

(92%)

Phenylmagnesium
bromide

Biphenyl

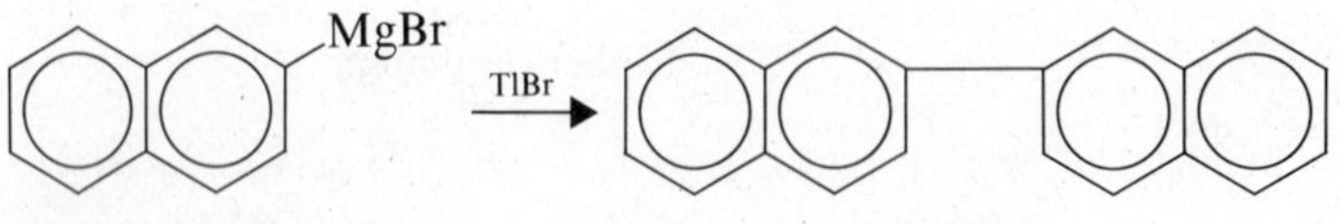

2-Naphthylmagnesium
bromide

(84%)
2,2'-Binaphthyl

Reactions with Compounds Containing Acidic Hydrogens

Grignard reagents and organolithium compounds are very strong bases. They react with compounds that have a hydrogen more acidic than the hydrogens of the hydrocarbon from which the Grignard reagent or organolithium is derived. We can understand how these reactions occur if we represent the Grignard reagent and organolithium compounds in the following ways.

$$\overset{\delta-}{R}:\overset{\delta+}{MgX} \quad \text{and} \quad \overset{\delta-}{R}:\overset{\delta+}{Li}$$

When we do this we can see that the reactions of Grignard reagents with water and alcohols (below) are nothing more than acid-base reactions; they lead to the formation of the weaker conjugate acid and weaker conjugate base. The Grignard reagent behaves as though it were the anion of an alkane, *a carbanion.*

$$\overset{\delta-}{R}:\overset{\delta+}{MgX} + \overset{\cdot\cdot}{H}:\overset{\cdot\cdot}{O}H \longrightarrow R:H + \overset{-}{H}\overset{\cdot\cdot}{O}: + Mg^{+2} + X^-$$

| Grignard reagent (stronger base) | Water (stronger acid) | Alkane (weaker acid) | Hydroxide ion (weaker base) | | |

$$\overset{\delta-}{R}:\overset{\delta+}{MgX} + \overset{\cdot\cdot}{H}:\overset{\cdot\cdot}{O}R \longrightarrow R:H + \overset{-}{R}\overset{\cdot\cdot}{O}: + Mg^{+2} + X^-$$

| Grignard reagent (stronger base) | Alcohol (stronger acid) | Alkane (weaker acid) | Alkoxide ion (weaker base) | | |

Problem 15.4

Write similar equations for the reactions that take place when *n*-butyllithium is treated with (a) water, (b) ethanol. Designate the stronger and weaker acids and the stronger and weaker bases.

Problem 15.5

Assuming you have *tert*-butyl bromide, magnesium, dry ether, and deuterium oxide (D_2O) available, show how you might synthesize the following deuterium-labeled alkane.

$$\underset{\underset{D}{|}}{\overset{\overset{CH_3}{|}}{CH_3CCH_3}}$$

Grignard reagents and organolithium compounds are capable of abstracting hydrogens that are much less acidic than those of water and alcohols. Grignard reagents and organolithium compounds react with the terminal hydrogens of 1-alkynes, for example, and this is a useful method for the preparation of alkynyl magnesium halides and alkynyl lithium compounds. These reactions are also acid-base reactions.

$$RC\equiv CH + \overset{\delta-}{R'}:\overset{\delta+}{MgX} \longrightarrow RC\equiv\overset{\delta-}{C}:\overset{\delta+}{MgX} + R':H$$

| Terminal alkyne (stronger acid) | Grignard reagent (stronger base) | Alkynyl magnesium halide (weaker base) | Alkane (weaker acid) |

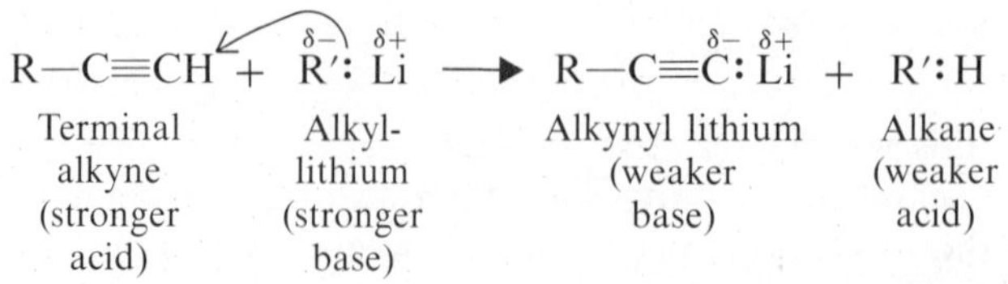

$$R\!-\!C\!\equiv\!CH + \overset{\delta-}{R'}\!:\!\overset{\delta+}{Li} \longrightarrow R\!-\!C\!\equiv\!\overset{\delta-}{C}\!:\!\overset{\delta+}{Li} + R'\!:\!H$$

| Terminal alkyne (stronger acid) | Alkyllithium (stronger base) | Alkynyl lithium (weaker base) | Alkane (weaker acid) |

The fact that these reactions go to completion is not surprising when we recall that the acidity constants of alkanes are of the order of $10^{-42} - 10^{-45}$, while those of terminal alkynes are 10^{-26} (Table 2.4, p. 68).

Grignard reagents are not only strong bases, they are also *powerful nucleophiles*. We will see in later chapters, when we study these reactions in detail, that reactions in which Grignard reagents act as nucleophiles are by far the most important. At this point, let us consider general examples that illustrate the ability of a Grignard reagent to act as a nucleophile by attacking saturated and unsaturated carbons.

Reactions of Grignard Reagents with Ethylene Oxide

An example of a reaction in which a Grignard reagent carries out a nucleophilic attack at a saturated carbon can be seen in its reaction with ethylene oxide. These reactions take the general form shown below and they give us a convenient synthesis of primary alcohols.

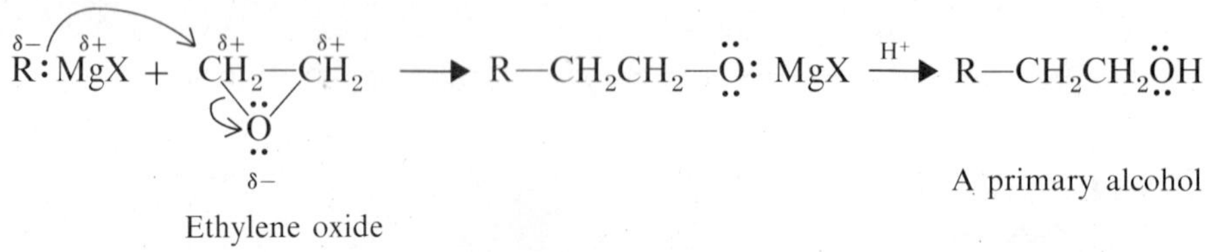

$$\overset{\delta-}{R}\!:\!MgX + \overset{\delta+}{CH_2}\!-\!\overset{\delta+}{CH_2} \longrightarrow R\!-\!CH_2CH_2\!-\!\overset{..}{\underset{..}{O}}\!:\!MgX \overset{H^+}{\longrightarrow} R\!-\!CH_2CH_2\overset{..}{\underset{..}{O}}H$$

$$\underset{\underset{\delta-}{\overset{..}{\underset{..}{O}}}}{}$$

Ethylene oxide   A primary alcohol

The nucleophilic alkyl group of the Grignard reagent attacks the partially positive carbon of the ethylene oxide ring. The highly strained ring opens, and the reaction leads to the formation of the salt of a primary alcohol. Subsequent acidification of the mixture yields the alcohol.

Reactions of Grignard Reagents with Carbonyl Compounds and Carbon Dioxide

From a synthetic point of view, the most important reactions of Grignard reagents and organolithium compounds are those in which Grignard reagents act as nucleophiles and attack an unsaturated carbon—*especially the carbon of a carbonyl group*. We will study these reactions in detail in Chapters 16 and 18. But we can give the general form of the reaction here.

The carbon atom of a carbonyl group is positively charged because of the inductive effect of the oxygen and because of the resonance contribution of the second structure shown below.

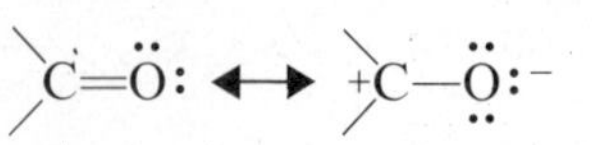

$$\overset{}{\underset{}{>}}C\!=\!\overset{..}{O}\!: \longleftrightarrow {}^+\overset{}{\underset{}{>}}C\!-\!\overset{..}{\underset{..}{O}}\!:^-$$

Resonance structures of the carbonyl group

Compounds that contain carbonyl groups are, therefore, highly susceptible to nucleophilic attack. Grignard reagents react with carbonyl compounds (aldehydes and ketones) in the following way:

$$\overset{\delta-}{R}\!:\!\overset{\delta+}{MgX} + \underset{}{\overset{}{C}}\!=\!\ddot{O}\!: \longrightarrow R-\overset{|}{\underset{|}{C}}-\ddot{\underset{\cdot\cdot}{O}}\!: MgX$$

Such a reaction is of great synthetic importance because it creates a new carbon-carbon single bond, a single bond between the carbon of the organic group of the Grignard reagent and the carbonyl carbon of the carbonyl compound. Acidification of the salts that are formed gives us a synthetic method for the preparation of alcohols.

$$R-\overset{|}{\underset{|}{C}}-\ddot{\underset{\cdot\cdot}{O}}\!: MgX \xrightarrow{HX} R-\overset{|}{\underset{|}{C}}-OH + MgX_2$$

An alcohol

Grignard reagents react in a similar way with the carbon of carbon dioxide. Acidification of the salts that are formed in these reactions gives us a synthesis of *carboxylic acids*. (We will study this reaction in Section 19.3).

$$\overset{\delta-}{R}\!:\!\overset{\delta+}{MgX} + \underset{\underset{\ddot{O}:}{\|}}{\overset{\ddot{O}:}{\|}}C \longrightarrow R-\overset{\ddot{O}:}{\underset{}{\overset{\|}{C}}}-\ddot{O}\!: MgX \xrightarrow{HX} \overset{O}{\overset{\|}{R}COH} + MgX_2$$

15.8 METALLOCENES: ORGANOMETALLIC SANDWICH COMPOUNDS

Cyclopentadiene reacts with phenylmagnesium bromide to give the Grignard reagent of cyclopentadiene. This reaction is not unusual, for it is simply another acid-base reaction like those we saw earlier. The methylene hydrogens of cyclopentadiene

| Cyclopentadiene | Phenylmagnesium bromide | | Cyclopentadienylmagnesium bromide | Benzene |

are much more acidic than the hydrogens of benzene and, therefore, the reaction goes to completion. (The methylene hydrogens of cyclopentadiene are acidic relative to ordinary methylene hydrogens because the cyclopentadienide anion is aromatic, cf., Section 12.7.)

When the Grignard reagent of cyclopentadiene is treated with ferrous chloride a reaction takes place that produces a highly unusual product, called *ferrocene*.

Ferrocene
(71% overall yield
from cyclopentadiene)

Ferrocene is an orange solid with a melting point of 174°. It is a highly stable compound; ferrocene can be sublimed at 100° and is not damaged when heated to 400°.

Many studies, including X-ray analysis, show that ferrocene is a compound in which the iron atom is located between two cyclopentadienide rings.

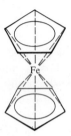

The carbon-carbon bond distances are all 1.40 Å and the carbon-iron bond distances are all 2.04 Å. Because of their structures, molecules like ferrocene have been called "sandwich" compounds.

The carbon-iron bonding in ferrocene results from overlap between the inner lobes of the p orbitals of the cyclopentadienide anions and $3d$ orbitals of the iron atom. Studies have shown, moreover, that this bonding is such that the rings of ferrocene are capable of essentially free rotation about an axis that passes through the iron atom and that is perpendicular to the rings.

Ferrocene is an *aromatic compound*. It undergoes a number of electrophilic aromatic substitution reactions, including sulfonation and Friedel-Crafts acylation.

The discovery of ferrocene (in 1951) was followed by the preparation of a number of similar aromatic compounds. These compounds, as a class, are called *metallocenes*.* Metallocenes with five-, six-, seven-, and even eight-membered rings have been synthesized from metals as diverse as manganese, cobalt, nickel, chromium, and uranium.

"Half-sandwich" compounds have been prepared through the use of metal carbonyls. Several are shown below.

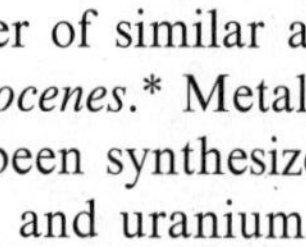

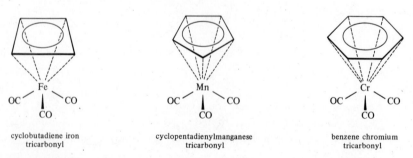

cyclobutadiene iron tricarbonyl

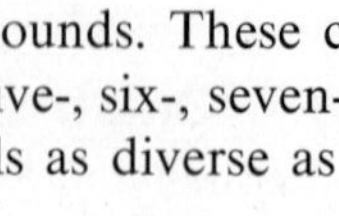

cyclopentadienylmanganese tricarbonyl

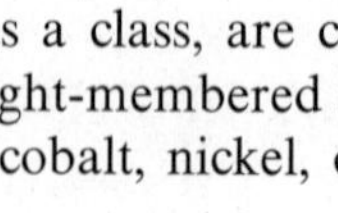

benzene chromium tricarbonyl

Although cyclobutadiene itself is *not* stable, the cyclobutadiene iron tricarbonyl complex is.

---

* Professors Ernest O. Fischer (of the Technical University, Munich) and Geoffrey Wilkinson (of Imperial College, London) received the Nobel Prize in 1973 for their pioneering work (performed independently) on the chemistry of organometallic sandwich compounds—or metallocenes.

15.9 ORGANIC HALIDES AND ORGANOMETALLIC COMPOUNDS IN THE ENVIRONMENT

Organic Halides

Since the discovery of the insecticidal properties of DDT in 1942, vast quantities of chlorinated hydrocarbons have been sprayed over the surface of the earth in an effort to destroy insects. These efforts initially met with incredible success in ridding large areas of the earth of disease-carrying insects, particularly those of typhus and malaria. As time has passed, however, we have begun to understand that this prodigious use of chlorinated hydrocarbons has not been without harmful—indeed tragic—side effects. Chlorinated hydrocarbons are usually highly stable compounds and they are only slowly destroyed by natural processes in the environment. As a result, many chloro-organic insecticides will remain in the environment for years. These highly stable pesticides are called "hard" pesticides.

Chlorohydrocarbons are also fat soluble and thus, they tend to accumulate in the fatty tissues of most animals. The food chain that runs from plankta to small fish to larger fish to birds and to larger animals including man tends to magnify the concentrations of chloro-organic compounds at each step.

The chlorohydrocarbon DDT is prepared from inexpensive starting materials, chlorobenzene and trichloroacetaldehyde. The reaction is catalyzed by acid.

DDT
1,1,1-Trichloro
2,2-bis(*p*-chlorophenyl)
ethane

In nature the principal decomposition product of DDT is DDE.

DDE
1,1-Dichloro-2,2-bis(*p*-chloro-
phenyl) ethene

Estimates indicate that nearly one billion pounds of DDE are now spread throughout the world ecosystem. One pronounced environmental effect of DDE has been in its action on egg-shell formation of many birds. DDE inhibits the enzyme *carbonic anhydrase* that controls the calcium supply for shell formation. As a consequence, the shells are often very fragile and do not survive until hatching occurs. During the past 10 years the populations of eagles, falcons, and hawks have dropped dramatically. There can be little doubt that DDE has been primarily responsible.

DDE also accumulates in the fatty tissues of man. Although man appears to have a short-range tolerance to moderate DDE levels, the long-range effects are far from certain.

Other hard insecticides are Aldrin, Dieldrin, and Chlordan. Aldrin is manufactured through the Diels-Alder reaction of hexachlorocyclopentadiene and nor-bornadiene.

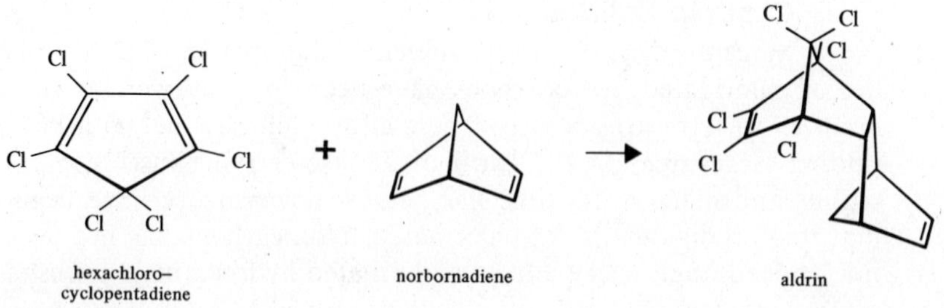

hexachloro–
cyclopentadiene

norbornadiene

aldrin

Chlordan is made by adding chlorine to the unsubstituted double bond of the Diels-Alder adduct obtained from hexachlorocyclopentadiene and cyclopentadiene. Dieldrin is made by converting an Aldrin double bond to an epoxide.

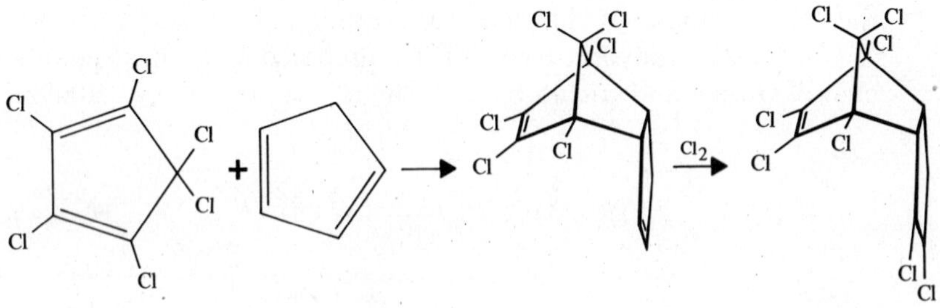

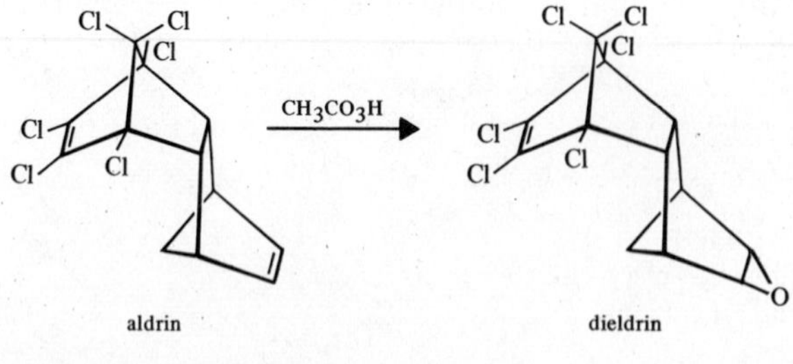

chlordan

aldrin

dieldrin

Lindane is another hard insecticide.

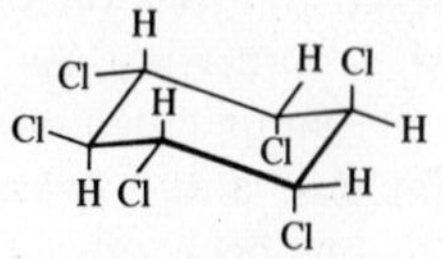

lindane

Other chlorinated organic compounds have been used extensively as herbicides. Two examples are 2,4-D and 2,4,5-T shown on the next page.

2,4-D
2,4-Dichlorophenoxy-
acetic acid

2,4,5-T
2,4,5-Trichlorophenoxy-
acetic acid

Enormous quantities of these two compounds were used as defoliants in the jungles of Indochina during the Vietnam war. The compound called 2,4,5-T has been shown to be a teratogen (a fetus-deforming agent). The teratogenic effect of 2,4,5-T may be a result of an impurity present in commercial 2,4,5-T, the compound 2,3,7,8-tetrachlorodibenzodioxin. (2,3,7,8-Tetrachlorodibenzodioxin is also highly toxic; it is many times as toxic, for example, as cyanide ion.)

2,3,7,8-Tetrachlorodibenzodioxin

Organometallic Compounds

With few exceptions, organometallic compounds are toxic. This toxicity varies greatly depending on the nature of the organometallic compound and the identity of the metal. Organic compounds of arsenic, antimony, lead, thallium, and mercury are toxic because the metal ions, themselves, are toxic. Certain organic derivatives of silicon are toxic even though silicon and most of its inorganic compounds are nontoxic.

Early in this century the recognition of the biocidal effects of organoarsenic compounds led Paul Ehrlich to his pioneering work in chemotherapy. Ehrlich sought compounds (which he called "magic bullets") that would show greater toxicity toward disease-causing microorganisms than they would toward their hosts. Ehrlich's research led to the development of Salvarsan and Neosalvarsan, two organoarsenic compounds that were used successfully in the treatment of diseases caused by spirochetes (e.g., syphilis) and trypanosomes (e.g., sleeping sickness). Salvarsan and Neosalvarsan are no longer used in the treatment of these diseases; they have been displaced by safer and more effective antibiotics. Ehrlich's research, however, initiated the field of chemotherapy (cf. Section 21.10).

Many microorganisms actually synthesize organometallic compounds and this discovery has an alarming ecological aspect. Mercury metal is toxic, but mercury metal is also unreactive. In the past untold tons of mercury metal have been disposed of by simply dumping it into lakes and streams and since mercury is toxic, many bacteria protect themselves from its effect by converting mercury metal to methylmercury ions (CH_3Hg^+) and to gaseous dimethylmercury $((CH_3)_2Hg)$. These organic mercury compounds are passed up the food chain (with modification) through fish to man where methylmercury ions act as a deadly nerve poison. Between 1953 and 1964, 116 people in Minamata, Japan, were poisoned by eating fish containing methyl mercury compounds. Arsenic is also methylated by organisms to the poisonous dimethyl arsine, $(CH_3)_2AsH$.

Ironically, chlorinated hydrocarbons appear to inhibit the biological reactions that bring about mercury methylation. Lakes polluted with organo chlorine pesticides show significantly lower mercury methylation. While this particular interaction of two pollutants may, in a certain sense, be beneficial, it is also instructive of the complexity of the environmental problems that we may face.

Tetraethyllead and other alkyllead compounds have been used as antiknock agents in gasoline since 1923. In the years since then, more than one trillion pounds of lead have been introduced into the atmosphere. In the northern hemisphere, gasoline burning alone has spread about 10 mg of lead on each square meter of the earth's surface. In highly industrialized areas the amount of lead per square meter is probably several hundred times higher. Because of the well-known toxicity of lead these facts must also be of great concern.

Additional Problems

15.6
Show how you might prepare 2-bromobutane from
(a) 2-Butanol, $CH_3CH_2CHOHCH_3$
(b) 1-Butanol, $CH_3CH_2CH_2CH_2OH$
(c) 1-Butene
(d) 1-Butyne

15.7
Show how you might prepare 1-bromobutane from each of the compounds listed in problem 15.6.

15.8
Show how you might carry out the following transformations.
(a) Cyclohexanol ⟶ chlorocyclohexane
(b) Cyclohexene ⟶ chlorocyclohexane
(c) 1-Methylcyclohexene ⟶ 1-bromo-1-methylcyclohexane
(d) 1-Methylcyclohexene ⟶ 2-bromo-1-methylcyclohexane
(e) 1-Bromo-1-methylcyclohexane ⟶ 1-deuterio-1-methylcyclohexane

15.9
Show how you might synthesize each of the following compounds from toluene.
(a) p-Bromotoluene
(b) p-Iodotoluene
(c) p-Tolylmagnesium bromide

(d) 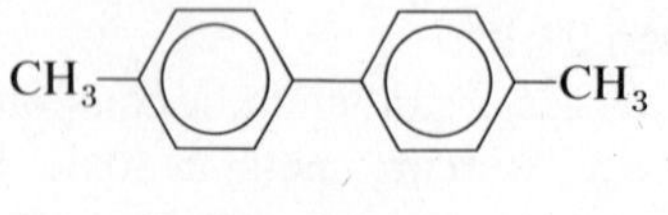 $CH_3-\langle\bigcirc\rangle-\langle\bigcirc\rangle-CH_3$

(e) 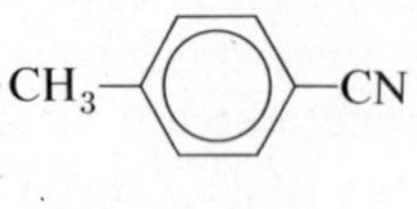 $CH_3-\langle\bigcirc\rangle-CN$

(f) 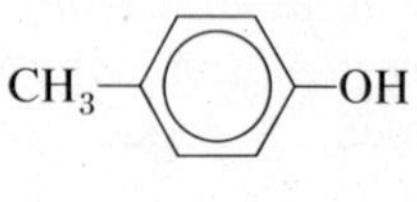 $CH_3-\langle\bigcirc\rangle-OH$

(g) 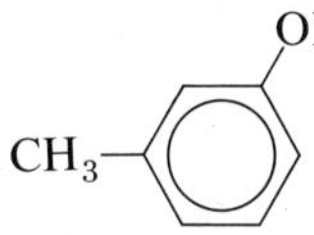 CH_3— (benzene ring with OH)

(h) 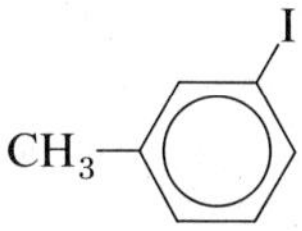 CH_3— (benzene ring with I)

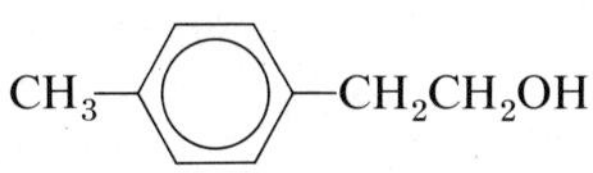

(i) CH_3—(benzene ring)—CH_2CH_2OH

(j) 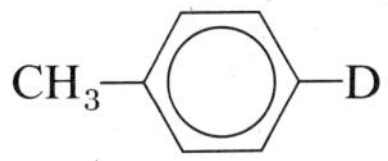 CH_3—(benzene ring)—D

15.10

acetyl group de activating

When ferrocene is acetylated twice in sequential fashion, one acetyl group becomes attached to each ring (rather than both acetyl groups to the same ring). How can you account for this?

15.11

How might you carry out the following transformations?
(a) Bromocyclopentane $\longrightarrow$ methylcyclopentane
(b) 3-Bromocyclopentene $\longrightarrow$ 3-methylcyclopentene
(c) Allyl bromide $\longrightarrow$ 1-pentene
(d) (E)-2-Iodo-2-butene $\longrightarrow$ (E)-3-methyl-2-heptene

15.12

What products would you expect from the following reactions?
(a) Phenyllithium + acetic acid $\longrightarrow$
(b) Phenyllithium + methyl alcohol $\longrightarrow$
(c) Methylmagnesium bromide + ammonia $\longrightarrow$
(d) (Four moles) methylmagnesium bromide + $SiCl_4$ $\longrightarrow$
(e) (Three moles) phenylmagnesium bromide + PCl_3 $\longrightarrow$
(f) (Two moles) ethylmagnesium bromide + $CdCl_2$ $\longrightarrow$
(g) Phenylmagnesium bromide + CO_2 $\xrightarrow{\quad}$
$\overline{(2)H_3O^+}$

(h) Phenylmagnesium bromide + $H-\overset{\overset{\displaystyle O}{\|}}{C}-H$ $\xrightarrow[(2)H_3O^+]{}$

15.13

When 1,2-dibromoethane is added to magnesium in ether a gas is evolved and the ether solution is found not to contain a Grignard reagent. The gas that is evolved decolorizes bromine in carbon tetrachloride. (a) What is the gas? (b) Suggest a mechanism for its formation.

15.14

Propose simple chemical tests that could be used to distinguish between the following pairs of compounds.

(a) Allyl bromide and *n*-propyl bromide
(b) Benzyl bromide and *p*-bromotoluene
(c) Vinyl chloride and benzyl chloride
(d) Phenyllithium and diphenylmercury
(e) Aldrin and Chlordan

15.15

The mechanism for the formation of DDT from chlorobenzene and trichloroacetaldehyde in sulfuric acid involves two electrophilic aromatic substitution reactions. In the first electrophilic substitution reaction the electrophile is protonated trichloroacetaldehyde. In the second the electrophile is a carbocation. Propose a mechanism for the formation of DDT.

15.16

What kind of reaction is involved in the conversion of DDT to DDE?

15.17

Compounds **A** and **B** are isomers with the molecular formula $C_4H_6Cl_2$. Both compounds decolorize bromine in carbon tetrachloride. The pmr spectrum of **A** shows only two signals: a singlet at $\delta 4.25$ and a singlet at $\delta 5.35$; the signal area ratio is 2:1. The pmr spectrum of **B** shows a singlet at $\delta 2.2$, a doublet at $\delta 4.15$, and a triplet at $\delta 5.7$; the signal area ratios are 3:2:1. Propose structures for **A** and **B**.

15.18

Propose a structure for compound **C**, $C_4H_8Cl_2$, whose pmr spectrum is given in Fig. 15.1.

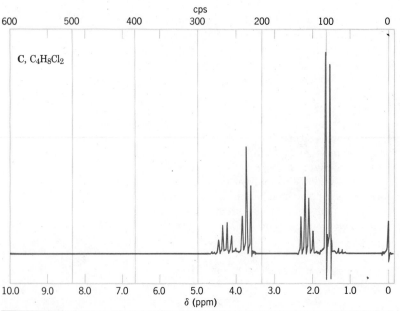

FIG. 15.1

*The pmr spectrum of compound **C**, problem 15.18. (Spectrum courtesy of Aldrich Chemical Co., Milwaukee, Wis.)*

15.19

The pmr spectrum of compound **D**, C_4H_7Cl, is given in Fig. 15.2. Compound **D** decolorizes bromine in carbon tetrachloride and gives a precipitate when treated with alcoholic $AgNO_3$. Propose a structure for **D**.

15.20
Compound **E**, C_4H_6, reacts with one mole of chlorine to yield a mixture of isomeric compounds **F** and **G** with molecular formula $C_4H_6Cl_2$. Compounds **F** and **G** react with chlorine to yield the same compound **H**, $C_4H_6Cl_4$. The pmr spectrum of compound **H** is shown in Fig. 15.3. Propose structures for **E**, **F**, **G**, and **H**.

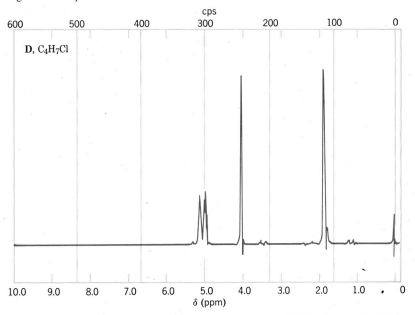

FIG. 15.2

*The pmr spectrum of compound **D**, problem 15.19. (Spectrum courtesy of Aldrich Chemical Co., Milwaukee, Wis.)*

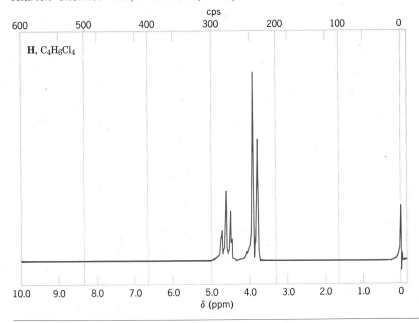

FIG. 15.3

*The pmr spectrum of compound **H**, problem 15.20. (Spectrum courtesy of Aldrich Chemical Co., Milwaukee, Wis.)*

16 ALCOHOLS, PHENOLS, AND ETHERS

16.1 STRUCTURE AND NOMENCLATURE

Alcohols are compounds in which a hydroxyl group is attached to a *saturated* carbon.* The saturated carbon may be a carbon of a simple alkyl group:

$$CH_3OH \qquad CH_3CH_2OH \qquad CH_3\underset{\underset{\displaystyle OH}{|}}{C}HCH_3$$

Methanol   Ethanol   2-Propanol
(methyl alcohol)   (ethyl alcohol)   (isopropyl alcohol)

The carbon may be a saturated carbon of an alkenyl or alkynyl group.

$$CH_2{=}CHCH_2OH \qquad H{-}C{\equiv}CCH_2OH$$

2-Propenol   2-Propynol
(allyl alcohol)   (propargyl alcohol)

Or, the carbon may be a saturated carbon that is attached to an aromatic ring.

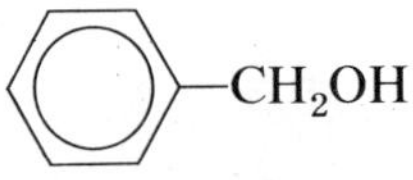

Benzyl alcohol

Compounds that have a hydroxyl group attached directly to a benzene ring are called *phenols*.

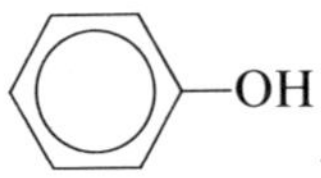

Phenol    *p*-Cresol
                   (a phenol)

Compounds that have a hydroxyl group attached to a polycyclic benzenoid ring are similar to phenols, but they are called, naphthols and phenanthrols, for example.

---

*Compounds in which a hydroxyl group is attached to an unsaturated carbon of a double bond (i.e., $C{=}C$) are called enols, cf. Section 18.8.

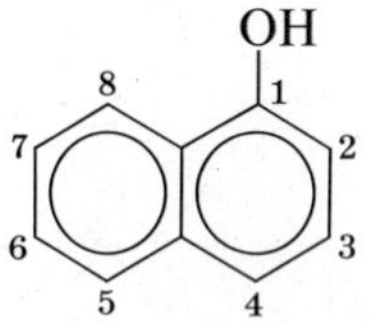

1-Naphthol
(α-naphthol)

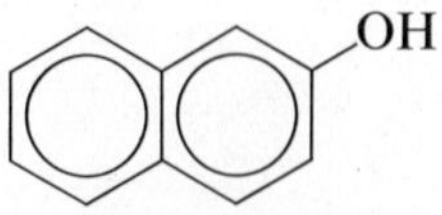

2-Naphthol
(β-naphthol)

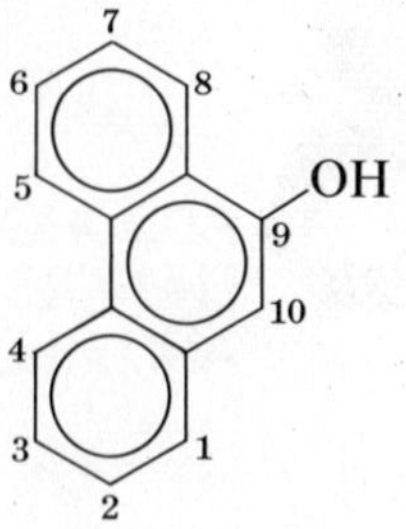

9-Phenanthrol

Ethers differ from alcohols and phenols in that the oxygen atom of an ether is bonded to two carbon atoms. The hydrocarbon groups may be alkyl, alkenyl, alkynyl, or aryl. Several examples are shown below.

$$CH_3CH_2-O-CH_2CH_3$$
Diethyl ether

$$CH_2=CHCH_2-O-CH_3$$
Methyl allyl ether

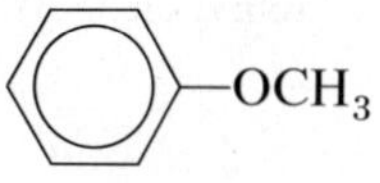

—OCH_3

Anisole

Nomenclature of Alcohols

Simple alcohols are often called by their *common* names. We have seen many examples already. In addition to *methyl alcohol, ethyl alcohol, isopropyl alcohol,* and *allyl alcohol,* for example, there are several others including the following:

$$CH_3CH_2CH_2OH$$
n-Propyl alcohol

$$CH_3CH_2CH_2CH_2OH$$
n-Butyl alcohol

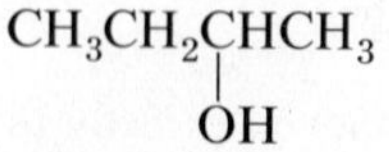

$$CH_3CH_2\underset{\underset{\displaystyle OH}{|}}{C}HCH_3$$
sec-Butyl alcohol

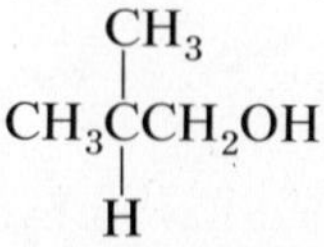

$$CH_3\overset{\overset{\displaystyle CH_3}{|}}{\underset{\underset{\displaystyle H}{|}}{C}}CH_2OH$$
Isobutyl alcohol

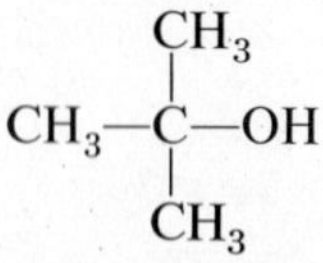

$$CH_3-\overset{\overset{\displaystyle CH_3}{|}}{\underset{\underset{\displaystyle CH_3}{|}}{C}}-OH$$
tert-Butyl alcohol

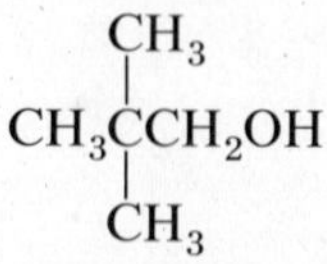

$$CH_3\overset{\overset{\displaystyle CH_3}{|}}{\underset{\underset{\displaystyle CH_3}{|}}{C}}CH_2OH$$
Neopentyl alcohol

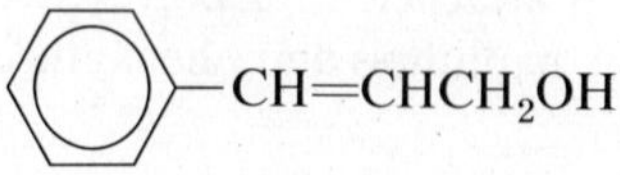

—$CH=CHCH_2OH$
Cinnamyl alcohol

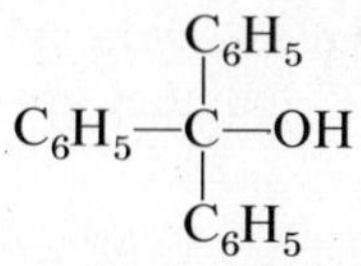

$$C_6H_5-\overset{\overset{\displaystyle C_6H_5}{|}}{\underset{\underset{\displaystyle C_6H_5}{|}}{C}}-OH$$
Triphenylcarbinol

The IUPAC rules for naming alcohols are as follows:

1. Select the longest continuous carbon chain *to which the hydroxyl is directly attached*. Change the name of the alkane corresponding to this chain by dropping the final *e* and adding *ol*. This gives the base name of the alcohol.
2. Number the longest continuous carbon chain so as to give the carbon bearing the hydroxyl group the lowest possible number. Indicate the position of the hydroxyl group by using this number; indicate the positions of other substituents (or multiple bonds) by using the numbers corresponding to their position along the carbon chain.

The examples below show how these rules are applied.

$$\overset{3}{C}H_3\overset{2}{C}H_2\overset{1}{C}H_2OH$$

1-Propanol

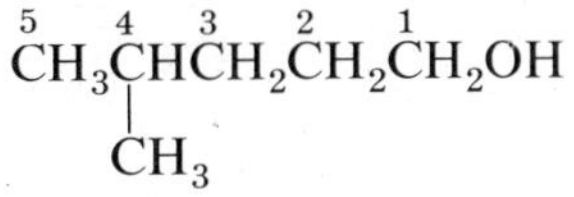

$$\overset{5}{C}H_3\overset{4}{C}HCH_2\overset{3}{C}H_2\overset{2}{C}H_2\overset{1}{C}H_2OH$$
$$\overset{|}{C}H_3$$

4-Methyl-1-pentanol

$$\overset{3}{C}lCH_2\overset{2}{C}H_2\overset{1}{C}H_2OH$$

3-Chloro-1-propanol

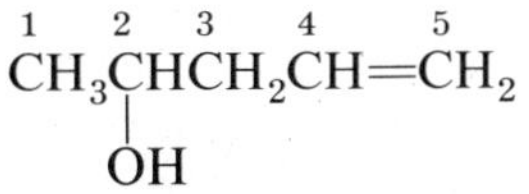

$$\overset{1}{C}H_3\overset{2}{C}HCH_2\overset{4}{C}H=\overset{5}{C}H_2$$
$$\overset{|}{O}H$$

4-Penten-2-ol

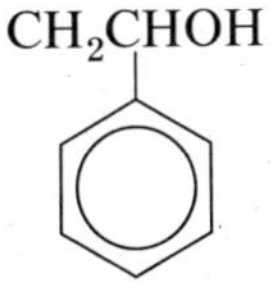

$$CH_2CHOH$$

1-Phenylethanol

—$$CH_2CH_2OH$$

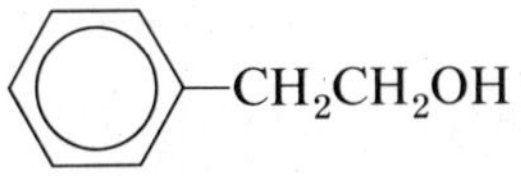

2-Phenylethanol

Alcohols containing two hydroxyl groups are commonly called glycols. In the IUPAC system they are named as *diols*.

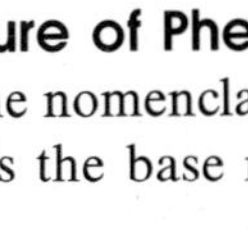

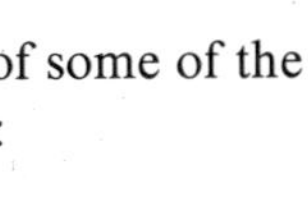

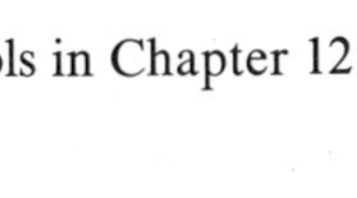

$$\underset{OH\ OH}{CH_2CH_2}$$

Common: Ethylene glycol
IUPAC: 1,2-Ethanediol

$$\underset{OH\,OH}{CH_3CHCH_2}$$

Propylene glycol
1,2-Propanediol

$$\underset{OH\quad OH}{CH_2CH_2CH_2}$$

Trimethylene glycol
1,3-Propanediol

Nomenclature of Phenols

We studied the nomenclature of some of the phenols in Chapter 12. In many compounds *phenol* is the base name:

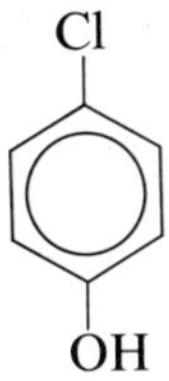

Cl

OH

p-Chlorophenol

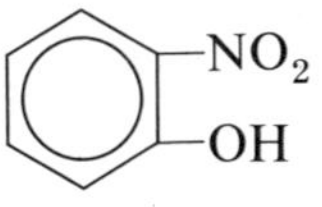

—$$NO_2$$
—OH

o-Nitrophenol

Br

OH

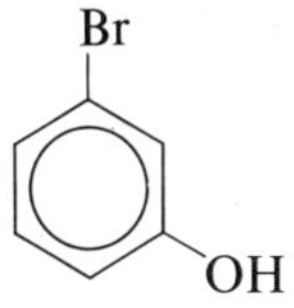

m-Bromophenol

The methylphenols, however, are called *cresols*:

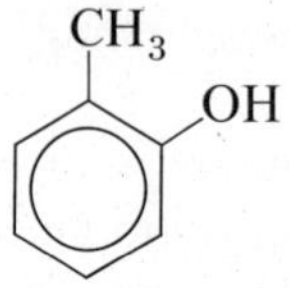

o-Cresol 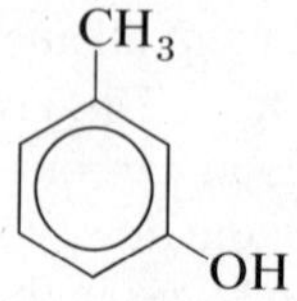   m-Cresol 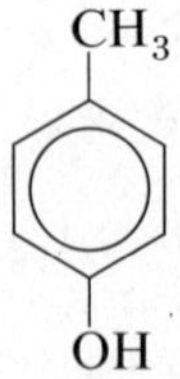   p-Cresol

The dihydroxybenzenes have special names:

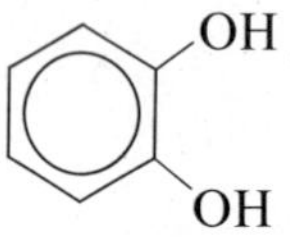

Catechol 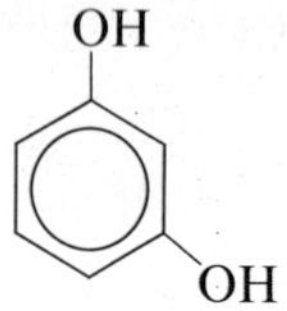   Resorcinol 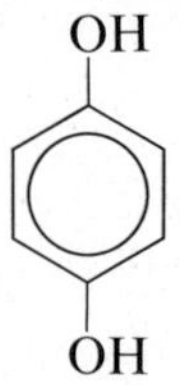   Hydroquinone

Nomenclature of Ethers

Ethers are frequently given common names and to do so is quite easy. One simply names both groups that are attached to the oxygen:

$CH_3OCH_2CH_3$ $CH_3CH_2OCH_2CH_3$
Methyl ethyl ether   Diethyl ether

$$C_6H_5O\overset{\displaystyle CH_3}{\underset{\displaystyle CH_3}{\overset{|}{\underset{|}{C}}}}-CH_3 \qquad CH_2{=}CHOCH{=}CH_2$$

Phenyl *tert*-butyl ether   Divinyl ether

IUPAC names are seldom used for simple ethers. IUPAC names are used for complicated ethers, however, and for compounds with more than one ether linkage. In the IUPAC system ethers are named as alkoxyalkanes, alkoxyalkenes, and alkoxyarenes.

$$CH_3\underset{\displaystyle OCH_3}{\overset{|}{C}}HCH_2CH_2CH_3 \qquad CH_3O{-}\bigcirc{-}CH_3$$

2-Methoxypentane   p-Methoxytoluene

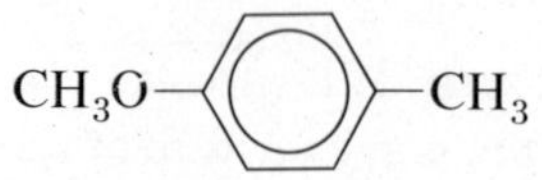

$CH_3OCH_2CH_2OCH_3$ $CH_3OCH{=}CHCH_3$
1,2-Dimethoxyethane   1-Methoxy-1-propene

Problem 16.1

Give appropriate names for all of the alcohols and ethers with the formulas (a) $C_5H_{12}O$, (b) $C_5H_{10}O$, and (c) $C_6H_{14}O$.

16.2 PHYSICAL PROPERTIES OF ALCOHOLS, PHENOLS, AND ETHERS

When one compares the boiling points of alcohols and phenols with the boiling points of hydrocarbons of roughly the same molecular weight, one is struck by the fact that alcohols and phenols boil at much higher temperatures. We can see these contrasts in Table 16.1.

TABLE 16.1 Comparison of Boiling Points of Alcohols, Phenols, and Hydrocarbons of Approximately the Same Molecular Weight

COMPOUND	MOL WT	bp °C (1 atm)
CH_3OH	32.04	65
CH_3CH_3	30.07	−88.6
CH_3CH_2OH	46.07	78.5
$CH_3CH_2CH_3$	44.09	−44.5
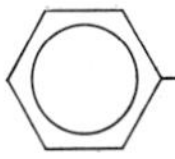 ⬡—OH	94.11	182
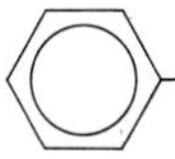 ⬡—CH_3	92.13	110.6

The boiling point of methanol is more than 150° higher than that of ethane even though the two compounds have approximately the same molecular weight. Ethanol boils 123° higher than propane and phenol boils almost 70° higher than toluene.

On the other hand, when we compare the boiling points of ethers with those of hydrocarbons of roughly the same molecular weight, the differences are not at all large (Table 16.2). Diethyl ether and pentane, for example, have almost the same boiling point.

TABLE 16.2 A Comparison of Boiling Points of Ethers and Hydrocarbons of Approximately the Same Molecular Weight

COMPOUND	MOL WT	bp °C (1 atm)
CH_3OCH_3	46.07	−25
$CH_3CH_2CH_3$	44.09	−44.5
$CH_3CH_2OCH_2CH_3$	74.12	34.6
$CH_3CH_2CH_2CH_2CH_3$	72.15	36
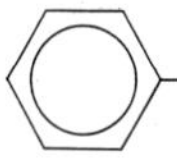 ⬡—OCH_3	108.13	158.3
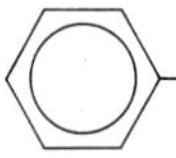 ⬡—CH_2CH_3	106.17	136

We can account for these contrasts and similarities on the basis of a single phenomenon: the presence or absence of strong *hydrogen bonds*.

Alcohols and phenols have a hydrogen atom attached to an oxygen atom. Because oxygen is highly electronegative, alcohols and phenols are capable of forming strong hydrogen bonds. Hydrogen bonding holds alcohol and phenol molecules together (causes them to be associated); alcohols and phenols, as a result, have boiling points that correspond to those of compounds with much greater molecular weight.

Hydrogen bonding   Hydrogen bonding
in an alcohol   in a phenol

TABLE 16.3 Physical Properties of Alcohols

COMPOUND	NAME	mp °C	bp °C (1 atm.)	DENSITY AT 20°	WATER SOLUBILITY g/100 g H_2O
Monohydroxy Alcohols					
CH_3OH	Methyl	−97	64.7	0.792	∞
CH_3CH_2OH	Ethyl	−114	78.3	0.789	∞
$CH_3CH_2CH_2OH$	n-Propyl	−126	97.2	0.804	∞
$CH_3CH(OH)CH_3$	Isopropyl	−88	82.3	0.786	∞
$CH_3CH_2CH_2CH_2OH$	n-Butyl	−90	117.7	0.810	7.9
$CH_3CH(CH_3)CH_2OH$	Isobutyl	−108	108.0	0.802	10.0
$CH_3CH_2CH(OH)CH_3$	sec-Butyl	−114	99.5	0.808	12.5
$(CH_3)_3COH$	tert-Butyl	25	82.5	0.789	∞
$CH_3(CH_2)_3CH_2OH$	n-Pentyl	−78.5	138.0	0.817	2.4
$CH_3(CH_2)_4CH_2OH$	n-Hexyl	−52	156.5	0.819	0.6
$CH_3(CH_2)_5CH_2OH$	n-Heptyl	−34	176	0.822	0.2
$CH_3(CH_2)_6CH_2OH$	n-Octyl	−15	195	0.825	0.05
$CH_3(CH_2)_7CH_2OH$	n-Nonyl		212	0.827	—
$CH_3(CH_2)_8CH_2OH$	n-Decyl	6	228	0.829	—
$CH_2{=}CHCH_2OH$	Allyl	−129	97	0.855	∞
$(CH_2)_4CHOH$	Cyclopentyl		140	0.949	—
$(CH_2)_5CHOH$	Cyclohexyl	24	161.5	0.962	3.6
$C_6H_5CH_2OH$	Benzyl	−15	205	1.046	4
Diols and Triols					
CH_2OHCH_2OH	Ethylene glycol	−16	197	1.113	∞
$CH_3CHOHCH_2OH$	Propylene glycol		187	1.040	∞
$CH_2OHCH_2CH_2OH$	Trimethylene glycol		215	1.060	∞
$CH_2OHCHOHCH_2OH$	Glycerol	18	290	1.261	∞

Ethers, by contrast, do not have a hydrogen attached to a strongly electronegative element and in this respect ethers are like hydrocarbons; ethers are not able to associate with each other through hydrogen bonding. Although ethers are more polar than hydrocarbons, their boiling points are comparable.

Ethers, however, *are* able to form hydrogen bonds with compounds such as water. Ethers, therefore, show water solubilities that are similar to those of alcohols of the same molecular weight and that are very different from the water solubilities of hydrocarbons.

Hydrogen bonding   Hydrogen bonding
between an ether and water   between an alcohol and water

Diethyl ether and 1-butanol, for example, have the same water-solubility, approximately 8.0g/100 ml at room temperature. Pentane, by contrast, is virtually insoluble in water.

Methyl alcohol, ethyl alcohol, both propyl alcohols, and *tert*-butyl alcohol are miscible with water in all proportions (Table 16.3). The remaining butyl alcohols show water solubilities between 7.9 and 12.5g/100 ml. The water solubility of alcohols gradually decreases as the hydrocarbon portion of the molecule lengthens; long chain alcohols are more "alkanelike" and are, therefore, less like water.

Phenols also show appreciable water solubilities and apparently form even stronger hydrogen bonds than alcohols. Phenol, itself, has a water solubility of 9.3g/100 ml at 20°, whereas the solubility of cyclohexanol is only 3.6g/100 ml.

The physical properties of a number of alcohols, phenols, and ethers are presented in Tables 16.3, 16.4, and 16.5.

TABLE 16.4 Physical Properties of Ethers

NAME	FORMULA	mp °C	bp °C	DENSITY 20°
Dimethyl ether	CH_3OCH_3	−140	−24.9	0.661
Methyl ethyl ether	$CH_3OCH_2CH_3$		7.9	0.697
Diethyl ether	$CH_3CH_2OCH_2CH_3$	−116	34.6	0.714
Di-*n*-propyl ether	$(CH_3CH_2CH_2)_2O$	−122	90.5	0.736
Diisopropyl ether	$(CH_3)_2CHOCH(CH_3)_2$		68	0.735
Di-*n*-butyl ether	$(CH_3CH_2CH_2CH_2)_2O$	−95	141	0.769
1,2-Dimethoxyethane	$CH_3OCH_2CH_2OCH_3$		83	0.863
Tetrahydrofuran	$(CH_2)_4O$	−108	65.4	0.888
Dioxane		11	101	
Anisole		−37.3	154	0.994

TABLE 16.5 Physical Properties of Phenols

NAME	FORMULA	mp °C	bp °C	WATER SOLUBILITY g/100 g H_2O
Phenol	C_6H_5OH	43	181	9.3
o-Cresol	$o\text{-}CH_3C_6H_4OH$	30	191	2.5
m-Cresol	$m\text{-}CH_3C_6H_4OH$	11	201	2.6
p-Cresol	$p\text{-}CH_3C_6H_4OH$	35.5	201	2.3
o-Chlorophenol	$o\text{-}ClC_6H_4OH$	8	176	2.8
m-Chlorophenol	$m\text{-}ClC_6H_4OH$	33	214	2.6
p-Chlorophenol	$p\text{-}ClC_6H_4OH$	43	220	2.7
o-Nitrophenol	$o\text{-}O_2NC_6H_4OH$	45	217	0.2
m-Nitrophenol	$m\text{-}O_2NC_6H_4OH$	96		1.4
p-Nitrophenol	$p\text{-}O_2NC_6H_4OH$	114		1.7
2,4-Dinitrophenol	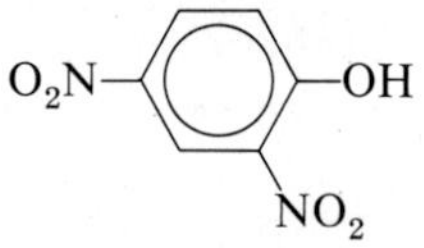	113		0.6
2,4,6-Trinitrophenol (picric acid)	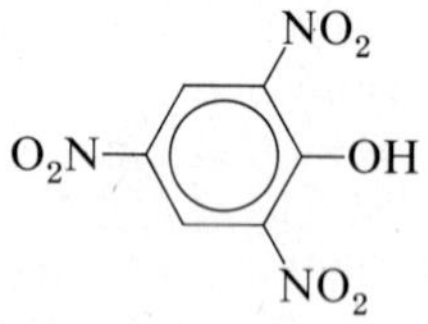	122		1.4

Problem 16.2

How can you account for the fact that the boiling point of ethylene glycol is much higher than that of either n-propyl alcohol or isopropyl alcohol even though all three compounds have roughly the same molecular weight?

16.3 IMPORTANT ALCOHOLS AND ETHERS

Methyl Alcohol

At one time, most methyl alcohol was produced by the destructive distillation of wood (i.e., heating wood to a high temperature in the absence of air). It was because of this method of preparation that methyl alcohol came to be called "wood alcohol." Today, most methyl alcohol is prepared by the catalytic hydrogenation of carbon monoxide. This reaction takes place under high pressure and a temperature of 300° to 400°.

$$CO + H_2 \xrightarrow[\substack{200\text{-}300 \text{ atm.} \\ ZnO\text{--}CrO_3}]{300\text{-}400°} CH_3OH$$

Methyl alcohol is highly toxic. Ingestion of even small quantities of methyl alcohol can cause blindness; large quantities cause death. Methyl alcohol poisoning can also occur by inhalation of the vapors or by prolonged exposure to the skin.

Ethyl Alcohol

Ethyl alcohol can be made by the fermentation of sugars and it is the alcohol of alcoholic beverages. The synthesis of ethyl alcohol in the form of wine by the fermentation of the sugars of fruit juices was probably man's first accomplishment in the field of organic synthesis. Sugars from a wide variety of sources can be used in the preparation of alcoholic beverages. Often, these sugars are from grains and it is this that accounts for ethyl alcohol having the synonym "grain alcohol."

Fermentation is usually carried out by adding yeast to a mixture of sugars and water. Yeast contains enzymes that promote a long series of reactions that ultimately convert a simple sugar ($C_6H_{12}O_6$) to ethyl alcohol and carbon dioxide.

$$C_6H_{12}O_6 \xrightarrow{\text{yeast}} 2CH_3CH_2OH + 2CO_2$$
$$(\sim 95\% \text{ yield})$$

Fermentation alone does not produce beverages with an alcohol content greater than 12 to 15%. To produce beverages of higher alcohol content the aqueous solution must be distilled. Brandy, whiskey, and vodka are produced in this way. The "proof" of an alcoholic beverage is simply twice the percentage of alcohol (by volume). One hundred proof whiskey is 50% alcohol. The flavors of the various distilled liquors result from other organic compounds that distill with the alcohol and water.

Distillation of a solution of ethyl alcohol and water will not yield ethyl alcohol more concentrated than 95%. A mixture of 95% ethyl alcohol and 5% water boils at a lower temperature (78.15°) than either pure ethyl alcohol (bp 78.3°) or pure water (bp 100°). Such a mixture is an example of *azeotrope*.* Pure ethyl alcohol can be prepared by adding benzene to the mixture of 95% ethyl alcohol and water and then distilling this solution. Benzene forms a new azeotrope with ethyl alcohol and water that is 7.5% water. This azeotrope boils at 64.9° and allows removal of the water (along with some ethyl alcohol). Eventually pure ethyl alcohol distills over. Pure ethyl alcohol is called *absolute alcohol*.

Ethyl alcohol is quite cheap but when it is used for beverages it is highly taxed. (The tax is greater than $20 per gallon in most states.) Federal law requires that most ethyl alcohol used for scientific and industrial purposes be adulterated or "denatured" to make it undrinkable. A variety of denaturants are used including methyl alcohol.

$$CH_2{=}CH_2 + H_2O \xrightarrow{\text{acid}} CH_3CH_2OH$$

Ethyl alcohol is an important industrial chemical. Most ethyl alcohol for industrial purposes is produced by the acid-catalyzed hydration of ethene.

Ethyl alcohol is a *hypnotic* (sleep producer). It depresses activity in the upper brain even though it gives the illusion of being a stimulant. Ethyl alcohol is also toxic but it is much less toxic than methyl alcohol. In rats the lethal dose of ethyl alcohol is 13.7 g per kilogram of body weight.

---

* Azeotropes can also have a boiling point that is higher than that of either of the pure components.

Isopropyl Alcohol

This alcohol is more toxic than ethyl alcohol. Isopropyl alcohol is commonly used as rubbing alcohol.

Ethylene Glycol

Ethylene glycol has a low molecular weight, a high boiling point, and is miscible with water in all proportions. These properties make ethylene glycol an ideal automobile antifreeze. Much ethylene glycol is sold for this purpose under a variety of trade names.

Diethyl Ether

Diethyl ether is a very low boiling, highly flammable compound. Care should always be taken when diethyl ether is used in the laboratory because open flames or sparks from light switches can cause explosive combustion of mixtures of diethyl ether and air.

Most ethers react slowly with oxygen in the air to form organic peroxides. These peroxides, which may accumulate in ethers that have been left standing for long periods in contact with air, are dangerously explosive and may detonate without warning when ether solutions are distilled to near dryness. Since ethers are used frequently in extractions, one should take care to test for and decompose any peroxides present in the ether before a distillation is carried out. (Consult a laboratory manual for instructions.)

Diethyl ether was first employed as a surgical anesthetic by C. W. Long of Jefferson, Georgia, in 1842. Long's use of diethyl ether was not published, but shortly thereafter, diethyl ether was introduced into surgical use at the Massachusetts General Hospital in Boston by J. C. Warren.

The idea of chemical anesthesia dates from 1799 with Sir Humphrey Davy's discovery of the anesthetic properties of nitrous oxide (laughing gas). During the nineteenth century chloroform was also used as an anesthetic; Queen Victoria, for example, gave birth to her children under chloroform anesthesia. Chloroform is not used as an anesthetic any longer because its use causes liver damage. (It did not, however, appear to shorten Queen Victoria's life appreciably.)

Two popular modern anesthetics are divinyl ether (vinethene) and cyclopropane. Cyclopropane acts very rapidly producing unconsciousness within seconds. Like diethyl ether, divinyl ether and cyclopropane are highly flammable and produce explosive mixtures in air. Thus, special skills are required in the application of these compounds.

16.4 PREPARATION OF ALCOHOLS

We have already studied three methods for the synthesis of alcohols: hydration of alkenes, oxymercuration-demercuration, and hydroboration-oxidation.

Hydration of Alkenes (discussed in Section 6.3)

Alkenes add water in the presence of an acid catalyst. The addition follows Markovnikov's rule; thus, except for the hydration of ethylene, the reaction produces secondary and tertiary alcohols. The reaction is reversible and the mechanism for the hydration of an alkene is simply the reverse of that for the dehydration of an alcohol (p. 172).

Problem 16.3

Show how you would prepare each of the following alcohols by acid-catalyzed hydration of the appropriate alkene.

    (a) *tert*-Butyl alcohol   (c) Cyclopentanol

    (b) 2-Hexanol   (d) 1-Methylcyclohexanol

Problem 16.4

When 3,3-dimethyl-1-butene is subjected to acid-catalyzed hydration the major product is 2,3-dimethyl-2-butanol. How can you explain this?

Oxymercuration-Demercuration (discussed in Section 6.4)

Addition of water to an alkene can be achieved indirectly through oxymercuration-demercuration. In the first step of this synthesis an alkene is allowed to react with mercuric acetate in a water-tetrahydrofuran solution. This step produces a hydroxyalkylmercury compound (below) in which mercury is bonded to the least substituted carbon. In the second step, sodium borohydride reduction of this intermediate results in replacement of the mercury substituent by hydrogen. The net result of the two steps is Markovnikov addition of —H and —OH. Both steps can be carried out rapidly and the yields obtained are usually high. An example is the following synthesis of 2-pentanol.

$$CH_3CH_2CH_2CH{=}CH_2 \xrightarrow[\substack{THF-H_2O\\(15\ sec.)}]{Hg(OAc)_2} \underset{\substack{(93\%)\\ \text{Hydroxyalkylmercury compound}}}{CH_3CH_2CH_2\underset{OH}{\underset{|}{CH}}{-}\underset{HgOAc}{\underset{|}{CH_2}}} \xrightarrow[\substack{OH^-\\(1\ hr)}]{NaBH_4} \underset{\substack{\text{2-pentanol}\\(93\%)}}{CH_3CH_2CH_2\underset{OH}{\underset{|}{CH}}CH_3}$$

Problem 16.5

Show how you might employ the oxymercuration-demercuration synthesis to prepare each of the following alcohols from an appropriate alkene.

    (a) 2-Hexanol   (d) 2-Phenyl-2-propanol

    (b) 1-Methylcyclohexanol

    (c) 2,4,4-Trimethyl-2-pentanol

Hydroboration-Oxidation (discussed in Section 6.5)

Hydroboration-oxidation gives us a method for adding water to a double bond in a manner opposite to that of acid-catalyzed hydration or oxymercuration-demercuration. The reaction takes place in two steps: in the first step boron hydride adds to the double bond; in the second step oxidation with hydrogen peroxide in aqueous base replaces the boron group with a hydroxyl group.

$$3R-CH=CH_2 + \frac{1}{2}(BH_3)_2 \longrightarrow (R-CH_2CH_2)_3B$$

$$(R-CH_2CH_2)_3B \xrightarrow[OH^-,\ H_2O]{H_2O_2} 3R-CH_2CH_2OH + B(OH)_3$$

The stereochemistry of hydroboration-oxidation is a net *syn* addition of —H and —OH.

Problem 16.6

Show how you would utilize the hydroboration-oxidation procedure to prepare each of the following alcohols.

(a) 3,3-Dimethyl-1-butanol
(b) 1-Hexanol
(c) 2-Phenylethanol
(d) *trans*-2-Methylcyclohexanol

Reduction of Carbonyl Compounds

Primary and secondary alcohols can be synthesized by the reduction of a variety of compounds that contain the carbonyl ($\diagdown$C$=$O) group. Several general examples are shown below.

$$\underset{\text{Carboxylic acid}}{R-\overset{\overset{\textstyle O}{\|}}{C}-OH} \xrightarrow{\text{(H)}} \underset{\text{1° Alcohol}}{R-CH_2OH}$$

$$\underset{\text{Ester}}{R-\overset{\overset{\textstyle O}{\|}}{C}-OR'} \xrightarrow{\text{(H)}} \underset{\text{1° Alcohol}}{R-CH_2OH}\ (+\ R'OH)$$

$$\underset{\text{Aldehyde}}{R-\overset{\overset{\textstyle O}{\|}}{C}-H} \xrightarrow{\text{(H)}} \underset{\text{1° Alcohol}}{R-CH_2OH}$$

$$\underset{\text{Ketone}}{R-\overset{\overset{\textstyle O}{\|}}{C}-R'} \xrightarrow{\text{(H)}} \underset{\text{2° Alcohol}}{R-\underset{\underset{\textstyle OH}{|}}{C}H-R'}$$

Reductions of carboxylic acids are the most difficult and prior to 1946, direct reduction of acids was not possible. However, the discovery in 1946 of the powerful reducing agent, *lithium aluminum hydride,* gave organic chemists a method for reducing acids to primary alcohols in excellent yields. Two examples (p. 587) are the lithium aluminum hydride reductions of acetic acid and 2,2-dimethylpropanoic acid.

$$\underset{\substack{O \\ \parallel}}{CH_3\overset{O}{\overset{\parallel}{C}}-OH} + \underset{\substack{Lithium \\ aluminum \\ hydride}}{LiAlH_4} \xrightarrow[(2)H_2O]{Et_2O} \underset{\substack{Ethyl\ alcohol \\ (100\%)}}{CH_3CH_2OH}$$

Acetic acid

$$\underset{\substack{2,2\text{-Dimethylpropanoic} \\ acid}}{CH_3-\overset{CH_3}{\underset{CH_3}{\overset{|}{\underset{|}{C}}}}-COOH} + LiAlH_4 \xrightarrow[(2)H_2O]{Et_2O} \underset{\substack{Neopentyl\ alcohol \\ (92\%)}}{CH_3\overset{CH_3}{\underset{CH_3}{\overset{|}{\underset{|}{C}}}}-CH_2OH}$$

Esters can be reduced in several ways: through the use of sodium in alcohol, by high pressure hydrogenation (a reaction preferred for industrial processes and often referred to as "hydrogenolysis") or through the use of lithium aluminum hydride.

$$R-\overset{O}{\overset{\parallel}{C}}OC_2H_5 \xrightarrow[C_2H_5OH]{Na} R-CH_2OH + C_2H_5OH$$

$$R-\overset{O}{\overset{\parallel}{C}}OR' + H_2 \xrightarrow[\substack{175° \\ 5000\ psi}]{CuO \cdot CuCr_2O_4} RCH_2OH + R'OH$$

$$R-\overset{O}{\overset{\parallel}{C}}OR' + LiAlH_4 \xrightarrow[(2)H_2O]{Et_2O} RCH_2OH + R'OH$$

The last method is the one most commonly used now in small scale laboratory synthesis.

Aldehydes and ketones can also be reduced to alcohols with hydrogen and a metal catalyst, with sodium in alcohol, and with lithium aluminum hydride. The reagent most often used however, is the reducing agent sodium borohydride, $NaBH_4$.

$$\underset{Butanal}{CH_3CH_2CH_2\overset{O}{\overset{\parallel}{C}}H} + NaBH_4 \xrightarrow{H_2O} \underset{\substack{1\text{-Butanol} \\ (85\%)}}{CH_3CH_2CH_2CH_2OH}$$

$$\underset{2\text{-Butanone}}{CH_3CH_2\underset{\overset{\parallel}{O}}{C}CH_3} + NaBH_4 \xrightarrow{H_2O} \underset{\substack{2\text{-Butanol} \\ (87\%)}}{CH_3CH_2\underset{\overset{|}{OH}}{C}HCH_3}$$

The key step in the reduction of a carbonyl compound by either lithium aluminum hydride or sodium borohydride is the transfer of a _hydride ion_ from the metal to the carbonyl carbon. In this transfer the hydride ion acts as a _nucleophile_. Since boron and aluminum compounds are Lewis acids they react as electrophiles at the carbonyl oxygen and facilitate the hydride transfer. The mechanism for the reduction of a ketone by sodium borohydride is illustrated below.

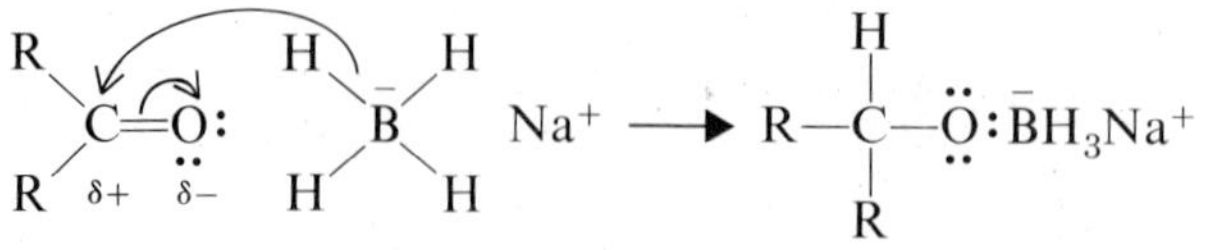

This step is then repeated until all hydrogens attached to boron have been transferred. The boron complex decomposes in water to form the secondary alcohol.

$$(R_2CHO)_4B^- \ Na^+ + 3H_2O \longrightarrow 4R\underset{\underset{OH}{|}}{C}HR + NaH_2BO_3$$

Sodium borohydride is a milder reducing agent than lithium aluminum hydride. While lithium aluminum hydride will reduce acids, esters, aldehydes, and ketones; sodium borohydride will reduce only aldehydes and ketones. Lithium aluminum hydride *reacts violently with water* and therefore reductions with lithium aluminum hydride must be carried out in anhydrous ether solutions. (Water is added cautiously after the reaction is over to decompose the aluminum complex.) Sodium borohydride reductions, by contrast, can be carried out in water or alcohol solutions.

Problem 16.7

Which reducing agent ($NaBH_4$ or $LiAlH_4$) would you use to carry out each of the following transformations?

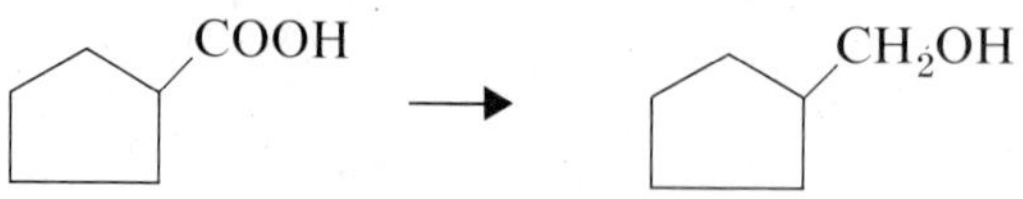

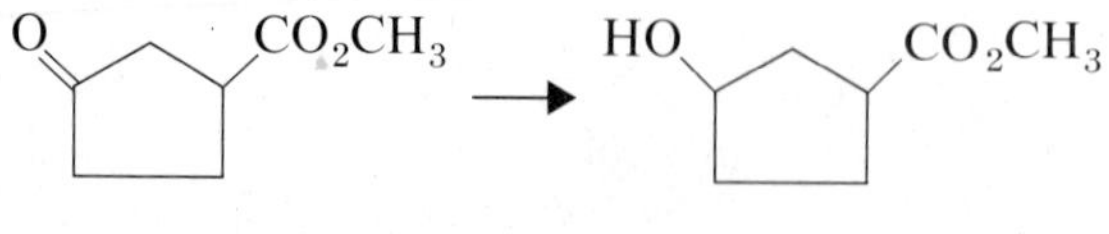

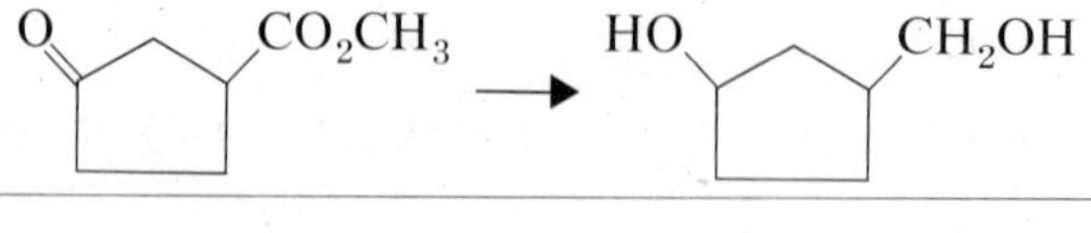

16.5 ALCOHOLS FROM GRIGNARD REAGENTS

Grignard reagents react with carbonyl compounds to produce alcohols. The reaction takes place in the following way:

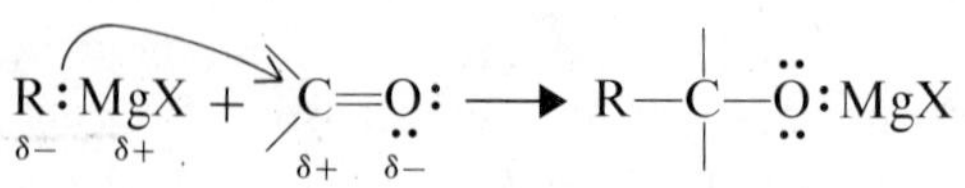

This reaction is very much like the reaction of sodium borohydride that we saw in Section 16.4: Sodium borohydride reacts by transferring a hydride ion to the carbonyl carbon; a Grignard reagent reacts by transferring a *carbanion*.

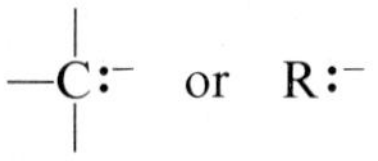

$$—\overset{|}{\underset{|}{C}}:^{-} \quad \text{or} \quad R:^{-}$$

A carbanion

In the sodium borohydride reaction, boron acts as a Lewis acid and associates with the oxygen atom of the carbonyl group. In the Grignard reaction magnesium acts in the same way.

The product formed when a Grignard reagent adds to a carbonyl group is an alkoxide ion $—\overset{|}{\underset{|}{C}}—\overset{..}{\underset{..}{O}}:^{-}$; that is associated with $Mg^{+2}X^{-}$. When water or dilute acid is added to the reaction mixture after the Grignard addition is over, an acid-base reaction takes place to produce an alcohol.

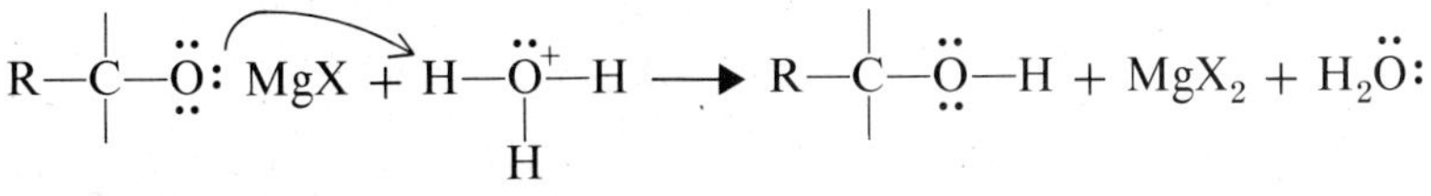

$$R—\overset{|}{\underset{|}{C}}—\overset{..}{\underset{..}{O}}:MgX + H—\overset{+}{\underset{|}{\overset{..}{O}}}—H \longrightarrow R—\overset{|}{\underset{|}{C}}—\overset{..}{\underset{..}{O}}—H + MgX_2 + H_2\overset{..}{\underset{..}{O}}:$$
$$\phantom{R—\overset{|}{\underset{|}{C}}—\overset{..}{\underset{..}{O}}:MgX + H—}\overset{|}{H}$$

Magnesium halide X^{-} Alcohol
alkoxide

Grignard additions to carbonyl compounds are important because they can be used to prepare primary, secondary, or tertiary alcohols.

1. A Grignard reagent reacts with formaldehyde, for example, to give a primary alcohol.

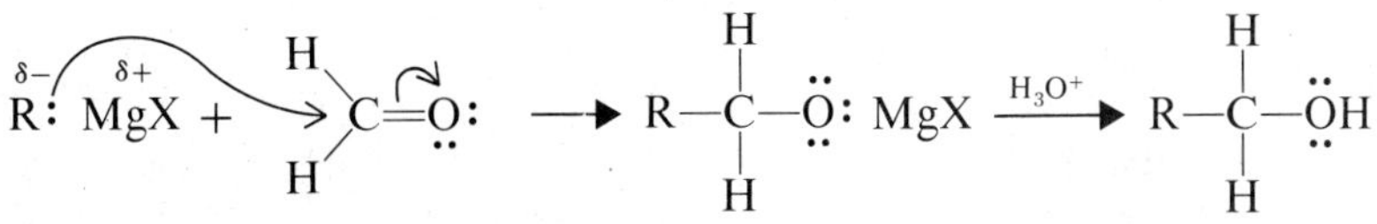

Formaldehyde   1° Alcohol

2. Grignard reagents react with higher aldehydes to give secondary alcohols.

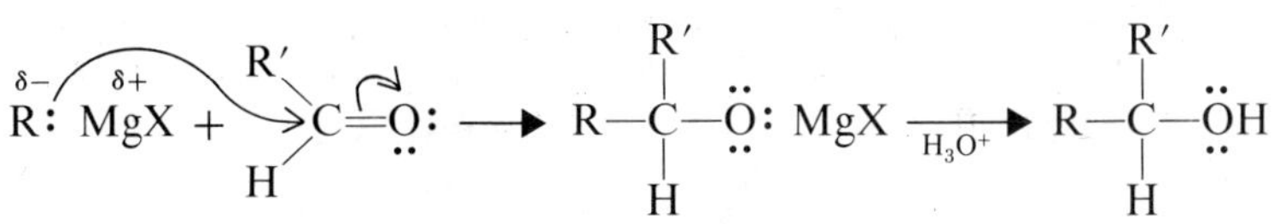

Higher   2° Alcohol
aldehyde

3. And, Grignard reagents react with ketones to give tertiary alcohols.

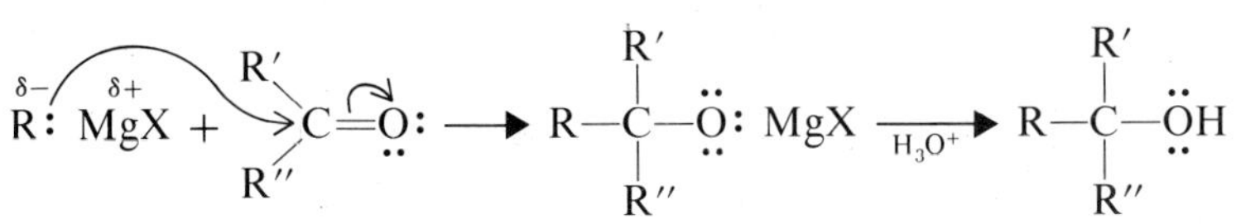

Ketone   3° Alcohol

Specific examples of these reactions are shown below.

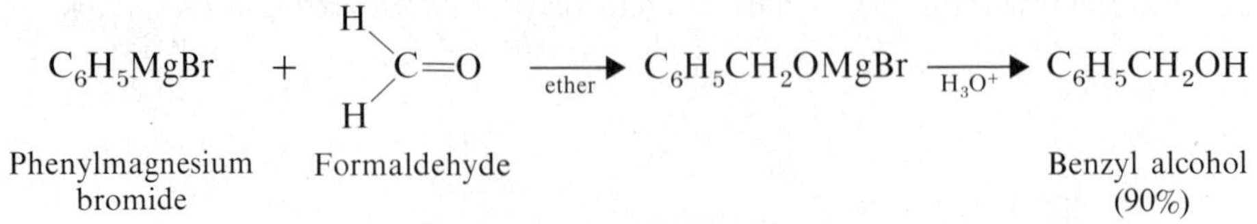

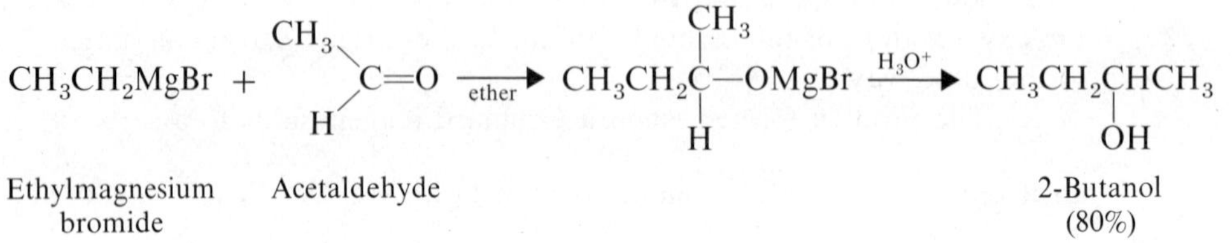

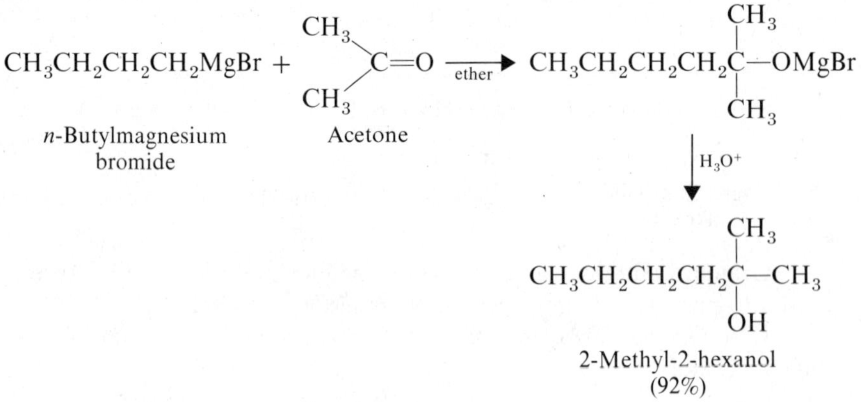

$$C_6H_5MgBr \quad + \quad \begin{matrix} H \\ \diagdown \\ \diagup \\ H \end{matrix} C=O \quad \xrightarrow{\text{ether}} \quad C_6H_5CH_2OMgBr \quad \xrightarrow{H_3O^+} \quad C_6H_5CH_2OH$$

Phenylmagnesium bromide	Formaldehyde		Benzyl alcohol (90%)

$$CH_3CH_2MgBr + \begin{matrix} CH_3 \\ \diagdown \\ \diagup \\ H \end{matrix} C=O \xrightarrow{\text{ether}} \begin{matrix} CH_3 \\ | \\ CH_3CH_2C{-}OMgBr \\ | \\ H \end{matrix} \xrightarrow{H_3O^+} \begin{matrix} CH_3CH_2CHCH_3 \\ | \\ OH \end{matrix}$$

Ethylmagnesium bromide	Acetaldehyde		2-Butanol (80%)

$$CH_3CH_2CH_2CH_2MgBr + \begin{matrix} CH_3 \\ \diagdown \\ \diagup \\ CH_3 \end{matrix} C=O \xrightarrow{\text{ether}} \begin{matrix} CH_3 \\ | \\ CH_3CH_2CH_2CH_2C{-}OMgBr \\ | \\ CH_3 \end{matrix}$$

n-Butylmagnesium   Acetone
bromide

$$\Big\downarrow H_3O^+$$

$$\begin{matrix} CH_3 \\ | \\ CH_3CH_2CH_2CH_2C{-}CH_3 \\ | \\ OH \end{matrix}$$

2-Methyl-2-hexanol
(92%)

4. Grignard reagents also add to the carbonyl group of esters. The initial product that forms from the addition is unstable and it loses magnesium alkoxide to form a ketone. Ketones are more reactive toward Grignard reagents than esters, so as soon as a molecule of the ketone is formed in the reaction mixture, it reacts with a second mole of the Grignard reagent. After hydrolysis, the overall product is a tertiary alcohol with two identical alkyl groups, groups that correspond to the alkyl portion of the Grignard reagent.

$$R{:}\ MgX + \begin{matrix} R' \\ \diagdown \\ \diagup \\ R''\ddot{O} \end{matrix} C{=}\ddot{O}{:} \longrightarrow \left[\begin{matrix} R' \\ | \\ R{-}C{-}\ddot{O}{-}MgX \\ | \\ {:}\ddot{O}{-}R'' \end{matrix} \right] \xrightarrow{-R''OMgX}$$

Ester

$$\left[\begin{matrix} R' \\ \diagdown \\ \diagup \\ R \end{matrix} C{=}\ddot{O}{:} \right] \xrightarrow{RMgX} \begin{matrix} R' \\ | \\ R{-}C{-}\ddot{O}MgX \\ | \\ R \end{matrix} \xrightarrow{H_3O^+} \begin{matrix} R' \\ | \\ R{-}C{-}OH \\ | \\ R \end{matrix}$$

3° Alcohol

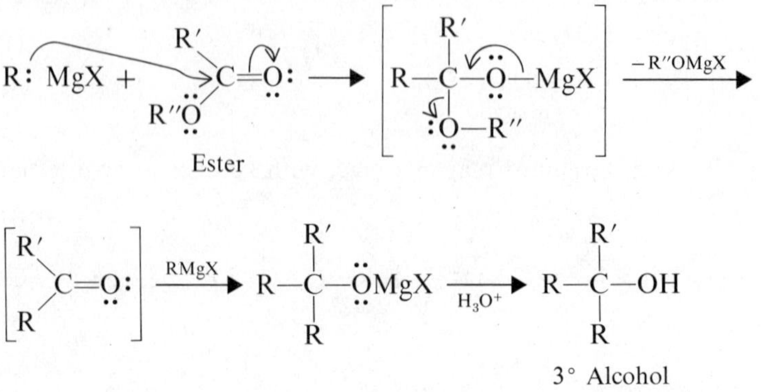

A specific example of this reaction is shown below.

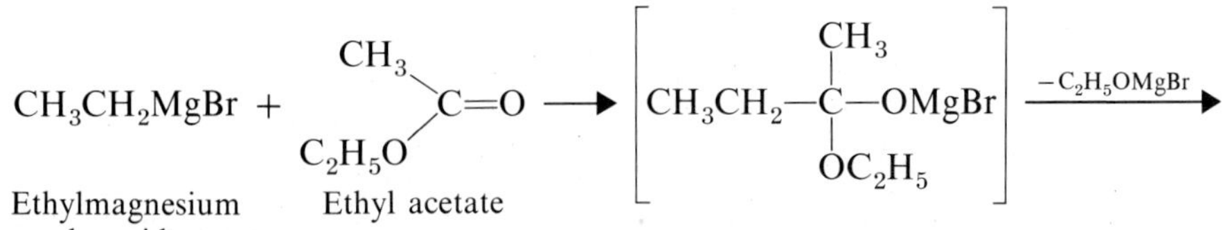

$$CH_3CH_2MgBr + \begin{matrix} CH_3 \\ C=O \\ C_2H_5O \end{matrix} \longrightarrow \left[\begin{matrix} CH_3 \\ | \\ CH_3CH_2-C-OMgBr \\ | \\ OC_2H_5 \end{matrix} \right] \xrightarrow{-C_2H_5OMgBr}$$

Ethylmagnesium   Ethyl acetate
   bromide

$$\left[\begin{matrix} CH_3 \\ C=O \\ CH_3CH_2 \end{matrix} \right] \xrightarrow{CH_3CH_2MgBr} \begin{matrix} CH_3 \\ | \\ CH_3CH_2C-CH_2CH_3 \\ | \\ OMgBr \end{matrix} \xrightarrow{H_3O^+} \begin{matrix} CH_3 \\ | \\ CH_3CH_2CCH_2CH_3 \\ | \\ OH \end{matrix}$$

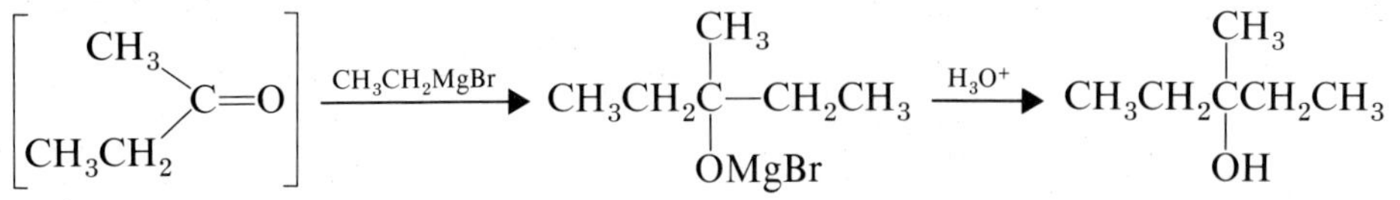

3-Methyl-3-pentanol
(67%)

Grignard reagents also react with epoxides to form alcohols by opening the three-membered ring (Section 15.7). After hydrolysis the product is a primary alcohol with two carbons more than the organic group of the Grignard reagent.

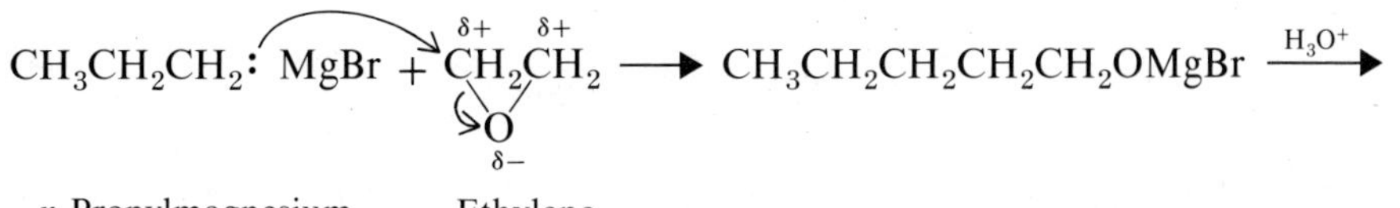

$$CH_3CH_2CH_2 \overset{\frown}{:} MgBr + \overset{\delta+}{CH_2}\overset{\delta+}{CH_2} \longrightarrow CH_3CH_2CH_2CH_2CH_2OMgBr \xrightarrow{H_3O^+}$$
$$\underset{\delta-}{\overset{\diagdown O \diagup}{}}$$

n-Propylmagnesium   Ethylene
     bromide   oxide

$$CH_3CH_2CH_2CH_2CH_2OH$$
1-Pentanol
(76%)

These examples begin to show us that by using the Grignard synthesis skillfully we can synthesize almost any alcohol we wish. In planning a Grignard synthesis we must simply choose the correct Grignard reagent and the correct aldehyde, ketone, ester, or epoxide. We can do this by examining the alcohol we wish to prepare and by paying special attention to the groups attached to the carbon bearing the —OH group. Many times there may be more than one way of carrying out the synthesis. In these cases our final choice will probably be dictated by the availability of starting compounds. Let us consider two examples.

Suppose we want to prepare 3-phenyl-3-pentanol. We examine its structure and we see that groups attached to the carbon bearing the —OH are a *phenyl group*

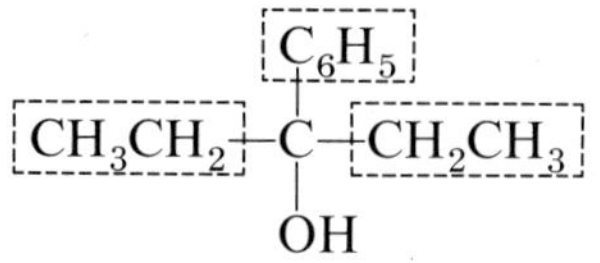

$$\begin{matrix} C_6H_5 \\ | \\ CH_3CH_2-C-CH_2CH_3 \\ | \\ OH \end{matrix}$$

3-Phenyl-3-pentanol

and *two ethyl groups*. This means that we can synthesize this compound in three different ways.

1. We can use a ketone with two ethyl groups (3-pentanone) and allow it to react with phenylmagnesium bromide:

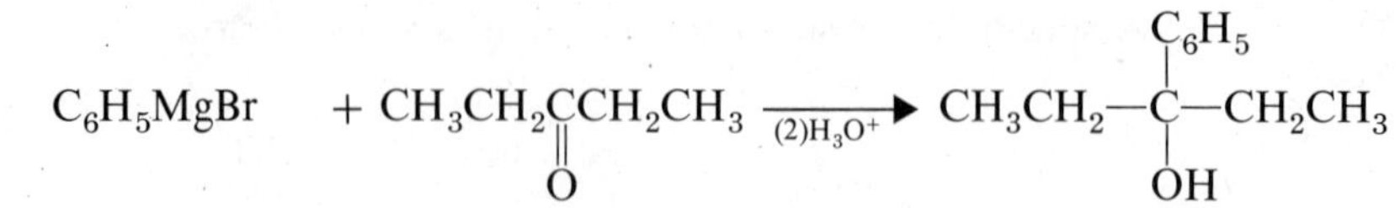

$$C_6H_5MgBr \quad + CH_3CH_2\underset{\underset{O}{\|}}{C}CH_2CH_3 \xrightarrow[(2)H_3O^+]{} CH_3CH_2-\underset{\underset{OH}{|}}{\overset{\overset{C_6H_5}{|}}{C}}-CH_2CH_3$$

Phenylmagnesium   3-Pentanone   3-Phenyl-3-pentanol
bromide

2. We can use a ketone containing an ethyl group and a phenyl group (phenyl ethyl ketone) and allow it to react with ethylmagnesium bromide:

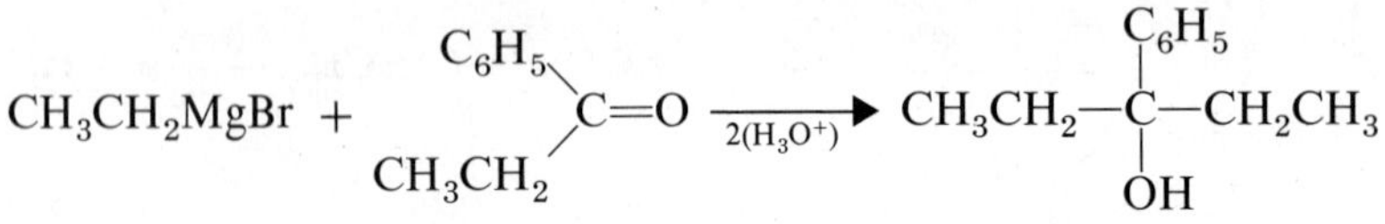

$$CH_3CH_2MgBr \; + \; \underset{CH_3CH_2}{\overset{C_6H_5}{>}}C{=}O \xrightarrow[2(H_3O^+)]{} CH_3CH_2-\underset{\underset{OH}{|}}{\overset{\overset{C_6H_5}{|}}{C}}-CH_2CH_3$$

Ethylmagnesium   Phenyl ethyl   3-Phenyl-3-pentanol
bromide   ketone

3. Or, we can use an ester of benzoic acid and allow it to react with two moles of ethylmagnesium bromide:

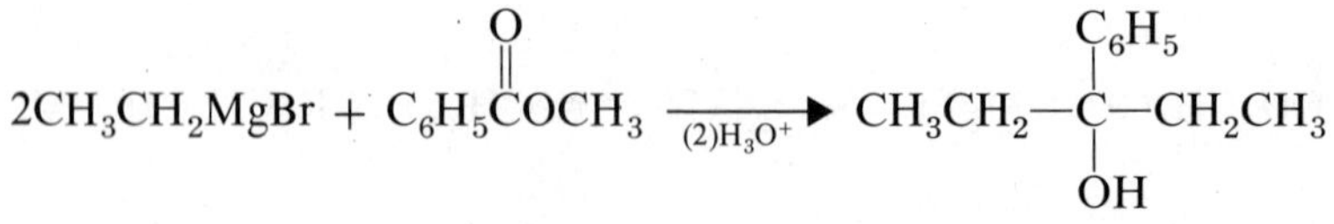

$$2CH_3CH_2MgBr + C_6H_5\underset{\underset{O}{\|}}{\overset{\overset{O}{\|}}{C}}OCH_3 \xrightarrow[(2)H_3O^+]{} CH_3CH_2-\underset{\underset{OH}{|}}{\overset{\overset{C_6H_5}{|}}{C}}-CH_2CH_3$$

Ethylmagnesium   Methyl   3-Phenyl-3-pentanol
bromide   benzoate

All of these methods will be likely to give us our desired compound in yields greater than 80%.

As another example, let us consider the problem of synthesizing 2-phenyl-ethanol.

$C_6H_5CH_2CH_2OH$

2-Phenylethanol

Here our choice of routes is somewhat more limited because the product that we desire is a primary alcohol. Still, however, two routes are open to us. We can allow benzylmagnesium bromide to react with formaldehyde, or we can allow phenyl-magnesium bromide to react with ethylene oxide.

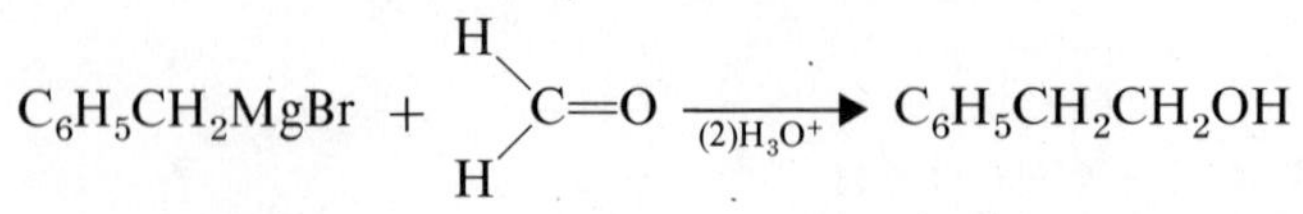

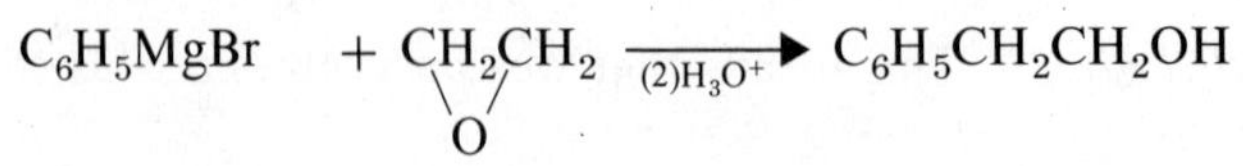

$$C_6H_5CH_2MgBr \; + \; \underset{H}{\overset{H}{>}}C{=}O \xrightarrow[(2)H_3O^+]{} C_6H_5CH_2CH_2OH$$

Benzylmagnesium   Formalde-   2-Phenylethanol
bromide   hyde

$$C_6H_5MgBr \quad + \; \underset{\underset{O}{\diagdown\diagup}}{CH_2CH_2} \xrightarrow[(2)H_3O^+]{} C_6H_5CH_2CH_2OH$$

Phenylmagnesium   Ethylene   2-Phenylethanol
bromide   oxide

Problem 16.8

Show how Grignard reactions could be used to synthesize each of the following compounds.

(a) *tert*-Butyl alcohol (two ways)

(b) 2-Pentanol (two ways)

(c) $\underset{\underset{OH}{|}}{\overset{\overset{CH_3}{|}}{C_6H_5C}}CH_2CH_3$ (three ways)

(d) 1-Hexanol (two ways)

Limitations of the Grignard Synthesis

While the Grignard synthesis is one of the most versatile of all general synthetic procedures, it is not without its limitations. Most of these limitations arise from the very feature of the Grignard reagent that makes it so useful, its *extraordinary reactivity as a nucleophile and a base.*

The Grignard reagent is a very powerful base; it is effectively a carbanion. Thus, it is not possible to prepare a Grignard reagent from an organic group that contains an *acidic hydrogen;* and by an acidic hydrogen we mean any hydrogen more acidic than the hydrogens of an alkane or alkene. We cannot, for example, prepare a Grignard reagent from a compound containing an O—H group, an N—H group, an —SH group, a —COOH group, or an —SO₃H group. If we were to attempt to prepare a Grignard reagent from an organic halide containing any of these groups, the formation of the Grignard reagent would simply fail to take place. (Even if a Grignard reagent were to form, it would immediately react with the acidic group.)

Grignard reactions are so sensitive to acidic compounds that when we prepare a Grignard reagent we must take special care to exclude moisture from our apparatus, and we must use anhydrous ether as our solvent.

Even acetylenic hydrogens are acidic enough to react with Grignard reagents. This is a limitation that we can use, however. We can make acetylenic Grignard reagents by allowing terminal alkynes to react with alkyl Grignard reagents (cf. Section 15.7). We can then use these acetylenic Grignard reagents to carry out other syntheses. For example:

$$C_6H_5C\equiv CH + C_2H_5MgBr \longrightarrow C_6H_5C\equiv CMgBr + C_2H_6\uparrow$$

$$C_6H_5C\equiv CMgBr + C_2H_5\overset{\overset{O}{\|}}{C}H \xrightarrow[(2)H^+]{} C_6H_5C\equiv C-\underset{\underset{OH}{|}}{C}HC_2H_5$$

(52%)

When we plan Grignard syntheses we must also take care not to plan a reaction in which a Grignard reagent is treated with an aldehyde, ketone, or ester that contains an acidic group. (Other than when we deliberately let it react with a terminal alkyne.) If we were to do this, the Grignard reagent would simply react as a base with the acidic hydrogen rather than react at the carbonyl or epoxide carbon as a nucleophile. If we were to treat 4-hydroxy-2-butanone with methylmagnesium bromide, for example, the following reaction would take place first,

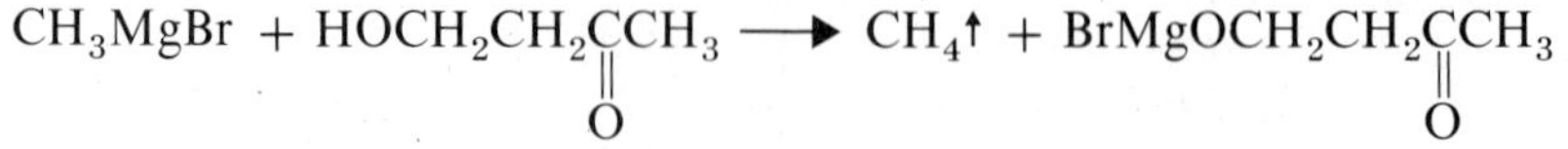

$$CH_3MgBr + HOCH_2CH_2CCH_3 \longrightarrow CH_4\uparrow + BrMgOCH_2CH_2CCH_3$$
$$\qquad\qquad\qquad\quad \overset{\|}{O} \qquad\qquad\qquad\qquad\qquad\qquad\quad \overset{\|}{O}$$

rather than

$$\qquad\qquad\qquad\qquad\qquad\qquad\qquad\qquad\qquad\overset{\displaystyle CH_3}{|}$$
$$CH_3MgBr + HOCH_2CH_2CCH_3 \;\;\overset{\times}{\longrightarrow}\;\; HOCH_2CH_2\overset{|}{C}\!-\!CH_3$$
$$\qquad\qquad\qquad\quad \overset{\|}{O} \qquad\qquad\qquad\qquad\qquad \overset{|}{OMgBr}$$

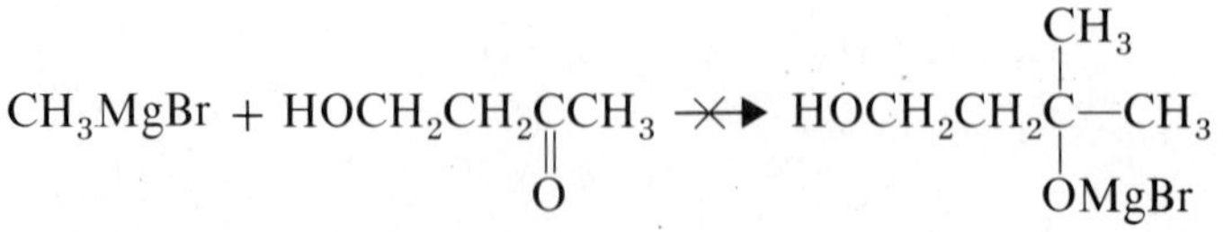

Since Grignard reagents are powerful nucleophiles we cannot prepare a Grignard reagent from any organic halide that contains a carbonyl, epoxy, or cyano (—CN) group.* If we were to attempt to carry out this kind of reaction, any Grignard reagent that formed would only react with the unreacted starting material.

16.6 REACTIONS OF ALCOHOLS

We can classify the reactions of alcohols into two general groups: (1) those reactions that take place with cleavage of the O—H bond, and (2) those reactions which take place with cleavage of the C—O bond.

$$-\overset{|}{\underset{|}{C}}\!-\!O\!\!\mid\!\!H \qquad\qquad -\overset{|}{\underset{|}{C}}\!\!\mid\!\!O\!-\!H$$

O—H bond cleavage   C—O bond cleavage

We begin our study of the reactions of alcohols with those reactions in which the O—H bond is broken.

Reactions Involving O—H Bond Breaking

Alcohols as Acids. Alcohols are weak acids. The acidity constants (K_a's) of most alcohols are of the order of 10^{-18}. This means that alcohols are slightly weaker acids than water ($K_a \sim 10^{-16}$), but they are much stronger acids than terminal alkynes ($K_a \sim 10^{-26}$) or ammonia ($K_a \sim 10^{-36}$). Alcohols are, of course, very much stronger acids than alkanes ($K_a \sim 10^{-42}$).

The conjugate base of an alcohol is an *alkoxide ion*. Because an alcohol is a weaker acid than water the alkoxide ion is a stronger base than the hydroxide ion. This means that if we add an alcohol to an aqueous solution of sodium hydroxide the position of equilibrium of the acid-base reaction (shown below) will favor hydroxide ions and alcohol molecules. Very little alkoxide ion will be present in the solution.

$$:\!\overset{\cdot\cdot}{\underset{\cdot\cdot}{O}}H^- \;\; + \;\; R\overset{\cdot\cdot}{O}H \;\; \underset{\longrightarrow}{\overset{\longleftarrow}{\;\;\;\;}} \;\; R\overset{\cdot\cdot}{\underset{\cdot\cdot}{O}}\!:^- \;\; + \;\; H\overset{\cdot\cdot}{O}H$$

Hydroxide ion   Alcohol   Alkoxide ion   Water
(weaker base)   (weaker acid)   (stronger base) (stronger acid)

* Reactions of cyano compounds will be discussed in Chapter 19.

On the other hand, since ammonia is a much weaker acid than an alcohol, the amide ion is a much stronger base than the alkoxide ion. If we were to add an alcohol to a solution of sodium amide in liquid ammonia, the position of equilibrium would favor the formation of alkoxide ions and ammonia (from the amide ion).

$$:\ddot{N}H_2^- \quad + \quad R\ddot{O}H \quad \rightleftharpoons \quad R\ddot{O}:^- \quad + \quad :NH_3$$

Amide ion	Alcohol	Alkoxide ion	Ammonia
(stronger base)	(stronger acid)	(weaker base)	(weaker acid)

Problem 16.9

Write equations for the acid-base reactions that would occur if ethanol were added to ether solutions of each of the following compounds. In each equation label the stronger acid, the stronger base, and so on.

(a) $CH_3C\equiv CNa$

(b) $CH_3CH_2CH_2CH_2Li$

(c) CH_3CH_2MgBr

Sodium and potassium alkoxides are often used as bases in organic syntheses. We use alkoxides when we carry out reactions that require stronger bases than hydroxide ion, but do not require exceptionally powerful bases such as the amide ion or the anion of an alkane. We also use alkoxide ions when (for reasons of solubility) we need to carry out a reaction in an alcohol solvent rather than in water.

Alcohol solutions of sodium and potassium alkoxides can be prepared by adding the alkali metal—sodium or potassium—to an alcohol. Sodium ethoxide in ethanol and potassium *tert*-butoxide in *tert*-butyl alcohol are often prepared in this way. The reactions that take place are oxidation-reduction reactions; the highly electropositive metals displace hydrogen from the alcohols.

$$CH_3CH_2\ddot{O}H + Na \longrightarrow CH_3CH_2\ddot{O}:Na + \frac{1}{2}H_2$$
(excess)   Sodium ethoxide (in ethanol)

$$CH_3-\underset{\underset{CH_3}{|}}{\overset{\overset{CH_3}{|}}{C}}-\ddot{O}H + K \longrightarrow CH_3-\underset{\underset{CH_3}{|}}{\overset{\overset{CH_3}{|}}{C}}-\ddot{O}:K + \frac{1}{2}H_2$$
(excess)   Potassium *tert*-butoxide (in *tert*-butanol)

Potassium *tert*-butoxide in *tert*-butyl alcohol is a stronger base than sodium ethoxide in ethanol.

Ester Formation. We have already seen examples of reactions in which alcohols react with acyl chlorides or acid anhydrides to form esters of carboxylic acids. We will study the properties of carboxylic acid esters in detail in Chapter 19.

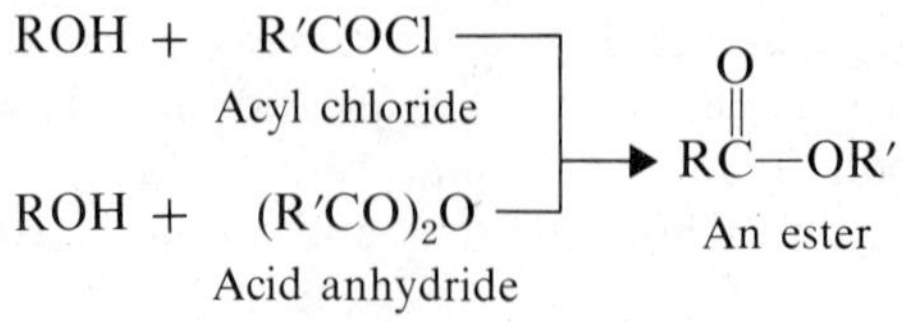

$$ROH + R'COCl$$
Acyl chloride

$$ROH + (R'CO)_2O$$
Acid anhydride

$$\begin{array}{c} O \\ \parallel \\ RC-OR' \end{array}$$
An ester

Alcohols also form esters when they react with inorganic derivatives. The more important esters of this type are those known as *sulfonates*. Ethyl alcohol, for example, reacts with methanesulfonyl chloride to form *ethyl methanesulfonate* and with *p*-toluenesulfonyl chloride to form *ethyl p-toluenesulfonate*.

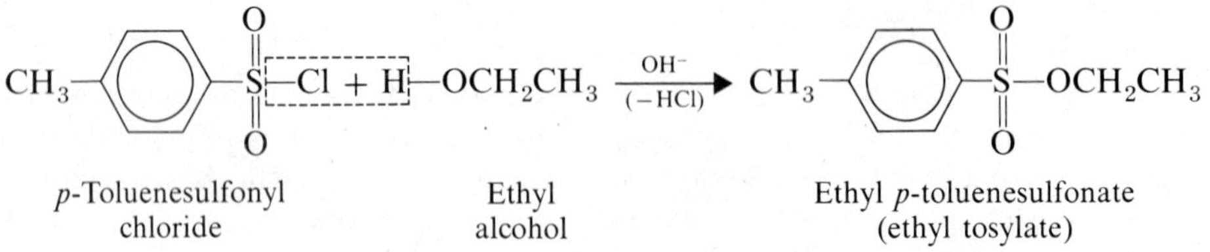

$$\underset{\substack{\text{Methanesulfonyl}\\\text{chloride}}}{CH_3S-Cl} + \underset{\substack{\text{Ethyl}\\\text{alcohol}}}{H-OCH_2CH_3} \xrightarrow[(-HCl)]{OH^-} \underset{\substack{\text{Ethyl methanesulfonate}\\\text{(ethyl mesylate)}}}{CH_3S-OCH_2CH_3}$$

$$\underset{\substack{\text{p-Toluenesulfonyl}\\\text{chloride}}}{CH_3-\langle\rangle-S-Cl} + \underset{\substack{\text{Ethyl}\\\text{alcohol}}}{H-OCH_2CH_3} \xrightarrow[(-HCl)]{OH^-} \underset{\substack{\text{Ethyl p-toluenesulfonate}\\\text{(ethyl tosylate)}}}{CH_3-\langle\rangle-S-OCH_2CH_3}$$

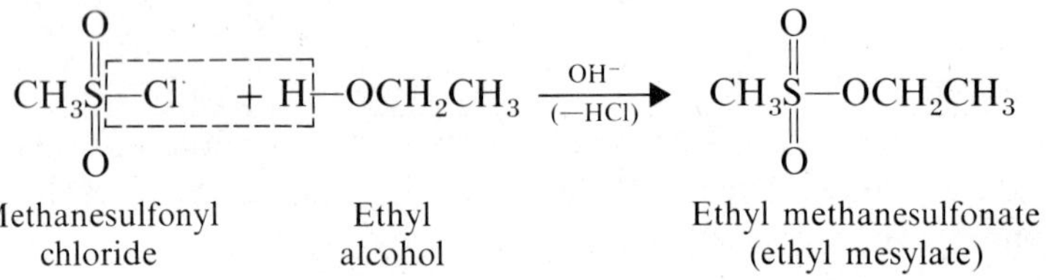

Sulfonyl chlorides are usually prepared by treating sulfonic acids with phosphorus pentachloride.

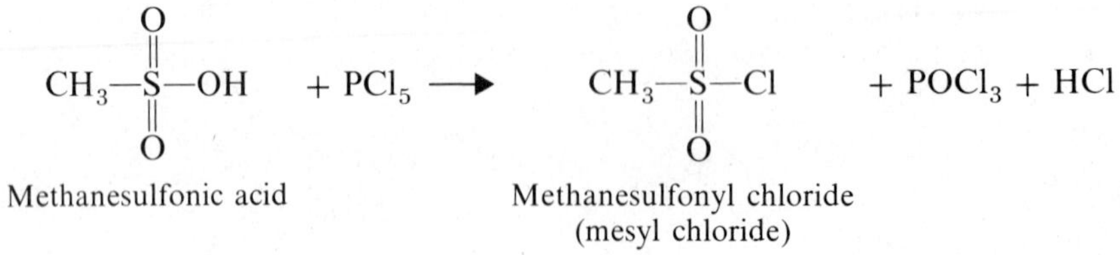

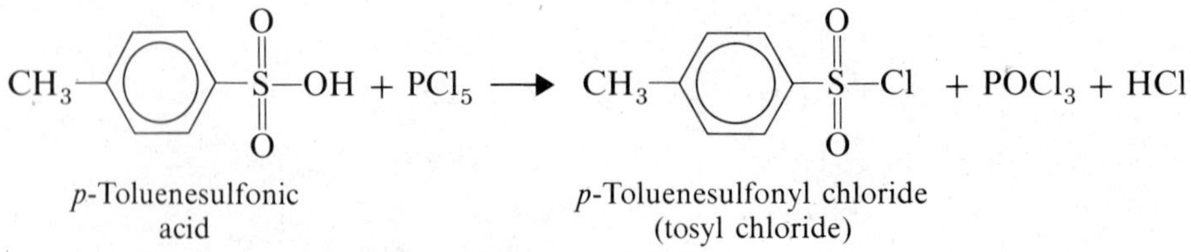

$$\underset{\substack{\text{Methanesulfonic acid}}}{CH_3-S-OH} + PCl_5 \longrightarrow \underset{\substack{\text{Methanesulfonyl chloride}\\\text{(mesyl chloride)}}}{CH_3-S-Cl} + POCl_3 + HCl$$

$$\underset{\substack{\text{p-Toluenesulfonic}\\\text{acid}}}{CH_3-\langle\rangle-S-OH} + PCl_5 \longrightarrow \underset{\substack{\text{p-Toluenesulfonyl chloride}\\\text{(tosyl chloride)}}}{CH_3-\langle\rangle-S-Cl} + POCl_3 + HCl$$

Methanesulfonyl chloride and *p*-toluenesulfonyl chloride are used so often that organic chemists have shortened their rather long names to "mesyl chloride" and "tosyl chloride," respectively. The methanesulfonyl group is often called a "mesyl" group and the *p*-toluenesulfonyl group is called a "tosyl" group. Methane sulfonates are known as "mesylates" and *p*-toluenesulfonates are known as "tosylates."

$$CH_3-\overset{\overset{O}{\|}}{\underset{\underset{O}{\|}}{S}}-$$

or Ms—
the mesyl group

$$CH_3-\overset{\overset{O}{\|}}{\underset{\underset{O}{\|}}{S}}-$$

or Ts—
the tosyl group

$$CH_3-\overset{\overset{O}{\|}}{\underset{\underset{O}{\|}}{S}}-OR \quad \text{or MsOR}$$

An alkyl mesylate

$$CH_3-\overset{\overset{O}{\|}}{\underset{\underset{O}{\|}}{S}}-OR \quad \text{or TsOR}$$

An alkyl tosylate

Problem 16.10

Starting with benzene or toluene and any necessary alcohols or inorganic reagents, show how you would prepare each of the following sulfonates.
(a) Methyl benzenesulfonate
(b) Isopropyl *p*-toluenesulfonate (isopropyl tosylate)
(c) *tert*-Butyl *p*-bromobenzenesulfonate (*tert*-butyl brosylate)

Alkyl *sulfonates*, differ from alkyl or dialkyl *sulfates* (Section 6.3). Sulfonates are esters of sulfonic acids, RSO_2OH, whereas sulfates are esters of sulfuric acid.

$$CH_3-\overset{\overset{O}{\|}}{\underset{\underset{O}{\|}}{S}}-OCH_3 \qquad CH_3O-\overset{\overset{O}{\|}}{\underset{\underset{O}{\|}}{S}}-OH \qquad CH_3O-\overset{\overset{O}{\|}}{\underset{\underset{O}{\|}}{S}}-OCH_3$$

Methyl methanesulfonate   Methyl hydrogen sulfate   Dimethyl sulfate

Alkyl sulfonates contain one carbon-sulfur bond and one carbon-oxygen-sulfur linkage, whereas alkyl sulfates and dialkylsulfates contain only carbon-oxygen-sulfur linkages. Alkyl and dialkyl sulfates are made by treating alkenes or alcohols with sulfuric acid. Dimethyl sulfate, for example, can be obtained by distilling a mixture of methyl alcohol and sulfuric acid at reduced pressure. Organic sulfonates and organic sulfates are frequently used in nucleophilic substitution reactions because

$$R-O^- + CH_3-O-\overset{\overset{O}{\|}}{\underset{\underset{O}{\|}}{S}}-CH_3 \longrightarrow R-O-CH_3 + {}^-O-\overset{\overset{O}{\|}}{\underset{\underset{O}{\|}}{S}}-CH_3$$

An alkoxide ion   Methyl methanesulfonate   A methyl ether   Mesylate ion

$$R-O^- + CH_3O-\overset{\overset{O}{\|}}{\underset{\underset{O}{\|}}{S}}-OCH_3 \longrightarrow R-OCH_3 + {}^-O-\overset{\overset{O}{\|}}{\underset{\underset{O}{\|}}{S}}-OCH_3$$

An alkoxide ion   Dimethyl sulfate   A methyl ether   Methyl sulfate ion

Good leaving groups

sulfonate ions and sulfate ions are good *leaving groups* (Chapter 17). Alkoxide ions, for example, react with organic sulfonates and organic sulfates to form ethers. In these reactions, the alkoxide ion acts as a nucleophile.

Problem 16.11

What products would be formed when (a) methyl methanesulfonate and (b) dimethyl sulfate react with hydroxide ion?

Oxidation of Alcohols. We saw, in Section 2.10, that primary alcohols can be oxidized to aldehydes and carboxylic acids.

$$R-CH_2OH \xrightarrow{(O)} \underset{\text{Aldehyde}}{R-\overset{\displaystyle O}{\overset{\|}{C}}-H} \xrightarrow{(O)} \underset{\text{Carboxylic acid}}{R-\overset{\displaystyle O}{\overset{\|}{C}}-OH}$$

1° Alcohol

Since the oxidation of aldehydes to carboxylic acids usually takes place with milder oxidizing agents than those required to oxidize primary alcohols to aldehydes, it is difficult to stop the oxidation at the aldehyde stage. One way of doing this, however, is to remove the aldehyde as soon as it is formed. This can often be done because aldehydes have lower boiling points than carboxylic acids (why?), and they can be distilled from the reaction mixture as they are formed. Another way to oxidize primary alcohols to aldehydes is through the use of special oxidizing agents. We will see examples of how this can be done later in this section.

Secondary alcohols can be oxidized to ketones. The reaction usually stops at the ketone stage because further oxidation requires the breaking of a carbon-carbon bond.

Tertiary alcohols can be oxidized but only under very forcing conditions. Tertiary alcohols are not easily oxidized because their oxidation requires the breaking of a carbon-carbon bond. Oxidations of tertiary alcohols, when they do occur, are of little synthetic utility.

$$R-\overset{\displaystyle R}{\underset{\displaystyle R}{\overset{\|}{C}}}-OH \qquad \textit{Difficult to oxidize}$$

3° Alcohol

The relative ease of oxidation of primary and secondary alcohols compared with difficulty in oxidizing tertiary alcohols forms the basis for the chromic acid test we gave earlier (Section 9.14).

Primary alcohols can be oxidized to carboxylic acids with potassium permanganate. The reaction is usually carried out in basic aqueous solution and MnO_2 precipitates as the oxidation takes place. After the oxidation is complete, filtration allows removal of the MnO_2 and acidification of the filtrate gives the carboxylic acid.

$$R-CH_2OH + KMnO_4 \xrightarrow[\substack{H_2O \\ \text{heat}}]{OH^-} RCOO^-K^+ + MnO_2$$

$$\downarrow H^+$$

$$RCOOH$$

A variety of oxidizing agents have been used to oxidize secondary alcohols to ketones. The most commonly used reagent is chromic acid, H_2CrO_4. Chromic acid is usually prepared *in situ* by adding chromic oxide (CrO_3) or sodium dichromate ($Na_2Cr_2O_7$) to sulfuric acid. Oxidations of secondary alcohols are generally carried out in acetic acid solutions. The balanced equation is shown below.

$$3 \begin{array}{c} R \\ \diagdown \\ CHOH \\ \diagup \\ R \end{array} + 2H_2CrO_4 + 6H^+ \longrightarrow 3 \begin{array}{c} R \\ \diagdown \\ C{=}O \\ \diagup \\ R \end{array} + 2Cr^{+++} + 8H_2O$$

As chromic acid oxidizes the alcohol to the ketone, chromium is reduced from the +6 oxidation state (H_2CrO_4) to the +3 oxidation state (Cr^{+++}).* Chromic acid oxidations of secondary alcohols generally give ketones in excellent yields. Two specific examples are the oxidation of 2-octanol to 2-octanone and the oxidation of (−)-menthol to (−)-menthone.

$$CH_3CH(CH_2)_5CH_3 + H_2CrO_4 \xrightarrow[\text{reflux}]{} CH_3C(CH_2)_5CH_3 + Cr^{+++}$$
$$\underset{\text{OH}}{|} \qquad\qquad\qquad \underset{\text{O}}{\|}$$

2-Octanol   2-Octanone
(96%)

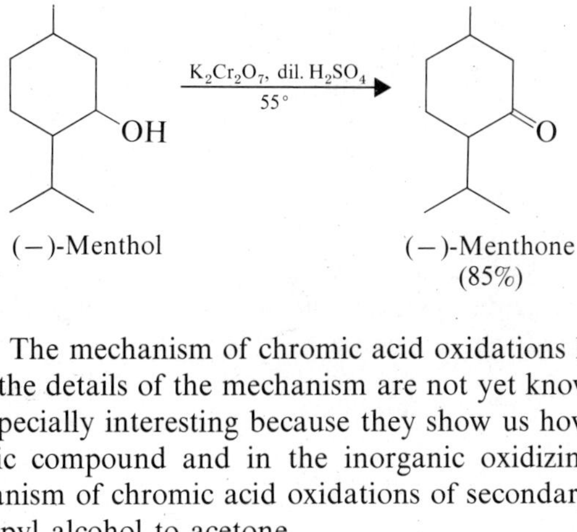

(−)-Menthol   (−)-Menthone
(85%)

The mechanism of chromic acid oxidations has been investigated. Although all of the details of the mechanism are not yet known, those details that are known are especially interesting because they show us how oxidation states change in the organic compound and in the inorganic oxidizing agent. We can illustrate the mechanism of chromic acid oxidations of secondary alcohols with the oxidation of isopropyl alcohol to acetone.

In step 1, isopropyl alcohol reacts with chromic acid to produce a *chromate ester*. No change in oxidation state occurs in this step.

Step 1

Isopropyl alcohol   Chromic acid Cr^{VI} Isopropyl chromate Cr^{VI} (a chromate ester)

---

* It is the color change that accompanies this change in oxidation state that allows chromic acid to be used as a test for primary and secondary alcohols.

Step 2 is the slowest step. In it, the chromate ester loses a proton and an $HCrO_3^-$ ion to produce acetone. This step may occur in two ways: the chromate ester may transfer a proton to water, or the proton may be transferred in a cyclic mechanism.

Step 2a

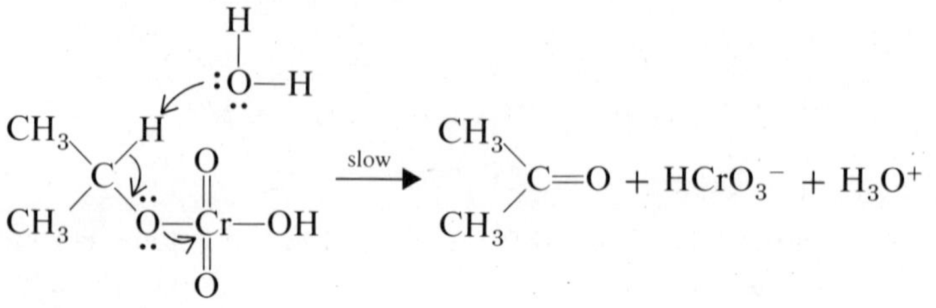

Isopropyl chromate
Cr^{VI}

Acetone Cr^{IV}

Step 2b

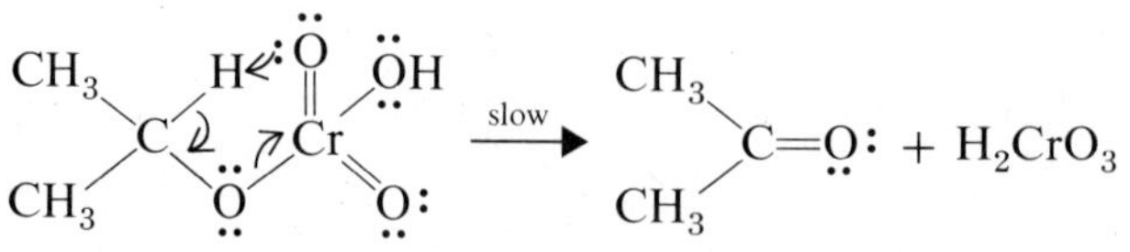

Isopropyl chromate
Cr^{VI}

Acetone   · Cr^{IV}

In either mechanism, the oxidation state of the alcohol increases to that of the ketone and the oxidation state of chromium changes from Cr^{VI} (of the ester) to Cr^{IV} (of H_2CrO_3). Current evidence suggests that chromium IV reacts with a Cr^{VI} species to form two equivalents of a Cr^V derivative, which also oxidizes alcohols to ketones. Subsequent oxidation of isopropyl alcohol by Cr^{VI} or Cr^V takes place until chromium is eventually reduced to Cr^{III} (Cr^{+3}).

Chromic acid usually oxidizes primary alcohols all the way to carboxylic acids. A complex of chromic oxide and pyridine, however, will oxidize a primary alcohol to an aldehyde and stop at that stage.

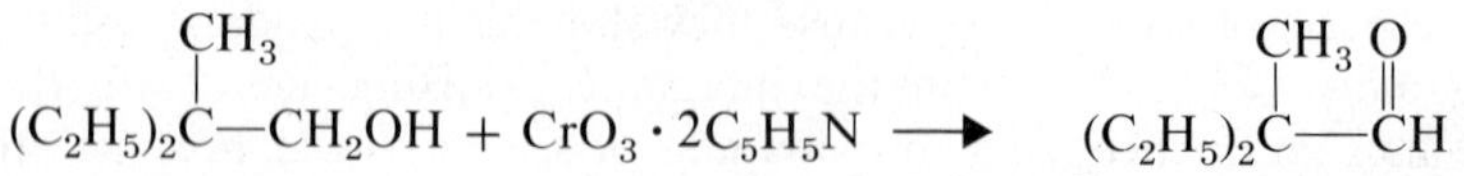

$$(C_2H_5)_2\overset{\overset{CH_3}{|}}{C}-CH_2OH + CrO_3 \cdot 2C_5H_5N \longrightarrow (C_2H_5)_2\overset{\overset{CH_3}{|}}{C}-\overset{\overset{O}{||}}{C}H$$

2-Methyl-2-ethyl-1-
butanol

2-Methyl-2-ethylbutanal

The chromic oxide-pyridine complex does not attack double bonds as the following example shows.

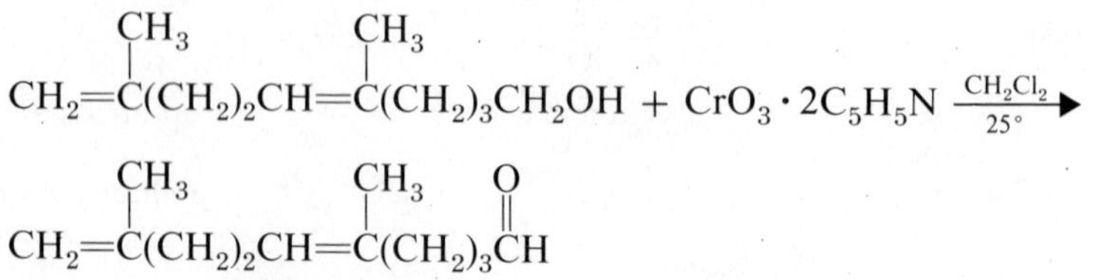

$$CH_2=\overset{\overset{CH_3}{|}}{C}(CH_2)_2CH=\overset{\overset{CH_3}{|}}{C}(CH_2)_3CH_2OH + CrO_3 \cdot 2C_5H_5N \xrightarrow[25°]{CH_2Cl_2}$$

$$CH_2=\overset{\overset{CH_3}{|}}{C}(CH_2)_2CH=\overset{\overset{CH_3}{|}}{C}(CH_2)_3\overset{\overset{O}{||}}{C}H$$

(92%)

Reactions Involving C—O Bond Breaking

There are several reactions of alcohols occurring with cleavage of the carbon-oxygen that we have seen before.

Dehydration of Alcohols (discussed in Section 5.8). When alcohols are heated with strong acids they undergo the elimination of water (dehydration) and form alkenes. The mechanism for this reaction is given in Section 5.8. Two specific examples of alcohol dehydrations are shown below.

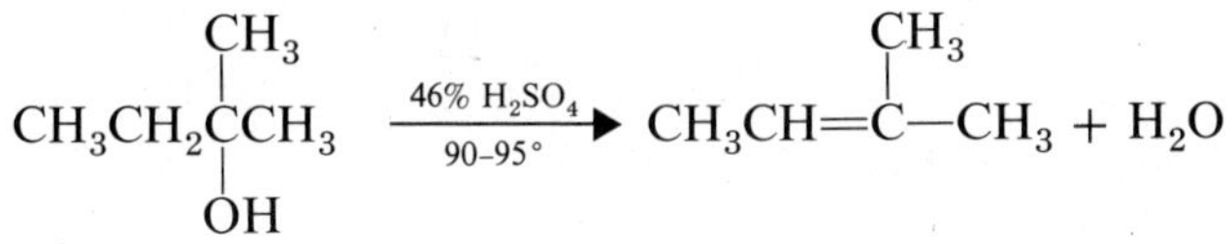

$$\underset{\substack{\text{2-Methyl-2-butanol}}}{\underset{\substack{|\\ \text{OH}}}{\underset{\substack{|\\ \text{CH}_3}}{\text{CH}_3\text{CH}_2\overset{\displaystyle \text{CH}_3}{\underset{}{\text{C}}}\text{CH}_3}}} \xrightarrow[90\text{–}95°]{46\% \text{ H}_2\text{SO}_4} \underset{\substack{\text{2-Methyl-2-butene}\\(84\%)}}{\text{CH}_3\text{CH}=\overset{\displaystyle \text{CH}_3}{\text{C}}-\text{CH}_3 + \text{H}_2\text{O}}$$

$$\underset{\substack{\text{2-Pentanol}}}{\underset{\substack{|\\ \text{OH}}}{\text{CH}_3\text{CH}_2\text{CH}_2\text{CHCH}_3}} \xrightarrow[90\text{–}95°]{62\% \text{ H}_2\text{SO}_4} \underset{\substack{\text{2-Pentene}\\(80\%)}}{\text{CH}_3\text{CH}_2\text{CH}=\text{CHCH}_3 + \text{H}_2\text{O}}$$

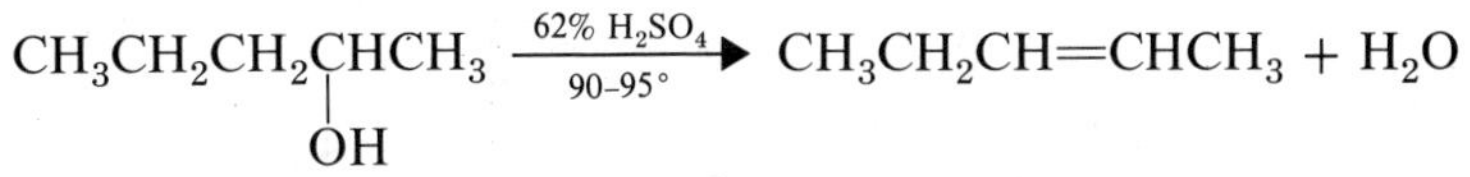

As these examples demonstrate, alcohol dehydrations generally produce the more highly substituted and, thus, *the more stable* alkene. The dehydration of 2-methyl-2-butanol produces, mainly, 2-methyl-2-butene rather than 2-methyl-1-butene, and the dehydration of 2-pentanol yields, mainly, 2-pentene rather than 1-pentene.

Problem 16.12

(a) Which 2-pentene isomer would you expect to obtain in the greatest amount from the dehydration of 2-pentanol, *cis*-2-pentene or *trans*-2-pentene? (b) Dehydration of 1-phenyl-2-propanol yields mainly *trans* (and some *cis*)-1-phenylpropene. The reaction produces practically no 3-phenylpropene. Explain.

Problem 16.13

When 3,3-dimethyl-2-butanol is treated with 85% phosphoric acid the following products are obtained.

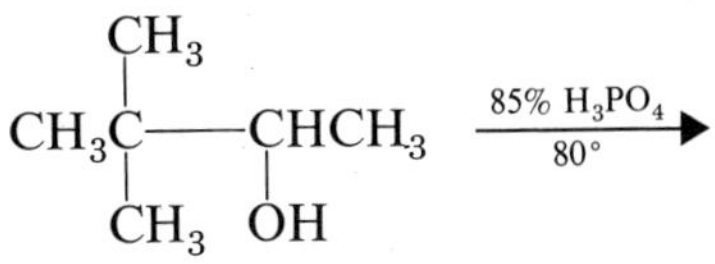

$$\underset{\substack{\text{3,3-Dimethyl-2-butanol}}}{\underset{\substack{|\quad\;\, |\\ \text{CH}_3\;\text{OH}}}{\overset{\substack{\text{CH}_3\\|}}{\text{CH}_3\text{C}}-\text{CHCH}_3}} \xrightarrow[80°]{85\% \text{ H}_3\text{PO}_4}$$

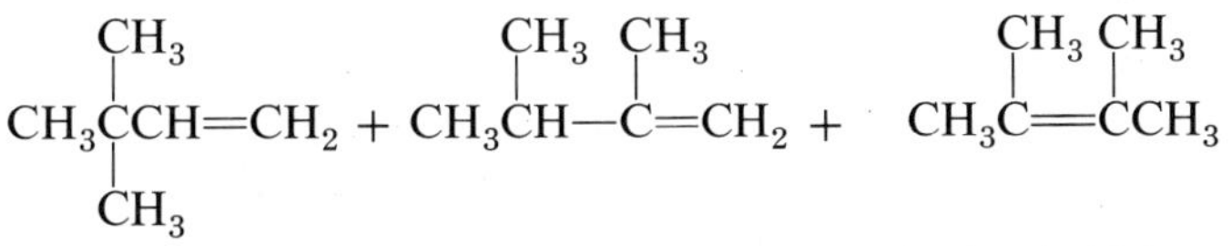

$$\underset{\substack{\text{3,3-Dimethyl-1-}\\ \text{butene}\\(0.4\%)}}{\underset{\substack{|\\ \text{CH}_3}}{\overset{\substack{\text{CH}_3\\|}}{\text{CH}_3\text{C}}\text{CH}=\text{CH}_2}} + \underset{\substack{\text{2,3-Dimethyl-1-}\\ \text{butene}\\(20\%)}}{\overset{\substack{\text{CH}_3\;\text{CH}_3\\|\quad\;\; |}}{\text{CH}_3\text{CH}-\text{C}=\text{CH}_2}} + \underset{\substack{\text{2,3-Dimethyl-2-butene}\\(80\%)}}{\overset{\substack{\text{CH}_3\;\text{CH}_3\\|\quad\;\; |}}{\text{CH}_3\text{C}=\text{CCH}_3}}$$

(a) Write a mechanism that accounts for the formation of each product. (b) Why is 2,3-dimethyl-2-butene the major product?

Reaction with Hydrogen Halides. We saw in Section 15.2 that alcohols react with the hydrogen halides (HI, HBr, and HCl) to form alkyl halides. This reaction also involves cleavage of the carbon-oxygen bond of the alcohol.

$$R\!-\!\!\mid\!\!OH + HX \longrightarrow R\!-\!X + H_2O$$

The order of reactivity of the hydrogen halides is HI > HBr > HCl (HF is generally unreactive), and the order of reactivity of alcohols is benzyl and allyl > 3° > 2° > 1°.

The reaction is *acid catalyzed;* it does not proceed at an appreciable rate unless a strong acid is present. The strong acid may be the hydrogen halide itself, or it may be concentrated sulfuric acid that has been added to the mixture. Primary and secondary alcohols are often converted to alkyl iodides and bromides by allowing them to react with a mixture of a sodium halide and concentrated sulfuric acid. This not only generates the hydrogen halide in the mixture (*in situ*); it also provides the acid catalyst.

$$ROH + NaX \xrightarrow{H_2SO_4} RX + NaHSO_4 + H_2O$$

Hydrogen chloride, does not react with primary and secondary alcohols unless zinc chloride or some similar Lewis acid is added to the reaction mixture as well. Zinc chloride, a good Lewis acid, acts as an acid catalyst in this reaction. It forms a complex with the alcohol through association with an unshared pair of electrons on the oxygen atom. This weakens the carbon-oxygen bond and in the attack by the chloride ion it breaks to give the alkyl halide.

$$R\!-\!\ddot{O}\!: + ZnCl_2 \rightleftharpoons R\!-\!\overset{+}{\ddot{O}}\!\!-\!\bar{Z}nCl_2$$
$$\quad\;\; | \qquad\qquad\qquad\quad |$$
$$\quad\;\; H \qquad\qquad\qquad\quad H$$

$$:\!\ddot{Cl}\!:^- + R\!-\!\overset{+}{\ddot{O}}\!\!-\!\bar{Z}nCl_2 \longrightarrow :\!\ddot{Cl}\!-\!R + [Zn(OH)Cl_2]^-$$
$$\qquad\qquad\quad |$$
$$\qquad\qquad\quad H$$

$$[Zn(OH)Cl_2]^- + H^+ \rightleftharpoons ZnCl_2 + H_2O$$

Tertiary alcohols, however, react rapidly with concentrated hydrochloric acid alone at room temperature.

$$\underset{\substack{\text{tert-Butyl} \\ \text{alcohol}}}{CH_3\!-\!\underset{\underset{CH_3}{|}}{\overset{\overset{CH_3}{|}}{C}}\!-\!OH} + HCl_{(conc.)} \xrightarrow{25°} \underset{\substack{\text{tert-Butyl} \\ \text{chloride} \\ (94\%)}}{CH_3\!-\!\underset{\underset{CH_3}{|}}{\overset{\overset{CH_3}{|}}{C}}\!-\!Cl} + H_2O$$

The reactions of secondary, tertiary, allyl, and benzyl alcohols appear to proceed through a mechanism involving the formation of carbocations. We can illustrate this mechanism with the reaction of *tert*-butyl alcohol and hydrochloric acid.

The first two steps are the same as the mechanism for the dehydration of an alcohol (Section 5.8). The alcohol accepts a proton and then the protonated alcohol dissociates to form a carbocation and water.

Step 1

$$CH_3-\underset{\underset{CH_3}{|}}{\overset{\overset{CH_3}{|}}{C}}-\overset{..}{\underset{..}{O}}-H + H-\overset{+}{\underset{H}{\overset{..}{O}}}-H \underset{}{\overset{fast}{\rightleftharpoons}} CH_3\underset{\underset{CH_3}{|}}{\overset{\overset{CH_3}{|}}{C}}-\overset{H}{\overset{+}{O}}-H + \overset{..}{\underset{H}{\overset{..}{O}}}-H$$

Step 2

$$CH_3-\underset{\underset{CH_3}{|}}{\overset{\overset{CH_3}{|}}{C}}-\underset{..+}{O}-H \overset{slow}{\rightleftharpoons} CH_3-\underset{\underset{CH_3}{|}}{\overset{\overset{CH_3}{|}}{C}}{}^+ + \overset{H}{\underset{..}{O}}-H$$

In step 3 the mechanisms for the dehydration of an alcohol and the formation of an alkyl halide differ. In dehydration reactions the carbocation loses a proton to form an alkene. In alkyl halide formation, the carbocation reacts with a nucleophile, *a halide ion*.

Step 3

$$CH_3-\underset{\underset{CH_3}{|}}{\overset{\overset{CH_3}{|}}{C}}{}^+ + \overset{..}{\underset{..}{Cl}}{:}^- \overset{fast}{\rightleftharpoons} CH_3-\underset{\underset{CH_3}{|}}{\overset{\overset{CH_3}{|}}{C}}-\overset{..}{\underset{..}{Cl}}{:}$$

How can we account for the different course of these two reactions?

When we dehydrate alcohols we usually carry out the reaction in concentrated sulfuric acid. The only nucleophiles present in this reaction mixture are water and hydrogen sulfate ($HSO_4{}^-$) ions. Both are poor nucleophiles but, more importantly, both are present in low concentrations. Under these conditions, the highly reactive carbonium ion stabilizes itself by losing a proton and becoming an alkene.

> In the reverse reaction, that is, the hydration of an alkene (Section 6.3) the carbonium ion *does* react with a nucleophile. It reacts with water. Alkene hydrations are carried out in dilute sulfuric acid where the water concentration is high. In some instances, too, carbocations may react with $HSO_4{}^-$ ions or with sulfuric acid, itself. When they do they form alkylsulfate esters, $R-OSO_2OH$.

When we convert an alcohol to an alkyl halide, we carry out the reaction in the presence of acid and *in the presence of halide ions*. Halide ions are good nucleophiles (much stronger nucleophiles than water), and since they are present in high concentration, most of the carbocations stabilize themselves by accepting a halide-ion electron pair.

These two reactions, dehydration and alkyl halide formation, also furnish us with another example of the competition between nucleophilic substitution reactions and elimination reactions (cf. Section 9.10). The dehydration of an alcohol is an elimination reaction. The conversion of an alcohol to an alkyl halide is a nucleophilic substitution reaction that proceeds via a carbocation. Very often, in conversions of alcohols to alkyl halides, we find that the reaction is accompanied by the formation of some alkene (i.e., by elimination). Thus, not all of the carbocations react with nucleophiles; some stabilize themselves by losing protons.

Not all acid-catalyzed conversions of alcohols to alkyl halides proceed through the formation of carbocations. Primary alcohols and methyl alcohol apparently react through a mechanism in which the carbon bears only a partial positive charge:

$$:\ddot{X}:^- \ + \ ^{\delta+}\underset{|}{\overset{|}{C}}\!\!-\!\!\overset{H}{\underset{..}{\overset{|}{O}}}\!\!-\!\!H^{\delta+} \ \longrightarrow \ :\ddot{X}\!\!-\!\!\underset{|}{\overset{|}{C}}\!- \ + \ :\overset{H}{\underset{..}{\overset{|}{O}}}\!\!-\!\!H$$

(Protonated 1°
or methyl
alcohol)

In these reactions the function of the acid is to produce a protonated alcohol (or a related Lewis acid complex if a catalyst such as $ZnCl_2$ is used). The carbon-oxygen bond of the protonated alcohol is weakened but not completely broken. However, because the bond is weakened the halide ion is able to displace a molecule of water from carbon and form an alkyl halide.

Although halide ions (particularly iodide and bromide ions) are strong nucleophiles they are not strong enough to carry out substitution reactions with alcohols themselves. That is, reactions of the following type do not occur to any appreciable extent.

$$:\ddot{Br}:^- \ + \ \underset{|}{\overset{|}{C}}\!\!-\!\!\ddot{O}H \ \longrightarrow \ :\ddot{Br}\!\!-\!\!\underset{|}{\overset{|}{C}}\!- \ + \ ^-:\ddot{O}H$$

They do not occur because such reactions would require the breaking of the strong carbon-oxygen bond of the alcohol.

> The reverse reaction, that is, the reaction of an alkyl halide with hydroxide ion, does occur and is another method for the synthesis of alcohols. We will see this reaction in Chapter 17.

We can see now why the reactions of alcohols with hydrogen halides are acid catalyzed. With allyl, benzyl, tertiary, and secondary alcohols the function of the acid is to produce a carbocation. With methyl alcohol and primary alcohols the function of the acid is to produce an oxonium ion in which the carbon-oxygen bond is sufficiently weakened to allow the halide ion to displace a molecule of water.

As we might expect, many reactions of alcohols with hydrogen halides, particularly those in which carbocations are formed, are accompanied by rearrangements.

Reactions with Phosphorus Halides and Thionyl Chloride (discussed in Section 15.2). Alcohols are also converted to alkyl halides by the action of phosphorus halides PBr_3 and PCl_3.

$$3R\!-\!OH + PBr_3 \ \longrightarrow \ 3R\!-\!Br + P(OH)_3$$
$$3R\!-\!OH + PCl_3 \ \longrightarrow \ 3R\!-\!Cl + P(OH)_3$$

Both of these reactions proceed with cleavage of the carbon-oxygen bond of the alcohol. However, unlike the reactions of alcohols with acids, the reactions of alcohols with phosphorus halides usually do not involve carbocations and usually do not occur with rearrangement of the carbon skeleton. For this reason phosphorus halides are often preferred as reagents for transforming alcohols to the corresponding alkyl halide. In the reactions of alcohols with phosphorus halides, the carbon-oxygen bond is weakened through the formation of a protonated *phosphite ester*. The general mechanism is shown below.

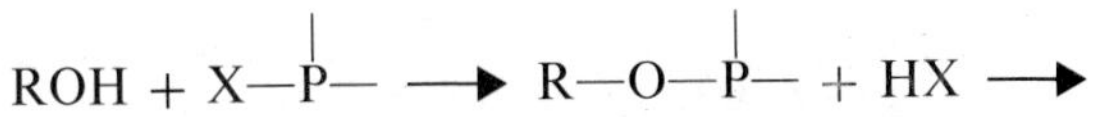

$$ROH + X-\overset{|}{\underset{|}{P}}- \longrightarrow R-O-\overset{|}{\underset{|}{P}}- + HX \longrightarrow$$

A phosphite
ester

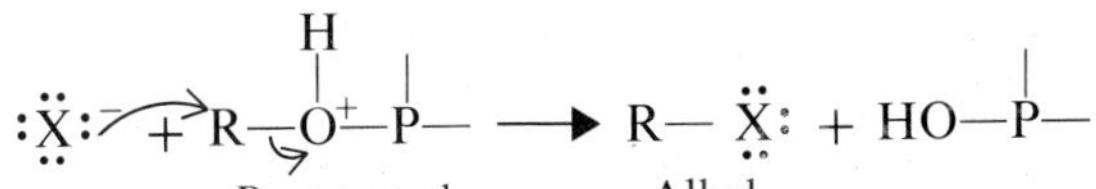

$$:\ddot{X}: + R-\overset{|}{\underset{|}{\overset{+}{O}}}-\overset{|}{\underset{|}{P}}- \longrightarrow R-\ddot{\ddot{X}}: + HO-\overset{|}{\underset{|}{P}}-$$

Protonated   Alkyl
phosphite   halide
ester

Thionyl chloride can also be used to convert alcohols to alkyl chlorides (usually without rearrangement of the carbon skeleton). This reaction, too, ultimately involves cleavage of the carbon-oxygen bond. The first step in the reaction is the formation of a *sulfite* ester (an alkyl chlorosulfite).

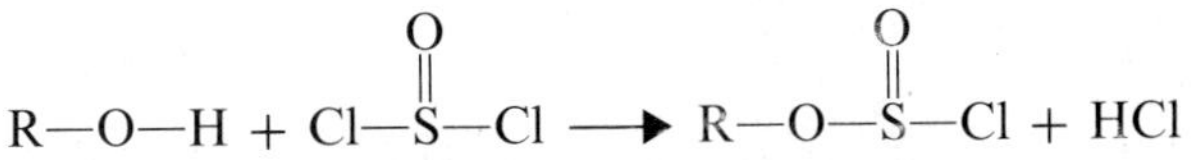

$$R-O-H + Cl-\overset{O}{\overset{||}{S}}-Cl \longrightarrow R-O-\overset{O}{\overset{||}{S}}-Cl + HCl$$

Thionyl   Alkyl chloro-
chloride   sulfite
(an ester)

Then the alkyl chlorosulfite loses SO_2, and an alkyl chloride is formed. It is in this step that the carbon-oxygen bond breaks:

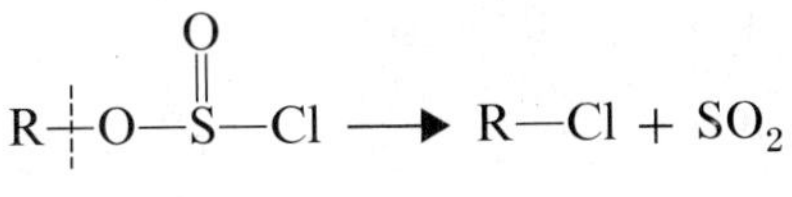

$$R\overset{O}{\underset{|}{+}}O-\overset{||}{S}-Cl \longrightarrow R-Cl + SO_2$$

16.7 POLYHYDROXY ALCOHOLS

Compounds containing two hydroxyl groups are called "glycols" or diols. The simplest possible diol is the unstable compound methylene glycol or methanediol.

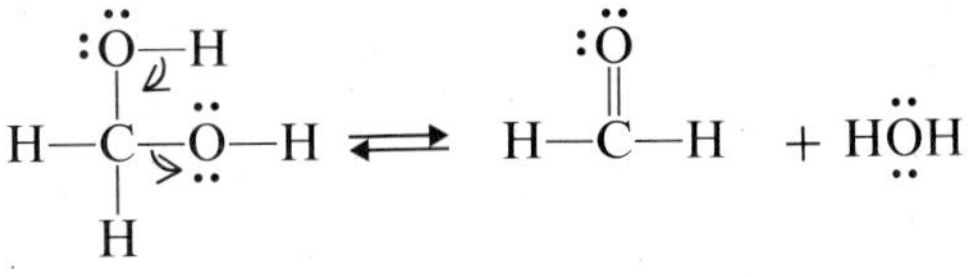

$$H-\overset{\overset{\displaystyle :\ddot{O}-H}{|}}{\underset{\overset{\displaystyle |}{H}}{C}}-\ddot{O}-H \rightleftharpoons H-\overset{\overset{\displaystyle :\ddot{O}}{||}}{C}-H + H\ddot{O}H$$

Methanediol   Formaldehyde

Methanediol is a *gem*-diol (a diol that has both hydroxyl groups attached to the same carbon). Most *gem*-diols are unstable except in an aqueous solution. When the water is removed, the *gem*-diol dehydrates and forms an aldehyde or ketone. When methanediol dehydrates it produces formaldehyde.

Gem-diols with strong electron-withdrawing groups can usually be isolated. One example is chloral hydrate (2,2,2-trichloroethanediol). Chloral hydrate is used as a sleep-inducing drug.

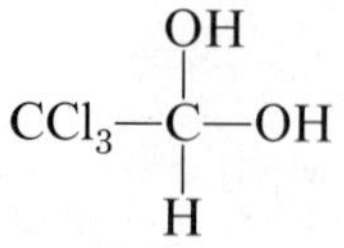

$$CCl_3-\overset{\displaystyle OH}{\underset{\displaystyle H}{\overset{|}{\underset{|}{C}}}}-OH$$

Chloral hydrate
(mp 51.7°)

Vic-diols are much more stable than *gem*-diols. *Vic*-diols can be prepared by the hydroxylation of alkenes (cf. Section 6.7 and Section 6.8).

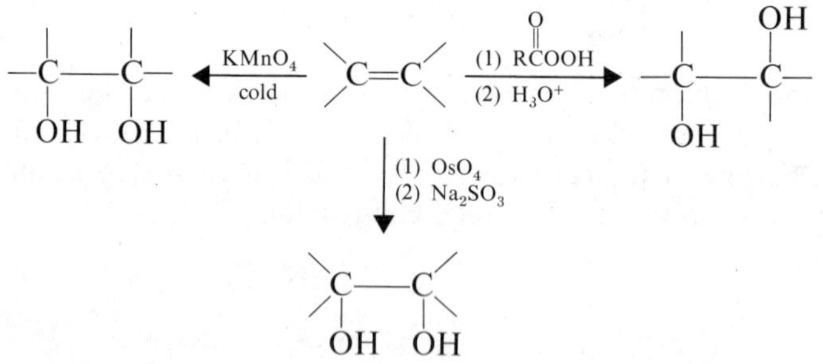

An important triol is the compound glycerol (1,2,3-propanetriol).

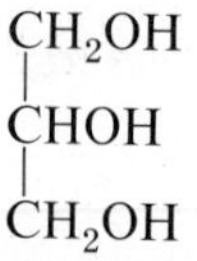

$$\begin{array}{l} CH_2OH \\ | \\ CHOH \\ | \\ CH_2OH \end{array}$$

Glycerol

Glycerol esters are, as we will see, very important compounds in biochemistry. Glycerol, itself, is a viscous hygroscopic liquid that is nontoxic. It is often used as a moistening agent in food, tobacco, and cosmetics.

The ester formed when glycerol reacts with nitric acid is the well-known explosive nitroglycerin.

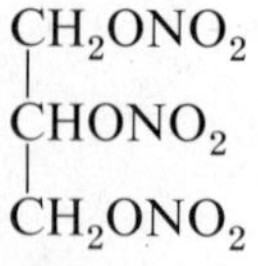

$$\begin{array}{l} CH_2ONO_2 \\ | \\ CHONO_2 \\ | \\ CH_2ONO_2 \end{array}$$

Glyceryl trinitrate or "nitroglycerin"

Nitroglycerin is very sensitive to shock but becomes much more stable and, therefore, safer when it is absorbed by sawdust or diatomaceous earth. In this form nitroglycerin is called "dynamite." Dynamite was invented by the Swedish industrial chemist Alfred Nobel.

In 1895 Nobel established a trust fund for the purpose of awarding annual prizes for exceptional contributions to the fields of chemistry, physics, medicine, and literature and to the cause of world peace. Prizes have been awarded since 1900. The first recipient of the Nobel Prize for chemistry was J. H. van't Hoff (p. 231). Marie Sklodowska Curie, a Polish chemist who worked in France, won two Nobel Prizes for science; one for her work in physics (1903) and one for her work in chemistry (1911). Professor Linus Pauling of the California Institute of Technology is the only person to have won two Nobel Prizes in distinctly separate areas. Pauling won the Nobel Prize for his contributions to chemistry in 1954, and for his contributions toward world peace in 1962.

16.8 ETHERS

Synthesis of Ethers

We have already studied two methods for synthesizing ethers.

1. Ethers can be prepared by solvomercuration-demercuration (Section 6.4) using an alcohol as the solvent in the first (solvomercuration) stage.

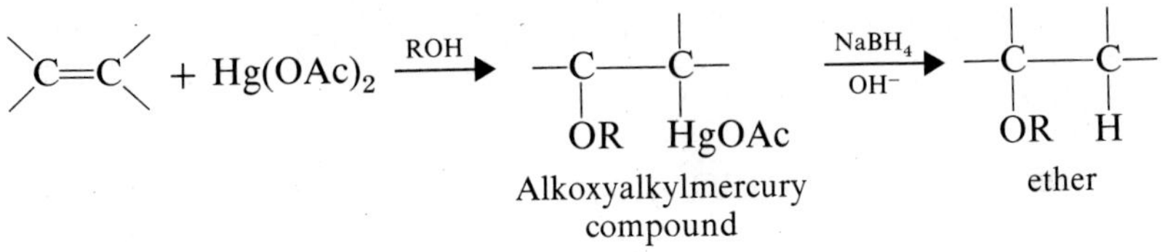

Alkoxyalkylmercury
compound

ether

2. Cyclic ethers with three-membered rings (epoxides) can be synthesized by treating an alkene with a peracid (Section 6.7).

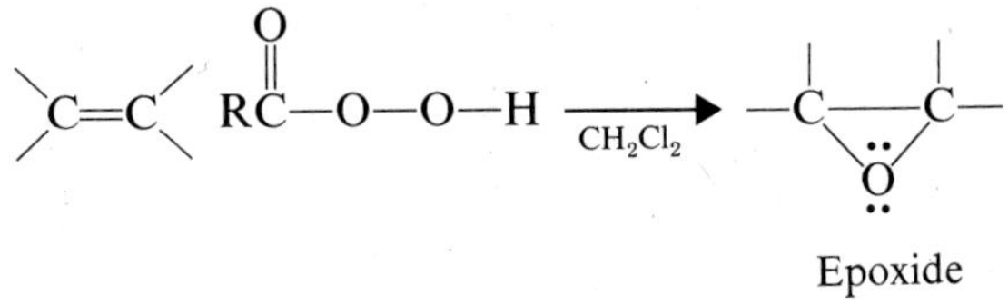

Epoxide

3. Ethers can also be prepared from simple alcohols by heating them with sulfuric acid. Ether formation takes place at a *lower* temperature than that required for dehydration to an alkene.

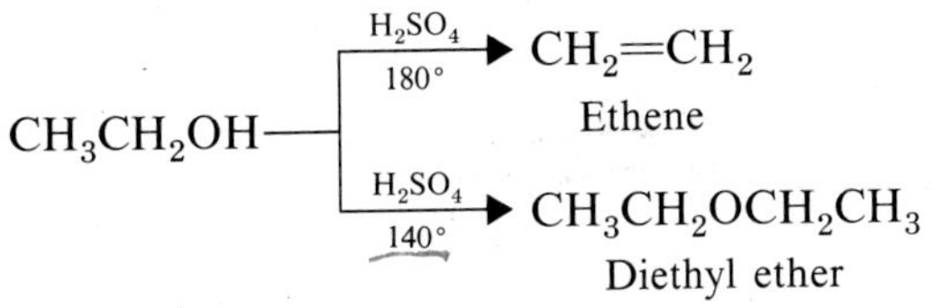

Primary alcohols react by forming an alkyl sulfate first.

$$CH_3CH_2OH + HOSO_3H \rightleftharpoons CH_3CH_2OSO_3H$$
Ethyl hydrogen sulfate

The alkyl hydrogen sulfate then reacts with an alcohol in a nucleophilic substitution reaction.

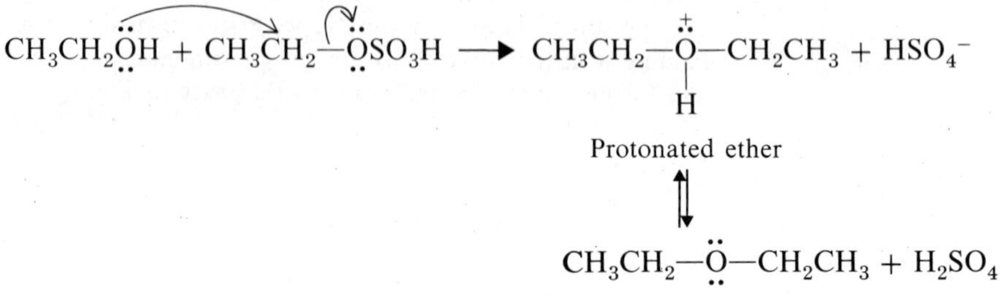

$$CH_3CH_2\ddot{O}H + CH_3CH_2-\ddot{O}SO_3H \longrightarrow CH_3CH_2-\overset{+}{\underset{H}{\ddot{O}}}-CH_2CH_3 + HSO_4^-$$

Protonated ether

$$CH_3CH_2-\ddot{O}-CH_2CH_3 + H_2SO_4$$

We see from the mechanism given above that the nucleophile, ethanol, displaces a hydrogen sulfate ion from ethyl hydrogen sulfate. This produces a *protonated ether*. The protonated ether, however, is in equilibrium with the ether itself, and, at the temperature at which the reaction is carried out (140°), the ether distills out of the reaction mixture.

Secondary alcohols appear to react first with concentrated H_2SO_4 to form carbocations and these then react with another molecule of the alcohol.

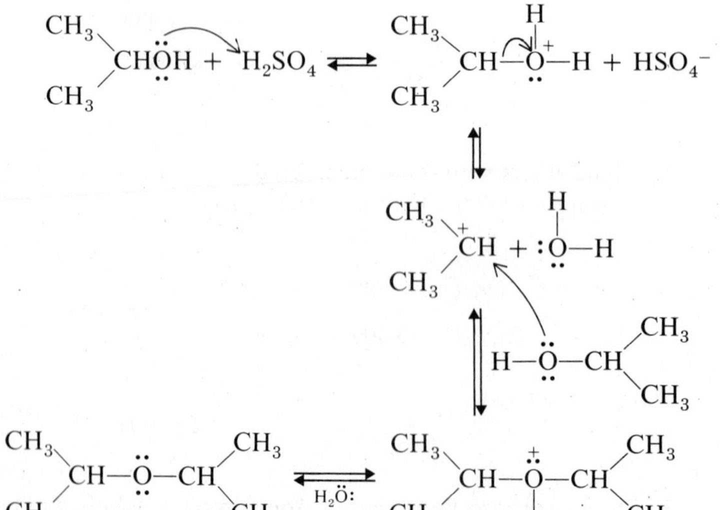

$$\underset{CH_3}{\overset{CH_3}{>}}CH\ddot{O}H + H_2SO_4 \rightleftharpoons \underset{CH_3}{\overset{CH_3}{>}}CH-\overset{H}{\underset{}{\overset{|}{\ddot{O}}}}{}^{+}-H + HSO_4^-$$

$$\underset{CH_3}{\overset{CH_3}{>}}\overset{+}{CH} + :\overset{H}{\underset{}{\overset{|}{\ddot{O}}}}-H$$

$$H-\ddot{O}-CH\underset{CH_3}{\overset{CH_3}{<}}$$

$$\underset{CH_3}{\overset{CH_3}{>}}CH-\ddot{O}-CH\underset{CH_3}{\overset{CH_3}{<}} \underset{H_2\ddot{O}:}{\rightleftharpoons} \underset{CH_3}{\overset{CH_3}{>}}CH-\overset{+}{\underset{H}{\ddot{O}}}-CH\underset{CH_3}{\overset{CH_3}{<}}$$

$$+ H_3\ddot{O}:^+$$

Heating tertiary alcohols with concentrated sulfuric acid does not lead to ditertiary ethers. This reaction is probably prevented by the steric crowding that would result in the ether. When tertiary alcohols are heated with sulfuric acid, dehydration to an alkene takes place instead.

Unsymmetrical ethers are not usually prepared by acid-catalyzed dehydration of alcohols because the reaction can lead to a mixture of products.

$$
\begin{array}{c}
\text{ROR} \\
+ \\
\text{ROH} + \text{R'OH} \underset{\text{H}_2\text{SO}_4}{\rightleftharpoons} \text{ROR'} + \text{H}_2\text{O} \\
+ \\
\text{R'OR'}
\end{array}
$$

Unsymmetrical ethers can be prepared from alkenes using the solvomercuration-demercuration reaction with an alcohol as the solvent (Section 6.4). Unsymmetrical ethers can also be prepared through a method known as the *Williamson synthesis.*

4. The Williamson synthesis is a nucleophilic substitution reaction. (We will discuss its mechanism and limitations in Chapter 17.) The synthesis consists of a reaction between a sodium alkoxide and an alkyl halide, alkyl sulfonate, or alkyl sulfate.

$$\text{RO}^-\text{Na}^+ + \text{R'}\!-\!\text{X} \longrightarrow \text{RO}\!-\!\text{R'} + \text{NaX}$$

$$(\text{X} = \text{I, Br, OSO}_2\text{R''}, \text{ or OSO}_3\text{R''})$$

A specific example of the Williamson synthesis is shown below.

$$\text{CH}_3\text{CH}_2\text{CH}_2\text{OH} + \text{Na} \longrightarrow \text{CH}_3\text{CH}_2\text{CH}_2\ddot{\text{O}}\!:^- \text{Na}^+ + \tfrac{1}{2}\text{H}_2$$

n-Propyl alcohol   Sodium *n*-propoxide

$$\Big\downarrow \text{CH}_3\text{CH}_2\text{I}$$

$$\text{CH}_3\text{CH}_2\text{OCH}_2\text{CH}_2\text{CH}_3 + \text{Na}^+\text{I}^-$$

Ethyl *n*-propyl ether
(70%)

Reactions of Ethers

Dialkyl ethers react with very few reagents. The only reactive sites that a dialkyl ether presents to another reactive substance are the C—H bonds of the alkyl groups and the —Ö— group of the ether linkage.

Ethers are like alkanes in that they undergo free-radical substitution reactions but these are of little synthetic importance.

The oxygen of the ether linkage makes ethers basic. Ethers can react with proton donors or Lewis acids to form *oxonium salts.*

$$\text{CH}_3\text{CH}_2\ddot{\text{O}}\text{CH}_2\text{CH}_3 + \text{HBr} \rightleftharpoons \text{CH}_3\text{CH}_2\!-\!\overset{+}{\underset{\underset{\text{H}}{|}}{\ddot{\text{O}}}}\!-\!\text{CH}_2\text{CH}_3 \ \text{Br}^-$$

An oxonium salt

$$\text{CH}_3\text{CH}_2\ddot{\text{O}}\text{CH}_2\text{CH}_3 + \text{BF}_3 \rightleftharpoons \text{CH}_3\text{CH}_2\!-\!\overset{+}{\underset{\underset{\underline{\text{BF}}_3}{|}}{\ddot{\text{O}}}}\!-\!\text{CH}_2\text{CH}_3$$

(an oxonium salt)
Boron trifluoride etherate

The oxonium salt formed when diethyl ether reacts with boron trifluoride is stable enough to allow its distillation at low pressures (bp 48° at 10 Torrs).

Heating ethers with very strong acids (HI, HBr, and H_2SO_4) causes them to undergo reactions in which the carbon oxygen bond breaks. Diethyl ether, for example, reacts with hot concentrated hydrobromic acid to give two moles of ethyl bromide.

$$CH_3CH_2OCH_2CH_3 + HBr \longrightarrow 2CH_3CH_2Br + H_2O$$

The mechanism for this reaction begins with formation of an oxonium ion. A nucleophilic substitution reaction then produces ethyl alcohol and ethyl bromide.

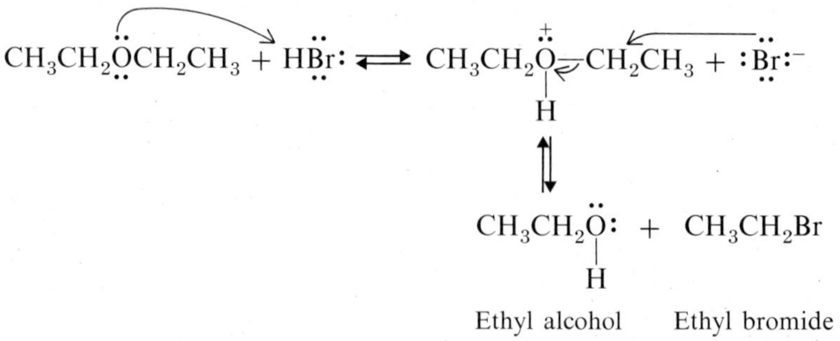

$$CH_3CH_2\overset{..}{\underset{..}{O}}CH_2CH_3 + H\overset{..}{\underset{..}{Br}}: \rightleftharpoons CH_3CH_2\overset{\overset{+}{}}{O} \rightleftharpoons CH_2CH_3 + :\overset{..}{\underset{..}{Br}}:^-$$
$$\underset{H}{|}$$

$$CH_3CH_2\overset{..}{O}: + CH_3CH_2Br$$
$$\underset{H}{|}$$

Ethyl alcohol   Ethyl bromide

Finally, ethyl alcohol can react with HBr to form a second mole of ethyl bromide.

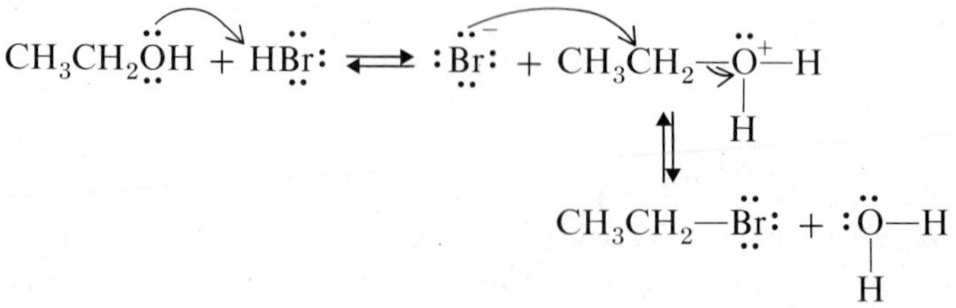

$$CH_3CH_2\overset{..}{O}H + H\overset{..}{\underset{..}{Br}}: \rightleftharpoons :\overset{..}{\underset{..}{Br}}: + CH_3CH_2 \rightarrow \overset{+}{O}-H$$
$$\underset{H}{|}$$

$$CH_3CH_2-\overset{..}{\underset{..}{Br}}: + :\overset{..}{O}-H$$
$$\underset{H}{|}$$

16.9 SYNTHESIS OF PHENOLS

Phenols can be prepared from aryl thallium ditrifluoroacetates through their reaction

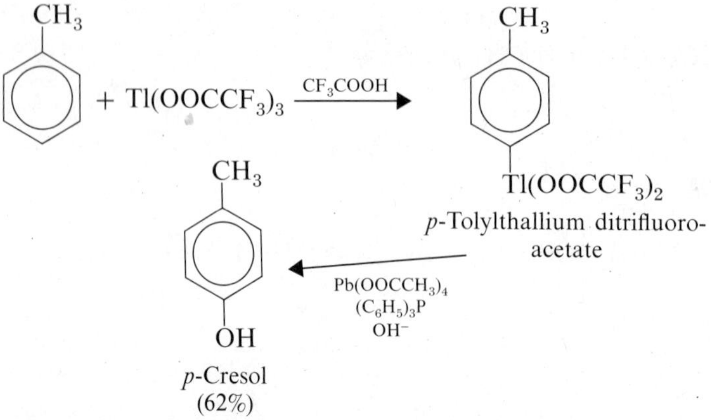

p-Tolylthallium ditrifluoro-
acetate

p-Cresol
(62%)

with lead tetraacetate and triphenylphosphine (Section 15.3). An example is the synthesis of *p*-cresol from toluene.

An older method for the preparation of phenols is through the fusion of sodium arylsulfonates with alkali metal hydroxides. This method works quite well for the preparation of *p*-cresol as the following example shows. However, the conditions required to bring about the reaction are so vigorous that its use in the preparation of many phenols is limited.

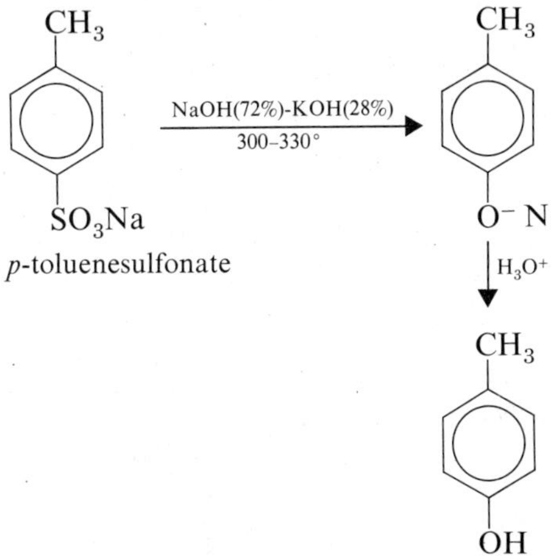

Sodium *p*-toluenesulfonate

p-Cresol
(63–70% overall from
sodium *p*-toluene sulfonate)

We will see a method for the preparation of phenols from diazonium salts in Chapter 21 that involves much milder reaction conditions. In Chapter 17 we will see how phenols can be prepared through nucleophilic substitution reactions.

16.10 REACTIONS OF PHENOLS

Phenols as Acids

Although phenols are structurally similar to alcohols, they are much stronger acids. The acidity constants of most alcohols are of the order of 10^{-18}. However, as we see in Table 16.6 the acidity constants of phenols are of the order of 10^{-11} or greater. We also see in Table 16.6, that strong electron-withdrawing groups on the benzene ring increase the acidity of phenols. For example, 2,4-dinitrophenol ($K_a = 1.1 \times 10^{-4}$) is a stronger acid than acetic acid and 2,4,6-trinitrophenol ($K_a = 4.2 \times 10^{-1}$) is almost as strong as some mineral acids. (The compound 2,4,6-trinitrophenol is called *picric acid*.)

We can explain both effects—the greater acidity of phenols relative to alcohols and the greater acidity of phenols with electron-withdrawing groups—through the application of resonance theory. Before we do this, however, it will be helpful if we reconsider acid-base theory in general terms.

Acid-base reactions are reactions that are under *equilibrium control*. That is, they are reversible reactions and, thus, the relative amounts of conjugate acid and base formed are governed by the position of an equilibrium. We have seen other

TABLE 16.6 The Acidity Constants of Phenols

NAME	FORMULA	ACIDITY CONSTANT K_a (in H_2O at 25°)
Phenol	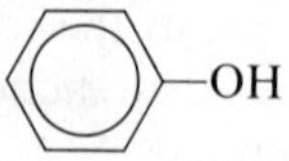	1.3×10^{-10}
o-Cresol	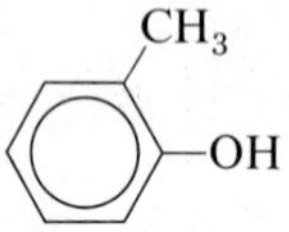	6.3×10^{-11}
m-Cresol	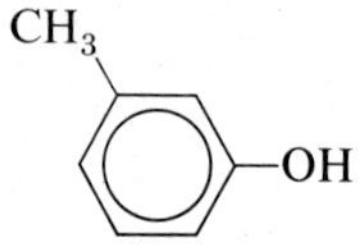	9.8×10^{-11}
p-Cresol		6.7×10^{-11}
o-Chlorophenol	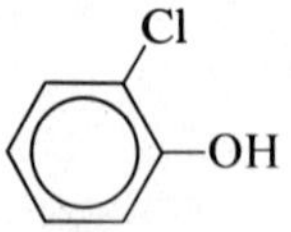	7.7×10^{-9}
m-Chlorophenol	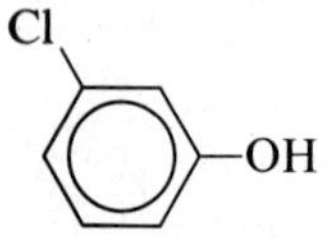	1.6×10^{-9}
p-Chlorophenol		6.3×10^{-10}
o-Nitrophenol	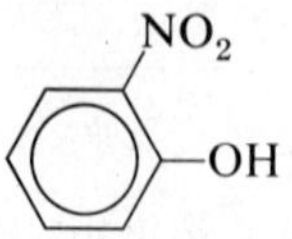	6.8×10^{-8}
m-Nitrophenol	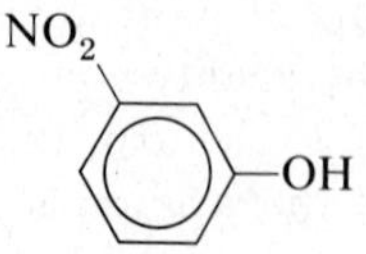	5.3×10^{-9}
p-Nitrophenol	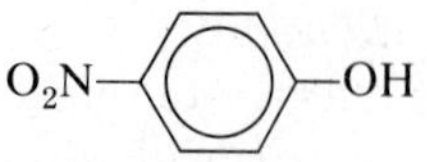	7×10^{-8}
2,4-Dinitrophenol	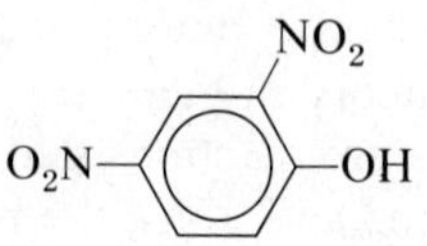	1.1×10^{-4}

TABLE 16.6 The Acidity Constants of Phenols (Continued)

NAME	FORMULA	ACIDITY CONSTANT K_a (in H_2O at 25°)
2,4,6-Trinitrophenol (picric acid)		4.2×10^{-1}
1-Naphthol		4.9×10^{-10}
2-Naphthol		2.8×10^{-10}

reactions that result from equilibrium control; for example, 1,4-additions to conjugated dienes (p. 359) and meta-thallation reactions of benzene compounds (p. 557). We have also seen that we can account for reactions that are under equilibrium control by concerning ourselves with the relative energies of the products and the reactants, rather than with relative magnitudes of energies of activation. The reason for this is simple; in reactions under equilibrium control both the products and the reactants have sufficient energy to surmount the energy barrier between them.

Thus, with acid-base reactions, we can assess the effects of variations in structure on K_a by estimating the energy difference between the acid and its conjugate base. Structural factors that stabilize the conjugate base, $A{:}^-$, more than they stabilize the acid, HA, cause the value of K_a to be larger. (HA will be, therefore, a stronger acid). Conversely, structural factors that stabilize the acid more than they do the conjugate base cause the value of K_a to be smaller. (In this case HA will be a weaker acid.)

Resonance effects are often an important factor in acid-base reactions, and resonance effects usually stabilize the conjugate base more than they stabilize the acid. We can see a vivid example of this if we return now to one of our original tasks and account for the greater acidity of a phenol relative to that of an alcohol.

Let us compare two *superficially* similar compounds, cyclohexanol and phenol.

Cyclohexanol
$K_a \simeq 10^{-18}$

Phenol
$K_a = 1.3 \times 10^{-10}$

Although phenol is a weak acid when compared with a carboxylic acid such as acetic acid ($K_a = 10^{-5}$), phenol is a much stronger acid than cyclohexanol (by a factor of almost 10^8). Phenol, for example, will dissolve in aqueous sodium hydroxide whereas cyclohexanol will not. The hydroxide ion is strong enough as a base so that the following reaction goes to completion and produces water-soluble sodium phenoxide.

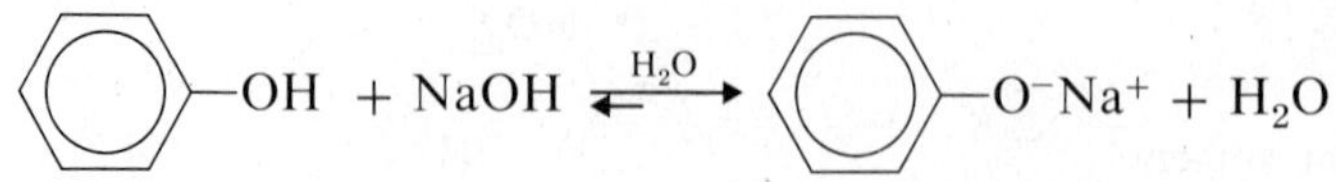

(slightly soluble)   (soluble)

The corresponding reaction of cyclohexanol with aqueous sodium hydroxide does not occur to any appreciable extent.

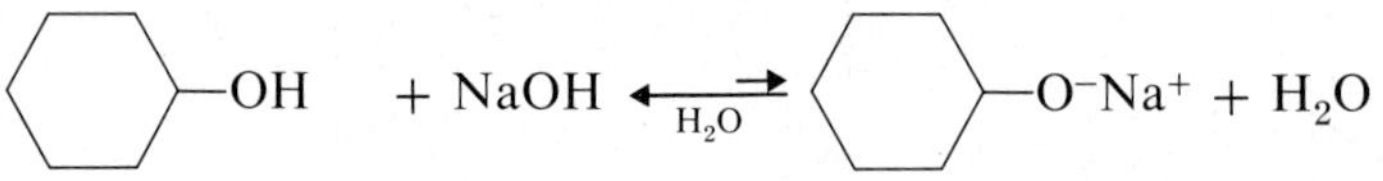

(very slightly soluble)   (soluble)

We can write resonance structures for both phenol and the phenoxide ion such as those shown below. (No analogous resonance structures are possible for cyclohexanol, of course.)

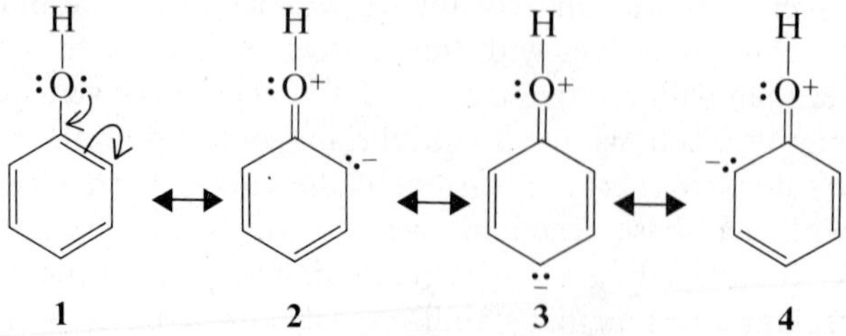

Resonance structures for phenol

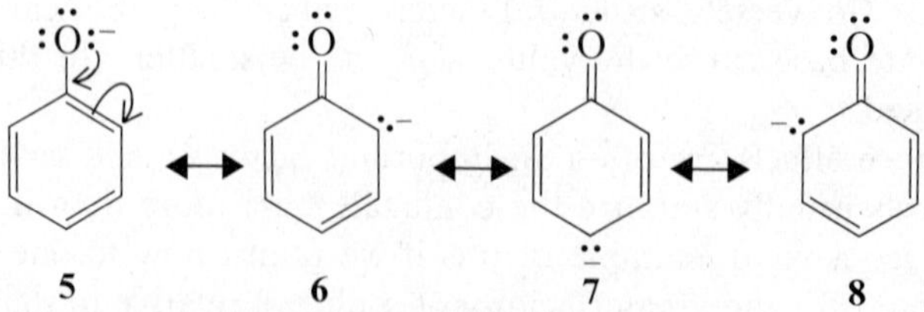

Resonance structures for the phenoxide ion

We see that with the exception of the Kekulé structure, the resonance structures for phenol (**2–4**) require separation of opposite charges while the corresponding resonance structures (**6–8**) for the phenoxide ion do not. *Energy is required to separate opposite charges,* and therefore we can conclude that resonance makes a *smaller* stabilizing contribution to phenol than it does to the phenoxide ion. That is, *resonance stabilizes the conjugate base* (the phenoxide ion) *more than it does the acid* (phenol).

Resonance stabilization of the phenoxide ion is particularly important because the negative charge is *delocalized* over the benzene ring. It is not localized on the oxygen atom as it would be in the anion of cyclohexanol:

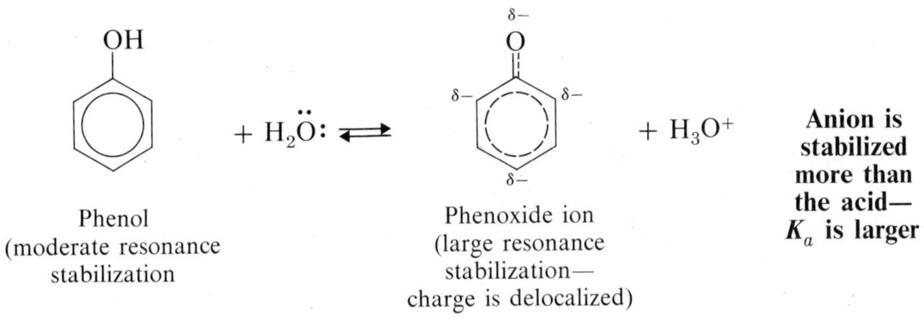

Phenol
(moderate resonance
stabilization

Phenoxide ion
(large resonance
stabilization—
charge is delocalized)

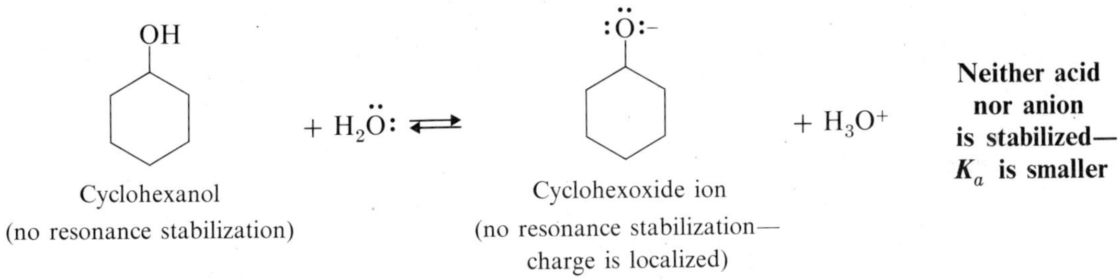

**Anion is
stabilized
more than
the acid—
K_a is larger**

Cyclohexanol
(no resonance stabilization)

Cyclohexoxide ion
(no resonance stabilization—
charge is localized)

**Neither acid
nor anion
is stabilized—
K_a is smaller**

Since resonance stabilizes the phenoxide ion more than it does phenol itself, ionization of phenol is a less endothermic reaction than the corresponding ionization of cyclohexanol (Fig. 16.1). The equilibrium between phenol and its conjugate base will, as a consequence, favor the formation of the conjugate base and the hydronium ion to a greater extent than the corresponding equilibrium of cyclohexanol (where neither the alcohol nor the alkoxide ion is resonance stabilized). Both compounds are weak acids, but, of the two, phenol is the stronger.

In the analysis given here, we have made our comparisons solely on the basis of enthalpy changes (ΔH's) for the two reactions. Such an analysis contains an oversimplification. In actuality, the equilibrium constant for a reaction is directly related to the standard free-energy change for the reaction, ΔG^0 (p. 288). The equation that relates these two quantities is

$$\log K_{eq} = -\frac{\Delta G^0}{2.303RT}$$

where R is the gas constant and T is the absolute temperature.

Since ΔG^0 is related both to the enthalpy change ΔH^0 and to the entropy change ΔS^0, that is,

$$\Delta G^0 = \Delta H^0 - T\Delta S^0$$

entropy effects can be important. In the two reactions considered here, the entropy changes have been of the same order of magnitude and have, therefore, canceled out. In other reactions, however, this may not be the case and, in fact, entropy changes may be the predominant influence.

Problem 16.14

The carbon-oxygen bond of phenol is much stronger than that of an alcohol. Phenol, for example, is not converted to bromobenzene when it is refluxed with concentrated

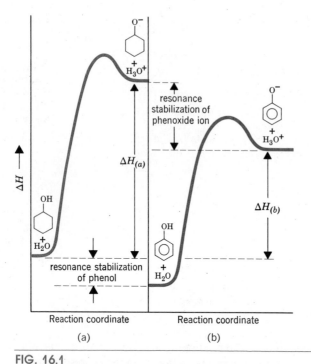

FIG. 16.1

Potential energy diagrams for the ionization of cyclo-hexanol (a) and phenol (b). Resonance stabilizes the phenoxide ion to a greater extent than it does phenol, itself; thus, the ionization of phenol is a less endothermic reaction than the ionization of cyclo-hexanol, that is, $\Delta H_{(a)} > \Delta H_{(b)}$. Phenol, therefore, is the stronger acid.

hydrobromic acid. Similar treatment of cyclohexanol, however, does give bromo-cyclohexane.

$$\bigcirc\text{—OH} + \text{HBr} \xrightarrow{\text{reflux}} \quad \text{No reaction}$$

$$\bigcirc\text{—OH} + \text{HBr} \xrightarrow{\text{reflux}} \bigcirc\text{—Br} + \text{H}_2\text{O}$$

Although resonance structures **2–4** do not make as great a contribution to the phenol hybrid as the corresponding structures (**6–8**) do to the phenoxide ion, structures **2–4** do help us understand why the carbon-oxygen bond of phenols is very strong. Explain.

If we examine Table 16.6 we see that phenols having electron withdrawing groups (Cl— or NO_2—) attached to the benzene ring are more acidic than phenol itself. On the other hand, those phenols bearing electron-releasing groups (i.e., CH_3—) are less acidic than phenol. We can account for this trend by examining both the resonance and the inductive effect that these groups have on the phenoxide ion.

Methyl groups are electron releasing. Because of this the benzene ring of the cresol anions can not delocalize a negative charge as effectively as the benzene ring of the phenoxide ion.

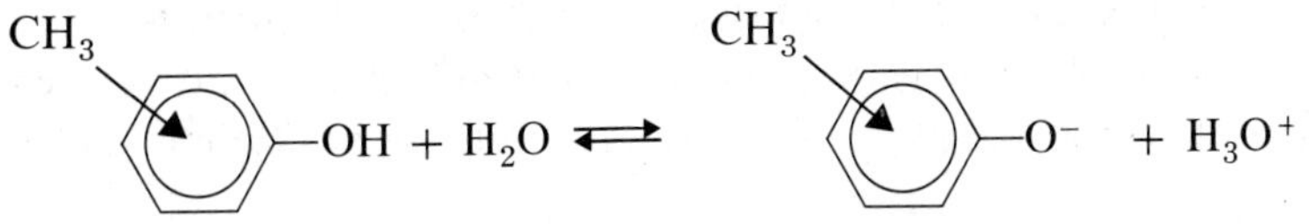

Electron-releasing group inhibits delocalization of the charge

Thus, the anion (the conjugate base) of a cresol is not stabilized by resonance to as great an extent as is the phenoxide ion. The cresols, consequently, are less acidic than phenol.

Electron-withdrawing groups, by contrast, assist in delocalization of the negative charge of phenol anions. Electron-withdrawing groups, therefore, help *stabilize* the anion and *increase* the acidity of the phenol.

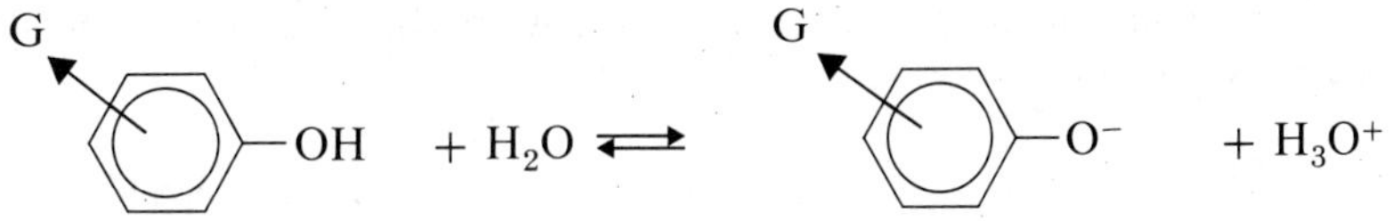

G = electron-withdrawing group (Cl or NO$_2$)

Electron-withdrawing group assists delocalization of charge

Nitro groups are particularly effective in delocalizing the negative charge when they are in an ortho or para position relative to the —OH group. In such compounds we find that resonance effects are important. We can see how if we examine the resonance structures for *p*-nitrophenoxide ion written below.

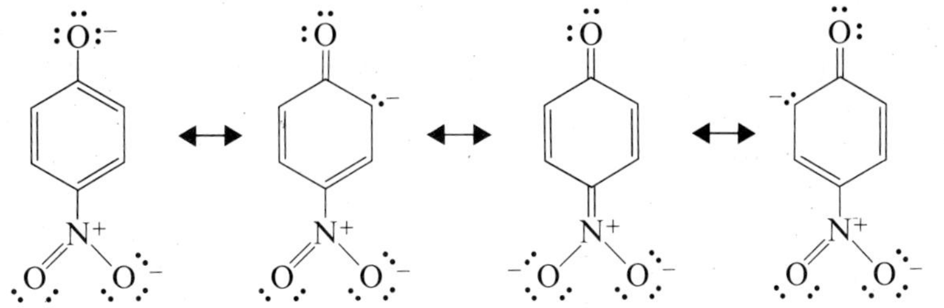

Exceptionally stable

One of these resonance structures is *exceptionally stable* because electronegative oxygens of the nitro group accommodate all the negative charge.

Problem 16.15

The unusually great acidity of 2,4,6-trinitrophenol ($K_a = 4.2 \times 10^{-1}$) can be accounted for because three exceptionally stable structures contribute to its anion. Write formulas for these three structures.

Other Reactions of the O—H Group of Phenols

Phenols react with carboxylic acid anhydrides and acid chlorides to form esters. These reactions are quite similar to those of alcohols (Section 16.6).

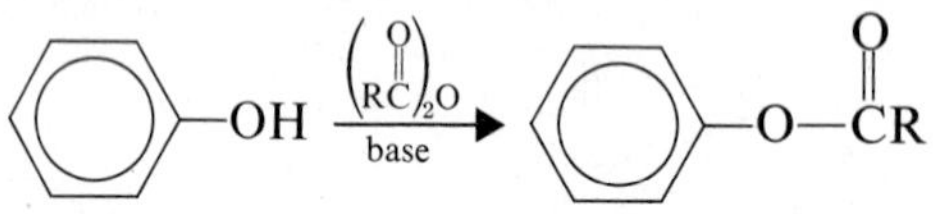

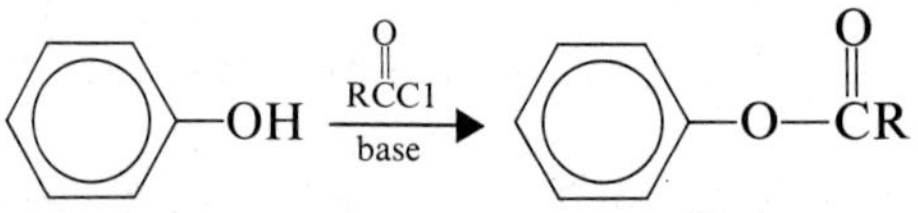

Phenols can be converted to ethers through the Williamson synthesis. Since phenols are more acidic than alcohols they can be converted to sodium phenoxides through the use of sodium hydroxide (rather than metallic sodium, the reagent used to convert alcohols to alkoxide ions). A specific example is the synthesis of anisole from phenol.

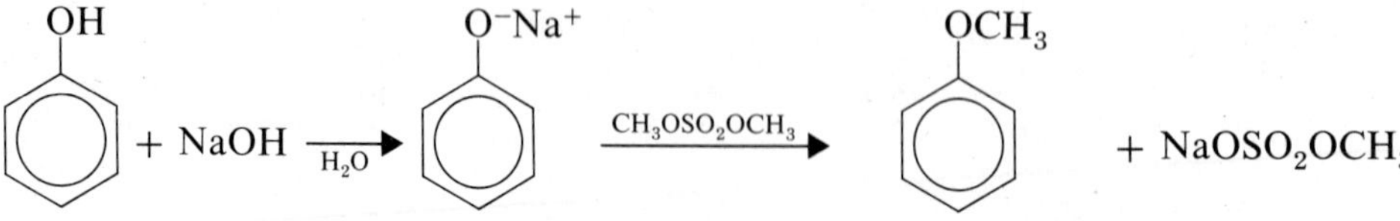

(72–75% overall)

When alkyl aryl ethers react with strong acids such as HI and HBr, the reaction produces an alkyl halide and a phenol. The phenol does not react further to produce an aryl halide because the carbon-oxygen bond is very strong (cf. problem 16.14).

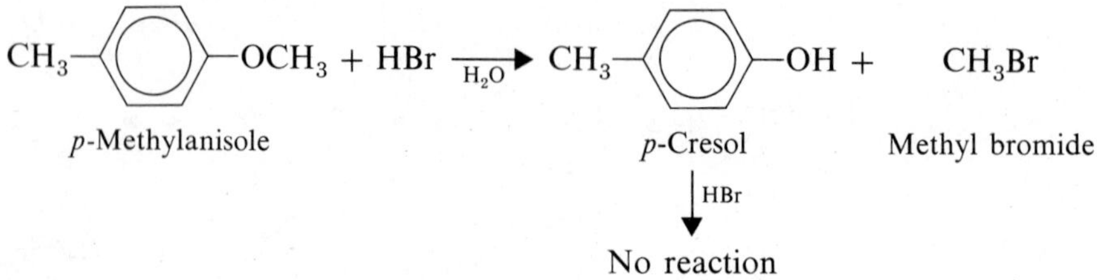

p-Methylanisole   p-Cresol   Methyl bromide

No reaction

Reactions of the Benzene Ring of Phenols

The hydroxyl group is a powerful activating group—and an ortho-para director—in electrophilic aromatic substitutions. Phenol itself reacts with bromine in aqueous solution to yield 2,4,6-tribromophenol in nearly quantitative yield.

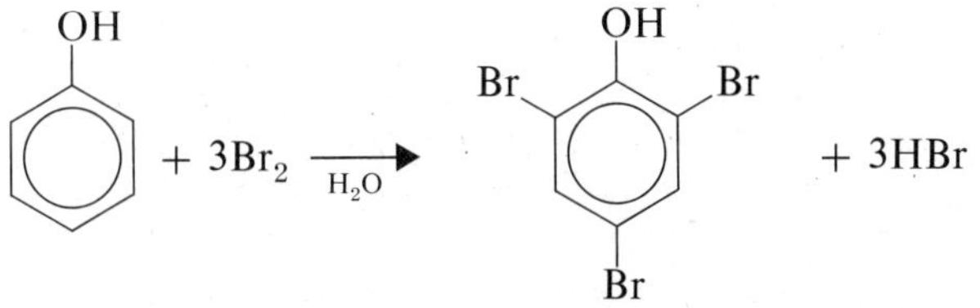

2,4,6-Tribromophenol
($\sim$100%)

Monobromination of phenol can be achieved by carrying out the reaction in carbon disulfide at a low temperature. The major product is the para isomer.

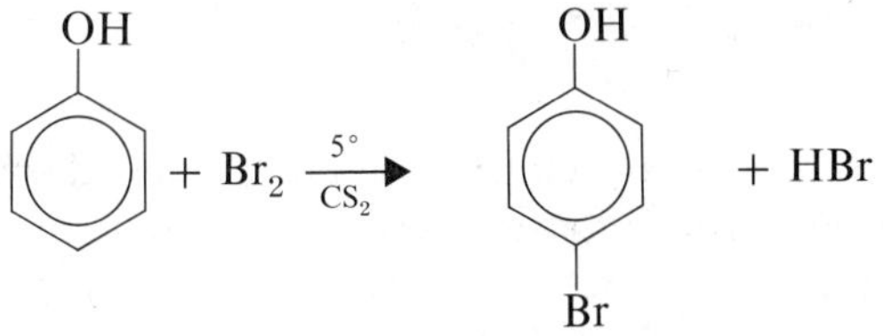

p-Bromophenol
(80–84%)

Phenol reacts with dilute nitric acid to yield a mixture of o- and p-nitrophenol. Although the yield is relatively low (because of oxidation of the ring) the ortho and para isomers can be separated by steam distillation. o-Nitrophenol is the more volatile isomer because its hydrogen bonding is intramolecular (below) (its effective molecular weight is lower). Thus, o-nitrophenol passes over with the steam, and p-nitrophenol remains in the distillation flask.

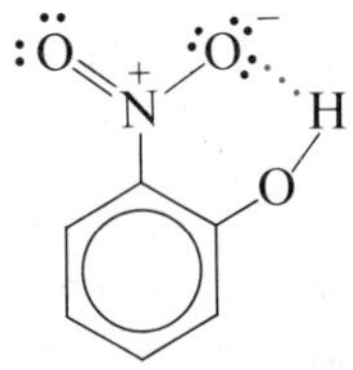

o-Nitrophenol
(more volatile because of
intramolecular hydrogen bonding)

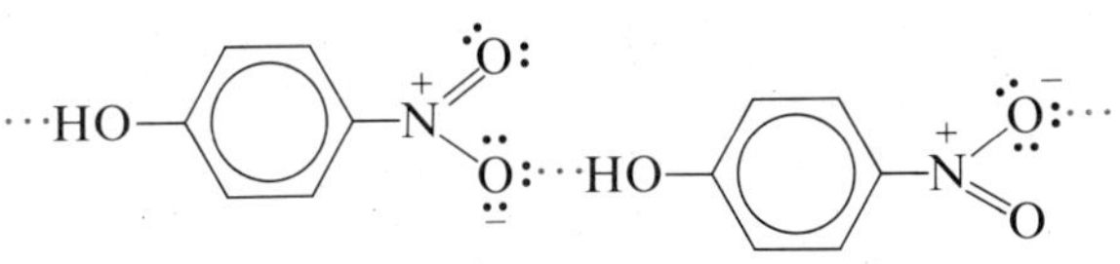

p-Nitrophenol
(less volatile because of intermolecular
hydrogen bonding)

The phenoxide ion is even more susceptible to electrophilic aromatic substitution than phenol itself. (Why?) Use is made of the high reactivity of the phenoxide ring in a reaction called the *Kolbe reaction*. In the Kolbe reaction carbon dioxide acts as the electrophile.

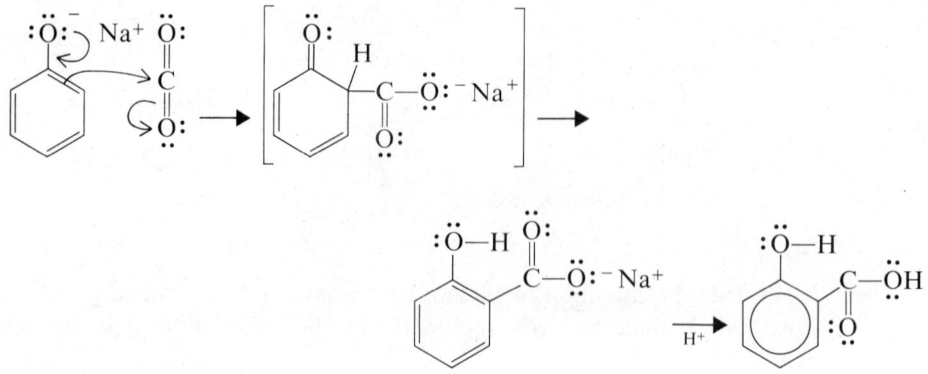

Salicylic acid

The reaction is usually carried out by allowing sodium phenoxide to absorb carbon dioxide (sodium phenyl carbonate is formed) and then heating the product to 125° under a pressure of several atmospheres of carbon dioxide. Subsequent acidification of the mixture produces *salicylic acid.*

Salicylic acid reacts with acetic anhydride to produce man's most widely used pain reliever—*aspirin.*

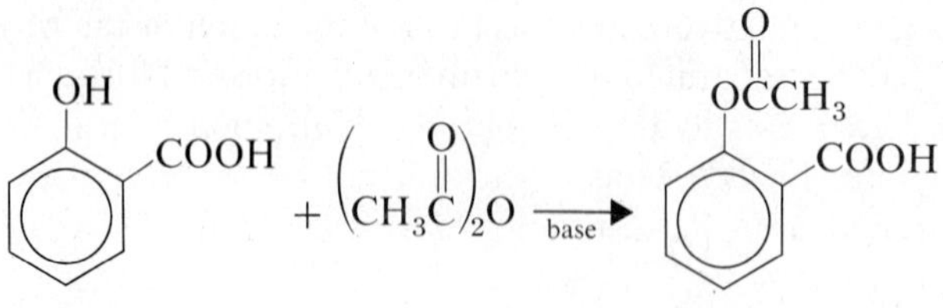

Salicylic
acid

Acetylsalicylic acid
(aspirin)

16.11 THIOLS, THIOETHERS, AND THIOPHENOLS

Sulfur is directly below oxygen in Group VI of the Periodic Table and, as we might expect, there are sulfur counterparts of the oxygen compounds that we have studied in this chapter.

The sulfur counterpart of an alcohol is called a *thiol* or a *mercaptan.* The name mercaptan comes from the Latin, *mercurium captans,* meaning "capturing mercury." Mercaptans react with mercuric ions and the ions of other heavy metals to form precipitates. The compound, CH_2CHCH_2OH, known as British Anti-Lewisite
    $\qquad\qquad\qquad\qquad\quad$ | $\quad$ |
    $\qquad\qquad\qquad\qquad\quad$ SH $\,$ SH
(BAL), was developed as an antidote for poisonous arsenic compounds used as war gases. BAL is also an effective antidote for mercury poisoning.

Several simple thiols are shown below and on the next page.

$$CH_3CH_2SH \qquad\qquad CH_3CH_2CH_2SH$$

Ethanethiol $\qquad\qquad\qquad$ Propanethiol
(ethyl mercaptan) $\qquad\qquad$ (propyl mercaptan)

$$\underset{\substack{\text{3-Methyl-1-butanethiol}\\\text{(isopentyl mercaptan)}}}{\overset{\overset{\displaystyle CH_3}{|}}{CH_3CHCH_2CH_2SH}} \qquad \underset{\substack{\text{2-Propenethiol}\\\text{(allyl mercaptan)}}}{CH_2{=}CHCH_2SH}$$

Compounds of sulfur, generally, and the low-molecular-weight thiols, in particular, are noted for their disagreeable odors. Anyone who has passed anywhere near a general chemistry laboratory in which hydrogen sulfide (H_2S) was being used has noticed the strong odor of that substance—the odor of rotten eggs. Another sulfur compound, 3-methyl-1-butanethiol is one unpleasant constituent of the liquid that skunks use as a defensive weapon. Propanethiol evolves from freshly chopped onions, and allyl mercaptan is one of the compounds responsible for the odor and flavor of garlic.

Aside from their odors, analogous sulfur and oxygen compounds show other chemical differences. These differences arise largely from the following features of sulfur compounds.

1. The sulfur atom is larger and more polarizable than the oxygen atom. As a result, sulfur compounds are more powerful nucleophiles and compounds containing —SH groups are stronger acids than their oxygen analogs. The ethyl mercaptide ion ($CH_3CH_2\ddot{S}:^-$), for example, is a much stronger nucleophile when it reacts at carbon atoms than is the ethoxide ion ($CH_3CH_2\ddot{O}:^-$). On the other hand since ethanol is a weaker acid than ethanethiol, the ethoxide ion is the stronger base.

2. The bond-dissociation energy ($\sim$80 kcal/mole) of the S—H bond of thiols is much less than that ($\sim$110 kcal/mole) of the O—H bond of alcohols. The weakness of the S—H bond allows thiols to undergo an oxidative coupling reaction when they react with mild oxidizing agents; the product is a disulfide:

$$\underset{\text{A thiol}}{2RS{-}H} + H_2O_2 \longrightarrow \underset{\text{A disulfide}}{RS{-}SR} + 2H_2O$$

Alcohols do not undergo an analogous reaction. When alcohols are treated with oxidizing agents, oxidation takes place at the weaker C—H bond ($\sim$85 kcal/mole) rather than at the strong O—H bond.

$$\underset{\overset{|}{H}}{\overset{|}{-}\overset{|}{C}{-}O{-}H} \xrightarrow{\text{[O]}} -\overset{|}{\underset{\displaystyle \cdot}{C}}{-}O{-}H$$

3. The sulfur atom has vacant $3d$ orbitals; these assist in delocalizing electrons (and, thus, negative charge) on adjacent atoms. This means that hydrogens on carbons that are adjacent to a thioether group are more acidic than those of oxygen ethers. Thioanisole, for example, reacts with *n*-butyllithium in the following way.

$$\langle \bigcirc \rangle{-}SCH_3 + C_4H_9{:}^- \ Li^+ \longrightarrow \langle \bigcirc \rangle{-}SCH_2{:}^- \ Li^+ + C_4H_{10}$$

The anion is stabilized by resonance structures that utilize the sulfur d orbitals:

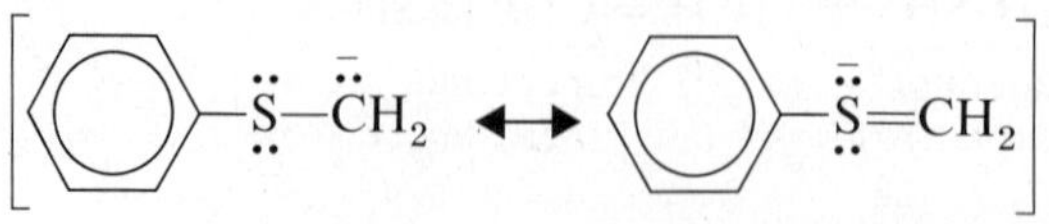

$$\left[\bigcirc\!\!-\ddot{\underset{..}{S}}\!-\bar{\ddot{C}}H_2 \longleftrightarrow \bigcirc\!\!-\ddot{\underset{..}{S}}\!=\!CH_2 \right]$$

Anisole does not undergo an analogous reaction because oxygen is not able to provide this type of resonance stabilization of the anion.

The $\diagdown\!S\!O\!\diagup$ group of sulfoxides and the positive sulfur of sulfonium ions are even more effective in delocalizing negative charge on an adjacent atom:

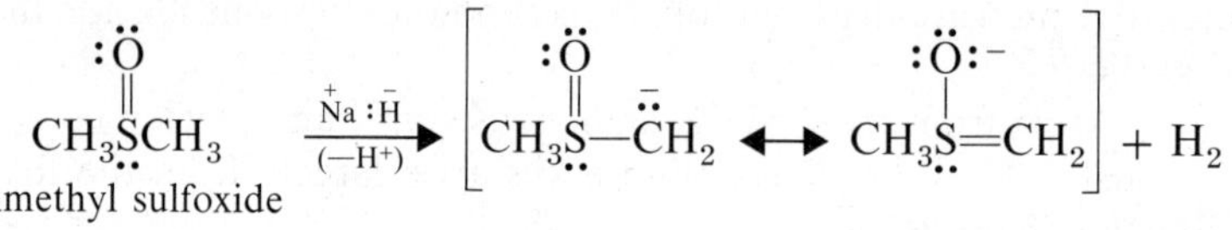

$$\underset{\text{dimethyl sulfoxide}}{CH_3\overset{:\ddot{O}}{\underset{..}{\overset{\|}{S}}}CH_3} \xrightarrow[\;(-H^+)\;]{\overset{+}{Na}\,:\bar{H}} \left[CH_3\overset{:\ddot{O}}{\underset{..}{\overset{\|}{S}}}\!-\!\bar{\ddot{C}}H_2 \longleftrightarrow CH_3\overset{:\ddot{O}:^-}{\underset{..}{S}}\!=\!CH_2 \right] + H_2$$

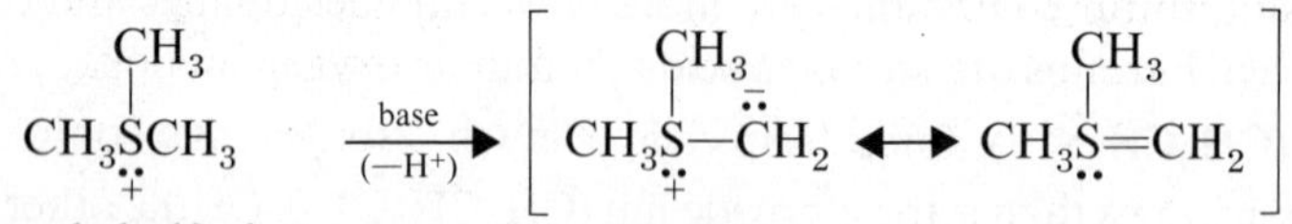

$$\underset{\substack{\text{Trimethylsulfonium}\\\text{bromide}}}{CH_3\overset{CH_3}{\underset{..}{\overset{|}{\underset{+}{S}}}}CH_3} \xrightarrow[\;(-H^+)\;]{\text{base}} \underset{\text{An ylide*}}{\left[CH_3\overset{CH_3}{\underset{+}{\overset{|}{\underset{..}{S}}}}\!-\!\bar{\ddot{C}}H_2 \longleftrightarrow CH_3\overset{CH_3}{\underset{..}{\overset{|}{S}}}\!=\!CH_2 \right]}$$

The anions formed in the reactions given above are of synthetic use. They can be used to synthesize epoxides, for example (cf. Section 18.11).

Preparation of Thiols

Alkyl bromides and iodides react with potassium hydrogen sulfide to form thiols. (Potassium hydrogen sulfide can be generated by passing gaseous H_2S into an alcoholic solution of potassium hydroxide.)

$$R\!-\!Br + KOH + \underset{\text{(excess)}}{H_2S} \xrightarrow[\text{heat}]{C_2H_5OH} RSH + KBr + H_2O$$

The thiol that forms is sufficiently acidic to form an alkyl sulfide ion in the presence of potassium hydroxide. Thus, if excess H_2S is not employed in the reaction, the major product of the reaction will be a thioether. The thioether results from the following reactions:

$$R\!-\!SH + KOH \longrightarrow R\!-\!\ddot{\underset{..}{S}}\!:^- K^+$$

$$R\!-\!\ddot{\underset{..}{S}}\!:^- K^+ + R\!-\!\ddot{\underset{..}{Br}}: \longrightarrow R\!-\!\ddot{\underset{..}{S}}\!-\!R + KBr$$

Alkyl halides also react with thiourea to form (stable) S-alkylisothiouronium salts. These can be used to prepare thiols.

*An ylide is a neutral molecule that can be represented by a resonance structure having a negative carbon directly attached to a positive heteroatom (i.e., sulfur, phosphorous, etc.)

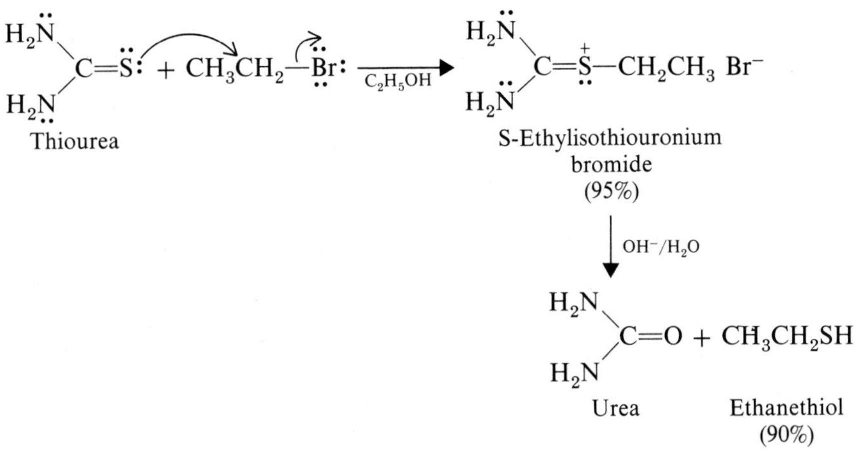

$$\underset{\text{Thiourea}}{\overset{H_2\ddot{N}}{\underset{H_2\ddot{N}}{\diagdown}}\hspace{-0.5em}C=\ddot{S}:} + CH_3CH_2\!-\!\ddot{\underset{\cdot\cdot}{B}}r: \xrightarrow{C_2H_5OH} \underset{\substack{\text{S-Ethylisothiouronium}\\ \text{bromide}\\ (95\%)}}{\overset{H_2\ddot{N}}{\underset{H_2\ddot{N}}{\diagdown}}\hspace{-0.5em}C=\overset{+}{\underset{\cdot\cdot}{S}}\!-\!CH_2CH_3 \; Br^-}$$

$$\downarrow{\scriptstyle OH^-/H_2O}$$

$$\underset{\text{Urea}}{\overset{H_2N}{\underset{H_2N}{\diagdown}}\hspace{-0.5em}C=O} + \underset{\substack{\text{Ethanethiol}\\ (90\%)}}{CH_3CH_2SH}$$

Physical Properties of Thiols

Thiols form very weak hydrogen bonds; their hydrogen bonds are not nearly as strong as those of alcohols. Because of this, low-molecular-weight thiols have lower boiling points than corresponding alcohols. Ethanethiol, for example, boils more than 40° lower than ethanol (37° versus 78°). The relative weakness of S—H hydrogen bonds is also evident when we compare the boiling points of ethanethiol and dimethylsulfide:

$$\underset{\text{bp } 37^\circ}{CH_3CH_2SH} \qquad \underset{\text{bp } 38^\circ}{CH_3SCH_3}$$

Physical properties of several thiols are given in Table 16.7.

TABLE 16.7 Physical Properties of Thiols

COMPOUND	STRUCTURE	MP C°	BP C°
Methanethiol	CH_3SH	-123	6
Ethanethiol	CH_3CH_2SH	-144	37
1-Propanethiol	$CH_3CH_2CH_2SH$	-113	67
2-Propanethiol	$(CH_3)_2CHSH$	-131	58
1-Butanethiol	$CH_3(CH_2)_2CH_2SH$	-116	98

Thiols and Disulfides in Biochemistry

Thiols and disulfides are important compounds in living cells, and there are many biochemical oxidation-reduction reactions in which they are interconverted.

$$2RSH \underset{[H]}{\overset{[O]}{\rightleftharpoons}} R\!-\!S\!-\!S\!-\!R$$

Lipoic acid, for example, an important cofactor in biological oxidations undergoes this oxidation-reduction reaction.

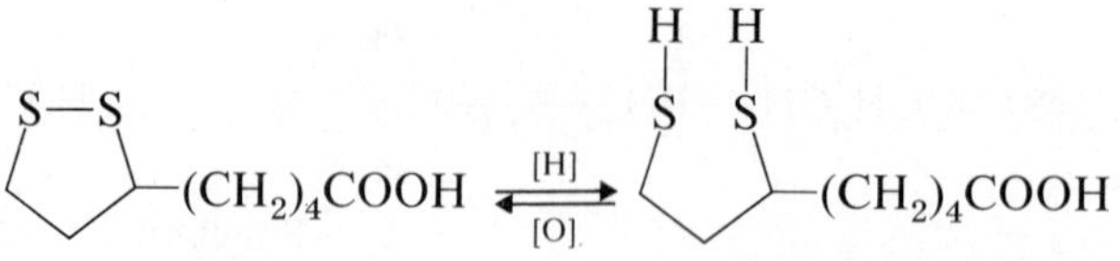

Lipoic acid   Dihydrolipoic acid

The amino acids *cysteine* and *cystine* are interconverted in a similar way.

$$2HOOC\underset{\displaystyle NH_2}{\underset{|}{C}}HCH_2SH \underset{[H]}{\overset{[O]}{\rightleftharpoons}} HOOC\underset{\displaystyle NH_2}{\underset{|}{C}}HCH_2S{-}SCH_2\underset{\displaystyle NH_2}{\underset{|}{C}}HCOOH$$

Cysteine   Cystine

As we will see later, the disulfide linkages of cystine units are important in determining the overall shapes of protein molecules.

Additional Problems

16.16

Show how each of the following transformations could be carried out.

(a) Styrene ⟶ 1-phenylethanol (two ways)
(b) Styrene ⟶ 2-phenylethanol
(c) Styrene ⟶ 1-methoxy-1-phenylethane
(d) 1-Phenylethanol ⟶ 1-phenylethyl ethyl ether
(e) Phenylacetic acid ($C_6H_5CH_2COOH$) ⟶ 2-phenylethanol
(f) Phenyl methyl ketone ($C_6H_5COCH_3$) ⟶ 1-phenylethanol
(g) Toluene ⟶ 2-phenylethanol
(h) Benzene ⟶ 2-phenylethanol
(i) Methyl phenylacetate ($C_6H_5CH_2CO_2CH_3$) ⟶ 2-phenylethanol

16.17

Show how 1-butanol could be transformed into each of the following compounds. (You may use any necessary inorganic reagents and you need not show the synthesis of a particular compound more than once.)

(a) 1-Butene
(b) 2-Butanol
(c) 2-Butanone ($CH_3COCH_2CH_3$)
(d) 1-Bromobutane
(e) 2-Bromobutane
(f) 1-Pentanol
(g) 1-Hexene
(h) 3-Methyl-3-heptanol
(i) 1-Butanal ($CH_3CH_2CH_2CHO$)
(j) 4-Octanol
(k) 3-Methyl-4-heptanol
(l) Pentanoic acid ($CH_3CH_2CH_2CH_2COOH$)
(m) *n*-Butyl *sec*-butylether (three ways)
(n) Di-*n*-butyl ether (two ways)
(o) *n*-Butyllithium
(p) *n*-Octane

16.18
Give structures and names for the compounds that would be formed when 1-propanol is treated with each of the following reagents.
(a) Sodium metal
(b) Sodium metal then 1-bromobutane
(c) Methanesulfonyl chloride
(d) *p*-Toluenesulfonyl chloride
(e) Acetyl chloride
(f) Hot basic $KMnO_4$
(g) Phosphorus trichloride
(h) Thionyl chloride
(i) Sulfuric acid at 140°
(j) Refluxing concentrated hydrobromic acid
(k) 2-Methylpropene and mercuric acetate, followed by treatment with $NaBH_4$ and OH^-
(l) Benzene and BF_3

16.19
Give structures and names for the compounds that would be formed when 2-propanol is treated with each of the reagents given in problem 16.18.

16.20
What products would be obtained from each of the following acid-base reactions?
(a) Sodium ethoxide in ethanol + phenol $\longrightarrow$
(b) Phenylmagnesium bromide in ether + ethanol $\longrightarrow$
(c) Ethanethiol + aqueous sodium hydroxide $\longrightarrow$
(d) Ethanol + CH_3CH_2SNa $\rightleftharpoons$
(e) Phenol + aqueous sodium hydroxide $\longrightarrow$
(f) Sodium phenoxide + aqueous hydrochloric acid $\longrightarrow$
(g) Sodium ethoxide in ethanol + H_2O $\longrightarrow$

16.21
The acidity constant for the initial ionization of carbonic acid (H_2CO_3 + H_2O $\rightleftharpoons$ HCO_3^- + H_3O^+) is 4.3×10^{-7}. Which of the following compounds would you expect to dissolve in aqueous sodium bicarbonate (aq. $NaHCO_3$)?
(a) Phenol
(b) *p*-Cresol
(c) *o*-Chlorophenol
(d) 2,4-Dinitrophenol
(e) 2,4,6-Trinitrophenol
(f) Benzoic acid ($K_a = 6.4 \times 10^{-5}$)

16.22
What compounds would you expect to be formed when each of the following ethers is refluxed with excess concentrated hydrobromic acid?
(a) Methyl ethyl ether
(b) Phenyl ethyl ether
(c) Tetrahydrofuran

(d) Dioxane $\left(\text{i.e.,} \right.$ 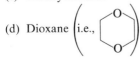 $\left. \right)$

16.23
(a) Give two methods for synthesizing *p*-ethylphenol from ethylbenzene. (b) How would you synthesize *m*-ethylphenol from ethylbenzene?

16.24

Complete the following equations.

(a) $CH_3CH\!\!-\!\!CH_2$ + H_3O^+ $\xrightarrow[H_2O]{}$

(b) CH_3CH_2OH + CH_2CH_2 $\xrightarrow[OH^-]{}$

(c) C_6H_5OH + CH_2CH_2 $\xrightarrow[OH^-]{}$

(d) 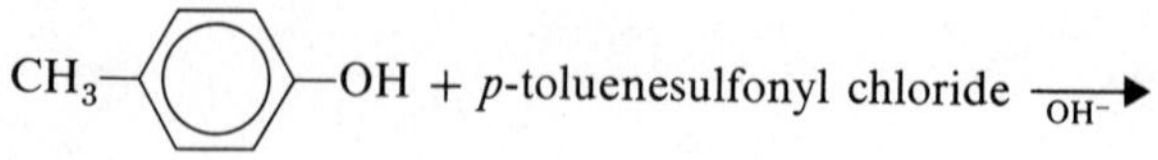 $CH_3\!\!-\!\!\langle \bigcirc \rangle\!\!-\!\!OH$ + p-toluenesulfonyl chloride $\xrightarrow[OH^-]{}$

(e) Cyclohexanol + acetic anhydride $\longrightarrow$

(f) Phenol + phthalic anhydride $\longrightarrow$

(g) p-Cresol + Br_2 $\xrightarrow{H_2O}$

(h) Benzyl bromide + thiourea $\longrightarrow$

(i) Product of (h) + OH^-/H_2O $\longrightarrow$

(j) Product of (i) + H_2O_2 $\longrightarrow$

(k) Product of (i) + NaOH $\longrightarrow$

(l) Product of (k) + benzyl bromide $\longrightarrow$

(m) Benzyl alcohol + $KMnO_4$ $\xrightarrow[heat]{OH^-}$

(n) Benzyl alcohol + $CrO_3 \cdot 2C_5H_5N$ $\xrightarrow[25°]{CH_2Cl_2}$

(o) Benzyl alcohol + Na $\longrightarrow$

(p) Product of (o) + $CH_3OSO_2OCH_3$ $\longrightarrow$

(q) cis-3-Hexene + $C_6H_5\overset{\overset{\displaystyle O}{\|}}{C}OOH$ $\longrightarrow$

(r) Product of (q) + H_3O^+/H_2O $\longrightarrow$

16.25

Show how you might prepare each of the following alcohols through a Grignard synthesis. (Assume that you have available any necessary organic halides, aldehydes, ketones, esters, and epoxides as well as any necessary inorganic reagents.)

(a) $CH_3CH_2\overset{\overset{\displaystyle CH_3}{|}}{\underset{\underset{\displaystyle CH_3}{|}}{C}}OH$ (three ways)

(b) 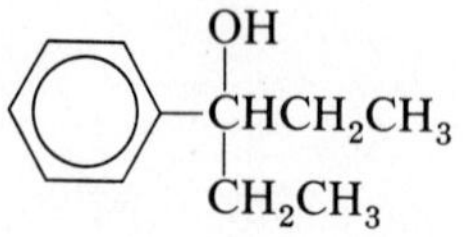 $\langle \bigcirc \rangle\!\!-\!\!\overset{\overset{\displaystyle OH}{|}}{\underset{\underset{\displaystyle CH_2CH_3}{|}}{C}}HCH_2CH_3$ (three ways)

(c)

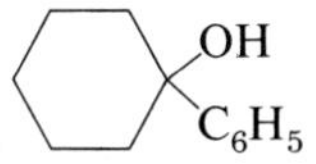

OH
C$_6$H$_5$

(d) 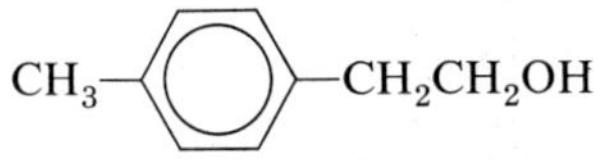 CH$_3$—⟨◯⟩—CH$_2$CH$_2$OH

(e) 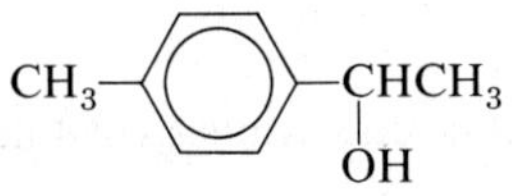 CH$_3$—⟨◯⟩—CHCH$_3$ (two ways)
                        |
                       OH

16.26

Describe a simple chemical test that could be used to distinguish between each of the following pairs of compounds.

(a) *p*-Cresol and benzyl alcohol
(b) Cyclohexanol and cyclohexane
(c) Cyclohexanol and cyclohexene
(d) Allyl propyl ether and dipropyl ether
(e) Anisole and *p*-cresol
(f) Picric acid and 2,4,6-trimethylphenol

16.27

Write a mechanism that accounts for the following reaction.

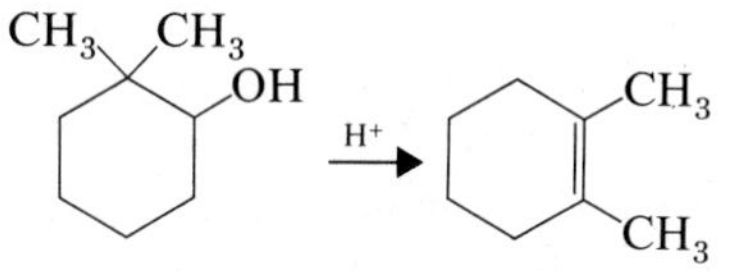

CH$_3$ CH$_3$
      OH   CH$_3$
              —H+→
                            CH$_3$

16.28

Thymol (below) can be obtained from thyme oil. Thymol is an effective disinfectant and is used in many antiseptic preparations. (a) Suggest a synthesis of thymol from *m*-cresol and propylene. (b) Suggest a method for transforming thymol into menthol (p. 599).

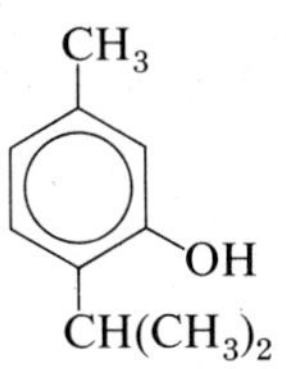

CH$_3$

OH
CH(CH$_3$)$_2$
Thymol

16.29

Carvacrol is another naturally occurring phenol and it is an isomer of thymol (problem 16.28). Carvacrol can be synthesized from *p*-cymene (*p*-isopropyltoluene) by ring sulfonation and alkali fusion of the sulfonic acid. Explain why this synthesis yields mainly carvacrol and very little thymol.

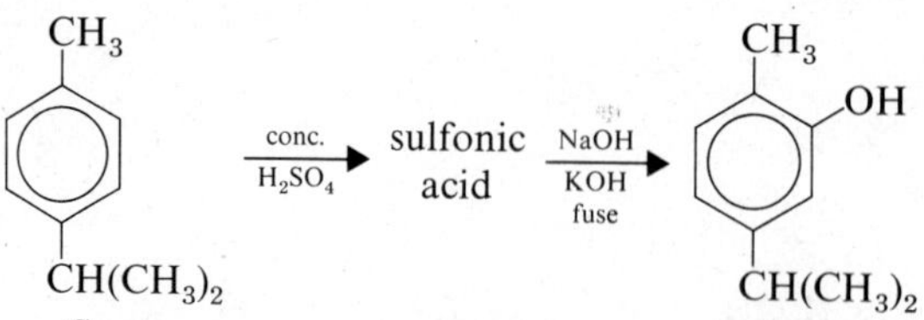
CH_3
conc.
H_2SO_4
sulfonic acid
NaOH
KOH
fuse
CH_3
OH
$CH(CH_3)_2$
$CH(CH_3)_2$
p-Cymene
Carvacrol

16.30

A widely used synthetic antiseptic is 4-*hexylresorcinol.* Suggest a synthesis of 4-hexylresorcinol from resorcinol and hexanoic acid.

16.31

Anethole (below) is the chief component of anise oil. Suggest a synthesis of anethole from anisole and propanoic acid.

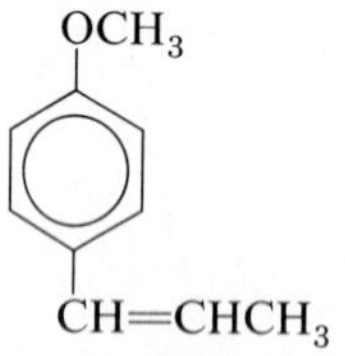
OCH_3
$CH{=}CHCH_3$
Anethole

16.32

Allyl disulfide, $CH_2{=}CHCH_2S{-}SCH_2CH{=}CH_2$, is another important component of oil of garlic. Suggest a synthesis of allyl disulfide starting with allyl bromide.

16.33

A compound X, $C_{10}H_{14}O$, dissolves in aqueous sodium hydroxide but is insoluble in aqueous sodium bicarbonate. Compound X reacts with bromine in water to yield a dibromo derivative, $C_{10}H_{12}Br_2O$. The 3000–4000 cm^{-1} region of the infrared spectrum of X shows a broad peak centered at 3250 cm^{-1}; the 680–840 cm^{-1} region shows a peak at 830 cm^{-1}. The pmr spectrum of X gives the following:

Singlet	$\delta 1.3$ (9H)
Singlet	$\delta 4.9$ (1H)
Multiplet	$\delta 7.0$ (4H)

What is the structure of X?

16.34

The infrared and pmr spectra of compound Y, $C_9H_{12}O$, are given in Fig. 16.2. Propose a structure for Y.

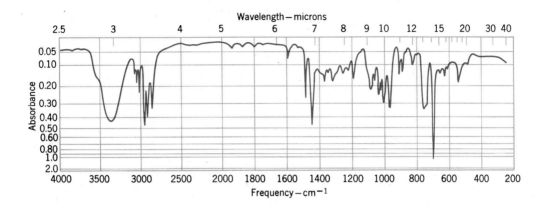

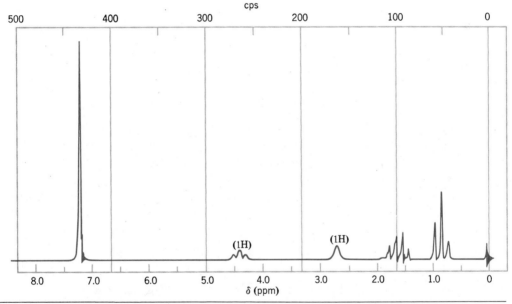

FIG. 16.2

The infrared and pmr spectra of compound Y, problem 16.34. (Infrared spectrum courtesy of Sadtler Research Laboratories, Philadelphia, Pa. The pmr spectrum courtesy of Varian Associates, Palo Alto, Calif.)

16.35

Starting with allyl alcohol, outline a synthesis of British Anti-Lewisite, $CH_2SHCHSHCH_2OH$

16.36

The widely used antioxidant and food preservative called **BHA** (**B**utylated **H**ydroxy **A**nisole) is actually a mixture of 2-*tert*-butyl-4-methoxyphenol and 3-*tert*-butyl-4-methoxyphenol. **BHA** is synthesized from *p*-methoxyphenol and 2-methylpropene. (a) Suggest how this is done. (b) Another widely used antioxidant is **BHT** (**B**utylated **H**ydroxy **T**oluene). **BHT** is actually 2,6-di-*tert*-butyl-4-methylphenol and the raw materials used in its production are *p*-cresol and 2-methylpropene. What reaction is used here?

* 16.37

Compound **Z**, $C_5H_{10}O$, decolorizes bromine in carbon tetrachloride. The infrared spectrum of

Z shows a broad peak in the 3200–3600 cm⁻¹ region. The pmr spectrum of **Z** is given in Fig. 16.3. Propose a structure for **Z**.

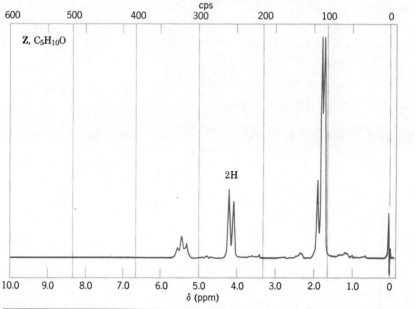

FIG. 16.3

*The pmr spectrum of compound **Z**, problem 16.37. (Spectrum courtesy of Aldrich Chemical Co., Milwaukee, Wis.)*

* **16.38**

A synthesis of lipoic acid (p. 624) is outlined below. Supply the missing reagents and intermediates.

$$\underset{\underset{O}{\parallel}}{Cl-C}(CH_2)_4CO_2C_2H_5 \xrightarrow[\text{AlCl}_3]{CH_2=CH_2} \text{(a)} \ C_{10}H_{17}ClO_3 \xrightarrow{\text{NaBH}_4}$$

$$\underset{\underset{OH}{|}}{ClCH_2CH_2CH}(CH_2)_4CO_2C_2H_5 \xrightarrow{\text{(b)}} \underset{\underset{Cl}{|}}{ClCH_2CH_2CH}(CH_2)_4CO_2C_2H_5$$

$$\xrightarrow[\text{(d)}]{\text{(c)}} \underset{\underset{SCH_2C_6H_5}{|}}{C_6H_5CH_2SCH_2CH_2CH}(CH_2)_4CO_2H \xrightarrow[\text{(2) H}^+]{\text{(1) Na, NH}_3}$$

$$\text{(e)} \ C_8H_{16}S_2O_2 \xrightarrow{O_2} \text{lipoic acid}$$

* 16.39

A Grignard reagent that is a key intermediate in an industrial synthesis of Vitamin A (p. 703) can be prepared in the following way:

$$HC\equiv CLi + CH_2=CHCCH_3 \xrightarrow[(2)\ NH_4^+]{} A\ (C_6H_8O) \xrightarrow{H^+}$$

(the second reactant bears a C=O above the central carbon)

$$B,\ HOCH_2CH=\overset{\displaystyle CH_3}{C}—C\equiv CH \xrightarrow{2C_2H_5MgBr} C\ (C_6H_6Mg_2Br_2O)$$

(a) What are the structures of compounds **A** and **C**?

(b) The acid-catalyzed rearrangement of **A** to **B** takes place very readily. What two factors account for this?

* 16.40

The remaining steps in the industrial synthesis of Vitamin A (as its acetate) are as follows: The Grignard reagent **C** from problem 16.39 is allowed to react with the aldehyde shown below. (See problems 18.12 and 20.33 for its synthesis.)

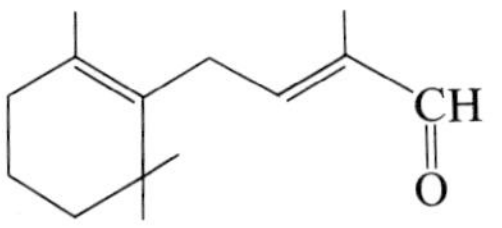

After acidification, the product obtained from this step is a diol **D**. Selective hydrogenation of the triple bond of **D** using Lindlar's catalyst yields **E** $(C_{20}H_{32}O_2)$. Treating **E** with one mole of acetic anhydride yields a monoacetate, **F**, and dehydration of **F** yields Vitamin A acetate. What are the structures of **D–F**?

* 16.41

Heating acetone with an excess of phenol in the presence of hydrogen chloride is the basis for an industrial process used in the manufacture of a compound called "bisphenol A." (Bisphenol A is used in the manufacture of epoxy resins and a polymer called "Lexan," cf. problem 22.21.) Bisphenol A has the molecular formula $C_{15}H_{16}O_2$ and the reactions involved in its formation are similar to those involved in the synthesis of DDT (problem 15.15). Write out these reactions and give the structure of bisphenol A.

* 16.42

One chemical-warfare agent used in World War I is a powerful vesicant called "mustard gas." (The name comes from its mustardlike odor; mustard gas, however, is not a gas but a high-boiling liquid that was dispersed as a mist of tiny droplets.) Mustard gas can be synthesized from ethylene oxide in the manner shown below. Outline the reactions involved.

$$2CH_2\!-\!CH_2 + H_2S \longrightarrow C_4H_{10}SO_2 \xrightarrow[ZnCl_2]{HCl} C_4H_8SCl_2$$
$$\overset{\displaystyle O}{}\qquad\qquad\qquad\qquad\qquad\text{"mustard gas"}$$

17 NUCLEOPHILIC SUBSTITUTION AND ELIMINATION REACTIONS

17.1 NUCLEOPHILIC SUBSTITUTION AT A SATURATED CARBON

We have now seen a number of reactions in which a nucleophile (i.e., a reagent that can donate an electron pair) replaces some other group (a leaving group) at a saturated carbon. In Chapter 10, for example, we saw that alkynes can be synthesized by the following nucleophilic substitution reaction:

$$R-C\equiv C:^- + R'-CH_2-\overset{..}{\underset{..}{Br}}: \xrightarrow[\text{substitution}]{\text{nucleophilic}} R-C\equiv C-CH_2R' + :\overset{..}{\underset{..}{Br}}:^-$$

Nucleophile   Leaving group

The nucleophile, in this reaction, is an acetylide ion (a carbanion) and the leaving group is a bromide ion.

In Chapter 16 we saw several other examples of nucleophilic substitution at a saturated carbon. The Williamson synthesis of ethers is a nucleophilic substitution reaction:

$$R-\overset{..}{\underset{..}{O}}:^- + R'-CH_2-\overset{..}{\underset{..}{X}}: \longrightarrow R-\overset{..}{\underset{..}{O}}-CH_2R' + :\overset{..}{\underset{..}{X}}:^-$$

Nucleophile   Leaving group

$$\text{(phenoxide)} \overset{..}{\underset{..}{O}}:^- + CH_3-O-\underset{\underset{O}{\parallel}}{\overset{\overset{O}{\parallel}}{S}}-O-CH_3 \longrightarrow \text{(phenyl)}-\overset{..}{\underset{..}{O}}CH_3 + {}^-O-\underset{\underset{O}{\parallel}}{\overset{\overset{O}{\parallel}}{S}}-O-CH_3$$

Nucleophile   Leaving group

In these two examples the nucleophiles are alkoxide ions and phenoxide ions, and the leaving groups are halide ions and methyl sulfate ions.

In the examples of nucleophilic substitution reactions that we have just examined, the nucleophiles and the leaving groups have been negatively charged ions. This is not always the case, however. The reaction of a primary alcohol with hydrobromic acid is an example of a reaction in which the leaving group is an electrically neutral molecule (water).

$$:\overset{..}{\underset{..}{Br}}:^- + R-CH_2-\overset{+}{\underset{\underset{H}{|}}{\overset{..}{O}}}-H \longrightarrow R-CH_2-\overset{..}{\underset{..}{Br}}: + :\overset{..}{\underset{\underset{H}{|}}{O}}-H$$

Nucleophile   Protonated 1° alcohol   Leaving group

In this chapter we will see a number of reactions in which the nucleophiles are electrically neutral. Two examples are shown below.

$$R_3N: \quad + CH_3-\overset{..}{\underset{..}{Br}}: \longrightarrow R_3\overset{+}{N}-CH_3 + :\overset{..}{\underset{..}{Br}}:^-$$

Nucleophile   (4° ammonium   Leaving
(3° amine)   salt)   group

$$H_2\overset{..}{O}: + R-CH_2-\overset{..}{\underset{..}{O}}-Ts \longrightarrow R-CH_2-\overset{+}{\underset{\underset{H}{|}}{\overset{..}{O}}}-H + {}^-:\overset{..}{\underset{..}{O}}Ts$$

Nucleo-   Alkyl tosylate   (Protonated   Leaving
phile   alcohol)   group

$$Ts = -SO_2-\langle\bigcirc\rangle-CH_3$$

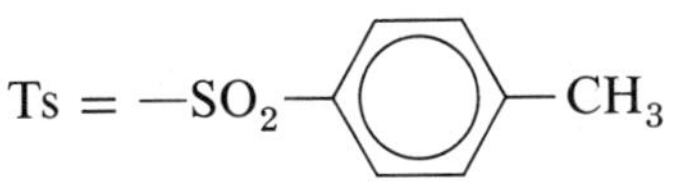

Nucleophilic substitution reactions, therefore, fall into one of the following categories,

$$Nu:^- + R-L \longrightarrow Nu-R + :L^-$$
$$Nu: + R-L \longrightarrow Nu^+-R + :L^-$$
$$Nu:^- + R-L^+ \longrightarrow Nu-R + :L$$
$$Nu: + R-L^+ \longrightarrow Nu^+-R + :L$$

where $Nu:$ or $Nu:^-$ represent nucleophiles and $L:$ and $L:^-$ represent leaving groups.

Since nucleophilic substitution reactions take place with a wide range of nucleophiles and leaving groups, they are of considerable synthetic importance. Some of the synthetic applications of nucleophilic substitution reactions are listed in Table 17.1.

When nucleophilic substitutions take place at primary, secondary, or methyl carbons, the yields of the reactions can be quite high. (We will see later that when substitution is attempted at a tertiary carbon the yields are very low.) Several specific examples of syntheses carried out using nucleophilic substitution reactions are the following.

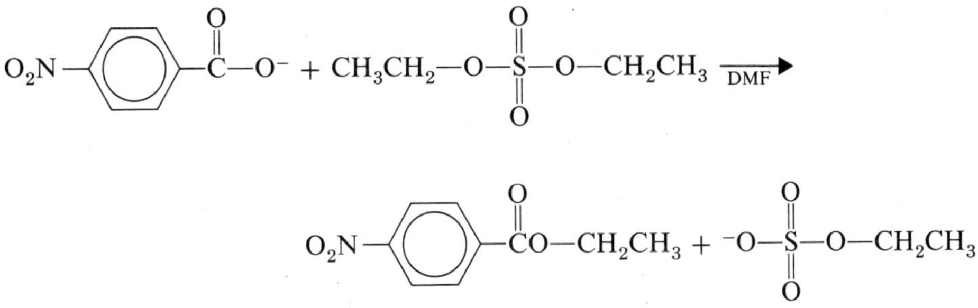

$$O_2N-\langle\bigcirc\rangle-\overset{\overset{O}{\|}}{C}-O^- + CH_3CH_2-O-\overset{\overset{O}{\|}}{\underset{\underset{O}{\|}}{S}}-O-CH_2CH_3 \xrightarrow{DMF}$$

$$O_2N-\langle\bigcirc\rangle-\overset{\overset{O}{\|}}{C}O-CH_2CH_3 + {}^-O-\overset{\overset{O}{\|}}{\underset{\underset{O}{\|}}{S}}-O-CH_2CH_3$$

(88%)

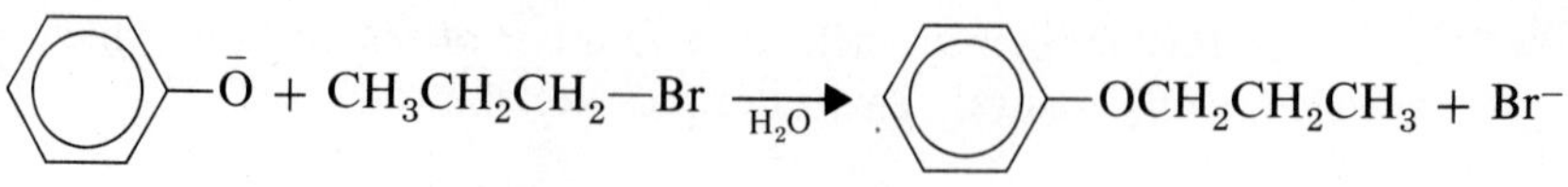

$$C_6H_5\text{—}\bar{O} + CH_3CH_2CH_2\text{—}Br \xrightarrow[\text{H}_2\text{O}]{} C_6H_5\text{—}OCH_2CH_2CH_3 + Br^-$$

(63%)

$$\underset{\underset{CH_3}{|}}{C_6H_{13}CH}\text{—}OTs + Br^- \xrightarrow[\underset{\overset{||}{O}}{CH_3SCH_3}]{} \underset{\underset{CH_3}{|}}{C_6H_{13}CH}\text{—}Br + \bar{O}Ts$$

(52%)

$$HC\equiv CCH_2CH_2\text{—}OTs + I^- \xrightarrow[\underset{\overset{||}{O}}{CH_3CCH_3}]{} HC\equiv CCH_2CH_2\text{—}I + \bar{O}Ts$$

(64%)

$$CH_3(CH_2)_{10}CH_2\text{—}Br + CN^- \xrightarrow[\text{aq. ethanol}]{} CH_3(CH_2)_{10}CH_2\text{—}CN + Br^-$$

(90%)

Table 17.1 Synthetically Important Nucleophilic Substitution Reactions

NUCLEOPHILE		SUBSTRATE ($-L = -\ddot{X}:$, $-OSO_2O\ddot{R}$, $-OTs$, etc.)	PRODUCT	SYNTHESIS OF
$H\ddot{O}:^-$ Hydroxide ion	+	R—L	$\xrightarrow[(-L^-)]{}$ R—$\ddot{O}H$ ⎤	
$H_2\ddot{O}:$	+	R—L	$\xrightarrow[(-HL)]{}$ R—$\ddot{O}H$ ⎦	Alcohols
$R\ddot{O}:^-$ Alkoxide ion	+	R—L	$\xrightarrow[(-L^-)]{}$ R—$\ddot{O}$—R ⎤	
R—$\ddot{O}H$ alcohol	+	R—L	$\xrightarrow[(-HL)]{}$ R—$\ddot{O}$—R ⎦	Ethers
$H\ddot{S}:^-$ Hydrosulfide ion	+	R—L	$\xrightarrow[(-L^-)]{}$ R—$\ddot{S}H$	Thiols
$:NH_3$ Ammonia	+	R—L	$\xrightarrow[(-HL)]{}$ R—$\ddot{N}H_2$	1° Amine
$R\ddot{N}H_2$ 1° Amine	+	R—L	$\xrightarrow[(-HL)]{}$ R—$\underset{\underset{H}{\mid}}{N}$—R	2° Amine
$R_2\ddot{N}H$ 2° Amine	+	R—L	$\xrightarrow[(-HL)]{}$ $R_3\ddot{N}$	3° Amine

Table 17.1 (continued) Synthetically Important Nucleophilic Substitution Reactions

NUCLEOPHILE		SUBSTRATE ($-L = -\ddot{X}:$, $-OSO_2O\ddot{R}$, $-OTs$, etc.)	PRODUCT	SYNTHESIS OF
$R_3N:$ 3° Amine	+	$R-L$	$\longrightarrow$ $R_4\overset{+}{N}\ L^-$	4° Ammonium salt
N_3^- Azide ion	+	$R-L$	$\xrightarrow[(-L^-)]{}$ $R-N_3$	Alkyl azide
I^- Iodide ion	+	$R-L$	$\xrightarrow[(-L^-)]{}$ $R-I$	Alkyl iodide
$\overset{\overset{O}{\|\|}}{R C} O^-$ Carboxylate ion	+	$R-L$	$\xrightarrow[(-L^-)]{}$ $\overset{\overset{O}{\|\|}}{R C} OR$	Ester
$\overset{\overset{O}{\|\|}}{R-C}-OH$ Carboxylic acid	+	$R-L$	$\xrightarrow[(-HL)]{}$ $\overset{\overset{O}{\|\|}}{R C}-OR$	
CN^- Cyanide ion	+	$R-L$	$\xrightarrow[(-L^-)]{}$ $R-CN$	Nitrile
$(C_2H_5O\overset{\overset{O}{\|\|}}{C})_2CH:^-$ Carbanion	+	$R-L$	$\xrightarrow[(-L^-)]{}$ $(C_2H_5O\overset{\overset{O}{\|\|}}{C})_2CH-R$	Substituted Malonic ester (Section 20.5)
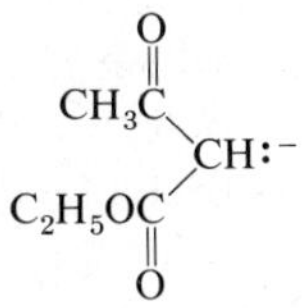 $\overset{CH_3\overset{O}{\overset{\|\|}{C}}}{\underset{C_2H_5O\underset{O}{\overset{\|\|}{C}}}{}}CH:^-$ Carbanion	+	$R-L$	$\xrightarrow[(-L^-)]{}$ $\overset{CH_3\overset{O}{\overset{\|\|}{C}}}{\underset{C_2H_5O\underset{O}{\overset{\|\|}{C}}}{}}CH-R$	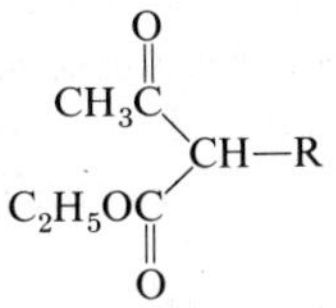 Substituted acetoacetic ester (Section 20.4)

Problem 17.1

What products would you expect from each of the following nucleophilic substitution reactions?

(a) $NaOH + CH_3-O\overset{\overset{O}{\|\|}}{\underset{\|\|}{\underset{O}{S}}}-CH_3 \longrightarrow$

(b) $(CH_3)_3N: + C_6H_5CH_2-Br \longrightarrow$

(c) $NaN_3 + CH_3CH_2-O\overset{\overset{O}{\|}}{\underset{\underset{O}{\|}}{S}}-\bigcirc-CH_3 \longrightarrow$

(d) $NaI + C_6H_5CH_2-Cl \longrightarrow$

(e) $CH_3COONa + CH_3\overset{|}{\underset{Br}{C}}HCH_3 \longrightarrow$

(f) $\overset{C_2H_5O\overset{O}{\overset{\|}{C}}}{\underset{C_2H_5O\underset{O}{\underset{\|}{C}}}{}}CH{:}^- \overset{+}{Na} + CH_3CH_2-Br \longrightarrow$

(g) $CH_3CH_2MgX + CH_2{=}CHCH_2-Br \longrightarrow$

17.2 MOLECULARITY OF NUCLEOPHILIC SUBSTITUTION REACTIONS: REACTION KINETICS

Of primary importance in formulating a mechanism for a chemical reaction is knowledge of a quantity called the *molecularity of the reaction*. The molecularity of a reaction is the number of molecules or ions that participate in the transition state of the rate-limiting step. If only a single species (molecule or ion) is involved in the transition state of the rate-limiting step, the reaction is said to be *unimolecular*. If two species are involved the reaction is said to be *bimolecular*, and if three are involved, the reaction is *termolecular*.

One of the best ways to determine the molecularity of a chemical reaction is through a study of reaction *kinetics*. That is, we measure the way in which *the rate of the reaction varies as we vary the concentration of each reactant*. Before we give specific examples of how this is done let us review briefly the factors that affect reaction rates.

We saw earlier (p. 131) that the collision theory of reaction rates tells us that the overall rate of a chemical reaction depends on the three factors: *the collision frequency, the energy factor, and the probability factor*.

Reaction rate	=	collision frequency	×	energy factor	×	probability factor
(the number of molecules in each unit volume that are converted from starting material to product in each second)		(the number of collisions between reactant molecules occurring in each unit volume per second)		(the fraction of collisions having energy greater than E_{act})		(the fraction of collisions having the proper orientation)

Let us assume that we are studying a reaction of the type:

$$Nu{:}^- + R-L \longrightarrow Nu-R + L{:}^-$$

Let us assume, further, that we carry out the same reaction several times *at the same temperature,* but with *different initial concentrations of Nu:⁻ and R—L.* In each case we will measure the rate at which the reaction takes place by measuring the rate *soon after each reaction begins.** We can do this in several ways. We can withdraw small samples from the reaction mixture at specific intervals of time and determine the concentrations of reactants or products. The rate of the reaction will be equal to the rate at which Nu:⁻ or R—L disappear from the reaction mixture or to the rate at which L:⁻ and Nu—R are formed in the reaction mixture. Just which quantity we determine is often a matter of experimental convenience. We can, for example, measure the rate of disappearance of halide ions by precipitating them with silver nitrate and weighing the precipitate. We can measure the rate of disappearance of hydroxide ions by titrating the samples with acid.

We can also follow the rate of many reactions spectroscopically. Spectroscopic measurements are particularly effective when the reactants and products absorb radiation in different regions; for then, we can simply follow the rate at which a particular spectroscopic peak appears in or disappears from the reaction mixture. Since the peak intensity is proportional to the concentration of the species absorbing radiation, we can easily obtain the rate of conversion of starting material to product.

Now let us suppose that by using one of these methods we measure the rates of several reactions of a specific Nu:⁻ and a specific R—L with different initial concentrations of Nu:⁻ and R—L. According to the collision theory, the different rates that we obtain must arise from differing collision frequencies. They cannot arise from different energy factors because we carried out all of the reactions at the same temperature. They cannot arise from different probability factors because the reacting species are the same in each reaction and the fraction of collisions having the proper orientation will be unchanged.

If we find that the reaction rate is directly proportional to the concentration of Nu:⁻ and to the concentration of R—L, it is reasonable to assume that Nu:⁻ and R—L collide with each other in order to bring about a reaction and *that both Nu:⁻ and R—L are involved in the transition state of the rate-limiting step.* We can assume, therefore, that the reaction is *bimolecular.*

We can describe this bimolecular reaction with equations. Since the rate is directly proportional to the concentration of Nu:⁻ and to the concentration of R—L, we can write the following proportionality:

Rate α [Nu:⁻] [R—L]

We can make this proportionality into an equation—called the *reaction rate equation*—by including a constant, k (called the *rate constant*).

Rate $= k$[Nu:⁻] [R—L]

A reaction rate equation like this is said to be *first order* with respect to [Nu:⁻] and [R—L]. That is, the reaction rate is proportional to [Nu:⁻] raised to the first power and to [R—L] raised to the first power. However, since the reaction rate is proportional to two concentrations, that is, [Nu:⁻] and [R—L], the reaction rate is

---

* We want to obtain *the initial reaction rate* because as the reaction progresses, the concentrations of the reactants will decrease.

said to be *second order* overall.* The kinds of numbers that we might get from a kinetic study of this reaction are shown in Table 17.2.

TABLE 17.2 Hypothetical Second-Order Reaction

$$Nu\overset{-}{:} + R-L \longrightarrow R-Nu + L^-$$

EXPT. NO.	INITIAL CONCENTRATION		INITIAL RATE OF FORMATION OF R—Nu OR L$^-$
	Nu$\overset{-}{:}$	R—L	
(1)	1 mole/liter	1 mole/liter	0.01 mole/liter sec^{-1}
(2)	2 mole/liter	1 mole/liter	0.02 mole/liter sec^{-1}
(3)	1 mole/liter	2 mole/liter	0.02 mole/liter sec^{-1}
(4)	2 mole/liter	2 mole/liter	0.04 mole/liter sec^{-1}

17.3 KINETICS OF NUCLEOPHILIC SUBSTITUTION REACTIONS AT A SATURATED CARBON

Let us now examine two actual nucleophilic substitution reactions: (1) the reaction of ethyl chloride with sodium hydroxide in aqueous ethanol and (2) the reaction of diphenylchloromethane with sodium fluoride in liquid sulfur dioxide.

$$(1)\ NaOH + CH_3CH_2Cl \xrightarrow[\substack{80\%\ C_2H_5OH \\ 55°}]{20\%\ H_2O} CH_3CH_2OH + NaCl$$

$$(2)\ NaF + C_6H_5\overset{\overset{\displaystyle H}{|}}{\underset{\underset{\displaystyle C_6H_5}{|}}{C}}-Cl \xrightarrow[-10°]{liq.\ SO_2} C_6H_5\overset{\overset{\displaystyle H}{|}}{\underset{\underset{\displaystyle C_6H_5}{|}}{C}}-F + NaCl$$

The S$_N$2 Reaction

The rate expression for the reaction of ethyl chloride with sodium hydroxide is *second order*. Experiments have shown that the reaction rate depends both on the concentration of ethyl chloride and on the concentration of hydroxide ion.

$$Rate = k[CH_3CH_2Cl][OH^-]$$

It has been found that if one *doubles* the concentration of ethyl chloride while keeping the hydroxide ion concentration constant, the rate of the reaction *doubles*. It has

---

* In general the overall order of a reaction is equal to the sum of the exponents a and b in the rate equation.

$$Rate = k[A]^a[B]^b$$

If in some other reaction, for example, we found that the

$$Rate = k[A]^2[B]$$

then we would say that the reaction rate is second order with respect to [A], first order with respect to [B], and third-order overall.

also been found that if one *doubles* the concentration of hydroxide ion while keeping the concentration of ethyl chloride constant, the rate *doubles*. And, finally, if one simultaneously *doubles both concentrations* (hydroxide ion and ethyl chloride) the rate *quadruples*.*

We can conclude, therefore, that the transition state for the rate limiting step of the ethyl chloride reaction involves both a hydroxide ion and an ethyl chloride molecule and that the reaction is *bimolecular*. We call this type of reaction an S_N2 reaction, meaning Substitution, Nucleophilic, bimolecular.

The S_N1 Reaction

When the kinetics of the reaction of diphenylchloromethane with sodium fluoride was studied it was found that the rate expression is *first order*.

$$\text{Rate} = k[(C_6H_5)_2CHCl]$$

That is, at a constant temperature the rate of the reaction is dependent *only* on the concentration of diphenylchloromethane and it is independent of the concentration of fluoride ion. In other words, when one doubles the concentration of diphenylchloromethane one finds that the rate of the reaction doubles. But, when one doubles or even triples the concentration of sodium fluoride the reaction rate does not change appreciably.

We conclude, therefore, that only diphenylchloromethane is involved in the transition state for the rate-limiting step and that this reaction is *unimolecular*. We call this type of reaction an S_N1 reaction (Substitution, Nucleophilic unimolecular).

17.4 STEREOCHEMISTRY AND MECHANISMS OF THE S_N2 REACTION

The bimolecular nucleophilic substitution reaction (the S_N2 reaction) occurs with inversion of configuration of the carbon that undergoes substitution. We cannot observe an inversion of configuration when the substrate for the reaction is an achiral molecule such as ethyl chloride, of course. But, we can detect an inversion of configuration when the S_N2 reaction takes place with a chiral substrate.

A compound that contains one chiral center and, therefore, exists as a pair of enantiomers is 2-bromooctane. These enantiomers have been obtained separately and are known to have the configurations and rotations shown below.

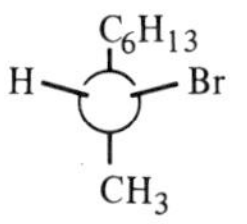

R-(−)-2-bromooctane
[α] = −34.6°

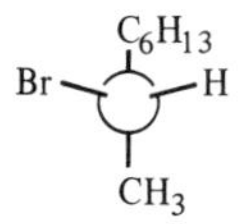

S-(+)-2-bromooctane
[α] = +34.6°

The alcohol, 2-octanol is also chiral. The configurations and rotations of the 2-octanol enantiomers have also been determined:

*In all of these experiments the temperature is the same.

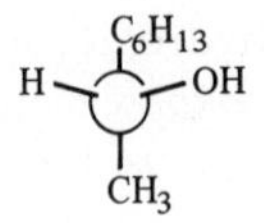

R-(−)-2-octanol
[α] = −9.9°

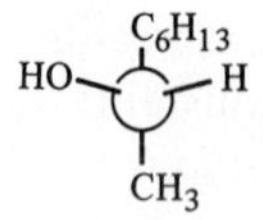

S-(+)-2-octanol
[α] = +9.9°

When R-(−)-2-bromooctane reacts with sodium hydroxide, the product that is obtained from the reaction is (S)-(+)-2-octanol. The reaction (below) is second order and it takes place with *complete inversion of configuration.*

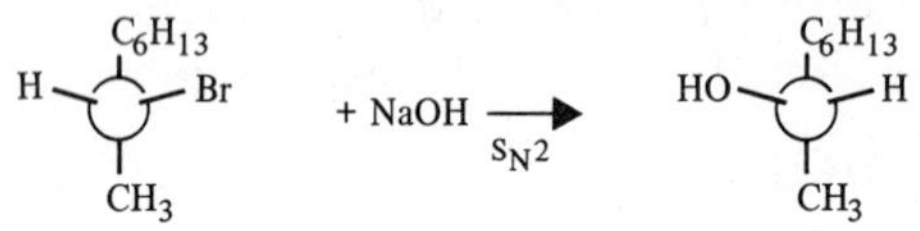

R-(−)-2-bromooctane
[α] = −34.6°
optical purity = 100%

S-(+)-2-octanol
[α] = +9.9°
optical purity = 100%

We also observe an inversion of configuration in the S_N2 reactions of cyclic molecules. When *cis*-3-methylcyclopentyl bromide, for example, reacts with cyanide ion in an S_N2 reaction, the product is *trans*-3-methylcyclopentyl cyanide.

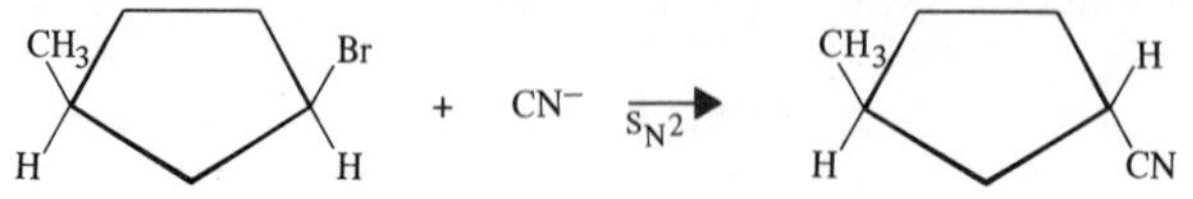

cis-3-methylcyclopentyl
bromide

trans-3-methylcyclopentyl
cyanide

Once again, we observe that the carbon atom that is the object of the substitution reaction has its configuration inverted.

How can we account for these results in terms of a reaction mechanism? One possible mechanism—one that is based on a mechanism proposed in the 1930s by Sir Christopher Ingold* of the University College, London—is outlined below.

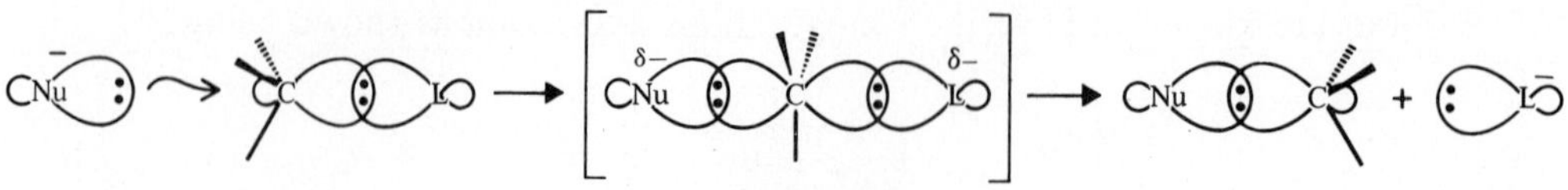

transition state

According to this mechanism the nucleophile approaches the carbon bearing the leaving group from the back side, that is, from the side directly opposite the leaving group. The orbital that contains the electron pair of the nucleophile begins to overlap with the small back lobe of the carbon sp^3-orbital. As the reaction progresses the lobe

---

* Ingold and his co-workers were the pioneers in this field. Their work provided the foundation on which our understanding of nucleophilic substitution and elimination is built.

between the nucleophile and carbon grows and the lobe between the carbon and the leaving group shrinks. As this happens the leaving group is pushed away and the configuration of the carbon atom inverts. (The configuration of the carbon atom inverts in much the same way as the configuration of an umbrella inverts when it is caught by a strong wind.)

The Ingold mechanism is a one-step displacement mechanism:

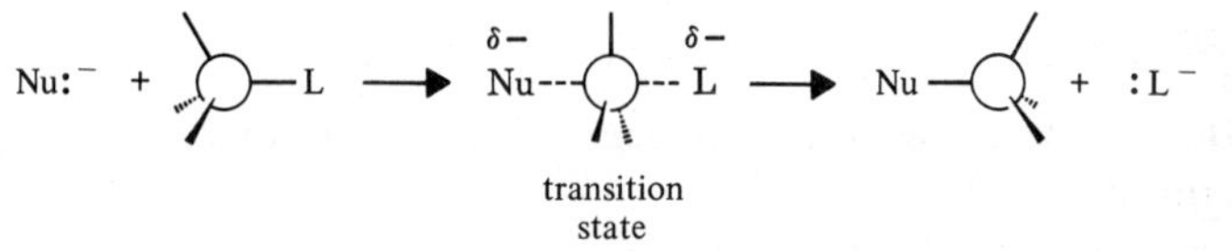

between the nucleophile and the leaving group are
partially bonded to the carbon undergoing inversion. Since this transition state involves both the nucleophile and the substrate, this mechanism accounts for the second-order reaction kinetics that we observe. Since the Ingold mechanism also requires inversion of configuration of the carbon at which displacement takes place, it also explains the stereochemistry of the S_N2 reaction.

For many years the Ingold mechanism was widely accepted as the only mechanism for bimolecular nucleophilic substitution reactions. In recent years, however, evidence has been advanced to support another possible mechanism: one involving *ion pairs*. The ion pair mechanism seems to be particularly important in those nucleophilic substitution reactions called *solvolysis reactions;* however, it may operate in other S_N2 reactions. *A solvolysis reaction is one in which the nucleophile is a molecule of the solvent* (solvent + *lysis:* cleavage by solvent).

> Solvolysis reactions are often described as being pseudo-first order since, by convention, the concentration of the solvent is not included in the rate expression. Even if it were included, we would not be able to detect a variation in rate with a change in concentration of the solvent since the solvent concentration is always very large and is, therefore, essentially constant. As we will see, however, some solvolysis reactions are actually bimolecular nucleophilic substitution reactions and occur with complete inversion of configuration.
>
> The first clear demonstration of ion pair involvement in solvolysis reactions was given by Professor S. Winstein (of the University of California, Los Angeles).

Two examples of solvolysis reactions for which ion pair mechanisms have been proposed are the hydrolysis and acetolysis of 2-octyl sulfonates.

Hydrolysis of a 2-Octyl Sulfonate

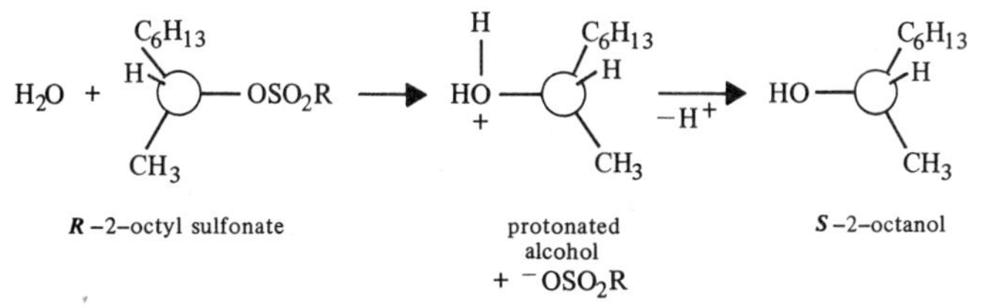

Acetolysis of a 2-Octyl Sulfonate

R-2-octyl sulfonate   protonated acetate + $^-OSO_2R$ S-2-octyl acetate

Both of these reactions have been shown to take place with complete inversion of configuration.

On the basis of kinetic evidence (beyond the scope of our treatment here) it has been proposed that these reactions take place with the rapid but reversible formation of an *"intimate"* (or tight) ion pair followed by a slow reaction with a solvent molecule. An intimate ion pair is an ionic intermediate in which the cation and anion are in close proximity and are not separated by solvent molecules.

Sol = H— or CH_3C—   intimate ion pair

(retention of configuration)   (inversion of configuration)

The intimate ion pair retains the configuration of the original sulfonate. However, when it reacts with the solvent, the displacement takes place from the backside and produces an inversion of configuration. In all likelihood, inversion occurs because

FIG. 17.1

Potential energy diagrams for the one-step mechanism for an S_N2 reaction
(a), *and for the intimate ion-pair mechanism* (b).

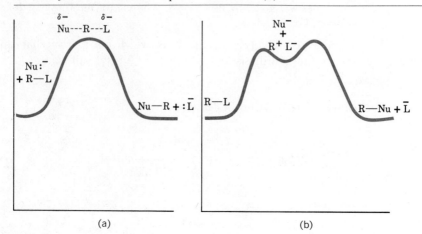

(a)   (b)

the intimate ion pair still has partial bonding between the carbon and the leaving group, and attack of the nucleophile must occur from the back side in the same manner as for an S_N2 reaction of a neutral compound.

The difference between the Ingold mechanism and the ion pair mechanism is most apparent in their potential energy diagrams (Fig. 17.1). In the one-step displacement mechanism there is a single transition state; in the ion pair mechanism there are two transition states and the ion pair occupies an energy minimum lying between them.

At the time of this writing there is not enough evidence available to distinguish between these two mechanistic possibilities for most S_N2 reactions. However, with either mechanism the stereochemistry of S_N2 reactions is clear: S_N2 reactions, whether they take place in one step or through the formation of ion pairs, *occur with inversion of configuration.*

17.5 STEREOCHEMISTRY AND MECHANISMS OF THE S_N1 REACTION

When 1-phenylethyl chloride reacts with water in aqueous acetone, that is,

$$C_6H_5\underset{\underset{Cl}{|}}{C}HCH_3 + H_2O \xrightarrow[(-HCl)]{acetone} C_6H_5\underset{\underset{OH}{|}}{C}HCH_3$$

the reaction rate is first order; it depends only on the concentration of 1-phenyl-ethylchloride and is independent of the concentration of water.

$$Rate = k\ [C_6H_5\underset{\underset{Cl}{|}}{C}HCH_3]$$

The stereochemistry of the reaction is shown below.

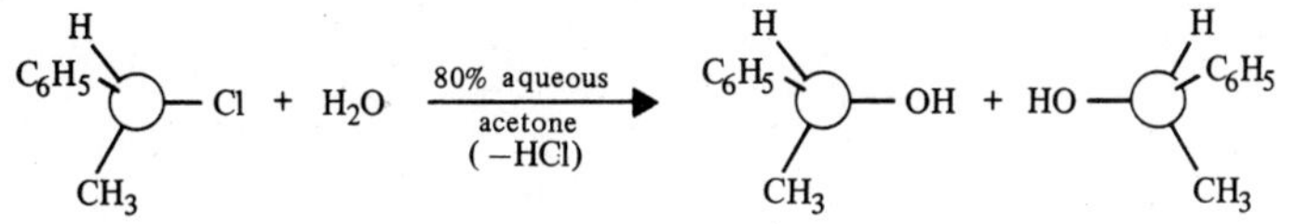

S–1–phenylethyl chloride (optically pure)

S –1–phenylethanol   R–1–phenylethanol

98% racemization
2% net inversion

We see from the equation (above) that the reaction of 1-phenylethyl chloride of the **S** configuration gives a product, 1-phenylethanol, of which 51% has the opposite (**R**) configuration and 49% has the same (**S**) configuration. In other words, 51% of the 1-phenylethyl chloride molecules have had their configuration inverted by the reaction, while the remainder (49%) have retained their original configuration. We describe this situation by saying that the reaction has taken place with 98% *racemization* and 2% *net* inversion.

We can account for the fact that this reaction is first order if we assume that the rate-limiting step (or slow step) for the reaction involves the organic halide alone. A general mechanism is the following.

Step 1 $R-X \xrightarrow{\text{slow}} R^+ + X^-$

Step 2 $R^+ + H_2O \xrightarrow{\text{fast}} R\overset{+}{O}H_2$

Step 3 $R\overset{+}{O}H_2 \xrightarrow[-H^+]{\text{fast}} ROH$

Step 1, the formation of a carbocation, is the slow step. Step 2 is a rapid reaction of the carbocation with water and step 3 is the rapid loss of a proton.

Since step 1 involves the organic halide alone (we are, for the moment, neglecting the involvement of solvent molecules) the overall rate of the reaction must correspond to the rate of this step,

Rate = $k[RX]$

and the reaction as a whole must show first-order kinetics.

Problem 17.2

Write a mechanism that will account for the fact that the reaction of diphenylchloro-methane and sodium fluoride (p. 636) gives a first-order rate equation.

A more detailed mechanism (shown below) illustrates one way in which we can account for the overall stereochemistry of the 1-phenylethyl chloride reaction.

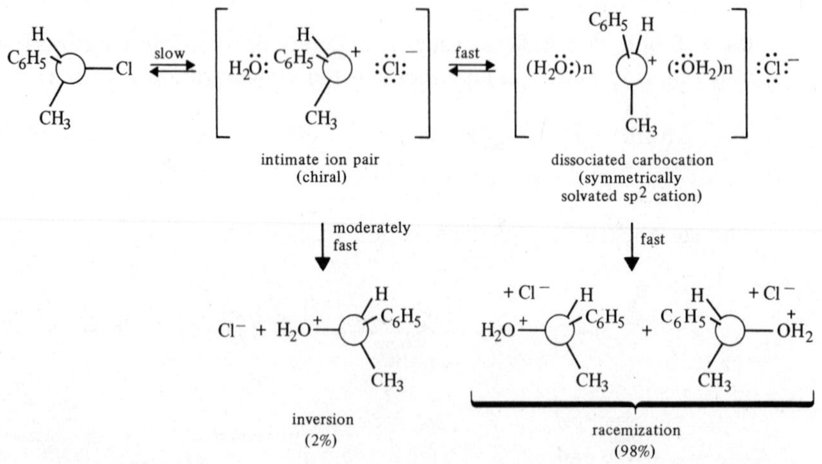

Here we see the important part played by solvent molecules and we also see the formation of two different cationic intermediates: an intimate ion pair and a more dissociated carbocation. The intimate ion pairs are formed first and they are solvated only on the back side. A relatively small number of the intimate ion pairs collapse to give an inverted product. However, since the developing cation in this reaction is a stable *benzyl* carbocation, most of the intimate ion pairs survive long enough to become more dissociated. The positive carbons of the dissociated carbocations are sp^2 hybridized and they are solvated on both the front and back sides. They react equally rapidly with water molecules at either face to give a racemic modification of the protonated alcohol.

There is also evidence, however, suggesting that a third type of cationic intermediate called a "solvent-separated" ion pair may intervene between the intimate ion pair and the dissociated carbocation in reactions of this type. A solvent-separated ion pair, as its name suggests, is one in which a molecule of solvent is situated between the carbocation and the anion. There is also evidence that solvent-

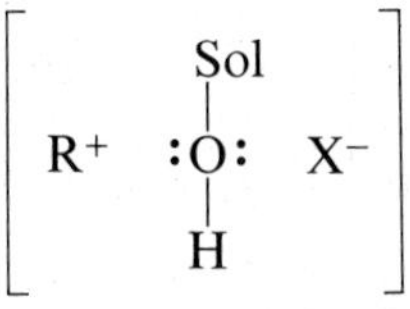

$$
\left[
\begin{array}{ccc}
& \overset{\displaystyle \text{Sol}}{\underset{\displaystyle \text{H}}{\overset{|}{\underset{|}{\text{:O:}}}}} & \\
R^+ & & X^-
\end{array}
\right]
$$

A solvent-separated
ion pair

separated ion pairs sometimes preferentially react with the intervening solvent molecule by a mechanism that takes place with *retention of configuration:*

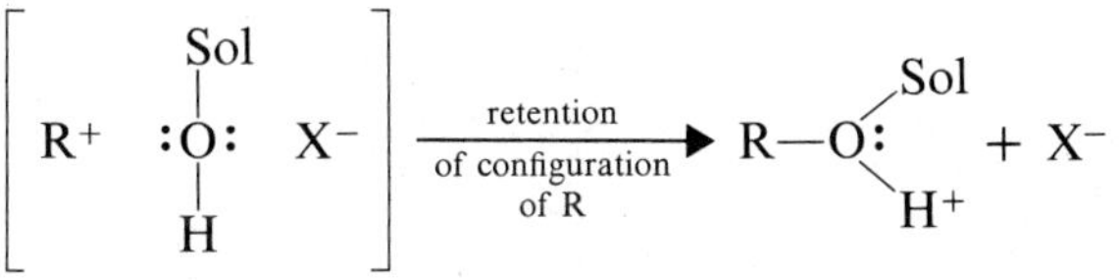

$$
\left[
\begin{array}{ccc}
& \overset{\displaystyle \text{Sol}}{\underset{\displaystyle \text{H}}{\overset{|}{\underset{|}{\text{:O:}}}}} & \\
R^+ & & X^-
\end{array}
\right]
\xrightarrow[\substack{\text{of configuration} \\ \text{of R}}]{\text{retention}}
R{-}O\underset{\displaystyle H^+}{\overset{\displaystyle Sol}{:}} + X^-
$$

Thus, it is possible that not all of the racemic product comes from dissociated carbocations and that more than 2% of the reaction may take place through intimate ion pairs. Part of the racemic product may result from a balanced collapse of intimate ion pairs (with inversion) and of solvent-separated ion pairs (with retention).

17.6 SUMMARY OF S$_N$ REACTION MECHANISMS AT A SATURATED CARBON

Nucleophilic substitution reactions may very well take place through a spectrum of mechanisms ranging from a one-step displacement mechanism at one end and a mechanism involving fully dissociated carbocations at the other. Intervening between these limits may be mechanisms involving at least two kinds of ion pairs, intimate ion pairs and solvent-separated ion pairs. We can represent this spectrum in the following way.

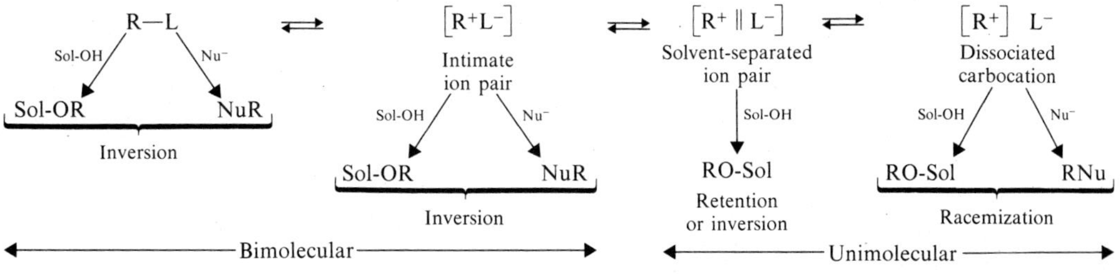

Whether a reaction gives a first-order or second-order rate equation will depend on just which of these mechanisms operates. We will obtain a second-order rate equation for those reactions that take place by a one-step displacement mechanism or through an intimate ion pair since, in these instances, the rate-limiting step is bimolecular. We will describe these reactions in our subsequent discussions as being bimolecular or S_N2 reactions.*

We will obtain a first-order rate equation for those reactions that involve dissociated carbocations or solvent separated ion pairs, for in these reactions the rate-limiting step is unimolecular. In future discussions we will describe these reactions as being unimolecular or S_N1.

> The only exception to these generalizations is a solvolysis reaction. A solvolysis reaction is *bimolecular* because the transition state of its rate-limiting step involves two species: the substrate and the solvent. A solvolysis reaction, however, will show *pseudo first-order kinetics* because the solvent concentration is very large and is, consequently, essentially constant.

The stereochemical possibilities for nucleophilic substitution reactions are summarized in Table 17.3.

17.7 FACTORS AFFECTING THE RATES OF S_N1 AND S_N2 REACTIONS AT A SATURATED CARBON

Experiments have shown that a number of factors affect the relative rates of unimolecular and bimolecular nucleophilic substitution reactions. The most important factors are:

1. The structure of the carbon bearing the leaving group.
2. The concentration and reactivity of the nucleophile (for bimolecular reactions only).
3. The nature of the leaving group.
4. The polarity and nature of the solvent.

---

* The mechanism involving an intimate ion pair is truly a bimolecular reaction since the transition state of the rate-limiting step,

$$R^+L^- + Nu\!:^- \longrightarrow NuR + L^-$$

or

$$R^+L^- + Sol\text{-}OH \longrightarrow Sol\!-\!\overset{+}{\underset{H}{O}}\!-\!R + L^-$$

involves two species: the intimate ion pair and the nucleophile or solvent.

TABLE 17.3 The Stereochemistry of Nucleophilic Substitution Reactions

MOLECULARITY	SUBSTRATE	REPRESENTATION	STEREOCHEMISTRY
S$_N$2	R—L Alkyl halide, tosylate, etc.	R—L	Nucleophilic attack by the solvent or the nucleophile from the back side gives an inverted product by a one-step displacement mechanism.
S$_N$2	R—L Intimate ion pair	[R$^+$L$^-$]	Nucleophilic attack by the solvent or the nucleophile from the back side gives an inverted product.
S$_N$1	R$^+$ O—H (S) L$^-$ Solvent-separated ion pair	[R$^+$‖L$^-$]	Nucleophilic attack by the solvent from front side occurs with retention of configuration. (Attack by another nucleophile may occur with inversion.)
S$_N$1	(O—H/S)$_n$ R$^+$ (O—H/S)$_n$ L$^-$ Dissociated carbocation	[R$^+$] L$^-$	Nucleophilic attack by the solvent or the nucleophile from the front or back- side gives a racemic product.

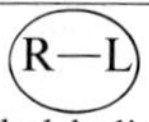

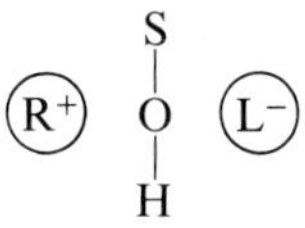

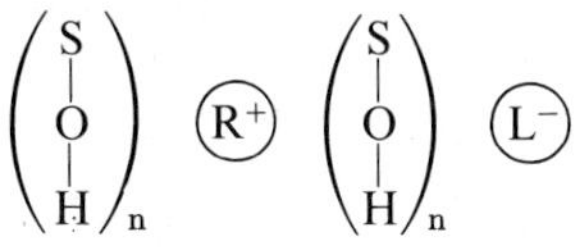

The relative rates at which S$_N$1 and S$_N$2 reactions take place will ultimately determine just which reaction pathway a given nucleophile and substrate will follow. Since the stereochemical outcome of the two pathways is different, it is of considerable importance in planning experiments to take these factors into account.

The Effect of the Nature of the Carbon Bearing the Leaving Group

S$_N$2 Reactions. Simple alkyl halides show the following general order of reactivity in S$_N$2 reactions:

Methyl > primary > secondary > (tertiary)

TABLE 17.4 Relative Rates of Reactions of Alkyl Halides in S_N2 Reactions

SUBSTITUENT	COMPOUND	RELATIVE RATE
Benzyl	$C_6H_5CH_2X$	120
Allyl	$CH_2=CHCH_2X$	40
Methyl	CH_3-X	30
1°	CH_3CH_2-X	1
2°	$(CH_3)_2CHX$	0.02
Neopentyl	$(CH_3)_3CCH_2X$	0.00001
3°	$(CH_3)_3CX$	~0

Simple allylic and benzylic halides are quite reactive in S_N2 reactions and are generally more reactive than primary halides and even methyl halides. Neopentyl halides, even though they are primary halides are very unreactive. Table 17.4 gives the relative rates for examples of each type.

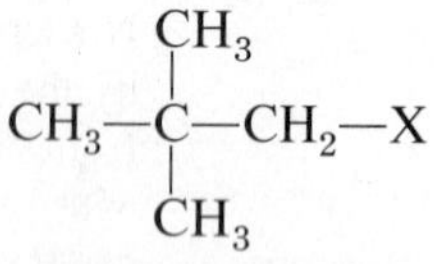

$$CH_3-\underset{\underset{CH_3}{|}}{\overset{\overset{CH_3}{|}}{C}}-CH_2-X$$

A neopentyl halide

The general order of reactivity for S_N2 reactions arises primarily from steric effects.* An S_N2 reaction requires an approach by the nucleophile to a distance within bonding range of the carbon undergoing attack. Because of this, bulky substituents on or near the carbon have a dramatic inhibiting effect (Fig. 17.2). Of the simple alkyl halides, methyl halides react most rapidly in S_N2 reactions because only three small hydrogens interfere with the approaching nucleophile (Fig. 17.2). Neopentyl and tertiary halides are the least reactive because bulky groups present a strong hindrance to the approaching nucleophile. (Tertiary substrates, for all practical purposes, do not react by an S_N2 mechanism.)

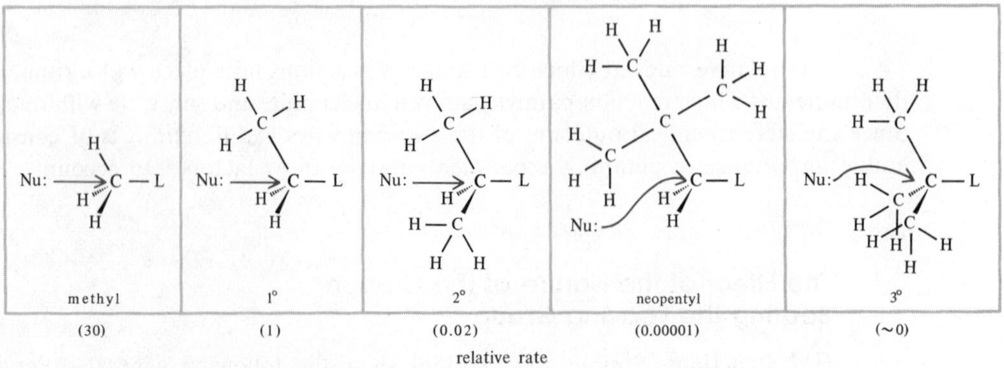

relative rate

FIG. 17.2
Steric effects in the S_N2 reaction.

*The high reactivity of benzyl and allyl halides may also result from an electronic factor.

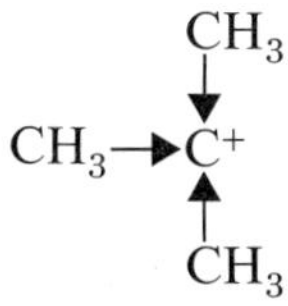

3° Carbocation
Stabilized by electron-release of three alkyl groups

$CH_2=CH-\overset{+}{C}H_2 \longleftrightarrow \overset{+}{C}H_2-CH=CH_2$

Allyl carbocation
Stabilized by resonance

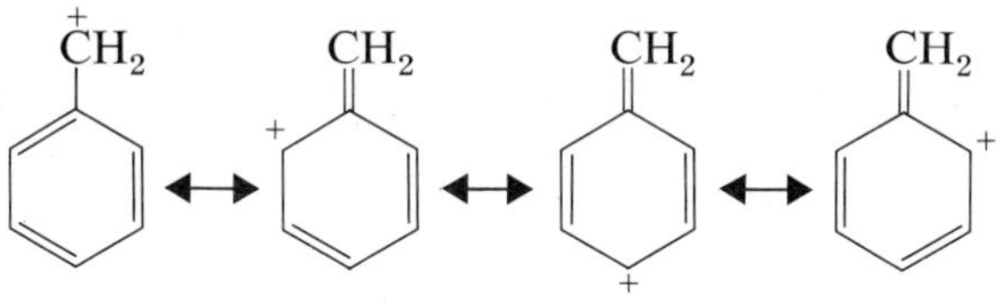

Benzyl carbocation
Stabilized by resonance

S_N1 Reactions. *The primary factor that determines the reactivity of organic substrates in an S_N1 reaction is an electronic factor.* Except for those reactions that take place in strongly acidic media, the only organic compounds that undergo reaction by an S_N1 path are those that are capable of forming relatively stable carbocations. An S_N1 reaction, therefore, seems to be limited to those compounds in which the leaving group is attached to a *tertiary, allylic,* or *benzylic carbon* or to some other stabilizing group.

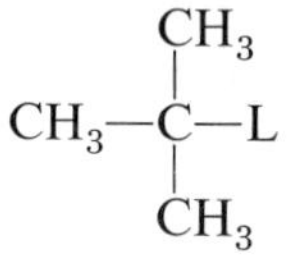

$CH_2=CHCH_2L$ $C_6H_5CH_2L$

Compounds of this type are capable of reacting by an S_N1 path in ordinary solvents. They react by forming relatively stable carbocations

We see then that simple allyl and benzyl substrates can react by either an S_N2 mechanism or an S_N1 mechanism, whereas tertiary substrates react only by an S_N1 mechanism. This is perfectly consistent with our theory of reactivity. Simple allyl and benzyl substrates are sterically similar to primary substrates. They have only one group attached to the carbon that undergoes nucleophilic attack, and this group does not severely hinder the approach of a nucleophile. Therefore, simple allyl and benzyl substrates are capable of reacting through an S_N2 mechanism. They can also react by an S_N1 mechanism because they are capable of forming relatively stable carbocations. Tertiary substrates react readily by an S_N1 pathway and for the same reason. Tertiary halides, however, do not react by an S_N2 mechanism because the three groups attached to the carbon bearing the leaving group prevent the nucleophile from coming within bonding distance.

If we consider only those reactions occurring in nonacidic media, we can summarize the effect of the carbon bearing the leaving group in the following way.

These substrates give only S_N2 reactions	These substrates give only S_N1 reactions
$CH_3{-}L$ $R{-}CH_2{-}L$ $R{-}\underset{\underset{R}{\vert}}{CH}{-}L$	$R{-}\underset{\underset{R}{\vert}}{\overset{\overset{R}{\vert}}{C}}{-}L$

These substrates may give either S_N1 or S_N2 reactions	These substrates give only S_N1 reactions
$Ar{-}CH_2{-}L$ $Ar{-}\underset{\underset{R}{\vert}}{CH}{-}L$	$Ar{-}\underset{\underset{R}{\vert}}{\overset{\overset{R}{\vert}}{C}}{-}L$
$CH_2{=}CHCH_2{-}L$ $CH_2{=}CH\underset{\underset{R}{\vert}}{CH}{-}L$	$CH_2{=}CH{-}\underset{\underset{R}{\vert}}{\overset{\overset{R}{\vert}}{C}}{-}L$

Problem 17.3

Primary halides of the type, $ROCH_2X$ apparently undergo S_N1 type reactions, while most primary halides do not. Can you propose a resonance explanation that will account for this?

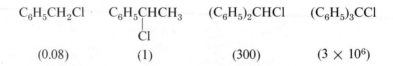

Problem 17.4

Each of the following chlorides undergo solvolysis in ethanol at the relative rates given in parentheses. How can you explain these results?

$$C_6H_5CH_2Cl \quad C_6H_5\underset{\underset{Cl}{\vert}}{CH}CH_3 \quad (C_6H_5)_2CHCl \quad (C_6H_5)_3CCl$$

$$(0.08) \qquad\qquad (1) \qquad\qquad (300) \qquad\qquad (3 \times 10^6)$$

Problem 17.5

Bridged cyclic compounds like those shown below are extremely *unreactive* in S_N2 reactions.

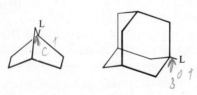

(a) Give a reason that will explain this. (b) How can you explain the fact that compounds of this type are also more unreactive in S_N1 reactions than similar noncyclic compounds? (Consider the fact that carbonium ions are generally sp^2 hybridized.) *carbo cannot be* sp^2 *hyb*

The Effect of the Concentration and Strength of the Nucleophile

The rates of S_N1 reactions are unaffected by either the concentration or the identity of the nucleophile. The rates of S_N2 reactions, however, depend *both* on the concentration *and* the identity of the attacking nucleophile. We saw in Section 17.3 how increasing the concentration of the nucleophile increases the rate of an S_N2 reaction. We can now examine how the rate of an S_N2 reaction depends on the identity of the nucleophile.

We describe nucleophiles as being *strong* or *weak*. When we do this we are really describing their relative reactivities in S_N2 reactions. A strong nucleophile is one that reacts rapidly with a given substrate. A weak nucleophile is one that reacts slowly with the same substrate under the same reaction conditions.

The relative strengths of nucleophiles can be correlated with several structural features:

1. We find that a negatively charged nucleophile is always a stronger nucleophile than its conjugate acid. Thus OH^- is a stronger nucleophile than H_2O, RO^- is stronger than ROH, RS^- is stronger than RSH.

2. When we examine a group of nucleophiles in which the nucleophilic atom is the same, relative nucleophilicities are found to parallel basicities. Oxygen compounds, for example, show the following order of reactivity:

$$RO^- > HO^- > ArO^- > RCOO^- > ROH > H_2O$$

This is also their order of basicity. (An alkoxide ion is a stronger base than a hydroxide ion, a hydroxide ion is a stronger base than a phenoxide ion, and so on.)

3. The relative strength of nucleophiles does not always parallel basicity, however. When we examine the relative nucleophilicity of compounds within the same group of the periodic table we find that in solvents such as alcohols and water the nucleophile with the larger nucleophilic atom is stronger. Thiols (R—SH) are stronger nucleophiles than alcohols (ROH); RS^- ions are stronger than RO^- ions; and the halide ions show the following order.

$$I^- > Br^- > Cl^- > F^-$$

This effect has been ascribed to the relative polarizability of the nucleophilic atoms. The outermost electrons of larger atoms are more loosely held by the nucleus and thus they are more easily distorted as they attack a positively charged atom. This cannot be an entire explanation, however, especially for the halide ions as we shall see in (4) below.

4. In a solvent, such as *dimethylformamide,*

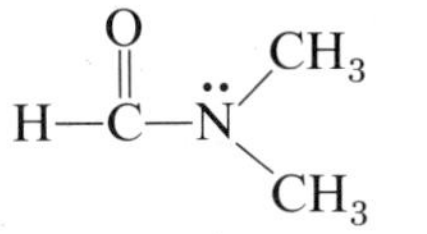

N,N-Dimethylformamide
(DMF)

the relative order of reactivity of chloride, bromide and iodide ions is

$$Cl^- > Br^- > I^-$$

This is the opposite of their strength as nucleophiles in alcohol or water solutions.

This effect seems to be related to the ability of the particular solvent to *solvate anions*. Alcohols and water are called *protic* solvents because they have a hydrogen that is attached to a strongly electronegative atom (oxygen). Alcohols and water are very effective in solvating anions because these hydrogens can form strong hydrogen bonds to the unshared electron pairs of anions. Smaller anions seem to be more effectively solvated (and thus, stabilized) by protic solvents than larger ones. This may be because with smaller anions the negative charge is more concentrated. Thus, in alcohol or water the smaller Cl⁻ ion has a more effective barrier of stabilizing solvent molecules surrounding it than the larger I⁻ ion. Therefore, in protic solvents the Cl⁻ ion is less reactive and is the weaker nucleophile.

Aprotic solvents are those solvents that do not have a hydrogen that is attached to a strongly electronegative element. Most aprotic solvents (benzene, the alkanes, and so on) are relatively nonpolar; they do not dissolve ionic compounds, and are not used as solvents for most S_N2 reactions. In recent years, however, a number of *highly polar aprotic solvents* have been discovered. Dimethylformamide is one example of a highly polar aprotic solvent; dimethylsulfoxide and dimethylacetamide are two others.

$$CH_3-\overset{\overset{\displaystyle O}{\|}}{S}-CH_3 \qquad CH_3\overset{\overset{\displaystyle O}{\|}}{C}N\overset{\diagup CH_3}{\diagdown CH_3}$$

Dimethylsulfoxide (DMSO) Dimethylacetamide (DMA)

All three solvents (DMF, DMSO, and DMA) are capable of dissolving ionic compounds and they solvate cations very well. However, they do not solvate anions to any appreciable extent. In these solvents anions are unencumbered by a layer of solvent molecules and, therefore, they are poorly stabilized by solvation. These "naked" anions, are highly reactive both *as bases and as nucleophiles*. The unsolvated Cl⁻ ion is more basic than the unsolvated I⁻ ion; thus, in the aprotic solvent, dimethylformamide, chloride ions are stronger nucleophiles.

S_N2 reactions, in general, are strongly favored by the use of polar aprotic solvents.

Problem 17.6

Which would you expect to be the stronger nucleophile: (a) The amide ion (NH_2^-) or ammonia (NH_3)? (b) The phenoxide ion or the *p*-nitrophenoxide ion?

Problem 17.7

How can you account for the fact that the S_N2 reaction of isopropyl iodide and quinuclidine takes place at a rate 700 times faster than the corresponding reaction of isopropyl iodide and triethylamine even though both nucleophiles are tertiary amines. (Hint: consider steric factors in the attacking nucleophiles.)

$$\text{Quinuclidine} \quad N: + \underset{CH_3}{\overset{CH_3}{\diagdown}}CHI \xrightarrow{\text{faster}} \overset{+}{N}-CH(CH_3)_2I^-$$

Quinuclidine

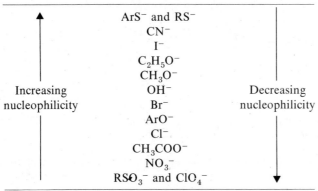

Triethylamine

The relative strengths of a number of common nucleophiles in protic solvents are given in Table 17.5.

TABLE 17.5 Relative Nucleophilic Strengths of Nucleophiles (in protic solvents)

Increasing nucleophilicity	ArS⁻ and RS⁻ CN⁻ I⁻ $C_2H_5O^-$ CH_3O^- OH⁻ Br⁻ ArO⁻ Cl⁻ CH_3COO^- NO_3^- RSO_3^- and ClO_4^-	Decreasing nucleophilicity

The Nature of the Leaving Group

The best leaving groups are those that give the most stable molecule or ion after they depart. This means that in general, the less basic the group is, the more easily it departs from carbon. Water, for example, is a better leaving group than hydroxide ion. As a consequence, alcohols do not react with halide ions in neutral media, for under these conditions the leaving group would be a hydroxide ion. Alcohols react readily with halide ions in acidic media, however. In acidic media the leaving group is water.

$$R—OH + X^- \not\longrightarrow R—X + OH^-$$

(this reaction does not take place because the leaving group is a strongly basic hydroxide ion)

$$R—\overset{+}{\underset{H}{\ddot{O}H}} + X^- \longrightarrow R—X + H_2\ddot{O}:$$

(this reaction takes place because the leaving group is a weak base).

Problem 17.8

Account for the fact that ethers are cleaved readily when they are heated with strong acids but are not cleaved in weak acids or in neutral media.

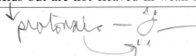

In addition to water and alcohols, other good leaving groups are halide ions ($I^- > Br^- > Cl^- > F^-$), sulfonate ions, sulfate ions, and carboxylate ions.

The Polarity of the Solvent

Polar solvents generally solvate and stabilize ions better than nonpolar solvents. (This is true of aprotic solvents, even though they only solvate cations.) Since nearly all nucleophilic substitution reactions involve ions as either reactants or products, the polarity of the medium in which the reaction takes place can strongly affect the rate of the reaction. Not all nucleophilic substitution reactions are affected in the same way, however. The rates of S_N1 reactions of the type,

$$S_N1 \quad R{-}L \longrightarrow R^+ + L^-$$

are greatly increased by the use of protic solvents of high polarity. Highly polar solvents not only solvate the ions produced by this reaction, they also stabilize the transition state in which the ions are developing.

$$\overset{\delta+}{R}{-}{-}{-}{-}{-}{-}{-}{-}{-}\overset{\delta-}{L}$$

Charge generation
in the transition state

They stabilize this transition state because it involves separated positive and negative charges.

The rates of S_N1 reactions of the type,

$$S_N1 \quad R{-}L^+ \longrightarrow R^+ + L$$

are slightly *decreased* by the use of solvents of high polarity. The reason for this effect is a *dispersal of charge* in the transition state.

$$\overset{\delta+}{R}{-}{-}{-}{-}{-}{-}{-}{-}{-}\overset{\delta+}{L}$$

Dispersal of charge
in the transition state

Highly polar solvents are better able to solvate and, thus, stabilize the reactant $R{-}L^+$ (in which the positive charge is concentrated) than they are able to solvate and stabilize the transition state. This causes a higher energy of activation and thus a slower reaction rate.

S_N2 reactions of the types illustrated below show a decrease in reaction rate in highly polar protic solvents.

All of these S_N2 reactions take place more slowly in highly polar protic solvents

$$Nu{:}^- + R{-}L \longrightarrow \overset{\delta-}{Nu}{-}{-}{-}\overset{\delta-}{R}{-}{-}{-}L \longrightarrow Nu{-}R + L^-$$
charge is dispersed

$$Nu{:} + R{-}L^+ \longrightarrow \overset{\delta+}{Nu}{-}{-}{-}\overset{\delta+}{R}{-}{-}{-}L \longrightarrow Nu^{\pm}{-}R + L$$
charge is dispersed

$$Nu{:}^- + R{-}L^+ \longrightarrow \overset{\delta-}{Nu}{-}{-}{-}\overset{\delta+}{R}{-}{-}{-}L \longrightarrow Nu{-}R + L$$
charge is decreased

In each of these reactions the charges of the transition state relative to those of the reactants are either dispersed or decreased.

The only S$_N$2 reactions that show rate increases in polar protic solvents are those in which the reactants do not bear charges at all:

$$\text{Nu:} + \text{R—L} \longrightarrow \overset{\delta+}{\text{Nu---R}}\overset{\delta-}{\text{---L}} \longrightarrow \text{Nu}^{\pm}\text{R} + \text{L}^-$$

charge separation

In reactions of this type, highly polar solvents stabilize the transition state because it involves charge separation (i.e., developing ions). Highly polar solvents, however, do not affect the relative energies of the reactants appreciably. The net result is a lower energy of activation and a faster reaction when the reaction is carried out in a highly polar solvent.

A good indication of a solvent's polarity is a quantity called the *dielectric constant*. The greater the dielectric constant, the greater is the solvents polarity and the greater is its ability to solvate ions and developing ions. The dielectric constants of some common protic solvents are shown in Table 17.6.

Table 17.7 gives the dielectric constants of several common aprotic solvents. Polar aprotic solvents accelerate the rates of all S$_N$2 reactions because the nucleophiles are unsolvated and, therefore, are highly reactive.

Problem 17.9

Each of the following reactions can be carried out in mixtures of water and ethanol. State whether you would expect the reaction rate to *increase* or *decrease* as the percentage of water in the reaction mixture is increased.

(a) $HO^- + CH_3Br \longrightarrow CH_3OH + Br^-$

(b) $NH_3 + CH_3CH_2I \longrightarrow CH_3CH_2\overset{+}{N}H_3 + I^-$

(c) $Br^- + CH_3\overset{+}{O}(CH_3)_2 \longrightarrow CH_3Br + CH_3OCH_3$

(d) $CH_3\underset{\underset{\displaystyle CH_3}{|}}{\overset{\overset{\displaystyle CH_3}{|}}{C}}Br + OH^- \longrightarrow CH_3\underset{\underset{\displaystyle CH_3}{|}}{\overset{\overset{\displaystyle CH_3}{|}}{C}}OH + Br^-$

(e) $OH^- + CH_2{=}CHCH_2Br \xrightarrow{\text{S}_\text{N}2} CH_2{=}CHCH_2OH + Br^-$

(f) $H_2O + CH_2{=}CHCH_2Br \xrightarrow{\text{S}_\text{N}1} CH_2{=}CHCH_2\overset{+}{O}H_2 + Br^-$

Problem 17.10

Would you expect the nucleophilic substitution reaction between ethyl bromide and sodium cyanide to take place faster in dimethylsulfoxide or in ethanol? Explain your answer.

TABLE 17.6 Dielectric Constants of Common Protic Solvents

	SOLVENT	FORMULA	DIELECTRIC CONSTANT
	Water	H_2O	80
	Formic acid	$\overset{\overset{O}{\parallel}}{H}COH$	59
Increasing solvent polarity	Methanol	CH_3OH	33
	Ethanol	CH_3CH_2OH	24
	Acetic acid	$CH_3\overset{\overset{O}{\parallel}}{C}OH$	6

TABLE 17.7 Dielectric Constants of Common Aprotic Solvents

	SOLVENT	FORMULA	DIELECTRIC CONSTANT
	Dimethylsulfoxide	$(CH_3)_2S{=}O$	45
	N,N-dimethylformamide	$(CH_3)_2N\overset{\overset{O}{\parallel}}{C}H$	38
	Acetonitrile	CH_3CN	38
	Nitrobenzene	$C_6H_5NO_2$	36
Increasing solvent polarity	Acetone	$CH_3\overset{\overset{O}{\parallel}}{C}CH_3$	25
	Diethyl ether	$(CH_3CH_2)_2O$	4
	Benzene	C_6H_6	3
	Pentane	$CH_3(CH_2)_3CH_3$	2

17.8 NEIGHBORING GROUP PARTICIPATION IN NUCLEOPHILIC SUBSTITUTION REACTIONS

Not all nucleophilic substitution reactions take place with racemization or with inversion of configuration. Some nucleophilic substitution reactions take place with overall *retention of configuration*.

One factor that can lead to retention of configuration in a nucleophilic substitution reaction is a phenomenon known as *neighboring group participation*. Let us see how this factor operates by examining the stereochemistry of two reactions in which 2-bromopropanoic acid is converted to lactic acid.

$$\underset{\substack{| \\ Br}}{CH_3CHCOOH} \longrightarrow \underset{\substack{| \\ OH}}{CH_3CHCOOH}$$

2-Bromopropanoic   Lactic acid
acid

When *S*-2-bromopropanoic acid is treated with concentrated sodium hydroxide the reaction is *bimolecular* and it takes place with *inversion of configuration*. This, of course, is the normal stereochemical result for an S_N2 reaction.

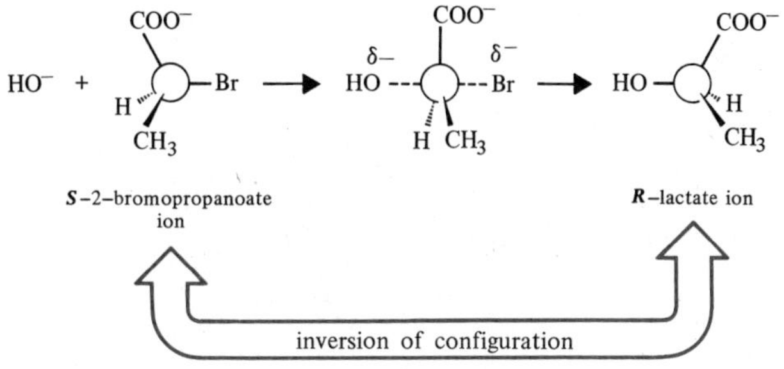

S-2-bromopropanoate ion

R-lactate ion

inversion of configuration

However, when the same reaction is carried out with a low hydroxide ion concentration and in the presence of Ag_2O, it takes place with an overall *retention of configuration*. In this case, the mechanism for the reaction involves the participation of the carboxylate group. In step 1 (below) an oxygen of the carboxylate group attacks the chiral carbon from the back side and displaces the bromine. (Silver ion aids in this process in much the same way that protonation assists the ionization of an alcohol.) The configuration of the chiral carbon inverts in step 1 and a cyclic ester called an α-lactone forms.

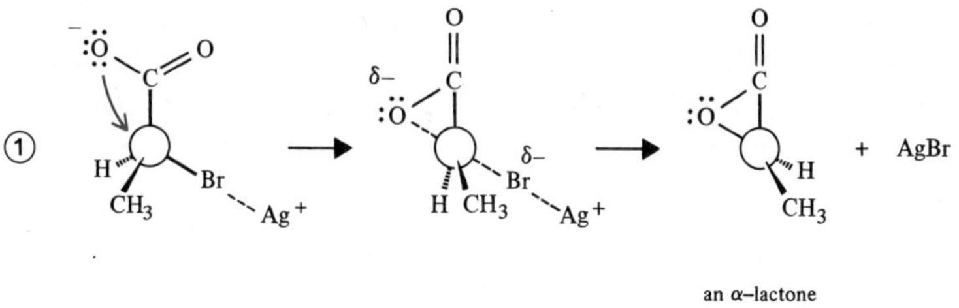

① + AgBr

an α-lactone

The highly strained three-membered ring of the α-lactone opens when it is attacked by a water molecule in step 2. *This step also takes place with an inversion of configuration.*

②

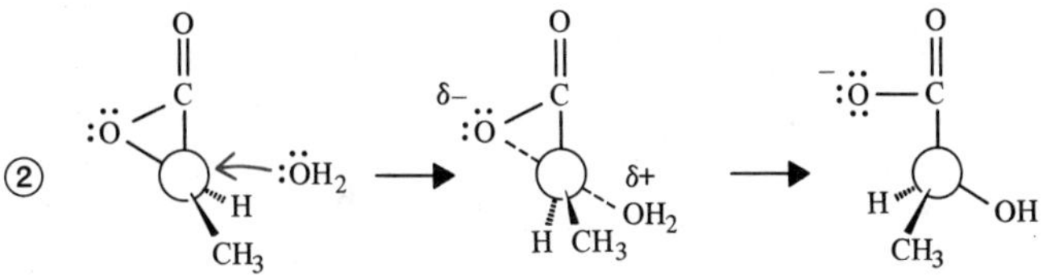

The net result of two inversions (in steps 1 and 2) is an overall *retention of configuration.*

Problem 17.11

The phenomenon of configuration inversion in a chemical reaction was discovered in 1896 by Paul von Walden. (Configuration inversions are still called Walden inversions in his honor.) Walden's proof of configuration inversion was based on the following cycle.

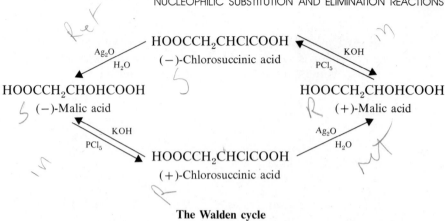

The Walden cycle

(a) Basing your answer on the preceding discussion, which reactions of the Walden cycle are likely to take place with overall inversion of configuration and which are likely to occur with overall retention of configuration? (b) Malic acid with a negative optical rotation is now known to have the S configuration. What are the configurations of the other compounds in the Walden cycle? (c) Walden also found that when (+)-malic acid is treated with thionyl chloride (rather than PCl_5) the product of the reaction is (+)-chlorosuccinic acid. How can you explain this result? (d) Assuming that the reaction of (−)-malic acid and thionyl chloride has the same stereochemistry, outline a Walden cycle based on the use of thionyl chloride instead of PCl_5.

Neighboring group participation can also lead to *cyclization reactions*. Epoxides, for example, can be prepared from 2-bromo-alcohols by treating them with sodium hydroxide. This reaction involves the following steps.

$$
\underset{\substack{| \\ :\ddot{Br}:}}{R-CH-CHR'} \;\underset{\ce{OH-}}{\rightleftharpoons}\; \underset{\substack{| \\ \ddot{:}\ddot{Br}:}}{\overset{:\ddot{O}:^-}{R-CH-CH-R'}} \;\longrightarrow\; \underset{}{\overset{:\ddot{O}:}{RCH-CHR'}} + :\ddot{Br}:^-
$$

Problem 17.12

(a) How would you synthesize a 2-halo-alcohol (a halohydrin) from an alkene? (b) Show how you could use this method to synthesize propylene oxide from propylene. (c) *Trans*-2-chlorocyclohexanol reacts readily with sodium hydroxide to yield cyclohexene oxide. *Cis*-2-chlorocyclohexanol does not undergo this reaction, however. How can you account for this?

When neighboring group participation occurs during the rate-determining step for a reaction, the rate is often markedly accelerated. This effect, called *anchimeric assistance* (Gr. *anchi* + *meros,* meaning neighboring parts) can be seen in the relative rates of S_N1 solvolysis reactions of isobutyl chloride and 2-phenyl-1-chloropropane. When 2-phenyl-1-chloropropane undergoes S_N1 solvolysis it does so much more rapidly than isobutyl chloride. The phenyl group is thought to assist in the ionization step by stabilizing the transition state leading to the phenonium ion

$$CH_3CHCH_2Cl$$
with CH_3 above

Isobutyl chloride

$$CH_3CHCH_2Cl$$
with phenyl ring above

2-Phenyl-1-chloropropane

intermediate below. The methyl group of isobutyl chloride is apparently unable to provide a similar kind of assistance when it undergoes solvolysis.

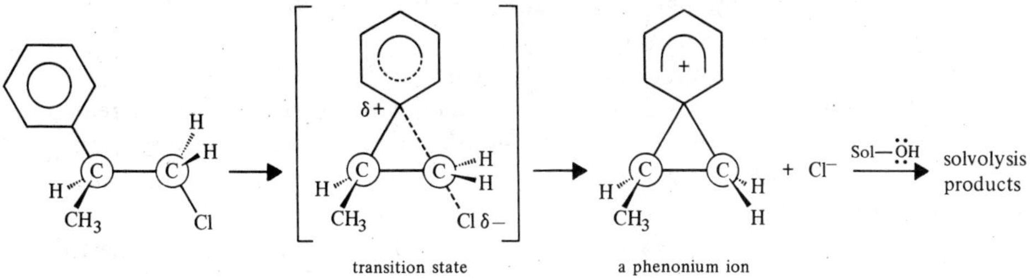

transition state   a phenonium ion

Problem 17.13

The phenonium ion formed as an intermediate in this reaction (above) strongly resembles the σ complexes we saw in electrophilic substitution reactions. What relative order of reactivity would you expect the following compounds to show: 2-phenyl-1-chloropropane, 2-(*p*-nitrophenyl)-1-chloropropane; 2-(*p*-hydroxyphenyl)-1-chloropropane; 2-(*p*-tolyl)-1-chloropropane?

Neighboring group participation and anchimeric assistance are important in many reactions that are catalyzed by enzymes.

17.9 1,2 ELIMINATION REACTIONS

All nucleophiles are also potential bases and all bases are potential nucleophiles. It should not be surprising, then, that nucleophilic substitution reactions and elimination reactions often compete with each other. We have seen, for example, that an alkyl halide can react with a hydroxide ion to form an alcohol. This is a substitution reaction.

$$RCH_2CH_2X + OH^- \xrightarrow{\text{substitution}} RCH_2CH_2OH + X^-$$
an alcohol

But we have also seen that an alkyl halide can react with a hydroxide ion by dehydrohalogenation to form an alkene. This reaction is a 1,2 elimination (or a β-elimination).

$$R\overset{2}{C}H_2\overset{1}{C}H_2X + OH^- \longrightarrow RCH{=}CH_2 + H_2O + X^-$$
with β under position 2 and α under position 1

We have already studied many examples of 1,2 elimination reactions. In addition to the preparation of alkenes and alkynes by dehydrohalogenation (Sections

5.12 and 9.11), we saw that alkenes can be prepared by dehydration (Sections 5.8 and 16.6). We also saw an example of the competition between substitution and elimination reactions in the reactions of sodium acetylides with alkyl halides in Section 9.10.

Elimination reactions can take place by a variety of mechanisms. The mechanistic paths followed by a particular substrate and base depend on a number of factors, many of which are the same as those for substitution reactions. The most important factors are:

1. The structure of the substrate undergoing elimination.
2. The strength, concentration, and structure of the base.
3. The polarity of the solvent.
4. The temperature of the reaction.

Since these factors influence the mechanistic paths for a given reaction, they also influence the relative percentages of substitution and elimination products, the orientation of eliminations, the likelihood of rearrangements, and the stereochemistry of the elimination reaction.

Elimination reactions that involve loss of a proton and the formation of an alkene can be formulated in the general way shown below:

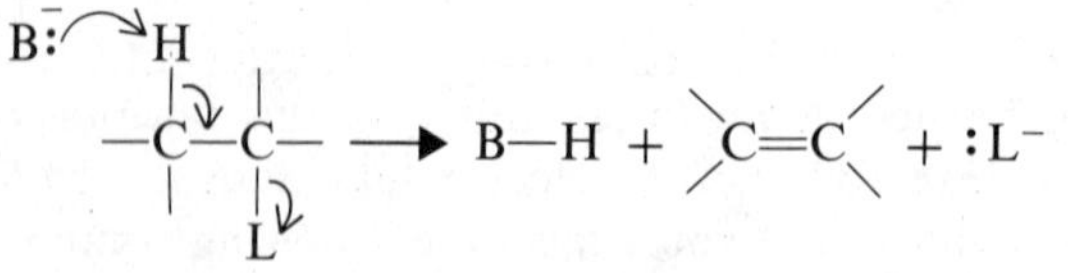

$$\text{B:}^- \overset{\frown}{\quad} \text{H}$$
$$-\overset{|}{\underset{|}{C}}-\overset{|}{\underset{|}{C}}- \longrightarrow \text{B}-\text{H} + \underset{/}{\overset{\backslash}{\quad}}C=C\overset{/}{\underset{\backslash}{\quad}} + \text{:L}^-$$
$$\overset{|}{\underset{L}{}}$$

This formulation is intended to show *only the flow of electrons* and does not imply that the bonds are made and broken at the same time (i.e., that the reaction is concerted). Indeed, for certain elimination reactions the question of just when bonds are broken and formed (and to what extent) is still a matter of some controversy among chemists who study their reaction mechanisms.

There is evidence indicating that the mechanisms of elimination reactions can vary from those in which the initial step is the formation of a bond between the base and the proton (a carbanion elimination) to those in which the initial step is the cleavage of the bond between the leaving group and the carbon (a carbocation mechanism). In between these two extremes lie the possibilities of mechanisms involving ion pairs and a mechanism in which the bonds are broken and formed at the same time (a concerted mechanism).

Carbanion Mechanisms

Carbanion mechanisms appear to operate when the β hydrogens are appreciably acidic, a situation that occurs when the β carbon of the substrate bears a strong electron-withdrawing group. The elimination reaction (p. 659) that occurs when a 2-methoxy-1-nitroethyl compound is treated with methoxide ion is an example. The reaction shows first-order reaction kinetics, and the rate depends only on the concentration of the anion. This mechanism is called an (*E1*) *anion* mechanism (i.e., **E**limination, unimolecular, anion) because the rate limiting step is unimolecular.

However, some reactions with carbanionlike mechanisms may show second-order kinetics when the β-hydrogen is less acidic, and some carbanion-type reactions apparently proceed through the formation of ion pairs.

$$CH_3\ddot{O}:^- + H-\underset{\underset{\overset{|}{N^+}}{|}}{\overset{|}{C}}-\underset{|}{\overset{|}{C}}-OCH_3 \xrightarrow{\text{fast}} CH_3\ddot{O}-H + :\underset{\underset{\overset{|}{N^+}}{|}}{\overset{|}{C}}-\underset{|}{\overset{|}{C}}-\ddot{O}CH_3 \longleftrightarrow \underset{\underset{\overset{|}{N^+}}{|}}{\overset{|}{C}}-\underset{|}{\overset{|}{C}}-\ddot{O}CH_3$$

Resonance
stabilized anion

$$\text{Rate} = k\left[\underset{\underset{NO_2}{|}}{\overset{\overset{\displaystyle-}{\cdot\cdot}}{C}}-\underset{|}{\overset{|}{C}}-OCH_3\right]$$

$$\downarrow \text{slow}$$

$$\underset{O_2N}{}\overset{}{C}=C\overset{}{} + {}^-:\ddot{O}CH_3$$

Carbocation Mechanisms

Elimination reactions in which the initial step involves cleavage of the bond between the leaving group and its carbon may lead to the formation of fully dissociated carbocations. These reactions are called, simply, E1 reactions (**E**limination, **u**nimolecular).

E1 reactions always accompany S_N1 reactions because the initial step for both reactions is the same. When *tert*-butyl chloride is treated with 80% aqueous ethanol at 25°, for example, the reaction produces the *substitution products* in 83% yield and the elimination product (2-methylpropene) in 17% yield.

$$CH_3\underset{\underset{CH_3}{|}}{\overset{\overset{CH_3}{|}}{C}}-Cl \xrightarrow[\substack{20\% \ H_2O \\ 25°}]{80\% \ C_2H_5OH}$$

$$\xrightarrow{S_N1} CH_3\underset{\underset{CH_3}{|}}{\overset{\overset{CH_3}{|}}{C}}-OH + CH_3\underset{\underset{CH_3}{|}}{\overset{\overset{CH_3}{|}}{C}}-OCH_2CH_3$$

(83%)

$$\xrightarrow{E1} CH_2=C\overset{\diagup CH_3}{\diagdown CH_3}$$

(17%)

The initial step for both reactions is the formation of a *tert*-butyl cation (or an ion pair). This is also the rate-limiting step for both reactions.

$$CH_3-\underset{\underset{CH_3}{|}}{\overset{\overset{CH_3}{|}}{C}}-\ddot{C}l: \xrightarrow{\text{slow}} CH_3\underset{\underset{CH_3}{|}}{\overset{\overset{CH_3}{|}}{C}}^+ + :\ddot{C}l:^-$$

(solvated)   (solvated)

Whether substitution takes place depends on the next step (the fast step). If a solvent molecule reacts as a nucleophile at the positive carbon of the *tert*-butyl cation the product is ethyl *tert*-butyl ether or *tert*-butyl alcohol and the reaction is S_N1.

$$CH_3-\overset{\underset{\displaystyle CH_3}{|}}{\overset{\displaystyle CH_3}{C^+}}\,\,Sol-\overset{..}{\underset{..}{O}}H \longrightarrow CH_3\overset{\underset{\displaystyle CH_3}{|}}{\overset{\displaystyle CH_3}{C}}-\overset{Sol}{\underset{H}{\overset{|}{\overset{..}{O}:}}} \rightleftharpoons CH_3\overset{\underset{\displaystyle CH_3}{|}}{\overset{\displaystyle CH_3}{C}}-O-Sol + H^+$$

(Sol = C_2H_5— or H—)

If, however, a solvent molecule acts as a base and accepts one of the β protons, the product is 2-methylpropene and the reaction is E1.

$$Sol-\overset{..}{\underset{H}{\overset{|}{O}:}}\,\,H-CH_2\!-\!\overset{\underset{\displaystyle CH_3}{|}}{\overset{\displaystyle CH_3}{C^+}} \longrightarrow Sol-\overset{..}{\underset{H}{\overset{|}{O}^\pm}}H + CH_2\!=\!C\overset{\displaystyle CH_3}{\underset{\displaystyle CH_3}{<}}$$

In most reactions of this type the S_N1 reaction is favored over the E1 reaction. Increasing the temperature of the reaction, however, favors reaction by the E1 mechanism at the expense of the S_N1 mechanism.

Since the E1 reaction and the S_N1 reaction proceed through the formation of a common intermediate, E1 reactions show some of the same reactivity effects as the S_N1 reaction. E1 reactions are favored with substrates that can form stable carbocations; they are also favored by the use of weak nucleophiles (bases) and they are generally favored by the use of polar solvents.

When the carbocation formed in an E1 reaction can lose a proton in more than one way, the product formed is usually the *most highly* substituted alkene. An example of this behavior can be seen in the solvolysis reaction of 2-bromo-2-methylbutane.

$$CH_3CH_2\overset{\underset{\displaystyle Br}{|}}{\overset{\displaystyle CH_3}{C}}CH_3 \xrightarrow[25°]{C_2H_5OH} CH_3CH_2\overset{\underset{\displaystyle +}{}}{\overset{\displaystyle CH_3}{\underset{|}{C}}}-CH_3 \xrightarrow[(-H^+)]{C_2H_5OH}$$

2-Methyl-2-butene (a trisubstituted alkene) (27%)

2-Methyl-1-butene (a disubstituted alkene) (7%)

Substitution product (66%)

This behavior was first formulated into a rule in 1875 by the Russian chemist Alexander Saytzeff. Thus, when an elimination reaction produces the most highly substituted alkene, it is said to proceed with *Saytzeff orientation* (or in accord with Saytzeff's rule).*

---

* Like Markovnikov, Saytzeff originally formulated his rule in terms of the number of hydrogens on each carbon atom. Saytzeff stated that in elimination reactions a hydrogen is preferentially lost from the adjacent carbon that is poorer in hydrogen. Markovnikov stated (p. 51) that in addition reactions a hydrogen is added to the carbon with the greater number of hydrogens. However, Professor Louis Fieser of Harvard University pointed out, several years ago, that both rules were anticipated by St. Matthew: "unto him that hath

We account for Saytzeff's rule on the basis of the transition state theory and the relative stability of alkenes. (Trisubstituted alkenes, we recall, are more stable than disubstituted alkenes, and disubstituted alkenes are more stable than monosubstituted alkenes.) Since the transition state leading to a particular alkene has a partially developed double bond, the transition state will be stabilized by the same factors that stabilize the alkene. The reaction with a transition state leading to the more stable (more highly substituted) alkene will have a lower energy of activation and will, therefore, occur faster.

Some cationic eliminations appear to occur through the formation of ion pairs as well as through fully dissociated carbocations. And, as we might expect, cationic eliminations often take place with rearrangement of the carbon skeleton. An example can be seen in the solvolysis reaction of neopentyl bromide below. Here both the substitution reaction and the elimination reaction give a rearranged product.

$$CH_3\underset{\underset{CH_3}{|}}{\overset{\overset{CH_3}{|}}{C}}CH_2Br \xrightarrow{C_2H_5OH} CH_3\underset{\underset{CH_3}{|}}{C}{=}CHCH_3 + CH_3\underset{\underset{CH_3}{|}}{\overset{\overset{OC_2H_5}{|}}{C}}CH_2CH_3$$

2-Methyl-2-butene   2-Ethoxy-2-methylbutane

Problem 17.14

The solvolysis reaction of neopentyl bromide given above takes place extremely slowly. What structural feature accounts for this?

Cationic eliminations are also common when tertiary and secondary alcohols are heated with strong acids (p. 172). In these reactions, carbocations are formed by the loss of a molecule of water from a protonated alcohol. Rearrangements are also frequent occurrences in these reactions.

The Concerted E2 Reaction

When 2-phenylethyl halides ($C_6H_5CH_2CH_2X$) are treated with ethoxide ion, the elimination reactions that take place proceed through transition states in which cleavage has occurred to some degree in both the C—H and C—X bonds. A general mechanism for these reactions is given below.

$$C_2H_5\ddot{O}{:}^- \quad\overset{\frown}{H} \atop C_6H_5{-}CH{-}CH_2 \longrightarrow \left[\begin{array}{c} \overset{\delta-}{C_2H_5\ddot{O}{-}{-}{-}H} \\ C_6H_5CH{=}CH_2 \\ {:}\ddot{X}{:}^{\delta-} \end{array} \right] \longrightarrow$$

Transition state

$$C_2H_5\ddot{O}H + C_6H_5CH{=}CH_2 + {:}\ddot{X}{:}^-$$

What is some of the evidence for this mechanism?

even more shall be added, but from him that hath not shall be taken away even that which he hath" (Matt. XXV, 29).

1. *The rate expressions for the reactions are second order.* The reaction rates depend both on the concentration of the 2-phenylethyl halide and the concentration of the ethoxide ion; doubling either concentration doubles the rate of a given reaction. (Because the rate expressions are second order, the mechanism is described as an E2 mechanism (i.e., Elimination, bimolecular).

2. *The reactions show a sizable hydrogen-deuterium isotope effect.* When eliminations are carried out with 2-phenylethyl halides containing deuterium at the 2 position, the reaction rates are considerably slower than the corresponding reactions of compounds containing only ordinary hydrogen. Carbon-deuterium bonds are known to be broken more slowly than carbon-hydrogen bonds and this phenomenon is known as an *isotope effect* (cf. p. 500). Thus, an isotope effect of this kind is consistent with a *mechanism that involves carbon-hydrogen bond cleavage in the transition state.*

3. *The reactions show a substantial element effect.* The relative strength of carbon-halogen bonds is:

$$-\overset{|}{\underset{|}{C}}-F > -\overset{|}{\underset{|}{C}}-Cl > -\overset{|}{\underset{|}{C}}-Br > -\overset{|}{\underset{|}{C}}-I$$

The relative reactivity of a 2-phenylethyl halide toward ethoxide ion is just the opposite:

$$C_6H_5CH_2CH_2I > C_6H_5CH_2CH_2Br > C_6H_5CH_2CH_2Cl > C_6H_5CH_2CH_2F$$

An element effect of this type is consistent with a *transition state in which carbon-halogen bond breaking is occurring.* Thus, 2-phenylethyl iodide reacts fastest because the carbon-iodine bond is weakest; 2-phenylethyl fluoride reacts slowest because the carbon-fluorine bond is strongest.

4. When the reaction is carried out in the presence of C_2H_5OD, none of the unreacted 2-phenylethyl halide recovered from the reaction is found to contain deuterium. This is strong evidence against a mechanism involving the reversible formation of a carbanion as the initial step; that is,

$$C_6H_5CH_2CH_2X \underset{C_2H_5OH}{\overset{C_2H_5O^-}{\rightleftharpoons}} C_6H_5\overset{..}{C}HCH_2X \longrightarrow C_6H_5CH{=}CH_2 + X^-$$

For, if such a mechanism operated, some deuterium would certainly be found in the unreacted starting material when the reaction is carried out in C_2H_5OD.

$$C_6H_5CH_2CH_2X + \overset{C_2H_5O^-}{\longrightarrow} C_6H_5\overset{..}{C}HCH_2X \overset{C_2H_5OD}{\longrightarrow} C_6H_5CHDCH_2X$$

$$\downarrow$$

$$C_6H_5CH{=}CH_2 + X^-$$

5. E2 reactions, generally, are not accompanied by rearrangements. This is consistent with a mechanism that does not involve the formation of carbocations.

Substitution and Elimination in E2 Reactions. Since E2 reactions are carried out with a high concentration of a strong base (and thus a high concentration of a strong nucleophile), substitution reactions by an S_N2 path often compete with the elimination reaction. When the nucleophile (base) attacks a β-hydrogen, elimination occurs. When the nucleophile attacks the carbon bearing the leaving group substitution results.

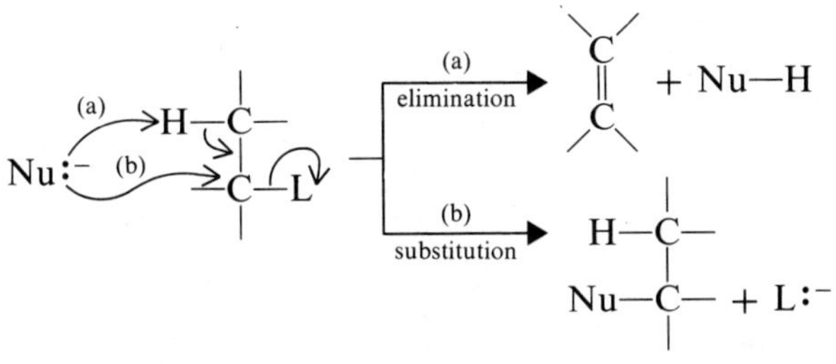

When the substrate is a primary halide and the base is ethoxide ion, substitution is highly favored.

$$C_2H_5O^-Na^+ + CH_3CH_2Br \xrightarrow[\substack{55^\circ \\ (-NaBr)}]{C_2H_5OH} \underset{(90\%)}{CH_3CH_2OCH_2CH_3} + \underset{(10\%)}{CH_2{=}CH_2}$$

With secondary halides, however, the elimination reaction is favored.

$$C_2H_5O^-Na^+ + CH_3\underset{\substack{| \\ Br}}{C}HCH_3 \xrightarrow[\substack{55^\circ \\ (-NaBr)}]{C_2H_5OH} \underset{(21\%)}{CH_3\underset{\substack{| \\ O \\ | \\ C_2H_5}}{C}HCH_3} + \underset{(79\%)}{CH_2{=}CHCH_3}$$

And with tertiary halides the elimination reaction is highly favored, especially when the reaction is carried out at higher temperatures.

$$C_2H_5O^-Na^+ + CH_3\overset{\substack{CH_3 \\ |}}{\underset{\substack{| \\ Br}}{C}}CH_3 \xrightarrow[\substack{25^\circ \\ (-NaBr)}]{C_2H_5OH} \underset{(9\%)}{CH_3\overset{\substack{CH_3 \\ |}}{\underset{\substack{| \\ O \\ | \\ C_2H_5}}{C}}CH_3} + \underset{(91\%)}{CH_2{=}\overset{\substack{CH_3 \\ |}}{C}CH_3}$$

$$C_2H_5O^-Na^+ + CH_3\overset{\substack{CH_3 \\ |}}{\underset{\substack{| \\ Br}}{C}}CH_3 \xrightarrow[\substack{55^\circ \\ (-NaBr)}]{C_2H_5OH} \underset{(100\%)}{CH_2{=}\overset{\substack{CH_3 \\ |}}{C}CH_3}$$

Increasing the reaction temperature is only one way of favorably influencing an elimination reaction of an alkyl halide. Another way is to use a strong sterically hindered base such as the *tert*-butoxide ion. The bulky methyl groups of the *tert*-butoxide ion appear to inhibit its reacting by substitution, so elimination reactions take precedence. We can see an example of this effect in the two reactions shown below. The relatively unhindered methoxide ion reacts with *n*-octadecyl bromide primarily by *substitution;* the bulky *tert*-butoxide ion gives mainly *elimination.**

$$CH_3O^- + CH_3(CH_2)_{15}CH_2CH_2\text{---}Br \xrightarrow[\text{reflux}]{CH_3OH}$$

$$CH_3(CH_2)_{15}CH\!=\!CH_2 + CH_3(CH_2)_{15}CH_2CH_2OCH_3$$
$$\qquad\qquad (1\%) \qquad\qquad\qquad\qquad (99\%)$$

$$\underset{\underset{CH_3}{|}}{\overset{\overset{CH_3}{|}}{CH_3\text{---}C\text{---}O^-}} + CH_3(CH_2)_{15}CH_2CH_2\text{---}Br \xrightarrow[40°]{(CH_3)_3COH}$$

$$CH_3(CH_2)_{15}CH\!=\!CH_2 + CH_3(CH_2)_{15}CH_2CH_2\text{---}O\text{---}\underset{\underset{CH_3}{|}}{\overset{\overset{CH_3}{|}}{C}}\text{---}CH_3$$

$$\qquad (85\%) \qquad\qquad\qquad\qquad\qquad (15\%)$$

The steric bulk of the *tert*-butoxide ion also influences the orientation of elimination reactions, as the following examples illustrate.

$$CH_3CH_2O^- + CH_3CH_2\text{---}\underset{\underset{CH_3}{|}}{\overset{\overset{CH_3}{|}}{C}}\text{---}Br \xrightarrow[CH_3CH_2OH]{70°} CH_3CH\!=\!C\!\!\underset{CH_3}{\overset{CH_3}{<}} + CH_3CH_2C\!\!\underset{CH_3}{\overset{CH_2}{<}}$$

$$\qquad\qquad\qquad\qquad\qquad\qquad\qquad (69\%) \qquad\quad (31\%)$$

$$CH_3\text{---}\underset{\underset{CH_3}{|}}{\overset{\overset{CH_3}{|}}{C}}\text{---}O^- + CH_3CH_2\text{---}\underset{\underset{CH_3}{|}}{\overset{\overset{CH_3}{|}}{C}}\text{---}Br \xrightarrow[(CH_3)_3COH]{75°} CH_3CH\!=\!C\!\!\underset{CH_3}{\overset{CH_3}{<}} + CH_3CH_2C\!\!\underset{CH_3}{\overset{CH_2}{<}}$$

$$\qquad\qquad\qquad\qquad\qquad\qquad\qquad\qquad (27.5\%) \qquad\quad (72.5\%)$$

Here we see that the major product of the *tert*-butoxide-promoted elimination is the *least* highly substituted alkene. An elimination reaction that proceeds in this manner is said to follow a *Hofmann orientation* (p. 667), rather than the Saytzeff orientation. The tendency of the *tert*-butoxide ion to give the least substituted alkene is probably related to its steric bulk and its relative inability to abstract internal hydrogens.

Stereochemistry of E2 Reactions

E2 reactions are usually stereospecific. With most open-chain compounds E2 reactions take place from a transition state in which the proton being eliminated and the leaving group are *anti* to each other.

---

* This effect does not occur when the substrate is *n*-octadecyl tosylate. Even when the base is potassium *tert*-butoxide, *n*-octadecyl tosylate gives a 99% yield of the substitution product. The reason for this difference in the behavior of bromides and tosylates is not understood but it can be very important in organic syntheses.

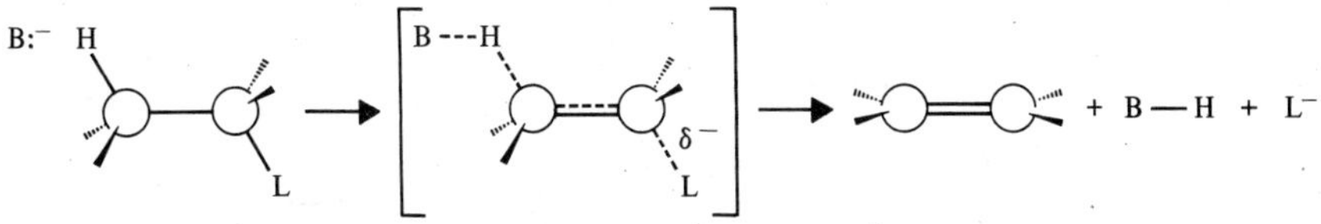

anti relation
between proton
and leaving group

One explanation that has been advanced to account for a preferred anti orientation is that it allows the large lobe from the C—H bond to interact with the small lobe of the C—L bond (below). In this sense then, the E2 reaction resembles the S_N2 reaction in that an electron pair of the C—H bond displaces the leaving group from the back side.

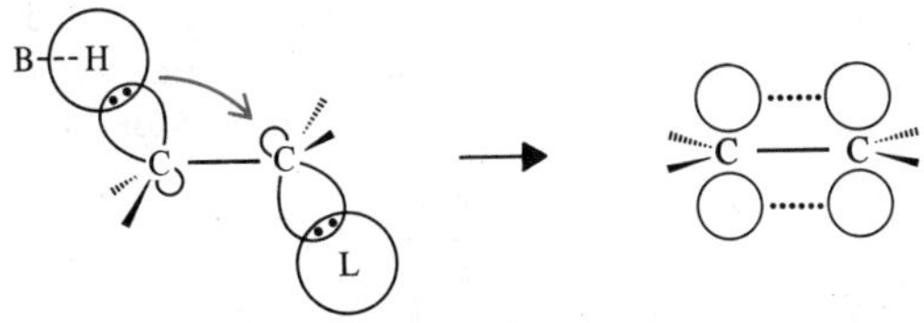

Problem 17.15

When the deuterium-labeled isomer below (*erythro*-2-bromobutane-3-*d*) undergoes elimination, the reaction yields *trans*-2-butene and *cis*-2-butene-2-*d* (as well as some 1-butene-3-*d*). The reaction does not yield *cis*-2-butene or *trans*-2-butene-2-*d*.

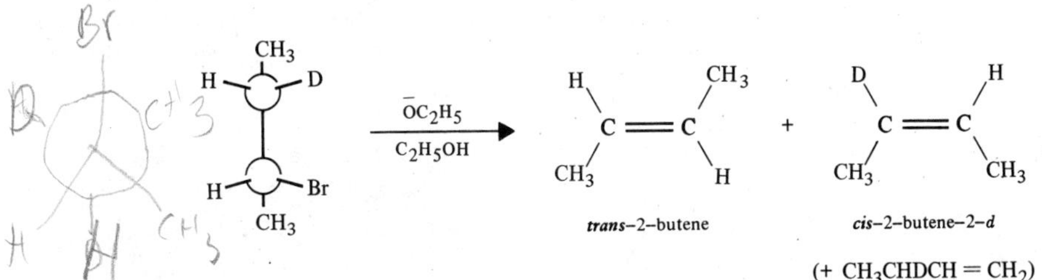

erythro–2–bromobutane–3–*d*

trans–2–butene

cis–2–butene–2–*d*

(+ $CH_3CHDCH = CH_2$)

but no

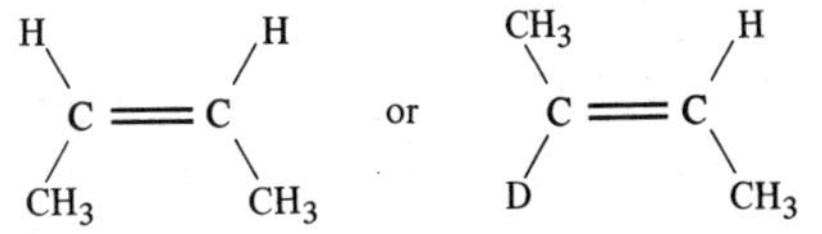

cis–2–butene

trans–2–butene–2–*d*

How can you explain these results?

With cyclic compounds the stereochemistry of elimination is more complex. *Cis*-2-phenylcyclohexyl tosylate, **1**, undergoes elimination at a rate that has been estimated as being 10^4 times faster than *trans*-2-phenyl-cyclohexyl tosylate, **2**. This evidence indicates a preference for an *anti* relation between the proton at C-2 and the leaving group, because **1** can assume an anti conformation easily and **2** cannot.

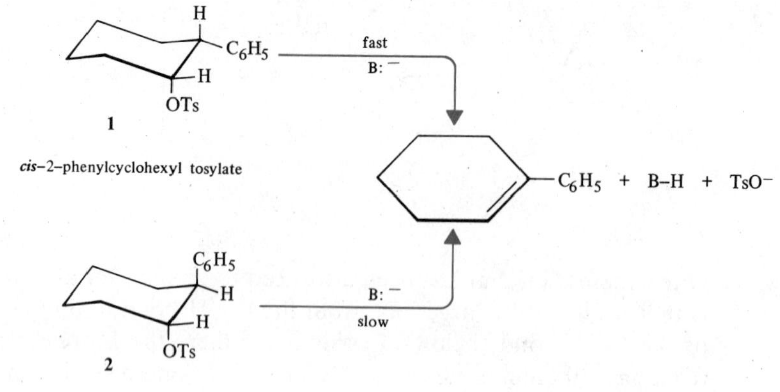

1

cis–2–phenylcyclohexyl tosylate

2

trans–2–phenylcyclohexyl tosylate

On the other hand when the deuterium-labeled compound **7** undergoes elimination, 94% of the product contains no deuterium. This cyclic system is rigid and thus a *syn* elimination must take place.

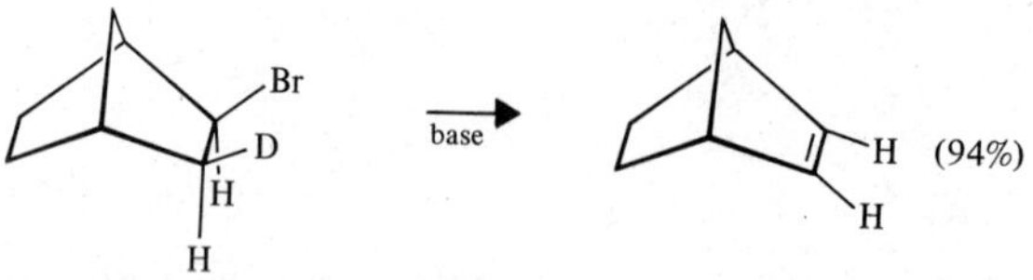

7   (no deuterium)

Results such as this have led to the suggestion that elimination reactions *only require a transition state in which the proton and the leaving group are coplanar*. This can be achieved in either an *anti coplanar* relation or a *syn coplanar* relation.

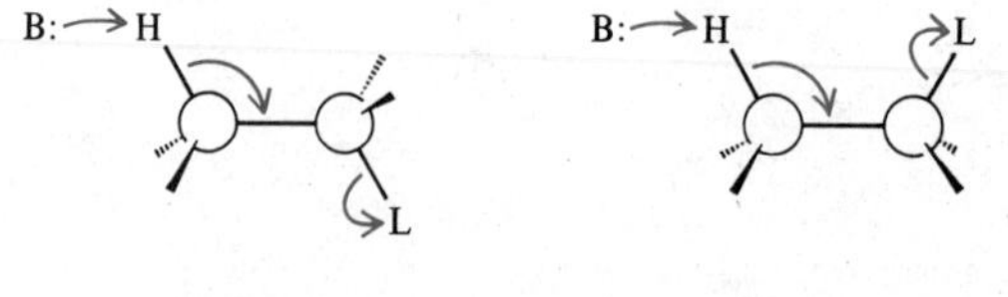

anti coplanar
transition state

syn coplanar
transition state

Thus, open-chain compounds apparently proceed through an anti coplanar transition state because this avoids a high energy eclipsed conformation. Cyclic molecules such as **7**, however, undergo elimination from a syn coplanar transition state because the molecular structure is such that it requires an eclipsed conformation.

The Hofmann Elimination

All of the elimination reactions that we have described so far have involved electrically neutral substrates. However, a number of eliminations are known in which the substrate bears a positive charge. One of the most important of these is the elimination reaction that takes place when a quaternary ammonium hydroxide (p. 667) is heated. The products are an alkene, water, and a tertiary amine.

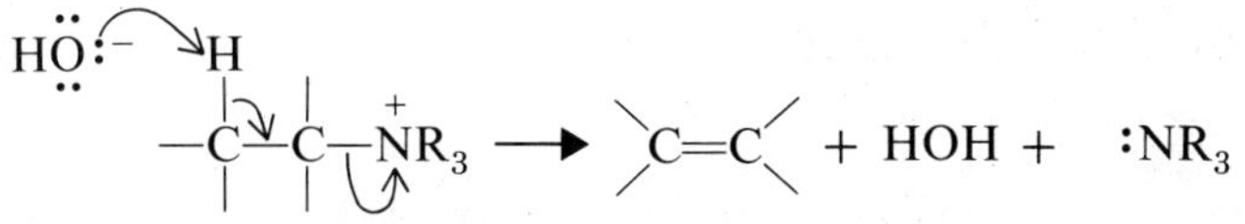

$$HO:^- \curvearrow H$$

$$-\overset{|}{\underset{|}{C}} \curvearrow \overset{|}{\underset{|}{C}} - \overset{+}{N}R_3 \longrightarrow \ \ \underset{/}{\overset{\backslash}{C}} = \overset{/}{\underset{\backslash}{C}} \ + HOH + \ :NR_3$$

A quaternary
ammonium hydroxide $\longrightarrow$ An alkene + Water + A tertiary amine

This elimination reaction was discovered in 1851 by August W. von Hofmann and has since come to bear his name.

Quaternary ammonium hydroxides can be prepared from quaternary ammonium halides in aqueous solution through the use of silver oxide.

$$2RCH_2CH_2\overset{+}{N}(CH_3)_3 \ X^- + Ag_2O + H_2O \longrightarrow 2RCH_2CH_2\overset{+}{N}(CH_3)_3 \ OH^- + 2AgX\downarrow$$

A quaternary ammonium
halide

Silver halide precipitates from the solution and can be removed by filtration. The quaternary ammonium hydroxide can then be obtained by evaporation of the water.

Problem 17.16

Quaternary ammonium halides can be prepared through a variety of substitution reactions using amines as nucleophiles. Give structures for the products A through C in each of the following examples.

(a) $C_2H_5Br + \ :N(CH_3)_3 \longrightarrow$ A

            Trimethyl amine   Ethyltrimethylammonium bromide

(b) $CH_3CH_2CH_2\overset{..}{N}CH_3 + CH_3I \longrightarrow$ B

                 $\overset{|}{CH_3}$

         Dimethylpropylamine   Trimethylpropylammonium iodide

(c) 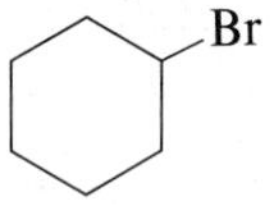   [cyclohexyl]—Br $+ (CH_3)_3N: \longrightarrow$ C

         Cyclohexyl   Trimethyl   Trimethylcyclohexylammonium
         bromide   amine   bromide

While most eliminations involving neutral substrates tend to follow the *Saytzeff rule* (p. 660), eliminations with charged substrates tend to follow the *Hofmann rule* and yield mainly the least substituted alkene. We can see an example of this behavior if we compare the reactions given below.

$$C_2H_5O^-Na^+ + CH_3CH_2\overset{|}{\underset{Br}{C}}HCH_3 \xrightarrow[25°]{C_2H_5OH}$$

$$CH_3CH=CHCH_3 + CH_3CH_2CH=CH_2 + NaBr + C_2H_5OH$$
$$(75\%) \qquad\qquad (25\%)$$

$$CH_3CH_2\underset{\underset{+}{\overset{|}{N(CH_3)_3}}}{CH}CH_3 \quad \overset{-}{O}H \xrightarrow[150°]{}$$

$$CH_3CH=CHCH_3 + CH_3CH_2CH=CH_2 + (CH_3)_3N\text{:} + H_2O$$
$$\underset{(5\%)}{} \qquad\qquad \underset{(95\%)}{}$$

$$CH_3CH_2\underset{\underset{+}{\overset{|}{S(CH_3)_2}}}{CH}CH_3 \quad \overset{-}{O}C_2H_5 \longrightarrow$$

$$CH_3CH=CHCH_3 + CH_3CH_2CH=CH_2 + (CH_3)_2S + C_2H_5OH$$
$$\underset{(26\%)}{} \qquad\qquad \underset{(74\%)}{}$$

The precise mechanistic reasons for these differences are complex and are not yet fully understood. One possible explanation is that the transition states of elimination reactions with charged substrates have considerable carbanion character. This means that these transition states show little resemblance to the final alkene product and, thus, are not stabilized appreciably by a developing double bond.

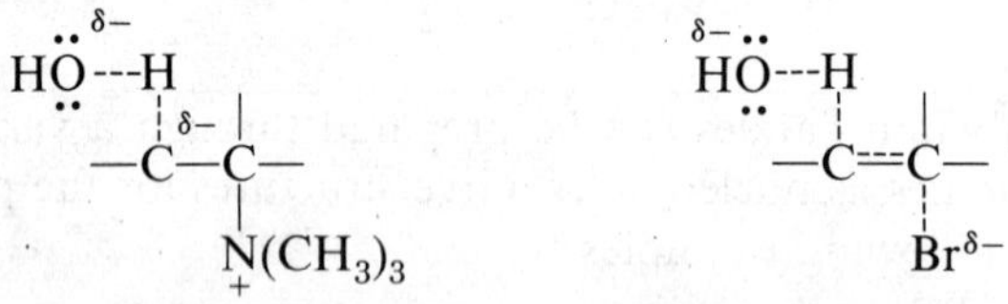

Carbanionlike transition state
(gives Hofmann orientation)

Alkenelike transition state
(gives Saytzeff orientation)

With a charged substrate, the base attacks the most acidic hydrogen instead. Primary hydrogens are more acidic because their carbon bears only one electron-releasing group.*

An elegant application of the Hofmann elimination can be seen in Richard Willstätter's synthesis of cyclooctatetraene (p. 669) from pseudopelletierine, a substance that can be obtained from the bark of the pomegranate tree. This synthesis was achieved in 1911, and in it Willstätter utilized *four* Hofmann eliminations. Even though most steps gave yields greater than 90%, the overall yield was only 3%. Still,

* It was the Hofmann elimination that led Hofmann to propose his rule, but not all Hofmann eliminations give a Hofmann orientation. The Hofmann elimination of the cyclohexyl derivative below gives mainly a Saytzeff orientation.

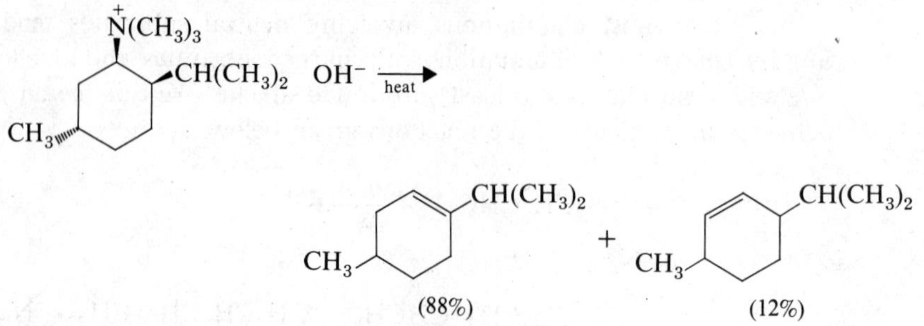

$$(88\%) \qquad\qquad\qquad (12\%)$$

This is possibly a gentle reminder of Nature's reluctance to be categorized.

Willstätter was able to obtain enough cyclooctatetraene to show that it was not aromatic. As we noted in Chapter 12, this was an important step in our understanding of aromatic compounds.

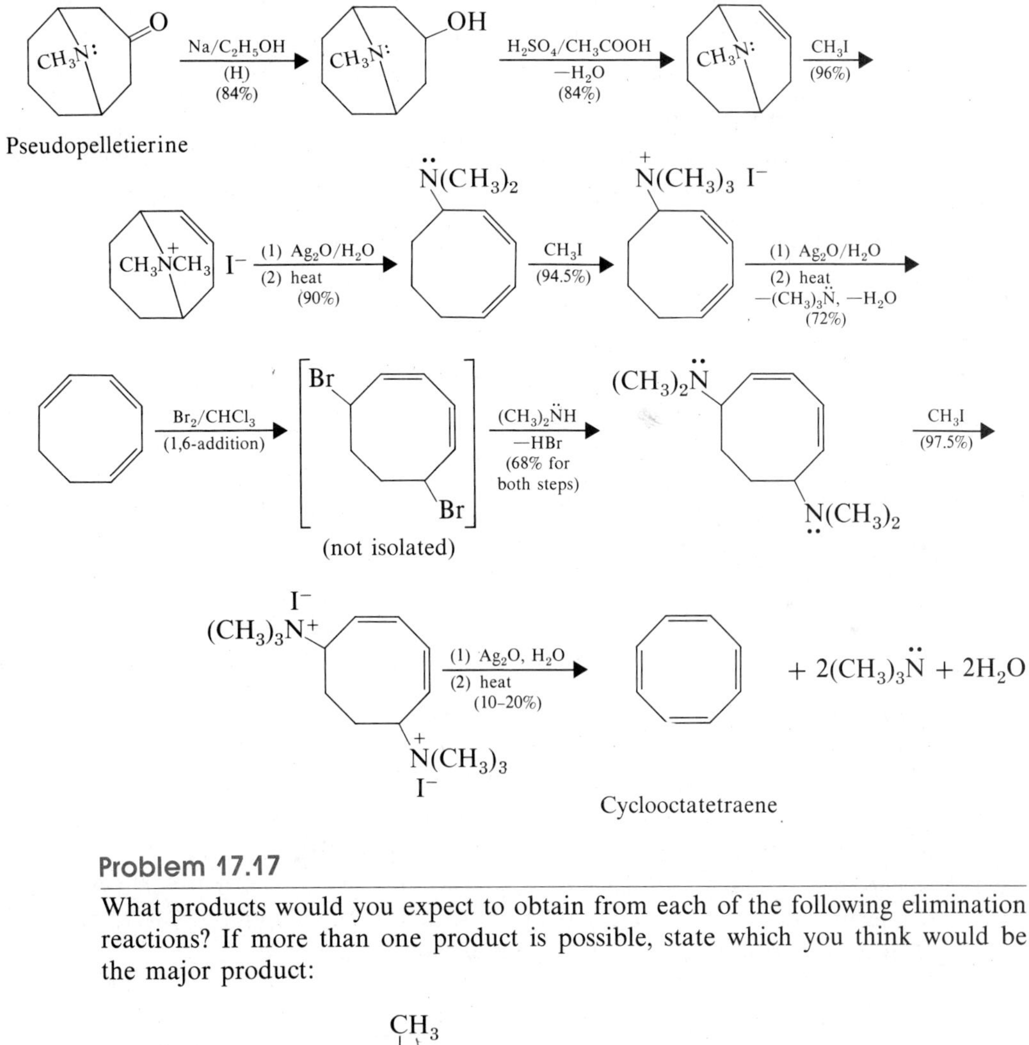

Pseudopelletierine

Cyclooctatetraene

Problem 17.17

What products would you expect to obtain from each of the following elimination reactions? If more than one product is possible, state which you think would be the major product:

(a) 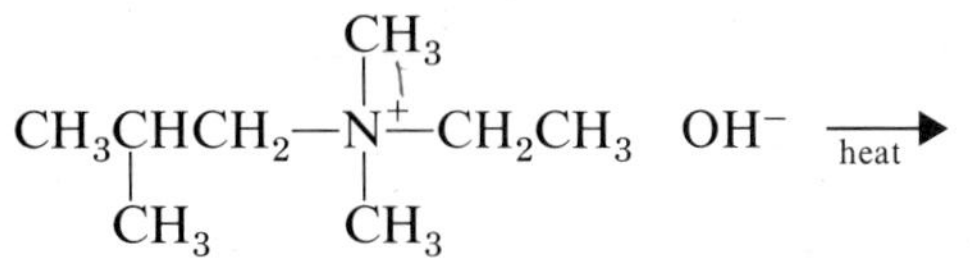 CH_3CHCH_2—$\overset{+}{\underset{\underset{CH_3}{|}}{N}}$—$CH_2CH_3$ OH^- $\xrightarrow{\text{heat}}$
$\quad\quad\underset{CH_3}{|}\quad\overset{CH_3}{|}$

(b) 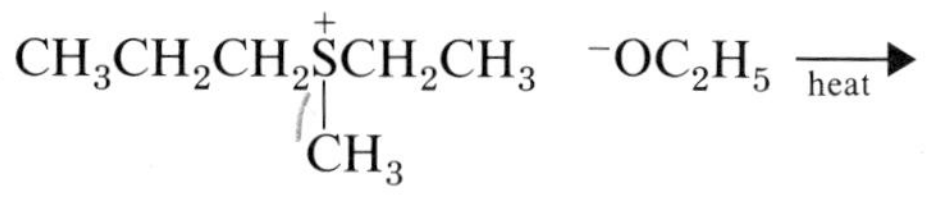 $CH_3CH_2CH_2\overset{+}{\underset{\underset{CH_3}{|}}{S}}CH_2CH_3$ $^-OC_2H_5$ $\xrightarrow{\text{heat}}$

(c) 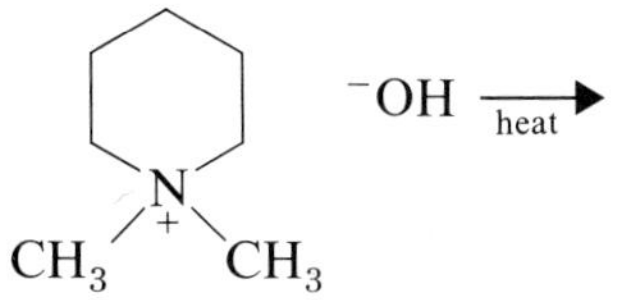 ^-OH $\xrightarrow{\text{heat}}$

(d) Product of (c) $\xrightarrow[\substack{(2)\ Ag_2O/H_2O \\ (3)\ heat}]{(1)\ CH_3I}$

17.10 INTRAMOLECULAR ELIMINATIONS

There are several elimination reactions in which the leaving group and the base are a part of the same molecule. These eliminations are *thermal eliminations;* they require only the application of heat. We will discuss two important examples: acetate pyrolysis and the Cope elimination.

Acetate Pyrolysis

When acetate esters of alcohols with a β-hydrogen are heated to a temperature of 200–500° acetic acid is eliminated and an alkene is formed.

$$CH_3CH_2CH_2CH_2CH_2OCCH_3 \xrightarrow{470°} CH_3CH_2CH_2CH=CH_2 + HOCCH_3$$

1-Pentyl acetate	1-Pentene	Acetic acid

The reaction appears to take place through a cyclic transition state in which the carbonyl oxygen of the acetate group acts as a base. Because the transition state is a six-membered ring, acetate pyrolyses are, of necessity, syn eliminations.

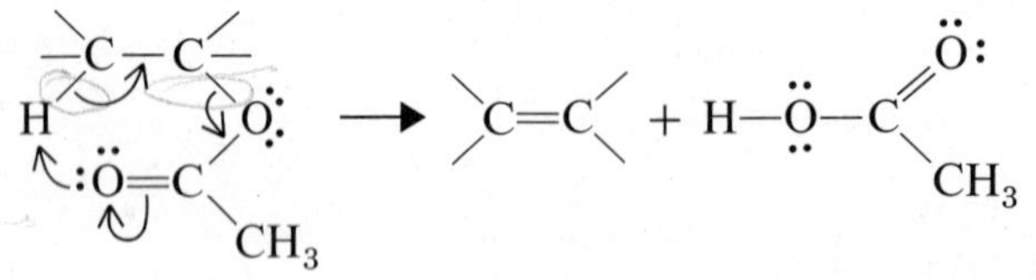

Alkene	Acetic acid

A cyclic anti elimination does not occur because the transition state would be highly strained and, thus, of very high energy. (A cyclic anti elimination would require a transition state that resembles a six-membered ring with a *trans* double bond.)

Problem 17.18

When the diastereomeric acetates, A and B, are heated they both yield *trans*-1,2-diphenylethene. The major product from A contains deuterium; however, that from B does not. Write transition states that explain these results.

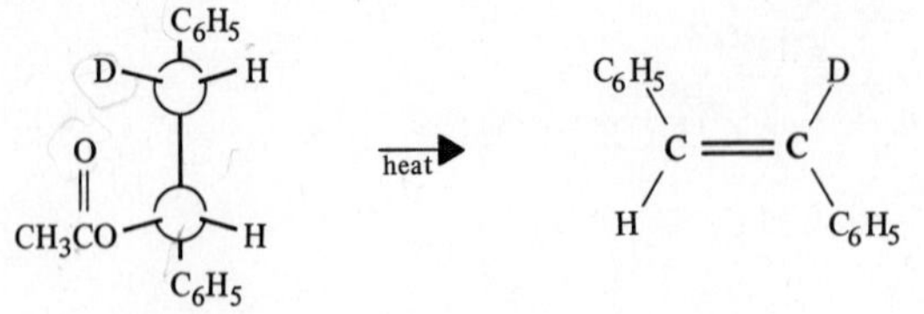

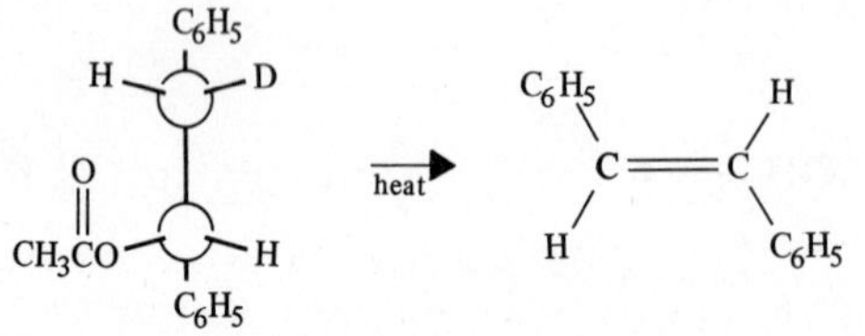

The Cope Elimination

Tertiary amine N-oxides undergo the elimination of a dialkylhydroxylamine when they are heated. The temperatures required, however, are considerably lower than those required for acetate pyrolysis.

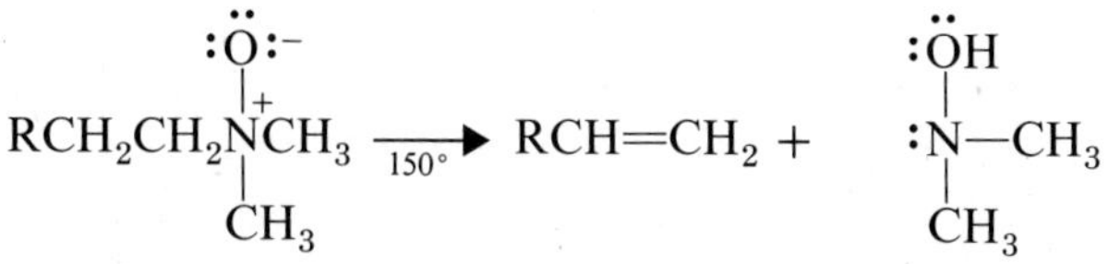

$$RCH_2CH_2\overset{\overset{\displaystyle :\ddot{O}:^-}{|}}{\underset{\underset{\displaystyle CH_3}{|}}{\overset{+}{N}}}CH_3 \xrightarrow{150°} RCH{=}CH_2 + \quad :\overset{\overset{\displaystyle :\ddot{O}H}{|}}{\underset{\underset{\displaystyle CH_3}{|}}{N}}{-}CH_3$$

A tertiary amine   An alkene   Dimethylhydroxyl
    N-oxide   amine

The Cope elimination is also a syn elimination and proceeds through a cyclic transition state:

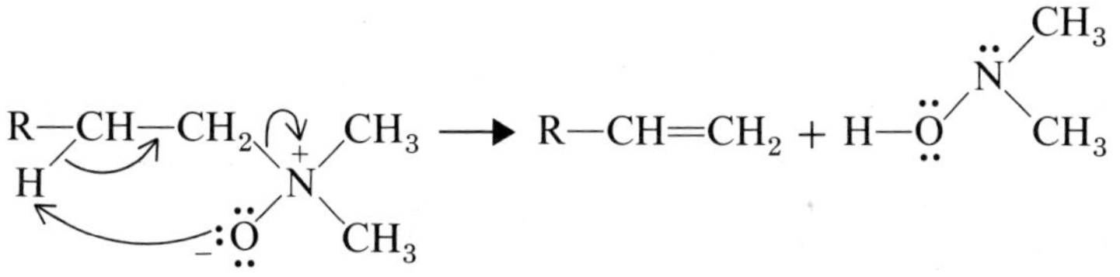

$$R{-}CH{-}CH_2 \cdots CH_3 \longrightarrow R{-}CH{=}CH_2 + H{-}\ddot{O} \overset{\displaystyle \ddot{N} \overset{\nearrow CH_3}{\searrow CH_3}}{}$$

Tertiary amine N-oxides are easily prepared by treating tertiary amines with hydrogen peroxide:

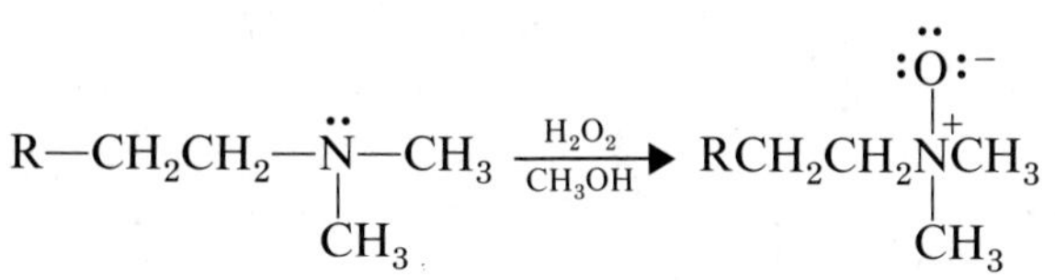

$$R{-}CH_2CH_2{-}\overset{\overset{\displaystyle ..}{}}{\underset{\underset{\displaystyle CH_3}{|}}{N}}{-}CH_3 \xrightarrow[CH_3OH]{H_2O_2} RCH_2CH_2\overset{\overset{\displaystyle :\ddot{O}:^-}{|}}{\underset{\underset{\displaystyle CH_3}{|}}{\overset{+}{N}}}CH_3$$

17.11 NUCLEOPHILIC SUBSTITUTION AND ELIMINATION REACTIONS AT UNSATURATED CARBONS

We have seen in earlier chapters that benzene compounds and alkenes are susceptible to electrophilic attack at their unsaturated carbons. Benzene compounds react primarily by substitution,

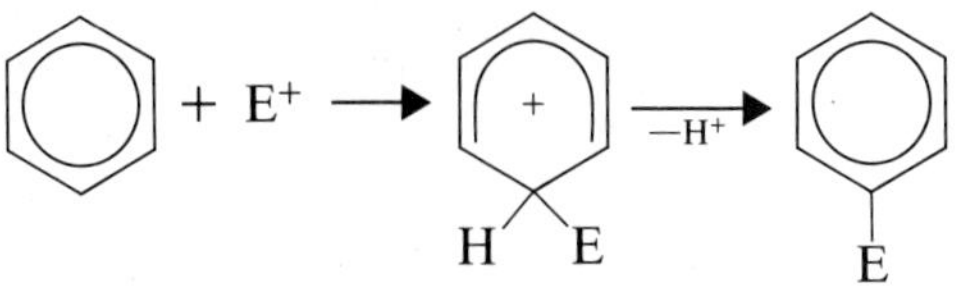

$$\bigcirc + E^+ \longrightarrow \overset{+}{\bigcirc}\underset{H\quad E}{} \xrightarrow{-H^+} \bigcirc\underset{E}{}$$

and alkenes react primarily by addition.

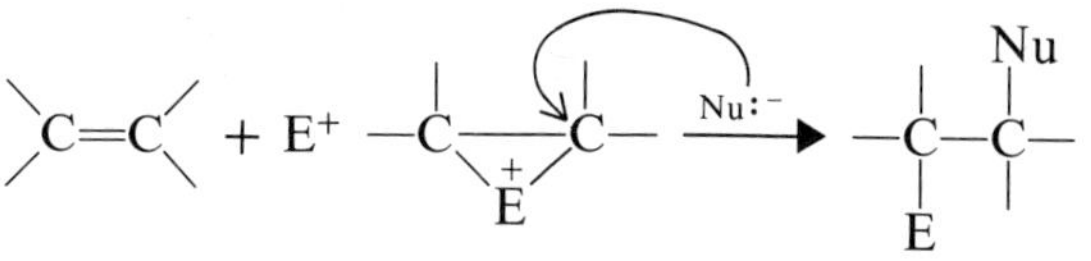

$$\overset{\diagdown}{\diagup}C{=}C\overset{\diagup}{\diagdown} + E^+ \quad {-}\overset{|}{C}\underset{\overset{+}{E}}{-}\overset{|}{C}{-} \xrightarrow{Nu:^-} {-}\overset{|}{\underset{E}{C}}\overset{\overset{Nu}{|}}{\underset{|}{C}}{-}$$

Most benzene compounds and *most* alkenes, however, are unreactive toward nucleophilic attack at their unsaturated carbons. Chlorobenzene, for example, can

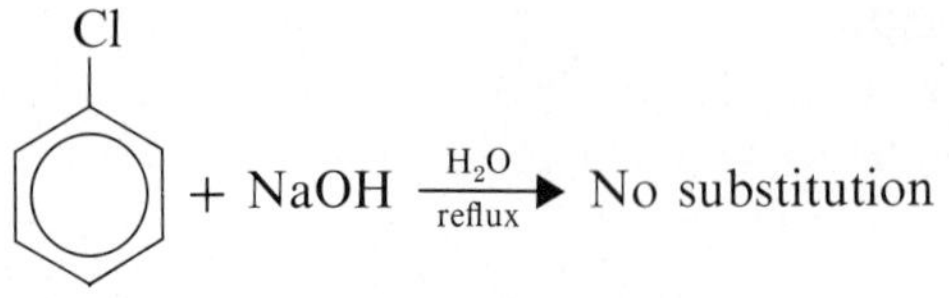

$$\text{(chlorobenzene)} + NaOH \xrightarrow[\text{reflux}]{H_2O} \text{No substitution}$$

be boiled with sodium hydroxide, for days and no detectable amount of phenol (or phenoxide) is produced.* Similarly, when vinyl chloride is heated with sodium hydroxide no substitution occurs:

$$CH_2{=}CHCl + NaOH \xrightarrow[\text{reflux}]{H_2O} \text{No substitution}$$

We can understand this lack of reactivity on the basis of two factors. (1) The unsaturated carbons of benzene compounds and alkenes are especially electron rich because of the electrons of their π bonds. Thus, we would not expect an electron-rich nucleophile to attack an electron-rich carbon in an S_N2 fashion. (2) Phenyl and vinyl cations are very unstable; less stable even than primary carbocations. We would not expect, therefore, for reactions to take place by an S_N1 path either.

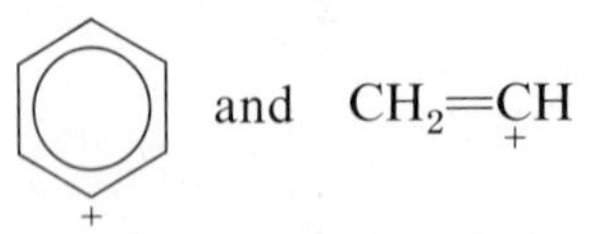

and $\quad CH_2{=}\overset{+}{CH}$

Highly unstable carbocations

Bimolecular Nucleophilic Aromatic Substitution Reactions

Some nucleophilic substitution reactions at unsaturated carbons *do* occur readily, however. But they only happen when an electronic factor makes the unsaturated carbon susceptible to nucleophilic attack. With benzene compounds, nucleophilic substitution can occur when strong electron-withdrawing groups are ortho or para to the leaving group. Examples are reactions of the compounds shown below.

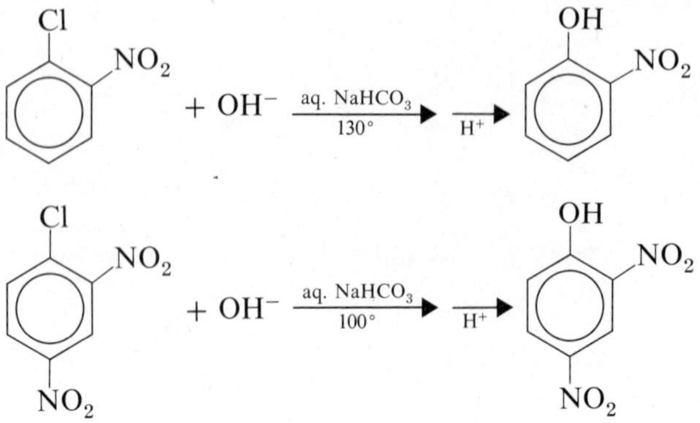

* Phenoxide ion is formed at very high temperatures (370°) and under pressure (cf. p. 674).

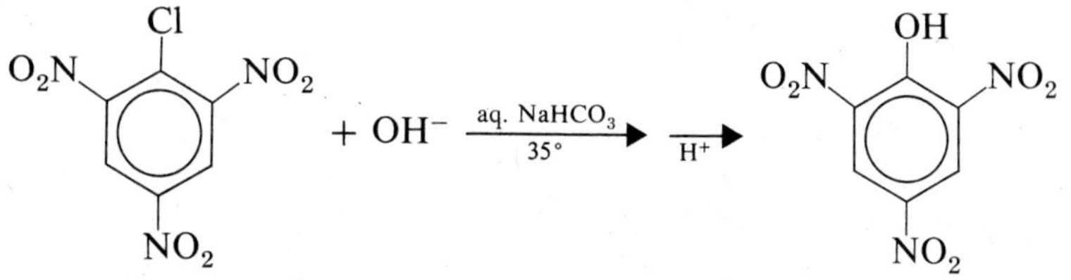

We also see in these examples that the temperature required to bring about the reaction is related to the number of ortho or para nitro groups. Of the three compounds, o-nitrochlorobenzene requires the highest temperature (p-nitrochlorobenzene reacts at 130° as well) and 2,4,6 trinitrochlorobenzene requires the lowest temperature.

A meta-nitro group does not produce a similar activating effect. For example, m-nitrochlorobenzene gives no reaction when it is heated with an amine at 180–190°.

The mechanism that operates in the nucleophilic substitution reactions of ortho and para nitrobenzene compounds is *bimolecular*. It involves the formation of an intermediate, called a Meisenheimer complex, in which both the nucleophile and the leaving group are bonded to carbon.

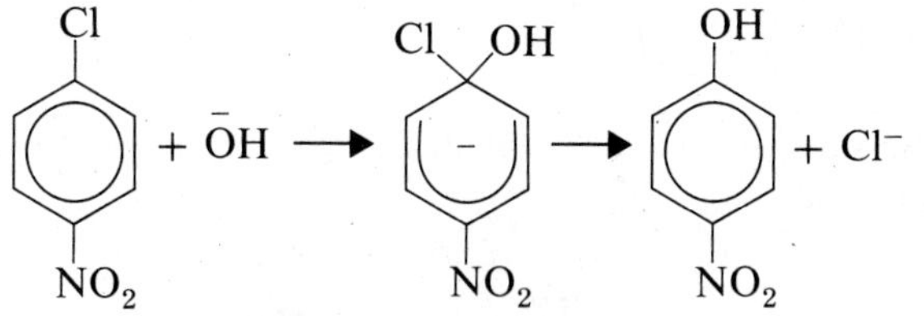

Meisenheimer complex

The Meisenheimer complex, except for its charge, bears a strong resemblance to the σ complexes that we encountered in electrophilic aromatic substitution reactions.

In electrophilic aromatic substitutions, σ complexes are stabilized by *ortho* and *para electron-releasing groups*. In nucleophilic aromatic substitutions, Meisenheimer complexes are stabilized by *ortho* and *para electron-withdrawing groups*. If we examine the following resonance structures, we can see how.

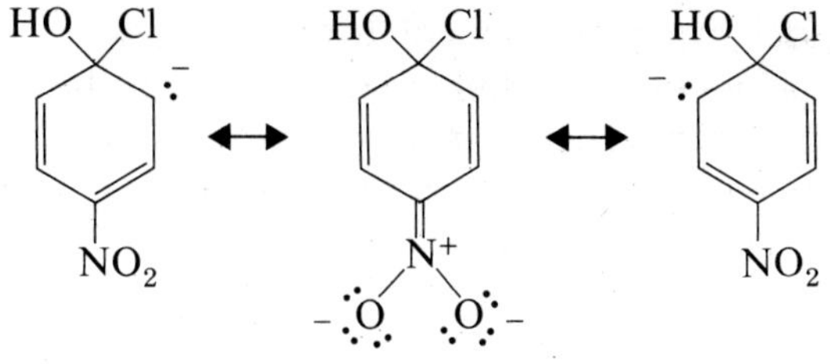

Especially stable
(negative charge is on oxygen)

Problem 17.19

What products would you expect from each of the following bimolecular nucleophilic substitution reactions?

(a) p-Nitrochlorobenzene + CH_3ONa $\xrightarrow{100°}$

(b) o-Nitrochlorobenzene + CH_3NH_2 $\xrightarrow[160°]{C_2H_5OH}$

(c) 2,4-Dinitrochlorobenzene + $C_6H_5NH_2$ $\xrightarrow{95°}$

Vinyl halides also undergo bimolecular nucleophilic substitution when they are activated by a strong electron-withdrawing group. An example is shown below.

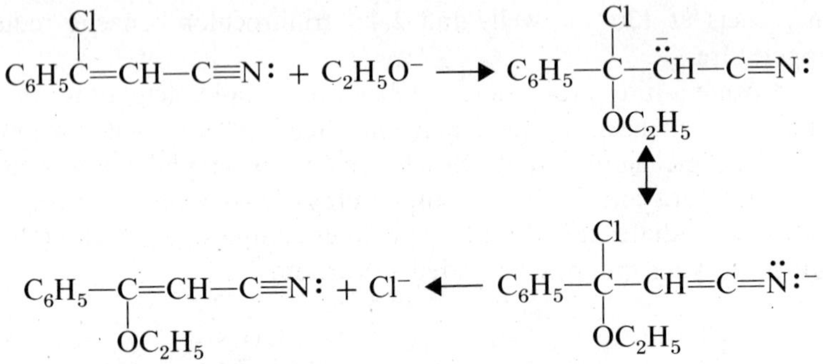

$$C_6H_5\overset{\underset{|}{Cl}}{C}=CH-C\equiv N\colon + C_2H_5O^- \longrightarrow C_6H_5-\overset{\underset{|}{Cl}}{C}-\overset{\cdot\cdot}{C}H-C\equiv N\colon$$
$$\underset{OC_2H_5}{\big|}$$

$$\updownarrow$$

$$C_6H_5-\underset{\underset{|}{OC_2H_5}}{C}=CH-C\equiv N\colon + Cl^- \longleftarrow C_6H_5-\underset{\underset{|}{OC_2H_5}}{\overset{\overset{|}{Cl}}{C}}-CH=C=\overset{\cdot\cdot}{N}\colon^-$$

Nucleophilic substitution reactions at the unsaturated carbon of a carbonyl group are very common.

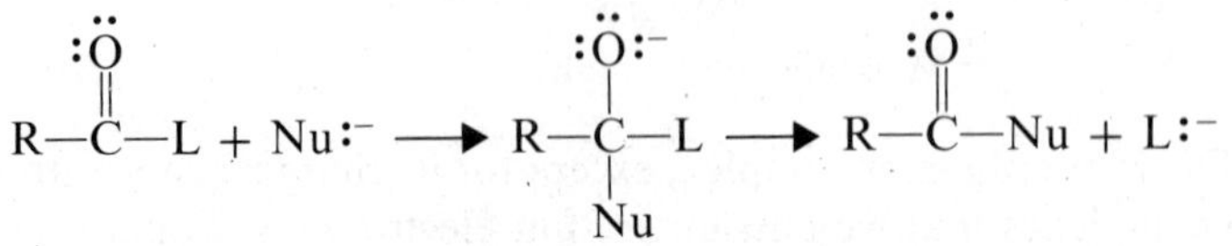

$$\overset{:\overset{\cdot\cdot}{O}}{\underset{}{\underset{}{R-\overset{\|}{C}-L}}} + Nu\colon^- \longrightarrow R-\underset{\underset{Nu}{\big|}}{\overset{:\overset{\cdot\cdot}{O}\colon^-}{C}}-L \longrightarrow \overset{:\overset{\cdot\cdot}{O}}{\underset{}{R-\overset{\|}{C}-Nu}} + L\colon^-$$

We will see a number of examples of this important reaction in Chapter 19.

Nucleophilic Aromatic Substitution Through an Elimination-Addition Mechanism: Benzyne

Although aryl halides such as chlorobenzene and bromobenzene do not react with most nucleophiles under ordinary circumstances, they do react under forcing conditions. Chlorobenzene can be converted to phenol by heating it with aqueous sodium hydroxide in a pressurized reactor at 340°

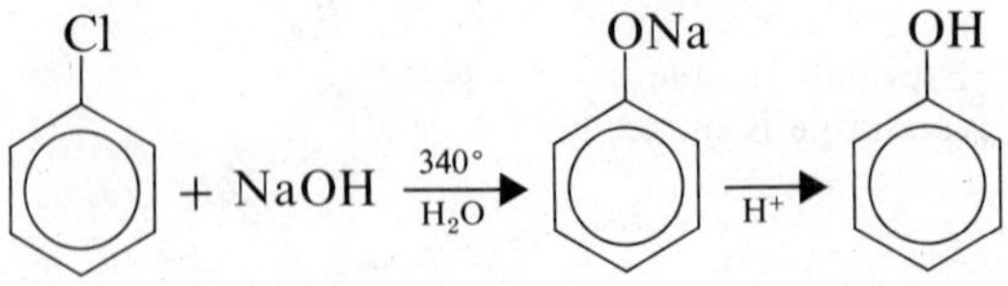

phenol

Bromobenzene reacts with the very powerful base, NH_2^-, in liquid ammonia:

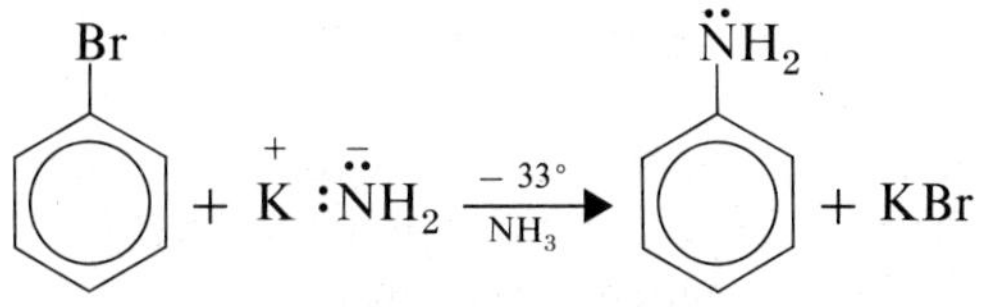

aniline

These reactions do not take place through a bimolecular substitution reaction; they take place through an *elimination-addition mechanism* that involves the formation of an unusual intermediate called *benzyne*. We can illustrate this mechanism with the reaction of bromobenzene and amide ion.

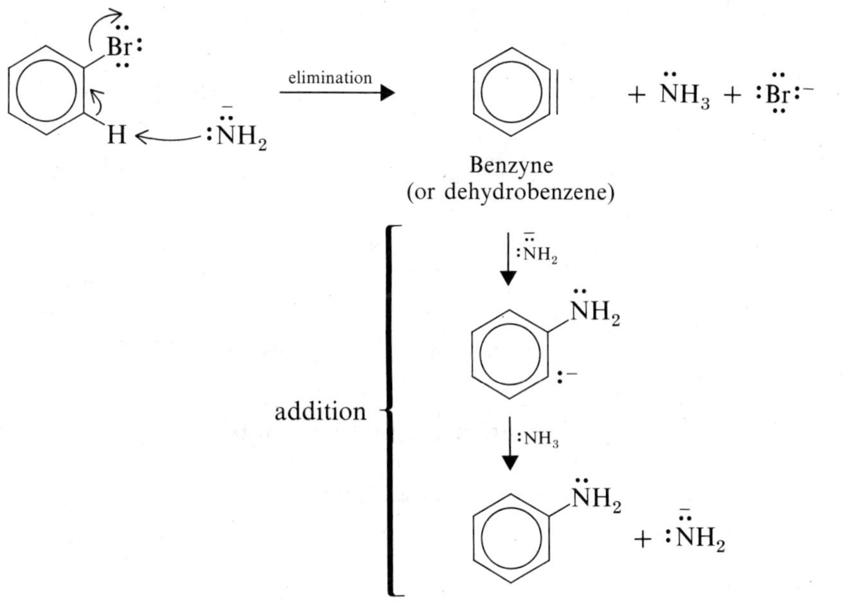

Benzyne
(or dehydrobenzene)

addition

In the first step (above), the amide ion initiates an elimination reaction by abstracting an ortho proton. This elimination reaction produces the highly unstable, and thus, highly reactive *benzyne*. Benzyne then reacts with any available nucleophile (in this case with amide ion) by a two step addition reaction to produce aniline.

The nature of benzyne, itself, will become clearer if we examine the orbital diagram below.

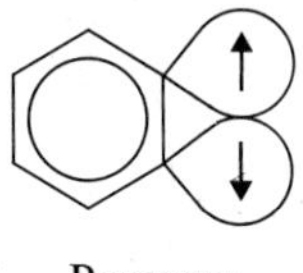

Benzyne

The extra bond in benzyne results from the overlap of sp^2 orbitals on adjacent carbons of the ring. The axes of these sp^2 orbitals lie in the same plane as that of the ring, and consequently, they do not overlap with the π orbitals of the aromatic system. They do not appreciably disturb the aromatic system and they do not make an appreciable resonance contribution to it. The extra bond is weak. Even though the ring hexagon is probably somewhat distorted in order to bring the sp^2 orbitals closer together, overlap between them is not large. Benzyne, as a result, is highly unstable and highly reactive.

What, then, is some of the evidence for the existence of benzyne and for an elimination-addition mechanism in some nucleophilic aromatic substitutions?

The first piece of clear-cut evidence was an experiment done by Professor J. D. Roberts in 1953; one that marked the beginning of benzyne chemistry. Roberts showed that when ^{14}C-labeled bromobenzene is treated with amide ion in liquid ammonia the aniline that is produced has the label equally divided between the 1 and 2 positions. This result is consistent with the elimination-addition mechanism below but is, of course, not at all consistent with a bimolecular displacement.

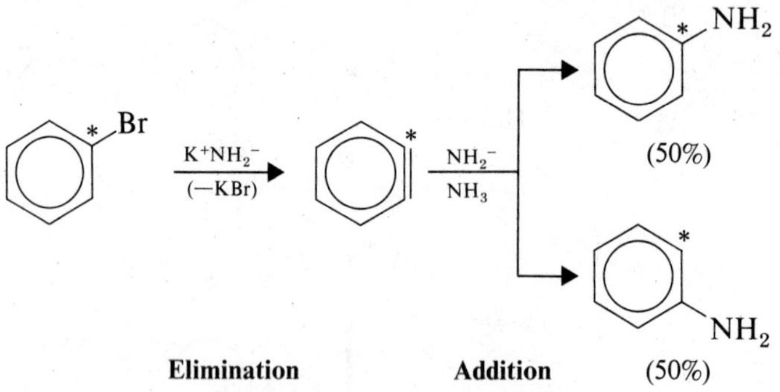

Elimination   **Addition**   (50%)

An even more striking illustration can be seen in the reaction below. When the ortho-derivative **1** is treated with sodium amide, the only organic product obtained is *meta*-trifluoromethylaniline.

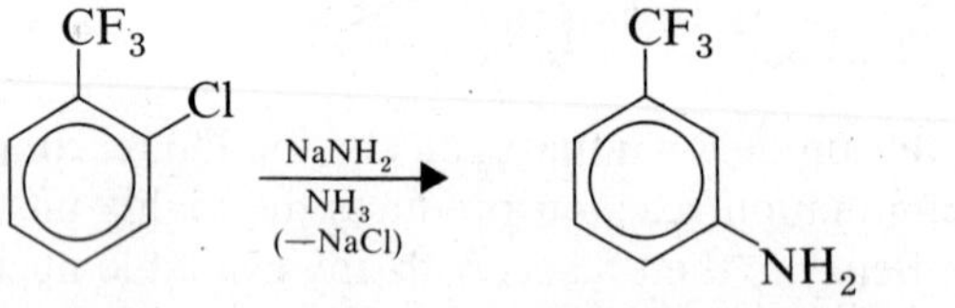

1   *m*-Trifluoromethylaniline

This result can also be explained by an addition-elimination mechanism. The first step produces the benzyne **2**:

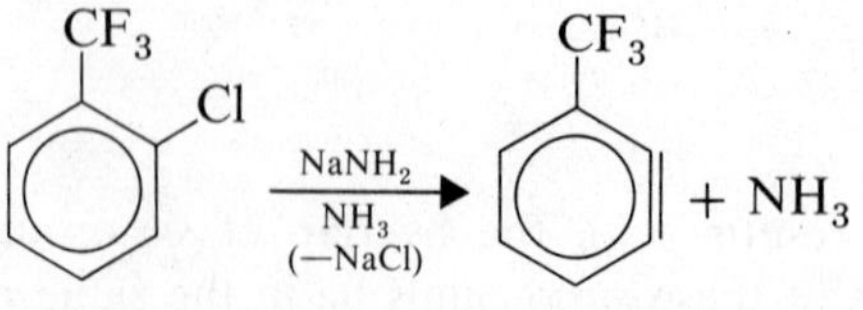

1   **2**

This benzyne then adds an amide ion in the way that produces, the more stable carbanion **3** rather than the less stable carbanion **4**.

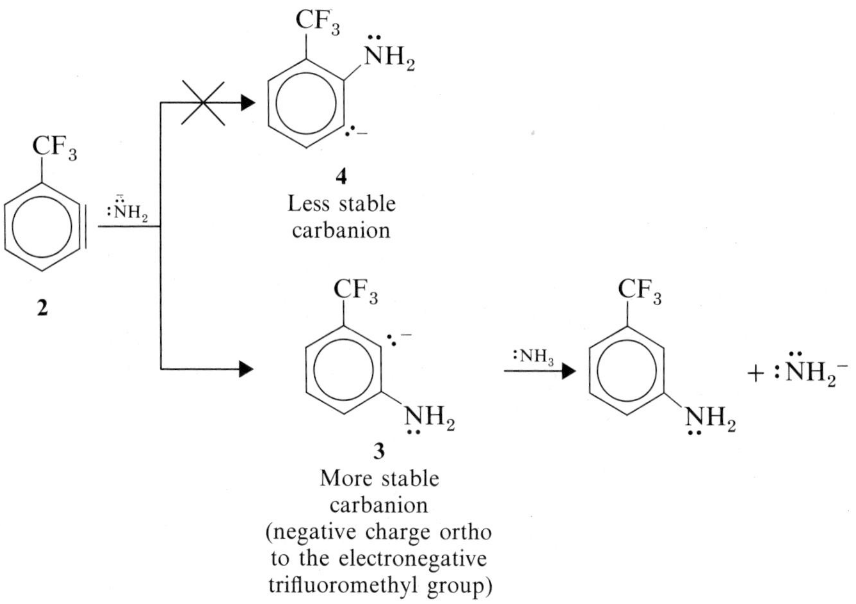

4
Less stable
carbanion

3
More stable
carbanion
(negative charge ortho
to the electronegative
trifluoromethyl group)

Carbanion **3** then accepts a proton from ammonia to form *m*-trifluoromethylaniline.

Carbanion **3** is more stable than **4** because the carbon bearing the negative charge is closer to the highly electronegative trifluoromethyl group. The trifluoromethyl group stabilizes the negative charge through its inductive effect. (Resonance effects are not important here because the sp^2 orbital that contains the electron pair does not overlap with the π orbitals of the aromatic system.)

Benzyne intermediates have been "trapped" through the use of Diels-Alder reactions. When benzyne is generated in the presence of the diene, *furan,* the product is a Diels-Alder-adduct.

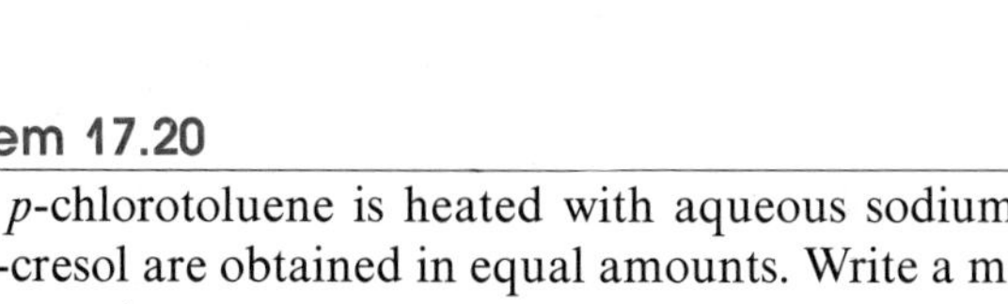

Benzyne
(generated by
an elimination
reaction)

Diels-Alder adduct

Problem 17.20

When *p*-chlorotoluene is heated with aqueous sodium hydroxide at 340°, *p*-cresol and *m*-cresol are obtained in equal amounts. Write a mechanism that would account for this result.

Problem 17.21

When 2-bromo-3-methylanisole is treated with amide ion in liquid ammonia no substitution takes place. This has been interpreted as providing evidence for the elimination-addition mechanism. Explain.

Additional Problems

17.22

When 2-iodooctane reacts with radioactive iodide ions ($^*I^-$) the starting materials and the products are chemically equivalent.

$$C_6H_{13}\underset{\displaystyle I}{\underset{|}{C}}HCH_3 + {}^*I^- \longrightarrow C_6H_{13}\underset{\displaystyle {}^*I}{\underset{|}{C}}HCH_3 + I^-$$

When the reaction is carried out with optically active 2-iodooctane, the rate of racemization is found to be exactly *twice* the rate of incorporation of radioactive iodine. This experiment is considered to provide convincing proof that an S_N2 reaction takes place with inversion of configuration. (a) How is this true? (b) What would you expect the relative rates of racemization and incorporation of radioactive iodine to be if the reaction took place through the formation of an achiral intermediate such as a carbocation?

17.23

(a) How can you account for the following relative S_N1 reaction rates?

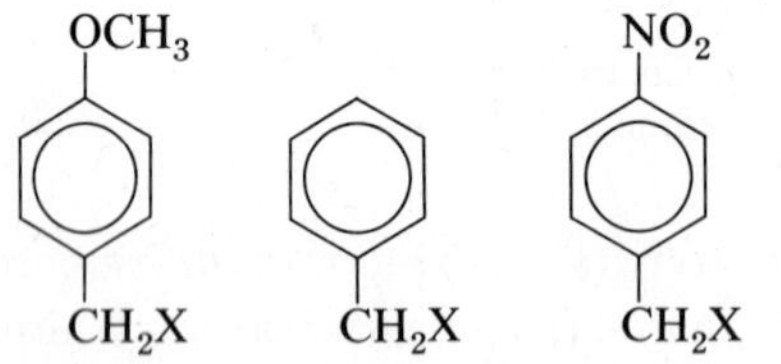

(5×10^3) (1) (5×10^{-3})

(b) How can you explain the fact that a *m*-methoxybenzyl halide undergoes S_N1 reaction at about the same rate as an unsubstituted benzyl halide?

17.24

The two syntheses of 1-phenyl-2-ethoxypropane shown below give products with opposite optical rotations.

$$\underset{\alpha = +33.0°}{C_6H_5CH_2\underset{\displaystyle OH}{\underset{|}{C}}HCH_3} \xrightarrow{K} \text{potassium salt} \xrightarrow{C_2H_5Br} \underset{\alpha = +23.5°}{C_6H_5CH_2\underset{\displaystyle OC_2H_5}{\underset{|}{C}}HCH_3}$$

$$\downarrow \text{TsCl/base}$$

$$C_6H_5CH_2\underset{\displaystyle OTs}{\underset{|}{C}}HCH_3 \xrightarrow[K_2CO_3]{C_2H_5OH} \underset{\alpha = -19.9°}{C_6H_5CH_2\underset{\displaystyle OC_2H_5}{\underset{|}{C}}HCH_3}$$

How can you explain this?

17.25

Explain the different results obtained from the two reactions shown below.

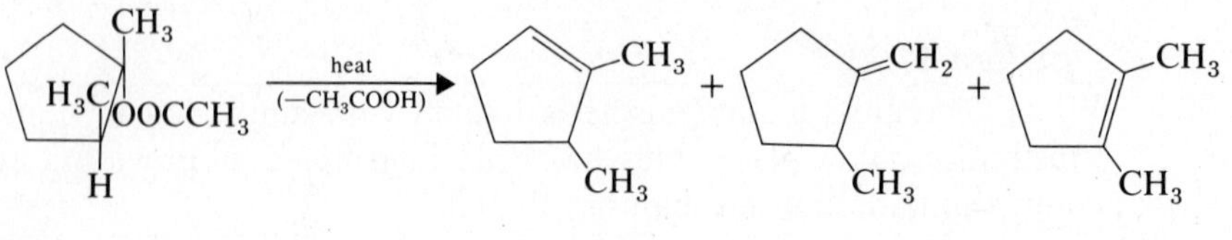

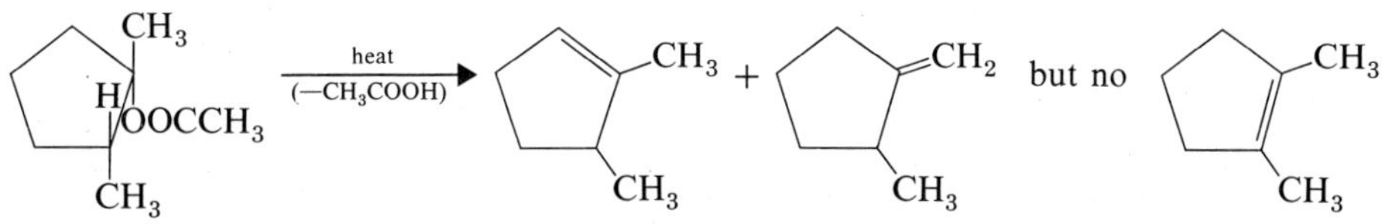

17.26

When neopentyl iodide is treated with silver acetate in acetic acid the product that is obtained from the reaction is *tert*-pentyl acetate.

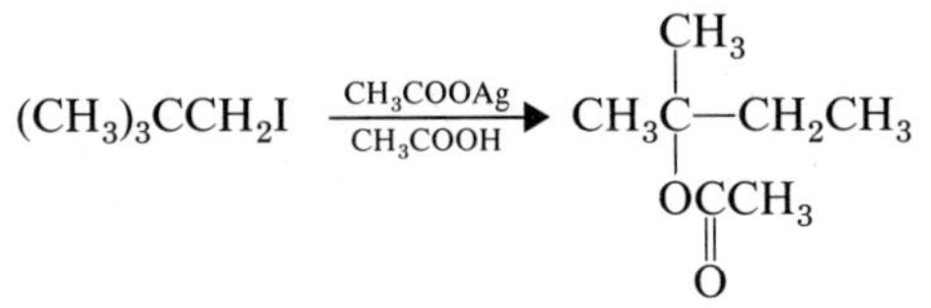

Neopentyl iodide   *tert*-Pentyl acetate

Propose a mechanism for this reaction.

17.27

Secondary halides can be converted to alcohols of the opposite configuration through: (1) an S_N2 reaction with sodium hydroxide,

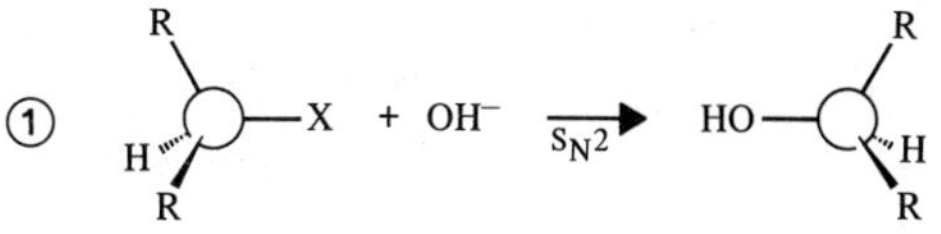

or (2) by an S_N2 reaction with sodium acetate followed by hydrolysis of the ester.

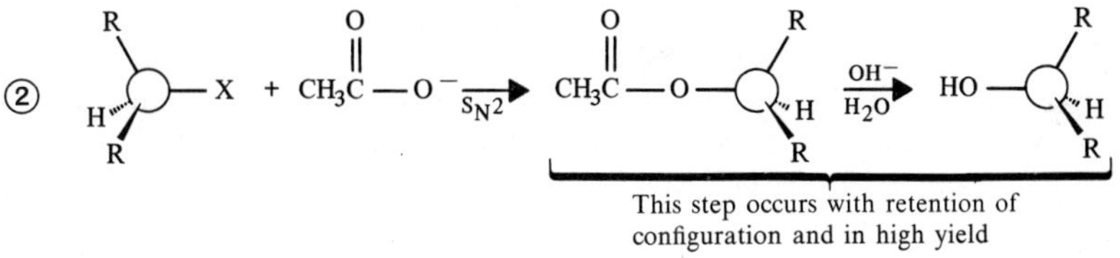

This step occurs with retention of
configuration and in high yield

The second method usually gives a much higher overall yield of the alcohol. Why?

17.28

Write a mechanism that explains the formation of tetrahydrofuran from the reaction of 4-chloro-1-butanol and aqueous sodium hydroxide.

17.29

(a) Consider the general problem of converting a tertiary alkyl halide to an alkene, for example, the conversion of *tert*-butyl chloride to 2-methylpropene. What experimental conditions would you choose to insure that elimination is favored over substitution? (b) Consider the opposite problem, that of carrying out a substitution reaction on a tertiary alkyl halide. Use as your example the conversion of *tert*-butyl chloride to ethyl *tert*-butyl ether. What experimental conditions would you employ to insure the highest possible yield of the ether?

17.30

Show how you might use nucleophilic substitution reactions to prepare each of the following compounds from appropriate alkyl halides.

(a) $C_6H_5CH_2CN$ (a nitrile)
(b) $CH_3CH_2CH_2SH$ (a thiol)
(c) $CH_3CH_2CH_2SCH_3$ (a thioether)
(d) $C_6H_5CH_2OCH_2CH_3$ (an ether)
(e) $CH_3CH_2CH_2N_3$ (an azide)

(f) $CH_3CH_2\overset{\overset{\displaystyle O}{\|}}{C}OCH_2C_6H_5$ (an ester)

(g)
$$
\begin{array}{c}
\ce{CH2-CH2}\\
| \qquad \diagdown\\
\ce{CH2} \qquad \ce{S}\\
| \qquad \diagup\\
\ce{CH2-CH2}
\end{array}
\text{ (a cyclic thioether)}
$$

(h) $C_6H_5CH_2P(C_6H_5)_3^+$ Br^- (a phosphonium salt) .

17.31

Although $(C_6H_5)_2CHCl$ is a secondary halide, it undergoes S_N1 solvolysis in 80% ethanol at a rate that is 1300 times faster than for the tertiary halide, $(CH_3)_3CCl$. Explain.

17.32

Many S_N2 reactions of alkyl chlorides and alkyl bromides are catalyzed by the addition of sodium or potassium iodide. For example, the hydrolysis of methyl bromide takes place much faster in the presence of sodium iodide. Explain.

17.33

When Grignard reagents are prepared from allyl halides, that is,

$$RCH{=}CHCH_2X + Mg \xrightarrow{\text{ether}} RCH{=}CHCH_2MgX$$

unavoidable by-products of the reactions are compounds with the formula $RCH{=}CHCH_2CH_2CH{=}CHR$. By-product formation is especially prevalent when concentrated solutions are used. Explain.

17.34

Treating propene oxide with sodium ethoxide in ethanol gives primarily 1-ethoxy-2-propanol and very little 2-ethoxy-1-propanol. What factor accounts for this?

* **17.35**

When sodium ethoxide reacts with epichlorohydrin, labeled with ^{14}C as shown in **I** below, the major product is an epoxide bearing the label as in **II**. Provide an explanation.

$$\underset{\textbf{I}}{\underset{\text{Epichlorohydrin}}{Cl{-}CH_2{-}\underset{\diagdown O \diagup}{CH}{-}\overset{*}{C}H_2}} \xrightarrow{\text{NaOC}_2\text{H}_5} \underset{\textbf{II}}{C_2H_5O\overset{*}{C}H_2{-}\underset{\diagdown O \diagup}{CH}{-}CH_2}$$

* **17.36**

Explain the following experimental observations: diphenylchloromethane, $(C_6H_5)_2CHCl$, reacts with sodium azide, NaN_3, in aqueous acetone to give a mixture of $(C_6H_5)_2CHN_3$ and $(C_6H_5)_2CHOH$. The rate of this reaction is unaffected by increasing the concentration of NaN_3. When diphenylbromomethane, $(C_6H_5)_2CHBr$, reacts with NaN_3 under the same conditions, the reaction takes place faster but the ratio of products $[(C_6H_5)_2CHN_3$: $(C_6H_5)_2CHOH]$ is the same.

* 17.37

The relative rates of ethanolysis of several primary alkyl halides are as follows: CH_3CH_2Br, 1.0; $CH_3CH_2CH_2Br$, 0.28; $(CH_3)_2CHCH_2Br$, 0.030; $(CH_3)_3CCH_2Br$, 0.0000042. (a) Is this reaction likely to be S_N1 or S_N2? (b) What factor will account for these relative reactivities?

* 17.38

In contrast to S_N2 reactions, S_N1 reactions show relatively little nucleophile selectivity. That is, when more than one nucleophile is present in the reaction medium, S_N1 reactions show only a slight tendency to discriminate between weak nucleophiles and strong nucleophiles, whereas S_N2 reactions show a marked tendency to discriminate. (a) Provide an explanation for this behavior. (b) Show how your answer accounts for the fact that $CH_3CH_2CH_2CH_2Cl$ reacts with 0.01 M NaCN in ethanol to yield primarily $CH_3CH_2CH_2CH_2CN$ whereas under the same conditions, $(CH_3)_3CCl$ reacts to give primarily $(CH_3)_3COCH_2CH_3$.

* 17.39

In 1949, D. J. Cram published the first of a series of papers on the solvolysis of 1-methyl-2-phenylpropyl tosylates, **A** and **C**. These reactions displayed a remarkable stereospecificity: when the optically active tosylate **A** was heated in acetic acid, the reaction yielded almost exclusively the optically active acetate **B**. On the other hand, heating the optically active tosylate **C** in acetic acid gave the racemic acetate **D**, **E**. Provide an explanation for these results.

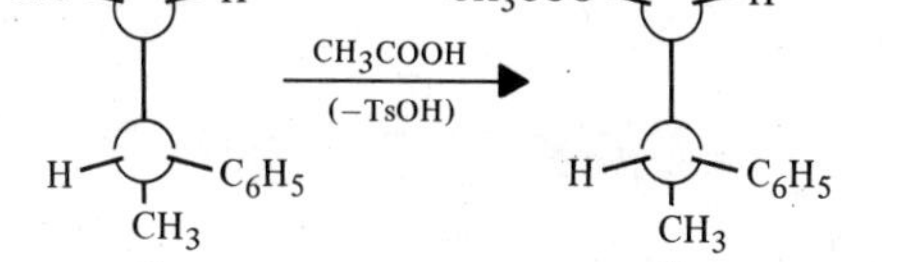

Optically active

Racemic modification

18 ALDEHYDES AND KETONES

18.1 INTRODUCTION

The carbonyl group, $\diagdown$C=O, is one of the most important functional groups in organic chemistry.

Although we have had some experience with carbonyl compounds in earlier chapters we will now consider their chemistry in detail. In this chapter we will study aldehydes—compounds in which the carbonyl group is bonded to carbon and hydrogen—and ketones—compounds in which the carbonyl group is bonded to two carbons. In Chapter 19 we will study compounds in which the carbonyl group is bonded to an oxygen (acids, esters, anhydrides), to a nitrogen (amides), and to halogens (acyl halides).

18.2 NOMENCLATURE OF ALDEHYDES AND KETONES

In the IUPAC system we name aliphatic aldehydes by replacing the final *e* of the name of the corresponding alkane with *al*. Since the aldehyde group must be at the end of the chain of carbons there is no need to designate its position. When other substituents are present, however, we assume that the carbonyl group occupies position -1. (Many aldehydes also have common names; these are given below in parentheses.)

$$\begin{array}{ccc} \overset{O}{\underset{\|}{\text{H}-\text{C}-\text{H}}} & \overset{O}{\underset{\|}{\text{CH}_3\text{C}-\text{H}}} & \overset{O}{\underset{\|}{\text{CH}_3\text{CH}_2\text{C}-\text{H}}} \\ \text{Methanal} & \text{Ethanal} & \text{Propanal} \\ \text{(formaldehyde)} & \text{(acetaldehyde)} & \text{(propionaldehyde)} \end{array}$$

$$\begin{array}{cc} \overset{O}{\underset{\|}{\text{CH}_3\text{CH}_2\text{CH}_2\text{C}-\text{H}}} & \overset{O}{\underset{\|}{\text{CH}_3\underset{\underset{\text{CH}_3}{|}}{\text{CH}}\text{C}-\text{H}}} \\ \text{Butanal} & \text{2-Methylpropanal} \\ \text{(butyraldehyde)} & \text{(isobutyraldehyde)} \end{array}$$

$$\begin{array}{cc} \overset{O}{\underset{\|}{\text{ClCH}_2\text{CH}_2\text{CH}_2\text{CH}_2\text{C}-\text{H}}} & \overset{O}{\underset{\|}{\text{C}_6\text{H}_5\text{CH}_2\text{C}-\text{H}}} \\ \text{5-Chloropentanal} & \text{Phenylethanal} \\ & \text{(phenylacetaldehyde)} \end{array}$$

When the carbonyl group is attached to an aromatic ring we name the compound as a benzaldehyde, tolualdehyde, naphthaldehyde, and so on.

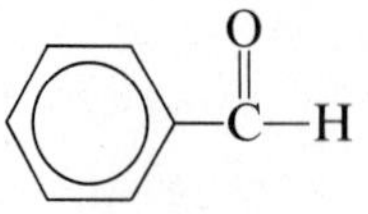

Benzaldehyde

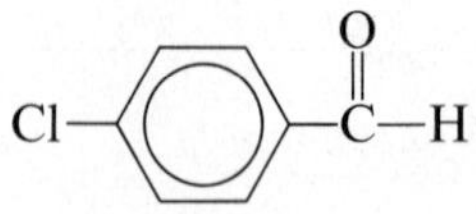

p-Chloro-benzaldehyde

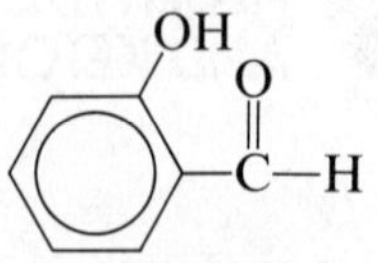

o-Hydroxybenzaldehyde
(salicylaldehyde)

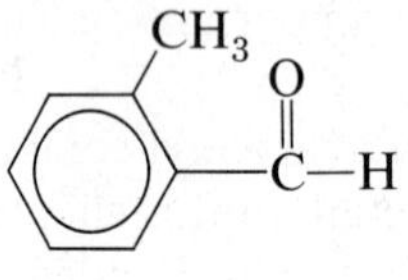

o-Tolualdehyde

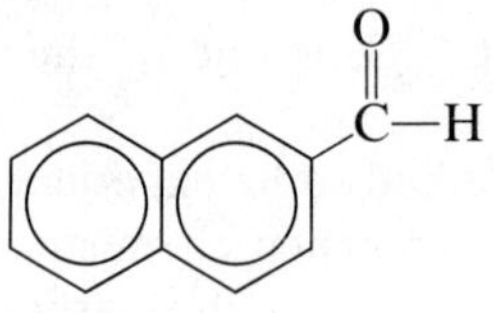

2-Naphthaldehyde

We name aliphatic ketones by replacing the final *e* of the name of the corresponding alkane with *one*. We then number the chain in the way that gives the carbonyl carbon the lower possible number and we use this number to designate its position. (No number is necessary for propanone and butanone. Why?)

$$CH_3CCH_3 \qquad CH_3CH_2CCH_3 \qquad CH_3CCH_2CH_2CH_3$$

Propanone
(acetone)

Butanone
(methyl ethyl ketone)

2-Pentanone
(methyl propyl ketone)

Common names for ketones (in parentheses above and below) are obtained simply by naming the two groups attached to the carbonyl group and adding the word *ketone*.

Some aromatic ketones have special names.

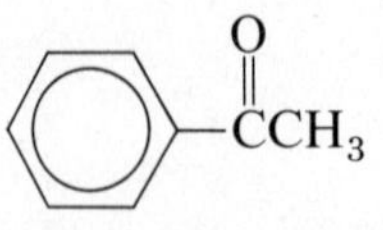

Acetophenone
(methyl phenyl ketone)

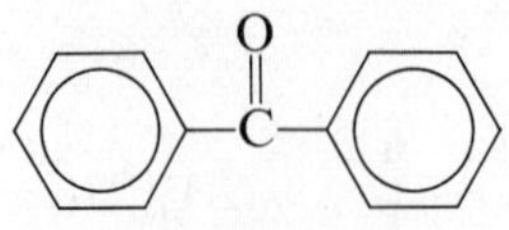

Benzophenone
(diphenyl ketone)

When we find it necessary to name the $-\overset{O}{\overset{\|}{C}}H$ group as a substituent, we call it the *formyl group*. When $R\overset{O}{\overset{\|}{C}}-$ groups are named as substituents, they are called *acyl groups*.

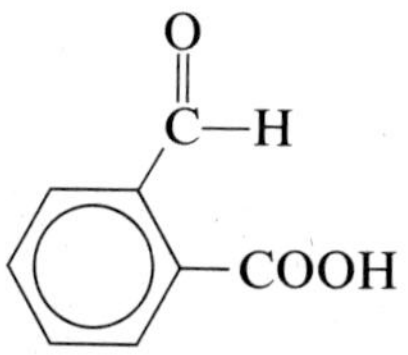

o-Formylbenzoic acid 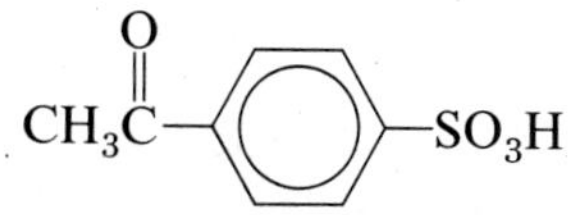   *p*-Acetylbenzenesulfonic acid

Problem 18.1

(a) Give IUPAC names for the seven aldehyde and ketone isomers with the formula $C_5H_{10}O$. (b) How many aldehydes and ketones contain a benzene ring and have the formula C_8H_8O? (c) What are their names?

18.3 PHYSICAL PROPERTIES

The carbonyl group is a polar group; thus, aldehydes and ketones have higher boiling points than hydrocarbons of the same molecular weight. However, since aldehydes and ketones cannot form strong hydrogen bonds to each other, they have lower boiling points than corresponding alcohols.

$CH_3CH_2CH_3$	$CH_3CH_2\overset{\text{O}}{\overset{\|}{C}}H$	$CH_3\overset{\text{O}}{\overset{\|}{C}}CH_3$	$CH_3CH_2CH_2OH$
Propane	Propanal	Acetone	Propanol
bp $-44.5°$	$49°$	$56.2°$	$97.1°$

Problem 18.2

Which compound in each pair listed below would you expect to have the higher boiling point? (Answer this problem without consulting tables.)
    (a) Pentanal or 1-pentanol
    (b) 2-Pentanone or 2-pentanol
    (c) *n*-Pentane or pentanal
    (d) Acetophenone or 2-phenylethánol
    (e) Benzaldehyde or benzyl alcohol

The carbonyl oxygen allows aldehydes and ketones to form strong hydrogen bonds to water. As a result, low-molecular-weight aldehydes and ketones show appreciable water solubilities. Acetone and acetaldehyde are soluble in water in all proportions.

Table 18.1 lists the physical properties of a number of common aldehydes and ketones.

A number of aromatic aldehydes obtained from natural sources have very pleasant fragrances. Some of these are shown on page 684.

TABLE 18.1 Physical Properties of Aldehydes and Ketones

FORMULA	NAME	mp °C	bp °C	WATER SOLUBILITY
HCHO	Formaldehyde	−92	−21	Very sol.
CH_3CHO	Acetaldehyde	−125	21	∞
CH_3CH_2CHO	Propanal	−81	49	Very sol.
$CH_3(CH_2)_2CHO$	Butanal	−99	76	Sol.
$CH_3(CH_2)_3CHO$	Pentanal	−91.5	102	Sl. sol.
$CH_3(CH_2)_4CHO$	Hexanal	−56	128	Sl. sol.
C_6H_5CHO	Benzaldehyde	−57	178	Sl. sol.
$C_6H_5CH_2CHO$	Phenylacetaldehyde	33	193	Sl. sol.
CH_3COCH_3	Acetone	−95	56.1	∞
$CH_3COCH_2CH_3$	Butanone	−86	79.6	Very sol.
$CH_3COCH_2CH_2CH_3$	2-Pentanone	−78	102	Sol.
$CH_3CH_2COCH_2CH_3$	3-Pentanone	−42	102	Sol.
$C_6H_5COCH_3$	Acetophenone	21	202	Insol.
$C_6H_5COC_6H_5$	Benzophenone	48	306	Insol.

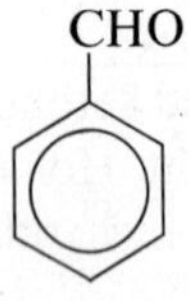

Benzaldehyde
(from bitter almonds)

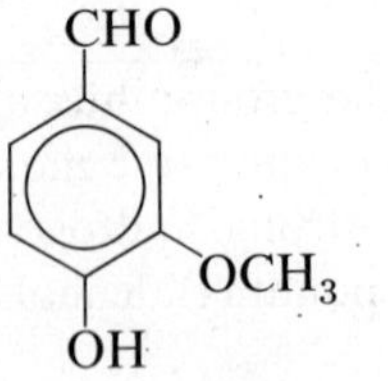

Vanillin
(from vanilla bean)

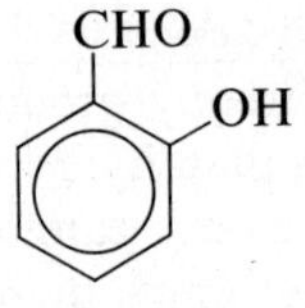

Salicylaldehyde
(from meadow sweet)

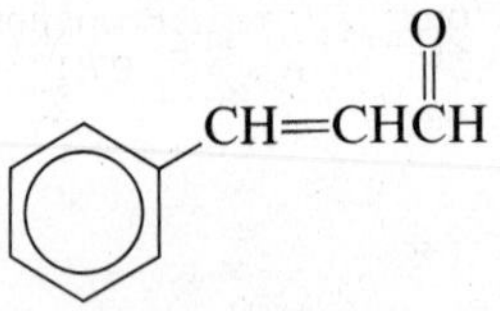

Cinnamaldehyde
(from cinnamon)

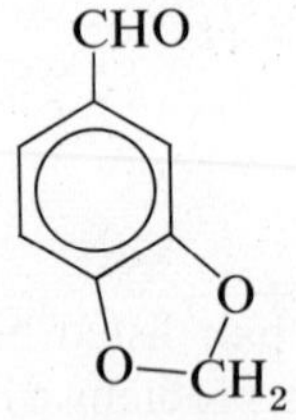

Piperonal
(made from safrole,
odor of heliotrope)

Spectroscopic Properties of Aldehydes and Ketones

Carbonyl groups of aldehydes and ketones give rise to very strong C=O stretching bands in the 1665–1780 cm^{-1} region of the *infrared spectrum*. The exact location of the peak (p. 685) depends on the structure of the aldehyde or ketone.

The CHO group of aldehydes also gives two weak bands in the 2700–2775 and 2820–2900 cm^{-1} region of the infrared spectrum.

The aldehydic proton gives a peak far downfield ($\delta = 9$–10) in *pmr spectra*.

The carbonyl groups of saturated aldehydes and ketones give a weak absorption band in the *ultraviolet region* between 270–300 nm. This band is shifted to longer wavelengths (300–350 nm) when the carbonyl group is conjugated with a double bond.

C=O STRETCHING FREQUENCIES			
R—CHO	$1720-1740$ cm^{-1}	RCOR	$1705-1720$ cm^{-1}
Ar—CHO	$1695-1715$ cm^{-1}	ArCOR	$1680-1700$ cm^{-1}
$-\overset{\mid}{C}=\overset{\mid}{C}-CHO$	$1680-1690$ cm^{-1}	$-\overset{\mid}{C}=\overset{\mid}{C}-COR$	$1665-1680$ cm^{-1}
		Cyclohexanones	$1665-1685$ cm^{-1}
		Cyclopentanones	$1708-1725$ cm^{-1}

18.4 PREPARATION OF ALDEHYDES

Aldehydes can be prepared by several methods that involve oxidation or reduction. However, since aldehydes are easily oxidized and reduced, we must use special reagents or techniques.

1. Preparation of aldehydes by oxidation of primary alcohols. Primary alcohols can be oxidized to aldehydes. A reagent commonly used for this purpose is a complex formed when chromic oxide (or $Na_2Cr_2O_7$) reacts with pyridine.

$$C_6H_{13}CH_2OH + CrO_3\text{-}C_5H_5N \xrightarrow{CH_2Cl_2} C_6H_{13}CHO$$

1-Heptanol   Chromic oxide-pyridine   Heptanal (93%)

$$C_6H_5CH=CHCH_2OH \xrightarrow[\substack{C_5H_5\overset{+}{N}HCl \\ C_5H_5N}]{Na_2Cr_2O_7} C_6H_5CH=CHCHO$$

Cinnamyl alcohol   Cinnamaldehyde (80%)

2. Preparation of aldehydes by oxidation of methylbenzenes. Strong oxidizing agents oxidize methylbenzenes to benzoic acids (p. 490). However, it is possible to halt the oxidation at the aldehyde stage by using reagents that convert the methyl

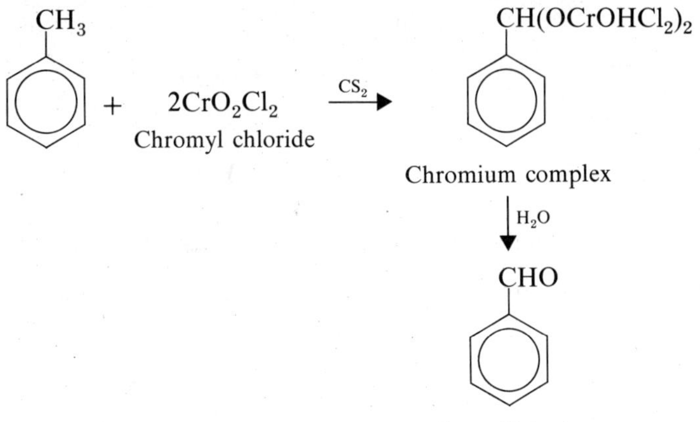

CH$_3$ + 2CrO$_2$Cl$_2$ $\xrightarrow{CS_2}$ CH(OCrOHCl$_2$)$_2$

Chromyl chloride   Chromium complex

$\downarrow$ H$_2$O

CHO

Benzaldehyde (90%)

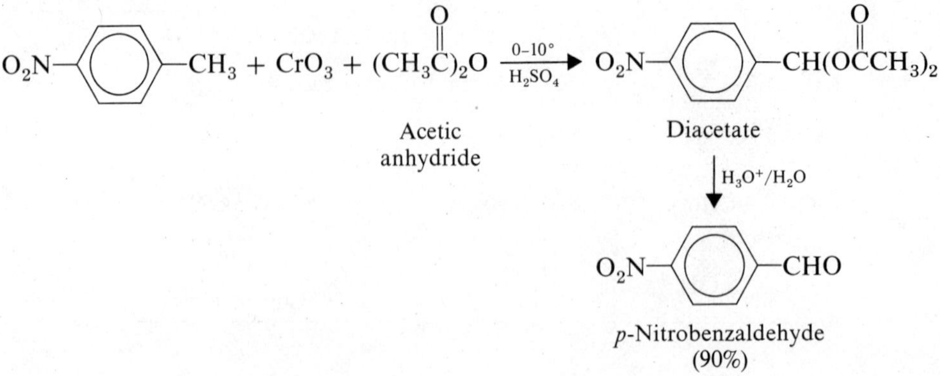

$$O_2N-\langle\bigcirc\rangle-CH_3 + CrO_3 + (CH_3\overset{O}{\overset{\|}{C}})_2O \xrightarrow[H_2SO_4]{0-10°} O_2N-\langle\bigcirc\rangle-CH(O\overset{O}{\overset{\|}{C}}CH_3)_2$$

Acetic
anhydride

Diacetate

$$\downarrow H_3O^+/H_2O$$

$$O_2N-\langle\bigcirc\rangle-CHO$$

p-Nitrobenzaldehyde
(90%)

group to an intermediate that is difficult to oxidize further. Two methods, based on the use of chromium compounds are illustrated above and on page 685. In the first method, chromyl chloride oxidizes the methyl group to a chromium complex that is hydrolyzed to an aldehyde when it is treated with water. In the second method, the oxidation is carried out in acetic anhydride (as the solvent) using chromic oxide. This method yields a diacetate that can be hydrolyzed to an aldehyde with aqueous acid.

3. Preparation of aldehydes by reduction of acid derivatives. A number of special methods can be used to convert acid derivatives to aldehydes. Acyl chlorides, for example, can be reduced to aldehydes by using hydrogen and a palladium catalyst that has been treated with sulfur. This technique, called the *Rosenmund reduction,* usually gives aldehydes in excellent yields.

$$R-\overset{O}{\overset{\|}{C}}-OH \xrightarrow{SOCl_2} R-\overset{O}{\overset{\|}{C}}-Cl \xrightarrow[Pd(S)]{H_2} R-\overset{O}{\overset{\|}{C}}-H$$
(80–90%)

Acid chlorides can also be reduced to aldehydes by treating them with lithium tri-*tert*-butoxyaluminum hydride, LiAlH[OC(CH$_3$)$_3$]$_3$.

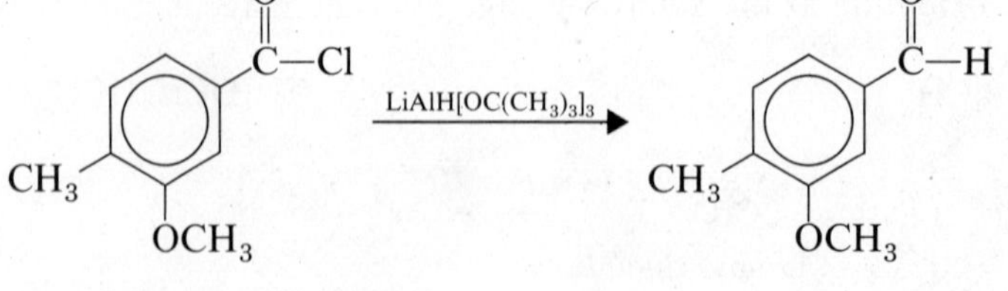

$$\xrightarrow{LiAlH[OC(CH_3)_3]_3}$$

18.5 PREPARATION OF KETONES

We have seen two methods for the preparation of ketones in earlier chapters.

1. Preparation of ketones through Friedel-Crafts acylation reactions (discussed in Section 13.7).

$$ArH + R-\overset{O}{\overset{\|}{C}}-Cl \xrightarrow{AlCl_3} Ar-\overset{O}{\overset{\|}{C}}-R$$
An aryl alkyl
ketone

or

$$ArH + Ar\overset{\overset{\displaystyle O}{\|}}{-C}-Cl \xrightarrow{AlCl_3} Ar\overset{\overset{\displaystyle O}{\|}}{-C}-Ar$$

A diaryl ketone

2. Preparation of ketones by oxidation of secondary alcohols. (discussed in Section 16.6).

$$R\overset{\overset{\displaystyle OH}{|}}{-CH}-R' \xrightarrow{(O)} R\overset{\overset{\displaystyle O}{\|}}{-C}-R'$$

Two other methods for the preparation of ketones are based on the use of organometallic compounds.

3. Preparation of ketones through the reaction of organocadmium compounds with acid chlorides. Grignard reagents are too reactive to be used to prepare ketones. We saw in Section 16.5 that Grignard reagents react with esters to yield tertiary alcohols. Grignard reagents also react with acid chlorides in the same general way.

$$R\overset{\overset{\displaystyle O}{\|}}{-C}-Cl + 2R'MgX \xrightarrow[(2)HOH]{} R\overset{\overset{\displaystyle OH}{|}}{\underset{\underset{\displaystyle R'}{|}}{C}}-R' + MgXCl + MgXOH$$

However, if a Grignard reagent is first converted to a dialkylcadmium by treating it with anhydrous cadmium chloride, subsequent treatment with an acid chloride gives a ketone in good yield.

$$2RMgX + CdCl_2 \longrightarrow R_2Cd \xrightarrow{2R'\overset{\overset{\displaystyle O}{\|}}{-C}-Cl} 2R\overset{\overset{\displaystyle O}{\|}}{-C}-R'$$

Dialkylcadmium
+
2MgXCl

This synthesis succeeds because cadmium reagents are too unreactive to attack the ketone that is produced. Two examples of this method are shown below.

$$2(CH_3)_2CHMgBr + CdCl_2 \xrightarrow{ether} [(CH_3)_2CH]_2Cd \quad \curlyvee \quad 2mgbrcl$$

$$\xrightarrow[ether]{2CH_3CH_2\overset{\overset{\displaystyle O}{\|}}{C}Cl}$$

$$2(CH_3)_2CH\overset{\overset{\displaystyle O}{\|}}{C}CH_2CH_3$$

(60%)

$$2CH_3MgI + CdCl_2 \xrightarrow{ether} (CH_3)_2Cd \xrightarrow{2C_2H_5O\overset{\overset{\displaystyle O}{\|}}{C}(CH_2)_8\overset{\overset{\displaystyle O}{\|}}{C}Cl} 2C_2H_5O\overset{\overset{\displaystyle O}{\|}}{C}(CH_2)_8\overset{\overset{\displaystyle O}{\|}}{C}CH_3$$

(84%)

4. Preparation of ketones from lithium dialkylcuprates. When an ether solution of a lithium dialkylcuprate is treated with an acid chloride at $-78°$, the product that forms is a ketone. This ketone synthesis is a variation of the Corey-House alkane synthesis (Section 3.13).

$$\underset{\text{Lithium dialkylcuprate}}{R_2CuLi} + R'{-}\overset{\overset{\displaystyle O}{\|}}{C}{-}Cl \longrightarrow R'{-}\overset{\overset{\displaystyle O}{\|}}{C}{-}R$$

Two specific examples are the preparations of methyl cyclohexyl ketone and 2,2-dimethyl-3-heptanone.

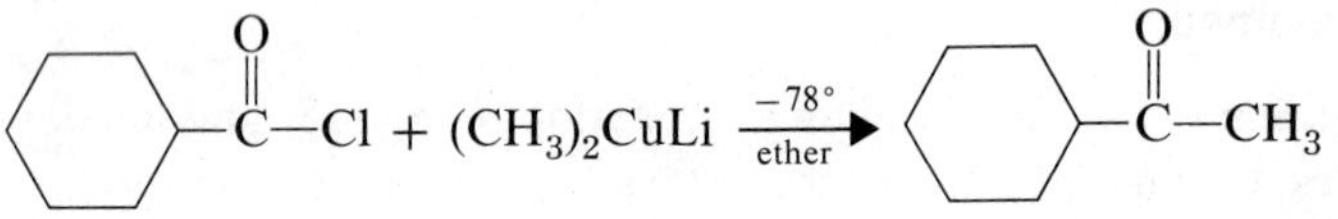

(81%)

$$CH_3\underset{\underset{\displaystyle CH_3}{|}}{\overset{\overset{\displaystyle CH_3}{|}}{C}}{-}\overset{\overset{\displaystyle O}{\|}}{C}{-}Cl + (CH_3CH_2CH_2CH_2)_2CuLi \xrightarrow[\text{ether}]{-78°} CH_3\underset{\underset{\displaystyle CH_3}{|}}{\overset{\overset{\displaystyle CH_3}{|}}{C}}{-}\overset{\overset{\displaystyle O}{\|}}{C}{-}CH_2CH_2CH_2CH_3$$

(90%)

Problem 18.3

What reagents would you use to carry out each of the following reactions?

(a) Benzene $\longrightarrow$ bromobenzene $\longrightarrow$ phenylmagnesium bromide $\longrightarrow$ benzyl alcohol $\longrightarrow$ benzaldehyde

(b) Toluene $\longrightarrow$ benzoic acid $\longrightarrow$ benzoyl chloride $\xrightarrow[\text{(two ways)}]{}$ benzaldehyde

(c) *p*-Bromotoluene $\longrightarrow$ [chromium complex] $\xrightarrow{H_2O}$ *p*-bromobenzaldehyde

(d) *m*-Chlorotoluene $\longrightarrow$ [a diacetate] $\xrightarrow[H_2O]{H^+}$ *m*-chlorobenzaldehyde

(e) 1-Phenylethanol $\longrightarrow$ acetophenone

(f) Benzene $\longrightarrow$ acetophenone

(g) Benzoyl chloride $\xrightarrow[\text{(two ways)}]{}$ acetophenone

18.6 GENERAL CONSIDERATION OF THE REACTIONS OF CARBONYL COMPOUNDS

Before we begin a detailed study of carbonyl compounds it will be helpful if we examine the structures of carbonyl compounds in a general way. As we do this we will find that certain structural features of carbonyl compounds underlie—*and thus unify*—most of the reactions that we will discuss in this chapter and the next.

Structure of the Carbonyl Group

The carbonyl carbon is sp^2 hybridized; thus it, and the three atoms attached to it, lie in the same plane. The bond angles between the three attached atoms are what we would expect of a trigonal coplanar structure: they are approximately 120°.

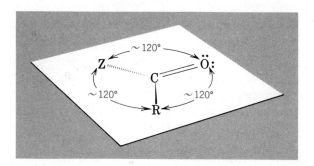

The carbonyl carbon bears a substantial partial *positive* charge; the carbonyl oxygen bears a substantial partial *negative* charge.

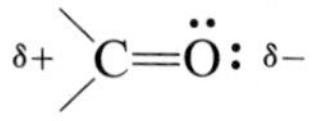

$$\delta+ \quad \overset{\diagdown}{\underset{\diagup}{C}}=\overset{..}{O}: \,\delta-$$

This charge distribution arises from two effects: (1) the inductive effect of the electronegative oxygen and (2) the resonance contribution of the second structure shown below.

$$\overset{\diagdown}{\underset{\diagup}{C}}=\overset{..}{O}: \quad \longleftrightarrow \quad \overset{\diagdown}{\underset{\diagup}{C}}{}^+\!\!-\overset{..}{\underset{..}{O}}:^-$$

Resonance structures for the carbonyl group

Nucleophilic Addition to the Carbon-Oxygen Double Bond

One highly characteristic reaction of aldehydes and ketones is a *nucleophilic addition* to the carbon-oxygen double bond. Aldehydes and ketones are especially susceptible to nucleophilic addition because of the structural features that we have just mentioned. The trigonal coplanar arrangement of groups around the carbonyl carbon means that the carbonyl carbon is relatively open to attack from above or below. The positive charge on the carbonyl carbon means that it is especially susceptible to attack by a nucleophile. The negative charge on the carbonyl oxygen means that nucleophilic addition is susceptible to acid-catalysis. We can visualize nucleophilic addition to the carbon-oxygen double bond occurring in two general ways:

1. In the presence of a reagent that consists of a strong nucleophile and a weak electrophile, addition will usually take place in the following way.

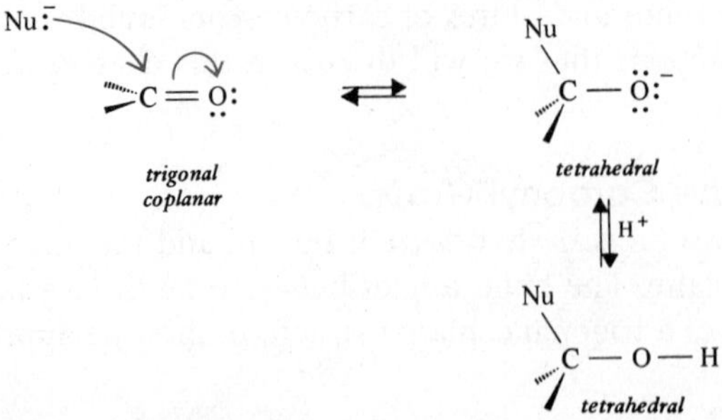

$$Nu:^- \quad \overset{\frown}{\underset{\diagup}{{}_{''''}C}} = \overset{..}{\underset{..}{O}}: \quad \rightleftharpoons \quad \overset{Nu}{\underset{\diagup}{\overset{\diagdown}{\underset{''}{}}}} \overset{|}{\underset{|}{C}} - \overset{..}{\underset{..}{O}}:^-$$

trigonal
coplanar   *tetrahedral*

$$\updownarrow H^+$$

$$\overset{Nu}{\underset{\diagup}{\overset{\diagdown}{\underset{''}{}}}} \overset{|}{\underset{|}{C}} - O - H$$

tetrahedral

In this type of addition the nucleophile uses its electron pair to form a bond to the carbonyl carbon. As this happens an electron pair of the carbon-oxygen π bond shifts out to the carbonyl oxygen and the hybridization state of the carbon changes from sp^2 to sp^3. The important aspect of this step is the ability of the carbonyl oxygen to accommodate the electron pair of the carbon-oxygen double bond.

In the second step the oxygen associates with an electrophile (usually a proton). This happens because the oxygen is now much more basic; it carries a full negative charge, and it resembles the oxygen of an alkoxide ion. (In some reactions the oxygen of the carbonyl group actually becomes the oxygen of an alkoxide ion.)

2. A second general mechanism that operates in nucleophilic additions to carbon-oxygen double bonds is an acid-catalyzed mechanism:

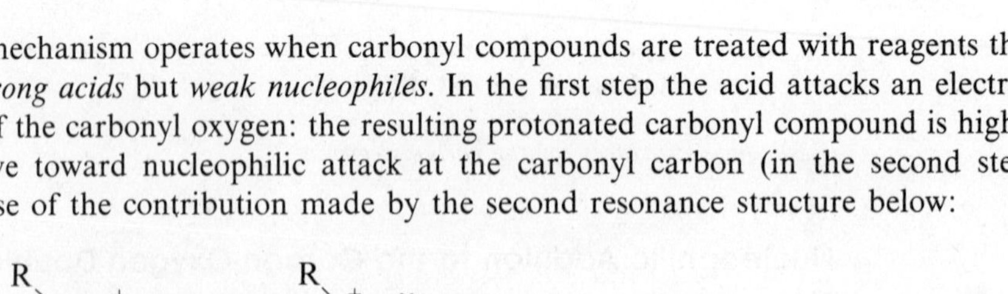

$$\overset{R'_{''''}}{\underset{R}{\diagup}}C = \overset{..}{\underset{..}{O}}: \; + \; H^+ \; \rightleftharpoons \; \overset{R'_{''''}}{\underset{R}{\diagup}}C = \overset{\overset{+}{O}H}{\underset{..}{}} \; \longleftrightarrow \; \overset{R'_{''''}}{\underset{R}{\diagup}}\overset{}{\underset{+}{C}} - \overset{..}{\underset{..}{O}}H$$

(or a Lewis acid)

$$\overset{R'_{''''}}{\underset{R}{\diagup}}\overset{\frown}{C} = \overset{}{\underset{+}{O}} - H \; + \; Nu:^- \; \rightleftharpoons \; \overset{Nu}{\underset{\underset{R}{\diagup}{R'_{''''}}}{\overset{\diagdown}{\underset{}{}}}}C - \overset{..}{\underset{..}{O}} - H$$

This mechanism operates when carbonyl compounds are treated with reagents that are *strong acids* but *weak nucleophiles.* In the first step the acid attacks an electron pair of the carbonyl oxygen: the resulting protonated carbonyl compound is highly reactive toward nucleophilic attack at the carbonyl carbon (in the second step) because of the contribution made by the second resonance structure below:

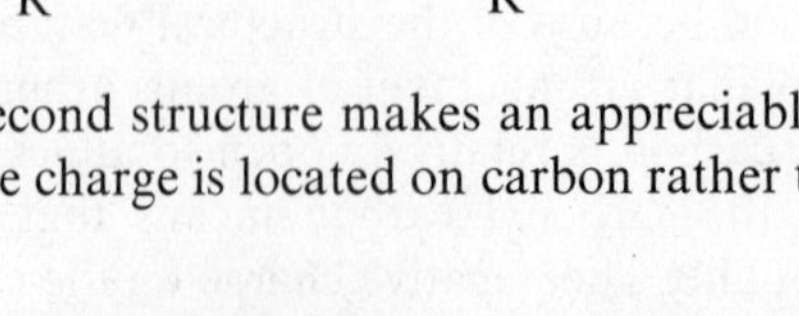

$$\overset{R}{\underset{R}{\diagdown\diagup}}C = \overset{+}{\underset{..}{O}} - H \; \longleftrightarrow \; \overset{R}{\underset{R}{\diagdown\diagup}}\overset{+}{C} - \overset{..}{\underset{..}{O}} - H$$

This second structure makes an appreciable contribution to the hybrid because the positive charge is located on carbon rather than on the more electronegative oxygen.

Reversibility of Nucleophilic Additions to the Carbon-Oxygen Double Bond

Most nucleophilic additions to carbon-oxygen double bonds are reversible; the overall results of these reactions will depend, therefore, on the position of an equilibrium. This behavior contrasts markedly with most electrophilic additions to carbon-carbon double bonds and with nucleophilic substitutions at saturated carbons. The latter reactions are essentially irreversible, and overall results are dictated by relative reaction rates.

Subsequent Reactions of Addition Products

Nucleophilic addition to a carbon oxygen double bond may lead to a product that is stable under the reaction conditions that we employ. If this is the case we are then able to isolate products with the general structure shown below:

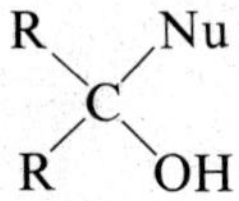

$$\begin{array}{ccc} R & & Nu \\ & \diagdown C \diagup & \\ R & & OH \end{array}$$

In other reactions the product formed initially may be unstable and may spontaneously undergo additional reactions. Even if the initial addition product is stable, however, we may deliberately bring about a subsequent reaction by changing the reaction conditions. When we begin our study of specific reactions we will see that one common subsequent reaction is an *elimination reaction.*

Problem 18.4

The reaction of an aldehyde or ketone with a Grignard reagent (p. 588) is a nucleophilic addition to the carbon-oxygen double bond. (a) What is the nucleophile? (b) The magnesium portion of the Grignard reagent plays an important part in this reaction. What is its function? (c) What product is formed initially? (d) What product forms when water is added?

Problem 18.5

The reactions of aldehydes and ketones with $LiAlH_4$ and $NaBH_4$ are nucleophilic additions to the carbonyl group. What is the nucleophile in these reactions?

18.7 THE ADDITION OF HYDROGEN CYANIDE AND SODIUM BISULFITE

Hydrogen Cyanide Addition

Hydrogen cyanide adds to the carbonyl groups of aldehydes and most ketones to form compounds called *cyanohydrins.**

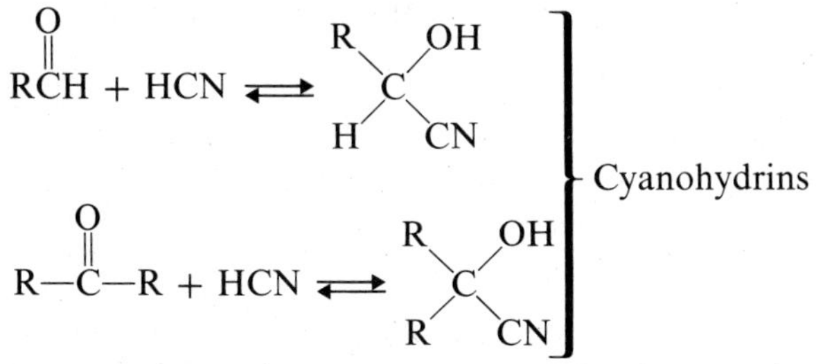

$$\begin{array}{c} O \\ \| \\ RCH + HCN \end{array} \rightleftharpoons \left.\begin{array}{c} R \diagdown C \diagup OH \\ H \diagup \diagdown CN \\[2em] R \diagdown C \diagup OH \\ R \diagup \diagdown CN \end{array}\right\} \text{Cyanohydrins}$$

$$\begin{array}{c} O \\ \| \\ R-C-R + HCN \end{array}$$

*Ketones in which the carbonyl group is highly hindered do not undergo this reaction.

Hydrogen cyanide is a very weak acid but the cyanide ion is a strong nucleophile. Therefore, cyanohydrin formation is initiated by a nucleophilic attack by the cyanide ion on the carbonyl carbon.

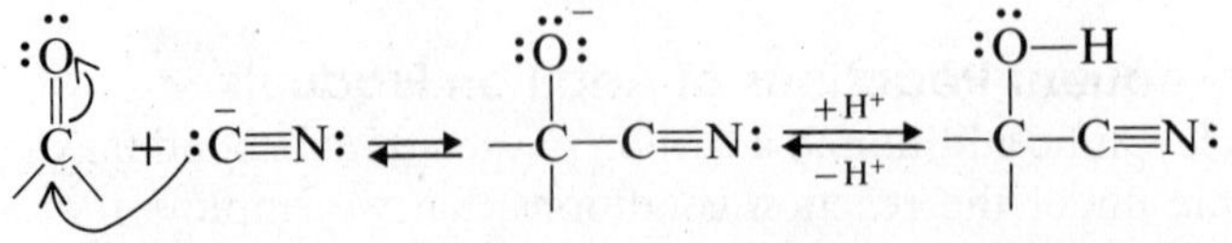

$$
\overset{\overset{\displaystyle ::\!O}{\|}}{C} + :\!\bar{C}{\equiv}N: \rightleftharpoons -\overset{\overset{\displaystyle :\ddot{O}:^-}{|}}{\underset{|}{C}}-C{\equiv}N: \underset{-H^+}{\overset{+H^+}{\rightleftharpoons}} -\overset{\overset{\displaystyle :\ddot{O}-H}{|}}{\underset{|}{C}}-C{\equiv}N:
$$

Liquid hydrogen cyanide can be used for this reaction, but since HCN is very toxic and volatile, it is safer to generate HCN in the reaction mixture. This can be done by treating the aldehyde or ketone with aqueous sodium cyanide and then by adding sulfuric acid slowly to the reaction mixture. *Even with this procedure, however, great care must be taken and the reaction must be carried out in a fume hood.* An example of the use of this procedure is the synthesis of acetone cyanohydrin that follows. (We will see later that a still-safer method for the synthesis of cyanohydrins is through the use of sodium bisulfite addition products.)

Cyanohydrins are useful intermediates in organic synthesis. Depending on the conditions used, acidic hydrolysis converts cyanohydrins to α-hydroxyacids or to α,β-unsaturated acids:

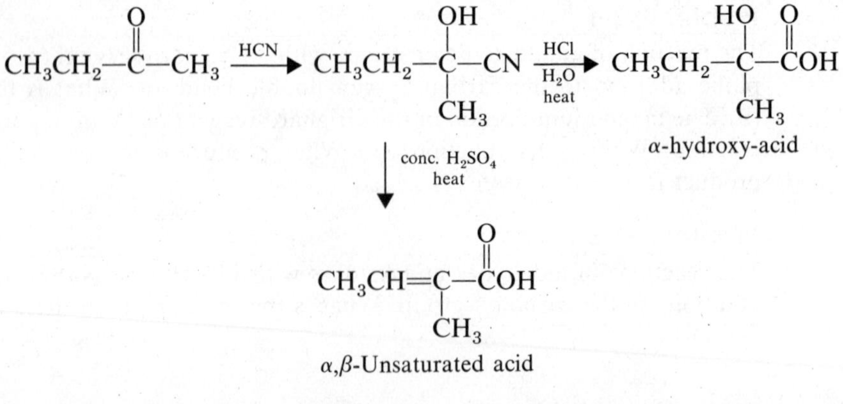

α-hydroxy-acid

α,β-Unsaturated acid

Treating acetone cyanohydrin with acid and methanol converts it to methyl methacrylate. Methyl methacrylate is the starting material for the synthesis of the polymer known as Plexiglas or Lucite (Section 8.3).

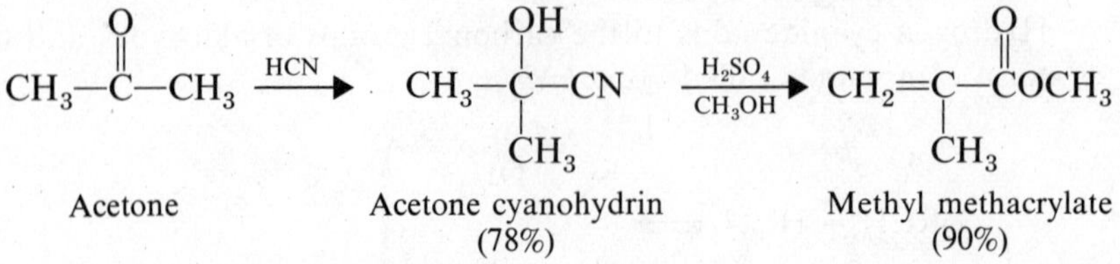

Acetone   Acetone cyanohydrin   Methyl methacrylate
                         (78%)   (90%)

Treating the cyanohydrin obtained from cyclohexanone with lithium aluminum hydride gives a β-amino alcohol.

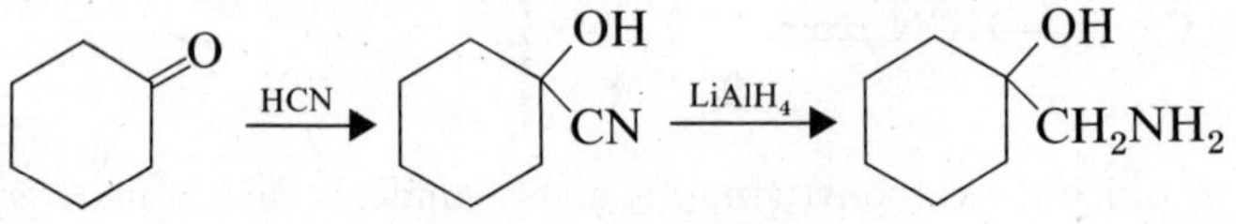

Sodium Bisulfite Addition

Sodium bisulfite ($NaHSO_3$) adds to a carbonyl group in a way that is very similar to the addition of HCN.

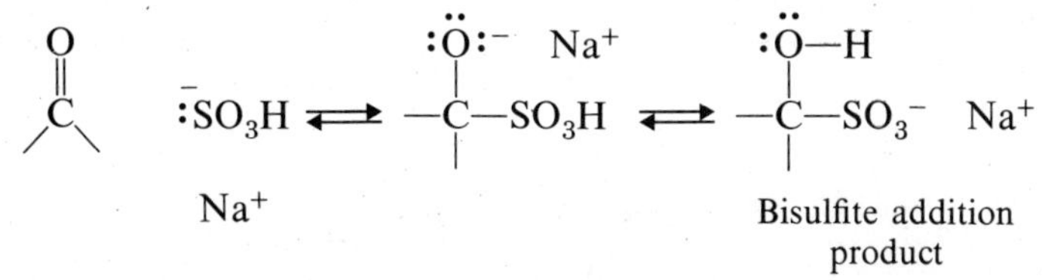

$$
\underset{Na^+}{\overset{O}{\underset{\|}{C}}} \quad :\!SO_3H^- \rightleftharpoons \overset{:\overset{..}{O}:^- \ Na^+}{\underset{|}{-C-SO_3H}} \rightleftharpoons \overset{:\overset{..}{O}-H}{\underset{|}{-C-SO_3^-}} \ Na^+
$$

Bisulfite addition product

This reaction takes place with aldehydes and some ketones. Most aldehydes react with one equivalent of sodium bisulfite to give the addition product in 70–90% yield. Under the same conditions, methyl ketones give yields varying from 12–56%. Most higher ketones do not give bisulfite addition products in appreciable amounts because the addition is very sensitive to steric hindrance. However, since the reaction is an equilibrium, yields from aldehydes and methyl ketones can be improved by using an excess of sodium bisulfite.

Because bisulfite addition compounds are crystalline salts, a bisulfite addition reaction is often used in separating aldehydes and methyl ketones from other substances. Since bisulfite addition is reversible, the aldehyde or methyl ketone can be regenerated after a separation has been made. This can be done by adding either an acid or a base.

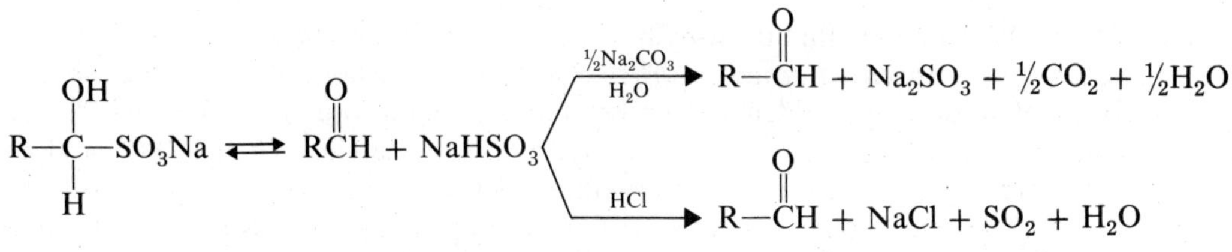

$$
\overset{OH}{\underset{H}{\underset{|}{R-C-SO_3Na}}} \rightleftharpoons R\overset{O}{\overset{\|}{CH}} + NaHSO_3
\begin{cases}
\xrightarrow[H_2O]{\frac{1}{2}Na_2CO_3} R-\overset{O}{\overset{\|}{CH}} + Na_2SO_3 + \frac{1}{2}CO_2 + \frac{1}{2}H_2O \\
\xrightarrow{HCl} R-\overset{O}{\overset{\|}{CH}} + NaCl + SO_2 + H_2O
\end{cases}
$$

We can also employ bisulfite addition compounds to synthesize cyanohydrins in a procedure that avoids the dangerous addition of acid to solutions of sodium cyanide. The small amount of bisulfite ion in equilibrium with the addition compound serves as the acid instead. Thus, we need add only sodium cyanide.

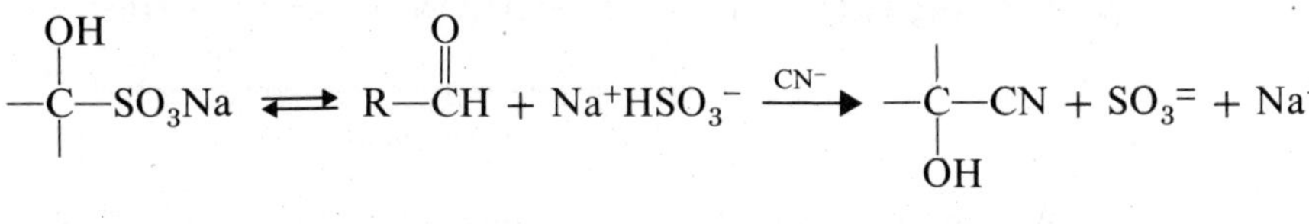

$$
\underset{|}{\overset{OH}{\underset{|}{-C-SO_3Na}}} \rightleftharpoons R-\overset{O}{\overset{\|}{CH}} + Na^+HSO_3^- \xrightarrow{CN^-} \underset{OH}{\overset{|}{\underset{|}{-C-CN}}} + SO_3^= + Na^+
$$

An example that illustrates this method and that also illustrates the hydrolysis of a cyanohydrin to an α-hydroxyacid is the reaction shown below.

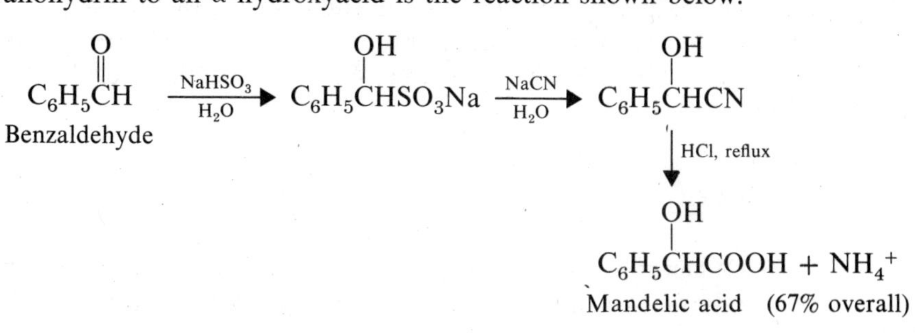

$$
\underset{\text{Benzaldehyde}}{C_6H_5\overset{O}{\overset{\|}{CH}}} \xrightarrow[H_2O]{NaHSO_3} C_6H_5\overset{OH}{\overset{|}{CHSO_3Na}} \xrightarrow[H_2O]{NaCN} C_6H_5\overset{OH}{\overset{|}{CHCN}}
$$

$$
\downarrow \text{HCl, reflux}
$$

$$
\underset{\text{Mandelic acid \quad (67\% overall)}}{C_6H_5\overset{OH}{\overset{|}{CHCOOH}} + NH_4^+}
$$

Problem 18.6

(a) Show how you might prepare lactic acid from acetaldehyde through a cyano-hydrin intermediate. (b) What stereoisomeric form of lactic acid would you expect to obtain?

18.8 KETO-ENOL TAUTOMERISM

The Acidity of the α-Hydrogen of Carbonyl Compounds

In addition to a propensity toward nucleophilic attack at the carbonyl carbon a second characteristic of carbonyl compounds is an unusual acidity of hydrogens on carbons adjacent to the carbonyl group. (These hydrogens are usually called the α-hydrogens and the carbon to which they are attached is called the α-carbon.)

$$R-\overset{\overset{\displaystyle :\overset{..}{O}}{\|}}{C}-\overset{\overset{\displaystyle \alpha}{|}}{\underset{\underset{\displaystyle H}{|}}{C}}-\overset{\overset{\displaystyle \beta}{|}}{\underset{\underset{\displaystyle H}{|}}{C}}-$$

This hydrogen is unusually acidic   *This hydrogen is not*

When we say that the α-hydrogens are acidic, *we mean that they are unusually acidic for hydrogens attached to carbon.* The acidity constants (K_a's) for the α-hydrogens of most simple aldehydes or ketones are of the order of 10^{-19}–10^{-20}. This means that they are more acidic than hydrogens of acetylene ($K_a = 10^{-25}$) and are far more acidic than the hydrogens of ethene ($K_a = 10^{-36}$) or of ethane ($K_a = 10^{-42}$).

The acidity of the α-hydrogens of carbonyl compounds is a result of a phenomenon that we encountered in our study of phenols: when a carbonyl compound loses an α-proton the anion that is produced is *stabilized by resonance.*

$$R-\overset{\overset{\displaystyle :\overset{..}{O}}{\|}}{C}-\overset{\overset{\displaystyle H}{|}}{\underset{\underset{}{}}{C}}HR' \overset{B:^-}{\longrightarrow} R-\overset{\overset{\displaystyle :\overset{..}{O}}{\|}}{C}-\overset{..}{C}HR \longleftrightarrow R-\overset{\overset{\displaystyle :\overset{..}{O}:^-}{|}}{C}=CHR$$

$$\underbrace{\qquad\qquad\qquad\qquad\qquad}_{}$$
$$\text{\textbf{A}}\qquad\qquad\qquad\qquad\text{\textbf{B}}$$

Resonance-stabilized anion

We see (above) that two resonance structures, **A** and **B**, can be written for the anion. In structure **A** the negative charge is on carbon and in structure **B** the negative charge is on oxygen. We would expect structure **B** to make a large contribution to the hybrid because oxygen, being highly electronegative, is better able to accommodate the negative charge. We can depict the hybrid in the following way.

$$R-\overset{\overset{\displaystyle O^{\delta-}}{\|}}{C}\overset{\displaystyle \delta-}{=\!=\!=}CHR$$

When this resonance-stabilized anion accepts a proton it can do so in either of two ways: it can accept the proton at carbon to form the original carbonyl compound (in the *keto* form), or it may accept the proton at oxygen to form an *enol*.

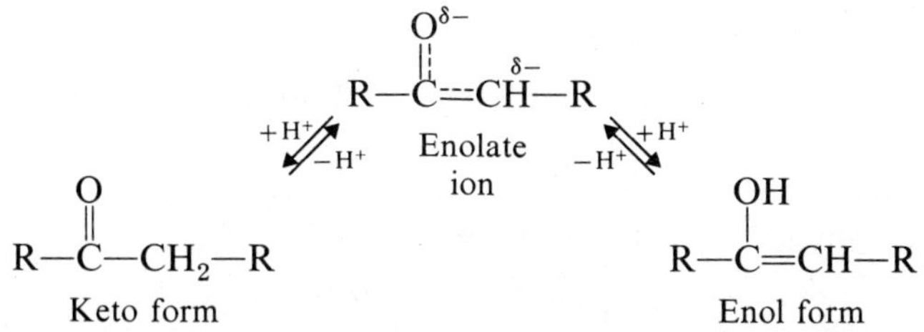

$$
R-\overset{\overset{\displaystyle O^{\delta-}}{\|}}{C}=\overset{\delta-}{C}H-R
$$

Enolate ion

$$
R-\overset{\overset{\displaystyle O}{\|}}{C}-CH_2-R
$$

Keto form

$$
R-\overset{\overset{\displaystyle OH}{|}}{C}=CH-R
$$

Enol form

Both of these reactions are, of course, reversible. Because the same resonance stabilized anion can be obtained by *removal of a proton from an enol* it is often called an *enolate ion*.

Keto and Enol Tautomers

The keto and enol forms of carbonyl compounds are structural isomers. However, the two forms are easily interconverted in the presence of traces of acids and bases. Chemists, therefore, use a special term to describe this type of structural isomerism. Interconvertible keto and enol forms are said to be *tautomers* and their interconversion is called *tautomerization*.

Keto-enol tautomerism is only one kind of tautomerism – a kind that should be called *proton tautomerism* since the two forms differ in the location of a proton.

Cyclooctatetraene exists in equilibrium with the bicyclic form shown below.

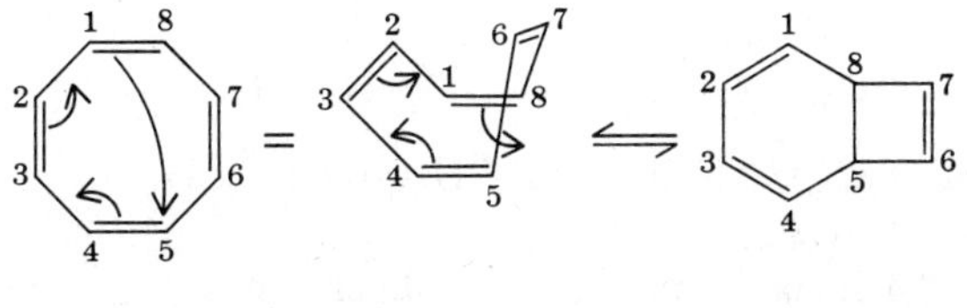

These two compounds are also tautomers since they are *easily interconverted structural isomers*. This kind of tautomerism is called *valence-bond tautomerism* since the two forms differ in the locations of covalent bonds.

One point needs to be emphasized: tautomers are not resonance structures. While tautomers differ in the location of their electrons, *they also differ in the location of their nuclei*. In the enol form of a carbonyl compound the proton is located next to the oxygen (within bonding distance); in the keto form the proton is next to carbon. In cyclooctatetraene, carbons 5 and 8 are relatively far apart; in its valence-bond tautomer they are within a normal carbon-carbon bond distance.

Under most circumstances, we encounter keto-enol tautomers in a state of equilibrium. (The surfaces of ordinary laboratory glassware are able to catalyze the interconversion and establish the equilibrium.) For simple monocarbonyl compounds such as acetone and acetaldehyde the amount of the enol form present at equilibrium is *very small*. In acetone it is much less than 1%; in acetaldehyde the enol concentration is too small to be detected. The greater stability of the keto forms of monocarbonyl compounds (below) can be related to the greater strength of the carbon-oxygen π bond compared to the carbon-carbon π bond ($\sim$87 kcal/mole versus $\sim$60 kcal/mole).

	Keto form		Enol form

Acetaldehyde:

$$CH_3\overset{\overset{\displaystyle O}{\|}}{C}H \rightleftharpoons CH_2{=}\overset{\overset{\displaystyle OH}{|}}{C}H$$

($\sim$100%)　　　(extremely small)

Acetone:

$$CH_3\overset{\overset{\displaystyle O}{\|}}{C}CH_3 \rightleftharpoons CH_2{=}\overset{\overset{\displaystyle OH}{|}}{C}CH_3$$

(>99%)　　　(1.5×10^{-4})　　.0015

Cyclohexanone:

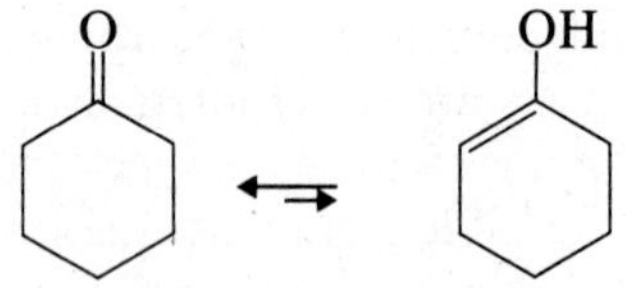

(98.8%)　　　(1.2%)

With compounds having two carbonyl groups separated by one saturated carbon (called β-dicarbonyl compounds), the amount of enol present at equilibrium is far higher. For example, 2,4-pentanedione exists in the enol form to an extent of 76%.

$$CH_3\overset{\overset{\displaystyle O}{\|}}{C}CH_2\overset{\overset{\displaystyle O}{\|}}{C}CH_3 \rightleftharpoons CH_3\overset{\overset{\displaystyle OH}{|}}{C}{=}CH\overset{\overset{\displaystyle O}{\|}}{C}CH_3$$

(24%)　　　　　　(76%)
2,4-Pentanedione　　Enol form

The greater stability of the enol form of β-dicarbonyl compounds can be attributed to stability gained through internal hydrogen bonding in a cyclic form.

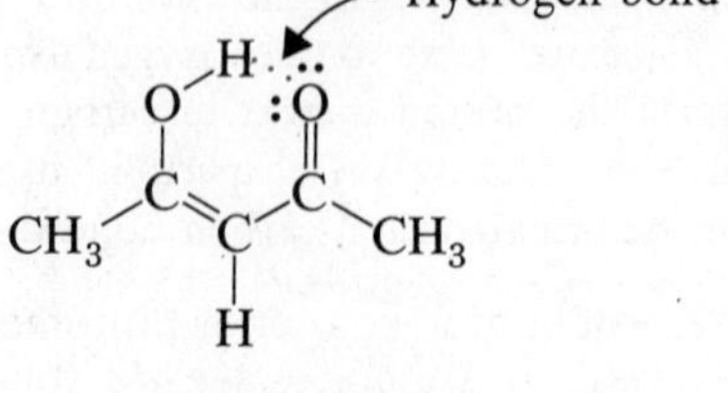

Enols in Biochemistry

An important reaction in the synthesis and degradation of sugars is one in which a three-carbon *ketone* and a three-carbon *aldehyde* are interconverted.

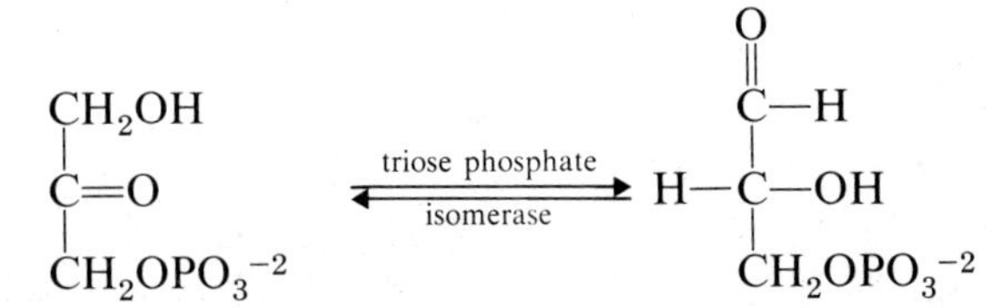

$$
\begin{array}{ccc}
CH_2OH & & O \\
| & & \| \\
C{=}O & \underset{\text{isomerase}}{\overset{\text{triose phosphate}}{\rightleftharpoons}} & C{-}H \\
| & & | \\
CH_2OPO_3^{-2} & & H{-}C{-}OH \\
& & | \\
& & CH_2OPO_3^{-2} \\
\text{Dihydroxyacetone} & & (R)\text{-Glyceraldehyde} \\
\text{phosphate} & & \text{3-phosphate}
\end{array}
$$

This remarkable reaction is catalyzed by an enzyme called *triose phosphate isomerase*. Although all the details of the enzymatic reaction are not fully understood, there is substantial evidence that the interconversion proceeds through the enol intermediate (actually an *enediol*) shown below:

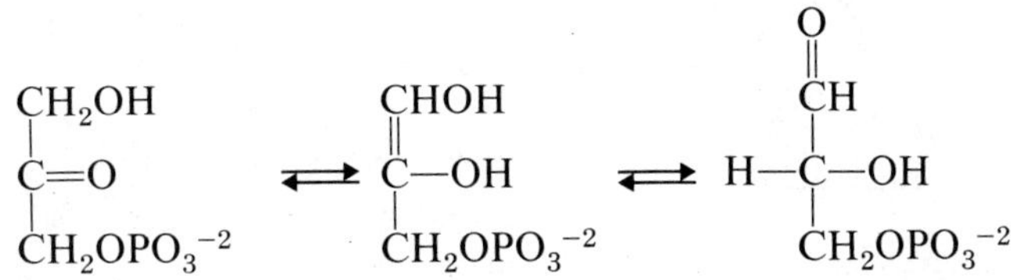

$$
\begin{array}{ccccc}
CH_2OH & & CHOH & & O \\
| & & \| & & \| \\
C{=}O & \rightleftharpoons & C{-}OH & \rightleftharpoons & CH \\
| & & | & & | \\
CH_2OPO_3^{-2} & & CH_2OPO_3^{-2} & & H{-}C{-}OH \\
& & & & | \\
& & & & CH_2OPO_3^{-2} \\
\text{Dihydroxyacetone} & & \text{``enediol''} & & (R)\text{-Glyceraldehyde} \\
\text{phosphate} & & \text{3-Phosphate} & & \text{3-phosphate}
\end{array}
$$

Triose phosphate isomerase apparently binds the enediol 3-phosphate in manner similar to that shown in Fig. 18.1. Then, if the enediol 3-phosphate accepts a proton at carbon-1, the reaction produces dihydroxyacetone phosphate. If the enediol 3-phosphate accepts a proton at carbon-2, the product is (R)-glyceraldehyde 3-phosphate. (This overall reaction is *stereoselective*. Achiral dihydroxyacetone phosphate interconverts only with the R-enantiomer of glyceraldehyde 3-phosphate.)

We will see one biochemical fate of dihydroxyacetone phosphate and (R)-glyceraldehyde 3-phosphate in Section 18.9.

18.9 THE ALDOL ADDITION: THE ADDITION OF ENOLATE IONS TO ALDEHYDES AND KETONES

When acetaldehyde reacts with dilute sodium hydroxide at room temperature (or below) dimerization takes place and 3-hydroxybutanal forms. Since 3-hydroxy-

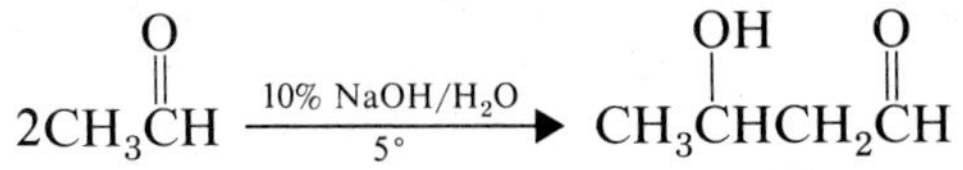

$$
\begin{array}{ccc}
& O & \quad OH \quad O \\
& \| & \quad | \quad\quad \| \\
2CH_3CH & \xrightarrow[5°]{10\% \text{ NaOH/H}_2\text{O}} & CH_3CHCH_2CH \\
& & (50\%) \\
& & \text{3-Hydroxybutanal} \\
& & \text{``aldol''}
\end{array}
$$

butanal is both an *ald*ehyde and an alcoh*ol* it has been given the common name "aldol" and reactions of this general type have come to be known as *aldol additions* (or *aldol condensations*).

The mechanism for the aldol addition is particularly interesting because this single reaction illustrates two important characteristics of carbonyl compounds: the

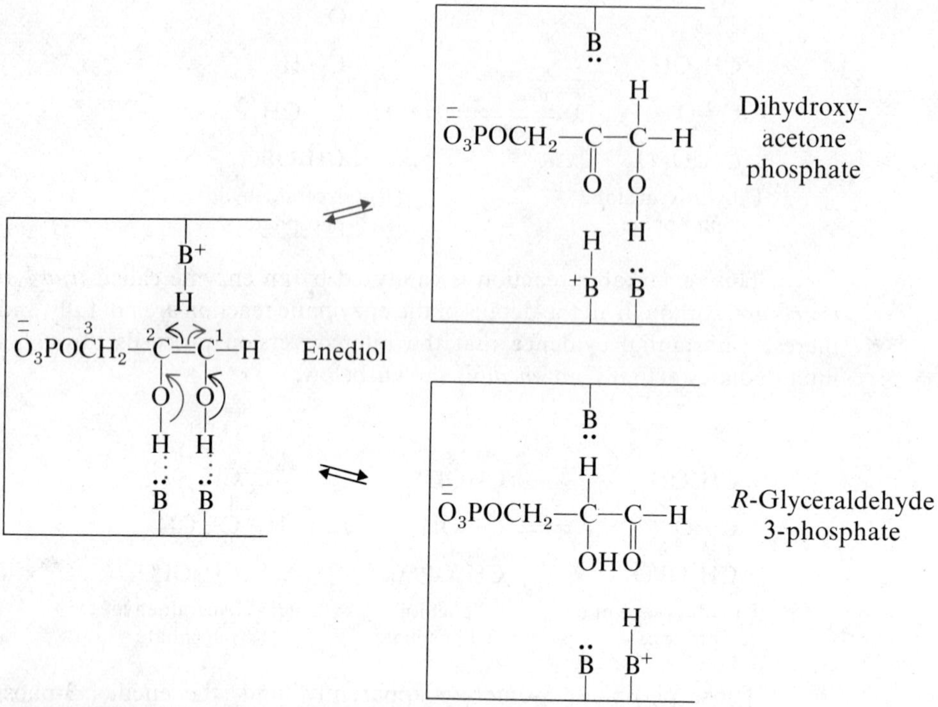

FIG. 18.1

A proposed mechanism for triose phosphate isomerase interconversion of dihydroxy-acetone phosphate and R-glyceraldehyde 3-phosphate. (Colored arrows show the electron flow when protonation occurs at carbon-1.)

acidity of their α-hydrogens and the tendency of their carbonyl groups to undergo nucleophilic addition.

In the first step of the aldol addition the base (hydroxide ion) abstracts a proton from the α-carbon of one molecule of acetaldehyde. This step yields the resonance stabilized enolate ion:

Step 1 $HO:^- + H-CH_2CH=O \rightleftharpoons HOH + \left[:CH_2-CH=O \leftrightarrow CH_2=CH-O:^- \right]$

Enolate ion

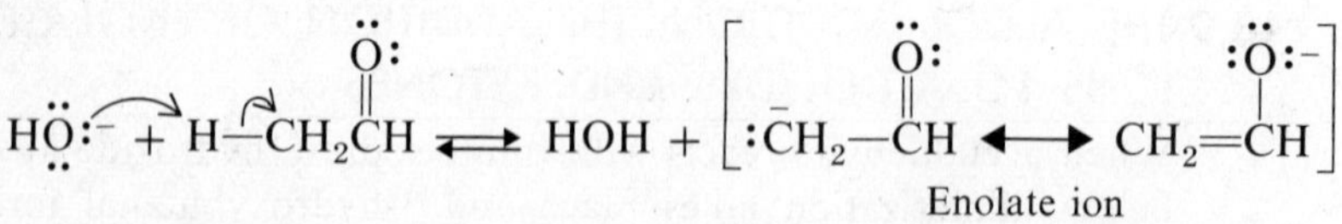

In the second step, the enolate ion acts as a nucleophile (actually as a carbanion) and attacks the carbonyl carbon of a second molecule of acetaldehyde. This step gives an alkoxide ion.

Step 2 $CH_3CH=O + :CH_2-CH=O \rightleftharpoons CH_3CHCH_2CH=O$

An alkoxide ion

$CH_2=CH-O:^-$

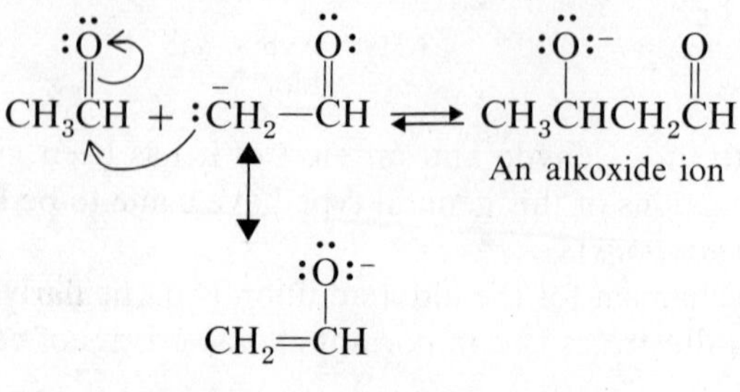

In the third step, the alkoxide ion abstracts a proton from water to form aldol. This step takes place because the alkoxide ion is a stronger base than a hydroxide ion.

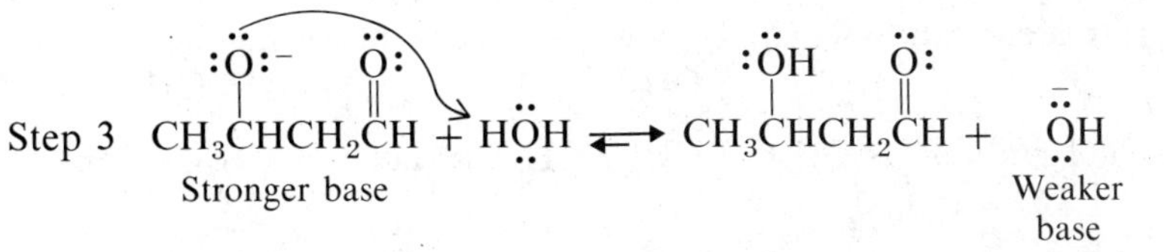

Step 3 CH_3CHCH_2CH + HÖH $\rightleftharpoons$ CH_3CHCH_2CH +   ÖH

Stronger base   Weaker base

If the basic reaction mixture containing the aldol is heated, dehydration takes place and crotonaldehyde (2-butenal) is formed. Dehydration occurs readily because of the acidity of the remaining α-hydrogens and because the product contains conjugated double bonds.

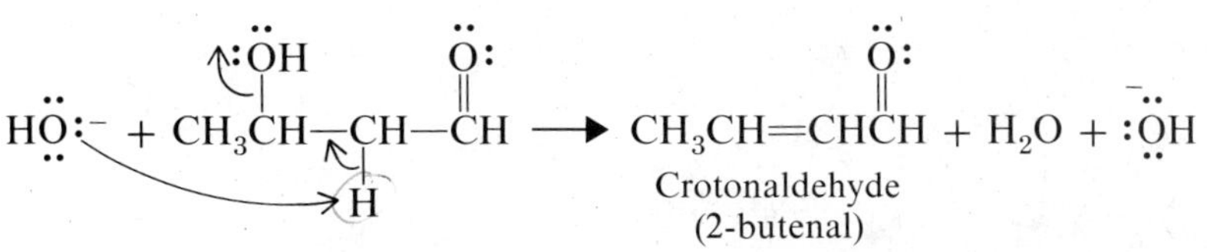

HÖ:⁻ + CH_3CH—CH—CH $\longrightarrow$ CH_3CH=CHCH + H_2O + :ÖH

Crotonaldehyde
(2-butenal)

In some aldol reactions, dehydration occurs so readily that we cannot isolate the product in the aldol form; we obtain the derived *enal* instead. An aldol *condensation* occurs rather than an aldol *addition*.

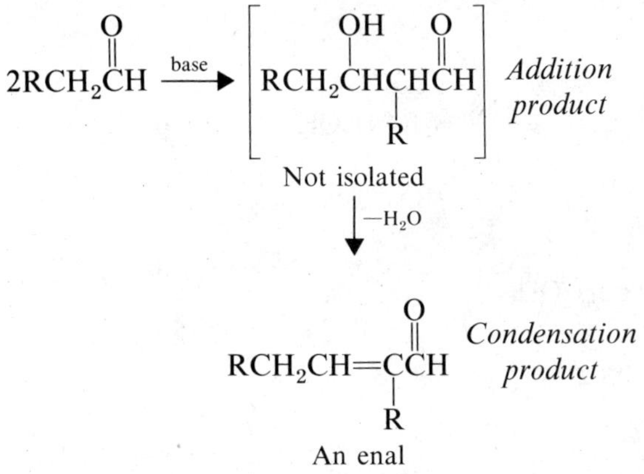

$2RCH_2CH$ $\xrightarrow{\text{base}}$ $\left[RCH_2CHCHCH \atop R \right]$ *Addition product*

Not isolated

$\downarrow$ —H_2O

RCH_2CH=CCH   *Condensation product*
R

An enal

The aldol addition or condensation is a general reaction of aldehydes that possess an α-hydrogen. Propanal, for example, reacts with aqueous sodium hydroxide to give 3-hydroxy-2-methylpentanal.

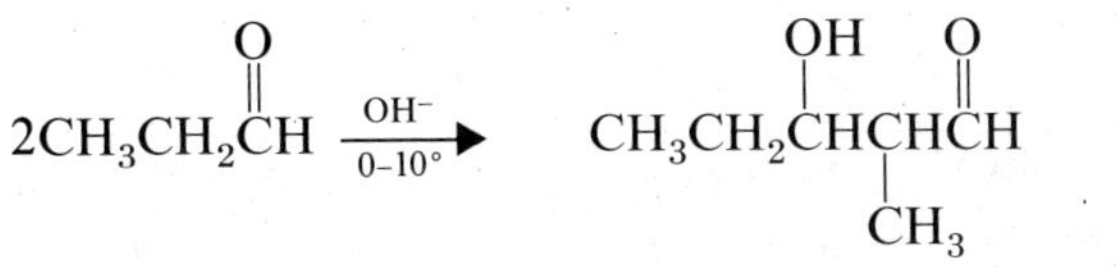

$2CH_3CH_2CH$ $\xrightarrow[0-10°]{OH⁻}$ $CH_3CH_2CHCHCH$
CH_3

3-Hydroxy-2-methylpentanal
(55–60%)

Problem 18.7

(a) Show all steps in the aldol addition that occurs when propanal ($CH_3CH_2\overset{\overset{\displaystyle O}{\|}}{C}H$) is treated with base. (b) How can you account for the fact that the product of the aldol addition is $CH_3CH_2\overset{\overset{\displaystyle OH}{|}}{C}H\overset{\underset{\displaystyle CH_3}{|}}{C}H\overset{\overset{\displaystyle O}{\|}}{C}H$ and not $CH_3CH_2\overset{\overset{\displaystyle OH}{|}}{C}HCH_2CH_2\overset{\overset{\displaystyle O}{\|}}{C}H$? (c) What product would be formed if the reaction mixture were heated?

The aldol addition is important in organic synthesis because it gives us a method for introducing a carbon-carbon bond between two smaller molecules. Because aldol products contain two functional groups, —OH and —CHO, we can use them to carry out a number of subsequent reactions. Examples are shown below.

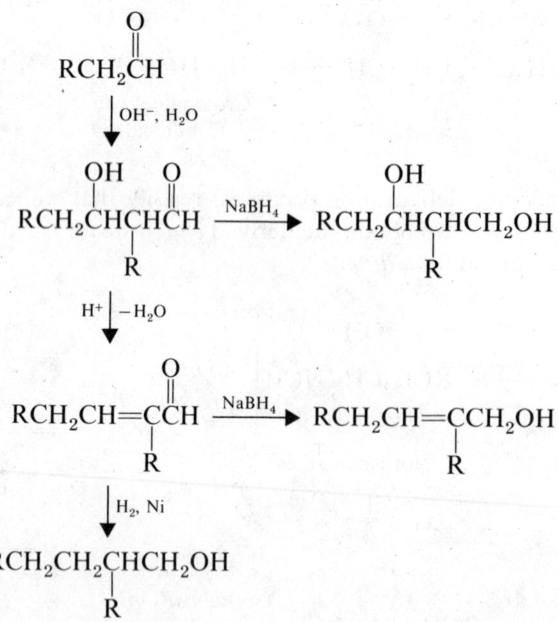

$$RCH_2\overset{\overset{\displaystyle O}{\|}}{C}H$$

$$\downarrow OH^-,\ H_2O$$

$$RCH_2\overset{\overset{\displaystyle OH}{|}}{C}H\underset{\underset{\displaystyle R}{|}}{C}H\overset{\overset{\displaystyle O}{\|}}{C}H \xrightarrow{NaBH_4} RCH_2\overset{\overset{\displaystyle OH}{|}}{C}H\underset{\underset{\displaystyle R}{|}}{C}HCH_2OH$$

$$\downarrow H^+\ -H_2O$$

$$RCH_2CH{=}\underset{\underset{\displaystyle R}{|}}{C}\overset{\overset{\displaystyle O}{\|}}{C}H \xrightarrow{NaBH_4} RCH_2CH{=}\underset{\underset{\displaystyle R}{|}}{C}CH_2OH$$

$$\downarrow H_2,\ Ni$$

$$RCH_2CH_2\underset{\underset{\displaystyle R}{|}}{C}HCH_2OH$$

Problem 18.8

Show how each of the following products could be synthesized from butanal.
  (a) 3-Hydroxy-2-ethylhexanal
  (b) 2-Ethyl-2-hexen-1-ol
  (c) 2-Ethyl-1-hexanol
  (d) 2-Ethyl-1,3-hexanediol (the insect repellent "6-12")

Problem 18.9

Most simple ketones do not undergo aldol additions because the equilibrium is unfavorable. Acetone can be induced to react if the reaction is carried out in a special apparatus that allows the addition product to be removed from contact with the base as soon as it forms. (a) Write a mechanism for an aldol-type addition of acetone in base. (b) What compound would you obtain when the aldol product is dehydrated?

Crossed Aldol Additions

"Mixed" or "crossed" aldol additions are of little synthetic importance if both reactants have α-hydrogens because these reactions give a complex mixture of products. If, for example, we were to carry out a crossed aldol addition using acetaldehyde and propanal we might expect to obtain at least four products.

$$
\underset{\substack{\text{O} \\ \|}}{CH_3CH} + \underset{\substack{\text{O} \\ \|}}{CH_3CH_2CH} \xrightarrow{\text{base}} \underset{\substack{\text{OH} \quad \text{O} \\ | \qquad \|}}{CH_3CHCH_2CH}
$$

3-Hydroxybutanal
(from two moles of acetaldehyde)

+

$$
\underset{\substack{\text{OH} \quad \text{O} \\ | \qquad \|}}{CH_3CHCHCHCH}
$$
$$
\underset{|}{CH_3}
$$

2-Methyl-3-hydroxypentanal
(from two moles of propanal)

+

$$
\underset{\substack{\text{OH} \quad \text{O} \\ | \qquad \|}}{CH_3CHCHCH} \qquad \text{and} \qquad \underset{\substack{\text{OH} \quad \text{O} \\ | \qquad \|}}{CH_3CH_2CHCH_2CH}
$$
$$
\underset{|}{CH_3}
$$

2-Methyl-3-hydroxy-
butanal

3-Hydroxypentanal

(from one mole of acetaldehyde and one mole of propanal)

Problem 18.10

We have already seen how the first two products (above) are formed. Write mechanisms that illustrate the formation of 2-methyl-3-hydroxybutanal and 3-hydroxypentanal.

Crossed aldol reactions *are* practical, however, when one reactant does not have an α-hydrogen and, thus, cannot undergo self-condensation. We can avoid other side reactions by placing this component in base and by slowly adding the reactant with an α-hydrogen to the mixture. Under these conditions the concentration of the reactant with an α-hydrogen will always be low and most of it will be present as an enolate ion. The main reaction that will take place is one between this enolate ion and the component that has no α-hydrogen. The examples listed in Table 18.2 illustrate this technique.

As the examples in Table 18.2 also show, the crossed aldol reaction is often accompanied by dehydration. Whether or not dehydration occurs can, at times, be determined by our choice of reaction conditions.

TABLE 18.2 Crossed Aldol Condensations

THIS REACTANT WITH NO α-HYDROGEN IS PLACED IN BASE	THIS REACTANT WITH AN α-HYDROGEN IS ADDED SLOWLY	PRODUCT
$\underset{\text{Benzaldehyde}}{C_6H_5\overset{\displaystyle O}{\overset{\|}{C}}H}$	$+ \quad \underset{\text{Propionaldehyde}}{CH_3CH_2\overset{\displaystyle O}{\overset{\|}{C}}H} \quad \xrightarrow[10°]{OH^-}$	$\underset{\substack{(68\%)\\ \text{2-Methyl-3-phenyl-}\\ \text{2-propenal}\\ (\alpha\text{-methylcinnamaldehyde})}}{C_6H_5CH=\overset{\displaystyle CH_3}{\underset{}{C}}-\overset{\displaystyle O}{\overset{\|}{C}}H}$
$\underset{\text{Benzaldehyde}}{C_6H_5\overset{\displaystyle O}{\overset{\|}{C}}H}$	$+ \quad \underset{\text{Phenylacetaldehyde}}{C_6H_5CH_2\overset{\displaystyle O}{\overset{\|}{C}}H} \quad \xrightarrow[20°]{OH^-}$	$\underset{\text{2,3-Diphenyl-2-propenal}}{C_6H_5CH=\underset{\displaystyle C_6H_5}{\overset{\displaystyle O}{\overset{\|}{C}}}CH}$
$\underset{\text{Formaldehyde}}{H\overset{\displaystyle O}{\overset{\|}{C}}H}$	$+ \quad \underset{\text{2-Methylpropanal}}{CH_3\overset{\displaystyle CH_3}{\underset{}{C}}H-\overset{\displaystyle O}{\overset{\|}{C}}H} \quad \xrightarrow[40°]{\text{dil. } Na_2CO_3}$	$\underset{\substack{\text{3-Hydroxy-2,2-}\\ \text{dimethylpropanal}\\ (>64\%)}}{CH_3-\overset{\displaystyle CH_3}{\underset{\displaystyle CH_2OH}{C}}-\overset{\displaystyle O}{\overset{\|}{C}}H}$

Problem 18.11

When excess formaldehyde in basic solution is treated with acetaldehyde, the following reaction takes place.

$$H\overset{\displaystyle O}{\overset{\|}{C}}H + CH_3\overset{\displaystyle O}{\overset{\|}{C}}H \xrightarrow[40°]{\text{dil. } Na_2CO_3} HOCH_2-\overset{\displaystyle CH_2OH}{\underset{\displaystyle CH_2OH}{C}}-CHO \qquad (82\%)$$

Write a mechanism that accounts for the formation of the product.

Ketones can be used as one component in crossed aldol reactions that are called *Claisen-Schmidt* reactions. Two examples are shown below.

$$C_6H_5\overset{\displaystyle O}{\overset{\|}{C}}H + CH_3\overset{\displaystyle O}{\overset{\|}{C}}CH_3 \xrightarrow[100°]{OH^-} \underset{\substack{\text{4-Phenyl-3-buten-2-one}\\ \text{(benzalacetone)}\\ (70\%)}}{C_6H_5CH=CH\overset{\displaystyle O}{\overset{\|}{C}}CH_3}$$

$$\underset{\text{C}_6\text{H}_5\overset{\text{O}}{\overset{\|}{\text{C}}}\text{H}}{} + \underset{\text{CH}_3\overset{\text{O}}{\overset{\|}{\text{C}}}\text{C}_6\text{H}_5}{} \xrightarrow[20°]{\text{OH}^-} \underset{\text{C}_6\text{H}_5\text{CH}=\text{CH}\overset{\text{O}}{\overset{\|}{\text{C}}}\text{C}_6\text{H}_5}{}$$

1,3-Diphenyl-2-propenone
(benzalacetophenone)
(85%)

An important step in the commercial synthesis of Vitamin A makes use of a Claisen-Schmidt reaction between citral and acetone:

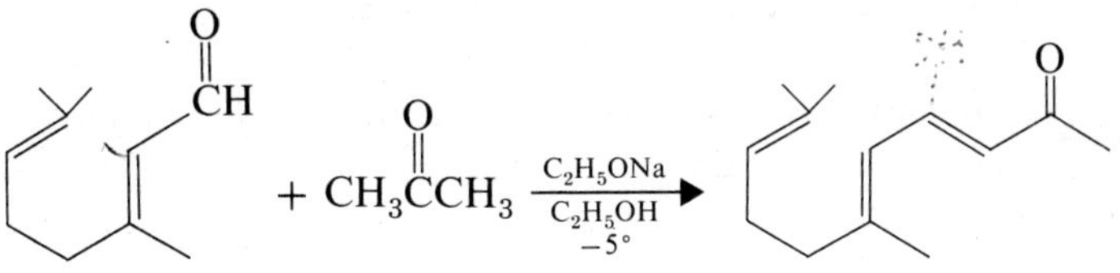

Citral   + $CH_3\overset{\text{O}}{\overset{\|}{C}}CH_3$ $\xrightarrow[\substack{\text{C}_2\text{H}_5\text{OH} \\ -5°}]{\text{C}_2\text{H}_5\text{ONa}}$ Pseudoionone
(49%)

Citral is a naturally occurring aldehyde that can be obtained from lemongrass oil.

Problem 18.12

When pseudoionone is treated with BF_3 in acetic acid, ring closure takes place and α- and β-ionone are produced. This is the next step in the Vitamin A synthesis.

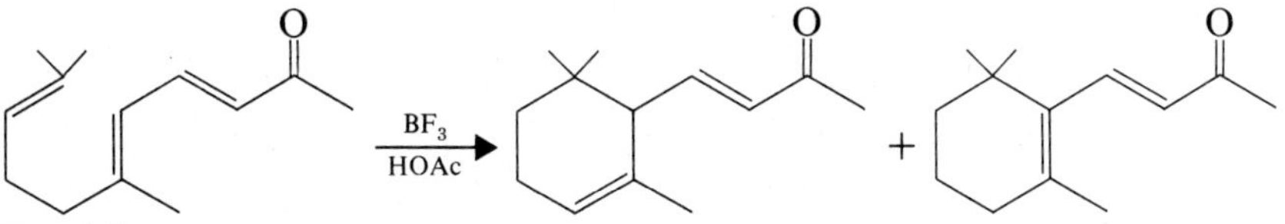

Pseudoionone $\xrightarrow[\text{HOAc}]{\text{BF}_3}$ α-Ionone + β-Ionone

(a) Write mechanisms that explain the formation of α- and β-ionone. (b) β-Ionone is the major product. How can you explain this? (c) Which ionone would you expect to absorb at longer wavelengths in the visible-ultraviolet region? Why?

Aldol Additions in Biochemistry

A number of aldol-type additions take place in the living cells of plants and animals. One *in vivo* reaction that is an important step in the synthesis of glucose involves a crossed aldol addition between R-glyceraldehyde 3-phosphate and di-hydroxyacetone phosphate (cf. p. 697) to produce D-fructose-1,6-diphosphate.* The enzyme that catalyzes this reaction has the appropriate name, *aldolase.* The reaction is completely stereospecific; it produces only D-fructose-1,6-diphosphate, although three other stereoisomers are theoretically possible.

The aldolase reaction, like other aldol additions is reversible. Thus, the cell can use the reaction to synthesize glucose (via fructose 1,6-diphosphate) or it can use the reaction to degrade glucose (via glyceraldehyde 3-phosphate and dihydroxy-

---

*We will discuss the meaning of the designation D in Chapter 24.

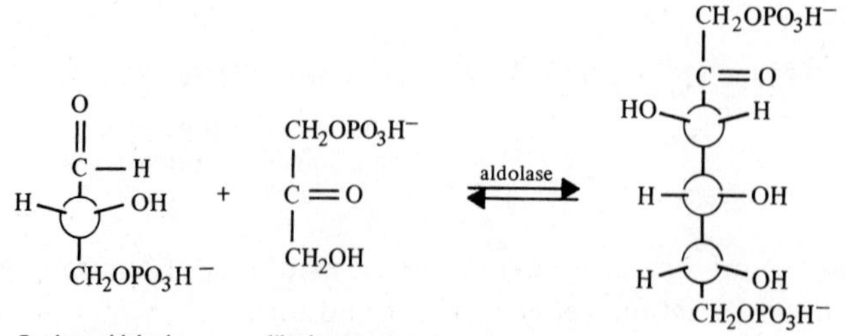

R–glyceraldehyde 3–phosphate   dihydroxyacetone phosphate   D–fructose 1,6–diphosphate

acetone phosphate) to carbon dioxide and water. Since glucose has far greater chemical potential energy than carbon-dioxide and water, glucose *synthesis* gives the cell a method for storing energy. Glucose *degradation* gives the cell a source of energy.

18.10 THE CANNIZZARO REACTION

When aldehydes *that have no α-hydrogen* are placed in concentrated alkali they undergo a reaction known as the Cannizzaro reaction. The products of the Cannizzaro reaction are an alcohol and the salt of a carboxylic acid. Three examples of Cannizzaro reactions are the following.

$$2C_6H_5-\overset{O}{\overset{||}{C}}-H \xrightarrow{60\% \text{ KOH}} C_6H_5CH_2OH + C_6H_5\overset{O}{\overset{||}{C}}-O^-$$
(80-90% yield)

$$2R-\overset{R}{\underset{R}{\overset{|}{C}}}-\overset{O}{\overset{||}{CH}} \xrightarrow{Ca(OH)_2} R-\overset{R}{\underset{R}{\overset{|}{C}}}-CH_2OH + R-\overset{R}{\underset{R}{\overset{|}{C}}}-\overset{O}{\overset{||}{C}}-O^-$$

$$2HCH \xrightarrow{50\% \text{ NaOH}} CH_3OH + H\overset{O}{\overset{||}{C}}-O^-Na^+$$

The Cannizzaro reaction is an *oxidation-reduction reaction* with the aldehyde acting both as an oxidizing agent and as a reducing agent. That is, one mole of the aldehyde reduces another to an alcohol; in this process the first aldehyde is oxidized to a carboxylate ion.

A likely mechanism for the Cannizzaro reaction is the following.

Step 1 $Ar-\overset{\overset{\cdot\cdot}{O}}{\overset{||}{C}}-H + {}^-\!:\!\overset{\cdot\cdot}{O}H \rightleftharpoons Ar-\overset{:\overset{\cdot\cdot}{O}:^-}{\underset{H-\overset{\cdot\cdot}{O}:}{\overset{|}{C}}}-H$

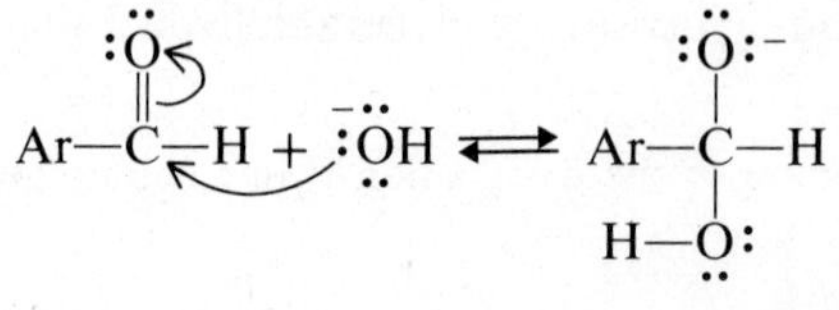

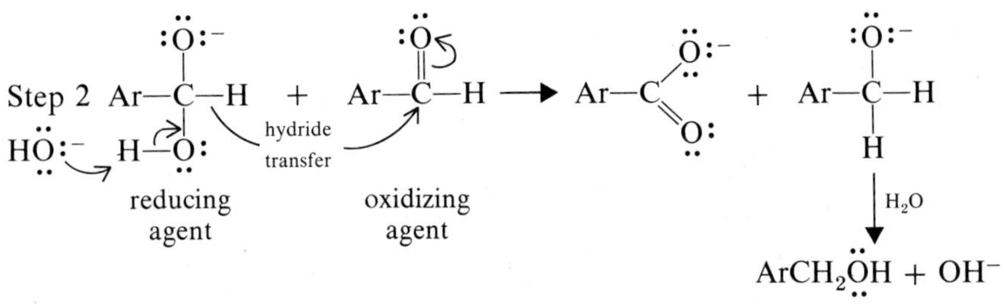

Step 2 $\text{Ar}-\overset{\overset{\displaystyle ..}{O:^-}}{\underset{\overset{\displaystyle ..}{HO:^-}}{\underset{|}{C}}}-H$ + $\text{Ar}-\overset{\overset{\displaystyle ..}{O}}{C}-H$ ⟶ $\text{Ar}-C\overset{\overset{\displaystyle ..}{O:^-}}{\underset{\overset{\displaystyle ..}{O:}}{}}$ + $\text{Ar}-\overset{\overset{\displaystyle ..}{O:^-}}{\underset{\overset{|}{H}}{C}}-H$

reducing agent   oxidizing agent

$\downarrow H_2O$

$ArCH_2\overset{..}{\underset{..}{O}}H + OH^-$

In step 1 a hydroxide ion attacks the carbonyl carbon of one aldehyde molecule and in step 2, oxidation-reduction takes place through a transfer of a hydride ion. (In step 2 the Cannizzaro reaction resembles $LiAlH_4$ and $NaBH_4$ reductions of aldehydes and ketones cf. p. 587.)

Cannizzaro reactions themselves are of little synthetic use since most aldehydes are "expensive" reducing agents. However, crossed Cannizzaro reactions, using inexpensive formaldehyde as one component (and in excess), are practical. Formaldehyde has the advantage, too, that it will preferentially accept the hydroxide ion in step 1 (why?) and thus act as the reducing agent in step 2. The reaction shown below illustrates this method.

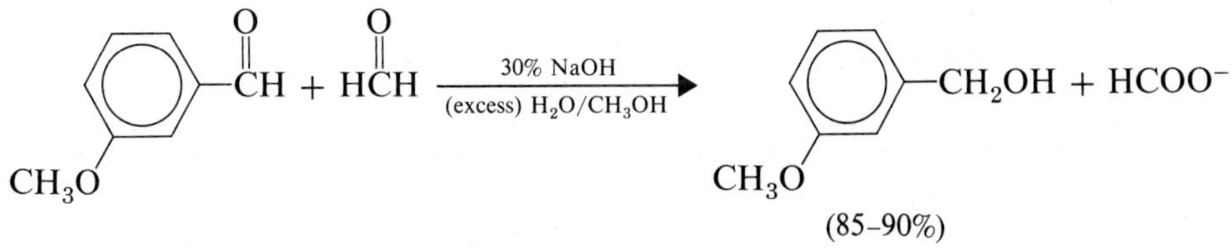

$(85-90\%)$

Problem 18.13

When acetaldehyde is treated with four equivalents of formaldehyde in aqueous calcium hydroxide (at 15–45°) the product of the reaction is pentaerythritol, $HOCH_2C(CH_2OH)_3$ (71% yield). Using your answer to problem 18.11, show how pentaerythritol forms.

18.11 THE ADDITION OF YLIDES

The Wittig Reaction

Aldehydes and ketones react with phosphorus ylides to yield *alkenes* and triphenylphosphine oxide.

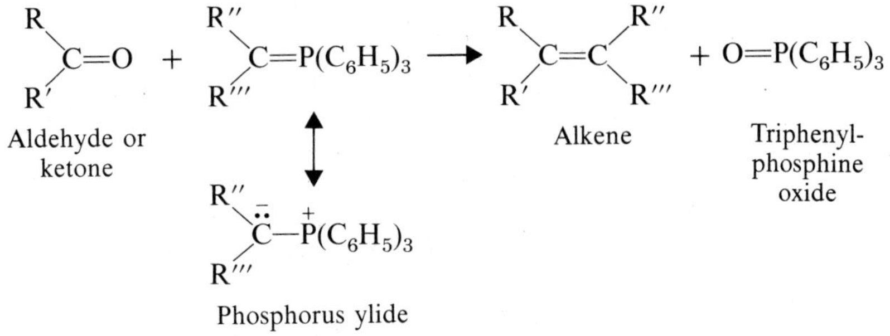

This reaction, known as the *Wittig reaction**, has proved to be a valuable method for synthesizing alkenes. The Wittig reaction is applicable to a wide variety of compounds and it gives a great advantage over most other alkene syntheses in that no ambiguity exists as to the location of the double bond in the product.

Phosphorus ylides are easily prepared from triphenylphosphine and alkyl halides. Their preparation involves two reactions:

General Reaction

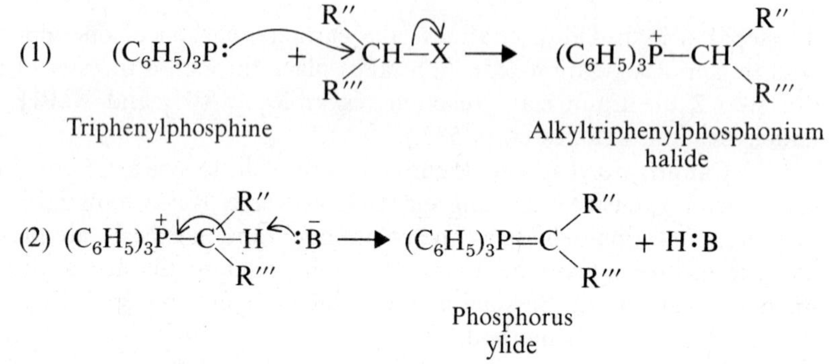

(1) $(C_6H_5)_3P:$ + $\overset{R''}{\underset{R'''}{>}}CH-X$ ⟶ $(C_6H_5)_3\overset{+}{P}-CH\overset{R''}{\underset{R'''}{<}}$ X^-

Triphenylphosphine   Alkyltriphenylphosphonium
halide

(2) $(C_6H_5)_3\overset{+}{P}-\overset{R''}{\underset{R'''}{C}}-H$ $:\bar{B}$ ⟶ $(C_6H_5)_3P=\overset{R''}{\underset{R'''}{C}}$ $+$ $H:B$

Phosphorus
ylide

Specific Example

(1) $(C_6H_5)_3P:$ + CH_3Br $\xrightarrow{C_6H_6}$ $(C_6H_5)_3\overset{+}{P}-CH_3$ Br^-
(89%)
Methyltriphenylphosphonium
bromide

(2) $(C_6H_5)_3\overset{+}{P}-CH_3$ + $:\!\bar{C}H_2SOCH_3$ $\xrightarrow{CH_3SOCH_3}$ $(C_6H_5)_3P=CH_2$ + CH_3SOCH_3
  Br^- $\quad\quad$ Na^+ $\quad\quad\quad\quad\quad\quad\quad\quad\quad\quad\quad$ + NaBr

The first reaction is a nucleophilic substitution reaction. Triphenylphosphine acts as a nucleophile and displaces a halide ion from the alkyl halide to give an alkyltriphenylphosphonium salt. The second reaction is an acid-base reaction. A strong base (usually an alkyllithium, or Na^+ $\bar{C}H_2SOCH_3$) removes a proton from the carbon that is attached to phosphorus to give the ylide.

Phosphorus ylides can be represented as a hybrid of the two resonance structures shown below. It is the contribution made to the hybrid by the second

$(C_6H_5)_3P=\overset{R''}{\underset{R'''}{C}}$ ⟷ $(C_6H_5)_3\overset{+}{P}-\overset{R''}{\underset{R'''}{\bar{C}}:}$

structure that explains the reaction of the ylide with an aldehyde or ketone; the ylide acts as a nucleophile—in effect as a carbanion—and attacks the carbonyl carbon. This step yields a charge separated intermediate called a *betaine* that (often spontaneously) loses triphenylphosphine oxide to give the alkene.

---

* Discovered in 1954 by Professor Georg Wittig, then at the University of Tübingen.

General Mechanism

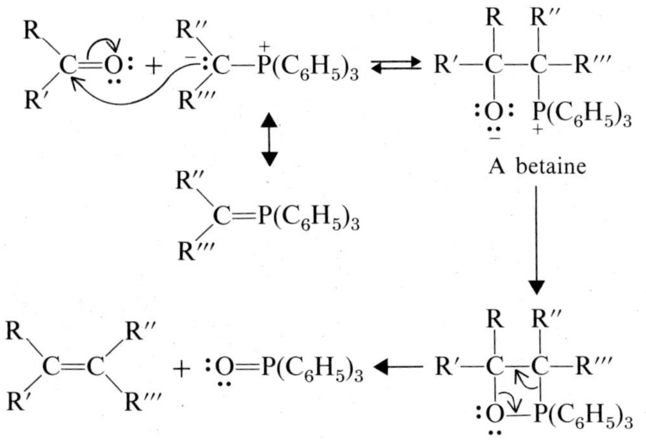

A betaine

Specific Example

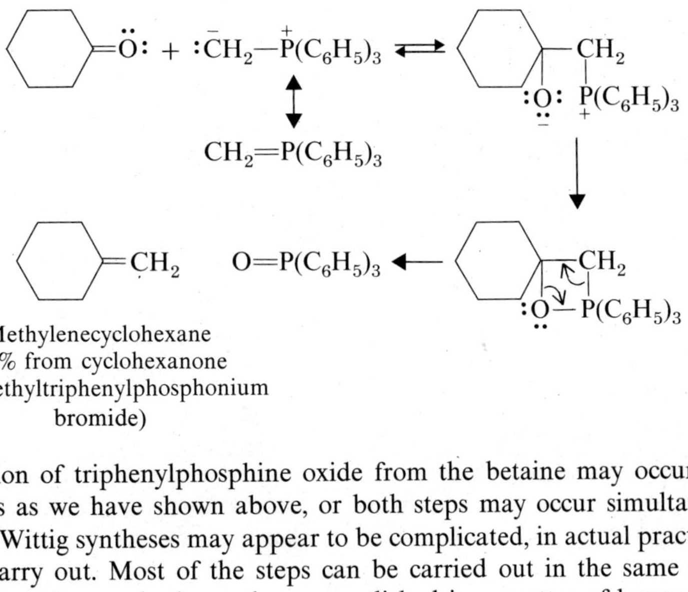

Methylenecyclohexane
(86% from cyclohexanone
and methyltriphenylphosphonium
bromide)

The elimination of triphenylphosphine oxide from the betaine may occur in two separate steps as we have shown above, or both steps may occur simultaneously.

While Wittig syntheses may appear to be complicated, in actual practice they are easy to carry out. Most of the steps can be carried out in the same reaction vessel and the entire synthesis can be accomplished in a matter of hours.

The Wittig reaction is most often employed to introduce a $=CH_2$ group into a compound using $CH_2=P(C_6H_5)_3$ as the ylide. However, the Wittig reaction can be used in other ways as the following examples show. Methyl and primary halides give higher yields of alkyltriphenylphosphonium salts (the first step) than secondary halides (why?). The last step of a Wittig synthesis can lead to a mixture of *cis* and *trans* isomers.

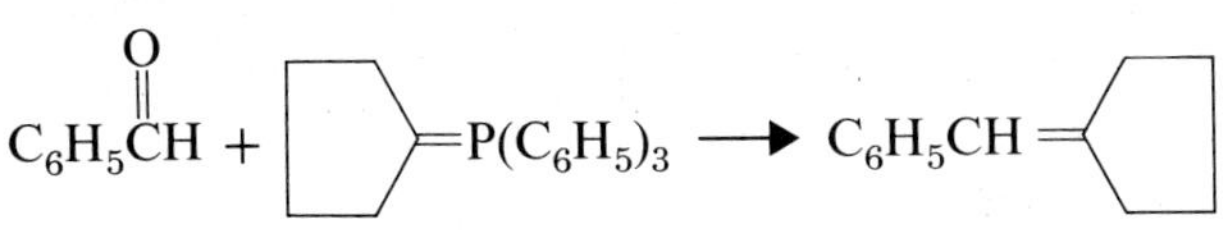

(65%)

$$\underset{O}{\overset{\parallel}{CH_3C}H} + CH_3(CH_2)_4CH=P(C_6H_5)_3 \longrightarrow CH_3CH=CH(CH_2)_4CH_3 + (C_6H_5)_3P=O$$

(69%)
(cis and trans)

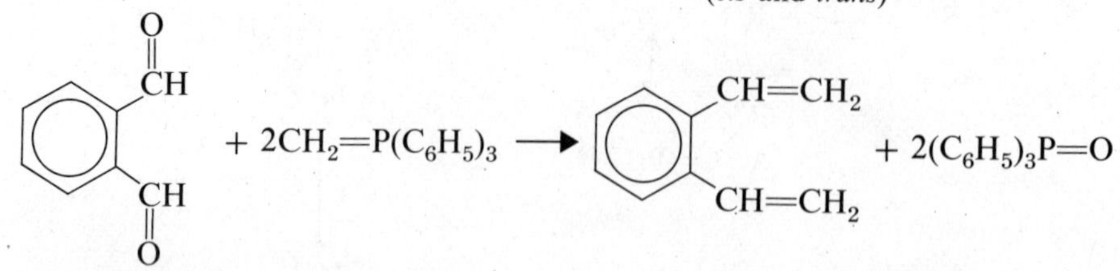

o-Phthalaldehyde

(85%)
o-Divinylbenzene

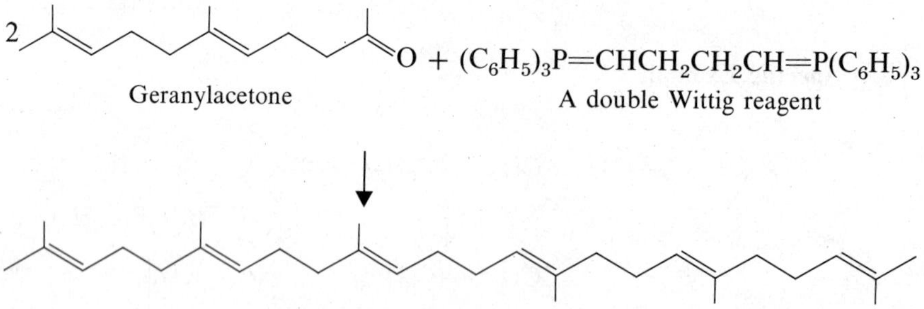

Geranylacetone + $(C_6H_5)_3P=CHCH_2CH_2CH=P(C_6H_5)_3$
A double Wittig reagent

Squalene
(12.5%)

Problem 18.14

Starting with triphenylphosphine and the appropriate alkyl halide, show how you would prepare each of the (different) ylides illustrated above.

Problem 18.15

In addition to triphenylphosphine assume that you have available as starting materials any necessary aldehydes, ketones and organic halides. Show how you might synthesize each of the following alkenes using the Wittig reaction.

(a) $C_6H_5CH=\overset{\overset{\displaystyle CH_3}{|}}{C}-CH_3$

(b) $C_6H_5\underset{\underset{\displaystyle CH_3}{|}}{C}=CH_2$

(c) $C_6H_5\underset{\underset{\displaystyle CH_3}{|}}{C}=CHCH_3$

(d) $\overset{\displaystyle CH_3}{\underset{\displaystyle CH_3}{>}}C=CH_2$

(e)

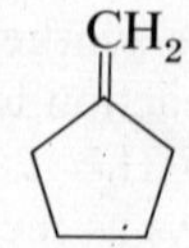

(f) $CH_3CH_2CH=\overset{\overset{\displaystyle CH_3}{|}}{C}CH_2CH_3$

(g) $C_6H_5CH=CHCH=CH_2$

(h) $C_6H_5CH=CHC_6H_5$

Problem 18.16

Triphenylphosphine can be used to convert epoxides to alkenes, for example,

$$C_6H_5CH\overset{\overset{\ddot{\ddot{O}}}{\frown}}{\;}CHCH_3 + (C_6H_5)_3P: \longrightarrow C_6H_5CH=CHCH_3 + (C_6H_5)_3PO$$

Propose a likely mechanism for this reaction.

Problem 18.17

The Wittig reaction can be used in the synthesis of aldehydes, for example,

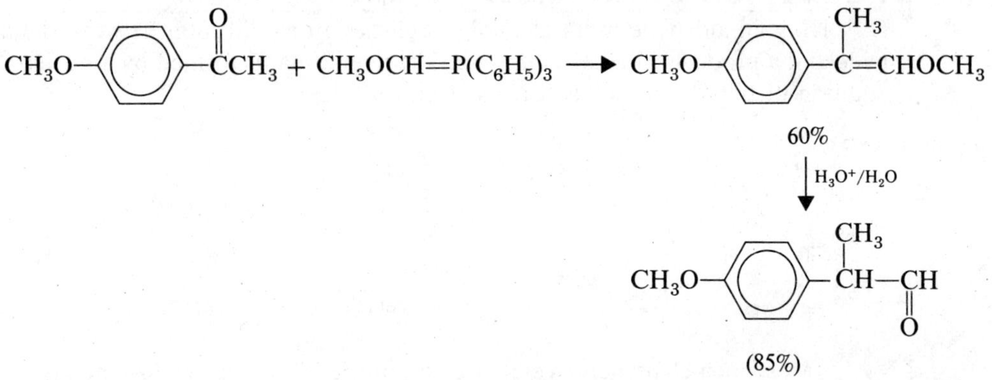

$$CH_3O\text{—}\underset{}{\bigcirc}\text{—}\overset{O}{\overset{\|}{C}}CH_3 + CH_3OCH=P(C_6H_5)_3 \longrightarrow CH_3O\text{—}\bigcirc\text{—}\overset{CH_3}{\underset{}{C}}=CHOCH_3$$

60%

$\downarrow$ H$_3$O$^+$/H$_2$O

$$CH_3O\text{—}\bigcirc\text{—}\overset{CH_3}{\underset{}{CH}}\text{—}\overset{}{\underset{\overset{\|}{O}}{CH}}$$

(85%)

(a) How would you prepare $CH_3OCH=P(C_6H_5)_3$? (b) Why does the second reaction (above) yield an aldehyde? (c) How would you use this method to prepare ⬡CHO from cyclohexanone?

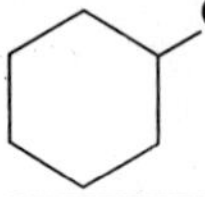

The Addition of Sulfur Ylides

Sulfur ylides also react as nucleophiles at the carbonyl carbon of aldehydes and ketones. The betaine that forms usually decomposes to an *epoxide* rather than to an alkene.

$$(CH_3)_2\overset{+}{\underset{\ddot{}}{S}}\text{—}CH_3 \xrightarrow[\underset{0°}{CH_3SOCH_3}]{Na^+\bar{C}H_2SOCH_3} [(CH_3)_2\overset{+}{\underset{\ddot{}}{S}}\text{—}\overset{\ddot{}}{C}H_2 \longleftrightarrow (CH_3)_2\overset{}{\ddot{S}}=CH_2]$$

Resonance stabilized sulfur ylide

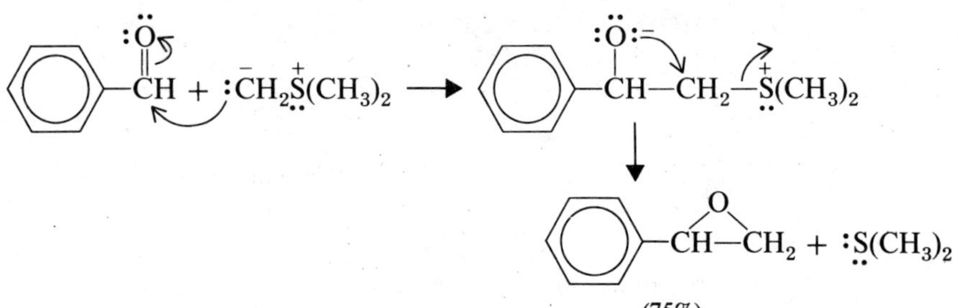

Reaction with benzaldehyde gives styrene oxide (75%) plus :S(CH$_3$)$_2$.

Problem 18.18

Show how you might use a sulfur ylide to prepare

(a) 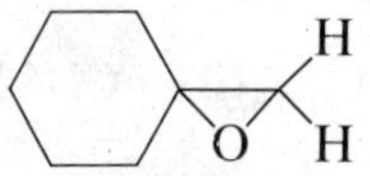 [structure: cyclohexane spiro epoxide with H, O, H]

(b) 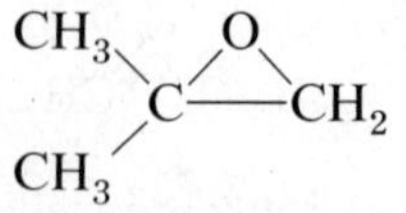 [structure: CH_3, CH_3, C, O, CH_2 epoxide]

18.12 THE ADDITION OF ALCOHOLS: ACETALS AND KETALS

Dissolving an aldehyde in an alcohol establishes an equilibrium between the aldehyde and a product called a *hemiacetal*. The hemiacetal is formed by a nucleophilic addition of the alcohol to the carbonyl group:

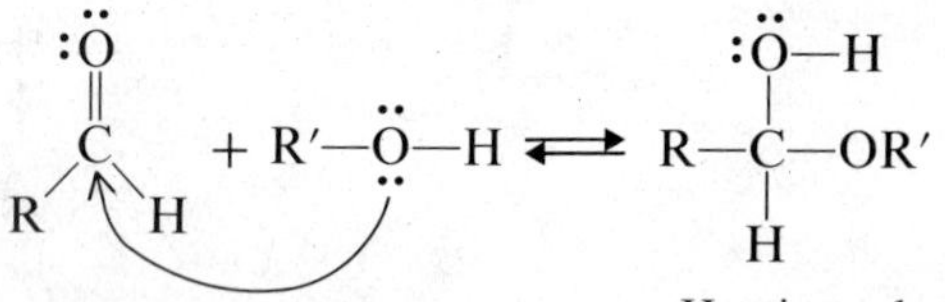

$$\underset{R}{\overset{:\ddot{O}:}{\underset{H}{C}}} + R'-\ddot{O}-H \rightleftharpoons R-\underset{H}{\overset{:\ddot{O}-H}{C}}-OR'$$

Hemiacetal

Most open-chain hemiacetals are not sufficiently stable to allow their isolation. Cyclic hemiacetals with five- or six-membered rings, however, are usually much more stable:

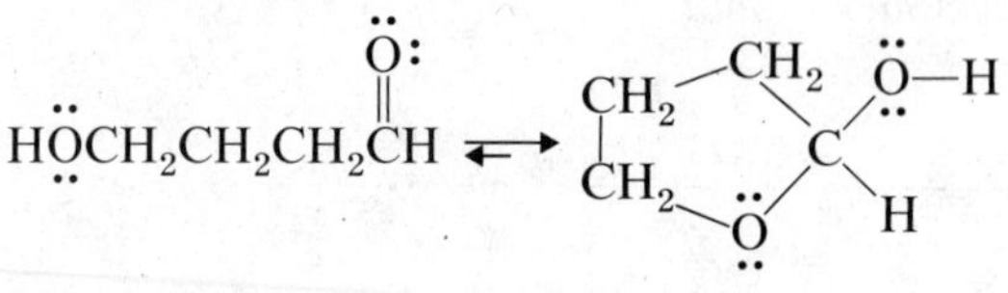

$$HOCH_2CH_2CH_2CH \rightleftharpoons [cyclic structure]$$

Most simple sugars (Chapter 24) exist, primarily, in a cyclic hemiacetal form.

[structure: A hexose (a sugar)] $\rightleftharpoons$ [structure: Cyclic (hemiacetal) form of a hexose with Hemiacetal group labeled]

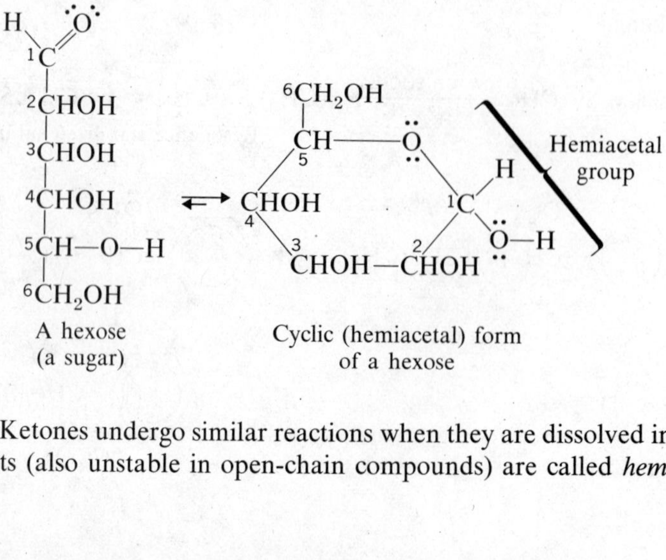

A hexose
(a sugar)

Cyclic (hemiacetal) form
of a hexose

Ketones undergo similar reactions when they are dissolved in an alcohol. The products (also unstable in open-chain compounds) are called *hemiketals*.

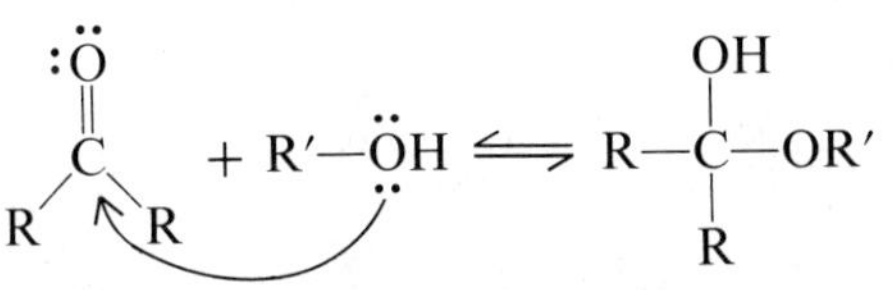

A hemiketal

If we take an alcohol solution of an aldehyde and pass into it a small amount of gaseous HCl, a second reaction takes place. The hemiacetal reacts with a second mole of the alcohol and produces an *acetal*.

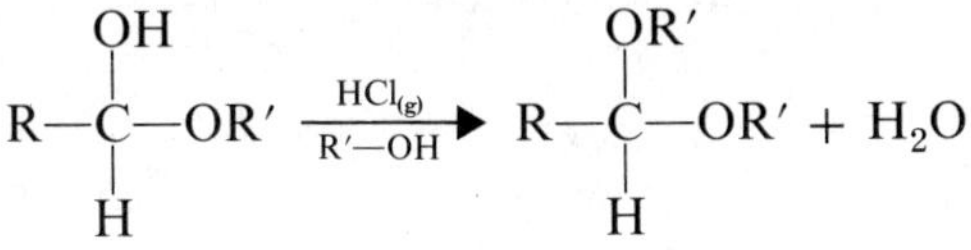

Hemiacetal   An acetal

The mechanism for acetal formation involves an acid-catalyzed elimination of water followed by a second *addition* of the alcohol.

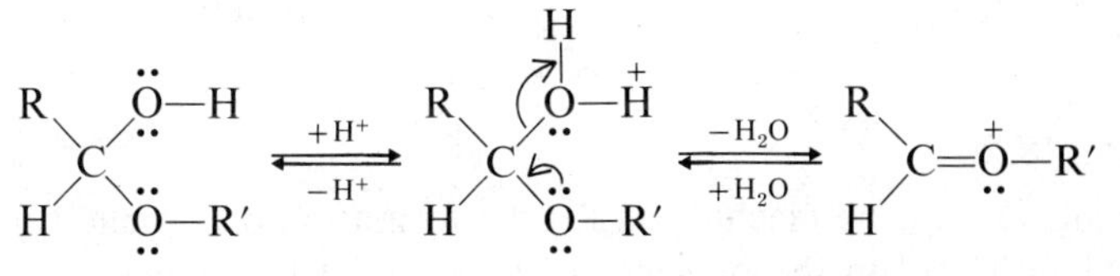

Hemiacetal

Acid-catalyzed elimination of water

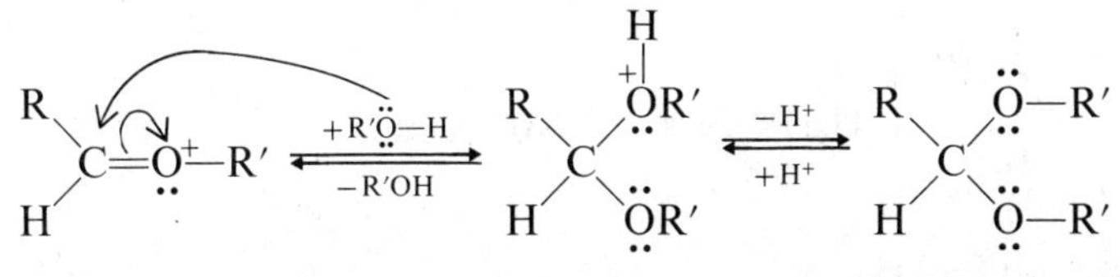

Acetal

Addition of a second mole of the alcohol

Problem 18.19

(a) Write another resonance structure for the intermediate, RCH=O⁺R. (b) Which structure would you expect to make a larger contribution to the hybrid? (c) Why?

    All steps in the formation of an acetal from an aldehyde are reversible. If we dissolve an aldehyde in a large excess of an anhydrous alcohol and add a small amount of an anhydrous acid (i.e., gaseous HCl or conc. H_2SO_4), the equilibrium will strongly favor the formation of an acetal. After the equilibrium is established we can isolate the acetal by evaporating the excess alcohol.

    If we then place the acetal in water and add a small amount of acid, all of the steps reverse and under these conditions (an excess of water) the equilibrium favors the formation of the aldehyde. The acetal undergoes *hydrolysis*.

$$\underset{\text{Acetal}}{\overset{R\diagdown\diagup OR'}{\underset{H\diagup\diagdown OR'}{C}}} + H_2O \underset{\substack{(several \\ steps)}}{\overset{H^+}{\rightleftharpoons}} \underset{\text{Aldehyde}}{R-\overset{\overset{\displaystyle O}{\|}}{C}-H} + 2R'OH$$

Ketal formation is not favored when ketones are treated with simple alcohols and gaseous HCl. Cyclic ketal formation *is* favored, however, when a ketone is treated with an excess of a glycol and a trace of acid.

$$\underset{R\diagup}{\overset{R\diagdown}{C}}=O + \underset{(excess)}{\overset{HOCH_2}{\underset{HOCH_2}{|}}} \overset{H^+}{\rightleftharpoons} \underset{\text{Cyclic ketal}}{\overset{R\diagdown\diagup O-CH_2}{\underset{R\diagup\diagdown O-CH_2}{C}|}}$$

This reaction, too, can be reversed by treating the ketal with aqueous acid.

$$\underset{R\diagup\diagdown O-CH_2}{\overset{R\diagdown\diagup O-CH_2}{C}|} + H_2O \overset{H^+}{\underset{}{\rightleftharpoons}} \underset{R\diagup}{\overset{R\diagdown}{C}}=O + \underset{CH_2OH}{\overset{CH_2OH}{|}}$$

Problem 18.20

Outline all steps in the mechanism for the formation of a cyclic ketal from acetone and ethylene glycol in the presence of gaseous HCl.

Although acetals and cyclic ketals are hydrolyzed to aldehydes and ketones in aqueous acid, *they are stable in aqueous base.*

$$\underset{H\diagup\diagdown OR'}{\overset{R\diagdown\diagup OR'}{C}} + H_2O \overset{OH^-}{\longrightarrow} \quad \text{No reaction}$$

$$\underset{R\diagup\diagdown O-CH_2}{\overset{R\diagdown\diagup O-CH_2}{C}|} + H_2O \overset{OH^-}{\longrightarrow} \quad \text{No reaction}$$

Because of this property, acetals and ketals give us a convenient method for protecting aldehyde and ketone groups from undesired reactions in basic solutions. (Acetals and ketals are really *gem*-diethers and, like ethers, they are relatively unreactive.) We can convert an aldehyde or ketone to an acetal or cyclic ketal, carry out a reaction on some other part of the molecule, and then hydrolyze the acetal or ketal with aqueous acid.

As an example, let us consider the problem of converting

to

A   **B**

Keto groups are more easily reduced than ester groups. Any reducing agent (i.e., $LiAlH_4$ or H_2/Ni) that will reduce the ester group of **A** will reduce the keto group as well. But, if we "protect" the keto group by converting it to a cyclic ketal, we can reduce the ester group without affecting the cyclic ketal. After we finish the ester reduction we can hydrolyze the cyclic ketal and obtain our desired product, **B.**

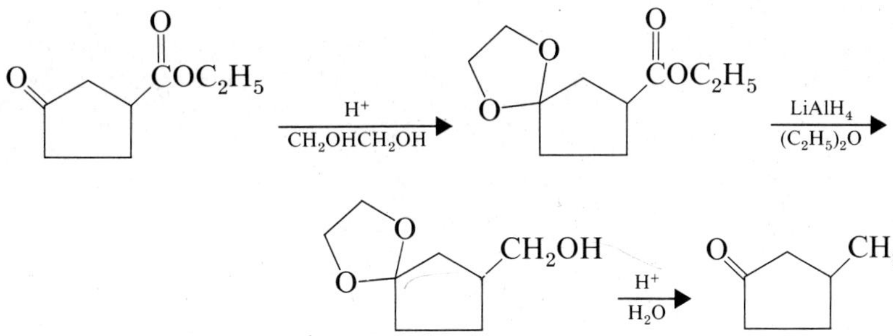

Problem 18.21

(a) Show how you might use a cyclic ketal in carrying out the following transformation.

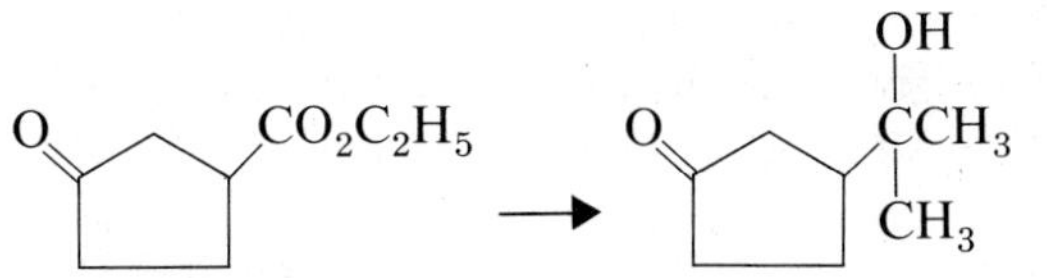

(b) Why would a direct addition of methylmagnesium bromide to **A** fail to give **C?**

Cyclic ketal formation can also be used to protect hydroxyl groups from undesired reactions. The synthesis of vitamin C from (−) sorbose provides an especially interesting application of this technique and, at the same time, it illustrates a number of other familiar reactions.

The starting compound, (−)-sorbose is a pentahydroxyketone. (It is also a simple sugar that is almost as sweet as ordinary table sugar.) In solution (−)-sorbose exists as an equilibrium mixture of an open-chain form, **1a,** and cyclic form, **1b.**

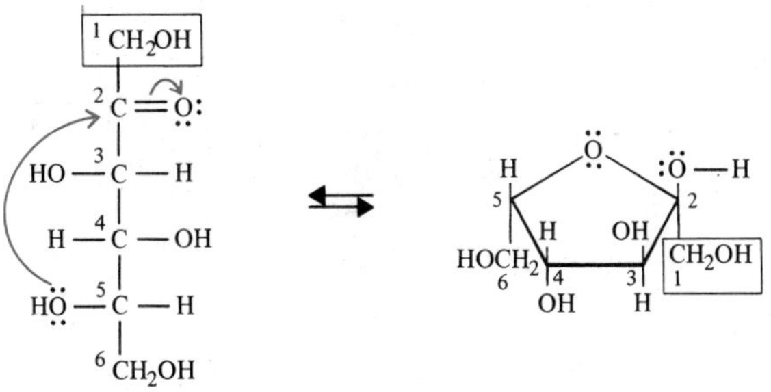

1a
(−)−sorbose
(open chain form)

1b
(−)−sorbose
(cyclic hemiketal form)

The vitamin C synthesis requires that the —CH$_2$OH group at position 1 of (−)-sorbose be oxidized to a —COOH group with potassium permanganate. However, direct treatment of (−)-sorbose with KMnO$_4$ not only causes oxidation of the alcohol group at position 1, it causes oxidation of the other four alcohol groups as well; carbon-carbon bonds break and the whole molecule falls apart.

However, the cyclic form of (−)-sorbose reacts with two moles of acetone (in the presence of a trace of acid) to form a *tricyclic* ketal, **2**. The hydroxyl groups of the cyclic hemiketal form of (−)-sorbose are properly positioned for this reaction to take place. (There are actually three cyclic ketal linkages in **2**. Can you find them?)

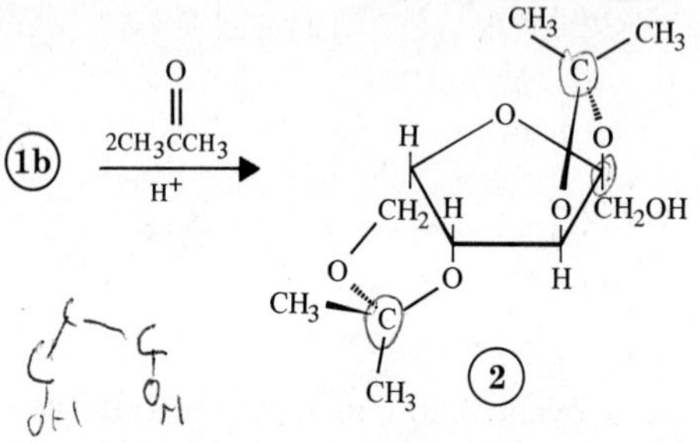

Compound **2** has only one free —CH$_2$OH group: *the one that we need to oxidize*. All of the others are now protected in ketal linkages. Oxidation of **2** with potassium permanganate gives **3**; and hydrolysis of **3** with aqueous acid gives **4**.

The remaining steps in the synthesis of vitamin C are outlined below.

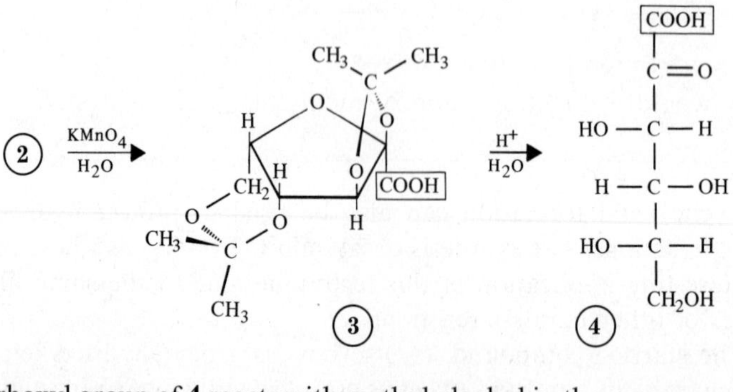

The carboxyl group of **4** reacts with methyl alcohol in the presence of acid to produce a methyl ester, **5**. Compound **5** enolizes when it is treated with base to give **6** and when **6** is heated it cyclizes to Vitamin C.

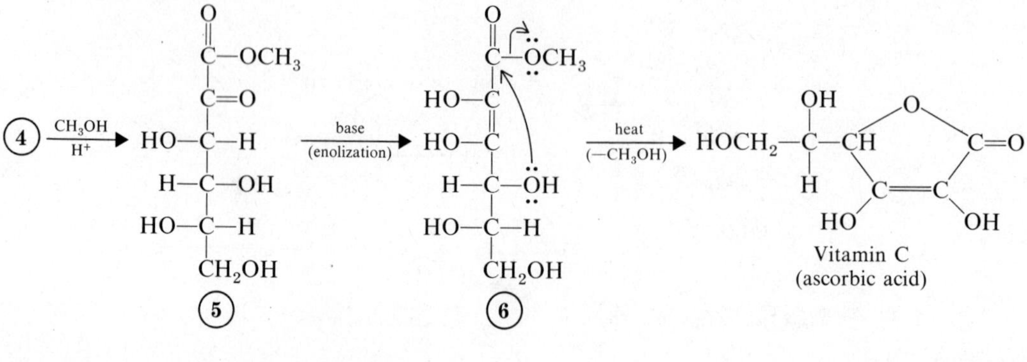

Thioacetals and Thioketals

Aldehydes and ketones react with thiols to form *thioacetals* and *thioketals*.

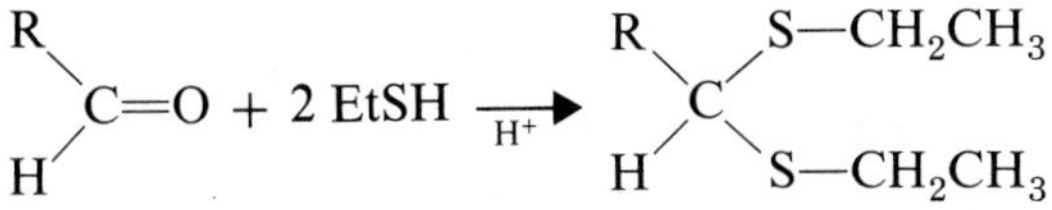

$$\begin{array}{c}R\\\\H\end{array}\!\!\!\!\diagdown\!\!\!\!C\!\!=\!\!O + 2\ EtSH \xrightarrow{H^+}\ \begin{array}{c}R\\\\H\end{array}\!\!\!\!\diagdown\!\!\!\!C\!\!\!\!\begin{array}{c}S\!\!-\!\!CH_2CH_3\\\\S\!\!-\!\!CH_2CH_3\end{array}$$

Thioacetal

$$\begin{array}{c}R\\\\R\end{array}\!\!\!\!\diagdown\!\!\!\!C\!\!=\!\!O + HSCH_2CH_2SH \xrightarrow{BF_3}\ \begin{array}{c}R\\\\R\end{array}\!\!\!\!\diagdown\!\!\!\!C\!\!\!\!\begin{array}{c}S\!\!-\!\!CH_2\\|\\S\!\!-\!\!CH_2\end{array}$$

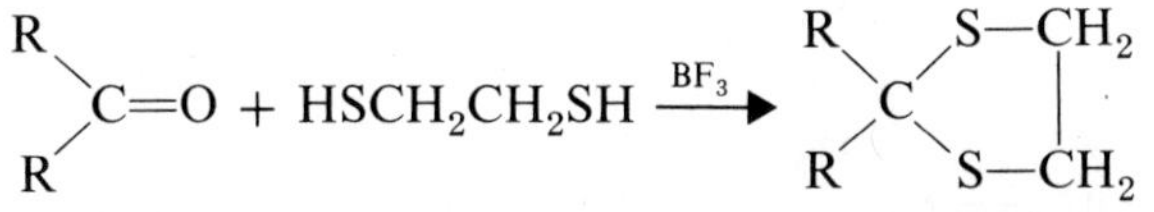

Cyclic thioketal

Thioacetals and thioketals are important in organic synthesis because they react with Raney nickel to yield hydrocarbons.* These reactions, (i.e., thioacetal

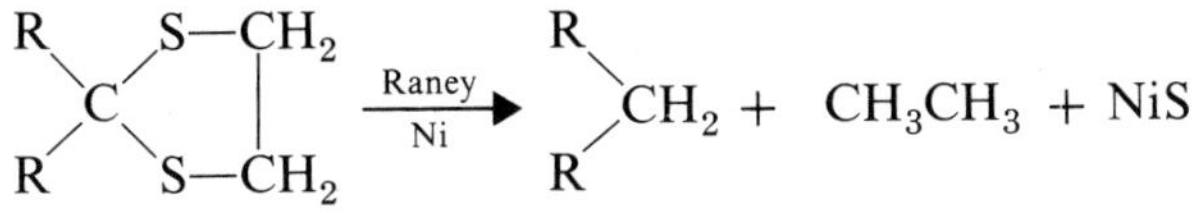

$$\begin{array}{c}R\\\\R\end{array}\!\!\!\!\diagdown\!\!\!\!C\!\!\!\!\begin{array}{c}S\!\!-\!\!CH_2\\|\\S\!\!-\!\!CH_2\end{array} \xrightarrow[Ni]{Raney}\ \begin{array}{c}R\\\\R\end{array}\!\!\!\!\diagdown\!\!\!\!CH_2 +\ CH_3CH_3 + NiS$$

or thioketal formation and subsequent "desulfurization") give us an additional method (cf. pp. 462 and 718) for converting carbonyl groups of aldehydes and ketones to —CH$_2$— groups.

Problem 18.22

Show how you might use thioacetal or thioketal formation and Raney nickel desulfurization to convert: (a) cyclohexanone to cyclohexane; (b) $CH_3\overset{\overset{\displaystyle O}{\displaystyle \|}}{C}CH_2CH_2CO_2C_2H_5$ to $CH_3CH_2CH_2CH_2CO_2C_2H_5$.

18.13 THE ADDITION OF DERIVATIVES OF AMMONIA

Aldehydes and ketones react with a number of derivatives of ammonia in the general way shown below.

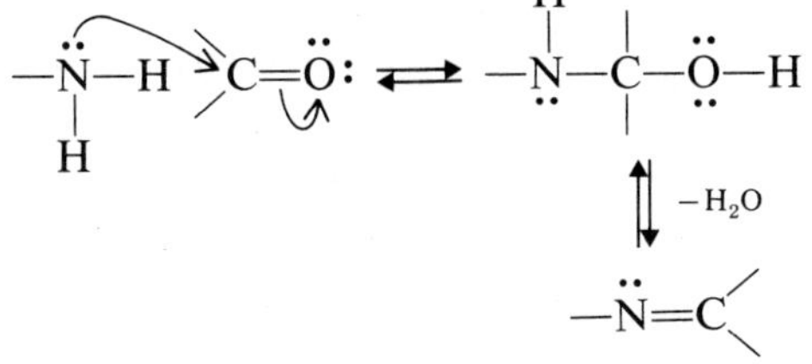

$$-\overset{..}{\underset{|\\H}{N}}\!\!-\!\!H\ \ \overset{\frown}{\diagup}\ \ \overset{\curvearrowleft}{C}\!\!=\!\!\overset{..}{\underset{..}{O}}: \ \rightleftharpoons\ -\overset{\overset{\displaystyle H}{\displaystyle |}}{\underset{..}{N}}\!\!-\!\!\overset{|}{\underset{|}{C}}\!\!-\!\!\overset{..}{\underset{..}{O}}\!\!-\!\!H$$

$$\Updownarrow\ _{-H_2O}$$

$$-\overset{..}{N}\!\!=\!\!C\diagdown$$

Table 18.3 lists several important examples of this general reaction.

* Raney nickel is a special nickel catalyst that contains adsorbed hydrogen.

TABLE 18.3 Reaction of Aldehydes and Ketones with Derivatives of Ammonia

1. Reactions with Hydroxylamine
   General Reaction

$$\text{C=O} + \text{H}_2\text{N---OH} \longrightarrow \text{C=N---OH} + \text{H}_2\text{O}$$

Aldehyde or   Hydroxylamine   An oxime
ketone

   Specific Example

$$\begin{array}{c}\text{CH}_3 \\ \diagdown \\ \text{H}\end{array}\text{C=O} + \text{H}_2\text{NOH} \xrightarrow[\text{base}]{\text{weak}} \begin{array}{c}\text{CH}_3 \\ \diagdown \\ \text{H}\end{array}\text{C=NOH}$$

Acetaldehyde   Acetaldoxime

2. Reactions with Hydrazine, Phenylhydrazine, and 2,4-Dinitrophenylhydrazine
   General Reactions

Aldehyde or
ketone

$$\text{C=O} + \text{H}_2\text{NNH}_2 \longrightarrow \text{C=NNH}_2 + \text{H}_2\text{O}$$

Hydrazine   A hydrazone

$$\text{C=O} + \text{H}_2\text{NNHC}_6\text{H}_5 \longrightarrow \text{C=NNHC}_6\text{H}_5 + \text{H}_2\text{O}$$

Phenylhydrazine   A phenylhydrazone

$$\text{C=O} + \text{H}_2\text{NNH}\!\!-\!\!\bigcirc\!\!-\!\!\text{NO}_2 \longrightarrow \text{C=NNH}\!\!-\!\!\bigcirc\!\!-\!\!\text{NO}_2 + \text{H}_2\text{O}$$

$$\qquad\qquad\qquad \text{NO}_2 \qquad\qquad\qquad\qquad \text{NO}_2$$

2,4-Dinitrophenyl-   A 2,4-Dinitrophenyl-
hydrazine   hydrazone

   Specific Examples

$$\begin{array}{c}\text{C}_6\text{H}_5 \\ \diagdown \\ \text{CH}_3\text{CH}_2\end{array}\text{C=O} + \text{H}_2\text{NNH}_2 \xrightarrow{\text{heat}} \begin{array}{c}\text{C}_6\text{H}_5 \\ \diagdown \\ \text{CH}_3\text{CH}_2\end{array}\text{C=NNH}_2 + \text{H}_2\text{O}$$

Propiophenone   Propiophenone hydrazone

$$\begin{array}{c}\text{C}_6\text{H}_5 \\ \diagdown \\ \text{CH}_3\end{array}\text{C=O} + \text{H}_2\text{NNHC}_6\text{H}_5 \xrightarrow[\text{CH}_3\text{COOH}]{\text{H}_3\text{O}^+} \begin{array}{c}\text{C}_6\text{H}_5 \\ \diagdown \\ \text{CH}_3\end{array}\text{C=NNHC}_6\text{H}_5 + \text{H}_2\text{O}$$

Acetophenone   Acetophenone phenylhydrazone

$$\begin{array}{c}\text{C}_6\text{H}_5 \\ \diagdown \\ \text{H}\end{array}\text{C=O} + \text{H}_2\text{NNH}\!\!-\!\!\bigcirc\!\!-\!\!\text{NO}_2 \xrightarrow[\substack{\text{C}_2\text{H}_5\text{OH} \\ \text{H}_2\text{O}}]{\text{HCl}} \begin{array}{c}\text{C}_6\text{H}_5 \\ \diagdown \\ \text{H}\end{array}\text{C=NNH}\!\!-\!\!\bigcirc\!\!-\!\!\text{NO}_2 + \text{H}_2\text{O}$$

$$\qquad\qquad\qquad\qquad \text{NO}_2 \qquad\qquad\qquad\qquad\qquad \text{NO}_2$$

Benzaldehyde   Benzaldehyde
                                        2,4-dinitrophenylhydrazone

TABLE 18.3 Reaction of Aldehydes and Ketones with Derivatives of Ammonia (Continued)

3. Reactions with Semicarbazide
   General Reaction

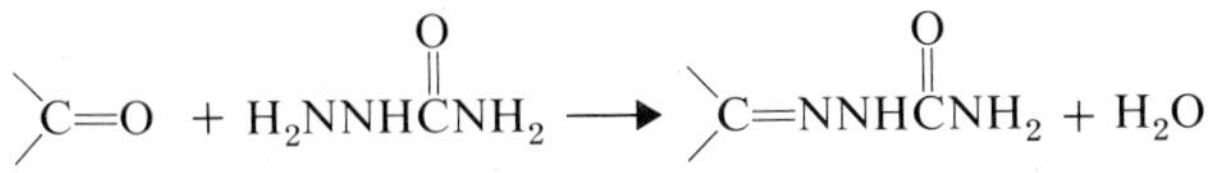

Aldehyde or ketone	Semicarbazide	A semicarbazone

Specific Example

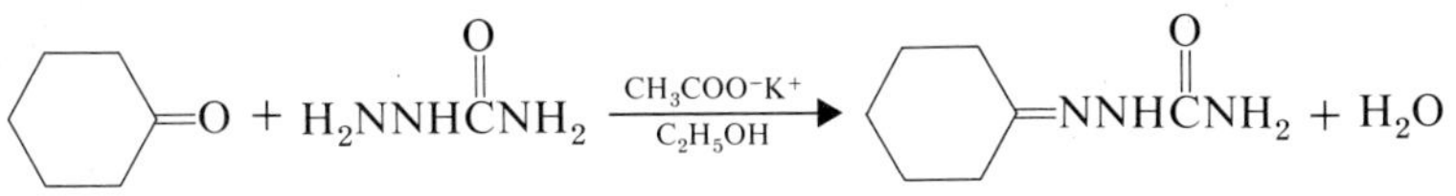

Cyclohexanone semicarbazone

2,4-Dinitrophenylhydrazones, Semicarbazones, and Oximes

The reactions of aldehydes and ketones with 2,4-dinitrophenylhydrazine, semicarbazide, and hydroxylamine are often used for purposes of identification. The products of the reactions, that is, 2,4-dinitrophenylhydrazones, semicarbazones, and oximes, are usually relatively insoluble and they have sharp characteristic melting points (cf. Table 18.4).

Imines and Enamines

Aldehydes and ketones react with primary amines to form *imines*. (Imines are also called "Schiff bases.")

Imines are important in many biochemical reactions because many enzymes use an —NH_2 group of an amino acid to react with an aldehyde or ketone to form

TABLE 18.4 Derivatives of Aldehydes and Ketones

ALDEHYDE OR KETONE	mp OF 2,4-DINITRO-PHENYLHYDRAZONE	mp OF SEMICARBAZONE	mp OF OXIME
Acetaldehyde	168.5°	162°	46.5°
Acetone	128	187 dec.	61
Benzaldehyde			130
o-Tolualdehyde	195	208	49
m-Tolualdehyde	211	213	60
p-Tolualdehyde	239	221	79
Phenylacetaldehyde	121	156	103

$$\underset{\text{Acetaldehyde}}{\overset{\displaystyle H}{\underset{|}{CH_3C{=}O}}} + \underset{\underset{\text{Methylamine}}{}}{H_2\overset{..}{N}{-}CH_3} \xrightarrow[\underset{(-H_2O)}{Na_2SO_4}]{\text{ether}} \underset{\underset{\underset{\text{(an imine)}}{\text{Acetaldimine}}}{(40\%)}}{\overset{\displaystyle H}{\underset{|}{CH_3C{=}\overset{..}{N}CH_3}}}$$

an imine linkage. We saw an example of this in the reaction of rhodopsin with all-*trans*-retinal (Section 11.1).

Imines are also formed as intermediates in a useful synthesis of amines (Section 21.4).

Aldehydes and ketones react with secondary amines to form *enamines* (*ene* + *amine*):

$$\underset{\underset{\text{Piperidine}}{}}{\overset{\displaystyle CH_3}{\underset{\underset{\displaystyle H}{|}}{CH_3CHC{=}O}}} + H{-}\overset{..}{N}\hexagon \xrightarrow[\underset{(-H_2O)}{0°}]{K_2CO_3} \underset{\underset{\underset{\text{An enamine}}{(75\%)}}{}}{\overset{\displaystyle CH_3}{\underset{|}{CH_3C{=}CH{-}\overset{..}{N}}}}\hexagon$$

Enamines are useful synthetic intermediates. We will discuss their chemistry in Section 20.3.

Hydrazones: The Wolff-Kishner Reduction

Hydrazones are occasionally used to identify aldehydes and ketones. But, unlike 2,4-dinitrophenylhydrazones, the melting points of simple hydrazones are often low. Hydrazones, however, do form the basis for a useful method by which carbonyl groups of aldehydes and ketones are transformed into $-CH_2-$ groups, called the *Wolff-Kishner reduction:*

$$\underset{\underset{\text{or ketone}}{\text{Aldehyde}}}{\overset{\diagdown}{\underset{\diagup}{C}}{=}\overset{..}{O}{:}} + H_2N{-}NH_2 \xrightarrow[\text{heat}]{\text{base}} [\underset{\text{Hydrazone}}{\overset{\diagdown}{\underset{\diagup}{C}}{=}N{-}NH_2}]$$

$$\downarrow$$

$$\overset{\diagdown}{\underset{\diagup}{C}}H_2 + N_2$$

Specific Example

$$\hexagon{-}\overset{\overset{\displaystyle O}{\|}}{C}CH_2CH_3 + H_2NNH_2 \xrightarrow[\underset{200°}{\text{Triethyleneglycol}}]{\text{NaOH}} \hexagon{-}CH_2CH_2CH_3$$

$$(82\%)$$

The Wolff-Kishner reduction is a useful complement to the Clemmensen reduction (p. 462) and the reduction of thioacetals, (p. 715) because all three reactions convert $\overset{\diagdown}{\underset{\diagup}{C}}{=}O$ groups into $-CH_2-$ groups. The Clemmensen reduction takes place in strongly acidic media and can be used for those compounds that are sensitive to base. The Wolff-Kishner reduction takes place in strongly basic solutions and can be used for those compounds that are sensitive to acid. The reduction of thioacetals takes place in neutral solution and can be used for compounds that are sensitive to both acids and bases.

18.14 HALOGENATION OF KETONES

Ketones that have an α-hydrogen react readily with halogens by substitution. The rates of these halogenation reactions *accelerate when acids or bases are added and substitution takes place almost exclusively at the α-carbon:*

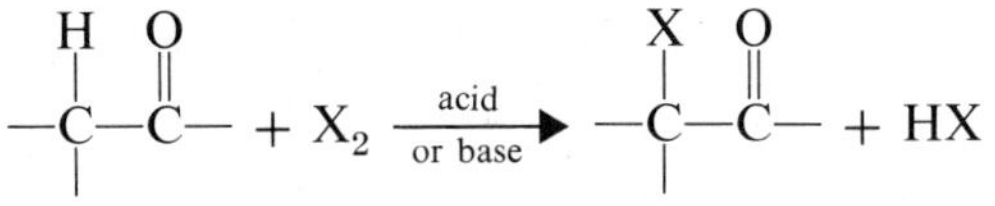

This behavior of ketones can be accounted for in terms of two related properties that we have already encountered: the acidity of the α-hydrogens of ketones and the tendency of ketones to form enols.

Base-Promoted Halogenation

In the presence of bases, halogenation takes place through the slow formation of an enolate ion followed by a rapid reaction with halogen.

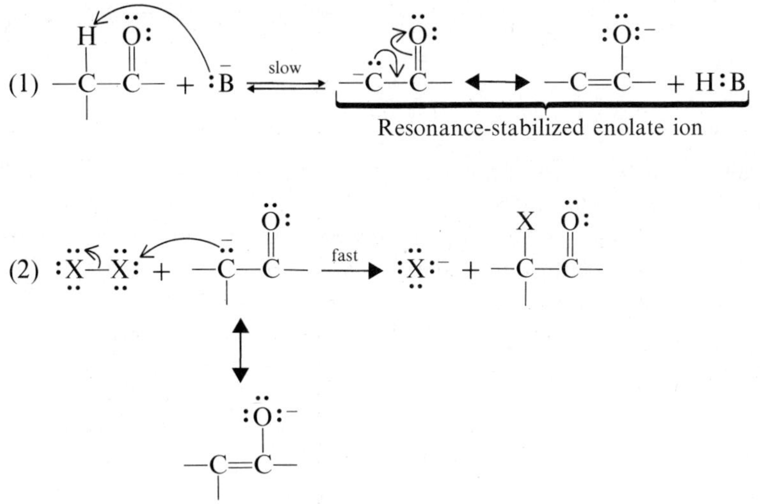

Resonance-stabilized enolate ion

Acid-Catalyzed Halogenation

In the presence of acids, halogenation takes place through the formation of an enol.

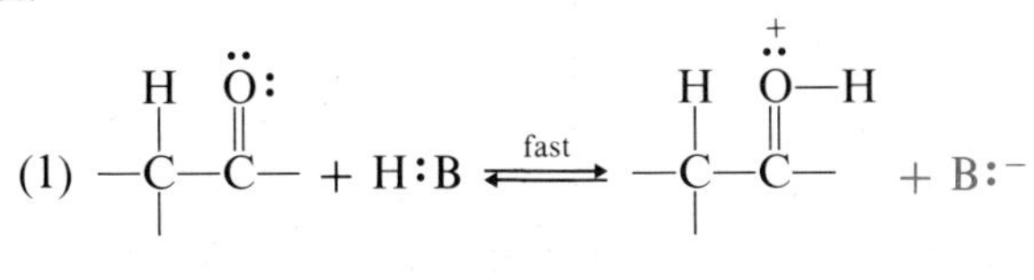

(2) $\overset{-}{B:}$ + $-\overset{|}{\underset{|}{C}}-\overset{H\quad \overset{+}{O}-H}{\underset{||}{C}}$ $\underset{\text{slow}}{\rightleftharpoons}$ H:B + $-\overset{|}{C}=\overset{O-H}{\underset{|}{C}}-$

enol

The enol then reacts with halogen to give an α-haloketone.

(3) $:\overset{..}{\underset{..}{X}}-\overset{..}{\underset{..}{X}}:$ + $-\overset{|}{C}=\overset{\overset{..}{O}-H}{\underset{|}{C}}-$ $\xrightarrow{\text{fast}}$ $:\overset{..}{\underset{..}{X}}:^-$ + $-\overset{X}{\underset{|}{C}}-\overset{\overset{+}{O}-H}{\underset{|}{C}}-$

(4) $-\overset{X}{\underset{|}{C}}-\overset{\overset{+}{O}-H}{\underset{|}{C}}-$ + $:\overset{..}{\underset{..}{X}}:^-$ $\underset{}{\overset{\text{fast}}{\rightleftharpoons}}$ $-\overset{X}{\underset{|}{C}}-\overset{\overset{..}{O}:}{\underset{|}{C}}-$ + $H:\overset{..}{\underset{..}{X}}:$

Part of the evidence that supports these mechanisms comes from studies of reaction kinetics. Both base-promoted and acid-catalyzed halogenations of ketones *show initial rates that are independent of the halogen concentration.* The mechanisms that we have written are in accord with this observation: in both instances the slow step of the mechanism occurs prior to the intervention of the halogen.

Problem 18.23

Why do we say that the first reaction is "base promoted" rather than "base catalyzed"?

Problem 18.24

Additional evidence for the halogenation mechanisms that we presented above come from the following facts:
   (a) Optically active phenyl *sec*-butyl ketone undergoes racemization at exactly the same rate that it undergoes base-promoted or acid-catalyzed halogenation.
   (b) Phenyl *sec*-butyl ketone undergoes acid-catalyzed iodination at the same rate that it undergoes acid-catalyzed bromination.
   (c) Phenyl *sec*-butyl ketone undergoes hydrogen-deuterium exchange in the presence of OD^- and D_2O. Exchange takes place only at the α-carbon and the rate of exchange is equal to the rate of base-promoted bromination.

Phenyl *sec*-butyl ketone $\xrightarrow[D_2O]{OD^-}$ Deuterium-labeled phenyl *sec*-butyl ketone

Explain how each of these observations support the mechanisms that we presented.

The Haloform Reaction

When methyl ketones react with halogens in the presence of base, multiple halogenations tend to occur at the carbon of the methyl group. Multiple halogena-

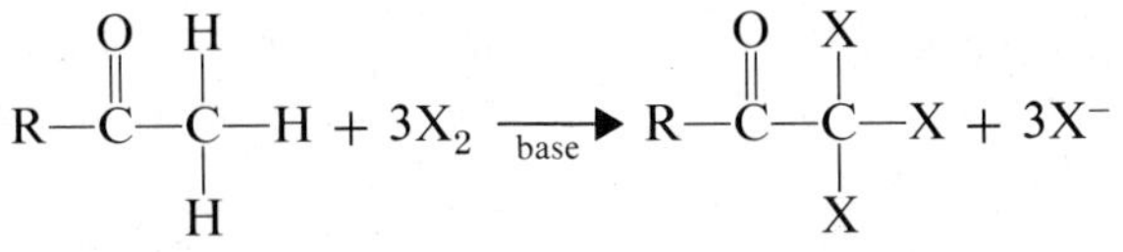

$$R-\overset{\overset{\displaystyle O}{\|}}{C}-\overset{\overset{\displaystyle H}{|}}{\underset{\underset{\displaystyle H}{|}}{C}}-H + 3X_2 \xrightarrow[\text{base}]{} R-\overset{\overset{\displaystyle O}{\|}}{C}-\overset{\overset{\displaystyle X}{|}}{\underset{\underset{\displaystyle X}{|}}{C}}-X + 3X^-$$

tions occur because substitution of the first halogen makes the remaining α-hydrogens on the methyl carbon more acidic. It does this through its inductive effect.

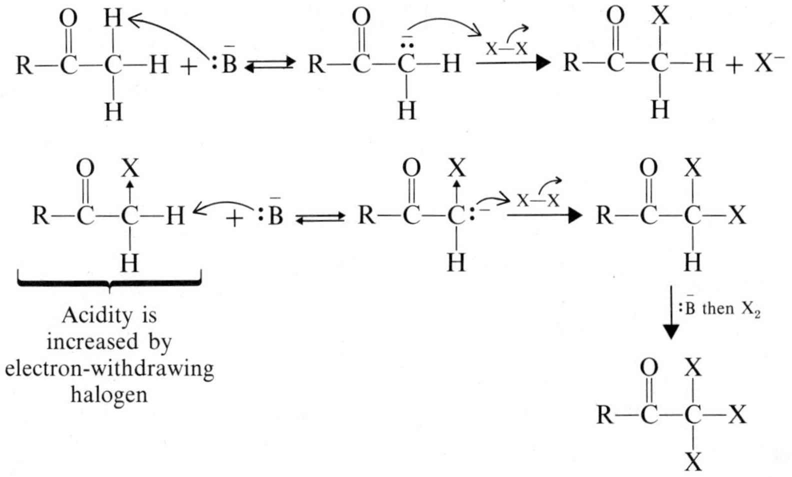

Acidity is
increased by
electron-withdrawing
halogen

When methyl ketones react with halogens in aqueous sodium hydroxide (i.e., in *hypohalite solutions*) an additional reaction takes place. Hydroxide ion attacks the carbonyl carbon of the trihaloketone and causes a cleavage of the carbon-carbon bond between the carbonyl group and the trihalomethyl group. This ultimately produces a carboxylate ion and a *haloform* (i.e., either $CHCl_3$, $CHBr_3$ or CHI_3). The

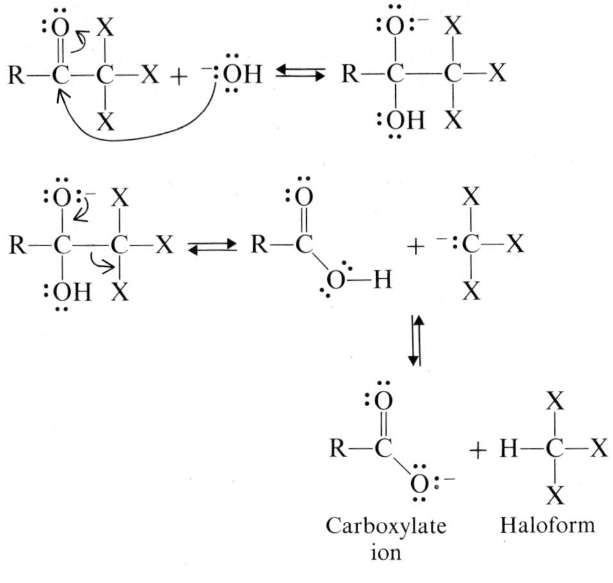

Carboxylate   Haloform
ion

initial step (p. 721) is a nucleophilic attack by hydroxide ion on the carbonyl carbon. In the next step carbon-carbon bond cleavage occurs and the haloform anion, $:CX_3^-$, departs. This step can occur because the haloform anion is unusually stable; its negative charge is dispersed by the three electronegative halogens. Finally, a proton transfer takes place between the carboxylic acid and the haloform anion.

The haloform reaction is of synthetic utility as a means of converting methyl ketones to carboxylic acids:

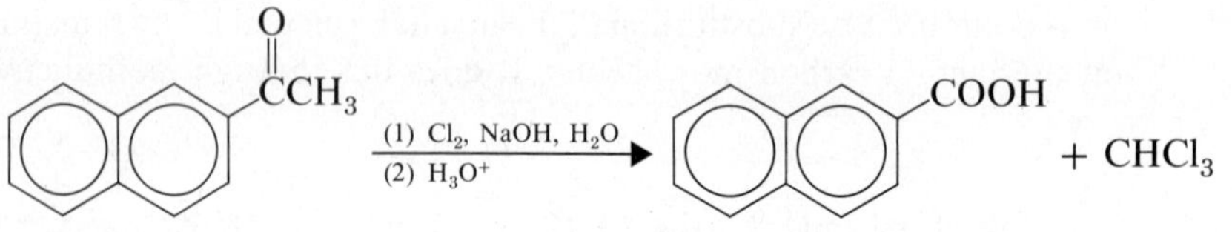

2-Acetylnaphthalene → 2-Naphthoic acid (87%) + $CHCl_3$

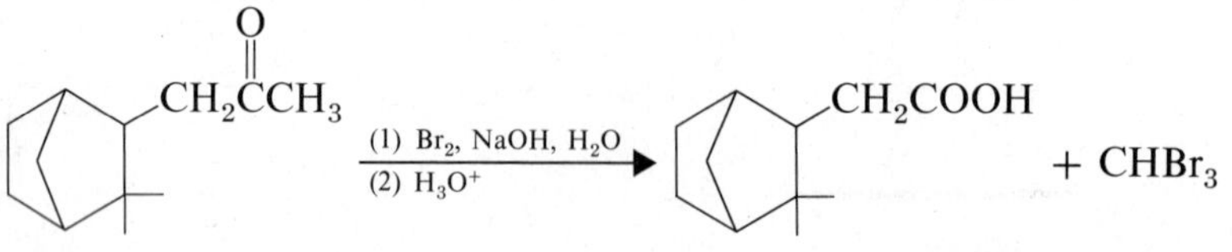

(1) Br_2, NaOH, H_2O (2) H_3O^+ → CH_2COOH + $CHBr_3$

$$CH_3C{=}CHCCH_3 \xrightarrow[\text{(2) } H_3O^+]{\text{(1) } Cl_2,\ NaOH,\ H_2O} CH_3C{=}CHCOOH + CHCl_3$$

(50%)

When the haloform reaction is used in synthesis, chlorine and bromine are most commonly used as the halogen component. Chloroform ($CHCl_3$) and bromoform ($CHBr_3$) are both liquids and are easily separated from the acid.

The Iodoform Reaction

The haloform reaction using iodine and aqueous sodium hydroxide is called the *iodoform reaction*. The iodoform reaction is frequently used in structure determinations because it allows identification of the two groups shown below:

$$-\overset{\underset{\|}{O}}{C}{-}CH_3 \quad \text{and} \quad -\overset{\underset{|}{OH}}{C}H{-}CH_3$$

Compounds containing either of these groups react with iodine in sodium hydroxide to give bright yellow precipitates of *iodoform* (CHI_3, mp 119°). Compounds containing the $-CHOHCH_3$ group give a positive iodoform test because they are first oxidized to methyl ketones:

$$-\overset{\underset{|}{OH}}{C}HCH_3 + I_2 + 2OH^- \longrightarrow -\overset{\underset{\|}{O}}{C}CH_3 + 2I^- + H_2O$$

Methyl ketones then react with iodine and hydroxide ion to produce iodoform:

$$-\underset{\underset{O}{\|}}{C}-CH_3 + 3I_2 + 3OH^- \longrightarrow -\underset{\underset{O}{\|}}{C}-CI_3 + 3I^- + 3H_2O$$

$$-\underset{\underset{O}{\|}}{C}-CI_3 + OH^- \longrightarrow -\underset{\underset{O}{\|}}{C}-O^- + CHI_3\downarrow$$

Yellow
precipitate

The group to which the —COCH$_3$ or —CHOHCH$_3$ function is attached can be aryl, alkyl, or hydrogen. Thus, even ethyl alcohol and acetaldehyde give positive iodoform tests.

$$CH_3-\underset{\underset{}{\overset{\overset{O}{\|}}{C}}}{}-H \xrightarrow[\text{NaOH/H}_2\text{O}]{\text{I}_2\ (excess)} H\overset{\overset{O}{\|}}{C}-O^- + CHI_3$$

Acetaldehyde

$$CH_3-\underset{\underset{H}{|}}{\overset{\overset{OH}{|}}{C}}-H \xrightarrow[\text{NaOH/H}_2\text{O}]{\text{I}_2\ (excess)} H\overset{\overset{O}{\|}}{C}-O^- + CHI_3$$

Ethyl alcohol

Problem 18.25

Which of the following compounds would give a positive iodoform test?

(a) Acetone   (f) 1-Phenylethanol
(b) Acetophenone   (g) 2-Phenylethanol
(c) Pentanal   (h) 2-Butanol
(d) 2-Pentanone   (i) 1-Acetylnaphthalene
(e) 3-Pentanone   (j) 3-Pentanol

18.15 α,β-UNSATURATED ALDEHYDES AND KETONES

When α,β-unsaturated aldehydes and ketones react with nucleophilic reagents they may do so in two ways: they may react by a *simple addition*, that is, one in which

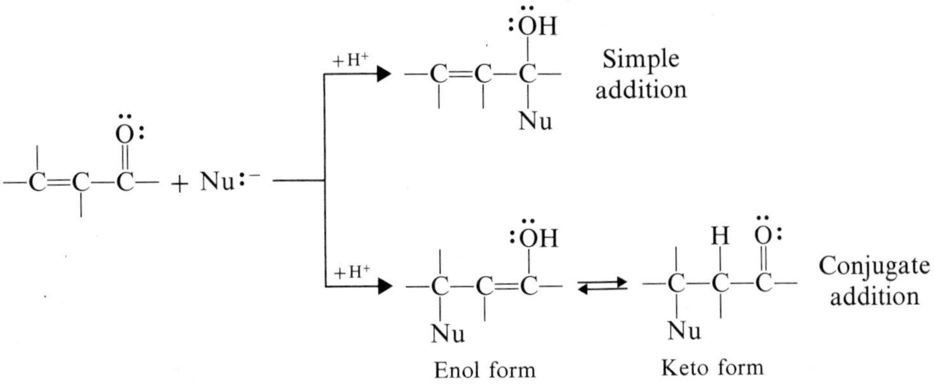

Simple addition

Conjugate addition

Enol form   Keto form

the nucleophile adds across the double bond of the carbonyl group; or they may react by a *conjugate addition,* one that resembles the 1,4-addition reactions of conjugated dienes (p. 358).

In many instances both modes of addition occur in the same reaction mixture. As examples, let us consider the Grignard reactions shown below.

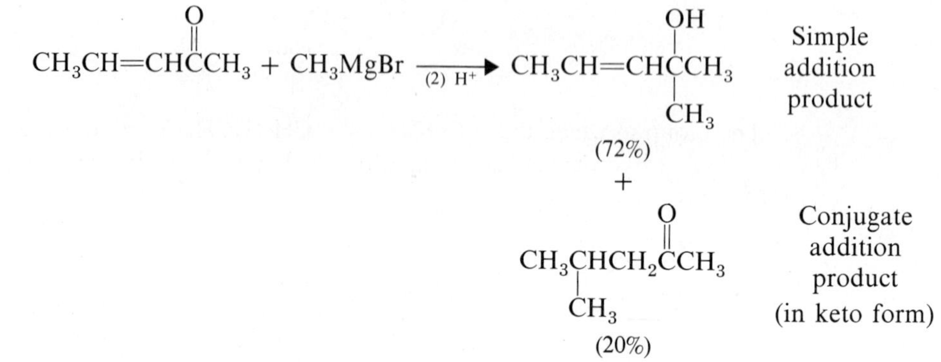

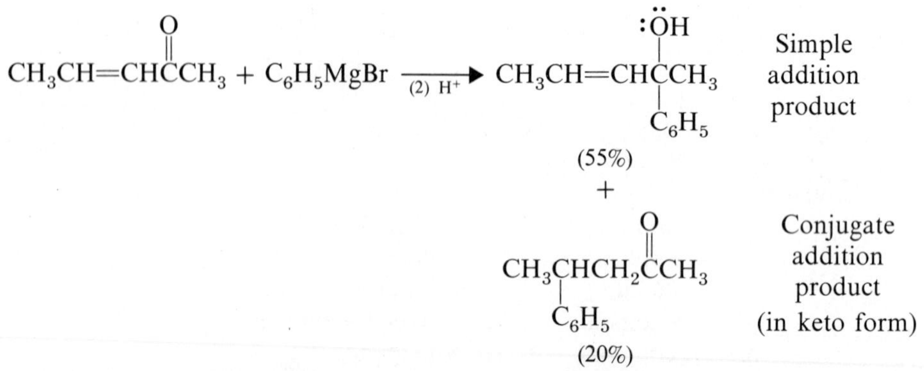

$$\text{CH}_3\text{CH}=\text{CHCCH}_3 + \text{CH}_3\text{MgBr} \xrightarrow[\text{(2) H}^+]{} \underset{\underset{\text{CH}_3}{|}}{\text{CH}_3\text{CH}=\text{CHCCH}_3}$$

Simple addition product

(72%)

+

$$\underset{\underset{\text{CH}_3}{|}}{\text{CH}_3\text{CHCH}_2\text{CCH}_3}$$

Conjugate addition product (in keto form)

(20%)

$$\text{CH}_3\text{CH}=\text{CHCCH}_3 + \text{C}_6\text{H}_5\text{MgBr} \xrightarrow[\text{(2) H}^+]{} \underset{\underset{\text{C}_6\text{H}_5}{|}}{\text{CH}_3\text{CH}=\text{CHCCH}_3}$$

Simple addition product

(55%)

+

$$\underset{\underset{\text{C}_6\text{H}_5}{|}}{\text{CH}_3\text{CHCH}_2\text{CCH}_3}$$

Conjugate addition product (in keto form)

(20%)

In each of these examples we see that simple addition is favored, but that conjugate addition accounts for a substantial amount of the product.

If we examine the resonance structures that contribute to the overall hybrid for an α,β-unsaturated aldehyde or ketone (below) we will be in a better position to understand these reactions.

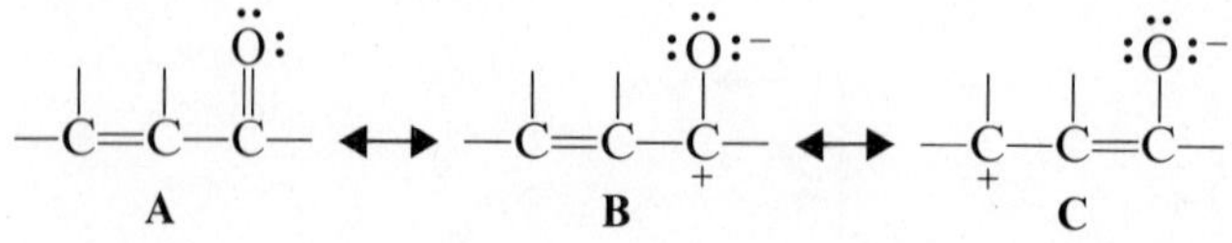

Although structures **B** and **C** involve separated charges they make a significant contribution to the hybrid because, in each, the negative charge is carried by electronegative oxygen. Structures **B** and **C** not only indicate that the oxygen of the hybrid should bear a partial negative charge; but they also indicate that *both the carbonyl carbon and the β-carbon should bear a partial positive charge.* They indicate that we should represent the hybrid in the following way:

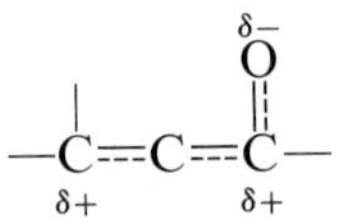

$$-\overset{|}{C}=\overset{|}{C}=\overset{\overset{\overset{\delta-}{O}}{\|}}{\underset{\delta+}{C}}-$$

This structure tells us that we should expect an electrophilic reagent to attack the carbonyl oxygen and a nucleophilic reagent to attack either the carbonyl carbon or the β-carbon.

This is exactly what happens in the Grignard reactions that we saw earlier. The electrophilic magnesium attacks the carbonyl oxygen; the nucleophilic carbon of the Grignard reagent attacks either the carbonyl carbon or the β-carbon.

Simple Addition

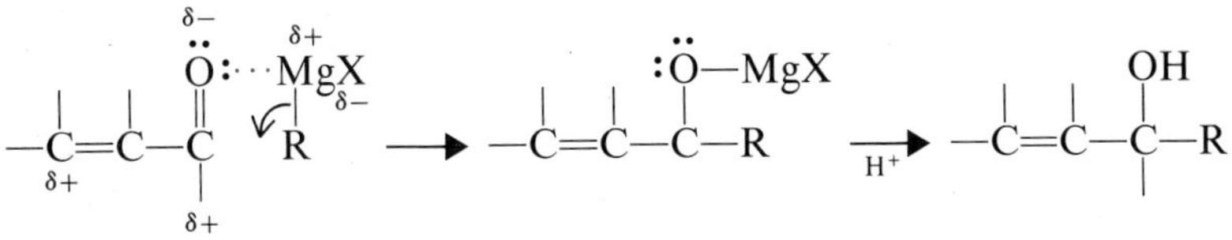

Conjugate Addition

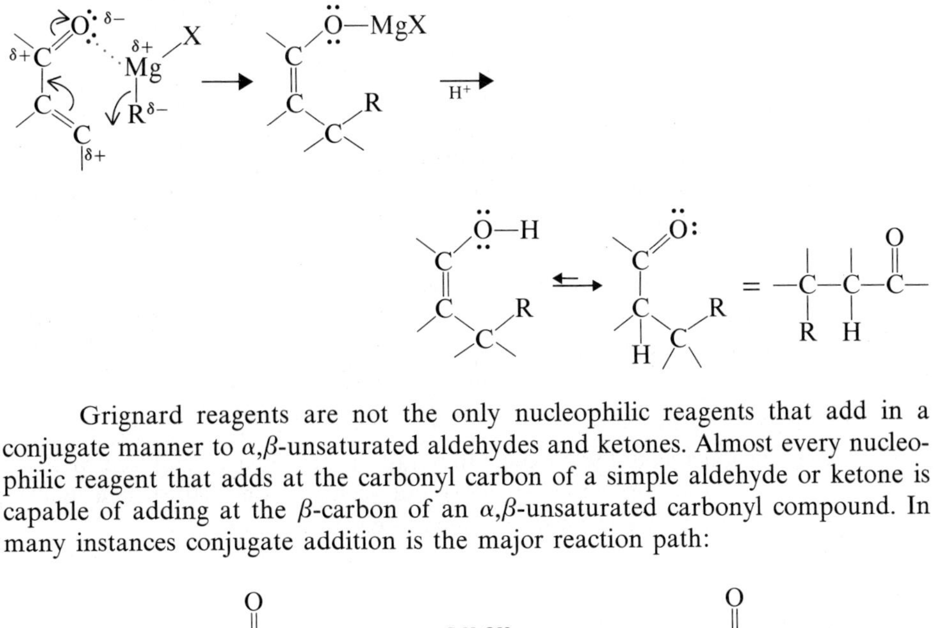

Grignard reagents are not the only nucleophilic reagents that add in a conjugate manner to α,β-unsaturated aldehydes and ketones. Almost every nucleophilic reagent that adds at the carbonyl carbon of a simple aldehyde or ketone is capable of adding at the β-carbon of an α,β-unsaturated carbonyl compound. In many instances conjugate addition is the major reaction path:

$$C_6H_5CH{=}CHCC_6H_5 + CN^- \xrightarrow[CH_3COOH]{C_2H_5OH} C_6H_5CH{-}CH_2CC_6H_5$$

with the product bearing $\overset{O}{\overset{\|}{}}$ on the carbonyl and CN on the CH carbon.

(95%)

$$CH_3C{=}CHCCH_3 + CH_3NH_2 \xrightarrow[H_2O]{} CH_3C{-}CH_2CCH_3$$

with CH_3 and $\overset{O}{\overset{\|}{}}$ groups; product bearing CH_3, CH_3NH, and $\overset{O}{\overset{\|}{}}CCH_3$.

(75%)

The sequence shown below illustrates how a conjugate aldol addition followed by a simple aldol condensation may be used to build one ring on to another. This procedure is known as the *Robinson annellation* ("ring forming") reaction.

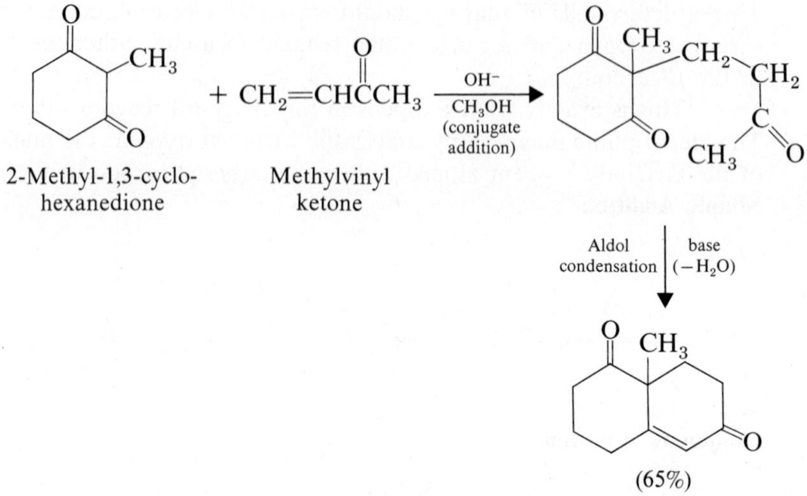

2-Methyl-1,3-cyclo-
hexanedione

Methylvinyl
ketone

Aldol condensation | base (−H₂O)

(65%)

Problem 18.26

(a) Propose step-by-step mechanisms for both steps of the Robinson annellation sequence shown above. (b) Would you expect 2-methyl-1,3-cyclohexanedione to be more or less acidic than cyclohexanone? Explain your answer.

Conjugate nucleophilic additions to α,β-unsaturated carbonyl compounds are known generally as *Michael reactions*. We will see other examples of Michael reactions in later chapters.

Problem 18.27

What product would you expect to obtain from the base-catalyzed Michael reaction of (a) benzalacetophenone (p. 703) and acetophenone? (b) of benzalacetophenone and cyclopentadiene? Show all steps in each mechanism.

Problem 18.28

When acrolein reacts with hydrazine, the product is a dihydropyrazole:

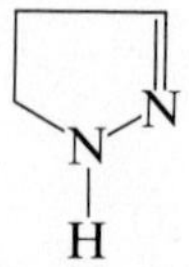

Acrolein   Hydrazine   A dihydropyrazole

Suggest a mechanism that explains this reaction.

18.16 CHEMICAL TESTS FOR ALDEHYDES AND KETONES

In addition to the iodoform reaction (p. 722) a variety of other chemical methods can be used in elucidating the structures of aldehydes and ketones.

Aldehydes and ketones can be differentiated from noncarbonyl compounds through their reactions with derivatives of ammonia. Semicarbazide, 2,4-dinitrophenylhydrazine, and hydroxylamine react with aldehydes and ketones to form precipitates. Semicarbazones and oximes are usually colorless, while 2,4-dinitrophenyl hydrazones are usually orange. The melting point of these derivatives can also be used in identifying specific aldehydes and ketones.

The ease with which aldehydes undergo oxidation provides a useful test that differentiates aldehydes from most ketones.

Tollens' (or silver mirror) Test

Mixing aqueous silver nitrate with aqueous ammonia produces a solution known as Tollens' reagent. The reagent contains the silver diammine ion, $Ag(NH_3)_2^+$. Although this ion is a very weak oxidizing agent it will oxidize aldehydes to carboxylate ions. As it does this, silver is reduced from the $+1$ oxidation state (of $Ag(NH_3)_2^+$) to metallic silver. If the rate of reaction is slow and the walls of the vessel are clean, metallic silver deposits on the walls of the test tube as a mirror; if not, it deposits as gray to black precipitate. Tollens' reagent gives a negative result with all ketones except α-hydroxyketones.

$$\underset{\text{Aldehyde}}{R-\overset{\displaystyle O}{\overset{\|}{C}}-H} + Ag(NH_3)_2^+ \xrightarrow{\text{aq. } NH_3} R-\overset{\displaystyle O}{\overset{\|}{C}}-\bar{O} + \underset{\substack{\text{Silver} \\ \text{mirror}}}{Ag\downarrow}$$

$$\underset{\alpha\text{-Hydroxyketone}}{R-\overset{\displaystyle O}{\overset{\|}{C}}-\overset{\displaystyle OH}{\overset{|}{C}}HR} + Ag(NH_3)_2^+ \xrightarrow{\text{aq. } NH_3} R-\overset{\displaystyle O}{\overset{\|}{C}}-\overset{\displaystyle O}{\overset{\|}{C}}R + \underset{\substack{\text{Silver} \\ \text{mirror}}}{Ag\downarrow}$$

$$\underset{\text{Ketone}}{R-\overset{\displaystyle O}{\overset{\|}{C}}-R} + Ag(NH_3)_2^+ \xrightarrow{\text{aq. } NH_3} \text{No reaction}$$

18.17 SUMMARY OF THE ADDITION REACTIONS OF ALDEHYDES AND KETONES

Table 18.5 summarizes the nucleophilic addition reactions of aldehydes and ketones that occur at the carbonyl carbon.

TABLE 18.5 Nucleophilic Addition Reactions of Aldehydes and Ketones

(1) Addition of Organometallic Compounds
Grignard Reagents (Section 16.5)
General Reaction

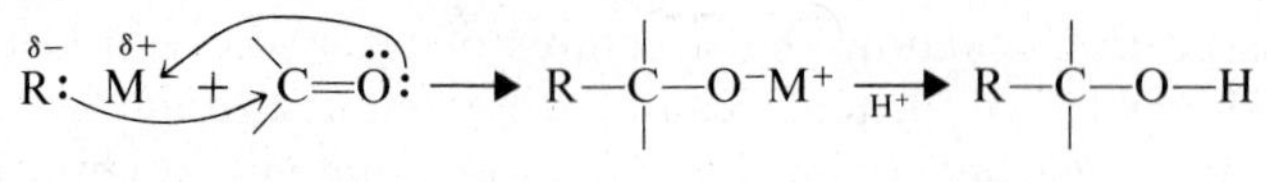

$$\overset{\delta-}{R:}\overset{\delta+}{M} + \underset{}{\overset{}{C}}=\overset{\cdot\cdot}{O}: \longrightarrow R-\overset{|}{\underset{|}{C}}-O^-M^+ \xrightarrow[H^+]{} R-\overset{|}{\underset{|}{C}}-O-H$$

Specific Examples

$$CH_3CH_2MgBr + CH_3\overset{O}{\overset{||}{C}}-H \xrightarrow[(2)H^+]{} CH_3CH_2\overset{OH}{\underset{}{CHCH_3}}$$
(67%)

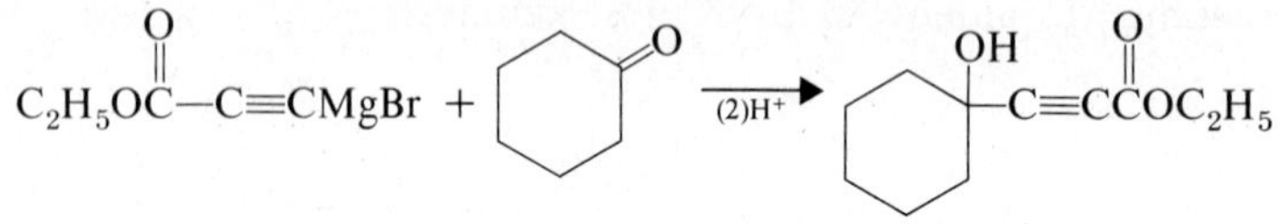

$$C_2H_5O\overset{O}{\overset{||}{C}}-C\equiv CMgBr + \text{(cyclohexanone)} \xrightarrow[(2)H^+]{} \text{(cyclohexane ring)}\overset{OH}{\underset{}{}}-C\equiv C\overset{O}{\overset{||}{C}}OC_2H_5$$
(67%)

(2) Addition of Hydride Ion (Section 16.4)
General Reaction

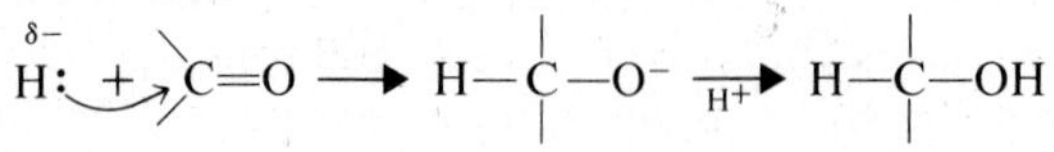

$$\overset{\delta-}{H:} + \overset{}{C}=O \longrightarrow H-\overset{|}{\underset{|}{C}}-O^- \xrightarrow[H^+]{} H-\overset{|}{\underset{|}{C}}-OH$$

Specific Examples Using Metal Hydrides (Section 16.4)

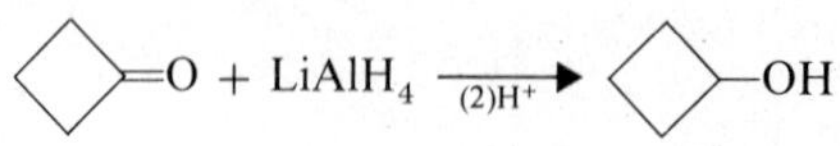

$$\text{(cyclobutane)}=O + LiAlH_4 \xrightarrow[(2)H^+]{} \text{(cyclobutane)}-OH$$
(90%)

$$CH_3\overset{O}{\overset{||}{C}}CH_2CH_2CH_3 + NaBH_4 \xrightarrow[OH^-]{CH_3OH} CH_3\overset{OH}{\underset{}{CHCH_2CH_2CH_3}}$$
(100%)

The Cannizzaro Reaction (Section 18.10)

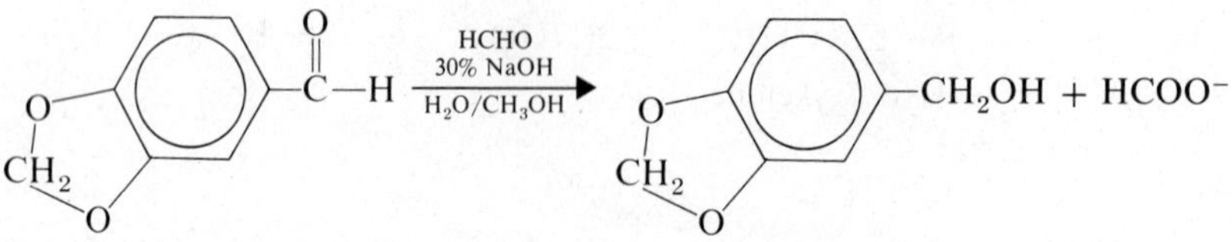

$$\text{(benzodioxole)}\overset{O}{\overset{||}{C}}-H \xrightarrow[H_2O/CH_3OH]{\substack{HCHO \\ 30\% NaOH}} \text{(benzodioxole)}-CH_2OH + HCOO^-$$
(85–90%)

(3) Addition of Hydrogen Cyanide and Sodium Bisulfite (Section 18.7)
General Reaction

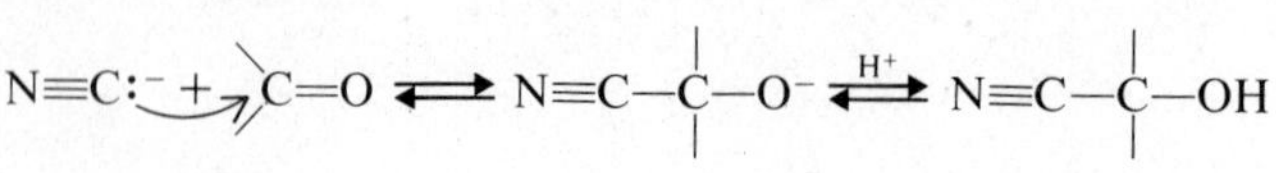

$$N\equiv C:^- + \overset{}{C}=O \rightleftharpoons N\equiv C-\overset{|}{\underset{|}{C}}-O^- \overset{H^+}{\rightleftharpoons} N\equiv C-\overset{|}{\underset{|}{C}}-OH$$

Specific Example

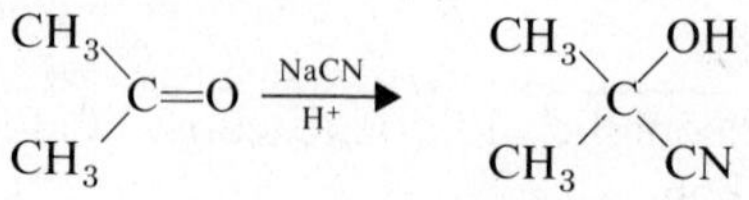

$$\overset{CH_3}{\underset{CH_3}{>}}C=O \xrightarrow[H^+]{NaCN} \overset{CH_3}{\underset{CH_3}{>}}\overset{OH}{\underset{CN}{C}}$$
(77–78%)

Acetone cyanohydrin

TABLE 18.5 Nucleophilic Addition Reactions of Aldehydes and Ketones (Continued)

General Reaction

$$\text{NaHSO}_3 + \text{C=O} \rightleftharpoons \text{C} \begin{smallmatrix} \text{SO}_3^- \ \text{Na}^+ \\ \\ \text{OH} \end{smallmatrix}$$

Specific Example

$$\text{CH}_3\text{C=O} + \text{NaHSO}_3 \rightleftharpoons \text{CH}_3\text{C} \begin{smallmatrix} \text{SO}_3^- \ \text{Na}^+ \\ \\ \text{OH} \end{smallmatrix} \text{H}$$

(88%)

(4) The Aldol Addition. The Addition of Enolate Ions (Section 18.9)

General Reaction

$$-\text{C}=\text{C}- + \text{C=O} \rightleftharpoons -\overset{\text{O}}{\overset{\|}{\text{C}}}-\text{C}-\text{C}-\text{O}^- \overset{\text{H}^+}{\rightleftharpoons} -\overset{\text{O}}{\overset{\|}{\text{C}}}-\text{C}-\text{C}-\text{OH}$$

Specific Example

$$\overset{\text{O}}{\overset{\|}{\text{HCCH}_3}} + \text{OH}^- \rightleftharpoons \text{H}-\overset{\text{O}^-}{\overset{\|}{\text{C}}}=\text{CH}_2 \overset{\text{CH}_3\text{CH, H}_2\text{O}}{\rightleftharpoons} \overset{\text{O}}{\overset{\|}{\text{HC}}}-\text{CH}_2\overset{\text{OH}}{\overset{|}{\text{C}}}\text{CH}_3 \;\; \overset{|}{\text{H}}$$

(5) Addition of Ylides (Section 18.11)

The Wittig Reaction

$$\text{Ar}_3\text{P}=\text{C}- + \text{C=O} \rightleftharpoons -\overset{|}{\text{C}}-\overset{|}{\text{C}}- \overset{-\text{Ar}_3\text{PO}}{\longrightarrow} \text{C=C} \\ \text{Ar}_3\overset{}{\text{P}}_+ \;\; \text{O}_-$$

The Addition of Sulfur Ylides

$$\text{R}_2\text{S}=\text{CH}_2 + \text{C=O} \longrightarrow -\overset{|}{\text{C}}-\overset{|}{\text{C}}- \overset{-\text{R}_2\text{S}}{\longrightarrow} -\overset{|}{\text{C}}-\overset{|}{\text{C}}- \\ \text{R}_2\overset{}{\text{S}}_+ \;\; \text{O}_- \qquad\qquad \text{O}$$

(6) Addition of Alcohols (Section 18.12)

General Reaction

$$\text{R}-\overset{..}{\underset{..}{\text{O}}}-\text{H} + \text{C=O} \rightleftharpoons \text{R}-\text{O}-\overset{|}{\text{C}}-\text{OH} \overset{\text{ROH}}{\underset{\text{H}^+}{\longrightarrow}} \text{R}-\text{O}-\overset{|}{\text{C}}-\text{O}-\text{R}$$

Hemiacetal   Acetal or
or hemiketal   ketal

Specific Example

$$\text{C}_2\text{H}_5\text{OH} + \overset{\text{O}}{\overset{\|}{\text{CH}_3\text{CH}}} \rightleftharpoons \text{C}_2\text{H}_5\text{O}-\overset{\text{CH}_3}{\underset{\text{H}}{\text{C}}}-\text{OH} \overset{\text{C}_2\text{H}_5\text{OH}}{\underset{\text{H}^+}{\longrightarrow}} \text{C}_2\text{H}_5\text{O}\overset{\text{CH}_3}{\underset{\text{H}}{\text{C}}}-\text{OC}_2\text{H}_5$$

TABLE 18.5 Nucleophilic Addition Reactions of Aldehydes and Ketones (Continued)

(7) Addition of Derivatives of Ammonia (Section 18.13)

General Reaction

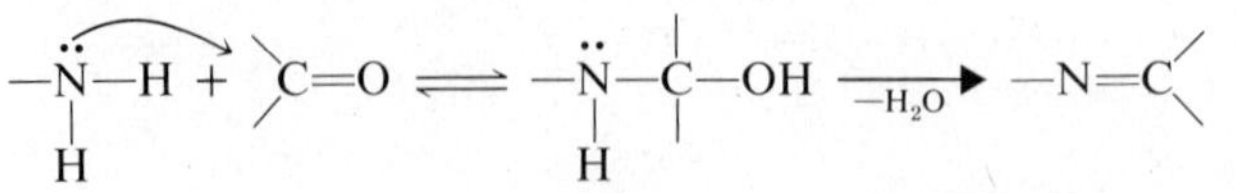

$$-\ddot{N}-H + \; \rangle C=O \;\rightleftharpoons\; -\ddot{N}-\overset{|}{\underset{|}{C}}-OH \xrightarrow[-H_2O]{} -N=C\langle$$
$$\;\;\;\; H \qquad\qquad\qquad H$$

Specific Examples

$$\begin{array}{c} O \\ \parallel \\ CH_3CH \end{array} + NH_2OH \longrightarrow CH_3CH=NOH$$
$$\text{Acetaldoxime}$$

$$\begin{array}{c} O \\ \parallel \\ C_6H_5CH \end{array} + H_2NNHC_6H_5 \longrightarrow C_6H_5CH=NNHC_6H_5$$
$$\text{Benzaldehyde phenylhydrazone}$$

Not only do α,β-unsaturated aldehydes and ketones undergo all of these reactions, they also undergo nucleophilic addition at their β-carbons (Section 18.15).

General Reaction

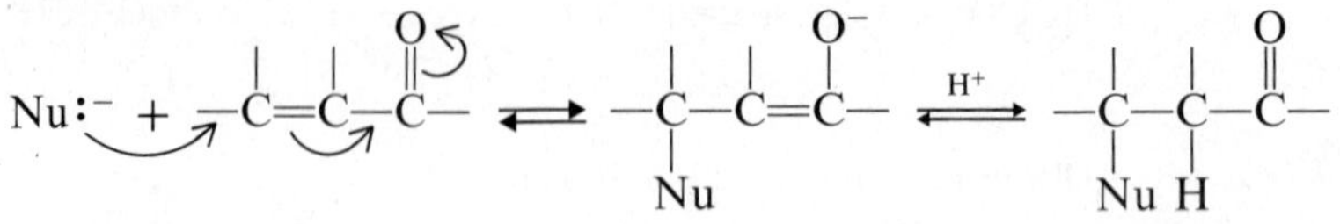

$$Nu{:}^- + \; \overset{}{\underset{}{-C=C-\overset{O}{\overset{\parallel}{C}}-}} \rightleftharpoons \; -\overset{O^-}{\underset{Nu}{\overset{|}{C}-C=C-}} \xrightarrow{H^+} -\overset{O}{\underset{Nu\;H}{\overset{\parallel}{C}-C-C-}}$$

Additional Problems

18.29
Give structural formulas for each of the following compounds.
(a) Formaldehyde
(b) Acetaldehyde
(c) Phenylacetaldehyde
(d) Acetone
(e) Methyl ethyl ketone
(f) Acetophenone
(g) Benzalacetone
(h) Benzalacetophenone
(i) Benzophenone
(j) Salicylaldehyde
(k) Piperonal
(l) Vanillin

18.30
Write structural formulas for the products formed when propanal reacts with each of the following reagents.
(a) $NaBH_4$ in aqueous NaOH
(b) C_6H_5MgBr then H_2O
(c) $LiAlH_4$ then H_2O
(d) $NaHSO_3$
(e) $NaHSO_3$ then NaCN
(f) OH^-, H_2O
(g) OH^-, H_2O then heat
(h) H_2 and Pt
(i) $HOCH_2CH_2OH$ and H^+
(j) $CH_3CH=P(C_6H_5)_3$
(k) Br_2 in Acetic acid
(l) $Ag (NH_3)_2^+$
(m) Hydroxylamine
(n) Semicarbazide
(o) Phenylhydrazine
(p) Cold dilute $KMnO_4$
(q) $HSCH_2CH_2SH$, H^+
(r) $HSCH_2CH_2SH$, H^+, then Raney nickel

18.31
Give structural formulas for the products formed (if any) from the reaction of acetone with each reagent in problem 18.30.

18.32

What products would be obtained from each of the following reactions of p-tolualdehyde?

(a) p-Tolualdehyde + acetaldehyde $\xrightarrow{OH^-}$

(b) p-Tolualdehyde $\xrightarrow[\text{NaOH}]{\text{conc.}}$

(c) p-Tolualdehyde + formaldehyde $\xrightarrow[\text{NaOH}]{\text{conc.}}$

(d) p-Tolualdehyde + cold dilute $KMnO_4 \longrightarrow$

(e) p-Tolualdehyde + $KMnO_4 \xrightarrow{\text{heat}}$

(f) p-Tolualdehyde + $CH_2{=}P(C_6H_5)_3 \longrightarrow$

18.33

What products would be obtained from each of the following reactions of acetophenone?

(a) Acetophenone + $HNO_3 \xrightarrow{H_2SO_4}$

(b) Acetophenone + $Cl_{2(\text{excess})} \xrightarrow{OH^-}$

(c) Acetophenone + $CH_2{=}P(C_6H_5)_3 \longrightarrow$

(d) Acetophenone + $NaBH_4 \xrightarrow[OH^-]{H_2O}$

(e) Acetophenone + $C_6H_5MgBr \xrightarrow{(2)\ H_2O}$

18.34

(a) Give three methods for synthesizing phenyl n-propyl ketone from benzene and any other needed reagents. (b) Give three methods for transforming phenyl n-propyl ketone into n-butylbenzene.

18.35

Review the mechanism given for the Cannizzaro reaction on p. 704. (a) What products would you expect to obtain when benzaldehyde is treated with concentrated NaOD in D_2O? (b) What products would you expect to obtain when benzaldehyde is treated with NaOH in H_2O? (c) Would these experiments help verify the proposed mechanism? How?

18.36

Outline simple chemical tests that would distinguish between each of the following:
(a) Benzaldehyde and benzyl alcohol
(b) Hexanal and 2-hexanone
(c) 2-Hexanone and 3-hexanone
(d) 2-Hexanol and 2-hexanone
(e) 2-Hexanol and 3-hexanol
(f) 3-Hexanone and 3-hexanol
(g) Benzalacetophenone and benzophenone
(h) 1-Phenylethanol and 2-phenylethanol
(i) Pentanal and diethyl ether

(j) $\underset{\text{O}}{\overset{\text{O}}{CH_3\overset{\|}{C}CH_2\overset{\|}{C}CH_3}}$ and $\underset{\text{OH}\quad\text{O}}{CH_3\overset{|}{C}{=}CH\overset{\|}{C}CH_3}$

(k) 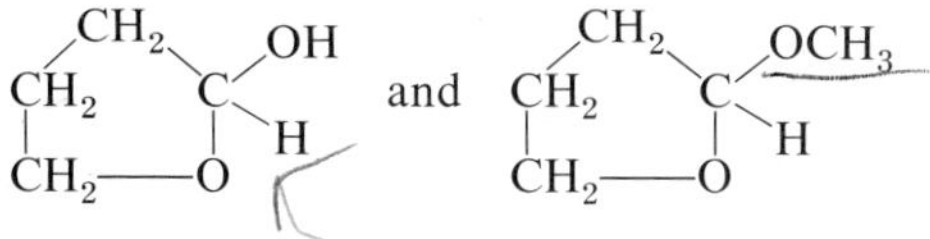 $\underset{\overset{|}{CH_2}\overset{|}{\underset{|}{CH_2}}}{\overset{CH_2}{\underset{}{}}}\underset{\overset{|}{O}}{\overset{OH}{\underset{H}{C}}}$ and $\underset{\overset{|}{CH_2}\overset{|}{\underset{|}{CH_2}}}{\overset{CH_2}{\underset{}{}}}\underset{\overset{|}{O}}{\overset{OCH_3}{\underset{H}{C}}}$

18.37

(a) Infrared spectroscopy gives an easy method for deciding whether the product obtained from the addition of a Grignard reagent to an α,β-unsaturated ketone is the simple addition product or the conjugate addition product. Explain. (What peak or peaks would you look for?) (b) How might you follow the rate of the following reaction using ultraviolet spectroscopy?

$$(CH_3)_2C{=}CHCCH_3 + CH_3NH_2 \xrightarrow{H_2O} (CH_3)_2CCH_2CCH_3$$

with carbonyl O groups, and CH_3NH substituent.

18.38

The hydrogens of the γ-carbon of crotonaldehyde are appreciably acidic ($K_a \simeq 10^{-20}$). (a) Write resonance structures that will explain this fact.

$$\overset{\gamma}{C}H_3\overset{\beta}{C}H{=}\overset{\alpha}{C}HCHO$$
Crotonaldehyde

(b) Write a mechanism that accounts for the following reaction.

$$C_6H_5CH{=}CHCHO + CH_3CH{=}CHCHO \xrightarrow[C_2H_5OH]{base} C_6H_5(CH{=}CH)_3CHO$$
(87%)

18.39

What reagents would you use to bring about each step of the following syntheses?

(a)

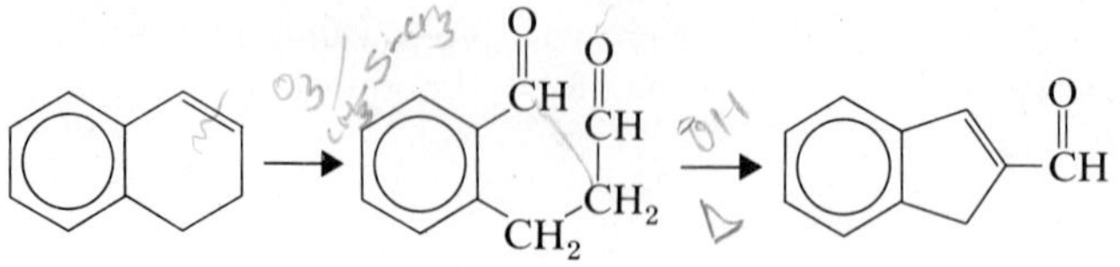

(b)

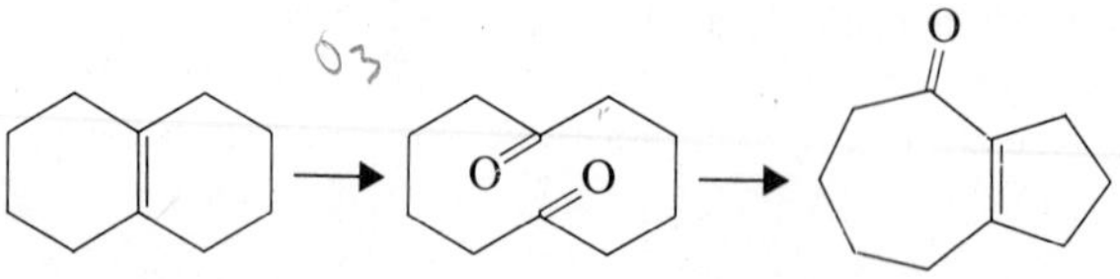

(c)

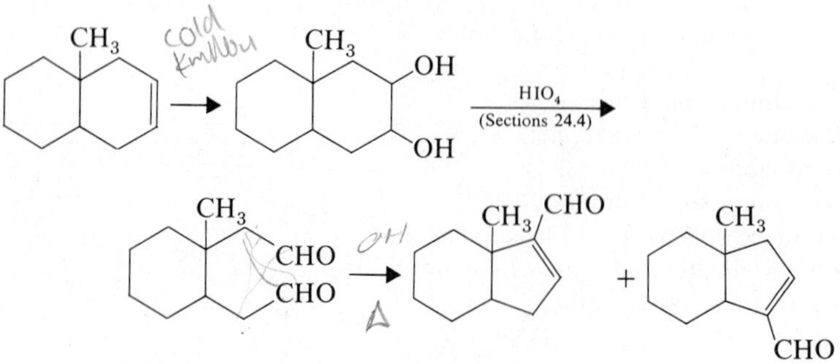

(d)

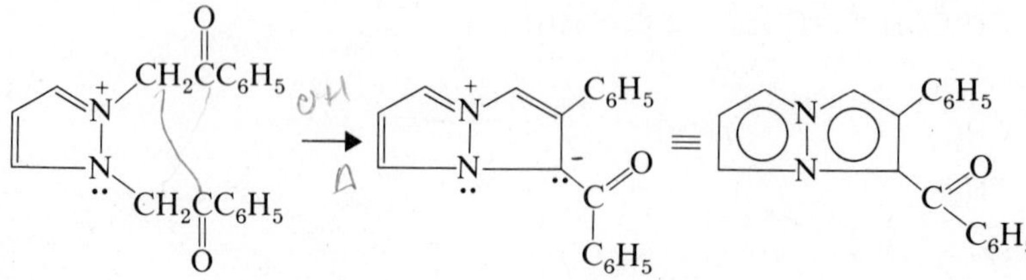

18.40
The α-hydrogens of nitroalkanes (i.e., $R{-}CH_2NO_2$) are appreciably acidic ($K_a = \sim 10^{-10}$). (a) Write resonance structures for the anion of nitromethane that will account for the acidity of nitromethane. (b) Assuming that you have available aldehydes, ketones, and nitroalkanes, show how you might synthesize each of the following.
(a) $HOCH_2CH_2NO_2$
(b) $C_6H_5CH{=}CHNO_2$
(c) $C_6H_5CH{=}C(CH_3)NO_2$

18.41
A useful synthesis of sesquiterpene ketones, called *cyperones,* was accomplished through a modification of the Robinson annellation procedure shown below.

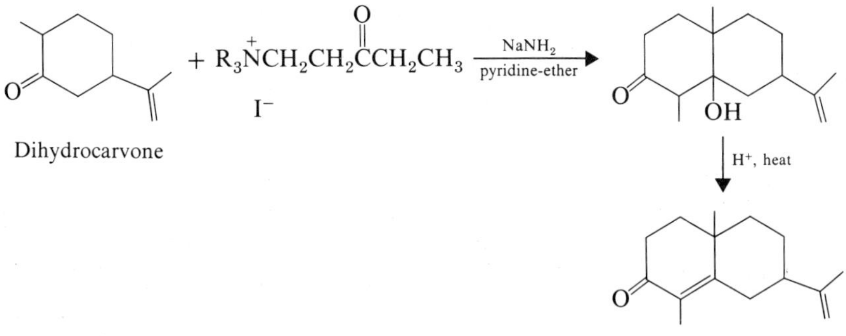

Dihydrocarvone

A cyperone

Write a mechanism that accounts for each step of this synthesis.

18.42
Dimethyl sulfoxide reacts with methyl iodide to give trimethylsulfoxonium iodide.Trimethylsulfoxonium iodide forms an ylide when it is treated with sodium hydride.

$$CH_3\ddot{S}(=O)CH_3 + CH_3{-}I \longrightarrow CH_3\overset{+}{S}(=O)(CH_3)CH_3\,I^- \xrightarrow{NaH} CH_2{=}S(=O)(CH_3)CH_3 + NaI + H_2$$

Trimethylsulfoxonium iodide — Ylide

Professor E. J. Corey of Harvard University has shown that this ylide reacts with ketones to give epoxides, that is,

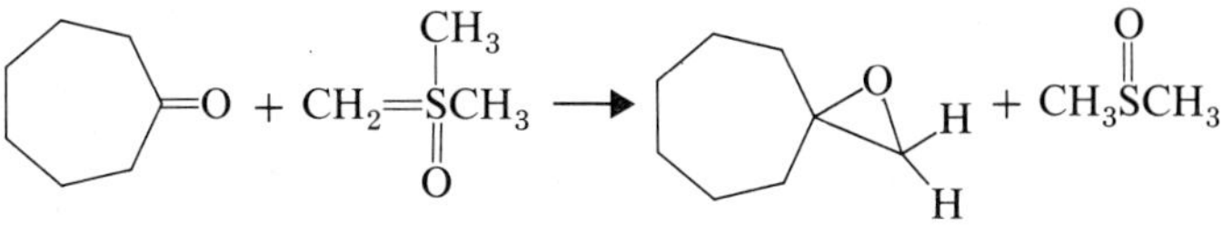

(71%)

Propose a mechanism for this epoxide synthesis.

18.43
(a) A compound U ($C_9H_{10}O$) gives a negative iodoform test. The infrared spectrum of U shows a strong absorption peak at 1690 cm^{-1}. The pmr spectrum of U gives the following:

Triplet $\delta 1.2$ (3H)
Quartet $\delta 3.0$ (2H)
Multiplet $\delta 7.7$ (5H)

What is the structure of U?

(b) A compound V is an isomer of U. Compound V gives a positive iodoform test; its infrared spectrum shows a strong peak at 1705 cm^{-1}. The pmr spectrum of V gives the following:

Singlet $\delta 2.0$ (3H)
Singlet $\delta 3.5$ (2H)
Multiplet $\delta 7.1$ (5H)

What is the structure of V?

18.44

Compounds W and X are isomers; they have the molecular formula C_9H_8O. The infrared spectra of both compounds show a strong absorption band near 1715 cm^{-1}. Oxidation of either compound with hot, basic potassium permanganate followed by acidification yields phthalic acid. The pmr spectrum of W shows a multiplet at $\delta 7.3$ and a singlet at $\delta 3.4$. The pmr spectrum of X shows a multiplet at $\delta 7.5$, a triplet at $\delta 3.1$, and a triplet at $\delta 2.5$. Propose structures for W and X.

18.45

Compounds Y and Z are isomers with the molecular formula $C_{10}H_{12}O$. The infrared spectra of both compounds show a strong absorption band near 1710 cm^{-1}. The pmr spectra of Y and Z are given in Fig. 18.2 and Fig. 18.3. Propose structures for Y and Z.

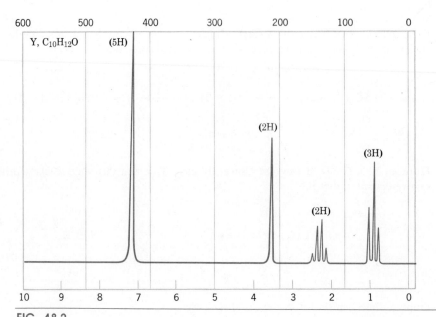

FIG. 18.2
The pmr spectrum of compound Y, problem 18.45. (Courtesy Aldrich Chemical Co., Milwaukee, Wis.)

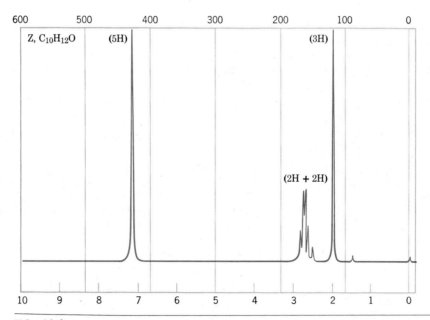

FIG. 18.3

The pmr spectrum of compound Z, problem 18.45. (Courtesy Aldrich Chemical Co., Milwaukee, Wis.)

18.46

Compound **A** has the molecular formula $C_6H_{12}O_3$ and shows a strong infrared absorption peak at 1710 cm^{-1}. When treated with iodine in aqueous sodium hydroxide **A** gives a yellow precipitate. When **A** is treated with Tollens' reagent no reaction occurs; however if **A** is treated first with water containing a drop of sulfuric acid and then treated with Tollens' reagent, a silver mirror forms in the test tube. Compound **A** shows the following pmr spectrum.

> singlet $\delta 2.1$
> doublet $\delta 2.6$
> singlet $\delta 3.2$ (6H)
> triplet $\delta 4.7$

Write the structure for **A.**

* 18.47

3-Hydroxybenzaldehyde undergoes a Cannizzaro reaction readily; 2-hydroxybenzaldehyde and 4-hydroxybenzaldehyde, however, fail to react. Explain.

* 18.48

When semicarbazide, $H_2NNHCONH_2$, reacts with a ketone (or an aldehyde) to form a semicarbazone (Sect. 18.13) only one nitrogen of semicarbazide acts as a nucleophile and attacks the carbonyl carbon of the ketone. The product of the reaction, consequently, is $R_2C=NNHCONH_2$ rather than $R_2C=NCONHNH_2$. What factor accounts for the fact that two nitrogens of semicarbazide are relatively nonnucleophilic?

* **18.49**

Treating a solution of *cis*-1-decalone (below) with base causes an isomerization to take place. When the system reaches equilibrium the solution is found to contain about 95% *trans*-1-decalone and about 5% *cis*-1-decalone. Explain.

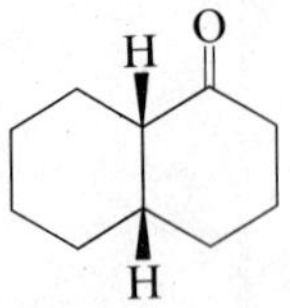

Cis-l-decalone

* **18.50**

Dihydropyran (below) reacts readily with an alcohol in the presence of a trace of anhydrous HCl or H_2SO_4, to form a tetrahydropyranyl ether.

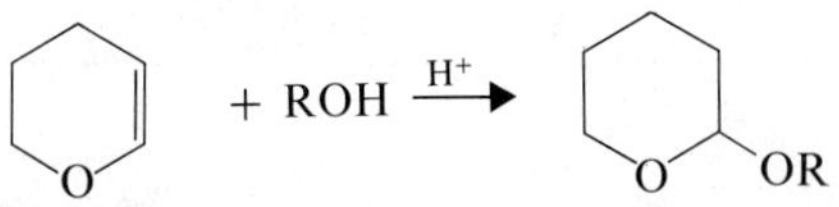

Dihydropyran   + ROH $\xrightarrow{H^+}$ Tetrahydropyranyl ether

(a) Write a plausible mechanism for this reaction. (b) Tetrahydropyranyl ethers are stable in aqueous base but hydrolyze rapidly in aqueous acid to yield the original alcohol and another compound. Explain. (What is the other compound?) (c) The tetrahydropyranyl group can be used as a "protecting" group for alcohols and phenols. Show how you might use it in a synthesis of $HOCH_2CH_2CH_2CH_2\overset{\underset{\textstyle CH_3}{|}}{\underset{\underset{\textstyle CH_3}{|}}{C}}OH$ starting with $HOCH_2CH_2CH_2CH_2Cl$.

CARBOXYLIC ACIDS AND THEIR DERIVATIVES: NUCLEOPHILIC SUBSTITUTION AT ACYL CARBON

19.1 INTRODUCTION

The carboxyl group, $-\overset{\overset{\displaystyle O}{\|}}{C}OH$, is one of the most widely occurring functional groups in chemistry and biochemistry. Not only are carboxylic acids themselves important, but the carboxyl group is the parent group of a large family of related compounds (Table 19.1).

All of these carboxylic acid derivatives contain the acyl group:

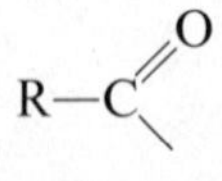

$$R-C\overset{\displaystyle O}{\diagup}$$

An acyl group

As a result, they are often called *acyl compounds*.

TABLE 19.1 Carboxylic Acid Derivatives

STRUCTURE	NAME
$R-\overset{\overset{\displaystyle O}{\|}}{C}-Cl$	Acyl (or acid) chloride
$R-\overset{\overset{\displaystyle O}{\|}}{C}-O-\overset{\overset{\displaystyle O}{\|}}{C}-R$	Acid anhydride
$R-\overset{\overset{\displaystyle O}{\|}}{C}-O-R'$	Ester
$R-\overset{\overset{\displaystyle O}{\|}}{C}-NH_2$	
$R-\overset{\overset{\displaystyle O}{\|}}{C}-NHR'$	Amides
$R-\overset{\overset{\displaystyle O}{\|}}{C}-NR_2$	
$R-\overset{\overset{\displaystyle O}{\|}}{C}-SR'$	Thiol ester

19.2 NOMENCLATURE AND PHYSICAL PROPERTIES

Carboxylic Acids

IUPAC names for carboxylic acids are obtained by dropping the final e of the name of the alkane corresponding to the longest chain in the acid and by adding *oic acid*. The examples listed below illustrate how this is done.

$$\underset{\text{(formic acid)}}{\underset{\text{acid}}{\underset{\text{Methanoic}}{\overset{\overset{\text{O}}{\|}}{\text{HCOH}}}}} \quad \underset{\text{(acetic acid)}}{\underset{\text{acid}}{\underset{\text{Ethanoic}}{\overset{\overset{\text{O}}{\|}}{\text{CH}_3\text{COH}}}}} \quad \underset{\text{(propionic acid)}}{\underset{\text{acid}}{\underset{\text{Propanoic}}{\overset{\overset{\text{O}}{\|}}{\text{CH}_3\text{CH}_2\text{COH}}}}} \quad \underset{\text{(butyric acid)}}{\underset{\text{acid}}{\underset{\text{Butanoic}}{\overset{\overset{\text{O}}{\|}}{\text{CH}_3\text{CH}_2\text{CH}_2\text{COH}}}}}$$

$$\underset{\text{4-Methylhexanoic acid}}{\overset{\overset{\text{O}}{\|}}{\text{CH}_3\text{CH}_2\text{CHCH}_2\text{CH}_2\text{COH}}} \quad \underset{\text{4-Hexenoic acid}}{\overset{\overset{\text{O}}{\|}}{\text{CH}_3\text{CH}{=}\text{CHCH}_2\text{CH}_2\text{COH}}}$$
$$\underset{\text{CH}_3}{}$$

Many carboxylic acids have common names that are derived from Latin or Greek words that indicate one of their natural sources (Table 19.2). Methanoic acid is called formic acid (from the Latin, *formica,* or ant). Ethanoic acid is called acetic acid (from the Latin, *acetum,* or vinegar). Butanoic acid is one compound responsible for the odor of rancid butter, thus its common name is butyric acid (from the Latin, *butyrum,* or butter). Hexanoic acid is one compound associated with the odor of goats, hence its common name, caproic acid (from the Latin *caper,* or goat). Pentanoic acid, with odor in between that of rancid butter and goats, is appropriately named valeric acid (from the Latin, *valerum,* to be strong). Octadecanoic acid takes its common name, stearic acid, from the Greek word, *stear,* for tallow.

Most of these common names have been with us for a long time and some are likely to remain in common usage for even longer, so it is helpful to be familiar with them. In this text we will always refer to methanoic acid and ethanoic acid as formic acid and acetic acid. However, in almost all other instances we will use IUPAC names.

Carboxylic acids are polar substances. They can form strong hydrogen bonds with each other and with water. As a result, carboxylic acids generally have high boiling points, and low-molecular-weight carboxylic acids show appreciable water solubility. The first four carboxylic acids (Table 19.2) are miscible with water in all proportions. As the length of the carbon chain increases, water solubility declines.

Carboxylate Salts

Salts of carboxylic acids are named as *-ates* in common nomenclature and as *-oates* in the IUPAC system. Thus, CH_3COONa is sodium acetate or sodium ethanoate. Other examples are:

$$\underset{\text{Sodium }p\text{-nitrobenzoate}}{\text{O}_2\text{N}{-}\!\!\bigcirc\!\!{-}\overset{\overset{\text{O}}{\|}}{\text{C}}{-}\text{O}^-\text{Na}^+} \qquad \underset{\underset{\text{Potassium 3-methylpentanoate}}{\text{CH}_3}}{\text{CH}_3\text{CH}_2\text{CHCH}_2\overset{\overset{\text{O}}{\|}}{\text{C}}{-}\text{O}^-\text{K}^+}$$

TABLE 19.2 Carboxylic Acids

STRUCTURE	IUPAC NAME	COMMON NAME	mp °C	bp °C	WATER SOL. (g/100 g H_2O) 25°	K_a (at 25°)
HCOOH	Methanoic acid	Formic acid	8	100.5		1.77×10^{-4}
CH_3COOH	Ethanoic acid	Acetic acid	16.6	118		1.76×10^{-5}
CH_3CH_2COOH	Propanoic acid	Propionic acid	−21	141		1.34×10^{-5}
$CH_3(CH_2)_2COOH$	Butanoic acid	Butyric acid	− 6	164		1.54×10^{-5}
$CH_3(CH_2)_3COOH$	Pentanoic acid	Valeric acid	−34	187	4.97	1.52×10^{-5}
$CH_3(CH_2)_4COOH$	Hexanoic acid	Caproic acid	− 3	205	1.08	1.31×10^{-5}
$CH_3(CH_2)_6COOH$	Octanoic acid	Caprylic acid	16	239	0.07	1.28×10^{-5}
$CH_3(CH_2)_8COOH$	Decanoic acid	Capric acid	31	269	0.015	1.43×10^{-5}
$CH_3(CH_2)_{10}COOH$	Dodecanoic acid	Lauric acid	44		0.006	
$CH_3(CH_2)_{12}COOH$	Tetradecanoic acid	Myristic acid	54		0.002	
$CH_3(CH_2)_{14}COOH$	Hexadecanoic acid	Palmitic acid	63		0.0007	
$CH_3(CH_2)_{16}COOH$	Octadecanoic acid	Stearic acid	70		0.0003	
$CH_2ClCOOH$	Chloroacetic acid		63	189	Very sol.	1.40×10^{-3}
$CHCl_2COOH$	Dichloroacetic acid		10.8	192		3.32×10^{-2}
CCl_3COOH	Trichloroacetic acid		56.3	198	Very sol.	2.00×10^{-1}
$CH_3CHClCOOH$	2-Chloropropanoic acid			186		1.47×10^{-2}
CH_2ClCH_2COOH	3-Chloropropanoic acid		61	204	Sol.	1.47×10^{-3}
C_6H_5COOH	Benzoic acid		122	250	0.34	6.46×10^{-5}
$p\text{-}CH_3C_6H_4COOH$	p-Toluic acid		180	275	0.03	4.33×10^{-5}
$p\text{-}ClC_6H_4COOH$	p-Chlorobenzoic acid		242		0.009	1.04×10^{-4}
$p\text{-}NO_2C_6H_4COOH$	p-Nitrobenzoic acid		242		0.03	3.93×10^{-4}
	1-Naphthoic acid		160		Insol.	2.00×10^{-4}
	2-Naphthoic acid		185		Insol.	6.80×10^{-5}

Sodium and potassium salts of most carboxylic acids are readily soluble in water. This is true even of the long-chain carboxylic acids. Sodium or potassium salts of long-chain carboxylic acids are the major ingredients of soap (cf., p. 883).

Acidity of Carboxylic Acids

When we examine the acidity constants of the carboxylic acids listed in Table 19.2 we find that most unsubstituted carboxylic acids have K_a's in the range of 10^{-4} to 10^{-5}. This means that carboxylic acids react readily with aqueous solutions of sodium hydroxide and sodium bicarbonate to form soluble sodium salts. We can use solubility tests, therefore, to distinguish water-insoluble carboxylic acids from water-insoluble phenols and alcohols. Water-insoluble carboxylic acids will dissolve in either aqueous sodium hydroxide or aqueous sodium bicarbonate:

Benzoic acid   Sodium benzoate
(water insoluble)   (water soluble)

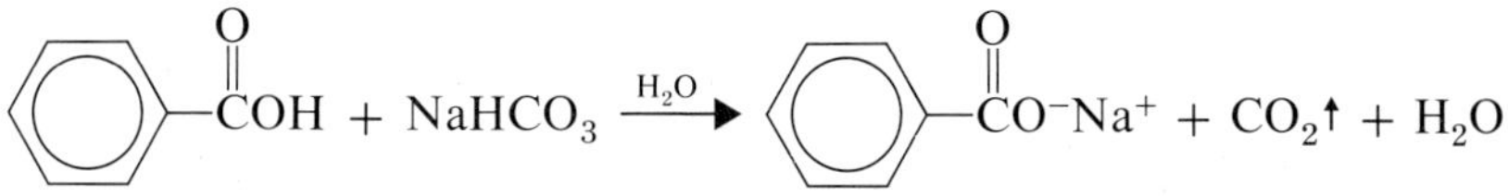

(water insoluble) (water soluble)

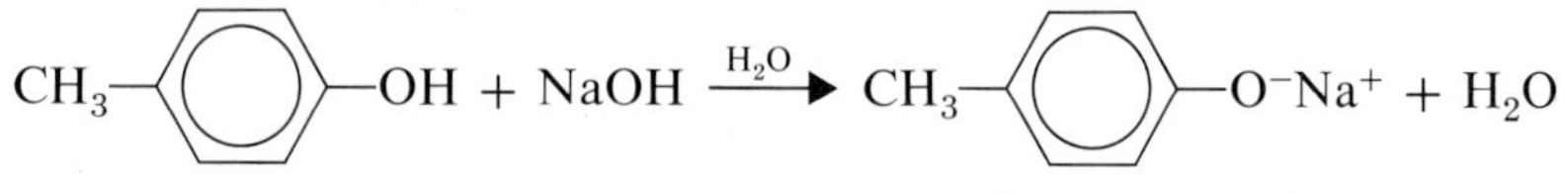

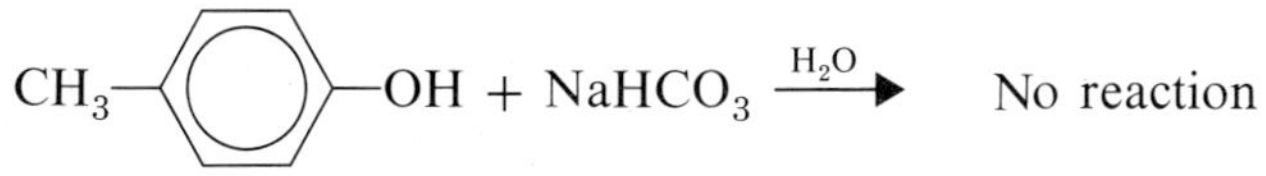

Water-insoluble phenols dissolve in aqueous sodium hydroxide but (except for the nitrophenols) do not dissolve in aqueous sodium bicarbonate:

p-Cresol (water insoluble) Sodium *p*-cresoxide (water soluble)

(water insoluble)

Water-insoluble alcohols do not dissolve in either aqueous sodium hydroxide or sodium bicarbonate.

Problem 19.1

(a) When excess carbon dioxide is passed into a solution of sodium benzoate and sodium *p*-cresoxide in aqueous sodium hydroxide, *p*-cresol separates from the solution (as an oil) but sodium benzoate remains in solution. Write equations that will provide an explanation for this. (b) Given the following reagents—aqueous sodium hydroxide, aqueous hydrochloric acid, ether, and carbon dioxide—explain how you would separate a mixture of benzoic acid, *p*-cresol, and cyclohexanol.

We also see, in Table 19.2, that carboxylic acids having electron-withdrawing groups are stronger than unsubstituted acids. The chloroacetic acids, for example, show the following order of acidities:

	$Cl_3C{-}COOH$	>	$Cl_2CH{-}COOH$	>	$ClCH_2{-}COOH$	>	$H_3C{-}COOH$
K_a:	2.0×10^{-1}		3.32×10^{-2}		1.40×10^{-3}		1.76×10^{-5}

With aliphatic acids this acid-strengthening effect of electron-withdrawing groups is primarily an *inductive effect,* and it decreases as the number of σ bonds between the electron-withdrawing group and the carboxyl group increases.

	$CH_3CHClCOOH$	$ClCH_2CH_2COOH$
K_a:	1.47×10^{-2}	1.47×10^{-3}

Electron-withdrawing groups increase the acidity of aliphatic acids because they stabilize the negatively charged carboxylate ion.

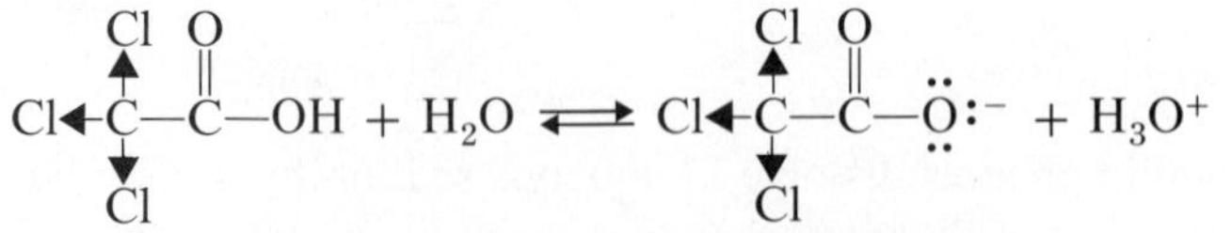

Stabilized by
dispersal of charge

Electron-withdrawing groups also increase the acidity of aromatic acids:

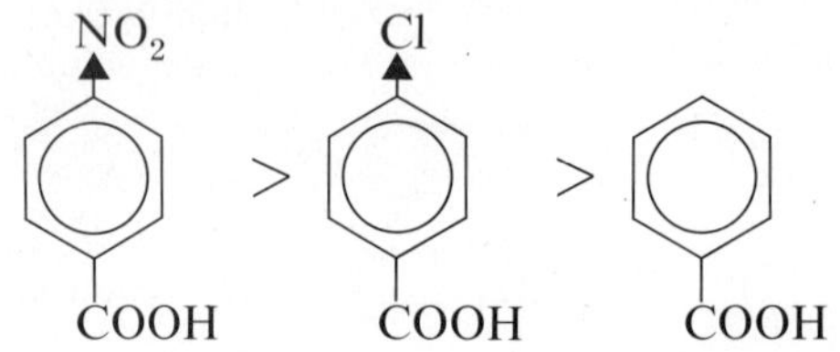

K_a: 3.93×10^{-4} 1.04×10^{-4} 6.46×10^{-5}

With the *p*-nitro group, inductive and *resonance effects* are important in stabilizing the carboxylate ion.

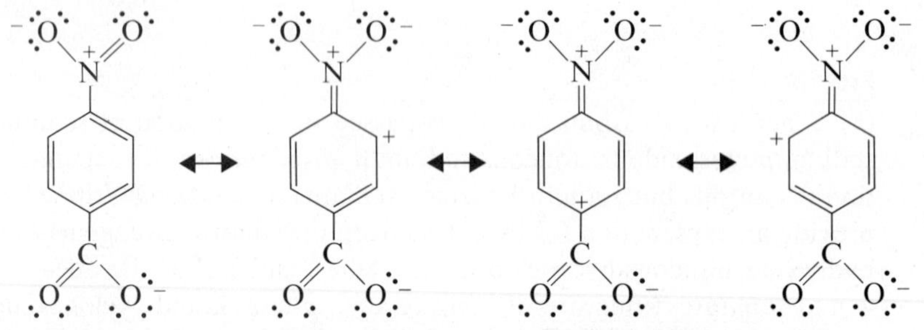

Resonance structures for the p-nitrobenzoate ion

Problem 19.2

Which acid of each pair shown below would you expect to be stronger?
   (a) CH_3COOH or CH_2FCOOH
   (b) CH_2FCOOH or $CH_2ClCOOH$
   (c) $CH_2ClCOOH$ or $CH_2BrCOOH$
   (d) $CH_2ClCH_2CH_2COOH$ or $CH_3CHClCH_2COOH$
   (e) $CH_3CH_2CHClCOOH$ or $CH_3CHClCH_2COOH$

   (f)

   (g)

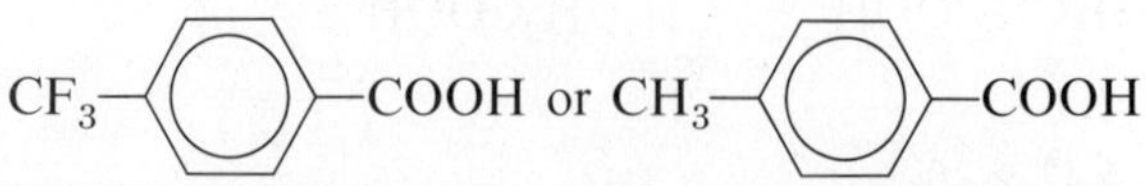

TABLE 19.3 Dicarboxylic Acids

| STRUCTURE | NAME | mp °C | K_a AT 25° | |
			K_1	K_2
HOOC—COOH	Oxalic acid	189 (dec.)	5.9×10^{-2}	6.4×10^{-5}
HOOCCH$_2$COOH	Malonic acid	136	1.4×10^{-4}	2.0×10^{-6}
HOOC(CH$_2$)$_2$COOH	Succinic acid	182	6.9×10^{-5}	2.5×10^{-6}
HOOC(CH$_2$)$_3$COOH	Glutaric acid	98	4.6×10^{-5}	3.5×10^{-6}
HOOC(CH$_2$)$_4$COOH	Adipic acid	153	3.7×10^{-5}	2.4×10^{-6}
cis-HOOC—CH=CH—COOH	Maleic acid	131	1.4×10^{-2}	8.0×10^{-7}
trans-HOOC—CH=CH—COOH	Fumaric acid	307	9.3×10^{-5}	3.6×10^{-5}
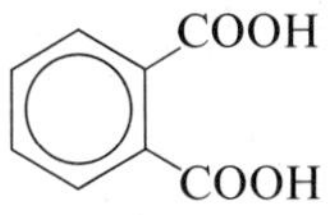	Phthalic acid	231	1.3×10^{-4}	3.9×10^{-6}
	Isophthalic acid	345	2.9×10^{-4}	2.5×10^{-6}
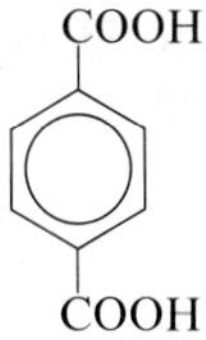	Terephthalic acid	Sublimes	3.1×10^{-4}	3.5×10^{-5}

Dicarboxylic Acids

Dicarboxylic acids are named as *alkanedioic acids* in the IUPAC system. Most simple dicarboxylic acids have common names (Table 19.3) and these are the names that we will use.

Problem 19.3

Suggest explanations for the following facts.

(a) K_1 for all of the dicarboxylic acids in Table 19.3 are higher than the K_a for monocarboxylic acids with the same number of carbons.

(b) The difference between K_1 and K_2 for dicarboxylic acids of type HOOC(CH$_2$)$_n$COOH decreases as n increases.

Esters

The names of esters are derived from the names of the alcohol portion (with the ending -*yl*) and the acid portion (with the ending -*ate* or -*oate*). The portion of the name derived from the alcohol comes first.

$$CH_3\overset{O}{\overset{\|}{C}}-OCH_2CH_3 \qquad CH_3CH_2\overset{O}{\overset{\|}{C}}-O\overset{CH_3}{\underset{CH_3}{\overset{|}{\underset{|}{C}}}}-CH_3$$

Ethyl acetate or   *tert*-Butyl propanoate
ethyl ethanoate

$$Cl-\overset{O}{\underset{}{\bigcirc}}-\overset{O}{\overset{\|}{C}}OCH_3 \qquad CH_3\overset{O}{\overset{\|}{C}}OCH{=}CH_2$$

Methyl *p*-chlorobenzoate   Vinyl acetate or
vinyl ethanoate

$$CH_3CH_2O\overset{O}{\overset{\|}{C}}-\overset{O}{\overset{\|}{C}}OCH_2CH_3 \qquad CH_3CH_2O\overset{O}{\overset{\|}{C}}CH_2\overset{O}{\overset{\|}{C}}OCH_2CH_3$$

Diethyl oxalate   Diethyl malonate

Esters are polar compounds but they cannot form strong hydrogen bonds to each other. As a result, esters have boiling points that are lower than those of acids and alcohols of comparable molecular weight. The boiling points (Table 19.4) of esters are about the same as those of comparable aldehydes and ketones. Esters have lower water solubilities than acids and alcohols.

Unlike the low molecular weight acids, esters usually have pleasant odors. The odors of some esters resemble those of fruits and these esters are used in the manufacture of synthetic flavors:

$$CH_3\overset{O}{\overset{\|}{C}}OCH_2CH_2\underset{\underset{CH_3}{|}}{CH}CH_3 \qquad CH_3CH_2CH_2CH_2\overset{O}{\overset{\|}{C}}-OCH_2CH_2\underset{\underset{CH_3}{|}}{CH}CH_3$$

Isopentyl acetate   Isopentyl pentanoate
(used in synthetic banana flavor)   (used in synthetic apple flavor)

TABLE 19.4 Carboxylic Acid Esters

NAME	STRUCTURE	mp °C	bp °C	WATER SOLUBILITY (g./100 g. at 20° C)
Methyl formate	$HCOOCH_3$	−99	31.5	Very sol.
Ethyl formate	$HCOOCH_2CH_3$	−79	54	
Methyl acetate	CH_3COOCH_3	−99	57	24.4
Ethyl acetate	$CH_3COOCH_2CH_3$	−82	77	7.39 (25° C)
n-Propyl acetate	$CH_3COOCH_2CH_2CH_3$	−93	102	1.89
n-Butyl acetate	$CH_3COOCH_2(CH_2)_2CH_3$	−78	125	1.0 (22° C)
Ethyl propanoate	$CH_3CH_2COOCH_2CH_3$	−73	99	1.75
Ethyl butanoate	$CH_3(CH_2)_2COOCH_2CH_3$	−93	120	0.51
Ethyl pentanoate	$CH_3(CH_2)_3COOCH_2CH_3$	−91	145	0.22
Ethyl hexanoate	$CH_3(CH_2)_4COOCH_2CH_3$	−68	168	0.063
Methyl benzoate	$C_6H_5COOCH_3$	−12	199	Insol.
Ethyl benzoate	$C_6H_5COOCH_2CH_3$	−35	213	Insol.
Phenyl acetate	$CH_3COOC_6H_5$		196	Insol.
Methyl salicylate	$o\text{-}HOC_6H_4COOCH_3$	− 9	223	Insol.

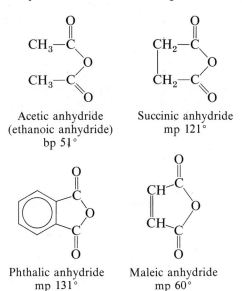

$$CH_3(CH_2)_2\overset{\overset{\displaystyle O}{\|}}{C}OCH_2CH_2CH_2CH_3$$

Butyl butanoate
(used in synthetic pineapple
flavor)

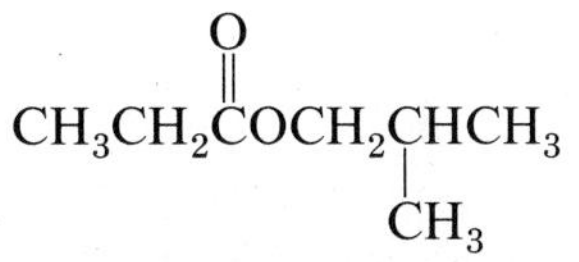

$$CH_3CH_2\overset{\overset{\displaystyle O}{\|}}{C}OCH_2\underset{\underset{\displaystyle CH_3}{|}}{C}HCH_3$$

Isobutyl propanoate
(used in synthetic rum flavor)

Glyceryl trimyristate is the major constituent of nutmeg and methyl salicylate is chief component of oil of wintergreen.

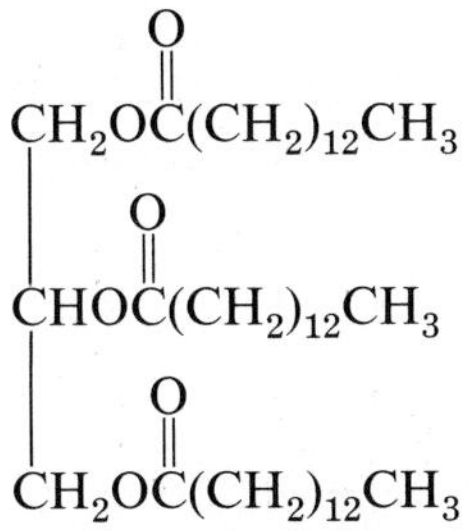

$$CH_2O\overset{\overset{\displaystyle O}{\|}}{C}(CH_2)_{12}CH_3$$
$$CHO\overset{\overset{\displaystyle O}{\|}}{C}(CH_2)_{12}CH_3$$
$$CH_2O\overset{\overset{\displaystyle O}{\|}}{C}(CH_2)_{12}CH_3$$

Glyceryl trimyristate
(nutmeg)

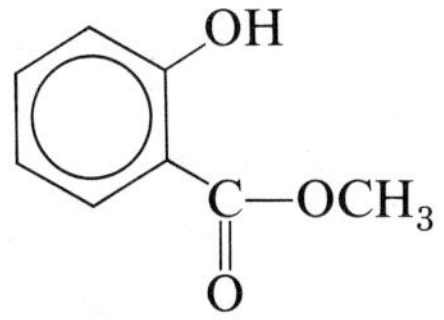

Methyl salicylate
(oil of wintergreen)

Carboxylic Acid Anhydrides

Most anhydrides are named by dropping the word *acid* from the name of the carboxylic acid and then adding the word *anhydride*.

Acetic anhydride
(ethanoic anhydride)
bp 51°

Succinic anhydride
mp 121°

Phthalic anhydride
mp 131°

Maleic anhydride
mp 60°

Acyl Chlorides

Acyl chlorides are often called *acid chlorides*. They are named by dropping -*ic* from the name of the acid and then adding *yl chloride*. Examples are:

$$CH_3\overset{\overset{\displaystyle O}{\|}}{C}-Cl \qquad CH_3CH_2\overset{\overset{\displaystyle O}{\|}}{C}-Cl \qquad C_6H_5\overset{\overset{\displaystyle O}{\|}}{C}-Cl$$

Acetyl chloride   Propanoyl chloride   Benzoyl chloride
(ethanoyl chloride)   mp −94°, bp 80°   mp −1°, bp 197°
mp −112°, bp 51°

Acyl chlorides and carboxylic anhydrides have boiling points in the same range as esters of comparable molecular weight.

Amides

Amides that have no substituent on nitrogen are named by dropping *-ic* from the common name of the acid (or *-oic* from the IUPAC name) and then adding *-amide*.

$$CH_3\overset{\overset{\displaystyle O}{\|}}{C}-NH_2 \qquad CH_3CH_2\overset{\overset{\displaystyle O}{\|}}{C}-NH_2 \qquad \bigcirc\!\!-\overset{\overset{\displaystyle O}{\|}}{C}-NH_2$$

Acetamide or   Propanamide   Benzamide
ethanamide   mp 79°, bp 213°   mp 130°, bp 290°
mp 82°, bp 221°

Substituents on the nitrogen atom of amides are named as alkyl groups and the named substituent is prefaced by *N-*, or *N,N-*. Examples are:

$$CH_3\overset{\overset{\displaystyle O}{\|}}{C}-N\overset{\displaystyle CH_3}{\underset{\displaystyle CH_3}{}} \qquad CH_3\overset{\overset{\displaystyle O}{\|}}{C}-NHC_2H_5$$

N,N-Dimethylacetamide   *N*-Ethylacetamide
bp 185°   bp 205°

Amides with one (or no) substituent on nitrogen are able to form strong hydrogen bonds to each other and have, consequently, high melting points and boiling points. *N,N*-disubstituted amides cannot form strong hydrogen bonds to each other; they have lower melting points and boiling points.

$$R-\overset{\overset{\displaystyle \ddot{O}:\text{---}H-\overset{\displaystyle R'}{\underset{}{N}}-\overset{\overset{\displaystyle O}{\|}}{C}-R}{\underset{\displaystyle NH}{\underset{\displaystyle |}{C}}}$$

Amide-hydrogen bonding

Spectroscopic Properties of Acyl Compounds

Infrared spectroscopy is of considerable importance in determining structures of carboxylic acids and their derivatives. The C=O stretching band is one of the most prominent in their infrared spectra since it is always a strong band. The C=O

TABLE 19.5 Carbonyl Stretching Absorptions of Acyl Compounds

TYPE OF COMPOUND	FREQUENCY RANGE, cm^{-1}
Carboxylic Acids	
R—COOH	1700–1725
$-\overset{\mid}{C}=C-COOH$	1690–1715
ArCOOH	1680–1700
Acid Anhydrides	
$R-\overset{O}{\overset{\|}{C}}-O-\overset{O}{\overset{\|}{C}}-R$	1800–1850 and 1740–1790
$Ar-\overset{O}{\overset{\|}{C}}-O-\overset{O}{\overset{\|}{C}}-Ar$	1780–1830 and 1730–1770
Acyl Chlorides	
$R-\overset{O}{\overset{\|}{C}}Cl$ and $Ar-\overset{O}{\overset{\|}{C}}Cl$	1780–1850
Esters	
$R-\overset{O}{\overset{\|}{C}}-OR$	1735–1750
$Ar-\overset{O}{\overset{\|}{C}}-OR$	1715–1730
Amides	
$R\overset{O}{\overset{\|}{C}}-NH_2$, $R\overset{O}{\overset{\|}{C}}NHR$ and $R\overset{O}{\overset{\|}{C}}NR_2$	1630–1690
Carboxylate Ions	
$RCOO^-$	1550–1630

stretching band, moreover, occurs at different frequencies for acids, esters, and amides, for example. And the precise location of this band is often helpful in structure determination. Table 19.5 gives the location of this band for most acyl compounds.

The hydroxyl groups of carboxylic acids also give rise to a broad peak in the 2500 to 2700-cm^{-1} region arising from O—H stretching vibrations. The N—H stretching vibrations of amides absorb between 3140 and 3500 cm^{-1}.

The acidic protons of carboxylic acids usually absorb very far downfield (δ10–12) in their *pmr spectra*.

19.3 PREPARATION OF CARBOXYLIC ACIDS

Most of the methods for the preparation of carboxylic acids are familiar ones.

1. **By oxidation of alkenes** (discussed in Section 6.8).

$$R—CH{=}CHR' \xrightarrow[\text{heat}]{\text{KMnO}_4} RCOOH + R'COOH$$

2. **By oxidation of aldehydes and primary alcohols** (discussed in Sections 18.16 and 16.6).

$$R—CHO \xrightarrow[\substack{\text{or}\\ \text{Ag(NH}_3)_2{}^+}]{\text{dil. KMnO}_4} RCOOH$$

$$RCH_2OH \xrightarrow[\substack{\text{heat}\\ \text{(2) H}^+}]{\text{KMnO}_4,\ \text{OH}^-} RCOOH$$

3. **By oxidation of alkylbenzenes** (discussed in Section 13.10).

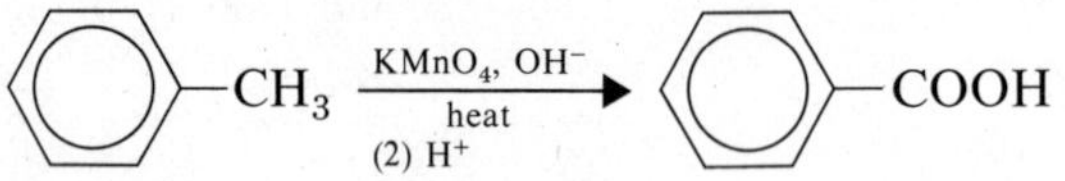

$$\text{〈〉}—CH_3 \xrightarrow[\substack{\text{heat}\\ \text{(2) H}^+}]{\text{KMnO}_4,\ \text{OH}^-} \text{〈〉}—COOH$$

4. **By oxidation of methyl ketones** (discussed in Section 18.14).

$$R—\overset{\overset{\displaystyle O}{\|}}{C}—CH_3 \xrightarrow[\text{(2) H}^+]{\text{(1) X}_2/\text{NaOH}} R—\overset{\overset{\displaystyle O}{\|}}{C}OH + CHX_3$$

5. **By hydrolysis of cyanohydrins and other nitriles.** We saw, in Section 18.7, that aldehydes and ketones can be converted to cyanohydrins; and that these can be hydrolyzed to α-hydroxy acids.

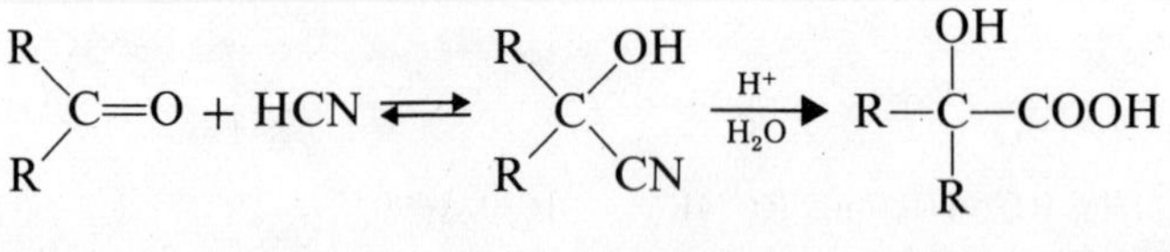

$$\underset{R}{\overset{R}{\diagdown}}C{=}O + HCN \rightleftharpoons \underset{R}{\overset{R}{\diagup}}\overset{OH}{\underset{CN}{C}} \xrightarrow[\text{H}_2\text{O}]{\text{H}^+} R—\overset{\overset{\displaystyle OH}{|}}{\underset{\underset{\displaystyle R}{|}}{C}}—COOH$$

Nitriles can also be prepared through nucleophilic substitution reactions of alkyl halides and sodium cyanide. Hydrolysis of the nitrile yields a carboxylic acid:

General Reaction

$$R—CH_2X + CN^- \longrightarrow RCH_2CN \xrightarrow[\substack{\text{H}_2\text{O}\\ \text{heat}}]{\text{H}^+} RCH_2COOH + NH_4{}^+$$

$$\xrightarrow[\substack{\text{H}_2\text{O}\\ \text{heat}}]{\text{OH}^-} RCH_2COO^- + NH_3$$

Specific Examples

$$HOCH_2CH_2Cl \xrightarrow[\text{(80\%)}]{\text{NaCN}} \underset{\substack{\text{2-Hydroxy-}\\ \text{propanenitrile}}}{HOCH_2CH_2CN} \xrightarrow[\text{(2) H}_3\text{O}^+]{\text{(1) OH}^-,\ \text{H}_2\text{O}} \underset{\substack{\text{3-Hydroxypropanoic}\\ \text{acid}}}{HOCH_2CH_2COOH}$$

$$\text{(75–80\%)}$$

$$BrCH_2CH_2CH_2Br \xrightarrow[(77-86\%)]{NaCN} NCCH_2CH_2CH_2CN \xrightarrow[(83-85\%)]{H_3O^+} HOOCCH_2CH_2CH_2COOH$$

1,3-Pentanedi-
nitrile

Glutaric acid

This synthetic method is generally limited to the use of *primary alkyl* halides. The cyanide ion is a relatively strong base and the use of secondary and tertiary alkyl halides leads, primarily, to an alkene (through elimination) rather than to a nitrile (through substitution). Aryl halides (except for those with ortho and para nitro groups) do not react with sodium cyanide.

6. **By carbonation of Grignard reagents.** Grignard reagents react with carbon dioxide to yield magnesium carboxylates (Section 15.7). Acidification of the magnesium carboxylate produces a carboxylic acid.

$$R-X + Mg \xrightarrow{ether} RMgX \xrightarrow{CO_2} RCOOMgX \xrightarrow{H^+} RCOOH$$

or

$$Ar-Br + Mg \xrightarrow{ether} ArMgBr \xrightarrow{CO_2} ArCOOMgBr \xrightarrow{H^+} ArCOOH$$

This synthesis of carboxylic acids is applicable to primary, secondary, tertiary, allyl, benzyl, and aryl halides.

CH₃ group on carbon:

$$CH_3-\underset{\underset{CH_3}{|}}{\overset{\overset{CH_3}{|}}{C}}-Cl \xrightarrow{Mg \atop ether} CH_3\underset{\underset{CH_3}{|}}{\overset{\overset{CH_3}{|}}{C}}MgCl \xrightarrow[(2)\ H_2O]{(1)\ CO_2} CH_3\underset{\underset{CH_3}{|}}{\overset{\overset{CH_3}{|}}{C}}COOH$$

tert-Butyl chloride

2,2-Dimethylpropanoic acid
(79–80% overall)

$$CH_3CH_2CH_2CH_2Cl \xrightarrow{Mg \atop ether} CH_3CH_2CH_2CH_2MgCl \xrightarrow[(2)\ H_2O]{(1)\ CO_2} CH_3CH_2CH_2CH_2COOH$$

n-Butyl chloride

Pentanoic acid
(80% overall)

Benzoic acid
(85%)

Problem 19.4

Show how you would prepare each of the following carboxylic acids through a Grignard synthesis.

(a) Phenylacetic acid
(b) 2,2-Dimethylpentanoic acid
(c) 3-Butenoic acid
(d) *p*-Toluic acid
(e) Hexanoic acid

Problem 19.5

(a) Which of the carboxylic acids in problem 19.4 could be prepared by a nitrile synthesis as well? (b) Which synthesis, Grignard or nitrile, would you choose to prepare HOCH$_2$CH$_2$CH$_2$CH$_2$COOH from HOCH$_2$CH$_2$CH$_2$CH$_2$Br? Why?

19.4 NUCLEOPHILIC SUBSTITUTIONS AT ACYL CARBON

In our study of carbonyl compounds in the last chapter we saw that a characteristic reaction of aldehydes and ketones is one of *nucleophilic addition* to the carbon oxygen double bond.

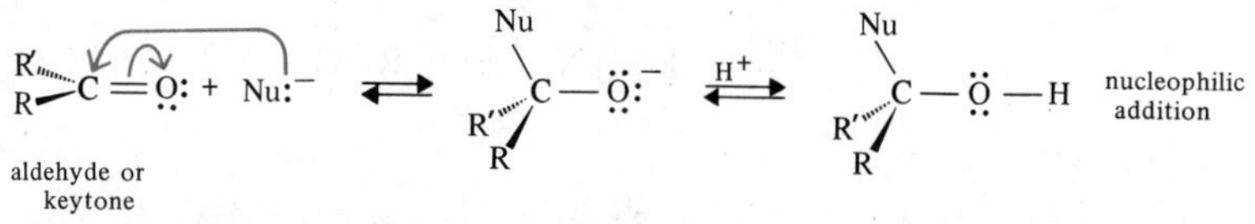

As we study carboxylic acids and their derivatives in this chapter we will find that their reactions are characterized by *nucleophilic substitution* reactions that take place at their acyl (carbonyl) carbons. We will encounter many reactions of the general type shown below:

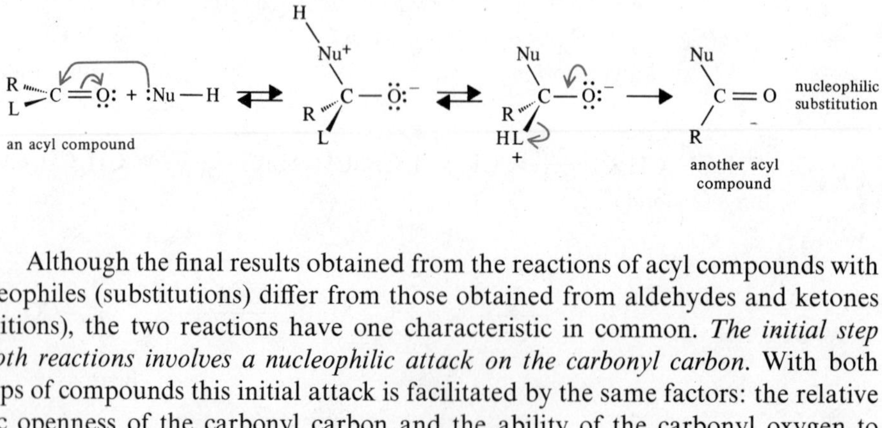

Although the final results obtained from the reactions of acyl compounds with nucleophiles (substitutions) differ from those obtained from aldehydes and ketones (additions), the two reactions have one characteristic in common. *The initial step in both reactions involves a nucleophilic attack on the carbonyl carbon.* With both groups of compounds this initial attack is facilitated by the same factors: the relative steric openness of the carbonyl carbon and the ability of the carbonyl oxygen to accommodate an electron pair of the carbon-oxygen double bond.

It is after the initial nucleophilic attack has taken place that the two reactions differ. The tetrahedral intermediate formed from an aldehyde or ketone usually accepts a proton to form an *addition product*. By contrast, the intermediate formed from an acyl compound usually ejects a leaving group; this leads to regeneration of the carbon-oxygen double bond and a *substitution product*.

Acyl compounds react as they do because they all have good potential leaving groups attached to the carbonyl carbon: an acyl chloride, for example, generally reacts by losing *a chloride ion*—a very weak base—and thus, a very good leaving group.

General Reaction

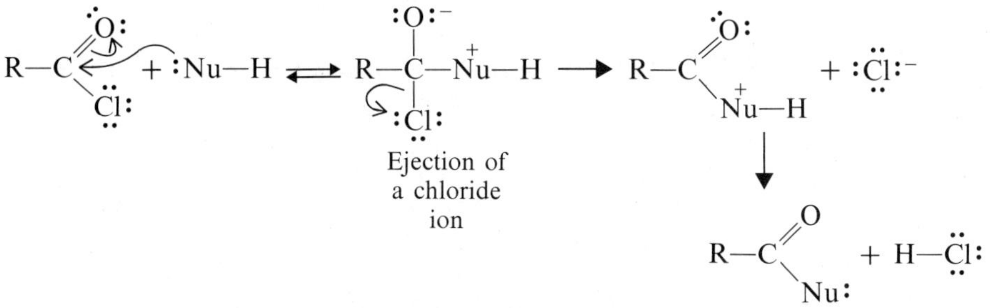

Example: The reaction of an acyl chloride with water.

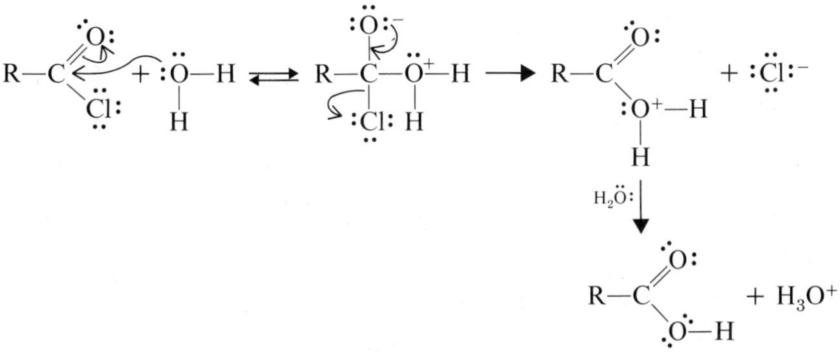

An acid anhydride generally reacts by losing *a carboxylate ion* or a molecule of a *carboxylic acid*—both are weak bases and good leaving groups.

General Reaction

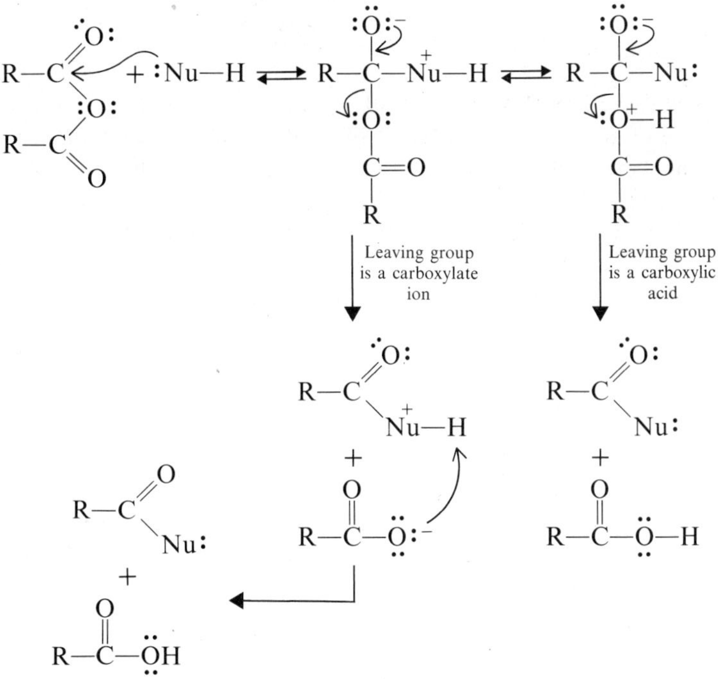

Example: The reaction of a carboxylic acid anhydride with an alcohol.

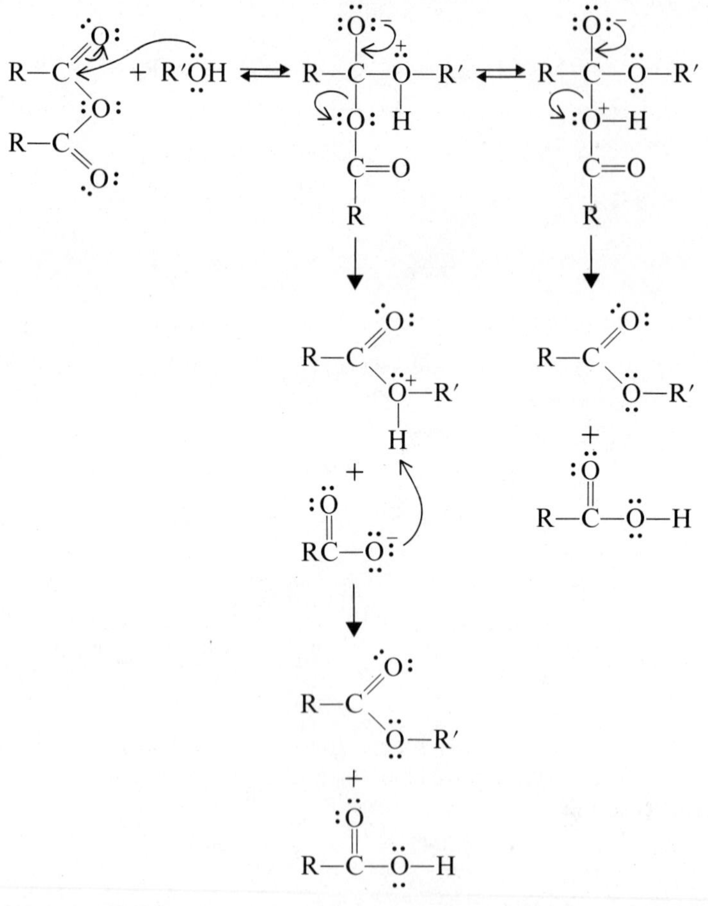

As we will see later, esters generally undergo nucleophilic substitution by losing a molecule of an *alcohol,* acids react by losing a molecule of *water,* and amides react by losing a molecule of *ammonia* or *amine.* All of the molecules lost in these instances are weak bases and are good leaving groups.

For an aldehyde or ketone to react by substitution the tetrahedral intermediate would have to eject a hydride ion ($H:^-$) or an alkanide ion ($R:^-$). Both are *very powerful bases* and both are, therefore, *very poor leaving groups.*

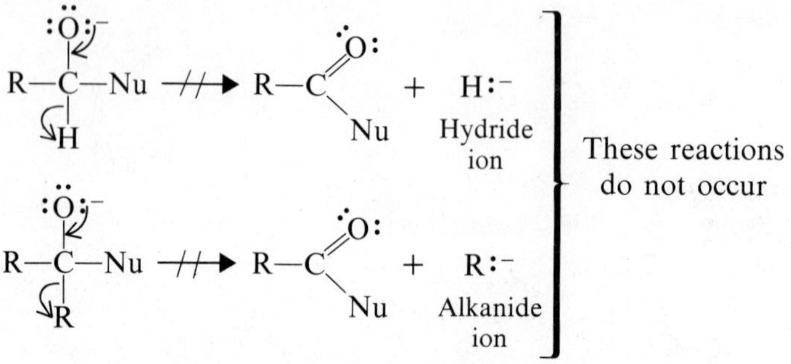

Relative Reactivity of Acyl Compounds

Of the acid derivatives that we will study in this chapter, acyl chlorides are the most reactive toward nucleophilic substitution and amides are the least reactive. In general, the overall order of reactivity is:

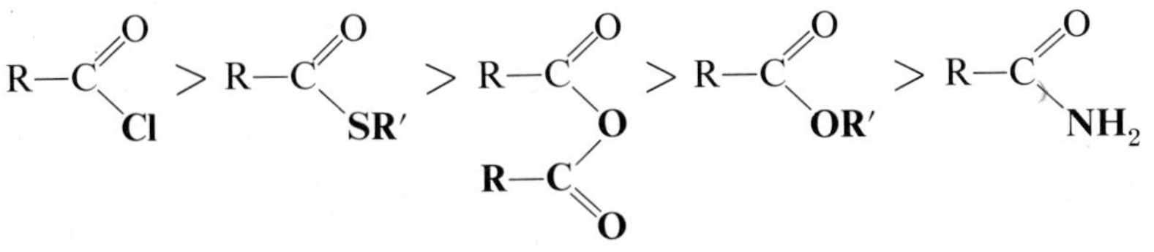

| Acyl chloride | Thiol ester | Acid anhydride | Ester | Amide |

We can account for this overall order of reactivity by taking into account three factors: (1) the basicities of the leaving groups, (2) resonance effects, and (3) inductive effects.

Let us look at the effect of the leaving groups first. The general order of reactivity of acid derivatives roughly parallels the basicities of the potential leaving groups. Chloride ions are the weakest bases and acyl chlorides are the most reactive. Amines are the strongest bases and amides are the least reactive.

All of the acid derivatives have an atom with an unshared electron pair adjacent to the carbonyl group. As a result of this, resonance contributions of the following kind can be important.

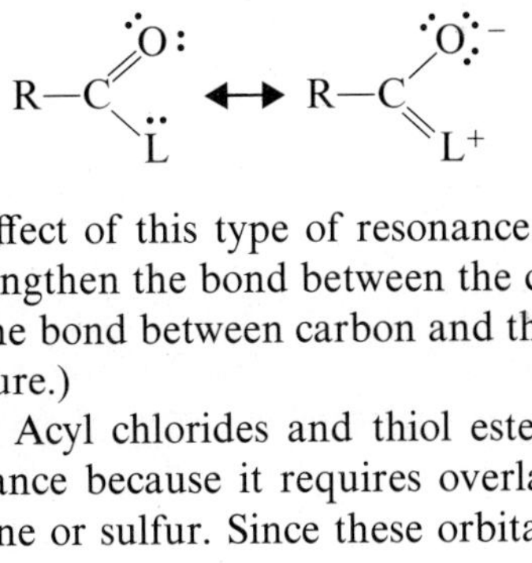

The effect of this type of resonance will be to stabilize the group as a whole and to strengthen the bond between the carbonyl carbon and the leaving group. (Notice that the bond between carbon and the leaving group is a double bond in the second structure.)

Acyl chlorides and thiol esters are probably least affected by this kind of resonance because it requires overlap of a $2p$-carbon orbital with a $3p$ orbital of chlorine or sulfur. Since these orbitals are of different sizes, overlap between them is not large.

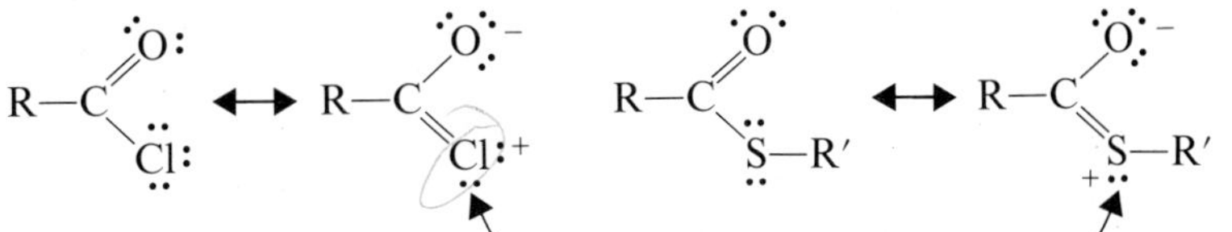

These structures do not make a large contribution because of ineffective orbital overlap

Thus, on the basis of resonance, we would expect acyl chlorides and thiol esters to be the least stabilized and the most reactive. The added inductive effect of the strongly electronegative chlorine explains why acyl chlorides are the most reactive of all.

With esters and amides, resonance effects (below) play a much larger part because overlap between a $2p$ orbital of carbon and a $2p$ orbital of oxygen or nitrogen is more favorable.

$$R-C \overset{\ddot{O}:}{\underset{\ddot{O}R}{}} \longleftrightarrow R-C \overset{\ddot{O}:^-}{\underset{\overset{+}{O}R}{}} \qquad R-C \overset{\ddot{O}:}{\underset{\ddot{N}H_2}{}} \longleftrightarrow R-C \overset{\ddot{O}:^-}{\underset{\overset{+}{N}H_2}{}}$$

As a result, ester and amide groups are stabilized to a considerable extent, and are, therefore, the least reactive. The greater inductive effect of oxygen versus nitrogen is in accord with the greater reactivity of an ester versus an amide.

With acid anhydrides, orbital overlap is favorable but the stabilization of a particular carbonyl group is less than that of an ester because the resonance effect is shared between two carbonyl groups.

$$
\begin{array}{ccc}
R-C \overset{\ddot{O}:}{} & R-C \overset{\ddot{O}:^-}{} & R-C \overset{\ddot{O}:}{} \\
:\ddot{O}: \longleftrightarrow & \overset{+}{O}: \longleftrightarrow & \overset{+}{O}: \\
R-C \overset{}{} & R-C \overset{}{} & R-C \overset{}{} \\
.\ddot{O}: & \ddot{O}: & \ddot{O}:^-
\end{array}
$$

Synthesis of Acid Derivatives

As we begin now to explore the syntheses of carboxylic acid derivatives we will find that in many instances one acid derivative can be synthesized through a nucleophilic substitution reaction of another. The order of reactivities that we have presented gives us a clue as to which syntheses are practical and which are not. In general, *less reactive acyl compounds can be synthesized from more reactive ones, but the reverse is usually difficult* and, when possible, requires special conditions and/or a catalyst.

19.5 SYNTHESIS OF ACYL CHLORIDES

Since acyl chlorides are the most reactive of the acid derivatives, we must use highly reactive substances to prepare them. We use other acid chlorides, *the acid chlorides of inorganic acids:* We use **PCl$_5$** (an acid chloride of phosphoric acid), **PCl$_3$** (an acid chloride of phosphorous acid), and **SOCl$_2$** (an acid chloride of sulfurous acid).

All of these reagents react with carboxylic acids to give acyl chlorides in good yield.

General Reactions

$$\underset{\text{Thionyl chloride}}{\overset{O}{\underset{\|}{R}COH} + SOCl_2} \longrightarrow R-\overset{O}{\overset{\|}{C}}-Cl + SO_2 + HCl$$

$$3RCOH + PCl_3 \longrightarrow 3RCCl + H_3PO_3$$

Phosphorus
trichloride'

$$RCOH + PCl_5 \longrightarrow RCCl + POCl_3 + HCl$$

Phosphorus
pentachloride

Specific Examples

$$C_6H_5COH + SOCl_2 \longrightarrow C_6H_5CCl + SO_2 + HCl$$

Benzoic acid   Benzoyl chloride
(91%)

$$3CH_3COH + PCl_3 \longrightarrow 3CH_3CCl + H_3PO_3$$

Acetic acid   Acetyl chloride
(67%)

$$CH_3(CH_2)_6COH + PCl_5 \longrightarrow CH_3(CH_2)_6CCl + POCl_3 + HCl$$

Octanoic acid   Octanoyl chloride
(82%)

These reactions all involve nucleophilic substitutions by chloride ion on a highly reactive intermediate: an acylchlorosulfite, an acylchlorophosphite, or an acylchlorophosphate. Thionyl chloride, for example, reacts with a carboxylic acid in the following way.

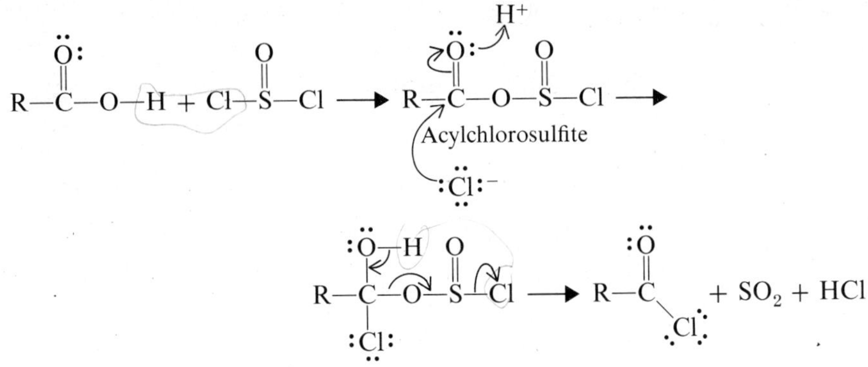

19.6 SYNTHESIS OF CARBOXYLIC ACID ANHYDRIDES

Carboxylic acids react with acyl chlorides in the presence of pyridine to give carboxylic acid anhydrides.

$$\underset{\substack{\parallel \\ O}}{R-C}-OH + \underset{\substack{\parallel \\ O}}{R'-C}-Cl + \underset{N}{\bigcirc} \longrightarrow \underset{\substack{\parallel \\ O}}{R-C}-O-\underset{\substack{\parallel \\ O}}{C}-R' + \underset{\substack{N^+ \\ H}}{\bigcirc} Cl^-$$

This is the most frequently used laboratory method for the preparation of anhydrides. The method is quite general and can be used to prepare mixed anhydrides (R ≠ R′) or simple anhydrides (R = R′).

Sodium salts of carboxylic acids also react with acyl chlorides to give anhydrides:

$$\underset{\substack{\parallel \\ O}}{R-C}-O^-Na^+ + \underset{\substack{\parallel \\ O}}{R'-C}-Cl \longrightarrow \underset{\substack{\parallel \\ O}}{R-C}-O-\underset{\substack{\parallel \\ O}}{C}-R' + Na^+Cl^-$$

In this reaction a carboxylate ion acts as a nucleophile and brings about a nucleophilic substitution reaction at the acyl carbon of the acyl chloride.

Acetic anhydride is an important industrial reagent. One industrial synthesis of acetic anhydride begins with dehydration of acetic acid at 700° to a highly reactive compound called *ketene*.

$$CH_3C\underset{OH}{\overset{O}{\diagdown}} \xrightarrow[700°]{AlPO_4} CH_2=C=O + H_2O$$

Acetic acid   Ketene

Ketene is then allowed to react with acetic acid; this reaction gives acetic anhydride.

$$CH_2=C=O + \underset{\substack{\parallel \\ O}}{CH_3C}-OH \longrightarrow \underset{\substack{\parallel \\ O}}{CH_3C}-O-\underset{\substack{\parallel \\ O}}{C}CH_3$$

Ketene   Acetic acid   Acetic anhydride

Acetic anhydride can be used as a dehydrating agent in the preparation of other anhydrides:

$$2C_6H_5\overset{O}{\overset{\parallel}{C}}OH + (CH_3\overset{O}{\overset{\parallel}{C}})_2O \underset{}{\overset{heat}{\rightleftharpoons}} (C_6H_5\overset{O}{\overset{\parallel}{C}})_2O + 2CH_3\overset{O}{\overset{\parallel}{C}}OH$$

Benzoic acid   Acetic   Benzoic   Acetic
               anhydride   anhydride   acid
                            (74%)

This synthesis is successful because acetic acid distills out as the reaction takes place. This shifts the equilibrium to the right.

Cyclic anhydrides can often be prepared by simply heating the appropriate dicarboxylic acid. This is true, however, only when anhydride formation leads to a five- or six-membered ring.

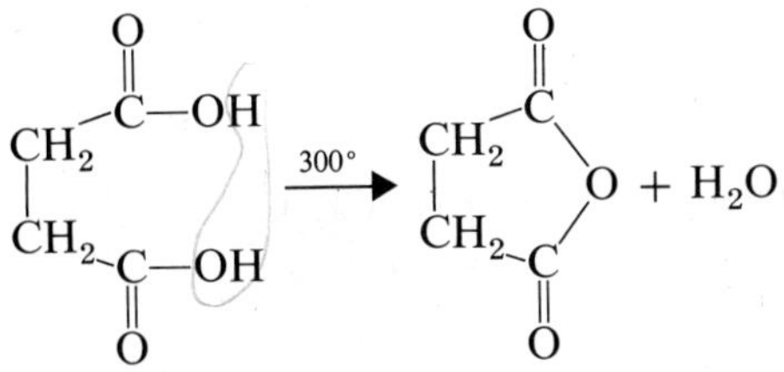

Succinic
anhydride

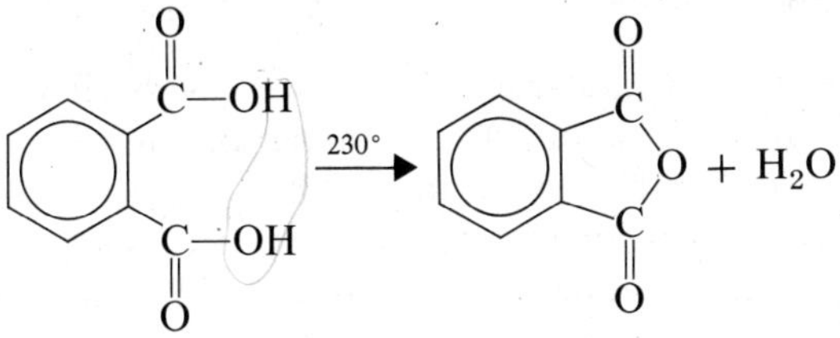

(~100%)

Problem 19.6

Give an equation showing a different method for the preparation of each of the
following anhydrides.

(a) $CH_3COCC_6H_5$

(b) $CH_3(CH_2)_4COC(CH_2)_4CH_3$
(c) Glutaric anhydride

Problem 19.7

When maleic acid is heated to 200° it loses water and becomes maleic anhydride.
Fumaric acid, a diastereomer of maleic acid, requires a much higher temperature
before it dehydrates; when it does it also yields maleic anhydride. Explain.

19.7 ESTERS

Synthesis of Esters: Esterification

Carboxylic acids react with alcohols to form esters through a condensation
reaction known as *esterification:*
General Reaction

$$R-\overset{O}{\underset{\|}{C}}-OH + R'-OH \underset{}{\overset{H^+}{\rightleftarrows}} R-\overset{O}{\underset{\|}{C}}-OR' + H_2O$$

Specific Examples

$$CH_3\overset{O}{\overset{||}{C}}OH + CH_3CH_2OH \underset{}{\overset{H^+}{\rightleftharpoons}} CH_3\overset{O}{\overset{||}{C}}OCH_2CH_3 + H_2O$$

Acetic acid   Ethyl alcohol   Ethyl acetate

$$C_6H_5\overset{O}{\overset{||}{C}}OH + CH_3OH \underset{}{\overset{H^+}{\rightleftharpoons}} C_6H_5\overset{O}{\overset{||}{C}}OCH_3 + H_2O$$

Benzoic acid   Methyl alcohol   Methyl benzoate

Esterification reactions are acid catalyzed. They proceed very slowly in the absence of strong acids, but reach equilibrium within a matter of a few hours when an acid and an alcohol are refluxed with a small amount of concentrated sulfuric acid or hydrogen chloride. Since the position of equilibrium controls the amount of the ester formed, the use of an excess of either the carboxylic acid or the alcohol increases the yield. Just which component we choose to use in excess will depend on its availability and cost. The yield of an esterification reaction can also be increased by removing water from the reaction mixture as it is formed.

When benzoic acid reacts with methanol that has been labeled with ^{18}O, the labeled oxygen appears in the ester: This result reveals just which bonds break in the esterification.

$$C_6H_5\overset{O}{\overset{||}{C}}{+}OH + CH_3O{+}H \underset{}{\overset{H^+}{\rightleftharpoons}} C_6H_5\overset{O}{\overset{||}{C}}{-}OCH_3 + H_2O$$

The results of the labeling experiment and the fact that esterifications are acid catalyzed are both consistent with the mechanism shown below. This mechanism is typical of acid-catalyzed nucleophilic substitution reactions at acyl carbons.

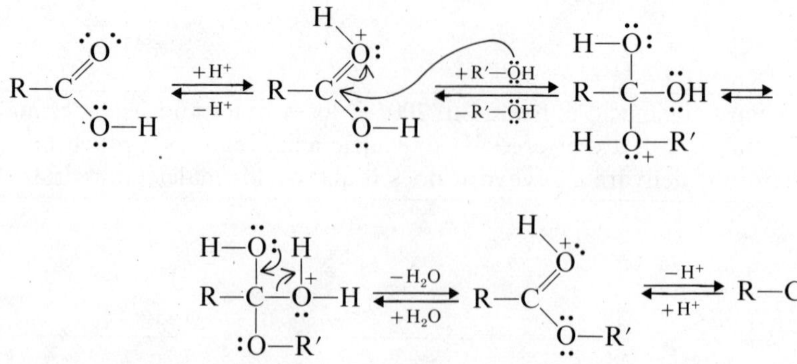

If we follow the forward reactions in the mechanism above we have the mechanism for the *acid-catalyzed esterification of an acid*. If, however, we follow the reverse reactions we have the mechanism for the *acid-catalyzed hydrolysis of an ester:*

$$R\overset{O}{\overset{||}{-C}}{-}OR' + H_2O \underset{}{\overset{H_3O^+}{\rightleftharpoons}} R\overset{O}{\overset{||}{-C}}{-}OH + R'{-}OH$$

Just which result we will obtain will depend on the reaction conditions we choose. If we want to esterify an acid we use an excess of the alcohol and if possible remove the water as it is formed. If we want to hydrolyze an ester we use a large excess of water; that is, we reflux the ester with dilute aqueous HCl or dilute aqueous H_2SO_4.

Problem 19.8

Where would you expect to find the labeled oxygen if you carried out an acid-catalyzed hydrolysis of methyl benzoate in ^{18}O-labeled water?

Steric factors strongly affect the reaction rates of acid-catalyzed hydrolyses of esters. The presence of large groups near the reaction site, whether in the alcohol component or the acid component, slows both reactions markedly. Tertiary alcohols, for example, react so slowly in esterification reactions that they usually undergo elimination instead. The dibromobenzoic acid shown below cannot be esterified through treatment with an alcohol and an acid catalyst; and methyl-2,6-dibromo-benzoate cannot be hydrolyzed with aqueous acid. (Special techniques are required to bring about both reactions.)

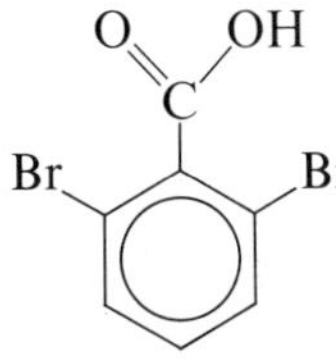

Bulky ortho substituents prevent acid-catalyzed esterification

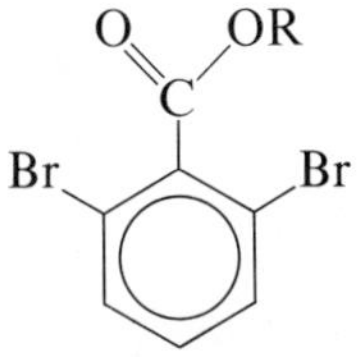

Bulky ortho substituents prevent acid-catalyzed hydrolysis

Esters can also be synthesized through the reaction of acid chlorides with alcohols. Since acid chlorides are much more reactive toward nucleophilic substitution than carboxylic acids, the reaction of an acid chloride and an alcohol occurs rapidly and does not require an acid catalyst. Pyridine is usually added to the reaction mixture to react with the HCl that forms.

General Reaction

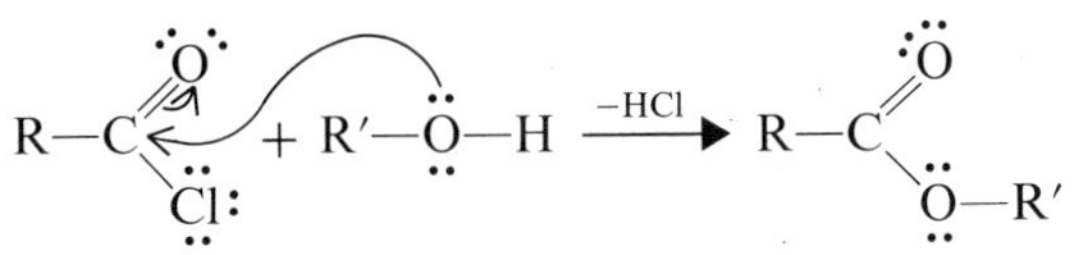

Specific Examples

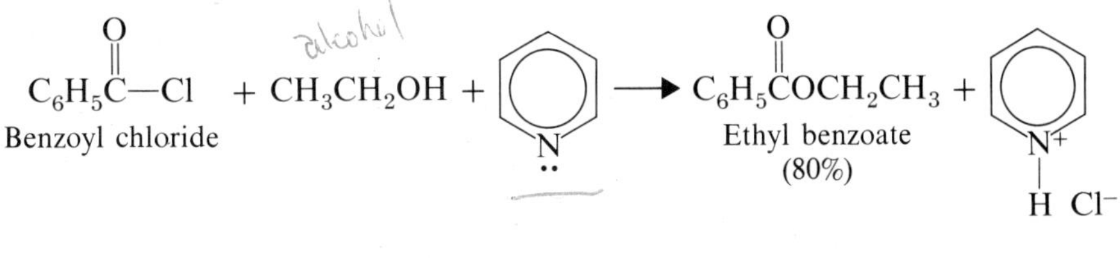

$$C_6H_5\overset{O}{\overset{\|}{C}}-Cl \quad + \quad CH_3CH_2OH \quad + \quad \longrightarrow \quad C_6H_5\overset{O}{\overset{\|}{C}}OCH_2CH_3 \quad +$$

Benzoyl chloride   alcohol

Ethyl benzoate (80%)

H Cl⁻

$$\underset{\text{Malonyl chloride}}{\text{Cl}-\overset{\text{O}}{\overset{\|}{\text{C}}}\text{CH}_2\overset{\text{O}}{\overset{\|}{\text{C}}}-\text{Cl}} + 2\text{CH}_3\underset{\underset{\text{CH}_3}{|}}{\overset{\overset{\text{CH}_3}{|}}{\text{C}}}-\text{OH} + 2 \underset{\text{N}}{\bigcirc} \longrightarrow$$

$$\underset{\underset{\text{Di-}\textit{tert}\text{-butyl malonate}}{(83\%)}}{\text{CH}_3\underset{\underset{\text{CH}_3}{|}}{\overset{\overset{\text{CH}_3}{|}}{\text{C}}}-\text{OCCH}_2\text{CO}-\underset{\underset{\text{CH}_3}{|}}{\overset{\overset{\text{CH}_3}{|}}{\text{C}}}\text{CH}_3} + 2 \underset{\underset{\text{H}}{\overset{\text{N}^+}{\bigcirc}}}{} \text{Cl}^-$$

Acid anhydrides also react with alcohols to form esters in the absence of an acid catalyst.

General Reaction

$$\underset{\text{RC}}{\overset{\text{O}}{\underset{\text{O}}{\text{RC}}}} \overset{\ddot{\text{O}}}{\underset{\ddot{\text{O}}}{}} + \text{R}'-\ddot{\text{O}}\text{H} \xrightarrow[-\text{RCOH}]{} \text{RC}\overset{\ddot{\text{O}}}{\underset{\ddot{\text{O}}-\text{R}'}{}}$$

Specific Example

$$\underset{\underset{\text{anhydride}}{\text{Acetic}}}{(\text{CH}_3\overset{\text{O}}{\overset{\|}{\text{C}}})_2\text{O}} + \underset{\underset{\text{alcohol}}{\text{Benzyl}}}{\text{C}_6\text{H}_5\text{CH}_2\text{OH}} \longrightarrow \underset{\text{Benzyl acetate}}{\text{CH}_3\overset{\text{O}}{\overset{\|}{\text{C}}}\text{OCH}_2\text{C}_6\text{H}_5} + \text{CH}_3\text{COOH}$$

Cyclic anhydrides react with one mole of an alcohol to form a compound that is both *an ester and an acid*. This reaction is often used in the resolution of enantiomeric alcohols (p. 271).

$$\underset{\underset{\text{anhydride}}{\text{Phthalic}}}{\bigcirc} + \underset{\underset{\underset{\text{sec-Butyl alcohol}}{}}{\text{OH}}}{\text{CH}_3\text{CHCH}_2\text{CH}_3} \xrightarrow{110°} \underset{\underset{(97\%)}{\underset{\text{phthalate}}{\textit{sec}\text{-Butyl hydrogen}}}}{\bigcirc}$$

Problem 19.9

Esters can also be synthesized by a transesterification reaction, that is,

$$\underset{\underset{\text{ester}}{\text{High boiling}}}{\text{R}-\overset{\text{O}}{\overset{\|}{\text{C}}}-\text{OR}'} + \underset{\underset{\text{alcohol}}{\text{High boiling}}}{\text{R}''-\text{OH}} \underset{}{\overset{\text{H}^+}{\rightleftarrows}} \underset{\underset{\text{ester}}{\text{Higher boiling}}}{\text{RC}-\text{OR}''} + \underset{\underset{\text{alcohol}}{\text{Low boiling}}}{\text{R}'-\text{OH}}$$

In this procedure we shift the equilibrium to the right by allowing the low boiling alcohol to distill out of the reaction mixture. The mechanism for transesterification is similar to that for an acid-catalyzed esterification (or an acid-catalyzed ester hydrolysis). Write a mechanism for the transesterification reaction shown below.

$$CH_2{=}CHCOCH_3 + CH_3CH_2CH_2CH_2OH \overset{H^+}{\rightleftharpoons}$$
Methyl acrylate   n-Butyl alcohol

$$CH_2{=}CHCOCH_2CH_2CH_2CH_3 + CH_3OH$$
n-Butyl acrylate   Methyl
(94%)   alcohol

Base-Promoted Hydrolysis of Esters: Saponification

Esters not only undergo acid hydrolysis, they also undergo *base-promoted hydrolysis*. Base-promoted hydrolysis is sometimes called *saponification*. Refluxing an ester with aqueous sodium hydroxide, for example, produces an alcohol and the sodium salt of the acid:

$$RC{-}OR' + NaOH \overset{H_2O}{\longrightarrow} R{-}C{-}O^-Na^+ + R'OH$$
Ester   Sodium carbox-   Alcohol
ylate

The carboxylate ion is very unreactive toward nucleophilic substitution because it is negatively charged. Base-promoted hydrolysis of an ester, as a result, is an essentially irreversible reaction.

The mechanism for the base-promoted hydrolysis of an ester also involves a nucleophilic substitution at the acyl carbon:

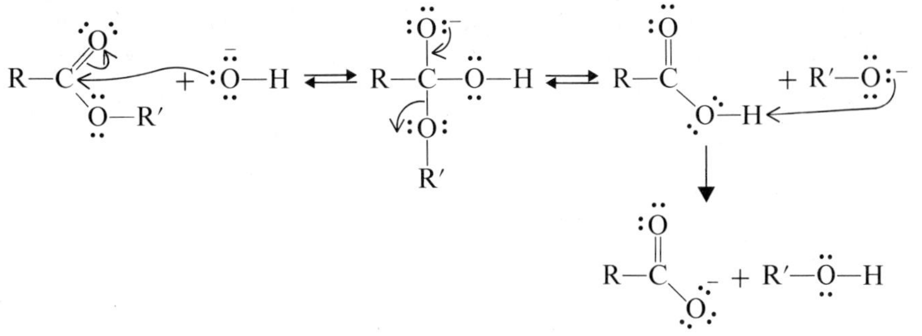

Part of the evidence that nucleophilic attack occurs at the acyl carbon comes from studies in which esters of chiral alcohols were subjected to base-promoted hydrolysis. The reaction of an ester with hydroxide ion could conceivably occur in two ways: reaction could take place through a nucleophilic substitution at the acyl carbon of the acid component (path A, p. 760) or through a nucleophilic substitution at the alkyl carbon of the alcohol portion (path B). Reaction by path A should lead to retention of configuration in the alcohol. Reaction by path B should lead to an inversion of configuration of the alcohol. *Inversion of configuration is almost never observed.* In almost every instance basic hydrolysis of an ester of a chiral alcohol proceeds with *retention of configuration.*

Path A: Nucleophilic substitution at the acyl carbon

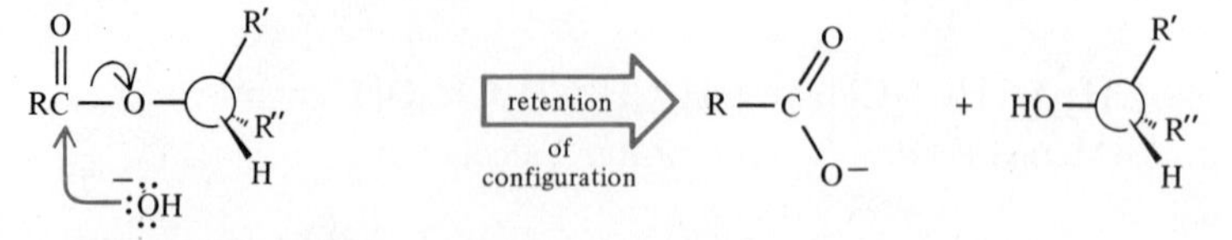

Path B: Nucleophilic substitution at the alkyl carbon

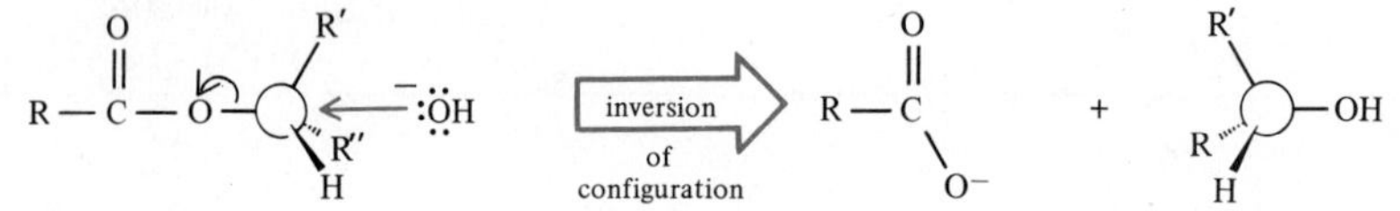

(*this reaction seldom occurs with esters of carboxylic acids*)

Although nucleophilic attack at the alkyl carbon seldom occurs with esters of carboxylic acids, it is the preferred mode of attack with esters of sulfonic acids (p. 597).

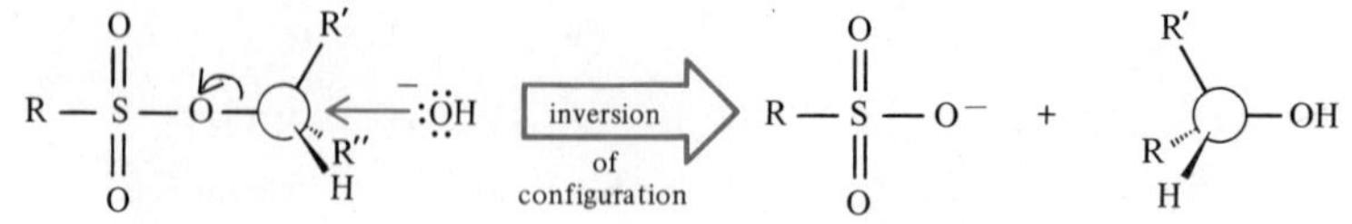

an alkyl sulfonate

Problem 19.10

(a) Write stereochemical formulas for compounds **A–F.**

1. *cis*-3-Methylcyclopentanol + $C_6H_5SO_2Cl \longrightarrow$ **A**

$$\xrightarrow[\text{heat}]{^-\text{OH}} \textbf{B} + C_6H_5SO_3^-$$

2. *cis*-3-Methylcyclopentanol + $C_6H_5\overset{\overset{\displaystyle O}{\|}}{C}{-}Cl \longrightarrow$ **C**

$$\xrightarrow[\text{reflux}]{^-\text{OH}} \textbf{D} + C_6H_5CO_2^-$$

3. *R*-2-Bromooctane + $CH_3COO^-Na^+ \longrightarrow$ **E** + NaBr

$$\downarrow \begin{array}{l} OH^-, H_2O \\ \text{(reflux)} \end{array}$$

F

4. *R*-2-Bromooctane + $OH^- \xrightarrow{\text{alcohol}}$ **F** + Br$^-$

(b) Which of the last two methods, (3) or (4), would you expect to give a higher yield of **F?** Why?

Problem 19.11

Base-promoted hydrolysis of methyl mesitoate occurs through an attack on the alcohol carbon instead of the acyl carbon.

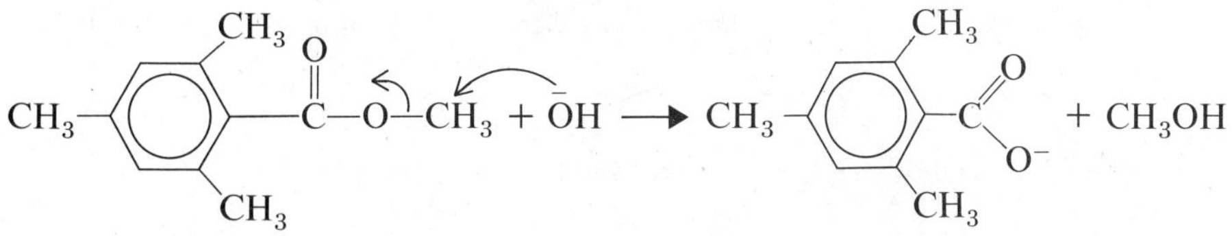

Methyl mesitoate

(a) Can you suggest a reason that will account for this unusual behavior? (b) Suggest an experiment with labeled compounds that would confirm this mode of attack.

Lactones

Carboxylic acids that have a hydroxyl group on a γ- or δ- carbon undergo an intramolecular esterification reaction and yield cyclic esters known as γ- or δ-*lactones*.

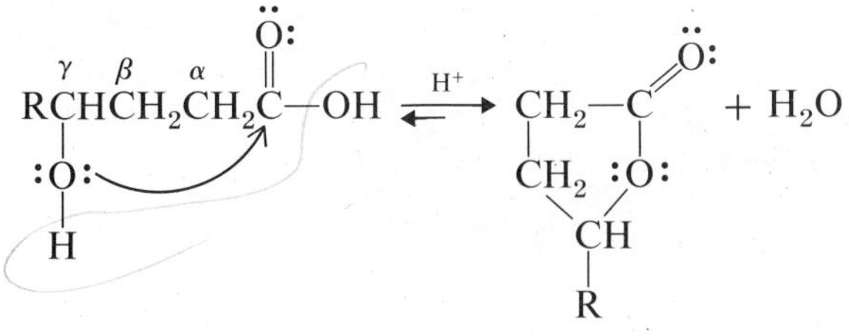

A γ-hydroxy acid   A γ-lactone

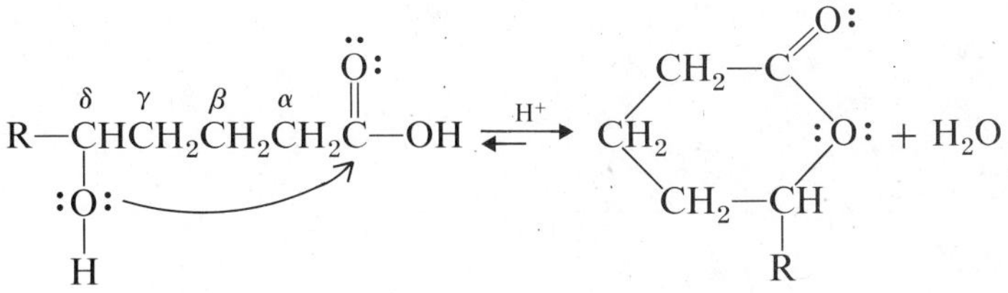

A δ-hydroxy acid   A δ-lactone

Lactones are hydrolyzed by aqueous base just as other esters are. Acidification of the sodium salt, however, may lead spontaneously back to the γ- or δ-lactone, particularly if excess acid is used.

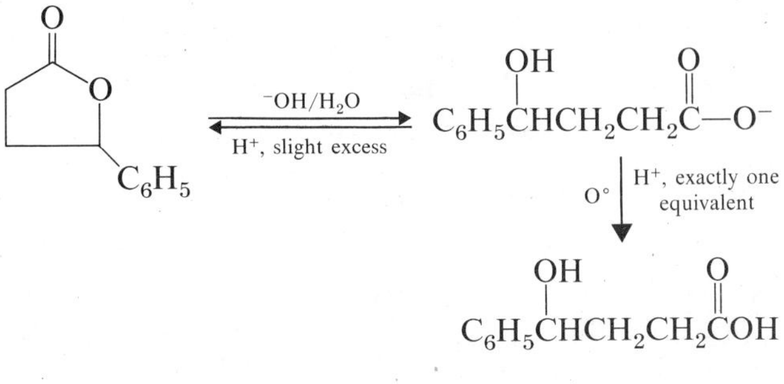

Many lactones occur in nature. Almost all are γ- or δ-lactones; that is, almost all contain five- or six-membered rings.

Beta lactones (lactones with four-membered rings) have been detected as intermediates in some reactions. They are highly reactive, however. If one attempts to prepare a β-lactone from a β-hydroxy acid, dehydration usually occurs instead:

$$\underset{\substack{\text{OH}\\ \\ \beta\text{-Hydroxy acid}}}{\text{RCHCH}_2\overset{\displaystyle O}{\overset{\|}{\text{C}}}\text{OH}} \xrightarrow[\substack{\text{or}\\ \text{acid}}]{\text{heat}} \underset{\substack{\alpha,\beta\text{-Unsaturated}\\ \text{acid}}}{\text{RCH}=\text{CH}\overset{\displaystyle O}{\overset{\|}{\text{C}}}\text{OH}} + \text{H}_2\text{O}$$

$$\xrightarrow[]{\text{heat}} \times \xrightarrow[]{} \underset{\substack{\beta\text{-Lactone}\\ \text{(does not form)}}}{\begin{array}{c}\text{RCH}-\text{CH}_2\\ \big| \qquad \big|\\ \text{O}-\text{C}\\ \diagdown_{\text{O}}\end{array}}$$

When α-hydroxy acids are heated they form cyclic diesters called *lactides*.

$$2\,\underset{\text{OH}}{\text{RCH}\overset{\displaystyle O}{\overset{\|}{\text{C}}}\text{OH}} \xrightarrow{\text{heat}} \underset{\text{A lactide}}{\begin{array}{c}\overset{\displaystyle O}{\overset{\|}{\text{C}}}-\text{O}\\ \diagup \qquad \diagdown\\ \text{R}-\text{CH} \qquad \text{CH}-\text{R}\\ \diagdown \qquad \diagup\\ \text{O}-\underset{\displaystyle\underset{\|}{\text{O}}}{\text{C}}\end{array}}$$

Alpha lactones also occur as intermediates in some reactions (cf. p. 655).

19.8 AMIDES

Synthesis of Amides

Amides can be prepared in a variety of ways starting with acyl chlorides, acid anhydrides, esters, carboxylic acids, and carboxylate salts. All of these methods involve nucleophilic substitution reactions by ammonia or an amine at an acyl carbon. As we might expect, acid chlorides are the most reactive and carboxylate ions are the least.

Amides from Acyl Chlorides

Primary amines, secondary amines, and ammonia all react rapidly with acid chlorides to form amides:*

---

* Acyl chlorides also react with tertiary amines by a nucleophilic substitution reaction. The acyl ammonium ion that forms, however, is not stable in the presence of water or any

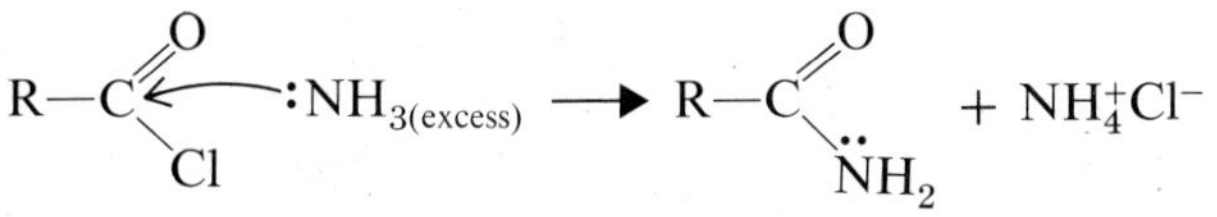

$$\underset{\text{Ammonia}}{R-\overset{O}{\underset{Cl}{C}}\ \ :NH_{3(\text{excess})}} \longrightarrow \underset{\text{An amide}}{R-\overset{O}{\underset{\ddot{N}H_2}{C}}} + NH_4^+Cl^-$$

$$R-\overset{O}{\underset{Cl}{C}}\ \ R'\ddot{N}H_{2(\text{excess})} \longrightarrow R-\overset{O}{\underset{\ddot{N}HR'}{C}} + RNH_3^+Cl^-$$

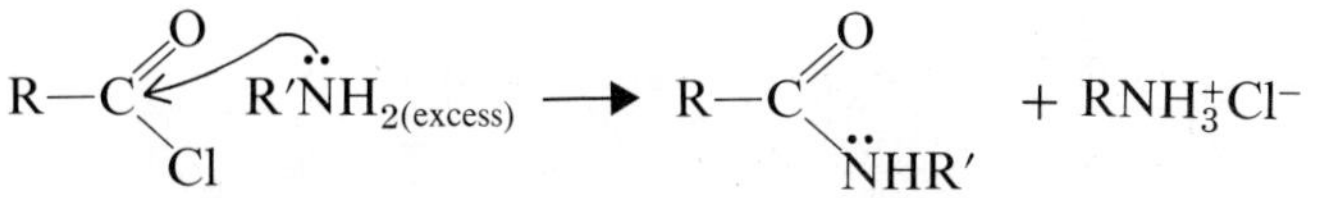

An *N*-substituted
amide

$$R-\overset{O}{\underset{Cl}{C}}\ \ R'-\underset{R''}{\ddot{N}H}_{(\text{excess})} \longrightarrow R-\overset{O}{\underset{\underset{R''}{N-R'}}{C}} + R'NH_2^+Cl^-$$

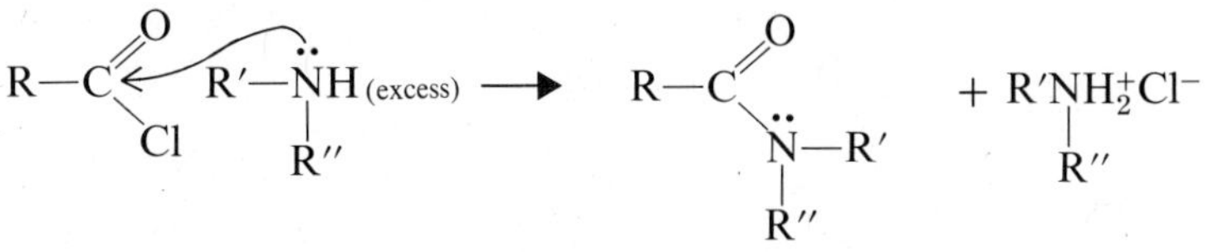

An *N,N*-disubstituted
amide

Since acyl chlorides are easily prepared from carboxylic acids, this is one of the most widely used laboratory methods for the synthesis of amides. The reaction between the acyl chloride and the amine (or ammonia) usually takes place at room temperature (or below) and produces the amide in high yield.

$$\underset{\underset{CH_3}{\overset{\displaystyle CH_2}{|}}}{CH_3CH_2\overset{O}{\overset{\|}{CH}C}-Cl} + 2NH_3 \xrightarrow{\text{benzene}} \underset{\underset{CH_3}{\overset{\displaystyle CH_2}{|}}}{CH_3CH_2\overset{O}{\overset{\|}{CH}C}-NH_2} + NH_4Cl$$

(91%)

$$CH_2{=}CH\overset{O}{\overset{\|}{C}}{-}Cl + 2NH_3 \longrightarrow CH_2{=}CH\overset{O}{\overset{\|}{C}}{-}NH_2 + NH_4Cl$$

(80%)

hydroxylic solvent.

$$\underset{\substack{\text{Acyl chloride}}}{R-\overset{O}{\underset{Cl}{C}}} + \underset{\substack{\text{3° Amine}}}{R_3N\!:} \longrightarrow \underset{\substack{\text{Acyl ammonium ion}}}{R-\overset{O}{\overset{\|}{C}}-\overset{+}{N}R_3\ Cl^-}$$

$$\downarrow\ {\scriptstyle H_2O}$$

$$R-\overset{O}{\overset{\|}{C}}OH + H\overset{+}{N}R_3\ Cl^-$$

Acyl pyridinium ions are probably involved as intermediates in those reactions of acyl chlorides that are carried out in the presence of pyridine.

Amides from Carboxylic Acid Anhydrides

Acid anhydrides react with ammonia and with primary and secondary amines and form amides through reactions that are analogous to those of acyl chlorides.

$$\left(\underset{\substack{\| \\ O}}{RC}\right)_2 O + \ddot{N}H_3 \longrightarrow \underset{\substack{\| \\ O}}{RC}-\ddot{N}H_2 + RCOOH$$

$$\left(\underset{\substack{\| \\ O}}{RC}\right)_2 O + R'-\ddot{N}H_2 \longrightarrow \underset{\substack{\| \\ O}}{RC}-\ddot{N}H-R' + RCOOH$$

$$\left(\underset{\substack{\| \\ O}}{RC}\right)_2 O + \underset{\substack{| \\ R''}}{R'-\ddot{N}H} \longrightarrow \underset{\substack{\| \quad | \\ O \quad R''}}{RC-\ddot{N}-R'} + RCOOH$$

Cyclic anhydrides react with ammonia or an amine in the same general way as acyclic anhydrides; however, the reaction produces a product that is both an amide and an ammonium salt. Acidifying the ammonium salt gives a compound that is both an amide and an acid:

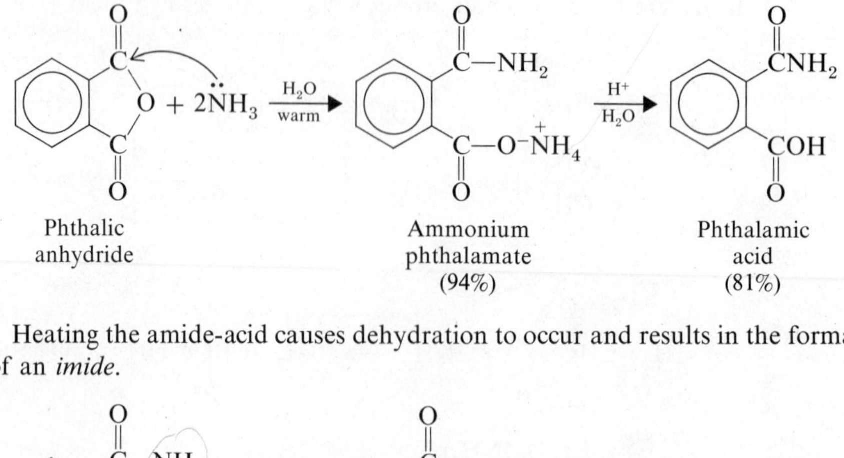

Phthalic anhydride | Ammonium phthalamate (94%) | Phthalamic acid (81%)

Heating the amide-acid causes dehydration to occur and results in the formation of an *imide*.

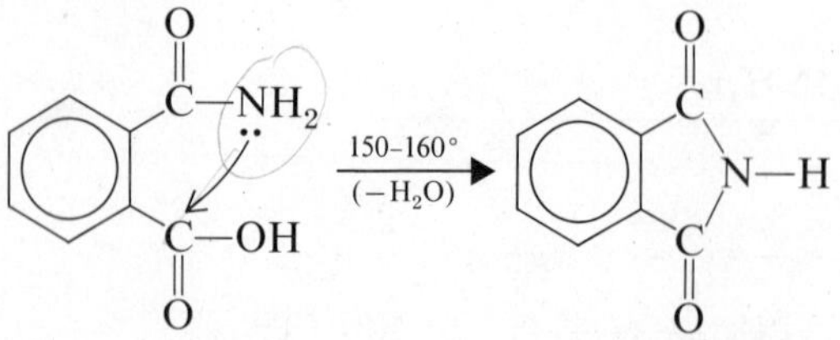

Phthalamic acid | Phthalimide (~100%)

Amides from Esters

Esters undergo nucleophilic substitution at their acyl carbons when they are treated with ammonia or with primary and secondary amines. These reactions take place more slowly than those of acyl chlorides and anhydrides, but they are synthetically useful.

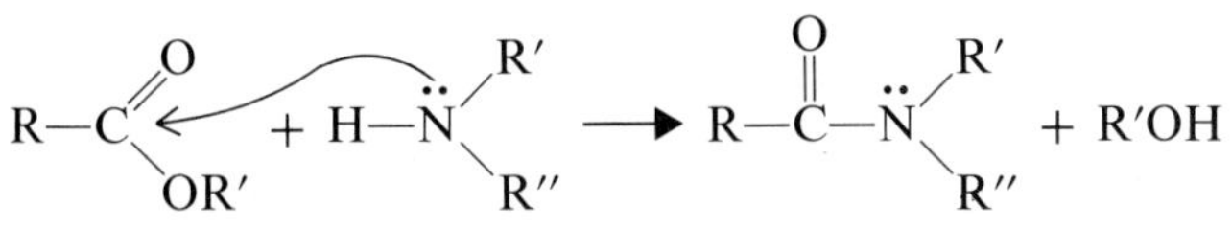

$$R-C{\overset{O}{<}}_{OR'} + H-{\overset{..}{N}}{\overset{R'}{<}}_{R''} \longrightarrow R-C{\overset{O}{<}}-{\overset{..}{N}}{\overset{R'}{<}}_{R''} + R'OH$$

R' and/or R''
may be H

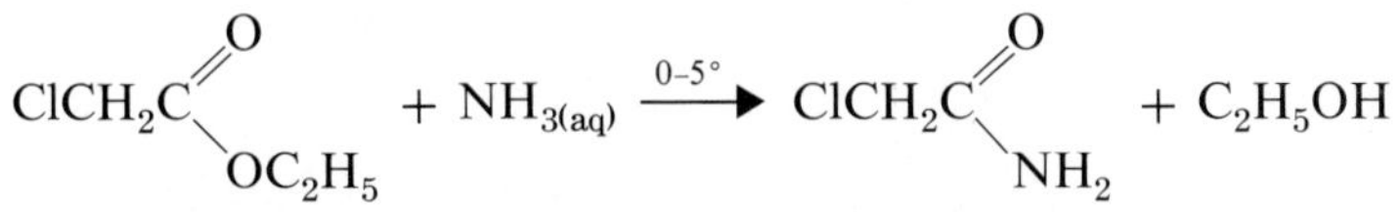

$$ClCH_2C{\overset{O}{<}}_{OC_2H_5} + NH_{3(aq)} \xrightarrow{0-5°} ClCH_2C{\overset{O}{<}}_{NH_2} + C_2H_5OH$$

Ethyl chloroacetate   Chloroacetamide
                                  (62–87%)

Amides from Carboxylic Acids and Ammonium Carboxylates

Carboxylic acids react with aqueous ammonia to form ammonium salts.

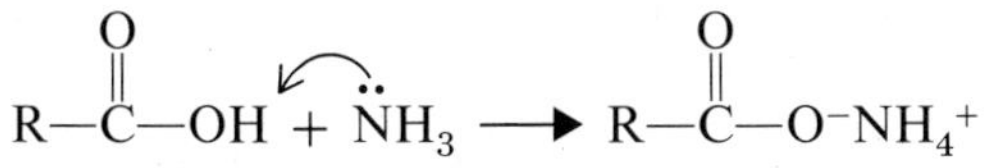

$$R-{\overset{O}{\overset{||}{C}}}-OH + {\overset{..}{N}}H_3 \longrightarrow R-{\overset{O}{\overset{||}{C}}}-O^-NH_4^+$$

An ammonium
carboxylate

Because of the low reactivity of the carboxylate ion toward nucleophilic substitution, further reaction does not usually take place in aqueous solution. However, if we evaporate the water and subsequently heat the dry salt, dehydration produces an amide.

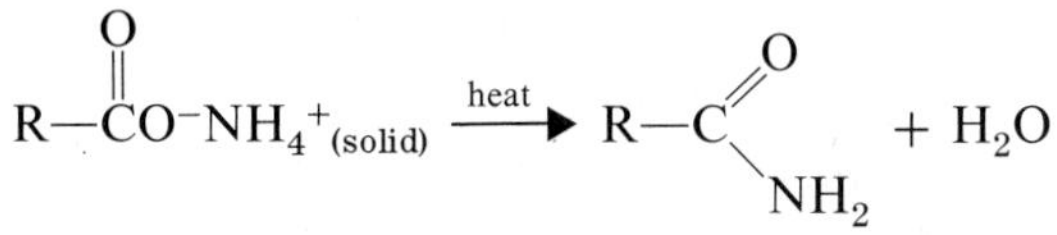

$$R-{\overset{O}{\overset{||}{C}}}O^-NH_4^+{}_{(solid)} \xrightarrow{heat} R-C{\overset{O}{<}}_{NH_2} + H_2O$$

An alternate method for synthesizing amides from acids involves adding solid ammonium carbonate to an excess of the acid and then slowly distilling the mixture.

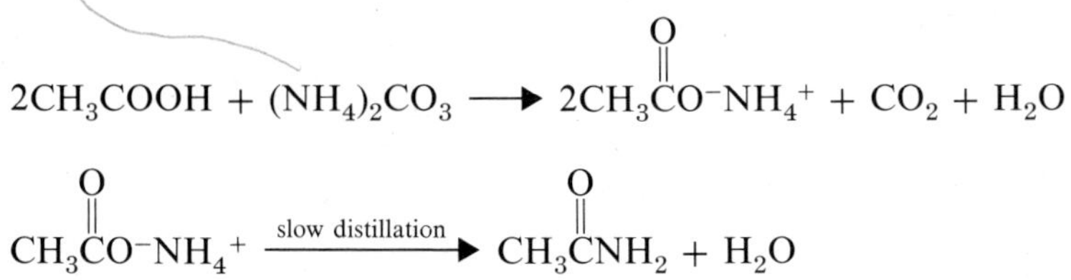

$$2CH_3COOH + (NH_4)_2CO_3 \longrightarrow 2CH_3{\overset{O}{\overset{||}{C}}}O^-NH_4^+ + CO_2 + H_2O$$

$$CH_3{\overset{O}{\overset{||}{C}}}O^-NH_4^+ \xrightarrow{slow\ distillation} CH_3{\overset{O}{\overset{||}{C}}}NH_2 + H_2O$$

Acetamide
(87–90%)

Amides are of great importance in biochemistry. The linkages that join individual amino acids together to form proteins are, primarily, amide linkages (p. 63). As a consequence, much research has been done to find new and mild ways for amide synthesis. One especially useful reagent is the compound dicyclohexyl-carbodiimide, $C_6H_{13}N=C=N-C_6H_{13}$. Dicyclohexylcarbodiimide promotes amide formation by reacting with the carboxyl group of an acid and activating it toward nucleophilic substitution.

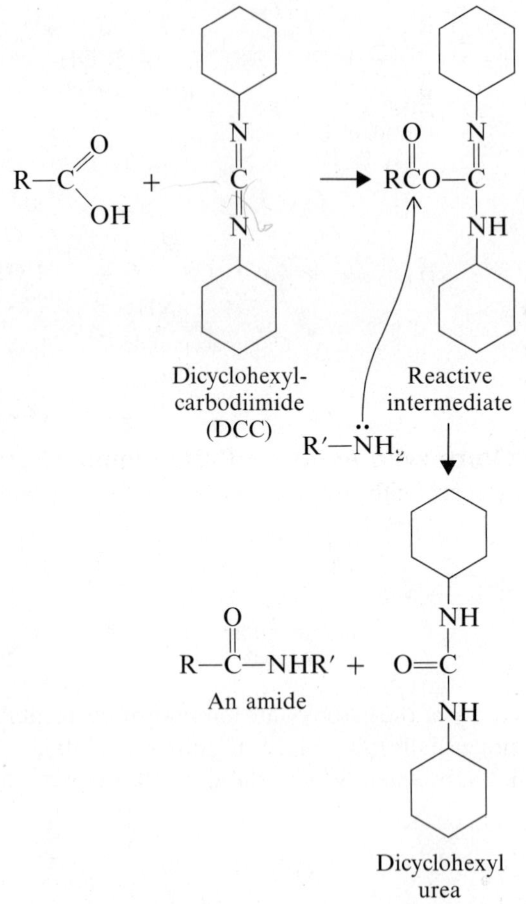

Dicyclohexyl-
carbodiimide
(DCC)

Reactive
intermediate

$R'-\ddot{N}H_2$

R—C—NHR′ +
An amide

Dicyclohexyl
urea

The intermediate in this synthesis does not need to be isolated, and both steps take place at room temperature. Amides are produced in very high yield. In Chapter 25 we will see how dicyclohexylcarbodiimide can be used in an automated synthesis of proteins.

Hydrolysis of Amides

Amides undergo hydrolysis when they are heated with aqueous acid or aqueous base.

Acidic Hydrolysis

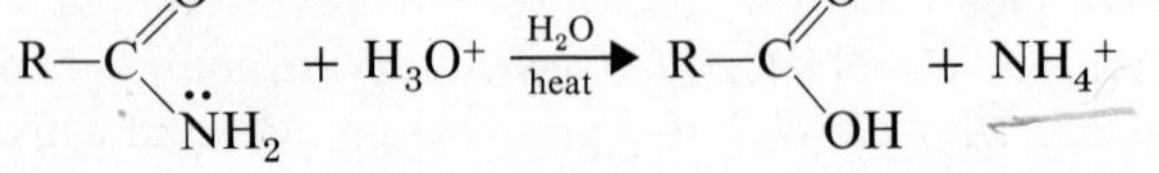

Basic Hydrolysis

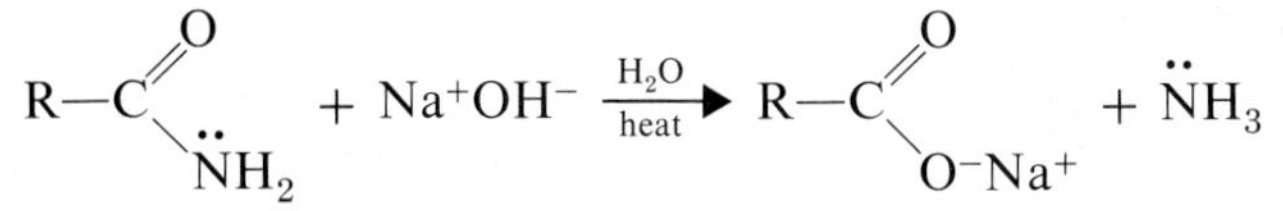

$$R-\overset{\overset{\textstyle O}{\|}}{\underset{\ddot{N}H_2}{C}} + Na^+OH^- \xrightarrow[\text{heat}]{H_2O} R-\overset{\overset{\textstyle O}{\|}}{\underset{O^-Na^+}{C}} + \ddot{N}H_3$$

N-substituted amides and *N,N*-disubstituted amides also undergo hydrolysis in aqueous acid or base. Amide hydrolysis by either method takes place more slowly than the corresponding hydrolysis of an ester. Thus, amide hydrolyses generally require more forcing conditions.

The mechanism for acid hydrolysis of an amide is similar to that given on p. 756 for the acid hydrolysis of an ester. Water acts as a nucleophile and attacks the protonated amide. The leaving group in the acidic hydrolysis of amide is ammonia (or an amine).

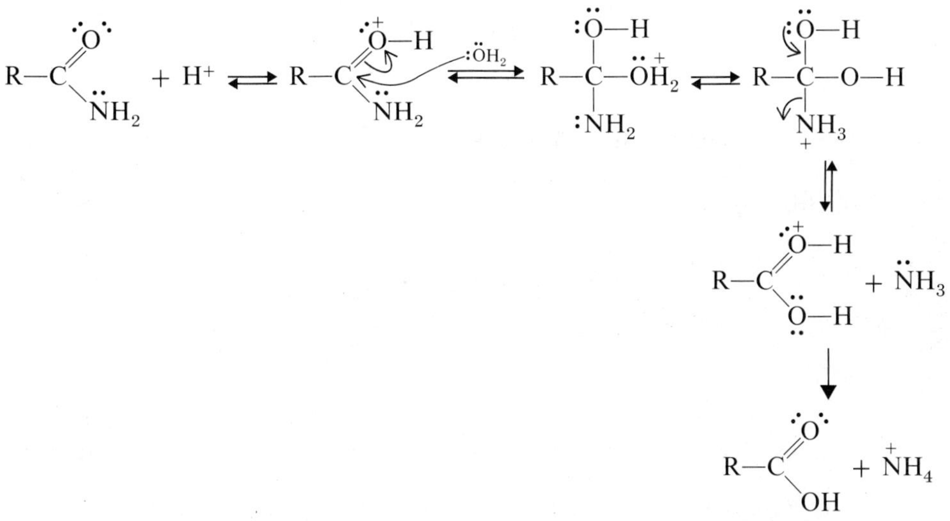

There is evidence that in basic hydrolyses of amides, hydroxide ions act both as nucleophiles and as bases. In the first step (below) a hydroxide ion attacks the acyl carbon of the amide. In the second step, a hydroxide ion removes a proton to give a dianion. In the final step the dianion loses a molecule of ammonia (or an amine); this step is synchronized with a proton transfer from water.

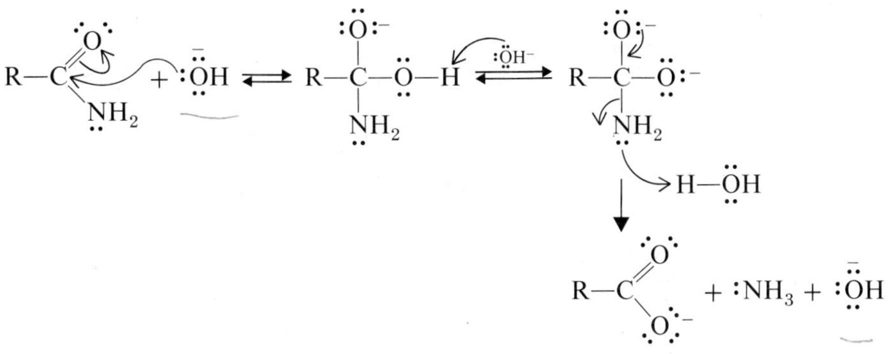

Problem 19.12

What products would you obtain from acidic and basic hydrolysis of each of the following amides.

(a) *N*,*N*-diethylbenzamide

(b)

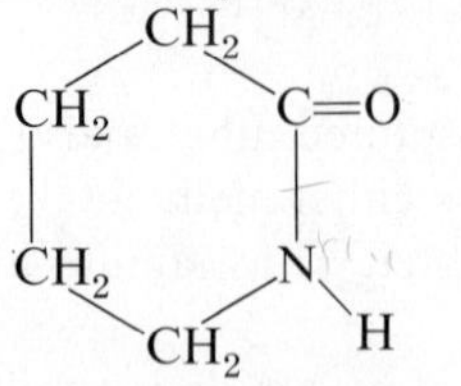

(c) 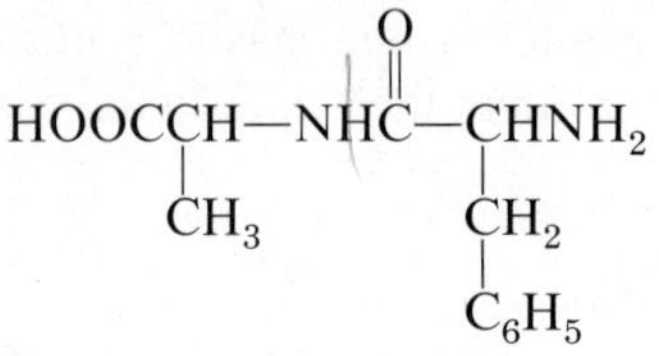 $\underset{\underset{CH_3}{|}}{HOOCCH}-\underset{}{NHC}-\underset{\underset{\underset{C_6H_5}{|}}{CH_2}}{CHNH_2}$ (a dipeptide)

Dehydration of Amides

Amides react with phosphorus pentoxide (P_2O_5) or with boiling acetic anhydride to form nitriles.

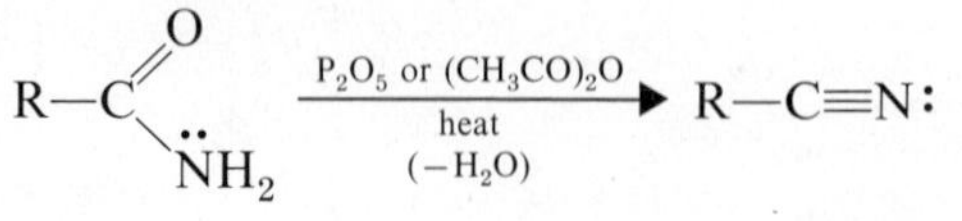

$$R-\overset{O}{\overset{\|}{C}}\underset{\ddot{N}H_2}{} \xrightarrow[\substack{heat \\ (-H_2O)}]{P_2O_5 \text{ or } (CH_3CO)_2O} R-C\equiv N\colon$$

A nitrile

This is a useful synthetic method for preparing nitriles that are not available through nucleophilic substitution reactions between alkyl halides and cyanide ion.

Problem 19.13

(a) Show all steps in the synthesis of $(CH_3)_3CCN$ from $(CH_3)_3CCOOH$. (b) What product would you expect to obtain if you attempted to synthesize $(CH_3)_3CCN$ using the following reaction?

$$(CH_3)_3C-Br + CN^- \longrightarrow$$

Lactams

Cyclic amides are called lactams. The size of the lactam ring is designated through the use of Greek letters in a way that is analogous to lactone nomenclature.

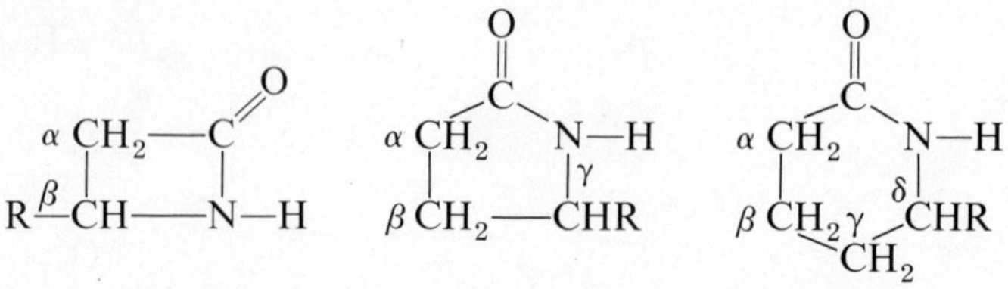

A β-lactam   A γ-lactam   A δ-lactam

Gamma-lactams and δ-lactams often form spontaneously from γ- and δ-amino acids. Beta-lactams, however, are highly reactive; their strained four-membered rings open easily in the presence of nucleophilic reagents. The penicillin antibiotics (below) contain a β-lactam ring.

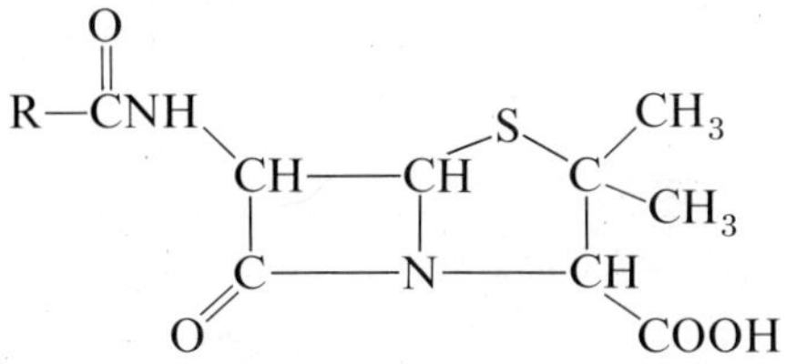

R = $C_6H_5CH_2$— (penicillin G)

R = C_6H_5CH— (ampicillin)
　　　|
　　　NH_2

R = $C_6H_5OCH_2$— (penicillin V)

The penicillins apparently act by interfering with the synthesis of the bacterial cell walls. It is thought that they do this by reacting with an amino group of an essential enzyme of the cell wall biosynthetic pathway. This reaction, which involves ring opening of the β-lactam and acylation of the amino group, inactivates the enzyme.

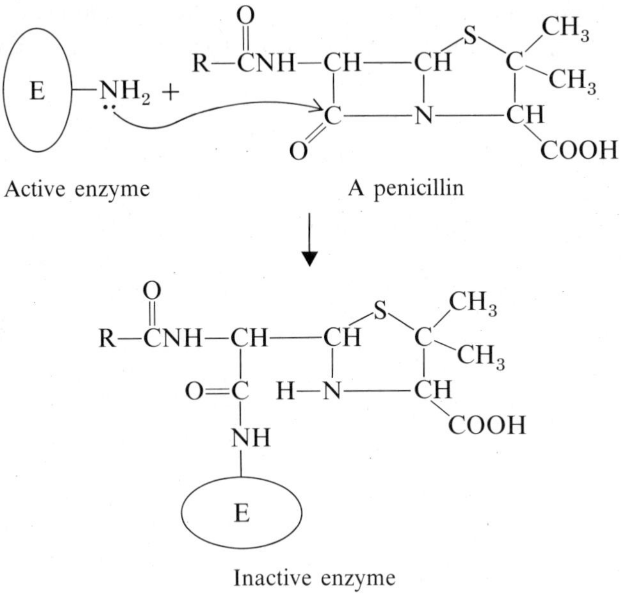

Active enzyme　　　　　A penicillin

Inactive enzyme

19.9 α-HALO ACIDS: THE HELL-VOLHARD-ZELINSKI REACTION

Aliphatic carboxylic acids react with bromine or chlorine in the presence of phosphorus (or a phosphorus halide) to give α-halo acids through a reaction known as the Hell-Volhard-Zelinski reaction.

General Reaction

$$R—CH_2COOH \xrightarrow[P]{X_2} \underset{\underset{X}{|}}{R}CHCOOH$$

α-Halo acid

Specific Examples

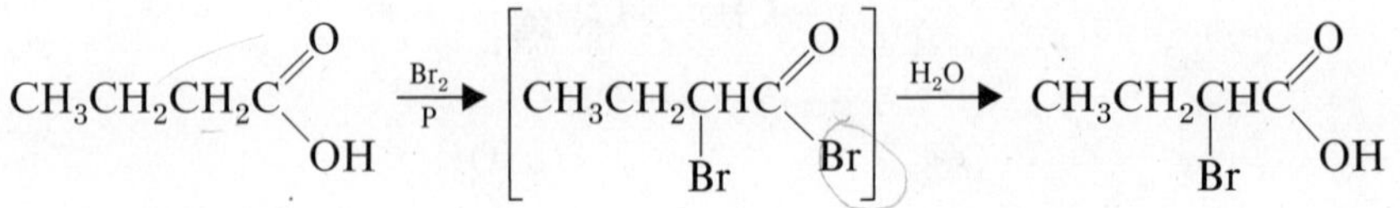

Butanoic acid   2-Bromobutanoic acid
(77%)

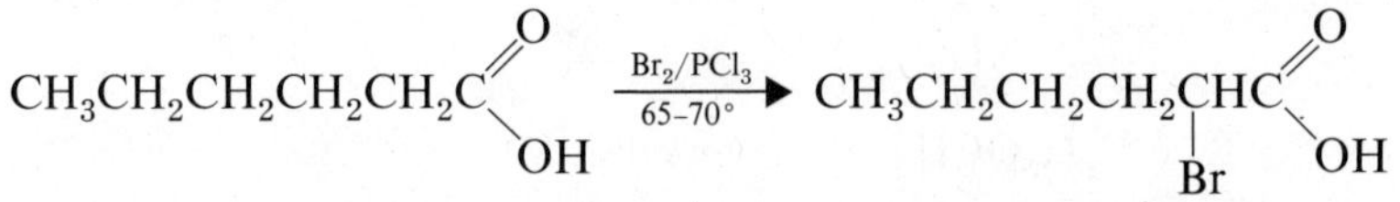

Hexanoic acid   2-Bromohexanoic acid
(83–89%)

Halogenation occurs specifically at the α-carbon. If more than one mole of bromine or chlorine is used in the reaction, the products obtained are α,α-dihalo acids or α,α,α-trihalo acids.

The mechanism for the Hell-Volhard-Zelinski reaction involves the formation of an enol from an acyl halide. Enol formation accounts for specific α-halogenation:

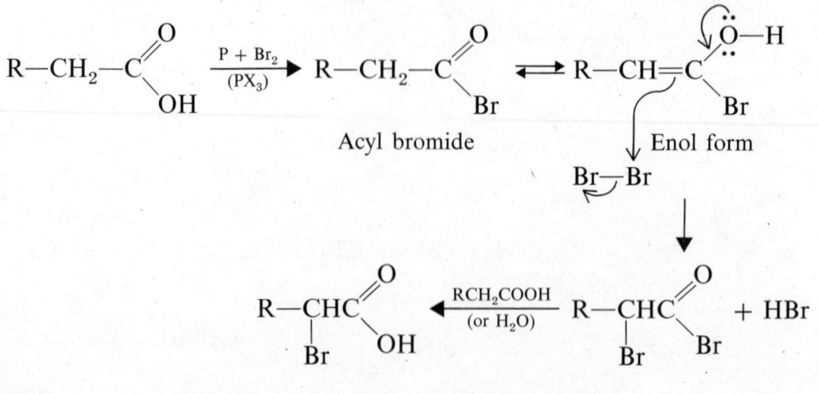

Acyl bromide   Enol form

Alpha-halo acids are important synthetic intermediates because they are capable of reacting with a variety of nucleophiles:

Conversion to α-Hydroxy Acids

$$\underset{\underset{X}{|}}{R}—CHCOOH \xrightarrow[\text{(2) H}^+]{\text{OH}^-} \underset{\underset{OH}{|}}{R}—CHCOOH + X^-$$

α-Halo acid   α-Hydroxy acid

Example:

$$CH_3CH_2\underset{\underset{Br}{|}}{C}HCOOH \xrightarrow[100°]{K_2CO_3, H_2O} CH_3CH_2\underset{\underset{OH}{|}}{C}HCOOH$$

2-Hydroxybutanoic acid
(69%)

Conversion to α-Amino Acids

$$R-\underset{\underset{X}{|}}{C}HCOOH + 2NH_3 \longrightarrow R\underset{\underset{NH_2}{|}}{C}HCOOH + NH_4X$$

α-Halo
acid

Example:

$$\underset{\underset{Br}{|}}{C}H_2COOH + NH_{3(excess)} \longrightarrow \underset{\underset{NH_2}{|}}{C}H_2COOH + NH_4Br$$

α-Aminoacetic acid
(glycine)
(60–64%)

19.10 THIOL ESTERS

Thiol esters can be prepared through reactions of a thiol with an acid chloride.

$$R-\overset{O}{\underset{Cl}{C}} + R'-SH \longrightarrow R-\overset{O}{\underset{S-R'}{C}} + HCl$$

Thiol ester

$$CH_3\overset{O}{\underset{Cl}{C}} + CH_3SH \xrightarrow{pyridine} CH_3\overset{O}{\underset{SCH_3}{C}} + \underset{\underset{H^+ \ Cl^-}{|}}{\underset{N}{\bigcirc}}$$

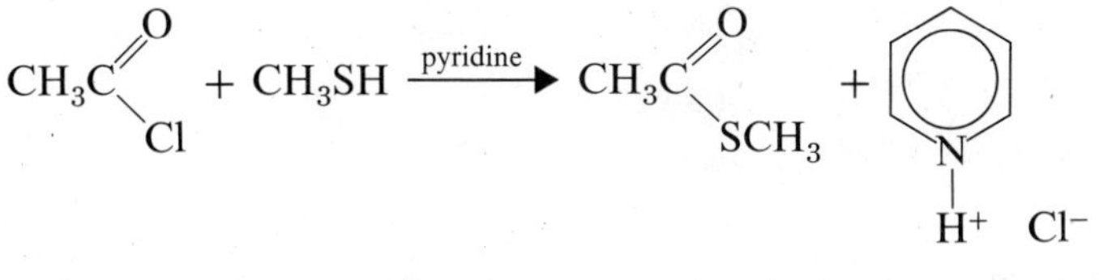

Although thiol esters are not often used in laboratory syntheses, they are of great importance in syntheses that occur within living cells. One of the important thiol esters in biochemistry is a compound called "acetylcoenzyme A."

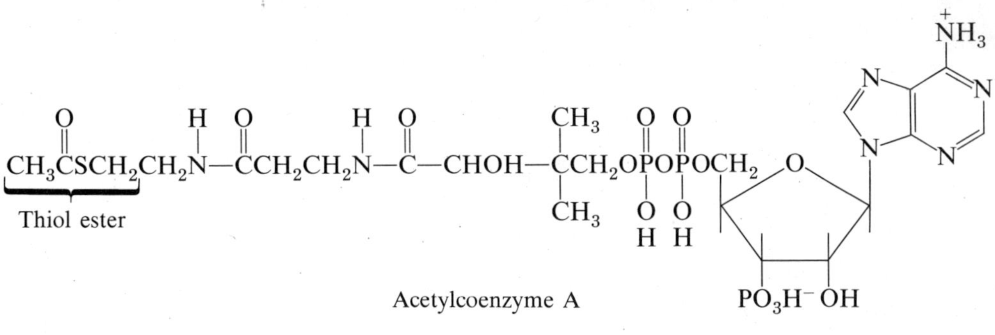

$$\underbrace{CH_3\overset{O}{C}SCH_2CH_2}_{\text{Thiol ester}}N-\overset{H}{\underset{}{}}\overset{O}{\underset{}{C}}CH_2CH_2N-\overset{H}{\underset{}{}}\overset{O}{\underset{}{C}}-CHOH-\underset{\underset{CH_3}{|}}{\overset{CH_3}{\underset{}{C}}}CH_2O\overset{O}{\underset{O}{P}}O\overset{O}{\underset{O}{P}}OCH_2 ...$$

Acetylcoenzyme A

The important part of this rather complicated structure is the thiol ester at the beginning of the chain; because of this, acetylcoenzyme A is usually abbreviated:

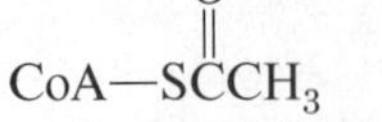

$$\text{CoA}-\overset{\overset{\displaystyle O}{\parallel}}{\text{SC}}\text{CH}_3$$

and coenzyme A, itself, is abbreviated:

CoA—SH

In certain biochemical reactions, an *acyl*coenzyme A operates as an *acylating agent;* it transfers an acyl group to another nucleophile in a reaction that involves a nucleophilic attack at the acyl carbon of the thiol ester. An example is the reaction shown below:

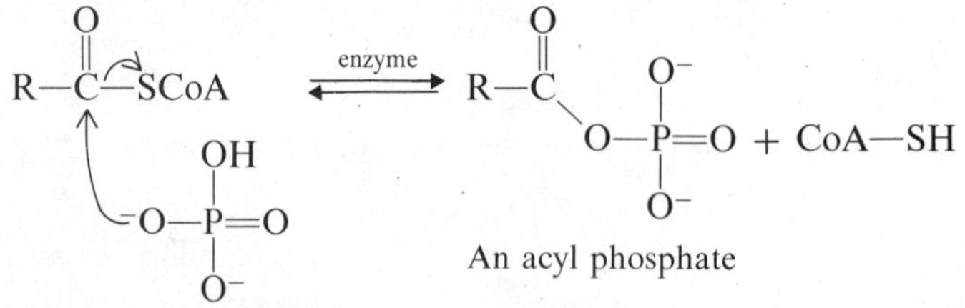

An acyl phosphate

This reaction is catalyzed by an enzyme called *phosphotransacetylase.*

The α-hydrogens of the acetyl group of acetylcoenzyme A are appreciably acidic. Acetylcoenzyme A, as a result, also functions as an *alkylating agent.* Acetyl-coenzyme A, for example, reacts with oxaloacetate ion to form citrate ion in a reaction that resembles an aldol addition.

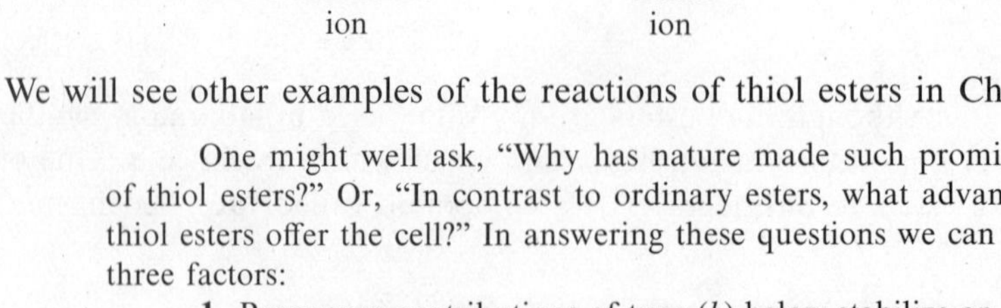

Oxaloacetate
ion

Citrate
ion

We will see other examples of the reactions of thiol esters in Chapter 23.

One might well ask, "Why has nature made such prominent use of thiol esters?" Or, "In contrast to ordinary esters, what advantages do thiol esters offer the cell?" In answering these questions we can consider three factors:

1. Resonance contributions of type (*b*) below stabilize an ordinary ester and make the carbonyl group less suceptible to nucleophilic attack.

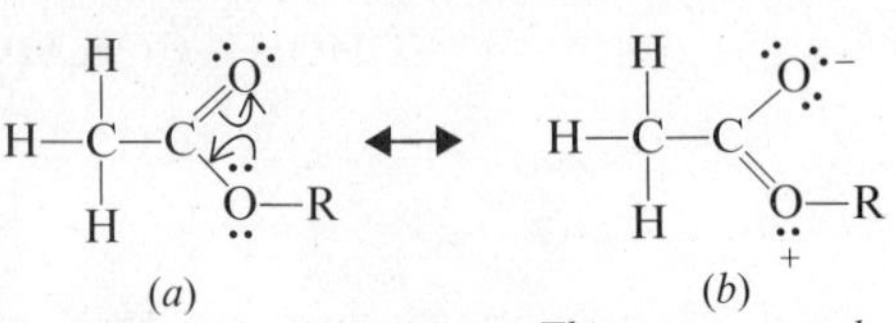

(*a*)   (*b*)

*This structure makes
an important contribution*

By contrast, thiol esters are not as effectively stabilized by a similar resonance contribution because structure (d) below requires overlap between the $3p$ orbital of sulfur and a $2p$ orbital of carbon. Since this overlap is not large, resonance stabilization by (d) is not as effective. Structure (e) does, however, make an important contribution; one that makes the carbonyl group more susceptible to nucleophilic attack.

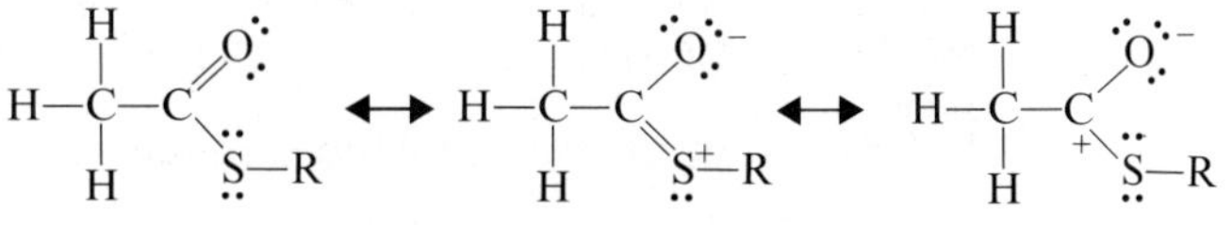

(c)

(d)
This structure is not an important contributor

(e)
This structure makes the carbonyl carbon susceptible to nucleophilic attack

2. A resonance contribution from the similar structure (g) below makes the α-hydrogens of thiol esters more acidic than those of ordinary esters.

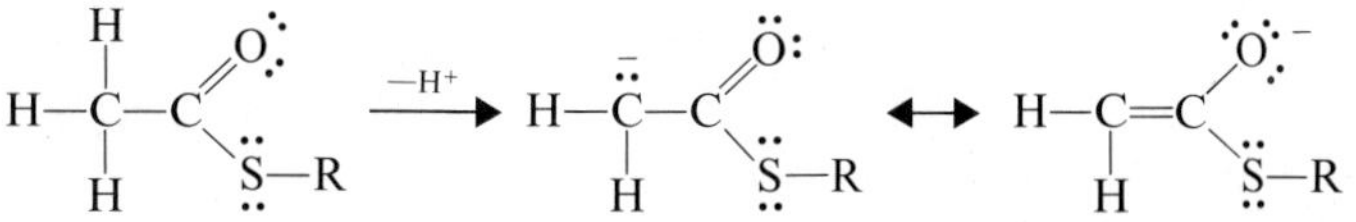

(f)

(g)
This structure effectively stabilizes the anion of a thiol ester

3. The carbon-sulfur bond of a thiol ester is weaker than the carbon-oxygen bond of an ordinary ester and $^-\!:\!\overset{..}{\underset{..}{S}}\!-\!R$ is a better leaving group than $^-\!:\!\overset{..}{\underset{..}{O}}R$.

Factors 1 and 3 make thiol esters effective *acylating agents;* factor 2 makes them effective *alkylating* agents. Thus, we should not be surprised when we encounter reactions like the one shown below:

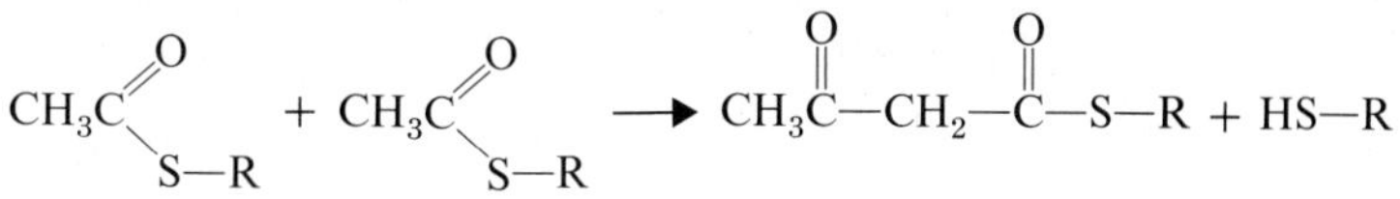

In this reaction one mole of a thiol ester acts as an acylating agent and the other acts as an alkylating agent (cf., p. 906).

19.11 SUMMARY OF THE REACTIONS OF CARBOXYLIC ACIDS AND THEIR DERIVATIVES

The reactions of carboxylic acids and their derivatives are summarized below.

Reactions of Carboxylic Acids

1. As acids (discussed in Section 19.2).

$$RCOOH + NaOH \longrightarrow RCOO^-Na^+ + H_2O$$
$$RCOOH + NaHCO_3 \longrightarrow RCOO^-Na^+ + H_2O + CO_2$$

2. Reduction (discussed in Section 16.4).

$$RCOOH + LiAlH_4 \xrightarrow[\text{(2) } H_2O]{} RCH_2OH$$

3. Conversion to acid chlorides (discussed in Section 19.5).

$$RCOOH + SOCl_2 \longrightarrow RCOCl + SO_2 + HCl$$
$$3RCOOH + PCl_3 \longrightarrow 3RCOCl + H_3PO_3$$
$$RCOOH + PCl_5 \longrightarrow RCOCl + POCl_3 + HCl$$

4. Conversion to acid anhydrides (discussed in Section 19.6).

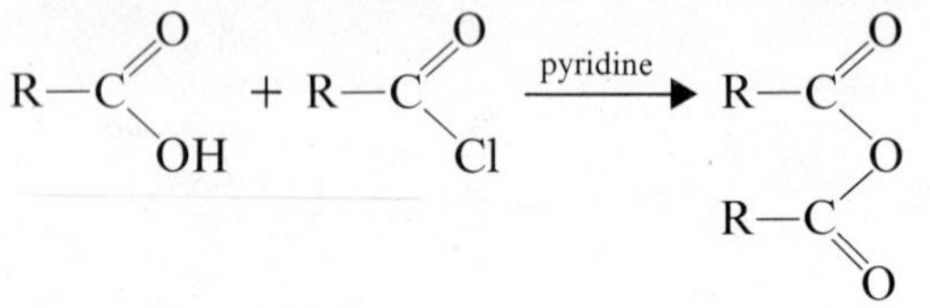

or

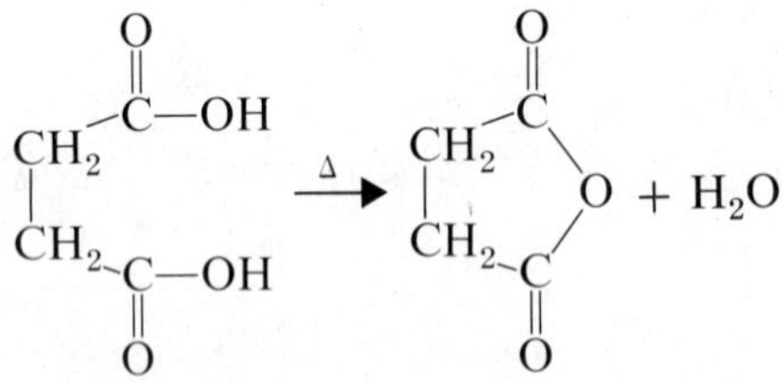

5. Conversion to esters (discussed in Section 19.7).

$$R{-}\overset{O}{\overset{\|}{C}}OH + R'{-}OH \underset{}{\overset{H^+}{\rightleftharpoons}} R{-}\overset{O}{\overset{\|}{C}}{-}OR' + H_2O$$

or

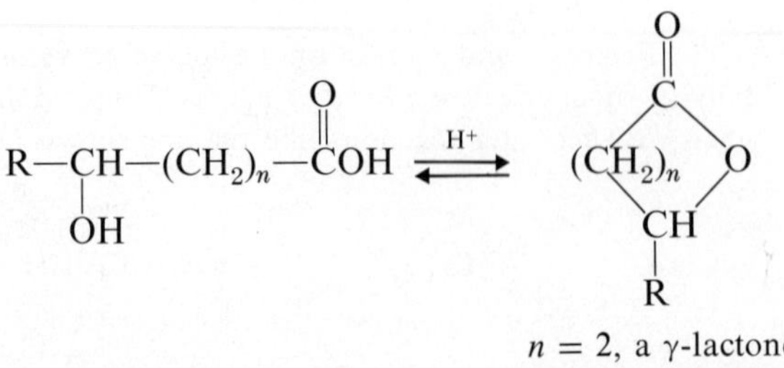

$n = 2$, a γ-lactone
$n = 3$, a δ-lactone

6. Conversion to amides and imides (discussed in Section 19.8).

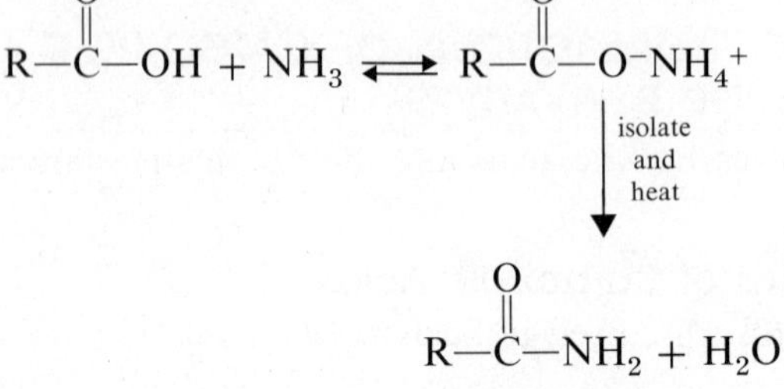

$$R{-}\overset{O}{\overset{\|}{C}}{-}OH + NH_3 \rightleftharpoons R{-}\overset{O}{\overset{\|}{C}}{-}O^-NH_4^+$$

isolate
and
heat

$$R{-}\overset{O}{\overset{\|}{C}}{-}NH_2 + H_2O$$

An amide

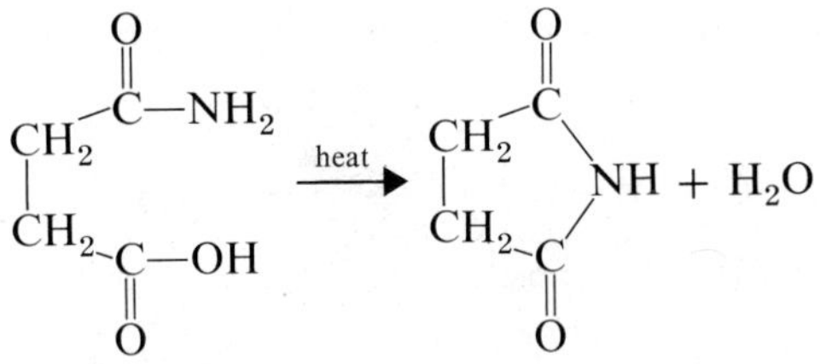

$$\underset{\underset{\substack{CH_2 \\ | \\ CH_2}}{}}{\overset{\overset{O}{\parallel}}{C}-NH_2} \quad \xrightarrow{\text{heat}} \quad \underset{\underset{CH_2}{\overset{CH_2}{}}}{\overset{\overset{O}{\parallel}}{C}} \underset{\overset{\parallel}{O}}{C} NH + H_2O$$

An imide

7. Conversion into lactams (discussed in Section 19.8).

$$R-\underset{\underset{NH_2}{|}}{CH}(CH_2)_n\overset{\overset{O}{\parallel}}{C}OH \quad \xrightarrow{\text{heat}} \quad \underset{\underset{(CH_2)_n}{}}{RCH}\underset{}{\overset{\overset{O}{\parallel}}{C}}NH + H_2O$$

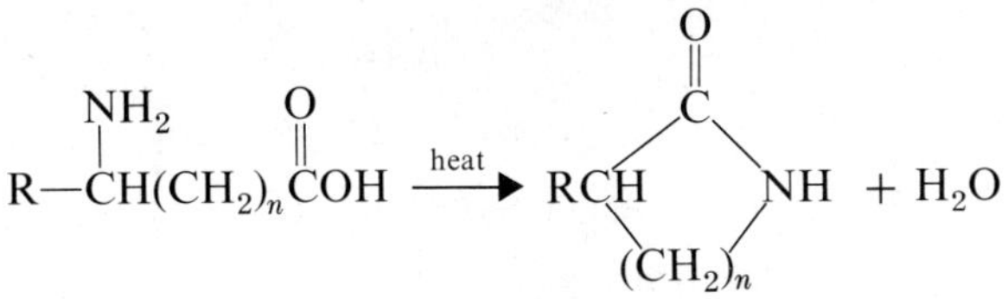

$n = 2$, a γ-lactam
$n = 3$, a δ-lactam

8. Alpha halogenation (discussed in Section 19.9).

$$R-CH_2COOH + X_2 \xrightarrow{P} R-\underset{\underset{X}{|}}{CH}COOH$$

$X_2 = Cl_2$ or Br_2

Reactions of Acyl Chlorides

1. Conversion into acids (discussed in Section 19.4).

$$R-\overset{\overset{O}{\parallel}}{C}-Cl + H_2O \longrightarrow R-\overset{\overset{O}{\parallel}}{C}-OH + HCl$$

2. Conversion into anhydrides (discussed in Section 19.6).

$$R-\overset{\overset{O}{\parallel}}{C}-Cl + R-\overset{\overset{O}{\parallel}}{C}-O^- \longrightarrow R-\overset{\overset{O}{\parallel}}{C}-O-\overset{\overset{O}{\parallel}}{C}-R + Cl^-$$

3. Conversion into esters (discussed in Section 19.7).

$$R-\overset{\overset{O}{\parallel}}{C}-Cl + R'-OH \xrightarrow{\text{pyridine}} R-\overset{\overset{O}{\parallel}}{C}-OR'$$

4. Conversion into amides (discussed in Section 19.8).

$$R-\overset{\overset{O}{\parallel}}{C}-Cl + NH_{3(excess)} \longrightarrow R-\overset{\overset{O}{\parallel}}{C}-NH_2 + NH_4Cl$$

$$R-\overset{\overset{O}{\parallel}}{C}-Cl + NH_2R_{(excess)} \longrightarrow R-\overset{\overset{O}{\parallel}}{C}-NHR + RNH_3Cl$$

$$R-\overset{\overset{O}{\parallel}}{C}-Cl + NHR_{2(excess)} \longrightarrow R-\overset{\overset{O}{\parallel}}{C}-NR_2 + R_2NH_2Cl$$

5. Conversion into thiol esters (discussed in Section 19.10).

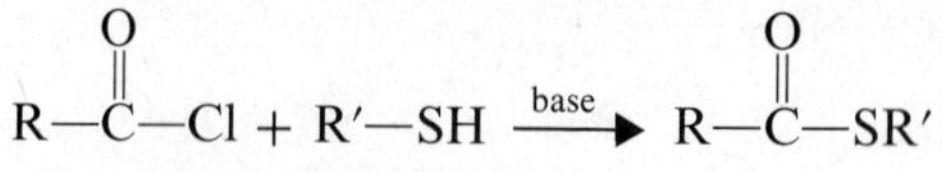

$$\underset{\underset{}{}}{R-\overset{\overset{O}{\parallel}}{C}-Cl} + R'-SH \xrightarrow{base} R-\overset{\overset{O}{\parallel}}{C}-SR'$$

6. Conversion into ketones.

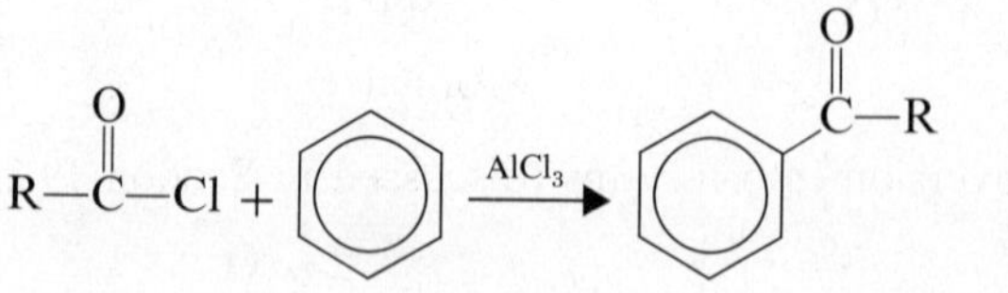

$$R-\overset{\overset{O}{\parallel}}{C}-Cl + \bigcirc \xrightarrow{AlCl_3} \bigcirc\overset{\overset{O}{\parallel}}{C}-R \qquad \text{(discussed in Section 13.7)}$$

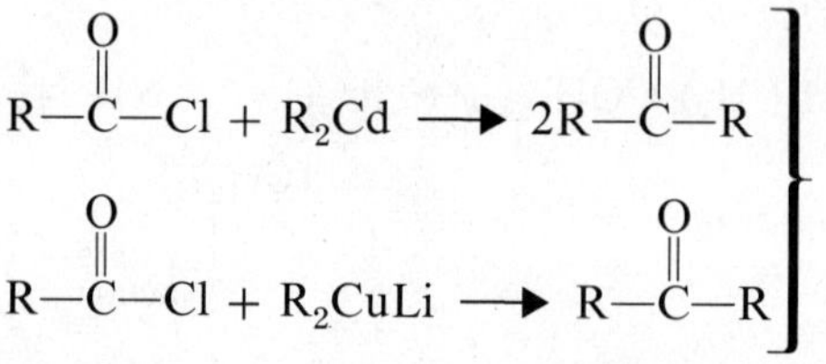

$$\left.\begin{array}{l} R-\overset{\overset{O}{\parallel}}{C}-Cl + R_2Cd \longrightarrow 2R-\overset{\overset{O}{\parallel}}{C}-R \\[2em] R-\overset{\overset{O}{\parallel}}{C}-Cl + R_2CuLi \longrightarrow R-\overset{\overset{O}{\parallel}}{C}-R \end{array}\right\} \text{(discussed in Section 18.5)}$$

7. Conversion into aldehydes (discussed in Section 18.4).

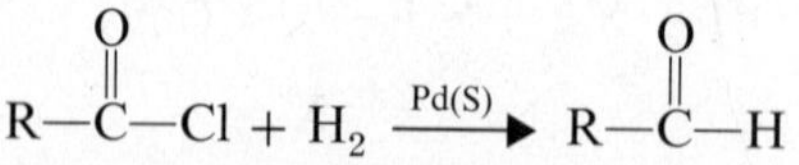

$$R-\overset{\overset{O}{\parallel}}{C}-Cl + H_2 \xrightarrow{Pd(S)} R-\overset{\overset{O}{\parallel}}{C}-H$$

8. Reaction with carbanions (discussed in Section 20.4).

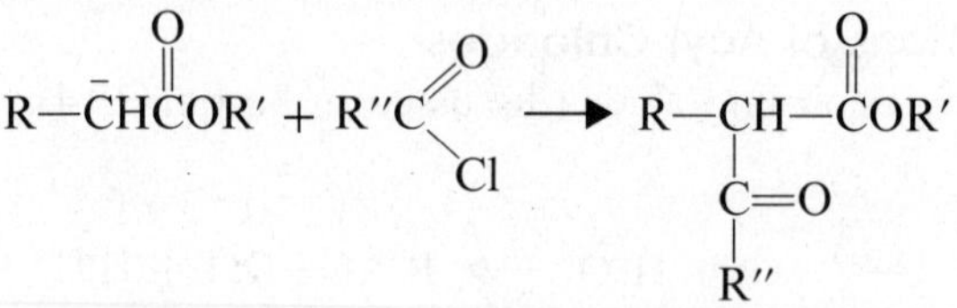

$$R-\overset{\overset{\bar{}}{}}{\underset{}{C}}H\overset{\overset{O}{\parallel}}{C}OR' + R''\overset{\overset{O}{\diagup}}{\underset{\diagdown Cl}{C}} \longrightarrow R-\underset{\underset{\underset{R''}{|}}{\underset{C=O}{|}}}{CH}-\overset{\overset{O}{\parallel}}{C}OR'$$

Reactions of Acid Anhydrides

1. Conversion into acids.

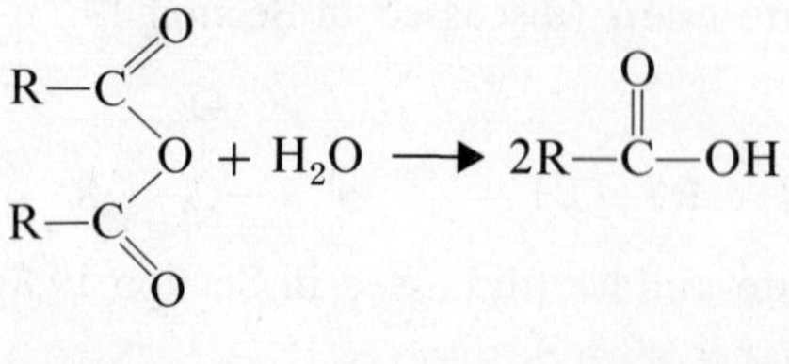

$$\underset{R-C\diagdown_{O}^{\diagup O}}{\overset{R-C\diagup^{\diagdown O}}{}} + H_2O \longrightarrow 2R-\overset{\overset{O}{\parallel}}{C}-OH$$

2. Conversion into esters (discussed in Section 19.7).

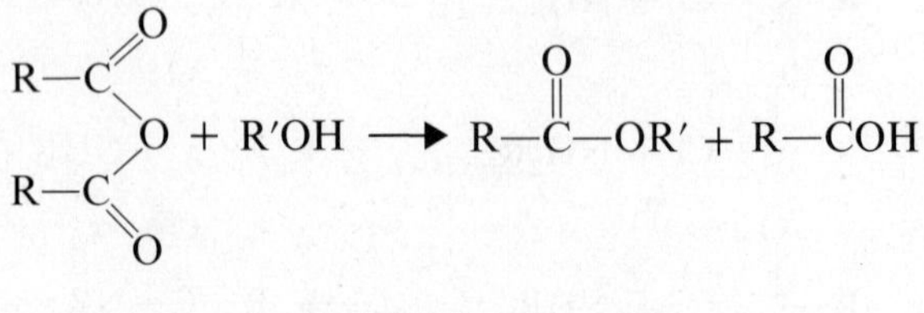

$$\underset{R-C\diagdown_{O}^{\diagup O}}{\overset{R-C\diagup^{\diagdown O}}{}} + R'OH \longrightarrow R-\overset{\overset{O}{\parallel}}{C}-OR' + R-\overset{\overset{O}{\parallel}}{C}OH$$

or

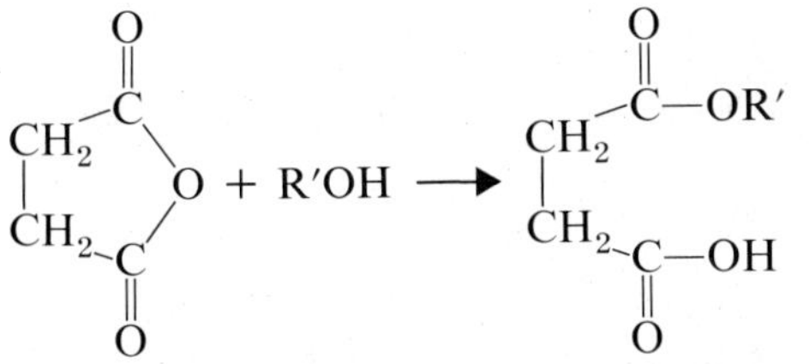

$$
\underset{\substack{| \\ CH_2 \diagdown C \diagup O}}{\overset{\substack{O \\ \| \\ C \\ CH_2}}{}} + R'OH \longrightarrow \underset{\substack{| \\ CH_2 {-} C {-} OH \\ \| \\ O}}{\overset{\substack{O \\ \| \\ C {-} OR' \\ CH_2}}{}}
$$

3. Conversion into amides and imides (discussed in Section 19.8).

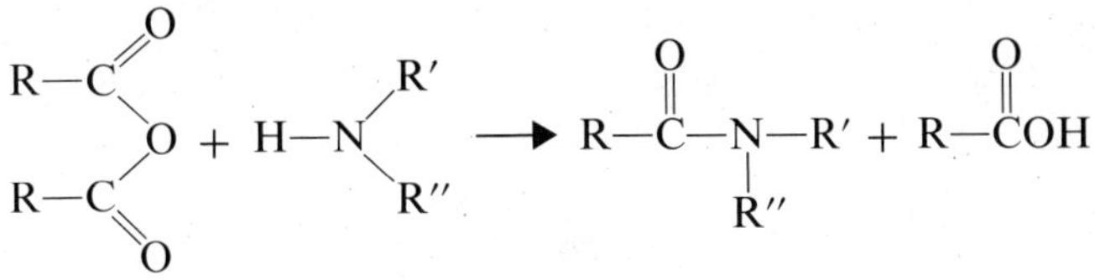

$$
\underset{\substack{R - C \\ \| \\ O}}{\overset{\substack{O \\ \| \\ R - C}}{}} O + H - N \underset{R''}{\overset{R'}{}} \longrightarrow \underset{\substack{| \\ R''}}{\overset{\substack{O \\ \| \\ R - C - N - R'}}{}} + R - \overset{O}{\overset{\|}{C}}OH
$$

R' and/or R'' may be H

or

$$
\underset{\substack{| \\ CH_2 \diagdown C \diagup O}}{\overset{\substack{O \\ \| \\ C \\ CH_2}}{}} + R' - NH_2 \longrightarrow \underset{\substack{| \\ CH_2 - C - OH \\ \| \\ O}}{\overset{\substack{O \\ \| \\ C - NHR' \\ CH_2}}{}} \xrightarrow{heat} \underset{\substack{| \\ CH_2 \diagdown C \\ \| \\ O}}{\overset{\substack{O \\ \| \\ C \\ CH_2}}{}} N - R'
$$

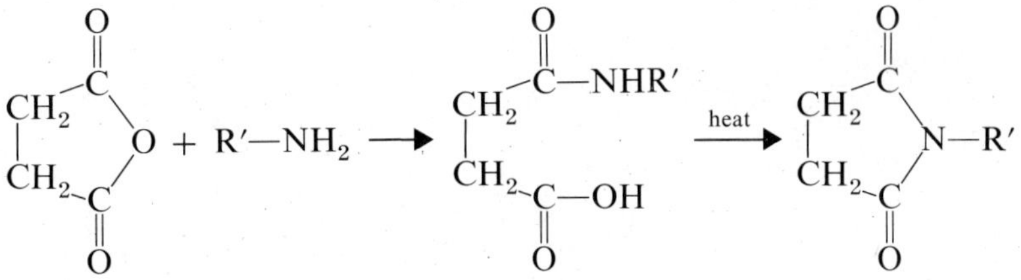

R' may be H

4. Conversion into ketones (discussed in Section 13.7).

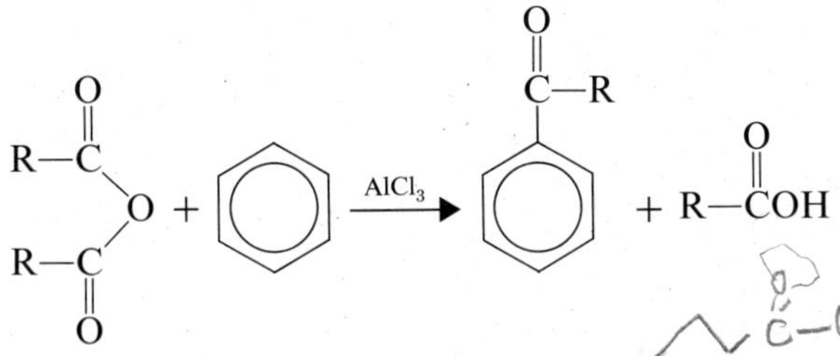

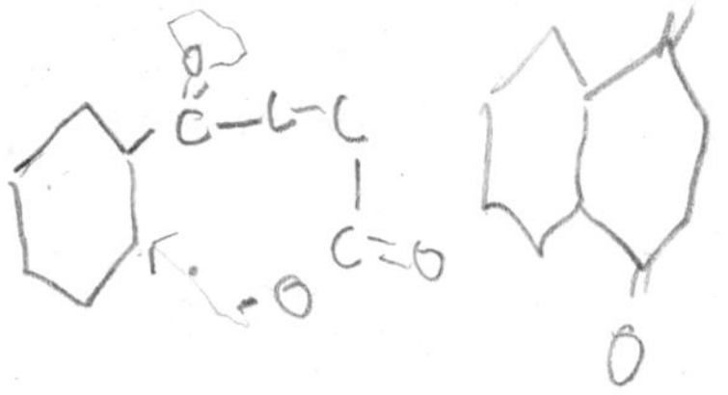

$$
\underset{\substack{R - C \\ \| \\ O}}{\overset{\substack{O \\ \| \\ R - C}}{}} O + \text{(benzene ring)} \xrightarrow{AlCl_3} \text{(benzene ring)} \overset{\substack{O \\ \| \\ C - R}}{} + R - \overset{O}{\overset{\|}{C}}OH
$$

Reactions of Esters

1. Hydrolysis (discussed in Section 19.7).

$$
R - \overset{O}{\overset{\|}{C}} - O - R' + H_2O \xrightarrow{H^+} R - \overset{O}{\overset{\|}{C}} - OH + R' - OH
$$

$$
R - \overset{O}{\overset{\|}{C}} - O - R' + OH^- \xrightarrow[H_2O]{} R - \overset{O}{\overset{\|}{C}} - O^- + R' - OH
$$

2. Conversion into other esters: transesterification (discussed in Section 19.7).

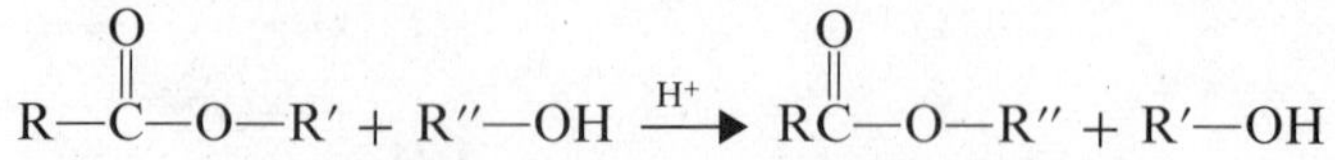

$$\underset{\substack{\| \\ O}}{R-C-O-R'} + R''-OH \xrightarrow{H^+} \underset{\substack{\| \\ O}}{RC-O-R''} + R'-OH$$

3. Conversion into amides (discussed in Section 19.8).

$$\underset{\substack{\| \\ O}}{R-C-OR'} + HN\diagdown \begin{matrix} R'' \\ R''' \end{matrix} \longrightarrow \underset{\substack{\| \\ O}}{R-CN}\diagdown \begin{matrix} R'' \\ R''' \end{matrix} + R'-OH$$

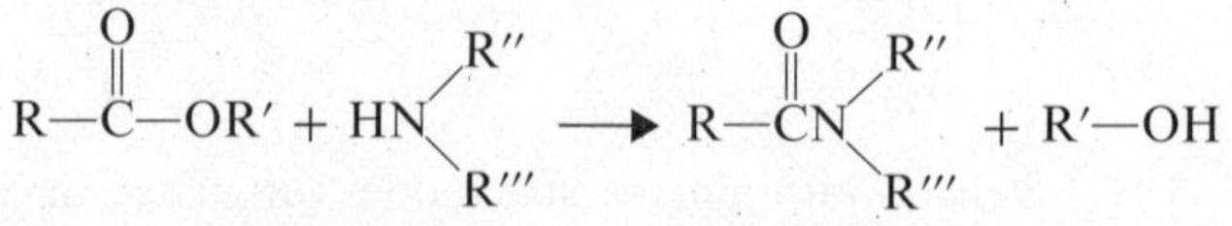

R''— and/or R'''— may be H—

4. Reaction with Grignard reagents (discussed in Section 16.5).

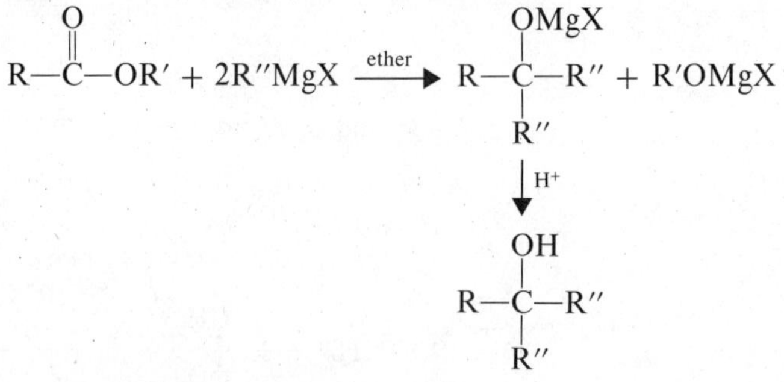

$$\underset{\substack{\| \\ O}}{R-C-OR'} + 2R''MgX \xrightarrow{ether} \underset{\substack{| \\ R''}}{R-\overset{OMgX}{\underset{|}{C}}-R''} + R'OMgX$$

$$\downarrow H^+$$

$$\underset{\substack{| \\ R''}}{R-\overset{OH}{\underset{|}{C}}-R''}$$

5. Reduction (discussed in Section 16.4).

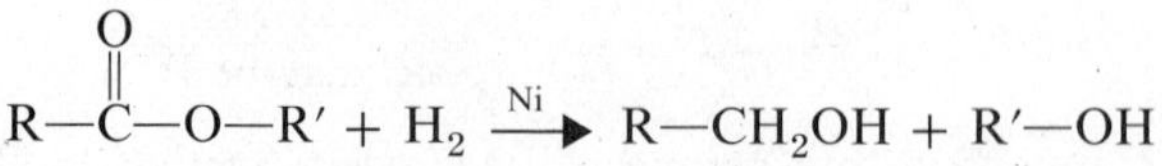

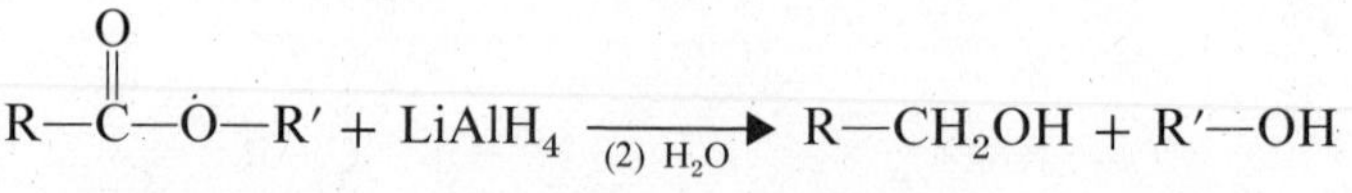

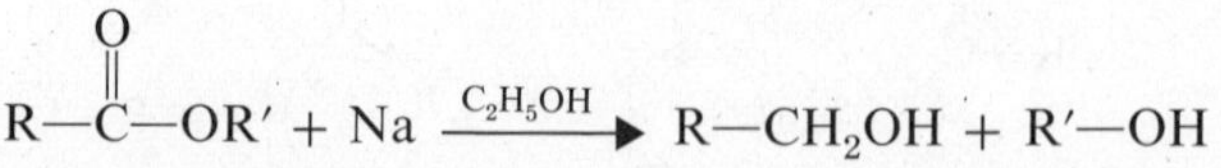

$$\underset{\substack{\| \\ O}}{R-C-O-R'} + H_2 \xrightarrow{Ni} R-CH_2OH + R'-OH$$

$$\underset{\substack{\| \\ O}}{R-C-O-R'} + LiAlH_4 \xrightarrow[(2)\ H_2O]{} R-CH_2OH + R'-OH$$

$$\underset{\substack{\| \\ O}}{R-C-OR'} + Na \xrightarrow{C_2H_5OH} R-CH_2OH + R'-OH$$

6. Reaction with carbanions: Claisen condensation (discussed in Section 20.2).

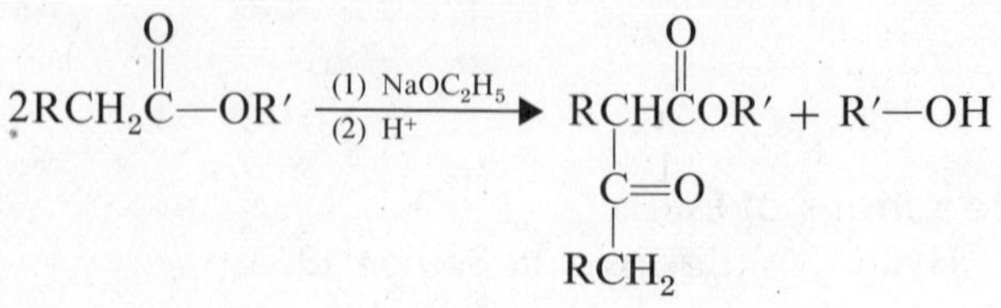

$$2R CH_2 \underset{\substack{\| \\ O}}{C-OR'} \xrightarrow[(2)\ H^+]{(1)\ NaOC_2H_5} \underset{\substack{| \\ C=O \\ | \\ RCH_2}}{R CH \underset{\substack{\| \\ O}}{COR'}} + R'-OH$$

Reactions of Amides

1. Hydrolysis (discussed in Section 19.8).

$$\underset{\substack{| \\ R''}}{R-\underset{\substack{\| \\ O}}{C}-NR'} + H_3O^+ \xrightarrow{H_2O} R-\underset{\substack{\| \\ O}}{C}-OH + \underset{\substack{| \\ R''}}{R'-NH_2^+}$$

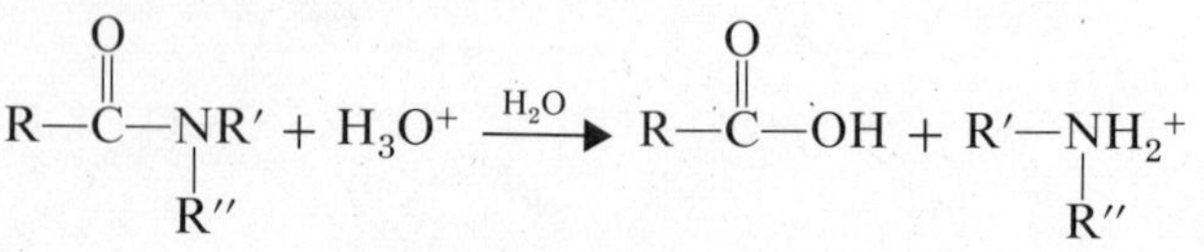

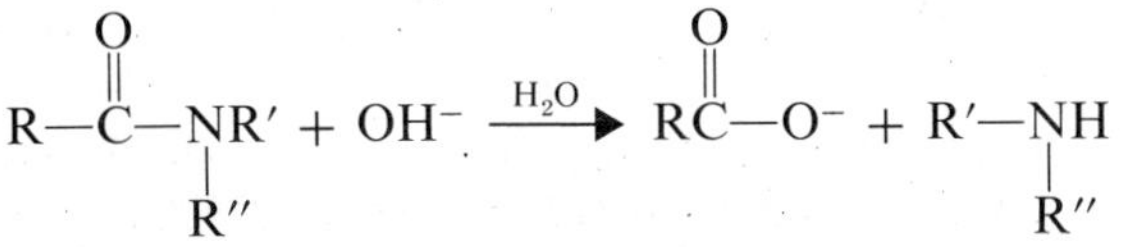

$$R-\overset{\overset{\displaystyle O}{\|}}{C}-\underset{\underset{\displaystyle R''}{|}}{N}R' + OH^- \xrightarrow{H_2O} R\overset{\overset{\displaystyle O}{\|}}{C}-O^- + R'-\underset{\underset{\displaystyle R''}{|}}{N}H$$

R— and/or R'— may be H—

2. Conversion into nitriles: dehydration (discussed in Section 19.8).

$$R-\overset{\overset{\displaystyle O}{\|}}{C}NH_2 \xrightarrow[\substack{heat \\ (-H_2O)}]{P_2O_5} R-C\equiv N$$

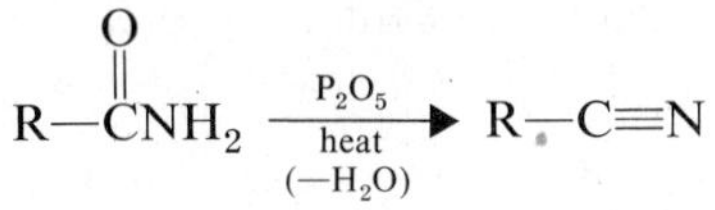

3. Conversion into imides (discussed in Section 19.8).

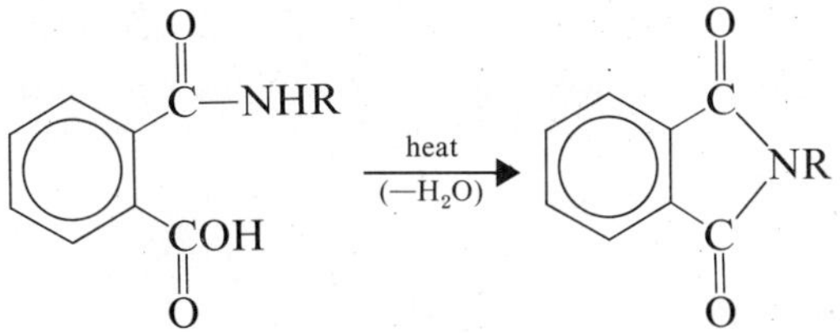

4. Conversion into amines (discussed in Section 21.4).

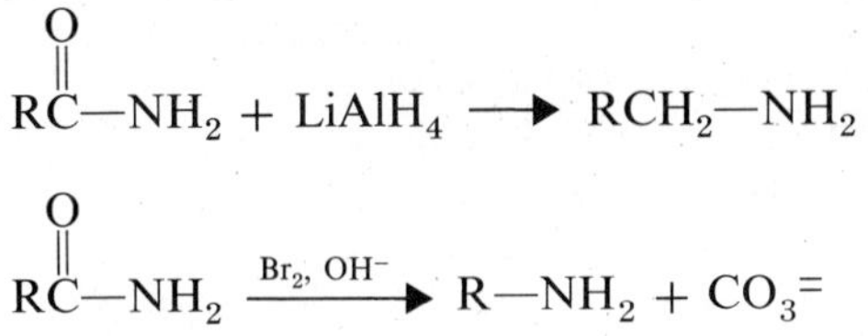

$$R\overset{\overset{\displaystyle O}{\|}}{C}-NH_2 + LiAlH_4 \longrightarrow RCH_2-NH_2$$

$$R\overset{\overset{\displaystyle O}{\|}}{C}-NH_2 \xrightarrow{Br_2, OH^-} R-NH_2 + CO_3^=$$

19.12 CHEMICAL TESTS FOR ACYL COMPOUNDS

Carboxylic acids are weak acids and their acidity helps us to recognize them. Aqueous solutions of water-soluble carboxylic acids give an acid test with blue litmus paper; water-insoluble carboxylic acids dissolve in aqueous sodium hydroxide and aqueous sodium bicarbonate. The latter reagent helps us distinguish carboxylic acids from most phenols. Except for the di- and trinitrophenols, phenols do not dissolve in aqueous sodium bicarbonate: carboxylic acids not only dissolve in aqueous sodium bicarbonate, they cause the evolution of carbon dioxide.

It is often helpful to determine the equivalent weight of a carboxylic acid by titrating a measured quantity of the acid with a standard solution of sodium hydroxide. For a monocarboxylic acid, the equivalent weight equals the molecular weight; for a dicarboxylic acid, the equivalent weight is one-half the molecular weight, and so on.

All acid derivatives can be hydrolyzed to carboxylic acids. The conditions required to bring about hydrolysis vary greatly, with acyl chlorides being the easiest to hydrolyze and amides being the most difficult.

Acyl chlorides hydrolyze in water and thus give a precipitate when treated with aqueous silver nitrate. Acid anhydrides dissolve when heated briefly with aqueous sodium hydroxide.

Esters and amides hydrolyze slowly when they are refluxed with sodium hydroxide: esters produce a carboxylate ion and an alcohol; amides produce a carboxylate ion and an amine or ammonia. The hydrolysis products, the acid and the alcohol or amine can be isolated and identified. Since base-promoted hydrolysis of an unsubstituted amide produces ammonia, this can often be detected by holding moist red litmus in the vapors above the reaction mixture.

Base-promoted hydrolysis of an ester consumes one mole of hydroxide ion for each mole of the ester. It is often convenient, therefore, to carry out the hydrolysis quantitatively.

$$\underset{\text{1 mole}}{\text{RCOOR}'} + \underset{\text{1 mole}}{\text{OH}^-} \longrightarrow \text{RCOO}^- + \text{R}'\text{OH}$$

This allows us to determine the *equivalent weight* of the ester. We can do this by hydrolyzing a known weight of the ester with an excess of a standard solution of sodium hydroxide. After the hydrolysis is complete, we can titrate the excess sodium hydroxide with a standard acid and determine the number of moles of hydroxide ion that were consumed in the hydrolysis reaction.

Additional Problems

19.14

Write structural formulas for each of the following compounds.

(a) Hexanoic acid
(b) Hexanamide
(c) N-Ethylhexanamide
(d) N,N-Diethylhexanamide
(e) 3-Hexenoic acid
(f) 2-Methyl-4-hexenoic acid
(g) Hexanedioic acid
(h) Phthalic acid
(i) Isophthalic acid
(j) Terephthalic acid
(k) Diethyl oxalate
(l) Diethyl adipate
(m) Isobutyl propanoate
(n) 2-Naphthoic acid
(o) Maleic acid
(p) Malic acid
(q) Fumaric acid
(r) Succinic acid
(s) Succinimide
(t) Malonic acid
(u) Diethyl malonate

19.15

Give IUPAC or common names for each of the following compounds.

(a) C_6H_5COOH
(b) C_6H_5COCl
(c) $C_6H_5CONH_2$
(d) $(C_6H_5CO)_2O$
(e) $C_6H_5COOCH_2C_6H_5$
(f) $C_6H_5COOC_6H_5$
(g) $CH_3COOCH(CH_3)_2$
(h) $CH_3CON(CH_3)_2$
(i) CH_3CN

(j)

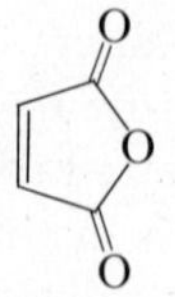

(k)

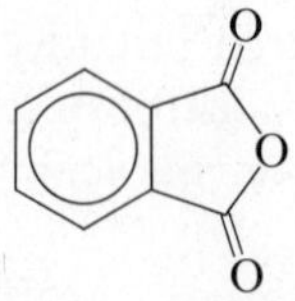

(l)

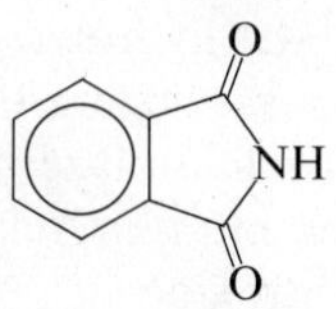

(m) 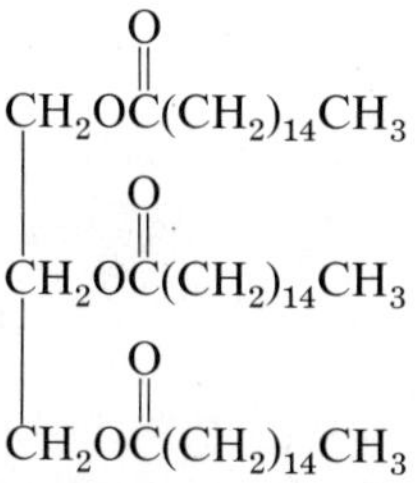 $CH_2OC(CH_2)_{14}CH_3$

$CH_2OC(CH_2)_{14}CH_3$

$CH_2OC(CH_2)_{14}CH_3$

(n) $HOOCCH_2CCOOH$

(o)

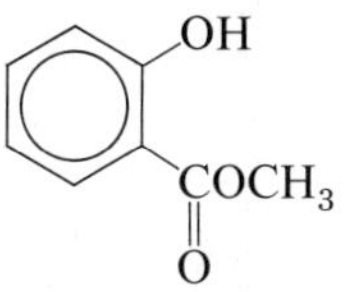

19.16
Show how benzoic acid can be synthesized from each of the following compounds.
(a) Bromobenzene   (e) Benzaldehyde
(b) Toluene   (f) Stryene
(c) Benzonitrile, C_6H_5CN (g) Benzyl alcohol
(d) Acetophenone

19.17
Show how phenylacetic acid can be prepared from
(a) Phenylacetaldehyde
(b) Benzyl bromide (two ways)

19.18
Show how pentanoic acid can be prepared from
(a) 1-Pentanol
(b) 1-Bromobutane (two ways)
(c) 2-Hexanone
(d) 5-Decene
(e) 1-Pentanal

19.19
What products would you expect to obtain when acetyl chloride reacts with each of the following?
(a) H_2O
(b) $AgNO_3/H_2O$
(c) $CH_3(CH_2)_2CH_2OH$ and pyridine
(d) $NH_{3(excess)}$
(e) $C_6H_5CH_3$ and $AlCl_3$
(f) H_2 and Pd(S)
(g) $(CH_3)_2CuLi$
(h) $(CH_3CH_2)_2Cd$

(i) $CH_3NH_{2(excess)}$
(j) $C_6H_5NH_{2(excess)}$
(k) $(CH_3)_2NH_{(excess)}$
(l) CH_3CH_2SH + pyridine
(m) $CH_3COO^-Na^+$
(n) CH_3COOH and pyridine
(o) Phenol and pyridine

19.20
What products would you expect to obtain when acetic anhydride reacts with each of the following?
(a) $NH_{3(excess)}$
(b) H_2O
(c) $CH_3CH_2CH_2OH$
(d) C_6H_6 + $AlCl_3$
(e) $CH_3CH_2NH_{2(excess)}$
(f) $(CH_3CH_2)_2NH_{(excess)}$

19.21
What products would you expect to obtain when succinic anhydride reacts with each of the reagents given in problem 19.20?

19.22
Show how you might carry out the following transformations.

(a) Succinic anhydride ⟶

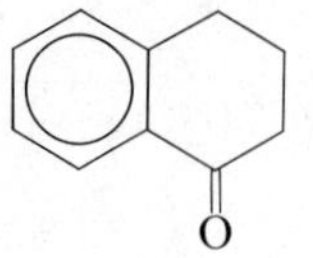

(b)

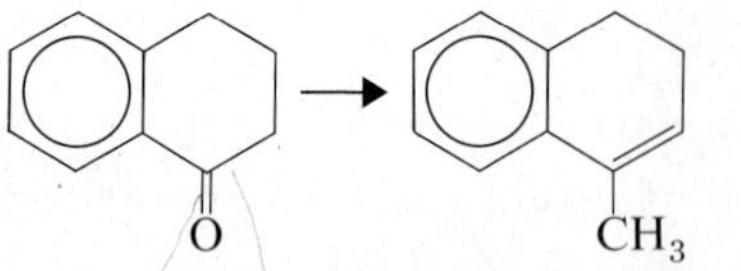

(c)

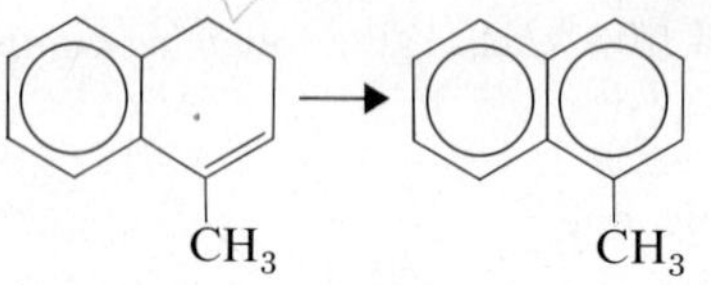

(d) Phthalic anhydride ⟶

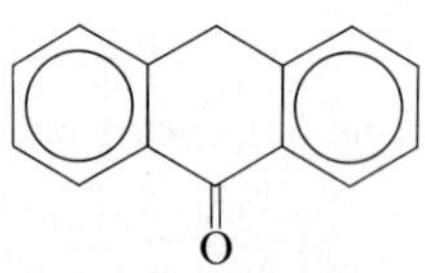

(e) Phthalic anhydride ⟶ *N*-methylphthalimide

(f) Maleic anhydride ⟶

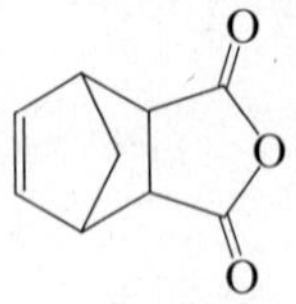

(g) Maleic anhydride ⟶

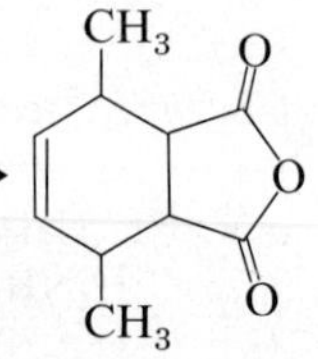

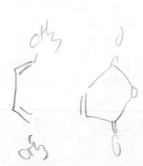

19.23

What products would you expect to obtain when ethyl propanoate reacts with each of the following?

(a) H_3O^+, H_2O (d) CH_3NH_2
(b) OH^-, H_2O (e) $LiAlH_4$, then H_2O
(c) 1-octanol, HCl   (f) C_6H_5MgBr, then H_2O

19.24

What products would you expect to obtain when propanamide reacts with each of the following?

(a) H_3O^+, H_2O
(b) OH^-, H_2O
(c) P_2O_5 and heat

19.25

Outline a simple chemical test that would serve to distinguish between

(a) Benzoic acid and methyl benzoate
(b) Benzoic acid and benzoyl chloride

(c) Benzoic acid and benzamide
(d) Benzoic acid and *p*-cresol
(e) Ethyl benzoate and benzamide
(f) Benzoic acid and cinnamic acid
(g) Ethyl benzoate and benzoyl chloride
(h) α-Chlorobutanoic acid and butanoic acid

19.26
What products would you expect to obtain when each of the following compounds is heated?
(a) 4-Hydroxybutanoic acid
(b) 3-Hydroxybutanoic acid
(c) 2-Hydroxybutanoic acid
(d) Glutaric acid

(e) $CH_3CHCH_2CH_2CH_2\overset{\overset{\displaystyle O}{\|}}{C}OH$
 $\quad\quad\;\;|$
 $\quad\quad\; NH_2$

(f)
$$\text{C}_6\text{H}_4\left(\overset{\overset{\displaystyle O}{\|}}{\text{C}}-\text{OH}\right)(\text{CH}_2\text{NH}_2)$$

19.27
Give stereochemical formulas for compounds **A–Q.**

(a) R-(−)-2-Butanol $\xrightarrow[\text{pyridine}]{\text{TsCl}}$ **A** $\xrightarrow{\text{CN}^-}$ **B**(C_5H_9N) $\xrightarrow[\text{H}_2\text{O}]{\text{H}_2\text{SO}_4}$

$\quad\quad\quad\quad\quad\quad\quad\quad\quad\quad\quad$ (+)-**C**$(C_5H_{10}O_2)$ $\xrightarrow{\text{LiAlH}_4}$ (−)-**D**$(C_5H_{12}O)$

(b) R-(−)-2-Butanol $\xrightarrow[\text{pyridine}]{\text{PBr}_3}$ **E**(C_4H_9Br) $\xrightarrow{\text{CN}^-}$

$\quad\quad\quad\quad\quad$ **F**(C_5H_9N) $\xrightarrow[\text{H}_2\text{O}]{\text{H}_2\text{SO}_4}$ (−)-**C**$(C_5H_{10}O_2)$ $\xrightarrow{\text{LiAlH}_4}$ (+)-**D**$(C_5H_{12}O)$

(c) **A** $\xrightarrow{\text{CH}_3\text{COO}^-}$ **G**$(C_6H_{12}O_2)$ $\xrightarrow{\text{OH}^-}$ (+)-**H**$(C_4H_{10}O)$ + CH_3COO^-

(d) (−)-**D** $\xrightarrow{\text{PBr}_3}$ **J**$(C_5H_{11}Br)$ $\xrightarrow[\text{ether}]{\text{Mg}}$ **K**$(C_5H_{11}MgBr)$ $\xrightarrow[\text{(2) H}^+]{\text{(1) CO}_2}$ **L**$(C_6H_{12}O_2)$

(e) R-(+)-Glyceraldehyde $\xrightarrow{\text{HCN}}$ $\underbrace{\text{**M**}(C_4H_7NO_3) + \text{**N**}(C_4H_7NO_3)}$

$\quad\quad\quad\quad\quad\quad\quad\quad\quad\quad$ Diastereomers, separated
$\quad\quad\quad\quad\quad\quad\quad\quad\quad\quad$ by fractional crystallization

(f) **M** $\xrightarrow[\text{H}_2\text{O}]{\text{H}_2\text{SO}_4}$ **P**$(C_4H_8O_5)$ $\xrightarrow[\text{HNO}_3]{\text{(O)}}$ *meso*-tartaric acid

(g) **N** $\xrightarrow[\text{H}_2\text{O}]{\text{H}_2\text{SO}_4}$ **Q**$(C_4H_8O_5)$ $\xrightarrow[\text{HNO}_3]{\text{(O)}}$ (−)-tartaric acid

19.28
(a) (±)-Pantetheine and (±)-pantothenic acid, important intermediates in the synthesis of coenzyme **A**, were prepared by the following route. Give structures for compounds **A–D.**

$$\underset{\underset{\displaystyle CH_3}{|}}{CH_3CHCHO} + H\overset{\overset{\displaystyle O}{\|}}{C}H \xrightarrow[\text{H}_2\text{O}]{\text{K}_2\text{CO}_3} (\pm)\text{-}\textbf{A}(C_5H_{10}O_2) \xrightarrow[\text{(2) KCN}]{\text{(1) NaHSO}_3}$$

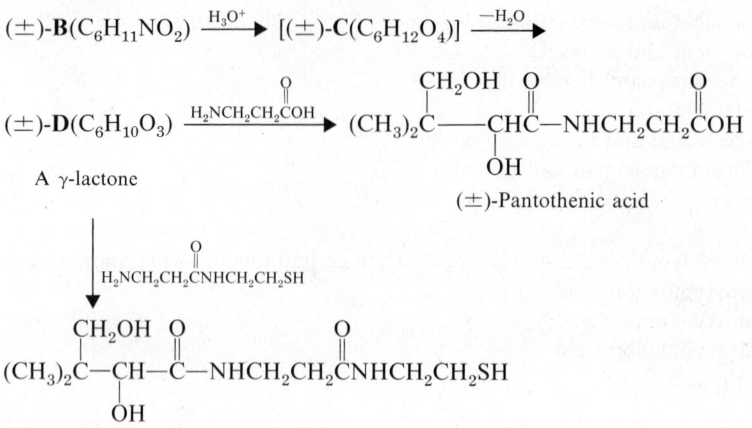

$$(\pm)\text{-}\mathbf{B}(C_6H_{11}NO_2) \xrightarrow{H_3O^+} [(\pm)\text{-}\mathbf{C}(C_6H_{12}O_4)] \xrightarrow{-H_2O}$$

$$(\pm)\text{-}\mathbf{D}(C_6H_{10}O_3) \xrightarrow{H_2NCH_2CH_2COH} $$

A γ-lactone

(±)-Pantothenic acid

(±)-Pantetheine

(b) The γ-lactone, (±) **D** can be resolved. If the (−)-γ-lactone is used in the last step, the pantotheine that is obtained is identical with that obtained naturally. The (−)-γ-lactone has the R configuration. What is the stereochemistry of naturally occurring pantotheine? (c) What products would you expect to obtain when (±)-pantotheine is refluxed with aqueous sodium hydroxide?

19.29

The infrared and pmr spectra of phenacetin ($C_{10}H_{13}NO_2$) are given in Fig. 19.1. Phenacetin is an analgesic and anti-pyretic compound, and is the P of A-P-C tablets (**A**spirin-**P**henacetin-**C**affeine). When phenacetin is refluxed with aqueous sodium hydroxide it yields phenetidine ($C_8H_{11}NO$) and sodium acetate. Propose structures for phenacetin and phenetidine.

19.30

Given below are the pmr spectra and carbonyl absorption peaks of five acyl compounds. Propose structures for each.

	pmr spectrum		IR spectrum
(a) $C_8H_{14}O_4$	Triplet	$\delta1.2$ (6H)	1740 cm^{-1}
	Singlet	$\delta2.5$ (4H)	
	Quartet	$\delta4.1$ (4H)	
(b) $C_{11}H_{14}O_2$	Doublet	$\delta1.0$ (6H)	1720 cm^{-1}
	Multiplet	$\delta2.1$ (1H)	
	Doublet	$\delta4.1$ (2H)	
	Multiplet	$\delta7.8$ (5H)	
(c) $C_{10}H_{12}O_2$	Triplet	$\delta1.2$ (3H)	1740 cm^{-1}
	Singlet	$\delta3.5$ (2H)	
	Quartet	$\delta4.1$ (2H)	
	Multiplet	$\delta7.3$ (5H)	
(d) $C_2H_2Cl_2O_2$	Singlet	$\delta6.0$	Broad peak 2500–2700 cm^{-1}
	Singlet	$\delta11.70$	1705 cm^{-1}
(e) $C_4H_7ClO_2$	Triplet	$\delta1.3$	1745 cm^{-1}
	Singlet	$\delta4.0$	
	Quartet	$\delta4.2$	

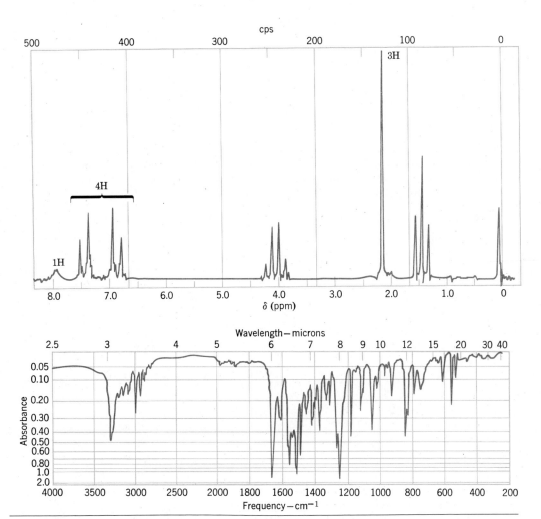

FIG. 19.1

The pmr and infrared spectra of phenacetin. (The pmr spectrum, courtesy Varian Associate, Palo Alto, Calif. Infrared spectrum, courtesy Sadtler Research Laboratories, Philadelphia, Pa.)

19.31

The active ingredient of the insect repellent "Off" is *N,N*-diethyl-*m*-toluamide, *m*-CH$_3$C$_6$H$_4$CON(CH$_2$CH$_3$)$_2$. Outline a synthesis of this compound starting with *meta*-toluic acid.

19.32

Amides are much weaker bases than corresponding amines. For example, most water-insoluble amines, RNH$_2$, will dissolve in dilute aqueous acids (e.g., aqueous HCl, H$_2$SO$_4$, etc.) by forming water-soluble alkylammonium salts, RNH$_3^+$X$^-$. Corresponding amides, RCONH$_2$, *do not dissolve in dilute aqueous acids*, however. Propose an explanation for the much lower basicity of amides when compared to amines.

19.33

While amides are much less basic than amines, they are much stronger acids. Amides have acidity constants (K_a's) in the range $10^{-14} - 10^{-16}$, while for amines, $K_a = 10^{-33} - 10^{-35}$. (a) What factor accounts for the much greater acidity of amides? (b) *Imides,* that is, compounds with the structure $(RC)_2NH$, are even stronger acids than amides. For imides $K_a = 10^{-9} - 10^{-10}$, and, as a consequence, water-insoluble imides dissolve in aqueous NaOH by forming soluble sodium salts. What extra factor accounts for the greater acidity of imides?

* **19.34**

Alkyl thiolacetates, $CH_3CSCH_2CH_2R$, can be prepared by a peroxide-initiated reaction between thiolacetic acid, CH_3CSH, and an alkene, $CH_2{=}CHR$. (a) Outline a reasonable mechanism for this reaction. (b) Show how you might use this reaction in a synthesis of 3-methyl-2-butanethiol from 2-methyl-2-butene.

* **19.35**

On heating, cis-4-hydroxycyclohexanecarboxylic acid forms a lactone but trans-4-hydroxy-cyclohexanecarboxylic acid does not. Explain.

* **19.36**

R-(+)-Glyceraldehyde can be transformed into (+)-malic acid by the following synthetic route. Give stereochemical structures for all of the intermediates.

$$R\text{-}(+)\text{-Glyceraldehyde} \xrightarrow[\text{oxidation}]{Br_2, H_2O} (-)\text{-glyceric acid} \xrightarrow{PBr_3}$$

$$(-)\text{-3-bromo-2-hydroxypropanoic acid} \xrightarrow{NaCN} C_4H_5NO_3 \xrightarrow[\text{heat}]{H_3O^+} (+)\text{-malic acid}$$

* **19.37**

R-(+)-Glyceraldehyde can also be transformed into (−)-malic acid. This synthesis begins with the conversion of R-(+)-glyceraldehyde into (−)-tartaric acid as shown in problem 19.27 parts (e) and (g). Then (−)-tartaric acid is allowed to react with phosphorus tribromide in order to replace one alcoholic —OH group with —Br. This step takes place with inversion of configuration at the carbon that undergoes attack. Treating the product of this reaction with zinc and acid produces (−)-malic acid. (a) Outline all steps in this synthesis by writing stereochemical structures for each intermediate. (b) The step in which (−)-tartaric acid is treated with phosphorus tribromide produces only one stereoisomer even though there are two replaceable —OH groups. How is this possible? (c) Suppose that the step in which (−)-tartaric acid is treated with phosphorus tribromide had taken place with "mixed" stereochemistry—with both inversion and retention at the carbon under attack. How many stereoisomers would have been produced? (d) What difference would this have made to the overall outcome of the synthesis?

* **19.38**

Cantharidin is a powerful vesicant that can be isolated from dried beetles (*Cantharis vesicatoria* or "spanish fly"). Outlined below is the stereospecific synthesis of cantharidin reported by Professor Gilbert Stork of Columbia University in 1953. Supply the missing reagents (a)–(o).

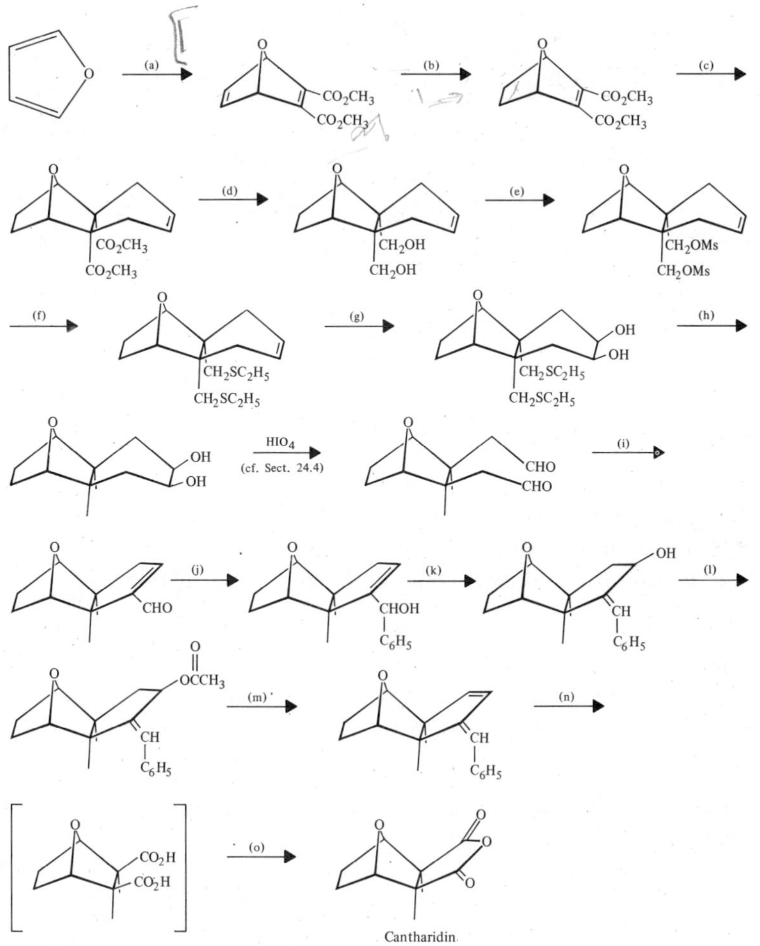

Cantharidin

* **19.39**

Examine the structure of cantharidin (problem 19.38) carefully and (a) suggest a possible two-step synthesis of cantharidin starting with furan (p. 436). (b) F. von Bruchhausen and H. W. Bersch at the University of Münster attempted this two-step synthesis in 1928 only a few months after Diels and Alder published their first paper describing their new diene addition and found that the expected addition failed to take place. Von Bruchhausen and Bersch also found that although cantharidin is stable at relatively high temperatures, heating cantharidin with a palladium catalyst causes cantharidin to decompose. They identified furan and dimethylmaleic anhydride among the decomposition products. What has happened in the decomposition and what does this suggest about why the first step of their attempted synthesis failed?

20

SYNTHESIS AND REACTIONS OF β-DICARBONYL COMPOUNDS

20.1 INTRODUCTION

Compounds having two carbonyl groups separated by an intervening carbon, that is,

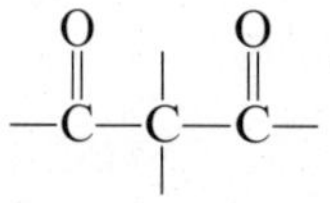

$$\underset{\underset{\displaystyle |}{\overset{\displaystyle \|}{-C}}}{\overset{\displaystyle O}{}}-\underset{\underset{\displaystyle |}{}}{\overset{}{C}}-\underset{\underset{\displaystyle |}{}}{\overset{\displaystyle \overset{\displaystyle O}{\|}}{C}}-$$

are called β-dicarbonyl compounds and these compounds are highly versatile reagents for organic synthesis. In this chapter we will explore some of the methods for preparing β-dicarbonyl compounds and some of their important reactions.

20.2 THE CLAISEN CONDENSATION: THE SYNTHESIS OF β-KETO ESTERS

When ethyl acetate reacts with sodium ethoxide, it undergoes *a condensation reaction;* after acidification, the product that we obtain is the β-keto ester, ethyl acetoacetate (commonly called, *acetoacetic ester*).

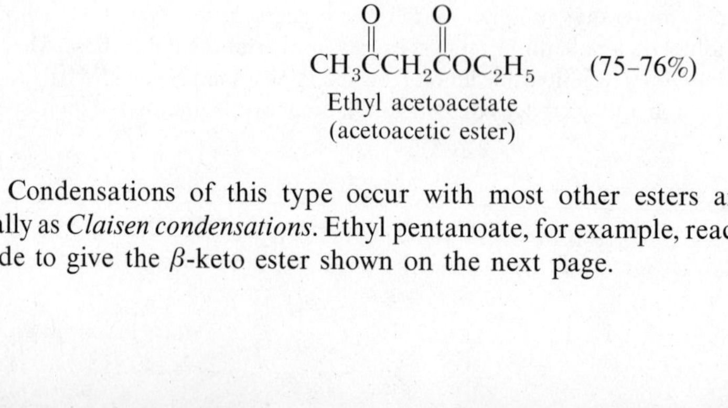

$$2CH_3\overset{O}{\overset{\|}{C}}OC_2H_5 \xrightarrow{\text{NaOC}_2\text{H}_5} \left[CH_3\overset{O}{\overset{\|}{C}}\overset{\displaystyle}{C}H\overset{O}{\overset{\|}{C}}OC_2H_5 \right] + C_2H_5OH$$

$$\underset{\substack{\text{Na}^+ \\ \text{Sodioacetoacetic} \\ \text{ester}}}{} \qquad \underset{\substack{\text{(removed by} \\ \text{distillation)}}}{}$$

$$\downarrow \text{HCl}$$

$$CH_3\overset{O}{\overset{\|}{C}}CH_2\overset{O}{\overset{\|}{C}}OC_2H_5 \qquad (75\text{--}76\%)$$

Ethyl acetoacetate
(acetoacetic ester)

Condensations of this type occur with most other esters and are known generally as *Claisen condensations*. Ethyl pentanoate, for example, reacts with sodium ethoxide to give the β-keto ester shown on the next page.

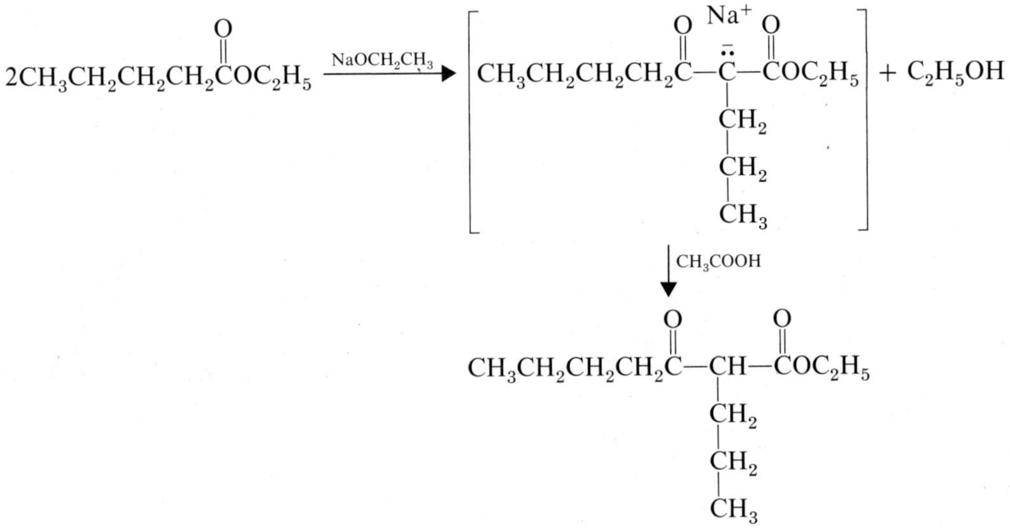

$$2CH_3CH_2CH_2CH_2\overset{O}{\overset{\|}{C}}OC_2H_5 \xrightarrow{NaOCH_2CH_3} \left[CH_3CH_2CH_2CH_2\overset{O}{\overset{\|}{C}}-\overset{Na^+ \ O}{\overset{\ddot{\underset{}{C}}}{\underset{\underset{\underset{CH_3}{|}}{\underset{CH_2}{|}}}{\overset{|}{CH_2}}}}-\overset{O}{\overset{\|}{C}}OC_2H_5 \right] + C_2H_5OH$$

$$\downarrow CH_3COOH$$

$$CH_3CH_2CH_2CH_2\overset{O}{\overset{\|}{C}}-\underset{\underset{\underset{CH_3}{|}}{\underset{CH_2}{|}}}{\overset{|}{\underset{CH_2}{\underset{|}{CH}}}}-\overset{O}{\overset{\|}{C}}OC_2H_5$$

(77%)

If we look closely at these two examples we can see that, overall, both reactions involve a condensation in which one ester loses an α-hydrogen and the other loses an ethoxide ion; that is,

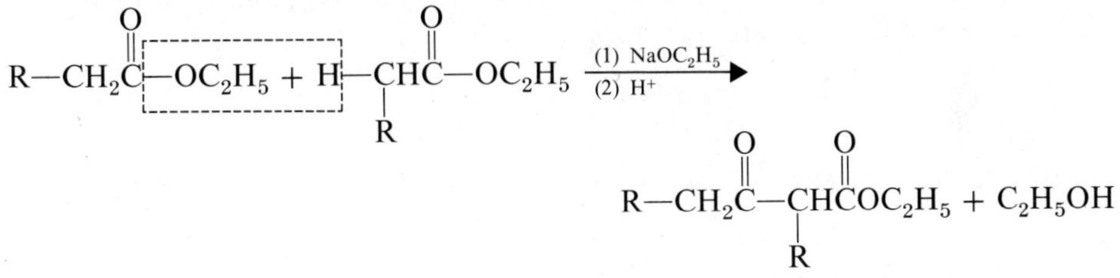

$$R-CH_2\overset{O}{\overset{\|}{C}}\boxed{-OC_2H_5} + \boxed{H}-\overset{O}{\overset{\|}{\underset{R}{C}H}}\overset{O}{\overset{\|}{C}}-OC_2H_5 \xrightarrow[(2) \ H^+]{(1) \ NaOC_2H_5}$$

$$R-CH_2\overset{O}{\overset{\|}{C}}-\underset{R}{\overset{O}{\overset{\|}{C}H}}COC_2H_5 + C_2H_5OH$$

A β-keto ester

We can understand how this happens if we examine the reaction mechanism.

The first step of a Claisen condensation resembles that of an aldol addition (p. 697). Ethoxide ion abstracts an α-proton from the ester. Although the α-protons of an ester are not as acidic as those of aldehydes and ketones, the enolate anion that forms is stabilized by resonance in a similar way.

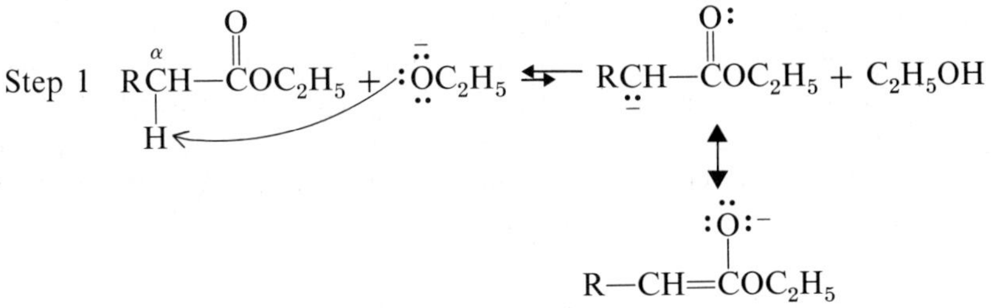

$$\text{Step 1} \quad \overset{\alpha}{R\underset{\underset{H}{|}}{C}H}-\overset{O}{\overset{\|}{C}}OC_2H_5 + :\overset{\_}{\underset{\ddot{}}{O}}C_2H_5 \Longleftrightarrow RCH-\overset{\overset{\ddot{\ddot{O}}:}{\|}}{C}OC_2H_5 + C_2H_5OH$$

$$\updownarrow$$

$$R-CH=\overset{:\ddot{O}:^-}{\underset{}{C}}OC_2H_5$$

In the second step the enolate anion attacks the carbonyl carbon of a second molecule of the ester. It is at this point that the Claisen condensation and the aldol addition *differ*, and they differ in an understandable way: in the aldol reaction nucleophilic attack leads to *addition;* in the Claisen condensation it leads to *substitution.*

Step 2 $RCH_2C \overset{\displaystyle \overset{..}{\overset{..}{O}:}}{\underset{\displaystyle OC_2H_5}{\big\langle}} + \ ^-:CH-COC_2H_5 \ \rightleftharpoons \ RCH_2C-CH-COC_2H_5$

with the intermediate:

$$RCH_2\overset{\displaystyle :O:^-}{\underset{\displaystyle C_2H_5\overset{..}{O}:}{C}}-\underset{\displaystyle R}{CH}-\overset{\displaystyle \overset{..}{O}:}{C}OC_2H_5$$

$$RCH_2\overset{\displaystyle :O}{C}-\underset{\displaystyle R}{CH}-\overset{\displaystyle \overset{..}{O}}{C}OC_2H_5$$

$$+ \ ^-:\overset{..}{\underset{..}{O}}C_2H_5$$

Although the products of this second step are a β-keto ester and ethoxide ion, all of the equilibria up to this point have been unfavorable. Very little product would be formed if this were the last step in the reaction.

The final step of a Claisen condensation is an acid-base reaction that takes place between ethoxide ion and the β-ketoester. *The position of equilibrium for this step is favorable,* and we can make it even more favorable by distilling ethanol from the reaction mixture as it forms.

Step 3 $RCH_2\overset{\displaystyle O}{C}-\underset{\displaystyle R}{\overset{\displaystyle H}{C}}-\overset{\displaystyle O}{C}OC_2H_5 + \ :\overset{..}{\underset{..}{O}}C_2H_5 \ \rightleftharpoons$

   β-Keto ester   Ethoxide ion
   (stronger acid)   (stronger base)

$$RCH_2\overset{\displaystyle :O}{C}-\underset{\displaystyle R}{\overset{..}{C}:}-\overset{\displaystyle :O}{C}OC_2H_5 + \ C_2H_5OH$$

   β-Keto ester anion   Ethanol
   (weaker base)   (weaker acid)

Beta-keto esters are stronger acids than ethanol; they react with ethoxide ion to produce ethanol and anions of β-keto esters. (It is this that pulls the equilibrium to the right.) Beta-keto esters are exceptionally acidic (much more acidic than ordinary esters) because their enolate anions are highly resonance stabilized; their negative charge is delocalized into two carbonyl groups:

$$RCH_2-\overset{\displaystyle \overset{..}{O}:}{C}-\underset{\displaystyle R}{\overset{..}{C}}-\overset{\displaystyle \overset{..}{O}:}{C}OC_2H_5 \longleftrightarrow RCH_2-\overset{\displaystyle :\overset{..}{O}:^-}{C}=\underset{\displaystyle R}{C}-\overset{\displaystyle \overset{..}{O}:}{C}OC_2H_5 \longleftrightarrow RCH_2-\overset{\displaystyle :\overset{..}{O}}{C}-\underset{\displaystyle R}{C}=\overset{\displaystyle :\overset{..}{O}:^-}{C}OC_2H_5$$

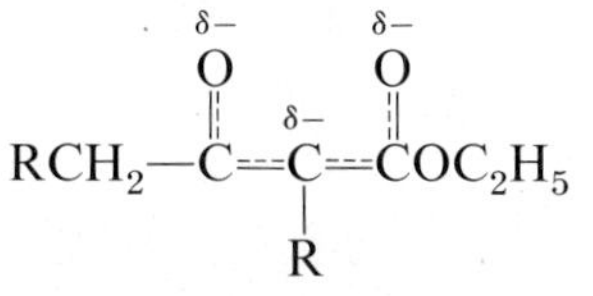

$$\overset{\delta-}{\underset{R}{\underset{|}{RCH_2-\overset{O}{\overset{||}{C}}\overset{\delta-}{=\!=\!=}\overset{\delta-}{C}\overset{O}{\overset{||}{=\!=\!=}COC_2H_5}}}}$$

Resonance hybrid

After steps 1–3 of a Claisen condensation have taken place, we add an acid to the reaction mixture. This brings about a rapid protonation of the anion and produces the β-keto ester as an equilibrium mixture of its keto and enol forms.

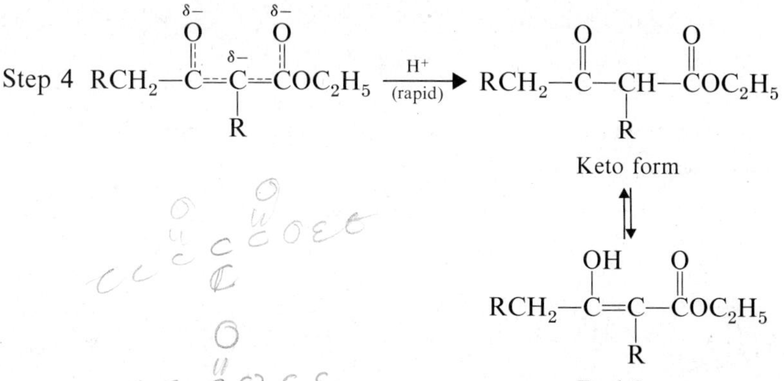

Step 4 $RCH_2-\overset{O}{\overset{||}{C}}\overset{\delta-}{=}\overset{\delta-}{\underset{R}{\underset{|}{C}}}\overset{O}{\overset{||}{=}COC_2H_5} \xrightarrow[\text{(rapid)}]{H^+} RCH_2-\overset{O}{\overset{||}{C}}-\overset{}{\underset{R}{\underset{|}{CH}}}-\overset{O}{\overset{||}{C}}OC_2H_5$

Keto form

$$RCH_2-\overset{OH}{\overset{|}{C}}=\overset{}{\underset{R}{\underset{|}{C}}}-\overset{O}{\overset{||}{C}}OC_2H_5$$

Enol form

Problem 20.1

(a) Write a mechanism for all steps of the Claisen condensation that takes place when ethyl propanoate reacts with ethoxide ion. (b) What products form when the reaction mixture is acidified?

When diethyl adipate is heated with sodium ethoxide, subsequent acidification of the reaction mixture gives 2-carbethoxycyclopentanone.

$$C_2H_5O\overset{O}{\overset{||}{C}}(CH_2)_4\overset{O}{\overset{||}{C}}OC_2H_5 \xrightarrow[\text{(2) H}^+]{\text{(1) NaOC}_2H_5}$$

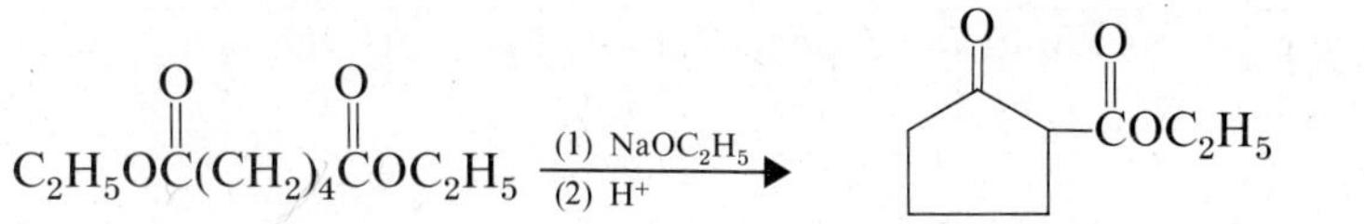

Diethyl adipate

2-Carbethoxycyclopentanone
(74–81%)

The mechanism for this reaction, called the *Dieckmann condensation,* is very similar to that of the Claisen condensation.

Problem 20.2

(a) Show all steps in the mechanism for the Dieckmann condensation. (b) What product would you expect from a Dieckmann condensation of diethyl heptanedioate, $C_2H_5O\overset{O}{\overset{||}{C}}(CH_2)_5\overset{O}{\overset{||}{C}}OC_2H_5$? (c) Can you account for the fact that diethyl glutarate does not undergo a Dieckmann condensation?

Crossed Claisen Condensations

Crossed Claisen condensations (like crossed aldol condensations) are possible when one ester component has no α-hydrogens and is, therefore, unable to undergo self-condensation. Ethyl benzoate, for example, condenses with ethyl acetate to give ethyl benzoylacetate.

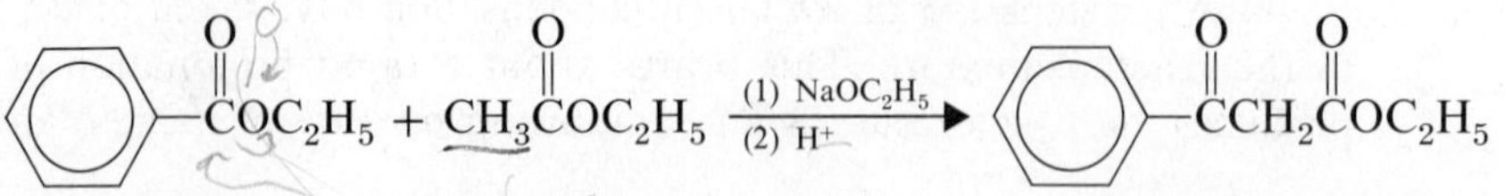

Ethyl benzoate   Ethyl benzoylacetate (60%)

Ethyl phenylacetate condenses with diethyl carbonate to give diethyl phenyl-malonate.

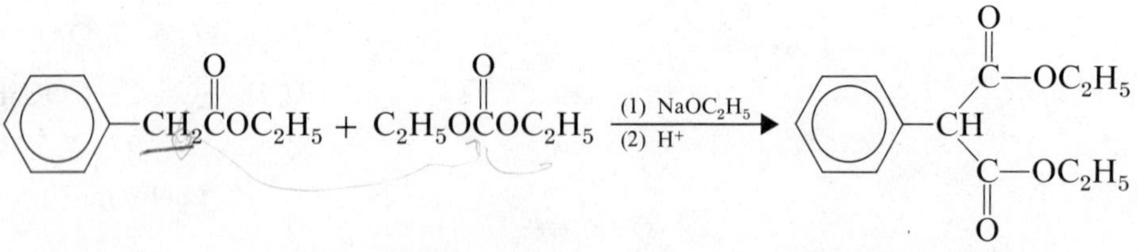

Ethyl phenylacetate   Diethyl carbonate   Diethyl phenylmalonate (65%)

Problem 20.3

Write mechanisms that account for the products that are formed in the two crossed Claisen condensations illustrated above.

Problem 20.4

What products would you expect to obtain from each of the following crossed Claisen condensations?

(a) Ethyl propanoate + diethyl oxalate $\xrightarrow[\text{(2) H}^+]{\text{(1) NaOCH}_2\text{CH}_3}$

(b) Ethyl acetate + ethyl formate $\xrightarrow[\text{(2) H}^+]{\text{(1) NaOCH}_2\text{CH}_3}$

Esters that have only one α-hydrogen can not be converted to β-keto esters through the use of sodium ethoxide. (Why not?) However, they can be converted to β-keto esters through reactions that utilize very strong bases. The strong base converts the ester to its enolate anion in nearly quantitative amounts; this allows us to *acylate* the enolate anion by treating it with an acyl chloride or an ester. An example of this technique that makes use of the very powerful base, sodium tri-phenylmethide, is shown at the top of the next page.

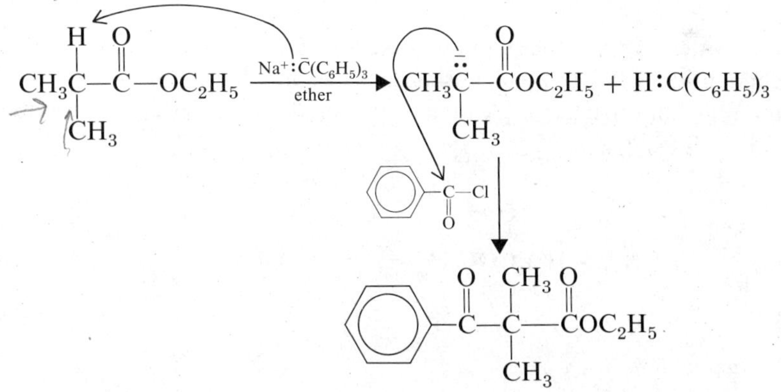

Ethyl 2,2-dimethyl-2-benzoylacetate
(79%)

Acylation of Other Carbanions

Enolate anions derived from ketones also react with esters in nucleophilic substitution reactions that resemble Claisen condensations:

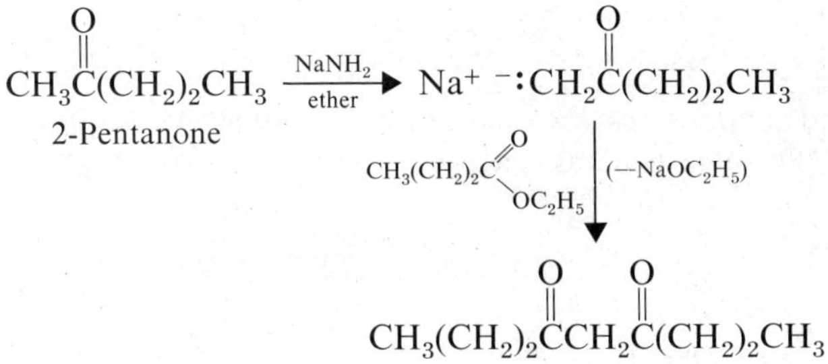

2-Pentanone

4,6-Nonanedione
(76%)

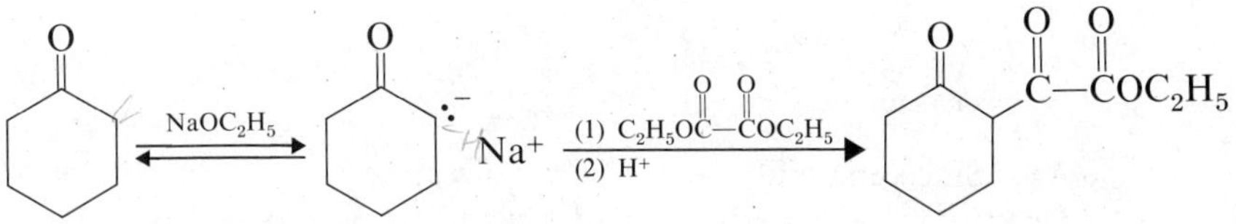

(67%)

Problem 20.5

Show how you might synthesize each of the following compounds using, as your starting materials, esters, ketones, acyl halides, and so on.

(a) 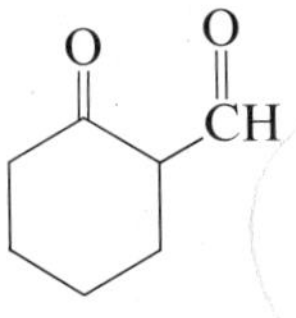 (b) 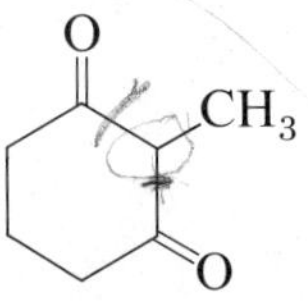 (c)

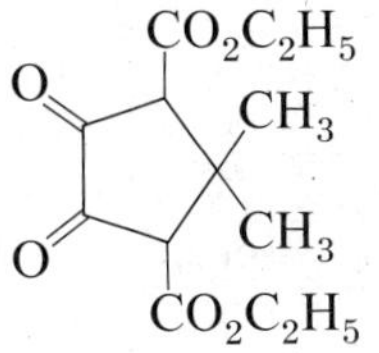

Problem 20.6

Keto esters are capable of undergoing cyclization reactions similar to the Dieckmann condensation. Write mechanisms that account for the products formed in each of the following reactions. [Also account for the ring size of the product in part (b).]

(a) $CH_3C(CH_2)_4COC_2H_5 \xrightarrow[\text{(2) H}^+]{\text{(1) NaOC}_2\text{H}_5}$

2-Acetylcyclopentanone

(b) $CH_3C(CH_2)_3COC_2H_5 \xrightarrow[\text{(2) H}^+]{\text{(1) NaOC}_2\text{H}_5}$

Cyclohexan-1,3-dione

20.3 ACYLATION OF ENAMINES: SYNTHESIS OF β-DIKETONES

Aldehydes and ketones react with secondary amines to form compounds called *enamines* (p. 717). The general reaction for enamine formation can be written as follows:

$$
\underset{\substack{\text{Aldehyde}\\\text{or ketone}}}{\overset{\ddot{\text{O}}}{\underset{H}{\overset{\|}{-\text{C}-\text{C}}}}} + \underset{2° \text{ Amine}}{H\ddot{\text{N}}-R} \rightleftharpoons -\overset{\text{OH}}{\underset{H}{\overset{|}{\text{C}}-\overset{|}{\text{C}}-\ddot{\text{N}}}}\overset{R}{\underset{R}{}} \rightleftharpoons \underset{\text{Enamine}}{-\text{C}=\text{C}\underset{R}{\overset{\ddot{\text{N}}\,R}{}}} + H_2O
$$

Since enamine formation requires the loss of a molecule of water, enamine preparations are usually carried out in a way that allows water to be removed as an azeotrope or with a drying agent. This drives the reversible reaction to completion. Enamine formation is also catalyzed by the presence of a trace of an acid. The secondary amines most commonly used to prepare enamines are cyclic amines such as pyrrolidine, piperidine, and morpholine.

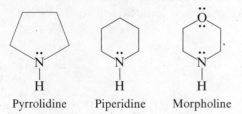

Pyrrolidine   Piperidine   Morpholine

Cyclohexanone, for example, reacts with pyrrolidine in the following way.

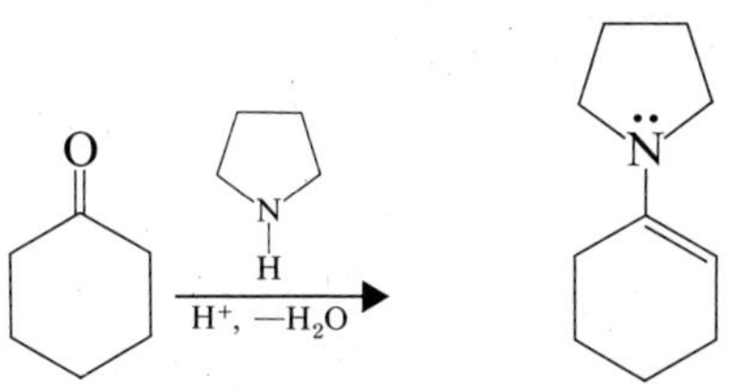

1-Pyrrolidinocyclohexene
(an enamine)

Enamines are good nucleophiles and an examination of the resonance structures below will show us that we should expect enamines to have both a nucleophilic nitrogen and a *nucleophilic carbon*.

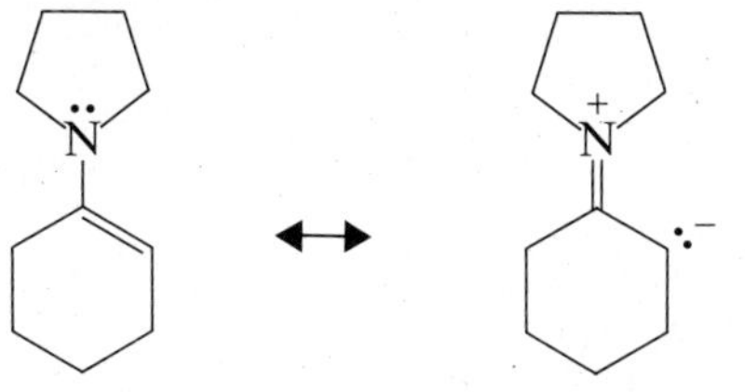

*Contribution made
by this structure
confers nucleophil-
icity on nitrogen*

*Contribution made
by this structure
confers nucleophil-
icity on carbon and
decreases nucleophilicity
of nitrogen*

It is the nucleophilicity of the carbon of enamines that makes them particularly useful reagents in organic synthesis.

When an enamine reacts with an acyl halide or an acid anhydride the product that one obtains is the *C*-acylated compound. The iminium ion that forms, hydrolyzes when water is added during the reaction workup and the overall reaction provides a synthesis of β-diketones.

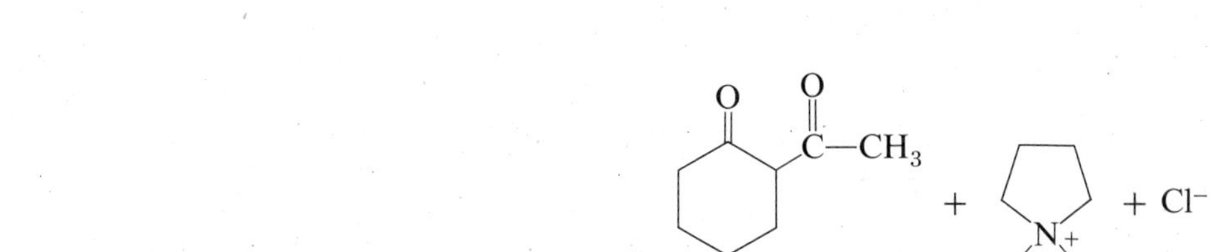

Iminium salt

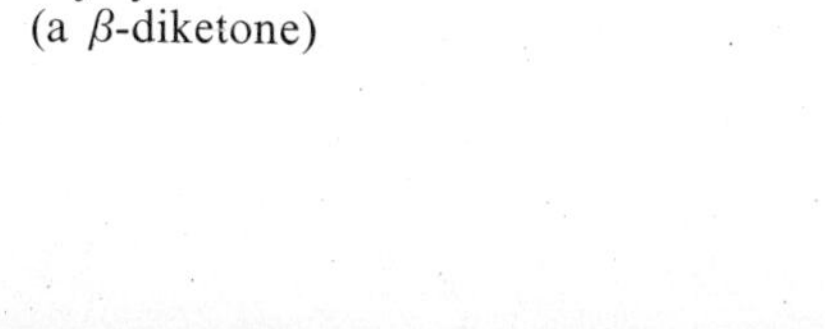

2-Acetylcyclohexanone
(a β-diketone)

Although *N*-acylation may occur in this synthesis, the *N*-acyl product is unstable and can act as an acylating agent itself.

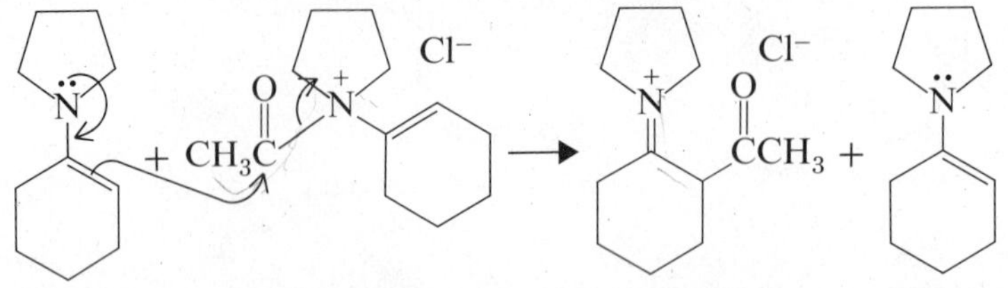

| Enamine | *N*-Acylated enamine | *C*-Acylated iminium salt | Enamine |

As a consequence, the yields of *C*-acylated products are generally high.

Problem 20.7

One mole of a tertiary amine is generally added to an enamine acylation reaction to neutralize any acid that might be formed. (a) Show how the iminium salt formed in the acylation reaction might react with the enamine to generate hydrogen chloride. (b) What effect would this acid-base reaction have on the yield if the tertiary amine were not present?

Enamines can be alkylated as well as acylated. While alkylation of an enamine may lead to the formation of a considerable amount of *N*-alkylated product, heating the *N*-alkylated product often converts it to a *C*-alkyl compound. This is particularly true when the alkyl halide is an allyl or benzyl halide.

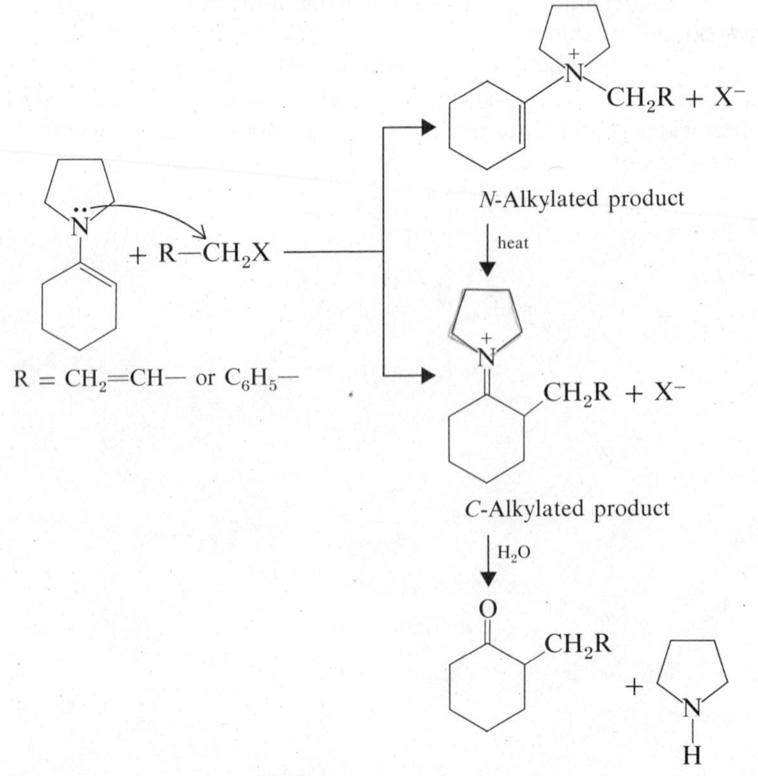

N-Alkylated product

C-Alkylated product

R = CH$_2$=CH— or C$_6$H$_5$—

Enamine alkylations are S_N2 reactions; thus, when we choose our alkylating agents we are usually restricted to the use of methyl, primary, allyl, and benzyl halides. Alpha-haloesters can also be used as the alkylating agents and this reaction provides a convenient synthesis of γ-ketoesters:

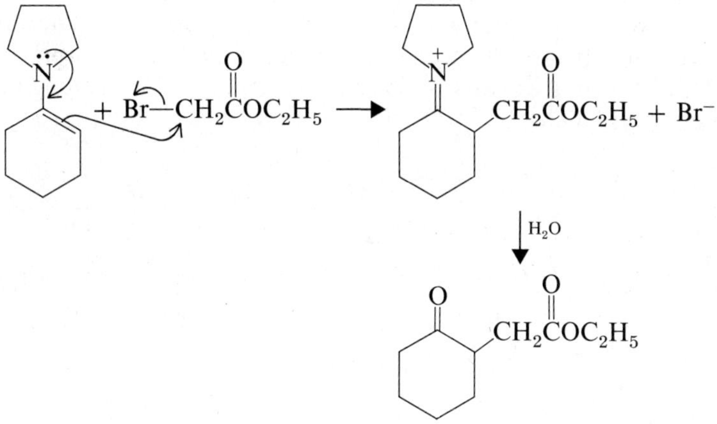

(75%)
(A γ-ketoester)

Problem 20.8

Show how you could employ enamines in syntheses of the following compounds.

(a) 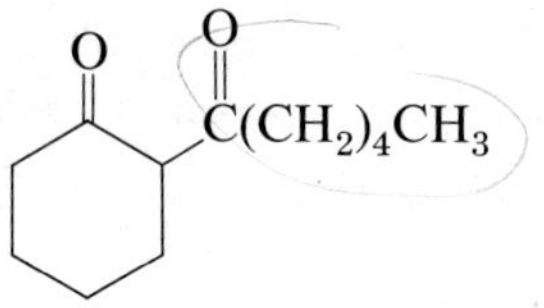

(c) 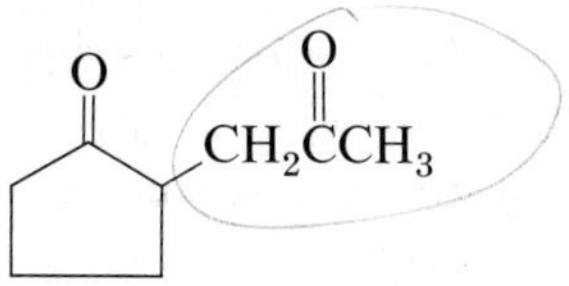

(b) 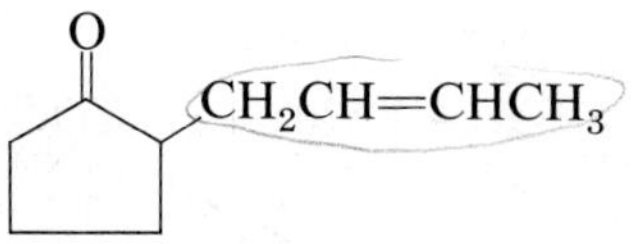

(d) 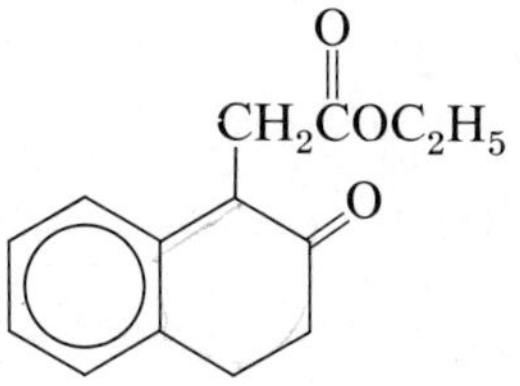

20.4 THE ACETOACETIC ESTER SYNTHESIS

Acetoacetic esters are useful reagents for the preparation of methyl ketones of the types shown below:

$$CH_3—\overset{O}{\overset{\|}{C}}—CH_2—R \quad \text{or} \quad CH_3—\overset{O}{\overset{\|}{C}}—\underset{R}{\underset{|}{C}H}—R$$

Monosubstituted acetone — Disubstituted acetone

Two factors make such syntheses practical. (1) The methylene protons of an aceto-acetic ester are appreciably acidic and (2) acetoacetic acids (i.e., β-ketoacids) de-carboxylate readily.

As we have seen (Section 20.2) the methylene protons of acetoacetic ester (below) are more acidic than the −OH proton of ethanol because they are located between two carbonyl groups and yield a highly stabilized enolate anion. This means that we can convert acetoacetic ester to an enolate anion using sodium ethoxide as a base. We can then carry out an alkylation reaction by treating the enolate anion with an alkyl halide:

$$CH_3\overset{\overset{\displaystyle \ddot{O}:}{\|}}{C}-CH_2-\overset{\overset{\displaystyle \ddot{O}:}{\|}}{C}OC_2H_5 + C_2H_5O^-Na^+ \rightleftharpoons CH_3\overset{\overset{\displaystyle \ddot{O}:}{\|}}{C}-\overset{\overset{\displaystyle Na^+}{\underset{\displaystyle \cdot\cdot}{C}}}{H}-\overset{\overset{\displaystyle \ddot{O}:}{\|}}{C}-OC_2H_5 + C_2H_5OH$$

Acetoacetic ester   Sodium ethoxide   Sodioacetoacetic ester

ß kdi

$$\big\downarrow R-X$$

$$CH_3\overset{\overset{\displaystyle \ddot{O}:}{\|}}{C}-\underset{\underset{\displaystyle R}{|}}{C}H-\overset{\overset{\displaystyle \ddot{O}:}{\|}}{C}-OC_2H_5 + NaX$$

Monoalkylacetoacetic ester

The monoalkylacetoacetic ester still has one appreciably acidic hydrogen and, if we desire, we can carry out a second alkylation reaction. Since the monoalkyl-acetoacetic ester is somewhat less acidic than acetoacetic ester itself (why?), it is usually helpful to use a stronger base.

$$CH_3\overset{\overset{\displaystyle O}{\|}}{C}-\underset{\underset{\displaystyle R}{|}}{C}H-\overset{\overset{\displaystyle O}{\|}}{C}-OC_2H_5 + (CH_3)_3CO^-K^+ \rightleftharpoons CH_3\overset{\overset{\displaystyle O}{\|}}{C}-\overset{\overset{\displaystyle K^+}{\underset{\displaystyle R}{\underset{\displaystyle |}{\underset{\cdot\cdot}{C}}}}}{}-\overset{\overset{\displaystyle O}{\|}}{C}OC_2H_5 + (CH_3)_3COH$$

Monoalkylacetoacetic ester   Potassium tert-butoxide

$$\big\downarrow R'-X$$

$$CH_3\overset{\overset{\displaystyle O}{\|}}{C}-\underset{\underset{\displaystyle R}{|}}{\overset{\overset{\displaystyle R'}{|}}{C}}-\overset{\overset{\displaystyle O}{\|}}{C}-OC_2H_5 + KX$$

Dialkylacetoacetic ester

If our goal is the preparation of a monosubstituted acetone we carry out only one alkylation reaction. We then hydrolyze the monoalkylacetoacetic ester using dilute sodium or potassium hydroxide. Subsequent acidification of the reaction mixture gives an alkylacetoacetic acid and heating this β-ketoacid to 100° brings about decarboxylation.

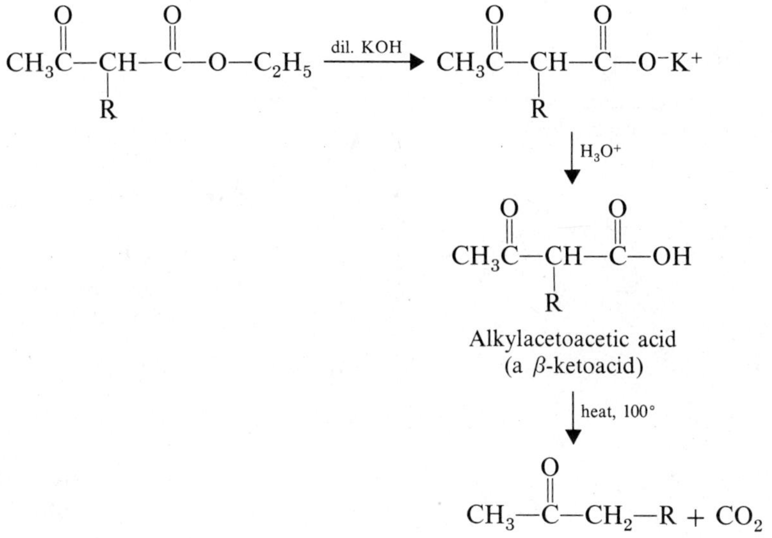

$$CH_3\overset{O}{\overset{\|}{C}}-CH-\overset{O}{\overset{\|}{C}}-O-C_2H_5 \xrightarrow{\text{dil. KOH}} CH_3\overset{O}{\overset{\|}{C}}-CH-\overset{O}{\overset{\|}{C}}-O^-K^+$$
$$\qquad\qquad\overset{|}{R} \qquad\qquad\qquad\qquad\qquad\qquad\overset{|}{R}$$

$$\Big\downarrow H_3O^+$$

$$CH_3\overset{O}{\overset{\|}{C}}-CH-\overset{O}{\overset{\|}{C}}-OH$$
$$\qquad\qquad\overset{|}{R}$$

Alkylacetoacetic acid
(a β-ketoacid)

$$\Big\downarrow \text{heat, }100°$$

$$CH_3-\overset{O}{\overset{\|}{C}}-CH_2-R + CO_2$$

Decarboxylations of β-ketoacids occur at relatively low temperatures for two reasons.

(1) When the carboxylate ion decarboxylates, it forms a resonance stabilized anion:

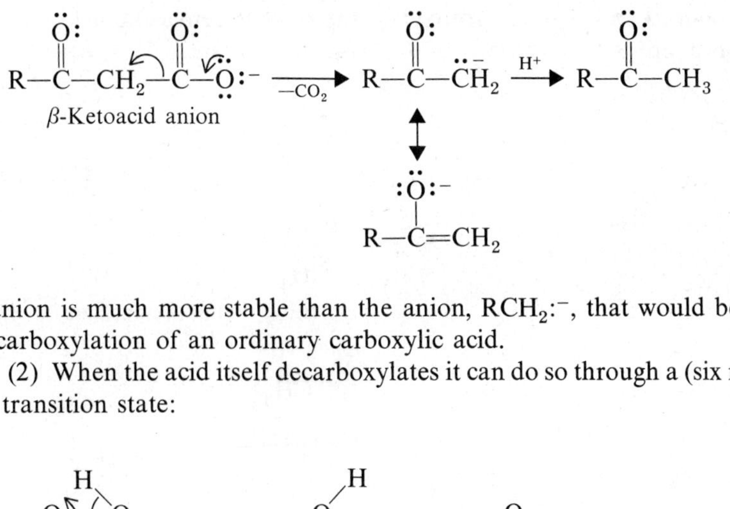

$$R-\overset{\overset{..}{O}:}{\overset{\|}{C}}-CH_2-\overset{\overset{..}{O}:}{\overset{\|}{C}}-\overset{..}{O}:^- \xrightarrow{-CO_2} R-\overset{\overset{..}{O}:}{\overset{\|}{C}}-\overset{..}{C}H_2 \xrightarrow{H^+} R-\overset{\overset{..}{O}:}{\overset{\|}{C}}-CH_3$$

β-Ketoacid anion

$$\updownarrow$$

$$R-\overset{:\overset{..}{O}:^-}{\overset{|}{C}}=CH_2$$

This anion is much more stable than the anion, RCH_2:$^-$, that would be produced by decarboxylation of an ordinary carboxylic acid.

(2) When the acid itself decarboxylates it can do so through a (six membered) cyclic transition state:

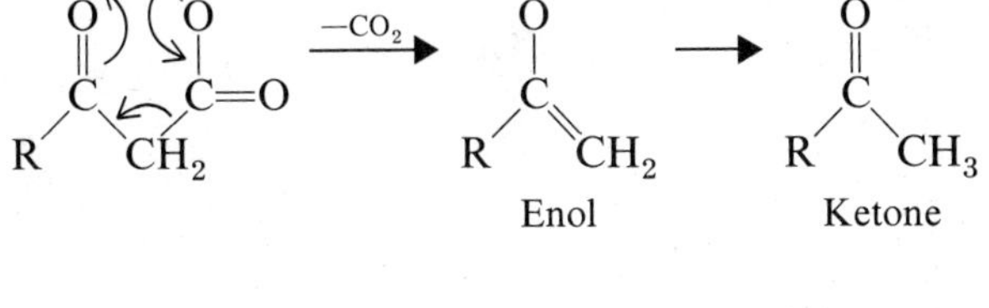

This produces an enol directly and avoids an anionic intermediate. The enol then tautomerizes to a methyl ketone.

A specific example of an acetoacetic ester synthesis of a monosubstituted acetone is the synthesis of 2-heptanone shown on page 798.

$$\underset{\substack{\text{Ethyl acetoacetate}\\ \text{(acetoacetic ester)}}}{\overset{\overset{\displaystyle O}{\parallel}\quad\;\;\overset{\displaystyle O}{\parallel}}{CH_3C-CH_2-COC_2H_5}} \xrightarrow[\text{(2) } CH_3CH_2CH_2CH_2Br]{\text{(1) } NaOC_2H_5/C_2H_5OH} \underset{\substack{\text{Ethyl } n\text{-butylacetoacetate}\\ \text{(69–72\%)}}}{\overset{\overset{\displaystyle O}{\parallel}\;\;\;\;\overset{\displaystyle O}{\parallel}}{CH_3C-\underset{\substack{|\\CH_2\\|\\CH_2\\|\\CH_2\\|\\CH_3}}{CH}-COC_2H_5}} \xrightarrow[\text{(2) } H_3O^+]{\text{(1) dil. NaOH}}$$

$$\underset{}{\overset{\overset{\displaystyle O}{\parallel}\;\;\;\;\overset{\displaystyle O}{\parallel}}{CH_3C-\underset{\substack{|\\CH_2\\|\\CH_2\\|\\CH_2\\|\\CH_3}}{CH}-C-OH}} \xrightarrow[-CO_2]{\text{heat}} \underset{\substack{\text{2-Heptanone}\\ \text{(52–61\% overall from}\\ \text{ethyl acetoacetate)}}}{\overset{\overset{\displaystyle O}{\parallel}}{CH_3C-CH_2CH_2CH_2CH_2CH_3}}$$

If our goal is the preparation of a disubstituted acetone, we carry out two successive alkylation reactions; we hydrolyze the dialkylacetoacetic ester that is produced and then decarboxylate the dialkylacetoacetic acid. An example of this procedure is the synthesis of 3-butyl-2-heptanone.

$$\underset{}{\overset{\overset{\displaystyle O}{\parallel}\;\overset{\displaystyle O}{\parallel}}{CH_3CCH_2COC_2H_5}} \xrightarrow[\text{(2) } CH_3CH_2CH_2CH_2Br]{\text{(1) } NaOC_2H_5/C_2H_5OH} \underset{\substack{\text{Ethyl } n\text{-butylacetoacetate}\\ \text{(69–72\%)}}}{\overset{\overset{\displaystyle O}{\parallel}\;\overset{\displaystyle O}{\parallel}}{CH_3CCHCOC_2H_5}\atop\underset{CH_3}{|\\(CH_2)_3\\|}} \xrightarrow[\text{(2) } CH_3CH_2CH_2CH_2Br]{\text{(1) } (CH_3)_3COK/(CH_3)_3COH}$$

$$\underset{\substack{\text{Ethyl di-}n\text{-butylacetoacetate}\\ \text{(77\%)}}}{\overset{\overset{\displaystyle O}{\parallel}}{CH_3C}-\overset{\overset{\displaystyle CH_3}{|}\;\overset{\displaystyle (CH_2)_3}{|}}{\underset{\substack{|\\(CH_2)_3\\|\\CH_3}}{C}}-COOC_2H_5} \xrightarrow[\text{(2) } H_3O^+]{\text{(1) dil. NaOH}} \overset{\overset{\displaystyle O}{\parallel}}{CH_3C}-\overset{\overset{\displaystyle CH_3}{|}\;\overset{\displaystyle (CH_2)_3}{|}}{\underset{\substack{|\\(CH_2)_3\\|\\CH_3}}{C}}-COOH \xrightarrow[-CO_2]{\text{heat}} \underset{\substack{\text{3-Butyl-2-heptanone}}}{\overset{\overset{\displaystyle O}{\parallel}}{CH_3C}-\underset{\substack{|\\(CH_2)_3\\|\\CH_3}}{CH(CH_2)_3CH_3}}$$

Although both alkylations in the example given above were carried out with the same alkyl halide, we could have used different alkyl halides if our synthesis had required it.

Problem 20.9

(a) To what general class of reactions does the alkylation step of an acetoacetic ester synthesis belong? (b) Occasional side products of alkylations of sodioacetoacetic esters are compounds with the following general structure.

$$\overset{..}{R\overset{..}{O}:} \quad \overset{..}{\overset{..}{O}:}$$
$$CH_3C\!\!=\!\!CHCOC_2H_5$$

Explain how these are formed.

Problem 20.10

Show how you would use the acetoacetic ester synthesis to prepare each of the following:
  (a) 2-Pentanone
  (b) 3-Propyl-2-hexanone
  (c) 4-Phenyl-2-butanone

Problem 20.11

Attempts to bring about monoalkylation of an acetoacetic ester are sometimes complicated by competing dialkylation reactions even when only one mole of sodium ethoxide and one mole of alkyl halide are employed. (a) Write a mechanism showing how dialkylation might occur when ethyl acetoacetate is treated with one mole of sodium ethoxide and one mole of methyl iodide. (b) Does the greater acidity of ethyl acetoacetate versus that of ethyl methylacetoacetate favor monoalkylation or dialkylation? (c) Dialkylations are more common when methyl and ethyl halides are employed than when larger alkyl halides are used. What factor accounts for this? (d) Suppose you wanted to prepare 3-methyl-2-hexanone. Which alkylation would you carry out first to ensure the best yield?

Problem 20.12

The acetoacetic ester synthesis generally gives best yields when primary halides are used in the alkylation step. Secondary halides give low yields and tertiary halides give practically no alkylation product at all. (a) Explain. (b) What products would you expect from the reaction of sodioacetoacetic ester and *tert*-butyl bromide? (c) Bromobenzene cannot be used as an alkylating agent in an acetoacetic ester synthesis. Why not?

Problem 20.13

Since the products obtained from Claisen condensations are β-ketoesters, subsequent hydrolysis and decarboxylation of these products gives a general method for the synthesis of ketones. Show how you would employ this technique in a synthesis of 4-heptanone.

The acetoacetic ester synthesis can also be carried out using haloesters and haloketones. The use of a α-haloester provides a convenient synthesis of γ-ketoacids:

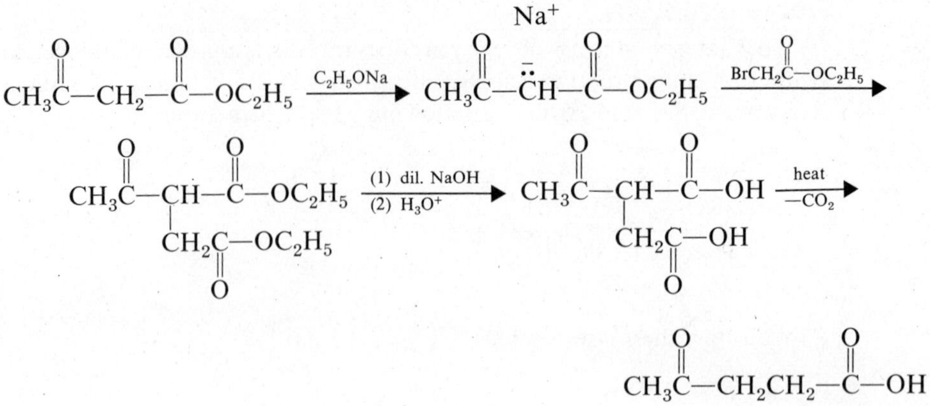

$$CH_3\overset{O}{\overset{\|}{C}}-CH_2-\overset{O}{\overset{\|}{C}}-OC_2H_5 \xrightarrow{C_2H_5ONa} CH_3\overset{O}{\overset{\|}{C}}-\overset{..}{C}H-\overset{O}{\overset{\|}{C}}-OC_2H_5 \;\; Na^+ \xrightarrow{BrCH_2\overset{O}{\overset{\|}{C}}-OC_2H_5}$$

$$CH_3\overset{O}{\overset{\|}{C}}-\underset{\underset{\underset{O}{\|}}{\underset{CH_2\overset{}{C}-OC_2H_5}{|}}}{CH}-\overset{O}{\overset{\|}{C}}-OC_2H_5 \xrightarrow[(2)\ H_3O^+]{(1)\ dil.\ NaOH} CH_3\overset{O}{\overset{\|}{C}}-\underset{\underset{\underset{O}{\|}}{\underset{CH_2\overset{}{C}-OH}{|}}}{CH}-\overset{O}{\overset{\|}{C}}-OH \xrightarrow[-CO_2]{heat}$$

$$CH_3\overset{O}{\overset{\|}{C}}-CH_2CH_2-\overset{O}{\overset{\|}{C}}-OH$$
$$\gamma\text{-Ketovaleric acid}$$

The use of an α-haloketone in an acetoacetic ester synthesis provides a general method for preparing γ-diketones:

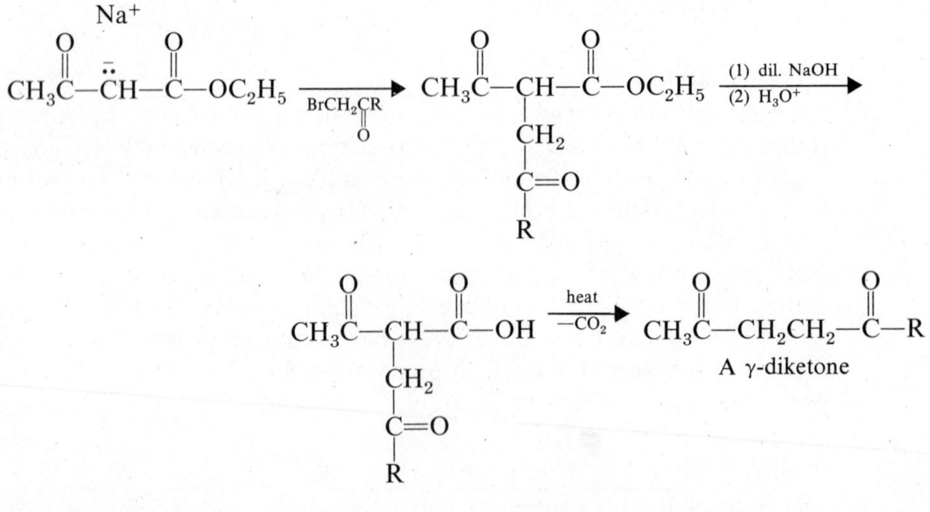

$$CH_3\overset{O}{\overset{\|}{C}}-\overset{..}{C}H-\overset{O}{\overset{\|}{C}}-OC_2H_5 \;\; Na^+ \xrightarrow{BrCH_2\overset{}{C}R\underset{O}{\|}} CH_3\overset{O}{\overset{\|}{C}}-\underset{\underset{\underset{R}{|}}{\underset{\underset{C=O}{|}}{\underset{CH_2}{|}}}}{CH}-\overset{O}{\overset{\|}{C}}-OC_2H_5 \xrightarrow[(2)\ H_3O^+]{(1)\ dil.\ NaOH}$$

$$CH_3\overset{O}{\overset{\|}{C}}-\underset{\underset{\underset{R}{|}}{\underset{\underset{C=O}{|}}{\underset{CH_2}{|}}}}{CH}-\overset{O}{\overset{\|}{C}}-OH \xrightarrow[-CO_2]{heat} CH_3\overset{O}{\overset{\|}{C}}-CH_2CH_2-\overset{O}{\overset{\|}{C}}-R$$
$$A\ \gamma\text{-diketone}$$

Problem 20.14

In the synthesis of the ketoacid given above, the dicarboxylic acid decarboxylates in a specific way; it gives

$$CH_3\overset{O}{\overset{\|}{C}}CH_2CH_2\overset{O}{\overset{\|}{C}}OH \text{ rather than } CH_3\overset{O}{\overset{\|}{C}}\underset{\underset{CH_3}{|}}{C}H\overset{O}{\overset{\|}{C}}OH. \text{ Explain.}$$

Problem 20.15

How would you use the acetoacetic ester synthesis to prepare the following?

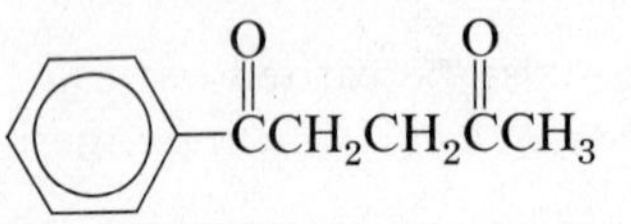

$$\text{⬡}-\overset{O}{\overset{\|}{C}}CH_2CH_2\overset{O}{\overset{\|}{C}}CH_3$$

Anions obtained from acetoacetic esters undergo acylation when they are treated with acyl chlorides or acid anhydrides. Since both of these acylating reagents react with alcohols, acylation reactions are carried out best in aprotic solvents. Sodium hydride can be used to generate the enolate anion.

$$CH_3-\overset{\overset{O}{\|}}{C}-CH_2-\overset{\overset{O}{\|}}{C}-OC_2H_5 \xrightarrow[\text{aprotic solvent}]{Na^+:H^-}$$

$$\overset{Na^+}{\underset{}{CH_3-\overset{\overset{O}{\|}}{C}-\overset{..}{\overset{-}{C}}H-\overset{\overset{O}{\|}}{C}-OC_2H_5}} + H_2 \xrightarrow[(-NaCl)]{R\overset{O}{\overset{\|}{C}}Cl} CH_3-\overset{\overset{O}{\|}}{C}-\underset{\underset{R}{\underset{\|}{C=O}}}{CH}-\overset{\overset{O}{\|}}{C}-OC_2H_5 \xrightarrow[(2)\ H_3O^+]{(1)\ dil.\ NaOH}$$

$$CH_3-\overset{\overset{O}{\|}}{C}-\underset{\underset{R}{\underset{\|}{C=O}}}{CH}-\overset{\overset{O}{\|}}{C}-OH \xrightarrow[-CO_2]{heat} \underset{\text{A }\beta\text{-diketone}}{CH_3-\overset{\overset{O}{\|}}{C}-CH_2-\overset{\overset{O}{\|}}{C}-R}$$

Acylations of acetoacetic esters, followed by hydrolysis and decarboxylation give us a method for preparing β-diketones.

Problem 20.16

How would you use the acetoacetic ester synthesis to prepare the following?

$$\underset{}{\bigcirc\!\!\!\!-\overset{\overset{O}{\|}}{C}CH_2\overset{\overset{O}{\|}}{C}CH_3}$$

One further variation of the acetoacetic ester synthesis involves the conversion of an acetoacetic ester to a resonance-stabilized *dianion* by using a very strong base such as potassium amide in liquid ammonia.

$$CH_3-\overset{\overset{..}{\overset{O}{\|}}{}}{C}-CH_2-\overset{\overset{..}{\overset{O}{\|}}{}}{C}-OC_2H_5 \xrightarrow[\text{liq. :NH}_3]{2K^+:\overset{..}{N}H_2^-} \left[^-:CH_2-\overset{\overset{..}{\overset{O}{\|}}{}}{C}-\overset{\overset{-}{..}}{C}H-\overset{\overset{..}{\overset{O}{\|}}{}}{C}-\overset{..}{\overset{..}{O}}C_2H_5 \right] 2K^+$$

$$\updownarrow$$

$$\left[CH_2=\underset{\underset{:\overset{..}{O}:}{|}}{C}-CH=\underset{\underset{\overset{..}{O}:^-}{|}}{C}-\overset{..}{\overset{..}{O}}C_2H_5 \right] 2K^+$$

$$\updownarrow$$

etc.

When this dianion is treated with one mole of an alkyl or acyl halide it undergoes alkylation or acylation at its terminal carbon rather than at its interior one. This orientation of the alkylation reaction apparently results from the greater basicity (and thus nucleophilicity) of the terminal carbanion. After monoalkylation or monoacylation has taken place, the anion that remains can be neutralized by adding ammonium chloride.

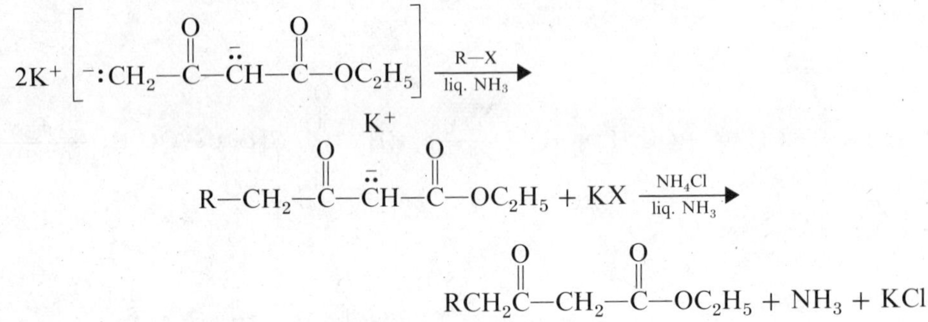

$$2K^+ \left[^-:CH_2-\overset{O}{\underset{\|}{C}}-\overset{..}{\underset{}{\overset{-}{C}H}}-\overset{O}{\underset{\|}{C}}-OC_2H_5 \right] \xrightarrow[\text{liq. NH}_3]{R-X}$$

$$\overset{K^+}{\underset{}{}}$$

$$R-CH_2-\overset{O}{\underset{\|}{C}}-\overset{..}{\underset{}{\overset{-}{C}H}}-\overset{O}{\underset{\|}{C}}-OC_2H_5 + KX \xrightarrow[\text{liq. NH}_3]{NH_4Cl}$$

$$RCH_2\overset{O}{\underset{\|}{C}}-CH_2-\overset{O}{\underset{\|}{C}}-OC_2H_5 + NH_3 + KCl$$

Problem 20.17

Show how you could use ethyl acetoacetate in a synthesis of

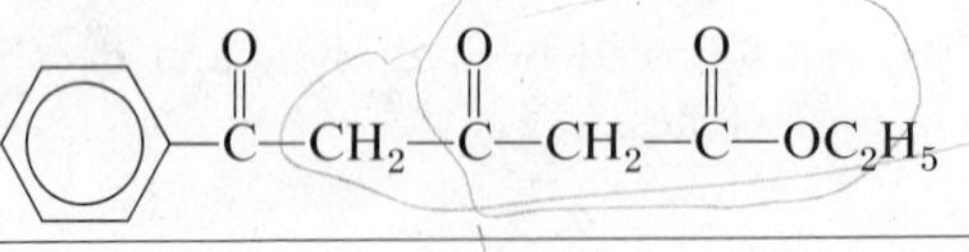

$$\bigcirc-\overset{O}{\underset{\|}{C}}-CH_2-\overset{O}{\underset{\|}{C}}-CH_2-\overset{O}{\underset{\|}{C}}-OC_2H_5$$

20.5 THE MALONIC ESTER SYNTHESIS

A useful counterpart of the acetoacetic ester synthesis—one that allows the synthesis of *mono-* and *disubstituted acetic acids*—is called the *malonic ester synthesis*.

The malonic ester synthesis resembles the acetoacetic ester synthesis in several respects.

1. Diethyl malonate, the starting compound, forms a highly stable enolate ion:

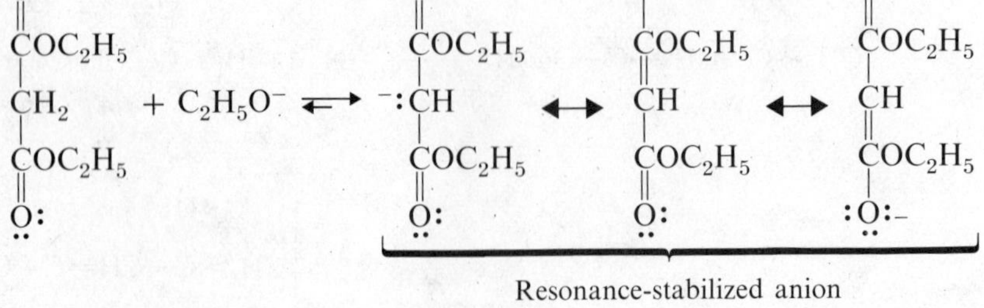

$$\overset{\overset{..}{O}:}{\underset{}{\|}}$$
$$COC_2H_5$$
$$CH_2 \quad + C_2H_5O^- \rightleftharpoons {}^-:CH \quad \leftrightarrow \quad CH \quad \leftrightarrow \quad CH$$
$$COC_2H_5$$
$$\overset{}{\underset{..}{O}:}$$

Resonance-stabilized anion

2. This enolate ion can be alkylated,

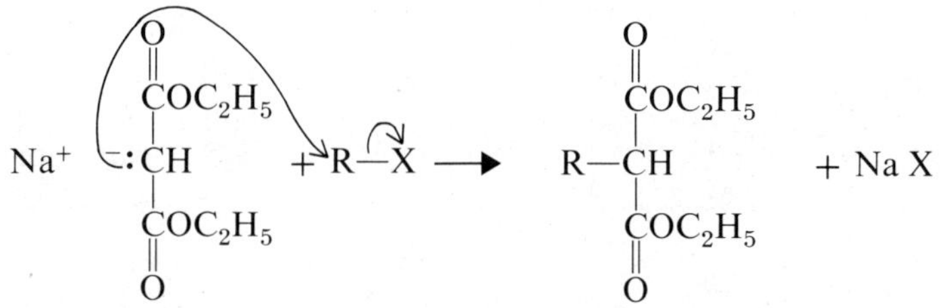

Sodiomalonic ester → Monoalkylmalonic ester + Na X

and the product can be alkylated again if our synthesis requires it:

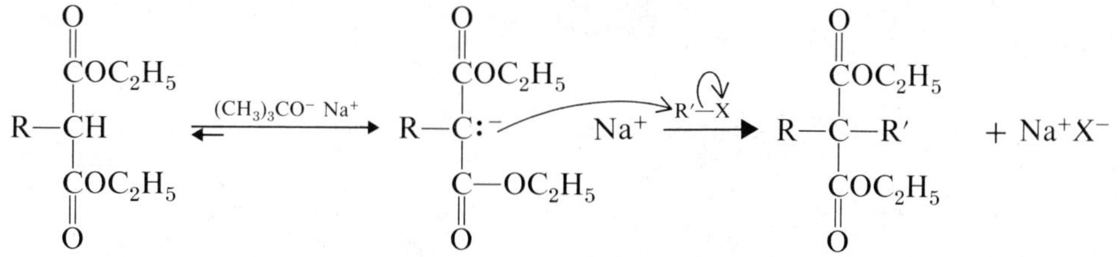

Dialkylmalonic ester

3. The mono- or dialkylmalonic ester can then be hydrolyzed to a mono- or dialkylmalonic acid, and substituted malonic acids decarboxylate readily. Decarboxylation gives a mono- or disubstituted acetic acid.

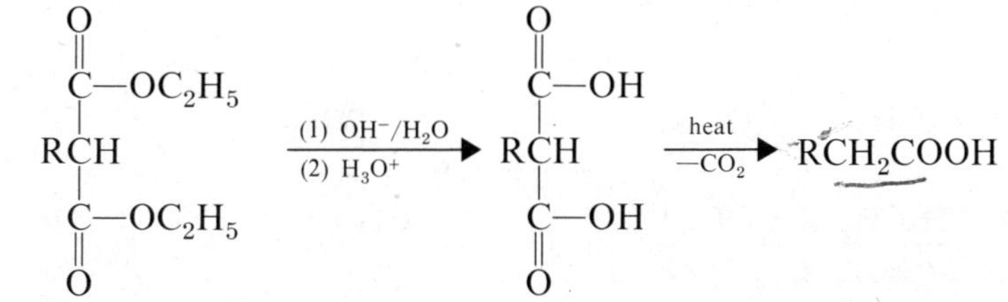

Monoalkylmalonic ester → Monoalkyl-acetic acid

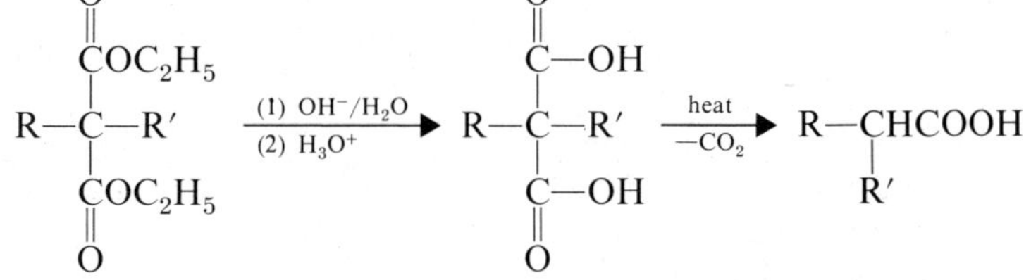

Dialkylmalonic ester → Dialkylacetic acid

Two specific examples of the malonic ester synthesis are the syntheses of hexanoic acid and 2-ethylpentanoic acid given on the next page.

$$\text{O} \atop \underset{\underset{\text{O}}{\overset{|}{\underset{\text{COC}_2\text{H}_5}{|}}}}{\overset{\text{COC}_2\text{H}_5}{\overset{|}{\underset{|}{\text{CH}_2}}}} \quad + \text{ NaOC}_2\text{H}_5 + \text{CH}_3\text{CH}_2\text{CH}_2\text{CH}_2\text{Br} \xrightarrow{\text{reflux}}$$

$$\underset{\substack{\\ \text{Diethyl }n\text{-butylmalonate}\\ (80\text{–}90\%)}}{\text{CH}_3\text{CH}_2\text{CH}_2\text{CH}_2\overset{\overset{\overset{\text{O}}{\|}}{\text{COC}_2\text{H}_5}}{\underset{\underset{\text{O}}{\|}}{\underset{\text{COC}_2\text{H}_5}{\text{CH}}}}} \xrightarrow[\text{(2) dil. H}_2\text{SO}_4, \text{ reflux}]{\text{(1) 50\% KOH, reflux}} \underset{\substack{\text{Hexanoic acid}\\ (75\%)}}{\text{CH}_3\text{CH}_2\text{CH}_2\text{CH}_2\text{CH}_2\text{COOH}}$$

$$\underset{\underset{\text{O}}{\overset{|}{\underset{\text{COC}_2\text{H}_5}{|}}}}{\overset{\overset{\text{O}}{\|}}{\overset{\text{COC}_2\text{H}_5}{\overset{|}{\underset{|}{\text{CH}_2}}}}} \xrightarrow[\text{CH}_3\text{CH}_2\text{I}]{\text{NaOC}_2\text{H}_5} \underset{\text{Diethyl ethylmalonate}}{\text{CH}_3\text{CH}_2\overset{\overset{\overset{\text{O}}{\|}}{\text{COC}_2\text{H}_5}}{\underset{\underset{\text{O}}{\|}}{\underset{\text{COC}_2\text{H}_5}{\text{CH}}}}} \xrightarrow[\text{CH}_3\text{CH}_2\text{CH}_2\text{Br}]{\text{NaOC}_2\text{H}_5}$$

$$\underset{\text{CH}_3\text{CH}_2\text{CH}_2}{\overset{\text{CH}_3\text{CH}_2}{\diagdown}} \text{C} \underset{\text{COC}_2\text{H}_5}{\overset{\text{COC}_2\text{H}_5}{\diagup}} \xrightarrow[\text{(2) H}_3\text{O}^+]{\text{(1) OH}^-/\text{H}_2\text{O}} \underset{\substack{\\ \text{Ethylpropylmalonic acid}}}{\overset{\text{CH}_3\text{CH}_2}{\underset{\text{CH}_3\text{CH}_2\text{CH}_2}{\diagup}} \text{C} \overset{\text{C}-\text{OH}}{\underset{\text{C}-\text{OH}}{\diagdown}}} \xrightarrow[180°]{\text{heat}}$$

$$\underset{\substack{\\ \text{2-Ethylpentanoic acid}}}{\text{CH}_3\text{CH}_2\text{CH}_2\underset{\underset{\text{CH}_3}{\overset{|}{\underset{|}{\text{CH}_2}}}}{\text{CHCOOH}}}$$

Problem 20.18

Outline all steps in a malonic ester synthesis of each of the following:

  (a) Pentanoic acid
  (b) 2-Methylpentanoic acid
  (c) 4-Methylpentanoic acid

Two variations of the malonic ester synthesis make use of dihaloalkanes. In the first of these, two moles of sodiomalonic ester are allowed to react with a dihaloalkane. Two consecutive alkylations occur giving a tetraester; hydrolysis and decarboxylation of the tetraester yields a dicarboxylic acid. An example is the synthesis of glutaric acid:

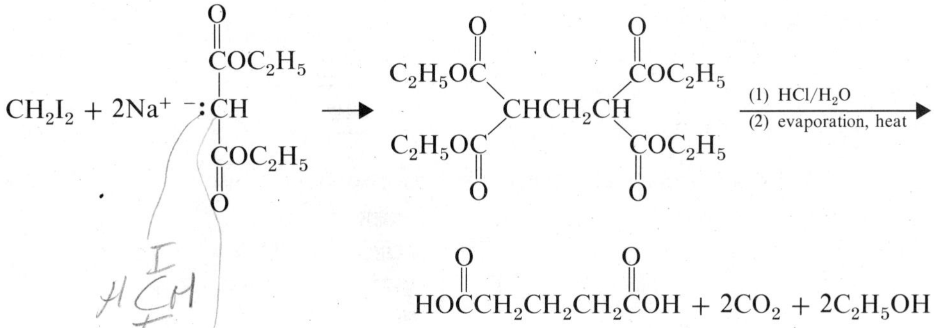

$$CH_2I_2 + 2Na^+ \ ^-:CH \begin{smallmatrix} \overset{O}{\overset{\|}{COC_2H_5}} \\ \\ \underset{O}{\overset{\|}{COC_2H_5}} \end{smallmatrix} \longrightarrow$$

$$\begin{matrix} C_2H_5O\overset{O}{\overset{\|}{C}} & & & \overset{O}{\overset{\|}{COC_2H_5}} \\ & CHCH_2CH & \\ C_2H_5O\underset{O}{\overset{\|}{C}} & & & \underset{O}{\overset{\|}{COC_2H_5}} \end{matrix} \xrightarrow[\text{(2) evaporation, heat}]{\text{(1) HCl/H}_2\text{O}}$$

$$HO\overset{O}{\overset{\|}{C}}CH_2CH_2CH_2\overset{O}{\overset{\|}{C}}OH + 2CO_2 + 2C_2H_5OH$$

Glutaric acid
(80% from tetraester)

In a second variation one mole of sodiomalonic ester is allowed to react with one mole of a dihaloalkane. This gives a haloalkylmalonic ester, which when treated with sodium ethoxide undergoes an internal alkylation reaction. This method has been used to prepare three, four, five, and six-membered rings. An example is the synthesis of cyclobutanecarboxylic acid.

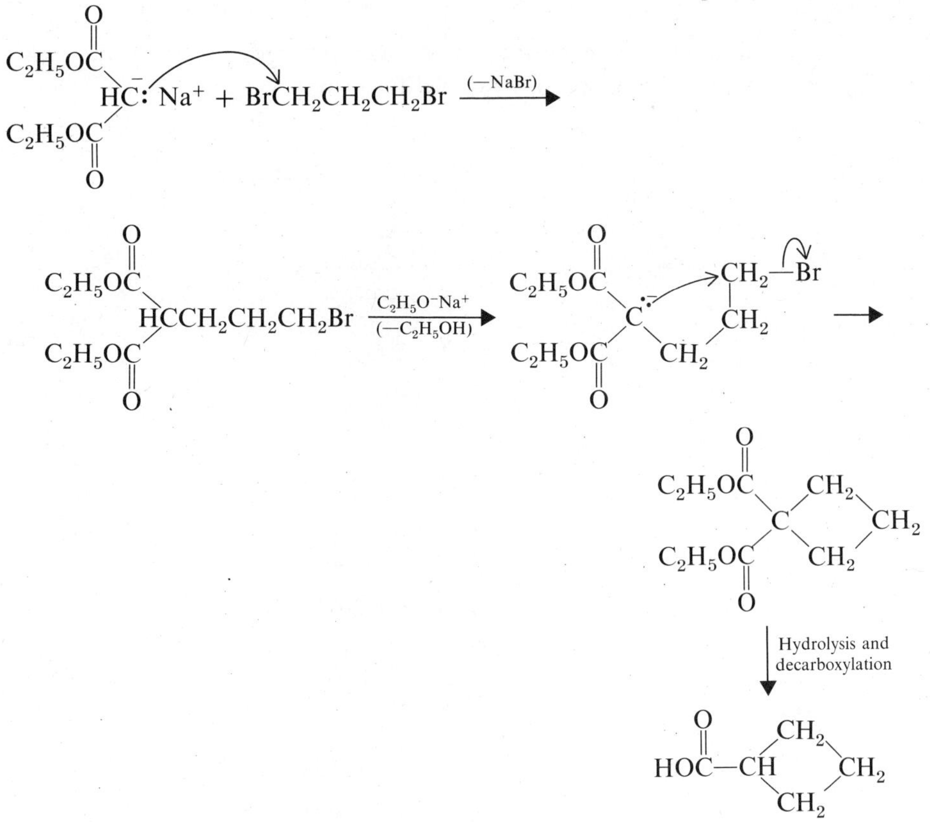

$$\begin{matrix} C_2H_5O\overset{O}{\overset{\|}{C}} \\ HC:^- Na^+ + BrCH_2CH_2CH_2Br \xrightarrow{(-NaBr)} \\ C_2H_5O\underset{O}{\overset{\|}{C}} \end{matrix}$$

$$\begin{matrix} C_2H_5O\overset{O}{\overset{\|}{C}} \\ HCCH_2CH_2CH_2Br \\ C_2H_5O\underset{O}{\overset{\|}{C}} \end{matrix} \xrightarrow[(-C_2H_5OH)]{C_2H_5O^-Na^+} \begin{matrix} C_2H_5O\overset{O}{\overset{\|}{C}} & CH_2-Br \\ \overset{..}{C} & \\ C_2H_5O\underset{O}{\overset{\|}{C}} & CH_2 & CH_2 \end{matrix} \longrightarrow$$

$$\begin{matrix} & \overset{O}{\overset{\|}{}} & \\ C_2H_5O\overset{}{C} & & CH_2 \\ & C & CH_2 \\ C_2H_5O\underset{}{C} & & CH_2 \\ & \underset{O}{\overset{}{}} & \end{matrix}$$

↓ Hydrolysis and decarboxylation

$$HO\overset{O}{\overset{\|}{C}}-CH \begin{matrix} CH_2 \\ \\ CH_2 \end{matrix} CH_2$$

Cyclobutanecarboxylic acid

20.6 FURTHER REACTIONS OF ACTIVE METHYLENE COMPOUNDS

Because of the acidity of their methylene hydrogens, malonic and acetoacetic esters are often called "active methylene compounds." Another useful active methylene compound is ethyl cyanoacetate.

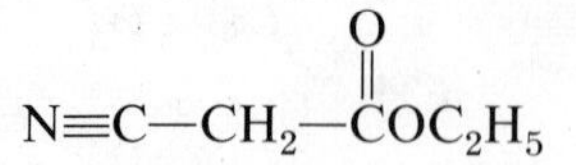

$$N \equiv C - CH_2 - \overset{\overset{O}{\|}}{C}OC_2H_5$$

Ethyl cyanoacetate also reacts with base to yield a resonance stabilized anion.

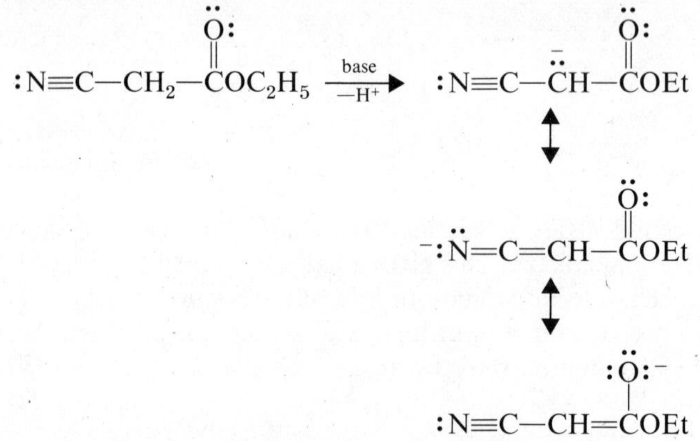

$$:N \equiv C - CH_2 - \overset{\overset{\ddot{O}:}{\|}}{C}OC_2H_5 \xrightarrow[-H^+]{base} :N \equiv C - \overset{\ddot{C}}{C}H - \overset{\overset{\ddot{O}:}{\|}}{C}OEt$$

$$\updownarrow$$

$$-:\ddot{N} = C = CH - \overset{\overset{\ddot{O}:}{\|}}{C}OEt$$

$$\updownarrow$$

$$:N \equiv C - CH = \overset{:\ddot{O}:^-}{C}OEt$$

And ethyl cyanoacetate anions also undergo alkylation reactions. They can be dialkylated with isopropyl iodide, for example.

$$\begin{array}{c} CH_3 \\ \diagdown CHI \\ CH_3 \end{array} + \begin{array}{c} \overset{O}{\|} \\ COC_2H_5 \\ | \\ CH_2 \\ | \\ CN \end{array} \xrightarrow[C_2H_5OH]{C_2H_5ONa} \begin{array}{c} CH_3 \\ \diagdown \\ CH_3 \end{array} \overset{CO_2C_2H_5}{\underset{CN}{CHCH}} \xrightarrow[CH_3]{C_2H_5ONa/C_2H_5OH} \overset{CH_3}{\underset{CH_3}{\diagdown CHI}}$$

(63%)

$$\begin{array}{c} CH_3 \\ \diagdown \\ CH_3 \end{array} CH - \overset{CO_2C_2H_5}{\underset{CN}{\overset{|}{C}}} - CH \begin{array}{c} CH_3 \\ \diagup \\ CH_3 \end{array}$$

(95%)

The Knoevenagel Condensation

Active methylene compounds undergo condensation reactions with aldehydes and ketones known as Knoevenagel condensations. These condensations are catalyzed by very weak bases. Two examples are the following:

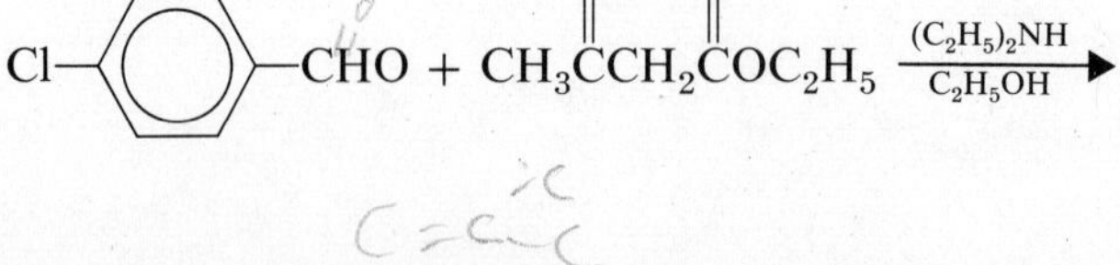

$$Cl - \left\langle \bigcirc \right\rangle - CHO + CH_3\overset{\overset{O}{\|}}{C}CH_2\overset{\overset{O}{\|}}{C}OC_2H_5 \xrightarrow[C_2H_5OH]{(C_2H_5)_2NH}$$

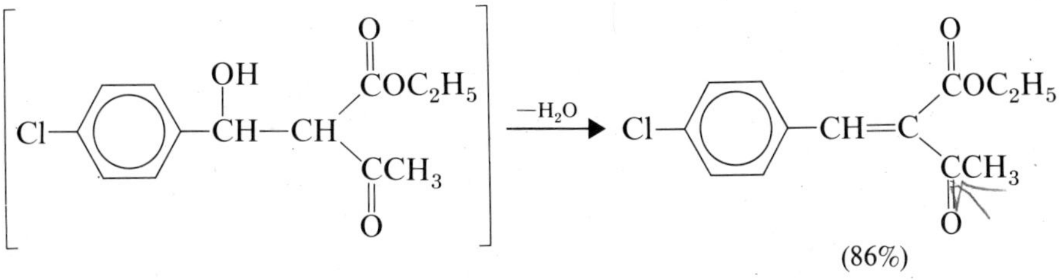

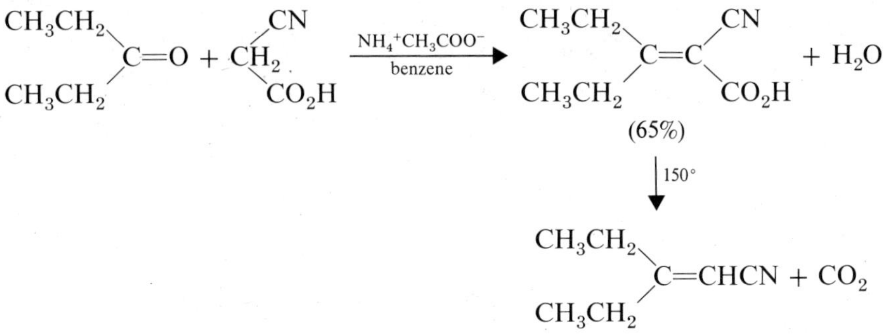

(86%)

(65%)

$150°$

When Knoevenagel condensations take place with malonic acid, decarboxylation usually occurs spontaneously. (These reactions are sometimes called the Doebner modification of the Knoevenagel condensation.)

(80%)

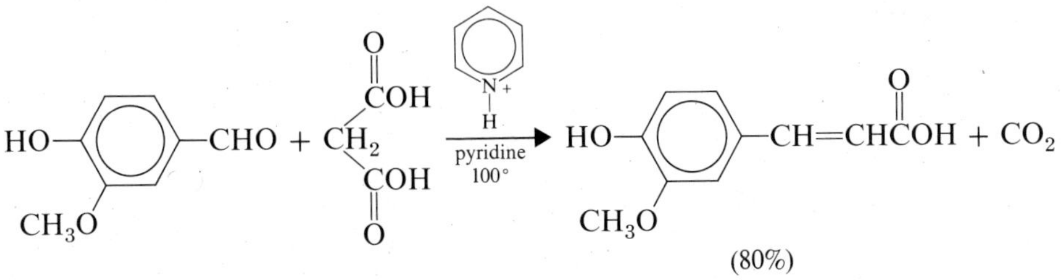

Michael Additions

Active methylene compounds also undergo conjugate additions to α,β-unsaturated carbonyl compounds (cf. p. 726). These reactions are known as Michael additions. An example is the following:

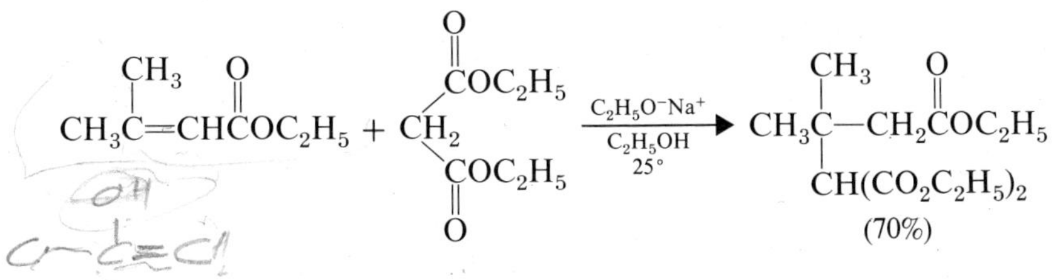

(70%)

The mechanism for this reaction involves (1) formation of an anion from the active methylene compound,

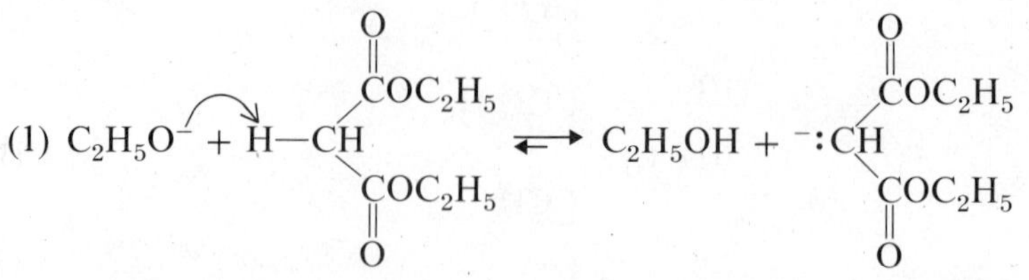

(2) conjugate addition of the anion to the α,β-unsaturated ester, followed by (3) the acceptance of a proton.

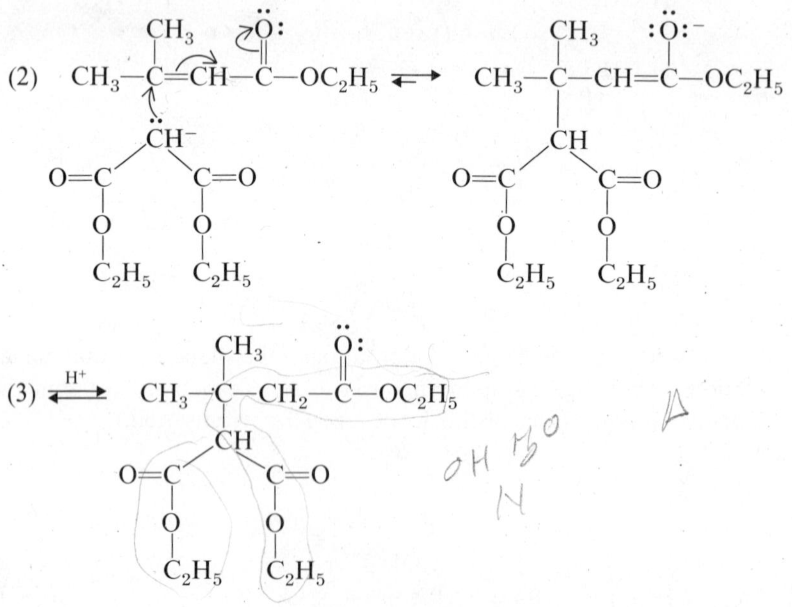

Problem 20.19

How would you prepare $HO\overset{O}{\overset{\|}{C}}CH_2\overset{CH_3}{\underset{CH_3}{C}}CH_2\overset{O}{\overset{\|}{C}}OH$ from the product of the Michael addition given above?

Michael additions take place with a variety of other reagents; these include acetylenic esters and α,β-unsaturated nitriles:

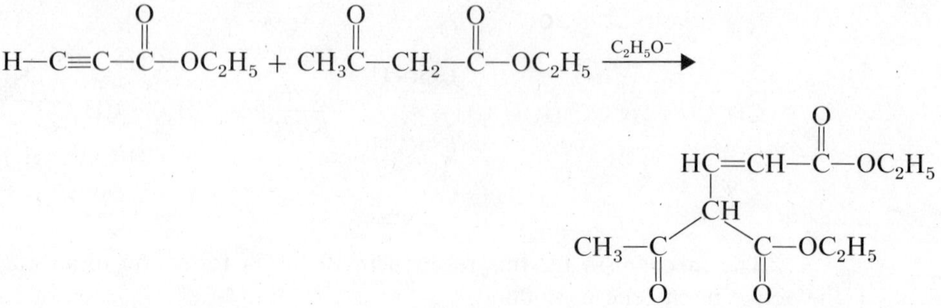

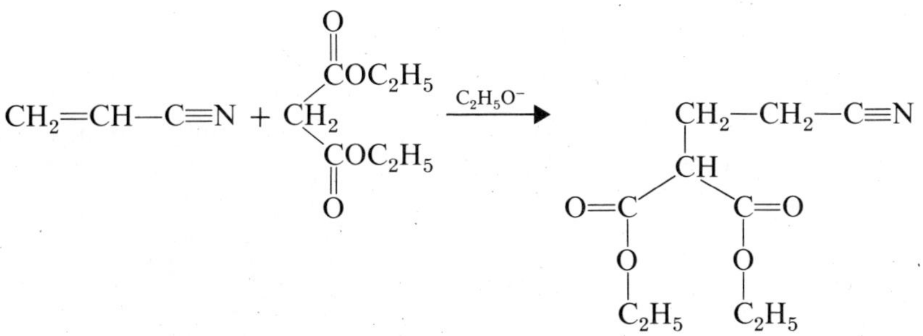

$$CH_2=CH-C\equiv N + \underset{\underset{O}{\overset{||}{\underset{COC_2H_5}{\overset{O}{\overset{||}{COC_2H_5}}}}}}{CH_2} \xrightarrow{C_2H_5O^-} \begin{array}{c} CH_2-CH_2-C\equiv N \\ | \\ CH \\ O=C \qquad C=O \\ | \qquad\qquad | \\ O \qquad\qquad O \\ | \qquad\qquad | \\ C_2H_5 \qquad C_2H_5 \end{array}$$

Enamines can also be used in Michael additions. The example given below is especially interesting because it illustrates how the use of an optically active amine can exert overall *steric control* on the reaction.

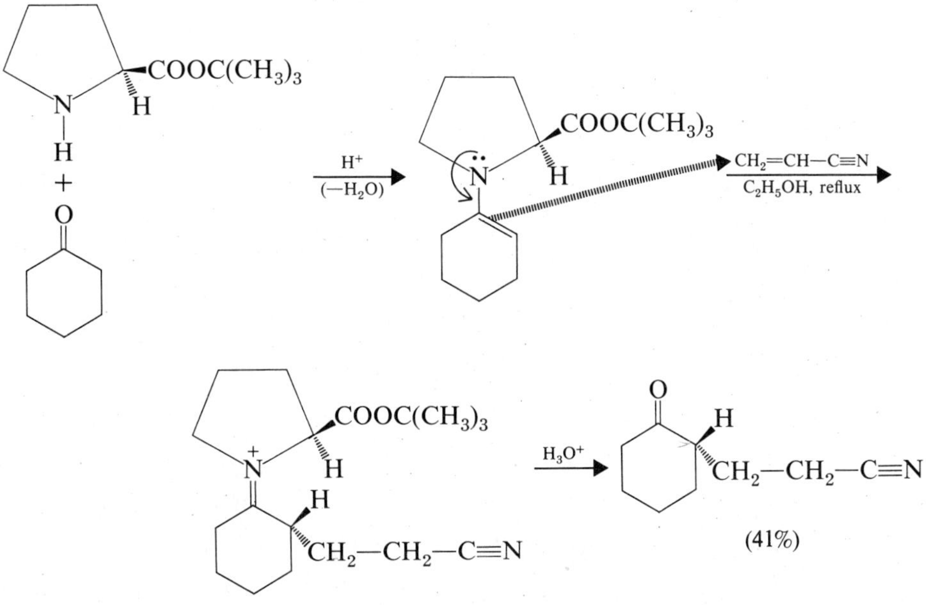

In this reaction the bulky *tert*-butyl ester group attached to the pyrrolidine ring apparently hinders the reaction at one face (the top) of the enamine. As a result, the reaction occurs preferentially at the other face (the bottom) and gives an optically active product.

The Mannich Reaction

Active methylene compounds react with formaldehyde and a primary or secondary amine to yield compounds called Mannich bases. An example is the reaction of acetone, formaldehyde, and diethylamine shown below.

$$CH_3-\overset{O}{\overset{||}{C}}-CH_3 + H-\overset{O}{\overset{||}{C}}-H + (C_2H_5)_2NH \xrightarrow{HCl} CH_3-\overset{O}{\overset{||}{C}}-CH_2-CH_2-N(C_2H_5)_2 + H_2O$$

(a Mannich base)

The Mannich reaction apparently proceeds through a variety of mechanisms depending on the reactants and the conditions that are employed. One mechanism that appears to operate in neutral or acidic media involves (1) initial reaction of the secondary amine with formaldehyde to yield an iminium ion and (2) subsequent reaction of the iminium ion with the enol form of the active methylene compound.

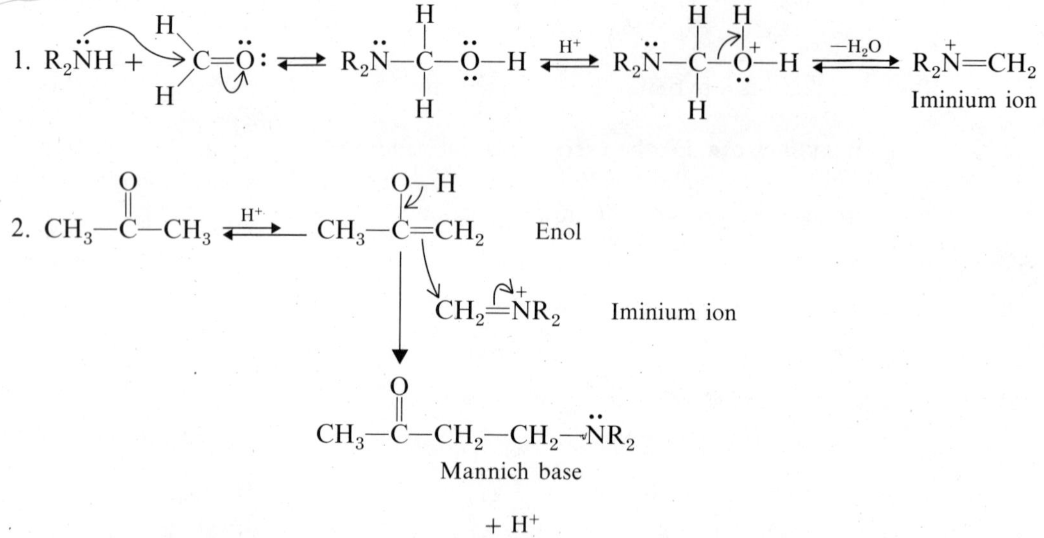

1. $R_2\ddot{N}H + \underset{H}{\overset{H}{>}}C=\ddot{O}: \rightleftharpoons R_2\ddot{N}-\overset{H}{\underset{H}{C}}-\ddot{O}-H \overset{H^+}{\rightleftharpoons} R_2\ddot{N}-\overset{H}{\underset{H}{C}}-\overset{H}{\underset{\ddot{O}}{O}}-H \overset{-H_2O}{\rightleftharpoons} R_2\overset{+}{N}=CH_2$

   Iminium ion

2. $CH_3-\overset{O}{\overset{\|}{C}}-CH_3 \overset{H^+}{\rightleftharpoons} CH_3-\overset{O-H}{\overset{\|}{C}}=CH_2$ Enol

   $\overset{+}{CH_2}=NR_2$ Iminium ion

   $CH_3-\overset{O}{\overset{\|}{C}}-CH_2-CH_2-\ddot{N}R_2$
   Mannich base

   $+ H^+$

Problem 20.20

Outline reasonable mechanisms that account for the products of the following Mannich reactions:

(a) 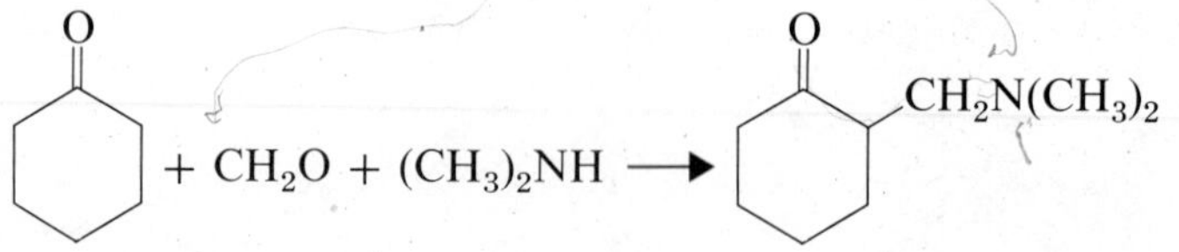   (cyclohexanone) $+ CH_2O + (CH_3)_2NH \longrightarrow$ (2-(dimethylaminomethyl)cyclohexanone $CH_2N(CH_3)_2$)

(b) 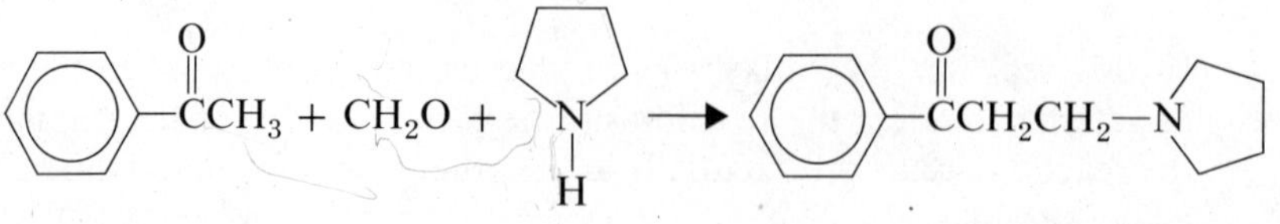   (Ph)$-\overset{O}{\overset{\|}{C}}CH_3 + CH_2O +$ (pyrrolidine $\overset{N}{\underset{H}{}}$) $\longrightarrow$ (Ph)$-\overset{O}{\overset{\|}{C}}CH_2CH_2-N$(pyrrolidine)

(c)   OH (4-methylphenol) $+ CH_2O + (CH_3)_2NH \longrightarrow$ OH $(CH_3)_2NCH_2$ ... $CH_2N(CH_3)_2$ / CH_3

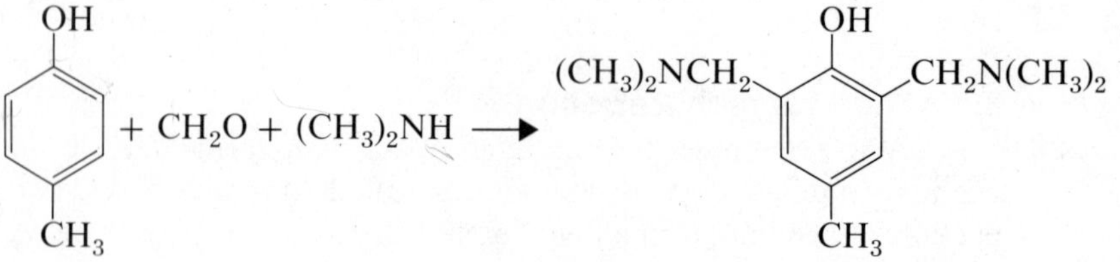

20.7 BARBITURATES

In the presence of sodium ethoxide, diethyl malonate reacts with urea to form a compound called barbituric acid.

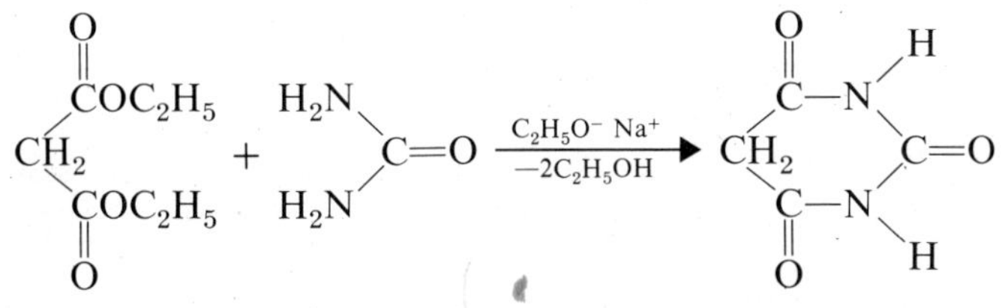

Barbituric acid

Barbituric acid is a pyrimidine (cf. p. 817) and it exists in several tautomeric forms.

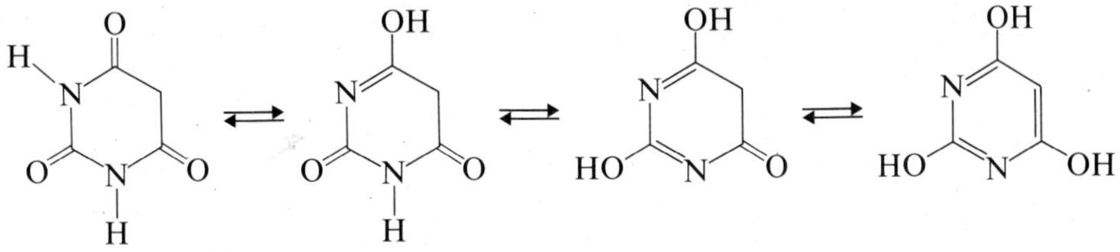

As its name suggests barbituric acid is a moderately strong acid, stronger even than acetic acid.* Its anion is highly resonance stabilized.

Salts of barbituric acid derivatives are *barbiturates*. Barbituric acid derivatives and barbiturates have been used in medicine as soporifics (sleep producers) since 1903. One of the earliest barbiturates introduced into medical use is the compound veronal (diethylbarbituric acid). Veronal is usually used as its sodium salt.

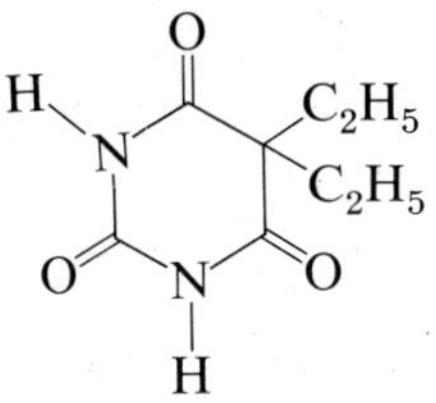

Veronal
(diethylbarbital)

Although barbiturates are very effective soporifics, their use is also hazardous. They are addictive, and overdosage, often with fatal results, is common.

Additional Problems

20.21
Show all steps in the following syntheses. You may use any other needed reagents but should begin with the compound given. You need not repeat steps carried out in earlier parts of this exercise.

(a) 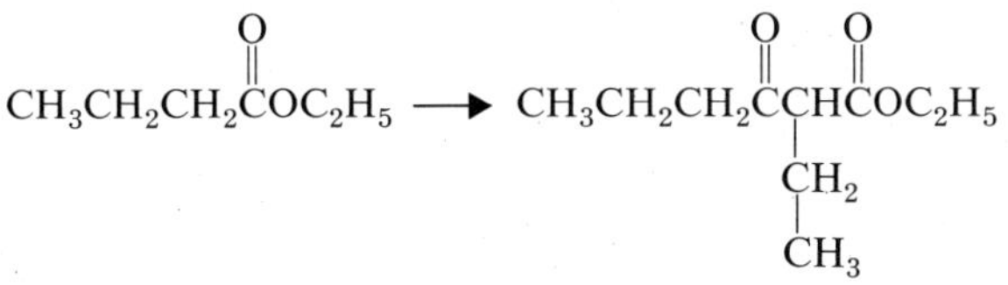 $CH_3CH_2CH_2COC_2H_5 \longrightarrow CH_3CH_2CH_2CCHCOC_2H_5$
                                                                    CH_2
                                                                    CH_3

* Barbituric acid was given its name by Adolf von Baeyer in 1864. The barbituric part of the name is thought to have been motivated by Baeyer's gallantry toward a friend named Barbara.

(b) $CH_3CH_2CH_2\overset{O}{\overset{\|}{C}}OC_2H_5 \longrightarrow CH_3CH_2CH_2\overset{O}{\overset{\|}{C}}CH_2CH_2CH_3$

(c) $C_6H_5CH_2\overset{O}{\overset{\|}{C}}OC_2H_5 \longrightarrow C_6H_5\overset{CH_3}{\overset{|}{C}H}COOH$

(d) $CH_3CH_2CH_2\overset{O}{\overset{\|}{C}}OC_2H_5 \longrightarrow CH_3CH_2\overset{\overset{O}{\overset{\|}{C}}OC_2H_5}{\underset{\underset{O\ \ O}{C-CO C_2H_5}}{C}}$

(e) $CH_3CH_2CH_2\overset{O}{\overset{\|}{C}}OC_2H_5 \longrightarrow CH_3CH_2CH_2\overset{O\ O}{\overset{\|\ \|}{C}C}OC_2H_5$

(f) $C_6H_5CH_2\overset{O}{\overset{\|}{C}}OC_2H_5 \longrightarrow C_6H_5\overset{O}{\overset{\|}{C}}HCOC_2H_5$
$\qquad\qquad\qquad\qquad\qquad\qquad \underset{\underset{O}{CH}}{|}$

(g)

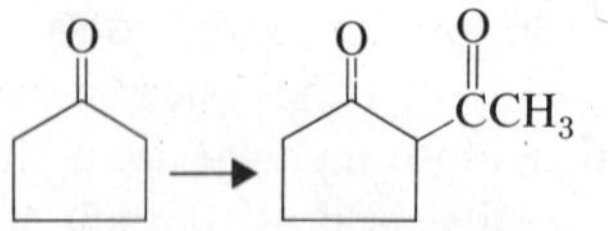

(h)

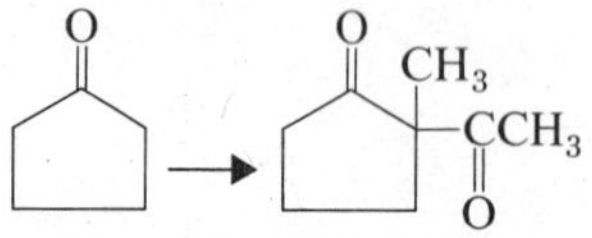

(i)

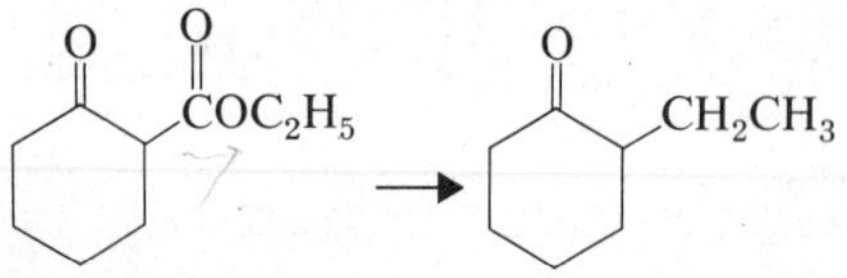

20.22
Outline syntheses of each of the following from acetoacetic ester and any other required reagents.
(a) Methyl *tert*-butyl ketone
(b) 2-Hexanone
(c) 2,5-Hexanedione
(d) 4-Hydroxypentanoic acid
(e) 2-Ethyl-1,3-butanediol
(f) 1-Phenyl-1,3-butanediol
(g) 1-Phenyl-1,3,5-pentanetriol

20.23
Outline syntheses of each of the following from diethyl malonate and any other required reagents.
(a) 2-Methylbutanoic acid
(b) 4-Methyl-1-pentanol
(c) $CH_3CH_2\overset{}{\underset{\underset{CH_2OH}{|}}{C}}HCH_2OH$

(d) $HOCH_2CH_2CH_2CH_2OH$

20.24

The synthesis of cyclobutanecarboxylic acid given on p. 805 was first carried out by William Perkin, Jr. in 1883, and it represented the first synthesis of an organic compound with a ring smaller than six carbons. (There was a general feeling at the time that such compounds would be too unstable to exist.) Earlier in 1883, Perkin reported what he mistakenly believed to be a cyclobutane derivative obtained from the reaction of acetoacetic ester and 1,3-dibromopropane. The reaction that Perkin had expected to take place was the following:

$$BrCH_2CH_2CH_2Br + CH_3\overset{O}{\underset{\|}{C}}CH_2\overset{O}{\underset{\|}{C}}OC_2H_5 \xrightarrow{C_2H_5ONa}$$

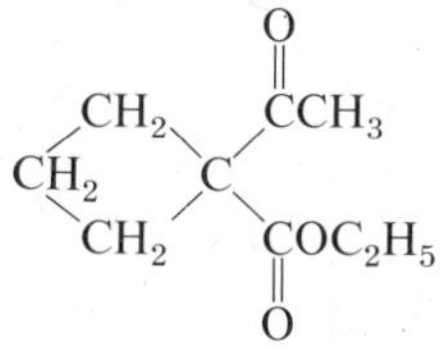

$$\begin{array}{c} CH_2 \\ CH_2 \quad C \\ CH_2 \end{array} \begin{array}{c} \overset{O}{\underset{\|}{C}}CH_3 \\ \overset{}{C}OC_2H_5 \\ \overset{\|}{O} \end{array}$$

The molecular formula for his product agreed with the formulation given above and alkaline hydrolysis and acidification gave a nicely crystalline acid (also having the expected molecular formula). The acid, however, was quite stable to heat and resisted decarboxylation. Perkin later found that both the ester and the acid contained six-membered rings (five carbons and one oxygen). Recall the charge distribution in the enolate ion obtained from acetoacetic ester and propose structures for Perkin's ester and acid.

20.25

(a) In 1884 Perkin achieved a successful synthesis of cyclopropanecarboxylic acid from sodiomalonic ester and 1,2-dibromoethane. Outline the reactions involved in this synthesis. (b) In 1885 Perkin synthesized five-membered carbocyclic compounds **D** and **E** in the following way:

$$2Na^+ : \bar{C}H(CO_2C_2H_5)_2 + BrCH_2CH_2CH_2Br \longrightarrow A\ (C_{17}H_{28}O_8) \xrightarrow{2C_2H_5O^-Na^+} \xrightarrow{Br_2}$$

$$B\ (C_{17}H_{26}O_8) \xrightarrow[(2)\ H_3O^+]{(1)\ OH^-/H_2O} C\ (C_9H_{10}O_8) \xrightarrow{heat} D\ (C_7H_{10}O_4) + E\ (C_7H_{10}O_4)$$

D and **E** are diastereomers; **D** can be resolved into enantiomeric forms while **E** cannot. What are the structures of **A, B, C, D,** and **E**? (c) Ten years later Perkin was able to synthesize 1,4-dibromobutane; he later used this and diethyl malonate to prepare cyclopentanecarboxylic acid. Show the reactions involved.

20.26

Thiolesters of the type RCH_2$\overset{O}{\underset{\|}{C}}$SCH_2CH_3 are more reactive in Claisen condensations than are ordinary esters. How can you account for this?

[handwritten: very acidic   good leaving group]

20.27

Write mechanisms that account for the products of the following reactions.

(a) $C_6H_5CH{=}CH\overset{O}{\underset{\|}{C}}OC_2H_5 + CH_2(\overset{O}{\underset{\|}{C}}OC_2H_5)_2 \xrightarrow{NaOCH_2CH_3}$

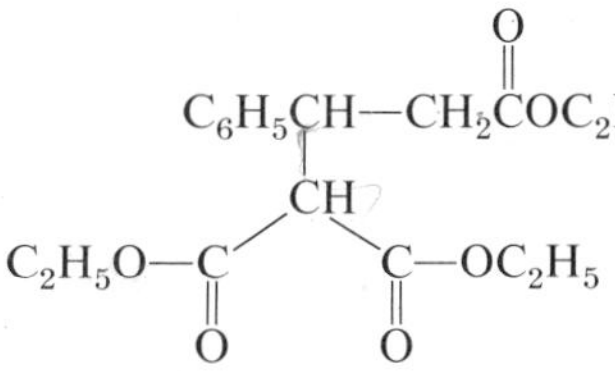

$$\begin{array}{c} C_6H_5\underset{\underset{\displaystyle CH}{|}}{CH}{-}CH_2\overset{O}{\underset{\|}{C}}OC_2H_5 \\ \\ C_2H_5O{-}\underset{\underset{\displaystyle O}{\|}}{C} \qquad \underset{\underset{\displaystyle O}{\|}}{C}{-}OC_2H_5 \end{array}$$

(b) 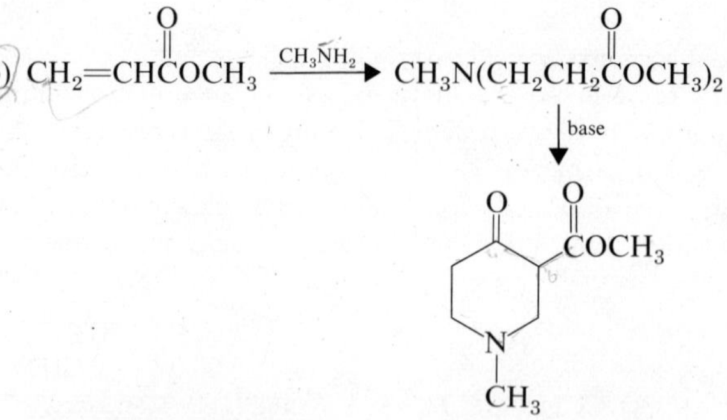 $CH_2=CHCOCH_3 \xrightarrow{CH_3NH_2} CH_3N(CH_2CH_2COCH_3)_2$

(c)

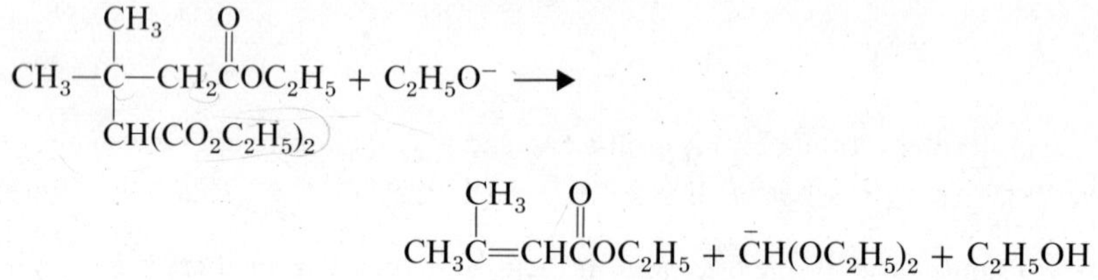

(d)

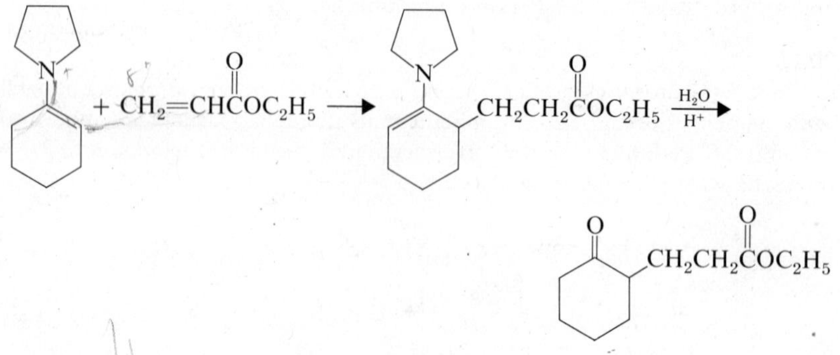

(e)

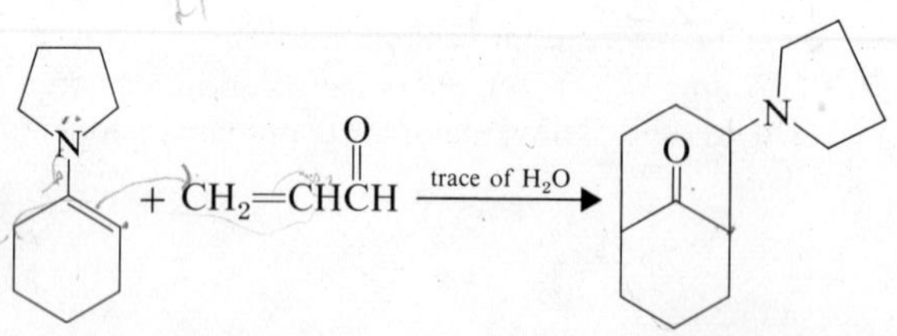

20.28

Knoevenagel condensations in which the active methylene compound is a β-keto ester or a β-diketone often yield products that result from one molecule of aldehyde or ketone and two molecules of the active methylene component, for example,

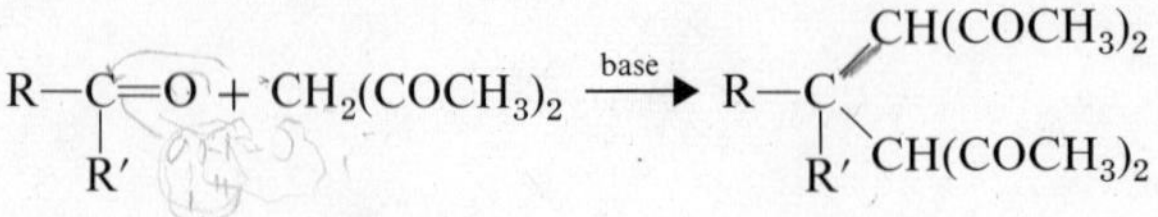

Suggest a reasonable mechanism that will account for the formation of these products.

20.29

Thymine (below) is one of the heterocyclic bases found in DNA. Starting with ethyl propanoate and using any other needed reagents show how you might synthesize thymine.

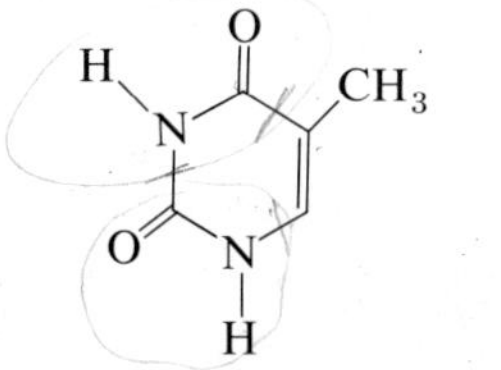

20.30

The mandibular glands of queen bees secrete a fluid that contains a remarkable compound known as "queen substance." When even an exceedingly small amount of the queen substance is transferred to worker bees it inhibits the development of their ovaries and prevents the workers from rearing new queens. Queen substance, a monocarboxylic acid with the molecular formula $C_{10}H_{16}O_3$, has been synthesized by the following route.

Cycloheptanone $\xrightarrow[\text{(2) } H_3O^+]{\text{(1) } CH_3MgI}$ **A** $(C_8H_{16}O)$ $\xrightarrow{H^+,\ \text{heat}}$ **B** (C_8H_{14}) $\xrightarrow[\text{(2) } Zn,\ H_2O]{\text{(1) } O_3}$

C $(C_8H_{14}O_2)$ $\xrightarrow[\text{pyridine}]{CH_2(CO_2H)_2}$ Queen Substance $(C_{10}H_{16}O_3)$

On catalytic hydrogenation queen substance yields compound **D**, which, on treatment with iodine in sodium hydroxide and subsequent acidification, yields a dicarboxylic acid **E**, that is,

Queen substance $\xrightarrow[Pd]{H_2}$ **D** $(C_{10}H_{18}O_3)$ $\xrightarrow[\text{(2) } H_3O^+]{\text{(1) } I_2/NaOH}$ **E** $(C_9H_{16}O_4)$

Provide structures for the queen substance and compounds **A–E**.

20.31

Linalool, an essential oil (p. 403) that can be isolated from a variety of plants, is 3,7-dimethyl-1,6-octadien-3-ol. Linalool is used in making perfumes and it can be synthesized in the following way:

$$CH_2{=}C{-}CH{=}CH_2 + HBr \longrightarrow \mathbf{F}\,(C_5H_9Br) \xrightarrow[\text{ester}]{\text{Sodioacetoacetic}}$$
$$|\ CH_3$$

G $(C_{11}H_{18}O_3)$ $\xrightarrow[\text{(2) } H_3O^+,\ \text{(3) heat}]{\text{(1) dil. NaOH}}$ **H** $(C_8H_{14}O)$ $\xrightarrow[\text{(2) } H_3O^+]{\text{(1) } LiC{\equiv}CH}$

I $(C_{10}H_{16}O)$ $\xrightarrow[\substack{\text{Lindlar's} \\ \text{catalyst}}]{H_2}$ Linalool

Outline the reactions involved. (Hint: compound **F** is the more stable isomer capable of being produced in the first step.)

20.32

Compound **J**, a compound with two four-membered rings, has been synthesized by the route shown below. Outline the steps that are involved.

$$NaCH(CO_2C_2H_5)_2 + BrCH_2CH_2CH_2Br \longrightarrow [C_{10}H_{17}BrO_4]$$

$$\xrightarrow{\text{NaOC}_2\text{H}_5} C_{10}H_{16}O_4 \xrightarrow{\text{LiAlH}_4} C_6H_{12}O_2 \xrightarrow{\text{HBr}}$$

$$C_6H_{10}Br_2 \xrightarrow[\text{2 NaOC}_2\text{H}_5]{\text{CH}_2(\text{CO}_2\text{C}_2\text{H}_5)_2} C_{13}H_{20}O_4 \xrightarrow[\text{(2) H}_3\text{O}^+]{\text{(1) OH}^-,\ \text{H}_2\text{O}}$$

$$C_9H_{12}O_4 \xrightarrow{\text{heat}} J(C_8H_{12}O_2) + CO_2$$

* 20.33

When an aldehyde or a ketone is condensed with ethyl α-chloroacetate in the presence of sodium ethoxide, the product is an α,β-epoxy ester called a *glycidic ester*. The synthesis is called the Darzens condensation.

$$\underset{\overset{|}{\text{R}}}{\overset{\overset{\text{R}'}{|}}{\text{C}}}{=}\text{O} + \text{ClCH}_2\text{COOC}_2\text{H}_5 \xrightarrow{\text{C}_2\text{H}_5\text{ONa}} \underset{\text{R}}{\overset{\text{R}'}{\text{C}}}\!\!-\!\!\underset{\text{O}}{\text{CHCOOC}_2\text{H}_5} + \text{NaCl} + \text{C}_2\text{H}_5\text{OH}$$

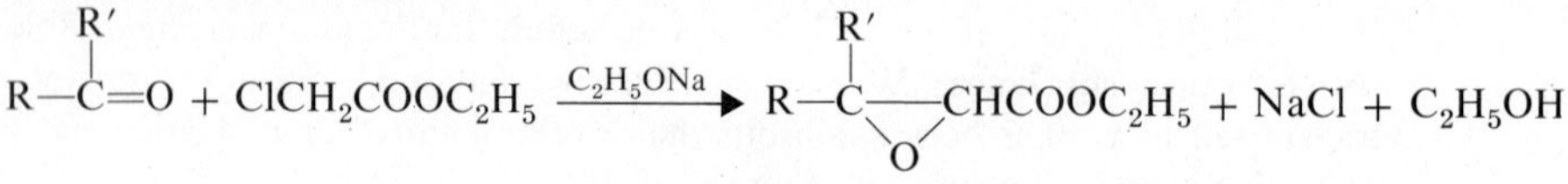

A glycidic ester

(a) Outline a reasonable mechanism for the Darzens condensation. (b) Hydrolysis of the epoxy ester leads to an epoxy acid that, on heating with pyridine, furnishes an aldehyde.

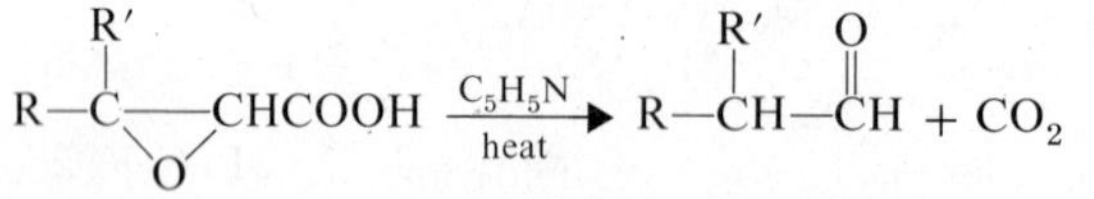

$$\underset{\text{R}}{\overset{\text{R}'}{\text{C}}}\!\!-\!\!\underset{\text{O}}{\text{CHCOOH}} \xrightarrow[\text{heat}]{\text{C}_5\text{H}_5\text{N}} \text{R}\!-\!\overset{\overset{\text{R}'}{|}}{\text{CH}}\!-\!\overset{\overset{\text{O}}{\|}}{\text{CH}} + CO_2$$

What is happening here? (c) Starting with β-ionone (p. 703) show how you might synthesize the aldehyde shown below. (This aldehyde is another intermediate in the industrial synthesis of Vitamin A that we began describing in problem 16.39.)

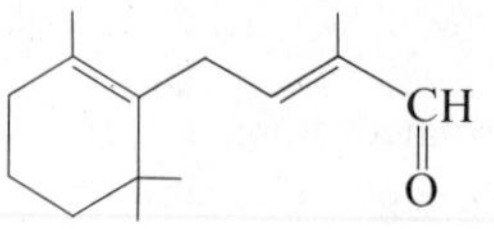

* 20.34

The *Perkin condensation* is an aldol-type condensation in which an aromatic aldehyde, ArCHO, reacts with a carboxylic acid anhydride, $(\text{RCH}_2\text{CO})_2\text{O}$, to give an α,β-unsaturated acid, ArCH=CRCOOH. The catalyst that is usually employed is the potassium salt of the carboxylic acid, RCH_2COOK. (a) Outline the Perkin condensation that takes place when benzaldehyde reacts with propanoic anhydride in the presence of potassium propanoate. (b) How would you use a Perkin condensation to prepare *p*-chlorocinnamic acid, $p\text{-ClC}_6\text{H}_4\text{CH}{=}\text{CHCOOH}$?

* 20.35

(+)-Fenchone is a terpenoid that can be isolated from fennel oil. ($\pm$) Fenchone has been synthesized through the route shown below. Supply the missing intermediates and reagents.

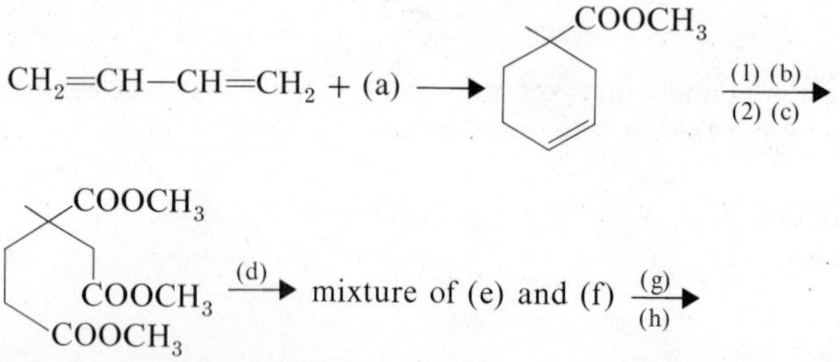

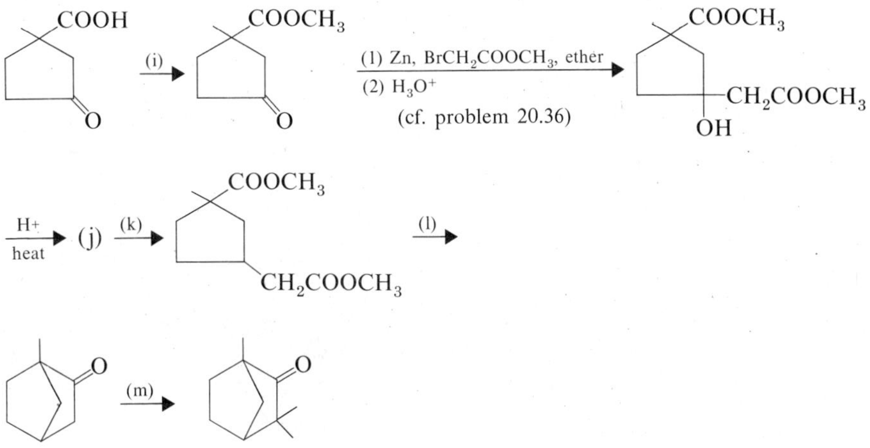

(±)-Fenchone

* 20.36

One step in the preceding problem involves the following reaction.

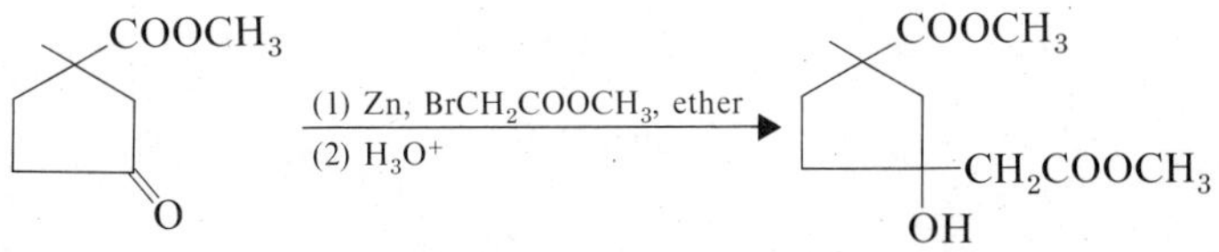

This is an example of the *Reformatsky reaction*. In general terms, the Reformatsky reaction involves treating an α-bromo ester with zinc in the presence of an aldehyde or ketone; the product (after acidification) is a β-hydroxy ester. (a) What general synthesis does this resemble? (b) What kind of reagent is likely to be an intermediate in the Reformatsky reaction? (c) The success of the Reformatsky reaction depends on the selectivity of this reagent. Explain. (d) What factor accounts for this selectivity? (e) Show how you might use the Reformatsky reaction in a synthesis of $C_6H_5CH_2CHCOOCH_3$ starting with benzaldehyde.

$$\underset{CH_3}{\quad}$$

* 20.37

Using as your starting compounds benzaldehyde, acetone, diethyl malonate, ethanol, and any needed inorganic reagents show how you might synthesize the compound shown below.

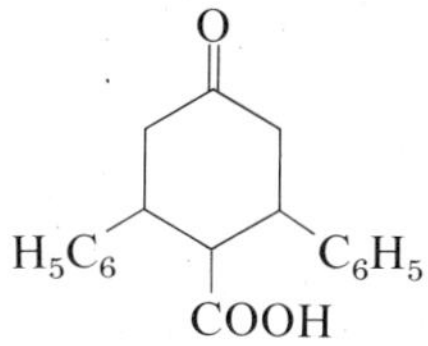

21 AMINES

21.1 NOMENCLATURE

In common nomenclature most primary amines are named by specifying the organic group attached to nitrogen and then adding the ending -amine:

Primary Amines

$$CH_3\ddot{N}H_2$$

Methylamine

$$CH_3CH_2\ddot{N}H_2$$

Ethylamine

$$CH_3CHCH_2\ddot{N}H_2$$
$$|$$
$$CH_3$$

Isobutylamine

$$CH_3$$
$$|$$
$$CH_3C-\ddot{N}H_2$$
$$|$$
$$CH_3$$

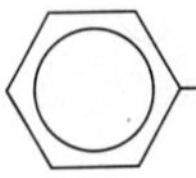

tert-Butylamine

$$\bigcirc\!-CH_2\ddot{N}H_2$$

Benzylamine

$$CH_2=CHCH_2\ddot{N}H_2$$

Allylamine

Most secondary and tertiary amines are named in the same general way; we either designate the organic groups individually if they are different, or use the prefixes di- or tri- if they are the same.

Secondary Amines

$$CH_3\ddot{N}HCH_2CH_3$$

Methylethylamine

$$(CH_3CH_2)_2\ddot{N}H$$

Diethylamine

$$CH_3$$
$$|$$
$$(CH_3CH_2CH)_2\ddot{N}H$$

Di-sec-butylamine

$$C_6H_5CH_2\ddot{N}HCH_2CH_3$$

Ethylbenzylamine

Tertiary Amines

$$(CH_3CH_2)_3\ddot{N}$$

Triethylamine

$$CH_2CH_3$$
$$|$$
$$CH_3\ddot{N}CH_2CH_2CH_3$$

Methylethylpropylamine

Salts of amines and quaternary ammonium compounds are named by adding -*ammonium* and then the name of the anion.

Quaternary Ammonium Compounds and Amine Salts

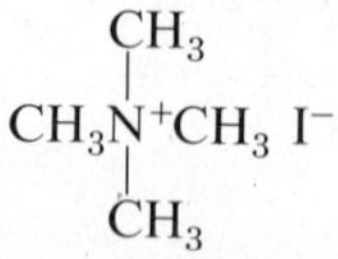

$$CH_3\overset{\displaystyle CH_3}{\underset{\displaystyle CH_3}{N^+}}CH_3 \; I^-$$

Tetramethylammonium iodide
(a quaternary ammonium compound)

$$CH_3\overset{\displaystyle CH_3}{\underset{\displaystyle CH_3}{N^\pm}}H \; Br^-$$

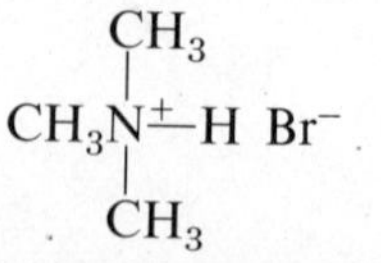

Trimethylammonium bromide
(a tertiary amine salt)

$$(CH_3CH_2\overset{+}{N}H_3)_2 \; SO_4^{=}$$

Ethylammonium sulfate
(a primary amine salt)

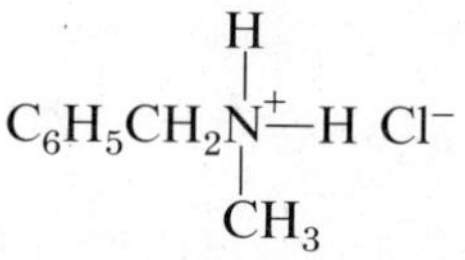

$$C_6H_5CH_2\overset{\displaystyle H}{\underset{\displaystyle CH_3}{N^\pm}}H \; Cl^-$$

Methylbenzylammonium chloride
(a secondary amine salt)

In the IUPAC system, the —NH_2 group is named as the *amino* group. We often use this system for naming complicated amines.

$$H_2\ddot{N}CH_2CH_2OH$$

2-Aminoethanol

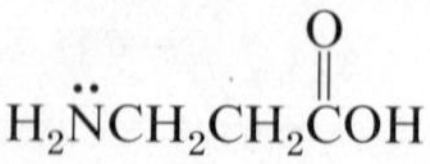

$$H_2\ddot{N}CH_2CH_2\overset{O}{\overset{\|}{C}}OH$$

3-Aminopropanoic acid

$$CH_3\underset{\displaystyle :NHCH_3}{CH}CH_2OH$$

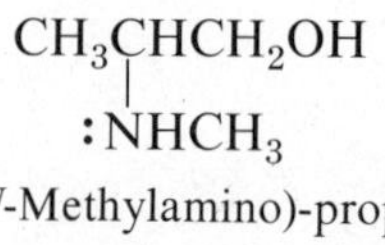

2-(*N*-Methylamino)-propanol

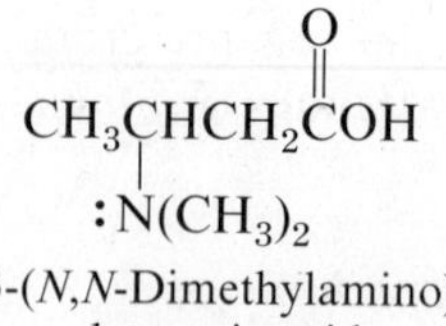

$$CH_3\underset{\displaystyle :N(CH_3)_2}{CH}CH_2\overset{O}{\overset{\|}{C}}OH$$

3-(*N*,*N*-Dimethylamino)-butanoic acid

As two examples above illustrate, whenever an H of the amino (—NH_2) group is replaced by some other group, the name of that other group preceded by a capital N appears in the name of the amine. The capital N means that the group is attached to nitrogen rather than to some carbon in the parent chain of the amine.

The more important *aromatic* and *heterocyclic* amines have the following names:

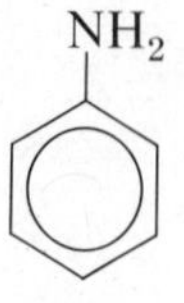

Aniline

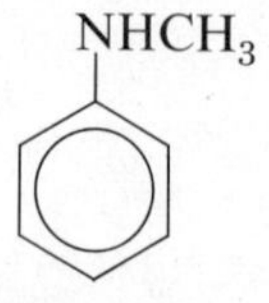

N-Methylaniline

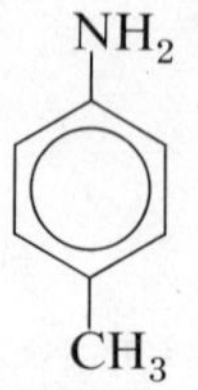

p-Toluidine

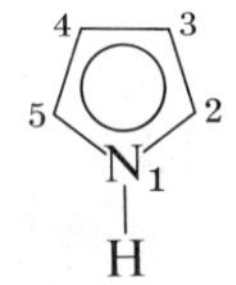

Pyrrole 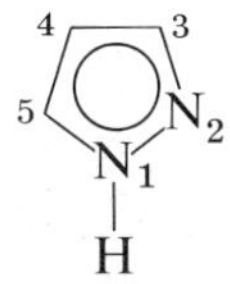 Pyrazole 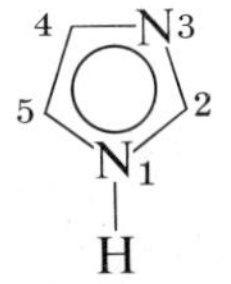 Imidazole 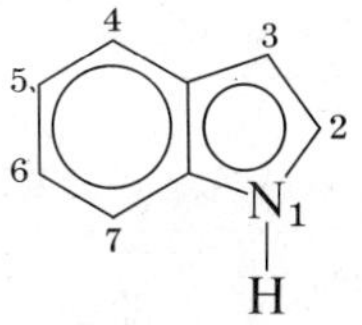 Indole

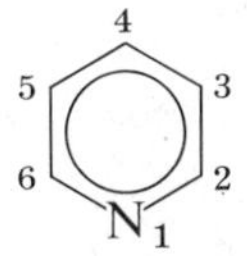

Pyridine 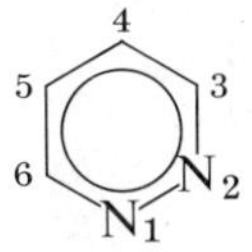 Pyridazine 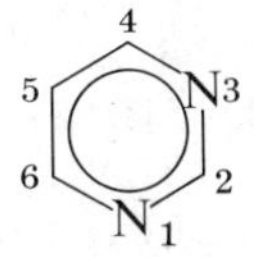 Pyrimidine 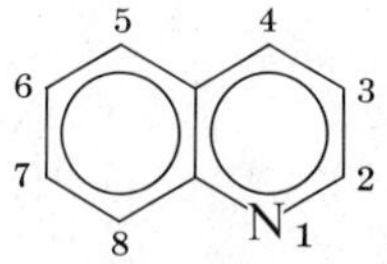 Quinoline

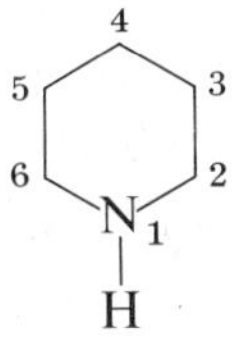

Piperidine 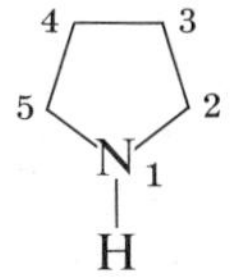 Pyrrolidine 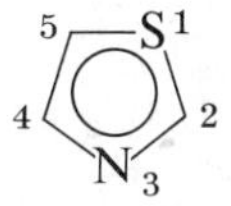 Thiazole 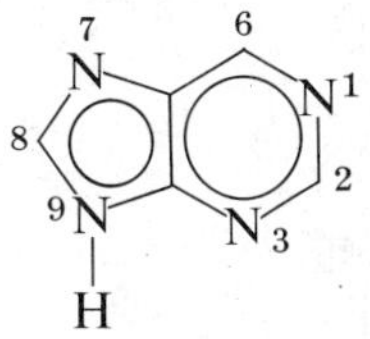 Purine

21.2 PHYSICAL PROPERTIES OF AMINES

Amines are moderately polar substances; they have boiling points that are higher than those of alkanes but generally lower than alcohols. Molecules of primary and secondary amines can form strong hydrogen bonds to each other and to water. Molecules of tertiary amines cannot form hydrogen bonds to each other but they can form hydrogen bonds to water. As a result, tertiary amines generally boil at lower temperatures than primary and secondary amines of comparable molecular weight, but all low-molecular-weight amines are very water soluble.

Table 21.1 lists the physical properties of some common amines.

21.3 BASICITY OF AMINES: AMINE SALTS

Amines are relatively weak bases. They are stronger bases than water but are far weaker bases than hydroxide ions, alkoxide ions, and so forth.

A convenient expression for relating basicities is a quantity called the *basicity constant*, K_b. When an amine dissolves in water, the following equilibrium is established,

$$\overset{..}{R}NH_2 + H_2O \rightleftarrows RNH_3^+ + OH^-$$

and K_b is given by the expression:

$$K_b = \frac{[RNH_3^+][OH^-]}{[RNH_2]}$$

The larger the value of K_b, the greater is the tendency of the amine to accept a proton from water and, thus, the greater will be the concentration of RNH_3^+ and

TABLE 21.1 Physical Properties of Amines

NAME	STRUCTURE	mp °C	bp °C	WATER SOLUBILITY	K_b (25°)
Primary Amines					
Methylamine	CH_3NH_2	−94	−6	Very sol.	4.4×10^{-4}
Ethylamine	$CH_3CH_2NH_2$	−84	17	Very sol.	5.6×10^{-4}
Propylamine	$CH_3CH_2CH_2NH_2$	−83	49	Very sol.	4.7×10^{-4}
Isopropylamine	$(CH_3)_2CHNH_2$	−101	33	Very sol.	5.3×10^{-4}
Butylamine	$CH_3(CH_2)_2CH_2NH_2$	−51	78	Very sol.	4.1×10^{-4}
Isobutylamine	$(CH_3)_2CHCH_2NH_2$	−86	68	Very sol.	3.1×10^{-4}
sec-Butylamine	$CH_3CH_2CH(CH_3)NH_2$	−104	63	Very sol.	3.6×10^{-4}
tert-Butylamine	$(CH_3)_3CNH_2$	−68	45	Very sol.	2.8×10^{-4}
Cyclohexylamine	Cyclo-$C_6H_{11}NH_2$		134	Sl. sol.	4.4×10^{-4}
Benzylamine	$C_6H_5CH_2NH_2$		185		2.0×10^{-5}
Aniline	$C_6H_5NH_2$	−6	184	3.7 g/100 g	3.8×10^{-10}
p-Toluidine	p-$CH_3C_6H_4NH_2$	44	200	Sl. sol.	1.2×10^{-9}
p-Anisidine	p-$CH_3OC_6H_4NH_2$	57	244	V. sl. sol.	2.0×10^{-9}
p-Chloroaniline	p-Cl—$C_6H_4NH_2$	70	232	Insol.	1.0×10^{-10}
p-Nitroaniline	p-$NO_2C_6H_4NH_2$	148	232	Insol.	1.0×10^{-13}
Secondary Amines					
Dimethylamine	$(CH_3)_2NH$	−96	7	Very sol.	5.2×10^{-4}
Diethylamine	$(CH_3CH_2)_2NH$	−48	56	Very sol.	9.6×10^{-4}
Dipropylamine	$(CH_3CH_2CH_2)_2NH$	−40	110	Very sol.	9.5×10^{-4}
N-Methylaniline	$C_6H_5NHCH_3$	−57	196	Sl. sol.	5.0×10^{-10}
Diphenylamine	$(C_6H_5)_2NH$	53	302	Insol.	6.0×10^{-14}
Tertiary Amines					
Trimethylamine	$(CH_3)_3N$	−117	3.5	Very sol.	5.0×10^{-5}
Triethylamine	$(CH_3CH_2)_3N$	−115	90	14 g/100 g	5.7×10^{-4}
Tripropylamine	$(CH_3CH_2CH_2)_3N$	−90	156	Sl. sol.	4.4×10^{-4}
N,N-Dimethylaniline	$C_6H_5N(CH_3)_2$	3	194	Sl. sol.	11.5×10^{-10}

OH$^-$ in the solution. Larger values of K_b, therefore, are associated with those amines that are stronger bases and smaller values of K_b are associated with those amines that are weaker bases.

The basicity constant of ammonia at 25°C is 1.8×10^{-5}.

$$\ddot{N}H_3 + H_2O \rightleftharpoons \overset{+}{N}H_4 + OH^-$$

$$K_b = 1.8 \times 10^{-5} = \frac{[NH_4^+][OH^-]}{[NH_3]}$$

When we examine the basicity constants of the amines given in Table 21.1, we see that most aliphatic primary amines (e.g., methylamine, ethylamine) are somewhat stronger bases than ammonia:

	$\ddot{N}H_3$	$CH_3\ddot{N}H_2$	$CH_3CH_2\ddot{N}H_2$	$CH_3CH_2CH_2\ddot{N}H_2$
K_b	1.8×10^{-5}	4.4×10^{-4}	5.6×10^{-4}	4.7×10^{-4}

We can account for this on the basis of the electron-releasing ability of an alkyl group. An alkyl group releases electrons, and it *stabilizes* the alkyl ammonium ion that results from the acid-base reaction *by dispersing its positive charge*. It stabilizes the alkyl ammonium ion to a greater extent than it stabilizes the amine.

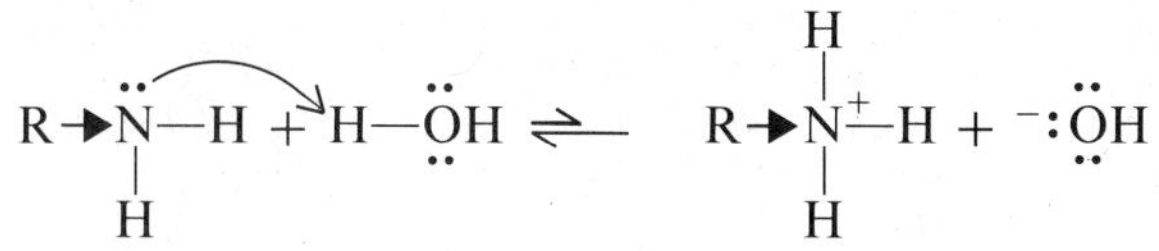

$$R \rightarrow \overset{..}{\underset{\overset{|}{H}}{N}} - H + H - \overset{..}{\underset{..}{O}}H \rightleftharpoons R \rightarrow \overset{\overset{H}{|}}{\underset{\overset{|}{H}}{N^+}} - H + {}^- : \overset{..}{\underset{..}{O}}H$$

By releasing electrons, R-
stabilizes the alkyl
ammonium ion
through dispersal of charge

When we examine the basicity constants of the aromatic amines (e.g., aniline, *p*-toluidine) in Table 21.1, we see that they are much weaker bases than ammonia.

	NH_3	$C_6H_5NH_2$	$p\text{-}CH_3C_6H_4NH_2$
K_b	1.8×10^{-5}	3.8×10^{-10}	1.2×10^{-9}

We can account for this effect on the basis of resonance contributions to the overall hybrid of an aromatic amine. For aniline, the following contributors are important.

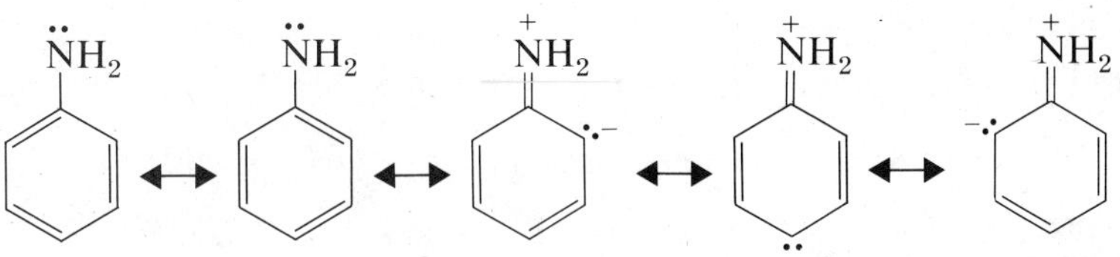

Structures **1** and **2** are the Kekulé structures that contribute to any benzene derivative. Structures **3–5**, however, *delocalize* the unshared electron pair of the nitrogen over the ortho and para positions of the ring. This delocalization of the electron pair makes it less available to a proton but, more importantly, *delocalization of the electron pair stabilizes aniline.*

When aniline accepts a proton it becomes an anilinium ion.

$$C_6H_5\overset{..}{N}H_2 + H_2O \rightleftharpoons C_6H_5\overset{+}{N}H_3 + {}^-\overset{}{O}H$$

Anilinium
ion

Since the electron pair of nitrogen accepts the proton, we are able to write only *two* resonance structures for the anilinium ion—the two Kekulé structures:

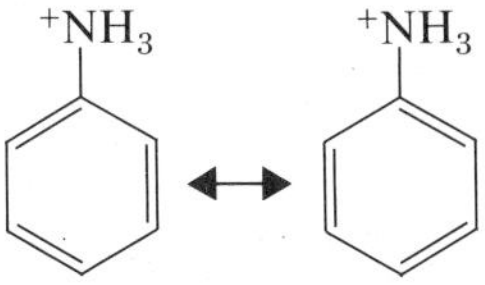

Structures corresponding to **3–5** are not possible for the anilinium ion and, consequently, resonance does not stabilize the anilinium ion to as great an extent as it does aniline itself. This greater stabilization of the reactant (aniline) when compared

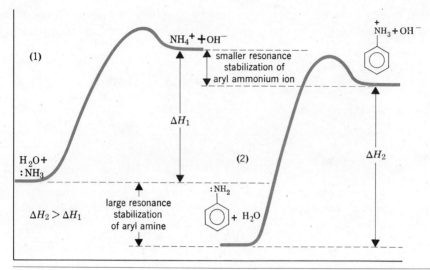

FIG. 21.1

Potential energy diagram for the reaction of NH_3 with H_2O (1), and for the reaction of an arylamine with H_2O (2).

to that of the product (anilinium ion) means that ΔH for the reaction,

$$\text{aniline} + H_2O \longrightarrow \text{anilinium ion} + OH^- \text{ (Fig. 21.1)},$$

will be a larger positive quantity than that for the reaction,

$$\text{ammonia} + H_2O \longrightarrow \text{ammonium ion} + OH^-$$

Aniline, as a result, is the weaker base.

Problem 21.1

Diphenylamine, $(C_6H_5)_2NH$, is a far weaker base ($K_b = 6 \times 10^{-14}$) than aniline ($K_b = 3.8 \times 10^{-10}$). Explain this, using resonance structures.

Problem 21.2

p-Toluidine and *p*-anisidine (see Table 21.1) are stronger bases than aniline, whereas *p*-chloroaniline and *p*-nitroaniline are weaker bases. Suggest explanations.

Amine Salts

When primary, secondary, and tertiary amines react with acids, they form amine salts:

$$CH_3CH_2\overset{\cdot\cdot}{N}H_2 + HCl \xrightarrow{H_2O} CH_3CH_2\overset{+}{N}H_3 \ Cl^-$$
Ethylammonium chloride
(an amine salt)

$$(CH_3CH_2)_2\overset{\cdot\cdot}{N}H + HBr \xrightarrow{H_2O} (CH_3CH_2)_2\overset{+}{N}H_2 \ Br^-$$
Diethylammonium bromide

$$(CH_3CH_2)_3\overset{..}{N} + HI \xrightarrow{H_2O} (CH_3CH_2)_3\overset{+}{N}H \quad I^-$$
Triethylammonium iodide

Quaternary ammonium halides do not undergo such a reaction for they are salts already and they do not have an unshared electron pair.

$$(CH_3CH_2)_4\overset{+}{N} \quad Br^-$$
Tetraethylammonium bromide
(an amine salt—does not
undergo reaction with acid)

Quaternary ammonium hydroxides, however, are strong bases. As solids, or in solution, they consist *entirely* of quaternary ammonium cations (R_4N^+) and hydroxide ions, OH^-; they are, therefore, strong bases—as strong as sodium or potassium hydroxide. Quaternary ammonium hydroxides react with acids to form quaternary ammonium salts:

$$(CH_3)_4\overset{+}{N} OH^- + HCl \longrightarrow (CH_3)_4\overset{+}{N} Cl^- + H_2O$$

Almost all alkylammonium chlorides, bromides, iodides, and sulfates are water soluble. Thus, a water-insoluble primary, secondary, or tertiary amine will dissolve in aqueous HCl, HBr, HI, or H_2SO_4. Acid solubility provides a convenient

$$\overset{|}{\underset{/}{-}N}: \quad + \quad \begin{matrix} HX \\ (or \ H_2SO_4) \end{matrix} \longrightarrow \overset{|}{\underset{/}{-}N}-H \quad \begin{matrix} X^- \\ (or \ HSO_4^-) \end{matrix}$$

Water-insoluble   Water-soluble
amine   salt

chemical method for distinguishing amines from nonbasic compounds that are water insoluble. Acid solubility also gives us a useful method for separating amines from nonbasic compounds that are water insoluble.

Problem 21.3
Outline a procedure for separating cyclohexylamine from cyclohexane using dilute HCl, aqueous NaOH, and ether.

Problem 21.4
Outline a procedure for separating a mixture of benzoic acid, *p*-cresol, aniline, and benzene using acids, bases, and organic solvents.

21.4 PREPARATION OF AMINES

1. Through Nucleophilic Substitution Reactions
Primary amine salts can be prepared from ammonia and alkyl halides through nucleophilic substitution reactions. Subsequent treatment of the resulting ammonium salts with base give primary amines.

$$\overset{..}{N}H_3 + R{-}X \longrightarrow R{-}NH_3{}^+X^- \xrightarrow{\ OH^-\ } RNH_2$$

This method is often of limited synthetic application because multiple alkylations tend to occur. When ethyl bromide reacts with ammonia, for example, the ethylammonium bromide that is produced initially can react with ammonia to liberate ethylamine. Ethylamine can then compete with ammonia and react with ethyl bromide to give diethylammonium bromide. Repetitions of acid-base and alkylation reactions ultimately produce some tertiary amines and even some quaternary ammonium salts if the alkyl halide is present in excess.

$$C_2H_5Br + \overset{..}{N}H_3 \longrightarrow C_2H_5NH_3{}^+Br^-$$

$$C_2H_5NH_3{}^+Br^- + NH_3 \rightleftharpoons C_2H_5\overset{..}{N}H_2 + NH_4{}^+Br^-$$

$$C_2H_5\overset{..}{N}H_2 + C_2H_5Br \longrightarrow (C_2H_5)_2NH_2{}^+Br^-$$

$$(C_2H_5)_2NH_2{}^+Br^- + \overset{..}{N}H_3 \rightleftharpoons (C_2H_5)_2\overset{..}{N}H + NH_4{}^+Br^-$$

$$(C_2H_5)_2\overset{..}{N}H + C_2H_5Br \longrightarrow (C_2H_5)_3NH^+Br^-$$

Multiple alkylations can be minimized by using a large excess of ammonia. (Why?) An example of this technique can be seen in the synthesis of alanine from 2-bromopropanoic acid:

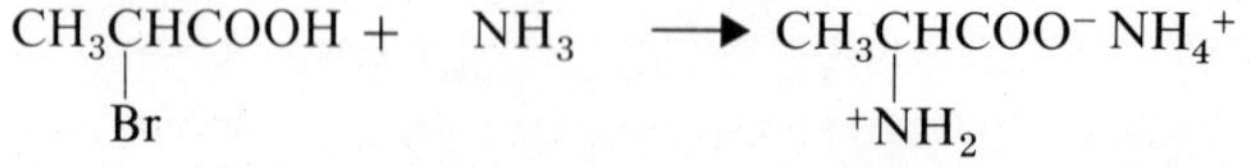

$$\underset{\substack{| \\ Br \\ (1\ mole)}}{CH_3CHCOOH} + \underset{(70\ moles)}{NH_3} \longrightarrow \underset{\substack{| \\ {}^+NH_2 \\ Alanine \\ (65\text{--}70\%)}}{CH_3CHCOO^- NH_4{}^+}$$

Nucleophilic substitution reactions can also be carried out using primary, secondary, and tertiary amines:

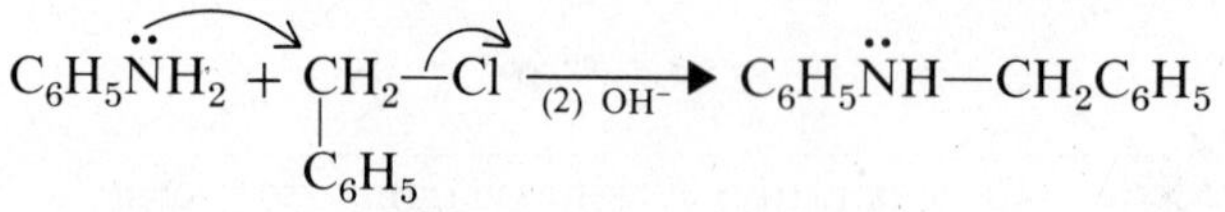

$$\underset{(4\ moles)}{C_6H_5\overset{..}{N}H_2} + \underset{\substack{| \\ C_6H_5 \\ (1\ mole)}}{CH_2{-}Cl} \xrightarrow{(2)\ OH^-} \underset{(96\%)}{C_6H_5\overset{..}{N}H{-}CH_2C_6H_5}$$

$$(CH_3)_2\overset{..}{N}H + \underset{NO_2}{\overset{Br}{\bigcirc}} \xrightarrow{NaHCO_3} \underset{\substack{NO_2 \\ (94\text{--}97\%)}}{\overset{\overset{..}{N}(CH_3)_2}{\bigcirc}}$$

$$\bigcirc N{:} + CH_3{-}I \xrightarrow{CH_3CN} \underset{(95\%)}{\bigcirc \overset{+}{N}{-}CH_3\ \ I^-}$$

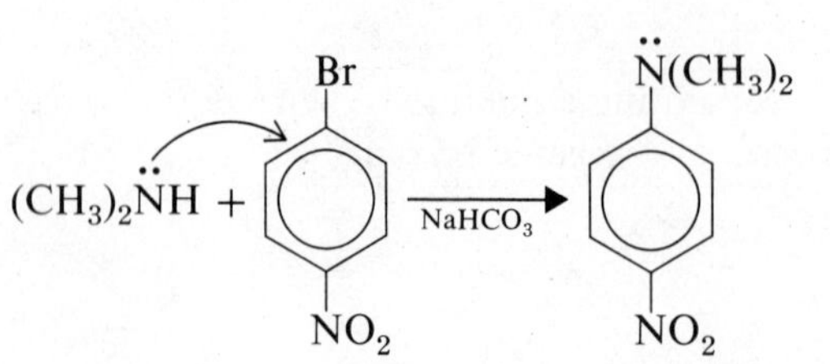

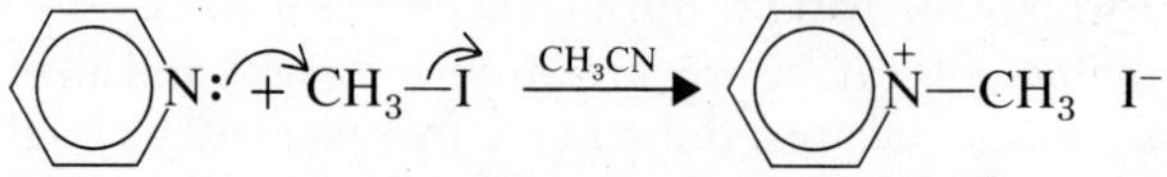

Potassium phthalimide (below) can be used to prepare primary amines through a method known as the *Gabriel synthesis*. This synthesis avoids the complications of multiple alkylations that occur when alkyl halides are treated with ammonia:

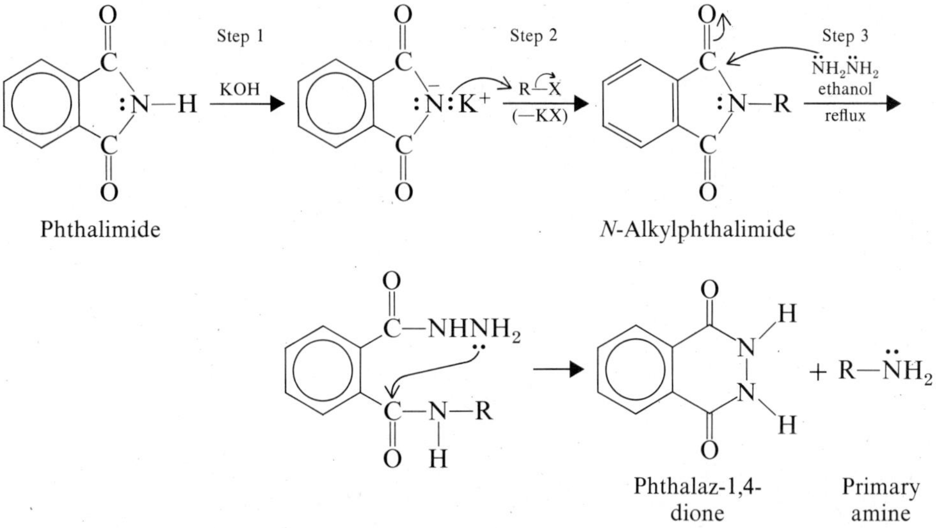

Phthalimide   N-Alkylphthalimide

Phthalaz-1,4-   Primary
dione   amine

Phthalimide is quite acidic ($Ka \simeq 10^{-9}$); it can be converted to potassium phthalimide by potassium hydroxide (step 1). The phthalimide anion is a strong nucleophile and (in step 2) it reacts with an alkyl halide to give an N-alkylphthalimide. Treating the N-alkylphthalimide with hydrazine (NH_2NH_2) in refluxing ethanol (step 3) gives a primary amine and a phthalaz-1,4-dione.

Syntheses of amines using nucleophilic substitution reactions are, as we might expect, restricted to the use of methyl, primary, and secondary alkyl halides, and to aryl halides activated by strong *o*- or *p*-electron-withdrawing groups. The use of tertiary halides leads almost exclusively to elimination reactions.

Problem 21.5

(a) Write resonance structures for the phthalimide anion that will account for the acidity of phthalimide. (b) Would you expect phthalimide to be more or less acidic than benzamide? Why?

Problem 21.6

Outline a preparation of benzylamine using the Gabriel synthesis.

2. Preparation of Amines Through Reduction of Nitro Compounds

The most widely used method for preparing aromatic amines involves nitration of the ring and subsequent reduction of the nitro group to an amino group.

$$Ar-H \xrightarrow[H_2SO_4]{HNO_3} Ar-NO_2 \xrightarrow{(H)} Ar-NH_2$$

We studied ring nitration in Chapter 13 and saw there that it is applicable to a wide variety of aromatic compounds. Reduction of the nitro group can also be carried out in a number of ways. The most frequently used methods are based on catalytic hydrogenation, or treatment of the nitro compound with acid and iron, zinc, or tin, or a metal salt such as $SnCl_2$. Several specific examples are shown below.

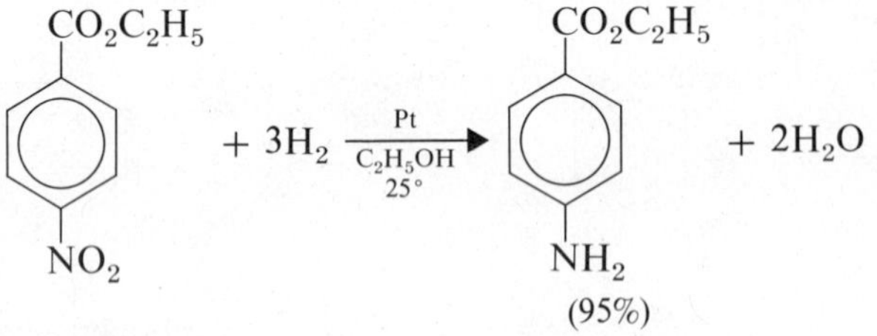

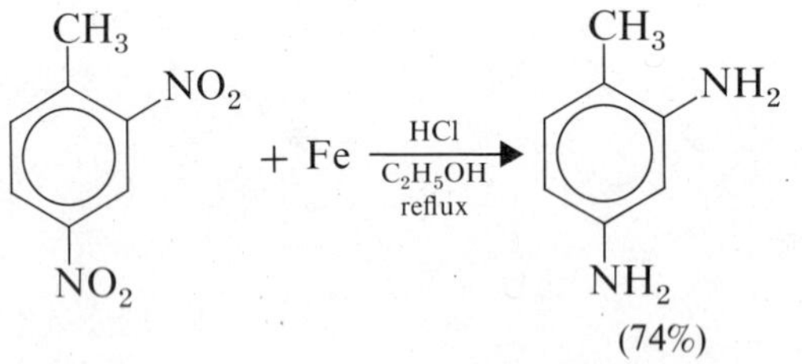

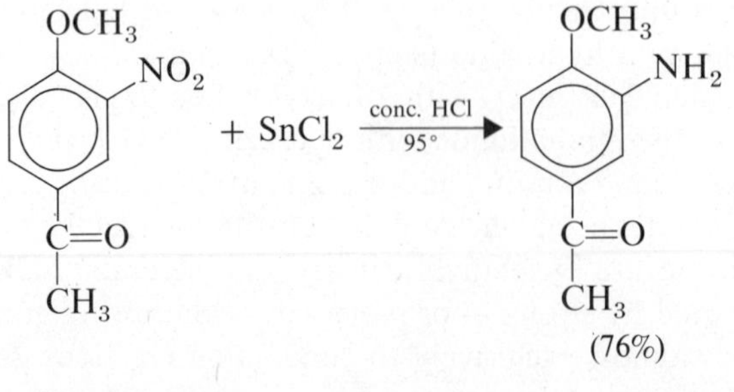

Selective reduction of one nitro group of a dinitro compound can often be achieved through the use of hydrogen sulfide in aqueous (or alcoholic) ammonia:

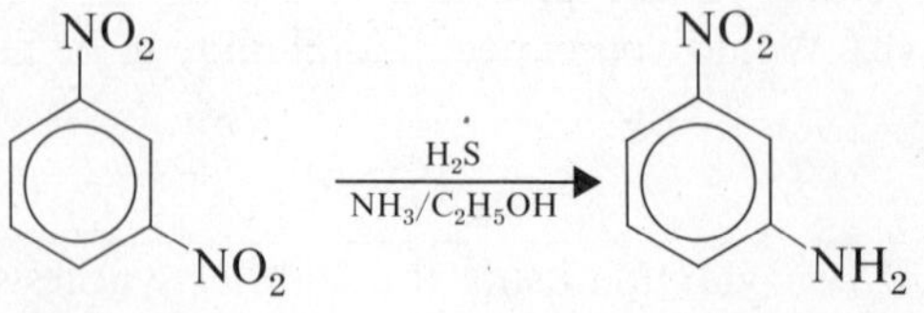

m-Dinitrobenzene   m-Nitroaniline
(70–80%)

When this method is used, the amount of the hydrogen sulfide must be carefully measured because the use of an excess may result in the reduction of more than one nitro group.

It is not always possible to predict just which nitro group will be reduced, however. Treating 2,4-dinitrotoluene with hydrogen sulfide and ammonia, results in reduction of the 4-nitro group:

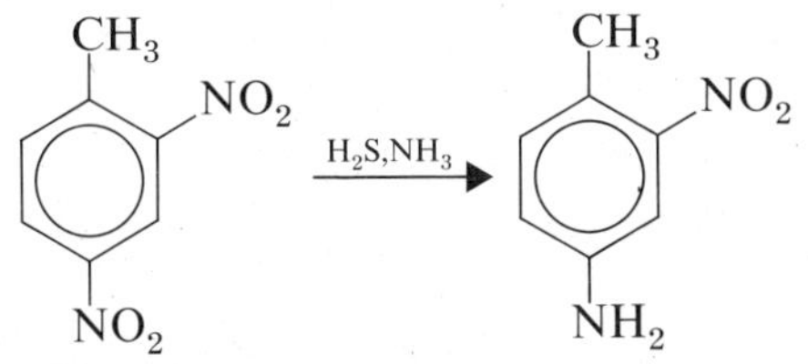

On the other hand, monoreduction of 2,4-dinitroaniline causes reduction of the 2-nitro group:

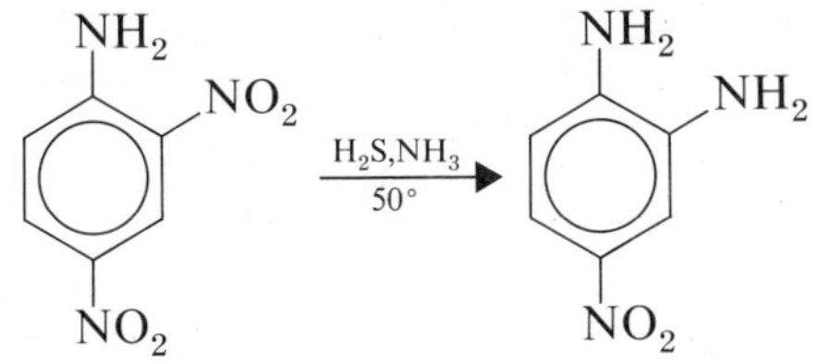

(60%)

3. Preparation of Amines Through Reductive Amination

Aldehydes and ketones can be converted to primary amines through catalytic or chemical reduction in the presence of ammonia. This process, called *reductive amination,* proceeds through the formation of an imine.

General Reaction

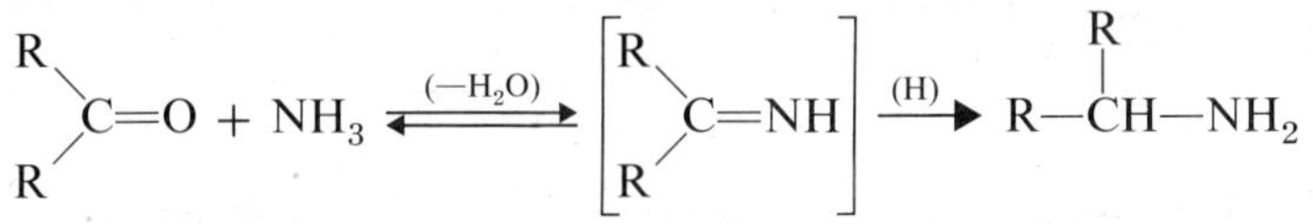

Aldehyde or ketone — Imine — A 1° amine

Specific Examples

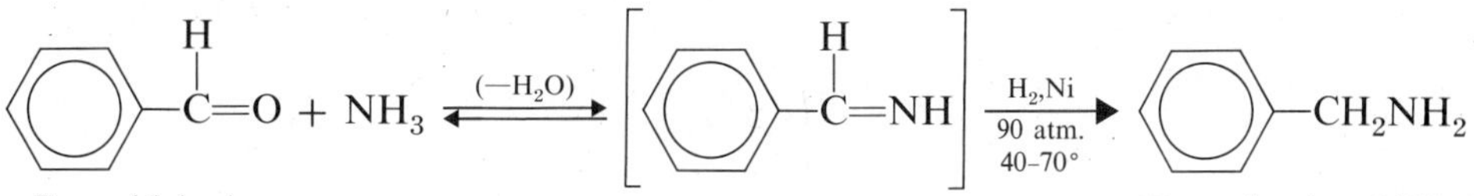

Benzaldehyde — Benzylamine (89%)

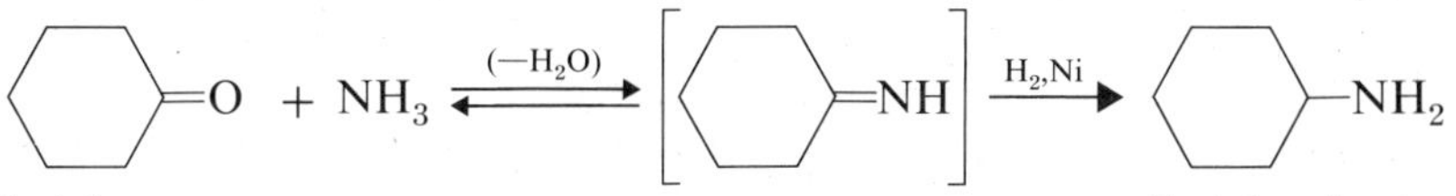

Cyclohexanone — Cyclohexylamine

Secondary amines can be prepared through reductive amination of an aldehyde or ketone in the presence of a primary amine.

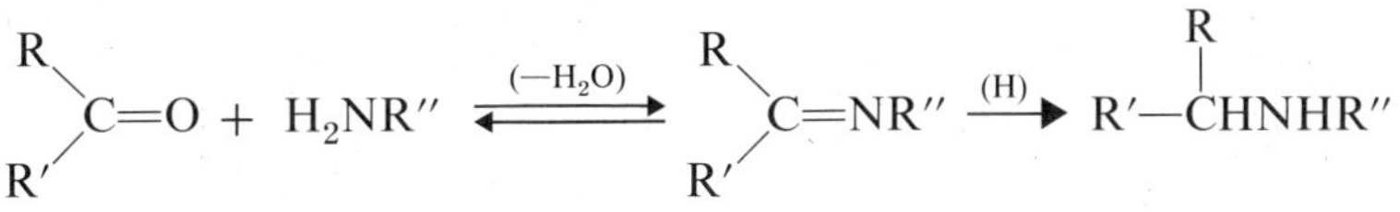

Aldehyde or ketone — 1° Amine — An imine — 2° Amine

An example is the synthesis of ethylbenzylamine from benzaldehyde and ethylamine:

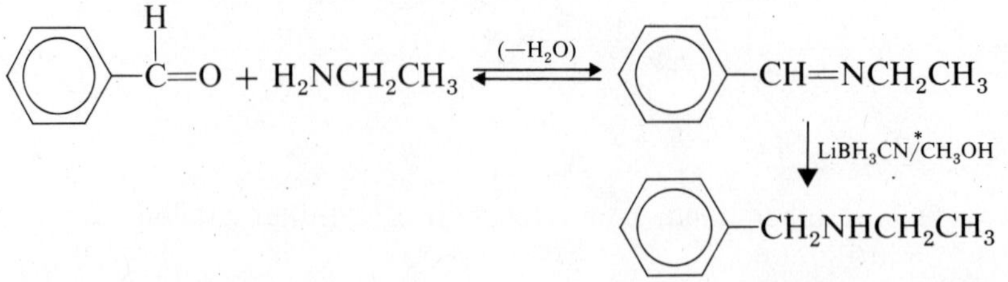

(72%)

Problem 21.7

Show how you might prepare each of the following amines through reductive amination.

    (a) $CH_3(CH_2)_3CH_2NH_2$

    (b) $C_6H_5CHCH_3$

          NH_2

    (c) $CH_3(CH_2)_4CH_2NHC_6H_5$

Problem 21.8

Reductive amination of a ketone is almost always a better method for preparation

$$R'$$
$$|$$

of amines of the type $RCHNH_2$, than treatment of an alkyl halide with ammonia. Why would this be true?

4. Preparation of Amines Through Reduction of Amides and Nitriles

    Amides and nitriles can be reduced to amines. Reduction of a nitrile yields a primary amine; reduction of an amide can yield a primary, secondary or tertiary amine.

$$R-C\equiv N \xrightarrow{(H)} RCH_2NH_2$$

Nitrile   1° Amine

(If R′ and R″ = H, the product is a 1° amine,
if R′ = H, the product is a 2° amine.)

    Nitriles can be reduced with hydrogen and a catalyst or by chemical reducing agents. Lithium aluminum hydride is often used as a chemical reducing agent for nitriles and amides. Several specific examples are listed on page 827.

---

*A reducing agent similar to $LiBH_4$. $LiBH_3CN$ (and $NaBH_3CN$) reduce iminium groups more rapidly than they reduce carbonyl groups.

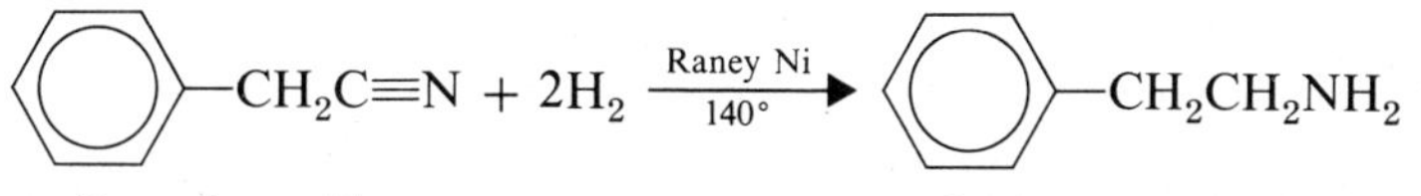

Benzyl cyanide $\xrightarrow[\text{140°}]{\text{Raney Ni}}$ 2-Phenylethylamine (71%)

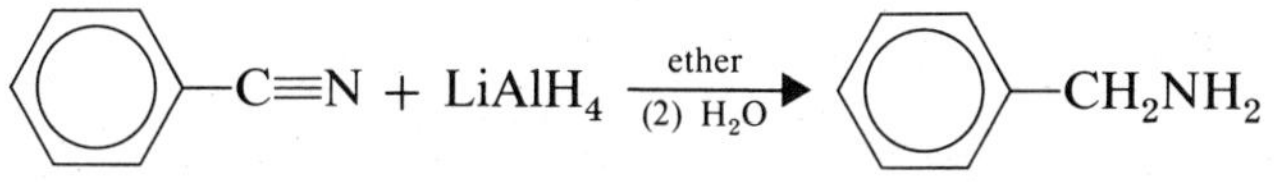

Benzonitrile + LiAlH$_4$ $\xrightarrow[\text{(2) H}_2\text{O}]{\text{ether}}$ Benzylamine (72%)

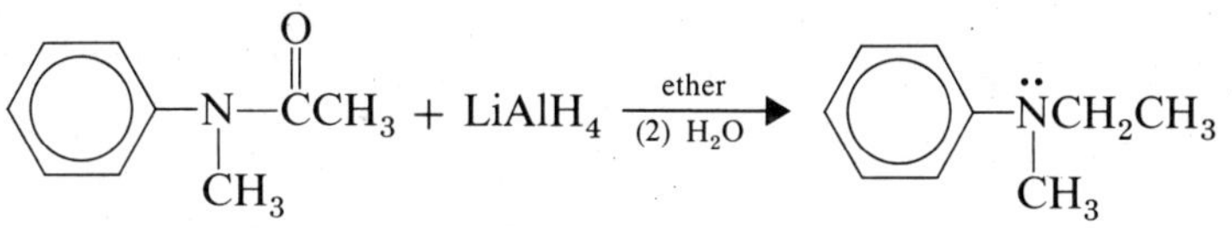

N-Methylacetanilide + LiAlH$_4$ $\xrightarrow[\text{(2) H}_2\text{O}]{\text{ether}}$ N-Ethyl-N-methyl aniline

Problem 21.9

Show how you might utilize the reduction of an amide or a nitrile to carry out each of the following transformations.

(a) Benzoic acid $\longrightarrow$ N-ethy-N-benzylamine
(b) 1-Bromopentane $\longrightarrow$ 1-hexylamine
(c) Propanoic acid $\longrightarrow$ tripropylamine

5. Preparation of Amines Through the Hofmann Degradation of Amides

Amides react with solutions of bromine or chlorine in sodium hydroxide to yield amines through a reaction known as the *Hofmann degradation*:

$$R-\overset{\overset{\displaystyle O}{\|}}{C}NH_2 + Br_2 + 4NaOH \xrightarrow{H_2O} RNH_2 + 2NaBr + Na_2CO_3 + 2H_2O$$

From the equation above we can see that the carbonyl carbon of the amide is lost (as $CO_3^=$) and that the R group of the amide becomes attached to the nitrogen of the amine.

The mechanism for this interesting reaction involves the following steps. In step 1 the amide undergoes a base-promoted *N-bromination*. In step 2 the *N*-bromo amide reacts with base to yield an anion which in step 3 simultaneously rearranges and loses a bromide ion to give an *isocyanate*. Finally, in step 4, the isocyanate undergoes hydrolysis to yield an amine and $CO_3^=$ ion.

Step 1

N-Bromo amide

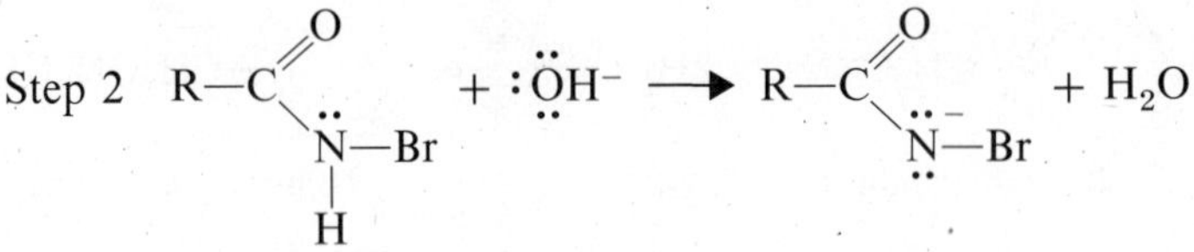

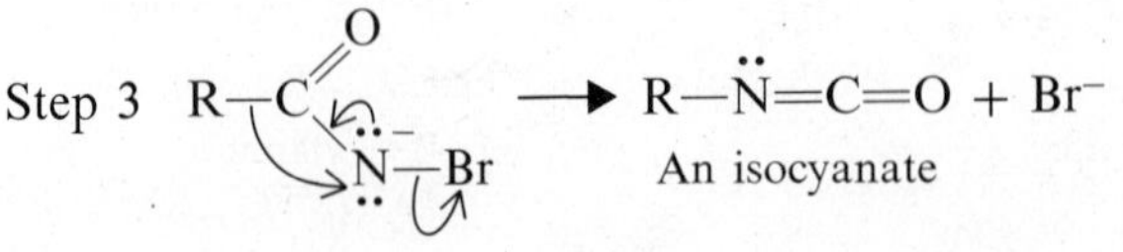

Step 2 $R-C\overset{O}{\underset{\underset{H}{\overset{|}{N}}-Br}{\diagdown}}$ $+ :\ddot{O}H^-$ $\longrightarrow$ $R-C\overset{O}{\underset{\overset{..}{\underset{..}{N}}-Br}{\diagdown}}$ $+ H_2O$

Step 3 $R-C\overset{O}{\underset{\overset{..}{N}-Br}{\diagdown}}$ $\longrightarrow$ $R-\ddot{N}=C=O + Br^-$

An isocyanate

Step 4 $R-\ddot{N}=C=O + 2O\overset{-}{H}$ $\longrightarrow$ $R-NH_2 + CO_3^=$

Studies of Hofmann degradations of optically active amides in which the chiral carbon is directly attached to the carbonyl group have shown that they occur with *retention of configuration*. Thus, the R group migrates to nitrogen with its electrons, *but without inversion*.

With lower-molecular-weight amides, Hofmann degradations using bromine in sodium hydroxide give excellent yields as the first two examples shown below illustrate. With high-molecular-weight amides, the yields are not as good but can be improved through the use of bromine and sodium methoxide in methanol.

$$CH_3CH_2CH_2CH_2CH_2\overset{O}{\overset{||}{C}}NH_2 \xrightarrow[H_2O]{Br_2,OH^-} CH_3CH_2CH_2CH_2CH_2NH_2$$

Hexanamide   *n*-Pentylamine
(88%)

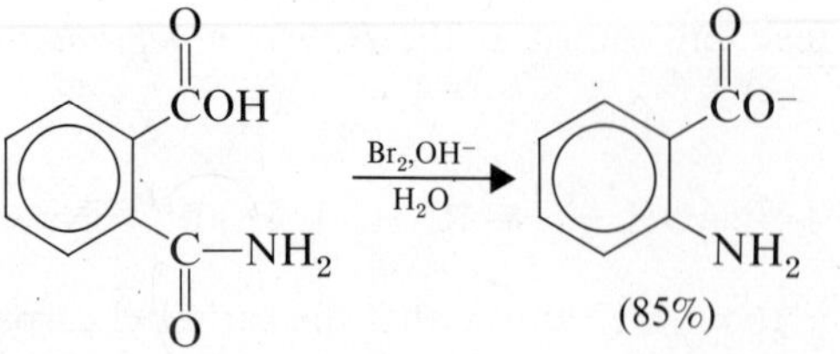

CH_3(CH_2)_{14}$\overset{O}{\overset{||}{C}}$NH_2 $\xrightarrow[CH_3OH]{Br_2,CH_3O^-}$ CH_3(CH_2)_{13}CH_2NH_2
(100%)

Problem 21.10

Using a different method for each part, but taking care in each case to select the *best* method, show how each of the following transformations might be accomplished.

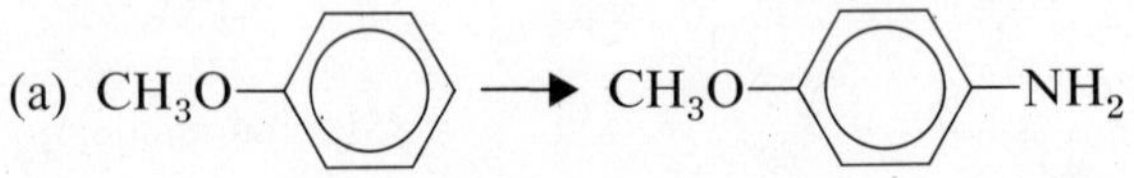

(a) CH_3O— ⟨○⟩ $\longrightarrow$ CH_3O— ⟨○⟩ —NH_2

(b) 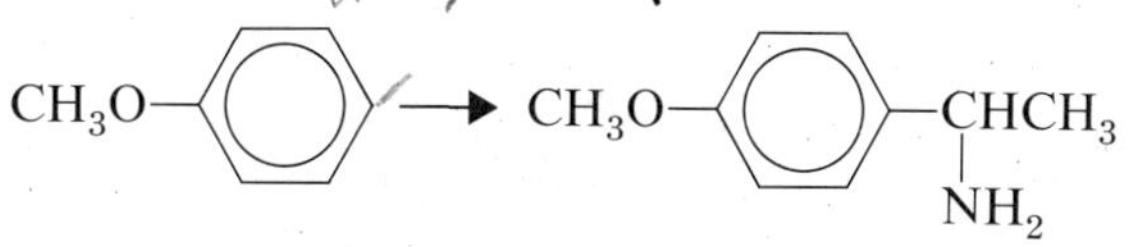 CH$_3$O—⟨◯⟩— $\longrightarrow$ CH$_3$O—⟨◯⟩—CHCH$_3$
   |
   NH$_2$

(c) 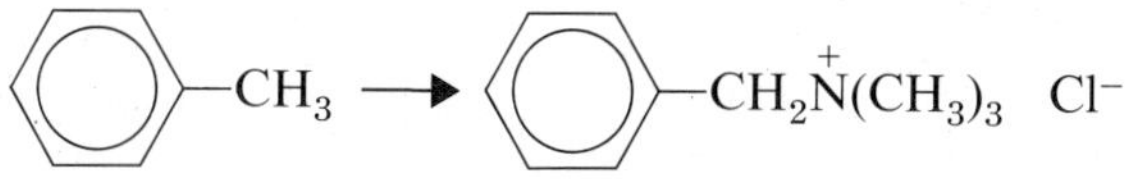 ⟨◯⟩—CH$_3$ $\longrightarrow$ ⟨◯⟩—CH$_2\overset{+}{N}$(CH$_3$)$_3$ Cl$^-$

(d) 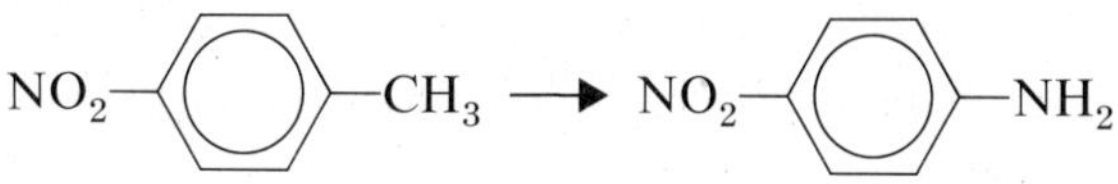 NO$_2$—⟨◯⟩—CH$_3$ $\longrightarrow$ NO$_2$—⟨◯⟩—NH$_2$

(e) 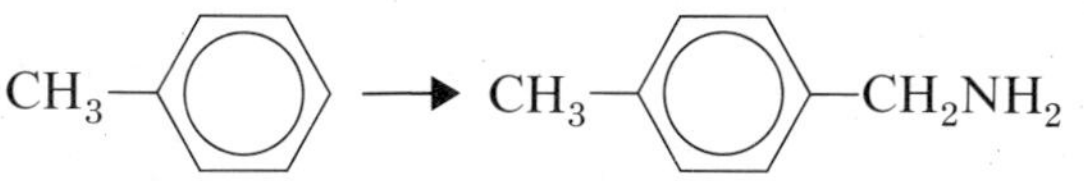 CH$_3$—⟨◯⟩ $\longrightarrow$ CH$_3$—⟨◯⟩—CH$_2$NH$_2$

21.5 REACTIONS OF AMINES

We have encountered a number of important reactions of amines in earlier sections of this book. We saw reactions in which primary, secondary, and tertiary amines act *as bases* in Section 21.3. We saw their reactions as *nucleophiles* in *alkylation reactions* in Chapter 17 and Section 21.4, and as *nucleophiles* in *acylation reactions* in Chapter 19. In Section 20.3, we found that enamines undergo *C*-acylation and *C*-alkylation and in Chapter 13 we saw that an amino group on an aromatic ring acts as a powerful *activating group* and as an *ortho-para director*.

The structural feature of amines that underlies all of these reactions and that forms a basis for our understanding of most of amine chemistry is the ability of nitrogen to share an electron pair:

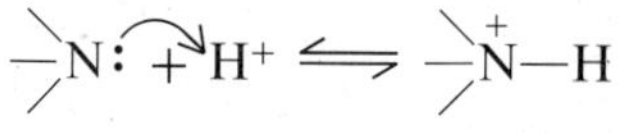

$$\overset{|}{\underset{|}{N}}\!:\curvearrowright\!H^+ \rightleftharpoons \overset{|}{\underset{|}{\overset{+}{N}}}\!\!-\!H$$

An amine acting as a base

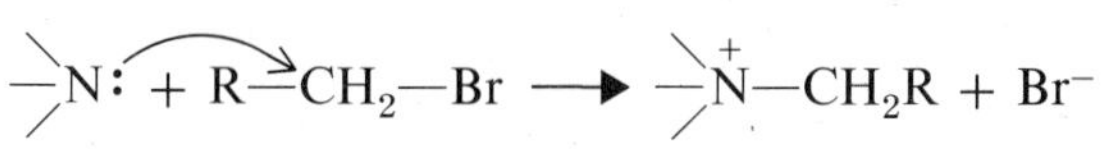

$$\overset{|}{\underset{|}{N}}\!:\; +\; R\!\!\curvearrowright\!\!CH_2\!-\!Br \longrightarrow \overset{|}{\underset{|}{\overset{+}{N}}}\!\!-\!CH_2R + Br^-$$

An amine acting as a nucleophile in an alkylation reaction

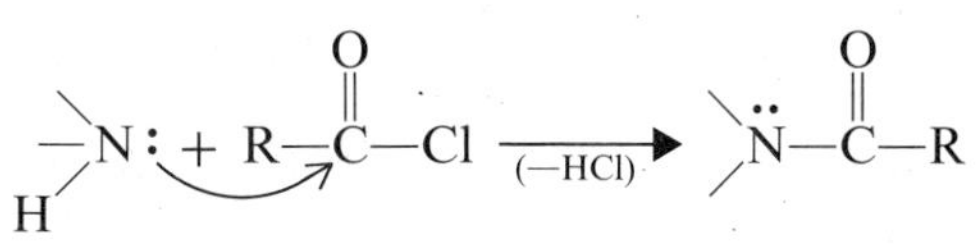

$$\underset{H}{\overset{|}{N}}\!:\; +\; R\!\!\curvearrowright\!\!\overset{\overset{\displaystyle O}{\|}}{C}\!-\!Cl \xrightarrow[(-HCl)]{} \overset{|}{\underset{}{\overset{..}{N}}}\!-\!\overset{\overset{\displaystyle O}{\|}}{C}\!-\!R$$

An amine acting as a nucleophile in an acylation reaction

In the examples given above the amine acts as a nucleophile by donating its electron pair to an electrophilic reagent. In the examples on page 830, resonance contributions involving the nitrogen electron pair make *carbon* atoms nucleophilic.

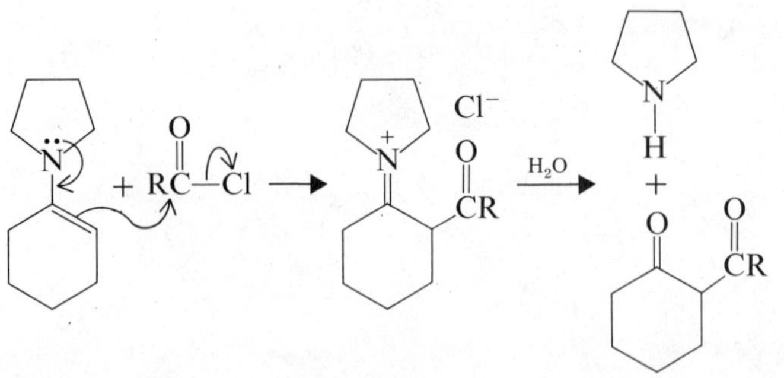

An enamine acting as a carbon nucleophile in an acylation reaction

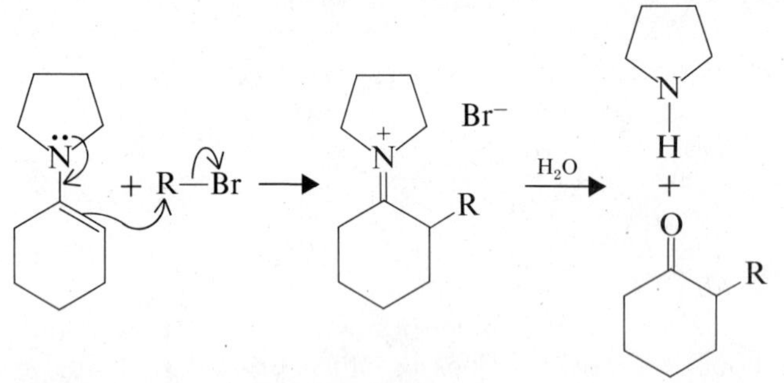

An enamine acting as a carbon nucleophile in an alkylation reaction

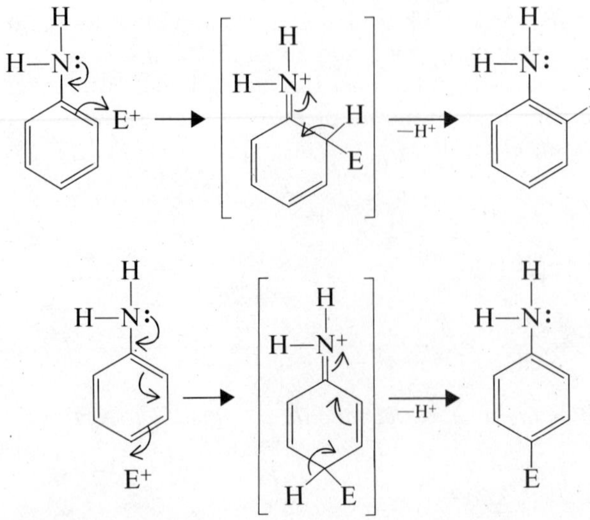

The amino group acting as an activating group and as an ortho-para director in electrophilic aromatic substitution

Problem 21.11

Review the chemistry of amines given in earlier sections and provide a specific example of each of the six reactions illustrated above and on page 829.

21.6 REACTIONS OF AMINES WITH NITROUS ACID

Nitrous acid, HONO, is a weak, unstable acid. It is usually prepared *in situ* by treating sodium nitrite ($NaNO_2$) with an aqueous solution of a strong acid:

$$HCl_{(aq)} + NaNO_{2(aq)} \longrightarrow HONO_{(aq)} + NaCl_{(aq)}$$
$$H_2SO_4 + 2NaNO_{2(aq)} \longrightarrow 2HONO_{(aq)} + Na_2SO_{4(aq)}$$

Nitrous acid reacts with all classes of amines. The products that we obtain from these reactions depend on whether the amine is primary, secondary, or tertiary and whether the amine is aliphatic or aromatic.

Reactions of Primary Aliphatic Amines with Nitrous Acid

Primary aliphatic amines react with nitrous acid through a reaction called *diazotization* to yield highly unstable aliphatic *diazonium salts*. Even at low temperatures, *aliphatic* diazonium salts decompose spontaneously by losing nitrogen to form carbocations. The carbocations go on to produce mixtures of alkenes, alcohols, and alkyl halides.

General Reaction

$$R\!-\!NH_2 \ + NaNO_2 + HX \xrightarrow[\text{H}_2\text{O}]{\text{(HONO)}} \left[R\!-\!\overset{+}{N}\!\equiv\!N\!:\ X^- \right]$$

1° Aliphatic   Aliphatic diazonium salt
amine   (*highly unstable*)

$$\downarrow {-N_2\ (\text{i.e., }:N\equiv N:)}$$

$$R^+ + X^-$$

$$\downarrow$$

Alkenes, alcohols, alkyl halides

Specific Examples

$$CH_3CH_2CH_2CH_2NH_2 + NaNO_2 + HCl \xrightarrow{\text{H}_2\text{O}} \left[CH_3CH_2CH_2CH_2\!-\!\overset{+}{N}\!\equiv\!N\!: \right]$$

$$\xrightarrow{-N_2} CH_3CH_2CH_2CH_2{}^+\!\!-\!\!$$

$$CH_2\!=\!CHCH_2CH_3 + CH_3CH\!=\!CHCH_3 + CH_3CH_2CH_2CH_2OH +$$

1-Butene	2-Butene	1-Butanol
(26%)	(10%)	(25%)

$$\underset{\overset{|}{OH}}{CH_3CH_2CHCH_3} + CH_3CH_2CH_2CH_2Cl + \underset{\overset{|}{Cl}}{CH_3CH_2CHCH_3}$$

2-Butanol	1-Chlorobutane	2-Chlorobutane
(13%)	(5%)	(3%)

Diazotizations of primary aliphatic amines are of little synthetic importance because they yield such a complex mixture of products. Diazotizations of primary aliphatic amines are used in some analytical procedures, however, because the evolution of nitrogen is quantitative. They can also be used to generate and thus study the behavior of carbocations in water, acetic acid, and other solvents.

Problem 21.12

Write mechanisms that account for the formation of each product in the reaction given above from the *n*-butyl cation.

Reactions of Primary Aryl Amines with Nitrous Acid

Primary aromatic amines react with nitrous acid to give arenediazonium salts. While arenediazonium salts are unstable, they are far more stable than aliphatic diazonium salts; they do not decompose at an appreciable rate when the temperature of the reaction is kept below 5°C.

$$Ar-NH_2 \ + \ NaNO_2 \ + \ 2HX \ \longrightarrow \ Ar-\overset{+}{N}\equiv N \text{:} \ X^- \ + \ NaX \ + \ 2H_2O$$

Primary aryl
amine

Arenediazonium
salt
(*stable if kept below*
5°C)

Diazotization reactions of primary aromatic amines are of considerable synthetic importance because the diazonium group, $-\overset{+}{N}\equiv N\text{:}$, can be replaced by a variety of other functional groups. We will examine these reactions in Section 21.7.

Reactions of Secondary Amines with Nitrous Acid

Aromatic and aliphatic secondary amines react with nitrous acid to yield *N*-nitrosoamines. *N*-Nitrosoamines usually separate from the reaction mixture as oily yellow liquids.

General Reaction

$$R_2NH \quad + \ HONO \ \longrightarrow \ R_2N-N=O \ + \ H_2O$$

Secondary aliphatic
amine

N-Nitrosoamine

$$ArNHR \quad + \ HONO \ \longrightarrow \ Ar-\underset{\underset{R}{|}}{\overset{\overset{O}{\|}}{N}}-R \ + \ H_2O$$

Secondary aromatic
amine

N-Nitrosoamine

Specific Examples

$$(CH_3)_2\overset{..}{N}H \ + HCl + NaNO_2 \ \xrightarrow[\text{H}_2\text{O}]{\text{(HONO)}} \ (CH_3)_2\overset{..}{N}-\overset{..}{N}=O$$

Dimethylamine

N-Nitrosodimethyl-
amine
(a yellow oil)

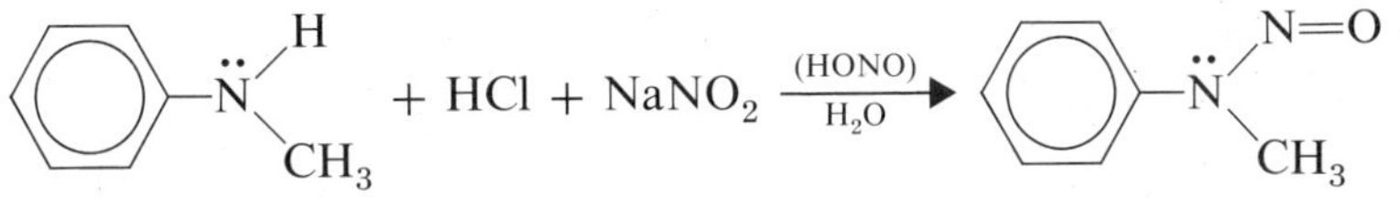

N-Methylaniline

N-Nitroso-N-methyl-
aniline
(87–93%, a yellow oil)

Reactions of Tertiary Amines with Nitrous Acid

When an aliphatic tertiary amine reacts with nitrous acid, an equilibrium is established between the tertiary amine, its salt, and an N-nitrosoammonium compound.

$$2R_3N: \quad + HX + NaNO_2 \rightleftharpoons R_3\overset{+}{N}H \ X^- + \ R_3\overset{+}{N}-N=O \ X^-$$

Tertiary aliphatic   Amine salt   N-Nitrosoammonium
amine   compound

While N-nitrosoammonium compounds are stable at low temperatures, at higher temperatures and in aqueous acid they decompose to produce aldehydes. These reactions are of little synthetic importance, however. Tertiary aromatic amines react with nitrous acid to form C-nitroso aromatic compounds. Nitrosation takes place almost exclusively at the para position if it is open; and, if not, at the ortho position.

General Reaction

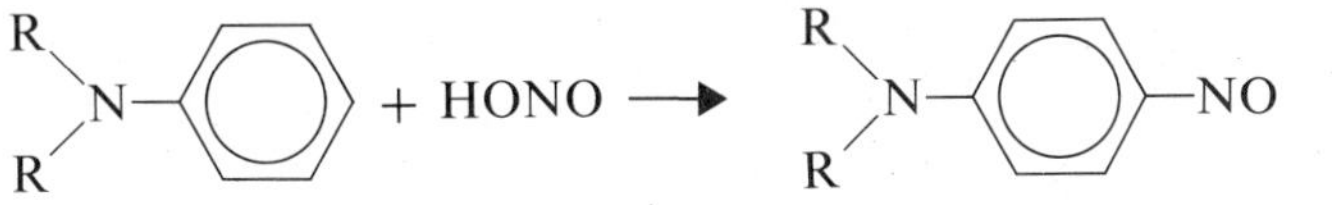

Tertiary aromatic
amine

p-Nitroso-N,N-dialkylaniline

Specific Example

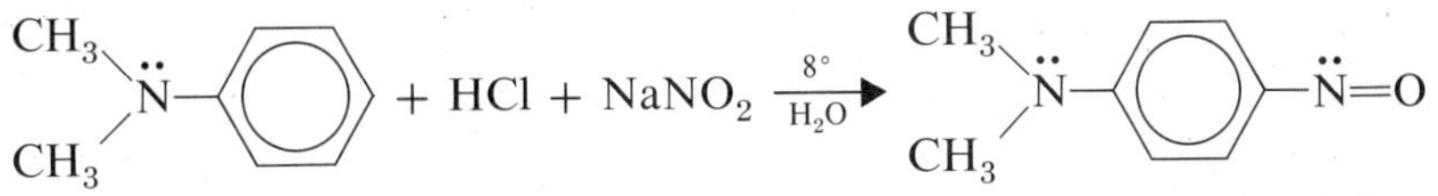

p-Nitroso-N,N-dimethyl-
aniline
(80–90%)

Problem 21.13

Para nitrosation of N,N-dimethylaniline (C-nitrosation) is believed to take place through an electrophilic attack of NO$^+$ ions. (a) Show how NO$^+$ ions might be formed in an aqueous solution of NaNO$_2$ and HCl. (b) Write a mechanism for p-nitrosation of N,N-dimethylaniline. (c) Tertiary aromatic amines and phenols undergo C-nitrosation reaction, whereas most benzene compounds do not. How can you account for this?

21.7 REPLACEMENT REACTIONS OF ARENEDIAZONIUM SALTS

Diazonium salts are highly useful synthetic intermediates in the chemistry of aromatic compounds because the diazonium group can be replaced by a number of other groups, including —F, —Cl, —Br, —I, —CN, —OH, and —H.

Diazonium salts are almost always prepared by diazotizing primary aromatic amines. Primary aromatic amines can be synthesized through reduction of nitro compounds that are readily available through direct nitration reactions. Thus, the diazonium salt is an important link in a series of very useful synthetic chains:

Syntheses Using Diazonium Salts

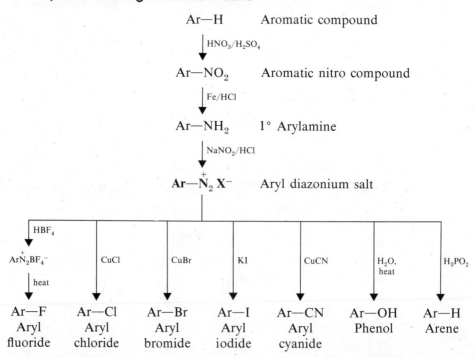

Most arenediazonium salts are unstable at temperatures above 5–10°C, and many explode when dry. Fortunately, however, most of the replacement reactions of diazonium salts do not require isolation of the diazonium salt. We simply add another reagent (CuCl, CuBr, KI, and so on) to the reaction mixture, gently warm the solution, and the replacement reaction (accompanied by the evolution of nitrogen) takes place.

Only in the replacement of the diazonium group by —F, need we isolate a diazonium salt. We do this by adding HBF_4 to the reaction mixture; this causes the sparingly soluble and reasonably stable arenediazonium fluoroborate, ArN_2^+ BF_4^-, to precipitate.

The Sandmeyer Reaction: Replacement of the Diazonium Group by —Cl, —Br, or —CN

Arenediazonium salts react with cuprous chloride, curpous bromide, and cuprous cyanide to give products in which the diazonium group has been replaced by —Cl, —Br, and —CN, respectively. These reactions are known generally as *Sandmeyer reactions*. Several specific examples follow on the next page.

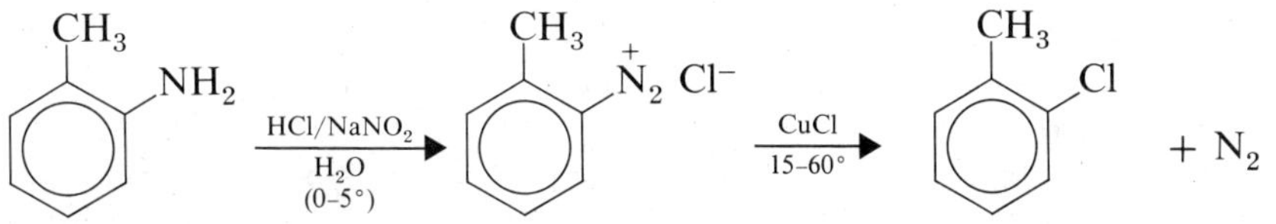

o-Toluidine → *o*-Chlorotoluene
(74–79% overall)

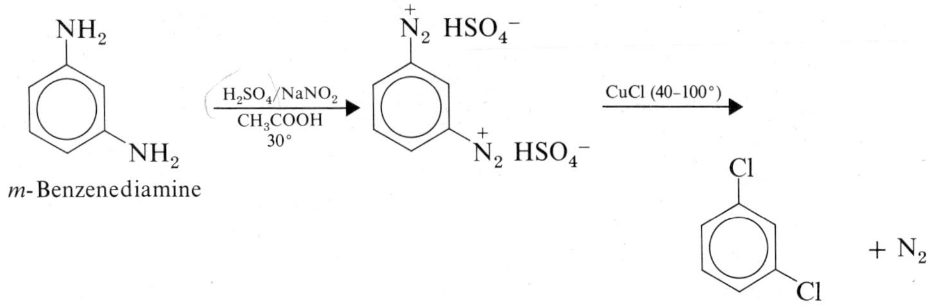

m-Benzenediamine → *m*-Dichlorobenzene
(70% overall)

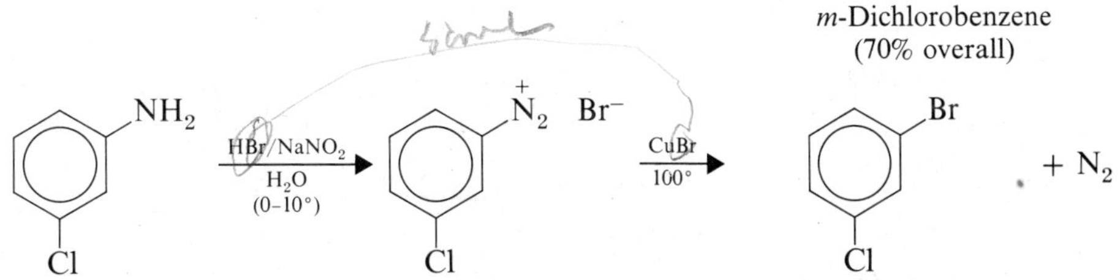

m-Chloroaniline → *m*-Bromochlorobenzene
(70% overall)

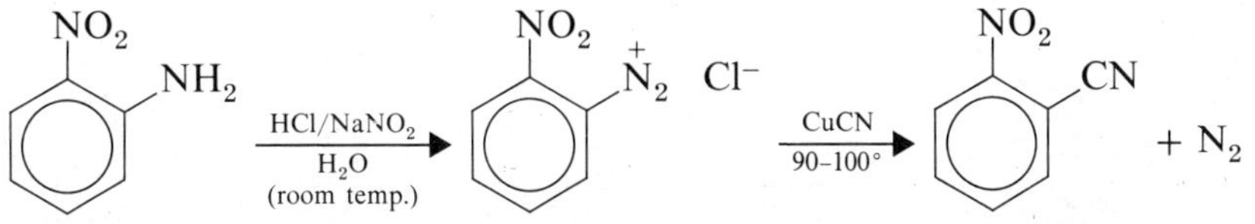

o-Nitroaniline → *o*-Nitrobenzo-
nitrile
(65% overall)

Replacement by —I

Arenediazonium salts react with potassium iodide to give products in which the diazonium group has been replaced by —I. An example is the synthesis of *p*-iodonitrobenzene:

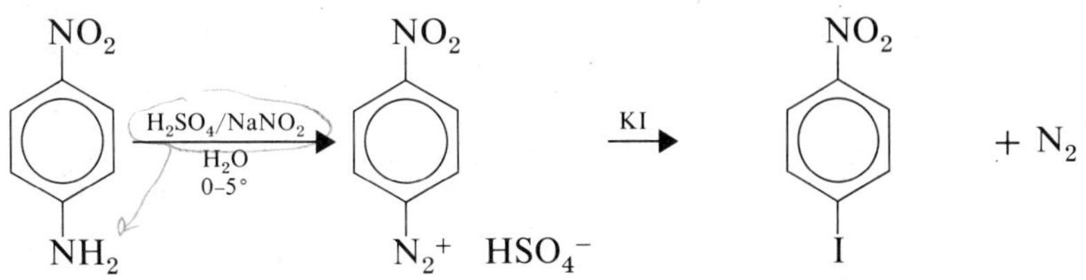

p-Iodonitrobenzene
(81% overall)

Aryl iodides can also be prepared through thallation reactions, cf. p. 553.

Replacement by —F

The diazonium group can be replaced by fluorine by treating the diazonium salt with fluoroboric acid (HBF_4). The diazonium fluoroborate that precipitates is isolated, dried, and heated until decomposition occurs. An aryl fluoride is produced.

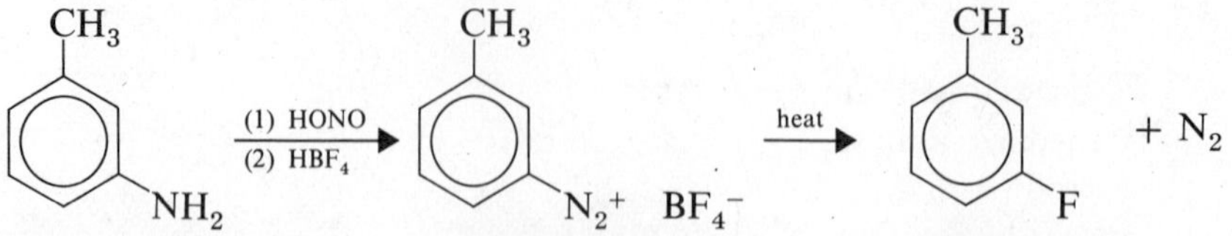

m-Toluidine   *m*-Toluenediazonium fluoroborate (79%)   *m*-Fluorotoluene (69%)

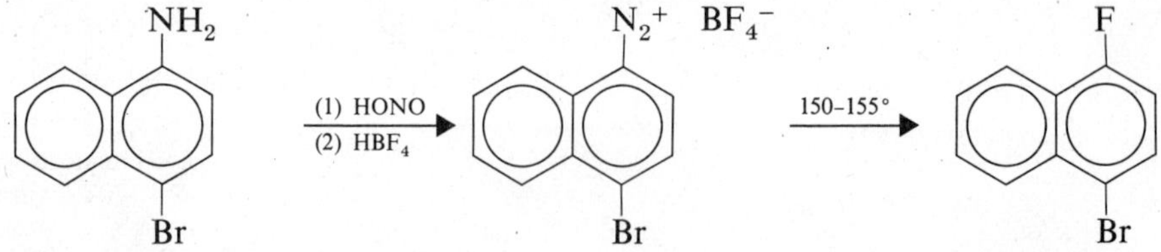

4-Bromo-1-naphthylamine   (isolate and dry)   1-Fluoro-4-bromo-naphthalene (97% overall)

Replacement by —OH

The diazonium group can be replaced by a hydroxyl group by simply acidifying the reaction mixture strongly and heating it:

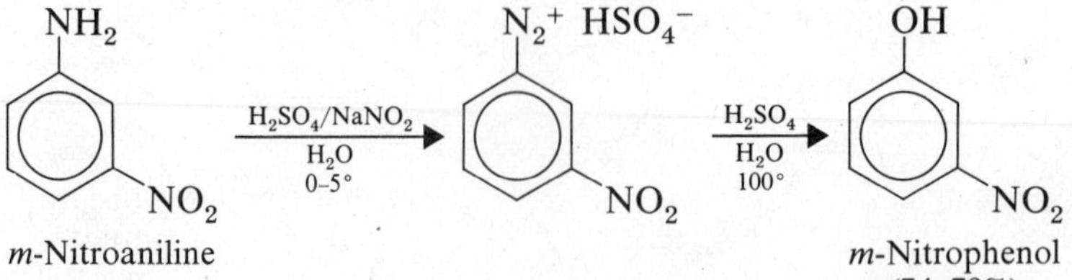

m-Nitroaniline   *m*-Nitrophenol (74–79%)

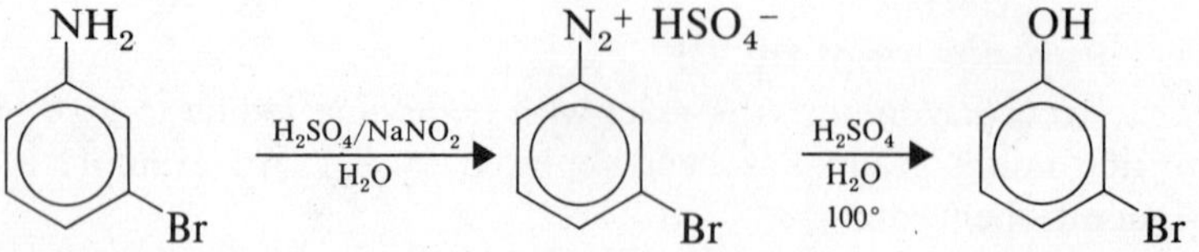

m-Bromoaniline   *m*-Bromophenol (78% overall)

Problem 21.14

In the preceding examples of diazonium reactions we have illustrated syntheses beginning with the compounds (a)-(f) below. Show how you might prepare each of these compounds from benzene.

    (a) *m*-Diaminobenzene

    (b) *m*-Nitroaniline

(c) *m*-Chloroaniline

(d) *m*-Bromoaniline

(e) *o*-Nitroaniline

(f) *p*-Nitroaniline

Replacement by —H

Arenediazonium salts react with hypophosphorus acid (H_3PO_2) or with ethanol to yield products in which the diazonium group has been replaced by —H.

Since we usually begin a synthesis using diazonium salts by nitrating an aromatic compound, that is, replacing —H by —NO_2 and then by —NH_2, it may seem strange that we would ever want to replace a diazonium group by —H. However, replacement of the diazonium group by —H can be a useful reaction: we can introduce an amino group into an aromatic ring in order to influence the orientation of a subsequent reaction. Later we can remove the amino group (i.e., carry out a *deamination* reaction) by diazotizing it and treating the diazonium salt with H_3PO_2 or ethanol.

Let us assume, for example, that we need to prepare 1,3,5-tribromobenzene:

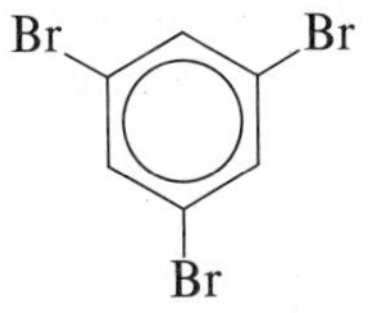

1,3,5-Tribromobenzene

Direct bromination of benzene will not give us this product because bromine is an ortho-para director:

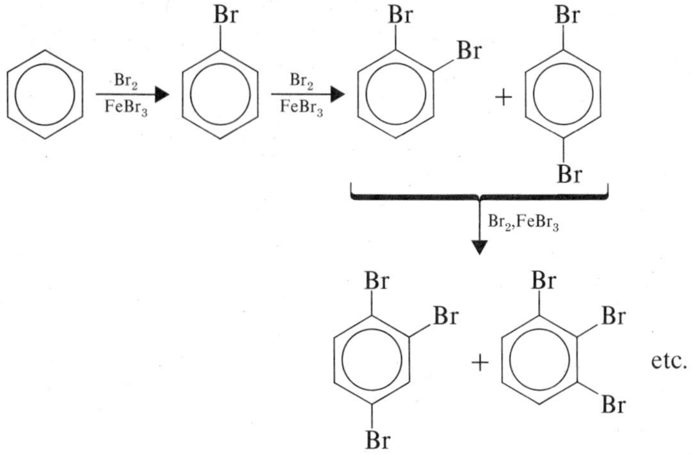

If, however, we begin our synthesis with aniline we can take advantage of the activating and directing effect of the amino group:

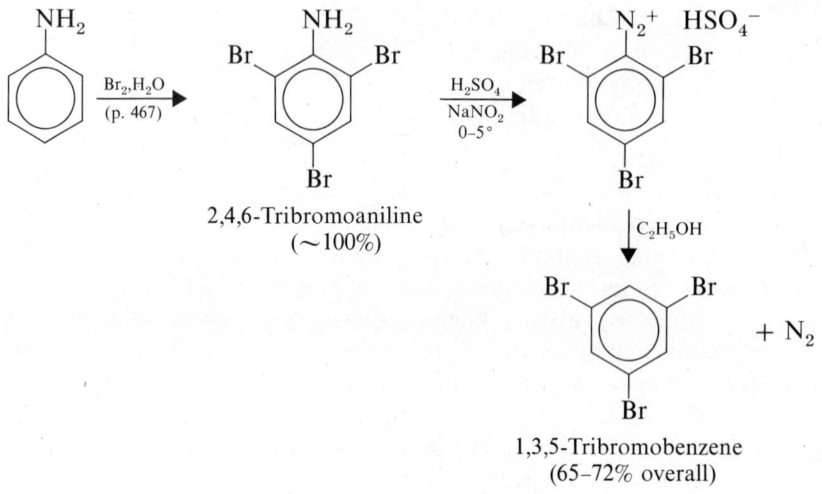

NH2
Br2,H2O
(p. 467)
NH2
Br
Br
Br
2,4,6-Tribromoaniline
(~100%)
H2SO4
NaNO2
0–5°
N2+ HSO4−
Br
Br
Br
C2H5OH
Br
Br
Br
+ N2
1,3,5-Tribromobenzene
(65–72% overall)

We can see another example of the usefulness of a deamination reaction in the synthesis of *m*-bromotoluene below. We cannot prepare *m*-bromotoluene by direct bromination of toluene or by a Friedel-Crafts alkylation of bromobenzene because both reactions give *o*- and *p*-bromotoluene. (Both CH_3— and Br— are ortho-para directors.) However, if we begin with *p*-toluidine (by nitrating toluene, separating the *p* isomer, and reducing the nitro group) we can carry out the following sequence of reactions and obtain *m*-bromotoluene in good yield.

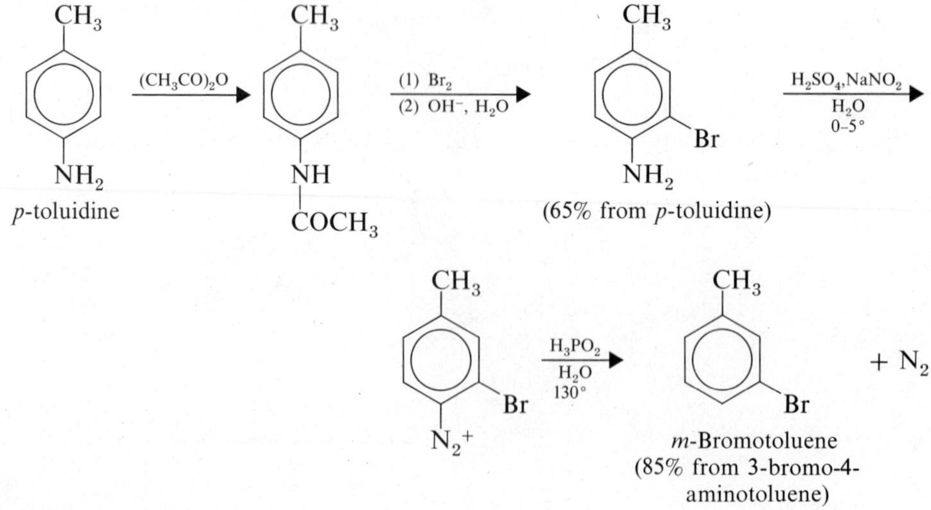

CH3
NH2
p-toluidine
(CH3CO)2O
CH3
NH
COCH3
(1) Br2
(2) OH−, H2O
CH3
Br
NH2
(65% from p-toluidine)
H2SO4,NaNO2
H2O
0–5°
CH3
Br
N2+
H3PO2
H2O
130°
CH3
Br
+ N2
m-Bromotoluene
(85% from 3-bromo-4-aminotoluene)

Problem 21.15

Suggest how you might modify the synthesis given above in order to prepare 3,5-dibromotoluene.

Problem 21.16

Although direct bromination of toluene gives a mixture of *o*- and *p*-bromotoluene, the mixture is difficult to separate because the boiling points of *o*-bromotoluene and *p*-bromotoluene are almost the same (182° and 185°, respectively). Ortho and para nitrotoluene, however, can be separated easily (bp *o*-nitrotoluene, 222°, bp *p*-nitrotoluene 238°). Suggest syntheses of *o*- and *p*-bromotoluene beginning with the nitration of toluene.

Problem 21.17

(a) On p. 836 we showed a synthesis of *m*-fluorotoluene starting with *m*-toluidine. How would you prepare *m*-toluidine from toluene? (b) How would you prepare *m*-chlorotoluene? (c) *m*-Bromotoluene? (d) *m*-Iodotoluene? (e) *m*-Cyanotoluene? (f) *m*-Toluic acid?

Problem 21.18

Starting with *p*-nitroaniline [problem 21.14 (f)] show how you might synthesize 1,2,3-tribromobenzene.

21.8 COUPLING REACTIONS OF DIAZONIUM SALTS

Diazonium salts are weak electrophiles; they react with highly reactive aromatic compounds to yield *azo* compounds. This electrophilic aromatic substitution reaction is often called a *diazo coupling reaction*.

General Reaction

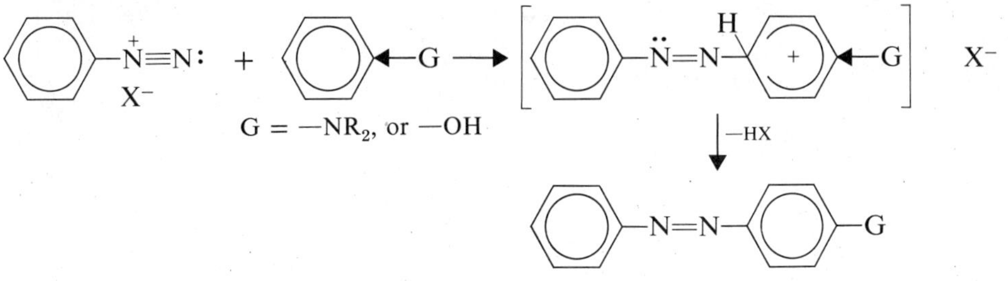

$G = -NR_2$, or $-OH$

An *azo compound*

Specific Examples

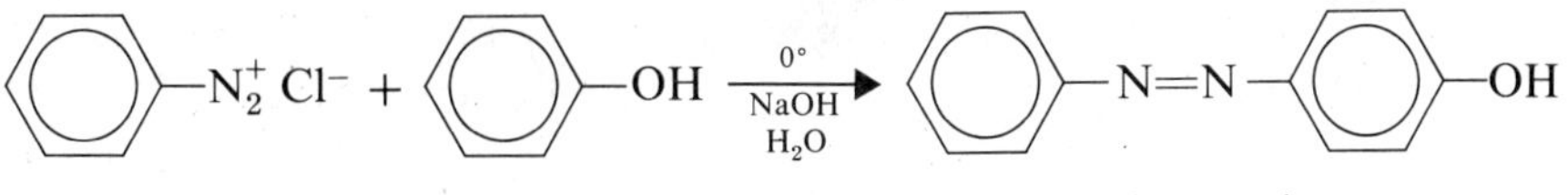

| Benzenediazonium chloride | Phenol | *p*-Hydroxyazobenzene (orange solid) |

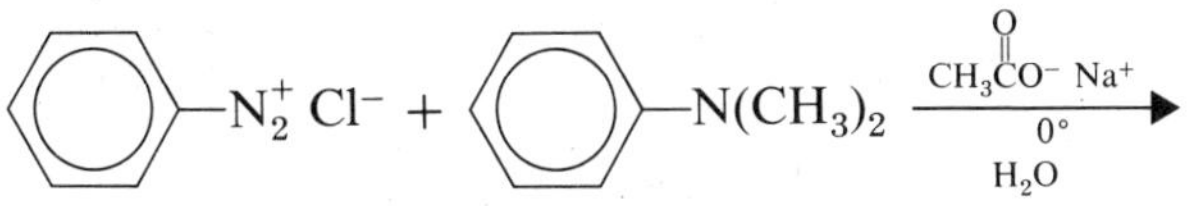

Benzenediazonium chloride   N,N-dimethyl-aniline

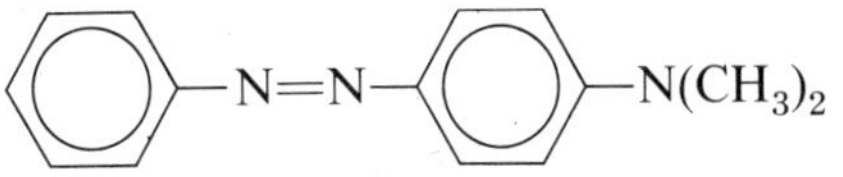

p-Dimethylaminoazobenzene (yellow solid)

Coupling reactions between diazonium salts and phenols take place most rapidly in *slightly* alkaline solution. Under these conditions an appreciable amount of the phenol is present as a phenoxide ion, ArO⁻, and phenoxide ions are even more reactive toward electrophilic substitution than are phenols themselves. (Why?) If the solution is too alkaline, however, (pH > 10) the diazonium salt itself reacts with hydroxide ion to form an unreactive diazohydroxide or diazotate ion:

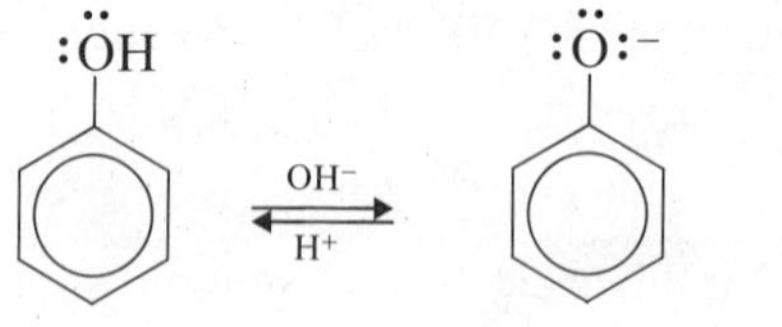

Phenol
(couples slowly)

Phenoxide ion
(couples rapidly)

$$\text{Ar—N} \equiv \text{N:} \xrightleftharpoons[\text{H}^+]{\text{OH}^-} \text{Ar—N} = \ddot{\text{N}}\text{—OH} \xrightleftharpoons[\text{H}^+]{\text{OH}^-} \text{Ar—N} = \ddot{\text{N}}\text{—}\ddot{\text{O}}\text{:}^-$$

Diazonium
ion
(couples)

Diazohydroxide
(does not couple)

Diazotate ion
(does not couple)

Coupling reactions of amines take place most rapidly in slightly acidic solutions (pH 5–7). Under these conditions the concentration of the diazonium salt is at a maximum; at the same time an excessive amount of the amine has not been converted to an unreactive amine salt:

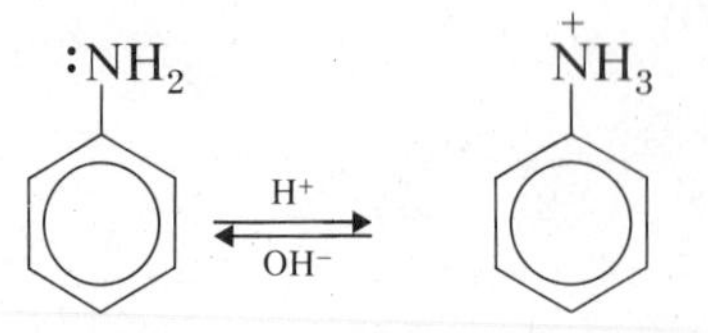

Amine
(couples)

Amine salt
(does not couple)

If the pH of the solution is lower than pH 5 the rate of amine coupling is slow.

With phenols and aniline derivatives, coupling takes place almost exclusively at the para position if it is open. If it is not, coupling takes place at the ortho position.

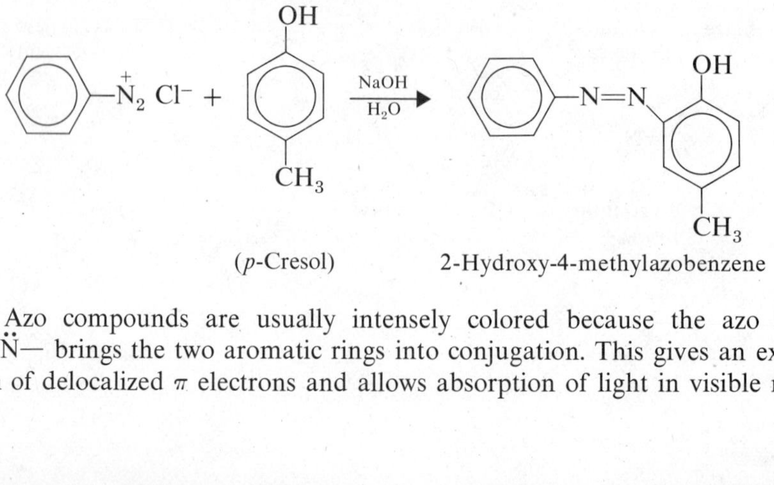

(*p*-Cresol)

2-Hydroxy-4-methylazobenzene

Azo compounds are usually intensely colored because the azo linkage —$\ddot{\text{N}}$=$\ddot{\text{N}}$— brings the two aromatic rings into conjugation. This gives an extended system of delocalized π electrons and allows absorption of light in visible regions.

Azo compounds, because of their intense colors, and because they can be synthesized from relatively inexpensive compounds, are used extensively as *dyes*.

Azo dyes almost always contain one or more $-SO_3^- Na^+$ groups to confer water solubility on the dye and assist in binding the dye to the surfaces of polar fibers (wool, cotton, or nylon). Many dyes are based on coupling reactions of naphthylamines and naphthols. "H-acid" (below) is a particularly versatile component in dye manufacture; not only does it contain sulfonic acid groups, but it can couple in two different ways depending on the pH of the medium.

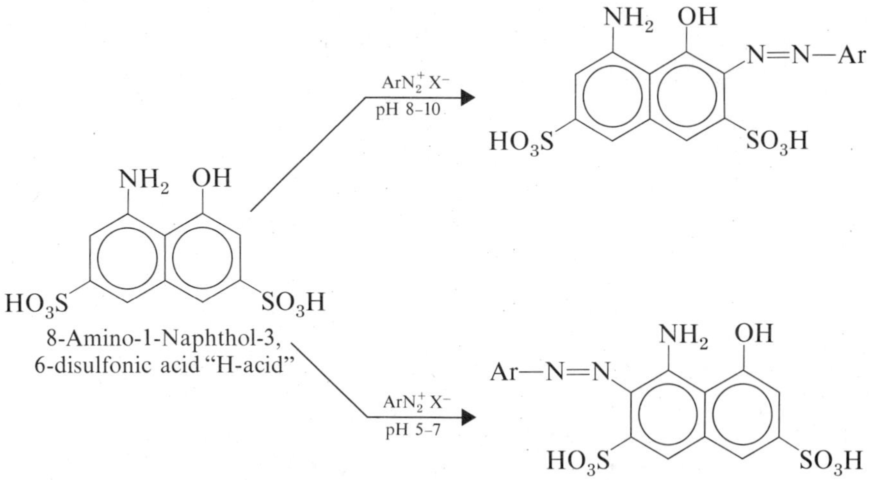

8-Amino-1-Naphthol-3, 6-disulfonic acid "H-acid"

Diamine Green B, a dye introduced into commercial use in 1891, is based on a coupling reaction using H-acid.

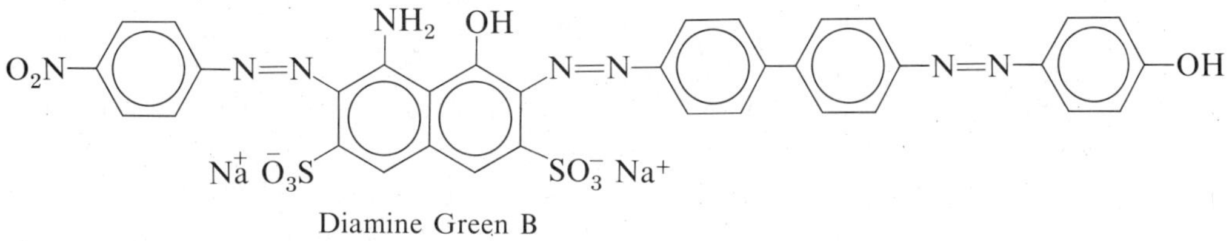

Diamine Green B

Orange II, a dye introduced even earlier (1876) is based on β-naphthol.

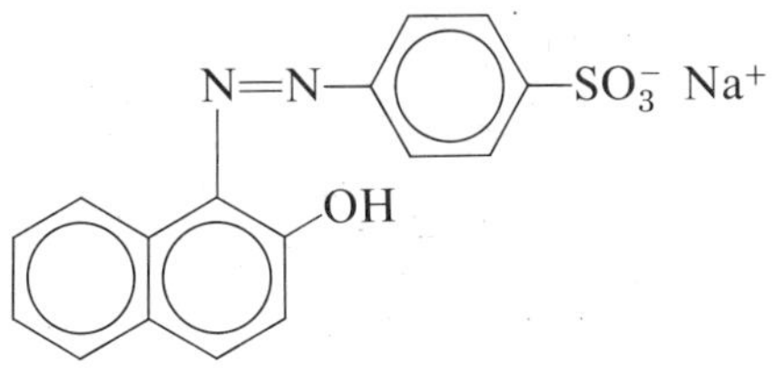

Orange II

Problem 21.19

Outline a synthesis of Orange II from β-naphthol and *p*-aminobenzene sulfonic acid.

Problem 21.20

Outline a synthesis of Diamine Green B using, as starting materials, phenol, aniline, H-acid, and benzidine.

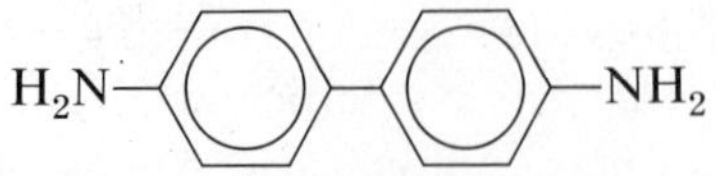

$$H_2N-\langle\bigcirc\rangle-\langle\bigcirc\rangle-NH_2$$

Benzidine

21.9 REACTIONS OF AMINES WITH SULFONYL CHLORIDES

Primary and secondary amines react with sulfonyl chlorides to form *sulfonamides*.

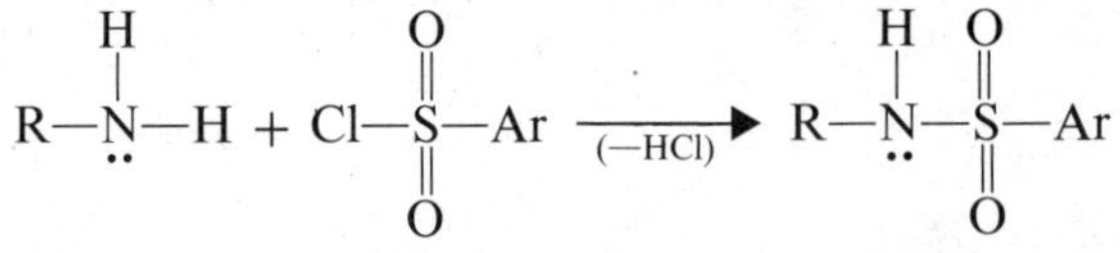

1° Amine	Sulfonyl chloride		N-Substituted sulfonamide

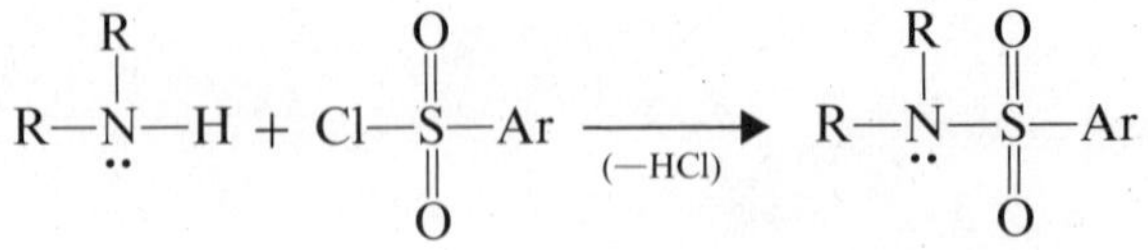

2° Amine		N,N-Disubstituted sulfonamide

Sulfonamide formation is the basis for a chemical test, called the Hinsberg test, that we can use to demonstrate whether an amine is primary, secondary, or tertiary. We carry out a Hinsberg test in two steps. First, we shake a mixture containing a small amount of the amine and benzene sulfonyl chloride in *excess* potassium hydroxide. Next, we allow time for a reaction to take place, and then acidify the reaction mixture. Each type of amine, 1°, 2°, or 3°, will give us a different *visible* result after each of these two stages of the test.

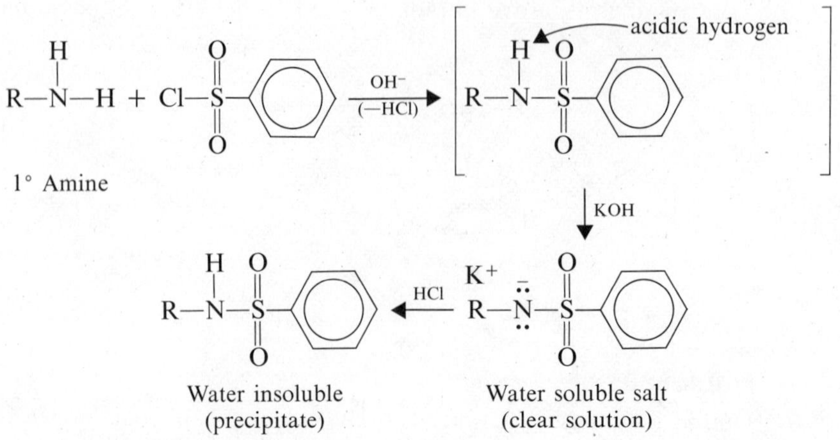

1° Amine

acidic hydrogen

Water insoluble (precipitate)	Water soluble salt (clear solution)

Primary amines react with benzenesulfonyl chloride to form *N*-substituted benzenesulfonamides. These, in turn, undergo acid-base reactions with the excess potassium hydroxide to form water-soluble potassium salts. (These reactions take place because the hydrogen attached to nitrogen is made acidic by the strongly electron-withdrawing —SO$_2$— group.) At this stage our test tube will contain a clear solution. Acidification of this solution will, in the next stage, cause the water insoluble *N*-substituted sulfonamide to precipitate.

Secondary amines react with benzenesulfonyl chloride in aqueous potassium hydroxide to form insoluble *N,N*-disubstituted sulfonamides that precipitate after the first stage. *N,N*-disubstituted sulfonamides do not dissolve in aqueous potassium hydroxide because they do not have an acidic hydrogen. Acidification of the mixture obtained from a secondary amine produces no visible result—the *N,N*-disubstituted sulfonamide remains as a precipitate.

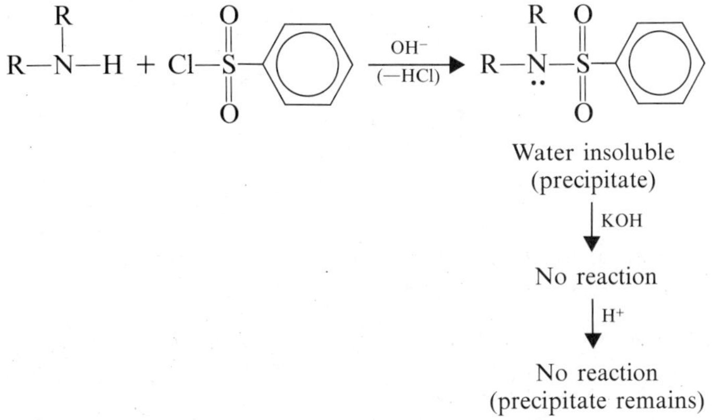

Tertiary amines do not react with benzenesulfonyl chloride. If the tertiary amine is water insoluble, no change will take place in the mixture as we shake it with benzenesulfonyl chloride and aqueous KOH. When we acidify the mixture the tertiary amine will dissolve because it will form a water-soluble salt.

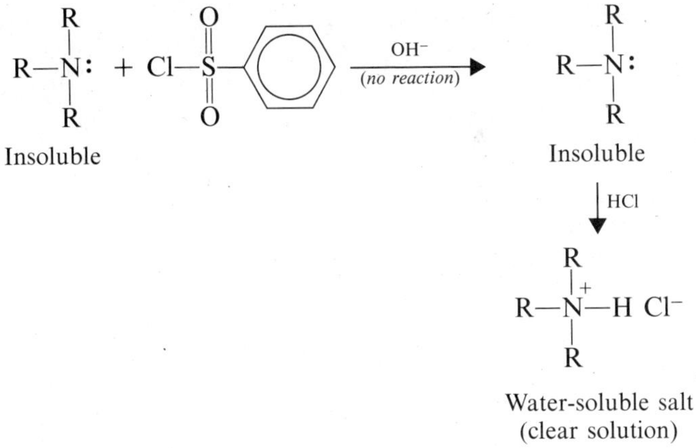

Problem 21.21

Write equations for the reactions that would occur when each of the following amines is treated with benzenesulfonyl chloride in aqueous KOH and when the resulting solution or mixture is subsequently acidified. Tell exactly what you would observe at each stage.

(a) Cyclohexylamine

(b) *N*-Methylcyclohexylamine

(c) *N,N*-Dimethylcyclohexylamine

Problem 21.22

An amine **A** has the molecular formula C_7H_9N. **A** reacts with benzenesulfonyl chloride in aqueous potassium hydroxide to give a clear solution; acidification of the solution gives a precipitate. When **A** is treated with $NaNO_2$ and HCl at 0–5°C, and then with 2-naphthol; an intensely colored compound is formed. **A** gives a single strong absorption peak in the 680–840 cm^{-1} region at 815 cm^{-1}. What is the structure of **A**?

21.10 THE SULFA DRUGS: SULFANILAMIDE

Chemotherapy

Chemotherapy may be defined as using chemical agents to selectively destroy infectious organisms without simultaneously destroying the host. Although it may be difficult to believe, in this age of "wonder drugs," chemotherapy is a relatively modern phenomenon. Prior to 1900 only three specific chemical remedies were known: mercury (for syphilis—but often with disastrous results), cinchona bark (for malaria), and ipecacaunha (for dysentery).

Modern chemotherapy began with the work of Paul Ehrlich early in this century and particularly with his discovery in 1909 of Salvarsan as a remedy for syphilis (p. 571). Ehrlich invented the term "chemotherapy" and in his research sought for what he called "magic bullets," that is, chemicals that would be toxic to infectious microorganisms but harmless to humans.

As a medical student, Ehrlich had been impressed with the ability of certain dyes to stain tissues selectively. Working on the idea that "staining" was a result of a chemical reaction between the tissue and the dye, Ehrlich sought dyes with selective affinities for microorganisms. He hoped that in this way he might find a dye that could be modified so as to render it specifically lethal to microorganisms.

Sulfa Drugs

Between 1909 and 1935, tens of thousands of chemicals, including many dyes, were tested by Ehrlich and others in a search for such "magic bullets." Very few compounds, however, were found to have any promising effect. Then, in 1935, an amazing event happened. The daughter of Gerhard Domagk, a doctor employed by a German dye manufacturer, contracted a streptococcal infection from a pin prick. As his daughter neared death, Domagk decided to give her an oral dose of a dye called Prontosil. Prontosil had been developed at Domagk's firm (I.G. Farbenindustrie) and tests with mice had shown that Prontosil inhibited the growth of streptococci. Within a short time the little girl recovered. Domagk's gamble not only saved his daughter's life, but it also initiated a new and spectacularly productive phase in modern chemotherapy.*

A year later, in 1936, Ernest Fourneau of the Pasteur Institute in Paris demonstrated that Prontosil breaks down in the human body to produce sulfanilamide, and that sulfanilamide is the actual active agent against streptococci.

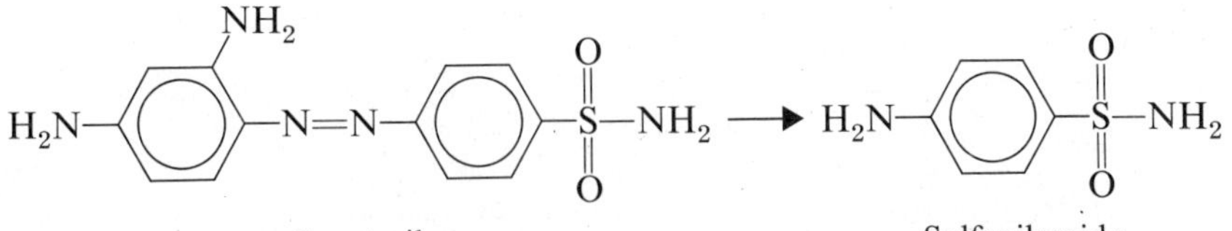

Prontosil   Sulfanilamide

Fourneau's announcement of this result set in motion a search for other chemicals (related to sulfanilamide) that might have even better chemotherapeutic effects. Literally thousands of chemical variations were played on the sulfanilamide theme; the structure of sulfanilamide was varied in almost every imaginable way. The best therapeutic results were obtained from compounds in which one hydrogen of the —SO_2NH_2 group was replaced by some other group, usually a heterocyclic amine. Among the most successful variations were the compounds shown below.

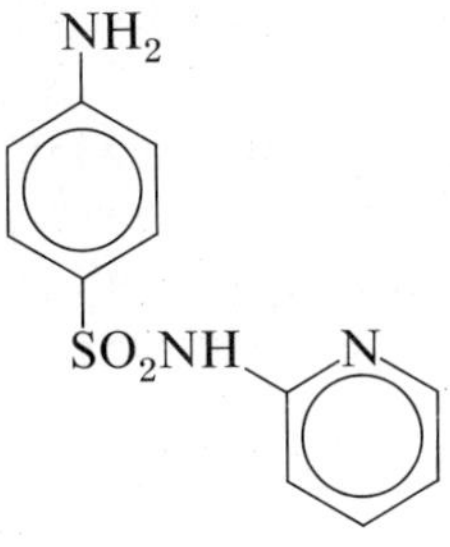

Sulfapyridine 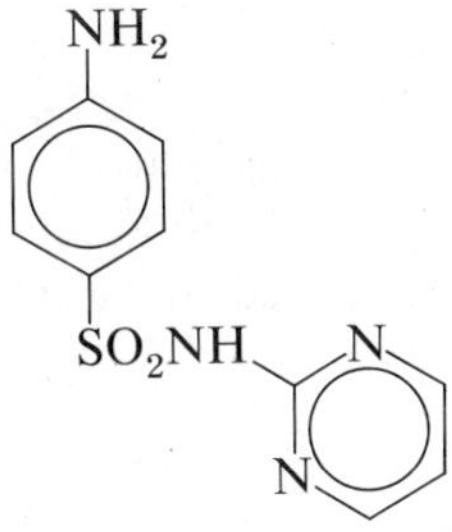   Sulfadiazine

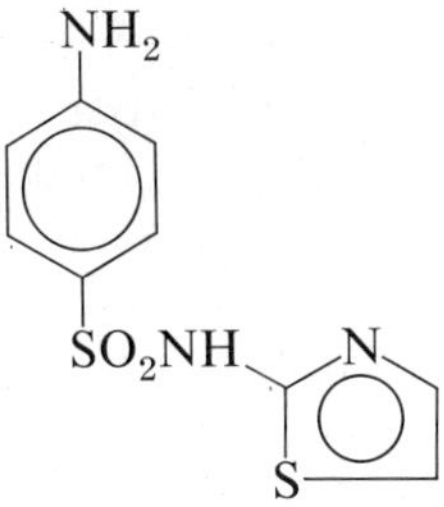

Sulfathiazole 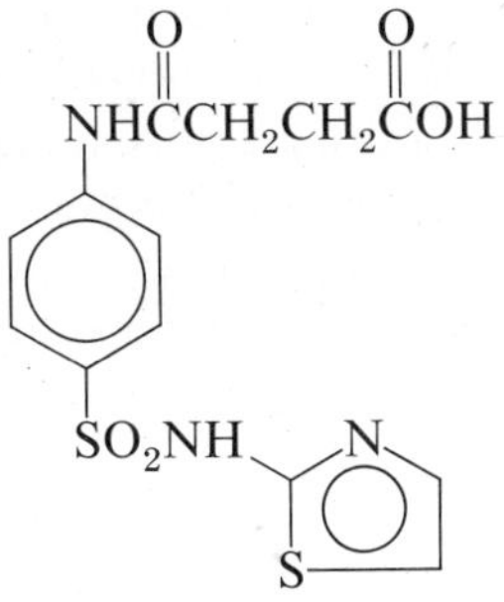   Succinoylsulfathiazole 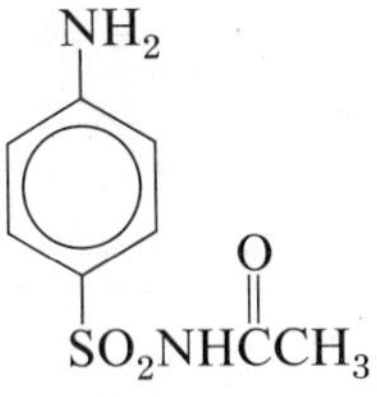   Sulfacetamide

Sulfapyridine was shown to be effective against pneumonia in 1938. (Prior to that time pneumonia epidemics had brought death to tens of thousands.) Sulfacetamide was used successfully in treating urinary tract infections in 1941. Succinoylsulfathiazole and the related compound phthalylsulfathiazole were used as chemo-

---

*Domagk was awarded the Nobel Prize for medicine in 1939.

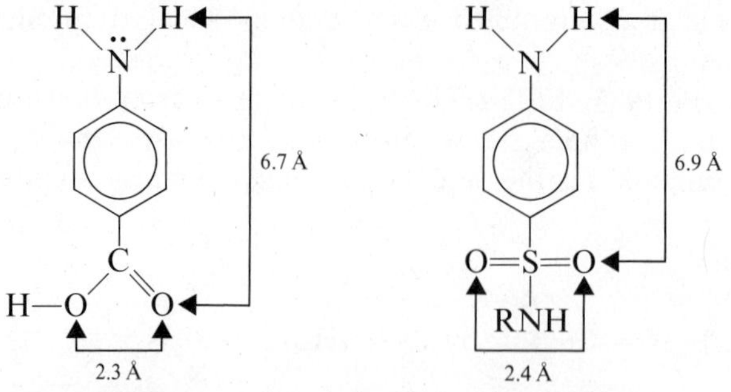

FIG. 21.2
The structural similarity of p-*aminobenzoic acid and a sulfanilamide. (From A. Korolkovas,* Essentials of Molecular Pharmacology, *Wiley, New York, 1970, p. 105. Used with permission.)*

therapeutic agents against infections of the gastrointestinal tract beginning in 1942. (Both compounds are slowly hydrolyzed internally to sulfathiazole.) Sulfathiazole saved the lives of countless wounded soldiers during World War II.

In 1940 a discovery by D. D. Woods laid the groundwork for our understanding of how the sulfa drugs work. Woods observed that the inhibition of growth of certain microorganisms by sulfanilamide is competitively overcome by p-aminobenzoic acid. Woods noticed the structural similarity between the two compounds (Fig. 21.2) and reasoned that the two compounds compete with each other in some essential metabolic process.

Essential Nutrients and Antimetabolites

All higher animals and many microorganisms lack the biochemical ability to synthesize certain essential organic compounds. These essential nutrients include vitamins, certain amino acids, unsaturated carboxylic acids, purines, and pyrimidines. The aromatic amine, p-aminobenzoic acid, is an essential nutrient for those bacteria that are sensitive to sulfanilamide therapy. Enzymes within these bacteria use p-aminobenzoic acid to synthesize another essential compound called *folic acid.*

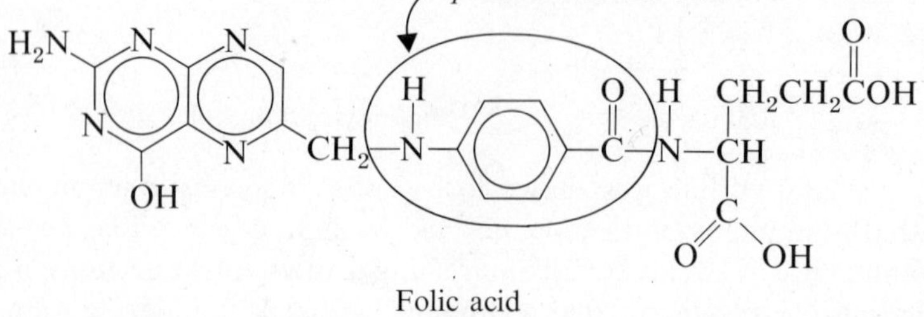

Folic acid

Chemicals that inhibit the growth of microbes are called *antimetabolites.* The sulfanilamides are antimetabolites for those bacteria that require p-aminobenzoic acid. The sulfanilamides apparently inhibit those enzymic steps of the bacteria that are involved in the synthesis of folic acid. The bacterial enzymes are apparently unable to distinguish between a molecule of a sulfanilamide and a molecule of p-aminobenzoic acid; thus, sulfanilamide "inhibits" the bacterial enzyme. Because the microorganism is unable to synthesize enough folic acid when sulfanilamide is present, it dies. Humans are unaffected by sulfanilamide therapy because we derive our folic acid from dietary sources (folic acid is a vitamin) and do not synthesize it from p-aminobenzoic acid.

The discovery of the mode of action of the sulfanilamides has led to the discovery of many new and effective antimetabolites. One example is *Methotrexate*, a derivative of folic acid that has been used successfully in treating certain carcinomas:

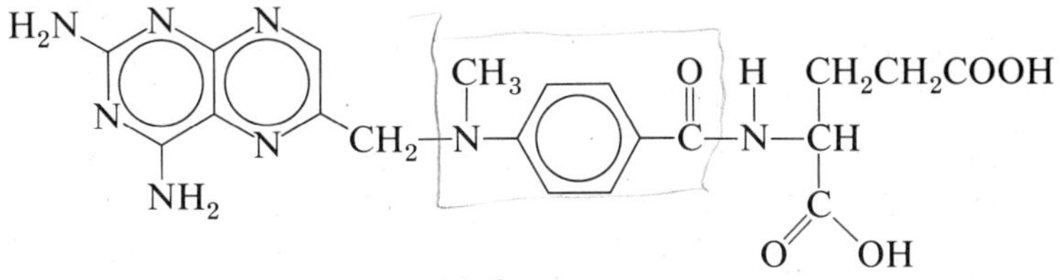

Methotrexate

Methotrexate, by virtue of its resemblance to folic acid, can enter into some of the same reactions as folic acid, but it cannot serve the same function, particularly in important reactions involved in cell division. Although methotrexate is toxic to all dividing cells, those cells that divide most rapidly—*cancer cells*—are most vulnerable to its effect.

Synthesis of Sulfa Drugs

Sulfanilamides can be synthesized from aniline through the sequence of reactions shown below.

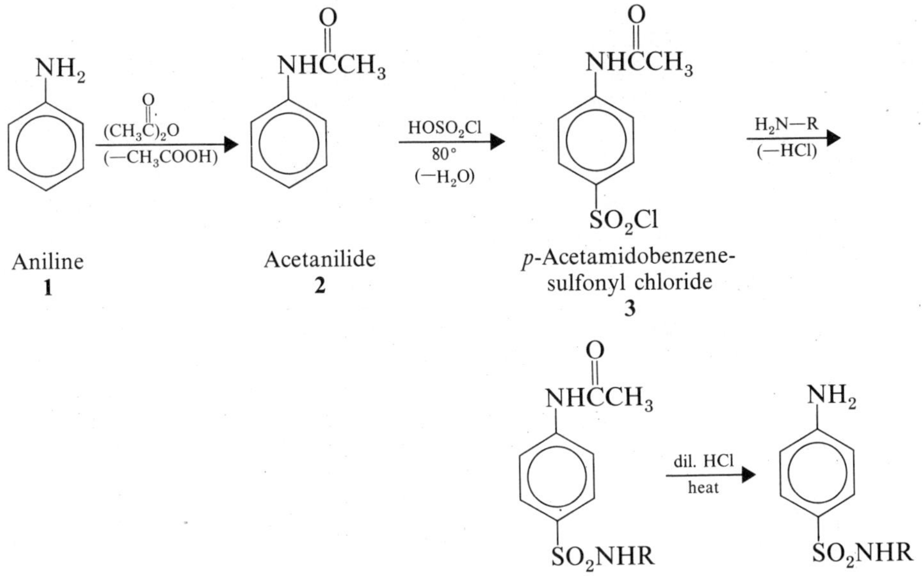

Acetylation of aniline produces acetanilide, **2**. Treatment of **2** with chloro-sulfonic acid brings about an electrophilic aromatic substitution reaction and yields *p*-acetamidobenzenesulfonyl chloride, **3**. Addition of ammonia or a secondary amine gives the diamide, **4**. (**4** is an amide of both a carboxylic acid and a sulfonic acid.) Finally, refluxing **4** with dilute hydrochloric acid selectively hydrolyzes the carboxa-mide linkage and produces sulfanilamide.

Problem 21.23

(a) Starting with aniline and assuming that you have 2-aminothiazole available, show how you would synthesize sulfathiazole. (b) How would you convert sulfa-thiazole to succinoylsulfathiazole? (c) To phthalylsulfathiazole?

2-Aminothiazole

21.11 ANALYSIS OF AMINES

Chemical Analysis

Amines are characterized by their basicity, and thus, by their ability to dissolve in dilute aqueous acid (p. 820). Primary, secondary, and tertiary amines can be distinguished from each other on the basis of the Hinsberg test (p. 842). Primary aromatic amines are often detected through diazonium salt formation and subse-quent coupling with β-naphthol to form a brightly colored azo dye (p. 841).

Spectroscopic Analysis

The *infrared spectra* of primary and secondary amines are characterized by absorption bands in the 3300–3500 cm^{-1} region that arise from N—H stretching vibration. Primary amines give two bands in this region; secondary amines generally give only one. Absorption bands arising from C—N stretching vibrations of aliphatic amines occur in the 1020–1220 cm^{-1} region but are usually weak and difficult to identify. Aromatic amines generally give a strong C=N stretching band in the 1250–1360 cm^{-1} region.

The *pmr spectra* of primary and secondary amines show N—H proton ab-sorptions in region 1–5 δ. These peaks are sometimes difficult to identify and are best detected by proton counting.

Additional Problems

21.24

Write structural formulas for each of the following compounds.

(a) Methylbenzylamine
(b) Triisopropylamine
(c) *N*-Methyl-*N*-ethylaniline
(d) *m*-Toluidine
(e) 2-Methylpyrrole
(f) *N*-Ethylpiperidine
(g) *N*-Ethylpyridinium bromide
(h) 3-Pyridinecarboxylic acid

(i) Indole
(j) Acetanilide
(k) Dimethylammonium chloride
(l) 2-Methylimidazole
(m) Sulfapyridine
(n) Tetrapropylammonium chloride

(o) Pyrrolidine
(p) N,N-Dimethyl-p-toluidine
(q) p-Anisidine
(r) Benzidine
(s) p-Aminobenzoic acid
(t) Histidine

21.25

Give common or IUPAC names for each of the following compounds.

(a) $CH_3CH_2CH_2NH_2$

(b) $C_6H_5NHCH_3$

(c) $(CH_3)_2CH\overset{+}{N}(CH_3)_3$ I^-

(d) o-$CH_3C_6H_4NH_2$

(e) o-$CH_3OC_6H_4NH_2$

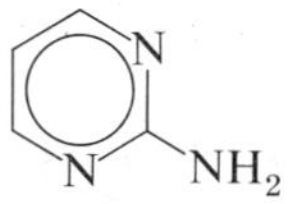

(h) $C_6H_5CH_2NH_3^+$ Cl^-

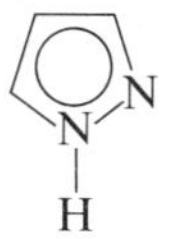

(i) $C_6H_5N(CH_2CH_2CH_3)_2$

(j) $C_6H_5SO_2NH_2$

(k) $CH_3NH_3^+$ CH_3COO^-

(l) $HOCH_2CH_2CH_2NH_2$

(f)

(g)

(m)

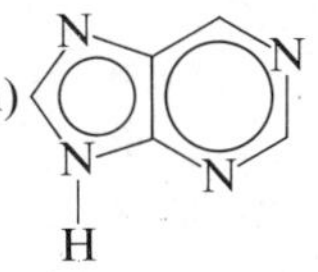

(n)

21.26

Show how you might prepare benzylamine from each of the following compounds.

(a) Benzonitrile
(b) Benzamide
(c) Benzyl bromide (two ways)
(d) Benzyl tosylate

(e) Benzaldehyde
(f) Phenylnitromethane
(g) Phenylacetamide

21.27

Show how you might prepare aniline from each of the following compounds.

(a) Benzene   (b) Bromobenzene   (c) Benzamide

21.28

Show how you might synthesize each of the following compounds from n-butyl alcohol.

(a) n-Butylamine (free of 2° and 3° amines)
(b) n-Pentylamine

(c) n-Propylamine
(d) Methyl n-butylamine

21.29

Give structures for compounds **A-F**:

N-Methylpiperidine + CH_3I $\longrightarrow$ **A** $(C_7H_{16}NI)$ $\xrightarrow[H_2O]{Ag_2O}$ **B** $(C_7H_{17}NO)$ $\xrightarrow[(-H_2O)]{heat}$

C $(C_7H_{15}N)$ $\xrightarrow{CH_3I}$ **D** $(C_8H_{18}NI)$ $\xrightarrow[H_2O]{Ag_2O}$ **E** $(C_8H_{19}NO)$ $\xrightarrow{heat}$

F (C_5H_8) + H_2O + $(CH_3)_3N$

21.30

Show how you might convert aniline into each of the following compounds. (You need not repeat steps carried out in earlier parts of this problem.)

(a) Acetanilide
(b) *N*-Phenylphthalimide
(c) *p*-Nitroaniline
(d) Sulfanilamide
(e) *N,N*-Dimethylaniline
(f) Fluorobenzene
(g) Chlorobenzene
(h) Bromobenzene

(i) Iodobenzene
(j) Benzonitrile
(k) Benzoic acid
(l) Phenol
(m) Benzene
(n) *p*-Hydroxyazobenzene
(o) *p*-Dimethylaminoazobenzene

21.31
What products would you expect to be formed when each of the following amines reacts with aqueous sodium nitrite and hydrochloric acid?
(a) Propylamine
(b) Dipropylamine
(c) *N*-Propylaniline
(d) *N,N*-dipropylaniline
(e) *p*-Propylaniline

21.32
(a) What products would you expect to be formed when each of the amines in the preceding problem reacts with benzenesulfonyl chloride and excess aqueous potassium hydroxide? (b) What would you observe in each reaction? (c) What would you observe when the resulting solution or mixture is acidified?

21.33
(a) What product would you expect to obtain from the reaction of piperidine with aqueous sodium nitrite and hydrochloric acid? (b) From the reaction of piperidine and benzenesulfonyl chloride in excess aqueous potassium hydroxide?

21.34
Give structures for the products of each of the following reactions.
(a) Ethylamine + benzoyl chloride $\longrightarrow$

(b) Methylamine + acetic anhydride $\longrightarrow$

(c) Methylamine + succinic anhydride $\longrightarrow$

(d) Product of (c) $\xrightarrow{\text{heat}}$

(e) Pyrrolidine + phthalic anhydride $\longrightarrow$

(f) Pyrrole + acetic anhydride $\longrightarrow$

(g) Aniline + propanoyl chloride $\longrightarrow$

(h) Tetraethylammonium hydroxide $\xrightarrow{\text{heat}}$

(i) *m*-Dinitrobenzene + H_2S $\xrightarrow[\text{C}_2\text{H}_5\text{OH}]{\text{NH}_3}$

(j) *p*-Toluidine + $Br_{2(\text{excess})}$ $\xrightarrow[\text{H}_2\text{O}]{}$

21.35
Starting with benzene or toluene outline syntheses of the following compounds using diazonium salts as intermediates. (You need not repeat syntheses carried out in earlier parts of this problem.)
(a) *o*-Cresol
(b) *m*-Cresol
(c) *p*-Cresol
(d) *m*-Dichlorobenzene
(e) *m*-Dicyanobenzene
(f) *m*-Iodophenol
(g) *m*-Bromobenzonitrile
(h) 3,5-Dibromonitrobenzene

(i) 3,5-Dibromoaniline
(j) 3,4,5-Tribromophenol
(k) 3,4,5-Tribromobenzonitrile
(l) 2,6-Dibromobenzoic acid
(m) 1,3-Dibromo-2-iodobenzene
(n) 2-Nitro-4-bromotoluene
(o) 3-Nitro-4-methylphenol

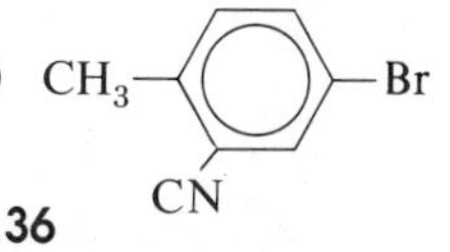

(q)

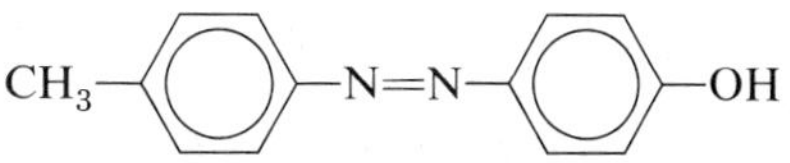

(r)

(p)

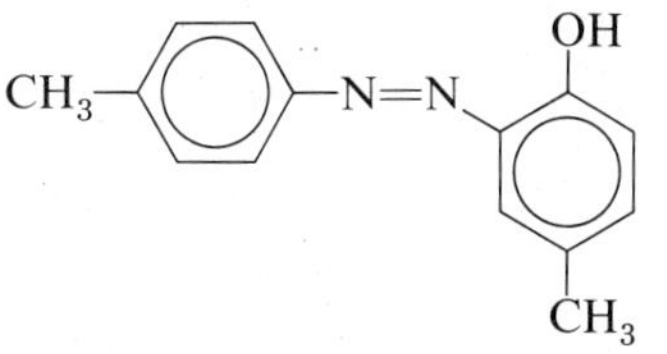

21.36
Write equations for simple chemical tests that would distinguish between
(a) Benzylamine and benzamide
(b) Allylamine and propylamine
(c) *p*-Toluidine and *N*-methylaniline
(d) Cyclohexylamine and piperidine
(e) Pyridine and benzene
(f) Cyclohexylamine and aniline
(g) Triethylamine and diethylamine
(h) Tripropylammonium chloride and tetrapropylammonium chloride
(i) Tetrapropylammonium chloride and tetrapropylammonium hydroxide

21.37
Describe with equations how you might separate a mixture of aniline, *p*-cresol, benzoic acid, and toluene using ordinary laboratory reagents.

21.38
Give structures for compounds **A-H**.

(a) 2,5-Hexanedione + $(NH_4)_2CO_3 \xrightarrow{100°}$ **A** (C_6H_9N), a pyrrole

(b) $CH_3\overset{O}{\overset{\|}{C}}CH_2NH_2$ + acetone $\xrightarrow{\text{base}}$ **B** (C_6H_9N), an isomer of **A**

(c) $CH_3NHNH_2 + (CH_3O)_2CHCH_2CH(OCH_3)_2 \xrightarrow[H_2O]{H^+}$ **C** ($C_4H_6N_2$), a pyrazole

(d) 2,5-Hexanedione + hydrazine $\xrightarrow{\text{heat}}$

D ($C_6H_{10}N_2$), a dihydropyridazine $\xrightarrow{O_2}$ **E** ($C_6H_8N_2$), a pyridazine

(e) Aniline + $CH_2{=}CH\overset{O}{\overset{\|}{C}}CH_3 \xrightarrow[\text{FeCl}_3]{\text{ZnCl}_2}$ **F** ($C_{10}H_9N$), a quinoline

(f) 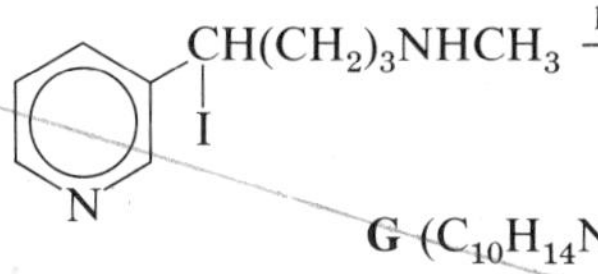 $\xrightarrow{\text{heat}} \xrightarrow{OH^-}$

G ($C_{10}H_{14}N_2$), nicotine $\xrightarrow[\text{(2) H}^+]{\text{(1) KMnO}_4,\ \text{OH}^-}$ **H** ($C_6H_5NO_2$), nicotinic acid

21.39
Show how you might synthesize β-aminopropionic acid, $H_2NCH_2CH_2COOH$, from succinic anhydride (β-aminopropionic acid is used in the synthesis of pantothenic acid; cf. problem 19.28).

21.40

Basing your answer on the theory of antimetabolite activity presented on p. 846, which of the following compounds would you expect to be *inactive* as inhibitors of folic acid biosynthesis?

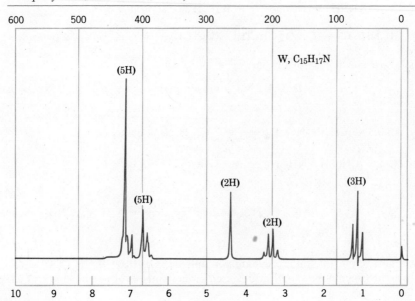

FIG. 21.3

The pmr spectrum of W (problem 21.42). (Courtesy Aldrich Chemical Company Inc., Milwaukee, Wis.)

21.41

A commercial synthesis of folic acid consists of heating the three compounds listed below with aqueous sodium bicarbonate. Propose reasonable mechanisms for the reactions that lead to folic acid.

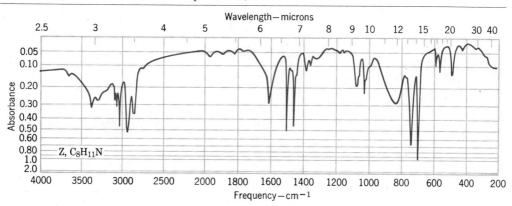

Folic acid
(~10%)

FIG. 21.4

Infrared and pmr spectra for compound Z, problem 21.43. (Courtesy Sadtler Research Laboratories, Inc., Philadelphia, Pa.)

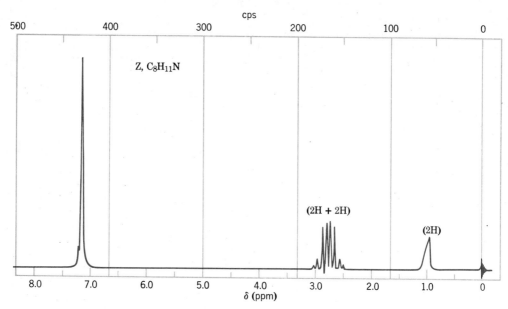

21.42

When compound W, $C_{15}H_{17}N$, is treated with benzenesulfonyl chloride and aqueous potassium hydroxide no apparent change occurs. Acidification of this mixture gives a clear solution. The pmr spectrum of W is shown in Fig. 21.3 page 852. Propose a structure for W.

21.43

Propose structures for compounds X, Y, and Z.

$$X(C_7H_7Br) \xrightarrow{NaCN} Y(C_8H_7N) \xrightarrow{LiAlH_4} Z(C_8H_{11}N)$$

The pmr spectrum of X gives two signals, a multiplet at $\delta 7.3$ (5H) and a singlet at $\delta 4.25$ (2H); the 680–840 cm^{-1} region of the infrared spectrum of X shows a peak at 690 cm^{-1} and 770 cm^{-1}. The pmr spectrum of Y is similar to that of X: multiplet $\delta 7.3$ (5H), singlet $\delta 3.7$ (2H). The pmr and infrared spectra of Z are shown in Fig. 21.4 page 853.

* 21.44

Using reactions that we have studied in this chapter, propose a mechanism that accounts for the following reaction.

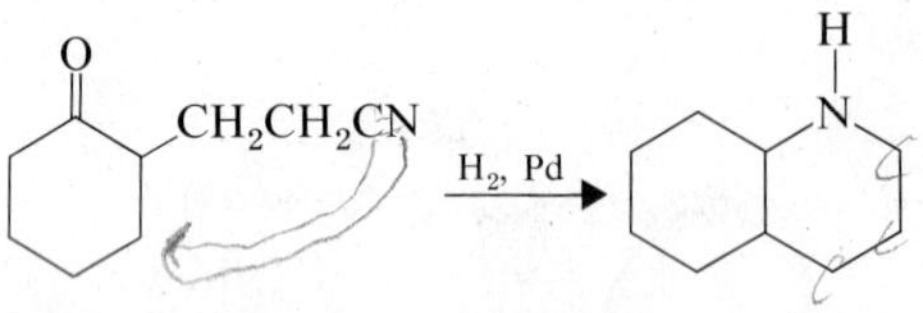

* 21.45

Antipyrine is a compound that has been used as an analgesic and to reduce fevers. It can be synthesized from ethyl acetoacetate and phenylhydrazine through the following steps.

$$CH_3\overset{O}{\overset{\|}{C}}CH_2COOC_2H_5 + C_6H_5NHNH_2 \xrightarrow{heat} C_{12}H_{16}N_2O_2 \xrightarrow[heating]{further}$$

$$C_{10}H_{10}N_2O \xrightarrow[NaOH]{CH_3OSO_2OCH_3} \text{Antipyrine, } C_{11}H_{12}N_2O$$

Outline the reactions that take place.

22

SPECIAL TOPICS III

REACTIONS OF HETEROCYCLIC AMINES
ALKALOIDS
CONDENSATION POLYMERS

22.1 REACTIONS OF HETEROCYCLIC AMINES

Heterocyclic amines undergo many reactions that are similar to those of the amines that we have studied in earlier chapters.

Heterocyclic Amines as Bases

Nonaromatic heterocyclic amines have basicity constants that are approximately the same as those of acyclic amines.

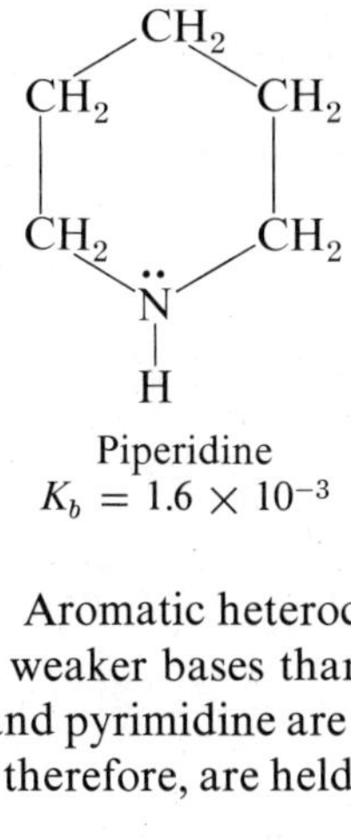

Piperidine	Pyrrolidine	Diethylamine
$K_b = 1.6 \times 10^{-3}$	$K_b = 1.3 \times 10^{-3}$	$K_b = 9.6 \times 10^{-4}$

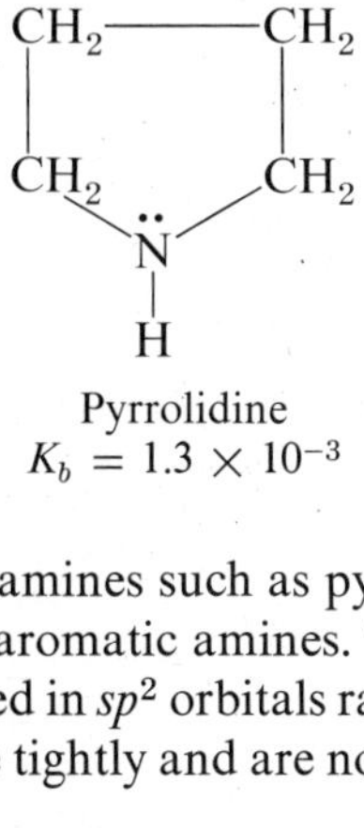

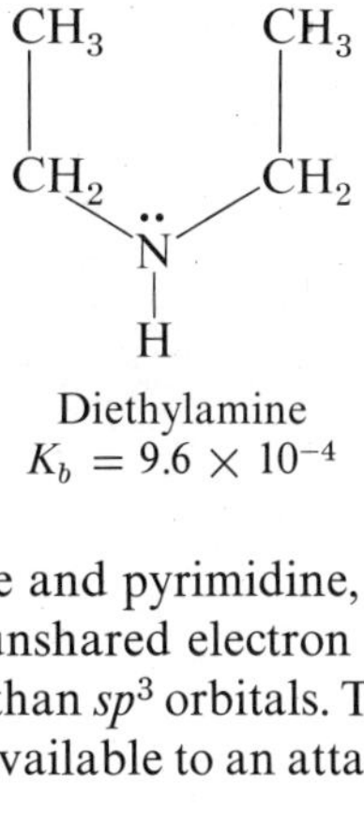

Aromatic heterocyclic amines such as pyridine and pyrimidine, however, are much weaker bases than nonaromatic amines. The unshared electron pairs of pyridine and pyrimidine are located in sp^2 orbitals rather than sp^3 orbitals. These electron pairs, therefore, are held more tightly and are not as available to an attacking proton.

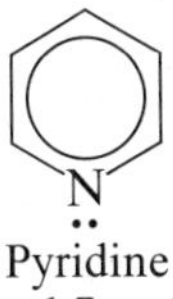

Pyridine	Pyrimidine
$K_b = 1.7 \times 10^{-9}$	$K_b = 5 \times 10^{-12}$

Pyrrole is a very weak base because its electron pair is a part of the aromatic sextet (p. 436). For pyrrole to accept a proton, a highly endothermic reaction must occur because in it the resonance energy of the pyrrole ring is lost.

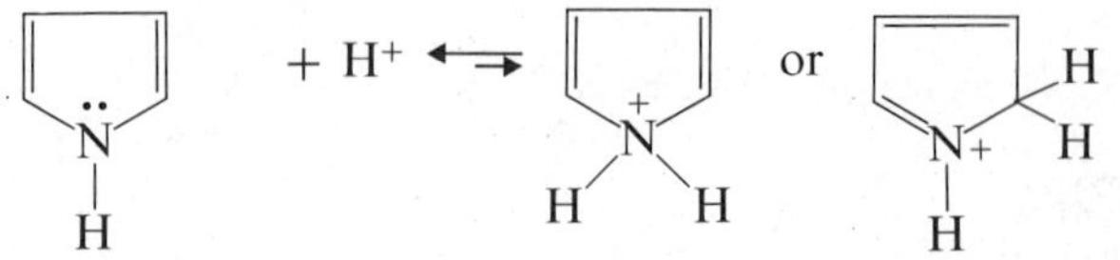

Pyrrole
$K_b = 2.5 \times 10^{-14}$
(an aromatic ring
of six π electrons)

(nonaromatic rings)

Imidazole has two nitrogens: the electron pair of one nitrogen is like that of pyrrole; it is a part of the aromatic sextet and, therefore, does not accept a proton readily. The other electron pair is much more available to an attacking proton and, as a result, imidazole is a much stronger base than pyrrole. It is an even stronger base than pyridine.

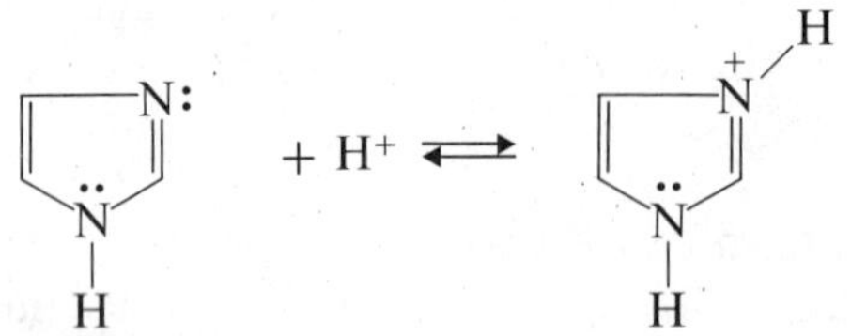

Imidazole
$K_b = 1.6 \times 10^{-7}$
(an aromatic ring)

(an aromatic ring)

The imidazole ring occurs naturally in the amino acid, histidine, an important constituent of proteins.

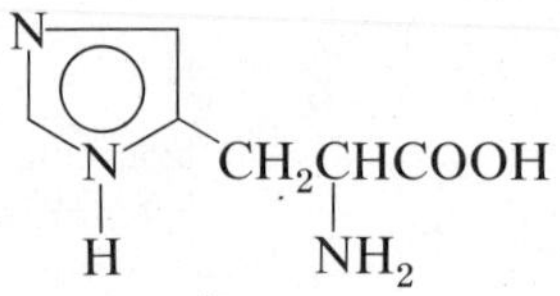

Histidine

Heterocyclic Amines as Nucleophiles in Alkylation and Acylation Reactions

Most heterocyclic amines undergo alkylation and acylation reactions in much the same way as acyclic amines.

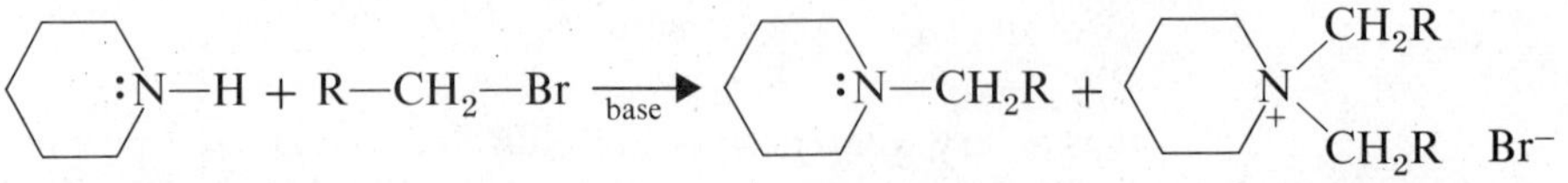

Piperidine

N-Alkylpiperidine

N,N-Dialkylpiper-
idium bromide

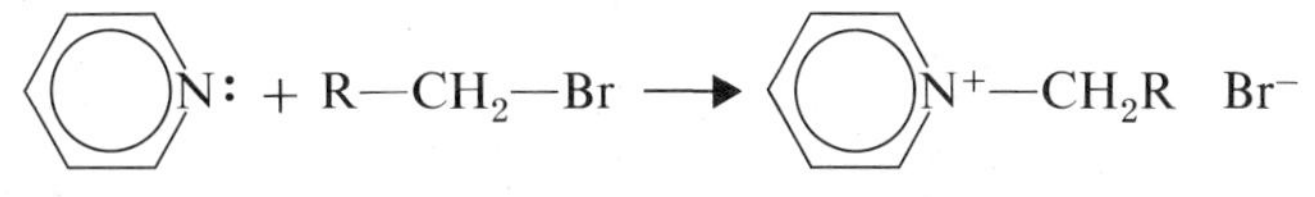

Pyridine

N-Alkylpyridinium
bromide

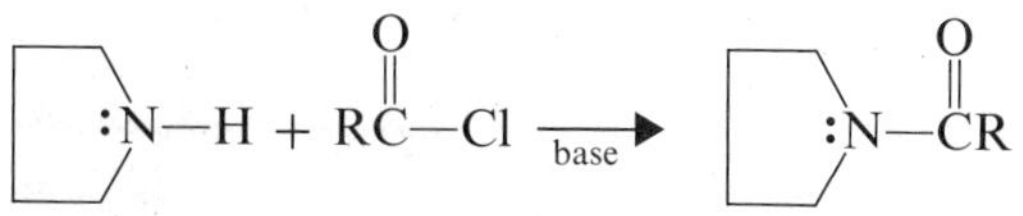

Pyrrolidine

N-Acylpyrrolidine
(an amide)

Problem 22.1

What products would you expect to obtain from the following reactions?
- (a) Piperidine + acetic anhydride ⟶
- (b) Pyridine + methyl iodide ⟶
- (c) Pyrrolidine + phthalic anhydride ⟶
- (d) Pyrrolidine + (excess) methyl iodide $\xrightarrow{\text{(base)}}$
- (e) Product of (d) + Ag_2OH_2O, then heat ⟶

Electrophilic Substitution Reactions of Aromatic Heterocyclic Amines

Pyrrole is highly reactive toward electrophilic substitution and substitution takes place primarily at position 2.

General Reaction

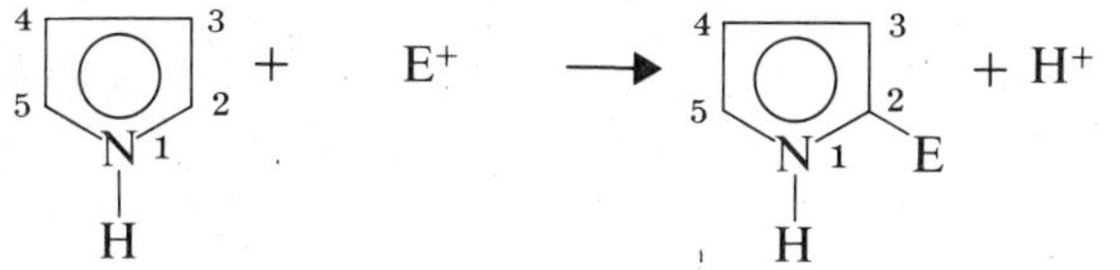

Pyrrole   Electrophile   2-Substituted pyrrole

Specific Example

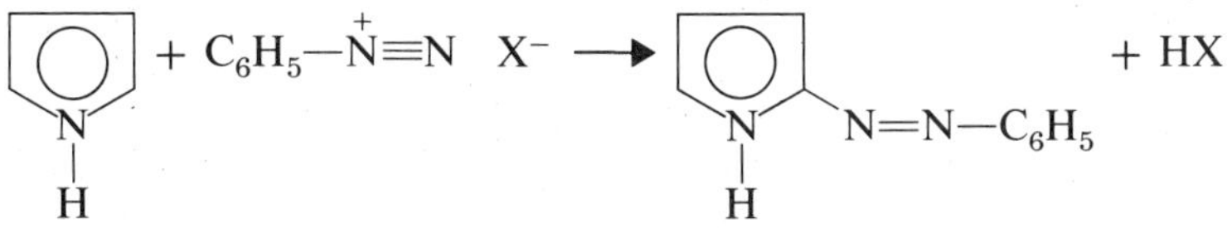

We can understand why electrophilic substitution at the 2 position is preferred if we examine the resonance structures shown on the next page.

Substitution at the 2 position of pyrrole

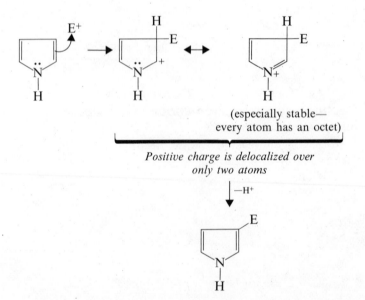

(especially stable—
every atom has an octet)

Positive charge is delocalized over three atoms

Substitution at the 3 position of pyrrole

(especially stable—
every atom has an octet)

*Positive charge is delocalized over
only two atoms*

We see that while an especially stable structure contributes to the hybrid for both intermediates, the intermediate arising from attack at the 2 position is stabilized by one additional resonance structure, and the positive charge is delocalized over three atoms rather than two. This means that this intermediate is more stable and that attack at the 2 position has a lower activation energy.

Pyridine is much less reactive toward electrophilic substitution than benzene. Pyridine does not undergo Friedel-Crafts acylation or alkylation; it does not couple with diazonium compounds. Bromination of pyridine can be accomplished but only in the vapor phase at 200° where a free-radical mechanism may operate. Nitration and sulfonation also require forcing conditions. Electrophilic substitution, when it occurs, nearly always takes place at the 3 position.

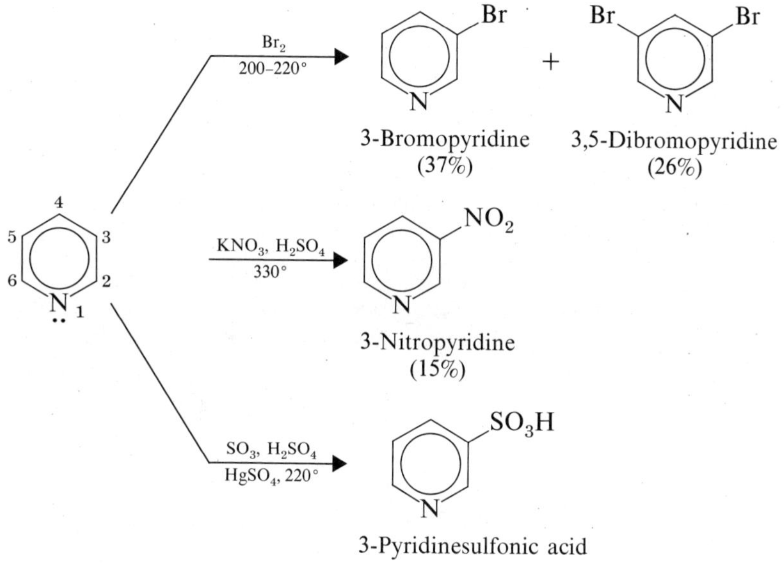

3-Bromopyridine
(37%)

3,5-Dibromopyridine
(26%)

3-Nitropyridine
(15%)

3-Pyridinesulfonic acid

We can, in part, attribute the lower reactivity of pyridine (when compared to benzene) to the difference in electronegativity of nitrogen when compared to carbon. Nitrogen, being more electronegative, is less able to accommodate the electron deficiency that characterizes the transition state leading to the positively charged σ complex in electrophilic substitution.

Pyridine

*Transition state
is of higher energy
because of greater
electronegativity
of nitrogen*

σ Complex

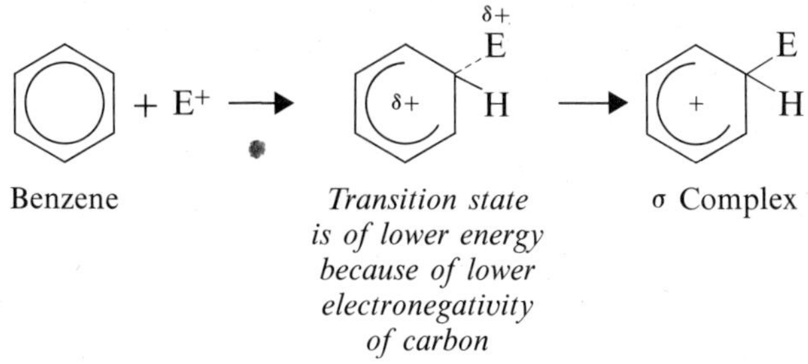

Benzene

*Transition state
is of lower energy
because of lower
electronegativity
of carbon*

σ Complex

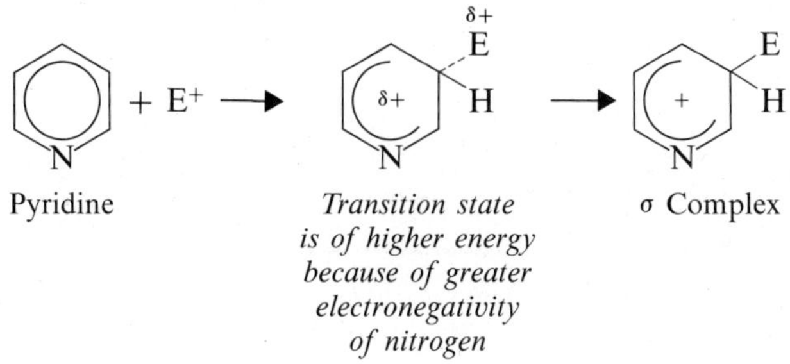

The low reactivity of pyridine toward electrophilic substitution also arises from the fact that pyridine is converted initially to a pyridinium ion by an electrophile or proton.

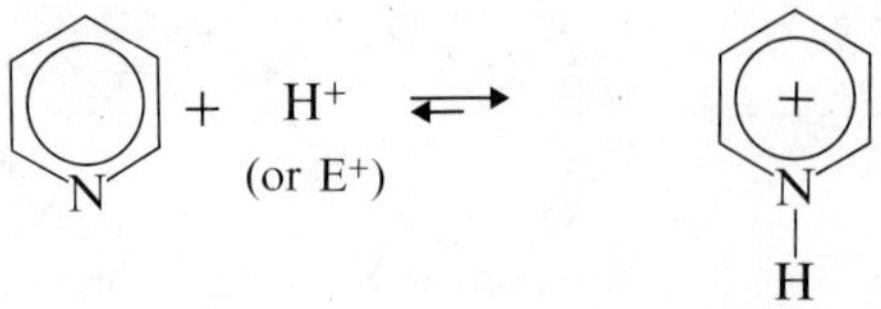

Pyridinium ion
(*highly unreactive because
of positive charge*)

Electrophilic attack at the 4 position (or the 2 position) is unfavorable because an especially unstable resonance structure contributes to the intermediate hybrid.

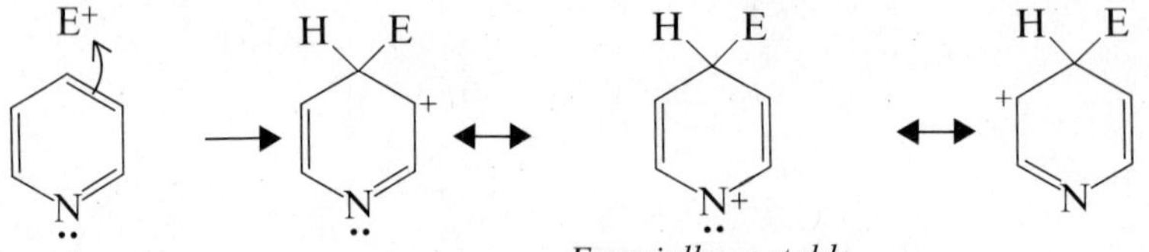

*Especially unstable,
nitrogen has a sextet
and a positive charge*

Similar resonance structures can be written for attack at the 2 position.

No especially unstable *or stable* structure contributes to the hybrid arising from attack at the 3 position; as a result, attack at the 3 position is preferred but occurs slowly.

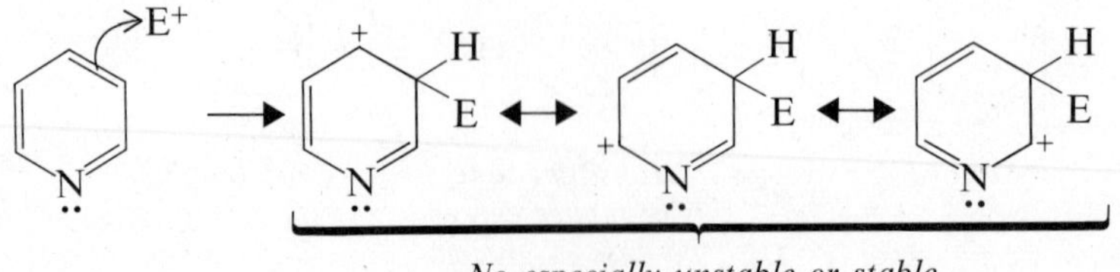

*No especially unstable or stable
structure contributes to the hybrid*

Pyrimidine is even less reactive toward electrophilic substitution than pyridine. (Why?) When electrophilic substitution takes place, it occurs at the 5 position.

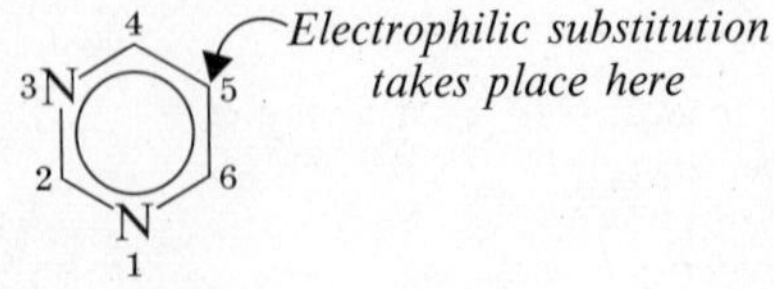

*Electrophilic substitution
takes place here*

Pyrimidine

Imidazole is much more susceptible to electrophilic substitution than pyridine or pyrimidine, but is less reactive than pyrrole. Imidazoles with 1 substituents undergo electrophilic substitution at the 4 position.

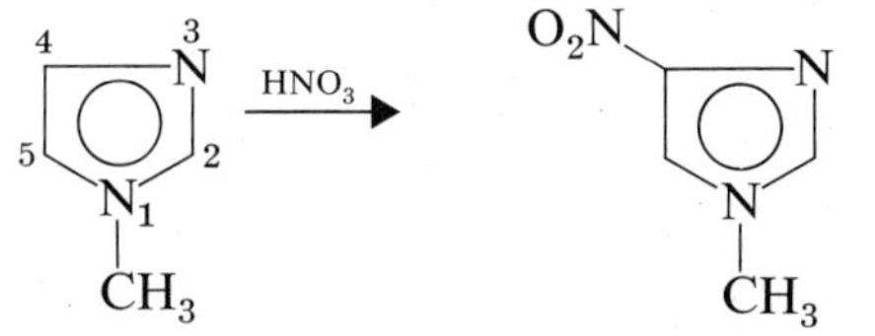

1-Methyl-4-nitroimidazole

Imidazole, itself, undergoes electrophilic substitution in a similar fashion. Tautomerism, however, makes the 4 and 5 positions equivalent.

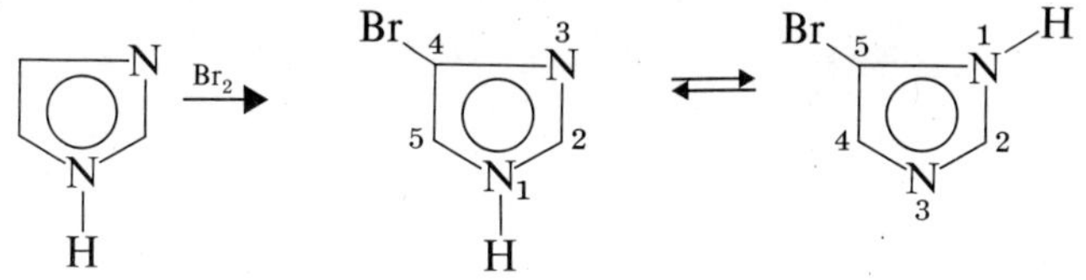

4-(5)-Bromoimidazole

Problem 22.2

Both pyrrole and imidazole are weak acids; they react with strong bases to form anions:

Pyrrole anion   Imidazole anion

(a) These anions resemble a carbocyclic anion that we have studied before. What is it?
(b) Write resonance structures that account for the stabilities of pyrrole and imidazole anions.

Nucleophilic Substitutions of Pyridine

In its reactions, the pyridine ring resembles a benzene ring with a strong electron-withdrawing group; pyridine is unreactive toward electrophilic substitution but reactive toward nucleophilic substitution.

In the previous section we compared the reactivity of pyridine and benzene toward electrophilic substitution and there we attributed pyridine's lower reactivity to the greater electronegativity of its ring nitrogen. Nitrogen, we said, because it is more electronegative than carbon, is less able to accommodate the electron deficiency that characterizes the transition state of the rate-limiting step in electrophilic aromatic substitution. On the other hand, nitrogen's greater electronegativity makes it *more* able to accommodate the excess *negative* charge that an aromatic ring must accept in *nucleophilic substitution*.

Pyridine reacts with sodium amide, for example, to form 2-aminopyridine: In this remarkable reaction (called the Chichibabin Reaction) amide ion, $:NH_2^-$, displaces a hydride ion, $H:^-$.

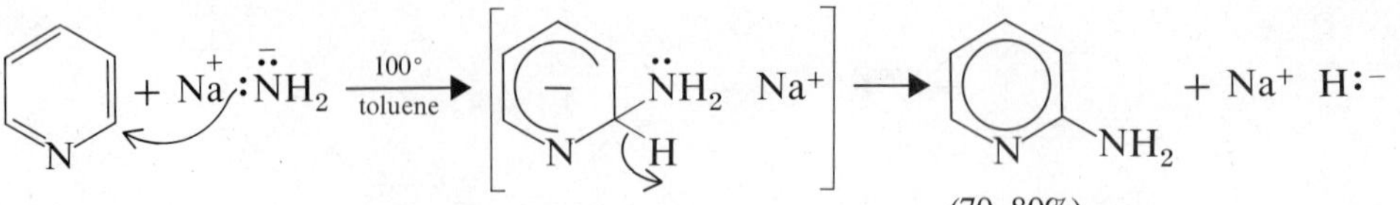

(70–80%)

If we examine the resonance structures that contribute to the intermediate in this reaction we will be able to see how the ring nitrogen accommodates the negative charge:

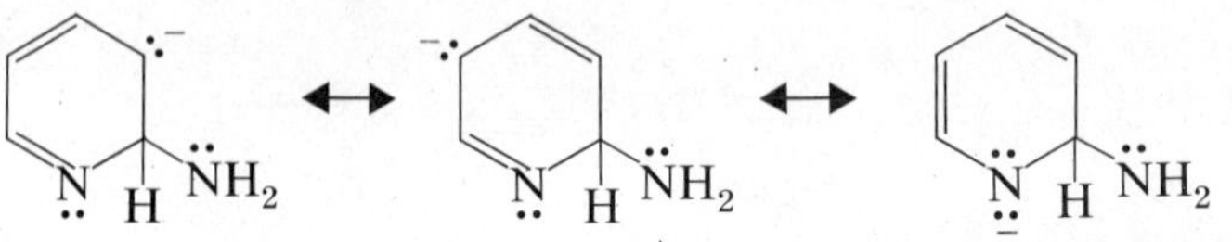

Especially stable because negative charge is on electronegative nitrogen

In the next step the intermediate loses a hydride ion and becomes 2-aminopyridine.*

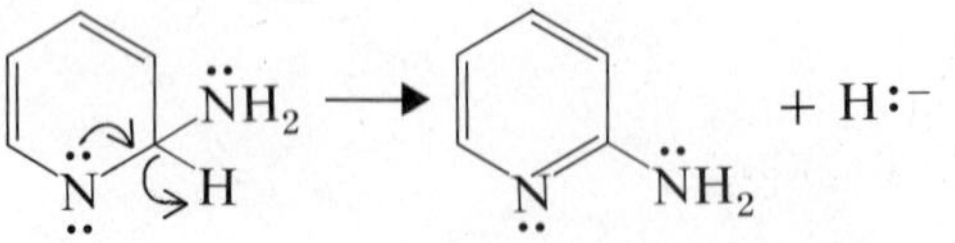

Pyridine undergoes similar nucleophilic substitution reactions with phenyl lithium, butyl lithium and potassium hydroxide.

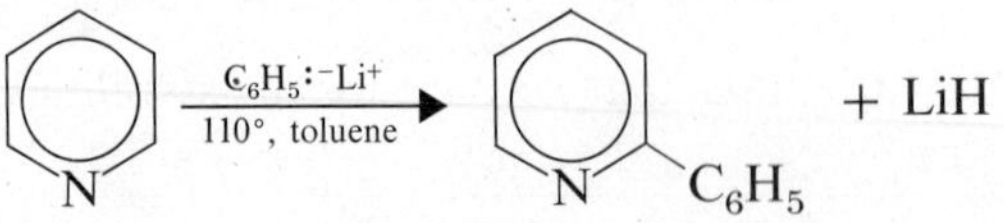

2-Phenylpyridine

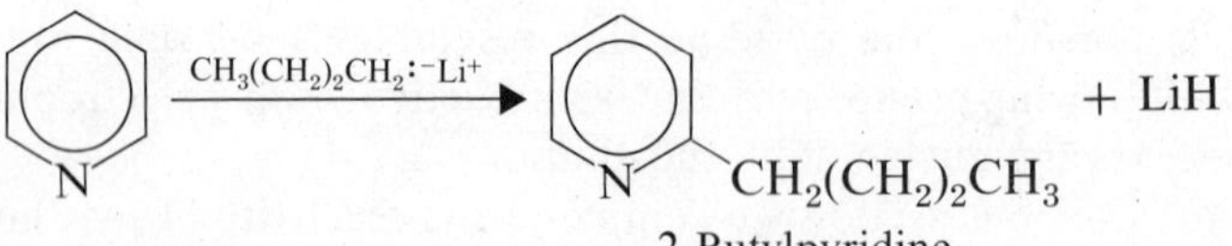

2-Butylpyridine

---

* In actual practice, a subsequent reaction occurs; 2-aminopyridine reacts with the sodium hydride to produce a sodioderivative:

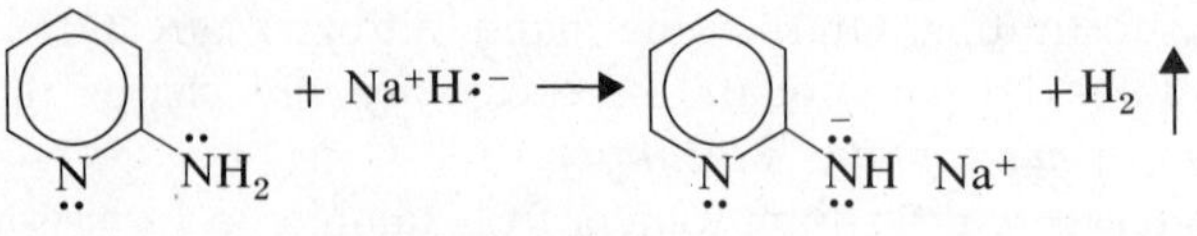

When the reaction is over, the addition of cold water to the reaction mixture converts the sodioderivative to 2-aminopyridine.

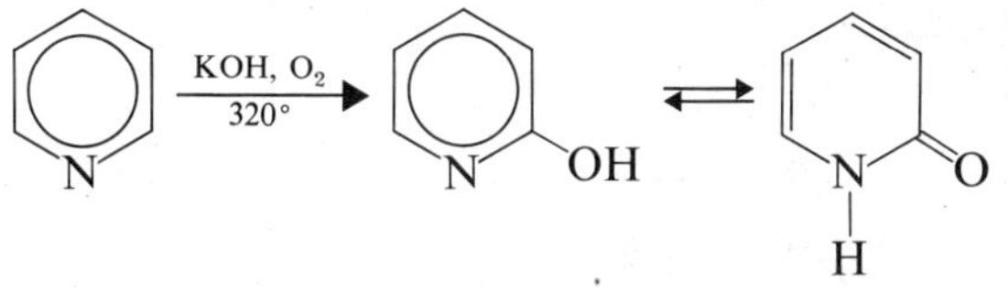

2-Pyridinol   2-Pyridone
(50%)

Nucleophilic Additions to Pyridinium Ions

Pyridinium ions are especially susceptible to nucleophilic attack at the 2 or 4 position because of the contributions of the resonance forms shown below.

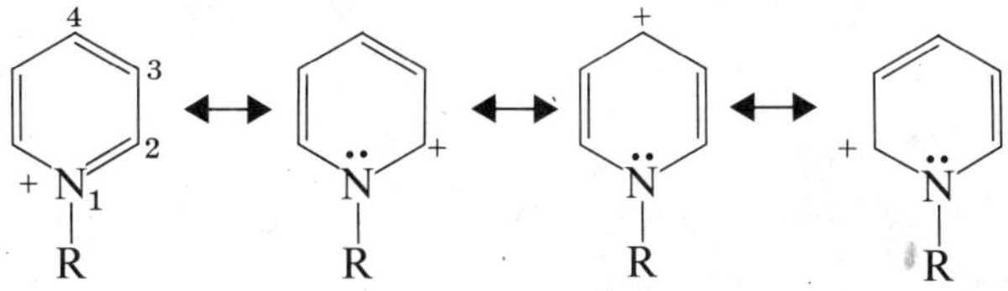

N-alkylpyridinium halides, for example, react with hydroxide ions primarily at position 2; this causes the formation of an addition product called a *pseudo base:*

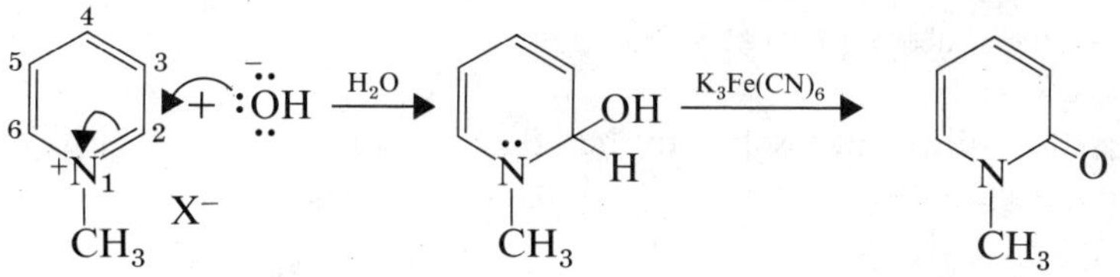

Pseudo base   *N*-Methylpyridone
(65–70%)

Oxidation of the pseudo base with potassium ferricyanide (above) produces an *N*-alkylpyridone.

Nucleophilic additions to pyridinium ions, especially the addition of *hydride ions,* have been of considerable interest to chemists because these reactions resemble the biological reduction of the important coenzyme, nicotinamide adenine dinucleotide, NAD$^+$ (p. 439).

A number of model reactions have been carried out in connection with these studies. Treating an *N*-alkylpyridinium ion with sodium borohydride, for example, brings about hydride addition, but addition occurs at position 2 and is usually accompanied by over reduction:

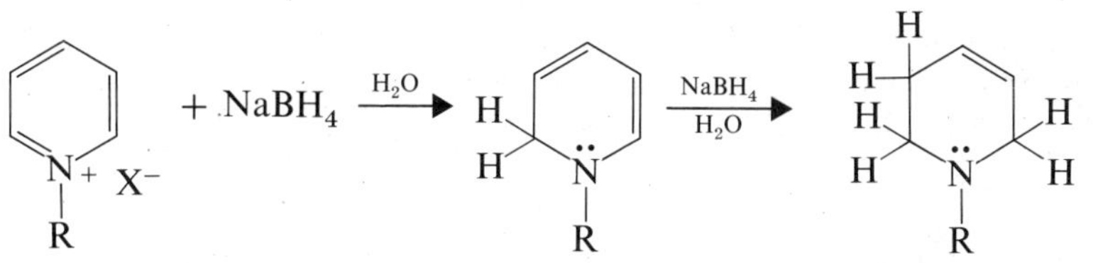

N-Alkyl-   1,2-Dihydro-   A 1,2,3,6-Tetrahydro-
pyridinium   pyridine   pyridine
iodide

Treating a pyridinium ion with basic sodium dithionite ($Na_2S_2O_4$), however, brings about specific addition to position 4:

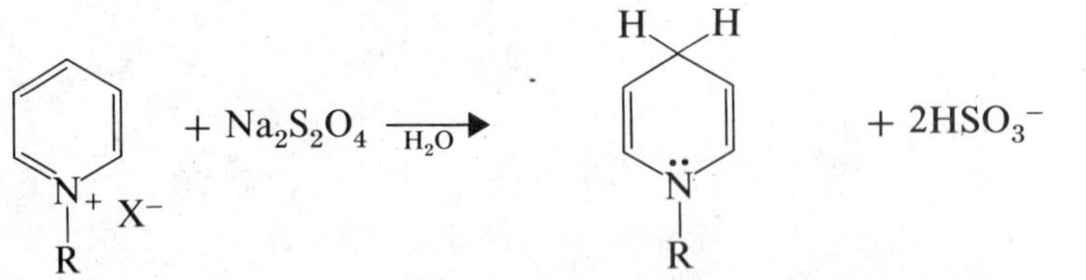

A 1,4-dihydropyridine

Sodium dithionite in aqueous base also reduces NAD^+ to NADH. The NADH

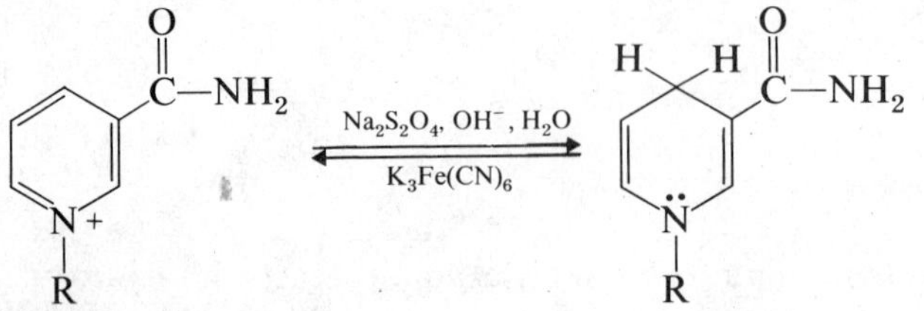

NAD$^+$
(see p. 439 for
the structure of R)

NADH

formed by dithionite reduction has been shown to be biologically active and can be oxidized to NAD^+ with potassium ferricyanide.

Problem 22.3

An alternative mechanism to the one given for the amination of pyridine on p. 862, involves a "pyridyne" intermediate, that is,

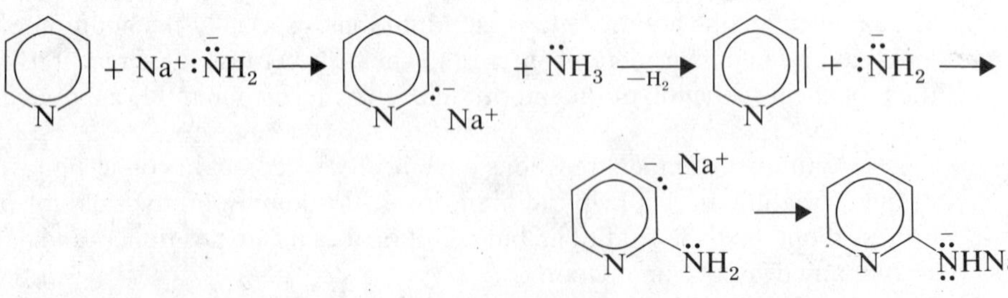

This mechanism was disallowed on the basis of an experiment in which 3-deuterio-pyridine was allowed to react with sodium amide. Consider the fate of deuterium in both mechanisms and explain.

Problem 22.4

2-Halopyridines undergo nucleophilic substitution much more readily than pyridine itself. What factor accounts for this?

22.2 ALKALOIDS

Extracting the bark, roots, leaves, berries, and fruits of plants often yields nitrogen-containing bases called *alkaloids*. The name alkaloid comes from the fact that these substances are "alkalilike;" that is, since alkaloids are amines they often react with acids to yield soluble salts. The nitrogen atoms of most alkaloids are present in heterocyclic rings. In a few instances, however, nitrogen may be present as a primary amine or as a quaternary ammonium group.

When administered to animals most alkaloids produce striking physiological effects and these effects *vary greatly* from alkaloid to alkaloid. Some alkaloids stimulate the central nervous system, others cause paralysis; some alkaloids elevate blood pressure, others lower it. Certain alkaloids act as pain relievers; others act as tranquilizers; still others act against infectious microorganisms. Most alkaloids are toxic when their dosage is large enough, and with some this dosage is very small. In spite of this, many alkaloids find use in medicine.

Systematic names are seldom used for alkaloids and their common names have a variety of origins. In many instances the common name reflects the botanical source of the compound. The alkaloid, strychnine, for example, comes from the seeds of the *strychnos* plants. In other instances the names are more whimsical: the name of the opium alkaloid, morphine, comes from Morpheus, the ancient Greek god of dreams; the name of the tobacco alkaloid, nicotine, comes from Nicot, an early French ambassador who sent tobacco seeds to France. The one characteristic that alkaloid names have in common is the ending -ine, reflecting the fact that they are all amines.

Alkaloids have been of interest to chemists for nearly two centuries and in that time thousands of alkaloids have been isolated. Most of these have had their structures determined through the application of chemical and physical methods and in many instances these structures have been confirmed by independent synthesis. A complete account of the chemistry of the alkaloids would (and does) occupy volumes; here we have space to consider only a few representative examples.

Alkaloids Containing a Pyridine or Reduced Pyridine Ring

The predominant alkaloid of the tobacco plant is nicotine:

Nicotine   Nicotinic acid

In very small doses nicotine acts as a stimulant, but in larger doses it causes depression, nausea, and vomiting. In still larger doses it is a violent poison. Nicotine salts are used as insecticides.

Oxidation of nicotine with concentrated nitric acid produces pyridine-3-carboxylic acid—a compound that is called *nicotinic acid*. While the consumption of nicotine is of no benefit to humans, nicotinic acid is a vitamin; it is incorporated into the important coenzyme, nicotinamide adenine dinucleotide.

Problem 22.5

Nicotine has been synthesized by the route shown below. All of the steps involve reactions that we have seen before. Suggest reagents that could be used for each.

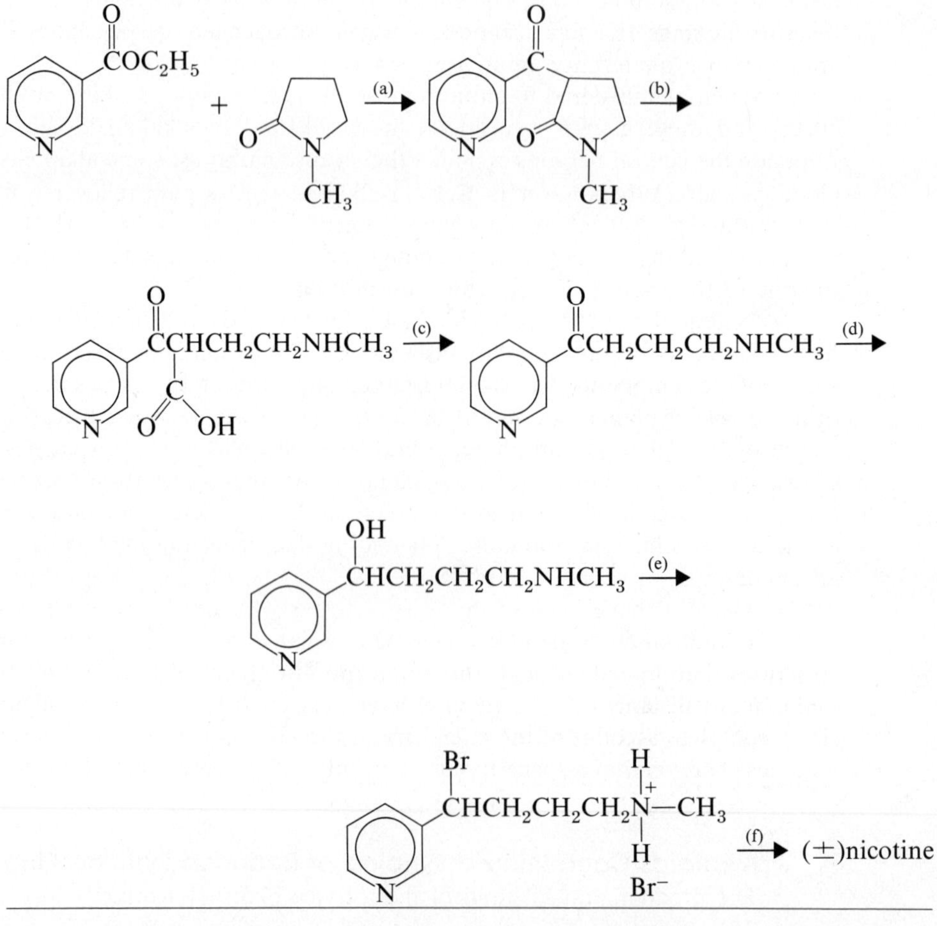

There are a number of alkaloids that contain a piperidine ring. These include coniine (from the water hemlock), atropine (from *Atropa belladonna* and other genera of the plant family Solanaceae), and cocaine (from *Erythroxylon coca*).

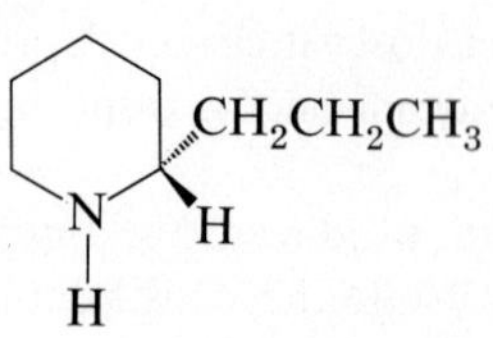

Coniine
((−)-2-propylpiperidine)

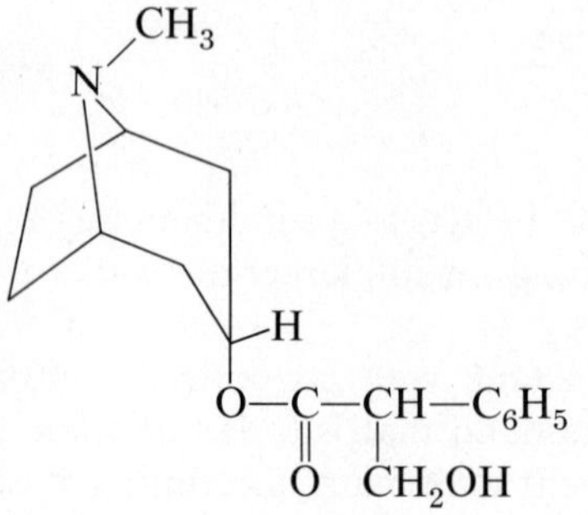

Atropine

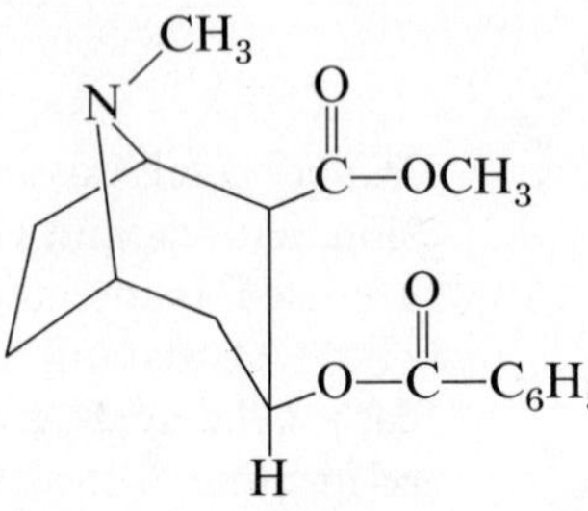

Cocaine

Coniine is highly toxic; its ingestion may cause weakness, drowsiness, nausea, labored respiration, paralysis, and death. Coniine was the toxic substance of the "hemlock" used in the execution of Socrates.

In small doses cocaine decreases fatigue, increases mental activity, and gives a general feeling of well-being. Prolonged use of cocaine, however, leads to psychical addiction and to periods of deep depression. Cocaine is also a local anesthetic and, for a time, it was used medically in that capacity. When its tendency to cause addiction was recognized, efforts were made to develop other local anesthetics. This led, in 1905, to the synthesis of novocaine, a compound that has some of the same structural features as cocaine (i.e., its benzoate ester and tertiary amine groups).

$$CH_3CH_2 \diagdown \atop CH_3CH_2 \diagup N-CH_2CH_2-O-\overset{\overset{\textstyle O}{\|}}{C}-\langle \bigcirc \rangle -NH_2$$

Novocaine

Atropine is an intense poison. In dilute solutions (0.5-1.0%) it is used to dilate the pupil of the eye in ophthalmic examinations. Alkaloids related to atropine are contained in the 12-hour continuous-release capsules used to relieve symptoms of the common cold.

Problem 22.6

The principal alkaloid of *Atropa belladonna* is the optically active alkaloid *hyoscyamine*. During its isolation hyoscyamine is often racemized by bases to optically inactive atropine. (a) What chiral center is likely to be involved in the racemization? (b) In hyoscyamine this chiral center has the *S* configuration. Write a three-dimensional structure for hyoscyamine.

Problem 22.7

Hydrolysis of atropine gives tropine and (±) tropic acid. (a) What are their structures? (b) Even though tropine has chiral carbons, it is optically inactive. Explain. (c) An isomeric form of tropine called ψ-tropine has also been prepared by heating tropine with base. ψ-Tropine is also optically inactive. What is its structure?

Problem 22.8

In 1891, G. Merling transformed tropine (cf. problem 22.7) into 1,3,5-cycloheptatriene (tropylidene) through the following sequence of reactions.

Tropine $(C_8H_{15}NO)$ $\xrightarrow{-H_2O}$ $C_8H_{13}N$ $\xrightarrow{CH_3I}$ $C_9H_{16}NI$ $\xrightarrow[\text{(2) heat}]{\text{(1) Ag}_2\text{O/H}_2\text{O}}$

$C_9H_{15}N$ $\xrightarrow{CH_3I}$ $C_{10}H_{18}NI$ $\xrightarrow[\text{(2) heat}]{\text{(1) Ag}_2\text{O/H}_2\text{O}}$ 1,3,5-cycloheptatriene $+ (CH_3)_3N + H_2O$

Write out all of the reactions that take place.

Problem 22.9

Many alkaloids appear to be synthesized in plants through reactions that resemble the Mannich reaction (p. 809). Recognition of this (by R. Robinson in 1917) led to a synthesis of tropinone that takes place under "physiological conditions," that is, at room temperature and at pH values near neutrality. This synthesis is shown below. Propose reasonable mechanisms that account for the overall course of the reaction.

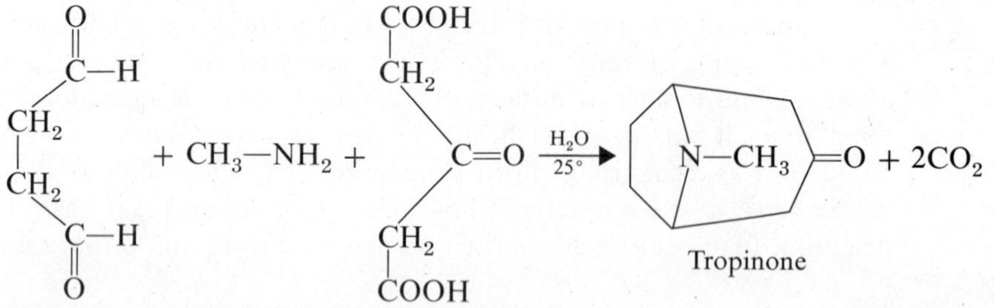

Alkaloids Containing an Isoquinoline or Reduced Isoquinoline Ring

Papaverine, morphine, and codeine are all alkaloids obtained from the opium poppy, *Papaver somniferum*.

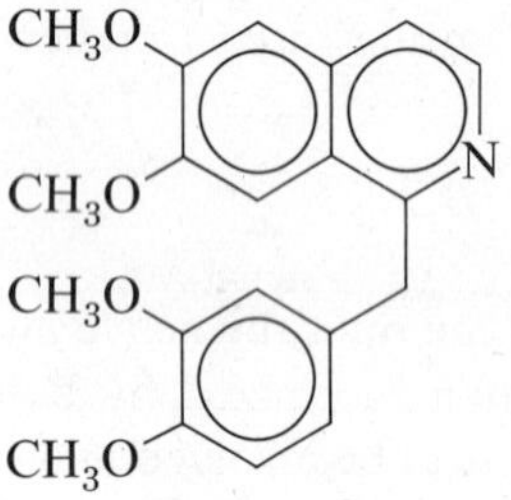

Papaverine

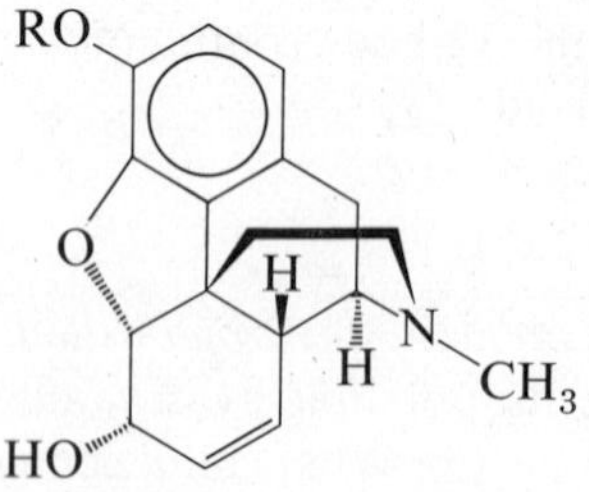

Morphine (R = H)
Codeine (R = CH$_3$)

Opium has been used since earliest recorded history. Morphine was first isolated from opium in 1803 and its isolation represented one of the first instances of the purification of the active principle of a drug. One hundred and twenty years were to pass, however, before the complicated structure of morphine was deduced, and its final confirmation through independent synthesis (by Professor Marshall Gates of the University of Rochester) did not take place until 1952.

Morphine is one of the most potent analgesics known and it is still used extensively in medicine to relieve pain, especially "deep" pain. Its greatest drawbacks, however, are its tendencies to lead to addiction and to depress respiration. These disadvantages have brought about a search for morphinelike compounds that do not have these disadvantages. One of the newest candidates is the compound pentazocine. Pentazocine is a highly effective analgesic and it is nonaddictive; unfortunately however, like morphine, it depresses respiration.

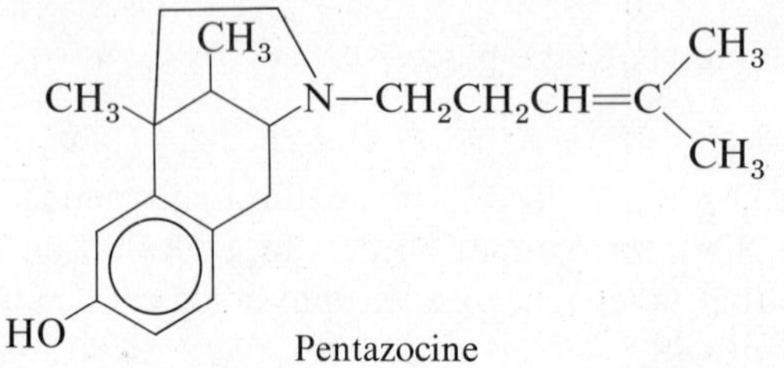

Pentazocine

Problem 22.10

Papaverine has been synthesized by the following route:

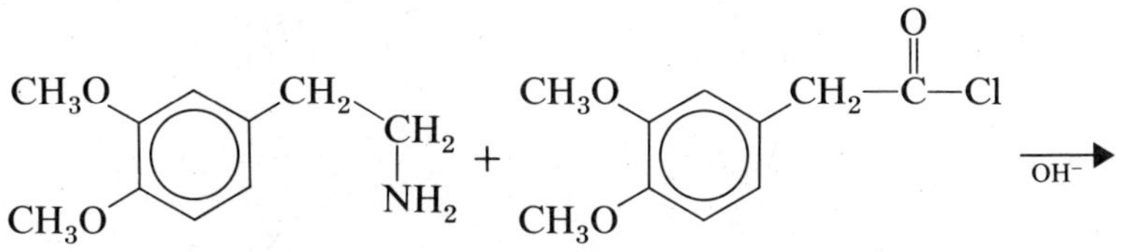

$$C_{20}H_{25}NO_5 \xrightarrow[\substack{\text{heat} \\ (-H_2O)}]{P_2O_5} \text{Dihydropapaverine} \xrightarrow[\substack{\text{heat} \\ (-H_2)}]{Pd} \text{Papaverine}$$

Outline the reactions involved.

Problem 22.11

One of the important steps in the synthesis of morphine involved the following transformation:

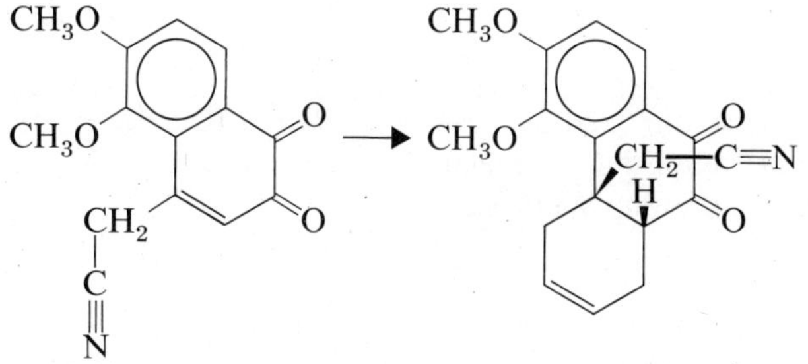

Suggest how this step was accomplished.

Problem 22.12

When morphine reacts with two moles of acetic anhydride it is transformed into the highly addictive narcotic, heroin. What is the structure of heroin?

Alkaloids Containing Indole or Reduced Indole Rings

A large number of alkaloids are derivatives of an indole ring system. These range from the relatively simple *gramine* to the highly complicated structures of *strychnine* and *reserpine*.

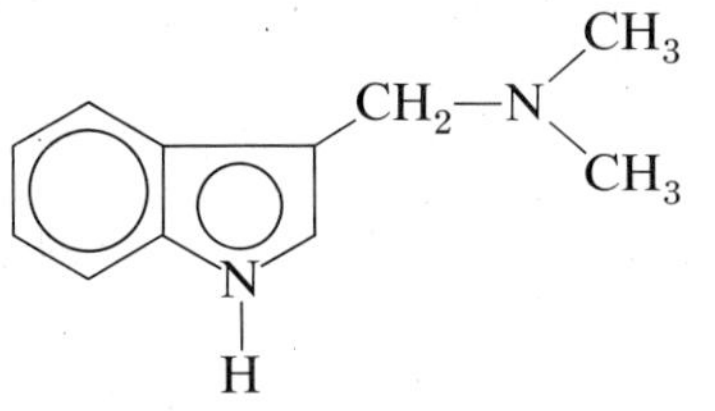

Gramine

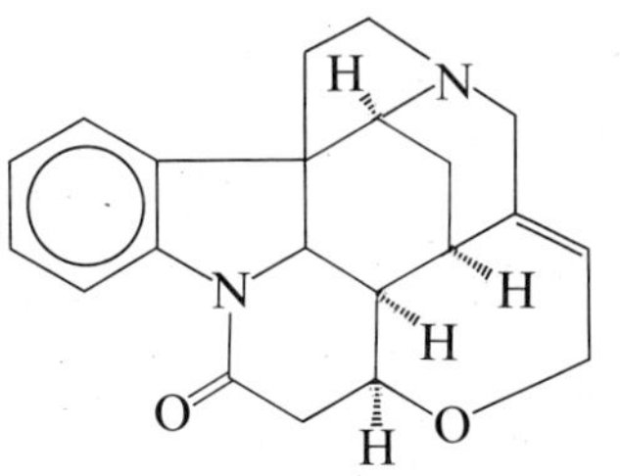

Strychnine

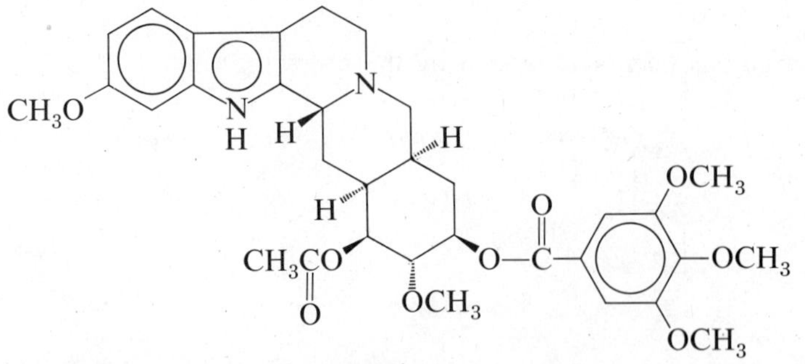

Reserpine

Gramine can be obtained from chlorophyll-deficient mutants of barley. Strychnine, a very bitter and highly poisonous compound, comes from the seeds of *Strychnos nux-vomica.* Strychnine is a central nervous system stimulant and has been used medically (in low dosage) to counteract poisoning by central nervous system depressants. Reserpine can be obtained from the Indian snake root *Rauwolfia serpentina,* a plant that has been used in native medicine for centuries. Reserpine is used in modern medicine as a tranquilizer and as an agent to lower blood pressure.

Problem 22.13

Gramine has been synthesized by heating a mixture of indole, formaldehyde, and dimethylamine. (a) What general reaction is involved here? (b) Outline a reasonable mechanism for the gramine synthesis.

Biosynthesis of Alkaloids

The primary starting materials for alkaloid synthesis in plants appear to be α-amino acids. More than 20 different α-amino acids occur naturally and they are also the main building blocks for proteins (Chapter 25). Two amino acids, in particular, are important in alkaloid biosynthesis. These are tyrosine and tryptophan:

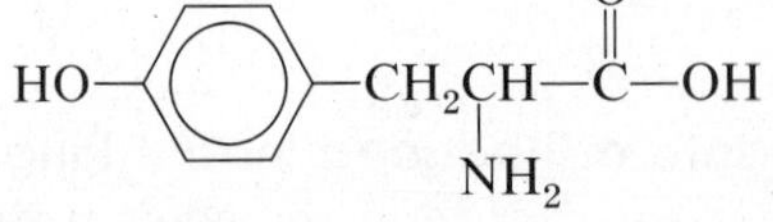

Tyrosine 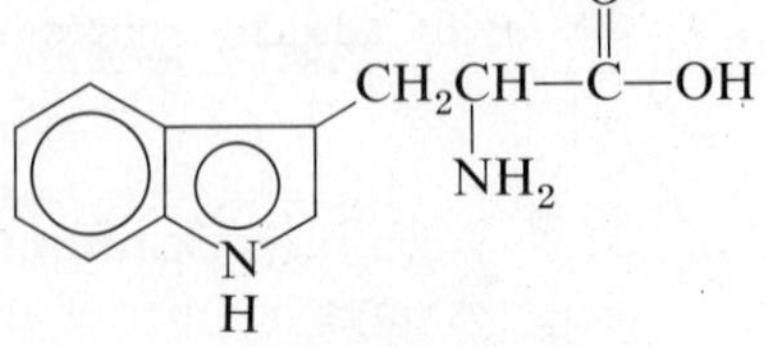   Tryptophan

Two general reactions appear to be of central importance in alkaloid biosynthesis—the Mannich reaction and the oxidative coupling of phenols. We studied the Mannich reaction earlier in Section 20.6 (cf. also problem 22.9). The oxidative coupling of phenols is a free-radical process that is catalyzed by enzymes in plants and that can also be carried out (usually less successfully) in the laboratory. A simple formulation of an oxidative phenol coupling is outlined below using phenol itself as the starting compound. Loss of an electron and a proton from phenol leads to a resonance-stabilized free radical. Two free radicals can then undergo coupling in a variety of ways:

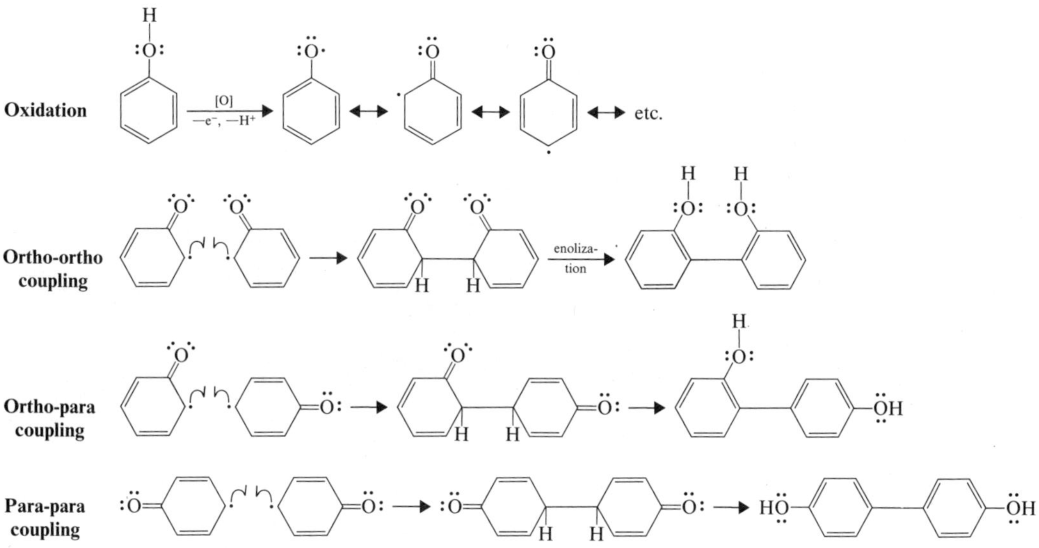

Oxidation

Ortho-ortho coupling

Ortho-para coupling

Para-para coupling

The biosynthetic route that the opium poppy uses to synthesize morphine is now known. Most of the morphine molecule, it turns out, is constructed from two molecules of tyrosine.

The synthesis begins with the oxidation of tyrosine to 3,4-dihydroxyphenylalanine:

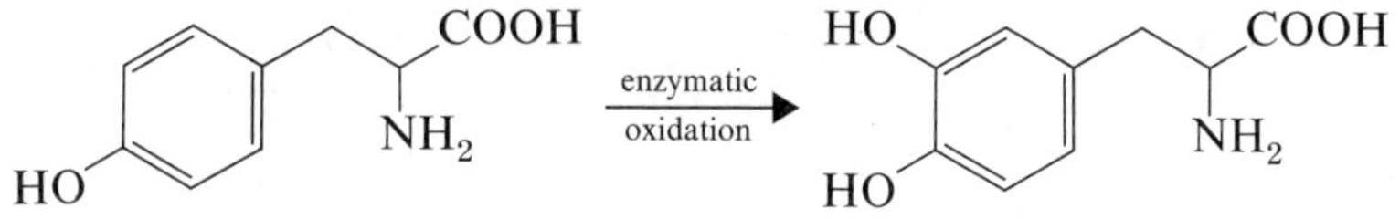

Tyrosine   3,4-Dihydroxyphenylalanine

Further enzyme-catalyzed reactions transform 3,4-dihydroxyphenylalanine into 3,4-dihydroxyphenylpyruvic acid, **I**, and into 3,4-dihydroxyphenylethylamine **II** (Fig. 22.1). These two molecules then react in a Mannich-type condensation to yield norlaudanosoline, **III.**

Methylation of norlaudanosoline at two of its —OH groups and at its —N—H group yields reticuline, **IV***a* (Fig. 22.2). A reticuline molecule can be twisted into conformation **IV***b*; one that allows an ortho-para oxidative phenolic coupling to take place yielding salutaridine. Reduction of salutaridine produces salutaridinol. Then salutaridinol is transformed into thebaine. (In this highly unusual step the oxygen bridge is installed through a reaction that is accompanied by *the displacement of a hydroxide ion.*) Finally, several additional enzymatic reactions transform thebaine into morphine.

Problem 22.14

There is considerable evidence that oxidative phenol couplings are important in the biosynthesis of bulbocapnine and glaucine (p. 874). (These two alkaloids have what is called an *apomorphine* ring system.) Both compounds appear to arise from reticuline. Show the type of oxidative phenol coupling that is involved in each biosynthesis. (Assume that methylation of —OH groups in both alkaloids and synthesis of the —O—CH$_2$—O— bridge in bulbocapnine occur after the oxidative phenol couplings.)

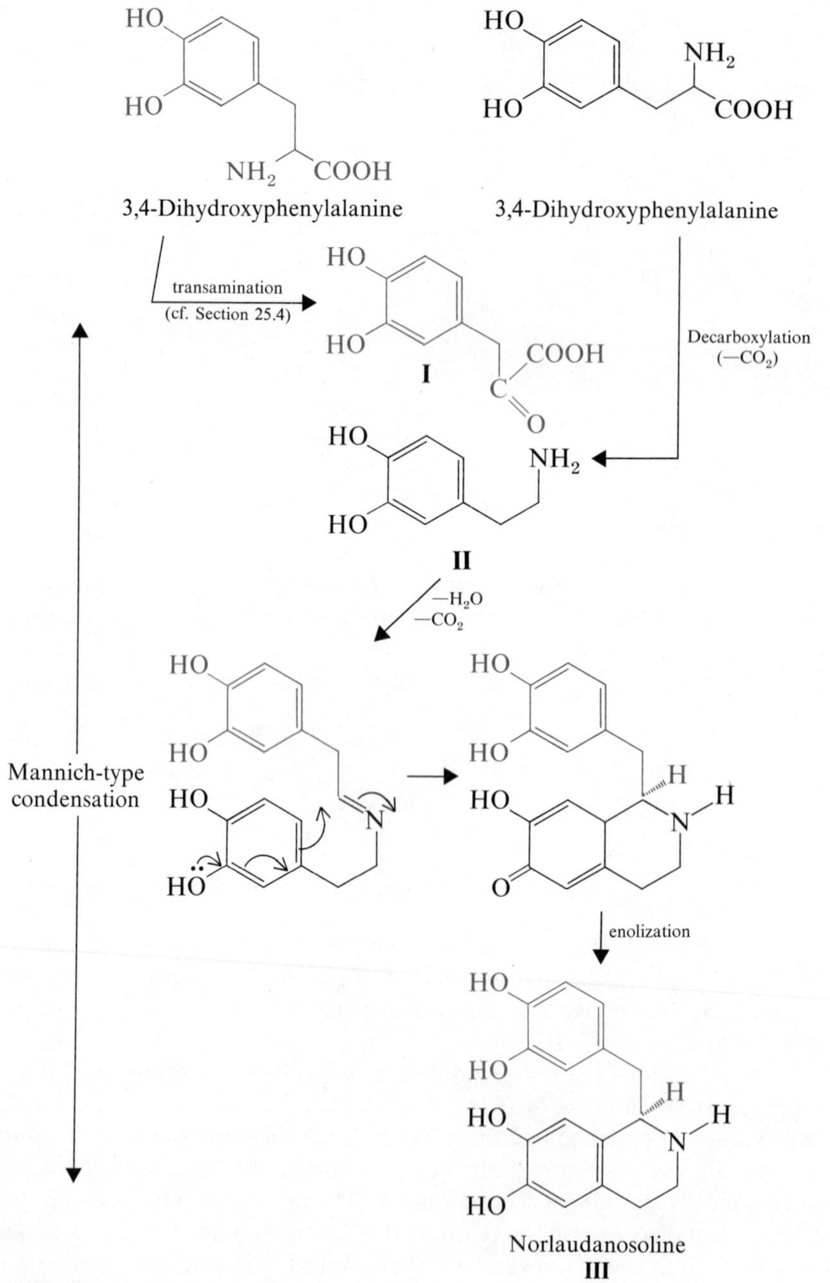

FIG. 22.1
The biosynthesis of norlaudanosoline from two moles of 3,4-dihydroxy-phenylalanine.

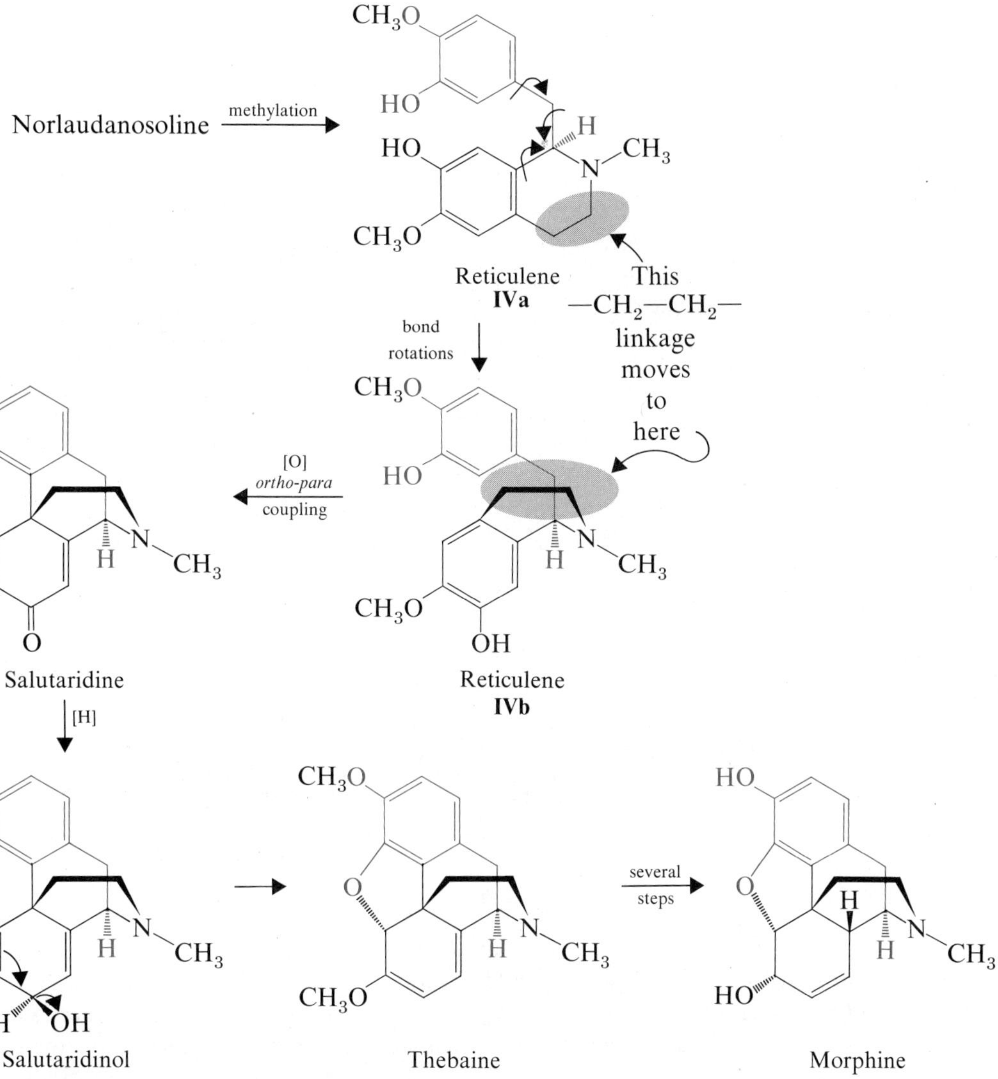

FIG. 22.2
The biosynthesis of morphine from norlaudanosoline.

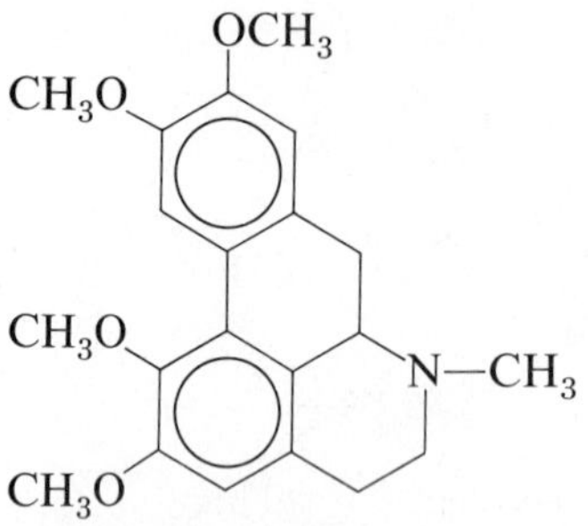

Glaucine

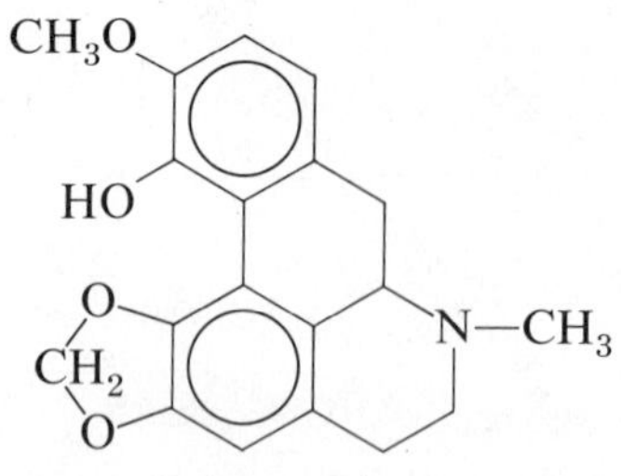

Bulbocapnine

Problem 22.15

Much of the evidence for plant biosynthetic pathways comes from experiments in which labeled compounds are fed to the plant. Later, intermediates and alkaloids are isolated from the plant and the locations of the labeled atoms are determined. An example is an experiment in which tyrosine labeled at the α-carbon was administered to opium poppies (*Papaver somniferum*); the papaverine isolated from the poppies had the labeled atoms at positions 1 and 3:

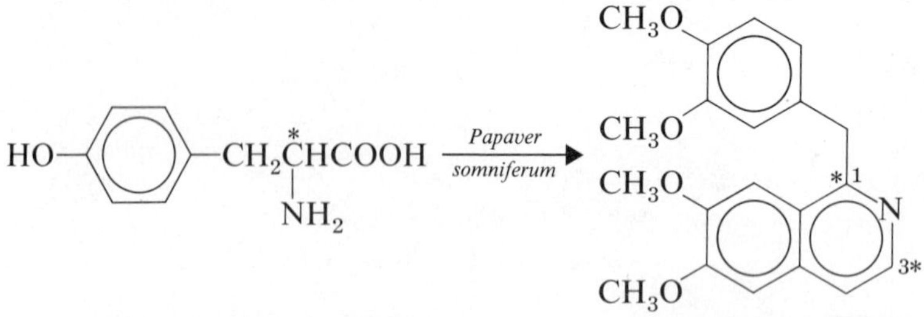

Labeled tyrosine

Labeled papaverine

(a) Is this evidence consistent with a biosynthesis of papaverine from tyrosine (i.e., with tyrosine leading to norlaudanosoline as in Fig. 22.1) and then a conversion of norlaudanosoline to papaverine? (b) Experiments using labeled norlaudanosoline show that opium poppies efficiently convert it into papaverine. Does this support the pathway outlined in (a)? (c) What major steps must take place in the conversion of norlaudanosoline to papaverine?

Problem 22.16

Harmine is an alkaloid isolated from *Peganum harmala L.* When tryptophan and pyruvic acid labeled in the positions shown below were fed to the plant, the harmine produced had the labeling pattern indicated.

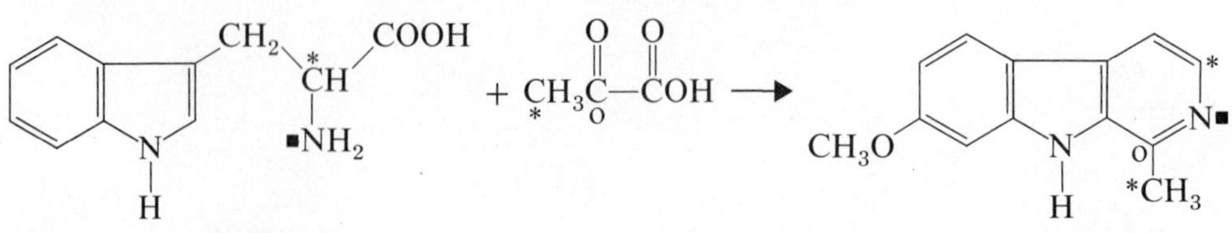

Tryptophan
* and o are ^{14}C labels
■ is an ^{15}N label

Pyruvic acid

Harmine

Show how these results are consistent with the following pathway: (1) decarboxylation of tryptophan (to tryptamine), (2) a Mannich-type condensation of tryptamine and pyruvic acid, then (3) dehydrogenation, (4) hydroxylation, and (5) methylation.

22.3 CONDENSATION POLYMERS

We saw, in Section 8.3, that large molecules with many repeating subunits—called *polymers*—can be prepared through addition reactions of alkenes. These polymers, we noted, are called *addition polymers*.

Another broad group of polymers are those called *condensation polymers*. These polymers, as their name suggests, are prepared through condensation reactions—reactions in which monomeric subunits are joined through intermolecular eliminations of small molecules such as water or alcohols. Among the most important condensation polymers are those called *polyamides* and *polyesters*.

Polyamides

Silk and wool are two naturally occurring polymers that man has used for centuries to fabricate articles of clothing. Silk and wool are examples of that group of organic compounds that we call *proteins*—a group of compounds that we will discuss in detail in Chapter 25. At this point we need only to notice (below) that the repeating subunits of proteins are derived from α-amino acids and that these subunits are joined by amide linkages. Proteins, therefore, are polyamides.

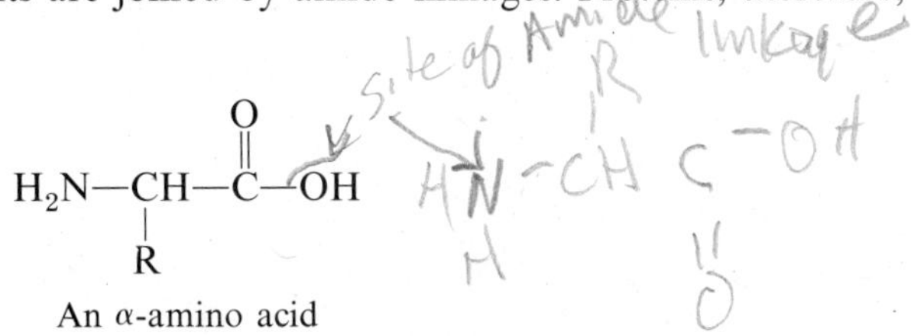

$$H_2N-CH-\overset{\overset{O}{\|}}{C}-OH$$
$$R$$

An α-amino acid

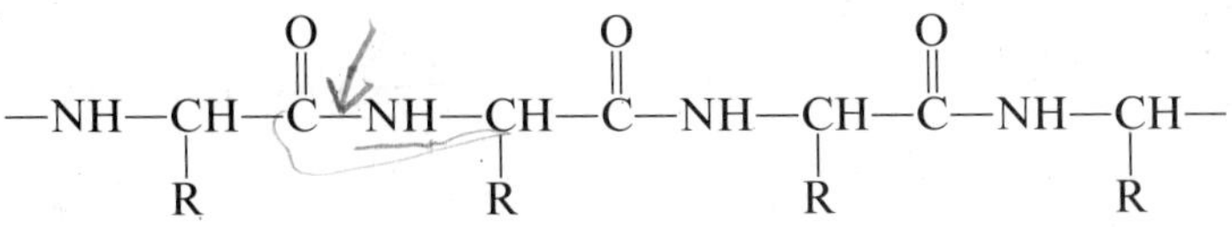

A portion of a polyamide chain as it might occur in a protein

It was a search for a synthetic material with properties similar to that of silk that led to the discovery of synthetic polyamide called Nylon.

One of the most important Nylons, called *Nylon 6,6*, can be prepared from the six-carbon dicarboxylic acid, adipic acid, and the six-carbon diamine, hexamethylenediamine. In the commercial process these two compounds are allowed to react in equimolar proportions in order to produce a 1:1 salt,

$$n\text{HO}\overset{\overset{\displaystyle O}{\|}}{C}-(CH_2)_4-\overset{\overset{\displaystyle O}{\|}}{C}OH + n\text{H}_2N-(CH_2)_6-NH_2 \longrightarrow$$

Adipic acid   Hexamethylenediamine

$$n\left[^-O\overset{\overset{\displaystyle O}{\|}}{C}-(CH_2)_4-\overset{\overset{\displaystyle O}{\|}}{C}-O^- \quad H_3\overset{+}{N}-(CH_2)_6-\overset{+}{N}H_3\right] \xrightarrow[\text{(polymerization)}]{\text{heat}}$$

1:1 salt (Nylon salt)

$$^-O\overset{\overset{\displaystyle O}{\|}}{C}-(CH_2)_4-\overset{\overset{\displaystyle O}{\|}}{C}-\left[NH-(CH_2)_6-NH-\overset{\overset{\displaystyle O}{\|}}{C}-(CH_2)_4-\overset{\overset{\displaystyle O}{\|}}{C}\right]_{n-1}-NH-(CH_2)_6-NH_3^+ + 2n\ H_2O$$

Nylon 6,6
(a polyamide)

Then, heating the 1:1 salt (Nylon salt) to a temperature of 270°C at a pressure of 250 pounds per square inch causes a polymerization to take place. Water molecules are lost as condensation reactions occur between $-\overset{\overset{\displaystyle O}{\|}}{C}-O^-$ and $-NH_3^+$ groups of the salt; this results in the formation of a polyamide.

    The Nylon 6,6 produced in this way has a molecular weight of ~10,000, has a melting point of ~250°, and can be spun into fibers from a melt. The fibers are then stretched to about four times their original length. This orients the linear polyamide molecules so that they are parallel to the fiber axis and allows hydrogen bonds to form between $-NH-$ and $>C=O$ groups on adjacent chains. Called "cold drawing," this stretching process greatly increases the fiber strength.

    Another type of Nylon, Nylon 6, can be prepared by a ring-opening polymerization of ε-caprolactam:

ε-Caprolactam
(a cyclic amide)

$$-NH-\left[\overset{\overset{\displaystyle O}{\|}}{C}(CH_2)_5-NH-\overset{\overset{\displaystyle O}{\|}}{C}-(CH_2)_5-NH\right]_n\overset{\overset{\displaystyle O}{\|}}{C}-$$

Nylon 6

    In this process ε-caprolactam is allowed to react with water; this converts some of it to ε-aminocaproic acid. Then heating this mixture to 250° drives off water as ε-caprolactam and ε-aminocaproic acid react to produce the polyamide. Nylon 6 can also be converted into fibers by melt spinning.

Problem 22.17

The raw materials for the production of Nylon 6, can be obtained in several ways as indicated on the next page. Give equations for each synthesis of adipic acid of hexamethylenediamine.

(a) Cyclohexanone $\xrightarrow{(O)}$ adipic acid

(b) Adipic acid $\xrightarrow{2NH_3}$ a salt $\xrightarrow{heat}$ $C_6H_{12}N_2O_2$ $\xrightarrow[\text{catalyst}]{350°}$

$\qquad\qquad\qquad\qquad$ $C_6H_8N_2$ $\xrightarrow[\text{catalyst}]{4H_2}$ hexamethylenediamine

(c) 1,3-Butadiene $\xrightarrow{Cl_2}$ $C_4H_6Cl_2$ $\xrightarrow{2NaCN}$ $C_6H_6N_2$ $\xrightarrow[\text{Ni}]{H_2}$

$\qquad\qquad\qquad\qquad$ $C_6H_8N_2$ $\xrightarrow[\text{catalyst}]{4H_2}$ hexamethylenediamine

(d) Tetrahydrofuran $\xrightarrow{2HCl}$ $C_4H_8Cl_2$ $\xrightarrow{2NaCN}$ $C_6H_8N_2$ $\xrightarrow[\text{catalyst}]{4H_2}$

$\qquad\qquad\qquad\qquad\qquad\qquad\qquad\qquad$ hexamethylenediamine

Polyesters

One of the most important polyesters is poly(ethylene terephthalate), a polymer that is marketed under the names *Dacron, Terylene,* and *Mylar.*

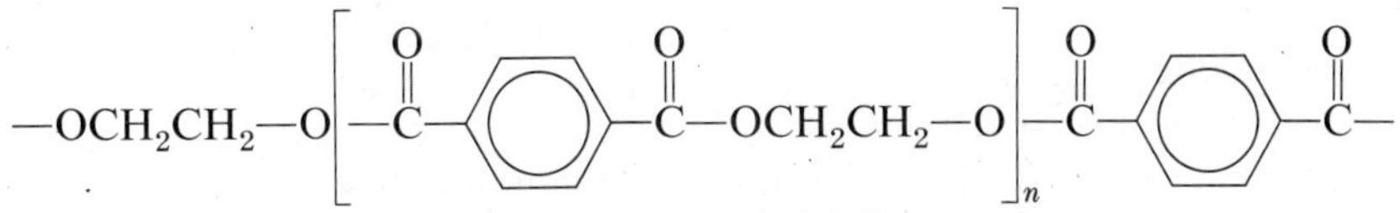

Poly(ethylene terephthalate)
(Dacron, Terylene, or Mylar)

Although one can obtain poly(ethylene terephthalate) by a direct acid-catalyzed esterification of ethylene glycol and terephthalic acid, the polymer produced in this way is of low molecular weight and is not generally useful.

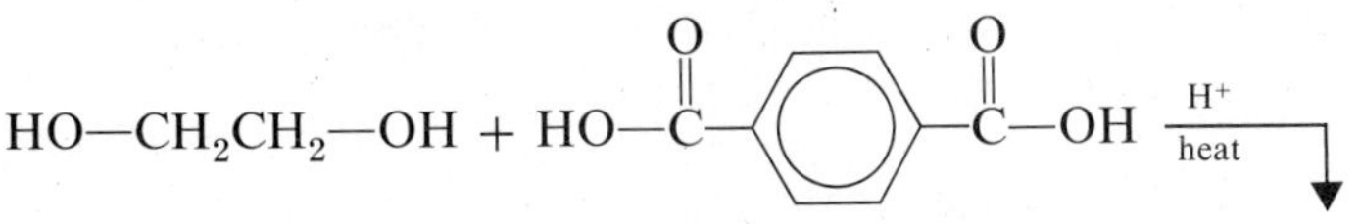

Ethylene glycol $\qquad\qquad$ Terephthalic acid

Poly(ethylene terephthalate)
of low molecular weight $+ H_2O$

Moreover, high temperatures are required to drive off the water produced by the esterification and this causes the polymer to undergo some decomposition.

A much better method for synthesizing poly(ethylene terephthalate) is based on transesterification reactions—reactions in which one ester is converted into another. One commercial synthesis utilizes two transesterifications. In the first (p. 878), dimethyl terephthalate and excess ethylene glycol are heated to 200° in the presence of a basic catalyst. Distillation of the mixture results in the loss of methanol (bp 78°) and the formation of a new ester, one formed from two moles of ethylene glycol and one mole of terephthalic acid. When this new ester is heated to a higher temperature (~280°), ethylene glycol distills and polymerization takes place.

First Transesterification

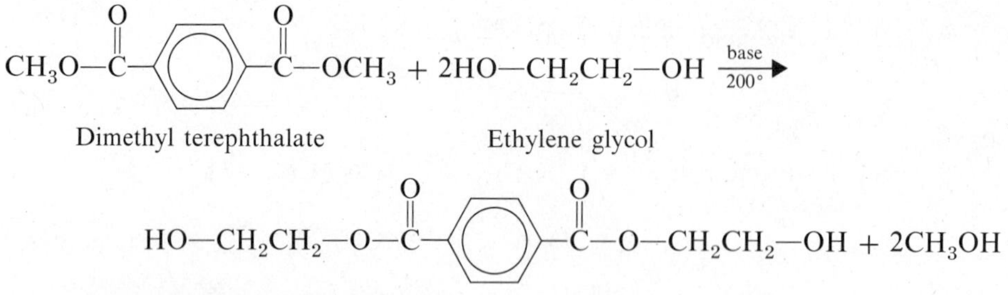

$$CH_3O-\overset{\overset{\displaystyle O}{\|}}{C}-\bigcirc-\overset{\overset{\displaystyle O}{\|}}{C}-OCH_3 + 2HO-CH_2CH_2-OH \xrightarrow[200°]{base}$$

Dimethyl terephthalate   Ethylene glycol

$$HO-CH_2CH_2-O-\overset{\overset{\displaystyle O}{\|}}{C}-\bigcirc-\overset{\overset{\displaystyle O}{\|}}{C}-O-CH_2CH_2-OH + 2CH_3OH$$

Second Transesterification

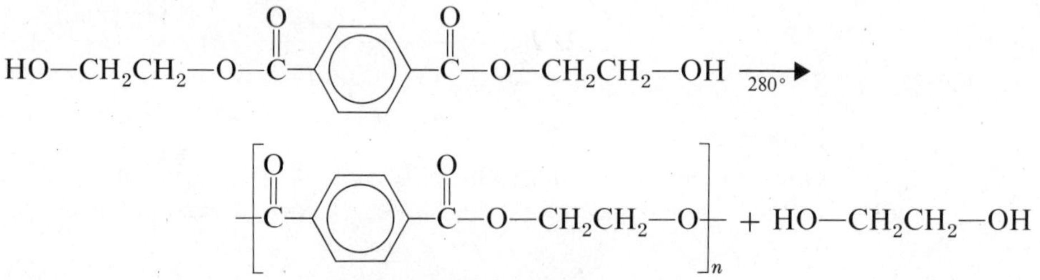

$$HO-CH_2CH_2-O-\overset{\overset{\displaystyle O}{\|}}{C}-\bigcirc-\overset{\overset{\displaystyle O}{\|}}{C}-O-CH_2CH_2-OH \xrightarrow{280°}$$

$$\left[\overset{\overset{\displaystyle O}{\|}}{C}-\bigcirc-\overset{\overset{\displaystyle O}{\|}}{C}-O-CH_2CH_2-O\right]_n + HO-CH_2CH_2-OH$$

Poly(ethylene terephthalate)

The poly(ethylene terephthalate) that results from the second transesterification melts at ∼270°. It can be melt spun into fibers to produce *Dacron* or *Terylene;* it can also be made into a film, in which form it is marketed as *Mylar.*

Problem 22.18

Transesterification reactions are catalyzed by either acids or bases. Using the transesterification reaction that takes place when dimethyl terephthalate is heated with ethylene glycol as an example, outline reasonable mechanisms for (a) the base-catalyzed reaction and (b) for the acid-catalyzed reaction.

Problem 22.19

Kodel (below) is another polyester that enjoys wide commercial use.

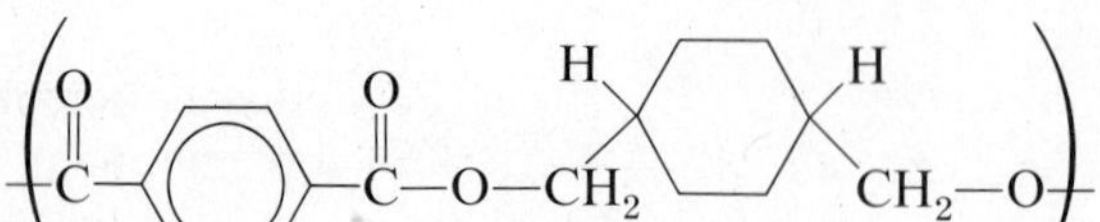

$$\left(\overset{\overset{\displaystyle O}{\|}}{C}-\bigcirc-\overset{\overset{\displaystyle O}{\|}}{C}-O-CH_2\overset{H}{\underset{}{\bigcirc}}\overset{H}{CH_2}-O\right)_n$$

Kodel

Kodel is also produced through a transesterification reaction. (a) What methyl ester and what alcohol are required for the synthesis of *Kodel?* (b) The alcohol can be prepared from dimethyl terephthalate. How might this be done?

Problem 22.20

Heating phthalic anhydride and glycerol together yields a polyester called *Glyptal.* Glyptal is especially rigid because the polymer chains are "cross linked." Write a portion of the structure of *Glyptal* and show how cross linking occurs.

Problem 22.21

Lexan, a high molecular weight "polycarbonate," is manufactured by mixing "bisphenol A" with phosgene in the presence of pyridine. Suggest a structure for Lexan.

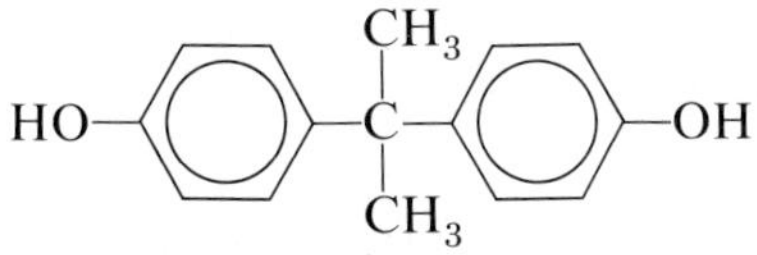

Bisphenol A 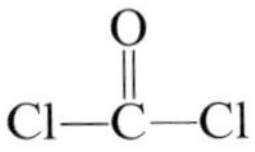   Phosgene

* Problem 22.22

The familiar "epoxy resins" or "epoxy glues" usually consist of two components that are sometimes labeled "resin" and "hardener." The resin is manufactured by allowing bisphenol A (problem 22.21) to react with an excess of epichlorohydrin, CH_2—$CHCH_2Cl$, in the presence of a base until a low molecular weight polymer is obtained. (a) What is a likely structure for this polymer and (b) what is the purpose of using an excess of epichlorohydrin? The hardener is usually an amine such as $H_2NCH_2CH_2NHCH_2CH_2NH_2$. (c) What reaction takes place when the resin and hardener are mixed?

23
SPECIAL TOPICS IV
LIPIDS

23.1 INTRODUCTION

When plant or animal tissues are extracted with a nonpolar solvent, (e.g., ether, chloroform, benzene, or an alkane) a portion of the material usually dissolves. The components of this soluble fraction are called *lipids*.*

Lipids include a wide variety of structural types including the following:
Carboxylic acids (or "fatty" acids)
Triacylglycerols (or neutral fats)
Phospholipids
Glycolipids
Waxes
Terpenes
Steroids

We discussed the chemistry of terpenes earlier as a special topic (Section 11.4). We can now examine the properties of other members of the lipid group.

23.2 FATTY ACIDS AND TRIACYLGLYCEROLS

Only a small portion of the total lipid fraction consists of free carboxylic acids. Most of the carboxylic acids in the lipid fraction are found as *esters of glycerol*, that is, as *triacylglycerols*.† The most common are *triacylglycerols of long-chain carboxylic acids* (Fig. 23.1).

Triacylglycerols are the oils and fats of plant or animal origin. They include such common substances as peanut oil, olive oil, soybean oil, corn oil, linseed oil,

*Lipids are defined in terms of the physical operation that we use to isolate them. In this respect, the definition of a lipid differs from that of proteins (Chapter 25) or carbohydrates (Chapter 24). The latter are defined on the basis of their structures.

†In older literature triacylglycerols were referred to as triglycerides or simply as glycerides.

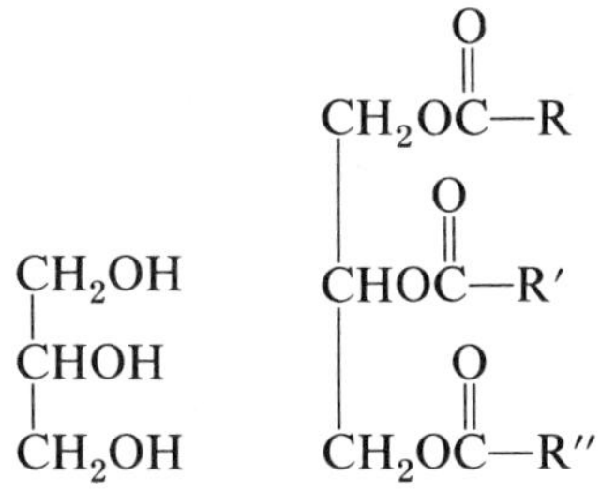

$$
\begin{array}{cc}
& \begin{array}{c} O \\ \parallel \\ CH_2OC-R \\ | \quad O \\ \quad \parallel \\ CHOC-R' \\ | \quad O \\ \quad \parallel \\ CH_2OC-R'' \end{array}
\end{array}
$$

$$
\begin{array}{c}
CH_2OH \\ | \\ CHOH \\ | \\ CH_2OH \\ (a)
\end{array}
$$

(b)

FIG. 23.1
(a) *Glycerol* (b) *A triacylglycerol. R, R', and R" are usually long-chain* ($\geq C_{14}$) *alkyl groups. R, R', and R" may also contain one or more carbon-carbon double bonds. In a typical triacylglycerol, R, R', and R" are all different.*

butter, lard, and tallow. Those triacylglycerols that are liquids at room temperature are generally known as *oils;* those that are solids are usually called *fats.*

The carboxylic acids that are obtained by hydrolysis of naturally occurring fats and oils usually have unbranched chains with an even number of carbon atoms. (We will see, later, that these facts give us important clues as to how they are synthesized in plants and animals.) The most common carboxylic acids obtained from fats and oils are the C_{14}-, C_{16}-, and C_{18}- acids shown in Table 23.1.

In addition to most of the common fatty acids given in Table 23.1, the hydrolysis of butter gives small amounts of saturated even-numbered carboxylic acids in the C_4-C_{12} range. These are butyric (butanoic), caproic (hexanoic), caprylic (octanoic), capric (decanoic), and lauric (dodecanoic) acids. The hydrolysis of coconut oil also gives short-chain carboxylic acids and a large amount of lauric acid.

Other less common fatty acids and their sources are given in Table 23.2. Table 23.3 gives the fatty acid composition of a number of common fats and oils.

TABLE 23.1 Common Fatty Acids

Saturated Carboxylic Acids	mp °C
$CH_3(CH_2)_{12}COOH$ Myristic acid (tetradecanoic acid)	54
$CH_3(CH_2)_{14}COOH$ Palmitic acid (hexadecanoic acid)	63
$CH_2(CH_2)_{16}COOH$ Stearic acid (octadecanoic acid)	70

Unsaturated Carboxylic Acids

$$CH_3(CH_2)_5 \overset{\displaystyle }{\underset{H}{C}} = \overset{\displaystyle (CH_2)_7COOH}{\underset{H}{C}}$$

Palmitoleic acid
(*cis*-9-hexadecenoic acid)   32

$$CH_3(CH_2)_7 \overset{\displaystyle }{\underset{H}{C}} = \overset{\displaystyle (CH_2)_7COOH}{\underset{H}{C}}$$

Oleic acid
(*cis*-9-octadecenoic acid)   4

$$CH_3(CH_2)_4 \overset{}{\underset{H}{C}} = \overset{CH_2}{\underset{H \; H}{C}} \; \overset{}{\underset{H}{C}} = \overset{(CH_2)_7COOH}{\underset{H}{C}}$$

Linoleic acid
(*cis, cis* 9,12-octadecadienoic acid)   −5

$$CH_3CH_2 \overset{}{\underset{H}{C}} = \overset{CH_2}{\underset{H \; H}{C}} \; \overset{}{\underset{H \; H}{C}} = \overset{CH_2}{\underset{}{C}} \; \overset{}{\underset{H \; H}{C}} = \overset{(CH_2)_7COOH}{\underset{H}{C}}$$

Linolenic acid
(*cis, cis, cis* 9,12,15-octadecatrienoic acid)   −11

TABLE 23.2 Some Less Common Fatty Acids

NUMBER OF CARBONS	NAME	STRUCTURE	SOURCE
18	Eleostearic acid	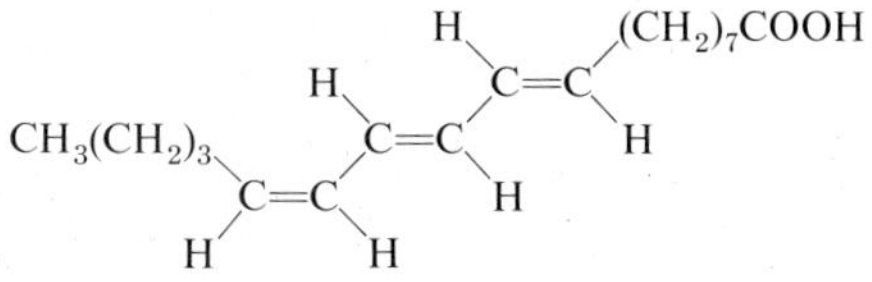	The main (80%) fatty acid obtained by hydrolysis of tung oil
18	Ricinoleic acid	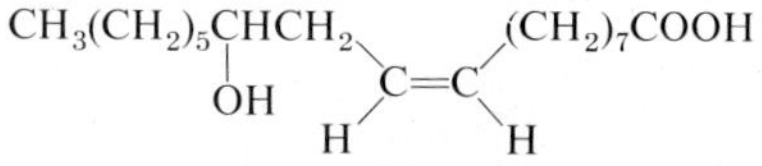 $CH_3(CH_2)_5CHCH_2$ with OH group, $C=C$, $(CH_2)_7COOH$, H, H	The main (80%) fatty acid obtained by hydrolysis of castor oil
18	Chaulmoogric acid	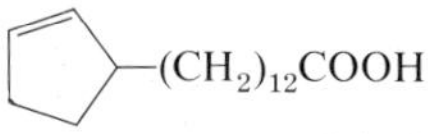 (cyclopentene)—$(CH_2)_{12}COOH$	From oil of *Hydnocarpus kuazii* (used in the treatment of leprosy)
19	Sterculic acid	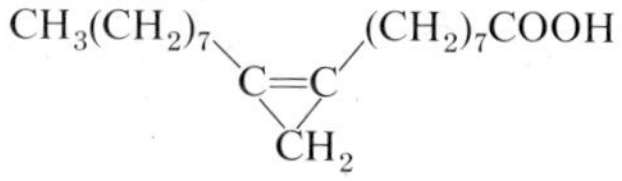 $CH_3(CH_2)_7$ $C=C$ $(CH_2)_7COOH$ with CH_2	From kernel oil of *Sterculia foetida*
19	Lactobacillic acid	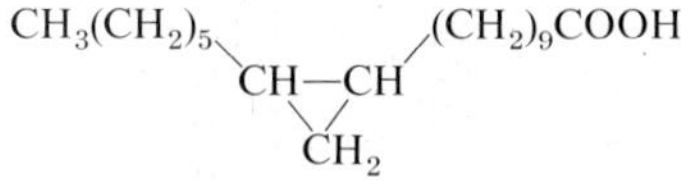 $CH_3(CH_2)_5$ $CH—CH$ $(CH_2)_9COOH$ with CH_2	From phospholipids of bacteria
19	Tuberculostearic acid	$CH_3(CH_2)_7CH(CH_2)_8COOH$ with CH_3	From tubercule bacilli
20	Arachidonic acid	$CH_3(CH_2)_4(CH=CHCH_2)_4(CH_2)_2COOH$	From human fat (0.3–1.0%) liver, lecithins
22	Cetoleic acid	$CH_3(CH_2)_9CH=CH(CH_2)_9COOH$	From fish oils
22	Erucic acid	$CH_3(CH_2)_7CH=CH(CH_2)_{11}COOH$ (*cis*)	From seed oils of rape, wallflower nasturtium, and mustard
24	Nervonic acid	$CH_3(CH_2)_7CH=CH(CH_2)_{13}COOH$ (*cis*)	From fish oils and brain tissues
27	Mycolipenic acid	$CH_3(CH_2)_{17}CHCH_2CHCH=C—COOH$ with CH_3, CH_3, CH_3	From tubercule bacilli

Hydrogenation of Triacylglycerols

We see from the data given in Table 23.3 that most oils are made up of high percentages of unsaturated fatty acids. The fact that oils have lower melting points than fats is apparently related to this factor because hydrogenation of an oil usually produces a solid fat.

TABLE 23.3 Fatty Acid Composition Obtained by Hydrolysis of Common Fats and Oils*

AVERAGE COMPOSITION OF FATTY ACIDS (mole %)

FAT OR OIL	SATURATED								UNSATURATED			
	C_4 BUTYRIC ACID	C_6 CAPROIC ACID	C_8 CAPRYLIC ACID	C_{10} CAPRIC ACID	C_{12} LAURIC ACID	C_{14} MYRISTIC ACID	C_{16} PALMITIC ACID	C_{18} STEARIC ACID	C_{16} PALMIT-OLEIC ACID	C_{18} OLEIC ACID	C_{18} LINOLEIC ACID	C_{18} LINOLENIC ACID
Animal Fats												
Butter	3–4	1–2	0–1	2–3	2–5	8–15	25–29	9–12	4–6	18–33	2–4	
Lard						1–2	25–30	12–18	4–6	48–60	6–12	0–1
Beef tallow						2–5	24–34	15–30		35–45	1–3	0–1
Vegetable Oils												
Olive						0–1	5–15	1–4		67–84	8–12	
Peanut							7–12	2–6		30–60	20–38	
Corn						1–2	7–11	3–4	1–2	25–35	50–60	
Cottonseed						1–2	18–25	1–2	1–3	17–38	45–55	
Soybean						1–2	6–10	2–4		20–30	50–58	5–10
Linseed							4–7	2–4		14–30	14–25	45–60
Coconut		0–1	5–7	7–9	40–50	15–20	9–12	2–4	0–1	6–9	0–1	
Marine Oils												
Cod liver						5–7	8–10	0–1	18–22	27–33	27–32	

*Data adapted from John R. Holum, *Introduction to Organic and Biological Chemistry*, Wiley, New York, (1969), p. 300, and from *Biology Data Book*, Philip L. Altman and Dorothy S. Ditmer, eds., Federation of American Societies for Experimental Biology, Washington, D.C., (1964).

$$\begin{array}{l}
\overset{\displaystyle O}{\underset{\displaystyle \|}{}}\\
CH_2OC(CH_2)_7CH{=}CH(CH_2)_7CH_3 \qquad \text{(from oleic acid)}\\[4pt]
\overset{\displaystyle O}{\underset{\displaystyle \|}{}}\\
CHOC(CH_2)_7CH{=}CHCH_2CH{=}CH(CH_2)_4CH_3\\[4pt]
\overset{\displaystyle O}{\underset{\displaystyle \|}{}}\\
CH_2OC(CH_2)_7CH{=}CHCH_2CH{=}CH(CH_2)_4CH_3
\end{array}$$

(from linoleic acid)

A triacylglycerol of a typical vegetable oil
(a liquid)

$$\xrightarrow[\text{Ni}]{\text{H}_2 \text{ (high pressure)}}$$

$$\begin{array}{l}
\overset{\displaystyle O}{\underset{\displaystyle \|}{}}\\
CH_2OC(CH_2)_{16}CH_3\\[4pt]
\overset{\displaystyle O}{\underset{\displaystyle \|}{}}\\
CHOC(CH_2)_{16}CH_3\\[4pt]
\overset{\displaystyle O}{\underset{\displaystyle \|}{}}\\
CH_2OC(CH_2)_{16}CH_3
\end{array}$$

Glyceryl tristearate
(a solid fat)

Commercial cooking fats such as Crisco and Spry, for example, are manufactured in just this way. Vegetable oils are hydrogenated until a semisolid of an appealing consistency is obtained. Complete hydrogenation is avoided because a completely saturated triacylglycerol is very hard and brittle.

Saponification of Triacylglycerols: Soaps

Alkaline hydrolysis (i.e., "saponification") of triacylglycerols produces glycerol and a mixture of salts of long-chain carboxylic acids:

$$\begin{array}{l}
\overset{\displaystyle O}{\underset{\displaystyle \|}{}}\\
CH_2OCR\\[4pt]
\overset{\displaystyle O}{\underset{\displaystyle \|}{}}\\
CHOCR' \quad + 3NaOH\\[4pt]
\overset{\displaystyle O}{\underset{\displaystyle \|}{}}\\
CH_2OCR''
\end{array}
\longrightarrow
\begin{array}{l}
CH_2OH \; +\\[4pt]
\\
CHOH\\[4pt]
\\
CH_2OH
\end{array}
\quad
\begin{array}{l}
\overset{\displaystyle O}{\underset{\displaystyle \|}{}}\\
RCO^-\;\;Na^+\\[4pt]
\overset{\displaystyle O}{\underset{\displaystyle \|}{}}\\
R'CO^-\;\;Na^+\\[4pt]
\overset{\displaystyle O}{\underset{\displaystyle \|}{}}\\
R''CO^-\;\;Na^+
\end{array}$$

Glycerol   Sodium carboxylates
"soap"

These salts of long-chain carboxylic acids are *soaps* and this is exactly the way in which most soaps are manufactured. Fats and oils are boiled in aqueous sodium hydroxide until hydrolysis is complete. Adding sodium chloride to the mixture then causes the soap to precipitate. (After the soap has been separated, glycerol can be isolated from the aqueous phase by distillation.) Crude soaps are

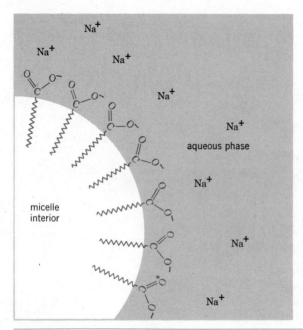

FIG. 23.2
A portion of a soap micelle.

usually purified by several reprecipitations. Perfumes can be added if a toilet soap is the desired product; sand, sodium carbonate, and other fillers can be added to make a scouring soap; and air can be blown into the molten soap if the manufacturer wants to market a soap that floats.

The sodium salts of long-chain carboxylic acids (soaps) are almost completely miscible with water. However, they do not dissolve as we might expect, that is, as individual ions. Except in very dilute solutions, soaps exist as *micelles* (Fig. 23.2). Soap micelles are usually spherical clusters of carboxylate ions that are dispersed throughout the aqueous phase. The carboxylate ions are packed together with their negatively charged (and thus, *polar*) carboxylate groups at the surface and with their nonpolar hydrocarbon chains on the interior. The sodium ions are scattered throughout the aqueous phase as individual solvated ions.

Micelle formation accounts for the fact that soaps dissolve in water. The nonpolar (and thus, *hydrophobic*) alkyl chains of the soap remain in a nonpolar environment—in the interior of the micelle. The polar (and therefore, *hydrophilic*) carboxylate groups are exposed to a polar environment—that of the aqueous phase. Because the surfaces of the micelles are negatively charged, individual micelles repel each other and remain dispersed throughout the aqueous phase.

Soaps serve their function as "dirt removers" in a similar way. Most dirt particles (on the skin, for example) become surrounded by a layer of an oil or fat. Water molecules alone are unable to disperse these greasy globules because they are unable to penetrate the oily layer and separate the individual particles from each other or from the surface to which they are stuck. Soap solutions, however, *are* able to separate the individual particles because their hydrocarbon chains can "dissolve" in the oily layer (Fig. 23.3). As this happens each individual particle

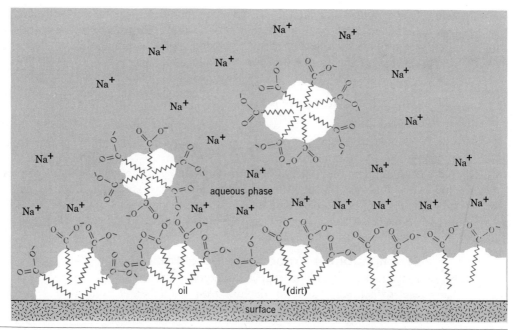

FIG. 23.3
Dispersal of oil-coated dirt particles by a soap.

develops an outer layer of carboxylate ions and presents the aqueous phase with a much more compatible exterior—a polar surface. The individual globules now repel each other and thus, become dispersed throughout the aqueous phase. Shortly thereafter they make their way down the drain.

Synthetic detergents (Fig. 23.4) function in the same way as soaps; they have long nonpolar alkane chains with polar groups at the end. The polar groups of most synthetic detergents are sodium sulfonates or sodium sulfates.

Synthetic detergents offer an advantage over soaps; they function well in "hard" water, that is, water containing Ca^{++}, Fe^{++}, Fe^{+++}, and Mg^{++} ions. Calcium, iron, and magnesium salts of alkane sulfonic acids and alkyl hydrogen sulfates are largely water soluble and, thus, synthetic detergents remain in solution. Soaps, by contrast, form precipitates—the ring around the bathtub—when they are used in hard water.

$$CH_3(CH_2)_nCH_2OSO_2O^- \quad Na^+$$
Sodium alkyl sulfate

$$CH_3(CH_2)_nCH_2SO_2O^- \quad Na^+$$
Sodium alkanesulfonate

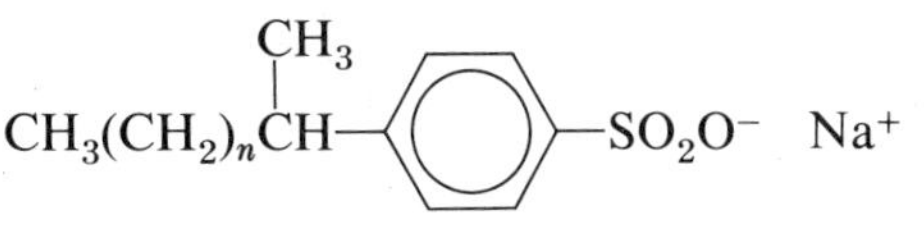

$$CH_3(CH_2)_n\overset{\displaystyle CH_3}{\underset{\displaystyle |}{CH}}\!\!-\!\!\bigcirc\!\!-\!\!SO_2O^- \quad Na^+$$
Sodium alkylbenzenesulfonate

FIG. 23.4
Typical synthetic detergents.

One serious disadvantage of some synthetic detergents is that they are not "biodegradable," that is, they are not broken down into harmless chemicals by microorganisms in the soil. For many years detergents were manufactured with the general structures shown below.

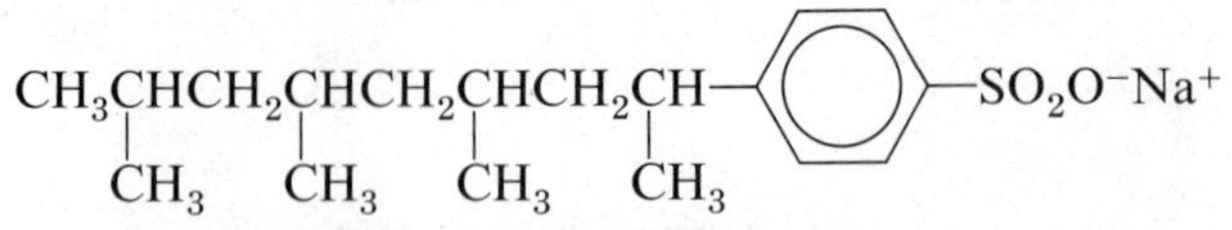

$$CH_3CHCH_2CHCH_2CHCH_2CH—⟨◯⟩—SO_2O^-Na^+$$
$$\qquad\;|\qquad\;|\qquad\;|\qquad\;|$$
$$\qquad CH_3\quad CH_3\quad CH_3\quad CH_3$$

Tetrapropylene-based sodium alkylbenzene sulfonate

Soil bacteria are unable to degrade tetrapropylene-based detergents to any appreciable extent and, as a result, they accumulated in the ground and surface water supplies. The use of these detergents has now been banned. The detergents shown in Fig. 23.4 are all completely biodegradable or largely so.

Reactions of the Carboxyl Group of Fatty Acids

Fatty acids, as we might expect, undergo typical reactions of carboxylic acids. They react with LiAlH$_4$ to form alcohols with alcohols and mineral acid to form esters, and with thionyl chloride to form acyl chlorides:

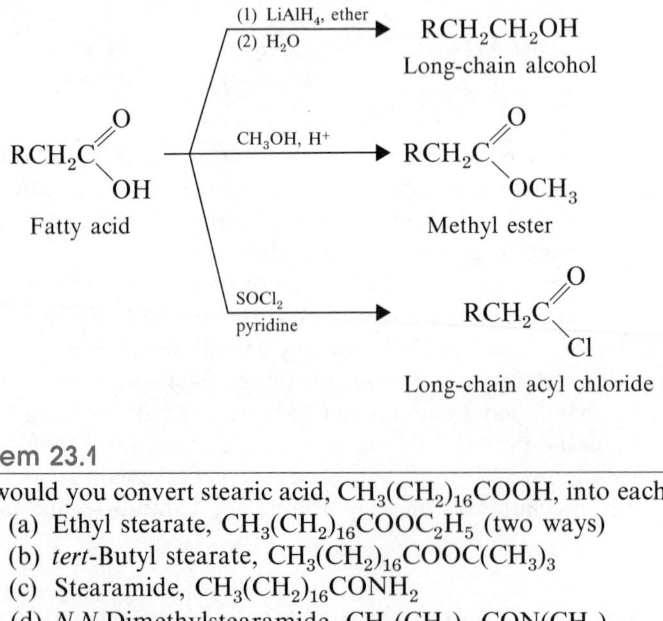

$$\xrightarrow[\text{(2) H}_2\text{O}]{\text{(1) LiAlH}_4,\ \text{ether}} RCH_2CH_2OH$$

Long-chain alcohol

$$RCH_2C\overset{O}{\underset{OH}{\big\diagup}}$$

Fatty acid

$$\xrightarrow{\text{CH}_3\text{OH, H}^+} RCH_2C\overset{O}{\underset{OCH_3}{\big\diagup}}$$

Methyl ester

$$\xrightarrow[\text{pyridine}]{\text{SOCl}_2} RCH_2C\overset{O}{\underset{Cl}{\big\diagup}}$$

Long-chain acyl chloride

Problem 23.1

How would you convert stearic acid, $CH_3(CH_2)_{16}COOH$, into each of the following?
- (a) Ethyl stearate, $CH_3(CH_2)_{16}COOC_2H_5$ (two ways)
- (b) *tert*-Butyl stearate, $CH_3(CH_2)_{16}COOC(CH_3)_3$
- (c) Stearamide, $CH_3(CH_2)_{16}CONH_2$
- (d) *N,N*-Dimethylstearamide, $CH_3(CH_2)_{16}CON(CH_3)_2$
- (e) *n*-Octadecylamine, $CH_3(CH_2)_{16}CH_2NH_2$
- (f) *n*-Heptadecylamine, $CH_3(CH_2)_{15}CH_2NH_2$
- (g) Octadecanal, $CH_3(CH_2)_{16}CHO$

- (h) 1-Octadecyl stearate, $CH_3(CH_2)_{16}\overset{O}{\overset{\|}{C}}OCH_2(CH_2)_{16}CH_3$
- (i) 1-Octadecanol, $CH_3(CH_2)_{16}CH_2OH$ (two ways)

- (j) 2-Nonadecanone, $CH_3(CH_2)_{16}\overset{O}{\overset{\|}{C}}CH_3$

(k) 1-Bromooctadecane, $CH_3(CH_2)_{16}CH_2Br$

(l) Nonadecanoic acid, $CH_3(CH_2)_{16}CH_2COOH$

Reactions of the Alkyl Chain of Saturated Fatty Acids

Fatty acids are like other carboxylic acids in that they undergo specific α-halogenation when they are treated with bromine or chlorine in the presence of phosphorus. This is the familiar Hell-Volhard-Zelinski reaction.

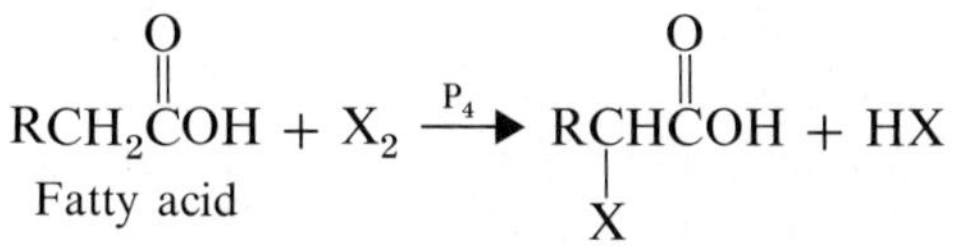

$$\underset{\text{Fatty acid}}{RCH_2\overset{\overset{\displaystyle O}{\|}}{C}OH} + X_2 \xrightarrow{P_4} R\underset{\underset{\displaystyle X}{|}}{CH}\overset{\overset{\displaystyle O}{\|}}{C}OH + HX$$

Problem 23.2

How would you transform myristic acid into each of the following?

(a) $CH_3(CH_2)_{11}\underset{\underset{\displaystyle Br}{|}}{CH}COOH$

(c) $CH_3(CH_2)_{11}\underset{\underset{\displaystyle CN}{|}}{CH}COOH$

(b) $CH_3(CH_2)_{11}\underset{\underset{\displaystyle OH}{|}}{CH}COOH$

(d) $CH_3(CH_2)_{11}\underset{\underset{\displaystyle NH_2}{|}}{CH}COOH$

Reactions of the Alkenyl Chain of Unsaturated Fatty Acids

The double bonds of the carbon chains of fatty acids undergo characteristic alkene addition reactions:

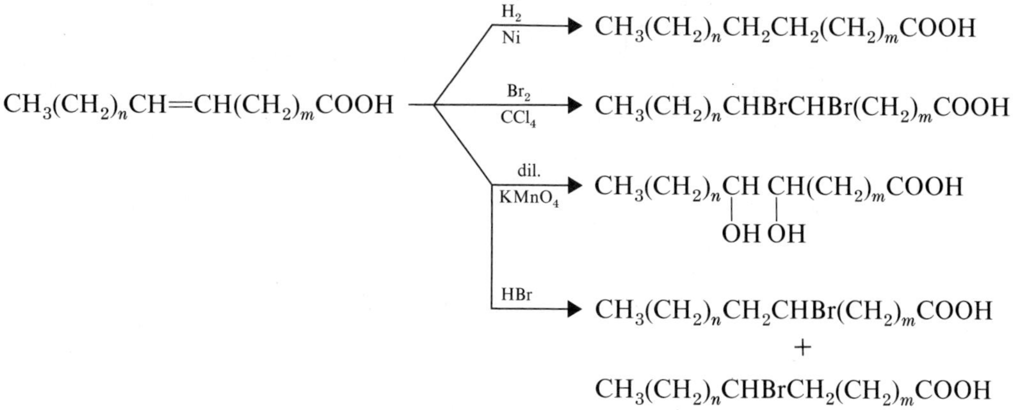

$$CH_3(CH_2)_nCH{=}CH(CH_2)_mCOOH$$

$\xrightarrow[Ni]{H_2}$ $CH_3(CH_2)_nCH_2CH_2(CH_2)_mCOOH$

$\xrightarrow[CCl_4]{Br_2}$ $CH_3(CH_2)_nCHBrCHBr(CH_2)_mCOOH$

$\xrightarrow[KMnO_4]{\text{dil.}}$ $CH_3(CH_2)_n\underset{\underset{\displaystyle OH}{|}}{CH}\underset{\underset{\displaystyle OH}{|}}{CH}(CH_2)_mCOOH$

$\xrightarrow{HBr}$ $CH_3(CH_2)_nCH_2CHBr(CH_2)_mCOOH$

$+$

$CH_3(CH_2)_nCHBrCH_2(CH_2)_mCOOH$

Problem 23.3

Using palmitoleic acid as an example and neglecting stereochemistry, illustrate each of the following reactions of the double bond.

(a) Addition of iodine   (c) Hydroxylation

(b) Addition of hydrogen   (d) Addition of HCl

Problem 23.4

(a) How many stereoisomers are theoretically possible when bromine adds to the double bond of palmitoleic acid? (b) In actuality the addition of bromine yields primarily one set of enantiomers, $(\pm)$-*threo*-9,10-dibromohexadecanoic acid. The addition of bromine is an anti addition to the double bond (i.e., it apparently takes place through a bromonium ion intermediate). Taking into account the *cis* stereochemistry of the double bond of palmitoleic acid and the stereochemistry of the bromine addition, write three-dimensional structures for the $(\pm)$-*threo*-9,10-dibromohexadecanoic acids.

Oxidations of unsaturated fatty acids are often used to locate double bonds. Direct oxidative cleavage of a double bond with $KMnO_4$ in acetone sometimes leads to overoxidation; as a result, some of the carboxylic acids that one obtains from the reaction mixture may have had their chains shortened. A much better procedure, one that gives unambiguous results because overoxidation is avoided, involves (1) initial hydroxylation of the double bond using osmium tetroxide, dilute aqueous potassium permanganate, or a peracid; (2) subsequent cleavage of the glycol (p. 927) using periodic acid or lead tetraacetate. When this procedure is employed, the two carbons of the double bond are oxidized to the aldehyde stage. The following example is an illustration.

$$CH_3(CH_2)_7 \quad (CH_2)_7COOH$$
$$\underset{H}{\overset{}{C}} = \underset{H}{\overset{}{C}} \quad \xrightarrow[\text{(2) } H_3O^+]{\overset{O}{\overset{\parallel}{HCOOH}}} \quad CH_3(CH_2)_7 - \underset{\underset{OH}{|}}{CH} - \underset{\underset{OH}{|}}{CH} - (CH_2)_7COOH$$

Oleic acid

$(\pm)$-*threo*-9,10-Dihydroxyocta-
decanoic acids

$\downarrow HIO_4$ (c.f. Section 24.4.)

$$CH_3(CH_2)_7\overset{O}{\overset{\parallel}{C}}H \quad + \quad H\overset{O}{\overset{\parallel}{C}}(CH_2)_7COOH$$

Nonanal
(Pelargonic
aldehyde)
(89%)

Nonanaloic acid
(azelaic half aldehyde)
(76%)

Problem 23.5

When oleic acid is hydroxylated using dilute aqueous permanganate or osmium tetraoxide, the products obtained are $(\pm)$-*erythro*-9,10-dihydroxyoctadecanoic acids, diastereomers of the *threo*-9,10-dihydroxyoctadecanoic acids above. Recall the stereochemistry of hydroxylation reactions (pp. 215, 264) and give stereochemical formulas for *threo*- and *erythro*-9,10-dihydroxyoctadecanoic acids.

Problem 23.6

When oleic acid is heated to 180–200° (in the presence of a small amount of selenium) an equilibrium is established between oleic acid (33%) and an isomeric compound called elaidic acid (67%). Suggest a possible structure for elaidic acid.

Problem 23.7

Gadoleic acid, $C_{20}H_{38}O_2$, a fatty acid that can be isolated from cod-liver oil, can be cleaved by hydroxylation and subsequent treatment with periodic acid to $CH_3(CH_2)_9CHO$ and $OHC(CH_2)_7COOH$. (a) What two stereoisomeric structures are possible for gadoleic acid? (b) What spectroscopic technique would make possible a decision as to the actual structure of gadoleic acid? (c) What peaks would you look for?

Problem 23.8

Vaccenic acid, a structural isomer of oleic acid, has been synthesized through the following reaction sequence:

$$\text{1-Octyne} + NaNH_2 \xrightarrow[NH_3]{} \textbf{A} \ (C_8H_{13}Na) \xrightarrow{ICH_2(CH_2)_7CH_2Cl} \textbf{B} \ (C_{17}H_{31}Cl) \xrightarrow{NaCN}$$

$$\textbf{C} \ (C_{18}H_{31}N) \xrightarrow{KOH, H_2O} \textbf{D} \ (C_{18}H_{31}O_2K) \xrightarrow{H_3O^+} \textbf{E} \ (C_{18}H_{32}O_2) \xrightarrow{H_2, Pd}$$

$$\text{vaccenic acid} \ (C_{18}H_{34}O_2).$$

Propose a structure for vaccenic acid and for the intermediates **A–E**.

Problem 23.9

ω-Fluorooleic acid can be isolated from a shrub, *Dechapetalum toxicarium,* that grows in Sierra Leone. The compound is highly toxic to warm-blooded animals; it has found use as an arrow poison in tribal warfare, in poisoning enemy water supplies, and by witch doctors "for terrorizing the native population." Powdered fruit of the plant has been used as a rat poison, hence ω-fluorooleic acid has the common name "ratsbane." A synthesis of ω-fluorooleic acid is outlined below. Give structures for compounds **F** through **I**.

$$\text{1-Bromo-8-fluorooctane} + \text{sodium acetylide} \longrightarrow \textbf{F} \ (C_{10}H_{17}F) \xrightarrow[\text{(2) I(CH}_2)_7\text{Cl}]{\text{(1) NaNH}_2}$$

$$\textbf{G} \ (C_{17}H_{30}FCl) \xrightarrow[DMSO]{NaCN} \textbf{H} \ (C_{18}H_{30}NF) \xrightarrow[\text{(2) H}^+]{\text{(1) KOH}} \textbf{I} \ (C_{18}H_{31}O_2F) \xrightarrow[Pd—BaSO_4]{H_2}$$

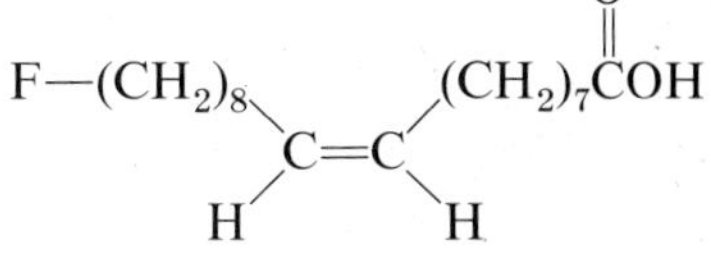

$$F-(CH_2)_8 \underset{H}{\overset{}{}} C=C \underset{H}{\overset{}{}} (CH_2)_7\overset{\overset{\displaystyle O}{\|}}{C}OH$$

ω-Fluorooleic acid
(46% yield, overall)

Biological Function of Triacylglycerols

The primary function of the triacyglycerols of mammals is as a source of chemical energy. When triacylglycerols are converted to carbon dioxide and water by biochemical reactions (i.e., when triacylglycerols are "metabolized") they yield more than twice as many kilocalories per gram as do carbohydrates or proteins. This is largely because of the high proportion of carbon-hydrogen bonds per molecule.

Triacylglycerols are distributed throughout nearly all types of body cells but they are stored primarily as "body fat" in certain depots of specialized connective tissue known as *adipose tissue.*

The saturated triacylglycerols of the body can be synthesized from all three major foodstuffs: proteins, carbohydrates, and fats or oils. Certain polyunsaturated fatty acids, however, are essential in the diets of higher animals.

23.3 PHOSPHOLIPIDS

Another large class of lipids are those called *phospholipids.* Most phospholipids are structurally derived from a glycerol derivative known as a *phosphatidic acid.* In a phosphatidic acid, two hydroxyl groups of glycerol are ester linked to fatty acids and one terminal hydroxyl group is ester linked to *phosphoric acid.*

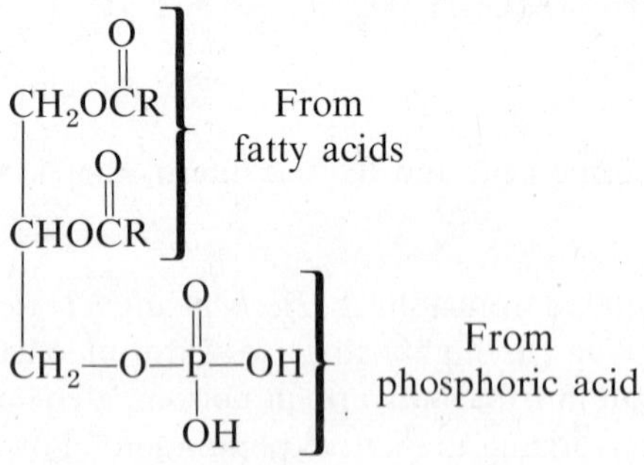

$$
\begin{array}{l}
\left. \begin{array}{l}
\text{CH}_2\text{OCR} \\[4pt]
\quad\quad\ \ \overset{\text{O}}{\overset{\|}{} } \\[-6pt]
\text{CHOCR}
\end{array} \right\} \quad \begin{array}{l}\text{From}\\ \text{fatty acids}\end{array} \\[18pt]
\left. \text{CH}_2\!-\!\text{O}\!-\!\overset{\overset{\text{O}}{\|}}{\underset{\text{OH}}{\text{P}}}\!-\!\text{OH} \right\} \quad \begin{array}{l}\text{From}\\ \text{phosphoric acid}\end{array}
\end{array}
$$

A phosphatidic acid
(a diacylglycerophosphate)

Phosphatides

In *phosphatides* the phosphate group of a phosphatidic acid is bound through another ester linkage to one of the nitrogen-containing compounds shown below.

$$\text{HOCH}_2\text{CH}_2\overset{+}{\text{N}}(\text{CH}_3)_3 \ \ \text{OH}^- \qquad \text{HOCH}_2\text{CH}_2\text{NH}_2$$

Choline   Ethanolamine

$$\text{HOCH}_2\underset{\overset{|}{\text{COO}^-}}{\text{CHNH}_3{}^+}$$

L-Serine

The most important phosphatides are the *lecithins,* the *cephalins, phosphatidyl serines,* and the *plasmalogens* (a phosphatidyl derivative). Their general structures are shown in Table 23.4.

Phosphatides resemble soaps and detergents in that they are molecules having both polar and nonpolar groups (Fig. 23.5a). Like soaps and detergents, too, phosphatides "dissolve" in aqueous media by forming micelles. There is evidence that in biological systems the preferred micelles consist of three-dimensional arrays of "stacked" bimolecular micelles (Fig. 23.5b).

The hydrophilic and hydrophobic portions of phophatides make them per-

TABLE 23.4 Phosphatides

Lecithins

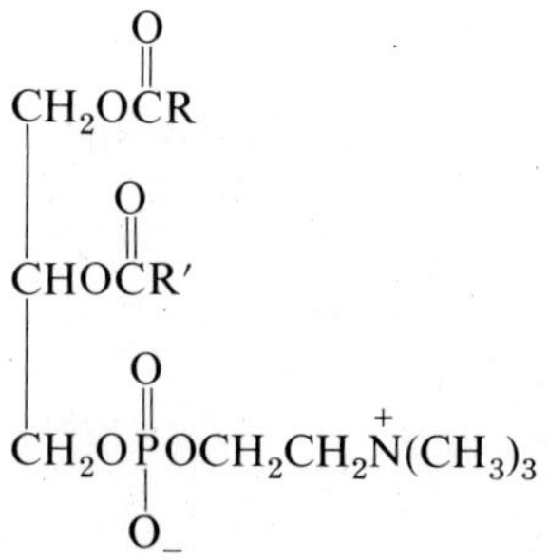

$$CH_2OCR \quad (\overset{O}{\overset{\|}{})}$$

$$|$$

$$CHOCR' \quad (\overset{O}{\overset{\|}{})}$$

$$|$$

$$CH_2O\overset{O}{\overset{\|}{P}}OCH_2CH_2\overset{+}{N}(CH_3)_3 \quad \text{(from choline)}$$

$$\underset{O_-}{|}$$

R is saturated and R' is unsaturated

Cephalins

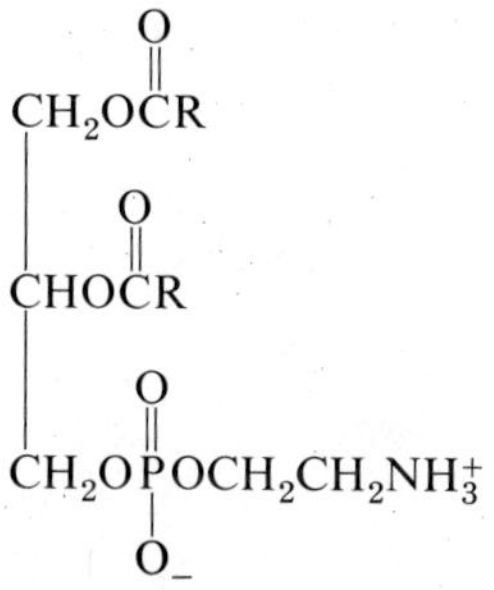

$$CH_2OCR \quad (\overset{O}{\overset{\|}{})}$$

$$|$$

$$CHOCR \quad (\overset{O}{\overset{\|}{})}$$

$$|$$

$$CH_2O\overset{O}{\overset{\|}{P}}OCH_2CH_2NH_3^+ \quad \text{(from ethanolamine)}$$

$$\underset{O_-}{|}$$

Phosphatidyl Serines

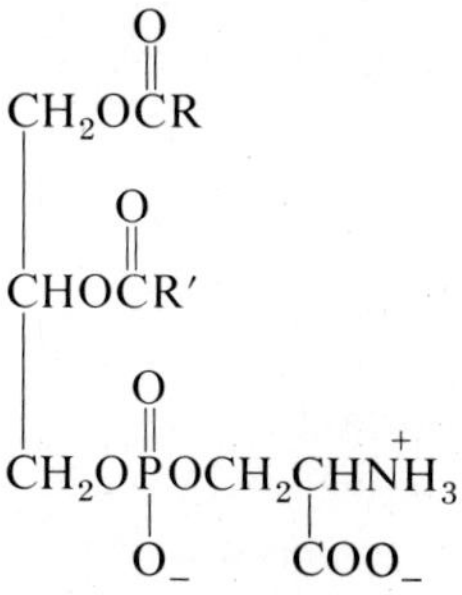

$$CH_2OCR \quad (\overset{O}{\overset{\|}{})}$$

$$|$$

$$CHOCR' \quad (\overset{O}{\overset{\|}{})}$$

$$|$$

$$CH_2O\overset{O}{\overset{\|}{P}}OCH_2\overset{+}{C}HNH_3 \quad \text{(from L-serine)}$$

$$\underset{O_-}{|} \quad \underset{COO_-}{|}$$

R and R' are like those of lecithins

Plasmalogens

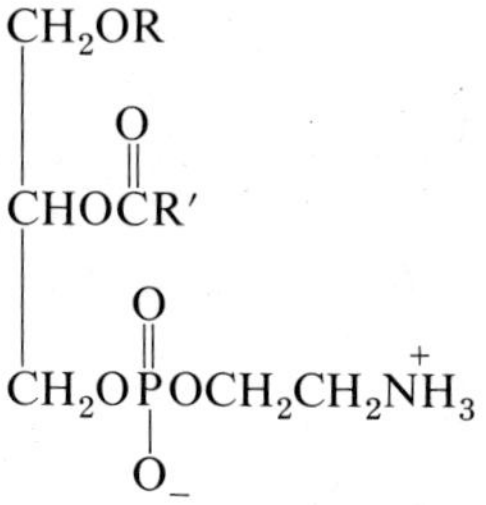

$$CH_2OR$$

$$|$$

$$CHOCR' \quad (\overset{O}{\overset{\|}{})}$$

$$|$$

$$CH_2O\overset{O}{\overset{\|}{P}}OCH_2CH_2\overset{+}{N}H_3 \quad \text{(from ethanolamine)}$$

$$\underset{O_-}{|}$$

$$\text{or } \overset{+}{N}(CH_3)_3 \quad \text{(from choline)}$$

R is $-CH=CH(CH_2)_nCH_3$ (from an α,β-unsaturated ether)

R' is that of an unsaturated fatty acid

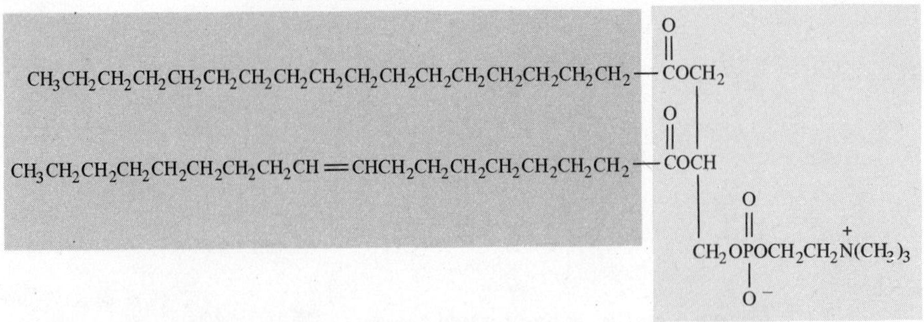

non–polar group   polar group

$CH_3CH_2CH_2CH_2CH_2CH_2CH_2CH_2CH_2CH_2CH_2CH_2CH_2CH_2CH_2CH_2CH_2$—$\overset{\overset{O}{\|}}{C}OCH_2$

$CH_3CH_2CH_2CH_2CH_2CH_2CH_2CH_2CH$=$CHCH_2CH_2CH_2CH_2CH_2CH_2CH_2$—$\overset{\overset{O}{\|}}{C}OCH$

$CH_2O\overset{\overset{O}{\|}}{P}OCH_2CH_2\overset{+}{N}(CH_2)_3$
$\underset{O^-}{|}$

(a)

$(CH_3)_3\overset{+}{N}CH_2CH_2O\overset{\overset{O}{\|}}{P}OCH_2$
$\underset{O^-}{|}$

$CHO\overset{\overset{O}{\|}}{C}$ /\/\/\/\/\/\/\/\/\ $\overset{\overset{O}{\|}}{C}OCH_2$

/\/\/\/\/\/\/\/\

$CH_2O\overset{\overset{O}{\|}}{C}$ /\/\/\/\/\/\/\/\ $\overset{\overset{O}{\|}}{C}OCH$

/\/\/\/\/\/\/\

$CH_2O\overset{\overset{O}{\|}}{P}OCH_2CH_2\overset{+}{N}(CH_3)_3$
$\underset{O^-}{|}$

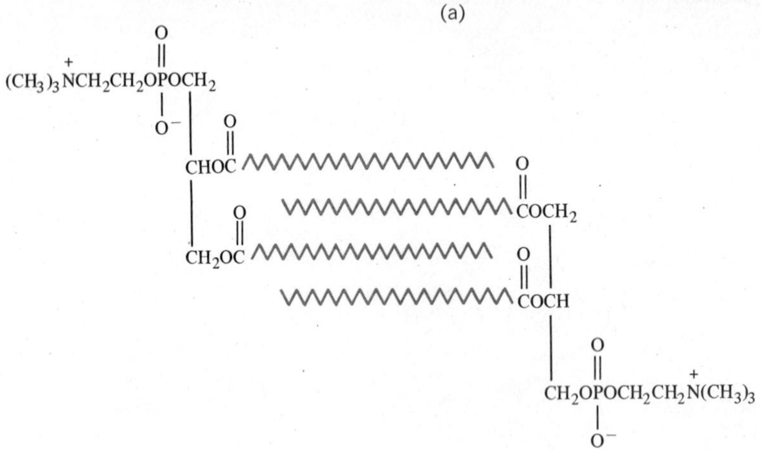

(b)

FIG. 23.5
(a) *Polar and nonpolar sections of a phosphatide.* (b) *A bimolecular phosphatide micelle.*

fectly suited for one of their most important biological functions: they form a portion of a structural unit that creates an interface between an organic and an aqueous environment. This is particularly true in cell walls and membranes where phosphatides are often found associated with proteins. Phosphatides also appear to be an essential factor in the formation of blood clots.

Problem 23.10

Under suitable conditions all of the ester (and ether) linkages of a phosphatide can be hydrolyzed. What organic compounds would you expect to obtain from the complete hydrolysis of (a) a lecithin, (b) an ethanolamine-based cephalin, (c) a choline-based plasmalogen. [Note: pay particular attention to the fate of the α,β-unsaturated ether in part (c)]

Derivatives of Sphingosine

Another important group of lipids are those that are derived from *sphingosine*, called *sphingolipids*. Two sphingolipids, a typical *sphingomyelin* and a typical *cerebroside*, are shown on the next page.

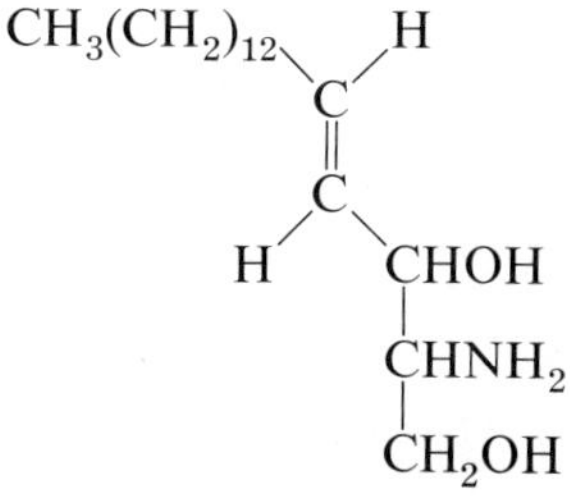

A sphingosine

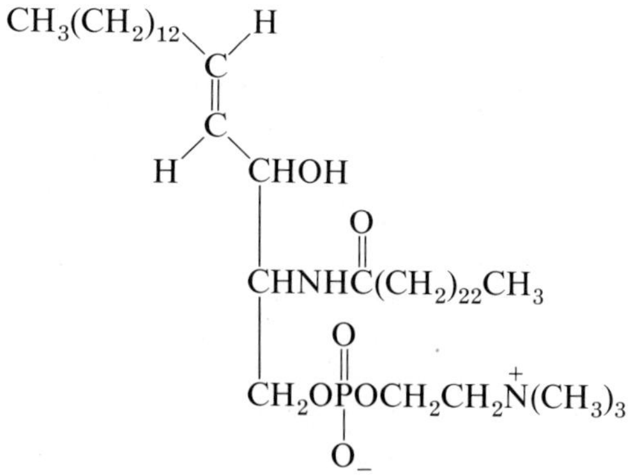

A sphingomyelin
(a sphingolipid)

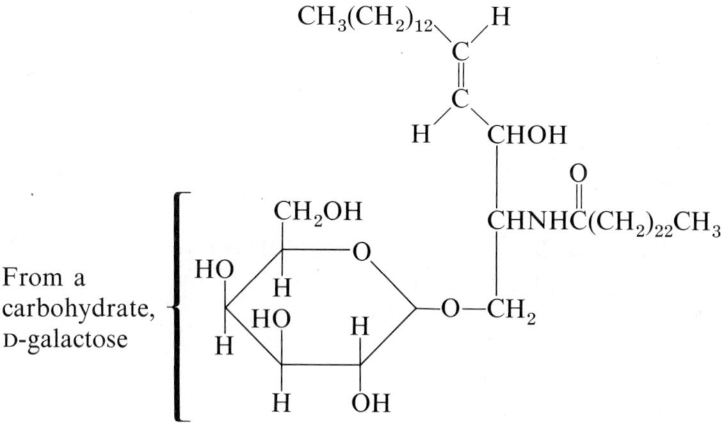

From a carbohydrate, D-galactose

A cerebroside
(also a sphingolipid)

Sphingomyelins are phosphatides. On hydrolysis they yield a sphingosine, choline, phosphoric acid, and a 24-carbon fatty acid called lignoceric acid. In a sphingomyelin this last component is bound in an *amide* linkage to the —NH$_2$ group of a sphingosine. The sphingolipids do not yield glycerol when they are hydrolyzed.

The cerebroside that we have shown above is an example of a *glycolipid*. Glycolipids have a polar group that is based on the presence of a *carbohydrate*. They do not yield phosphoric acid or choline when they are hydrolyzed.

The sphingolipids, together with proteins and polysaccharides, make up *myelin,* the protective coating that encloses nerve fibers or *axons*. The axons of nerve cells carry electrical nerve impulses; myelin.has been described as having a function relative to the axon similar to that of the insulation on an ordinary electric wire.

23.4 WAXES

Most waxes are esters of long-chain fatty acids and long-chain alcohols. Waxes are found as protective coatings on the skin, fur, or feathers of animals and on the leaves and fruits of plants. Several esters isolated from waxes are the following.

$$\text{CH}_3(\text{CH}_2)_{14}\overset{\overset{\text{O}}{\|}}{\text{C}}\text{OCH}_2(\text{CH}_2)_{14}\text{CH}_3$$

Cetyl palmitate
(from spermaceti)

$$\text{CH}_3(\text{CH}_2)_n\overset{\overset{\text{O}}{\|}}{\text{C}}\text{OCH}_2(\text{CH}_2)_m\text{CH}_3$$

$n = 24$ and 26; $m = 28$ and 30
(from beeswax)

$$\text{HOCH}_2(\text{CH}_2)_n\overset{\overset{\text{O}}{\|}}{\text{C}}-\text{OCH}_2(\text{CH}_2)_m\text{CH}_3$$

$n = 16\text{--}28$; $m = 30$ and 32
(from carnauba wax)

23.5 STEROIDS

The lipid fractions obtained from plants and animals contain an important group of compounds known as *steroids*. Many steroids are important "biological regulators" and steroids nearly always show dramatic physiological effects when they are administered to living organisms. Among these important compounds are male and female sex hormones, adrenocortical hormones, D vitamins, the bile acids, and certain cardiac poisons.

Structure and Systematic Nomenclature of Steroids

As we saw earlier (p. 113), steroids are derivatives of the perhydrocyclopentano-phenanthrene ring system shown below.

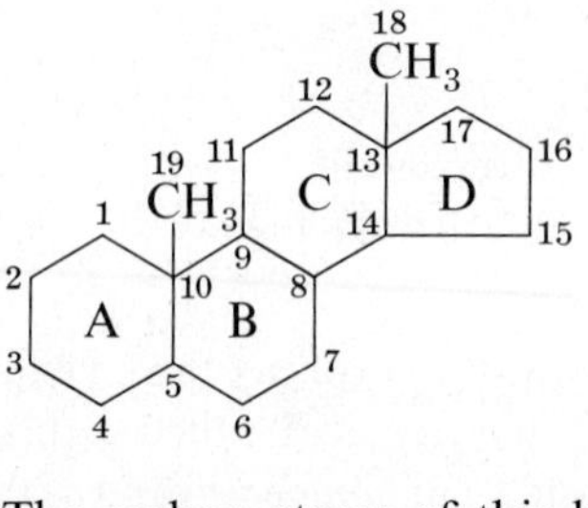

The carbon atoms of this basic ring system are numbered in the way that we have indicated above. The four rings are designated with letters.

In most steroids the B,C and C,D ring junctions are *trans*. The A,B ring junction, however, may be either *cis* or *trans* and this gives rise to two general groups of steroids having the three dimensional structures shown in Fig. 23.6.

Problem 23.11

Draw the two basic ring systems given in Fig. 23.6 for the 5α and 5β series showing all hydrogens of the cyclohexane rings. Label each hydrogen as to whether it is axial or equatorial.

The methyl groups that are attached at points of ring junction (i.e., those numbered 18 and 19) are called *angular methyl groups* and they serve as important reference points for stereochemical designations. The angular methyl groups protrude above the general plane of the ring system when it is written in the manner shown in Fig. 23.6. By convention, other groups that lie on the same general side of the

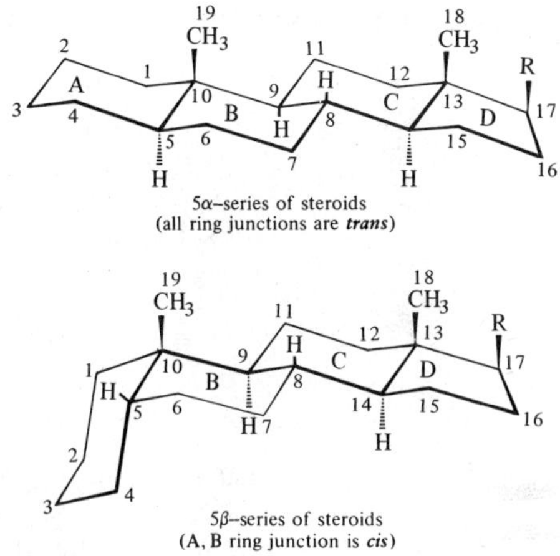

FIG. 23.6
The basic ring systems of the 5α- and 5β-series of steroids.

molecule as the angular methyl groups (i.e., on the top side) are designated as β *substituents;* those that lie generally on the bottom (i.e., are *trans* to the angular methyl groups) are designated as α *substituents.* When these designations are applied to the hydrogen atom at position 5, the ring system in which the A,B ring junction is *trans* becomes the 5α series; and the ring system in which the A,B ring junction is *cis* becomes the 5β series.

In systematic nomenclature the nature of the R group at position 17 determines (primarily) the base name of an individual steroid. These names are derived from the steroid hydrocarbon names given in Table 23.5.

TABLE 23.5 Names of Steroid Hydrocarbons

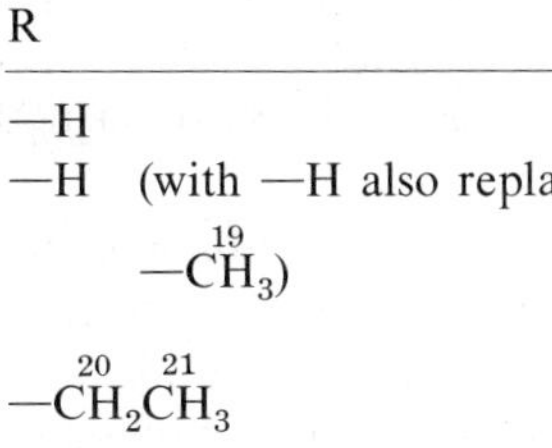

R	NAME
—H	Androstane
—H (with —H also replacing $\overset{19}{-CH_3}$)	Estrane
$\overset{20}{-CH_2}\overset{21}{CH_3}$	Pregnane
$-\overset{20}{C}H\overset{22}{CH_2}\overset{23}{CH_2}\overset{24}{CH_3}$ $\underset{21}{CH_3}$	Cholane
$-\overset{20}{C}H\overset{22}{CH_2}\overset{23}{CH_2}\overset{24}{CH_2}\overset{25}{CH}\overset{26}{CH_3}$ $\underset{21}{CH_3} \qquad \underset{27}{CH_3}$	Cholestane

Two examples that illustrate the way these base names are used are shown below.

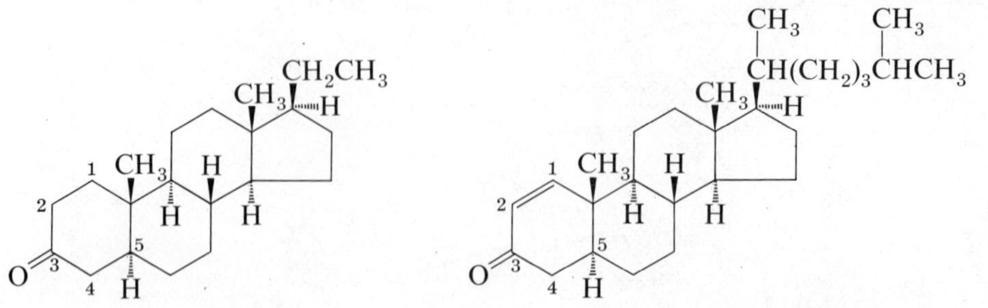

5α-Pregnan-3-one

5α-Cholest-1-en-3-one

We will see that many steroids also have common names and that the names of the steroid hydrocarbons given in Table 23.5 are derived from these common names.

Problem 23.12

(a) Androsterone, a secondary male sex hormone, has the systematic name, 5α-androstan-3α-ol-17-one. Give a three-dimensional formula for androsterone. (b) Norethynodrel, a synthetic steroid that has been widely used in oral contraceptives, has the systematic name, 17-α-ethynyl-17β-hydroxy-5(10)-estren-3-one. Give a three-dimensional formula for norethynodrel.

Cholesterol

Cholesterol (p. 897), one of the most widely occurring steroids, can be isolated by extraction of nearly all animal tissues. Human gallstones are a particularly rich source.

Although cholesterol was isolated as early as 1770, the last stereochemical details of its structure were not known until 1955. Two German chemists, Adolf Windaus and Heinrich Wieland, were responsible for outlining the structure of cholesterol; they received the Nobel Prize for their work in 1928.*

Part of the difficulty in assigning an absolute structure to cholesterol can be appreciated by noticing that cholesterol contains *eight* chiral centers. This means that there are 2^8 or 256 possible stereoisomeric forms of the basic structure, *only one of which is cholesterol.*

Cholesterol occurs widely in the body, but not all of the biological functions of cholesterol are yet known. Cholesterol is known to serve as an intermediate in the biosynthesis of steroid hormones and of bile acids but far more cholesterol is present in the body than is needed for these purposes. High levels of blood cholesterol have been implicated in the development of arteriosclerosis (hardening of the arteries) and in heart attacks that occur when cholesterol-containing plaques block

---

*The original cholesterol structure proposed by Windaus and Wieland was incorrect. This became evident in 1932 as a result of x-ray diffraction studies done by the British physicist, J. D. Bernal. By the end of 1932, however, English scientists, and Wieland himself, using Bernal's results, were able to outline the correct structure of cholesterol.

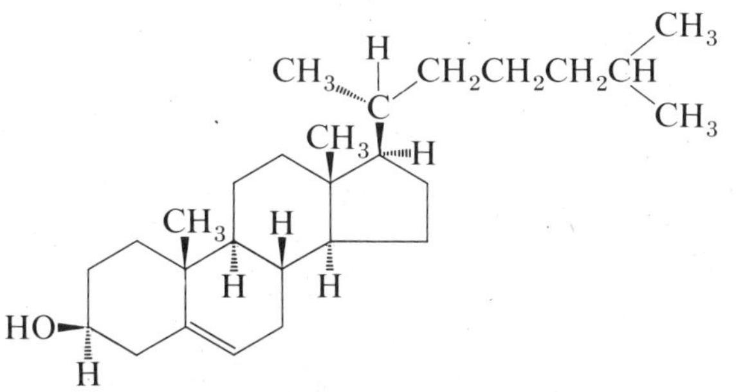

Absolute configuration of cholesterol
(5-cholesten-3β-ol)

arteries of the heart. Considerable research is being carried out in the area of cholesterol metabolism with the hope of finding ways of minimizing cholesterol levels through the use of dietary adjustments or drugs.

Sex Hormones

The sex hormones can be classified into three major groups: (1) the female sex hormones, or *estrogens,* (2) the male sex hormones, or *androgens,* and (3) the pregnancy hormones, or *progestins.*

The first sex hormone to be isolated was the estrogen, *estrone.* Working independently, Adolf Butenandt (in Germany) and Edward Doisy (in the USA) isolated estrone from the urine of pregnant women. They published their discovery in 1929. Later, Doisy was able to isolate the much more potent estrogen, *estradiol* (the true female sex hormone). In this research Doisy had to extract *four tons* of sow ovaries in order to obtain just *12 milligrams* of estradiol.

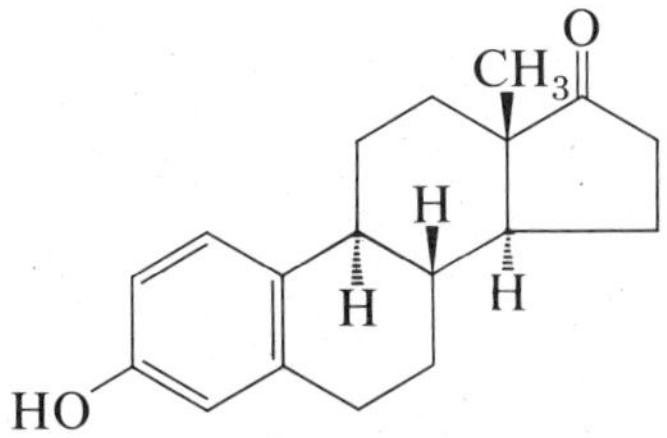

Estrone
(3-hydroxy-1,3,5(10)-estra-
trien-17-one)

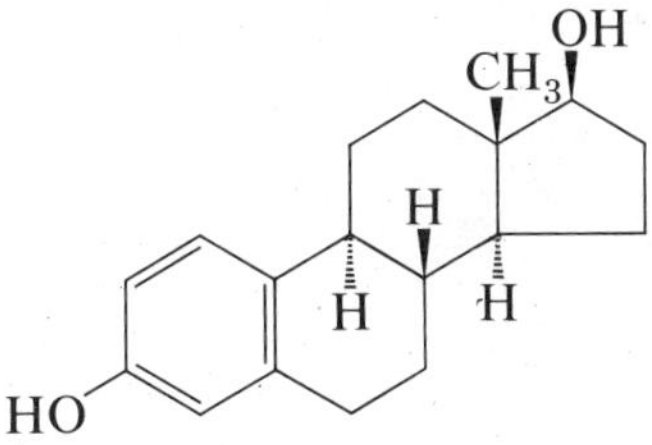

Estradiol
(3,17β-dihydroxyestra-1,3,5(10)-
triene)

Estradiol is secreted by the ovaries and it promotes the development of the secondary female characteristics that appear at the onset of puberty. Estrogens also stimulate the development of the mammary glands during pregnancy and have been shown to induce estrus (heat) in laboratory animals.

In 1931 Butenandt and Kurt Tscherning isolated the first androgen, *andros-terone.* They were able to obtain 15 milligrams of this hormone by extracting approximately 15,000 liters of male urine. Soon afterwards (in 1935), Ernest Laqueur (in Holland) isolated another male sex hormone, *testosterone,* from steer testes. It soon became clear that testosterone is the true male sex hormone and that androste-rone is a metabolized form of testosterone that is excreted in the urine.

Androsterone
(5α-androstan-3α-ol-17-one)

Testosterone
(17β-hydroxy-4-androsten-3-one)

Testosterone is secreted by the testes. It is the hormone that promotes the development of secondary male characteristics: the growth of facial and body hair; the deepening of the voice; muscular development; and the maturation of the male sex organs.

Testosterone and estradiol, then, are the chemical compounds from which "maleness" and "femaleness" are derived. It is especially interesting to examine their structural formulas and see how very slightly these two compounds differ. Testosterone has an angular methyl group at the A,B ring junction that is missing in estradiol. Ring A of estradiol is a benzene ring and, as a result, estradiol is a phenol. Ring A of testosterone contains an α,β-unsaturated keto group.

Problem 23.13

The estrogens (estrone and estradiol) are easily separated from the androgens (androsterone and testosterone) on the basis of one of their chemical properties. What is the property and how could such a separation be accomplished?

Progesterone
(4-pregnene-3.20-dione)

Progesterone is the most important *progestin* (pregnancy hormone). After ovulation occurs, the remnant of the ruptured ovarian follicle (called the *corpus luteum*) begins to secrete progesterone. This hormone prepares the lining of the uterus for implantation of the fertilized ovum, and continued progesterone secretion is necessary for the completion of pregnancy. (Progesterone is secreted by the placenta after secretion by the corpus luteum declines.)

Progesterone *also suppresses ovulation,* and it is the chemical agent that apparently accounts for the fact that pregnant women do not conceive again while pregnant. It was this observation that led to the search for synthetic progestins that could be used as oral contraceptives. (Progesterone, itself, requires very large doses to be effective in suppressing ovulation when taken orally.) A number of such compounds have been developed and are now widely used. In addition to norethynodrel (cf. problem 23.12), another widely used synthetic progestin is its double bond isomer, *norethinodrone.*

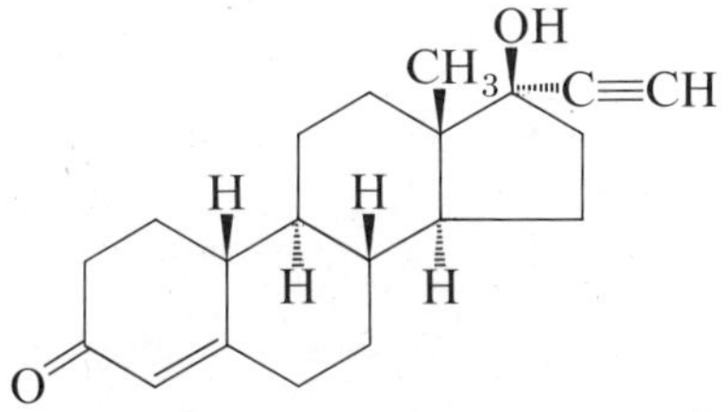

Norethinodrone
(17α-ethynyl-17β-hydroxy-4-estren-3-one)

Synthetic estrogens have also been developed and these are often used in oral contraceptives in combination with synthetic progestins. A very potent synthetic estrogen is the compound called *ethynylestradiol* or *novestrol.*

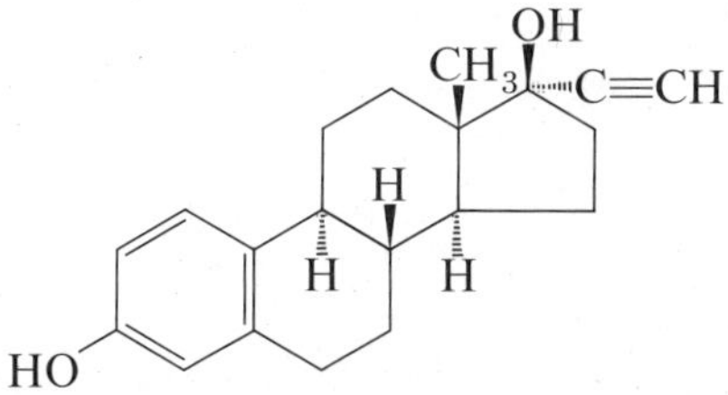

Ethynylestradiol
(17α-ethynyl-1,3,5(10)-estratriene-3,17β-diol)

Adrenocortical Hormones

At least 28 different hormones have been isolated from the adrenal cortex. Included in this group are the two steroids shown below.

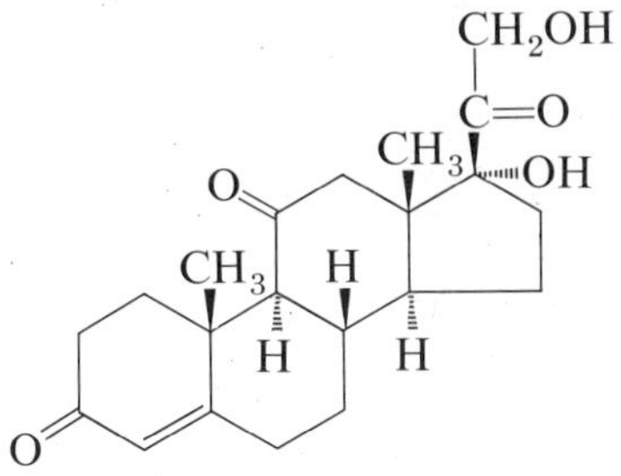

Cortisone
(17α,21-dihydroxypregn-4-ene-
3,11,20-trione)

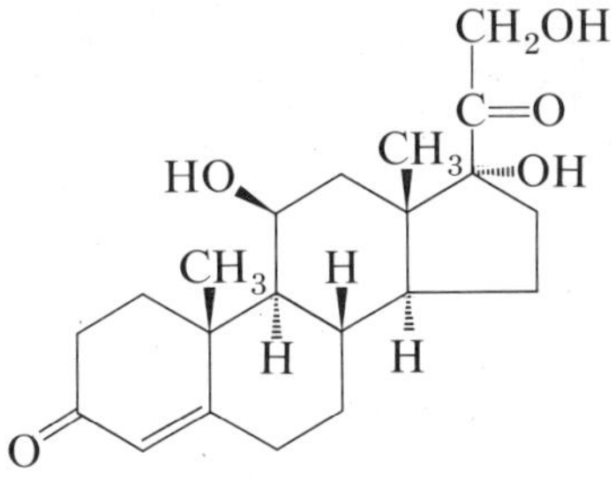

Cortisol
(11β,17α,21-trihydroxypregn-4-ene-
3,20-dione)

Most of the adrenocortical steroids are characterized by an oxygen function at position 11 (a keto group in cortisone, for example, and a β-hydroxyl in cortisol). Cortisol is the major hormone synthesized by the human adrenal cortex.

The adrenocortical steroids are apparently involved in the regulation of a large number of biological activities including carbohydrate, protein, and lipid metabolism, water and electrolyte balance, and reactions to allergic and inflammatory phenomena. Recognition of the antiinflammatory effect of cortisone and its usefulness in the treatment of rheumatoid arthritis in 1949, has led to extensive research in this area. Many 11-oxygenated steroids are now used in the treatment of a variety of disorders ranging from Addison's disease, to asthma, and to skin inflammations.

D Vitamins

The demonstration, in 1919, that sunlight exerted a curative effect on children suffering from rickets—a childhood disease characterized by poor bone growth—began a long search for a chemical explanation. Soon thereafter, it was discovered that irradiation of certain foodstuffs increased their antirachitic properties and, in 1930, the search led to a steroid that can be isolated from yeast, called *ergosterol*. Irradiation of ergosterol was found to produce a highly active material. In 1932, Windaus and his co-workers in Germany demonstrated that this highly active substance was vitamin D_2. The photochemical reaction that takes place is one in which the dienoid ring B of ergosterol opens to produce a conjugated triene:

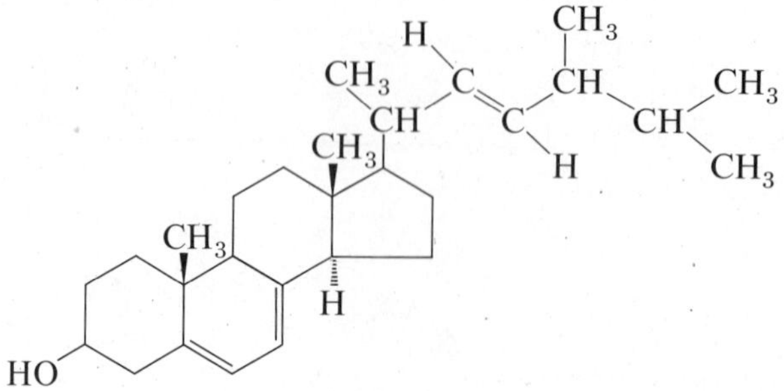

Ergosterol

ultraviolet light, room temperature

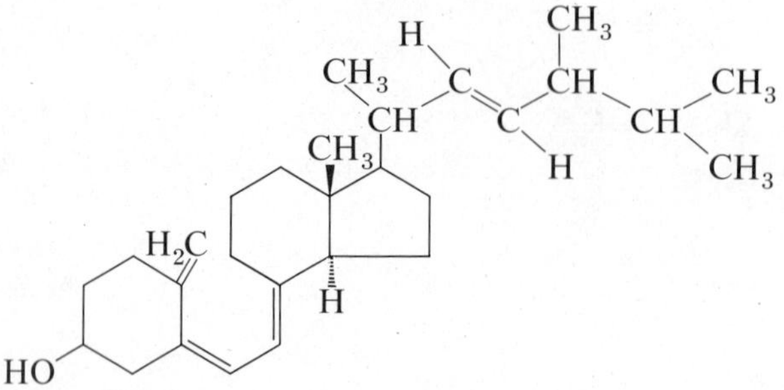

Vitamin D_2

Other Steroids

The structures, sources, and physiological properties of a number of other important steroids are given in Table 23.6.

TABLE 23.6 Other Important Steroids

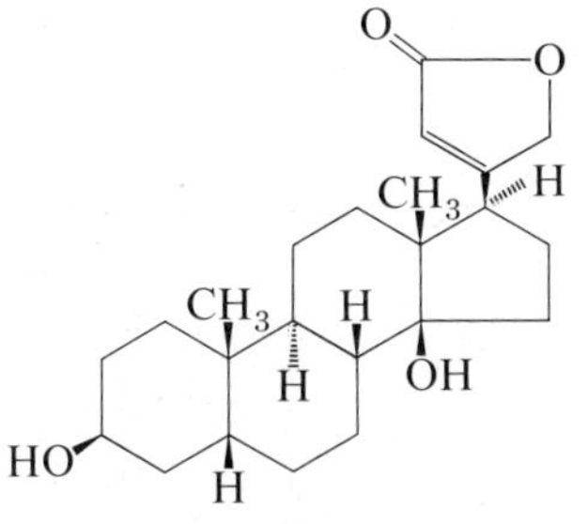

Digitoxigenin

Digitoxigenin is a cardiac aglycone that can be isolated by hydrolysis of digitalis, a pharmaceutical that has been used in treating heart disease since 1785. In digitalis, sugar molecules are joined in acetal linkages to the 3—OH of the steroid. In small doses digitalis strengthens the heart muscle, in larger doses it is a powerful heart poison.

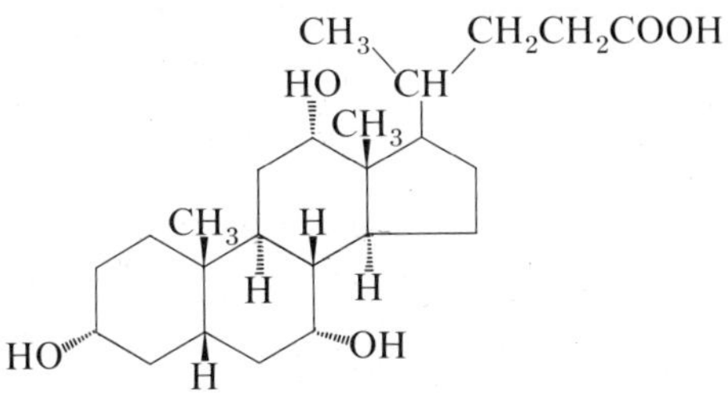

Cholic acid

Cholic acid is the most abundant acid obtained from the hydrolysis of human or ox bile. Bile is produced by the liver and stored in gall bladder. When secreted into the small intestine, bile emulsifies lipids by acting as a soap. This aids in the digestive process.

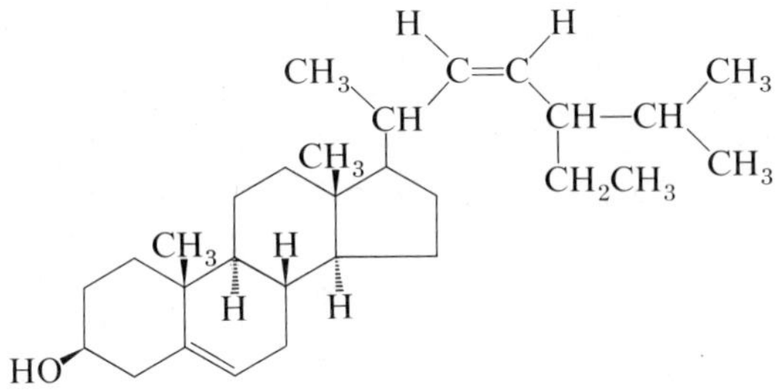

Stigmasterol

Stigmasterol is a widely occurring plant steroid that is obtained commercially from soybean oil.

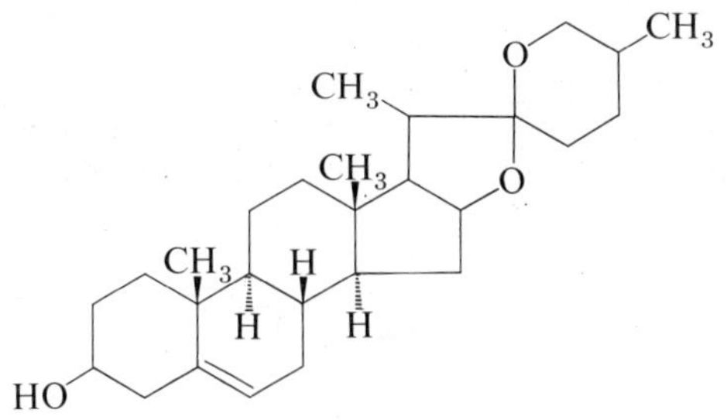

Diosgenin

Diosgenin, obtained from a particular species of yams, is used as the starting material for a commercial synthesis of cortisone.

Reactions of Steroids

Steroids undergo all of the reactions that we might expect of molecules containing double bonds, hydroxyl groups, keto groups, and so on. While the stereochemistry of steroid reactions is often quite complex, it is many times strongly influenced by the steric hindrance presented at the β face of the molecule by the angular methyl groups. Thus, many reagents react preferentially at the relatively unhindered α face. This is particularly true when the reaction takes place at a functional group very near an angular methyl group and when the attacking reagent is bulky. Examples that illustrate this tendency are shown below.

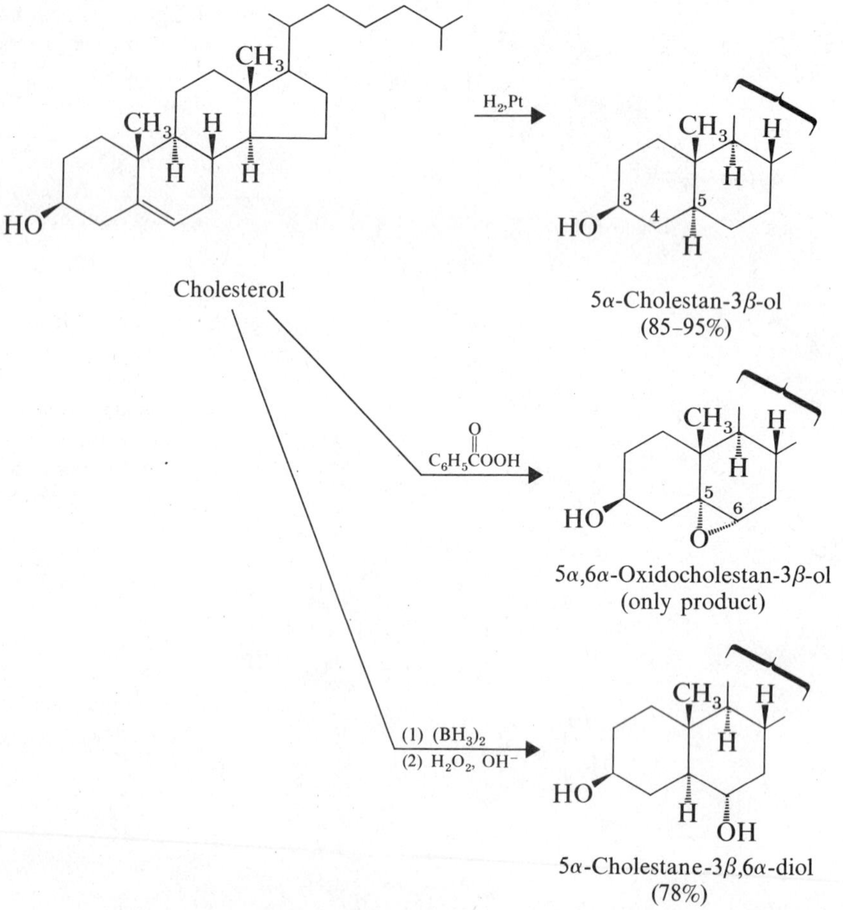

Cholesterol

5α-Cholestan-3β-ol
(85–95%)

5α,6α-Oxidocholestan-3β-ol
(only product)

5α-Cholestane-3β,6α-diol
(78%)

When the epoxide ring of 5α,6α-oxidocholestan-3β-ol (above) is opened, attack by chloride ion must occur from the β face but it takes place at the more open 6 position.

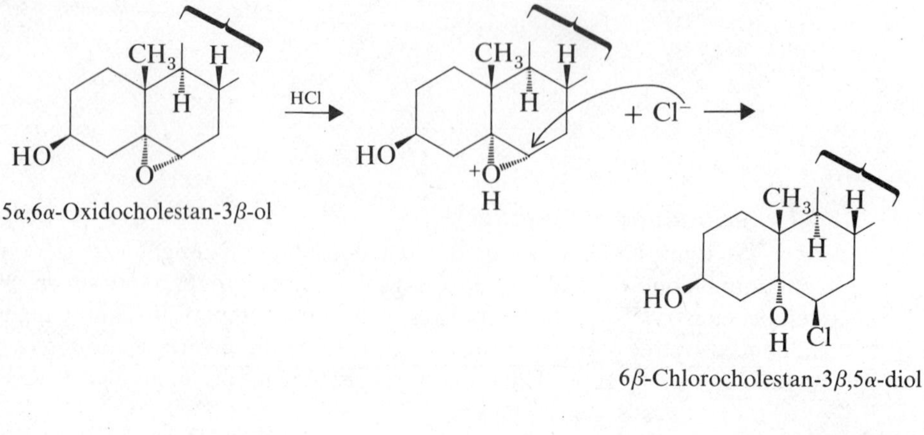

5α,6α-Oxidocholestan-3β-ol

6β-Chlorocholestan-3β,5α-diol

Problem 23.14

Show how you might convert cholesterol into each of the following compounds.

(a) 5α,6β-Dibromocholestan-3β-ol
(b) Cholestane-3β,5α,6β-triol
(c) 5α-Cholestan-3-one
(d) 6α-Deuterio-5α-cholestan-3β-ol
(e) 6β-Bromocholestane-3β-5α-diol

Problem 23.15

Give structural formulas and names for compounds **A** and **B**.

$$5\alpha\text{-Cholest-2-ene} \xrightarrow{\text{C}_6\text{H}_5\overset{\overset{\displaystyle O}{\|}}{\text{C}}\text{OOH}} \textbf{A} \text{ (an epoxide)} \xrightarrow{\text{HBr}} \textbf{B}$$

[Hint: **B** is not the most stable stereoisomer]

Problem 23.16

When cholesterol is treated with bromine in carbon tetrachloride the initial product that forms is the 5α,6β-dibromocholestan-3β-ol. If this product is allowed to stand in chloroform solution it slowly comes into equilibrium with the 5β,6α-dibromide. (The 5β,6α-dibromide predominates at equilibrium; the mixture contains 85% of the 5β,6α-dibromocholestan-3β-ol and only 15% of 5α,6β-dibromocholestan-3β-ol.) Consider conformational effects and account for this phenomenon.

The relative openness of equatorial groups (when compared to axial groups) also influences the stereochemical course of steroid reactions. When 5α-cholestane-3β,7α-diol (below) is treated with excess ethyl chloroformate ($\text{C}_2\text{H}_5\text{OCOCl}$), only the equatorial 3β-hydroxyl becomes esterified. The axial 7α-hydroxyl is unaffected by the reaction.

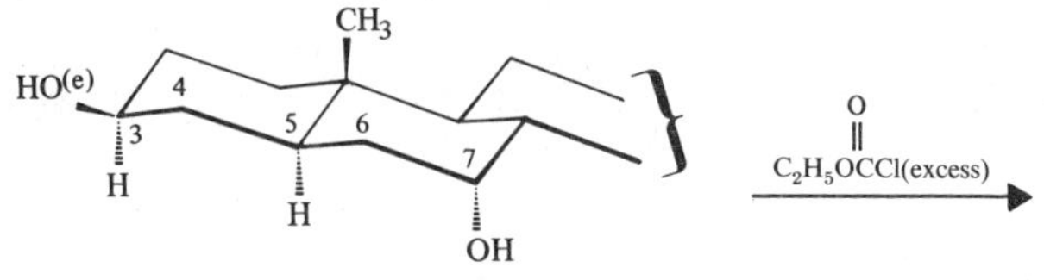

5α–cholestane –3β, 7α–diol

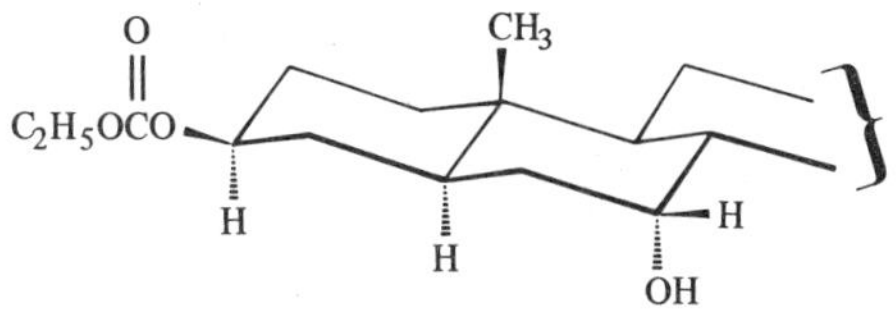

$$\xrightarrow{\underset{\text{C}_2\text{H}_5\text{OCCl(excess)}}{\overset{\overset{\displaystyle O}{\|}}{}}}$$

(only product)

By contrast, treating 5α-cholestane-3β,7β-diol with excess ethyl chloroformate leads to esterification of both hydroxyl groups. In this instance both groups are equatorial.

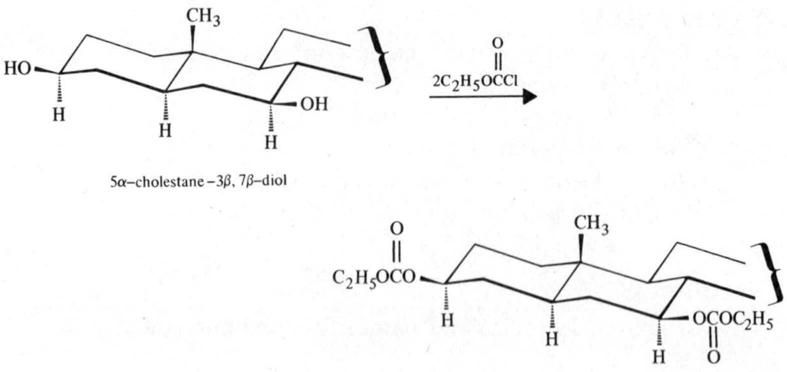

5α–cholestane –3β, 7β–diol

23.6 LIPID BIOSYNTHESIS

Biosynthesis of Fatty Acids

The fact that most naturally occurring fatty acids are made up of an even number of carbon atoms suggests the possibility that they are assembled from two carbon units. The idea that these units might be acetate (CH_3COO^-) units was put forth as early as 1893. Many years later when radioactively labeled compounds became available, it became possible to test and confirm this hypothesis.

When an animal is fed acetic acid labeled with carbon-14 at the carboxyl group, the fatty acids that the animal synthesizes contains the label at alternate carbons beginning with the carboxyl carbon:

$$CH_3\overset{*}{C}OOH \qquad CH_3CH_2CH_2\overset{*}{C}H_2CH_2\overset{*}{C}H_2CH_2\overset{*}{C}H_2CH_2\overset{*}{C}H_2CH_2\overset{*}{C}H_2CH_2\overset{*}{C}H_2CH_2\overset{*}{C}OOH$$

Feeding carboxyl-labeled acetic acid ($C^* = {}^{14}C$)

Yields palmitic acid labeled at these positions

Conversely, feeding acetic acid labeled at the methyl carbon yields a fatty acid labeled at the other set of alternate carbons:

$$\overset{*}{C}H_3COOH \qquad \overset{*}{C}H_3CH_2\overset{*}{C}H_2CH_2\overset{*}{C}H_2CH_2\overset{*}{C}H_2CH_2\overset{*}{C}H_2CH_2\overset{*}{C}H_2CH_2\overset{*}{C}H_2CH_2COOH$$

Feeding methyl-labeled acetic acid

Yields palmitic acid labeled at these positions

The biosynthesis of fatty acids is now known to begin with acetyl coenzyme A:

$$CH_3\overset{O}{\overset{||}{C}}-S-CoA$$

The acetyl portion of acetyl coenzyme A can be synthesized in the cell from acetic acid; it can also be synthesized from carbohydrates, proteins, and other fats.

$$
\underset{\substack{CH_3\overset{\displaystyle O}{\overset{\|}{C}}OH}}{}
$$

$$
\begin{array}{l}
CH_3\overset{O}{\overset{\|}{C}}OH \\
\text{Carbohydrates} \\
\text{Proteins} \\
\text{Fats}
\end{array}
\xrightarrow[\text{CoA—SH}]{}
\underset{\text{Acetyl coenzyme A}}{CH_3\overset{O}{\overset{\|}{C}}S\text{—CoA}}
$$

Although the methyl group of acetyl coenzyme A is already activated toward condensation reactions by virtue of its being a part of a thiol-ester (Section 19.10); nature activates it again by converting it to *malonyl coenzyme A*.

$$
\underset{\text{Acetyl CoA}}{CH_3\overset{O}{\overset{\|}{C}}S\text{—CoA}} + CO_2
\underset{\text{acetyl CoA carboxylase*}}{\rightleftarrows}
\underset{\text{Malonyl CoA}}{HO\overset{O}{\overset{\|}{C}}CH_2\overset{O}{\overset{\|}{C}}S\text{—CoA}}
$$

The next steps in fatty acid synthesis involve the transfer of acyl groups of malonyl CoA and acetyl coenzyme A to the thiol group of a coenzyme called *acyl carrier protein* or ACP—SH.

$$
\underset{\text{Malonyl CoA}}{HO\overset{O}{\overset{\|}{C}}CH_2\overset{O}{\overset{\|}{C}}S\text{—CoA}} + ACP\text{—SH}
\rightleftarrows
\underset{\text{Malonyl—S—ACP}}{HO\overset{O}{\overset{\|}{C}}CH_2\overset{O}{\overset{\|}{C}}S\text{—ACP}} + CoA\text{—SH}
$$

$$
\underset{\text{Acetyl CoA}}{CH_3\overset{O}{\overset{\|}{C}}S\text{—CoA}} + ACP\text{—SH}
\rightleftarrows
\underset{\text{Acetyl—S—ACP}}{CH_3\overset{O}{\overset{\|}{C}}S\text{—ACP}} + CoA\text{—SH}
$$

Acetyl-S-ACP and malonyl-S-ACP then condense with each other to form acetoacetyl-S-ACP:

$$
\underset{\text{Acetyl-S-ACP}}{CH_3\overset{O}{\overset{\|}{C}}S\text{—ACP}} + \underset{\text{Malonyl-S-ACP}}{HO\overset{O}{\overset{\|}{C}}CH_2\overset{O}{\overset{\|}{C}}S\text{—ACP}}
\rightleftarrows
$$

$$
\underset{\text{Acetoacetyl-S-ACP}}{CH_3\overset{O}{\overset{\|}{C}}CH_2\overset{O}{\overset{\|}{C}}S\text{—ACP}} + CO_2 + ACP\text{—SH}
$$

The molecule of CO_2 that is lost in this reaction is the same molecule that was incorporated into malonyl CoA in the acetyl CoA carboxylase reaction.

This remarkable reaction bears a strong resemblance to the malonic ester syntheses that we saw earlier (Section 20.5) and it deserves special comment. One can imagine, for example, a more economical synthesis of acetoacetyl-S-ACP, that is, a simple condensation between two moles of acetyl-S-ACP.

---

* This step also requires one mole of adenosine triphosphate (p. 914) and an enzyme that transfers the carbon dioxide.

$$\underset{}{\overset{O}{\underset{\|}{CH_3CS}}}-ACP + \underset{}{\overset{O}{\underset{\|}{CH_3CS}}}-ACP \rightleftarrows \underset{}{\overset{O \quad \; O}{\underset{\| \quad \|}{CH_3CCH_2CS}}}-ACP + ACP-SH$$

Studies of this last reaction, however, have revealed that it is highly *endothermic* and that the position of equilibrium lies very far to the left. By contrast, the condensation of acetyl-S-ACP and malonyl-S-ACP is highly exothermic and the position of equilibrium lies far to the right. The favorable thermodynamics of the condensation reaction utilizing malonyl-S-ACP comes about because *the reaction also produces a highly stable substance: carbon dioxide.* Thus, decarboxylation of the malonyl group provides the condensation reaction with thermodynamic assistance.

The next three steps in fatty acid synthesis transform the acetoacetyl group of acetoacetyl-S-ACP into a butyryl (butanoyl) group. These steps involve (1) reduction of the keto group (utilizing NADPH* as the reducing agent), (2) dehydration of an alcohol, and (3) reduction of a double bond (again utilizing NADPH).

Reduction of Keto Group

$$\underset{\text{Acetoacetyl-S-ACP}}{\overset{O \quad \; O}{\underset{\| \quad \|}{CH_3CCH_2CS}}}-ACP + NADPH + H^+ \rightleftarrows \underset{\beta\text{-Hydroxybutyryl-S-ACP}}{\overset{OH \quad \; O}{\underset{| \quad \| }{CH_3CHCH_2CS}}}-ACP + NADP^+$$

Dehydration of Alcohol

$$\underset{\beta\text{-Hydroxybutyryl-S-ACP}}{\overset{OH \quad \; O}{\underset{| \quad \| }{CH_3CHCH_2CS}}}-ACP \rightleftarrows \underset{\text{Crotonyl-S-ACP}}{\overset{O}{\underset{\|}{CH_3CH{=}CHCS}}}-ACP + H_2O$$

Reduction of Double Bond

$$\underset{\text{Crotonyl-S-ACP}}{\overset{O}{\underset{\|}{CH_3CH{=}CHCS}}}-ACP + NADPH + H^+ \rightleftarrows \underset{\text{Butyryl-S-ACP}}{\overset{O}{\underset{\|}{CH_3CH_2CH_2CS}}}-ACP + NADP^+$$

These steps complete one cycle of the overall fatty acid synthesis. Their net result is the conversion of two acetate units into the four-carbon butyrate unit of butyryl-S-ACP. (This conversion requires, of course, the crucial intervention of a molecule of carbon dioxide.) At this point, another cycle begins and the chain is lengthened by two more carbons:

Condensation

$$\underset{\text{(four carbons)}}{\overset{O}{\underset{\|}{CH_3CH_2CH_2CS}}}-ACP + \underset{}{\overset{O \quad \; O}{\underset{\| \quad \|}{HOCCH_2CS}}}-ACP \longrightarrow$$

$$\underset{}{\overset{O \quad \quad \; O}{\underset{\| \quad \quad \|}{CH_3CH_2CH_2CCH_2CS}}}-ACP + CO_2 + ACP-SH$$

*NADPH is *nicotinamide adenine dinucleotide phosphate,* a coenzyme that is very similar in structure and function to NADH p. 439.

Reduction

$$CH_3CH_2CH_2\overset{\overset{\displaystyle O}{\|}}{C}CH_2\overset{\overset{\displaystyle O}{\|}}{C}S-ACP \xrightarrow{\quad\text{NADPH} \quad \text{NADP}^+ \atop +H^+\quad\quad}$$

$$CH_3CH_2CH_2\overset{\overset{\displaystyle OH}{|}}{C}HCH_2\overset{\overset{\displaystyle O}{\|}}{C}S-ACP$$

Dehydration

$$CH_3CH_2CH_2\overset{\overset{\displaystyle OH}{|}}{C}HCH_2\overset{\overset{\displaystyle O}{\|}}{C}S-ACP \xrightarrow{\quad H_2O \quad} CH_3CH_2CH_2CH=CH\overset{\overset{\displaystyle O}{\|}}{C}S-ACP$$

Reduction

$$CH_3CH_2CH_2CH=CH\overset{\overset{\displaystyle O}{\|}}{C}S-ACP \xrightarrow{\quad\text{NADPH} \quad \text{NADP}^+ \atop +H^+\quad\quad}$$

$$CH_3CH_2CH_2CH_2CH_2\overset{\overset{\displaystyle O}{\|}}{C}S-ACP$$
(six carbons)

Subsequent turns of the cycle continue to lengthen the chain by two carbon units until a long-chain fatty acid is produced. The overall equation for the synthesis of palmitic acid, for example, can be written as follows:

$$CH_3\overset{\overset{\displaystyle O}{\|}}{C}S-CoA + 7HO\overset{\overset{\displaystyle O}{\|}}{C}CH_2\overset{\overset{\displaystyle O}{\|}}{C}S-CoA + 14NADH + 14H^+ \longrightarrow$$

$$CH_3(CH_2)_{14}COOH + 7CO_2 + 8CoA-SH + 14NAD^+ + 6H_2O$$

One of the most remarkable aspects of fatty acid synthesis is that the entire cycle appears to be carried out by a complex of enzymes that are clustered into a single unit. The molecular weight of this cluster of proteins, *called fatty acid synthetase,* has been estimated as 2,300,000.* The synthesis begins with a single molecule of acetyl-S-ACP serving as a primer. Then, in the synthesis of palmitic acid, for example, successive condensations of seven molecules of malonyl-S-ACP occur with each condensation followed by reduction, dehydration, and reduction. All of these steps, which result in the synthesis of a 16-carbon chain, take place before the fatty acid is released from the enzyme cluster.

The acyl carrier protein has been isolated and purified; its molecular weight is approximately 10,000. The protein contains a chain of groups called a *phosopho-pantetheine group* that is identical to that of coenzyme A (p. 771). In ACP this chain is attached to a protein (rather than to an adenosine phosphate as it is in coenzyme A):

---

* As isolated from yeast cells. Fatty acid synthetases from different sources have different molecular weights; that from pigeon liver, for example, has a molecular weight of 450,000.

$$\overbrace{}^{\text{Pantothenic acid}} \qquad \overbrace{}^{\beta\text{-Aminoethanethiol}}$$

Protein—O—P—O—CH$_2$—C—CHOH—C—NH—CH$_2$—CH$_2$—C—NH—CH$_2$—CH$_2$—SH

$$\underbrace{}_{\substack{\text{Phosphopantetheine group} \\ \text{20.2 Å}}}$$

The length of the phosphopantetheine group is 20.2 Å, and it has been postulated that it acts as a "swinging arm" in transferring the growing acyl chain from one enzyme of the cluster to the next (Fig. 23.7).

Biosynthesis of Steroids

We saw in Section 11.4, that the five-carbon compound, 3-isopentenylpyrophosphate, is the actual "isoprene unit" that nature uses in constructing terpenoids and carotenoids. We can now extend that biosynthetic pathway in two directions. We can show how 3-isopentenylpyrophosphate (like the fatty acids) is ultimately derived from acetate units and how cholesterol, the precursor of most of the important steroids, is synthesized from 3-isopentenylpyrophosphate.

In the 1940s, Konrad Bloch of Harvard University used labeling experiments to demonstrate that all of the carbon atoms of cholesterol can be derived from acetic acid. Using *methyl-labeled* acetic acid, for example, Bloch found the following label distribution in the cholesterol that was synthesized.

FIG. 23.7
The phosphopantetheine group as a swinging arm in the fatty acid synthetase complex. (From A.L. Lehninger, Biochemistry, *Worth Publishers, Inc., New York, (1970), p. 519. Used with permission.)*

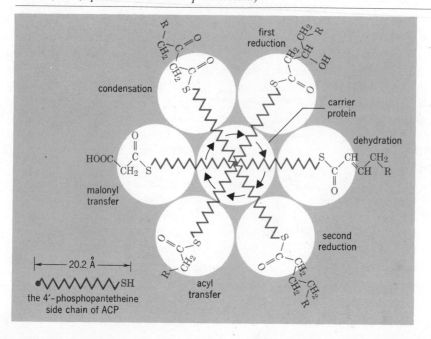

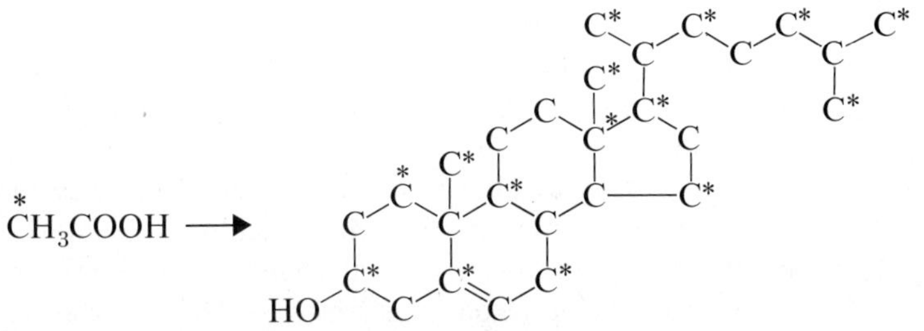

$$\overset{*}{C}H_3COOH \longrightarrow$$

Bloch also found that feeding *carboxyl-labeled* acetic acid led to incorporation of the label into all of the other carbons of cholesterol (the unstarred carbons of the formula given above).

Subsequent research by a number of investigators has shown that 3-isopentenylpyrophosphate is synthesized from acetate units through the following sequence of reactions:

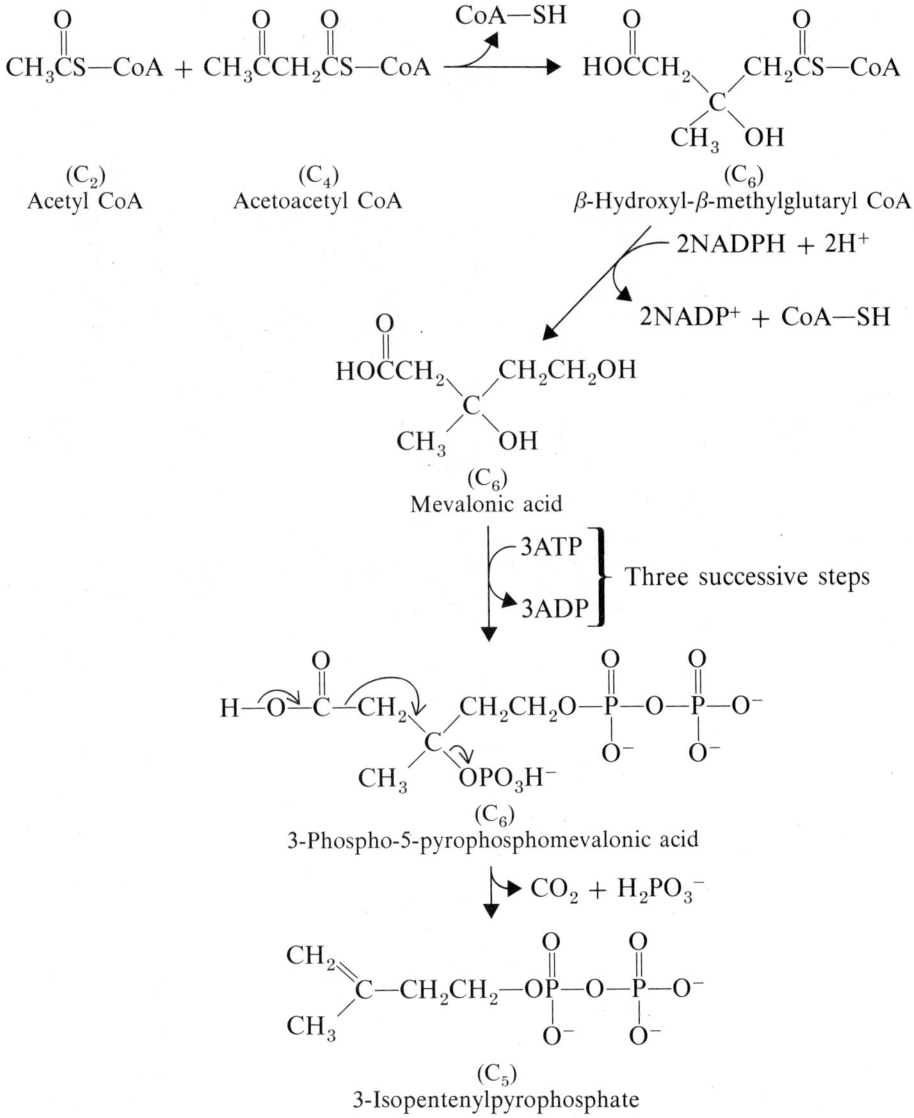

(C_2)
Acetyl CoA

(C_4)
Acetoacetyl CoA

(C_6)
β-Hydroxyl-β-methylglutaryl CoA

$-$ 2NADPH + 2H$^+$

2NADP$^+$ + CoA—SH

(C_6)
Mevalonic acid

$-$ 3ATP
$\longrightarrow$ 3ADP $\Big\}$ Three successive steps

(C_6)
3-Phospho-5-pyrophosphomevalonic acid

$CO_2 + H_2PO_3^-$

(C_5)
3-Isopentenylpyrophosphate

The first step of this synthetic pathway is straightforward. Acetyl CoA (from one mole of acetate) and acetoacetyl CoA (from two moles of acetate) condense to form the six-carbon compound, β-hydroxy-β-methylglutaryl CoA. This step is followed by an enzymatic reduction of the thiol ester group of β-hydroxy-β-methylglutaryl CoA to the primary alcohol of mevalonic acid. The key to understanding this pathway was in the discovery that mevalonic acid was an intermediate and that this six-carbon compound could be transformed into the five-carbon 3-isopentenylpyrophosphate by successive phosphorylations and decarboxylation.

Then 3-isopentenylpyrophosphate (3-IPP) isomerizes to produce an equilibrium mixture that contains 2-isopentenylpyrophosphate (2-IPP), and these two compounds condense to form geranylpyrophosphate, a 10-carbon compound. Geranylpyrophosphate subsequently condenses with another mole of 3-isopentenylpyrophosphate to form farnesylpyrophosphate, a 15-carbon compound. (Geranylpyrophosphate and farnesylpyrophosphate are the precursors of the mono- and sesquiterpenes, p. 408.)

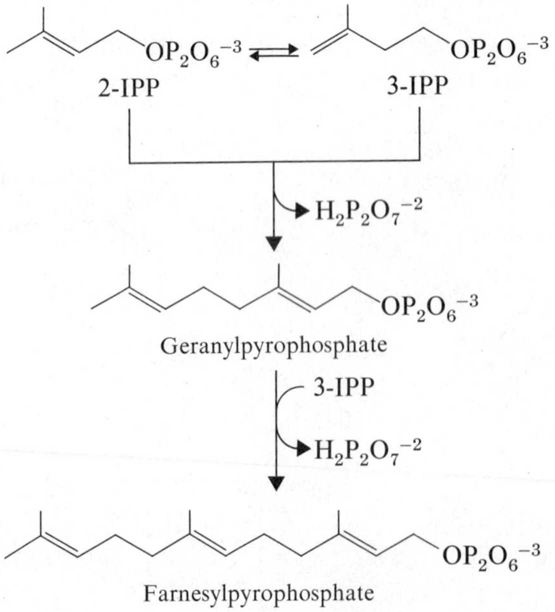

2-IPP   3-IPP

Geranylpyrophosphate

Farnesylpyrophosphate

Two moles of farnesylpyrophosphate then undergo a reductive condensation to produce squalene.

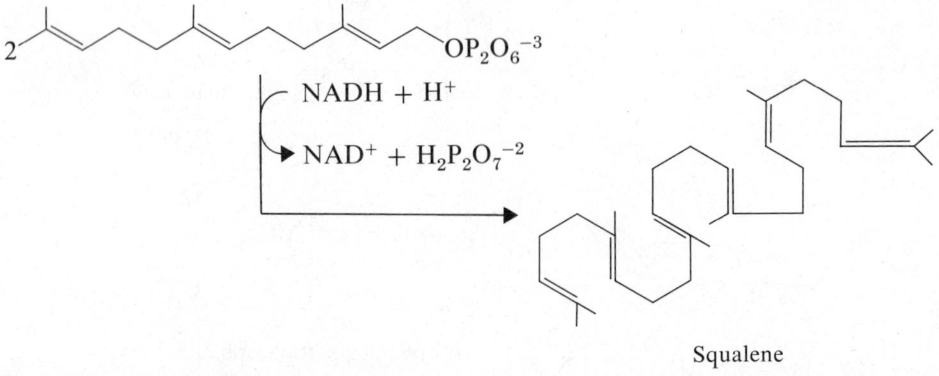

Squalene

Squalene is the direct precursor of cholesterol. Oxidation of squalene yields squalene 2,3-epoxide, which undergoes a remarkable series of ring closures accompanied by concerted methyl and hydride migrations to yield lanosterol (cf. p. 223). Lanosterol is then converted to cholesterol through a series of enzyme-catalyzed reactions.

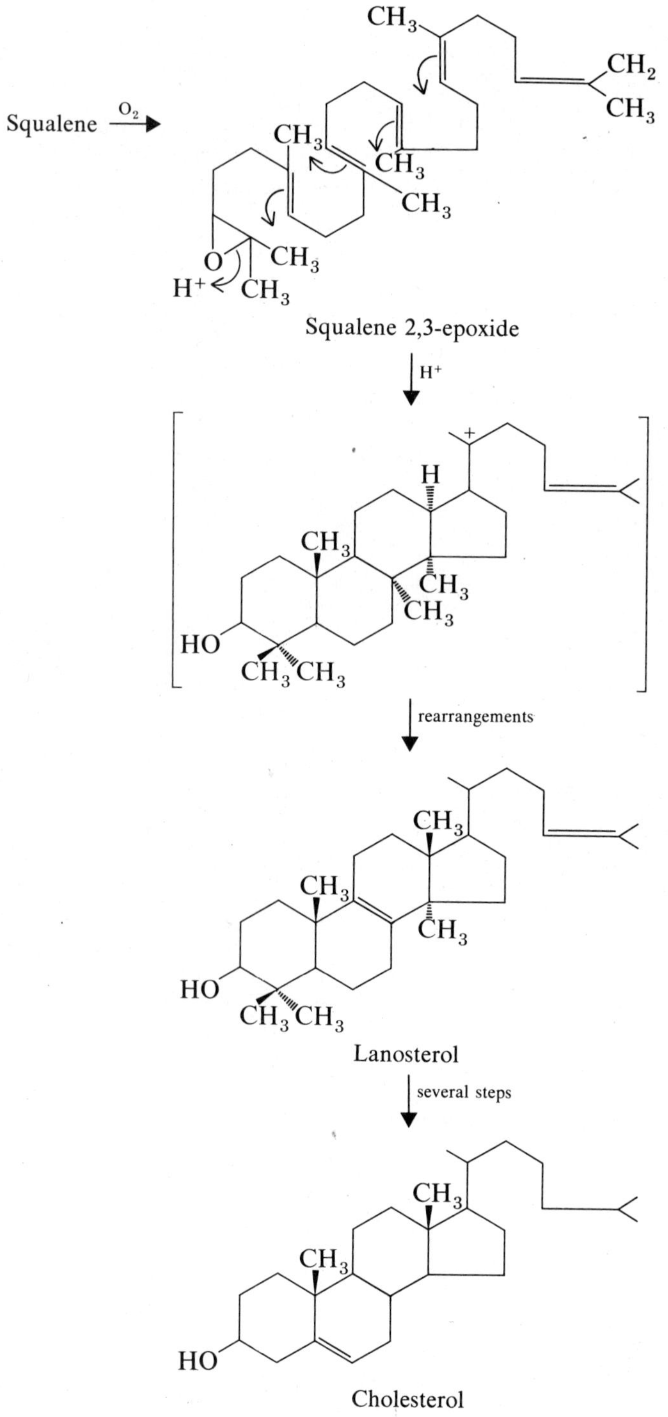

Squalene $\xrightarrow{O_2}$

Squalene 2,3-epoxide

Lanosterol

Cholesterol

* Problem 23.17

One of the first laboratory syntheses of cholesterol was achieved by Professor R. B. Woodward and his students at Harvard University in 1951. Many of the steps of this synthesis are outlined below. Supply the missing reagents.

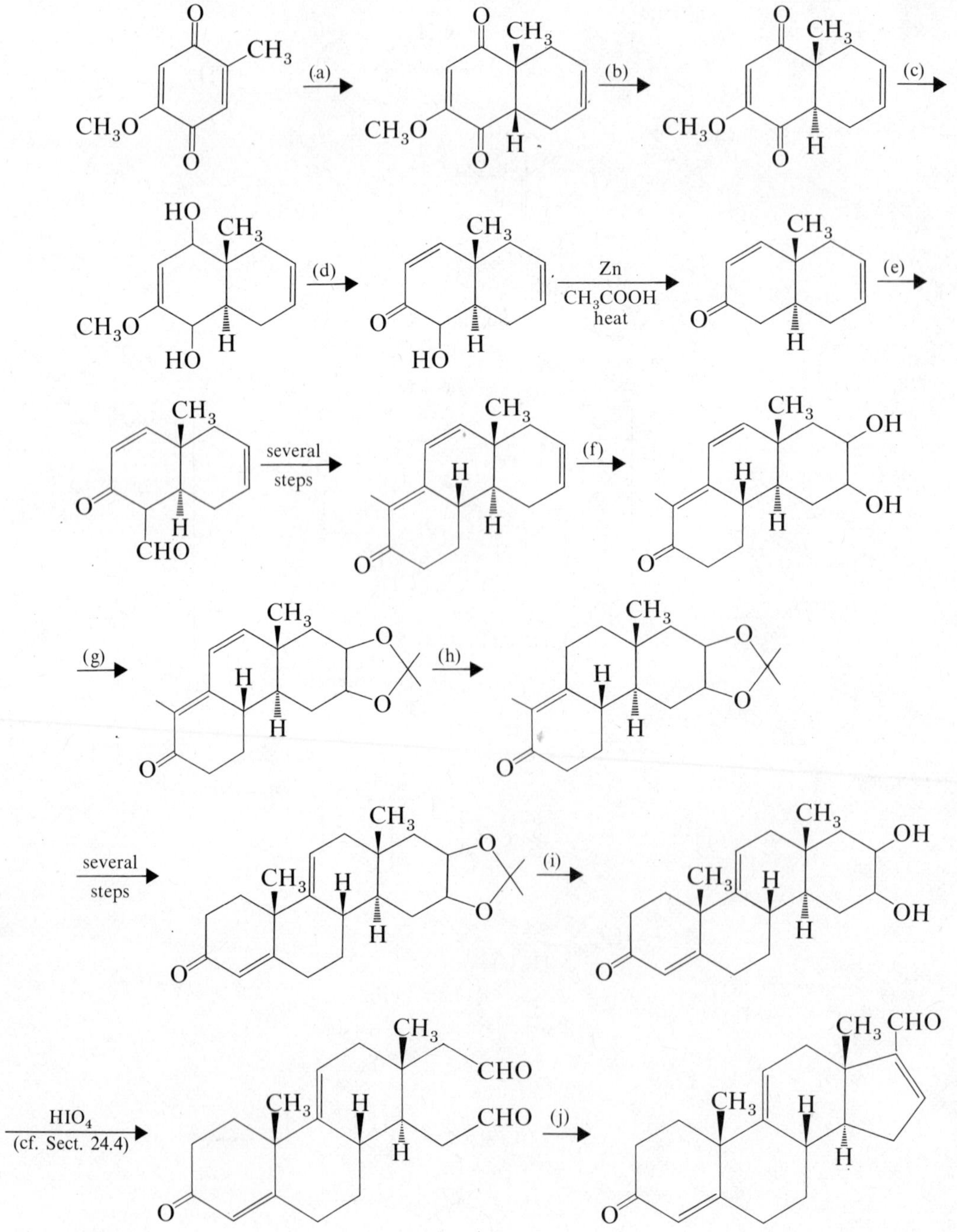

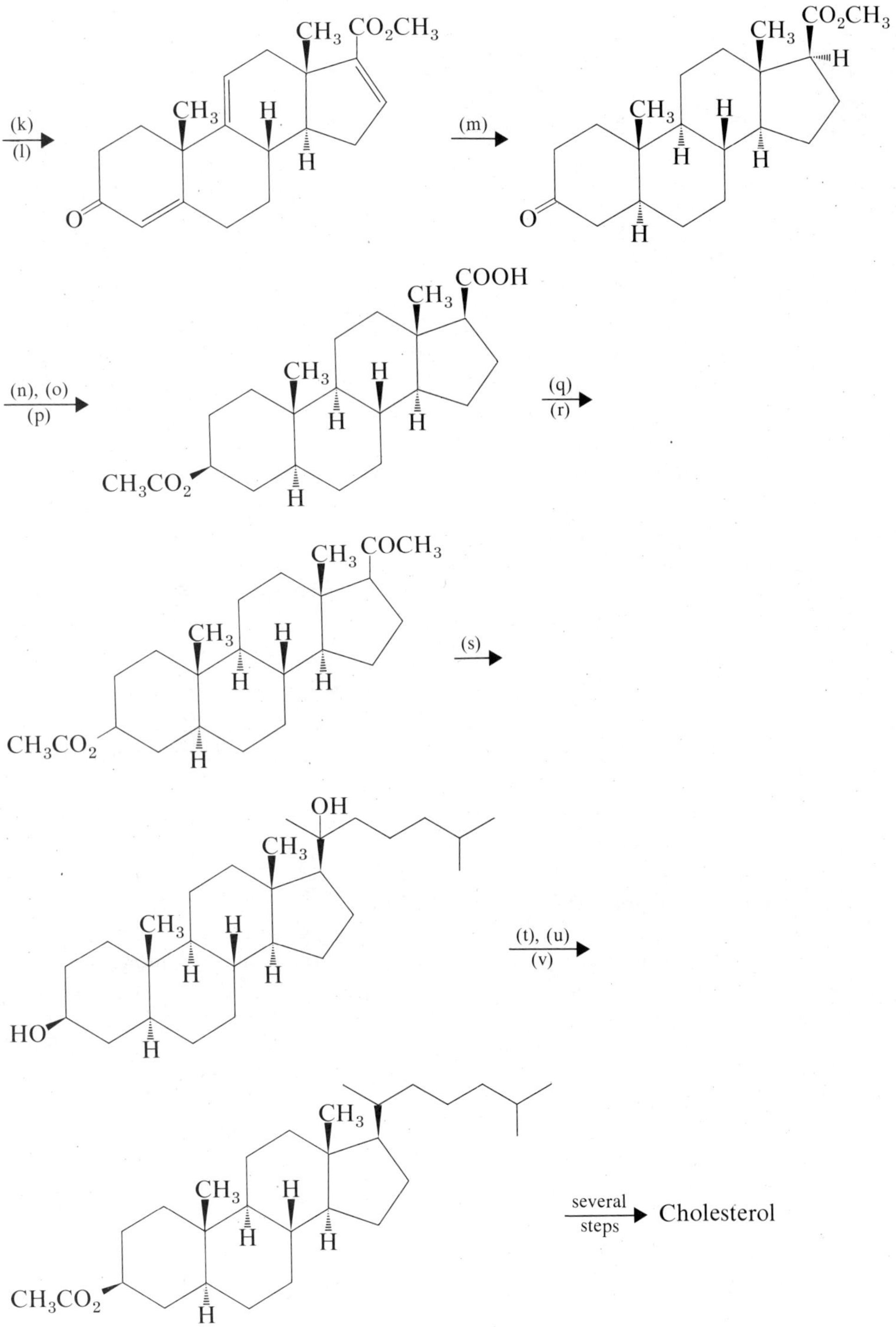
(k)
(l)
CH_3 CO_2CH_3
CH_3
H
H
O
(m)
CH_3 CO_2CH_3
H
CH_3
H
H
H
H
O
H
(n), (o)
(p)
CH_3 COOH
CH_3
H
H
H
CH_3CO_2
H
(q)
(r)
CH_3 $COCH_3$
CH_3
H
H
H
CH_3CO_2
H
(s)
OH
CH_3
CH_3
H
H
H
HO
H
(t), (u)
(v)
CH_3
CH_3
H
H
H
CH_3CO_2
H
several
steps
Cholesterol

24 CARBOHYDRATES

24.1 INTRODUCTION

Classification of Carbohydrates

The group of compounds known as carbohydrates received their general name because of early observations that they often have the formula $C_x(H_2O)_y$—that is, they appear to be "hydrates of carbon." Simple carbohydrates are also known as sugars or saccharides (L. *saccharum,* sugar) and the ending of the names of most sugars is *ose.* Thus, we have such names as *sucrose* for ordinary table sugar, *glucose* for blood sugar, *maltose* for malt sugar, and so on.

Carbohydrates are usually defined as *polyhydroxy aldehydes and ketones or substances that hydrolyze to yield such compounds.* Although this definition draws attention to the important functional groups of carbohydrates, it is not entirely satisfactory. We will later find that because carbohydrates contain $\diagdown C{=}O$ groups and —OH groups, they exist, primarily, as *hemiacetals* and *acetals* or as *hemiketals* and *ketals.*

The simplest carbohydrates, those that cannot be cleaved into smaller, simpler carbohydrates by hydrolysis, are called *monosaccharides.* Carbohydrates that undergo hydrolysis to produce only two moles of a monosaccharide are called *disaccharides;* those that yield three moles of a monosaccharide are called *trisaccharides;* and so on. (Carbohydrates that hydrolyze to yield 2 to 10 moles of a monosaccharide are sometimes called *oligosaccharides.*) Carbohydrates that yield a large number of moles of monosaccharide, that is, > 10, are known as *polysaccharides.*

Maltose and sucrose are examples of disaccharides; they are monosaccharide dimers. On hydrolysis, maltose yields two moles of the monosaccharide, glucose; sucrose undergoes hydrolysis to yield one mole of glucose and one mole of the monosaccharide, fructose. Starch and cellulose are examples of polysaccharides; they are both glucose polymers. Hydrolysis of either starch or cellulose yields a large number of glucose units.

$$\text{One mole of maltose} \xrightarrow[\text{H}^+]{\text{H}_2\text{O}} \text{two moles of glucose}$$
$$\text{(a disaccharide)} \qquad \text{(a monosaccharide)}$$

$$\text{One mole of sucrose} \xrightarrow[\text{H}^+]{\text{H}_2\text{O}} \text{one mole of glucose} + \text{one mole of fructose}$$
$$\text{(a disaccharide)} \qquad \text{(monosaccharides)}$$

$$\begin{array}{c}\text{One mole of starch}\\ \text{or}\\ \text{one mole of cellulose}\\ \text{(polysaccharides)}\end{array} \xrightarrow[\text{H}^+]{\text{H}_2\text{O}} \text{many moles of glucose}$$

Carbohydrates are the most abundant organic constituents of plants and animals. They not only serve as an important source of chemical energy for living organisms (sugars and starches are important in this respect), but in plants and in some animals they serve as important constituents of supporting tissues (this is the primary function of the cellulose found in wood, cotton, and flax, for example).

We encounter carbohydrates at almost every turn of our daily lives. The paper on which this book is written is largely cellulose; so too is the cotton of our clothes and the wood of our houses. The flour from which we make bread is mainly starch, and starch is also a major constituent of many other foodstuffs, such as potatoes, rice, beans, corn, and peas.

Photosynthesis and Carbohydrate Metabolism

Carbohydrates are synthesized in green plants through a process called *photosynthesis;* a process that uses solar energy to reduce, or "fix," carbon dioxide. The overall equation for photosynthesis can be written as follows:

$$x CO_2 + y H_2O + \text{solar energy} \longrightarrow \underset{\text{Carbohydrate}}{C_x(H_2O)_y} + x O_2$$

Many individual enzyme-catalyzed reactions take place in the general photosynthetic process and not all are fully understood. We know, however, that photosynthesis begins with the absorption of light by the important green pigment of plants, chlorophyll (Fig. 24.1). The green color of chlorophyll and, therefore, its ability to absorb sunlight in the visible region, is due primarily to its extended conjugated system. As photons of sunlight are trapped by chlorophyll, energy

FIG. 24.1

*Chlorophyll-*a. (*The structure of chlorophyll-*a *was established largely through the work of H. Fischer, R. Willstätter, and J. B. Conant. A synthesis of chlorophyll-*a *from simple organic compounds was achieved by R. B. Woodward in 1960.*)

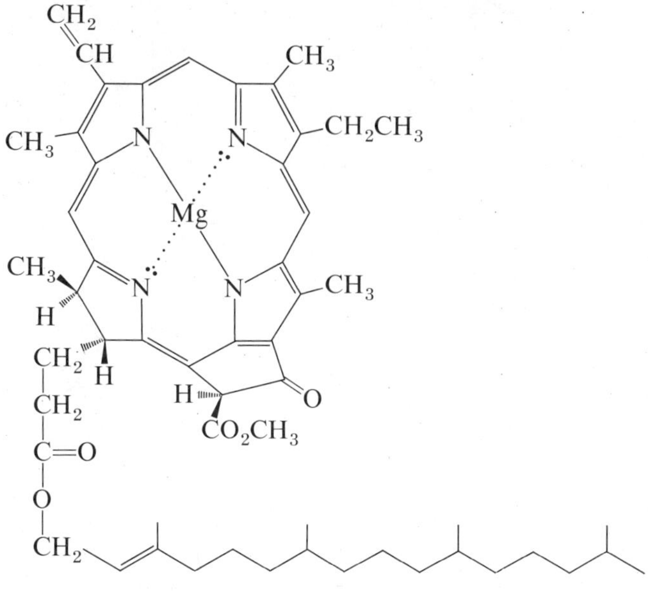

becomes available to the plant in a chemical form that can be used to carry out the reactions that reduce carbon dioxide to carbohydrates and oxidize water to oxygen.

Carbohydrates act as a major chemical repository for solar energy. Their energy is released when animals or plants metabolize carbohydrates to carbon dioxide and water.

$$C_x(H_2O)_y + xO_2 \longrightarrow xCO_2 + yH_2O + \text{energy}$$

The metabolism of carbohydrates also takes place through a series of enzyme-catalyzed reactions in which each energy-yielding step is an oxidation (or the consequence of an oxidation).

Although some of the energy released in the oxidation of carbohydrates is inevitably converted to heat, much of it is conserved in a new chemical form through reactions that are coupled to the synthesis of adenosine triphosphate (ATP) from adenosine diphosphate (ADP) and inorganic phosphate (P_i) (Fig. 24.2). The phos-

FIG. 24.2

The synthesis of adenosine triphosphate from adenosine diphosphate and (hydrogen) phosphate ion. This reaction takes place in all living organisms, and adenosine triphosphate is the major compound into which the chemical energy released by biological oxidations is transformed.

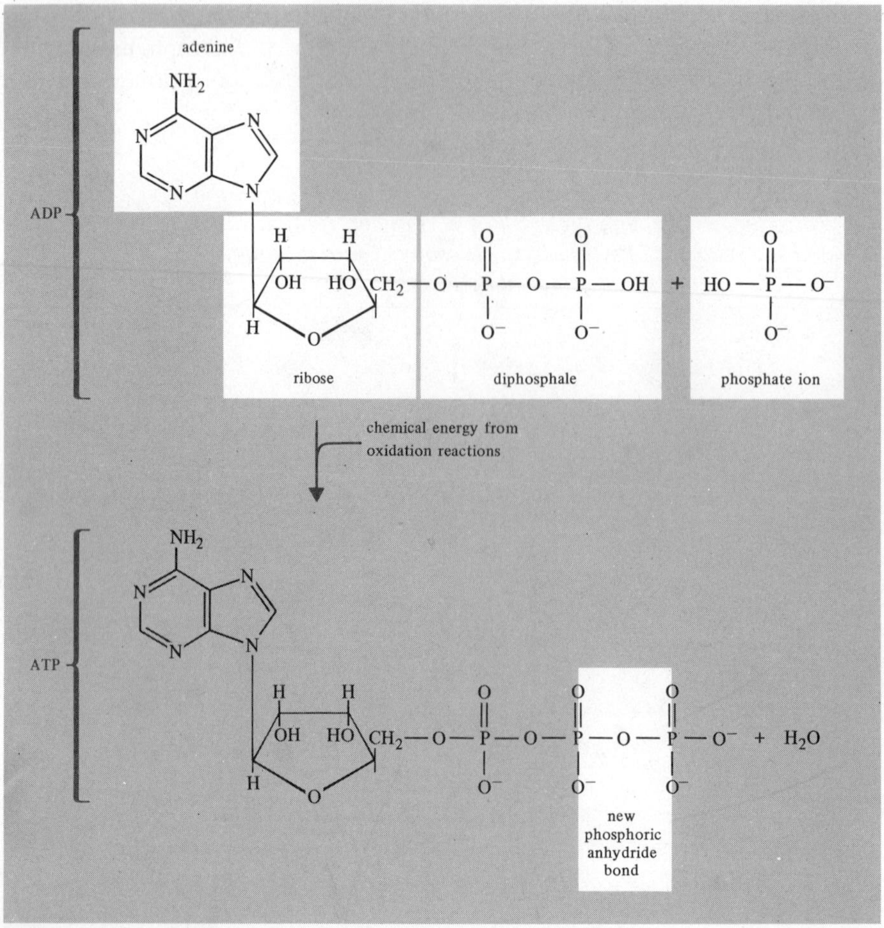

phoric acid anhydride bond that forms between the terminal phosphate group of ADP and the phosphate ion becomes another repository of chemical energy. Plants and animals can use the conserved energy of ATP to carry out all of their energy-requiring processes; the contraction of a muscle, the synthesis of a macromolecule, and so on. When the energy contained in ATP is utilized, a coupled reaction takes place in which ATP is hydrolyzed:

$$\text{ATP} + \text{H}_2\text{O} \xrightarrow[\text{energy}]{} \text{ADP} + \text{P}_i$$

or a new anhydride linkage is created:

$$\underset{\text{O}}{\overset{\overset{\displaystyle O}{\|}}{\text{R}-\text{C}-\text{OH}}} + \text{ATP} \longrightarrow \text{R}-\overset{\overset{\displaystyle O}{\|}}{\text{C}}-\text{O}-\overset{\overset{\displaystyle O}{\|}}{\underset{\underset{\displaystyle O^-}{|}}{\text{P}}}-\text{O}^- + \text{ADP}$$

Acyl phosphate

24.2 MONOSACCHARIDES

Classification of Monosaccharides

Monosaccharides are classified according to (1) the number of carbon atoms present in the molecule and (2) whether they contain an aldehyde or keto group. Thus, a monosaccharide containing three carbon atoms is called a *triose;* one containing four carbons is called a *tetrose;* one containing five carbons is a *pentose;* and one containing six carbons is a *hexose.* A monosaccharide containing an aldehyde group is called an *aldose;* and one containing a keto group is called a *ketose.* These two classifications are frequently combined: a four-carbon aldose, for example, is called an *aldotetrose,* a five-carbon ketose is called a *ketopentose.*

$$
\begin{array}{ll}
\text{CHO} & \text{CH}_2\text{OH} \\
| & | \\
\text{CHOH} & \text{C}=\text{O} \\
| & | \\
\text{CHOH} & \text{CHOH} \\
| & | \\
\text{CH}_2\text{OH} & \text{CHOH} \\
& | \\
& \text{CH}_2\text{OH}
\end{array}
$$

An aldotetrose   A ketopentose

Problem 24.1

How many chiral carbons are contained by the (a) aldotetrose and (b) ketopentose given above? (c) How many stereoisomers would you expect from each general structure?

D and L Designations of Monosaccharides

The simplest monosaccharides are the compounds glyceraldehyde and dihydroxyacetone (p. 916). Of these two compounds, only glyceraldehyde contains a chiral carbon.

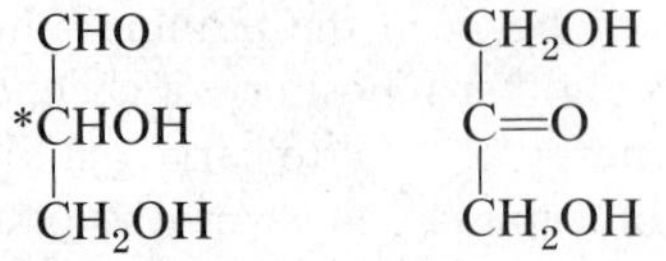

$$
\begin{array}{cc}
\text{CHO} & \text{CH}_2\text{OH} \\
| & | \\
{}^*\text{CHOH} & \text{C}{=}\text{O} \\
| & | \\
\text{CH}_2\text{OH} & \text{CH}_2\text{OH} \\
\text{Glyceraldehyde} & \text{Dihydroxyacetone} \\
\text{(an aldotriose)} & \text{(a ketotriose)}
\end{array}
$$

Glyceraldehyde exists, therefore, in two enantiomeric forms which are known to have the absolute configurations shown below.

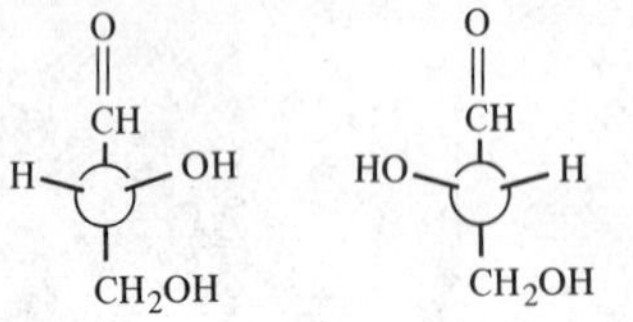

(+)-glyceraldehyde   (−)-glyceraldehyde

We saw on p. 267 that according to the Cahn-Ingold-Prelog convention, (+)-glyceraldehyde should be designated R-(+)-glyceraldehyde and (−)-glyceraldehyde should be designated S-(−)-glyceraldehyde.

Early in this century, before the absolute configurations of any organic compounds were known, another system of stereochemical designations was introduced. According to this system (first suggested by M. A. Rosanoff in 1906), (+)-glyceraldehyde is designated D-(+)-glyceraldehyde and (−)-glyceraldehyde is designated L-(−)-glyceraldehyde. These two compounds, moreover, serve as configurational standards for all monosaccharides: a monosaccharide *whose highest-numbered chiral carbon* has the same configuration as D-(+)-glyceraldehyde is designated as a D sugar; one whose highest-numbered chiral carbon has the same configuration as L-glyceraldehyde is designated as an L sugar.

D and L designations are like R and S designations in that they are not necessarily related to the optical rotations of the sugars to which they are applied.

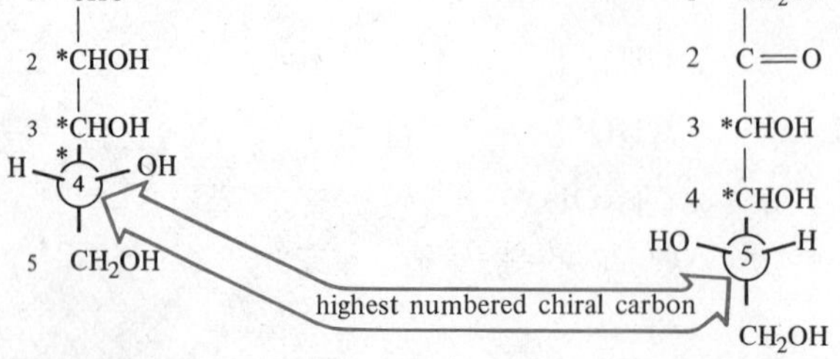

a D–aldopentose   an–L-ketohexose

Thus, one may encounter other sugars that are D-(+)- or D-(−)- and that are L-(+) or L-(−)-.

The D-L system of stereochemical designations is thoroughly entrenched in the literature of carbohydrate chemistry and even though it has the disadvantage of specifying the configuration of only one chiral center—that of the highest-numbered chiral center—we will employ the D-L system in our designations of carbohydrates.

Problem 24.2

Write three-dimensional formulas for each aldotetrose and ketopentose isomer in problem 24.1 and designate each as to whether it is a D or L sugar.

Structural Formulas for Monosaccharides

Later in this chapter we will see how the great carbohydrate chemist, Emil Fischer,* was able to establish the stereochemical configuration of the aldohexose D-(+)-glucose, the most abundant monosaccharide. In the meantime, however, we can use D-(+)-glucose as an example illustrating the various ways of presenting the structures of monosaccharides.

Fischer represented the structure of D-(+)-glucose with the cross formulation, **1**, in Fig. 24.3. This type of formulation is now called a Fischer projection formula and is still a useful one for carbohydrates. *When we use Fischer projection formulas, however, we must not* (in our mind's eye) *remove them from the plane of the page in order to test their superposability.* In terms of more familiar formulations, the Fischer projection formula translates into formulas **2** and **3**.†

In IUPAC nomenclature and using the Cahn-Ingold-Prelog system of stereochemical designations, the open-chain form of D-(+)-glucose is 2*R*, 3*S*, 4*R*, 5*R* -2,3,4,5,6-pentahydroxyhexanal.

Although many of the properties of D-(+)-glucose can be explained in terms of an open-chain structure (**1**, **2**, or **3**), there is a considerable body of evidence indicating that the open-chain structure exists, primarily, in equilibrium with two cyclic forms. These can be represented by structures **4** and **5** or **6** and **7**. The cyclic forms of D-(+)-glucose are *hemiacetals* that are formed by an intramolecular reaction of the —OH group at C-5 with the aldehyde group. Cyclization creates a new chiral carbon at C-1 and this explains how two cyclic forms are possible. These two cyclic forms are *diastereomers* that differ only in the configuration of C-1; in carbohydrate chemistry diastereomers of this type are called *anomers* and C-1 is called the *anomeric carbon.*

Structures **4** and **5** for the glucose anomer are called Haworth formulas‡ and although they do not give an accurate picture of the shape of the six-membered

---

* Emil Fischer (1852–1919) was professor of organic chemistry at the University of Berlin. In addition to monumental work in the field of carbohydrate chemistry, where Fischer and his co-workers established the configuration of most of the monosaccharides, Fischer also made important contributions to studies of amino acids, proteins, purines, indoles, and to stereochemistry generally. As a graduate student Fischer discovered phenylhydrazine, a reagent that was highly important in his later work with carbohydrates. Fischer was the second recipient (in 1902) of the Nobel Prize for Chemistry.

† The meaning of formulas **1, 2,** and **3** can be seen best through the use of molecular models: we first construct a chain of six-carbon atoms with the —CHO group at the top and a —CH_2OH group at the bottom. We then bring the CH_2OH group up behind the chain until it almost touches the —CHO group. Holding this model so that the —CHO and —CH_2OH group are directed generally away from us, we then begin placing —H and —OH groups on each of the four remaining carbons. The —OH group of carbon-2 is placed on the right; that of carbon-3 on the left; and those of carbons-4 and -5 on the right.

‡ After the English chemist W. N. Haworth (1883–1950) who, in 1926, along with E. L. Hirst, demonstrated that the cyclic form of glucose acetals consists of a six-membered ring. Haworth received the Nobel Prize for his work in carbohydrate chemistry in 1937.

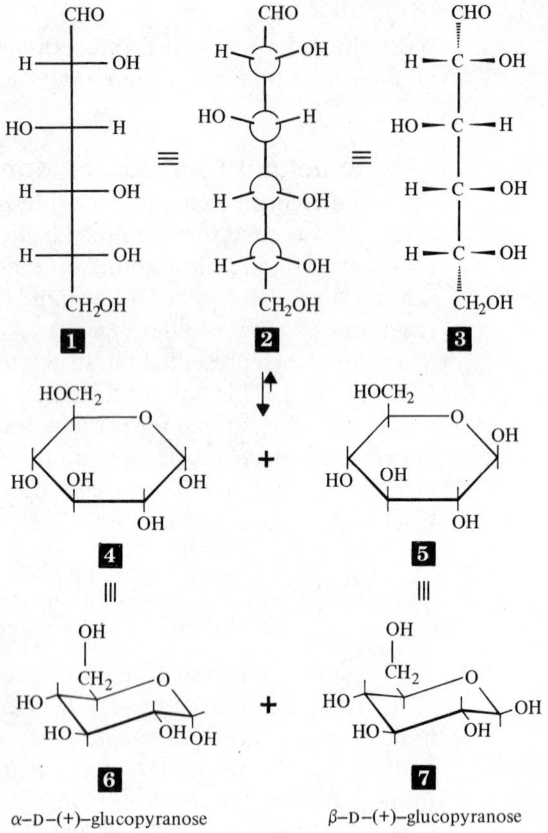

FIG. 24.3
(a) *1–3 are formulas used for the open-chain structure of* D-(+)-*glucose.* (b) *4–7 are formulas used for the two cyclic hemiacetal forms of* D-(+)-*glucose.*

α–D–(+)–glucopyranose   β–D–(+)–glucopyranose

ring, they have many practical uses. Figure 24.4 demonstrates how the representation of each chiral carbon of the open-chain form can be correlated with its representation in the Haworth formula.

The two glucose anomers are designated as an α *anomer* or a β *anomer* depending on the location of the —OH group of C-1. When we draw the cyclic forms of a D sugar in the orientation shown in Figs. 24.3 or 24.4, the α anomer has the —OH *down* and the β anomer has the —OH *up*.

Studies of the structures of the cyclic hemiacetal forms of D-(+)-glucose using X-ray analysis have demonstrated that the actual conformations of the rings are the chair forms represented by formulas **6** and **7** in Fig. 24.3. This is exactly what we would expect from our studies of the conformations of cyclohexane (Chapter 3), and it is especially interesting to notice that in the β anomer of D-glucose all of the large substituents, —OH or —CH₂OH, are equatorial. In the α anomer the only bulky axial substituent is the —OH at C-1.

It is convenient at times to represent the cyclic structures of a monosaccharide without specifying whether the configuration of the anomeric carbon is α or β. When we do this we will use formulas such as the following.

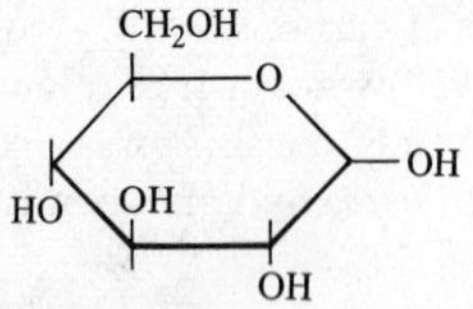

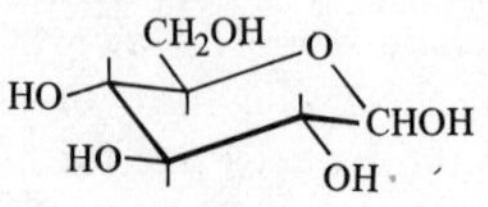

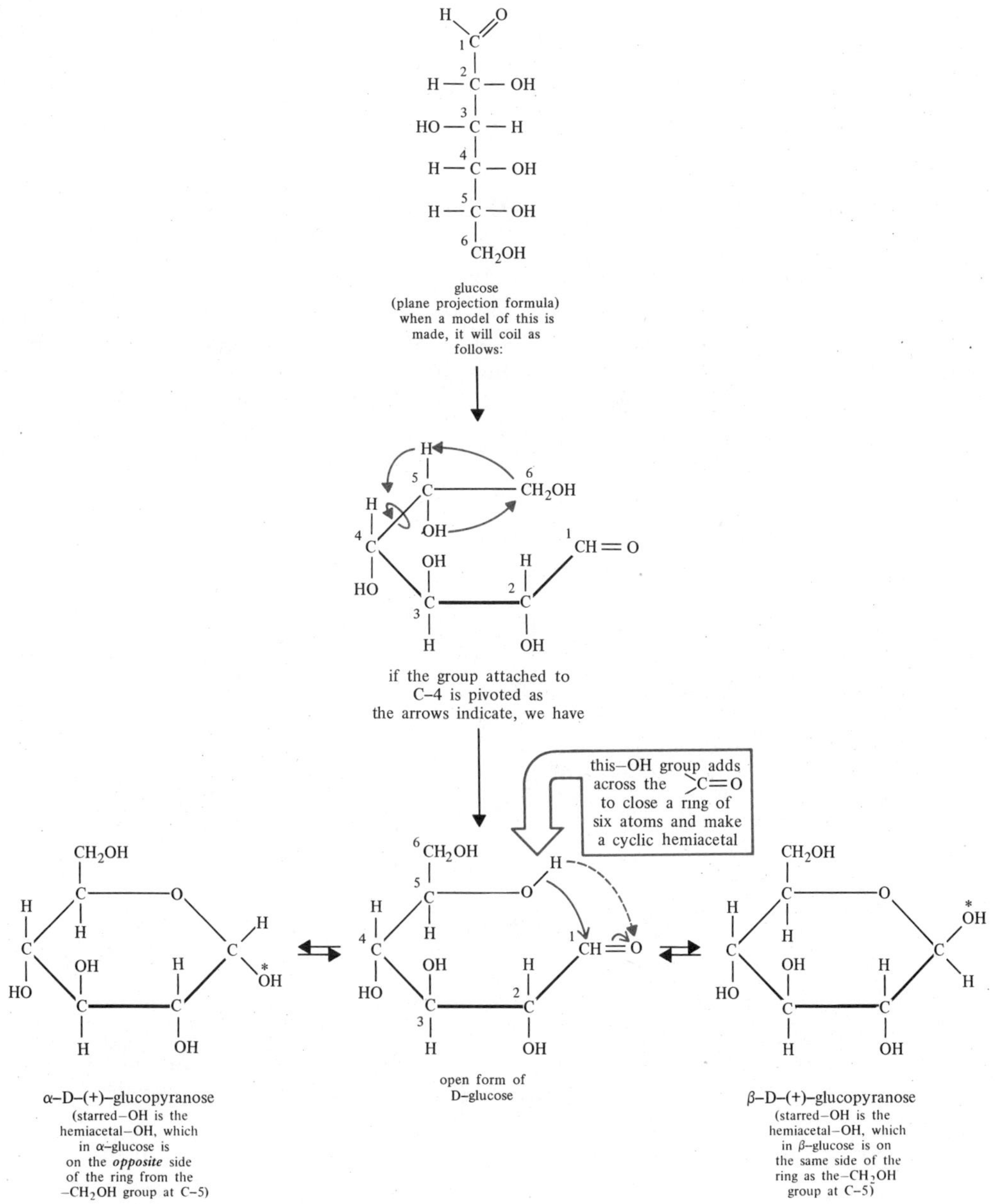

FIG. 24.4
The Haworth formulas for the cyclic hemiacetal forms of D-(+)-*glucose and their relation to the open-chain polyhydroxy aldehyde structure.* (*From John R. Holum,* Introduction to Organic and Biological Chemistry, *Wiley, New York, (1969), p. 282. Used by permission.*)

Not all carbohydrates exist in equilibrium with six-membered hemiacetal rings; in several instances the ring is five membered. (Even glucose exists, to a small extent in equilibrium with five-membered hemiacetal rings.) Because of this, a system of nomenclature has been introduced to allow designation of the ring size: if the monosaccharide ring is six membered, the compound is called a *pyranose;* if the ring is five membered the compound is designated as a *furanose*.* Thus, the full name of compound **4** (or **6**) is α-D-(+)-glucopyranose, while that of **5** (or **7**) is β-D-(+)-glucopyranose.

24.3 MUTAROTATION AND GLYCOSIDE FORMATION

Part of the evidence for the cyclic hemiacetal structure for D-(+)-glucose comes from experiments in which both forms have been isolated. Ordinary D-(+)-glucose has a melting point of 146°. However, when D-(+)-glucose is crystallized by evaporating an aqueous solution kept above 98°, a second form of D-(+)-glucose with a melting point of 150° can be obtained. When the optical rotations of these two forms are measured they are found to be significantly different, but when an aqueous solution of either form is allowed to stand, its rotation changes—the specific rotation of one form decreases and the rotation of the other increases—*until both solutions show the same value*. A solution of ordinary D-(+)-glucose (mp 146°) has an initial specific rotation of +112° but ultimately the specific rotation of this solution falls to +52.7°. A solution of the second form of D-(+)-glucose (mp 150°) has an initial specific rotation of +19°; but on standing, the specific rotation of this solution rises to +52.7°. This phenomenon is called *mutarotation*.

The explanation for mutarotation lies in the existence of an equilibrium between the open-chain form of D-(+)-glucose and the α and β forms of the cyclic hemiacetals.

α–D–(+)–glucopyranose
(mp 146°, $[\alpha]_D^{25}$ = +112°)

β–D–(+)–glucopyranose
(mp 150°, $[\alpha]_D^{25}$ = +18.7°)

X-ray analysis has confirmed that ordinary D-(+)-glucose has the α configuration at the anomeric carbon and that the higher melting form has the β configuration.

There is also evidence indicating that the amount of open-chain D-(+)-glucose in solution is very small. Solutions of D-(+)-glucose give no observable ultraviolet absorption band for a carbonyl group, and solutions of D-(+)-glucose give a negative

---

* These names come from the names of the oxygen heterocycles *pyran* and *furan* + *ose*.

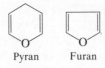

Pyran   Furan

test with Schiff's reagent—a special reagent that requires a relatively high concentration of a free aldehyde group (rather than a hemiacetal) in order to give a positive test.

Assuming that the concentration of the open-chain form is negligible, one can calculate the percentages of the α and β anomers present at equilibrium. These percentages, 36% α anomer and 64% β anomer, are in accord with a greater stability of β-D-($+$)-glacopyranose. This is what we might expect on the basis of its having an equatorial anomeric hydroxyl:

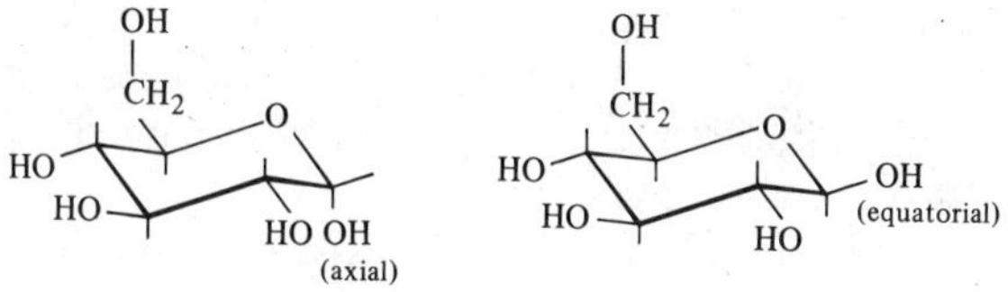

α–D–(+)–glucopyranose
(36% at equilibrium)

β–D–(+)–glucopyranose
(64% at equilibrium)

The β anomer of a pyranose is not always the more stable, however. With D-mannose the equilibrium favors the α anomer and this result—termed an *anomeric effect*—has been ascribed to an unfavorable interaction between the dipoles of the equatorial hydroxyl and the ring oxygen in the β anomer. The α anomer allows for maximum separation of the dipoles.

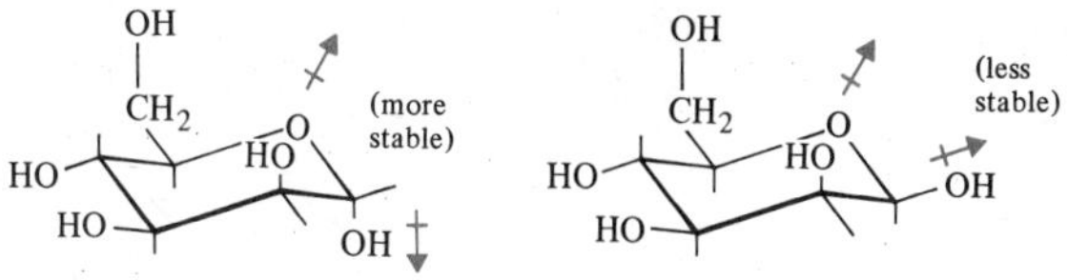

α–D–mannopyranose
(69% at equilibrium)

β–D–mannopyranose
(31% at equilibrium)

The anomeric effect is particularly important when the oxygen substituent at C-1 is an —OCH_3 or —$\overset{\overset{\displaystyle O}{\displaystyle \|}}{OCCH_3}$ group: in such compounds the α anomer is almost always the more stable.

When a small amount of gaseous hydrogen chloride is passed into a solution of D-($+$)-glucose in methanol, a reaction takes place that results in the formation of anomeric methyl *acetals:*

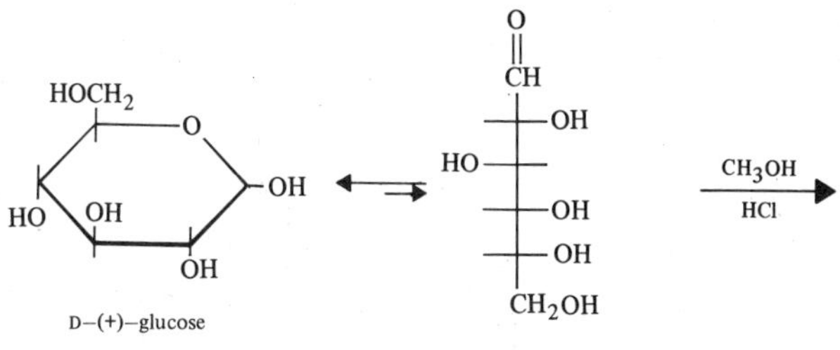

D–(+)–glucose

methyl α–D–glucopyranoside
(mp 165°, $[\alpha]_D^{25} = +158°$)

methyl β–D–glucopyranoside
(mp 107°, $[\alpha]_D^{25} = -33°$)

Carbohydrate acetals, generally, are called *glycosides* and an acetal of glucose is called a *glucoside*. (Acetals of mannose are *mannosides,* ketals of fructose are *fructosides,* and so on.) The methyl D-glucosides have been shown to have six-membered rings (p. 929 and 938) so they are properly named methyl α-D-gluco-pyranoside and methyl β-D-glucopyranoside.

Since glycosides are acetals, they do not exist in equilibrium with an open-chain form in neutral or basic aqueous media. Under these conditions (that is, in neutral or basic solutions) glycosides do not show mutarotation.

Problem 24.3

Glycosides *do* exhibit mutarotation in aqueous acid. Explain.

Problem 24.4

(a) Write conformational formulas for methyl α-D-glucopyranoside and (b) methyl β-D-glucopyranoside.

Problem 24.5

Write Haworth and conformational formulas for methyl α-D-mannopyranoside.

24.4 OXIDATION REACTIONS OF MONOSACCHARIDES

A number of oxidizing agents are used to identify functional groups of carbohydrates and in elucidating their structures. The most important are (1) Benedict's or Tollens' reagent, (2) bromine water, (3) nitric acid, and (4) periodic acid. Each of these reagents produces a different and usually specific effect when it is allowed to react with a monosaccharide; we should now examine what these are.

Benedict's or Tollens' Reagent: Reducing Sugars

Benedict's reagent (an alkaline solution containing a cupric citrate complex ion) and Tollens' solution ($Ag(NH_3)_2{}^+OH^-$) oxidize and thus give positive tests with aldehydes *and with α-hydroxy ketones.* Both reagents, therefore, give positive tests with aldoses *and ketoses.* This is true even though aldoses and ketoses exist primarily as cyclic hemiacetals.

We studied the use of Tollens' silver mirror test in Chapter 19. Benedict's solution (and the related Fehling's solution that contains a cupric tartrate complex ion) give brick-red precipitates of Cu_2O when they oxidize an aldehyde or α-hydroxy-ketone. Since the solutions of cupric tartrates and citrates are blue, the appearance of a brick-red precipitate is a vivid and unmistakable indication of a positive test.

$$
\begin{array}{c}
\overset{\textstyle O}{\underset{\textstyle \|}{}}\\
R-CH\\
\text{(aldehyde}\\
\text{or aldose)}
\end{array}
$$

$$
\underset{\substack{\text{Benedict's}\\\text{solution}\\\text{(blue)}}}{Cu^{++}\ (\text{complex})}\ +\quad
\underset{\substack{\text{(α-hydroxy ketone}\\\text{or ketose)}}}{\overset{\textstyle \text{or}}{R-\underset{\textstyle \underset{\textstyle O}{\|}}{C}CH_2OH}}\quad
\longrightarrow\quad
\underset{\text{(brick red)}}{Cu_2O}\ +\ \text{oxidation products}
$$

Sugars that give positive tests with Tollens' or Benedict's solutions are known as *reducing sugars,* and all carbohydrates that contain a *hemiacetal group* or a *hemiketal group* give positive tests. In aqueous solution these hemiacetals or hemiketals exist at equilibrium with relatively small, but not insignificant, amounts of noncyclic aldehydes or α-hydroxy ketones. It is the latter that undergo the oxidation, perturbing the equilibrium to produce more aldehyde or α-hydroxy ketone, which undergoes oxidation and so forth, until the test reagent is exhausted.

Carbohydrates that contain only acetal or ketal groups do not give positive tests with Benedict's or Tollens' solution and they are called *nonreducing sugars.* Acetals or ketals do not exist in equilibrium with aldehydes or α-hydroxy ketones in the basic aqueous media of the test reagents.

Reducing Sugar   Nonreducing Sugar

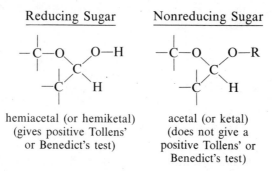

hemiacetal (or hemiketal)   acetal (or ketal)
(gives positive Tollens'   (does not give a
or Benedict's test)   positive Tollens' or
     Benedict's test)

Problem 24.6

How might you distinguish between α-D-glucopyranose (i.e., D-glucose) and methyl α-D-glucopyranoside?

Although Benedict's and Tollens' reagent have some use as diagnostic tools (Benedict's solution is often used in quantitative determinations of glucose in blood), none of these reagents is useful as a preparative reagent in carbohydrate oxidations. Oxidations with both reagents take place in alkaline solution, *and in alkaline solution sugars undergo a complex series of reactions that lead to isomerizations and fragmentations.*

Problem 24.7

If D-glucose is treated with aqueous calcium hydroxide and the solution is allowed to stand for several days, a mixture of products results, including D-mannose, D-fructose, glycolic aldehyde, and D-erythrose.

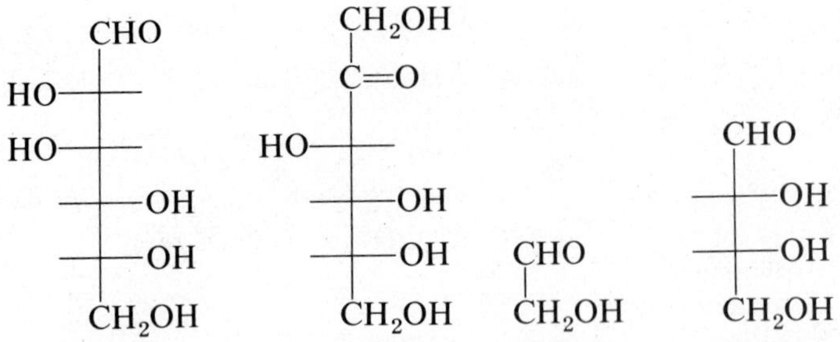

CHO — HO— — HO— — —OH — —OH — CH$_2$OH
D-Mannose

CH$_2$OH — C=O — HO— — —OH — —OH — CH$_2$OH
D-Fructose

CHO — CH$_2$OH
Glycolic aldehyde

CHO — —OH — —OH — CH$_2$OH
D-Erythrose

D-mannose results from the reversible formation of an enolate ion; D-fructose results from an enediol; and glycolic aldehyde and D-erythrose result from a *reverse* aldol addition. Using the open-chain forms of all the sugars involved, write mechanisms that explain the formation of each product.

Bromine Water: The Synthesis of Aldonic Acids

Monosaccharides do not undergo isomerization and fragmentation reactions in mildly acidic solution. Thus, a useful oxidizing reagent for preparative purposes is bromine in water (pH = 6.0). Bromine water is a general reagent that brings about the conversion of an aldose to an *aldonic acid*. That is, it selectively oxidizes the —CHO group to a —COOH group:

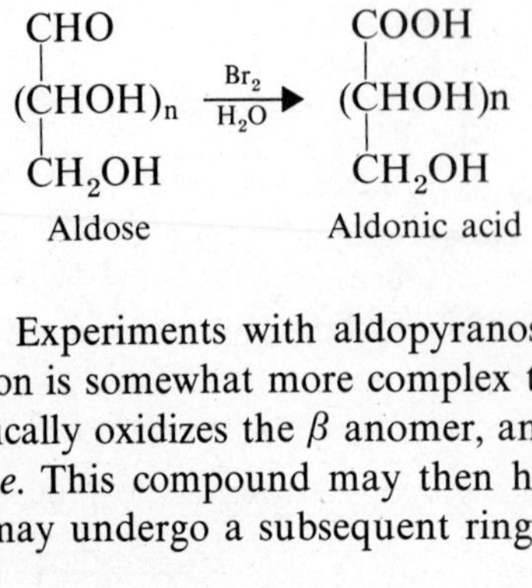

$$\underset{\text{Aldose}}{\overset{\text{CHO}}{\underset{\text{CH}_2\text{OH}}{(\text{CHOH})_n}}} \xrightarrow[\text{H}_2\text{O}]{\text{Br}_2} \underset{\text{Aldonic acid}}{\overset{\text{COOH}}{\underset{\text{CH}_2\text{OH}}{(\text{CHOH})n}}}$$

Experiments with aldopyranoses have shown that the actual course of the reaction is somewhat more complex than we have indicated above. Bromine water specifically oxidizes the β anomer, and the initial product that forms is a δ-*aldonolactone*. This compound may then hydrolyze to an aldonic acid, and the aldonic acid may undergo a subsequent ring closure to form a γ-*aldonolactone*.

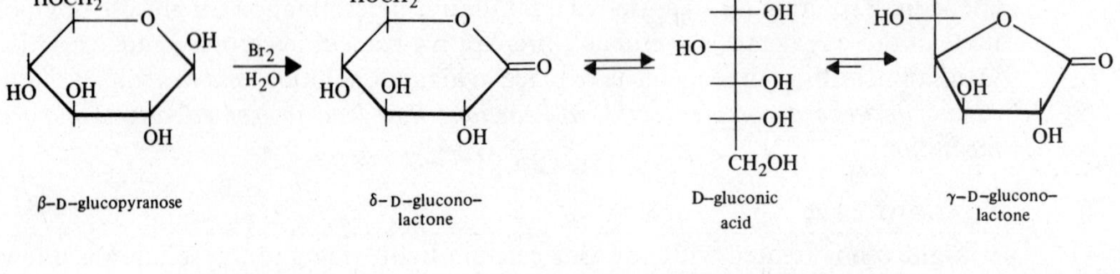

β-D-glucopyranose $\quad$ δ-D-glucono- lactone $\quad$ D-gluconic acid $\quad$ γ-D-glucono- lactone

Although α-aldopyranoses also undergo this oxidation, they react much more slowly. The rate of oxidation of β-D-glucopyranose, for example, is 250 times as fast as that of α-D-glucopyranose. The very slow rate of oxidation of α-D-glucopyranose reflects its conversion to the β anomer (followed by oxidation) rather than direct oxidation of the α anomer.

A mechanism that accounts for this behavior requires attack by Br^+ at the anomeric oxygen followed by loss of HBr from an intermediate in which the —H and —OBr groups are *antiparallel* to each other. Beta-anomers can achieve this stereochemistry easily.

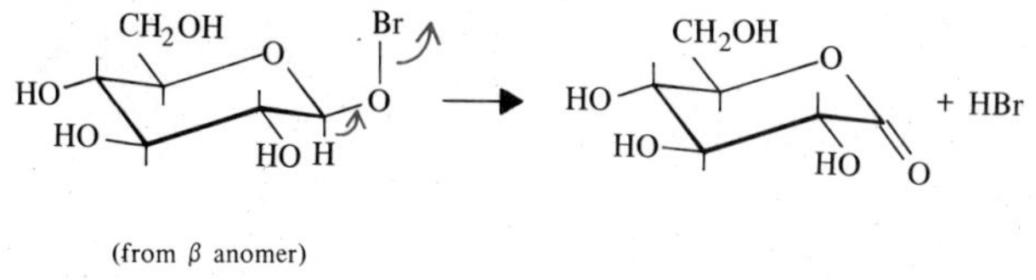

(from β anomer)

Antiparallel Arrangement

With α anomers the axial hydrogens at the 3 and 5 positions hinder an antiparallel arrangement:

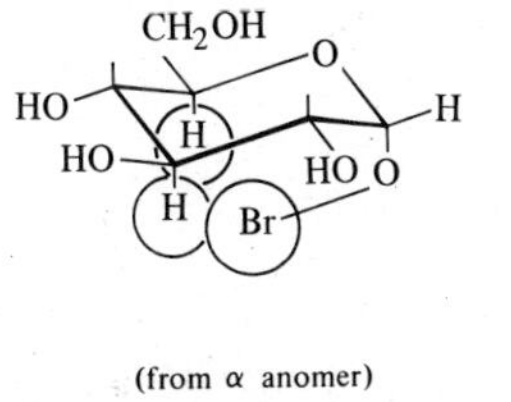

(from α anomer)

Antiparallel arrangement is hindered

Problem 24.8

Write an equation to illustrate the bromine-water oxidation of β-D-mannopyranose (cf. page 921). The name of the aldonic acid is D-mannonic acid and the names of the aldonolactones are δ- and γ-D-mannonolactone.

Nitric Acid Oxidation: Aldaric Acids

Dilute nitric acid—a stronger oxidizing agent than bromine water—oxidizes both the —CHO group and the terminal —CH_2OH group of an aldose to —COOH groups. These dicarboxylic acids are known as *aldaric acids*.

$$
\begin{array}{c}
CHO \\
| \\
(CHOH)n \\
| \\
CH_2OH
\end{array}
\xrightarrow{HNO_3}
\begin{array}{c}
COOH \\
| \\
(CHOH)n \\
| \\
COOH
\end{array}
$$

Aldose   Aldaric acid

It is not known whether a lactone is an intermediate in the oxidation of an aldose to an aldaric acid; however, aldaric acids form γ and δ lactones readily.

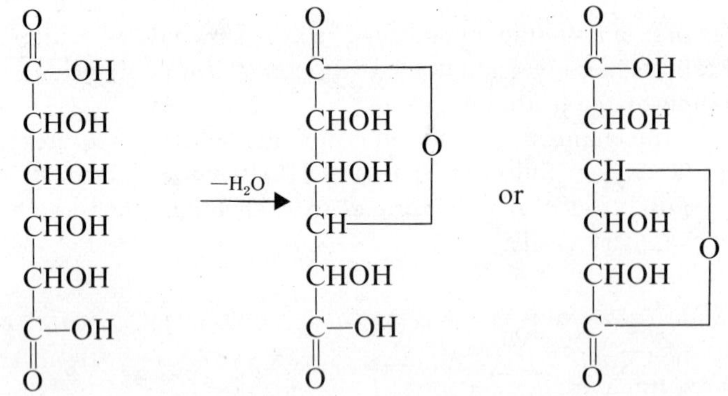

Aldaric acid
(from an aldohexose)

γ-Lactones of an aldaric acid

The aldaric acid obtained from D-glucose is called D-glucaric acid.*

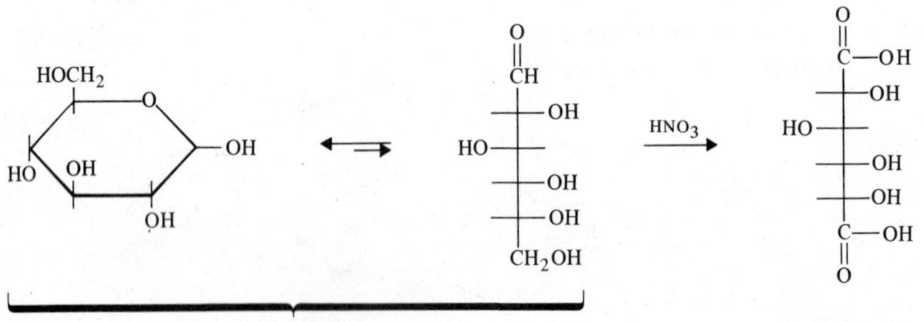

D-glucose

D-glucaric acid

Problem 24.9

(a) Would you expect D-glucaric acid to be optically active? (b) Write the open-chain structure for the aldaric acid (mannaric acid) that would be obtained by nitric acid oxidation of D-mannose. (c) Would you expect it to be optically active? (d) What aldaric acid would you expect to obtain from D-erythrose (cf. problem 24.7)? (e) Would it show optical activity? (f) D-Threose, a diastereomer of D-erythrose, yields an optically active aldaric acid when it is subjected to nitric acid oxidation. Write Fischer projection formulas for D-threose and its nitric acid oxidation product. (g) What are the names of the aldaric acids obtained from D-erythrose and D-threose? (See p. 270.)

Problem 24.10

D-Glucaric acid undergoes lactonization to yield two different γ lactones. What are their structures?

Periodate Oxidations: Oxidative Cleavage of Polyhydroxy Compounds

Compounds that have hydroxyl groups on adjacent atoms undergo oxidative cleavage when they are treated with aqueous periodic acid, HIO_4. The reaction

* Older terms for an aldaric acid are a *glycaric* acid or a *saccharic* acid.

breaks carbon-carbon bonds and produces carbonyl compounds (aldehydes, ketones, or acids). The stoichiometry of the reaction is:

$$-\overset{|}{\underset{|}{C}}-OH$$
$$\text{-----|------- } + HIO_4 \longrightarrow 2-\overset{|}{C}=O + HIO_3 + H_2O$$
$$-\overset{|}{\underset{|}{C}}-OH$$

Since the reaction usually takes place in quantitative yield, valuable information can often be gained by measuring the number of moles of periodic acid that are consumed in the reaction as well as by identifying the carbonyl products.*

Periodate oxidations are thought to take place through a cyclic intermediate:

$$-\overset{|}{\underset{|}{C}}-OH \quad \xrightarrow{H_2O} \quad \cdots \quad \longrightarrow \quad -\overset{|}{C}=O \quad + IO_3^-$$
$$-\overset{|}{\underset{|}{C}}-OH \quad + IO_4^- \qquad \qquad \qquad -\overset{|}{C}=O$$

Before we discuss the use of periodic acid in carbohydrate chemistry we should illustrate the course of the reaction with several simple examples. Notice in these periodate oxidations that *for every C—C bond broken, a C—O bond is formed at each carbon.*

1. When 1,2-propanediol is subjected to this oxidative cleavage the products are formaldehyde and acetaldehyde.

$$\begin{array}{ccc} H & & H \\ | & & | \\ H-\overset{|}{C}-OH & & H-C=O \quad \text{(formaldehyde)} \\ \text{-----|------} \; + IO_4^- \longrightarrow & & + \\ H-\overset{|}{C}-OH & & H-C=O \quad \text{(acetaldehyde)} \\ | & & | \\ CH_3 & & CH_3 \end{array}$$

1,2-Propanediol

2. When three or more —CHOH groups are contiguous, the internal ones

$$\begin{array}{ccc} & & H \\ H & & | \\ | & & H-C=O \\ H-\overset{|}{C}-OH & & + \qquad \text{(formaldehyde)} \\ \text{-----|------} & & O \\ & & \| \\ H-\overset{|}{C}-OH + 2IO_4^- \longrightarrow & H-C-OH \quad \text{(formic acid)} \\ \text{-----|------} & & + \\ H-\overset{|}{C}-OH & & H-C=O \quad \text{(formaldehyde)} \\ | & & | \\ H & & H \end{array}$$

---

*The reagent, lead tetraacetate ($Pb(OOCCH_3)_4$), brings about cleavage reactions similar to those of periodic acid. The two reagents are complementary; periodic acid works well in aqueous solutions and lead tetraacetate gives good results in organic solvents.

are obtained as *formic acid*. Periodate oxidation of glycerol, for example, gives two moles of formaldehyde and one mole of formic acid.

3. Oxidative cleavage also takes place when an —OH group is adjacent to the carbonyl group of an aldehyde or ketone. Glyceraldehyde yields two moles of formic acid and one mole of formaldehyde,

$$
\begin{array}{ccc}
\overset{\displaystyle O}{\overset{\displaystyle \|}{\underset{|}{C}}}\text{—H} & & \overset{\displaystyle O}{\overset{\displaystyle \|}{H\text{—}C}}\text{—OH} \quad \text{(formic acid)} \\
\text{------|------} & & + \\
H\text{—}\underset{|}{C}\text{—OH} \;+\; 2IO_4^- \longrightarrow & \overset{\displaystyle O}{\overset{\displaystyle \|}{H\text{—}C}}\text{—OH} \quad \text{(formic acid)} \\
\text{------|------} & & + \\
H\text{—}\underset{|}{C}\text{—OH} & & H\text{—}\underset{|}{C}\text{=O} \quad \text{(formaldehyde)} \\
H & & H
\end{array}
$$

Glyceraldehyde

while dihydroxyacetone gives two moles of formaldehyde and one mole of carbon dioxide.

$$
\begin{array}{ccc}
H & & H \\
H\text{—}\underset{|}{C}\text{—OH} & & H\text{—}\underset{|}{C}\text{=O} \quad \text{(formaldehyde)} \\
\text{------|------} & & + \\
\underset{|}{C}\text{=O} \;+\; 2IO_4^- \longrightarrow & O\text{=}C\text{=O} \quad \text{(carbon dioxide)} \\
\text{------|------} & & + \\
H\text{—}\underset{|}{C}\text{—OH} & & H\text{—}\underset{|}{C}\text{=O} \quad \text{(formaldehyde)} \\
H & & H
\end{array}
$$

4. Periodic acid does not cleave compounds in which the hydroxyl groups are separated by an intervening —CH_2— group, nor those in which a hydroxyl group is adjacent to an ether or acetal function.

$$
\begin{array}{l}
CH_2OH \\
| \\
CH_2 \quad + \; IO_4^- \longrightarrow \; \text{No reaction} \\
| \\
CH_2OH
\end{array}
$$

$$
\begin{array}{l}
CH_2OCH_3 \\
| \\
CHOH \quad + \; IO_4^- \longrightarrow \; \text{No reaction} \\
| \\
CH_2R
\end{array}
$$

Problem 24.11

What products would you expect to be formed when each of the following compounds is treated with an appropriate amount of periodic acid? How many moles of HIO_4 would be consumed in each case?

(a) 2,3-Butanediol

(b) 1,2,3-Butanetriol

(c) $CH_2OHCHOHCH(OCH_3)_2$

(d) $CH_2OHCHOHCOCH_3$

(e) $CH_3COCHOHCOCH_3$

(f) *cis*-Cyclopentanediol

(g) $CH_3\overset{\overset{\displaystyle CH_3}{|}}{\underset{\underset{\displaystyle OH}{|}}{C}}-\overset{}{\underset{\underset{\displaystyle OH}{|}}{CH_2}}$

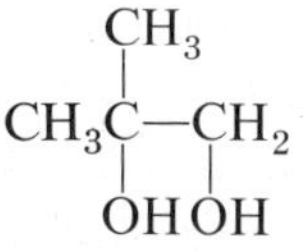

(h) D-Erythrose

Problem 24.12

Show how periodic acid could be used to distinguish between an aldohexose and a ketohexose. What products would you obtain from each and how many moles of HIO_4 would be consumed?

In 1936 E. L. Jackson and C. S. Hudson used periodate oxidations in an elegant procedure for determining the ring size of glycosides. Their method also permits us to demonstrate that different glycosides of the α-D-hexose series have the same configurations at C-1 and at C-5. This method is outlined in Fig. 24.5.

FIG. 24.5

Jackson and Hudson's method. Placing the —OCH₃ *group on the right in structure **1** indicates that the glycoside is an α anomer. The configuration of C-2, C-3, and C-4 is not specified, thus, **1** is any α-methyl-D-hexopyranoside.*

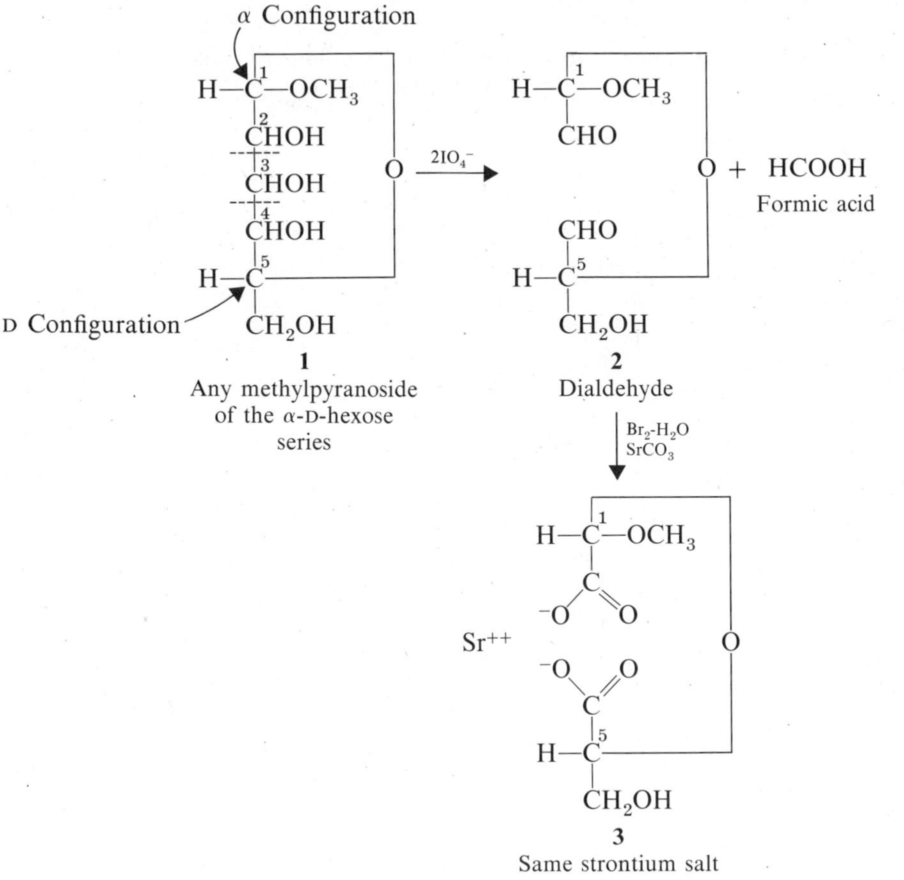

The fact that the methyl glycoside, **1**, consumes two moles of periodate ion and produces one mole of formic acid demonstrates that the glycoside contains a six-membered ring. The oxidation also eliminates three centers of chirality in **1**. (C-2 and C-4 are oxidized to aldehyde groups and C-3 is eliminated as formic acid.) The fact that all methyl pyranosides of the α-D-hexose series yield the same dialdehyde, **2**, as can be shown by conversion to the same strontium salt, **3**, demonstrates that they all have the same configuration at C-1 and C-5.

Problem 24.13

When methyl furanosides of the α-D-pentose series are subjected to Jackson and Hudson's procedure they, too, yield the same dialdehyde, **2**, and strontium salt, **3**. (a) Outline the reactions involved. (b) In what respect would you obtain a different result from applying Jackson and Hudson's method to a methyl-α-D-pentofuranoside and a methyl-α-D-hexopyranoside?

Problem 24.14

(a) What are compounds **4** and **5** below?

$$\text{Strontium salt } \mathbf{3} \xrightarrow[\text{H}_2\text{O}]{\text{H}^+} \begin{cases} \mathbf{4} \ (C_2H_2O_3) \\ \mathbf{5} \ (C_3H_6O_4) \end{cases}$$

(b) Bromine-water oxidation of (+)-glyceraldehyde also produces compound **5**. What is the special significance of this fact?

24.5 REDUCTION OF MONOSACCHARIDES: ALDITOLS

Aldoses (and ketoses) can be reduced with sodium borohydride to compounds called *alditols*.

$$\begin{array}{c} \text{CHO} \\ | \\ \text{(CHOH)}_n \\ | \\ \text{CH}_2\text{OH} \\ \text{Aldose} \end{array} \xrightarrow[\text{H}_2/\text{Pt}]{\underset{\text{or}}{\text{NaBH}_4}} \begin{array}{c} \text{CH}_2\text{OH} \\ | \\ \text{(CHOH)}_n \\ | \\ \text{CH}_2\text{OH} \\ \text{Alditol} \end{array}$$

Reduction of D-glucose, for example, yields D-glucitol.

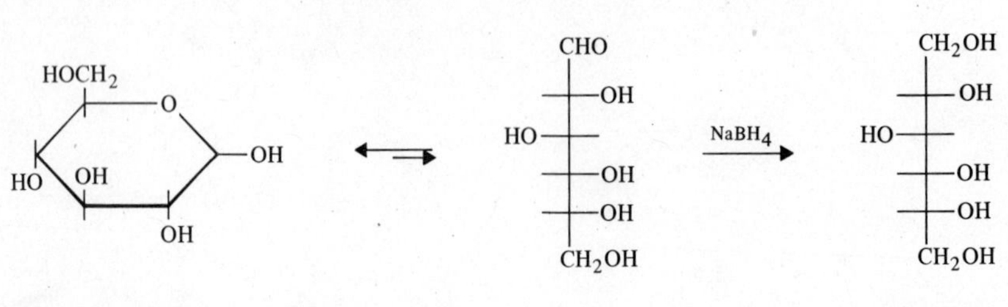

D–glucitol

Problem 24.15

(a) Would you expect D-glucitol to be optically active? (b) Write Fisher projection formulas for all of the D-aldohexoses that would yield *optically inactive alditols*.

24.6 REACTIONS OF MONOSACCHARIDES WITH PHENYLHYDRAZINE: OSAZONES

The aldehyde group of an aldose reacts with such carbonyl reagents as hydroxylamine and phenylhydrazine. With hydroxylamine, the product is the expected one: an oxime. With phenylhydrazine, however, three moles of phenylhydrazine are consumed and a second phenylhydrazone group is introduced at C-2. The product is called an *phenylosazone*.

$$
\begin{array}{c}
\text{H} \\
| \\
\text{C=O} \\
| \\
\text{CHOH} \\
| \\
(\text{CHOH})_n \\
| \\
\text{CH}_2\text{OH}
\end{array}
\; + \; 3\text{C}_6\text{H}_5\text{NHNH}_2 \longrightarrow
\begin{array}{c}
\text{H} \\
| \\
\text{C=NNHC}_6\text{H}_5 \\
| \\
\text{C=NNHC}_6\text{H}_5 \\
| \\
(\text{CHOH})n \\
| \\
\text{CH}_2\text{OH}
\end{array}
\; + \; \text{C}_6\text{H}_5\text{NH}_2 + \text{NH}_3 + \text{H}_2\text{O}
$$

Aldose   Phenylosazone

Although the mechanism for osazone formation is not known with certainty, it probably depends on a series of reactions in which $\diagdown\text{C=N}-$ behaves very much like $\diagdown\text{C=O}$.

$$
\begin{array}{c}
\text{H}^+ \\
\text{CH=N}-\text{NHC}_6\text{H}_5 \\
\text{H}-\text{C}-\text{OH} \\
|
\end{array}
\rightleftharpoons
\begin{array}{c}
\text{H}\quad\text{H} \\
\text{CH}-\text{N}-\text{N}-\text{C}_6\text{H}_5 \\
\text{C}-\text{O}-\text{H} \\
|
\end{array}
\xrightarrow[\text{C}_6\text{H}_5\text{NH}_2]{}
$$

$$
\begin{array}{c}
\text{CH=NH} \\
| \\
\text{C=O} \\
|
\end{array}
\xrightarrow{2\text{C}_6\text{H}_5\text{NHNH}_2}
\begin{array}{c}
\text{CH=NNHC}_6\text{H}_5 \\
| \\
\text{C=NNHC}_6\text{H}_5 \\
|
\end{array}
\; + \; \text{NH}_3 + \text{H}_2\text{O}
$$

Osazone formation results in a loss of chirality at C-2 but does not affect other chiral centers; D-glucose and D-mannose, for example, yield the same phenylosazone:

$$
\begin{array}{c}
\text{CHO} \\
\text{—OH} \\
\text{HO—} \\
\text{—OH} \\
\text{—OH} \\
\text{CH}_2\text{OH}
\end{array}
\xrightarrow{\text{C}_6\text{H}_5\text{NHNH}_2}
\begin{array}{c}
\text{CH=NNHC}_6\text{H}_5 \\
\text{C=NNHC}_6\text{H}_5 \\
\text{HO—} \\
\text{—OH} \\
\text{—OH} \\
\text{CH}_2\text{OH}
\end{array}
\xleftarrow{\text{C}_6\text{H}_5\text{NHNH}_2}
\begin{array}{c}
\text{CHO} \\
\text{HO—} \\
\text{HO—} \\
\text{—OH} \\
\text{—OH} \\
\text{CH}_2\text{OH}
\end{array}
$$

D-Glucose   Same phenylosazone   D-Mannose

This experiment (first done by Emil Fischer) establishes that D-glucose and D-mannose have the same configuration about C-3, C-4, and C-5. Diastereomeric aldoses (such as D-glucose and D-mannose) that differ only in configuration at C-2 are called *epimers*.*

Problem 24.16

Although D-fructose is not an epimer of D-glucose or D-mannose (D-fructose is a ketohexose), D-fructose does yield the same phenylosazone. (a) Using Fischer projection formulas, write an equation for this reaction. (b) What information about the stereochemistry of D-fructose does this experiment yield?

24.7 SYNTHESIS AND DEGRADATION OF MONOSACCHARIDES

Kiliani-Fischer Synthesis

In 1886 Heinrich Kiliani discovered that an aldose can be converted to the epimeric aldonic acids of next higher carbon number through the addition of hydrogen cyanide and subsequent hydrolysis of the epimeric cyanohydrins. Fischer later extended this method by showing that aldonolactones obtained from the aldonic acids can be reduced to aldoses. Today this method for lengthening the carbon chain of an aldose is called the Kiliani-Fischer synthesis.

We can illustrate the Kiliani-Fischer synthesis with the synthesis of D-threose and D-erythrose (aldotetroses) from D-glyceraldehyde (an aldotriose) on page 933.

Addition of hydrogen cyanide to glyceraldehyde produces two epimeric cyanohydrins because the reaction creates a new chiral center. The cyanohydrins can be separated easily (since they are diastereomers) and each can be converted to an aldose through hydrolysis, acidification, lactonization, and reduction. One cyanohydrin ultimately yields D-(−)-erythrose and the other yields D-(−)-threose.

We can be sure that the aldotetroses that we obtain from this Kiliani-Fischer synthesis are both D sugars because the starting compound is D-glyceraldehyde and its chiral carbon is unaffected by the synthesis. On the basis of the Kiliani-Fischer synthesis we cannot know just which aldotetrose has both —OH groups on the right and which has the top —OH on the left. However, if we oxidize both aldotetroses to aldaric acids, one [D-(−)-erythrose] will yield an *optically inactive* product while the other [D-(−)-threose] will yield a product that is *optically active* (cf. problem 24.9).

Problem 24.17

(a) What are the structures of L-(+)-threose and L-(+)-erythrose? (b) What aldotriose would you use to prepare them in a Kiliani-Fischer synthesis?

Problem 24.18

(a) Outline a Kiliani-Fischer synthesis of epimeric aldopentoses starting with D-(−)-erythrose (use Fischer projection formulas). (b) The two epimeric aldopentoses that one obtains are D-(−)-arabinose and D-(−)-ribose. Nitric acid oxidation

---

* The term epimer has taken on a broader meaning and is now often applied to any pair of diastereomers that differ only in the configuration of a single carbon.

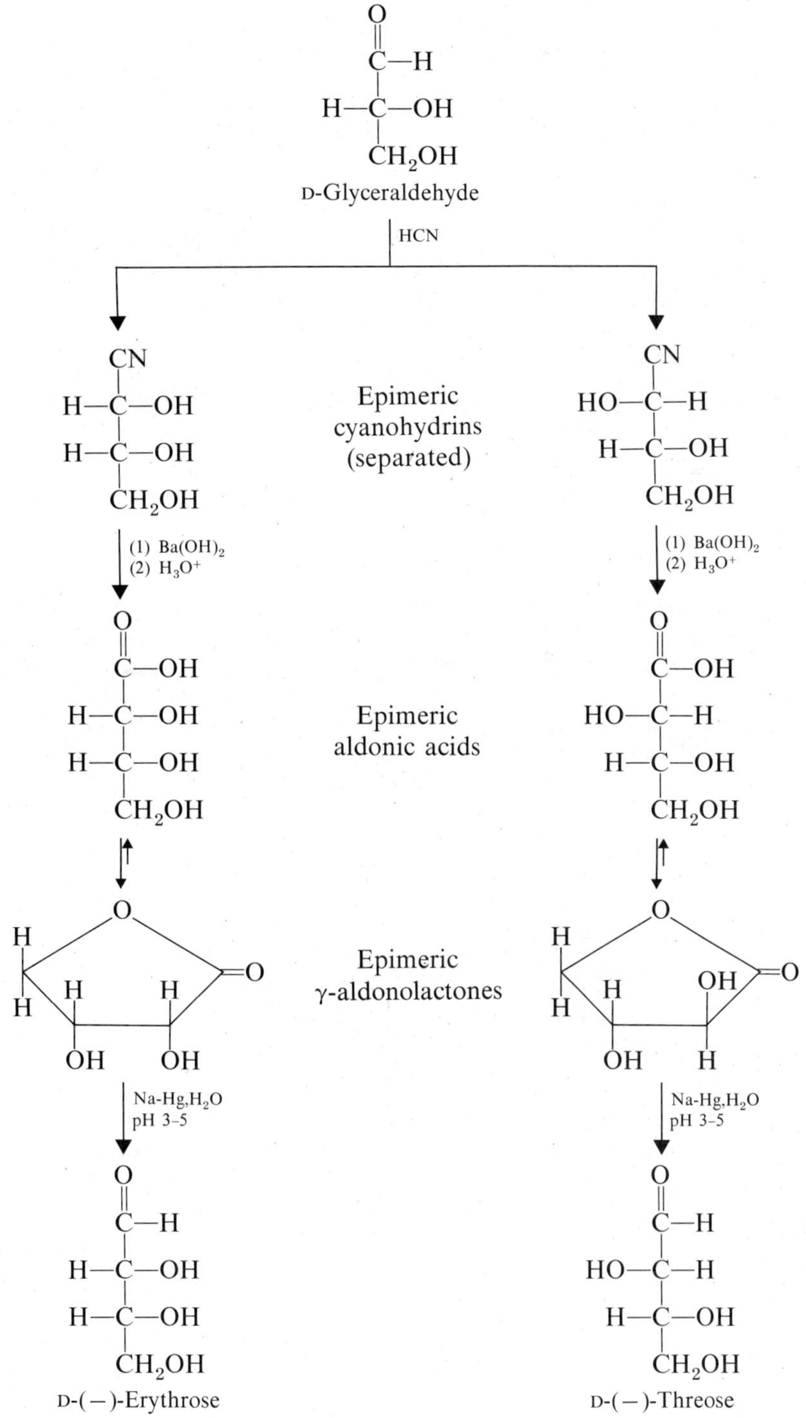

D-Glyceraldehyde

HCN

Epimeric cyanohydrins (separated)

(1) Ba(OH)$_2$
(2) H$_3$O$^+$

Epimeric aldonic acids

Epimeric γ-aldonolactones

Na-Hg,H$_2$O
pH 3–5

D-(−)-Erythrose

D-(−)-Threose

of D-(−)-ribose yields an optically inactive aldaric acid, while similar oxidation of D-(−)-arabinose yields an optically active product. On the basis of this information alone, which Fischer projection formula represents D-(−)-arabinose and which represents D-(−)-ribose?

Problem 24.19

Subjecting D-(−)-threose to a Kiliani-Fischer synthesis yields two other epimeric aldopentoses, D-(+)-xylose and D-(−)-lyxose. D-(+)-xylose can be oxidized (with nitric acid) to an optically inactive aldaric acid, while similar oxidation of D-(−)-lyxose gives an optically active product. What are the structures of D-(+)-xylose and D-(−)-lyxose?

Problem 24.20

There are a total of eight aldopentoses. In problems 24.18 and 24.19 you have arrived at the structures of four. What are the names and structures of the four that remain?

The Ruff Degradation

Just as the Kiliani-Fischer synthesis can be used to lengthen the chain of an aldose by one carbon, the Ruff degradation* can be used to shorten the chain by a similar unit. The Ruff degradation involves (1) oxidation of the aldose to an aldonic acid using bromine water and (2) oxidative decarboxylation of the aldonic acid to the next lower aldose using hydrogen peroxide and ferric sulfate. D-(−)-ribose, for example, can be degraded to D-(−)-erythrose:

D-(−)-Ribose   D-Ribonic acid   D-(−)-Erythrose

Problem 24.21

The aldohexose, D-(+)-galactose can be obtained by hydrolysis of *lactose*, a disaccharide found in milk. When D-(+)-galactose is treated with nitric acid it yields an optically inactive aldaric acid. When D-(+)-galactose is subjected to a Ruff degradation it yields D-(−)-lyxose (cf. problem 24.19). Using only these data, write the Fischer projection formula for D-(+)-galactose.

24.8 THE D-FAMILY OF ALDOSES

The Ruff degradation and the Kiliani-Fischer synthesis allow us to place all of the aldoses into families or "family trees" based on their relation to D- or L-glyceraldehyde. Such a tree is constructed in Fig. 24.6, and includes the structures of the D-aldohexoses, **1–8**.

---

* Developed by Otto Ruff, a German chemist, 1871–1939.

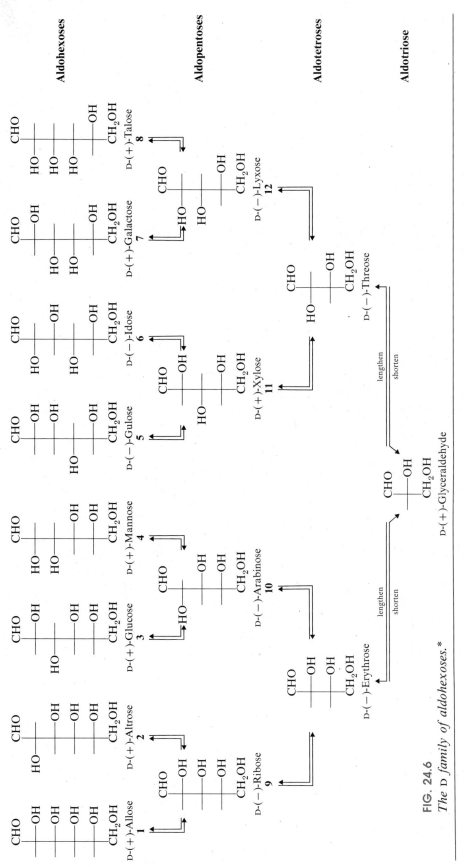

FIG. 24.6
*The D family of aldohexoses.**

* A useful mnemonic for the D-aldohexoses: "all altruists gladly make gum in gallon tanks. Write the names in a line and above each write CH_2OH. Then, for C_5 write OH to the right all the way across. For C_4 write OH to the right four times, then H; for C_3, OH, OH, then H, H, and repeat; C_2, alternate OH and H to the right." (L. F. Fieser and Mary Fieser, *Organic Chemistry*, Reinhold Publishing Corp., New York, 1956, p. 359.)

Most, but not all, of the naturally occurring aldoses belong to the D family with D-(+)-glucose being by far the most common. D-(+)-Galactose can be obtained from milk sugar (lactose); but L-(−)-galactose occurs in a polysaccharide obtained from the vineyard snail, *helix pomatia*. L-(+)-Arabinose is found widely, but D-(−)-arabinose is scarce, being found only in certain bacteria and sponges. Threose, lyxose, gulose, and allose do not occur naturally but one or both forms (D or L) of each have been synthesized.

24.9 FISCHER'S PROOF OF THE CONFIGURATION OF D-(+)-GLUCOSE

Emil Fischer began his work on the stereochemistry of (+)-glucose in 1888, only 12 years after van't Hoff and LeBel had made their proposal concerning the tetrahedral structure of carbon. Only a small body of data was available to Fischer at the beginning: only a few monosaccharides were known, including (+)-glucose, (+)-arabinose, and (+)-mannose. [(+)-Mannose had just been synthesized by Fischer.] The sugars (+)-glucose and (+)-mannose were known to be aldohexoses; (+)-arabinose was known to be an aldopentose.

Since an aldohexose has four chiral carbons, 2^4 (or 16) stereoisomers are possible—*one of which is* (+)-*glucose*. Fischer arbitrarily decided to limit his attention to the eight structures with the D configuration given in Fig. 24.6 (structures **1-8**). Fischer realized that he would be unable to differentiate between enantiomeric configurations because methods for determining the absolute configuration of organic compounds had not been developed. It was not until 1949, when Bijvoet (p. 268) determined the absolute configuration of D-(+)-tartaric acid (and, hence, D-(+)-glyceraldehyde) that Fischer's arbitrary assignment of (+)-glucose to the D family was known to be correct.

Fischer's assignment of structure **3** to (+)-glucose was based on the following reasoning.

1. Nitric acid oxidation of (+)-glucose gives an optically active aldaric acid. This eliminates structures **1** and **7** from consideration because both compounds would yield *meso*-aldaric acids.

2. *Degradation of* (+)-*glucose gives* (−)-*arabinose, and nitric acid oxidation of* (−)-*arabinose gives an optically active aldaric acid.* This means that (−)-arabinose cannot have configurations **9** or **11** and must have either structure **10** or **12**. It also establishes that (+)-glucose cannot have configuration **2, 5,** or **6**. This leaves structures **3, 4,** and **8** as possibilities for (+)-glucose.

3. A Kiliani-Fischer synthesis beginning with (−)-arabinose gives (+)-glucose and (+)-mannose; nitric acid oxidation of (+)-mannose gives an optically active aldaric acid. This together with the fact that (+)-glucose yields a different but also optically active aldaric acid establishes structure **10** as the structure of (−)-arabinose and eliminates structure **8** as a possible structure for (+)-glucose. Had (−)-arabinose been represented by structure **12** a Kiliani-Fischer synthesis would have given the two aldohexoses, **7** and **8**, one of which (**7**) would yield an optically inactive aldaric acid on nitric acid oxidation.

Two structures now remain, **3** and **4**; one structure represents (+)-glucose and one represents (+)-mannose. Fischer realized that (+)-glucose and (+)-mannose were epimeric (at C-2) but a decision as to which compound was represented by which structure was most difficult.

With brilliant logic, however, Fischer realized that if (+)-glucose has structure
3, *one other sugar,* L-gulose (structure **13** below), would yield the same aldaric acid
on nitric acid oxidation. On the other hand, if (+)-glucose has structure **4** its aldaric
acid cannot be obtained by oxidation of another aldohexose.

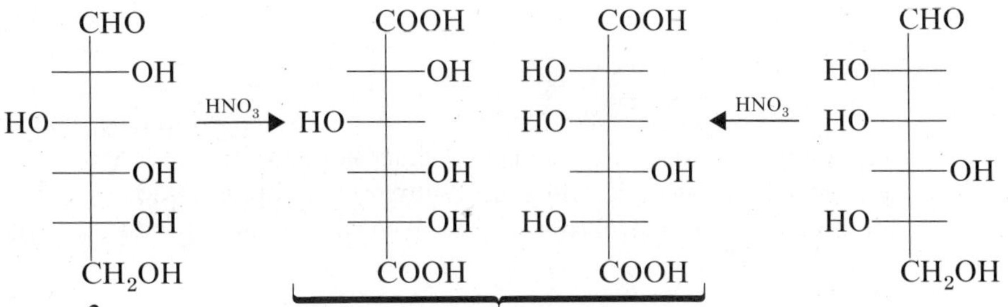

These are identical
(they can be superimposed by
rotating one structure 180°)

13
L-(+)-Gulose
(enantiomer of **5**)

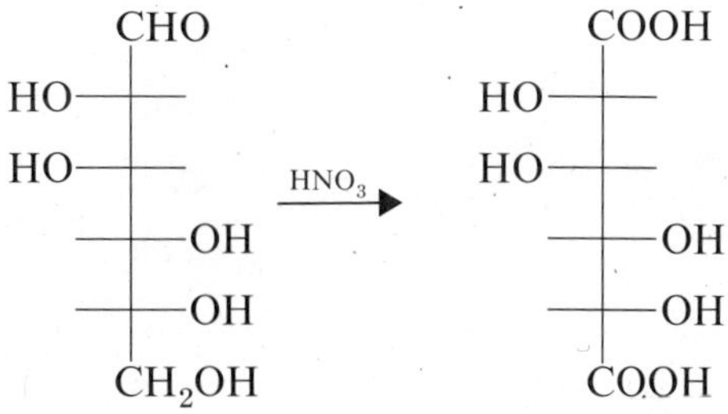

4

(No other aldohexose,
D or L, will yield
this aldaric acid)

The key compound, L-gulose, was unknown—so Fischer set out to synthesize
it.

4. *Fischer synthesized* L-(+)-*gulose,* **13**, *and showed that on oxidation it gave
the same aldaric acid as* (+)-*glucose.* This confirmed structure **3** as the correct
structure of (+)-glucose and established **4** as the structure of (+)-mannose. (It also
proved the structure of L-(+)-gulose.)

Fischer's synthesis of L-(+)-gulose began with one of the γ-lactones of
D-glucaric acid (cf., problem 24.10) and was carried out as follows:

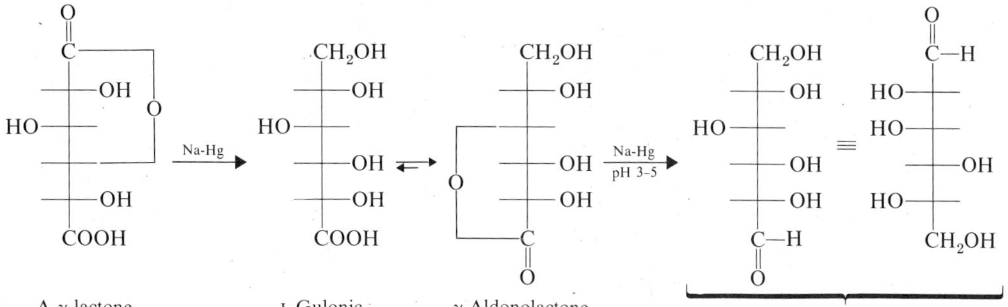

A γ-lactone
of D-glucaric
acid

L-Gulonic
acid

γ-Aldonolactone

L-(+)-Gulose

Problem 24.22

What product would result if the other γ-lactone of D-glucaric acid (problem 24.10) were subjected to the reaction sequence given above?

24.10 METHYLATION OF MONOSACCHARIDES

Since methyl glycosides are acetals they do not undergo isomerization or fragmentation in basic media. Thus, a methyl glucoside can be converted to a pentamethyl derivative by treating it with excess dimethyl sulfate in aqueous sodium hydroxide. Subsequent acid hydrolysis of the glycosidic methyl produces a tetramethyl-D-glucose.

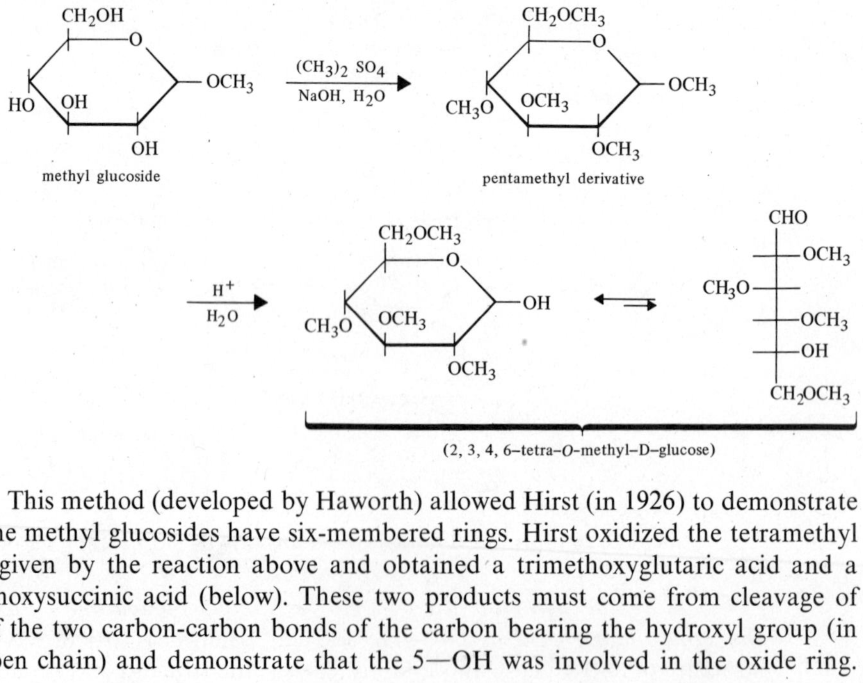

methyl glucoside

pentamethyl derivative

(2, 3, 4, 6–tetra–O–methyl–D–glucose)

This method (developed by Haworth) allowed Hirst (in 1926) to demonstrate that the methyl glucosides have six-membered rings. Hirst oxidized the tetramethyl ether given by the reaction above and obtained a trimethoxyglutaric acid and a dimethoxysuccinic acid (below). These two products must come from cleavage of one of the two carbon-carbon bonds of the carbon bearing the hydroxyl group (in the open chain) and demonstrate that the 5—OH was involved in the oxide ring.

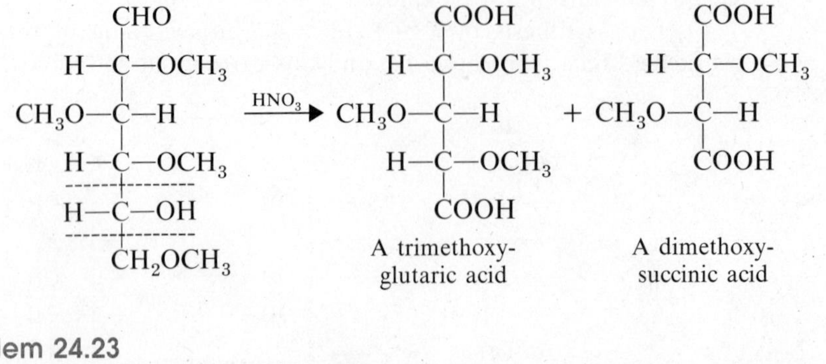

A trimethoxy-glutaric acid

A dimethoxy-succinic acid

Problem 24.23

What products would have been obtained from the Haworth-Hirst procedure if the methyl glucoside had been a furanoside?

Methylation procedures, as we will see in the next section, are especially important in deciphering the structures of disaccharides.

24.11 DISACCHARIDES

Sucrose

Ordinary table sugar is a disaccharide called *sucrose*. Sucrose, the most widely occurring disaccharide, is found in all photosynthetic plants and is obtained commercially from sugar cane or sugar beets. Sucrose has the structure shown in Fig. 24.7.

The structure of sucrose is based on the following evidence.

1. Sucrose has the molecular formula $C_{12}H_{22}O_{11}$.

2. Acid hydrolysis of sucrose yields one mole of D-glucose and one mole of D-fructose.

3. Sucrose is a nonreducing sugar; it gives negative tests with Benedict's and Tollens' solution. Sucrose does not form an osazone and does not undergo mutarotation. These facts mean that neither the glucose nor the fructose portions of sucrose has a hemiacetal or hemiketal group. Thus, the two hexoses must have a glycoside linkage that involves C-1 of glucose and C-2 of fructose; for only in this way will both carbonyl groups be present as full acetals (or glycosides).

4. The stereochemistry of the glycoside linkages can be inferred from experiments done with enzymes. Sucrose is hydrolyzed by an α-*glucosidase* obtained from yeast but not by β-glucosidases. This indicates *an α configuration at the glucoside portion*. Sucrose is also hydrolyzed by *sucrase,* an enzyme known to hydrolyze β-fructofuranosides but not α-fructofuranosides. This indicates *a β configuration at the fructoside portion.*

5. Methylation of sucrose gives an octamethyl derivative that, on hydrolysis, gives 2,3,4,6-tetra-*O*-methylglucose and 1,3,4,6-tetra-*O*-methylfructose. The identities of these two products demonstrate that the glucose portion is a *pyranoside* and that the fructose portion is a *furanoside*.

FIG. 24.7
Two representations of the formula for (+)-sucrose (α-D-gluco-pyranosyl β-D-fructofuranoside).

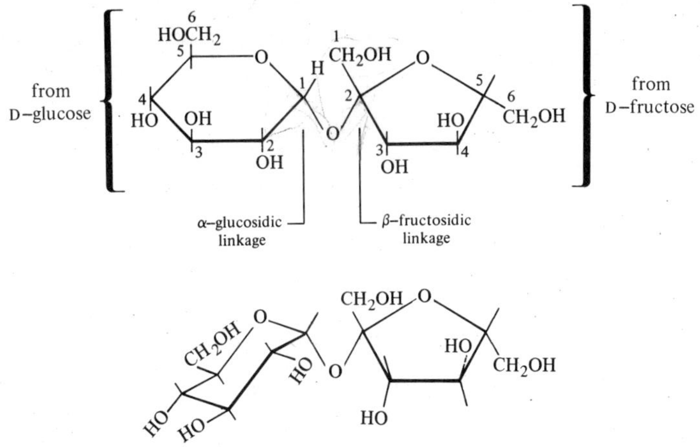

The structure of sucrose has been confirmed by X-ray analysis and by an unambiguous synthesis.

Maltose

When starch (p. 943) is hydrolyzed by the enzyme *diastase,* one product is a disaccharide known as *maltose* (Fig. 24.8).

1. When maltose is subjected to acid hydrolysis it yields two moles of D-(+)-glucose.

2. Unlike sucrose, *maltose is a reducing sugar;* it gives positive tests with Fehling's, Benedict's, and Tollens' solutions. Maltose also reacts with phenylhydrazine to form a monophenylosazone (i.e., it incorporates two molecules of phenylhydrazine).

3. Maltose exists in two anomeric forms; α-maltose, $[\alpha]_D^{25} = +168°$, and β-maltose, $[\alpha]_D^{25} = +112°$. The maltose anomers undergo mutarotation to yield an equilibrium mixture, $[\alpha]_D^{25} = +136°$.

Facts **2** and **3** demonstrate that one of the glucose residues of maltose is present in a hemiacetal form; the other, therefore, must be present as a glucoside. The configuration of this glucosidic linkage can be inferred as α, because maltose is hydrolyzed by α-glucosidases and not by β-glucosidases.

4. Maltose reacts with bromine water to form a monocarboxylic acid, maltonic acid (Fig. 24.9a). This fact, too, is consistent with the presence of only one hemiacetal group.

5. Methylation of maltonic acid followed by hydrolysis gives 2,3,4,6-tetra-*O*-methyl-D-glucose and 2,3,5,6-tetra-*O*-methyl-D-gluconic acid. That the first product has a free —OH at C-5 indicates that the nonreducing glucose portion is present as a pyranoside; that the second product, 2,3,5,6-tetra-*O*-methyl-D-gluconic acid, has a free —OH at C-4 indicates that this position was involved in a glycosidic linkage with the nonreducing glucose.

Only the size of the reducing glucose ring needs to be determined.

6. Methylation of maltose itself, followed by hydrolysis (Fig. 24.9b), gives

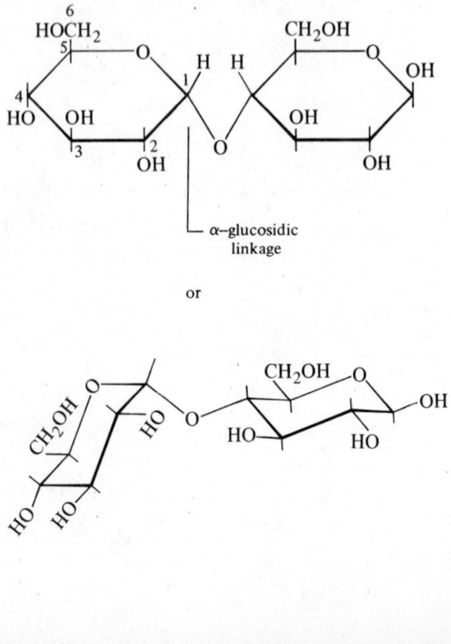

FIG. 24.8

Two representations of the structure of the β anomer of (+)-maltose, 4-O-(α-D-glucopyranosyl) β-D-glucopyranose.

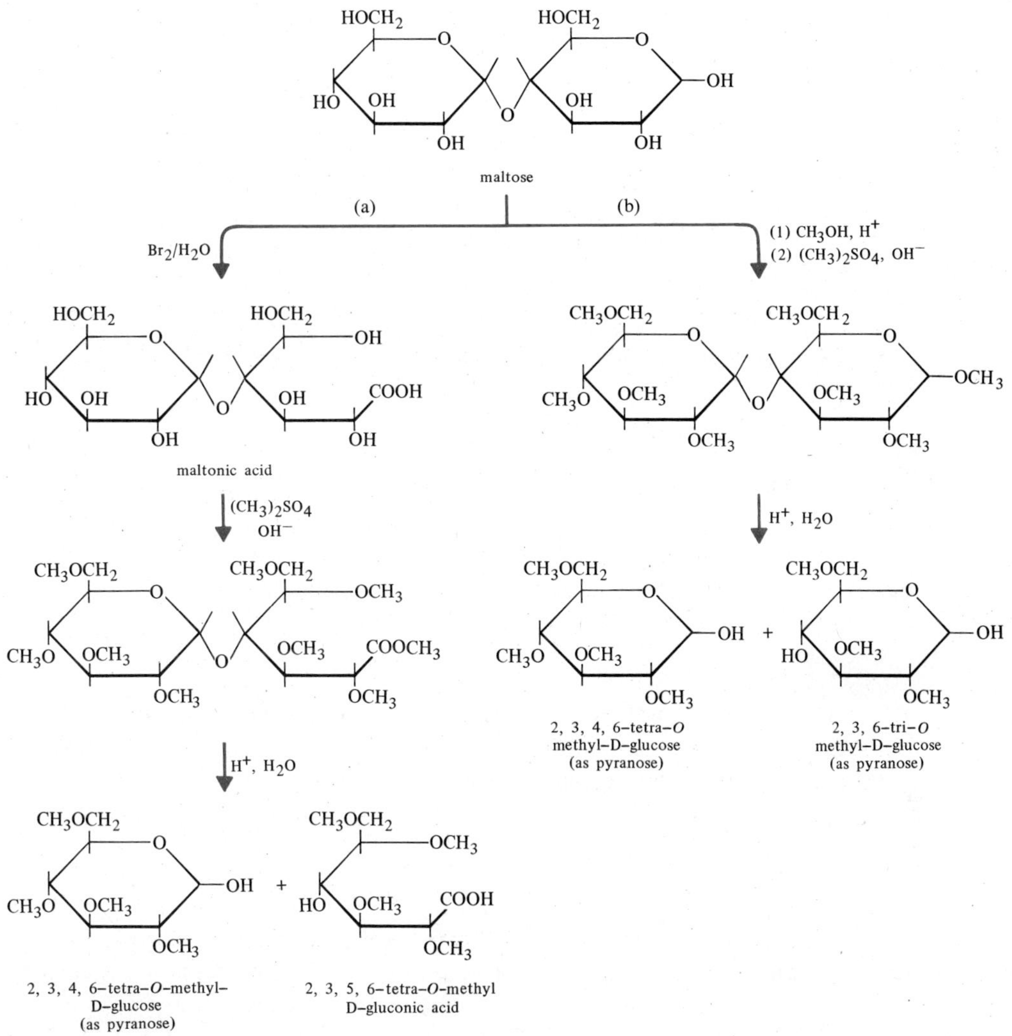

FIG. 24.9
(a) *Oxidation of maltose to maltonic acid followed by methylation and hydrolysis.* (b) *Methylation and subsequent hydrolysis of maltose itself.*

2,3,4,6-tetra-O-methyl-D-glucose and 2,3,6-tri-O-methyl-D-glucose. The free —OH at C-5 in the latter product indicates that it must have been involved in the oxide ring and that the reducing glucose is present as a *pyranose*.

Cellobiose

Partial hydrolysis of cellulose gives the disaccharide, cellobiose, $C_{12}H_{22}O_{11}$ (Fig. 24.10). Cellobiose resembles maltose in every respect except one: the configuration of its glycosidic linkage.

Cellobiose, like maltose, is a reducing sugar that, on acid hydrolysis, yields two moles of D-glucose. Cellobiose also undergoes mutarotation and forms a phenyl-

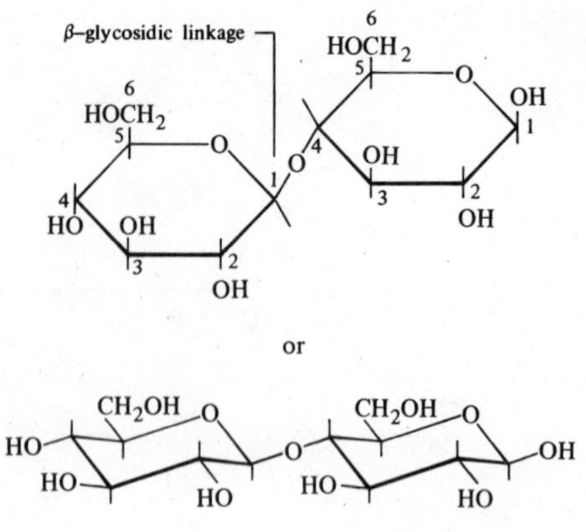

FIG. 24.10
Two representations of the β anomer of cel-lobiose, 4-O-(β-D-glucopyranosyl) β-D-glu-copyranose.

osazone. Methylation studies show that C-1 of one glucose unit is connected in glycosidic linkage with C-4 of the other and that both rings are six membered. Unlike maltose, however, cellobiose is hydrolyzed by *β-glucosidases* and not by α-glucosidases: this indicates that the glycosidic linkage in cellobiose is β (Fig. 24.10).

Lactose

Lactose (Fig. 24.11) is a disaccharide that is present in the milk of humans, cows, and almost all other mammals. Lactose is a reducing sugar that hydrolyzes to yield glucose and galactose; the glycosidic linkage is β.

24.12 POLYSACCHARIDES

Three important polysaccharides, all of which are polymers of D-glucose, are starch, glycogen, and cellulose. Starch is the principal food reserve of plants; glycogen

FIG. 24.11
Two representations of the β anomer of lactose, 4-O-(β-D-galac-topyranosyl) β-D-glucopyranose.

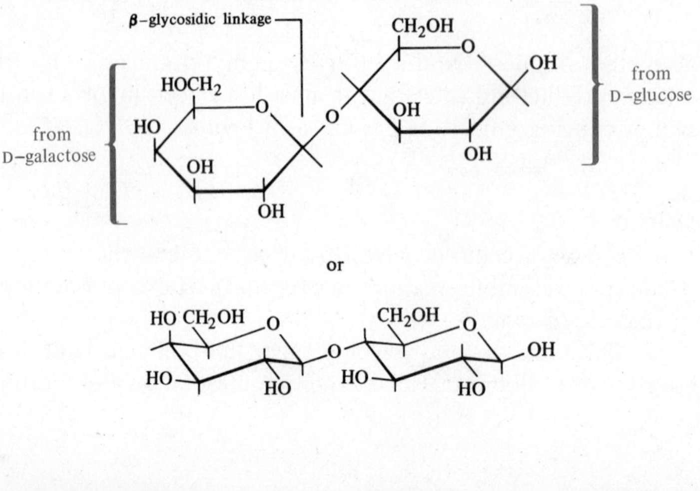

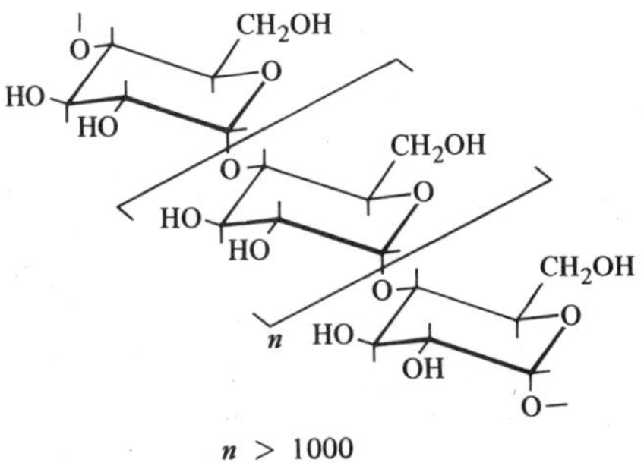

FIG. 24.12
Partial structure of amylose, an unbranched polymer of D-glucose connected in α, 1:4 glycosidic linkages.

functions as a carbohydrate reserve for animals; and cellulose serves as structural material in plants. As we examine the structures of these three polysaccharides we will be able to see how each is especially suited for its function.

Starch

Starch occurs as microscopic granules in the roots, tubers, and seeds of plants. Corn, potatoes, wheat, and rice are important commercial sources. Heating starch with water causes the granules to swell and produce a colloidal suspension from which two major components can be isolated. One fraction is called *amylose* and the other *amylopectin*. Most starches yield 10–20% amylose and 80–90% amylopectin.

Physical measurements show that amylose typically consists of more than 1000 D-glucopyranoside units *connected in α linkages* between C-1 of one unit and C-4 of the next (Fig. 24.12). Thus, in the ring size of its glucose units and in the configuration of the glycosidic linkages between them, amylose resembles maltose.

Chains of D-glucose units with α-glycosidic linkages such as those of amylose tend to assume a helical arrangement. This results in a compact shape for the amylose molecule even though its molecular weight is quite large (150,000–600,000).

Amylopectin has a structure similar to that of amylose (i.e., α, 1:4 links), with the exception that in amylopectin the chains are branched. Branching takes place between C-6 of one glucose unit and C-1 of another and occurs at intervals of 20–25 glucose units (Fig. 24.13). Physical measurements indicate that amylopectin has a molecular weight of one to six million; thus amylopectin consists of hundreds of interconnecting chains of 20–25 glucose units each.

Glycogen

Glycogen has a structure very much like that of amylopectin; however, in glycogen the chains are much more highly branched. Methylation and hydrolysis of glycogen indicates that there is one end group for every 10 to 12 glucose units; branches may occur as often as every 6. Glycogen has a very high molecular weight. Studies of glycogens isolated under conditions that minimize the likelihood of hydrolysis indicate molecular weights as high as 100 million.

The size and structure of glycogen beautifully suit its function as reserve carbohydrate for animals. First, its size makes it too large to diffuse across cell membranes; thus, glycogen remains inside the cell where it is needed as an energy source. Second, because glycogen incorporates tens of thousands of glucose units

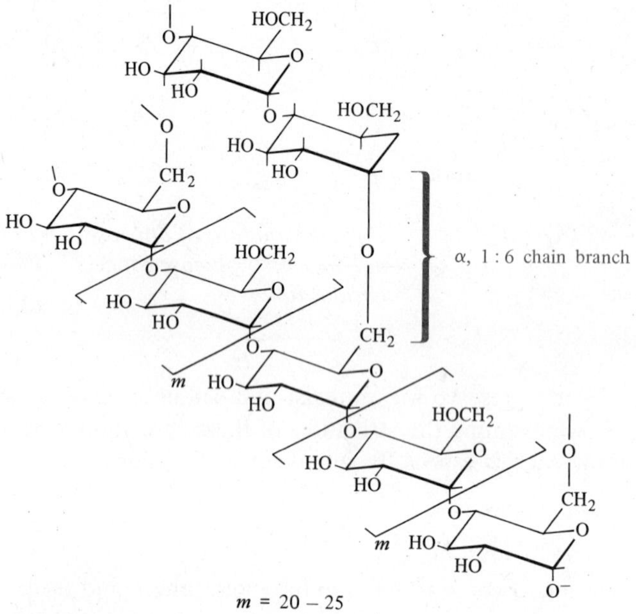

FIG. 24.13
Partial structure of amylopectin.

in a single molecule it solves an important osmotic problem for the cell. Were so many glucose units present in the cell as individual molecules, the osmotic pressure within the cell would be enormous—so large that the cell membrane would almost certainly break.* Finally, the localization of glucose units within a large, highly branched structure simplifies one of the cell's logistical problems: that of having a ready source of glucose when cellular glucose concentrations are low and of being able to store glucose rapidly when cellular glucose concentrations are high. There are enzymes within the cell that catalyze the reactions by which glucose units are detached from (or attached to) glycogen. These enzymes operate at end groups by hydrolyzing (or forming) $\alpha,1:4$ glycosidic linkages. Because glycogen is so highly branched, a very large number of end groups are available at which these enzymes can operate. At the same time the overall concentration of glycogen (in moles per liter) is quite low because of its enormous molecular weight.

Amylopectin presumably serves a similar function in plants. The fact that amylopectin is less highly branched than glycogen is, however, not a serious disadvantage. Plants have a much lower metabolic rate than animals—and plants, of course, do not require sudden bursts of energy.

Animals also store energy as fats (triacylglycerols) as well as glycogen. Fats,

---

* The phenomenon of osmotic pressure occurs whenever two solutions of different concentrations are separated by a membrane that will allow penetration (osmosis) of the solvent but not of the solute. The osmotic pressure, π, on one side of the membrane is related to the number of moles of solute particles, n, the volume of the solution, V, and RT (the gas constant times the absolute temperature):

$$\pi V = n\,RT$$

because they are more highly reduced, are capable of furnishing much more energy; metabolism of a typical fatty acid, for example, liberates more than twice as much energy per carbon as glucose or glycogen. Why then, we might ask, has Nature developed two different energy repositories? Glucose (from glycogen) is readily available and is highly water soluble.* Glucose, as a result, diffuses rapidly through the aqueous medium of the cell and serves as an ideal source of "ready energy." Long-chain fatty acids, by contrast, are almost insoluble in water and their concentration inside the cell could never be very high. They would be a poor source of energy if the cell were in an energy pinch. On the other hand, fatty acids, (as triacylglycerols) because of their caloric richness are an excellent energy repository for long-term energy storage.

Cellulose

When we examine the structure of cellulose we find another example of a polysaccharide in which nature has arranged monomeric glucose units in a manner that suits its function. Cellulose contains D-glucopyranoside units linked in 1:4 fashion in very long unbranched chains. Unlike starch and glycogen, however, the linkages in cellulose are *β-glycosidic linkages* (Fig. 24.14). This configuration of the anomeric carbons of cellulose makes cellulose chains essentially linear; they do not tend to coil into helical structures as do glucose polymers when linked in an α–1:4 manner.

The linear arrangement of β-linked glucose units in cellulose presents a uniform distribution of —OH groups on the outside of each chain. When two or more cellulose chains make contact, the hydroxyl groups are ideally situated to "zip" the chains together by forming hydrogen bonds. Zipping many cellulose chains together in this way gives a highly insoluble, rigid, and fibrous polymer that is ideal as cell-wall material for plants.

This special property of cellulose chains, we should emphasize, is not just a result of β,1:4 glycosidic linkages; it is also a consequence of the precise stereochemistry of D-glucose at each chiral carbon. Were D-galactose or D-allose units linked in a similar fashion, they almost certainly would not give rise to a polymer with properties like cellulose. Thus, we get another glimpse of why D-glucose occupies such a special position in the chemistry of plants and animals. Not only is it the most stable aldohexose (because it can exist in a chair conformation that allows all of its bulky groups to occupy equatorial positions); but its special stereochemistry

* Glucose is actually liberated as glucose-6-phosphate, which is also water soluble.

FIG. 24.14
A portion of a cellulose chain.

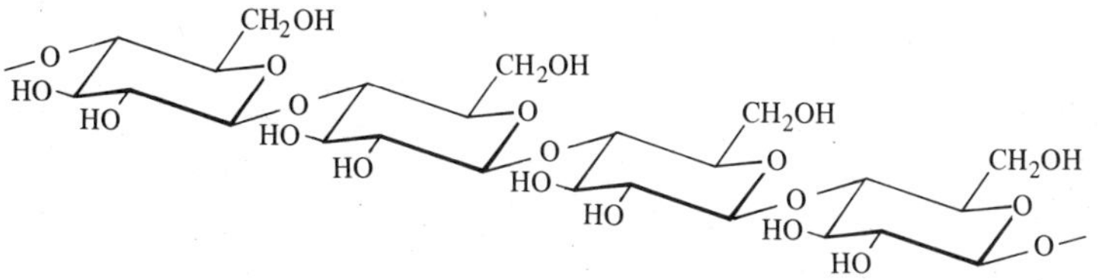

also allows it to form helical structures when α linked as in starches, and rigid linear structures when β linked as in cellulose.*

Perhaps we should ask ourselves one other question: Why has Nature "chosen" D-(+)-glucose for its special role rather than L-(−)-glucose, its mirror image? Here an answer cannot be given with any certainty. The selection of D-(+)-glucose may simply have been a random event early in the course of the evolution of enzyme catalysts. Once this selection was made, however, the chirality of the active sites of the enzymes involved would retain a bias toward D-(+)-glucose and against L-(−)-glucose (because of the improper fit of the latter). Once introduced, this bias would be perpetuated and extended to other catalysts.

Finally, when we speak of Nature selecting or choosing a particular molecule for a given function, we do not mean to imply that evolution operates on a molecular level. Evolution, of course, takes place at the level of organism populations and molecules are selected only in the sense that their use gives the organism an increased likelihood of surviving and procreating.

Additional Problems

24.24
Give appropriate structural formulas to illustrate each of the following:

(a) An aldopentose
(b) A ketohexose
(c) An L-monosaccharide
(d) A glycoside
(e) An aldonic acid
(f) An aldaric acid
(g) An aldonolactone
(h) A pyranose
(i) A furanose
(j) A reducing sugar
(k) A pyranoside
(l) A furanoside
(m) Epimers
(n) Anomers
(o) A phenylosazone
(p) A disaccharide
(q) A polysaccharide
(r) A nonreducing sugar

24.25
Draw Haworth and conformational formulas for each of the following:
(a) α-D-Allopyranose
(b) Methyl β-D-allopyranoside
(c) Methyl 2,3,4,6-tetra-O-methyl-β-D-allopyranoside

24.26
Draw structures for furanose and pyranose forms of D-ribose. Show how you could use periodate oxidation to distinguish between a methyl ribofuranoside and a methyl ribopyranoside.

24.27
One reference book lists D-mannose as being dextrorotatory; another lists it as being levorotatory. Both references are correct. Explain.

---

*Another interesting and important fact about cellulose: the digestive enzymes of humans cannot attack its β-1,4 linkages. Hence, cellulose cannot serve as a food source for humans, as can starch. Cows and termites, however, can use cellulose (of grass and wood) as a food source because symbiotic bacteria in their digestive systems furnish β-glucosidases.

24.28

The starting material for the commercial synthesis of vitamin C (p. 713) is L-sorbose (below); it can be synthesized from D-glucose through the following reaction sequence:

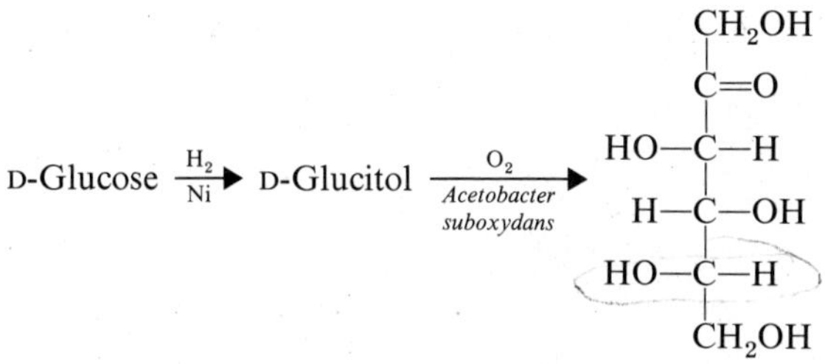

$$
\text{D-Glucose} \xrightarrow[\text{Ni}]{\text{H}_2} \text{D-Glucitol} \xrightarrow[\substack{\textit{Acetobacter} \\ \textit{suboxydans}}]{\text{O}_2}
$$

CH₂OH
|
C=O
|
HO—C—H
|
H—C—OH
|
HO—C—H   *C₂ actually*
|
CH₂OH

L-Sorbose

The second step of this sequence illustrates the use of a bacterial oxidation; the microorganism, *Acetobacter suboxydans,* accomplishes this step in 90% yield. The overall result of the synthesis is the transformation of a D-aldohexose, D-glucose, into an L-ketohexose, L-sorbose. What does this mean about the specificity of the bacterial oxidation?

24.29

What two aldoses would yield the same phenylosazone as L-sorbose (problem 24.28)?

24.30

In addition to fructose (problem 24.16) and sorbose (problem 24.28) there are two other 2-ketohexoses, *psicose* and *tagatose.* D-Psicose yields the same phenylosazone as D-allose (or D-altrose); D-tagatose yields the same osazone as D-galactose (or D-talose). What are the structures of D-psicose and D-tagatose?

24.31

A, B, and **C** are three aldohexoses. **A** and **B** yield the same optically active alditol when they are reduced with hydrogen and a catalyst; **A,** and **B** yield different phenylosazones when treated with phenylhydrazine. **B** and **C** give the same phenylosazone but different alditols. Assuming that all are D-sugars, give names and structures for **A, B,** and **C.**

24.32

Although monosaccharides undergo complex isomerizations in base (cf., problem 24.7), aldonic acids are epimerized specifically at C-2 when they are heated with pyridine. Show how you could make use of this reaction in a synthesis of D-mannose from D-glucose.

24.33

(a) The most stable conformation of most aldopyranoses is one in which the largest group—the —CH₂OH group—is equatorial. However, D-idopyranose exists primarily in a conformation with an axial —CH₂OH group. Write formulas for the two chair conformations of α-D-idopyranose (one with the —CH₂OH group axial and one with the —CH₂OH group equatorial) and provide an explanation.

24.34

(a) Heating D-altrose with dilute acid produces a nonreducing *anhydro sugar,* C₆H₁₀O₅. Methylation of the anhydro sugar followed by acid hydrolysis yields 2,3,4-tri-O-methyl-D-altrose. The formation of the anhydro sugar takes place through a chair conformation of β-D-altropyranose in which the —CH₂OH group is axial. What is the structure of the anhydro sugar and how is it formed? (b) D-Glucose also forms an anhydro sugar but the conditions required are much more drastic than for the corresponding reaction of D-altrose. Explain.

24.35

Although mutarotation of the D-glucopyranoses is catalyzed by either acids or bases, one of the most effective catalysts is 2-hydroxypyridine:

2-Hydroxypyridine

Write a mechanism that accounts for this.

24.36

Show how the following experimental evidence can be used to deduce the structure of lactose (p. 942).
1. Acid hydrolysis of lactose, $C_{12}H_{22}O_{11}$, gives equimolar quantities of D-glucose and D-galactose. Lactose undergoes a similar hydrolysis in the presence of a *β-galactosidase*.
2. Lactose is a reducing sugar and forms a phenylosazone; it also undergoes mutarotation.
3. Oxidation of lactose with bromine water followed by hydrolysis with dilute acid gives D-galactose and D-gluconic acid.
4. Bromine water oxidation of lactose followed by methylation and hydrolysis gives 2,3,5,6-tetra-O-methyl-D-δ-gluconolactone and 2,3,4,6-tetra-O-methyl-D-galactose.
5. Methylation and hydrolysis of lactose gives 2,3,6-tri-O-methyl-D-glucose and 2,3,4,6-tetra-O-methyl-D-galactose.

24.37

Deduce the structure of the disaccharide, *melibiose,* from the following data.
1. Melibiose is a reducing sugar that undergoes mutarotation and forms a phenylosazone.
2. Hydrolysis of melibiose with acid or with an *α-galactosidase* gives D-galactose and D-glucose.
3. Bromine-water oxidation of melibiose gives *melibionic acid.* Hydrolysis of melibionic acid gives D-galactose and D-gluconic acid. Methylation of melibionic acid followed by hydrolysis gives 2,3,4,6-tetra-O-methyl-D-galactose and 2,3,4,5-tetra-O-methyl-D-gluconic acid.
4. Methylation and hydrolysis of melibiose gives 2,3,4,6-tetra-O-methyl-D-galactose and 2,3,4-tri-O-methyl-D-glucose.

24.38

Trehalose is a disaccharide that·can be obtained from yeasts, fungi, sea urchins, algae, and insects. Deduce the structure of trehalose from the following information.
1. Acid hydrolysis of trehalose yields only D-glucose.
2. Trehalose is hydrolyzed by α-glucosidases but not by β-glucosidases.
3. Trehalose is a nonreducing sugar; it does not mutarotate, form a phenylosazone, or react with bromine water.
4. Methylation of trehalose followed by hydrolysis yields two moles of 2,3,4,6-tetra-O-methyl-D-glucose.

24.39

Outline chemical tests that will distinguish between each of the following:
(a) D-Glucose and D-glucitol
(b) D-Glucitol and D-glucaric acid
(c) D-Glucose and D-fructose
(d) D-Glucose and D-galactose
(e) Sucrose and maltose
(f) Maltose and maltonic acid

(g) Methyl β-D-glucopyranoside and 2,3,4,6-tetra-O-methyl-β-D-glucopyranose
(h) Methyl α-D-ribofuranoside (**I**) and methyl 2-deoxy-α-D-ribofuranoside (**II**)

I   **II**

* 24.40

A group of oligosaccharides, called *Schardinger dextrins* can be isolated from *Bacillus macerans* when the bacillus is grown on a medium rich in amylose. These oligosaccharides are all *nonreducing*. A typical Schardinger dextrin undergoes hydrolysis when treated with an acid or an α-glucosidase to yield six, seven, or eight molecules of D-glucose. Complete methylation of a Schardinger dextrin followed by acid hydrolysis yields only 2,3,6,tri-O-methyl-D-glucose. Propose a general structure for a Schardinger dextrin.

* 24.41

Isomaltose is a disaccharide that can be obtained by enzymic hydrolysis of amylopectin. Deduce the structure of isomaltose from the following data:
(1) Hydrolysis of one mole of isomaltose by acid or by an α-glucosidase gives two moles of D-glucose.
(2) Isomaltose is a reducing sugar.
(3) Isomaltose is oxidized by bromine water to isomaltonic acid. Methylation of isomaltonic acid and subsequent hydrolysis yields 2,3,4,6-tetra-O-methyl-D-glucose and 2,3,4,5-tetra-O-methyl-D-gluconic acid.
(4) Methylation of isomaltose itself followed by hydrolysis gives 2,3,4,6-tetra-O-methyl-D-glucose and 2,3,4-tri-O-methyl-D-glucose.

* 24.42

Stachyose occurs in the roots of several species of plants. Deduce the structure of stachyose from the following data:
(1) Acidic hydrolysis of one mole of stachyose yields two moles of D-galactose, one mole of D-glucose, and one mole of D-fructose.
(2) Stachyose is a nonreducing sugar.
(3) Treating stachyose with an α-galactosidase produces a mixture containing D-galactose, sucrose, and a nonreducing trisaccharide called *raffinose*.
(4) Acidic hydrolysis of raffinose gives D-glucose, D-fructose, and D-galactose. Treating raffinose with an α-galactosidase yields D-galactose and sucrose. Treating raffinose with invertase (an enzyme that hydrolyzes sucrose) yields fructose and *melibiose* (cf. problem 24.37).
(5) Methylation of stachyose followed by hydrolysis yields 2,3,4,6-tetra-O-methyl-D-galactose, 2,3,4-tri-O-methyl-D-galactose, 2,3,4-tri-O-methyl-D-glucose, and 1,3,4,6-tetra-O-methyl-D-fructose.

* **24.43**

Arbutin, a compound that can be isolated from the leaves of bearberry, cranberry, and pear trees, has the molecular formula $C_{12}H_{16}O_7$. When arbutin is treated with aqueous acid or with a β-glucosidase, the reaction produces D-glucose and a compound **X** with the molecular formula $C_6H_6O_2$. The pmr spectrum of compound **X** consists of two singlets, one at $\delta 6.8$ (4H) and one at $\delta 7.9$ (2H). Methylation of arbutin followed by acidic hydrolysis yields 2,3,4,6-tetra-O-methyl-D-glucose and a compound **Y**, $C_7H_8O_2$. Compound **Y** is soluble in dilute aqueous NaOH but is insoluble in aqueous $NaHCO_3$. The pmr spectrum of **Y** shows a singlet at $\delta 3.9$ (3H), a singlet at $\delta 4.8$ (1H) and a multiplet (that resembles a singlet) at $\delta 6.8$ (4H). Treating compound **Y** with aqueous NaOH and $(CH_3)_2SO_4$ produces compound **Z**, $C_8H_{10}O_2$. The pmr spectrum of **Z** consists of two singlets, one at $\delta 3.75$ (6H) and one at $\delta 6.8$ (4H). Propose structures for arbutin and for compounds **X**, **Y**, and **Z**.

* **24.44**

D-Glucose reacts with acetone in the presence of sulfuric acid to yield a compound with the molecular formula $C_{12}H_{19}O_6$ which has been given the common name "diacetone glucose." D-Galactose undergoes a similar reaction to yield an isomeric product. "Diacetone glucose" has three five-membered rings; the corresponding compound obtained from D-galactose has two five-membered rings and a six-membered ring. (a) Write structures for these two compounds and (b) explain their formation. (Hint: use models.)

25 AMINO ACIDS AND PROTEINS

25.1 INTRODUCTION

The three important groups of biological polymers are polysaccharides, proteins, and nucleic acids. We studied polysaccharides in Chapter 24 and saw that they function primarily as energy reserves and, in plants, as structural materials. When we study nucleic acids in Chapter 26 we will find that they serve two major purposes: storage and transmission of information. Of the three groups of biopolymers, proteins have the most diverse functions. As enzymes and hormones, proteins catalyze and regulate the reactions that occur in the body; as muscles and tendons they provide the body with the means for movement; as skin and hair they give it an outer covering; as hemoglobins they transfer all-important oxygen to its most remote corners; as antibodies they provide it with a means of protection against disease; and in combination with other substances in bone they provide it with structural support.

Given such diversity of functions we should not be surprised to find that proteins come in all sizes and shapes. By the standard of most of the molecules we have studied even small proteins have very high molecular weights. Lysozyme, an enzyme, is a relatively small protein and yet its molecular weight is 14,600. The molecular weights of most proteins are much larger. Their shapes cover a range from the globular proteins such as lysozyme and hemoglobin to the helical coils of α-keratin (hair, nails, wool) and the pleated sheets of silk fibroin.

And yet, in spite of such diversity of size, shape, and function, all proteins have common features that allow us to decipher their structures and understand their properties. Proteins are *polyamides* (cf. Section 22.3) and their monomeric units are amino acids. Although hydrolysis of naturally occurring proteins may yield up to 22 different amino acids, all amino acids have two important structural features in common: They are all α-amino acids and, with the exception of glycine (whose molecules are achiral), almost all naturally occurring amino acids have the L configuration at the α-carbon.* That is, they have the same relative configuration as L-glyceraldehyde:

$$
\begin{array}{ccc}
\underset{R}{\overset{COOH}{H_2N-\overset{|}{\underset{|}{C}}-H}} & \underset{CH_2OH}{\overset{CHO}{HO-\overset{|}{\underset{|}{C}}-H}} & \underset{NH_2}{\overset{CH_2COOH}{|}}
\end{array}
$$

an L–α–amino–acid   L–glyceraldehyde   glycine

* Some D-amino acids have been obtained from the material comprising the cell walls of bacteria, and by hydrolysis of certain antibiotics.

TABLE 25.1 L-Amino Acids Found in Proteins

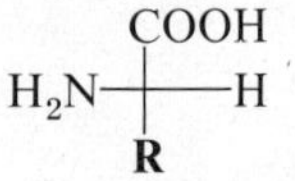

STRUCTURE OF **R**	NAME	ABBREVIATION
R group is neutral		
$-H$	Glycine	Gly
$-CH_3$	Alanine	Ala
$-CH(CH_3)_2$	Valine[e]	Val
$-CH_2CH(CH_3)_2$	Leucine[e]	Leu
$-CHCH_2CH_3$ $\quad$ CH_3	Isoleucine[e]	Ile
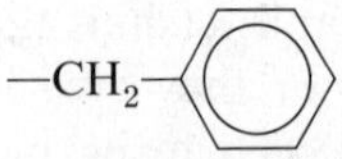 $-CH_2-\bigcirc$	Phenylalanine[e]	Phe
$-CH_2CONH_2$	Asparagine	Asn
$-CH_2CH_2CONH_2$	Glutamine	Gln
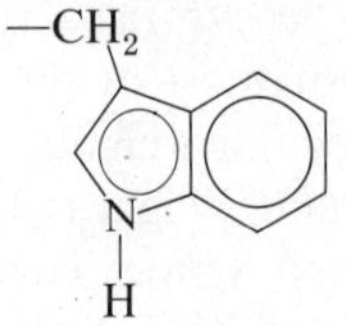 $-CH_2$ (indole)	Tryptophan[e]	Trp
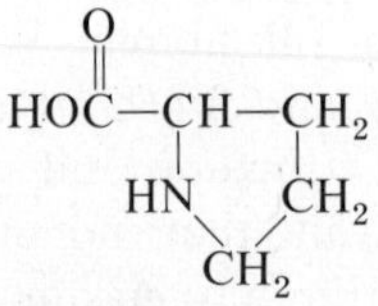 (proline complete structure)	Proline	Pro
R contains an —OH group		
$-CH_2OH$	Serine	Ser
$-CHOH$ $\quad$ CH_3	Threonine[e]	Thr
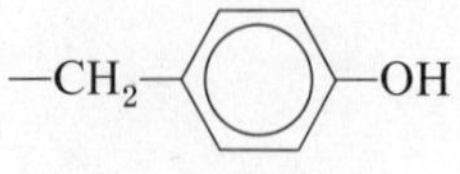 $-CH_2-\bigcirc-OH$	Tyrosine	Tyr
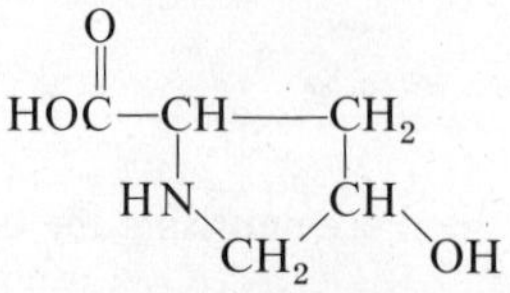 (hydroxyproline complete structure)	Hydroxyproline	Hyp

TABLE 25.1 L-Amino Acids Found in Proteins (Continued)

STRUCTURE OF **R**	NAME	ABBREVIATION
R contains sulfur		
$-CH_2SH$	Cysteine	Cys
$-CH_2-S$ $\quad\quad\mid$ $-CH_2-S$	Cystine	Cys-Cys
$-CH_2CH_2SCH_3$	Methionine[e]	Met
R contains a carboxyl group		
$-CH_2COOH$	Aspartic acid	Asp
$-CH_2CH_2COOH$	Glutamic acid	Glu
R contains a basic amino group		
$-CH_2CH_2CH_2CH_2NH_2$	Lysine[e]	Lys
$\qquad\qquad\qquad\overset{NH}{\overset{\|}{}}$ $-CH_2CH_2CH_2NH-C-NH_2$	Arginine	Arg
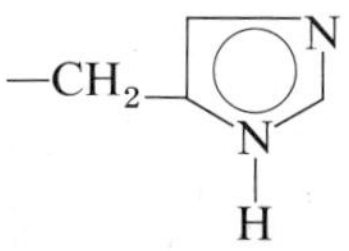 $-CH_2$ (imidazole ring with N, N—H)	Histidine	His

e = essential amino acids

25.2 AMINO ACIDS

Structures and Names

The 22 α-amino acids that can be obtained from proteins can be subdivided into five different groups on the basis of the structures of their R groups. These are given in Table 25.1.

Essential Amino Acids

Amino acids can be synthesized by all living organisms. Many higher species, however, are deficient in their ability to synthesize all of the amino acids they need for their proteins. Thus these higher organisms require certain amino acids as a part of their diet. For humans there are eight essential amino acids; these are designated with the superscript e in Table 25.1.

Amino Acids as Dipolar Ions

Since amino acids contain both a basic group ($-NH_2$) and an acidic group ($-COOH$), they are amphoteric compounds. In the dry solid state, amino acids exist as *dipolar ions,* a form in which the carboxyl group is present as a carboxylate ion,

—COO⁻, and the amine group is present as an ammonium group, —NH₃⁺. In aqueous solution, an equilibrium exists between the dipolar ion and the anionic and cationic forms of an amino acid.

$$H_2NCHCOO^- \underset{OH^-}{\overset{H_3O^+}{\rightleftharpoons}} \overset{+}{H_3N}-CHCOO^- \underset{OH^-}{\overset{H_3O^+}{\rightleftharpoons}} \overset{+}{H_3N}-CHCOOH$$

:---:		:---:		:---:
R		R		R
Anionic		Dipolar		Cationic
form		ion		form

The position of equilibrium depends on the pH of the solution and the nature of the amino acid. In strongly acidic solutions all amino acids are present primarily as cations; in strongly basic solutions they are present as anions. At some intermediate pH, called the *isoelectric point,* the concentration of the dipolar ion is at its maximum, and the concentrations of the anions and cations are equal. At this pH there is no net migration of the amino acid when placed in an electric field.

An amino acid with a neutral *R* group is somewhat more acidic than it is basic; as a result, its isoelectric point occurs at a pH slightly lower than that of a neutral aqueous solution (pH 7). The isoelectric point of glycine, for example, is at pH 6.1, that of phenylalanine is at pH 6.3, of cystine at pH 5.0.

If the side chain of the amino acid gives the amino acid an extra basic amino group the pH of the isoelectric point is much higher. Lysine, for example, has its isoelectric point at pH 9.7. If the side chain gives the amino acid an extra carboxyl group, the isoelectric point is lower. Aspartic acid has its isoelectric point at pH 2.7.

Problem 25.1

Write structures for the form of lysine that would predominate (a) in strongly acid solution and (b) in strongly basic solution. (c) At its isoelectric point the predominant form of lysine is the dipolar ion,

$$\overset{+}{H_3N}CH_2CH_2CH_2CH_2CHCOO^-$$
$$|$$
$$NH_2$$

rather than

$$H_2NCH_2CH_2CH_2CH_2CHCOO^-$$
$$|$$
$$^+NH_3$$

Explain.

Problem 25.2

What form of glutamic acid would you expect to predominate in (a) strongly acid solution? (b) Strongly basic solution? (c) At its isoelectric point (pH 3.2)? (d) The isoelectric point of glutamine (pH 5.65) is considerably higher than that of glutamic acid. Explain.

25.3 LABORATORY SYNTHESIS OF α-AMINO ACIDS

A variety of methods have been developed for the laboratory synthesis of α-amino acids. We will describe here three general methods, all of which are based on reactions we have seen before.

Direct Ammonolysis of an α-Halo Acid

$$R—CH_2COOH \xrightarrow[X_2]{P} RCHCOOH \xrightarrow{NH_{3(excess)}} R—CHCOO^-$$
$$\underset{X}{|} \qquad\qquad\qquad \underset{\overset{+}{N}H_3}{|}$$

This method is probably used least often because yields tend to be poor. We saw an example of this method on p. 822.

From Potassium Phthalimide

This method is a modification of the Gabriel synthesis of amines (p. 823). The yields are usually high and the products are easily purified.

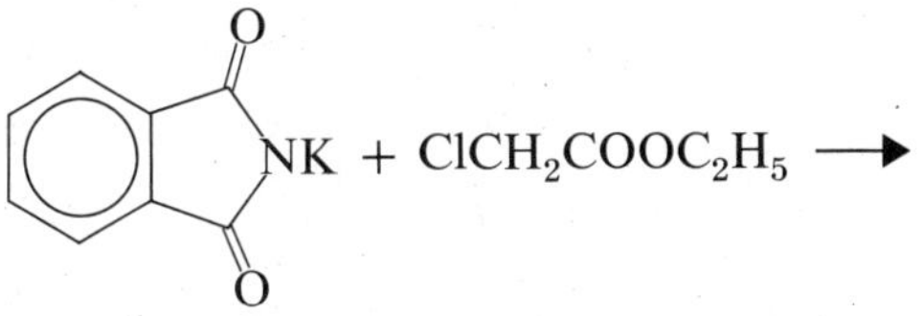

Potassium Ethyl chloroacetate
phthalimide

(97%)   Glycine Phthalic
          (85%)   acid

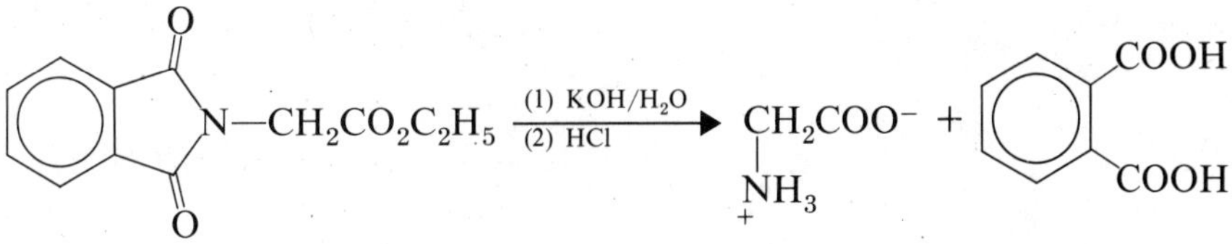

From Amido Malonic Esters and Imido Malonic Esters

Diethyl aminomalonate can be prepared from diethyl malonate through the following reactions.

$$CH_2(CO_2C_2H_5)_2 + CH_3(CH_2)_3ONO \xrightarrow[(2)\ H_2SO_4]{(1)\ 0-20°} HON{=}C(CO_2C_2H_5)_2 + CH_3(CH_2)_3OH$$

Diethyl malonate   *n*-Butyl   Diethyl
       nitrite   oximinomalonate
               (87%)

$$\downarrow 2H_2/Ni$$

$$H_2NCH(CO_2C_2H_5)_2$$
Diethyl aminomalonate
(65%)

Converting the amino group of diethyl aminomalonate to an amide gives diethyl benzamidomalonate, a very useful reagent for amino acid synthesis.

$$H_2NCH(CO_2C_2H_5)_2 + C_6H_5COCl \xrightarrow{pyridine} C_6H_5CONHCH(CO_2C_2H_5)_2$$

Diethyl benzamidomalonate
(90%)

An example of its utility is the following synthesis of DL-leucine.

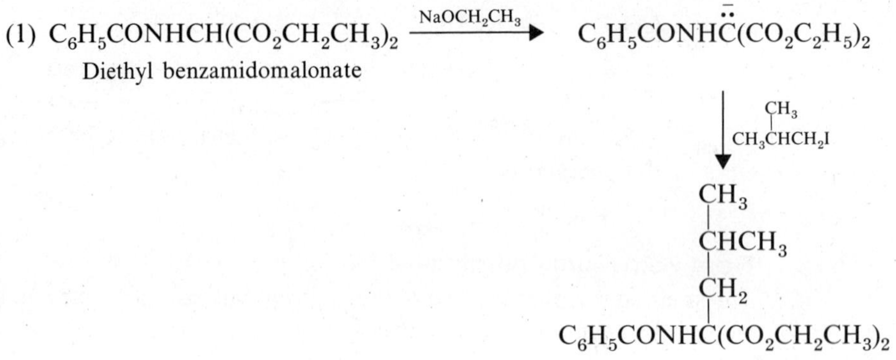

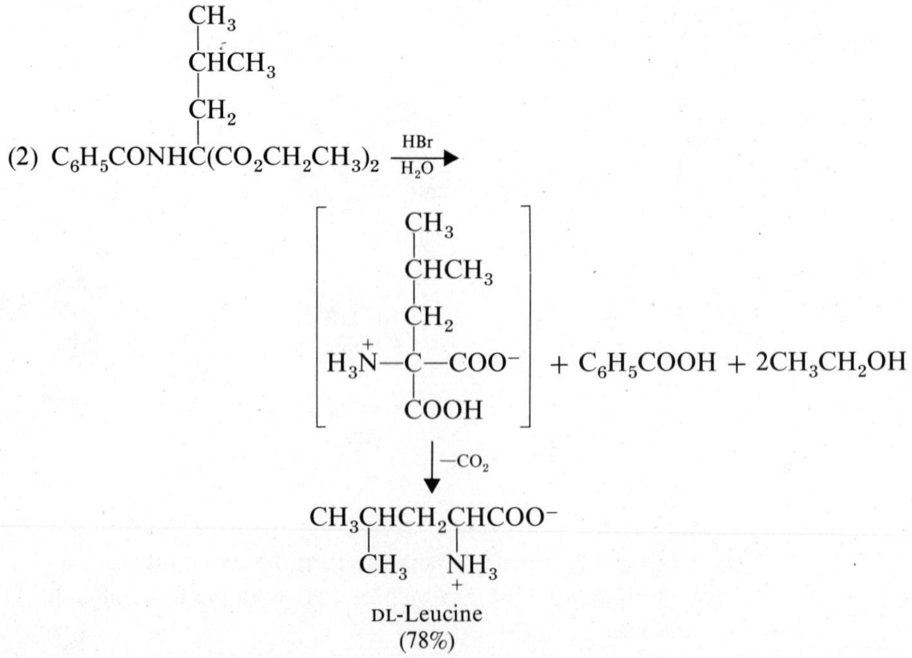

(1) $C_6H_5CONHCH(CO_2CH_2CH_3)_2$ $\xrightarrow{NaOCH_2CH_3}$ $C_6H_5CONH\overset{..}{C}(CO_2C_2H_5)_2$

Diethyl benzamidomalonate

$$\downarrow \overset{CH_3}{CH_3CHCH_2I}$$

$$\begin{array}{c} CH_3 \\ | \\ CHCH_3 \\ | \\ CH_2 \\ | \\ C_6H_5CONH\overset{}{C}(CO_2CH_2CH_3)_2 \end{array}$$

(2)
$$\begin{array}{c} CH_3 \\ | \\ CHCH_3 \\ | \\ CH_2 \\ | \\ \end{array}$$
$C_6H_5CONH\overset{}{C}(CO_2CH_2CH_3)_2$ $\xrightarrow[H_2O]{HBr}$

$$\left[\begin{array}{c} CH_3 \\ | \\ CHCH_3 \\ | \\ CH_2 \\ | \\ H_3\overset{+}{N}-C-COO^- \\ | \\ COOH \end{array} \right] + C_6H_5COOH + 2CH_3CH_2OH$$

$$\downarrow -CO_2$$

$$\begin{array}{c} CH_3CHCH_2CHCOO^- \\ | \quad\quad | \\ CH_3 \quad N\overset{}{H}_3 \\ \quad\quad + \end{array}$$

DL-Leucine
(78%)

In the first step of the synthesis diethyl benzamidomalonate reacts with sodium ethoxide to give an enolate ion; this then reacts with isobutyl iodide in an alkylation reaction. In the second step acid hydrolysis cleaves the amide and ester linkages and the malonic acid that is produced decarboxylates spontaneously to give DL-leucine.

A variation of this procedure uses potassium phthalimide and bromomalonic ester to prepare an *imido* malonic ester. This method is illustrated with a synthesis of methionine.

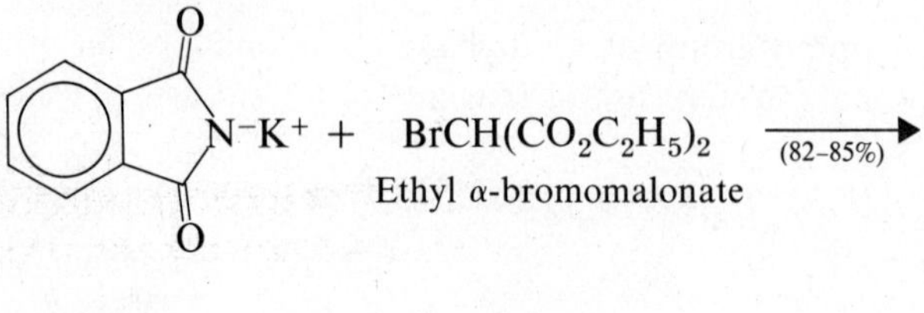

N^-K^+ + $BrCH(CO_2C_2H_5)_2$ $\xrightarrow{(82-85\%)}$

Ethyl α-bromomalonate

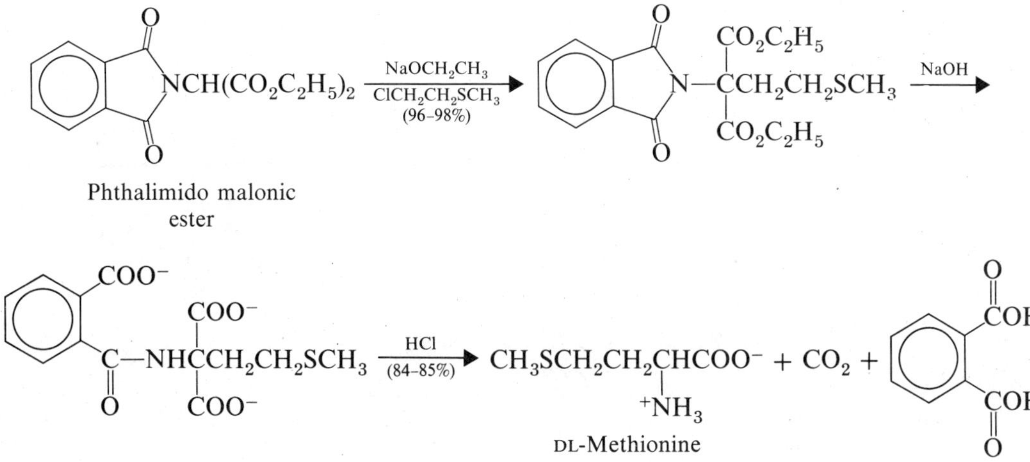

Phthalimido malonic
ester

DL-Methionine

Problem 25.3

The following amino acids have been synthesized from diethyl benzamidomalonate and the appropriate alkyl halide. Outline each method. (The percentages in parentheses are the yields actually obtained.)

      (a) DL-Phenylalanine (90%)
      (b) DL-Aspartic acid   (62%)
      (c) DL-Valine   (71%)

Problem 25.4

Starting with ethyl α-bromomalonate and potassium phthalimide and using any other necessary reagents show how you might synthesize:

      (a) DL-Leucine
      (b) DL-Alanine
      (c) DL-Phenylalanine

Resolution of DL-Amino Acids

With the exception of glycine, which has no chiral carbon, the amino acids that are produced by the methods we have outlined are all produced as racemic modifications. In order to obtain the naturally occurring L-amino acid we must, of course, resolve the racemic modifications. This can be done in a variety of ways including the methods outlined in Section 7.12.

One especially interesting method for resolving amino acids is based on the use of enzymes called *deacylases*. These enzymes catalyze the hydrolysis of *N-acyl-amino acids* in living organisms. Since the active site of the enzyme is chiral it hydrolyzes only *N*-acylamino acids of the L configuration. When it is exposed to a racemic modification of *N*-acylamino acids, only the L-amino acid is affected and the products, as a result, are separated easily.

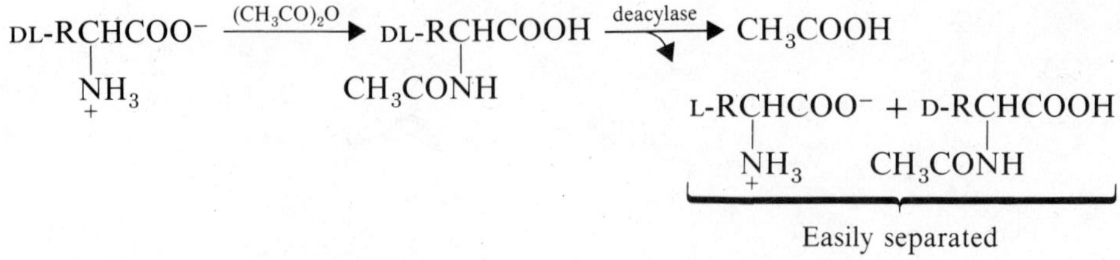

$$\underset{\underset{+}{NH_3}}{\text{DL-RCHCOO}^-} \xrightarrow{\text{(CH}_3\text{CO)}_2\text{O}} \underset{CH_3CONH}{\text{DL-RCHCOOH}} \xrightarrow{\text{deacylase}} CH_3COOH$$

$$\underset{\underset{+}{NH_3} \qquad CH_3CONH}{\text{L-RCHCOO}^- + \text{D-RCHCOOH}}$$

Easily separated

25.4 BIOSYNTHESIS OF AMINO ACIDS

Two of the most important methods used by living cells for the synthesis of amino acids are described below.

Reductive Amination

This biosynthesis bears a remarkable resemblance to one laboratory method that we have seen for the synthesis of amines (p. 825). In enzymatic reductive amination, α-ketoglutaric acid combines with ammonia in the presence of the reducing agent, NADH. The product is L-glutamic acid.

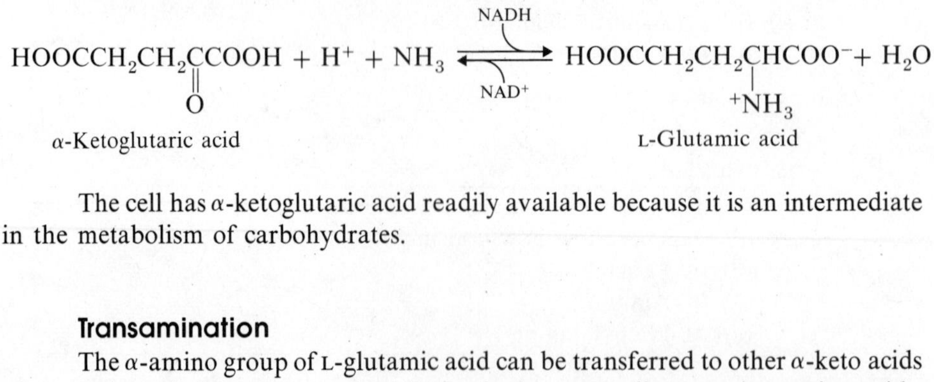

$$\underset{\alpha\text{-Ketoglutaric acid}}{\underset{O}{\overset{\parallel}{HOOCCH_2CH_2CCOOH}}} + H^+ + NH_3 \underset{\text{NAD}^+}{\overset{\text{NADH}}{\rightleftharpoons}} \underset{L\text{-Glutamic acid}}{\underset{^+NH_3}{HOOCCH_2CH_2CHCOO^-}} + H_2O$$

The cell has α-ketoglutaric acid readily available because it is an intermediate in the metabolism of carbohydrates.

Transamination

The α-amino group of L-glutamic acid can be transferred to other α-keto acids whose carbon chains correspond to those of other naturally occurring amino acids. These reactions are catalyzed by enzymes called *transaminases*. An example is the biosynthesis of L-aspartic acid from L-glutamic acid and oxaloacetic acid.

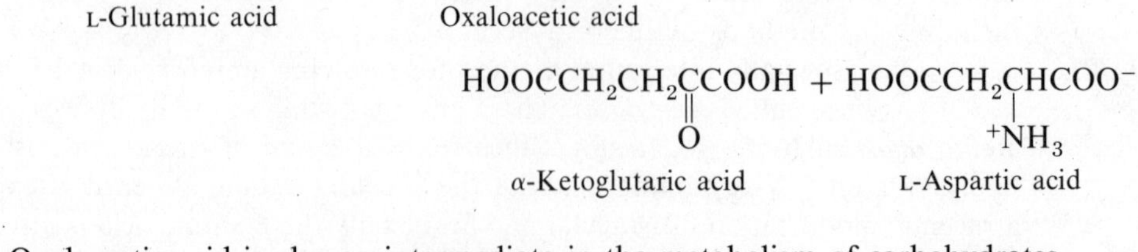

$$\underset{L\text{-Glutamic acid}}{\underset{\underset{+}{NH_3}}{HOOCCH_2CH_2CHCOO^-}} + \underset{\text{Oxaloacetic acid}}{\underset{O}{\overset{\parallel}{HOOCCH_2CCOOH}}} \overset{\text{transaminase}}{\rightleftharpoons}$$

$$\underset{\alpha\text{-Ketoglutaric acid}}{\underset{O}{\overset{\parallel}{HOOCCH_2CH_2CCOOH}}} + \underset{L\text{-Aspartic acid}}{\underset{^+NH_3}{HOOCCH_2CHCOO^-}}$$

Oxaloacetic acid is also an intermediate in the metabolism of carbohydrates.

25.5 ANALYSIS OF AMINO ACID MIXTURES

The amide linkages that join α-amino acids in proteins are commonly called *peptide linkages* and α-amino acid polymers with molecular weights less than 10,000 are usually called *polypeptides*. Those with molecular weights greater than 10,000 are called proteins. This division is quite arbitrary and the two terms are sometimes interchanged. In actuality, both proteins and polypeptides are *polyamides*.

We can represent the structures of polypeptides using the symbols for the amino acids. The dipeptide glycylvaline, for example, is represented as Gly · Val, and the dipeptide valylglycine as Val · Gly. In each case the amino acid whose carboxyl group is involved in an amide linkage is placed first.

$$NH_2CH_2\overset{\displaystyle O}{\overset{\|}{C}}-NHCHCOH \qquad NH_2CHC-NHCH_2COH$$

Glycylvaline
(Gly · Val)

Valylglycine
(Val · Gly)

The tripeptide glycylvalylphenylalanine can be represented in the following way.

$$NH_2CH_2\overset{\displaystyle O}{\overset{\|}{C}}-NHCHC-NHCHCOH$$

Glycylvalylphenylalanine
(Gly · Val · Phe)

When a protein or polypeptide is refluxed with $6N$-hydrochloric acid for 24 hours, hydrolysis of all of the amide linkages usually takes place and this produces a mixture of amino acids. One of the first tasks that we face when we attempt to determine the structure of a polypeptide or protein is the separation and identification of the individual amino acids in such a mixture. Since as many as 22 different amino acids may be present, this could be a formidable task if we are restricted to conventional methods.

Fortunately, techniques have been developed, based on the principle of elution chromatography, that simplify this problem immensely and even allow it to be automated. Automatic amino acid analyzers were developed at the Rockefeller Institute in 1950 and have since become commercially available. They are based on the use of insoluble polymers, containing sulfonate groups called *cation-exchange resins* (Fig. 25.1).

If an acidic solution containing a mixture of amino acids is passed through a column packed with a cation exchange resin the amino acids will be adsorbed by the resin because of attractive forces between the negatively charged sulfonate groups and the positively charged amino acids. The strength of the adsorption will vary with the basicity of the individual amino acids; those that are most basic will

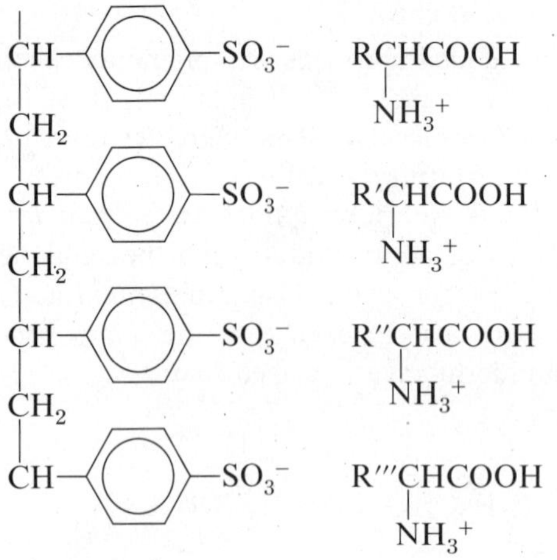

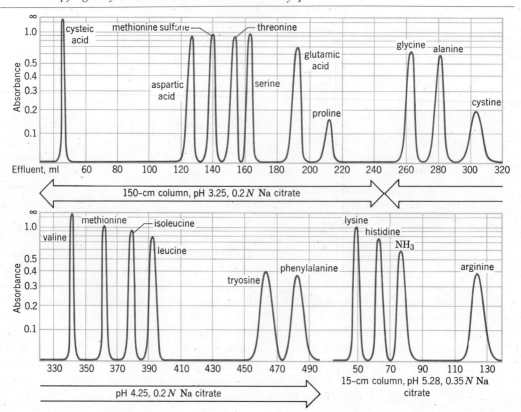

FIG. 25.1

A section of a cation-exchange resin with adsorbed amino acids.

FIG. 25.2

Typical result given by an automatic amino acid analyzer. [Reprinted with permission from D. H. Spackman, W. H. Stein, and S. Moore, Anal. Chem. 30, *1190 (1958). Copyright by the American Chemical Society.]*

be held most strongly. If the column is then eluted with a buffered solution at a given pH the individual amino acids will move down the column at different rates and ultimately become separated. At the end of the column the eluate is allowed to mix with *ninhydrin,* a reagent that reacts with the amino acids to give derivatives with an intense purple color (λ_{max} 570 nm). The amino acid analyzer is designed so that it can measure the absorbance of the eluate (at 570 nm) continuously and record this absorbance as a function of the volume of the effluent.

A typical graph obtained from an automatic amino acid analyzer is shown in Fig. 25.2. When the procedure is standardized the positions of the peaks are characteristic of the individual amino acids and the areas under the peaks correspond to their relative amounts.

25.6 AMINO ACID SEQUENCE OF PROTEINS AND POLYPEPTIDES

Once we have determined the amino acid composition of a protein or a polypeptide we should then determine its molecular weight. A variety of methods are available for doing this, including chemical methods, ultracentrifugation, light scattering, osmotic pressure and X-ray diffraction. With the molecular weight and amino acid composition we will now be able to calculate the "molecular formula" of the protein; that is, we will know how many of each type of amino acid are present as *amino acid residues* (i.e., RCHCO—) in each protein molecule. Unfortunately, however,

$$-\text{NH}$$

we have only begun our task of determining its structure. The next step is a formidable one, indeed: we must determine the order in which the amino acids are connected; that is, we must determine the *covalent structure of the polyamide.*

A simple tripeptide composed of 3 different amino acids can have 6 different amino acid sequences; a tetrapeptide composed of 4 different amino acids can have as many as 24. For a protein with a molecular weight of 10,000 or more, composed of 20 different amino acids, the number of possibilities approaches infinity.

In spite of this, a number of methods have been developed that allow the amino acid sequences to be determined and these, as we shall see, have been applied with amazing success. In our discussion here we will limit our attention to two methods that illustrate how sequence determinations can be done: *terminal residue analysis* and *partial hydrolysis.*

Terminal Residue Analysis

One of the ends of a polypeptide chain terminates in an amino acid residue that has a free —NH_2 group, the other terminates in an amino acid residue with a free —COOH group. These two amino acids are called the *N-terminal* residue and the *C-terminal* residue, respectively.

$$\text{H}_2\text{N}-\underset{\underset{\text{R}}{|}}{\text{CHCO}} \left(\text{NH}\underset{\underset{\text{R}}{|}}{\text{CHCO}}\right) \text{NH}\underset{\underset{\text{R}}{|}}{\text{CHCOOH}}$$

N-Terminal   *C*-Terminal
residue   residue

One very useful method for determining the *N*-terminal amino acid residue, called the *Sanger method*, is based on the use of 2,4-dinitrofluorobenzene (DNFB).* When a polypeptide is treated with DNFB in mildly basic solution, a nucleophilic aromatic substitution reaction takes place involving the free amino group of the *N*-terminal residue. Subsequent hydrolysis of the polypeptide gives a mixture of amino acids in which the *N*-terminal amino acid bears a label, *the 2,4-dinitrophenyl group*. As a result, after separating this amino acid from the mixture, we can identify it.

$$O_2N-\langle\bigcirc\rangle-F + H_2\ddot{N}CHCO-NHCHCO\sim \text{ etc.} \xrightarrow{HCO_3^-}$$

NO₂ — R — R'

2,4-Dinitrofluoro-
benzene
(DNFB) Polypeptide

$$O_2N-\langle\bigcirc\rangle-NHCHCO-NHCHCO\sim \xrightarrow{H_3O^+}$$

NO₂ — R — R'

Labeled polypeptide

$$O_2N-\langle\bigcirc\rangle-NHCHCOOH + H_3\overset{+}{N}CHCOO^-$$

NO₂ — R — R'

Labeled *N*-terminal amino   Mixture of
acid   amino acids

separate and identify

Problem 25.5

The electron-withdrawing property of the 2,4-dinitrophenyl group makes separation of the labeled amino acid very easy. Suggest how this is done.

Of course, 2,4-dinitrofluorobenzene will react with any free amino group present in a polypeptide including the ε-amino group of lysine. But, only the *N*-terminal amino acid residue will bear the label at the α-amino group.

A second method of *N*-terminal analysis is the *Edman degradation* (developed by Pehr Edman of the University of Lund, Sweden). This method offers an advantage over the Sanger method in that it removes the *N*-terminal residue and leaves the remainder of the peptide chain intact. The Edman degradation is based on a labeling reaction between the *N*-terminal amino group and phenyl isothiocyanate, $C_6H_5N=C=S$ (p. 961). When the labeled polypeptide is treated with acid, the *N*-terminal amino acid residue splits off as a phenylthiohydantoin. The product can be identified by comparison with phenylthiohydantoins prepared from standard amino acids.

* This method was introduced by Frederick Sanger of Cambridge University in 1945. Sanger made extensive use of this procedure in his determination of the amino acid sequence of insulin.

$$\bigcirc\!\!-N\!\!=\!\!C\!\!=\!\!S + H_2\overset{..}{N}CHCO\!\!-\!\!NHCHCO\!\!\sim \quad \text{etc.}$$
$$\underset{R}{} \qquad \underset{R'}{}$$

$$\downarrow \; OH^-, \; pH9$$

$$\bigcirc\!\!-NH\!\!-\!\!\overset{S}{\overset{\|}{C}}\!\!-\!\!NHCHCO\!\!-\!\!NHCHCO\!\!\sim \quad \text{etc.}$$
$$\underset{R}{} \qquad \underset{R'}{}$$

Labeled polypeptide

$$\downarrow \; H^+$$

$$\bigcirc\!\!-\!\!N \underset{\underset{O}{\overset{\|}{C}}\!\!-\!\!\underset{R}{CH}}{\overset{\overset{S}{\overset{\|}{C}}}{\diagup \diagdown}} NH \quad + \quad H_2NCHCO\!\!\sim$$
$$\underset{R'}{}$$

Phenylthiohydantoin	Polypeptide with one
	less amino acid residue

The polypeptide that remains after the first Edman degradation can be submitted to another degradation to identify the next amino acid in the sequence and this process has even been automated. Unfortunately, Edman degradations cannot be repeated indefinitely. As residues are successively removed, amino acids formed by hydrolysis during the acid treatment accumulate in the reaction and interfere with the procedure. The Edman degradation, however, has been successfully applied to polypeptides with as many as 50 amino acid residues.

C-terminal residues can be identified through the use of digestive enzymes called *carboxypeptidases*. These enzymes specifically hydrolyze the amide bond of the amino acid residue containing a free —COOH group, liberating it as a free amino acid. A carboxypeptidase, however, will continue to attack the polypeptide chain that remains, successively lopping off C-terminal residues. As a consequence, it is necessary to follow the amino acids released as a function of time. The procedure can only be applied to a limited amino acid sequence, for, at best, after a time the situation becomes too confused to sort out.

Consider the polypeptide sequences shown in Fig. 25.3. Carboxypeptidase hydrolysis might give a result like that shown in Fig. 25.3a when the amino acids are released at the same rate. If the amino acids are released at different rates, however, we might obtain a result such as that shown in Fig. 25.3b; in this case we would even have difficulty in deciding which amino acid was the original C-terminal amino acid.

Problem 25.6

(a) Write a reaction showing how 2,4-dinitrofluorobenzene could be used to identify the N-terminal amino acid of Val · Ala · Gly. (b) What products would you expect (after hydrolysis) when Val · Lys · Gly is treated with 2,4-dinitrofluorobenzene?

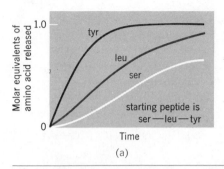
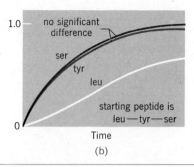

FIG. 25.3

The release by carboxypeptidase of amino acids from a peptide having the C-terminal sequence shown. (a) All bonds cleaved at the same rate. (b) Ser removed slowly; tyr cleaved rapidly; leu cleaved at the same rate as tyr. (From Robert Barker, Organic Chemistry of Biological Compounds, © *1971. Reprinted by permission of Prentice-Hall, Inc.*

Problem 25.7

Write the reactions involved in a sequential Edman degradation of Met · Ile · Arg.

Partial Hydrolysis

Sequential analysis using the Edman degradation or carboxypeptidase becomes impractical with proteins or polypeptides of appreciable size. Fortunately, however, we can resort to another technique; that of *partial hydrolysis*. Using dilute acids or enzymes we attempt to break the polypeptide chain into small fragments, ones that we can identify using DNFB or the Edman degradation. Then we examine the structures of these smaller fragments looking for points of overlap and attempt to piece together the amino acid sequence of the original polypeptide.

Consider a simple example: we are given a pentapeptide known to contain valine (two residues), leucine (one residue), histidine (one residue) and phenylalanine (one residue). With this information, we can write the "molecular formula" of the protein in the following way, using commas to indicate that the sequence is unknown.

Val$_2$, Leu, His, Phe

Then let us assume that by using 2,4-dinitrofluorobenzene and carboxypeptidase we discover that valine and leucine are the *N*-terminal and *C*-terminal residues, respectively. This gives us:

Val(Val, His, Phe) Leu

But, the sequence of the three nonterminal amino acids is still unknown.

We then subject the pentapeptide to partial acid hydrolysis and obtain the following dipeptides.

Val · His + His · Val + Val · Phe + Phe · Leu

The points of overlap (i.e., His, Val, and Phe) tell us that the original pentapeptide must have been:

Val · His · Val · Phe · Leu

Two enzymes are also frequently used to affect partial hydrolyses: *trypsin* preferentially hydrolyzes peptide bonds in which the carboxyl group is a part of a lysine or arginine residue. Chymotrypsin preferentially hydrolyzes peptide bonds at the carboxyl groups of phenylalanine, tyrosine, and tryptophan. It will also attack the peptide bonds at the carboxyl groups of leucine, methionine, asparagine, and glutamine.

Problem 25.8

Glutathione is a tripeptide found in most living cells. Partial acid hydrolysis of glutathione yields two dipeptides Cys · Gly and one composed of Glu and Cys. When this second dipeptide was treated with DNFB, acid hydrolysis gave N-labeled Glu. (a) Based on this information alone, what structures are possible for glutathione? (b) Synthetic experiments have shown that the second dipeptide has the following structure.

$$\overset{+}{H_3}NCHCH_2CH_2CONHCHCOO^-$$
$$\underset{\displaystyle COO^-}{}\underset{\displaystyle CH_2SH}{}$$

What is the structure of glutathione?

Problem 25.9

Give the amino acid sequence of the following polypeptides using only the data given by partial acidic hydrolysis.

(a) Ser, Hyp, Pro, Thr $\xrightarrow[H_2O]{H^+}$ Ser · Thr · + Thr · Hyp + Pro · Ser

(b) Ala, Arg, Cys, Val, Leu $\xrightarrow[H_2O]{H^+}$

Ala · Cys · + Cys · Arg + Arg · Val + Leu · Ala

25.7 PRIMARY STRUCTURES OF POLYPEPTIDES AND PROTEINS

The covalent structure of a protein or polypeptide is called its *primary structure*. By using the techniques we described in the previous sections, chemists have had remarkable success in determining the primary structures of polypeptides and proteins. The compounds described in the following pages are important examples.

Oxytocin and Vasopressin

Oxytocin and vasopressin (Fig. 25.4) are two rather small polypeptides with strikingly similar structures (where oxytocin has leucine, vasopressin has arginine and where oxytocin has isoleucine, vasopressin has phenylalanine). In spite of the similarity of their amino acid sequences these two polypeptides have quite different physiological effects. Oxytocin occurs only in the female of a species and stimulates uterine contractions during childbirth. Vasopressin occurs in males and females; it causes contraction of peripheral blood vessels and an increase in blood pressure. Its major function, however, is as an *antidiuretic;* physiologists often refer to vasopressin as *antidiuretic hormone.*

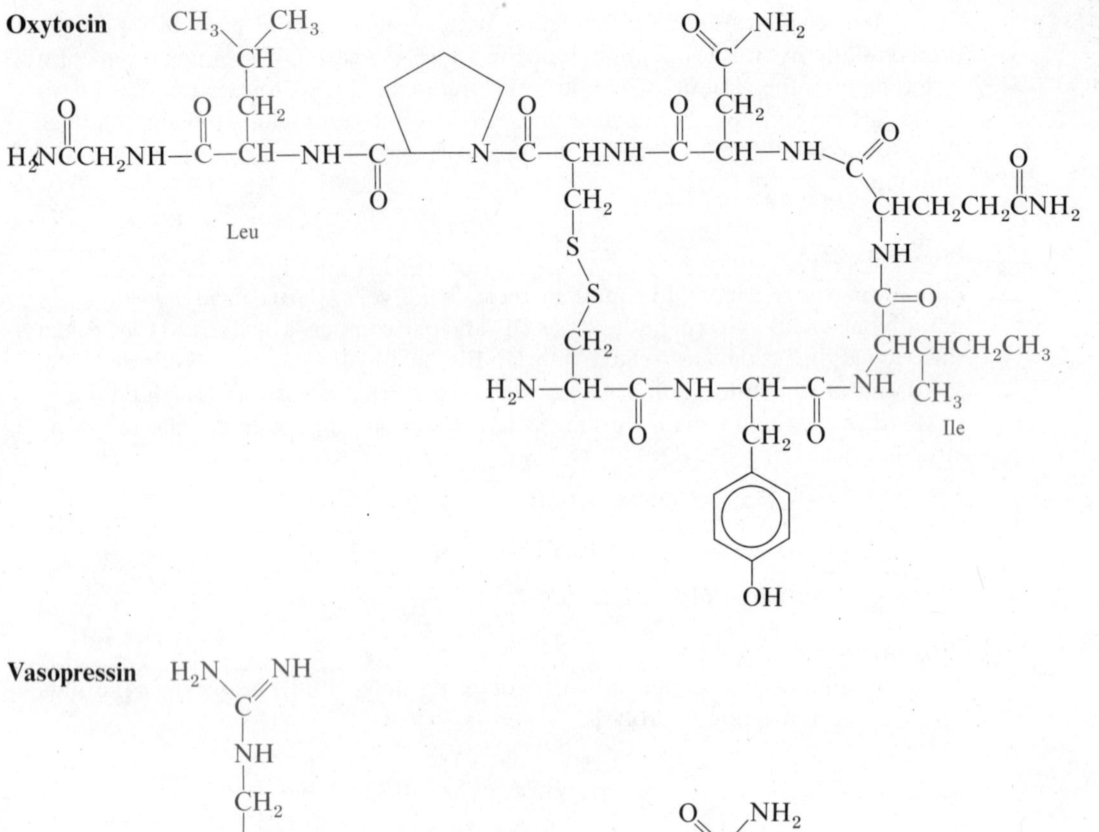

Oxytocin

Vasopressin

FIG. 25.4
The structures of oxytocin and vasopressin.

The structures of oxytocin and vasopressin also illustrate the importance of the disulfide linkage between cysteine residues in the overall primary structure of a polypeptide. In these two molecules this disulfide linkage leads to a cyclic structure.*

Problem 25.10

Treating oxytocin with certain reducing agents (e.g., sodium in liquid ammonia) brings about a single chemical change that can be reversed by air oxidation. What chemical changes are involved?

Insulin

Insulin is a hormone secreted by the pancreas that regulates glucose metabolism. Insulin deficiency in humans cause diabetes mellitus.

The amino acid sequence of bovine insulin (Fig. 25.5) was determined by Sanger in 1953. (Sanger received the Nobel Prize in 1958.) Bovine insulin has a total of 51 amino acid residues in two polypeptide chains, called the A and B chains. These chains are joined by two disulfide linkages. The A chain contains an additional disulfide linkage between cysteine residues at positions 6 and 11.

Since Sanger's determination of the amino acid sequence of bovine insulin, the amino acid sequences of insulin from a number of species have become known. Those from vertebrates all have an A chain of 21 residues and a B chain of 30. The A chains of insulins from humans, pigs, dogs, rabbits, and sperm whales are the same. The B chains of cows, pigs, dogs, goats, sperm whales, and horses are identical. The B chains of insulins from humans and elephants are also identical.

The principal differences in the insulin A chains occur at positions 8, 9, and 10. In humans these are,

Thr · Ser · Ile

in cows,

Ala · Ser · Val

and in sheep,

Ala · Gly · Val

Similar studies have been carried out with other proteins. These have been very useful in constructing a phylogenetic tree that allows an estimation to be made of the probable times of divergence of the major genera and species of living organisms. As we might expect, the number of amino acid residue differences is proportional to their phylogenetic differences and to the time of divergence. One protein that has been studied extensively in this regard is *cytochrome c,* a protein important in biological oxidations. Cytochrome c, isolated from monkeys shows only one different residue from that obtained from humans; that obtained from horses, however, differs by 12 residues. Cytochrome-c molecules isolated from vertebrates generally differ by as many as 43–48 residues from those isolated from a microorganism such as yeast.

---

* Vincent du Vigneaud of Cornell Medical College synthesized oxytocin and vasopressin in 1953; he received the Nobel Prize in 1955.

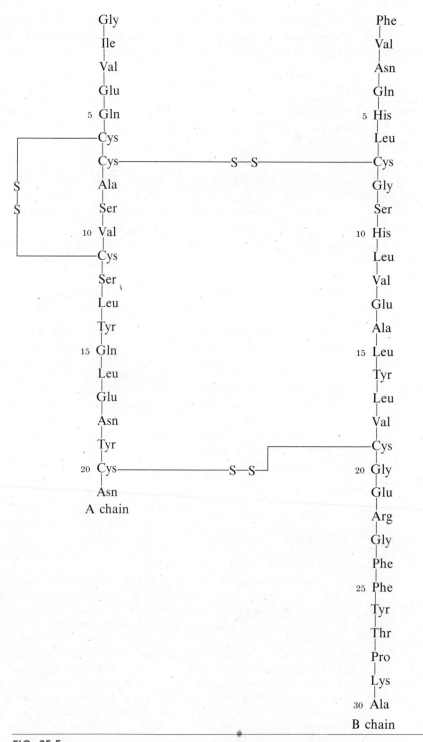

FIG. 25.5

The amino acid sequence of bovine insulin. (From A.L. Lehninger, Biochemistry, *Worth Publishers, Inc., New York, 1970, p. 92. Used with permission.)*

Other Polypeptides and Proteins

Successful sequential analyses have been achieved with a number of other polypeptides and proteins including:

1. Bovine *ribonuclease*. This enzyme, which catalyzes the hydrolysis of ribonucleic acids (p. 998), has a single chain of 124 amino acid residues and four intrachain disulfide linkages.

2. Human *hemoglobin*. There are four peptide chains in this important oxygen-carrying protein. Two identical α chains have 141 residues each, and two identical β chains have 146 residues. The genetically based disease, sickle cell anemia, has been shown to result from a single amino acid error in the β chain. In normal hemoglobin, position -6 has a glutamic acid residue, in sickle cell hemoglobin position -6 is occupied by valine.

> Red blood cells (erythrocytes) containing hemoglobin with this amino acid residue error tend to become crescent shaped ("sickle") when the partial pressure of oxygen is low, as it is in venous blood. These distorted cells are more difficult for the heart to pump through small capillaries. They may even block capillaries by clumping together; at other times the red cells may even split open.
>
> Children who inherit this genetic trait from both parents suffer from a severe form of the disease and usually do not live past the age of two. Children who inherit the disease from only one parent generally have a much milder form.
>
> Sickle cell anemia arose among the populations of central and western Africa where, ironically, it may have had a beneficial effect. People with a mild form of the disease are far less susceptible to malaria than those with normal hemoglobin. Malaria, a disease caused by an infectious microorganism, is especially prevalent in central and western Africa.
>
> Mutational changes such as those that give rise to sickle cell anemia are very common. Approximately 150 different types of mutant hemoglobin have been detected in humans; fortunately, most are harmless.

3. Bovine *trypsinogen* and *chymotrypsinogen*. These two enzyme precursors have single chains of 229 and 245 residues, respectively.

4. *Gamma globulin*. This immunoprotein has a total of 1320 amino acid residues in four chains. Two chains have 214 residues each; the other two have 446.

25.8 PROTEIN AND POLYPEPTIDE SYNTHESIS

We saw in Chapter 19 that the synthesis of an amide linkage is a relatively simple one. We must first "activate" the carboxyl group of an acid by converting it to an anhydride or acid chloride and then allow it to react with an amine:

$$\underset{\text{Anhydride}}{R-\overset{\overset{O}{\|}}{C}-O-\overset{\overset{O}{\|}}{C}-R} + \underset{\text{Amine}}{R'-NH_2} \longrightarrow \underset{\text{Amide}}{R-\overset{\overset{O}{\|}}{C}-NHR'} + R-COOH$$

The problem becomes somewhat more complicated, however, when both the acid group and the amino group are present in the same molecule as they are in an amino acid and, especially, when our goal is the synthesis of a naturally occurring polyamide where the sequence of different amino acids is all important. Let us consider, as an example, the synthesis of the simple dipeptide alanylglycine, Ala · Gly. We might first activate the carboxyl group of alanine by converting it to an acid chloride and then we might allow it to react with glycine. Unfortunately, however, we cannot prevent alanyl chloride from reacting with itself. So our reaction would yield not only Ala · Gly but also Ala · Ala. It could also lead to Ala · Ala · Ala and Ala · Ala · Gly, and so on. The yield of our desired product would be low and we would also have a difficult problem separating the dipeptides, tripeptides, and so on.

$$
\underset{\substack{\displaystyle | \\ +NH_3 \\ \text{Ala}}}{CH_3CHCO^-} \xrightarrow[\text{(2) } H_3NCH_2COO^-]{\text{(1) SOCl}_2 \quad} \underset{\substack{\displaystyle | \\ NH_2 \\ \text{Ala · Gly}}}{CH_3CHCNHCH_2COOH} + \underset{\substack{\displaystyle | \quad | \\ NH_2 \;\; CH_3 \\ \text{Ala · Ala}}}{CH_3CHCNHCHCOOH} +
$$

$$
\underset{\substack{\displaystyle | \quad\quad | \\ NH_2 \;\; CH_3 \;\; CH_3 \\ \text{Ala · Ala · Ala}}}{CH_3CHCNHCHCNHCHCOOH} + \underset{\substack{\displaystyle | \quad\quad | \\ NH_2 \;\; CH_3 \\ \text{Ala · Ala · Gly}}}{CH_3CHCNHCHCNHCH_2COOH}
$$

Protecting Groups

The solution to this problem is to "protect" the amino group of the first amino acid before we activate it and allow it to react with the second. By protecting the amino group we mean that we must convert it to some other group of low nucleophilicity—*one that will not react with a reactive acyl derivative.* The protecting group must be carefully chosen because after we have synthesized the amide linkage between the first amino acid and the second we will want to be able to remove the protecting group without disturbing the new amide bond.

A number of reagents have been developed to meet these requirements. Two that are often used are *benzyl chloroformate* and *tert-butyloxycarbonyl azide.*

$$
\underset{\text{Benzyl chloroformate}}{C_6H_5CH_2-O-\overset{\displaystyle O}{\overset{\|}{C}}-Cl} \qquad \underset{\textit{tert}\text{-Butyloxycarbonyl azide}}{(CH_3)_3C-O-\overset{\displaystyle O}{\overset{\|}{C}}-N_3}
$$

Both reagents react with amino groups (p. 969) to form derivatives that are unreactive toward further acylation. Both derivatives, however, are of a type that allow removal of the protecting group under conditions that do not affect peptide bonds. The benzyloxycarbonyl group (abbreviated Z-) can be removed by catalytic hydrogenation or by treating the derivative with cold HBr in acetic acid. The *tert*-butyloxycarbonyl group (abbreviated Boc-) can be removed through treatment with HCl or CF_3COOH in acetic acid.

Benzyloxycarbonyl Group

$$\text{Benzyl chloroformate} \quad \langle\bigcirc\rangle-CH_2O\overset{O}{\overset{\|}{C}}-Cl + H_2N-R \xrightarrow[25°]{OH^-} \langle\bigcirc\rangle-CH_2O\overset{O}{\overset{\|}{C}}-NH-R + Cl^-$$

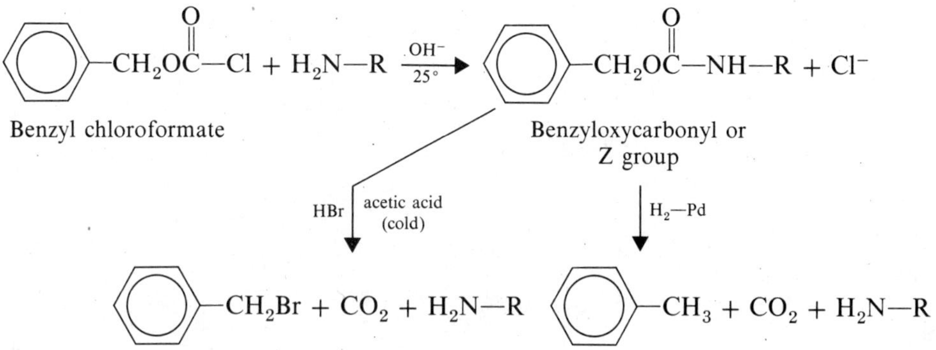

Benzyl chloroformate

Benzyloxycarbonyl or
Z group

$$\xrightarrow[\text{(cold)}]{\text{HBr} \quad \text{acetic acid}} \quad \langle\bigcirc\rangle-CH_2Br + CO_2 + H_2N-R$$

$$\xrightarrow{H_2-Pd} \quad \langle\bigcirc\rangle-CH_3 + CO_2 + H_2N-R$$

tert-Butyloxycarbonyl Group

$$(CH_3)_3C-O\overset{O}{\overset{\|}{C}}-N_3 + H_2N-R \xrightarrow[25°]{base} \underbrace{(CH_3)_3C-O\overset{O}{\overset{\|}{C}}-NHR}_{} + N_3^-$$

tert-Butoxycarbonyl
azide

tert-Butyloxycarbonyl
or Boc group

$$\xrightarrow[\text{acetic acid, 25°}]{\text{HCl or CF}_3\text{COOH} \atop \text{in}} \quad (CH_3)_2C=CH_2 + CO_2 + H_2N-R$$

The easy removal of both groups (Z- and Boc-) in acidic media results from the exceptional stability of the carbocations that are formed initially. The benzyloxycarbonyl group gives a *benzyl cation;* the *tert*-butyloxycarbonyl group yields, initially, a *tert-butyl cation.*

Removal of the benzyloxycarbonyl group with hydrogen and a catalyst depends on the fact that benzyl-oxygen bonds are weak and are subject to hydrogenolysis at low temperatures.

$$C_6H_5CH_2-O\overset{O}{\overset{\|}{C}}R \xrightarrow[25°]{H_2,Pd} C_6H_5CH_3 + HO\overset{O}{\overset{\|}{C}}R$$

A benzyl ester

Activation of the Carboxyl Group

Perhaps the most obvious way to activate a carboxyl group is to convert it to an acyl chloride. This method was used in early peptide syntheses but acyl chlorides are actually more reactive than necessary and, as a result, their use leads to complicating side reactions. A much better method is to convert the carboxyl group of the "protected" amino acid to a mixed anhydride using ethyl chloroform-

ate, $Cl-\overset{O}{\overset{\|}{C}}-OC_2H_5$

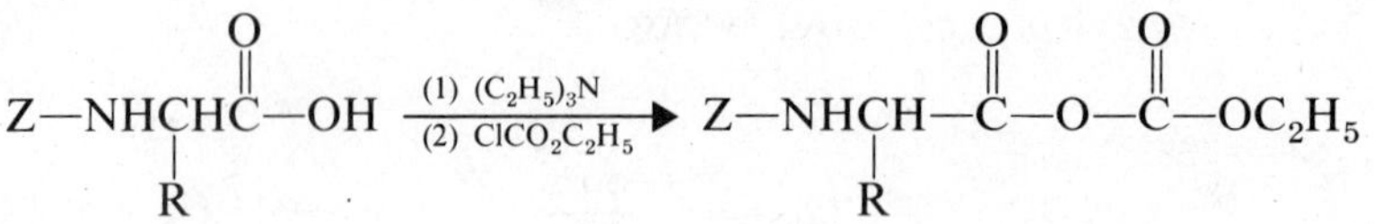

$$Z-NHCH\underset{R}{\overset{\overset{\displaystyle O}{\|}}{C}}-OH \xrightarrow[\text{(2) ClCO}_2\text{C}_2\text{H}_5]{\text{(1) (C}_2\text{H}_5)_3\text{N}} Z-NHCH\underset{R}{-}\overset{\overset{\displaystyle O}{\|}}{C}-O-\overset{\overset{\displaystyle O}{\|}}{C}-OC_2H_5$$

"Mixed anhydride"

The mixed anhydride can then be used to acylate another amino acid and form a peptide linkage.

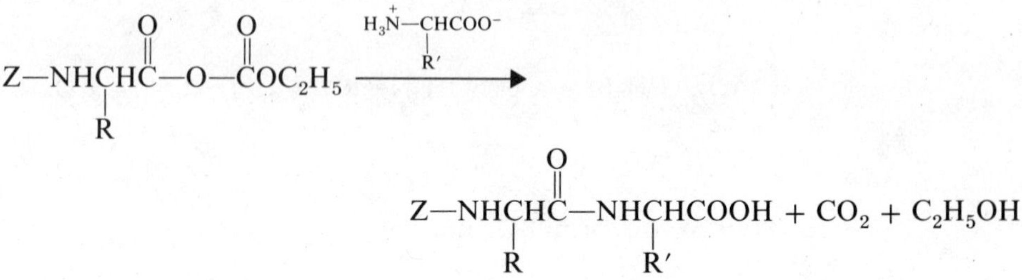

$$Z-NHCH\underset{R}{\overset{\overset{\displaystyle O}{\|}}{C}}-O-\overset{\overset{\displaystyle O}{\|}}{C}OC_2H_5 \xrightarrow{\overset{+}{H_3}N-\underset{R'}{CHCOO^-}}$$

$$Z-NHCH\underset{R}{\overset{\overset{\displaystyle O}{\|}}{C}}-NHCHCOOH \underset{R'}{} + CO_2 + C_2H_5OH$$

Peptide Synthesis

Let us examine now how we might use these reagents in the preparation of the simple dipeptide, Ala·Leu. The principles involved here can, of course, be extended to the synthesis of much longer polypeptide chains.

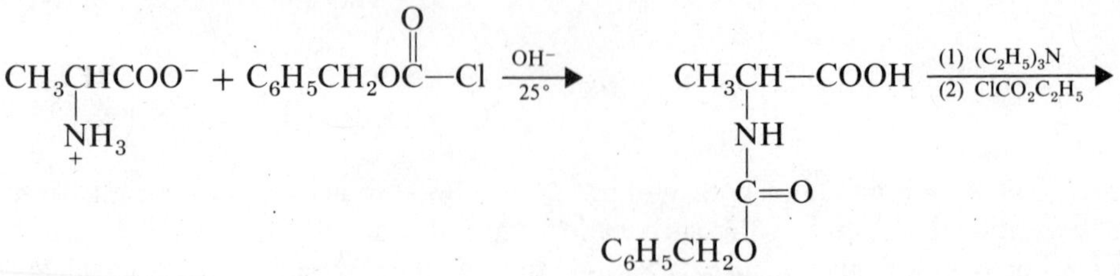

$$\underset{\underset{+}{\overset{|}{N}H_3}}{CH_3CHCOO^-} + C_6H_5CH_2O\overset{\overset{\displaystyle O}{\|}}{C}-Cl \xrightarrow[25°]{OH^-} CH_3CH-COOH \xrightarrow[\text{(2) ClCO}_2\text{C}_2\text{H}_5]{\text{(1) (C}_2\text{H}_5)_3\text{N}}$$

Ala   Benzyl chloro-   Z-Ala
formate

$$CH_3CH-\overset{\overset{\displaystyle O}{\|}}{C}-O\overset{\overset{\displaystyle O}{\|}}{C}OC_2H_5 \xrightarrow[CO_2 + C_2H_5OH]{\overset{\overset{+}{N}H_3}{(CH_3)_2CHCH_2CHCOO^-} \atop \text{Leu}}$$

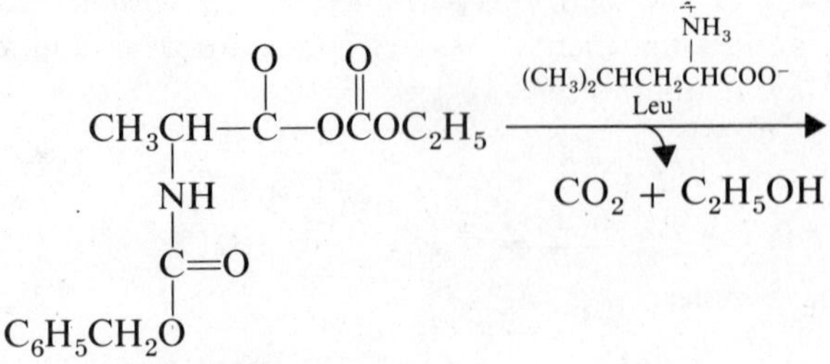

Mixed anhydride
of Z-Ala

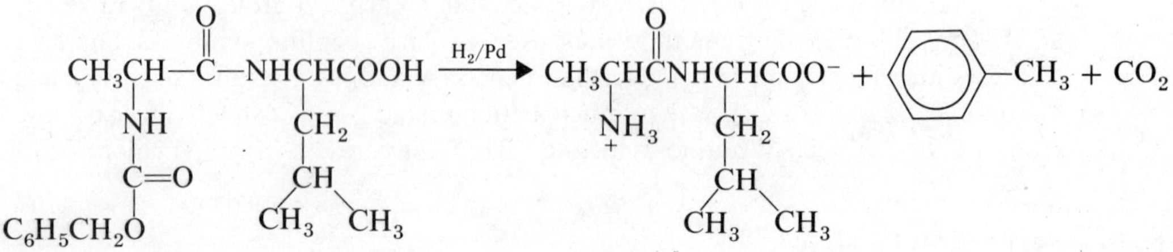

$$CH_3CH-\overset{\overset{\displaystyle O}{\|}}{C}-NHCHCOOH \xrightarrow{H_2/Pd} CH_3CH\overset{\overset{\displaystyle O}{\|}}{C}NHCHCOO^- + \bigcirc-CH_3 + CO_2$$

Z-Ala·Leu   Ala·Leu

Problem 25.11
_____
Show all steps in the synthesis of Gly·Val·Ala using the *tert*-butyloxycarbonyl (Boc-) group as a protecting group.

Problem 25.12
_____
The synthesis of a polypeptide containing lysine requires the protection of both amino groups. (a) Show how you might do this in a synthesis of Lys·Ile using the benzyloxycarbonyl group as a protecting group. (b) The benzyloxycarbonyl group

$$\overset{\text{NH}}{\underset{\|}{}}$$

can also be used to protect the guanido group, —NHC—NH$_2$, of arginine. Show a synthesis of Arg·Ala.

Problem 25.13
_____
The terminal carboxyl groups of glutamic acid and aspartic acid are often protected through their conversion to benzyl esters. What mild method could be used for removal of this protecting group?

Automated Peptide Synthesis

Although the methods that we have described thus far have been used to synthesize a number of polypeptides including ones as large as insulin, they are extremely time consuming. One must isolate and purify the product at almost every stage. Thus, a real advance in peptide synthesis came with the development by R. B. Merrifield (at Rockefeller University) of a procedure for automating peptide synthesis.

The Merrifield method is based on the use of a polystyrene resin similar to the one we saw on p. 958, *but one that contains —CH$_2$Cl groups* instead of sulfonic acid groups. This resin is used in the form of small beads and is insoluble in most solvents.

The first step in automated peptide synthesis (Fig. 25.6) involves a reaction that attaches the first protected amino acid residue to the resin beads. After this step is complete the protecting group is removed and the next amino acid (also protected) is condensed with the first using dicyclohexyl carbodiimide (p. 766) to activate its carboxyl group. Then removal of the protecting group of the second residue readies the resin-dipeptide for the next step.

The great advantage of this procedure is that purification of the resin with its attached polypeptide can be carried out at each stage by simply washing the resin with an appropriate solvent. Impurities, because they are not attached to the insoluble resin are simply carried away by the solvent. In the automated procedure each cycle of the "protein making machine" requires only four hours and attaches one new amino acid residue.

The Merrifield technique has been applied successfully to the synthesis of ribonuclease, a protein with 124 amino acid residues. The synthesis involved 369 chemical reactions and 11,931 automated steps—all were carried out without isolating an intermediate. The synthetic ribonuclease not only had the same physical characteristics of the natural enzyme; it possessed the biological activity as well.

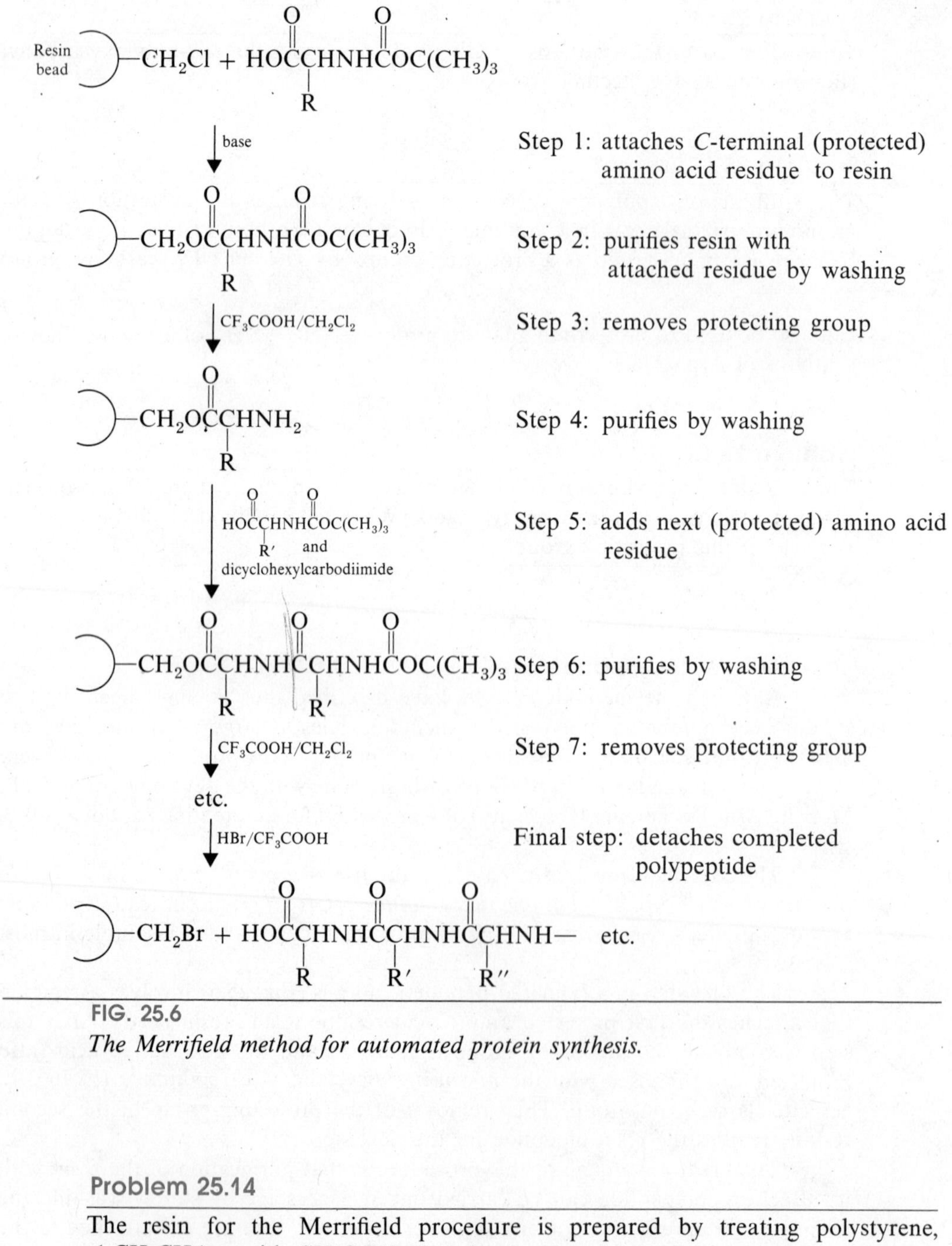

Resin bead $-CH_2Cl$ + $HOC\overset{O}{\underset{}{\parallel}}CHNH\overset{O}{\underset{}{\parallel}}COC(CH_3)_3$
            |
            R

↓ base

Step 1: attaches *C*-terminal (protected) amino acid residue to resin

$-CH_2O\overset{O}{\underset{}{\parallel}}C\underset{R}{\overset{}{\mid}}CHNH\overset{O}{\underset{}{\parallel}}COC(CH_3)_3$

Step 2: purifies resin with attached residue by washing

↓ CF_3COOH/CH_2Cl_2

Step 3: removes protecting group

$-CH_2O\overset{O}{\underset{}{\parallel}}C\underset{R}{\overset{}{\mid}}CHNH_2$

Step 4: purifies by washing

↓ $HOC\overset{O}{\underset{}{\parallel}}CHNH\overset{O}{\underset{}{\parallel}}COC(CH_3)_3$
  $\underset{R'}{}$ and
  dicyclohexylcarbodiimide

Step 5: adds next (protected) amino acid residue

$-CH_2O\overset{O}{\underset{R}{\parallel}}CCHNH\overset{O}{\underset{R'}{\parallel}}CCHNH\overset{O}{\underset{}{\parallel}}COC(CH_3)_3$

Step 6: purifies by washing

↓ CF_3COOH/CH_2Cl_2

Step 7: removes protecting group

etc.

↓ HBr/CF_3COOH

Final step: detaches completed polypeptide

$-CH_2Br$ + $HOC\overset{O}{\underset{R}{\parallel}}CCHNH\overset{O}{\underset{R'}{\parallel}}CCHNH\overset{O}{\underset{R''}{\parallel}}CCHNH-$ etc.

FIG. 25.6

The Merrifield method for automated protein synthesis.

Problem 25.14

The resin for the Merrifield procedure is prepared by treating polystyrene, $+CH_2CH+_n$, with CH_3OCH_2Cl and a Lewis acid catalyst. (a) What reaction is
            |
            C_6H_5
involved? (b) After purification, the completed polypeptide or protein can be detached from the resin by treating it with HBr in trifluoroacetic acid under conditions mild enough not to affect the amide linkages. What structural feature of the resin makes this possible?

Problem 25.15

Outline the steps in the synthesis of Ala · Phe · Lys using the Merrifield procedure.

25.9 SECONDARY AND TERTIARY STRUCTURE OF PROTEINS

We have seen how amide and disulfide linkages constitute the covalent or *primary structure* of proteins. Of equal importance in understanding how proteins function is knowledge of the way in which the peptide chains are arranged in three dimensions. Involved here are the secondary and tertiary structures of proteins.

Secondary Structure

The secondary structure of a protein is derived from the ways in which a polypeptide chain can interact with itself through the formation of hydrogen bonds.

The major experimental technique that has been used in elucidating the secondary structures of proteins is X-ray analysis.

When X-rays pass through a crystalline substance they produce diffraction patterns. Analysis of these patterns often yield indications of a regular repetition of particular structural units with certain specific distances between them, called *repeat distances*. The complete X-ray analysis of a molecule as complex as a protein often takes years of painstaking work. Nonetheless, many X-ray analyses have been done and they have revealed that the polypeptide chain of a natural protein can interact with itself in two major ways; through formation of a *pleated sheet* and an *α helix*.*

To understand how these interactions occur let us look first at what X-ray analysis has revealed about the geometry of the peptide bond itself. Peptide bonds tend to assume a geometry such that six atoms of the amide linkage are coplanar (Fig. 25.7). The carbon-nitrogen bond of the amide linkage is unusually short, indicating that resonance contributions of the type shown below are important.

$$\overset{\cdot\cdot}{\underset{/}{\overset{\backslash}{N}}}-C\overset{\overset{O}{\parallel}}{\underset{\backslash}{}} \longleftrightarrow \overset{+}{\underset{/}{\overset{\backslash}{N}}}=C\overset{\overset{O^-}{}}{\underset{\backslash}{}}$$

---

* Pioneers in the X-ray analysis of proteins were two American scientists, Linus Pauling (p. 607) and Robert B. Corey. Beginning in 1939, Pauling and Corey initiated a long series of studies of the conformations of peptide chains. At first they used crystals of single amino acids, then dipeptides and tripeptides, and so on. Moving on to larger and larger molecules and using the precisely constructed molecular models, they were able to understand the secondary structures of proteins for the first time.

FIG. 25.7

The geometry and bond distance of the peptide linkage. The six enclosed atoms tend to be coplanar and assume a 'transoid' arrangement.

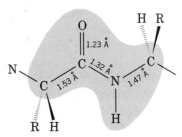

The carbon-nitrogen bond, consequently, has considerable double bond character (~40%) and rotations of groups about this bond are severely hindered.

Rotations of groups attached to the amide nitrogen and the carbonyl carbon are relatively free, however, and these rotations allow peptide chains to form different conformations.

The transoid arrangement of groups around the relatively rigid amide bond would cause the R groups to alternate from side to side of a single fully extended peptide chain:

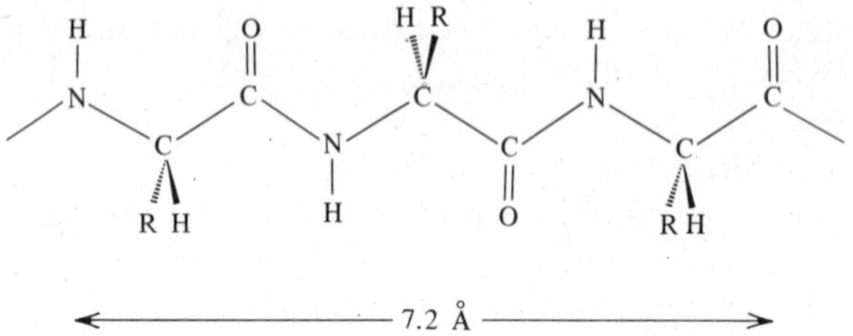

Calculations show that such a polypeptide chain would have a repeat distance (i.e., distance between alternating units) of 7.2 Å.

Fully extended polypeptide chains could conceivably form a flat-sheet structure with each alternating amino acid in each chain forming two hydrogen bonds with an amino acid in the adjacent chain:

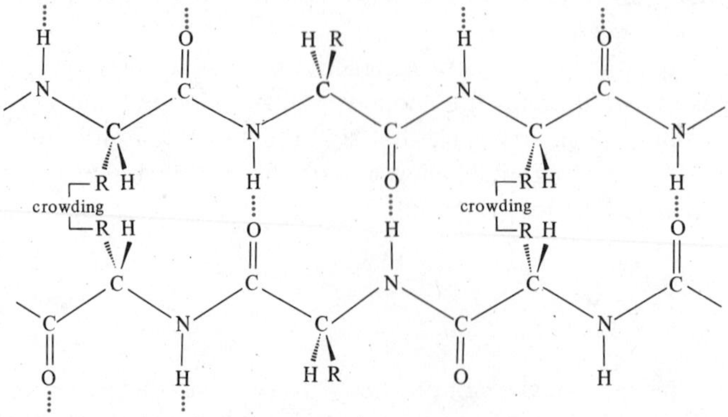

hypothetical flat–sheet structure

This structure does not exist in naturally occurring proteins because of the crowding that would exist between R groups. If such a structure did exist, it would have the same repeat distance as the fully extended peptide chain, that is, 7.2 Å.

Slight rotations of bonds, however, can transform a flat-sheet structure into what is called the *pleated-sheet* or *β configuration*. The pleated-sheet structure gives small and medium sized R groups room enough to avoid van der Waals repulsions and is the predominant structure of silk fibroin (48% glycine and 38% serine and alanine residues). The pleated-sheet structure has a slightly shorter repeat distance, 7.0 Å, than the flat sheet.

Of far more importance in naturally occurring proteins is the structure called the α helix (Fig. 25.8). This structure is a right-handed helix with 3.6 amino acid

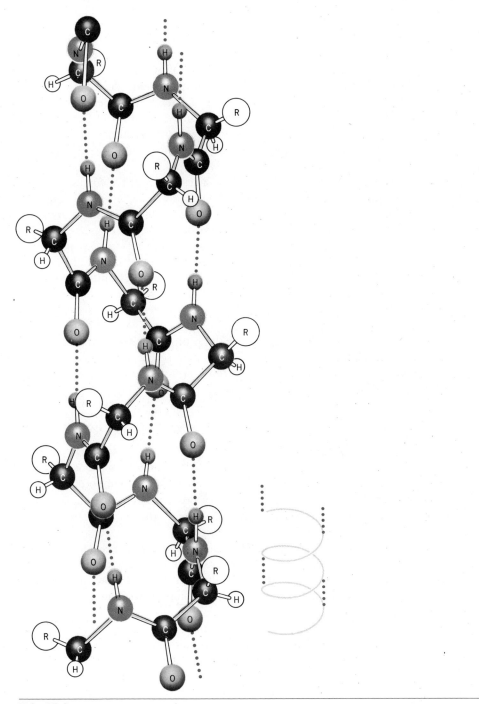

FIG. 25.8
*A representation of the α-helical structure of a polypep-
tide. Hydrogen bonds are denoted by dotted lines.*

residues per turn. Each amide group in the chain is bonded to an amide group at a distance of three amino acid residues in either direction and the R groups all extend away from the axis of the helix. The repeat distance of the α helix is 1.5 Å.

The α-helical structure is found in many proteins; it is the predominant structure of the polypeptide chains of fibrous proteins such as *myosin,* the protein of muscle, and of *α-keratin,* the protein of hair, unstretched wool, and nails.

Not all peptide chains can exist in an α-helical form. Certain peptide chains assume what is called a *random coil arrangement,* a structure that is flexible, changing, and statistically random. Synthetic polylysine, for example, exists as a random coil and does not normally form an α helix. At pH 7 the ϵ-amino groups of the lysine residues are positively charged and, as a result, repulsive forces between them are so large they overcome any stabilization that would be gained through hydrogen bond formation of an α helix. At pH 12, however, the ϵ-amino groups are uncharged and polylysine spontaneously forms an α helix.

The presence of proline or hydroxyproline residues in polypeptide chains produces another striking effect: because the nitrogen atoms of these amino acids are part of five-membered rings, the groups attached by the nitrogen—α-carbon bond cannot rotate enough to allow for an α-helical structure. Wherever proline or hydroxyproline occur in a peptide chain their presence causes a kink or bend and interrupts the α helix.

The polypeptide chains of globular proteins such as hemoglobin, ribonuclease, α-chymotrypsin, and lysozyme (p. 977) contain segments of α helix and segments of random coil. Proline and hydroxyproline are often found in those regions of the structure where the conformation changes.

Tertiary Structure

The tertiary structure of a protein is its three-dimensional shape that arises from foldings of its polypeptide chains. These foldings do not occur randomly: under the proper environmental conditions they occur in one particular way—a way that is characteristic of a particular protein and one that is often highly important to its function.

A variety of forces are involved in the folding process including the disulfide bonds of the primary structure. One characteristic of most proteins is that the folding takes place in such a way as to expose the maximum number of polar (hydrophilic) groups to the aqueous environment and enclose a maximum number of nonpolar (hydrophobic) groups within its interior.

The soluble globular proteins tend to be much more highly folded than fibrous proteins. However, fibrous proteins also have a tertiary structure; the α-helical strands of α-keratin, for example, are wound together into a "super helix." This super helix has a repeat distance of 5.1 Å units indicating that the super helix makes one complete turn for each 35 turns of the α helix. The tertiary structure does not end here, however. Even the super helixes can be wound together to give a ropelike structure of seven strands.

Myoglobin (Fig. 25.9) and hemoglobin (Sec. 25.11) were the first proteins (in 1957 and 1959) to be subjected to a completely successful X-ray analysis. This work was accomplished by J. C. Kendrew and Max Perutz at Cambridge University

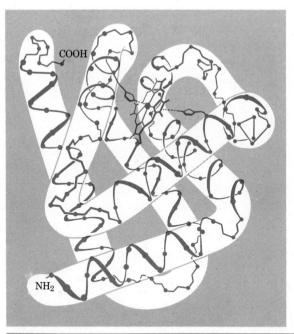

FIG. 25.9

The three-dimensional structure of myoglobin. (From R.E. Dickerson, The Proteins II, *H. Neurath, ed., Academic Press, New York, 1964, p. 634. Used with permission.)*

in England. (They received the Nobel Prize in 1962.) Since then a number of other proteins including lysozyme, ribonuclease, and α-chymotrypsin have yielded to complete structural analysis.

25.10 LYSOZYME: MODE OF ACTION OF AN ENZYME

Lysozyme is made up of 129 amino acid residues (Fig. 25.10). Three short segments of the chain between residues 5–15, 24–34, and 88–96 have the structure of an α helix; the residues between 41–45 and 50–54 form pleated sheets, and a hairpin turn occurs at residues 46–49. The remaining polypeptide segments of lysozyme have a random coil arrangement.

The discovery of lysozyme is an interesting story in itself:

"One day in 1922 Alexander Fleming was suffering from a cold. This is not unusual in London, but Fleming was a most unusual man and he took advantage of the cold in a characteristic way. He allowed a few drops of his nasal mucus to fall on a culture of bacteria he was working with and then put the plate to one side to see what would happen. Imagine his excitement when he discovered some time later that the bacteria near the mucus had dissolved away. For a while he thought his ambition of finding a universal antibiotic had been realized. In a burst of activity he quickly established that the antibacterial action of the mucus was due

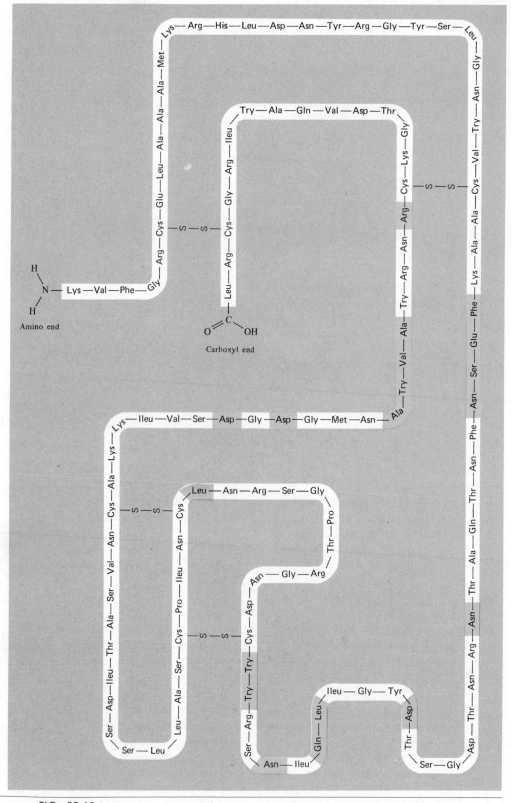

FIG. 25.10
The covalent structure of lysozyme. The amino acids that line the active site of lysozyme are shown in color.

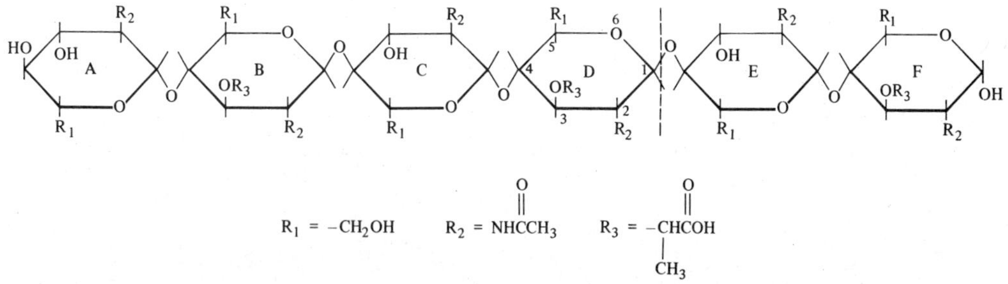

$$R_1 = -CH_2OH \qquad R_2 = NHCCH_3 \qquad R_3 = -CHCOH$$

FIG. 25.11

A hexasaccharide that has the general structure as the cell wall polysaccharide on which lysozyme acts. Two different amino sugars are present: rings A, C, and E are derived from a monosaccharide called N-acetylglucosamine; rings B, D, and F are derived from a monosaccharide called N-acetylmuramic acid. When lysozyme acts on this oligosaccharide, hydrolysis takes place and it results in cleavage of the glycosidic linkage between rings D and E.

to the presence of an enzyme; he called this substance lysozyme because of its capacity to lyse, or dissolve the bacterial cells. Lysozyme was soon discovered in many tissues and secretions of the human body, in plants and most plentifully of all in the white of an egg. Unfortunately Fleming found that it is not effective against the most harmful bacteria. He had to wait seven years before a strangely similar experiment revealed the existence of a genuinely effective antibiotic: penicillin."

This story was related by Professor David C. Phillips of Oxford University who many years later used X-ray analysis to discover the three-dimensional structure of lysozyme.*

Phillips' X-ray diffraction studies of lysozyme are especially interesting because they have also revealed important information about how this enzyme acts on its substrate. Lysozyme's substrate is a polysaccharide of amino sugars that makes up part of the bacterial cell wall. An oligosaccharide that has the same general structure as the cell wall polysaccharide is shown in Fig. 25.11.

By using oligosaccharides (made up of *N*-acetylglucosamine units only) on which lysozyme acts very slowly, Phillips and his co-workers were able to discover how the substrate fits into the enzyme's active site. This site is a deep cleft in the lysozyme structure (Fig. 25.12*a*). The oligosaccharide is held in this cleft by hydrogen bonds and as the enzyme binds the substrate two important changes take place: the cleft in the enzyme closes slightly and ring D of the oligosaccharide is "flattened" out of its stable chair conformation. This flattening causes atoms 1, 2, 5, and 6 of ring D to become coplanar; it also distorts ring D in such a way as to make the glycosidic linkage between it and ring E more suceptible to hydrolysis.†

Hydrolysis of the glycosidic linkage probably takes the course illustrated in Fig. 25.12*b*. The carboxyl group of glutamic acid (residue number 35) donates a proton to the oxygen between rings D and E. Protonation leads to cleavage of the

---

* Quotation from David C. Phillips, "The Three-Dimensional Structures of an Enzyme Molecule," Copyright © (1966) by Scientific American, Inc. All rights reserved.

† R. H. Lemieux and G. Huber of the National Research Council of Canada have shown that when an aldohexose is converted to a carbocation the ring of the carbocation assumes just this flattened conformation.

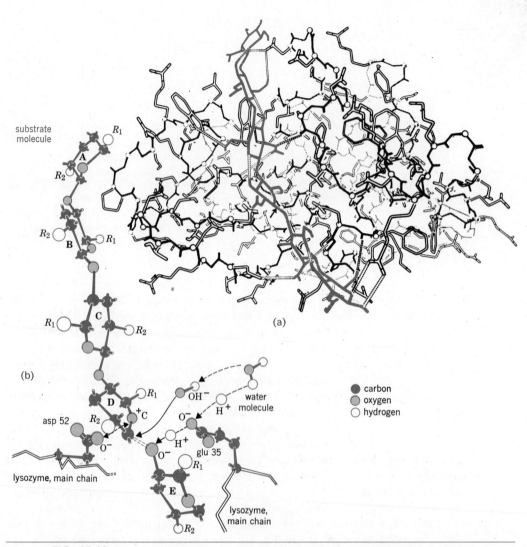

substrate
molecule

R_1

A

R_2

R_2 B R_1

R_1 C R_2

(a)

(b)

R_1

D

OH⁻

water
molecule

H⁺

asp 52 R_2 ⁺C

O⁻

H⁺

glu 35

R_1

O⁻ O⁻

lysozyme, main chain

R_1

E

lysozyme,
main chain

R_2

● carbon
● oxygen
○ hydrogen

FIG. 25.12

(a) *This drawing shows the backbones of the lysozyme-substrate complex. The substrate (in this drawing a hexasaccharide) fits into a cleft in the lysozyme structure and is held in place by hydrogen bonds. As lysozyme binds the oligosaccharide, the cleft in its structure closes slightly. (Adapted with permission from* Atlas of Protein Sequence and Structure *1969, ed. Margaret O. Dayoff, National Biomedical Research Foundation, Washington, D. C. 1969. The drawing was made by Irving Geis based on his perspective painting of the molecule which appeared in* Scientific American, *November 1966. The painting was made of an actual model assembled at the Royal Institution, London, by D. C. Phillips and his colleagues, based on their X-ray crystallography results.*) (b) *A possible mechanism for lysozyme action. This drawing shows an expanded portion of the figure above and illustrates how hydrolysis of the acetal linkage between rings D and E of the substrate may occur. Glutamic acid (residue 35) donates a proton to the interring oxygen. This causes the formation of a carbocation that is stabilized by the carboxylate ion aspartic acid (residue 52). A water molecule supplies an* OH^- *to the carbonium ion and* H^+ *to glutamic acid. (Adapted with permission from "The Three-Dimensional Structures of an Enzyme Molecule," by David C. Phillips. Copyright ©* *Nov. 1966 by Scientific American, Inc., All rights reserved.*)

glycosidic link and to the formation of a carbocation at C—1 of ring D. This carbocation is stabilized by the negatively charged carboxylate group of aspartic acid (residue number 52) which lies in close proximity. A water molecule diffuses in and supplies an OH⁻ ion to the carbonium ion and a proton to replace that lost by glutamic acid.

When the polysaccharide is a part of a bacterial cell wall, lysozyme probably first attaches itself to the cell wall by hydrogen bonds. After hydrolysis has taken place lysozyme falls away leaving behind a bacterium with a punctured cell wall.

25.11 HEMOGLOBIN: A CONJUGATED PROTEIN

Some proteins contain as a part of their structure a nonprotein group called a *prosthetic group*. An example is the oxygen-carrying protein, hemoglobin. Each of the four polypeptide chains of hemoglobin is bound to a prosthetic group called *heme* (Fig. 25.13). The four polypeptide chains of hemoglobin are wound in such a way as to give hemoglobin a roughly spherical shape. Moreover, each heme group lies in a crevice with the hydrophobic vinyl groups of its porphyrin structure surrounded by side chains of hydrophobic amino residues. The two propanoate side chains of heme lie near positively charged amino groups of lysine and arginine residues.

The iron of the heme group is in the +2 (ferrous) oxidation state and it forms a coordinate bond to a nitrogen of the imidazole group of histidine of the polypeptide chain. This leaves one valence of the ferrous ion free to combine with oxygen as shown on the next page.

FIG. 25.13

The structure of heme, the prosthetic group of hemo-globin. Heme has a structure similar to that of chlorophyll (p. 913) in that each is derived from a heterocyclic, porphyrin, ring. The iron of heme is in the ferrous (+2) oxidation state.

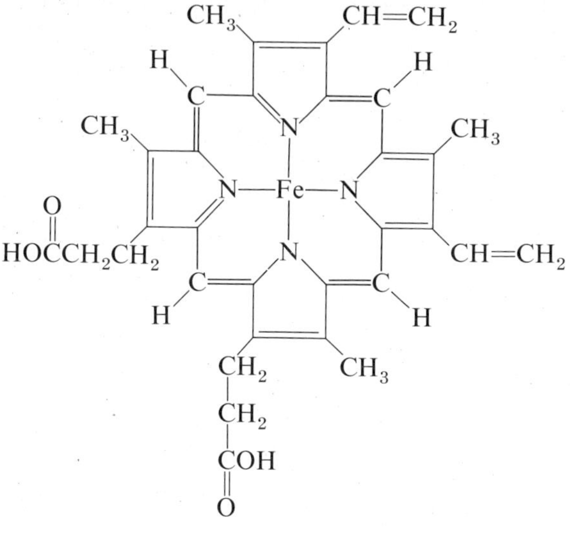

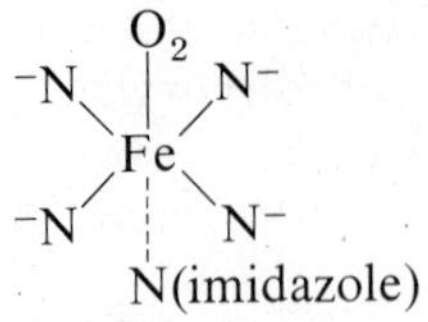

O_2

$-N$ — $N-$

Fe

$-N$ — $N-$

N(imidazole)

A portion of oxygenated
hemoglobin

The fact that the ferrous ion of the heme group combines with oxygen is not particularly remarkable; many similar compounds do the same thing. What is remarkable about hemoglobin is that when the heme combines with oxygen the ferrous ion does not become readily oxidized to the ferric state. Studies with model heme compounds in water, for example, show that they undergo a rapid combination with oxygen but they also undergo a rapid oxidation of the iron from Fe^{+2} to Fe^{+3}. When these same compounds are embedded in the hydrophobic environment of a polystyrene resin, however, the iron is easily oxygenated and deoxygenated and this occurs *with no change in oxidation state of iron*. In this respect, it is especially interesting to note that X-ray studies of hemoglobin have revealed that the polypeptide chains provide each heme group with a similar hydrophobic environment.

Additional Problems

25.16
(a) Which amino acids in Table 25.1 have more than one chiral carbon? (b) Write projection formulas for the isomers of each of these amino acids that would have the L configuration at the α-carbon. (c) What kind of isomers have you drawn in each case?

25.17
(a) Which amino acid in Table 25.1 would react with nitrous acid (i.e., a solution of $NaNO_2$ and HCl) to yield lactic acid? (b) All of the amino acids in Table 25.1 liberate nitrogen when they are treated with nitrous acid except two; which are these? (c) What product would you expect to obtain from treating tyrosine with excess bromine water? (d) What product would you expect to be formed in the reaction of phenylalanine with ethanol in presence of hydrogen chloride? (e) What product would you expect from the reaction of alanine and benzoyl chloride in aqueous base?

25.18
(a) On the basis of the following sequence of reactions Emil Fischer was able to show that (−)serine and L-(+)-alanine have the same configuration. Write projection formulas for the intermediates **A–C**.

$(-)$serine $\xrightarrow[\text{CH}_3\text{OH}]{\text{HCl}}$ **A** $(C_4H_{10}ClNO_3)$ $\xrightarrow{\text{PCl}_5}$ **B** $(C_4H_9Cl_2NO_2)$ $\xrightarrow[\text{(2) OH}^-]{\text{(1) H}_3\text{O}^+\text{, H}_2\text{O, heat}}$

C $(C_3H_6ClNO_2)$ $\xrightarrow[\text{dil. H}_3\text{O}^+]{\text{Na-Hg}}$ L-(+)-alanine

(b) The configuration of L-(−)-cysteine can be related to that of L-(−)-serine through the following reactions. Write projection formulas for **D** and **E**.

B [from part (a) above] $\xrightarrow{\text{OH}^-}$ **D** $(C_4H_8ClNO_2)$ $\xrightarrow{\text{NaSH}}$

E $(C_4H_9NO_2S)$ $\xrightarrow[\text{(2) OH}^-]{\text{(1) H}_3\text{O}^+\text{, H}_2\text{O, heat}}$ L-(−)-cysteine

(c) The configuration of L-(−)-asparagine can be related to that of L-(−)-serine in the following way. What is the structure of F?

$$\text{L-(−)-asparagine} \xrightarrow[\substack{\text{Hofmann}\\\text{degradation}}]{\text{NaOBr/OH}^-} \textbf{F} \ (C_3H_7N_2O_2)$$

$$\textbf{C} \text{ [from part (a)]} \xrightarrow{\text{NH}_3}$$

25.19

(a) DL-Glutamic acid has been synthesized from diethyl acetamidomalonate in the following way: outline the reactions involved.

$$\underset{\substack{\text{Diethyl acetamido-}\\\text{malonate}}}{\overset{\displaystyle O}{\overset{\|}{CH_3CNHCH(CO_2C_2H_5)_2}}} + CH_2{=}CH{-}C{\equiv}N \xrightarrow[\substack{C_2H_5OH\\(95\% \text{ yield})}]{\text{NaOC}_2H_5}$$

$$\textbf{G} \ (C_{12}H_{18}N_2O_5) \xrightarrow[\substack{\text{reflux 6 hr.}\\(66\% \text{ yield})}]{\text{conc. HCl}} \text{DL-Glutamic acid}$$

(b) Compound **G** has also been used to prepare the amino acid, DL-ornithine through the route shown below.

$$\textbf{G} \ (C_{12}H_{18}N_2O_5) \xrightarrow[\substack{68°, \ 1000 \text{ psi}\\(90\% \text{ yield})}]{\text{H}_2/\text{Ni}} \textbf{H} \ (C_{10}H_{16}N_2O_4, \text{ a } \delta\text{-lactam}) \xrightarrow[\substack{\text{reflux 4 hr.}\\(97\% \text{ yield})}]{\text{conc. HCl}}$$

$$\text{DL-Ornithine hydrochloride, } C_5H_{13}ClN_2O_2$$

L-Ornithine is a naturally occurring amino acid but does not occur in proteins. In one metabolic pathway L-ornithine serves as a precursor for L-arginine.

25.20

The *Strecker synthesis* of DL-alanine is outlined below:

$$\underset{\text{Acetaldehyde}}{CH_3CHO} \xrightarrow[\text{HCN}]{\text{NH}_3} \underset{\underset{NH_2}{|}}{CH_3CHC{\equiv}N} \xrightarrow{\text{H}_3O^+} \underset{\underset{\text{DL-Alanine}}{\underset{^+NH_3}{|}}}{CH_3CHCOO^-}$$

(a) Outline a Strecker synthesis of DL-phenylalanine. (b) DL-methionine can also be synthesized by a Strecker synthesis. The required starting aldehyde can be prepared from acrolein, $CH_2{=}CHCHO$, and methanethiol, CH_3SH. Outline all steps in this synthesis of DL-methionine.

25.21

Transamination reactions (p. 956) involving L-glutamic acid provide animals with pathways for redistributing amino groups. These pathways are important because in a given meal an animal may eat protein that, when hydrolyzed, provides it with a mixture of amino acids quite different from that which is optimal for its metabolism. It might, for example, ingest protein rich in phenylalanine but poor in aspartic acid. Assuming that α-ketoglutaric acid and oxaloacetic acid are available in the cell as a result of carbohydrate metabolism, show how two transamination reactions (the first synthesizing glutamic acid) would result in a net synthesis of aspartic acid from phenylalanine.

25.22

Bradykinin is a nonapeptide released by blood plasma globulins in response to a wasp sting. It is a very potent pain-causing agent. Its molecular formula is Arg_2, Gly, Phe_2, $Pro_3 \cdot$ Ser. The use of 2,4-dinitrofluorobenzene and carboxypeptidase show that both terminal residues are arginine. Partial acid hydrolysis of Bradykinin gives the following di- and tripeptides:

$$Phe \cdot Ser + Pro \cdot Gly \cdot Phe + Pro \cdot Pro + Ser \cdot Pro \cdot Phe + Phe \cdot Arg + Arg \cdot Pro$$

What is the amino acid sequence of Bradykinin?

25.23

Complete hydrolysis of a heptapeptide showed that it had the following molecular formula:

$$Ala_2, Glu, Leu, Lys, Phe, Val$$

Deduce the amino acid sequence of this heptapeptide from the following data.
1. Treatment of the heptapeptide with 2,4-dinitrofluorobenzene followed by incomplete hydrolysis gave, among other products: Val labeled at the α-amine group, lysine labeled at the ϵ-amino group, and a dipeptide, DNP—Val $\cdot$ Leu (DNP = 2,4-Dinitrophenyl-).
2. Hydrolysis of the heptapeptide with carboxypeptidase gives an initial high concentration of alanine, followed by a rising concentration of glutamic acid.
3. Partial enzymatic hydrolysis of the heptapeptide gave a dipeptide, **A,** and a tripeptide, **B.**

    (a) Treatment of **A** with 2,4-dinitrofluorobenzene followed by hydrolysis gave DNP-labeled leucine and lysine labeled only at the ϵ-amino group.

    (b) Complete hydrolysis of **B** gave phenylalanine, glutamic acid, and alanine. When **B** was allowed to react with carboxypeptidase the solution showed an initial high concentration of glutamic acid. Treatment of **B** with 2,4-dinitrofluorobenzene followed by hydrolysis gave labeled phenylalanine.

25.24

Synthetic polyglutamic acid exists as an α helix in solution at pH 2–3. When the pH of such a solution is gradually raised through the addition of base, a dramatic change in optical rotation takes place at pH 5. This change has been associated with the unfolding of the α helix and the formation of a random coil. What structural feature of polyglutamic acid, and what chemical change, can you suggest as an explanation for this transformation?

* 25.25

Part of the evidence for restricted rotation about the carbon-nitrogen bond in a peptide linkage (see pp. 973–974) comes from pmr studies done with simple amides. For example, at room temperature and with the instrument operating at 60 MHz, the pmr spectrum of N,N-dimethylformamide, $(CH_3)_2NCHO$, shows a doublet at $\delta 2.80 \, (3H)$, a doublet at $\delta 2.95 \, (3H)$ and a multiplet at $\delta 8.05 \, (1H)$. When the spectrum is determined at lower magnetic field strength (i.e., with the instrument operating at 30 MHz) the doublets are found to have shifted so that the distance (in Hertz) that separates one doublet from the other is smaller. When the temperature at which the spectrum is determined is raised, the doublets persist until a temperature of 111° is reached, then the doublets coalesce to become a single signal. Explain in detail how these observations are consistent with the existence of a relatively large barrier to rotation about the carbon-nitrogen bond of N,N-dimethylformamide.

26

SPECIAL TOPICS V
NUCLEIC ACIDS:
PROTEIN SYNTHESIS

"... I cannot help wondering whether some day an enthusiastic scientist will christen his new born twins Adenine and Thymine."

F. H. C. Crick*

26.1 INTRODUCTION

The molecules that preserve hereditary information and that transcribe and translate that information in a way that allows the synthesis of all the varied enzymes of the cell are the nucleic acids, deoxyribonucleic acid (DNA) and ribonucleic acid (RNA). These biological polymers are sometimes found associated with proteins and in this form they are known as *nucleoproteins*.

It has been from studies of nucleic acids themselves that has come much of our knowledge of how genetic information is preserved, how it is passed on to succeeding generations of the organism, and how it is transformed into the working parts of the cell. For these reasons we will focus our attention on the structures and properties of nucleic acids and of their components, *nucleotides* and *nucleosides*.

26.2 NUCLEOTIDES AND NUCLEOSIDES

Mild degradations of nucleic acids yield their monomeric units, compounds that are called *nucleotides*. A general formula for a nucleotide and the specific structure of one nucleotide, called adenylic acid, are shown in Fig. 26.1.

Complete hydrolysis of a nucleotide furnishes:

1. A heterocyclic base, either a purine or pyrimidine.
2. A five-carbon monosaccharide, either D-ribose or 2-deoxy-D-ribose.
3. A phosphate ion.

The central portion of the nucleotide is the monosaccharide and it is always present as a five-membered ring, that is, as a furanoside. The heterocyclic base of a nucleotide is attached through an *N*-glycosidic linkage to C-1' of the ribose or deoxyribose unit and this linkage is always β. The phosphate group of a nucleotide is present as a phosphate ester and it may be attached at C-5' or C-3'. (In nucleotides the carbons of the monosaccharide portion are designated with primed numbers, i.e., 1', 2', 3', and so on.)

Removal of the phosphate group of a nucleotide converts it to a compound known as a *nucleoside*. The nucleosides that can be obtained from DNA all contain

---

* Who along with J. D. Watson and Maurice Wilkins shared the Nobel Prize in 1962 for their proposal of (and evidence for) the double helix structure of DNA. (Taken from F. H. C. Crick, "The Structure of the Hereditary Material," *Scientific American,* October 1954.)

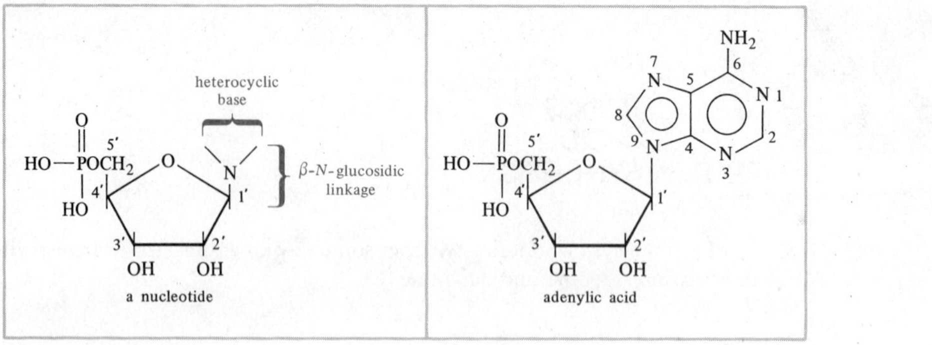

(a)   (b)

FIG. 26.1

(a) *General structure of a nucleotide obtained from RNA. The heterocyclic base is a purine or pyrimidine. In nucleotides obtained from DNA, the sugar component is 2'-deoxyribose, that is, the —OH at position 2' is replaced by —H. The phosphate group of the nucleotide shown above is attached to the 5'-carbon; it may also be attached to the 3'-carbon. The heterocyclic base is always attached through a β-glycosidic linkage at C-1'.* (b) *Adenylic acid, a typical nucleotide.*

2-deoxy-D-ribose as their sugar component and one of four heterocyclic bases, either adenine, guanine, cytosine, or thymine:

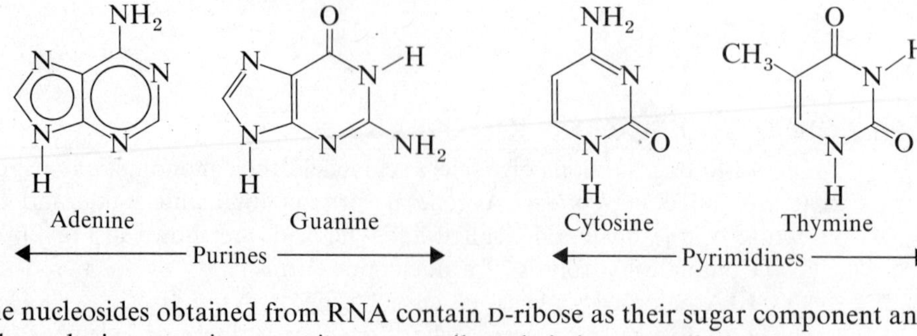

The nucleosides obtained from RNA contain D-ribose as their sugar component and either adenine, guanine, cytosine, or uracil as their heterocyclic base.*

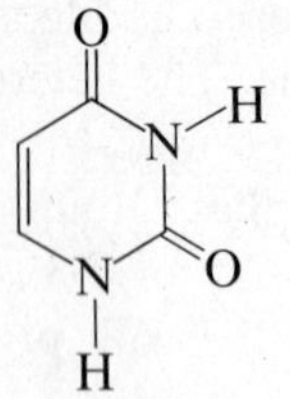

Uracil
(a pyrimidine)

* Notice that in an RNA nucleoside (or nucleotide) uracil replaces thymine. (Some nucleosides obtained from specialized forms of RNA may also contain other, but similar, purines and pyrimidines.)

The heterocyclic bases obtained from nucleosides are capable of existing in more than one tautomeric form. The forms that we have shown (p. 986) are the predominant forms that the bases assume when they are present in nucleic acids.

Problem 26.1

Write the structures of the other tautomeric forms of adenine, guanine, cytosine, thymine, and uracil.

The names and structures of the nucleosides found in DNA are shown in Fig. 26.2; those found in RNA are given in Fig. 26.3.

Problem 26.2

The nucleosides shown in Figs. 26.2 and 26.3 are stable in dilute base. In dilute acid, however, they undergo rapid hydrolysis yielding a sugar (deoxyribose or ribose) and a heterocyclic base. (a) What structural feature of the nucleoside accounts for this behavior? (b) Propose a reasonable mechanism for the hydrolysis.

Nucleotides are named in several ways. Adenylic acid (Fig. 26.1), for example, is sometimes called 5′-adenylic acid in order to designate the position of the phosphate group; it is also called adenosine 5′-phosphate, or simply adenosine mono-

FIG. 26.2
Nucleosides that can be obtained from DNA. (Undesignated bonds are to —H.)

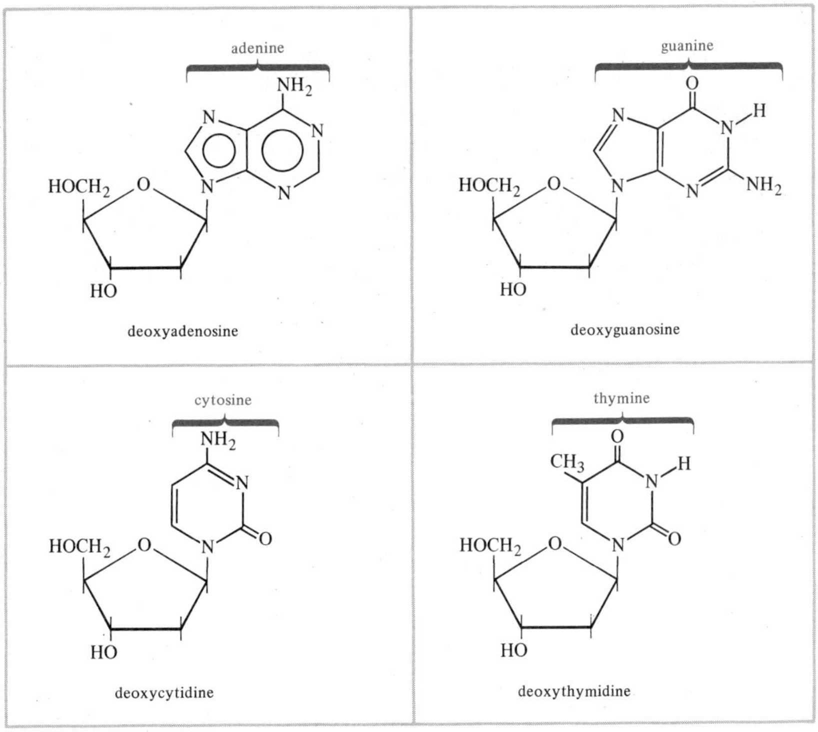

adenine

deoxyadenosine

guanine

deoxyguanosine

cytosine

deoxycytidine

thymine

deoxythymidine

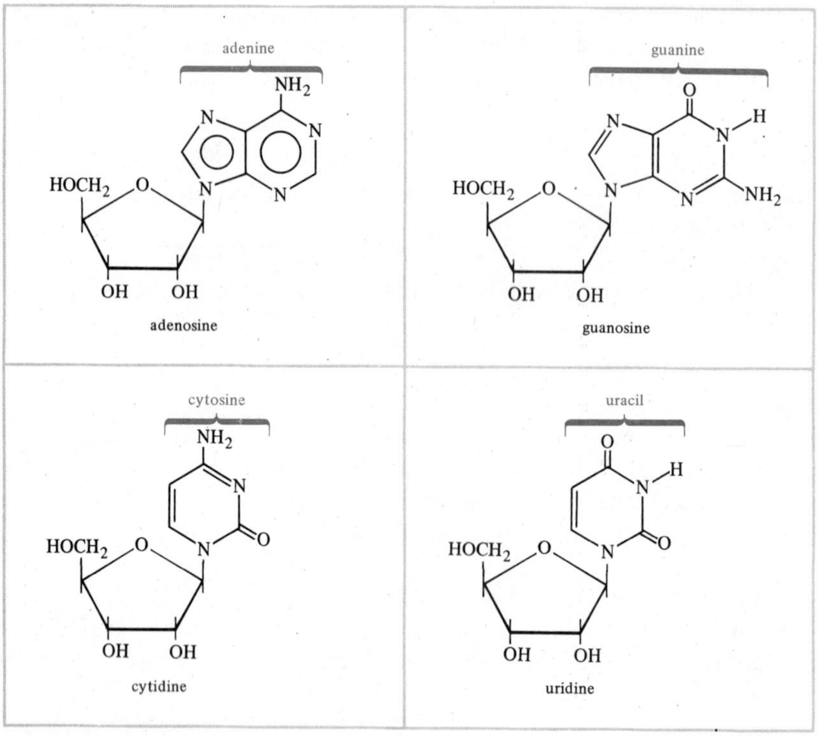

FIG. 26.3
Nucleosides that can be obtained from RNA. (Undesignated bonds are to
—H.)

phosphate (AMP). Uridylic acid is called 5′-uridylic acid, uridine 5′-phosphate, or uridine monophosphate (UMP), and so on.

Nucleosides and nucleotides are found in places other than as part of the structure of DNA and RNA. We have seen, for example, that adenosine units are part of the structures of the important coenzymes, NAD (p. 439) and coenzyme A (p. 771). The 5′-triphosphate of adenosine is, of course, the important energy source, ATP (p. 914). The compound called, 3′,5′-cyclic adenylic acid (Fig. 26.4) is an important regulator of hormone activity. Cells synthesize this compound from ATP through the action of an enzyme, *adenyl cyclase*. In the laboratory, 3′,5′-cyclic adenylic acid

FIG. 26.4
3′,5′-Cyclic adenylic acid and its
laboratory and biosynthesis.

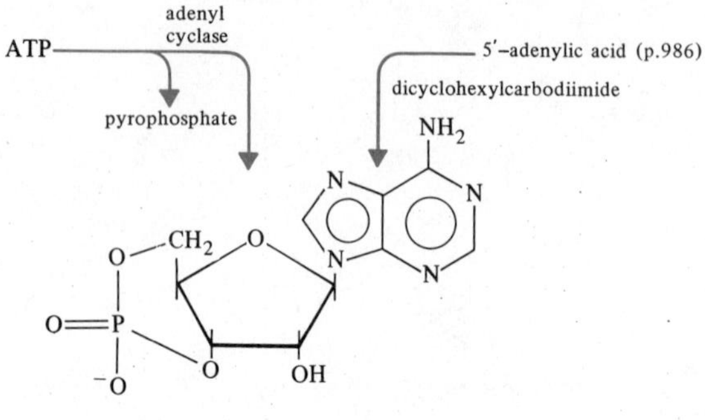

can be prepared through dehydration of 5′-adenylic acid with dicyclohexyl carbodi-imide.

Problem 26.3

When 3′,5′-cyclic adenylic acid is treated with aqueous sodium hydroxide the major product that is obtained is 3′-adenylic acid (adenosine 3′-phosphate) rather than 5′-adenylic acid. Suggest an explanation that accounts for the course of this reaction.

26.3 LABORATORY SYNTHESIS OF NUCLEOSIDES AND NUCLEOTIDES

A variety of methods have been developed for the synthesis of nucleosides. One technique uses reactions that assemble the nucleoside from suitably activated and protected ribose derivatives and heterocyclic bases. An example is the following synthesis of adenosine.

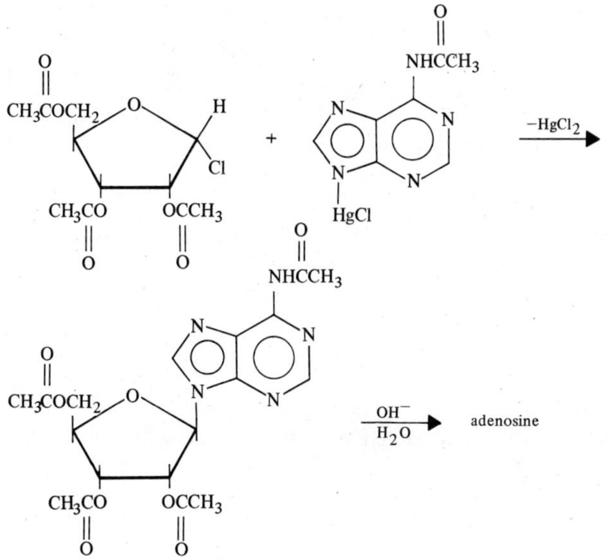

Another technique involves formation of the heterocyclic base on a protected *N*-aminoribose derivative:

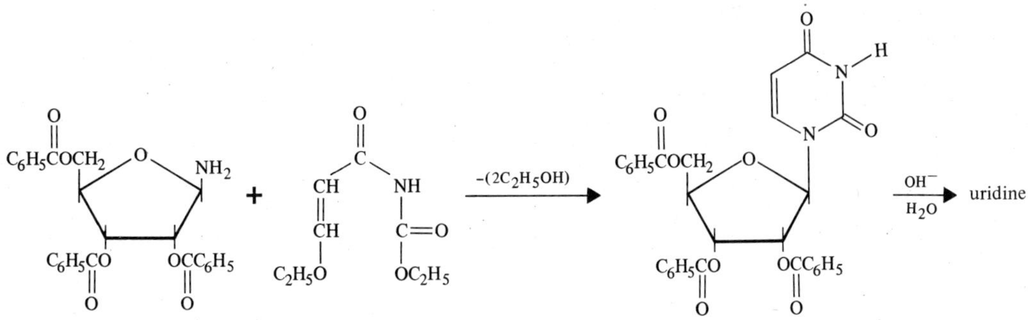

2,3,5–tri–*O*–benzoyl–
β–D–ribofuranosylamine

β–ethoxy–*N*–ethoxy–
carbonylacrylamide

Problem 26.4

Basing your answer on reactions that you have seen before, propose a likely mechanism for the condensation reaction given at the bottom of page 989.

Still a third technique involves the synthesis of a nucleoside with a substituent in the heterocyclic ring that can be replaced with other groups. This method has been used extensively to synthesize unusual nucleosides, ones that do not necessarily occur naturally. An example (below) makes use of a 6-chloropurine derivative obtained from the appropriate ribofuranosyl chloride and chloromercuripurine.

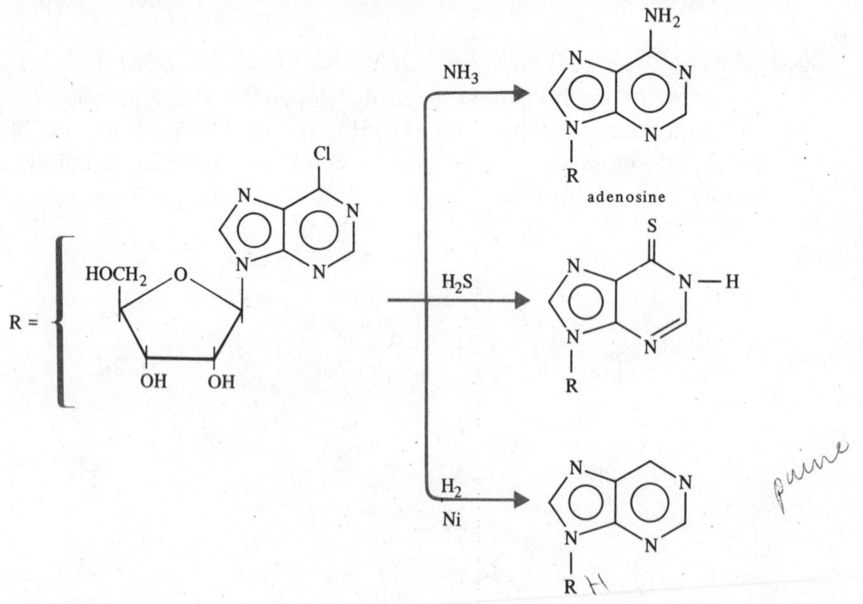

Numerous phosphorylating agents have been used to convert nucleosides to nucleotides. One of the most useful is dibenzyl phosphochloridate.

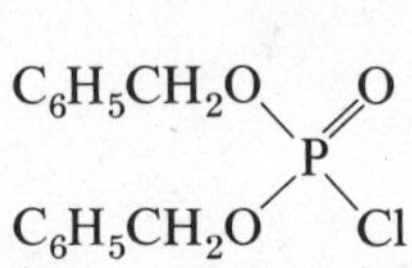

Dibenzyl phosphochloridate

Specific phosphorylation of the 5'—OH can be achieved if the 2' and 3' OH groups of the nucleoside are protected by an isopropylidene group (below).

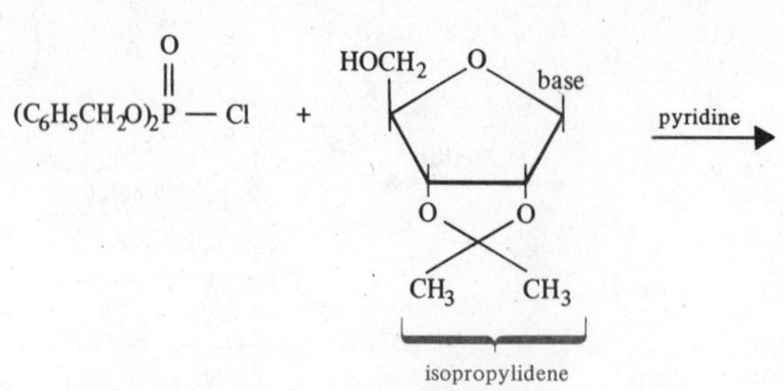

isopropylidene
protecting group

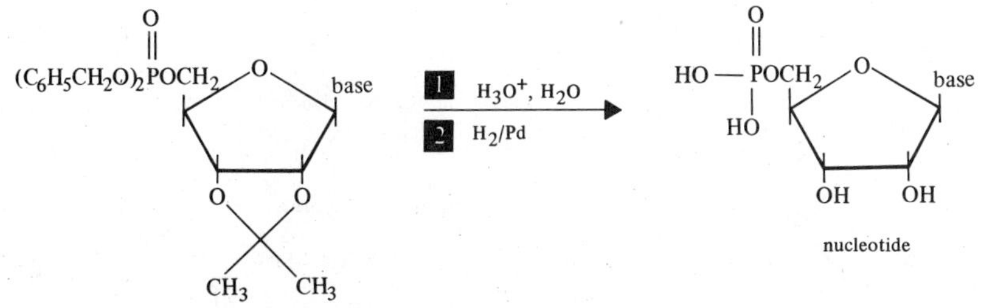

nucleotide

Mild acid hydrolysis removes the isopropylidene group, and hydrogenolysis cleaves the benzyl phosphate bonds.

Problem 26.5

(a) What kind of linkage is involved in the isopropylidene protected nucleoside and why is it susceptible to mild acid hydrolysis? (b) How might such a protecting group be installed?

26.4 DEOXYRIBONUCLEIC ACID: DNA

Primary Structure

Nucleotides bear the same relation to a nucleic acid that amino acids do to a protein; they are its monomeric units. The connecting links in proteins are amide groups, in nucleic acids they are phosphate ester linkages. Phosphate esters link the 3'-hydroxyl of one ribose (or deoxyribose) with the 5'-hydroxyl of another. This makes the nucleic acid a long unbranched chain with a "backbone" of sugar and phosphate units with heterocyclic bases protruding from the chain at regular intervals (Fig. 26.5).

It is, as we will see, the *base sequence* along the chain of DNA that contains the encoded genetic information. This sequence of bases can be determined through techniques based on selective enzymatic hydrolyses and the actual base sequences have been worked out for a number of smaller nucleic acids.

Secondary Structure

It was the now-classic proposal of Watson and Crick (made in 1953 and verified shortly thereafter by the X-ray analysis of Wilkins) that gave a model for the secondary structure of DNA. The secondary structure of DNA is especially important because it enables us to understand how the genetic information is preserved, how it can be passed on during the process of cell division, and how it can be transcribed to provide a template for protein synthesis.

Of prime importance to Watson and Crick's proposal was an earlier observation by E. Chargaff that certain regularities can be seen in the percentages of heterocyclic bases obtained from the DNA of a variety of species. Table 26.1 gives results that are typical of those that can be obtained.

Chargaff pointed out that for all species examined:

1. The total mole percentage of purines is approximately equal to that of the pyrimidines, that is, $(\%G + \%A)/(\%C + \%T) \simeq 1$.

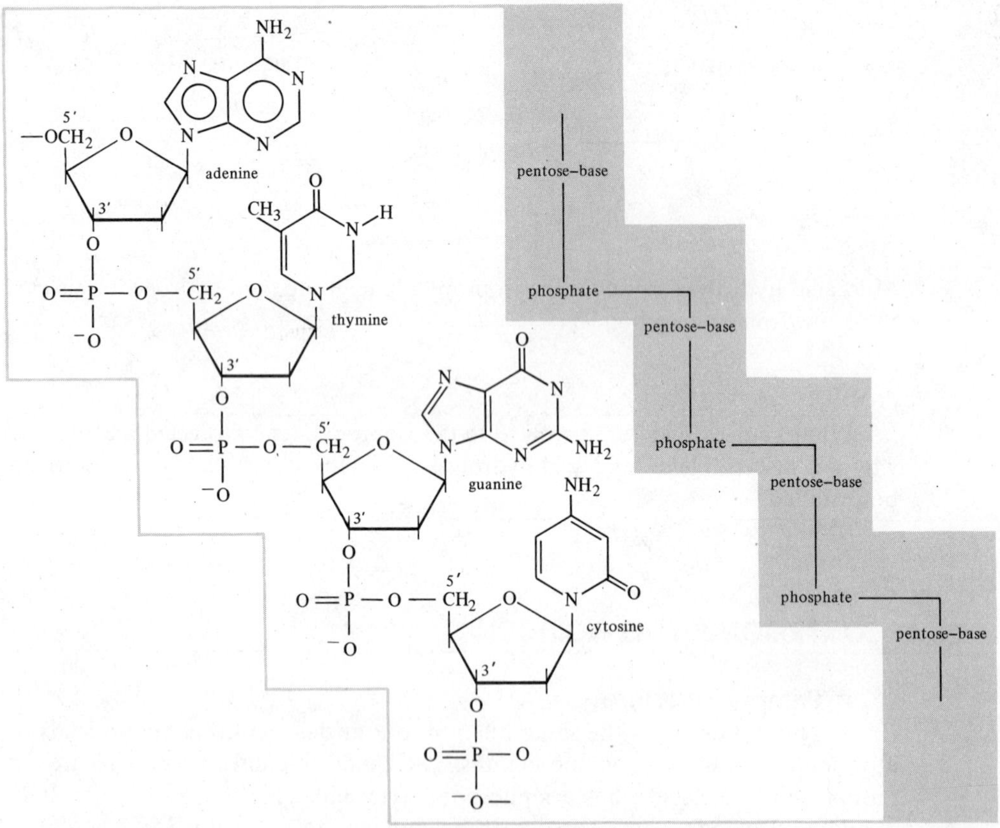

FIG. 26.5
Hypothetical segment of a single DNA chain showing how phosphate ester groups link the 3' and 5'-hydroxyls of deoxyribose units. RNA has a similar structure with two exceptions: a hydroxyl replaces hydrogen at the 2'-position of each ribose unit and uracil replaces thymine.

TABLE 26.1 DNA Composition of Various Species

	BASE PROPORTIONS, MOLES %							
SPECIES	G	A	C	T	$\dfrac{G+A}{C+T}$	$\dfrac{A+T}{G+C}$	$\dfrac{A}{T}$	$\dfrac{G}{C}$
Sarcina lutea	37.1	13.4	37.1	12.4	1.02	0.35	1.08	1.00
Escherichia coli K12	24.9	26.0	25.2	23.9	1.08	1.00	1.09	0.99
Wheat germ	22.7	27.3	22.8*	27.1	1.00	1.19	1.01	1.00
Bovine thymus	21.5	28.2	22.5*	27.8	0.96	1.27	1.01	0.96
Staphylococcus aureus	21.0	30.8	19.0	29.2	1.11	1.50	1.05	1.11
Human thymus	19.9	30.9	19.8	29.4	1.01	1.52	1.05	1.01
Human liver	19.5	30.3	19.9	30.3	0.98	1.54	1.00	0.98

*Cytosine + methylcytosine.

2. The mole percentage of adenine is nearly equal to that of thymine (i.e., %A/%T ≃ 1) and that the mole percentage of guanine is nearly equal to that of cytosine (i.e., %G/%C ≃ 1).

Chargaff also noted that the ratio that varies from species to species is the ratio (%A + %T)/(%G + %C). He noted, moreover, that while this ratio is characteristic of the DNA of a given species, it is the same for DNA obtained from different tissues of the same animal, and does not vary appreciably with the age or conditions of growth of individual organisms within the same species.

Watson and Crick also had X-ray data that gave them the bond distances and angles of the purines and pyrimidines of model compounds; in addition they had results from Wilkins that indicated an unusually long repeat distance, 34 Å, in natural DNA.

Reasoning from these data, Watson and Crick proposed a double helix as a model for the secondary structure of DNA. According to this model, two nucleic acid chains are held together by hydrogen bonds between base pairs on opposite strands. This double chain is wound into a helix with both chains sharing the same axis. The base pairs are on the inside of the helix and the sugar-phosphate backbone on the outside (Fig. 26.6). The pitch of the helix is such that 10 successive nucleotide pairs give rise to one complete turn in 34 Å (the repeat distance). The exterior width of the spiral is ~20 Å and the internal distance between 1'-positions of ribose units on opposite chains is ~11 Å.

Using molecular scale models Watson and Crick observed that the internal distance of the double helix is such that it allows only a purine-pyrimidine type of hydrogen bonding between base pairs. Purine-purine base pairs do not occur because they would be too large to fit, and pyrimidine-pyrimidine base pairs do not occur because they would be too far apart to form effective hydrogen bonds.

Watson and Crick went one crucial step further in their proposal. Assuming that the oxygen-containing heterocyclic bases existed in keto forms (p. 811) they argued that base pairing through hydrogen bonds can only occur in a specific way:

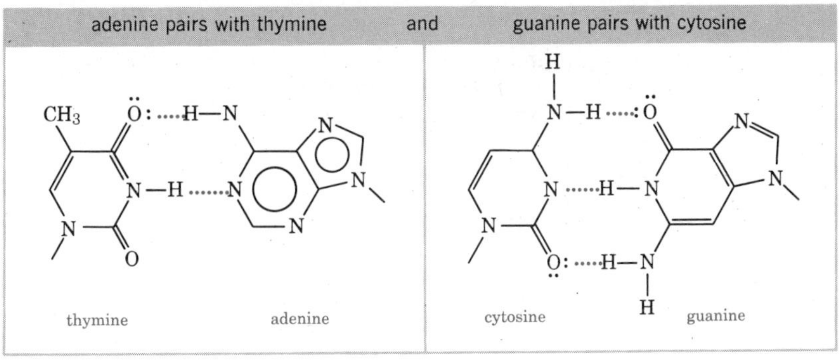

The bond distances of these base pairs are shown in Fig. 26.7.

Specific base pairing of this kind is consistent with Chargaff's finding that %A/%T ≃ 1 and that %G/%C ≃ 1.

Specific base pairing also means that the two chains of DNA are complementary. Wherever adenine appears in one chain, thymine must appear opposite it in the other; wherever cytosine appears in one chain, guanine must appear in

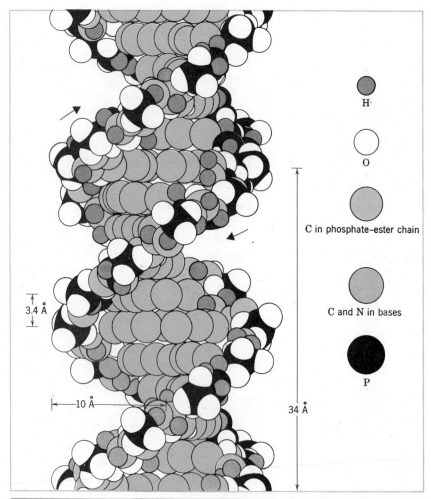

FIG. 26.6
A molecular model of a portion of the DNA double helix. (From *Chemistry and Biochemistry: A Comprehensive Introduction* by A. L. Neal. Copyright © 1971 by McGraw-Hill Inc. Used with permission of McGraw-Hill Book Company.)

the other. Thus, a segment of a double chain might, if it were linear, resemble the following:

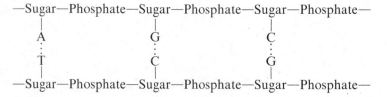

In Watson and Crick's own words, "The phosphate-sugar backbone of our model is completely regular, but any sequence of the pairs of bases can fit into the structure. It follows that, in a long molecule, many different permutations are possible, and it therefore seems likely that the precise sequence of the bases is the

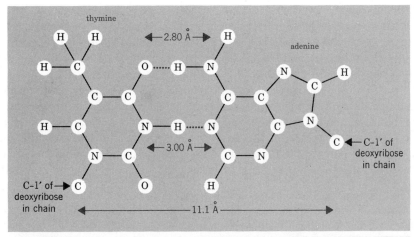

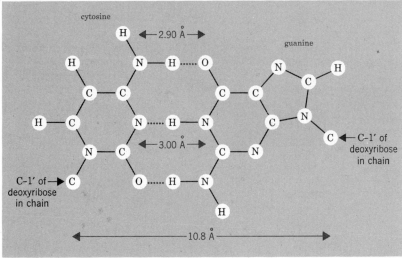

FIG. 26.7
Dimensions of thymine-adenine and cytosine-guanine base pairs. The dimensions are such that they allow the formation of strong hydrogen bonds and also allow the base pairs to fit inside the two phosphate-ribose chains of the double helix. [*Adapted from L. Pauling and R.B. Corey,* Arch. Biochem. Biophys., 65, *164 (1956).*]

code which carries the genetic information. If the actual order of the bases on one of the pair of chains were given, one could write down the exact order of the bases on the other one, because of the specific pairing. Thus, one chain is, as it were, the complement of the other, and it is this feature which suggests how the deoxyribonucleic acid molecule might replicate itself."

Replication of DNA

The Watson-Crick proposal gave, for the first time, a model that permitted an understanding of how the genetic information of a cell might be passed on to daughter cells at the time of cell division. Again, in their own words:

"Previous discussions of self-duplication have usually involved the concept of a template or mould. Either the template or mould was supposed to copy itself directly or it was to produce a 'negative,' which in turn was to act as a template and produce the original 'positive' once again. In no case has it been explained in detail how it would do this in terms of atoms and molecules."

"Now our model for deoxyribonucleic acid is, in effect, a *pair* of templates, each of which is complementary to the other. We imagine that prior to duplication the hydrogen bonds are broken, and the two chains unwind and separate. Each chain then acts as a template for the formation on to itself of a new companion chain, so that eventually we shall have two pairs of chains, where we had only one before. Moreover, the sequence of pairs of bases will have been duplicated exactly." *

* This quotation and the previous ones were taken from J. D. Watson and F. H. C. Crick, "Genetical Implications of the Structure of Deoxyribonucleic Acid," *Nature, 171* (1953), pps. 965, 966. Used with permission.

FIG. 26.8

Replication of DNA. The double strand unwinds from one end and complementary strands are formed along each chain.

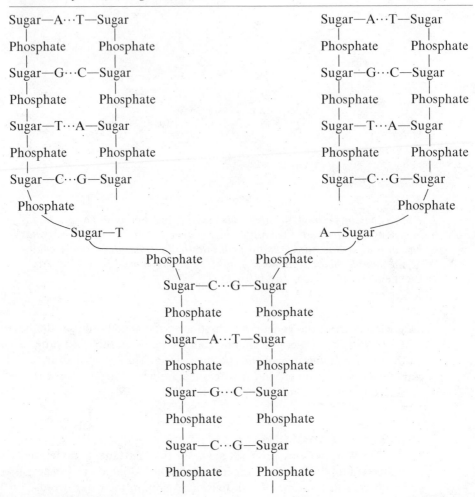

Thus, the genetic information is encoded in the pattern of bases (i.e., adenine, thymine, guanine, and cytosine), along the complementary DNA chains and all the directions necessary for the replication of a cell are written in the four-letters A,G,T,C. There are according to genetic calculations, approximately 1500 base pairs in a single gene. Four different bases give, therefore, a possibility of 4^{1500} different gene isomers. According to Watson, this number is so large that it is larger than all the different genes that have existed since life appeared on this planet.

Since the original publications of Watson and Crick, studies have been made that suggest that the chains of DNA do not unwind completely before replication occurs. Instead, the chains begin unwinding at one end and the complementary strands are formed as the unwinding takes place. An illustration of how this might occur is given in Fig. 26.8.

Problem 26.6

(a) There are approximately six billion base pairs in the DNA of a single human cell. Assuming that this DNA exists as a double helix calculate the length of all the DNA contained in a human cell. (b) The weight of DNA in a single human cell is 6×10^{-12} g. Assuming that the earth's population is about three billion we can conclude that all of the genetic information that gave rise to all human beings now alive was once contained in the DNA of a corresponding number of fertilized ova. What is the total weight of this DNA? (The volume that this DNA would occupy is approximately that of a raindrop, yet if the individual molecules were laid end to end they would stretch to the moon and back almost eight times.)

Problem 26.7

(a) The most stable tautomeric form of guanine is the lactam form below. This is the form that is normally present in DNA and, as we have seen, it pairs specifically with cytosine. If guanine tautomerizes to the abnormal lactim form, it pairs with thymine instead. Write structural formulas showing the hydrogen bonds in this abnormal base pair.

Lactam form
of guanine

Lactim form
of guanine

(b) Improper base pairings that result from tautomerizations occurring during the process of DNA replication have been suggested as a source of spontaneous mutations. We saw in part (a) that if a tautomerization of guanine occurred at the proper moment it could lead to the introduction of thymine (instead of cytosine) into its complementary DNA chain. What error would this new DNA chain introduce into *its* complementary strand during the next replication even if no further tautomerizations take place?

Problem 26.8

Mutations can also be caused chemically, and nitrous acid is one of the most potent chemical *mutagens*. One explanation that has been suggested for the mutagenic effect of nitrous acid is the deamination reactions that it produces with purines and pyrimidines bearing amino groups. When, for example, an adenine-containing nucleotide is treated with nitrous acid, it is converted to a hypoxanthine derivative:

Adenine
nucleotide

Hypoxanthine
nucleotide

(a) Basing your answer on reactions you have seen before, what are likely intermediates in the adenine ⟶ hypoxanthine interconversion? (b) Adenine normally pairs with thymine in DNA, but hypoxanthine pairs with cytosine. Show the hydrogen bonds of a hypoxanthine-cytosine base pair. (c) Show what errors an adenine ⟶ hypoxanthine interconversion would generate in DNA through two replications.

26.5 RNA AND PROTEIN SYNTHESIS

Soon after the Watson-Crick hypothesis was published, scientists began to extend it to yield what Crick has called "the central dogma of molecular genetics." This dogma states that genetic information flows from:

$$DNA \longrightarrow RNA \longrightarrow proteins$$

The synthesis of proteins is, of course, all important to a cell's function because proteins (as enzymes) catalyze all its reactions. Even the very primitive cells of bacteria require as many as 3000 different enzymes. This means that the DNA molecules of these cells must contain a corresponding number of genes to direct the synthesis of these proteins. A gene is that segment of the DNA molecule that contains the information necessary to direct the synthesis of one protein (or one polypeptide).

DNA is found primarily in the nucleus of the cell; protein synthesis takes place primarily in that part of the cell called the *cytoplasm*. Protein synthesis requires that two major processes take place; the first takes place in the cell nucleus, the second in the cytoplasm. The first process is *transcription,* a process in which the genetic message is transcribed on to a form of RNA called messenger RNA (*m*RNA). The second process involves two other forms of RNA, called ribosomal RNA (*r*RNA) and transfer RNA (*t*RNA).

Messenger RNA Synthesis—Transcription

Protein synthesis begins in the cell nucleus with the synthesis of messenger RNA. Part of the DNA double helix unwinds sufficiently to expose on a single chain

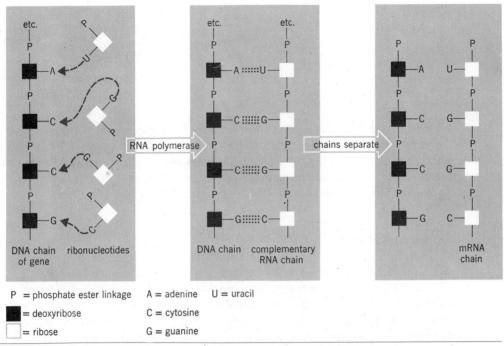

P = phosphate ester linkage   A = adenine   U = uracil

■ = deoxyribose   C = cytosine

☐ = ribose   G = guanine

FIG. 26.9

Transcription of the genetic code from DNA to messenger RNA.

a portion corresponding to at least one gene. Ribonucleotides, present in the cell nucleus, assemble along the exposed DNA chain pairing with the bases of DNA. The pairing patterns are the same as those in DNA with the exception that in RNA uracil replaces thymine. The ribonucleotide units of messenger RNA are joined into a chain by an enzyme called RNA *polymerase*. This process is illustrated in Fig. 26.9.

Problem 26.9

Write structural formulas showing how the keto form of uracil in messenger RNA can pair with adenine in DNA through hydrogen bond formation.

After messenger RNA has been synthesized in the cell nucleus it migrates into the cytoplasm where, as we will see, it acts as a template for protein synthesis.

Ribosomes—rRNA

Scattered throughout the cytoplasm of most cells are small bodies called ribosomes. Ribosomes of *E. Coli.* for example, are about 180 Å in diameter and are composed of approximately 60% RNA (ribosomal RNA) and 40% protein. They apparently exist as two associated subunits called the 50S and 30S subunits (Fig. 26.10); together they form a 70S ribosome.* Although the ribosomes are at the site

---

* S stands for svedberg unit; it is used in describing the behavior of proteins in an ultracentrifuge.

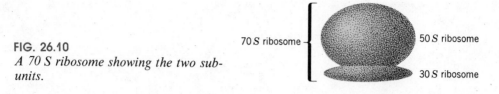

FIG. 26.10
A 70 S ribosome showing the two sub-units.

70 *S* ribosome

50 *S* ribosome

30 *S* ribosome

of protein synthesis, ribosomal RNA itself does not direct protein synthesis. Instead, a number of ribosomes become attached to a chain of messenger RNA and form what is called a *polysome*. It is along the polysome—with messenger RNA acting as the template—that protein synthesis takes place. One of the functions of ribosomal RNA is to bind the ribosome to the messenger RNA chain.

Transfer RNA

Transfer RNA has a very low molecular weight when compared to that of messenger RNA or ribosomal RNA. Transfer RNA, consequently, is much more soluble than messenger RNA or ribosomal RNA and is sometimes referred to as soluble RNA. The function of transfer RNA is to transport amino acids to specific areas of the messenger RNA of the polysome. There are, therefore, at least 20 different forms of transfer RNA, one for each of the 20 amino acids that are incorporated into proteins.*

The structures of a number of transfer RNA's have been determined. They are composed of a relatively small number of nucleotide units (70–90 units) folded into several loops or arms through base pairing along the chain (Fig. 26.11). One arm always terminates in the sequence cytosine-cytosine-adenine. It is to this arm that a specific amino acid becomes attached *through an ester* linkage to the 3′—OH of the terminal adenine. This attachment reaction is catalyzed by an enzyme that is specific for the transfer RNA and for the amino acid. The specificity may grow out of the enzymes ability to recognize base sequences along other arms of the transfer RNA.

At the loop of still another arm is a specific sequence of bases, called the *anticodon*. The anticodon is highly important because it allows the transfer RNA to bind with a specific site—called the *codon*—of messenger RNA. This binding between bases of the codon and anticodon is governed by what is called *the genetic code*.

The Genetic Code

The genetic message that specifies a particular amino acid is encoded as a sequence of three bases in messenger RNA. We can easily see why at least three bases are required to carry this message. There are 20 different amino acids used in protein synthesis but there are only four different bases in messenger RNA. If only two bases were used, there would be only 4^2 or 16 possible combinations, a number too small to accommodate all of the possible amino acids. However, with

---

* Although proteins are composed of 22 different amino acids, protein synthesis requires only 20. Proline is converted to hydroxyproline and cysteine is converted to cystine after synthesis of the polypeptide chain has taken place.

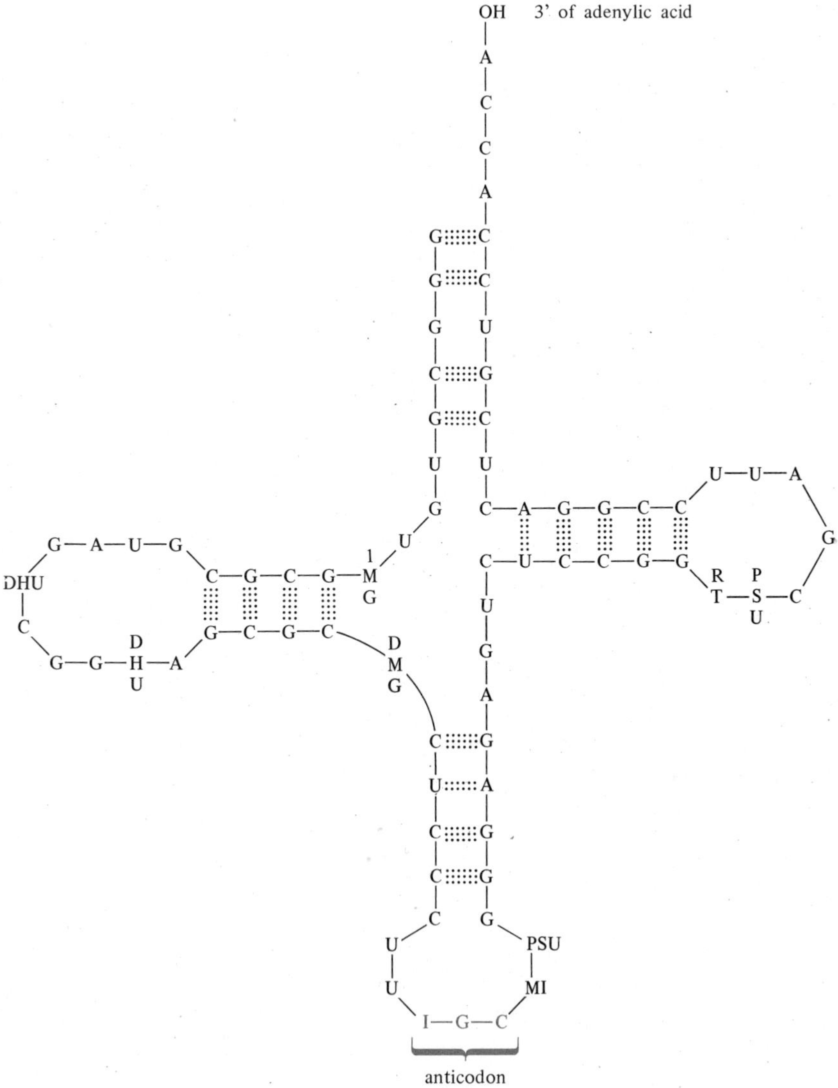

OH   3' of adenylic acid

FIG. 26.11

Structure of a transfer RNA isolated from yeast that has the specific function of transferring alanine residues. Transfer RNA's often contain unusual nucleosides. PSU = pseudouridine, RT = ribothymidine, MI = 1 methylinosine, I = inosine, DMG = N-(2)-methylguanosine, DHU = 4,5-dihydrouridine, 1 MG = 1-methylguanosine.

anticodon

a three-base code, 4^3 or 64 different sequences are possible. This is far more than are needed and it allows for multiple ways of specifying an amino acid. It also allows for sequences that punctuate protein synthesis, sequences that say, in effect, "start here" and "end here."

The messenger RNA code for the 20 amino acids has been deciphered and the base sequences that specify each amino acid are given in Table 26.2.

TABLE 26.2 The Messenger RNA Genetic Code

AMINO ACID	BASE SEQUENCE	AMINO ACID	BASE SEQUENCE	AMINO ACID	BASE SEQUENCE
Ala	GCA	His	CAC	Ser	AGC
	GCC		CAU		AGU
	GCG				UCA
	GCU	Ile	AUA		UCG
			AUC		UCC
Arg	AGA		AUU		UCU
	AGG				
	CGA	Leu	CUA	Thr	ACA
	CGC		CUC		ACC
	CGG		CUG		ACG
	CGU		CUU		ACU
			UUA		
Asn	AAC		UUG	Tyr	UGG
	AAU				UAC
		Lys	AAA		UAU
Asp	GAC		AAG		
	CAU			Val	GUA
					GUG
Cys	UGC	Met	AUG		GUC
	UGU				GUU
		Phe	UUU		
			UUC	Chain initiation:	
Gln	CAA				
	CAG	Pro	CCA	Met$_{formyl}$	AUG
			CCC		
Glu	GAA		CCG	Chain termination:	
	GAG		CCU		
					UAA
Gly	GGA				UAG
	GGC				
	GGG				
	GGU				

Both methionine (Met) and *N*-formyl methionine (Met$_{formyl}$) have the same messenger RNA code (AUG); however, *N*-formylmethionine (below) is carried by

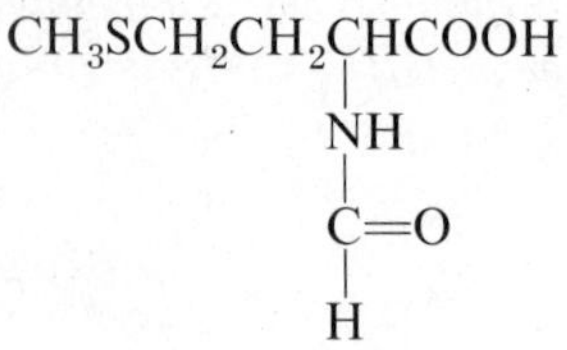

$$CH_3SCH_2CH_2CHCOOH$$
$$|$$
$$NH$$
$$|$$
$$C{=}O$$
$$|$$
$$H$$

N-Formylmethionine

a different transfer RNA from that which carries methionine. *N*-formylmethionine appears to be the first amino acid incorporated into the chain of all proteins and the transfer RNA that carries Met$_{formyl}$ appears to be the punctuation mark that says "start here." Before the polypeptide synthesis is complete, N-formylmethionine is removed from the protein chain by an enzymatic hydrolysis.

We are now in a position to see how the synthesis of a hypothetical polypeptide might take place. Let us imagine that a long strand of messenger RNA is in the cytoplasm of a cell and that it is in contact with ribosomes. Also in the cytoplasm are the 20 different amino acids, each acylated to its own specific transfer RNA.

Now (Fig. 26.12), a transfer RNA bearing Met_{formyl} uses its anticodon (UAC) to associate with the proper codon (AUG) on that portion of messenger RNA that is in contact with a ribosome. The next triplet of bases on this particular messenger RNA chain is AAA; this is the codon that specifies lysine. A lysyl-transfer RNA with the anticodon UUU attaches itself to this site. The two amino acids, Met_{formyl} and Lys, are now in the proper position for an enzyme to join them in peptide linkage. After this happens, the ribosome moves down the chain so that it is in contact with the next codon. This one, GUA, specifies valine. A transfer RNA bearing valine (and with the anticodon CAU) binds itself to this site. Another enzymatic reaction takes place attaching valine to the polypeptide chain. Then the whole process repeats itself again and again; the ribosome moves along the messenger RNA chain, other transfer RNA's move up with their amino acids, new peptide bonds are formed, and the polypeptide chain grows. At some point an enzymatic reaction removes Met_{formyl} from the beginning of the chain. Finally, when the chain is the proper length the ribosome reaches a punctuation mark, UAA, saying "stop here." The ribosome separates from the messenger RNA chain and so, too, does the protein.

Even before the polypeptide chain is fully grown, it begins to form its own specific secondary and tertiary structure (Fig. 26.13). This happens because its primary structure is correct—its amino acids are ordered in just the right way. Hydrogen bonds form giving rise to specific segments of α helix, pleated sheet, and random coil. Then the whole thing folds and bends; enzymes install disulfide linkages, so that when the chain is fully grown, the whole protein has just the shape it needs to do its job.

If this protein happens to be lysozyme it has a deep cleft, or jaw, where a specific polysaccharide fits. And if it is lysozyme, and a certain bacterium wanders by, that jaw begins to work; it bites its first polysaccharide in half.

In the meantime other ribosomes nearer the beginning of the messenger RNA chain are already moving along, each one synthesizing another molecule of the polypeptide. The time required to synthesize a protein depends, of course, on the number of amino residues it contains, but indications are that each ribosome can cause 150 peptide bonds to be formed each minute. Thus, a protein, such as lysozyme, with 129 amino acid residues requires less than a minute for its synthesis. However, if four ribosomes are working their way along a single messenger RNA chain, the polysome can produce a lysozyme molecule every 13 seconds.

But why, we might ask, is all this protein synthesis necessary—particularly in a fully grown organism? The answer is that proteins are not permanent; they do not get themselves synthesized once and remain intact in the cell for the lifetime of the organism. They are synthesized when and where they are needed. Then they are taken apart, back to amino acids; enzymes disassemble enzymes. Some amino acids get metabolized for energy; others—new ones—come in from the food we have eaten and the whole process begins again.

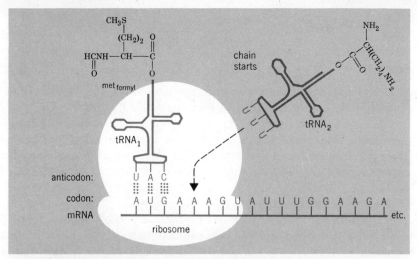

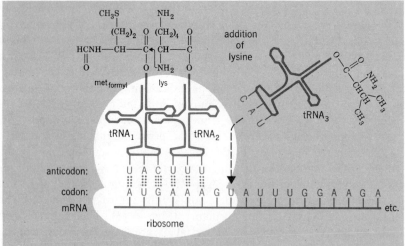

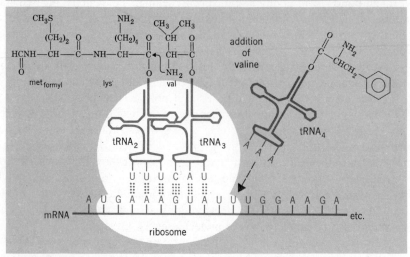

FIG. 26.12

Step-by-step growth of a polypeptide chain with messenger RNA acting as a template. Transfer RNA's carry amino acid residue to the site of messenger RNA that is in contact with a ribosome. Codon-anticodon pairing occurs between messenger RNA and transfer RNA at the ribosomal surface. An enzymatic reaction joins the amino acid residues through an amide linkage. After the first amide bond is formed the ribosome moves to the next codon on messenger RNA. A new transfer RNA arrives, pairs, and transfers its amino acid residues to the growing peptide chain, and so on.

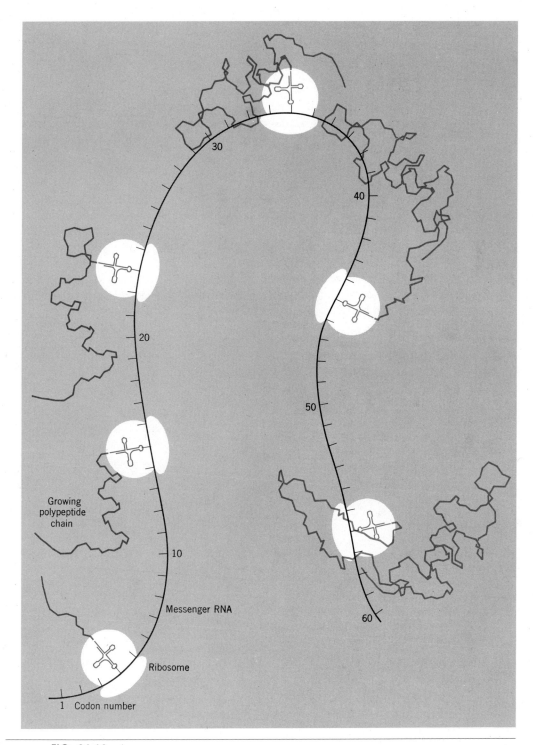

FIG. 26.13
The folding of a protein molecule as it is synthesized. [*Adapted with permission from D. C. Phillips, "The Three-Dimensional Structure of an Enzyme Molecule,"* Copyright © 1966 Scientific American, Inc. All rights reserved.]

Problem 26.10

A segment of DNA has the following sequence of bases:

$$...A\ C\ C\ C\ C\ C\ A\ A\ A\ A\ T\ G\ T\ C\ G...$$

(a) What sequence of bases would appear in messenger RNA transcribed from this segment? (b) Assume that the first base in this messenger RNA is the beginning of a codon. What order of amino acids would be translated into a polypeptide synthesized along this segment? (c) Give anticodons for each transfer RNA associated with the translation in part (b).

Problem 26.11

(a) Using the first codon given for each amino acid in Table 26.2 write the base sequence of messenger RNA that would translate the synthesis of the following pentapeptide:

$$Arg \cdot Ile \cdot Cys \cdot Tyr \cdot Val$$

(b) What base sequence in DNA would transcribe a synthesis of the messenger RNA? (c) What anticodons would appear in the transfer RNA's involved in the pentapeptide synthesis?

Problem 26.12

Explain how an error of a single base in each strand of DNA could bring about the amino acid residue error that causes sickle cell anemia (p. 967).

APPENDIX
MASS SPECTROSCOPY

A.1 THE MASS SPECTROMETER

In a mass spectrometer (Fig. A.1) molecules in the gaseous state under low pressure are bombarded with a beam of high-energy electrons. The energy of the beam of electrons is usually 70 eV (electron volts) and one of the things this bombardment can do is dislodge one of the electrons of the molecule and produce a positively charged ion called *the molecular ion.*

$$\underset{\substack{Molecule}}{M} \quad + \quad \underset{\substack{High\text{-}energy \\ electron}}{e^-} \quad \longrightarrow \quad \underset{\substack{Molecular \\ ion}}{M^{\ddagger}} \quad + \; 2e^-$$

The molecular ion is not only a cation, but because it contains an odd number of electrons, it also is a free radical. Thus it belongs to a general group of ions called *radical cations.* If, for example, the molecule under bombardment is a molecule of ammonia the following reaction will take place.

$$\underset{\substack{H}}{H:\overset{..}{\underset{..}{N}}:H} + e^- \longrightarrow \left[\underset{\substack{H}}{H:\overset{.}{\underset{..}{N}}:H} \right]^+ + 2e^-$$

molecular ion, M$\overset{.}{^+}$
(a radical cation)

An electron beam with an energy of 70 electron volts ($\sim$1600 kcal/mole) not only dislodges electrons from molecules, producing molecular ions, it also imparts to the molecular ions considerable surplus energy. Not all molecular ions will have the same amount of surplus energy, but for most, the surplus will be far in excess of that required to break covalent bonds (50–100 kcal/mole). Thus, soon after they are formed, most molecular ions literally fly apart—they undergo *fragmentation.* Fragmentation can take place in a variety of ways depending on the nature of the particular molecular ion and as we will see later, the way a molecular ion fragments can give us highly useful information about the structure of a complex molecule. Even with a relatively simple molecule like ammonia, however, fragmentation can produce several new cations. The molecular ion can eject a hydrogen atom, for example, and produce the cation, NH_2^+.

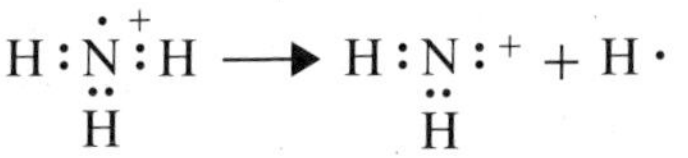

$$\underset{\substack{H}}{H:\overset{.\;+}{\underset{..}{N}}:H} \longrightarrow \underset{\substack{H}}{H:\overset{..}{\underset{..}{N}}:^+} + H \cdot$$

This NH_2^+ cation can then lose a hydrogen atom to produce $NH^{\ddagger}$, which can lead, in turn, to N^+.

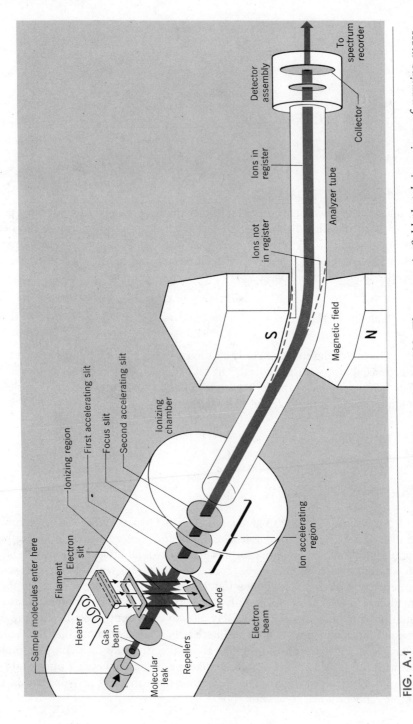

FIG. A.1
Mass spectrometer. Schematic diagram of CEG model 21-103. The magnetic field that brings ions of varying mass-to-charge ratios into register is perpendicular to the page. (From John R. Holum, Organic Chemistry: A Brief Course, Wiley, New York, 1975. Used with permission.)

$$\text{H}\!:\!\underset{\displaystyle \overset{\textstyle \cdot\cdot}{\text{H}}}{\text{N}}\!:^{+} \longrightarrow \text{H}\!:\!\underset{\displaystyle \cdot}{\text{N}}\!:^{+} + \text{H}\cdot$$

$$\text{H}\!:\!\underset{\displaystyle \cdot}{\text{N}}\!:^{+} \longrightarrow \;:\!\text{N}\!:^{+} + \text{H}\cdot$$

The mass spectrometer then *sorts* these cations on the basis of their mass/charge or m/e ratio. Since for all practical purposes the charge on all of the ions is $+1$, this amounts to sorting them on the basis of their mass. The conventional mass spectrometer does this by accelerating the ions through a series of slits and then sending the ion beam into a curved tube (see Fig. A.1 again). This curved tube passes through a variable magnetic field and the magnetic field exerts an influence on the moving ions. Depending on its strength at a given moment, the magnetic field will cause ions with a particular m/e ratio to follow a curved path that exactly matches the curvature of the tube. These ions are said to be "in register." Because they are in register, these ions pass through another slit and impinge on an ion collector where the intensity of the ion beam is measured electronically. The intensity of the beam is simply a measure of the relative abundance of the ions with a particular m/e ratio. Some mass spectrometers are so sensitive that they can detect the arrival of a *single ion*.

The actual sorting of ions takes place in the magnetic field and this sorting takes place because laws of physics govern the paths followed by charged particles when they move through magnetic fields. Generally speaking, a magnetic field such as this, will cause ions moving through it to move in a path that represents part of a circle. The radius of curvature of this circular path (r, cm) is related to the m/e ratio of the ions, to the strength of the magnetic field ($\mathbf{H}$, gauss) and to the accelerating voltage (V, volts). The equation that describes the relationship is,

$$\frac{m}{e} = \frac{4.82 \times 10^{-5}r^2\mathbf{H}^2}{V}$$

We can see that if r and V are made to be constant, then,

$$\frac{m}{e} \propto \mathbf{H}^2$$

What this equation tells us is that if we keep the accelerating voltage constant and progressively increase the magnetic field, ions whose m/e ratios are progressively larger will travel in a circular path of radius r that exactly matches that of the curved tube. That is, ions with progressively increasing m/e ratios will be "in register" and thus will be detected at the ion collector. Since, as we said earlier, the charge on nearly all of the ions is unity, this means that *ions of progressively increasing mass arrive at the collector and are detected.*

What we have described is called "magnetic focusing" (or "magnetic scanning") and all of this is done automatically by the mass spectrometer. The spectrometer displays the results by plotting a series of peaks of varying intensity in which each peak corresponds to ions of a particular m/e ratio. This display (Fig. A.2) is one form of a *mass spectrum*.

> Ion sorting can also be done with "electrical focusing." In this technique, the magnetic field is held constant and the accelerating voltage is varied. Both methods, of course, accomplish the same thing, and some high resolution mass spectrometers employ both techniques.

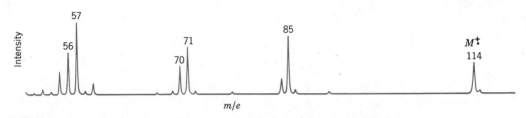

FIG. A.2

A portion of the mass spectrum of n-*octane.*

To summarize: a mass spectrometer bombards organic molecules with a beam of high-energy electrons causing them to ionize and fragment. It then separates the resulting mixture of ions on the basis of their mass/charge ratio and records the relative abundance of each ionic fragment. It displays this result as a plot of ion intensity versus m/e.

A.2 THE MASS SPECTRUM

Mass spectra are usually published as bar graphs or in tabular form. Figure A.3 shows us how this is done with the mass spectrum of ammonia. In either presentation, the most intense peak—called the *base peak*—is assigned the arbitrary value of 100%. The intensities of all other peaks are given proportionate values, as percentages of the base peak.

The masses of the ions given in a mass spectrum are those that we would calculate for the ion by assigning the constituent atoms *masses rounded off to the nearest whole number*. For the commonly encountered atoms the nearest whole number masses are:

$$H = 1$$
$$C = 12$$
$$N = 14$$
$$O = 16$$
$$F = 19$$

In the mass spectrum of ammonia we see peaks at $m/e = 14, 15, 16,$ and 17. These correspond to the molecular ion and to the fragments we saw earlier.

$$NH_3 \xrightarrow{-e^-} [NH_3]^{+} \xrightarrow{-H\cdot} [NH_2]^{+} \xrightarrow{-H\cdot} [NH]^{+} \xrightarrow{-H\cdot} [N]^{+}$$
$$m/e \quad = \quad 17 \qquad\qquad 16 \qquad\quad 15 \qquad\quad 14$$
$$\text{(molecular ion)}$$

By convention we express,

$$H:\overset{\cdot}{\underset{\overset{\cdot\cdot}{H}}{N}}:H^+ \quad\quad \text{as} \quad\quad [NH_3]^{+}$$

$$H:\underset{\overset{\cdot\cdot}{H}}{N}:^{+} \quad\quad \text{as} \quad\quad [NH_2]^{+}$$

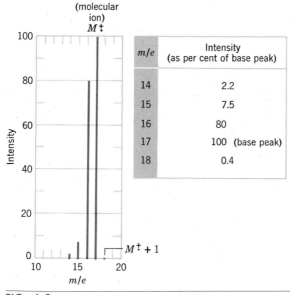

FIG. A.3

The mass spectrum of NH₃ presented as a bar graph and in tabular form.

$$H:\underset{\cdot\cdot}{N}:^{+} \quad \text{as} \quad [NH]_\cdot^{+}$$

and $:N:^{+}$ as $[N]^{+}$

In the case of ammonia, the base peak is the peak due to the molecular ion. This is not always the case, however; in many of the spectra that we will see later the base peak (the most intense peak) will be at an m/e value different from that of the molecular ion. This happens because in many instances the molecular ion fragments so rapidly that some other ion at a smaller m/e value becomes the most intense peak. In a few cases the molecular ion peak is extremely small, and sometimes it is absent altogether.

One other feature in the spectrum of ammonia requires explanation: the small peak that occurs at m/e 18. In the bar graph we have labeled this peak $M^{+} + 1$ to indicate that it is one mass unit greater than the molecular ion. The $M^{+} + 1$ peak appears in the spectrum because most elements (e.g., nitrogen and hydrogen) have more than one naturally occurring isotope (Table A.1). Although most of the NH_3 molecules in a sample of ammonia are composed of $^{14}N^1H_3$, a small but detectable fraction of molecules will be composed of $^{15}N^1H_3$. (A very tiny fraction of molecules will also be composed of $^{14}N^1H_2{}^2H$.) These molecules—$^{15}N^1H_3$ or $^{14}N^1H_2{}^2H$—will produce molecular ions at m/e 18, i.e., at $M^{+} + 1$.

The spectrum of ammonia begins to show us with a simple example, how the masses (or m/e's) of individual ions can give us information about the composition of the ions and how this information can allow us to arrive at possible structures for a compound. Problems A.1–A.3 will allow us further practice with this technique.

TABLE A.1 Principal Stable Isotopes of
Common Elements[a]

ELEMENT	MOST COMMON ISOTOPE	NATURAL ABUNDANCE OF OTHER ISOTOPES (BASED ON 100 ATOMS OF MOST COMMON ISOTOPE)			
Carbon	^{12}C 100	^{13}C	1.08		
Hydrogen	^{1}H 100	^{2}H	0.016		
Nitrogen	^{14}N 100	^{15}N	0.38		
Oxygen	^{16}O 100	^{17}O	0.04	^{18}O	0.20
Fluorine	^{19}F 100				
Sulfur	^{32}S 100	^{33}S	0.78	^{34}S	4.40
Chlorine	^{35}Cl 100	^{37}Cl	32.5		
Bromine	^{79}Br 100	^{81}Br	98.0		
Iodine	^{127}I 100				

[a] Data obtained from R. M. Silverstein, G. C. Bassler, and T. C. Morrill, *Spectrometric Identification of Organic Compounds*, 3rd Ed., Wiley, N.Y. 1974, p. 13.

Problem A.1

Propose a structure for the compound whose mass spectrum is given in Fig. A.4 and make reasonable assignments for each peak.

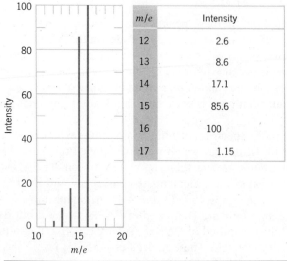

m/e	Intensity
12	2.6
13	8.6
14	17.1
15	85.6
16	100
17	1.15

FIG. A.4
Mass spectrum for problem A.1.

Problem A.2

Propose a structure for the compound whose mass spectrum is given in Fig. A.5 and make reasonable assignments for each peak.

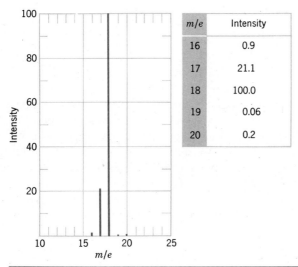

m/e	Intensity
16	0.9
17	21.1
18	100.0
19	0.06
20	0.2

FIG. A.5
Mass spectrum for problem A.2.

Problem A.3

The compound whose mass spectrum is given in Fig. A.6 contains three elements, one of which is fluorine. Propose a structure for the compound and make reasonable assignments for each peak.

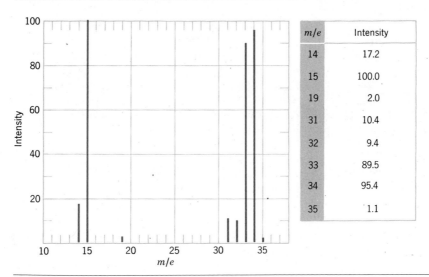

m/e	Intensity
14	17.2
15	100.0
19	2.0
31	10.4
32	9.4
33	89.5
34	95.4
35	1.1

FIG. A.6
Mass spectrum for problem A.3.

A.3 DETERMINATION OF MOLECULAR FORMULAS AND MOLECULAR WEIGHTS

The Molecular Ion and Isotopic Peaks

Look at Table A.1 for a moment. We see that most of the common elements found in organic compounds have naturally occurring *heavier* isotopes. For three of the elements—carbon, hydrogen, and nitrogen—the principal heavier isotope is one mass unit greater than the most common isotope. The presence of these elements in a compound will give rise to a small isotopic peak one unit greater than the molecular ion—at $M^{\ddagger} + 1$. For four of the elements—oxygen, sulfur, chlorine, and bromine—the principal heavier isotope is two mass units greater than the most common isotope. The presence of these elements in a compound give rise to an isotopic peak at $M^{\ddagger} + 2$.

$M^{\ddagger} + 1$ **Elements:** C, H, N
$M^{\ddagger} + 2$ **Elements:** O, S, Br, Cl

Isotopic peaks give us one method for determining molecular formulas. To understand how this can be done, let us begin by noticing that the isotope abundances in Table A.1 are based on 100 atoms of the normal isotope. Now let us suppose, as an example, that we have 100 molecules of methane, CH_4. On the average there will be 1.08 molecules that contain ^{13}C and 4×0.016 molecules that contain ^{2}H. Altogether then, these heavier isotopes should contribute an $M^{\ddagger} + 1$ peak whose intensity is $\sim 1.14\%$ of the intensity of the molecular ion.

$$1.08 + 4\,(0.016) \simeq 1.14\%$$

Notice that the result of this calculation correlates well with the intensity of the $M^{+} + 1$ peak in the actual spectrum of methane given in Fig. A.4.

Problem A.4

What relative intensity would you expect for the $M^{\ddagger} + 1$ peak of NH_3? Compare your answer with the spectrum in Fig. A.3.

Problem A.5

What relative intensities would you expect for the $M^{\ddagger} + 1$ and $M^{\ddagger} + 2$ peaks in water? Compare your answer with the spectrum in Fig. A.5.

As the number of atoms in a molecule increases, calculations like this become more and more complex and time consuming. Fortunately, however, these calculations can be done readily with computers, and tables are now available that give relative values for the $M^{\ddagger} + 1$ and $M^{\ddagger} + 2$ peaks for all combinations of common elements with molecular formulas up to mass 500.* Part of the data obtained from one of these tables is given in Table A.2.

* See J. H. Beynon and A. E. Williams, *Mass and Abundance Tables for Use in Mass Spectrometry,* Elsevier, Amsterdam, 1963; J. H. Beynon, *Mass Spectrometry and Its Application to Organic Chemistry,* Elsevier, Amsterdam, 1960; or R. M. Silverstein, G. C. Bassler, and T. C. Morrill, *Spectrometric Identification of Organic Compounds,* 3rd Ed., Wiley, New York, 1974.

TABLE A.2 Relative Intensities of $M^+ + 1$ and $M^+ + 2$ Peaks for Various Combinations of C, H, N, and O for Mass 72 and 73

M^+	FORMULAS	PERCENT OF M^+ INTENSITY		M^+	FORMULAS	PERCENT OF M^+ INTENSITY	
		$M^+ + 1$	$M^+ + 2$			$M^+ + 1$	$M^+ + 2$
72	CH_2N_3O	2.30	0.22	73	CHN_2O_2	1.94	0.41
	CH_4N_4	2.67	0.03		CH_3N_3O	2.31	0.22
	$C_2H_2NO_2$	2.65	0.42		CH_5N_4	2.69	0.03
	$C_2H_4N_2O$	3.03	0.23		C_2HO_3	2.30	0.62
	$C_2H_6N_3$	3.40	0.04		$C_2H_3NO_2$	2.67	0.42
	$C_3H_4O_2$	3.38	0.44		$C_2H_5N_2O$	3.04	0.23
	C_3H_6NO	3.76	0.25		$C_2H_7N_3$	3.42	0.04
	$C_3H_8N_2$	4.13	0.07		$C_3H_5O_2$	3.40	0.44
	C_4H_8O	4.49	0.28		C_3H_7NO	3.77	0.25
	$C_4H_{10}N$	4.86	0.09		$C_3H_9N_2$	4.15	0.07
	C_5H_{12}	5.60	0.13		C_4H_9O	4.51	0.28
					$C_4H_{11}N$	4.88	0.10
					C_6H	6.50	0.18

Data from J. H. Beynon, *Mass Spectrometry and Its Application to Organic Chemistry,* Elsevier, Amsterdam, 1960.

Let us use the data in Table A.2 to determine a molecular formula for the compound whose mass spectrum is given in Fig. A.7. First, we notice that the mass of the molecular ion is 72. This tells us, right away, the molecular weight of the compound rounded off to the nearest mass unit. We also see in the left-hand column of Fig. A.7 that the intensities of the peaks are expressed in the normal way—as

m/e	Intensity (as per cent of base peak)	m/e	Intensity (as per cent of M^+)
27	59.0	72 M^+	100.0
28	15.0	73 $M^+ + 1$	4.5
29	54.0	74 $M^+ + 2$	0.3
39	23.0		Recalculated to base on M^+
41	60.0		
42	12.0		
43	79.0		
44	100.0 (base)		
72	73.0 M^+		
73	3.3		
74	0.2		

FIG. A.7

Mass spectrum of an unknown compound.

percentages of the most intense peak, in this instance, the peak at m/e 44. The data for the $M^{\ddagger} + 1$ and $M^{\ddagger} + 2$ peaks in Table A.2, however, are calculated as percentages of the $M^{\ddagger}$ intensity. This means that we need to recalculate the intensities of the peaks in our spectrum at m/e 72, 73, and 74 by basing them on m/e 72 = 100. These results are shown below (and in the right-hand column of Fig. A.7).

m/e	% of $M^{\ddagger}$	
72	100	$M^{\ddagger}$
73	4.5	$M^{\ddagger} + 1$
74	0.3	$M^{\ddagger} + 2$

The low intensity of the $M^{\ddagger} + 2$ peak disallows the possibility of a compound containing sulfur, chlorine, or bromine (see Table A.1). Thus we are justified in assuming that the compound consists of carbon, hydrogen, nitrogen, and oxygen and we are justified in using Table A.2. If we now look under m/e 72, we see that only one formula gives $M^{\ddagger} + 1$ and $M^{\ddagger} + 2$ peaks with intensities close to that of our compound—the formula C_4H_8O. This, then, must be the molecular formula of our compound.

Problem A.6

(a) Write structural formulas for at least 14 stable compounds that have the formula C_4H_8O. (b) The infrared spectrum of the unknown compound shows a strong peak near 1730 cm^{-1}. Which structures now remain as possible formulas for the compound? (We continue with this compound in problem A.15.)

Problem A.7

Use the mass spectral data given in Fig. A.8 to determine the molecular formula for the compound. (The pmr spectrum of this compound is the subject of problem 25.25.)

m/e	INTENSITY (AS PERCENT OF BASE PEAK)
14	8.0
15	38.6
18	16.3
28	39.7
29	23.4
42	46.6
43	10.7
44	100.0 (Base)
73	86.1 $M^{\ddagger}$
74	3.2
75	0.2

FIG. A.8
Mass spectrum for problem A.7.

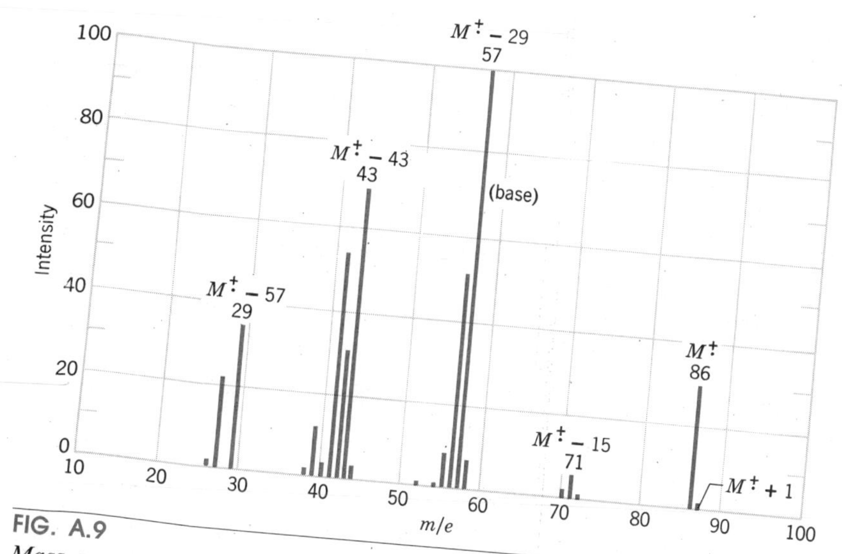

FIG. A.9

Mass spectrum of n-*hexane.*

Figure A.9 shows us the kind of fragmentation a longer chain alkane can undergo. The example here is *n*-hexane and we see a reasonably intense molecular ion at m/e 86 accompanied by a small $M^{\ddagger} + 1$ peak. There is also a smaller peak at m/e 71 ($M^{\ddagger} - 15$) corresponding to the loss of $\cdot CH_3$ and the base peak is at m/e 57 ($M^{\ddagger} - 29$) corresponding to the loss of $\cdot CH_2CH_3$. The other prominent peaks are at m/e 43 ($M^{\ddagger} - 43$) and m/e 29 ($M^{\ddagger} - 57$) corresponding to the loss of $\cdot CH_2CH_2CH_3$ and $\cdot CH_2CH_2CH_2CH_3$, respectively. The important fragmentations are just the ones we would expect:

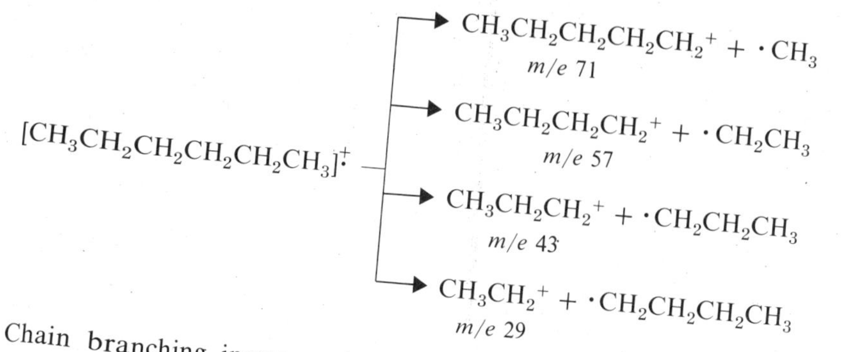

$$[CH_3CH_2CH_2CH_2CH_2CH_3]^{\ddagger}$$

$$\longrightarrow CH_3CH_2CH_2CH_2CH_2^+ + \cdot CH_3$$
$$m/e\ 71$$

$$\longrightarrow CH_3CH_2CH_2CH_2^+ + \cdot CH_2CH_3$$
$$m/e\ 57$$

$$\longrightarrow CH_3CH_2CH_2^+ + \cdot CH_2CH_2CH_3$$
$$m/e\ 43$$

$$\longrightarrow CH_3CH_2^+ + \cdot CH_2CH_2CH_2CH_3$$
$$m/e\ 29$$

Chain branching increases the likelihood of cleavage at a branch point because a more stable carbocation can result. When we compare the mass spectrum of 2-methylbutane (Fig. A.10) with the spectrum of *n*-hexane we see a much more intense peak at $M^{\ddagger} - 15$. Loss of a methyl radical from 2-methylbutane can give a secondary carbocation:

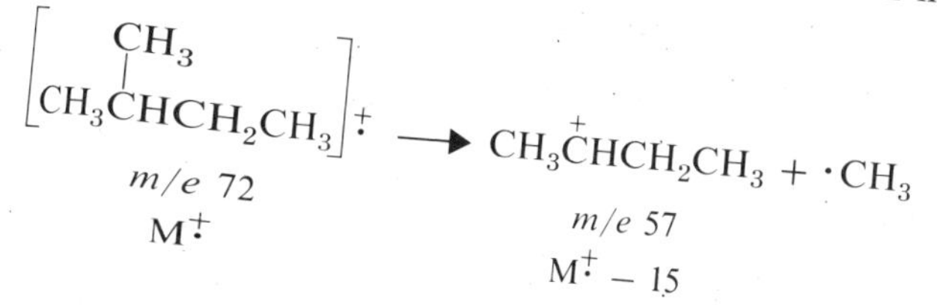

$$\begin{bmatrix} CH_3 \\ | \\ CH_3CHCH_2CH_3 \end{bmatrix}^{\ddagger} \longrightarrow CH_3\overset{+}{C}HCH_2CH_3 + \cdot CH_3$$

$$\begin{array}{cc} m/e\ 72 & m/e\ 57 \\ M^{\ddagger} & M^{\ddagger} - 15 \end{array}$$

Problem A.8

The peaks due the chlorine and bromine isotopes are especially helpful in recognizing compounds that contain these elements. (a) What factor makes this possible? (b) What approximate intensities would you expect for the $M^{\ddagger}$ and $M^{\ddagger} + 2$ peaks of CH_3Cl? (c) For the $M^{\ddagger}$ and $M^{\ddagger} + 2$ peaks of CH_3Br? (d) An organic compound gives an $M^{\ddagger}$ peak at m/e 122 and a peak of nearly equal intensity at m/e 124. What is a likely molecular formula for the compound?

High-Resolution Mass Spectroscopy

All of the spectra that we have described so far were determined on what are called "low-resolution" mass spectrometers. These spectrometers, as we noted earlier, measure m/e values to the nearest whole-number mass unit. Most laboratories are equipped with this type of mass spectrometer.

Some laboratories, however, are equipped with the more expensive "high-resolution" mass spectrometers. These spectrometers can measure m/e values to three or four decimal places and thus they provide an extremely accurate method for determining molecular weights. And, because molecular weights can be measured so accurately, these spectrometers also allow us to determine molecular formulas.

The determination of a molecular formula by an accurate measurement of a molecular weight is possible because the actual masses of atomic particles (nuclides) are not integers (see Table A.3). Consider, as examples, the three molecules, O_2, N_2H_4, and CH_3OH. The actual atomic masses of the molecules are all different.

$$^{16}O_2 = 2(15.9949) = 31.9898$$
$$N_2H_4 = 2(14.0031) + 4(1.00783) = 32.0375$$
$$CH_4O = 12.0000 + 4(1.00783) + 15.9949 = 32.0262$$

TABLE A.3 Exact Masses of Nuclides

ISOTOPE	MASS
^1H	1.00783
^2H	2.01410
^{12}C	12.00000 (std)
^{13}C	13.00336
^{14}N	14.0031
^{15}N	15.0001
^{16}O	15.9949
^{17}O	16.9991
^{18}O	17.9992
^{19}F	18.9984
^{32}S	31.9721
^{33}S	32.9715
^{34}S	33.9679
^{35}Cl	34.9689
^{37}Cl	36.9659
^{79}Br	78.9183
^{81}Br	80.9163
^{127}I	126.9045

High-resolution mass spectrometers are available that are capable of measuring mass with an accuracy of 1 part in 40,000. Thus, such a spectrometer can easily distinguish between these three molecules and tell us the molecular formula.

A.4 FRAGMENTATION

In most instances the molecular ion is a highly energetic species and in the case of a complex molecule a great many things can happen to it. The molecular ion can break apart in a variety of ways and the fragments that are produced can then undergo further fragmentation and so on. In a certain sense mass spectroscopy is a "brute force" technique. Striking an organic molecule with 70 eV electrons is a little like firing a howitzer at a house made of matchsticks. That fragmentation takes place in any sort of predictable way is truly remarkable—and yet it does. Many of the same factors that govern ordinary chemical reactions seem to apply to fragmentation processes and many of the principles that we have learned about the relative stabilities of carbocations, free radicals, and molecules will help us to make some sense out of what takes place. And, as we learn something about what kind of fragmentations to expect, we will be much better able to use mass spectra as an aid in determining the structures of organic molecules.

We cannot, of course, in the limited space that we have here, look at these processes in great detail, but we can examine some of the more important ones.

As we begin, keep two important principles in mind. (1) The reactions that take place in a mass spectrometer are *unimolecular*—that is, they involve only a *single* molecular fragment. This is true because the pressure in a mass spectrometer is kept so low ($\sim 10^{-6}$ torr) that reactions requiring bimolecular collisions do not occur. (2) The relative ion abundances, as measured by ion intensities, are extremely important. We will see that the appearance of certain prominent peaks in the spectrum gives us important information about the structures of the fragments produced and about their original locations in the molecule.

Fragmentation by Cleavage of a Single Bond

One important type of fragmentation is the simple cleavage of a single bond. With a radical cation this cleavage can take place in at least two ways; each way produces a *cation* and a *free radical*. With the molecular ion obtained from propane, for example, the two possibilities are:

$$CH_3CH_2\overset{+}{\cdot}CH_3 \longrightarrow \underset{m/e\ 29}{CH_3CH_2^+} + \cdot CH_3$$

$$CH_3CH_2\overset{+}{\cdot}CH_3 \longrightarrow CH_3CH_2\cdot + \underset{m/e\ 15}{^+CH_3}$$

These two modes of cleavage do not take place equally, however. While the relative abundance of ions produced by such a cleavage is influenced both by the stability of the carbocation and by the stability of the free radical; *the carbocation's stability is more important*.* In the spectrum of propane the peak at m/e 29 ($CH_3CH_2^+$) is the

*This can be demonstrated through thermochemical calculations that we cannot go into here. The interested student is referred to F. W. McLafferty, *Interpretation of Mass Spectra*, 2nd Ed., W. A. Benjamin, Reading, Mass., 1973, p. 41 and pp. 210–211.

TABLE A.4 Ionization Potentials of Selected Molecules

COMPOUND	IONIZATION POTENTIAL (ELECTRON VOLTS)
$CH_3(CH_2)_3NH_2$	8.7
C_6H_6	9.2
C_2H_4	10.5
CH_3OH	10.8
C_2H_6	11.5
CH_4	12.7

most intense peak; the peak at m/e 15 (CH_3^+) has an intensity of only 5.6%. Th reflects the greater stability of $CH_3CH_2^+$ when compared to CH_3^+.

Fragmentation Equations

Before we go further, we need to examine some of the conventions that used in writing equations for fragmentation reactions. In the two equation cleavage of the single bond of propane that we have just written, we have loca the odd electron and the charge on one of the carbon-carbon sigma bonds molecular ion. When we write structures this way, the choice of just where to l the odd electron and the charge is sometimes arbitrary. When possible, howe write the structure showing the molecular ion that would result from the rem one of the most loosely held electrons of the original molecule. Just which e these are can usually be estimated from ionization potentials (Table A ionization potential of a molecule is the amount of energy (in eV) required t an electron from the molecule.] As we might expect, ionization potential that the nonbonding electrons of nitrogen and oxygen and the pi electrons and aromatic molecules are held more loosely than the electrons of localizir and carbon-hydrogen sigma bonds. Thus the convention of the molecule contains electron and charge is especially applicable when the molecule contains only ca nitrogen, double bond, or aromatic ring. If the molecule contains a great many of th and carbon-hydrogen sigma bonds, and if it contains charge is so arbi choice of where to localize the odd electron and the charge is so arbi impractical. In these instances we usually resort to another conventio formula for the radical cation in brackets and place the odd elect outside. Using this convention we would write the two fragmentati propane in the following way.

$$[CH_3CH_2CH_3]^{\ddagger} \longrightarrow CH_3CH_2^+ + \cdot CH_3$$

$$[CH_3CH_2CH_3]^{\ddagger} \longrightarrow CH_3CH_2\cdot + CH_3^+$$

Problem A.9

The most intense peak in the mass spectrum of 2,2-dimethylbut (a) What carbocation does this peak represent? (b) Using the co just described, write an equation that shows how this carbo molecular ion.

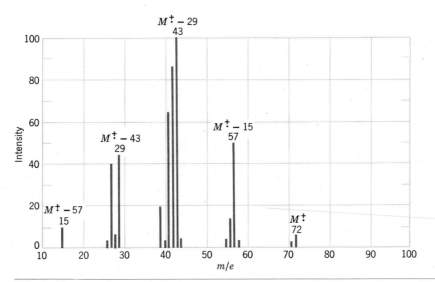

FIG. A.10
The mass spectrum of 2-methylbutane.

whereas with *n*-hexane, loss of a methyl radical can yield only a primary cation.

With 2,2-dimethylpropane (Fig. A.11), this effect is even more dramatic. Loss of a methyl radical by the molecular ion produces a *tertiary* carbocation and this reaction takes place so readily that virtually none of the molecular ions survive long enough to be detected.

$$
\begin{bmatrix} & CH_3 & \\ CH_3-&\overset{|}{\underset{|}{C}}-CH_3 \\ & CH_3 & \end{bmatrix}^{\ddot{+}} \longrightarrow CH_3-\overset{CH_3}{\underset{CH_3}{\overset{|}{\underset{|}{C^+}}}} + \cdot CH_3
$$

$$
\begin{array}{cc}
m/e\ 72 & m/e\ 57 \\
M^{\ddot{+}} & M^{\ddot{+}} - 15
\end{array}
$$

Problem A.10

In contrast to 2-methylbutane and 2,2-dimethylpropane the mass spectrum of 3-methylpentane (not given) has a peak of very low intensity at $M^{\ddot{+}} - 15$. It has a peak of very high intensity at $M^{\ddot{+}} - 29$, however. Explain.

Carbocations stabilized by resonance are usually also prominent in mass spectra. Several ways that resonance-stabilized carbocations can be produced are outlined below.

(1) Alkenes frequently undergo fragmentations that yield allylic cations.

$$CH_2 \overset{+\cdot}{\frown} CH \overset{\smile}{\frown} CH_2 : R \longrightarrow \overset{+}{C}H_2 - CH = CH_2 + \cdot R$$
$$m/e\ 41$$

(2) Carbon-carbon bonds next to an atom with an unshared electron pair usually break readily because the resulting carbocation is resonance stabilized.

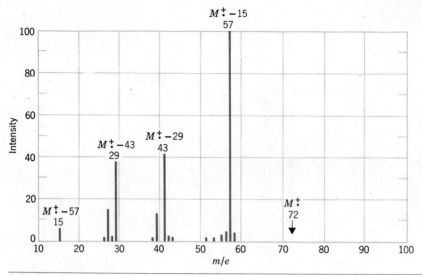

FIG. A.11

Mass spectrum of 2,2-dimethylpropane.

$$R-\overset{+\bullet}{Z}\overset{\frown}{}CH_2:CH_3 \longrightarrow R-\overset{+}{Z}=CH_2 + \cdot CH_3$$

$$\updownarrow$$

$$R-\overset{\bullet\bullet}{Z}-\overset{+}{C}H_2$$

Z = N, O, or S; R may also be H.

(3) Carbon-carbon bonds next to the carbonyl group of an aldehyde or ketone break readily because resonance-stabilized acylium ions are produced

$$\overset{R}{\underset{R'}{\diagdown}}\overset{\bullet\bullet}{C}\overset{\frown}{=}\overset{\bullet+}{O}: \longrightarrow R'-C\overset{+}{\equiv}\overset{\bullet\bullet}{O}: + R\cdot$$

$$\updownarrow$$

$$R'-\overset{+}{C}=\overset{\bullet\bullet}{O}:$$

or

$$\overset{R}{\underset{R'}{\diagdown}}\overset{\bullet\bullet}{C}\overset{\frown}{=}\overset{\bullet+}{O}: \longrightarrow R-C\overset{+}{\equiv}\overset{\bullet\bullet}{O}: + R'\cdot$$

$$\updownarrow$$

$$R-\overset{+}{C}=\overset{\bullet\bullet}{O}:$$

(4) Alkyl-substituted benzenes undergo loss of a hydrogen atom or methyl group to yield the highly stable tropylium ion (cf. Sect. 12.6). This fragmentation gives a prominent peak (sometimes the base peak) at m/e 91.

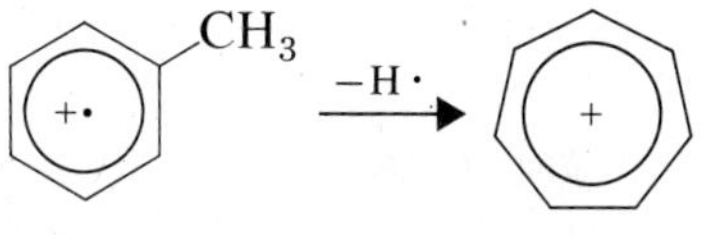

m/e 91

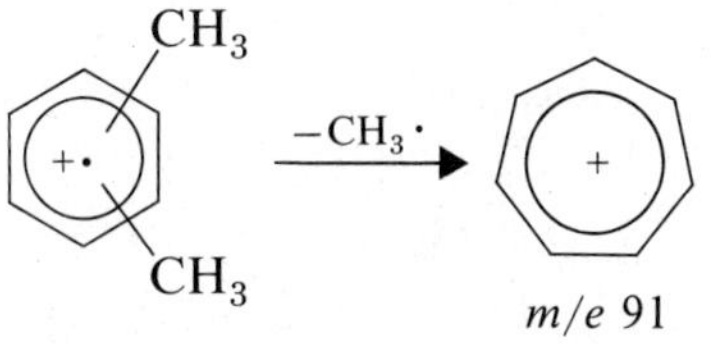

m/e 91

(5) Substituted benzenes also lose their substituent and yield a phenyl cation at m/e 77.

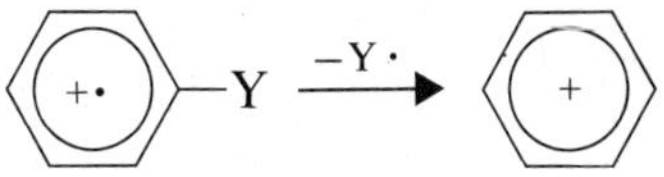

m/e 77

$$Y = \text{halogen}, -NO_2, -\overset{\overset{\textstyle O}{\|}}{C}R, -R, \text{ etc.}$$

Problem A.11

The mass spectrum of 4-methyl-1-hexene (not given) shows intense peaks at m/e 57 and m/e 41. What fragmentation reactions account for these peaks?

Problem A.12

Explain the following observations that can be made about the mass spectra of alcohols:

(a) The molecular ion peak of a primary or secondary alcohol is very small; with a tertiary alcohol it is usually undetectable. (b) Primary alcohols show a prominent peak at m/e 31. (c) Secondary alcohols usually give prominent peaks at m/e 45, 59, 73 and so forth. (d) Tertiary alcohols have prominent peaks at m/e 59, 73, 87 and so forth.

Problem A.13

The mass spectra of isopropyl butyl ether and propyl butyl ether are given in Figs. A.12 and A.13. (a) Which spectrum represents which ether? (b) Explain your choice.

The Nitrogen Rule

For most of the elements found in organic compounds, there is a correspondence between the most common *valence* of the element and the *mass* of its most abundant isotope—either *both are even numbers* or *both are odd numbers*. For carbon, for example, the valence is 4 and the mass 12; for hydrogen the valence is 1, the mass 1; for oxygen the valence is 2 and the mass 16. The only important exception is the element nitrogen; here the valence is odd (3) and the mass even (14).

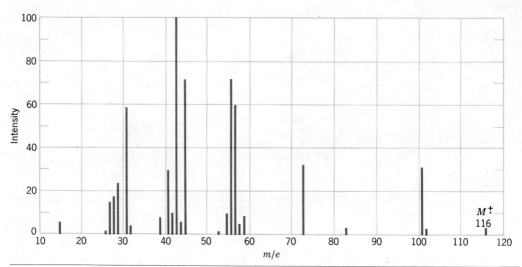

FIG. A.12
Mass spectrum for problem A.13.

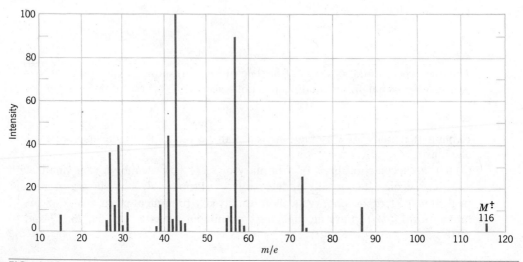

FIG. A.13
Mass spectrum for problem A.13.

These numbers lead to an important rule called the **nitrogen rule:** *Compounds containing an even number of nitrogen atoms (zero is an even number)* will give a molecular ion with an even mass number.

Problem A.14

Demonstrate the validity of the nitrogen rule by showing how it applies to the molecular ions obtained from the following: (a) C_6H_6O (b) C_2H_5Cl (c) C_6H_{12} (d) C_3H_9N (e) CN_2H_6 (f) C_2H_5NO.

Fragmentation by Cleavage of Two Bonds

Many peaks in mass spectra can be explained by fragmentation reactions that involve the breaking of two covalent bonds. When a radical-cation undergoes this type of fragmentation the products are *a new radical-cation* and *a neutral molecule.* Some important examples are the following.

(1) Alcohols frequently show a prominent peak at $M^{\ddagger} - 18$. This corresponds to the loss of a molecule of water.

$$R-CH-CH_2 \longrightarrow R-CH^{\cdot\ddagger}CH_2 + H-O-H$$
$$M^{\ddagger} \qquad\qquad M^{\ddagger} - 18$$

or

$$[R-CH_2-CH_2-OH]^{\ddagger} \longrightarrow [R-CH=CH_2]^{\ddagger} + H_2O$$
$$M^{\ddagger} \qquad\qquad M. - 18$$

(2) Cycloalkenes can undergo a retro-Diels-Alder reaction that produces an alkene and an alkadiene radical cation.

$$\left[\bighexagon\right]^{\ddagger} \longrightarrow \left[\bigpentagon\right]^{\ddagger} + \begin{array}{c} CH_2 \\ \| \\ CH_2 \end{array}$$

(3) Carbonyl compounds with a hydrogen on their γ-carbon undergo a fragmentation called the *McLafferty rearrangement.*

$$\left[\begin{array}{c} O \quad H \\ \| \quad \diagdown \\ Y-C \quad CHR \\ \diagdown \quad \diagup \\ CH_2 \quad CH_2 \end{array}\right]^{\ddagger} \longrightarrow \left[\begin{array}{c} H \\ O \diagdown \\ \| \\ Y-C \\ \diagdown \\ CH_2 \end{array}\right]^{\ddagger} + RCH=CH_2$$

Y may be R, H, OR, OH, etc.

In addition to these reactions, we frequently find peaks in mass spectra that result from the elimination of other small stable neutral molecules, for example, H_2, NH_3, CO, HCN, H_2S, alcohols, and alkenes.

Additional Problems

A.15

Reconsider problem A.6 and the spectrum given in Fig. A.7. Important clues to the structure of this compound are the peaks at m/e 44 (the base peak) and m/e 29. Propose a structure for the compound and write fragmentation equations showing how these peaks arise.

A.16

The homologous series of primary amines, $CH_3(CH_2)_nNH_2$, from CH_3NH_2 to $CH_3(CH_2)_{13}NH_2$ all have their base (largest) peak at m/e 30. What ion does this peak represent and how is it formed?

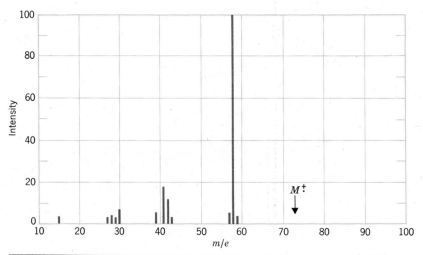

FIG. A.14
*Mass spectrum of compound **A** (problem A.17).*

A.17

The mass spectrum of compound **A** is given in Fig. A.14. The pmr spectrum of **A** consists of two singlets with area ratios of 9 : 2. The larger singlet is at δ1.2, the smaller one at δ1.3. Propose a structure for compound **A.**

A.18

The mass spectrum of compound **B** is given in Fig. A.15. The infrared spectrum of **B** shows a broad peak between 3200–3600 cm⁻¹. The pmr spectrum of **B** shows the following peaks: a triplet at δ0.9, a singlet at δ1.1, and a quartet at δ1.6. The area ratios of these peaks is 3 : 7 : 2 respectively. Propose a structure for **B.**

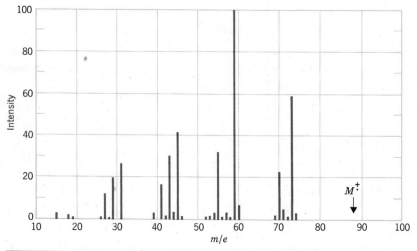

FIG. A.15
*Mass spectrum of compound **B** (problem A.18).*

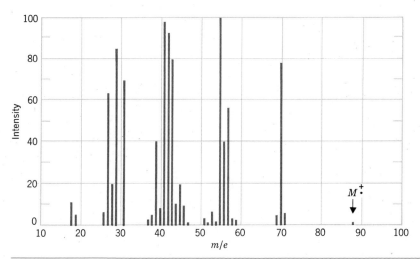

FIG. A.16
Mass spectrum of compound C (problem A.19).

A.19

The mass spectrum of compound **C** is given in Fig. A.16. Compound **C** is an isomer of **B** and the infrared spectrum of **C** also shows a broad peak in the 3200–3600 region. The pmr spectrum of **C** is given in Fig. A.17. Propose a structure for **C**.

A.20

The mass spectrum of compound **D** is given in Fig. A.18. **D** shows a strong infrared peak at 1710 cm⁻¹. The pmr spectrum of **D** is given in Fig. A.19. Propose a structure for **D**.

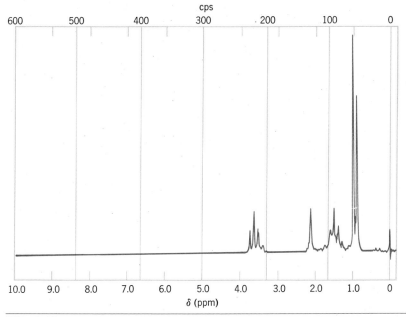

FIG. A.17
The pmr spectrum of compound C. (Courtesy of Aldrich Chemical Co., Milwaukee, Wis.)

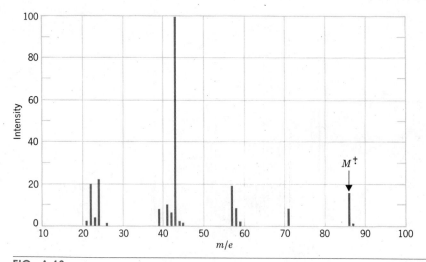

FIG. A.18
*The mass spectrum of compound **D** (problem A.20).*

A.21
Propose a structure for compound **E** whose mass spectrum is given in Fig. A.20.

A.22
Propose a structure for compound **F** whose mass spectrum is given in Fig. A.21 and whose pmr spectrum is given in Fig. A.22.

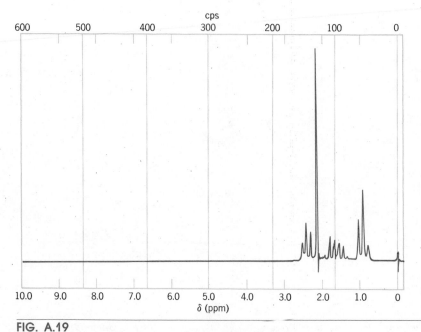

FIG. A.19
*The pmr spectrum of compound **D**. (Courtesy of Aldrich Chemical Co. Milwaukee, Wis.)*

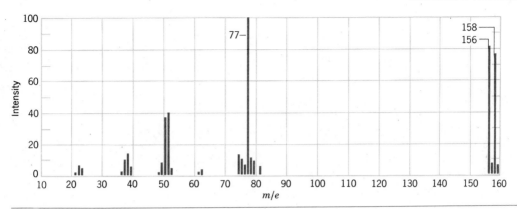

FIG. A.20

The mass spectrum of compound **E.**

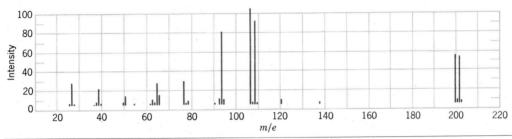

FIG. A.21

The mass spectrum of compound **F.** (*Adapted with permission from E. Stenhagen, S. Abrahamsson, and F. W. McLafferty,* Registry of Mass Spectral Data, *Vol. II, Wiley, New York, 1974, p. 992.*)

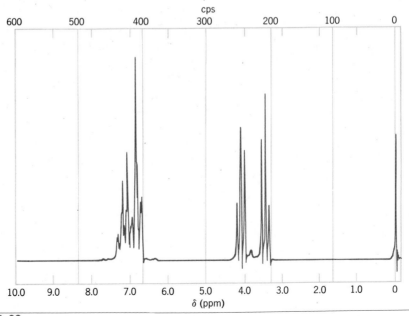

FIG. A.22

The pmr spectrum of compound **F.** (*Courtesy of Aldrich Chemical Co., Milwaukee, Wis.*)

BIBLIOGRAPHY OF SUGGESTED READINGS

CHAPTER 1

O. T. Benfey, *From Vital Force to Structural Formulas,* Houghton Mifflin, Boston, 1964.

R. J. Gillespie, "The Electron-Pair Repulsion Model for Molecular Geometry," *J. Chem. Educ., 47,* 18 (1970).

P. E. Verkade, "August Kekulé," *Proc. Chem. Soc.,* 205 (1958).

L. Pauling, *The Nature of the Chemical Bond,* 3rd ed., Cornell University Press, Ithaca, N.Y., 1960.

G. W. Wheland, *Resonance in Organic Chemistry,* Wiley, New York, 1955.

J. B. Hendrickson, D. J. Cram, and G. S. Hammond, *Organic Chemistry,* McGraw-Hill, New York, 1970, Chapter 2.

CHAPTER 2

O. T. Benfey, *The Names and Structures of Organic Compounds,* Wiley, New York, 1966.

J. D. Roberts, R. Stewart, and M. C. Caserio, *Organic Chemistry: Methane to Macromolecules,* Benjamin, New York, 1971, Chapter 2.

CHAPTER 3

G. W. Wheland, *Advanced Organic Chemistry,* 3rd ed., Wiley, New York, 1960.

Lloyd N. Ferguson, "Ring Strain and Reactivity of Alicycles," *J. Chem. Educ., 47,* 46 (1970).

C. A. Coulson, *Valence,* Oxford University Press, New York, 1952, Chapter VIII.

J. March, *Advanced Organic Chemistry,* McGraw-Hill, New York, 1968, pp. 101–109, 117–120.

G. H. Posner, "Substitution Reactions Using Organocopper Reagents," *Organic Reactions,* Vol. 22, Wiley, New York, 1975.

CHAPTER 4

S. W. Benson, "Bond Energies," *J. Chem. Educ., 42,* 502 (1965).

R. T. Morrison and R. N. Boyd, *Organic Chemistry,* 3rd ed., Allyn and Bacon, Boston, 1973, Chapter 2.

W. A. Pryor, *Free Radicals,* McGraw-Hill, New York, 1965.

E. S. Huyser, *Free-Radical Chain Reactions,* Wiley, New York, 1970.

C. Walling, *Free Radicals in Solution,* Wiley, New York, 1957.

W. A. Pryor, *Introduction to Free Radical Chemistry,* Prentice-Hall, Englewood Cliffs, N.J., 1965.

CHAPTER 5

L. F. Fieser and M. Fieser, *Advanced Organic Chemistry,* Reinhold, New York, 1961, Chapter 5.

G. A. Olah and P. v. R. Schleyer, eds., *Carbonium Ions,* Wiley, New York, 1968.

M. Orchin and H. H. Jaffé, *The Importance of Antibonding Orbitals,* Houghton Mifflin, Boston, 1967.

CHAPTER 6

O. T. Benfey, *Introduction to Organic Reaction Mechanisms,* McGraw-Hill, New York, 1970, Chapter 5.

W. H. Saunders, *Ionic Aliphatic Reactions,* Prentice-Hall, Englewood Cliffs, N.J., 1965, Chapter 2.

H. C. Brown, *Hydroboration,* Benjamin, New York, 1962.

H. C. Brown and P. J. Geoghegan, Jr., "Solvomercuration-Demercuration. I.," *J. Org. Chem., 35,* 1844 (1970).

W. S. Johnson, "Non-enzymic Biogenetic-like Olefinic Cyclizations," *Accounts Chem. Res., 1,* 1 (1968).

J. G. MacConnell and Robert M. Silverstein, "Recent Results in Insect Pheromone Chemistry," *Angew. Chem. internat Edit., 12,* 644 (1973).

CHAPTER 7

E. L. Eliel, *Elements of Stereochemistry,* Wiley, New York, 1969.

K. Mislow, *Introduction to Stereochemistry,* Benjamin, New York, 1965.

D. F. Mowery, Jr., "Criteria for Optical Activity in Organic Molecules," *J. Chem. Educ., 46,* 269 (1969).

D. Whittaker, *Stereochemistry and Mechanism,* Clarendon Press, Oxford, 1973, Chapters 1, 2, and 5.

CHAPTER 8

R. H. Maybury, "The Language of Quantum Mechanics," *J. Chem. Educ., 39,* 367 (1962).

J. Hine, *Divalent Carbon,* Ronald Press, New York, 1964.

G. L. Closs, "Structures of Carbenes and the Stereochemistry of Carbene Additions to Olefins," *Topics in Stereochemistry,* Vol. 3, Wiley, New York, 1968.

W. E. Parham and E. E. Schweizer, "Halocyclopropanes from Halocarbenes," *Organic Reactions,* Vol. 13, Wiley, New York, 1963.

H. E. Simmons, T. L. Cairns, S. A. Vladuchick, and C. M. Hoiness, "Cyclopropanes from Unsaturated Compounds, Methylene Iodide, and Zinc-Copper Couple," *Organic Reactions,* Vol. 20, Wiley, New York, 1973.

F. W. Billmeyer, *Textbook of Polymer Science,* 2nd ed., Wiley, New York, 1971.

L. R. G. Treloar, *Introduction to Polymer Science,* Springer-Verlag, New York, 1970.

CHAPTER 9

L. F. Fieser and M. Fieser, *Advanced Organic Chemistry,* Reinhold, New York, 1961, Chapter 6.

T. F. Rutledge, *Acetylenic Compounds: Preparation and Substitution Reactions,* Reinhold, New York, 1968.

T. F. Rutledge, *Acetylenes and Allenes: Addition Cyclization and Polymerization Reactions,* Reinhold, New York, 1969.

R. L. Shriner, R. C. Fuson, and D. Y. Curtin, *Systematic Identification of Organic Compounds,* Wiley, New York, 1964.

T. L. Jacobs, "The Synthesis of Acetylenes," *Organic Reactions,* Vol. 5, Wiley, New York, 1949.

CHAPTER 10

A. Liberles, *Introduction to Molecular Orbital Theory,* Holt, Rinehart and Winston, New York, 1966.

M. Orchin and H. H. Jaffé, *The Importance of Antibonding Orbitals,* Houghton Mifflin, 1967.

J. Sauer, "Diels Alder Reactions, Part I." *Angew. Chem. internat Edit., 5,* 211 (1966); "Part II," *Ibid., 6,* 16 (1967).

CHAPTER 11

R. Hubbard and A. Kropf, "Molecular Isomers in Vision," *Bio-organic Chemistry: Readings from Scientific American,* M. Calvin and M. Jorgenson, eds., W. H. Freeman, San Francisco, 1968.

R. H. Johnson and T. P. Williams, "Action of Light upon the Visual Pigment Rhodopsin," *J. Chem. Educ., 47,* 736 (1970).

E. L. Menger, ed., "Special Issue on the Chemistry of Vision," *Accounts Chem. Res., 8,* (3), 81–112, 1975.

R. B. Woodward and R. Hoffman, *The Conservation of Orbital Symmetry,* Academic Press, New York, 1970.

C. R. Noller, *Chemistry of Organic Compounds,* Saunders, Philadelphia, 1965, Chapter 42.

J. W. Cornforth, "Terpenoid Biosynthesis," *Chemistry in Britain, 4,* 102 (1968).

CHAPTER 12

G. M. Badger, *Aromatic Character and Aromaticity,* Cambridge University Press, 1969.

R. Breslow, "Antiaromaticity," *Accounts Chem. Res., 6,* 393 (1973).

F. Sondheimer, "The Annulenes," *Accounts Chem. Res., 5,* 81 (1972).

CHAPTER 13

L. M. Stock, *Aromatic Substitution Reactions,* Prentice-Hall, Englewood Cliffs, N.J., 1968.

J. March, *Advanced Organic Chemistry,* McGraw-Hill, New York, 1968, Chapter 11.

G. A. Olah, *Friedel-Crafts and Related Reactions,* Vol. I, Wiley, New York, 1963.

W. R. Dolbier, Jr., "Electrophilic Additions to Alkenes," *J. Chem. Educ., 46,* 342 (1969).

CHAPTER 14

R. M. Silverstein and G. C. Bassler, *Spectrometric Identification of Organic Compounds,* Wiley, New York, 1967.

J. R. Dyer, *Applications of Absorption Spectroscopy of Organic Compounds,* Prentice-Hall, Englewood Cliffs, N.J., 1965.

J. D. Roberts, *Nuclear Magnetic Resonance,* McGraw-Hill, New York, 1959.

CHAPTER 15

E. C. Taylor and Alexander McKillop, "Thallium in Organic Synthesis," *Accounts Chem. Res., 3,* 338 (1970).

J. S. Thayer, "Some Biological Aspects of Organometallic Chemistry," *J. Chem. Educ., 48,* 806 (1971).

J. D. Roberts and M. C. Caserio, *Basic Principles of Organic Chemistry,* Benjamin, New York, 1965, Chapter 12.

CHAPTER 16

S. Patai, ed., *Chemistry of the Hydroxyl Group,* Wiley, New York, 1971.

S. Patai, ed., *Chemistry of the Ether Linkage,* Wiley, New York, 1967.

L. B. Clapp, *The Chemistry of the OH Group,* Prentice-Hall, Englewood Cliffs, N.J., 1967.

CHAPTER 17

C. K. Ingold, *Structure and Mechanism in Organic Chemistry,* 2nd ed., Cornell University Press, Ithaca, N.Y., 1969, Chapters 7 and 9.

D. J. Raber and J. M. Harris, "Nucleophilic Substitution Reactions at Secondary Carbon Atoms," *J. Chem. Educ., 49,* 60 (1972).

R. A. Sneen, "Organic Ion Pairs as Intermediates in Nucleophilic Substitution and Elimination Reactions," *Accounts Chem. Res., 6,* 46 (1973).

F. G. Bordwell, "How Common are Base-Initiated, Concerted 1,2-Eliminations?" *Accounts Chem. Res., 5,* 374 (1972).

J. F. Bunnett, "The Base-Catalyzed Halogen Dance, and Other Reactions of Aryl Halides," *Accounts Chem. Res., 5,* 139 (1972).

CHAPTER 18

C. A. Buehler and D. E. Pearson, *Survey of Organic Synthesis,* Wiley, New York, 1970.

H. O. House, *Modern Synthetic Reactions,* 2nd ed., Benjamin, New York, 1972.

S. Patai, ed., *The Chemistry of the Carbonyl Group,* Vol. 1, Wiley, New York, 1966.

S. Patai and J. Zabicky, eds., *The Chemistry of the Carbonyl Group,* Vol. 2, Wiley, New York, 1970.

A. J. Nielson and W. J. Houlihan, "The Aldol Condensation," *Organic Reactions,* Vol. 16, Wiley, New York, 1968.

E. Vedejs, "Clemmensen Reduction of Ketones," *Organic Reactions,* Vol. 22, Wiley, New York, 1975.

M. W. Rathke, "The Reformatsky Reaction," *Organic Reactions,* Vol. 22, Wiley, New York, 1975.

CHAPTER 19

S. Patai, ed., *The Chemistry of Carboxylic Acids and Esters,* Wiley, New York, 1969.

L. F. Fieser and M. Fieser, *Advanced Organic Chemistry,* Reinhold, New York, 1961, Chapters 11, 23, and 24.

S. Patai, ed., *The Chemistry of Amides,* Wiley, New York, 1969.

C. D. Gutsche, *The Chemistry of Carbonyl Compounds,* Prentice-Hall, Englewood Cliffs, N.J., 1967.

CHAPTER 20

C. R. Hauser and B. E. Hudson, "The Acetoacetic Ester Condensation and Certain Related Reactions," *Organic Reactions,* Vol. 1, Wiley, New York, 1942.

H. O. House, *Modern Synthetic Reactions,* Benjamin, New York, 1965, Chapters 7 and 9.

W. McCrae, *Basic Organic Reactions,* Heyden and Son, Ltd., London, 1973, Chapters 3 and 4.

J. P. Schaefer and J. J. Bloomfield, "The Dieckmann Condensation," *Organic Reactions,* Vol. 15, Wiley, New York, 1967.

G. Jones, "The Knovenagel Condensation," *Organic Reactions,* Vol. 15, Wiley, New York, 1967.

T. M. Harris and C. M. Harris, "The γ-Alkylation and γ-Arylation of Dianions of β-Dicarbonyl Compounds," *Organic Reactions,* Vol. 17, Wiley, New York, 1969.

A. G. Cook, *Enamines: Synthesis, Structure, and Reactions,* Dekker, New York, 1969.

CHAPTER 21

S. Patai, ed., *The Chemistry of the Amino Group,* Wiley, New York, 1968.

L. F. Fieser and M. Fieser, *Advanced Organic Chemistry,* Reinhold, New York, 1961, Chapters 14 and 21.

H. K. Porter, "The Zinin Reduction of Nitroarenes," *Organic Reactions,* Vol. 20, Wiley, New York, 1973.

H. Zollinger, *Diazo and Azo Chemistry,* Wiley, New York, 1961.

CHAPTER 22

L. A. Paquette, *Principles of Modern Heterocyclic Chemistry,* Benjamin, New York, 1968.

G. A. Swan, *An Introduction to Alkaloids,* Wiley, New York, 1967.

T. A. Geissman and D. H. G. Crout, *Organic Chemistry of Secondary Plant Metabolism,* Freeman, Cooper and Co., San Francisco, 1969, Chapters 16 to 19.

J. K. Stille, *Industrial Organic Chemistry,* Prentice-Hall, Englewood Cliffs, N.J., 1968.

CHAPTER 23

L. F. Fieser and M. Fieser, *Advanced Organic Chemistry,* Reinhold, New York, 1961, Chapter 30.

L. F. Fieser, "Steroids," *Bio-organic Chemistry: Readings from Scientific American,* Freeman, San Francisco, 1968.

A. White, P. Handler, E. L. Smith, *Principles of Biochemistry,* McGraw-Hill, New York, 1964, Chapters 5 and 6.

A. L. Lehninger, *Biochemistry,* Worth, New York, 1970, Chapter 23.

E. E. Conn and P. K. Stumpf, *Outlines of Biochemistry,* 3rd ed., Wiley, New York, 1972, Chapters 3 and 12.

J. R. Hanson, *Introduction to Steroid Chemistry,* Pergamon Press, New York, 1968.

CHAPTER 24

L. F. Fieser and M. Fieser, *Advanced Organic Chemistry,* Reinhold, New York, 1961. Chapter 29.

C. R. Noller, *Chemistry of Organic Compounds,* Saunders, New York, 1965, Chapter 18.

D. E. Green and R. F. Goldberger, *Molecular Insights into the Living Process,* Academic Press, 1967, Chapters 2 and 3.

C. S. Hudson, "Emil Fischer's Discovery of the Configuration of Glucose," *J. Chem. Educ., 18,* 353 (1941).

R. Barker, *Organic Chemistry of Biological Compounds,* Prentice-Hall, Englewood Cliffs, N.J., 1971, Chapter 5.

CHAPTER 25

C. R. Noller, *Chemistry of Organic Compounds,* Saunders, New York, 1965, Chapter 19.

L. F. Fieser and M. Fieser, *Advanced Organic Chemistry,* Reinhold, New York, 1961, Chapter 31.

J. R. Holum, *Organic Chemistry: A Brief Course,* Wiley, New York, 1975, Chapter 15.

The following articles from, *Bio-organic Chemistry: Readings from Scientific American,* M. Calvin and M. J. Jorgenson, eds., Freeman, San Francisco, 1968:

P. Doty, "Proteins," p. 15.

W. H. Stein and S. Moore, "The Chemical Structure of Proteins," p. 23.

E. O. P. Thompson, "The Insulin Molecule," p. 34.

M. F. Perutz, "The Hemoglobin Molecule," p. 41.

E. Zuckerkandl, "The Evolution of Hemoglobin," p. 53.

D. C. Phillips, "The Three-Dimensional Structure of an Enzyme Molecule," p. 67.

L. Pauling, R. B. Corey, and R. Hayward, "Structures of Protein Molecules," *Scientific American,* October 1954, p. 54.

H. D. Law, *The Organic Chemistry of Peptides,* Wiley, New York, 1970.

D. E. Green and R. F. Goldberger, *Molecular Insights into the Living Process,* Academic Press, 1967, Chapters 4 and 5.

E. E. Conn and P. K. Stumpf, *Outlines of Biochemistry,* 3rd ed., Wiley, New York, 1972, Chapter 4.

R. E. Dickerson and I. Geis, *The Structure and Action of Proteins,* Harper and Row, New York, 1969.

CHAPTER 26

R. Barker, *Organic Chemistry of Biological Compounds,* Prentice-Hall, Englewood Cliffs, N.J., 1971, Chapter 8.

E. E. Conn and P. K. Stumpf, *Outlines of Biochemistry,* 2nd ed., Wiley, New York, 1972, Chapters 5, 18, and 19.

J. D. Watson, *Molecular Biology of the Gene,* 2nd ed., Benjamin, New York, 1970.

The following articles in *Bio-organic Chemistry: Readings from Scientific American,* M. Calvin and M. J. Jorgenson, eds., Freeman, San Francisco, 1968:

F. H. C. Crick, "The Structure of the Hereditary Material," p. 75.

R. W. Holley, "The Nucleotide Sequence of a Nucleic Acid," p. 82.

ANSWERS TO SELECTED PROBLEMS

CHAPTER 1

1.5 (a) $\underset{+\quad+}{\overset{H-Br;}{\longrightarrow}}$ (b) $\underset{+\quad+}{\overset{I-Cl;}{\longrightarrow}}$ (c) H_2, $\mu=0$;

(d) Cl_2, $\mu=0$.

1.6 (a) (b) (d) (e), tetrahedral; (c) trigonal planar; (f) linear.

1.7 The bond moments cancel.

1.9 Trigonal planar structure causes bond moments to cancel.

1.14 (a) and (d), (e) and (f).

1.15 (a) C, 40.0%; H, 6.66%; O, 53.3%; (b) C, 32.0%; H, 6.66%; N, 18.7%, O, 42.7%; (c) C, 12.8%; H, 1.78%; Br, 85.4%.

1.19 (a) NH_2; (b) CH; (c) C_2H_4O.

1.20 (b) $C_3H_3N_3$.

1.21 CH_2O.

1.22 C_2H_4.

1.24 $C_{11}H_{12}Cl_2N_2O_5$.

CHAPTER 2

2.1 (a) two; (b) one; (c) CCl_3CH_3 and $CHCl_2CH_2Cl$; (d) yes; (e) two; (f) CCl_3CH_2Cl and $CHCl_2CHCl_2$; (g) one.

2.5 (a) $CH_3CCl(CH_3)CH_3$; (b) $CH_3CHICH_2CH_3$.

2.7 (a) yes; (b) $CH_2=CClCH_3$ and $CH_3CCl_2CH_3$.

2.8 (a) ethyne + 2HCl; (b) ethene + Cl_2; (c) ethene + Br_2; (d) ethyne + 2Br_2; (e) ethyne + HCl, then, product + HBr.

2.10 Molecules of trimethylamine cannot form hydrogen bonds to each other whereas molecules of n-propylamine can.

2.11 (a) alkyne; (d) aldehyde.

2.13 (b) substitution; (f) condensation.

2.16 (c) tertiary; (e) secondary.

2.17 (a) secondary; (c) tertiary.

2.20 (b) ethylene glycol; (f) propionic acid.

2.22 (b) $CH_3CH_2CH_2OH$; (e) $CH_3CH_2CH_2X$; (l) $CH_3N(CH_3)CH_2CH_3$.

2.25 ester.

CHAPTER 3

3.7 (a) $CH_3CHClCHClCH_2CH_3$; (k) $CH_3CH(CH_3)CH_2CH_2CH_2Cl$; (m) $CH_3C(CH_3)_2CH_2Cl$; (n) $CH_3CH(CH_3)CH_2CH_3$.

3.8 (a) 3,4-dimethylhexane; (f) cyclopentylcyclopentane.

3.10 (a) neopentane (or 2,2-dimethylpropane); (d) cyclopentane.

3.11 (a) 2,4-dimethyl-3-isopropylpentane; (f) 5,5-dibutylnonane.

3.12 (a) neopentane.

3.15 (d) chloroethane.

3.18 (c) *trans*-1,4-dimethylcyclohexane.

3.21 (c) $CH_2=CH(CH_2)_3CH_3 + H_2 \xrightarrow{Pt}$
$CH_3(CH_2)_4CH_3$.

CHAPTER 4

4.1 (a) -102 kcal/mole; (d) $+13$ kcal/mole.

4.6 (c) $\Delta H = +88$ kcal/mole; $E_{act} = +88$ kcal/mole.

4.8 Their hydrogens are all equivalent; replacing any one yields the same product.

4.9 (b) $\Delta H = -8.5$ kcal/mole.

4.12 $\underset{\cdot}{CH_3}\overset{CH_3}{\underset{|}{C}}CH_2CH_3 > CH_3\overset{CH_3}{\underset{|}{CH}}CHCH_3 >$

$\cdot CH_2\overset{CH_3}{\underset{|}{CH}}CH_2CH_3 \simeq CH_3\overset{CH_3}{\underset{|}{CH}}CH_2CH_2\cdot$

4.18 The mechanism involves protolysis of a carbon-carbon bond of isobutane.

4.20 (a) $CH_3-H \longrightarrow CH_3\cdot + H\cdot$,
$E_{act} = \Delta H = +104$ kcal/mole
$CH_3CH_2-H \longrightarrow CH_3CH_2\cdot + H\cdot$,
$E_{act} = \Delta H = +98$ kcal/mole
Since the E_{act} for the ethane C—H cracking reaction is lower, the ethane reaction occurs at a lower temperature;
(b) $CH_3CH_3 \longrightarrow 2CH_3\cdot$, $E_{act} = \Delta H = +88$ kcal/mole. This reaction has a lower E_{act} than the ethane C—H cracking reaction given in part (a).
(c) $CH_3CH_2-CH_2CH_3 \longrightarrow 2CH_3CH_2\cdot$,
$E_{act} = \Delta H = +82$ kcal/mole

$$CH_3CH_2CH_2—CH_3 \longrightarrow$$
$$CH_3CH_2CH_2 \cdot + CH_3 \cdot,$$
$$E_{act} = \Delta H = +85 \text{ kcal/mole.}$$

CHAPTER 5

5.2 (a) isobutane; (b) $4CO_2 + 4H_2O$; (c) yes; (d) cyclobutane and methyl-cyclopropane; (e) yes.

5.4 (a) 2-methyl-2-butene; (d) 4-methylcyclohexene.

5.8 Add a drop of Br_2 in CCl_4 to a small amount of each isomer. Bromine will add to the double bond of 1-hexene producing a colorless solution. No reaction will take place with cyclohexane as long as the mixture is not heated and is protected from light; the red-brown bromine color, consequently, will persist.

5.9 (c), (g), (h), and (l) can exist as *cis-trans* isomers.

5.12 (a) *cis-trans* isomerization. This happens because at 300° the molecules have enough energy to surmount the rotational barrier of the carbon-carbon double bond. (b) *trans*-2-butene because it is more stable.

5.16 (a) *cis*-1,2-dimethylcyclopentane; (b) *cis*-1,2-dimethylcyclohexane; (c) *cis*-1,2-dideuteriocyclohexane.

CHAPTER 6

6.1 1-iodo-2-chloropropane.

6.6 (c) cyclohexene + $Hg(OAc)_2$ + CH_3OH, then $NaBH_4$ + OH^-.

6.8 (c) 1-methylcyclopentene + $(BH_3)_2$, then CH_3COOD.

6.10 (c) 3-methyl-1-pentene.

6.13 (a) 1,2-dibromobutane; (f) CH_3CH_2OH; (j) cyclopentene; (m) 2-chlorohexane; (q) $2CH_3CH_2COO^-$

6.18 2-methylpropene > propene > ethene.

6.21 4-methylcyclohexene.

6.28

$$\overset{\overset{\displaystyle CH_3}{|}}{CH_3C}=CHCH_2CH_2\overset{\overset{\displaystyle CH_3}{|}}{C}=CHCH_2CH_2\overset{\overset{\displaystyle CH_2}{||}}{C}CH=CH_2.$$

CHAPTER 7

7.1 chiral: (a), (e), (f), (g), (h); achiral: (b), (c), (d).

7.6 (b), (c), (d).

7.11 75% (S)-(+)-2-butanol and 25% R-(−)-2-butanol.

7.13 (a) diastereomers; (b) diastereomers; (c) diastereomers; (e) yes; (f) no.

7.14 (a) and (b).

7.17 (a) *trans*-1,2-dibromocyclopentane; (b) as a racemic modification; (c) *cis*-1,2-dibromocyclopentane.

7.24 They are diastereomers.

7.27 (a) enantiomers; (d) diastereomers; (g) two molecules of the same compound; (j) enantiomers; (n) structural isomers.

7.31 (b) and (c).

7.33 (a) (3S,6S)-3,6-dimethyloctane; (b) yes.

CHAPTER 8

8.1 (a) 0.91 mole/liter; (b) 0.50 mole/liter; (c) 0.0010 mole/liter.

8.4 Head-to-tail polymerization involves the formation of a more stable free radical.

8.9 Addition of singlet methylene to each face of *trans*-2-butene occurs at the same rate.

8.11 ethane, isobutane, butane, 2,3-dimethyl-butane, 2-methylpentane, hexane.

CHAPTER 9

9.2 (a) $HC≡CNa + NH_3$; (d) $HC≡CH + CH_3CH_2ONa$.

9.4 (c) 2-butyne + $(BD_3)_2$, then CH_3COOD; (d) 1-butyne + Sia_2BH, then CH_3COOD.

9.7 (b) cyclooctyne.

9.9 (a) and (j).

9.12 (a) $CH_3CH_2CH_2CBr=CHBr$; (d) $CH_3CH_2CH_2CH=CHBr$; (j) $CH_3CH_2CH_2C≡CCH_3$; (k) $CH_3CH_2CH_2C≡C—C≡$
$$CCH_2CH_2CH_3;$$
(n) $CH_3CH_2CH_2COOH + CO_2$.

9.14 (a) 1-pentene + Br_2, then $2NaNH_2$ and heat; (e) 1-bromopropane + $HC≡CNa$.

9.15 (a) Br_2/CCl_4 or dil. $KMnO_4$; (e) sodium fusion, then HNO_3, then $AgNO_3$.

CHAPTER 10

10.1 (a) $^{14}CH_2=CH—CH_2—X +$
$X—^{14}CH_2—CH=CH_2$; (c) in equal amounts.

10.4 (a) $CH_3CH_2CHClCH=CHCH_3$ +
$CH_3CHClCH=CHCH_2CH_3$;

(b) $CH_3CHClCH=CHCH_3$ +
$CH_3CH=CHCHClCH_3$, possibly,
$CH_2ClCH=CHCH_2CH_3$ and
$CH_2=CHCHClCH_2CH_3$.

10.10 (a) 1,4-dibromobutane + 2KOH, alcohol, and heat; (g) $HC\equiv CCH=CH_2$ + Sia_2BH, then CH_3COOH, or H_2 + P-2 catalyst.

10.13 (a) 1-butene + N-bromosuccinimide, then KOH in alcohol and heat;
(e) cyclopentane + Br_2, hν, then KOH in alcohol and heat, then N-bromosuccinimide.

10.15 (a) $Ag(NH_3)_2OH$; (c) CrO_3 in H_2SO_4;
(e) $AgNO_3$ in alcohol.

10.21 **A** is 1,3-cyclohexadiene;
B is 1,4-cyclohexadiene.

10.25 This is another example of rate versus equilibrium control of a reaction. The *endo* adduct, **G,** is formed faster, and at the lower temperature it is the major product. The *exo* adduct, **H,** is more stable, and at the higher temperature it is the major product.

CHAPTER 11

11.2 *trans*-3,4-dimethylcyclobutene.
11.4 *trans,trans*-2,4-hexadiene $\xrightarrow{h\nu}$
cis-3,4-dimethylcyclobutene $\xrightarrow{heat}$
cis,trans-2,4-hexadiene.

11.6 (a) conrotatory, heat; (b) conrotatory, heat; (c) disrotatory, light.

11.10 (a) 4n + 2 where n = 0; (b) allylic;
(c) conrotatory, since the cyclopropyl anion is a 4n π-electron system (where n = 1).

11.12 **1** is *cis,trans*-cyclonona-1,3-diene;
2 is *cis,cis*-cyclonona-1,3-diene.

11.20 (a) $CH_3\overset{\overset{\displaystyle O}{\|}}{C}CH_3$ +

$H\overset{\overset{\displaystyle O}{\|}}{C}CH_2CH_2-\overset{\overset{\displaystyle O}{\|}}{C}-\overset{\overset{\displaystyle O}{\|}}{C}H$ + 2H$\overset{\overset{\displaystyle O}{\|}}{C}$H;

(d) $CH_3\overset{\overset{\displaystyle O}{\|}}{C}CH_3$ +

$H\overset{\overset{\displaystyle O}{\|}}{C}CH_2CH_2\overset{\overset{\displaystyle O}{\|}}{C}CH_3$ + H$\overset{\overset{\displaystyle O}{\|}}{C}CH_2OH$.

CHAPTER 12

12.1 (a) compound (d) only; (b) none.
12.5 Tropylium bromide is a largely ionic compound consisting of the cycloheptatrienyl (tropylium) cation and a bromide anion.
12.14 (a) *p*-dibromobenzene, mp +87°;
(b) *o*-dibromobenzene, mp +6°;
(c) *m*-dibromobenzene, mp −7°.
12.22 (a) *o*-diaminobenzene, mp +104°;
m-diaminobenzene, mp +63°;
p-diaminobenzene, mp +142°.

CHAPTER 13

13.3 The following reaction takes place:
$2Fe + 3X_2 \longrightarrow 2FeX_3$.
13.4 $HO-NO_2 + HO-NO_2 \rightleftharpoons$
$H_2O^+-NO_2 + {}^-O-NO_2$
$H_2O^+-NO_2 \rightleftharpoons H_2O + \overset{+}{N}O_2$.
13.9 (a) benzene +

$CH_3CH_2CH_2CH_2CH_2\overset{\overset{\displaystyle O}{\|}}{C}Cl \xrightarrow{AlCl_3}$

$C_6H_5\overset{\overset{\displaystyle O}{\|}}{C}CH_2CH_2CH_2CH_2CH_3 \xrightarrow[HCl]{Zn(Hg)}$

$C_6H_5CH_2CH_2CH_2CH_2CH_2CH_3$.

13.11 (a) At the lower temperature the reaction is rate controlled; at the higher temperature it is equilibrium controlled. (b) *p*-toluenesulfonic acid.

13.14 (a) $CH_2ClCH_2\overset{+}{N}(CH_3)_3Cl^-$; (b) slower;
(c) The positive nitrogen makes it strongly electron withdrawing; (d) The structures are similar to those for $C_6H_5CF_3$ on pages 475 and 476.

13.26 Introduce the chlorine into the benzene ring first, otherwise the double bond will undergo addition of chlorine when ring chlorination is attempted.

13.28 (a) 2-methyl-5-acetylbenzenesulfonic acid;
(c) 2,4-dimethoxynitrobenzene;
(e) 3-nitro-4-hydroxybenzenesulfonic acid;
(g) *m*-chlorobenzotrichloride.

13.29 (a) toluene, $KMnO_4$, OH^-, heat; then H_3O^+; then Cl_2, $FeCl_3$.
(c) aniline, CH_3COCl; then conc. H_2SO_4; then Br_2, $FeBr_3$; then dilute H_2SO_4.

(f) toluene, CH_3COCl, $AlCl_3$; then isolate *ortho* isomer.

(g) product of (f), Br_2, OH^-; then H_3O^+.

(i) toluene, HNO_3, H_2SO_4; then isolate *para* isomer, then Br_2, $FeBr_3$.

13.31 p-$NO_2C_6H_4O$—$\overset{\overset{\displaystyle O}{\|}}{C}C_6H_5$ and

o-$NO_2C_6H_4O$—$\overset{\overset{\displaystyle O}{\|}}{C}C_6H_5$.

CHAPTER 14

14.2 (a) one; (b) two; (c) one; (d) three; (e) two; (f) three.

14.7 a doublet (3H) downfield; a quartet (1H) upfield.

14.8 (a) CH_3CHICH_3; (b) CH_3CHBr_2; (c) $CH_2ClCH_2CH_2Cl$.

14.10 (a) $C_6H_5CH(CH_3)_2$;
(b) $C_6H_5CH(NH_2)CH_3$;
(c)

14.17 A, *o*-bromotoluene; **B,** *p*-bromotoluene; **C,** *m*-bromotoluene; **D,** benzyl bromide.

14.19 phenylacetylene.

14.22

F

14.25 G, $CH_3CH_2CHBrCH_3$; **H,** CH_2=$CBrCH_2Br$.

CHAPTER 15

15.2 Complex formation between the carbonyl oxygen and thallium trifluoroacetate does not allow easy delivery of the thallium to the *ortho* position because the intermediate would resemble an eight-membered ring. Thallation, consequently, takes place at the *para* position.

15.3 (a) *tert*-butylbenzene, Br_2, and $Tl(OOCCF_3)_3$
(d) propylbenzene, $Tl(OOCCF_3)_3$, 25°; then KI, H_2O.
(g) ethylbenzene, $Tl(OOCCF_3)_3$, 25°; then KCN, H_2O and $h\nu$.

15.5 $(CH_3)_3CBr + Mg \xrightarrow{\text{ether}}$
$(CH_3)_3CMgBr \xrightarrow{D_2O} (CH_3)_3CD$.

15.10 Acetylation deactivates the ring toward further electrophilic aromatic substitution.

15.14 (a) Br_2/CCl_4 or $KMnO_4/H_2O$;
(c) alcoholic $AgNO_3$; (e) Br_2/CCl_4 or $KMnO_4/H_2O$ (one double bond of both compounds is apparently unreactive toward electrophilic addition).

CHAPTER 16

16.7 (a) $LiAlH_4$; (b) $NaBH_4$; (c) $LiAlH_4$.

16.11 (a) $CH_3OH + {}^-OSO_2CH_3$;
(b) $CH_3OH + {}^-OSO_2OCH_3$.

16.16 (a) C_6H_5CH=CH_2, H_2O, H^+, and heat; or C_6H_5CH=CH_2, $Hg(OOCCH_3)_2$, H_2O, then $NaBH_4$, OH^-; (e) $C_6H_5CH_2COOH$, $LiAlH_4$, ether; (h) C_6H_6, Br_2, $FeBr_3$; then Mg, Et_2O; then ethylene oxide; then H_2O.

16.18 (a) $CH_3CH_2CH_2ONa$;
(e) $CH_3CH_2CH_2OOCCH_3$;
(i) $(CH_3CH_2CH_2)_2O$;
(l) $C_6H_5CH(CH_3)_2 + C_6H_5CH_2CH_2CH_3$.

16.21 (d), (e), and (f).

16.26 (a) aqueous NaOH; (c) Br_2/CCl_4, or $KMnO_4/H_2O$, or CrO_3/H_2SO_4;
(e) aqueous NaOH.

16.31

16.33 *p-tert*-butylphenol.

16.34 1-phenyl-1-propanol.

CHAPTER 17

17.1 (a) $CH_3OH + NaOMs$;
(c) $CH_3CH_2N_3 + NaOTs$;
(e) $CH_3COOCH(CH_3)_2 + NaBr$
(g) $CH_3CH_2CH_2CH$=$CH_2 + MgBrX$.

17.5 (a) Steric hindrance prevents nucleophilic attack from the back-side;
(b) The ring structures prevent the formation of a coplanar arrangement of groups around the positive carbon of the carbocation.

17.8 In strong acid the leaving group is ROH; in a neutral or weakly acidic solution the leaving group would have to be RO⁻.

17.9 (a) decrease; (c) decrease; (e) decrease.

17.13 2-(*p*-hydroxyphenyl)-1-chloropropane > 2-(*p*-tolyl)-1-chloropropane > 2-phenyl-1-chloropropane > 2-(*p*-nitrophenyl)-1-chloropropane.

17.19 (a) [structure with OCH₃ and NO₂ on benzene ring] (b) [structure with NO₂ and NHCH₃ on benzene ring]

(c) [structure with NHC₆H₅, NO₂, and NO₂ on benzene ring]

17.22 (a) If every substitution involves an inversion, then racemization will be complete when only half the substrate has incorporated radioactive iodine. Thus, the rate of racemization *should be twice the rate of incorporation of radioactive iodine.* (b) If an achiral intermediate such as a carbocation ion were involved, one would expect the rate of racemization to equal the rate of incorporation of radioactive iodine.

CHAPTER 18

18.2 (a) 1-pentanol; (c) pentanal; (e) benzyl alcohol.

18.5 a hydride ion.

18.15 (a) $C_6H_5CH_2Br + (C_6H_5)_3P$, then strong base, then CH_3COCH_3.
(c) $CH_3CH_2Br + (C_6H_5)_3P$, then strong base, then $C_6H_5COCH_3$.
(e) $CH_3I + (C_6H_5)_3P$, then strong base, then cyclopentanone.
(g) $CH_2{=}CHCH_2Br + (C_6H_5)_3P$, then strong base, then C_6H_5CHO.

18.19 (a) $R\overset{+}{\underset{\cdot\cdot}{C}}H{-}\overset{\cdot\cdot}{\underset{\cdot\cdot}{O}}{-}R$; (b) $RCH{=}\overset{+}{\underset{\cdot\cdot}{O}}{-}R$;
(c) $RCH{=}\overset{+}{\underset{\cdot\cdot}{O}}{-}R$ because the carbon and oxygen atoms both have an octet of electrons.

18.23 Because base is consumed as the reaction takes place.

18.25 (a), (b), (d), (f), (h), and (i).

18.30 (a) $CH_3CH_2CH_2OH$;
(c) $CH_3CH_2CH_2OH$;
(e) $CH_3CH_2CH(OH)CN$;
(g) $CH_3CH_2CH{=}C(CH_3)CHO$;
(j) $CH_3CH_2CH{=}CHCH_3$;
(l) $CH_3CH_2COO^-$;
(n) $CH_3CH_2CH{=}NNHCONH_2$;
(p) CH_3CH_2COOH.

18.36 (a) Tollens' reagent; (c) I_2/NaOH;
(e) I_2/NaOH; (g) Br_2/CCl_4;
(i) Tollens' reagent; (k) Tollens' reagent.

CHAPTER 19

19.2 (a) CH_2FCOOH; (c) $CH_2ClCOOH$;
(e) $CH_3CH_2CHClCOOH$;
(g) $CF_3{-}$[benzene ring]${-}COOH$.

19.4 (a) $C_6H_5CH_2Br + Mg$ + ether, then CO_2, then H_3O^+.
(c) $CH_2{=}CHCH_2Br + Mg$ + ether, then CO_2, then H_3O^+.

19.5 (a), (c), and (e).

19.8 in the carboxyl group of benzoic acid.

19.13 (a) $(CH_3)_3CCOOH + SOCl_2$, then NH_3, then P_2O_5, heat.
(b) $CH_2{=}\underset{\underset{CH_3}{|}}{C}{-}CH_3$.

19.19 (a) CH_3COOH;
(c) $CH_3COOCH_2(CH_2)_2CH_3$;
(e) $p\text{-}CH_3COC_6H_4CH_3 + o\text{-}CH_3COC_6H_4CH_3$;
(g) CH_3COCH_3; (i) $CH_3CONHCH_3$;
(k) $CH_3CON(CH_3)_2$;
(m) $(CH_3CO)_2O$;
(o) $CH_3COOC_6H_5$.

19.25 (a) $NaHCO_3/H_2O$; (c) $NaHCO_3/H_2O$;
(e) OH^-/H_2O, heat, detect NH_3 with litmus paper; (g) $AgNO_3$/alcohol.

19.30 (a) diethyl succinate; (c) ethyl phenylacetate; (e) ethyl chloroacetate.

CHAPTER 20

20.4 (a) $CH_3CHCOCOOC_2H_5$;
$\quad\quad\quad\quad |$
$\quad\quad\quad\quad CO_2C_2H_5$

$\quad\quad\quad\quad O$
$\quad\quad\quad\quad \|$
(b) $HCCH_2CO_2C_2H_5$

20.9 (a) Nucleophilic substitution, S_N2

(b) O-alkylation that results from the oxygen of the enolate ion acting as a nucleophile.

20.12 (a) Reactivity is the same as with any S_N2 reaction. With primary halides substitution is highly favored, with secondary halides elimination competes with substitution, and with tertiary halides elimination is the exclusive course of the reaction. (b) acetoacetic ester and 2-methyl propene; (c) Bromobenzene is unreactive toward nucleophilic substitution.

20.25 **D** is *trans*-1,2-cyclopentanedicarboxylic acid, **E** is *cis*-1,2-cyclopentanedicarboxylic acid.

CHAPTER 21

21.7 (a) $CH_3(CH_2)_3CHO + NH_3 \xrightarrow{H_2,\ Ni}$
$\quad\quad\quad\quad CH_3(CH_2)_3CH_2NH_2$

(c) $CH_3(CH_2)_4CHO + C_6H_5NH_2 \xrightarrow{H_2,\ Ni}$
$\quad\quad\quad\quad CH_3(CH_2)_4CH_2NHC_6H_5$

21.8 The reaction of a secondary halide with ammonia is almost always accompanied by some elimination.

21.10 (a) anisole + HNO_3 + H_2SO_4; then Fe + HCl   (b) anisole + CH_3COCl + $AlCl_3$, then NH_3 + H_2 + Ni;   (c) toluene + Cl_2 and light, then $(CH_3)_3N$; (d)p-nitrotoluene + $KMnO_4$ + OH^-; then H_3O^+, then $SOCl_2$ followed by NH_3; then NaOBr (Br_2 in NaOH); (e) toluene + $Tl(OOCCF_3)_3$, CF_3COOH, then KCN,-H_2O,$h\nu$, then $LiAlH_4$.

21.18 p-nitroaniline + Br_2 + Fe, followed by $HCl/NaNO_2$ followed by CuBr, then Fe/HCl, then $HCl/NaNO_2$ followed by H_3PO_2.

21.42 **W** is N-benzyl-N-ethylaniline.

CHAPTER 22

22.4 Halide ions are much better leaving groups than hydride ions, thus 2-halo-pyridines undergo nucleophilic substitution much more readily than pyridine itself.

22.11 By carrying out a Diels-Alder reaction using 1,3-butadiene as the diene.

22.14 Para-ortho coupling of reticuline yields the precursor of glaucine; ortho-ortho coupling gives the precursor of bulbocapnine.

CHAPTER 23

23.1 (a) C_2H_5OH, H^+, heat or $SOCl_2$, then C_2H_5OH

(d) $SOCl_2$, then $(CH_3)_2NH$

(g) $SOCl_2$, then $LiAlH[OCC(CH_3)_3]$

(j) $SOCl_2$, then $(CH_3)_2CuLi$

23.6 Elaidic acid is *trans*-9-octadecenoic acid.

23.8 **A** is $CH_3(CH_2)_5C{\equiv}CNa$, **B** is $CH_3(CH_2)_5C{\equiv}CCH_2(CH_2)_7CH_2Cl$, **C** is $CH_3(CH_2)_5C{\equiv}CCH_2(CH_2)_7CH_2CN$, **E** is $CH_3(CH_2)_5C{\equiv}CCH_2(CH_2)_7CH_2COOH$. Vaccenic acid is

$$CH_3(CH_2)_5 \quad\quad\quad (CH_2)_9COOH$$
$$\diagdown\quad\quad\diagup$$
$$C{=}C$$
$$\diagup\quad\quad\diagdown$$
$$H\quad\quad\quad\quad H$$

23.16 The $5\alpha,6\beta$-dibromocholestan-3β-ol that forms initially is unstable because both bromines are axial and because steric repulsions occur between the C-10 methyl and the bromine at C-6. Isomerization takes place to produce the more stable $5\beta,6\alpha$-dibromocholestan-3β-ol.

CHAPTER 24

24.1 (a) two; (b) two; (c) four

24.3 Acid catalyzes hydrolysis of the glycosidic (acetal) group.

24.11 (a) $2CH_3CHO$, one mole HIO_4

(b) $HCHO + HCOOH + CH_3CHO$, two moles HIO_4

(c) $HCHO + OHCCH(OCH_3)_2$, one mole HIO_4

(d) $HCHO + HCOOH + CH_3COOH$, two moles HIO_4

(e) $2CH_3COOH + HCOOH$, two moles, HIO_4

24.22 D-(+)-glucose

24.27 One anomeric form of D-mannose is dextrorotatory, ($[\alpha]_D = +29.3°$), the other is levorotatory ($[\alpha]_D = -17.0°$).

24.28 The microorganism selectively oxidizes the—CHOH-group of D-glucitol that corresponds to C-5 of D-glucose.

24.31 **A** is D-altrose, **B** is D-talose, **C** is D-galactose

CHAPTER 25

25.1 (a) $\overset{+}{H_3}NCH_2CH_2CH_2CH_2CHCOOH$
$\overset{|}{{}^+NH_3}$

(b) $H_2NCH_2CH_2CH_2CH_2CHCOO^-$
$\overset{|}{NH_2}$

(c) The α-amino group is less basic because of the proximity of the electron-withdrawing carboxylate group.

25.5 The labeled amino acid no longer has a basic —NH_2 group; it is, therefore, insoluble in aqueous acid.

25.8
$\overset{+}{H_3}NCHCH_2CH_2CONHCHCONHCH_2COOH$
$\overset{|}{COO^-} \qquad\qquad \overset{|}{CH_2SH}$

25.22 Arg·Pro·Pro·Gly·Phe·Ser·Pro·Phe·Arg

25.23 Val·Leu·Lys·Phe·Ala·Glu·Ala

CHAPTER 26

26.2 The N-glycosidic linkage (like an O-glycosidic linkage) is stable in aqueous base but is susceptible to acid hydrolysis. (b) a mechanism analogous to that given on page 711 for the hydrolysis of an acetal.

26.3 The reaction presumably occurs through an S_N2 mechanism with hydroxide ion acting as a nucleophile. Attack occurs preferentially at the primary 5'-carbon rather than the secondary 3'-carbon.

26.6 (a) approximately two meters; (b) 0.018 g.

26.10 (a) U G G|G G G|U U U|U A C|A G C
  mRNA

(b)· Tyr · Gly · Phe · Tyr · Ser
  amino acids

(c) A C C|C C C|A A A|A U G|U C G
  anticodons

26.12 A change from C-T-T to C-A-T or a change from C-T-C to C-A-C.

INDEX

Page numbers printed in **boldface** refer to tables of physical properties. Those preceded by an A refer to the appendix.

The Modern Periodic Table of the Elements

atomic number → 1 **H** ← atomic mass 1.0079

PERIODS

	IA	IIA	IIIB	IVB	VB	VIB	VIIB	VIII	VIII	VIII	IB	IIB	IIIA	IVA	VA	VIA	VIIA	O
1	1 **H** 1.00797																	2 **He** 4.00260
2	3 **Li** 6.941	4 **Be** 9.01218											5 **B** 10.81	6 **C** 12.01115	7 **N** 14.0067	8 **O** 15.9994	9 **F** 18.99840	10 **Ne** 20.179
3	11 **Na** 22.98977	12 **Mg** 24.305											13 **Al** 26.98154	14 **Si** 28.086†	15 **P** 30.97376	16 **S** 32.06	17 **Cl** 35.453	18 **Ar** 39.948
4	19 **K** 39.098	20 **Ca** 40.08	21 **Sc** 44.9559	22 **Ti** 47.90	23 **V** 50.9414	24 **Cr** 51.996	25 **Mn** 54.9380	26 **Fe** 55.847	27 **Co** 58.9332	28 **Ni** 58.71	29 **Cu** 63.546	30 **Zn** 65.38	31 **Ga** 69.72	32 **Ge** 72.59	33 **As** 74.9216	34 **Se** 78.96	35 **Br** 79.904	36 **Kr** 83.80
5	37 **Rb** 85.4678	38 **Sr** 87.62	39 **Y** 88.9059	40 **Zr** 91.22	41 **Nb** 92.9064	42 **Mo** 95.94	43 **Tc** 98.9062	44 **Ru** 101.07	45 **Rh** 102.9055	46 **Pd** 106.4	47 **Ag** 107.868	48 **Cd** 112.40	49 **In** 114.82	50 **Sn** 118.69	51 **Sb** 121.75	52 **Te** 127.60	53 **I** 126.9045	54 **Xe** 131.30
6	55 **Cs** 132.9054	56 **Ba** 137.34	57 *****La** 138.9055	72 **Hf** 178.49	73 **Ta** 180.9479	74 **W** 183.85	75 **Re** 186.2	76 **Os** 190.2	77 **Ir** 192.22	78 **Pt** 195.09	79 **Au** 196.9665	80 **Hg** 200.59	81 **Tl** 204.37	82 **Pb** 207.19	83 **Bi** 208.9804	84 **Po** (210)	85 **At** (210)	86 **Rn** (222)
7	87 **Fr** (223)	88 **Ra** 226.0254	89 †**Ac** (227)	104 **Ku** (261)	105 **Ha** (260)													

*
58 **Ce** 140.12	59 **Pr** 140.9077	60 **Nd** 144.24	61 **Pm** (147)	62 **Sm** 150.4	63 **Eu** 151.96	64 **Gd** 157.25	65 **Tb** 158.9254	66 **Dy** 162.50	67 **Ho** 164.9304	68 **Er** 167.26	69 **Tm** 168.9342	70 **Yb** 173.04	71 **Lu** 174.97

†
90 **Th** 232.0381	91 **Pa** 231.0359	92 **U** 238.029	93 **Np** 237.0482	94 **Pu** (244)	95 **Am** (243)	96 **Cm** (247)	97 **Bk** (247)	98 **Cf** (251)	99 **Es** (254)	100 **Fm** (257)	101 **Md** (258)	102 **No** (255)	103 **Lr** (256)